Advanced Foundation Engineering

Advanced Foundation Engineering

Principles, Performance and Prospects

Moses Karakouzian
Department of Civil and Environmental Engineering and Construction
University of Nevada, Las Vegas (UNLV)

Abolfazl Eslami
Department of Civil and Environmental Engineering
Amirkabir University of Technology (AUT), Tehran

WILEY

This edition first published 2026

Registered Offices
John Wiley & Sons, Inc., 111 River Street, Hoboken, NJ 07030, USA
John Wiley & Sons Ltd. New Era House, 8 Oldlands Way, Bognor Regis, West Sussex, PO22 9NQ, UK

For details of our global editorial offices, customer services, and more information about Wiley products visit us at www.wiley.com.

The manufacturer's authorized representative according to the EU General Product Safety Regulation is Wiley-VCH GmbH, Boschstr. 12, 69469 Weinheim, Germany, e-mail: Product_Safety@wiley.com.

Wiley also publishes its books in a variety of electronic formats and by print-on-demand. Some content that appears in standard print versions of this book may not be available in other formats.

Library of Congress Cataloging-in-Publication Data Applied for

Hardback ISBN: 9781394203451

Cover Design: Wiley
Cover Image: © jovannig/stock.adobe.com
Back Cover Image: Courtesy of Moses Karakouzian

Set in 9.5/12.5pt STIXTwoText by Straive, Chennai, India

Contents

Preface

Authors

Welcome to a unique exploration of Foundation Engineering in this comprehensive volume titled "*Advanced Foundations Engineering: Principles, Performance and Prospects.*" The "bedrock" of this book is the extensive collective experience of its authors, whose academic pursuits, groundbreaking research endeavors, and practical consulting engagements span more than seven decades that uniquely positioning them to distill a wealth of insights into this comprehensive volume. This substantial tenure reflects the depth of their individual contributions as well as their shared commitment to advancing the field of foundation engineering. As stewards of this dynamic discipline, the authors provide a narrative of foundation engineering that reflects the history and the future trajectory of the field, offering readers insights drawn from the ever-expanding wealth of knowledge. This wealth of experience enriches every page, providing readers with a nuanced understanding of foundation engineering that goes beyond conventional wisdom. As leaders in the field, the authors bring a unique perspective, blending theoretical mastery with practical insights gained from tackling real-world challenges, and providing the readers with an understanding of the intricacies that define modern foundation engineering. This collaborative effort aims to serve as a guide for those seeking a deep understanding and insight into the complexities of foundation engineering, fostering excellence in the next generation of professionals.

Response to Global Public and Private Demand for New Infrastructure

Against the backdrop of an unprecedented population growth, increased urbanization, and rapidly evolving and transforming world, the demand for new infrastructure is more pronounced than ever. The pressing need for housing, transportation, energy, and various other amenities has created a unique set of challenges that necessitate a departure from conventional practices, and resulted in major changes in the way we approach construction projects. Accordingly, infrastructure development is no longer confined to easily accessible or geotechnically favorable locations. Instead, there is an urgent need to explore and utilize sites that were once considered impractical, unsuitable, and unprofitable, due to challenging terrain, unsuitable soil conditions, remoteness, or lack of accessibility. The traditional reluctance to utilize such sites is being replaced by drive to exploit their potential. Difficult terrains, characterized by rugged landscapes, steep slopes, or adverse environmental conditions, are now viewed not as obstacles but as opportunities for innovative engineering solutions. Areas with historically unsuitable soil conditions are being reevaluated with advancements in geotechnical engineering, allowing for the development of new methods to minimize challenges such as poor bearing capacity or high susceptibility to settlement. Previously isolated areas are now being connected through innovative transportation solutions, reducing remoteness and facilitating responsible development. By embracing difficult terrains, reevaluating soil conditions, and extending development to remote and less accessible areas, the engineering community is rising to the challenge of creating infrastructure that not only meets the needs of a growing population but does so in a manner that is sustainable, inclusive, and forward-thinking.

This book contextualizes foundation engineering within the broader scope of global infrastructure development, addressing the challenges and opportunities that arise from this surge in demand for infrastructure in the public and private sectors. By examining the challenges and opportunities inherent in this demand, readers will be able to contribute meaningfully and holistically to the development of sustainable, resilient, and innovative foundations that underpin the structures shaping our world.

Interactive Nature of Foundation Design

Foundation design is not a static process; it demands an interactive and adaptive approach that considers a myriad of factors. This book stresses the interactivity inherent in foundation design, emphasizing the dynamic relationship between design principles and the unique challenges posed by diverse geological and environmental contexts. It guides readers through the intricacies of customizing foundation designs to specific site conditions and demands of each project transcending conventional wisdom and avoiding one-size-fits-all solutions. By fostering an understanding of the dynamic relationship between foundations and their environments, readers are equipped to address the complexities inherent in designing resilient structures that stand the test of time.

Laboratory and In Situ Testing of Soils for Geotechnical Engineering Properties

An in-depth and thorough understanding of soil behavior is fundamental for making informed decisions in foundation engineering. This book places a strong emphasis on the important role played by laboratory and in situ testing in acquiring essential geotechnical engineering soil properties, with a specific focus on mechanical properties like strength and stiffness. Laboratory testing provides controlled conditions to analyze soil samples, yielding valuable insights into their mechanical behavior. It enables the determination of parameters like shear strength, compressibility, and consolidation characteristics, forming the foundation for precise design calculations. While laboratory testing is adequate for certain projects such as earth dams and slopes, the complexity and variability inherent in foundation engineering demand a more comprehensive approach. The authors contend that relying solely on laboratory testing may oversimplify the real-world conditions. Accordingly, the combination of laboratory and in situ testing becomes particularly crucial for foundations and structures where the soil and structure interact. In situ testing captures the actual response of soil under existing field conditions, offering a more realistic understanding of its behavior. This is vital for foundations, where the interaction with the surrounding soil is paramount. The book explores various in situ testing techniques, such as cone penetration tests, pressuremeter tests, and seismic methods, providing readers with a comprehensive toolkit for field assessments. Through detailed discussions and practical examples, the book provides readers with insights into the synergistic use of laboratory and in situ testing and holistic approach to soil characterization. This knowledge ensures that foundation selection, analyses, and designs are based on accurate understanding of the soil's mechanical properties in both controlled laboratory settings and the real-world environment.

Innovative Foundations

As the built environment continues to evolve, so do the demands on innovative foundation designs. This section of the book discusses innovative foundation solutions that push the boundaries of conventional practice by presenting advances in semi-deep or intermediate foundations in contrast to shallow and deep foundations, and advances in soil stabilization and ground improvement techniques. The exploration of advances in semi-deep or intermediate foundations represents a groundbreaking departure from conventional practices in foundation engineering. While shallow foundations bear loads near the surface and deep foundations penetrate deeper into the ground, semi-deep alternatives occupy a middle ground, offering a unique set of advantages. This section introduces and elaborates on innovative techniques such as skirted, floating, shell, bucket, and spudcan hybrid foundations, that combine elements of both shallow and deep foundation design principles. Additionally, this section discusses advances in soil stabilization and ground improvement techniques such as earthworks densification and compaction, physico-chemical modifications, hydraulic modifications, and soil reinforcement, to mention a few. By presenting these advancements through insightful discussions and practical examples, the book aims to equip engineers with an understanding of the semi-deep or intermediate foundations, and soil stabilization and ground improvement techniques, encouraging them to embrace inventive solutions that address the unique challenges posed by different geological and environmental contexts.

Important and Dominant Role of Deep Foundations

Deep foundations have become indispensable in construction of the foundations of heavy structures such as tall buildings, bridges and offshore wind turbines, particular for those structures exposed to combined vertical, moment, and horizontal (VMH) loading. Accordingly, a considerable portion (18 out of a total of 27 chapters) of this book, is devoted to all aspects related to deep foundations discourse including analysis (VMH loading), design (performance based), pile groups, construction, innovations, semi-deep foundations, in situ site investigation, capacity and integrity testing, performance under cyclic loading, performance under extreme conditions, data mining, numerical analysis, uncertainty and reliability, Load and Resistance Factored Design (LRFD), damage assessment and repair, reuse of existing foundations, sustainability, value engineering and optimization, selection of foundations. Also included in the book is an extensive discussion on CPT and CPTu techniques (simulating pile driving) as versatile tools for in situ data acquisition to complement deep foundation analysis, design and installation.

Whether faced with soft, compressible cohesive soils; loose and liquefiable sands and silts; problematic deposits; or challenging geological formations, deep foundations exhibit a remarkable functionality, providing exceptional performance regarding load bearing, deformation control, and general stability. Deep foundations can be constructed from mini to mega scale as single entities or as a group, with or without rafts, offering a range of design possibilities tailored to the unique demands of each project, accommodating diverse structural requirements. The extensive and comprehensive interaction between deep foundations and the surrounding confining soil delivers substantial load-bearing capacity, controls deformations and settlements, and enhances the resistance and resilience to nonvertical forces, i.e., cyclic, seismic and wind loads, providing a level of stability that is unmatched.

While addressing the practical aspects, the significance of deep foundations extends to economic and environmental considerations. By optimizing the consumption of construction materials and minimizing the necessity for extensive excavation, deep foundations contribute to more sustainable construction practices. Furthermore, the longevity and durability provided by deep foundations reduce the need for future maintenance and repair, presenting a cost-effective and environmentally conscious solution.

Testing of Shallow and Deep Foundations

Shallow foundations are commonly utilized for light structures at sites having competent soil conditions. The performance of these foundations is critical to the overall stability of a project. This book presents both bench-scale and field testing methodologies for shallow foundations, offering a comprehensive guide to evaluating and optimizing their design. By examining real-world scenarios and outcomes, readers will develop a profound understanding of the factors influencing shallow foundation behavior, enabling them to make informed decisions, ensuring the reliability and safety of shallow foundations.

For deep foundations, field testing is indispensable for assessing performance and mitigating potential risks. Recognizing the critical role of field testing of deep foundations, this book dedicates a comprehensive discussion exploring various techniques employed in the field such as static and dynamic load testing, cross-hole sonic logging (CSL) testing, etc. Through a mix of theoretical frameworks and practical considerations, engineers will be equipped with the skills needed to design and assess deep foundations with confidence. Moreover, through detailed discussions and case studies, engineers will gain the knowledge needed to deal with the challenges associated with designing and evaluating deep foundations in various geological settings.

Data Mining in Design and Analysis of Foundations

The importance of data mining in the design and analysis of foundations cannot be overstated, as it serves as a transformative tool in extracting valuable insights from vast datasets, thereby enhancing the precision and efficiency of geotechnical engineering practice, and facilitating informed decision-making through optimal design strategies. This book recognizes the groundbreaking role of data mining in the design and analysis of foundations. In current foundation engineering practice, projects generate an ever-expanding volume of data stemming from sources such as geotechnical laboratory and in situ testing and investigations, subsurface exploration, field testing of foundations, construction methodologies, and real-time performance monitoring. Data mining techniques enable engineers to sift through this wealth of information, identifying patterns, correlations, and valuable hidden relationships that may elude conventional analytical approaches. By employing advanced algorithms, these techniques enable the extraction of information from vast datasets, facilitating a deeper understanding of soil-structure interaction and foundation behavior, and identifying critical factors influencing performance and longevity. Whether it is analyzing soil composition, assessing load-bearing capacities, or predicting settlement patterns, data mining helps geotechnical engineers to make informed decisions, thereby contributing to the development of safer, more resilient, and economically viable foundation solutions. The integration of data mining in foundation design creates opportunities for innovation and continuous improvement in the field, reinforcing the role of empirical insights in shaping the future of geotechnical engineering.

Reliability and Uncertainty Analysis and LRFD Design

In an era where precision is paramount, this book places a strong emphasis on and a rigorous examination of reliability and uncertainty analysis in foundation engineering with a specific focus on LRFD. The adoption of LRFD principles ensures that foundation designs are not just safe but optimized for performance and efficiency. By equipping readers with tools and methodologies to handle the complexities of uncertain conditions, it allows the practitioners to make informed decisions in the face of complex, unpredictable and often ambiguous conditions, ensuring robustness of foundation designs and the reliability, resilience of foundation structures.

Numerical Methods and Tools

Modern foundation engineering relies heavily on and is inseparable from advanced numerical methods and tools. This book provides a detailed exploration of state-of-the-art numerical techniques, demystifying them by offering readers a comprehensive understanding of their use and application. With practical examples and step-by-step explanations, readers, whether novices or seasoned professionals, will gain proficiency in leveraging numerical methods to effectively apply these tools and enhance the precision and efficiency of their analysis and design processes.

Damage and Repair of Foundations

Foundations, despite their robust nature, are not immune to the forces of time and external factors. In the face of unpredictable forces and environmental stressors, foundations are susceptible to damages that require careful consideration and effective repair strategies. This book dedicates significant attention to the often-overlooked aspects of damage and repair, offering through comprehensive discussions, detailed case studies, and practical insights, a holistic exploration of strategies to diagnose, assess, mitigate, and rehabilitate foundation damages. The goal is not just to design foundations that withstand external pressures but to equip engineers with the tools to ensure the longevity and resilience of structures in the face of unforeseen challenges through effective damage repair.

Sustainability

Sustainability is not a buzzword; it is a mandate for responsible engineering. The imperative of sustainable practices in engineering is a central theme of this book. It not only emphasizes the principles of sustainability in foundation engineering but also provides practical guidance on how to design foundations that meet performance criteria and contribute to a more sustainable built environment. By integrating sustainability into foundational design, this book equips readers to be stewards of both structural integrity and environmental responsibility. By incorporating ecological considerations into foundation design, this book advocates for responsible practices that leave a positive imprint on the world. From environmentally conscious material choices to energy-efficient design strategies, this section will provide the engineers with the knowledge needed to build a greener future.

Reuse of Existing Foundations

In the pursuit of sustainable development, the reuse of existing foundations is a compelling and critical consideration. This book promotes adapting and optimizing existing foundations for new structures, minimizing environmental impact and maximizing resource efficiency. It explores the complexities of assessing, adapting, and optimizing existing foundations for new challenges. Through practical guidance and illustrative examples, readers will learn to exploit the potential use of existing foundations. By addressing the environmental and economic implications of reusing foundations, it provides readers with valuable insights into minimizing the ecological footprint of construction projects. This section provides guidelines for repurposing foundations, breathing new life into structures of the past while satisfying contemporary sustainability goals.

Value Engineering and Optimization

Striking a balance between performance and cost requires strategic decision-making. This book introduces readers to value engineering and optimization, providing tools and methodologies to make informed choices that align with project goals and constraints. By fostering a mindset of efficiency without compromise, it allows the engineers to find the equilibrium between functionality, sustainability, and cost-effectiveness. By incorporating these principles into the foundation design process, practitioners can achieve optimal outcomes in a holistic and efficient manner.

Selection of Foundation – Project-Based Design Approach

Foundation selection is not a one-size-fits-all undertaking. It demands an understanding of the specific project requirements and site conditions. This book emphasizes the importance of a project-based design approach in making informed decisions regarding the selection of foundations with the understanding that each construction project is inherently unique, demanding a tailored approach to foundation selection. "Project-based design" refers to an approach in engineering and other fields where the design process is tailored specifically to the unique requirements and constraints of a particular project. Instead of applying a standardized or generic design, project-based design takes into account the specific characteristics, goals, and challenges of the project at hand. Recognizing the interplay between structures, geological conditions, and environmental factors, the authors provide 24 key considerations that guide the foundation selection process. These considerations include the type of structure being built, the loads the foundation will need to support, the geological and geotechnical conditions of the site, environmental and sustainability requirements, economic viability, and others. Collectively, these considerations provide a comprehensive framework for engineers to follow. By emphasizing a project-centric approach, this book provides the readers with the knowledge and tools necessary to examine the complex interplay of factors influencing foundation selection, ensuring that each project benefits from a tailored and optimized foundation solution.

Numerous Case Studies

Real-world challenges demand pragmatic, practical, and workable solutions. This book is replete with numerous case studies that bridge theory and application, offering readers a firsthand look at the complexities and successes of foundation engineering. Each case study provides a roadmap for approaching a myriad of scenarios and cultivating the problem-solving skills crucial for success in the field. From challenging geological conditions to unexpected construction complications, these case studies, drawn from real-world projects, offer readers a unique opportunity to explore and examine the complexities of diverse scenarios. By presenting a spectrum of challenges and solutions, this book offers a hands-on, experiential learning opportunity.

Our Challenge to Students, Academicians, and Professionals

Our book aspires to issue a bold challenge to students, academicians, and professionals in the field of foundation engineering. By presenting groundbreaking insights, innovative approaches, and forward-thinking concepts, we seek to inspire them to embark on a journey of exploration, pushing the boundaries of traditional practices. The goal is to encourage the creation of novel prospects within foundation engineering, sparking a paradigm shift that fosters creativity, resilience, and transformative advancements in the industry. Through this challenge, we aim to catalyze a collective effort towards shaping a dynamic and progressive future for foundation engineering professionals.

March 5, 2026
Las Vegas, NV

Moses Karakouzian
Department of Civil and Environmental Engineering and Construction
University of Nevada, Las Vegas (UNLV)

Abolfazl Eslami
Department of Civil and Environmental Engineering
Amirkabir University of Technology (AUT), Tehran

Acknowledgments

We want to take a moment to express our gratitude to the many people and organizations who made this book possible. This comprehensive work on advanced foundation engineering is the culmination of collective efforts, and we extend our sincere gratitude to all those who have contributed to its realization and provided support and encouragement.

The academic community forms the bedrock of knowledge. We acknowledge our fellow academicians and researchers for your dedication to advancing knowledge in the field of foundation engineering paving the way for our work, and we are honored to be a part of this community. Our interactions with our colleagues and university professors, have been intellectually stimulating and influential in shaping our combined experience of 70 years in academia, research, and consulting. Their commitment to fostering an environment of intellectual curiosity has left an indelible mark on the ideas explored in this book. We also acknowledge the invaluable insights and experiences shared by professional practitioners. Their perspectives from the field have contributed to shaping the content of this book, and have added a practical dimension to the theoretical discussions, creating a well-rounded resource for both academics and practitioners.

The success of this book owes much to the invaluable contributions of Dr. Amirhossein Ebrahimipour and Dr. Sara Heidarie Golafzani. Their expertise, insights, and commitment to advancing foundation engineering have been instrumental in shaping the content and depth of this work. Their dedication to excellence has truly elevated the quality of this project.

Undertaking a project of this magnitude required a significant network of support and contributors who were involved in various capacities and played a vital role. In this context, we acknowledge Sajjad Shirani, Mehrdad Karami, Donya Afshar, Mahdi Shakeran, and Amir Nikouei for their indispensable diligence in research assistance, logistical support, and administrative tasks. Their efforts have eased the complexities of the journey, rendering this venture more manageable. Moreover, we acknowledge Ms. Shelby Smith and Ms. Meagan Madriaga-Hopkins for their tireless editing of major parts of the book. Furthermore, we extend our appreciation for the cooperation of Hasan Moghadasi, Arash Tadayon, Pouya Bahrami, Ali Mousavi Noqondar, Parisa Khodabakhsh, Mohammadreza Pordel, Dorsa Shadlou, Mohammad Esmailzade, Mostafa Nazmi, and Ramin Ezzatdoost. Their collaborative spirit and contributions have been invaluable to the success of this project.

We acknowledge the energy and enthusiasm of current and past graduate students and fellows, Dr. Richard Stone, Dr. Visar Fahrangi, Dr. Mohammad Nazari Sharabian, Dr. Negasi Gebremichael, Dr. Rouzbeh Afshar, Dr. Mahmoud Motahari, Dr. Mehdi Khalili, Dr. Kazem Jadidi, Dr. Yahia Iskander, Dr. Meysam Imani, Dr. Davood Akbarimehr, Dr. Ebrahim Shojaei, Dr. Asgar Nasiri, Dr. Masoud Nobahar, Dr. Seyyed Mohammad Fattahi, Mr. Hasan Baranizadeh, Mr. Makan Eslami, Mr. Kaamyar Eslami, Mr. Barmak Biroun Hosseini, Dr. Sudip Basack, Dr. Ghritartha Goswami, Mr. Ashkan Saradar, and Mr. Mohammad Moein, who have injected vitality into this project. Their questions, discussions, and collaborative spirit have been invigorating. Their involvement is a testament to the importance of nurturing the next generation of scholars and practitioners.

We devote our heartfelt gratitude for the guidance and constructive support from our pioneers, Dr. Bengt H. Fellenius and late Dr. Richard Campanella. Their scholarly input, insightful critiques, and encouragement have significantly enriched the principles and backbone of this book. Their mentorship has been a source of inspiration throughout this challenging yet rewarding process.

We are deeply grateful to our respective Civil and Environmental Engineering Departments at the University of Nevada, Las Vegas (UNLV), and Amirkabir University of Technology (AUT), Tehran, for their unwavering support throughout the process of writing this book. The generous financial assistance by the Civil and Environmental Engineering Department at UNLV has been instrumental in bringing this project to fruition, enabling us to conduct necessary research and access valuable resources. Moreover, we extend our sincere appreciation to the Civil and Environmental Engineering Department at AUT for granting a sabbatical leave to Dr. Abolfazl Eslami, providing him with the dedicated time and space needed to focus on this endeavor, as well as for sharing and granting the use of the comprehensive AUT: Geo-CPT & Pile Database, which significantly enhanced the content and depth of this work. These gestures reflect our respective departments' commitment to fostering research and academic pursuits. The collaborative spirit within each department has been a source of inspiration, and we are thankful for the open exchange of ideas and insights that have contributed to the overall quality of this book.

A special acknowledgment is extended to the staff of the Clark County Building Department, Las Vegas, Nevada, for their support in providing us with technical information and data on major projects in the area. The information provided insights into real-world challenges and practical considerations that have grounded the theoretical aspects of this book in the realities of professional practice.

We acknowledge Dr. James Bristow for his support and guidance in facilitating in expanding our network of professionals, which has been a guiding force in this endeavor. His expertise and mentorship have shaped the trajectory of this work, and we are grateful for the opportunity to benefit from his wealth of knowledge and experience. We also acknowledge Dr. Mandro M. Eslami for his support in brainstorming for ideas for the organization of this book and for his guidance and help in logistical and administrative support, facilitating the smooth progress of our work.

We express our appreciation to Wiley, the publisher, for their commitment to disseminating scholarly works. Their professionalism, editorial support, and dedication have been crucial in delivering high-quality publications and bringing this book to fruition.

To the readers, users, students, academicians, and professionals alike, we appreciate you choosing this book. We hope it provides you with a valuable resource of useful information and practical knowledge. We are very optimistic that this book will serve as a compelling challenge to students, academicians, and professionals in the field of foundation engineering. By probing into innovative methodologies, cutting-edge research, and emerging trends, we aspire to motivate you to explore novel prospects within the realm of foundation engineering. We believe that the content presented herein will kindle fresh perspectives, encouraging you to push boundaries and contribute to the evolution of this critical discipline. Accepting and confronting the challenges outlined in this book has the potential to reshape the future landscape of foundation engineering practice. Thus, your interest in our accomplishment gives our work purpose, and we are truly grateful for your readership.

Last but certainly not least, we extend heartfelt thanks to our families for their unwavering support and understanding. Their encouragement and patience have sustained us through the peaks and valleys of this journey. This work is a testament to the collective strength of our familial bonds.

We extend our sincere apologies for any oversight in the acknowledgment section of this book. It is not our intention to omit any names, and we deeply regret any disappointment or inconvenience caused. Your support and contributions are invaluable to us, and we want to express our gratitude to each person who played a role in the creation of this work. We appreciate your understanding and assure you that your contributions are not overlooked or underestimated. In the upcoming editions, we will make the necessary amendments to ensure that proper recognition is given to all those who deserve it.

Each individual and entity mentioned here has played a distinctive role in the creation of this book, and we are deeply grateful for the collaborative spirit that has shaped its development.

We would like to express our deepest gratitude to our wives, Kohar and Zoya, whose unwavering love, patience, and encouragement sustained us throughout the five-year journey of writing this book. The long hours of research, writing, revisions, and coordination required for a project of this scope inevitably demanded time and energy that might otherwise have been spent with our families. Their understanding, quiet strength, and constant support made this work possible. Their encouragement during moments of fatigue and their belief in the value of this endeavor provided us with the motivation to persevere. We dedicate our heartfelt thanks to them for their companionship, sacrifice, and enduring inspiration during the creation of this book.

Thank you all for being a part of this journey. Your support has made this book a reality.

March 5, 2026

Warm regards,
Moses Karakouzian and Abolfazl Eslami

Topic A

Background

Section I

Context

1

An Introduction to Foundation Engineering

Chapter Highlights

- Introduction to Foundation Engineering
- Geosystems and Foundation Systems
- Relevant Disciplines
- Realm of Foundation Engineering
  - Technical
  - Constructional and Practical
  - Feasibility and Durability
- Challenging Conditions
  - Superstructure
  - Site Investigation
  - Subsoil
  - Environmental and Circumferential
  - Construction and Material
  - Sustainability Aspect
- Book Organization
  - I: Context
  - II: Geotechnical Engineering
  - III: Spread Footings and Mats
  - IV: Ground Modifications
  - V: Skirted, Floating, and Shell Foundations
  - VI: Single Piles and Drilled Shafts
  - VII: Group Actions and Extreme Conditions
  - VIII: Testing and Numerical Approaches
  - IX: Trends and Prospects
  - X: Sustainable Development

1.1 Introduction

Foundation elements or systems represent essential structural components responsible for transferring various combinations of loads from a superstructure to the underlying geological materials. These foundation units may either independently support the imposed loads or work in conjunction with other elements, such as basement walls, floors, or slabs.

The foundation plays a pivotal role in mitigating the significantly concentrated stresses within structural components like walls, columns, or piers, where compression stresses typically range from 10 to 200 MPa. Its primary function is to distribute

Advanced Foundation Engineering: Principles, Performance and Prospects, First Edition. Moses Karakouzian and Abolfazl Eslami.

these stresses to the subsoil while ensuring that they remain within the acceptable bearing stress limits of 0.05–0.5 MPa. In this context, examples of footings or foundations can include road pavements, steel column baseplates, and even the roots of trees and plants.

Figure 1.1 illustrates a range of common structures, including skyscrapers, bridges, wind turbines, storage facilities, airport runways, and offshore platforms. Figure 1.2 showcases various types of transitional elements or units, which serve as foundations in diverse engineering applications.

Figure 1.1 Common types of structures in civil engineering (Sergii Figurnyi / Adobe Stock Photos. Kansai Design Co. Ltd. Lukasz Z / Adobe Systems Incorporated).

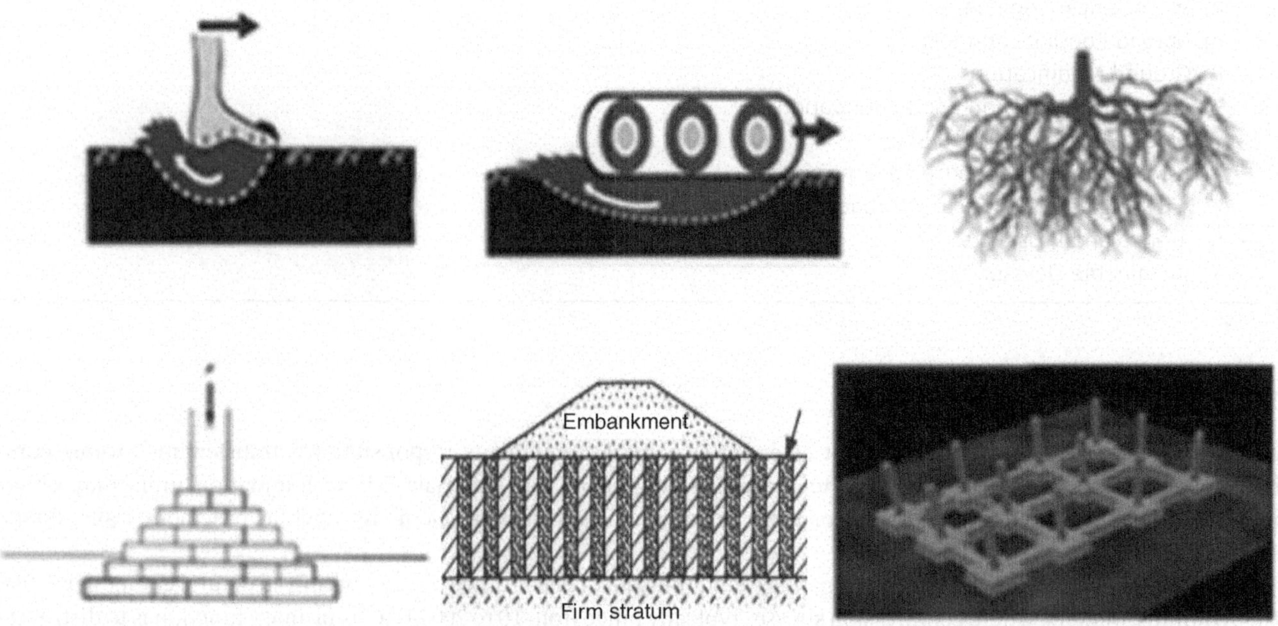

Figure 1.2 Different units acting as foundation elements.

1.2 Geosystems and Foundations

1.2.1 Geosystems

Geosystems represent a combination of various natural and artificial elements and materials assembled to meet specific needs. Within this context, there is a growing preference for utilizing a higher proportion of geomaterials and minimizing the use of artificial materials, aligning with principles of value engineering and sustainable development. Geosystems encompass a wide spectrum within the fields of geotechnical and geological engineering, environmental geotechnics, geohydrology, and applied geophysics. Figure 1.3 illustrates several exemplary geosystems.

1.2.2 Foundation Systems

Foundations are typically categorized into four main groups based on the ratio of embedment depth to breadth. These categories include shallow foundations, shallow foundations on stabilized soil, semi-deep foundations, and deep foundations. Shallow and deep foundations are considered the primary foundation types, while the other two fall into the intermediate category. Figure 1.4 provides an illustration of these various foundation types.

Despite commonly focusing on specific components like the foundation when discussing substructures, it is important to recognize that the substructure can be viewed as a comprehensive "foundation system." This foundation system encompasses not only the embedded portion of the structure but also the surrounding soil and other structural elements such as walls, horizontal diaphragms, and the lower part of the structure that interfaces with the ground. In essence, the

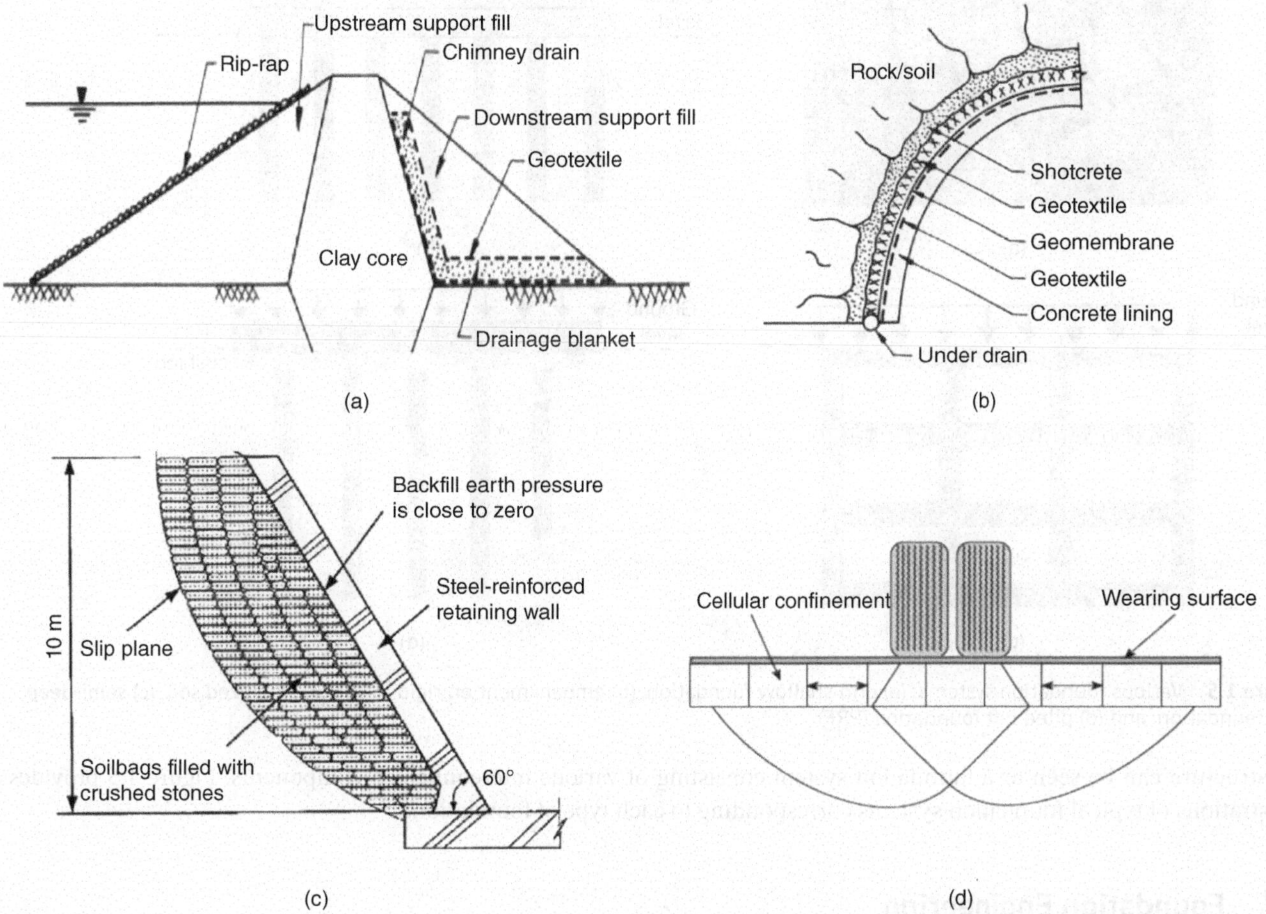

Figure 1.3 Various geosystems: (a) cross-section of the earth dam with geotextile filters, (b) cross-section of a tunnel vault showing the general arrangement of the lining system, (c) application of sandbags for soil slope stability, and (d) application of geocell for road pavement.

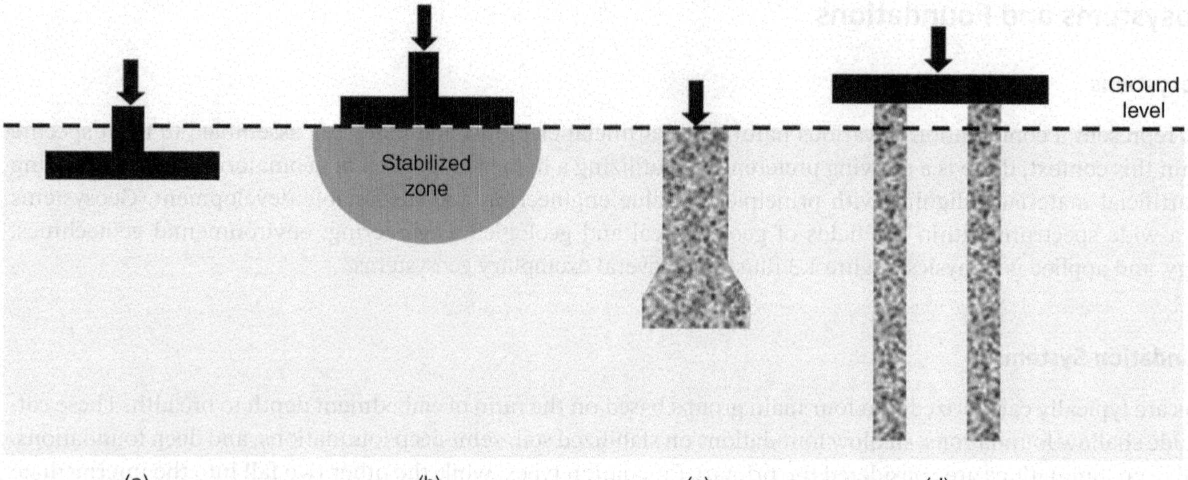

Figure 1.4 Different foundation types: (a) shallow, (b) shallow on stabilized soil, (c) semi-deep, and (d) deep.

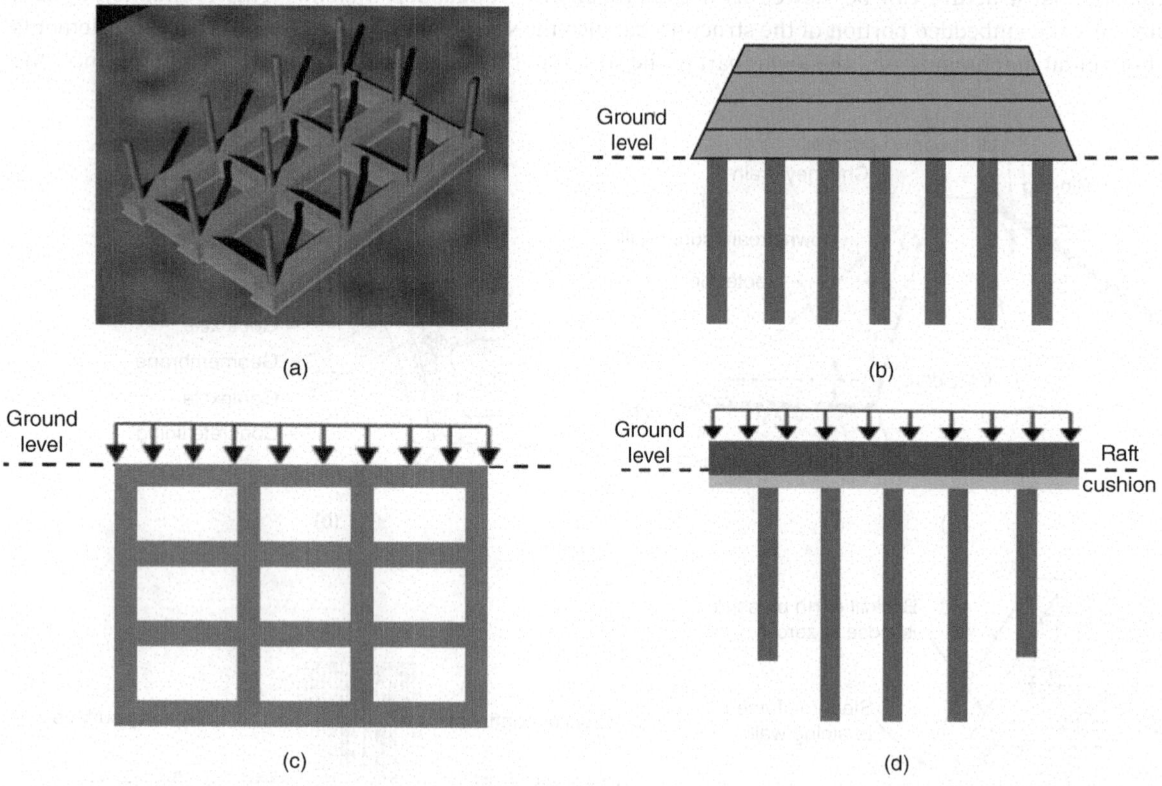

Figure 1.5 Various foundation systems: (a) grid shallow foundation, (b) embankment on rigid intrusion stabilized soil, (c) semi-deep box foundation, and (d) piled raft foundation (PRF).

substructure can be seen as a foundation system consisting of various interconnected components. Figure 1.5 provides illustrations of typical foundation systems corresponding to each type of foundation.

1.3 Foundation Engineering

Foundation engineering is a specialized field within geotechnical engineering that combines principles of soil mechanics, structural engineering, and project-specific serviceability requirements to design and build foundations for various types of structures, including those onshore, offshore, and inland. Unlike a routine technical procedure, foundation engineering is often seen as more of an artistic approach because well-designed and constructed foundation systems exhibit optimal performance throughout a project's life span. The primary objectives of a foundation engineer encompass creating foundation systems that are technically sound, safe, serviceable, constructible, cost-effective, and efficient in supporting superstructures. Foundation engineering should be viewed as a multidisciplinary endeavor that involves a collaborative and knowledge-driven process among structural, geotechnical, and construction engineers.

The foundation of a structure plays a pivotal role in addressing several critical factors, including the origin and characteristics of the applied loads and the capacity of any civil engineering project to accommodate foundation movements. It is important to recognize that any footing unit or foundation system essentially functions as a structural component responsible for the secure transfer of loads from the superstructure to the underlying subsoil. Therefore, a thorough understanding of internal stress distribution and the selection of appropriate materials are crucial elements in the design process.

From a geotechnical engineering perspective, every type of foundation engages with the surrounding ground. Consequently, the design process should take into account the engineering properties and behavior of the adjacent geomaterials.

In essence, while the construction materials and equipment are undeniably significant aspects of foundation construction, purposeful consideration of the interaction with geomaterials and the performance of the surrounding soil, especially in challenging conditions, can lead to an efficient and cost-effective practice.

1.4 Realm of Foundation Engineering

Efficient foundation engineering practice encompasses a range of technical, practical, and economic considerations, with a primary focus on performance-based design. These considerations can be summarized as follows:

1) **Bearing Capacity:** This aspect evaluates the ability of the underlying soil or rock to support loads without experiencing failure. It depends on factors such as shear strength properties of geomaterials, foundation geometry, embedment depth, and loading conditions. Various equations and guidelines are available in foundation design references to assess bearing capacity.
2) **Serviceability (Settlement and Torsion):** Serviceability concerns encompass deformation and durability controls in foundation design. Key deformations to consider include settlement, heave, angular distortion, tilt, lateral displacement, and vibration. While exceeding serviceability limits may not lead to structural failure, it can result in increased maintenance costs, aesthetic issues, reduced functionality, and environmental concerns.
3) **Structural Design:** The structural design of foundations addresses stress distribution within foundation elements due to externally applied forces from the superstructure and the reaction from the supporting soil. In modern reinforced concrete foundations, this involves determining internal stresses caused by flexural and shear forces and designing appropriate concrete mixes and steel reinforcements. The structural design may follow different approaches, such as ASD (Allowable Stress Design), USD (Ultimate Stress Design), or LRFD (Load and Resistance Factor Design), depending on foundation type, materials, and regional codes.
4) **Stability Control:** Stability considerations pertain to ensuring the foundation remains stable and secure under various loading conditions and soil behavior scenarios. It involves assessing factors like sliding, overturning, and settlement-induced tilting of foundation elements.
5) **Model or Full-Scale Testing:** Testing, whether at model or full scale, provides valuable insights into foundation behavior and helps validate design assumptions. These tests can involve physical models, laboratory experiments, or field load tests to assess actual performance.
6) **Constructional Aspects:** Practical aspects of construction, including equipment, methods, and construction sequencing, are critical for executing the foundation design successfully. Efficient construction practices can significantly impact project costs and timelines.
7) **Durability:** Foundation durability is essential for long-term performance. It involves selecting materials and construction techniques that can withstand environmental factors and maintain their structural integrity over time.
8) **Economic Requirements:** Balancing technical and safety considerations with economic feasibility is crucial in foundation engineering. Cost-effective solutions that meet performance requirements are sought while ensuring the project remains financially viable.

In summary, foundation engineering involves a holistic approach that addresses safety, serviceability, and economic factors. It requires a thorough understanding of the underlying geomaterials, structural behavior, and construction techniques. Settlement analysis, in particular, is a complex aspect of foundation engineering, necessitating comprehensive data, informed analysis, and sound engineering judgment to ensure tolerable settlement effects on both the superstructure and substructure.

As Fellenius (2023) aptly emphasizes, the analysis and design of foundations constitute an iterative process, wherein imposed loads, corresponding settlements, and foundation geometry are interdependent factors influenced by geotechnical capacity, structural capacity, and settlement requirements.

Stability control becomes paramount when foundations are subjected to atypical forces beyond vertical loads. These forces, including horizontal, bending, uplift forces, soil and water lateral pressures, seismic and wind loads, as well as potential pullouts due to vessel or vehicle collisions, require careful consideration. Stability checks encompass the assessment of sliding, overturning, overall slope stability, and the interface tension between the foundation and underlying soil or rock. Any unsatisfactory stability conditions can result in significant damage and failure, affecting both the foundations and the structures they support.

Foundation testing, whether at a model scale or full scale in situ, during the safety design phase, proves highly advantageous. This is especially relevant for extensive projects involving multiple piles or footings, particularly on challenging sites, as it helps validate that the actual foundation behavior aligns with design assumptions. Such testing is typically conducted using full, medium, or even small-scale prototypes. Deviations from expected behavior may necessitate revisions to the foundation design. However, it is crucial to acknowledge that foundation testing addresses individual elements of the substructure system. The interaction of foundation system elements, such as piles and single footings, or combinations like rafts and piles, constitutes a hybrid behavior. Consequently, overall foundation behavior may not be directly inferred from foundation test results without accounting for foundation–soil interaction effects.

In line with Coduto et al. (2016), **durability requirements** linked to serviceability limits for foundations ensure that foundations can withstand various forms of physical, chemical, and biological deterioration. Over the structure's design life, foundations are expected to maintain their performance. Durability limits often relate to issues like corrosion or other time-dependent weakening processes. Therefore, it is imperative to protect foundation materials, whether concrete, steel, or wood, against environmental threats, including soil, water, and atmospheric conditions.

Constructability performance requirements are closely tied to the practical aspects of construction and must align with typical contractor practices. This means that the foundation system should be implementable without requiring extraordinary methods or specialized equipment. Decision-making in this regard hinges on various factors, including project site characteristics, foundation type, nearshore or offshore environments, ground conditions, groundwater levels, available installation equipment, specified budget, project timeline, and proximity to adjacent structures or utilities. Hence, it is crucial to recognize that foundation engineering and design are multidisciplinary endeavors that rely on collaboration and knowledge-sharing.

While foundation design necessitates a higher degree of conservatism compared to superstructure design, this conservatism must be judiciously justified. Excessive conservatism can lead to overconsumption of materials and equipment, protracted construction timelines, and unwarranted project costs. Therefore, efficient foundation design, coupled with appropriate conservatism, should incorporate technical and practical engineering features that align with the principles of value engineering.

1.5 Challenging Conditions and Role of Foundation Engineer

The selection of foundations for a project hinges on several key determinants, broadly categorized as follows:

1) **Superstructure and Loading:** The nature and magnitude of the superstructure, including the applied loads, significantly influence foundation selection.
2) **Site Investigation and Characterization:** Comprehensive site investigation and accurate characterization of the subsurface conditions are essential in making informed foundation decisions.
3) **Subsoil Condition:** Understanding the properties and behavior of the underlying subsoil is crucial for designing foundations that can withstand various geotechnical challenges.
4) **Environmental and Circumferential Conditions:** Environmental factors and surrounding conditions, such as seismic activity, liquefaction potential, and proximity to adjacent structures, play a pivotal role in foundation choices.

5) **Construction Requirements and Material Availability:** Practical considerations, including construction methods, material availability, and equipment, impact foundation feasibility.
6) **Sustainable Development Principles:** The principles of sustainable development emphasize environmentally responsible and resource-efficient foundation systems that contribute to energy efficiency and minimal environmental impact.

Challenging conditions arise when one or more of these factors deviate significantly from the norm. For instance, challenging conditions may involve problematic surface soil near a site's surface when erecting a tall and heavy structure. Another example relates to offshore projects with substantial lateral and moment loads that necessitate rigorous stability measures. Other challenges are high liquefaction potential and seismic activity, which are common abnormal conditions that warrant careful planning. Similarly, urban construction requires diligent attention due to the presence of nearby structures.

Geotechnical engineering encounters inherent uncertainties arising from natural geomaterials, measurement processes, and the estimation of design geotechnical parameters. These uncertainties can magnify the risks in geotechnical designs. Here, extensive site investigation and characterization become paramount. The application of in situ testing, particularly cone penetration tests (CPT and CPTu), provides a wealth of accurate and continuous data, reducing uncertainty and enhancing reliability.

Another challenge foundation engineers face is to consider sustainability as a factor in the selection and design of foundation systems. A sustainable foundation system encompasses both the structure and the implementation of environmentally responsible and resource-efficient practices throughout the entire foundation life cycle. This life cycle includes planning, design, construction, operation, maintenance, renovation, and eventual demolition. Sustainable foundation systems for buildings aim not only to be energy-efficient but also environmentally friendly. The evaluation of the sustainability of different foundation systems takes into account three key pillars of sustainability: environmental (planet), economic (profit), and social (people). Adhering to the principles of sustainable development, future construction endeavors should prioritize reducing environmental pollution and ensuring the efficient utilization of material resources. Sustainable foundation systems aim to optimize the consumption of materials, ground, and energy while promoting substantial interactions between foundations and the surrounding soil. This approach leads to reduced environmental impact and more efficient structural design.

In challenging conditions, foundation engineers must excel in selecting and designing the most efficient foundation system. This often involves combining typical foundations and developing innovative hybrid foundations. An example of such conditions is seen in offshore wind turbine (OWT) installations. These structures face unique challenges due to their geometry and complex loads, which are a mixture of cyclic and dynamic components. Apart from the self-weight of the whole system, four main lateral loads are acting on an OWT structure: wind, wave, 1P (rotor frequency), and 2P/3P (blade-passing frequency) loads (Figure 1.6a) (Bhattacharya 2019). The previously mentioned loads can be combined and presented as equivalent horizontal load (H), overturning moment (M), and self-weight of the system (V).

Offshore wind turbines (OWTs) installation is a unique type of structure due to its geometry and the loads acting on it, which are a mixture of cyclic and dynamic components. Apart from the self-weight of the whole system, four main lateral loads are acting on an OWT structure: wind, wave, 1P (rotor frequency), and 2P/3P (blade-passing frequency) loads (Figure 1.6a) (Bhattacharya 2019). The previously mentioned loads can be combined and presented as equivalent horizontal load (H), overturning moment (M), and self-weight of the system (V). Since dominant loads in an OWT are horizontal load and overturning moment, sufficient embedment, as well as breadth of foundation, is required to provide favorable stability of the whole system. Also, practical and economic requirements must be satisfied. Figure 1.6b illustrates typical foundations for OWTs, such as monopile and bucket. Among hybrid foundation systems that are applicable, particularly for offshore wind turbines, hybrid suction bucket foundations (HSBF), and monopole footing can be named.

The concept of HSBF combines a skirted mat and one or more suction caissons with a skirt length that extends significantly beyond the mat skirt. This results in a significant contribution of the caisson to the overall foundation capacity. Figure 1.6c depicts a schematic view of this concept (Gaudin et al. 2011).

Another new hybrid foundation, named the monopile-footing composite foundation, combines the monopile and circular footing foundations. In comparison with the traditional monopile foundation, the hybrid foundation provides higher bearing capacity, especially against lateral loads and moment, and more stiffness. This foundation system is schematically shown in Figure 1.6d.

The following paragraphs provide several examples of challenging conditions that were addressed by the foundation engineer.

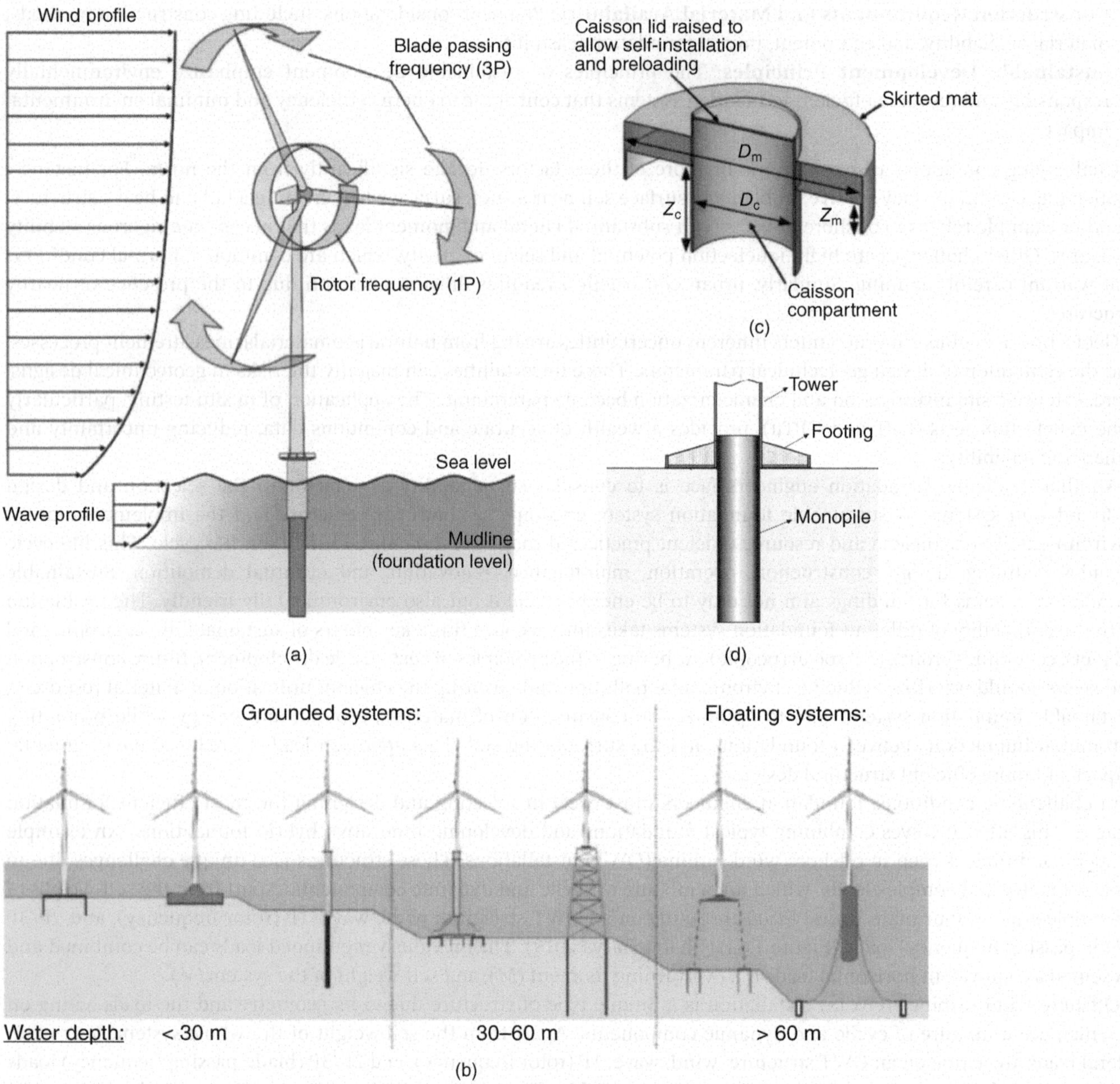

Figure 1.6 (a) Realistic distribution of main loads imposing on OWTs, (b) typical foundations for offshore wind turbines (Bhattacharya 2019 / With permission of John Wiley & Sons), (c) hybrid suction bucket foundations (HSBF) (Gaudin et al. 2011 / International Society of Offshore and Polar Engineers), and (d) monopole combined with footing (Anastasopoulos and Theofilou 2015 / with permission of Elsevier).

1.5.1 Torre Latino Americana in Mexico City

Figure 1.7 provides an example of using combined foundation systems: the floating raft foundation in conjunction with deep foundations and the piled raft combined with basement shear walls.

In a floating box foundation, a substantial portion of the superstructure's load is counterbalanced through buoyancy, with deep foundations introduced to mitigate excessive settlement concerns. It is noteworthy that in certain scenarios where conventional bottom-up construction poses challenges, a top-down construction approach may be recommended. A prominent example of this combined approach is the Torre Latino Americana in Mexico City. This location is known for its Mexico City Clay, which is prone to significant consolidation settlements, often measuring several meters. In this project, the foundation was strategically placed at a depth of 13 m, effectively equating to half of the building's weight. The introduction of deep foundations was instrumental in transferring the loads to deeper and more robust soil layers, thereby ensuring controlled

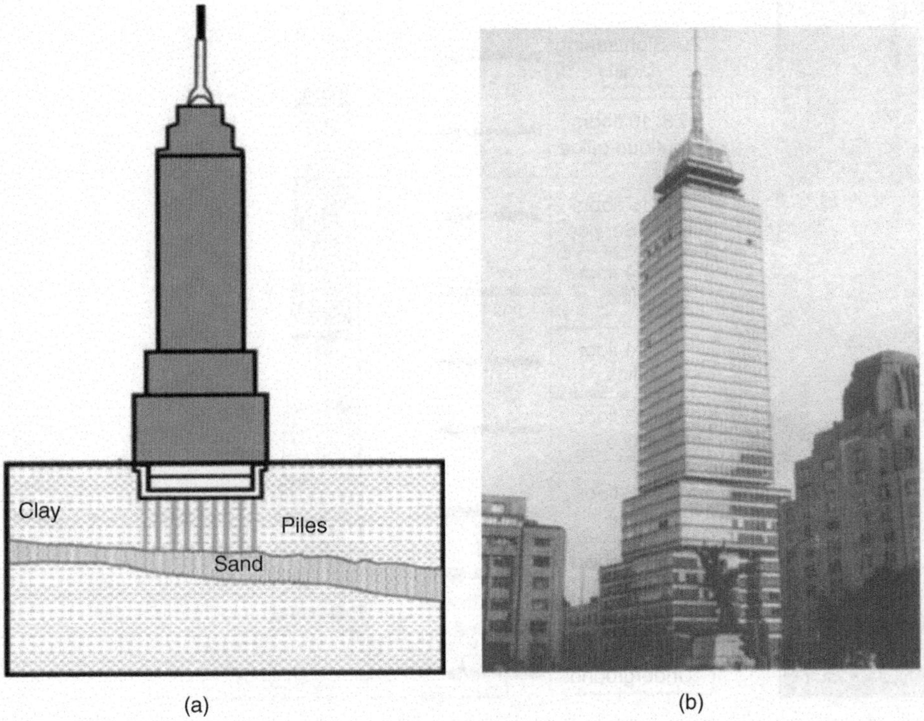

Figure 1.7 Torre Latino Americana: (a) schematic view and foundation system (Brûlé et al. 2019 / Walter de Gruyter GmbH / CC BY 4.0) and (b) actual view (Zeevaert 1957 / Emerald Group Publishing Limited).

settlement levels (Zeevaert 1957). Figure 1.7 offers both a schematic representation of this foundation system and an actual view of the building, providing a clear visualization of the engineering solution employed.

1.5.2 Shanghai Tower

Piled raft foundations are applicable in cases where high bearing capacity and low settlement are required, especially for tall buildings where considerable lateral loads due to wind or earthquake loads, along with an extraordinary vertical load of structural weight, resulting in a combination of loads (VMH). This load combination, which usually is not unique, is one of the main concerns in the design of foundation systems. In some cases, rigid shear walls of basements are connected to the piled raft, aiming to decrease both maximum and differential settlements by increasing the rigidity of the foundation. This combination is recognized as a successful foundation system for high-rise buildings underlain by soft soil. Figure 1.8 depicts views of the superstructure and its combined foundation system. The Shanghai Tower is supported on a piled raft combined with shear walls. Shear walls with a 5-story height are constructed on a 6 m thickness raft supported on about 950 bored deep foundations (Zhu et al. 2012).

1.5.3 Artificial Islands

Soil improvement practices are viewed as complementary solutions alongside foundation implementations. Over time, ground modification methods have evolved from limited depth and scope to encompass more extensive and profound applications, as exemplified by the creation of artificial islands in recent decades. These man-made reclaimed areas, typically situated in nearshore regions, serve various purposes such as residential, industrial, transportation, and leisure. In the construction of artificial islands, two primary stages of soil improvement are typically undertaken: first in the seabed, followed by enhancements in the reclaimed zone.

Many artificial islands with various applications have emerged all over the world. Chubu Centrair (see Figure 1.9a), Kansai Airports in Japan, Rize–Artvin Airport in Turkey, and Dalian Jinzhouwan Airport in China can be listed as artificial islands. Some artificial islands are recognized as historical landmarks, like Treasure Island in San Francisco, USA, near the Golden Gate Bridge (see Figure 1.9b), and for leisure and residential purposes, such as Palm artificial islands in Dubai, UAE.

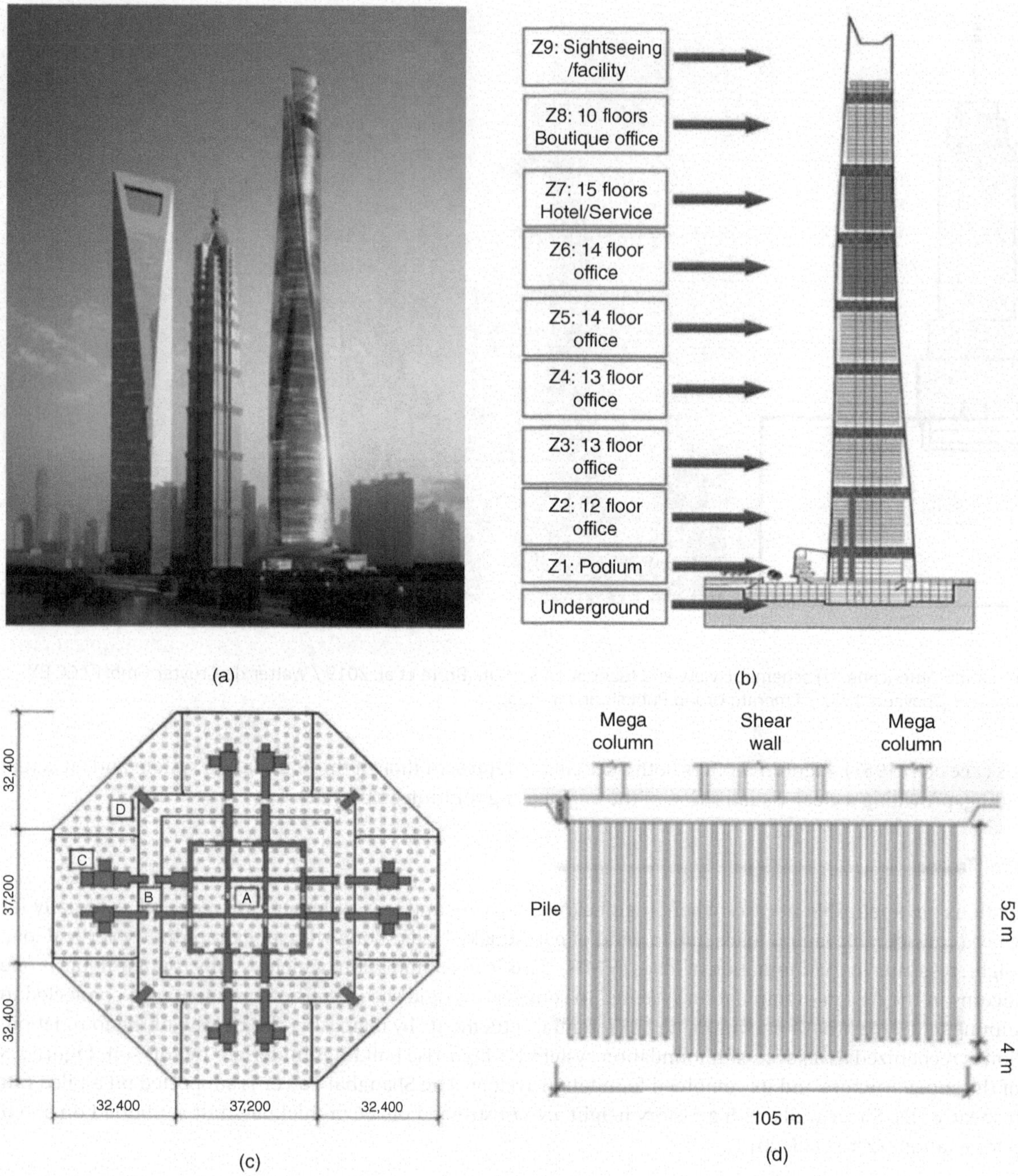

Figure 1.8 Shanghai Tower: (a) perspective view, (b) elevation view (Wang et al. 2021 / with permission of John Wiley & Sons), (c) foundation system plan (Zhao and Liu 2017 / with permission of Korea Institute of Science and Technology Information), and (d) foundation system section.

1.5.4 Retrofitting and Reinforcing

The annual reports on foundation failures, often resulting from uncertainty sources such as extreme events, heavy loading, or human errors during design and construction phases, highlight the advantages of retrofitting and reinforcing existing foundation systems rather than opting for complete demolition and reconstruction. This approach is particularly significant in the context of sustainable development principles. An illustrative example of retrofitting is provided in the case of a concrete pedestrian bridge that failed during a flood event, as documented by Eslami et al. (2016).

In certain scenarios, micropiles are employed to anchor the foundation shaft into the underlying formation, as depicted in Figure 1.10a. This involves drilling the shaft, placing the micropile inside it, and extending it to the desired depth in the

Figure 1.9 Various artificial islands: (a) Chubu Centrair Airport in Japan, Wikipedia (2023a) (Behbeh / Wikimedia Commons / CC BY SA 3.0) and (b) Treasure Island (San Francisco, USA), Wikipedia (2023a) (USDA / Wikimedia Commons / Public Domain).

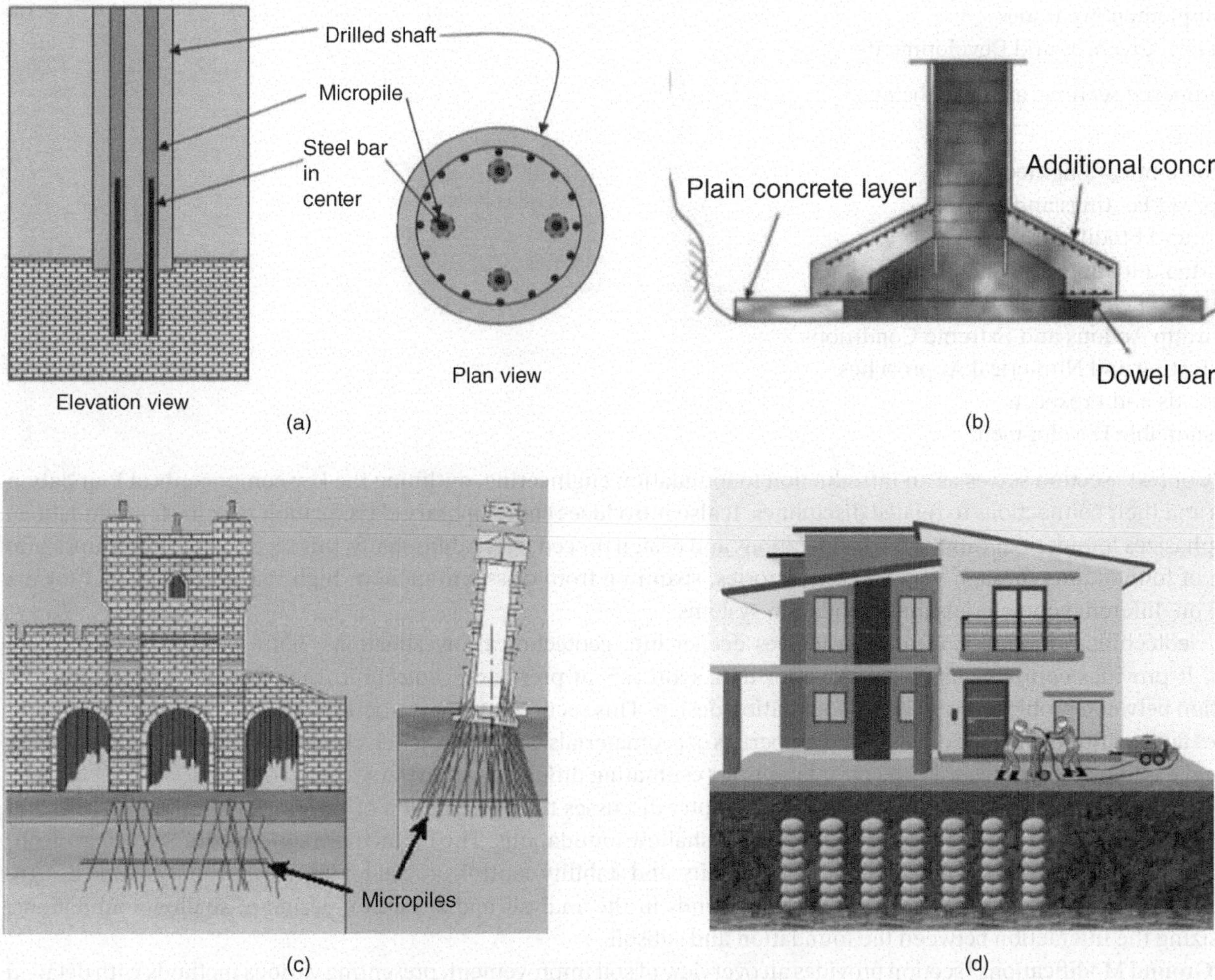

Figure 1.10 (a) Micropiles drilled through an existing drilled shaft (Brown et al. 2010 / U.S. Department of Transportation / Public Domain), (b) increasing the dimensions of an existing footing (Bhattacharjee 2018 / Emerald Publishing Limited), (c) application of micropiles for retrofit, and (d) grouting improvement for leveling and strengthening.

substrate to enhance the structural integrity of the pile. Access to the top of the shaft is a prerequisite for implementing this method (Brown et al. 2010). Another approach to augment the size of an existing footing is illustrated in Figure 1.10b, where both the footing's dimensions and the attached column are simultaneously enlarged.

Figure 1.10c demonstrates the use of micropiles to strengthen existing foundations, particularly in the case of historical buildings. Grouting, depicted in Figure 1.10d, is another common method employed for foundation leveling in adjacent construction scenarios or for enhancing the geotechnical performance of current foundations.

1.6 Book Organization

The current book is organized in the form of 6 topics, 10 sections, and 27 chapters. The topics are as follows:

- A: Background
- B: Shallow Foundations
- C: Semi-Deep Foundations
- D: Deep Foundations
- E: Complementary Issues
- F: Recent Advances and Developments

The proposed sections are listed below:

- I: Context
- II: Geotechnical Engineering
- III: Spread Footings and Mats
- IV: Ground Modifications
- V: Skirted, Floating, and Shell Foundations
- VI: Single Piles and Drilled Shafts
- VII: Group Actions and Extreme Conditions
- VIII: Testing and Numerical Approaches
- IX: Trends and Prospects
- X: Sustainable Development

The "Context" section serves as an introduction to foundation engineering, outlining the key components of foundation systems and their connections to related disciplines. It also introduces contemporary classification systems for foundations and emphasizes foundational analysis considerations and design procedures. Additionally, this section offers a glimpse into the role of foundations through various case histories, spanning from classic to modern high-rise structures and towers situated on different geomaterials and foundation systems.

The "Geotechnical Engineering" section delves deeper into geotechnical investigations, both in terms of depth and breadth. It provides comprehensive guidance for data synthesis in preparing geotechnical reports and underscores the interaction between problematic soils and foundation design. This section further introduces soil and rock parameters and discusses factors influencing the engineering properties of geomaterials. Moreover, it delves into Soil Behavior Classification (SBC) and explores various records and correlations for estimating different parameters.

The "Spread Footings and Mats" section's initial chapter discusses the construction of various foundations on different geomaterials, including nontraditional and specialized shallow foundations. The following chapter addresses key geotechnical requirements, such as bearing capacity, serviceability, and stability control, crucial for foundation analysis and design. The final chapter in this section explores emerging trends in the analysis and structural design of shallow foundations, emphasizing the interaction between the foundation and subsoil.

The "Ground Modifications" section provides an overview of soil improvement, presenting various methods with detailed concepts and calculation procedures through practical examples and case studies. It also focuses on the geotechnical behavior of shallow foundations on stabilized and improved grounds, highlighting primary modification approaches.

The "Skirted, Floating, and Shell Foundations" section is dedicated to semi-deep foundations, featuring diverse types and construction techniques. It also covers semi-deep foundations widely employed in offshore projects, including skirted and bucket foundations. The section then delves into floating foundations, encompassing design and construction criteria, and concludes with an exploration of construction trends and geotechnical and structural design principles for shell foundations.

The "Single Piles and Drilled Shafts" section shifts the focus to single piles and drilled shafts, detailing construction and installation methods and equipment for various types, including specialized ones. It outlines general algorithms for pile design and differentiates pile load transfer and failure mechanisms around the pile toe, utilizing static analysis and in situ testing. This section also addresses performance under lateral loading and structural considerations.

The "Group Actions and Extreme Conditions" section delves into the complexities of group actions in deep foundations. It explores pile groups and piled raft foundations (PRFs), elucidating the fundamentals of bearing capacity under vertical, lateral, and combined loading, along with load–displacement behavior via case studies. The final part of this section addresses performance under extreme conditions, including cyclic and seismic loading, as well as geohazards like liquefaction and scouring.

The "Testing and Numerical Approaches" section introduces two complementary topics: full-scale load testing and numerical analysis. Full-scale load tests offer an efficient approach to assess foundation performance, while model-scale testing plays a vital role in research and practical studies. Numerical analysis is presented as a versatile tool for complex and time-consuming investigations in foundation engineering.

The "Trends and Prospects" section centers on recent trends and prospects in foundation engineering, covering topics such as uncertainty and reliability, the LRFD approach, relevant codes, and data mining-based design. It underscores the importance of considering uncertainty sources in analysis and design and highlights the pivotal role of databases in geotechnical engineering.

The concluding section (Sustainable Development) of the book explores sustainable development through three distinct parts: foundation damages and prevention and repair techniques, foundation selection and value engineering, and reusing existing foundations after demolishing a building. This section provides valuable insights into mitigating foundation issues, optimizing foundation selection, and maximizing the utilization of existing foundations when buildings are demolished.

1.7 Major Points

1) Foundations are structural elements responsible for transferring the imposed loads from a superstructure to the underlying ground.
2) Geosystems represent a fusion of various natural and artificial components and materials assembled to meet specific needs. They encompass a wide spectrum of applications in geotechnical and geological engineering, environmental geotechnics, geohydrology, and applied geophysics.
3) Despite the conventional notion that foundations are limited to the transitional portion of the substructure, it is essential to recognize that the substructure itself can be conceptualized as a comprehensive "foundation system." Such a system encompasses not only the embedded portion of the structure but also the surrounding soil in the lateral and underlying regions of the foundation unit.
4) The primary objective of foundation engineering is to design and execute technically sound, construction-feasible, and cost-effective foundations that offer dependable support for superstructures. This field draws knowledge and expertise from geotechnical engineering, structural engineering, and construction engineering.
5) Key requirements in the analysis and design of foundations include evaluating bearing capacity, ensuring serviceability (addressing settlement and torsion concerns), conducting structural design, verifying stability control, performing full or model scale testing, considering construction-related aspects, assessing durability, and meeting economic constraints.
6) The selection of foundations for a project hinges on several determinant factors, encompassing superstructure and loading, site investigation and characterization, subsoil conditions, environmental and circumferential circumstances, construction requirements, material availability, and sustainable development principles. Challenging conditions arise when one or a combination of these factors deviates significantly from the norm. In such cases, a comprehensive assessment by the foundation engineer that accounts for all these factors is indispensable.
7) The present book is structured around 6 overarching topics, each divided into 10 sections and comprising a total of 27 chapters. The topics cover background information, shallow foundations, semi-deep foundations, deep foundations, supplementary topics, as well as recent advances and developments. The sections delve into various aspects, including the context, geotechnical engineering, spread footings and mats, ground modifications, skirted, floating, and shell foundations, single piles and drilled shafts, group actions and extreme conditions, testing and numerical approaches, trends and prospects, and sustainable development.

References

Anastasopoulos, I. and Theofilou, M. (2015). Hybrid foundation for offshore wind turbines: environmental and seismic loading. *Soil Dynamics and Earthquake Engineering* 80: 192–209.

Bhattacharjee, J. (2018). Deterioration of concrete structures along with case studies in India. *Proceedings of the Institution of Civil Engineers – Forensic Engineering* 171 (2): 80–90. https://doi.org/10.1680/jfoen.17.00010.

Bhattacharya, S. (2019). *Design of Foundations for Offshore Wind Turbines*. Wiley.

Brown, D.A., Turner, J.P., and Castelli, R.J. (2010). Drilled shafts: construction procedures and LRFD design methods. *GEOTECHNICAL ENGINEERING CIRCULAR NO. 10*. Federal Highway Administration (FHWA).

Brûlé, S., Enoch, S., and Guenneau, S. (2019). Role of nanophotonics in the birth of seismic megastructures. *Nanophotonics* 8 (10): 1591–1605.

Coduto, D.P., Kitch, W.A., and Yeung, M.R. (2016). *Foundation Design Principles and Practices*, 3e. Upper Saddle River, NJ: Prentice Hall, Inc.

Eslami, A., Heidarie Golafzani, S., and Jamshidi Chenari, R. (2016). Assessment of Babolsar concrete pedestrian bridge failure for 1964 flood event and retrofitting practice. *Engineering Failure Analysis* 68: 101–112.

Fellenius, B.H. (2023). Basics of foundation design, Electronic Edition. www.Fellenius.net, 432 p (accessed 12 April 2023).

Gaudin, C., Mohr, H., Cassidy, M., et al. (2011). Centrifuge experiments of a hybrid foundation under combined loading. *Conference: International Society of Offshore and Polar Engineers (ISOPE)*, Maui, Hawaii, USA (19–24 June 2011).

Wang, L., Zhao, X., and Liu, S. (2021). A study of differential foundation settlement of piled raft and its effect on the long-term vertical shortening of super-tall buildings. *The Structural Design of Tall and Special Buildings* 30: 1–19. e1832. https://doi.org/10.1002/tal.1832.

Wikipedia (2023a). Chubu Centrair International Airport, https://en.wikipedia.org/wiki/Chubu_Centrair_International_Airport.

Wikipedia (2023b). Treasure Island, San Francisco, https://en.wikipedia.org/wiki/Treasure_Island,_San_Francisco.

Zeevaert, L. (1957). Foundation design and behavior of tower Latino Americana in Mexico City. *Géotechnique* 7: 115–133.

Zhao, X. and Liu, S. (2017). Foundation differential settlement included time-dependent elevation control for super tall structures. *International Journal of High-Rise Buildings* 6 (1): 83–89.

Zhu, Y., Poon, D., Zuo, S., and Fu, G. (2012). Structural design challenges of Shanghai Tower. *CTBUH 2012 9th World Congress*, Shanghai (19–21 September 2012). Chicago: CTBUH.

2

Foundation Systems – Forms and Function

Chapter Highlights

- Process for Engineering of Foundations: Data, Design, and Construction
- Foundation Categorization: Geometry and Construction Aspects
- Intermediate Alternatives
  - Semi-deep Foundations
  - Ground Modification
- Foundation Form Aspects
- Insights on Foundation Systems: Feature and Function
- Overall Procedure for Foundation Design
- Case Histories
  - Horns Rev Offshore Wind Farm, North Sea, Denmark
  - Al Faisaliah Building, Riyadh, Saudi Arabia
  - Milad Tower, Tehran, Iran
  - Torre Latino Americana, Mexico City, Mexico
  - Shanghai Tower, Shanghai, China
  - Oil Storage Tanks, Poti, Georgia

2.1 Process for Engineering of Foundations: Data, Design, and Construction

To ensure proper foundation performance, Burland et al. (2012) have advocated for a cyclic perspective pattern and data feedback approach, as illustrated in Figure 2.1b. This stands in contrast to the conventional, linear rule (Figure 2.1a), which is often insufficient. Recent developments in geotechnical databases have yielded several advantages, including design optimization, the evaluation of current methods, the development of new approaches, and the completion of geotechnical interpretations. Eslami and Heidarie (2021) have introduced a novel approach known as the relevant databased approach for deep foundations. This proposed algorithm aids in comparing the characteristics and conditions of a specific foundation with similar cases stored in the database through screening. Consequently, it facilitates the decision-making process regarding the priority of design methods. Furthermore, in cases where data availability is limited, the approach allows for the replication or judicious selection of necessary information based on available similar data. Ultimately, this approach leads to the utilization of site-specific methods that adhere to a performance-based design while also meeting the requirements of value engineering, assessed through various criteria. A detailed exploration of this proposed approach is presented in Figure 2.1c and will be further discussed in Chapter 25.

2.2 Foundation Categories: Geometry and Construction Aspects

Foundations can be categorized into two main types: shallow and deep foundations. This classification is based on factors such as their plan geometry, embedment depth, load transfer mechanism, and whether they support the structure

Advanced Foundation Engineering: Principles, Performance and Prospects, First Edition. Moses Karakouzian and Abolfazl Eslami.

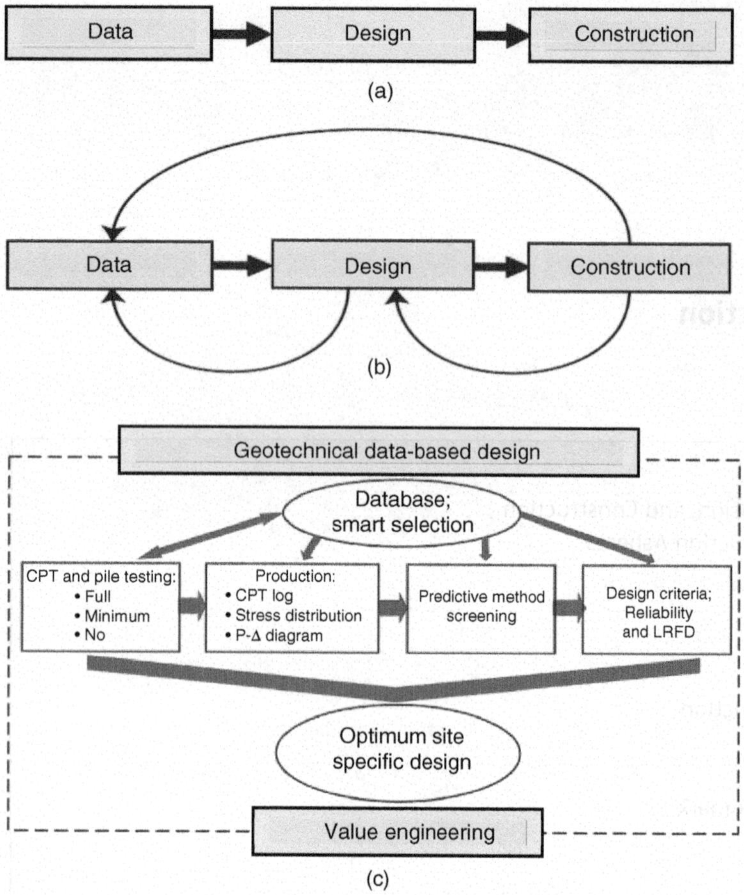

Figure 2.1 Data, design and construction relation: (a) conventional approach; (b) interactive trend; (c) databased approach.

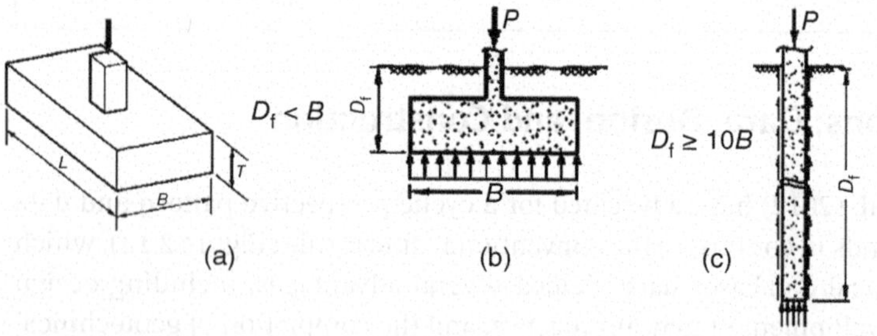

Figure 2.2 (a) General foundation dimensions, (b) schematic view of shallow foundation, and (c) schematic view of deep foundation.

individually or as part of a group. Figure 2.2a shows the typical dimensions of a foundation, including breadth (B), length (L), and thickness (T). Additionally, Figure 2.2b and 2.2c provide schematic representations of the two different foundation types and illustrate the concept of embedment depth (D_f).

Shallow foundations are typically situated close to or at the ground surface, and their primary function is to transfer loads directly to the soil immediately beneath them. The distribution of normal reaction forces relies on the contact area between the foundation's bottom and the soil. Among shallow foundations, the most commonly employed type is the single foundation, which ensures that transmitted loads are spread over a sufficient soil area to maintain soil stress within acceptable limits. Shallow foundations find extensive use for supporting structures with light to moderate loads when the soil conditions are relatively favorable. Single footings are typically employed to support individual columns, and in cases where multiple closely spaced columns are needed to support a wall, they can be combined into continuous or strip footings.

When two-directional strip foundations are used, they form grid foundations, which offer more efficient load distribution compared to a group of individual spread foundations. When applied loads increase or when weaker surface soil conditions

Figure 2.3 Instances of prevalent shallow foundations: (a) combined, (b) grid, (c) attached single foundation, and (d) mat foundation.

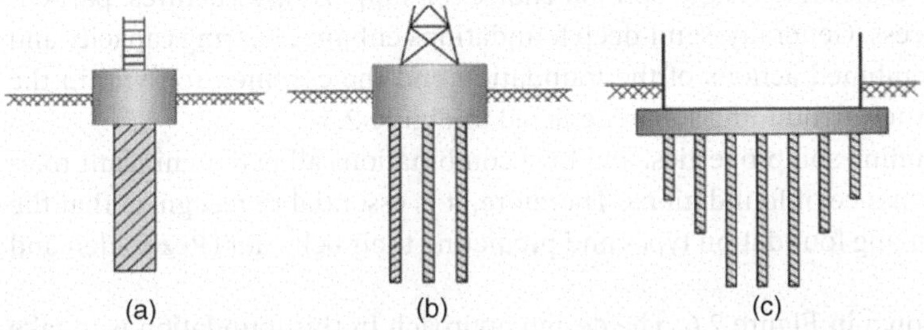

Figure 2.4 Different applications of piles: (a) single shaft or caisson, (b) pile group, and (c) piled raft foundation (PRF).

are present, single and combined footings may merge to cover the majority of the superstructure's footprint, resulting in mat or raft foundations. Compared to other types of shallow foundations, raft foundations offer several advantages, including ease of construction, structural continuity, enhanced rigidity, basement waterproofing, resistance against instability conditions, and the ability to distribute loads over a larger area. Figure 2.3 provides an illustration of various commonly used types of shallow foundations.

Deep foundations, often referred to as piles, serve to transfer applied loads to deeper layers of soil through reactions at their toes and shafts. These structural elements can be either precast components driven into the ground or constructed in place. The utilization of piles becomes imperative in various scenarios, including supporting heavy structures, high-rise buildings, both onshore and offshore structures, bypassing existing weak and compressible soil deposits at or near the ground surface, and ensuring stability of the structure. Given that soil-bearing capacity and stiffness typically increase with depth, a hybrid approach combining piles and the surrounding soil can harness substantial bearing capacity, resulting in minimal settlement and heightened stability. Figure 2.4 provides a visual representation of a typical application of deep foundations. Piles play a crucial role in preventing excessive settlement or lateral movement of structures by securely transmitting the superstructural loads through weak or problematic soil layers into the more competent underlying soils and rocks (Randolph 2003).

2.3 Intermediate Alternatives: Semi-deep Foundations

Shallow foundations are designed to transmit superstructure loads to the underlying soil layers at a relatively small depth (D_f), as compared to the foundation's width (B). They are the preferred foundation type for residential and low-rise buildings having low-level loads. However, they are known to have several shortcomings, including limited bearing capacity, significant settlement, and reduced stability, especially under high-level loads imposed by high-rise buildings, towers, bridges, and marine structures. Accordingly, deep foundations ($D_f/B > 10$) have become the preferred choice for supporting heavy, tall, and critically important structures. However, the widespread application of deep foundations is constrained by factors

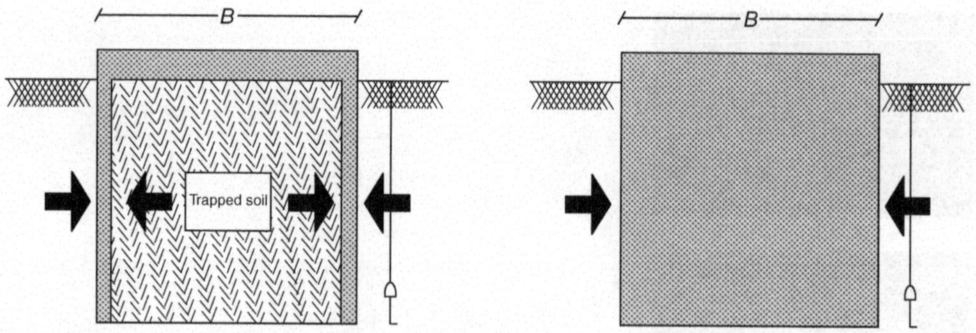

Figure 2.5 Hybrid confinement of soil and foundation.

such as construction equipment, cost, implementation time, and environmental considerations associated with methods like drilled shafts and pile driving. In response, two intermediate types of foundations have emerged as alternatives to shallow and deep foundations. These are semi-deep foundations and foundations based on ground improvements.

Semi-deep foundations represent a distinct foundation category, notable for their performance characteristics. This relatively recent concept is gaining momentum as an increasingly optimal choice for supporting structures, particularly in situations where depths are challenging to access. Generally, semi-deep foundations enhance bearing capacity and improve load–displacement behavior through the combined actions of the foundation and the confined soil within the foundation, or the interaction of the foundation and the surrounding soil as depicted in Figure 2.5.

The shape, dimensions, embedment depth, surrounding soil properties, and load combinations all play significant roles in influencing the geotechnical and structural performance of foundations. Therefore, it is essential to recognize that the D/B factor should not be the sole criterion for categorizing foundation types and predicting their behavior (Rezazadeh and Eslami 2017).

An example of a novel semi-deep foundation is shown in Figure 2.6. The design approach in this foundation is to take advantage of a pair of the near-surface competent caliche layers. The competent caliche layers are bonded to the top and bottom of a continuous flight auger pile as a composite foundation system named caliche stiffened pile (CSP). The key objective is to optimize the required pile length in a cost-effective approach without ameliorating soil properties (Stone et al. 2023).

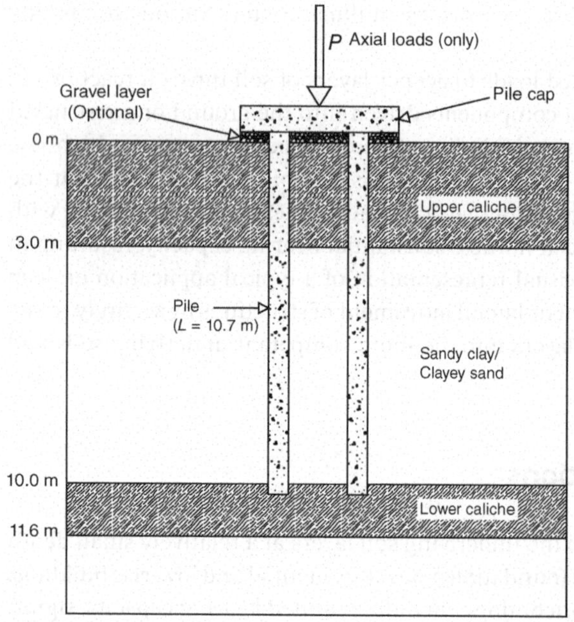

Figure 2.6 Caliche stiffened pile foundation system (Stone et al. 2023 / Canadian Science Publishing).

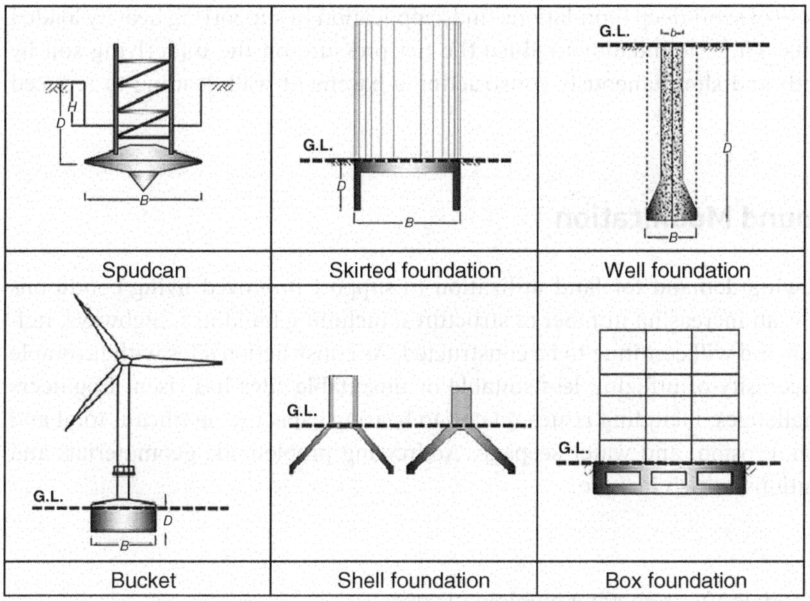

Figure 2.7 Various types of semi-deep foundations (Eslami et al. 2020 / with permission of Elsevier).

Semi-deep foundations are emerging as an increasingly optimal choice for supporting structures, particularly in situations where accessing significant depths is challenging. As shown in Figure 2.7, various types of semi-deep foundations include:

- Spudcan
- Skirted foundation
- Well foundation
- Bucket foundation or suction caisson
- Shell foundation
- Floating and box foundation

Following are short descriptions of some of the semi-deep foundations. Expanded discussions of semi-deep foundations are presented in Chapters 10–13.

- **Spudcan Foundations:** Spudcans are the prevalent foundations for jack-up platforms. These spudcans are typically circular in plan and feature shallow conical undersides, typically inclined at angles ranging from 15° to 30° to the horizontal.
- **Skirted Foundations:** Skirted foundations are continuous circular foundations with thin skirts encircling their perimeter. These skirts serve multiple functions, including confining soft surface soils, transmitting loads to the more stable underlying layers to resist lateral loads and moments, and enhancing protection against piping and scour. They also prevent soil beneath the foundation from squeezing out and protect against potential damage during adjacent construction excavations. The ease of installation makes them economically attractive, especially in areas with challenges related to pile construction. Additionally, they can transmit loads through end resistance at the tip level and shaft friction alongside the skirt (Rezazadeh and Eslami 2017).
- **Well Foundations:** Well foundations resemble under-reamed drilled shafts. When dealing with homogeneous stiff clay, hardpan, or soft cohesive rock at shallow depths, under-reamed foundations offer a cost-effective solution.
- **Bucket Foundation (Suction Caisson):** Bucket foundations serve as cost-effective substructures for wind turbines and offshore oil and gas facilities, providing embedded support.
- **Shell Foundations:** Certain types of foundation structures, such as the inverted arc foundation or brick arch foundations, can be considered as a form of shell foundation. The inverted arc foundation, which resembles an upside-down shell or arch, can be regarded as a shell foundation due to its curved shape and the forces it generates within the structure. Shells can provide structural strength, load distribution, and stability to the foundation system, particularly in challenging soil conditions or areas prone to settlement.

- **Floating and Box Foundations:** This category of semi-deep foundations finds application in supporting heavily loaded structures situated on soft and weak deposits. These foundations reduce the net pressure on the underlying soil by increasing the foundation's embedment depth and simultaneously constructing a basement wall, leading to reduced settlement.

2.4 Intermediate Alternatives: Ground Modification

With the ongoing urbanization, there is a growing demand for land utilization to support improved living conditions and transportation infrastructure. Consequently, an increasing number of structures, including buildings, highways, railways, tunnels, levees, and earth dams, have been and will continue to be constructed. As construction sites with favorable geotechnical conditions become scarcer, the necessity of utilizing less suitable or unsuitable sites has risen. Engineers are confronted with mounting geotechnical challenges, including issues related to bearing capacity, significant total and differential settlements, instability, liquefaction, erosion, and water seepage. Addressing problematic geomaterials and geotechnical conditions requires a range of solutions, which include:

- Avoiding and altering the site.
- Designing superstructures accordingly.
- Replacing better and non-problematic geomaterials after removing problematic ones.
- Improving geomaterial properties and geotechnical conditions.

When superstructures are built on the ground, there are four foundation options for bearing on the natural ground: replaced surface deposits, compacted/consolidated layer, composite ground, and piles to deeper layers.

Ground modifications are performed for the following partial list of purposes:

- Increasing bearing capacity.
- Decreasing erosion potential and subsequent distortion of the structure.
- Reducing compressibility and minimizing settlements.
- Controlling shrinkage, swell, and reducing permeability.
- Redirecting seepage to eliminate or decrease water pressure.
- Preventing deleterious physical or chemical attack.
- Minimizing liquefaction potential susceptibility.
- Minimizing the effect of variability of soils.

Different methods and techniques are widely used for ground modifications, which can be classified into:

- Earthworks
- Densification
- Physical and chemical modification
- Hydraulic control for ground modification
- Reinforcement
- New trends; bioremediation (microorganisms and plant engineering)

Alternatively, soil improvement methods can also be classified into several categories based on different criteria, including the depth of enhancement (shallow and deep methods), the timing in relation to construction (pre-construction, during construction, and post-construction methods), time-dependent procedures, installation techniques (extrusion and intrusion), and the speed of operation (fast and slow methods).

In the following sections, we will provide brief descriptions of various categories of ground modification methods. For more detailed information on soil improvement methods, please refer to Chapter 8.

2.4.1 Earthworks

Traditional ground improvement methods often involve over-excavation and replacement, a practice that remains widely used. This approach centers on the removal of problematic geomaterials, replacing them with more stable fills, typically utilizing materials like rock, gravel, or sand.

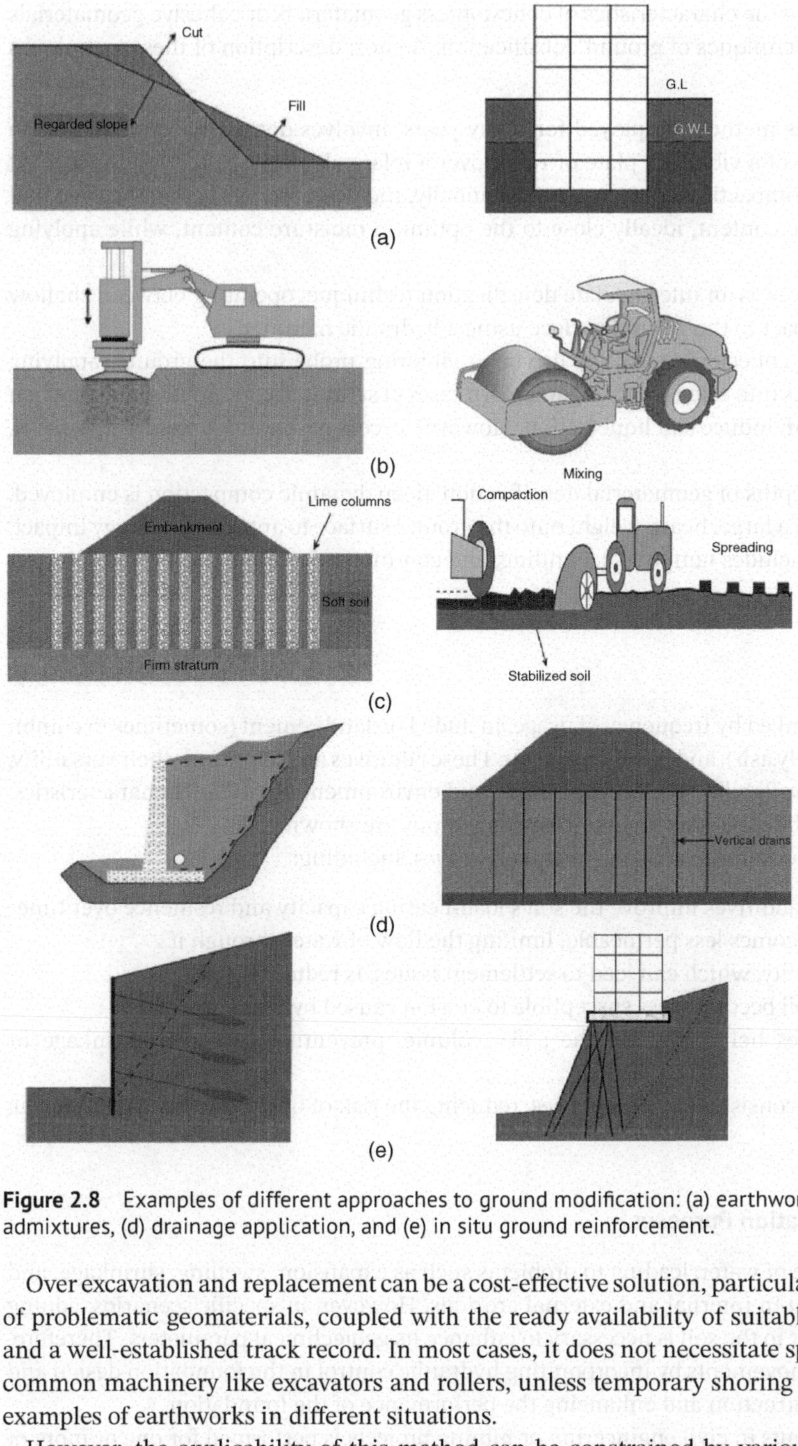

Figure 2.8 Examples of different approaches to ground modification: (a) earthwork, (b) compaction, (c) ground modification with admixtures, (d) drainage application, and (e) in situ ground reinforcement.

Over excavation and replacement can be a cost-effective solution, particularly when dealing with limited areas and depths of problematic geomaterials, coupled with the ready availability of suitable fill materials. It boasts simplicity, reliability, and a well-established track record. In most cases, it does not necessitate specialized contractors or equipment, except for common machinery like excavators and rollers, unless temporary shoring and dewatering are needed. Figure 2.8a shows examples of earthworks in different situations.

However, the applicability of this method can be constrained by various site-specific conditions. Factors such as the need for extensive excavation, a high groundwater table, the presence of existing on-site or nearby structures and utility lines, restricted access for trucks on the site, long-distance hauling of fill materials, challenges associated with disposing of problematic geomaterials, and project timelines can all impact its feasibility.

2.4.2 Densification

Both shallow and deep densification methods have gained widespread use for improving geomaterial properties near the surface and at greater depths. These techniques involve applying vibrations, pressure, kneading, and impact to the ground surface. They have proven effective in enhancing the characteristics of cohesionless geomaterials or cohesive geomaterials with low plasticity. Figure 2.8b shows typical techniques of ground densification. A short description of these techniques is provided below.

- **Routine Plate or Roller Compaction:** This method, employed for many years, involves densifying geomaterials to shallow depths through the repeated passage of a vibratory plate or roller over a relatively thin lift. To achieve optimal densification, it is crucial to select suitable compaction equipment. Additionally, the geomaterials should be prepared at the appropriate lift thickness and moisture content, ideally close to the optimum moisture content, while applying sufficient compaction energy.
- **Rapid Impact Compaction (RIC):** RIC serves as an intermediate densification technique, operating between shallow and deep densification. It swiftly delivers impact to the ground surface using a hydraulic hammer.
- **Vibro Compaction:** This method compacts cohesionless soil by driving a vibrating probe into the ground, applying lateral vibratory forces that reorganize particles into a denser arrangement. In cases of saturated cohesionless soil or when water is injected into the ground, vibration can induce soil liquefaction. However, excess pore water pressure dissipates, leading to soil densification.
- **Deep Dynamic Compaction:** For greater depths of geomaterial densification, deep dynamic compaction is employed. This process involves the repeated dropping of a large, heavy weight onto the ground surface to apply high-energy impact. Dynamic replacement, a related technique, includes tamping, backfilling, and continuous tamping until stone columns are formed.

2.4.3 Physical and Chemical Stabilization

The most commonly used artificial additives, ranked by frequency of usage, include Portland cement (sometimes in combination with fly ash), lime (often combined with fly ash), and bitumen and tar. These additives are favored for their versatility, wide applicability to various soil types, ready availability, cost-effectiveness, and environmentally friendly characteristics. In Figure 2.8c, examples of physical ground stabilization using lime-treated columns are shown.

The purpose of mixing these additives into the soil is to achieve several objectives, including:

- **Enhancing Strength and Durability:** The additives improve the soil's load-bearing capacity and resilience over time.
- **Reducing Permeability:** The treated soil becomes less permeable, limiting the flow of water through it.
- **Minimizing Deformability:** Soil deformability, which can lead to settlement issues, is reduced.
- **Enhancing Erosion Resistance:** Treated soil becomes less susceptible to erosion caused by wind or water.
- **Providing Volume Stability:** The additives help maintain the soil's volume, preventing excessive shrinkage or expansion.
- **Controlling Variability:** They ensure more consistent soil properties, reducing the risk of unexpected behavior during construction or over time.

2.4.4 Hydraulic Control for Ground Modification Purposes

Soil-related issues often stem from the presence of water, leading to problems such as expansion, swelling, shrinkage, and liquefaction. Additionally, water flow can result in internal and external erosions. However, in specific scenarios, adding controlled and predetermined amounts of water to the soil is necessary to enhance its geotechnical parameters. Therefore, it is crucial to control the groundwater and its movements by incorporating hydraulic control in the foundation design and construction for the purpose of facilitating construction and enhancing the performance of the foundation.

Controlling the groundwater and its movements in civil engineering or mining projects is performed for one or more of the following reasons:

- Provide a dry working area.
- Stabilize constructed or natural slopes.
- Decrease lateral pressures on foundations or retaining structures.
- Decrease the compressibility of granular soils.
- Increase the bearing capacity of foundations.

- Enhance the workability or hauling characteristics of borrow materials.
- Prevent liquefaction potential during earthquakes.
- Eliminate soil particle movement by groundwater.
- Prevent surface erosion.
- Prevent or decrease damages due to frost heave.

Drains serve multiple purposes in various applications. Figure 2.8d shows an example of horizontal drainage application at the back of a retaining wall and vertical drains under an embankment. Drains serve multiple purposes in various applications. Here are a few examples:

- **Retaining Wall Drains:** These drains are designed to alleviate lateral earth pressures on the facing of retaining walls, reducing the risk of wall failure.
- **Chimney Drains:** Typically installed in dams or levees, chimney drains help lower water heads between upper and lower streams, contributing to stability and safety.
- **Vertical Drains:** These drains are employed to expedite the dissipation of excess pore water pressure within soft soils, aiding in soil consolidation and strengthening.

2.4.5 Reinforcement

In situ ground reinforcement is employed to stabilize unstable ground caused by natural or human-induced changes in geotechnical conditions. Various methods, such as ground anchors, soil nails, micro-piles, and slope-stabilizing piles, are utilized for in situ ground reinforcement to address initial challenges, as depicted in Figure 2.8e. Moreover, steel or plastic geogrids are used to reinforce the soil, with the reinforcement's effectiveness governed by factors like tensile strength and bond development with the surrounding soil. The presence of reinforcement significantly enhances the mechanical properties of granular soil, and different modes of failure can be observed based on the confining pressure.

2.4.6 New Trends; Bioremediation and Plant Engineering

The concept of utilizing biological processes for soil improvement, known as the bio-mediated ground modification technique, has shown significant potential in geotechnical engineering applications, both in terms of performance and environmental sustainability. Bio-mediated soil improvement generally involves the biochemical reaction within a soil mass, resulting in the formation of calcite precipitate to modify the soil's engineering properties. Microbially induced calcite precipitation (MICP) is an interdisciplinary field of study combining engineering, chemistry, and microbiology. MICP can be applied to improve the efficiency of in situ bioremediation (Rajasekar et al. 2017). Encouraging outcomes from studies have highlighted the substantial potential for broader applications within geotechnical engineering techniques.

Bio-mediated ground modification represents an innovative and novel approach that can effectively mitigate issues like liquefaction and landslides in loose sand, which can otherwise lead to foundation deformation or failure. This promising technique has demonstrated its effectiveness in various applications, including enhancing shear strength, reducing soil permeability, improving the strength and durability of concrete and mortar, remediating cracks in buildings, enhancing soil engineering properties, and cementation of sand columns (Umar et al. 2016).

Plant engineering or stabilization involves the use of vegetation, such as grass, bushes, and trees, either alone or in conjunction with inert materials like steel, concrete, and rocks, to enhance soil stability. Plant engineering techniques can be categorized based on site conditions and intended purposes, encompassing surface protection for erosion control, stabilization using plants, and methods that combine plants with inert materials (Punetha et al. 2019).

Utilizing vegetation for slope stabilization, particularly shallow surface materials, has some significant advantages, including:

- Interception of rainfall by foliage (including evaporative losses).
- Reduction of soil moisture and increase of soil suction by uptake from plant root systems and transpiration.
- Physical soil reinforcement by root systems.
- Reduction of sloughing and loss of loose surface materials due to "catchment" by shrubs and trees.
- Stabilization by buttressing and arching between adjacent trees.

2.5 Foundation Form Aspects

Structures can be classified in various ways, including by type, structural system, application, material, element, and form. When it comes to categorizing structures based on their form, one prevalent approach involves dividing them into four general categories. The first category encompasses one-dimensional structures, which include items such as ropes, cables, struts, columns, beams, and arches. The second category encompasses two-dimensional structures, comprising membranes, plates, slabs, shells, vaults, domes, synclastic, and anticlastic forms. The third category comprises three-dimensional structures, which include solid masses. The final category encompasses structures that do not fit into any of the previous categories; these are often combinations of the aforementioned structural forms.

Foundations, like structures, can also be categorized based on their geometry. Three primary foundation categories are examined: one-dimensional (1D) or linear foundations, two-dimensional (2D) or planar foundations, and three-dimensional (3D) or volumetric foundations (Ebrahimipour et al. 2023; Eslami 2022). Further details are provided below.

- **One-Dimensional or Linear Foundations:** These foundations have one dimension that is significantly larger than the other two dimensions. Typically, they transmit loads along their central axis to deeper layers of the subsoil. Due to their substantial depth, concerns about bearing capacity, stability, or settlement are minimal. However, construction and economic challenges may arise. Notable examples of this foundation category include various types of single piles, single piers and caissons, and micro-piles. Figure 2.9a illustrates an example of this foundation type.
- **Two-Dimensional or Planar Foundations:** In these foundations, one dimension (thickness) is smaller than the other two dimensions, giving them a plate-like shape. These foundations can be implemented horizontally, and in some cases, vertically or at an inclined orientation. Consequently, they are divided into two subcategories: horizontal planar foundations and non-horizontal planar foundations. Mats represent a common example of the first subcategory, while shell foundations align with the characteristics of the second. These footings typically require more detailed monitoring of stability and settlement. However, they offer sufficient bearing capacity and reasonable construction costs. Figure 2.9b provides an illustration of this foundation type.
- **Three-Dimensional or Volumetric (Massive) Foundations:** These foundations are significant in all three dimensions and exhibit behavior akin to hybrid structures. There is substantial interaction between the soil and the foundation in these cases. Consequently, they possess a high-bearing capacity, reduced concerns regarding stability, and limited settlement due to their strong connection with the soil. Examples of such foundations include piled raft foundations (PRFs) and box foundations. Figure 2.9c provides visual representations of this foundation category.

Figure 2.10 illustrates the functioning of the aforementioned foundation systems under applied loads. Linear foundations, shown in Figure 2.10a, primarily bear loads through shaft (side) and toe resistance, making them ideal for transferring axial loads. In planar foundations, base resistance is a crucial component. Shell foundations, considered planar, leverage their geometric form to enhance the bearing process and improve performance, as depicted in Figure 2.10b. Volumetric

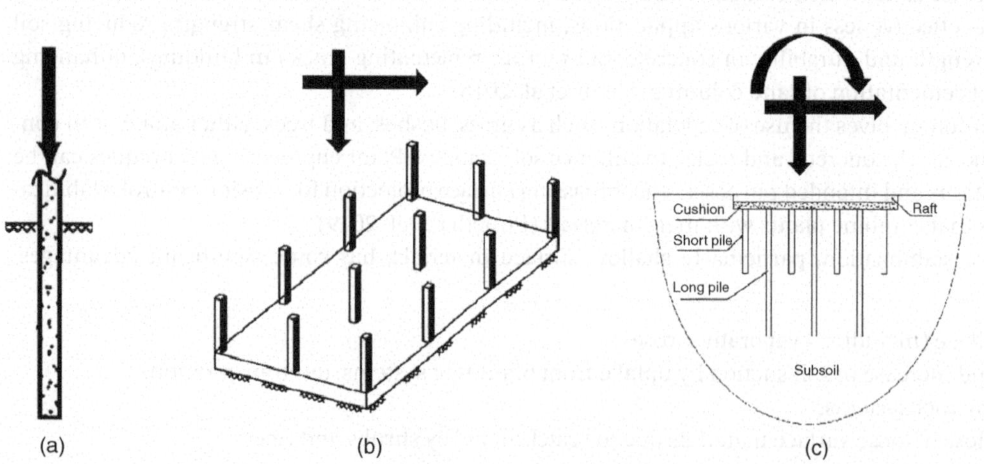

Figure 2.9 Typical categories of foundations based on form and prevalent applied loads: (a) one dimensional, (b) two dimensional, and (c) three dimensional.

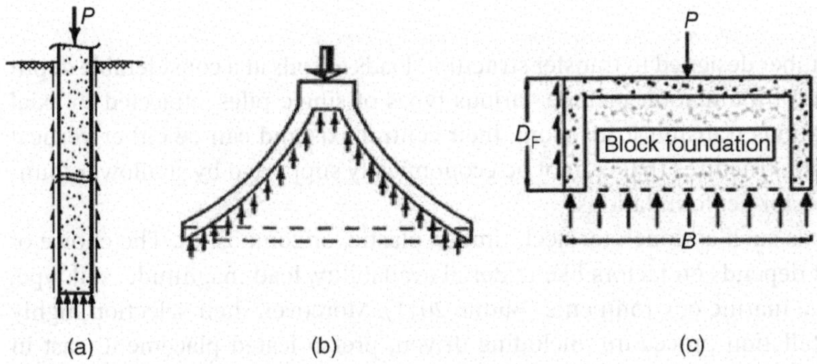

Figure 2.10 Function of various types of foundations regarding form and load transfer: (a) linear foundation, (b) planar foundation, and (c) volumetric (massive) foundation.

or massive foundations, forming a geosystem (a combination of soil and foundation unit), resist a combination of loads through toe and shaft resistance, Figure 2.10c. The interactive performance of various foundation elements, forming a cohesive block, ensures effective support under diverse superstructure and soil conditions.

2.6 Insights on Foundation Systems; Features and Functions

From a structural system perspective, structures can be categorized into different types, including form-act, vector-act, section-act, surface-act, and hybrid systems. Following are short descriptions of each of these systems:

- **Form-Act Structure Systems:** These systems are characterized by their non-rigidity and flexibility, taking on various shapes and being anchored by fixed ends. They support both themselves and the surrounding space. Ancestors of form-act structures include vertical hanger cables, which transmit loads directly to their point of suspension, and vertical columns that transfer loads in the opposite direction to their base point.
- **Vector-Act Structure Systems:** In this system, structures consist of short, solid, straight-line elements that can only transmit forces along their length due to their small cross-sectional area relative to their length. Trusses are prime examples of vector-act structures.
- **Section-Act Structure Systems:** These systems involve linear elements that are straight and maintain their length. If these elements possess material strength, they can serve structural purposes. With compressive strength, they function as compression members, with tensile strength as tension members, and if they also exhibit definite bending rigidity, they can be employed as beams.
- **Surface-Act Structure Systems:** Structural surfaces are used to create mechanisms that redirect forces. The key characteristics of surface-act structures are the structural continuity of elements in two axes, providing resistance against compressive, tensile, and shear stresses. Proper shaping of these surfaces redirects acting forces and distributes them evenly as small unit stresses across the surface.
- **Hybrid Structure Systems:** These systems arise when two distinct structural systems with similar bearing functions combine to form a single operational structure with new mechanical properties. The precondition for a hybrid system is that the two parent systems are equally capable of bearing loads (Engel 2013).

Just as structures can be categorized based on their structural systems, foundations can also be classified similarly. There are five primary categories of foundations:

- **Section-Act Foundations:** These comprise foundations like mats or grid foundations.
- **Vector-Act Foundations:** Examples include single piles and micro-piles.
- **Surface-Act Foundations:** Shell foundations are an instance of this category.
- **Block Act (Hybrid) Foundations:** This group includes bucket foundations, box foundations, pile raft foundations, and pile group foundations.

The following subsections describe each of these foundation systems.

2.6.1 Vector-Act Foundations

A vector-act foundation is a slender structural member designed to transfer structural loads to soils at a considerable depth beneath a structure's base. Examples of this type of foundation include various types of single piles subjected to axial loading, as well as micro-piles. Vector-act foundations transmit loads along their central axis and can be either vertical or inclined. They are typically employed to support structures that cannot be economically supported by shallower foundations. Figure 2.11 provides visual examples of vector-act foundations.

Vector-act foundations are crafted from materials such as concrete, steel, timber, plastic, or composites. The choice of vector-act foundation type for a particular project depends on factors like material availability, load magnitude, soil type, and environmental conditions, including corrosive marine environments (Budho 2011). Moreover, their selection highly depends on the categorization based on the installation procedure, including driven, prefabricated placement, cast in place (bored), drilled displacement (DD), bottom driving, continuous flight auger (CFA), and implemented via torque or vibration. More details about the types and construction of these foundations are discussed in Chapter 14.

In cases where upper soil layers are compressible and weak, vector-act foundations provide increased bearing capacity by transmitting loads to deeper, more robust soil layers. This, in turn, reduces surface layer stress and minimizes settlement, making it feasible to undertake substantial projects on weaker soils.

In some cases, the foundation is subjected to uplifting forces, and vector-act foundations are commonly used to resist the uplifting force due to their higher depth and significant mobilized shaft resistance riverine or onshore structures rely on. Bridge abutments and piers are mainly supported on vector-act foundations due to scouring phenomena to avoid the risk of losing the bearing capacity, excessive settlement, and instability during flooding. One of the major benefits of these foundations is their effective application in both onshore and offshore projects.

In cases where existing active or difficult soils as well as the risk of liquefaction in the surface layers are high, the use of vector-act foundations, along with crossing the layer is highly well-practiced. Suppose the superstructure is high, and the effect of lateral loads, such as earthquakes and wind loads, is noticeable. Vector-act foundations increase the resistant moments and improve the structure stability against sliding, uplift, and overturning loads. If the structure is under horizontal forces, vector-act foundations resist bending, while supporting the vertical loads. This condition is usually encountered in designing and constructing geostructures and foundations of tall buildings subjected to extreme winds or earthquake forces (Das and Sivakugan 2019).

Nevertheless, despite their widespread use in substantial, tall, and significant structures, the utilization of vector-act foundations is curtailed by factors such as construction equipment requirements, costs, implementation time, and environmental considerations. The performance of these foundations has been evaluated comprehensively by Fellenius (2023).

2.6.2 Section-Act Foundations

These foundations encompass various types of shallow foundations, including spread footings, grid foundations, and at the ultimate condition, mat or raft foundations. They rely on the reaction of the subsoil to support the impending loads and are among the most commonly used foundation types worldwide. Figure 2.12 provides visual examples of section-act foundations.

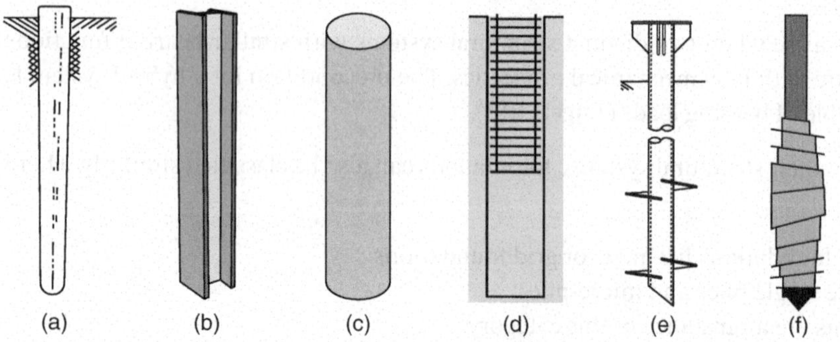

Figure 2.11 Examples of vector-act foundations: (a) wooden pile, (b) steel H-pile, (c) pipe pile (hollow or composite), (d) drilled shaft, (e) helical pile, and (f) Omega DDP.

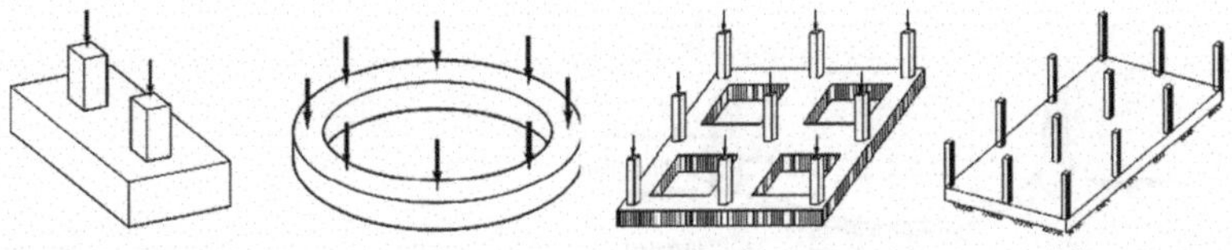

Figure 2.12 Typical section-act foundations: (a) ring, (b) combined, (c) grid, and (d) mat.

One of the most considerable parameters that affect a section-act foundation's performance is its rigidity. Fully flexible foundations are bent to achieve a uniform pressure under them, while completely rigid foundations settle uniformly, but have a nonuniform distribution of stress. Based on some characteristics of a foundation, it can be considered rigid or flexible. Increasing the thickness of the foundation, the elasticity modulus of the concrete (E_c), the rigidity of the superstructure, and the foundation's second moment of area (I_F) lead to more rigidity. On the other hand, increasing the coefficient of subgrade reaction (K_s), the distance between two consequent columns, elasticity modules of the subsoil (E_s), and the difference between loads of the consequent columns underlying by the foundation lead to less rigidity (Eslami 2013).

For determining internal forces, the governing factors are the applied loading pattern, foundation rigidity combined with superstructure plus underneath soil condition. This complex pattern requires advanced tools and methods especially realizing soil–structure interaction through numerical modeling. Despite all the advantages these foundations have, large section-act poses problems in analysis, design, and ultimately, implementation. The analysis methods and designs of these bases are not ordinary and straightforward.

Furthermore, it should be mentioned that the settlement, either immediately or through consolidation of these foundations, is significant due to their large dimensions. In addition, because large section-act foundations stress a relatively large volume of influenced subsoil, they have more settlement than other foundations. The settlement estimation of large section-act foundations is more complicated than the determination of their bearing capacity. The settlement depends on the foundation rigidity, foundation type, soil type, soil homogeneity, groundwater condition, and construction. Previously published works on large section-act foundations have stressed concerns about the temperature buildup due to large volumes of concrete hydration (Hemsley 2000).

2.6.3 Surface-Act Foundations

Surface-act foundations, often referred to as "shell foundations," come in various shapes, including inverted cones and semispheres, which will be explained further in Chapter 13.

From a mathematical and geometrical perspective, surface-act foundations are surfaces that do not lie on the same plane. However, in engineering terms, they possess a structural design that derives its load-bearing capacity from its shape rather than its mass. Much like the functionality of a shell, a surface-act foundation transfers the load to the ground primarily due to its specific shape, not the mass or volume it encompasses. Consequently, the design of these foundations allows for maximum performance to be achieved using minimal materials, resulting in an economic advantage. Furthermore, the geometry of these foundations is strategically designed to internally distribute incoming loads, thereby increasing the load transfer area to the soil beneath. This design feature enhances bearing capacity and effectively manages settlement, further reinforcing their economic efficiency. Consequently, these foundations have become an ideal choice in the field of foundation engineering (Kurian 2006).

Surface-act foundations exhibit great diversity in their shapes. Some of the more commonly encountered shapes for these foundations are cone or inverted cone, inverted dome, hyperbolic paraboloid, elliptic paraboloid, folded plate, and combinations of these forms. Figure 2.13 provides schematic representations of these various foundation shapes.

2.6.4 Block-Act (Hybrid) Foundations

These foundations represent comprehensive solutions in terms of soil and foundation interaction, especially in challenging environments. Notably, semi-deep foundations tailored for marine applications, such as spudcans, buckets, and skirted foundations, have gained prominence. In inland projects, box foundations, PRFs, and foundations on massive, stabilized grounds are common choices.

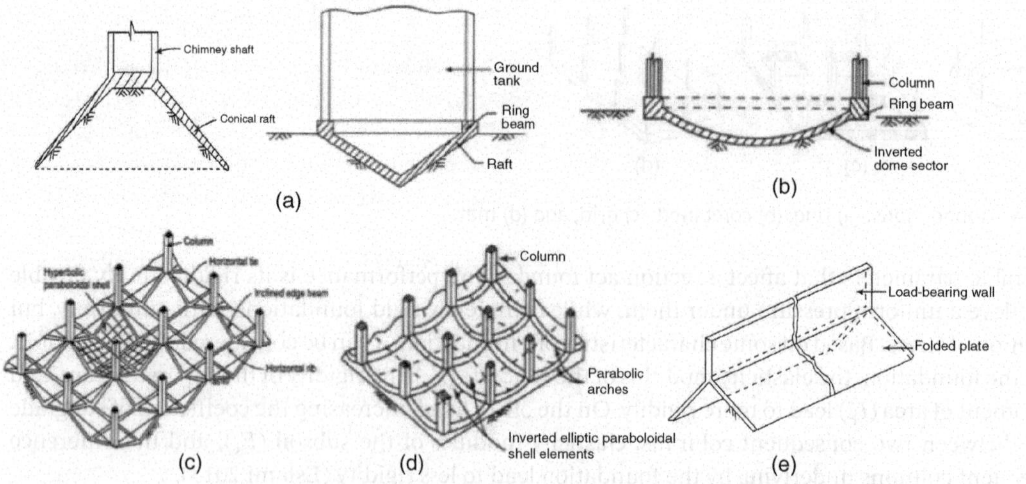

Figure 2.13 Different shapes of surface-act foundations: (a) cone and inverted cone, (b) inverted dome, (c) hyperbolic paraboloid, (d) elliptic paraboloid, and (e) folded plate (Kurian 2006 / Alpha Science International Limited).

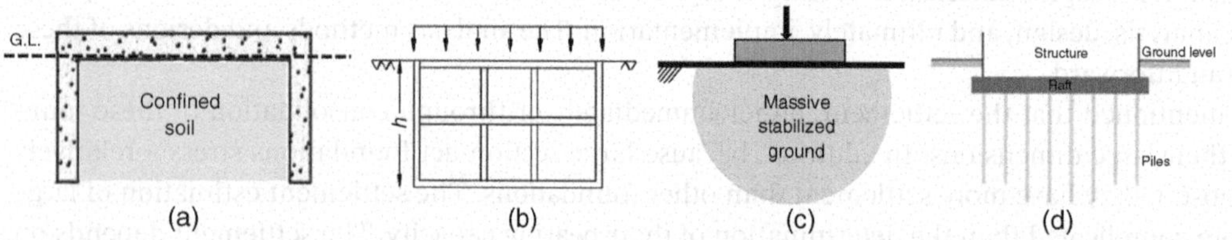

Figure 2.14 Various types of hybrid foundations: (a) skirted foundation, (b) box foundation, (c) massive stabilized ground, and (d) piled raft foundation.

Bucket foundations serve as embedded foundations and cost-efficient substructures for wind turbines and offshore oil and gas facilities (Ibsen et al. 2015). An illustration of bucket foundations is provided in Figure 2.14a.

Skirted foundations are continuous semi-deep circular foundations featuring thin skirts encircling their circumference. The primary function of these skirts is to confine soft surface soils and efficiently transfer loads, ensuring resistance against sliding, overturning, piping, and scour. They achieve load transfer through end resistance at the base level and shaft friction alongside the skirt (Rezazadeh and Eslami 2017).

Box foundations are employed in scenarios where weak layers extend to significant depths, making deep foundations unfeasible. Placing the foundation at lower depths and excavating the soil below it counteracts the superstructure's gravity load, minimizing settlement. For massive structures of utmost importance, the foundation system must possess exceptional flexural stiffness. This can be achieved by combining floor and ceiling slabs with walls both around and within the basement. An example of a box foundation is depicted in Figure 2.14b.

In some instances, a retrofitted or reinforced soil mass can function as a block-act foundation. Techniques like deep soil mixing, grouting, or implementing stone columns enhance subsoil-bearing capacity or reduce settlement, forming a cohesive block foundation, as schematically shown in Figure 2.14c.

Another example of these foundations is PRFs, consisting of a raft foundation and a group of piles. The raft behaves akin to section-act foundations, while the piles act as vector-act foundations. In cases where the failure mechanism involves blockage, the entire mass, including piles and interlocked soil, acts as a unified unit, significantly contributing to the required stability. Figure 2.14d illustrates a schematic of a PRF.

2.7 Overall Procedure for Foundation Design

The process of foundation design is well established, and generally involves the following steps (Poulos et al. 2017; Viggiani et al. 2014):

1) Compiling the geological and hydrogeological evidence and information about the site area.

2) Planning and implementation of the site investigation to recognize site stratigraphy and variability.
3) In situ testing to recognize the appropriate engineering properties of the proposed layers. Laboratory testing supplements the in situ testing and obtains more detailed information on the layer behavior than may be possible within in situ testing.
4) Creation of a geotechnical model for the site, incorporating the key strata and their engineering properties. In some cases, where ground conditions are variable, several models for different site parts may be necessary to consider the site variability.
5) Synthesis of geological and geotechnical evidence; available information on the subsoil; planning, execution, and interpretation of the in situ testing site and laboratory investigations on the subsoil or underlain layers; development of a geotechnical characterization of the subsoil.
6) Preliminary assessment of foundation requirements and foundation choice, based upon a combination of experience and relatively simple analysis and design methods. In this assessment, considerable simplification of both the geotechnical profile(s) and the structural loading is necessary (bearing capacity criteria).
7) Reorganization of the design, based on more accurate representations of the structural layout, applied loadings, ground conditions, settlement prediction, differential rotation, and other displacements. Check these items with admissibility and function of structures from this stage and beyond; close interaction with the structural designer is essential for favored foundation design.
8) Detailed design joined with the structural engineer. Based on the applied load, and foundation base, as well as the amount of rigidity or flexibility, the internal stresses are calculated, and as a result, suitable material is allocated.
9) Definition of the installation methods and preparation of the technical specifications.
10) Evaluation of the cost to assist in the choice between possible alternative solutions.
11) In situ foundation testing at or before this stage is highly desirable, if not essential, to represent that the actual foundation behavior is consistent with the design presumptions; this commonly takes the form of testing of prototype or near-prototype piles or footings. If the behavior deviates from what is expected, then the foundation design may need to be revised. Such a revision may be either positive (a decrease in foundation requirements) or negative (a rise in foundation requirements). In order to make this decision, the foundation engineer must know that foundation testing involves only the foundation system elements. The foundation system, piles, and raft within the system will interact. Thus, the overall foundation behavior may not be assessed directly from the foundation test results without considering the foundation–soil interaction effects.
12) Surveillance of the performance of the building during and after construction. At the very least, settlements at several locations around the foundation should be monitored. Ideally, some of the footings, piles, and raft sections should also be monitored to measure load sharing among the foundation elements or contact pressure. Such surveillance is becoming more acceptable as a routine procedure for high-rise buildings, but not always for more conventional structures. As with any application of the observational method (OM), if the measured performance significantly violates the design expectations, then a contingency plan should be performed to address such departures. It should be noted that departures may involve settlements and differential settlements that are more significant than expected or smaller than expected.

2.8 Case Histories

Case History 2.1

Monopiles of Horns Rev 1 (Vector-Act Foundation System)

The wind turbine in question is part of the Horns Rev Offshore Wind Farm, constructed in 2003 and situated in the North Sea west of Esbjerg, Denmark (Augustesen et al. 2009; Jalbi et al. 2017). This wind farm comprises 80 turbines with a tower height of 60 m and a rotor diameter of 80 m. The turbines are evenly spaced in a regular quadrilateral grid, with distances of 560 m (seven times the rotor diameter) in both the east–west and north–south directions, resulting in a total park area of approximately 20 km^2. A comprehensive overview is provided in Figure 2.15a.

(Continued)

Case History 2.1 (Continued)

(a) (b)

Figure 2.15 The Horns Rev 1 Offshore Wind Farm: (a) general view and (b) schematic of monopole and soil profile (Jalbi et al. 2017 / with permission of Elsevier).

The turbine foundations are steel monopiles with outer diameters typically measuring 4 m and lengths ranging from 30 to 32.7 m. Monopiles are widely recognized as vector-act foundation systems. A transition piece, featuring outer and inner diameters of 4.24 and 4.15 m, respectively, connects the tower to the monopole, as depicted in Figure 2.15b. To characterize the soil strata, a site investigation program was conducted, involving boreholes, cone penetration tests, and laboratory triaxial tests. The soil profile primarily consists of deep sand, divided into six sub-layers. The upper 6.5 m below the seabed comprises sand, followed by a 7.5 m layer of silty sand. From 14 to 18.2 m below the seabed, silt/sand, including organic material, is found, followed by a layer of sand.

Case History 2.2

Al Faisaliah Building, Riyadh, Saudi Arabia (Section-Act Foundation System)

The Al Faisaliah building complex is one of the prominent projects in recent decades, in which a broad mat foundation was used (Hemsley 2000). The Al Faisaliah building complex main feature in Riyadh center is a 40-story office tower (Figure 2.16). A section through the building and a plan of the foundation are shown in Figure 1.20. The 3–4 m thick mat foundation has an area of 1600 m^2 in the plan, giving a volume of 6000 m^3. The tower has a height-to-breadth ratio of 5.8. The raft supports a total load of just over 1.1 GN. The tower tapers from 45 m square to zero at a height of 263 m.

The superficial ground deposits comprise 1–6 m of silty sand and gravel overlying Jurassic limestone (Arab and Jubaila formations). The limestone was assessed as having an unconfined compressive strength of approximately 20 MPa, a rock quality designation (RQD) of 25–50%, and a Young modulus (E) of approximately 2 GPa. For design purposes, an allowable net-bearing pressure of 850 kPa was selected for the core foundation to minimize settlement, and a 33% overstress was permitted under wind loads. These values were predicted to produce 10–20 mm settlements beneath the columns and 20–30 mm under the core, with the majority occurring during construction. Although the rock could safely support substantially higher pressures, the values adopted were selected to satisfy settlement serviceability criteria. One of the main challenges and concerns through the construction process was the temperature buildup due to large volumes of concrete hydration along with local hot weather.

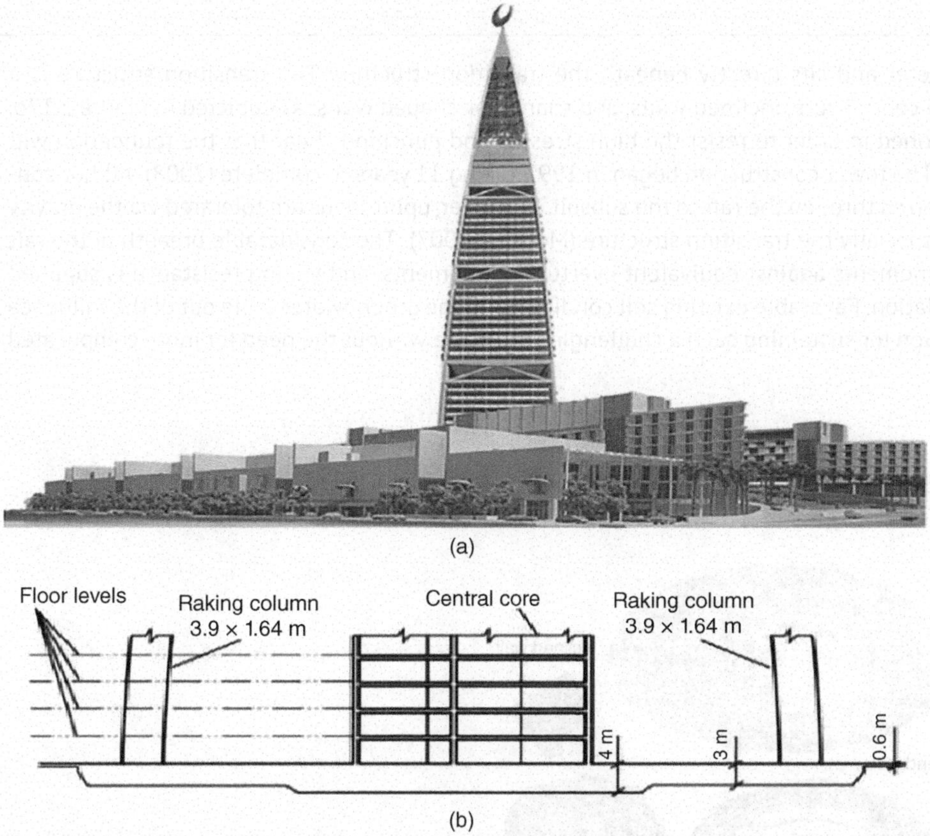

Figure 2.16 Al Faisaliah building complex: (a) computer simulation and (b) foundation section (Hemsley 2000 / with permission of Thomas Telford Ltd.).

Case History 2.3

Milad Tower, Tehran, Iran (Surface-Act Foundation System)

Milad Tower, with a height of 435 m, located in the northwest of Tehran, the capital city of Iran, is one of the highest telecommunication towers worldwide (Figure 2.17a). This multipurpose tower provides communication, broadcasting, wireless access, Aerology, and traffic control. It comprises the foundation, lobby, main shaft, head structure, and antenna mast. The main shaft comprises a central shaft and four legs. The central shaft is comprised of two nested octagons. With varying hollow trapezoidal sections, the legs disappear at the level of 240 m, so that the primary shaft has a diameter of 28 m at zero levels and diminishes to 17 m at the level of 240 m. The central shaft prolongs up to the level of 315 m. The tower head structure prolongs from the level of 247.5–315 m. A steel 120 m high antenna mast prolongs from the top of the head structure up to the level of 435 m (Zafarani et al. 2008).

Geotechnical site investigations have shown that the existing deposit is composed of a variety of different soils, mainly GC-SC and CL, based on the unified method. The bearing capacity after imposing required safety factors and realizing the allowed settlement of the tower is equal to 490 kPa.

The concrete shaft carries most of the gravitational and lateral loads of the structure. It is 315 m high, and consists of four main tapered trapezoidal walls, and two octagonal shapes connected by several walls. The octagons are post-tensioned in order to increase the bending capacity and stiffness of the structure to reduce the deflections. The foundation and lobby area consists of a circular foundation and a transition structure. The circular mat foundation

(Continued)

Case History 2.3 (Continued)

is approximately 66 m in diameter and sits directly beneath the transition structure. The transition structure is a pyramid shape, and comprises a central core, inclined walls, and triangular-shaped walls, as depicted in Figure 2.17b. The structure is also post-tensioned in order to resist the high stresses and punching shear that the foundation will experience during its life span. The tower construction began in 1997, taking 11 years to complete (2008). All horizontal and vertical loads are transferred through the raft to the subsoil. Moreover, uplift loads are tolerated via the gravity weight of the foundation and especially the transition structure (Mohseni 2007). The considerable breadth of the raft provides resistant overturning moments against equivalent overturning moments, and sliding resistance is supplied through the weight of the foundation. Favorable existing soil condition and the groundwater table out of the influence zone contributes to the foundation for sustaining such a challenging structure without the need for more complicated and massive foundations.

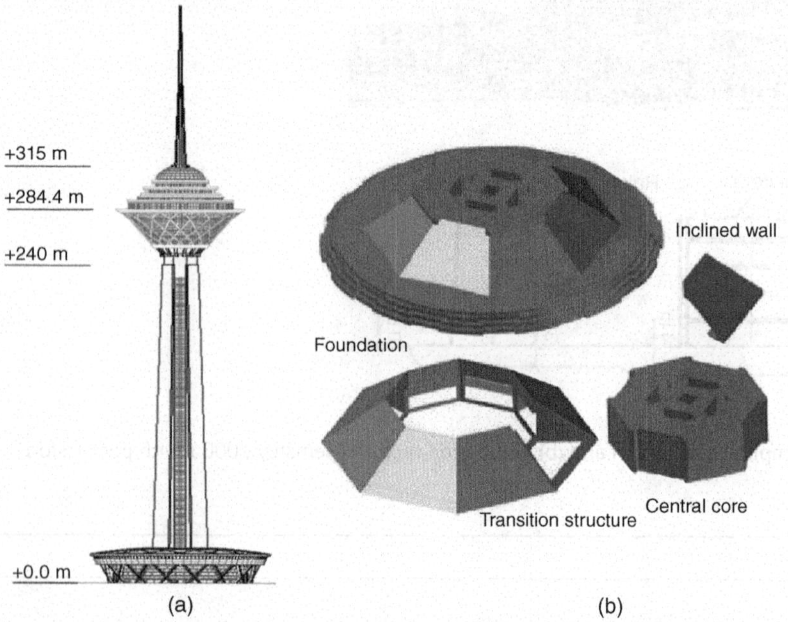

Figure 2.17 The Milad Tower: (a) completed form (Zafarani et al. 2008 / Hamid Zafarani) and (b) the transition structure and foundation (Rezayibana 2020 / International Journal of Engineering / CC BY 4.0).

The transition structure rises from below the ground level – 14.00 m to the level of 0.00. After 130,000 m of excavation, concrete and foundation operations began. The removed soil weight is approximately equal to the tower's total weight, and the volume of concreting for the foundation and transition structure is approximately 21,000 m^3.

Case History 2.4

Torre Latino Americana, Mexico City, Mexico (Hybrid Foundation System)

The 43-story Torre Latino Americana construction in Mexico City was a vitally important milestone in floating foundation technology (Zeevaert 1957; Coduto et al. 2016). It is significant because of the tough soils there. Figure 2.18 shows the tower's general view. The soil profile is generally as follows:

- **0–5.5 m Depth:** Old fill that includes Aztec artifacts; groundwater table at 2 m.
- **5.5–9.1 m Depth:** Becarra sediments – interbedded sands, silts, and clays.

Figure 2.18 General view of Torre Latino Americana Tower and (Coduto et al. 2016 / with permission of Pearson Education).

- **9.1–33.5 m Depth**: Tacubaya clays – soft volcanic clay; moisture content = 100–400%, $C_c = 8$; $S_u = 35–70$ kPa.
- **33.5–70.0 m Depth**: Tarango sands and clays – harder and stronger deposits; much less compressible than the Tacubaya clays.

Due to very soft and compressible Tacubaya clays at depths of 9.1–33.5 m, a considerable settlement was observed in the buildings around the tower. For the Palace of Fine Arts, located across the street from the tower and with similar characteristics, a settlement of 3 m was observed from 1904 to 1962; it continues to settle at a rate of 0.5 in (12 mm) per year. Based on Holtz et al. (2023), the compression index (C_c) for Mexico City clay is in the range of 7–10, which is near to problematic organic clays ($10 \leq C_c \leq 15$) and has a meaningful difference in comparison to typical clays including Chicago silty clay ($0.15 \leq C_c \leq 0.3$) and Boston blue clay ($0.3 \leq C_c \leq 0.5$).

To avoid these large settlements and provide safety and good service, the designers decided to implement a PRF system at a depth of 13 m. This project is a successful example of using a floating foundation, or equating soil removed with part of the building weight. By performing this method, half of the weight of the structure was replaced with the removed soil. Therefore, to transfer the load and sew the foundation to deeper layers, a series of driven piles were used under the raft, the end of which was extended to a dense layer of Tarango sand.

By installing piles under the foundation, a hybrid raft-pile foundation was obtained, a successful system of foundations for reducing total and nonuniform settlements. Interesting aspects of this project are the importance of geotechnical studies in identifying subsurface layers, the presence of very problematic soil at a depth of 10 m below the ground, the optimal aspects of using floating foundations, and the importance of end-bearing piles in controlling the settlement.

Case History 2.5

Shanghai Tower, Shanghai, China (Hybrid Foundation System)

The Shanghai Tower is a mega-tall skyscraper in Lujiazui, Pudong, Shanghai (Figure 2.19a) (Wang et al. 2021). It is considered the second-tallest building in the world after Burj Khalifa. The height of the tower is 632 m. It consists of a 124-story tower, a 7-story podium, and a 5-story basement (El Gendy et al. 2018).

(Continued)

Case History 2.5 (Continued)

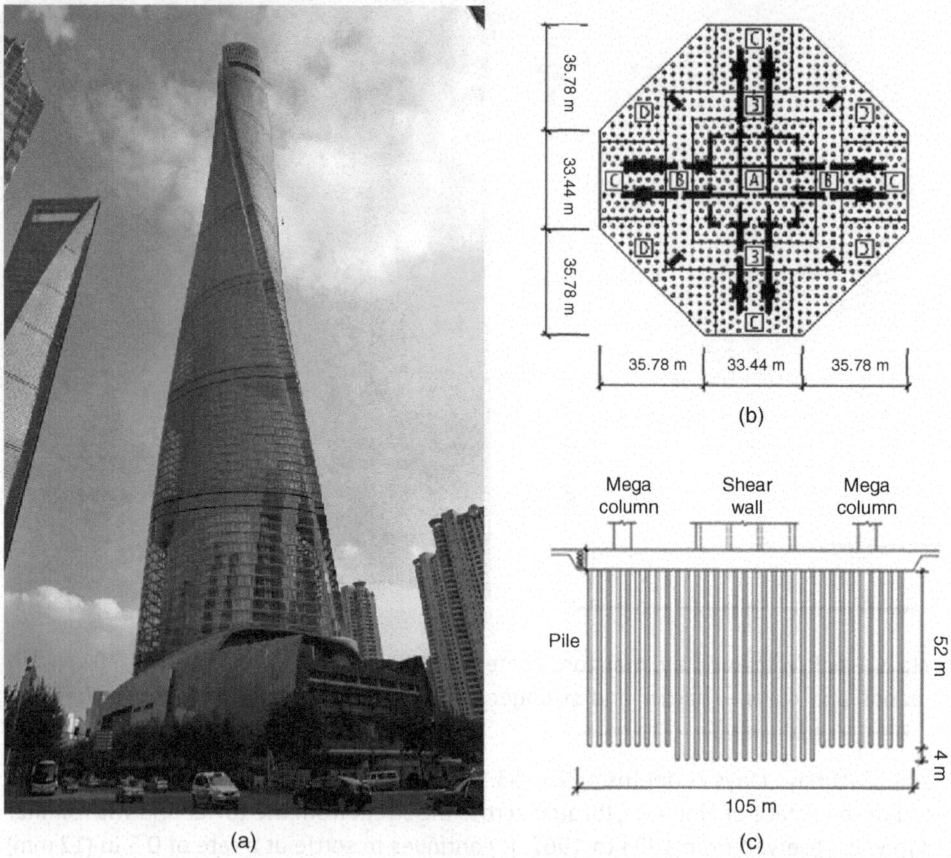

Figure 2.19 Shanghai Tower: (a) general view (Baycrest / Wikimedia Commons / CC BY-SA 2.5), (b) foundation layout, plan view, and (c) foundation layout, front view (Wang et al. 2021 / John Wiley & Sons).

The tower is supported by a PRF in which the thickness of the raft under the tower is 6 m, and the area of the raft is 8945 m^2 (Figure 2.19b). About 950 bored piles with a diameter of 1 m and a pile space of 3 m are arranged uniformly. As shown in Figure 2.19c, the effective lengths of bored piles vary from area to area. More specifically, it is 56 m in Area A but 52 m in other areas. However, the bearing capacities of the pile are the same among the whole raft. The PRF is embedded in the site with the soil condition of class IV. From the geotechnical site investigation, it is indicated that the subsoil consists of clay, silty clay, and sand, underlain by a completely decomposed granite, which can be divided into 9 layers and 14 sublayers. The raft is supported by the top layer, and the pile ending is beard by the fifth, seventh, or ninth layers (Wang et al. 2021).

Case History 2.6

Oil Storage Tanks, Poti, Georgia (Hybrid Foundation System)

This project includes the installation of eight large diameter storage tanks of oil and handling appurtenances for the onshore reservoir facility to be built in two phases in the Port of Poti, the oldest and the most prominent gate of Georgia connecting Europe and Central Asia, which is located on the Eastern side of the Black Sea in Georgia (Duzceer 2003). Figure 2.20a depicts a view of the mentioned tanks. The layout plan of the oil storage tanks and structures constructed in Phase I is shown in Figure 2.20b. Two of the tanks have a diameter of 28.50 m. The diameters of the third and fourth

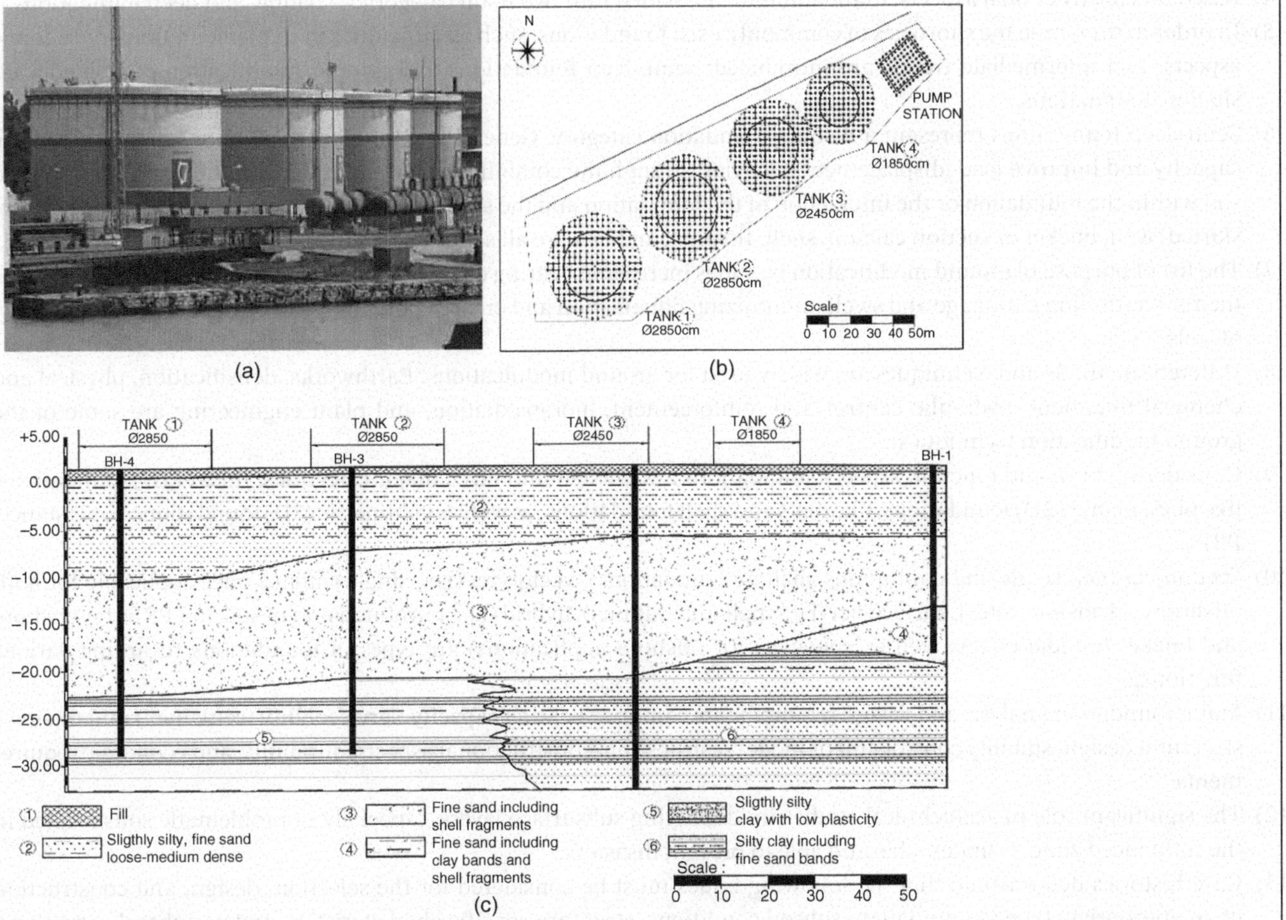

Figure 2.20 Oil tank farm of Poti, (a) tanks view (Scientific Research Firm Gamma 2009 / with permission of Gamma Scientific Inc.), (b) plan of tanks and the underlying, and (c) subsoil profile (Duzceer 2003).

tanks are 24.50 m and 18.50 m, respectively. All the tanks have a height of 18 m; the maximum allowable height for oil storage tanks has been specified by API 650 (1984) (Duzceer 2003).

The tank subsoil conditions typically consist of 1.5 m fill overlying 20 m thick loose to average dense silty sand. Silty sand is underlain by 15 m thick medium-stiff to stiff clay. Groundwater is encountered at a depth of 1.5 m. The general soil profile is shown in Figure 2.20c. To improve the region soil, stone columns with a diameter of 1 m, a distance from the center to the center of 2.5 m, and a length of 20 m have been used. This configuration ultimately leads to a replacement area ratio of 12.5–16.5%. A square pattern is used in the implementation of the stone columns. The tanks were subjected to hydrostatic testing according to the project's technical specifications. Each tank was filled with seawater in four stages up to a maximum capacity of 180 kPa. The water level was kept constant for 12 hours at each loading stage, and finally, the settlement of the tanks was measured at specific points that were previously instrumented (Duzceer 2003).

2.9 Major Points

1) Structural elements that transfer upcoming loads of superstructure to the ground are called foundations.
2) Foundation engineering mainly aims to design and implement technically sound, construction-feasible, and economic foundations that provide reliable support for the superstructure.
3) Foundation engineering comprises knowledge and skills from geotechnical engineering, structural engineering, and construction engineering.

4) Based on constructional aspects, foundations are classified into two main categories: shallow and deep foundations.
5) In order to overcome the shortages of commonly used foundations, such as difficulties in implementation or economic aspects, two intermediate trends are introduced: semi-deep foundations and ground modification combined with shallow foundations.
6) Semi-deep foundations represent a distinct foundation category. Generally, semi-deep foundations enhance bearing capacity and improve load–displacement behavior through the combined actions of the foundation and the confined soil within the foundation or the interaction of the foundation and the surrounding soil. Foundations named spudcan, skirted, well, bucket or suction caisson, shell, floating, and box are all semi-deep foundations.
7) The list of purpose of ground modification includes increasing bearing capacity, decreasing compressibility and settlements, controlling shrinkage and swell, minimizing liquefaction and erosion potential, and minimizing the variability of soils.
8) Different methods and techniques are widely used for ground modifications. Earthworks, densification, physical and chemical treatment, hydraulic control, soil reinforcement, bioremediation, and plant engineering are some of the ground modification techniques.
9) Considering form and function aspects, foundations are divided into three major categories: linear (1D) foundations, like piles; planar (2D) foundations, like mats and shell foundations, as well as volumetric (3D) foundations, for instance, PRFs.
10) Section-act (i.e., single and continuous shallow foundations), vector-act (i.e., single (mono) piles applicable for both offshore and onshore sites), surface-act (i.e., shell foundations), hybrid (i.e., retrofitted mas of soil and PRFs for onshore and bucket foundations as well as spudcans for offshore applications) are types of foundations regarding various functions.
11) Major foundation analysis and design requirements comprise bearing capacity, serviceability (settlement and torsion), structural design, stability control, full or model scale testing, constructional aspects, durability, and economic requirements.
12) The significant role of geotechnical studies in identifying subsurface layers, especially in problematic soils located in the influenced zone, is undeniable and able to prevent disasters.
13) Case histories demonstrate that the following issues must be considered for the selection, design, and construction of an appropriate type of foundation: subsoil conditions, structure-specific characteristics, and peripheral conditions, along with the relevant available databases.
14) Overall, in realizing geotechnical and practical requirements in the selection of foundation systems, three primary influential factors are load combinations, which are determined by the type and characteristics of the superstructure, foundation system type (linear, planar, and volumetric) as well as soil type and strength (very soft to very hard or very loose to very dense). In this regard, a comprehensive and thoughtful assessment, which includes all the abovementioned factors, is essential.

References

Augustesen, A.H., Brødbæk, K.T., Møller, M. et al. (2009). Numerical modelling of large-diameter steel piles at horns rev. In: *Proceedings of the Twelfth International Conference on Civil, Structural and Environmental Engineering*, Civil-Comp Proceedings No. 91 (ed. B.H.V. Topping, L.F.C. Neves, and R.C. Barros). Scotland: Computing Civil-Comp Press.

Budho, M. (2011). *Soil Mechanics and Foundations*, 3e. New Jersey, USA: Wiley.

Burland, J., Chapman, T., Skinner, H.D., and Brown, M. (2012). *ICE Manual of Geotechnical Engineering Volume 2: Geotechnical Design, Construction, and Verification*. Institution of Civil Engineers.

Coduto, D.P., Kitch, W.A., and Yeung, M.R. (2016). *Foundation Design Principles and Practices*, 3e. Upper Saddle River, NJ: Prentice Hall, Inc.

Das, B.M. and Sivakugan, N. (2019). *Principles of Foundation Engineering*, 9e. Boston: Cengage Learning.

Duzceer, R. (2003). Ground improvement of oil storage tanks using stone columns. *The 12th Pan American Conference on Soil Mechanics and Geotechnical Engineering*, Cambridge, USA (22–26 June 2003).

Ebrahimipour, A., Eslami, A., Imani, M., and Imam, R. (2023). Foundation systems performance; technical and practical aspects. *Proceedings of 13th International Congress on Civil Engineering*, Tehran, Iran (17–19 October 2023).

El Gendy, M., El Gendy, A., and El Gendy, O. (2018). Analysis of piled raft of Shanghai Tower in Shanghai by the program ELPLA. GEOTEC Software Inc. PO Box 14001 Richmond Road PO, Calgary AB, Canada T3E 7Y7.

Engel, H. (2013). *Structure Systems*, 5e. Germany: Hatje Cantz.

Eslami, A. (2013). *Foundation Engineering, Design, and Construction*, 2e. Tehran: Building and Housing Research Center.

Eslami, A. (2022). Foundation systems for tall buildings. Key Lecture for Construction Engineering Organization, Iran (15 June 2022). https://abolfazleslami.com/.

Eslami, A. and Heidarie Golafzani, S. (2021). Relevant data-based approach upon reliable safety factor for pile axial capacity. *Marine Georesources and Geotechnology* 39 (11): 1373–1386.

Eslami, A., Moshfeghi, S., Molaabasi, H., and Eslami, M. (2020). *Piezocone and Cone Penetration Test (CPTu and CPT) Applications in Foundation Engineering*, 1e. Elsevier.

Fellenius, B.H. (2023). Basics of foundation design. Electronic Edition. www.Fellenius.net, 432 p.

Hemsley, J.A. (2000). *Design Applications of Raft Foundations*, 1e. London: Thomas Telford.

Holtz, R.D., Kovacs, W.D., and Sheahan, T.C. (2023). *An Introduction to Geotechnical Engineering*, 3e. Pearson.

Ibsen, L.B., Barari, A., and Larsen, K.A. (2015). Effect of embedment on the plastic behavior of bucket foundations. *Journal of Waterway, Port, Coastal, and Ocean Engineering* 141 (6): 1–9.

Jalbi, S., Shadlou, M., and Bhattacharya, S. (2017). Practical method to estimate foundation stiffness for design of offshore wind turbines. In: *Wind Energy Engineering*, 329–352.

Kurian, N.P. (2006). *Shell Foundations: Geometry, Analysis, Design, and Construction*. Alpha Science International Limited.

Mohseni, A. (2007). Technical and practical characteristics of Milad tower in Tehran. http://abdolazim.blogfa.com/post/130 (accessed 25 April 2023).

Poulos, A., Brown, C., McCulloch, D., and Cole, J. (2017). Context-aware augmented reality object commands. U.S. Patent No. 9,791,921. Washington, DC: U.S. Patent and Trademark Office.

Punetha, P., Samanta, M., and Sarkar, S. (2019). Bioengineering as an effective and ecofriendly soil slope stabilization method: a review. In: *Landslides: Theory, Practice, and Modelling. Advances in natural and Technological Hazards Research*, vol. 50 (ed. S. Pradhan, V. Vishal, and T. Singh), 201–224. Cham: Springer.

Rajasekar, A., Moy Ch.K.S., and Wilkinson, S. (2017). MICP and advances towards eco-friendly and economical applications. *IOP Conference Series: Earth and Environmental Science* 78: 012016.

Randolph, M.F. (2003). Science and empiricism in pile foundation design. *Géotechnique* 53 (10): 847–875.

Rezayibana, B. (2020). The effect of soil type on seismic response of tall telecommunication towers with random vibration analysis. *International Journal of Engineering (IJE), IJE Transactions C: Aspects* 33 (3): 419–426.

Rezazadeh, S. and Eslami, A. (2017). Bearing capacity of semi-deep skirted foundations on clay using stress characteristics and finite element analyses. *Marine Georesources and Geotechnology* 36: 625–639.

Scientific Research Firm Gamma (2009). Environmental Audit Report for Current Operations of Poti Sea Port.

Stone, R.C., Farhangi, V., Fatahi, B., and Karakouzian, M. (2023). A novel short pile foundation system bonded to highly cemented layers for settlement control. *Canadian Geotechnical Journal* 60 (9): 1332–1351. https://doi.org/10.1139/cgj-2020-0710.

Umar, M., Kassim Kh.A., and Chiet, K.T.P. (2016). Biological process of soil improvement in civil engineering: a review. *Journal of Rock Mechanics and Geotechnical Engineering* 8: 767–774.

Viggiani, C., Mandolini, A., and Russo, G. (2014). *Piles and Pile Foundations*. CRC Press.

Wang, L., Zhao, X., and Liu, S. (2021). A study of differential foundation settlement of piled raft and its effect on the long-term vertical shortening of super-tall buildings. *Struct Design Tall Spec Build* 30: 1–19.

Zafarani, H., Ghorbani-Tanha, A.K., Rahimian, M., and Noorzad, A. (2008). Seismic response analysis of Milad tower in Tehran, Iran, under site-specific simulated ground motions. *The 14th World Conference on Earthquake Engineering*, Beijing, China (12–17 October 2008).

Zeevaert, L. (1957). Foundation Design and Behavior of Tower Latino Americana in Mexico City. *Géotechnique* 7: 115–133.

Section II

Geotechnical Engineering

3

Geotechnical Site Investigation

Chapter Highlights

- An Introduction to Geotechnical Engineering
- Missions, Activities, and Liabilities of a Geotechnical Engineer
- Geotechnical Investigations Extensions and Stages
- Source of Data
- Geotechnical Testing for Foundations; In situ and Laboratory Tests
- Geotechnical Reports for Foundation Engineering
- Problematic Geomaterials; Properties and Identification
- Case Histories
  - Torre Latino Americana
  - Bolivian Experimental Site
  - Amir Abaad Tank Farm

3.1 Geotechnical Engineering

Geotechnical engineering is the systematic application of methods and techniques that enable construction in, on, or with geomaterials. The design of structures involving soil or rock relies heavily on the engineering properties of these geomaterials. Geotechnical engineering practices encompass various activities, including site investigations to assess geomaterial suitability for construction. Geotechnical engineers also monitor and control groundwater levels and flow. At times, they assess and remediate contaminated soils. Practical applications of geotechnical engineering include foundation engineering, underground structures, ground excavations, slope stability analyses, dams, natural or artificial fills, and roadway and airfield pavements.

Despite significant advancements, many geotechnical engineering calculation methods remain approximate due to the inherent inhomogeneity of geomaterials and varying environmental conditions. Soil and rock are more sensitive to environmental factors than prefabricated building materials like concrete and steel. Therefore, it is crucial to gain comprehensive knowledge of the interactions between geomaterials and the environment to assess geotechnical engineering behavior accurately and predict project performance.

Figure 3.1 illustrates the interaction and overlap between geotechnical engineering and various engineering branches. Geotechnical engineering activities are integral to a team, collaborating with professions such as structural engineering and hydraulics engineering. The final design of any project emerges from the cooperation of these disciplines.

The application, significance, and functionality of a structure, coupled with the widespread use of natural geomaterials in construction projects, demand a thorough investigation, identification, collection, and evaluation of geomaterial engineering properties. This necessity has given rise to the field of "geotechnical site investigation" within geotechnical engineering practice. Geotechnical site engineering encompasses the following aspects:

- Identifying and characterizing the location, depth, and thickness of soil and rock layers.
- Recovering samples for observation, laboratory testing, and assessment.

Advanced Foundation Engineering: Principles, Performance and Prospects, First Edition. Moses Karakouzian and Abolfazl Eslami.

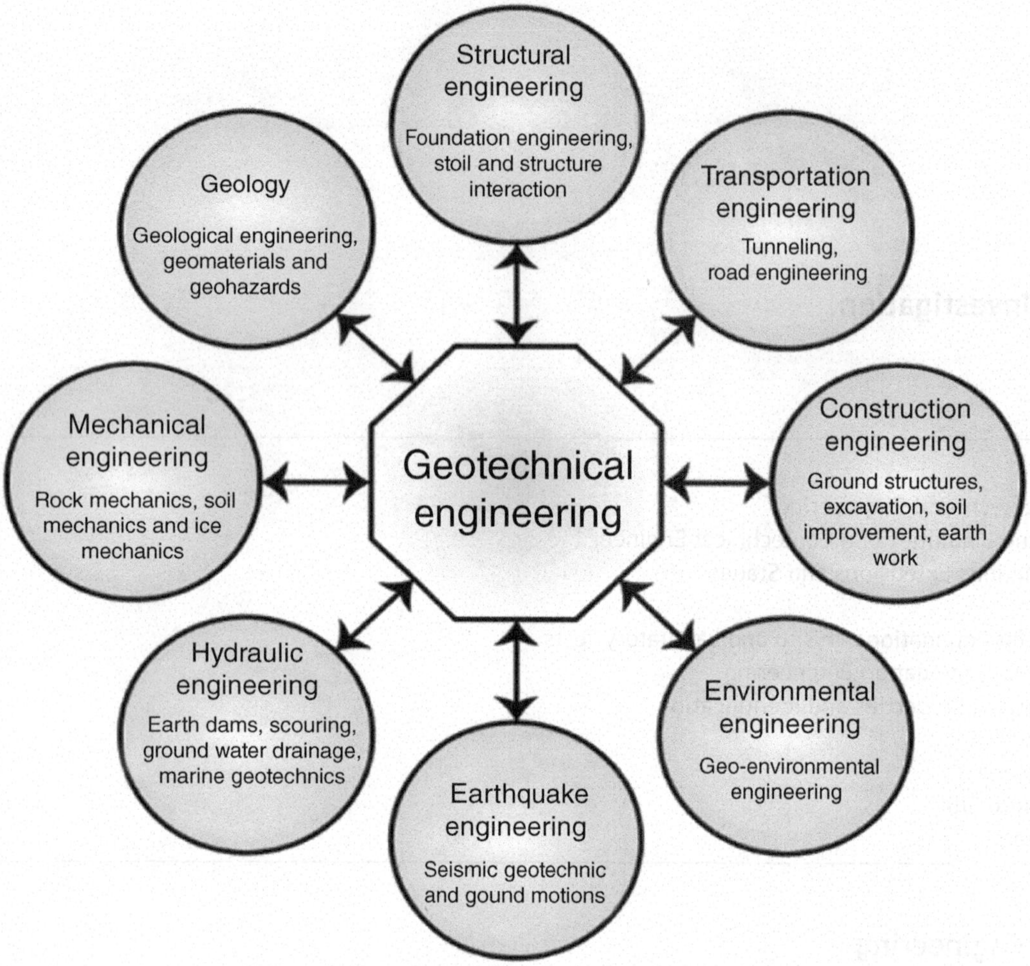

Figure 3.1 Demonstration of the overlap of geotechnical engineering with other disciplines.

- Analyzing groundwater conditions and their potential impact on projects.
- Determining the bearing capacity of geomaterials to support imposed loads safely.
- Investigating potential effects and complications arising from construction, earthworks, or excavation activities.
- Selecting and designing appropriate foundation systems for a given structure.
- Identifying the need for slope stabilization and addressing natural and artificial deposits.
- Evaluating the requirement for utilizing ground or underground structures and their possible designs.
- Assessing site response to potential natural or artificial geohazards, such as seismic loads.

In contrast to small-scale projects like residential houses or two to three-story commercial buildings, large-scale projects such as tall buildings, bridges, and power plants require extensive geotechnical investigations. These investigations involve high costs for in situ studies and laboratory testing. On project sites where problematic geomaterials such as collapsible silts, liquefiable sands, or expansive clays are present, geotechnical challenges can significantly impact the project's feasibility, both from economic and technical perspectives.

In many projects, the geotechnical engineer plays a crucial role in both the design and construction phases. During the construction phase, the geotechnical engineer's services may include:

- **Assessing Geomaterial Conditions:** Evaluate the conditions of geomaterials observed in site investigations and laboratory testing, comparing them with the expected conditions during the design phase. Significant changes in the site's conditions may necessitate modifications to the initial design when unexpected variations are encountered during the construction phase.

- **Monitoring Actual Performance:** Compare the actual performance of the project with the predicted performance in the design stage. Special equipment is employed for monitoring and recording different performance measures, such as displacements and groundwater levels. This practice, known as the observational method (OM), allows adjustments to be made based on actual conditions encountered during construction.
- **Quality Assurance:** Conduct quality assurance checks and provide various geotechnical services throughout the construction phase, particularly for foundations, ensuring their stability and integrity. These services continue until the final construction phases. Furthermore, geotechnical engineers may provide ongoing monitoring services for project sites prone to long-term settlement. Such sites require continuous monitoring for several months to several years after construction.

3.2 Geotechnical Investigations Extensions

Geotechnical investigations are instrumental in the design and construction of diverse projects. This section presents an overview of the assessments, objectives, and applications of geotechnical investigations.

3.2.1 Assessments of Geotechnical Investigations

Prior to conducting any in situ investigation or field testing, the design team must compile essential data for the proposed development. This includes:

- **Location and Project Type:** Understanding the specific location and type of project is crucial.
- **Ground Leveling and Existing Topography:** Analyzing the existing ground levels and topographical features provides valuable context.
- **Accessibility to Roads and Nearby Facilities:** Assessing transportation access and nearby facilities is essential for logistics.
- **Geometrical Dimensions in Plan and Height:** Determining the project's layout and vertical dimensions aids in accurate design.
- **Method of Construction:** Understanding the construction techniques to be employed is vital for assessing the site's requirements.
- **Anticipated Load Combinations Imposed from the Structure:** Evaluating expected loads on the structure is fundamental to ensure proper foundation design.
- **Evidence of Former or In-Progress Development:** Studying past or ongoing developments in the area offers insights into the geological history of the site.

One of the initial and fundamental phases in project design is geotechnical site investigation, involving the assessment of subsurface soil and rock characteristics. This assessment can be conducted through various approaches, including:

- **Remote Sensing:** Utilizing regular aerial photographs to gain an overview of the site.
- **Data Analysis:** Investigating available data from geological and geographical maps, geotechnical reports, and geological hazard maps.
- **Site Survey:** Conducting on-site surveys and visual observations to understand surface conditions.
- **Local Experience:** Leveraging local knowledge and insights from ongoing construction operations.
- **Geophysical and Seismic Evaluations:** Employing geophysical and seismic methods to understand subsurface conditions.
- **Drilling, Sampling, and Groundwater Assessment:** Performing drilling to obtain soil samples and assess groundwater levels.
- **In Situ and Laboratory Testing:** Conducting tests on-site and in the laboratory to analyze soil properties.

- **Environmental Evaluation:** Assessing environmental factors that might impact the site.
- **Instrumentation and Monitoring:** Installing instruments for real-time monitoring of site behavior and geotechnical parameters.

3.2.2 Objectives of Geotechnical Investigations

In any civil project, the primary objectives include durability, engineering performance, acceptable serviceability, and economic efficiency. Achieving an optimal foundation system design necessitates a combination of engineering expertise, extensive geotechnical experience, and the utilization of advanced analytical tools and theories available in the field.

3.2.3 Applications of Geotechnical Investigations

Following are the applications of geotechnical investigations.

- Conducting a comprehensive site evaluation for the planned project's construction.
- Assessing the strength and stiffness properties of underlying soil and rock.
- Determining soil compressibility characteristics.
- Selecting the most suitable foundation system.
- Evaluating the need for ground modification, if necessary.
- Choosing the optimal construction method for the foundation.
- Identifying potential borrow pit locations.
- Exploring the efficient integration of fabricated materials with natural resources.
- Anticipating structural performance under static, dynamic, and seismic loads during and after construction.
- Monitoring foundation performance and its interaction with the main structure.
- Conducting back analysis and design optimization.
- Identifying potential ground issues, such as instability or geohazards.
- Recognizing potential geo-environmental problems and proposing appropriate remedial measures.
- Assessing the project's technical, constructional, and economic design aspects.

3.3 Stages for Geotechnical Investigations

Commonly, a site investigation program is carried out according to the following stages:

1) Data Collection and Literature Review
2) Site Visit and Nondestructive Testing
3) Drilling, Sampling, and In Situ Testing
4) Laboratory Testing on Collected Samples
5) Data Analysis and Synthesis, and Geotechnical Investigation Report
6) Instrumentation, Surveillance, and Monitoring

Stage I – Data Collection and Literature Review: In any site investigation program, the key steps include gathering site plan information, specifying the locations of structural elements like foundations, columns, and walls, determining the applied loads from the superstructure including type, size, height, number of stories, and underground floors. Additionally, it involves identifying the structure's importance and limitations concerning serviceability aspects such as anticipated displacement and settlement. The investigation also entails reviewing previous geotechnical site investigation reports for local projects, evaluating the condition of existing structures, examining topographic, geological, seismic hazard, and shear wave velocity maps, aerial photographs, and local agency reports. Moreover, it involves assessing hydrological and climatological data and considering the conditions of any existing mines adjacent to the project site.

Stage II – Site Visit and Nondestructive Testing: Preliminary investigations primarily involve conducting a site visit and nondestructive testing to create geological and topographical site maps. This phase includes collecting data on former constructions and evaluating the performance of existing structures. Geomaterial conditions of the project site are assessed, considering the appropriateness of drilling equipment. Additionally, potential geohazards like earthquakes, landslides, and subsidence are analyzed, along with possible issues during field operation and construction. At this stage, verifying the collected data and information with the current site condition is crucial.

Stage III – Drilling, Sampling, and In Situ Testing: Although nondestructive methods offer valuable insights into subsurface conditions, the geotechnical investigation must progress through field operations involving activities like digging trenches, excavating galleries, and drilling. Sampling and in situ tests are conducted to provide detailed and precise subsurface profiling, including layer thickness, the extent of soil and rock layers, and groundwater fluctuations.

Stage IV – Laboratory Testing on Collected Samples: During this stage, the collected samples are taken to a laboratory for comprehensive testing. Basic and advanced laboratory tests are conducted to assess the soil and rock's key properties. These studies primarily focus on the geomaterials' indices, strength, stiffness, compressibility, and durability characteristics.

Stage V – Data Analysis and Synthesis, and Geotechnical Investigation Report: During this stage, geotechnical investigation reports are prepared to provide a detailed description of the subsurface profiles, site exploration methods, in situ and laboratory tests, obtained results, groundwater conditions, and any information regarding problematic geomaterials or unusual situations that may arise during the construction phase or in the long-term. The reports synthesize information from field exploration, in situ and laboratory tests, along with a review of potential geohazard scenarios. They also include recommendations for foundation design based on the collected data and analyses.

Stage VI – Instrumentation, Surveillance, and Monitoring: Upon completing site investigations and entering the construction phase, it is crucial to employ instrumentation to monitor geotechnical-structural interactions, aiding in the anticipation of potential scenarios during construction. Additionally, information from adjacent projects can offer insights into the general subsurface conditions. The depth of geotechnical investigations conducted in the preceding stages is determined by the project's significance, subsurface conditions, available prior information, and the expertise and judgment of the geotechnical engineer.

3.4 Extent of Geotechnical Site Investigation for Foundation Engineering

While there are no specific guidelines for the extent of geotechnical site investigations, the number and depth of borings are contingent upon the project's type, significance, site conditions, geological complexities, available equipment, local expertise, and engineering judgment. Generally, as the foundation load increases, the distance between borings should decrease, and the depth should increase. The selection of the number, location, and depth of borings is crucial for a comprehensive understanding of the subsurface profile within the structure's influence zone. During the initial investigation phase, considering a smaller number of borings with wider spacing is practical. In this stage, where detailed structural requirements and designs are unavailable, drilling deeper than the minimum required depth can provide a more accurate understanding of soil and bedrock stratification. However, for subsequent and final investigations, it becomes necessary to increase both the number and depth of borings.

3.4.1 Boring Numbers and Spacing

The selection of boring and sampling points on the site plan depends on factors such as the project type, applied loads, subsurface conditions, significant changes in soil layers, and the specific phase of the investigation.

For projects like roads, a boring spacing of about 500–1000 m is adequate in typical flat ground conditions. For structures such as dams and power stations, a more common practice is to have a boring spacing of approximately 20–100 m. In the case of structures like piers or abutments with a width less than or equal to 30 m, it is recommended to have a minimum of one exploration boring per substructure. Additional borings are necessary if there are more noticeable variations in subsurface conditions horizontally than vertically. For ordinary buildings or special projects like antenna towers with vertical structures, one boring is insufficient. In these cases, it is advisable to consider at least three borings, although the number can be reduced to two if the subsurface conditions are assessed as uniform. Relevant organizations and regulations provide recommendations for the number, spacing, and depth of borings tailored to different applications such as buildings, towers, industrial facilities, bridges, and storages.

3.4.2 Depth of Investigations

A geotechnical exploration should be conducted within the specific area where construction operations or structures will influence the ground. The following recommendations can be applied in this context:

According to the criteria outlined by the American Society of Civil Engineers (ASCE 1993) and the Canadian Engineering Manual (CFEM 2006), the minimum required depth is determined by the larger value between two specific criteria: a depth at which the net stress due to the structure reaches either 5% of the soil effective stress (σ_z) or 10% of the applied pressure from the foundation to the sub-ground (q_{applied}). If noticeably dense soil layers or rock are found at a shallower depth than determined by this criterion, the exploration depth can be adjusted, allowing for a shallower boring depth. Additionally, it is crucial to account for the increase in soil net stress due to the influence of adjacent structures when calculating the boring depth. The additional stresses induced by nearby structures might necessitate an increase in the minimum boring depth. For deep foundations, the required exploration boring depth is the larger value between 6 m and two times the maximum dimension of the foundation group below the expected tip of the foundation (AASHTO 2014).

Taking into account the 2:1 stress distribution with depth and assuming the total area of all the individual footings is represented by a virtual mat foundation with dimensions $B \times L$ (where B and L are the equivalent mat foundation's width and length), derived from stress overlaps of individual footings, the minimum boring depth can be calculated using Equation (3.1) based on the 5% criterion.

$$0.05\gamma H = \frac{q_{\text{avg}} \times B.L}{(B+H)(L+H)} \tag{3.1}$$

where q_{avg} is the average pressure from the structure, and H is the minimum boring depth based on the 5% effective stress criteria.

The extent of investigations on the plan can be calculated by Equations 3.2 and 3.3.

$$\frac{L'}{L} = 1 + \frac{H}{B} \tag{3.2}$$

$$\frac{B'}{B} = 1 + \frac{H}{B} \tag{3.3}$$

where L' and B' are the length and width of the investigation area, respectively.

In a situation in which the groundwater level is near the ground surface, the buoyant/submerged unit weight (γ') must be used in calculations, and consequently, the boring depth increases.

As mentioned above, the values of boring depth (H), width (B'), and length (L') of the investigation area can be calculated based on the width (B) and length (L) of the foundation and the average applied pressure (q_{avg}) (Tables 3.1 and 3.2; Figures 3.2 and 3.3).

Overall, the guidelines presented in Table 3.3 can be considered the minimum investigation depths for common structures placed on spread footings or mat foundations (Coduto 2001).

Table 3.1 *H/B* values for calculating the minimum required boring depth based on 5% effective stress criteria ($\gamma_{\text{avg}} = 16\ \text{kN/m}^3$).

	L/B						
P_{avg}/B	1	2	3	4	6	8	10
0.05	0.345	0.381	0.396	0.405	0.414	0.419	0.422
0.10	0.533	0.601	0.633	0.652	0.673	0.684	0.692
0.15	0.672	0.767	0.814	0.842	0.874	0.892	0.904
0.20	0.785	0.905	0.964	1.000	1.043	1.068	1.083
0.30	0.969	1.128	1.210	1.261	1.323	1.359	1.383
0.40	1.120	1.309	1.411	1.476	1.555	1.602	1.633
0.50	1.243	1.465	1.584	1.660	1.756	1.813	1.851
0.60	1.354	1.602	1.737	1.825	1.934	2.000	2.050
0.70	1.454	1.725	1.874	1.972	2.095	2.171	2.222
0.80	1.545	1.838	2.000	2.108	2.244	2.328	2.386
0.90	1.629	1.942	2.117	2.233	2.382	2.474	2.537
1.00	1.707	2.040	2.226	2.351	2.511	2.611	2.680

Table 3.2 *H/B* values for calculating the minimum required boring depth based on 5% effective stress criteria ($\gamma_{avg} = 18\ kN/m^3$).

	L/B						
P_{avg}/B	1	2	3	4	6	8	10
0.05	0.320	0.350	0.364	0.371	0.379	0.383	0.386
0.10	0.497	0.558	0.586	0.603	0.622	0.632	0.638
0.15	0.629	0.716	0.757	0.783	0.811	0.827	0.838
0.20	0.737	0.846	0.900	0.933	0.971	0.993	1.010
0.30	0.912	1.059	1.134	1.181	1.237	1.269	1.290
0.40	1.060	1.233	1.330	1.385	1.457	1.499	1.527
0.50	1.175	1.381	1.490	1.561	1.647	1.699	1.733
0.60	1.281	1.512	1.637	1.717	1.817	1.877	1.918
0.70	1.380	1.630	1.768	1.858	1.971	2.040	2.086
0.80	1.464	1.738	1.889	1.988	2.113	2.189	2.241
0.90	1.550	1.838	2.000	2.108	2.244	2.328	2.386
1.00	1.620	1.930	2.105	2.220	2.367	2.458	2.521

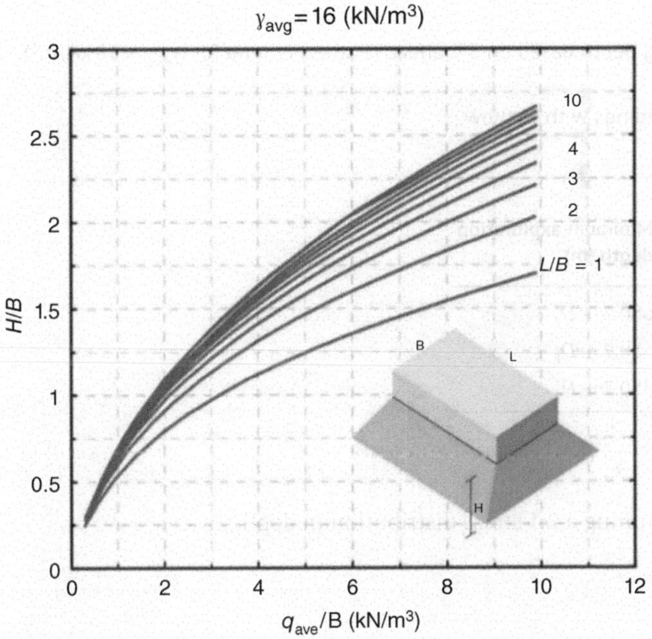

Figure 3.2 *H/B* values for calculating the minimum required boring depth based on 5% effective stress criteria for ($\gamma_{avg} = 16\ kN/m^3$).

Example 3.1 Considering a shallow foundation system for a 10-story building with dimensions of 15 × 15 m. if the soil unit weight is 1.8 ton/m^3, the groundwater elevation is 25 m below the ground level, and the average load for each story is considered 1 ton/m^2, calculate the appropriate extent, in the plan and depth, for investigation borings.

Step 1: Determining the average applied pressure (q_{avg}):

$$q_{avg} = 10 \times 1 = 10\ \text{ton/m}^2$$

Step 2: Calculating the appropriate boring depth:

$$\frac{L}{B} = \frac{15}{15} = 1, \quad \frac{q_{avg}}{B} = \frac{10}{15} = 0.67\ \text{ton/m}^2 \rightarrow \frac{H}{B} = 1.3, \quad H = 1.3 \times 15 = 19.5\ \text{m}$$

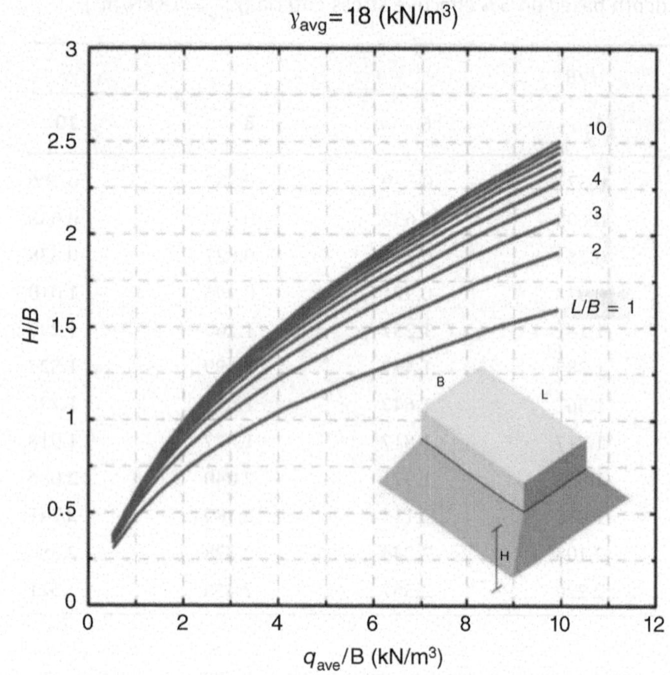

Figure 3.3 *H/B* values for calculating the minimum required boring depth based on 5% effective stress criteria for ($\gamma_{avg} = 18$ kN/m^3).

Table 3.3 Guidelines for the depth of investigation borings for buildings with shallow foundations (Sowers 1979/Pearson Education).

Subsurface conditions	Structure footprint area (m^2) for each exploratory boring	Minimum exploration depth (m)
Poor quality and/or erratic	100–300	$6S0.7 + D_f$
Average	200–400	$5S0.7 + D_f$
High-quality and uniform	300–1000	$3S0.7 + D_f$

*S: number of stories; *D_f: anticipated depth of the foundation (m).

Step 3: Calculating the investigation extent in the plan by assuming a 2:1 stress distribution in depth:

$$\frac{B'}{B} = 1 + \frac{H}{B} = 1 + \frac{19.5}{15} = 2.3 \rightarrow B' = 34.5\text{ m}$$
$$\frac{L'}{L} = 1 + \frac{H}{L} = 1 + \frac{19.5}{15} = 2.3 \rightarrow L' = 34.5\text{ m}$$

According to Table 3.3, the following equation can be used to calculate the boring depth:

$$d = AS^{0.7} + D_f$$

Assuming a high-quality condition for subsoil and 1 m for embedment depth, according to Table 3.3, the exploratory depth of 16 m agrees with the one calculated in Step 2.

Example 3.2 Considering a shallow foundation system for a 25-story building with the dimensions of 15 × 20 m. if the soil unit weight is 1.8 ton/m^3, the groundwater elevation is 25 m, and the average load for each story is considered 1 ton/m^2, what would be a proper investigation depth based on the following three methods?

a) 5% effective stress criterion
b) 10% of the transferred stress from the foundation to the subsoil

a) $q_{avg} = 25 \times 1 = 25\ \text{ton/m}^2, 0.05 \times (H \times 1.8) = \frac{25 \times 15 \times 20}{(15+H)(20+H)} \Rightarrow H = 33\ \text{m}$

b) $0.1 \times q_{avg} = \frac{q_{avg} \times 15 \times 2\ 0}{(15+H)(20+H)} \Rightarrow H = 37\ \text{m}$

3.5 Sources of Geotechnical Data

Geotechnical engineers primarily work with natural geomaterials, which exhibit uncertain and variable properties. Due to this inherent spatial variability, geotechnical engineers strive to minimize uncertainties to design secure foundations. This is accomplished by utilizing geotechnical data from various sources and tests, as listed below. To minimize uncertainties effectively, it is essential to incorporate diverse sets of measured and collected data. This overarching objective can be attained through different sources of geotechnical data, obtained via the following approaches:

1) Maps and technical literature review
2) Site visit
3) On situ testing
4) Geophysical exploration
5) Drilling
6) Sampling
7) Laboratory tests
8) In situ penetration tests
9) Physical modeling or Medium Scale Testing (Bench Scale)
10) Full-size field tests
11) Instrumentation and monitoring

Each of these categories is briefly reviewed in the following sections.

3.5.1 Maps and Technical Literature Review

Remote sensing constitutes the initial step in site investigation. In geotechnical exploration, a variety of pertinent maps are employed, including geological, geographical, topographical, and hydrological maps. Geohazard maps, such as those detailing liquefaction, landslides, and fault lines, are also valuable resources.

In addition to a site-specific literature review, any literature, maps, or site investigation reports accessible for adjacent sites or local projects prove invaluable. These resources serve as primary tools, enhancing the assessment of regional geology, soil composition, and bedrock stratification. They provide engineers with essential insights for planning new site investigations effectively.

3.5.2 Site Visit

A site visit or field exploration, involving a visual assessment of the project site, plays a crucial role in the preliminary investigation of any project. Several essential pieces of data and information can be gathered through a site visit, including:

- Surface geomaterials conditions.
- Surface water and drainage conditions.
- Performance of existing structures.
- Interaction with adjacent structures.
- The occurrence of any potential geohazards.
- Road access for future operations (e.g., site investigation and construction).

3.5.3 On Situ Testing

Following the site visit, various visual and manual tests should be conducted to validate the observations and develop a comprehensive and optimized site investigation program. The following topics are relevant to in situ visual and manual testing:

- Color and smell of surface geomaterials.
- Texture and fabric.
- Initial evaluation of soil gradation and plasticity.
- Moisture condition.
- Soil strength near the surface, which can be determined by a pocket penetrometer or a movable vehicle.
- Any indications of chemical activities, considerable erosion, or corrosion phenomena.

3.5.4 Geophysical Exploration

Traditional investigation methods like excavations of test pits, wells, trenches, galleries, and borings cover only a limited area of the project site. The results from these investigations do not fully represent the soil properties across the entire project site and should be interpreted cautiously. Geophysical exploration or nondestructive tests (NDT) can offer detailed and vital information over a larger portion of the site without disturbing the soil. Geophysical tests are particularly valuable for providing rapid three-dimensional (3D) coverage of large areas. However, geophysical explorations are cost-effective when the project site is sufficiently large, and other methods like penetration tests become impractical and uneconomical. It is important to note that nondestructive tests do not directly measure soil engineering properties such as plasticity, compressibility, shear strength, or grain size distribution. Their primary application lies in determining the depth and location of bedrock in structures like marine installations or dams. For instance, geophysical tests can assess soil layer characteristics and stratification.

There are three major categories of geophysical tests: seismic tests, electrical tests, and magnetic tests. Seismic tests rely on the differential seismic wave velocity in different soil types. Soil types are distinguished by measuring the wave velocity resulting from stimulation. Additionally, wave reflection allows for the determination of the location and depth of the bedrock. Electrical tests are based on differences in electrical conductivity or resistance in various soil types. This method distinguishes soil types by measuring apparent electric resistance through an electrical current. The magnetic approach, although more limited compared to other methods, relies on variations in the Earth's magnetic field. Generally, various geophysical parameters, such as ground resistivity and shear wave velocity, are measured alongside certain penetration tests to elucidate subsurface conditions. Table 3.4 outlines some common geophysical methods and their applications.

3.5.5 Drilling

The office work, literature review, site visit, and on-site testing are followed by field operations, which include horizontal excavations (galleries), vertical drilling boreholes (BH), test pits (TP), etc. The equipment used for excavations and borings varies based on the width, depth, and type of geomaterial formation. Boring equipment examples include hand-operated augers, portable power-operated vehicles, and heavier drilling rigs.

3.5.6 Sampling

The main objectives of excavation and boring operations are to obtain soil and rock samples for conducting laboratory tests and characterizing the subsurface soil profile. Two approaches are employed for sampling:

1) **Undisturbed Sampling:** These samples maintain soil texture, structure, natural stress states, and physical parameters such as unit weight and moisture, similar to in situ conditions.
2) **Disturbed Sampling:** These samples reflect conditions where the properties of soil have changed during drilling and sampling procedures.

Special efforts must be made during the sampling procedure to preserve the in situ soil structure using appropriate equipment. Attention should be given to the following factors to minimize disturbance:

- **Elimination of Vertical and Horizontal Stresses:** Ensure stresses within the soil skeleton are eliminated.
- **Volume Changes:** Minimize changes due to stress reduction.
- **Shearing and Compressibility:** Prevent shearing and compressibility during drilling and sampling.

Table 3.4 Some common geophysical investigation methods (Hunt 2005/Taylor & Francis Group).

Method	Applications	Comments
Seismic refraction (from surface)	Evaluate layer depths and velocities, land, or water geologic sections interpreted.	Most appropriate if velocities increase with depth and rock surface is regular.
Seismic direct (uphole, downhole, crosshole)	Obtain velocities for specific layer; dynamic properties; rock mass quality.	Requires drill holes. Crosshole yields the best results. Expensive.
Seismic reflection	General subsurface section is depicted. Water bodies yield clearest sections.	Land results are difficult to interpret. Velocities not obtained. Layer depth comps require other data.
Electrical resistivity	Locate saltwater boundaries, clean granular, and clay layer, bedrock depth, and underground mines by measured anomalies.	Difficult to interpret. Depends on wide variations. Probe configurations vary. No engineering properties were achieved.
Electrical conductivity	Obtain subsurface sections to find contaminant plumes by measured anomalies.	Difficult to interpret. Depends on wide variations. No engineering properties were achieved.
Gravimeter	Detect faults, domes, intrusions, cavities, and buried valleys by measured anomalies.	Accurate surface elevations are needed. Not commonly used. Measures density differences.
Magnetometer	Mineral prospecting, location of large igneous masses.	Commonly not used in engineering or groundwater studies.
Ground-probing radar	General subsurface section is depicted. Most useful to show buried pipe, bedrock, and voids boulders.	Interpretation difficult. Limited to shallow depths. No engineering properties were achieved.
Thermography	Shallow subsurface section is depicted. Useful for water pipeline leaks.	Interpretation difficult. Limited to shallow depths. No engineering properties were achieved.

- **Friction:** Reduce friction between soil sample and sampling apparatus.
- **Evaporation:** Prevent soil moisture evaporation, which can alter the sample's water content.
- **Air Bubbles:** Eliminate air bubble pressure in the soil.
- **Climatic Effects:** Account for changes in climatic conditions.
- **Handling:** Avoid shaking of the soil sample during drilling and transfer to the laboratory.

In practice, specific sampling methods like block sampling and thin-wall coring pipes can minimize disturbances during cohesive soil sampling. However, for non-cohesive soils, additional measures like ground freezing are necessary. Refer to Table 3.5 for suitable sampling procedures for different soil types.

3.5.7 Laboratory Tests

Soil and rock samples are subjected to laboratory tests to assess their fundamental, index, and engineering properties. The samples necessary for these tests must be meticulously obtained during field operations and stored appropriately for optimal results. In certain tests, disturbed samples can also be utilized to identify geotechnical properties unaffected by disturbances, such as Atterberg limits and natural water content. Various laboratory tests are typically categorized and may include:

- Classification and index tests, such as density and water content, gradation (sieve analysis, hydrometer test, and wash test), and Atterberg limits.
- Compressibility tests, including ramming tests, consolidation, inflation, and CBR (California Bearing Ratio).
- Water flow tests, such as permeability, capillary, and sand boil.

Table 3.5 Some sampling techniques (adapted from Sabatini et al. 2002; Loehr et al. 2016).

Sampler	Disturbed/Undisturbed	Appropriate soil types	Method of penetration
Split-barrel (split spoon)	Disturbed	Sands, silts, clays	Hammer driven
Thin-walled shelby tube	Undisturbed	Clays, silts, fine-grained soils, clayey sands	Mechanically pushed
Continuous push	Partially undisturbed	Sands, silts, and clays	Hydraulic push with plastic lining
Piston	Undisturbed	Silts and clays	Hydraulic push
Pitcher	Undisturbed	Stiff to hard clay, silt, sand, partially weathered rock, and frozen or resin-impregnated granular soil	Rotation and hydraulic pressure
Denison	Undisturbed	Stiff to hard clay, silt, sand, and partially weathered rock	Rotation and hydraulic pressure
Modified California	Disturbed	Sands, silts, clays, and gravels	Hammer driven (large split spoon)
Continuous Auger	Disturbed	Cohesive soils	Drilling w/Hollow Stem Augers
Bulk	Disturbed	Gravels, sands, silts, and clays	Hand tools, bucket augering
Block	Undisturbed	Cohesive soils and frozen or resin impregnated granular soil	Hand tools

- Chemical tests to identify sulfate and chloride contents, water and soil pH, corrosivity, and electrical resistivity.
- Shear strength tests, such as simple shear, direct shear, uniaxial, and triaxial tests
- Expansion and collapse tests.
- Dynamic tests, including resonant column, cyclic triaxial, and cyclic simple shear tests.

Although laboratory tests provide various platforms considering drainage and consolidation conditions, sample size, static or cyclic loading, and stress path for different analyses, they come with limitations such as sample size and modeling original in situ stress states. To address some of these constraints, physical modeling (medium- or full-scale testing), and in-field testing (e.g., in situ penetration tests) can be utilized as compensatory measures.

3.5.8 In Situ Penetration Tests (Please see Section 3.6)

3.5.9 Physical Modeling or Medium-Scale Testing (Bench Scale)

Due to the challenges and complexities involved in conducting full-scale physical modeling tests for geo-structures, geotechnical engineers often opt for constructing small- or medium-scale prototype models in the laboratory. These models allow them to simulate field conditions, replicating the physical and mechanical performance of the structure and its environmental factors. Specifically, these physical models analyze the performance of footings and piles in the laboratory. Small-scale piles or footings are employed to ensure that laboratory conditions closely mirror in situ conditions, including aspects like aspect ratio, state parameters, confining stresses, and other geotechnical factors.

Various apparatuses are utilized for physical modeling, such as simple chambers (1g), frustum confining vessels (FCV), centrifuge apparatus (ng), and calibration chambers. These tools help investigate the performance of different foundation systems in controlled laboratory environments. Consequently, the use of scaled-up model testing for foundations has witnessed a significant increase.

3.5.10 Full-Scale Tests

For most projects, particularly in geotechnical engineering applications like excavations, pile groups, combined foundations, airports, and dams, conducting full-scale testing can be intricate, expensive, and time-consuming. However, in specific cases involving multiple geo-structural units, full-scale testing is essential, especially for individual elements like anchors, single piles, and single footings. These tests are often carried out following established standard procedures, providing crucial performance data for various foundation loadings.

Full-scale testing is indispensable for investigating and quantifying the desired function of the foundation. It allows structural elements to be tested under real operational and environmental conditions, offering valuable insights through direct measurements rather than relying on indirect investigations. Consequently, this approach provides precise and reliable information, enhancing the overall understanding of foundation behavior.

3.5.11 Instrumentation, Surveillance, and Monitoring

As previously discussed, the inherent heterogeneity of geomaterials leads to complex behavior compared to engineered materials like steel and concrete. Consequently, evaluating the actual properties and performance of geomaterials involves challenges and uncertainties. In geotechnical engineering, instrumentation and monitoring play a crucial role. Utilizing various instruments to monitor the performance of geotechnical structures and measure stresses and displacements at different construction and serviceability stages is essential. Specialized sensitive equipment is installed to monitor these structures, and the recorded data is digitized using transducers.

Monitoring provides essential and detailed information for both design and practical applications. Essentially, these sophisticated instruments serve as surveillance tools, assessing geomaterial performance during and after construction. The necessity of instrumentation and monitoring can be attributed to the following reasons:

- **Cost Reduction:** Appropriate instrumentation helps mitigate design risks and potential hazards in sites with complex conditions, such as problematic soils, leading to lower risks and cost savings.
- **Safety Factor Reduction:** Proper instrumentation reduces the need for assumptions in selecting design parameters, resulting in lower safety factors and reduced expenses.
- **Design Modification:** In cases where existing design parameters are insufficient, data from instrumentation and monitoring during the project's initial phases can aid in determining necessary design modifications.
- **Safety:** Instrumentation can provide advance warnings of potential geohazards, particularly in slope stability projects, enhancing overall safety measures.
- **Operation Procedure Control:** Sophisticated instruments offer valuable data on the effectiveness of operations like ground modification, enabling precise control over procedures. For instance, the number of drains employed can be optimized based on this data.

3.6 In Situ Penetration Tests

In many instances, traditional geotechnical site investigation methods involving boring, sampling, transferring samples to the laboratory, and conducting tests can be cumbersome, expensive, and time-intensive processes. Consequently, bringing test equipment to the project site has become a preferred approach. Furthermore, advancements in in situ testing technology have led to a more detailed understanding of geomaterial behavior.

Since the beginning of the new century, significant strides have been made in the field of in situ testing, particularly in identifying soil stiffness and strength parameters. In situ tests have found extensive application in non-cohesive soils where sampling is challenging. However, it is important to recognize that in situ tests also have limitations, such as uncontrolled drainage conditions and stress paths.

Therefore, a comprehensive evaluation of geomaterial properties necessitates the concurrent use of both in situ and laboratory tests. In fact, in situ and laboratory tests are complementary, and conducting them simultaneously yields more reliable data and parameters for design purposes.

The in situ tests can be used to directly evaluate subsurface conditions, geomaterial engineering properties, and characteristics. The following in situ tests are more common in geotechnical practice:

- SPT: Standard Penetration Test
- CPT: Cone Penetration Test
- CPTu: Piezocone Penetration Test
- DMT: Dilatometer Test
- PMT: Pressure Meter Test
- VST: Vane Shear Test
- PLT: Plate Load Test
- BSPT: Becker Penetration Test

Performing SPT, VST, and PMT requires drilling a borehole employing a suitable rotary drilling rig and equipment. However, specific types of VST and PMT (i.e., fully displacement), where the drilling rod is directly pushed to the desired depth, can be conducted without boreholes. Conversely, DMT, CPT, and CPTu do not require a borehole because they utilize direct penetration technology. Table 3.6 outlines the suitable range of in situ tests based on various soil properties.

The in situ tests can be divided into two main categories:

1) Logging tests (e.g., CPT) are basically penetration tests and provide continuous records of in-depth soil measurements. Logging tests are commonly repeatable, fast, and economical.
2) Specific tests (e.g., PMT) are generally used to identify specific soil parameters such as deformation modulus or shear strength. Specific tests are usually more expensive and time-consuming compared to logging tests.

It is preferable to use both logging tests and specific tests to achieve more reliable data. Table 3.7 presents the general procedure and major records of commonly used in situ tests for foundation engineering. It aims to assess the role and application of dominant in situ tests via a few case studies at the end of this chapter.

3.7 Variability from Laboratory and In Situ Testing

It is essential to acknowledge the limitations of in situ testing equipment in geotechnical investigations. Moreover, measurement errors should be considered in the context of soil variability. SPT exhibits high variability as an in situ test, while dilatometer and electrical cone penetration tests demonstrate the lowest variability. The variabilities of laboratory tests and in situ tests are presented in Tables 3.8 and 3.9, respectively. In clays, the variability of Atterberg and shear strength tests is high; however, soil density can be evaluated with high accuracy.

3.8 Geotechnical Reports

Upon completing site investigations, it is the geotechnical engineer's responsibility to synthesize and interpret the information, construct a site geology model, and purpose appropriate geomaterial parameters for design.

Typically, there are two types of reports in geotechnical engineering practice (Brown et al. 2010): geotechnical site investigation and geotechnical design report.

A geotechnical site investigation report must include the in situ investigations and laboratory tests carried out and present the resulting data. Typically, this type of report does not include analysis and interpretation of the subsurface conditions or any design recommendations. Table 3.10 represents different categories of information that should typically be included in a geotechnical investigation report.

A geotechnical design report, also known as a foundation report, typically includes an evaluation of actual subsurface conditions and potential geohazards, as well as geotechnical analyses and recommendations for designing the appropriate foundation system with any related earth structures. Geotechnical design reports must characterize different soil and rock units from an engineering point of view and propose appropriate design parameters for them. This requires a summary of all resulting data and their analyses to verify the suggested design parameters. Groundwater elevation and its fluctuation, which are significant for both design and construction procedures, must be accurately evaluated. Any design issue, specifically the foundation axial resistance, settlement analysis, group behavior, and response to lateral loading, should be mentioned regarding the recommended methodologies.

It is necessary to provide a site location map on a regional map for reference. Exact locations of all site project drilling, sampling, in situ tests, and any other type of exploratory investigations must be indicated clearly on the site map. Preferably, the site plan should be a well-detailed topographical map with elevation contours and an adequately established benchmark. Moreover, site location plans can be shown directly on aerial photographs that show true North (*N*) direction.

Geotechnical reports are usually accompanied by representing subsurface conditions obtained from the result of in situ and laboratory tests. Longitudinal profiles are commonly created along the alignment of the project, and for critical locations with special conditions, a limited number of transverse profiles may be included. Subsurface profiles, with an understanding of the geological setting and proper engineering judgment, assist the geotechnical engineer in interpreting subsurface conditions between investigation sites.

Table 3.6 Various types of in situ tests and their applications (Lunne et al. 1997/Taylor & Francis Group).

Group	Device	Soil parameters													Ground type						
		Soil type	Profile	u	*Φ'	S_u	I_D	m_v	c_v	k	G_0	σ_h	OCR	σ–ϵ	Hard rock	Soft rock	Gravel	Sand	Silt	Clay	Peat
Penetrometers	Dynamic	C	B	—	C	C	C	—	—	—	C	—	C	—	—	C	B	A	B	B	B
	Mechanical	B	A/B	—	C	C	B	C	—	—	C	C	C	—	—	C	C	A	A	A	A
	Electric (CPT)	B	A	—	C	B	A/B	C	—	—	B	B/C	B	—	—	C	C	A	A	A	A
	Piezocone (CPTU)	A	A	A	B	B	A/B	B	A/B	B	B	B/C	B	C	—	C	—	A	A	A	A
	Seismic (SCPT/SCPTU)	A	A	A	B	A/B	A/B	B	A/B	B	A	B	B	B	—	C	—	A	A	A	A
	Flat dilatometer (DMT)	B	A	C	B	B	C	B	—	—	B	B	B	C	C	C	—	A	A	A	A
	Standard penetration test (SPT)	A	B	—	C	C	B	—	—	—	C	—	C	—	—	C	B	A	A	A	A
	Resistivity probe	B	B	—	B	C	A	C	—	—	—	—	—	—	—	C	—	A	A	A	A
Pressure-meters	Pre-bored (PBP)	B	B	—	C	B	C	B	C	—	B	C	C	C	A	A	B	B	B	A	B
	Self-boring (SBP)	B	B	A(1)	B	B	B	B	A(1)	B	A(2)	A/B	B	A/B(2)	—	B	—	B	B	A	B
	Full displacement (FDP)	B	B	—	C	B	C	C	C	—	A(2)	C	C	C	—	C	—	B	B	A	A
Others	Vane	B	C	—	—	A	—	—	—	—	—	—	B/C	B	—	—	—	—	—	A	B
	Plate load	C	—	—	C	B	B	B	C	C	A	C	B	B	B	A	B	B	B	A	A
	Screw plate	C	C	—	C	B	B	B	C	C	A	C	B	—	—	—	—	A	A	A	A
	Borehole permeability	C	—	A	—	—	—	—	B	A	—	—	—	—	A	A	A	A	A	A	B
	Hydraulic fracture	—	—	B	—	—	—	—	C	C	—	B	—	—	B	—	—	—	—	A	C
	Crosshole/downhole/surface seismic	C	C	—	—	—	—	—	—	—	A	—	B	—	A	A	A	A	A	A	A

Applicability: A = high, B = moderate, C = low, — = none.
*Φ' = Will depend on soil type: (1) = Only when pore pressure sensor fitted; (2) = Only when displacement sensor fitted; u = in situ static pore pressure; Φ' = effective internal friction angle; S_u = undrained shear strength; I_D = density index; m_v = constrained modulus; c_v = coefficient of consolidation; k = coefficient of permeability; G_0 = shear modulus at small strains; σ_h = horizontal stress; OCR = over consolidation ratio; σ–ϵ = stress–strain relationship.

Table 3.7 Testing procedures and records presentation of commonly used in situ tests.

Test and standard	Suitable soils	Testing procedure		Records: Graphical presentation	Records: Numerical quantity
Standard Penetration Test (SPT) ASTM D1586	Almost all soil types	• Insert an SPT split-spoon sampler into the hole. • Raise a 63.5-kg hammer to a distance of 760 mm and allow it to fall. • Repeat this process until the sampler has penetrated 450 mm. • Record the number of hammer blows required for each 150-mm interval for three intervals.	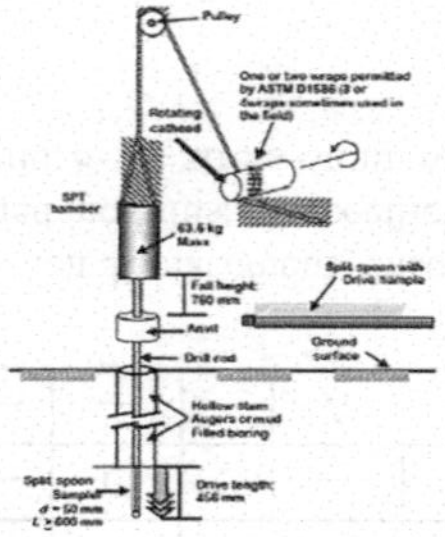	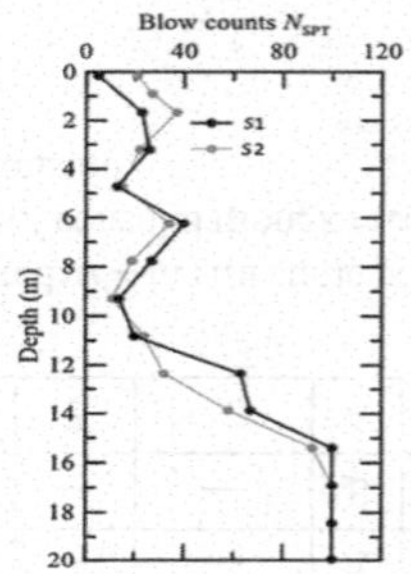	*N*: The number of blows required to achieve a penetration of 12 in. (300 mm)
Vane Shear Test (VST) ASTM D2573	Soft to medium fine-grained soils	• Insert a metal vane into the soil. • Rotate the vane at a rate of 6°/min until the soil fails in shear. • Record the maximum torque during the rotation.	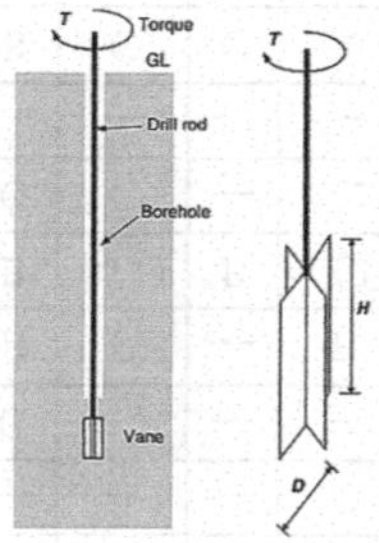	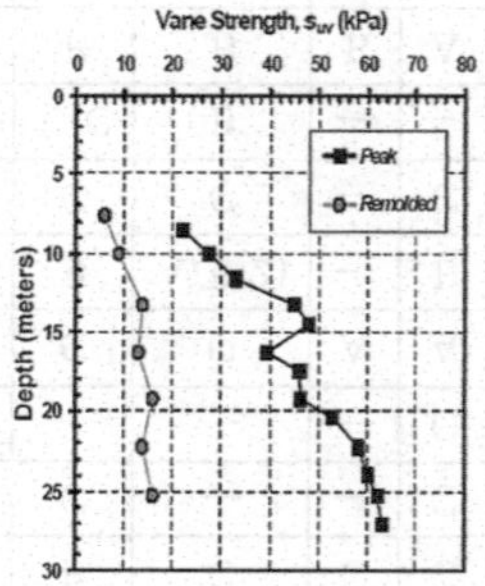	S_u: Undrained shear strength $S_u = \frac{6T_{max}}{7\pi D^3}$
Cone Penetration Test (CPT) ASTM D3441	All except gravels	• The cone is pushed continuously by a hydraulic jack at a speed of 20 mm/s. • The readings are taken at every 10–50 mm.	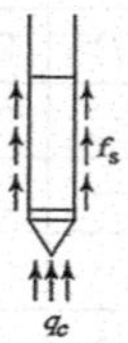	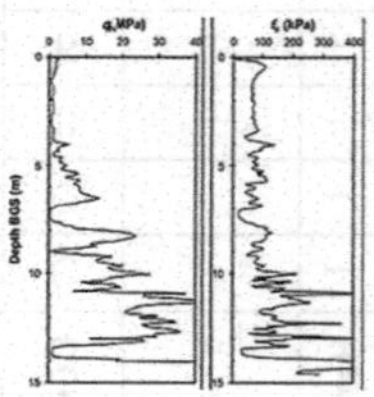	q_c: Cone point resistance f_s: Sleeve friction
Piezocone Penetration Test (CPTu) ASTM D6067	Fine-grained (very soft to hard) Granular (loose to medium-dense)	• Probe is pushed into the ground at a speed of 2 cm/s to reach a predefined depth or the probe cannot be pushed any further due to obstacles.	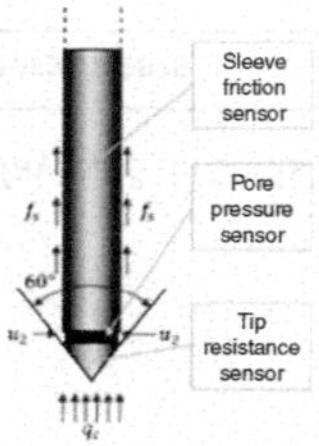	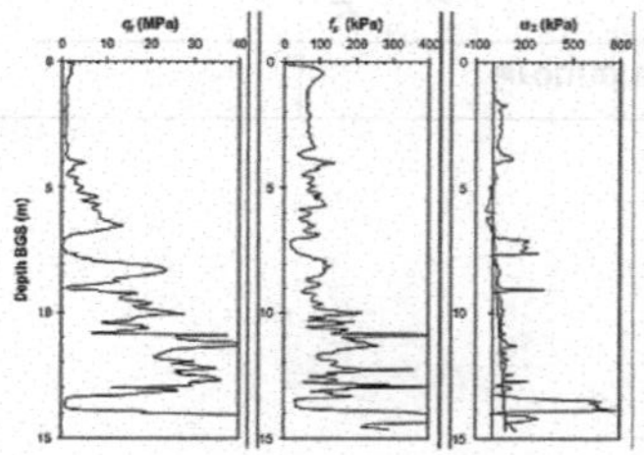	q_c: Cone point resistance f_s: Sleeve friction u_2: Pore pressure at cone shoulder

Dilatometer Test (DMT) ASTM D6635	Almost all soil types	• A tapered stainless-steel blade with an 18-degree wedge tip that is pushed vertically into the ground at 8-in. (200 mm) intervals at a rate of (0.8-in./s) (20 mm/s). • The blade is connected to a readout pressure gauge at the ground surface via a special air tubing with internal wire pre-threaded through push rods. • A 2.4 in. (60 mm) diameter flexible steel membrane located on one side of the blade is inflated pneumatically.	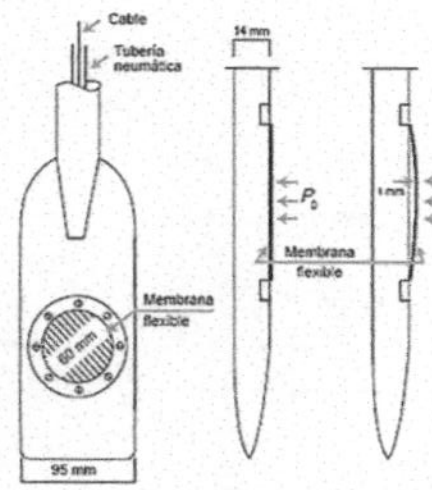	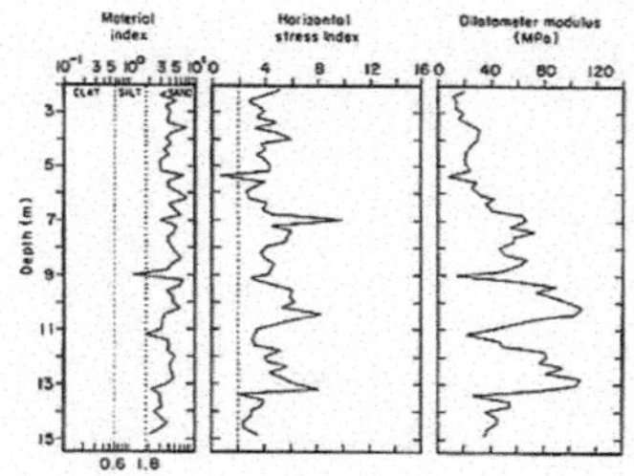	I_D: Material index K_D: Horizontal stress index E_D: Dilatometer modulus
Pressure meter Test (PMT) ASTM D4719	Almost all types of soils and soft rock	• The test is conducted in a pre-bored hole with a diameter that is between 1.03 and 1.2 times the nominal diameter of the probe. • The most common probe has a diameter of 58 mm and a length of 420 mm. • Probe cells can be expanded by either liquid or gas. • Pressure is applied in increments and the new volume of the cell is measured. The process is continued until the soil fails or the device's pressure limit is reached.	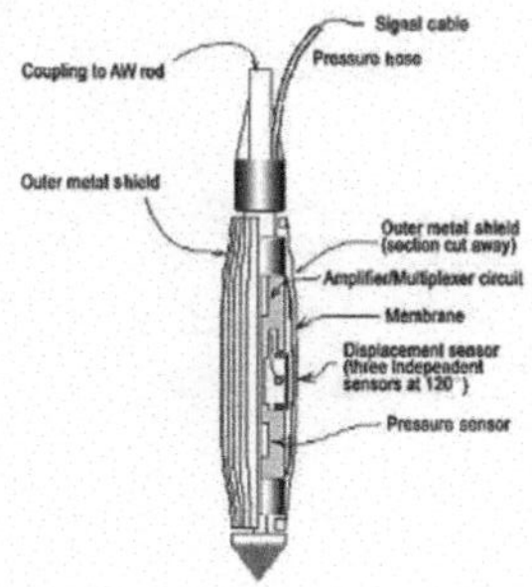	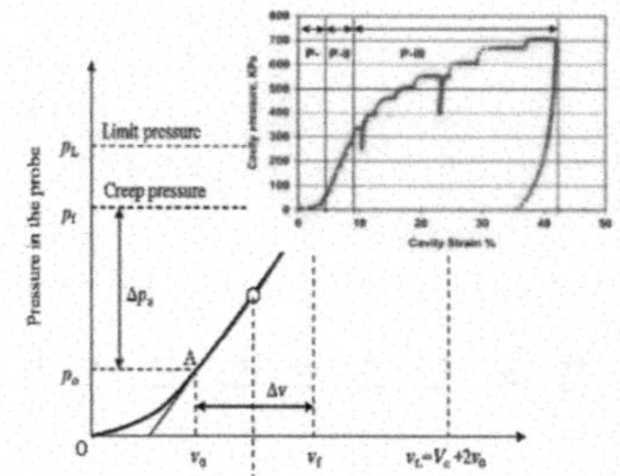	E_p: Pressuremeter modulus P_f: Creep, or yield pressure P_l: Limit pressure G: Shear Modulus
Becker Penetration Test (BPT)	Sands through boulders	• Using a small diesel pile-driving hammer and percussion action to drive a 135–230 mm (5.5–9.0 in.) diameter double-wall steel casing into the ground. • The cuttings are sent to the top by blowing air through the casing.	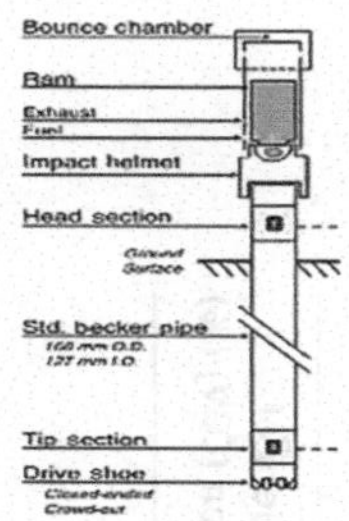	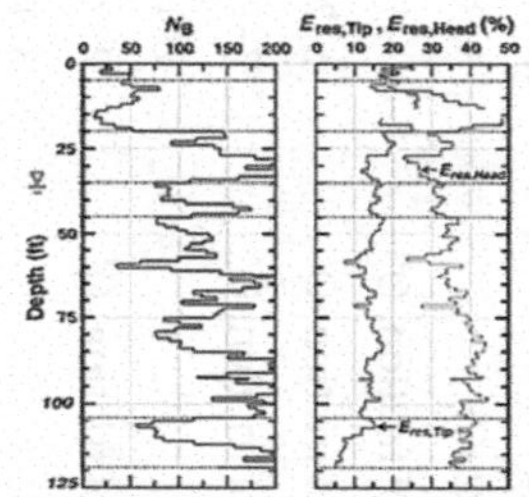	N_B: The number of blows required to advance the casing 300 mm (1 ft) E_{res}: Residual energies measured at head and tip sections
Plate Load Test (PLT) ASTM D1194	Almost all soil types	• Typical sizes of plates range from 0.15 m (6 in.) to 0.91 m (36 in.). • Plate rests uniformly on the soil; arrange a loading system to provide a sufficient reaction; and provide a means for measuring plate deformation.	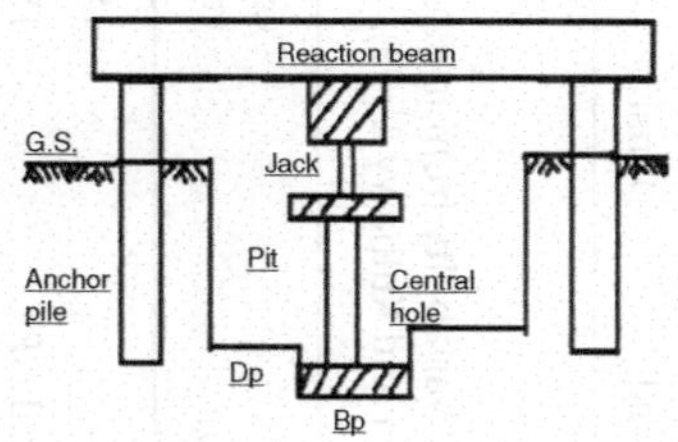	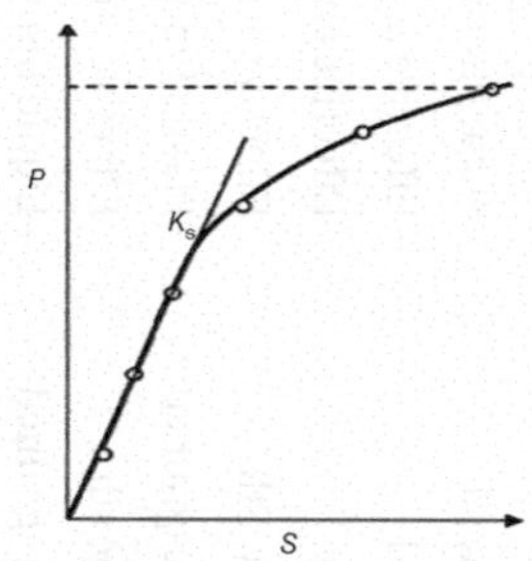	K_s: Subgrade reaction E_s: Elastic modulus Q_U: Ultimate bearing capacity of shallow foundation

Table 3.8 Variability from laboratory tests (adapted from Phoon and Kulhawy 1999).

		Coefficient of variation (COV) (%)	
Soil type	**Property**	**Range**	**Mean**
Fine-grained	Plasticity index	5–51	24
Clay, silt	Effective angle of friction, Ø′	7–56	24
Clay, silt	Shear strength, Cu	19–20	20
Clay, silt	Shear strength, Cu	8–38	19
Sand	Effective angle of friction, Ø′	13–14	14
Clay	Effective angle of friction, Ø′	6–22	14
Clay, Silt	Effective angle of friction, Ø′	3–29	13
Fine-grained	Plastic limit	7–18	10
Clay, silt	Effective angle of friction, Ø′	2–22	8
Fine-grained	Liquid limit	3–11	7
Fine-grained	Density	1–2	1

Table 3.9 Variability from in situ tests (adapted from Phoon and Kulhawy 1999).

Test	Coefficient of variation (COV) (%)
SPT	15–45
Mechanical CPT	15–25
SBP	15–25
VST	10–20
PBP	10–20
Electric CPT	5–15
DMT	5–15

Traditional in situ and laboratory tests have significantly contributed to refining design procedures over the last century. However, recent advancements in geophysical testing, analytical and numerical modeling, and the availability of geotechnical databanks have further enhanced foundation design procedures (Figure 3.4).

Regardless of the approach chosen, engineering judgment and experience must play a pivotal role in the selection of the final foundation design.

3.9 Problematic Soils

Certainly, some soils pose challenges in geotechnical engineering due to their tendencies to collapse, expand, disperse, lack strength significantly, undergo substantial settlement, or exhibit corrosive properties. These characteristics can be attributed to their mineralogy, fabric, composition, or the nature of their pore fluids. This phenomenon is particularly noticeable in fine-grained soil types with potential for chemical and electrical activities. Problematic soils encompass various types, including normally consolidated, sensitive, expansive, dispersive clay, collapsible soils, organic soils and peat, frozen deposits, liquefiable sands, and non-engineered fills. Abnormal activities in these soils can lead to significant

Table 3.10 Typical information included in a geotechnical investigation report (Brown et al. 2010/U.S. Department of Transportation/Public Domain).

Background information	Overview of the project (bridge, retaining wall, structure, or other facilities)
	General site conditions (topography, geology, ground cover, drainage, accessibility)
	Specific methods employed for site investigation
Scope of site investigation	Plans showing exact locations of all drillings, test pits, and in situ tests
	Number, locations, and depths of all borings and in situ tests
	Types and frequency of samples obtained; standards used
	Types and numbers of laboratory tests; standards used
	Subcontractors operating the work and dates of work
	Systems used for soil and rock classification
Data presentation	Logs of all borings and test pits
	Water level readings and other groundwater data collected
	Data tabulations and plots from each in situ test hole
	Results of geophysical testing
	Summary tables and datasheets for conducted lab tests
	Rock core photographs
	Geologic mapping data sheets and summary plots
	Subsurface profiles from field and lab test data; statistical summaries
	Existing information from previous investigations (boring logs, data)

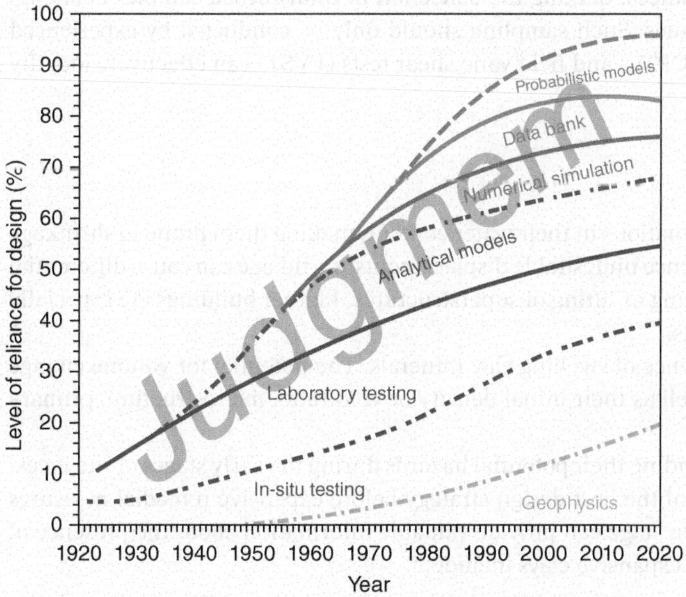

Figure 3.4 Evolution of site geotechnical characterization (adapted from Lacasse 1985).

damage to superstructures and financial losses. According to Coduto et al. (2016), the annual damages in the USA due to project malfunctions on challenging soils surpass the hazards of tornadoes, floods, and earthquakes. Several approaches can be employed to address problematic geomaterials:

1) Avoiding construction on the site.
2) Designing structures accordingly.
3) Improving geomaterial engineering properties using modification techniques.
4) Removing problematic geomaterials and replacing them with non-problematic geomaterials.

3.9.1 Soft Clay (Normally Consolidated Clays)

Clays with soft to medium consistency are prevalent across extensive regions and have undergone consolidation solely under their own weight. The application of additional overburden can lead to substantial long-term settlement, potentially reaching up to 30% of the original height. Without mitigation measures and appropriate precautions, structures may displace or settle unexpectedly, causing uncontrolled and undesired deformations. The extent and approximate settlement rate can be forecasted through analyses based on consolidation or oedometer test results. Furthermore, amendments are necessary to replicate laboratory findings accurately in real-world field conditions.

3.9.2 Sensitive Clays

Sensitive clays exhibit a remolded strength significantly lower, often a quarter or less, than their undisturbed strength. Typically, these clays have in situ water contents equal to or greater than their liquid limits (LL), indicating their presence. When foundations are placed on deposits consolidated by the weight of eroded materials in a project site, applied loads might generate excessive shearing stresses under foundation systems, leading to potential catastrophic failure. Deep excavations in sensitive clays can be perilous due to their substantial loss in shear strength. This effect is pronounced during pile driving in these soils, resulting in a drastic strength reduction, facilitating pile foundation construction. However, this phenomenon usually resolves within days or weeks, restoring the original strength.

Assessing the physical properties of sensitive clays requires laboratory tests on undisturbed and remolded samples. Notably, these clays are highly sensitive to minor disturbances, making the collection of undisturbed samples challenging, demanding special equipment and advanced techniques. Such sampling should only be conducted by experienced personnel. Additionally, cone penetration tests (CPT and CPTu) and field vane shear tests (FVST) can effectively identify these surprising soils in the field.

3.9.3 Expansive Soils

Certain clays undergo gradual volume changes due to fluctuations in their water content, making them prone to shrinkage or swelling. Structures built on expansive clays may experience undesirable displacements; shrinkage can cause differential settlement, while pressures from swelling can lead to heaving or lifting of superstructures. Lighter buildings are especially susceptible to failures induced by these ground movements.

Expansive clays are primarily characterized by the presence of swelling clay minerals. The potential for volume change depends on the type and quantity of these minerals, as well as their initial density or void ratio, microstructure, primary water content, and applied vertical stress.

Identifying the presence of expansive soils and understanding their potential hazards during the early stages of site investigations is crucial. This awareness ensures the selection of the best design strategy before expensive remedial measures are taken. Field observations in the initial site investigation stage can provide valuable information about the presence of expansive clays and their potential hazards. Some signs of expansive clays include:

Soil Characteristics

- Spacing and width of wide or deep shrinkage cracks.
- High dry strength and low wet strength – high plasticity soil.
- Stickiness and low trafficability when wet.
- Shear surfaces have a glazed or shiny appearance.

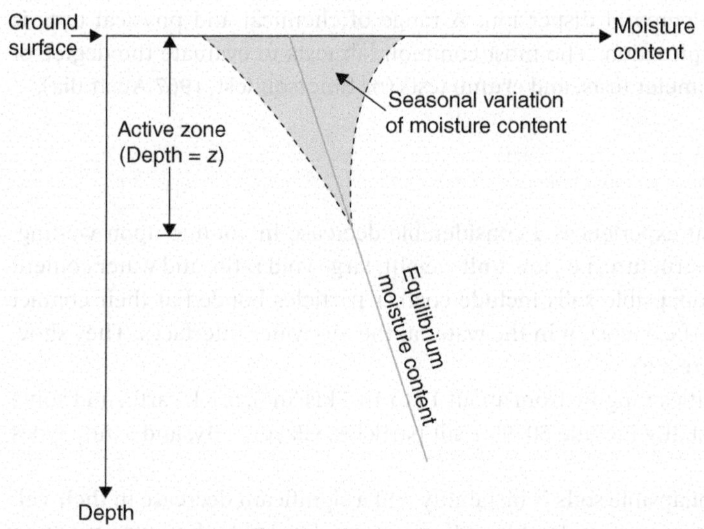

Figure 3.5 Definition of the active zone (Das and Sivakugan 2019 / Cengage Learning).

Geology and Topography

- Undulating topography.
- Evidence of low permeability by surface drainage and infiltration features.

Environmental Conditions

- Vegetation type
- Climate

Methods of determining volume changes in expansive clays can be grouped into (i) empirical (determining swelling potential from liquid and plastic limits, activity, void ratio, and natural water content), (ii) soil suction, and (iii) oedometer methods. Moreover, due to expansive clays being sensitive to water, it is vital to identify the depth at which periodic moisture changes occur (Active Zone). The moisture content fluctuation is highest near the ground surface and decreases with depth (Figure 3.5).

3.9.4 Dispersive Clays

Certain types of clay are susceptible to soil dispersion, a process in which soil aggregates break down into very small particles in the presence of water. The distinction between dispersive clays and normal clays lies in their electrochemical properties. The key factor governing a clay's susceptibility to dispersion is the concentration of dissolved sodium cations (Na^+) in the pore water compared to other major basic cations such as magnesium (Mg^+) and calcium (Ca^+). Sodium increases the diffused double water layer thickness around individual clay particles, reducing the attractive force between particles and making detachment easier. Flow-through cracks in dispersive clays can lead to rapid erosion and crack enlargement. Dispersion can occur due to chemical or physical reactions. Key factors in initiating internal erosion include hydraulic gradient, gap-graded soil, and overburden pressure.

Dispersive soils are unsuitable as fill material in constructing hydraulic earth structures or as foundation material. Utilizing dispersive clays in hydraulic structures like embankment dams can lead to catastrophic failures if not identified and used correctly. Failures in dispersive clay dams caused by internal erosion are a common occurrence.

Some visual identifiers of the existence of dispersive clays are:

- Deep cuts due to erosion in steep slopes.
- Caves and tunnels similar to limestone-dissolved karsts.
- Natural water canals with sharp turns.
- Sinkholes.
- Muddy water that never settles.

There are also some laboratory tests to identify the degree of dispersion. A range of chemical and physical tests is suggested, rather than relying on a single analysis interpretation. The most common lab tests to evaluate the degree of dispersion are pinhole tests, chemical tests, double hydrometer tests, and crumb tests (or Emerson test, 1967 Australia).

3.9.5 Collapsible Soils

Collapsible soils are defined as any unsaturated soil that experiences a considerable decrease in volume upon wetting, additional loading, or both. Collapsible soils have a loose structure, i.e., low unit weight, large void ratio, and water content far less than saturation. Commonly, the structures of collapsible soils include coarser particles bonded at their contact points by the finer fraction (silt or clay) or possibly by surface tension in the water at the air–water interfaces. They show relatively high apparent strength in their dry states (Figure 3.6).

Soils such as loess (void ratio >1.0 and in situ dry densities ranging from about 10 to 16.5 kN/m^3), brick earth, and some windblown silts may be prone to collapse. These soils usually include 50–90% silt particles. Clayey, silty, and sandy types tend to fall into the silty category.

One of the main critical problems encountered with collapsible soils is instability and a significant decrease in their volume due to small changes in the soil water content, resulting in considerable settlement and damage to the superstructure. Jennings and Knight (1975) have proposed an approach to illustrate the collapse potential of a soil: an un-disturbed soil sample has been taken at its natural moisture content in a consolidation ring. Then, step loads are exerted on the sample up to a pressure of 200 kN/m^2. The specimen is then saturated for 24 hours. In fact, this test defines the void ratios e_1 and e_2 prior to and following the flood, respectively (Figure 3.7).

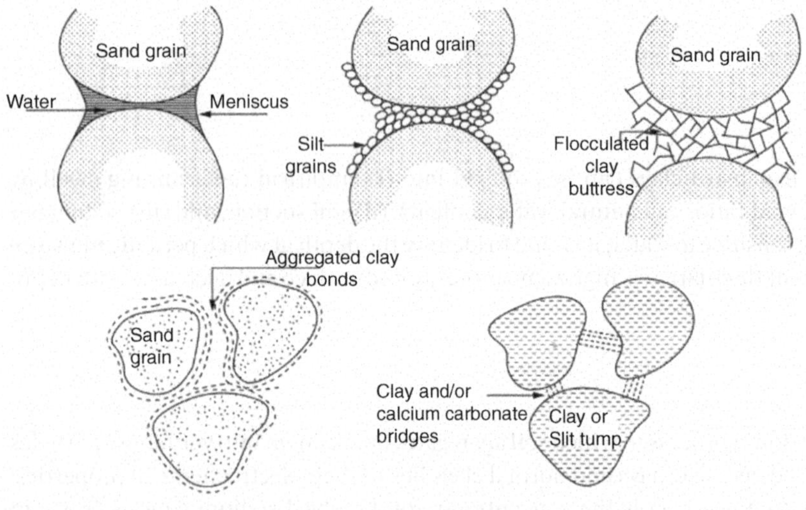

Figure 3.6 Typical bonding arrangement formed in collapsible soils (John et al. 2012 / Emerald Publishing Limited).

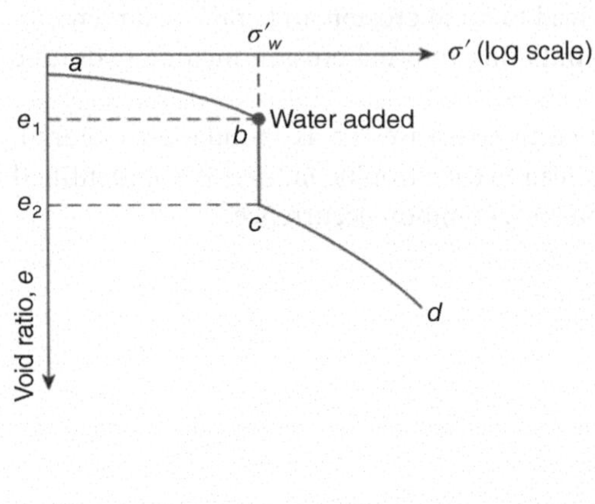

Figure 3.7 Void ratio-log presser curve for a collapsing soil (Das and Sivakugan 2019 / Cengage Learning).

In order to provide an economical and efficient engineering solution, four basic steps must be undertaken when dealing with collapsible soils (after Popescu 1986):

1) **Identification:** Determine whether a collapsible soil exists.
2) **Classification:** If a collapsible soil exists, how significant is it?
3) **Quantification:** Assess the degree of collapse that will occur.
4) **Evaluation:** Assess the design options.

3.9.6 Organic Soils and Peat

Peat and organic soils are unconsolidated surficial deposits characterized by extreme wetness, softness, and compressibility. They can also form a stratum beneath other surficial deposits. The term "peat" refers to highly organic soils primarily derived from plant remains. Peats are highly compressible surficial deposits with a high content of noncrystalline colloid (humus), constituting the subsurface of wetland systems. Although plant fibers may be visible, they might not be apparent in advanced stages of decomposition. In geotechnical engineering, soils with organic contents exceeding 20% are considered organic soils, and peat specifically refers to organic soil with an organic content exceeding 75% (Table 3.11).

Dealing with peat and organic soils poses challenges in sampling, in situ testing, settlement, stability, stabilization, and construction due to their high compressibility and low shear strength. Recognizing these issues during the initial site investigation is crucial. Peat and organic soils are significantly weaker and more compressible than inorganic soils, making them unsuitable for most geotechnical engineering projects. Typically, the standard practice involves avoiding these soils, removing them, or using deep foundations (piles) as a solution.

Organic soils and peat have specific characteristics that differentiate them from most mineral soils and require special construction considerations. These special characteristics include (Huat et al. 2014):

1) High water content (moisture up to 1500%).
2) Low shear strength (5–20 kPa).
3) High compressibility (Figure 3.8).
4) A high degree of spatial variability.

Table 3.11 Soil types based on organic content (ASTM D4427).

Basic soil type	Description	Organic content (%)
Clay or silt or sand	Slightly organic	2–20
Organic soil	—	25–75
Peat	—	>75

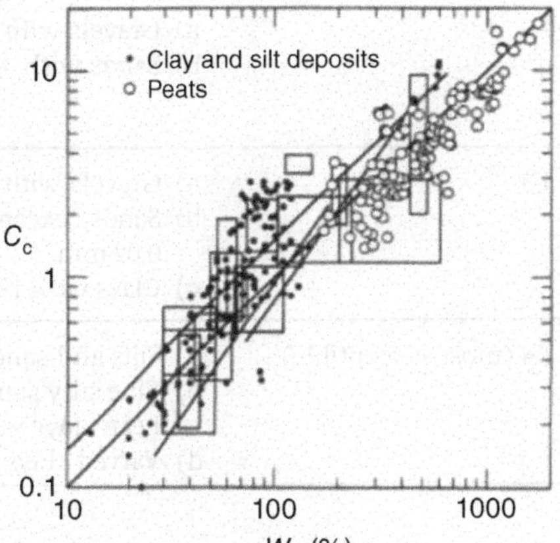

Figure 3.8 Relation between natural water content and compressibility coefficient of silt, clay, and organic soils (Mesri et al. 1997 / American Society of Civil Engineers).

5) Potential for progressive decomposition as a consequence of changing environmental conditions.
6) High permeability compared with clay.
7) High charge and high specific surface area of the colloidal fraction of peat.

3.9.7 Frozen Soils

Frost heave is a common foundation problem in areas with frozen ground phenomena, where winter temperatures rarely exceed the freezing point, and summer temperatures are just warm enough to cause thawing near the ground surface. The ground undergoes considerable disturbance due to the expansion of growing ice bodies (water expands by approximately 9% when it freezes) or excess pore water pressures in confined water-bearing lenses.

To design foundation systems in areas susceptible to frost heave, it is crucial to evaluate the depth of frost penetration (the greatest depth to which the ground might become frozen) and the soil frost susceptibility (Table 3.12). Foundations must be constructed at a depth below the frost penetration level to protect them from the effects of frost.

3.9.8 Liquefiable Soils

Liquefaction is a phenomenon linked to zero effective stress caused by elevated pore water pressure resulting from the tendency of sand structures to densify under cyclic loading. A characteristic feature of liquefaction events is the generation of excess pore pressure under undrained loading conditions. In this scenario, rapid compaction of sand is not possible due to the relative incompressibility of water. Instead, excess pore water pressure builds up, and its magnitude increases with the duration of cyclic loading. Typically, these additional pressures begin to dissipate only after the ground stops shaking. Liquefaction occurs in loose and saturated sand deposits. In the case of clays, cyclic softening leads to reduced strength and stiffness. Under seismic conditions, there is a potential for compression in all types of sand, whether loose, dense, dry, partially saturated, or fully saturated, due to the imposition of dynamic compaction.

The factors that influence the onset and progression of liquefaction include:

1) Seismological factors such as the cycling load duration, frequency, and amplitude of the cyclic shear stress.
2) Site-specific factors such as depth and thickness of different soil strata, the grain size distribution and relative density of the soil, and groundwater table elevation.

The stiffness and strength of soils significantly decrease when liquefaction occurs, leading the soil to lose its passive particle grid structure. Structures with foundations on liquefied soils may experience abrupt loss of support, causing severe

Table 3.12 Frost susceptibility of different soil types (adapted from Coduto 2001).

Group	Soil types	USCS group symbols
F1 (least susceptible)	Gravels with 3–10% finer than 0.02 mm	GW, GP, GW-GM, GP-GM
F2	a) Gravels with 10–20% finer than 0.02 mm b) Sands with 3–15% finer than 0.02 mm	GM, GW-GM, GP-GM SW, SP, SM, SW-SM, SP-SM
F3	a) Gravels with more than 20% finer than 0.02 mm b) Sands, except very fine silty sands, with more than 15% finer than 0.02 mm c) Clays with IP 7 12, except varved clays	GM, GC SM, SC CL, CH
F4 (most susceptible)	a) Silts and sandy silts b) Fine silty sands with more than 15% finer than 0.02 mm c) Lean clays with IP <12 d) Varved clays and other fine-grained, banded sediments	GM, GC, SM, CL, CL-ML

(a) (b)

Figure 3.9 Seismic compression and liquefaction-induced failures: (a) lateral spreading (USGS/EERI), (b) heave (floatation) (wordpress org).

settlement, tilting, and structural damage. Seismic compression and liquefaction can also lead to landslides, lateral spreads, and failures of waterfront retaining structures, among other issues (see Figure 3.9).

3.9.9 Non-engineered Fills

The term "fill" refers to material deposited by human activities. The material may be natural or have been modified before deposition, and include industrial waste, domestic waste, building waste, slag, mining and quarry waste, and soil waste.

The term "non-engineered fill" also known as "uncontrolled fill" denotes fill that undergoes no specific treatment and compaction measures during or after deposition. Typically, non-engineered fills have low bearing capacities, settle unevenly, and undergo significant displacement due to factors unrelated to applied loading (refer to Table 3.13). Some fills may also pose additional risks, such as contamination, gas emissions, and spontaneous combustion.

Table 3.13 Possible self-weight settlement of different fill materials (with permission of Leach and Goodger 1991).

Material	Potential self-weight settlement: depth of fill (%)
Well-compacted and well-graded sand and gravel	0.5
Well-compacted shale, chalk, and rock-fills	0.5
Medium-compacted rock-fills	1.0
Well-compacted clay	0.5
Lightly compacted clay and chalk	1.5
Lightly compacted clay placed in deep layers	1.0–2.0
Nominally compacted opencast backfill	1.2
Uncompacted sand	3.5
Poorly compacted chalk	1.0
Uncompacted (pumped) clay fill	12.0
Well-compacted mixed refuse	30.0
Well-controlled domestic refuse placed in layers and well compacted	10.0
Opencast backfill compacted by scarpers	0.6–0.8

Table 3.14 Factors influencing fill behavior (Eslami et al. 2015/Springer Nature).

Factor	Description	Effects
Method of deposition and degree of compaction	Thin layers with heavily compaction or thick layers with low or no compaction efforts	Low shear strength, especially, in fine-grained materials; non-compacted materials will sustain large settlements
Nature of materials	Homogeneity, chemical composition, organic content	Non-homogeneity causes differential settlements; seepage of chemical waste.
Particle size distribution	Coarse grained, fine grained, or mixed	Controls pervious, so the short- and long-time shear strength of materials and type of settlements.
Age	Time that has elapsed since placement	In fine-grained materials, especially landfills, settlement is a function of time.
Boundaries	Surface and depth boundaries	Shallow, medium, or deep fills control the total and differential settlements.
Water table	Water table and its fluctuations	Inundation may cause large settlements; drawdown water table causes settlement in loose fills.
Stress history	Last buildings, preloading	Pre-consolidation reduces fills settlement.

The engineering properties and behavior of non-engineered fill are influenced by various factors such as the composition of the deposited material, deposition method, compaction level, particle size distribution, age, boundaries, water table, and stress history, as detailed in Table 3.14. Project sites with existing non-engineered fills often require treatment and modification to mitigate potential deformations caused by self-weight or foundation loads. The degree of improvement necessary depends on the structure's function, load, and geometry, as well as the characteristics and age of the fill material. When borrowed fills are used as construction material beneath foundations, they must be placed and compacted according to engineering designs.

3.10 Case Histories

Case History 3.1

Torre Latino Americana, Mexico City, Mexico

As presented in Chapter 1, Torre Latino Americana in Mexico City is an important milestone in compensated pile-supported mat technology, as shown in Figure 3.10a (Coduto et al. 2016). Its challenging soil conditions are described in Figure 3.10b.

Due to the highly compressible Tacubaya clays, significant settlements were observed in other structures near this building. For instance, the Palace of Fine Arts, located across the street from the Tower, settled over 3 m from 1904 to 1962 and continues to settle at a rate of 12 mm/year (White 1962).

To mitigate settlement, geotechnical engineers excavated to a depth of 13 m and drove to the Tarango sands to create a compensated pile-supported mat foundation. The removal of this soil compensated for about half the building's weight. Consequently, the building loads the deeper soils and has settled. However, this settlement is by design, and it has approximately matched that of the surrounding ground. This project serves as an exemplary case study underscoring the importance of thorough site investigation. Figure 3.11a and b illustrate the conventional soil profile of the site and the in situ test records conducted in Mexico City.

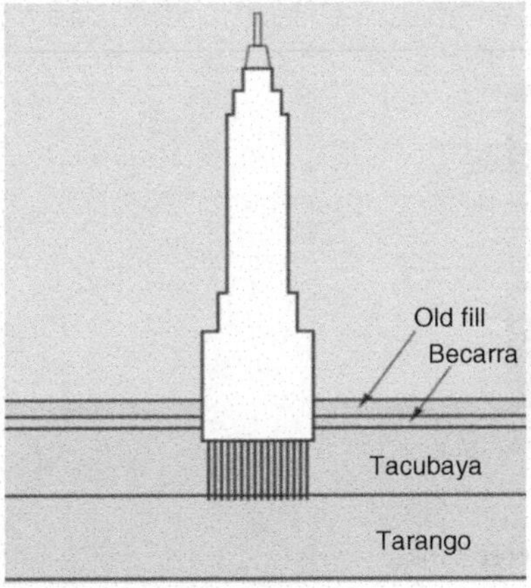

(a)

Depth (m)	Material
0–5.5	Old fill that includes Aztec artifacts. Groundwater table at 2 m
5.5–9.1	Becarra sediments–Interbedded sands, silts, and clays
9.1–33.5	Tacubaya clays–Soft volcanic clay; moisture content = 100–400, $C_c/(1 + e)$ = 0.80; S_u = 35–70 kPa
33.5–70	Tarango sands and clays–Harder and stronger deposits; much less compressible than the Tacubaya clays

(b)

Figure 3.10 Torre Latino Americana soil profile (a) graphical and (b) tabular.

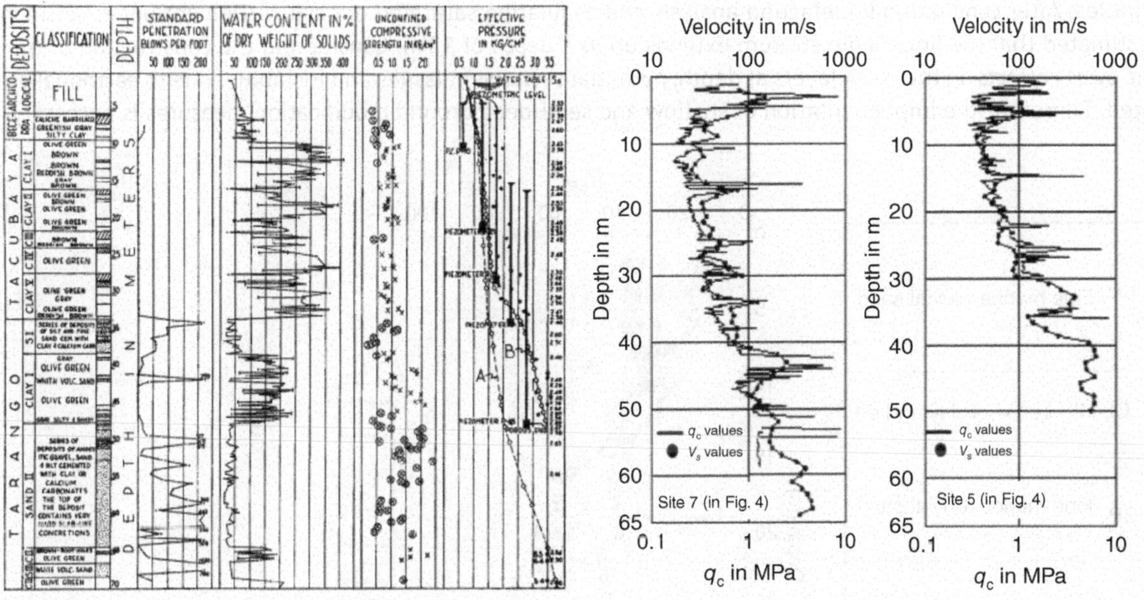

Figure 3.11 (a) SPT result of Torre Latino Americana (Zeevaert 1957 / Emerald Publishing Limited) and (b) Stratigraphic characteristics of Mexico City soil deposits (Romo and Silvis 2003 / with permission of Elsevier).

Case History 3.2

Site Investigation and Ground Modification Assessment of Amir Abaad Port Tank Farm (Courtesy of Sham-e Consulting Engineering Co. 2015)

This project involved soil improvement practices and the construction of a terminal for oil storage and product exports, situated in Amir Abaad Port on the northern coast of Iran along the Caspian Sea. The site covers an area of 29,773 m^2, with an estimated storage capacity of 24,000 tons of oil products. Figure 3.12a and b depict an aerial photo of the site during construction and a view of the completed tanks, respectively Eslami et al. (2025).

(Continued)

Case History 3.2 (Continued)

Figure 3.12 Amir Abaad Port tank farm: (a) aerial photo during construction and (b) completed tanks.

Based on excavated boreholes at the site, the subsoil consists of fine marine alluvial sand with small amounts of silt, extending to an average depth of 20 m. Clay and silt lenses with a maximum thickness of 1 m are observed at depths ranging from 8.5 to 15.5 m. Figure 3.13a and b illustrate the general soil profile of the site and SPT records from various boreholes. After conducting liquefaction analysis and evaluating saturated deposit settlement under seismic loads, it is estimated that the liquefiable stratum extends up to a depth of 12 m. Considering the construction of the tanks and the stress effects on the weak layers at depth, potential settlement issues and instability due to earthquakes are anticipated. Therefore, the implementation of shallow and semi-deep ground modification measures is necessary.

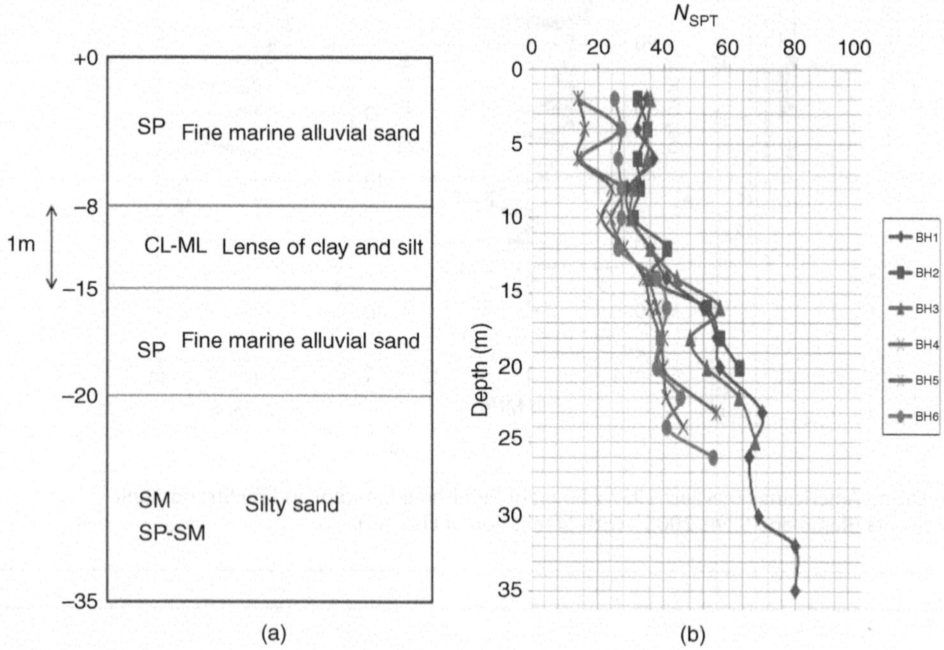

Figure 3.13 Amir Abaad Port tank farm: (a) general soil profile and (b) SPT records.

Various soil improvement methods were evaluated considering technical, practical, time, and economic aspects. Ultimately, deep dynamic compaction (DDC) was selected as the optimal and most feasible approach. Tampers weighing 11 and 23 tons were dropped from a height between 18 and 22 m using a 165-tons crane to initiate the soil improvement process. The procedure was followed by leveling and shallow compaction. The effectiveness of the ground modification was assessed through plate load tests (PLT) and standard penetration tests (SPT). A comparison of the results before and after ground modification is illustrated in Figure 3.14a and b, respectively, clearly demonstrating the effectiveness of the soil improvement practice.

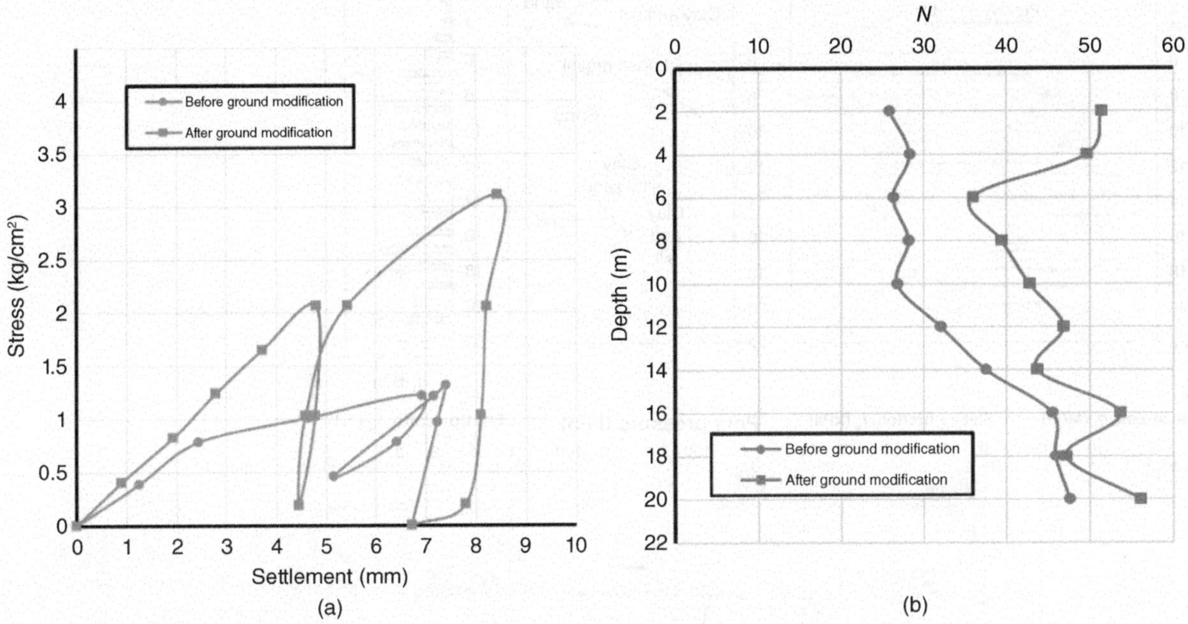

Figure 3.14 Soil improvement practice assessment via in situ test records: (a) PLT and (b) SPT.

Case History 3.3

Bolivian Experimental Site for Testing

The Bolivian Experimental Site for Testing (B.E.S.T) serves the purpose of providing meticulously documented geotechnical information from a location where various types of pile tests have been conducted (Fellenius et al. 2017). The findings of these field studies were presented at the 3rd International Conference on Deep Foundations in Bolivia (C.F.P.B).

In the upper 10–20 m of the site's profile, there are normally consolidated layers comprising clays, silts, and sands in various combinations and thicknesses. The top 5–6 m consist of loose silt and sand, followed by a 6- to 7-m layer of compact silt and sand. At approximately 11 m, there is a 1-m thick layer of soft silty clay, succeeded by a 1-m thick layer of compact sand. Below the 12-m mark, the profile alternates between layers of compact to dense silty sand and layers of loose sand, both approximately 2 m thick. The groundwater table at the site fluctuates seasonally between the ground surface and a depth of about 0.5 m. The site's geotechnical conditions were thoroughly examined using traditional in situ and laboratory techniques. Figure 3.15a–c display the results of laboratory tests, standard penetration tests (SPT), cone penetration test with pore pressure measurement (CPTu), and results from the flat dilatometer test (DMT).

(Continued)

Case History 3.3 (Continued)

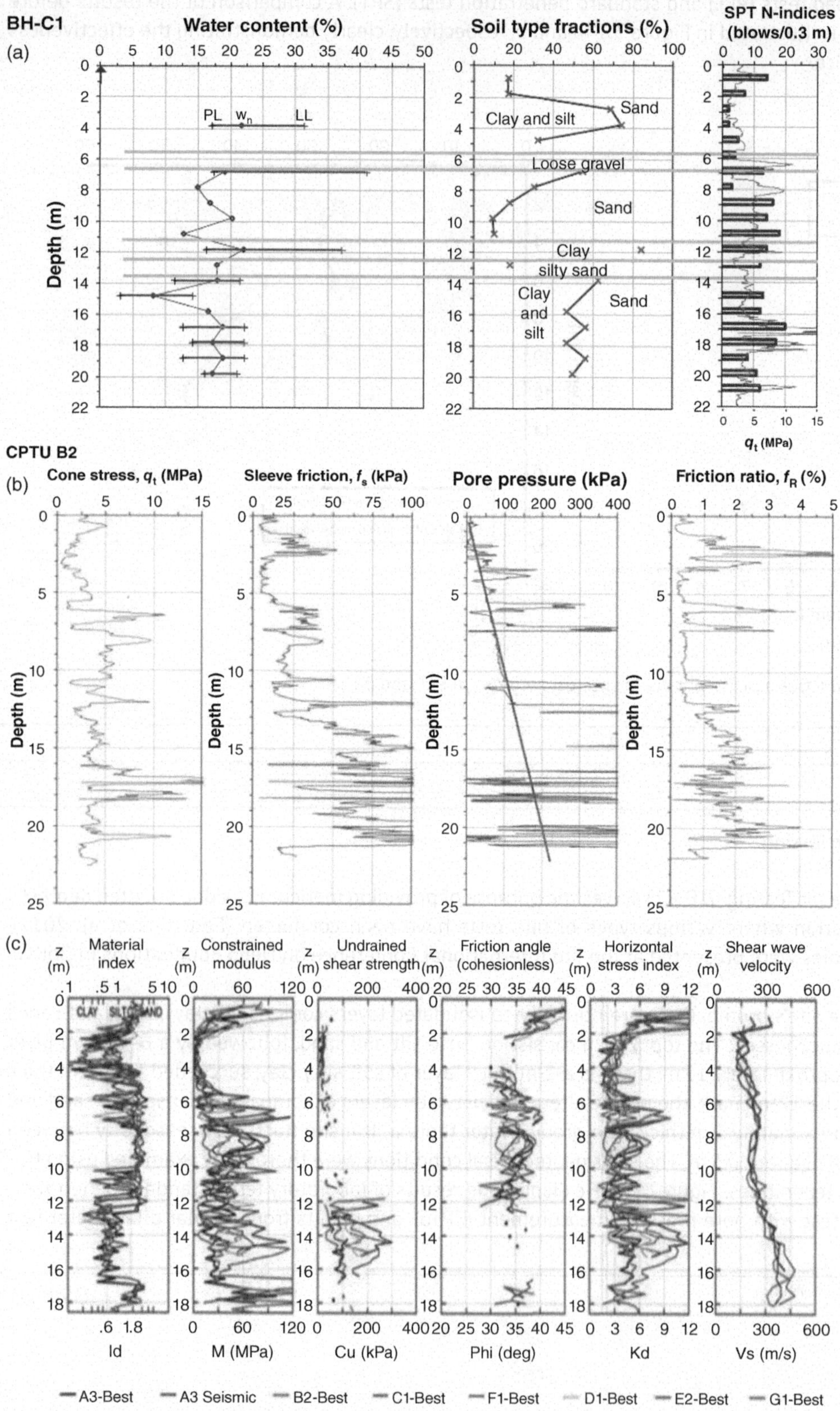

Figure 3.15 Bolivian Experimental Site, an example of site investigation data: (a) borehole and SPT, (b) CPTu, and (c) DMT compilation (Fellenius et al. 2017 / International Conference on Soil Mechanics and Geotechnical Engineering).

3.11 Major Points

1) Soil and rock, being natural materials, exhibit varying properties influenced by local environmental conditions. Therefore, their engineering characteristics need to be individually explored and defined for each project, a process known as site investigation.
2) The initial phase of a site investigation program involves data collection and a comprehensive literature review.
3) Typical site investigations for standard projects involve infield assessments, boring, sampling, and both in situ and laboratory testing. For projects with unusual subsurface conditions, a more in-depth geotechnical exploration and knowledge-based analysis are imperative, contingent upon the project's significance.
4) The selection of the location, number, and depth of boring and sampling points depends on factors such as the structure's importance, soil and bedrock stratification, applied loads, and the specific phase of the investigation.
5) Samples obtained from the site are subjected to laboratory tests, enabling soil classification and determination of critical parameters such as strength and compressibility.
6) To accurately mimic the physical and mechanical behavior of the structure and its environment, medium-scale or full-scale physical modeling tests in the field are essential, providing insights into real-world conditions.
7) In situ tests are instrumental in delineating subsurface profiles and soil engineering parameters. These tests are particularly valuable for soils that are challenging to sample conventionally.
8) Problematic geomaterials encompass soils and rocks susceptible to collapse, expansion, dispersion, excessive settlement, or even failure under relatively low stress. Identifying these materials and assessing the extent of the issues are crucial aspects of any site investigation project.
9) Site investigation and testing efforts yield a vast amount of data. Geotechnical engineers play a pivotal role in synthesizing and interpreting this information. They are responsible for selecting appropriate geomaterial design parameters and conducting engineering analyses for the design process.
10) During data collection and synthesis, it is crucial to consider soil variability, limitations of testing equipment, and potential measurement errors to ensure accurate and reliable results.

3.12 Questions and Problems

3.12.1 Explanatory

1 What are the main tasks of a geotechnical engineer in a project's reconnaissance and consultation phase?

2 Explain the process of planning a geotechnical site investigation.

3 List the important points of instrumentation and monitoring in geotechnical design.

4 How does one evaluate soil and bedrock characteristics in a site investigation?

5 Name visual identifiers of the existence of dispersive clays and explain them.

6 How can it be known if a project site has collapsible soil via traffic trespassing?

7 What are the main features and problems of organic soils from a geotechnical point of view?

8 What is a "Non-Engineered Fill?" Name the main problems one encounters with it and explain why.

3.12.2 Conceptual

1 Explain the significance of in situ testing in foundation engineering.

2 A project site has a 5 m thick soft clay layer near the ground surface. List possible problems and how to evaluate them on this site.

3 Based on Examples 3.1 and 3.2, discuss the factors that are the most influential on the required depth of boreholes.

4 Explain why we must be concerned about unsaturated soils in a project site.

5 Explain CPT and SPT applications in granular soils and compare these two methods' advantages and disadvantages.

6 Examine the advantages of using SCPTu over the other penetration tests.

7 Compare and contrast FVST and CPTu for recognition of clays.

8 Describe the evolution of site geotechnical characterization over the past 100 years based on Figure 3.4. What do you think about future practical approaches?

9 Explain the role of site investigation in a foundation system selection for the Torre Latino Americana project (Case History 3.1).

10 Compare and contrast the results of the site investigation for the Bolivian experimental site (Case History 3.3).

3.12.3 Problems

1 Consider a shallow foundation system for (a) an 8-story and (b) a 32-story building with the same 20×40 m dimensions. If the soil unit weight is 1.7 ton/m^3, the groundwater elevation is 20 m, and the average load for each story is considered 1 ton/m^2 in both cases, calculate the appropriate extent in the plan and depth for investigation borings for both projects.

2 Figure 3.11a shows the SPT result in the Torre Latino Americana project site (Case History 3.1), and Figure 3.11b represents the typical CPT results in Mexico City. Check the compatibility of these results with each other.

3 Consider the given CPT-log (Golden Ears Bridge, Cone Tec, Vancouver, BC) and evaluate the most appropriate implementation depth of piles toes at this site.

- $Z = 1–10$ m
- $Z = 10–20$ m
- $Z = 20–30$ m

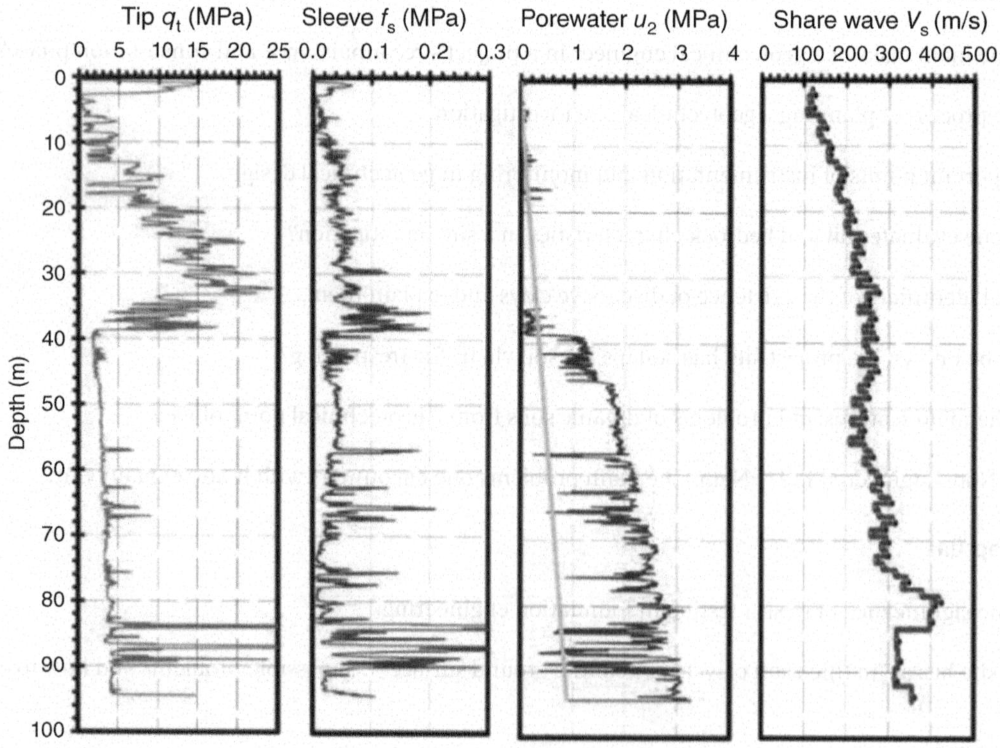

4 Four fills with the same height of 10 m but different materials and conditions are listed in in the table below. Answer the following questions:
   a) Which fill has experienced most of its possible settlement? Explain.
   b) Calculate the self-weight settlement of each fill in the next 20 years from the current situation. Explain the difference between their long-term settlement.
   c) Do any of these fills have a critical condition? Explain why and suggest proper measures to reduce each fills' long-term settlement.

Fill	Material
A	Well-compacted sand and gravel
B	Uncompacted sand
C	Uncompacted clay
D	Well-compacted mixed refuse

5 Compare the parameters related to the stiffness and soil strength before and after soil improvement practices in Case History 3.2 (the Amir Abaad Tank Farm project).

References

AASHTO (American Association of State Highway and Transportation Officials) (2014). *AASHTO LRFD Bridge Design Specifications*, 7e, Customary U.S. Units. Washington, DC: AASHTO.

ASCE (1993). Bearing capacity of soils. Technical Engineering and Design Guides. US Corps of Engineers, No. 7.

Brown, D.A., Turner, J.P., and Castelli, R.J. (2010). *Drilled Shafts: Construction Procedures and LRFD Design Methods*. US Department of Transportation, Federal Highway Administration (FHWA).

Canadian Geotechnical Society (2006). *Canadian Foundation Engineering Manual, CFEM*, vol. 4, 488. Vancouver: BiTech Publishers.

Coduto, D.P. (2001). *Foundation Design: Principles and Practices*, 2e. Upper Saddle River, NJ: Prentice Hall, Inc.

Coduto, D.P., Kitch, W.A., and Yeung, M.R. (2016). *Foundation Design Principles and Practices*, 3e. Upper Saddle River, NJ: Prentice Hall, Inc.

Das, B.M. and Sivakugan, N. (2019). *Principles of Foundation Engineering*, 9e. Boston: Cengage Learning.

Eslami, A., Ahmadnezhad, M., Kenarsari, A.E., and Rezazadeh, S. (2015). Experimental study on stiffness properties of non-engineered clay and granular fills. *Arabian Journal of Geosciences* 8 (5): 3065–3075.

Eslami, A., Ebrahimipour, A., Fattahi, S.M., et al. (2025). Sustainable ground improvement and hybrid foundation for tank farm on liquefiable coastal deposit: case study. *Marine Georesources and Geotechnology* 43 (2). https://doi.org/10.1080/1064119X.2025.2471825.

Fellenius, B.H., Terceros H.M., and Massarsch, K.R. (2017). Bolivian experimental site for testing. *Presentation of field testing programme. 3rd Bolivian International Conference on Deep Foundations*, Santa Cruz de la Sierra, Bolivia (27–29 April), Vol. 2, pp. 3–31.

Holtz, R.D., Kovacs, W.D., and Sheahan, T.C. (2023). *An Introduction to Geotechnical Engineering*, 3e. Pearson.

Huat, B.B., Prasad, A., Asadi, A., and Kazemian, S. (2014). *Geotechnics of organic soils and peat*. CRC Press, Taylor & Francis Group.

Hunt, R.E. (2005). *Geotechnical Engineering Investigation Handbook*, 2e. CRC Press.

Jennings, J.E. and Knight, K. (1975). The additional settlement of foundation due to collapse of sandy soils on wetting. In: *4th International Conference on Soil Mechanics and Foundation Engineering*. London.

John, B., Tim, C., and Hilary, S. (2012). *ICE Manual of Geotechnical Engineering*. Thomas Telford Ltd.

Lacasse, S. (1985). Design parameters of clays from in-situ and lab tests. *Proc. Symposium on New Concepts in Geotechnical Engineering*, Rio de Janeiro; also, Norwegian Geotechnical Inst. Report No. 52155-50, Oslo.

Leach, B.A. and Goodger, H.K. (1991). Building on derelict land. In: *CIRIA Special Publication 76*. London.

Loehr, J. E., Lutenegger, A., Rosenblad, B. L., et al. (2016). Geotechnical site characterization. Geotechnical Engineering Circular No. 5 (No. FHWA-NHI-16-072). National Highway Institute (US).

Lunne, T., Robertson, P.K., and Powell, J.J. (1997). *Cone Penetration Testing in Geotechnical Practice*. London: Blackie Academic and Professional.

Mesri, G., Stark, T., Ajlouni, M., and Chen, C. (1997). Secondary compression of peat with or without surcharging. *Journal of Geotechnical and Geoenvironmental Engineering* 123: 411–421. https://doi.org/10.1061/(ASCE)1090-0241(1997)123:5(411).

NCHRP (2019). Guide specification for the service life design of highway bridges: project 12–108. Modjesky & Masters for National Cooperative Highway Research Program, Washington DC (October 2019).

Phoon, K.K. and Kulhawy, F.H. (1999). Characterization of geotechnical variability. *Canadian Geotechnical Journal* 36 (4): 612–624.

Popescu, M.E. (1986). A comparison between the behaviour of swelling and of collapsing soils. *Engineering Geology (Amsterdam).* 23 (2): 145–163.

Romo, M. and Silvia, G. (2003). Neurofuzzy mapping of CPT values into soil dynamic properties. *Soil Dynamics and Earthquake Engineering* 23: 473–482. https://doi.org/10.1016/S0267-7261(03)00066-6.

Sabatini, P.J., Bachus, R.C., Mayne, P.W., et al. (2002). Evaluation of soil and rock properties. Geotechnical Engineering Circular No. 5 (*No. FHWA-IF-02-034*).

Sham-e Consulting Engineering Co. (2015). Technical report of site investigation and soil improvement practice assessment of oil tank farm of Amir Abaad port for Arad Omran Caspian Company.

Sowers, G.F. (1979). Introductory soil mechanics and foundations. *Geotechnical Engineering* 92: 114–117.

White, R.E. (1962). Caissons and Cofferdams," Chapter 10 in. In: *Foundation Engineering* (ed. G.A. Leonards). McGraw Hill.

Zeevaert, L. (1957). Foundation design and behavior of Tower Latino Americana in Mexico City. *Géotechnique* 7: 115–133.

4

Soil and Rock Parameters for Foundations

Chapter Highlights

- Soil Parameters for Foundation Engineering
- Factors Affecting Engineering Properties of Soils
- Soil Identification via Visual and Manual
- Soil Classification Based on Laboratory Element Testing
- Soil Behavior Classification (SBC) Through In Situ Testing Records
- Correlations for Strength and Stiffness Parameters
- Relations for Relevant Parameters
- Case Histories
  - Lake Urmia
  - Bothkennar Test Site
  - Explosive Compaction Assessment

4.1 Introduction

Understanding the properties and behaviors of geomaterials is a highly intricate task due to their nonhomogeneous and nonuniform nature concerning physical and chemical compositions. The composition of geomaterials is influenced by various factors, including their source, weathering processes, age, drainage, degree of saturation, sedimentation processes, and encountered geological phenomena. Therefore, foundation engineers must acknowledge the inherent variability in geomaterial properties in their design and construction recommendations.

Typically, soil parameters and properties used for designing foundation systems are obtained through laboratory and in situ field testing. However, it is essential to recognize the limitations of data obtained from laboratory testing, as it may not fully represent the in situ soil properties. These limitations are primarily due to sample disturbance, challenges in replicating in situ conditions in the laboratory, and scale effects. Consequently, it is advisable to complement laboratory tests with in situ testing, especially for large, critical, and sensitive projects where the repercussions of foundation failure could be catastrophic.

Given the challenges associated with collecting intact specimens, application of confining stresses mimicking the state of stress of the specimen source, and minimizing scale effects in laboratory testing procedures, in situ geotechnical investigations and testing are preferred. They offer direct and indirect applications in determining settlement, bearing capacity, and stability issues. By relying on in situ testing methods, the accuracy and reliability of assessments regarding settlement, bearing capacity, and stability can be significantly enhanced.

Key soil parameters used in foundation engineering are:

- Unit weight (γ)
- Internal friction angle (Ø)
- Relative density (D_r)

Advanced Foundation Engineering: Principles, Performance and Prospects, First Edition. Moses Karakouzian and Abolfazl Eslami.

Table 4.1 Factors affecting engineering properties of soils

Factor	Description
Quality of the clay	The term "clay" is used to define any cohesive soil comprising sufficient clay minerals present that drying produces shrinkage with the formation of cracks or fissures.
Natural cementation and aging	The aging process appears to raise the soil cementation by a variable amount. The effect of aging and cementation in sand is not nearly as pronounced as that of clay, but still the effect, as a statistical accumulation from a very large number of grain contacts, can be significant for foundation design.
Over-consolidation	It has been recognized that prior stresses on a soil element lead to a stress history that is retained by the soil structure until a new stress state exceeds the maximum previous one. The behavior of over-consolidated soils under new loads is dissimilar to that of normally consolidated soils, so it is essential to be able to recognize the occurrence, especially for cohesive soils.
Mode of deposit formation	Soil deposits that have been transported via water or wind tend to be made up of small grain sizes, and initially, to be somewhat loose with large void ratios.
	Soil deposits developed via a glacier transporting agent tend to be more varied in composition. These deposits may contain large sand or clay lenses.
	The major consideration for residual soil deposits is the amount of rainfall that has taken place. Large amounts of surface water tend to leach materials from the upper zones to greater depths. A resulting stratum of fine particles at some depth can affect the site's strength and settlement characteristics.
Water content	A rise in the pore pressure in any soil will decrease the shear strength. For cohesionless soils, a sufficient rise can reduce the shear strength to zero, and the end result is a viscous fluid.
	Also, Matric suction is a soil parameter that influences the behavior of unsaturated soils in terms of shear strength and permeability. A rise in water content tends to reduce the shear strength of cohesive soils.

- Compressibility (C_c)
- Stiffness (E)
- Undrained shear strength (S_u)
- Constrained modulus (m_v)
- Over Consolidation Ratio (OCR)

Other soil parameters occasionally used in foundation engineering are:

- Shear wave velocity (V_s)
- Small strain shear modulus (G_0)
- Permeability (k)
- Sensitivity (S_t)

Most of the factors that influence the engineering properties of soils are the result of geological processes that occur over extended time periods. Table 4.1 provides a comprehensive list of the key factors that have a significant impact on these properties.

4.2 Basic Soil Mechanics for Foundation Engineering

To analyze how soil responds to foundation loads, geotechnical engineers need to examine its physical composition, mechanical properties, hydrological characteristics, and occasionally, its chemical properties. The subsequent paragraphs delve into these aspects.

4.2.1 Phase Diagrams: Parameters and Relations

To establish the physical and mechanical properties of soils it is essential to define specific terms and understand the relationships between them. Typically, a soil mass can be represented schematically, as illustrated in Figure 4.1, depicting

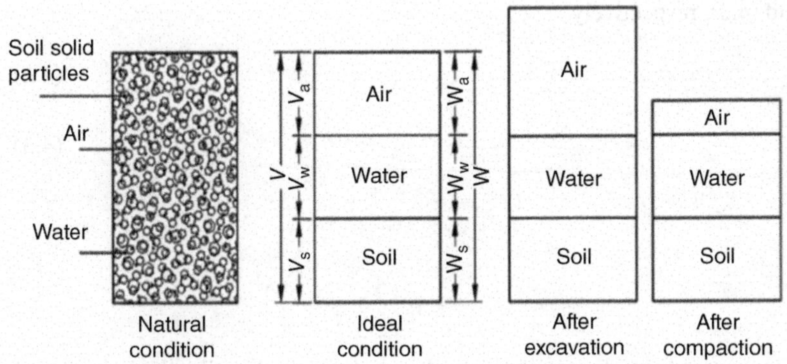

Figure 4.1 Various phases in a mass of soil and relevant weight and volumes.

its natural state, ideal state, state after excavation, and state after compaction. A soil mass comprises solid particles, water (liquids), and air, each with its own weight and volume.

The volume of soil in its natural state remains fixed, but excavation and boring can lead to expansion and increased volume. Conversely, in certain civil projects or ground modification practices, compaction results in a reduction in volume. Table 4.2 outlines the definitions and typical range of phase parameters associated with soil constituents.

The in situ unit weight is a critical parameter, particularly for coarse-grained soils. It plays a crucial role in evaluating compressibility, settlement, and liquefaction potential. In the case of sands, relative density (D_r) serves as an important indicator of their liquefaction and compaction potential. The relative density (D_r) can be defined using Equation 4.1:

$$D_r = \frac{e_{max} - e}{e_{max} - e_{min}} \tag{4.1}$$

where,

e_{max} = the maximum void ratio,
e_{min} = the minimum void ratio,
e = the current void ratio

Table 4.2 Definition and typical range of phase parameters.

No.	Phase parameters	Equation	Typical variation range
1	Void ratio	$e = \frac{V_v}{V_s}$	0.1–1.5
2	Porosity	$n = \frac{V_v}{V} = \frac{e}{1+e}$	9–60%
3	Water content	$\omega = \frac{W_w}{W_s} \times 100$	3–70%
4	Degree of saturation	$S_r = \frac{V_w}{V_v} \times 100$	2–100%
5	Specific gravity	$G_s = \frac{W_s}{V_s \times \gamma_w}$	2.6–2.8
6	Wet density	$\gamma_{wet} = \gamma = \frac{W}{V}$	1.1–2.1 ton/m^3
7	Dry density	$\gamma_d = \frac{W_{dry}}{V}$	1–1.9 ton/m^3
8	Saturated density	$\gamma_{sat} = \frac{W_{sat}}{V}$	1.2–2.3 ton/m^3
9	Submerged density	$\gamma' = \gamma_{sat} - \gamma_w$	0.4–1.1
10	Air content	$A = \frac{V_a}{V} \times 100$	5–30%

e_{max} and e_{min} are calculated by Equations 4.2 and (4.3), respectively.

$$e_{max} = \frac{G_s \gamma_w}{\gamma_{d,min}} - 1 \tag{4.2}$$

$$e_{min} = \frac{G_s \gamma_w}{\gamma_{d,max}} - 1 \tag{4.3}$$

where,

$\gamma_{d,min}$ = the minimum dried unit weight,
$\gamma_{d,max}$ = the maximum dried unit weight,
G_s = specific gravity.

Relative density (D_r) is classified as very loose, loose, medium dense, dense, and very dense to describe the behavior of coarse-grained soils, particularly sand deposits, as provided in Table 4.3.

When the water content of fine-grained soils increases, it causes a transformation in their volume and state, transitioning from a solid to a liquid state. The specific water content at which this change occurs, resulting in a shift from solid to semisolid, is referred to as the shrinkage limit (SL). Similarly, the water content that marks the boundary between the plastic and semisolid states of the soil is known as the plastic limit (PL). On the other hand, the liquid limit (LL) corresponds to the water content at which the soil initiates a liquid state. These limits – shrinkage limit, plastic limit, and liquid limit – are illustrated in Figure 4.2.

The strength and compressibility (deformation) of soils can be loosely correlated with the different states illustrated in Figure 4.2. In the liquid state, the soil exhibits the lowest strength and experiences the largest deformation. Conversely, in the solid state, the soil displays the highest strength and undergoes minimal deformation (Budho 2011). The liquidity index (LI) is an Atterberg limit used to estimate the soil's strength and compressibility. It is expressed by Equation 4.4:

$$\text{LI} = \frac{\omega - \text{PL}}{\text{LL} - \text{PL}} \tag{4.4}$$

Table 4.3 Coarse-grained soils density classification based on D_r.

D_r (%)	0–15	15–35	35–65	35–80	80–100
Density	Very loose	Loose	Medium dense	Dense	Very dense

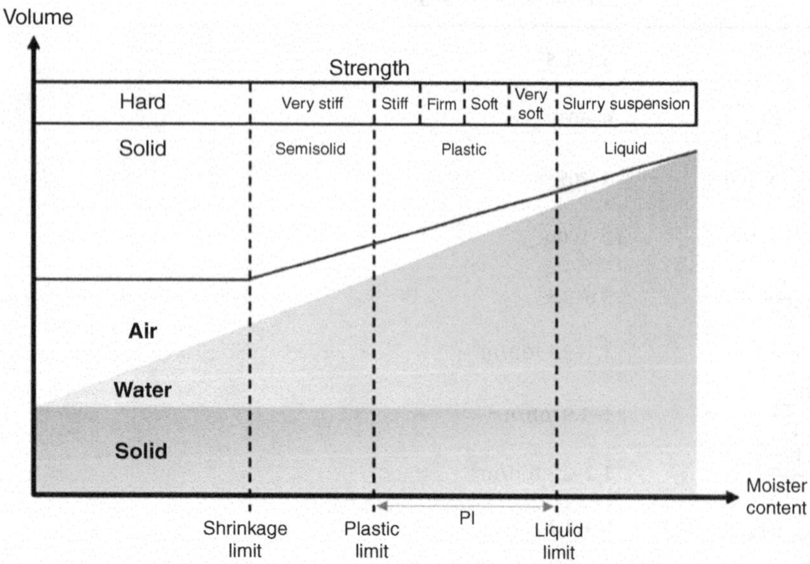

Figure 4.2 Variation of soil volume and different moisture content.

where

PL = plastic limit,
LL = liquid limit,
ω = water content

Table 4.4 presents the description of the strength of fine-grained soils based on the liquidity index.

Skempton (1953) proposed a relationship between the plasticity index (PI), which is the difference between the liquid limit and plastic limit of soils, and the clay fraction content. According to Skempton, for soils with a specific mineralogy, the plasticity index is linearly correlated to the amount of clay fraction present. He introduced the term "activity" (A) to describe the influence of clay fractions on the plasticity index. Equation 4.5 is commonly used to calculate the activity parameter (A). Based on the liquidity index (LI), different soil types can be described. A description of various soils categorized by their liquidity index is provided in Table 4.5.

$$A = \frac{\text{PI}}{\text{Clay fraction (\%)}} \tag{4.5}$$

As discussed and explained in the current section, it is important to note that the physical, chemical, and electrical properties of fine-grained soils are significantly influenced by the soil water content as well as the amount of the fines.

4.2.2 Geostatic Stress

Geostatic stresses originate from the gravitational force exerted on a mass of soil. The geostatic stress can be calculated via the following equation:

$$\sigma_z = \sum \gamma h \tag{4.6}$$

where,

σ_z = geostatic vertical total stress,
γ = total unit weight of soil layer,
h = thickness of the soil layer

Effective stress is defined as the stress exerted by the solid soil particles through particle-to-particle contact. According to this definition, a portion of the total vertical stress is carried by the solid soil particles, while the remaining portion is carried by the pore water.

$$\sigma_z = \sigma'_z + u \tag{4.7}$$

Table 4.4 Description of the strength of fine-grained soils based on LI (Budho 2011 New Jersey).

Values of LI	Description of soil strength
LI < 0	Semisolid state – high strength, brittle (sudden) fracture is expected
0 < LI < 1	Plastic state – intermediate strength, soil deforms like a plastic material
LI > 1	Liquid state – low strength, soil deforms like a viscous fluid

Table 4.5 Activity of clay-rich soils (Budho 2011 New Jersey).

Description	Activity, A
Inactive	<0.75
Normal	0.75–1.25
Active	1.25–2
Highly active	>6

where

σ_z = vertical total stress,
σ'_z = effective stress,
u = pore water pressure

The water pressure in the soil is calculated using the following equation:

$$u = u_h + u_e \pm u_i \tag{4.8}$$

where

u_h = hydrostatic water pressure = $Z_w \cdot \gamma_w$,
Z_w = the water depth from ground level,
γ_w = water unit weight,
u_e = excess pore water pressure,
u_i = depending on upward or downward water gradient = $i\gamma_w$,
i = hydraulic gradient ($\Delta h/l$)

4.2.3 Shear Strength of Soils

The majority of foundation failures occur when the shear stresses generated by foundation loads and construction activities surpass the inherent shear strength of the soil. The determination of stress increase in the soil caused by foundation loads and construction activities is a crucial responsibility of geotechnical engineers.

In geotechnical practice, the Mohr–Coulomb failure criterion is widely employed to assess the shear strength of soils. Generally, the failure criterion is based on the following equation:

$$\tau = c + \sigma' \tan \varphi \tag{4.9}$$

where

τ = shear strength,
c = effective or total cohesion,
φ = effective or total friction angle,
σ' = effective or mean confining stress acting on the shear surface

Figure 4.3a and b illustrates stress circles at yield, touching the Coulomb envelope for effective strength parameters (φ' and c') and undrained strength parameters ($\varphi = 0$, $c = S_u$), respectively.

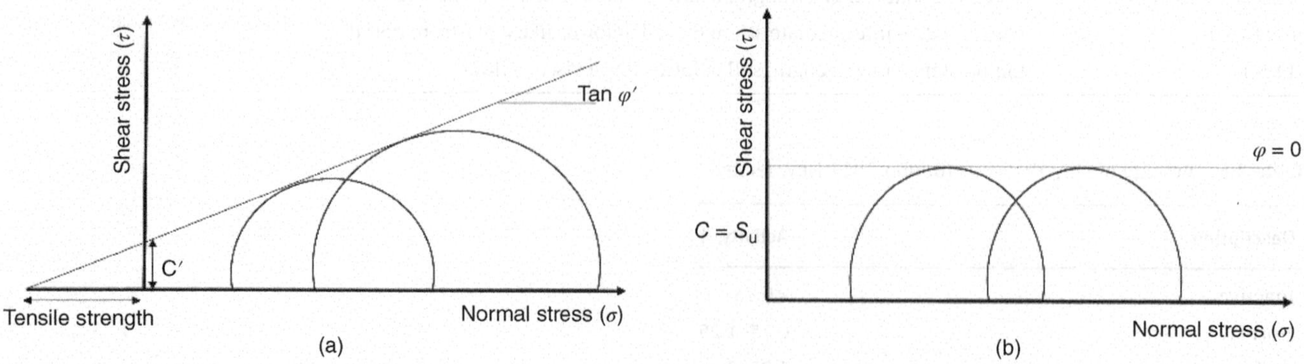

Figure 4.3 Stress circles at yield; one touches Coulomb envelope: (a) effective strength parameters and (b) undrained strength parameters.

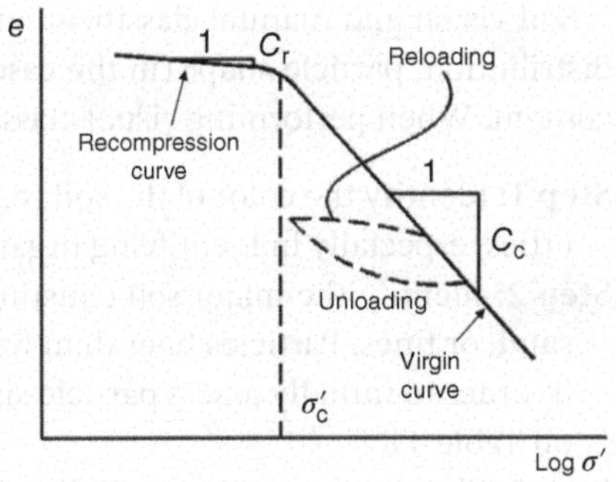

Figure 4.4 e-log σ' formulation of the soil stress–strain curve.

4.2.4 Consolidation Settlement

Consolidation settlement is a phenomenon that involves the time-dependent alteration in the volume of voids within the soil due to an increase in effective vertical stress (σ'_z) and the dissipation of excess pore pressure. In saturated soils, where soil particles and fluids within the voids are considered incompressible, volume changes occur when pore water is squeezed out from the soil matrix in response to external loading. The dissipation of the increased pore water pressure is a gradual process that takes time. This time-dependent process is known as consolidation. Consolidation settlements primarily occur in fine-grained saturated soils.

The *e*-log-*p* method is a commonly used analytical approach for estimating consolidation settlement. In this method, the stress–strain behavior of the soil is represented by two straight lines, as illustrated in Figure 4.4. The pre-consolidation stress (σ'_c) refers to the maximum effective stress that the soil has experienced throughout its geological history. When the stress is below the pre-consolidation stress, the soil is relatively stiff, and its strain–stress relationship follows the recompression curve (utilizing the parameter C_r in the calculation). On the other hand, when the stress exceeds the pre-consolidation stress, the soil becomes more compressible, and its strain–stress behavior follows the virgin curve (utilizing the parameter C_c in the calculation).

4.3 Visual and Manual Soil Identification

The first and vital phase of any geotechnical engineering project is understanding and characterizing the subsoil conditions. Identifying the soil type, whether it is classified in general terms as coarse (e.g., gravel) or fine (e.g., clay), provides essential insights into potential challenges during excavation or other construction activities. Having a systematic soil identification procedure, even in the initial planning stages, is advantageous for making informed design and construction decisions.

The following are activities for classifying the soil layers at a project site and their functions (Bowles and Guo 1996):

1) **Using Other Databases for Prediction:** Soil classification enables the engineer to utilize existing databases or literature to predict the performance of foundations in similar soil conditions. By matching the soil classification with known case studies or empirical data, the engineer can estimate the behavior of foundations and make informed design decisions.
2) **Creating a Local Success/Failure Database:** Classifying soil layers helps in establishing a local database specific to the project site. This database records the performance of foundations in similar soil conditions, including both successful and failed cases. It allows future engineers to learn from past experiences and make more accurate predictions and decisions for similar projects in the same region.
3) **Maintaining a Permanent Record:** Soil classification serves the purpose of creating a permanent record that can be referenced by future professionals. This record ensures that the soil conditions and their classification are documented, providing valuable information for subsequent investigations, assessments, or design modifications.
4) **Contributing to the General Body of Knowledge:** By classifying soil layers and documenting the findings, foundation engineers can contribute to the collective knowledge in the field. This information can be shared through journal papers or conference presentations, allowing the broader geotechnical community to benefit from the insights and experiences gained in specific projects. It helps in advancing the understanding of foundation behavior and enhancing the common terminology used in geotechnical engineering.

Soil visual and manual classification relies on various features, including color, major and minor constituents, particle distribution, particle shape (in the case of coarse-grained soils), plasticity (in the case of fine-grained soils), and moisture content. When performing visual classification of soil, the following steps can be followed:

Step 1: Identify the color of the soil (e.g., brown, gray, brownish-gray). The color provides an indication of likely soil properties, especially in identifying organic soils (Table 4.6).

Step 2: Identify the major soil constituents using Table 4.7 as coarse gravel, fine gravel, coarse sand, medium sand, fine sand, or fines. Particles finer than fine sand cannot be discerned with the naked eye at a distance of 20 cm. Additionally, in order to initially assess particle sizes, sedimentation tests can be conducted, and the results can be interpreted based on Table 4.8.

Step 3: Identify the moisture condition as dry, moist, or wet, in accordance with the criteria in Table 4.9

Step 4: If the major soil constituent is sand or gravel, identify particle distribution, particle shape, and consistency.

- **Particle Distribution:** Soil can be categorized as either well-graded (containing a wide range of particle sizes) or poorly graded (composed of particles of similar size).

Table 4.6 Effect of color (Look 2007/Taylor & Francis Group).

Color effect	Significance
Light to dark	Increasing moisture content; dry soils are generally lighter than wet soils
Black or dark shades of brown and grey	Organic matter likely
Bright shades of brown, grey, red, yellow, and whites	Inorganic soils
Mottled colors	Poor drainage
Red, yellow–brown	Presence of iron oxides

Table 4.7 Grain size distribution.

Soil constituent	Size limits	Familiar example
Boulder	305 mm or more	Larger than a basketball
Cobbles	76–305 mm	Grapefruit size
Coarse gravel	19–76 mm	Orange or lemon size
Fine gravel	4.75–19 mm	Grape or pea size
Coarse sand	2–4.75 mm	Resembles rocksalt
Medium sand	0.42–2 mm	Resembles sugar or table salt
Fine sand	0.075–0.42 mm	Resembles powdered sugar
Fines	Less than 0.0075 mm	—

Table 4.8 Sedimentation tests for initial assessment of particle sizes (Look 2007 / With permission of Taylor & Francis Group).

Description	Approximate time for particles to settle in 100 mm of water
Coarse sand	1 s
Fine sand	10 s
Silt	1–10 min
Clay	1 h

Table 4.9 Criteria for describing moisture conditions (with permission of ASTM).

Description	Criteria
Dry	Absence of moisture, dusty, dry to the touch
Moist	Damp, but no visible water
Wet	Visible free water; usually, soil is below the water table

Table 4.10 Criteria for describing angularity of coarse-grained particles (with permission of ASTM).

Description	Criteria
Angular	Particles have sharp edges and relatively plane sides with unpolished surfaces.
Subangular	Particles are similar to angular description but have rounded edges.
Subrounded	Particles have nearly plane sides but have well-rounded corners and edges.
Rounded	Particles have smoothly curved sides and no edges.

Table 4.11 Consistency of non-cohesive soil (adapted from Look 2007).

Term	Symbol	Field assessment	
Very loose	VL	50 mm peg easily driven	Foot imprints easily
Loose	L	12 mm reinforcing bar easily pushed by hand	Shovels easily
Medium dense	MD	12 mm bar needs a hammer to drive >200 mm	Shoveling difficult
Dense	D	50 mm peg hard to drive 12 mm bar needs a hammer to drive <200 mm	Needs a pick for excavation
Very dense	VD	12 mm bar needs a hammer to drive <60 mm	Picking difficult
Cemented	C	12 mm bar needs a hammer to drive <20 mm	Cemented, indurated, or large-size particles

Table 4.12 Criteria for describing dry strength (with permission of ASTM).

Description	Criteria
None	The dry specimen crumbles into powder with mere pressure of handling.
Low	The dry specimen crumbles into powder with some finger pressure.
Medium	The dry specimen breaks into pieces or crumbles with considerable finger pressure.
High	The dry specimen cannot be broken with finger pressure. The specimen will break into pieces between the thumb and a hard surface.
Very high	The dry specimen cannot be broken between the thumb and a hard surface.

- **Particle Shape:** Particle shape, whether angular, subangular, subrounded, or rounded, can be determined using Table 4.10.
- **Soil Consistency:** Field assessments of soil consistency often rely on tactile perception. While a pocket penetrometer can quantify values, it has limitations related to scale effects, conversions, the sample used, and the soil type (refer to Table 4.11).

Step 5: If the major soil constituents are fines, perform the following tests:

- **Dry Strength Test:** Mold a sample into a 25 mm size ball and let it dry. Test the strength of the dry sample by crushing it between the fingers. Describe the strength as none, low, medium, high, or very high depending on the test results as shown in Table 4.12.

- **Dilatancy Test:** Choose an adequate amount of material from the specimen to mold into a ball approximately 12 mm in diameter. Mold the material, adding water if necessary, until it achieves a soft, nonsticky consistency. Shape the soil into a smooth ball in the palm of one hand using a knife blade or a small spatula. Shake the soil ball horizontally by striking the side of one hand vigorously against the other hand several times. Observe the reaction of water appearing on the surface of the soil. Squeeze the sample by closing the hand around it or pinching the soil between the fingers. Note the reaction, categorizing it as none, slow, or rapid based on the criteria in Table 4.13. The reaction refers to the speed at which water appears during shaking and disappears during squeezing.
- **Consistency Test:** Try to crush a small amount of soil by pressing it between your thumb and forefinger or by squeezing it in the palm of your hand. Describe the consistency as very soft, soft, firm, hard, or very hard in accordance with the criteria in Table 4.14.
- **Plasticity (or Toughness) Test (ASTM):** Roll the samples into a thread about 3 mm in diameter. Fold the thread and reroll it repeatedly until the thread crumbles at a diameter of 3 mm. Note (i) the pressure required to roll the thread when it is near crumbling; (ii) whether it can support its own weight; (iii) whether it can be molded back into a coherent mass; and (iv) whether it is tough during kneading. Describe the toughness according to the criteria in Table 4.15. A low to medium toughness, along with nonplastic to low plasticity, is an indication that the soil is silty; otherwise, the soil is clayey.
- **Plasticity Test (Holtz et al. 2023):** Mold a soil specimen at its natural water content into a lump and roll it out by hand on a smooth surface or between the palms into the smallest thread possible without causing excessive cracking or crumbling. The smaller the thread that can be rolled out, the more plastic the soil is (Table 4.16).
- Also, aforementioned soil tests applying fine-grained soils are compared and contrasted as presented in Table 4.17.

Table 4.13 Criteria for describing dilatancy (with permission of ASTM).

Description	Criteria
None	No visible change in the specimen.
Slow	Water appears slowly on the surface of the specimen during shaking and does not disappear or disappears slowly upon squeezing.
Rapid	Water appears quickly on the surface of the specimen during shaking and disappears quickly upon squeezing.

Table 4.14 Consistency of non-cohesive soil (adapted from Look 2007).

Term	Symbol	Field assessment	Thump pressure penetration
Very soft	VS	Exudes between fingers when squeezed	>25 mm
Soft	S	Can be molded by light finger pressure	>10 mm
Firm	F	Can be molded by strong finger pressure	<10 mm
Stiff	St	Cannot be molded by fingers; can be indented by thumb pressure	<5 mm
Very stiff	Vst	Can be indented by a thumbnail	<1 mm
Hard	H	Difficult to be indented by thumbnail	~0 mm

Table 4.15 Criteria for describing toughness (with permission of ASTM).

Description	Criteria
Low	Only slight pressure is required to roll the thread near the plastic limit. The thread and the lump are weak and soft.
Medium	Medium pressure is required to roll the thread to near the plastic limit. The thread and the lump have medium stiffness.
High	Considerable pressure is required to roll the thread to near the plastic limit. The thread and the lump have very high stiffness.

Table 4.16 Manual plasticity test results interpretation (Holtz et al. 2023/Pearson Education).

Description	Soil type	Smallest diameter of thread
Nonplastic	Silt	None
Slight	Clayey silt	6.0 mm
Low	Silt and clay	3.0 mm
Medium	Clay and silt	1.6 mm
High	Silty clay	0.8 mm
Very high	Clay	0.4 mm

Table 4.17 Field determination of soil components (adapted from Hunt 2005).

Component	Characteristic	Determination
Silt	Strength	Low when air-dried; crumbles easily.
	Dilatancy test	Mixed with water to thick paste consistency; appears wet and shiny when shaken in palm of hand, but when palm is cupped and sample squeezed, surface immediately dulls and dries.
	Dispersion test	Mixed with water in container: particles settle out in 1/4 to 1 h.
	Thread test	Rolls into thin threads in wet state, but threads break when picked up by one end.
Clay	Strength	High when air-dried; breaks with difficulty.
	Plasticity	When mixed with water to form paste and squeezed in hand, specimen merely deforms, and surface does not change in appearances.
	Dispersion test	Remains in suspension from several hours to several days in container.
	Thread test	Can be rolled into fine threads that remain intact; fineness depends on clay content and mineralogy.
Organic soil	Adhesion	Sticky and greasy feel when smeared between fingers.
	Strength	Relatively high when air-dried.
	Odor	Decayed organic matter; gases
	Organic matter	Root fibers, etc.
	Shrinkage	Very high

4.4 Soil Classification Based on Laboratory Results

Soil classification, from an engineering perspective, is carried out based on soil engineering properties that are relevant to their use as building materials or for foundation support. Soil classification systems categorize soil into groups and subgroups based on shared engineering properties, including grain-size distribution, liquid limit, shrinkage limit, and plastic limit. Modern engineering classification systems aim to facilitate a seamless transition from field observations and in situ test results to fundamental predictions of soil engineering properties and behaviors.

Various soil classification systems rely on grain size, physical characteristics, and soil composition. Soil components, defined by grain size, typically encompass boulders, cobbles, gravel, sand, silt, and clay. Table 4.18 presents several commonly used soil classification systems. In addition to laboratory test-based classifications, the last entry in Table 4.18 outlines a field identification-based system of classification.

The Unified Soil Classification System (USCS) is the most widely used geotechnical engineering classification system for soils in North America. It comprises three main classification groups: (i) coarse-grained soils, (ii) fine-grained soils, and (iii) highly organic soils. The USCS further subdivides these three major soil classes for greater clarity. Gravels and sands are differentiated based on grain size, while clays and silts are distinguished by their Atterberg limits. The classification

Table 4.18 Field determination of soil components (Hunt 2005/Taylor & Francis Group).

System	Grain diameter (mm)
	0.0006, 0.002, 0.006, 0.02, 0.06, 0.2, 0.6, 2.0, 4.76, 19, 76
M.I.T and British Standards Institute	Clay; Silt (f, m, c); Sand (f, m, c); Gravel
	0.001, 0.005, 0.074, 0.25, 2.0, 9, 24, 76
American Association of State Highway Officials (AASHO)	Colloids; Clay; Silt; Sand (f, c); Gravel (f, m, c); Boulders
	0.002, 0.05, 0.25, 0.5, 2.0, 76
U.S. Dept. of Agriculture (USDA)	Clay; Silt; Sand (vf, f, m, c, vc); Gravel (f, m); Cobbles
Unified Soil Classification System (USDR, USAEC)	Clay and Silt; Sand (f, m, c); Gravel (f, c); Cobbles
	No. 200, 40, 10, 4, ¾ in., 3 in.
	Grain diameter in U.S. standard sieve sizes
	No. 200, 60, 30, 10, 3⁄16, 3⁄8 in., 1.0 in., 3 in.
American Society for Engineering Education (ASEE) (Burmister)	Clay and Silt (Silt f, c); Sand (f, m, c); Gravel (f, m, c)
Field Identification	Not discernible; Hand lens; Visible to eye; Measurable

system categorizes some soils as "well-graded" and others as "poorly-graded," and it also separates "high-plasticity" soils from "low-plasticity" soils. Moderately organic soils are considered subdivisions of silts and clays, and their distinction from inorganic soils is based on changes in their plastic properties upon drying. Table 4.19 provides a detailed explanation of the major and minor soil categories using gradation curve analysis and the plasticity chart.

The graphical curves are illustrated in Figure 4.5a for various coarse-grained soils in gradation curves and fine-grained soils in the plasticity chart, as indicated in Figure 4.5b.

The soil characteristics have been well-depicted in Table 4.20 particularly in construction and feasibility for foundation engineering through divisions and classifications based on USCS.

Table 4.19 Unified soil classification system (with permission of ASTM).

Major Division			Symbol	Typical Names	Laboratory Classification Criteria		
Coarse-grained soils (more than half of the material is larger than no. 200 sieve size)	Gravels (more than half of the coarse fraction is larger than No. 4 sieve size)	Clean gravel (little or no fines)	GW	Well-graded gravels, gravel-sand mixtures, little or no fine	Determine percentages of sand gravel from grain-size curve. Depending on percentage of fines (fraction smaller than No. 200 sieve size). Coarse organized – grained soils are classified as follows: Less than 5 % GW, GP, SW, SP; More than 12 % GM, GC, SM, SC; 5 to 12 % Borderline cases requiring dual symbols	$C_u = \frac{D_{60}}{D_{10}} > 4$; $1 < C_c = \frac{(D_{30})^2}{D_{10} \times D_{60}} < 3$	
			GP	Poorly graded gravels, gravel-sand mixtures, little or no fines		Not meeting all gradation requirement for GW	
		gravels with fines (appreciable amount of fines)	GM	Silty gravels, gravel-sand clay mixtures		Atterberg limits below "A" line or PI<4	Above "A" line with 4<PI<7 are border line cases requiring use of dual symbols
			GC	Clay gravels, gravel sands, little or no fines		Atterberg limits below "A" line or PI>7	
	Sands (more than half of the coarse fraction is smaller than no. 4 sieve size)	Clean sands (little or no fines)	SW	Well-graded sands, gravelly sands, little or no fines		$C_u = \frac{D_{60}}{D_{10}} > 4$; $1 < C_c = \frac{(D_{30})^2}{D_{10} \times D_{60}} < 3$	
			SP	Poorly graded sands, gravelly sands, little or no fines		Not meeting all gradation requirements for GW	
		Sands with fines (appreciable amount of fines)	SM	Silty sands, sand-silt mixtures		Atterberg limits above "A" line or PI<4	Limits plotting in hatched zone with 4<PI<7 are borderline cases requiring use of dual symbols
			SC	Clayey sands, sand-clay mixtures		Atterberg limits above "A" line or PI<7	
Fine-grained soils (more than half of the material is smaller than no. 200 sieve)	Silts and clays (liquid limit < 50)		ML	Inorganic silts and very fine, rock flour; silty or clayey fine sands; or clayey silts with slight plasticity	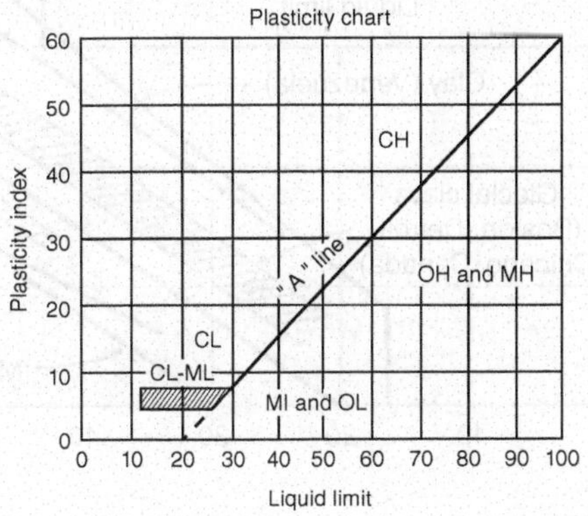 A line: PI = 0.73 × (LL – 20)		
			CL	Inorganic clays of low to medium plasticity; gravelly clays, sandy clays, silty clays, lean clays			
			OL	Organic silts and organic silty clays of low plasticity			
	Silts and clays (liquid limit > 50)		MH	Inorganic silts; micaceous or diatomaceous fine sandy or silty soils; elastic silts			
			CH	Inorganic clays of high plasticity; fat clays			
	Highly organic soil		Pt	Peat and other highly organic soils			

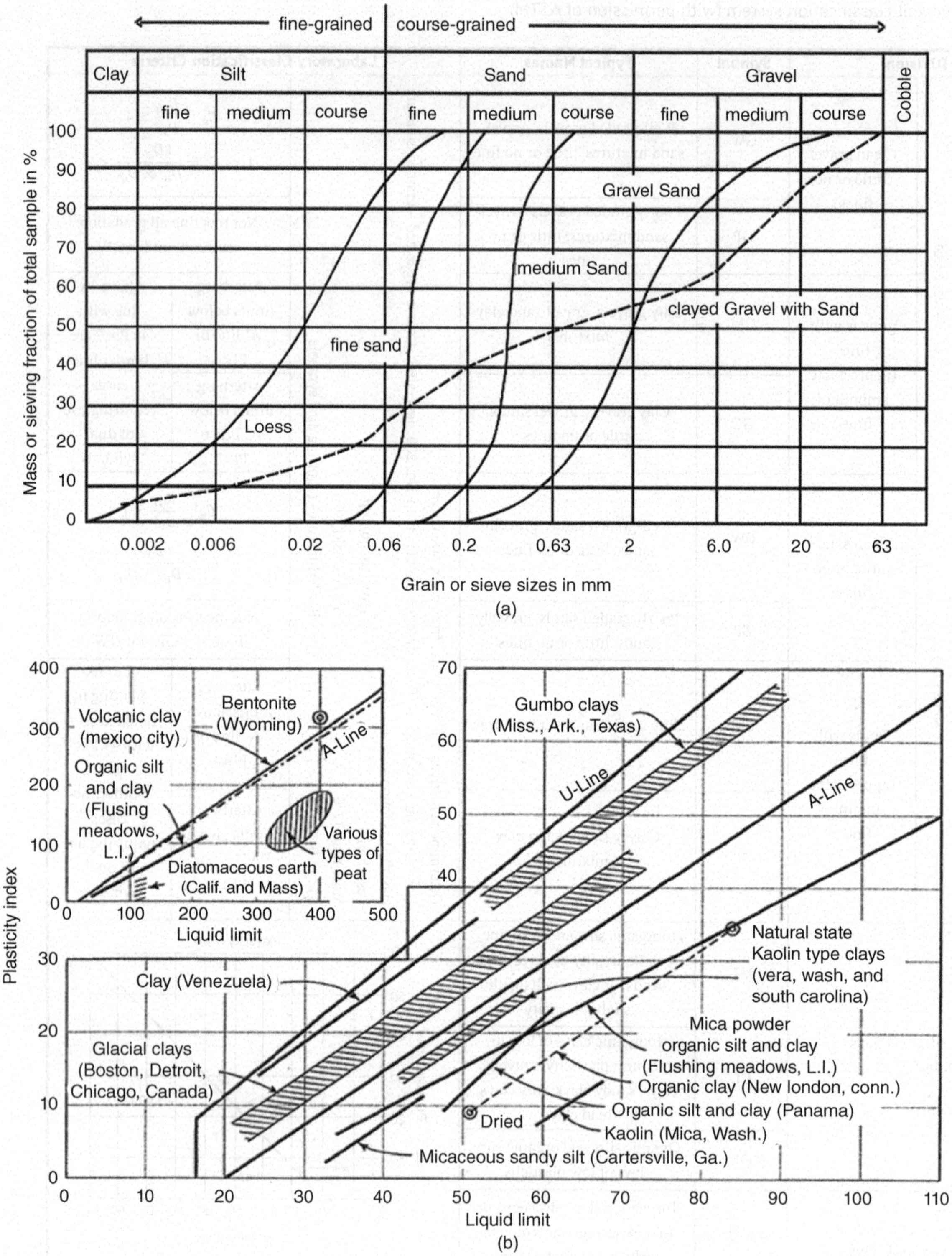

Figure 4.5 (a) Grain size distribution curves (sieve curves) for different soil types and (b) relation between liquid limit and plasticity index for typical soils (Casagrande 1948/with permission of American Society of Civil Engineers).

Table 4.20 Soil characteristics based on experiences and practices (with permission of Holtz et al. 2023).

Major division	Symbol	Unit dry density (g/cm^3)	CBR	Subgrade modulus (kg/cm^3)	Permeability when compacted	Potential frost action	Shear strength	Compressibility	Workability as a construction material	Desirability for using as a foundation
Gravels and gravelly soils	GW	2.00–2.24	40–80	8.3–13.8	Pervio us	None to very slight	Excellent	Negligible	Excellent	Excellent
	GP	1.76–2.24	30–60	8.3–13.8	Very previous	None to very slight	Good	Negligible	Good	Good
	GM(d)	2.00–2.32	40–60	8.3–13.8	Semi-pervious to impervious	Slight to medium	Good	Negligible	Good	Fair to good
	GM(ud)	1.84–2.16	20–30	5.5–13.8	Semi- pervious to impervious	Slight to medium	Good	Negligible	Good	Fair to good
	GC	2.08–2.32	20–40	5.5–13.8	Impervious	Slight to medium	Fair to good	Very low	Good	Fair
Sands and sandy soils	SW	1.76–2.08	20–40	5.5–11.1	Pervious	None to very slight	Excellent	Negligible	Excellent	Good to excellent
	SP	1.68–2.16	10–40	4.2–11.1	Pervious	None to very slight	Good	Very low	Fair	Fair to good
	SM(d)	1.92–2.16	15–40	4.2–11.1	Semi- pervious to impervious	Slight to high	Good	Low	Fair	Fair
	SM(ud)	1.60–2.08	10–20	2.8–8.3	Semi- pervious to impervious	Slight to high	Good	Low	Fair	Fair
	SC	1.60–2.16	5–20	2.8–8.3	Impervious	Slight to high	Fair to good	Low	Good	Fair to good
Silts and clays with LL < 50	ML	1.44–2.08	$\leqslant$15	2.8–5.5	Semi- pervious to impervious	Medium to very high	Fair	Medium	Fair	Poor to fair
	CL	1.44–2.08	$\leqslant$15	1.4–4.2	Impervious	Medium to very high	Fair	Medium	Fair to good	Poor to fair
	OL	1.44–1.68	$\leqslant$5	1.4–2.8	Semi- pervious to impervious	Medium to very high	Poor	Medium	Fair	Poor
Silts and clays with LL < 50	MH	1.28–1.68	$\leqslant$10	1.4–2.8	Semi- pervious to impervious	Medium to very high	Poor to fair	High	Poor	Very poor
	CH	1.44–1.84	$\leqslant$15	1.4–4.2	Impervious	Medium	Poor	High	Poor	Very poor
	OH	1.28–1.76	$\leqslant$5	0.7–2.8	Impervious	Medium	Poor	High	Poor	Very poor
Highly organic soils	Pt	—	—	—	—	Slight	—	—	Not suitable	Not suitable

a) When compacted and saturated.

4.5 Soil Behavior Classification Based on In Situ Tests

In every foundation design, it is essential to understand the pertinent soil properties and parameters that govern soil behavior. Soil behavior is influenced by various factors, including particle size, shape, mineral composition, pore fluid characteristics, particle packing, stress history, and other factors.

Due to recent advances in in situ test procedures that have increased the reliability of the in situ test results, one of the main applications of in situ tests today is to determine the soil profile and classify its behavior.

4.5.1 Soil Classification Based on Standard Penetration Test Blow Counts

The standard penetration test (SPT) is a test conducted during a boring in the field to measure the approximate soil resistance to the penetration of a split-spoon sampler at various depths below the ground surface. An advantage is that SPT has been widely used for a long time, resulting in significant experience. It is also relatively simple and economical. While there is some correlation of the *N* value with the engineering parameters of soil (Table 4.21), the test results are affected by many factors, and there are many sources of errors.

Table 4.21 Strength from SPT on different types of soils (adapted from Look 2007).

Material	Description	SPT (*N* Value)	Corrected SPT ((No)60)
Clay	Very soft	≤2	—
	Soft	2–5	—
	Firm	5–10	—
	Stiff	10–20	—
	Very stiff	20–40	—
	Hard	>40	—
Clean sand	Very loose	≤4	≤3
	Loose	4–10	3–8
	Medium dense	10–30	8–25
	Dense	30–50	25–42
	Very dense	>50	>42
Fine sand	Very loose	—	≤3
	Loose	—	3–7
	Medium dense	—	7–23
	Dense	—	23–40
	Very dense	—	>40
Medium sand	Very loose	—	≤3
	Loose	—	3–8
	Medium dense	—	8–25
	Dense	—	25–43
	Very dense	—	>43
Coarse sand	Very loose	—	≤3
	Loose	—	3–8
	Medium dense	—	8–27
	Dense	—	27–47
	Very dense	—	>47

4.5.2 Soil Classification Based on FVST Records

The results of standard penetration tests (SPT) in fine-grained soils do not accurately predict the undrained shear strength due to the development of high pore-water pressure during the SPT test. However, field vane shear test (FVST) or small-scale VST can provide more accurate estimates of undrained shear strength the SPT tests. The general correlation between the consistency and undrained shear strength is shown in Table 4.22.

4.5.3 Soil Classification Based on CPT Data

One of the main purposes of performing a cone penetrometer test (CPT) in a site investigation project is to identify the subsurface soil. One of the first attempts to classify soils based on CPT results was conducted by Begemann (1965), who showed that the type of soil is a function of the ratio between the cone resistance and the sleeve friction. In addition, Schmertmann (1978), Campanella (1986a, b), and Olsen and Mitchell (1995) used cone resistance (q_c) or normalized cone resistance, the friction ratio (F_r), and the proposed Soil Classification Chart, which is shown in Figure 4.6a–c.

The development of the piezocone allowed the additional measurement of pore pressure during the test, which allows for better soil classification. The additional parameter used for soil classification is the pore pressure ratio B_q, which can be calculated by Equation 4.10. Senneset et al., (1989) proposed a graph in this regard, as shown in Figure 4.6d.

$$B_q = \frac{u_2 - u_0}{(q_t - \sigma_0')} \tag{4.10}$$

where

u_0 = pore pressure due to groundwater table,
u_2 = pore pressure measured at the shoulder,
σ_0' = total overburden stress,
q_t = corrected tip resistance

Corrected tip resistance (q_t) can be calculated by Equation 4.11:

$$q_t = q_c - u_2(1 - a_n) \tag{4.11}$$

where

q_c = Cone resistance,
a_n = net area ratio, approximately equal to the ratio of shaft cross-section and cone cross-section areas, usually measured in a calibration cell (Lunne et al. 1997).

Over the years, many researchers have proposed different soil classification methods and charts based on either CPT or CPTu results. They include Jones and Rust (1982), Jefferies and Davies (1991), and Ramsey (2002).

Jefferies and Davies (1993) introduced the soil behavior index (I_{SBT}), which describes the boundaries between different soil types and can be calculated by Equation 4.12:

$$I_{SBT} = \left[\left(3.47 - \log\left(\frac{q_c}{p_a}\right)\right)^2 + (\log R_f + 1.22)^2\right]^{0.5} \tag{4.12}$$

Table 4.22 General correlation between consistency and undrained shear strength (adapted from Das 2006).

Consistency	S_u (kN/m²)
Very soft	0–25
Soft	25–50
Medium	50–100
Hard	100–200
Very hard	200–400

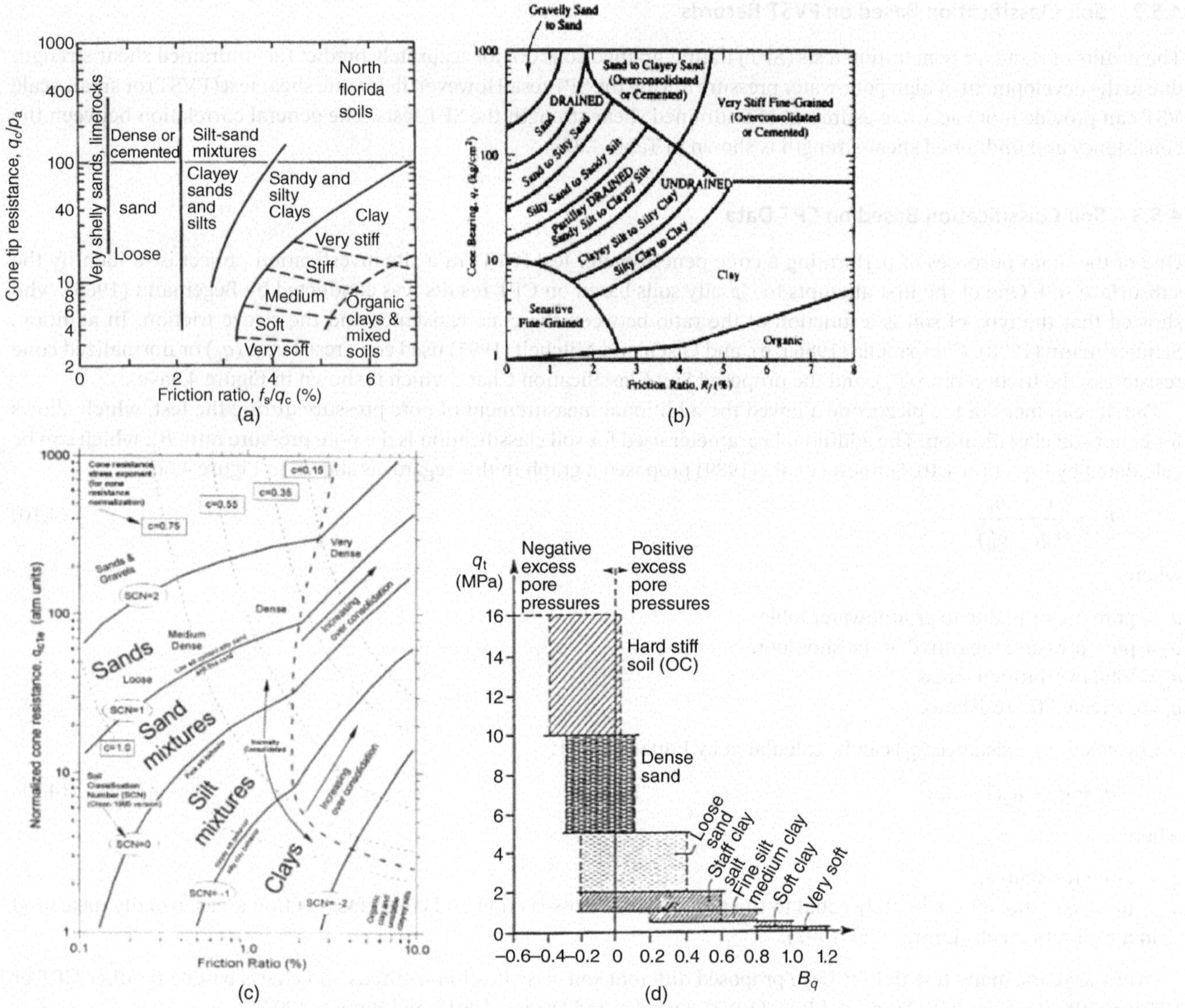

Figure 4.6 Some of pioneer soil profiling charts: (a) Schmertmann (1978)/U.S. Department of Transportation / Public Domain, (b) Adapted from Campanella (1986a, b), (c) Adapted from Senneset and Janbu (1989), and (d) Adapted from Olsen and Mitchell (1995).

where

R_f = friction ratio,

q_c = cone resistance

The values of I_c defined on the boundaries of various soil types on the Robertson (1990) classification chart is depicted in Figure 4.7 (Table 4.23).

Robertson (1990) proposed a development of the Robertson et al. (1986) profiling chart, shown in Figure 4.8, plotting a normalized cone resistance, against a normalized friction ratio in a cone resistance chart. The accompanying pore pressure ratio chart plots the normalized cone resistance against the pore pressure ratio.

Robertson (2016) presented an update of some popular CPT-based SBT classification systems. The presented updates use behavior-based descriptions and comprise a method to identify the existence of microstructure in soils. Examples of updated classification systems are depicted in Figure 4.9.

Eslami and Fellenius (1997) compiled a database consisting of CPT and CPTu data linked with results of boring, sampling, laboratory testing, and soil characteristics. The data points were plotted in a profiling chart, and envelopes were drawn enclosing five soil types. The envelopes are shown in Figure 4.10a. Plotting an "effective" cone resistance defined by Equation 4.13 was found to provide a more consistent envelope delineation than a plot of only the cone resistance.

$$q_E = q_t - u_2 \tag{4.13}$$

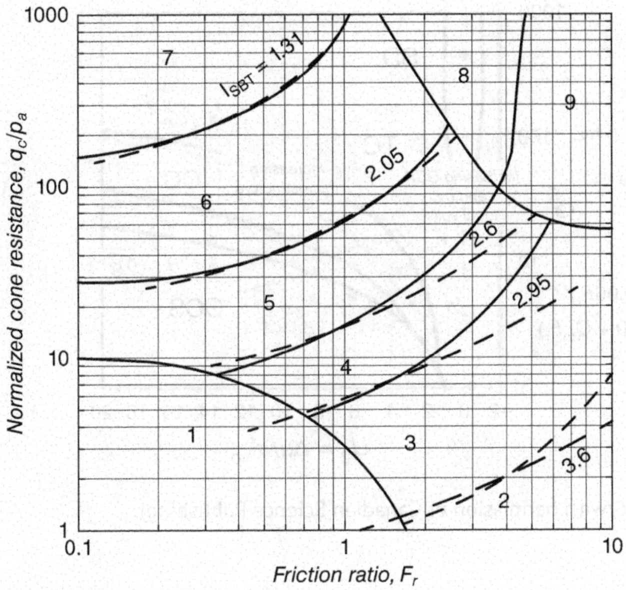

Figure 4.7 Values of I_c on the Robertson (1990) classification chart (adapted from Robertson (1990); Jefferies and Davies 1991).

Table 4.23 Soil classification based on I_{SBC} (Li et al. 2016/with permission of Emerald Group Publishing Ltd.).

Soil classification	Zone	CPTu index (I_{SBC})
Gravelly sands	7	$I_{SBC} < 1.25$
Sands – clean sand to silty sand	6	$1.25 < I_{SBC} < 1.90$
Sand mixture – silty sand to sandy silt	5	$1.90 < I_{SBC} < 2.54$
Silt mixture – clayey silt to silty clay	4	$2.54 < I_{SBC} < 2.82$
Clays	3	$2.82 < I_{SBC} < 3.22$
Organic clays	2	$3.22 < I_{SBC}$

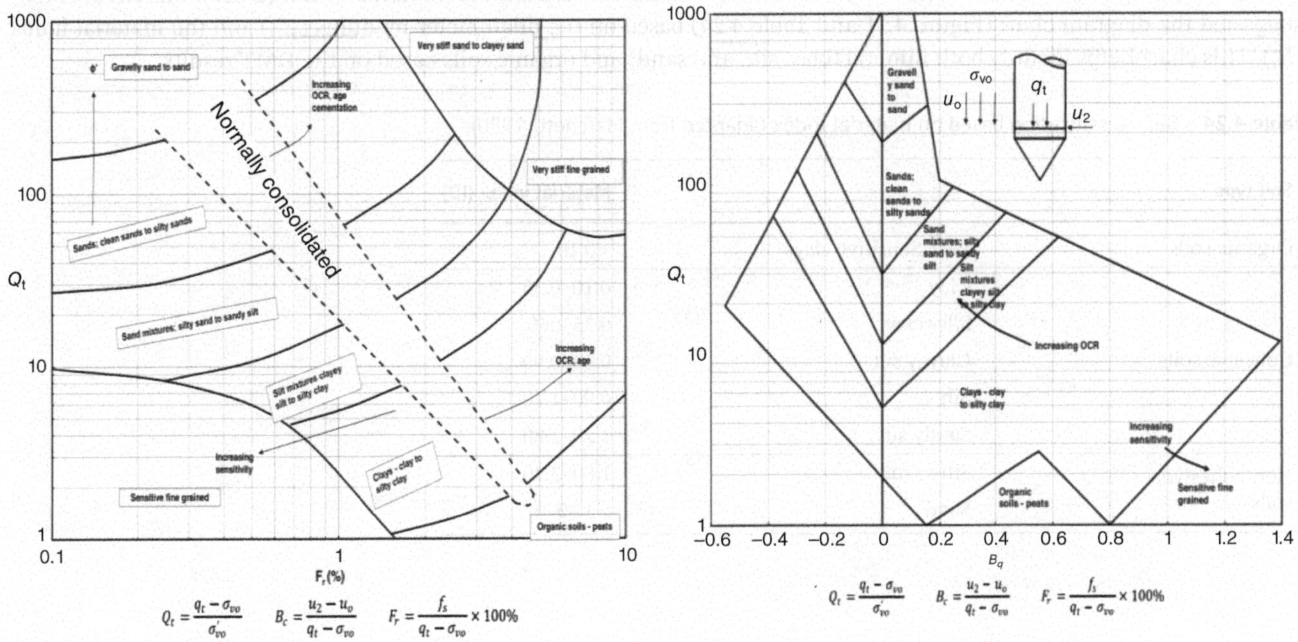

Figure 4.8 Proposed soil behavior type classification system from CPT/CPTu data (adapted from Robertson 1990).

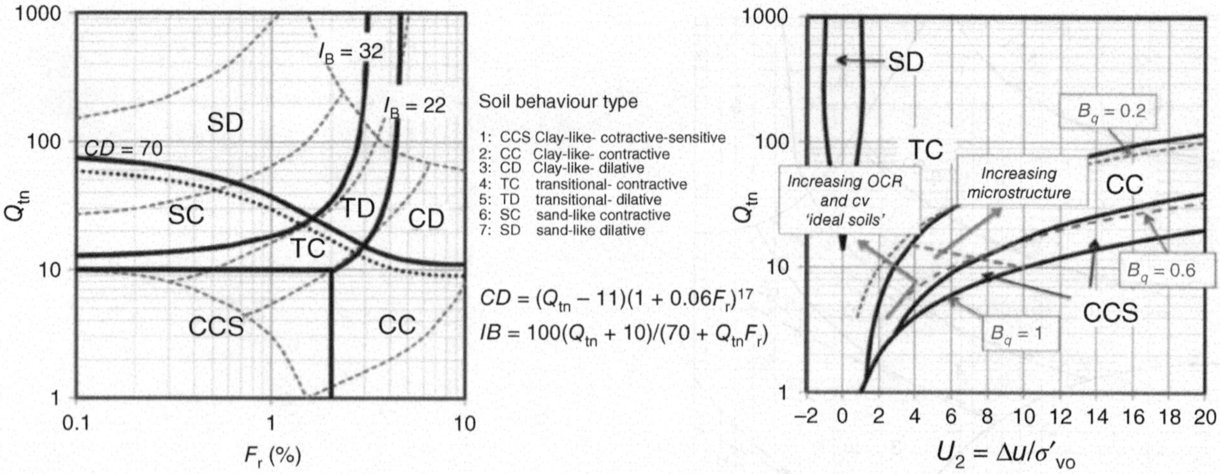

Figure 4.9 Examples of updated classification charts (Robertson 2016/with permission of Canadian Science Publishing).

where

q_E = effective cone resistance,
q_t = cone resistance corrected for pore water pressure on the shoulder (Equation 4.11),
u_2 = pore pressure measured at cone shoulder

A new format of classification chart, i.e., triangular form, is proposed by Eslami et al. (2015) containing Cone Tip Resistance (q_c), Sleeve Friction (f_s), and Pore Pressure (u_2). Eslami et al. (2023) developed a triangular classification chart for deltaic soils by compiling a marine database of the CPTu soundings and their soil profiling in their vicinity. The proposed chart is presented in Figure 4.10b. Overall, a qualitative chart is suggested for the general behavior classification of soils, dividing them into five generic zones (types), as illustrated in Figure 4.10c.

4.5.4 Soil Classification Based on DMT Records

In recent years, static probing by dilatometer penetration tests (DMT) has been a widely used method of in situ investigation to provide the information needed by civil engineers for design, construction, securing construction permits, and operation control. The diagram charts are commonly used for soil classification from the dilatometer test (DMT). Marchetti (1980) suggested the diagram chart (Figure 4.11 and Table 4.24) based on the dilatometer modulus (E_D) and the material index (I_D). This chart helps identify both mineral (clay, silt, and sand) and organic soils based on the DMT results.

Table 4.24 Soil classification based on material index (adapted from Marchetti 1980).

Soil type		**Material index (ID)**
Organic soils	Peat/Sensitive clays	<0.10
	Clay	0.10–0.35
	Silty clay	0.35–060
Cohesive soils	Clayey silt	0.60–0.90
	Silt	0.90–1.20
	Sandy silt	1.20–1.80
Non-cohesive soils	Silty sand	1.80–3.30
	Sand	>3.30

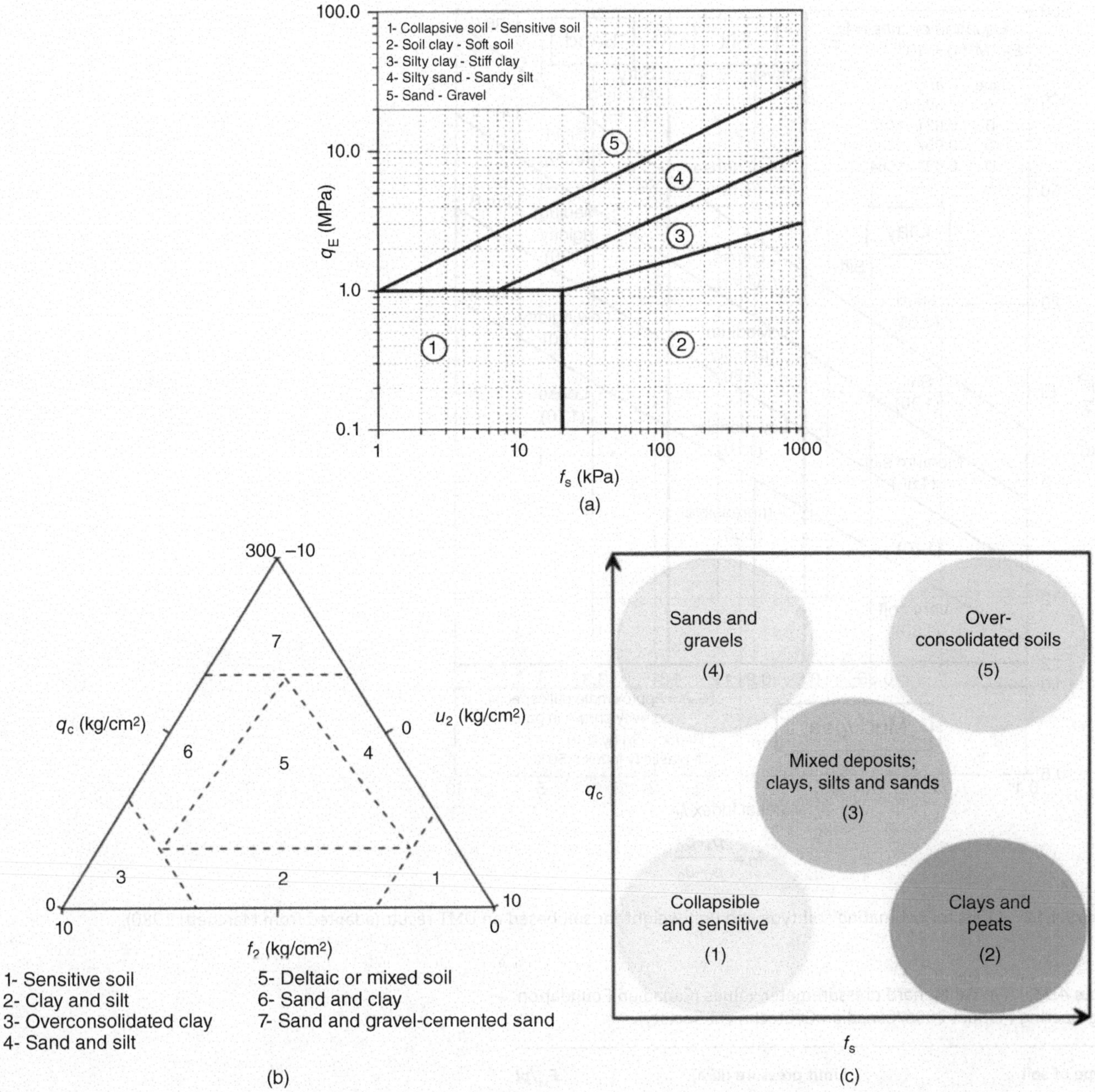

Figure 4.10 (a) classification chart based on effective cone resistance and shaft resistance (Eslami and Fellenius 1997 / With permission of Canadian Science Publishing), (b) enhanced triangular charts for deltaic soils (Eslami et al. 2023 with permission of ELSEVIER), and (c) general qualitative classification chart.

4.5.5 Soil Classification Based on PMT Measurements

The borehole pressuremeter test (PMT) has been widely used in France since it was first developed by Menard in 1956. It is not common to use PMT for soil classification, but as shown in Table 4.25, typical Menard pressuremeter values can be used to identify soil types.

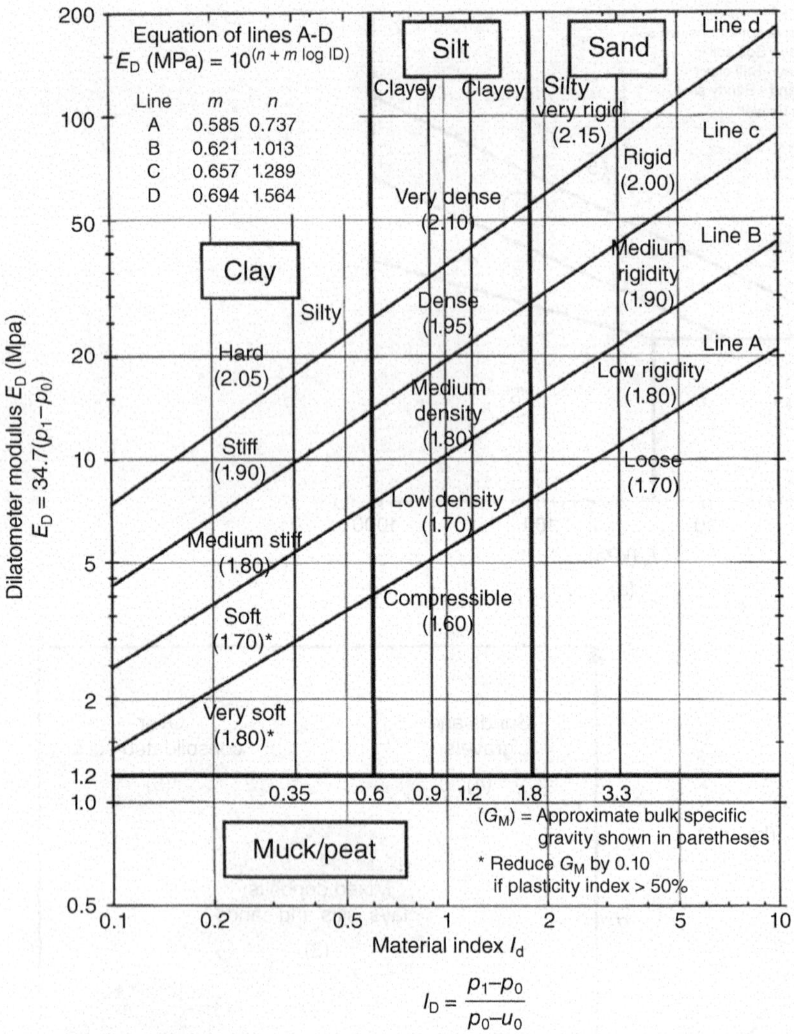

Figure 4.11 Chart for estimating soil type and unit weight for soil based on DMT result (adapted from Marchetti 1980).

Table 4.25 Typical Menard pressuremeter values (Canadian Foundation Engineering Manual 2006/Canadian Geotechnical Society).

Type of soil	Limit pressure (kPa)	E_m/pl
Soft clay	50–300	10
Firm clay	300–800	10
Stiff clay	600–2500	15
Loose silty sand	100–500	5
Silt	200–1500	8
Sand and gravel	1200–5000	7
Till	1000–5000	8
Old fill	400–1000	12
Recent fill	50–300	12

Example 4.1 Determine the behavioral classification of soil in the range of 4–7 m using given SPT, PMT, and DMT results, if E_{PMT} in the range of 4–7 m is equal to 7000 kPa.

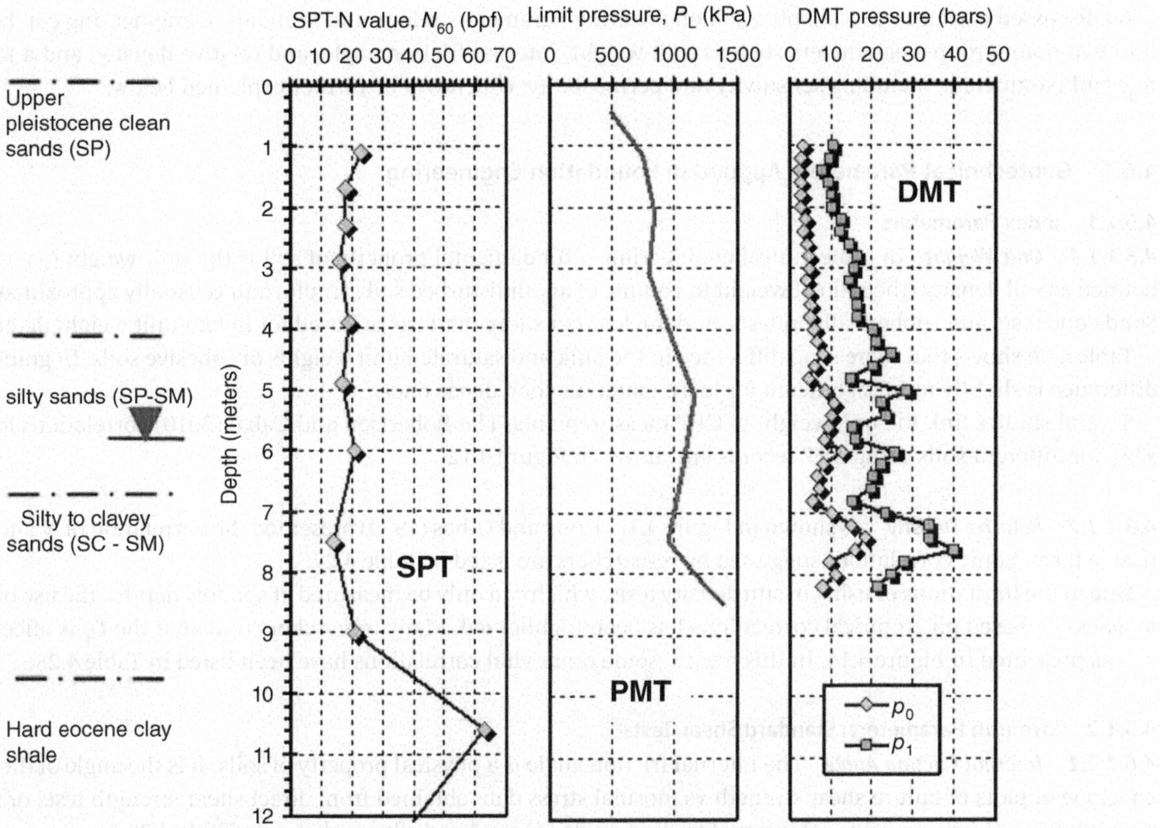

Depth (m)	*N*-SPT	SPT-soil type (Look 2007)	Pl (kPa)-PMT	E_{PMT}/pl	PMT-soil type (CFEM 2006)	Pressure-DMT (bars) p_0	p_1	u_0		I_D	DMT-soil type (Marchetti 1980)
4–5	18	Stiff clay/Medium dense Clean Sand	1000	7	Silty sand	9	22	0	45.11	1.44	Dense sandy silt
5–6	19	Stiff clay/Medium dense Clean Sand	1150	6	Silty sand	10	26	0	55.52	1.6	Very dense sandy silt
6–7	18	Stiff clay/Medium dense Clean Sand	1050	6.67	Silty sand	9.5	20	0.0981	36.43	1.11	Dense sandy silt

Comparing the results with the given soil profile show that, generally, the in situ tests predict the soil type well and with minor errors. In this regard, PMT and DMT are more accurate than SPT records since SPT is not typically used for subsoil classification and identification.

4.6 Soil Geotechnical Parameters: Determination and Correlations

Each geotechnical report includes a description of the subsurface soil stratification and estimates of geotechnical parameter values (referred to as soil engineering parameters) for each soil layer identified.

The primary purpose of this section is to present commonly used correlations employed by geotechnical practitioners to assess design parameters crucial in geotechnical activities, particularly foundation design. Empiricism plays a significant role in geotechnical engineering. These correlations serve as an initial assessment and can be compared against laboratory and in situ test results. The literature offers a wide range of diagrams and empirical equations, which are developed based on historical field and laboratory databases, as well as professional experience.

As discussed earlier in this chapter, soil engineering parameters relevant to foundation engineering can be categorized into two main groups: parameters such as unit weight, internal friction angle, and relative density; and a few additional relevant parameters, including sensitivity and permeability, which will be further explained below.

4.6.1 Geotechnical Parameters Applied in Foundation Engineering

4.6.1.1 Index Parameters

4.6.1.1.1 Unit Weight In geotechnical engineering, a fundamental property of soil is the unit weight (γ). Unit weight is defined as soil density (the ratio of weight to volume of an undisturbed soil sample) and is usually approximated in the lab. Sand cone tests and rubber balloon tests, and nuclear densitometer tests are used for in situ unit weight determinations.

Table 4.26 shows that there is no difference in the bulk and saturated unit weights of cohesive soils. In granular soils, the difference is slightly more significant for loose materials than dense ones.

Several studies link the unit weight to CPT measurements. The Robertson and Cabal (2010) correlations for calculating γ/γ_w for different soils using CPT records are shown in Figure 4.12.

4.6.1.1.2 Relative Density As shown in Figure 4.13, Holtz and Gibbs (1979) presented the correlation of N and D_r in a more usable form. Some correlations suggested by researchers are listed in Table 4.27.

Due to the limitations of using in situ density tests, which can only be measured at shallow depths, the use of CPT records to assess D_r based on empirical correlations has been highlighted. Many researchers note that the D_r is affected by q_c and σ_v', as presented in Figure 4.14. In this regard, some other vital correlations have been listed in Table 4.28.

4.6.1.2 Strength Parameters: Standard Shear Tests

4.6.1.2.1 Internal Friction Angle The internal friction angle is a physical property of soils. It is the angle of the shear failure envelope of plots of failure shear strength vs. normal stress data obtained from direct shear strength tests or triaxial shear strength tests. Common values of internal friction angle for sand and silt are shown in Table 4.29.

Table 4.26 Typical values for bulk and saturated unit weights (Das and Sivakugan 2016/Cengage Learning).

Soil type		Bulk unit weight (kN/m³)		Saturated unit weight (kN/m³)	
		Loose	Dense	Loose	Dense
Granular soils	Gravel	16	18	20	21
	Well-graded sand and gravel	19	21	21.5	23
	Coarse or medium sand	16.5	18.5	20	21.5
	Well graded sand	18	21	20.5	22.5
	Fine or silty sand	17	19	20	21.5
	Rock fill	15	17.5	19.5	21
	Brick hardcore	13	17.5	16.5	19
	Slag fill	12	15	18	20
	Ash fill	6.5	10	13	15
Cohesive soils	Peat	12		12	
	Organic clay	15		15	
	Soft clay	17		17	
	Firm clay	18		18	
	Stiff clay	19		19	
	Hard clay	20		20	
	Stiff or hard glacial clay	21		21	

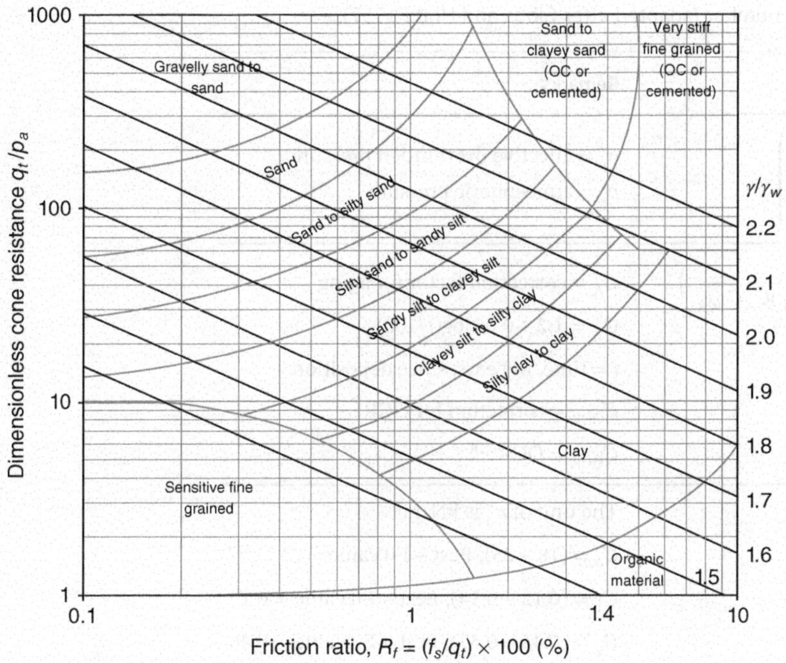

Figure 4.12 Normalized unit weight and R_f (%) (adapted from Robertson and Cabal 2010).

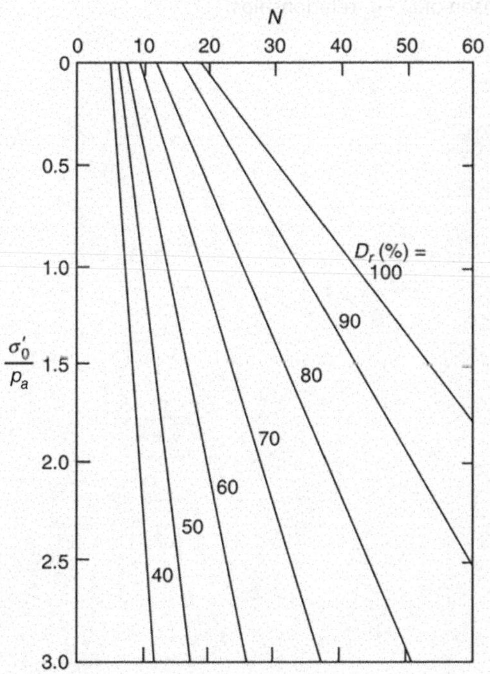

Figure 4.13 Variation of N with σ_0/p_a and D_r (adapted after Gibbs and Gibbs 1979).

Coarse-grained soils shear strength is commonly presented in the form of the friction angle, ϕ. Some researchers have proposed friction angles in sands based on one of the following methods:

- Cavity expansion
- Bearing capacity
- Empirically based on calibration chamber tests

Table 4.27 Proposed correlation for D_r based on SPT number (adapted after Gibbs and Holtz 1957).

References	Proposed correlation	Remarks
Meyerhof (1957)	$D_r = 20.4\left(\frac{N}{0.7+\left(\frac{\sigma'_o}{p_a}\right)}\right)^{0.5}$	σ'_o = effective overburden pressure p_a = atmospheric pressure
Kulhawy and Mayne (1990)	$D_r = \left(\frac{(N_1)_{60}}{(60+25\log D_{50})C_A C_{OCR}}\right)^{0.5}$	C_A = correction factor for aging $C_A = 1.2 + 0.05\log(t/100)$ t = time, in years, since deposition C_{OCR} = correction for OCR $C_{OCR} = OCR^{0.18}$
Yoshida et al. (1988)	$D_r = C_0(\sigma'_o)^{-C_1}N^{C_2}$	The unit of σ'_o is kN/m^2 $C_0 = (18-25)$, Best – FitValue = 25 $C_1 = (0.12-0.14)$, Best – FitValue = 0.12 $C_2 = (0.44-0.57)$, Best – FitValue = 0.46

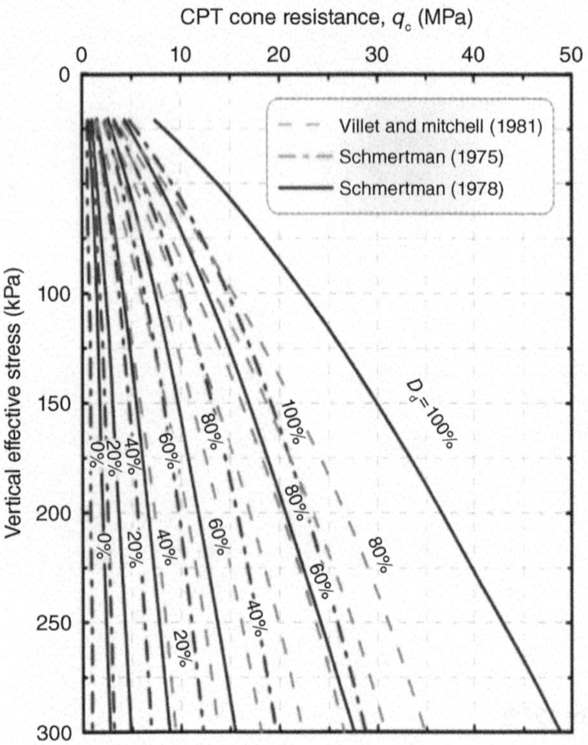

Figure 4.14 Comparison of D_r–q_c relationships.

The friction angle (ϕ) of granular soils increases with the grains' angularity, surface roughness, and relative density. Poorly graded granular soils typically have lower ϕ than well-graded ones. Figures 4.15 and 4.16 show the friction angles determined from different granular soils' triaxial compression tests with no plastic fines.

Schmertmann (1975) proposed a correlation between effective overburden pressure σ_0, N, and ϕ, and this correlation is shown in Figure 4.17. Some other correlations are listed in Table 4.30.

Table 4.28 Correlations predicting D_r from CPT records.

References	Proposed correlation	Remarks
Baldi et al. (1986)	$D_r = \frac{1}{C_2}\ln\left(\frac{q_c}{C_0(\sigma_v')^{0.55}}\right)$	C_0 and C_2: soil constants, $C_0 = 157$ and $C_2 = 2.41$ normally consolidated sand. q_c and σ_v' are in the kPa unit
Jamiolkowski et al. (2001)	$D_r = 26.8\ln\frac{q_c/p_a}{(\sigma_v'/p_a)^{0.5}} - b_x$	$b_x = 52.5$ for *high* compressibility sands $b_x = 67.5$ for *medium* compressibility sands $b_x = 82.5$ for *low* compressibility sands
Kulhawy and Mayne (1990)	$D_r = \frac{Q_{cn}}{305Q_cQ_{OCR}}$	$Q_{cn} = q_t/p_a/(\sigma_v'/p_a)^{0.5}$ Q_c= Compressibility factor (0.91 for high, 1.0 for medium, and 1.09 for low). Q_{OCR} = Over consolidation factor, $OCR^{0.18}$
Mayne (2007)	$D_r = 100\left(0.268\ln\left(\frac{q_t/p_a}{\sqrt{\sigma_v'/p_a}}\right) - 0.675\right)$	q_c and σ_v' are in kPa unit

Table 4.29 Common amount of friction angle for sand and silt (adapted from Das 2006).

Soil type		ϕ
Angular sand	Loose	27–30
	Medium	30–35
	Dense	35–38
Rounded sand	Loose	30–35
	Medium	35–40
	Dense	40–45
Sandy gravel		34–48
Silt		26–35

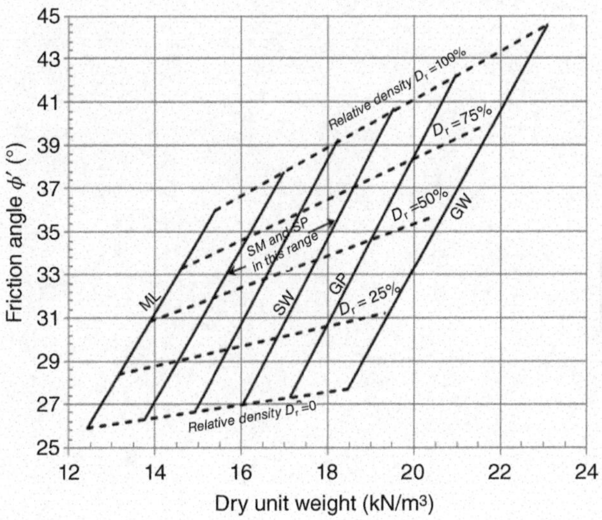

Figure 4.15 Friction angles of granular soils (adapted from U.S. Navy 1982).

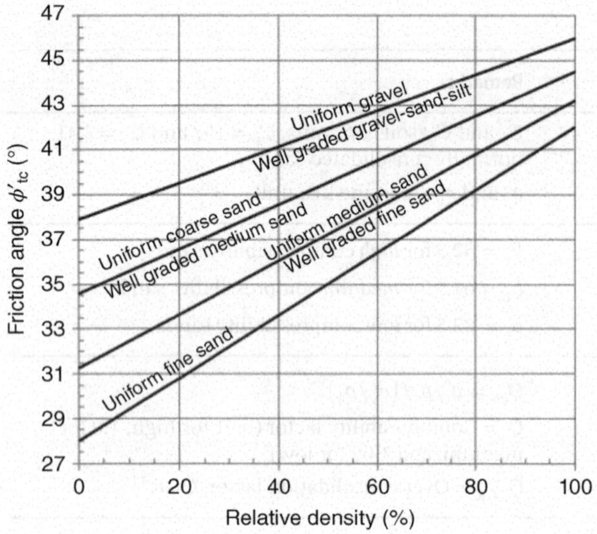

Figure 4.16 $Ø'_{tc}$ – D_r relations (adapted from Schmertmann 1978).

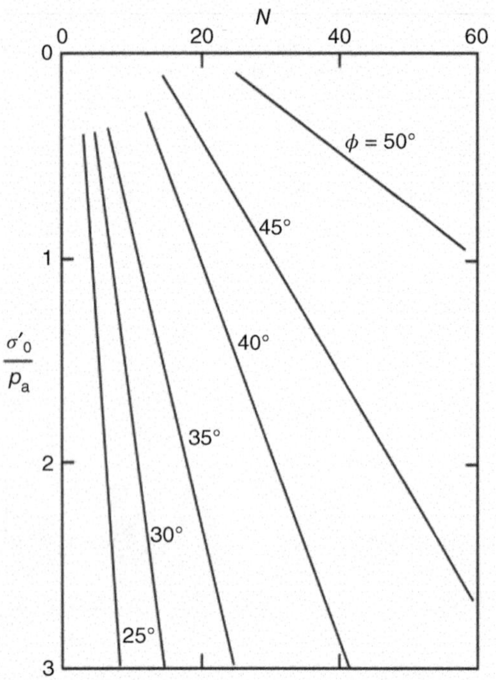

Figure 4.17 Variation of Ø with SPT number and σ'_o/p_a (adapted after Schmertmann 1975).

Table 4.30 Variation of Ø with *N*.

References	Proposed correlation
Meyerhof (1959)	$\varphi = 28 + 0.15\left[20.4\left(\dfrac{N}{0.7 + (\sigma'_o/p_a)}\right)\right]^{0.5}$
Wolff (1989)	$\varphi = 27.1 + 0.3N_1 - 0.00054N_1^2$
Kulhawy and Mayne (1990)	$\varphi = \tan^{-1}\left[\dfrac{N}{12.2 + 20.3\,(\sigma'_o/p_a)}\right]^{0.34}$
Hatanaka and Uchida (1996)	$\varphi = \sqrt{20(N_1)_{60}} + 20$

Bergdahl et al. (1993) proposed a correlation of q_c and relative density with friction angles for cohesionless or mixed soils as shown in Table 4.31. Mitchell and Durgunoglu (1983) suggested a correlation between cone tip resistance and internal friction angle as depicted in Figure 4.18.

Other researchers (e.g., Meyerhof (1974), Robertson et al. (1986), Kulhawy and Mayne (1990), and Uzielli et al. (2013)) introduced equations to predict Ø as summarized in Table 4.32. Moreover, Mayne (2014) proposed correlations to predict the friction angle based on empirical correlation and case histories.

By applying two basic equations on the determination of a deep foundation's bearing capacity, one for corrected q_t (q_E) tip and the other for f_s, and using the effective bearing capacity instead of the total stress, a dual equation system with two unknowns can be achieved under static loading conditions as follows:

$$CN_c + qN_q + 0.5\gamma BN_\gamma = q_E = q_t - u_2 \tag{4.14}$$

$$C + \sigma' \tan(\delta) = f_s \tag{4.15}$$

Eslami and Mohammadi (2016) present approaches to predict φ and C of marine deposits from CPTu records based on bearing capacity and cavity expansion theories. They summarize the soil classification chart in five categorized areas based on CPTu test results. For specified q_t and F_r (friction ratio), the soil type can be determined in every depth interval.

Table 4.31 Correlation of q_c and relative density with friction angle for cohesionless or mixed soils (after Bergdahl et al. 1993).

Relative density	q_c (MPa)	ϕ
Very weak	0.0–2.5	29–32
Weak	2.5–5.0	32–35
Medium	5.0–10.0	35–37
Large	10.0–20.0	37–40
Very large	>20.0	40–42

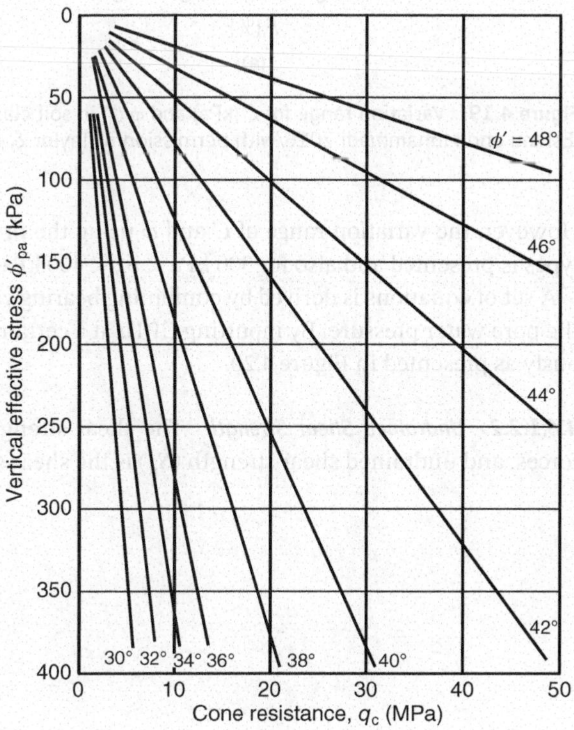

Figure 4.18 Friction angle changes and cone tip resistance (Mitchell and Durgunoglu 1983/American Society of Civil Engineers).

Table 4.32 Proposed correlations for friction angles based on CPT results.

References	Correlations	Soil type and remarks
Meyerhof (1974)	$\varphi = \tan^{-1}(q_c/0.5N_q)$	Sand q_c (MPa)
Robertson et al. (1986)	$\varphi = \tan^{-1}\left[0.1 + 0.38\log\left(q_c/\sigma_v'\right)\right]$	Sand
Kulhawy and Mayne (1990)	$\varphi = 17.6 + 11\log\left(q_c/\sqrt{100\sigma_v'}\right)$	Sand σ_v' and q_c are in kPa unit
Uzielli et al. (2013)	$\varphi = 25\left(q_c/\sqrt{100\sigma_v'}\right)^{0.1}$	Sand σ_v' and q_c are in kPa unit
Robertson and Cabal (2010)	$\tan\varphi' = \frac{1}{2.68}\left(\log\left(q_c/\sigma_v'\right) + 0.29\right)$	Uncemented, unaged, moderately compressible quartz sands
Mayne (2014)	$\varphi = 29.5B_q^{0.121}[0.256 + 0.33B_q + \log Q_t]$	Cohesive soils $Q_t = \frac{q_t - \sigma_v}{\sigma_v'}$ $B_q = \frac{u_2 - u_0}{q_t - \sigma_v}$

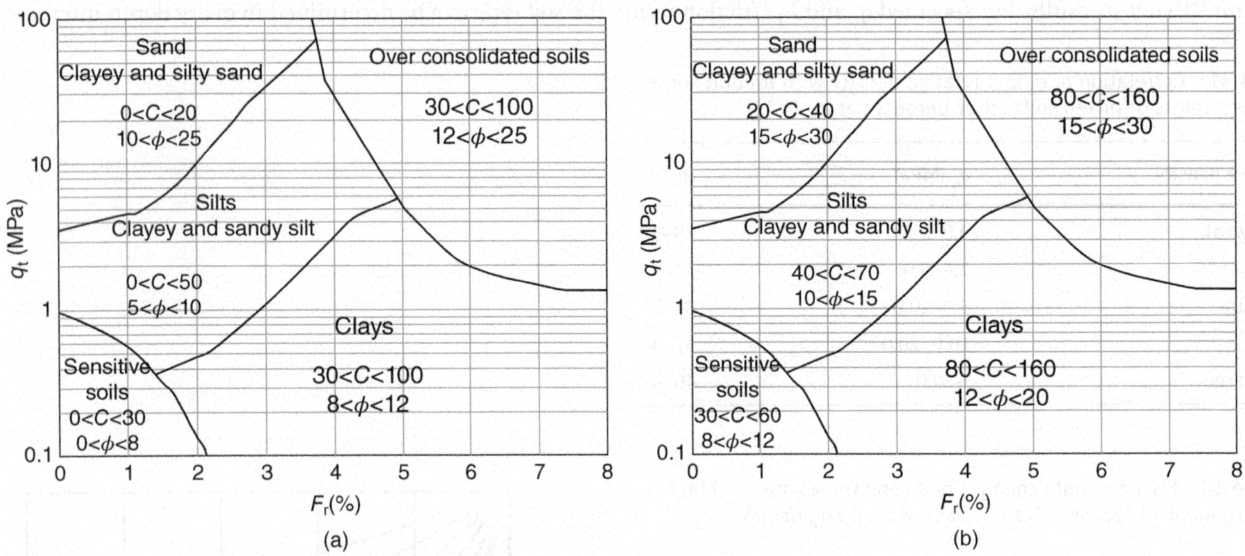

Figure 4.19 Variation range for C (kPa) and ϕ (°) in soil classification chart: (a) 200 kPa $< \sigma_v' <$ 300 kPa, (b) 300 kPa $< \sigma_v' <$ 400 kPa (Eslami and Mohammadi 2016/with permission of Taylor & Francis).

However, the variation range of C and φ using the proposed method at 200 kPa $< \sigma_v' <$ 300 kPa into five categorized soil types is presented and also for 300 kPa $< \sigma_v' <$ 400 kPa in Figure 4.19.

A set of equations is derived by combining bearing capacity theories and direct shear modes of failure while considering the pore water pressure. By inputting CPTu at a certain depth, soil shear strength parameters can be calculated simultaneously as presented in Figure 4.20.

4.6.1.2.2 Undrained Shear Strength The shear strength of soil is the maximum internal resistance to applied shearing forces, and undrained shear strength (S_u) is the shear strength of soil when sheared at a constant volume.

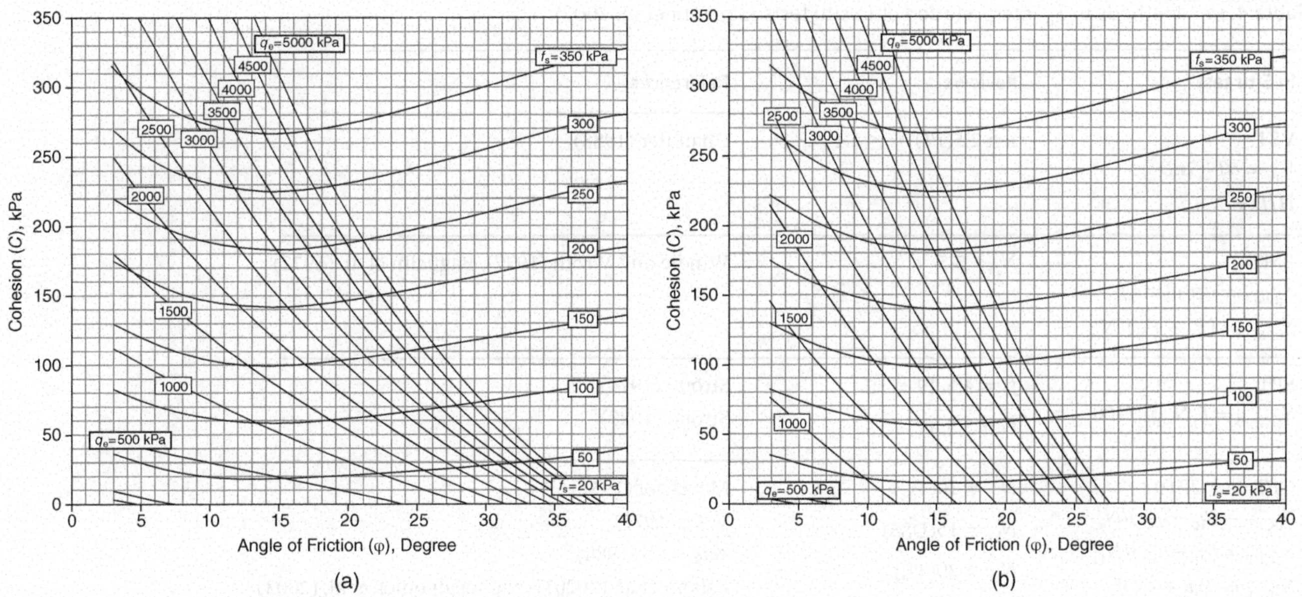

Figure 4.20 Graphic estimation of C and φ: (a) $\sigma'_v = 50$ kPa, (b) $\sigma'_v = 200$ kPa (Eslami and Mohammadi 2016/with permission of Taylor & Francis).

Table 4.33 Approximate variation of consistency, *N*, and undrained cohesion of clay Terzaghi and Peck (1967) / With permission of John Wiley & Sons.

Consistency	*N*	C_u(kN/m^2)
Very soft	0–2	<15
Soft	2–4	12–25
Medium	4–8	25–50
Stiff	8–15	50–100
Very Stiff	15–30	100–200
Hard	>30	>200

There are some proposed correlations between the SPT results and the shear strength of the soil. Terzaghi and Peck suggest approximate consistency, corresponding *N* value, and undrained cohesion (c_u) of clay soils (1967), provided in Table 4.33. These values need to be used with care.

- Beside the soil profile classification, one of the main engineering parameters of fine-grained soil, derived from CPT, is the undrained shear strength (S_u) of clayey soil. S_u cannot be directly measured by the CPT, yet it can be obtained with empirical correlations and must be calibrated by laboratory tests such as triaxial compression or direct shear tests. Several researchers investigated the relationship between the S_u–CPT parameters. Some methods in S_u interpretation of in situ tests are listed in Table 4.34.

Table 4.34 Methods in S_u interpretation of in situ tests (Sabatini et al. 2002).

In Situ test	Reviews	References
VST: $S_{uv} = 6T/(7\pi D^3)$ $H/D = 2$	$\mu \cong 2.5(PI)^{-0.3} \leq 1.1$	Chandler (1988)
PMT: $S_{uPMT} = dp/d(\ln\varepsilon_v)$ $S_{uPMT} = (P_L - P_O)/N_c$	$N_c = 5.5$	Windle and Wroth (1977), Baguelin et al. (1972)
SPT: $S_{u(N60)} = f_1 N_{60} P_a/100$	$f_1 = 4.5$, PI = 50 $f_1 = 5.5$, PI = 15	Stroud (1974) Stroud (1989)
CPT and CPTu: $S_{uCPT} = (q_c - \sigma_{vo})/N_c$ $S_{uCPT} = (q_t - \sigma_{vo})/N_{kt}$ $S_{uCPT} = (0.8 - 1.2)f_s$	$N_{kt} = 10$(TC) $N_{kt} = 15$(DSS) $N_{kt} = 20$(TE)	Meyerhof (1951) Vesic (1977) Aas et al. (1988) Eslami et al. (2020) Anagnostopoulos et al. (2003)
DMT: $S_{uDMT} = 0.22\sigma'_{vo}(0.5K_D)^{1.25}$ $S_{uDMT} = (P_o - u_o)/10$ $S_{uDMT} = d_s\sigma'_{vo}(0.5K_D)^{1.25}$	TC: $d_s = 0.20$ VST: $d_s = 0.19$ DSS: $d_s = 0.14$	Marchetti (1980) Schmertmann (1991) Lacasse and Lunne (1988)
PLT: $S_{uPLT} = q_{ult}/6.18$	—	Meyerhof (1951)

Example 4.2 The given figure shows the CPT results from a project site. Determine the required items.

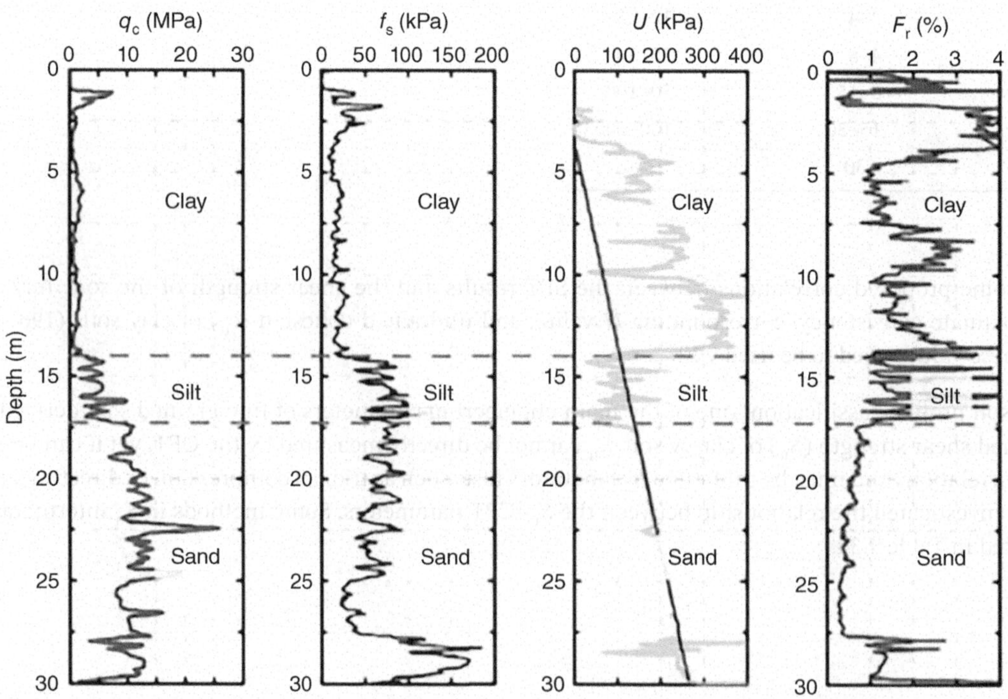

a) Calculate the relative density of the soil from a depth of 20–25 m, assuming $\sigma_v' = 160$ kPa.
q_c, and u can be considered 14 MPa and 180 KPa, respectively.
Using the proposed correlation by Mayne (2007):

$$q_t = q_c + u(1-a) \xrightarrow{\text{by assuming } a=0.6} q_t = 14{,}000 + 180(1-0.6) = 14{,}072 \text{ kPa}$$

$$D_r = 100\left(0.268\ln\left(\frac{q_t/p_a}{\sqrt{\sigma_v'/p_a}}\right) - 0.675\right)$$

$$D_r = 100\left(0.268\ln\left(\frac{140.72}{\sqrt{1.6}}\right) - 0.675\right) = 58.7\%$$

b) Calculate the friction angle of the soil from a depth of 20–25 m, assuming $\sigma_v' = 160$ kPa.
q_c can be considered 14 MPa.
Based on Kulhawy and Mayne (1990):

$$\varphi = 17.6 + 11\text{Log}\left(\frac{q_c}{\sqrt{100\sigma_v'}}\right) \rightarrow \varphi = 17.6 + 11\text{Log}\left(\frac{14{,}000}{\sqrt{16{,}000}}\right) = 40$$

Based on Uzielli et al. (2013):

$$\varphi = 25\left(\frac{q_c}{\sqrt{100\sigma_v'}}\right)^{0.1} \rightarrow \varphi = 25\left(\frac{14{,}000}{\sqrt{16{,}000}}\right)^{0.1} = 40$$

c) Calculate undrained shear strength 0–14 m depth (Assume $\sigma_v = 110$ kPa).

$$S_u = \frac{(q_c - \sigma_v)}{N_c},\quad q_c = 2.5\text{ MPa},\ N_c = 20,\ \sigma_v = 110\text{ kPa} \Rightarrow S_u = \frac{(2500-110)}{20} = 119.5\text{ kPa}$$

4.6.1.3 Stiffness and Compressibility Parameters

4.6.1.3.1 Stiffness Soil modulus is recognized as a complex parameter in geotechnical engineering due to the various conditions involved. The Young's modulus (E) is a soil deformability parameter that relates to the linear elastic behavior of geotechnical materials. However, it is important to note that the soil modulus cannot be equated to the slope of the stress–strain curve, as geo-materials do not exhibit ideal linear elastic behavior. This implies that the soil modulus is stress-dependent and undergoes changes with varying stress levels. Therefore, it must be considered as a parameter that is influenced by stress.

When a foundation is subjected to loading on a normally consolidated deposit, the secant modulus (E_s) can be utilized to estimate settlement. To obtain E_s, a slope is drawn from the origin to an arbitrary point on the stress–strain curve. Additionally, the tangent slope (E_t), associated with the tangent modulus, can be employed to assess progressive settlement when incremental loads are applied. Table 4.35 presents suggested relationships between elastic modulus and the Cone Penetration Test (CPT) results.

As Baldi et al. (1986) suggested that the stiffness of fine-grained soil is affected by different factors, including the mineralogy, grain size, particle shape, grain fabric, drainage conditions, mean effective stress, and stress–strain behavior. In a given soil, cone tip resistance is mainly controlled by the void ratio and the effective stress conditions. Some methods in E interpretation of in situ tests are listed in Table 4.36.

4.6.1.3.2 Constrained Modulus The constrained modulus of clays can be calculated in the laboratory using the consolidation test:

$$M = \frac{1}{m_v} \tag{4.16}$$

where

m_v = the coefficient of volume compressibility.

Table 4.35 Drained elastic modulus for soils using CPT records (adapted from Look 2007).

Soil		q_c (MPa)	Drained elastic modulus (MPa)	E_t From CPT
Sand	Very loose	<2.5	<10	$(2-4)q_c$
	Loose	2.5–5	10–20	
	Medium dense	5–10	20–30	$(3-7)q_c$
	Dense	10–20	30–60	$(6-10)q_c$
	Very dense	>20	>60	
Clay	Very soft	<2	<5	$(1-2)q_c$
	Soft	2–5	5–15	$(1-2)q_c$
	Medium	5–10	15–25	$(2-4)q_c$
	Hard	10–20	>25	$(3-8)q_c$

Table 4.36 Elastic modulus parameter of in situ tests (adapted from Bowles 1997).

Soil type	CPT	SPT
Sand	$E_s=(2-4)q_u$ $=8000q_u$ - - - - - - - - - - - $E_s=1.2\,(3D_r^2+2)\cdot q_c$ $E_s=(1+D_r^2)\cdot q_c$	$E_s=500(N+15)$ $=7000\sqrt{N}$ $=6000N$ - - - - - - - - - - - - - - $E_s=(15{,}000-22{,}000)\cdot \ln N$
Saturated sand	$E_s=F\cdot q_c$ $e=1.0\quad F=3.5$ $e=0.6\quad F=7.0$	$E_s=250(N+15)$
OCR sand	$E_s=(6-30)q_c$	$E_s=40{,}000+1050\cdot N$ $E_{s(OCR)}\approx E_{s(NC)}\sqrt{OCR}$
Sandy gravel	—	$E_s=1200(N+6)$ $=600(N+6)\quad N\le 15$ $=600(N+6)+2000\quad N>15$
Clay sand	$E_s=(3-6)q_c$	$E_s=320(N+15)$
Silty sand	$E_s=(1-2)q_c$ $q_c<2500$ kPa $E_s'=2.5q_c$ $2500<q_c<5000$ kPa $E_s'=4q_c+5000$	$E_s=300(N+6)$
Soft clay	$E_s=(3-8)q_c$	—

Lunne et al. (1997) assert that most correlations between penetrometer test results and drained constrained modulus are based on the tangent modulus from oedometer tests. The correlations are typically represented as:

$$M=\alpha q_c \tag{4.17}$$

where,

$M=$ the constrained modulus,
$\alpha=$ the constrained modulus factor. Relations between α and q_c for cohesive soils are summarized in Table 4.37.

Table 4.37 Constrained modulus coefficient for cohesive soils (adapted from Sanglerat 1972; after Frank and Magnan 1995).

Soil type	q_c (MPa)	α
Low plasticity clay (CL)	$q_c < 0.7$	$3 < \alpha < 8$
	$0.7 < q_c < 2$	$1 < \alpha < 5$
	$q_c > 2$	$1 < \alpha < 2.5$
Silt, low liquid limit (ML)	$q_c < 2$	$3 < \alpha < 6$
	$q_c > 2$	$1 < \alpha < 2$
High plasticity clay (CH)/ silt, high liquid limit (MH)	$q_c < 2$	$2 < \alpha < 6$
	$q_c > 2$	$1 < \alpha < 2$
Organic silt (OL)	$q_c < 1.2$	$2 < \alpha < 8$
Organic clay (OH)/Peat (Pt) with $50 < w < 100$		$1.5 < \alpha < 4$
Organic clay (OH)/Peat (Pt) with $100 < w < 200$	$q_c < 0.7$	$1 < \alpha < 1.5$
Organic clay (OH)/Peat (Pt) $w > 200$		$\alpha < 4$

4.6.1.3.3 Over Consolidation Ratio In geotechnical engineering, in order to describe the stress history condition of a soil deposit, the over-consolidation ratio (OCR) is introduced as Equation 4.18:

$$\text{OCR} = \frac{\text{maximum past effective vertical stress}}{\text{current effective vertical stress}} = \frac{\sigma'_p}{\sigma'_v} \tag{4.18}$$

The value of σ'_p can be determined by the odometer test. However, conducting odometer tests is time-consuming, especially the complicated ones that use the stress path approach. Additionally, several factors can affect the test results, including sample disturbance and the interpretation methods selected to find the maximum past consolidation pressure. Hence, it is suggested that in situ tests be carried out as an efficient and economical way to estimate OCR from field data.

There are several existing methods based on in situ test data for assessing the OCR. Table 4.38 and Figure 4.21 show some proposed SPT and CPT correlations for OCR. Additionally, Figure 4.22 shows some other correlations for other in situ tests, such as DMT and VST, based on data collected by Kulhawy and Mayne (1990).

Table 4.38 Proposed correlations for OCR.

References	Correlations	Remarks
Mayne and Kemper (1988)	$\text{OCR} = 0.37\left({}^{(q_c-\sigma_v)}/_{\sigma'_v}\right)^{1.01}$	—
Mayne and Kemper (1988)	$\text{OCR} = 0.193\left({}^{N}/_{\sigma'_o}\right)^{0.689}$	σ'_o is in MN/m^2
Trevor and Mayne (2004)	$\text{OCR} = 2(0.029 + 0.409M)\left[\frac{1}{1.95M+1}\left(\frac{q_t - u_2}{\sigma'_v}\right)\right]^{\frac{1}{\theta}}$	$\theta = 0.8 - 0.9$ $M = \frac{6\sin\varphi'}{3 - \sin\varphi'}$
Mayne (2007)	$\sigma'_p = k(q_t - \sigma_v)$	k: Pre-consolidation cone factor with an expected range of 0.2–0.5
Robertson (2009)	$\text{OCR} = 0.25Q_t^{1.25}$	—
Robertson (2012)	$\text{OCR} = (2.625 + 1.75\log F_r)^{-1.25}Q_t^{1.25}$	—
Chanmee et al. (2017)	$\text{OCR} = k\left(\frac{q_t - \sigma_v}{\text{OCR}\sigma'_v}\right)$	Under consolidated deposits $k = 0.14$–0.4

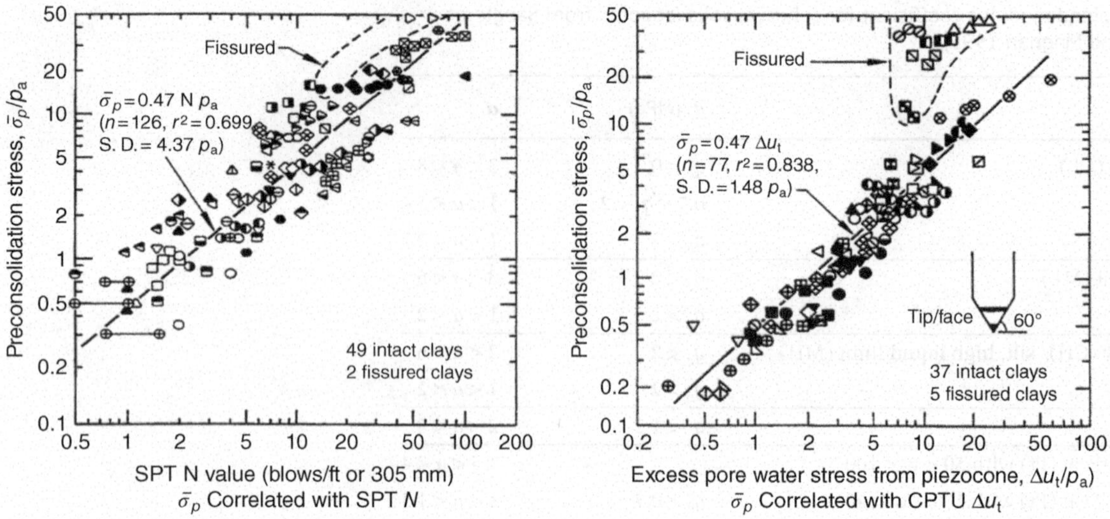

Figure 4.21 SPT and CPT correlation with OCR (Kulhawy and Mayne 1990/Electric Power Research Institute).

Table 4.39 Empirical correlation for the compression ratio (Ameratunga et al. 2016 / With permission of Springer Nature).

Soil type	Correlation for C_r	References
Marine clays of Southeast Asia	$0.0043w_n$	Azzouz et al. (1976)
	0.0045LL	Balasubramaniam and Brenner (1981)
Bangkok clays	0.00.463LL − 0.013	Balasubramaniam and Brenner (1981)
	$0.00566w_n - 0.037$	
French clays	$0.0039w_n - 0.013$	Balasubramaniam and Brenner (1981)
Indiana clays	$0.0249 + 0.003w_n$	Lo and Lovell (1982)
	0.0294 + 0.00238LL	
	$0.0125 + 0.152e_0$	
	$0.2037(e_0 - 0.2465)$	Goldberg et al. (1979)
Clays from Greece and parts of the US	0.002(LL + 9)	Azzouz et al. (1976)
	$0.14(e_0 + 0.007)$	
	$0.003(w_n + 7)$	
	$0.126(e_0 + 0.003\text{LL} - 0.06)$	
Chicago clays	$0.208e_0 + 0.0083$	Azzouz et al. (1976)
Inorganic and organic clays and silty soils	$0.156e_0 + 0.0107$	Elnaggar and Krizek (1970)

4.6.1.3.4 Consolidation Indexes Selected empirical correlations for the compression ratio as compiled by Djoenaidi (1985) are listed in Table 4.39.

The slope of the consolidation line in the e-log σ_v' space is known as the compression index (C_c). When the clay is normally consolidated, the e- σ_v' values will be located on this line, irrespective of the stress level. When clay is normally consolidated, C_c is a measure of its stiffness and a key parameter in evaluating the final settlements due to consolidation. It is usually related to the initial void ratio (e_0), in situ natural water content (w_n), liquid limit (LL), or plasticity index (PI). Some of the suggested empirical correlations for C_c are listed in Table 4.40.

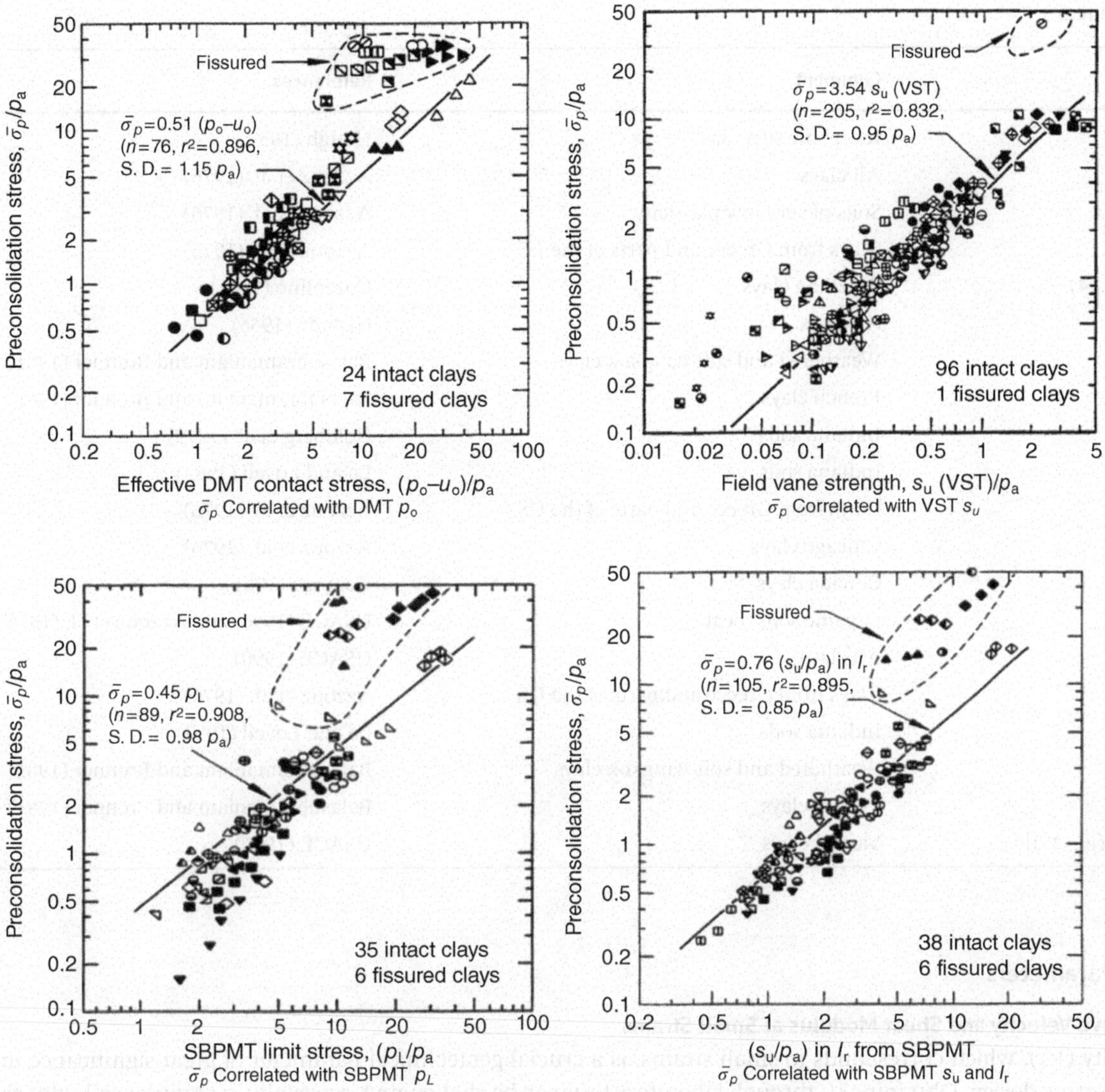

Figure 4.22 DMT, VST, and SBPMT correlation with OCR (Kulhawy and Mayne 1990).

Table 4.40 Empirical correlations for C_c (Ameratunga et al. 2016 / With permission of Springer Nature).

Correlation for C_c	Comment	References
0.009(LL – 10)	Undisturbed clay of sensitivity less than 4. Reliability ± 30%	Terzaghi and Peck (1948)
0.007(LL – 10)	Remolded clay	Skempton (1944)
0.0046(LL – 9)	Sao Paulo, Brazil clays	Cozzolino (1961)
0.0186(LL – 30)	Soft silty Brazilian clays	Cozzolino (1961)
0.01(LL – 13)	All clays	USACE (1990)
0.008(LL – 8.2)	Indiana soils	Lo and Lovell (1982)
0.21 + 0.008LL	Weathered and soft Bangkok clays	Balasubramaniam and Brenner (1981)

(Continued)

Table 4.40 (Continued)

Correlation for C_c	Comment	References
$0.3(e_0 - 0.27)$	Inorganic silty clay	Hough (1957)
$1.15(e_0 - 0.35)$	All clays	Azzouz et al. (1976)
$0.75(e_0 - 0.50)$	Soils of very low plasticity	Azzouz et al. (1976)
$0.4(e_0 - 0.25)$	Clays from Greece and parts of the US	Azzouz et al. (1976)
$0.256 + 0.43(e_0 - 0.84)$	Brazilian clays	Cozzolino (1961)
$0.54(e_0 - 0.35)$	All clays	Nishida (1956)
$0.22 + 0.29e_0$	Weathered and soft Bangkok clays	Balasubramaniam and Brenner (1981)
$0.575e_0 - 0.241$	French clays	Balasubramaniam and Brenner (1981)
$0.5363(e_0 - 0.411)$	Indiana soils	Goldberg et al. (1979)
$0.496e_0 - 0.195$	Indiana soils	Lo and Lovell (1982)
$0.40(e_0 - 0.25)$	Clays from Greece and parts of the US	Azzouz et al. (1976)
$0.01w_n$	Chicago clays	Azzouz et al. (1976)
$0.01w_n$	Canada clays	Koppula (1981)
$0.0115w_n$	Organic soils, peat	USACE (1990) and Azzouz et al. (1976)
$0.012w_n$	All clays	USACE (1990)
$0.01(w_n - 5)$	Clays from Greece and parts of the US	Azzouz et al. (1976)
$0.0126w_n - 0.162$	Indiana soils	Lo and Lovell (1982)
$0.008w_n + 0.20$	Weathered and soft Bangkok clays	Balasubramaniam and Brenner (1981)
$0.0147w_n - 0.213$	French clays	Balasubramaniam and Brenner (1981)
$(1 + e_0)[0.1 + 0.006(w_n\text{-}25)]$	Varved clays	USACE (1990)

4.6.2 Related Parameters

4.6.2.1 Shear Wave Velocity and Shear Modulus at Small Strains

Shear wave velocity (V_s), which corresponds to small strains, is a crucial geotechnical parameter of great significance in earthquake engineering design. Obtaining V_s through laboratory tests can be challenging, especially in granular soils, due to difficulties in collecting undisturbed samples. As a result, in situ seismic tests are preferred for determining V_s. Techniques such as down-hole testing, cross-hole testing, and surface wave velocity measurements are employed to establish a shear wave velocity profile with depth.

Numerous correlations have been proposed to estimate shear wave velocity based on results from Standard Penetration Tests (SPT) and Cone Penetration Tests (CPT) for various soil types. The correlation functions utilize input parameters such as q_c, f_s, N, and σ'_v, and can be in linear, logarithmic, or power form. Tables 4.41 and 4.42 provide a compilation of some of these correlations.

It is generally accepted that the best way of assessing small strain (<0.001%) shear modulus (G_0) is to use shear wave velocity ($G_0 = \rho V_s^2$). However, in situ penetration tests are mainly carried out to assess G_0 using empirical correlations, unless project-specific detailed investigations are performed. Some of the correlations for G_0 prediction based on CPT records are presented in Table 4.43.

4.6.2.2 Sensitivity

Sensitivity (S_t) is defined as the ratio of peak undisturbed shear strength to remolded shear strength (Equation 4.19). Many researchers have reported that the shear strength of remolded samples is affected by skin friction, f_s.

$$S_t = \frac{S_u}{S_{u(\text{remolded})}} = \frac{(q_t - \sigma_v)}{N_{kt}}\left(\frac{1}{f_s}\right) \qquad (4.19)$$

Table 4.41 Some correlations between v_s (m/s) and N (Ameratunga et al. 2016 / With permission of Springer Nature).

Source	Soil type	V_s (m/s)
Imai (1977)	All soils	$91N^{0.337}$
	Sand	$80.6N^{0.331}$
	Clay	$80.2N^{0.292}$
Ohta and Goto (1978)	All soils	$85.35N^{0.348}$
Seed and Idriss (1981)	All soils	$61.4N^{0.5}$
Sykora and Stokoe (1983)	Sand	$100.5N^{0.29}$
Okamoto et al. (1989)	Sand	$125N^{0.3}$
Pitilakis et al. (1999)	Sand	$145N^{0.178}$
	Clay	$132N^{0.271}$
Kiku et al. (2001)	All soils	$68.3N^{0.292}$
Jafari et al. (2002)	Sand	$22N^{0.77}$
	Clay	$27N^{0.73}$
Hasancebi and Ulusay (2007)	All soils	$99N^{0.309}$
	Sand	$90.82N^{0.319}$
	Clay	$97.89N^{0.269}$
Dikmen (2009)	All soils	$58N^{0.39}$
	Sand	$73N^{0.33}$
	Silt	$60N^{0.36}$
	Clay	$44N^{0.48}$

Table 4.42 Empirical correlations between V_s and CPT data (adapted from Ameratunga et al. 2016).

References	Proposed correlation (m/s)	Soil type	Units of parameters	
			q_c	f_s
Hegazy and Mayne (1995)	$V_s = 12.02(q_c)0.319(f_s) - 0.0466$	Sand	kPa	kPa
	$V_s = 3.18(q_c)0.549(f_s)\ 0.025$	Clay	kPa	kPa
Mayne and Rix (1995)	$V_s = 1.75(q_c)0.627$	Clay	kPa	kPa
Iyisan (1996)	$V_s = 55.3(q_c)0.377$	Clay	kgf/cm²	—
	$V_s = 218 + 0.70\ q_c$	Sand	kgf/cm²	—
Piratheepan (2002)	$V_s = 25.3(q_c)0.163(f_s)\ 0.029D0.115$	Sand	kPa	kPa
	$V_s = 11.9(q_c)0.269(f_s)\ 0.108D0.127$	Clay	kPa	kPa
Tun (2003)	$V_s = 109.29+52.674\ln(q_c)$	All	MPa	MPa
Madiai and Simoni (2004)	(1) $V_s = 211(q_c)0.23$ (2) $V_s = 155(q_c)0.29(f_s) - 0.10$	All	MPa	MPa
Mayne (2006)	$V_s = 18.5 + 118.8\log(f_s)$	All	—	kPa
Paoletti et al. (2010)	$V_s = 50[(q_c/p_a)0.43 - 3]$	Sand	kPa	—
MolaAbasi et al. (2015)	$V_s = 100[1.36 - 0.35f_s + 0.15q_c - 0.05f_s2 - 0.018q_c2 + 0.39f_s\ q_c]$	Clay	MPa	MPa
	$V_s = 100[1.73 + 2.74f_s + 0.03q_c - 4.015f_s2 - 0.00026q_c2 + 0.007f_s q_c]$	Sand	MPa	MPa
	$V_s = 100[1.47 + 2.07f_s + 0.10q_c + 9.50f_s2 - 0.0023q_c2 - 0.034f_s\ q_c]$	Mixed	MPa	MPa
	$V_s = 100[1.40 + 1.59f_s + 0.09q_c - 1.33f_s2 - 0.002q_c2 + 0.05f_s\ q_c]$	All	MPa	MPa

D: depth in meter.

Table 4.43 Empirical correlations for G_0 based on q_c.

References	Correlations	Remarks
Mayne and Rix (1995)	$G_{max} = 99.5{p_a}^{0.305}(q_c)^{{e_0}^{\frac{0.695}{1.13}}}$	Cohesive soils, e_0 = initial void ratio
Eslaamizaad and Robertson (1997)	$\left(\frac{G_0}{q_c}\right) = 1634\left(\frac{q_c}{\sqrt{\sigma'_v}}\right)^{-0.75}$	Cohesionless soils
Schnaid (2009)	$G_0 = b(q_c \sigma'_v p_a)^{0.3}$	Cohesionless soils, b = 280 and 110 for an upper and lower bond.

Units (kPa).

Table 4.44 Permeability from CPT results (Robertson 1990).

Soil behavior type	Soil permeability (m/s)
Sensitive fine-grained	3×10^{-9} to 3×10^{-8}
Organic soils-peats	1×10^{-8} to 1×10^{-6}
Clays-clay to silty clay	1×10^{-10} to 1×10^{-7}
Silt mixtures clayey silt to silty clay	3×10^{-9} to 1×10^{-7}
Sand mixtures; silty sand to sandy silt	1×10^{-7} to 1×10^{-5}
Sands; clean sands to silty sands	1×10^{-5} to 1×10^{-3}
Gravelly sand to sand	1×10^{-3} to 1
Very stiff sand to clayey sand	1×10^{-8} to 1×10^{-6}
Very stiff fine-grained	1×10^{-9} to 1×10^{-7}

Schmertmann's (1978) suggested Equation 4.20 to determine the shear strength of remolded soils.

$$S_t = \frac{N_s}{R_f} \tag{4.20}$$

where

R_f = the friction ratio,
N_s = a constant.

4.6.2.3 Permeability

Robertson (1990) proposes a relationship between permeability (k) and the soil behavior type for CPT as shown in Table 4.44. Robertson and Cabal (2010) suggest Equation 4.21 to evaluate the average permeability (k_{ave}):

$$\begin{cases} k = 10^{(0.952-3.04I_c)} & 1 < I_c \le 3.27 \\ k = 10^{(-4.52-1.37I_c)} & 3.27 < I_c \le 4 \end{cases} \tag{4.21}$$

where

$$I_c = \sqrt{(3.47 - \log Q_t)^2 + (\log F_r + 1.22)^2} \tag{4.22}$$

4.7 Rock Properties and Classification

In rock engineering, there are many rock mass classification systems to quantify the quality of rock masses by taking into account the properties of the intact rock along with the discontinuities. One of the most commonly used systems is the Geomechanics Classification or Rock Mass Rating System (RMR) (Bieniawski 1989), which gives indices of rock mass quality called RMR.

Table 4.45 Field estimates of uniaxial compressive strength (after Hoek 2007/Evert Hoek).

Grade[a)]	Term	Uniaxial compressive strength (MPa)	Point load index[b)] (MPa)	Field estimate of strength	Examples
R6	Extremely strong	>250	>10	Specimen can only be chipped with a geological hammer	Fresh basalt, chert, diabase, gneiss, granite, quartzite
R5	Very strong	100–250	4–10	Specimen requires many strong blows of a geological hammer to fracture it	Amphibolite, sandstone, basalt, gabbro, gneiss, granodiorite, limestone, marble, rhyolite, tuff
R4	Strong	50–100	2–4	Specimen requires more than one blow of a geological hammer to fracture it	Limestone, marble, phyllite, sandstone, schist, shale
R3	Medium strong	25–50	1–2	It cannot be scraped or peeled with a pocket knife, specimen can be fractured with a single blow from a geological hammer	Claystone, coal, concrete, schist, shale, siltstone
R2	Weak	5–25	—	It can be peeled with a pocket knife with difficulty, shallow indentation made by a firm blow with the point of a geological hammer	Chalk, rocksalt, potash
R1	Very weak	1–5	—	Crumbles under firm blows with point of a geological hammer, can be peeled by a pocket knife	Highly weathered or altered rock
R0	Extremely weak	0.25–1	—	Indented by thumbnail	Stiff fault gouge

a) Grade according to Brown (1981).
b) Point load tests on rocks with a uniaxial compressive strength below 25 MPa are likely to yield ambiguous results.

The RMR of a rock mass is the sum of the ratings given to the rock mass with respect to the following six parameters:

1) Uniaxial compressive strength of the intact rock (σ_{ci})
2) Rock Quality Designation (RQD)
3) Spacing of discontinuities
4) Condition of discontinuities
5) Groundwater conditions
6) Orientation of discontinuities

4.7.1 Uniaxial Compressive Strength of the Intact Rock

The uniaxial or unconfined compressive strength of the intact rock is the maximum axial stress that an unconfined cylindrical intact rock core can take before failing, which is only a property of the intact rock. Table 4.45 gives estimates of σ_{ci} for different rocks (Hoek 2007).

4.7.2 Rock Quality Designation

The Rock Quality Designation, RQD, is obtained from core log data and is an index of rock mass quality. The RQD is a modified core recovery, defined as the percentage of intact core pieces longer than 100 mm in the total length of the core. The core should be drilled with a double-tube core barrel and be at least NW size (54.7 mm in diameter).

$$\text{RQD} = \left(\frac{\sum \text{Lengh of core pieces} > 10\text{ cm}}{\text{Total length of core run}} \right) \times 100\% \tag{4.23}$$

4.7.3 Rock Mass Rating System

Tables 4.46–4.48 show ratings based on the main rock parameters determined in the field.

Table 4.46 Rock mass rating system – part A (Bieniawski 1989 / With permission of John Wiley & Sons).

		A. Classification parameters and their rating							
1	Strength of intact rock	Point-load strength index (MPa)	>10	4–10	2–4	1–2	Or this low range, unconfined compressive test is prefer0072ed		
		Unconfined compressive strength (MPa)	>250	100–250	50–100	25–50	5–25	1–5	<1
		Rating	15	12	7	4	2	1	0
2	Drill core quality RQD (%)	Value	90–100	75–90	50–75	25–50	<25		
		Rating	20	17	13	8	3		
3	Spacing of discontinuities (m)	Value	>2	0.6–2	0.2–0.6	0.06–0.2	<0.06		
		Rating	20	15	10	8	5		
4	Conditions of discontinuities	Value	Very rough surfaces, not continuous, no separation, unweathered wall rock	Slightly rough surfaces, separation <1 mm, slightly weathered walls	Slightly rough surfaces, separation <1 mm, highly weathered walls	Slickensided surfaces or gouge <5 mm thick or separation 1–5 mm continuous	Soft gouge >5 mm thick or separation >5 mm continuous		
		Rating	30	25	20	10	0		
5	Groundwater	Inflow per 10 m tunnel length (l/min)	None or	<10	10–25	25–125	>125		
		Ratio of joint water pressure to major principal stress	0	<0.1	0.1–0.2	0.2–0.5	>0.5		
		General conditions	Completely dry	Damp	Wet	Dripping	Flowing		
		Rating	15	10	7	4	0		

Table 4.47 Rock mass rating system – part B (Bieniawski 1989 / With permission of John Wiley & Sons).

B. Rating adjustment for joint orientations						
Strike and dip orientations of discontinuities		**Very favorable**	**Favorable**	**Very favorable**	**Unfavorable**	**Very favorable**
Ratings	Tunnels and mines	0	−2	−5	−10	−12
	Foundations	0	−2	−7	−15	−25
	Slopes	0	−5	−25	−50	−60

Table 4.48 Rock mass rating system – part C (Bieniawski 1989 / With permission of John Wiley & Sons).

C. Rock mass classes and corresponding design parameters and engineering properties					
Class No.	**I**	**II**	**III**	**IV**	**V**
RMR	100–81	80–61	60–41	40–21	<21
Description	Very good	Good	Fair	Poor	Very poor
Average stand-up time	20 yr for 15 m span	1 yr for 10 m span	1 wk for 5 m span	10 h for 2.5 m span	30 min for 1 m span
Cohesion of rock mass (MPa)	>0.4	0.3–0.4	0.2–0.3	0.1–0.2	<0.1
Internal friction angle of rock mass (°)	>45	35–45	25–35	15–25	<15

Example 4.3 The CPT record of a site is given in graphical and tabular format. Based on SBC charts, The soil type from the ground level up to a depth of 6 m is sand. Lower than the depth of 6 m is clay. Determine the required parameters.

Depth (m)	q_c (MPa)	f_s (MPa)	u (MPa)	R_f(%)
1–2	20.25	0.14	0.01	0.42
2–3	5.73	0.04	0.01	0.64
3–4	11.89	0.08	0.005	0.79
4–5	11.73	0.12	0.01	0.86
5–6	10.47	0.05	0.02	0.64
6–7	34.78	0.23	0.11	0.72
7–8	0.84	0.02	0.24	5
8–9	0.83	0.019	0.24	4.2
9–10	0.83	0.02	0.19	5.16
10–11	0.83	0.02	0.28	3.7
11–12	1.01	0.019	0.31	3.14
12–13	0.83	0.01	0.31	3.16
13–14	1.19	0.02	0.3	3.07
14–15	1.19	0.018	0.35	2.79
15–16	1.014	0.016	0.34	2.97
16–17	1.37	0.018	0.36	2.5
17–18	1.37	0.014	0.47	2.42

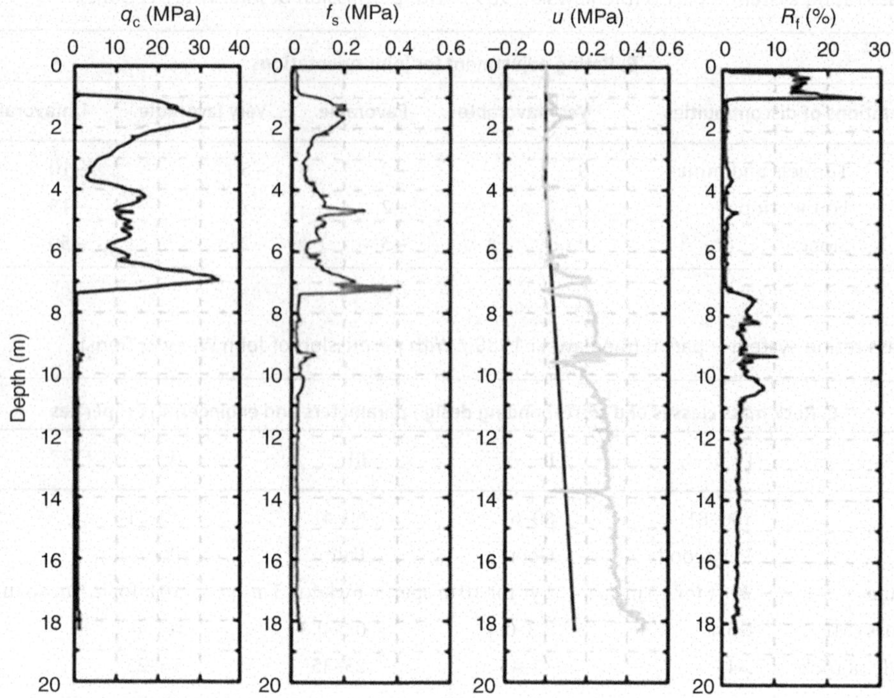

a) Plot the profile of the unit weight and elasticity modulus with depth for both layers.

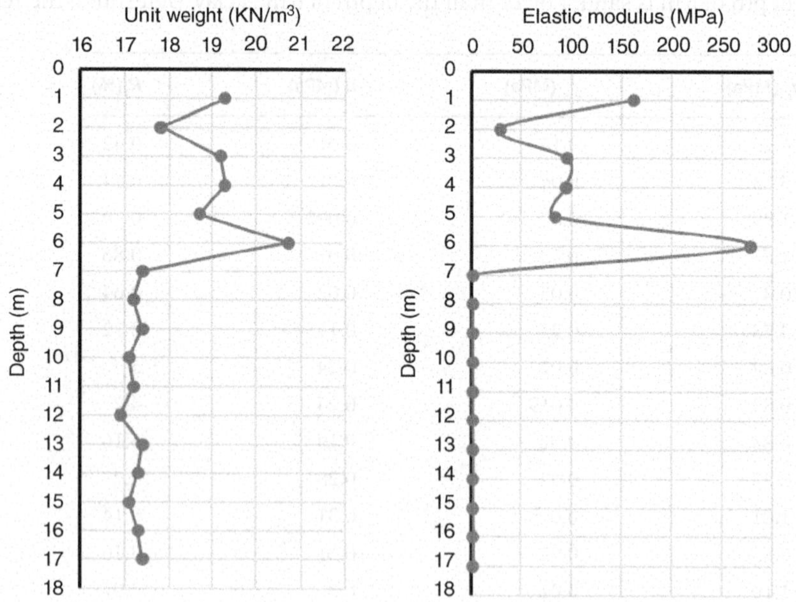

b) Plot he profile of relative density and internal friction angle with depth for the sand layer.

$$q_t = q_c + u(1 - a) \rightarrow \text{by assuming } a = 0.6$$

For depths 1–2 m $q_t = 20.25 + 0.02(1 - 0.6) = 20.258$ (MPa)

$$D_r = 100\left(0.268 \ln\left(\frac{q_t/p_a}{\sqrt{\sigma'_v/p_a}}\right) - 0.675\right)$$

For depths 2–3 m $D_r = 100\left(0.268\ln\left(\frac{57.34}{\sqrt{0.371}}\right) - 0.675\right) = 54\%$

For Sand layer: $\varphi = 17.6 + 11\text{Log}\left(q_c/\sqrt{100\sigma_v'}\right)$

For depth 3–4 m $\varphi = 17.6 + 11\text{Log}(11{,}890/\sqrt{5130}) = 42$

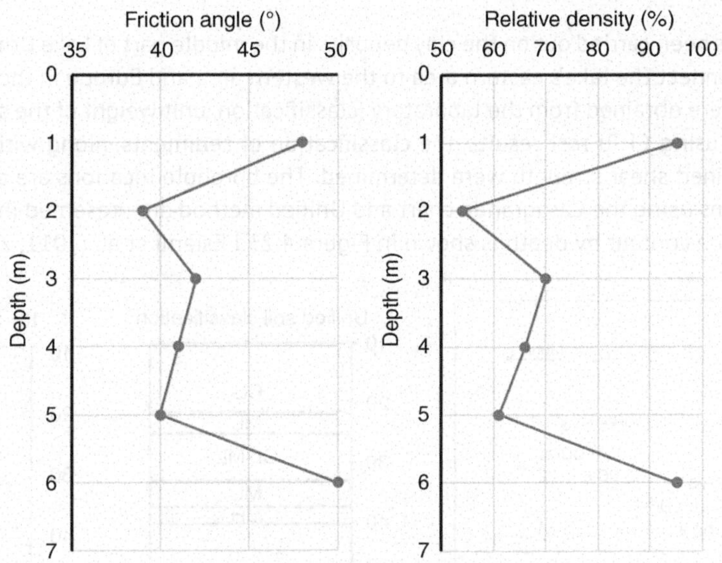

c) Plot the profile of undrained shear strength and sensitivity with depth for the clay layer.

$$S_{uCPT} = \frac{(q_c - \sigma_{vo})}{N_c}$$

For depths 17–18 m $S_{uCPT} = (1370 - 304.7)/20 = 53.26$

$$S_t = \frac{S_u}{S_{u(remolded)}} = \frac{(q_t - \sigma_v)}{N_{kt}}\left(\frac{1}{f_s}\right)$$

For 17–18 m $S_t = \frac{S_u}{S_{u(remolded)}} = \frac{(1558-304.7)}{15}\left(\frac{1}{14}\right) = 5.97$

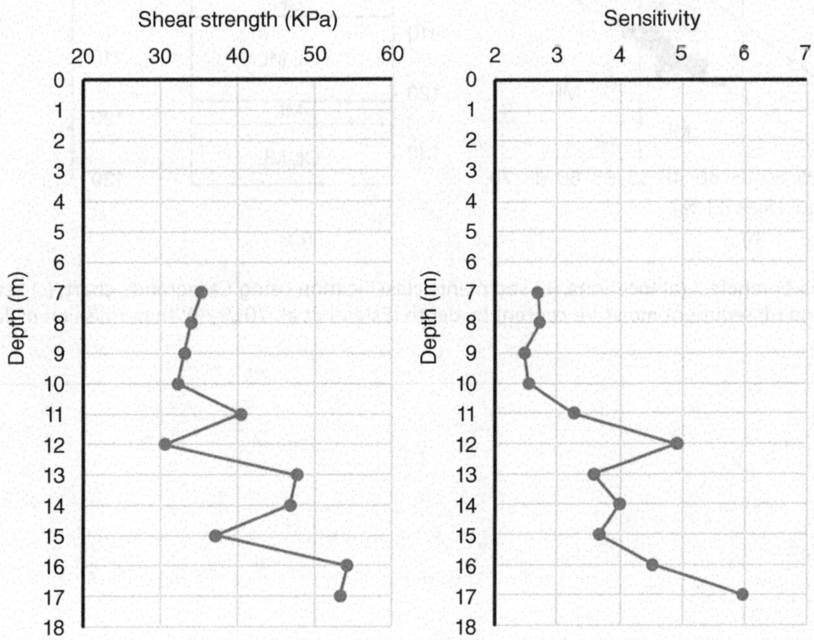

4.8 Case Histories

Case History 4.1

Geotechnical Site Characterization of the Lake Urmia

A comprehensive study has been carried out on the clay deposits in the middle part of Lake Urmia. A causeway over this lake was constructed to connect the lake's eastern area to the western area and Europe in the middle part of the lake. Geotechnical properties were obtained from the laboratory (classification, unit weight of the soil, and undrained shear strength) tests. Moreover, using CPTu test results, the classification of sediments, along with the unit weight of the soil, sensitivity, and undrained shear strength were determined. The borehole locations are depicted in Figure 4.23a. The sediment classifications using the Casagrande chart and Unified method are presented in Figure 4.23b and c. The profile of sediment moisture content by depth is shown in Figure 4.23d Eslami et al. (2011, 2019).

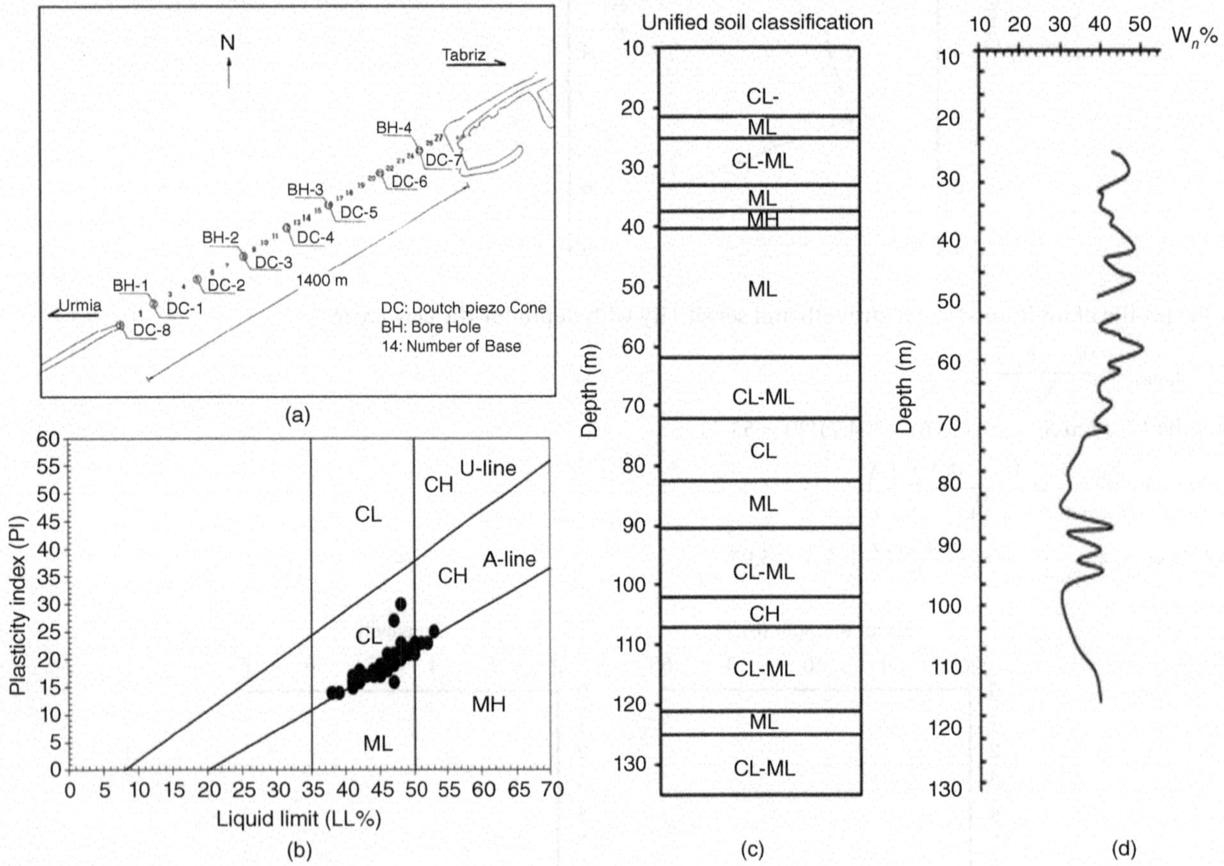

Figure 4.23 Lake Urmia boreholes: (a) locations, (b) sediments classification using Casagrande chart, (c) sediments classification using Unified method, and (d) sediment moisture content by depth (Eslami et al. 2019 / With permission of Taylor & Francis Group).

CPTu records as well as the soil profile of one typical borehole and sensitivity log are presented in Figure 4.24.

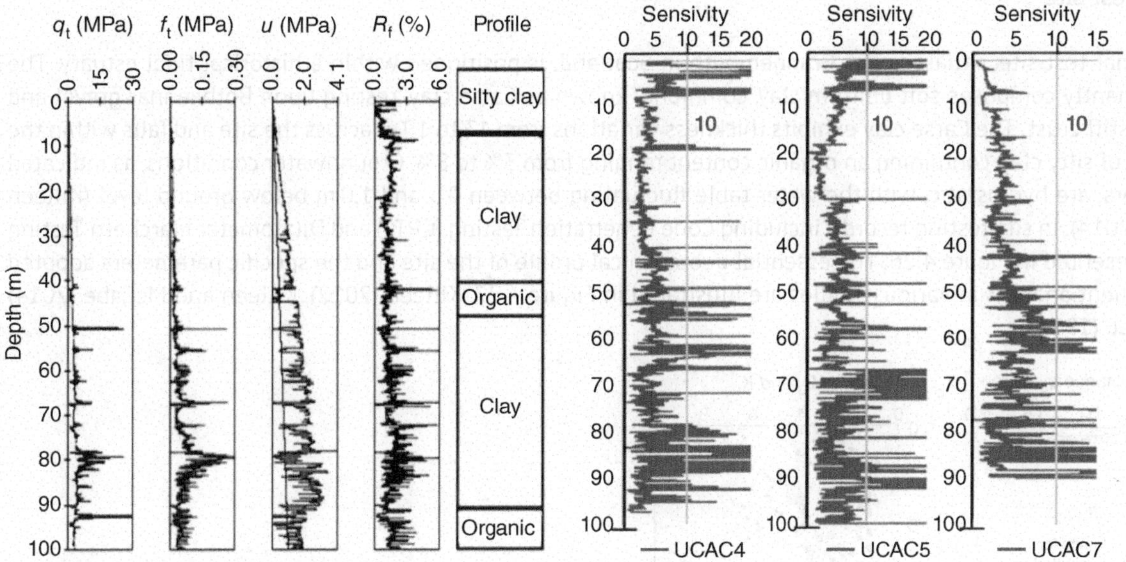

Figure 4.24 Typical CPTu records of lake Urmia (adapted from Eslami et al. 2011).

Soil behavior classification using Robertson (1990) and Eslami and Fellenius (1997) are illustrated in Figure 4.25a and b, respectively.

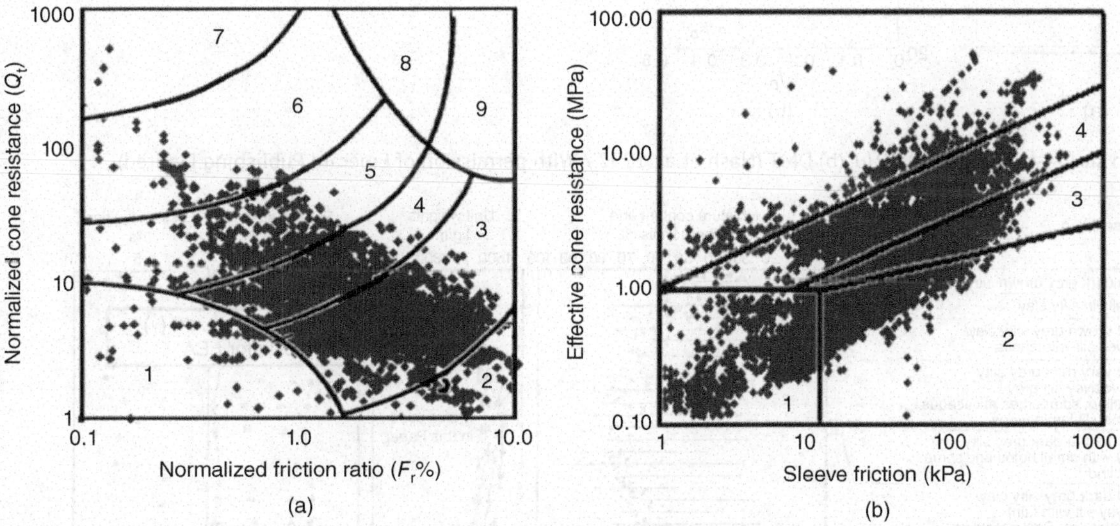

Figure 4.25 (a) soil behavior classification using Robertson (1990) method, (b) soil behavior classification using the Eslami and Fellenius (1997) method (Eslami et al. 2011/Elsevier/CC BY 4.0).

Case History 4.2

Bothkennar Test Site

The Bothkennar test site, situated near Grangemouth in Scotland, is positioned within a historical tidal estuary. The site predominantly comprises soft uniform clay, commonly known as Carse clay, resting upon Bothkennar gravel and covered by a stiff crust. The Carse clay exhibits thickness variations from 13 to 19 m across the site and falls within the classification of silty clay, containing an organic content ranging from 3% to 8%. Groundwater conditions, as indicated by piezometers, are hydrostatic, with the water table fluctuating between 0.5 and 1.0 m below ground level (Killeen and McCabe 2014). In situ testing records, including Cone Penetration Testing (CPTu) and Dilatometer Marchetti Testing (DMT), are presented in Figure 4.26. The essential geotechnical profile of the site and the specific parameters adopted for finite element analysis in various studies are illustrated in Figure 4.27, Killeen (2012), Killeen and McCabe (2014) and Nash et al. (1992).

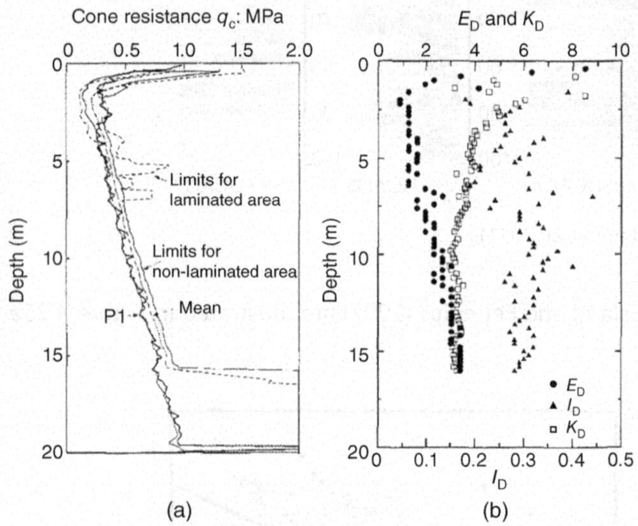

Figure 4.26 In situ tests records: (a) CPTu, (b) DMT (Nash et al. 1992 / With permission of Emerald Publishing Limited).

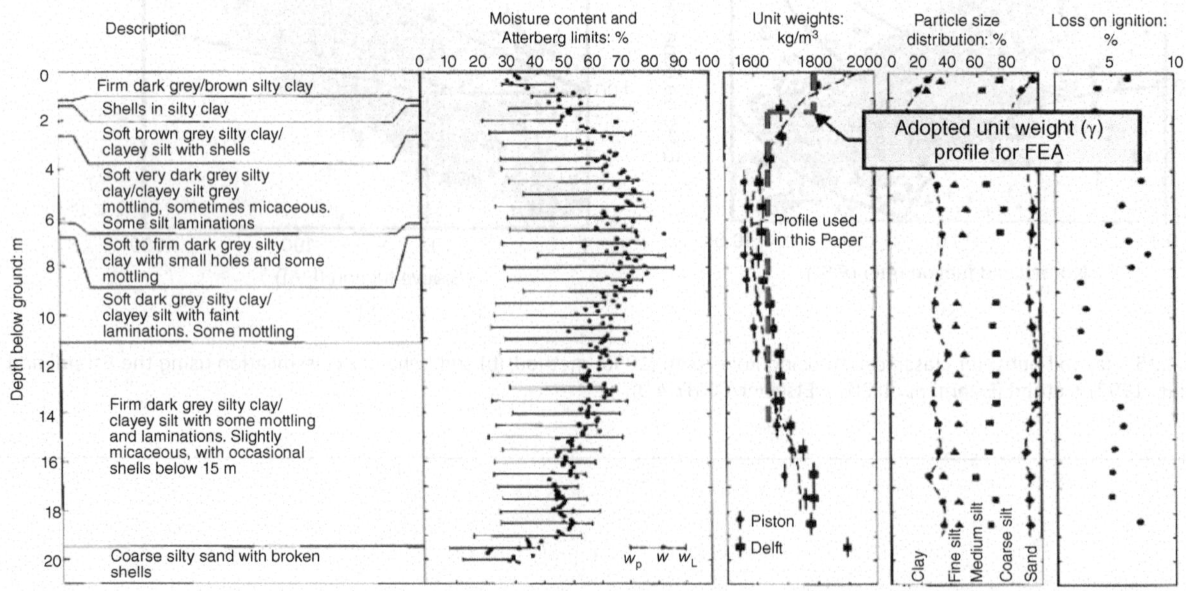

Figure 4.27 Key geotechnical profile of the site and adopted parameters for AFE (Killeen 2012/Killeen).

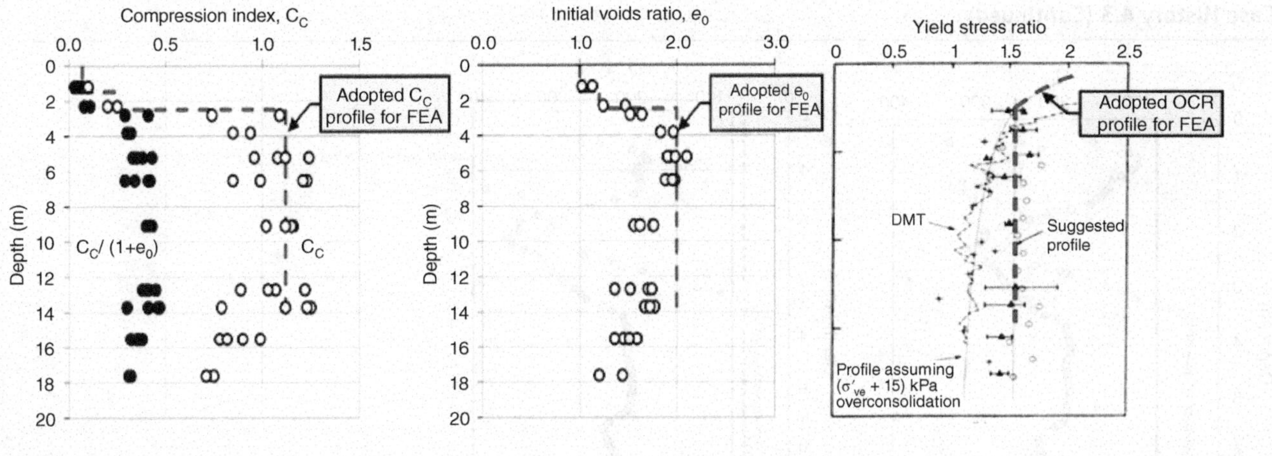

Figure 4.27 *(Continued)*

Case History 4.3

Explosive Compaction Application for Liquifiable Sands

Explosive compaction (EC) or Blast densification (BD) has been realized as an efficient technique to improve soil and mitigate the liquefaction potential in loose saturated sands. Due to its continuous and precise records, Piezocone (CPTu) is the most applicable in situ test in geotechnical practice to evaluate liquefaction potential. Eslami (2015) compiled a data bank including eight case histories in different locations to investigate EC effects on the mitigation of loose sands' instability. The site's geomaterials were in the category of fine to medium sand, silty sand, and a mixture of sand and gravel with relative density between 30% and 60% and a thickness of 5–40 m. Four CPT-based criteria were used, including the cyclic stress ratio approach, cone tip resistance (q_c) variations before and after modification, Q_{tn} and q_{c1N}, and soil behavior classification charts. Figure 4.28 depicts the variation of Q_{tn} and q_{c1N} before and after compaction. Analyses show that, due to EC, the soil state changes from loose to dense, the sand's contractive behavior changes to dilative, and the liquefaction potential diminishes. Also, by using soil behavior classification charts pre and post-explosion, improved soils are not in the liquefiable zone anymore as illustrated in Figure 4.29.

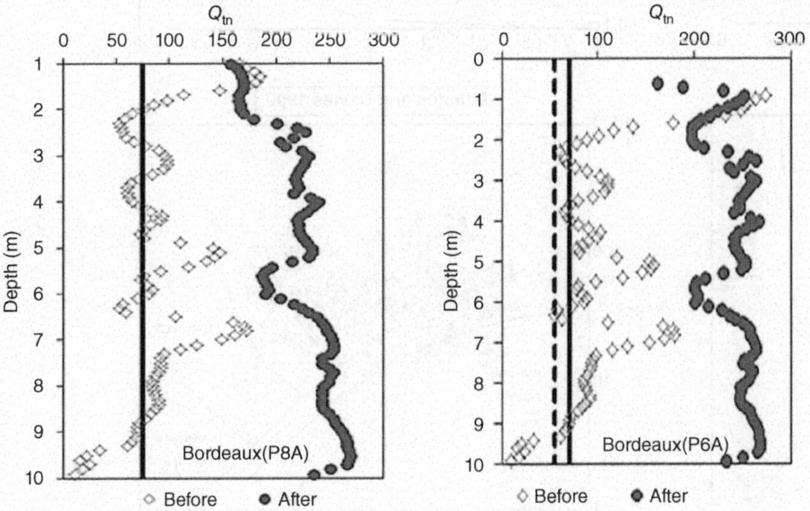

Figure 4.28 Variation of Q_{tn} and q_{c1N} before and after compaction (Eslami 2015 / With permission of Springer Nature).

(Continued)

Case History 4.3 (Continued)

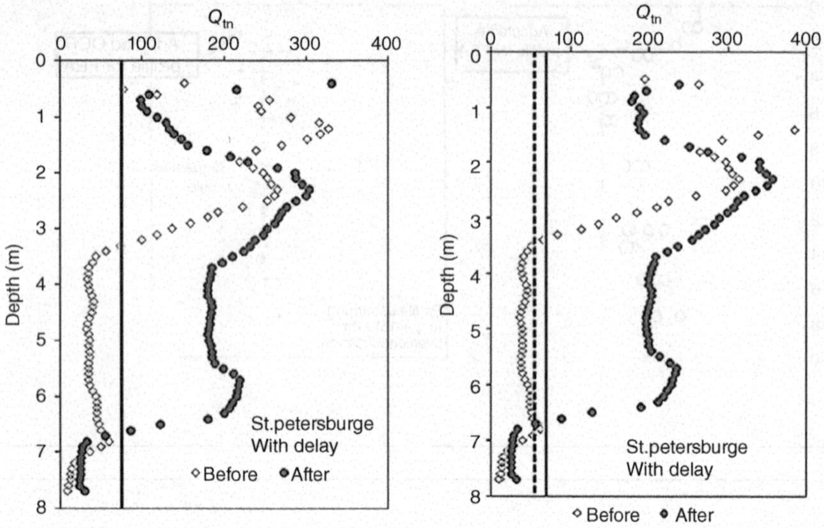

Figure 4.28 *(Continued)*

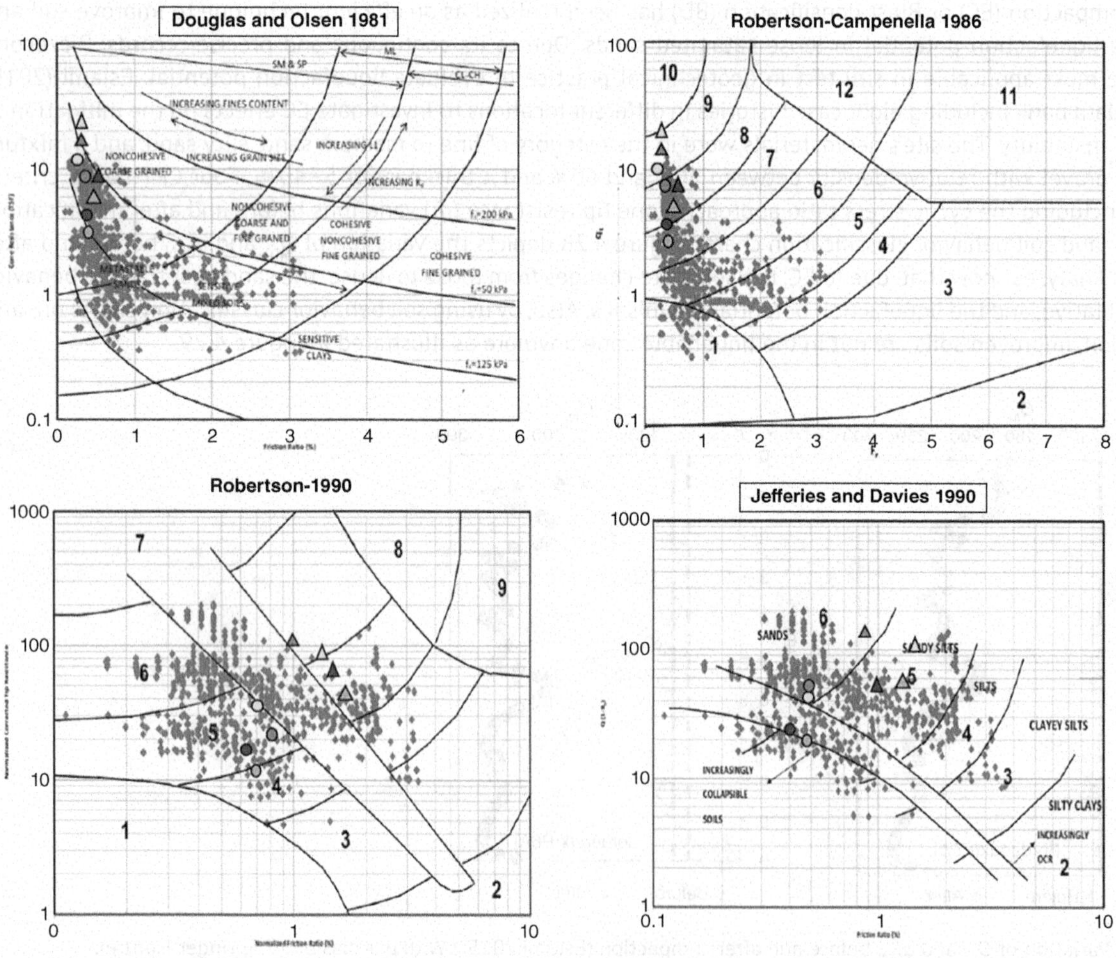

Figure 4.29 CPT-based classification charts for soil behavior assessment pre and post-explosion (Eslami 2015 / With permission of Springer Nature).

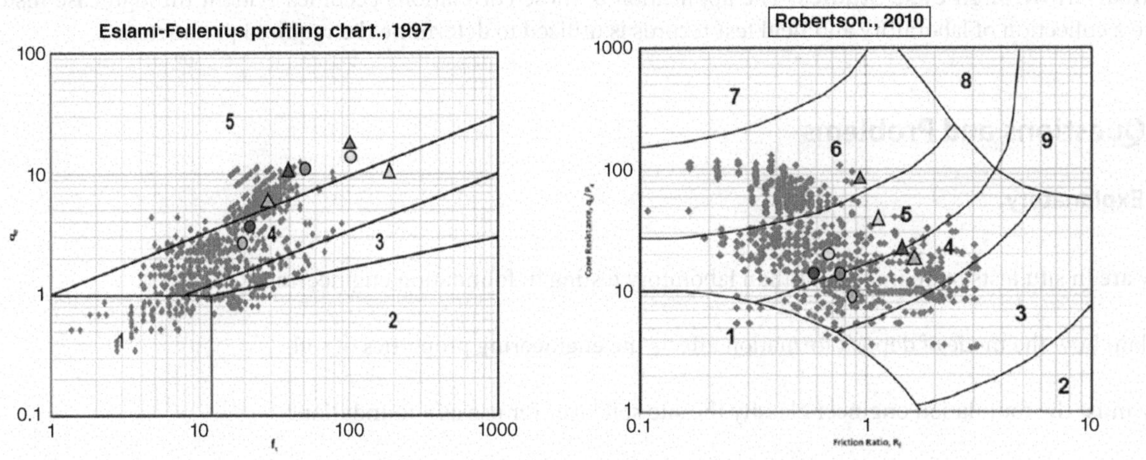

Jebba Dam(Zone I) after explosion	▲	Jebba Dam(Zone I) before explosion	●
Jebba Dam(Test) after explosion	△	Jebba Dam(Test) before explosion	○
St. Petersburg I (Without delay) after explosion	▲	St. Petersburg I (Without delay) before explosion	●
St. Petersburg II (With delay) after explosion	▲	St. Petersburg II (With delay) before explosion	●

Figure 4.29 *(Continued)*

4.9 Major Points

1) **Soil Stratification in Geotechnical Design:** Geotechnical designs necessitate determining subsurface soil stratification and estimating geotechnical parameter values (soil engineering parameters) for each individual soil layer.
2) **Factors Influencing Soil Properties:** Time-dependent geological processes significantly impact soil engineering properties. Crucial factors include clay quality, natural cementation, aging, over-consolidation, mode of deposit formation, and water content.
3) **Understanding Soil Phase Relations:** To analyze the physical properties and mechanical behaviors of soils in foundation engineering, understanding phase relations – defined relationships encompassing physical, chemical, and electrical properties – is essential. Fine-grained soil properties strongly depend on water content and fine content.
4) **Subsurface Soil Identification:** The initial step in any geotechnical project involves identifying and describing subsurface soil conditions. Employing systematic approaches, including visual and manual methods, is advantageous even during project planning stages.
5) **Soil Classification in Engineering:** Soil classification in engineering is based on properties relevant to their use as construction materials or foundation support. Modern classification systems categorize soils based on properties like grain-size distribution, liquid limit, and plastic limit. These systems facilitate a smooth transition from field observations to predictions of soil behaviors.
6) **Categorizing Soil Engineering Parameters:** Soil engineering parameters in foundation engineering fall into two groups: applied parameters and related parameters.
7) **Subcategories of Applied Parameters:** Applied parameters are further divided into subgroups, including index parameters (e.g., unit weight and relative density), strength parameters (e.g., undrained shear strength and internal friction angle), and stiffness and compressibility parameters (e.g., over-consolidation ratio and constrained modulus).
8) **Ensuring Parameter Accuracy:** Given soil variability, geotechnical engineers must compare parameters obtained from different methods with established criteria and data banks to ensure accuracy.
9) **Rock Mass Rating System (RMR) in Foundation Engineering:** The Rock Mass Rating System (RMR), a common rock classification system, can estimate rock parameters crucial in foundation engineering.

10) **Application through Case Studies:** The application of these correlations becomes evident through case histories, where a collection of laboratory and field test records is utilized to determine the required parameters.

4.10 Questions and Problems

4.10.1 Explanatory

1 Why are in situ tests more important than laboratory testing in foundation engineering?

2 Explain how the mode of deposit formation affects the engineering properties of soil.

3 Why must the foundation engineer classify the site soil layer for use as a foundation?

4 Discuss the dominant in situ tests for behavior classification of fine-grained soils.

5 Explain the role of void ratio and water content through relative density and liquidity index.

6 Name some well-documented research sites around the world. Is this important?

7 Explain the process of liquefaction mitigation via blast densification based on Case Case History 4.3.

8 Discuss the existing soil profile and adopted parameters of the Bothkennar site (Case 4.2).

9 List the parameters of the soil phase diagram and provide their definition.

10 Explain how to evaluate soil strength via visual and manual soil classification.

4.10.2 Conceptual

1 Classify the deposit presented in Example 4.2 based on three different soil behavior classification charts and compare the results.

2 Explain the main features of GW, SP, and ML soils based on the Unified Soil Classification System.

3 DMT is conducted on a project site. Classify a soil and estimate its unit weight if the dilatometer modulus and material index are 5 MPa and 1.6, respectively.

4 Explain the changes in parameters due to soil improvement practice, based on Case History 3.3 (explosive compaction assessment).

5 Discuss the trend and evolutions of Eslami et al. publications from the 1990s up to the 2020s.

6 What are the differences and similarities between the results of laboratory and in situ tests in Urmia Lake site characterization (Case 4.1)?

7 Classify the soil layers based on the given DMT record and draw the unit weight and soil undrained shear strength (S_u) variations with depth.

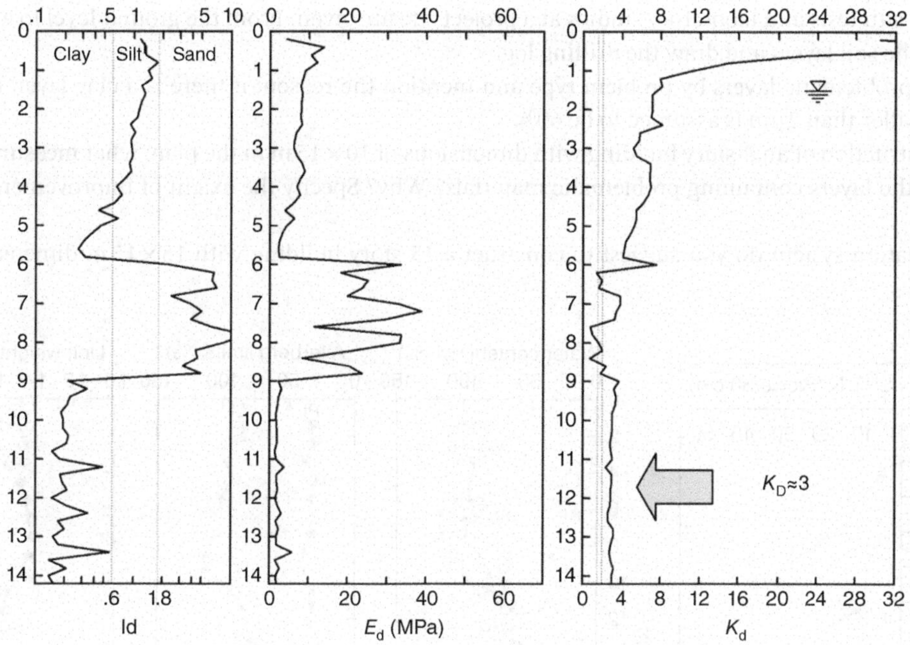

4.10.3 Problems

1 Resolve Example 4.2 using different methods or correlations compared to existing ones.

2 Based on the given CPTu records, classify the deposit in 5 m intervals based on two methods and then determine the key strength and stiffness parameters.

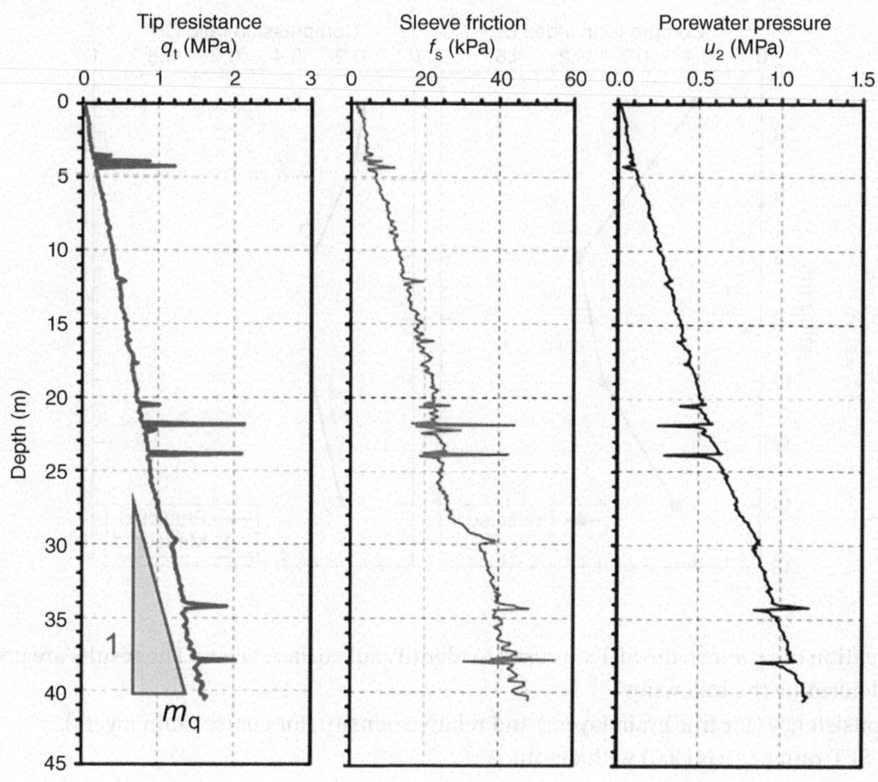

3 Results from in situ tests and laboratory studies at a project site are given. From the ground level to a depth of 15 m:

a) Determine the soil layers and draw the drilling log.
b) Specify the problematic layers by problem type and mention the reason; if there is a clay layer, the percentage of particles smaller than 2 μm is assumed to be 60%.
c) For the construction of an 8-story building with dimensions of 10 × 15 m in the plan, what measures do you suggest to improve the layers containing problematic materials? Why? Specify the extent of improvement in the plan and depth.
d) What foundation system do you suggest to construct a 15-story building with 15 × 15 m dimensions in the plan? Why?

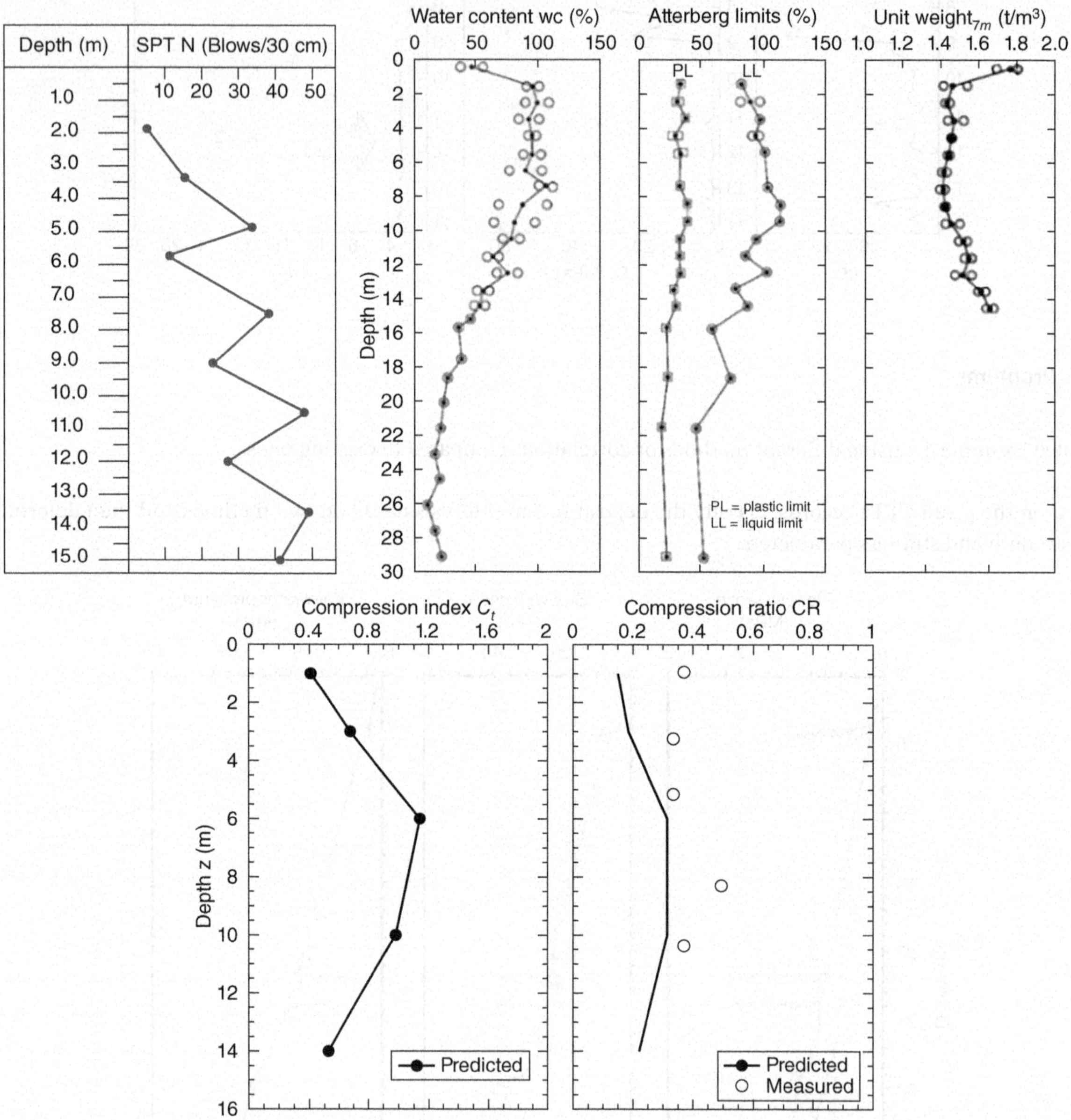

4 In the site investigation of a project, the SPT was used to identify subsurface layers. The results are presented. According to these results, determine the following:

a) The state of consistency (for fine grain layers) and relative density (for coarse grain layers);
b) The modified SPT number ((N1)60) with depth;

c) The soil's relative density and internal friction angle at a depth of 10 m from the ground surface;
d) The soil and OCR's undrained shear strength at depths of 25 and 50 m from the ground surface.

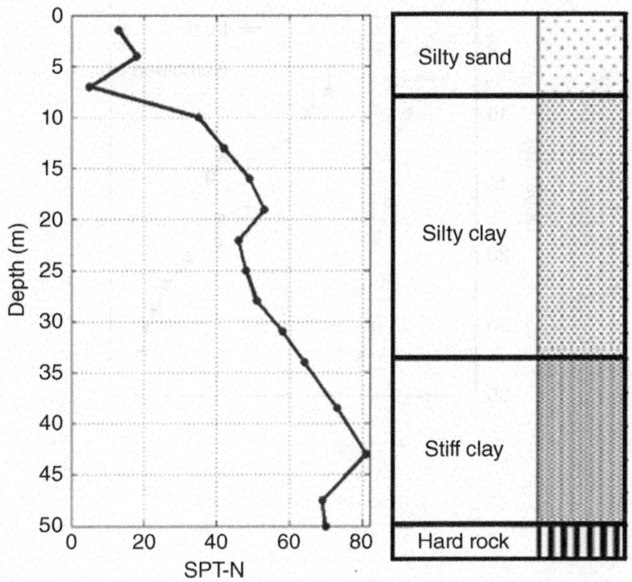

5 Calculate soil strength, stiffness, and related parameters for each layer based on SPT and CPT results given below; draw their variation with depth and compare the resulting parameters.

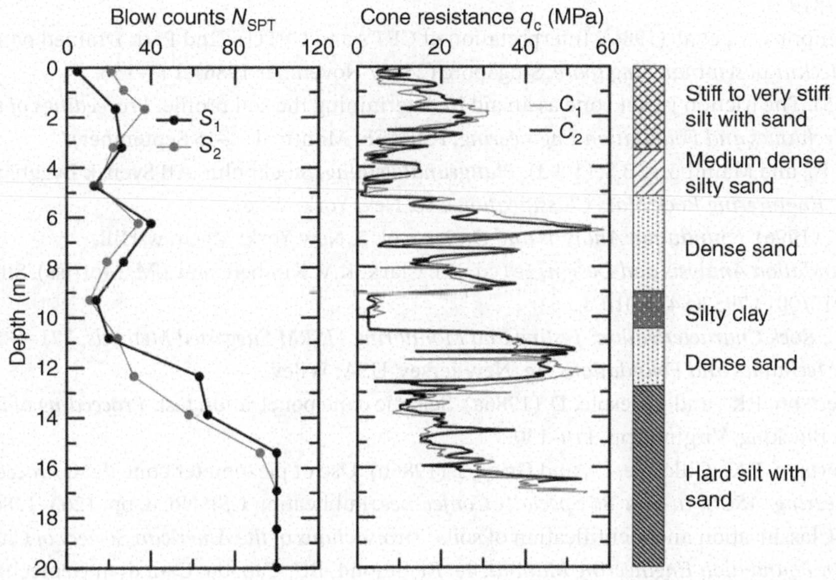

6 The result of a vane shear test (VST) at a project site is shown. The soil is normally consolidated, and the groundwater level is at a depth of 5 m from the ground surface. Given the overall diameter of the vane is 65 mm and its height is 130 mm:

- Determine the required torque to rotate the blades at depths of 10 m, 19 m, and 25 m.
- Plot the variation of soil sensitivity, shear strength and plasticity index with depth (consider the LL of the soil at each depth equal to the remolded strength + 10).
- Calculate the bearing capacity of a square foundation with a side of 5 m, which is implemented at a depth of 8 m.

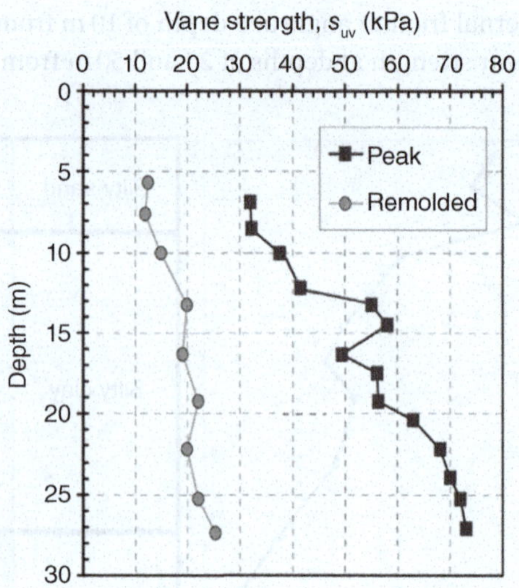

References

Ameratunga, J., Sivakugan, N., and Das, B.M. (2016). *Correlations of Soil and Rock Properties in Geotechnical Engineering*. Springer.

Anagnostopoulos, A., Koukis, G., Sabatakakis, N., and Tsiambaos, G. (2003). Empirical correlations of soil parameters based on cone penetration tests (CPT) for Greek soils. *Geotechnical and Geological Engineering* 21: 377–387. https://doi.org/10.1023/B:GEGE.0000006064.47819.1a.

Baldi, G., Bellotti, R., Ghionna, V., et al. (1986). Interpretation of CPT's and CPTU's. 2nd Part: Drained penetration. *Proceedings 4th international geotechnical seminar, Singapore*, Singapore (25–27 November 1986), 143–156.

Begemann, H.K.S. (1965). The friction jacket cone as an aid in determining the soil profile. *Proceedings of the 6th International Conference on Soil Mechanics and Foundation Engineering*, ICSMFE, Montreal (8–15 September).

Bergdahl, U., Ottosson, E., and Malmborg, B.S. (1993). *Plattgrundläggning*. Stockholm: AB Svensk Byggtjänst.

Bieniawski, Z.T. (1989). *Engineering Rock Mass Classifications*, 1e. New York: Wiley.

Bowles, J.E. and Guo, Y. (1996). *Foundation Analysis and Design*, vol. 5. New York: McGraw-Hill.

Bowles, J.E. (1997). *Foundation Analysis and Design*, 5e (ed. B.J. Clark, K.V. Kimbell, and J.M. Morriss). Singapore: McGraw-Hill. https://doi.org/10.1016/0013 7952(84)90010-3.

Brown, E.T. (ed.) (1981). *Rock Characterization, Testing and Monitoring - ISRM Suggested Methods*, 171–183. Oxford: Pergamon.

Budho, M. (2011). *Soil Mechanics and Foundations*, 3e. New Jersey, USA: Wiley.

Campanella, R.G., Robertson, P.K., and Gillespie, D. (1986a). Seismic cone penetration test. *Proceeding of in Situ 86, ASCE Specialty Conference*. Blacking, Virginia, pp. 116–130.

Campanella, R.G., Robertson, P.K., Gillespie, D., and Grieg, J. (1986b). Use of piezometer cone data. *Proceedings of American Society of Civil Engineering, ASCE, In-Situ '86 Specialty Conference*. Publication GSP No. 6, pp. 1263–1280.

Casagrande, A. (1948). Classification and identification of soils. *Transactions of the American Society of Civil Engineers* 113 (1).

CFEM (2006). *Canadian Foundation Engineering Manual*, 4e. Richmond, BC, Canada: Canadian Geotechnical Society.

Chanmee, N., Chai, J., Hino, T., and Wang, J. (2017). Methods for evaluating over consolidation ratio from piezocone sounding results. *Underground Space* 2 (3): 182–194.

Das, B.M. (2006). *Principles of Geotechnical Engineering*. Thomson Learning (KY).

Das, B.M. and Sivakugan, N. (2016). *Fundamentals of Geotechnical Engineering*, 5e. Boston, USA: Cengage Learning.

Djoenaidi, W.J. (1985). A compendium of soil properties and correlations. MEngSc thesis. University of Sydney, Australia.

Eslaamizaad, S. and Robertson, P.K. (1997). Evaluation of settlement of footings on sand from seismic in-situ tests. *Proceedings of the 50th Canadian Geotechnical Conference, Ottawa, Ontario*, vol. 2, pp. 755–764.

Eslami, A. (2015). Investigation of explosive compaction (EC) for liquefaction mitigation using CPT records. *Bulletin of Earthquake Engineering*.

Eslami, A. and Fellenius, B.H. (1997). Pile capacity by direct CPT and CPTu methods applied to 102 case histories. *Canadian Geotechnical Journal* 34 (6): 886–904. https://doi.org/10.1139/t97-056.

Eslami, A., Aflaki, E., and Hosseini, B. (2011). Evaluating CPT and CPTu based pile bearing capacity estimation methods using Urmiyeh Lake Causeway piling records. *Scientia Iranica* 18: 1009–1019. https://doi.org/10.1016/j.scient.2011.09.003.

Eslami, A., Alimirzaei, M., Aflaki, E., and MolaAbasi, H. (2015). Deltaic soil behavior classification using CPTu records—proposed approach and applied to fifty-four case histories. *Marine Georesources & Geotechnology* 35 (1): 62–79.

Eslami, A., Akbarimehr, D., Aflaki, E., and Hajitaheriha, M.M. (2019). Geotechnical site characterization of the Lake Urmia super-soft sediments using laboratory and CPTu records. *Marine Georesources and Geotechnology* 38 (10): 1223–1234.

Eslami, A., Golafzani, S.H., and Naghibi, M.H. (2023). Developed triangular charts; deltaic CPTu-based soil behavior classification using AUT: CPTu-Geo-Marine Database. *Probabilistic Engineering Mechanics* 71: 103380.

Eslami, A. and Mohammadi, A. (2016). Drained soil shear strength parameters from CPTu data for marine deposits by analytical model. *Ships and Offshore Structures* 11 (8).

Frank, R. and Magnan, J.P. (1995). Cone penetration testing in France: national report. *Proceedings CPT '95*, Linkoping, Sweden (4–5 October 1995). Swedish Geotechnical Society, 3, 147–156.

Gibbs, H.J. and Holtz, W.G. (1957). Research on determining the density of sand by spoon penetration testing. *Proceedings, 4th International Conference on Soil Mechanics And Foundation Engineering,* London (12–24 August 1957), I, 35–39.

Hatanaka, M. and Uchida, A. (1996). Empirical correlations between penetration resistance and internal friction angle of sandy soils. *Soils Found* 36 (4): 1–10.

Hoek, E. (2007). Practical rock engineering. A set of notes. http://www.rocscience.com.

Holtz, W.G. and Gibbs, H.J. (1979). Discussion of "SPT and relative density in coarse sand". *Journal of the Geotechnical Engineering Division* 105 (3): 439–441.

Holtz, R.D. and Kovacs, W.D. (1981). An Introduction to Geotechnical Engineering, 1e. Prentice-Hall, Inc.

Holtz, R.D., Kovacs, W.D., and Sheahan, T.C. (2023). *An Introduction to Geotechnical Engineering*, 3e. Pearson.

Hunt, R.E. (2005). *Geotechnical Engineering Investigation Handbook*, 2e. CRC Press. https://doi.org/10.1201/9781420039153.

Jamiolkowski, M., Lopresti, D.C.F., and Manassero, M. (2001). Evaluation of relative density and shear strength of sands from cone penetration test and flat dilatometer test. *Soil Behaviour, and Soft Ground Construction* (GSP 119), Boston, Massachusetts (5–6 October 2001). ASCE, Reston, 201–238.

Jefferies, M. and Davies, M. (1991). Soil classification by the cone penetration test: discussion. *Canadian Geotechnical Journal* 28 (1): 173–176.

Jefferies, M.G. and Davies, M.P. (1993). Use of CPTu to estimate equivalent SPT N60. *Geotechnical Testing Journal* 16 (4): 458–468. https://doi.org/10.1520/GTJ10286J.

Jones, G. A. and Rust, E. (1982). Piezometer penetration testing, CUPT. *Proceedings of the 2nd European Symposium on Penetration Testing*, ESOPT-2, 2, Amsterdam (24–27 May), 607–614.

Killeen, M.M. (2012). Numerical modelling of small groups of stone columns. PhD thesis. National University of Ireland, Galway.

Killeen, M.M. and McCabe, B.A. (2014). Settlement performance of pad footings on soft clay supported by stone columns: a numerical study. *Journal of Soils and Foundations* 54 (4): 760–776.

Kulhawy, F.H. and Mayne, P.W. (1990). Manual on estimating soil properties for foundation design (No. EPRI-EL-6800) Electric Power Research Institute, Palo Alto, CA (USA); Cornell University, Ithaca, NY (USA) (Geotechnical Engineering Group).

Li, J., Cassidy, M.J., Huang, J. et al. (2016). Probabilistic identification of soil stratification. *Géotechnique* 66 (1): 16–26.

Look, B.G. (2007). *Handbook of Geotechnical Investigation and Design Tables*, 1e. Taylor & Francis. https://doi.org/10.1201/9780203946602.

Lunne, T., Robertson, P.K., and Powell, J.J.M. (1997). *Cone Penetration Testing in Geotechnical Practice*. London: Blackie Academic and Professional/Chapman-Hall Publishers 312 p.

Marchetti, S. (1980). In situ tests by flat dilatometer. *Journal of the Geotechnical Engineering Division* 106: 299–321. https://doi.org/10.1061/AJGEB6.0000934.

Mayne, P.W. (2007). Cone penetration testing state-of-practice, NCHRP synthesis. Transportation Research Board Report Project, 20–05, 118 pages.

Mayne, P.W. (2014). Interpretation of geotechnical parameters from seismic piezocone tests. In: P.K. Robertson, K.I. Cabal (eds) *Proceedings 3rd International Symposium on Cone Penetrometer Testing*, CPT'14, Las Vegas (12–14 May 2014), 47–73.

Mayne, P.W. and Kemper, J.B. (1988). Profiling OCR in stiff clays by CPT and SPT. *Geotechnical Testing Journal* 11 (2): 139–147.

Meyerhof, G.G. (1959). Compaction of sands and the bearing capacity of piles. *Journal of the Soil Mechanics and Foundations Division* 85 (6): 1–29.

Meyerhof, G.G. (1974). Ultimate bearing capacity of footings on sand layer overlying clay. Canadian *Geotechnical Journal* 11 (2): 223–229.

Mitchell, J.K. and Durgunoglu, H.T. (1983). Cone resistance as measure of sand strength. *Journal of the Geotechnical Engineering Division* 104 (GT7): 995–1012.

Nash, D.F.T., Powell, J.J.M., and Lloyd, I.M. (1992). Initial investigations of the soft clay test site at Bothkennar. *Géotechnique* 42 (2): 163–181.

Olsen, R.S. and Mitchell, J.K. (1995). CPT stress normalization and prediction of soil classification. *International Symposium on Cone Penetration Testing*: CPT 95, Linkoping, Sweden (4–5 October 1995), 257–262.

Ramsey, N. (2002). A calibrated model for the interpretation of cone penetration tests (CPTs)s in North Sea quaternary soils. In: *Offshore Site Investigation and Geotechnics—Diversity and Sustainability*, 341–356. London: Society for Underwater Technology.

Robertson, P.K. (1990). Soil classification using the cone penetration test. *Canadian Geotechnical Journal* 27 (1): 151–158. https://doi.org/10.1139/t90-014.

Robertson, P.K. (2009). Interpretation of cone penetration tests – a unified approach. *Canadian Geotechnical Journal* 46: 1337–1355.

Robertson, P.K. (2012). Interpretation of in-situ tests – some insights. *J.K. Mitchell Lecture, Proceedings of ISC'4*, Recife, Brazil (18–21 September).

Robertson, P.K. (2016). Cone penetration test (CPT)-based soil behaviour type (SBT) classification system—an update. *Canadian Geotechnical Journal* 53 (12): 1910–1927.

Robertson, P.K. and Cabal, K.L. (2010). *Guide to Cone Penetration Testing for Geotechnical Engineering*. Gregg Drilling & Testing.

Robertson, P.K., Campanella, R.G., Gillespie, D., and Greig, J. (1986). Use of piezometer cone data. *Proceedings of the In-Situ'86 Use of In-situ testing in Geotechnical Engineering*, New York (23–25 June 1986). ASCE, Reston, Virginia, 1263–1280.

Sabatini, P.J., Bachus, R.C., Mayne, P.W., et al. (2002). Geotechnical Engineering Circular No. 5: Evaluation of soil and rock properties. Geotechnical Engineering Circular No. 5 (No. FHWA-IF-02-034).

Senneset, K., Sandven, R., and Janbu, N. (1989). Evaluation of Soil Parameters from Piezocone Tests. Transportation Research Record 1235. *National Research Council*, Washington, DC, pp. 24–37.

Sanglerat, G. (1972). *The Penetrometer and Soil Exploration*. Amsterdam: Elsevier Pub 488 pp.

Schmertmann, J.H. (1975). Measurement of in situ shear strength, Keynote lecture. *Proceedings of the conference on in-situ measurement of soil properties*, New York (1–4 June 1975). ASCE, 2, 57–138.

Schmertmann, J.H. (1978). Guidelines for cone penetration test performance and design. *Report FHWA-TS-78-209*. U.S. Dept of Transportation, Washington, 145 pp.

Schnaid, F. (2009). *In-situ Testing in Goemechanics* (329pp). London: Taylor and Francis.

Skempton, A. (1953). Soil mechanics in relation to geology. *Proceedings of the Yorkshire Geological Society* 29 (1): 33–62.

Sowers, G.F. (1979). *Introductory Soil Mechanics and Foundations, Geotechnical Engineering*, 4e. New York: Macmillan.

Terzaghi, K. and Peck, R.B. (1967). *Soil Mechanics in Engineering Practice*, 2e. New York: Wiley.

Trevor, F.A. and Mayne, P.W. (2004). Undrained shear strength and OCR of marine clays from piezocone test results. *Proceeding ISC-2 on Geotechnical and Geophysical Site Characterization* (ed. Viana da Fonseca, Mayne), 391–398. Rotterdam: Mill Press.

U.S. Navy (1982). *Soil Mechanics – Design Manual 7.1*. Washington, DC: Department of the Navy, Naval Facilities Engineering Command, U.S. Government Printing Office.

Uzielli, M., Mayne, P.W., and Cassidy, M.J. (2013). Probabilistic Assessment of Design Strengths for Sands from *In-Situ Testing Data in Modern Geotechnical Design Codes of Practice*, pp. 214–227.

Wolff, T.F. (1989). Pile capacity prediction using parameter functions. *Predicted and Observed Axial Behavior of Piles, Results of a Pile Prediction Symposium*, New York (25 June 1989). ASCE Geotechnical Special Publication 23, 96–106.

Topic B

Shallow Foundations

Section III

Spread Footings and Mats

5

Types and Construction Methods

Chapter Highlights

- Introduction
- Single Spread Footings
- Combined Foundations
  - Two Columns
  - Strip and Continuous
  - Grid
- Mat Foundations
- Uncommon Shallow Foundations
  - Grillage Foundations
  - Attached Single Foundations
  - Industrial Slabs
- Case History
  - Large Foundations Performance on Strong Soils
  - Adjacent Construction Complications
  - Le Royal Hotel complex, Amman, Jordan

5.1 Introduction of Shallow Foundations

Foundations serve as structures that transfer loads from superstructures to the subsoil. These structures can be made of diverse materials such as masonry, concrete, reinforced concrete, steel, wood, or even encased soil with geosynthetics. Designing a foundation necessitates in-depth knowledge of soil properties and their response to loads. This involves considering factors like surrounding conditions, construction methods, and potential future changes. The effectiveness of foundation design relies on the engineer's ability to analyze and integrate these factors.

Among the various types of foundations, shallow foundations are among the most common, especially in building projects (Figure 5.1). In these foundations, the depth (D_f) is less than the width ($D_f \leq B$). Shallow foundations are typically constructed after excavating and removing weathered soils while bypassing the frost depth and weak surface layers. These foundations are highly prevalent due to their cost-effectiveness and ease of construction. They are commonly used for small to medium-sized structures located on sites with moderately good soil conditions. Even large structures can utilize shallow foundations if the site has exceptionally good soil or shallow bedrock.

With structural and construction considerations, the following systems are common shallow foundations:

- Single spread footings
- Combined and strip (multiple columns, wall, and ring)
- Mat or raft

Advanced Foundation Engineering: Principles, Performance and Prospects, First Edition. Moses Karakouzian and Abolfazl Eslami.

Figure 5.1 Typical section of shallow foundations.

5.2 Single Spread Foundations (Footings)

A single footing can be defined as an enlarged base at the bottom of a column or wall, dispersing the applied loads across a sufficiently large soil area. Usually, each column and wall in a structure will have its dedicated footing, leading to potentially numerous individual footings in a typical structure. Single footing foundations support the weight of a column or wall and represent the simplest, most prevalent, and cost-effective types of foundations. Various factors can influence the classification of single foundations, including:

- Form in plan and section
- Material
- Construction methods

5.2.1 Form in Plan

Spread footings can be designed in various shapes and sizes in order to satisfy specific project requirements. These shapes include square, rectangular, and circular as shown in Figure 5.2. The shape selection depends on the column section and the load type, including axial force and the moment in one or two directions. Square-shaped footings typically support a single centrally located column. Rectangular spread footings are used when designing a square footing with a sufficiently large base area is not practical, or significant moments are applied in one direction only. Circular spread footings have a round footprint. Smaller circular footings are built by drilling with an auger. These foundations are commonly used in flagpoles and power transmission lines.

5.2.2 Form in Section

Depending on thickness or vertical cross-section, single foundations can be categorized as spread, stepped, or sloped. A simple (pad) footing consisting of a concrete flat plate or a slab (Figure 5.3a) with a thickness of more than 1 m is uneconomical. Stepped foundations provide the necessary thickness to resist punching shear, allow adequate column-to-foundation connectivity, and reduce the required reinforcing bars to resist flexural moments in the footing (see Figure 5.3b). Taper-haunch or sloped foundations can be an alternative to stepped foundations optimizing material consumption such as concrete and reinforcement. However, constructing the formwork for the sloped foundations can be more challenging compared to stepped foundations (Figure 5.3c).

5.2.3 Materials

Until the mid-nineteenth century, nearly all spread footings were constructed of masonry (Figure 5.4a and b). Dimension-stone footings were constructed from cut and dressed stones of specific sizes that were fit together with

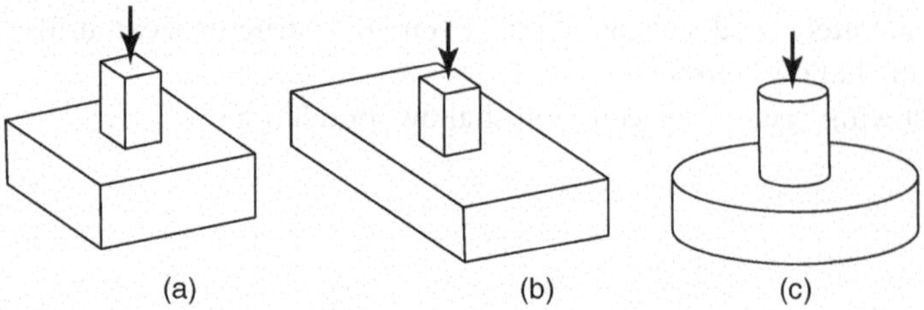

Figure 5.2 Different plans of footings: (a) square, (b) rectangular, and (c) circular.

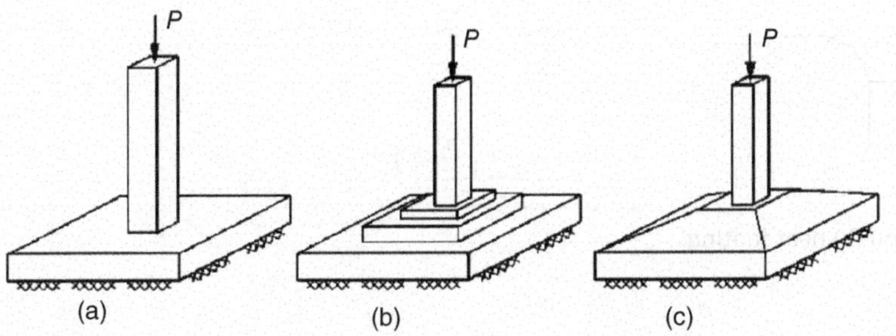

Figure 5.3 Different sections of footings: (a) simple, (b) stepped, and (c) sloped.

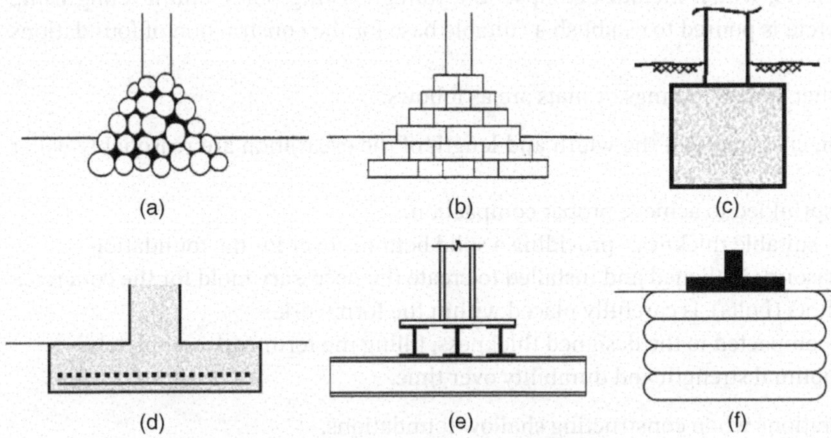

Figure 5.4 Different footings in regard to material: (a) rubble stone, (b) dimension-stone, (c) plain concrete, (d) reinforced concrete, (e) steel, and (f) pad on reinforced soil.

very small gaps. In contrast, rubble-stone footings were constructed from random-sized material joined with mortar (Peck et al. 1974). These footings had low tensile strength and required large height-to-width ratios to limit the flexural stresses and prevent tensile failures.

As larger structures became prevalent, it became necessary to provide shorter and lighter footings while maintaining the same base dimensions. This change needed structural materials capable of sustaining flexural stresses. Plain concrete was used for footing construction (Figure 5.4c).

Figure 5.4d shows a common reinforced concrete footing, which is very strong, durable, economical, and easy to construct. These footings are thinner than their old counterparts and do not require extensive excavations and intrusion into basements. Thus, almost all spread footings are now built from reinforced concrete. In some cases, footings made of steel (Figure 5.4e), as well as simple pads constructed on confined soils via geosynthetic layers (Figure 5.4f), have been used as shallow foundations.

5.2.4 Construction Methods and Embedment

The choice of construction procedure is heavily influenced by existing soil conditions, categorized as soft, medium, or hard. In instances where soft or loose soil does not meet geotechnical and structural requirements, soil stabilization and modification measures become necessary. Structural fill, involving the removal or replacement of problematic soils, can stabilize these grounds. When the soil possesses adequate strength and stiffness, and excavation is challenging, a spread foundation is placed directly on the ground surface. In low-lying areas susceptible to movement from sea or river water, or when the groundwater level is close to the surface, foundations are constructed at elevated levels. In such scenarios, full-depth formworks made of materials like steel, masonry, or wood are employed, referred to as formed footings (Figure 5.5a). In standard soil conditions, excavation is typically performed using machinery like backhoes or excavators, and concreting is done without any formwork, creating what is known as a neat footing (Figure 5.5b).

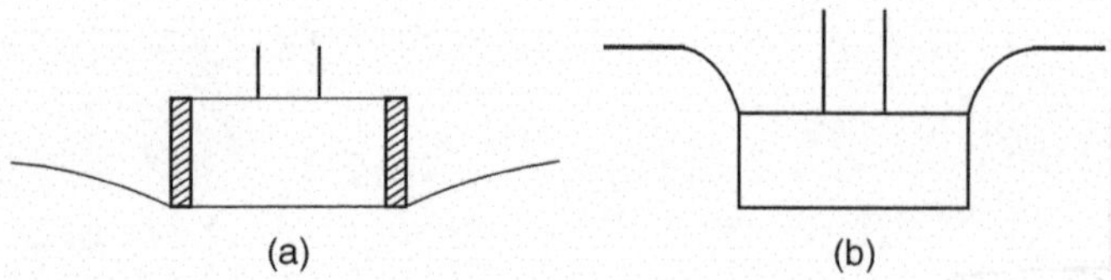

Figure 5.5 Methods of concreting: (a) formed footing and (b) neat footing.

5.2.5 Construction Considerations

To ensure the efficient and durable performance of the footing throughout the project's lifetime, specific special requirements must be considered. Typically, the initial step in spread footing construction involves the removal of any peaty and organic soils. The preparation phase follows, which includes compaction, filling existing holes, and leveling using coarse-grained soils. Subsequently, lean concrete is poured to establish a suitable base for the construction of foundations and structural elements.

The general steps in the construction of either spread footings or mats are as follows:

1) The area is excavated to the desired depth, ensuring that the width and length of the excavation are reasonably wider than the dimensions of the footing.
2) The subgrade is compacted, and water is sprinkled to achieve proper compaction.
3) A layer of lime or lean concrete is laid to a suitable thickness, providing a solid bottom cover for the foundation.
4) If formwork is required, steel, wood, or masonry is aligned and installed to create the necessary mold for the concrete.
5) Reinforcement, along with dowels or anchors (bolts), is carefully placed within the formwork.
6) Portland Cement Concrete is poured and compacted to the designed thickness, filling the formwork completely.
7) The concrete is properly cured to ensure optimal strength and durability over time.

The following are other important considerations when constructing shallow foundations.

5.2.5.1 Scour Potential

Erosion caused by intense water flow under extreme weather conditions can cause significant settlements and even structural collapse. Foundations for bridges and structures near rivers, channels, and floodplains must be built below anticipated scouring levels. However, if uncertainties exist regarding these scouring limits, alternative foundation types should be considered to ensure stability and safety.

5.2.5.2 Influence Depth and Frost Protection

Stresses exerted on the soil beneath spread footings diminish with depth below the footing. Normally, for square and rectangular footings, at a depth below two times the width of the footing, these stresses reduce to about ten percent of the stresses at the soil–footing interface. However, closely spaced footings can cause overlapping stresses, increasing the depth of the zone of the influence. In the case of continuous spread footings, the induced stresses decrease to about ten percent of the stresses at the soil–footing interface at a depth approximately four times the width of the footing. Spread footings must be designed and constructed below the frost depth to avoid frost-susceptible materials.

The dimensions of a footing significantly influence settlement and bearing capacity assessments. Generally, presumptive bearing capacities found in the literature are not contingent on width. Therefore, these presumptive capacities should be used cautiously and primarily for preliminary designs.

5.2.5.3 Continuity and Interactive Considerations

A foundation system consists of three components: the foundation, the superstructure elements (columns, walls, etc.) that are connected to the foundation, and the soil in which the foundation is constructed. For optimal and efficient performance of the foundation system it is important to have these three components connected and act as a unit. Additionally, two or more foundations can be connected by ties or strips for additional efficiency.

5.2.5.3.1 Combined in Plan In earthquake-affected areas, lateral movements of the earth can cause horizontal displacement, generating significant forces in the structural components of the foundation, and potentially resulting in damage to

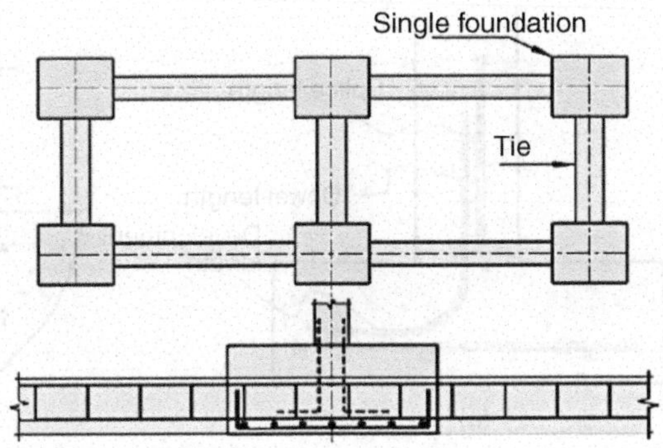

Figure 5.6 How to tie single foundations.

the structure of the foundations and excessive displacements. To prevent this, it is recommended to connect single foundations with reinforced concrete ties. In determining the tie section, the axial force affecting the ties due to lateral movement under earthquake conditions (moving away or approaching the foundations) must be considered. Accordingly, the ties between the two foundations should be calculated for an axial tensile force with a pressure of 10% of the most significant column load. Longitudinal bars must be continuous and extend at least as long as the development length. They should not be cut off on the foundation and must be continuously intersected (refer to Figure 5.6).

5.2.5.3.2 Surrounding by Ground The depth of shallow foundations significantly influences bearing capacity, serviceability, and stability, crucial aspects in geotechnical foundation design. Increasing the foundation depth and anchoring it in the ground establishes a more thorough interaction between the foundation and soil, resulting in higher bearing capacity and reduced settlement compared to foundations constructed at ground level. Ground-anchored foundations also offer greater resistance against uplift loads due to the soil weight on the foundation and an extended failure surface. Furthermore, incorporating the passive lateral load of the soil contributes to higher safety factors, meeting stability requirements (see Figure 5.7).

5.2.5.4 Connection with Superstructure

5.2.5.4.1 Concrete Columns The link between concrete columns and the foundation is established through reinforcing bars, commonly known as rebars, as shown in Figure 5.8a. These rebars transmit axial loads and moments from the columns to the foundation. In general, the foundation thickness is dictated by the required development lengths of the rebars within the foundation to accommodate the effective transfer of the axial loads and moments from the column.

5.2.5.4.2 Steel Columns The connection between steel columns and the foundation is achieved using anchor rods embedded in the foundation supporting the column loads as shown in Figure 5.8b. The anchor rods are bolted to steel plates welded to the bottom of the columns. The compressive axial loads from the columns are distributed by the plates beneath the columns and transferred to the foundations. The tensile axial loads and moments from the columns are transmitted to the foundation through bolted anchor rods. Additionally, shear loads from the columns are transferred to the foundation through bolted anchor rods, and friction develops between the plates and the foundation.

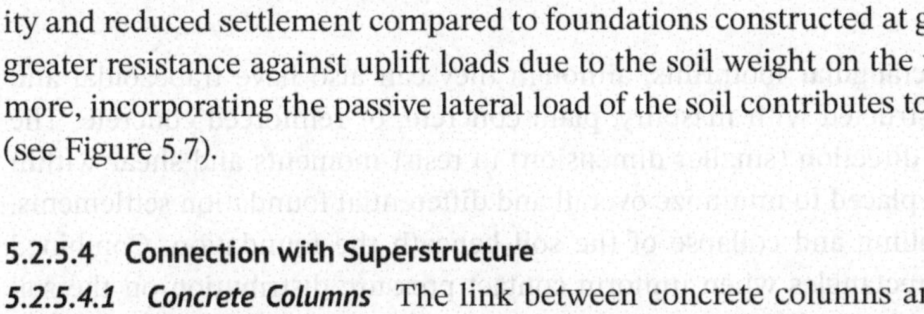

Figure 5.7 Different modes of foundations depth and comparison with foundation thickness.

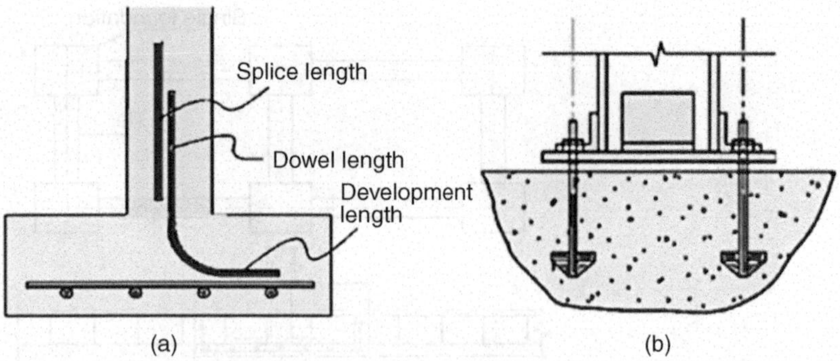

Figure 5.8 Connecting column to the foundation: (a) concrete and (b) steel.

5.3 Combined Foundations

Combined foundations carry the loads from more than one column. The design of combined foundations depends upon the number and layout of the columns the foundation carries. Common combined foundation configurations are as follows:

- Two-Column Foundations
- Strip and Continuous Foundations
- Grid Foundations

5.3.1 Two-Column Foundations

Two-column combined foundations usually have rectangular footprints, although they can also have trapezoidal and T-shaped footprints. These foundations may be constructed with masonry, plain concrete, or reinforced concrete. The primary reinforcements are placed in the transverse direction (smaller dimension) to resist moments and shear within the foundation. Longitudinal reinforcements are also placed to minimize overall and differential foundation settlements, as well as resist foundation movements due to swelling and collapse of the soil beneath the foundation. Combined foundation footprints (Figure 5.9a) are designed as rectangles when uniform contact pressure distribution on the soil beneath the foundation exists. In cases where significant differences between column loads exist, trapezoidal, and T-shaped foundation footprints are recommended to maintain uniform contact pressure distribution on the soil beneath the foundation and increase the rigidity of the foundation system to prevent differential settlements (Figure 5.9b and c). The construction and reinforcing process is less challenging in T-shaped foundations compared to those of trapezoidal

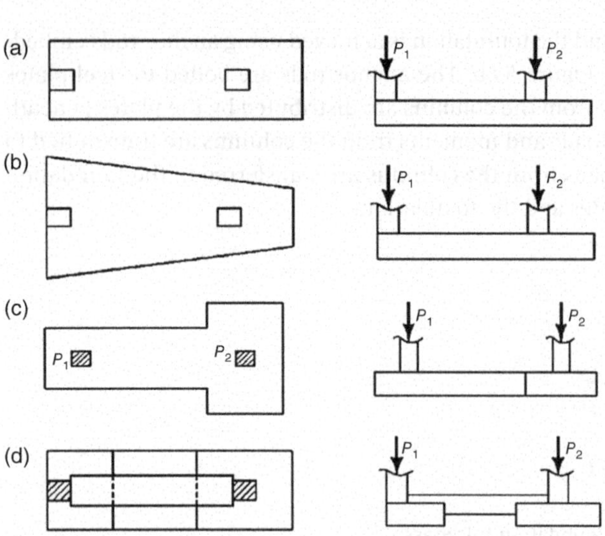

Figure 5.9 Section and plan of double-column combined foundations: (a) rectangular, (b) trapezoidal, (c) T-shaped, and (d) cantilever.

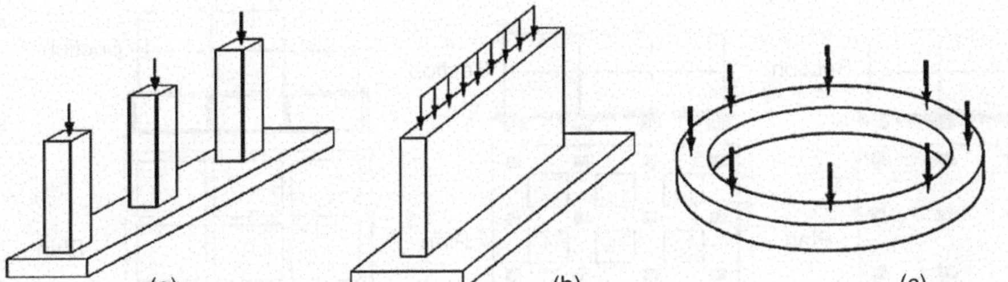

Figure 5.10 Types of strip and continuous foundations: (a) multicolumn, (b) wall, and (c) ring.

ones. Cantilever foundations comprise two separate footings attached via a strap beam. In this case, uniform soil pressure can be achieved by matching the center of the foundation footprints with the center of gravity of the column loads (Figure 5.9d).

Combined foundations are economical when:

- The columns are near one another.
- The column is near the property line. The combined foundation will allow the footing to be constructed within the property line, as well as prevent eccentric column loading.
- The dimension of one side of the footing is limited to lower values.

5.3.2 Strip and Continuous Foundations

Strip foundations are used to support the load of multiple adjacent columns, and their length is much greater than their width. An example of a typical strip foundation is shown in Figure 5.10a. Wall footings, shown in Figure 5.10b, are continuous foundations that support load-bearing walls. Ring spread footings are continuous footings that have been wrapped in a circular shape, as seen in Figure 5.10c. This type of foundation is commonly used to support the walls of aboveground circular storage tanks. In geotechnical assessments, tanks are often treated as circular foundations with diameters equal to the tank's diameter.

5.3.3 Grid Foundations

Grid foundations consist of longitudinal and transverse strip foundations that work together to transfer loads in two directions and adjust for differential settlements, particularly in loose soils. Figure 5.11 illustrates a schematic of a grid foundation.

5.4 Mat Foundations

One of the most commonly used types of shallow foundations is the mat foundation, also known as a raft foundation. A mat foundation is a large spread footing made of reinforced concrete that typically covers the entire footprint of the structure. Mat foundations are used to support loads of large buildings, silos, chimneys, towers, storage tanks, and large machines. Due to the large contact area of the mat foundation, the load is distributed over a larger area, resulting in lower soil stresses compared to other foundation types. This, in turn, reduces the risk of shear failure.

Figure 5.11 Schematic of grid foundation.

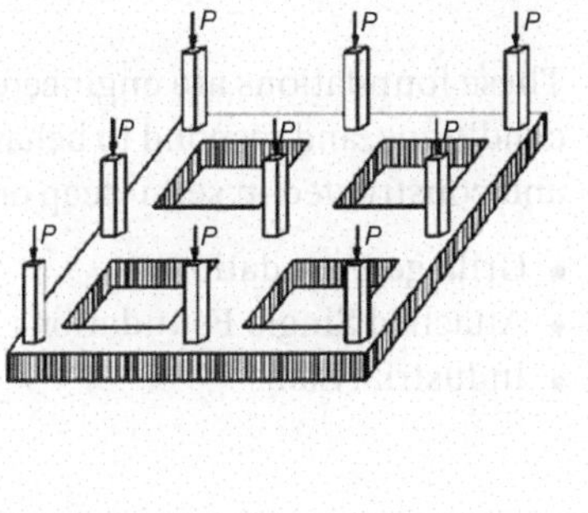

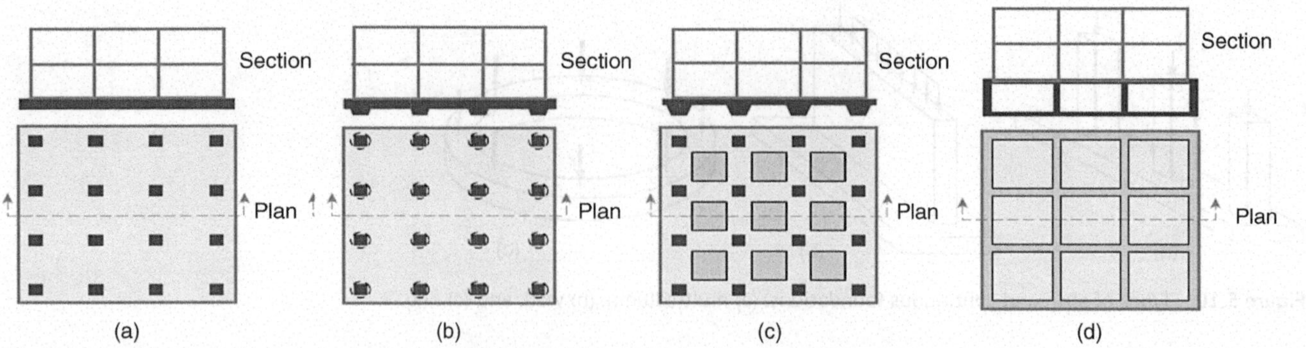

Figure 5.12 Types of mats: (a) flat plate, (b) flat plate thickened under columns, (c) beams and slab, and (d) slab with basement walls as a part of the mat.

Mat foundations are popular choices due to their straightforward construction process and effectiveness in supporting structures on weak soils at shallow depths. However, selecting the appropriate foundation system requires an evaluation of the soil conditions and load-bearing requirements, as well as a rigorous analysis of the settlement criteria.

Mat foundations are typically recommended in the following situations:

1) Low soil bearing capacity
2) The need to reduce foundation/soil contact pressure by distributing the loads over a large area
3) Relatively covering most of the total footprint of the structure by any other foundation type
4) Overlapping the footprint of individual footings as a result of placing columns or walls too closely to one another
5) High probability of having differential settlements when individual footings are used
6) Existence of highly compressible soil layers
7) Unrealistic use of any other kind of foundation

Depending on the soil conditions and the imposed loads, various types of mats may be used. The various types of mats are as follows:

Flat Plate Mats: One of the simplest types of mat foundations is the flat plate mat. It is suitable when columns and walls are uniformly placed at small intervals, and the loads are relatively small. Reinforcements are placed in both directions, with an additional reinforcement placed beneath load-bearing walls and columns. The thickness is typically limited to 300 mm for economic reasons (see Figure 5.12a).

Plate Thickened under Columns: Plate Thickened under Columns is a type of raft foundation used when the columns and load-bearing walls are subjected to heavier loads. In this type, the slab is thickened beneath the columns and walls, and more reinforcements are used to resist diagonal shear and moments (Figure 5.12b).

Two-way Beam and Slab: In this kind of mat foundation, beams are cast monolithically, with the mat slab connecting the columns and walls. This type of foundation is used when the distance between the columns and the walls is considerable, and their loads are unequal to each other (Figure 5.12c).

Rigid Frame Mats – (Slab with Basement Walls as a Part of the Mat Foundation): In this type of mat foundation, walls are constructed along the perimeter and interior of the footprint of the structure. These walls act as deep beams 90 cm in depth. The walls are sandwiched between two slabs: one at the bottom and the other on the top of the walls, forming a cellular mat foundation. This kind of mat is very rigid and economical (Figure 5.12d).

5.5 Uncommon Shallow Foundations

These foundations are engineered to endure nonstandard loading conditions from the superstructure, address specific soil conditions, and respond to behavioral demands from both the superstructure and substructure. They are typically designed and constructed in semi-deep or deep foundation excavations. Examples include:

- Grillage Foundations
- Attached Single Foundations
- Industrial Slabs

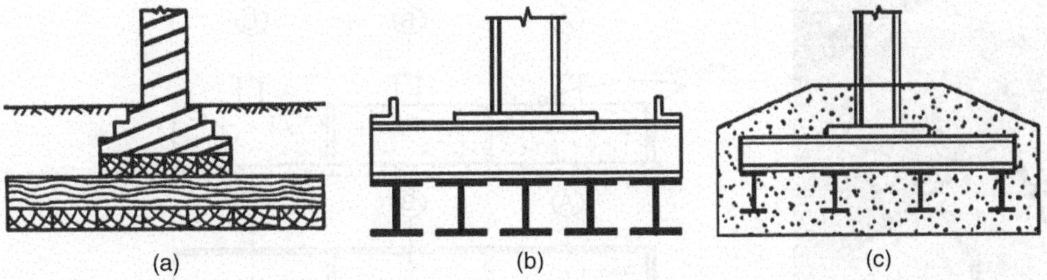

Figure 5.13 Grillage foundations: (a) timber, (b) steel, and (c) concrete frame.

5.5.1 Grillage Foundations

Grillage foundations are engineered for heavy superstructure loads and rocky bases, often chosen when reinforced concrete solutions are impractical due to excessive tensile, compressive, and shear stresses in the foundation system. They are designed to distribute loads over a larger area at the base of the structure. Based on the materials used, grillage foundations are categorized as steel or timber grillage foundations. Steel grillage foundations comprise wide-flange steel beams arranged in two welded or bolted layers perpendicular to each other (Figure 5.13b). To enhance protection and durability, concrete may be poured around the grillage (Figure 5.13c). For lighter loads or temporary constructions on loose soils, two layers of timber can be employed, covered with concrete (Figure 5.13a).

The method for constructing steel grillage foundations is as follows:

1) A trench is excavated to the designed width, length, and depth. The bottom of the trench is rammed and leveled.
2) A layer of lean cement concrete is laid and compacted.
3) Another layer of stronger cement concrete is poured and compacted providing a concrete bed.
4) Wide-flange steel beams arranged in two welded or bolted layers perpendicular to each other are placed over the concrete bed. The spaces between and around the steel beams are then filled with cement concrete.
5) The entire excavation is then filled with cement concrete. The structure in the form of a steel stanchion, column, and pier is then built on the grillage.

The method for constructing a timber grillage foundation is as follows:

1) Wooden planks are placed next to each other longitudinally on the prepared trench bed.
2) The planks should be placed on either side than the designed width of the base course of the structure.
3) On the bottom layer of planks, rectangular timber sections of proper size are placed from center to center.
4) Then, the top layer of planks, placed next to each other in width as same as that of the base course of the structure, is to be constructed on it.
5) On the prepared timber platform, the structure is built in the form of a timber column or even a masonry wall.

5.5.2 Attached Single Foundations

The analysis, design, and construction of single-footing foundations are simpler than those of mat foundations. Accordingly, single footings attached to each other at their bases can be used as an alternative to mat foundations. The single footings are separated by joints except at the bottom of the footings, where reinforcing steel can be provided to act as ties between the footings. The analysis and design are performed for each footing independently of other adjacent ones. Figure 5.14a and b shows typical perspectives, plans, and sections of the attached single foundations.

5.5.3 Industrial and Warehouse Slabs

A number of load-carrying structures may not be traditionally recognized as a shallow foundation, even though design processes and construction materials are very similar to those of shallow foundations. Industrial and warehouse slabs is an example of this category of foundations. In the past, foundation slabs for industrial facilities and warehouses were constructed without much consideration for the potential change of future use. However, the recent trend is to design and construct slabs for industrial and warehouse use considering potential future changes of use of the facility.

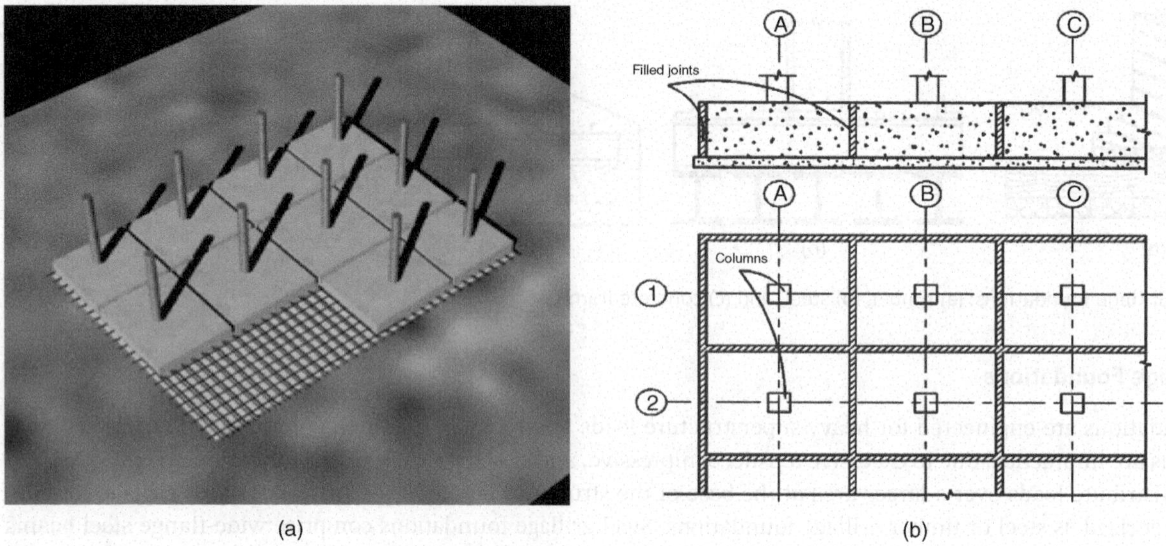

Figure 5.14 Attached are single foundations: (a) perspective and (b) section and plan.

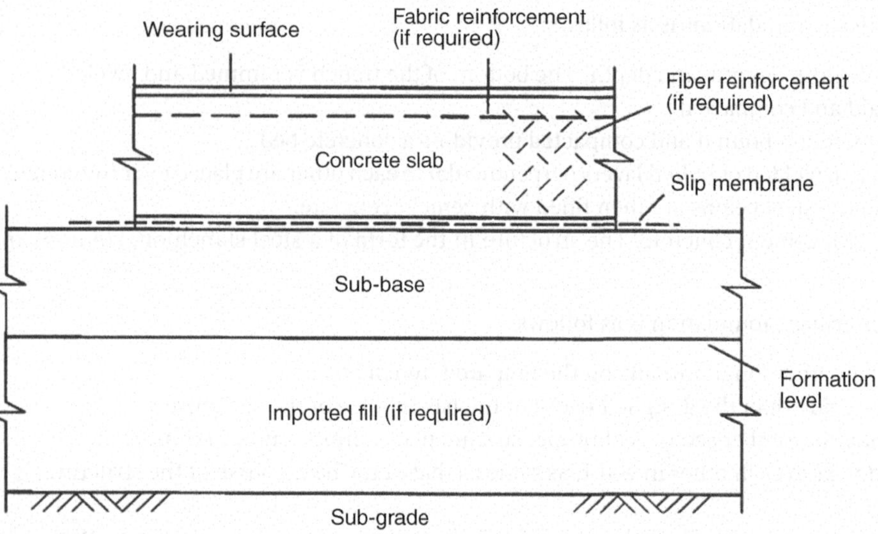

Figure 5.15 Main elements in concrete floor construction (Hemsley 2000 / With permission of Thomas Telford).

Accordingly, innovative design and construction methods of slabs have been introduced to provide flexibility for the requirements of different future uses of such facilities.

Figure 5.15 illustrates the essential foundation components required for the construction of concrete slabs for industrial and warehouse facilities. The quality and thickness of the concrete slab, its interaction with the subbase and subgrade, and the construction method all have a significant role in how well the foundation performs regarding its strength, stiffness, surface regularity, and slab surface abrasion resistance. The subbase impacts the foundation system's load-bearing capacity and reduces settlements. Following the excavation and backfilling with imported fill, the subbase is constructed above the compacted fill. The subbase is made up of imported well-graded engineered geomaterial that has been densely compacted to create a smooth and level working surface on which to build the concrete slab. Finally, a slab surface wearing course is applied as a sealant to resist chemical attack, minimize abrasion, and increase friction.

Cracking of the slab may result from excessive restraint to contraction brought on by concrete drying shrinkage. Cracking can be alleviated by providing a slip membrane and geotextile fabric layer between the concrete and the subbase reducing the friction between them. In addition to reducing friction, geotextile fabrics act as reinforcing the concrete thus improving the ability to bear loads.

5.6 Case Histories

Case History 5.1

Large Foundations Performance on Strong Soils

Several Canadian structures that were built on strong soils were instrumented to measure settlements throughout the 1970s and 1980s (Clark 1998). This was frequently done at the consultant's expense, when they wanted to verify the predicted settlements based on analyses.

The instrumentation work done gave insight into how very large buildings settle. In a few instances, the observations continued long after construction was completed. A summary of the settlement observations for structures on strong soils at the end of the construction period is shown in Table 5.1. Although the majority of these are in Alberta, there are also settlement documents from Toronto, Miami, Houston, and New York. Office buildings make up the majority of the case studies in Table 5.1, but two other structures with excellent total load and bearing pressure data have been converted for comparison by determining the number of stories that would correspond to the foundation load. They were the Canada Malting Building and the Calgary Tower. Table 5.1 shows that settlements are small and are comparable to structures that are nearly the same height. The effective use of shallow foundations on strong deposits is made clear by this issue.

Table 5.1 Summary of large foundations settlement on strong soils (Clark 1998 / With permission of Canadian Science Publishing).

Name and location of building	Foundation type	Below grade (m)	No. of stories	Settlement (mm)	Soil type
CN Tower, Edmonton	Raft	6.7–7.9	26	30	Clay till
A.G.T. Tower, Edmonton	Raft	13.7	34	48	Sand–clay till
Oxford Building, Edmonton	Raft	7.3–13.7	28	28	Sand–clay till
Omni International Hotel, Miami	Raft	—	26	30	Over consolidated soils
Sun Oil Building, Calgary	Spread footing–mat	6.1	32	25	Gravel, silt, and till
Scotia Centre, Calgary	Spread footing–mat	7.3	40	49	Silt till
First Canadian Centre, West Tower, Calgary	Raft	—	42	53	Till
Building A, Toronto	Raft	4.9	43	48	Clay-silt till
Bankers Hall, Calgary	Raft	18	47	55	Clay-silt till
Calgary Tower, Calgary	Ring	7	8	8	Shale
Canada Malting Building, Calgary	Raft	6	44	33	Till-shale

Case History 5.2

Complications Caused by Adjacent Construction (Eslami and Moghaddasi 2018)

One of the major considerations for design and serviceability criteria in urban areas is the effect of stress induced on the subsoil from adjacent constructions. If adequate prevention is not provided in the design, the overstress causes general or punching bearing capacity failure or differential settlement, and subsequently, causes damages to existing structures. When two foundations are adjacent to each other, stress bulbs overlap (Figure 5.16). As a result, the two foundations rotate and tilt toward each other. The rotation angle depends on the amount of the pressure overlap and soil conditions.

(Continued)

Case History 5.2 (Continued)

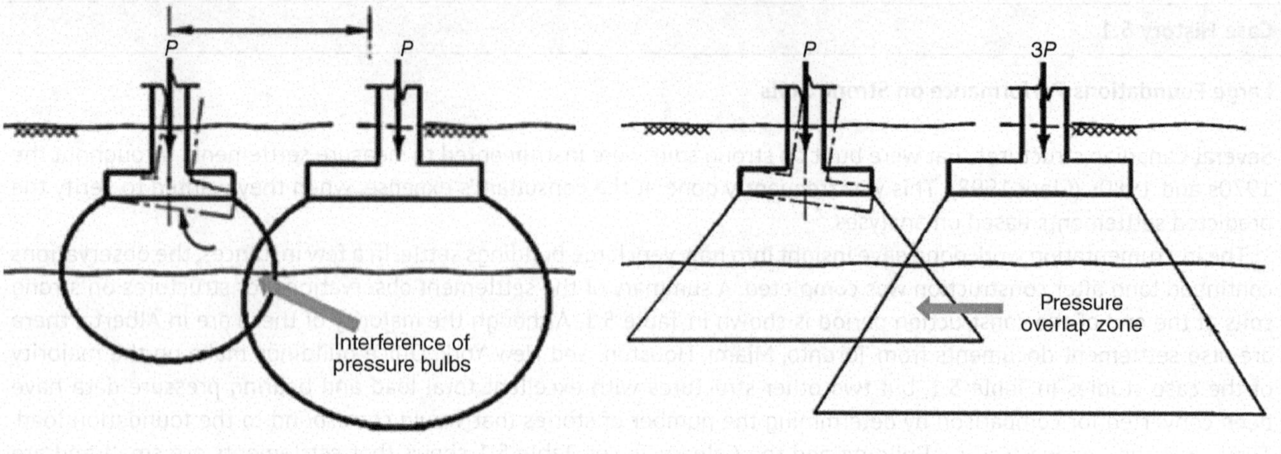

Figure 5.16 Pressure overlap zone for two adjacent foundations.

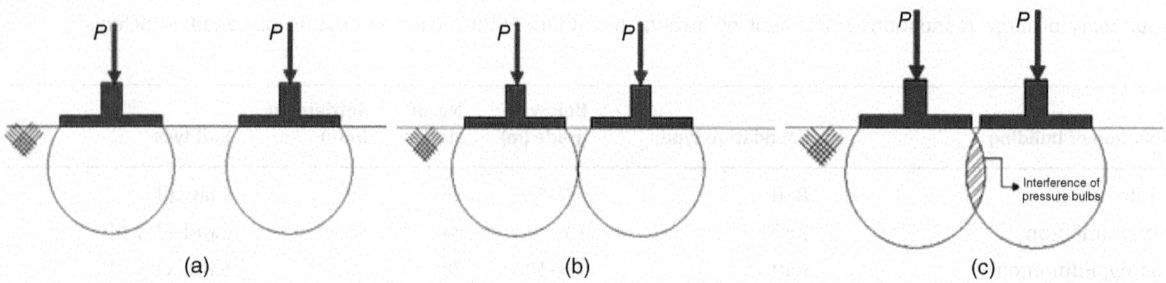

Figure 5.17 Comparison of pressure bulb interference for two adjacent foundations.

Figure 5.17 illustrates the varying interference of pressure bulbs resulting from the distance between two adjacent foundations. In Figure 5.17a, no interference occurs due to the significant distance between the foundations. Although the stress bulbs touch each other, the foundations do not experience any adjacent interactions. As the space between the foundations decreases, they come into contact, resulting in adjacent influence, as shown in Figure 5.17b. At a lower distance, the foundations experience even greater contact and are subject to greater adjacent influence, as shown in Figure 5.17c.

1) **Grain Silos:** Two-grain silos were built on Lake Agassiz clay in the Red River valley south of Winnipeg, Manitoba, Canada. These silos had ring foundations that were constructed too closely (Figure 5.18a). Since the silos were placed in saturated clay with high subsidence susceptibility and stress bulb interference (Figure 5.18b), nonuniform settlements occurred and eventually led to structural failure (Bozozuk 1976; Budho 2011; Dogangun et al. 2009).
2) **Santos Towers:** The structure in this case is located on the coastline of the city of Santos in Brazil. Due to the lack of suitable foundation selection in this area, until 1968, most structures were built on shallow foundations. According to the soil type (Figure 5.19a), which includes sand and marine clay, most of this city's structures face partial or large tilts. Nuncio Malzoni is a 17-story building built in 1967 on a shallow foundation. The depth of the foundation is 2 m, and the height of the building is 57 m. A short time after this building's construction, another similar building with a similar foundation was built on its left side. In 1968, the Nuncio Malzoni building underwent a 2.2-degree rotation (as shown in Figure 5.19b) with a maximum horizontal displacement of 2 m (Maffei and Gonçalves 2016; Pimenta et al. 1998).

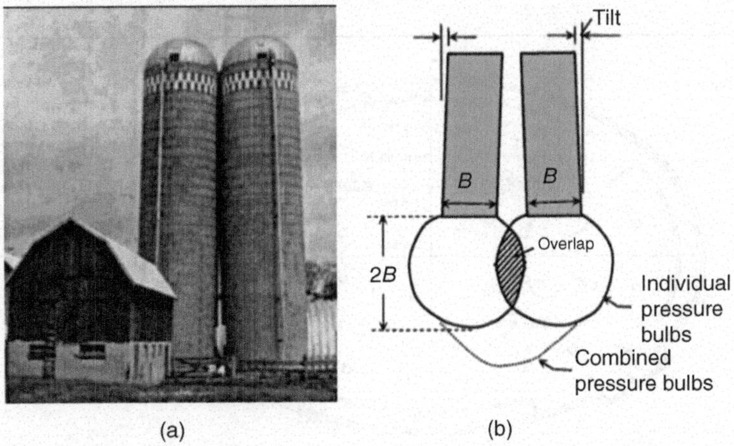

Figure 5.18 Adjacent silos: (a) Tilting and (b) Interference of pressure bulbs (Bozozuk 1976 / With permission of National Research Council Canada).

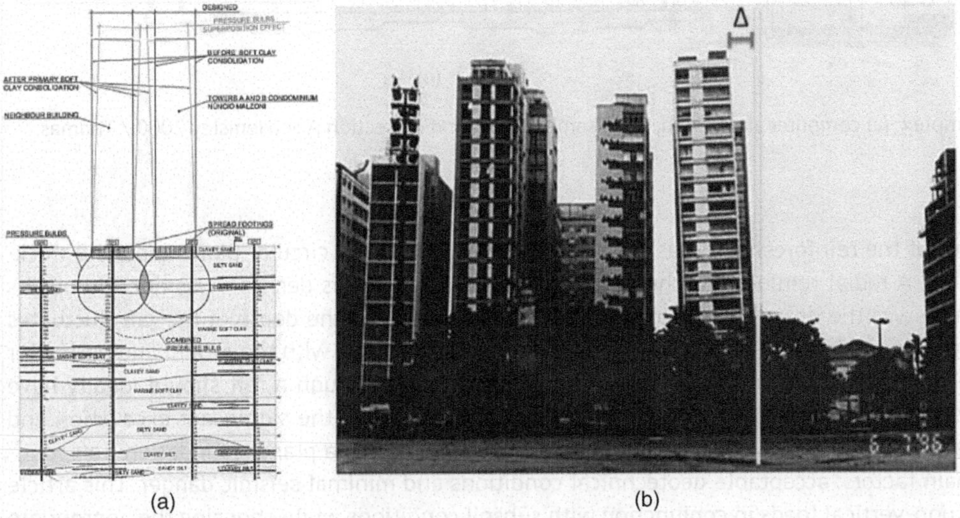

Figure 5.19 Santos towers: (a) Subsoil profile and superposition of pressure bulbs and (b) tilted tower (adapted from Maffei and Gonçalves 2016; Pimenta et al. 1998 / CIMNE).

Case History 5.3

Le Royal Hotel Complex, Amman, Jordan (Hemsley 2000)

Three levels of underground parking, in addition to a level for sports facilities and a further level for the mechanical plant, are present in the Le Royal Hotel complex. The raft founding level is 23 m below street level. Three important phases in the proposal gradually lower the ziggurat-shaped superstructure. The structure has an oval design (Figure 5.20b), and its major and minor axes measure 104 and 80 m, respectively, at ground level. Marlstones and silicified limestone from the Amman formation make up Amman's soil. The top 6 m of the rock are often decomposed; below this horizon, the rock becomes stronger, but because of the interbedded thin clay layers between the limestone, its bearing ability remains relatively consistent with depth. Based on the geotechnical ground study, an allowed bearing capacity of 650 kPa and a modulus of subgrade response of 80 MN/m^3 had been established. Considering that 460 kPa of overburden was removed, the bearing capacity seemed to be conservative.

(Continued)

Case History 5.3 (Continued)

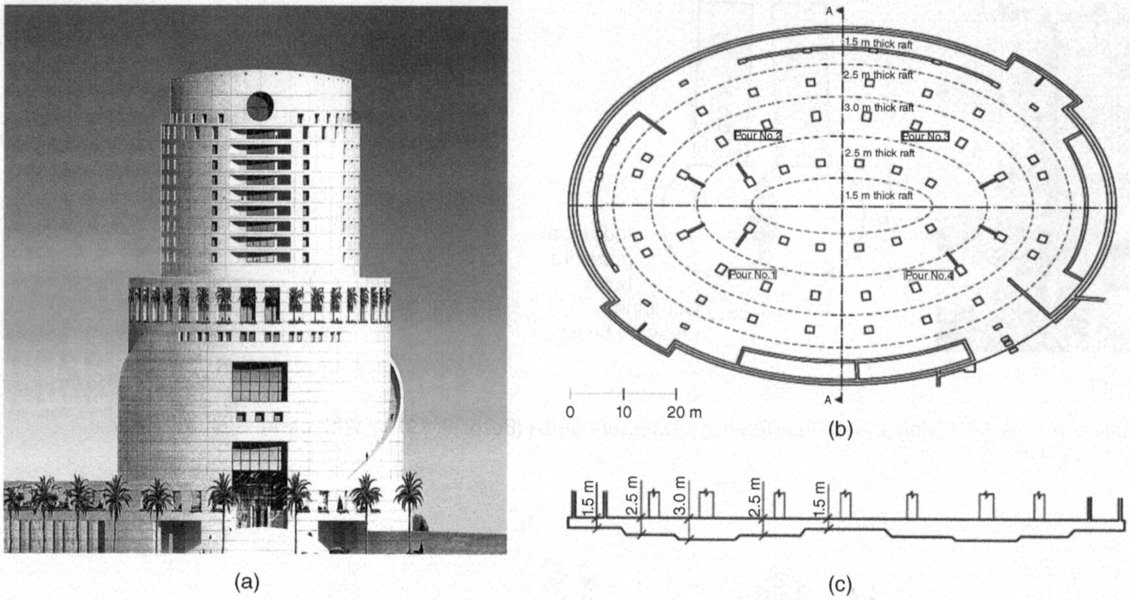

Figure 5.20 Le Royal Hotel Complex: (a) computer simulation, (b) basement plan, and (c) section A-A (Hemsley 2000 / Thomas Telford Ltd).

An orthogonal arrangement of the reinforcement was not possible due to the raft's circular plan shape and thickness fluctuations (Figure 5.20c). A radial reinforcing scheme was chosen, although this decision was not without its challenges. The punching shear was the determining element for raft thickness, and the depth of the raft fluctuates in a series of concentric circles (Figure 5.20c). The raft has a total plan area of 6480 m^2 with the atrium and perimeter zones having a minimum thickness of 1.5 m and a maximum thickness of 3 m. Although a raft should ideally have a uniform thickness, in this situation, a minimum volume method was chosen due to the significant excavation and concrete volume savings. Instead of a massively piled raft foundation or a pile group, a planar foundation (raft foundation) is used due to two main factors: acceptable geotechnical conditions and minimal seismic danger. This article makes clear the relevance of non-vertical loads in conjunction with subsoil conditions while choosing the appropriate foundation for a given project.

5.7 Major Points

1) One of the most typical foundations, particularly for building projects, is the shallow foundation. The depth (D_f) in these foundations is smaller than the width ($D_f < B$). Due to their inexpensive cost and simple construction, these foundations are the most common by far. They are most frequently employed in small to medium-sized structures on sites with fair to good soil conditions, but they can also be utilized on some major structures if they are built on sites with very good soil or shallow bedrock.
2) A single footing is an expanded section that sits under a column or wall and disperses the applied loads over a large enough region of soil. Typically, each column and wall has its own footings; therefore, a typical structure has hundreds of separate footings. The simplest, most prevalent, and most affordable forms are single foundations, which support columns.
3) Special requirements must be considered in the footing construction process for efficient and durable performance during the project's lifetime. These requirements include scour potential, influence depth and frost protection, continuity and interactive considerations, and connection with the superstructure.

4) Two columns, three columns, combined foundations, and eventually strip foundations, whose length is multiple times their breadth, are all variations of combined foundations, which support the weight of more than one column. Double-column foundations, strip and continuous foundations, and grid are often used as combined foundations.
5) Rafts are one of the most popular forms of shallow foundations (mats). A raft is a large spread foundation built of reinforced concrete that often encloses the whole footprint of the structure. Silos, chimneys, tower structures, storage tanks, and massive machinery are just a few examples of the types of constructions that rafts may support. Given the state of the soil and the applied load, a variety of mat types may be used. The many raft kinds include rigid frame, two-way beam, rigid plate, thicker plate beneath columns, and flat plate.
6) Uncommon shallow foundations are built on a semi-deep or deep surface and are developed through unconventional loading systems derived from the superstructure, the unique characteristics of the bed, and the preparation of special behavioral demands from the superstructure and substructure. A few of them are grillage foundations, attached single foundations, and industrial slabs.
7) Based on a few case studies, large foundations, i.e., mats in case of implementing on suitable and strong deposits have a satisfying geotechnical performance, which leads to their broad application in such conditions. It is important to note that complications resulting from the adjacent foundation constructions must be considered in the design process to prevent future failures and damages.

5.8 Questions and Problems

5.8.1 Explanatory

1 Define the difference between a square footing and a continuous foundation. When would each type be used?

2 Explain a structure for which that a ring footing would be appropriate.

3 Describe why a rectangular footing might be a good choice for supporting a column with a moment load in one direction.

4 What is the difference between mat and grid foundations?

5 Express the advantages and disadvantages of attached single foundations and grillage foundations.

6 What necessities lead to the use of combined foundations instead of single foundations?

7 What are the different types of raft foundations used in construction?

8 Explain the depth of influence, frost protection, and scour potential.

9 State the types of foundation construction methods.

10 What is the difference between steel and concrete columns connecting to foundations?

5.8.2 Conceptual

1 Suggest an appropriate foundation system for each of the projects below. Explain your reasoning.
(a) 5-story residential building on weak soil
(b) 20-story building on medium-strength soil
(c) cylindrical reservoir tank for oil
(d) beneath a retaining wall
(e) machinery with heavy loads on stiff soil

2 Considering the Le Royal Hotel complex, Amman, Jordan (Case History 5.3), as a successful example, what are your suggestions for engineers working on similar projects? Explain the lessons and experiences that can be gained through this case history.

3 Framed concrete sometimes surrounds the steel in grillage foundations. What do you think about the functions and behavior of concrete in these foundations?

4 Suggest some solutions to adjacent construction complications.

5 Discuss the geotechnical performance of large foundations on strong soils, using what you learn from Case History 5.1.

6 Compare four different types of two-column combined foundations realizing constructional and structural aspects.

7 Explain the role and requirements of continuity in foundation engineering.

8 Discuss the significance and role of foundation embedment depth based on technical aspects.

9 What are the alternatives to raft foundations?

References

Bowels, J.E. (1996). *Foundation Analysis and Design*, 8e. McGraw-Hill.

Bozozuk, M. (1976). *CBD-177 Tower Silo Foundations*. Canada: Canadian Building Digest, National Research Council.

Budho, M. (2011). *Soil Mechanics and Foundations*, 3e. Wiley.

CFEM (2006). *Canadian Foundation Engineering Manual*. British Columbia: Bitech Publishers LTD.

Clark, J.I. (1998). The settlement and bearing capacity of very large foundations on strong soils: 1996 R.M. Hardy keynote address. *Canadian Geotechnical Journal* 35: 131–145.

Coduto, D.P. (2016). *Foundation Design, Principle and Practices*, 3e. New Jersey: Prentice-Hall.

Das, B.M. (2016). *Principles of Foundation Engineering*, 8e. Cengage Learning.

Day, R.W. (2012). *Geotechnical Earthquake Engineering Handbook*, 2e. McGraw-Hill.

Dogangun, A., Karaca, Z., Durmus, A., and Sezen, H. (2009). Cause of damage and failures in silo structures. *Journal of Performance of Constructed Facilities* 23 (2): 65–71.

Eslami, A. (2013). *Foundation Engineering, Design, and Construction*, 2e. Tehran: Building and Housing Research Center.

Eslami, A. and Moghaddasi, H. (2018). Complications caused by neighboring and building construction in urban areas – three case studies. *Civil Engineering Research Journal* 5 (4): 00143–00152.

Eslami, A., Moshfeghi, S., Molaabasi, H., and Eslami, M. (2020). *Piezocone and Cone Penetration Test (CPTu and CPT) Applications in Foundation Engineering*, 1e. Elsevier.

Fellenius, B.H. (2023). Basics of foundation design, Electronic Edition. www.Fellenius.net, 432 p (accessed 6 March 2023).

FHWA (2016). Continuously Reinforced Concrete Pavement Manual; Guidelines for Design, Construction, Maintenance, and Rehabilitation. FHWA-HIF-16-026. Federal Highway Administration (FHWA).

Hemsley, J.A. (2000). *Design Applications of Raft Foundations*, 1e. London: Thomas Telford.

Maffei, C.E. and Gonçalves, H.H. (2016). Straightening two tilted buildings using two different techniques. *American Society of Civil Engineers (ASCE)* 1–8.

Peck, R.B. (1948). *History of Building Foundations in Chicago*, Bulletin Series No. 373. Urbana, IL: University of Illinois Engineering Experiment Station.

Peck, R.B., Hanson, W.E., and Thornburn, T.H. (1974). *Foundation Engineering*, 2e. New York: Wiley.

Pimenta, P.M., Maffei, C.E., Gonçalves, H.H., and Pauletti, R.M. (1998). A programming system for nonlinear dynamic and static analysis of tall buildings. In: *Computational Mechanics New Trends and Applications* (ed. S. Idelsohn, E. Oñate, and E. Dvorkin), 1–17. Barcelona, Spain: ©CIMNE.

Tomilson, M.J. (2001). *Foundation Design and Construction*, 7e. Prentice-Hall.

6

Geotechnical Design Aspects

Chapter Highlights

- Stresses in Soil
- Contact Pressure
- Determining Bearing Capacity
  - Base Resistance
  - Circumferential Resistance
  - In Situ Test Applications
  - Bearing Capacity of Rocks
- Settlement Estimation
  - Immediate Settlement
  - Consolidation Settlement
  - Creep Settlement
  - In Situ Test Applications
  - Differential Settlement and Distortion
- Stability Control
  - Sliding
  - Overturning
  - Uplift Capacity
  - Eccentricity
  - Buoyancy
- Case Histories
  - The Collapse of Fargo Grain Elevator (Bearing Capacity Failure)
  - El Sagrario Church and Metropolitan Cathedral, Mexico City (Differential Settlement)
  - Offshore and Onshore Load and Foundation (Exceptional Combined Loading)
  - Sloped Concrete Walls; New Approach (Stability Issue)

6.1 Stresses in Soil

Evaluating soil stresses is essential in foundation design and analysis. There are three sources that contribute to soil stresses. These sources are:

- Geostatic
- Lateral
- Induced

Advanced Foundation Engineering: Principles, Performance and Prospects, First Edition. Moses Karakouzian and Abolfazl Eslami.
 Published 2026 by John Wiley & Sons Ltd.

6.1.1 Geostatic Stress

The source of geostatic stresses is when gravity acts on soil mass, which is covered in Chapter 4.

6.1.2 Lateral Stress

The recognition of geostatic lateral stresses is crucial in various engineering applications. The following equation can be employed for this purpose:

$$\sigma'_h = K\sigma'_z \tag{6.1}$$

where

K = the coefficient of lateral earth pressure.

The K_0 value, with respect to the lateral displacement condition of soil, can be described as "at rest," indicating no significant movement. On the other hand, the active or passive states are associated with mobilized or resisting displacements, respectively.

In conventional conditions, the different K values can be calculated via the following equations:

$$K_0 = (1 - \sin\varphi) \tag{6.2}$$

$$K_0 = (1 - \sin\varphi)\text{OCR}^{\sin\varphi} \tag{6.3}$$

$$K_a = \tan^2\left(45 - \frac{\varphi}{2}\right) \tag{6.4}$$

$$K_p = \tan^2\left(45 + \frac{\varphi}{2}\right) \tag{6.5}$$

In conditions of soil–structure interaction due to construction activities, such as pile installation, the at-rest condition, K_0, is assumed.

6.1.3 Induced Stress

The induced or applied stresses are caused by external loads such as structural loads, fills, or any other type of surface loading. These types of stresses decrease with depth within the soil.

In rigid foundations, the contact pressure, q, is:

$$q = \frac{P}{A_f} \tag{6.6}$$

where

q = contact pressure
P = applied a concentrated vertical load
A_f = the foundation footprint area

An approximate way of calculating q_z, is the 2:1 method; accordingly, the q_z can be found for foundation $B \times L$, and contact pressure of q:

$$q_z = \frac{q \times B \times L}{(B + z)(L + z)} \tag{6.7}$$

Various theories, equations, and graphs are presented in foundation engineering to determine the I_z value, regarding the following equation:

$$\Delta\sigma_z = q_z = I_z \cdot q \tag{6.8}$$

Figure 6.1 depicts the dissipation of imposed pressure on the soil through the depth and by increasing the distance from the center of the loading area.

When the dimensions of the loading area are sufficiently large, there is no need to distribute or decrease the stresses with the depth. In such cases, the full value of the imposed stress is considered in the calculation of stress at every depth. Example 6.1 illustrates this concept.

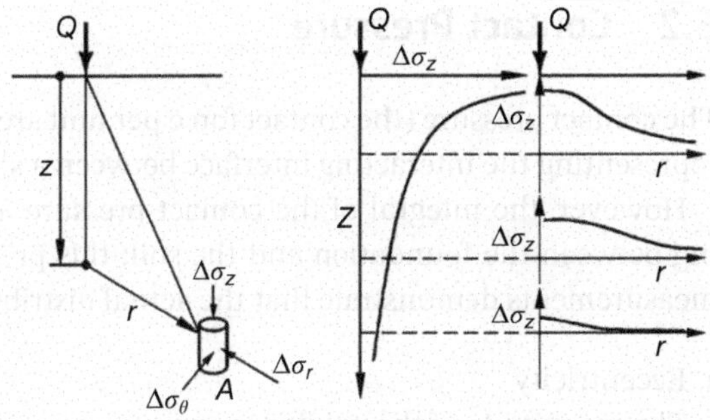

Figure 6.1 Distribution of stress at different depths.

Example 6.1 Consider a nonhomogeneous soil comprising a layer of 7 m sand underlying by clay. The GWT is located 3 m below the ground surface.

a) Plot the total and effective stress as well as hydrostatic water pressure up to $Z = 10$ m
b) Plot the effective stress before and after a single foundation (1×1 m) and a load of 10 tons
c) Plot the effective stress before and after an extended fill ($h = 3$ m) and a specific weight of 2 ton/m³

$$\gamma_{\text{sand-saturated}} = 1.9\ \text{ton/m}^3, \quad \gamma_{\text{sand-wet}} = 1.7\ \text{ton/m}^3, \quad \gamma_{\text{clay-saturated}} = 2\ \text{ton/m}^3$$

Sample calculation for $Z = 5$ m

$$\text{Total stress} = 3 \times 1.7 + 2 \times 1.9 = 8.9\ \text{ton/m}^2,$$

$$\text{Hydrostatic stress} = 2 \times 1 = 2\ \text{ton/m}^2,$$

$$\text{Effective stress} = 8.9 - 2 = 6.9\ \text{ton/m}^2$$

Single foundation: $\Delta\sigma = \frac{10}{(1+5)(1+5)} \cong 0.3\ \text{ton/m}^2 \rightarrow$ Effective stress $= 6.9 + 0.3 = 7.2\ \text{ton/m}^2$
Extended fill: $\Delta\sigma = 3 \times 2 = 6\ \text{ton/m}^2 \rightarrow$ Effective stress $= 6.9 + 6 = 12.9\ \text{ton/m}^2$

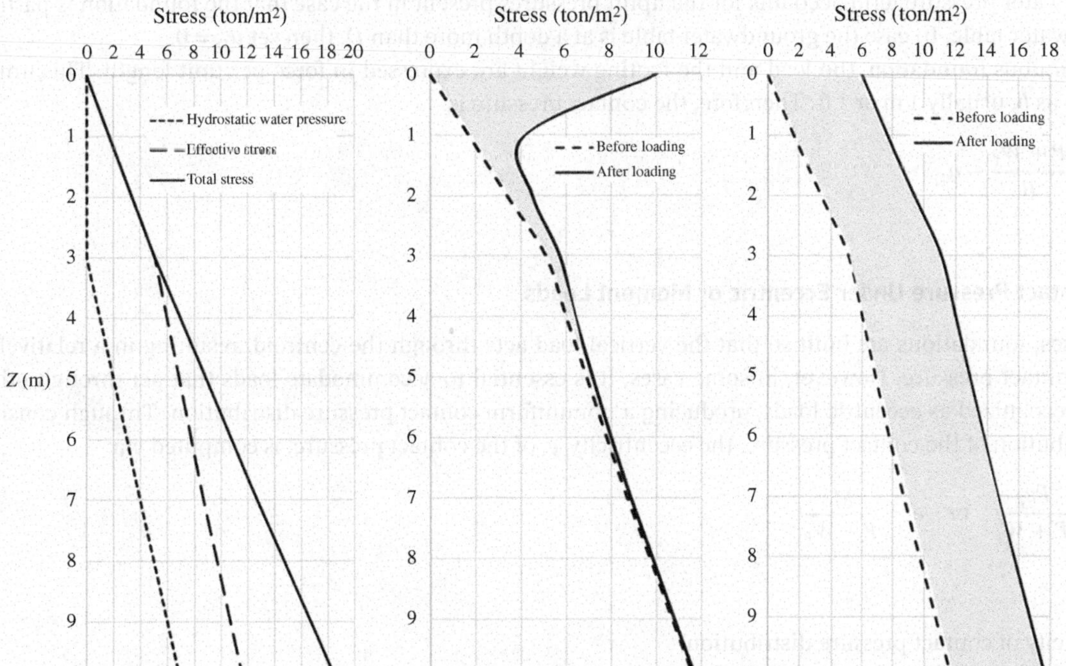

6.2 Contact Pressure

The contact pressure (the contact force per unit area between the foundation base and the subsoil) is an essential parameter representing the interacting interface between a shallow foundation and the underlying soil.

However, the integral of the contact pressure across a shallow foundation's base area must be equal to the force acting between the foundation and the soil; this pressure is not necessarily distributed evenly. Analytical studies and field measurements demonstrate that the actual distribution depends on various factors, comprising:

- Eccentricity
- The magnitude of the applied moment
- The structural rigidity of the foundation
- Stress–strain properties of the soil
- The roughness of the bottom of the foundation

6.2.1 Contact Pressure Under Axial Vertical Loads

The contact pressure beneath a shallow foundation is:

$$q = \frac{P + W_f}{A_f} - u_D \tag{6.9}$$

where

q = contact pressure
P = applied load
W_f = foundation weight + soil weight above the foundation
A_f = the foundation area
u_D = pore water pressure at the bottom of the foundation

The pore water pressure term accounts for the uplift pressures present in the case that the foundation is partially under the groundwater table. In case the groundwater table is at a depth more than D, then set $u_D = 0$.

In a continuous foundation, the load and the footing weight are expressed in force per unit length. This unit length is determined as b, usually 1 m or 1 ft. Therefore, the contact pressure is:

$$q = \frac{P + W_f}{B} - u_D \tag{6.10}$$

6.2.2 Contact Pressure Under Eccentric or Moment Loads

In most cases, foundations are built so that the vertical load acts through the centroid, resulting in a relatively uniform spread of contact pressure. However, in some cases, it is essential to accommodate loads that act through other points, which are recognized as eccentric loads, producing a nonuniform contact pressure distribution. Through considering the linear distribution of the contact pressure, the eccentricity, e, of the contact pressure, is computed via:

$$e = \frac{Pe_L}{P + W_f} \quad \text{or} \quad e = \frac{M}{P + W_f} \tag{6.11}$$

where

e = eccentricity of contact pressure distribution
e_L = eccentricity of the load

If the eccentric or moment loads occur only in the B direction, the contact pressure distribution is shown in Figure 6.2.

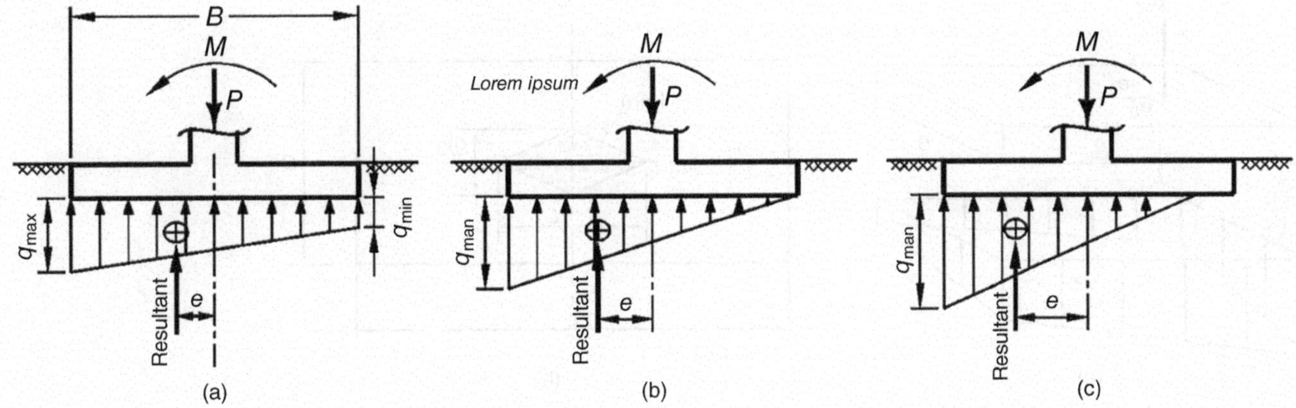

Figure 6.2 Distribution of contact pressure beneath footing with various eccentricities: (a) $e < B/6$, (b) $e = B/6$, and (c) $e > B/6$.

If $e \leq B/6$, the contact pressure distribution is trapezoidal, as shown in Figure 6.2a, and the minimum and maximum contact pressures on square or rectangular foundations are:

$$q_{\min} = \left(\frac{P + W_f}{A_f} - u_D\right)\left(1 - \frac{6e}{B}\right) \tag{6.12}$$

$$q_{\max} = \left(\frac{P + W_f}{A_f} - u_D\right)\left(1 + \frac{6e}{B}\right) \tag{6.13}$$

where

q_{min} = minimum contact pressure
q_{max} = maximum contact pressure

If $e = B/6$ (i.e., the resultant force acts at the third point of the foundation), $q_{min} = 0$, and the contact pressure distribution is triangular (Figure 6.2b). Consequently, as long as $e \leq B/6$, there will be some contact pressure distributed across the entire base area. However, when $e > B/6$, the contact pressure acts outside the third point, resulting in pressure distributions illustrated in Figure 6.2c. In theory, tension occurs on one side of the footing. However, since soil cannot withstand tension, the contact pressure on this side of the footing will disappear entirely. Furthermore, the elevated contact pressure on the opposite side may lead to significant settlement in that area. The actual behavior likely involves excessive tilting of the footing, which is undesirable. Consequently, to prevent this scenario, footings are designed with eccentric or moment loads, ensuring the following criterion is met:

$$e \leq B/6 \tag{6.14}$$

This criterion maintains compressive stresses along the entire base area. For rectangular foundations with the moment or eccentric load in the long direction, substitute L for B in Equation 6.14.

6.2.3 Contact Pressure Under Two-Way Eccentric or Moment Loads

If the resultant load acting on the base is eccentric in both the B and L directions, it must fall within the diamond-shaped kern (Figure 6.3b), for the contact pressure to be compressive along the entire base of the foundation. It falls within this kern only if the following condition is met:

$$\frac{6e_B}{B} + \frac{6e_L}{L} \leq 1.0 \tag{6.15}$$

where

e_B = eccentricity in the B direction
e_L = eccentricity in the L direction

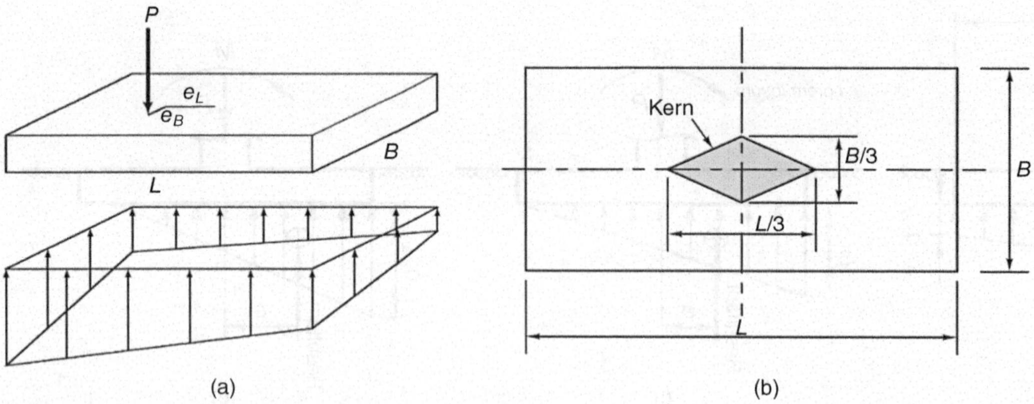

Figure 6.3 (a) Contact pressure distribution under a spread foundation, eccentric load in both directions and (b) to keep $q \geq 0$ along with the foundation base, the resultant force must be located in diamond-shaped kern.

If Equation 6.15 is satisfied, the magnitudes of q at the four corners of a square or shallow rectangular foundation are calculated via Equation 6.16 (Figure 6.3a):

$$q = \left(\frac{P + W_f}{BL} - u_D\right)\left(1 \pm \frac{6e_B}{B} \pm \frac{6e_L}{L}\right) \tag{6.16}$$

Example 6.2 A mat foundation weighing 8000 tons is supposed to support four-grain storage silos. The net weight of each silo is 3000 tons, and the weight of the internal load is 15,000 tons. The groundwater table is at great depth. Control the dimensions of the foundation in terms of one-way and two-way eccentricity and determine the minimum needed width to meet the requirements.

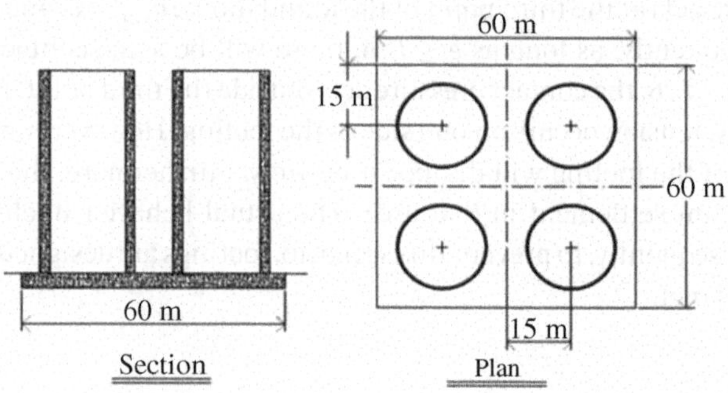

a) One-way eccentricity
   The most critical one-way eccentricity occurs when two adjacent silos are full, and the other two are empty.

$$P = 4 \times 3000 + 2 \times 15{,}000 = 42{,}000 \text{ tons}$$

$$M = 2 \times 15{,}000 \times 15 = 450{,}000 \text{ ton.m}$$

$$e = \frac{M}{P + W_f} = \frac{450{,}000}{42{,}000 + 8000} = 9 \text{ m} < \frac{B}{6} = 10 \text{ m O.K.}$$

b) Two-way eccentricity
   The most critical two-way eccentricity occurs when one silo is full, and the others are empty.

$$P = 4 \times 3000 + 1 \times 15{,}000 = 27{,}000 \text{ tons}$$

$$M_B = M_L = 1 \times 15{,}000 \times 15 = 225{,}000 \text{ ton.m}$$

$$e_B = e_L = \frac{M}{P + W_f} = \frac{225{,}000}{27{,}000 + 8000} = 6.4\ \text{m}$$

$$\frac{6e_B}{B} + \frac{6e_L}{L} = \frac{6 \times 6.43}{60} + \frac{6 \times 6.43}{60} = 1.3 \nleq 1$$

The dimensions of the foundation are not appropriate.

$$\frac{6 \times 6.43}{B} + \frac{6 \times 6.43}{B} = 1 \rightarrow B_{\min} = 77.2\ \text{m}$$

Consider $B = L = 77.5$ m

Conclusion

Although the dimensions of the foundation are satisfactory for one-way eccentricity, in the case of two-way eccentricity, the dimension should be increased to preclude any tensile stress beneath the foundation. In addition, the resultant force must be located within the kern.

6.3 Determining Bearing Capacity

Bearing capacity is one of the most essential requirements in foundation design. Ultimate bearing capacity is the highest foundation to soil interface pressure than can be tolerated by soil (q_{ult}). Safe bearing capacity (q_s) is achieved by dividing the ultimate bearing capacity (q_{ult}) by the factor of safety (FS). Safe bearing capacity (q_s) is used to determine the dimensions of the foundation.

$$q_s = \frac{q_{ult}}{FS} \tag{6.17}$$

To limit the uniform or nonuniform settlements of the foundation to an allowable level, the safe bearing capacity (q_s) is reduced to an allowable bearing value (q_a).

The factor of safety depends on various factors. Table 6.1 encapsulates some of the factors and their variation ranges.

There are five ways to designate the soils' bearing capacity:

1) Static analysis, based on analytical methods of subsoil soil rupture in the limit state and based on soil shear strength parameters, C, and φ
2) Complementary methods include using in situ test data such as CPT, PMT, VST, and SPT directly and indirectly
3) Presumed values recommended by codes and handbooks
4) Using the results of loading tests that directly designate the bearing capacity of the soil
5) Numerical analyzes that consider various mechanisms related to the limit state and failure use approximate solutions, based on finite element numerical methods or finite-difference

Bearing capacity computation includes two components. The first is the base resistance, which is also known as toe or tip resistance. The second is circumferential resistance, or side, as well as shaft resistance.

Table 6.1 Influential factors on bearing capacity and their variation range.

Factor	Range of variation
Soil type	Coarse-grained ⟷ fine-grained
Geotechnical site investigation	Extensive ⟷ limited
Variation of soil (in plan and profile)	Uniform ⟷ variable
Importance of structure and foundation failure consequences	Low ⟷ high
Possibility of extraordinary loads	Low ⟷ high
Factor of safety variation	2 ⟷ 4

6.3.1 Base Resistance

As the load on a foundation element gradually increases from an initial zero value, the element initially undergoes some degree of settlement. Once the load surpasses a certain threshold, a portion of the soil enters the plastic range, exhibiting plastic deformation. With further increases in load, these plastic zones within the soil mass may expand and extend towards a free boundary, typically the ground surface. At this stage, significant settlements can occur with minimal to no additional load increments. Bearing capacity failure of a foundation element is said to occur when it undergoes substantial settlements as described above (Salgado 2008).

General shear failure, local shear failure, and punching shear failure are three types of bearing capacity failures (Figure 6.4). A typical load–displacement curve for failure modes is also presented in Figure 6.4.

General shear failure is the most common failure mode observed in spread footings, particularly in relatively incompressible soils. General shear failure occurs in cases where stiff and saturated normally consolidated clays are subjected to rapid loading, and where undrained loading conditions prevail. The failure surface in such cases is well defined, and failure occurs abruptly. Adjacent to the spread foundation, a bulge may appear on the ground surface. It is important to note that while bulges may be visible on both sides of the footing, ultimate failure typically occurs on one side only, often accompanied by foundation rotation.

The opposite extreme is punching shear failure, which happens in very loose sands, or in a thin crust of firm soil underlain by a very weak soil, or in weak clays loaded under slowly drained conditions. The high compressibility of such soil profiles results in considerable settlements and poorly defined vertical shear surfaces. Little or no bulging happens at the ground surface, and failure develops gradually.

Local shear failure is an intermediate phenomenon between general shear failure and punching shear failure. In local shear failure, shear surfaces are well-defined beneath the spread footing and disappear near the ground surface. A small bulge may be noticed, but a significant settlement, perhaps on the order of half the footing width, is required prior to a precise shear surface forming near the ground. Sudden failure does not occur in contrast to the general shear failure case.

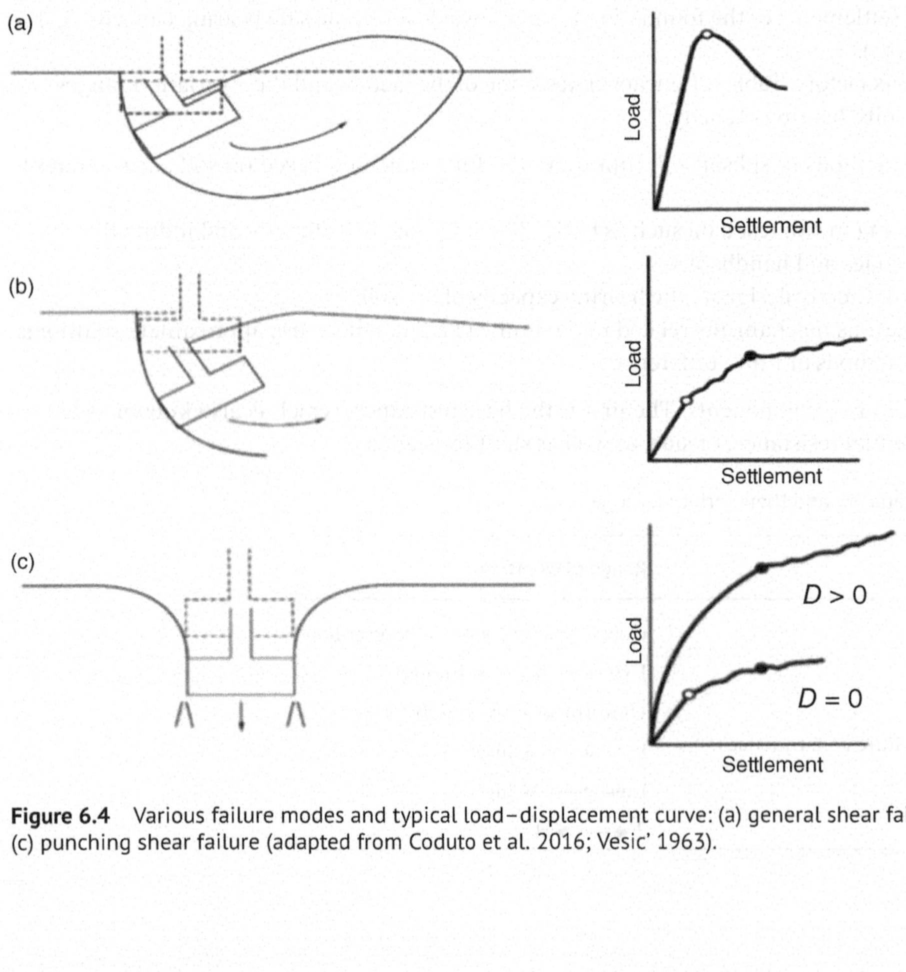

Figure 6.4 Various failure modes and typical load–displacement curve: (a) general shear failure, (b) local shear failure, and (c) punching shear failure (adapted from Coduto et al. 2016; Vesic' 1963).

6.3.1.1 Determination of Bearing Capacity Using Static Analysis

Terzaghi (1943) suggested a formula, which for the general shear failure mechanism beneath strip footings, is:

$$q_{\text{ult}} = CN_c + \overline{q}N_q + 0.5\gamma BN_\gamma \tag{6.18}$$

where

q_{ult} = ultimate bearing capacity
C = soil cohesion
$\overline{q}$ = surcharge around the foundation (γD_f)
γ = average effective unit weight of the soil beneath and around the foundation
B = foundation width
D_f = embedment depth of the foundation
N_c, N_q, N_γ = nondimensional factors as exponential functions of φ (see Table 6.2)
φ = soil internal friction angle

Terzaghi also recommended a revised formula for various shapes of foundations in case of punching failure occurring. Figure 6.5a shows the geometry of Terzaghi's bearing capacity formulas. Later on, Meyerhof (1956), Brinch Hansen (1970), and Vesić (1975) applied modifications for shape, depth, inclination, ground, and base factors to Terzaghi's original bearing capacity formula, which was conservative. A summary of these factors is presented in Table 6.3.

In common geotechnical engineering practice, the following equation is used (CFEM 2006):

$$q_{\text{ult}} = CN_c s_c i_c d_c + qN_q s_q i_q d_q + 0.5\gamma BN_\gamma s_\gamma i_\gamma d_\gamma \tag{6.19}$$

Table 6.2 Nondimensional factors for bearing capacity (adapted from CFEM 2006).

φ	N_c	N_q	N_γ	φ	N_c	N_q	N_γ
20	15	6	3	33	39	26	24
21	16	7	3	34	42	29	29
22	17	8	4	35	46	33	34
23	18	9	5	36	51	38	40
24	19	10	6	37	56	43	47
25	21	11	7	38	61	49	56
26	22	12	8	39	68	56	67
27	24	13	9	40	75	64	80
28	26	15	11	41	84	74	95
29	28	16	13	42	94	85	114
30	30	18	15	43	105	99	137
31	33	21	18	44	118	115	166
32	35	23	21	45	134	135	201

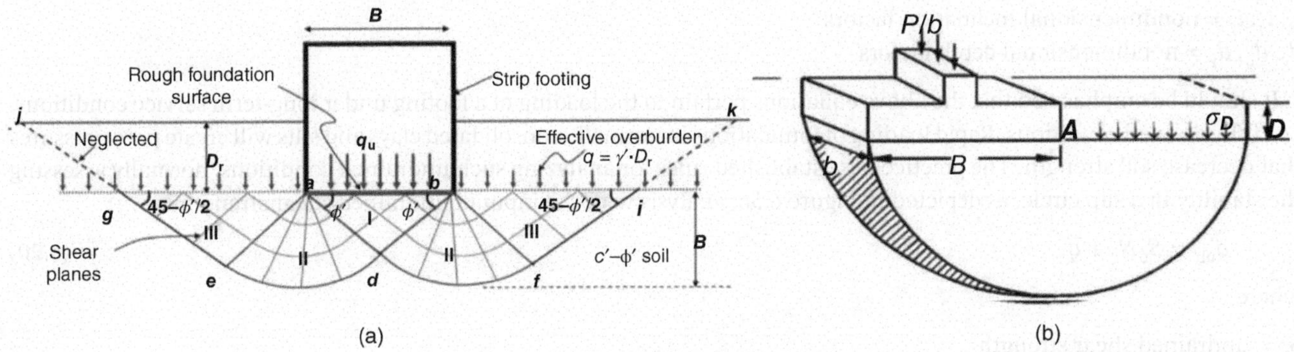

Figure 6.5 Schematic of shear failure zone: (a) drained condition and (b) undrained.

Table 6.3 Summary of different methods and their factors.

Factors	Meyerhof	Vesic'	Hansen
s_c	$1+0.2N_\phi B/L$	$1+N_q B/N_c L$	$1+N_q B/N_c L$
s_q	$1+0.1N_\phi B/L$	$1+\dfrac{B\tan\phi}{L}$	$1+\dfrac{B\sin\phi}{L}$
s_γ	$1+0.1N_\phi B/L$	$1-0.4B/L\geq 0.6$	$1-0.4B/L\geq 0.6$
i_c	$\left[1-\dfrac{\theta^\circ}{90}\right]^2$	$i_q-\dfrac{1-i_q}{N_q-1}$	$i_q-\dfrac{1-i_q}{N_q-1}$
i_q	$\left[1-\dfrac{\theta^\circ}{90}\right]^2$	$\left[1-\dfrac{H}{P+A_f c_a\cot\phi}\right]^m$	$\left[1-\dfrac{0.5H}{P+A_f c_a\cot\phi}\right]^{\alpha_1}$
i_γ	$\left[1-\dfrac{\theta^\circ}{\phi^\circ}\right]^2$	$\left[1-\dfrac{H}{P+A_f c_a\cot\phi}\right]^{m+1}>0$	$\left[1-\dfrac{0.7H}{P+A_f c_a\cot\phi}\right]^{\alpha_2}$
d_c	$1+0.2(N_\phi)^{0.5}D/B$	$1+0.4k$	$1+0.4k$
d_q	$1+0.1(N_\phi)^{0.5}D/B$	$1+2\tan\phi(1-\sin\phi)^2k$	$1+2\tan\phi(1-\sin\phi)^2k$
d_γ	$1+0.1(N_\phi)^{0.5}D/B$	1.0	1.0
N_c	$(N_q-1)\cot\phi$	$(N_q-1)\cot\phi$	$1.5(N_q-1)\tan\phi$
N_q	$N_\phi e^{\pi\tan\phi}$	$N_\phi e^{\pi\tan\phi}$	$N_\phi e^{\pi\tan\phi}$
N_γ	$(N_q-1)\tan(1.4\phi)$	$2(N_q+1)\tan\phi$	$1.5(N_q+1)\tan\phi$
Considerations	$N_\phi=\tan^2\left(45+\dfrac{\phi}{2}\right)$	$N_\phi=\tan^2\left(45+\dfrac{\phi}{2}\right)$ $\dfrac{D}{B}\leq 1\rightarrow k=\dfrac{D}{B}$ $\dfrac{D}{B}>1\rightarrow k=\tan^{-1}\left(\dfrac{D}{B}\right)$ $m=\dfrac{2+B/L}{1+B/L}$ A_f = Area of foundation P = Vertical force H = Horizontal force on foundation	$N_\phi=\tan^2\left(45+\dfrac{\phi}{2}\right)$ $\dfrac{D}{B}\leq 1\rightarrow k=\dfrac{D}{B}$ $\dfrac{D}{B}>1\rightarrow k=\tan^{-1}\left(\dfrac{D}{B}\right)$ $2\leq\alpha_1\leq 5$ $2\leq\alpha_2\leq 5$ A_f = Area of foundation P = Vertical force H = Horizontal force on foundation

where

s_c, s_q, s_γ = nondimensional shape factors
i_c, i_q, i_γ = nondimensional inclination factors
d_c, d_q, d_γ = nondimensional depth factors

It should be emphasized that the above equations pertain to the loading of a footing under long-term service conditions, i.e., fully drained conditions. Rapid loading of foundations in normally consolidated clays and silts will create pore pressures that decrease soil strength. The practice has established rules for analyzing such undrained conditions, normally assessing the stability in a slip circle, as depicted in Figure 6.5b; analysis with the input of undrained shear strength is:

$$q_{ult}=S_u N_c+\overline{q} \tag{6.20}$$

where

S_u = undrained shear strength
N_c = cohesion coefficient (5–6 for shallow foundations)

6.3.1.2 Comments on the Equations of Bearing Capacity

1) By increasing the soil cohesion parameter, the bearing capacity of the soil beneath the foundation increases.
2) Changes of N_c, N_q, and N_γ increase exponentially and ascendingly with the angle of friction, and therefore, bearing capacity also increases with an increasing angle of internal friction.
3) The bearing capacity of a foundation is directly influenced by the specific gravity of the surrounding soil and the bottom of the foundation. Additionally, variations in groundwater levels and changes in specific gravity can lead to a decrease or increase in the bearing capacity.
4) Increasing the depth of a foundation increases the surcharge ($\overline{q} = \gamma D_f$) and changes the failure path. As a result, it leads to an increase in the bearing capacity.
5) The width of the foundation increases the bearing capacity based on the bearing capacity's basic equation. However, Vesic (1975), has suggested limitations on its role. For widths exceeding 2 m, a reduction factor should be applied, which is partly related to N_γ:

$$r_\gamma = 1 - 0.25 \log \left(\frac{B}{2}\right) \tag{6.21}$$

6) Generally, the soil parameters such as C, φ, and γ, as well as the geometry of the foundation comprising B or L and D_f have significant roles in the determination of bearing capacity.
7) The zone of failure is usually $B/2 \tan \phi$ or $B/2 \tan(45 + \varphi/2)$, which is equal to B or $1.5B$. In case of weaker layers adjacent to the bottom of the foundation and in the influence zone of stress, the design should be done with high care and appropriate engineering judgment.
8) The relation between the angle of friction and main factors of bearing capacity is exponential, and interpolation, especially for an angle of friction more than 30°, leads to considerable errors, so it is essential to determine the factors from related equations or graphs directly.
9) In the case of a local shear failure mechanism, the N_q for more recent methods can be determined via:

$$N_q = (1 + \tan \phi')e^{\tan \phi'} \tan^2 \left(45 + \frac{\phi}{2}\right) \tag{6.22}$$

10) Other effective bearing capacity factors include multilayered soil, drainage, non-vertical, and dynamic loading conditions.
11) For strip foundations, plain strain conditions exist below the footing. Accordingly, the friction angle obtained from the triaxial test can be increased by up to 10%.
12) The equations that determine the bearing capacity, are functions of B (width), L (length) and D_f (depth) of the foundation. There are two common methods for using these equations: (i) Based on previous experiences, values of B, L and D_f are assumed, and the bearing capacity and the factor of safety calculated, and the safety factor is determined. (ii) The bearing capacity is calculated as a function of B, L, and D_f, and then the obtained equation is assessed parametrically.

6.3.2 Side or Circumferential Resistance

The bearing capacity is obtained from both circumferential resistance and base resistance sections in some foundations that have considerable embedment depths, such as semi-deep foundations and deep foundations. In such cases, the bearing capacity component due to the circumference of the foundation may have a more prominent role in the bearing capacity due to its considerable area. The frictional resistance mobilized at the soil-foundation side interface, called circumferential or side resistance, q_s. In cohesionless soils, q_s can be calculated by:

$$q_s = \sigma'_z \cdot K \cdot \tan \delta \tag{6.23}$$

where the K factor varies between K_0 and K_p depending on the nature of the lateral stress interaction of the soil and foundation. Usually, K is assumed to be equal to K_0. σ'_z is effective stress, and its average should be considered in each different layer. The δ usually ranges between $2/3\phi$ and $3/4\phi$.

In cohesive soils, the side resistance is often calculated by:

$$q_s = \alpha \cdot s_u \tag{6.24}$$

where

s_u = the undrained shear strength of the soil.
α = a dimension-less factor that is determined based on the value of s_u, and usually ranges between 0.5 and 1.

Example 6.3 Determine the ultimate bearing capacity of a foundation that has dimensions of $0.5 \times 2 \times 12$ m in three different executive conditions:

$$\gamma = 1.8\ \text{ton/m}^3, \quad S_u = z + 4\ (\text{ton/m}^2), \quad \alpha = 1, \quad N_c = 5, 9, \text{and } 12 \text{ (from shallow to deep conditions)}$$

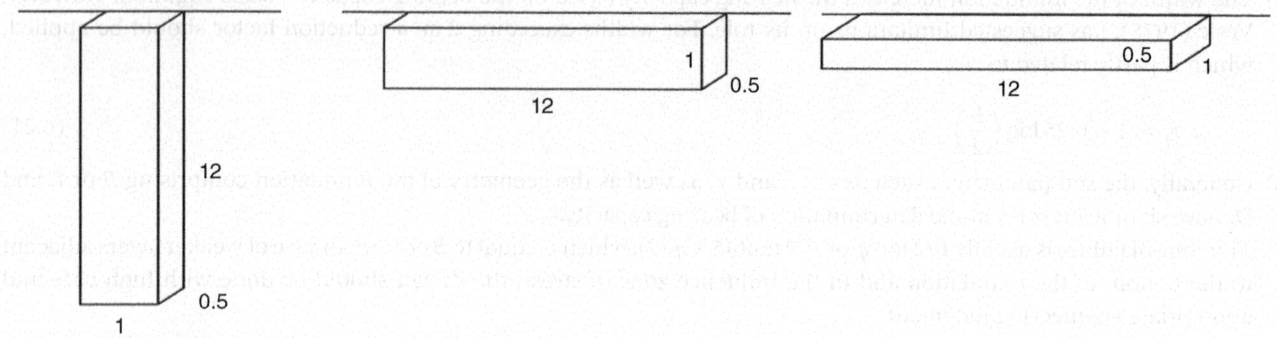

a) Shallow foundation

$$q_{ult} = S_u N_c + \overline{q}$$

$$z = 0.5\ \text{m} \rightarrow S_u = 0.5 + 4 = 4.5\ \text{ton/m}^2$$

$$q_{ult} = 4.5 \times 5 + 1.8 \times 0.5 = 23.4\ \text{ton/m}^2$$

$$P_{ult} = 23.4 \times 12 \times 1 = 280.8\ \text{tons}$$

b) Semi-deep foundation

$$q_{ult} = S_u N_c$$

$$z = 1\ \text{m} \rightarrow S_u = 1 + 4 = 5\ \text{ton/m}^2$$

$$q_{ult} = 5 \times 9 = 45\ \text{ton/m}^2$$

$$Q_{base} = 45 \times 12 \times 0.5 = 270\ \text{tons}$$

$$q_s = \alpha \cdot S_u$$

$$z = 0.5\ \text{m} \rightarrow S_u = 0.5 + 4 = 4.5\ \text{ton/m}^2$$

$$q_s = \alpha \cdot s_u = 1 \times 4.5 = 4.5\ \text{ton/m}^2$$

$$Q_{shaft} = 4.5 \times 2 \times (12 + 0.5) \times 1 = 112.5\ \text{tons}$$

$$Q_{total} = Q_{base} + Q_{shaft} = 270 + 112.5 = 382.5\ \text{tons}$$

c) Deep foundation

$$q_{\mathrm{ult}} = S_{\mathrm{u}} N_{\mathrm{c}}$$

$$z = 12\ \mathrm{m} \rightarrow S_{\mathrm{u}} = 12 + 4 = 16\ \mathrm{ton/m^2}$$

$$N_{\mathrm{c}} = 12$$

$$q_{\mathrm{ult}} = 16 \times 12 = 192\ \mathrm{ton/m^2}$$

$$Q_{\mathrm{base}} = 192 \times 1 \times 0.5 = 96\ \mathrm{tons}$$

$$q_{\mathrm{s}} = \alpha \cdot S_{\mathrm{u}}$$

$$z = 6\ \mathrm{m} \rightarrow S_{\mathrm{u}} = 6 + 4 = 10\ \mathrm{ton/m^2}$$

$$q_{\mathrm{s}} = \alpha \cdot S_{\mathrm{u}} = 1 \times 10 = 10\ \mathrm{ton/m^2}$$

$$Q_{\mathrm{shaft}} = 10 \times 2 \times (1 + 0.5) \times 12 = 360\ \mathrm{tons}$$

$$Q_{\mathrm{total}} = Q_{\mathrm{base}} + Q_{\mathrm{shaft}} = 96 + 360 = 456\ \mathrm{tons}$$

6.3.3 Conditions that Affect Bearing Capacity

In some conditions, modifications of the calculated bearing capacity are necessary. A few of these conditions are summarized as follows.

6.3.3.1 Groundwater Level Fluctuations

The presence of groundwater has the potential to decrease the cohesion (C) and increase the pore water pressure (u), which generally results in a reduction of the soil's bearing capacity. The ideal condition for achieving higher load-bearing capacity beneath a foundation is to ensure that the groundwater level does not intersect the failure zone, which extends vertically about B to $1.5B$ (B being the width of the foundation) below the foundation.

When calculating the long-term bearing capacity in the presence of high groundwater levels, the effective specific gravity of the soil is considered. Corrections are made to the basic equation of bearing capacity because the groundwater level determined during geotechnical investigations is not constant and typically varies. It is crucial to account for the possibility that the groundwater level may reach the failure zone beneath the foundation.

6.3.3.2 Layered Subsoil

Soils are often characterized by heterogeneity, meaning there can be different soil layers beneath a foundation with varying strength properties. When dealing with these layered soil conditions, it is important to consider mostly the impact of thick layers in the calculations, while ignoring the effect of thin layers. In such conditions, it becomes necessary to consider the weighted average of the strength properties of the different soil layers within the influence zone beneath the foundation to accurately determine the overall behavior and bearing capacity of the foundation.

$$C_{\mathrm{avg}} = \frac{C_1 H_1 + C_2 H_2 + \ldots + C_n H_n}{H_1 + H_2 + \ldots + H_n} \tag{6.25}$$

$$\phi_{\mathrm{avg}} = Arctan\left[\frac{\tan(\phi_1)H_1 + \tan(\phi_2)H_2 + \ldots + \tan(\phi_n)H_n}{H_1 + H_2 + \ldots + H_n}\right] \tag{6.26}$$

H, C, and ϕ are the thickness and shear parameters of each layer of soil within the influence zone. The influence zone can be determined via Equation 6.27, and is usually between B and $1.5B$ beneath the foundation.

$$Z_{\mathrm{inf}} = \frac{B}{2} \tan\left(45 + \frac{\phi}{2}\right) \tag{6.27}$$

In the case of more critical structures, more accurate analyses and procedures are required.

6.3.3.3 Foundation Under Eccentric Loads

In some cases, the foundation is under axial loads, and the bending moments are in one or two ways simultaneous, or the column would not be symmetric to the plan. In such cases, the dimension should be decreased considering $e_x = \frac{M_y}{P}$ and $e_y = \frac{M_x}{P}$

$$L' = L - 2e_y \tag{6.28}$$

$$B' = B - 2e_x \tag{6.29}$$

B and L are primary dimensions and B' and L' are reduced. Reduced dimensions (B' and L') must be used to calculate ultimate load and correction factors.

In this case, another approach is using reduction factors instead of reduced dimensions. The bearing capacity is calculated without considering the existence of moments or the eccentricity of vertical loads, and then the reduction factors are imposed.

$$q^*_{\text{ult}} = q_{\text{ult}} \times R_e \tag{6.30}$$

$$R_e = 1 - \frac{2e}{B} \quad \text{for cohesive soils} \tag{6.31}$$

$$R_e = 1 - \left(\frac{e}{B}\right)^{0.5} \quad \text{for non – cohesive soils} \tag{6.32}$$

q^*_{ult} is the reduced bearing capacity and is applicable for e/B between 0 and 0.3. In cases of eccentricity in both ways, the reduction factor is calculated via:

$$R_e = R_{e_x} \cdot R_{e_y} \tag{6.33}$$

Example 6.4 Determine the bearing capacity of given shallow foundation in three different conditions of the ground water table:

a) at the ground surface
b) at the foundation base
c) out of the influence zone

$$\varphi = 30°, \quad C = 0\,\text{kPa}, \quad \gamma_{\text{wet}} = 17\text{kN/m}^3, \quad \gamma_{\text{sat}} = 19\text{kN/m}^3, \quad \text{FS} = 3$$

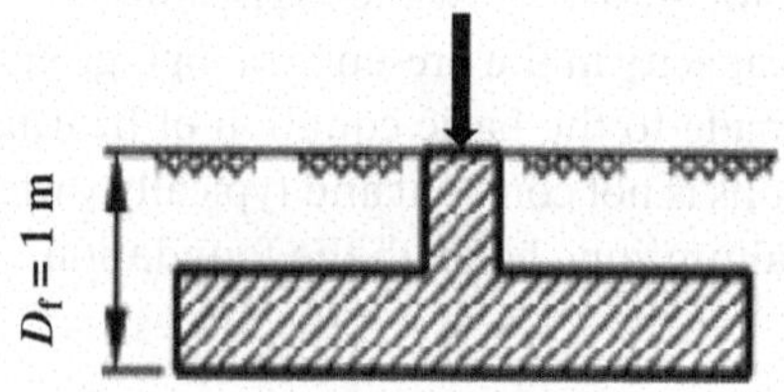

B × L = 2 × 2 m

$$C = 0 \rightarrow q_{\text{ult}} = qN_q s_q i_q d_q + 0.5\gamma B N_\gamma s_\gamma i_\gamma d_\gamma$$

$$\varphi = 30° \rightarrow N_q = 18, \quad N_\gamma = 15$$

$$i_q = i_\gamma = 1$$

$$\text{Meyerhof: } N_\phi = \tan^2\left(45 + \frac{\phi}{2}\right) = \tan^2\left(45 + \frac{30}{2}\right) = 3$$

$$s_q = s_\gamma = 1 + \frac{0.1 N_\phi B}{L} = 1 + 0.1 \times 3 \times \frac{2}{2} = 1.3$$

$$d_q = d_\gamma = 1 + \frac{0.1 (N_\phi)^{0.5} D}{\text{B}} = 1 + \frac{0.1 (3)^{0.5} 1}{2} = 1.087$$

a) $q_{ult} = 1 \times (19 - 10) \times 18 \times 1.3 \times 1 \times 1.087 + 0.5 \times (19 - 10) \times 2 \times 15 \times 1.3 \times 1 \times 1.087 = 419.7\,\text{kPa}, q_s = \frac{419.7}{3} = 139.9\,\text{kPa}$

b) $q_{ult} = 1 \times 17 \times 18 \times 1.3 \times 1 \times 1.087 + 0.5 \times (19 - 10) \times 2 \times 15 \times 1.3 \times 1 \times 1.087 = 623.2\,\text{kPa}, q_s = \frac{623.2}{3} = 207.7\,\text{kPa}$

c) $q_{ult} = 1 \times 17 \times 18 \times 1.3 \times 1 \times 1.087 + 0.5 \times 17 \times 2 \times 15 \times 1.3 \times 1 \times 1.087 = 792.7\,\text{kPa}, q_s = \frac{792.7}{3} = 264.2\,\text{kPa}$

6.3.4 In Situ Test Application in Determination of Bearing Capacity

In recent years, there has been a growing recognition of the value of in situ test methods as complementary tools to static analysis in determining bearing capacity. In situ testing techniques have witnessed increased utilization in geotechnical design, primarily due to the rapid advancement of in situ testing instruments. These advancements have facilitated a more accurate evaluation of the actual behavior of soil. This increased utilization of in situ testing techniques has been driven by a better understanding of the limitations and inadequacies of conventional laboratory testing methods. Researchers, such as Wroth (1988), have highlighted the need for improved evaluation techniques, leading to the greater adoption of in situ testing methods. The realization of these limitations has further emphasized the significance of incorporating in situ testing to enhance geotechnical design practices.

There are two approaches in estimating the bearing capacity of foundations via in situ testing:

1) **Direct Approach:** The bearing capacity is directly estimated from in situ testing records.
2) **Indirect Approach:** In situ testing records are initially used to estimate the soil strength parameters, such as cohesion (c), internal friction angle (ϕ), or any other pertinent soil parameters, such as density (γ) and Young's modulus (E_s); subsequently, the bearing capacity is calculated through static analysis methods. Due to the considerable uncertainties of static analysis methods, the direct methods are more common than the indirect approaches in geotechnical practice.

6.3.4.1 Direct Method: Standard Penetration Test

The Standard Penetration Test (SPT) is a crucial tool for determining the bearing capacity of both shallow and deep foundations. One of its key applications is in calculating the allowable bearing capacity of soils, which can be accomplished using Meyerhof relations and referring to Figure 6.6 (CFEM 1992). Figure 6.6 provide a graphical representation of the relationship between foundation dimensions, the N value obtained from SPT, and the allowable bearing capacity. These figures offer reliable values when the groundwater level is sufficiently low and does not intersect the failure zone beneath the foundation. However, caution should be exercised, as more than half of the obtained values should not be used when groundwater intrusion occurs. It is important to note that the reliability of the allowable bearing capacity is contingent upon the settlement being less than 25 mm. Furthermore, N values measured at different depths must be normalized to account for effective stress effects, and energy corrections should be applied to ensure accurate results.

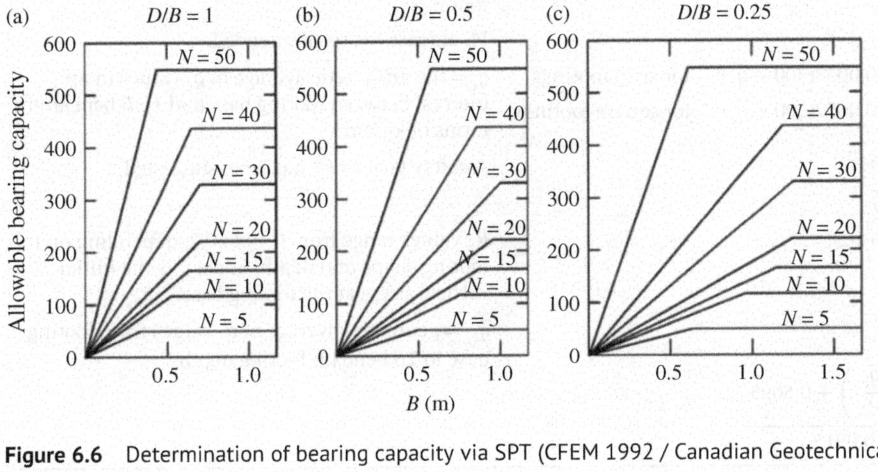

Figure 6.6 Determination of bearing capacity via SPT (CFEM 1992 / Canadian Geotechnical Society).

6.3.4.2 Direct Method: Vane Shear Test (VST)

Short-term conditions or a total stress analysis are considered for carrying capacity for foundations built on saturated clay soils. The vane shear test is used in soft clay soils to calculate bearing capacity in both the laboratory and the field. In this case, the bearing capacity will be calculated via the following equation:

$$q_{\text{ult}} = \mu S_{\text{u}} N_{\text{c}} \left(1 + 0.2\frac{D}{B}\right)\left(1 + 0.2\frac{B}{L}\right) + \sigma_0 \tag{6.34}$$

where

μ = reduction factor for VST, which can be determined via Figure 6.7
N_{c} = equals 5.7 in the Terzaghi method and 5.14 in other methods
σ_0 = the total stress at the bottom of the foundation

6.3.4.3 Direct Method: Cone Penetration Test (CPT and CPTu)

Direct CPT methods initiated theoretically or empirically relate the ultimate bearing capacity of soils, q_{ult} to the cone point resistance and q_c with some modification factors. Among various methods, some that geotechnical engineers commonly use are summarized in Table 6.4.

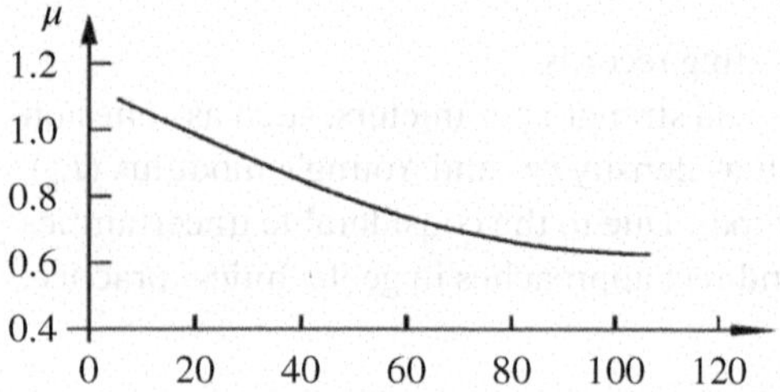

Figure 6.7 Determination of μ (CFEM 1992 / Canadian Geotechnical Society).

Table 6.4 Commonly used CPT-based methods for bearing capacity calculations.

References	Equations	Remarks
Schmertmann (1978)	$q_{\text{ult}} = \bar{q}N_q + 0.5\gamma BN_\gamma$ $N_q = N_\gamma = 1.25\sqrt{q_{c1} \times q_{c2}}$	q_{c1} = arithmetic average of q_c values in an interval between footing base and 0.5*B* beneath footing base. q_{c2} = arithmetic average of q_c values in an interval between 0.5*B* and 1.5*B* beneath footing base.
Meyerhof (1963)	$q_{\text{ult}} = \bar{q}_c\left(\frac{B}{12.2}\right)\left(1 + \frac{D_f}{B}\right)$	$\bar{q}_c$ = arithmetic average of q_c values in a zone including footing base and 1.5*B* beneath the footing. FS at least 3 is recommended.
Bowles (1996)	$q_{\text{ult}} = 28 - 0.0052(300 - \bar{q}_c)^{1.5}$, for strip footings $q_{\text{ult}} = 48 - 0.0052(300 - \bar{q}_c)^{1.5}$, for square footings	$\bar{q}_c$ = the arithmetic average of q_c values in an interval between footing base and 1.5*B* beneath in terms of kg/cm².
CFEM (2006)	$q_{\text{ult}} = 0.30\,\bar{q}_c$ $q_{\text{all}} = 0.10\,\bar{q}_c$	A safety factor of 3 has been suggested.
Tand et al. (1994)	$q_{\text{ult}} = R_k q_c + \sigma_{v0}$	R_k values range from 0.14 to 0.2, depending on the footing shape and depth, and σ_{v0} is the initial vertical stress at the footing base.
Eslami and Gholami (2006)	$q_{\text{ult}} = \bar{\alpha} \times \bar{q}_{cg}$ $\varphi = \dfrac{\log\left(\dfrac{\bar{q}_c}{\gamma' z}\right) + 0.5095}{0.0915}$	$\bar{q}_{cg}$ = geometric average of q_c values from footing base to 2*B* beneath footing depth.

Figure 6.8 Bearing capacity correlation factor for relating q_{ult} to $\overline{q}_{cg}$ (adapted from Eslami and Gholami 2006).

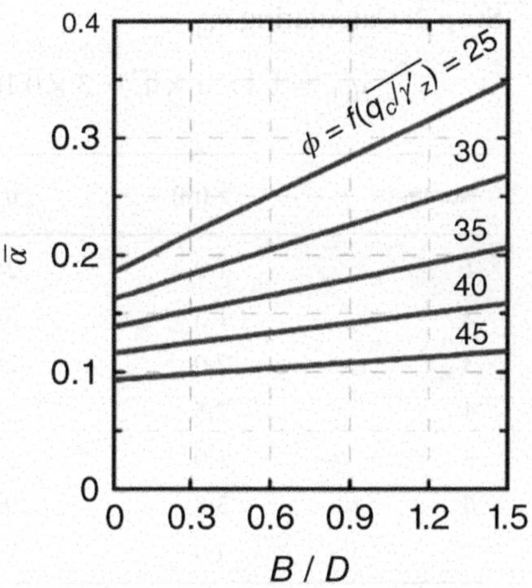

A summarized step-by-step procedure of the Eslami and Gholami (2006) method listed in the above table is as follows:

1) The zone located between the foundation base to 2*B* beneath can be divided into sublayers. The values of $\overline{q}_{cg}$ and $(\overline{q}_c/\gamma' z)_{cg}$ in this interval are calculated.
2) The average φ angle is obtained by using $(\overline{q}_c/\gamma' z)_{cg}$.
3) Based on D/B and φ values, from Figure 6.8, $\overline{\alpha}$ can be obtained.
4) Finally, the ultimate bearing capacity is calculated as: $q_{ult} = \overline{\alpha} \times \overline{q}_{cg}$.

Example 6.5 According to the given graph of CPT, find bearing capacity of the square shallow foundation if: $B = 3$ m, $D_f = 0.76$ m, GWT = 4.9 m, and $\gamma = 15.5$ kN/m^3.

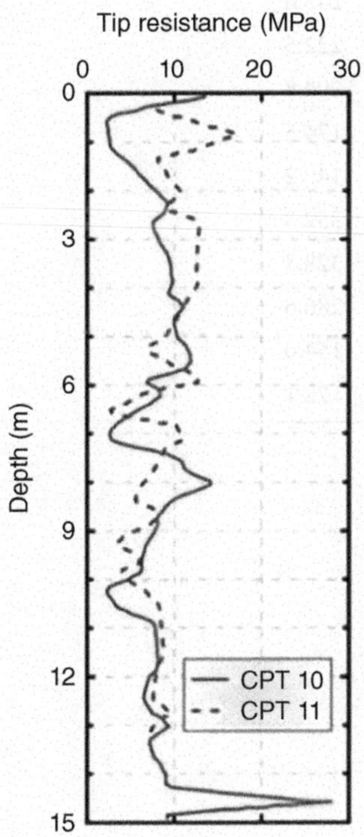

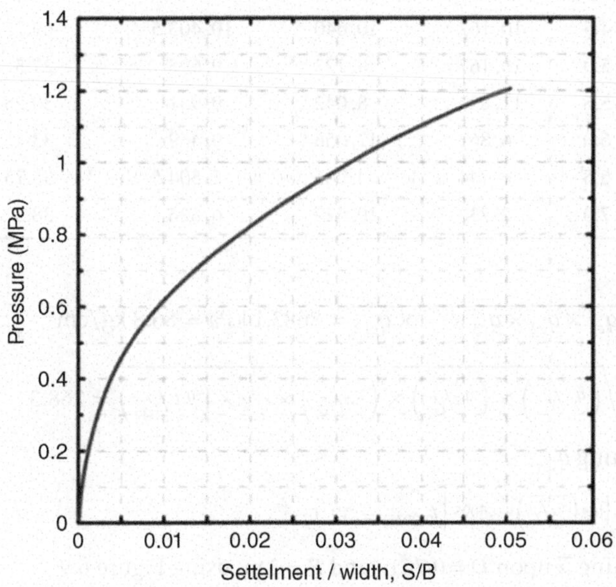

- See Table 6.4 – CFEM (2006)

Step 1: Computing $\overline{q_c}$ from the bottom level of the foundation down to 1*B* under the bottom level of the foundation

$$\overline{q_c} = 9.6 \text{ MPa}$$

Step 2: Calculating q_{ult}

$$q_{ult} = FS \times \alpha \times \overline{q_c} = 3 \times 0.10 \times 9.6 = 2.9\ \text{MPa}$$

Number	z (m)	q_{c-1} (kPa)	q_{c-2} (kPa)	q_{c-avg} (kPa)
1	1.0	2494	15,834	9,164
2	1.5	5445	8,751	7,098
3	2.0	7806	10,522	9,164
4	2.5	8161	10,876	9,518.5
5	3.0	8161	12,528	10,344.5
6	3.5	9577	12,647	11,112
7	4.0	9577	12,056	10,816.5

- Eslami and Gholami (2006) method:
Step 1: Calculating $\overline{q_{cg}}$ and $\overline{q_c/\gamma'_z}$ in the influence zone from the ground level down to 2*B* beneath the foundation.

Number	z(m)	q_{c-1} (kPa)	q_{c-2} (kPa)	q_{c-avg} (kPa)	Y'_z (kPa)	q_{c-avg}/Y'_z
1	0.5	2,258	9,105	5,681.5	7.75	733.1
2	1.0	2,494	15,834	9,164	15.5	591.2
3	1.5	5,445	8,751	7,098	23.25	305.3
4	2.0	7,806	10,522	9,164	31	295.6
5	2.5	8,161	10,876	9,518.5	38.75	245.6
6	3.0	8,161	12,528	10,344.5	46.5	222.5
7	3.5	9,577	12,647	11,112	54.25	204.8
8	4.0	9,577	12,056	10,816.5	62	174.5
9	4.5	10,167	10,640	10,403.5	69.75	149.2
10	5.0	10,167	9,223	9,695	27.5	352.5
11	5.5	11,820	8,042	9,931	30.25	328.3
12	6.0	6,862	12,056	9,459	33	286.6
13	6.5	6,744	4,265	5,504.5	35.75	154.0
14	7.0	2,730	10,522	6,626	38.5	172.1

$$\overline{q_{cg}} = \sqrt[12]{q_{c1} \times q_{c2} \times q_{c3} \times \ldots \times q_{c12}} = 8683.1\ \text{kPa} = 86.8\ \text{kg/cm}^2$$

$$\overline{q_c/\gamma'_z} = \sqrt[12]{\left(q_{c1}/\gamma'_{z_1}\right) \times \left(q_{c2}/\gamma'_{z_2}\right) \times \left(q_{c3}/\gamma'_{z_3}\right) \times \cdots \times \left(q_{c12}/\gamma'_{z_{12}}\right)} = 268.3$$

Step 2: Calculating q_{ult}

$$q_{ult} = \left[\log\left(\overline{q_c/\gamma'_z}\right) + 0.5095\right] / 0.0915 = 32.1$$

Step 3: Computing $\overline{\alpha}$ upon $D = 0.76$ m and $B = 3$ m, using Figure 6.9

$$\overline{\alpha} = 0.16$$

Step 4: Computing q_{ult}

$$q_{ult} = \overline{\alpha} \times \overline{q_{cg}} = 0.16 \times 86.8 = 13.9\ \text{kg/cm}^2 = 1.39\ \text{MPa}$$

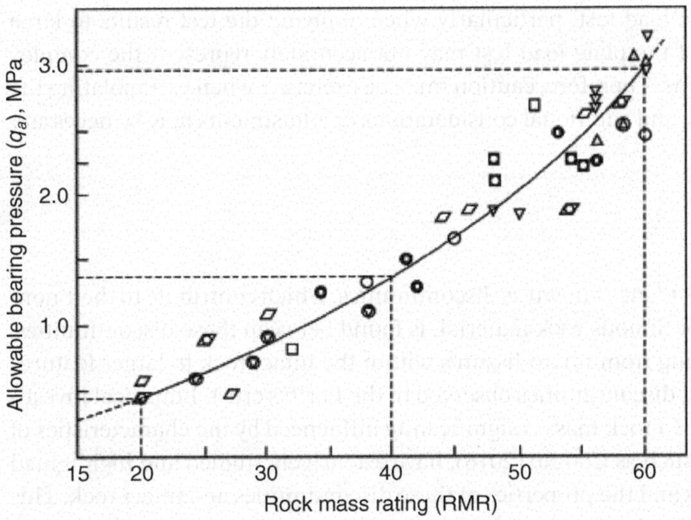

Figure 6.9 Allowable bearing pressure based on the rock mass rating and natural moisture content (Mehrotra 1992 / Mehrotra).

6.3.4.4 Direct Method: Pressure Meter Test (PMT)

The bearing capacity of shallow foundations can be calculated using the results of PMT via the following equation:

$$q_{\text{ult}} = K_g(p_l - p_0) + \sigma_0 \tag{6.35}$$

where

K_g = factor, which is a function of the geometry of the foundation and type of the soil; for square footings, it ranges between 0.8 and 3.5, and for strip footings, it ranges between 0.8 and 1.8
p_l = limit pressure
p_0 = the total horizontal stress at the bottom of the foundation
σ_0 = vertical surcharge at the level of the foundation bottom

6.3.4.5 Direct Method: Plate Load Test (PLT)

Plate Load Test (PLT) is highly advantageous for determining the bearing capacity of foundations due to its ability to simulate loading conditions that closely resemble those experienced by actual foundations. Consequently, its direct application for assessing bearing capacity is more prominent compared to other in situ tests. In cohesive soils or in soils where undrained conditions exist, the bearing capacity is not influenced by geometrical characteristics. Accordingly, the bearing capacity of both the plate and foundation can be considered equal. This means that the bearing capacity derived from the PLT can be directly applied to both the plate and foundation without differentiation (Equation 6.36).

$$q_{\text{ultP}} = q_{\text{ultF}} \tag{6.36}$$

where

q_{ultP} = Bearing capacity from the plate load test
q_{ultF} = Bearing capacity of foundation

However, for granular soils or where drained conditions exist, q_{ultF} is modified by dimensions of the plate and the foundation, Equation 6.37 is used:

$$q_{\text{ultF}} = q_{\text{ultP}}\frac{B_\text{F}}{B_\text{P}} \tag{6.37}$$

where

B_P = Width of plate
B_F = Width of foundation

It is crucial to acknowledge the limitations of the plate load test, particularly when applying the test results to large foundations, such as mat foundations. The dimensions of the plate load test may not accurately represent the complex behavior and load distribution patterns of larger foundations. Therefore, caution must be exercised when extrapolating the test records from plate load tests to large-scale foundations, and additional considerations or adjustments may be necessary to ensure reliable results.

6.3.5 Bearing Capacity of Rocks

Rocks are commonly characterized by the presence of weak planes known as discontinuities, which contribute to their non-continuous nature. The intact rock, which refers to the continuous rock material, is found between these discontinuities. Discontinuities can occur in various sets and scales, ranging from micro-fissures within the intact rock to larger features like fissures, joints, shears, and faults, which are the largest discontinuities observed in the Earth's crust. Empirical investigations have consistently demonstrated that the behavior of a rock mass is significantly influenced by the characteristics of both the discontinuities and the intact rock. Researchers, such as Coduto (2016), have extensively studied and highlighted the close relationship between the behavior of rock masses and the properties of these discontinuities and intact rock. This understanding of rock mass behavior plays a crucial role in various engineering applications related to rock mechanics and geotechnical engineering.

Constructing a foundation on rocks involves three key considerations. First, rocks possess a significantly higher ability and resistance to imposed loads compared to soil. Second, the presence of discontinuities and fragments in the rock mass can significantly reduce its resistance compared to intact rock. The compressive strength of intact rocks typically ranges from 50 to 2000 kg/cm^2. Accordingly, small individual foundations can effectively bear heavy loads. However, a single fissure in a specific direction within the bedrock can compromise the stability of the entire foundation. Third, construction on rocky terrain generally takes place at natural levels with minimal alterations, whereas in soil beds, the desired conditions can be achieved by leveling the bed. The inherent ability of rocks to withstand high shear, compressive, and tensile forces allows structures built on rocky foundations to exhibit greater safety and resistance to static loading compared to structures on soil. Overall, the favorable characteristics of rocks make them suitable for supporting structures subjected to significant loads, offering enhanced stability and durability compared to foundations on soil.

A rock mass is considered an intact rock when the distances between discontinuities are more than 0.3 m. Determination of the rock's allowable bearing capacity is done via determining the discontinuity's effects on the rock's behavior. The allowed bearing capacity can be calculated approximately through Equation 6.38:

$$q_a = K_{sp} q_{u-core} \tag{6.38}$$

where

q_a = allowed bearing capacity
q_{u-core} = average uniaxial compressive strength of rock
K_{sp} = empirical factor, which ranges between 0.1 and 0.4, determined using Table 6.5.

In some cases, the intact part of the rock is very weak, and the distances between discontinuities are very low or weathered. Usually, in such cases, rock can be considered a granular mass, and the equations of soil mechanics can be used to determine the allowed bearing capacity.

Based on Bowles (1996), the bearing capacity of rocks can be calculated via Equation 6.39, which Terzaghi gives, but the factors vary from the soil.

$$q_{ult} = CN_c s_c + \bar{q} N_q + 0.5 \gamma B N_\gamma s_\gamma \tag{6.39}$$

Table 6.5 K_{sp} factors regarding distances of discontinuities.

Relative distance	Distance (m)	K_{sp}
Relatively low	0.3–1	0.1
High	1–3	0.25
Very high	>3	0.4

Table 6.6 The shape factors regarding foundation type.

The factor	Strip	Round	Square
s_c	1	1.3	1.3
s_γ	1	0.6	0.8

$$N_c = 5\tan^4\left(45 + \frac{\varphi}{2}\right) \tag{6.40}$$

$$N_q = \tan^6\left(45 + \frac{\varphi}{2}\right) \tag{6.41}$$

$$N_\gamma = N_q + 1 \tag{6.42}$$

The factors of s_c and s_γ are determined based on the type of the foundation. Table 6.6 is used in this area. Ultimately, the calculated bearing capacity must be modified via the RQD factor.

$$q_{\text{ult}}{}' = q_{\text{ult}} \times \text{RQD}^2 \tag{6.43}$$

Figure 6.9 shows the allowable bearing pressure based on the rock mass rating (RMR) and natural moisture content.

6.3.6 Allowed Bearing Capacity, Recommendations of Codes and References

Constructional codes of different countries or design guidelines and manuals usually recommend some amount for the bearing capacity of the foundation. Table 6.7 presents a summary of the allowed bearing capacity of soils and rocks that are naturally existing and common for projects.

Case Study 6.1

Collapse of Fargo Grain Elevator (Coduto et al. 2016)

One of the most dramatic bearing capacity failures was the Fargo Grain Elevator collapse of 1955. This grain elevator was constructed near Fargo, North Dakota in 1954. It was a reinforced concrete structure comprising twenty cylindrical bins and other relevant structures, all supported on a 15.8 m wide, 66.4 m long, 0.71 m thick mat foundation.

The initial average net bearing pressure, represented as $q' = q - \sigma'_{z0}$, due to the empty structure's weight, was 76.1 kPa. As grain filling commenced in April 1955, q' began to rise. Notably, in such structures, the live load (grain) significantly surpasses the dead load. By mid-June, the average net bearing pressure had tripled, reaching 227 kPa. However, this escalating pressure led to rapid settlement, culminating in the elevator's collapse in the early hours of June 12, 1955. The failure resulted in a noticeable 2-m bulge.

Crucially, no geotechnical investigation preceded the elevator's construction. Nordlund and Deere (1970) conducted a thorough post-failure investigation. Their analysis revealed predominantly saturated clays with an undrained shear strength (S_u) ranging from 30 to 50 kPa. Bearing capacity analyses using this data indicated a net nominal unit bearing capacity of 197–312 kPa. This aligned well with the observed q' at the time of failure, measuring 228 kPa on average and 249 kPa at its maximum point.

The analysis of this failure underscored the trustworthiness of bearing capacity analyses. Basic exploration and testing would have revealed shear strength values that could have predicted this failure. If such evaluations had been conducted, and the design had incorporated an appropriate safety margin, this failure could have been averted. A parallel incident occurred in 1913 at a grain elevator near Winnipeg, Manitoba, roughly 320 km north of Fargo. Intriguingly, the findings of this failure investigation were published in 1953, two years prior to the Fargo incident. This case history serves as a classic example of engineers neglecting to glean lessons from past errors.

Table 6.7 Allowed bearing capacity of some typical rocks and soils.

Type of rock or soil	q_a (MPa)
Rocks	
Hard and non-weathered igneous rocks	10
Sandstone and hard limestones	4
Schists	3
Hard shales and soft sandstones[a)]	2
Soft shales[a)]	0.6–1
Soft limestones[a)]	0.6
Non-cohesive soils[b)]	
Dense gravel or dense gravel and sand	More than 0.6
Medium-dense gravel and sand	0.2–0.6
Loose gravel and sand	Less than 0.2
Dense sand	More than 0.3
Medium-dense sand	0.1–0.3
Loose sand ($N < 10$)	Less than 0.1
Cohesive soils[c)]	
Very stiff and hard clays	0.3–0.6
Stiff clays	0.15–0.3
Medium clays	0.075–0.15
Soft clays	Less than 0.075
Very soft clays	Not applicable
Organic soils	Not applicable
Non-engineered fills	Must be measured

a) There is a possibility of heaving due to excavation or unloading. Also softening occurs in water adsorption.
b) The abovementioned amounts are for conditions in which the breadth of the foundation is above 1 m, and the groundwater table is at least equal to B, lower than the bottom of the foundation.
c) These soils in long-term and saturated conditions are prone to consolidation settlement. Moreover, as a result of variation in moisture, shrinkage or swelling occurs.

6.4 Settlement Estimation

Settlement is the vertical deflection of soils under the foundation loading. The compressibility of soils involves load–displacement relationships; thus, the trend of stress–strain curves plus soil stiffness plays a vital role in settlement computations.

6.4.1 Stress Distribution Beneath the Foundations

Various relationships have been proposed to distribute stresses for different load types and foundation shapes, easily simplified using graphs. The stress points of the vertical section can be utilized for square and strip foundations based on Boussinesq relations. The relevant diagram is shown in Figure 6.10.

Observing the pressure isobars shows that, for square foundations at a depth of about $2B$–$3B$ and the horizontal distance B from the center of the foundation, the amount of stress has reached 10% of it at the bottom of the foundation. For settlement calculations, the zone beyond this area can be omitted. These dimensions can be used as the basis for depth and width of

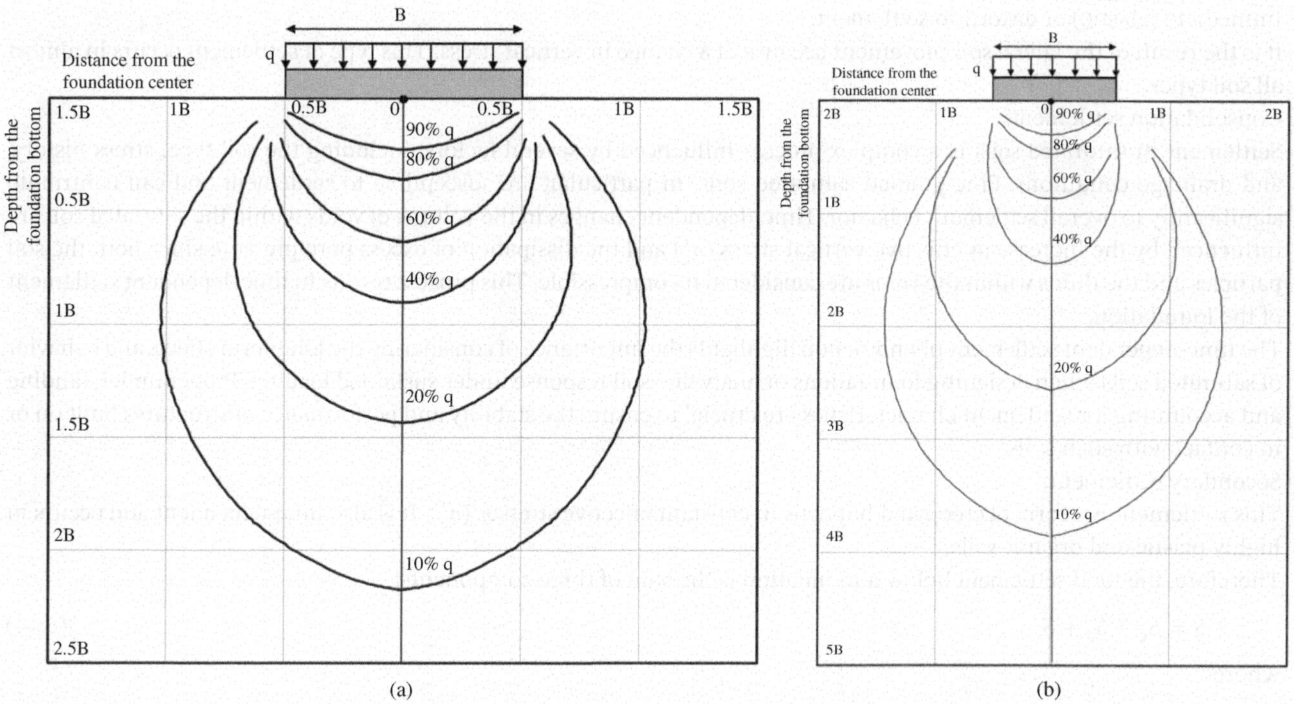

Figure 6.10 Pressure isobars regarding the Boussinesq equation for: (a) Square foundations; (b) strip foundations.

Figure 6.11 Comparison of compressible zone beneath a single foundation and a raft foundation under same stress.

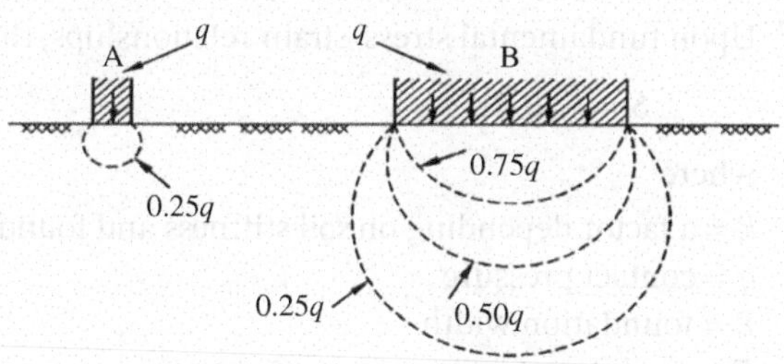

the geotechnical investigations, as well as in the determination of the proximity of two or more foundations. However, for strip foundations, at a depth of about $4B$ and the horizontal distance $1.5B$ from the center of the foundation, the amount of stress has reached 10% of it at the bottom of the foundation.

The zone of soil influenced by loading beneath a large foundation is significantly larger in volume compared to that of a smaller foundation experiencing the same intensity of contact pressure. As a result, the settlements observed under larger foundations are greater than those under smaller foundations when subjected to the same contact pressure. When multiple individual foundations are combined, their behavior resembles that of a raft foundation in terms of settlement. Figure 6.11 illustrates this concept. While raft foundations offer benefits such as enhanced bearing capacity, reduced nonuniform stresses, and resolution of local issues associated with individual foundations, it is important to note that settlements under the same stress are generally higher compared to ordinary single foundations. Nonuniform settlements can occur due to factors such as subsoil layering, variations in foundation patterns, uneven distribution of loads and stresses, and overlapping stresses from adjacent foundations. Additionally, differences in construction methods for different sections or other local factors should be carefully considered when assessing differential settlements. In summary, the occurrence of nonuniform settlements can be attributed to various factors, and their assessment requires careful consideration of subsoil conditions, foundation patterns, load distributions, and other local factors.

6.4.2 Different Settlement Types

There are several categories of foundations settlement. These are discussed below.

- Immediate (elastic) or distortion settlement:
It is the result of the lateral soil movement because of a change in vertical stress. This type of settlement occurs in almost all soil types.
- Consolidation settlement:
Settlement in saturated soils is a complex process influenced by several factors, including the soil type, stress history, and drainage conditions. Fine-grained saturated soils, in particular, are susceptible to settlement and can contribute significantly to overall settlement behavior. Time-dependent changes in the volume of voids within the saturated soil are influenced by the increase in effective vertical stress (σ_z') and the dissipation of excess pore pressure since both the soil particles and the fluids within the voids are considered incompressible. This process results in time dependent settlement of the foundation.
The time-dependent settlement phenomenon highlights the importance of considering the long-term effects and behavior of saturated soils when designing foundations or analyzing soil response under sustained loading. Proper understanding and accounting for settlement characteristics are crucial to ensure the stability and performance of structures built on or in contact with such soils.
- Secondary settlement:
This settlement is a form of creep and happens in constant effective stresses (σ_z'). It is also time-dependent and occurs in highly plastic and organic soils.
Therefore, the total settlement below a foundation is the sum of three components:

$$S = S_e + S_c + S_s \tag{6.44}$$

where

S_e = elastic settlement
S_c = consolidation settlement
S_s = secondary settlement

Upon fundamental stress–strain relationships, the settlement can be formulated as:

$$S = kqB/\mathrm{E} \tag{6.45}$$

where

k = a factor depending on soil stiffness and foundation rigidity
q = contact pressure
B = foundation width
E = modulus of soil

More investigations are required for k parameter ranges for different deposits, foundations, and settlement types.
However, two major approaches are recommended for settlement estimation:

– Modulus Based Method
In this approach, the vertical compression is:

$$\varepsilon_z = \frac{\Delta\sigma_z}{M} \tag{6.46}$$

where M is the constrained modulus (secant or tangent) equal to:

$$M = \frac{E(1-\nu)}{(1+\nu)(1-2\nu)} \tag{6.47}$$

Furthermore, the common equation for immediate and elastic settlement is:

$$S_d = \frac{qB(1-\nu^2)}{E_s}I \tag{6.48}$$

- e-log σ approach for consolidation settlement:

$$\varepsilon = \frac{\Delta e}{1+e_0} \tag{6.49}$$

$$\varepsilon_z = m_v \Delta\sigma' \cdot H; \quad \text{Or} \tag{6.50}$$

$$\varepsilon_z = \frac{C_c}{1+\varepsilon_0}\log\frac{\sigma_1'}{\sigma_0'} \tag{6.51}$$

where

C_c = compression index
m_v = coefficient of volume change
e_0 = initial void index
σ'_0 = initial vertical stress
σ'_1 = final vertical stress
H = compressive soil thickness or influence zone of the applied pressure

$$\Delta\sigma' = \sigma'_1 - \sigma'_0$$

Overall, the following factors determine the foundation settlement:

1) Foundation shape, depth, geometry, and rigidity
2) Soil stress–strain behavior
3) Stiffness-characters
4) Poisson effect
5) Stress distribution in depth
6) Drainage condition
7) Static or dynamic loading
8) Stress history
9) One-dimensional or three-dimensional deformation
10) Plasticity and organic content

Example 6.6 Determine the bearing capacity and estimate the settlement of the foundation system of a 40-story building in three conditions: (average load of each story equals to 1 ton/m^2)

a) Mat foundation, 25 × 25 m, on the ground
b) Mat foundation, 25 × 25 m, D_f = 15 m, floating foundation
c) Mat foundation, 25 × 25 m, D_f = 15 m, floating foundation in conjunction with massive soil improvement

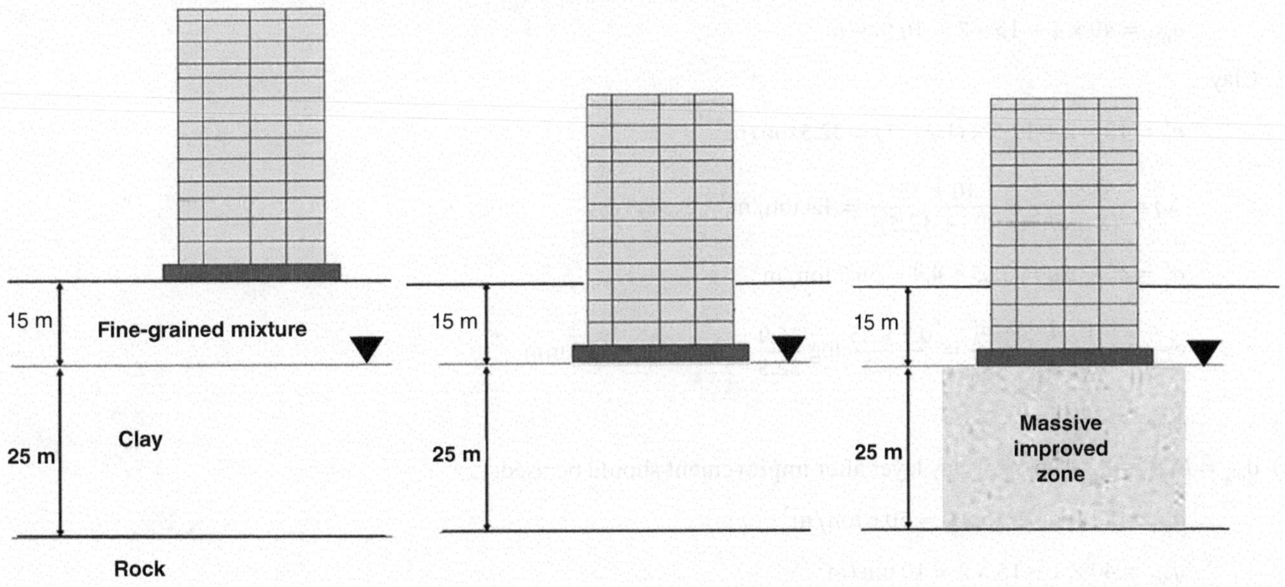

Soil layer	γ (ton/m^3)	e_0	C_c	K	$S_{u,avg}$ (kg/cm^2)	E (kg/cm^2)
Fine-grained mixture	2	—	—	0.4	0.5	200
Clay (pre Improvement)	1.2	2	0.7	—	0.25	—
Clay (Post Improvement)	1.5	2	0.22		0.4	

The influence zone for the determination of bearing capacity and estimation of the settlement are presumed B and $2B$, respectively:

a) $q_{ult} = N_c S_u + \sigma'_0$. The average of S_u between the ground surface to $Z = 40$ m should be used.

$$S_{uavg} = \frac{15 \times 0.5 + 10 \times 0.25}{25} = 0.4\ \text{kg/cm}^2 = 4\ \text{ton/m}^2$$

$$q_{ult} = 5.14 \times 4 + 0 = 20.6\ \text{ton/m}^2$$

$$q = 40 \times 1 = 40\ \text{ton/m}^2$$

Fine-grained mixture:

$$S_{\text{Fine-grained mixture}} = \frac{K \times q \times B}{E} = \frac{0.4 \times 40 \times 25}{200 \times 10} = 0.2\ \text{m} = 200\ \text{mm}$$

Clay:

$$\sigma'_0 = 15 \times 2 + 12.5 \times (1.2 - 1) = 32.5\ \text{ton/m}^2$$

$$\Delta q = \frac{q \times B^2}{(B + Z)^2} = \frac{40 \times 25^2}{(25 + 27.5)^2} = 9.1\ \text{ton/m}^2$$

$$\sigma'_1 = \sigma'_0 + \Delta q = 32.5 + 9.1 = 41.6\ \text{ton/m}^2$$

$$S_{\text{Clay}} = \frac{C_c H}{1 + e_0} \log \frac{\sigma'_1}{\sigma'_0} = \frac{0.7 \times 25}{1 + 2} \log \frac{41.6}{32.5} = 0.625\ \text{m} = 625\ \text{mm}$$

$$S_{\text{total}} = 200 + 625 = 825\ \text{mm}$$

b) $q_{ult} = N_c S_u + \sigma'_0$. The S_u of the clay layer should be used.

$$q_{ult} = 5.14 \times 2.5 + 15 \times 2 = 42.9\ \text{ton/m}^2$$

$$q_{net} = 40 \times 1 - 15 \times 2 = 10\ \text{ton/m}^2$$

Clay:

$$\sigma'_0 = 15 \times 2 + 12.5 \times (1.2 - 1) = 32.5\ \text{ton/m}^2$$

$$\Delta q = \frac{q \times B^2}{(B + Z)^2} = \frac{10 \times 25^2}{(25 + 12.5)^2} = 4.4\ \text{ton/m}^2$$

$$\sigma'_1 = \sigma'_0 + \Delta q = 32.5 + 4.4 = 36.9\ \text{ton/m}^2$$

$$S_{\text{Clay}} = \frac{C_c H}{1 + e_0} \log \frac{\sigma'_1}{\sigma'_0} = \frac{0.7 \times 25}{1 + 2} \log \frac{36.9}{32.5} = 0.322\ \text{m} = 322\ \text{mm}$$

$$S_{\text{total}} = 322\ \text{mm}$$

c) $q_{ult} = N_c S_u + \sigma'_0$. The S_u of clay layer after improvement should be used.

$$q_{ult} = 5.14 \times 4 + 15 \times 2 = 50.6\ \text{ton/m}^2$$

$$q_{net} = 40 \times 1 - 15 \times 2 = 10\ \text{ton/m}^2$$

Clay:

$$\sigma'_0 = 15 \times 2 + 12.5 \times (1.5 - 1) = 36.3\ \text{ton/m}^2$$

$$\Delta q = \frac{q \times B^2}{(B + Z)^2} = \frac{10 \times 25^2}{(25 + 12.5)^2} = 4.4\ \text{ton/m}^2$$

$$\sigma_1' = \sigma_0' + \Delta q = 36.3 + 4.4 = 40.7 \text{ ton/m}^2$$

$$S_{\text{Clay}} = \frac{C_c H}{1+e_0} \log \frac{\sigma_1'}{\sigma_0'} = \frac{0.22 \times 25}{1+2} \log \frac{40.7}{36.3} = 0.091 \text{ m} = 91 \text{ mm}$$

$$S_{\text{total}} = 91 \text{ mm}$$

6.4.2.1 Modified Equation for Settlement Estimation

In contrast to a surface footing, an embedded foundation exhibits several distinctive characteristics (Gazetas et al. 1985):

a) Soil stiffness generally increases with depth, meaning the footing loads are transmitted to a stiffer soil layer, resulting in reduced settlement, known as the Gibson effect. It is worth noting that Gazetas et al. (1985) omitted the Gibson effect for simplicity.
b) Normal soil stresses above the foundation level decrease settlement by enhancing confinement on the deforming half-space, referred to as the trench effect.
c) Part of the load on the footing might transfer through the sidewall, contingent on the shear resistance at the soil–wall interface. This load accommodation by side resistance decreases vertical settlement, termed the side wall-contact effect, supported by available analytical and numerical results.
   Gazetas et al. (1985) proposed the following equations for immediate settlement calculation of an arbitrarily shaped rigid footing embedded in deep homogeneous soil:

$$S_i = \frac{q_0 B(1-\nu^2)}{E_s} \cdot \mu_s \cdot \mu_{\text{emb}} \cdot \mu_{\text{wall}} \tag{6.52}$$

$$\mu_s = 0.45\left(\frac{A_b}{4L^2}\right)^{-0.38} \tag{6.53}$$

$$\mu_{\text{emb}} = 1 - 0.04\frac{D_f}{B}\left[1 + \frac{4}{3}\left(\frac{A_b}{4L^2}\right)\right] \tag{6.54}$$

$$\mu_{\text{wall}} = 1 - 0.16\left(\frac{A_w}{A_b}\right)^{0.54} \tag{6.55}$$

where

A_b = plan area of the foundation

$2L$ and $2B$ = the length and width of the rectangle circumscribed to the actual contact surface

A_w = the actual area of the sidewall in contact with the embedded portion of the footing

$A_b/4L^2$ is determined regarding the shape of the foundation, via Table 6.8

The geometry of the abovementioned equation is depicted in Figure 6.12.

Table 6.8 Determination of $A_b/4L^2$.

Shape of foundation	$A_b/4L^2$
Square	1
Rectangle	B/L
Circle	0.785
Strip	0

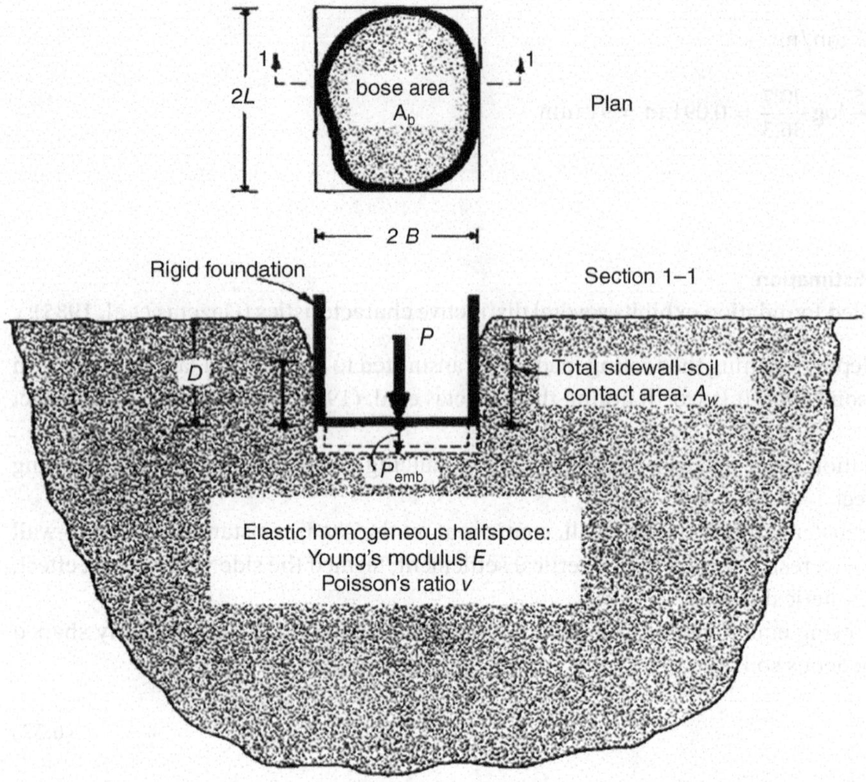

Figure 6.12 The geometry of determined foundation (Gazetas et al. 1985 / With permission of Emerald Publishing Limited).

Example 6.7 Estimate the total settlement for the square foundation with given characteristics and parameters. Assume that the weight of excavated soil is compensated with the foundation weight. The thickness of foundation is 1 m and the walls are connected to the foundation.

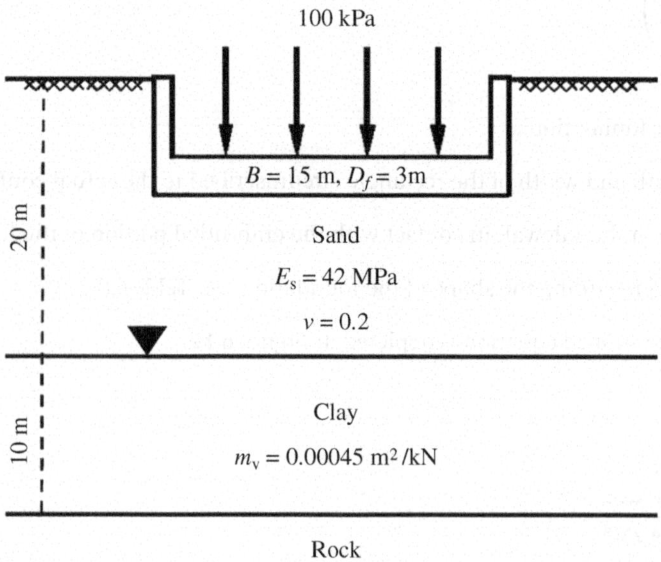

Immediate Settlement; Gazetas et al. (1985)

Method

$$\text{Square foundation: } \frac{A_b}{4L^2} = 1$$

$$\mu_s = 0.45\left(\frac{A_b}{4L^2}\right)^{-0.38} = 0.45(1)^{-0.38} = 0.45$$

$$\mu_{\text{emb}} = 1 - 0.04\frac{D_f}{B}\left[1 + \frac{4}{3}\left(\frac{A_b}{4L^2}\right)\right] = 1 - 0.04 \times \frac{3}{15} \times \left[1 + \frac{4}{3}(1)\right] = 0.981$$

$$\mu_{\text{wall}} = 1 - 0.16\left(\frac{A_w}{A_b}\right)^{0.54} = 1 - 0.16\left(\frac{4 \times 15 \times 3}{15 \times 15}\right)^{0.54} = 0.858$$

$$S_i = \frac{q_0 B(1 - \nu^2)}{E_s} \cdot \mu_s \cdot \mu_{\text{emb}} \cdot \mu_{\text{wall}} = \frac{100 \times 15 \times (1 - 0.2^2)}{42 \times 10^3} \times 0.45 \times 0.981 \times 0.858 = 0.013\text{ m} = 13\text{ mm}$$

Consolidation Settlement

$$\text{At the middle of the clay layer: } \Delta q = \frac{q \times B^2}{(B + Z)^2} = \frac{100 \times 15^2}{(15 + 22)^2} = 16.4\text{ kPa}$$

$$S_c = m_v \Delta\sigma' H = 0.00045 \times 16.4 \times 10 = 0.074\text{ m} = 74\text{ mm}$$

Total Settlement

$$S_{\text{total}} = S_i + S_c = 13 + 74 = 90\text{ mm}$$

Comments on the Equations of Settlement Estimation

1) Due to considerable differences in the settlement values obtained from various equations, the results should be used carefully, and the average of somc different methods should be used.
2) The data from the in situ test, as a complementary source, can be used to assess settlement.
3) The effect of the breadth of the foundation is considerable in the final value of the settlement. Although larger foundations provide higher bearing capacity, serviceability requirements must not be overlooked, and there should be a balance between major foundation design requirements.
4) There are two approaches in the case of multilayered soil. First, an average of layer can be used to estimate settlement. Another approach is calculating the settlement of each layer separately, and then a sum of these values is considered. The second approach is more preferred than the first one.

6.4.3 In Situ Tests Application in Settlement Estimation

In the estimation of foundation settlement under imposed loads using in situ testing, similar to the determination of bearing capacity, two significant trends can be observed: direct and indirect approaches. In the direct approach, settlement is directly estimated from in situ testing records. On the other hand, the indirect approach involves using in situ testing records to initially estimate certain soil parameters related to stiffness, particularly Young's modulus (E). Subsequently, the settlement is calculated using conventional methods. Both direct and indirect approaches provide valuable information for estimating foundation settlement. The choice of approach depends on factors such as the availability of suitable in situ testing methods, the desired accuracy of settlement estimation, and the specific project requirements.

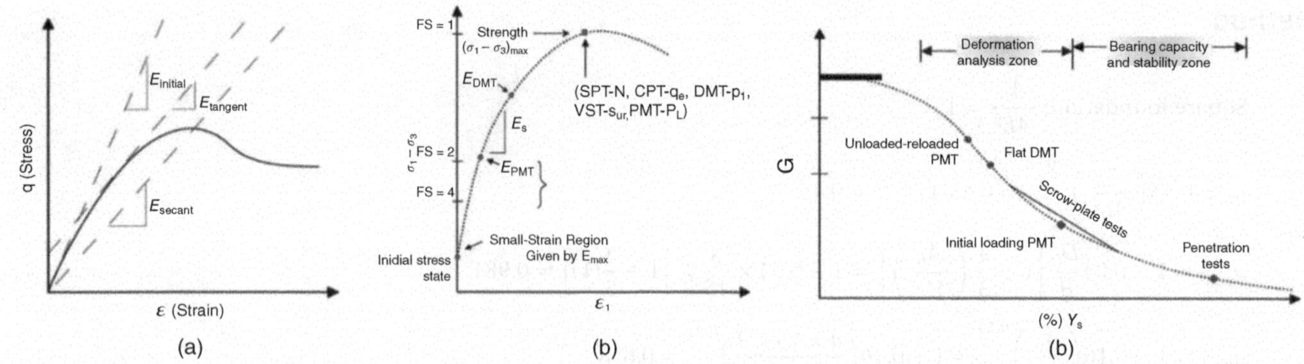

Figure 6.13 (a) q–ε curve acquired from a triaxial test (Eslami et al. 2019 / with permission of Elsevier), (b) ideal soil stress–strain relation and soil modulus at various strains, and (c) variation of shear modulus with shear strain (Sabatini et al. 2002 / U.S. DEPARTMENT OF TRANSPORTATION / Public Domain).

6.4.3.1 Determination of Soil Modulus

The determination of soil modulus in foundation engineering is a complex task due to the heterogeneity of the soil and various in situ conditions. The parameter Young's modulus (E_s) is commonly used to represent the deformability of soil and is associated with the linear elastic behavior of geotechnical materials. However, since geotechnical materials do not exhibit ideal linear elastic properties, the soil modulus does not directly correspond to the slope of the stress–strain (q–ε) curve. Instead, it is influenced by stress and strain magnitudes. Figure 6.13a illustrates the q–ε behavior observed in a triaxial test, highlighting the existence of multiple soil moduli. Figure 6.13b and c depict an ideal stress–strain relationship for soil and the variation of shear modulus with shear strain, respectively.

When drawing a slope from the origin to an arbitrary point in Figure 6.13a, a secant slope as secant modulus E_s can be used to predict the settlement of a footing loaded for the first time on a normally consolidated deposit. E_t, the tangent slope, relates to the tangent modulus and can determine incremental movements when an incremental load is applied.

6.4.3.2 Direct Methods: Application of SPT

Burland and Burbidge (1985) produced a more detailed method that uses a database of 100 case records of settlements on sands and gravels. For normally consolidated sand, the average settlement at the end of construction is calculated as:

$$S = f_s f_l f_t (qB^{0.7} I_c) \tag{6.56}$$

f_s, f_l, and f_t are correction factors, and I_c is a compression index that is related to the SPT N-value, as shown in Figure 6.14. The average N_{SPT} is taken over the depth of influence z_1, which is also obtained from the same figure in space (B plotted against z_1).

The authors concluded that the shape of the footing, the thickness of the compressive layer, and time also significantly affected predicted settlements. The correction factors express these effects. The shape factor for $L/B > 1$ is expressed as:

$$f_s = \left(\frac{1.25L/B}{(L/B) + 0.25} \right)^2 \tag{6.57}$$

For conditions in which the depth of influence of the applied pressure is greater than the thickness of the compressible material, H_s:

$$f_l = \frac{H_s}{z_1}\left(2 - \frac{H_s}{z_1}\right) \tag{6.58}$$

The time factor for estimating settlements more than three years after construction:

$$f_t = 1 + R_3 + R_t \log \frac{t}{3} \tag{6.59}$$

Conservative values of R_3 and R_t are 0.3 and 0.2, respectively, for static loading, and 0.7 and 0.8, respectively, for fluctuating loading. The effect of the previous over-consolidation of the soil is taken into account by expressing Equation 6.60 as:

$$S = (q - 2/3\sigma'_{v0})\, qB^{0.7} I_c \tag{6.60}$$

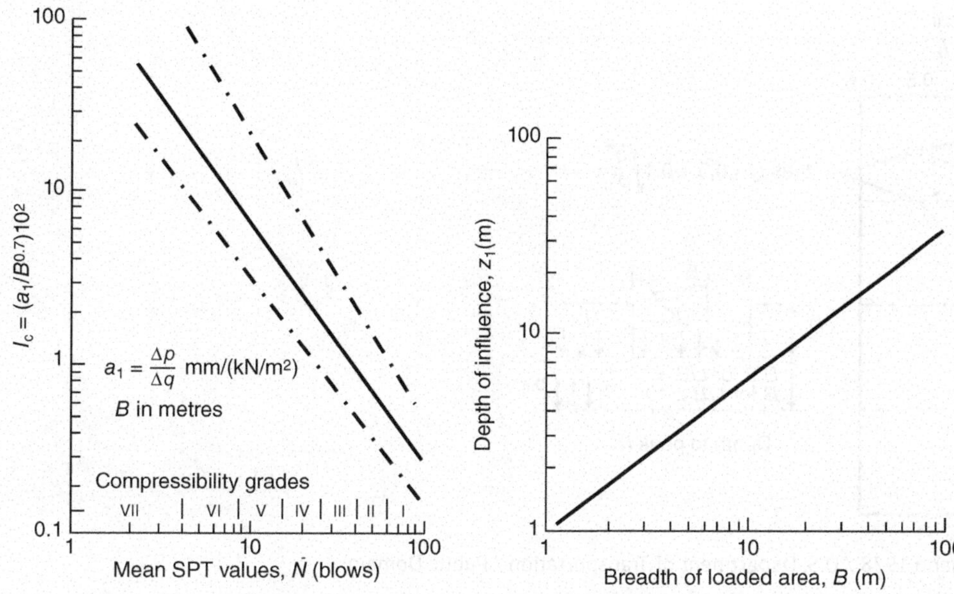

Figure 6.14 Prediction of settlements of spread footings on granular soils (Schnaid 2008 / With permission of Taylor & Francis Group).

where

σ'_{v0} = the maximum vertical effective stress at the foundation level.

Over-consolidation is known to have considerable effects on settlement, as well as for cases in which the stress increase remains below σ'_{v0}:

$$S = 1/3qB^{0.7}I_c \tag{6.61}$$

Interestingly, no corrections are made to the N-values concerning the effective overburden pressure (i.e., depth of footing and water table position). However, in deriving the compression index I_c, two cases may require the use of corrected N-values:

1) Very fine silty sand, below the water table:

$$N = 15 + 0.5\,(N_{measured} - 15) \tag{6.62}$$

2) Gravel or sandy gravel:

$$N = 1.25\,N_{measured} \tag{6.63}$$

6.4.3.3 Direct Methods: Application of CPT

Schmertmann (1978) presents Equation 6.64 to designate the settlement of sandy soils, via CPT results considering elasticity theory.

$$S = C_1C_2\Delta P\sum_{1}^{n}\frac{I_z}{C_3E}\Delta z \tag{6.64}$$

where

C_1 = the correction rate for the buried depth
C_2 = the correction rate of creep
C_3 = the correction rate for the foundation shape
ΔP = the net pressure applied to the ground by the foundation
I_z = the strain influence factor
Δz = the thickness of the examining layers
E = Young's modulus ($E = 2q_c$ here)

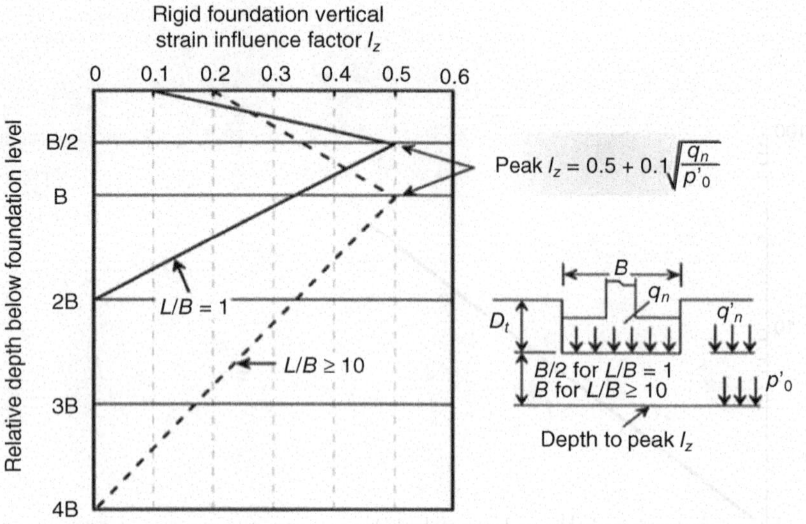

Figure 6.15 I_z diagrams (Schmertmann 1978 / U.S. Department of Transportation / Public Domain).

Figure 6.15 gives the vertical strain influence factor I_z distribution for square and strip foundations in case the ratio $L/B \geq 10$. Values for rectangular foundations for $L/B < 10$ can be calculated via interpolation.

The depths at which the maximum I_z occurs may be calculated as follows:

$$I_z = 0.5 + 0.1\sqrt{\frac{q_n}{p'_0}} \tag{6.65}$$

p'_0 is effective overburden pressure at depths $B/2$ and B for square and strip foundations, respectively. Furthermore, I_z is equal to 0.1 at the base and zero at depth $2B$ below the base for square footing, whereas for a strip foundation, it is 0.2 at the base and zero at depth $4B$.

Valikhah and Eslami (2019), building upon the work of Malekdoost and Eslami (2011), introduce a novel relationship connecting the stress exponent, modulus number, and CPT results. This relation is underpinned by a robust theoretical framework, enhancing the reliability of estimating shallow foundation settlement. Notably, their approach accounts for the scale effect, addressing differences between the influenced zone beneath the shallow foundation and the penetration cone.

In this proposed methodology, the scale effect is meticulously incorporated to provide a more realistic settlement estimation. Calibration investigations on various cases have revealed an optimal relationship that results in maximum accuracy. The terms in the relationship are: (i) α, constant coefficient, (ii) m, for estimating the modulus number, (iii) CPT diameter, b, typically 0.0357 or 0.0437 m for standard electronic penetrometers, and (iv) foundation width, B, measured in meters. Following is the relationship:

$$\alpha = 0.25b \times \left(\frac{2B+1}{3B}\right)^3 \tag{6.66}$$

$$m = \alpha \times q_t \tag{6.67}$$

q_t is corrected cone resistance for pore water pressure. The following equation has opted as a basis for the functional dependency of j and q_c:

$$j = \frac{q_c}{x + yq_c} \tag{6.68}$$

In this equation, q_c is in kPa, and two parameters of x and y are found by a study on the collected database as follows:

$$x = 0.02R_f + 0.5 \tag{6.69}$$

$$y = 7.53(\sigma'_0)^{-0.25} \tag{6.70}$$

R_f is friction ratio and σ'_0 is the initial effective soil stress in kPa at the middle of the target layer at which the settlement is estimated. The strain could be calculated via the:

$$\varepsilon = \frac{1}{2q_t j}\left[\left(\frac{\sigma'_0 + \Delta\sigma'}{100}\right)^j - \left(\frac{\sigma'_0}{100}\right)^j\right] \tag{6.71}$$

where

σ'_0 = initial effective stress in the middle of the investigated layer under in situ pressure.

$\Delta\sigma'$ indicates the increase of effective stress in the middle of the studied layer under loads of surcharges or applied stress.

The stress-influenced depth is considered as (2–3) × B beneath the footings for single foundations, (3–4) × B for the strip, and (1–1.5) × B for mat (B is the foundation width) (CFEM 2006).

The value of a settlement in the soil can be calculated from Equation 6.72:

$$\varepsilon = \frac{\Delta H}{H} \tag{6.72}$$

In this equation, ΔH is the settlement, and H is the thickness of the subject layer. The investigating zone can be divided into several thin layers to increase the accuracy of the calculations. Because CPT records, such as q_c and f_s, are measured every 20 mm depth interval, one can presume the layers have 20 mm thickness.

The proposed values of j applying slight changes are employed in the Campanella et al. (1986) and the Eslami–Fellenius (1997) charts as shown in Figure 6.16. Hence, the parameter of R_f must be replaced by f_s (sleeve friction) and q_E (effective cone resistance in Eslami–Fellenius [1997] method).

6.4.4 Differential Settlement and Distortion

Differential settlement, represented by δ_D, refers to the variation in settlement between two foundations or between two points on a foundation. Significant differential settlement can be problematic as it causes distortion in the structure, leading to challenges in terms of serviceability. There are numerous potential factors that can contribute to the occurrence of differential settlement and its effects. These factors are listed below.

- Variations in the soil profile
- Variations in the structural loads
- Design controlled by bearing capacity
- Construction tolerances
- Adjacent loading or unloading
- Groundwater table draw down
- Nonuniform construction

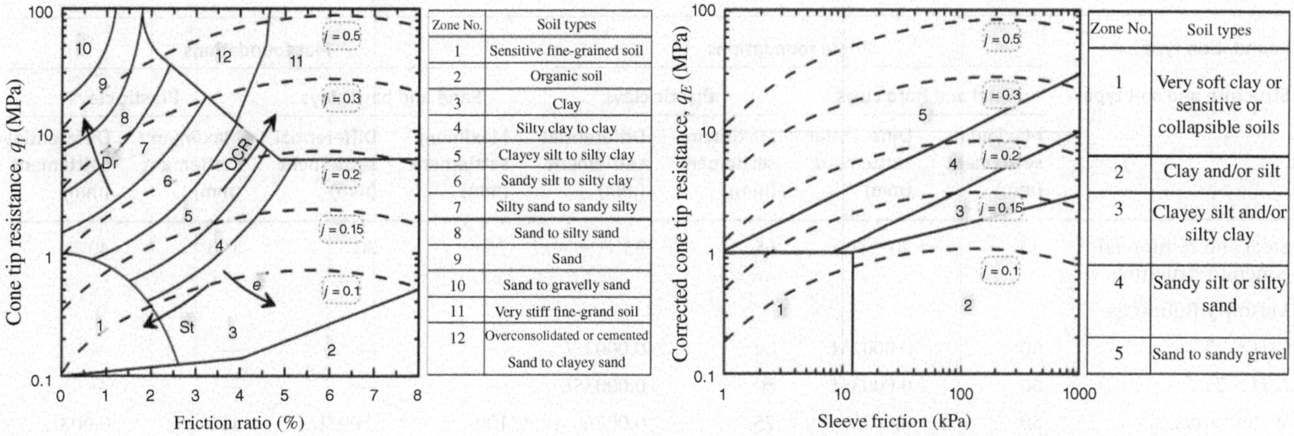

Zone No.	Soil types
1	Sensitive fine-grained soil
2	Organic soil
3	Clay
4	Silty clay to clay
5	Clayey silt to silty clay
6	Sandy silt to silty clay
7	Silty sand to sandy silty
8	Sand to silty sand
9	Sand
10	Sand to gravelly sand
11	Very stiff fine-grand soil
12	Overconsolidated or cemented Sand to clayey sand

Zone No.	Soil types
1	Very soft clay or sensitive or collapsible soils
2	Clay and/or silt
3	Clayey silt and/or silty clay
4	Sandy silt or silty sand
5	Sand to sandy gravel

Figure 6.16 Behavior classification charts with proposed "j" values (a) Campanella et al. (1986) with permission of American Society of Civil Engineering and (b) Eslami and Fellenius 1997 (Valikhah and Eslami 2019).

The rigidity of the frame of a structure has a major effect on differential settlement. In very flexible structures, each foundation acts almost independently from the foundations. Accordingly, the settlement of one foundation has almost no impact on the others. Although, much stiffer structures, probably due to shear walls or diagonal bracing, tends to smooth over differential settlement problems. In such cases when one foundation settles more than the others, a rigid structure will redirect some of its load to adjacent foundations, thus reducing the differential settlement (Coduto et al. 2016).

For the differential settlement δ_D for any two columns of a frame of the building, is the difference between the total settlements of the two columns. If L is the span or distance between the two columns, the angular distortion a is defined as:

$$\alpha = \frac{\delta_D}{L} \tag{6.73}$$

Differential settlement alone does not provide enough information to assess tolerable deflections. The serviceability and ultimate limit states of a structure are influenced by the distortion of the superstructure caused by foundation deflections. The same differential settlement causes different degrees of distortion for different spans. However, when two structural frames with comparable material properties experience the same angular distortion, a similar response can be expected in both frames. In other words, the frames should crack at approximately the same angular distortion. It is worth noting that the ratio of differential settlement to span, as expressed in Equation 6.73, represents the tangent of the angle. In most cases, this ratio is small enough that it can be approximated as equal to the angle itself. Figure 6.17 illustrates different scenarios relevant to serviceability, including uniform total settlement, nonuniform or differential total settlement, and distortion or tilting.

6.4.4.1 Allowable Limits

Predicted values of the total settlement, as well as the value of distortion, must be within the range of allowed values in order to satisfy the requirements of performance-based design. Tables 6.9 and 6.10 present a summary of values allowed for total as well as differential settlement and angular distortion, respectively.

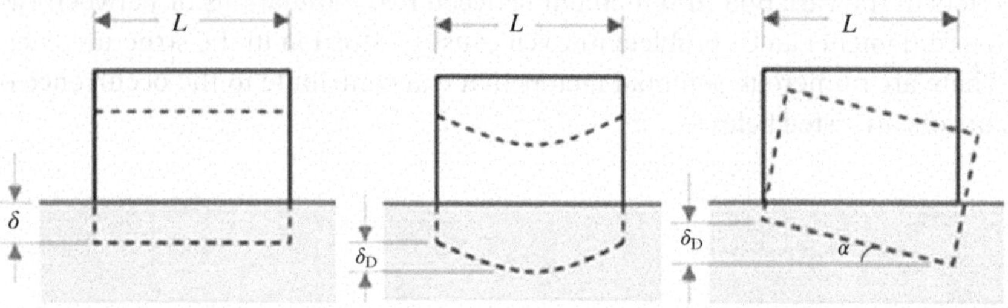

Figure 6.17 (a) Uniform total settlement, (b) differential total settlement, and (c) tilting.

Table 6.9 Allowed limits of different settlements.

Foundation type	Single foundations				Mat foundations			
Structure and soil type	Sand and hard clays		Plastic clays		Sand and hard clays		Plastic clays	
	Maximum settlement (mm)	Differential settlement (mm)	Maximum settlement (mm)	Differential settlement (mm)	Maximum settlement (mm)	Differential settlement (mm)	Maximum settlement (mm)	Differential settlement (mm)
Steel and reinforced concrete structures	50	20	65	25	75	30	100	40
Masonry Buildings								
$L/H \leq 25$	60	$0.00025L$	80	$0.00025L$	—	—	—	—
$L/H > 25$	60	$0.00035L$	80	$0.00035L$	—	—	—	—
Water tanks, silos, and chimneys	50	$0.002L$	75	$0.002L$	100	0.003L	125	0.003L

L: the deformed length of the wall or center to center of columns; H: the height of the wall from the foundation top.

Table 6.10 Allowable values of the angular distortion (Coduto et al. 2016 / Pearson Education).

Type of structure	$\alpha = \frac{\Delta w}{L}$
Steel tanks	1/25
Bridges with simply supported spans	1/125
Bridges with continuous spans	1/250
Buildings that are very tolerant of differential settlements, such as industrial buildings with corrugated steel siding and no sensitive interior finishes	1/250
Typical commercial and residential buildings	1/500
Overhead traveling crane rails	1/500
Buildings that are incredibly intolerant of differential settlement, such as those with a sensitive wall or floor finishes	1/1000
Machinery	1/1500
Buildings with unreinforced masonry load-bearing walls	
Length/height ≤ 3	1/2500
Length/height ≥ 5	1/1250

6.4.4.2 Induced Instability

The negative effects of nonuniform subsidence on a structure are more significant compared to uniform subsidence. In the case of uniform subsidence, buildings generally do not experience structural instability, but may experience serviceability issues. However, in cases of nonuniform subsidence where the tilt exceeds the allowable range, the structure becomes susceptible to damage or even collapse. It is therefore crucial to prevent excessive tilt and ensure that the structure remains within the acceptable limits to maintain its stability and integrity.

A parametric study has been conducted in this regard. For this study, a 10 m by 10 m square foundation for the building is assumed. Additionally, the height and weight of each floor of the building is assumed to be 3 m and 1 t/m^2, respectively. The studied parameters are Δ_{max}, the maximum lateral movement at the top of the structure as shown in Figure 6.18,

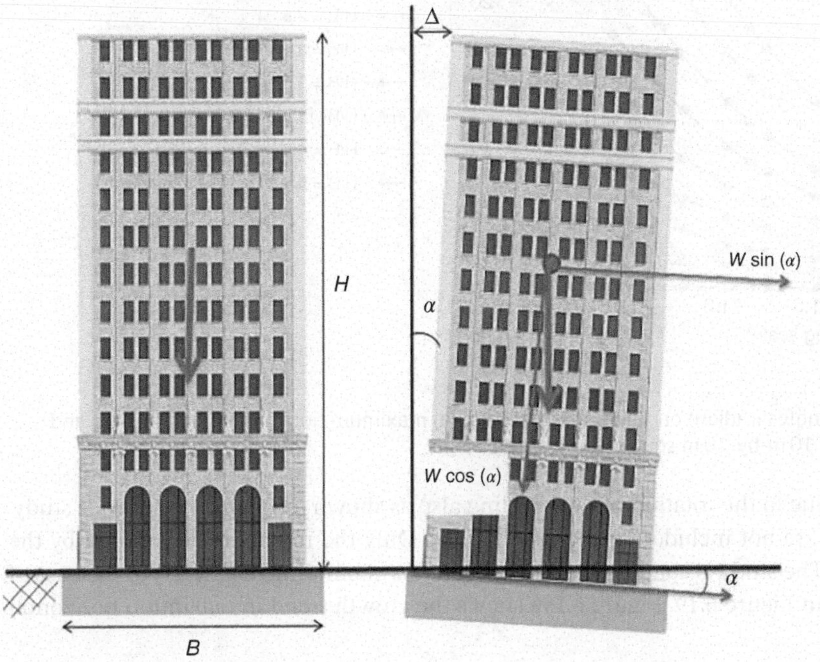

Figure 6.18 Schematic building rotation with resistance force and overturning force due to weight of the structure.

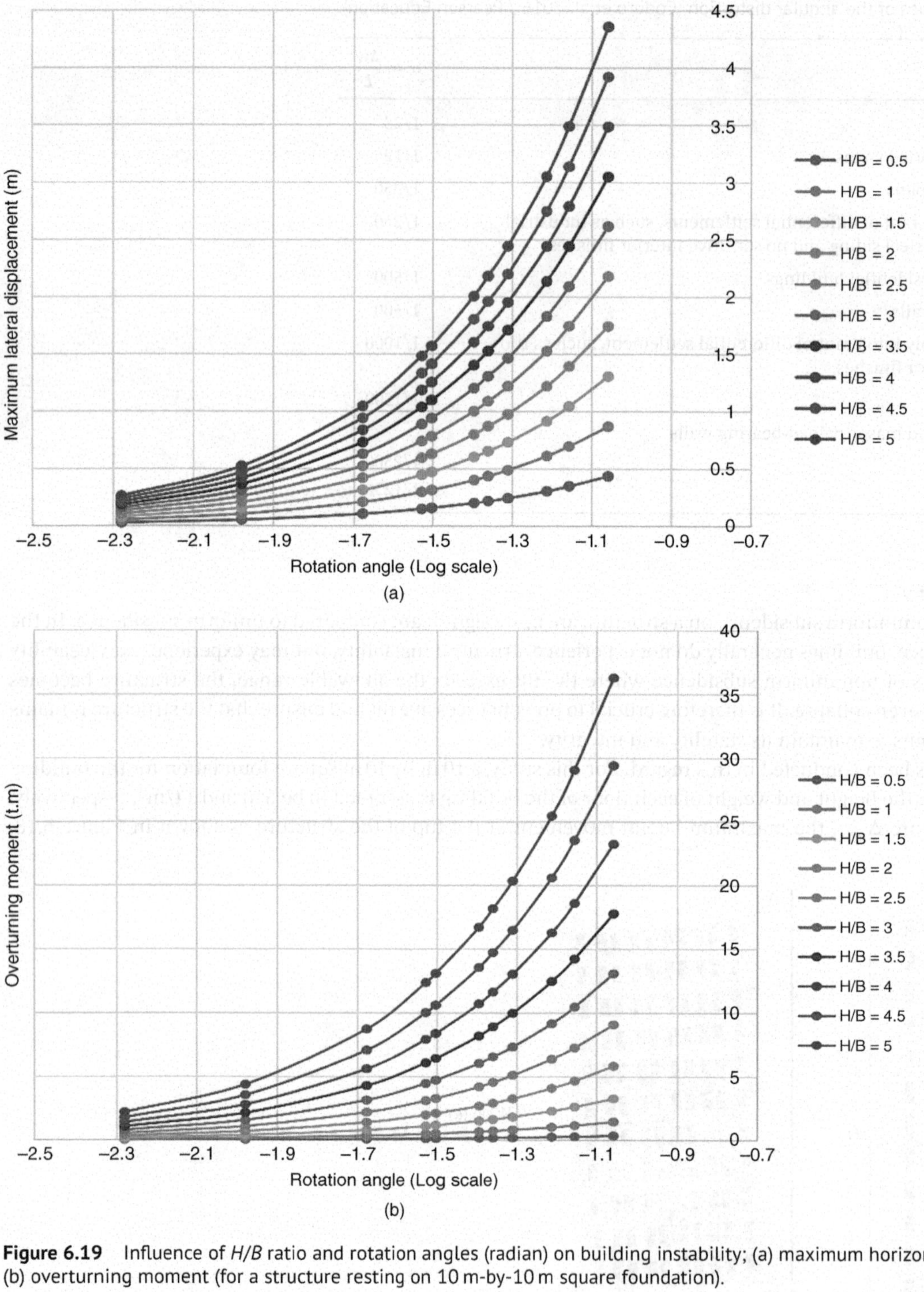

Figure 6.19 Influence of *H*/*B* ratio and rotation angles (radian) on building instability; (a) maximum horizontal displacement and (b) overturning moment (for a structure resting on 10 m-by-10 m square foundation).

and M_o, the overturning moment generated due to the rotation of the building also as shown in Figure 6.18. In this study, moments due to wind and earthquake loads are not included in the calculations. Only the moment, M_o, created by the weight of the rotated structure is considered. The study is conducted for varying ratios of building height (H) to the base of the square foundation (B) designated as H/B in Figure 6.19. Figure 6.19a shows the growth trend in maximum horizontal displacement, Δ_{max}, with increasing H/B ratio. Similarly, Figure 6.19b shows the upward trend of overturning moments, M_o, with increasing H/B ratio.

In Figure 6.18, overturning force component and maximum horizontal displacement due to the weight of the schematic rotated building are shown.

$$\Delta_{max} = H \cdot \sin(\alpha) \tag{6.74}$$

$$M_o = W \cdot \sin(\alpha) \cdot \frac{H}{2} \tag{6.75}$$

Case Study 6.2

El Sagrario Church and Metropolitan Cathedral, Mexico City: Differential Settlement (Shelley et al. 1997)

The cathedral area is 120 × 60 m, and the weight is approximately 1245 MN. A masonry platform was first constructed, acting as a raft with 80 × 140 m in plan dimensions. The average thickness is approximately 1.20 m, but it is thicker where the softer ground is found, indicating that differential settlement happened within the construction. The soil was reinforced via 22,500 wooden stakes, having a 20–25 cm diameter and a length between 2.2 and 4 m.

El Sagrario Church, built between 1749 and 1768, is located adjacent to the southeastern part of the cathedral, founded via the identical system as in the cathedral: short stakes (diameter = 10.0 cm; length = 1.8–2.0 m) were used to reinforce the soil, and later, a masonry platform was implemented to erect the footings, bearing the applied loads. Its western portion is supported by the cathedral's foundation raft and joined structurally along a common wall.

Differential settlements of the Metropolitan Cathedral initiate to happen in the earliest stages of construction. In 1666, differences in the elevations of 16 primary columns reached a maximum of 85 cm. Then, settlements and structural distortions accumulated; therefore, walls and columns lost them vertically, and fissures and cracks appeared. Towards the end of the nineteenth century, the differential settlement between the base of the western tower and the apse reached 150 cm. Settlements at the base of the western tower have been measured almost continually since 1907. Based on the measurements, it is known that over the first three decades of the century, settlement rates raised from a few centimeters per year to more than 20 cm/year, and the maximum settlement rate was observed in the 1940s when it exceeded 30 cm/year. In the early 1960s, the settlement rates decreased, but they increased again after the early 1980s. They now stand at about 10 cm/year in the central part of the city, and in its outskirts, mainly in its southern part, they exceed 30 cm/year. From 1907 to 1989, the base of the western tower settled more than 7.5 m concerning a fixed point on a hill near the city, outside the lake zone, which would not represent a problem had the settlings in the cathedral been uniform. In October 1989, the differential settlement between the base of the western tower and the apse reached 2.4 m, i.e., 90 cm were added within this century because of the effects of subsidence brought about by pumping from deep wells.

In 1972, the Sagrario Church's maximum differential settlement was 70 cm, and settlement rates during the church were 2.9 mm/month.

45 CPT tests were applied to determine the stratigraphy beneath the cathedral. The results of three CPT tests performed in the southern atrium are depicted in Figure 6.20. The archaeological fills are thicker, penetration resistance is larger and compressibility smaller than the tests performed on the western and eastern sides. The clay layers under the dome of the Sagrario Church have compressed more than 20.0 m, a decrease of about 50% from their initial thickness, where the depth of archaeological fills exceeds 23.0 m. The average phreatic level in 1952 in the cathedral was 2.8 m below the surface and 7.2 m in 1989 at the start of the project. Piezometric conditions have been modified during the project, primarily because of pumping.

(Continued)

Case Study 6.2 (Continued)

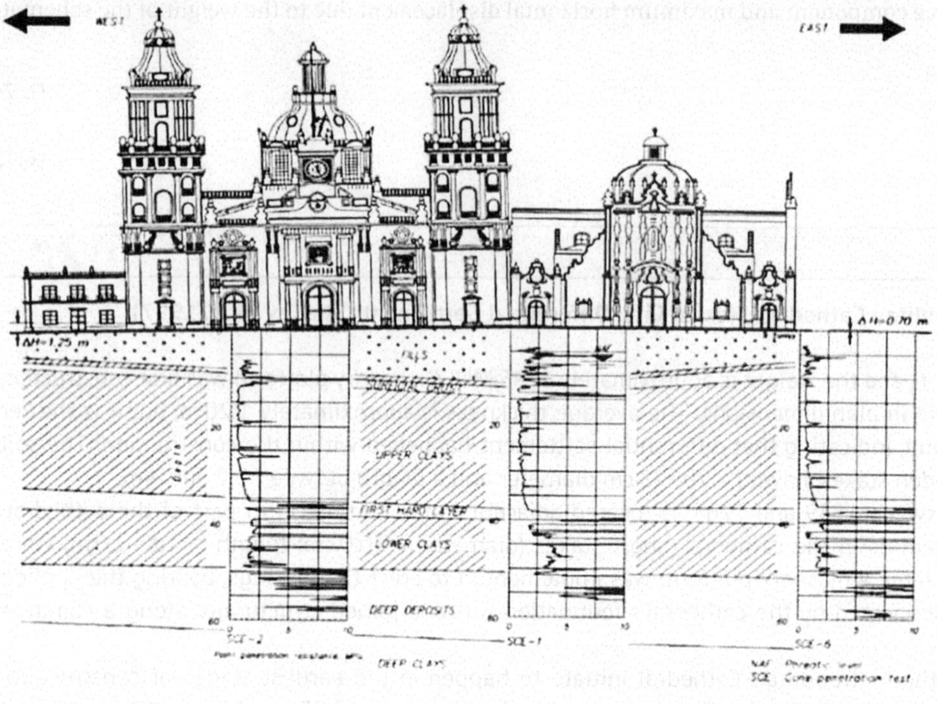

Figure 6.20 CPT tests and stratigraphy along an $E-W$ axis in the southern atrium (Shelley et al. 1997 / With permission of International Society for Soil Mechanics and Geotechnical Engineering).

A project was initiated to solve the problems of the cathedral and the Sagrario Church, applying the method of under excavation, chosen over other technically feasible but more costly underpinning projects requiring main structural modifications. It was concluded that the restoration of piezometric pressures during the perimeters of the two churches could only be recommended to complete others. Between August 1993 and December 1996, 2200 m^3 of soil were under excavation. Extensive instrumentation is used to monitor the structural response closely and modify the under-excavation program when essential. As recently as December 1996, the maximum corrective settlement was 68 cm. Under excavation has produced corrective changes in the inclinations and pre-existing fissures.

6.5 Stability Issues (External)

Stability of structures is an essential design consideration alongside bearing capacity and settlement calculations when it comes to foundation engineering. This aspect becomes particularly important when foundations are subjected to horizontal moment or uplift forces. Horizontal moments or uplift forces on the foundation of structures are created when (i) buildings are subjected to earthquake and wind loads, and (ii) bridges, retaining walls, marine structures, and dry ducks are subjected to wind loads and loads due to flowing water currents or buoyancy of water. These non-vertical loading conditions pose significant challenges for foundations design, especially for shallow foundations design. Therefore, it is crucial to accurately study and evaluate the effects of these forces on the stability of the structure (sliding, overturning, buoyancy, and uplift) from the foundation design point of view. Proper control and mitigation strategies must be implemented to ensure the stability of the foundation system.

6.5.1 Sliding

Spread footings resist applied shear loads in two ways: sliding friction along the footing base and passive earth pressure on the leading edge. Besides, an active earth pressure acts on the trailing edge. All these forces are shown in Figure 6.21. A summary of the required equations regarding ASD and LRFD methods is presented in Table 6.11.

Figure 6.21 Shear load acting on a spread footing.

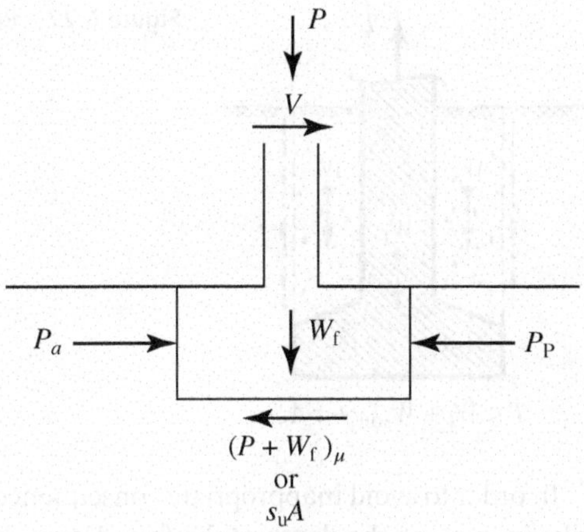

Table 6.11 Equations for controlling of sliding.

No	Approach	Soil type	Equation	Considerations
1	ASD	Cohesionless	$V_a = \dfrac{(P + W_f - u_D A)\mu + 0.5P_P}{FS} - P_A$	FS usually equals 1.5. $V \leq V_a$
2	ASD	Cohesive	$V_a = \dfrac{s_u A + 0.5 s_u BD}{FS}$	FS usually equals 2. $V \leq V_a$
3	LRFD	Cohesionless	$\sum \phi V_n = \phi_1(P_u + W_f - u_D A)\mu + \phi_2(0.5P_P)$	$V_u + 1.5\,P_A \leq \sum \phi V_n$
4	LRFD	Cohesive	$\sum \phi V_n = \phi_1 s_u A + \phi_2(0.5 s_u BD)$	$V_u \leq \sum \phi V_n$

where

V_a = allowable shear capacity
P = downward column load
V = ASD shear load
W_f = weight of footing
u_D = pore water pressure at foundation base
A = area of the footing
μ = interface coefficient between the footing and the soil; for footings cast on the ground, the concrete bonds with the soil at the interface, and sliding occurs in the soil just below the footing; in this case, we can simply assume $\mu = \tan \phi'$
P_P = total passive resultant force
P_A = total active resultant force
FS = factor of safety
P_u = minimum factored vertical column load
V_u = factored column shear load
ϕ_1 = resistance factor for sliding shear (0.8–0.9, for more details, see AASHTO 2012)
ϕ_2 = resistance factor for passive earth pressure (0.5, for more details, see AASHTO 2012)

6.5.2 Overturning

The factor of safety against overturning of the foundation can be calculated via the:

$$FS_{(overturning)} = \frac{\sum M_R}{\sum M_o} \tag{6.76}$$

where

$\sum M_o$ = sum of overturning moments about the center point

$\sum M_R$ = sum of resisting moments about the center point

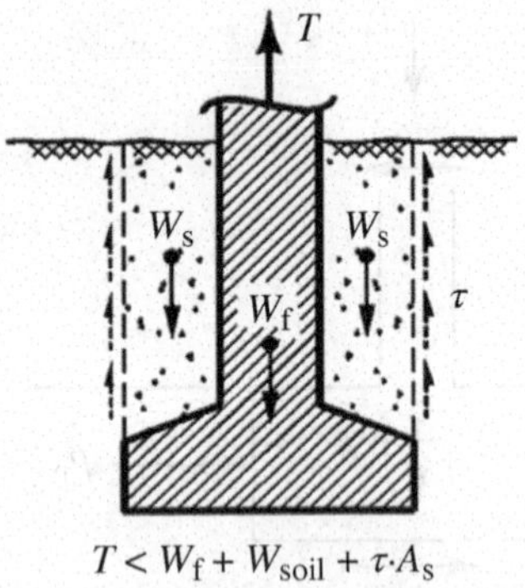

Figure 6.22 Resistant components against uplift loads.

In order to avoid inappropriate consequences of buildings overturning due to their foundations overturning, some actions like increasing the depth of the foundation, increasing the weight of the foundation, and using some elements to nail the foundation to the ground should be done.

6.5.3 Uplift Capacity

If the foundation is under tensile force T_u, for stability against uplift:

$$T_u \leq \frac{W_F + W_S + A_s\tau_V}{FS} \tag{6.77}$$

where

W_F = foundation weight
W_S = soil weight above the foundation
τ_V = vertical shear strength
FS = factor of safety equals 1.5 – 2
A_S = area of resistant shear forces around the foundation

Vertical shear strength in cohesionless soils can be calculated through $\tau_V = \sigma_z' \cdot K \cdot \tan\delta$. The K factor varies between K_0 and K_P. Moreover, the appropriate value depends on the soil and foundation interaction. Usually, K is assumed to be equal to K_0 and is calculated via $K_0 = 1 - \sin\phi$. The δ usually ranges between $2/3\phi$ and $3/4\phi$. In calculating the side resistance between two soil surfaces, $\tan\phi$ is used. In cohesive soils, $\tau_V = \alpha \cdot s_u$ is usually used.

In order to deal with large uplift forces, usually semi-deep and deep foundations are used. In the case of shallow foundations, they should be connected appropriately to adjacent foundations. Furthermore, to increase the shear strength and weight of the soil on the foundation's base, they are executed in the form of a pedestal with a larger diameter than the stem or body in the end.

The resistant components against uplift loads are depicted in Figure 6.22. The resistant zone of shear forces around the foundation is assumed to be conservatism and straightforward in the form of a cylinder.

6.5.4 Eccentricity

Stability requirements are essential when the eccentricity exceeds the proposed values, explained in detail in Sections 6.2.2 and 6.2.3.

Example 6.8 Calculate the safety factor against sliding and overturning for the given retaining wall. Also, check the eccentricity criteria and suggest solutions for any probable problem.

$$K_a = \tan^2\left(45 - \frac{\varphi}{2}\right) = \tan^2\left(45 - \frac{30}{2}\right) = \frac{1}{3} \qquad \varphi = 30°$$

$$K_p = \tan^2\left(45 + \frac{\varphi}{2}\right) = \tan^2\left(45 + \frac{30}{2}\right) = 3 \qquad \gamma_{soil} = 1.8\ \text{ton/m}^3$$

$$\gamma_{concrete} = 2.25\ \text{ton/m}^3$$

$$\tan\delta = 0.4$$

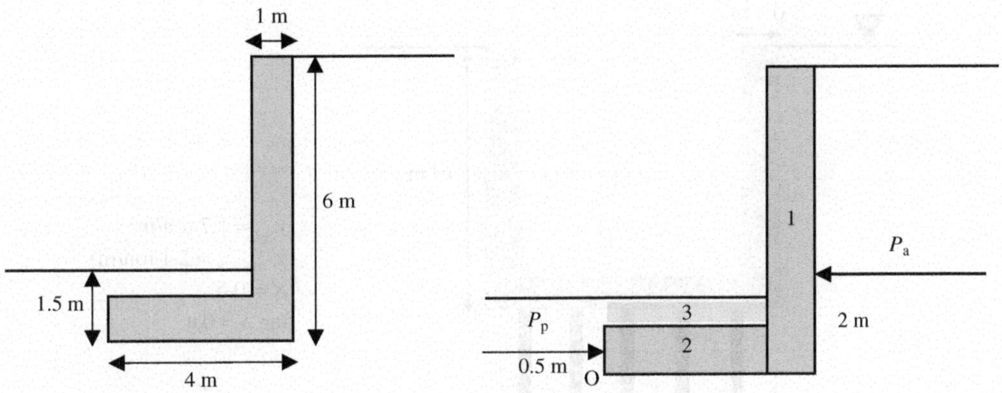

Force	Force magnitude per m	Force arm (m)	Moment magnitude per m (ton/m)
P_a	$\frac{1}{2}\times\frac{1}{3}\times 1.8\times 6^2 = 10.8$	2	21.6
	$\sum H = 10.8$		$\sum M_o = 21.6$
P_p	$\frac{1}{2}\times 3\times 1.8\times 1.5^2 = 6.1$	0.5	3.1
1	$6\times 1\times 2.25 = 13.5$	3.5	47.3
2	$3\times 1\times 2.25 = 6.8$	1.5	10.1
3	$0.5\times 3\times 1.8 = 2.7$	1.5	4.1
	$\sum W = 23$		$\sum M_R = 64.6$

Sliding Resistance

$$FS_{Sliding} = \frac{\sum W\times\tan\delta}{\sum H} = \frac{23\times 0.4}{10.8} = 0.9 \rightarrow \text{Not Acceptable}$$

Overturning Resistance

$$FS_{Overturning} = \frac{\sum M_R}{\sum M_o} = \frac{64.6}{21.6} = 3 \rightarrow \text{Acceptable}$$

Eccentricity

$$\sum M = \sum M_R - \sum M_o = 64.6 - 21.6 = 43\ \text{ton/m}$$

$$\overline{X} = \frac{\sum M}{\sum W} = \frac{43}{23} = 1.9\ \text{m}$$

$$e = \frac{B}{2} - \overline{X} = \frac{4}{2} - 1.9 = 0.1\ \text{m} < \frac{B}{6} = 0.7\ \text{m} \rightarrow \text{Acceptable}$$

Using a shear key or anchors is recommended to provide sufficient sliding resistance.

Example 6.9 A cubical concrete tank with 10 m dimensions and a 0.5 m wall thickness, is proposed to be built in a nearshore site. The site, in some cases, is prone to be fully saturated and, as a result, at risk of floatation for the empty tank. First, control the stability against uplift if the minimum required safety factor is 1.5. In case the minimum requirement is not met, determine the number of anchor piles with average tensile capacities of 15 tons that are needed to provide the stability of the tank. The other parameters are given below.

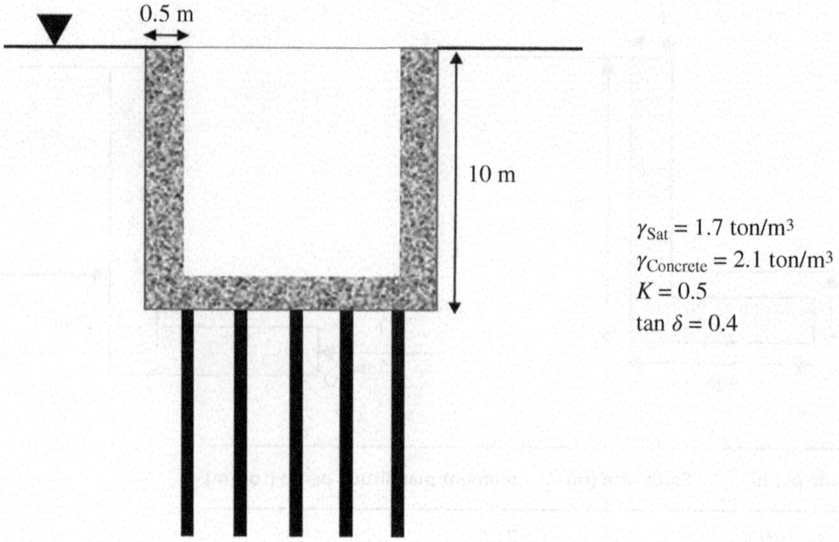

Controlling the Stability

$$W = V_{\text{Concrete}} \times \gamma_{\text{Concrete}} = (10^3 - 9^3) \times 2.1 = 569.1 \text{ tons}$$

$$q_s = \sigma'_z \cdot K \cdot \tan\delta = 10 \times (1.7 - 1) \times 0.5 \times 0.4 = 1.4 \text{ ton/m}^2$$

$$R_{\text{side}} = 1.4 \times 4 \times 10 \times 10 = 560 \text{ tons}$$

$$P_{\text{uplift}} = 10 \times 1 \times 10 \times 10 = 1000 \text{ tons}$$

$$\text{FS}_{\text{uplift}} = \frac{W + R_{\text{side}}}{P_{\text{uplift}}} = \frac{569.1 + 560}{1000} = 1.1 \not\geq 1.5$$

Number of Anchor Piles

$$\text{FS}_{\text{uplift}} = \frac{W + R_{\text{side}} + F_{\text{anchor}}}{P_{\text{uplift}}} \geq 1.5 \rightarrow \frac{569.1 + 560 + F_{\text{anchors}}}{1000} \geq 1.5 \rightarrow F_{\text{anchors}} \geq 370.9 \text{ tons}$$

$$\text{Number of anchor piles} \geq \frac{370.9}{15} = 24.7 \rightarrow 25 \text{ anchor piles are required.}$$

Case Study 6.3

Offshore and Onshore Loads and Foundation (Randolph et al. 2005)

The design of the shallow foundations of offshore structures requires more extensive site investigation and special procedures compared to those required for shallow foundations of onshore structures. This is primarily due to the increased size of the offshore structures they support and the harsher environmental conditions they face. Even smaller gravity base offshore structures can reach heights of 70 m, which is equivalent to an 18- or 20-story high-rise building, with a footprint measuring 50 m by 50 m. Larger offshore structures can exceed 400 m in height, requiring foundations with a plan area exceeding 15,000 m^2. In some cases, even a single bucket foundation can have a diameter of over 15 m. Apart from the sheer size of offshore structures, their foundations must withstand severe environmental forces, including wind, waves, and currents. In certain cold regions, ice forces also come into play. These environmental forces impose significant horizontal forces and moments on the foundations, which are not typically encountered onshore. A schematic depicting the loading conditions of a gravity base platform can be seen in Figure 6.23.

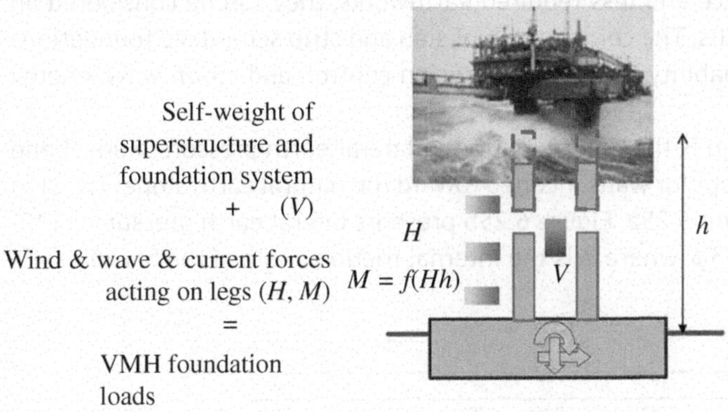

Figure 6.23 Combined loading of a gravity base platform (Randolph et al. 2005 / Millpress Science Publishers / IOS Press).

Design loads for an offshore gravity base structure and an onshore high-rise of similar height are compared in Figure 6.24. The self-weight of the two structures is moderately comparable, with the vertical design load of the offshore structure 35% larger than the onshore high-rise. However, the live loads are many times greater offshore than onshore; the horizontal design load is 1600% larger and the moment is 500% larger, which is reflected in an increased foundation area of around 150%. More essential than the absolute magnitude of the environmental loads is the ratio of horizontal load to moment. The offshore structure shown in Figure 6.24, M/HD is relatively low (approximately 0.35) compared to the onshore high-rise (approximately 1.6), indicating a high risk of sliding failure for offshore gravity base structures. Whether sliding, overturning, or vertical bearing capacity, limiting load is generally the critical design criteria for shallow offshore foundations. Tolerances of allowable deformation usually constrain allowable settlements of shallow offshore foundations to maintain the integrity of oil wells and pipelines, rather than platform stability. It is feasible to design offshore shallow foundations for settlements up to several meters.

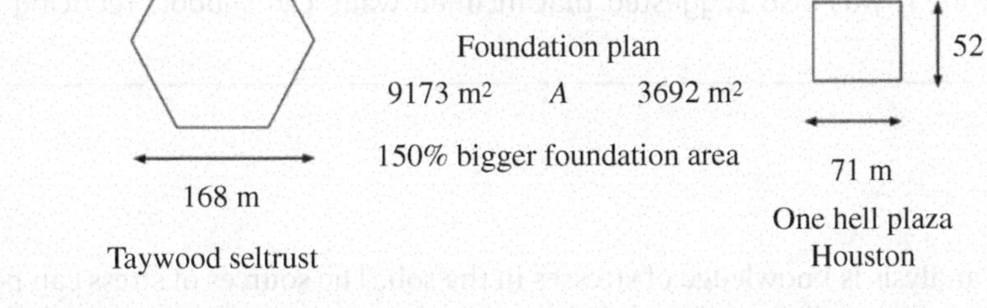

Figure 6.24 Comparison of offshore and onshore design loads (Randolph et al. 2005 / International Society for Soil Mechanics and Geotechnical Engineering).

Case Study 6.4

Sloped Concrete Walls; New Approach for Less Active and More Passive Force (Eslami and Eslami 2017)

Concrete seawalls are structures in coastal facilities for shore and slope protection. Owing to several advantages of sloped or inclined walls such as minimum soil disturbance and less required earthworks, they can be considered an appropriate alternative to vertical cantilever retaining walls. The combination of slab and strip semi-deep foundations instead of heel–toe slab foundations increases their capability for stability, erosion control, and storm wave energy dissipation.

One of the significant elements in retaining wall design is the determination of lateral earth pressure. Kerisel and Absi (1990) determined the lateral earth pressure coefficient for walls inclined toward the natural earth slope. The sign convention of Kerisel and Absi (1990) can be seen in Figure 6.25a. Figure 6.25b presents lateral earth pressure coefficients by Kerisel and Absi (1990) for a condition of ($\lambda = 2/3\phi$, where ϕ is the internal friction angle of the backfill soil).

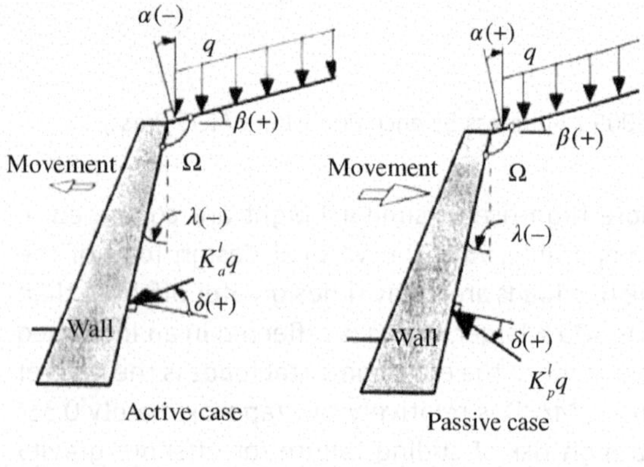

(a)

ϕ (°)		15	20	25	30	35	40	45
$\Omega = 90°$	K_a	0.542	0.447	0.369	0.304	0.249	0.206	0.165
	K_p	2.024	2.645	3.554	4.926	7.075	10.64	17.24
$\Omega = 95°$	K_a	0.517	0.420	0.340	0.274	0.221	0.177	0.139
	K_p	2.123	2.822	3.861	5.464	8.000	12.40	20.41
$\Omega = 100°$	K_a	0.494	0.394	0.314	0.248	0.196	0.153	0.116
	K_p	2.222	3.006	4.178	6.024	9.063	14.28	24.39
$\Omega = 105°$	K_a	0.471	0.370	0.289	0.1224	0.173	0.132	0.098
	K_p	2.331	3.202	4.545	6.667	10.20	16.57	29.41
$\Omega = 110°$	K_a	0.450	0.374	0.226	0.203	0.153	0.114	0.082
	K_p	2.445	3.417	4.918	7.353	11.54	19.23	34.48
$\Omega = 115°$	K_a	0.429	0.326	0.246	0.183	0.135	0.098	0.069
	K_p	2.557	3.636	5.346	8.130	12.99	22.22	41.67
$\Omega = 120°$	K_a	0.409	0.305	0.226	0.166	0.120	0.085	0.058
	K_p	2.681	3.876	5.803	9.010	14.78	25.42	50.00
$\Omega = 125°$	K_a	0.391	0.286	0.209	0.150	0.106	0.073	0.049
	K_p	2.809	4.126	6.263	10.00	16.57	29.70	58.82
$\Omega = 130°$	K_a	0.373	0.269	0.192	0.136	0.094	0.063	0.041
	K_p	2.941	4.405	6.802	10.99	18.87	34.48	71.43
$\Omega = 135°$	K_a	0.356	0.252	0.177	0.123	0.083	0.054	0.034
	K_p	3.086	4.687	7.389	12.19	21.28	40.00	83.33

(b)

Figure 6.25 Inclined walls (a) sign convention for determination of the lateral earth pressure coefficients and (b) lateral pressure coefficients A. Eslami et al 2008 / Journal of Advanced Materials in Engineering.

Following studies performed by these researchers and according to Boussinesq relations, by increasing the angle Ω, coefficient K_a decreases and K_p increases. It is clear from Figure 6.25b that although β is constant, an increase in λ leads to a decrease in K_a and an increase in K_p. The soil's lateral pressure decreases in active conditions and increases in passive conditions. For instance, for soils with $\phi = 20°$, when the backfill is horizontal, increasing the wall slope (λ) to 30° decreases K_a up to 32% and increases K_p up to 45%. This can be more notable for soils with higher internal friction angles.

Generally, the studies show that in static and seismic conditions, inclined walls are the superior option to use against internal instability and external instability factors. It was also suggested that inclined walls can support reducing operational time and cost.

6.6 Major Points

1) One of the major requirements for foundation analysis is knowledge of stresses in the soil. The sources of stress can be attributed to three kinds: geostatic, horizontal, and induced. The geostatic stresses are a result of the force of gravity acting directly on the soil mass. The geostatic horizontal stress is also vital for many engineering analyses. The

induced or applied stresses are caused by external loads, such as structural loads, fills, or any other type of surface loading.

2) The most critical parameter determining the interface between a shallow foundation and the soil that supports it is the contact pressure, which is the contact force per unit area between the bottom of the foundation and the subsoil. The value of contact pressure depends on factors comprising eccentricity, the magnitude of the applied moment, the structural rigidity of the foundation, stress–strain properties of the soil, and the roughness of the bottom of the foundation.
3) Efficient foundation engineering practice requires careful consideration of several key requirements, encompassing technical, practical, and economic aspects, with a primary focus on performance-based design. The following issues are generally regarded as major considerations in this field:
   - **Bearing Capacity:** Assessing the ability of the foundation to support the applied loads without excessive settlement or failure.
   - **Serviceability:** Addressing settlement and torsion to ensure the foundation meets acceptable performance criteria.
   - **Structural Design:** Designing the foundation to adequately distribute loads and withstand the anticipated forces.
   - **Stability Control:** Ensuring the stability of the foundation under various loading conditions and preventing undesired movements or failure.
   - **Full or Model Scale Testing:** Conducting tests, either at full scale or on scaled models, to validate design assumptions and assess the behavior of the foundation system.
   - **Constructional Aspects:** Considering practical aspects related to construction techniques, methods, and materials to ensure proper execution of the foundation.
   - **Durability:** Incorporating measures to ensure the longevity and resistance of the foundation against deterioration or degradation.
   - **Economic Aspects:** Balancing the technical requirements with cost-effectiveness and efficiency, optimizing the design and construction processes to achieve the desired performance within the available budget.

   By addressing these various aspects, foundation engineers strive to develop efficient and reliable foundation systems that meet the specific requirements of each project.
4) A foundation element is said to have failed in bearing capacity when it undergoes substantial settlements. The magnitude of the load causing the foundation element to fail in this manner, referred to as the ultimate bearing capacity, depends on a variety of factors, comprising the dimensions of the element, the depth at which it is implemented in the ground, and the shear strength of the soil. There are five ways to determine the soils' bearing capacity: static analysis, complementary methods including using in situ test data, directly and indirectly, presumed values recommended by codes and handbooks, and using the results of loading tests, modelling by numerical analyses.
5) Settlement is the vertical deflection of soils under the foundation loading. The compressibility of soils involves stress–strain relationships. In this regard, the trend of stress–strain curves plus soil stiffness plays a vital role in settlement computations. Generally, regarding static or dynamic loading by a foundation and consequent induced stresses distributed in depth, the types of settlement are recognized as immediate settlement, consolidation settlement, and secondary settlement.
6) Stability control is another primary requirement related to the geotechnical design of foundations, further than the determination of bearing capacity and settlement estimation. The resistance of a foundation against sliding, overturning, buoyancy, and uplift forces should be controlled and considered in this area.
7) Investigating past geotechnical failures is of utmost importance to prevent their recurrence, mitigate risks, and avoid potential loss of life and significant financial losses. A crucial lesson learned from these experiences is the recognition of subbase layers as a critical factor. Conducting thorough site investigations allows for more optimized designs and improved performance of substructures. Furthermore, the influence of the horizontal and uplift loads and moments on the dimensions and foundation behavior cannot be ignored and should be carefully considered during the design process. By incorporating these lessons into future projects, engineers can enhance the safety, efficiency, and overall success of geotechnical endeavors.

6.7 Questions and Problems

6.7.1 Explanatory

1 Name and explain each kind of stress in soil briefly.

2 What are the different conditions of the coefficient of lateral earth pressure determination?

3 List the factors that the value of contact pressure depends on.

4 How can the data from in situ tests be used to determine bearing capacity and settlement estimation?

5 Define the concept and the way of calculation for different settlement types.

6 What is a differential settlement, and what are its potential sources?

7 Name and explain the sources of resistance against uplift forces.

6.7.2 Conceptual

1 Explain the differences between a single foundation and a mat foundation, considering bearing capacity and serviceability requirements.

2 List some solutions to provide more resistance against horizontal and uplift loads in foundations.

3 What are the differences between offshore and onshore projects from the perspective of imposed loads and required foundation dimensions?

4 Discuss the reason for failure in the foundations of a bridge due to floods based on what you learned in this chapter.

5 Explain the complementary role of laboratory and in situ tests in determining bearing capacity and settlement estimation.

6 List the lessons that can be learned via the case histories explained in this chapter.

7 Assuming the same soil condition, compare the external stability for all four suggested retaining walls.

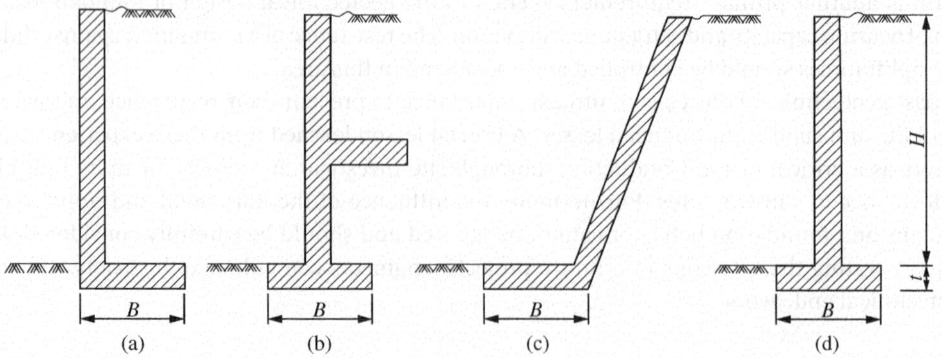

8 Compare and contrast the performance of a given inclined wall with a similar vertical wall. Also, discuss the role of the shear key in sliding and overturning resistance.

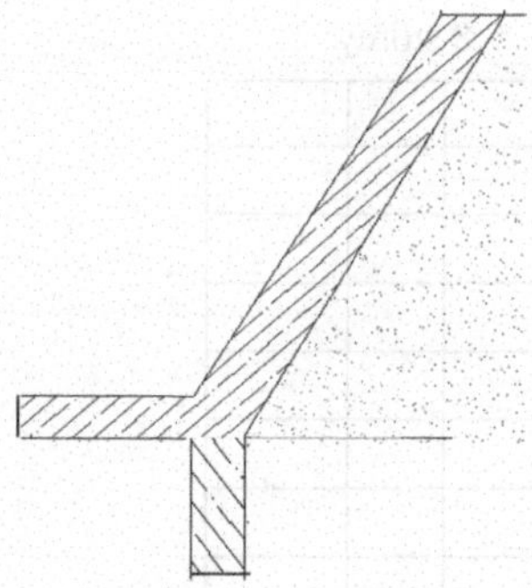

6.7.3 Problems

1 A mat foundation is supposed to withstand the load of five grain storage silos. The net weight of each silo is 3000 tons, and the weight of the inner load is 20,000 tons. The mat's weight is 7500 tons, and because the silos are loaded separately, they impose an eccentricity on the foundation. Control the requirements related to eccentric loads and specify the foundation's minimum width if required.

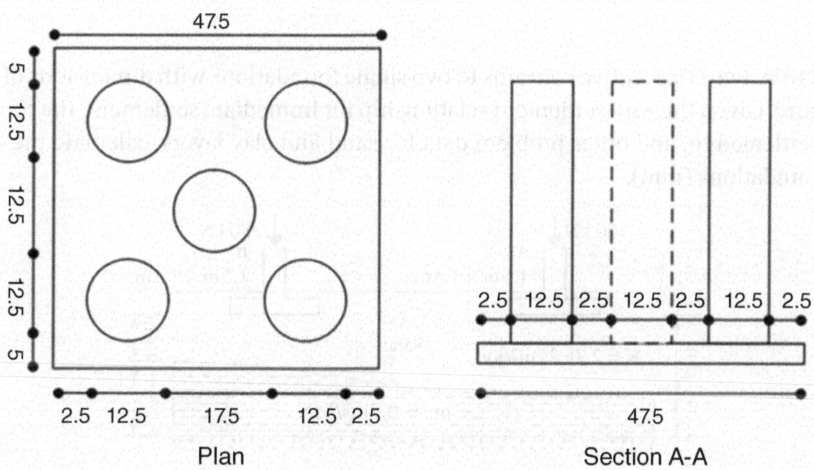

2 According to the figure below, a mat foundation 40 m wide, 50 m long, and 1.5 m thick transmits a total gravitational load of 50,500 tons to the ground through columns and walls. Calculate the average contact pressure and compare it with effective stress below the foundation before construction.
$\gamma_c = 2.4\,\text{t/m}^3$, $\gamma_w = 1\,\text{t/m}^3$, $\gamma = 1.8\,\text{t/m}^3$

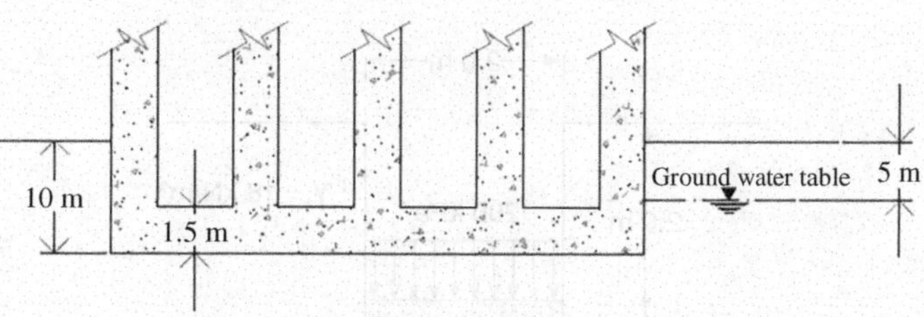

3 An eight-story building with 18×18 m dimensions is located in the plan as shown below, on the ground with soil layering, including a layer of sand and clay with specific parameters. The groundwater level is on the clay layer at a depth of 7 m above the ground, and $D_f = 2$ m. Determine the bearing capacity and estimate the settlement. Control the adequacy of foundation presuming FS = 3 and $q_{all} = 1.5\,\text{kg/cm}^2$.

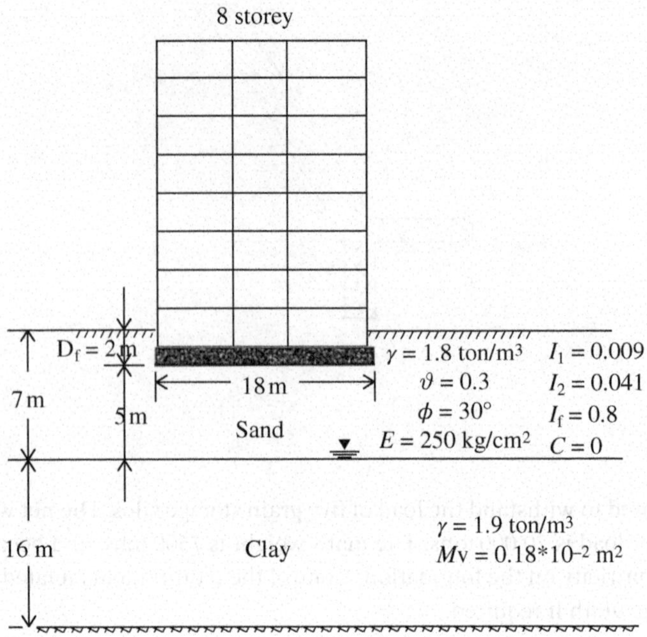

4 The load is applied from two consecutive columns to two single foundations with dimensions of 1.5 × 1.5 m, according to the following figure. Given the KqB settlement relationship for immediate settlement, the $S_c = m_v H \Delta\sigma$ relationship for consolidation settlement e, and other problem data for sand and clay layers, calculate the differential settlement between the two foundations (mm).

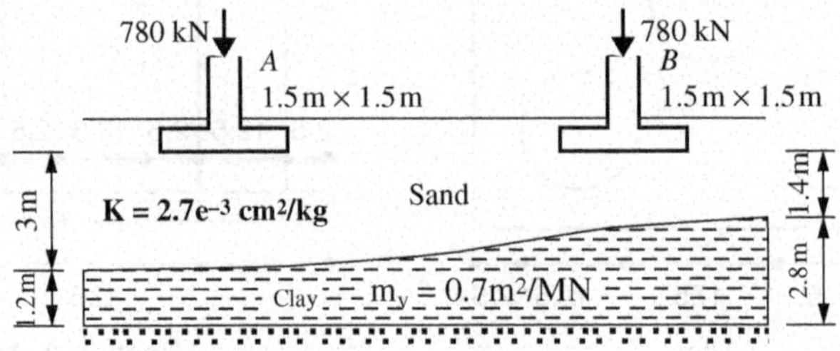

5 Redo Example 6.3 for the following dimensions: 6, 1.5, 0.5, and compare the results.

6 For the square foundation shown in the figure, calculate the amount of foundation settlement after five years, using the Schmertmann's method. The q_c value of the CPT test starts from 2.5 MPa in the first meter and increases by 0.5 MPa/m ($E = 2q_c$).

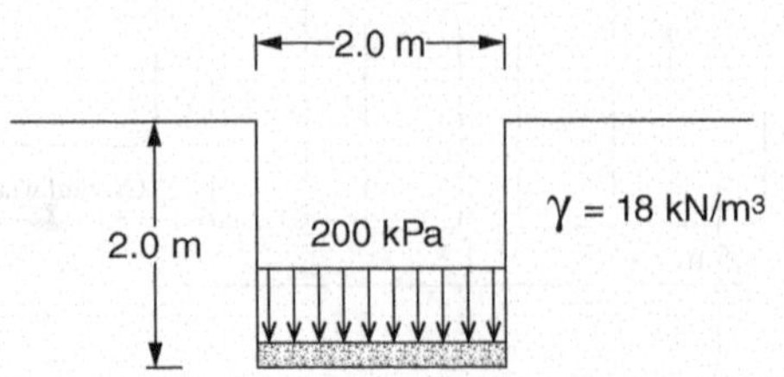

7 Redo Example 6.6 for settlement and assume Mexico City clay as the existing layer. Consider an amount of 7 and 0.7 for C_C before and after soil improvement practice. e_0 can be assumed equal to 6. Other parameters are as given in the example.

8 Consider a single foundation with given loads and soil characteristics. Determine the safety factor that can bear the incoming loads regarding bearing capacity and stability requirements.

$\phi = 34°, \quad \gamma_{soil} = 1.85\ t/m^3$

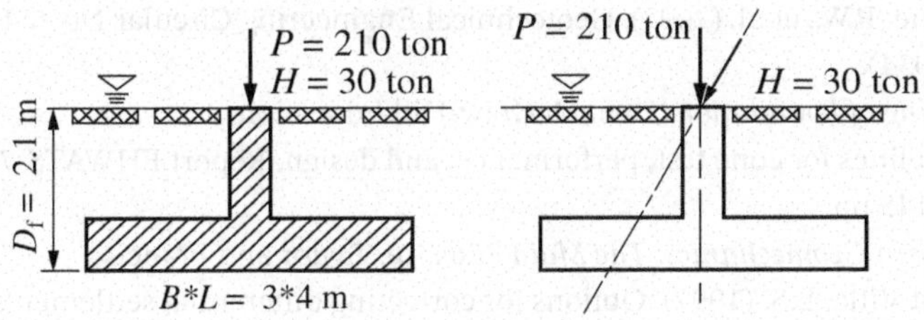

9 Redo Example 6.9, assuming a thickness of 75 cm for the tank and mobilization of 50% of side resistance. Compare the calculations and discuss the role of tank wall thickness and side resistance on resistance against uplift.

References

AASHTO (2012). *AASHTO LRFD Bridge Design Specifications*, 6ee. Washington, D.C.: American Association of State Highway and Transportation Officials.

Bowles, J.E. (1996). *Foundation Analysis and Design*, 8e. McGraw-Hill.

Brinch Hansen, J. (1970). A revised and extended formula for bearing capacity. *Bulletin of the Danish Geotechnical Institute* 28: 5–11.

Burland, J.B. and Burbidge, M.C. (1985). Settlement of foundations on sand and gravel. *Proceedings of Institution of Civil Engineers. Part 1: Design and Construction* 78: 1325–1381.

Campanella, R.G., Robertson, P.K., Gillespie, D., and Grieg, J. (1986). Use of piezometer cone data. Publication GSP No. 6. *Proceedings of American Society of Civil Engineering, ASCE, In-Situ 86 Specialty Conference*, pp. 1263–1280.

CFEM (1992). *Canadian Foundation Engineering Manual*. British Columbia: Bitech Publishers LTD.

CFEM (2006). *Canadian Foundation Engineering Manual*. British Columbia: Bitech Publishers LTD.

Coduto, D.P., Kitch, W.A., and Yeung, M.R. (2016). *Foundation Design Principles and Practices*, 3e. Upper Saddle River, NJ: Prentice Hall, Inc.

Eslami, M.M. and Eslami, A. (2017). Seawall case studies and failure analysis of sloped concrete walls under static and dynamic loads. *Marine Georesources and Geotechnology* 36 (3): 331–339.

Eslami, A. and Fellenius, B. (1997). Pile capacity by direct CPT and CPTu methods applied to 102 case histories. *Canadian Geotechnical Journal* 34: 886–904.

Eslami, A. and Gholami, M. (2006). Analytical model for the ultimate bearing capacity of foundations from cone resistance. *International Journal of Science and Technology* 13 (3): 223–233.

Eslami, A., Moshfeghi, S., Molaabasi, H., and Eslami, M. (2019). *Piezocone and Cone Penetration Test (CPTu and CPT) Applications in Foundation Engineering*, 1e. Elsevier.

Gazetas, G., Tassoulas, J.L., Dobry, R., and O'Rourke, M.J. (1985). Elastic settlement of arbitrarily shaped foundations embedded in half-space. *Géotechnique* 35 (3): 339–346.

Kerisel, J. and Absi, E. (1990). *Active and Passive Earth Pressure Tables*, 3ee. Rotterdam: A.A. Balkema.

Malekdoost, M. and Eslami, A. (2011). Application of CPT data for estimating foundations settlement. Sharif Civil. *Engineering Journal* 27 (2): 75e85. No. 1.

Mehrotra, V.K. (1992). Estimation of engineering parameters of rock mass. PhD thesis. Department of Civil Engineering, University of Roorkee, India.

Meyerhof, G.G. (1956). Penetration tests and bearing capacity of cohesionless soils. *ASCE Journal of the Soil Mechanics and Foundations Division* 82 (SM1): 1–19. (Reprinted in Meyerhof, 1982).

Meyerhof, G.G. (1963). Some recent research on the bearing capacity of foundations. *Canadian Geotechnical Journal* 1 (1): 16–26.

Nordlund, R.L. and Deere, D.U. (1970). Collapse of Fargo Grain Elevator. *ASCE Journal of the Soil Mechanics and Foundations Division* 96 (SM2): 585–607.

Randolph, M., Cassidy, M., Gourvenec, S., and Erbrich, C. (2005). Challenges of offshore geotechnical engineering. *Proceedings of the 16th International Conference on Soil Mechanics and Geotechnical Engineering* (12–16 September 2005). Rotterdam: Millpress.

Sabatini, P.J., Bachus, R.C., Mayne, P.W., et al. (2002). Geotechnical Engineering Circular No. 5: Evaluation of Soil and Rock Properties (No. FHWA-IF-02-034).

Salgado, R. (2008). *The Engineering of Foundations*, 1ee. McGraw-Hill Education.

Schmertmann, J.H. (1978). Guidelines for cone test, performance, and design. Report FHWATS-78209. *Federal Highway Administration*, Washington, 145 pp.

Schnaid, F. (2008). *In Situ Testing in Geomechanics: The Main Tests*, 1e. Taylor and Francis.

Shelley, E.O., Gonzalez, E.T., and Villa, E.S. (1997). Options for correcting differential settlements in Mexico City's Metropolitan Cathedral. *14th International Conference on Soil Mechanics and Foundation Engineering* (Hamburg) (6–12 September 1997). Rotterdam: Balkema, A.A.

Tand, K., Funegard, E., and Warden, P. (1994). Footing load test on sand. *Vertical and Horizontal Deformations of Foundations and Embankments* 8 (40): 164e178. ASCE Geotechnical Special Publication, 1.

Terzaghi, K. (1943). *Theoretical Soil Mechanics*. New York: Wiley.

Valikhah, F. and Eslami, A. (2019). CPT-based nonlinear stress–strain approach for evaluating foundation settlement: analytical and numerical analysis. *Arabian Journal for Science and Engineering* 1e6: 8819–8834.

Vesicá, A.S. (1963). Bearing capacity of deep foundations in sand. Highway Research Record, No. 39, 112–153, Highway Research Board, National Academy of Sciences, Washington, DC.

Vesicá, A.S. (1975). Bearing capacity of shallow foundations. In: *Foundation Engineering Handbook*, 1e (ed. H.F. Winterkorn and H.-Y. Fang), 121–147. New York: Van Nostrand Reinhold.

Wroth, C.P. (1988). Penetration testing-a more rigorous approach to interpretation. In: *Proceedings of the 1st International Symposium on Penetration Testing, ISOPT-1,* Orlando (20-24 March 1988), vol. 1, 303–311. Disney World.

7

Structural Analysis and Design

Chapter Highlights

- **Contact Pressure**
- **Subgrade Reaction Modulus**
- **Flexibility and Rigidity**
- **Comparative Study on Strip and Mat Foundation Examples**
- **Nonrigid Methods of Analysis**
- **Numerical Approach**
- **Structural Foundation Design**
  - Design for Shear (Two-Way and One-Way Shear)
  - Design for Flexure
  - Column Base Bearing
- **Routine Tabular and Graphical Approach**
- **Case Studies**
  - Al Faisaliah Building, Riyadh, Saudi Arabia (Hemsley 2000)
  - Shanghai Tower: Interactive Performance of Piled Raft Foundation and Basement Fin Walls (Sun et al. 2011; Zhu et al. 2012)

7.1 Introduction

The primary objective of structural design in foundation engineering is to ascertain the appropriate materials, such as concrete and steel, within the foundation structure to effectively withstand internal forces, including compression, tensile, and shear stresses. Once the footprint of the foundation is determined based on soil–structure interaction considerations satisfying bearing capacity and settlement requirements, the thickness of the foundation and how to reinforce it are some of the topics related to structural design.

The construction cost of foundations is usually between 5% and 15% of the total construction cost of any structure. With careful planning and proper selection of foundation type, significant savings can be made. The selection of the final design of foundations is done after comparing different alternatives in terms of performance (load bearing, tolerable settlements durability, stability, service), constructability, construction cost, and sustainability.

During foundation design, all forces that a structure may experience during its design life need to be considered. These forces generally include gravitational forces (dead and live loads) and lateral forces (wind, earthquake, and soil drift loads).

For the design of single foundations, it is assumed that they are rigid. This means that the flexural stiffness of the foundation is such that the soil-foundation contact pressure is independent of any changes in the footprint of the foundation and is considered to be linearly uniform or linearly nonuniform. However, in the case of other types of foundations, such as combined and raft foundations, the assumption of rigidity may not always be valid. In such cases, a more careful and detailed analysis is necessary to accurately assess the behavior and performance of the foundation system.

7.2 Principles of Internal Foundations Analysis

The purpose of analyzing foundations is to determine the internal forces used in structural design. Two main approaches are commonly employed for this purpose: the rigid method and the nonrigid (flexible) method. The rigid method is typically applied in the analysis of single footings. It assumes that the foundation is rigid and does not consider any flexibility. This approach provides simplified calculations for determining the internal forces. On the other hand, the nonrigid methods consider the flexibility of the foundation system. These methods consider the coefficient of subgrade reaction (k_s) in the analysis. They are used for analyzing foundations other than single footings, including flexible combined footings, continuous foundations, and rafts.

7.2.1 Soil-Foundation Contact Pressure

The soil-foundation contact pressure is a crucial factor as it represents the interaction between a shallow foundation and the underlying geomaterial, quantifying the applied force per unit area. Chapter 6 provides detailed information on this topic, covering various types of loading and applying principles of solid mechanics to determine the contact pressure.

According to the Theory of Elasticity, the stress distribution under symmetrically loaded footings is not uniform. The actual distribution of stress is dependent on both the flexibility of the footing and the underlying geomaterial. When footings are constructed on loose sand, the grains along the border tend to move laterally, while the internal soil is relatively constrained. This discrepancy produces a pressure diagram, shown qualitatively in Figure 7.1a. The theoretical pressure distribution shown in Figure 7.1b is for the general situation of rigid footings on any material. The high edge pressure may be explained by the fact that shear at the edge must occur before settlement. Due to the low rupture strength of soil and the fact that most footings are of intermediate rigidity, it is unlikely that substantial edge shear stresses would arise. The edge stress is also dependent on the compressible soil thickness, H, as seen in Figure 7.1b.

Due to the interaction of the footing rigidity with the soil type, condition, and timely reaction to stress, the pressure distribution under most footings will be relatively indefinite. As a result, it is customary to use the linear pressure distribution seen in Figure 7.1c under spread footings. The few documented field measurements show this assumption is reasonable (Bowles 1996).

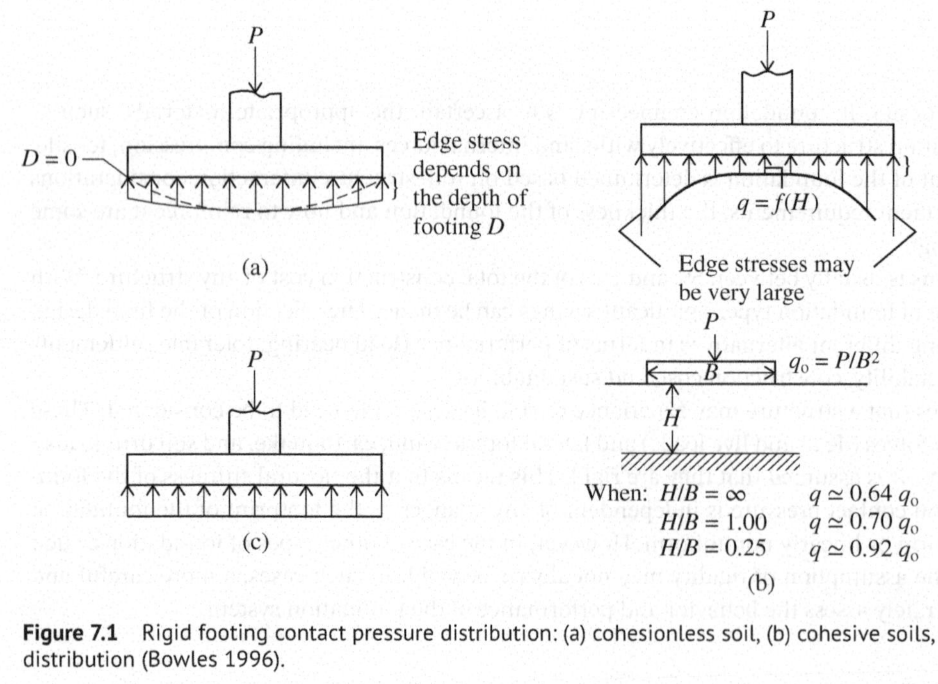

Figure 7.1 Rigid footing contact pressure distribution: (a) cohesionless soil, (b) cohesive soils, and (c) common assumed linear distribution (Bowles 1996).

7.2.2 Modulus of Subgrade Reaction

The modulus of subgrade reaction (k_s) is a theoretical relationship between soil pressure and deflection of the foundation. It is used for structural analysis of continuous footings, rafts, and a variety of pilings. When plate-load tests are conducted, the following basic equation is used (see Figure 7.2a):

$$k_s = \frac{q}{\delta} \tag{7.1}$$

where

q = bearing pressure
δ = settlement
k_s = subgrade reaction coefficient

Plots of q vs. δ from load tests produce the curves seen in Figure 7.2b qualitatively. If this sort of curve is used to determine k_s in the previous equation, the value is dependent on whether the modulus is tangent or secant and on the coordinates of q and δ.

Vesic' (1961) proposed Equation 7.2 to designate the subgrade reaction modulus:

$$k'_s = 0.65\sqrt[12]{\frac{E_s B^4}{E_f I_f}}\frac{E_s}{1-\mu^2} \tag{7.2}$$

where

E_s, E_f = soil and footing modulus
B, I_f = width and inertia moment of footing

k_s can be obtained based on k'_s via:

$$k_s = \frac{k'_s}{B} \tag{7.3}$$

For practical purposes, Equation 7.2 is simplified to:

$$k_s = \frac{E_s}{B(1-\mu^2)} \tag{7.4}$$

Table 7.1 can be used to estimate the value of k_s in order to ascertain the subgrade modules' order of magnitude. If a calculated value is more than two or three times the range shown in the table, the calculations should be rechecked for any extraordinary errors.

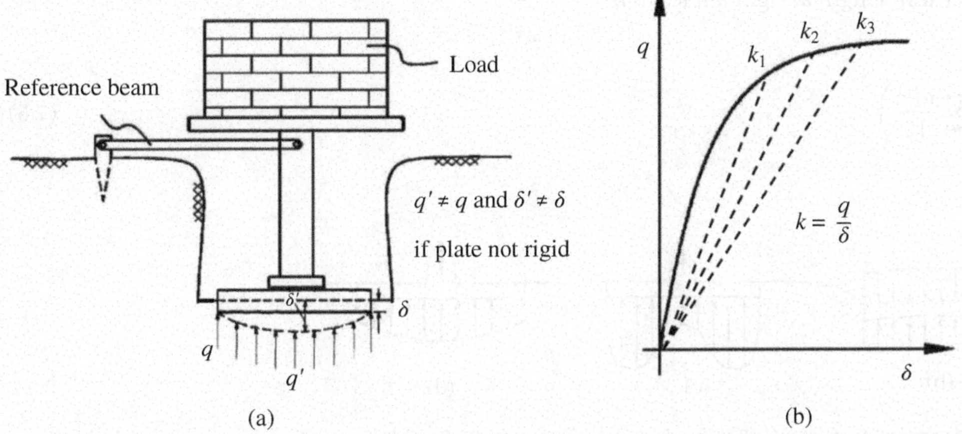

Figure 7.2 Subgrade reaction modulus determination. (a) Test set-up; (b) Stress-strain plot.

Table 7.1 Typical Range of Subgrade Reaction Modulus (Bowles 1996).

Soil	k_s (kN/m^3)
Loose sand	4,800–16,000
Medium dense sand	9,600–80,000
Dense sand	64,000–128,000
Clayey medium-dense sand	32,000–80,000
Silty medium-dense sand	24,000–48,000
Clayey soil:	
$q_a \leq 200$ kPa	12,000–24,000
$200 < q_a \leq 800$ kPa	24,000–48,000
$q_a > 800$ kPa	>48,000

Note: q_a = allowable bearing capacity = q_{ult}/FS.

7.2.3 Flexibility and Rigidity

The assumption of rigidity for single and combined two-column foundations is quite close to reality due to their typical dimensions. However, in the case of strip and raft foundations, assuming complete rigidity is not realistic and often results in excessive and sometimes inefficient designs. Figure 7.3 illustrates various stress distribution conditions based on the underlying deposit and the rigidity or flexibility of the foundation.

Since the rigidity and flexibility of foundations are influenced by multiple factors, it is essential to identify these influential factors first. This allows for the selection of appropriate analysis methods and the application of sound engineering judgment to ensure safe and optimal design. The following segment provides an overview of different methods for evaluating the rigidity of foundations. By employing these methods, engineers can accurately assess the behavior and performance of foundations, leading to more reliable and efficient designs.

Committee 336 of the American Concrete Institute proposed an approach to determine the rigidity of foundations in 1988. The rigidity factor K_r is determined as:

$$K_r = \frac{E' I_b}{E_s B^3} \tag{7.5}$$

where

E' = structure material elasticity modulus
E_s = soil elasticity modulus
B = foundation width
I_b = structure inertia moment per unit length at right angles to B

The term $E' I_b$ is defined as:

$$E' I_b = E' \left(I_F + \sum I_b' + \sum \frac{ah^3}{12} \right) \tag{7.6}$$

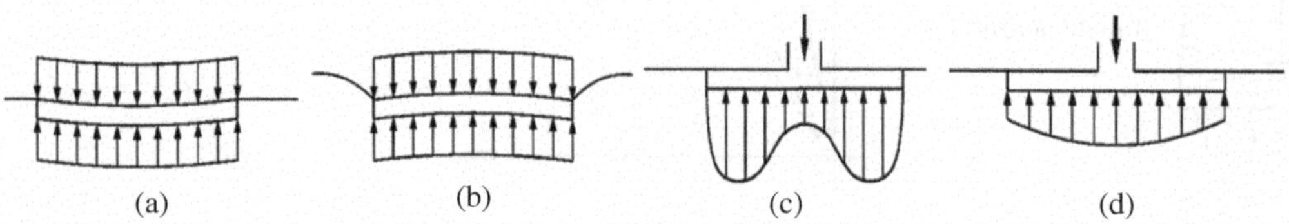

Figure 7.3 Stress distribution under spread footings: (a) flexible foundation on clay, (b) flexible foundation on sand, (c) rigid foundation on clay, and (d) rigid foundation on sand.

where

$E'I_b$ = superstructure and foundation flexural rigidity per unit length at right angles to B
$\sum E'I'_b$ = framed members flexural rigidity at right angles to B
$\sum E'\frac{ah^3}{12}$ = shear walls flexural rigidity
a = shear wall thickness
h = shear wall height
$E'I_F$ = foundation flexibility

Considering the value of K_r,

if $K_r > 0.5$, the foundation acts as a rigid one;
if $K_r < 0.5$, the foundation can be considered flexible (Das 2019).

If the following conditions are met, a strip footing is considered rigid:

1) The average length of two consequent spans is less than $1.75/\lambda$

$$\lambda = \sqrt[4]{\frac{K_s B}{4EI}}, \tag{7.7}$$

where

EI = foundation flexural rigidity
B = the breadth of the foundation
K_s = subgrade reaction coefficient

2) The length of the two spans and adjacent column loads should not differ by more than 20% of the larger value.
Considering L as the length of the strip, λL can be used to determine the rigidity status of the foundation.
If $\lambda L < \frac{\pi}{4}$, the foundation is rigid.
If $\lambda L > \pi$, the foundation is flexible.
Practically in all cases that $> \frac{\pi}{4}$, the flexible foundation equations can be used (Bowles 1996).
Another method to determine the rigidity of the foundation is defining the $\bar{k}$ factor. This factor is calculated as:

$$\bar{k} = \frac{1}{12}\left(\frac{E_c}{E_s}\right)\left(\frac{t}{L}\right)^3, \tag{7.8}$$

where

t = foundation thickness
L = foundation length in the direction of the determination of rigidity
E_c = concrete elasticity modulus
E_s = soil elasticity modulus

Based on the value of $\bar{k}$,

if $\bar{k} > 0.5$, the foundation can be considered rigid;
if $\bar{k} < 0.5$, the foundation can be considered rigid (Eslami 2013).

Example 7.1 Evaluate the rigidity or flexibility of the following foundations:

$$K_s = 30\frac{\text{MPa}}{\text{m}}, \quad E_s = 60\text{ MPa}, \quad f'_c = 25\text{ MPa}, \quad E_c = 23\text{ GPa}$$

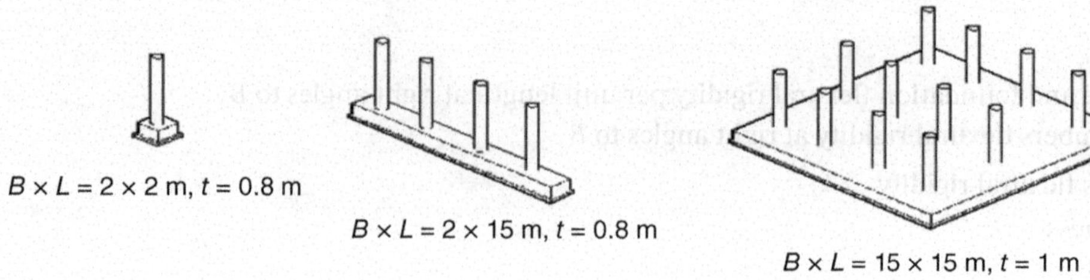

λL method

$$\text{Single: } \lambda = \sqrt[4]{\frac{K_s B}{4EI}} = \sqrt[4]{\frac{30 \times 2}{4 \times 23 \times 10^3 \times \frac{1}{12} \times 2 \times 0.8^3}} = 0.3 \rightarrow \lambda L = 0.6 < \frac{\pi}{4} \rightarrow \text{Rigid}$$

$$\text{Strip: } \lambda = \sqrt[4]{\frac{K_s B}{4EI}} = \sqrt[4]{\frac{30 \times 2}{4 \times 23 \times 10^3 \times \frac{1}{12} \times 2 \times 0.8^3}} = 0.3 \rightarrow \lambda L = 4.5 > \pi \rightarrow \text{Flexible}$$

$$\text{Mat: } \lambda = \sqrt[4]{\frac{K_s B}{4EI}} = \sqrt[4]{\frac{30 \times 15}{4 \times 23 \times 10^3 \times \frac{1}{12} \times 15 \times 1^3}} = 0.25 \rightarrow \lambda L = 3.75 > \pi \rightarrow \text{Flexible}$$

$\overline{k}$ method

$$\text{Single: } \overline{K} = \frac{1}{12}\left(\frac{E_c}{E_s}\right)\left(\frac{t}{L}\right)^3 = \frac{1}{12}\left(\frac{23 \times 10^3}{60}\right)\left(\frac{0.8}{2}\right)^3 = 2.04 > 0.5 \rightarrow \text{Rigid}$$

$$\text{Strip: } \overline{K} = \frac{1}{12}\left(\frac{E_c}{E_s}\right)\left(\frac{t}{L}\right)^3 = \frac{1}{12}\left(\frac{23 \times 10^3}{60}\right)\left(\frac{0.8}{15}\right)^3 = 0.005 < 0.5 \rightarrow \text{Flexible}$$

$$\text{Strip: } \overline{K} = \frac{1}{12}\left(\frac{E_c}{E_s}\right)\left(\frac{t}{L}\right)^3 = \frac{1}{12}\left(\frac{23 \times 10^3}{60}\right)\left(\frac{1}{15}\right)^3 = 0.0095 < 0.5 \rightarrow \text{Flexible}$$

7.2.4 Analysis of Grillage Foundations

Grillage foundations are used in cases where the loads from the superstructure are heavy and the ground conditions are hard or rocky. In situations where conventional foundation systems utilizing reinforced concrete are deemed insufficient in providing safety against tensile, compressive, and shear stresses, steel *I*-beams are utilized. The steel *I*-beams are arranged in two layers perpendicular to each other, surrounded by a concrete framework. In cases of lower loads or when steel *I*-beams are unavailable, temporary foundations can be constructed using two rows of orthogonal wooden beams stacked on top of each other and enclosed in concrete. The plan and section of the steel grillage foundations are depicted in Figure 7.4a,b.

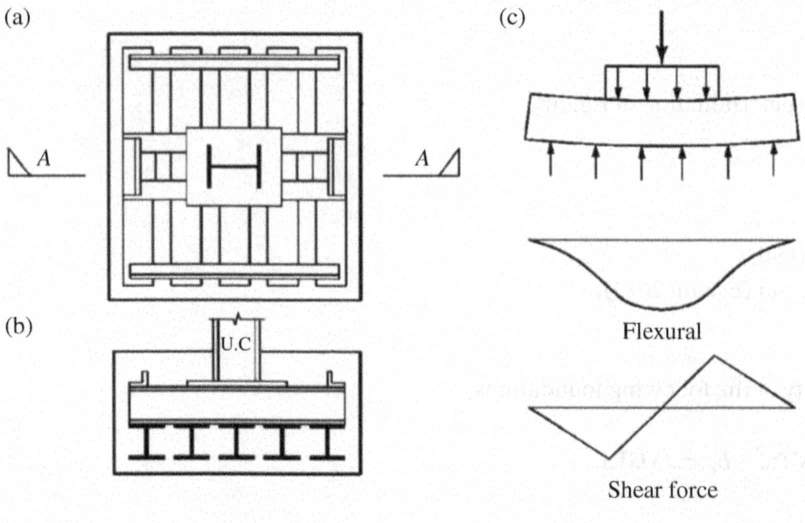

Figure 7.4 Grillage foundation: (a) plan, (b) section, and (c) typical flexural moment and shear force diagram.

The assumption of rigidity is commonly employed when analyzing these foundations. Figure 7.4c demonstrates the typical flexural moment and shear force diagrams of a grillage foundation.

Example 7.2 Rigidity evaluation and analysis of a grillage foundation:
The foundation is designed of IPB 500 profiles with a length of 3 m and a base plate with a breadth of 0.75 m. The load from the column is 15,000 kN, and the foundation is implemented on a bedrock that has an elasticity modulus of 500 MPa and a subgrade reaction modulus of 200 $\frac{\text{MPa}}{\text{m}}$. There are two profiles on the top layer, and the loads from the base plate and the lower profiles imposing on the upper profiles are distributed and uniform. It is assumed that the profiles work individually. The characteristics of the IPB500 profile are as follows:

$$I_x = I_y = 12{,}620\ \text{cm}^4,\quad S = 842\ \text{cm}^3,\quad A = 239\ \text{cm}^2,\quad h = 50\ \text{cm},\quad b = 30\ \text{cm}$$

$$E_{\text{steel}} = 200\ \text{GPa},\quad F_y = 240\ \text{MPa}$$

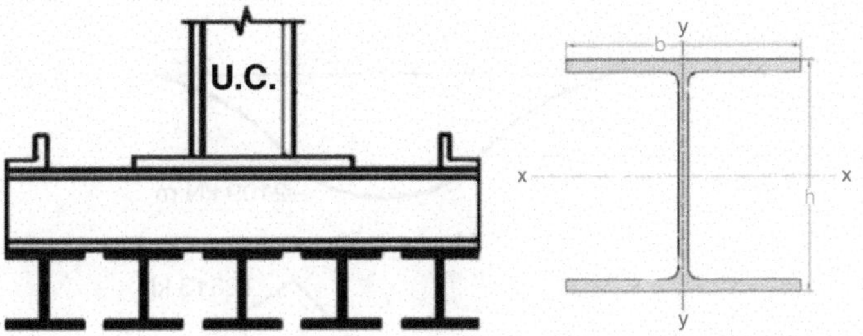

Rigidity Evaluation:

λL method

$$\lambda = \sqrt[4]{\frac{K_s B}{4EI}} = \sqrt[4]{\frac{200 \times 0.5}{4 \times 2 \times 10^5 \times 12{,}620 \times 10^{-8}}} = 0.998 \rightarrow \lambda L = 2.99 > \frac{\pi}{4} \rightarrow \text{Flexible}$$

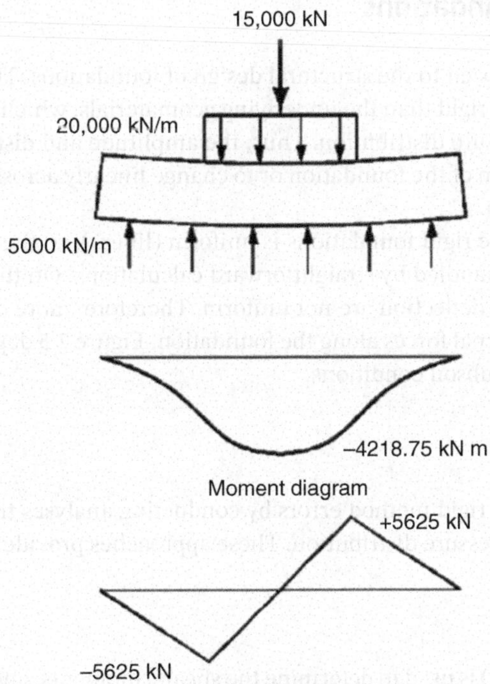

$\bar{k}$ method

$$\bar{K} = \frac{1}{12}\left(\frac{E_{steel}}{E_{soil}}\right)\left(\frac{t}{L}\right)^3 = \frac{1}{12}\left(\frac{2\times 10^5}{500}\right)\left(\frac{0.5}{3}\right)^3 = 0.154 < 0.5 \rightarrow \text{Flexible}$$

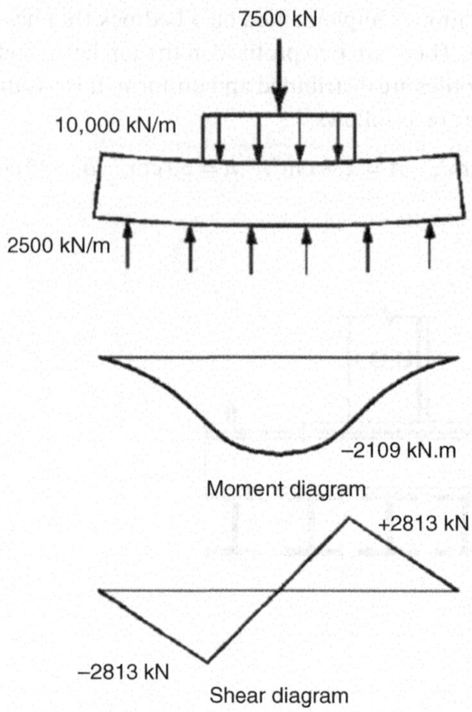

7.3 Analysis of Flexible Foundations

The rigid method is the simplest approach to the structural design of foundations. This approach is based on the assumption that the foundation is much more rigid than the underlying geomaterials, which implies that any distortions will have a negligible effect on the bearing pressure distribution. Thus, the amplitude and distribution of bearing pressure are considered uniform throughout the bottom of the foundation or to change linearly across the foundation based on the applied loads and the weight of the foundation.

The contact pressure (q) beneath the rigid foundations is uniform (linear), so the analysis and design of these foundations are not challenging and can be handled by straightforward calculations. On the other hand, for nonrigid or flexible foundations, the contact pressure and deflection are not uniform. Therefore, more complicated calculations are required due to variable k_s and subsequent internal forces along the foundation. Figure 7.5 depicts the bearing pressure distribution under a mat foundation for different subsoil conditions.

7.3.1 Simplified Classic Methods

These approaches compensate for the rigid method errors by conducting analyses that consider the foundation deformations and their effect on the bearing pressure distribution. These approaches provide more precise values for deformations and stresses.

7.3.1.1 Winkler Method

The "bed of springs" model (Figure 7.6) is used to determine the shears, moments, and deformations in the raft foundation, which serves as the basis for structural design. Because Winkler (1867) is credited with the first use of this approach to

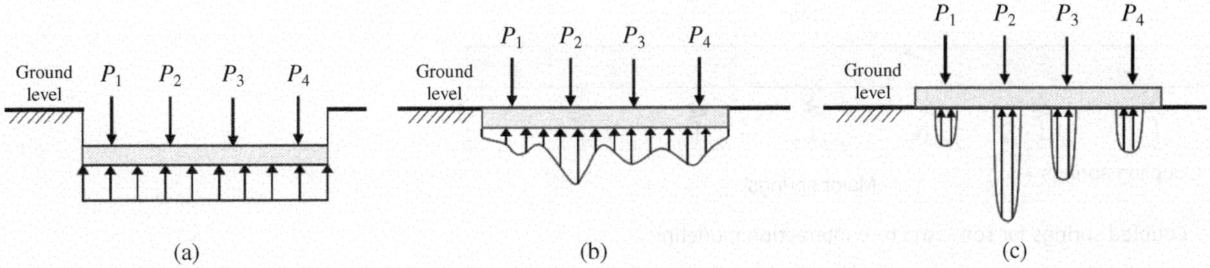

Figure 7.5 Bearing pressure distribution under mat foundation on various geomaterials: (a) soft, (b) stiff or compacted, (c) bedrock or very hard.

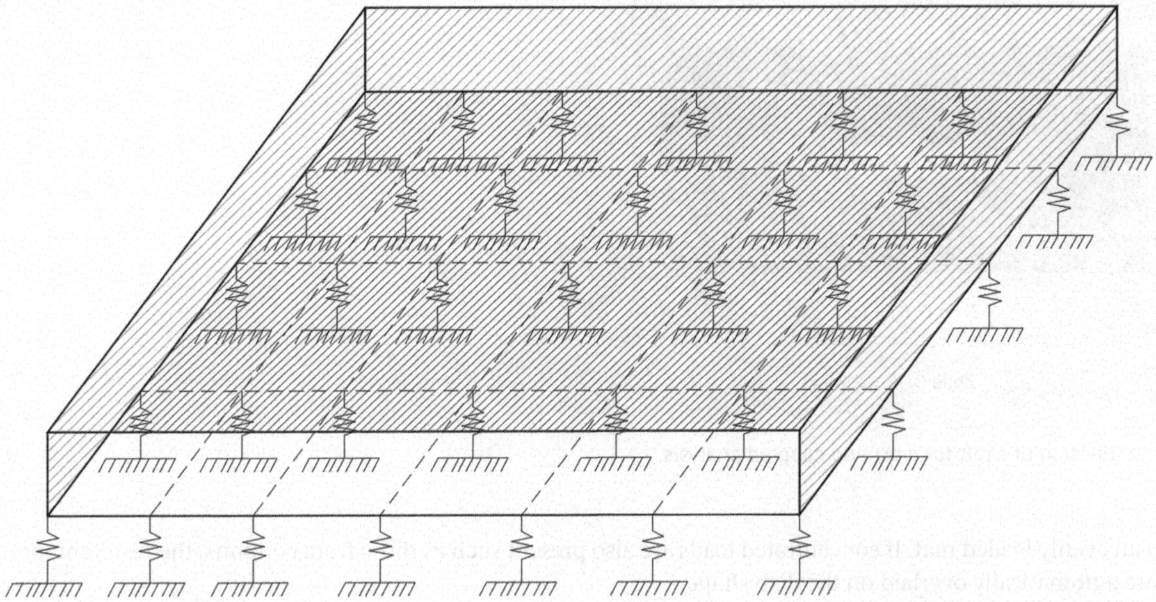

Figure 7.6 The "Spring Bed" analogy.

depict the interaction of soil and foundations, the analytical model is frequently referred to as a "Winkler foundation" or the "Winkler method." Additionally, it is referred to as a "beam on elastic foundation" analysis. This approach of describing bearing pressure is a subset of the analysis of the soil–structure interaction.

The Winkler method, in its classical form, assumes that each "spring" is linear and operates independently of the others and that all springs have the same k_s. This depiction achieves the intended result of raising the bearing pressure under the settlement-prone areas of the mat and represents a considerable advance over the rigid method. Closed-form solutions have been developed for a variety of shapes and loading conditions.

Unfortunately, the Winkler method's errors are often unconservative, resulting in an underestimate of the mat flexural stresses. Therefore, modifications are required.

7.3.1.2 Pseudo-coupled Method

The next step up from a Winkler analysis is to do a coupled analysis (Figure 7.7). Thus, the vertical springs and the evenly loaded mat cease to function separately. While this methodology is more precise than the Winkler method in theory, it is unclear how the coupling springs k_s values are selected.

The pseudo-coupled method (Horvath 1993; Liao 1991) seeks to solve the Winkler method's lack of coupling while avoiding the coupled method's analytical issues, which is accomplished via the use of springs that operate independently but have varying k_s values depending on their placement on the mat. To simulate the actual reaction of homogenous soil more precisely, springs at the mat's periphery should be stiffer than those in the center, resulting in the desired dish-shaped

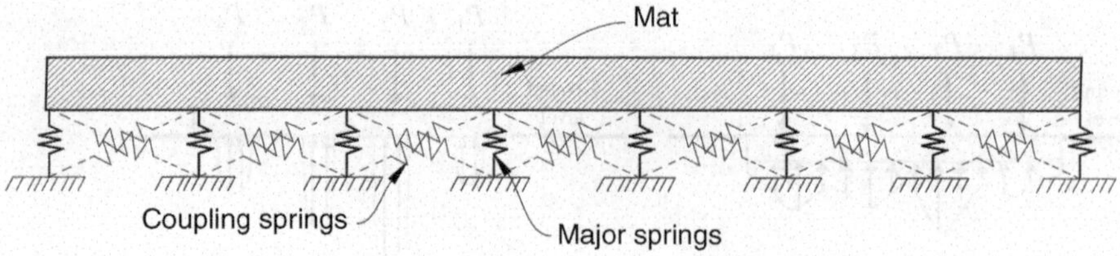

Figure 7.7 Coupled springs for soil–structure interaction modeling.

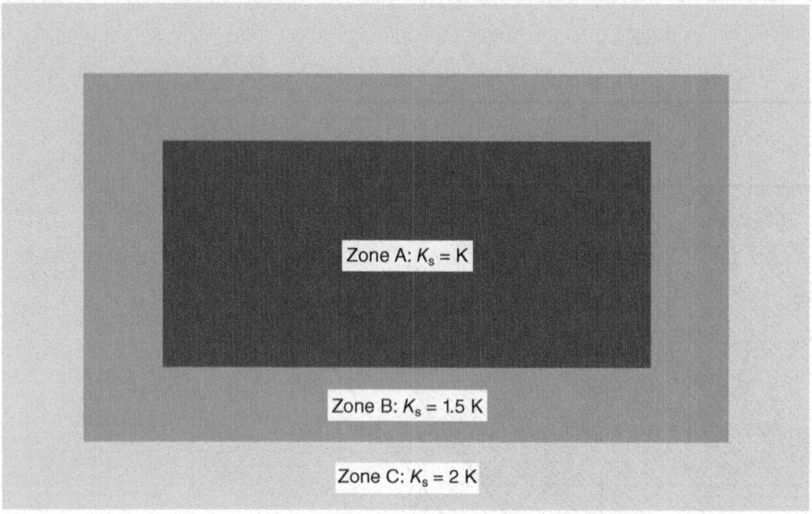

Figure 7.8 Typical division of a raft for a psuedo-coupled analysis.

deformation in an evenly loaded mat. If concentrated loads are also present such as those from columns, the resultant mat deformations are automatically overlaid on the dish shape.

According to model studies, satisfactory results are attained when the k_s values throughout the raft's perimeter are about double those in the center (ACI 1993), which may be accomplished in a number of ways, including the following:

1) Calculate the average value of k_s.
2) As illustrated in Figure 7.8, divide the mat into three or more concentric zones. The innermost zone should be 30–50% wider and 30–50% longer than the raft.
3) Each zone should be assigned a k_s value. These values should gradually grow out from the center so that the outermost zone's k_s is approximately double that of the innermost zone. Select the areas and k_s values so that the integral of the individual k_s values multiplied by their respective areas equals the average k_s multiplied by the overall area.
4) Utilize the Winkler "bed of springs" analysis to determine the shears, moments, and deformations of the raft.
5) Adjust the raft thickness and reinforce as necessary to provide the required strength and serviceability.

7.3.1.3 Beam on Elastic Foundation Concept

When the flexural rigidity of the footing is considered, a solution based on some sort of beam on an elastic foundation may be implemented (Figure 7.9). The primary differential equation is as follows:

$$EI\frac{d^4y}{dx^4} = q = -k_s'y \tag{7.9}$$

where

$k_s' = k_sB$
y = deflection of the foundation

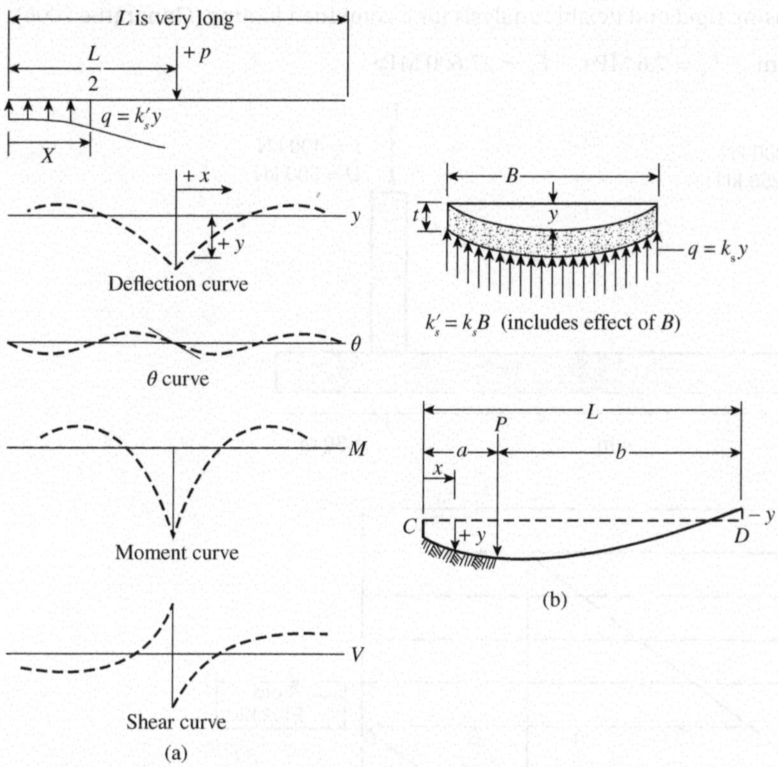

Figure 7.9 Beam on elastic foundation: (a) infinite length and (b) finite length (Bowles 1996).

Also, the slope of deflection (θ), bending moment (M), and shear force (Q) can be calculated via Equations 7.10–7.12. Table 7.2 contains closed-form solutions of the infinite beam on elastic foundations.

$$\theta = \frac{dy}{dx} \tag{7.10}$$

$$M = EI\frac{d^2y}{dx^2} \tag{7.11}$$

$$V = EI\frac{d^3y}{dx^3} \tag{7.12}$$

Table 7.2 Solutions in closed-form for an infinite beam on an elastic foundation (Bowles 1996).

Loading condition	y	θ	M	V
Concentrated load at end	$\frac{2V_1\lambda}{k'_s}D$	$\frac{-2V_1\lambda^2}{k'_s}A$	$\frac{-V_1}{\lambda}B$	$-V_1C$
Moment at end	$\frac{-2M_1\lambda^2}{k'_s}C$	$\frac{4M_1\lambda^3}{k'_s}D$	M_1A	$-2M_1\lambda B$
Concentrated load at center (+↓)	$\frac{P\lambda}{2k'_s}A$	$\frac{-P\lambda^2}{k'_s}B$	$\frac{P}{4\lambda}C$	$\frac{-P}{2}D$
Moment at center (+↷)	$\frac{M_0\lambda^2}{k'_s}B$	$\frac{M_0\lambda^3}{k'_s}C$	$\frac{M_0}{2}D$	$\frac{-M_0\lambda}{2}A$

$\lambda = \sqrt[4]{\frac{K_sB}{4EI}}$, $A = e^{-\lambda x}(\cos\lambda x + \sin\lambda x)$, $B = e^{-\lambda x}\sin\lambda x$, $C = e^{-\lambda x}(\cos\lambda x - \sin\lambda x)$, $D = e^{-\lambda x}\cos\lambda x$

Example 7.3 Shear and moment distribution using rigid and flexible analysis for a combined footing (Gunaratne 2006).

$B = 1.22$ m, $L = 6.08$ m, $B_{\text{Column}} = 0.3$ m, $E_s = 7.6$ MPa, $E_c = 27{,}600$ MPa

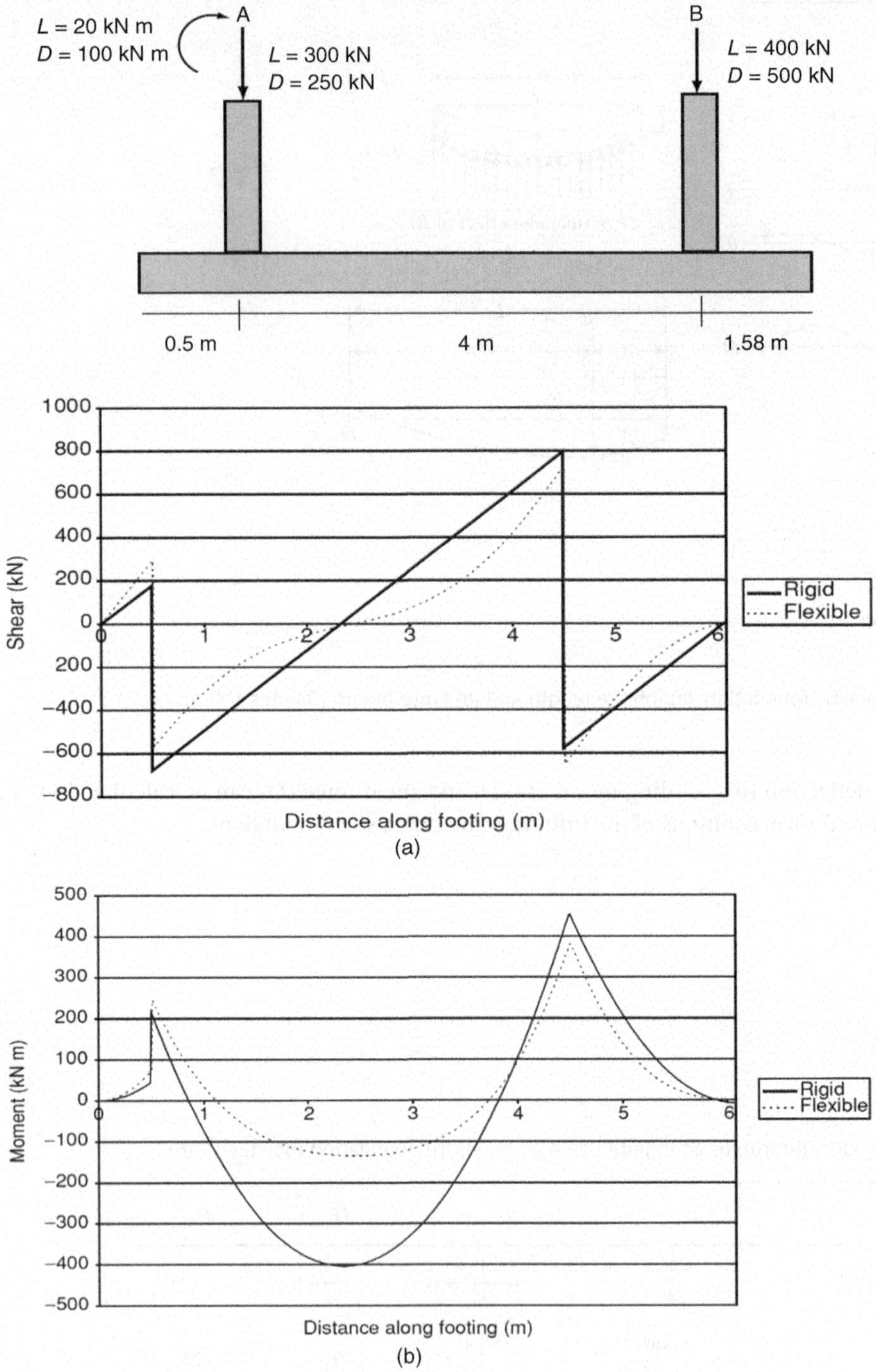

Mat (raft) footings may also be analyzed using the concept of slabs on elastic foundations. In polar coordinates, the governing differential equation for a free (unloaded) axisymmetrical structural plate or slab on an elastic foundation is as follows:

$$\frac{Eh^3}{12(1-\nu^2)}\left(\frac{d^4y}{dr^4}+\frac{2}{r}\frac{d^3y}{dr^3}-\frac{1}{r^2}\frac{d^2y}{dr^2}+\frac{1}{r^3}\frac{dy}{dr}\right)+k_s y = 0 \tag{7.13}$$

It is worth noting that the axisymmetric conditions eliminate the need for any θ terms in Equation 7.13.

Several investigators have proposed using the Winkler method to provide a closed-form solution for the basic differential equations for multiple loadings. These solutions have a number of significant disadvantages over the finite element solution provided in the next section, including the following:

1) Incorrect to take on the role of a weightless beam (since the weight will be a factor in case of footing tending to separate from the soil).
2) Challenging to eliminate the soil impact when the footing separates from the soil.
3) Difficult to compensate for boundary conditions with known rotation or deflection at specified sites.
4) Complex to impose different types of loads on a footing.
5) Challenging to alter footing properties along with member.
6) Hard to make a possible variation in subgrade reaction along the footing (Bowles 1996).

7.3.1.4 Finite Element Method (Structural Matrix Analysis Method)

This analytical technique partitions the soil into a network of tiny three-dimensional components, each with distinct engineering features and linked in a specific manner to the surrounding elements. After applying structural and gravitational stresses, the components are strained and deformed appropriately. This technique should accurately describe the foundation and allow for precise and economical design. This methodology should also not be confused with structural analysis techniques that represent the foundation with two-dimensional finite elements and the soil using Winkler springs.

The finite-element method (FEM) is the most efficient approach for handling problems involving beams on elastic foundations, but it needs a digital (or personal) computer. It is therefore straightforward when accounting for boundary conditions such as an unrotated or translated point, beam weight, and nonlinear soil effects.

The FEM is more adaptable than the finite-difference technique (FDM) in that it allows for the creation of equation models for individual elements and their usage in the beam model. All components must be identical in length and cross-section when using the finite-difference approach. The end element needs different equations rather than inner components, and modeling boundary conditions, as well as nonlinear soil effects, is challenging. The FDM initially had the benefit of using less computer memory because only the displacement is unknown at the node. This advantage was completely negated with the discovery of band matrix solution methods.

The footing must first be discretized into numerous one-dimensional (beam) components using the FEM approach. Figure 7.10 illustrates a typical discretization procedure used to prepare for load-deflection analysis.

Given that the beam is assumed to be resting on an elastic foundation, the modulus of vertical subgrade reaction k_s at any location (p) may be related to the deflection of the beam (y) at that point using the following formula:

$$p = k_s y \tag{7.14}$$

Thus, spring stiffness K_j (force/deflection) may be conveniently stated in terms of the modulus of the vertical subgrade reaction k_s. Furthermore, the dimensions of foundation (B and L) are as follows:

$$K_j = BLk_s \tag{7.15}$$

Using basic structural matrix analysis techniques, the element stiffness matrices may be combined to form the global stiffness matrix $[K]$. The spring stiffness K_j of each node may be added to the appropriate diagonal element of $[K]$ during the assembly process. Therefore:

$$[P] = [K][y], \tag{7.16}$$

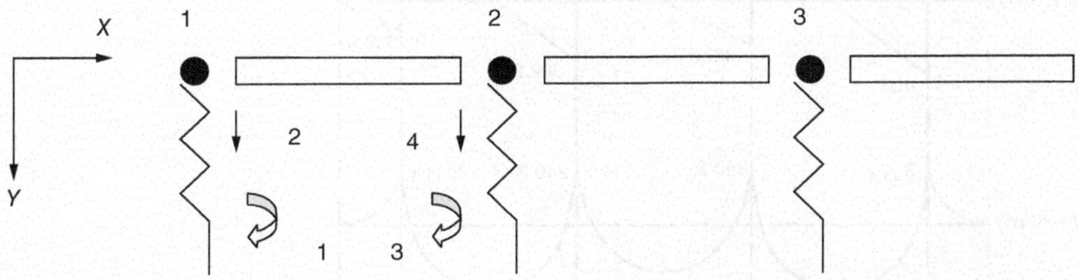

Figure 7.10 Typical discretization for stiffness analysis method (Gunaratne 2006 / With permission of Taylor & Francis Group).

where

$[P]$ = Load vector

$[y]$ = Deflection vector

Then, one may solve for and derive the global deflection vector using the global force vector. Finally, by inserting the nodal deflections into the individual element equations, the moments and shear forces inside the foundation may be computed.

Example 7.4 Analyze the given foundation in rigid and flexible conditions:

$f'_c = 21$ MPa, $f_y = 300$ MPa, $B = 1.25$ m, Column = 20 cm × 20 cm

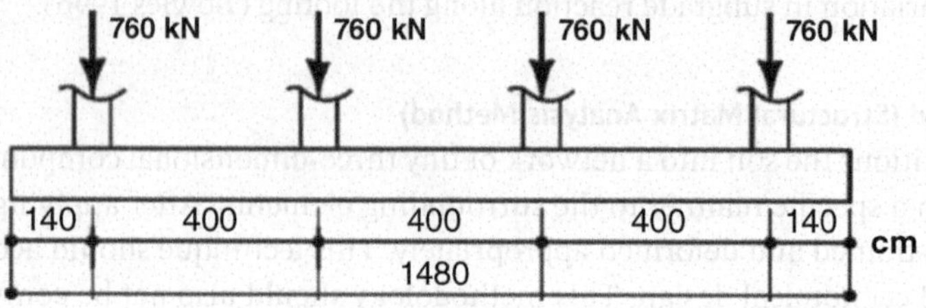

Rigid Condition ($K_s = 0.1\frac{\text{MPa}}{\text{m}}$)

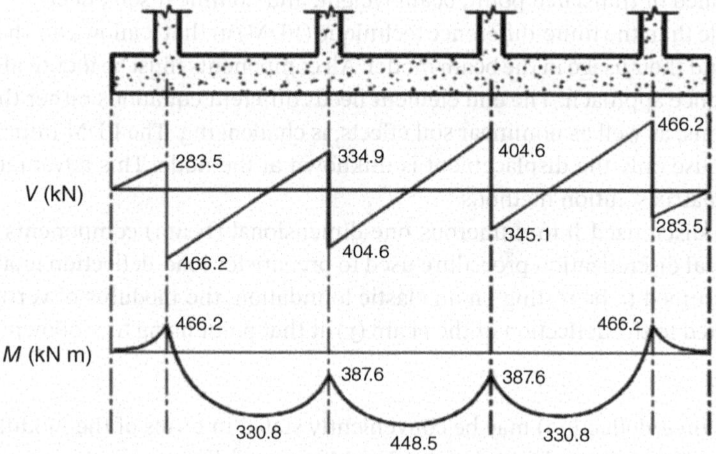

Flexible Condition ($K_s = 100\frac{\text{MPa}}{\text{m}}$) via FEM method

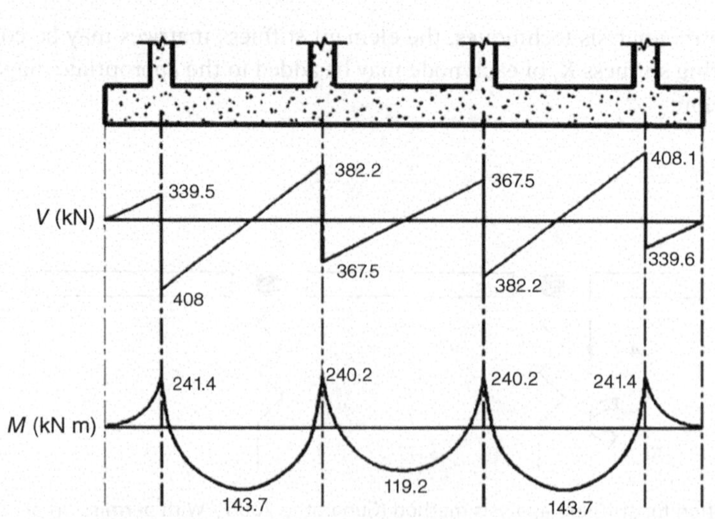

There is a considerable difference in the results of the comparison between flexible and rigid conditions. Therefore, for foundations, particularly strip and continuous, it is essential to implement a rigid and flexible analysis. Since simplified routine calculations can lead to incorrect results, using sophisticated analysis, a numerical approach, and supplementary software is recommended.

Example 7.5 Analyze one design strip (PQIK) of the given mat foundation in rigid and flexible conditions:

$$f_c' = 21\text{ MPa},\quad f_y = 300\text{ MPa},\quad \text{Column} = 40\text{ cm} \times 40\text{ cm},\quad T = 75\text{ cm}$$

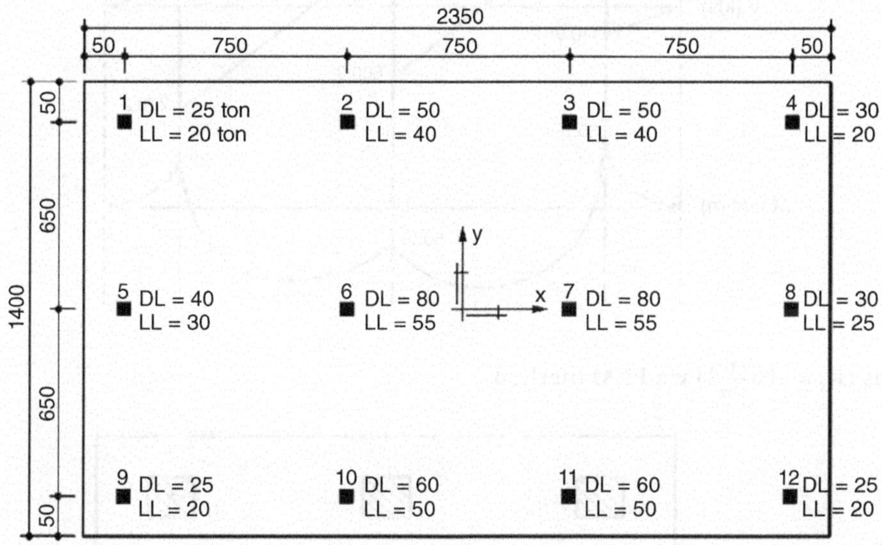

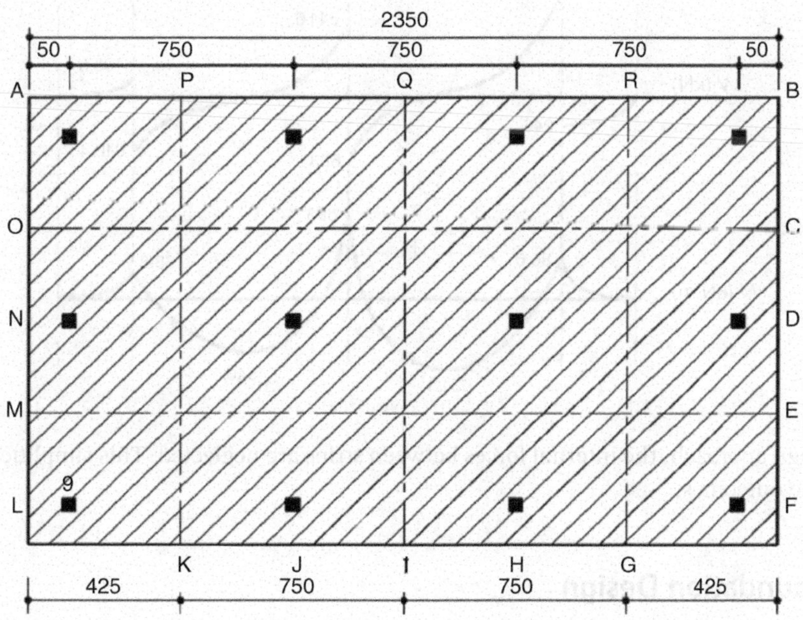

Rigid Conditions ($K_s = 0.2\frac{\text{MPa}}{\text{m}}$)

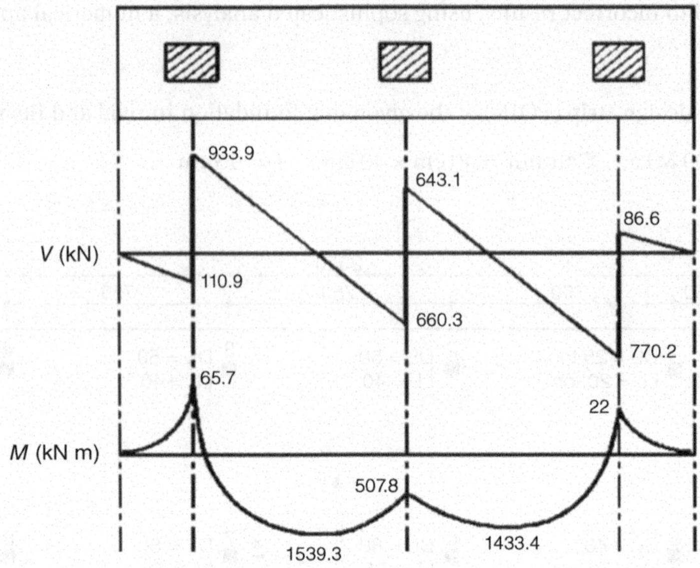

Flexible Conditions ($K_s = 100\frac{\text{MPa}}{\text{m}}$) via FEM method

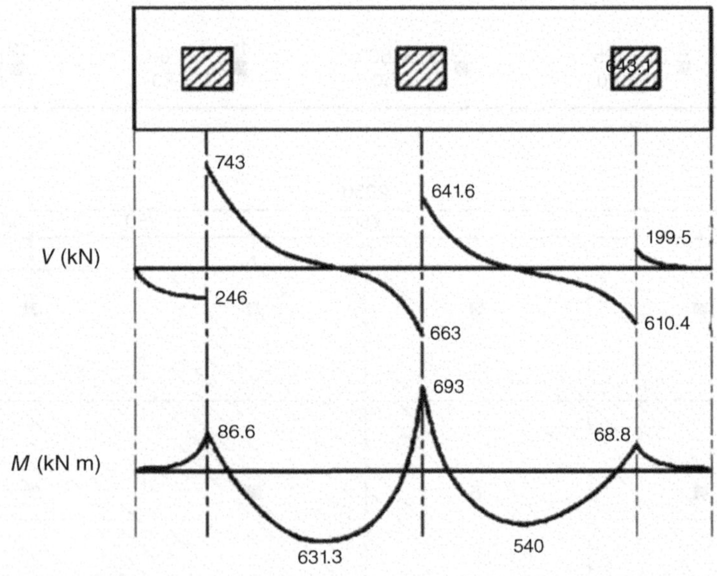

In using the strip design approach, the internal forces between strips are neglected. This simplified procedure is not true and can highly affect the outcomes.

7.4 Structural Foundation Design

Spread footings are structurally designed using the load and resistance factor design method (LRFD). The ACI 318 Building Code Requirements for Structural Concrete (ACI 2011) for buildings and the AASHTO LRFD Bridge Design Specifications (AASHTO 2012) for transportation-related structures are the two most often utilized codes in the United States.

All footings must be capable of supporting the structure imposed compressive loads. Additionally, footings may be needed to bear applied shear loads or moments. Figure 7.11 depicts two major failures of spread footings. Since structural design

Figure 7.11 Major structural failures of spread footings: (a) shear failure and (b) flexural failure.

is based on LRFD methods, these demands will be continually calculated as factored compressive load, P_u, factored shear load, V_u, and factored moment, M_u, using load combinations and load factors provided by the applicable structural code (e.g., ACI and AASHTO).

The structural capacity of a footing will be stated using LRFD methods as a resistance factor multiplied by the nominal capacity of the footing. Three capacities must be computed in general: the factored compressive capacity, ϕP_n, the factored shear capacity, ϕV_n, and the factored moment capacity, ϕM_n. By ensuring that the factored loads never exceed the factored resistance, a suitable degree of reliability is attained. Equations 7.17–7.19 express this requirement mathematically:

$$P_u \leq \phi P_n \tag{7.17}$$

where

$\phi = 0.65$ (ACI 2011) and 0.7 (AASHTO 2012);

$$V_u \leq \phi V_n \tag{7.18}$$

where

$\phi = 0.75$ (ACI 2011) and 0.9 (AASHTO 2012); and

$$M_u \leq \phi M_n, \tag{7.19}$$

where

$\phi = 0.9$ (ACI 2011) and 0.9 (AASHTO 2012).

The design process of footings can be summarized as follows:

1) The footing depth is determined regarding architectural requirements and soil conditions.
2) The footing width is determined based on geotechnical design.
3) The thickness of the footing is governed by shear demand. It may be essential to adjust the depth of the footings to provide enough coverage for the floor slabs.
4) Flexural demand dictates the reinforcing, size, and spacing of bars.
5) The details of the connection are governed by the type, shape, and design of the columns or walls.

The chapter design process will follow the ACI code, which is applicable to buildings, retaining walls, and similar structures. The method for designing footings for other structures such as bridges is similar, although the code requirements will vary.

7.5 Structural Design of Rigid Square Footings

This section discusses the structural design of square footings that support a single column in the center. Additional footing types are discussed in the following sections. The structural design process begins with an examination of the shear capability.

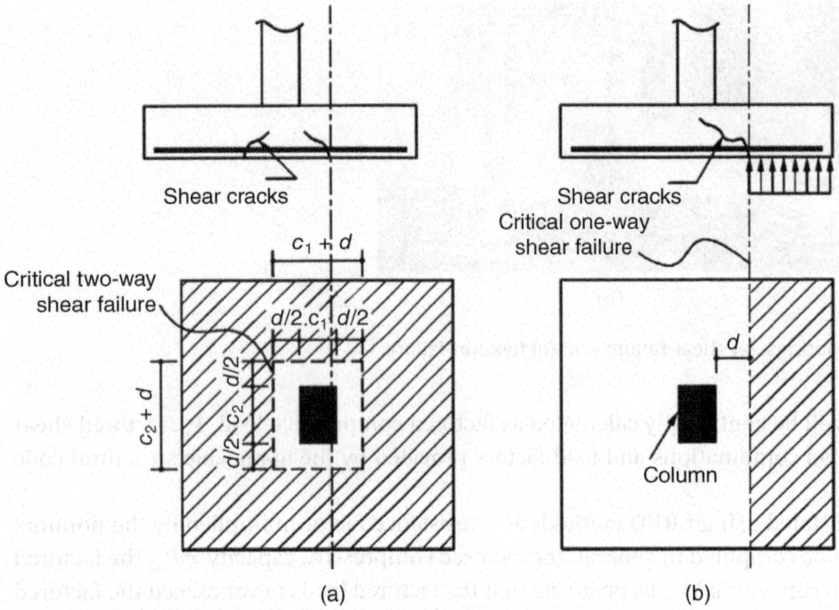

Figure 7.12 Shear failure modes: (a) one-way shear and (b) two-way shear.

7.5.1 Designing for Shear

ACI distinguishes two types of shear failure in reinforced concrete: one-way shear (beam shear or wide-beam shear) and two-way shear (diagonal tension shear or punching shear). These two kinds of failure correspond to the failures shown in Figure 7.12 when spread footings are used. Although the failure surfaces are inclined, as seen in Figure 7.11a, engineers simplify simulations via these idealized vertical surfaces.

7.5.1.1 Two-Way Shear

Generally, the foundation designer is supplied with the factored load in the column immediately above the footing. Factored normal and shear forces, P_u and V_u, as well as factored moment, M_u, may be included in the demand. The factored shear force acting on the footing's critical surface, V_{uc}, is calculated using the factored compressive column load, P_u, and the factored column moment, M_u. The factored column shear force has no effect on the critical section shear forces. The following equation may be used to compute the factored shear force on the critical surface:

$$V_{uc} = \frac{P_u}{4}\left[\frac{B^2 - (c+d)^2}{B^2}\right] + M_u\left[\frac{B^3 - (c+d)^3}{B^4}\right], \tag{7.20}$$

where

V_{uc} = factored shear force on the most critical face
P_u = factored normal column force
M_u = factored column moment
c = column width or diameter (for concrete columns) or base plate width (for steel columns)
d = effective footing depth
B = footing width

According to ACI, the nominal two-way shear capacity of square footings supporting square or circular columns positioned inside the footing (i.e., not on the edge or corner) is

$$V_{nc} = V_c = \frac{1}{3} b_0 d \sqrt{f'_c} \tag{7.21 SI}$$

$$V_{nc} = V_c = 4 b_0 d \sqrt{f'_c} \tag{7.21 English}$$

$$b_0 = c + d \tag{7.22}$$

where

V_{nc} = nominal two-way shear capacity on the critical section (kN, lb)
V_c = nominal two-way shear capacity of concrete (kN, lb)
b_0 = critical shear surface length = length of one face of inner block (mm, in.)
c = column width (mm, in.)
d = effective depth (mm, in.)
f'_c = 28-day compressive strength of concrete (MPa, lb/in.^{2})

Other requirements apply if the column is of a different form or is positioned near the footing edge or corner [ACI 11.11.2.1]. If the footing is built of prestressed concrete, additional requirements apply [ACI 11.11.2.2]; however, spread footings are rarely prestressed.

This analysis aims to determine the effective depth, d, that satisfies Equation 7.18. V_{uc} and V_{nc} are both dependent on the effective depth, d, but there is no straightforward solution. As a result, the following iterative procedure is required:

1) Assume that d has a trial value. Generally, a value close to the column width is a reasonable starting point. When choosing trial values of d, keep in mind that T must be a multiple of 3 in. or 100 mm, which means that only the matching values of d are worth consideration. With d_b equal to 25 mm (1 in.), the possible values of d are 200, 300, and 400 mm or 8, 11, 14, and 17 in.
2) Calculate V_{uc} and V_{nc} and verify that Equation 7.18 is fulfilled.
3) Repeat Steps 1 and 2 until the lowest d satisfying Equation 7.18 is found.
4) Calculate the footing thickness, T, using Equation 7.23 with d_b = 25 mm (1 in.). Calculate it as a multiple of 100 mm (3 in.). T must be at least 300 mm due to practical building considerations (12 in.).

$$d = T - 70\ \text{mm} - d_b \quad (7.23\ \text{SI})$$

$$d = T - 3 - d_b \quad (7.23\ \text{English})$$

7.5.1.2 One-Way Shear

Like two-way shear, one-way shear is analyzed by integrating the bearing stress for that footing section outside the critical surface. The critical surface for one-way shear is seen in Figure 7.11a.

The factored shear force on the critical surface can be computed through:

$$V_{uc} = \frac{P_u}{2}\left[\frac{B - c - 2d}{B}\right] + \frac{3M_u}{2B}\left[1 - \frac{(c+2d)^2}{B^2}\right] \quad (7.24)$$

where

V_{uc} = factored shear force on critical shear surfaces
B = footing width
c = column width
d = effective footing depth
P_u = factored column normal force
M_u = factored column moment

Based on ACI, the nominal one-way shear capacity is

$$V_{nc} = V_c = \frac{1}{6} b_w d \sqrt{f'_c} \quad (7.25\ \text{SI})$$

$$V_{nc} = V_c = 2 b_w d \sqrt{f'_c} \quad (7.25\ \text{English})$$

where

V_{nc} = nominal one-way shear capacity on the critical section (kN, lb)
V_c = nominal one-way shear capacity of concrete (kN, lb)
b_w = length of critical shear surface = B (mm, in.)
d = effective depth (mm, in.)
f'_c = 28-day compressive strength of concrete (MPa, lb/in.^{2})

Once again, when Equation 7.18 is met, the design is satisfactory. Both V_{uc} and V_{nc} are dependent on the effective depth, d, which must be found using Equations 7.24 and 7.25, as discussed in the section on two-way shear. The final design value for d is the greater of the one-way and two-way shear evaluations.

7.5.2 Flexural Design

Once the shear analysis is complete, the design process may proceed to the flexural analysis. Both the ACI and AASHTO codes have substantially identical flexural design specifications.

7.5.2.1 Flexural Design Principles

The basic design issue for flexural-loaded reinforced concrete members is as follows: given a calculated moment on the critical surface, M_{uc}, estimate the member's required cross-sectional dimensions and the required size and position of the reinforcing bars. Fortunately, flexural design in foundations is simpler than other structural components since some dimensions are dictated by geotechnical issues.

The amount of steel needed to resist flexure is proportional to the effective depth, d, the distance between the extreme compression fiber and the tension reinforcement's centroid, as seen in Figure 7.13.

As seen in Figure 7.13, the nominal moment capacity of a flexural component made of reinforced concrete with $f'_c \leq$ 30 MPa (4000 lb/in.^{2}) is

$$M_n = A_s f_y \left(d - \frac{a}{2}\right), \tag{7.26}$$

where a is the depth of the thrust block depicted in Figure 7.13.

$$a = \frac{\rho d f_y}{\beta_1 f'_c}, \tag{7.27}$$

where

β_1 = an empirical factor, equal to 0.85 for concrete with $f'_c \leq$ 30 MPa (4000 lb/in.^{2})
ρ = the steel ratio

$$\rho = \frac{A_s}{bd} \tag{7.28}$$

Setting $M_u = \phi M_n$, M_u = the factored moment at the section being analyzed, using $\beta_1 = 0.85$, and solving for A_s results in

$$A_s = \left(\frac{f'_c b_w}{1.176 f_y}\right)\left(d - \sqrt{d^2 - \frac{2.353 M_{uc}}{\phi f'_c b_w}}\right), \tag{7.29}$$

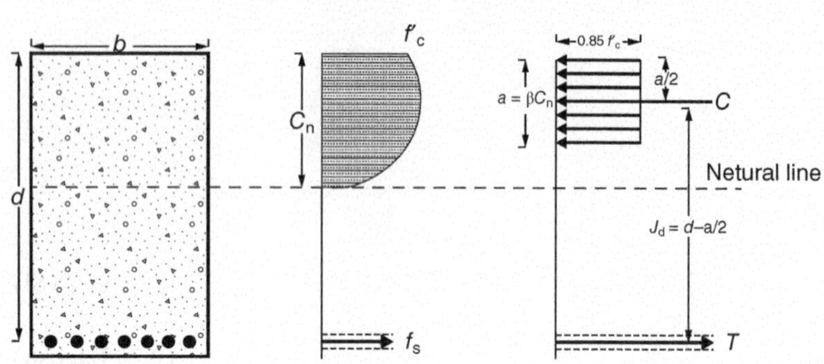

Figure 7.13 The reinforced concrete: (a) cross section, (b) actual stress distribution, and (c) equivalent stress distribution ($0.65 < \beta_1 < 0.85$).

where

A_s = reinforcing steel cross-sectional area
f'_c = concrete 28-day compressive strength
b_w = width of flexural member (in this case, this is the same as the footing width, B)
d = effective depth
ϕ = resistance factor = 0.9 for flexure [ACI 9.3.2]
M_{uc} = factored moment at the section being analyzed

The preceding analysis assumes that the flexural capacity is tension-controlled. If the tension zone is loaded to failure, the steel reinforcement will be placed before the concrete in the compression zone, guaranteeing that the footing will act as a ductile. A compatibility check must be done to ensure that this condition occurs.

The rebars must penetrate deeply enough into the concrete to provide adequate anchoring [ACI 15.6], referred to as the development length. With a clear gap of at least $2d_b$ between the bars and a concrete cover of at least d_b, the minimum needed development length, l_d, must be computed in accordance with [ACI 12.2.3].

7.5.2.2 Load Transfer from Column to Foundation

To ensure that the footing can support the axial column load, check its bearing capacity. This is especially likely to be a problem if the column has high compressive loads that might result in a footing failure resembling a loss of bearing capacity. Calculate the factored column load, P_u, and compare it to the nominal column bearing strength, P_{nb}, to rule out this possibility:

$$P_{nb} = 0.85 f'_c A_1 s \tag{7.30}$$

Then, determine whether the following statement is true:

$$P_u = \phi P_{nb} \tag{7.31}$$

where

P_u = factored column load
P_{nb} = nominal column bearing strength (i.e., bearing of column on top of footing)
f'_c = 28-day compressive strength of concrete used for the footing
$s = (A_2/A_1)^{0.5} \leq 2$ if the frustum in Figure 7.14 fits entirely within the footing (i.e., if $c + 4d \leq B$)
$s = 1$ if the frustum in Figure 7.14 does not fit entirely within the footing
A_1 = cross-sectional area of the column = c^2
A_2 = base area of the frustum $(c + 4d)^2$ as shown in Figure 7.14
c = column width or diameter
ϕ = resistance factor = 0.65

7.5.2.3 Application to Spread Footings

Spread footing bends in two perpendicular directions and may be constructed as a two-way slab using techniques like those used for a four-sided floor slab. For practical reasons, it is conventional to construct footings as though they were a one-way slab.

Once the amount of steel required to support the imposed load in one direction bending is determined, the identical steel area is placed in a perpendicular direction. In essence, the footing is reinforced twice, which is more than a rigorous two-way analysis requires.

7.5.2.4 Reinforcing Steel Area

In the design of flexural members, it is common to create a moment diagram and select sufficient steel for each section of the member. However, when dealing with isolated spread footings, the issue can be simplified due to the understanding that the

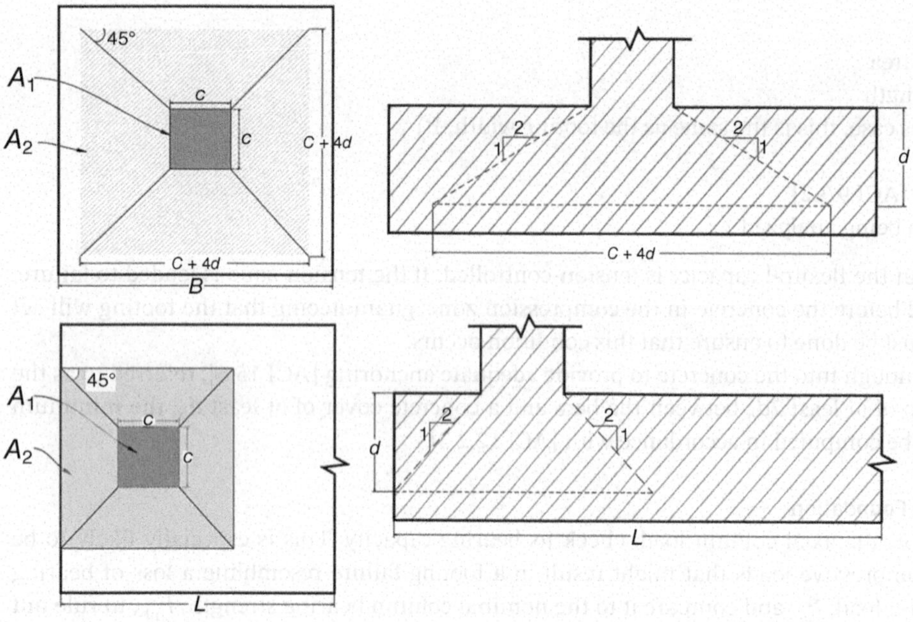

Figure 7.14 Application of a frustrum to find areas $\mathbf{A_1}$ and $\mathbf{A_2}$: (a) Single footing; (b) continues footing.

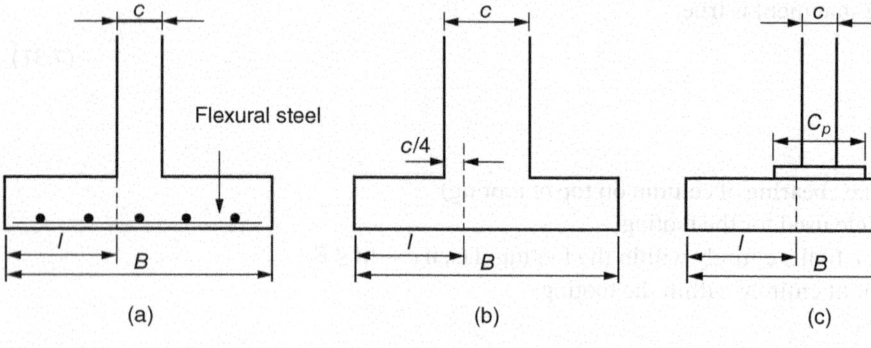

Figure 7.15 Bending critical section: (a) concrete column, (b) masonry column, and (c) steel column.

bending moment in the footing increases continuously from the edge to the center. Therefore, the critical section for bending should be positioned near the edge of the column. The exact location of this section is determined by the specifications of the column footing connection and varies depending on the type of column being connected to the footing. Figure 7.14 provides an illustration of the critical section placement for various column types. To simplify the calculations, a distance, denoted as "l" is designated between the critical section and the outer edge of the footing. This distance represents the cantilever distance. Figure 7.15 depicts the concept of the cantilevered distance. By considering this simplification, engineers can streamline the analysis and design process for isolated spread footings while still ensuring their structural integrity and performance.

The factored bending moment at the critical section, M_{uc}, is

$$M_{uc} = \frac{P_u l^2}{2B} + \frac{2M_u l}{B}, \tag{7.32}$$

where

P_u = factored normal force
M_u = factored moment
l = cantilever distance
B = footing width

Example 7.6 A 50 cm square reinforced concrete column supporting an 800 kN vertical dead weight and a 650 kN vertical live load must be supported on a square spread footing. The allowed bearing pressure based on bearing capacity has been found to be 0.3 MPa using ASD methods. According to a serviceability evaluation, the maximum bearing pressure that may be used to fulfill the settlement criteria is 0.19 MPa. The groundwater table is far below the footing bottom. Determine the needed width, B, thickness, T, and effective depth, d. Then design the required flexural steel.

Determination of Geometry
Because the footing is for a structure, the ACI code applies. The footing width must be selected using geotechnical analysis to ensure that the ultimate and serviceability limit states are not exceeded.

$$W_f = B^2 D\gamma_c = B^2(0.9\ \text{m})(24\ \text{kN/m}^3) = 21.6B^2$$

Because the allowable bearing pressures were estimated using ASD, the footing width must be computed using non-factored loads, and the allowable bearing stress is 0.3 MPa.

$$P = P_D + P_L + \ldots = 800\ \text{kN} + 650\ \text{kN} + 0 = 1450\ \text{kN}$$

$$B = \sqrt{\frac{P + W_f}{q_A + u_D}} = \sqrt{\frac{1450 + 21.6B^2}{300 + 0}} \rightarrow B = 2.28\ \text{m}$$

Then, depending on the serviceability limit state, we calculate the footing width. We must utilize allowable bearing stress of 0.19 MPa for this calculation.

$$B = \sqrt{\frac{P + W_f}{q_A + u_D}} = \sqrt{\frac{1450 + 21.6B^2}{190 + 0}} \rightarrow B = 2.93\ \text{m}$$

Settlement governs →use $B = 3$ m
We will now begin our structural design in order to calculate the thickness of the footing. Because this design is for a structure, ASCE 7 load factors apply:

$$P_u = 1.2\,P_D + 1.6\,P_L = 1.2 \times 800 + 1.6 \times 650 = 2000\ \text{kN}$$

Due to the large factored load and the fact that this is a large spread footing:

$$f_c' = 28\ \text{MPa} \quad \text{and} \quad f_y = 300\ \text{MPa}$$

Design for Shear
Because no moment or shear loads are applied, there is no need to verify for one-way shear. Find the necessary thickness using a two-way shear analysis.

Assume $T = 750$

$$d = T - 70\ \text{mm} - d_b = 750 - 70 - 25 = 655\ \text{mm}$$

$$V_{uc} = \frac{P_u}{4}\left[\frac{B^2 - (c+d)^2}{B^2}\right] + M_u\left[\frac{B^3 - (c+d)^3}{B^4}\right] = \frac{2000}{4}\left[\frac{3000^2 - (500 + 655)^2}{3000^2}\right] = 425.9\ \text{kN}$$

$$b_0 = c + d = 500 + 655 = 1155\ \text{mm}$$

$$V_{nc} = \frac{1}{3}b_0 d\sqrt{f_c'} = \frac{1}{3} \times 1155 \times 655 \times \sqrt{28} = 1334.4\ \text{kN}$$

$$\phi V_n = 0.75 \times 1334.4 = 1000.8\ \text{kN}$$

$$V_{uc} = 425.9\ \text{kN} < \phi V_n = 1000.8\ \text{kN accepatble}$$

Use $B = 3$ m; $T = 75$ cm; $d = 65.5$ cm

Design for flexure

$$l = \frac{B-c}{2} = \frac{3000-500}{2} = 1250\text{ mm} = 1.25\ m$$

$$M_{uc} = \frac{P_u l^2}{2B} + \frac{2M_u l}{B} = \frac{2000 \times 1.25^2}{2 \times 3} + 0 = 520.8\text{ kN m}$$

$$A_s = \left(\frac{f'_c b_w}{1.176 f_y}\right)\left(d - \sqrt{d^2 - \frac{2.353 M_{uc}}{\phi f'_c b_w}}\right) = \left(\frac{28000 \times 3}{1.176 \times 300000}\right)\left(0.655 - \sqrt{0.655^2 - \frac{2.353 \times 520.8}{0.9 \times 28000 \times 3}}\right)$$

$$= 0.002974\text{ m}^2 = 2974\text{ mm}^2$$

Assume 11 #19 bars ($A_s = 3124\text{ mm}^2$)

Control minimum steel:

$$A_s \geq 0.0018\, A_g$$

$$3124\text{ mm}^2 \geq 0.0018 \times 3000 \times 750$$

$$3124\text{ mm}^2 \ngeq 4050\text{ mm}^2 \text{ OK}$$

Use 15 #19 bars ($A_s = 4260\text{ mm}^2$)

$$4260\text{ mm}^2 \geq 0.0018 \times 3000 \times 750$$

$$4260\text{ mm}^2 \geq 4050\text{ mm}^2 \text{ OK}$$

Check ductility:

$$c_n = \frac{a}{\beta_1} = \frac{A_s f_y}{0.85^2 f'_c b} = \frac{4260 \times 300}{0.85^2 \times 28 \times 3000} = 21.06\text{ mm}$$

$$\varepsilon_t = 0.003\left(\frac{d - c_n}{c_n}\right) = 0.003\left(\frac{655 - 21.06}{21.06}\right) = 0.09 > 0.005 \text{ OK}$$

Development length must also be checked.

7.5.3 Tabular or Graphical Approach

Since single foundations are the most common type of foundations in construction projects, the design engineer is faced with the design of countless variations of these foundations. Thus, it is necessary to adopt a simple straightforward method to design them. Although simple computer programs or specialized software can be used to design foundations quickly, if one does not have access to a computer or software, there would be problems when designing foundations. The proposed approach for determining the geometry and required reinforcement of single foundations is depicted in Figure 7.16 and Tables 7.3–7.8. In cases where the table contents are zero, compressive reinforcements are required, and empty positions are determinants of very low effective depths. Furthermore, the minimum amount of effective depth is 15 cm.

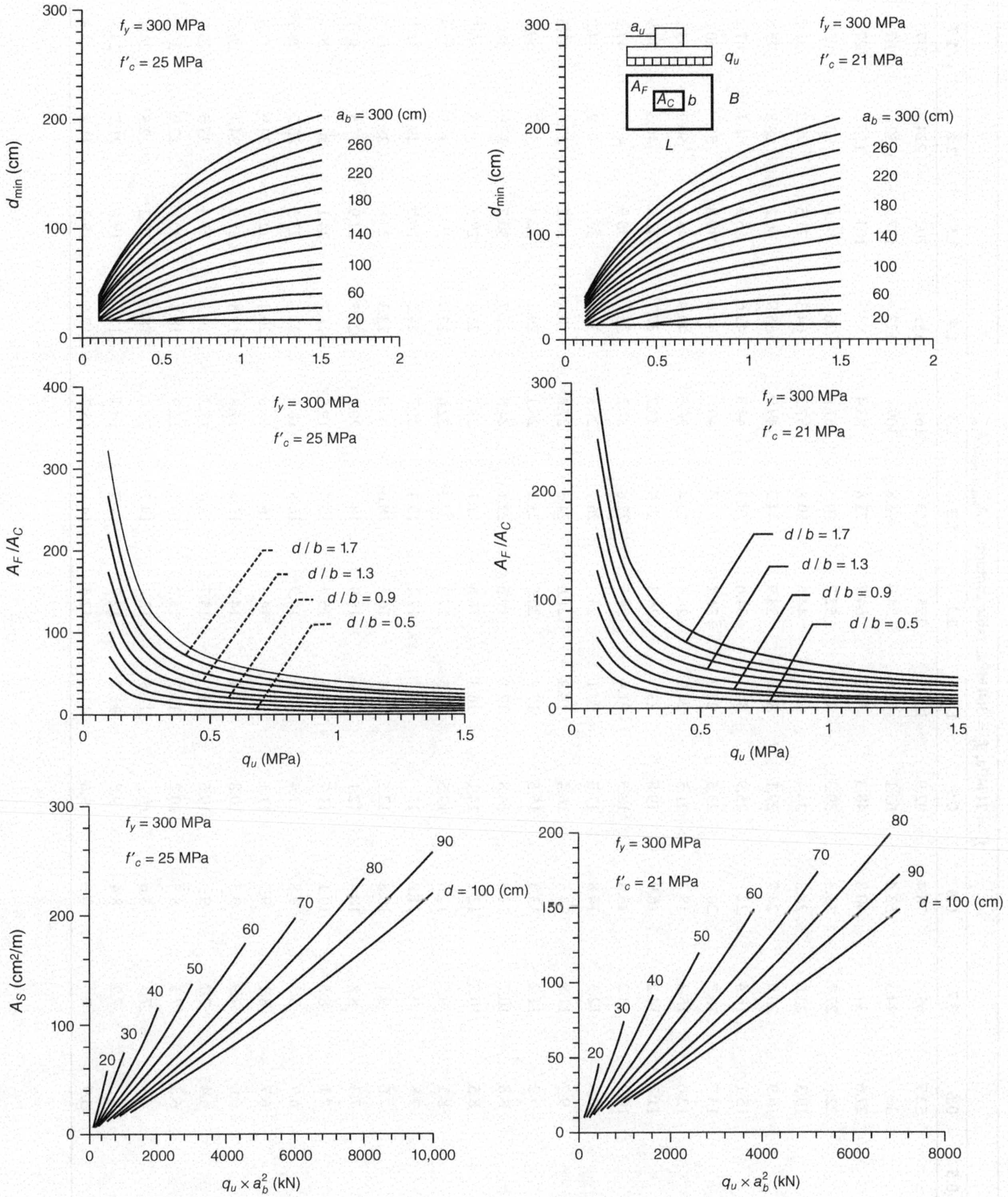

Figure 7.16 Graphs for determining the single foundations' thickness and reinforcement (Eslami 2013).

Table 7.3 Determining effective depth (*d*) based on two-way shear (Eslami 2013).

	$f'_c = 21$ MPa, $f_y = 300$ MPa, table contents $= A_{foundation}/A_{column}$												
q_u(MPa)	$d/b = 0.5$	0.6	0.7	0.8	0.9	1	1.1	1.2	1.3	1.4	1.5	1.6	1.7
0.1	41.4	52.7	65	78.4	92.9	109	125	143	161	181	202	224	247
0.15	28.4	36	44.3	53.4	63.1	73.6	84.8	96.8	109	123	137	152	167
0.2	21.8	27.6	34	40.8	48.3	56.2	64.7	73.8	83.4	93.5	104	115	127
0.25	17.9	22.6	27.7	33.3	39.3	45.8	52.7	60	67.8	76	84.6	93.7	103
0.3	15.3	19.3	23.6	28.3	33.4	38.8	44.6	50.8	57.3	64.3	71.5	79.2	87.2
0.35	13.4	16.9	20.6	24.7	29.1	33.8	38.9	44.2	49.9	55.9	66.2	68.8	75.8
0.4	12	15.1	18.4	22	25.9	30.1	34.6	39.3	44.3	49.6	55.2	61.1	67.2
0.45	11	13.7	16.7	20	23.5	27.2	31.2	35.5	40	44.8	49.8	55	60.6
0.5	10.1	12.6	15.3	18.3	21.5	24.9	28.5	32.4	36.5	40.9	45.4	50.2	55.2
0.55	9.4	11.7	14.2	16.9	19.8	23	26.3	29.9	33.7	37.7	41.9	46.3	50.9
0.6	8.8	10.9	13.2	15.8	18.5	21.4	24.5	27.8	31.3	35	38.9	43	47.2
0.65	8.3	10.3	12.5	14.8	17.3	20.1	23	26.1	29.3	32.8	36.4	40.2	44.2
0.7	7.8	9.7	11.8	14	16.4	18.9	21.6	24.5	27.6	30.8	34.2	37.8	41.5
0.75	7.5	9.2	11.2	13.3	15.5	17.9	20.5	23.2	26.1	29.2	32.4	35.7	39.3
0.8	7.1	8.8	10.7	12.6	14.8	17.1	19.5	22.1	24.8	27.7	30.7	33.9	37.3
0.85	6.9	8.5	10.2	12.1	14.1	16.3	18.6	21.1	23.7	26.4	29.3	32.3	35.5
0.9	6.6	8.1	9.8	11.6	13.5	15.6	17.8	20.2	22.6	25.3	28	30.9	33.9
0.95	6.4	7.8	9.4	11.2	13	15	17.1	19.4	21.7	24.2	26.9	29.6	32.5
1	6.2	7.6	9.1	10.8	12.5	14.4	16.5	18.6	20.9	23.3	25.8	28.5	31.3
1.05	6	7.3	8.8	10.4	12.1	13.9	15.9	18	20.2	22.5	24.9	27.5	30.1
1.1	5.8	7.1	8.5	10.1	11.7	13.5	15.4	17.4	19.5	21.7	24.1	26.5	29.1
1.15	5.7	6.9	8.3	9.8	11.4	13.1	14.9	16.8	18.9	21	23.3	25.7	28.1
1.2	5.5	6.7	8.1	9.5	11.1	12.7	14.5	16.3	18.3	20.4	22.6	24.9	27.3
1.25	5.4	6.6	7.9	9.3	10.8	12.4	14.1	15.9	17.8	19.8	21.9	24.1	26.5
1.3	5.3	6.4	7.7	9	10.5	12	13.7	15.4	17.3	19.3	21.3	23.5	25.7
1.35	5.2	6.3	7.5	8.8	10.2	11.7	13.3	15.1	16.9	18.8	20.8	22.9	25
1.4	5	6.1	7.3	8.6	10	11.5	13	14.7	16.4	18.3	20.2	22.3	24.4
1.45	5	6	7.2	8.4	9.8	11.2	12.7	14.3	16.1	17.9	19.8	21.7	23.8
1.5	4.9	5.9	7	8.3	9.6	11	12.5	14	15.7	17.5	19.3	21.2	23.3

Table 7.4 Determining effective depth (d) based on one-way shear (Eslami 2013).

	$f'_c = 21$ MPa, $f_y = 300$ MPa												
q_u(MPa)	a_b=20 cm	40	60	80	100	120	140	160	180	200	220	240	260
0.1	15	15	15	15	15	15.9	18.6	21.3	23.9	26.6	29.2	31.9	34.5
0.15	15	15	15	15	18.7	22.4	26.2	29.9	33.6	37.4	41.1	44.8	48.6
0.2	15	15	15	18.8	23.5	28.1	32.8	37.5	42.2	46.9	51.6	56.3	61
0.25	15	15	16.6	22.2	27.7	33.2	38.8	44.3	49.8	55.4	60.9	66.5	72
0.3	15	15	18.9	25.2	31.5	37.8	44.1	50.4	56.7	63	69.3	75.6	81.9
0.35	15	15	20.9	27.9	34.9	41.9	48.9	55.8	62.8	69.8	76.8	83.8	90.7
0.4	15	15.2	22.8	30.4	38	45.6	53.2	60.8	68.4	76	83.6	91.2	98.8
0.45	15	16.3	24.5	32.6	40.8	49	57.1	65.3	73.4	81.6	89.8	97.9	106
0.5	15	17.3	26	34.7	43.4	52	60.7	69.4	78.1	86.7	95.4	104	113
0.55	15	18.3	27.4	36.6	45.7	54.9	64	73.2	82.3	91.5	101	110	119
0.6	15	19.2	28.7	38.3	47.9	57.5	67	76.6	86.2	95.8	105	115	125
0.65	15	20	29.9	39.9	49.9	59.9	69.8	79.8	89.8	99.8	110	120	130
0.7	15	20.7	31	41.4	51.7	62.1	72.4	82.8	93.1	104	114	124	135
0.75	15	21.4	32.1	42.8	53.5	64.2	74.8	85.5	96.2	107	118	128	139
0.8	15	22	33	44.1	55.1	66.1	77.1	88.1	99.1	110	121	132	143
0.85	15	22.6	33.9	45.2	56.6	67.9	79.2	90.5	102	113	124	136	147
0.9	15	23.2	34.8	46.4	58	69.6	81.1	92.7	104	116	128	139	151
0.95	15	23.7	35.6	47.4	59.3	71.1	83	94.8	107	119	130	142	154
1	15	24.2	36.3	48.4	60.9	72.6	84.7	96.8	109	121	133	145	157
1.05	15	24.7	37	49.3	61.7	74	86.3	98.7	111	123	136	148	160
1.1	15	25.1	37.7	50.2	62.8	75.3	87.9	100	113	126	138	151	163
1.15	15	25.5	38.3	51	63.8	76.5	89.3	102	115	128	140	153	166
1.2	15	25.9	38.9	51.8	64.8	77.7	90.7	104	117	130	143	155	168
1.25	15	26.3	39.4	52.6	65.7	78.8	92	105	118	131	145	158	171
1.3	15	26.6	39.9	53.3	66.6	79.9	93.2	107	120	133	147	160	173
1.35	15	27	40.4	53.9	67.4	80.9	94.4	108	121	135	148	162	175
1.4	15	27.3	40.9	54.6	68.2	81.8	95.5	109	123	136	150	164	177
1.45	15	27.6	41.4	55.2	69	82.7	96.5	110	124	138	152	166	179
1.5	15	27.9	41.8	55.7	69.7	83.6	97.5	112	125	139	153	167	181

Table 7.5 Determining A_s in term of cm^2 for width of 1 m (Eslami 2013).

	$f'_c = 21$ MPa, $f_y = 300$ MPa													
$q_u a_b^2$ (kN)	$d = 15$ cm	20	25	30	35	40	45	50	55	60	65	70	75	80	85
50	6.4	4.7	5	6	7	8	9	10	11	12	13	14	15	16	17
60	7.7	5.7	5	6	7	8	9	10	11	12	13	14	15	16	17
70	9.1	6.7	5.3	6	7	8	9	10	11	12	13	14	15	16	17
80	10.5	7.7	6	6	7	8	9	10	11	12	13	14	15	16	17
90	11.9	8.6	6.8	6	7	8	9	10	11	12	13	14	15	16	17
100	13.3	9.7	7.6	6.3	7	8	9	10	11	12	13	14	15	16	17
120	16.3	11.7	9.2	7.6	7	8	9	10	11	12	13	14	15	16	17
140	19.4	13.8	10.8	8.9	7.5	8	9	10	11	12	13	14	15	16	17
160	22.6	15.9	12.4	10.2	8.6	8	9	10	11	12	13	14	15	16	17
180	26	18	14	11.5	9.8	8.5	9	10	11	12	13	14	15	16	17
200	29.6	20.2	15.6	12.8	10.9	9.4	9	10	11	12	13	14	15	16	17
300	0	32.1	24.2	19.6	16.5	14.3	12.6	11.3	11	12	13	14	15	16	17
400		45.9	33.4	26.7	22.4	19.3	17	15.2	13.8	12.7	13	14	15	16	17
500		0	43.4	34.1	28.4	24.4	21.4	19.1	17.3	15.8	14.5	14	15	16	17
600		0	54.4	42	34.6	29.6	25.9	23.1	20.9	19	17.5	16.2	15.1	16	17
700			0	50.3	41.1	35	30.5	27.2	24.5	22.3	20.5	18.9	17.6	16.5	17
800			0	59.2	47.8	40.5	35.2	31.3	28.1	25.6	23.5	21.7	20.2	18.9	17.7
900			0	68.8	54.8	46.5	40	35.4	31.9	29	26.6	24.5	22.8	21.3	20
1000			0	0	62.2	52	44.9	39.7	35.6	32.3	29.6	27.4	25.4	23.7	22.3
1200				0	78.2	64.2	55	48.4	43.3	39.2	35.8	33.1	30.7	28.6	26.9
1400				0	0	77.4	65.7	57.4	51.1	46.2	42.2	38.3	36	33.6	31.5
1600					0	91.8	76.9	66.7	59.2	53.4	48.6	44.7	41.4	38.6	36.2
1800					0	0	88.8	76.5	67.6	60.7	55.2	50.7	46.2	43.7	40.9
2000						0	102	86.7	76.2	68.3	61.9	56.8	52.5	48.8	45.6
2200						0	0	97.4	85.2	76	68.8	63	58.1	54	50.4

Table 7.6 Determining effective depth (*d*) based on two-way shear (Eslami 2013).

	$f'_c = 25$ MPa, $f_y = 300$ MPa, table contents = $A_{foundation}/A_{column}$												
q_u (MPa)	$d/b = 0.5$	0.6	0.7	0.8	0.9	1	1.1	1.2	1.3	1.4	1.5	1.6	1.7
0.1	45	57.3	70.7	85.3	101	118	136	155.3	175.7	197.2	219.9	243.8	268.8
0.15	30.7	39	48.1	57.9	68.6	80	92.2	105.1	118.9	133.4	148.7	164.8	181.6
0.2	23.6	29.9	36.8	44.3	52.3	61	70.2	80.1	90.5	101.5	113.1	125.3	138.1
0.25	19.3	24.4	30	36.1	42.6	49.6	57.1	65	73.4	82.3	91.7	101.6	111.9
0.3	16.5	20.8	25.5	30.6	36.1	42	48.3	55	62.1	69.6	77.5	85.8	94.5
0.35	14.5	18.2	22.3	26.7	31.4	36.6	42	47.8	54	60.5	67.3	74.5	82
0.4	12.9	16.2	19.8	23.8	28	32.5	37.3	42.4	47.9	53.6	59.7	66	72.7
0.45	11.7	14.7	18	21.5	25.3	29.3	33.7	38.3	43.2	48.3	53.7	59.4	65.4
0.5	10.8	13.5	16.5	19.7	23.1	26.8	30.7	34.9	39.4	44.1	49	54.2	59.6
0.55	10	12.5	15.2	18.2	21.3	24.7	28.3	32.2	36.3	40.6	45.1	49.9	54.8
0.6	9.4	11.7	14.2	16.9	19.8	23	26.3	29.9	33.7	37.7	41.9	46.3	50.9
0.65	8.8	11	13.3	15.9	18.6	21.5	24.7	28	31.5	35.2	39.1	43.2	47.5
0.7	8.4	10.4	12.6	15	17.5	20.3	23.2	26.3	29.6	33.1	36.8	40.6	44.7
0.75	7.9	9.9	11.9	14.2	16.6	19.2	22	24.9	28	31.3	34.7	38.4	42.2
0.8	7.6	9.4	11.4	13.5	15.8	18.2	20.9	23.6	26.6	29.7	33	36.4	40
0.85	7.3	9	10.9	12.9	15.1	17.4	19.9	22.5	25.3	28.3	31.4	34.6	38.1
0.9	7	8.6	10.4	12.4	14.4	16.7	19	21.6	24.2	27	30	33.1	36.3
0.95	6.7	8.3	10	11.9	13.9	16	18.3	20.7	23.2	25.9	28.7	31.7	34.8
1	6.5	8	9.7	11.4	13.4	15.4	17.6	19.9	22.3	24.9	27.6	30.5	33.4
1.05	6.3	7.8	9.3	11.1	12.9	14.9	16.9	19.2	21.5	24	26.6	29.3	32.2
1.1	6.1	7.5	9.1	10.7	12.5	14.4	16.4	18.5	20.8	23.2	25.7	28.3	31.1
1.15	6	7.3	8.8	10.4	12.1	13.9	15.9	17.9	20.1	22.4	24.8	27.4	30
1.2	5.8	7.1	8.5	10.1	11.7	13.5	15.4	17.4	19.5	21.7	24.1	26.5	29.1
1.25	5.7	6.9	8.3	9.8	11.4	13.1	14.9	16.9	18.9	21.1	23.3	25.7	28.2
1.3	5.5	6.8	8.1	9.6	11.1	12.8	14.5	16.4	18.4	20.5	22.7	25	27.4
1.35	5.4	6.6	7.9	9.3	10.8	12.4	14.2	16	17.9	19.9	22.1	24.3	26.7
1.4	5.3	6.5	7.7	9.1	10.6	12.1	13.8	15.6	17.5	19.4	21.5	23.7	26
1.45	5.2	6.3	7.6	8.9	10.3	11.9	13.5	15.2	17	19	21	23.1	25.3
1.5	5.1	6.2	7.4	8.7	10.1	11.6	13.2	14.9	16.6	18.5	20.5	22.6	24.7

Table 7.7 Determining effective depth (d) based on one-way shear (Eslami 2013).

	$f'_c = 25$ MPa, $f_y = 300$ MPa											
q_u(MPa)	$a_b = 20$ cm	40	60	80	100	120	140	160	180	200	220	240
0.1	15	15	15	15	15	15	17.2	19.7	22.2	24.6	27.1	29.5
0.15	15	15	15	15	17.4	20.9	24.4	27.8	31.3	24.8	38.3	41.7
0.2	15	15	15	17.5	21.9	26.3	30.7	35.1	39.5	43.8	48.2	52.6
0.25	15	15	15.6	20.8	26	31.2	36.4	41.6	46.8	52	57.2	62.4
0.3	15	15	17.8	23.7	29.6	35.6	41.5	47.4	53.3	59.3	65.2	71.1
0.35	15	15	19.8	26.4	32.9	39.5	46.1	52.7	59.3	65.9	72.5	79.1
0.4	15	15	21.6	28.8	36	43.2	50.3	57.5	64.7	71.9	79.1	86.3
0.45	15	15	23.2	31	38.7	46.5	54.2	61.9	69.7	77.4	85.2	92.9
0.5	15	16.5	24.7	33	41.2	49.5	57.7	66	74.2	82.5	90.7	99
0.55	15	17.4	26.1	34.9	43.6	52.3	61	69.7	78.4	87.1	95.9	105
0.6	15	18.3	27.4	36.6	45.7	54.9	64	73.2	82.3	91.4	101	110
0.65	15	19.1	28.6	38.2	47.7	57.3	66.8	76.3	85.9	95.4	105	115
0.7	15	19.8	29.7	39.7	49.6	59.5	69.4	79.3	89.2	99.1	109	119
0.75	15	20.5	30.8	41	51.3	61.5	71.8	82.1	92.3	103	113	123
0.8	15	21.2	31.7	42.3	52.9	63.5	74.1	84.6	95.2	106	116	127
0.85	15	21.8	32.6	43.5	54.4	65.3	76.2	87.1	97.9	109	120	131
0.9	15	22.3	33.5	44.7	55.8	67	78.1	89.3	101	112	123	134
0.95	15	22.9	34.3	45.7	57.1	68.6	80	91.4	103	114	126	137
1	15	23.4	35	46.7	58.4	70.1	81.8	93.4	105	117	129	140
1.05	15	23.8	35.7	47.7	59.6	71.5	83.4	95.3	107	119	131	143
1.1	15	24.3	36.4	48.6	60.7	72.8	85	97.1	109	121	134	146
1.15	15	24.7	37.1	49.4	61.8	74.1	86.5	98.8	111	124	136	148
1.2	15	25.1	37.7	50.2	62.8	75.3	87.9	100	113	126	138	151
1.25	15	25.5	38.2	51	63.7	76.4	89.2	102	115	127	140	153
1.3	15	25.8	38.8	51.7	64.6	77.5	90.4	103	116	129	142	155
1.35	15	26.2	39.3	52.4	65.5	78.6	91.6	105	118	131	144	157
1.4	15	26.5	39.8	53	66.3	79.5	92.8	106	119	133	146	159
1.45	15	26.8	40.2	53.6	67.1	80.5	93.9	107	121	134	148	161
1.5	15	27.1	40.7	54.2	67.8	81.4	94.9	109	122	136	149	163

Table 7.8 Determining A_s in term of cm^2 for width of 1m (Eslami 2013).

	$f'_c = 25$ MPa, $f_y = 300$ MPa														
$q_u a_b^2$ (kN)	$d = 15$ cm	20	25	30	35	40	45	50	55	60	65	70	75	80	85
50	6.4	4.7	5	6	7	8	9	10	11	12	13	14	15	16	17
60	7.7	5.7	5	6	7	8	9	10	11	12	13	14	15	16	17
70	9	6.6	5.3	6	7	8	9	10	11	12	13	14	15	16	17
80	10.4	7.6	6	6	7	8	9	10	11	12	13	14	15	16	17
90	11.8	8.6	6.8	6	7	8	9	10	11	12	13	14	15	16	17
100	13.2	9.6	7.6	6.3	7	8	9	10	11	12	13	14	15	16	17
120	16	11.6	9.1	7.5	7	8	9	10	11	12	13	14	15	16	17
140	19	13.6	10.7	8.8	7.5	8	9	10	11	12	13	14	15	16	17
160	22	15.7	12.3	10.1	8.6	8	9	10	11	12	13	14	15	16	17
180	25.2	17.8	13.9	11.4	9.7	8.5	9	10	11	12	13	14	15	16	17
200	28.9	19.9	15.5	12.7	10.8	9.4	9	10	11	12	13	14	15	16	17
300	0	31.2	23.8	19.4	16.4	14.2	12.6	11.3	11	12	13	14	15	16	17
400	0	43.8	32.6	26.3	22.2	19.2	16.9	15.1	13.7	12.5	13	14	15	16	17
500		58.3	42	33.5	28	24.2	21.3	19	17.2	15.7	14.5	14	15	16	17
600		0	52.1	41	34.1	29.3	25.7	23	20.8	18.9	17.4	16.1	15	16	17
700		0	63.1	48.8	40.3	34.5	30.2	27	24.3	22.2	20.4	18.9	17.6	16.4	17
800			0	57	46.7	39.8	34.8	31	27.9	25.5	23.4	21.6	20.1	18.8	17.7
900			0	65.7	53.4	45.3	39.5	35.1	31.6	28.8	26.4	24.4	22.7	21.2	19.9
1000			0	74.9	60.2	50.9	44.2	39.2	35.3	32.1	29.4	27.2	25.3	23.6	22.2
1200				0	74.8	62.4	53.9	47.7	42.7	38.8	35.6	32.8	30.5	28.5	26.7
1400				0	90.6	74.6	64	56.3	50.4	45.7	41.8	38.5	35.8	33.4	31.3
1600				0	0	87.6	74.6	65.3	58.2	52.6	48.1	44.3	41.1	38.3	35.9
1800					0	102	75.6	74.5	66.2	59.8	54.5	50.2	46.5	43.3	40.6
2000					0	117	97.1	84	74.5	67	61	56.1	51.9	48.4	45.3
2200					0	0	109	93.9	82.9	74.4	67.6	62.1	57.4	53.4	50
2400						0	122	104	91.6	82	74.4	68.2	63	58.6	54.8
2600						0	0	115	101	89.7	81.2	74.4	68.6	63.8	59.6
2800						0	0	126	110	97.6	88.2	80.6	74.3	69	64.5
3000						0	0	138	119	106	95.3	87	80.1	74.3	69.4

Example 7.7 Design a single foundation based on the given data via the graphical approach.

Live load = 900 kN	Dead load = 1200 kN
Column dimensions: 45 cm × 45 cm	$f'_c = 21$ MPa
Foundation dimensions: 280 cm × 280 cm	$f_y = 300$ MPa

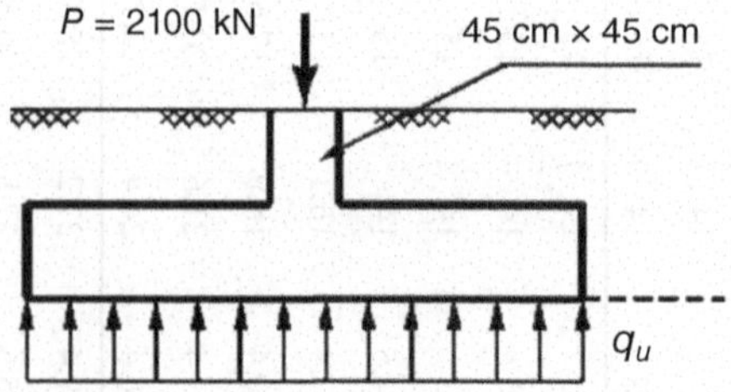

Determining the effective depth based on two-way shear:

$$\frac{A_F}{A_C} = \frac{280 \times 280}{45 \times 45} = 38.7, \ q_u = 0.41 \text{ MPa}$$

Based on Table 7.4 and using $\frac{A_F}{A_C} = 39.3, q_u = 0.4$ MPa

$$\frac{d}{b} = 1.2 \rightarrow d = 1.2 \times 45 = 54 \text{ cm}$$

Determining the effective depth based on one-way shear:

$$a_b = \frac{(280 - 45)}{2} = \frac{235}{2} = 117.5 \text{ cm}$$

Based on Table 7.5 and using $q_u = 0.4$ MPa

Δa_b	Δd
120–100	45.6–38
117.5–100	$X = ?$

$d = 38 + X = 38 + 6.5 = 44.65$ cm

Determining the required reinforcement:

$q_u \times a_b^2 = 4.09 \times 117.5^2 = 56467.6 \text{ kg} = 565$ kN
Based on Table 7.6 and using $q_u \times a_b^2 = 565$ kN, d = 55 cm

$q_u \times a_b^2$	A_s
60–50	20.9–17.3
56.5–50	$X = ?$

$A_s = 17.3 + X = 17.3 + 2.34 = 19.64 \text{ cm}^2$

Case History 7.1

Al Faisaliah Building, Riyadh, Saudi Arabia

The Al Faisaliah building complex main feature in Riyadh center is a 40-story office tower (Figure 7.17a) (Hemsley 2000). Figure 7.17b depicts a section through the building and the foundation's layout. The 3–4 m thick mat foundation has an area of 1600 m^2 in the plan, giving a volume of 6000 m^3. The tower has a height-to-breadth ratio of 5.8. The raft supports a total load of just over 1.1 GN. The tower tapers from a 45 m square to zero at a 263 m height.

Figure 7.17 Al Faisaliah Building complex, Riyadh, Saudi Arabia: (a) general view (swisshippo / Adobe Stock Photos) and (b) section of foundation (Hemsley 2000 / With permission of Thomas Telford).

(Continued)

Case History 7.1 (Continued)

The superficial ground deposits comprise 1–6 m of silty sand and gravel overlying Jurassic limestone (Arab and Jubaila formations). The limestone was assessed as having an unconfined compressive strength of approximately 20 MPa, a rock quality designation (RQD) of 25–50%, and Young's modulus (E) of approximately 2 GPa. For design purposes, an allowable net bearing pressure of 850 kPa was selected for the core foundation to minimize settlement, and a 33% overstress was permitted under wind loads. Pressures of 1000 kPa were permitted beneath the corner columns due to the pressure bulb's reduced size relative to its core. These values were predicted to produce 10–20 mm settlements beneath the columns and 20–30 mm under the core, most occurring during construction. Although the rock could safely support substantially higher pressures, the values adopted were selected to satisfy settlement serviceability criteria.

The allowable bearing pressures, together with the preliminary load take-down and wind loading calculations, were used for the initial sizing of the pad footings beneath the core and columns. Thicknesses were determined by considering the bending in the reinforced concrete footings on the assumption that they were rigid and gave a uniform distributed load on the underside; this resulted in an acceptable reinforcement percentage. By this method, thicknesses of 4 m for the tower core footing and 3 m for the column footings were adopted. Due to their proximity, footings to the core and the columns were linked together to form a cruciform shape, which further contributed to the stability of the structure.

The foundation loads were calculated for the final analysis and design for two combinations (dead plus live; dead, live, and wind) based on a 1 m thick slice through the tower. This two-dimensional slice through the tower foundations and supporting strata was analyzed to provide the predicted settlement profiles for each load combination, assuming symmetry for the tower axis. The settlement profile was determined using linear elastic finite elements, including the raft stiffness but ignoring the effect of the superstructure, using PLAXIS software.

A final analysis was conducted to design the orthogonal reinforcement after back-checking to confirm that the deflected raft form resembled that anticipated by the soil–structure interaction model. This reinforcement was frequently below the minimum standards, and in no case covered more than 1% of the cross-total section's area in both directions.

Case History 7.2

Shanghai Tower: Interactive Performance of Piled Raft Foundation and Basement Fin Walls (Sun et al. 2011; Zhu et al. 2012)

Several advantages can be obtained by setting fin walls in the basement of super tall buildings. The overall stiffness of the foundation could be improved and then differential foundation settlement could be reduced. On the other hand, basement shear stiffness could be effectively increased and the ground floor could be set as the lateral support with an appropriate percentage of basement fin walls. Besides, the fin walls can also increase punching resistance capacity for pile-raft foundation and can thus reduce the raft mat thickness (Sun et al. 2011).

There is a minimal ability to transfer gravity loads with little soil stiffness. Even though piles are concentrated under the core and super columns, pile group effects also contribute to the rather uniform settlement. Concrete fin walls five stories tall are installed at the basement levels to engage both core walls and super columns and more evenly transfer the tower load, reducing both total settlement and differential settlement. The walls have integrated steel plates to accommodate the significant forces being redistributed. The maximum predicted settlement is decreased by 20–30% and differential settlement is significantly decreased by these fin walls. The settlement contours with and without fin walls are depicted in Figure 7.18. The predicted tower peak settlement at five years is between 100 and 120 mm (Zhu et al. 2012). The foundation selection of the current project is an indicator of the superstructure's interactive role, foundation system, and underneath the soil in gaining efficient and favorable performance.

Sun et al. (2011) assesses the interactive performance of piled raft foundation and basement fin walls through finite element modeling. The structural responses of the two scenarios, with and without fin walls, are compared. Figure 7.19a

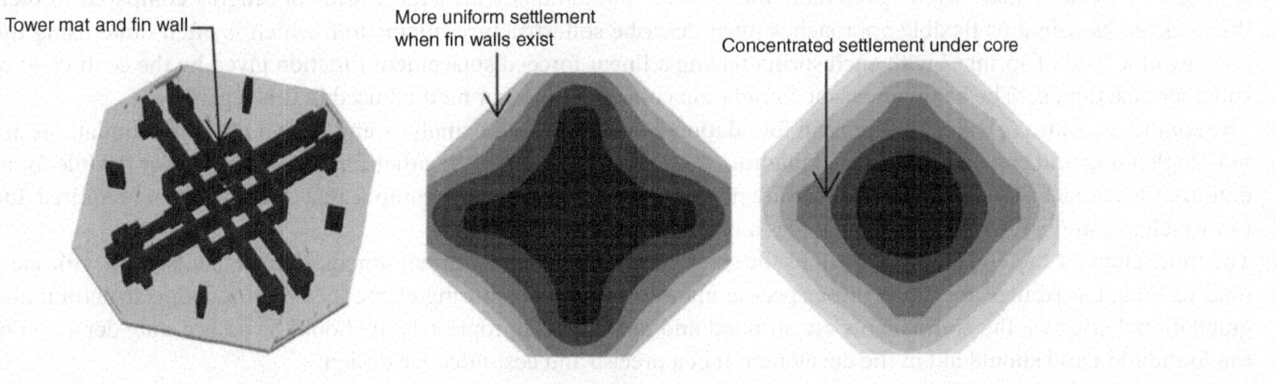

Figure 7.18 Tower mat and fin wall and its effects on settlement distribution (Zhu et al. 2012 / with permission of Council on Tall Buildings and Urban Habitat).

depicts the fin wall layout. Figure 7.19b shows that the maximum mat settlement occurs at the core wall area, while the minimum settlement occurs at the corner columns close to the edge of the mat. The mat settlement curve is in basin type. The arrangement of fin walls effectively improves the overall basement stiffness, and the differential settlement between the center and edge of the mat is significantly decreased. The differential settlement of the mat decreased from 64 to 34.7 mm. The differential settlement of the fin walls decreased from 20.3 to 5.2 mm.

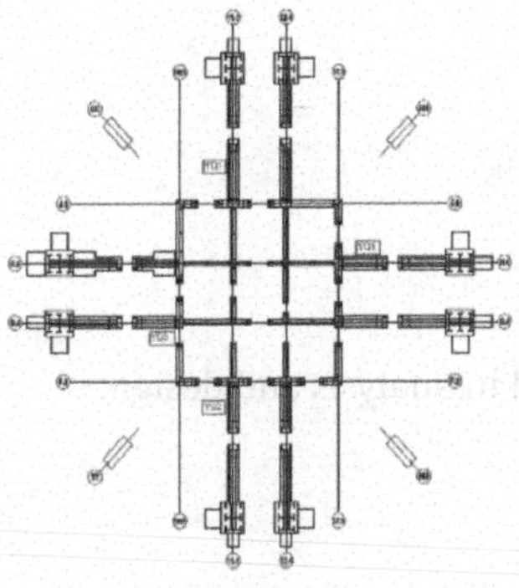

(a)

Position	Scenario 1 (with fin walls)	Scenario 2 (without fin walls)
Maximum settlement (core walls)	83.3	101.1
Minimum settlement (corner columns)	48.6	37.1
Differential settlement	34.7	64.0
Settlement of fin walls at core wall side	80.5	93.7
Settlement of fin walls at super column side	75.3	73.4
Differential settlement of fin walls	5.2	20.3

(b)

Figure 7.19 Shanghai Tower: (a) layout of the fin wall, (b) settlements of walls and columns (Sun et al. 2011 / with permission of Elsevier).

7.6 Major Points

1) Structural design consists of the analysis and design of the foundation, which first aims to determine the internal foundation materials to provide adequate resistance against internal forces. Then, the required thickness and the reinforcement of the foundation are determined.
2) The foundation analysis aims to determine the internal forces used in structural design. In this regard, two main approaches are used: the rigid method, which is usually used for single footings, and the nonrigid methods, which consider k_s in analysis and that are used for other foundations comprising flexible foundations like rafts.
3) The contact pressure (contact force per unit area) is the primary parameter that characterizes the interface between a shallow foundation and the underlying geomaterial.
4) The modulus of subgrade reaction is a relationship factor between soil pressure and deflection that is often utilized in foundation component structural analysis.
5) The rigidity assumption about single and combined two-column foundations is very close to reality due to their common dimensions. In the case of strip and raft foundations, the assumption of complete rigidity is not very close to reality and, as in many cases, leads to over-the-top and sometimes ineffective designs.

6) Nonrigid or flexible analyses are preferable for mats or foundations with large widths or lengths compared to their thicknesses. Nonrigid or flexible approaches must describe soil–structure interaction, which is often done using the analogy of a "bed of springs" with each spring having a linear force–displacement function given by the coefficient of subgrade reaction, k_s. The beam on elastic foundation concept is another method used in this regard.
7) The contact pressure (q) beneath the rigid foundation is uniform, so the analysis and design of these foundations are not challenging and can be handled by straightforward calculations. On the other hand, for nonrigid or flexible foundations, the contact pressure and deflection are not uniform; therefore, more complicated calculations are required due to variable, k_s, and subsequent internal forces along the foundation.
8) The finite element method (FEM) partitions the soil into a network of tiny three-dimensional elements, each with specified technical attributes and linked in a specific manner to the neighboring elements. After applying structural and gravitational stresses, the components are strained and deformed appropriately. It should be an accurate depiction of the foundation and should aid in the development of a precise and cost-effective design.
9) Spread footing structural design must account for shear and flexural failure modes as well as column base bearing failure. A shear failure occurs when the column or wall punches through the footing, while a flexural failure occurs when the footing's cantilever strength is inadequate.

7.7 Questions and Problems

7.7.1 Explanatory

1 Explain the contact pressure and the effect of foundation rigidity on it.

2 What is subgrade modulus reaction, and what is its application in foundation engineering?

3 Name and explain some factors that affect the rigidity of different foundations.

4 Distinguish between the real form of shear failure surfaces in footings and those utilized in analysis and design.

5 Explain the difference between thickness and effective depth.

6 Write and illustrate the fundamental equations for shear and flexural design.

7.7.2 Conceptual

1 Compare the closed-form solutions and finite element methods to analyze flexible foundations.

2 Generally, when designing reinforced concrete beams, we design for flexure first and then for shear. However, we design for shear first, followed by flexure with footings. Why?

3 What impact has the emergence of powerful but affordable digital computers had on the analysis and design of rafts? What changes are you anticipating as this trend continues?

4 How can a pile cap on four or five piles supporting one column in the center be analyzed based on single foundations concepts?

5 List the challenges in the design and construction of the Al Faisaliah Building (Case History 7.1) and suggest your solutions.

6 Discuss the role of superstructure elements on the rigidity of the foundation based on what you learned from Case History 7.2.

7.7.3 Problems

1 Evaluate the rigidity or flexibility of the given foundations:

$K_s = 24\frac{\text{MPa}}{\text{m}}$, $E_s = 72$ MPa, $f'_c = 21$ MPa, $E_c = 24$ GPa

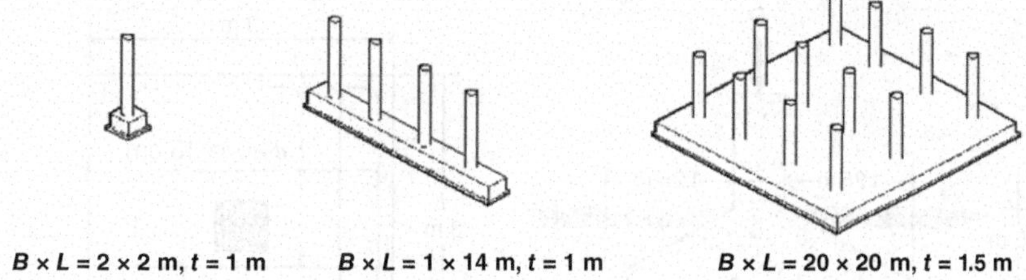

2 Rigidity evaluation and analysis of a grillage foundation:
The foundation is designed of IPB 500 Steel I-beam with a length of 2.5 m and a base plate with a breadth of 0.7 m. The load from the column is 20,000 kN, and the foundation is implemented on a bedrock that has an elasticity modulus of 600 MPa and a subgrade reaction modulus of 180 $\frac{\text{MPa}}{\text{m}}$. There are two profiles on the top layer, and the loads from the base plate and the lower profiles imposing on the upper profiles are distributed and uniform. It is assumed that the profiles work individually. The characteristic of the IPB500 Steel I-beam is as follows:

$$I_x = I_y = 12{,}620\ \text{cm}^4,\quad S = 842\ \text{cm}^3,\quad A = 239\ \text{cm}^2,\quad h = 50\ \text{cm},\quad b = 30\ \text{cm}$$

$$E_{steel} = 200\ \text{GPa}, F_y = 240\ \text{MPa}$$

3 Analyze the given combined foundation in rigid and flexible conditions.

$B = 1.4$ m, $L = 7.5$ m, $f'_c = 21$ MPa, $f_y = 300$ MPa,

Rigid Condition ($K_s = 0.1\frac{\text{MPa}}{\text{m}}$), Flexible Condition ($K_s = 100\frac{\text{MPa}}{\text{m}}$)

4 A 45 cm square reinforced concrete column supporting a 900 kN vertical dead weight and a 750 kN vertical live load must be supported on a square spread footing. The allowed bearing pressure based on bearing capacity has been found to be 0.25 MPa using ASD methods. According to a serviceability evaluation, the maximum bearing pressure that may be used to fulfill settlement criteria is 0.15 MPa. The groundwater table is far below the footing bottom. Determine the needed width, B, thickness, T, and effective depth, d. Then, design the required flexural steel.

5 As shown in the figure, a single foundation bears loads of a column with dimensions of 0.5 m × 0.5 m. Check the performance of the foundation for the following criteria:

a) Bearing capacity
b) Shear failure

c) Flexural failure

$$q_n = 0.3\,\text{MPa},\quad f_c' = 23\,\text{MPa},\quad f_y = 300\,\text{MPa},\quad D_{load} = 1500\,\text{kN},\quad L_{load} = 1000\,\text{kN},$$

$$\gamma_{soil} = 17.5\,\frac{\text{kN}}{\text{m}^3},\quad \gamma_c = 23.5\,\frac{\text{kN}}{\text{m}^3}$$

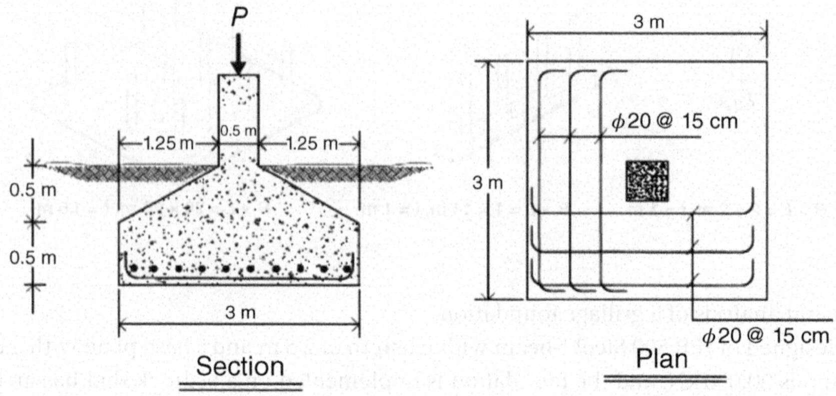

6 Design a single foundation based on the given data via the graphical (tabular) approach.

Live load = 1000 kN	Dead load = 1500 kN
Column dimensions: 50 × 50 cm	$f_c' = 21$ MPa
Foundation dimensions: 300 × 300 cm	$f_y = 300$ MPa

7 Repeat solving Example 7.6 via the tabular method (Assuming $f_c' = 25$ MPa).

References

AASHTO (2012). *AASHTO LRFD Bridge Design Specifications*, 6e. American Association of State Highway and Transportation Officials.

ACI (1993). *Suggested Analysis and Design Procedures for Combined Footings and Mats*, ACI 336.2 R- 88 (Reapproved 1993). American Concrete Institute.

ACI (2011). *Building Code Requirements for Structural Concrete (ACI 318-11)*. American Concrete Institute.

Bowels, J.E. (1996). *Foundation Analysis and Design*, 8e. McGraw-Hill.

Das, B. (2019). *Principles of foundation engineering*, 9e. Cengage Learning.

Eslami, A. (2013). *Foundation Engineering, Design, and Construction*, 2e. Building and Housing Research Center.

Gunaratne, M. (2006). *The Foundation Engineering Handbook*. CRC Press, Taylor & Francis Group.

Hemsley, J.A. (2000). *Design Applications of Raft Foundations*, 1e. Thomas Telford.

Horvath, J.S. (1993). *Subgrade Modeling for Soil-Structure Interaction Analysis of Horizontal Foundation Elements.* Manhattan College Research Report No. CE/ GE- 93-1. New York: Manhattan College.

Liao, S.S.C. (1991). *Estimating the Coefficient of Subgrade Reaction for Tunnel Design*. Internal Research Report. Parsons Brinkerhoff, Inc.

Sun, H.H., Zhao, X., Li, X.P., et al. (2011). Performance analysis of basement fin wall of the Shanghai Tower based on the interaction between pile-raft foundation and superstructure. *The Twelfth East Asia-Pacific Conference on Structural Engineering and Construction*, Hong Kong (26–28 January 2011). Netherlands: Procedia Engineering.

Winkler, E. (1867). *Die Lehre von Elastizität und Festigkeit (On Elasticity and Fixity)*. Prague: H. Dominicus (in German).

Zhu, Y., Poon, D., Zuo, S., and Fu, G. (2012). Structural design challenges of Shanghai Tower. CTBUH 2012 9th World Congress, Shanghai.

Section IV

Ground Modifications

8
Difficult Soil Improvement

Chapter Highlights

- **Background to Soil Improvement**
  - Reasons
  - Purpose
  - Benefits
- **Modification Techniques**
  - Earth Works
  - Densification and Compaction
  - Physicochemical Modifications
  - Hydraulic Modifications
  - Reinforcement
  - New Trends
- **Method Selection**
  - Technical
  - Practical
  - Economic
  - Environmental
- **Practical Examples**
- **Case Histories**
  - Application of Tire-Derived Aggregates as Fill Materials
  - Deep Dynamic Compaction Practice of Amir Abaad Port Tank Farm
  - Application of Vibro Stone Columns, Ghana
  - Geotechnical Aspects of Explosive Compaction
  - Sand Drains for Kansai Airport Artificial Island

8.1 Introduction

In recent years, the rapid growth of population and urban development has led to an increase in construction activities. However, these construction projects often encounter problematic soils, which can pose various technical and engineering challenges during both the construction phase and the operational life span of structures. Problematic soils are characterized by low bearing capacity and shear strength, among other undesirable properties. To address these issues, soil improvement techniques have been developed to enhance the performance of problematic soils. Soil improvement aims to mitigate the weaknesses of such soils by improving their bearing capacity, compressibility, subsidence potential, permeability, and swelling characteristics.

Advanced Foundation Engineering: Principles, Performance and Prospects, First Edition. Moses Karakouzian and Abolfazl Eslami.

8.2 Background

The first step in any construction project is to evaluate the soil properties and geological conditions of the site. If the soil is deemed unsuitable for construction due to its inadequate strength, appropriate soil improvement methods must be employed to enhance the soil conditions beneath the foundation. The selection of soil improvement techniques depends on various factors, including the soil condition and type, the nature of the construction project, environmental considerations, site-specific conditions, implementation constraints, project significance, time constraints, and economic considerations. Soil improvement methods can be employed either as a temporary measure to facilitate the construction of desired structures or as a permanent solution to enhance the physical and mechanical properties of the soil in the long term. The choice of the temporary or permanent solution and the appropriate technique will depend on the specific project requirements and the desired outcomes.

Overall, the benefits of soul modifications can be listed as follows:

- Improve soil strength properties;
- Reduce the amount of settlement and control the settlement;
- Increase bearing capacity;
- Improving soil permeability;
- Reduce costs and construction time;
- Consider the specific conditions of some industrial projects; and
- Reduce environmental damage.

8.3 Soil Improvement Methods

The selection of the appropriate soil improvement method is based on various factors, including the specific goals of improvement, site conditions, soil type, environmental considerations, economic factors, and project schedule. Soil improvement techniques can be classified based on different criteria, such as the loading process (static or dynamic), treatment location (internal or external), and implementation rate (slow or fast). Table 8.1 provides a classification of soil improvement techniques based on these factors. It is important to refer to the specific table mentioned (Table 8.1) to gain a detailed understanding of the classification and the corresponding soil improvement techniques under each category.

Ground modification methods can be classified into five general groups:

- Earthworks (Section 8.4)
- Densification and Compaction (Section 8.5)
- Physicochemical modifications (Section 8.6)
- Hydraulic Modifications (Section 8.7)
- Soil Reinforcement (Section 8.8)

8.4 Earth Works

Earthworks are ground modification measures that include (i) removing the existing problematic soil and replacing and compacting it with modified soil (cut and fill), (ii) unloading and floatation, and (iii) using lightweight materials.

8.4.1 Cut and Fill

For shallow layers of soft or loose soils with a thickness of less than 5 m, one cost-effective soil improvement method is to excavate and remove the problematic soil, replacing it with suitable materials, or treating and reusing the original soil after improvement.

In situations where the groundwater level is below the excavation floor, soft clay and loose sand can be compacted and, if necessary, mixed with lime, volcanic ash, or Portland cement. After treatment, these soils can be placed back in the excavated area. Soil displacement can also be utilized to improve soil resistance in certain cases. It is crucial to carefully design and manage the cut and fill operations to avoid the formation of failure surfaces and ensure the stability of the improved soil mass.

Table 8.1 Classification of ground improvement methods (after Chu et al. 2009).

Category	Method	Principle
A. Ground improvement without admixtures in non-cohesive soils or fill materials	A1. Dynamic compaction	Densification of granular soil by dropping a heavy weight from the air onto the ground.
	A2. Vibro compaction	Densification of granular soil using a vibratory probe inserted into the ground.
	A3. Explosive compaction	Shock waves and vibrations are generated by blasting to cause granular soil ground to settle through liquefaction or compaction.
	A4. Electric pulse compaction	Densification of granular soil using the shock waves and energy generated by electric pulse under ultra-high voltage.
	A5. Surface compaction (including rapid impact compaction)	Compaction of fill or ground at the surface or shallow depth using a variety of compaction machines.
B. Ground improvement without admixtures in cohesive soils	B1. Replacement/displacement (including load reduction using lightweight materials)	Removing bad soil by excavation or displacement and replacing it with good soil or rocks. Some lightweight materials may be used as backfill to reduce the load or earth pressure.
	B2. Preloading using fill (including the use of vertical drains)	Fill is applied and removed to pre-consolidate compressible soil so that its compressibility will be much reduced when future loads are applied.
	B3. Preloading using a vacuum (including combined fill and vacuum)	Vacuum pressure of up to 90 kPa is used to pre-consolidate compressible soil so that its compressibility will be much reduced when future loads are applied.
	B4. Dynamic consolidation with enhanced drainage (including the use of the vacuum)	Similar to dynamic compaction except vertical or horizontal drains (or together with vacuum) are used to dissipate pore pressures generated in the soil during compaction.
	B5. Electro-osmosis or electro-kinetic consolidation	DC current causes the water in soil or solutions to flow from anodes to cathodes, which are installed in the soil.
	B6. Thermal stabilization using heating or freezing	Changing the physical or mechanical properties of soil permanently or temporarily by heating or freezing the soil.
	B7. Hydro-blasting compaction	Collapsible soil (loess) is compacted by a combined wetting and deep explosion action along a borehole.
C. Ground improvement with admixtures or inclusions	C1. Vibro replacement or stone columns	Holes jetted into soft, fine-grained soil and backfilled with densely compacted gravel or sand to form columns.
	C2. Dynamic Replacement	Aggregates are driven into the soil by high energy dynamic impact to form columns. The backfill can be either sand, gravel, stones, or demolition debris.
	C3. Sand compaction piles	Sand is fed into the ground through a casing pipe and compacted by either vibration, dynamic impact, or static excitation to form columns.
	C4. Geotextile confined columns	Sand is fed into a closed bottom geotextile-lined cylindrical hole to form a column.
	C5. Rigid inclusions (or composite foundation)	Using piles, rigid or semi-rigid bodies or columns which are either premade or formed in situ to strengthen soft ground.
	C6. Geosynthetic reinforced column or pile-supported embankment	Using piles, rigid or semi-rigid columns or inclusions, and geosynthetic grids to enhance the stability and reduce the settlement of embankments.
	C7. Microbial methods	Using microbial materials to modify the soil to increase its strength or reduce its permeability.
	C8. Other methods	Unconventional methods, such as the formation of sand piles using blasting and the use of bamboo, timber, and other natural products.

(Continued)

Table 8.1 (Continued)

Category	Method	Principle
D. Ground improvement with grouting type admixtures	D1. Particulate grouting	Grout granular soil or cavities or fissures in soil or rock by injecting cement or other particulate grouts to either increase the strength or reduce the permeability of the soil or ground.
	D2. Chemical grouting	Solutions of two or more chemicals reacting in soil pores to form a gel or a solid precipitate to either increase the strength or reduce the permeability of soil or ground.
	D3. Mixing method (including premixing or deep mixing)	Treat the weak soil by mixing it with cement, lime, or other binders in situ using a mixing machine or before placement.
	D4. Jet grouting	High-speed jets at depth erode the soil and inject grout to form columns or panels.
	D5. Compaction grouting	Very stiff, mortar-like grout is injected into discrete soil zones and remains in a homogenous mass to densify loose soil or lift the settled ground.
	D6. Compensation grouting	Medium to high viscosity particulate suspensions are injected into the ground between a subsurface excavation and a structure in order to negate or reduce settlement of the structure due to ongoing excavation.
E. Earth reinforcement	E1. Geosynthetics or mechanically stabilized earth (MSE)	Use of the tensile strength of various steel or geosynthetic materials to enhance the soil's shear strength and the stability of roads, foundations, embankments, slopes, or retaining walls.
	E2. Ground anchors and soil nails	Use of the tensile strength of embedded nails or anchors to enhance the stability of slopes or retaining walls.
	E3. Biological methods using vegetation	Use of the vegetation roots for the stability of slopes.

8.4.2 Partially or Fully Compensated Foundations (Unloading and Floatation)

The reduction of the total settlement and the nonuniform settlement of a foundation using the reduction of the net load due to excavation is called the buoyancy principle. When part of the weight of the structure using the excavation is balanced, the building is called relatively floating, and when the total weight is balanced, the structure is called completely floating. A flotation scheme is proposed when the foundation is very compacted to a deep soil bed and when using deep foundations is costly. If underground floors are needed, floating is more important. Figure 8.1 shows this type of improvement.

8.4.3 Use of Lightweight Materials in Engineered Fills

Lightweight materials used in fills are specifically designed to reduce the weight of the fill material while still providing adequate load-bearing capacity. These materials are commonly used in applications where weight reduction is desired, such as in areas with poor soil conditions or where the load on the foundation needs to be minimized. Here are some examples of lightweight materials used in fills:

1) **Expanded Polystyrene (EPS) Geofoam:** EPS geofoam is a lightweight cellular plastic material that is commonly used in fill-in construction projects. It has a high strength-to-weight ratio, excellent compressive strength, and low thermal

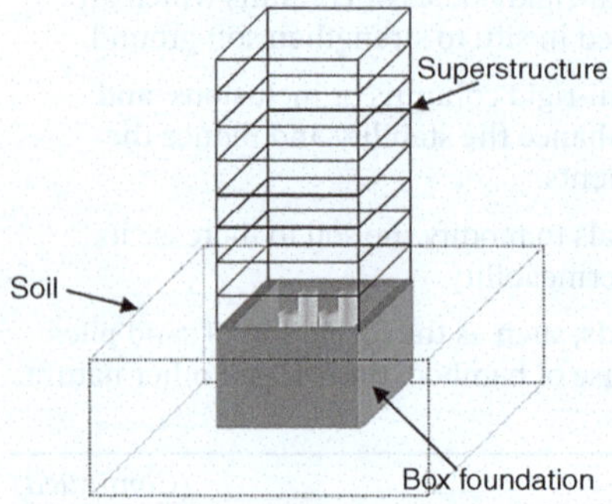

Figure 8.1 Using the floatation method in soil improvement.

conductivity. EPS geofoam blocks are lightweight and easy to handle, making them suitable for applications such as road embankments, bridge approaches, and lightweight backfill.

2) **Lightweight Aggregate Fill:** Lightweight aggregates, such as expanded clay, shale, or pumice, can be used as fill material to reduce the weight of the fill while providing stability. These aggregates are porous and have low unit weights compared to traditional aggregates. They are often used in applications like lightweight embankments, green roofs, and geotechnical fills.
3) **Cellular Concrete:** Cellular concrete, also known as foam concrete or aerated concrete, is a lightweight fill material made by introducing air or gas bubbles into a cementitious mixture. It has low density, good thermal insulation properties, and can be easily placed and compacted. Cellular concrete is commonly used in applications like lightweight fill for construction voids, slope stabilization, and backfilling utility trenches.
4) **Lightweight Expanded Clay Aggregate (LECA):** LECA is a lightweight aggregate made from expanded clay pellets. It is lightweight, porous, and provides good drainage properties. LECA is commonly used as lightweight fill material in applications such as retaining walls, landscaping, and sports field construction.
5) **Wood Chips or Mulch:** Wood chips or mulch can be used as lightweight fill material in certain applications. They are organic materials that have low density and can provide some degree of insulation and moisture retention. Wood chips or mulch are often used in landscaping projects, green roofs, and as temporary fill material.

These lightweight fill materials offer advantages such as reduced weight, improved load distribution, and ease of installation. However, it is important to consider the specific project requirements and consult with geotechnical engineers or experts to determine the most suitable lightweight fill material for the given application. Figure 8.2 shows photos of examples of lightweight material.

Figure 8.2 Different lightweight materials: (a) blast furnace slag (Derbyshire Aggregates Ltd), (b) geofoam, (c) expanded clay aggregate, and (d) waste rubber.

Case Study 8.1

Application of Tire-Derived Aggregates as Fill Materials

Tire-derived aggregates (TDA) have been examined as fill materials for highway embankments (Meles et al. 2013). Therefore, their performance and compression behavior, which has been employed for constructing an embankment, has been investigated for a highway test in Canada. This road had four test sections according to the passenger and light truck tires (PLTT), native soil as well as TDA–soil mixture, and off-the-road (OTR) truck tires, these types of tires are shown in Figure 8.3. Altogether, 25 settlement plates and 30 temperature probes have been placed in the road for monitoring the embankment. Moreover, the falling weight deflectometer (FWD) has been run under various levels of loads to determine the embankment materials' deflection behavior when the soil cover has been put on. According to the observations, FWD, and field data, PLTT has higher compressibility in comparison to TDA–soil and OTR mixture. In addition, the TDA–soil mix section exhibited the same function as the control section. Furthermore, each temperature probe did not exhibit any internal heating over the 8-month monitoring interval (Meles et al. 2013).

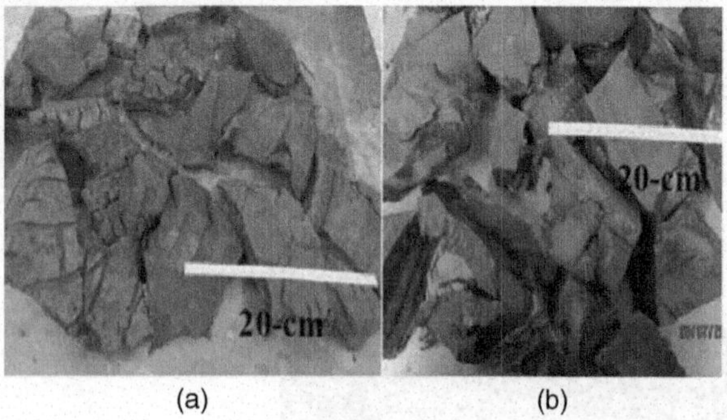

Figure 8.3 PLTT in (a) and OTR in (b) (Meles et al. 2013 / Taylor and Francis).

In this step, researchers fabricated a test embankment of four distinct sections consisting of OTR, PLTT, native soil, and a PLTT–soil mixture for an access road. Different geotechnical sensors have been used in the embankment to compare diverse sections according to the convenience in construction, immediate compression, susceptibility to interior stiffness and heating as well as field mixing of TDA and soil. A picture of this project is shown in Figure 8.4. General conclusions and observations are (Meles et al. 2013):

1) The test embankment has been fabricated with traditional construction instruments, facing no severe troubles.
2) According to the observations during the construction process, PLTT following the compaction had higher compressibility than OTR.
3) According to the unit weight that has been calculated based on the truck counts, OTR and PLTT have been lightweight so that the unit weight has been compressed and compacted.
4) Regarding the settlement measurement, TDA–soil and PLTT mixture sections showed the least and greatest settlements upon the completion of the construction process. Moreover, settlement measurements following the completion of the asphalt layers revealed that further time-dependent settlements are higher in the OTR section in comparison to the PLTT section.
5) According to the results, a mixture of TDA and soil showed the same function as the normal fill employed in the control section; nonetheless, it needs more effort to mix soil with TDA.
6) There has been not any evidence of internal heating through thermistors over the 8-month monitoring period for the abovementioned TDA sections.
7) According to the outputs from the FWD test, the PLTT section exhibited the greatest deflection under the different load levels exerted during the testing process.
8) It is possible to use stage paving to mitigate the probable distress, such as rutting or cracking, which influences the pavement's long-term performance.

Figure 8.4 Using TDA as a lightweight fill (Meles et al. 2013) material (Taylor and Francis).

8.5 Compaction and Densification

This category of soil improvement techniques includes (i) shallow compaction by rollers, (ii) dynamic compaction, and (iii) blasting.

8.5.1 Shallow Compaction by Rollers

The primary objectives of surface compaction are to enhance shear strength, reduce compressibility, decrease permeability, minimize liquefaction potential, control expansion and collapse, and improve overall stability. Achieving these goals is crucial for foundations and road construction. Rollers are extensively utilized for compacting surface layers. The effectiveness of rollers varies depending on their type and the level of compression they can achieve. Factors such as static weight, impact force, and vibration generation influence the performance of these construction machines. Roller weights range from 50 kg for manual compaction tools to as heavy as 60 tons for mechanized rollers. Apart from conventional rollers, specialized roller designs exist for specific purposes. Impact rollers and smart rollers are examples of such specialized types. One of the latest advancements in roller technology is the introduction of smart rollers, which offer programmable features for more precise and cost-effective compaction. Figure 8.5 illustrates an example of this innovative roller type.

8.5.2 Deep Dynamic Compaction and Rapid Impact Compaction

8.5.2.1 Deep Dynamic Compaction

Deep dynamic compaction is an effective soil improvement technique that enhances soil compaction by dropping steel or concrete weights. The weights used typically range from 5 to 40 tons and are dropped from heights of 10–40 m. The weights on the ground surface follow a specific pattern matching the plan and location of the foundations. This method ensures the targeted compaction of the soil in the designated areas.

The deep dynamic compaction method is a valuable technique used to improve problematic soils and enhance the overall performance of the foundations of various residential and office buildings, reservoirs, highways, railways, airports, and ports. By increasing the bearing capacity of the soil and reducing subsidence under the foundation, this method effectively mitigates the potential for liquefaction.

The deep dynamic compaction technique is commonly employed for the modification of various granular soils. However, it is important to note that it is not recommended for clay soils with a high plasticity index exceeding 8% and a high degree of saturation. Nonetheless, this technique has proven successful in improving fine-grained clay soils with a low plasticity index in several countries. When the groundwater level is high, typically within 2 m below the ground surface, the effectiveness of deep dynamic compaction is reduced due to the rapid drainage required for proper compaction. Generally, deep dynamic compaction is considered cost-effective for project areas larger than 5000 m^2. Figure 8.6 provides an illustration of the application of these improvements. The depth of improvement using the deep dynamic compaction (DDC) method can be estimated using the following empirical formula:

$$D_i = n_c\sqrt{W_t H_d} \tag{8.1}$$

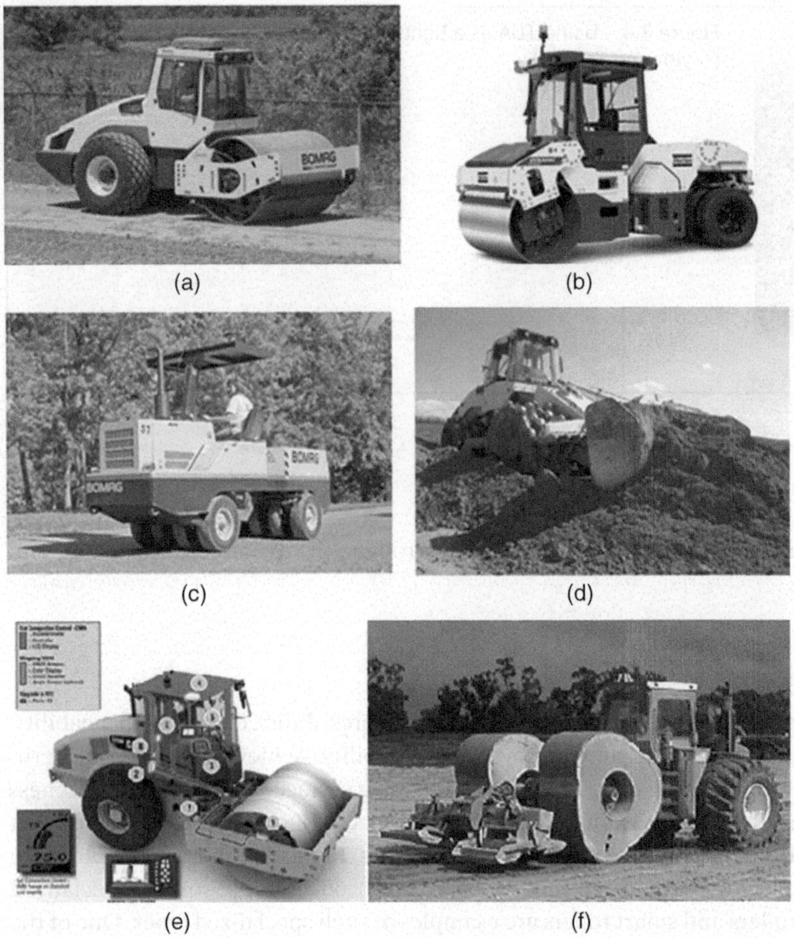

Figure 8.5 Different types of rollers. Typical shallow compaction rollers: (a) smooth drum roller, (b) combination roller, (c) pneumatic tire roller (BOMAG GmbH), (d) sheepsfoot / padfoot roller (BOMAG GmbH). Special shallow compaction rollers. (e) smart roller, and (f) impact roller.

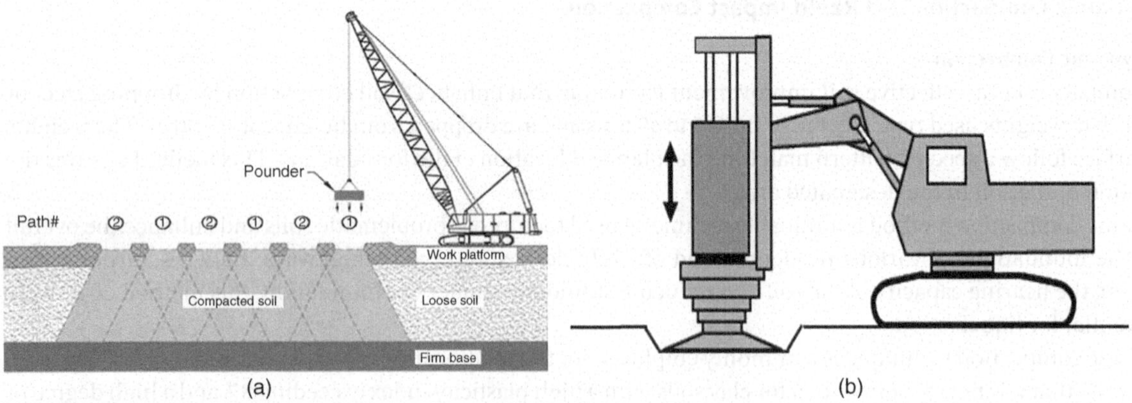

Figure 8.6 (a) Deep dynamic compaction and (b) Rapid impact compaction.

where

D_i = improvement depth (m)
W_t = tamper weight (tons)
H_d = drop height (m)
n_c = constant (Table 8.2)

Table 8.2 Recommended values for n_c (Han 2015 / With permission of John Wiley & Sons).

Soil type	Degree of saturation	n_c
Pervious soil deposits—granular soils	High	0.5
	Low	0.5–0.6
Semipervious deposits—primary silts with PI < 8	High	0.35–0.4
	Low	0.4–0.5
Semipervious deposits—primary clayey soils with PI > 8	High	Not recommended
	Low ($\omega < PL$)	0.35–0.4

PI = plasticity index, ω = moisture content, and PL = plastic limit. For $W_t H_d$ = 1–3 MJ/m^2 and a tamper drop using a single cable.

8.5.2.2 Rapid Impact Compaction

Rapid impact compaction is a high-frequency controlled energy compaction technique used to densify surface layers of soils (usually to a depth of 5–7 m) with a minimum impact on the immediate worksite environment. Figure 8.6b shows a schematic view of the rapid-impact compaction machine.

Example 8.1 DDC method via 80-ton tamper and 20 m drop height is used in a nearshore site. If the maximum required depth for improvement is 20 m, assess the sufficiency of the tamper weight and drop height combination for three different given conditions. If the current design is insufficient, suggest the appropriate combination to achieve the required depth.

- **Pervious Soil Deposits:** granular soils; high degree of saturation
- **Semipervious Deposits:** primary silts with PI < 8, low degree of saturation
- **Semipervious Deposits:** primary clayey soils with PI > 8, low degree of saturation

(a) Sufficiency Assessment:

Soil type	*W* (tons)	*H* (m)	$n_{c\text{-ave}}$	*D* (m)	Sufficiency
Granular soil	80	20	0.500	20	OK
Primary silts with PI < 8	80	20	0.450	18	Not OK
Primary clayey soil with PI > 8	80	20	0.375	15	Not OK

(b) Sufficiency Assessment, Without Changing the Drop Height:

Soil type	*H* (m)	*D* (m)	$n_{c\text{-ave}}$	*W* (tons)
Granular soil	20	20	0.500	80.000
Primary silts with PI < 8	20	20	0.450	98.765
Primary clayey soil with PI > 8	20	20	0.375	142.22

(c) Sufficiency Assessment, Without Changing the Tamper Weight:

Soil type	*W* (tons)	*D* (m)	$n_{c\text{-ave}}$	*H* (m)
Granular soil	80	20	0.500	20.0
Primary silts with PI < 8	80	20	0.450	24.7
Primary clayey soil with PI > 8	80	20	0.375	35.6

Case Study 8.2

Deep Dynamic Compaction Practice of Amir Abaad Port Tank Farm (Courtesy of Sham-e Consulting Engineering Co. 2015)

The Amir Abaad Port Tank Farm project, located in the north of Iran, covers an area of 29,773 m^2 and has a storage capacity of approximately 24,000 tons of oil products. Due to the construction of tanks and the potential impacts of stresses on weak layers at depth, as well as challenges related to settlement and seismic loads, it is essential to implement shallow and semi-deep soil improvement methods at the site. Several soil improvement methods for the site were considered, including deep dynamic compaction, drilled shaft, blasting, micro-piles, and vibro-compaction techniques. These methods were evaluated and compared based on technical, constructability, construction time, and economic criteria. After careful assessment of each method, deep dynamic compaction method was determined to be the most suitable method for the Amir Abaad Port Tank Farm project.

Figure 8.7 Soil improvement practice of Amir Abaad Port Tank Farm: (a) removing organic soils and leveling, (b) dropping the tampers, (c) craters resulted from tampers, (d) filling the craters, (e) inundating, and (f) shallow compaction and ironing (rollers).

For the deep dynamic compaction process, tampers with the weight of 11 and 23 tons and drop height between 18 and 22 m via a 165-ton crane were selected. The distance between square patterned drops is set to be 5 m. The process has been continued by leveling and shallow compaction. The effectiveness of ground modification is assessed through in situ tests. Figure 8.7 illustrates some of the soil improvement practice stages.

For more detailed information about this case, please refer to Case Study 3.2.

8.5.3 Stone Columns Stabilization and Compaction

This category of improvement techniques comprises in situ mixing of existing soil with additives like cement or lime or aggregates to form vertical columns in the soil.

8.5.3.1 Lime and Lime/Cement Columns

According to the studies, lime and lime/cement columns, wherein dry cement and quicklime are mixed in situ with soft soil are common for stabilizing silts, organic soil, and soft clay. Researchers use lime/cement and lime columns to:

- Increase the stability of roadways and embankments,
- Reduce settlements of roadways and embankments,
- Increasing the stability of trenches (water mains, heating ducts, and sewer lines),
- Stabilizing organic soils

Figure 8.8 depicts the installation of lime and lime/cement columns.

8.5.3.2 Vibro-compaction and Stone Columns

Vibro stone columns can be constructed using two methods: wet and dry. The wet top feed method, also known as displacement and replacement, involves the use of water jets to remove soft materials, stabilize the probe holes, and ensure proper placement of stone backfills around the vibrator tip. This method is widely used and considered the most cost-effective approach among deep vibratory techniques. However, the generation of spoil during the process can pose challenges for utilization in confined sites or environmentally sensitive areas. This method is shown in Figure 8.9.

The dry bottom feed method (displacement) approach utilizes a similar vibrator probe as the standard vibro replacement but adds a supply tube and hopper to directly feed the stone backfill to the vibrator tip. The bottom-feed vibro-replacement has been introduced as a thoroughly dry operation wherein the vibrator would remain in the ground over the course of the column construction procedure. Therefore, eliminating the flushing water removes the spoil and extends the site's ranges for treatment. Treatment may be done up to a depth that is not suppressed by the existence of groundwater. This method is shown in Figure 8.10.

8.5.3.3 Rammer Aggregate Piers – Geopier

The Geopier method, developed by Nathaniel S. Fox in 1984, is an advanced technology used to enhance soil-bearing capacity and reduce subsidence beneath surface foundations. Geopier has become a widely recognized and optimal solution

Figure 8.8 Installation of lime and lime / cement columns.

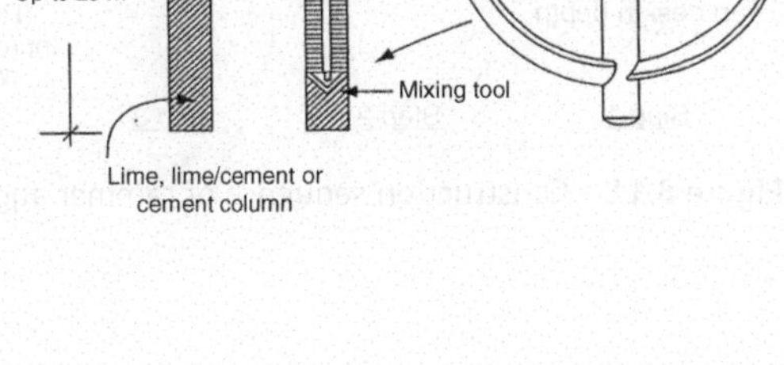

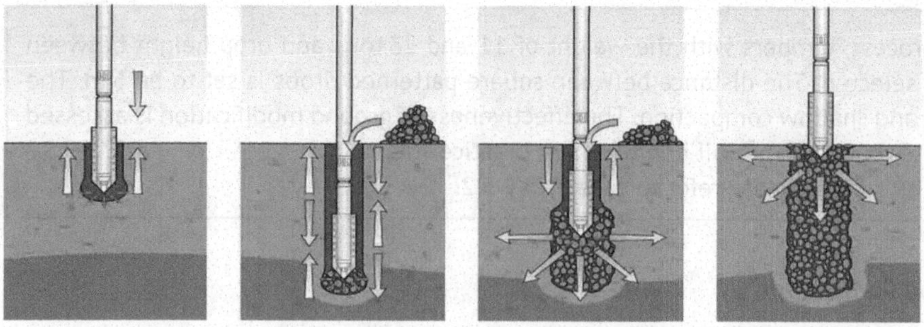

Figure 8.9 Schema of the wet, top feed procedure (Taube and Martin 2002).

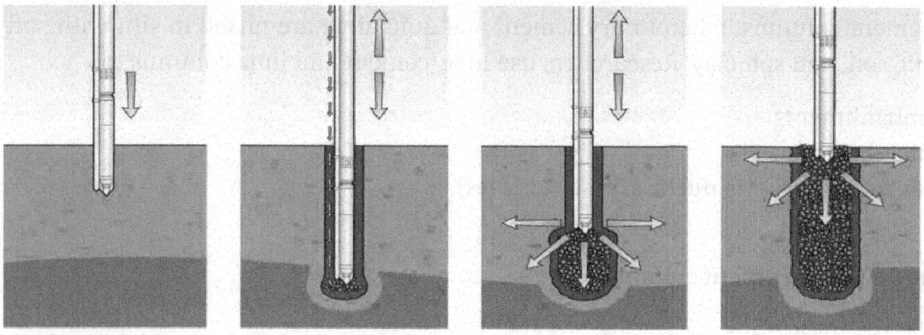

Figure 8.10 Schema of the dry-bottom-feed procedure (Taube and Martin 2002).

for soil improvement. It has been successfully applied in various construction projects, including embankment railroad tracks, reinforced earth walls, slope stabilization, and mitigating the risk of soil liquefaction. Geopiers offer a cost-effective and efficient approach to address soil-related challenges, providing stability and durability to structures built on weak or problematic soils. This method is shown in Figure 8.11.

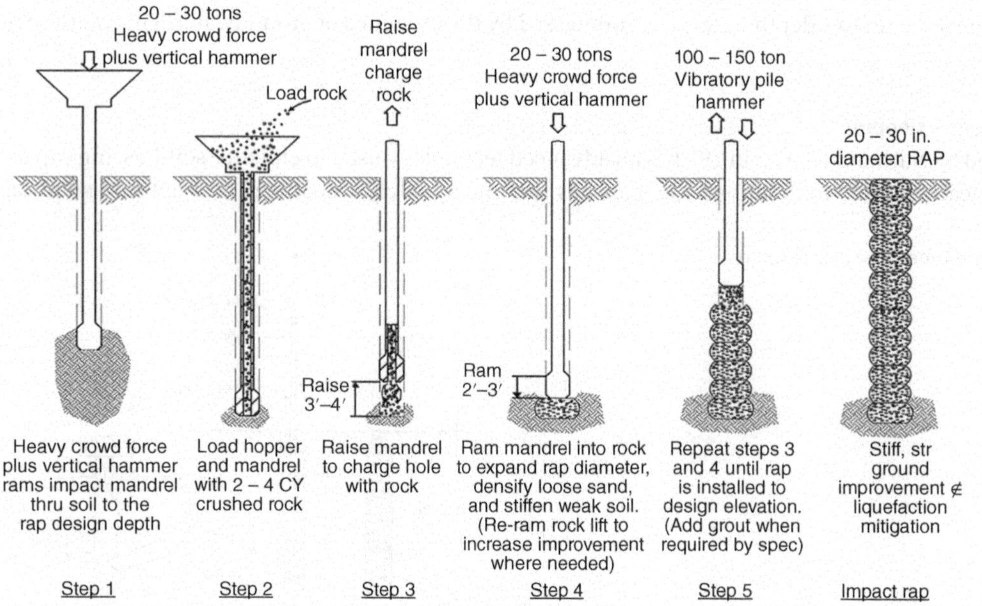

Figure 8.11 Construction sequence of rammer aggregate piers.

8.5.3.4 Deep Mixing

There is another type of stone column that is primarily executed utilizing the grouting procedure. Deep mixing (DM) can be named in this regard, which generally includes mixing in situ soil with a hardening agent.

8.5.3.5 Jet Grouting

In this method, the grout is used to erode the existing soil, mix it with grout, and then form hardened columns or walls. Grouting-based stone columns will be explained in Section 8.6.

8.5.4 Calculations for Bearing Capacity and Settlements of Soil–Column Composite

In practical applications, three common layout pattern configurations of soil columns are often used. Rectangular and triangular patterns are usually applied for most foundations, while radial or circular patterns are best suited for circular or ring foundations, such as tank foundations. The area replacement ratio, a_s (Equation 8.2), defined as the ratio of the cross-sectional area of all the soil columns to the area of the footprint of the building, is an important parameter that plays a crucial role in determining the settlement and bearing capacity of the treated ground. Equivalent strength and stiffness parameters for the treated ground are typically determined based on the area replacement ratio. Equation 8.3 provides a useful tool for calculating these parameters in relation to the area replacement ratio.

$$a_s = \frac{A_c}{A_e} = C\left(\frac{d_c}{s}\right)^2 \tag{8.2}$$

where

a_s = area replacement ration
A_c = column cross-sectional area
A_e = column tributary area
d_c = column diameter
s = center-to-center spacing between columns in a square or equilateral triangular pattern
C = 0.785 for a square pattern and 0.907 for an equilateral triangular pattern

$$f_{eq} = f_{column} a_s + f_{soil}(1 - a_s) \tag{8.3}$$

where

f_{eq} = equivalent parameter
f_{column} = parameter of column
f_{soil} = parameter of soil
a_s = area replacement ration

For a more accurate calculation, Equation (8.3) can be used for the internal friction angle.

$$\phi_{eq} = \arctan[a_s \tan \phi_c + (1 - a_s) \tan \phi_s] \tag{8.4}$$

where

ϕ_s = soil friction angle
ϕ_c = column friction angle
ϕ_{eq} = equivalent friction angle

Example 8.2 The rigid foundation of a commercial complex with dimensions of 20 m × 20 m is located on a layer of soft saturated soil. In order to modify the ground, 25% of the total surface area beneath the foundation is taken by the total cross-sectional area of all the stone columns installed as ground improvement. The ratio of the modulus of elasticity of the soil columns to that of the surrounding soil is 6. The maximum allowable settlement is 8 cm. Using equivalent parameters: Compare the bearing capacity of the foundation prior to and after ground modification.

a) Estimate the immediate settlement of the foundation prior to and before ground modification under the stress of 300 kPa beneath the foundation (K = 0.5)

Before stone implementation: $C = 25$ kPa, $\phi_s = 20°$, $E_s = 8$ MPa, $\gamma_{sat} = 19$ kN/m³

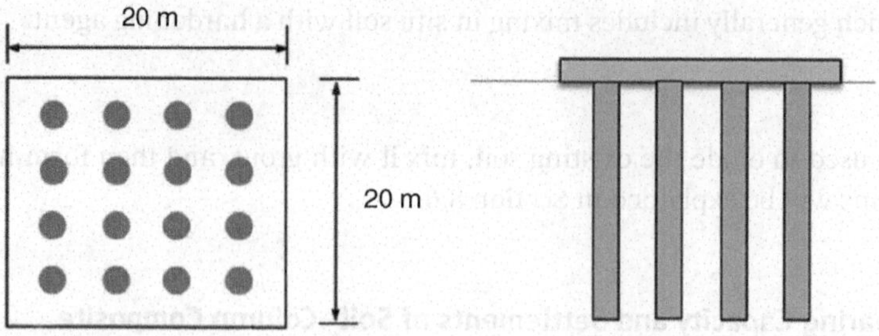

After stone implementation: $\phi_c = 40°$, $E_s = 20$ MPa

a) Before soil improvement,

Basic bearing capacity equation: $q_{ult} = CN_c + \bar{q}N_q + 0.5\gamma BN_\gamma$

$$\phi = 20° \rightarrow N_c = 17.7, N_q = 7.4, N_\gamma = 4.4$$

$$q_{ult} = 25 \times 17.7 + 0 \times N_q + 0.5 \times (19 - 10) \times 20 \times 4.4 = 838.5 \text{ kN}$$

After soil improvement,

$$\phi_{eq} = \arctan[a_s \tan \phi_c + (1 - a_s) \tan \phi_s] = \arctan[0.25 \times \tan 40 + (1 - 0.25) \times \tan 20] = 25.77°.$$

Conservatively: $\phi = 25° \rightarrow N_c = 25.1, N_q = 12.7, N_\gamma = 9.2$

$$q_{ult} = 25 \times 25.1 + 0 \times N_q + 0.5 \times (19 - 10) \times 20 \times 9.2 = 1455.5 \text{ kPa}$$

$$\Delta q_{ult} = \frac{1455.5 - 838.5}{838.5} = 0.736 = \text{a } 73.6\% \text{ increase in bearing capacity}$$

b) Before soil improvement:

$$S = \frac{kqB}{E} = \frac{0.5 \times 300 \times 20}{8 \times 10^3} = 0.375 \text{ m} = 37.5 \text{ cm} \nless 8\, cm \text{ Not OK}$$

After soil improvement:

$$E_{eq} = E_s(1 - a_s) + E_c a_s = 20 \times (1 - 0.25) + (6 \times 20) \times 0.25 = 45 \text{ Mpa}$$

$$S = \frac{kqB}{E} = \frac{0.5 \times 300 \times 20}{45 \times 10^3} = 0.067 \text{ m} = 6.7 \text{ cm} < 8 \text{ cm OK}$$

$$\Delta s = \frac{6.7 - 37.5}{37.5} = -0.821 = \text{a } 82.1\% \text{ decrease in settlement}$$

Case Study 8.3

Application of Vibro Stone Columns, Ghana

In this project, Vibro stone columns were used to support Ghana Oil Company (GOIL)'s MGO Storage Facility at the Port of Takoradi in Ghana. The development of the Takoradi Harbor in the western part of Ghana created an opportunity for GOIL to construct three additional fuel tanks in the port. A picture of the construction site is presented in Figure 8.12.

The challenge

The challenging conditions of saturated, soft clayey soil, limited space, high water table, and tight project schedules required the engineers to come up with innovative and cost-effective solutions. In such circumstances, vibro stone columns have demonstrated their effectiveness in accelerating consolidation settlement by up to 90% within a span of two weeks. Additionally, they help increase the cumulative bearing capacity, thereby limiting total and differential settlements. These columns have proven to be a reliable method for addressing the specific challenges posed by the project site.

Figure 8.12 Application of vibro stone columns in Ghana (Keller Africa/https://www.keller-africa.co.za/projects / last accessed on November 17, 2023).

The solution

For each of the three massive fuel storage tanks, vibro stone columns were designed and installed. 80 No. 800 mm diameter vibro stone columns were installed. Subsequently, the client expanded the scope by another 82 stone columns for their pump station foundations. A total of 322 no. top feed (wet method) vibro stone columns were installed to depths of approximately 8 m (https://www.keller-africa.co.za/projects).

8.5.5 Explosive Compaction or Blast Densification

Explosive compaction is typically limited to loose granular soils with a clay content of less than 5% or a silt content of less than 20%. The presence of clay pockets in sandy soils can reduce the effectiveness of the blasting operation. When immediate vibrations or shocks are applied to saturated loose granular soils, localized liquefaction can occur, along with the rearrangement of the soil grains. This leads to the transient transfer of load to the pore water and subsequent repositioning of the grains, resulting in a denser packing state. It is important to note that the weight of the soil contributes to this process.

Analyses have indicated that saturated conditions are more favorable for smoother propagation of shock waves during blasting compaction. In contrast, compaction through blasting in unsaturated soils can be somewhat unpredictable. Additionally, the presence of pore water serves as a lubricant, facilitating the rearrangement of soil grains. As the void spaces in the soil decrease due to compaction, a larger volume of water is displaced relative to the original void depth, resulting in densification. Therefore, the most efficient results of explosive compaction are typically observed in dry or well-saturated soils.

When the water volume in the soil is reduced from 100% saturation to the absorbed water content around the soil grains, the increased capillary tension between the grains can result in a lower density, particularly in finer soils. However, this effect of capillary tension can be mitigated by flooding or compacting the targeted area. Additionally, the presence of a small amount of gas in the soil can significantly dampen the shock wave generated by blasting. Figure 8.13 illustrates the propagation of the wave caused by blasting.

Furthermore, the distribution of charges, the timing of detonations between charges, and the size of the charges must be carefully considered to achieve the desired density of the materials and achieve satisfactory field test results. It has been observed that when there are minor delays (up to a few seconds) between the charges, compaction can extend deeper into a soil layer. However, larger delays between charges result in a higher degree of compaction than the minor delays. In all cases, the development of craters at the soil surface should be avoided. The amount of densification desired and the size of the charges are influenced by factors such as the type of soil, the position of the water table, dissipation of pore-water pressure, thickness of the soil layer, proximity to structures or bodies of water, spacing and overlapping effect of charges, and the type of explosives used. These factors need to be carefully considered to achieve effective and controlled compaction results.

Care must be taken to ensure that structures adjacent to the blasting site are not affected and that no large-scale slip or shear failure is induced. Excess pore pressure and settlement due to explosions are related to the ratio:

$$N_h = \frac{W^{\frac{1}{3}}}{R} \tag{8.5}$$

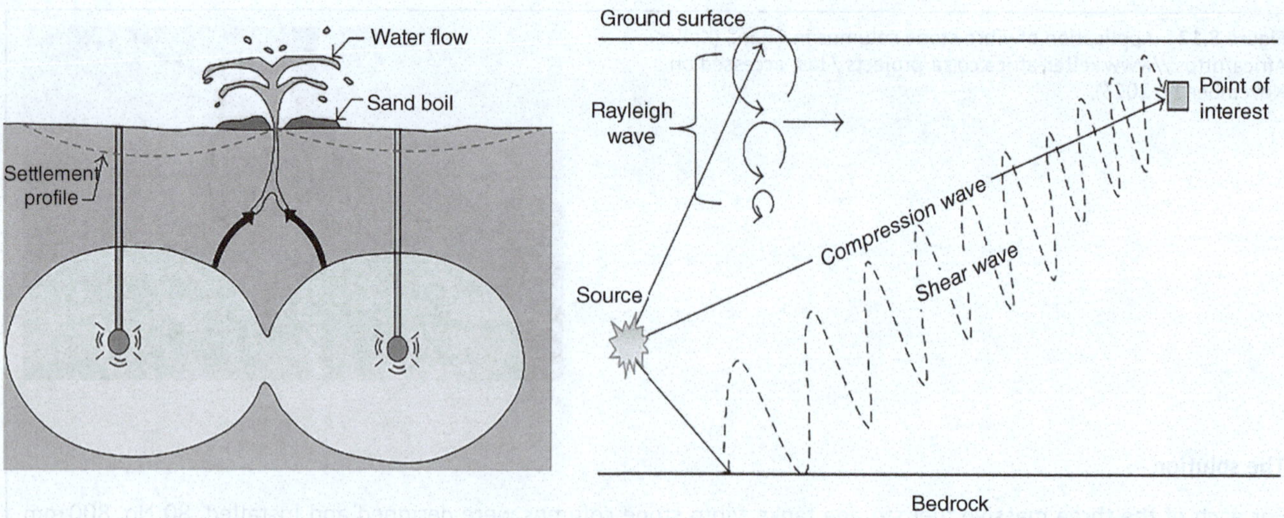

Figure 8.13 Wave propagation caused by explosion shock, soil liquefaction, and settlement (Shakeran et al. 2016 / Hindawi Publishing Corporation / CC BY 4.0).

where

N_h = Hopkinson number
W = weight of explosives, equivalent kilograms of TNT
R = radial distance from point of explosion, m

If N_h is less than the range of 0.09 to 0.15 little or no liquefaction is said to occur. This relationship can be used to estimate a safe distance from the explosion.

Tavakoli et al. (2014) proposed a novel approach to predict the ultimate settlement induced by explosive compaction based on powder factor, explosion phase number, and charge location depth. The average powder factor (PF_{avg}) has been used instead of the total applied energy. The ultimate settlement is as presented below:

$$\left[\frac{\Delta h}{H_0}\right](\%) = \frac{0.812\left(PF_{avg}^{0.71}\right)\left(1.285^{\log(N)}\right)}{d_{Tavg}^{0.18}} \tag{8.6}$$

where

Δh = settlement of improved soil
H_0 = initial thickness of the improved layer
N = explosive phase number
PF_{avg} = average powder factor (g/m^3) (Equation 8.6)
d_{Tavg} = average depth of charge location (m)

$$PF_{avg} = \frac{\sum PF_i}{N} \tag{8.7}$$

where

PF_i = powder factor in phase i

Case Study 8.4

Geotechnical Aspects of Explosive Compaction (Shakeran et al. 2016)

Shakeran et al. (2016), aiming to investigate the geotechnical aspects that play an important role in the performance of explosive compaction (EC), have compiled a database from thirteen-field tests and construction sites around the world, where EC has been successfully applied for modifying soil. This research focuses on the evaluation of grain size distribution and initial stability status of deposits besides changes in soil penetration resistance due to EC.

Generally, cone penetration test (CPT) can be used for the evaluation of EC before and after blasting (Figure 8.14). It is also common to perform CPT tests in different time intervals after the explosion because penetration resistance increases over time. Accordingly, in this study, the last reported CPT data has been reported for each site. A combination of CPT data and experimental equations may be used to determine other geotechnical parameters.

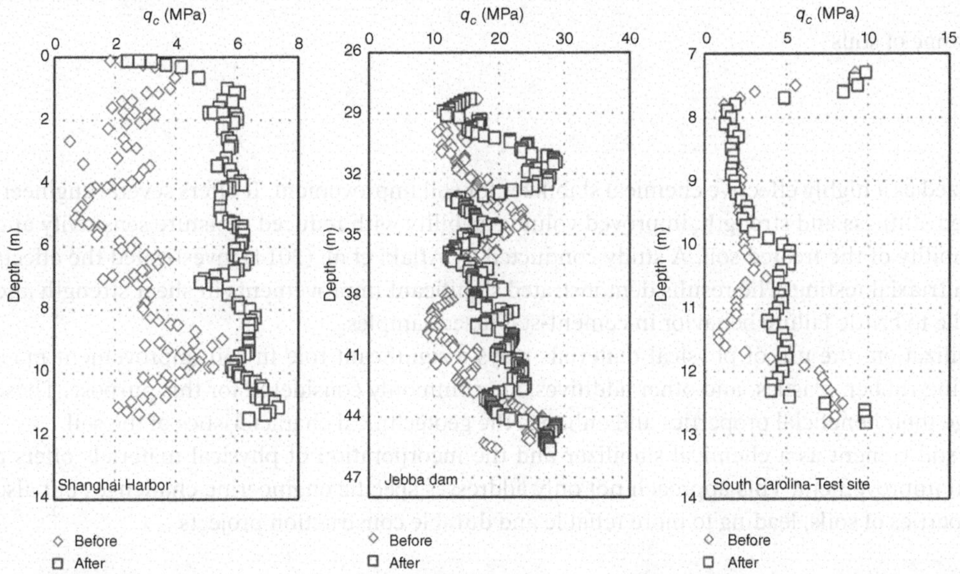

Figure 8.14 CPT records before and after EC in some sites in the database (Shakeran et al. 2016 / Hindawi Publishing-Corporation / CC BY 4.0).

A study of the grain size distribution of target deposits for improvement and design status of EC, in addition to analysis of CPT records before and after improvement, has been used and the following results are inferred for zoning:

- EC was performed successfully in a wide range of soil types, from coarse gravel with cobble or rubble with a diameter of 1 m to silty sands with 40% silt content and 10% clay. The best performance of EC was in alluvial sands with a fine content of less than 5% and hydraulically deposited layers with an initial relative density of 30–60%.
- The cone penetration test (CPT) can be used as an evaluation tool for EC along with the measurements of settlement. Analyses have shown that, through EC, the cone resistance of sand deposits easily increases up to 200% with a fine content of less than 5%, which would be followed by a 10–30% increase in the relative density of soil.
- In addition to the position of problematic deposits concerning grain size distribution and initial strength, other important parameters like depth of target layer, relative costs, and environmental limitations may play an important role in the selection of EC as an alternative for deep improvement among other methods.

8.6 Physical and Chemical Modification

These soil treatment methods include the application of additives, using heating or freezing, grouting, and soil mixing, as well as plant and bioengineering to enhance problematic soil properties.

8.6.1 Chemical and Physical Stabilization

The most important purpose of chemical additives is their ability to improve or control volume stability (control of inflation and shrinkage), resistance, and change of stress–strain and soil permeability properties. Popular inorganic materials applied for stabilizing soil include Portland cement and lime, and popular organic materials used include acrylamides, resins, polyesters, and bitumen.

Chemical additives play a crucial role in improving or controlling the volume stability (expansion and collapse), increasing bearing capacity, reducing settlements, stress–strain behavior, and reducing the permeability of soils. Inorganic materials like Portland cement and lime are commonly used for soil stabilization. Organic materials such as acrylamides, resins, polyesters, and bitumen are also widely employed for soil stabilization purposes. The selection of chemical additives depends on the specific requirements of the project and the characteristics of the soil being treated. The advantages of these chemical additives are:

- Applicable to a remarkable range of soils
- Abundantly available
- Affordable
- Environmentally friendly

Soil cement is widely recognized as a highly effective chemical stabilizer for soil improvement. It offers several engineering benefits, including enhanced stiffness and strength, improved volume stability with reduced moisture sensitivity and frost heave, and increased durability of the treated soil. A study conducted by Aflaki et al. (2014) investigated the effects of cement stabilization through triaxial testing. The results demonstrated significant improvements in shear strength and revealed a transition from ductile to brittle failure behavior in cement-stabilized samples.

In addition to chemical stabilization, the use of physical materials plays a significant role in soil improvement practices. Various materials, including rubber, plastics, and other additives, are commonly considered for this purpose. These materials are utilized to leverage their beneficial properties and enhance the geotechnical characteristics of the soil.

Overall, the combination of soil cement as a chemical stabilizer and the incorporation of physical materials offers a comprehensive approach to soil improvement. This approach not only addresses specific engineering challenges but also optimizes the geotechnical properties of soils, leading to more reliable and durable construction projects.

Example 8.3 In a cohesive soil mass with internal friction, by performing cement remediation operations, soil adhesion increases by 85%. The internal friction angle remained unchanged after the remediation, and the dry unit weight increased by 6%. If drilling without lateral restraint in this soil mass, what percentage of the tensile zone depth in the soil mass will change?

$$C_1 \rightarrow C_2 = 1.85\, C_1$$

$$\varphi_1 \rightarrow \varphi_2 = \varphi_1$$

$$\gamma_1 \rightarrow \gamma_2 = 1.06\, \gamma_1$$

$$k_{a1} = \tan^2\left(45 - \frac{\varphi_1}{2}\right) \rightarrow \varphi_2 = \varphi_1 \rightarrow k_{a1} = \tan^2\left(45 - \frac{\varphi_2}{2}\right) = k_{a2}$$

$$h_{\text{cr}} = ?\ \sigma_h = 0 \rightarrow k_a.\gamma.h_{\text{cr}} = 2c\sqrt{k_a}$$

$$\rightarrow h_{\text{cr}} = \frac{2c\sqrt{k_a}}{k_a.\gamma} \rightarrow h_{2_{\text{cr}}} = \frac{2c_2.\sqrt{k_{a_2}}}{k_{a_2}.\gamma_2}\ h_{2_{cr}} = \frac{2*1.85\ c_1\ \sqrt{k_{a_1}}}{k_{a_1}*1.06\ \gamma_1} = \frac{2c_1\ \sqrt{k_{a_1}}}{k_{a_1}*\gamma_1} * \frac{1.85}{1.06}$$

$$h_{\text{cr}_2} = 1.745\, h_{\text{cr}_1} \rightarrow \frac{h_{\text{cr}_2}}{h_{\text{cr}_1}} = 1.745$$

The depth of the tensile zone will increase by 74.5%.

Shallow stabilization using cement, lime, fly ash, slag, and bitumen is one of the most common ground improvement measures taken in different projects. Figure 8.15 presents some examples of this improvement method in actual sites.

8.6.2 Methods of Heating Soil In Situ (Heating, Freezing)

Heat treatment of a clayey soil at temperatures exceeding 400 °C results in pronounced permanent changes in its engineering properties. Heating a soil consumes a lot of energy. In an era of increasing energy costs and concerns about air pollution, large-scale heat treatment is unlikely to have much of a future, except in special circumstances. A special situation could arise where the soil already contains some fuel, either naturally or in the form of waste products. Heat treatment may also attract new interest in the production of bricklike building elements through medium-temperature

Figure 8.15 Shallow stabilization in actual cases.

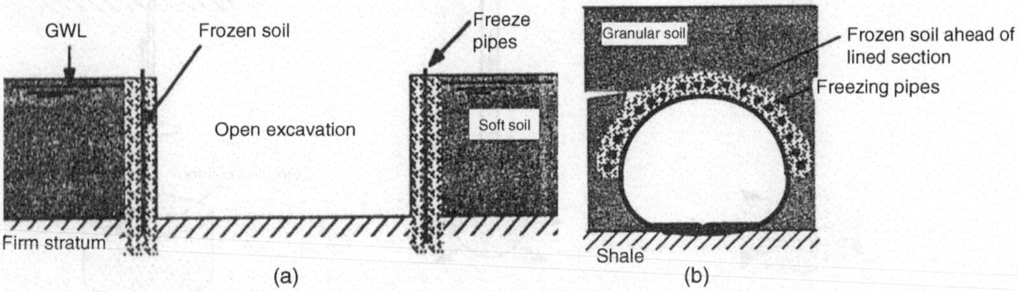

Figure 8.16 Ground freezing application: (a) for excavation and (b) for tunnel.

firing, possibly in combination with chemical stabilization. This type of improvement is shown for stabilizing an excavation as well as a tunnel in Figure 8.16.

Modifying ground through artificial freezing has great potential to solve difficult foundation problems because it promises a very high gain in strength and effective seepage control in a matter of hours or days after initiating the process. Typical applications are:

- Retaining structures for open excavations
- Lateral supports for sinking mine shafts
- Ground support and seepage control during tunneling
- Sealing of leaks in underground pipes
- Temporary stabilization of landslides
- Enabling undisturbed sampling in cohesionless soils

8.6.3 Grouting

Grouting is a process that involves injecting fluidized materials into voids or spaces within the ground or between structures and the ground. The primary objective of grouting is to enhance the strength, density, and permeability characteristics of the surrounding rock or soil. It serves as a means to create a more compact and less permeable medium. One of the key purposes of grouting is to fill voids that are inaccessible through other methods. By doing so, it prevents the transmission of stress between the ground and structures or within the ground itself. This helps in ensuring adequate stress transfer and stability, ultimately safeguarding the integrity of the structures and the surrounding ground. Overall, grouting plays a crucial role in improving the overall quality of the subsurface by filling voids, enhancing strength, and reducing permeability. It is a valuable technique employed in geotechnical engineering to address ground improvement and stabilization requirements.

Grouting is classified as bellows:

- Jet grouting (replacement),
- Grouting of voids,
- Compaction grouting (including slab-jacking),
- Displacement grouting,
- Penetration grouting (intrusion, permeation),
- Specific grouting applications and approaches like electro-grouting.

Grouting involves the use of various materials, which can be classified into suspensions, emulsions, and solutions. Suspensions consist of small solid particles dispersed in a liquid medium. Emulsions are two-phase mixtures consisting of colloidal droplets of liquid dispersed in another liquid. Solutions, on the other hand, are homogeneous mixtures of two or more substances in a liquid form. Different types of grouting methods are employed depending on the specific requirements of the project. The main types of grouting and their suitability for different applications are illustrated in Figure 8.17. Additionally, Figure 8.18 showcases various grouting methods and their respective applications. Figure 8.19 provides examples

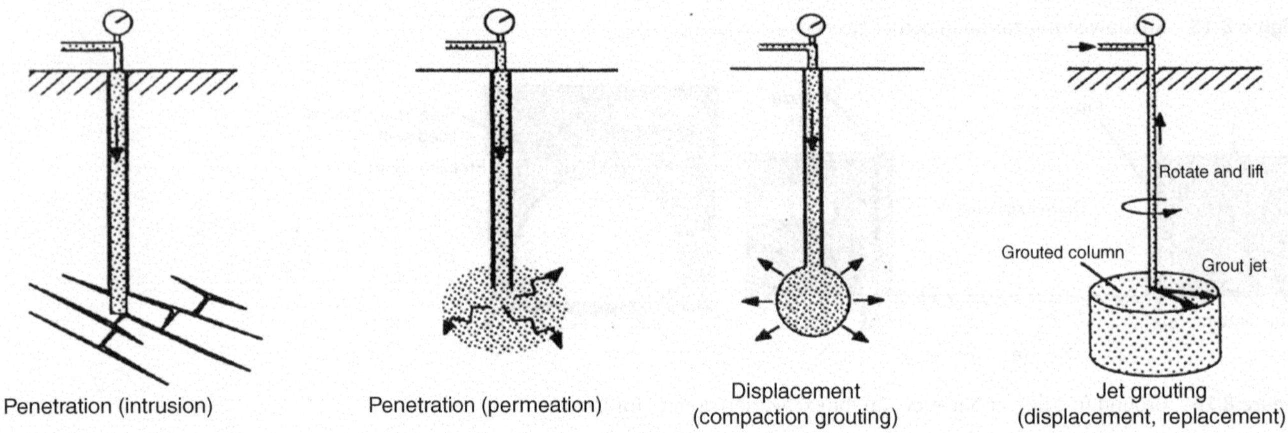

Figure 8.17 Types of grouting.

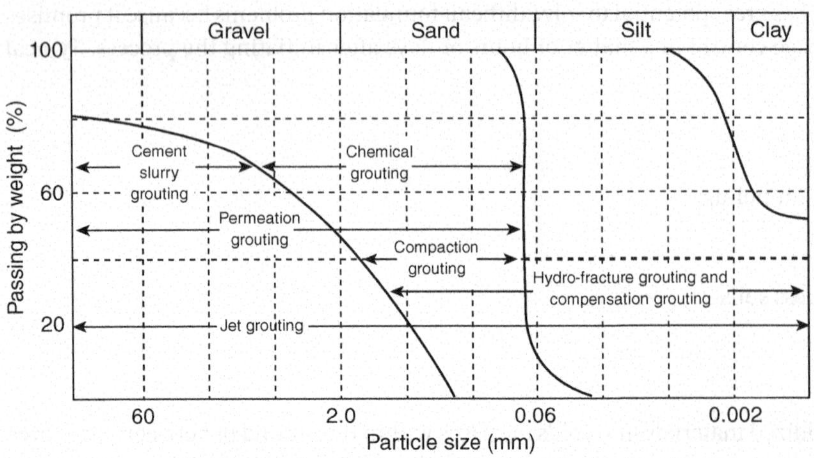

Figure 8.18 Suitability of various grouting approaches (data modified from Elias et al. (2006)).

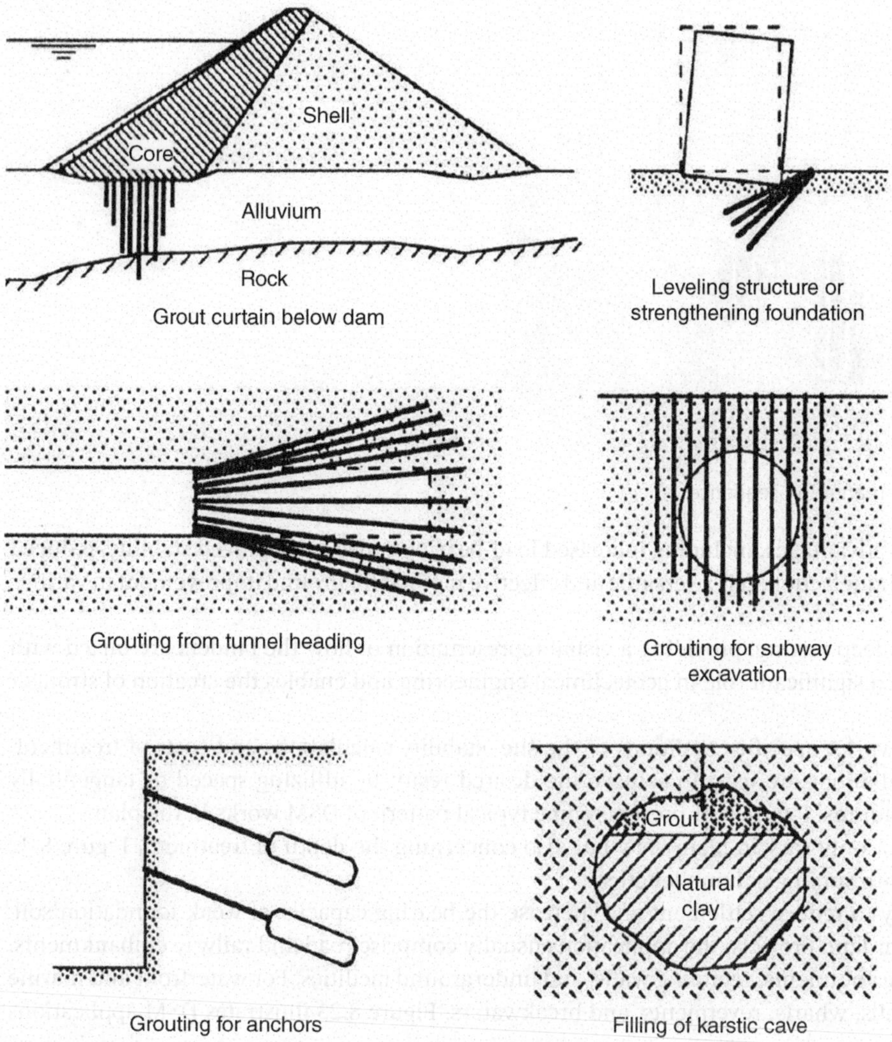

Figure 8.19 Typical applications of grouting.

of typical applications where the grouting process is employed. These applications demonstrate the wide range of uses for grouting in various construction and geotechnical projects.

Overall, the selection of grouting materials and methods depends on factors such as the nature of the ground, project requirements, and desired outcomes. These figures serve as valuable references in understanding the different aspects of grouting and its applications.

8.6.4 Deep Soil Mixing or Deep Mixing

Chemical agents, also known as binders, can be introduced into the ground to create improved geomaterials with enhanced strength and stiffness. This process involves the combination of the binders with existing geomaterials, resulting in the formation of hardened solid materials.

The soil mixing method is employed to achieve this improvement, using mechanical mixers or augers. This method can be applied near the surface, primarily for enhancing subgrade and base course materials, or at greater depths to create columns or walls. The technique of mixing binders with soils at significant depths, as depicted in Figure 8.20, is commonly referred to as deep mixing (DM) or deep soil mixing (DSM).

Deep mixing involves the mechanical blending of binders with the in situ soil, leading to the formation of a more uniform and improved soil mass. This process can be conducted using various techniques, such as column mixing or wall mixing, depending on the desired geotechnical objectives and site conditions.

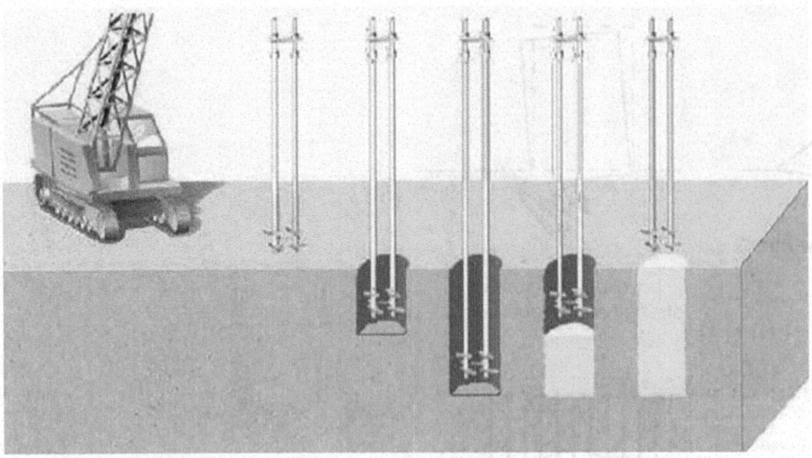

Figure 8.20 Deep soil mixing (DSM) implementation sequence.

The use of deep mixing offers several advantages, including increased load-bearing capacity, improved stability, reduced settlement, and enhanced resistance to liquefaction. It is a versatile and effective method for ground improvement in various geotechnical applications.

Figure 8.20 illustrates the process of deep mixing, providing a visual representation of how the binders are mixed with the soils at depth. This technique plays a significant role in geotechnical engineering and enables the creation of stronger and more stable ground conditions.

Depending on the purpose of DSM works, specific conditions of the site, stability calculations and costs of treatment, and different patterns of column installations are used to achieve the desired result by utilizing spaced or tangentially overlapping and single or combined columns. Figure 8.21 presents some typical pattern of DSM works in the plan.

Column installation patterns may not only vary in plain view but also concerning the depth of treatment. Figure 8.22 depicts examples of deep soil mixing with varying column lengths.

The purpose of using DSM is mainly to reduce settlement and increase the bearing capacity of weak foundation soil, as well as prevent sliding failure. For on-land projects, the applications usually comprise road and railway embankments, buildings, industrial halls, tanks, bridge abutments, retaining walls, and underground facilities. For waterfront and marine applications, they can include quay walls, wharfs, revetments, and breakwaters. Figure 8.23 illustrates DSM applications for foundation support.

8.6.5 Jet Grouting

Jet grouting is a special ground improvement technique between conventional grouting and deep mixing. Jet grouting utilizes grout (sometimes with water and/or air) to disturb the soil at depths and then mix the disturbed soil with grout to

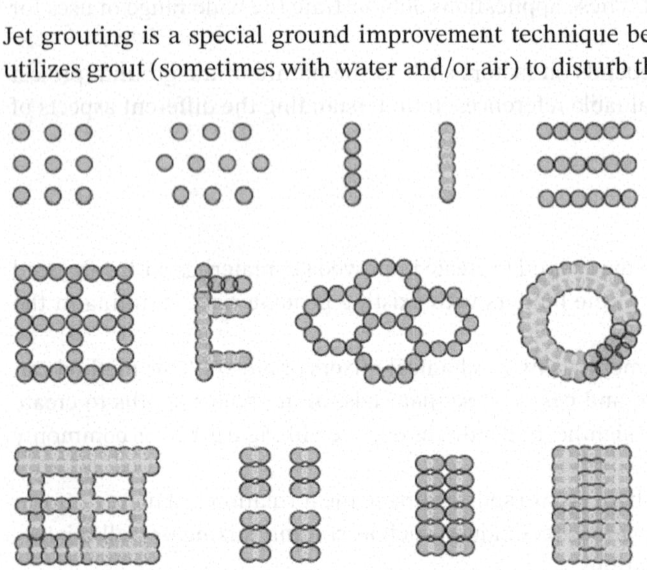

Figure 8.21 Various deep soil mixing (DSM) patterns.

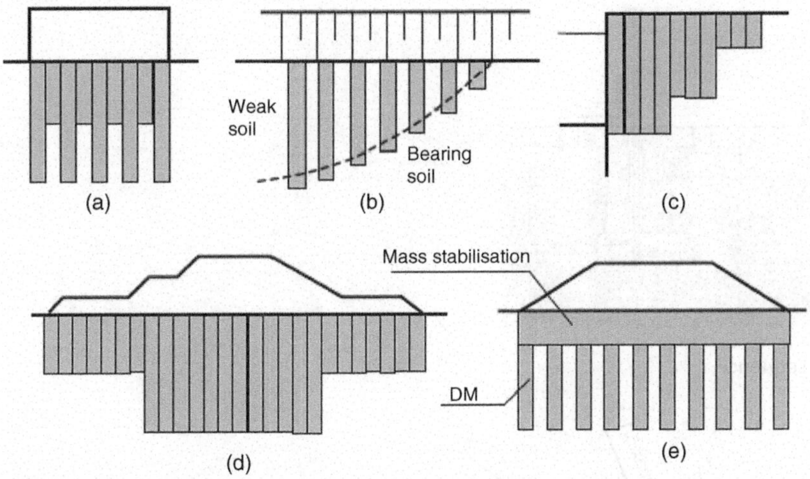

Figure 8.22 Applications of deep soil mixing (DSM): (a) combined short and long walls, (b) transition zone, (c) stepped columns, (d) embankment with berms, and (e) combined mass stabilization and DSM.

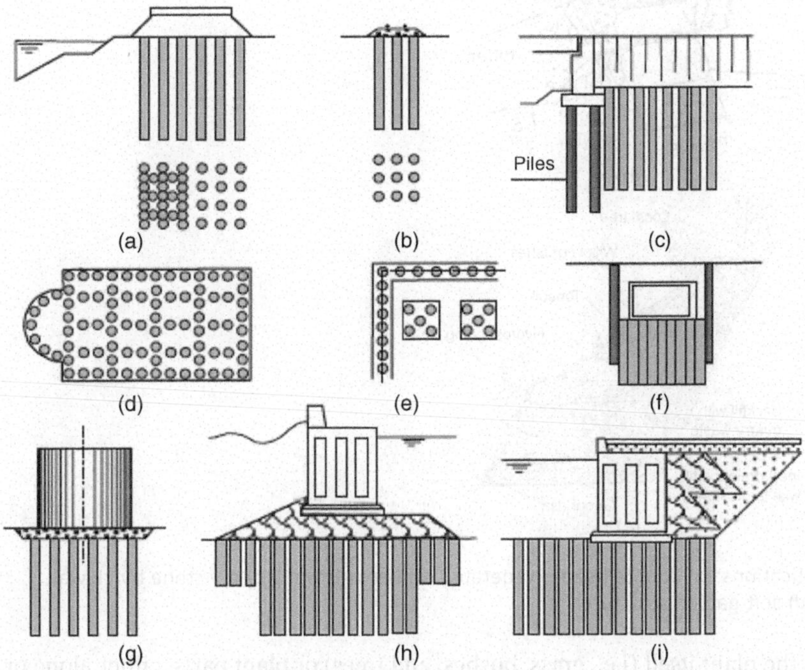

Figure 8.23 DSM applications for foundation support: (a) road embankment, (b) railway embankment, (c) bridge approach zone, (d) slab foundation, (e) strip and pad foundations, (f) culvert, (g) tank, (h) breakwater, and (i) quay wall.

form hardened columns or walls, as shown in Figure 8.24. Due to the hybrid nature of this technique, it is considered to be a deep mixing method by some authors and a grouting method by others.

8.6.6 New Trends: Plant Engineering and Bioengineering

Bioengineering constructions are the constructions wherein the live plants, associations of plants or plant materials are employed for structural purposes. They are frequently mixed with inert materials like gravel, stone, steel, wood, soil, or geosynthetics. Such approaches may lead to rich, live systems that retain self-balance without any artificial energy inputs. Therefore, a detailed knowledge of diverse construction approaches as well as proper assessment of the desired impacts and site conditions is essential in choosing the most efficient method.

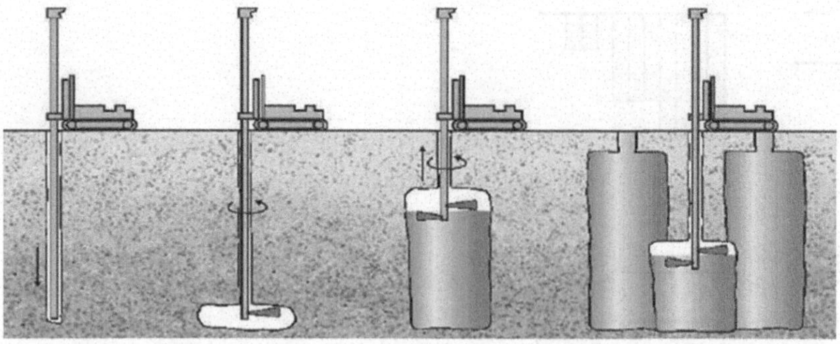

Figure 8.24 Jet grouting method implementation sequence.

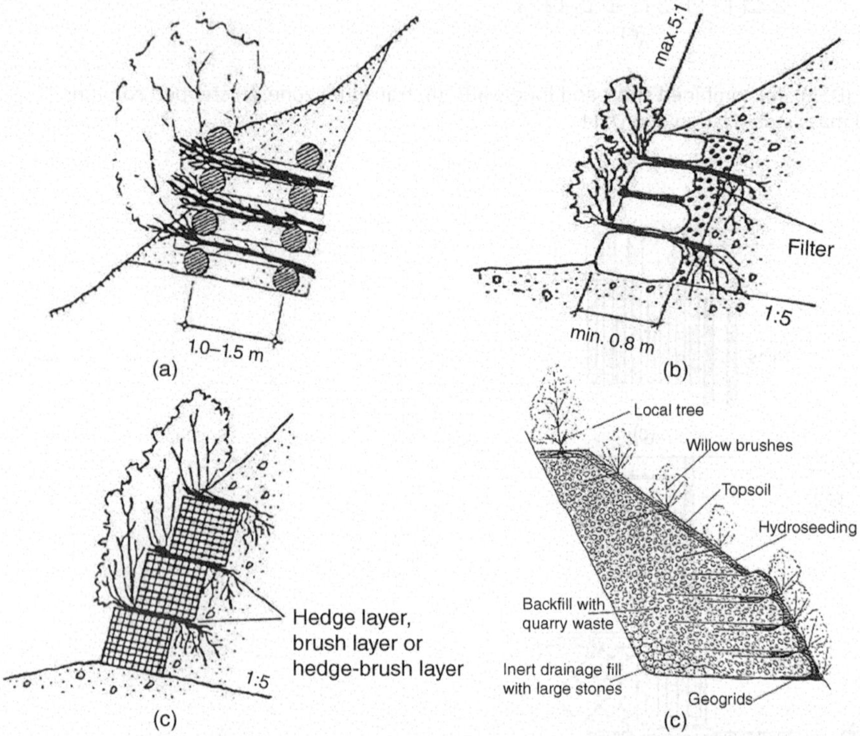

Figure 8.25 Illustrations of plant engineering applications: (a) double header vegetated timber crib wall, (b) dry stone block wall with live branches, (c) vegetated hard gabion, and (d) soft gabion structure.

Plant engineering or stabilization comprises the plant itself (i.e., grass, bushes, and trees) or plant parts, either alone or associated with inert materials such as steel, concrete, and rocks, to improve soil stability. Plant engineering techniques can be classified into various categories depending on the site conditions and intended use. These include surface protection or erosion control, stabilization using plants, and techniques involving both plants and inert materials (Punetha et al. 2019). The developing and growing of the roots remarkably elevate effectiveness of the stabilization. Some applications and types of this modification kind are presented in Figure 8.25.

The concept of using biological processes in soil improvement, known as the bio-mediated ground modification technique, has shown significant potential in geotechnical engineering applications. Bio-mediated soil improvement generally refers to the biochemical reaction within a soil mass to produce calcite precipitate to modify soil engineering properties. Microbially induced calcite precipitation (MICP) is an interdisciplinary field combining engineering, chemistry, and microbiology. MICP can be applied to improve the efficiency of in situ bioremediation (Rajasekar et al. 2017). Bio-mediated ground modification is considered an inventive and new approach that can prevent liquefaction and landslides in loose sand. The great promise of biological treatments has been demonstrated in various applications, including enhancing the shear strength and reducing the permeability of soils and cementation of the sand column (Umar et al. 2016).

8.7 Hydraulic Modification

Generally, all measures that are taken to control the water in a project are regarded as hydraulic modification practices. In this regard, the application of pumps for dewatering, using barriers against water seepage, and drains for accelerating the consolidation process can be listed.

8.7.1 Dewatering

8.7.1.1 Open Sumps and Ditches

This dewatering method is the cheapest and most common with the best performance in shallow excavations in dense uniformly graded coarse soils, rocks as well as in permeable soils that overlie the impermeable strata. Dewatering of clayey slopes may be accomplished by a combination of a toe drain and gravel-filled lateral slots.

8.7.1.2 Gravity Flow Wells

If the level of water in a borehole is lowered by steady pumping, groundwater will flow into the well under gravity until the phreatic surface, the levels at which the water pressure is atmospheric, has dropped to a new equilibrium position. This process can be used to lower the water level for construction purposes.

8.7.2 Drainage

Water is detrimental to the performance and life of roadways and earth structures in geotechnical engineering, and removing it prevents damage and failure, maintains performance, and prolongs the life of these structures. Different types of drains are used in this regard. Figure 8.26 illustrates different drainage systems.

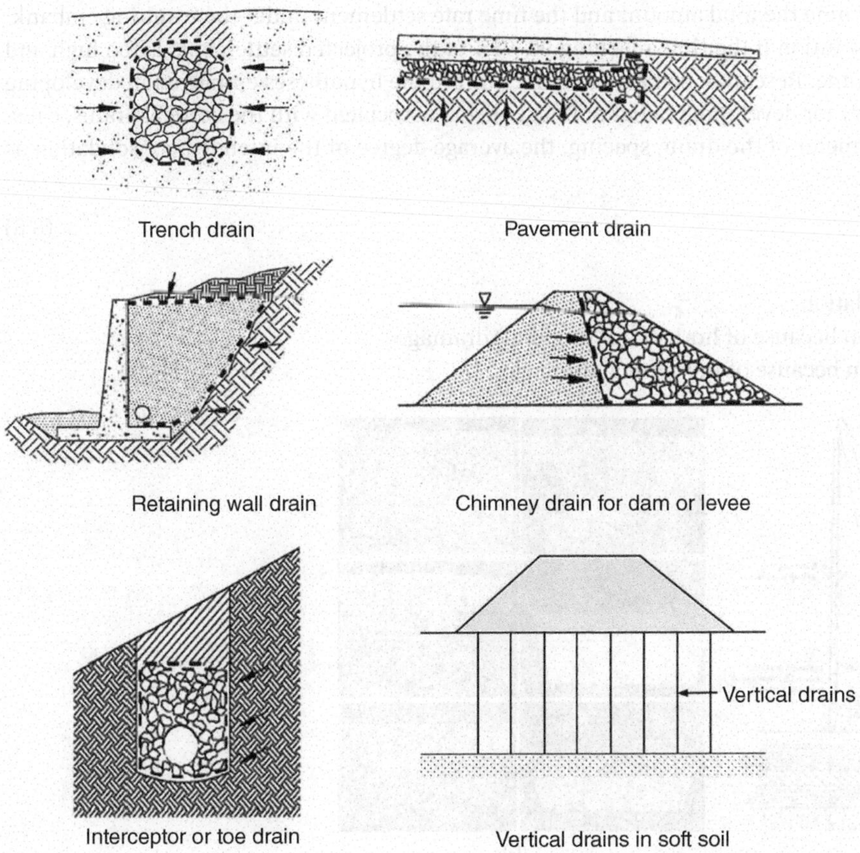

Figure 8.26 Different drainage systems (Han 2015 / With permission of John Wiley & Sons).

8.7.2.1 Filter Drains

It is possible to use the drainage system for controlling high-pore water pressures and high-pressure gradients. This kind of system involves synthetic filter materials, pipes, aggregate filters, and other conduits. Most drainage systems utilize the porous filter aggregates to collect water and direct it to the outlet channels.

8.7.2.2 Sand Drains

Studies have shown the specific efficiency of the sand drains in the stratified soils because of the greater permeability parallel to bedding. Researchers used sand drains in relation to dewatering systems to decline the level of groundwater in the stratified soils consisting of alternating layers of pervious and impervious soils. This system directs water from the more permeable layers to the dewatering screens situated in the opposite direction of the less permeable layers, which reduces dewatering costs. Moreover, sand drains would not be used in situations where loadings do not amount the pre-consolidation stress.

8.7.2.3 Band Drains

Wick drains can be best described as prefabricated vertical (PV) drains. Their primary use is accelerating the consolidation of the soft saturated compressible soils that are under loading. In simple terms, they largely decline the settlement time of embankment over the soft soil, and in this way, they speed up the rate of strength gain of the in situ soft soils. The effectiveness of a drain remarkably is dependent on its cross-sectional circumference instead of its cross-sectional area. Consequently, the band-shaped drains have higher efficiency in comparison to the drains with a circular cross-section. It should be noted that man-made fabric, in particular, polyethylene, polypropylene, polyester as well as PVC, have been increasingly employed for producing the band drains since the early 1970s. Now, there are several types of flat band drains; a majority of them contain a flat plastic core consisting of the drainage channels enclosed by a thin filter layer. This type of improvement is shown in Figure 8.27.

The principal purpose of any wick drain design is to choose the kind, spacing, as well as length of the drains to reach the intended degree of consolidation within a pre-defined timespan. The design procedure initiates with the settlement analyses without any wick drains to determine the total amount and the time rate settlement under the resulting embankment load. Wick drains may be a design solution if the time reaching 90–95% of the projected settlements is too high and goes beyond the contracted construction time. Researchers have therefore utilized the hypotheses employed in developing one-dimensional (1D) consolidation theory for developing radial drainage theory associated with the vertical drains, causing the association between time, the diameter of the drain, spacing, the average degree of the intended consolidation as well as coefficient of consolidation.

$$\overline{U} = 1 - (1 - \overline{U}_h)(1 - \overline{U}_v) \tag{8.8}$$

where

$\overline{U}$ = the total average degree of consolidation
$\overline{U}_h$ = the average degree of consolidation because of horizontal (or radial) drainage
$\overline{U}_v$ = the average degree of consolidation because of the vertical drainage

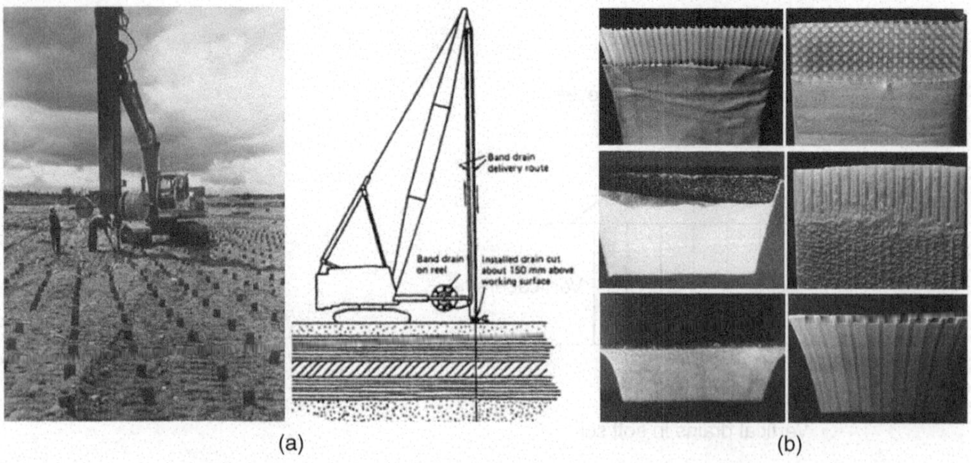

Figure 8.27 Band drains: (a) installation and (b) examples of different types.

Using an Idealistic solution (Barron's method) assumptions are:

(1) One-dimensional theory
(2) No drain resistance (infinite permeability of drain)
(3) No disturbance of soil (any consequence has been not observed on the soils permeability and consolidation features because of the drain installation)

The time necessary for achieving the desired level of consolidation (t) may be calculated as:

$$t = \frac{D^2}{8c_h} F(n) \ln\left(\frac{1}{1 - \overline{U}_h}\right) \tag{8.9}$$

where

D = the cylinder diameter under the drain influence (drain influence zone)
c_h = consolidation coefficient for horizontal drainage
$F(n)$ = drain spacing factor = $\ln\left(\frac{D}{d_w}\right) - \frac{3}{4}$(simplified)
d_w = the diameter of an equivalent circular drain

Another solution for calculating the timespan necessary to achieve the intended level of consolidation is the modified solution (Hansbo's method):

$$t = \frac{D^2}{8c_h}(F(n) + F_s + F_r) \ln\left(\frac{1}{1 - \overline{U}_h}\right) \tag{8.10}$$

where

$F(n)$ represents the drain spacing factor = $\ln\left(\frac{D}{d_w}\right) - \frac{3}{4}$
d_w implies the equivalent diameter = 0.5 (a + b), as depicted in Figure 8.28
F_s stands for the factor for soil disturbance; that is, $((k_h/k_s) - 1) \ln\left({}^{d_s}/_{d_w}\right)$
k_h refers to the permeability coefficient in the horizontal direction in an undisturbed soil
k_s is the permeability coefficient in the horizontal direction in the disturbed soils
d_s represents the diameter of the idealized disturbed zone surrounding the drain = (4 to 7) d_w
F_r = Factor of drain resistance $\pi Z(L - z)\left({}^{k_h}/_{q_w}\right)$
Z = distance below the top surface of the compressible soil layer
L = effective length of the drain, the drain length in case of drainage at just one end, half-length of the drain in case of the occurrence of drainage at the two ends
q_w = discharge capacity of the drain (at gradient = 1.0)

Example 8.4 Construction of a storage tank with a diameter of 39 m and a weight of 2500 tons is planned. The soil stratigraphy at the site is composed of an 8 m thick, soft clay layer resting on dense sand. Consolidation and permeability indexes are $2.7 \times 10^{-8}\ \frac{m^2}{s}$ and $6 \times 10^{-10}\ \frac{m}{s}$, respectively. The level of groundwater is near to the ground surface. Prefabricated vertical

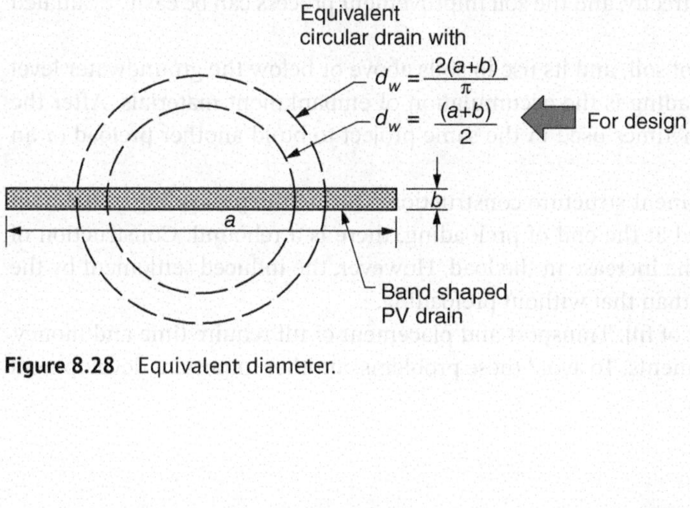

Figure 8.28 Equivalent diameter.

drains (PVD) are suggested to accelerate consolidation. The selected PVDs have 100 and 5 mm dimensions and penetrate to soft and sand layers. The allowed discharge capacity is equal 0.000686 $\frac{m^3}{s}$. It is assumed that there is no smear effect in the PDV installation process. Determine the required distance of PVDs in a square pattern for the condition that, after eight months of preloading, 80% of the settlement occurs.

$$t = 8 \text{ months} = 8 \times 30 \times 24 \times 3600 \text{ s}$$

$$T_v = \frac{C_v.t}{H_{dr}^2} = \frac{2.7 \times 10^{-8} \times 8 \times 30 \times 24 \times 3600}{\left(\frac{8}{2}\right)^2} = 0.0349$$

$$T_v = \frac{\pi}{4} U_v^{\ 2} \rightarrow U_v = 0.211 = 21.1\%$$

$$U = 1 - (1 - U_v)(1 - U_h) \rightarrow 0.8 = 1 - (1 - 0.211)(1 - U_h) \rightarrow U_h = 0.7465 = 74.65\%$$

Hansbo method

$$t = \frac{D^2}{8c_h}(F(n) + F_s + F_r) \ln\left(\frac{1}{1 - U_h}\right)$$

$$F_s = 0$$

$$F(n) = \ln\left(\frac{D}{d_w}\right) - \frac{3}{4}$$

$$d_w = \frac{100 + 5}{2} = 52.5 \text{ mm}$$

$$F_r = \frac{2\pi}{3} L^2 \frac{k_h}{q_w} \rightarrow F_r = \frac{2\pi}{3} \times 4^2 \times \frac{6 \times 10^{-10}}{0.000686} = 2.93 \times 10^{-5}$$

$$C_h = 2C_v = 2 \times 2.7 \times 10^{-8} = 5.4 \times 10^{-8}$$

$$8 \times 30 \times 24 \times 3600 = \frac{D^2}{8 \times 5.4 \times 10^{-8}} \left(\ln\left(\frac{D}{0.0525}\right) - \frac{3}{4} + 0 + 2.93 \times 10^{-5}\right) \ln\left(\frac{1}{1 - 0.7465}\right) \rightarrow D = 1.57 \text{ m}$$

$$D = 1.13S \rightarrow S = \frac{1.57}{1.13} = 1.389 \text{ m} \cong 1.39 \text{ m}$$

8.7.3 Preloading

Preloading is the temporary loading applied at the construction site of buildings and facilities in order to improve subsurface soils. This is usually done before the construction phase, but occasionally due to the construction of storage tanks. In principle, pre-construction works are required, even after the construction of the complete structure. It is possible that the materials can be used to apply the preload to the part of the final structure as in road infrastructures.

One of the biggest advantages of this method compared to other land improvement methods is its very low cost. The effects of this method can be evaluated immediately and directly, and the soil improvement process can be easily evaluated by measuring ground movements and pore pressures.

This method has been successfully applied to any type of soil, and its use in soils above or below the groundwater level has been successful. The most common method of preloading is the accumulation of embankment materials. After the preload is completed, these materials are moved and sometimes used in the same project to build another preload or an embankment.

Figure 8.29 illustrates fill preloading followed by permanent structure construction. During the preloading, settlement develops with loading and time. When the fill is removed at the end of preloading, there is a rebound. Construction of the permanent structure induces new settlement due to the increase in the load. However, the induced settlement by the structure after preloading is expected to be much smaller than that without preloading.

Fill preloading requires the use of a significant amount of fill. Transport and placement of fill require time and money. Sometimes, fill placement induces failure and large movements. To avoid these problems, another important development

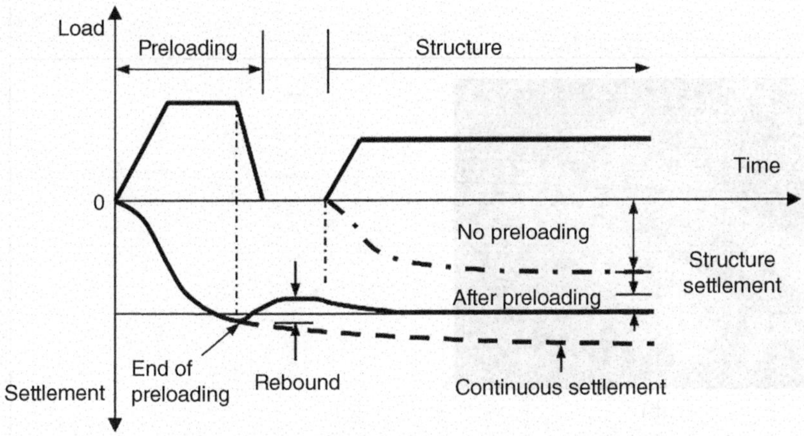

Figure 8.29 Fill preloading and application (Han 2015 / With permission of John Wiley & Sons).

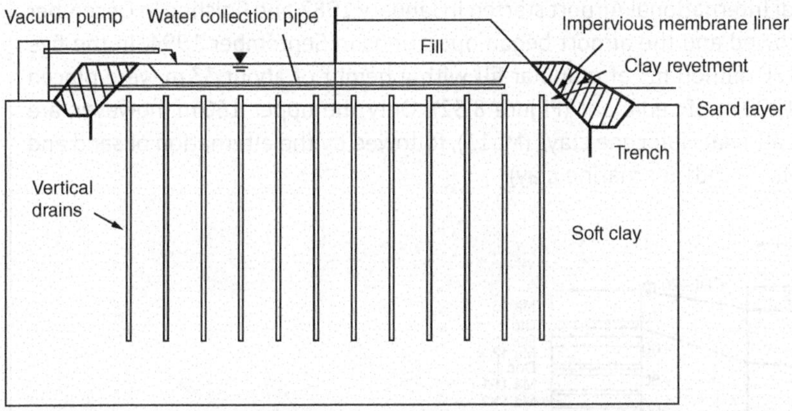

Figure 8.30 Vacuum preloading (Han 2015 / With permission of John Wiley & Sons).

of the preloading technology is the idea of vacuum preloading. Different from fill preloading, vacuum preloading utilizes atmospheric pressure on an air-tight system installed in the field by applying a vacuum inside the system (Figure 8.30).

Case Study 8.5

Sand Drains for Kansai Airport Artificial Island (Mesri and Funk 2014; Purzin et al. 2010)

The Kansai International Airport, located in Osaka Bay, was built in deep seawater ranging from 18 to 20 m to mitigate noise pollution and avoid land acquisition disputes. The construction of Island I, spanning 511 hectares, commenced in 1987, and its Runway I became operational in 1994. Island II, covering 545 hectares, began construction in 1999, and its Runway II became operational in 2007. This remarkable engineering feat involved the implementation of over 2.2 million vertical sand drains that fully penetrated the Holocene clay layer, ranging from 17.3 to 24.1 m thick, along with the placement of 430 million m^3 of fill material. The project received acclaim and was recognized as one of the "Monuments of the Millennium" by the American Society of Civil Engineers, honoring the ten civil engineering projects that had the greatest positive impact on twentieth-century life. It is worth noting that despite encountering geotechnical challenges during the initial phase of construction, the airport stands as an impressive achievement (Purzin et al. 2010). Aerial photos of the site are presented in Figure 8.31.

(Continued)

Case Study 8.5 (Continued)

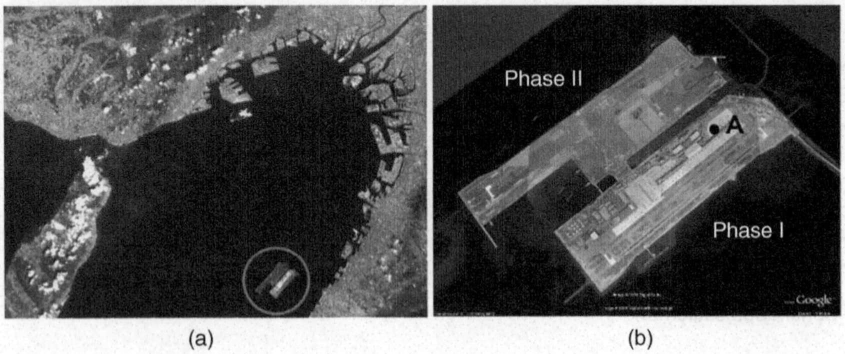

Figure 8.31 Kansai International Airport: (a) Osaka Bay and (b) the Phase I and II islands (Purzin et al. 2010 / Springer Nature).

Construction of the first phase island of Kansai International Airport started in January 1987 and finished in December 1991. Construction of the airport facilities followed and the airport began operations in September 1994. In the five years of the island's construction, more than 180 million m^3 of granular fill with a height of about 33 m were placed on the seabed, which consists of more than 1200 m of sediment (Figure 8.32). Only the upper 160 m, however, are considered compressible; the top 20 m are soft alluvial Holocene clays (Ma13), followed by the alternation of sand and clay layers of Pleistocene origins (Ma 7–12, "Ma" stands for marine clay).

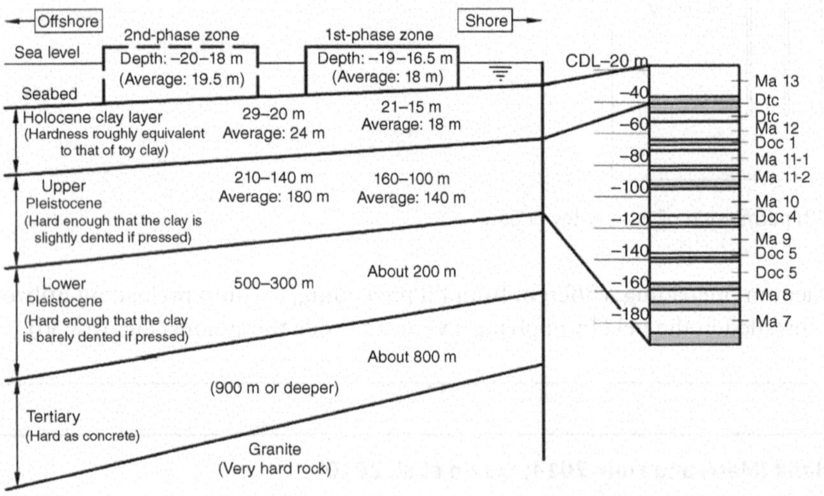

Figure 8.32 The soil profile of the seabed (Purzin et al. 2010 / With permission of Springer Nature).

The construction process included the following stages: first, the top 20 m of the seabed (Holocene clays) were treated within the design island area by installing vertical sand drains to accelerate compaction under the backfill. Next, the perimeter seawalls were built (Figure 8.33). Subsequently, the land reclamation took place in which the granular fill, taken from several excavations in the Osaka area (they practically leveled a couple of large hills), was placed within the seawalls up to a depth of about 3 m below the water level using bottom-dump barges. The final step was accomplished by employing four large barges, anchored inside the seawalls, which transferred the fill brought by the smaller barges from across the bay, to bring the island to the required 4 m above the water level. This height guarantees that the airport will not be affected by high tides brought by typhoons that hit the coast of Japan every September.

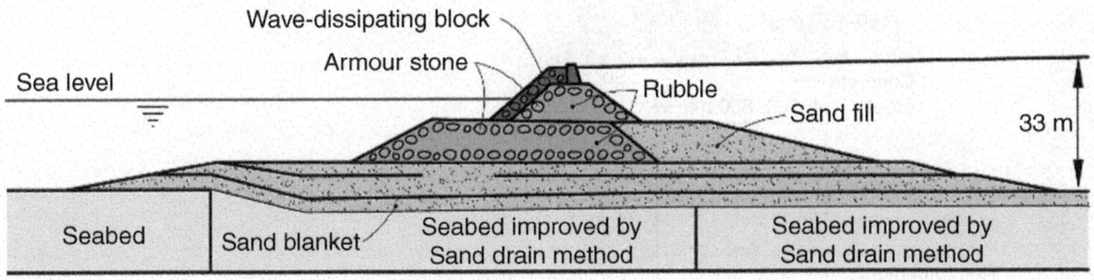

Figure 8.33 The cross-section of a seawall (Purzin et al. 2010 / With permission of Springer Nature).

8.8 Reinforcement

Soil reinforcement is among the most common ground modification techniques nowadays. Their performance is similar to concrete enhancement using reinforcement rebars. Soil reinforcement can be done in vertical, horizontal, and inclined conditions. Reinforcements, based on placement in or exterior to the stabilized soil mass, are divided into extrusions or intrusions.

8.8.1 Extrusions

Ground-water commonly results in a challenging problem during the excavation process. In fact instability can be observed in the case of the construction of deep excavations into the sandy ground. Therefore, these locations must be carefully addressed before and during the excavation process via dewatering and/or exclusion procedures. The economy of creating an obstacle to exclude the ground-water is dependent on the presence of an impermeable stratum under the excavation on order to form an efficient natural cut-off for the barriers. If there is no stratum, or if its placement is at too great a depth to be employed as a cut-off, upward seepage can occur, which results in instability at the excavation base. Types of extrusion techniques are:

- Sheet piling
- Contiguous bored piles
- Secant pile walls
- Bentonite cut-off walls
- Geomembrane barriers
- Concrete diaphragm walls
- Grout curtains and panels
- Ice walls

Figure 8.34 presents views of the steel and concrete sheet piles.

8.8.2 Intrusions

According to the research, reinforced earth has been considered one of the composite materials with the soil consisting of the reinforcing elements, so it usually involves strips of plastic geo-grids or galvanized steel. Its bond and tensile strength enhance the surrounding soil and govern the reinforcement effectiveness. Moreover, reinforcement largely ameliorates the mechanical features of the granular soil, and thus, two states of failure may occur with regard to the confining pressure, $\sigma 3$. It is possible to make the reinforcing elements from any materials that have an acceptable tensile strength of any shape and size, affording the friction surface required to prevent slippage and failure by pulling out. In addition, the materials must exhibit resistance to deterioration or corrosion due to other variables. Researchers have also employed aluminum or steel alloy strips, steel cables, wire mesh, polymeric geosynthetic materials, and glass-fiber reinforced plastic as the reinforcing elements. Figure 8.35 depicts typical types of intrusions. Types of intrusion techniques are:

- Reinforced soil
- Soil nailing

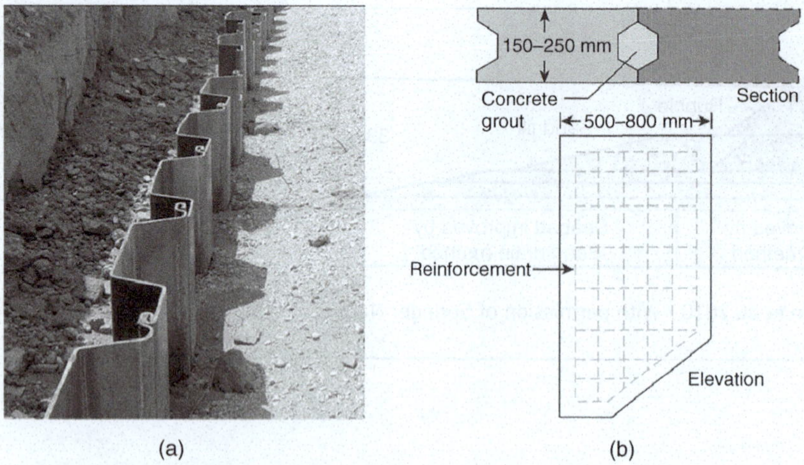

Figure 8.34 Sheet piles: (a) steel and (b) precast concrete (Das and Sivakugan 2019 / with permission of Cengage Learning).

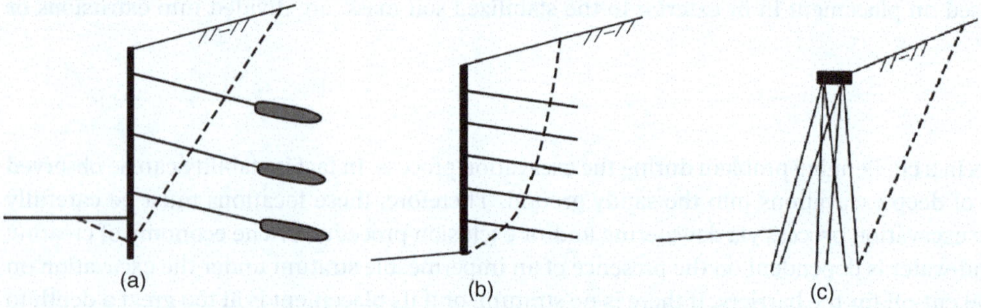

Figure 8.35 Some typical intrusions: (a) anchoring, (b) nailing, and (c) micro-piling (Han 2015 / With permission of John Wiley & Sons).

- Doweling
- Anchoring
- Reticulated micro-piling

Anchors are used to enhance the stability of slopes, excavations, and trenches in soils and rocks. The force provided by anchors is calculated through the difference between active and passive forces at the front and back of the wall. The friction resistant force of the anchor, is calculated through Equations 8.9 and 8.10.

For granular and non-cohesive soils:

$$T_R = \sigma_z' \times A_a \times L_a \times K_f \quad (8.11)$$

And, for cohesive soils:

$$T_R = \alpha S_u \times A_a \times L_a \quad (8.12)$$

where

A_a = side area of anchor in the unit length
L_a = anchor length
σ_z' = effective stress at the middle depth of the anchor (if the anchor is inclined)
α = reduction factor of undrained shear strength, ranging between 0.5 and 1 when S_u is between 40 and 150 kPa
K = factor related to soil type and density condition (Table 8.3)

Micro-piles are small piles with a diameter between 75 and 250 mm and are made of steel and concrete. They are usually cast in-place and have reinforcing rebars at the center. Micro-piles can be implemented in vertical or inclined directions and groups. They are generally applied to sustain superstructures and stabilize soil against displacement and instabilities. Figure 8.36 depicts the typical sequence of micro-pile construction.

Table 8.3 Values of *K* for calculating anchors force (CFEM 2006).

Soil type	Density condition		
	Loose	Medium	Dense
Silt	0.1	0.4	1
Fine Sand	0.2	0.6	1.5
Medium Sand	0.5	1.2	2
Coarse Sand, Gravel	1	2	3

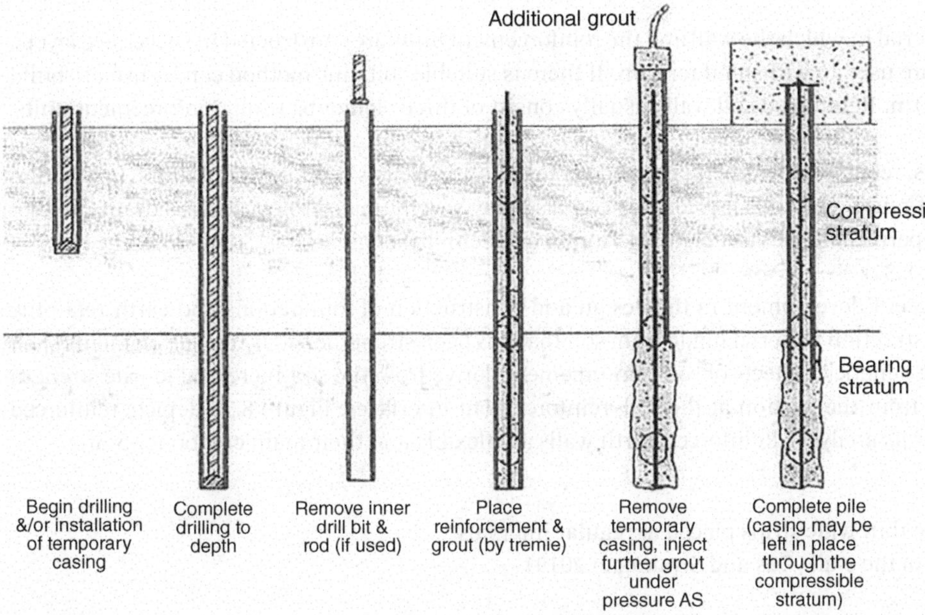

Figure 8.36 Micro-pile construction sequence (FHWA-2000 / U.S. Department of Transportation / Public Domain).

For design purposes, the common assumption is that the load is transferred through the side and the friction resistance between grout and soil without any contribution of base resistance. The allowed axial compressive or tensile capacity is calculated through the:

$$P_{G\text{-allowable}} = \frac{\alpha_{\text{bond-nominal-strength}}}{F.S.} \times \pi \times D_{\text{bond}} \times L_{\text{bond}} \tag{8.13}$$

where

$P_{G\text{-allowable}}$ = allowed axial geotechnical capacity
$F.S.$ = Factor of safety, equal 2.5
D_{bond} = diameter of bond area
L_{bond} = length of bond area
$\alpha_{\text{bond-nominal-strength}}$ = friction resistance when the grouting process is done while extracting the casing with the pressure suggested in Table 8.4.

8.8.3 Geosynthetics

One popular method of soil improvement is reinforcing the soil. This method works similarly to the improvement of concrete properties by reinforcement (Hajitaheriha et al. 2021). Soil reinforcement can be done vertically, horizontally, or obliquely by different elements. In addition to the use of piles, other major methods for soil reinforcement include the construction of stone and lime columns in the ground, the creation of micro-piles, restraint in the soil, and the use of the soil system. The arrangement and function of these methods differ.

Table 8.4 Values of bond strength for micro-piles' geotechnical capacity.

Soil type	$\alpha_{bond-nominal-strength}$ (kPa)
Soft fine-grained (silt and clay)	35–95
Stiff and dense fine-grained	70–190
Fine and relatively loose sands	70–100
Relatively fine and dense sands	120–360
Gravel	120–360

Reinforced soil is a composite material in which the wall and the reinforcement units are constructed in successive layers. The reinforcement units used in it are used to withstand tension. If there is suitable soil, this method can be used to build an embankment up to a height of 30 m. Reinforced soil walls usually consist of three elements: wall, reinforcement units, and soil.

Mesh, galvanized fibers and sheets, rebar, plastic, fiberglass, and geotextiles are used as reinforcement units. In addition to being used in the construction of reinforced soils, geotextiles can also play several major roles, such as drainage, erosion control, reinforcement, and impermeable membrane formation in geotechnical engineering. Some different types of geosynthetics are depicted in Figure 8.37.

The use of reinforced earth is a recent development in the design and construction of foundations and earth-retaining structures. Reinforced earth is a construction material made from soil that has been strengthened by tensile elements such as metal strips or geosynthetics. The beneficial effects of soil reinforcement derive from the soil-increased tensile strength and the shear resistance developed from the friction at the soil–reinforcement interfaces. Figure 8.38 depicts reinforced earth wall elements and the basis for its analysis. Reinforced earth walls are flexible and their main components are:

1) Backfill, which is a granular soil
2) Reinforcing strips (tie), which are thin wide strips placed at regular intervals
3) A cover or skin on the front face of the wall (Das and Sivakugan 2019)

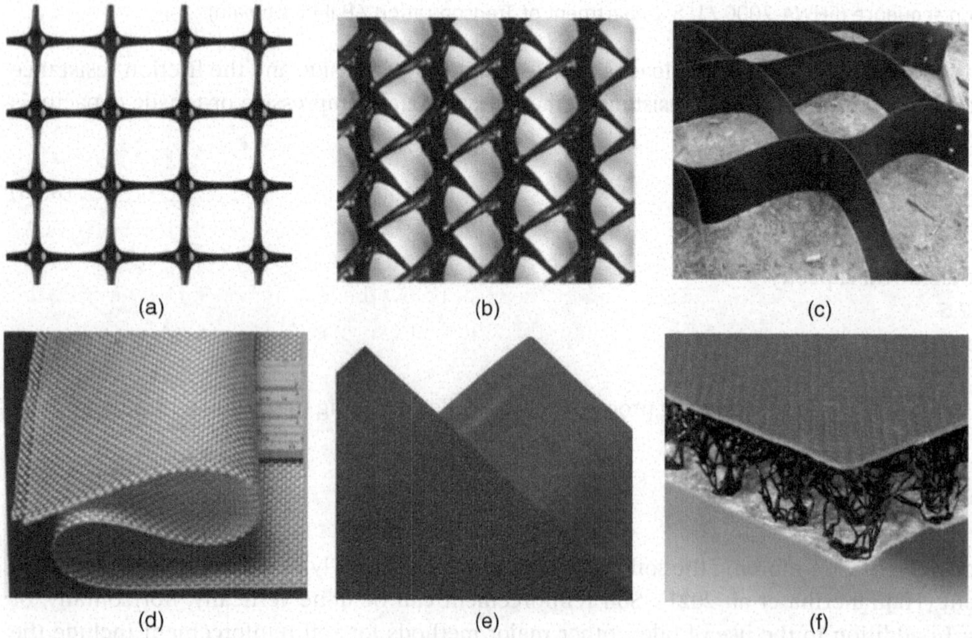

Figure 8.37 Some types of geosynthetics: (a) geogrid, (b) geonet, (c) geocell, (d) geotextile (woven), (e) geomembrane, and (f) geocomposite (Markiewicz et al. 2022 / MDPI / CC BY 4.0).

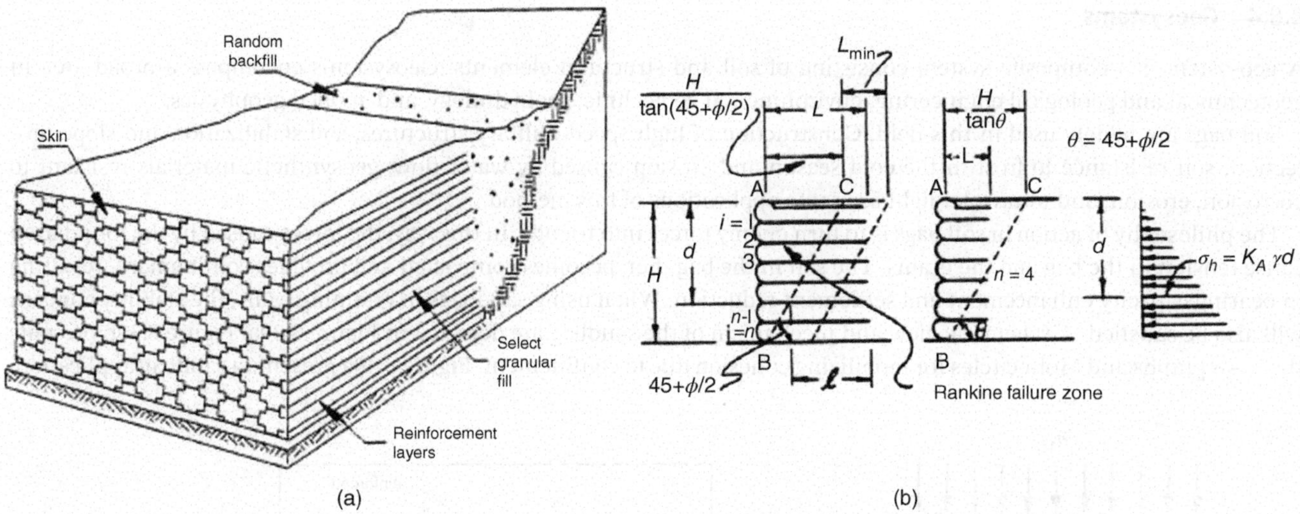

Figure 8.38 Reinforced earth wall: (a) elements (Robert Holtz 2001 / Robert Holtz) and (b) analysis bases.

Factor of Safety Against Tie Failure

$$FS = \frac{f_s \times w \times t}{K_a(\gamma z + q_s) \times s_y \times s_x} \tag{8.14}$$

where

s_y = horizontal distance between ties
s_x = vertical distance between ties
f_s = yield or breaking strength of the tie material
w = width of each tie (typically about 3 in.)
t = tie thickness. For metal strips extra thickness regarding the average annual corrosion rate during the structure's life span must be considered
z = the depth of purpose
γ = unit weight of soil

Total Length of Tie (Pullout Capacity)

$$FS = \frac{2w \times L_{min} \times (\gamma z + q_s) \times \tan\delta}{K_a(\gamma z + q_s) \times s_y \times s_x} = \frac{2w \times L_{min} \times \tan\delta}{K_a \times s_y \times s_x} \tag{8.15}$$

$$L_{min} = \frac{K_a \times s_y \times s_x \times R_s}{z \times w \times \tan\delta} \tag{8.16}$$

$$\sigma_d = K_a \times \gamma z \tag{8.17}$$

L_{min} is the length of the tie beyond the limits of the Rankine active failure zone, which makes an angle of $(45 + \frac{\phi}{2})$ with the horizontal direction. The length of ties within the failure zone (L_R) at the depth of z, considering H as the height of the wall is:

$$L_R = (H - z) \times \cot\left(45 + \frac{\phi}{2}\right) \tag{8.18}$$

Therefore, the total length of ties is calculated via the:

$$L = L_R + L_{min} \tag{8.19}$$

In design practice, the factor of safety for tie length and failure is considered in the range of 1.5–3.

8.8.4 Geosystems

A geosystem is a composite system consisting of soil and structural elements. Geosystems encompass a broad area in geotechnical and geological engineering, environmental geotechnic, geohydrology, and applied geophysics.

Soil bags are widely used in this field. Construction of high-speed military structures, soil stabilization and slope protection, soil resistance to frost in the cold season and erosion caused by water flow, geosynthetic materials resistant to corrosion, erosion, and ultraviolet light are some applications of this method.

The philosophy of action in soil bags is to turn enemy forces into friends. In this way, the forces caused by the foundation cause tension in the bag and the grains. The soil in the bag then becomes compacted and provides confinement, resulting in bearing capacity enhancement and settlement reduction. When using sandbags in retaining walls, the stability concern will also be satisfied. A schematic view and mechanism of the sandbag are depicted in Figure 8.39a. Figure 8.39b presents the $\tau - \sigma$ graphs and Mohr circles for mobilizing cohesion due to confinement. Figure 8.40a presents a sandbag application.

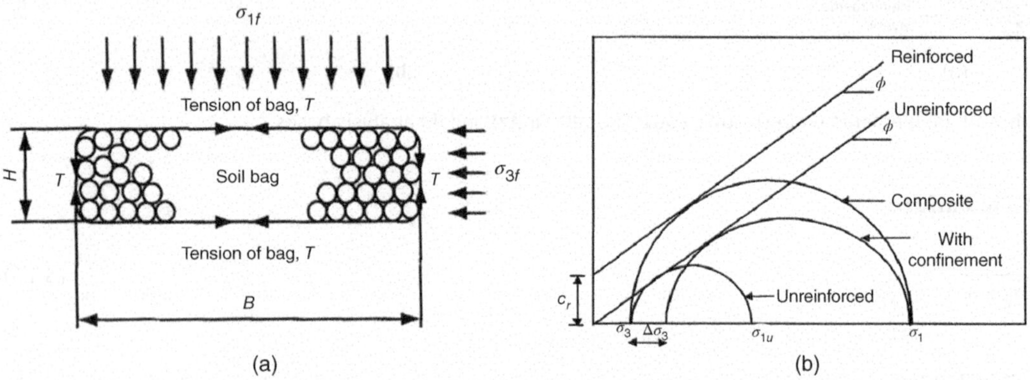

Figure 8.39 (a) Mechanism of a sandbag and (b) cohesion mobilization due to confinement.

Figure 8.40 Some typical geosystems: (a) sandbag, (b) crib wall (Planted Well), (c) gabion, and (d) geocell (vadimgouida / Adobestock).

For unreinforced confined conditions:

$$\sigma_1 = K_p(\sigma_3 + \Delta\sigma_3) \tag{8.20}$$

And, for the reinforced condition:

$$\sigma_r = \frac{1+\sin\phi}{1-\sin\phi}\sigma_3 + 2c_r\sqrt{\frac{1+\sin\phi}{1-\sin\phi}} = K_p\sigma_3 + 2c_r\sqrt{K_p} \tag{8.21}$$

$$c_r = \frac{\Delta\sigma_3}{2} \tag{8.22}$$

Crib walls, gabions, and geocells are other geosystems that are widely used in construction projects today. Crib walls have high flexibility in the face of high slope displacement. They are made element by element, and therefore, they are easy to transport and require little execution skills. They are beautiful in appearance, and as a result of their high permeability, the pressure of the pore-water pressure does not accumulate behind them. Figure 8.40b depicts an example of crib walls.

In gabions, broken rock fragments or river sand are placed inside metal baskets and used in execution. These types of geosystems have high flexibility and permeability and require little skill to implement, and as such, they have garnered attention in civil engineering. Figure 8.40c illustrates a gabion application.

Geocells are used in various applications such as rapid and temporary construction of roads in loose sand areas, stabilization of asphalt road infrastructure, stabilization of dirt road infrastructure, prevention of road collapse and separation. Figure 8.40d presents a typical geocell application.

8.9 Quality Control and Selection of Methods

Control and evaluation in civil engineering, geotechnical engineering, and especially in soil improvement are of significant importance. Various indicators are used in the assessment that soil settlement is of considerable importance in this regard. The percentage of moisture content and soil permeability are important parameters in the assessments.

The use of field experiments is also effective and practical in improvement control. In situ and on situ on-test concluding site density testing, SPT, CPT, PLT, PMT, and shear wave velocity evaluation can be used. Figure 8.41 shows the soil assessment of the site prior to and following the improved soil by in situ tests.

Practitioners employ field instrumentation like settlement platforms and gauges, inclinometers as well as piezometers for monitoring the functionality of the wick drains and controlling the surcharge and construction rate of the embankment. Importantly, designers and the instrumentation workforce must fully understand the given piece of instrumentation, which should be installed, and the strata wherein the instruments must be embedded. Furthermore, settlement measuring instruments, including deep settlement points, horizontal deflection tools, and platforms, are applied to measure the total amount and rate of consolidation. An inclinometer is utilized to measure the horizontal deflection and depth and as a warning against possible failures. Consequently, the pore-pressure instruments (piezometers) are employed for calculating the consolidation rate and excess build-up of the pore pressure, which implies possible failures.

The choice of improvement methods is of considerable importance in projects. Technical, construction, economical, durability, and geo-environmental criteria are effective in choosing the method. The decision to use a method should be accompanied by consideration of other methods of dealing with weak and compacted soils. Options vary greatly in impact, cost, schedule, and the space required. Therefore, the responsibility and influence of the geotechnical engineer at this stage of decision are crucial.

Whether the plan is a new or existing development, equipment and materials, availability, quantity and expected response to an earthquake or explosion, accuracy of predictions, and finally, cost should also be considered in choosing the method. The employer may also be inclined to one of a variety of methods for several reasons. The final selection steps of the improvement plan are as follows:

- Selection of usable improvement methods
- Proper development and evaluation of selected design methods
- Final design for one or more improvement methods
- Comparison of final designs and method selection
- Local experiment to confirm the effect of the selected method and to extend the method chosen.

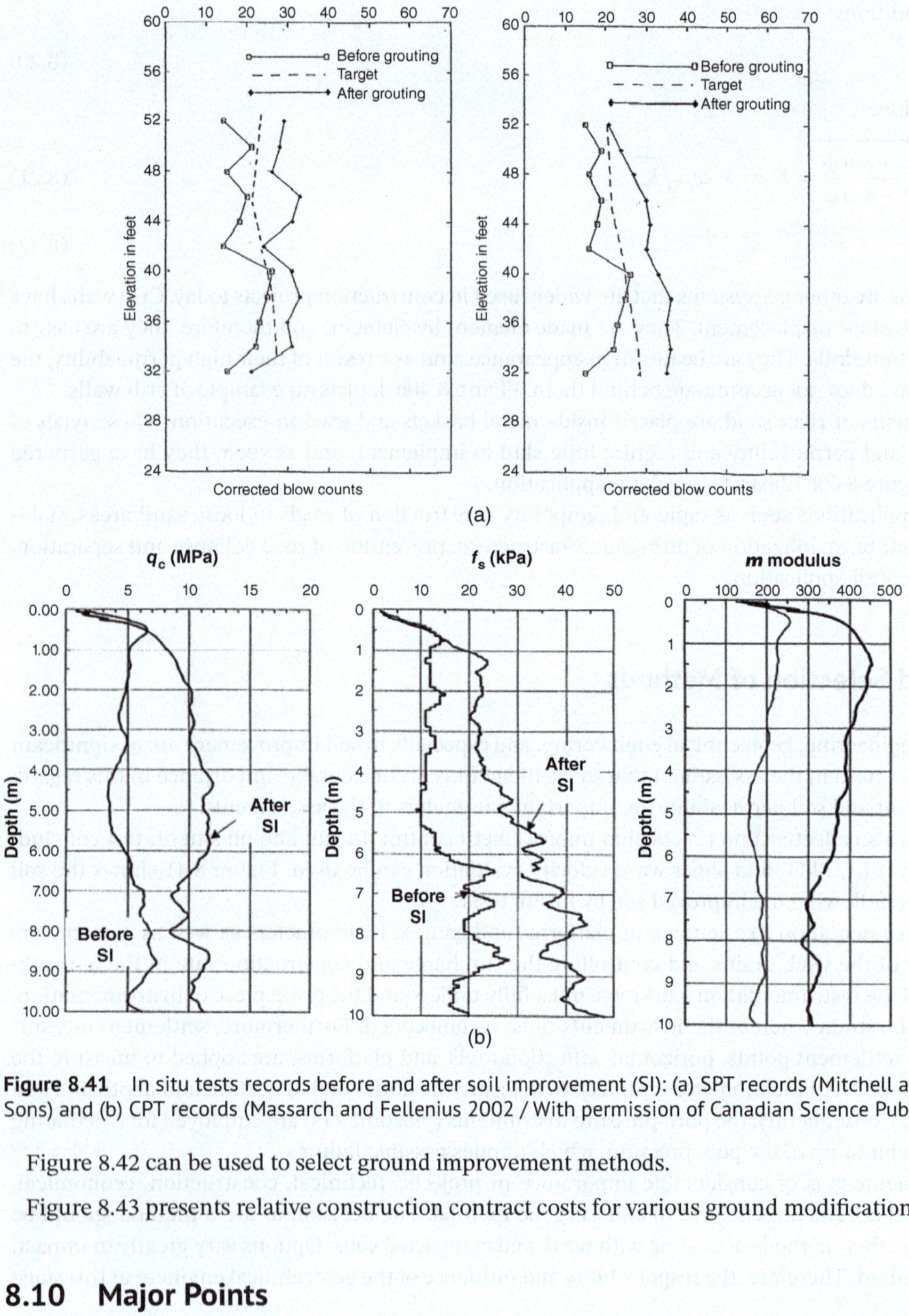

Figure 8.41 In situ tests records before and after soil improvement (SI): (a) SPT records (Mitchell and Wentz 1991 / John Wiley & Sons) and (b) CPT records (Massarch and Fellenius 2002 / With permission of Canadian Science Publishing).

Figure 8.42 can be used to select ground improvement methods.

Figure 8.43 presents relative construction contract costs for various ground modification approaches.

8.10 Major Points

1) There are several options in a project site with problematic geomaterials, such as avoiding construction on the site, planning the super-structures, eliminating and replacing the problematic geo-materials with the non-problematic geo-materials, and enhancing the geotechnical conditions and geo-material features.
2) Soil improvement from a geotechnical point of view is the modification of all soil features to improve its engineering functionality, like strength, lower permeability and compressibility, and more acceptable conditions of the groundwater.
3) There are several methods for soil improvement, and familiarity with these methods helps solve many foundation engineering problems. Generally, ground modification methods can be classified into five general groups: earthworks, densification and compaction, physicochemical modifications, hydraulic modifications, and reinforcement.

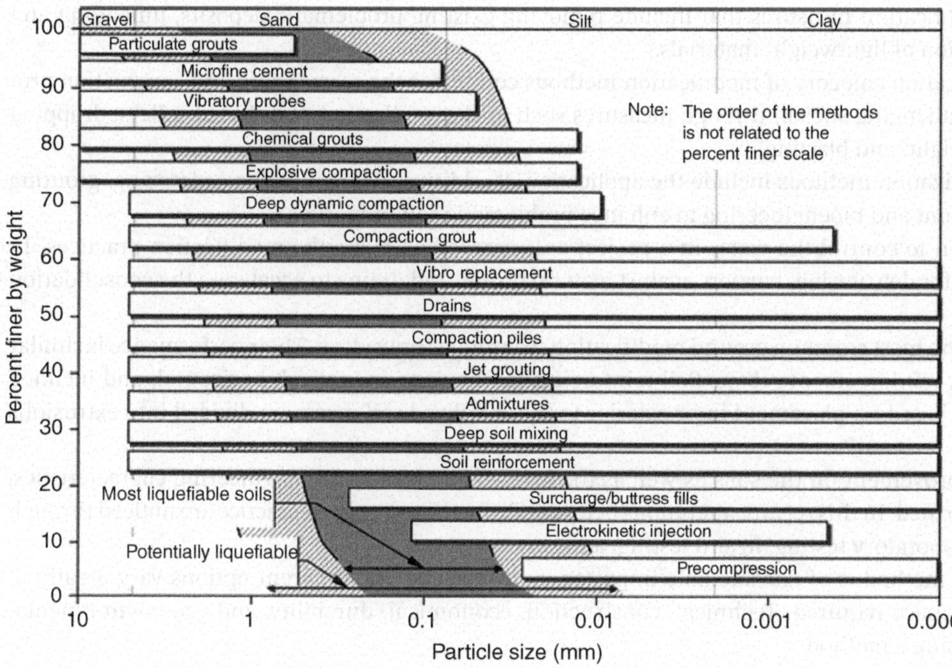

Figure 8.42 Improvement approaches of ground for various kinds of soils (adapted from Mitchell and Gallagher 1998).

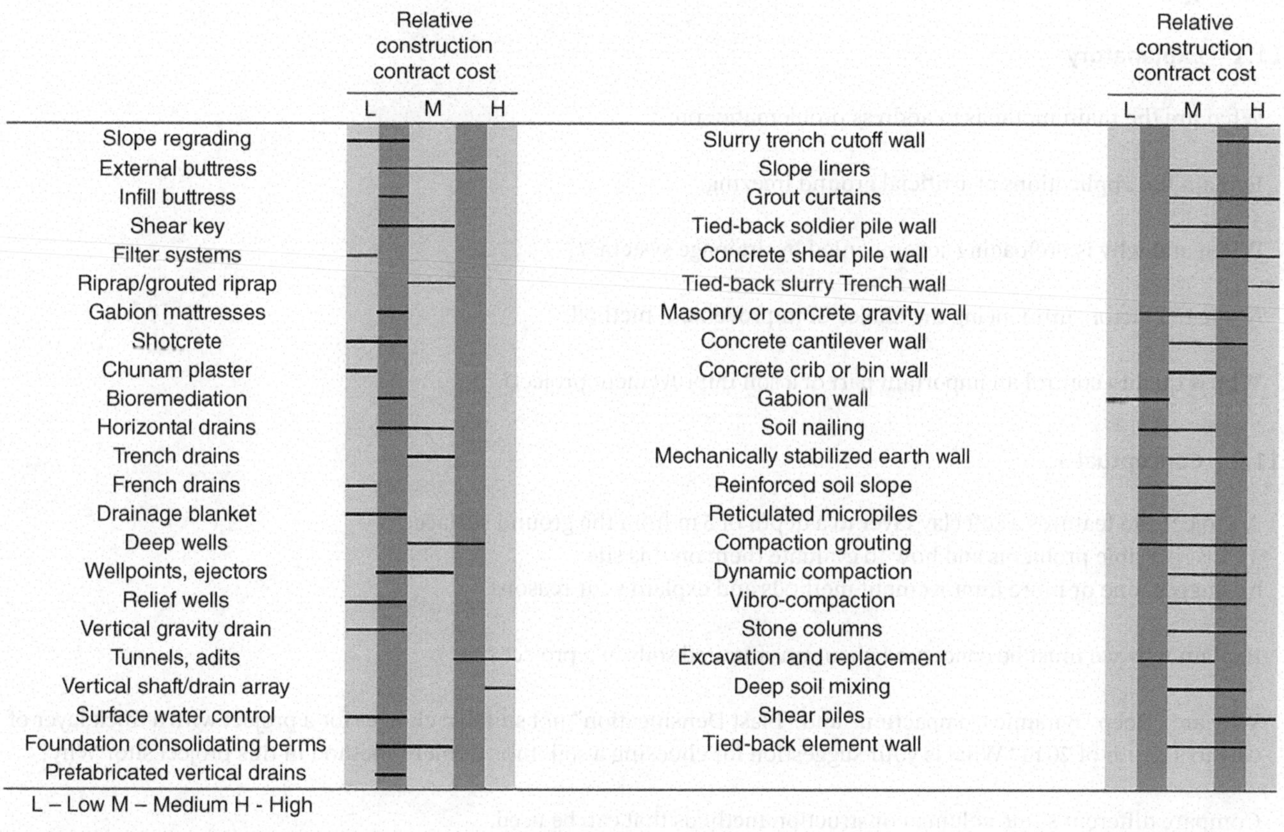

Figure 8.43 Relative construction contract costs for different improvement methods.

4) Earthworks are ground modification measures that include removing existing problematic deposits, unloading and floatation as well as application of lightweight materials.
5) The compaction and densification category of modification methods comprises the rearranging and compaction process of existing loose and challenging subsoil through measures such as the application of different rollers, dropping tampers from a particular height, and blasting.
6) Physical and chemical stabilization methods include the application of additives, using heating or freezing, grouting and soil mixing, as well as plant and bioengineering to enhance problematic soil properties.
7) Generally, all measures taken to control the water in a project are regarded as hydraulic modification practices. In this regard, applying pumps for dewatering, barriers against water seepage, and drains to accelerate the consolidation process can be listed.
8) Soil reinforcement is one of the most common ground modification techniques nowadays. Their performance is similar to enhancing concrete using reinforcement rebars. Soil reinforcing can be done in vertical, horizontal, and inclined conditions. Reinforcements, based on placement in or exterior to the stabilized soil mass, are divided into extrusions or intrusions.
9) To evaluate the effect of improvement on the soils as well as the level of changes in soil engineering characteristics, quality control must be performed. In this regard, alterations before and after improvement practice are noticed through tools such as sampling and laboratory testing, in situ testing, and surveying.
10) The choice of improvement method is of considerable importance in projects. The different options vary greatly in impact, cost, schedule, and space required. Technical, construction, economical, durability, and geo-environmental criteria are effective in choosing a method.

8.11 Questions and Problems

8.11.1 Explanatory

1 What are the main methods to address problematic soils?

2 Explain the applications of artificial ground freezing.

3 When and why is preloading accompanied by drainage systems?

4 Name the factors influencing the choice of improvement method.

5 Why is quality control an important part of a soil improvement project?

8.11.2 Conceptual

1 A project site features a soft clay layer to a depth of 5 m from the ground surface.
a) List possible problems and how to evaluate them on this site.
b) Suggest one or more improvement methods and explain your reasons.

2 Explain why we must be concerned about unsaturated soils in a project site.

3 Why are "Deep Dynamic Compaction" and "Blast Densification" not suitable choices for a project with a thick layer of clay to a depth of 20 m? What is your suggestion for choosing a soil improvement method in this project site? Why?

4 Compare different stone column construction methods that can be used.

5 The results of the CPT test in a project up to a depth of 30 m are similar to those shown in Figure 8.41.
a) Identify the layer or layers that may contain problematic soil by stating the reason.
b) Suggest one or more improvement methods for problem layers and explain your reasons.

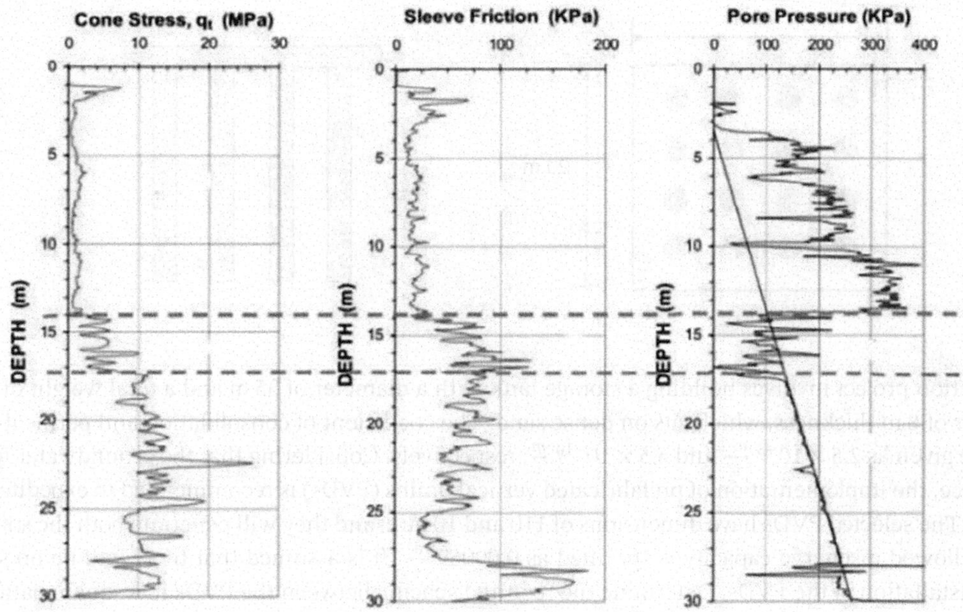

8.11.3 Problems

1 It is desirable to design a fully floating foundation for a structure weighing 7×10^3 tons.
a) Determine the depth of the excavation without considering the side friction of the excavation wall.
b) Determine the minimum undrained cohesion, c_{u2}, required for foundation to have a factor of safety of 2 against bearing capacity failure.

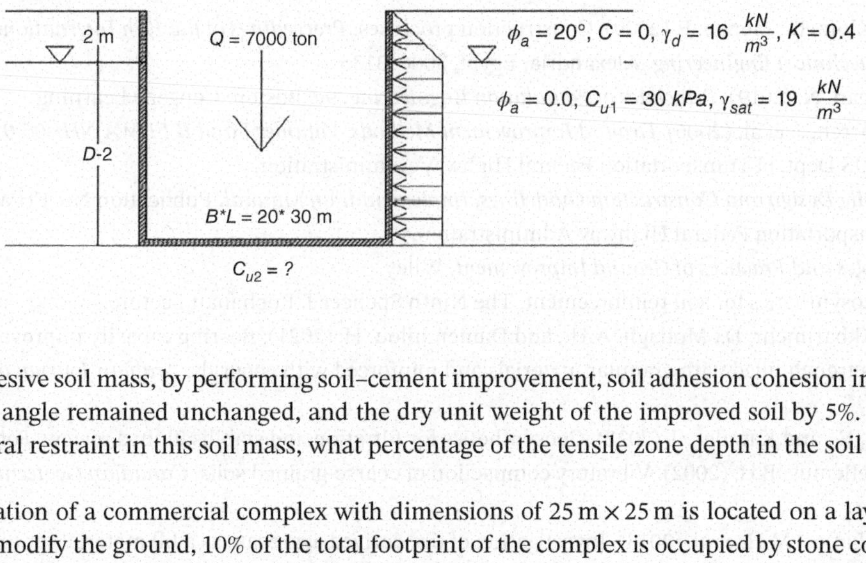

2 In a friction-cohesive soil mass, by performing soil–cement improvement, soil adhesion cohesion increased by 70%, the internal friction angle remained unchanged, and the dry unit weight of the improved soil by 5%. If drilling is carried out without lateral restraint in this soil mass, what percentage of the tensile zone depth in the soil mass will change?

3 The rigid foundation of a commercial complex with dimensions of 25 m × 25 m is located on a layer of soft saturated soil. In order to modify the ground, 10% of the total footprint of the complex is occupied by stone columns. The ratio of the modulus of elasticity the columns to that of the surrounding is expected to be 5. The maximum allowable settlement is 8 cm. Using equivalent parameters:
a) Compare the bearing capacity of the foundation prior to and after ground modification.
b) Estimate the immediate settlement of the foundation prior to and after ground modification under the stress of 320 kPa beneath the foundation ($K = 0.4$)
Before stones are implemented: $C = 30$ kPa, $\phi_s = 22°$, $E_s = 10$ MPa, $\gamma_{sat} = 19$ kN/m³
After stones are implemented: $\phi_c = 38°$, $E_s = 25$ MPa

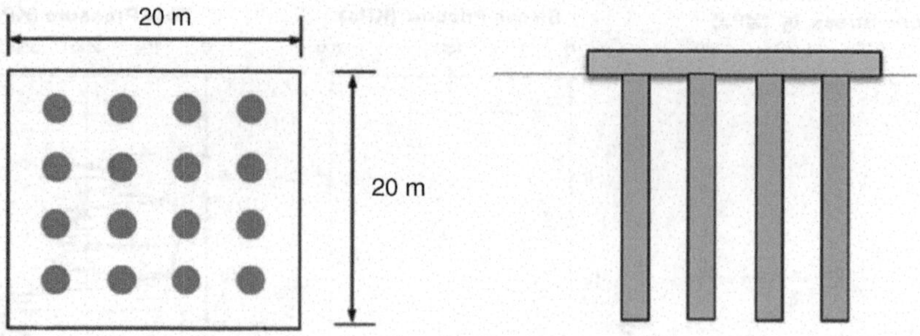

4 The construction project involves building a storage tank with a diameter of 35 m and a total weight of 3000 tons on a soft clay layer of 8 m thickness, which sits on dense sand. The coefficient of consolidation and permeability of the soft clay layer are given as $2.5 \times 10^{-8}\ \frac{m^2}{s}$ and $6.5 \times 10^{-10}\ \frac{m}{s}$, respectively. Considering that the groundwater level is near the ground surface, the implementation of prefabricated vertical drains (PVDs) is recommended to expedite the consolidation process. The selected PVDs have dimensions of 110 and 10 mm, and they will penetrate both the soft clay and sand layers. The allowed discharge capacity is specified as $0.00069\ \frac{m^3}{s}$. It is assumed that there will be no smearing effect during the installation of the PVDs. Determine the required spacing between the PVDs in a square pattern so that 80% of the settlement can be achieved after a nine-month preloading period.

References

Aflaki, E., Sedighi, P., and Eslami, A. (2014). Study on the failure behavior of three different stabilized problematic soils. *Scientia Iranica* 21 (4): 1231–1240.

CFEM (2006). *Canadian Foundation Engineering Manual*, 4e. Vancouver: Canadian Geotechnical Society, BiTech Publishers, 488 p.

Chu, J., Varaskin, S., Klotz, U., Menge, P. (2009). Construction processes. *Proceedings of the 17th International Conference on Soil Mechanics and Geotechnical Engineering*, Alexandria, Egypt, 3006–3135.

Das, B.M. and Sivakugan, N. (2019). *Principles of Foundation Engineering*, 9e. Boston: Cengage Learning.

Elias, V., Welsh, J., Warren, J. et al. (2006). *Ground Improvement Methods, Volumes I and II FHWA NHI-06-019 and FHWA NHI-06-020*. USA: US Dept. of Transportation Federal Highway Administration.

FHWA (2000). *Micropile Design and Construction Guidelines, Implementation Manual*. Publication No. FHWA-SA-97-070. US Department of Transportation Federal Highway Administration.

Han, J. (2015). *Principles and Practices of Ground Improvement*. Wiley.

Holtz, R.D. (2001). Geosynthetics for soil reinforcement. The Ninth Spencer J. Buchanan Lecture.

Hajitaheriha, M.M., Akbarimehr, D., Motlagh, A.H., and Damerchilou, H. (2021). Bearing capacity improvement of shallow foundations using a trench filled with granular materials and reinforced with geogrids. *Arabian Journal of Geosciences* 14 (15): 1–14.

Markiewicz, A., Koda, E., and Kawalec, J. (2022). Geosynthetics for filtration and stabilisation: a review. *Polymers* 14: 5492.

Massarch, K.R. and Fellenius, B.H. (2002). Vibratory compaction of coarse-grained soils. *Canadian Geotechnical Journal* 39: 695–709.

Meles, D., Bayat, A., Shafiee, M.H. et al. (2013). Investigation of tire derived aggregate as a fill material for highway embankment. *International Journal of Geotechnical Engineering* 8 (2): 182–190. https://doi.org/10.1179/1939787913y.0000000015.

Mesri, G. and Funk, J.R. (2014). Settlement of the Kansai International Airport Islands. *Journal of Geotechnical and Geoenvironmental Engineering* 141 (2): 1–16.

Mitchell, J.K. and Gallagher, P.M. (1998). *Engineering Guidelines on Ground Improvement for Civil Works Structures and Facilities*. Washington, DC: U.S. Army Corp of Engineering Division Directorate of Civil Works.

Mitchell, J.K. and Wentz, F.J. (1991). *Performance of Improved Ground During the Loma Prieta Earthquake*. Earthquake Engineering Research Center. Report no. UCB/EERC-91.

Punetha, P., Samanta, M., and Sarkar, S. (2019). Bioengineering as an effective and ecofriendly soil slope stabilization method: a review. In: *Landslides: Theory, Practice, and Modelling. Advances in Natural and Technological Hazards Research*, vol. 50 (ed. S. Pradhan, V. Vishal, and T. Singh). Cham: Springer.

Purzin, A.M., Alonso, E.E., and Pinyol, N.M. (2010). Unexpected excessive settlements: Kansai International Airport, Japan. In: *Geomechanics of Failures*, 23–43.

Rajasekar, A., Moy, C.K.S., and Wilkinson, S. (2017). MICP and advances towards eco-friendly and economical applications. *IOP Conference Series: Earth and Environmental Science* 78: 6.

Sham-e Consulting Engineering Co. (2015). Technical report of site investigation and soil improvement practice assessment of oil tank farm of Amir Abaad port for Arad Omran Caspian Company

Shakeran, M., Eslami, A., and Ahmadpour, M. (2016). Geotechnical aspects of explosive compaction. *Shock and Vibration* 2016: 1–14.

Taube, G & Martin, P.E., P.G. 2002, *Stone Columns for Industrial Fills*, Nicholson Construction Company, Cuddy, Pennsylvania.

Tavakoli, B., Shakeran, M., and Eslami, A. (2014). Settlement Prediction for Problematic Soils Improved by Explosive Compaction (EC) Method. VITAE-REVISTA DE LA FACULTAD DE QUIMICA FARMACEUTICA

Umar, M., Kassim, K.A., and Chiet, K.T.P. (2016). Biological process of soil improvement in civil engineering: a review. *Journal of Rock Mechanics and Geotechnical Engineering* 8: 767–774.

9

Foundations on Stabilized Soils

Chapter Highlights

- **Categories of Stabilized Soil Media under Foundations**
  - Mass Treatment
  - Rigid Intrusions
  - Reinforced Soil
- **Foundations on Massive Treated Soils**
- **Foundations on Rigid Intrusions (Stone Columns)**
  - Bearing Capacity
  - Settlement Estimation
  - Stability
- **Foundations on Reinforced Soils**
  - Bearing and Pullout Capacity
  - Confinement
- **Case Histories**
  - Dynamic Compaction at Dubai Palm Jumeira Sewage Treatment Plants [Massive Treated Soil]
  - Analytical and Numerical Investigation of Deformation Mechanism in Stone Column Supported Foundations [Rigid Intrusions]
  - Ground Improvement for Railway Embankment in Malaysia [Rigid Intrusions]
  - GRS Structures for Railways in Japan [Reinforced Soil]
  - Bearing Capacity of Geosynthetic Encased Stone Columns [Reinforced Rigid Intrusions]

9.1 Introduction

The application of ground modification techniques serves as an intermediate and complementary solution for both shallow and deep foundations. Soil improvement practices encompass a wide range of methods, with over 50 different approaches available that can be applied based on factors such as project type, soil type, and local requirements. One common application of soil improvement is to establish a suitable subbase for the construction of shallow foundations. From a foundation engineering perspective, soil improvement techniques can be broadly categorized into three groups: massive treatment, rigid intrusions, and reinforced soil.

Massive treatment refers to soil improvement techniques that involve modifying a large volume or mass of soil simultaneously. This category includes methods such as deep dynamic compaction (DDC), explosive compaction (EC), preloading with drainage, grouting, and massive soil mixing. Figure 9.1 illustrates examples of massive treatment, showcasing massive grouting and the application of vertical drains.

Rigid intrusions are employed when accessing deeper layers of soil proves challenging. These intrusions consist of vertical columns that improve bearing capacity, reduce settlement, and enhance stability. Examples of rigid intrusion methods include deep soil mixing, vibro-compaction stone columns, and lime/cement columns. Figure 9.2 illustrates different types of rigid intrusions.

Advanced Foundation Engineering: Principles, Performance and Prospects, First Edition. Moses Karakouzian and Abolfazl Eslami.

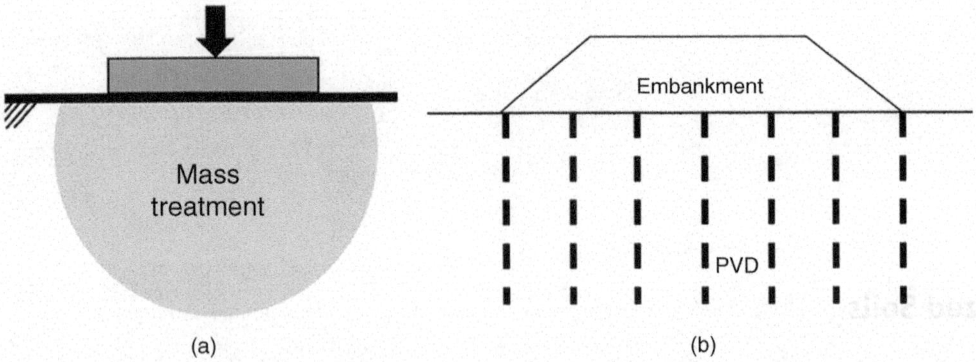

Figure 9.1 Massive treatment: (a) such as mass grouting and densification, (b) vertical drains (PVDs) and preloading.

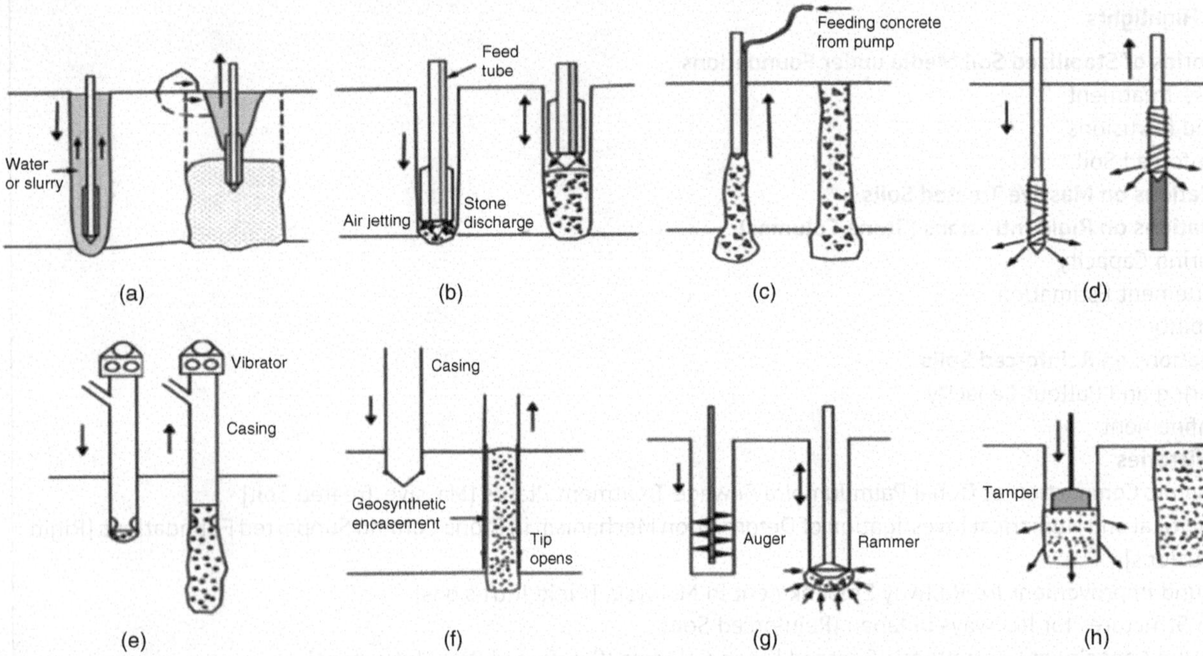

Figure 9.2 Various rigid intrusions: (a) vibro-replacement, (b) vibro-displacement, (c) vibro-concrete column, (d) controlled modulus(stiffness) column, (e) sand compaction column, (f) encased granular column, (g) rammed aggregate column, and (h) dynamic replacement (Han 2015 / John Wiley & Sons).

Reinforced soil techniques generally involve the removal of problematic soil and replacing it with superior granular soil. Reinforcements, such as geosynthetics or metallic strips, are then incorporated in layers. The contribution of the reinforcements is like that of the reinforcing bars in reinforced concrete. These geosystems serve the purpose of transferring and bearing loads in various applications, including foundations, walls, and slopes. Figure 9.3 provides visual examples of reinforced soil techniques, depicting the utilization of confined soil (like a sandbag mechanism) and layered geosynthetic reinforcement in foundation construction.

9.2 Foundations on Massive Treated Soil

The practice of massive treatment of soils involves modification of a large volume of soil extending from the ground surface to a depth of the zone of influence of the structure loads. Various techniques such as grouting, densification, and pre-consolidation can be employed to modify such soil masses. The geotechnical properties of the treated soil mass are significantly improved compared to the natural soil. For instance, grouting can lead to a substantial increase in soil cohesion, which positively affects the bearing capacity and stability of the soil. Densification methods often result in an increase in

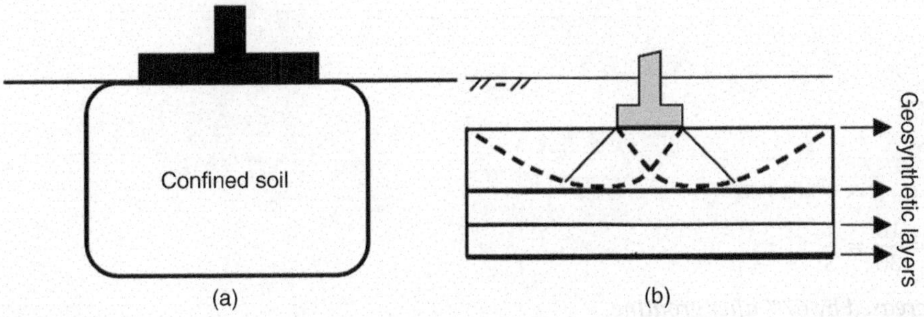

Figure 9.3 Foundation on reinforced soil: (a) confined soil (sandbag), (b) layered geosynthetic reinforcement.

the internal friction angle and modulus of elasticity of the soil. The internal friction angle plays a crucial role in bearing capacity and is essential for stability analysis of slopes. Moreover, a higher modulus of elasticity can lead to reduced foundation settlements. The following three examples (Examples 9.1–9.3) and Case Study 9.1 demonstrate the application of mass treatment methods.

Example 9.1 The results of a direct shear test in a site prior to and after grouting are shown in the graph. The elasticity modulus has increased from 10 to 30 MPa. Calculate and compare the desired parameters before and after ground modification:

- Bearing capacity of a strip foundation with a width of 1 m
- Foundation settlement
- Maximum depth that can be excavated without support (FS = 2)

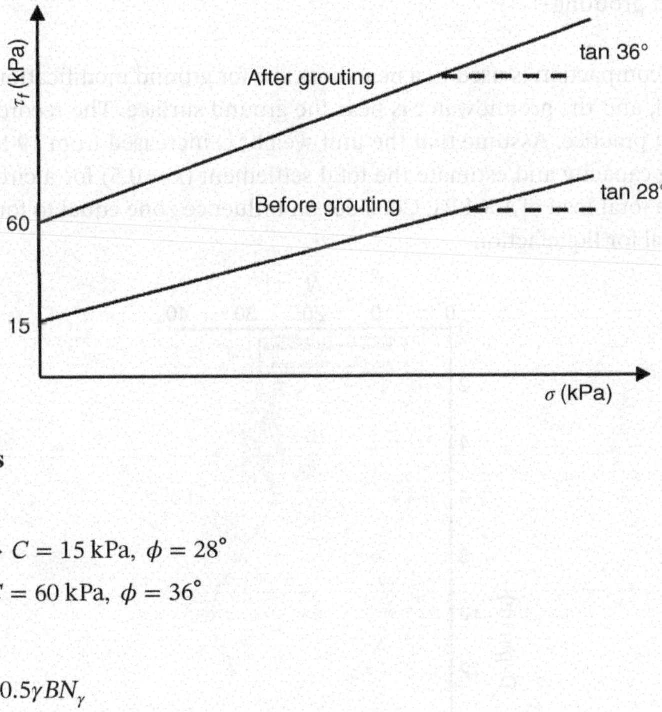

Shear strength parameters

$$\tau_f = \sigma \tan \phi + C$$

Prior to Grouting → $C = 15\,\text{kPa},\ \phi = 28°$

After Grouting → $C = 60\,\text{kPa},\ \phi = 36°$

Bearing capacity

$$q_{\text{ult}} = CN_c + qN_q + 0.5\gamma BN_\gamma$$

Assuming: $\gamma = 19\dfrac{\text{kN}}{\text{m}^3},\ D_f = 0\,\text{m}$

Prior to Grouting → $\phi = 28°$: $N_c = 26, N_\gamma = 11$

$$q_{\text{ult}} = 15 \times 26 + 0 + 0.5 \times 19 \times 1 \times 11 = 494.5\,\text{kPa}$$

After Grouting → $\phi = 36°$: $N_c = 51,\ N_\gamma = 40$

$$q_{\text{ult}} = 60 \times 51 + 0 + 0.5 \times 19 \times 1 \times 40 = 3440\,\text{kPa}$$

$$\frac{q_{ult_{\text{after grouting}}}}{q_{ult_{\text{before grouting}}}} = \frac{3440}{494.5} = 7$$

Settlement

$$S = \frac{kqB}{E}$$

$$\frac{S_{\text{after grouting}}}{S_{\text{before grouting}}} = \frac{E_{\text{before grouting}}}{E_{\text{after grouting}}} = \frac{10}{30} = \frac{1}{3}$$

The amount of settlement has been decreased by 67% after grouting.

Excavation depth without support

$$H_{cr} = \frac{4c}{\gamma\sqrt{K_a}}$$

$$K_a = \frac{1 - \sin\phi}{1 + \sin\phi}$$

$$\text{Prior to Grouting} \rightarrow K_a = \frac{1 - \sin 28}{1 + \sin 28} = 0.36$$

$$H_{\text{cr}} = \frac{4c}{\gamma\sqrt{K_a}} = \frac{4 \times 15}{19 \times \sqrt{0.36}} = 5.3\text{ m} \rightarrow H_{\text{cr-allowed}} = \frac{5.3}{2} = 2.65\text{ m}$$

$$\text{After Grouting} \rightarrow K_a = \frac{1 - \sin 36}{1 + \sin 36} = 0.26$$

$$H_{\text{cr}} = \frac{4c}{\gamma\sqrt{K_a}} = \frac{4 \times 60}{19 \times \sqrt{0.26}} = 24.8\text{ m} \rightarrow H_{\text{cr-allowed}} = \frac{24.8}{2} = 12.4\text{ m}$$

H_{cr} increases drastically after grouting.

Example 9.2 Deep dynamic compaction is used in a nearshore site for ground modification practice. The existing soil is mainly from cohesionless sand, and the groundwater is near the ground surface. The records of SPT are available before and after the soil improvement practice. Assume that the unit weight is increased from 19 to 20 $\frac{\text{kN}}{\text{m}^3}$ during improvement practice. Calculate the bearing capacity and estimate the total settlement ($k = 0.5$) for a circular foundation with a 15 m diameter and 1 m D_f, bearing a total load of 150 kPa. Consider an influence zone equal to foundation breadth in analysis. Also assess the general potential for liquefaction.

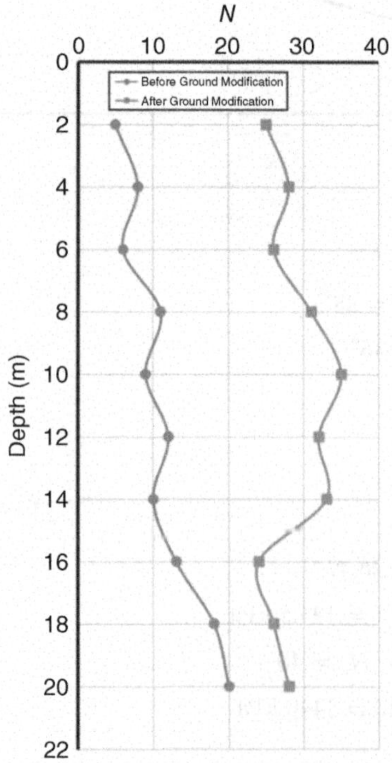

Before Soil Improvement

$N_{avg} = 8$

Wolff (1989): $\varphi = 27.1 + 0.3N_1 - 0.00054N_1^2 = 27.1 + (0.3 \times 8) - (0.00054 \times 8^2) = 29°$

Bowles (1997): $E_s = 500(N + 15) = 500 \times (8 + 15) = 11500\ \text{kPa} = 11.5\ \text{MPa}$

Bearing capacity

Basic equation: $q_{ult} = CN_c + \bar{q}N_q + 0.5\gamma BN_\gamma$

$\phi = 29° \rightarrow N_q = 16, N_\gamma = 13$

$q_{ult} = 0 + 1 \times (19 - 10) \times 16 + 0.5 \times (19 - 10) \times 15 \times 13 = 1021.5\ \text{kPa}$

Settlement

$$S = \frac{kqB}{E} = \frac{0.5 \times 150 \times 15}{11.5 \times 10^3} = 0.098\ \text{m} = 9.8\ \text{cm}$$

After Soil Improvement

$N_{avg} = 30$

Wolff (1989): $\varphi = 27.1 + 0.3N_1 - 0.00054N_1^2 = 27.1 + (0.3 \times 30) - (0.00054 \times 30^2) = 35°$

Bowles (1997): $E_s = 500(N + 15) = 500 \times (30 + 15) = 22500\ \text{kPa} = 22.5\ \text{MPa}$

Bearing capacity

Basic equation: $q_{ult} = CN_c + \bar{q}N_q + 0.5\gamma BN_\gamma$

$\phi = 35° \rightarrow N_q = 33, N_\gamma = 34$

$q_{ult} = 0 + 1 \times (20 - 10) \times 33 + 0.5 \times (20 - 10) \times 15 \times 34 = 2880\ \text{kPa}$

Settlement

$$S = \frac{kqB}{E} = \frac{0.5 \times 150 \times 15}{22.5 \times 10^3} = 0.05\ \text{m} = 5\ \text{cm}$$

$$\Delta q_{ult} = \frac{2880 - 1021.5}{1021.5} = 1.82 - \text{a } 182\% \text{ increase in bearing capacity}$$

$$\Delta s = \frac{5 - 9.8}{9.8} = -0.49 = \text{a } 49\% \text{ decrease in settlement}$$

Liquefaction potential assessment

Based on Table 4.21, for the corrected SPT N value of 8, the sand is recognized as loose, which has a high potential for liquefaction in case of full saturation and occurring repeated cyclic, dynamic, and high magnitude earthquake loads. After the soil improvement practice, the average *N* value rises to 30, which is considered dense, and therefore indicates that liquefaction is unlikely.

Example 9.3 A site consisting of mixed soil is treated massively via hydraulic modification. CPT records before and after the soil improvement practice are available. Assess the effect of ground modification practices on bearing capacity and settlement of given square foundation. Determine the bearing capacity in total stress analysis and use Gazetas's et al. (1985) method for settlement estimation. Assume that the given pressure is the net pressure, unit weight, and Poisson's ratio are constant during treatment process, and the influence zone is equal to the foundation breadth.

$$\gamma_{sat} = 19\ \frac{\text{kN}}{\text{m}^3}, \nu = 0.2, S_u = 1.2f_s, E_s = 2q_c$$

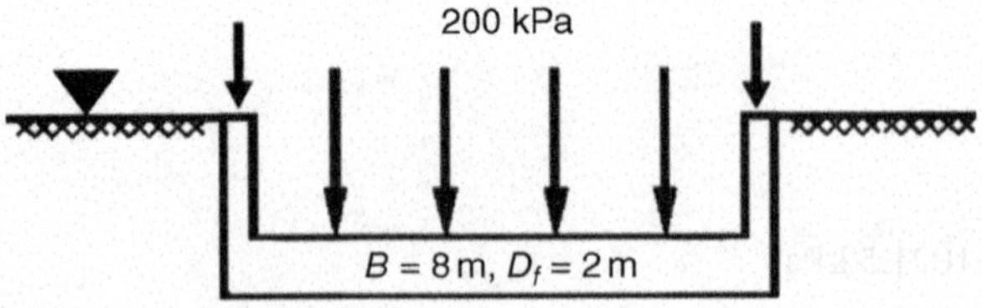

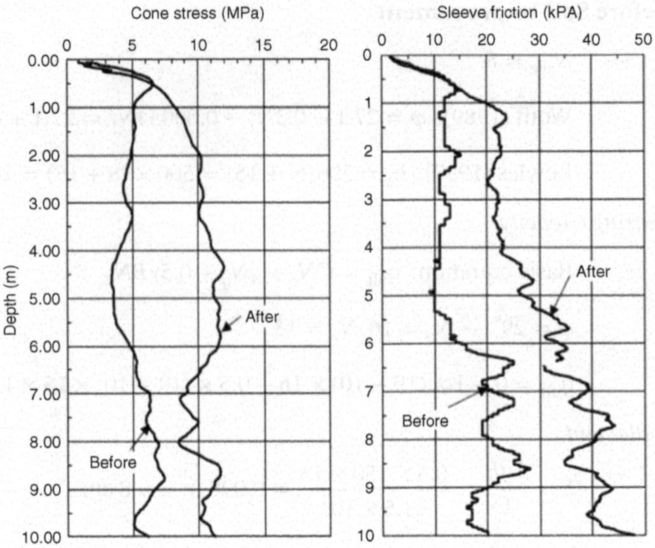

An average of CPT records from the ground surface to the depth of 10 m is used in the following calculations:

Before Soil Improvement

$$q_{c\text{-avg}} = 4\text{ MPa} \rightarrow E_s = 2 \times 4 = 8\text{ MPa}$$

$$f_{s\text{-avg}} = 15\text{ kPa} \rightarrow S_u = 1.2 \times 15 = 18\text{ kPa}$$

Bearing capacity

$$\text{Basic equation: } q_{\text{ult}} = S_u N_c + \overline{q}$$

$$q_{\text{ult}} = 18 \times 5.14 + 2 \times (19 - 10) = 110.5\text{ kPa}$$

Settlement: Gazetas et al. (1985) method

$$\text{Square foundation: } \frac{A_b}{4L^2} = 1$$

$$\mu_s = 0.45\left(\frac{A_b}{4L^2}\right)^{-0.38} = 0.45(1)^{-0.38} = 0.45$$

$$\mu_{\text{emb}} = 1 - 0.04\frac{D_f}{B}\left[1 + \frac{4}{3}\left(\frac{A_b}{4L^2}\right)\right] = 1 - 0.04 \times \frac{2}{8} \times \left[1 + \frac{4}{3}(1)\right] = 0.977$$

$$\mu_{\text{wall}} = 1 - 0.16\left(\frac{A_w}{A_b}\right)^{0.54} = 1 - 0.16\left(\frac{4 \times 8 \times 2}{8 \times 8}\right)^{0.54} = 0.84$$

$$S_i = \frac{q_0 B(1 - \nu^2)}{E_s} \cdot \mu_s \cdot \mu_{\text{emb}} \cdot \mu_{\text{wall}} = \frac{200 \times 8 \times (1 - 0.2^2)}{8 \times 10^3} \times 0.45 \times 0.977 \times 0.84 = 0.071\text{ m} = 71\text{ mm}$$

After Soil Improvement

$$q_{c\text{-avg}} = 10\text{ MPa} \rightarrow E_s = 2 \times 10 = 20\text{ MPa}$$

$$f_{s\text{-avg}} = 30\text{ kPa} \rightarrow S_u = 1.2 \times 30 = 36\text{ kPa}$$

Bearing capacity

$$\text{Basic equation: } q_{\text{ult}} = S_u N_c + \overline{q}$$

$$q_{\text{ult}} = 36 \times 5.14 + 2 \times (19 - 10) = 203\text{ kPa}$$

Settlement: Gazetas et al. (1985) method

$$\text{Square foundation: } \frac{A_b}{4L^2} = 1$$

$$\mu_s = 0.45\left(\frac{A_b}{4L^2}\right)^{-0.38} = 0.45(1)^{-0.38} = 0.45$$

$$\mu_{\text{emb}} = 1 - 0.04\frac{D_f}{B}\left[1 + \frac{4}{3}\left(\frac{A_b}{4L^2}\right)\right] = 1 - 0.04 \times \frac{2}{8} \times \left[1 + \frac{4}{3}(1)\right] = 0.977$$

$$\mu_{\text{wall}} = 1 - 0.16\left(\frac{A_w}{A_b}\right)^{0.54} = 1 - 0.16\left(\frac{4 \times 8 \times 2}{8 \times 8}\right)^{0.54} = 0.84$$

$$S_i = \frac{q_0 B(1 - \nu^2)}{E_s} \cdot \mu_s \cdot \mu_{\text{emb}} \cdot \mu_{\text{wall}} = \frac{200 \times 8 \times (1 - 0.2^2)}{20 \times 10^3} \times 0.45 \times 0.977 \times 0.84 = 0.028\text{ m} = 28\text{ mm}$$

$$\Delta q_{\text{ult}} = \frac{203 - 110.5}{110.5} = 0.84 = \text{an } 84\% \text{ increase in bearing capacity}$$

$$\Delta s = \frac{28 - 71}{71} = -0.61 = \text{a } 61\% \text{ decrease in settlement}$$

Case Study 9.1

Dynamic Compaction at Dubai's Palm Jumeira Sewage Treatment

Palm Jumeira is a set of artificial reclaimed islands off the coast of Dubai, United Arab Emirates (Hamidi and Varaksin 2015). It is the first of three internationally known palm-shaped reclamations, which inspired the concept of figure-shaped reclamation projects worldwide. The Palm consists of a tree trunk, a crown with seventeen fronds, three surrounding crescent islands forming an 11-km-long breakwater, and two comparable smaller islands on each side of the trunk in the shape of the logo: a palm. The island is 5 km by 5 km in size and has added around 78 km to Dubai's initial 72-km shoreline. Palm Jumeira was constructed using 94 million m^3 of sand and 7 million cubic meters of rock.

As shown in Figure 9.4, a design and construction contract were used to build sewage treatment plants (STPs) designated as Lots A-A and G-G on the topmost crescents of Palm Jumeira.

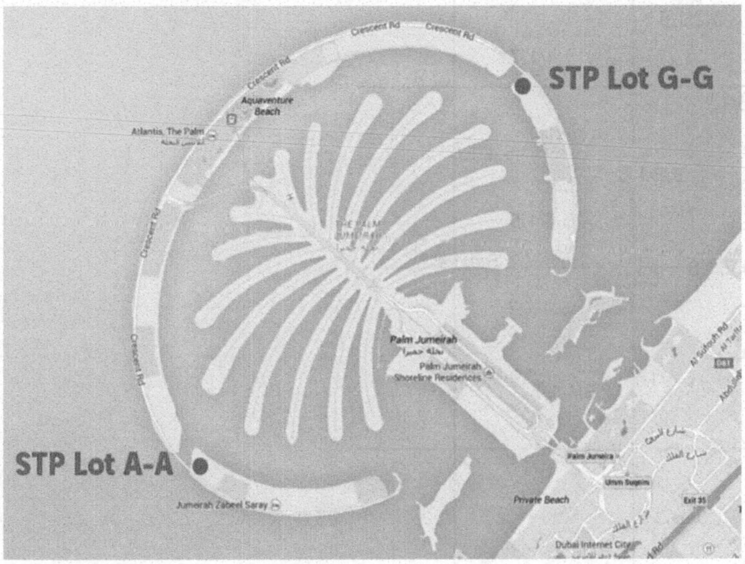

Figure 9.4 STPs locations on the island (Hamidi and Varaksin 2015 / with permission of Elsevier).

The reinforced concrete STP tanks have a diameter of 35.1 m. Each tank was designed with the expectation of being exposed to the following loads:

- Dead load of the reinforced concrete structure: 41 kPa
- Dead load of steel structures: 8 kPa

(Continued)

Case Study 9.1 (Continued)

- Live load of the liquid inside the tank: 71 kPa
- Total load: 120 kPa

Boreholes drilled on-site revealed that the surface crust of soil was typically dense, with SPT blow counts of up to 28; however, the deeper layers of soil were less dense with a minimum blow count of 5 in the upper 8 m of the reclamation. At depths higher than 8 m, the soil seemed to densify, with a minimum SPT blow count value of 18 and exceeding 50 at 13 m. The fines content (FC) of the soil in these two boreholes ranged between 2% and 10% in the top 13 m but climbed to 22% at a depth of 14.5 m. Additionally, the CPT measurements indicated that the first two meters of soil were made of very thick sand with q_c values as high as 25 MPa. The soil subsequently became loose, with q_c values as low as 3–4 MPa down to a depth of roughly 13 m, at which point good resistance was met and testing was discontinued.

Due to the unsatisfactory SPT and CPT findings, the project engineers determined that additional geotechnical procedures were required to guarantee the safe transfer of the tank loads to the underlying geomaterial. While piling was the primary method of supporting heavy structures on Palm Jumeira, it was estimated to be prohibitively expensive and difficult to construct piles on an island that had no connection to the mainland, relied on marine transportation, and lacked all necessary facilities and infrastructure at the time. Numerous ground improvement techniques were investigated to treat the loose hydraulic fill, including vibro-compaction and the use of stone columns; however, dynamic compaction and dynamic surcharging were ultimately applied.

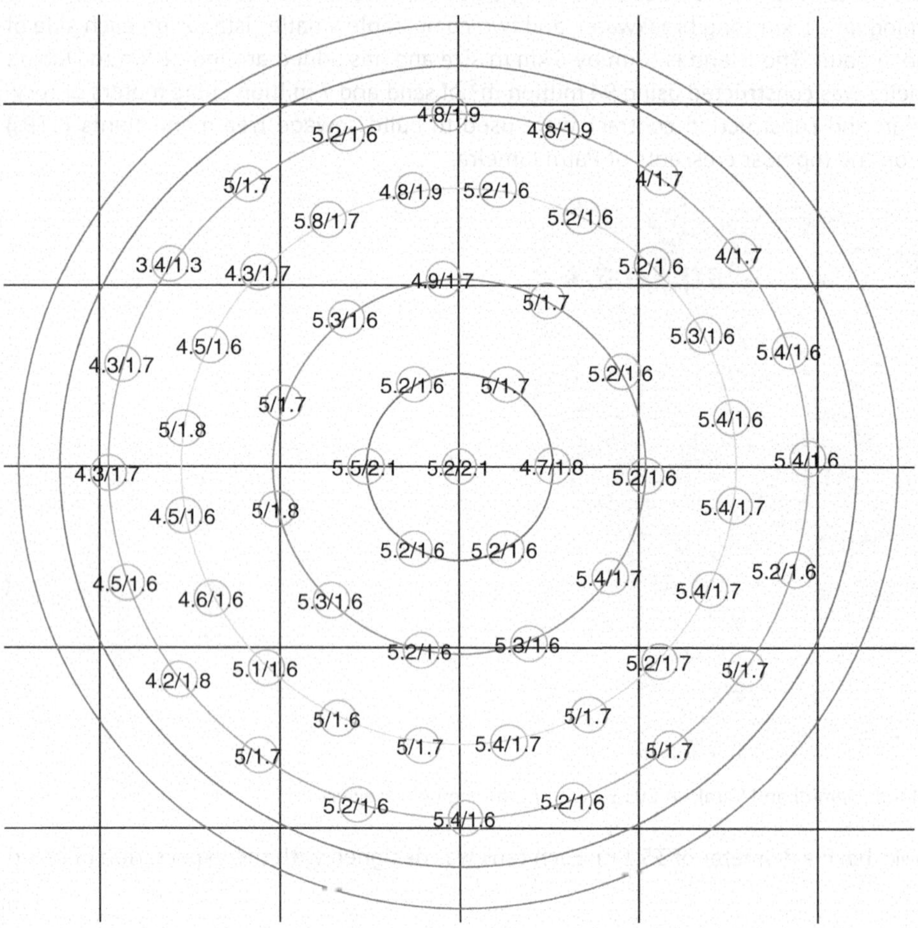

Figure 9.5 DC first phase crater-top diameter and depth (Hamidi and Varaksin 2015).

Within the first two stages of dynamic compaction, the pounder penetration and crater diameter for each print position were measured. The crater's average pounder penetration depth and upper diameters were 1.7 and 5 m during the first phase. These statistics were dramatically lowered in the second phase to 0.4 and 2.3 m, respectively. The diameters and depths of the crater tops during the first phase of dynamic compaction are depicted in Figure 9.5.

The ground level had decreased to +2.25 m DMD after dynamic compaction. With 150 m^3 of stone added, equating to a thickness of 0.13 m, it is possible to determine that the ground settles 0.68 m in addition to the settlements caused by dynamic surcharging. While the size of settlements produced by dynamic compaction seems to be much greater than those induced by dynamic surcharging, this does not diminish the significance of the dynamic surcharging impact, which was meant to minimize settlements in the deeper siltier strata.

After finishing the site's dynamic compaction and leveling, four PMTs were conducted to assess the ground conditions inside the improved region. P_{LM} is represented in Figure 9.6 before and after ground modification. The greatest improvement occurred at around 8 m deep, when, omitting the maximum values of one of the tests (A2), P_{LM} varied between roughly 2 and 4 MPa and E_M ranged between 23 and 30 MPa at approximately 4–8 m depth. Notably, improvement is seen at higher depths as a result of the combination of dynamic surcharging and pre-excavated dynamic compaction with a relatively high number of blows. Additionally, the lowest P_{LM} value is larger than 600 kPa after the modification, indicating that the young hydraulic fill is no longer vulnerable to creep as a result of its weight.

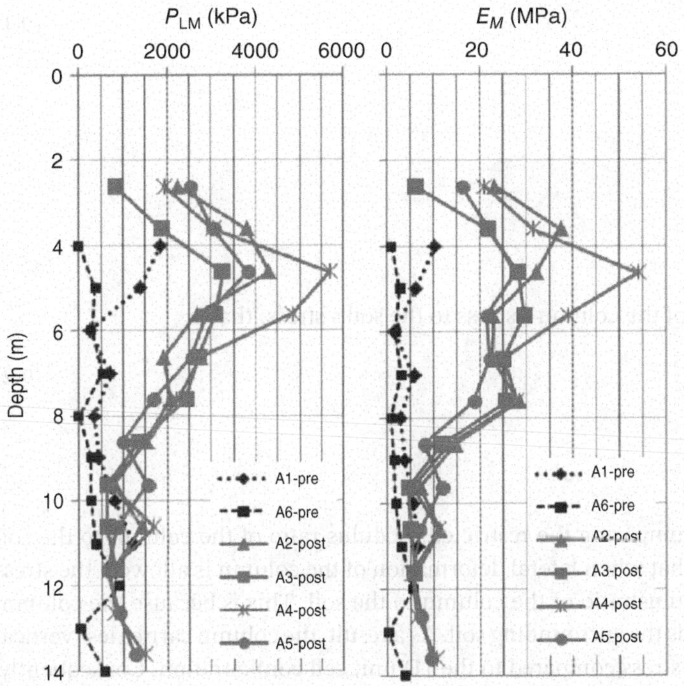

Figure 9.6 Pressuremeter records prior to and after ground modification (Hamidi and Varaksin 2015 / with permission of Elsevier).

9.3 Foundations on Rigid Intrusions (Stone Columns)

9.3.1 Introduction

Deep replacement procedures involve substantial modifications to the ground by excavating or displacing problematic geomaterials and replacing them with a more suitable or densified fill material or concrete in the form of columns. These columns, along with the surrounding soils form a composite ground or foundation that is capable of supporting vertical and lateral loads as well as moments. The composite ground or foundation enhances bearing capacity, reduces settlement, accelerates consolidation, and increases shear strength, which contributes to slope stability and resistance against liquefaction.

Granular columns have been widely utilized in geotechnical applications to serve various functions, including compaction, load bearing, reinforcement, stress distribution, and drainage. Compaction is often achieved during the installation of the columns, although its effectiveness may be limited in saturated and cohesive soils. Apart from compaction, the columns perform multiple functions throughout their service life, depending on the specific requirements of the project.

9.3.2 Principles

9.3.2.1 Unit Cells and Stress Concentration Ratio

The unit cell concept, comprising a single column and the surrounding soil, is commonly employed to analyze column-reinforced soft foundation soils. Previous theoretical advancements and analyses assumed that a one-dimensional (1D) unit cell restricts the lateral deformation of the column, as depicted in Figure 9.7a. However, it has been observed that increased lateral stress from the column impacts the excess pore water pressure variation in the surrounding soil. To account for the lateral displacement of the column, a modified unit cell is considered, as shown in Figure 9.7b, where the lateral displacement is represented as δ_h. This modified unit cell approach has been found to significantly influence the settlement and consolidation behavior of column-reinforced soft foundations.

When a 1D unit cell is subjected to equal vertical strain, the following equation is applicable:

$$\varepsilon_z = \frac{\Delta\sigma_c}{D_c} = \frac{\Delta\sigma_s}{D_s} \tag{9.1}$$

where

ε_z = vertical strain at a depth of z
$\Delta\sigma_c$ = vertical stress on the column
$\Delta\sigma_s$ = vertical stress on the soil
D_c = column constrained modulus
D_s = soil constrained modulus

The stress concentration ratio is expressed as the ratio of the column's stress to the soil's stress; that is

$$n_{1D} = \frac{\Delta\sigma_c}{\Delta\sigma_s} = \frac{D_c}{D_s} \tag{9.2}$$

where

n_{1D} = the stress concentration ratio under a 1D condition

The stress concentration ratio can be determined by comparing the restricted modulus ratio of the column to the soil under equal vertical strain. However, it should be noted that when lateral deformation of the column is allowed, the stress concentration ratio is no longer equal to the elastic modulus ratio of the column to the soil. This is because the column, experiencing concentrated stress, deforms laterally towards the surrounding soil. As a result, the column carries less vertical stress, while the surrounding soil bears a greater vertical stress compared to the 1D unit cell configuration. Consequently, when considering lateral column deformation, the stress concentration ratio is reduced compared to the case where no deformation is considered.

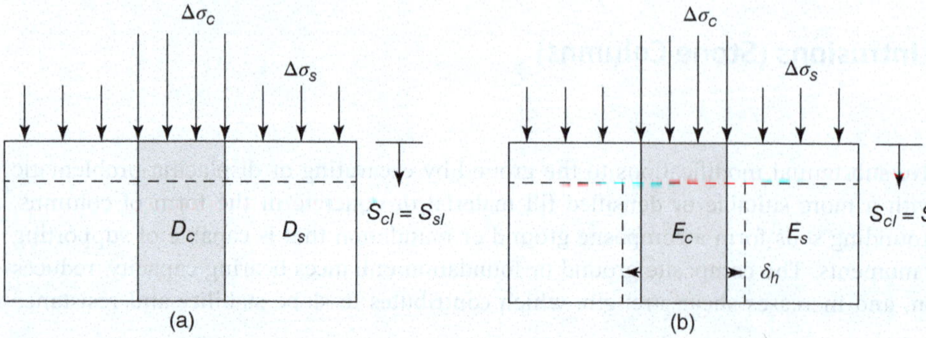

Figure 9.7 One-dimensional unit cells: (a) without and (b) with lateral deformation of a column (Han, 2015 / John Wiley & Sons).

9.3.2.2 Failure Modes

Columns constructed in soils are often used to support vertical loads. However, they are sometimes used to improve the shear resistance of soft soils. In addition to vertical loads, these columns are subjected to horizontal loads and stresses or movements under certain circumstances, such as inside excavations or beneath embankments. These columns fail in distinct ways under a variety of loading conditions.

Columns that experience vertical compression forces transfer the load to the surrounding soil through side friction or lateral confinement. Figure 9.8 illustrates the possible failure mechanisms in these columns. If the applied loads exceed the structural strength of the column, it may crush, especially in the case of brittle end-bearing concrete columns. Shear failure is commonly observed in granular columns. Short granular or concrete columns without an end-bearing layer are prone to punching failures. Bulging failure, which occurs predominantly at the top of granular columns in soft soils, is characterized by the column expanding to a diameter two to three times larger than the original column diameter. In flexible columns such as granular columns, there is a critical length beyond which no additional bearing capacity is provided. Typically, the critical length ranges between four and six times the column diameter.

Columns subjected to horizontal loading or movement provide resistance to shear, bending, or rotation. Shear failure is a potential failure mode in granular columns, while bending failure is more common in concrete columns, particularly if they are unreinforced. In soft soils, short concrete columns may experience rotation. To enhance the shear and bending resistance of columns, rigid inclusions can be incorporated within them. In the case of geosynthetic-encased columns, the burst of the geosynthetic reinforcement is another possible failure mechanism to consider.

When designing pile foundations, the piles bear all the loads, and the surrounding soils do not carry any load. However, in the case of composite foundations consisting of columns and soils, both the columns and soils share the applied loads and deformations. Existing theories for composite foundations have primarily focused on rigid loads and similar strain conditions. Under rigid loading conditions, the stress distribution on the columns and soil can be simplified, as illustrated in Figure 9.9.

The following relationship can be established upon the force equilibrium:

$$\Delta\sigma_z A = \Delta\sigma_s (A_e - A_c) + \Delta\sigma_c A_c \tag{9.3}$$

where

$\Delta\sigma_z$ = average vertical stress applied on the composite foundation
A_e = one column influence or tributary area
$\Delta\sigma_s$ = vertical stress on the soil
$\Delta\sigma_c$ = vertical stress on the column
A_c = the column cross-sectional area

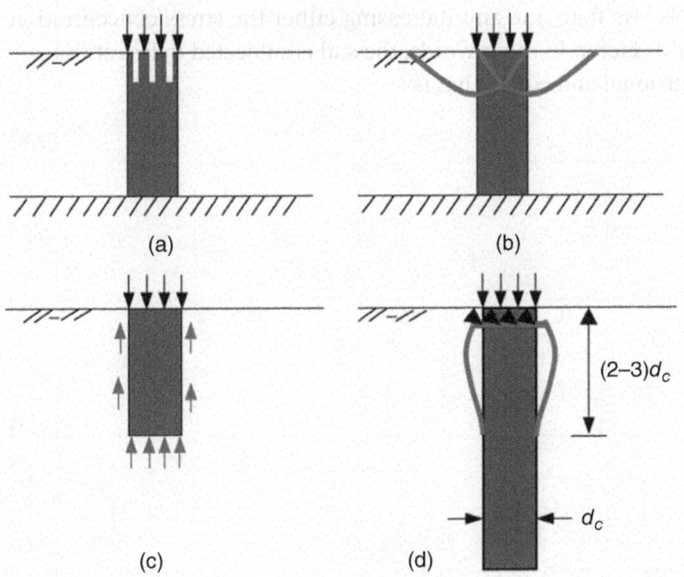

Figure 9.8 Individual columns failure modes for vertical loads: (a) crushing, (b) shear, (c) punching, and (d) bulging (Barksdale and Bachus 1983 / U.S. Department of Transportation / Public Domain).

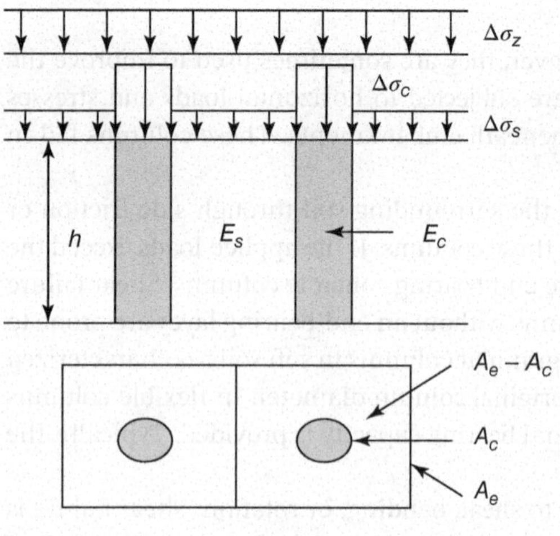

Figure 9.9 Stress distribution model (Han 2015 / John Wiley & Sons).

Dividing Equation (9.3) by A_e results in

$$\Delta\sigma_z = \Delta\sigma_s(1 - a_s) + \Delta\sigma_c a_s \tag{9.4}$$

where

a_s = the area replacement ratio (the ratio of the column cross-section area to the influence area)

Regarding the stress concentration ratio $\left(n = \frac{\Delta\sigma_c}{\Delta\sigma_s}\right)$, Equation (9.4) is rewriteable as

$$\Delta\sigma_z = [(1 - a_s) + na_s]\Delta\sigma_s = [1 + (n - 1)a_s]\Delta\sigma_s \tag{9.5}$$

The stress on the soil is

$$\Delta\sigma_s = \mu\Delta\sigma_z \tag{9.6}$$

$$\mu = \frac{1}{1 + (n - 1)a_s} \tag{9.7}$$

where

μ = the stress reduction factor

According to Equation (9.7), the stress reduction factor is less than one and increasing either the stress concentration ratio or the area replacement ratio results in a decrease in this factor. In other words, the soil is subjected to less stress.

Assume the column and soil deformations are one-dimensional and equal, that is

$$\varepsilon_c = \varepsilon_s = \varepsilon_z \tag{9.8}$$

where

ε_c = column vertical strain
ε_s = soil vertical strain
ε_z = average vertical strain

Dividing Equation (9.4) by Equation (9.8) leads to

$$E_{eq} = E_s(1 - a_s) + E_c a_s = [1 + (n - 1)a_s]E_s \tag{9.9}$$

where

E_{eq} = equivalent modulus of the composite foundation
E_s = soil modulus
E_c = column modulus

As presented in Chapter 8, equivalent strength and stiffness parameters can be used for geotechnical performance assessment.

Example 9.4 The rigid foundation of a commercial complex with dimensions of 15 m × 10 m is located on a layer of soft saturated soil. In order to modify the ground, 15% of the total surface beneath the foundation is replaced with stone columns, in which the column's ratio of the modulus of elasticity to the surrounding soil after improvement is 5, and the maximum allowed settlement is 7 cm. The net pressure is assumed to be 300 kPa. Using equivalent parameters:

a) Calculate the stress on the surrounding soil and the stress on stone columns.
b) Compare the bearing capacity of the foundation prior to and before ground modification.
c) Estimate the immediate settlement of foundation prior to and before ground modification under the stress of 300 kPa beneath the foundation ($K = 0.5$).

Before Improvement: $C = 20\ \text{kPa}, \phi_s = 20°, E_s = 5\ \text{MPa}, \gamma_{\text{sat}} = 19\ \frac{\text{kN}}{\text{m}^3}$

After Improvement: $\phi_c = 40°, E_s = 15\ \text{MPa}$

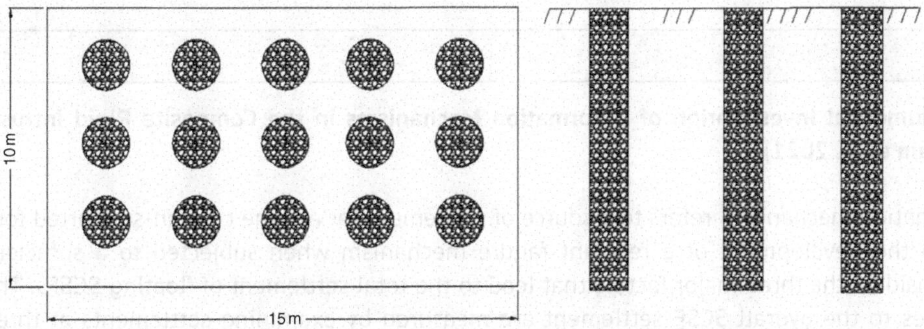

a) Given the assumption that the foundation is rigid:

$$\frac{A_c}{A} = \frac{15}{100} \rightarrow \frac{A}{A_c} = \frac{100}{15} \rightarrow \varepsilon_c = \varepsilon_s$$

The strain of the soil and the stone column is assumed to be equal:

$$A_c \cdot \sigma_c + A_s \cdot \sigma_s = \sigma \cdot A$$

$$\frac{A_c}{A} = \frac{15}{100} = \frac{A_c}{A_s + A_c} \rightarrow \frac{A_s}{A} = \frac{85}{100}$$

$$\sigma_c = \sigma_s \times \frac{E_C}{E_s} \rightarrow \sigma_c' = \sigma_s \times 5$$

$$0.15 \times \sigma_c + 0.85 \times \sigma_s = \sigma = 0.15 \times 5 \times \sigma_s + 0.85 \times \sigma_s = 300$$

$$\rightarrow \sigma_s = 187.5\ \text{kPa}$$

Stress on the surrounding soil

$$\rightarrow \sigma_c = \sigma_s \times 5 = 187.5 \times 5 = 937.5\ \text{kPa}$$

Stress on the stone columns

b) Before soil improvement,

Basic bearing capacity equation: $q_{\text{ult}} = CN_c + \overline{q}N_q + 0.5\gamma BN_\gamma$

$$\phi = 20° \rightarrow N_c = 15, N_q = 6, N_\gamma = 3$$

$$q_{\text{ult}} = 20 \times 15 + 0 \times N_q + 0.5 \times (19 - 10) \times 10 \times 3 = 435\ \text{kPa}$$

After soil improvement,

$$\phi_{\text{eq}} = \arctan[a_s \tan\phi_c + (1 - a_s)\tan\phi_s] = \arctan[0.15 \times \tan 40 + (1 - 0.15) \times \tan 20] = 23.52°.$$

Conservatively: $\phi = 23° \rightarrow N_c = 18, N_q = 9, N_\gamma = 5$

It is assumed that C is constant.

$$q_{\text{ult}} = 20 \times 18 + 0 \times N_q + 0.5 \times (19 - 10) \times 10 \times 5 = 585 \text{ kPa}$$

$$\Delta q_{\text{ult}} = \frac{585 - 435}{435} = 0.345 = \text{a } 34.5\% \text{ increase in bearing capacity}$$

c) Before soil improvement:

$$S = \frac{kqB}{E} = \frac{0.5 \times 300 \times 10}{5 \times 10^3} = 0.3 \text{ m} = 30 \text{ cm} \nless 7 \text{ cm Not OK}$$

After soil improvement:

$$E_{\text{eq}} = E_s(1 - a_s) + E_c a_s = 15 \times (1 - 0.15) + (5 \times 15) \times 0.15 = 24 \text{ MPa}$$

$$S = \frac{kqB}{E} = \frac{0.5 \times 300 \times 10}{24 \times 10^3} = 0.063 \text{ m} = 6.3 \text{ cm} < 7\text{cm OK}$$

$$\Delta s = \frac{6.3 - 30}{30} = -0.79 = \text{a } 79\% \text{ decrease in settlement}$$

Case Study 9.2

Analytical and Numerical Investigation of Deformation Mechanisms in the Composite Rigid Intrusions Supported Foundations (Imam et al. 2021)

The term "deformation mechanism" refers to a source of settlement for a stone column-supported foundation (SCSF) that fails due to the development of a relevant failure mechanism when subjected to a sufficiently great load. This research considers the three major factors that lead to the total settlement of floating SCSFs. The contributions of these processes to the overall SCSF settlement are measured by examining settlements at three typical stone column group locations.

As seen in Figure 9.10, the settlement of an SCSF derived from a unit-cell model may be the result of three components accumulating. The values of S_f, S_c, and S_s in this figure represent the vertical displacements of the raft foundation, the central point of the column toe (point A), and the point in mid-distance between neighboring columns at the column toe level (point B). The relative contribution of different settlement components to total raft settlement may be measured using the following equation:

$$\frac{(S_f - S_c)}{S_f} + \frac{(S_c - S_s)}{S_f} + \frac{(S_s)}{S_f} = \text{CCR} + \text{CPR} + \text{BPR} = 1 \tag{9.10}$$

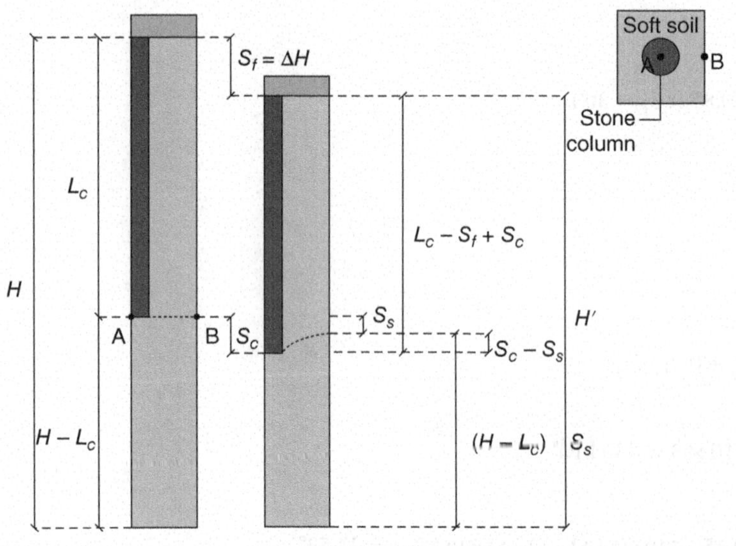

Figure 9.10 Configuration of stone column settlement in a unit cell model (Imam et al. 2021 / with permission of Elsevier).

The column compression ratio (CCR) and the individual column punching ratio (CPR) are the first two terms in Equation (9.10). The third element of Equation (9.10) takes the compression of the soil under the reinforced block into account and is referred to as the stone column group block punching ratio (BPR). The settlement contribution factors (SCF) are the ratios that describe the contributions of column compression, single-column punching, and reinforced block punching to the total settlement of the SCSF.

To determine the impact of each of these variables on the group's overall behavior, the stone column behavior chart (SCBC) chart is presented. This figure is intended to illustrate the relationship between the SCFs and the SCSFs' regulating deformation and failure processes.

The actual total settlement of an SCSF is often the result of a combination of many deformation mechanisms. Thus, the governing deformation processes of SCSFs are classified into seven distinct modes, including three primary mechanisms: block punching (B), column compression, including column bulging or shearing behavior (C), and individual column punching (P). The other four mechanisms are different combinations of two of the three primary modes, as well as a combination of all three primary modes: PC, BP, BC, and BPC. The current research identifies the deformation processes of stone column groups by settling three sample places around the core column. While the ruling mechanism of the whole group may be found by evaluating the mechanisms produced in all group locations, this may not occur at the center column's site.

The following procedures may be followed to recognize the deformation mechanism of a finite-sized stone column supported foundation:

1) Designate the raft, stone column, and soil settlements at three representative points for a representative unit cell model of the group through numerical or analytical methods (S_f, S_c, and S_s).
2) Designate SCFs for the unit cell models (CCR_{uc}, CPR_{uc}, and BPR_{uc}) via Equation (9.10).
3) Calculate SCFs for the finite-sized SCSF (CCR, CPR, and BPR) via

$$\frac{CCR}{CCR_{uc}} = f\left(\frac{B}{H}\right) = 2 - \left(\frac{9}{8} - \frac{9/8}{1 + (4 * B/H)^{1.5}}\right) \text{ for B/H} < 1.0 \text{ and } 1.0 \text{ for B/H} \geq 1.0 \quad (9.11)$$

$$\frac{BPR}{BPR_{uc}} = g\left(\frac{B}{H}\right) = 2 - \left(\frac{4.5}{3.5} - \frac{4.5/3.5}{1 + (4 * B/H)^{0.75}}\right) \text{ for B/H} < 1.33 \text{ and } 1.0 \text{ for B/H} \geq 1.33 \quad (9.12)$$

$$CPR = 1 - f\left(\frac{B}{H}\right) \cdot CCR_{uc} - g\left(\frac{B}{H}\right) \cdot BPR_{uc} \quad (9.13)$$

where

B = raft width

H = thickness of the compressible layer

4) Designate the mechanism of deformation by locating the associated status point on the SCBC (Figure 9.11).

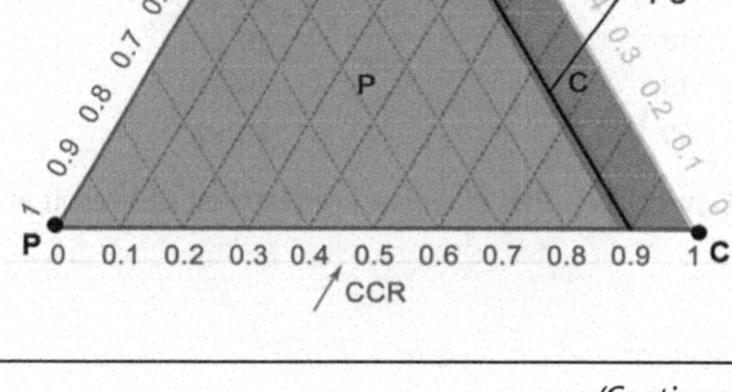

Figure 9.11 SCBC and regions representing various deformation mechanisms (Imam et al. 2021 / with permission of Elsevier).

(Continued)

Case Study 9.2 (Continued)

In general, a floating SCSF is most successful when "block punching" behavior prevails since all reinforced soils operate as a reasonably rigid block, and settlement occurs mostly as a result of soil compression under the reinforced zone. Thus, design parameters may be chosen in such a way that the deformation mechanism of the group of stone columns is oriented as closely as possible to point "B" on the SCBC.

Case Study 9.3

Ground Improvement for Railway Embankment in Malaysia (Arulrajah et al. 2009)

The 110-km electrified high-speed railway connects Rawang in Selangor to Bidor in the Perak state, Peninsular Malaysia. The geotechnical design of the project includes ground improvement to support the imposed dead and traffic loads for trains traveling up to 160 kph. The ground modification was necessary to assure the embankments' sufficient performance in terms of settlement and slope stability, as well as the project's completion within the specified project time.

The project's railway embankments ranged in height from 1 to 12 m. The embankment top had a minimum width of 14.9 m for embankments less than 10 m in height and a maximum width of 24.9 m for embankments more than ten meters in height. The soils discovered on the project site were a varied combination of extremely soft silts, clays, and loose sands that reached depths of up to 30 m. Regarding construction constraints, two approaches were required for the treatment process: (i) treatment of the entire width of the embankment in locations where the new alignments required the construction of two new tracks, and (ii) treatment in locations where a new track would be constructed first, and the existing live track would be rehabilitated later. The first stage would simply treat the breadth of the

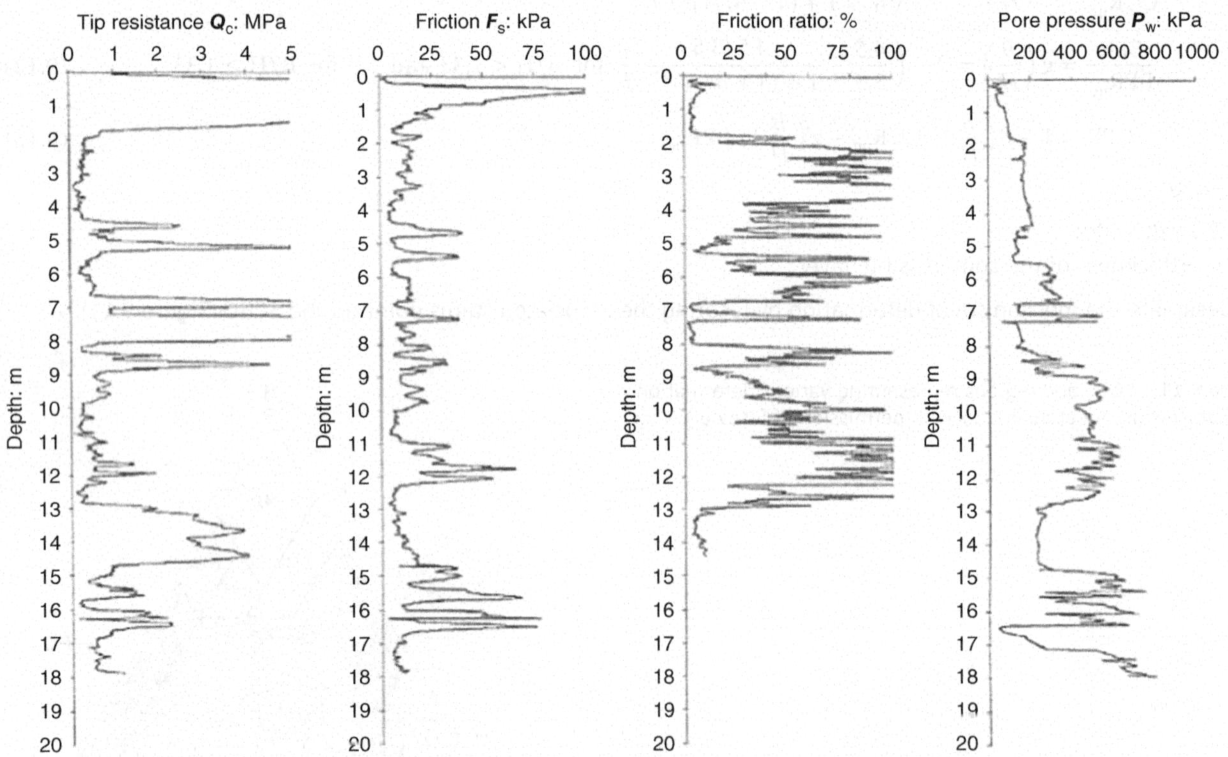

Figure 9.12 Plot of typical pretreatment CPT result at stone column location (Arulrajah et al, 2009 / Emerald Publishing Limited).

embankment under the new track. The second stage of treatment would involve work under the planned restored track, which would be completed after train operations were moved to the new live track.

Site surveys indicated a diverse variety of soil types along the railway from very soft silty clay or clayey silt to loose, silty, clayey sand. A common cone penetration test (CPT) plot at one of these stone column treatment locations is seen in Figure 9.12.

Stone columns were utilized to remediate soils over 14 km of the railway line's length. On the project site, about 1,100,000 linear meters of 0.8–1.0 m diameter stone columns were installed to depths of 6–30 m. The schematic diagram in Figure 9.13 illustrates the stone column treatment work completed in the project at the new alignment's places.

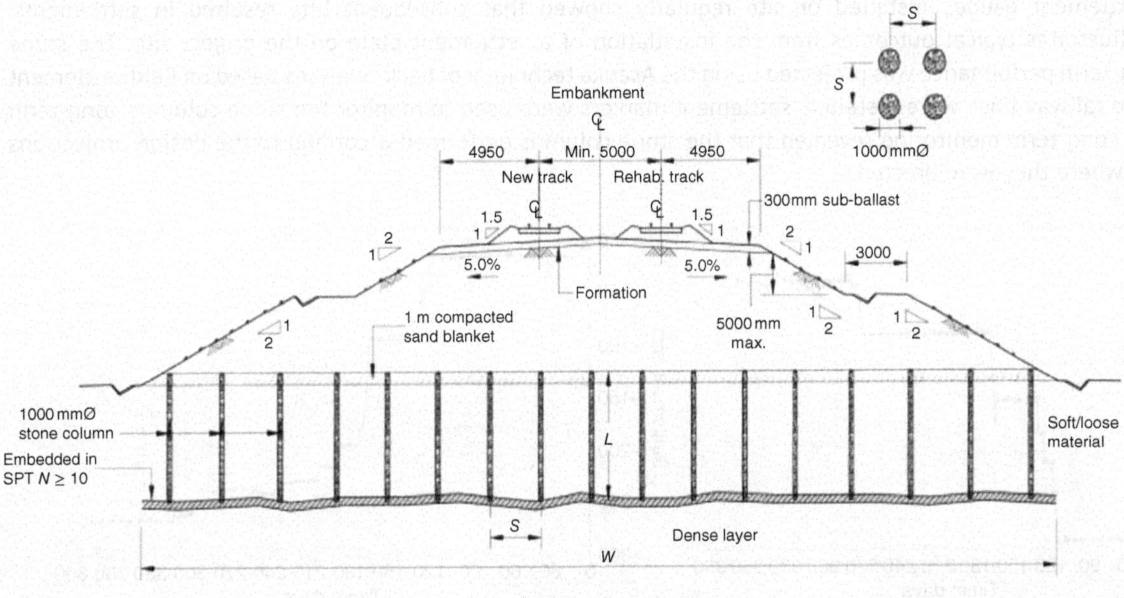

Figure 9.13 Schematic depicting the proposed treatment of stone columns for the new alignment, which will include two additional tracks (Arulrajah et al. 2009 / Emerald Publishing Limited).

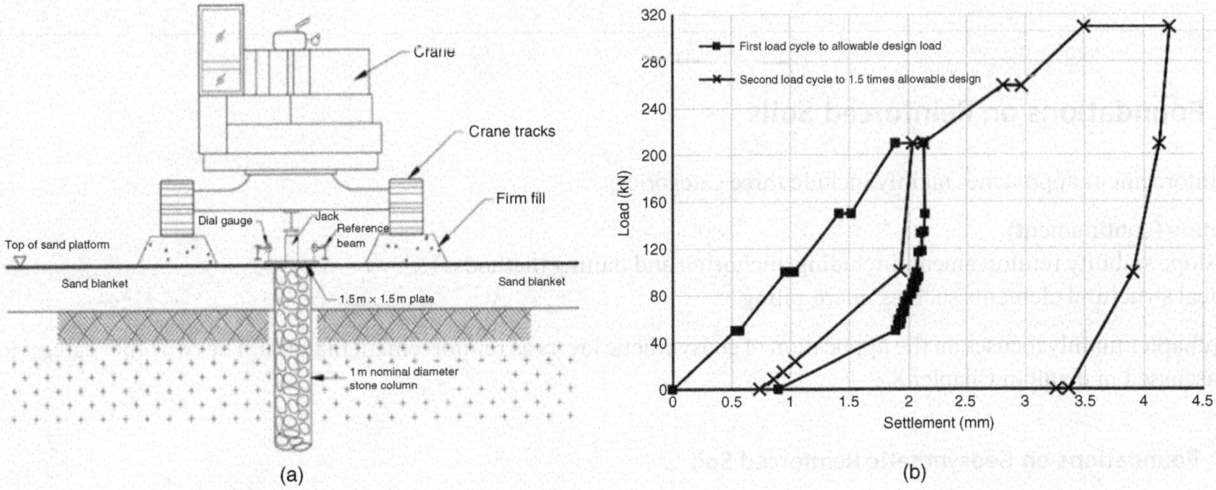

Figure 9.14 (a) Single column plate load test setup and (b) load–settlement curves for a single column plate load test (Arulrajah et al. 2009 / Emerald Publishing Limited).

(Continued)

Case Study 9.3 (Continued)

Following the installation of the stone columns, plate load tests on single columns or groups of four columns were conducted for acceptance reasons. The first cycle was conducted with the permissible design load applied and sustained for 24 hours. The second cycle was subjected to a maximum load of 150% of the design load. The load test acceptance criteria stated that the settlement should be no more than 50 mm below the permitted design load and no more than 80 mm below 150% of the allowable design load. The load test plate was 1.5 m × 1.5 m in size for a single column and 3 m × 3 m in size for a group of four columns. The schematic layout in Figure 9.14a illustrates the plate load test configuration for a single column load test. Figure 9.14b illustrates the results of a typical single-column PLT performed throughout the project.

Surface settlement gauges installed on-site regularly showed that subsequent lifts resulted in settlements. Figure 9.15 illustrates typical outcomes from the installation of a settlement plate on the project site. The stone columns' long-term performance was projected using the Asaoka technique of back-analyses based on field settlement data. After the railway lines were installed, settlement markers were used to monitor the stone columns' long-term performance. Long-term monitoring revealed that the stone columns performed according to the design projections in the region where they were erected.

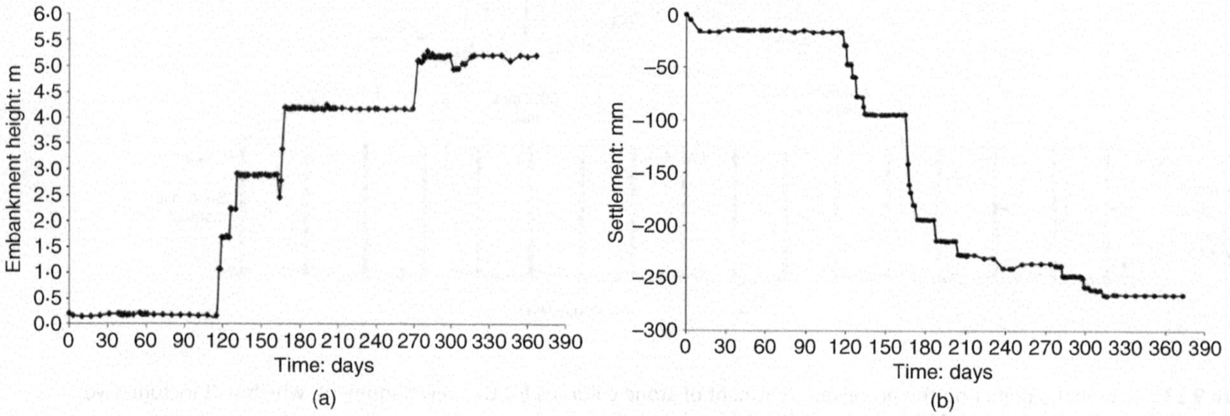

Figure 9.15 The settlement plate results installed in a stone column improved area (Arulrajah et al. 2009 / Emerald Publishing Limited).

9.4 Foundations on Reinforced Soils

Soil reinforcement approaches mainly include three categories:

- Sheeting (confinement).
- Soil slope stability reinforcement, including anchoring and nailing methods.
- Vertical structural elements such as micro-piling.

This chapter mainly focuses on the application of geosynthetic layers as reinforcement material. The two other categories were discussed in detail in Chapter 8.

9.4.1 Foundations on Geosynthetic Reinforced Soil

9.4.1.1 Introduction

In many projects, shallow footings are constructed on weak soils, resulting in low bearing capacity and significant settlement. This can lead to structural damage, reduced durability, and deterioration in performance. To address this

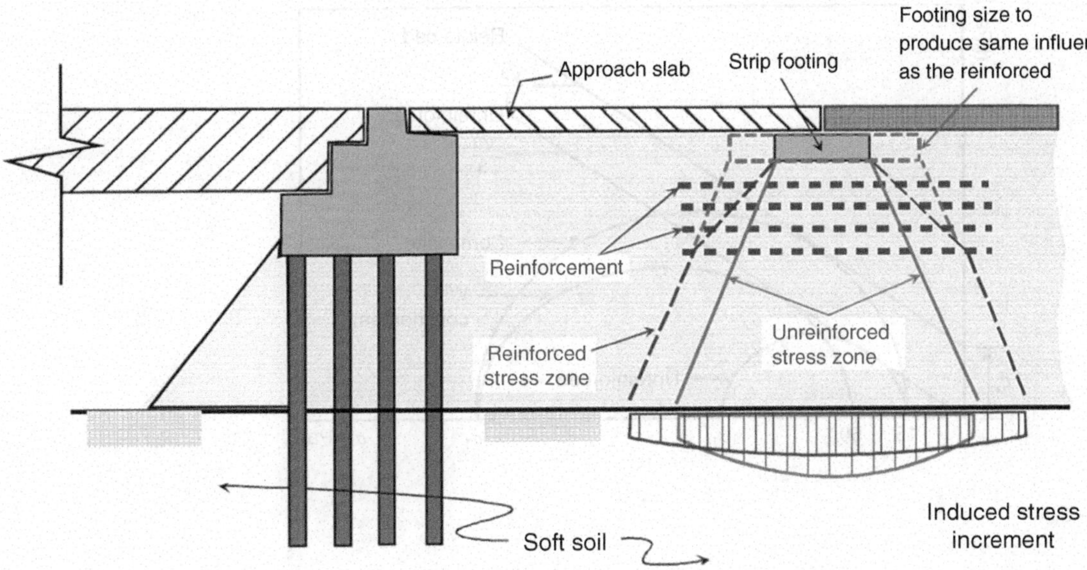

Figure 9.16 RSF applied to approach slab (Chen 2007 / with permission of Elsevier).

issue, traditional ground modification techniques involve replacing portions of the existing soil with stronger granular fill, increasing the width of the footing, or a combination of both. However, a more cost-effective alternative is the use of geosynthetic reinforced soils. This can be achieved by reinforcing the existing soil or by replacing it with compacted granular fill in combination with geosynthetics. The reinforced soil mass enhances the foundation's bearing capacity and provides a more uniform distribution of pressure, resulting in reduced settlement.

Geosynthetic reinforced soils are particularly suitable for the application in spread footings, including single and continuous foundations. Various types of reinforcements, such as geogrids, woven geotextiles, geocells, and metallic reinforcements, can be used to reinforce the soils. While metallic reinforcements (strips and meshes) have been used in laboratory research, they are less commonly used in practical applications. The fill materials commonly used in geosynthetic reinforced foundations include aggregates, sand, clay, and fly ash (Chen 2007). These materials provide the necessary strength and stability when combined with the geosynthetic reinforcements.

The application of geosynthetic reinforcement has been shown to enhance bearing capacity and reduce settlement in various investigations. These studies have primarily involved laboratory modeling and numerical analyses. The results consistently demonstrate that geosynthetic reinforcement is effective in improving the bearing capacity of the foundation. However, the reduction in settlement may be less pronounced compared to the enhancement in bearing capacity. Nonetheless, geosynthetic reinforcement still offers valuable benefits, including a reduction in the consumption of low-quality materials and an overall improvement in the performance of the foundation system.

One of the significant applications of reinforced soil foundations is in the design of approach slabs for highway bridges to minimize differential settlements developed between the roadway and the bridge slab. Differential settlements can cause the notorious bridge "bump" problem, which may result in unsafe driving conditions and damage to vehicles. One suggested solution is to use a rigid approach slab that distributes the traffic loads to two slab ends. To support this portion of the load, a shallow foundation is necessary at the approach slab ends. To achieve higher bearing capacity and minimize settlement, it is crucial to reinforce the underlying soil. This can be achieved through the application of reinforced soil techniques, which enhance the overall performance of the approach slab system (refer to Figure 9.16).

Example 9.5 Evaluate the effect of soil confinement on bearing capacity of a square footing with a 2 m width and a 1 m embedment depth based on given Mohr $\tau - \sigma$ of reinforced and unreinforced conditions. The unit weight is 18 $\frac{kN}{m^3}$, and groundwater is out of the influence zone.

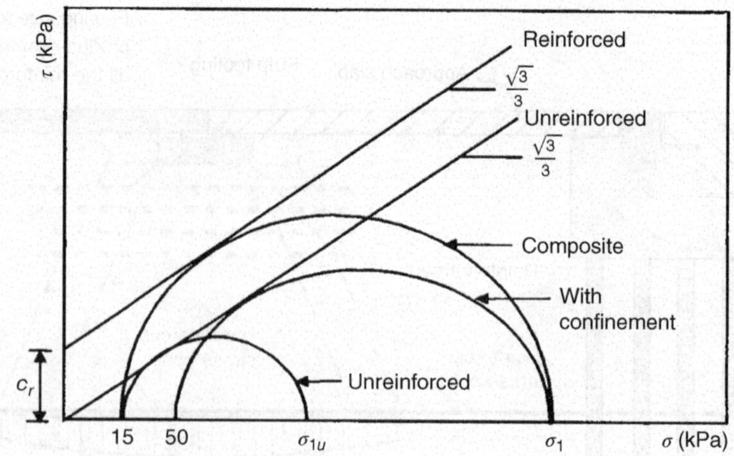

$$\tau_f = \sigma \tan \phi + C$$

Before Confinement

$$\phi = \tan^{-1}\left(\frac{\sqrt{3}}{3}\right) = 30°, C = 0\,\text{kPa}$$

$$q_{\text{ult}} = CN_c + qN_q + 0.5\gamma BN_\gamma$$

$$\phi = 30° \rightarrow N_c = 30, N_q = 18, N_\gamma = 15$$

$$q_{\text{ult}} = 0 + 18 \times 1 \times 18 + 0.5 \times 18 \times 2 \times 15 = 594\,\text{kPa}$$

After Confinement

$$\phi = \tan^{-1}\left(\frac{\sqrt{3}}{3}\right) = 30°$$

$$C = C_r = \frac{\Delta\sigma_3}{2}\sqrt{K_p}$$

$$K_p = \frac{1 + \sin\phi}{1 - \sin\phi} = \frac{1 + \sin 30}{1 - \sin 30} = 3 \rightarrow C_r = \frac{50 - 15}{2} \times \sqrt{3} = 30\,\text{kPa}$$

$$q_{\text{ult}} = CN_c + qN_q + 0.5\gamma BN_\gamma$$

$$\phi = 30° \rightarrow N_c = 30, N_q = 18, N_\gamma = 15$$

$$q_{\text{ult}} = 30 \times 30 + 18 \times 1 \times 18 + 0.5 \times 18 \times 2 \times 15 = 1494\,\text{kPa}$$

$$\frac{q_{\text{ult}_{\text{after confinement}}}}{q_{\text{ult}_{\text{before confinement}}}} = \frac{1494}{594} = 2.5$$

9.4.1.2 Principles

Different potential failure modes of geosynthetic reinforced foundations are depicted in Figure 9.17. As depicted in Figure 9.17a, shallow failure occurs when the distance from the base of the foundation to the uppermost reinforcement is considerable or the fill is weak. As illustrated in Figure 9.17b, interlayer failure can extend in considerable reinforcement spacing. In thick reinforced foundations with multiple reinforcement layers, a general failure involving one or more reinforcement layers may develop as depicted in Figure 9.17c. Stress at the footing base spreads to the underlying soil. In a thin reinforced zone and weak underlying soil, the distributed foundation may fail as a result of the low soil strength (Figure 9.17d). In the case of thin reinforced foundations but those that are wide and underlain by weak soil, the

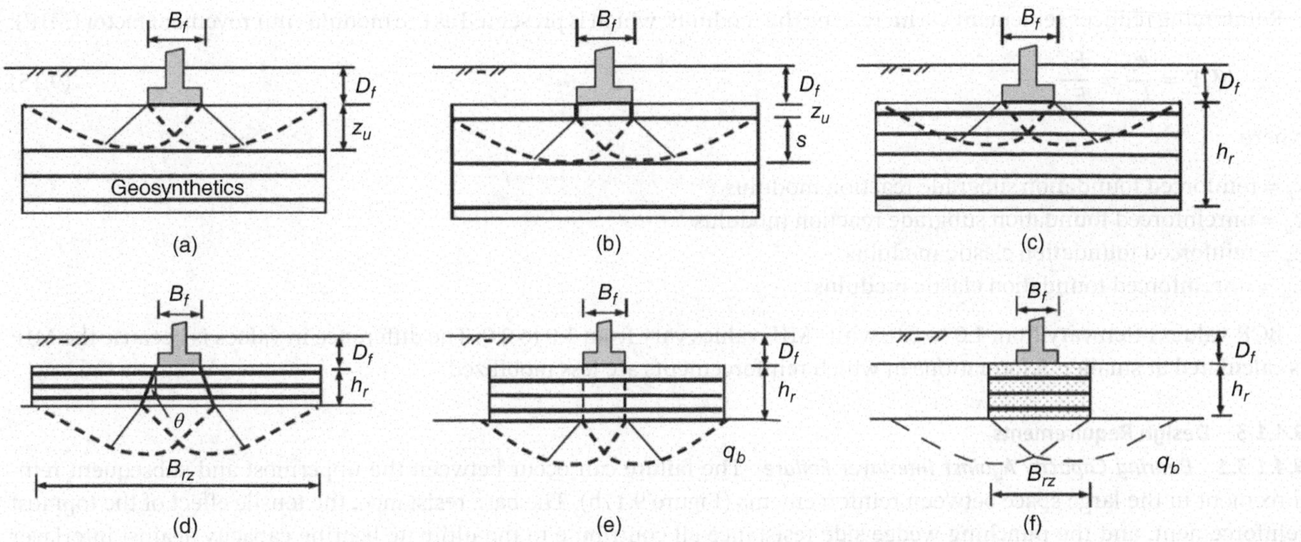

Figure 9.17 Different probable failure modes of foundations on geosynthetic reinforced soils: (a) shallow failure above the uppermost reinforcement, (b) interlayer failure between reinforcements, (c) general failure within the reinforced zone, (d) failure of the distributed foundation, (e) punching failure through the reinforced zone, and (f) reinforced zone punching failure (adapted from Wayne et al. 1998, Chen 2007, Han 2015).

footing may punch into the reinforced zone (Figure 9.17e). Figure 9.17f presents a reinforced zone punching through the underlying soil as a result of the low reinforced zone width and underlying weak soil.

Static plate loading tests are conducted on both unreinforced and reinforced foundations; representative results are shown in Figure 9.18. As seen in Figure 9.18, reinforcing enhances bearing capacity, which is frequently expressed as the bearing capacity ratio (BCR), expressed via:

$$\text{BCR} = \frac{q_{\text{ult},r}}{q_{\text{ult},u}}, \tag{9.14}$$

where

$q_{\text{ult},u}$ = unreinforced foundation ultimate bearing capacity
$q_{\text{ult},r}$ = reinforced foundation ultimate bearing capacity

Figure 9.18 Pressure vs. settlements of PLTs on unreinforced and reinforced foundations (Han 2015 / John Wiley & Sons).

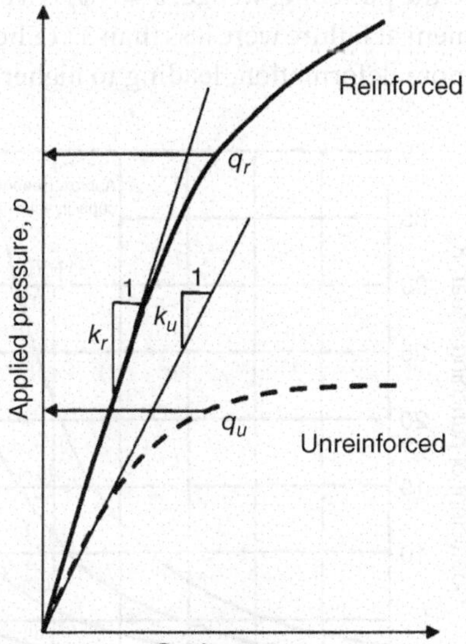

Reinforcing reduces settlement via increasing the modulus, which is presented as the modulus improvement factor (MIF):

$$\mathrm{MIF} = \frac{k_r}{k_u} = \frac{E_r}{E_u} \tag{9.15}$$

where

k_r = reinforced foundation subgrade reaction modulus
k_u = unreinforced foundation subgrade reaction modulus
E_r = reinforced foundation elastic modulus
E_u = unreinforced foundation elastic modulus

BCR values often vary from 1.0 to 3.0, while MIF values vary from 1.0 to 2.0. The difference in values is because the MIF is calculated at smaller deformations in which reinforcements are less mobilized.

9.4.1.3 Design Requirements

9.4.1.3.1 Bearing Capacity Against Interlayer Failure The failure can occur between the uppermost and subsequent reinforcement in the large space between reinforcements (Figure 9.17b). The base resistance, the tensile effect of the topmost reinforcement, and the punching wedge side resistance all contribute to the ultimate bearing capacity against interlayer failure:

$$q_{\mathrm{ult},r} = \mathrm{BCR} \cdot q_b + \frac{2C_1 c_a z_u}{B_f} + 2\gamma_r' z_u^2 \left(1 + \frac{2D_f}{z_u}\right) \frac{K_s \tan\phi_r}{B_f} + \frac{2C_1 T_a \tan\delta}{B_f} - \gamma_r' z_u \tag{9.16}$$

where

BCR = the footing bearing capacity ratio at the embedment depth of $D_f + z_u$
q_b = the unreinforced footing ultimate bearing capacity with the embedment depth of $D_f + z_u$
C_1 = 2 for square footing and 1 for continuous footing
ϕ_r = reinforced fill friction angle
γ_r' = reinforced fill effective unit weight
K_s = punching shear coefficient (Figure 9.19)
T_a = reinforcement allowable tensile strength (lesser tensile strength at a specific strain and reinforcement pullout capacity)
c_a = interface cohesion between punching wedge and reinforced fill
δ = interface friction angle between punching wedge and reinforced fill

Wayne et al. (1998) assumed full covering (wrapping) of the geosynthetic reinforcement at the punching wedge's edges (i.e., setting $\tan\delta = 1$ in Equation [9.16]). Chen (2007) ascribed the geosynthetic reinforcement to adding confining stress to the punching wedge; $\delta = \frac{2}{3}\phi_r$ is a common hypothesis. The majority of reported tensile strains in geosynthetic reinforcement at failure were less than 3% (Chen 2007), utilizing a 457 mm by 457 mm square plate size. Larger footings may generate more deformation, leading to higher strain in geosynthetic reinforcement. In the absence of field data, a tensile strain of

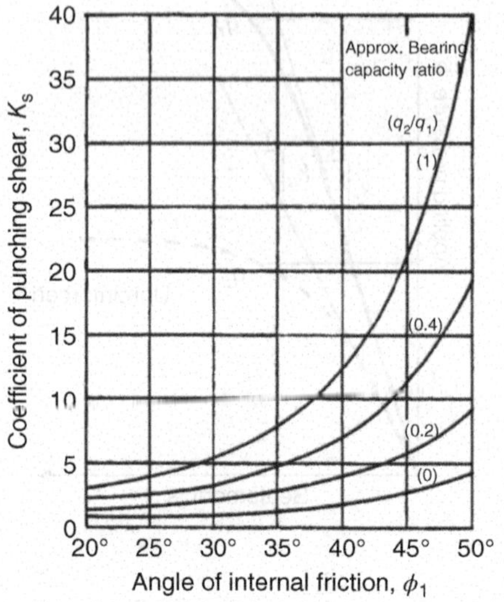

Figure 9.19 Punching shear coefficient (Meyerhof and Hanna 1978 / Canadian Science Publishing).

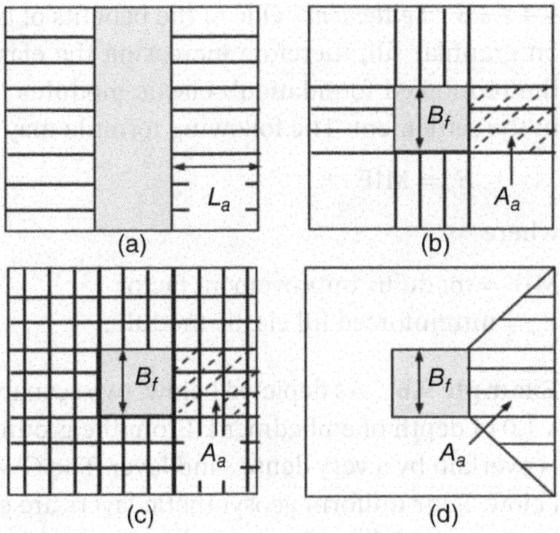

Figure 9.20 Pullout anchorage length or area: (a) continuous footing, (b) square footing and uniaxial reinforcement, (c) square footing and biaxial reinforcement, and (d) square footing and uniform reinforcement (Han 2015 / John Wiley & Sons).

3% in the reinforcement may be utilized to specify the ultimate bearing capacity of a reinforced foundation by restricting geosynthetic reinforcement overstress. At a specified strain, the tensile strength of geosynthetic reinforcement, ε_g, can be calculated via

$$T_g = J\varepsilon_g \tag{9.17}$$

where

J = the geosynthetic reinforcement tensile stiffness

The capacity of geosynthetic reinforcement on either side of the footing to pull out may be determined using:

$$T_{po} = 2C_i\left[c_r + \gamma_r'(D_f + z)\tan\phi_r\right]L_a \text{ (for a continuous footing)} \tag{9.18}$$

$$T_{po} = \frac{2C_i\left[c_r + \gamma_r'(D_f + z)\tan\phi_r\right]A_a}{B_f} \text{ (for a square footing)} \tag{9.19}$$

where z, L_a, and A_a are illustrated in Figure 9.20; and A_a can be designated through:

$$A_a = B_f\frac{B_{rz} - B_f}{2} \text{ (square footing with uniaxial and biaxial reinforcements)} \tag{9.20}$$

$$A_a = \frac{B_{rz}^2 - B_f^2}{4} \text{ (square footing with uniform reinforcements)} \tag{9.21}$$

A biaxial geogrid is strong in the lateral and longitudinal directions but weak in all others, particularly at 45° to the loading direction. Thus, conservatively, the contributions of a biaxial geogrid in directions other than the machine and cross-machine directions may be neglected. Since a triaxial geogrid has uniform tensile strength and stiffness in all directions, it may be used as a constant reinforcement.

When the spacing between the reinforcements is equal to or less than the distance to the highest reinforcement, the ultimate bearing capacity against interlayer failure is always greater than the bearing capacity against shallow failure above the uppermost reinforcement. Thus, if the ultimate bearing capacity against shallow failure is sufficient, no more analysis is necessary.

9.4.1.3.2 ***Bearing Capacity Against Punching Failure Through the Reinforced Zone*** In the case of a thin reinforced zone or weak underlying soil, punching failure into the reinforced zone will likely occur, as depicted in Figure 9.17e. This failure into the reinforced zone is a combination of the base resistance, reinforcement tensile effect, and the punching wedge side resistance:

$$q_{ult,r} = q_b + \frac{2C_1c_ah_r}{B_f} + C_1\gamma_r'h_r^2\left(1 + \frac{2D_f}{h_r}\right)\frac{K_s\tan\phi_r}{B_f} + \frac{2C_1T_a\tan\delta}{B_f} - \gamma_r'h_r \tag{9.22}$$

where

q_b = ultimate bearing capacity of the foundation with a width of B_f at a depth of $D_f + h_r$
T_a = total allowable tensile strength of reinforcement layers on one side of the footing

9.4.1.3.3 Settlement One of the benefits of geosynthetics in reinforced foundations is that they act as a lateral constraint on granular fill, therefore increasing the elastic modulus. According to laboratory and field plate loading experiments, the reinforced foundation's elastic modulus is greater than that of the unreinforced foundation, resulting in a decrease in the settlement. The following formula may be used to calculate the reinforced foundation's elastic modulus E_r:

$$E_r = \text{MIF} \cdot E_u \tag{9.23}$$

where

MIF = modulus improvement factor
E_u = unreinforced fill elastic modulus

Example 9.6 As depicted below, two scenarios are considered for a square concrete foundation with a 1.25 m width and a 1.0 m depth of embedment. From the ground surface, the site is comprised of a 7 m deep layer of soft clayey soil, which is overlain by a very dense sand layer. The GWT is located at a depth of 1 m, and the soft soil has the characteristics shown below. Four uniform geosynthetic layers are employed to increase the bearing capacity as seen in the figure. Geosynthetics have the following properties: at 3%, the secant modulus is 650 kN/m, whereas the interaction coefficient with fill is 0.9. The total load including foundation weight is 180 kN. (Floatation is not considered in calculation.) Consider 3 as the minimum required safety factor.

a) Determine the ultimate bearing capacity of the RSF if the punching failure through the reinforced foundation is controlling and compare it to the initial condition.
b) Estimate the consolidation settlement of the clay layer for both the initial and reinforced conditions. Consider 5.2 cm (2 in.) as the allowed limit.

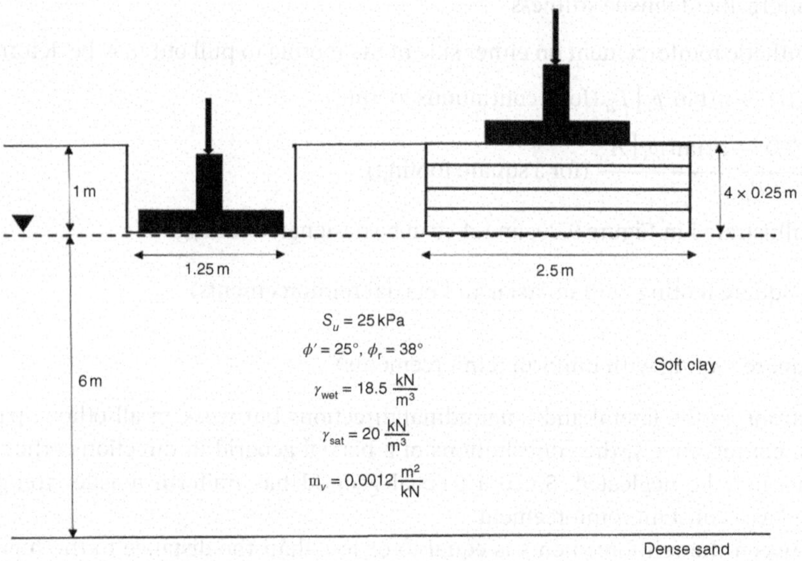

a) $p = \dfrac{P + W_f}{A_f} - u_D = \dfrac{180}{1.25 \times 1.25} - 0 = 115\ \text{kPa}$

Bearing capacity of natural soil

Short-term behavior: $q_{\text{ult}} = S_u N_c + \overline{q}$

$$q_{\text{ult}} = 25 \times 5.14 + 1 \times 18.5 = 147\ \text{kPa}$$

Long-term behavior: $q_{\text{ult}} = CN_c + \overline{q}N_q + 0.5\gamma BN_\gamma$

$$\phi = 25° \rightarrow N_c = 21, N_q = 11, N_\gamma = 7$$

$$q_{\text{ult}} = 0 + 1 \times 18.5 \times 11 + 0.5 \times (20 - 10) \times 1.25 \times 7 = 247\ \text{kPa}$$

Therefore, short-term behavior is controlling.

$$FS = \frac{147}{115} = 1.3 \ngtr 3 \text{ Not OK} \rightarrow \text{Therefore, reinforcing is essential.}$$

Bearing capacity of reinforced soil

$$A_a = \frac{\left(B_{rz}^2 - B_f^2\right)}{4} = \frac{(2.5^2 - 1.25^2)}{4} = 1.17\ \text{m}^2$$

The pullout capacities for four reinforcements are

$$T_{po} = \frac{2C_i\left[c_r + \gamma_r'(D_f + z)\tan\phi_r\right]A_a}{B_f}$$

$$T_{po1} = \frac{2\times 0.9\times[0 + (18.5)\times(0 + 0.25)\times\tan 38]\times 1.17}{1.25} = 6.1\ \frac{\text{kN}}{\text{m}}$$

$$T_{po2} = \frac{2\times 0.9\times[0 + (18.5)\times(0 + 0.5)\times\tan 38]\times 1.17}{1.25} = 12.2\frac{\text{kN}}{\text{m}}$$

$$T_{po3} = \frac{2\times 0.9\times[0 + (18.5)\times(0 + 0.75)\times\tan 38]\times 1.17}{1.25} = 18.3\ \frac{\text{kN}}{\text{m}}$$

$$T_{po4} = \frac{2\times 0.9\times[0 + (18.5)\times(0 + 1)\times\tan 38]\times 1.17}{1.25} = 24.4\ \frac{\text{kN}}{\text{m}}$$

$$T_g = J\varepsilon_g = 0.03\times 650 = 19.5\ \text{kN/m}$$

$$T_{a1} = 6.1\frac{\text{kN}}{\text{m}},\ T_{a2} = 12.2\frac{\text{kN}}{\text{m}},\ T_{a3} = 18.3\frac{\text{kN}}{\text{m}},\ \text{and } T_{a4} = 19.5\frac{\text{kN}}{\text{m}}.$$

Long term behavior: $q_b = 247$ kPa

$$q_{\text{ult},r} = q_b + \frac{2C_1c_ah_r}{B_f} + C_1\gamma_r'h_r^2\left(1 + \frac{2D_f}{h_r}\right)\frac{K_s\tan\phi_r}{B_f} + \frac{2C_1T_a\tan\delta}{B_f} - \gamma_r'h_r$$

$$q_{\text{ult},r} = 247 + \frac{2\times 2\times 0\times 1}{1.25} + 2\times(18.5)\times 1^2\times\left(1 + \frac{2\times 1}{1}\right)\times\frac{7\times\tan 38}{1.25}$$
$$+ \frac{2\times 2\times(6.1 + 12.2 + 18.3 + 19.5)\times\tan(0.67\times 38)}{1.25} - (18.5)\times 1 = 799\ \text{kPa}$$

The FS of the reinforced foundation is $FS_r = \frac{799}{115} = 6.9 > 3$ OK

b) *Settlement of foundation on natural soil*

At the middle of the clay layer: $\Delta q = \frac{q\times B^2}{(B+Z)^2} = \frac{180}{(1.25+3)^2} = 10$ kPa

$$S_c = m_v\Delta\sigma' H = 0.0012\times 10\times 6 = 0.072\ \text{m} = 72\ \text{mm} \nleq 52\ \text{mm Not OK}$$

Settlement of foundation on reinforced soil

At the middle of the clay layer: $\Delta q = \frac{q\times B^2}{(B+Z)^2} = \frac{180}{(2.5+3)^2} = 6$ kPa

$$S_c = m_v\Delta\sigma' H = 0.0012\times 6\times 6 = 0.043\ \text{m} = 43\ \text{mm} < 52\ \text{mm OK}$$

9.4.2 Application of Reinforcement for Soil Slope Stability

Geosynthetic and metallic reinforcements are used to reinforce the fill. While soil is strong in compression but weak in tension, reinforced soil is strong in both directions. Geosynthetics, geogrids, and geocells are used to reinforce slopes, embankments, earth-retaining walls, foundations, and highways. Metallic reinforcements such as steel strips and steel meshes are most often used in earth-retaining walls. Through anchoring, tensioned membranes, and lateral constraint through friction, interlocking, or close confinement, geosynthetics supply tension or confining stress to fill. One or more

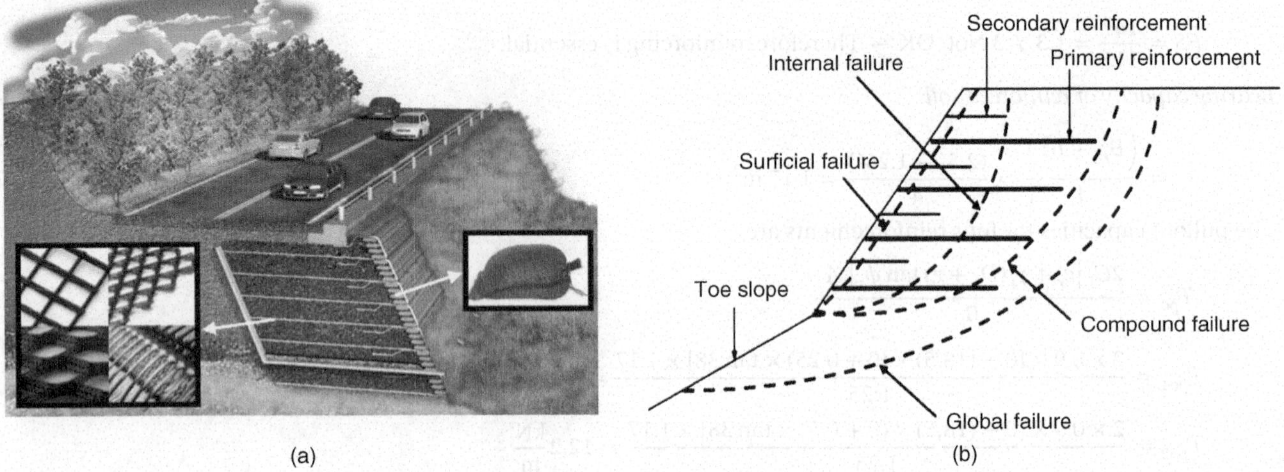

Figure 9.21 (a) Schematic view of geosynthetic-reinforced slope and (b) possible failure modes of a reinforced slope (figure courtesy of ACE Geosynthetics, used with permission).

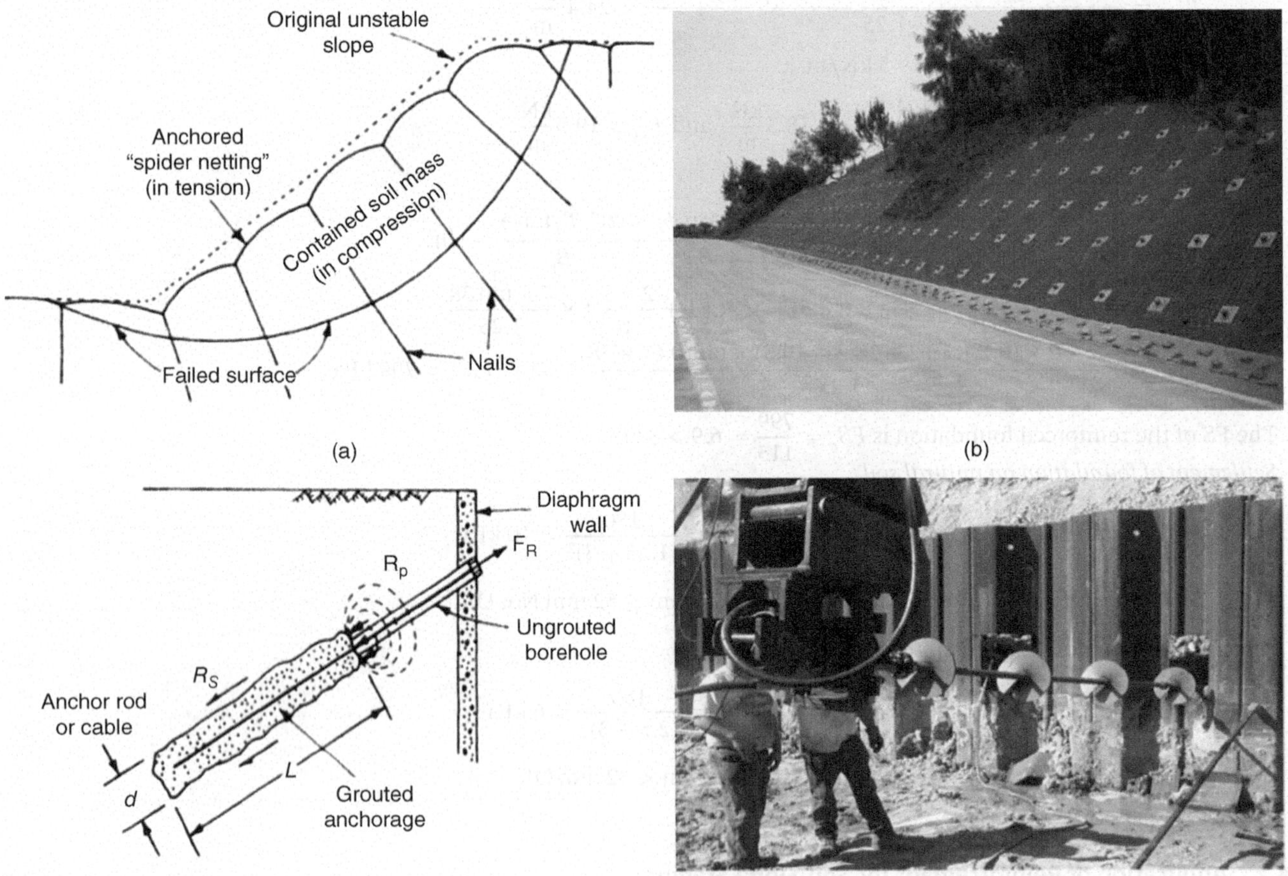

Figure 9.22 (a) Schematic cross-section of anchored geosynthetic "spider netting" used with soil nails to stabilize a slope, (b) actual application of soil nailing, (c) a stressed soil anchor, and (d) application of helical piles for stabilizing slopes and excavations.

mechanisms may be present in a single application. For example, anchoring is the major mechanism for reinforcing slopes and earth walls, whereas friction created by geotextiles, interlocking created by geogrids, and closed confinement created by geocells all contribute to the lateral restriction of granular fill in roadways. When differential settlement is severe, such as embankments over voids, the tensioned membrane effect may become critical in highways. Additionally, geosynthetic reinforcement may help prevent the formation of slip surfaces in reinforced foundations and alter the boundary conditions on top of weak foundations, hence increasing bearing capacity. The geosynthetic reinforced slope is depicted schematically in Figure 9.21a.

Four possible failure modes have been recognized for geosynthetic reinforced slopes: (i) surficial failure, (ii) internal failure, (iii) compound failure, and (iv) global failure (Figure 9.21b). In these conditions, above the slide surface, there is an active soil wedge; below the slip surface, there is a stable soil mass.

Other reinforcement elements are used to provide soil slope stability in which pullout capacity plays a significant role in geotechnical performance. Application of nails, anchors, and helical piles can be named in this regard. Figure 9.22 presents some typical applications and mechanisms of the abovementioned reinforcement elements.

Case Study 9.4

GRS Structures for Railways in Japan (Tatsuoka and Watanabe 2015)

Numerous geosynthetic reinforced soil (GRS) structures (retaining walls, box culverts, bridge abutments, and bridge systems) have become standard soil structures (retaining walls, box culverts, bridge abutments, and bridge systems) for Japan's high-speed train (HST) lines (named Shinkansen) and have been adopted in place of conventional structures.

The first HST line was the Tokaido Shinkansen, which connected Tokyo and Osaka and was inaugurated in 1964, just before the Tokyo Olympic Games. The overall length build (excluding two narrow-gauge lines) is 2765 km, which includes the following three generations: (i) Tokaido; (ii) Sanyo (the western section between Osaka and Okayama opened in 1972; the eastern section between Okayama and Hakata opened in 1975; the southern section of Tohoku between Tokyo and Morioka opened in 1982; and Joetsu, which opened in 1982; and (iii) anything built after 1982 (Figure 9.23a)

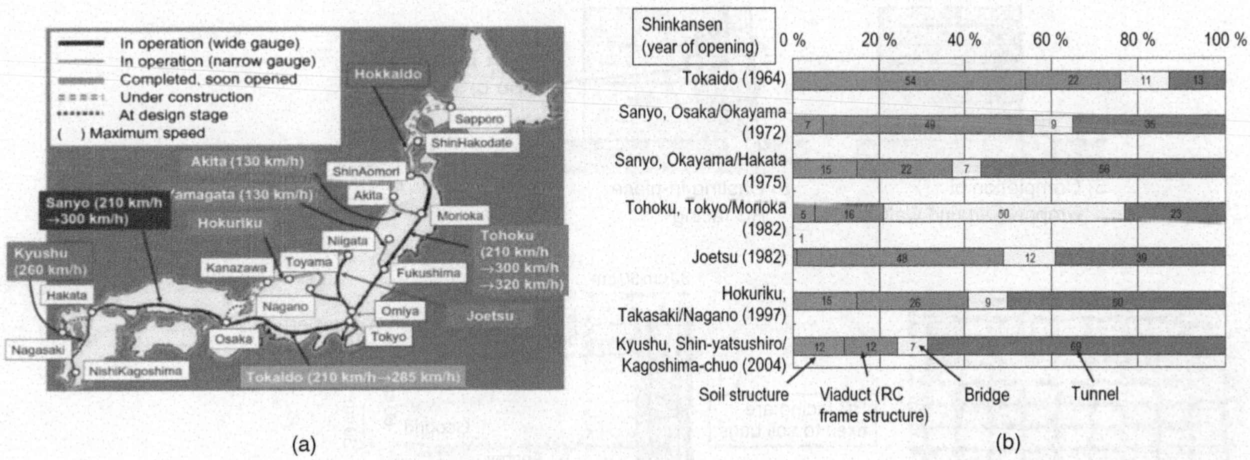

Figure 9.23 Shinkansen (a) network and (b) length ratios of different structure types (Tatsuoka and Watanabe 2015 / with permission of Elsevier).

For the first generation (Tokaido) grade separation, the embankment was built for more than half of the whole length (Figure 9.23b) in accordance with the design norm at the time. From the outset of train operations, considerable ongoing track maintenance was required to alleviate embankment settlement issues, notable bumps immediately after bridge abutments, and box culverts. With regard to the significant damage to railway embankments that happened

(Continued)

Case Study 9.4 (Continued)

after the Tokaido line's inauguration, various sections of the embankment were reinforced to provide sufficient stability during heavy rains and strong earthquakes.

While a mild sloping embankment has a larger base area than a reinforced concrete (RC) frame structure (such as a bridge), the embankment's construction cost per length is much lower. The experience gained during the construction of the HST line in Japan demonstrated that the construction cost (excluding land) is approximately 1/7 that of a viaduct and 1/15 that of a bridge; additionally, even when soil improvement becomes necessary, the cost is still 1/2 that of a viaduct and 1/5 that of a bridge. Additionally, by creating embankments in surrounding locations using material extracted from tunnels and the ground, the cost and environmental effect will be greatly reduced. Although, based on the lessons learned from the Tokaido Shinkansen embankment, the quantity of embankment and related retaining wall drastically decreased in the second generation (i.e., Sanyo, southern Tohoku, and Joetsu) (Figure 9.23b).

During the 1980s, the GRS retaining wall (RW) with a stage-constructed full-height rigid (FHR) face (Figure 9.24) was created. The first GRS RW railway of this kind was built in 1989. Other GRS RWs of this kind performed well during the 1995 Great Kobe earthquake, although several embankments and conventional-type RWs, as well as other forms of reinforced soil RWs, sustained major damage or collapsed. Numerous viaducts along the Sanyo Shinkansen were also heavily affected or collapsed. These occurrences revealed that the stability of these types of GRS RWs is much greater than that of embankments and conventional-type RWs under strong seismic loads such as those experienced during the earthquake.

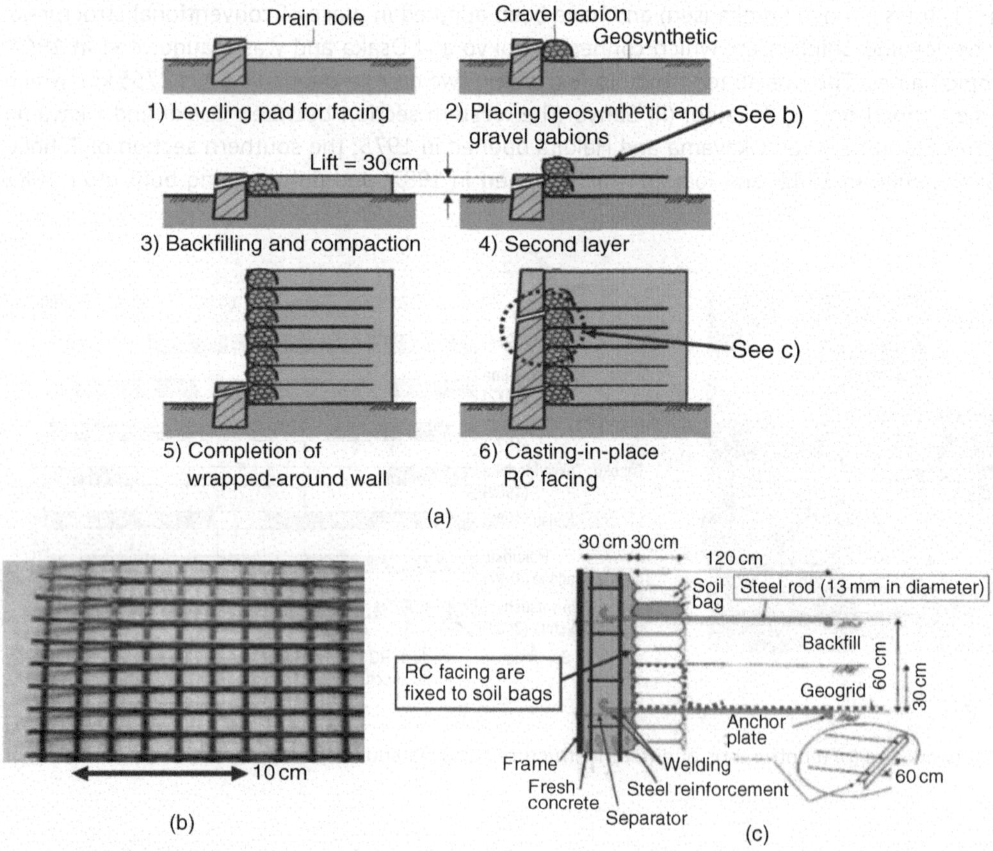

Figure 9.24 GRS RW with FHR facing: (a) construction procedure, (b) typical geogrid, and (c) facing construction (Tatsuoka et al. 1997 / Emerald Publishing Ltd.).

In the two decades following the 1995 Great Kobe earthquake, the design and construction strategy for soil structures on Japanese railroads has seen substantial changes. As Figure 9.25a indicates, there have been no difficulties with any of the GRS structures so far. Figure 9.25b illustrates the statistics. After being exposed to the 1995 Great

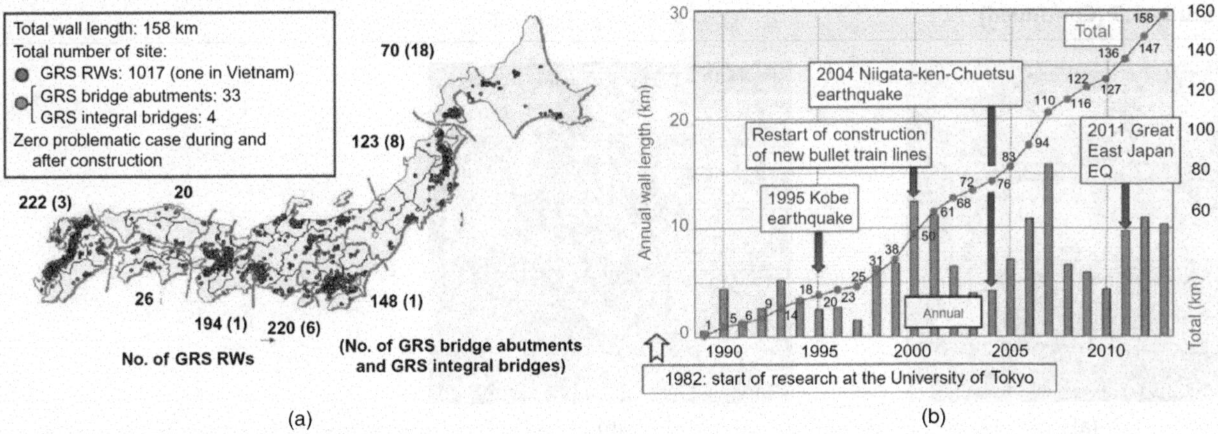

Figure 9.25 (a) Locations of GRS RWs with FHR facing [June 2014] and (b) annual and cumulative wall lengths (Tatsuoka and Watanabe 2015 / with permission of Elsevier).

Kobe earthquake and the 2011 Great East Japan earthquake, as well as heavy rains and flooding, the GRS technology has shown to be cost-effective, owing to its excellent resilience to these sorts of severe natural disasters.

These GRS structures were chosen because they are very cost-effective in comparison to conventional varieties; they need less construction and maintenance while providing more performance, including seismic stability. In comparison to conventional-type structures, the settlement in the backfill directly behind the facing (i.e., the bump) produced by long-term train loads and seismic loads becomes insignificant with GRS bridge abutments, GRS integrated bridges, and GRS box culverts.

Case Study 9.5

Bearing Capacity of Geosynthetic Encased Stone Columns (Ghazavi and Nazari Afshar 2013)

The use of stone columns in low-strength soil is an effective way to enhance the load-bearing capabilities of shallow footings. The ability of native soft soils to provide circumferential confinement is a major factor in stone column-bearing capacity. For certain large-body laboratory tests, Ghazavi and Nazari Afshar (2013) used stone columns with diameters of 60, 80, and 100 mm with a length-to-diameter ratio of 5. Additionally, experiments on groups of 60 mm-diameter stone columns were conducted to examine the impact of nearby columns on the reference loaded stone column. The goal of this study was to examine the efficacy of vertically enclosing different stone column diameters under the same conditions. Additionally, comparisons were made between reinforced and unreinforced vertically encased stone columns (VESC) encased in geotextiles, to explore the impact of reinforcing on the footing's load-carrying properties using vertical encased stone columns (VESC). Figure 9.26a,b depict the testing setup. The plan view of the test and the shape of the stone column after testing are shown in Figure 9.26c,d.

In general, the following inferences are derived from the test results:

1) Single stone columns are subject to the bulging failure mode. The bulging failure often happens between D and 2D from the head of the stone column. In the case of the stone column group, lateral deformation and bulging constituted the failure mode.
2) The use of OSCs enhances the ultimate load that soft soil can sustain. Utilizing vertical (VESCs) reinforcing material might further increase the ultimate load and stiffness of the treated soil.
3) Less lateral bulging occurs in VESCs than in OSCs as a result of the greater lateral confinement geosynthetic material offers.
4) The ultimate capacity and stiffness of stone columns improve with an increase in the length and strength of the reinforcing encasement.

(Continued)

Case Study 9.5 (Continued)

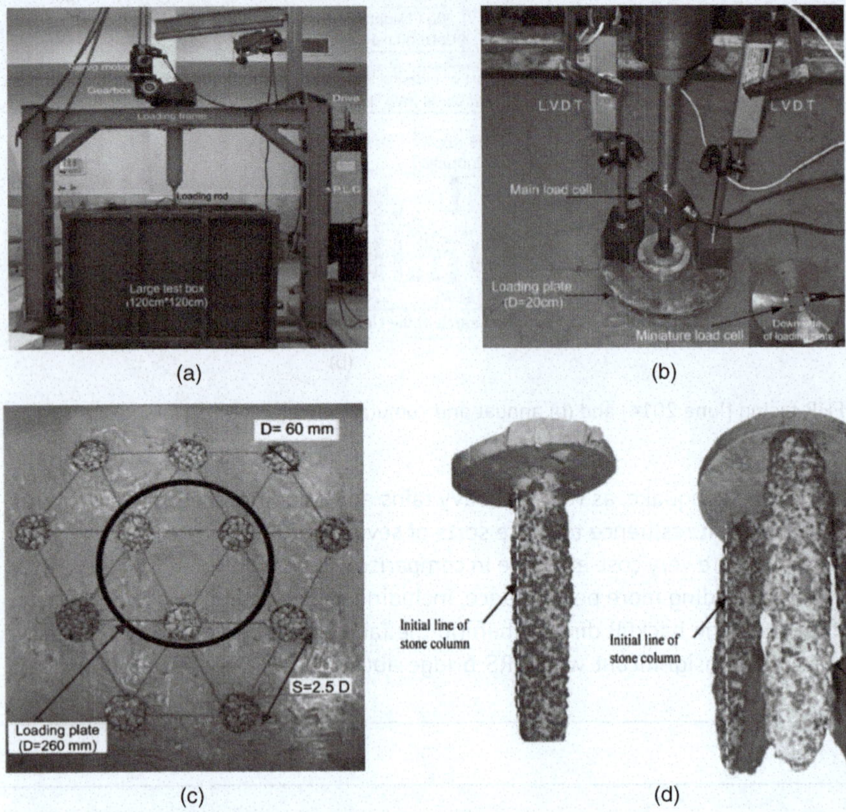

Figure 9.26 (a) Loading frame, (b) single stone column loading plate and data acquisition instruments, (c) plan view of group tests arrangement, and (d) stone column shape after testing (Ghazavi and Nazari Afshar 2013 / with permission of Elsevier).

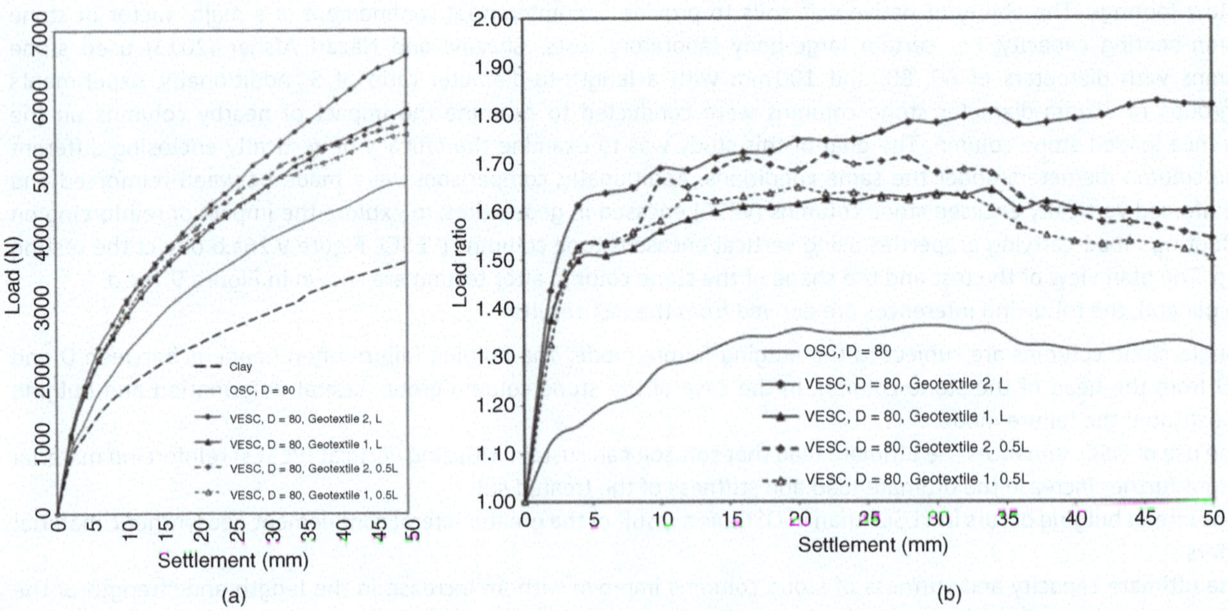

Figure 9.27 (a) Load–settlement variation of single stone column with diameters 80 mm, (b) Variation of load ratio vs. settlement for various stone columns with diameters of 80 mm (Ghazavi and Nazari Afshar 2013 / with permission of Elsevier).

5) Studies on a group of stone columns have demonstrated that the ultimate capacity of VESCs is higher than that of a group of OSCs.
6) VESCs have a greater stress concentration ratio value than OSCs. With an increase in the settlement and stone column diameter, this ratio falls.

Figure 9.27 displays two of the study's graphical results. The load–settlement behavior of isolated stone columns (OSCs and VESCs) with a diameter of 80 mm is shown in Figure 9.27a. The ultimate load-carrying capability of the soft soil has increased in all OSCs and VESCs. Additionally, the ultimate capacities of stone columns are increased by using geosynthetic material for vertical encasing. Because the settlement for stone columns has diameters of 80 mm and different tensile strengths for reinforcement materials, Figure 9.27b illustrates the variation of the load ratio (LR), which is the ratio of the ultimate load obtained from soft soil with no stone column to ultimate load obtained from reinforced soil with stone columns. The current study's full-length encasement of VESCs with stronger geotextiles has a maximum LR and a minimum LR for OSCs. Additionally, as the length and tensile strength of the reinforcing material in VESCs rise, the LR value also rises. This is because reinforcement material prevents bulging and gives columns lateral confinement.

9.4.3 Vertical Structural Elements

As discussed in Chapter 8, there are inclined and vertical structural reinforcement elements beside horizontal reinforcement elements. Micro-piles are among the reinforcement elements that are widely used for different applications.

Example 9.7 A storage tank with a 7.5 m diameter and 15 m height with a total pressure of 150 kPa, including the foundation weight, is constructed on a nearshore site. The existing soil is recognized as mixed deltaic soil and the embedment depth of the foundation is 1 m. Based on structural analysis, a horizontal load equal to 15% of the total vertical load acts at the midpoint of the storage height.

a) If the minimum required factor of safety is 3, control the sufficiency of the current foundation design for bearing the upcoming loads. Also, check the eccentricity.
b) If the minimum requirements are not satisfied, suggest a micro-pile design based on the given characteristics (FS = 2.5).
c) Discuss the different roles of the micro-piles.

$$\text{Soil: } C = 15\,\text{kPa},\ \phi' = 20^\circ,\ \gamma_{wet} = 18\,\frac{\text{kN}}{\text{m}^3},\ \gamma_{sat} = 19\,\frac{\text{kN}}{\text{m}^3}$$

$$\text{Micropile}: \alpha_{bond-nominal-strength} = 82\,\text{kPa},\ D_{bond} = 15\,\text{cm},\ L_{bond} = 10\,\text{m}$$

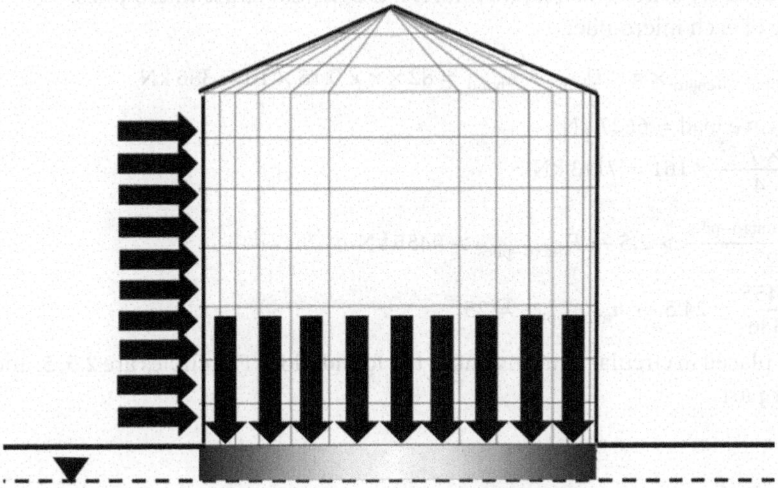

a) Total loads on the foundation surface are as follows

$$V = 150 \times \frac{\pi \times 7.5^2}{4} = 6627\ \text{kN}$$

$$H = 0.15 \times 6626.8 = 994\ \text{kN}$$

$$M = 994 \times \frac{15}{2} = 7455\ \text{kN m}$$

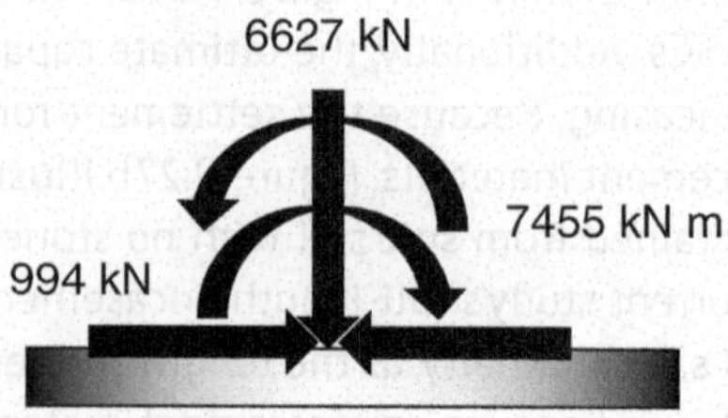

Checking eccentricity

$$e_x = e_y = \frac{M}{V} = \frac{7455}{6627} = 1.12\ \text{m}$$

$$\text{Circular foundation: } e \leq \frac{r}{4} \rightarrow e_x = e_y = 1.12\ \text{m} \nleq \frac{3.75}{4} = 0.94\ \text{m}$$

Therefore, the eccentricity requirement is not satisfied and appropriate measures must be conducted.

$$q_{\text{ult}} = CN_c + \overline{q}N_q + 0.5\gamma BN_\gamma$$

$$\phi = 20° \rightarrow N_c = 15, N_q = 6, N_\gamma = 3$$

$$q_{\text{ult}} = 15 \times 15 + 1 \times 18 \times 6 + 0.5 \times (19 - 10) \times 7.5 \times 3 = 434\ \text{kPa}$$

Reduction factors for both directions must be considered.

$$R_{\text{ex}} = R_{\text{ey}} = 1 - \left(\frac{e}{B}\right)^{0.5} = 1 - \left(\frac{1.12}{7.5}\right)^{0.5} = 0.61$$

$$R_e = R_{\text{ex}} \cdot R_{\text{ey}} = 0.61 \times 0.61 = 0.37$$

$$q^*_{\text{ult}} = q_{\text{ult}} \times R_e = 434 \times 0.37 = 161$$

$$\text{FS} = \frac{161}{150} = 1.07 \ngtr 2.5\ \text{Not OK}$$

The minimum factor of safety is not satisfied; therefore, it is essential to use micro-piles.

b) Geotechnical capacity of each micro-pile:

$$P_G = \alpha_{\text{bond-nominal-strength}} \times \pi \times D_{\text{bond}} \times L_{\text{bond}} = 82 \times \pi \times 0.15 \times 10 = 386\ \text{kN}$$

Total vertical compressive load = 6627 kN

$$\text{Load by the raft} = \frac{\pi \times 7.5^2}{4} \times 161 = 7113\ \text{kN}$$

$$\text{FS} = \frac{7113 + F_{\text{micro-piles}}}{6627} > 2.5 \rightarrow F_{\text{micro-piles}} > 9455\ \text{kN}$$

$$n_{\text{micro-piles}} > \frac{9455}{386} = 24.5 \rightarrow n_{\text{micro-piles}} = 25$$

Micro-piles should be placed in circular patterns under the foundation. The circles are 2.5, 5, and 7.5 m in diameter from the inside to the outer part.

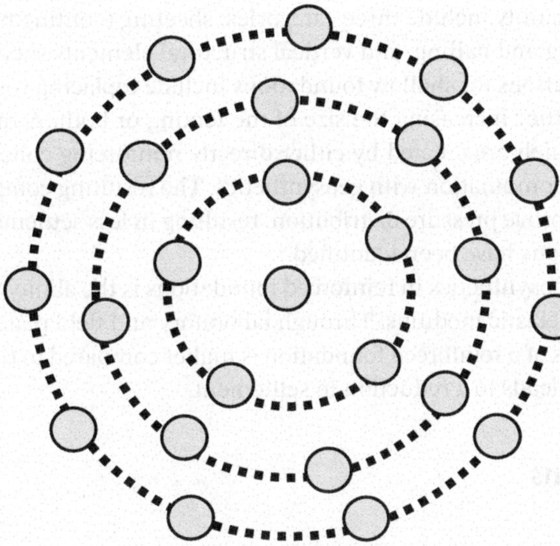

c) Micro-piles contribute to the geotechnical performance through three procedures of densification, grouting, and reinforcing. The implemented micro-piles overcome eccentricity (foundation lift-up) and also provide resistance against tensile loads. They also provide sufficient bearing capacity, either in compression or tension. Considering the high elasticity modulus of micro-piles compared to existing soil, the amount of settlement will be decreased. Also, any potential for collapse or liquefaction can be mitigated by applying driven steel tube pipes (micro-piles).

9.5 Major Points

1) Ground modification techniques play a crucial role in enhancing the performance of both shallow and deep foundations. These techniques serve as an intermediate and complementary solution in foundation engineering. Soil improvement methods can be broadly categorized into three groups based on their approach: massive treatment, rigid intrusions, and reinforced soil.
2) Massive treatment includes soil improvement techniques that modify a volume or mass of soil at the same time. In this regard, modification methods such as deep dynamic compaction (DDC), explosive compaction (EC), preloading along with drainage, grouting, and massive soil mixing can be named.
3) Rigid intrusions are used when accessing deeper layers is difficult. These elements are vertical columns that enhance the bearing capacity, reduce the settlement, and increase the stability. Improvement methods such as deep soil mixing, vibro-compaction stone columns, and lime/cement columns are rigid intrusions.
4) The objective of creating a composite ground or foundation is to achieve several improvements, including enhancing bearing capacity, minimizing and accelerating settlements, improving shear strength for slope stability, and increasing liquefaction resistance. This is accomplished by combining columns with the surrounding soils. In geotechnical applications, granular columns have proven to serve at least one of the following functions:
5) **Densification:** Granular columns are used to densify the soil, increasing its compaction and improving its load-bearing capacity.
6) **Load-bearing:** The columns contribute to bearing the applied loads, transferring them to the surrounding soil, and providing additional support.
7) **Reinforcing:** By incorporating granular columns, the soil is reinforced, enhancing its overall strength and stability.
8) **Stress distribution:** The columns help distribute stresses more uniformly throughout the composite ground, reducing localized stress concentrations.
9) **Drainage:** Granular columns can facilitate drainage within the soil, preventing the buildup of excess pore water pressure and improving the overall performance of the foundation.

   Depending on the specific project requirements and soil conditions, granular columns may be employed for one or more of these purposes. They offer versatility in geotechnical engineering, allowing engineers to tailor the composite ground to address the specific needs and challenges of a given project.

10) Soil reinforcement approaches mainly include three categories: sheeting (confinement), soil slope stability reinforcement, methods such as anchoring and nailing, and vertical structural elements such as micro-piling.
11) Typical ground modification solutions for shallow foundations include replacing some of the existing soil with a thick layer of stronger granular fill, either increasing the size of the footing or both. A more cost-effective option is to use geosynthetic reinforced soils, which are created by either directly reinforcing cohesive soils or by replacing existing soils with harder granular fill in combination with geosynthetics. The resulting composite zone will increase the foundation's bearing capacity and improve pressure distribution, resulting in less settlement. Six distinct types of failure for geosynthetic reinforced foundations have been identified.
12) One of the advantages of using geosynthetics in reinforced foundations is the ability to provide lateral constraint to the granular fill, which increases its elastic modulus. Through laboratory and field plate loading experiments, it has been observed that the elastic modulus of a reinforced foundation is higher compared to that of an unreinforced foundation. This increase in elastic modulus leads to a reduction in settlement.

9.6 Questions and Problems

9.6.1 Explanatory

1 What is the definition of mass-treated soil? Name some ground modification methods that are included in this category.

2 Explain and illustrate the components of a stone column-supported foundation.

3 Explain the relationship between the sandbag mechanisms and reinforced soil.

4 What are the effects of geosynthetics in the enhancement of foundations with geotechnical behavior?

5 Define different failure modes of geosynthetic reinforced foundations via suitable pictures.

6 Explain the critical failure mode in RSF and how to determine it.

7 Name and define the mechanisms that contribute to increasing the safety factor of a geosynthetic reinforced slope.

9.6.2 Conceptual

1 Name the most effective technique to raise the safety factor in geosynthetic reinforced slopes for global failure.

2 Consider the SCBC chart. Introduced by Imam et al. (2021), explain the procedure of failure mode determination.

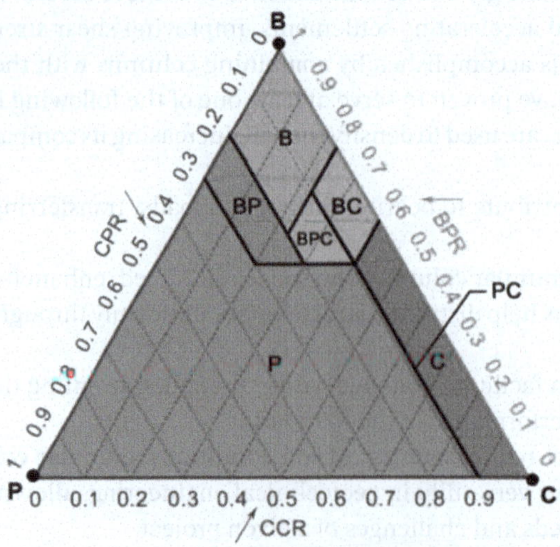

3 What are alternatives available to prevent the most critical failure mode, the punching of a footing through a geosynthetic reinforced foundation?

4 Compare the reinforcing method with other soil improvement methods that are available for shallow foundations.

5 Compare the static and dynamic behavior of conventional retaining walls and geosynthetic reinforced walls via the results of the investigations in Japan (Case Study 9.4).

6 Discuss the effects of encasement on the behavior of rigid intrusion composite foundations based on what you learned from Case Study 9.5.

7 Which lessons can be gained through soil improvement of the railway embankment in Malaysia via stone columns?

8 Considering the case study of STP tanks in Palm Jumeira, what are other suitable methods of soil improvement that could be used on this site? Compare these methods with the one applied.

9 Discuss the effect of an increase in lateral load on the number and performance of micro-piles in a project like the one presented in Example 9.7.

9.6.3 Problems

1 The direct shear test results in a site prior to and after grouting are shown in the graph. The elasticity modulus has increased from 15 to 40 MPa. Calculate and compare the desired parameters before and after ground modification:
- Bearing capacity of a 1.5 m-wide strip foundation.
- Settlement of foundation.

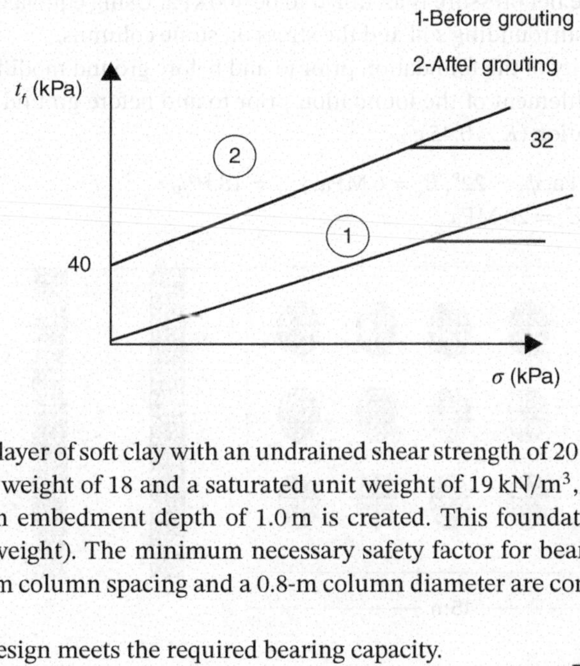

2 A site consists of a 20 m thick layer of soft clay with an undrained shear strength of 20 kPa and a 1.0 m deep groundwater table. The soil has a wet unit weight of 18 and a saturated unit weight of 19 kN/m^3, respectively. A square foundation with a width of 2.0 m and an embedment depth of 1.0 m is created. This foundation can support a load of 400 kN (including the foundation's weight). The minimum necessary safety factor for bearing failure is 2.5. Ten-meter-long granular columns with a 1.5-m column spacing and a 0.8-m column diameter are constructed in an equilateral triangle pattern.
a) Determine whether this design meets the required bearing capacity.
b) Calculate the percentage of reduction in the amount of settlement, considering $\frac{E_c}{E_s} = 10$.

3 A square concrete foundation with a 1.0 m width is planned with an embedment depth of 1.0 m. From the ground surface, the site is comprised of a 9 m deep layer of soft clayey soil, which is overlain by a very dense sand layer. The GWT is located at a depth of 0.6 m, and the soft soil has the given characteristics. Three uniform geosynthetic layers are employed to increase the bearing capacity, as seen in the figure. Geosynthetics have the following properties: at 3%, the secant modulus is 500 kN/m, whereas the interaction coefficient with fill is 0.9.

a) Determine the ultimate bearing capacity of the RSF and FS at the applied pressure if the punching failure through the reinforced foundation is controlling.
b) Calculate the MIF and percentage of reduction in the amount of settlement, considering the drained elastic modulus before and after soil reinforcement.

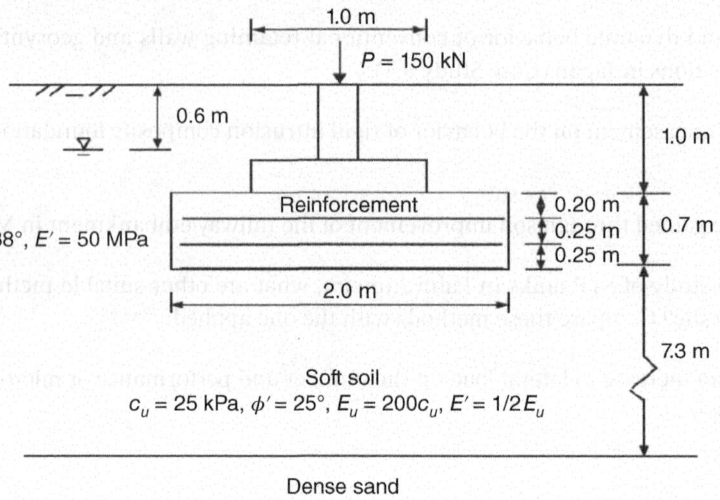

4 The rigid foundation of a commercial complex with dimensions of 15 m × 10 m is located on a layer of soft saturated soil. In order to modify the ground, 10% of the total surface beneath the foundation is replaced with stone columns, in which the column's ratio of the modulus of elasticity to the surrounding soil after improvement is 5, and the maximum allowed settlement is 8 cm. The net pressure is assumed to be 400 kPa. Using equivalent parameters:
a) Calculate the stress on the surrounding soil and the stress on stone columns.
b) Compare the bearing capacity of the foundation prior to and before ground modification.
c) Estimate the immediate settlement of the foundation prior to and before ground modification under the stress of 300 kPa beneath the foundation ($K = 0.45$).

Before Improvement: $C = 30$ kPa, $\phi_s = 22°$, $E_s = 6$ MPa, $\gamma_{sat} = 18$ kN/m³
After Improvement: $\phi_c = 38°$, $E_s = 20$ MPa

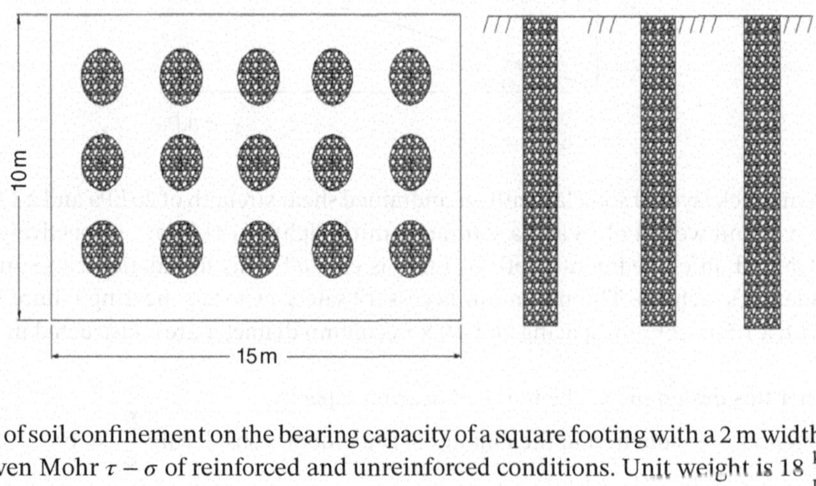

5 Evaluate the effect of soil confinement on the bearing capacity of a square footing with a 2 m width and a 1 m embedment depth based on given Mohr $\tau - \sigma$ of reinforced and unreinforced conditions. Unit weight is 18 $\frac{kN}{m^3}$, and groundwater is out of the influence zone.

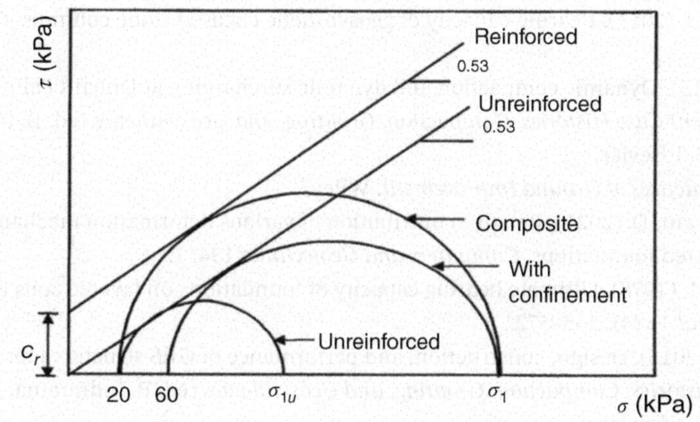

6 Resolve Example 9.3 using the given profile of the Janbu modulus number.

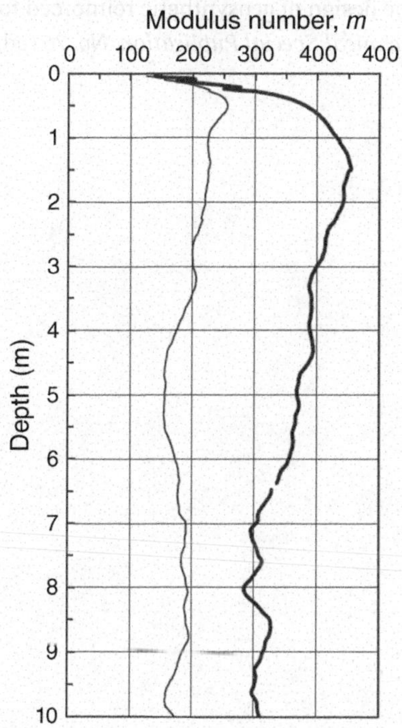

References

Arulrajah, A., Abdullah, A., Bo, M.W., and Bouazza, A. (2009). Ground improvement techniques for railway embankments. *Proceedings of the Institution of Civil Engineers, Ground Improvement* 162: 3–14.

Barksdale, R.D. and Bachus, R.C. (1983). Design and construction of stone columns, FHWA/R83/026.

Chen, Q. (2007). An experimental study on characteristics and behavior of reinforced soil foundation. Ph.D. Dissertation. Louisiana State University.

Gazetas, G., Tassoulas, J.L., Dobry, R., and O'Rourke, M.J. (1985). Elastic settlement of arbitrarily shaped foundations embedded in half-space. *Géotechnique* 35 (3): 339–346.

Ghazavi, M. and Nazari Afshar, J. (2013). Bearing capacity of geosynthetic encased stone columns. *Geotextiles and Geomembranes* 38: 26–36.

Hamidi, B. and Varaksin, S. (2015). Dynamic compaction and dynamic surcharging at Dubai's Palm Jumeira Sewage Treatment Plants. In: *Ground Improvement Case Histories: Compaction, Grouting, and Geosynthetics* (ed. B. Indraratna, J. Chu, and C. Rujikiatkamjorn), 309–335. Elsevier.

Han, J. (2015). *Principles and Practices of Ground Improvement*. Wiley.

Imam, R., Zarei, M., and Ghafarian, D. (2021). Relative contribution of various deformation mechanisms in the settlement of floating stone column-supported foundations. *Computers and Geotechnics* 134: 15.

Meyerhof, G.G. and Hanna, A.M. (1978). Ultimate bearing capacity of foundations on layered soils under inclined load. *Canadian Geotechnical Journal* 15 (4): 565–572.

Tatsuoka, F. and Watanabe, K. (2015). Design, construction, and performance of GRS structures for railways in Japan. In: *Ground Improvement Case Histories: Compaction, Grouting, and Geosynthetics* (ed. B. Indraratna, J. Chu, and C. Rujikiatkamjorn), 657–692. Elsevier.

Tatsuoka, F., Tateyama, M., Uchimura, T., and Koseki, J. (1997). Geosynthetic-reinforced soil retaining walls as important permanent structures. *Mercer Lecture, Geosynthetic International* 4 (2): 81–136.

Wayne, M.H., Han, J., and Akins, K. (1998). The design of geosynthetic reinforced foundations. In: *Design and Construction of Retaining Systems, ASCE Geo-Institute Geotechnical Special Publication, No. 76* (ed. J.J. Bowders et al.), 1–18. Reston, VA: ASCE Press.

Topic C

Semi-Deep Foundations

Section V

Skirted, Floating, and Shell Foundations

10

Types and Construction Approach

Chapter Highlight

- Introduction to Semi-deep Foundations
- Major Categories
  - Well Foundations
  - Skirted Foundations
  - Floating Foundations
  - Spudcan Foundations
  - Bucket Foundations
  - Shell Foundations
- Case Histories
  - Well Foundation for Transmission Towers
  - Floating Foundation of Savu Tower, Singapore (Wong et al. 1995)
  - Bucket Foundation of Dogger Bank Wind Farm, North Sea (Gres et al. 2016)
  - Hybrid Bucket Foundation (HBF) as a Foundation for Offshore Wind Turbines (OWT) (Zhang et al. 2016)
  - A Review of Offshore Foundations (Randolph et al. 2005)
  - Hypar Shell Foundation of St. Vincent's Chapel, Mexico (Melaragno 2012)

10.1 Introduction

Substructures are typically divided into two categories: shallow foundations and deep foundations. Shallow foundations are designed to transfer structural loads to the underlying soil layers at relatively small depths (D_f) compared to the foundation width (B). However, due to the rapid urbanization and the construction of heavy, tall, and important structures, such as bridges, marine structures, high-rise buildings, and towers, there has been an increased reliance on deep foundations ($D/B \geq 10$), such as piles, drilled piers, and shafts. However, the application of deep foundations can be limited by factors such as the availability of construction equipment, cost considerations, installation time, and environmental concerns associated with techniques like drilled shafts and pile driving (Elhakim 2005). As an alternative to shallow and deep foundations, there is a relatively new and novel type of foundation known as "semi-deep foundations," which is gaining popularity as an optimal alternative to both shallow and deep foundations. These foundations offer improved performance and are particularly suitable for supporting structures at accessible depths. Figure 10.1 illustrates the three types of foundations: shallow, semi-deep, and deep.

The depth-to-width ratio of semi-deep foundations can reach four in the different soil types. As a result, they can modify the bearing capacity and satisfy most demands of foundation design optimization. Semi-deep foundation is a very suitable alternative to shallow or deep foundations in situations where (i) deep problematic soil, such as loose deposits, exists beneath the foundation, (ii) it is not practical or possible to perform any soil improvement, and (iii) the cost of deep foundations would be excessive and unreasonable for the specific project (Rezazadeh and Eslami 2017a).

Advanced Foundation Engineering: Principles, Performance and Prospects, First Edition. Moses Karakouzian and Abolfazl Eslami.

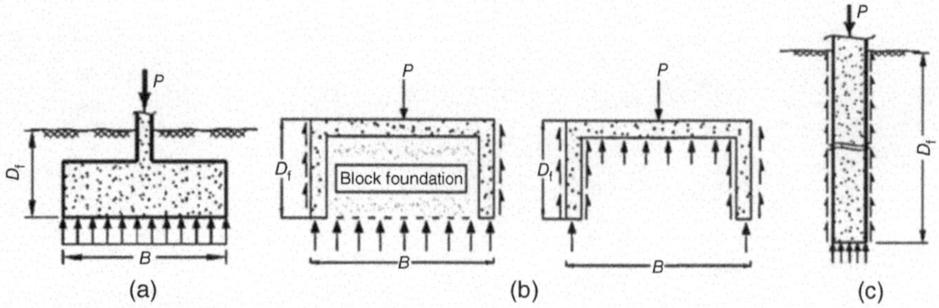

Figure 10.1 Main foundation categories: (a) shallow, (b) semi-deep, and (c) deep.

10.2 Geotechnical Background

A semi-deep foundation offers a solution that overcomes the limitations of other foundation types and can be a cost-effective and practical choice for various construction projects. By increasing the foundation depth, a semi-deep foundation provides the necessary stability and bearing capacity to withstand a combination of loading conditions. These foundations are particularly suitable for offshore and onshore projects where structures such as wind turbines, storage facilities, bridge piers, and other structures need to support significant and combined vertical, horizontal, and moment (VHM) loads. The concept of semi-deep foundations improves the performance of the foundation under these loading conditions, as depicted in Figure 10.2. By utilizing the advantages of semi-deep foundations, engineers can design and construct robust foundations that ensure the structural integrity and stability of the supported structures.

When designing foundations to withstand combined vertical, horizontal, and moment (VHM) loading, it is crucial to ensure the stability of the structure under varying loads and prevent excessive displacements. A successful design should guarantee the structure's safety even under maximum loads and ensure that the superstructure does not experience excessive movement. Bearing failure is typically evaluated using plasticity theory, while settlement is assessed using elasticity principles. Therefore, it is essential to find a design solution that satisfies both criteria by considering load–displacement curves or VHM failure envelopes. This process involves determining the most appropriate and cost-effective foundation configurations, as well as understanding the vertical bearing capacity. The shape, dimensions, and embedment depths of foundations, along with the characteristics of the surrounding soil and the applied load combinations, all influence the geotechnical and structural performance of foundations. Recent studies have shown that for foundations with significant width and depth-to-width ratios (D/B) ranging from 1 to 4, the kinematic mechanisms and failure modes differ between shallow and deep foundations. Accordingly, in semi-deep foundations, in addition to the D/B, other factors, such as the depth-to-width ratio, contact area, soil confinement, and the geometry and formation of a block foundation, all influence

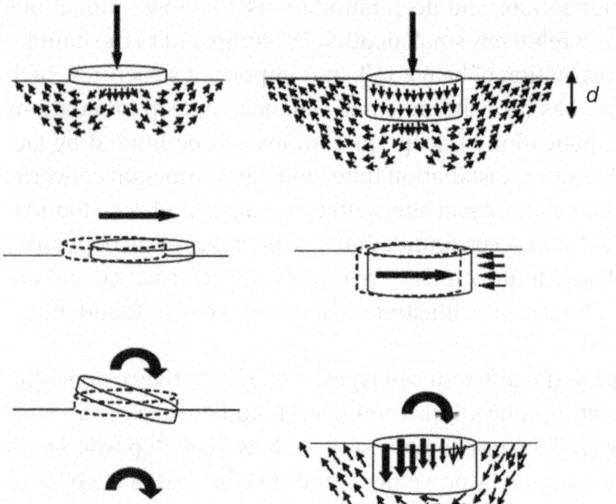

Figure 10.2 Effect of depth factor in enhancing the capacity of a foundation under VHM loading (Randolph et al. (2005) / Taylor & Francis Group).

the load–settlement behavior and must be considered in the design. Therefore, the D/B factor alone should not be the sole criterion for defining the foundation type (Yun and Bransby 2007).

In brief, the main specifications of a semi-deep foundation can be categorized as below:

1) Suitable choice for the foundation of structures sustaining large VHM loading, especially in marine projects facing wind, waves, and earthquake loads.
2) Possibility to use in loose deposits, like normally consolidated clay with unsatisfactory properties.
3) Foundation depth, shaft, toe, surrounding soils, and confined soil by the foundation significantly affect geotechnical performance improvement. It comprises settlement reduction, bearing capacity enhancement, and stability compared with shallow foundations.
4) The interaction of the block foundation with the surrounding soil (skin friction and passive lateral soil resistance) is crucial for stability against lateral and uplift loadings.

10.3 Categories

As depicted in Figure 10.3, the main categories of semi-deep foundations are:

- Well foundations
- Skirted foundations
- Floating foundations
- Bucket foundations or suction caissons
- Spudcans
- Shell foundations

10.4 Well Foundations

When dealing with challenging soil conditions where traditional ground modification methods are impractical, a well foundation can be a suitable choice. There are generally three types of well foundations based on the materials used. The first type involves physically and chemically treated materials, where a combination of geomaterial and a cohesive agent like cement and lime is used (Figure 10.4a). This approach enhances the properties of the foundation material. The second category consists of well foundations made from plain or reinforced concrete (Figure 10.4b). These foundations provide

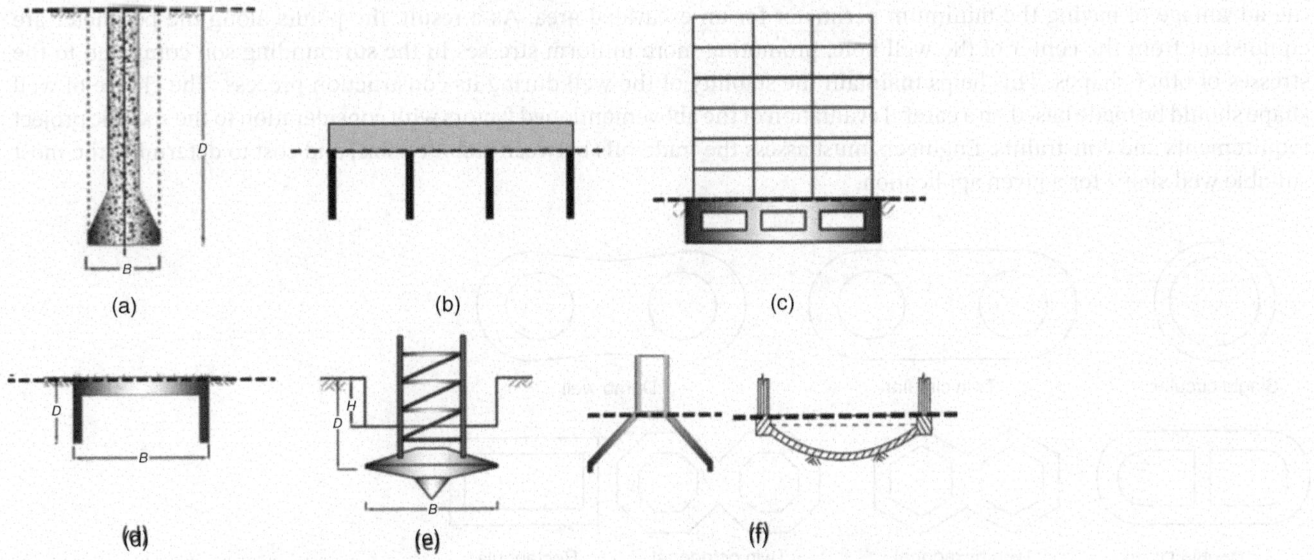

Figure 10.3 Various types of semi-deep foundations: (a) well foundation, (b) skirted foundation, (c) floating foundation, (d) bucket foundation, (e) spudcan, and (f) shell foundation (after Eslami et al. 2020) This date should be 2018.

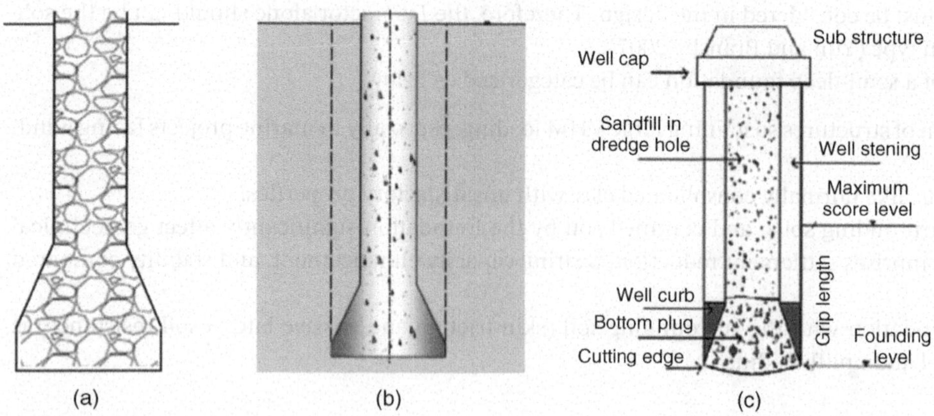

Figure 10.4 Different types of well foundations: (a) physically and chemically treated (mortar-treated geomaterial), (b) plain or reinforced concrete, and (c) sandfill.

strength and durability and are commonly used in various applications. The third type is sandfill well foundations. In this method, a hollow shaft or a box is submerged in the ground until it reaches a hard or dense layer in the soil (Gupta 2007). Creating a cutting edge at the end of the box facilitates its integration with the soft soil. Once the desired depth is reached, the materials inside the box are removed, and concrete is poured after placing reinforcement into it (Figure 10.4c). Each type of well foundation offers advantages depending on the specific soil conditions and project requirements. The choice of the appropriate well foundation type is crucial for achieving stability and ensuring the long-term performance of the structure.

In situations where the pull-out capacity of the foundation is a primary concern, a well foundation can be a simple and cost-effective solution. Due to its relatively low depth-to-width ratio (D/B), it is feasible to widen the end of the well. By widening the bottom portion of the well, the foundation can engage a larger soil area, thereby increasing the frictional resistance against pull-out forces. This utilization of the soil weight enhances the overall stability of the well foundation and improves its ability to resist upward or horizontal forces. Proper engineering analysis and design are essential to ensure the well foundation's stability and optimize its pull-out capacity based on the specific soil conditions and project requirements. Circular, Twin circular, Dumb well, Double-D, Twin hexagonal, Twin octagonal, and Rectangular are prevalent well formations depicted in Figure 10.5.

When selecting a well shape, several key factors should be considered. These include the size of the well and its stability, the cost of the excavation, and the vertical and horizontal forces and moments exerted on the well. A circular well offers the advantage of having the minimum perimeter for an excavated area. As a result, the points along the perimeter are equidistant from the center of the well hole, promoting more uniform stresses in the surrounding soil compared to the stresses of other shapes. This helps maintain the stability of the well during its construction process. The choice of well shape should be made based on a careful evaluation of the abovementioned factors with consideration to the specific project requirements and constraints. Engineers must assess the trade-offs between stability, size, and cost to determine the most suitable well shape for a given application.

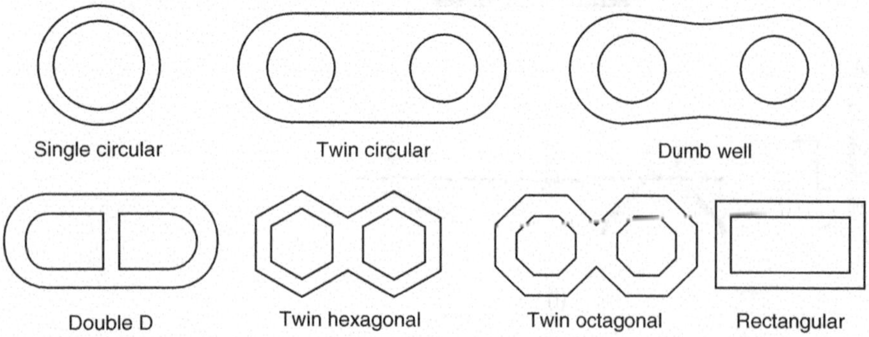

Figure 10.5 Different shapes of wells.

Case Study 10.1

Well Foundations for Transmission Towers

Transmission towers are tall structures, usually made from steel, constructed to support overhead power lines. These towers are built in various shapes and sizes. Typical heights range from 15 to 55 m. Figure 10.6 illustrates some typical transmission towers, including their height and typically carried voltage.

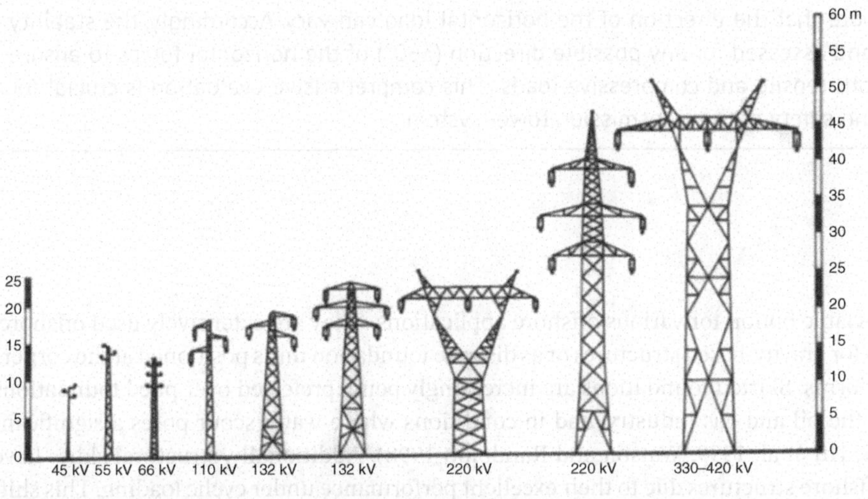

Figure 10.6 Typical transmission towers.

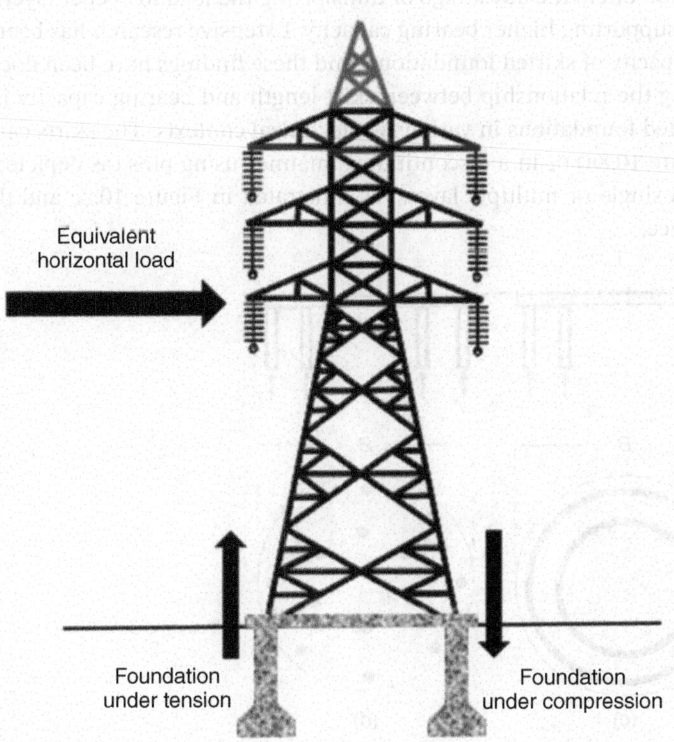

Figure 10.7 Typical transmission tower supported on belled well foundations.

(Continued)

Case Study 10.1 (Continued)

Transmission towers are typically lightweight structures as far as vertical loads are concerned. However, due to their significant height, they are subjected to high-magnitude horizontal loads, which necessitates ensuring their stability against overturning moments and tensile forces. Deep foundations are commonly used to achieve this stability. In cases where accessibility and transportation to the construction site pose challenges, belled well semi-deep foundations can be a suitable alternative. These foundations are designed to bear both compressive and tensile loads, providing the necessary stability for transmission towers. Figure 10.7 illustrates a typical transmission tower supported by belled well foundations. It is important to note that the direction of the horizontal load can vary. Accordingly, the stability of the foundation must be analyzed and assessed for any possible direction (360°) of the horizontal forces to ensure their favorable performance under both tensile and compressive loads. This comprehensive evaluation is crucial for maintaining the structural integrity and safety of the transmission tower system.

10.5 Skirted Foundations

Skirted foundations have emerged as a viable option for various offshore applications. They are extensively used offshore, either as standalone foundation systems for gravity-based structures or as discrete foundation units positioned at the corners of jacket structures and tension leg platforms. Skirted foundations are increasingly being preferred over piled foundations in offshore structures and facilities for the oil and gas industry, and in conditions where water scour poses a significant challenge (Bransby and Randolph 1998; Hu et al. 1999; Watson and Randolph 1998). Additionally, structural skirts have gained popularity in supporting large offshore structures due to their excellent performance under cyclic loading. This shift is driven by the potential cost savings achieved through reduced material usage and installation time. Structural skirts offer an alternative method to enhance bearing capacity and minimize footing settlement.

Compared to a surface foundation, the skirted foundation offers the advantage of transferring the load to deeper layers of soil, which are typically stronger and more capable of supporting higher bearing capacity. Extensive research has been conducted on the effects of skirt length on the bearing capacity of skirted foundations, and these findings have been documented in the literature (Tripathy 2013). Understanding the relationship between skirt length and bearing capacity is crucial in optimizing the design and performance of skirted foundations in various geotechnical contexts. The skirts can be implemented in a continuous form (as shown in Figure 10.8a) or in a discontinuous manner using pins (as depicted in Figure 10.8b). Furthermore, skirts can be installed in single or multiple layers, as illustrated in Figure 10.8c and d, respectively, providing flexibility in design and performance.

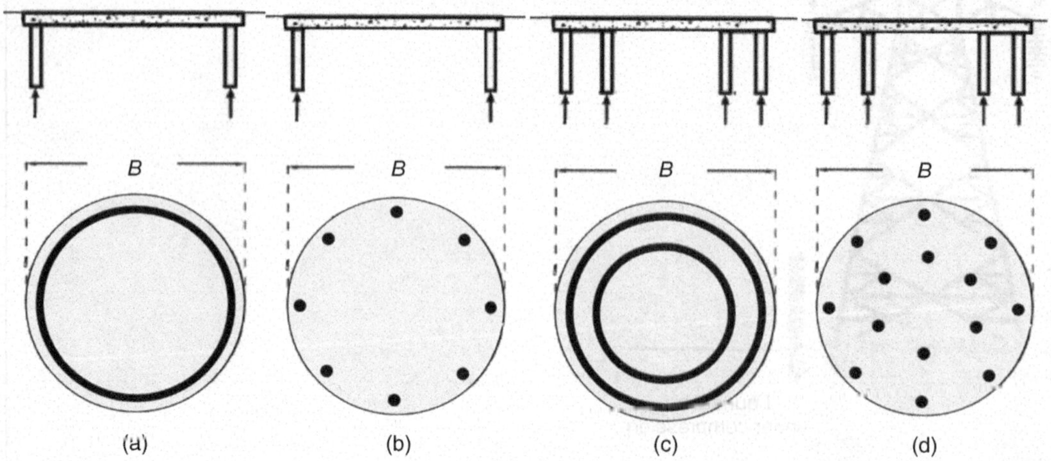

Figure 10.8 Skirt types: (a) continuous, (b) discontinuous, (c) multi continuous, and (d) multi discontinuous.

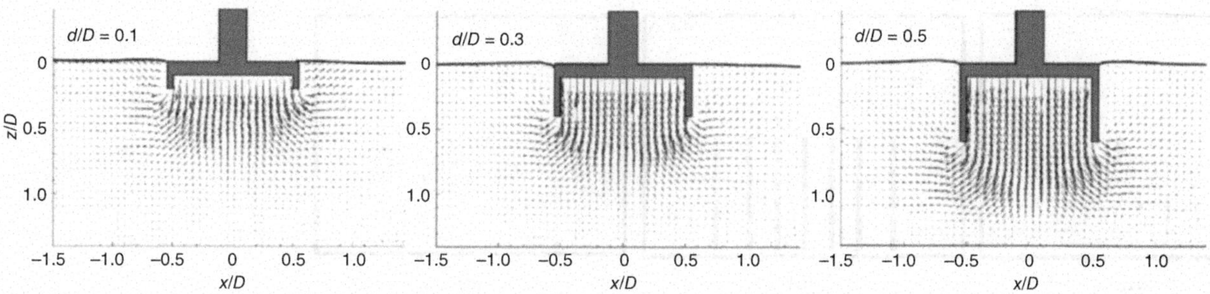

Figure 10.9 Normalized soil velocity vectors for foundations subjected to compression displacement (Mana et al. 2012 / Emerald Publishing Limited).

Skirted foundations are constructed by combining a shallow foundation, which has a certain width (B) and is located on the soil surface, with vertical skirts that extend both along the perimeter and internally within the raft foundation. These skirts are designed to penetrate the soil deposits to a depth that is typically twice the width of the foundation ($D/B = 2$). By incorporating the raft, skirts, and confined soil, the foundation forms a block structure that enhances its overall strength and stiffness, enabling it to efficiently bear the loads from the substructure (Byrne and Houlsby 2003; Tripathy 2013). This technique offers advantages such as not requiring soil excavation, making it suitable for construction in deep water. Figure 10.9 illustrates the effect of an increase in skirt depth on altering the soil movement and subsequently failure mechanism.

Al-Aghbari and Mohamedzein (2004a, b) conducted studies on the use of structural skirts to improve the performance of strip footings on sandy soils. Their findings indicated that structural skirts enhanced the bearing capacity and reduced settlement by up to three times. They also identified various factors that influence the improvement in bearing capacity achieved by using these skirts. To account for the combined effects of these factors, a parameter known as the skirt factor was introduced. It is suggested that this factor can be incorporated into the general equation for determining the ultimate bearing capacity of shallow foundations. By considering the skirt factor, engineers can better estimate the enhanced bearing capacity provided by structural skirts in their foundation design calculations.

The skirted foundation approach is based on attaching structural skirts to the edges of shallow foundations. This technique offers advantages such as not requiring soil excavation, making it suitable even in areas with a deep waters. The use of structural skirts to increase the effective depth of foundations has been employed for a considerable period, especially in marine environments where water scour is a concern. This improvement method does not require soil excavation; thus, it is not limited by the presence of a high water table. Structural skirts fixed to the edges of shallow foundations have been used for a significant time, principally to increase the effective depths of foundations in marine and other situations where water scour may be a problem (Bransby and Randolph 1998; Hu et al. 1999; Watson and Randolph 1998).

10.6 Floating Foundations

In recent years, the floating foundation technique has emerged as a promising method for foundation design and construction optimization. This approach involves excavating the soil and increasing the depth of the footing, resulting in improved performance in terms of strength and stiffness of the substructure system. Additionally, the floating foundation can be combined with a pile group for enhanced load-bearing capacity (Golder 1976). The floating foundation method is particularly suitable in two main scenarios. Firstly, it is applicable when the substructure requires high bending rigidity due to the significance and weight of the superstructure it supports. In such cases, the floating foundation can provide the necessary stability and support. Secondly, the floating foundation is well-suited for situations where the soil layer consists of deep soft deposits, which pose a high risk of settlement. In such conditions, the use of piles becomes necessary to mitigate settlement, and the floating foundation method can be a suitable alternative (see Figure 10.10). By implementing the floating foundation technique, engineers and designers can optimize foundation design, enhance structural performance, and address challenges posed by soft soil deposits and heavy superstructures.

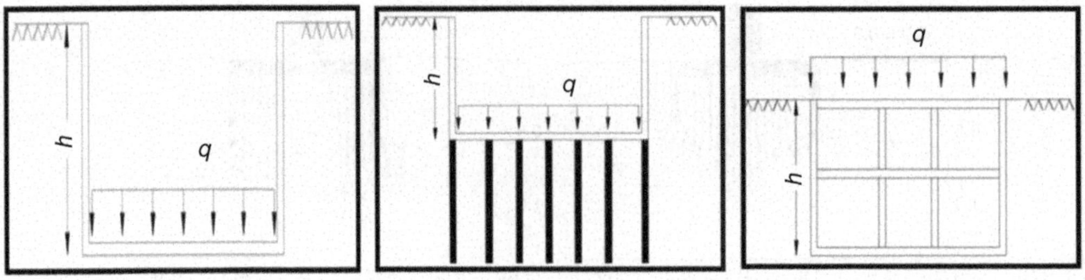

Figure 10.10 Types of compensated foundations: (a) raft, (b) PRF, and (c) box.

The concept of floating foundations can be traced back to the construction of Albion Mill in London around 1780. This early example of a floating foundation utilized approximately 50% floatation, meaning that the excavation reduced the increase in vertical effective stress $\Delta\sigma_z'$ by 50%. Analogous to a ship floating in water, the weight of the entire building can be counterbalanced by the weight of the excavated soil. In an ideal scenario where the dead and live loads from the structure remain constant, it would be theoretically possible to select an excavation depth that would result in $\Delta\sigma_z' = 0$. However, due to variations in live loads over time and fluctuations in groundwater table elevation, perfect balance between weights is not always achievable. The presence of groundwater creates an upward buoyancy force on submerged and dewatered basements, further reducing $\Delta\sigma_z'$. To ensure the stability and performance of a floating foundation, it is crucial to consider the worst possible loading conditions. Engineers must select an appropriate depth of excavation that guarantees $\Delta\sigma_z'$ is always greater than zero, thus maintaining an adequate level of vertical effective stress in the soil. This ensures that the foundation remains stable and can safely support the structure even under varying loads and groundwater conditions.

When analyzing floating foundations, conducting parametric studies is often beneficial to assess the effects of various parameters. Small changes in loads, groundwater levels, or other factors can sometimes have a significant impact on $\Delta\sigma_z'$. It is crucial to account for these potential variations to ensure the stability and performance of the foundation. The most significant problems with floating foundations usually occur during construction. The most critical phase is when the excavation is open, but construction has not yet begun. These excavations may lack stability and require special precautions to prevent failures or collapses.

In cases where the existing soil is not sufficiently stable, a top-down construction approach can be employed for the excavation. Unlike the traditional bottom-up method, top-down construction involves building permanent structures from the top downward, using deep excavations. Box foundations can increase rigidity, leading to reduced differential settlements. Additionally, locating the base level for seismic analysis at a deeper depth results in a significant decrease in shear forces and moments. These factors contribute to improved performance and stability of the foundation system.

Case Study 10.2

Floating Foundation of Savu Tower, Singapore

The Savu Tower is located in Singapore. This building has 33 floors with a height of 128 m, with 4 of the 33 floors located below the ground surface as substructure, as shown in Figure 10.11 (Wong et al. 1995). The average pressure applied by the main building of the tower is equal to 540 kPa, and the net pressure applied to the foundation is equal to 283 kPa. The retaining wall of the infrastructure includes piles. Four boreholes were used to obtain soil conditions to a depth of 40–70 m. The soil includes fine to coarse sand from the ground to a depth of about 4 m and Bouldery clay to a depth of more than 70 m. Bouldery clay is a hard, silty, and sandy clay. The groundwater level is at a depth of 2 m.

The foundation of the tower building is a raft-floating foundation. The dimensions of the foundation are 40 m by 68 m, and it has a thickness of 2.8 m, which is located at a depth of 16 m below the ground. The performance of the tower building foundation was observed after completing the foundation using instrumentation. The maximum settlement was 14 mm when the superstructure was completed.

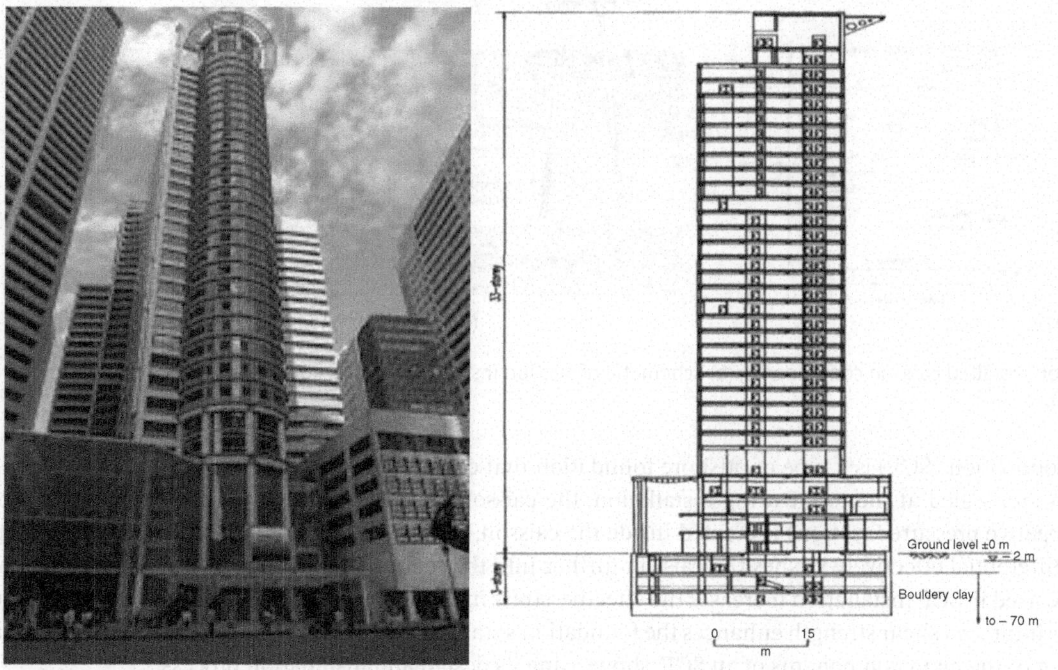

Figure 10.11 The Savu Tower infrastructure section (Wong et al. 1995).

10.7 Bucket Foundations

In recent years, offshore foundations have witnessed increased diversity, with the emergence of concrete or steel bucket foundations. These foundations serve as anchors for floating platforms or as the base for various sea bottom structures, typically in smaller sizes. Originally developed for the offshore oil and gas industry, bucket foundations have found applications in wind turbine construction as well. The loading conditions for these foundations involve a vertical load resulting from the slender construction, along with horizontal forces that generate significant overturning moments Barari (2012). Figure 10.12 provides a visual representation, showcasing both a schematic and an actual view of typical suction bucket foundations.

(a) (b)

Figure 10.12 Bucket foundation: (a) schematic view (Salem et al. 2021 / MDPI / CC BY 4.0), (b) actual view.

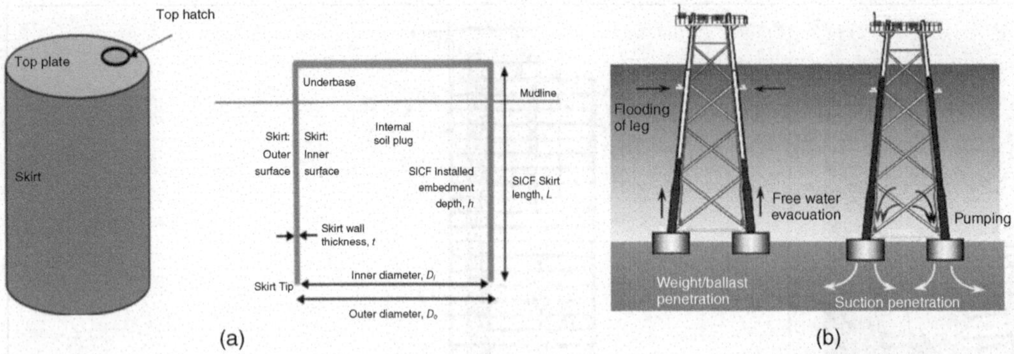

Figure 10.13 (a) Suction installed caisson components, (b) schematic of suction installation of bucket and caissons (adapted from Erbrich and Tjelta 1999).

A suction caisson foundation (SCF) is a type of offshore foundation that consists of a large steel cylinder. The cylinder is open at the bottom and sealed at the top. During installation, the caisson is initially penetrated the seabed under its own weight. Then, negative pressure (suction) is created inside the caisson, causing a pressure differential across the top plate. This pressure differential effectively pushes the caisson further into the seabed, embedding it to the desired depth. The suction technique used in SCF installation increases the effective stress in the surrounding soil, resulting in an increase in shear strength. This improved shear strength enhances the foundation's stability and load-bearing capacity. Figure 10.13a provides an illustration of the main components of an SCF, showcasing its design and installation process.

The diameter of SCFs can vary from 750 mm to a few meters. Contrary to pile foundations, SCFs have no difficulty and noise in installation. Moreover, due to their large diameters and burial skirts, these types of foundations can sustain enormous tensile and lateral loads.

During the installation of a suction caisson foundation (SCF), the foundation is gradually lowered to the seabed while the skirts penetrate the soil. The goal is to reach a sufficient depth where the soil can provide the necessary bearing capacity to support the self-weight of the caisson. Throughout this process, open valves in the lid allow water to escape. Once the self-weight skirt penetration has reached a point where a seal is formed between the skirt and the soil, the water inside the caisson is evacuated using a pumping system. By removing the water, a differential pressure (suction) is created across the caisson lid, which drives the caisson further into the seabed. In sandy soils, seepage flow develops around the caisson skirt. This seepage flow has the effect of reducing the effective stresses at the skirt tip (Andersen et al. 2008; Houlsby and Byrne 2005). This reduction in effective stress is crucial for a successful installation, particularly in dense sands where high skin friction is desired for stability. Figure 10.13b provides a schematic representation of this installation process, highlighting the water escape and suction-induced penetration of the caisson into the seabed.

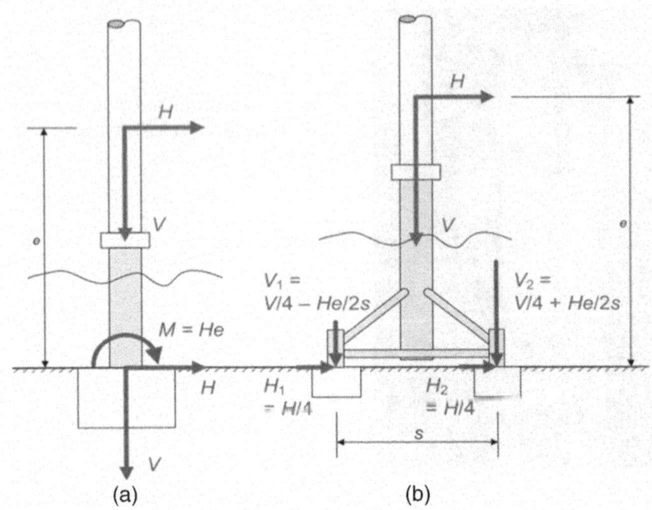

Figure 10.14 Typical structural configurations of SCF usage: (a) mono-caisson, (b) multi-caisson (assuming a tetrapod design).

Caisson foundations installed by suction are used with two main substructure configurations (Figure 10.14), as detailed below with their respective loading regimes.

- **Mono-caisson:** Where a single large diameter caisson (Figure 10.14a) supports the substructure. The critical design factors are typically the caisson response and capacity under the influence of overturning moments.
- **Multi-caisson:** Where three or four caissons typically support the substructure. Figure 10.14b shows a tetrapod-style substructure, but tubular jackets are more common. Typically, the caissons' vertical responses and capacities are the key design factors, specifically the "upwind" caisson, which may experience tension uplift.

In Figure 10.14, blue arrows are the resultant applied structural loads. Red arrows are the loads on the foundation (for case b, the loads are for a four-footed jacket, assuming equal distribution of horizontal load and no moment transferred to individual foundations).

Case Study 10.3

Bucket Foundation of Dogger Bank Wind Farm, North Sea (Gres et al. 2016)

Dogger Bank Wind Farm (DB) is a collection of offshore wind farms in the North Sea built 125–290 km off the east coast of Yorkshire, England. The meteorological mast is one of two masts supported by the Mono Bucket (MB) installed at the DB and is positioned west of the site. Figure 10.15a shows an on-site shot from the installation vessel. The MB is a welded steel foundation separated into four modules: shaft, webs, lid, and skirt. The foundation's main structure is made up of webs and shafts. The lid and skirt create horizontal and vertical bases, respectively. The MB is set up using a mix of self-weight and suction.

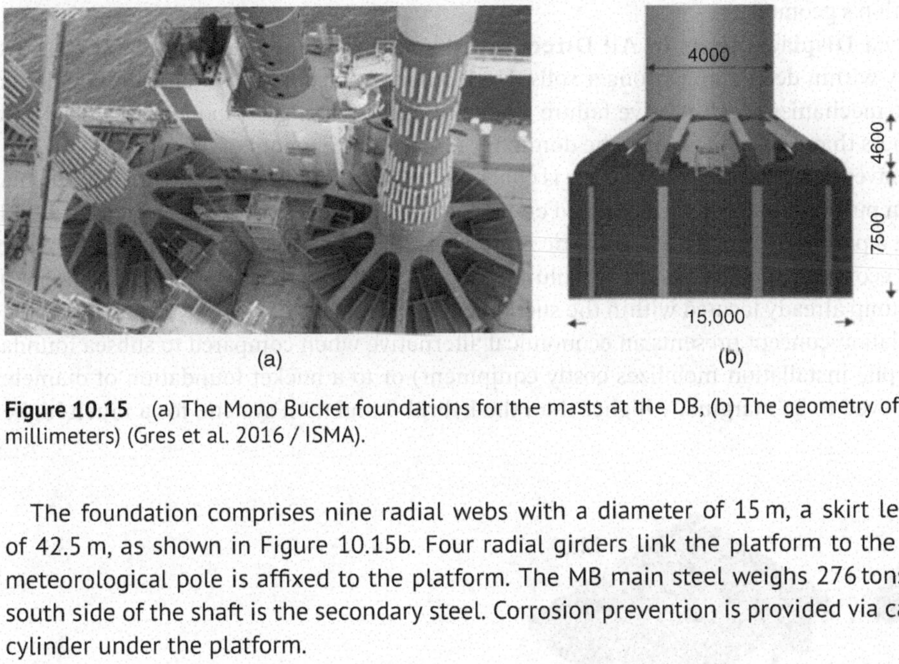

Figure 10.15 (a) The Mono Bucket foundations for the masts at the DB, (b) The geometry of the foundation (all dimensions in millimeters) (Gres et al. 2016 / ISMA).

The foundation comprises nine radial webs with a diameter of 15 m, a skirt length of 7.5 m, and a shaft length of 42.5 m, as shown in Figure 10.15b. Four radial girders link the platform to the MB shaft. A 91.5-m three-legged meteorological pole is affixed to the platform. The MB main steel weighs 276 tons. The boat landing affixed to the south side of the shaft is the secondary steel. Corrosion prevention is provided via cathodic protection attached to the cylinder under the platform.

10.7.1 Hybrid Suction Bucket Foundation

Subsea structures rest on the seabed and connect subsea wells to a leading platform via various flow lines, jumpers, and gathering lines. The wells, subsea vegetations, manifolds, sleds, flowlines, and gathering lines constitute a subsea system. They are usually founded on a mud mat foundation, which forms an integral part of the structure. The mud mat foundation is typically made from steel and may have dimensions ranging from 5 to 15 m in the plan. By nature, subsea structures have a low self-weight, and seabed vertical bearing capacity is typically not the governing design criterion. However, they often need to withstand large horizontal forces, sometimes from currents, but more often from loads applied by the connected equipment. One example is the loads generated by the thermal expansion of pipelines. As these loads are applied at various locations on the subsea structure, they usually result in the overturning moment, and torsion applied on the mud mat foundation. To enhance the horizontal capacity of the foundation, and to a lesser extent the moment capacity, different options can be considered. One option is the addition of a peripheral skirt to the mud mat foundation. This skirt can help increase the foundation's resistance to horizontal forces. Another option is the incorporation of piles at each corner of the mud mat foundation. This approach is typically chosen when minimizing foundation displacements, which is crucial from a serviceability standpoint (Guadin et al. 2011). By employing these measures, the subsea structures can effectively withstand the significant horizontal forces and moments exerted upon them, ensuring their stability and integrity in offshore environments.

The hybrid foundation system described herein proposes an alternative to increase conventional mud mat foundations' capacities and limit displacements in all degrees of freedom by equipping a skirted foundation with one (or multiple) suction caisson(s). Figure 10.16 presents the general concept of the hybrid suction bucket foundation.

In addition to increasing the combined bearing capacity, the caisson enables self-installation and offers the potential for suction-induced preloading by applying suction within the caisson compartment. A wide range of applications can be considered, notably in deep waters, where self-installation significantly reduces installation costs. It also offers potential in scenarios in which an increase of vertical bearing capacity by preloading is required. The expected benefits include (Guadin et al. 2011):

1) **Suction Installation:** The suction caisson can assist in the foundation installation, reducing the need for additional equipment. Current solutions for additional horizontal capacity often involve corner piles, which are expensive and require specialist operation. The suction installation is expected to be particularly advantageous in deep water.
2) **Additional Vertical, Horizontal, and Moment Capacities:** These are primarily attributed to the suction caisson's inclusion, which leads to (i) the mobilization of more profound and stronger soil and (ii) the potential change in failure mechanism related to the foundation's geometry.
3) **Increased Stiffness and Reduced Displacements in All Directions:** Limited settlement is expected as the caisson will mobilize bearing capacity within deeper and stronger soils. Horizontal displacements are also limited due to the possible change in the failure mechanism with passive failure (the dominant resisting component for the hybrid foundation) having a higher stiffness than shearing failure (the dominant resisting component of the skirted mat).
4) **Ease of Retrieval:** Retrieval of conventional subsea foundations is challenging due to the suction generated at the mat invert during pull-out, resulting in pull-out forces that can exceed crane capacity. While solutions are available to limit suction generation, such as using a perforated mat or an eccentric pull-out, reversing the pumps in a caisson appears to be a more straightforward and economical alternative. This solution only requires the mobilization of one crane of limited capacity, along with the pump already located within the suction caisson's lid.
5) **Reduced Cost:** The hybrid foundation concept presents an economical alternative when compared to subsea foundations reinforced by corner piles (pile installation mobilizes costly equipment) or to a bucket foundation of diameter and length, whose capacity is believed to be marginally higher than the hybrid foundation system, for a much higher manufacturing cost.

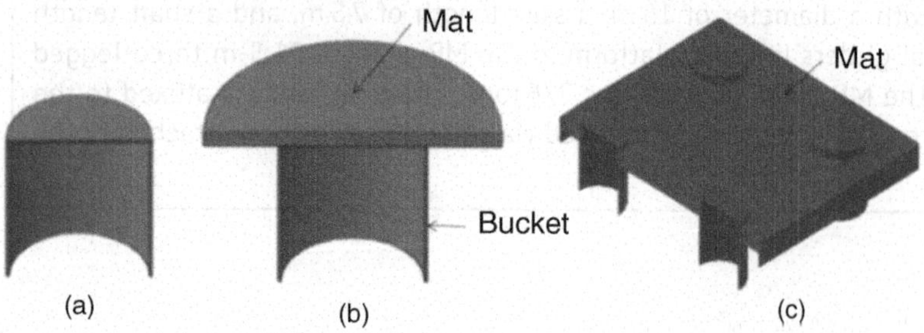

Figure 10.16 Bucket foundation types: (a) ordinary bucket foundation (OBF), (b) a combination of a mat foundation with an ordinary bucket (hybrid bucket foundation; HBF), and (c) multi buckets under square mat foundation (Kim et al. 2020).

Case Study 10.4

Hybrid Bucket Foundation (HBF) as a Foundation for Offshore Wind Turbines (OWT) (Zhang et al. 2016)

A mono-caisson foundation and a pre-stressed concrete arc, which act as a transmission structure, manufacture a shape of a typical hybrid bucket foundation (HBF) as shown in Figure 10.17. Implementing seven rooms with a honeycomb arrangement allows HBF more reliability during transfer and installation. In the presented case, a 3 MW HBF of 2100 tons with a remaining structure of 710 tons was selected as a prototype. The HBF model was made of steel on a scale of 1:10 and experimented on. The outer diameter (*D*), or experimented hybrid bucket, was about 9 m, and its skirt length (*L*) was 0.9 m. The inside of the bucket was divided into seven rooms by some plate stiffeners in which six circumferential rooms had the same size, and the middle one was larger in order to load HBF, a steel tube, as a part of the wind turbine tower, was connected to the top of the bucket reinforced by six triangular stiffener plates. The dimensions of the model are listed in Table 10.1.

Figure 10.17 The 2.5 MW offshore wind turbine supported by composite bucket foundation.

Table 10.1 Details of the model dimensions.

Parameter	Prototype	Model
Diameter D (m)	35	3.5
Skirt length L (m)	9	0.9
L/D	3.89	3.89
Skirt and bulkhead thickness t (m)	0.03	0.008
D/t	1166.7	437.5
Tip area (m^2)	6.447	0.1718
Net area (m^2)	955.178	9.4445
Total weight (kg)	2,810,000	2810.7

(Continued)

Case Study 10.4 (Continued)

Along the Jiangsu coast, a large artificial pool was excavated and used for HBF tests. The deposits in the test site were saturated silty clay brought from a wind farm located on the coast of the Jiangsu province of China. A tank was used to fill by soil layer by layer. The total height of the deposits was 1.2 m, and the top of the soil was leveled precisely. The soil was fully saturated for at least six months. The properties of the saturated clay were determined by soil laboratory tests. The soil's saturated density was 1806 kg/m^3, and the water content was 33.5%. Its liquid index and plasticity index were 11.75 and 1.19, respectively. The compressive modulus of the soil was 3.69 MPa, and its cohesion strength and internal friction angle were 11.84 kPa and 7.14°. As the skirt length of the HBF was 0.9 m, the sand specification below 1 m of the soil had no remarkable effect on the HBF technique's performance assessment.

10.8 Spudcan

Self-elevating mobile jack-up units are commonly used for offshore drilling, workovers on fixed platforms, and production support activities in shallow to moderate water depths. These units have a distinct design consisting of a buoyant triangular platform supported by three independent truss-work legs (Figure 10.18). The triangular platform serves as the main deck where equipment and personnel are located during operations. The weight of the deck and equipment is distributed fairly evenly to maintain stability. Each leg is equipped with a rack and pinion system, which allows the legs to be raised or lowered through the deck. This self-elevating feature enables the jack-up unit to adjust its height according to the water depth, providing a stable working platform above the water surface. By raising the legs, the jack-up unit can position the platform above the water, allowing drilling or other operations to be carried out. Once the operations are complete, the legs can be lowered to rest on the seabed, providing stability and support to the unit during stationary periods. The independent leg design allows the jack-up unit to operate in a wide range of water depths and adapt to uneven seabed conditions. It offers flexibility and mobility, making it a preferred choice for offshore activities in shallow to moderate water depths.

The foundations of independent-leg jack-up platforms, also known as spudcans, are designed to provide stability and support to the jack-up unit on the seabed. Spudcans have a roughly circular plan and resemble large, inverted cones. They typically have a shallow conical underside, sloping at an angle of around 15–30° to the horizontal, and feature a sharp protruding spigot. The size and shape of spudcans can vary depending on the manufacturer and the specific rig. In larger jack-up units, the diameter of the spudcans can exceed 20 m. Some spudcans may have noncircular shapes to optimize their performance and bearing capacity, as shown in Figure 10.19. Alternatively, some jack-ups use a mat support system that connects all of the legs and is comparable in size to the hull of the jack-up unit. These mats are designed to increase the bearing area of the foundations, which is particularly important in soft sediment conditions where soil stability is a concern. Jack-up units are available in different sizes to accommodate various water depths and operational requirements.

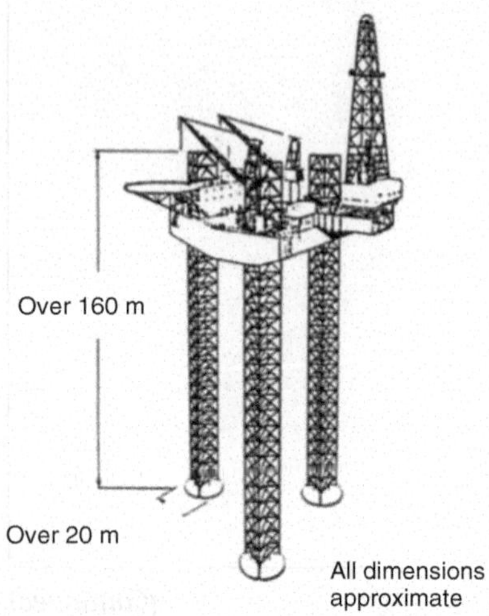

Figure 10.18 Typical jack-up unit used today (Cassidy et al. 2009, after Reardon 1986).

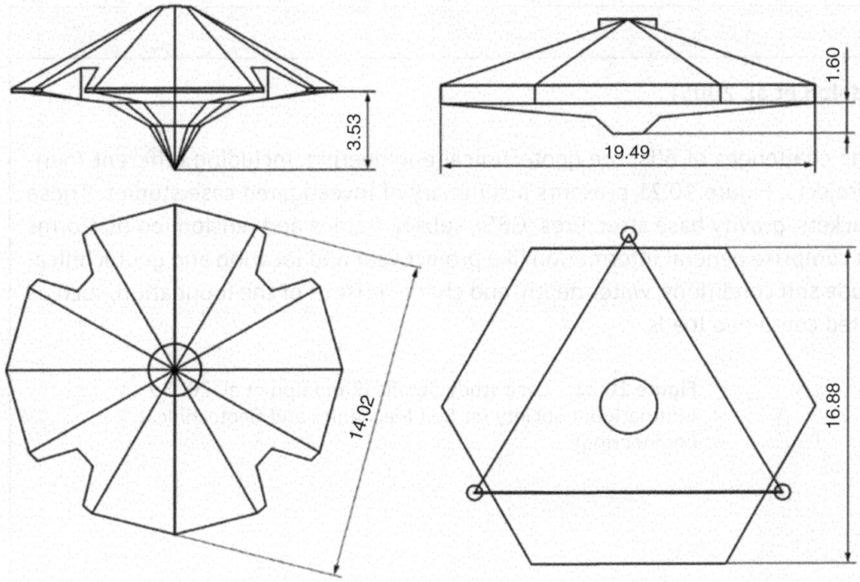

Figure 10.19 Typical noncircular spudcan configuration (ISO 2012).

Larger units may have leg lengths ranging from 100 to 170 m, while hull lengths typically range from 50 to 70 m (Randolph and Gourvence 2011). These dimensions are determined based on the specific design criteria and the intended deployment conditions. The foundation design of jack-up units, whether using spudcans or mat supports, is crucial for ensuring the stability, safety, and performance of the unit during offshore operations.

Jack-up units are highly valuable in the offshore industry due to their flexibility, cost-effectiveness, and self-installation capability. The process of deploying a jack-up unit involves towing it to the designated site while floating on its hull with the legs elevated out of the water, as shown in Figure 10.20a. Once the jack-up reaches the desired location, the legs are gradually lowered to the seabed. The legs continue to be jacked downward until sufficient bearing capacity is achieved, allowing the hull to be lifted clear of the water, as depicted in Figure 10.20b. This ensures that the unit is stable and securely positioned on the seabed. To further enhance the stability and performance of the spudcan foundations, a process called preload is often carried out. During preload, seawater is pumped into ballast tanks located in the hull of the jack-up. This additional weight applies a larger vertical load to the foundations, exceeding the expected service load. The purpose of this "proof test" is to ensure that the foundations can withstand the anticipated operational loads. Once the foundations have been adequately tested, the ballast tanks are emptied before the jack-up commences its operations, as shown in Figure 10.20c. The preload applied to the spudcan foundations typically ranges between 1.3 and 2 times the weight of the jack-up unit itself. This ensures that the foundations are sufficiently loaded and can safely support the structure during its operational phase (Randolph and Gourvence 2011). The ability of jack-up units to self-install and provide stable platforms for offshore operations has made them a preferred choice in the industry, especially in shallow to moderate water depths. They offer versatility, cost efficiency, and the ability to be deployed to various locations as needed.

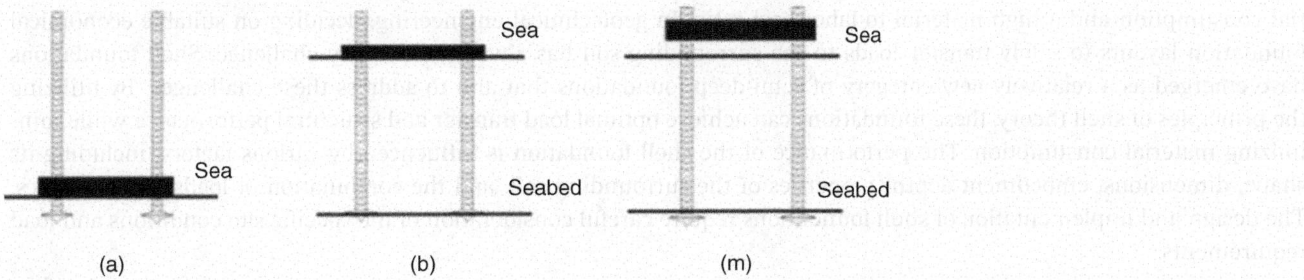

Figure 10.20 Jack-up installation procedure: (a) afloat, (b) installation and preload, and (c) operation (adapted from Randolph et al. 2005).

Case Study 10.5

A Review of Offshore Foundations (Randolph et al. 2005)

Randolph et al. (2005) have reviewed the challenges of offshore geotechnical engineering, including different foundation types implemented in offshore projects. Figure 10.21 presents a summary of investigated case studies. These foundations include concrete or steel buckets, gravity base structures (GBS), subsea frames and tension leg platforms (TLP) as shown in Figure 10.22. The data comprise general information like project year and location and geotechnical and circumferential conditions that include soil conditions, water depth, and characteristics of the foundation, such as foundation type, dimensions, and tolerated combined loads.

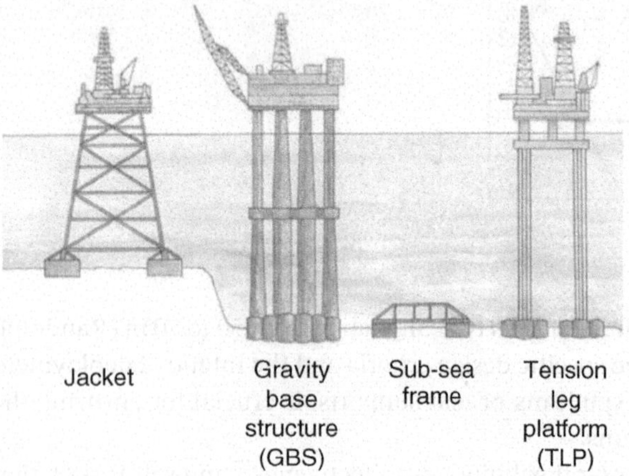

Figure 10.21 Case study details (Randolph et al. 2005 / International Society for Soil Mechanics and Geotechnical Engineering).

As mentioned before, spudcans are another offshore foundation that is widely implemented. Figure 10.23a presents the combined loading (VMH) acting on a spudcan and the effect of rotational fixity on its geotechnical performance. Similar to typical foundations, penetration or embedment depth is a determinant factor in geotechnical performance. Figure 10.23b illustrates the expansion of the combined loading yield surface with an increase in spudcan penetration.

10.9 Shell Foundations

Shell foundations belong to the relatively new category of "semi-deep foundations" (Tadayon et al. 2022). Figure 10.24 depicts some shell foundation applications. They are characterized by their geometric shape and efficient use of materials, offering structural integrity with minimal material consumption. Shell foundations have been recognized as a cost-effective solution due to their strength derived from their geometry rather than their mass. Additionally, shells foundations are regarded as a cost-effective solution because they offer the greatest structural integrity with the least amount of material consumption and a high material-to-labor cost ratio. In geotechnical engineering, deciding on suitable economical foundation layouts to safely transfer loads to the surrounding soil has always been a key challenge. Shell foundations have emerged as a relatively new category of semi-deep foundations that aim to address these challenges. By utilizing the principles of shell theory, these foundations can achieve optimal load transfer and structural performance while minimizing material consumption. The performance of the shell foundation is influenced by various factors, including its shape, dimensions, embedment depth, properties of the surrounding soil, and the combination of loads it experiences. The design and implementation of shell foundations require careful consideration of the specific site conditions and load requirements.

Project Year	Foundation type	Location	Water depth m	Soil conditions	Foundation plan	Foundation dimensions m, m²	Skirts d (d/D)	V ($-W$) MN	H MN	M MN	M/HD	Reference
Ekofisk Tank 1973	GBS	North Sea (Norway)	70	Dense sand to 26 m, layers stiff clay 16–18 m. >26 m hard clays and sands (s_u ~ 300 kPa)		A = 7390 D_{eqv} = 97	0.4 (0.004)	1900	786	28000	0.37	Clausen 1976 (p.400) Clausen 1976 (p.263) O'Reilly & Brown 1991 (p.124)
Beryl A 1975	Condeep	North Sea (Norway)	120	As at Ekofisk but sand layer only to 10 m		A = 6360 D_{eqv} = 90	4 (0.04)	1500	450	15000	0.37	Clausen 1976 (p.263)
Brent B 1975	Condeep	North Sea (Norway)	140	Stiff to hard clays (s_u ~ 300 kPa) with thin layers of dense sands to 45 m	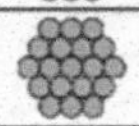	A = 6360 D_{eqv} = 90	4 (0.04)	2000	500	20000	0.44	O'Reilly & Brown 1991 (p.124)
Gullfaks C 1989	Deep skirted Condeep	North Sea (Norway)	220	Soft nc silty clays (s_u ~ 30 kPa) and silty clayey sands with dense sand layers (q_c ~ 4 MPa)		A = 16,000 D_{eqv} = 143 D_{cell} = 28 (t = 0.4)	22 (0.13)	5000	712	65440	0.64	Tjelta et al. 1990 Tjelta 1998
Snorre A 1991	TLP with concrete buckets	North Sea (Norway)	310	Very soft to soft nc clays (s_{um} = 0, k ~ 7 kPa/m to 20 m; 0 < q_c < 2 kPa @ 17 m)		A_{total} = 2724 D_{cell} = 17 (t = 0.35)	12 (0.7) (CFT, 0.4)	142 per CFT	21 per CFT	126 per CFT	0.20	Christophersen 1993 (p.435) Stove et al. 1992 (p.76)
Draupner E (Europipe) 1994	Jacket with steel buckets	North Sea (Norway)	70	22–25 m dense to very dense fine sand (q_c ~ 60 MPa) over stiff clay		A_{total} = 452 D = 12	6 (0.5)	57 per bucket	10	30	0.25	Bye et al. 1995 (p.870)
Sleipner SLT 1995	Jacket with steel buckets	North Sea (Norway)	70	As at Draupner E		A_{total} = 616 D = 14	5 (0.35)	134 per bucket	22	-	-	Bye et al. 1995 (p.876)
Troll A 1995	Deep skirted Condeep	North Sea (Norway)	305	Soft nc clays (s_{um} = 0, k ~ 3 kPa/m to 60 m; 0 < q_c < 1.5 kPa @ 40 m)	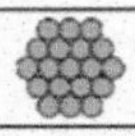	A = 16,596 D_{eqv} = 145 D_{cell} = 32	36 (0.25)	2353	512	94144	1.27	Andenaes et al. 1996 (p.62) Hansen et al. 1992 (p.924)
Wandoo 1997	GBS	NW Shelf W Australia	54	Thin layer dense calcareous sand (q_c ~ 3 MPa) over thick strong calcarenite (q_c ~ 30 MPa)		A = 7866 114 x 69 x 17 D_{eqv} = 100	0.3 (0.003)	755	165	7420	0.45	Humpheson. 1998 (p.362, 365)
Bayu-Undan 2003	Jacket with steel plates	Timor Sea N Australia	80	2 m very soft calcareous sandy silt over cemented calcarenite and limestone (q_c ~ 20 MPa)		A_{total} = 480 A_{plate} = 120 6 x 20	0.5 (0.04)	125 per plate	10	-	-	Neubecker & Erbrich 2004 (p.2)
Yolla 2004	Skirted GBS/jacket hybrid	Bass Strait S Australia	80	Firm calcareous sandy silt with very soft clay and sand layers (q_c ~ 2 MPa)		A = 2500 50 x 50 D_{equiv} = 56.4	5.5 (0.1)	-	-	-	-	

Figure 10.22 Applications of offshore foundations (NGI).

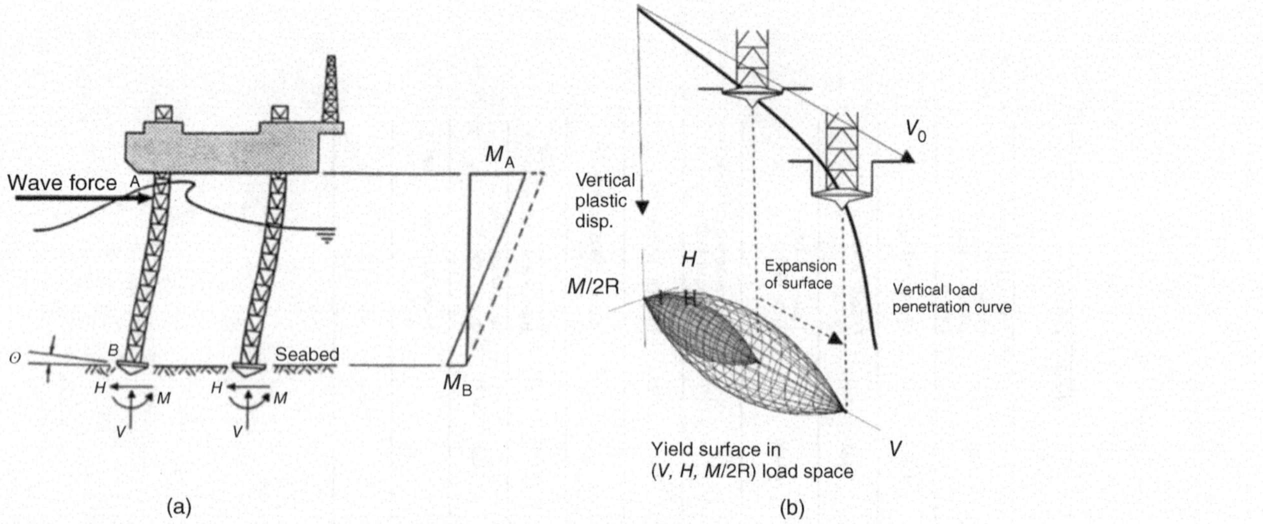

Figure 10.23 (a) The effect of rotational fixity on spudcan (Santa Maria 1988 / Santa Maria) and (b) expansion of the combined loading yield surface with spudcan penetration (Randolph et al. 2005 / International Society for Soil Mechanics and Geotechnical Engineering).

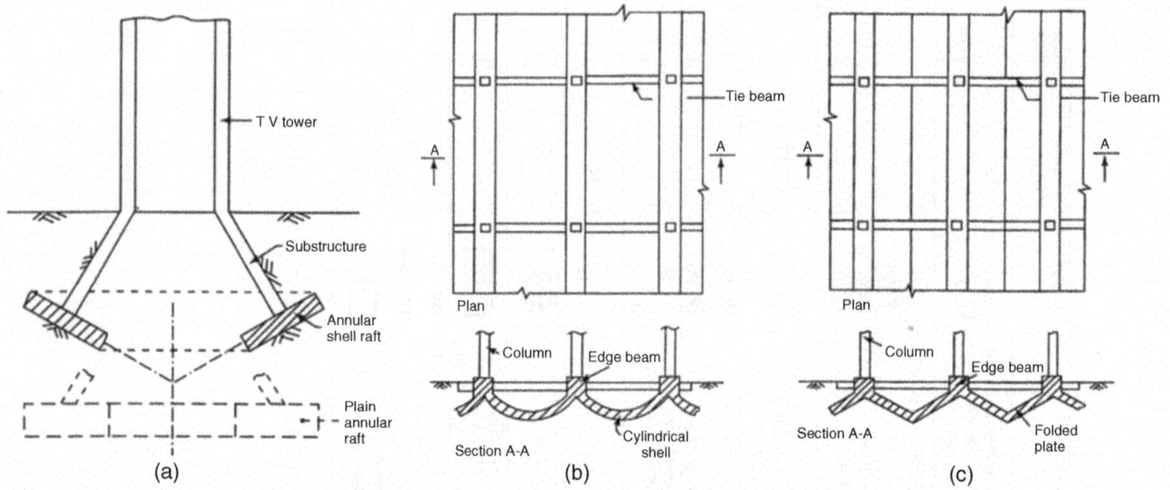

Figure 10.24 Some applications of shell foundations: (a) inverted cone ring raft, (b) cylindrical shell raft, and (c) folded plate strip raft (Kurian 2006 / Alpha Science International Limited).

In general, a foundation may be required to transfer vertical loads, horizontal loads, and moments to the soil. In contrast to footings and rafts, which resist the imposed load by bending (Figure 10.25a), piles, piers, and caissons are under direct compression (Figure 10.25b) or tension depending on the situation, but are subject to flexure as a result of horizontal load and moments. In this regard, shell foundations resist the loads via secondary flexure, as shown in Figure 10.25c. More details and explanations are presented in Chapter 13.

10.9.1 Construction

In general, shell foundation construction methods are divided into two types: in situ and prefabricated. In this section, most of the issues raised are related to the specific performance characteristics of these two types of shell foundations construction methods.

In the in situ construction method, the foundation is poured onto the soil cut and forms the desired shell's profile. For instance, in the case of axisymmetric shells like the cone and the inverted dome, both of which have a circular plan and exist as separate units, the correct profile can be obtained by rotating a suitable template around a vertical post placed in

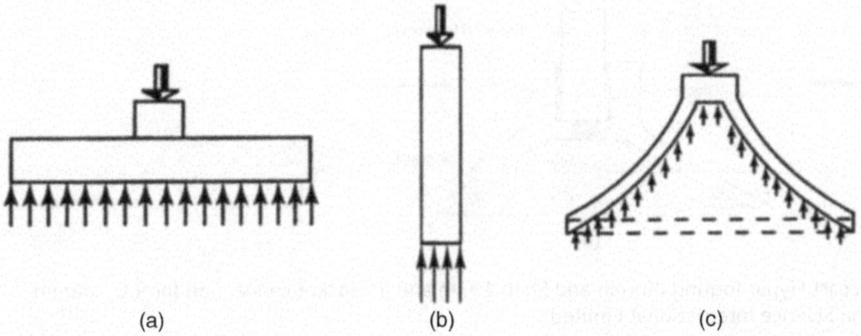

Figure 10.25 General structural action of foundations: (a) Footing (flexure), (b) Pier (axial compression), and (c) Shell foundation (axial forces with secondary flexure) (Kurian 2006 / Alpha Science International Limited).

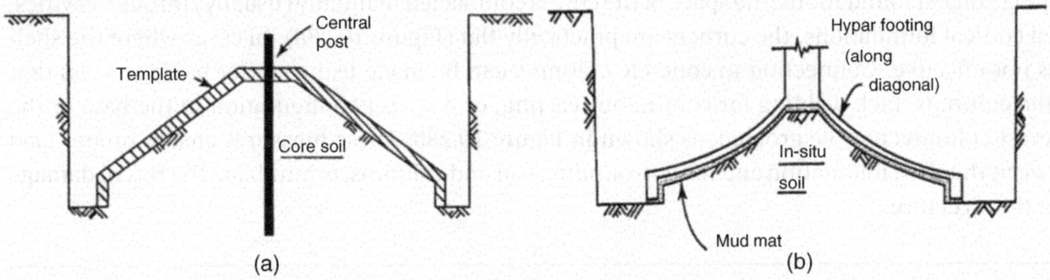

Figure 10.26 (a) Core profiling for cone, (b) in situ construction (Kurian 2006 / Alpha Science International Limited).

the center for the cone (Figure 10.26a). The same method can be used to execute other symmetrical shells, such as the inverted dome, the double cone folded shell, etc., by rotating the templates around the vertical central post. Several forms may be required during various stages of implementation, including soil profile creation, laying the mud mat, casting, and finishing the shell. The prepared soil surface is then covered with a thin coating (about 80 mm) of lean concrete (mud mat) or cement mortar (Figure 10.26b) (Kurian 2006).

In the construction of large shells, there is a need to use high-quality concrete and reduce the thickness of the shell to a minimum. To achieve this, prefabrication methods can be employed. Specifically, hypar (hyperbolic paraboloid) and conical footings can be prefabricated using inverted wooden molds. The molds for these footings are created by cutting and nailing plywood strips along the directions of the straight generating lines. A box is formed using wooden sides, and the inside of the box is filled with low-strength concrete. The concrete is then shaped to fit the desired pattern of the footing using cement paste (Figure 10.27) (Kurian 2006). This process allows for the creation of the desired shape and profile of the footing. By using prefabricated molds, the construction of large shells becomes more efficient. The use of wooden molds provides flexibility in shaping the concrete, allowing for precise replication of the desired pattern. Additionally, the use of high-quality concrete and the reduction of shell thickness contribute to the overall strength and stability of the foundation.

In both in situ and prefabricated methods, there should be no defects in any area of the ground contact surface. Because of load concentration on the point of the shell, there is a possibility of damage to the shell. Because there is not always full contact between the soil core and the foundation in the prefabricated state, the soil should not only be cut in the desired shape as in the in situ states. The prefabricated shell foundation should be constructed in a trench with a specified depth.

Figure 10.27 Inverted concrete mold (for cone) (Kurian 2006 / Alpha Science International Limited).

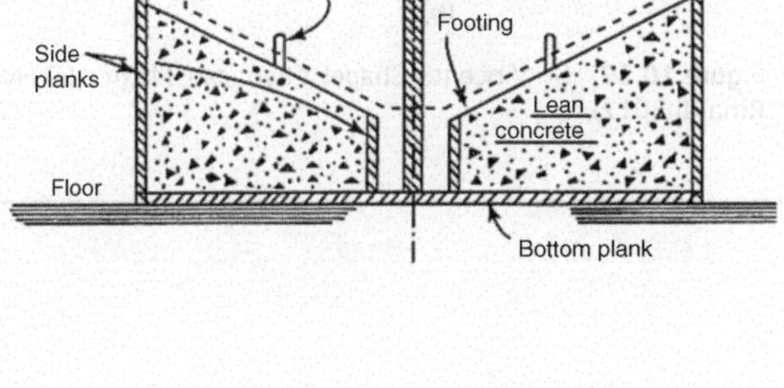

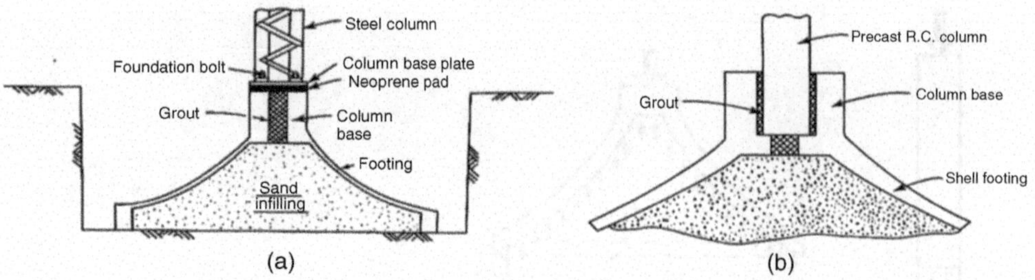

Figure 10.28 (a) Fixing of steel column to precast Hypar footing (Kurian and Shah 1974) and (b) socket connection for R.C. column with precast Hypar footing (Kurian 2006 / Alpha Science International Limited).

Dry sand should be poured to create a soil the core under the foundation which is then compacted so that the core below the foundation is suitable for load transfer. A thin layer of cement slurry may be injected, and the floor cavities of the columns will be closed. In steeply sloping conical foundations, the space of the core is compacted manually (usually) through cavities, but in Hypar and low-sloped conical foundations, the corners are practically flat (Figure 10.28a). In cases where the shell depth is large, this action is not effective. Connection to concrete columns can be made using starter bars (dowels) that protrude from the base of the columns, tack welding for continuous casting, or a socket configuration in the base of the columns into which the precast columns may be grouted, as shown in Figure 10.28b. These measures ensure proper load transfer and connection between the shell foundation and the surrounding soil and columns, minimizing the risk of damage and ensuring the stability of the structure.

Case Study 10.6

Hypar Shell Foundation of St. Vincent's Chapel, Mexico (Melaragno 2012)

In 1959–1960, Candela and Lopez built a chapel using three varieties of hyperbolic paraboloids joined together. Each hyperbolic paraboloid features a low peak projecting outward at the ends of the cantilevered members and a higher peak soaring vertically on a tilted axis. The overall shape of the building is virtually triangular (Figure 10.29).

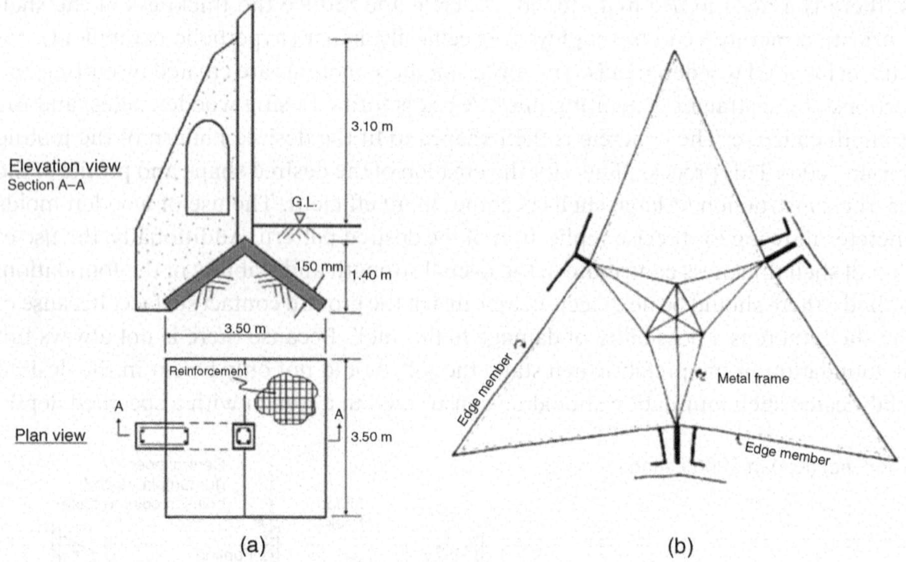

Figure 10.29 St. Vincent's Chapel, Coyoacan, Mexico (a) Hypar footing, (b) structural diagram (adapted from Melaragno 2012; Rinaldi 2012).

Only three concrete abutments serve as supports for the building. The cantilevered edges are 77 ft and 6 in. long, starting from the support.

1) Semi-deep foundations are a viable solution when soil improvement methods are not feasible, and using piles or shallower foundations would be costly or present geotechnical performance issues.
2) The geotechnical performance of semi-deep foundations is influenced by factors such as burial depth, shaft design, toe configuration, surrounding soils, and confined soil beneath the foundation. Compared to shallow foundations, semi-deep foundations offer benefits such as reduced settlement, enhanced bearing capacity, and improved stability. The combination of the skin and toe of a semi-deep foundation with the surrounding deposits and confined soil creates a block foundation, providing material efficiency and resistance against lateral loading and uplift. Examples of semi-deep foundation types include well, skirted, floating, bucket, spudcan, and shell foundations.
3) Well foundations are suitable for applications where both compression and tension capacity are important. A hollow shaft or box is submerged into the ground until it reaches a hard or dense layer in the soil. Well foundations are simple and cost-effective solutions, particularly when pull-out capacity is a primary concern. They are practical due to their relatively low depth ratio (*D*/*B*).
4) Skirted foundations are commonly used in offshore applications and offer cost savings by reducing material requirements and installation time. Structural skirts provide improved bearing capacity and reduced settlement for the foundation on the soil.
5) Floating foundations, where the burial depth of the footing is increased through soil excavation, have emerged as a promising foundation design and construction method. This approach improves the strength and stiffness of the substructure system. Floating foundations are advantageous in cases where high bending rigidity is required and deep soft deposits are present.
6) Concrete or steel bucket foundations are increasingly used as anchors for floating platforms or as foundations for various sea bottom structures. These foundations experience vertical loads due to their slender construction and are subjected to horizontal forces that induce significant overturning moments.
7) Spudcans, which serve as the foundations of independent leg jack-up platforms, are typically large inverted cones. They have a roughly circular plan and a shallow conical underside with a protruding spigot. Spudcans can vary in diameter and shape depending on the manufacturer and rig.
8) Shells are considered cost-effective solutions because they provide high structural integrity with minimal material consumption and a favorable material-to-labor cost ratio. In geotechnical engineering, selecting appropriate and economical foundation layouts for safely transferring loads to the surrounding soil has always been a challenge. In this context, shell foundations offer an appropriate solution for shallow foundations.

10.10 Major Points

1) Semi-deep foundations are a viable solution when soil improvement methods are not feasible, and using piles or shallower foundations would be costly or present geotechnical performance issues.
2) The geotechnical performance of semi-deep foundations is influenced by factors such as burial depth, shaft design, toe configuration, surrounding soils, and confined soil beneath the foundation. Compared to shallow foundations, semi-deep foundations offer benefits such as reduced settlement, enhanced bearing capacity, and improved stability. The combination of the skin and toe of a semi-deep foundation with the surrounding deposits and confined soil creates a block foundation, providing material efficiency and resistance against lateral loading and uplift. Examples of semi-deep foundation types include well, skirted, floating, bucket, spudcan, and shell foundations.
3) Well foundations are suitable for applications where both compression and tension capacity are important. A hollow shaft or box is submerged into the ground until it reaches a hard or dense layer in the soil. Well foundations are simple and cost-effective solutions, particularly when pull-out capacity is a primary concern. They are practical due to their relatively low depth ratio (*D*/*B*).

4) Skirted foundations are commonly used in offshore applications and offer cost savings by reducing material requirements and installation time. Structural skirts provide improved bearing capacity and reduced settlement for the foundation on the soil.
5) Floating foundations, where the burial depth of the footing is increased through soil excavation, have emerged as a promising foundation design and construction method. This approach improves the strength and stiffness of the substructure system. Floating foundations are advantageous in cases where high bending rigidity is required and deep soft deposits are present.
6) Concrete or steel bucket foundations are increasingly used as anchors for floating platforms or as foundations for various sea bottom structures. These foundations experience vertical loads due to their slender construction and are subjected to horizontal forces that induce significant overturning moments.
7) Spudcans, which serve as the foundations of independent leg jack-up platforms, are typically large inverted cones. They have a roughly circular plan and a shallow conical underside with a protruding spigot. Spudcans can vary in diameter and shape depending on the manufacturer and rig.
8) Shells are considered cost-effective solutions because they provide high structural integrity with minimal material consumption and a favorable material-to-labor cost ratio. In geotechnical engineering, selecting appropriate and economical foundation layouts for safely transferring loads to the surrounding soil has always been a challenge. In this context, shell foundations offer an appropriate solution for shallow foundations.

10.11 Questions

10.11.1 Explanatory

1 What is the definition of a semi-deep foundation in terms of dimensions and performance?

2 Which parts of a semi-deep foundation are active in the bearing foundation under VHM loading?

3 Name the main types of semi-deep foundations for marine structure application.

4 Explain how adding skirts to a shallow foundation can increase its vertical bearing capacity.

5 Explain the suction bucket foundation's mechanism of installation and how suction force facilitates the installation process.

6 List the benefits and application of shells in foundation engineering.

7 Compare and contrast the two typical construction methods of shell foundations.

8 Name and compare three different shapes of floating foundations.

9 Explain the spudcan installation process.

10 Explain the concept of HSBF by using appropriate figures.

10.11.2 Conceptual

1 Draw appropriate figures and explain the role of the foundation depth in the behavior of semi-deep foundations.

2 Explain the significant benefits of a semi-deep foundation in comparison with shallow/deep foundations.

3 Considering Case Study 10.1, suggest other solutions for the foundation of high transmission towers.

4 Based on Case Study 10.5, discuss the advantages and disadvantages of typically used offshore foundations.

5 Based on Case Studies 10.3 and 10.4, what are your suggestions for offshore wind turbine foundations in different conditions, including soil type and water depth?

6 Compare a shell foundation with a shallow and deep foundation regarding the load transfer mechanism via the given figure.

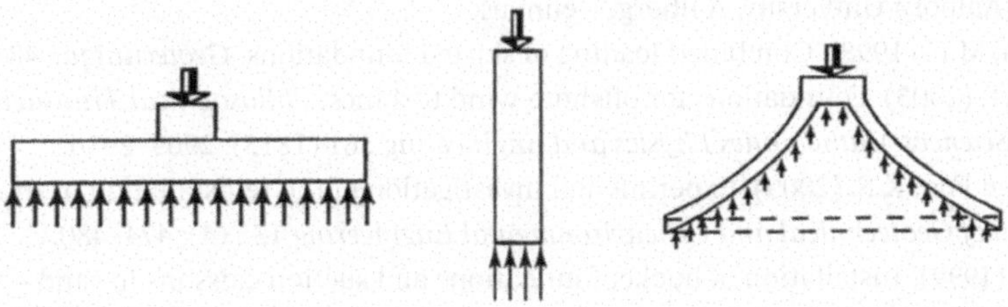

7 Discuss the role of foundation embedment depth in the geotechnical performance of marine foundations, especially under combined loading based on the below figure.

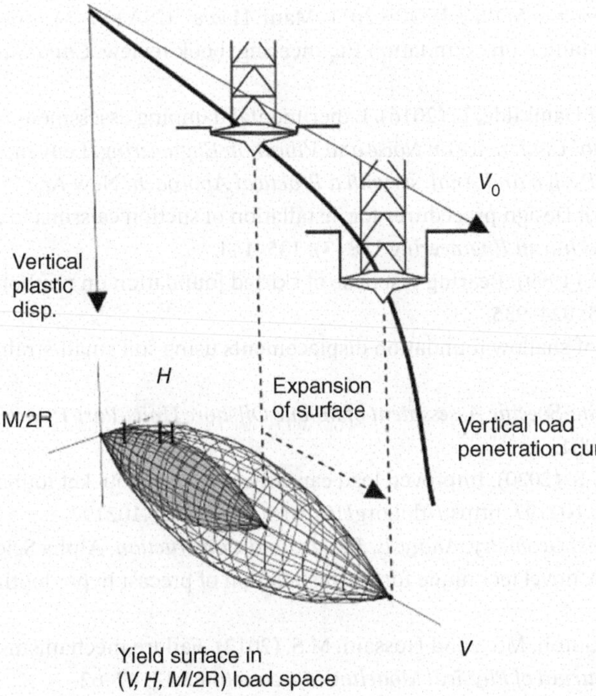

8 Compare and contrast the installation process of those semi-deep foundations that are implemented in offshore projects.

9 Compare shallow and semi-deep foundations implemented in marine projects regarding their function and construction (installation) procedures.

10 What are the advantages and disadvantages of semi-deep and deep foundations in offshore projects? Compare and contrast for different conditions.

References

Al-Aghbari, M.Y. and Mohamedzein, Y.E.A. (2004a). Bearing capacity of strip foundations with structural skirts. *Geotechnical and Geological Engineering* 21 (1): 43–57.

Al-Aghbari, M.Y. and Mohamedzein, Y.E.A. (2004b). Improving the performance of circular foundations using structural skirts. *Ground Improvement* 22 (1): 43–57.

Andersen, K.H., Jostad, P., and Dyvik, R. (2008). Penetration resistance of offshore skirted foundations and anchors in dense sand. *Geotechnical Geoenvironmental Engineering* 134: 106–116.

Barari, A. (2012). Characteristic behavior of bucket foundations. PhD thesis. Department of Civil Engineering, The Faculty of Engineering and Science, Aalborg University, Aalborg, Denmark.

Bransby, M.F. and Randolph, M.F. (1998). Combined loading of skirted foundations. *Geotechnique* 48 (5): 637–655.

Byrne, B.W. and Houlsby, G.T. (2003). Foundations for offshore wind turbines. *Philosophical Transactions: Mathematical, Physical and Engineering Sciences Mathematics Physics and Engineering* 361 (1813): 2909–2930.

Cassidy, M.J., Quah, C.K., and Foo, K.S. (2009). Experimental investigation of the reinstallation of spudcan footings close to existing footprints. *Journal of Geotechnical and Geoenvironmental Engineering* 135 (4): 474–480.

Erbrich, C.T. and Tjelta, T.I. (1999). Installation of bucket foundations and suction caissons in sand – Geotechnical performance. *Paper Presented at the Offshore Technology Conference*, Houston, Texas (3–6 May 1999), OTC-10990-MS. https://doi.org/10.4043/10990-MS.

Eslami, A., Moshfeghi, S., Molaabasi, H., and Eslami, M. (2020). *Piezocone and Cone Penetration Test (CPTu and CPT) Applications in Foundation Engineering*, 1e. Elsevier.

Gaudin, C. et al. (2011). Centrifuge experiments of a hybrid foundation under combined loading. *Proceedings International Society of Offshore and Polar Engineers (ISOPE)*, ISOPE 2011, Maui, Hawai, USA (19–24 June), 386–392.

Golder, H.Q. (1976). The Canadian Manual on Foundation Engineering: book review. *Canadian Geotechnical Journal* 13 (4): 499–504.

Gres, S., Fejerskov, M., Ibsen, L., and Damkilde, L. (2016). Experimental damping assessment of full scale offshore Mono Bucket foundation. *ISMA-The International Conference on Noise and Vibration Engineering*, Leuven, Belgium (19–21 September 2016).

Gupta, S.C. (2007). *Raft Foundation Design and Analysis with a Practical Approach*. New Age International.

Houlsby, G.T. and Byrne, B.W. (2005). Design procedures for installation of suction caissons in sand. *Proceedings of the Institution of Civil Engineers: Geotechnical Engineering* 158 (3): 135–144.

Hu, Y., Randolph, M., and Watson, P. (1999). Bearing response of skirted foundation on nonhomogeneous soil. *Geotechnical and Geoenvironmental Engineering* 125: 924–935.

In Elhakim, A.F. (2005). Evaluation of shallow foundation displacements using soil small-strain stiffness. *Journal of Materials Processing Technology* 1 (1): 1–8.

ISO (2012). *Natural Gas Industries-Site-Specific Assessment of Mobile Offshore Units-Part 1: Jack-Ups*. International Organization for Standardisation ISO: 19905-19901.

Kim, J.-H., Jeong, Y.-H., and Kim, D.-S. (2020). Improved load capacity of a hybrid bucket foundation for offshore structures in sand. *Applied Ocean Research* 101: 102197. https://doi.org/10.1016/j.apor.2020.102197.

Kurian, N.P. (2006). *Shell Foundations: Geometry, Analysis, Design, and Construction*. Alpha Science International, Limited.

Kurian, N.P. and Shah, S.H. (1974). A novel technique for the installation of precast hypar footings. *Bui. Int. Assn. Shell Spatial Structures* 56: 61–66.

Mana, D.S.K., Gourvence, S.M., Randolph, M.F., and Hossain, M.S. (2012). Failure mechanisms of skirted foundations in uplift and compression. *International Journal of Physical Modeling in Geotechnics* 12: 47–62.

Melaragno, M. (2012). *An Introduction to Shell Structures: The Art and Science of Vaulting*. Springer Science & Business Media.

Randolph, M. and Gourvence, S. (2011). *Offshore Geotechnical Engineering*, 559. Spon Press.

Randolph, M., Cassidy, M., Gourvenec, S., and Erbrich, C., (2005). Challenges of offshore geotechnical engineering. *Proceedings of the 16th International Conference on Soil Mechanics and Geotechnical Engineering* (12–16 September 2005). Rotterdam: Millpress.

Reardon, M.J. (1986). Review of the geotechnical aspects of jack-upunit operations. *Ground Engineering* 19 (7): 21–26.

Rezazadeh, S. and Eslami, A. (2017a). Bearing capacity of semi-deep skirted foundations on clay using stress characteristics and finite element analyses. *Marine Georesources and Geotechnology* 36 (6): 1–15.

Rinaldi, R. (2012). Inverted shell foundation performance in soil. PhD thesis. Department of Building, Civil & Environmental Engineering, Concordia University, Montreal, Quebec, Canada.

Salem, A., Jalbi, S., and Bhattacharya, S. (2021). Vertical stiffness functions of rigid skirted caissons supporting offshore wind turbines. *Journal of Marine Science and Engineering* 9: 573.

Santa Maria, P.E.L. (1988). Behaviour of footings for offshore structures under combined loading. DPhil thesis. University of Oxford.

Tadayon, A., Eslami, A., and Fakharian, K. (2022). Load–displacement behavior of semi-deep shell foundations on sand by experimental study. *International Journal of Civil Engineering* 20 (7): 839–855.

Tripathy, S. (2013). Load carrying capacity of skirted foundation on sand. Master thesis. Rourkela Institute, 211.

Watson, P.G. and Randolph, M.F. (1998). Skirted foundations in calcareous soil. *Geotechnical Engineering Journal, Proceedings of the Institute of Civil Engineering* 131: 171–179.

Wong, I.H., Ooi, I.K., and Brooms, B.B. (1995). Performance of raft foundations for high-rise buildings on the Bouldery clay in Singapore. *Canadian Geotechnical Journal* 219–236.

Yun, G. and Bransby, M.F. (2007). The undrained vertical bearing capacity of skirted foundations. *Soil and Foundations, Japanese Geotechnical Society, Elsevier* 47 (3): 493–505.

Zhang, P., Guo, Y., Liu, Y., and Ding, H. (2016). Experimental study on installation of hybrid bucket foundations for offshore wind turbines in silty clay. *Ocean Engineering* 114: 87–100. https://doi.org/10.1016/j.oceaneng.2016.01.025.

11

Skirted, Bucket, and Spudcan Foundations

Chapter Highlight

- Main Semi-Deep Foundation for Offshore applications
- Classic Approaches for Bearing Capacity Calculation
- VMH Approaches for Foundation Behavior Investigation
- 2D and 3D Correlations for Failure Envelope
- Comprehensive Examples
- Case studies
  - Numerical Modelling of Suction Bucket Foundation in Clay (Yin et al. 2020)
  - Centrifuge Study of Improving Load Capacity of Skirted Foundation by Hybrid Technique (Kim et al. 2020)
  - Bearing capacity of composite bucket foundations for offshore wind turbines in silty sand by Field Test. (Jia et al. 2018)
  - A new method to predict the bearing capacity–penetration curve of spudcans in multilayered clay soils (Mehralizadeh and Makarchian 2021)

11.1 Introduction

Semi-deep foundations have gained significant popularity in both offshore and onshore projects, particularly in ocean environments. They are widely utilized for establishing offshore wind farms and supporting jack-up platforms in the oil industry. The three main types of semi-deep foundations commonly employed are (a) skirted foundations, (b) bucket foundations, and (c) spudcan foundations. These semi-deep foundations offer advantages such as improved stability, increased load-bearing capacity, and reduced environmental impact compared to traditional deep foundations.

a) **Skirted Foundations:** Skirted foundations are commonly used in offshore structures. Before pile installation, rectangular footings are commonly used to temporarily support jacket platforms on the seafloor. These footings, equipped with skirted plates, are designed to bear vertical and horizontal loads from the upper structures. It is crucial to consider both the vertical bearing capacity and horizontal loading resistance in this context. Figure 11.1 illustrates the inclusion of shear keys or extended skirts beneath the foundation, which enhance the lateral resistance of the shallow concrete foundation.
b) **Bucket Foundations:** Bucket foundations are widely used in the offshore wind industry. They have a bucket-shaped structure that is placed on the seabed and then slowly penetrated into the soil. The bucket design helps to provide increased resistance against vertical and horizontal loads. Bucket foundations, also known as suction caissons, are typically installed using suction force. They resemble gravity-based foundations in shape but feature long skirts around their perimeter. A bucket foundation consists of a circular ribbed lid with a thin tubular skirt extending beneath it, resembling a bucket. These foundations are categorized as semi-deep foundations, as their diameter-to-skirt-length ratio (D/Z) ranges from 0.5 to 1, making them shorter than traditional piles (Bhattacharya 2019). The installation process of bucket foundations was discussed in detail in the previous chapter (Chapter 10). Figure 11.2 provides a visual representation of a bucket foundation.

Advanced Foundation Engineering: Principles, Performance and Prospects, First Edition. Moses Karakouzian and Abolfazl Eslami.

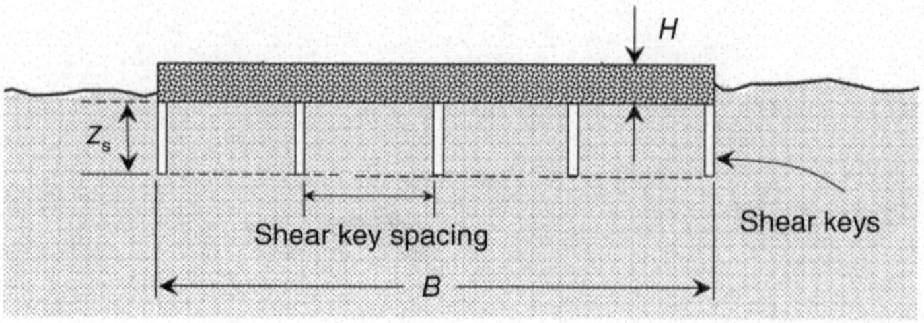

Figure 11.1 Details of skirted foundation (NAVFAC 2012 / Naval Facilities Engineering Systems Command / Public Domain).

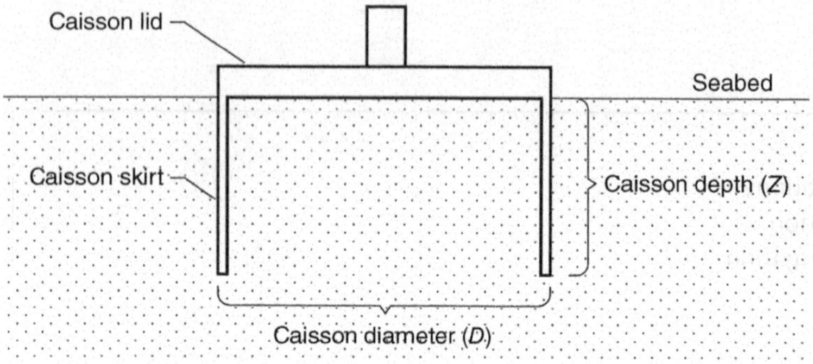

Figure 11.2 Typical scheme of bucket (caisson) foundation.

c) **Spudcan Foundations:** Spudcan foundations are primarily used in jack-up platforms for the oil and gas industry. They feature a large inverted conical baseplate known as a spudcan, which is lowered to the seabed. The spudcan helps to distribute the load and provides stability by increasing the bearing area thus supporting the individual legs of a jack-up platform. Figure 11.3 illustrates a schematic of an embedded spudcan foundation. Chapter 10 provides further details regarding the shapes and dimensions of spudcans. While various unconventional spudcan shapes are common in the offshore industry, for analytical purposes, spudcans are typically idealized as axisymmetric inverted conical footings. During penetration analysis, an equivalent conical angle is determined at each depth. The volume of the equivalent cone is equal to the volume of the penetrated portion of the spudcan (up to the largest cross-sectional area in the plan), and the planar area in contact with the soil remains consistent. Once the maximum spudcan diameter is reached, the equivalent conical footing remains constant (Randolph and Gourvenec 2011).

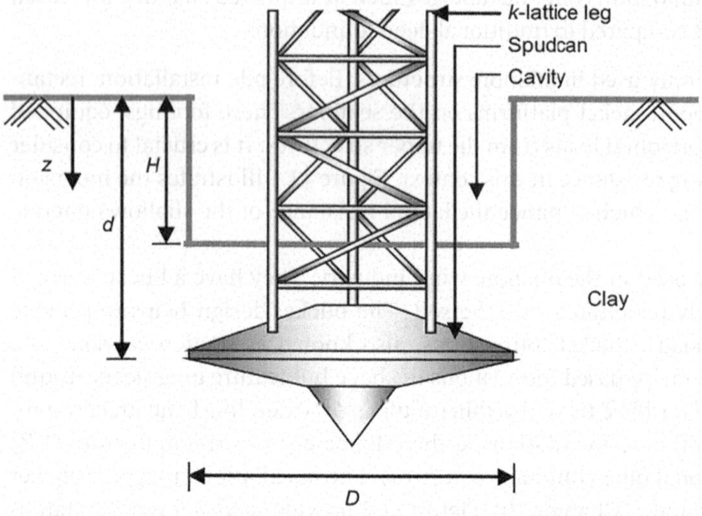

Figure 11.3 Schematic of embedded spudcan foundation showing open cavity (Hossain et al. 2005).

11.2 Bearing Capacity Determination: Classic and Modified Approaches

To determine the bearing capacity of a semi-deep foundation, sustained vertical (V), horizontal (H), and moment (M) combined loading (VMH) need to be considered. These combined loading is due to waves, sea currents, wind, and self-weight, which are common in offshore environments. Accordingly, determination of the bearing capacity of semi-deep foundations due to VMH loading requires special treatment instead of using traditional bearing capacity formulas with empirical modification developed for onshore foundations where the principal loading is in the vertical direction. The failure envelopes of the semi-deep foundations are mainly influenced by their geometry and embedment, as well as the strength properties of the soil. Almost all research that has been performed on semi-deep foundations has concentrated on determining the failure surface envelope under combined VMH loading. The following is a sample of the researchers and their research (analytical, numerical, and experimental) on the bearing capacity of semi-deep foundations under combined (VMH) loading: Upper and lower bound analyses by Bransby and Randolph (1999); Finite element analyses, and centrifuge tests by Yun and Bransby (2007); Skirted foundations in plane strain by Gourvenec (2008); and Finite element and upper bound plasticity analyses by Rezazadeh and Eslami (2017).

Just a note on the selection of soil strength properties in design for short-term vs. long-term static loadings. For **cohesive soils** such as clay, which fail before excess pore water pressures dissipate under static short-term loadings, undrained soil properties are used in the design. For static long-term loading conditions, the excess pore water pressures have dissipated, undrained soil properties are used in the design. However, static long-term bearing loading in cohesive soils is usually not the controlling design case compared to static short-term loading. For **cohesionless soils** like sand have high permeability, the excess pore pressures dissipate very rapidly. Consequently, bearing capacity failure in cohesionless soils typically occurs under drained conditions. Therefore, drained soil properties are utilized for both short-term and long-term design considerations. These principles and guidelines are outlined in the NAVFAC (Naval Facilities Engineering Command) manual, which provides valuable information for foundation design (NAVFAC 2012).

Details of bearing capacity calculations for each of these three types of semi-deep foundations are discussed in the next three sections (11.3, 11.4, and 11.5).

11.3 Bearing Capacity of Skirted Foundations

11.3.1 Vertical Bearing Capacity of Skirted Foundations (NAVFAC 2012)

11.3.1.1 Clay

For cohesive soil under undrained conditions, the maximum vertical load that a skirted foundation can support is calculated by Equation (11.1) according to the sketch in Figure 11.1.

$$V_0 = A'\{s_{uz}N_cK_c + \gamma_b D_f\} + (2B + 2L)(\min(D_f \cdot H + z_s))\frac{S_{ua}}{S_t} \tag{11.1}$$

where:

A' = effective base area of foundation depending on the load eccentricity

S_{uz} = undrained shear strength of cohesive soil averaged over the distance 0.7B below the foundation base

N_c = bearing capacity factor; for undrained failure $N_c = 5.14$

K_c = correction factor that accounts for load inclination, foundation shape, embedment depth, the inclination of the foundation base, and inclination of the ground. For a nearly square or round footing on a nearly level seafloor with a vertical load, $K_c = 1.2$

γ_b = buoyant unit weight of soil above the foundation base

D_f = depth of embedment of foundation

z_s = shear key height

B = base width

L = base length

H = base block height

S_{ua} = undrained strength averaged over the side soil contact zone

S_t = soil sensitivity; for a soft cohesive seafloor, $S_t \approx 3$

When a square or circular foundation on the seafloor surface sustains a concentrically vertical load, as well as seafloor and foundation are horizontal, Equation (11.1) can be summarized and rewritten as Equation (11.2).

$$V_0 = (6.17)As_{uz} \tag{11.2}$$

11.3.1.2 Sand

Generally, for sands, failure occurs under drained conditions. So, the maximum vertical bearing capacity of skirted foundation on sand soils is calculated by Equation (11.3)

$$V_0 = A'\gamma_b[D_f\{1+(N_qS_q-1)f_z\}+(B'/2)N_\gamma S_\gamma f_z]+(2B+2L)(\min(D_f\cdot H+z_s))\gamma_b tan\delta(D_f+\max(0\cdot D_f-H-z_s))/2 \tag{11.3}$$

where:

A' = effective base area of foundation depending on the load eccentricity
γ_b = buoyant unit weight of soil above the foundation base
D_f = depth of embedment of foundation
z_s = shear key height

$$N_q = e^{\pi tan\emptyset}\tan^2\left(45+\frac{\varphi}{2}\right);\ \text{(Meyerhof's 1963)} \tag{11.4}$$

$$N_\gamma = (N_q-1)\tan(1.4\emptyset);\ \text{(Meyerhof's 1963)} \tag{11.5}$$

$$s_q = s_\gamma = 1+0.1\times\tan^2(45+\varphi/2) \tag{11.6}$$

For a nearly square or round footing on a nearly level surface with a vertical load, $S_q = 1.0$, and S_γ can be assumed as 0.6.

f_z = depth attenuation factor
B' = effective base width of foundation depending on the load eccentricity
B = base width
L = base length
H = base block height
δ = effective friction angle alongside the footing = ϕ – 5 deg for rough-sided footings, = 0 for smooth-sided footings or where the soil is greatly disturbed
ϕ = soil friction angle

For a square or circular and horizontal base foundation on a horizontal seafloor surface, bearing capacity under concentrical vertical load can be determined by reduced Equation (11.3) as Equation (11.7).

$$V_0 = 0.3A\gamma_b BN_\gamma \tag{11.7}$$

11.3.2 Lateral Bearing Capacity of Skirted Foundations (NAVFAC 2012)

Lateral loading on foundations can be attributed to various factors, including a sloping seafloor, storm waves, current drag, and non-vertical line loading. To enhance the lateral bearing capacity, shear keys are incorporated, which push the failure surface deeper into the seafloor where the soils are stronger and offer greater resistance against lateral loading.

Figure 11.4 illustrates three potential failure modes for skirted foundations. When shear keys are designed to be adequately close together, sliding failure occurs at the base of the keys (Figure 11.4a). The lateral bearing capacity based on this failure mode will be discussed in the subsequent section. Shear key design primarily addresses the passive wedge failure mode (Figure 11.4c), while the occurrence of deep passive failure mode (Figure 11.4b) is rare, and therefore, it is not extensively covered in this chapter.

11.3.2.1 Clay

The maximum lateral load capacity, F_0 for a foundation on Clay soils under undrained conditions can be calculated by Equation (11.8)

$$F_0 = s_{uz}A + 2s_{ua}H_sL \tag{11.8}$$

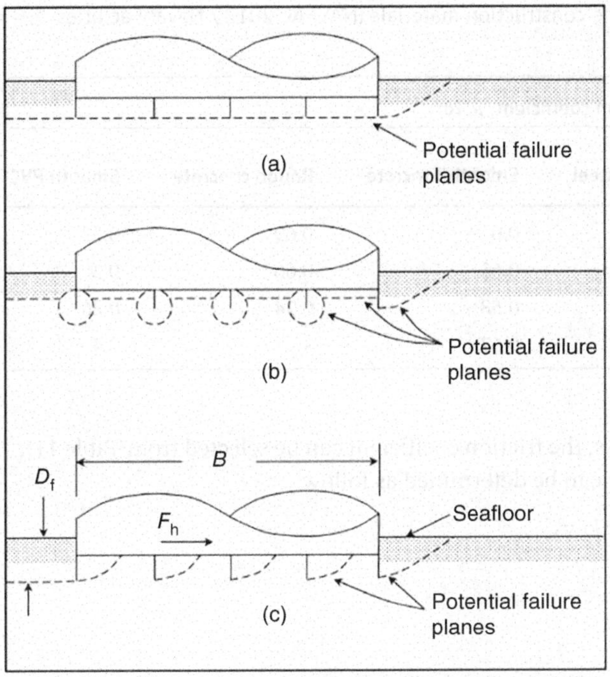

Figure 11.4 Possible failure modes when sliding resistance is exceeded: (a) sliding base failure, (b) deep passive failure, and (c) passive wedge failure (NAVFAC 2012 / Naval Facilities Engineering Systems Command / Public Domain).

where:

s_{uz} = undrained shear strength at the shear key tip
A = foundation base area
s_{ua} = undrained shear strength averaged over the side soil contact zone
H_s = side soil contact height = min (D_f, H + z_s)
D_f = depth of foundation base or skirt tip below the seafloor
H = height of foundation block
L = base length

Note: Generally, shear keys are used for resisting lateral loads. When shear keys do not exist, the lateral bearing capacity is limited to s_{uz}.A or F_n.μ whichever is lower (μ = coefficient of friction = 0.2 for clays).

11.3.2.2 Sand

For drained condition, lateral bearing capacity in the sand is calculated by Equation (11.9)

$$F_0 = \mu[(W_{bf} + W_{bst} + W_b - F_{ve})\cos\beta - F_h \sin\beta] + R_p \tag{11.9}$$

where:

μ = coefficient of friction between foundation base and soil or between soil and soil when shear keys cause this type of sliding failure
W_{bf} = buoyant weight of the foundation
W_{bst} = buoyant weight of the bottom-supported structure
W_b = buoyant weight of soil contained within the footing skirt = $\gamma_b A z_s$
A = footing base area enclosed by the footing skirt
z_s = height of footing shear key skirt
F_h = design environmental loading and mooring line loading in the horizontal direction (downslope is assumed positive)
F_{ve} = design environmental loading and mooring line loading in the vertical direction (upward is positive)
β = seafloor slope angle (deg)
R_p = passive soil resistance on the leading edge of the base and footing shear key skirt

Table 11.1 Coefficient of friction between cohesionless soils and marine construction materials (NAVFAC 2012 / Naval Facilities Engineering Systems Command / Public Domain).

Soil	Soil internal friction coefficient	Friction coefficient, μ, for:				
		Smooth steel	**Rough steel**	**Smooth concrete**	**Rough concrete**	**Smooth PVC**
Quartz Sand	0.67	0.27	0.6	0.6	0.69	0.33
Coralline Sand	0.67	0.2	0.63	0.63	0.66	0.2
Oolitic Sand	0.79	0.23	0.56	0.58	0.74	0.26
Foram Sand-Silt	0.64	0.4	0.66	0.67	–	0.4

Based on soil type and base material type and material roughness, the friction coefficient can be selected from Table 11.1. For conditions that there are no other references, the value of μ can be determined as follow:

$\mu = \tan(\varphi - 5\text{ deg})$ for rough steel or concrete base without shear keys
$\mu = \tan(\varphi)$ for a base with shear keys

where:

φ = soil friction angle (deg)

Note: In the case of a skirted foundation on sandy soils, increasing the weight of the foundation can often enhance the lateral bearing capacity. When the foundation is situated on a sloping surface, the downslope force induces sliding, which is accounted for by the sin β term in the relevant equations. However, it should be noted that certain factors like current scour or animal burrowing activity can lead to the removal of the passive wedge. Consequently, the contribution of the passive wedge is typically disregarded in such cases.

It is necessary to maintain stability against sliding that lateral bearing capacity be larger than the sum of forces causing sliding by a sufficient safety factor. The factor of safety for lateral sliding is determined by Equation (11.10)

$$F_s = \frac{F_0}{(W_{bf} + W_{bst} + W_b - F_{ve})\sin\beta + F_h\cos\beta} \tag{11.10}$$

As suggested by NAVFAC (2012), Safety Factor for long-term loading and short-term loading is 2 and 1.5 respectively. According to

The minimum foundation buoyant weight for skirted foundation ignoring the passive wedge effect can be determined from:

$$W_{bf} = \frac{(F_s + \mu tan\beta)F_h}{\mu - F_s tan\beta} + F_{ve} - W_{bst} - W_b \tag{11.11}$$

11.3.2.3 Clay-Long-Term Loading

After the dissipation of excess pore pressure in cohesive soil, the long-term lateral load capacity of a foundation on cohesive soil is based on failure analysis by drained properties of soil like sand with an additional contribution from c′ which is the effective soil cohesion. The lateral bearing capacity for a foundation on clay soil in the drained condition is determined by:

$$F_0 = C'A + \mu[(W_{bf} + W_{bst} + W_b - F_{ve})cos\beta - F_h sin\beta] \tag{11.12}$$

Therefore, the minimum buoyant weight of the foundation for this case is calculated by:

$$W_{bf} = \frac{(F_s + \mu tan\beta)F_h - \frac{C'A}{cos\beta}}{\mu - F_s tan\beta} + F_{ve} - W_{bst} - W_b \tag{11.13}$$

where

$\mu = \tan(\varphi' - 5\text{ deg})$ for rough steel or concrete base without shear keys
$\mu = \tan(\varphi')$ for a base with shear keys
φ' = drained soil friction angle (deg)

Normally, undrained analysis controls the design. On the other hand, in very few cases for instance very stiff clays, the long-term drained cohesive soil condition will yield lower capacity.

11.3.3 Depth and Spacing of Shear Keys (NAVFAC 2012)

Generally, the shear key height (zs) and spacing are controlled by the needed lateral bearing capacity of the skirted foundation. Recommended maximum depth of shear key for cohesive and cohesion less soil is 0.1B and 0.05B respectively B is the foundation width. For cohesionless soil, it is suggested that the height of the perimeter skirt (shear keys around the edge) will select around 0.1B aiming to control undermining of the foundation caused by scour and animal burrowing.

In cohesive soils and cohesionless soil, the minimum spacing of the shear key is recommended equally to 1zs and 2zs, respectively. By comparing the design load to the passive resistance provided per key, the number of shear keys (n) can be determined in each direction by:

$$n \geq \frac{[F_s F_{hp} + (W_{bf} + W_{bst}) \sin \beta]}{R_p} + 1 \tag{11.14}$$

where:

F_{hp} = resultant of applied load in the downslope direction
R_p = resistance developed by one key against movement along the longitudinal axis according to Equations 11.15 and 11.6

For cohesive soils, the R_p is calculated by:

$$R_p = \left[\frac{\gamma_b z_s^2}{2} + 2s_{ua} z_s\right] B \tag{11.15}$$

where:

S_{ua} = undrained shear strength averaged over the shear key height
B = dimension of foundation perpendicular to the movement direction

For cohesive soils, the R_p is calculated by:

$$R_p = \left[\frac{K_p \gamma_b z_s^2 B}{2}\right] \tag{11.16}$$

$$K_p = \tan^2(45 + \emptyset/2) \tag{11.17}$$

Example 11.1 Determining of vertical and lateral bearing capacity of square skirted foundations in drained sand and undrained clay based on the NAVFAC (2012) code is required.

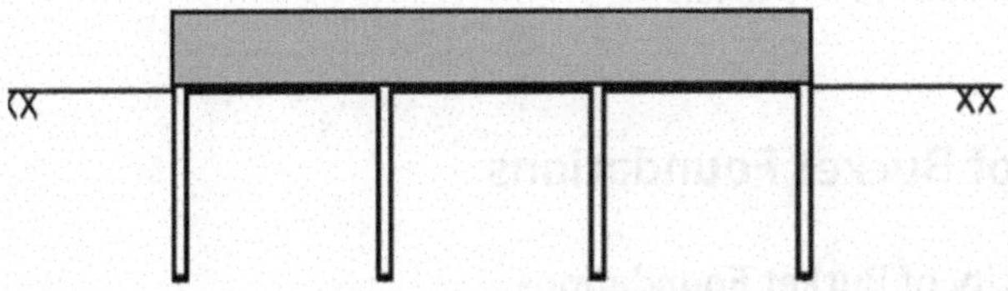

B = 15 m, H = 1m, Z_s = 3m
Sand: γ = 1.9 ton/m^3, ϕ = 30, f_z = 1
Clay: γ sat = 2 ton/m^3, S_t = 4, S_u = 30 kN/m^2
Concrete: γ = 2.4 ton/m^3; Consider the weight of the foundation as permanent vertical load.

Firstly, for vertical bearing capacity:

- **Drained Sand**

$$V_0 = A'\gamma_b[D_f\{1 + (N_q S_q - 1)f_z\} + (B'/2)N_\gamma S_\gamma f_z] + (2B + 2L)(\min(D_f \cdot H + z_s))\gamma_b \tan\delta(D_f + \max(0 \cdot D_f - H - z_s))/2$$

$$A' = A = 15 \times 15 = 225\ m^2;\ D_f = 3\ m$$

$$s_q = 1;\ s_\gamma = 0.6$$

$$N_q = e^{\pi \tan\emptyset} \tan^2(45 + \varphi/2) = e^{\pi \times \tan 30} \tan^2(60) = 18.4$$

$$N_\gamma = (N_q - 1)\tan(1.4\emptyset) = (18.4 - 1) \times \tan(42) = 15.66$$

$$V_0 = 225 \times (1.9 - 1)[3\{1 + (18.4 \times 1 - 1) \times 1\} + 7.5 \times 15.66 \times 0.6] + (30 + 30) \times 3 \times 0.9$$
$$\times \frac{\tan(30 - 5)(3 + 0)}{2} = 25448 + 113.3 = 25561.3\ ton$$

- **Undrained Clay**

$$V_0 = A'\{s_{uz} N_c K_c + \gamma_b D_f\} + (2B + 2L)(\min(D_f \cdot H + z_s))\frac{S_{ua}}{S_t}$$

$$\varphi = 0 \Rightarrow N_c = 5.14\ K_c = 1.2\ S_{uz} = S_{ua} = S_u = 30\ kPa = 3\ {}^{ton}/_{m^2}$$

$$V_0 = 225 \times \{3 \times 5.14 \times 1.2 + (2 - 1) \times 3\} + (30 + 30) \times (3)\frac{3}{4} = 4973.4\ ton$$

Secondly, for Lateral bearing capacity:

- **Drained Sand**

$$F_0 = \mu[(W_{bf} + W_{bst} + W_b - F_{ve})\cos\beta - F_h \sin\beta] + R_p$$

$$\mu = \tan(\varphi) = \tan(30) = 0.577;\ \beta = 0;\ F_{ve} = 0;\ W_{bst} = 0$$

$$W_{bf} = (2.4 - 1) \times 15 \times 15 \times 1 = 315\ ton$$

$$W_b = A\gamma_b z_s = (1.9 - 1) \times 15 \times 15 \times 3 = 607.5\ ton$$

$$K_p = \tan^2\left(45 + \frac{\varphi}{2}\right) = \tan^2(60) = 3;\ K_a = \tan^2\left(45 - \frac{\varphi}{2}\right) = \tan^2(30) = 0.33$$

$$R_p = \left[\frac{K_p \gamma_b z_s^2 B}{2}\right] = \left[\frac{3 \times 0.9 \times 3^2 \times 15}{2}\right] = 182.25\ ton$$

$$P_a = \left[\frac{K_a \gamma_b z_s^2 B}{2}\right] = \left[\frac{0.33 \times 0.9 \times 3^2 \times 15}{2}\right] = 20\ ton$$

$$F_0 = 0.577 \times [(315 + 607.5)] + 182.25 - 20 = 694.5\ ton$$

- **Undrained Clay**

$$F_0 = s_{uz}A + 2s_{ua}H_s L$$

$$F_0 = 3 \times 15 \times 15 + 2 \times 3 \times 3 \times 15 = 945\ ton$$

11.4 Bearing Capacity of Bucket Foundations

11.4.1 Vertical Bearing Capacity of Bucket Foundations

The bearing capacity of a bucket foundation under vertical load can be determined using two different methods, both of which yield similar results. In the case of a skirted foundation (Figure 11.5a), where the soil is trapped by the skirt creating a block, the total bearing capacity can be obtained by considering the sum of tip resistance and shaft friction. This approach takes into account the resistance provided by the soil at the tip of the foundation as well as the frictional resistance along the shaft. Alternatively, when considering internal soil flow within the bucket foundation (Figure 11.5b), the vertical bearing capacity can be determined by considering different factors such as tip resistance, raft resistance (at ground level), internal friction, and external friction. These factors collectively contribute to the overall capacity of the foundation. Both methods, whether based on the skirted approach or internal soil flow, lead to similar results when determining the vertical bearing capacity of a bucket foundation. The choice of method may depend on the specific characteristics of the foundation and the soil conditions at the site.

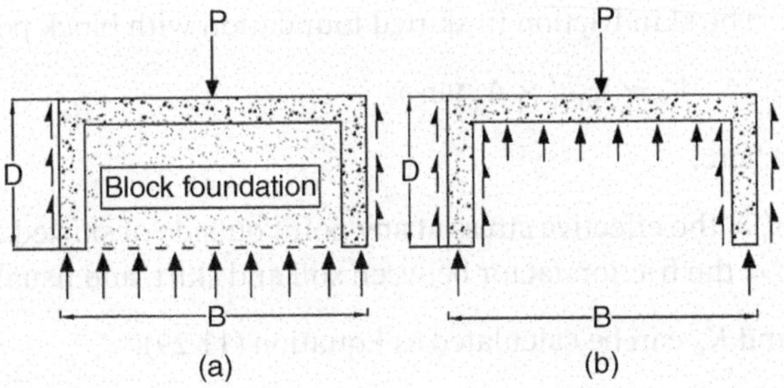

Figure 11.5 Two possible mechanisms for vertical bearing capacity: (a) block foundation and (b) internal soil flow (Rezazadeh and Eslami 2017 / Taylor & Francis Group).

11.4.1.1 Clay

For instance, for the skirted foundation in undrained uniform clay, the ultimate vertical capacity can be calculated by Equation (11.18).

$$V_0 = V_b + V_s \tag{11.18}$$

For the vertical loading, V_b is obtained by the standard end-bearing capacity formula, according to Equation (11.19)

$$V_b = N_c d_c s_c A S_u \tag{11.19}$$

where s_c equals 1.2 for a circular surface rough foundation; and $d_c = 1+n\,(D/B)$, where $0.2 < n < 0.4$ (Brinch Hansen 1970; Meyerhof 1963; Skempton 1951). V_b can be rewritten as Equation (11.20).

$$V_b = N_{cb} A S_u \tag{11.20}$$

where N_{cb} is the end bearing capacity factor for a rough-based and rough-sided circular skirted foundation, and for uniform clay, this factor depends on the embedment ratio.

For skirted foundations, undrained shear strength (S_u), interface roughness term (α), and the embedment ratio (D/B) play the main role in skin resistance as Equations (11.4a & b):

$$V_s = \alpha \cdot S_u \cdot \pi \cdot B \cdot D \tag{11.21a}$$

$$V_s = 4\alpha \cdot (D/B) A S_u \tag{11.21b}$$

The interface roughness factor is assumed as rough ($\alpha = 1$), and Equation (11.21b) can be rearranged as follows:

$$V_s = N_{cs} A S_u \tag{11.21c}$$

where,

N_{cs} = the shaft bearing capacity factor (4D/B), and becomes an enhancing part of the foundation total bearing capacity as the embedment ratio increases.

11.4.1.2 Sand

For sands, the basic equation for vertical bearing capacity is Equation (11.18). Considering block performance, based on Meyerhof (1963) classic formula for end-bearing capacity, V_b can be determined as Equation (11.22):

$$V_b = (\overline{q} N_q s_q d_q + 0.5B' N_\gamma s_\gamma d_\gamma) \times A_b \tag{11.22}$$

where the mentioned parameter in Equation (11.22) can be expressed as below:

$$\overline{q} = \gamma' D \tag{11.23}$$

$$N_q = e^{\pi tan\emptyset} \tan^2(45 + \varphi/2) \tag{11.24}$$

$$s_q = s_\gamma = 1 + 0.1 \times \tan^2(45 + \varphi/2) \tag{11.25}$$

$$d_q = d_\gamma = 1 + 0.1 \times \sqrt{\tan^2(45 + \varphi/2)} \times \frac{D}{B} \tag{11.26}$$

$$N_\gamma = (N_q - 1) \tan(1.4\emptyset) \tag{11.27}$$

The skin friction in skirted foundation with block performance in sands can be calculated by Equation (11.28):

$$V_s = K_0 \sigma_v' \times A_a \tan \delta \tag{11.28}$$

where,

σ_v' = the effective stress at any point outside of skirted foundation,
δ = the friction factor between soil and skirt, and usually taken equal to 2/3Ø,

and K_0 can be calculated as Equation (11.29):

$$K_0 = 1 - \sin Ø \tag{11.29}$$

Therefore, it can be seen that the embedment ratio (D/B) has an influential role in total bearing capacity, and increasing in this portion concludes in total bearing capacity.

11.4.2 Lateral Bearing Capacity of Bucket Foundations

11.4.2.1 Clay

For the horizontal bearing capacity of a circular skirted foundation in undrained saturated clay, there are three resistance factors, including: 1 – base cohesion; 2 – side cohesion; and 3 – soil resistance in the front of the skirted foundation. These factors are depicted in Figure 11.6.

The resistant force depicted in Figure 11.6 can be calculated as Equations (11.30a–11.30c):

$$F_1 = A_b S_u \tag{11.30a}$$

$$F_2 = 2BDS_u \tag{11.30b}$$

$$F_3 = N_c DBS_u \tag{11.30c}$$

where,

A_b = the base area of the foundation; for example, for a circular skirted foundation with block behavior, it is equal to $\pi \frac{D^2}{4}$.

11.4.2.2 Sand

As illustrated in Figure 11.6 the block behavior assumption can be considered for a skirted foundation under horizontal loading in the sand. Therefore, the resistant forces, F_1, F_2, F_3, are estimated by Equations (11.31a–11.31c):

$$F_1 = A_b \sigma_{v1}' \tan \delta \tag{11.31a}$$

$$F_2 = 2BDK_0 \sigma_v' \tan \delta \tag{11.31b}$$

$$F_3 = \frac{1}{2} K_p \gamma B D^2 \tag{11.31c}$$

where,

σ_{v1}' = the effective stress at the end of the skirted foundation, due to the weight of the foundation and vertical loading;

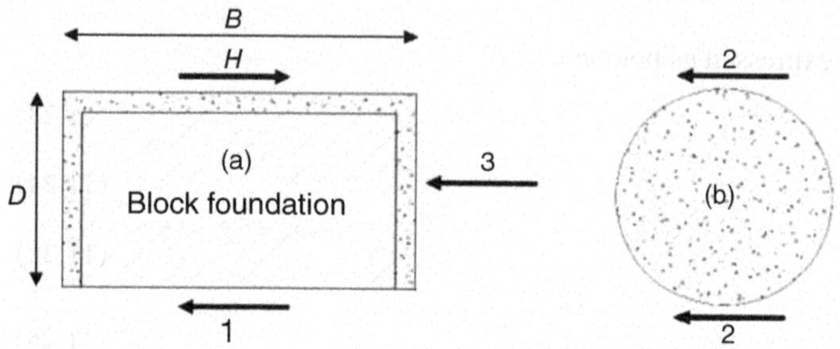

Figure 11.6 Possible resistance factors in the front of horizontal load implemented on a skirted foundation: (a) cross view and (b) top view.

K_p = passive lateral pressure coefficient, which can be calculated as Equation (11.32):

$$K_P = \frac{1 + \sin(\emptyset)}{1 - \sin(\emptyset)} \tag{11.32}$$

Example 11.2 The total vertical and horizontal bearing capacities for a skirted foundation with width (B) equal to 8 m and skirted length (D) equal to 4 m is requested. The foundation is implemented in uniformly saturated clay with $S_u = 50$ kPa. (Assumption of block foundation is acceptable).

Firstly, for vertical bearing capacity:

$$V_0 = V_b + V_s \quad V_b = N_{cb} A_b S_u \quad V_s = N_{cs} A_s S_u$$

$$A_b = \pi \times \frac{8^2}{4} = 50.26\, m^2; \quad A_s = \pi \times 8 \times 4 = 100.53\, m^2; \quad N_{cb} = N_c d_c s_c; \quad N_{cs} = 4\frac{D}{B} = 4 \times \frac{4}{8} = 0.5$$

$N_c = 9$; *According to Meyerhof* 1963.

$$d_c = 1 + n\frac{D}{B} = 1 + 0.3 \times \frac{4}{8} = 1.15$$

$$S_c = 1.2 \rightarrow N_{cb} = 9 \times 1.15 \times 1.2 = 12.42$$

$$V_b = 12.42 \times 50.26 \times 50 = 31211.46\, kN.$$

$$V_s = N_{cs} A_s S_u = 0.5 \times 100.53 \times 50 = 2513.25$$

$$V_0 = V_b + V_s \rightarrow V_0 = 31211.46 + 2513.25 = 33724.71\, kN$$

Secondly, for horizontal bearing capacity, V_H can be determined by summing up the resistant forces introduced in Figure 11.6 through Equations (11.30a–11.30c):

$$F_1 = A_b S_u = \pi \times \frac{8^2}{4} \times 50 = 2513.3\, kN$$

$$F_2 = 2BDS_u = 2 \times 8 \times 4 \times 50 = 3200\, kN$$

$$F_3 = N_c DBS_u = 5.14 \times 8 \times 4 \times 50 = 8224\, kN \cdot (N_c \textit{ is selected conservatively})$$

$$V_H = F_1 + F_2 + F_3 = 2513.3 + 3200 + 8224 = 13937.3\, kN$$

11.5 Bearing Capacity of Spudcans

11.5.1 Bearing Capacity of Spudcans Penetrated in Clay (ABS 2018)

Before a spudcan penetrates the seabed, it is crucial to calculate the bearing capacity at each depth of penetration and predict the load–penetration curve. Conventionally, the penetration of a spudcan is defined as the distance of the tip (lowest point on the spudcan) beneath the mudline. However, in bearing capacity analysis, it is often more convenient to use the lowest depth of the maximum plan area. The careful distinction between the two variables z and h is essential.

Most spudcans have a polygonal shape in plan view. To simplify the bearing capacity calculations, each plan section should be converted to an equivalent circle with a diameter, D, that encloses the same area as the polygon. This conversion allows for a consistent analysis of the foundation's bearing capacity. It is important to note that the displacement mechanism in the soil differs for a spudcan compared to more conventional foundations. As the spudcan is continuously pushed into the seabed, it displaces the soil along its path. Notably, after a certain depth of penetration, the soil flows back into the hole above the spudcan. This backflow of soil is an important feature to consider in the analysis. Equation (11.33) is the relationship between the normalized cavity depth (h_c/D) and the shear strength (S_{uH}/γ'D) at the backflow depth (hc) as shown in Figure 11.7. This equation helps account for the effects of soil displacement and backflow on the bearing capacity of the spudcan foundation.

$$\frac{h_c}{D} = \left(\frac{S_{uH}}{\gamma' D}\right)^{0.55} - \frac{1}{4}\left(\frac{S_{uH}}{\gamma' D}\right) \tag{11.33}$$

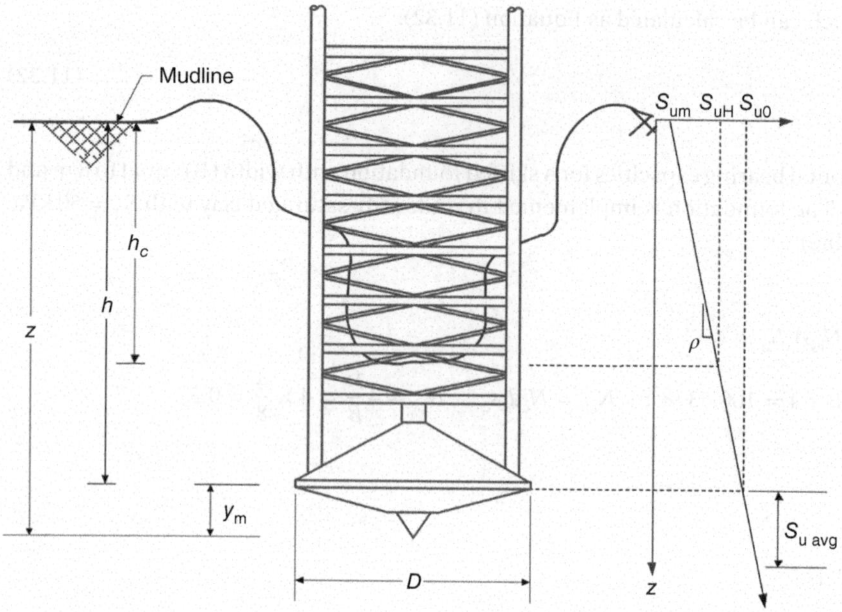

Figure 11.7 Important parameters for embedded spudcan with open soil cavity (ABS 2018 / American Bureau of Shipping).

Based on parameters introduced in Figure 11.7, the vertical bearing capacity of the spudcan foundation (V_0) in clay can be determined through Equations 11.34 to 11.36

- If $z \leq y_m$, then there is partial penetration and no backfill can occur:

$$V_0 = S_{u0}N_cA_{eff} + \gamma' V_c \tag{11.34}$$

where

S_{u0} = undrained shear strength at the lowest depth of the maximum plan area of spudcan
N_c = bearing capacity factor at shallow spudcan penetration prior to any backflow
γ' = submerged soil unit weight, in kN/m^3
D_{eff} = equivalent spudcan diameter at mudline, in m, presented in Figure 11.8
$A_{eff} = \frac{\pi D_{eff}^2}{4}$, in m^2
V_c = volume of the conical spudcan below mudline, in m^3

- If $z \geq y_m$ and $h \leq h_c$, then:

$$V_0 = S_{u0}N_cA + \gamma'(V + Ah) \tag{11.35}$$

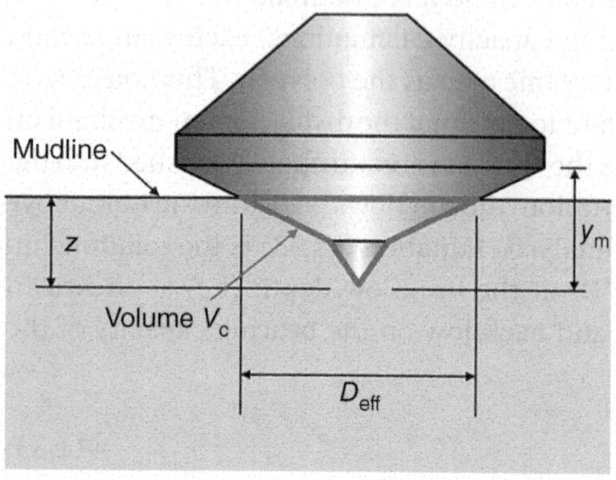

Figure 11.8 Definition of equivalent cone (ABS 2018 / American Bureau of Shipping).

where

A = spudcan maximum cross-sectional area, in m^2
V = volume of the conical spudcan below mudline when $z = y_m$, in m^3

- If $z \geq y_m$ and $h > h_c$, then:

$$V_0 = S_{u0}N_cA + \gamma'(V + Ah_c) \tag{11.36}$$

The above calculations give an approximation of bearing capacity to a depth where there is no backflow, named critical depth. Below the critical depth, there is full incremental backflow around the spudcan. This is clearly an idealization, as in reality, a more gradual transition will occur.

N_c can be determined by Equation (11.37). ABS (2018) presented a table for the estimation of N_c as a function of the con angle β, the dimensionless embedment depth h/D, the roughness factor α, ($0 \leq \alpha \leq 1$, a value of 0.5 is recommended in the absence of evidence that supports any other value), and the dimensionless measure of the rate of increase of strength with depth $\rho D/S_{um}$.

$$N_c = 10\left(1 + 0.065\frac{h}{D}\right) < 11.3 \tag{11.37}$$

11.5.2 Bearing Capacity of Spudcans Penetrated in Sand (ABS 2018)

After the widest cross-section of the spudcan penetrates below the ground surface, soil backflow starts immediately, and any unrestrained sand above the level of the spudcan flows freely into the hole above the spudcan. In this scenario, the backfill sand is typically assumed to rest at the "angle of repose," which can be approximated by the critical state friction angle φcv, as shown in Figure 11.9. However, it is worth noting that in cases where the spudcan has a small penetration depth in the sand, only partial penetration occurs, and backflow is not relevant. In such situations, the soil above the spudcan remains relatively undisturbed, and the analysis can focus on the bearing capacity and interaction of the spudcan with the surrounding soil without considering significant soil backflow.

- If $z \leq y_m$, then there is partial penetration and no backfill can occur and the Equation (11.38) should be used:

$$V_0 = \frac{1}{2}\gamma' D_{eff} N_\gamma F_{mob} A_{eff} + \gamma' V_c \tag{11.38}$$

- If $z > y_m$, then the bearing capacity is calculated by:

$$V_0 - F_{mob}\left(\frac{1}{2}\gamma' D_{eff} N_\gamma \xi_{h\gamma} + \gamma' h N_q \xi_{sq} \xi_{hq}\right) A + \gamma'(V - V_{soil}) \tag{11.39}$$

where

V_{soil} = volume of the backfill soil that rests on the spudcan, in m^3, presented in Figure 11.9
N_q, N_γ = bearing capacity factor, see Equations 11.24 & 11.27
ξ_{sq}, $\xi_{h\gamma}$ = empirical shape factors
ξ_{hq} = empirical depth factor
F_{mob} = mobilization factor

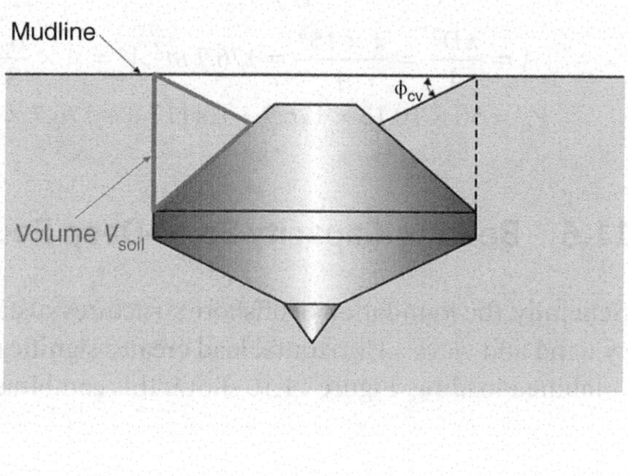

Figure 11.9 Definition of back flow parameters in sand (ABS 2018 / American Bureau of Shipping).

For estimation of the bearing capacity factor, N_γ, ABS (2018) suggested N_γ in terms of the cone apex angle, β, the roughness factor, α, ($0 \le \alpha \le 1$, a value of 0.5 is recommended in the absence of evidence that supports any other value), and the friction angle, ϕ'. The values are only presented for surface footings (h = 0). For h > 0, the value $\xi_{h\gamma} = 1.0$ is recommended (i.e., no adjustment should be made to this term to account for the depth of embedment).

To convert the plane strain condition to axial symmetry condition, The empirical shape factor $\xi_{sq} = 1 + \tan\phi'$ is applied. A depth factor $\xi_{hq} = 1 + 2\tan\phi'(1 - \sin\phi')2$ tan-1 (h/D) makes an increase in the contribution of this factor via depth.

The mobilized factor, F_{mob}, is used for the calculated resistance. ABS (2018) suggested values of 0.25 to 0.5 for this factor based on a back analysis of 10 case records, of which eight are from the North Sea Region and the others from the Gulf of Mexico and offshore Australia.

11.5.3 Correlations of Spudcan Capacity with Penetrometer Tests (ABS 2018)

When high-quality continuous data from penetration testing are available for instance piezocone penetration test, T-bar or ball penetrometer, the alternative method expressed in Equation (11.40) can be used for determining spudcan resistance in clay soils. Great caution must be practiced for use of the direct correlation and it is not recommended in the layered soil.

$$V_0 = \frac{N_c}{N_{kt}} q_{net} A + \gamma'(hA + V - V_{soil}) \tag{11.40}$$

where

q_{net} = net cone resistance, which is the measured cone resistance being corrected for the pore pressure effects and the total in situ vertical stress, in N/m².

Example 11.3 The total vertical bearing capacity for a spudcan foundation with width (D) equal to 15 m, the y_m equal to 2 m and embedded depth (h) equal to 3 m is requested. The foundation is implemented in uniformly saturated clay with $S_u = 50$ kPa and $\gamma' = 10$ kN/m³.

The N_c value for a typical deeply penetrated spudcan is approximately 9.0, and the N_{kt} value recommended for cone interpretation is approximately 13.5 to 18.6 based on the calibration method. The N_c/N_{kt} ratio varies between 0.48 and 0.67. The self-weight and buoyancy terms in the spudcan analysis have been included as $\gamma'(hA + V - V_{soil})$, in N. In the cases where the soil is sufficiently coarse partial drainage will occur during the penetrometer test, even though the spudcan penetration will be undrained. The N_c/N_{kt} ratio needs to be reduced to account for this effect.

The depth of the spudcan toe (z) is:

$$z = h + y_m = 3 + 2 = 5\,m > y_m$$

The back flow depth (h_c) should be calculated as follow:

$$\frac{h_c}{D} = \left(\frac{S_{uH}}{\gamma' D}\right)^{0.55} - \frac{1}{4}\frac{S_{uH}}{\gamma' D} = \left(\frac{50}{10 \times 15}\right)^{0.55} - \frac{1}{4} \times \frac{50}{10 \times 15} = 0.46 \rightarrow h_c = 0.46 \times 15 = 6.9\,m$$

According to values of h, z, h_c, and y_m, it can be con concluded that $z > y_m$ and $h < h_c$. So, Equation (11.35) is used for estimating the vertical bearing capacity as below:

$$V_0 = S_{u0} N_c A + \gamma'(V + Ah)$$

$$N_c = 10\left(1 + 0.065\frac{h}{D}\right) = 10\left(1 + 0.065 \times \frac{3}{15}\right) = 10.13$$

$$A = \frac{\pi D^2}{4} = \frac{\pi \times 15^2}{4} = 176.7\,m^2; V = A \times \frac{y_m}{3} = 176.7 \times \frac{2}{3} = 117.8\,m^3$$

$$V_0 = 50 \times 10.13 \times 176.7 + 10(117.8 + 176.7 \times 3) \cong 95977\,kN$$

11.6 Bearing Capacity Semi-Deep Foundations under Combined Loading (VMH)

Generally, the foundation of offshore structures sustains vertical load (V) due to self-weight and horizontal loads (H) caused by wind and waves. Horizontal load creates significant moment (M) on foundations; therefore, they are subjected to VMH combined loading. Figure 11.10 shows this combination of loads on the foundation of an offshore wind turbine.

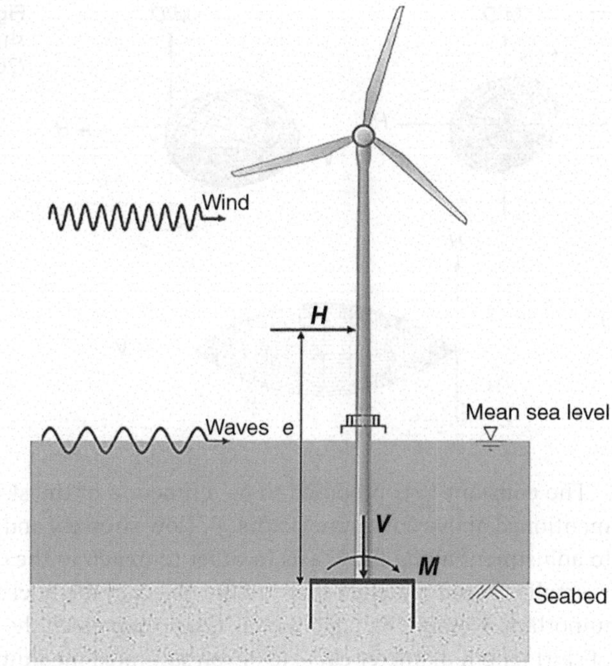

Figure 11.10 VMH loading for foundation of offshore wind turbine.

In recent years, researchers have focused on investigating the behavior of foundations under combined loads, including vertical, horizontal, and moment loading. The focus of their research has been using numerical simulations and laboratory studies to establish failure envelopes for foundations subjected to VMH loading. Several researchers have made significant contributions in presenting failure envelopes for skirted/hybrid foundations under VMH loading (Cheng et al. 2019; Ibsen et al. 2013; Kim et al. 2020; Rezazadeh and Eslami 2017; Villalobos et al. 2005; Xia et al. 2021; Zhang et al. 2010). The shape of the yield surface, characterized by hardening behavior, is a fundamental aspect of the proposed models for the failure envelope as reported in the literature. The vertical plastic settlement of the foundation plays a crucial role in controlling the shape of the yield surface. Empirical equations, such as Equation (11.41) proposed by Ibsen et al. (2013), are commonly used to describe the form of the yield surfaces:

$$f = \left(\frac{H}{h_0 V}\right)^2 + \left(\frac{M}{m_0 DV}\right)^2 - 2a\left(\frac{H}{h_0 V}\right)\left(\frac{M}{m_0 DV}\right) - F(V, V_t, V_{pre}) = 0 \tag{11.41}$$

where

f = a yield surface related to the shape of the failure surface,
D = the foundation diameter,
V_t = the capacity of the bucket foundation under tensile loading.

The general shape of the surface is obtained by the three parameters, h_0, m_0, and a, in the radial plane; the parameters h_0 and m_0 specify the size of the yield surface at the most comprehensive section of the surface, along the V-axis by HM = 0/V and MH = 0/DV, respectively, where HM = 0 is the value of H at the cross point with the M = 0 axis, and MH = 0 is the corresponding value for M. For determining the rotation ellipse in the radial planes, the eccentricity factor "a" is used. Figure 11.11 illustrates a complete shape of a rotated yield surface according to Equation (11.41) for shallow circular footing. For more detailed discussion on this topic, please refer to Ibsen et al. 2013.

There is a drawback according to the skirted foundation's behavior related to the presented yield surface correlation. Because of the tension-bearing capacity of the bucket foundation, its envelope surface has a cross point with V-axis at a negative value. In contrast, it will not intersect at zero vertical loads. Therefore, Villalobos et al. (2005) introduced a dimensionless constant, t_0, for modifying the yield surface. They finally proposed $F(V, V_t, V_{pre})$ as below:

$$F(V, V_t, V_{pre}) = \left(\frac{\beta_{12}}{(t_0+1)^{(\beta_1+\beta_2)}}\right) + \left(\frac{V}{V_{pre}} + t_0\right)^{2\beta_1}\left(1 - \frac{V}{V_{pre}}\right)^{2\beta_2} \tag{11.42}$$

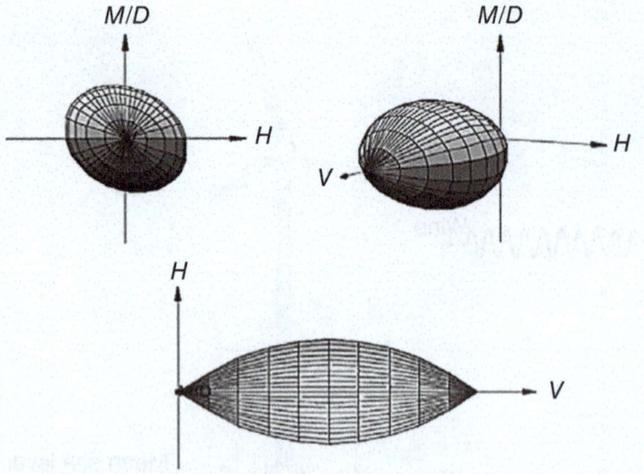

Figure 11.11 Illustration of yield envelope surface for a shallow footing shaped like a parabola and rotated ellipse, based on Byrne and Houlsby's (2003) expression.

The constant t_0 is proposed to be a function of the skirt length, D, relative to the diameter of the bucket (B), and the mentioned above curvature factors, β_1 (low stresses) and β_2 (high stresses), cause the parabolic shape of the yield surface to adjustment along the V axis in order to match to the experimental data. The self-weight of the wind turbine structure is small. Therefore, when they are the object of the bucket foundation, the tip of the yield surface at a low vertical load is important. Notably, Villalobos et al. (2005) expressed the yield surface function based on a bucket foundation with a ratio of skirt length diameter equal to 0.5 on saturated medium-dense sand.

Rezazadeh (2017) has done extensive research on the derivation of failure envelopes in 2D and 3D spaces for the behavior of a skirted foundation under H-V, M-V, H-M, and VMH combined loading in sand and undrained saturated clay by using numerical modeling. There are several factors, such as soil specifications, foundation dimensions, and load combinations, which can affect the size and shape of the failure envelope of a foundation in 2D/3D spaces. Following Rezazadeh and Eslami's (2017) research results, failure envelopes in 2D/3D spaces for clay and sand are presented, and the engineering applications of these surfaces will be discussed.

11.6.1 2D Failure Envelopes for Undrained Saturated Clay

11.6.1.1 H-V

In this space, the low boundary of a skirted foundation's bearing capacity under H-V combined loading can be described as Equation (11.43).

$$\frac{H}{H_{ult}} = 1 - 0.22\left(\frac{V}{V_{ult}}\right) - 0.47\left(\frac{V}{V_{ult}}\right)^2 - 0.21\left(\frac{V}{V_{ult}}\right)^3 \tag{11.43}$$

As Rezazadeh's and Eslami's (2017) research results demonstrate, with increases in the embedment depth ratio, D/B, and non-homogeneity coefficient $K = k.B/S_{um}$, where S_{um} = undrained shear strength at the ground surface; k = rate of increasing soil strength with depth, the failure surface of H/H_{ult}-V/V_{ult} moves slightly towards the inside, but the general shape is the same. Figure 11.12 illustrates the effect of D/B and K on the failure surface shape of a skirted foundation in an H-V loading space.

11.6.1.2 M-V

The shape of the yield surface for a skirted foundation in saturated clay with an undrained condition under M-V combination loads can be determined by Equation (11.44).

$$\frac{M}{M_{ult}} - 1 - 0.24\left(\frac{V}{V_{ult}}\right) - 0.89\left(\frac{V}{V_{ult}}\right)^2 + 0.13\left(\frac{V}{V_{ult}}\right)^3 \tag{11.44}$$

Comparison of Equation (11.44) with Equation (11.43) shows that the general shapes of the yield surfaces in H-V and M-V spaces are mostly the same, so increasing the embedment depth ratio and nonhomogeneity of clay can change the surface boundary curves, like the H-V surface.

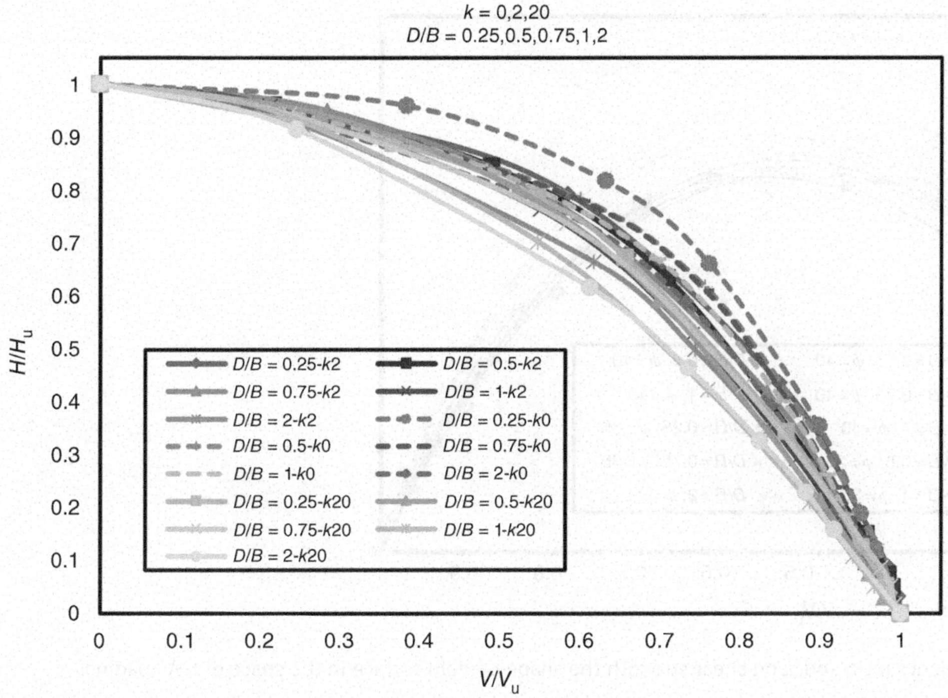

Figure 11.12 Effect of embedment depth ratio and non-homogeneity of clay on the shape of yield surface in the space of H-V loading (Rezazadeh 2017).

11.6.2 2D Failure Envelope for Sand

11.6.2.1 H-V

In the space of H-V combined loading for sand, the lower boundary of the skirted foundation bearing capacity can be evaluated by Equation (11.45).

$$\frac{H}{V_{ult}} = 0.02 + 0.64\left(\frac{V}{V_{ult}}\right) - 0.33\left(\frac{V}{V_{ult}}\right)^2 - 0.33\left(\frac{V}{V_{ult}}\right)^3 \tag{11.45}$$

As it is depicted in Figure 11.13 the increase of embedment depth ratio (D/B) has a negligible effect on the yield surface normalized by the ultimate vertical capacity (V_{ult}), as well as where the vertical load is more than half of the vertical bearing capacity; envelope curves are almost overlapping. On the other hand, where the embedment depth ratio (D/B) is enhanced, especially for D/B = 2, the size of the yield surface, especially for $V/V_{ult} < 0.5$, will expand.

11.6.2.2 M-V

Like H-V yield surfaces, in this space, normalizing V and M based on the vertical bearing capacity can make reasonable envelopes, which are shown in Figure 11.14. As presented, except for D/B = 2, all other curves have an almost similar shape. Therefore, the yield surface for a skirted foundation's bearing capacity in the sand under M-V combined loading can be conservatively described as Equation (11.46).

$$\frac{M}{BV_{ult}} = 0.007 + 0.34\left(\frac{V}{V_{ult}}\right) - 0.46\left(\frac{V}{V_{ult}}\right)^2 + 0.11\left(\frac{V}{V_{ult}}\right)^3 \tag{11.46}$$

11.6.3 3D Failure Pattern for VMH Loading

3D ultimate surfaces are utilized for quantifying of the effects of the different parameters, which can affect the shape and size of the envelope in a 3D space. Regarding the research conclusion reported by Rezazadeh and Eslami (2017), The effect of soil properties and foundation shape on the size and figure of a 3D envelope, which is normalized by ultimate capacity (V_{ult}, M_{ult}, H_{ult}), is negligible, whereas, in the sand, this similarity feature can be seen when normalization would be applied

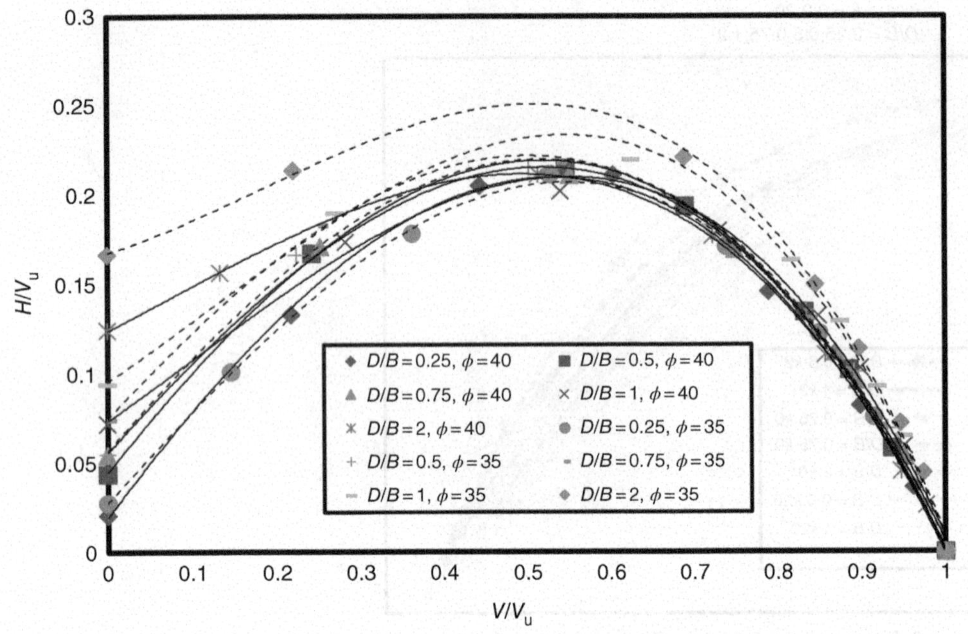

Figure 11.13 Effect of embedment depth ratio and sand shear strength the shape of yield surface in the space of H-V loading (Rezazadeh 2017).

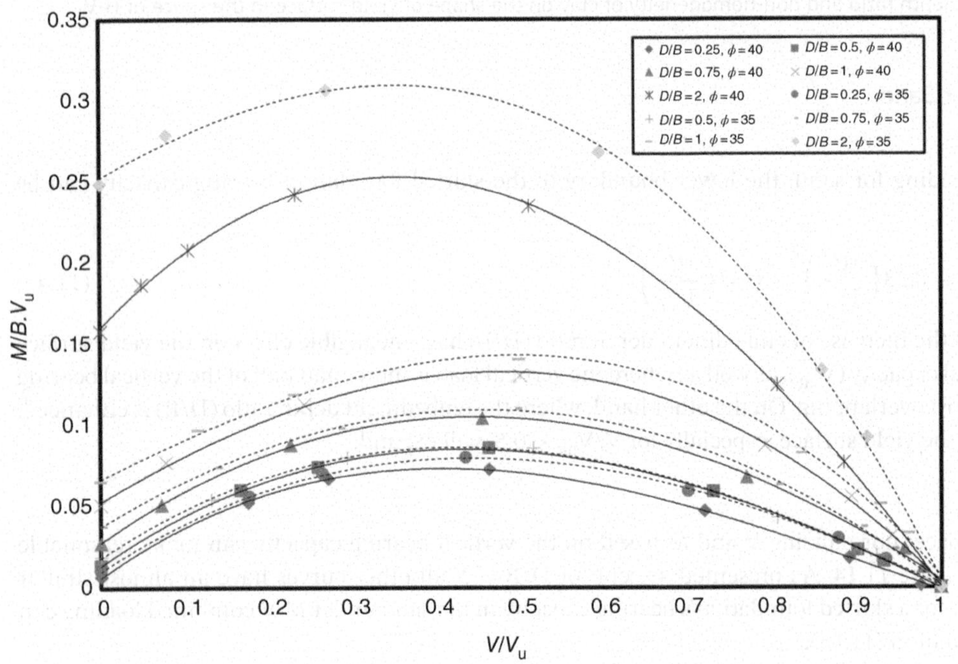

Figure 11.14 Effect of embedment depth ratio and sand shear strength the shape of yield surface in the space of M-V loading (Rezazadeh 2017).

by vertical ultimate bearing capacity (V_u). Figures 11.15 and 11.16 show the 3D ultimate surface for a specific embedment depth ratio (D/B = 0.25) for homogenous clay and medium sand, respectively.

There are three general parameters that have the main effects on the bearing capacity of skirted/hybrid foundations under combined loading (VMH), including:

- Shape of foundation
- Soil properties
- Quantity and combination of loading

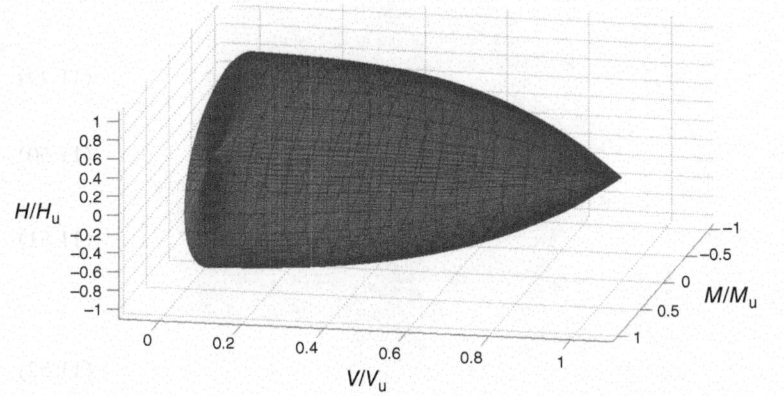

Figure 11.15 3D failure envelope in the homogenous clay for the skirted foundation with D/B = 0.25 (Rezazadeh 2017).

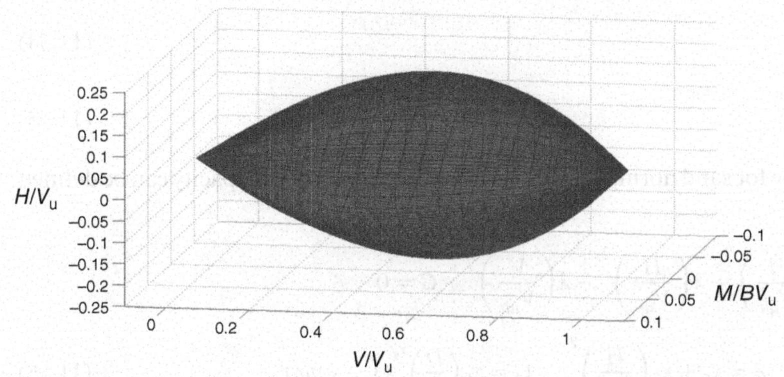

Figure 11.16 3D failure envelope in the medium sand for the skirted foundation with D/B = 0.25 (Rezazadeh 2017).

Regarding embedment depth ratio (D/B), in clay, ultimate envelopes related to M/M_{ult} -V/V_{ult} and V/V_{ult}-H/H_{ult} are shrinkage. With the increase in the D/B ratio. However, envelopes in the space of V/V_{ult}-H/H_{ult} enlarge, and the eccentricity increases. In sand, all curves of envelopes in the space of V/V_{ult}-M/BV_{ult}-H/V_{ult} enlarge by D/B increasing.

As for the role of soil specification, it can be said that with an increase in clay's non-homogeneity, the 3D ultimate surface in the space of V/V_{ult}-M/M_{ult}-II/H_{ult} becomes smaller. In addition, an increase in sand's strength/stiffness characteristic causes the 3D failure envelope in the space of V/V_{ult}-M/BV_{ult}- H/V_{ult} to be enlarged. Moreover, increasing the vertical load in clay makes the ultimate surface envelope small. On the other hand, when the vertical load increases to about half of the ultimate vertical capacity, the bearing capacity of the foundation is enhanced. Afterward, an increase in the vertical load causes a decrease in the skirted foundation's bearing capacity under VMH loading.

11.7 Main Correlations for Failure Pattern Under VMH Loading

Rezazadeh and Eslami 2017 utilized vast numerical modeling, along with the genetic algorithm optimization method, to excavate suitable coefficients for the best description of the failure envelope of skirted foundations under VMH loading.

The general function of the VMH failure envelope for clay normalized by ultimate bearing capacity can be defined as Equation (11.47):

$$F_{(VMH)} = \lambda\left(\frac{M}{M_{ult}}\right)^2 + \alpha\beta\left(\frac{M}{M_{ult}}\right)\left(\frac{H}{H_{ult}}\right) + \lambda\left(\frac{H}{H_{ult}}\right)^2 + \lambda\left(\frac{V}{V_{ult}}\right)^2 - \lambda = 0 \tag{11.47}$$

$$\alpha = x_1\left(\frac{D}{B}\right)^{x_2}, \beta = x_3\left(1 - \frac{V}{V_{ult}}\right)^{x_4}, \lambda = x_5 + x_6\left(\frac{H}{H_{ult}}\right)^2 \tag{11.48}$$

For homogeneous clay:

$$\alpha = -0.8958\left(\frac{D}{B}\right)^{9.2923} \tag{11.49}$$

$$\beta = 0.0375\left(1 - \frac{V}{V_{ult}}\right)^{0.3483} \tag{11.50}$$

$$\lambda = -0.0026 + 0.0551\left(\frac{H}{H_{ult}}\right)^{2} \tag{11.51}$$

For non-homogeneous clay:

$$\alpha = 0.0529\left(\frac{D}{B}\right)^{1.1493} \tag{11.52}$$

$$\beta = 0.2109 + 3.7042\left(\frac{V}{V_{ult}}\right)^{1.9729} \tag{11.53}$$

$$\lambda = -0.0017 + 0.0512\left(\frac{H}{V_{ult}}\right)^{2} \tag{11.54}$$

$$G = -0.13\left(\frac{D}{B}\right)^{0.8941}(1 - \sin\varphi)^{7.3528} \tag{11.55}$$

The general function of the VMH failure envelope for sand normalized by vertical ultimate bearing capacity can be defined as Equation (11.56):

$$F_{(VMH)} = \lambda\left(\frac{M}{B \cdot V_{ult}}\right)^{2} + \alpha\beta\left(\frac{M}{B \cdot V_{ult}}\right)\left(\frac{H}{V_{ult}}\right) + \lambda\left(\frac{H}{V_{ult}}\right)^{2} - \lambda\left(\frac{V}{V_{ult}}\right)^{2} - G = 0$$

$$\alpha = x_1\left(\frac{D}{B}\right)^{x_2}, \quad \beta = x_3 + x_4\left(\frac{V}{V_0}\right)^{x_5}, \quad \lambda = x_6 + x_7\left(\frac{H}{H_0}\right)^{2}, \quad G = x_8\left(\frac{D}{B}\right)^{x_9}(1 - sin\varphi)^{x_{10}} \tag{11.56}$$

For obtaining precise results, and for more confidence when one of the loads is zero, it is suggested to use Equations (11.57), (11.61), and (11.65) for the failure envelopes of M-H, V-H, and V-M.

$$F_{(VM)} = \left(\frac{M}{B \cdot V_{ult}}\right)^{2} + \alpha\left(\frac{M}{B \cdot V_{ult}}\right)\left(\frac{V}{V_{ult}}\right) + \beta\left(\frac{V}{V_{ult}}\right)^{2} - G = 0 \tag{11.57}$$

$$\alpha = -0.0017\left(\frac{D}{B}\right)^{8.0446} \tag{11.58}$$

$$\beta = -0.4319 + 0.4364\left(\frac{V}{V_{ult}}\right)^{1.006} \tag{11.59}$$

$$G = 0.4851\left(\frac{D}{B}\right)^{1.9671}(1 - \sin\varphi)^{5.0730} \tag{11.60}$$

$$F_{(VH)} = \left(\frac{H}{V_{ult}}\right)^{2} + \alpha\left(\frac{H}{V_{ult}}\right)\left(\frac{V}{V_{ult}}\right) + \beta\left(\frac{V}{V_{ult}}\right)^{2} - G = 0 \tag{11.61}$$

$$\alpha = 0.0031\left(\frac{D}{B}\right)^{9.2318} \tag{11.62}$$

$$\beta = -0.3076 + 0.3202\left(\frac{V}{V_{ult}}\right)^{0.6933} \tag{11.63}$$

$$G = 2.2855\left(\frac{D}{B}\right)^{0.4533}(1 - \sin\varphi)^{6.5116} \tag{11.64}$$

$$F_{(MH)} = \left(\frac{M}{B \cdot V_{ult}}\right)^{2} + \alpha\left(\frac{M}{B \cdot V_{ult}}\right)\left(\frac{H}{V_{ult}}\right) + \left(\frac{H}{V_{ult}}\right)^{2} - G = 0 \tag{11.65}$$

$$\alpha = 1.8842\left(\frac{D}{B}\right)^{0.3295} \tag{11.66}$$

$$G = 0.5775\left(\frac{D}{B}\right)^{1.6879}(1 - \sin\varphi)^{5.6642} \tag{11.67}$$

Example 11.4 Safety evaluation of skirted foundation of a storage tank in clay deposits

In this case study, a skirted foundation with B = 10 m and D = 5 m is selected for the foundation of a storage tank in saturated homogeneous clay, as depicted. The soil and loading properties are described below:

$S_u = 25$, $E_u = 10$ MPa; $V = 5$ MN, $H = 1.5$ MN, $M = 22.5$ MN.

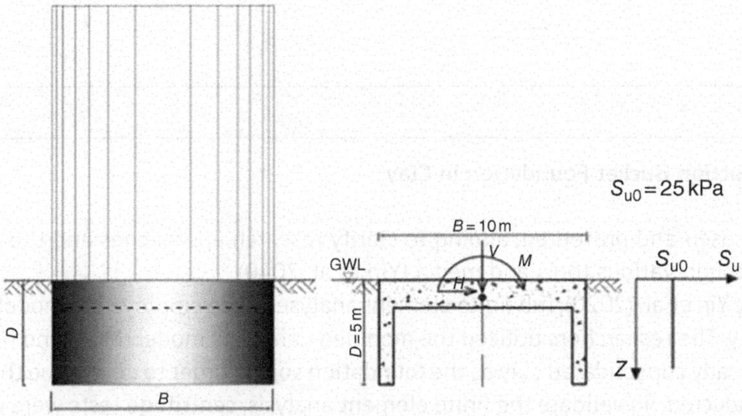

It is important to mention that it is possible to verify the safety of the foundation under VMH loading by the failure surface Equation (11.47), with coefficients defined by Equation (11.48) to Equation (11.51). Firstly, the ultimate bearing capacity of a skirted foundation under horizontal and moment loading should be determined by charts, determined by numerical analysis, which are illustrated below. Regarding these charts, it can be seen that the ultimate bearing capacity of a skirted foundation under pure horizontal/moment loads are 10.1 MN and 34 MN.m, respectively.

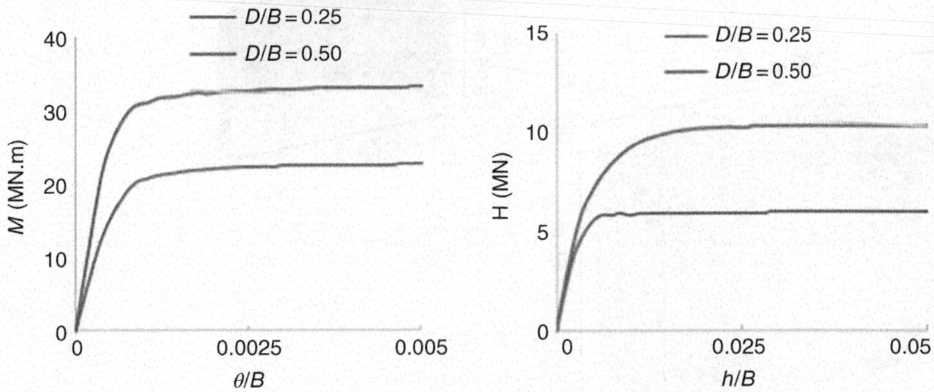

After calculating envelope coefficients based on shape, soil, and loading condition and replacing them in Equation (11.47), safety evaluation is accessible.

$$F_{(VMH)} = \lambda\left(\frac{M}{M_{ult}}\right)^2 + \alpha\beta\left(\frac{M}{M_{ult}}\right)\left(\frac{H}{H_{ult}}\right) + \lambda\left(\frac{H}{H_{ult}}\right)^2 + \lambda\left(\frac{V}{V_{ult}}\right)^2 - \lambda = 0$$

$$\alpha = -0.8958(0.5)^{9.2923} = -0.001428234$$

$$\beta = 0.0375\left(1 - \frac{5}{21.8}\right)^{0.3483} = 0.034247$$

$$\lambda = -0.0026 + 0.0551\left(\frac{1.5}{10.1}\right)^2 = -0.001385$$

$$F_{(VMH)} = -0.001385\left(\frac{22.5}{34}\right)^2 + (-0.001428234)(0.034247)\left(\frac{22.5}{34}\right)\left(\frac{1.5}{10.1}\right) + (-0.001385)\left(\frac{1.5}{10.1}\right)^2$$
$$+ (-0.001385)\left(\frac{5}{21.8}\right)^2 - (-0.001385) = 0.00067 > 0.$$

When F(VMH) is positive, it means that the equivalent point of VMH loads is outside of the failure envelope, and is not in safe status.

11.8 Case Studies

Case Study 11.1

Numerical Modeling of Suction Bucket Foundation in Clay

A few case studies are chosen and presented, aiming to clarify research approaches and the application of reviewed concepts and theories through various tools and media (Yin et al. 2020).

In a study conducted by Yin et al. (2020), full finite element analyses were employed to model the behavior of suction bucket foundations in clay. The researchers utilized the modified Cam-Clay model (MCC) and the hardening soil model (HS) to represent the normally consolidated clay as the foundation soil. In order to determine the necessary parameters, laboratory tests were conducted. To validate the finite element analysis, centrifuge tests were conducted. Additionally, the numerical analysis was extended using the radial displacement control method to investigate the failure envelope of the suction bucket foundation under combined vertical, horizontal, and moment loading (VMH loading). The results of this study provided new analytical formulations for the 3D failure patterns in the VMH loading space. The suction bucket foundation model was implemented in PLAXIS3D, as depicted in Figure 11.17. The embedment depth ratio (L/D ratio) was set to 0.5, with D being equal to 7.5 m.

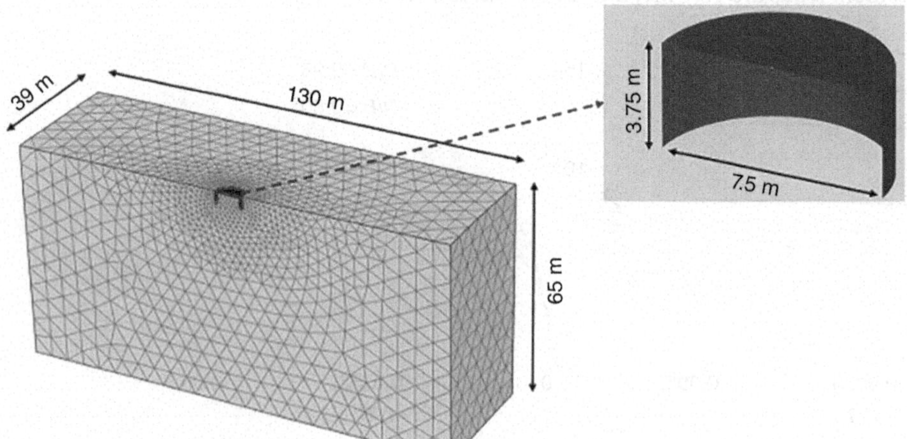

Figure 11.17 FEM model with geometries in PLAXIS 3D (Yin et al. 2020 / with permission of Elsevier).

The VMH loads are applied to the loading reference point (LRP) of the foundation. The weight of an offshore wind turbine (OWT) is typically light (around 6 MN) compared to the vertical bearing capacity of the OWT foundation. Therefore, the bearing capacity under horizontal and moment loading is critical.

Assumptions that researchers have taken for obtaining the failure push in the H-M space and main results are summarized below:

- Various displacements are applied to the LRP with a constant ratio between rotation and horizontal displacement (δ_θ/δ_u = constant).
- a shows the failure envelope in H-M space for two different utilized constitutive models. Figure 11.18 expressed the effect of vertical loading on the shape of the failure surface in the H-M plane.
- The effect of bending moment on the horizontal bearing capacity is remarkable.
- The incline ellipse is the best description for the figure of the suction bucket envelope in clay.
- The increase in pre-vertical load causes shrinkage of failure envelope size.

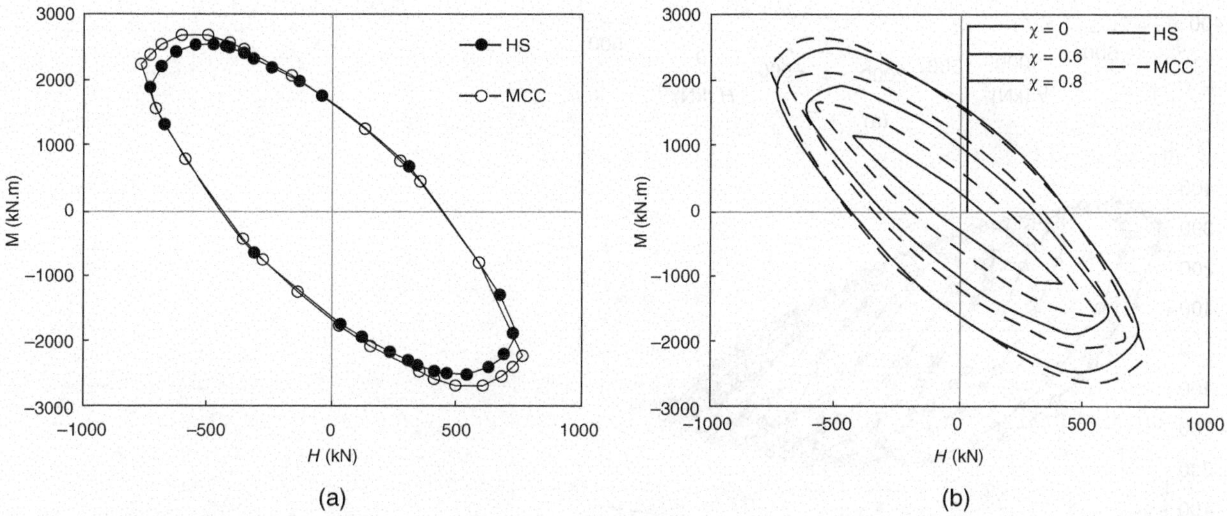

Figure 11.18 Failure envelopes in H-M plane at (a) $\chi = 0$ and (b) different vertical load levels based on HS and MCC models. χ = Vi/V0 (V0 is maximum vertical load) (Yin et al. 2020 / with permission of Elsevier).

Yin et al. combined the results in the H-M plane under different levels of vertical load and H-V plane, and they plotted the VMH 3D space envelope, which is depicted in Figure 11.19. The size of the inclined ellipse is controlled by vertical load. In this study, the researchers tried to obtain the best equations describing the failure surface of the suction caisson under VMH loading, and introduced their proposed Equations (11.68) and (11.69), based on Villalobos et al.'s (2009) formula.

$$F = \left(\frac{H}{h_0 V}\right)^2 + \left(\frac{M}{Dm_0 V_0}\right)^2 + 2e\left(\frac{H}{h_0 V_0}\right)\left(\frac{M}{Dm_0 V_0}\right) - F_V(V, V_0) = 0 \tag{11.68}$$

$$F_V(V, V_0) = \left(1 - \left(\frac{V}{V_0}\right)\right)\left(0.88\left(\frac{V}{V_0}\right)^2 + 1.08\left(\frac{V}{V_0}\right) + 1\right) \tag{11.69}$$

The parameters determined by adopting numerical results with the optimal curve are shown in Table 11.2.

(Continued)

Case Study 11.1 (Continued)

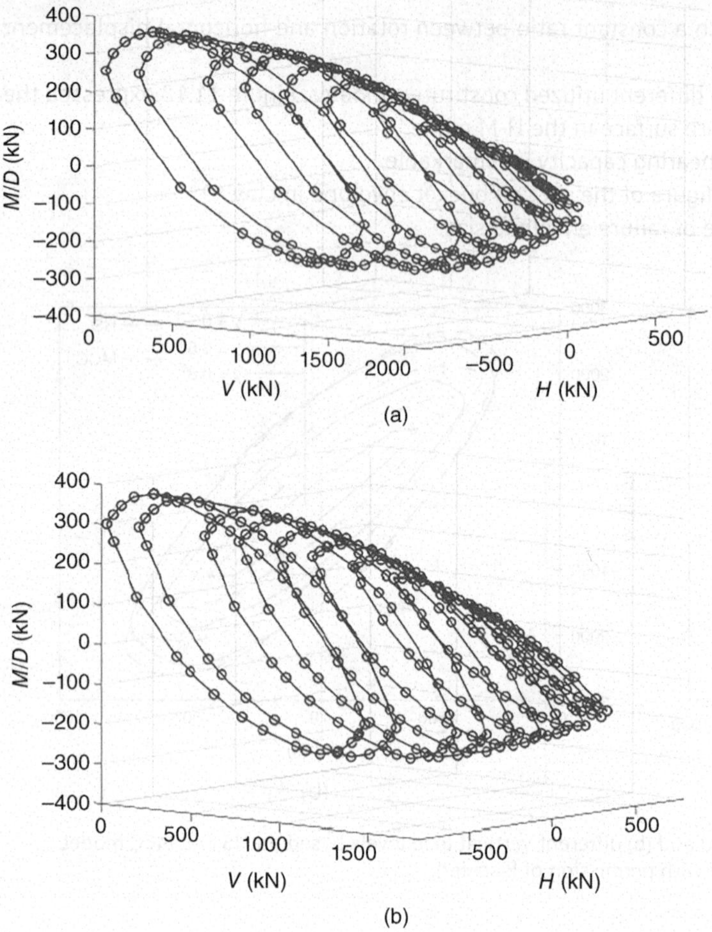

Figure 11.19 VMH failure envelope in 3D space using: (a) HS model and (b) MCC model (Yin et al. 2020 / with permission of Elsevier).

Table 11.2 Parameters of failure envelope for the suction bucket foundation (Yin et al. 2020/with permission of Elsevier).

Parameter	Name	Value (HS)	Value (MCC)
V_0	Vertical bearing capacity: kN	2150	1800
h_0	Dimension of failure surface (horizontal)	0.213	0.245
m_0	Dimension of failure surface (moment)	0.101	0.124
e	The eccentricity of the failure surface	0.768	0.80

Case Study 11.2

Centrifuge Study of Improving Load Capacity of Skirted Foundation by Hybrid Technique (Kim et al. 2020)

In recent years, the hybrid bucket foundation (HBF) design has been proposed for increasing the load capacity of ordinary bucket foundations (OBF); in this study, researchers investigated the comparison of HBFs and OBFs behavior under VMH loading by centrifuge (70 g) models The result confirmed the feasibility and superior performance of HBFs.

Pure vertical (V), Horizontal (H) loads and combined HM loading were applied to steel buckets. The schematic set of test, loading system, and HSBF's shape studied in this research are shown in Figure 11.20. Moreover, their characteristics are summarized in Table 11.3.

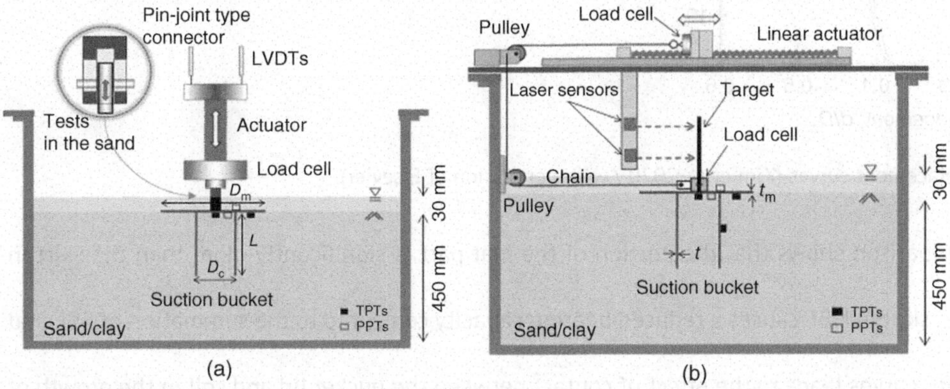

Figure 11.20 Scheme for vertical and horizontal loading: (a) vertical loading system and (b) horizontal loading system (Kim et al. 2020 / with permission of Elsevier).

Table 11.3 Characteristics of HBF and OBF model (adapted from Kim et al. 2020).

Parameter	Name	Model (mm)	Prototype (m)
D_c	Bucket diameter	71.4	5
D_m	Circular mat diameter	142.8	10
L	Skirt length	71.4	5
t	Tip thickness	3	0.02

The HBF comprises circular mats attached to a single bucket, where the ratio of mat diameter to bucket unit is two. The mat area added to the single bucket is around 60% of the total OBF surrounding area; besides, the HSBF submerge weight is 44% larger than the SBF one. However, due to the small portion of offshore wind turbine self-weight to vertical bearing capacity (V_y) of the foundation, the influence of this parameter is not notable. Researchers tried to probe the influence of HSBF's unit on the bearing capacity by testing a circular mat foundation (CMF) exclusively.

The selected soil was homogeneous silica sand with $\gamma' = 14.4\ \text{kN/m}^3$ ($D_r = 60\%$). For preparing the model, firstly, the soil was prepared to a depth of 450 mm. The deposits would be saturated by adding water from the surface and pre-spinning to g-level in the centrifuge before applying the loading test.

A remarkable result of this research, vertical load–displacement charts for different foundation models, are depicted in Figure 11.21. Major important conclusions by investigation of this chart can be as below:

- The charts have no peak value during the increase of load displacements due to the hardening property of silica sand.
- HBF vertical bearing capacity has far distance from other types of foundations.

(Continued)

Case Study 11.2 (Continued)

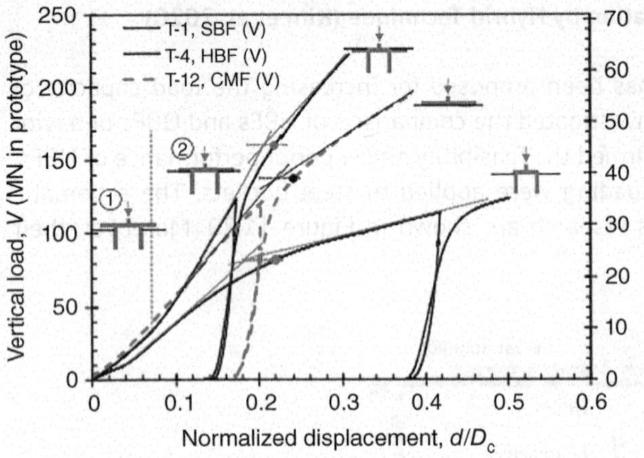

Figure 11.21 Vertical load–displacement curves (Kim et al. 2020 / with permission of Elsevier).

- The CMF and SBF curve comparison shows that the portion of the mat part is significantly more than the skirt in vertical bearing capacity.
- Interaction between mat and skirt in HBF causes a reduced bearing capacity compared to the summation of SBF and CMF curves.
- The S-shaped of SBF and HBF curves leads to the effect of contact between the bucket lid and soil in the growth of vertical bearing capacity sharply.

The horizontal load–rotation curves for SBF (T-2 and T-3) and HBF (T-5 and T-6) are depicted in Figure 11.22, where $M/HD_c = 0.1$ and V = self-weight. Two laser sensors were utilized for measuring the displacement for rotation angle calculation. The main conclusions that can be perceived from these curves are mentioned below:

- The hybrid technique modifies the horizontal capacity of SBF significantly simply because the mat part increased the contact area.
- The trend of the T-5 curve includes two curvature shapes, which is different from the behavior of T-6; its simple reason is that in T-5, the lid of HBF had a distance from the soil, whereas, in the T-6, there is complete contact between mat and soil.

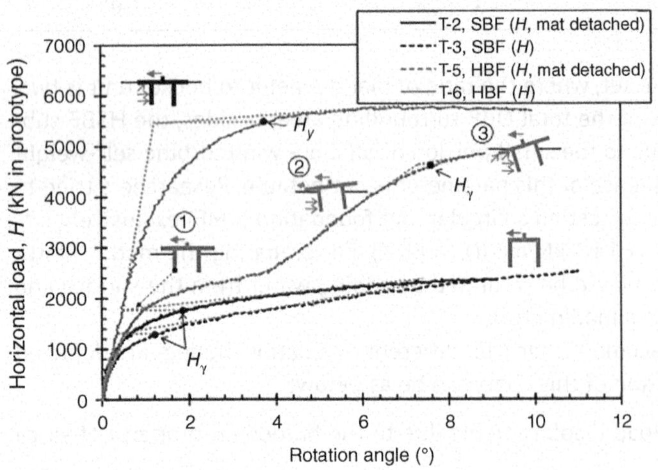

Figure 11.22 Horizontal load–rotation angle curves (in T-2 and T-5, the bottom of the bucket lid is partially detached from the ground surface) (Kim et al. 2020 / with permission of Elsevier).

- In the T-5 trend, three steps of rotational behavior could be seen: First, the mat was detached, and the skirt wall was a prominent role to resist horizontal load; second, by increasing the load, the foundation started to rotate, and mat edges intrude into moment resistant; and finally, the mat engaged to moment resistant completely by penetrating the ground, therefore the ultimate bearing capacity of HBF was reached.
- In the T-6, regarding full contact between mat and soil, the horizontal load steadily increased till bearing capacity was reached.

Kim and Kim (2019) reported that soil heaving during suction bucket installation could obstruct full penetration of the suction bucket foundation. Therefore, behavior like the T-5 _trend is possible in the prototype.

As shown in Figure 11.22, researchers used the secant line method proposed by Villalobos (2007) to determine the yield point and called H_y in the chart. H_y for T-6, T-5, T-3, and T-2 is 5450 kN, 4590 kN, 1280 kN, and 1770 kN, respectively, as it can be concluded that HBF bearing capacity is about 3.3 times greater than SBF. Moreover, one important piece of data inferred from curves presented in Figure 11.22 is the rotation angle corresponding with H_y. This parameter is 1.9°, 1.2°, 7.6°, and 3.8° for T-2 (SBF), T-3 (SBF), T-5 (HBF), and T-6 (HBF), respectively. Therefore, a notably higher rotation is required for HBF than SBF to reach H_y, and the hybrid foundation behaves stiffer against H-M combined loads.

The authors compared the bearing capacities of SBF and HBF foundations with different load eccentricities in the radial (M-H) plane presented in Figure 11.23. The expression proposed by Byrne and Houlsby (2003) and Larsen et al. (2013) seemed to be more conservative than SBF results, whereas those given by Achmus et al. (2013) overestimated the capacity of the SBF compared to the results of these tests.

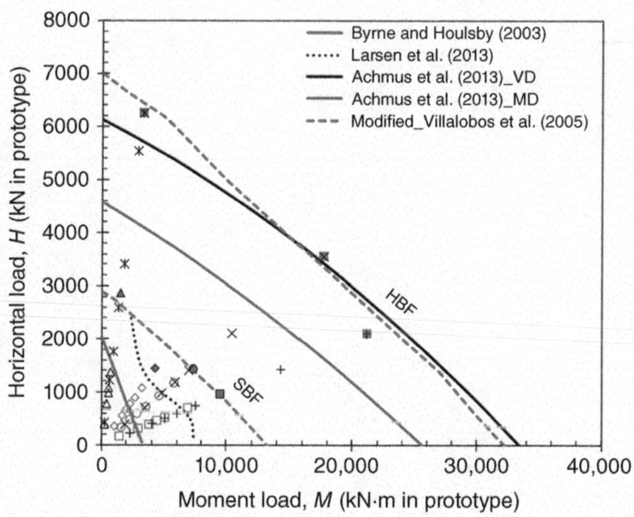

Figure 11.23 Comparison of predicted failure envelopes with test results (Kim et al. 2020 / with permission of Elsevier).

The authors used a modified general expression proposed by Villalobos et al. (2005) to describe the combined load capacity of bucket foundations under low vertical loads. This expression is presented by Equation (11.70), and authors calculated input parameters for SBF and HBF shown in Table 11.4.

Figure 11.24 shows the elliptical shape of the failure envelope in the radial (M-H) plane based on Equation (11.70) and the input parameter expressed in Table 11.4. It is clear that the failure surface for the HBF was higher than that for SBF, illustrating Hybrid technique increases bearing load capacity under combined VMH loading.

$$f = \left(\frac{H}{h_0 V_y}\right)^2 + \left(\frac{M}{m_0 D_c V_y}\right)^2 - 2a\left(\frac{H}{h_0 V_y}\right)\left(\frac{M}{m_0 D_c V_y}\right) - \left[\frac{\beta_{12}}{(t_0+1)^{(\beta_1+\beta_2)}}\right]^2 \left(\frac{V}{V_y}+t_0\right)^{2\beta_1}\left(1-\frac{V}{V_y}\right)^{2\beta_2} = 0 \quad (11.70)$$

$$\beta_{12} = \frac{(\beta_1+\beta_2)^{(\beta_1+\beta_2)}}{(\beta_1)^{\beta_1}(\beta_2)^{\beta_2}}; \quad t_0 = -\frac{V_t}{V_y}; \quad V_t = \text{Tensile Capacity of the Bucket}$$

(Continued)

Case Study 11.2 (Continued)

Table 11.4 Input parameters for the modified framework proposed by Villalobos et al. (2005) (adapted from Villalobos et al. 2005).

Inputs	Single bucket	Hybrid bucket
D_c	5 m	5 m
D_m	5 m	10 m
T	0.21m	0.21 m
V/V_y	0.018	0.0095
D_r	0.62%	0.62%
ϕ	41°	41°
$K\tan(\delta)$	0.34	0.34
β_1	0.84	0.84
β_2	1	1
β_{12}	3.56	3.56
h_0	0.15	0.15
m_0	0.135	0.135
a	−0.86	−0.86
t_0	0.0092	0.008

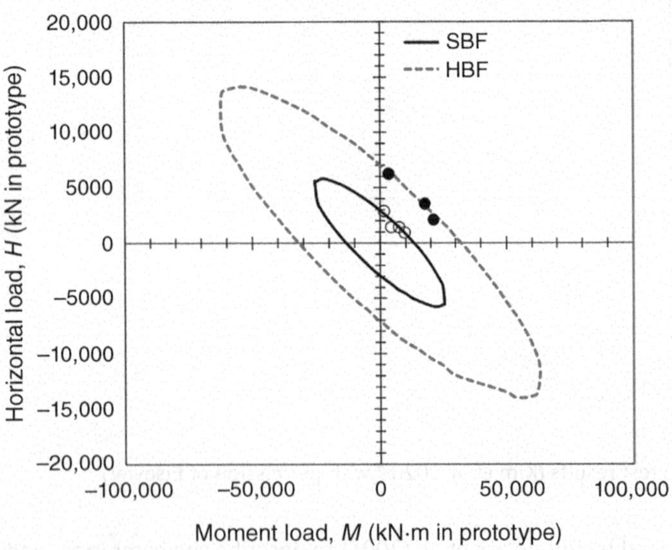

Figure 11.24 Failure envelopes of the SBF and HBF in the H–M plane (Kim et al. 2020 / with permission of Elsevier).

Case Study 11.3

Field Test for investigation of Bearing capacity of composite bucket foundations for offshore wind turbines in silty sand. (Jia et al. 2018)

Field tests were designed at a natural water pool with a water depth of about 0.4 m in saturated silty sand off the coast of Jiangsu to estimate the bearing capacity of a composite bucket foundation (CBF) with seven compartments. A part of the northeast coast of Jiangsu province in the Yellow Sea of China was selected for tests. Figure 11.25a shows soil grain size distribution. Moreover, before the CBF installation by suction forces, the cone penetration tests (CPTs) were conducted on the test shown in Figure 11.25b.

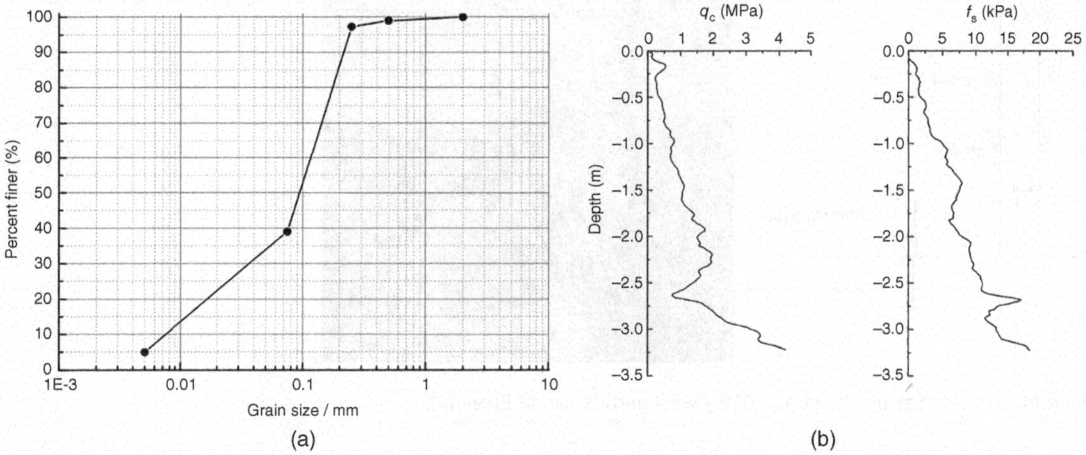

Figure 11.25 (a) typical grain size and (b) CPT records (Jia et al. 2018 / with permission of Elsevier).

In this study, CBF with a weight of 2100 t was selected as a prototype. The model scale of the CBF is 1:10, and the model of CBF is made of steel while the prototype structure is manufactured of reinforced concrete. The outer diameter (D) of the model was 3 m and the clear height of the outer and inner skirts (L) was equal to 0.9 m (Figure 11.26). For applying the horizontal load, a 5 m steel tube with an outer diameter of 0.66 m is used which was connected to the CBF top lid by welding and the height of implementation of the horizontal load was selected equal to 5 m with respect to the top lid of CBF. A force sensor was linked to a wire rope in order to tension force measurement. Lateral and vertical LVDTs were installed on the tube and top of the CBF respectively for displacement monitoring. The numbering and position of LVDTs are illustrated in Figure 11.27.

Figure 11.26 Model of composite bucket foundation (Jia et al. 2018 / with permission of Elsevier).

The main control parameters and related values of Two experiments (Test I and II) are listed in Table 11.5. The incremental load for each step was approximately 15% of the determined ultimate loading capacity (4 kN). Each load step was retained for a certain time until the displacement was often unchanged (especially in Test I), and the next load step was applied subsequently.

Figure 11.28 shows the comparison between the tests and numerical results. To calculate the load–displacement curves, Elastic module E (Young module) varying from 1Es to 5Es (compressive modulus) are utilized. The values based

(Continued)

Case Study 11.3 (Continued)

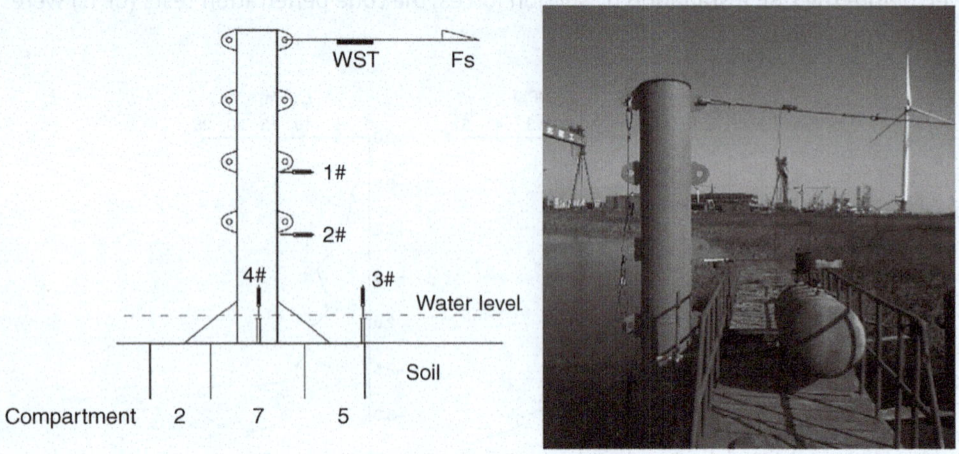

Figure 11.27 Horizontal load test set-up (Jia et al. 2018 / with permission of Elsevier).

Table 11.5 Main parameters of tests (adapted from Jia et al. 2018).

Test	Step number	Time (h)	Maximum load (kN)
I	11	9.08	40
II	8	4.07	32

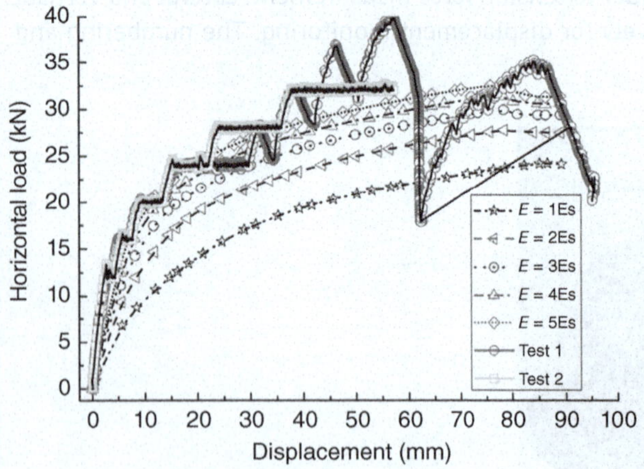

Figure 11.28 Load–displacement relationship curves of composite bucket foundation in field test and FEM model (Jia et al. 2018 / with permission of Elsevier).

on FEM analysis with E = 3–5Es agree well with those calculated via tests data before the displacement is less than 30 mm. Particularly, the trend of experimental results for the displacement less than 20 mm is almost the same with curved obtained by FEM model with E = 4Es. Whereas, for displacement above 40 mm, the curve obtained from the FEM model with E = 5Es presents a better tendency.

By the FEM model, the movement of the rotation center during loading can be monitored. As illustrated in Figure 11.29, by increasing loading till the ultimate state, the rotation center reaches a minimum on the left side of opposite to the loading direction. The position of the rotation center changes from near the center line of the Bucket foundation to the bottom of the bulkhead in the loading direction during an increase in the horizontal loads, the rotation center coordinate is approximately 0.45 R from the center line and with 0.8H beneath the top lid around the bulkhead in the loading direction when the horizontal load is reached the ultimate value.

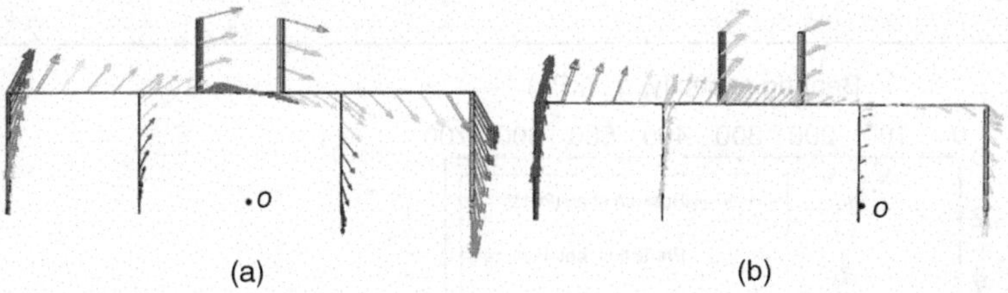

Figure 11.29 Deformation trend and rotation center position of CBF (O: rotation center): (a) initial loading and (b) ultimate loading (Jia et al. 2018 / with permission of Elsevier).

Case Study 11.4

A new method to predict the bearing capacity–penetration curve of spudcans in multilayered clay soils (Mehralizadeh and Makarchian 2021)

Mehralizadeh and Makarchian (2021) presented straightforward correlations for predicting the bearing capacity–penetration curve of the spudcan via numerical modeling and studying the determinant parameters of the bearing capacity of spudcan in multilayered clay soil.

The undrained shear strength of the layers is the most influential parameter on the bearing capacity in multilayered clays. In the current research the effects of the shear strength ratio of the first layer to the second layer ($\frac{S_{u1}}{S_{u2}}$) and the second layer to the third layer ($\frac{S_{u2}}{S_{u3}}$) were studied.

X, Y, and Z are dimensionless parameters that are introduced for normalizing the peak/initial resistance (q_p), the layer's interface resistance (q_{int}), and the resistance at the depth of 0.6H (q_{60}) according to base values (as depicted in Figure 11.30).

$$X = \frac{q_p}{q_{ave}} \text{ or } \frac{q_p}{q_1} \tag{11.71}$$

$$Y = \frac{q_{int}}{q_{ave}} \text{ or } \frac{q_{int}}{q_2} \tag{11.72}$$

$$Z = \frac{q_{60}}{q_{ave}} \text{ or } \frac{q_{60}}{q_3} \tag{11.73}$$

Figure 11.30 Schematic diagram of the uniform bearing capacity (profile A) and with the slight bearing capacity changes in depth (profile B) in single-layered, punch-through, and squeezing profiles in multilayered clays (Mehralizadeh and Makarchian 2021 / Taylor & Francis Group).

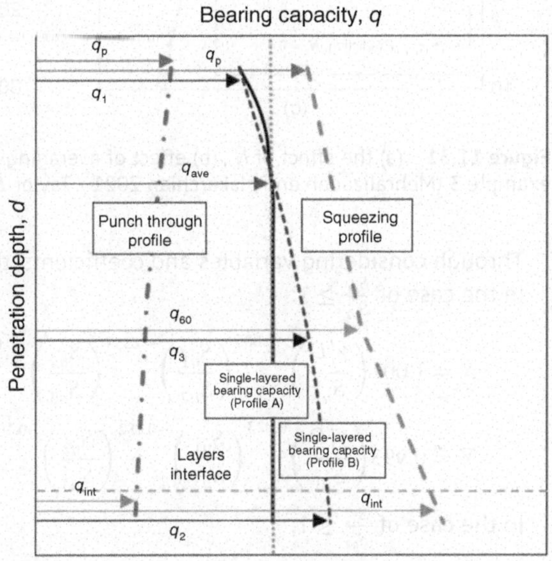

(Continued)

Case Study 11.4 (Continued)

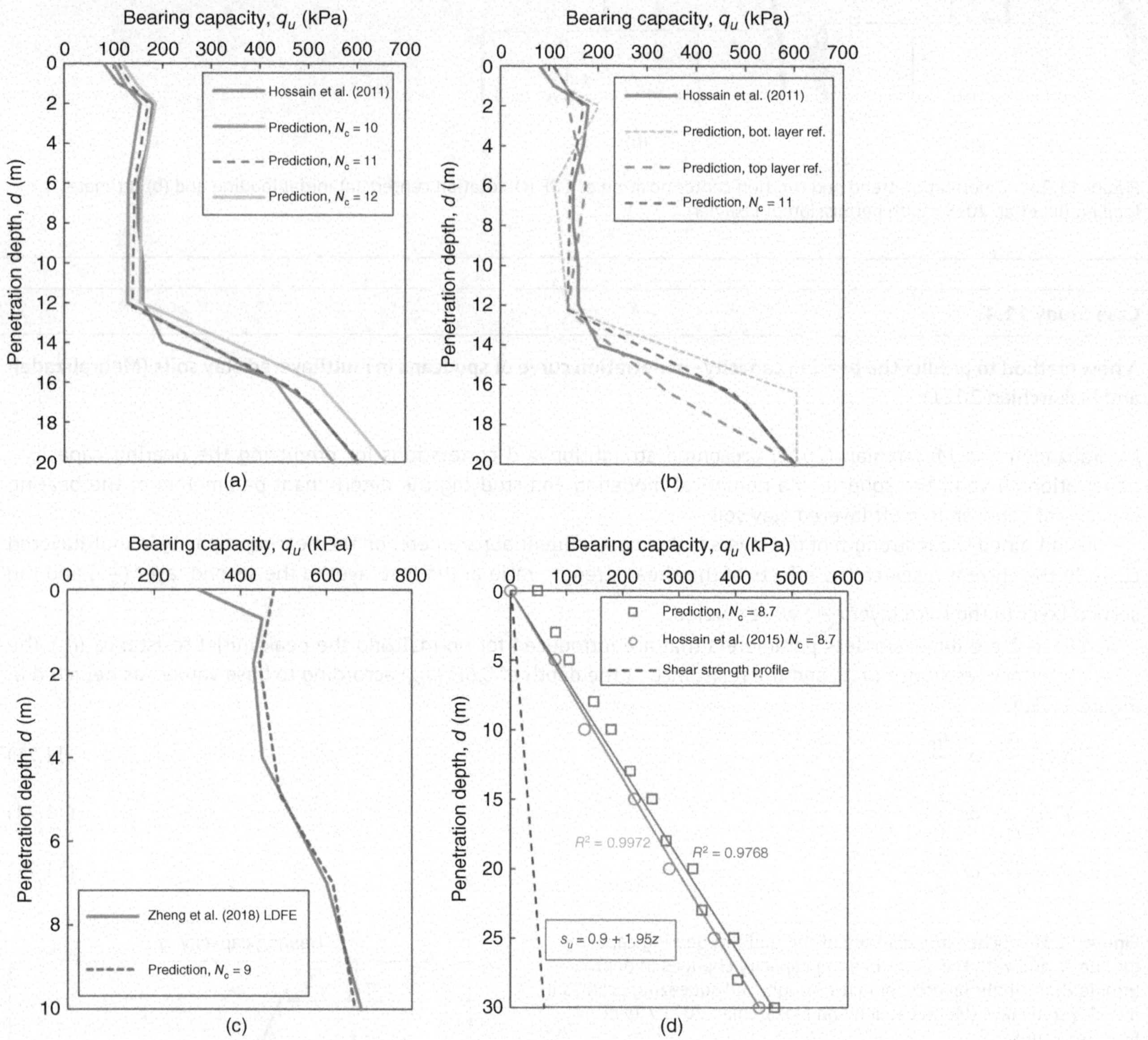

Figure 11.31 (a) the effect of N_c, (b) effect of averaging in example 1, (c) comparison for example 2, and (d) comparison for example 3 (Mehralizadeh and Makarchian 2021 / Taylor & Francis Group).

Through considering variables and coefficients, the following equations are obtained for prediction.

In the case of $\frac{S_{u1}}{S_{u2}} \geq 1$:

$$X = 1.001\left(\frac{\gamma_1' D}{S_{u1}}\right)^{-0.005}\left(\frac{S_{u1}}{S_{u2}}\right)^{-0.358}\left(\frac{S_{u2}}{S_{u3}}\right)^{-0.091}\left(\frac{H_1}{D}\right)^{0.355}\left(\frac{H_2}{D}\right)^{-0.024} \tag{11.74}$$

$$Y = 0.993\left(\frac{\gamma_1' D}{S_{u1}}\right)^{0.023}\left(\frac{S_{u1}}{S_{u2}}\right)^{-0.682}\left(\frac{S_{u2}}{S_{u3}}\right)^{-0.101}\left(\frac{H_1}{D}\right)^{0.043}\left(\frac{H_2}{D}\right)^{-0.159} \tag{11.75}$$

In the case of $\frac{S_{u1}}{S_{u2}} \leq 1$:

$$X = 1.000\left(\frac{S_{u1}}{S_{u2}}\right)^{-0.009}\left(\frac{S_{u2}}{S_{u3}}\right)^{0.006}\left(\frac{H_1}{D}\right)^{-0.016}\left(\frac{H_2}{D}\right)^{0.069} \tag{11.76}$$

$$Y = 0.995\left(\frac{\gamma_1' D}{S_{u1}}\right)^{-0.007}\left(\frac{S_{u1}}{S_{u2}}\right)^{-0.684}\left(\frac{S_{u2}}{S_{u3}}\right)^{-0.248}\left(\frac{H_1}{D}\right)^{-0.016}\left(\frac{H_2}{D}\right)^{0.358} \quad (11.77)$$

$$X \le Z = 1.01\left(\frac{\gamma_1' D}{S_{u1}}\right)^{0.013}\left(\frac{S_{u1}}{S_{u2}}\right)^{-0.204}\left(\frac{S_{u2}}{S_{u3}}\right)^{-0.033}\left(\frac{H_1}{D}\right)^{-0.583}\left(\frac{H_2}{D}\right)^{0.274} \le Y \quad (11.78)$$

The stages for predicting the bearing capacity–penetration curve for a spudcan in multilayered clay soils are listed below:

1) The information on layers is recorded and listed comprising layer thickness, effective unit weight and undrained shear strength.
2) The parameters of X, Y, and Z are calculated for each layer depending on $\frac{S_{u1}}{S_{u2}} \ge 1$ or $\frac{S_{u1}}{S_{u2}} \le 1$.
3) The average bearing capacity of the layer (q_{ave}) is calculated via the bearing capacity factor (N_c), without considering the effect of the underlying layers. Then, based on the values of the X, Y, and Z, the values of q_p, q_{int}, and q_{60} are calculated for the desired layer.
4) Based on the differences between the values of q_{int} of a layer with q_p of the underlying layer, the average of the two values should be used to determine the value of resistance at the interface of a layer. This was obtained practically.
5) In the last stage, the obtained points of q_p, q_{int}, and q_{60} and their relevant depth are connected to draw the bearing capacity–penetration curve of a spudcan.

Figure 11.31 presents the results of the comparison between the presented approach and previously published studies through three different examples.

11.9 Major Points

1) Skirted foundations, Bucket and spudcan foundations are the three main types of semi-deep foundations utilized for a wide range of onshore and offshore structures such as storage, oil industrial structures, and offshore wind turbines.
2) The combination of a shallow skirted foundation with a caisson makes a new type of semi-deep foundation, called a hybrid foundation. Nowadays, this new concept is considered for offshore substructures, especially offshore wind turbines (OWT), for its outstanding advantages described in the previous chapter. Applying the vertical (V), horizontal, (H), and moment (M) load combination to substructure due to waves, sea currents, wind, and self-weight is common in the offshore environment.
3) Bearing capacity of the Bucket foundation under vertical load can be determined via two different mechanisms. When soil is snared by a skirt, which means a block is created, the sum of tip resistance and shaft friction could be used for obtaining capacity. For internal soil flow, characters that make vertical resistance include tip, raft (at ground level), internal friction, and external friction.
4) In recent years, behavior investigations of foundations under combined loads, including vertical, horizontal, and moment, have been considered by researchers. Numerical and laboratory studies have been hired to develop a failure envelope for a foundation under VMH loading in this method. Providing a feasible foundation model under combined applied loads is the main challenge for incorporating structural analysis. The yield surface with hardening characteristics is the base of the proposed models in the literature.
5) Rezazadeh and Eslami (2017) have done extensive research on the derivation of the failure envelope in 2D and 3D spaces for the behavior of skirted foundations under H-V, M-V, H-M, and VMH combined loading in sand and undrained saturated clay by using numerical modeling. According to the authors' studies, there are several factors, such as soil specifications, foundation dimensions, and load combinations, which can affect the size and shape of the failure envelope of foundations in 2D/3D spaces.
6) Regarding embedment depth ratios (D/B) in clay, ultimate envelopes related to M/M_{ult} -V/V_{ult} and V/V_{ult} -H/H_{ult} are shrinkages, with the increase of the D/B ratio. However, envelopes in the space of V/V_{ult} -H/H_{ult} enlarge, and the eccentricity increases as well. In sand, all curves of envelopes in the space of V/V_{ult} -M/BV_{ult} -H/V_{ult} enlarge by D/B increasing.
7) As for the role of soil specification, it can be said that with the increase of non-homogeneity in clay, the 3D ultimate surface in the space of V/V_{ult}-M/M_{ult}-H/H_{ult} becomes smaller. Additionally, an increase in the strength/stiffness

characteristics of sand causes the 3D failure envelope in the space of V/V_{ult}-M/BV_{ult}-H/V_{ult} to be enlarged. Moreover, increasing the vertical load in clay makes the ultimate surface envelope small. On the other hand, when the vertical load increases to about half of the ultimate vertical capacity, the bearing capacity of the foundation enhances, and afterward, an increase in vertical load causes the bucket foundation bearing capacity to decrease under VMH loading.

8) The numerical study of the suction caissons under VMH loading in clay shows: (1) The effect of bending moment on the horizontal bearing capacity is remarkable; (2) Incline ellipse is the best description for the figure of the suction bucket envelope in clay; (3) The increase in pre-vertical load causes shrinkage of failure envelope size.
9) Centrifuge tests for assessing the effects of hybrid technology on the bearing capacity of suction are presented in the researched carried out by Kim et al. (2020). These tests show that the combination of a mat foundation with ordinary suction caissons can increase the horizontal bearing capacity of a simple bucket foundation; the average horizontal bearing capacity of the HBF (Hybrid Bucket Foundation) test specimens was 3.3 times greater than that of the SBF (Simple Bucket Foundation) test specimens.
10) The failure envelopes suggested by Villalobos et al. (2005) are able to express accurate test results for both simple bucket foundations and hybrid bucket foundations with appropriate input variables.
11) The rate of loading can affect the behavior of load–displacement of the bucket foundation. The comparison between the FEM model and field test in lateral load–displacement curves shows using a Young module (E) 5 times more than a compressive module (E_s) can lead to matching between the FEM model and Field test results.

11.10 Questions and Problems

11.10.1 Explanatory

1 What is the embedment ratio in a semi-deep foundation, and its normal range for this type of foundation?

2 Explain the difference between ordinary bucket foundations and hybrid bucket foundations.

3 Name the effective parameters on the bearing capacity of the skirted & bucket foundation in saturated clay and sand based on classical methods.

4 What are the parameters that control the shape of the 3D failure zone of a bucket foundation under VMH loading?

5 Regarding reviewed recent works, explain the effect of vertical loading on the bearing capacity of semi-deep foundations.

6 What is the normal range of skirt heights and their distances for skirted foundations?

7 Name major parameters that affected the bearing capacity of spudcans for long-term and short-term loading.

8 Explain the stages of predicting the bearing capacity – penetration curves of spudcans based on Case Study 11.4.

11.10.2 Conceptual

1 How can utilizing the hybrid technique enhance the bucket foundation bearing capacity under VMH loading?

2 Why is it essential to evaluate the safety of offshore structures foundations under combined VMH loading instead of using the classic bearing capacity formula for separated V, H, and M loading?

3 Regarding the second case study, explain the effect of mat–soil contact in the behavior of a hybrid suction bucket foundation.

4 Regarding the 3D failure zone formula, explain how the bearing capacity of the semi-deep foundation under VMH loading can be improved.

5 Regarding Figures 11.13 and 11.14, explain how vertical load can affect on bearing capacity of the bucket foundation in the M–H surface.

6 Discuss determinant factors on bearing capacity–penetration behavior of spudcans considering Case Study 11.4.

11.10.3 Problems

1 Resolve example 11.1 considering a skirted foundation installed in sand deposits with $\gamma' = 8\ \text{kN/m}^3$ and $\phi = 30°$ and the water level is at the ground surface.

2 Resolve example 11.3 considering the embedded depth of the spudcan is equal to 8 m.

3 Evaluate the safety situation of a wind turbine foundation in the sand, shown in the following schematic. Under combined loading, including H = 2 MN, V = 3 MN, M = 80 MN.m, due to the low weight of the turbine superstructure, the $V/V_{ult} = 0$ can be considered.

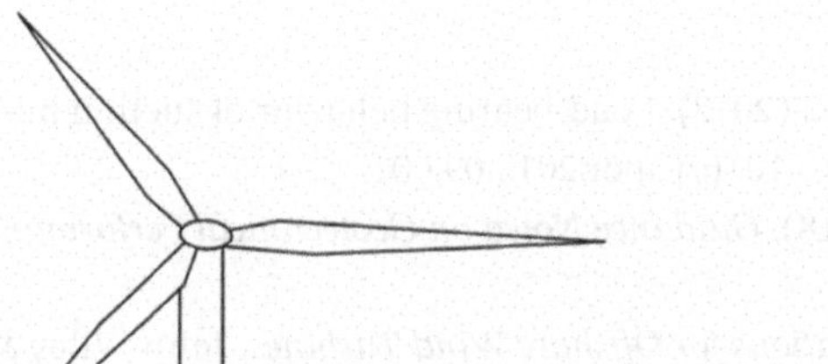

4 Via the classic methods presented in this chapter, determine the vertical and horizontal bearing capacity of a square skirted foundation with B = 6 m and D/B = 0.7 in (a) an undrained saturated clay with constant $S_u = 70$ kPa; and (b) in the sand with $\phi = 40°$ and $\gamma = 20\ \text{kN/m}^3$.

5 Evaluate the safety factor of the wind turbine foundation in the previous problem for D/B = 0.5,0.75,1 under VMH loading, and compare the results.

6 Evaluate the stability of the offshore jacket illustrated in the figure below. The substructure of this jacket comprises four skirted foundations, with D = 10 m and D/B = 0.25 penetrated in loose sand deposits with $\phi = 30°$. The distance between the foundations in the plan is about 60 m; the substructure of the jacket carries the self-weight of the jacket, W = 60 MN, and the wave design force, H = 12 MN, at the height of 80m above the mudline (use H-V envelope and correlations).

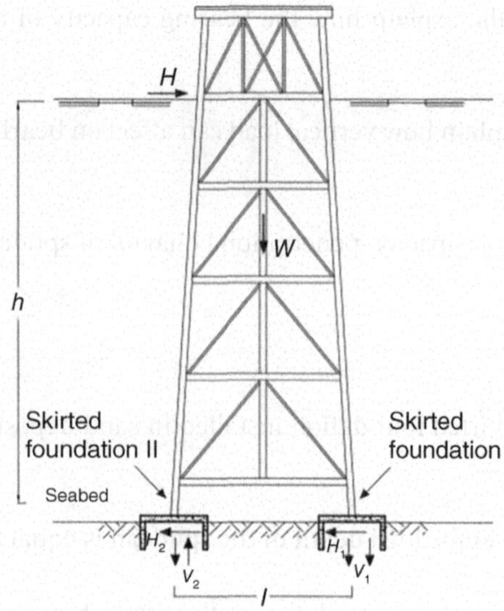

References

Achmus, M., Akdag, C.T., and Thieken, K. (2013). Load-bearing behavior of suction bucket foundations in sand. *Applied Ocean Research* 43: 157–165. https://doi.org/10.1016/j.apor.2013.09.001.

American Bureau of Shipping (ABS) (2018). *Guidance Noted on Geotechnical Performance of Spudcan Foundations*. Houston, TX: ABS.

Bhattacharya, S. (2019). *Design of Foundations for Offshore Wind Turbines*. John Wiley & Sons.

Bransby, M.F. and Randolph, M.F. (1999). The effect of embedment depth on the undrained response of skirted foundations to combined loading. *Soils and Foundations* 39 (4): 19–33.

Brinch Hansen, J. (1970). A revised and extended formula for bearing capacity. *The Danish Geotechnical Institute* 98: 5–11.

Byrne, B.W. and Houlsby, G.T. (2003). Foundations for offshore wind turbines. *Philosophical Transactions of the Royal Society of London* 361: 2909–2930.

Cheng, N., Gaudin, C., and Cassidy, M.J. (2019). Physical and numerical study of the combined bearing capacity of hybrid foundation systems. *Ocean Engineering* 179 (April 2017): 104–115. https://doi.org/10.1016/j.oceaneng.2019.03.002.

Gourvenec, S. (2008). Effect of embedment on the undrained capacity of shallow foundations under general loading. *Geotechnique* 58 (3): 177–185.

Hossain, M.S., Hu, Y., Randolph, M., and White, D. (2005). Limiting cavity depth for spudcan foundations penetrating clay. *Geotechnique* 55 (9): 679–690. https://doi.org/10.1680/geot.2005.55.9.679.

Ibsen, L., Barari, A., and Larsen, K. (2013). An experimental study of the drained capacity of bucket foundations for offshore applications. *International Conference on Case Histories in Geotechnical Engineering* 2: 1–6.

Jia, N., Zhang, P., Liu, Y., and Ding, H. (2018). Bearing capacity of composite bucket foundations for offshore wind turbines in silty sand. *Ocean Engineering* 151 (January): 1–11. https://doi.org/10.1016/j.oceaneng.2018.01.006.

Kim, J.H. and Kim, D. (2019). Soil displacement near a bucket foundation installed in sand by suction and jacking in a centrifuge. *Journal of Geotechnical and Geoenvironmental Engineering* 145 (11): https://doi.org/10.1061/(asce)gt.1943-5606.0002149.

Kim, J.H., Jeong, Y.H., and Kim, D.S. (2020). Improved load capacity of a hybrid bucket foundation for offshore structures in sand. *Applied Ocean Research* 101 (January): 102197. https://doi.org/10.1016/j.apor.2020.102197.

Larsen, K.A., Ibsen, L.B., and Barari, A. (2013). Modified expression for the failure criterion of bucket foundations subjected to combined loading. *Canadian Geotechnical Journal* 50 (12): 1250–1259. https://doi.org/10.1139/cgj-2012-0308.

Mehralizadeh, H. and Makarchian, M. (2021). A new method to predict the bearing capacity–penetration curve of spudcans in multi-layered clay soils. *Marine Georesources and Geotechnology* 40 (5): 511–522.

Meyerhof, G.G. (1963). Some recent research on the bearing capacity of foundations. *Canadian Geotechnical Journal* 1 (1): 16–26.

Naval Facilities Engineering Service Center (NAVFAC) (2012). Handbook for Marine Geotechnical Engineering. SP-2209-0CN, Port Hueneme, California.

Randolph, M. and Gourvenec, S. (2011). *Offshore Geotechnical Engineering*, 528. Taylor and Francis Group.

Rezazadeh, S. (2017). Bearing capacity analysis of skirted semi-deep foundations under combined loading. PhD thesis. Amirkabir University of Technology, p. 365.

Rezazadeh, S. and Eslami, A. (2017). Bearing capacity of semi-deep skirted foundations on clay using stress characteristics and finite element analyses. *Marine Georesources and Geotechnology* 36 (6): 1–15. https://doi.org/10.1080/1064119X.2017.1361488.

Skempton, A.W. (1951). The bearing capacity of clays. *Building Research Congress, London* (11–20 September 1951) 1: 180–189.

Villalobos, F. A. (2007). Model testing of foundations for offshore wind turbines. PhD thesis, University of Oxford, Oxford, England, 2007.

Villalobos, F.A., Byrne, B.W., and Houlsby, G.T. (2005). Moment loading of caissons installed in saturated sand. *Frontiers in Offshore Geotechnics, ISFOG 2005 – Proceedings of the 1st International Symposium on Frontiers in Offshore Geotechnics*, 411–416 (19–21 September 2005). London: Taylor & Francis Group. https://doi.org/10.1201/noe0415390637.ch41.

Xia, H., Zhou, X., Zhou, M. et al. (2021). Capacity of caissons in stiff-over-soft clay under combined V–H-M loadings. *Ocean Engineering* 229: 109007. https://doi.org/10.1016/j.oceaneng.2021.109007.

Yin, Z.Y., Teng, J.C., Li, Z., and Zheng, Y.Y. (2020). Modeling of suction bucket foundation in clay: from finite element analyses to macro-elements. *Ocean Engineering* 210 (May): 107577. https://doi.org/10.1016/j.oceaneng.2020.107577.

Yun, G. and Bransby, M.F. (2007). The horizontal-moment capacity of embedded foundations in undrained soil. *Canadian Geotechnical Journal* 44 (4): 409–424. https://doi.org/10.1139/T06-126.

Zhang, J.H., Chen, Z.Y., and Li, F. (2010). Three-dimensional limit analysis of suction bucket foundations. *Ocean Engineering* 37 (8–9): 790–799. https://doi.org/10.1016/j.oceaneng.2010.02.017.

12

Floating (Balanced) Foundations

Chapter Highlights

- **Background to Floating Foundations**
- **Principles and Functions**
  - Ship Buoyancy in Water
  - Buoyancy Properties
  - Compensated Foundations
- **Types of Floating Foundations**
- **Excavation Issues and Associated Difficulties**
- **Support Systems**
- **Design Aspects and Procedure**
- **Case Histories**
  - Liberty Mutual Insurance, The USA (Casagrande and Fadum 1944)
  - Piled Raft of Torre Latino Americana, Mexico City (Brûlé et al. 2019)
  - Hotel Meridien, Brazil (Hemsley 2000)
  - Top-down Construction Method for a Metro Station (Jeong and Kim 2018)
  - Albany Telephone Building, The USA (Glick 1936)
  - Raffles Complex, Singapore (Wong et al. 1995)
  - Culture Center, South Korea (Chung et al. 2005)

12.1 Background to Floating Foundations

Shallow foundations constructed on weak surface soils often lack sufficient bearing capacity, experience high settlement, and lack resistance against lateral loads. As a result, designers resort to considering alternative foundation types. Deep foundations are typically the alternative in foundation engineering practice. However, deep foundations can be costly and introduce construction challenges. Therefore, an intermediate approach, such as semi-deep foundations, may be considered. One type of semi-deep foundation is the "floating foundation," which involves excavating the soil and constructing a spread foundation at the bottom of the excavation. The weight of the excavated soil offsets part or the whole weight of the superstructure load. Another example of a semi-deep foundation is the well foundation. By excavating wells and reaching the dense layer, the desired bearing conditions can be achieved. Well foundations are also advantageous when dealing with interaction or dynamic forces, as the combined forces of foundation weight, soil shear strength, and arching effect provide resistance.

Advanced Foundation Engineering: Principles, Performance and Prospects, First Edition. Moses Karakouzian and Abolfazl Eslami.

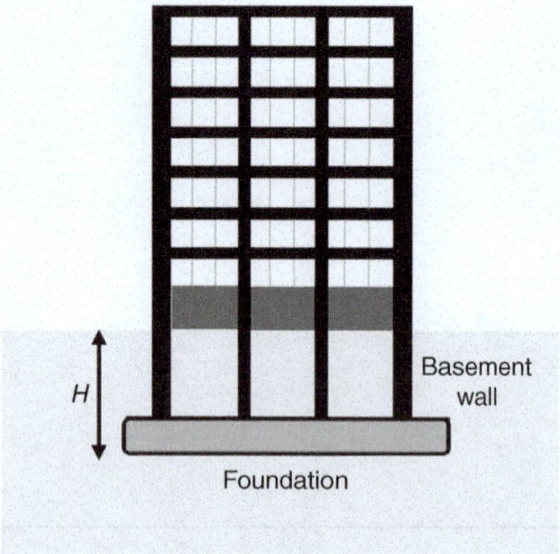

Figure 12.1 Excavation for floating foundation construction.

12.2 Floating Foundations Principles

Reducing the total and differential settlement is called the flotation principle by reducing the net load due to excavation. When part of the structure's weight is balanced using excavation, the structure is called relatively floating. When the entire structure's weight is balanced, the structure is completely buoyant (Brown 1971).

Floating foundations allow the weight of the building to roughly balance with the weight of excavated soil (with water) at the project site. It is important to note that the possibilities of complete equilibrium between the building weight and the excavated soil weight are low; as a result, floating foundations are frequently relatively buoyant.

Figure 12.1 shows an example of excavation for floating foundation construction. The excavation to a depth of $H = 5$ m for the basement relieves the soil at the foundation level of about 90 kN/m^2 of pressure. The weight of the substructure is about 30 kN/m^2, so a pressure of 60 kN/m^2 can be placed to implement the superstructure to balance the overburden's weight. The pressure of 60 kN/m^2 is approximately equivalent to a six-story building. Therefore, a 6-story building can float in soft and compressible soil without settlement. In practice, however, it is rarely possible to balance the loads completely (zero net loads on the soil). Factors in nature, such as groundwater level variations, affect the principle of buoyancy. Also, accurately estimating the live load is not possible in many cases. Another influential factor in the occurrence of floating foundation settlement is the reconsolidation of the soil, which has been heaved due to the removal of overburden pressure in excavation for the substructure (Tomlinson 2001).

Similarities have been used to define and understand floating foundations. The different definitions presented in this field are as follows:

- **The Similarity of Floating Foundations with Ship Buoyancy in Water:** The term "floating foundation" expresses a precise meaning; researchers often compare it to a ship in the water. For example, in connection with floating foundations, the building in the soil is similar to a ship floating in water (Fang 2001).
- **Buoyancy Properties of Floating Foundations:** The term "buoyancy" describes hydrostatic upward pressure due to the placement of foundations under the groundwater level in the soil, which means that the pressure from the structure balances the upward hydrostatic pressure (Fang 2001).
- **Compensated Foundations:** Used in defining these foundations, which means that the weight of the building is balanced by the soil weight removed.

The weight of the building should not be less than the upward hydrostatic pressure caused by groundwater. This force is equivalent to the size of the water, which is equivalent to the volume of the building under the groundwater level. If this happens (apart from the friction on the sides of the building), the building will move upwards. This is not likely in the building foundation but can occur in reservoirs for sewage works or swimming pools when they are empty of water.

12.3 Types of Floating Foundations

When the soil shear strength is too low to support the building, soil failure can occur. In such cases, where no strong layer exists at a reasonable depth and pile installation is not feasible, a floating foundation can be used to reduce the shear stress

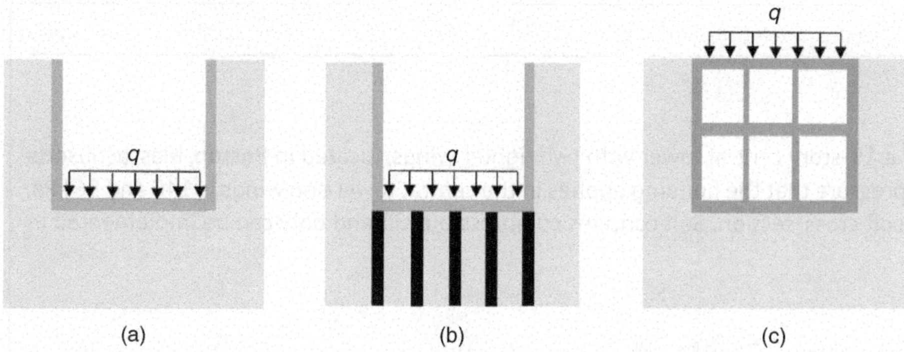

Figure 12.2 Different types of floating foundations in depth: (a) raft, (b) piled raft, and (c) box.

to an acceptable level. This approach also addresses the settlement issue. For buildings of any size, the excavation can be carried out to the maximum extent possible, necessitating a rigid raft or box-type foundation.

Floating foundations can be constructed using various methods, including raft foundations, upper slabs, middle walls, plain floating foundations, or piled raft foundations. Figure 12.2 provides examples of these construction types for floating foundations. The specific type of floating foundation chosen depends on factors such as soil conditions, building dimensions, and engineering requirements.

12.3.1 Compensated Raft

Buoyant raft (mat) foundations are primarily designed to protect the structure by reducing the net pressure caused by excavation, without utilizing the substructure space for architectural purposes. The raft is designed to be as light and rigid as possible, and the use of cells helps achieve this objective. However, this construction method limits the architectural utilization of the basement space. Alternatively, caissons can be used to implement buoyant raft foundations.

In the case of box foundations, they belong to the group of buoyant raft foundations, but their design aims to create additional usable space within the building for architectural purposes. These foundations help reduce the pressure on the soil caused by the overburden load, and the space created in the basement can be utilized as part of the building's functional areas. Box foundations offer opportunities for incorporating architectural features and maximizing the usable space within the structure.

12.3.1.1 Buoyant Caisson Raft

Caissons provide a method for implementing buoyant raft foundations. In this approach, the foundation is initially designed for shallow placement and then floated to the desired depth by removing soil from the created cells. Concrete is used to level the floors of the cells at the desired depth, and subsequently, the superstructure is constructed on top of them. Not all cells are excavated simultaneously, which ensures that there is always some soil overburden present to prevent ground heave in the excavated area. Figure 12.3 depicts an example of caisson implementation in a marine environment.

Figure 12.3 Implementation of the buoyant caisson raft (HCM Constructions).

Case Study 12.1

Liberty Mutual Insurance, The USA

The plan is 82 by 35 m and includes a 19-story central tower with two lighter wings, located in Boston, Massachusetts (Casagrande and Fadum 1944). The pressure that the building applies in the central tower and wings is 140 and 86 kPa, respectively. Figure 12.4 shows the soil cross-section. Soil contains compressible silt and only can be implemented in light buildings.

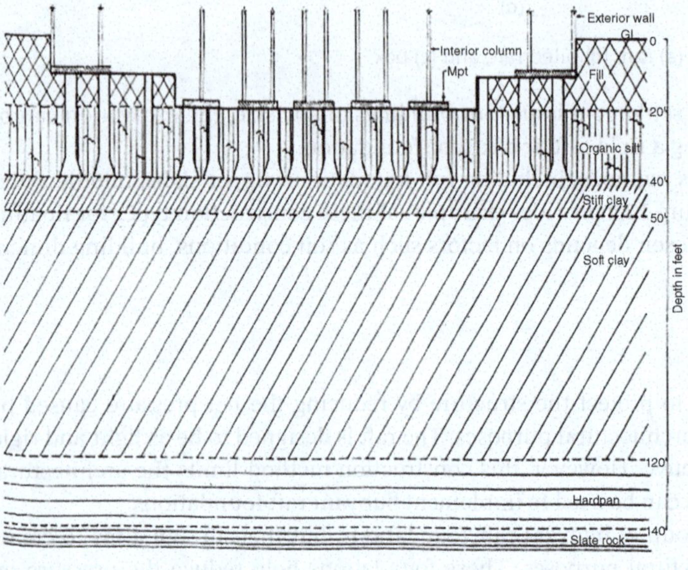

Figure 12.4 Liberty Mutual Insurance: cross-section and view of foundation and soil (Casagrande and Fadum 1944 / With permission of American Society of Civil Engineers).

The hard clay layer acts as a raft and spreads the load to the lower soft clay layer. Many buildings have withstood significant differential settlement due to the consolidation of the soft clay layer with a variable thickness or the different loads in different building areas due to the balance between structural and overburden pressure in buildings with basement floors.

Excavation equal to the superstructure weight is essential to eliminate or reduce asymmetric settlements. The net pressure on the soil was equivalent to 22.5 kPa. Therefore, the excavation depth under the central tower and the wings was 6 and 3 m. Loads of the columns were supported by floating foundations, which were placed on concrete foundations and extended to the surface of the hard clay layer to a depth of 12 m above the ground.

The total settlement was 51 mm after four years. There were no excavation problems in this case, and the excavation was done only to reduce the settlement. The foundation used in this sample is of the first type of floating foundation, in which the load enters the raft (hard clay), on top of which the settling layer was located.

12.3.2 Compensated Piled Raft

To enhance the geotechnical performance of raft foundations in terms of bearing capacity, settlement, and stability, deep foundations are often incorporated. This combined system is considered one of the most comprehensive foundation solutions. By utilizing deep foundations in conjunction with raft foundations, the load-bearing capacity of the foundation system is increased, settlement is reduced, and overall stability is improved. This approach allows for the efficient transfer of loads to deeper, more competent soil layers, providing a robust foundation solution for various types of structures.

Case Study 12.2

Piled Raft of Torre Latino Americana, Mexico City

As presented in Chapter 1, Torre Latino Americana is a 43-story (182 m) tower constructed on the soft challenging clay of Mexico City. The square base is 34.4 × 34.4 m, and the overall weight of the tower is around 23,500 tons. This project is one of the best examples of the application of floating foundations. A piled raft foundation system was implemented at a depth of 13 m, aiming to prevent extraordinary settlements, provide safety and favorable serviceability. By performing a floating foundation, or equating soil removed with part of the building weight, half of the weight of the structure was replaced with the removed soil. Therefore, to transfer the load and sew the foundation to deeper layers, a series of driven piles were used under the raft, the end of which was extended to a dense layer of Tarango sand. By installing piles under the foundation, a piled foundation was obtained, a successful system of foundations for reducing total and nonuniform settlements. Figure 12.5 depicts the schematic and actual view tower and its foundation system.

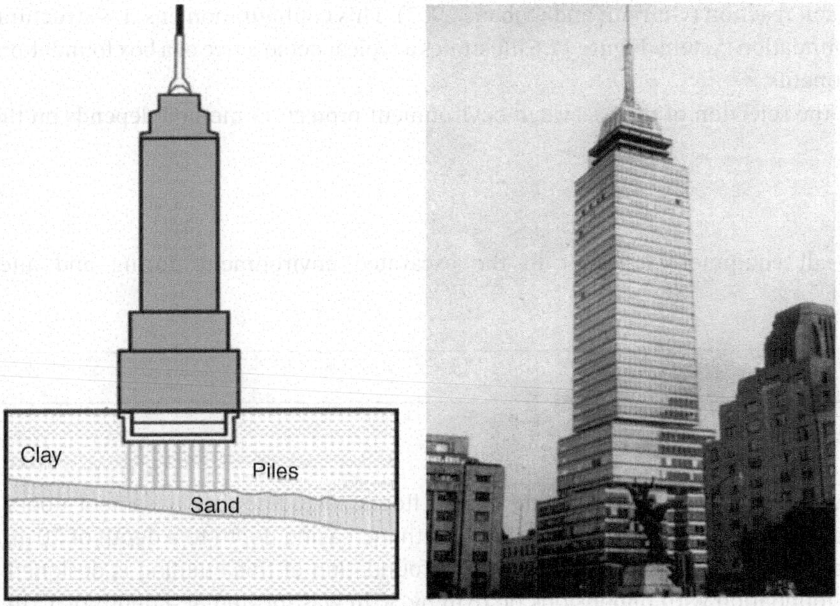

(a) (b)

Figure 12.5 Torre Latino Americana: (a) schematic view and piled raft foundation (Brûlé et al. 2019), (b) actual view (Zeevaert 1957/ICE Publishing).

12.3.3 Box Foundations

Box foundations offer the advantage of utilizing the substructure space for various purposes, such as warehouse storage or underground parking. To accommodate these functions, the basement floors of box foundations need to be spacious, without closely spaced walls or columns. The roof of the basement floor consists of slabs, which may or may not incorporate strong beams to provide the required rigidity. Box foundations are constructed using two horizontal diaphragms, namely the

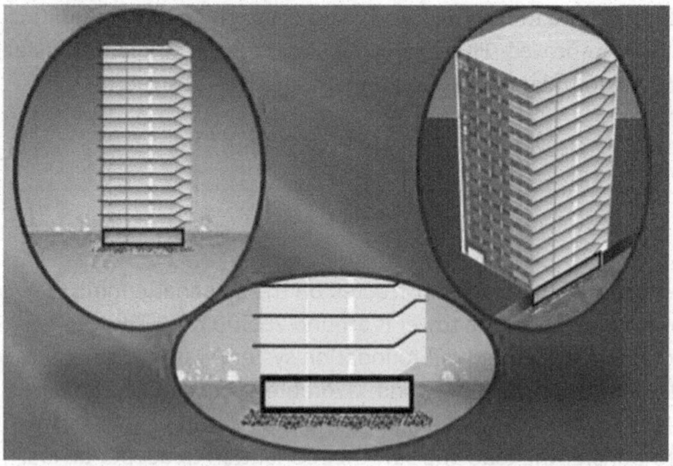

Figure 12.6 Typical perspective of box foundation.

upper and floor slabs, along with middle and perimeter walls. The upper and lower diaphragms are connected to the center walls, preventing rotational movements. The load-bearing base of the box foundation extends through the perimeter walls, minimizing the eccentricity of the overall soil reaction (Giuriani and Gubana 2007). This configuration ensures structural stability and efficient load transfer to the foundation system. Figure 12.6 illustrates a typical perspective of a box foundation, showcasing its distinct features and components.

In the construction of box foundations, the selection of the excavated environment protection method depends on the following parameters:

1) Foundation Depth
2) Soil conditions and groundwater level
3) Protection of buildings, roads, and all equipment adjacent to the excavated environment during and after implementation.

Case Study 12.3

Hotel Meridien, Brazil

The Meridien Hotel is located in Rio de Janeiro, Brazil. This building has 40 floors, including four basement floors (Hemsley 2000). The soil is fine coastal sand, the density of which increases with increasing depth to a depth of 20 m, the location of the bedrock. The groundwater level is at a depth of 2 m. The foundation of this building is designed based on a box foundation system. The foundation with dimensions of 26 m by 30 m was located at a depth of 13 m above the ground and was constructed as a box foundation at this depth. Boxes were used as water storage tanks. Figure 12.7 shows the view, plan, and cross-section of the infrastructure.

The total thickness of the foundation is 3 m, including the upper slabs and the floor with a thickness of 0.25 and 0.6, respectively, to withstand the applied pressure. The pressure applied by the foundation to the soil is about 500 kPa. This amount of pressure was allowed for the soil due to the amount of overburden pressure applied before excavation. The observed settlement was small and indicated the proper performance of the foundation. The infrastructure in this building has been implemented typically (from bottom to top) after excavation and protection of the excavation wall by the diaphragm wall. The measured settlement confirmed the effectiveness of this method.

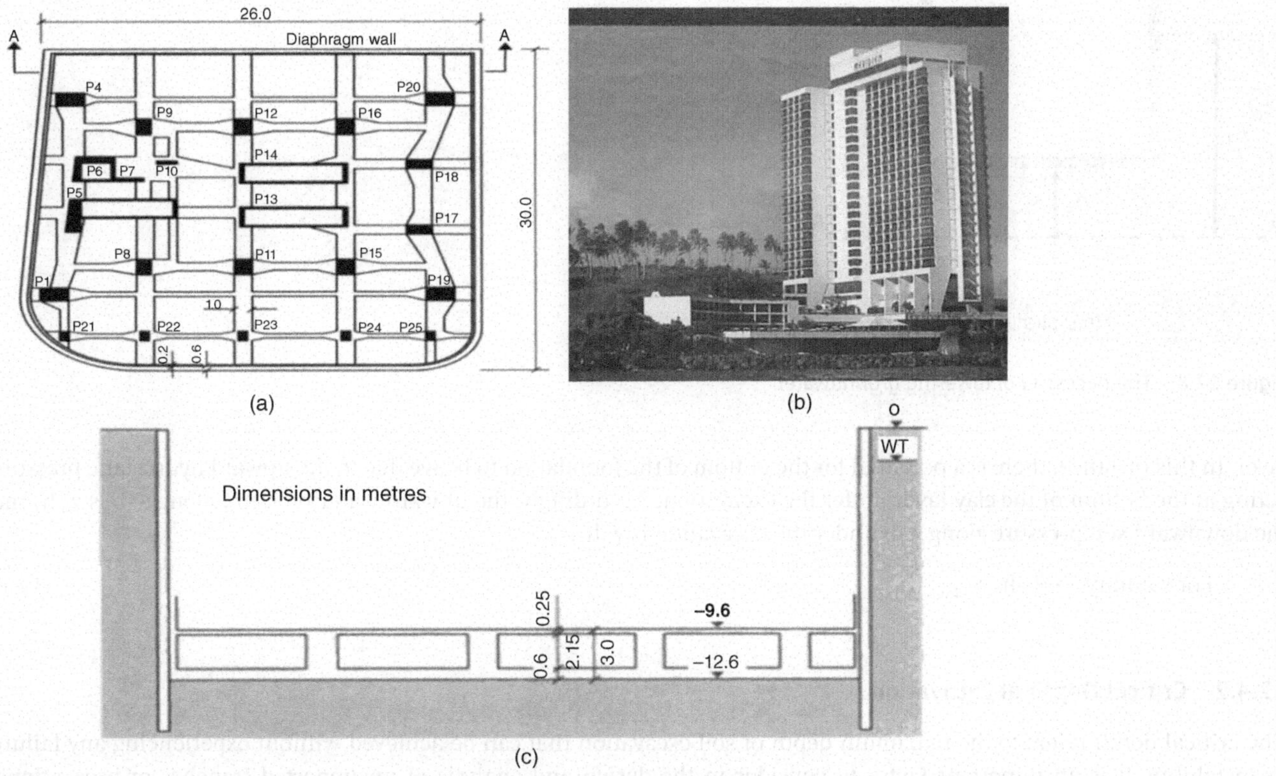

Figure 12.7 Meridian Hotel: (a) plan, (b) view, and (c) section A-A (Hemsley 2000 / With permission of Thomas Telford).

12.4 Excavation Issues and Associated Difficulties

Excavating the ground in the form of vertical cuts poses challenges, particularly in densely populated urban areas. It is crucial to carefully design and analyze such excavations, taking into consideration the geotechnical characteristics of the soil and various surrounding conditions, including superstructure loads, traffic loads, ground and surface water conditions, and neighboring buildings. Vertical cuts, especially slopes, are susceptible to instability, and it is essential to implement appropriate measures to prevent potential failures. Factors such as water seepage, dissipation of suction, and the passage of time can significantly influence the stability of the excavated slopes. These considerations become even more critical in the case of deep excavations, heavy traffic loads, and the presence of high and heavy structures in close proximity to the excavation area. By conducting thorough geotechnical assessments, considering proper drainage and soil stabilization techniques, and implementing effective slope protection measures, the risks associated with vertical cuts can be mitigated. Careful planning and execution are essential to ensure the safety and stability of the excavated slopes and surrounding areas.

12.4.1 Groundwater Lowering

Lowering of the groundwater level is needed when excavating in relatively impermeable clay soil below the groundwater level (The soil above the dotted line in Figure 12.8 is clay). The soil beneath the dotted line in Figure 12.8 is a permeable

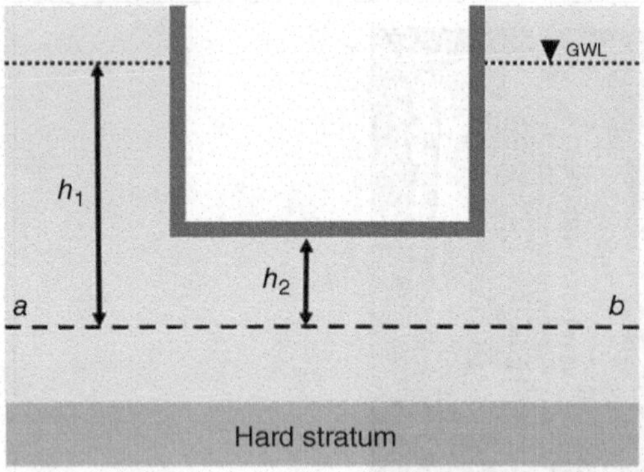

Figure 12.8 The necessity of lowering groundwater.

layer. In this instance, there is a potential for the bottom of the foundation to heave due to the upward hydrostatic pressure acting at the bottom of the clay layer under the excavation. Accordingly, the upward water pressure along a–b is $\gamma_w h_1$ and the downward soil pressure along a–b under the excavation is $\gamma .\ h_2$.

$$\text{For stability} \rightarrow \gamma \cdot h_2 > \gamma_w \cdot h_1 \tag{12.1}$$

12.4.2 Critical Depth of Excavation

The critical depth refers to the maximum depth of soil excavation that can be achieved without experiencing any failure or instability. It is an important factor to consider in the design and analysis of unsupported trenches or excavations. The critical depth is determined based on various geotechnical factors such as soil properties, groundwater conditions, slope stability, and the presence of any external loads or surcharges. It is crucial to ensure that the excavation does not exceed the critical depth to prevent potential failures such as slope collapse, soil movement, or structural instability.

To assess the critical depth, engineers perform stability analyses and consider factors such as soil shear strength, soil cohesion, angle of internal friction, groundwater pressure, and the potential for lateral soil movements. By evaluating these factors, the critical depth can be determined, and appropriate measures can be taken to ensure the stability and safety of the excavation. The critical depth analysis plays a vital role in excavation design and helps in determining the necessary support systems, shoring techniques, or reinforcement methods required to maintain the stability of the excavation and ensure the safety of personnel and adjacent structures.

The simple Equation 12.2 can be used to determine the critical depth for excavations in clay soils with different excavation length, width, and depth (see Figure 12.9) (Bjerrum and Eide 1956).

$$H_c = N_c \frac{S_u}{\gamma} \tag{12.2}$$

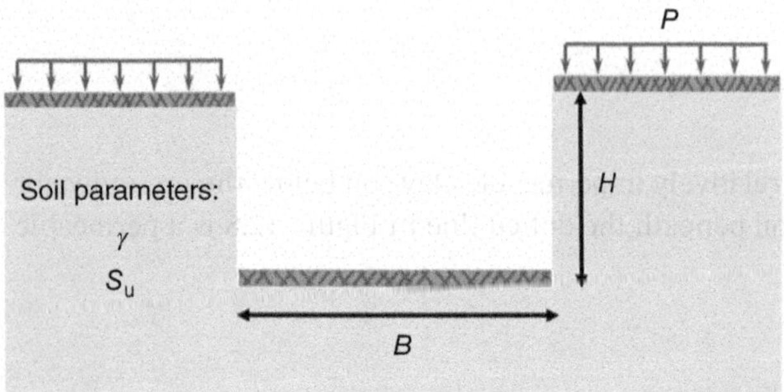

Figure 12.9 Parameters needed to calculate excavation critical height.

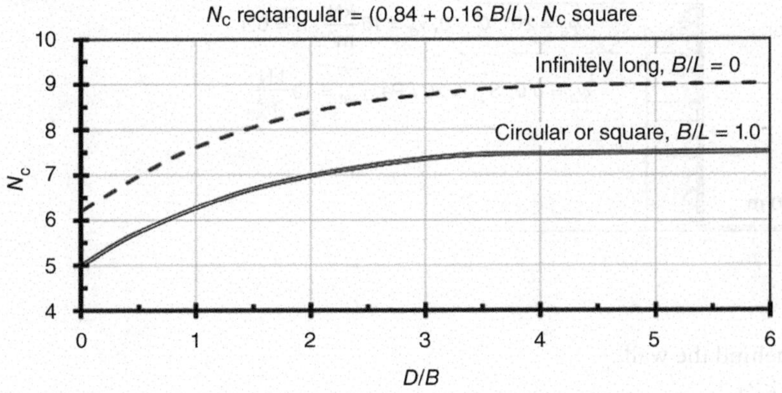

Figure 12.10 N_c values (Skempton 1951).

where

N_c = bearing capacity factor obtained from Figure 12.10.
γ = soil density
S_u = soil shear strength
H_c = critical height of excavation

Equation 12.3 provides the safety factor (F_s) against excavation failure in clay soils foe excavation depth of H (Bjerrum and Eide 1956).

$$F_s = N_c \frac{S_u}{\gamma \cdot H + P} \tag{12.3}$$

where,

P = overburden pressure,
F_s = The excavation failure safety factor varies between 1.35 and 1.5

This method determines the maximum excavation depth for design purposes. The safety factor tends to decrease in the following situations:

1. There is no possibility of damage in the adjacent excavation of buildings.
2. Clay is composed of stiff deposits.
3. Horizontal retaining wall inside the excavated area has high rigidity.

The factor of safety can be increased by filling the excavated space by water or bentonite. If the excavation is filled with water and bentonite, the value of the safety factor follows Equations 12.4 and 12.5:

$$F = \frac{N_c S_u}{H(\gamma - \gamma_w) + P} \tag{12.4}$$

$$F = \frac{N_c S_u}{H(\gamma - \gamma_m) + P} \tag{12.5}$$

where,

γ_m = Bentonite density
γ_w = water density

Example 12.1 It is desirable to design a floating foundation for a structure weighing 7×10^3 tons. Calculate the following:

a) Determining the critical height
b) Determining minimum undrained cohesion required for clay soil in order to safe load bearing with a factor of safety of 2

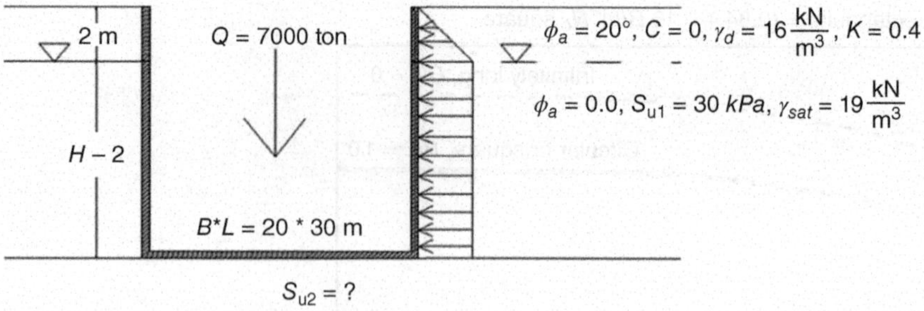

a) Calculation of wall resistance in the dry area behind the wall
At $z = 2\,m \Longrightarrow \sigma_h = k.\,\sigma_v = 2 \times 16 \times 0.4 = 12.8\,kPa$
Compressive force in this area per unit circumference of the wall:

$$12.8 \times \frac{2}{2} = 12.8\,kN/m$$

Given that the equivalent friction of the wall with the surrounding soil in that area is assumed to be 20 degrees, so the wall resistance in this area will be equal to:

$$F_{s1} = 12.8 \times \tan 20 \times P = 2 \times (20 + 30) \times 12.8 \times \tan 20 = 465.9\,kN$$

Calculation of wall resistance in the saturation zone with respect to the internal friction angle of zero

$$F_{s2} = S_{u1} \times (20 + 30) \times 2 \times (H - 2) = 3 \times 10^3(H - 2)\,kN$$

Total wall resistance $= F_s = F_{s1} + F_{s2} = 465.9 + 3 \times 10^3(H - 2)$
In order for the foundation to be floating, the load of the structure along with the weight of the foundation per unit area must be equal to the initial overhead stress before construction, so we have;
Initial overhead force = gravity load of the foundation taking into account the resistance of wall friction with the soil

$$7 \times 10^3 \times 9.81 - 465.9 - 3 \times 10^3(H - 2) = [2 \times 16 + (H - 2) \times 19] \times 20 \times 30 \Longrightarrow H = 5.4\,m$$

b) In order for the soil to be stable under this load, the stress on the soil must equal the soil's net bearing capacity.

$$q_{ult(net)} = S_u \cdot N_c + q' - q'$$

Assuming the application of the factor of safety of 2 in stability, we will therefore have:

$$\frac{q_{ult(net)}}{F_s} = S_{u2} \cdot \frac{N_c}{2} = [2 \times 16 + 3.4 \times 19]$$

Given the dependence of N_c on the B/L and D/B ratios of the corresponding graph, we estimate N_c for a rectangular foundation.

$$\frac{B}{L} = \frac{2}{3}\ \frac{D}{B} = \frac{5.4}{20} = 0.27$$

From the graph for the square foundation:

$$N_c(Square) \cong 5.2\quad N_c(rectangular) \cong 0.95$$

As a result, we will have:

$$S_{u2} = 38.6\,kPa$$

12.4.3 Upward Pressure Caused by Groundwater

Designing underground structures and substructures requires controlling the possibility of floating under upward hydrostatic forces. The flotation process follows Figure 12.11 due to the upward hydrostatic pressure at the basements of the buildings.

The weight of the completed structure is usually enough to overcome hydrostatic upward pressure in tall buildings with basements. However, the necessary measures to neutralize flotation are being developed during the construction stage of the structure. The hydrostatic upward pressure neutralization method selection depends on the soil conditions, the project's characteristics, and the construction method.

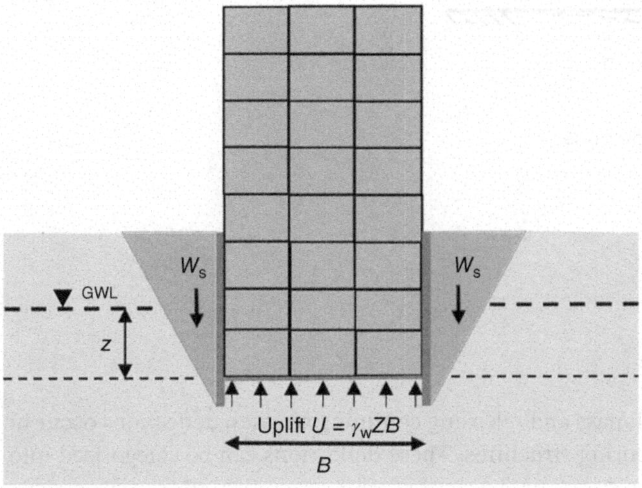

Figure 12.11 Hydrostatic upward pressure on the substructure floor.

It is necessary to prevent overturning and moving the substructure in water due to the upward groundwater pressure before entering the high superstructure loads to bear this pressure, which only occurs in saturated soils. Constant pumping can prevent this problem and other methods to keep groundwater levels low during construction.

After building the superstructure, to overcome the upward hydrostatic pressure, the resulting dead load is sufficient to balance the upward pressure. In some underground structures, there may be pure upward pressure due to the lightness of the superstructure. In this case, installing some anchors is necessary to prevent the overturning and moving of the substructure in water.

Figure 12.12 shows a sample of overcoming upward hydrostatic pressure. The increased dead load in the structure counteracts upward pressure, which leads to costly and heavy operation. If the soil surrounds the piles, as illustrated in Figure 12.12a friction piles most effectively. However, suppose the substructure is placed on the bedrock. In that case, the anchor bars or cables can be injected into the holes created in the bedrock, as shown in Figure 12.12b (Tomlinson 2001).

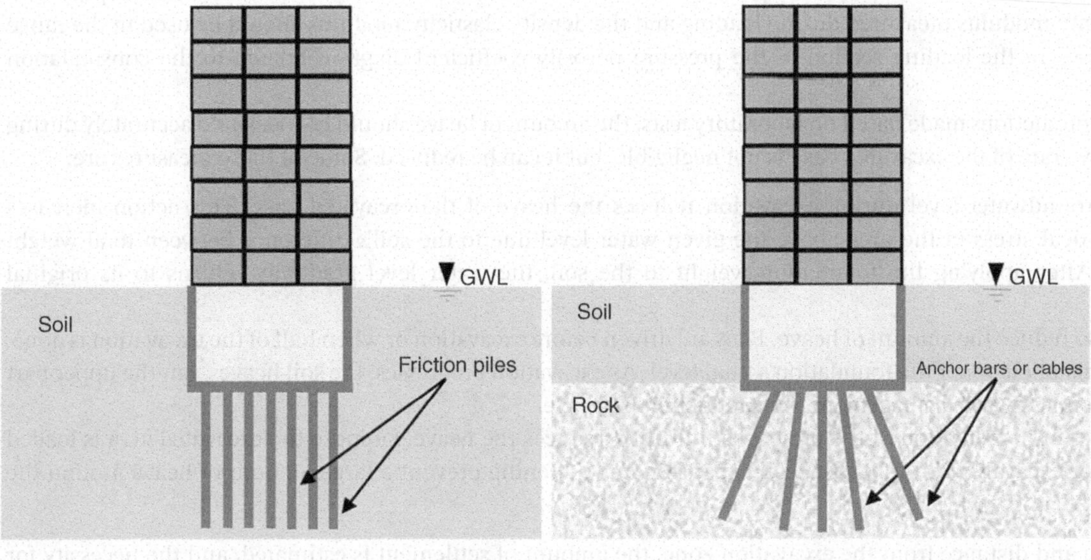

Figure 12.12 Overcoming upward hydrostatic pressure: (a) anchorage of floating structures in soil, (b) anchorage of floating structures to bedrock.

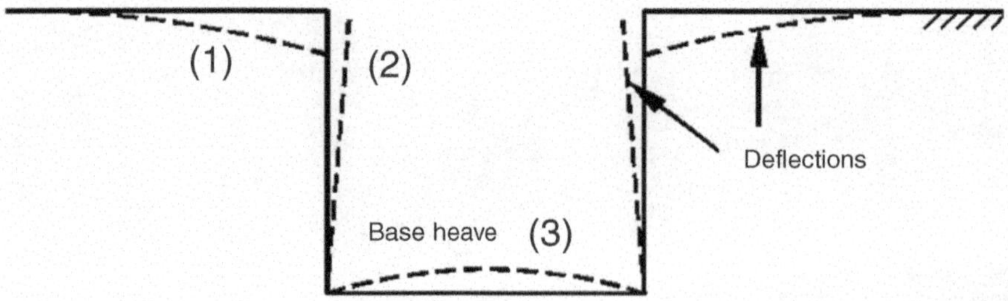

Figure 12.13 Excavation-induced deflections.

12.4.4 Deflections and Settlements

As a result of excavation, due to reasons such as extracting soil mass and releasing confining stresses, deflections occur in the zone of excavation, which can lead to instability in neighboring structures. These deflections can be categorized into three groups:

1) Settlement on the surface of adjacent grounds
2) Lateral deflection of excavation walls
3) Heave in the base of the excavation

Figure 12.13 depicts these three deflections schematically.

Lateral deflections of walls must be within allowed ranges that are determined in relevant codes and standards. It is essential to consider this issue in the analysis and design rigorously.

There are ways to reduce and control base heave. Three main reasons cause the excavation base heave (Fang 2001):

A) Elastic deformation of the excavation base soil
B) In the excavated environment, which remains open for a long time, there will be gradual soil heave and increased water due to the reduction of vertical pressure.
C) When the excavation depth approaches a critical amount, it is possible to settle the soil plastic settlement around the excavated site.

Although designers should avoid point C, its implementation is not as serious for buildings outside the adjacent excavation area in the second type of floating foundation. There is no need to balance the settlement of the surroundings. Before floating foundation construction and implementation, a laboratory test should evaluate the heave potential. However, the elasticity modulus measured during loading and the density elasticity modulus should be used in the range of appropriate stresses in the loading section of the pressure porosity coefficient diagram related to the consolidation tests.

In addition to the predictions made based on laboratory tests, the amount of heave should be measured accurately during excavation. The heave rate of the excavated base is not negligible, but it can be reduced. Some of these measures are:

A) Lowering the groundwater level during excavation reduces the heave of the excavated base. This action increases the effective vertical stress in the area above the given water level due to the soil's difference between total weight and buoyancy. After applying the foundation weight to the soil, the water level gradually returns to its original elevation.
B) Friction piles also reduce the amount of heave. Piles are driven before excavation or when half of the excavation is done. The pile's tip is installed under the foundation's final level. As excavation progresses, the soil heaves, but the upper part of the piles prevents upward displacement, thus reducing the heave.
C) Finally, the rapid implementation of the project significantly reduces the heave. Suppose the excavated area is loaded immediately after excavation. In that case, soil expansion and softening prevent a large amount of heave around the excavated base.

Based on soil type and distance from the excavation zone, the amount of settlement is estimated, and the necessity for strengthening the foundation and supporting the excavation is determined. This strengthening is relevant to soil type, the magnitude of loads imposed on the foundation, and their location. Based on CFEM (1992), three zones as depicted in Figure 12.14 must be considered for foundation strengthening.

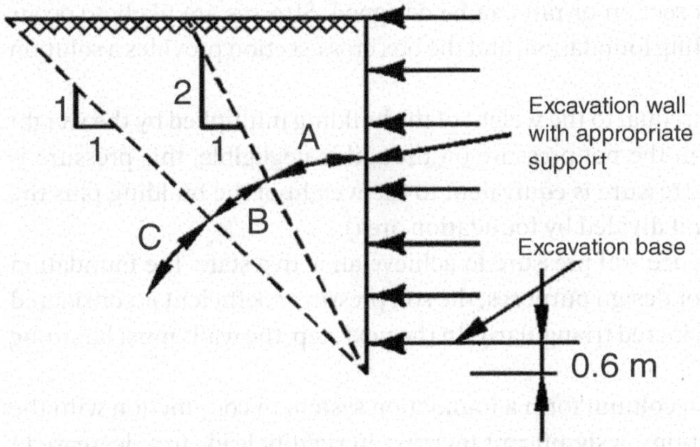

Figure 12.14 Zoning for strengthening of foundation adjacent to excavation (CFEM 1992 / BiTech Publishers).

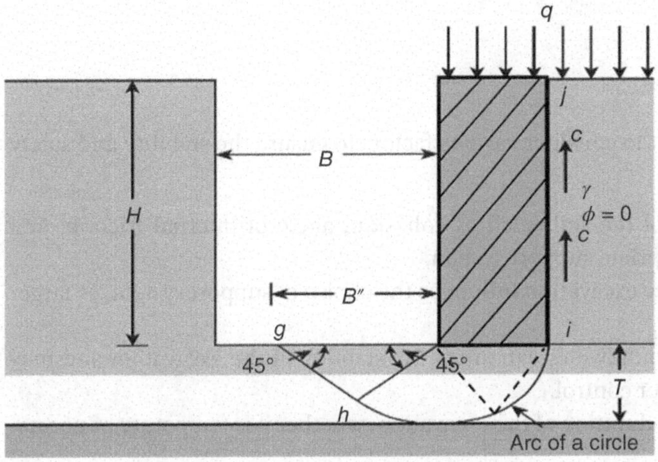

Figure 12.15 Base failure (Das and Sivakugan 2019 / With permission of Cengage Learning).

General recommendations are as follows:

Zone A: Usually there is a necessity to strengthen the foundations in this zone.
Zone B: The foundations in this zone may need to be strengthened based on structure type and loading condition. Horizontal and vertical pressure on the excavation walls due to foundations that are not strengthened must be considered.
Zone C: The foundations in this zone do not need to be strengthened but strengthening the foundations of other zones often must be continued to this zone. Horizontal pressure from strengthening elements of foundations is not considered usually.

12.4.5 Base Failure

By excavation to establish the foundation at the required depth, the vertical pressure on the soil decreases, meaning that the excavation base may expand or heave due to release from overburden pressure. In such cases, the soil of the wall acts as a column of load (the hatched column in Figure 12.15), and since the embedment depth is zero at one side, base failure occurs, which is shown for clay soil in Figure 12.15.

12.4.6 Foundation Systems Consideration

Although individual foundations can apply the principle of buoyancy to a building, this is unusual. Floating foundations are usually simple rafts and box cross-sections.

After estimating the pressure distribution on the soil, a box section or raft can be designed. Stresses are likely to occur during the flotation of the box section in the second type of floating foundation, and the box cross-section provides a solution by protecting it in the corners or only in the middle.

The foundation floor should be able to resist upward pressure equal to the weight of the building multiplied by the weight of the foundation divided by the foundation area. Even though the net pressure on the soil is negligible, this pressure is the total pressure on the soil, which might be significant (net pressure is equivalent to the weight of the building plus the weight of the foundation, minus the weight of the excavated soil divided by foundation area).

Since the foundation does not have enough flexibility to reduce soil pressure to achieve an active state, the foundation walls must be resistant to pressure in the rest state of the soil. For design purposes, the soil pressure coefficient is considered to be 0.5, and the overall pressure in the first step should be considered triangularly. In the next step, the walls must be strong enough to bear forces in both cases (Fang 2001).

Above all, a floating foundation in which basement walls and column form a foundation system in conjunction with the raft, perform more favorably than single rafts. In such foundations, a significant increase in rigidity leads to a decrease in differential settlements. Furthermore, the base level for seismic analysis will be located in depth, therefore considerable reduction in design shear forces and moments occur.

12.5 Supporting Systems

When selecting an excavation support system, it is important to consider various factors to ensure the stability and safety of the excavation. These factors include:

1) **Soil Type and Strength Parameters:** The properties of the soil, such as cohesion, angle of internal friction, and permeability, play a crucial role in determining the appropriate support system.
2) **Height and Width of Excavation:** The dimensions of the excavation influence the choice of support system, as larger excavations may require more robust support systems.
3) **Groundwater Condition:** The presence and level of groundwater can affect the stability of the excavation and may require additional measures for dewatering or groundwater control.
4) **Long-term or Short-term Serviceability:** The intended duration of the excavation, whether it is temporary or permanent, can impact the selection of the support system.
5) **Requirements of Deflection:** Some excavations may have specific deflection criteria that need to be met, such as restrictions on ground movement to protect adjacent structures.
6) **Adjacent Loading and Structures:** The proximity of existing structures or loads near the excavation site must be considered to prevent any adverse effects or damage during the excavation process.
7) **Local Tectonic and Seismic Conditions:** The geological and seismic conditions of the area can influence the choice of support system, particularly in regions prone to earthquakes or ground movements.

Figure 12.16 provides an overview of typical excavation support systems, including sloped excavation, sheet piling, diaphragm walls, bored pile walls, and top-down construction. Each system has its advantages and considerations, and the selection depends on the specific project requirements and site conditions.

12.5.1 Sloped (Stepped) Excavation

Sloped or stepped excavation is indeed one of the most economical methods for projects where there is sufficient space around the substructure to slope the walls of the excavated area and achieve a stable slope. This method is typically suitable when there are no issues with significant groundwater presence that could lead to erosion and slumping of slopes. Determining stable slopes for the excavated area involves considering factors such as the type of soil present. Different soil types have varying stability characteristics, and the slope design should be tailored accordingly. The duration for which the excavation area will remain open is also important, as longer durations may require more stable slopes to ensure safety. Additionally, the permitted degree of side slippage, or the allowable movement of the soil, should be considered in

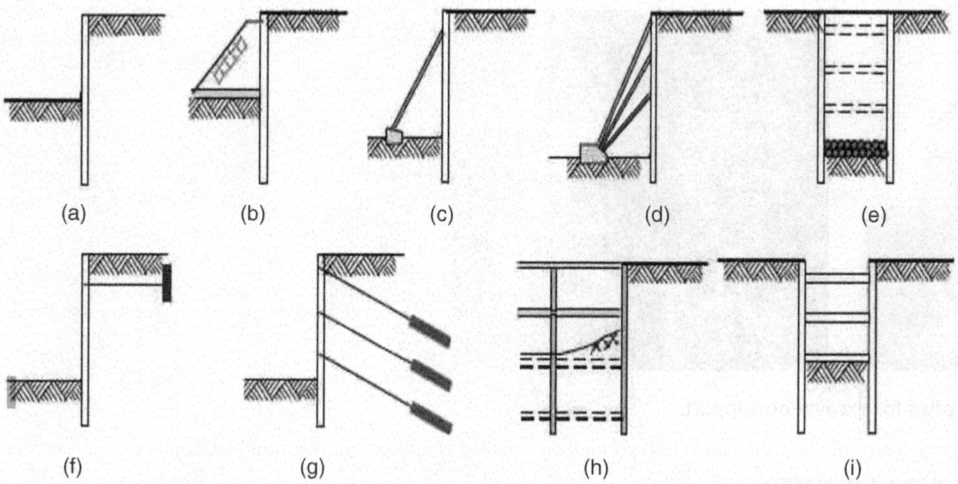

Figure 12.16 Typical systems for supporting the excavations: (a) cantilever wall, (b) berm and slab, (c) single-propped wall, (d) multi-propped wall, (e) soilcrete-slab, (f) tie-back wall, (g) ground anchors, (h) top-down construction, and (i) braced excavation.

determining the appropriate slope design. By taking these factors into account and implementing suitable slope designs, sloped or stepped excavation can provide a cost-effective solution for projects with favorable site conditions.

12.5.2 Sheet Piling

The sheet piling construction method is particularly suitable for projects where there is limited space around the excavation site, making it challenging to slope the ground. One advantage of sheet piling is that, if the soil conditions allow, the sheets can be removed and reused in other projects, making them a more cost-effective option compared to diaphragm walls. It is important to carefully investigate and control the effects of adjacent underground buildings, roads, or structures during the excavation process. Movements and excavations can have a significant impact on these structures. Excavation for box foundations, for example, can lead to settlement of the surrounding ground surface. This settlement can potentially cause damage to nearby buildings and result in cracks in drains and other services.

Several factors can contribute to settlement during excavation, including lateral displacement of the excavation face due to the yielding components of sheet piling, lateral displacement of the excavation under the influence of elastic displacement from the retaining wall after backfill completion, reduction of groundwater level due to pumping, and geohazards such as slope failure, landslides, heave, erosion, and other potential risks. Sheets are commonly used as temporary support systems during excavation.

They can be made of various materials, such as wood, concrete, or steel. In deep excavations, steel sheets are commonly employed. Figure 12.17 provides an example of a sheet pile wall, showcasing the use of steel sheets in the construction process.

12.5.3 Diaphragm Walls

After the construction of reinforced concrete guide walls and excavation of the trench between them, the installation of a diaphragm wall is carried out. This process involves temporarily protecting the diaphragm wall with bentonite slurry. To reinforce the diaphragm wall, steel reinforcement is placed at the desired level within the trench, followed by the pouring of bentonite replacement concrete. This process helps ensure the stability and strength of the diaphragm wall. In urban areas where concerns about vibration and noise are significant, diaphragm walls offer advantages over traditional sheet piling methods. Diaphragm walls effectively protect the surrounding environment during excavation activities. Figure 12.18 provides a visual representation of the diaphragm wall construction process. Diaphragm walls are particularly suitable for

Figure 12.17 Application of sheet piles for excavation support.

Figure 12.18 Implementation of underground retaining walls using diaphragm wall (Tomlinson 2001 / Pearson Education).

projects where obstacles, such as existing structures or utilities, hinder the use of sheet piling in the soil. Additionally, when there are limitations imposed by a high groundwater level or when adjacent foundations need protection against ground heave and soil erosion, diaphragm walls are a preferred choice. Overall, the use of diaphragm walls provides a reliable and efficient solution for excavation protection, ensuring the stability of the surrounding environment while facilitating construction activities in challenging urban settings.

12.5.4 Bored Pile Walls

In the contiguous bored pile system, piles can be installed in one or two rows, which can be secant or within a short distance of each other. Piles are first run one by one, and then the middle piles are installed. Using the secant bored pile method can solve the waterproofing problem. A secant pile wall is comprised of two drilled piles, one primary and the other secondary. Figure 12.19 shows a secant pile wall after excavation.

12.5.5 Top-down Construction

There is an increasing demand, particularly in densely populated urban areas, for improving construction methods that minimize the impact on adjacent structures during excavations. In response to this demand, the top-down construction technique has emerged as a popular solution, replacing the conventional bottom-up technique.

The top-down construction method offers several advantages in minimizing the impact on the surrounding environment and enabling deep excavations in narrow inner-city construction sites. It is particularly suitable for challenging soil conditions such as high groundwater levels and weak soils. The process begins with the installation of retaining walls along the outline of the structure. Diaphragm walls are commonly used for this purpose. These retaining walls serve as

Figure 12.19 Secant pile wall.

underground walls for the entire life span of the building. They provide stability to the excavated surface and prevent soil movement during construction. Once the retaining walls are in place, temporary columns are installed to support the structural loads during the excavation and construction phases. These columns ensure the stability and safety of the structure during the top-down construction process. With the temporary columns in place, ground floor slabs can be constructed over the excavated area. This allows for the simultaneous progress of construction activities both above and below the ground level. The top-down construction method allows for efficient utilization of the construction site and enables faster project completion compared to traditional bottom-up methods. By pre-installing the retaining walls and temporary columns, the top-down construction method minimizes the impact of noise and dust on the surrounding environment. It also provides better control over the excavation process and ensures the stability of the excavated area throughout construction. Figure 12.20 illustrates the top-down construction process, showcasing the sequence of installation of retaining walls, temporary columns, and ground floor slabs.

After constructing the first-floor slab, excavating under the first basement floor is done. The construction of the first basement floor is completed with the slab installation. When the first upper level is under construction, a similar procedure is carried out for the second basement level at the same time. This process continues until the target underground level is reached. Once the targeted underground excavation is complete, the final foundation (raft foundation) is installed. Construction of the superstructure continues until the target floor level is reached (Jeong and Kim 2018). This construction approach, known as the top-down construction method, offers advantages in terms of minimizing disruption to the surrounding environment, reducing construction time, and ensuring the stability of the excavated area throughout the construction process.

Case Study 12.4

Application of Top-down Construction Method for a Metro Station

Dadongmen metro station is the transfer station of Hefei Metro 1 and Metro 2, China (Jeong and Kim 2018). The transfer station is located at the intersection of Shengli Road and East of Changjiang Road, and the station of Metro 1 is located under the Shengli Road and the station of Metro 2 under the East of Changjiang Road. The transfer station is T-shape, and the station of Metro 1 is under the station of Metro 2. The excavation depth of Metro 1 is 31.7 m and 24.4 m for Metro 2.

All the stations of Metro 1 and Metro 2 are island platforms with a width of 14.0 m. The station of Metro 1 is a four-floor underground frame structure with three-span, and the length, width, and excavation depth of the standard section are 145.44 m, 23.5 m and 31.7 m, respectively. The maximum excavation depth of the station of Metro 1, the end well, is 33.1 m, the deepest station of Metro 1. Figure 12.21a is the elevation view of the Station of Metro 1.

(Continued)

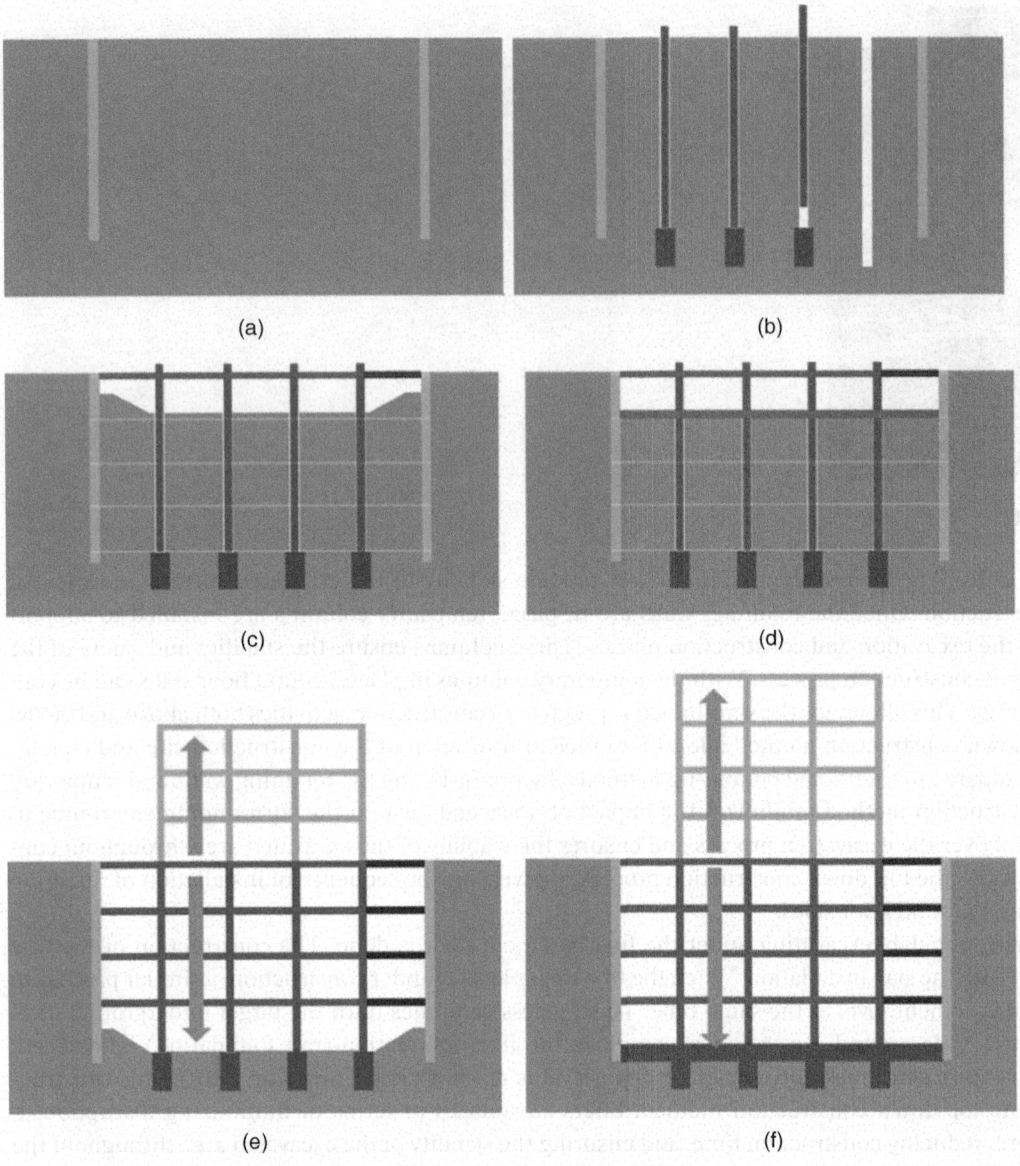

Figure 12.20 Schematic of top-down construction process: (a) installation of retaining walls, (b) installation of columns, (c) excavation of 1sr basement level, (d) construction of 1st basement level slab, (e) simultaneous construction, and (f) construction of base footing (Jeong and Kim 2018 / With permission of Springer Nature).

Case Study 12.4 (Continued)

The station of Metro 2 is a three-floor underground frame structure with a three-span, and the length, width, and excavation depth of the standard section are 220.5 m, 23.5 m, and 24.4 m, respectively. The excavation depth of the end well is 25.3 m. The intersection of the two stations is a three-floor underground frame structure with a multi-span. The total floor area of the transfer station is 34837 m^2. Figure 12.21b presents the soil profile of the site.

Considering the excavation depth and the surrounding, the top-down construction method is adopted for this project. The retaining structure of the station of Metro 1 is a diaphragm wall with a width of 1.2 m, and a depth of 40.7 m for the standard section (in reference to Figure 12.21a), 42.1 m for the end well. The retaining structure of the station of Metro 2 is a diaphragm wall with a width of 1.0 m, and a depth of 29.85 m for the standard section and 31.30 m for the end well. For the need for anti-float, there are many belled steel pipe piles in the transfer underground station. The underparts of the belled steel pipe piles are in the sandstone layer.

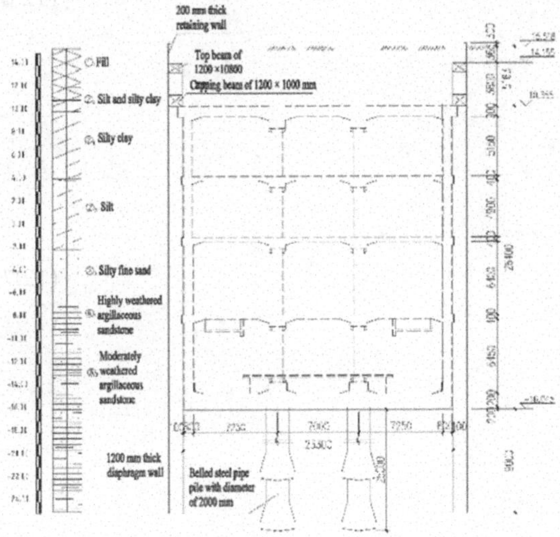

(a)

Name	Unit weight (kN/m^3)	Compression modulus (MPa)	CU: Cohesion (kPa)	CU: Friction angle (°)	Permeability coefficient (*m/d*): Horizontal	Permeability coefficient (*m/d*): Vertical
Fill	17.5	—	0	8	—	—
Silty clay	19.9	8	30	7	0.05	0.05
Silt	20.5	8.3	20	11	0.5	0.5
Silty fine sand	20.1	9.6	10	25	2.0	2.0
Silt and silty clay	17.6	3	5	6	0.01	0.01
Highly weathered argillaceous sandstone	21.0	30	35	30	0.2	0.2
Moderately weathered argillaceous sandstone	23.2	—	—	—	0.05	0.05

(b)

Figure 12.21 (a) The elevation view of the Station of Metro 1, (b) the soil profile of the site (adapted from Jeong and Kim 2018).

12.6 Floating Foundation Design

In general, floating foundations are designed for situations where the underlying soil layer is compressible (excessive settlements) and exhibits low bearing capacity. The objective is to improve the load-bearing capacity and control settlement.

12.6.1 Settlement Control

In the design of floating foundations, in a project in which the base failure does not occur and soil has an acceptable bearing capacity, the following items should be considered:

- Whether or not excavation is a significant problem.
- The compressible layers.
- The loading area of the building can be divided into the following two categories:

1. The building load should be carried out exactly at the foundation area. In this case, the settlement is calculated at ground level. If the settlement is more than the allowable amount, the stress must be reduced to reach the allowable settlement. The excavation depth will determine possible stress reduction.

2. A strong layer carries the building load in depth. In this case, the settlement under the building load on top of the strong layer should be calculated. If the calculated settlement is unacceptable, the stress should be reduced. The excavation depth will determine desired stress level. Also, determine how the load of the building is transferred to the strong layer (with pile).

Immediate settlement is determined by using equation (Equation 12.6). In this equation, the value of q equals to q_{net}. The soils encountered at the floating foundation depth are usually overconsolidated. Accordingly, in the equation of consolidation settlement (Equation 12.7), the C_r coefficient is used instead of C_c, which results in reduced consolidation settlement.

- Calculating the excavation base heave. If the value is acceptable, add it to the calculated settlement. The heave will generally be slight since the soil has an acceptable resistance in this type of floating foundation.

$$s = k\frac{qB}{E} \tag{12.6}$$

where,

k = a factor depending on soil stiffness and foundation rigidity,
q = contact pressure, $q_{net} = q - \gamma.\ D_f$
B = foundation width
E = modulus of soil

$$S = \frac{H_0 C_c}{1+e}\log\left(\frac{\sigma'_f}{\sigma'_0}\right) \tag{12.7}$$

where,

C_c = compression index,
H_0 = compressive soil thickness
e_0 = initial void index
σ'_0 = initial vertical stress
σ'_f = final vertical stress

12.6.2 Bearing Capacity Control

Excavation is a crucial aspect when dealing with floating foundations, as it can impact the stability and bearing capacity of the foundation. There are two main approaches for constructing floating foundations in relation to excavation. In the first approach, the foundation is made as a box section and floats in a suitable position in precast parts. In the second approach, retaining excavation is performed, and the foundation is constructed inside the excavated area.

Excavation is a crucial aspect when dealing with floating foundations, as it can impact the stability and bearing capacity of the foundation. There are two main approaches for constructing floating foundations in relation to excavation:

The following should be considered in the design of this type of floating foundations:

- The selection between the two types of floating foundation groups depends on factors such as cost and the safety of adjacent buildings. Here are the considerations for choosing between the two groups.
  (1) **Cost Considerations:** If there are no buildings near the excavated area, the first group (box section floating foundations) may be more cost-effective. This approach involves fabricating precast parts and positioning them in the desired location, which can be a cost-efficient method.
  (2) **Safety of Adjacent Buildings:** If there are buildings located near the excavated area, additional measures need to be taken to ensure their safety. In such cases, the second group (excavated floating foundations) is generally preferred. This approach involves constructing a retaining wall inside the excavated area to provide support and protection to the adjacent buildings.
- Calculating the critical depth for the excavation operation.
- Calculating the net stress on the soil at critical depth.
- The settlement should be calculated under net stress.

In the bearing capacity equation (Equation 12.8), by using the floating foundation, the value of foundation depth (D_f) is increased; therefore, the $\overline{q}$ term is increased; so, as a result, the bearing capacity increases.

$$q_{ult} = cN_c + \overline{q}N_q + 0.5\gamma BN_\gamma \tag{12.8}$$

where,

C = soil cohesion,
$\overline{q}$ = surcharge around the foundation (γD_f)
γ = average effective unit weight of the soil beneath and around the foundation
B = foundation width

D_f = embedment depth of the foundation
N_c, N_q, N_γ = nondimensional factors as exponential functions of φ

It is necessary to check whether the settlement is acceptable. If not, the following suggestions can reduce the amount of settlement:

In the first group:

1) Increase the area of the box plan.
2) Increase the excavation depth and fill the space below the box section using water and bentonite.

In the second group:

1) Increase the excavation area.
2) Transfer part of the load to the lower strong layers by piles.

Example 12.2 The 12-story building is located on a foundation measuring 20 x 20 m. According to the given information, it is desirable to determine the settlement in 2 cases:

a) Foundation on the ground level
b) Using the floating foundation at a depth of 2 m

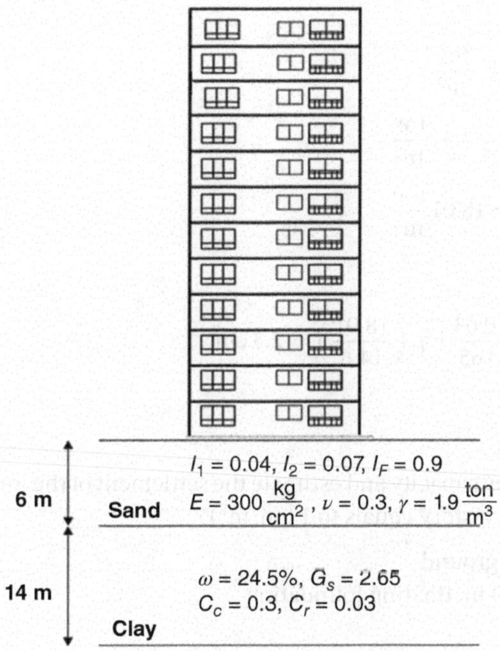

a)

$$q = 12\,\frac{ton}{m^2}$$

$$S = \frac{qB(1-\nu^2)}{E} I_0 I_F = \frac{12 \times 10 \times (1 - 0.3^2) \times 0.08 \times 0.9}{300} = 0.26\ \text{cm}$$

$$S = 4 \times 0.26 = 1.04\ \text{cm}$$

$$I_0 = I_1 + \frac{1 - 2\nu}{1 - \nu} I_2 = 0.08$$

$$\omega \cdot G_s = S_r \cdot e_0$$

$$0.245 \times 2.65 = 1 \times e_0$$

$$e_0 = 0.65$$

$$\gamma = \frac{G_s + e_0}{1 + e_0}\gamma_w = \frac{2.65 + 0.65}{1 + 0.65} \times 1 = 2\,\frac{ton}{m^3}$$

$$\sigma_0' = 1.9 \times 6 + 7 \times (2 - 1) = 18.3 \frac{\text{ton}}{\text{m}^2}$$

$$\Delta\sigma' = \frac{12 \times 20 \times 20}{(20 + 13)^2} = 4.41 \frac{\text{ton}}{\text{m}^2}$$

$$\sigma_f' = \sigma_0' + \Delta\sigma' = 18.3 + 4.41 = 22.8 \frac{\text{ton}}{\text{m}^2}$$

$$\Delta H = \frac{H_0 C_c}{1 + e} \log\left(\frac{\sigma_f'}{\sigma_0'}\right) = \frac{14 \times 0.3}{1 + 0.65} \log\left(\frac{22.8}{18.3}\right) = 23.8 \text{ cm}$$

$$S = 23.8 + 1.04 = 24.84 \text{ cm}$$

b)

$$q = 12 - 2 \times 1.9 = 8.2 \frac{\text{ton}}{\text{m}^2}$$

$$S = \frac{qB(1 - \nu^2)}{E} I_0 I_F = \frac{8.2 \times 10 \times (1 - 0.3^2) \times 0.08 \times 0.9}{300} = 0.17 \text{ cm}$$

$$I_0 = I_1 + \frac{1 - 2\nu}{1 - \nu} I_2 = 0.08$$

$$\sigma_c' = 18.3 \frac{\text{ton}}{\text{m}^2}$$

$$\sigma_0' = 4 \times 1.9 + 7 \times (2 - 1) = 14.6 \frac{\text{ton}}{\text{m}^2}$$

$$\Delta\sigma' = \frac{(12 - 2 \times 1.9) \times 20 \times 20}{(20 + 11)^2} = 3.41 \frac{\text{ton}}{\text{m}^2}$$

$$\sigma_f' = \sigma_0' + \Delta\sigma' = 14.6 + 3.41 = 18.01 \frac{\text{ton}}{\text{m}^2}$$

$$\sigma_0' < \sigma_f' < \sigma_c'$$

$$\Delta H = \frac{H_0 C_r}{1 + e} \log\left(\frac{\sigma_f'}{\sigma_0'}\right) = \frac{14 \times 0.03}{1 + 0.65} \log\left(\frac{18.01}{14.6}\right) = 2.3 \text{ cm}$$

$$S = 2.3 + 0.68 = 2.98 \text{ cm}$$

Example 12.3 Determine the bearing capacity and estimate the settlement of the foundation system of a 30-story building in two conditions: (average load of each story equals to 1 ton/m^2)

a) Mat foundation, 25 × 25 m, on the ground
b) Mat foundation, 25 × 25 m, $D_f = 10$ m, floating foundation

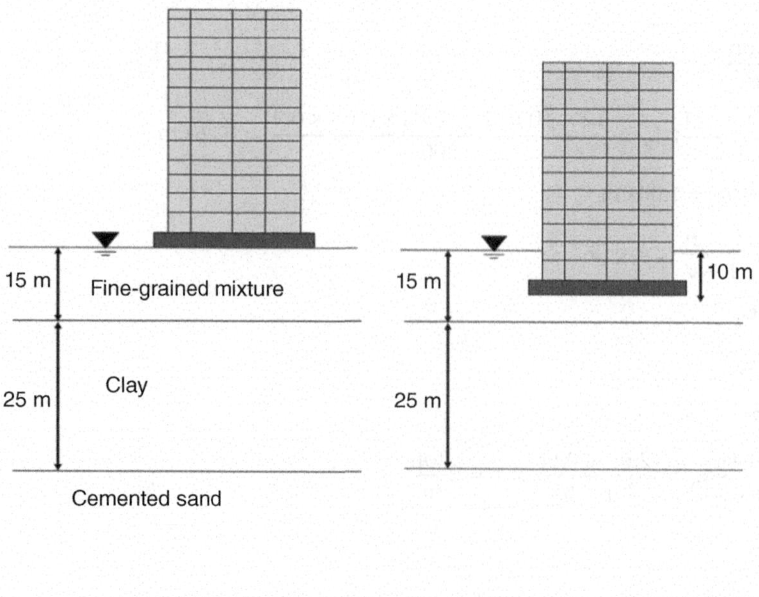

Soil Layer	γ (ton/m³)	K	$S_{u,avg}$ (kg/cm²)	C (kg/cm²)	φ	E (kg/cm²)
Fine-grained Mixture	2	0.4	0.5	–	–	150
Clay	1.2	0.2	0.25	–	–	50
Cemented Sand	2.2	0.15	–	0.5	36	1500

The influence zone used to determine bearing capacity and settlement estimation are presumed B and 2B:

a) $q_{ult} = N_c S_u + \sigma_0$, The average of S_u between the ground surface to Z = 25 m should be used.

$$S_{u_{avg}} = \frac{15 \times 0.5 + 10 \times 0.25}{25} = 0.4\frac{kg}{cm^2} = 4\frac{ton}{m^2}$$

$$q_{ult} = 5.14 \times 4 + 0 = 20.56\frac{ton}{m^2}$$

$$q = 30 \times 1 = 30\frac{ton}{m^2}$$

Fine-grained mixture:

$$S_{Fine-grained-mixture} = \frac{kqB}{E} = \frac{0.4 \times 30 \times 25}{150 \times 10} = 0.2\text{ m} = 200\text{ mm}$$

Clay:

$$\Delta q = \frac{q \times B^2}{(B+Z)^2} = \frac{30 \times 25^2}{(25+15)^2} = 11.72\frac{ton}{m^2}$$

$$S_{Clay} = \frac{kqB}{E} = \frac{0.2 \times 11.72 \times 25}{50 \times 10} = 0.117\text{ m} = 117\text{ mm}$$

Cemented Sand:

$$\Delta q = \frac{q \times B^2}{(B+Z)^2} = \frac{30 \times B^2}{(B+Z)^2} = 4.438\frac{ton}{m^2}$$

$$S_{cemented-sand} = \frac{kqB}{E} = \frac{0.15 \times 4.438 \times 25}{1500 \times 10} = 0.0011\text{ m} = 1.1\text{ mm}$$

$$S_{total} = 200 + 117 + 1.1 = 318.1\text{ mm}$$

b) $q_{ult} = N_c S_u + \sigma_0$, The average of S_u between the ground surface to Z = 10 to Z = 35 m should be used.

$$S_{u_{avg}} = \frac{5 \times 0.5 + 20 \times 0.25}{25} = 0.3\frac{kg}{cm^2} = 3\frac{ton}{m^2}$$

$$q_{ult} = 5.14 \times 3 + 10 \times (2-1) = 25.42\frac{ton}{m^2}$$

$$q = 30 \times 1 - 10 \times 2 = 10\frac{ton}{m^2}$$

Fine-grained mixture:

$$S_{Fine-grained-mixture} = \frac{kqB}{E} = \frac{0.4 \times 10 \times 25}{150 \times 10} = 0.067\text{ m} = 67\text{ mm}$$

Clay:

$$\Delta q = \frac{q \times B^2}{(B+Z)^2} = \frac{10 \times 25^2}{(25+5)^2} = 6.94\frac{ton}{m^2}$$

$$S_{Clay} = \frac{kqB}{E} = \frac{0.2 \times 6.94 \times 25}{50 \times 10} = 0.0694\text{ m} = 69.4\text{ mm}$$

Cemented Sand:

$$\Delta q = \frac{q \times B^2}{(B+Z)^2} = \frac{10 \times 25^2}{(25+30)^2} = 2.066 \frac{ton}{m^2}$$

$$S_{cemented-sand} = \frac{kqB}{E} = \frac{0.15 \times 2.066 \times 25}{1500 \times 10} = 0.0005\,m = 0.5\,mm$$

$$S_{total} = 67 + 69.4 + 0.5 = 136.9\,mm.$$

12.7 Case Studies

This section provides an overview of various types of floating foundations that have been implemented in practical cases. The geometric characteristics of the foundations, soil conditions, foundation depth, and superstructure pressures are discussed. The settlement measurements obtained through instrumentation in these practical cases are also presented.

Case Study 12.5

Albany Telephone Building, The USA

The Albany Telephone Building, constructed in 1929, has dimensions of 32.1 m by 52.5 m (Glick 1936). It consists of 11 floors in the superstructure and three floors in the infrastructure. The gravitational load acting on the foundation surface is measured at 215.5 kPa. Given the delicate nature of the telephone building, it is crucial to ensure that the foundation experiences no asymmetric settlement. Figure 12.22 illustrates the cross-section of the foundation.

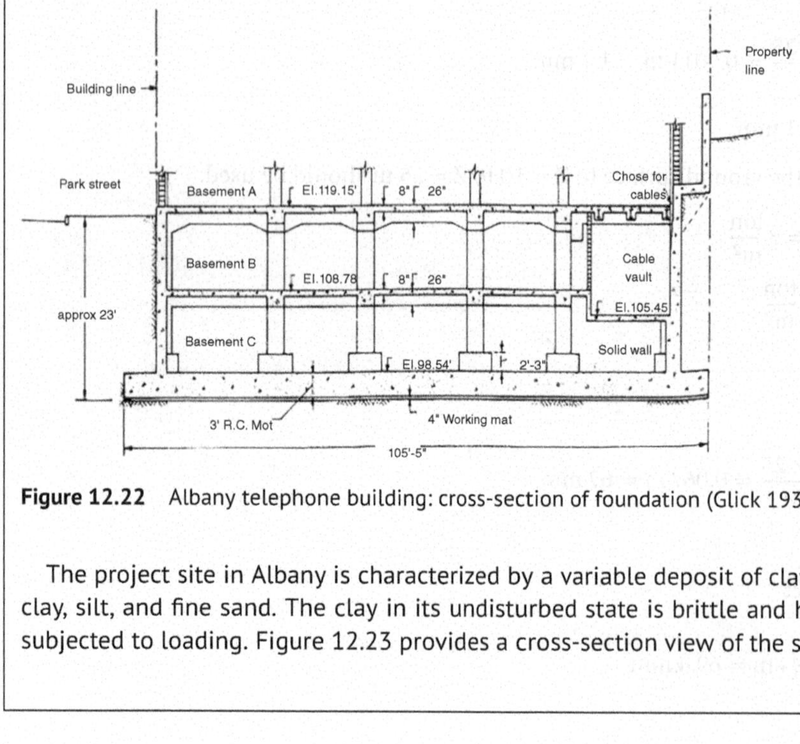

Figure 12.22 Albany telephone building: cross-section of foundation (Glick 1936 / ISSMGE).

The project site in Albany is characterized by a variable deposit of clay soil consisting of layers of plastic clay, silty clay, silt, and fine sand. The clay in its undisturbed state is brittle and hard, but it becomes soft and cohesive when subjected to loading. Figure 12.23 provides a cross-section view of the soil layers at the project site.

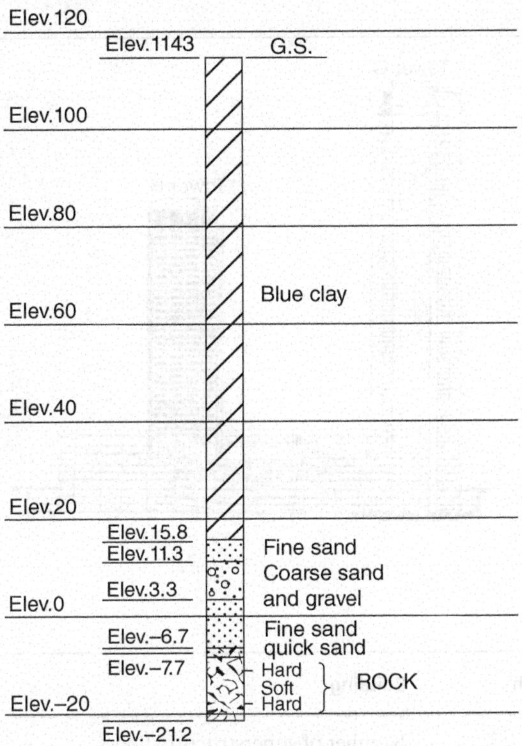

Figure 12.23 Albany telephone building soil cross-section (adapted from Glick 1936).

Due to the sensitivity of the soil, it was determined that friction piles would result in significant settlements. The alternative option of using piers to reach the bedrock was deemed costly. Therefore, a floating foundation in the form of a box with two basements was chosen as the foundation solution. The foundation design consists of a wall, a raft, and a basement slab for the infrastructure. A raft with a thickness of 0.91 m and a depth of 7 m was designed. The thickness of the basement slabs is 205 mm, with beams measuring 1015 mm wide and 660 mm thick placed between the columns in both directions. Taking into account the overburden reduction due to the 7-m excavation depth, the net pressure on the soil is reduced by 134 kPa. With the weight of the raft at 0.91 m thickness, the resulting net pressure is 100 kPa. Settlement analysis indicates that a net pressure of 100 kPa corresponds to an average settlement of approximately 178 mm. Considering the live load, one-third of the live load is factored into the settlement calculation along with the dead load. The net pressure on the soil for this combined load condition is 40 kPa, resulting in an estimated average settlement of 75 mm.

Case Study 12.6

Raffles Complex, Singapore

Raffles Complex consists of 4 tall buildings ranging from 28 to 73 floors, with three basement floors (Wong et al. 1995). The complex is located in Singapore. Its area is approximately 31,000 m^2. Each of the four towers in this complex is located separately on the floating foundation. A temporary retaining wall includes a sheet to protect the excavated area. The soil consists of medium-sized to coarse-grained sands to a depth of 14 m; the groundwater level is located at depths of 1.2 to 1.6 m. The settlements of the Complex were measured at different points during construction using instrumentation. The maximum settlements of the building related to the particular place were 6 to 10 mm. The maximum settlements of the buildings related to the north and south tower A, B, and C were measured as 21, 16, 27, and 48 mm, respectively (see Figure 12.24). Table 12.1 presents a summary of project data.

(Continued)

Case Study 12.6 (Continued)

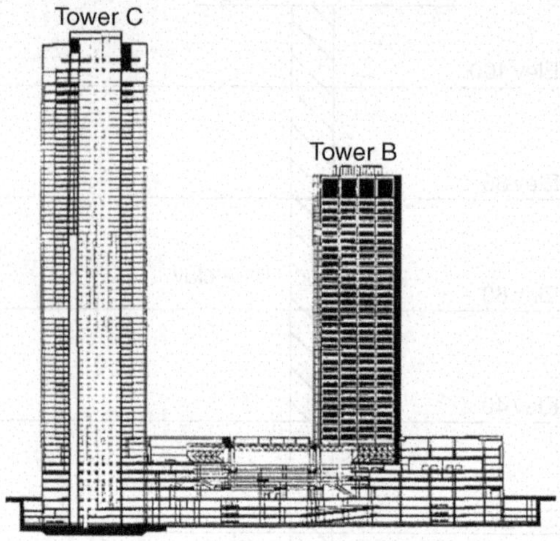

Figure 12.24 Raffles complex: section (Wong et al. 1995 / With permission of Canadian Science Publishing).

Table 12.1 Information about the Raffles Complex, Singapore.

Tower C	Tower B	Tower A South	Tower A North	Building
73	42	28	28	Number of superstructure floors
439	479	455	455	Equivalent pressure of the structure (kPa)
201	223	205	207	Net pressure applied (kPa)
1335	868	578	578	Total load (MN)
60.8×57.5	42×43	1680	37×37	Foundation area (m^2)
4.9	3	2.5	2.4	Thickness of foundation (m)
17.5	15.6	15.1	15	Depth of foundation (m)

Case Study 12.7

Culture Center, South Korea

The Cultural Center was built in Korea from 1995 to 2002 (Chung et al. 2005). Its total area with parking is 33,058 m^2, and the building occupies an area of 28,579 m^2. Figure 12.25 shows the plan of the Culture Center building. The different parts of the building had different weights, as given in Table 12.2.

The soil is composed of soft clay that contains layers of sand at different depths. Figure 12.26 shows the sampling and laboratory tests related to the soil. Due to the presence of soft soil deposits with high thickness at the project site, the use of deep foundations requires the installation to a depth of 20 to 70 m. As a result, using piles is costly. Therefore, the option of using piles was eliminated, and, in contrast, floating foundations were used along with soil reinforcement. However, this type of foundation was uncommon in Korea. Different parts of the building held different weights, so the excavation depth for the foundation varied.

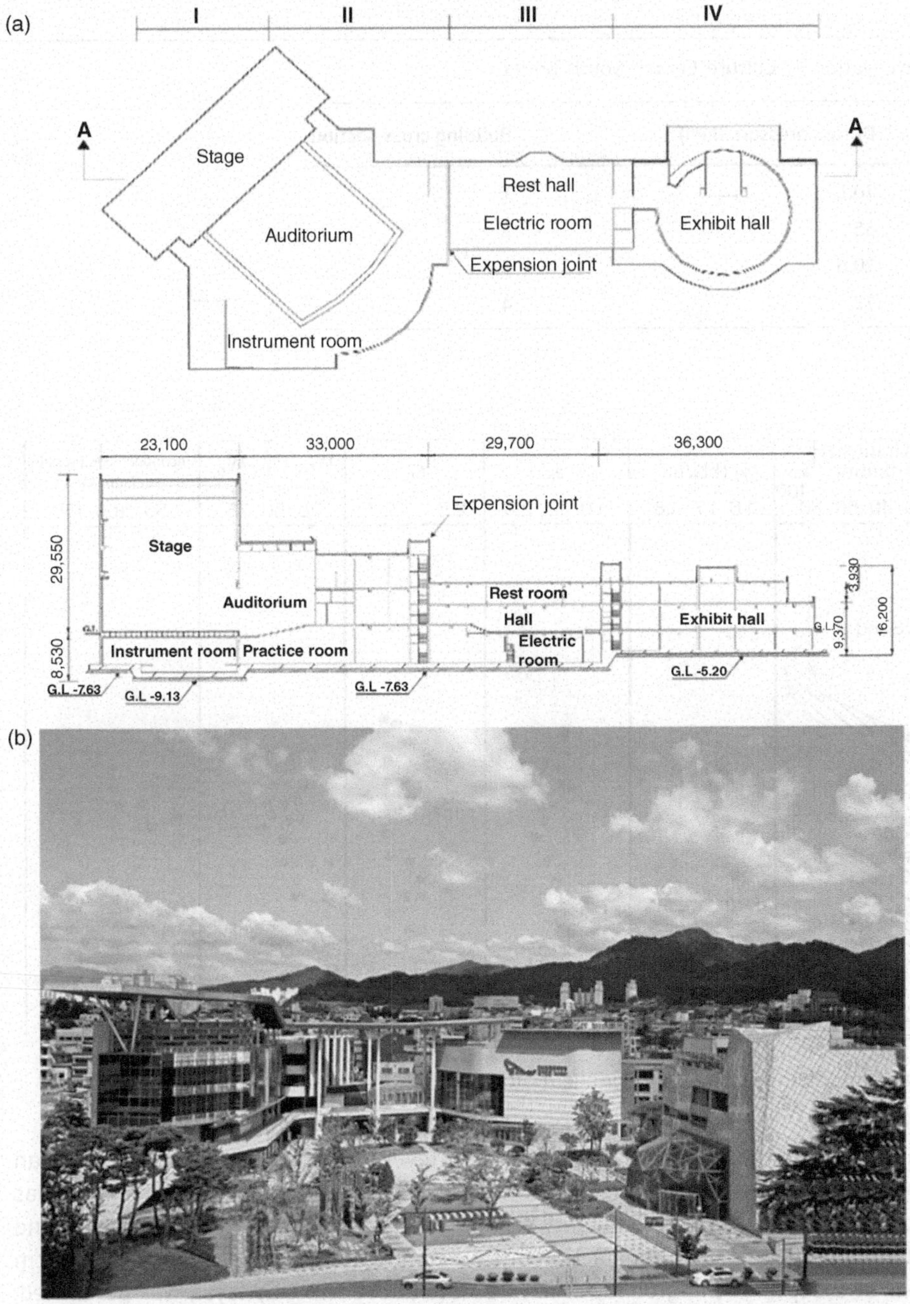

Figure 12.25 Culture center, South Korea: (a) plan and cross-section, (b) view (Chung et al. 2005 / With permission of Springer Nature).

(Continued)

Case Study 12.7 (Continued)

Table 12.2 Pressure and depth of excavation of Culture Center, South Korea.

Depth of excavation (m)	Design pressure (kPa)	Building cross-section
7.63	70.6	1
9.13	55	2
7.63	70.6	3
5.2	72	4

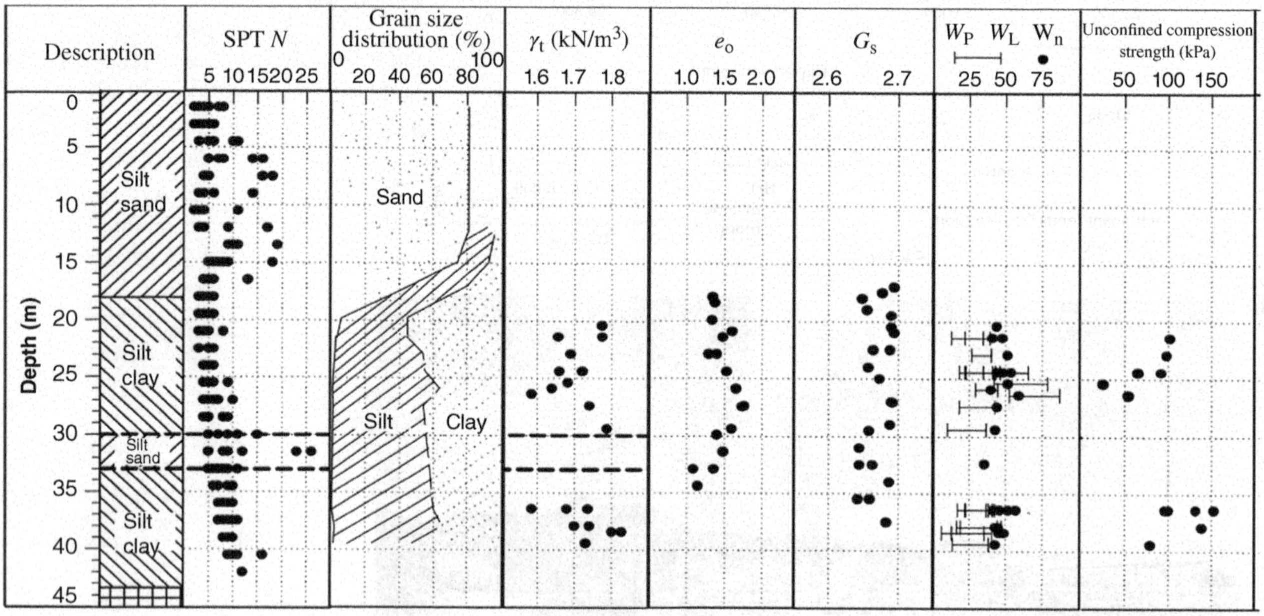

Figure 12.26 Practical soil profile diagrams (Chung et al. 2005 / Springer Nature).

Due to the presence of loose sand layers in the soil and high groundwater levels, liquefaction can occur during an earthquake. For this purpose, the soil–cement wall diameter of 55 cm to a depth of 22 m below the sand layer was made. This wall was implemented for two purposes: the temporary sealing wall for the excavated environment and protection against liquefaction. Also, to protect against liquefaction of soil–cement piles with a diameter of 100 cm at intervals of 3 m from each other at a depth of the ground, which was 2 m above the clay layer, was performed. In addition to protecting against liquefaction, piles effectively reduced total and differential settlements in the sand layer. A 60 cm diameter deep well was installed to control the groundwater level at 4 points. Total and differential settlements during construction were 65 and 46 mm, respectively. No settlement or cracking of the structure occurred three years after the completion of the building. Figure 12.27 depicts a schematic of the implemented system for groundwater control and liquefaction mitigation.

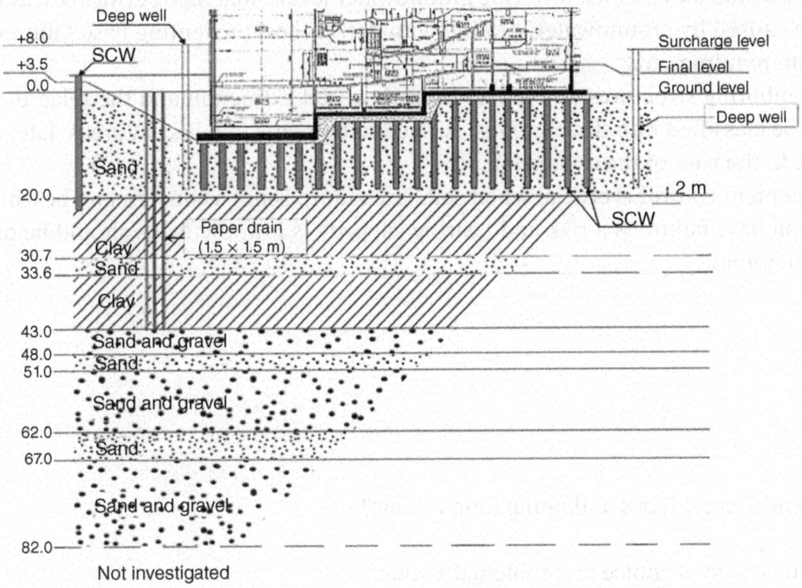

Figure 12.27 Equipment installed to control groundwater level and protect against liquefaction (Chung et al. 2005 / With permission of Springer Nature).

12.8 Major Points

1) Floating foundations are effective in reducing settlement by minimizing the stress transferred to the surrounding soil. They are suitable for heavily loaded structures resting on soft and low permeability soils. Caisson and box foundations are commonly used to create floating foundations that behave like buoyant rafts.
2) Three types of floating foundations are rafts, piled rafts, and box foundations. A compensated raft foundation is a type of mat foundation in which the net pressure on the foundation soil is lowered even more by increasing the foundation embedment depth and constructing a basement wall at the same time. Excavated soil partially or completely compensates for the structure weight, reducing subsequent settlements. The foundation is called a buoyant raft or a floating foundation when the structural load is entirely compensated.
3) Piled rafts are implemented for solving the problem of bearing capacity and settlement. Box foundations in which the walls and slabs of the basement are part of the foundation are applicable in cases where significant rigidity is required.
4) Floating foundations have the advantages of being utilized in low-bearing-capacity soils, reducing foundation settlement, stress reduction on the founding soil, and being used in high-moisture soils. Basement levels can be built on top of the foundation.
5) However, floating foundations also have several disadvantages. Repairing foundation slabs can be challenging due to the limited load capacity of the superstructure. The number of stories that can be constructed is limited, and material options are restricted to lightweight materials, requiring extra care and monitoring.
6) The selection of an excavation support system depends on various factors, including soil type and strength, excavation dimensions, groundwater conditions, long-term or short-term serviceability requirements, deflection criteria, adjacent structures, and local tectonic and seismic conditions. Common supporting systems include sloped excavations, sheet piling, diaphragm walls, bored pile walls, and top-down construction.
7) There is an increasing demand for construction methods that minimize impacts on adjacent structures, particularly in densely populated urban areas. As a result, the application of top-down construction techniques has become a preferred solution over conventional bottom-up techniques.

8) Challenges and considerations in excavation practices include lowering groundwater levels, managing critical excavation heights, addressing upward pressure caused by groundwater, dealing with deflections, preventing base failures, and carefully evaluating foundation system requirements.
9) Excavation can induce deflections in neighboring structures due to factors such as soil extraction and the release of confining stresses. These deflections can be classified into three groups: surface settlements in adjacent areas, lateral deflections of excavation walls, and heave in the base of the excavation.
10) In the design of floating foundations, settlement control is crucial when soil base failure is not a concern and bearing capacity is favorable. Conversely, when soil base failure is a risk and bearing capacity is low, the focus should be on managing the bearing capacity of the foundation.

12.9 Questions and Problems

12.9.1 Explanatory

1 What is the difference between the first and second types of floating foundations?

2 Describe why a floating foundation might be a good choice on problematic soils.

3 What similarities have been used to define and understand floating foundations?

4 Name and illustrate three types of floating foundations.

5 What are the advantages and disadvantages of floating foundations?

6 For what purposes are box foundations implemented?

7 Name and explain influential factors in the selection of an excavation support system.

8 List typical supporting systems that are implemented for excavations.

9 Illustrate different stages of the top-down construction approach.

10 What are the types of deflections induced by excavation?

12.9.2 Conceptual

1 What is the difference in design between the first and second types of floating foundations?

2 Why was a floating foundation used to build Liberty Mutual Insurance?

3 What procedures should be considered in designing the second type of floating foundations?

4 Explain the following terms:
- Caissons
- Compensated Foundation

5 Define the role of floatation in the geotechnical performance of a foundation system, considering the major requirements in foundation engineering through basic equations.

6 Why was a floating foundation used for the hotel Meridien and Raffles complex?

7 Explain how the implementation of floatation leads to a more sustainable design of foundations.

8 Illustrate practical solutions for base failure prevention.

9 Compare and contrast bottom-up and top-down construction methods. What are the advantages and disadvantages?

10 Consider Tower Latino Americana as a successful example of floating foundations application. List the recommendations and lessons for similar projects.

12.9.3 Problems

1 In cohesive soil, by performing cement improvement operations, shear strength increased by 70%, the internal friction angle remained unchanged, and the unit weight of soil increased by 10%. In this soil mass, what will be the percentage change of the critical height?

2 Golder and Crowch (1963) reported the Sugar Mill project, a building located in Guyana (South America) with heavy machinery and high-speed steam turbines. The weight of the building and machinery is about 22,000 kN. The dimensions of the plan are approximately 30.5 m by 18 m. The site soil consists of soft silty clay to lower depths and is typically consolidated. Its cohesion resistance gradually increases from 9.5 kPa at a depth of 2 m to 38 kPa at 20 m. The unit weight is assumed to be 16 $\frac{kN}{m^3}$. Determine the critical height of excavation in this project.

3 The 10-story building is located on a foundation measuring 15 × 15 m. According to the given information, it is desirable to determine the settlement in 2 cases:
a) Foundation on the ground level
b) Using the floating foundation at a depth of 2 m

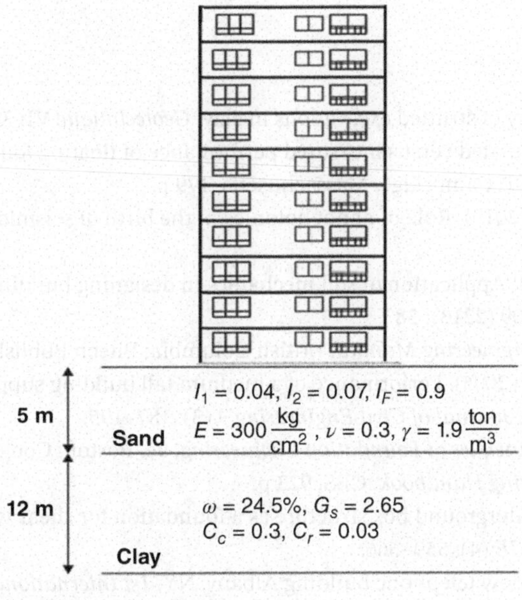

4 Determine the bearing capacity and estimate the settlement of the foundation system of a 25-story building in two conditions: (average load of each story equals 1 ton/m^2).
a) Mat foundation, 20 × 20 m, on the ground
b) Mat foundation, 20 × 20 m, D_f = 10 m, floating foundation

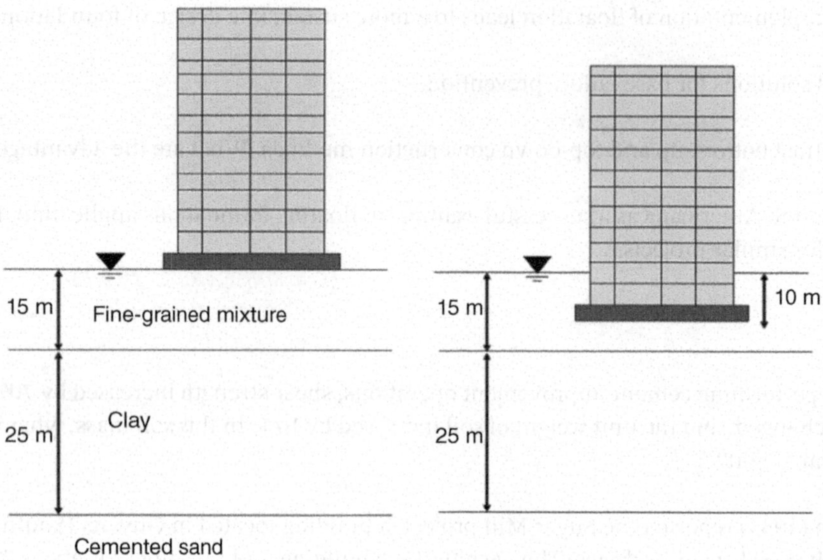

Soil Layer	γ (ton/m³)	K	$S_{u,avg}$ (kg/cm²)	C (kg/cm²)	φ	E (kg/cm²)
Fine-grained Mixture	1.9	0.35	0.45	–	–	180
Clay	1.4	0.25	0.30	–	–	60
Cemented Sand	2.1	0.1	–	0.6	35	1800

References

Bjerrum, L. and Eide, O. (1956). Stability of strutted excavations in clay. *Geotechnique* VI: 32.

Brown, T.L. (1971). Comparison of calculated versus measured performance of floating foundation at MIT, ScD thesis. Department of Civil Engineering, MIT, Cambridge, Massachusetts, 179 p.

Brûlé, S., Enoch, S., and Guenneau, S. (2019). Role of nanophotonics in the birth of seismic megastructures. *Nanophotonics* 8 (10): 1591–1605.

Casagrande, A. and Fadum, R.E. (1944). Application of soil mechanics in designing building foundations. *Transactions of the American Society of Civil Engineers* 109 (2213): 383.

CFEM (1992). *Canadian Foundation Engineering Manual.* British Columbia: Bitech Publishers LTD.

Chung, S.G., Kwag, S.H., and Lee, N.K. (2005). Performance of a medium tall building supported by a floating foundation on the Nakdong River deltaic deposit. *KSCE Journal of Civil Engineering* 9 (3): 187–195.

Das, B.M. and Sivakugan, N. (2019). *Principles of Foundation Engineering*, 9e. Boston: Cengage Learning.

Fang, H.Y. (2001). *Foundation Engineering Handbook*. CBS, 923 p.

Giuriani, E. and Gubana, A. (2007). Underground box structure as a foundation for shear walls in seismic resistant buildings. *Journal of Structural Engineering ASCE* (4): 559–566.

Glick, G.W. (1936). Foundations of the new telephone building Albany, NY. *1st International Conference on Soil Mechanics and Foundation Engineering (Harvard)*, Cambridge, Massachusetts (22–26 June 1936), 278. Graduate School of Engineering, Harvard University.

Golder, H.Q. and Crowch, E. (1963). Floating foundation for a factory building. Canadian Consulting Engineer.

Hemsley, J.A. (2000). *Design Applications of Raft Foundations.* Tomas Telford, 626 p.

Jeong, S. and Kim, D. (2018). Estimation of the load sharing ratio of pre-installed columns in top-down buildings on Korean rock. *KSCE Journal of Civil Engineering* 22 (12): 4852–4861.

Skempton, A. W. (1951). 'The bearing capacity of clays', Building research congress, London (11–20 September 1951), 1, pp. 180–189.

Tomlinson, M.J. (2001). *Foundation Design and Construction*, 7e. CEng FICE, FIStructE, 569 p.

Wong, I.H., Ooi, I.K., and Brooms, B.B. (1995). Performance of raft foundations for high-rise buildings on the Boulder Clay in Singapore. *Canadian Geotechnical Journal* 219–236.

Zeevaert, L. (1957). Foundation design and behavior of tower Latino Americana in Mexico City. *Géotechnique* 7: 115–133.

13

Shell Foundations

Chapter Highlights

- Introduction to Shell Foundations
- Overview of Shells in Civil Engineering
- Categories of Shell Foundations
- Geotechnical Design Approach
- Structural Design Approach
- Case Studies
  - Model Scale Study (Tadayon et al. 2022)
  - Numerical Assessment of Inverted Shell (Rinaldi 2012)
  - Field Study on Triangular Shells (Haut et al. 2016)
  - Sustainability and Reduction of Carbon Emission (Feickert 2022)
- Case Histories
  - T.V. Tower (Reiff and Burkhardt 2020)
  - The Nonoalco Tower (Enriquez and Fierro 1963)

13.1 Introduction

The shell structure refers to a thin, curved membrane or slab that serves both as a structural element and a protective covering. Shells are known for their strength and stiffness despite their thin and curved nature. Shell foundations offer an economical alternative to traditional foundations, particularly in situations where the soil is weak or problematic. The effectiveness of shell foundations relies on their geometric shape and continuous curvature.

The use of shells in construction has expanded in recent years, with various structures adopting thin shell elements. Figure 13.1 showcases examples of these structures. Shells can be found in a wide range of sizes, depending on their intended function. One notable example of a shell structure is the Hoover Dam in Las Vegas, USA, depicted in Figure 13.1d. As mentioned earlier, shells derive their strength from their unique shape. This allows them to achieve high structural performance with minimal material usage. This inherent efficiency contributes to the economic advantage of shell structures. Their versatility in accommodating various geometries and the flexibility of their application in architectural design further enhance their popularity in the construction industry (Kurian 2006). Additionally, shells are visually appealing due to their graceful and organic forms. The combination of their structural efficiency and aesthetic appeal makes shells a preferred choice in both construction and architectural projects. The ability to create striking and functional structures while optimizing material consumption is a significant advantage offered by shells.

The American Concrete Institute has defined a thin shell according to ACI (2008):

> "A three-dimensional spatial structure made up of one or more curved slabs or folded plates whose thicknesses are small compared to their other dimensions. Thin shells are characterized by their three-dimensional load-carrying behavior, which is determined by the geometry of their forms, by the manner in which they are supported, and by the nature of the applied load."

Advanced Foundation Engineering: Principles, Performance and Prospects, First Edition. Moses Karakouzian and Abolfazl Eslami.

Figure 13.1 Shell structures: (a) Sydney Opera House (Lorraine Murray 2023) (Bernard Spragg / Wikimedia Commons), (b) Kresge Auditorium (MIT building) (Beyond My Ken / Wikimedia Commons / CC BY-SA 4.0), (c) Savill Building (Wood Awards / https://www.woodawards.com/portfolio/savill-building/ / last accessed under Nov 20, 2023), and (d) Hoover Dam (Lynn Betts / Wikimedia Commons / Public Domain).

The behavior of a shell structure, such as a roof slab, differs from that of a plate element. While a plate element primarily bends under vertical loads, a shell structure resists these loads through internal forces within the membrane or surface, such as compression and tension. This characteristic enables shells to distribute and transfer loads more efficiently. In a traditional plate element, such as a simply supported beam or a roof slab with a span (L), the bending moment (M) increases with the square of the span ($M \propto L^2$). This means that as the distance between supports (L) increases, the bending moment and consequently the required thickness of the beam or slab increase significantly. However, in shell structures, the distribution of forces through the membrane or surface allows for greater load-bearing capacity and structural efficiency. Concrete, as a construction material, is most efficient under direct compression and least efficient under direct tension. In bending conditions, where both compression and tension forces are present, the efficiency of concrete lies between these two extremes. By incorporating reinforcement, such as steel bars, into the concrete shell, the tensile forces can be effectively resisted, making the design more economical and optimizing the use of materials (Kurian 2006). Figure 13.2 shows the usability of materials in different loads. The concept of utilizing the internal forces within the shell to resist loads and the efficiency of materials under different load conditions are important considerations in the design of shell structures.

Shells can achieve material savings compared to flat slabs, even though they have a larger surface area. This is because the curvature or creases in shells allow for a reduction in thickness while maintaining structural integrity. By distributing the loads efficiently through the shell's curved surface, the material requirements can be minimized. However, it is important to note that constructing shells may require more skill and expertise compared to flat slabs. The execution and construction process for shells can be more complex, especially for nonsymmetrical shapes or intricate designs. This may result in additional costs associated with labor and execution. The cost-effectiveness of shell structures depends on the balance between material savings and the additional costs of construction and labor. In cases where the savings in material costs outweigh the additional execution costs, constructing shells can lead to overall cost savings. The economic viability of shells

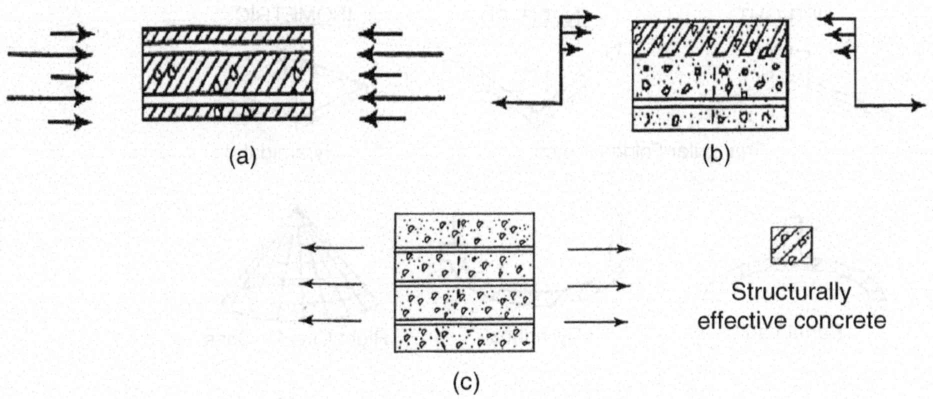

Figure 13.2 Efficiency of reinforced concrete sections: (a) direct compression, (b) bending, and (c) direct tension (Kurian 2006 / Alpha Science International Limited).

can vary depending on factors such as the cost-to-labor ratio in a particular country or region. It is also worth mentioning that the cost of materials and labor is not fixed and can vary over time and location (Kurian 2006). Therefore, economic considerations for shell structures should be evaluated based on the specific project circumstances and market conditions.

13.2 Overview of Shells in Civil Engineering

In general, shells are divided into three main categories based on their type and number of curvatures: (i) uncurved, (ii) single curved, and (iii) doubly curved. Figure 13.3 shows these different categories.

An uncurved shell, also known as a flat shell or plate, refers to a structure that is folded in an upright or inverted position without any significant curvature or radius of curvature. Single-curved shells have one set of curves in one direction. These shells do not have any Gaussian curvature. They can be formed by bending a flat plate in one direction. Examples of single-curved shells include cylindrical shells or barrel vaults. Doubly curved shells, on the other hand, have two sets of curves in two different directions. They are non-extensible, meaning that they cannot be flattened or unfolded into a flat shape without breaking. Doubly curved shells are naturally stronger and more rigid compared to single-curved shells. They can be further categorized into synclastic and anticlastic shells based on the direction of their curvature (Rinaldi 2012). Synclastic shells have two sets of curves that are concave in the same direction. Examples of synclastic shells include spherical shells and domes. Anticlastic shells have curvatures that are convex in opposite directions. Examples of anticlastic shells include hyperbolic paraboloids and saddle-shaped shells. The curvature of shells plays a significant role in their structural behavior and load-carrying capacity. The choice of single-curved or doubly curved shells depends on the specific design requirements, aesthetics, and structural performance desired for the project.

While shells have been commonly used in the construction of roofs and other above-ground structures, their use as foundation structures has been relatively limited. However, it is worth noting that certain types of foundation structures, such as the inverted arc foundation or brick arch foundations, can be considered as a form of shell foundation. The inverted arc foundation, which resembles an upside-down shell or arch, can be regarded as a shell foundation due to its curved shape and the forces it generates within the structure. Similarly, the use of brick arches as foundations in India, as depicted in Figure 13.4, demonstrates the application of shells in foundation construction (Kurian 2006). These historical examples highlight that the concept of utilizing shells in foundation structures is not entirely new. Shells can provide structural strength, load distribution, and stability to the foundation system, particularly in challenging soil conditions or areas prone to settlement. As engineering knowledge and construction techniques continue to evolve, there may be increased opportunities to explore the use of shells in innovative foundation designs. The advantages of shells, such as their efficient use of materials and structural integrity, can potentially offer benefits in terms of cost-effectiveness and long-term performance.

The economic factor is indeed a significant consideration in the use of shells as foundations. When compared to conventional plain foundations, shells can offer potential cost savings, especially in situations where there are higher loads or weaker soil conditions. The economic benefits of using shells as foundations become more pronounced when

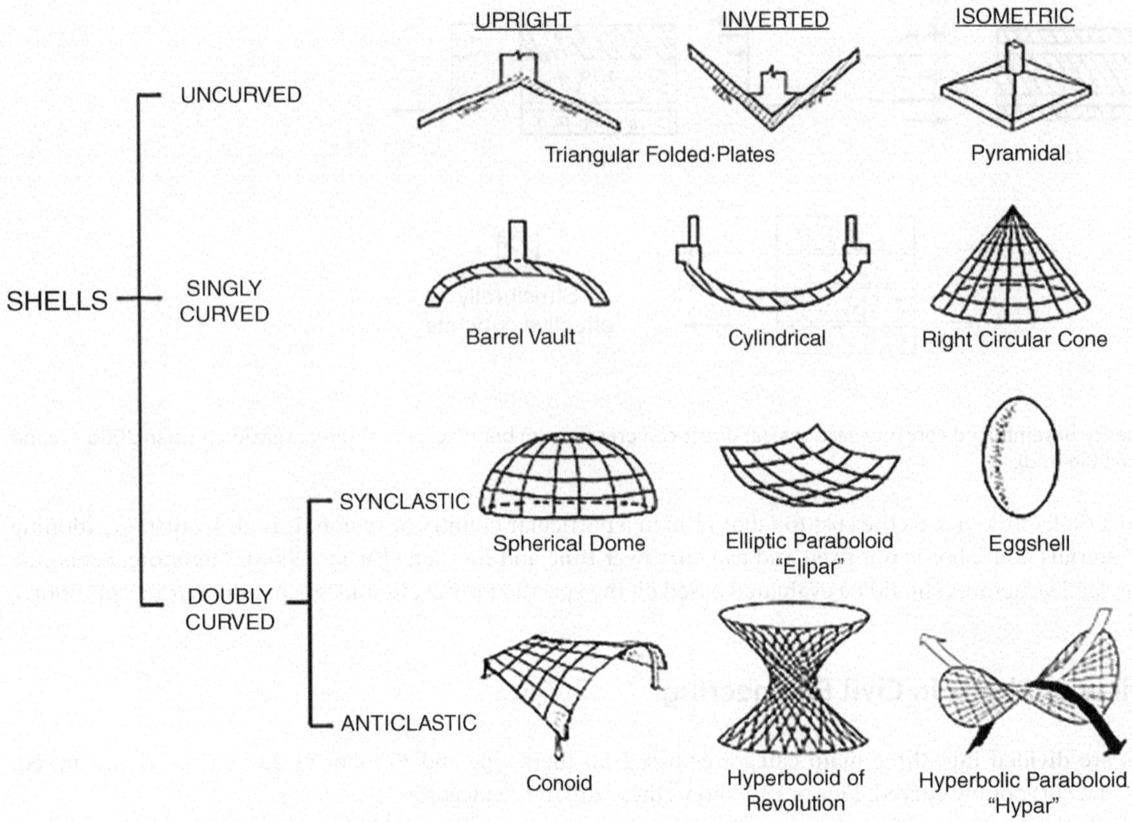

Figure 13.3 Shells classification (Rinaldi 2012 / Rinaldi).

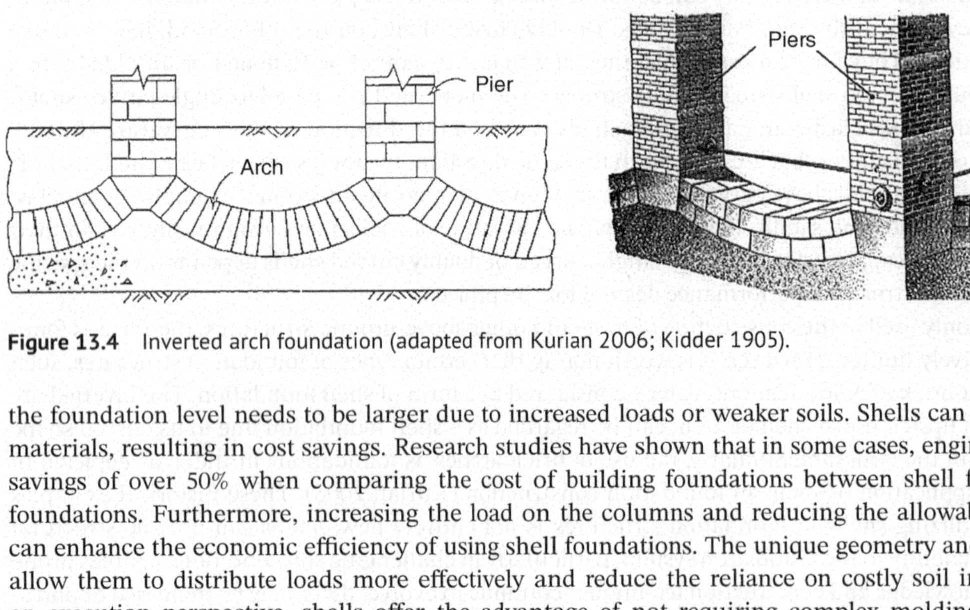

Figure 13.4 Inverted arch foundation (adapted from Kurian 2006; Kidder 1905).

the foundation level needs to be larger due to increased loads or weaker soils. Shells can provide a more efficient use of materials, resulting in cost savings. Research studies have shown that in some cases, engineers have estimated economic savings of over 50% when comparing the cost of building foundations between shell foundations and plain shallow foundations. Furthermore, increasing the load on the columns and reducing the allowable bearing capacity of the soil can enhance the economic efficiency of using shell foundations. The unique geometry and structural properties of shells allow them to distribute loads more effectively and reduce the reliance on costly soil improvement techniques. From an execution perspective, shells offer the advantage of not requiring complex molding processes during foundation construction (Kurian 2006). The main part of the foundation can be directly cast on the soil, forming the shell structure. This simplifies the construction process and can save time and labor costs. Different geometric shapes are employed when constructing shell foundations to suit specific requirements and site conditions. These shapes are selected based on their structural performance, load distribution capabilities, and overall suitability for the intended application.

13.3 Categories of Shell Foundations

Foundation systems can be categorized based on their load transfer mechanisms. The four main categories are vector-act, like single piles, section-act like mats, surface-act as shell foundations, block-act (hybrid), such as box foundations and pile groups. Shell foundations specifically fall under the category of surface-act foundations. Just like the structural shells, these foundations derive their load-bearing capacity from their shape rather than their mass. The unique shape of shell foundations allows them to efficiently transfer loads to the ground while utilizing the minimum amount of material. This characteristic of shells leads to economic advantages in terms of material consumption and overall cost-effectiveness. Surface-act foundations, including shell foundations, offer flexibility in terms of shape and dimensions. This flexibility allows them to be tailored as efficient and optimal solutions for various applications where section-act foundations are typically used. The commonly used shell foundations include conical, inverted dome, combined, hyperbolic paraboloid, elliptic paraboloid, and folded plate foundations. However, it is important to note that any foundation that can function similarly to a shell foundation, with a thickness smaller than its other dimensions, can be considered a type of shell foundation. This includes foundations like skirted foundations, which have a specific shape and behavior that allows them to act as surface-act foundations.

The selection of the most suitable foundation type is contingent upon the specific project requirements, site conditions, and load characteristics. Each foundation type has its own set of advantages and limitations, necessitating a comprehensive analysis to identify the most appropriate foundation system for a given project. Factors such as soil conditions, structural loads, and construction constraints must be carefully considered. By conducting a thorough analysis, engineers can determine the optimal foundation solution that ensures the safety, stability, and cost-effectiveness of the structure.

13.3.1 Cone

The simplest form of a shell foundation is a cone made of reinforced concrete, featuring axial symmetry. Its radial and circumferential reinforcements are designed similarly to a flat circular footing, but the construction process is more challenging. The thickness of the shell can either vary along its generator or remain uniform. This type of foundation is typically used as an individual footing due to its circular plan.

Conical foundations serve various purposes and are commonly employed in telecommunication towers, as depicted in Figure 13.6. Tall structures like towers often incorporate a conical substructure that connects the shaft to the annular raft at the bottom (Figure 13.5a). The annular raft carries the load and transmits it to the ground. These cones are often internally stiffened in different configurations to resist moments and shears (Figure 13.5b). Notable examples of such structures include the Moscow Tower (533 m) as well as towers in Berlin and Munich (Figure 13.5c).

In essence, these structures represent incomplete conical infrastructures that showcase the effectiveness of conical shells. Moreover, in large dimensions, conical foundations can be utilized in relatively tall structures such as chimney shafts. It is crucial for conical foundations to maintain perfect contact with the soil along their entire bottom surface. Figure 13.6a illustrates an inverted conical foundation suitable for structures like guyed masts. Additionally, on a larger scale, conical foundations can serve as the foundation for cylindrical structures such as ground storage tanks (as shown in Figure 13.6b) or overhead structures supported by a row of columns arranged in a circular fashion, such as water tanks (Kurian 2006).

13.3.2 Inverted Dome

In circular or overhead structures such as water tanks, a thin inverted dome is an economical option instead of a thick annular or circular raft foundation. The column's load transfer to the inverted dome is done through a ring beam (Figure 13.7).

13.3.3 Hyperbolic Paraboloid

The hyperbolic–parabolic shell, also known as the Hypar, combines aesthetics and flexibility in its design. It is characterized by a transition surface that moves a concave parabola over a convex parabola (Figure 13.8), resulting in a doubly curved

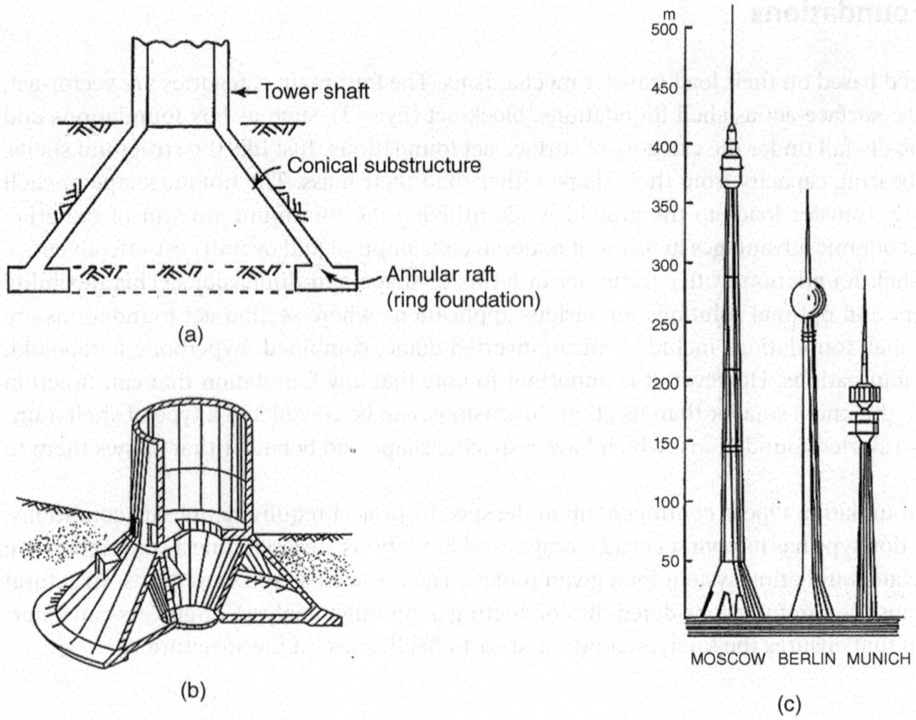

Figure 13.5 Conical substructure for telecommunication towers: (a) conical substructure for the tower, (b) stiffened conical substructure of the Stuttgart Television Tower (isometric view), and (c) television towers in Moscow, Berlin, and Munich (adapted from Leonhardt 1967).

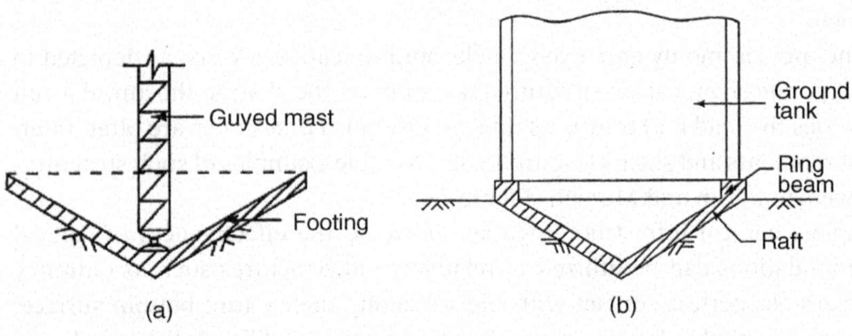

Figure 13.6 Inverted cone shell foundations: (a) footing for mast, (b) raft for the ground tank (Kurian 2006 / Alpha Science International Limited).

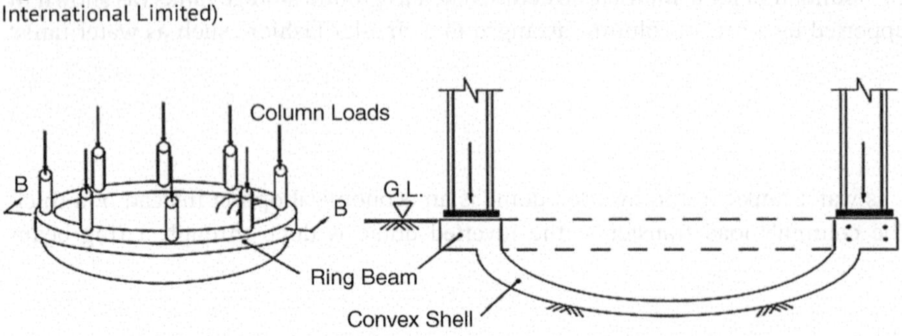

Figure 13.7 Inverted circular dome shell (Rinaldi 2012 / Rinaldi).

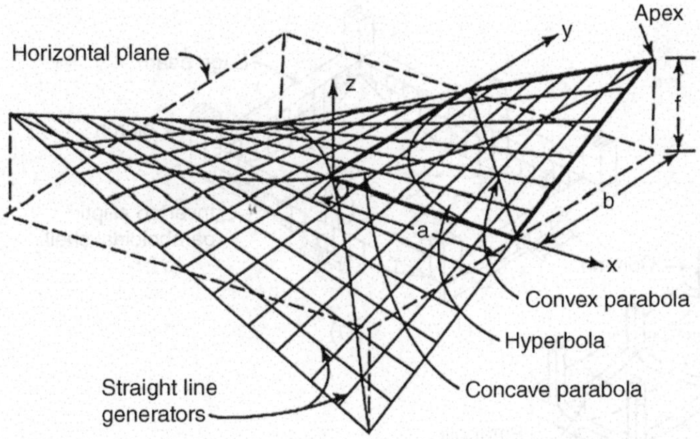

Figure 13.8 Geometry of the hyperbolic (Kurian 2006 / Alpha Science International Limited).

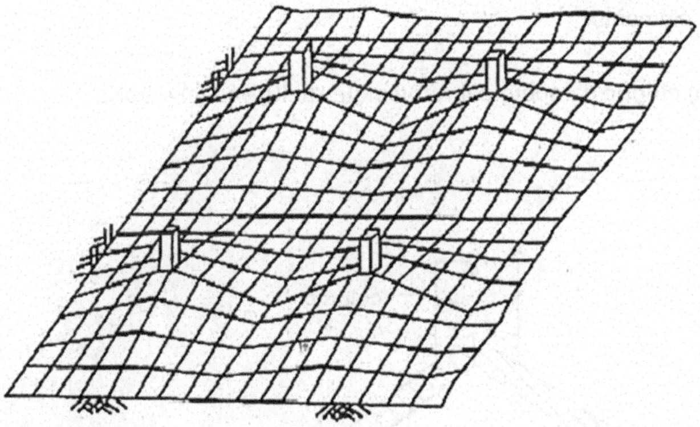

Figure 13.9 Hypar raft paraboloid (Kurian 2006 / Alpha Science International Limited).

anticlastic shape. The Hypar shell is both visually appealing and structurally efficient. The surface of the Hypar shell consists of two sets of perpendicular parabolas. In the case where the parabolas are identical, it is referred to as a rectangular hyperbolic–parabolic shell. In this type of Hypar shell, the surface is composed of straight lines at different angles that are parallel in space. However, in the plan view, these lines have perpendicular angles to each other, exhibiting the property of a straight line. This property indicates that the curve must be zero where different curves intersect in two directions. From an engineering perspective, the straight-line property of the Hypar shell is advantageous as it combines the benefits of both a shell and a flat surface. This property facilitates the formulation, reinforcement, and construction of the shell using concrete. The use of Hypar shells extends beyond single foundations and can also be employed as rafts for columns arranged in multiple rows (Figure 13.9) (Kurian 2006).

13.3.4 Elliptic Paraboloid

The elliptic paraboloid, also called the Ellpar, has a double synclastic curvature created by moving one parabola over another. Both parabolas in an Ellpar, unlike a Hypar, have a curvature in the same direction (Figure 13.10a). It is called an Ellpar due to its formation of an ellipse from the intersection of the horizontal plane with its shell.

An inverted Ellpar with edge beams can be used as a foundation for structures with a rectangular or square plan as an elliptical parabola (Figure 13.10b). It can also be extended as a raft for several columns (Figure 13.10c) (Kurian 2006).

13.3.5 Folded Plate

The most practical case of this type is a combination of a trapezoidal plate's four elements that create a pyramidal shape. It can also provide support for the column at its center. Since this type of foundation can have a rectangular or square plan, it can also combine multiple units instead of a raft foundation (Figure 13.11) (Kurian 2006).

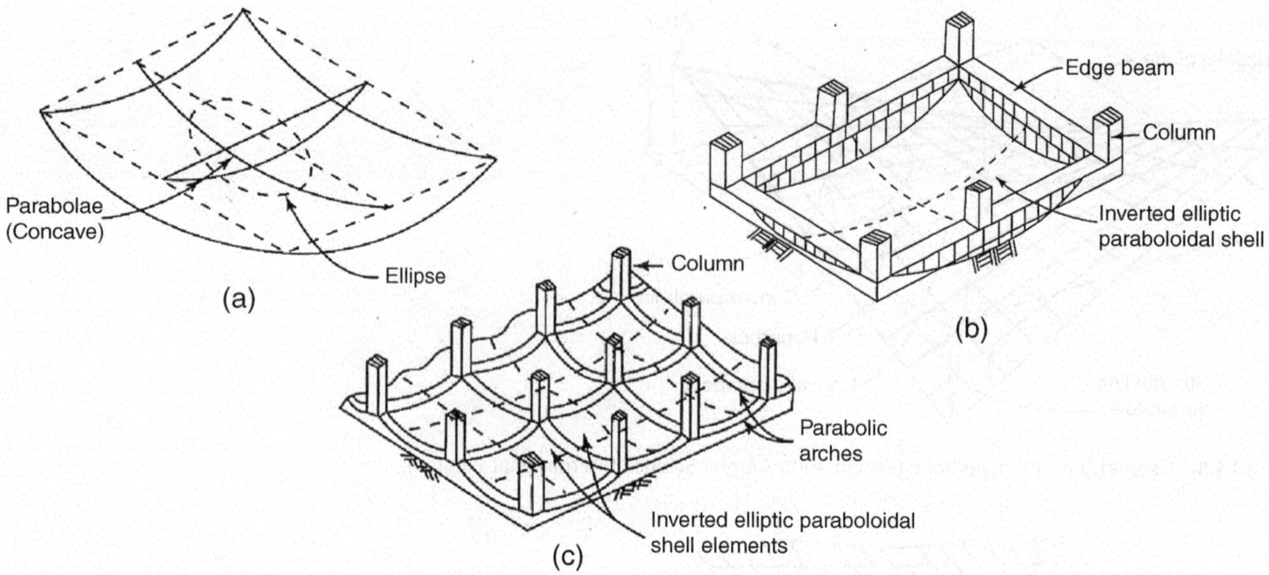

Figure 13.10 Elliptic paraboloidal shell rafts: (a) elliptic paraboloid, (b) elliptic paraboloidal raft (single shell), and (c) elliptic paraboloidal raft (multiple shell) (Kurian 2006).

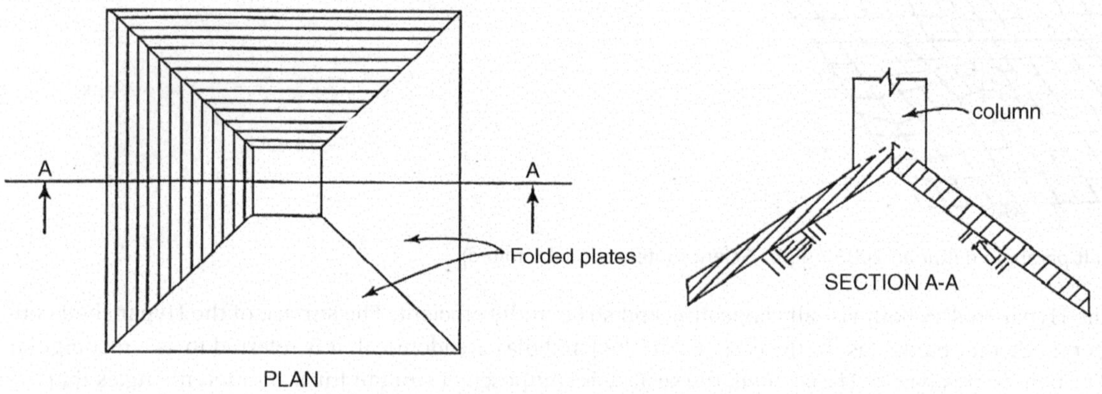

Figure 13.11 Folded plate footing (Kurian 2006 / Alpha Science International Limited).

13.3.6 Combined

The required level of contact between the shell foundation and the soil depends on two factors: bearing capacity and allowable settlement. To increase the contact surface area, the inverted dome can be combined with the frustum of an upright cone. In this configuration, the frustum compensates for the additional foundation area required if the load-bearing capacity of the soil exceeds the plan dimensions of the structure. It is important to note that the inner shell can be either full or partial. These combinations of inverted and upright cones are known as "combined shell foundations." They offer a combination of the characteristics of a shell and a folded plate. The corresponding plain counterpart, a thick annular raft, is referred to as a "double cone folded shell raft foundation" and is known by the same name (Figure 13.12).

13.4 Design Approaches

During the geotechnical design phase, the foundation plan dimensions are determined to meet the soil criteria of bearing capacity and settlement. The structural design phase focuses on accommodating the shear and flexural actions within the foundation structure. In the case of conical shell foundations, they are evaluated as individual footings designed to support

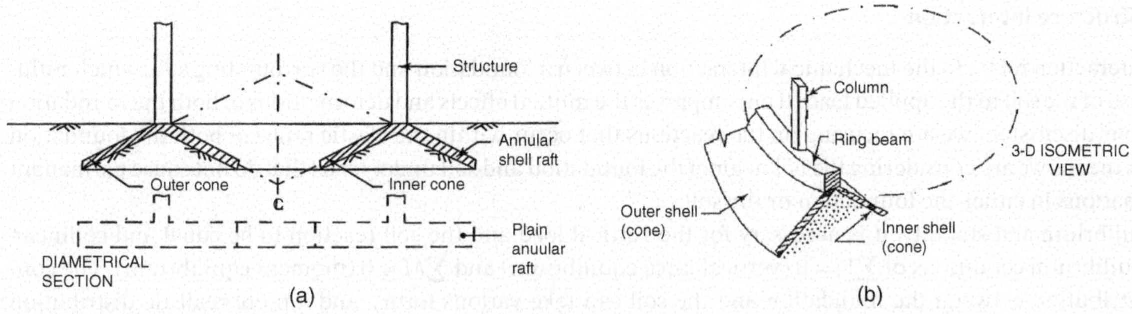

Figure 13.12 Double cone folded shell annular raft (Kurian 2006 / Alpha Science International Limited).

single columns. The structural analysis considers the specific loading conditions and assesses the capacity of the conical shell foundation to resist the applied forces and moments. Overall, the geotechnical and structural design phases work together to ensure that the conical shell foundation meets both the soil requirements and structural integrity necessary for a safe and reliable foundation system.

13.4.1 Basics of Design

The main components of a shell foundation include the girder, shell, and toe (edge) (Figure 13.13). The girder is usually known as the base of the column or the circular beam of the shell, which is the first element in absorbing the superstructure loads. In inverted shell foundations, edge beams (in linear load conditions) and annular beams (in nonlinear load conditions) act as a girder. The girder transmits the loads to the main element of the shell. The shell element acts as a wing as the main load transfer mechanism between the beam elements. Unlike roof structures, shell components in foundations must be reinforced to absorb tensile loads such as return load, and provide resistance to overturning, sliding, and lifting.

The girder, the middle beam, and the edge beam receive the main stresses directly from the applied load, while the shell absorbs the secondary stresses. Implementing edge beams and increasing the depth of their placement in shell foundations leads to improved transfer stress and increases their bearing capacity, which is especially important when there is a high-stress concentration around the geometry of the shell model (Rinaldi 2012). Finally, the components listed for the shell foundation are not necessarily named above and may have other components depending on their geometry or may not include the components listed above.

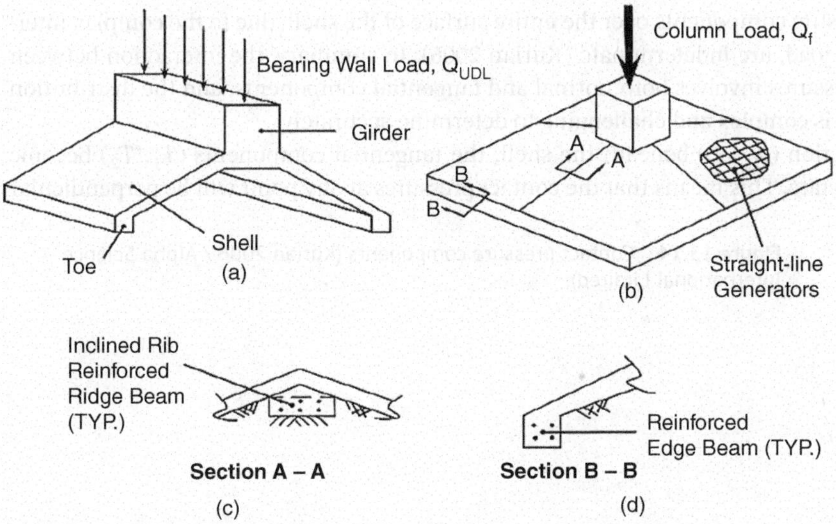

Figure 13.13 Shell components: (a) strip shell footing, (b) isolated shell footing, (c) ridge beam detail, and (d) edge beam detail (Rinaldi 2012 / Rinaldi).

13.4.2 Soil–Structure Interaction

Soil–structure interaction refers to the mechanical interaction between a foundation and the surrounding soil, which influences the response of the soil to the applied load. It encompasses the mutual effects and deformations of both the foundation and the soil. In our discussion, we are focusing on the reactions that occur within the elastic range of both the foundation and the soil. This means we are considering the behavior of the foundation and soil under loads that do not cause permanent or plastic deformations in either the foundation or the soil.

To ensure equilibrium and stability, it is necessary for the vertical load and the soil reaction to be equal and collinear, satisfying the equilibrium conditions of $\sum V = 0$ (vertical force equilibrium) and $\sum M = 0$ (moment equilibrium). The contact pressure distribution between the foundation and the soil can take various forms, and a more realistic distribution reflects the actual behavior of the soil–structure interaction. In structural design, it is important to consider the moment and shear values that result from the actual contact pressure distribution. This can be achieved through static indeterminate interaction analysis, which takes into account the nonuniform distribution of soil reactions. While conventional design approaches often assume a uniform average value for the contact pressure, known as rigid design, a more accurate and flexible design approach considers the actual interactive contact pressure distribution (Kurian 2006). For stable soil behavior under non-axial and monotonic loading conditions, it is beneficial to bring the contact pressure closer to a uniform distribution at the interface between the soil and the shell foundation. This helps to ensure a stable response and mitigate any potential uneven settlement or failure mechanisms.

13.4.3 Contact Pressure

The soil–structure interaction in shell foundations is more complex than in shallow foundations. When considering the contact between the shell and the soil at various points, the contact pressure generally consists of a normal component and a tangential component. The tangential component acts in the direction of lateral deformation of the shell at that particular point. The maximum value of the tangential component is limited by the coefficient of wall friction multiplied by the normal component, both acting on the same elemental surface.

For a shell with a regular geometric shape (such as a Hypar), each point on the surface has a normal vector and a tangential plane. There are two curves on the surface, one with maximum curvature and the other with minimum curvature. The planes containing these curves also have a normal vector perpendicular to each other, forming a right-angled frame. The decomposition of the tangential component of the contact pressure is done in the direction of the principal tangents, resulting in three components: N, T_1, and T_2 (as shown in Figure 13.14a). However, as the frame changes direction from one point to another while maintaining its orthogonal position, it is more appropriate to decompose the contact pressure into X, Y, and Z components, corresponding to the x, y, and z coordinate directions (as shown in Figure 13.14b). The actual values and distribution of these contact pressure components over the entire surface of the shell, due to the complex interaction between the shell and the soil under load, are indeterminate (Kurian 2006). In summary, the interaction between the shell and the soil in terms of contact pressures involves both normal and tangential components, and the distribution of these components over the shell's surface is complex and challenging to determine accurately.

In the case of a soft clay soil with no friction ($\varphi = 0$) beneath the shell, the tangential components (T_1, T_2) become unstable, resulting in an almost hydrostatic state. This means that the contact pressures at any point will be perpendicular

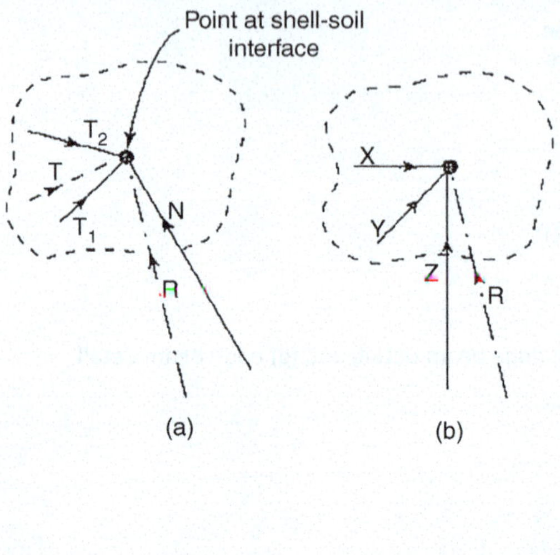

Figure 13.14 Contact pressure components (Kurian 2006 / Alpha Science International Limited).

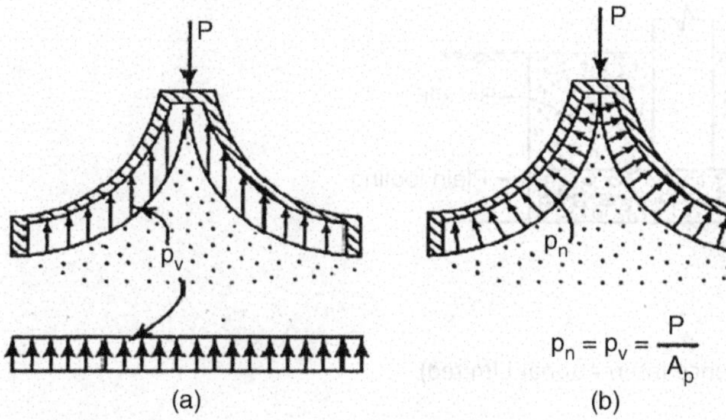

Figure 13.15 Simplified pressure distribution under shell foundation (a) frictional soil (sand), (b) cohesion soil (soft clay) (Kurian 2006 / Alpha Science International Limited).

to the surface of the shell. However, if the soil is frictional, such as sand, the tangential components will change the direction of the contact pressure, making it almost perpendicular to the ground surface. In this case, if the contact pressure is ideally vertical, it implies that $X = Y = 0$ and $R = Z$ everywhere. When assuming a uniform pressure distribution on the shell, a common assumption in the design of conventional shell foundations on frictional soils, the following Equation 13.1 is obtained (as shown in Figure 13.15a) (Kurian 2006).

$$R = P_v = \frac{P}{A_p} \tag{13.1}$$

where

R = resulting contact pressure
P_v = the vertical base pressure
P = applied load
A_p = cross-sectional area

Figure 13.15b shows that Equation 13.2 is obtained in a purely normal and uniform distribution. This relationship means that there will be a hydrostatic pressure distribution with the same intensity in all directions (Kurian 2006).

$$P_n = P_v = \frac{P}{A_p} \tag{13.2}$$

where

P_n = the normal pressure to the surface of the shell

13.5 Geotechnical Design Approach

Geotechnical design is based on the following criteria (Equation 13.3):

$$\text{Net loading intensity} \leq \text{Allowable soil pressure} \tag{13.3}$$

The formula above defines the allowable soil pressure, which is determined as the lesser value between the safe bearing pressure (related to bearing capacity) and the soil pressure for a specified acceptable settlement (related to settlement). The design criteria for bearing capacity and settlement are separately satisfied by considering the smaller value between the safe bearing pressure and the soil pressure corresponding to the permissible settlement.

When comparing shell foundations and plain foundations with equal horizontal cross-sectional area, there is little difference in the geotechnical design. However, in the design of shell foundations, the soil beneath the shell, known as the core soil, is included in the calculation of bearing capacity. This is done under the condition that the core soil is fully interacting with the foundation (as shown in Figure 13.16). To resist horizontal forces, the friction between the core soil and the shell foundation is typically greater than the friction between the concrete and the foundation soil in plain foundations (Kurian 2006).

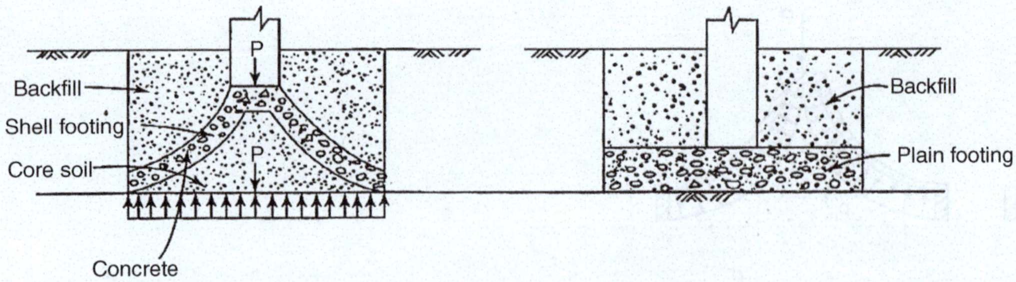

Figure 13.16 Shell footing and plain footing (Kurian 2006 / Alpha Science International Limited).

Agarwal and Gupta (1983) found an improved bearing capacity of shell footings in their research, attributing it to the increased soil-to-soil friction at the base of the shell footing. Szechy (1965) suggests that a downward recurved (concave) shape of the footing helps in achieving a wider load dispersion, thereby reducing settlement (as illustrated in Figure 13.17).

Kurian (2006) and Jayakrishna Devaki (2002) carried out an analytical study in which they change the geometry of the shell in order to better understand the reaction of soil under perfectly rigid shells, including bearing capacity and settlement. They experimented with different pure shell shapes using different ratios between the length of the base (a) and the rise of the shell (f) (see Figure 13.18b for dims). The bearing capacity of the shell increased for each of the 44 configurations tested as the rise (f) increased in comparison to the flat foundations for the same load in the upright configuration. When the rise to the half-base ratio (f/a) is equal to 1/2 and when it is equal to 1, the increase in bearing capacity varies from 14–34% to 20–42%. When (f/a) equals −1/2 for inverted shells, bearing capacity drops by 0–40%, and when (f/a) equals −1, it drops by 10–60% (Kurian 2006). In Table 13.1, a portion of these conclusions is displayed.

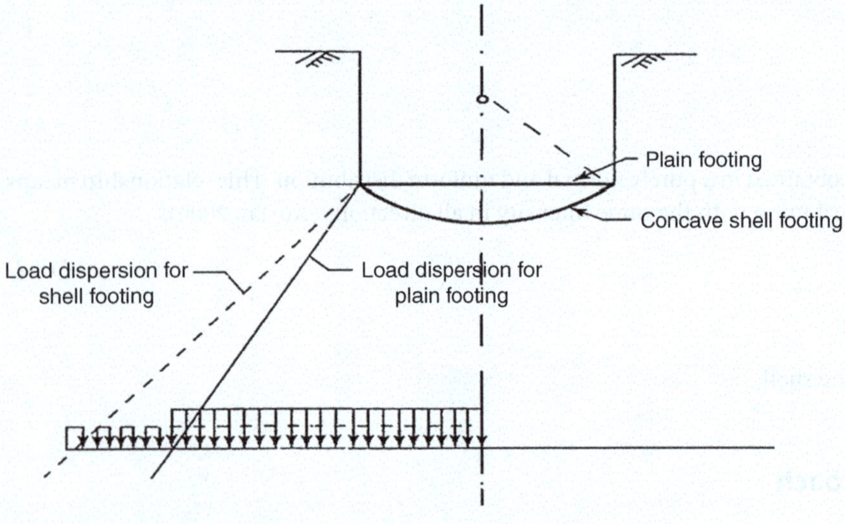

Figure 13.17 Load dispersion under plain and shell strip footings (Szechy 1965).

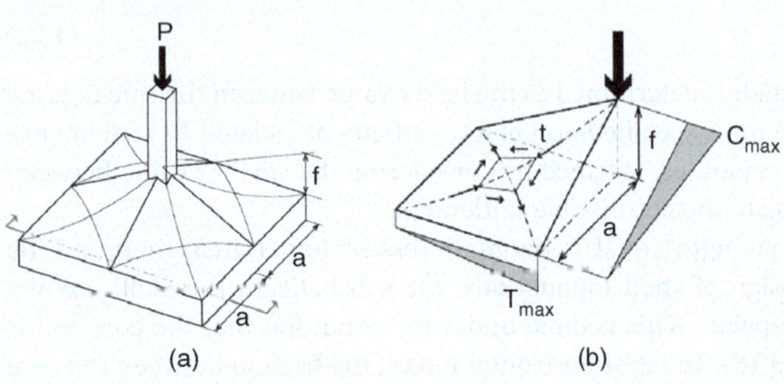

Figure 13.18 Hyperbolic paraboloid shell: (a) dimensions of shell footing, (b) section of shell footing with forces (Feickert 2022 / Springer Nature / CC BY 4.0).

Table 13.1 Bearing capacity and settlement of shell foundations under axisymmetric 3-D analyses (Jayakrishna Devaki 2002 / Jayakrishna Devaki).

Type of footing	Section	Plan	Rise to half-base ratio (f/a)	Bearing capacity (kPA)	Settlement at 1/3rd bearing capacity (mm)	Difference in bearing capacity from flat (%)
Flat square			0	148.14	0.450	0
Flat circular			0	103.04	0.171	0
Conical (upright)			+1/2	157.27	0.377	+34.47
Conical			+1	176.93	0.317	+41.75
Conical (inverted)			−1/2	73.60	0.308	+39.99
Conical			−1	73.60	0.384	−39.99
Hypar			+1/2	172.84	1.025	+14.28
Hypar			+1	185.19	0.552	+20.00
Hypar			−1/2	129.63	0.642	−12.49
Hypar			−1	98.77	1.453	−50.00

Example 13.1 A Hypar shell foundation with a 3.7 m width is placed on normally consolidated clayey soil. The applied net load is assumed to be 800 kN. For a 30° area of load dispersion, calculate:

- Δq at area 1 and 2 under the foundation.
- Total consolidation settlement of clayey layers.

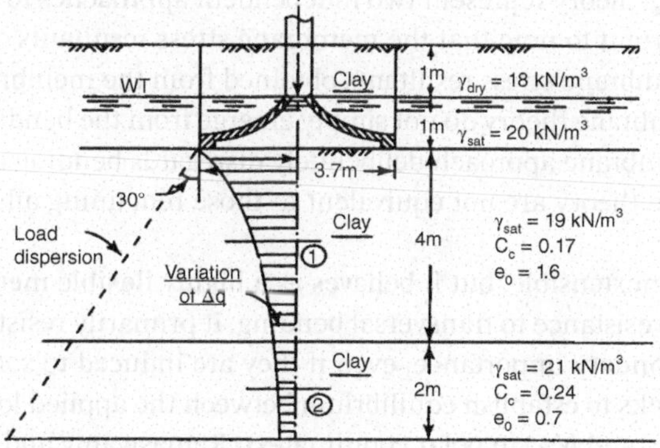

Δp for 30° dispersion:

Area at 1: $(3.7 + 2 \times 2 \times \tan 30°)^2 = 36\ \text{m}^2$

$$\Delta q = \frac{p_{\text{net}}}{A} = \frac{800}{36} = 22\ \text{kPa}$$

Area at 2: $(3.7 + 2 \times 5 \times \tan 30°)^2 = 90\ \text{m}^2$

$$\Delta q = \frac{p_{\text{net}}}{A} = \frac{800}{90} = 9\ \text{kPa}$$

Consolidation settlement:

At layer 1:

$$\sigma_0' = (1 \times 18) + (1 \times (20 - 10)) + (2 \times (19 - 10)) = 46\ \text{kPa}$$

$$s = \frac{H_0}{1 + e_0} \times C_c \times \log \frac{\sigma_1'}{\sigma_0'} = \frac{4}{1 + 1.6} \times 0.17 \times \log \frac{46 + 22}{46} = 44.5\ \text{mm}$$

At layer 2:

$$\sigma'_0 = (1 \times 18) + (1 \times (20 - 10)) + (4 \times (19 - 10)) + (1 \times (21 - 10)) = 75 \text{ kPa}$$

$$s = \frac{H_0}{1 + e_0} \times C_c \times \log \frac{\sigma'_1}{\sigma'_0} = \frac{2}{1 + 0.7} \times 0.24 \times \log \frac{75 + 9}{75} = 13.9 \text{ mm}$$

Total consolidation settlement: $44.5 + 13.9 = 58.4$ mm

13.6 Structural Design Approach

When analyzing shells, three theoretical approaches are commonly employed: membrane theory, bending theory, and ultimate strength theory. The membrane theory and bending theory are utilized for the working stress design of foundation shells, while the ultimate strength theory determines the maximum loads the shells can sustain under specific conditions. In the context of considering the soil reaction to be uniform or uniformly variable, and the alignment of the resultant of applied loads and moments with respect to the centroid of the contact area between the foundation and soil, the membrane theory is typically considered conventional.

Unlike plain structural slabs that resist applied loads through flexural action, shells rely on their geometry and the distribution of in-plane stresses to carry the loads, with minimal or no bending (Ramaswamy 1968). In the case of shells, if we consider a point (x, y) on the center surface of the shell (Figure 13.19), and assume that the shell primarily resists applied loads through in-plane stresses, the stress distribution at that point is characterized by three membrane stress resultants: the two "normal stress resultants," Nx and Ny, and the "membrane shear stress resultant," Nxy (or Nyx). These stress resultants are expressed per unit width, similar to plate analysis (Ramaswamy 1968).

In cases where the shell cannot resist loads solely through membrane stresses, the flexural mode becomes active, and bending stress resultants come into play. Therefore, in general, a total of eight stress resultants are present at any point within the shell, including the three membrane stress resultants and the five bending stress resultants. These bending stress resultants are obtained through a "bending analysis" of the shell.

The membrane theory and bending theory represent two independent approaches to shell analysis, each satisfying equilibrium in their own way. It is important to note that the membrane stress resultants obtained from the bending analysis may differ significantly from the membrane stress resultants obtained from the membrane analysis. The membrane stress resultants obtained through the membrane theory do not simply emerge from the bending theory by removing the bending part of the stress resultants. The membrane approach deliberately disregards bending effects, even though the membrane stress resultants from the membrane theory are not equivalent to those remaining after removing the bending part from the bending theory.

A shell can be considered highly inextensible, but it behaves as a highly flexible membrane with significant membrane rigidity. While it may have minimal resistance to transversal bending, it primarily resists applied loads through membrane stresses. Bending stresses are of secondary importance, even if they are induced to some extent. The membrane analysis is based on force equilibrium and seeks to establish equilibrium between the applied loads and the three membrane stress resultants N_x, N_y, and N_{xy} (or N_r, N_θ, and $N_{r\theta}$, in polar coordinates) (Ramaswamy and Keshava Rao 1961).

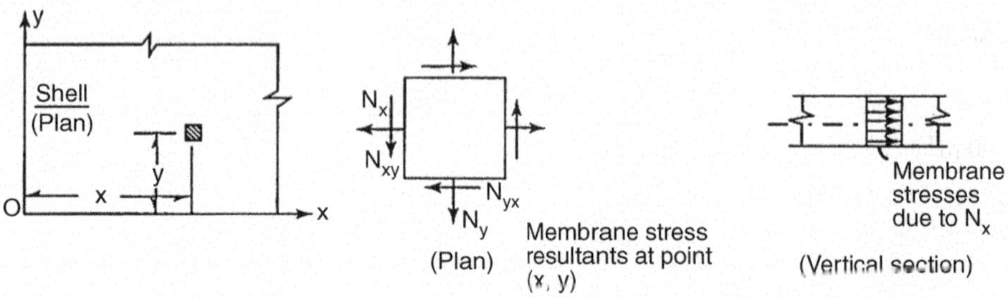

Figure 13.19 Normal and shear stresses (Kurian 2006 / Alpha Science International Limited).

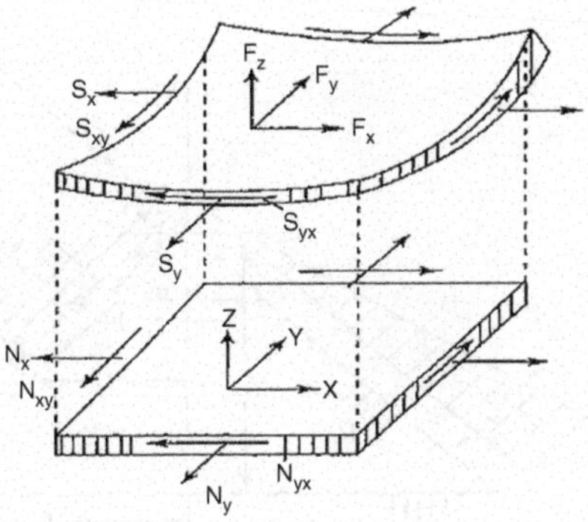

Figure 13.20 Real and pseudo stress resultants (Kurian 2006 / Alpha Science International Limited).

13.6.1 Real and Pseudo Stress Resultants

The introduction of projected or "pseudo" stress resultants, such as N_x, N_y, and N_{xy}, instead of the actual or "real" stress resultants, S_x, S_y, and S_{xy}, has proven to be useful in the membrane analysis of doubly curved shells like the Hypar and the Ellpar (Figure 13.20). In conjunction with the pseudo stress resultants, pseudo loads X, Y, and Z are introduced in place of the real loads F_x, F_y, and F_z in the coordinate directions (Figure 13.20). The pseudo stress resultants are defined in such a way that they exert a constant force within the corresponding coordinate direction on the projected sides, similar to how the real stress resultants act on specific sides. The horizontal components of the real forces acting on the sides of the actual element are equal to the pseudo forces acting on the corresponding sides of the projected element (Kurian 2006).

By introducing the notation $\frac{\partial_z}{\partial_x} = p$ and $\frac{\partial_z}{\partial_y} = q$ it is shown that:

$$S_x = N_x\sqrt{\frac{1+p^2}{1+q^2}} \tag{13.4}$$

$$S_y = N_y\sqrt{\frac{1+q^2}{1+p^2}} \tag{13.5}$$

$$S_{xy}(= S_{yx}) = N_{xy}(= N_{yx}) \tag{13.6}$$

where

S_x, S_x, and S_x = real stress resultant
N_x, N_y, and N_{xy} = pseudo stress resultant

The fictitious or pseudo loads (intensities) are so defined that:

Real load × actual surface area of the element = pseudo load × projected area of the element from which the equations are obtained:

$$X = F_x\sqrt{1+p^2+q^2} \tag{13.7}$$

$$Y = F_y\sqrt{1+p^2+q^2} \tag{13.8}$$

$$Z = F_z\sqrt{1+p^2+q^2} \tag{13.9}$$

where

F_x, F_y, and F_z = real loads
X, Y, and Z = pseudo loads

13.6.2 Membrane Stresses in a Conical Shell

In a conical shell subjected to uniform soil pressure, both vertical and normal, membrane stress resultants N_s, N_θ, and $N_{s\theta}$ develop (Figure 13.21). Figure 13.21 also illustrates the variation of the stress resultants N_s and N_θ along the generator of

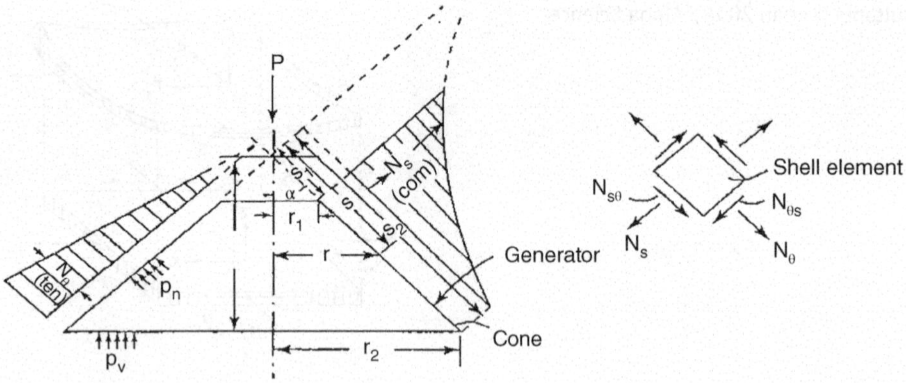

Figure 13.21 Membrane stress resultants in cone (Kurian 2006 / Alpha Science International Limited).

the shell, under uniform normal soil pressure. From Figure 13.21, it can be observed that under uniform normal contact pressure, the conical shell experiences meridional compression with maximum intensity at the top and decreasing to zero at the bottom (which represents the boundary condition). Additionally, there is hoop tension that decreases linearly upward from a maximum at the base, indicating an apparent "balance" between these two effects.

The cone-shaped footing should be designed to endure meridional compression and hoop tension caused by a uniform soil response acting within the vertical or normal direction, facing the soil underneath the foundation.

Stress resultants under vertical soil pressure:

$$N_s = -\frac{p_v}{2s} \tan \alpha \left[s_2^2 - s^2\right] \tag{13.10}$$

$$N_\theta = p_v s \frac{\alpha}{\cos \alpha} \tag{13.11}$$

$$N_{s\theta} = 0 \tag{13.12}$$

Stress resultants under normal soil pressure:

$$N_s = -\frac{p_n}{2s} \tan \alpha \left[s_2^2 - s^2\right] \tag{13.13}$$

$$N_\theta = p_n s \tan \alpha \tag{13.14}$$

$$N_{s\theta} = 0 \tag{13.15}$$

where

N_s = meridional compression per unit length
N_θ = hoop tension per unit length
$N_{s\theta}$ = shear stress resultant
α = half angle of the cone
S_2 = distance of the bottom of the cone from the apex
s = the slope length measured from the vertex
p_v = vertical base pressure
p_n = the normal pressure to the surface of the shell

Fixing the centerline triangle is the style method's initial and most vital stage (Figure 13.22). The size of the column outlines the truncation purpose of the cone. Calculating the value distance of column bases from the apex (S_1).

Based on the analysis, the design considerations for the conical shell can be summarized as follows:

1) **Meridional Compression:** The greatest meridional compression occurs at S_1, and the thickness of the shell at this point should be sufficient to resist the compression. From a practical perspective, a minimum thickness of 150 or 120 mm (depending on construction type) is recommended. Steel reinforcement should be provided at the top, radially distributed and curtailed as N_s reduces with distance. Each section should have a minimum nominal steel reinforcement of 0.5% of the gross section to ensure effectiveness against secondary bending.

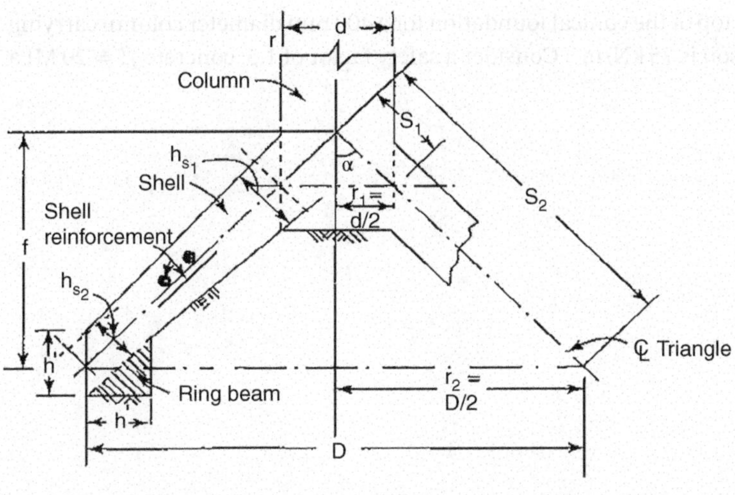

Figure 13.22 Dimensioning the conical footing (Kurian 2006 / Alpha Science International Limited).

2) **Hoop Tension:** Steel rings placed in the circumferential direction should fully resist the hoop tension. The hoop tension is greatest at the bottom and decreases linearly to a minimum at S_1. The spacing of hoop steel can be determined at these critical points and uniformly varied along the generator while satisfying the nominal steel demand. Sufficient hoops should be provided to tie the radial steel. The minimum concrete thickness at the top and bottom parts should be the maximum of: (i) thickness required to satisfy tension in the corresponding concrete section, (ii) thickness to limit the hoop steel to 5% of the section, and (iii) 150 or 120 mm, depending on construction technique.

The design thickness (h) at S_1 and S_2 is determined based on the higher values derived from considerations of meridional compression and hoop tension as described above. The footing profile can be completed by varying the thickness between S_1 and S_2 consistently and dropping verticals at S_1 and S_2.

This design approach ensures that the conical shell can effectively resist both meridional compression and hoop tension, providing structural stability and strength.

13.6.3 Limit State Design of Shell

Limit state design (LSD) of shell foundations must be done for factored loads obtained by multiplying characteristic loads by the proper partial safety factors, the latter depending on the type of load and the limits state being evaluated. This is true for any reinforced concrete structure. These factored loads produce the membrane stress resultants required for membrane design. The design strengths, also known as material properties, are obtained by multiplying the characteristic strengths of the materials (steel and concrete) by the proper partial safety factors. According to the limit state method, the design must satisfy the limit states of "collapse" and "serviceability." In the membrane mode, collapse includes direct compression or tension, as appropriate. The term "deflection" and "cracking" both fall under the purview of the term "serviceability limit state" with "cracking" being of particular significance to foundations due to the potential it presents for the infiltration of chemically aggressive groundwater and the subsequent risks of corrosion of the reinforcement and disturbance of concrete. The accompanying "short column formula" can be applied to compression, whether it be in a shell section or an edge beam (Kurian 2006).

$$V_u = 0.4 f'_c A_c + 0.67 f_y A_s \tag{13.16}$$

where

V_u = vertical compressive force
f'_c = characteristic strength of concrete
A_c = area of concrete
f_y = yield strength of steel
A_s = area of steel

Example 13.2 Calculate area of steel required at the top of the conical foundation for a 400 mm diameter column carrying the load of 1100 kN if the safe bearing capacity of the soil is 75 kN/m². Consider a safety factor of 1.5, concrete $f_c' = 20$ MPa and steel $f_y = 415$ MPa.

$$\frac{\pi D^2}{4} \times 75 = 1100\ \text{kN}$$

Diameter of cone bottom $D = 4.3$ m

Provide $D = 4.5$ m Radius = 2.25 m

Assume concreting slope of 1 to 2 horizontal

Rise = 2.25 ÷ 2 = 1.125 m

$$\tan\alpha = 2 \rightarrow \alpha = \frac{1}{2} \quad \text{central angle} = 63.44°$$

$$\sin\alpha = 0.894 \qquad \cos\alpha = 0.447$$

Since the column diameter is 400 mm → $S_1 = \sqrt{0.2^2 + 0.1^2} = 0.224$ m

$$S_2 = \sqrt{2.25^2 + 1.125^2} = 2.516\ \text{m}$$

$$P_n = P_v = \frac{1100}{\frac{\pi \times 4.5^2}{4}} = 69.2\ \frac{\text{kN}}{\text{m}^2}$$

$$P_u = 69.2 \times 1.5 = 103.8\ \frac{\text{kN}}{\text{m}^2}$$

Maximum compression per meter is at base of column at the top of the cone: $s = s_1$

$$N_s = \frac{103.8}{2 \times 0.224} \times \tan 63.44° \times [2.516^2 - 0.224^2] = 2910\ \frac{\text{kN}}{\text{m}}$$

Based on Limit State Design thickness h at 5% steel:

$0.95h \times 0.4 \times 20 + 0.05 \times 415 \times 0.67 = 2910 \rightarrow h = 135$ mm raise to 140 mm to adequate cover.

$$\text{Area of steel required} = 135 \times 0.05 = 6.75\ \frac{\text{mm}^2}{\text{m}}$$

13.7 Case Studies

Research on shell foundations can be divided into structural performance and geotechnical behavior. The first category focuses on studying the performance of shell structures related to membrane stresses, bending moments, shear strength, elastic deformation, and ultimate strength of the crust itself as a member of reinforced concrete structures. Compared to conventional shallow foundations, the research results on the performance of shell foundation structures can lead to economic benefits in concrete and steel rebars. The second category includes the geotechnical behavior of shell foundations to determine soil response in the form of subsidence, ultimate bearing capacity, contact pressure distribution, vertical and lateral stresses, and soil displacements.

13.7.1 Model Scale Study (Tadayon et al. 2022)

An experimental study on the geotechnical aspects of semi-deep shell foundations is presented in this article. In a 1 g chamber (Figure 13.23a), the behavior of physical models with various D/B ratios under axial pressure on the sand with $D_r = 29\%$, 58%, and 73% corresponding to $\phi = 30$, 35, and 40 have been studied. On the load–displacement relationship and

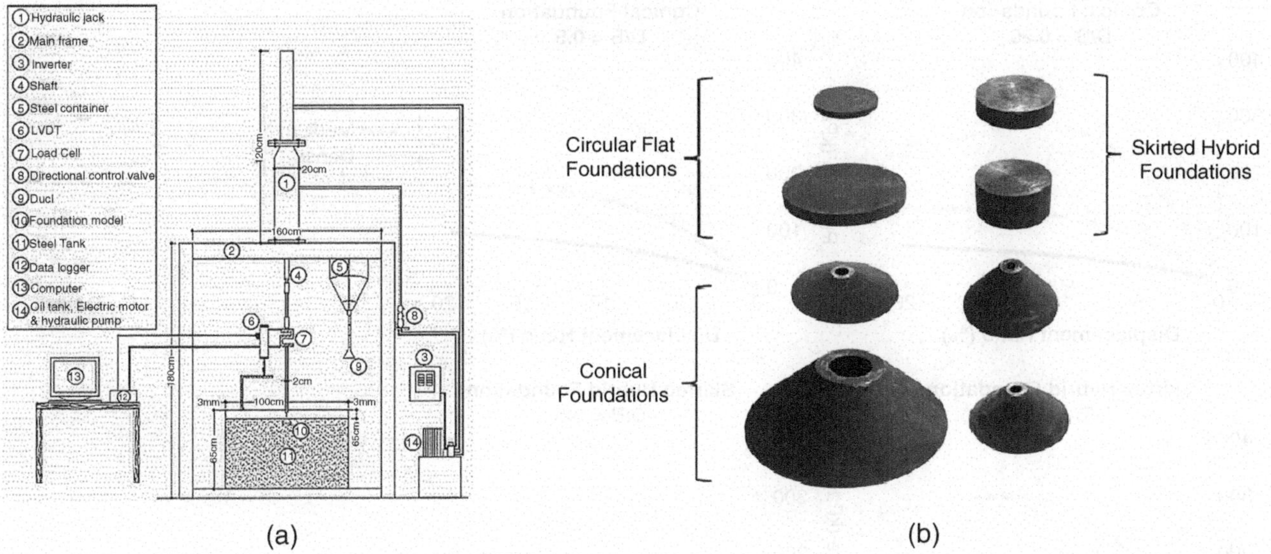

Figure 13.23 Overall view: (a) schematic layout of experimental test setup and (b) three model foundations (Tadayon et al. 2022 / Springer Nature).

ultimate bearing load (Q_u), the effects of shell design, foundation geometry, and soil characteristics have been researched. In test programs simulating the axisymmetric conditions, three kinds of model foundations: circular flat, skirted hybrid, and conical shell, were used on various sand states (Figure 13.23b).

The following conclusions are taken from this study:

- By enhancing soil properties, the bearing behavior of foundations is improved, resulting in a final bearing capacity of shell foundations on dense sand that is double that of medium sand and five to six times that of loose sand. For s/B less than about 11%, an increase in relative density in shallow depths ($D/B = 0.25$) causes an increase in the early slope of the normalized load–displacement relationship, which is followed by a decrease in slope and an increase in displacement ratio. Therefore, a turning point in the graphs of the Q/Q_u-displacement ratio is at a displacement of 8.8 mm. However, as depth increases ($D/B = 0.5$), nondimensional P–δ's slope decreases, causing the curves inverted (Figure 13.24).
- For both skirted hybrid and conical shell foundations, the initial slope of P–δ curves and the amount of movement necessary to mobilize bearing capacity in sand increase with increasing D/B. Due to the stiffer soil response, this rise can be seen more clearly in loose sand. For $D/B = 0.25$ and 0.5, respectively, the ratio of the ultimate load of semi-deep foundations to a circular flat one ranges from (1.2–1.5) to (1.5–2). The final bearing capacity of shell foundations increases to 1.3 when the D/B ratio is doubled from 0.25 to 0.5.
- The overall trend of the relationship between load and displacement and the mode of failure for all kinds of foundations depends on the state of the sand in a manner that strain hardening in loose and medium sand as well as strain softening in dense sand would occur. According to the curves' shapes, shear failure, local shear failure, and striking failure are each responsible for semi-deep sand shells with D_r values of 29%, 58%, and 73%, respectively.

13.7.2 Numerical Assessment of Inverted Shells (Rinaldi 2012)

Due to the soil's onerous nature and anisotropic characteristics, costly and sophisticated field testing, and/or arduous mathematical modeling are no longer viable. Using AutoCAD graphics software, prototypes of 24 suggested shell forms and 2 plain shapes were exhibited, including the simple, triangular, cylindrical, and combination shell shapes (Figure 13.25).

Based on the Mohr–Coulomb yield criteria, the elastic, perfectly plastic model was determined to be the most generally used soil model and is the one employed in this study. Using commercially available finite element software, PLAXIS, the shell footing models and sand bed medium would be modeled and assessed. It is assumed that the soil is homogeneous and of isotropic inconsistency. The analysis is built based on the elasticity theory. The produced and deformed forms of the mesh with pin fixities around the perimeter for an upright folded-plate shell are shown in Figure 13.26.

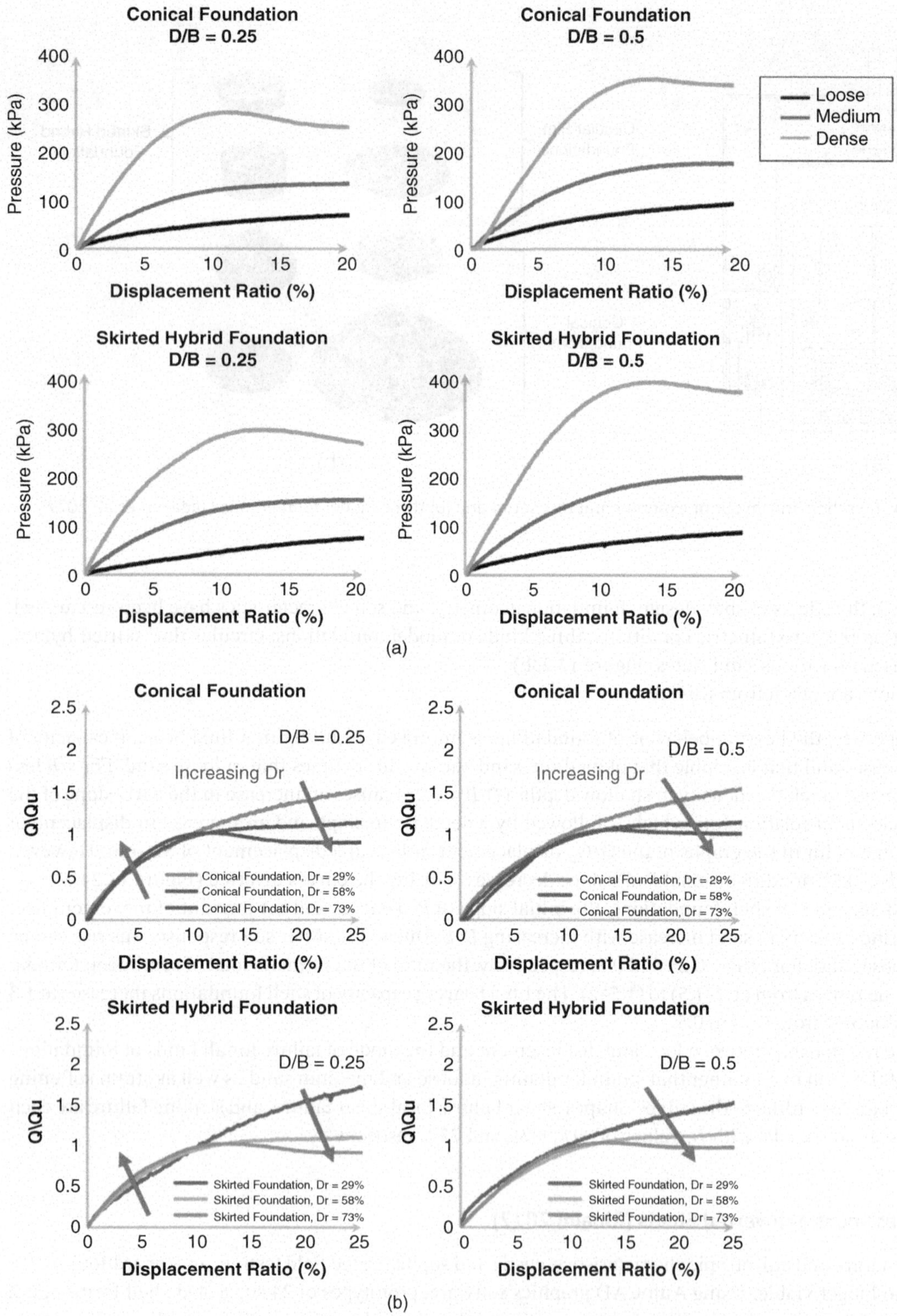

Figure 13.24 Conical and skirted hybrid foundations results: (a) pressure against displacement ratio on loose, medium, and dense sand and (b) normalized load–displacement (Tadayon et al. 2022 / Springer Nature).

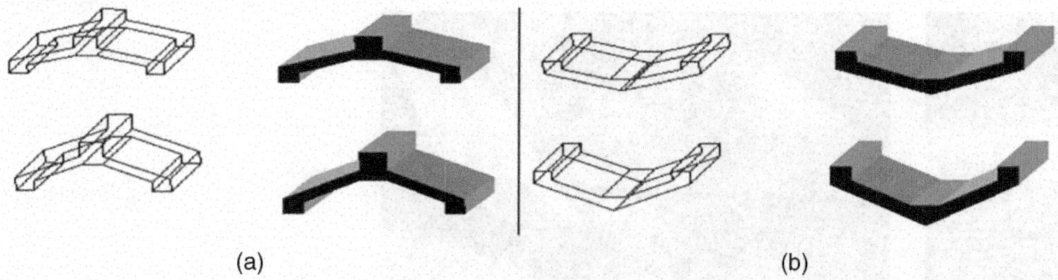

Figure 13.25 Triangular shell model rendering: (a) upright and (b) inverted (Rinaldi 2012 / Rinaldi).

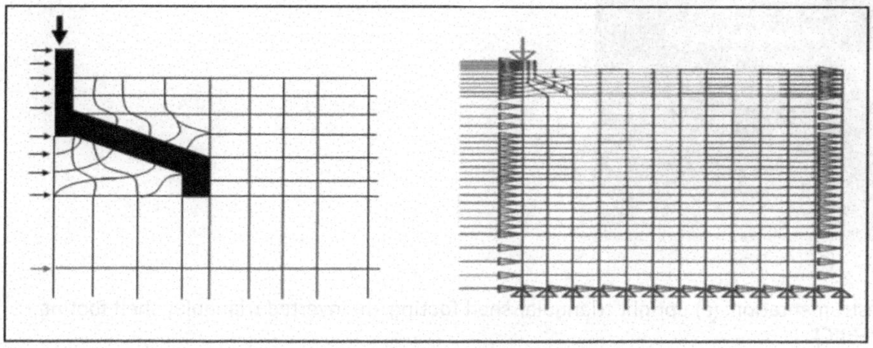

Figure 13.26 FEM mesh generation for embedded shell (Rinaldi 2012 / Rinaldi, R.).

Drained sand was chosen for the soil. A two-phase load plot was used to achieve the results. Construction of the footing and manipulation of the soil domain, including soil embedment, soil backfill, and soil border confinement, are all part of the initial phase. In the second phase, a uniform monotonic vertical force is given to the edge beams in inverted shells, as well as the vertical center column load in flat and upright shells. The footing-soil system was loaded in 2 kPa increments until the soil medium shear strength was reached.

From the numerical investigation, the soil response is generated for each shell tested. The following results were obtained:

- As with plain foundations, increased sand density increases capacity beneath shell footings.
- The effect of shell angle on footing capacity is a crucial determinant in performance, with a change in the shell angle of merely 18°, resulting in a 12% increase in load-carrying capacity.
- Increases in the depth of embedment from 0.50 to 0.75 indicated a 2% increase, implying that deeper shell footings had more capacity due to denser soil.

13.7.3 Field Study on Triangular Shells (Haut et al. 2016)

In this research, a finite element and field model test study on the performance of triangular shell footings were presented. Three different footing types were tested in the field for this study: a conventional flat footing, an "upright" triangular shell footing and an inverted triangular shell footing. Figure 13.27 depicts the experimental setup. Each footing had a plan size of 1 m × 1 m.

The inverted triangular shell footing demonstrated a higher load-carrying capacity than both the conventional flat strip footing and the "upright" triangular footing, as was to be anticipated (Figure 13.28).

Figure 13.29 displays the contact pressures that were measured using pressure cells that were set up in different places underneath the model footings. The edge part of the triangular shell footing was discovered to have a greater stress concentration. Stress was, however, more evenly dispersed across the shell contact surface for the inverted triangular shell.

Figure 13.27 Field model test set-up and instrumentation: (a) upright triangular shell footing, (b) inverted triangular shell footing, and (c) loading and testing (Haut et al. 2016/FEIIC).

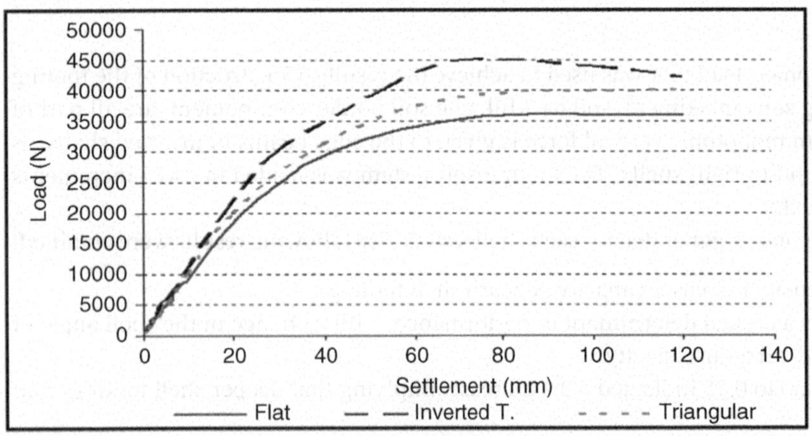

Figure 13.28 Measured load-settlement data of the field model footings (Haut et al. 2016).

The vertical stress of the inverted triangular shell footing was about 20% lower than the vertical stress of the erect triangular shell footing.

13.7.4 Sustainability and Reduction of Carbon Emission (Feickert 2022)

By 2060, it is anticipated that the floor area will have doubled due to the rising worldwide population. In addition, due to current construction methods, the building industry contributes 11% annually to global greenhouse gas emissions. Therefore, reducing embodied carbon will be crucial if global warming is to be kept to 1.5 °C above pre-industrial levels, and conventional building methods need to be rethought. This study addresses the need for a substantial rise in adequate housing due to rapid urbanization while reducing carbon emissions linked to reinforced concrete structural elements.

The term "embodied carbon (EC)" is frequently used to explain how building construction affects the environment. Embodied carbon is measured in carbon dioxide equivalence (CO_2e) units, which reflect the greenhouse gas emissions from

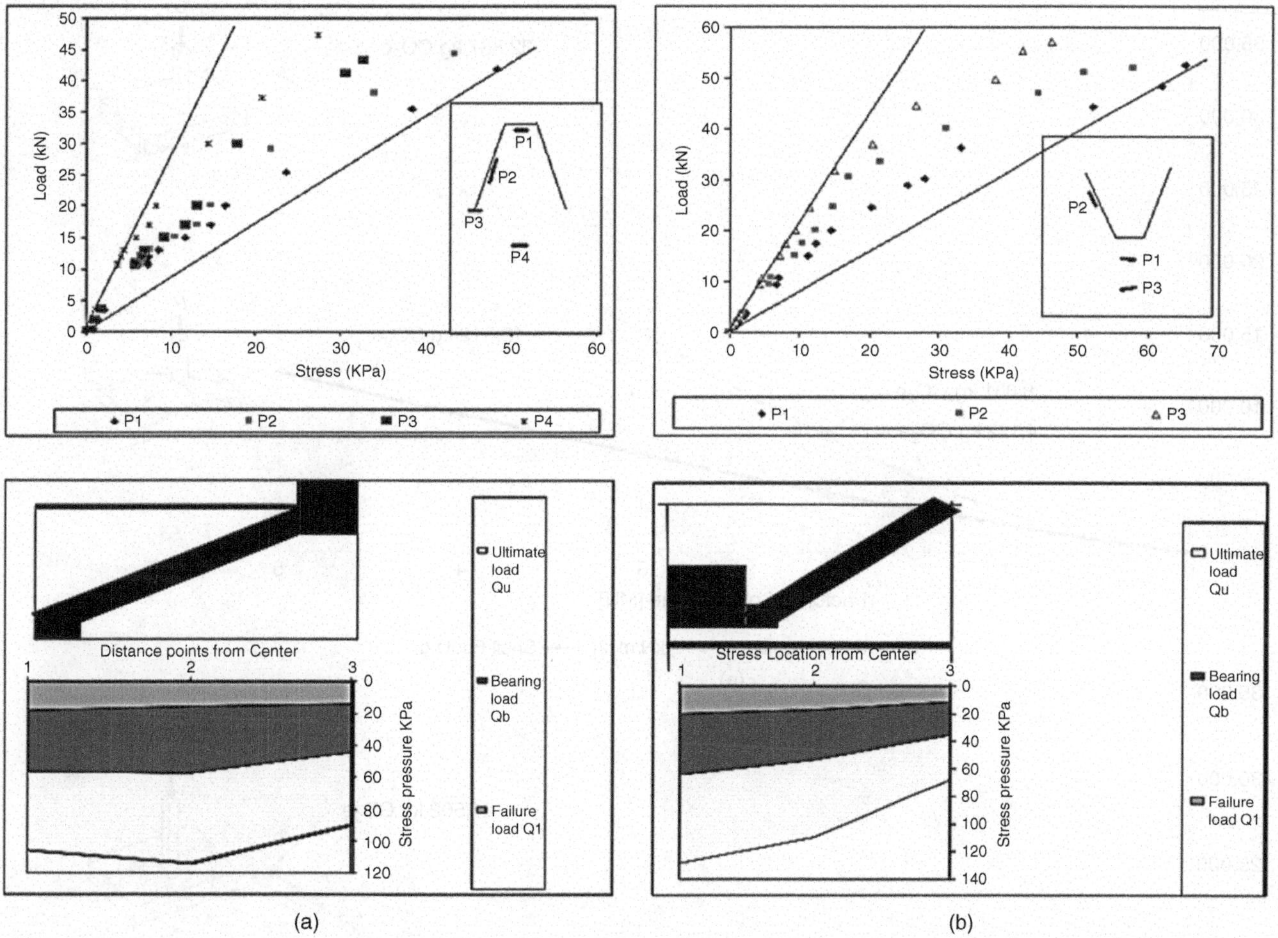

Figure 13.29 Stress distributions beneath the shell footings: (a) triangular shell, (b) inverted triangular shell (Haut et al. 2016).

different industrial processes' potential to cause global warming. As the major sources of embodied carbon in a building, the structural floor system, frame, and foundations reflect the systems with the greatest potential to reduce emissions. Material is used excessively at the expense of time in construction contexts where labor expenses are the determining factor.

Shells are better at distributing pressures through their cross-sections, which lowers the amount of material needed for structural support and, in turn, lowers their embodied carbon. For the same design load, the environmental effect of conventional prismatic foundations and shell foundations is assessed using existing analytical equations in a parametric design workflow. The embodied carbon in foundations is reduced by 48% by shells for a 2 MN column weight on clay soil. This method's applicability to different building typologies and site conditions is revealed by applying it methodically. In comparison to their prismatic counterparts, shells perform considerably better under high applied loads and in soils with low bearing capacity. The potential savings that could result from the shape optimization of numerous systems are then considered for foundations in the context of the entire building. The embodied carbon of a structure can be decreased by 72% by using shell foundations and shape-optimized floor slabs.

Even though the amount of savings is marginally reduced, shells continue to perform better than spread footings as the bearing capacity moves from clay to sand. In comparison to the spread footing, the shell footing saves 39% CO_2e for a 2 MN weight (3612 kg and 5945 kg, respectively) (Figure 13.30). The shell reduces carbon emissions from the spread footing by 56% for a 5 MN weight (10,807 kg CO_2e and 24,502 kg CO_2e, respectively) (Figure 13.30). Clay soils limit the extent of the savings, but they are still significant and demand investigation to learn how to be built.

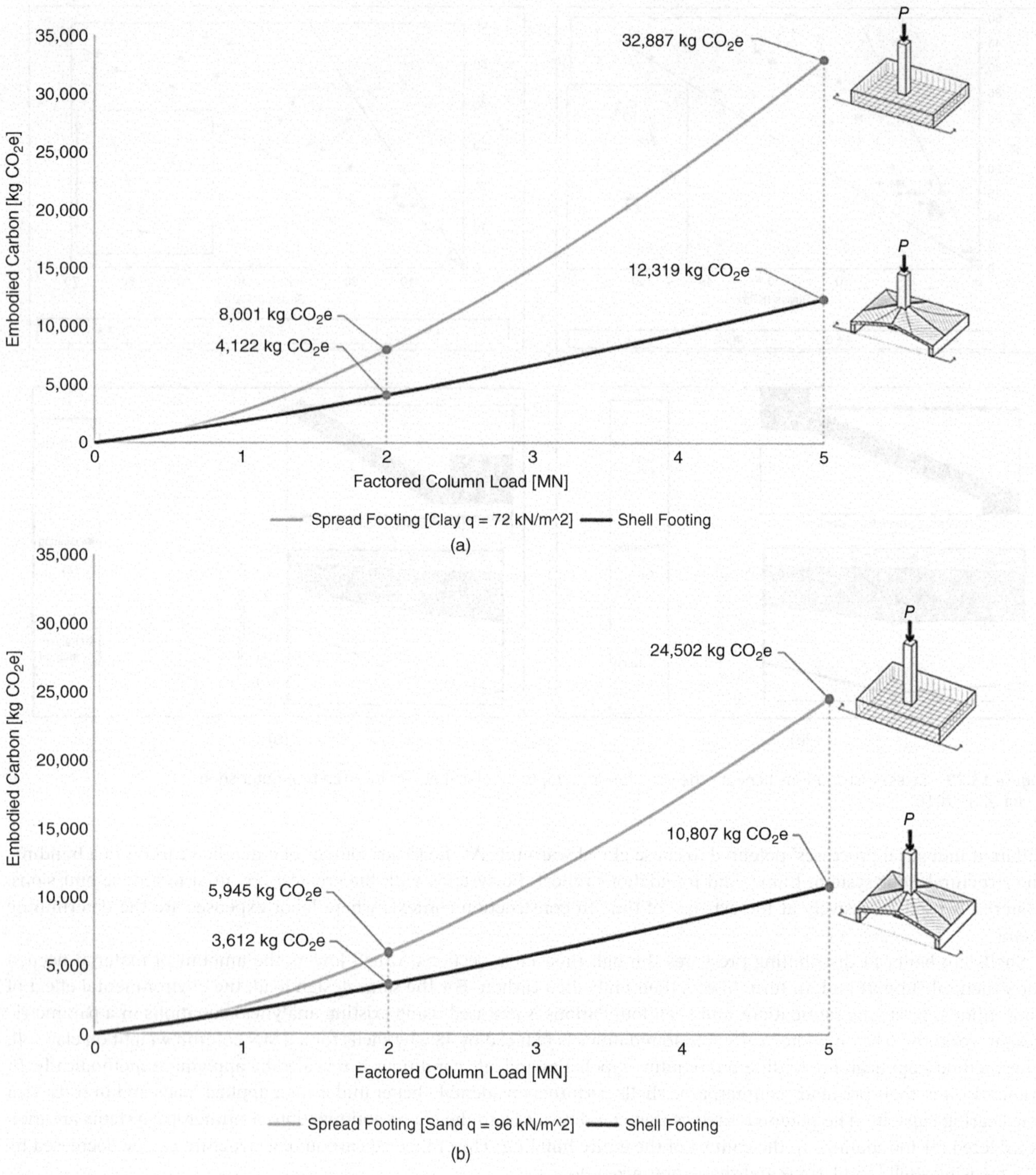

Figure 13.30 Embodied carbon for spread and shell footing: (a) on clay soil, (b) on sandy soil (Feickert 2022 / Springer Nature / CC BY 4.0).

13.8 Case Studies

Case Study 13.1

T.V. Tower, Germany (Reiff and Burkhardt 2020)

Due to its elevated location, Fritz Leonhardt suggested building a thin reinforced concrete tower rather than a steel lattice mast since it would be more visible from a distance. On June 10, 1954, the television tower's foundation stone ceremony took place. The innovative and aesthetically pleasing tower was opened on February 5, 1956, after 14 months of building and became a role model for the rest of the world.

The foundation body, the one-story entry building, the tower shaft, the so-called Korb (head or basket), and the antenna are all parts of the tower. The transmitter mast is included in the overall height, which is 217 m. The shaft has a diameter of 5.04 m under the tower head and 10.80 m at the foot of the tower. The platform's diameter is 15.10 m. The tower's base can be thought of as a unique engineering innovation. The base, which is entirely underground, is made up of two reinforced concrete truncated cones that are set against each other. They are supported by a 27 m diameter slab of pre-stressed, strengthened concrete. According to Fritz Leonhardt's account, this structure "appears like a spatial framework of great rigidity." A reinforced concrete slab 1-m-thick marks the end point of the foundation (Figure 13.31).

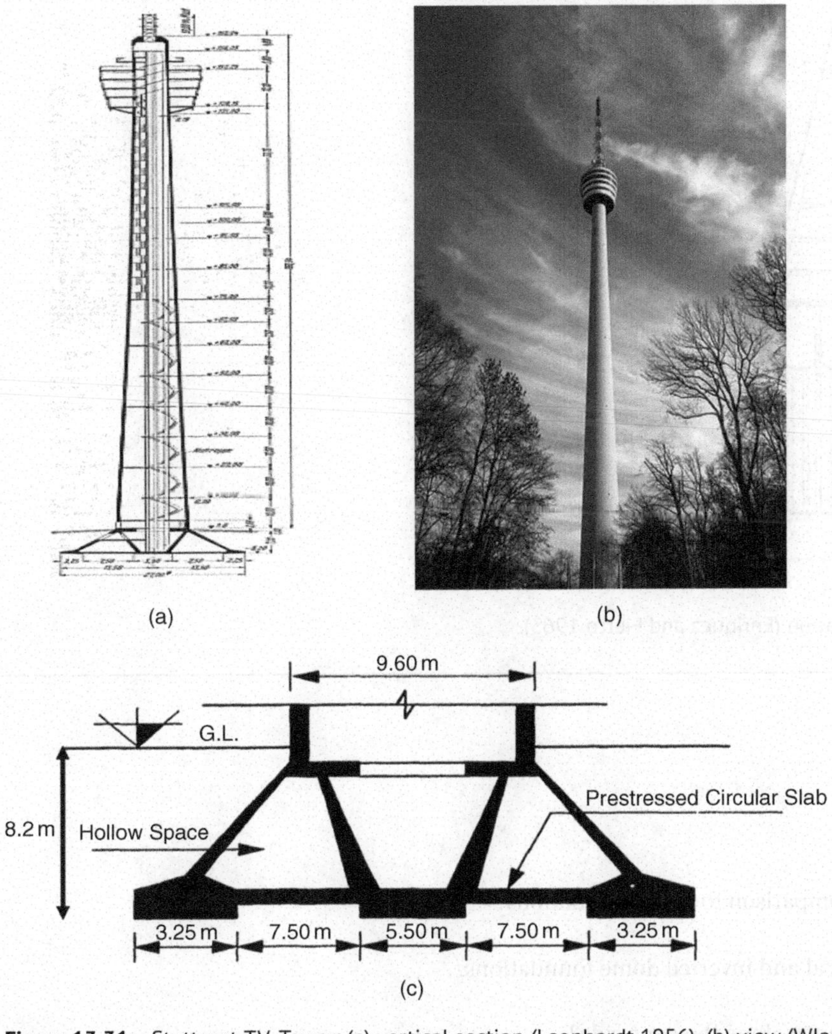

Figure 13.31 Stuttgart T.V. Tower: (a) vertical section (Leonhardt 1956), (b) view (Wladyslaw / Wikimedia Commons / CC BY-SA 3.0), (c) hollow conical shell foundation (Abdel-Rahman 1996 / Concordia University).

Case Study 13.2

The Nonoalco Tower, Mexico City (Enriquez and Fierro 1963)

One example of a highly ambitious development of a complicated surface-act foundation system was the foundation supporting the Nonoalco office tower, illustrated in Figure 13.32. The inverted barrel shells were linked by reinforced concrete girders and were supported overall by cast-in-place friction piles. This foundation system was estimated to have saved 50% of material costs compared to the conventional two-way slab foundation that was replaced. Moreover, the foundation system exhibited increased stiffness and reduced weight (Enriquez and Fierro 1963). Inverted barrel shells, recognized as surface-act foundation systems, provide more favorable performance via less consumption of materials, which is a usual benefit of the surface-act compared to the section-act foundation systems.

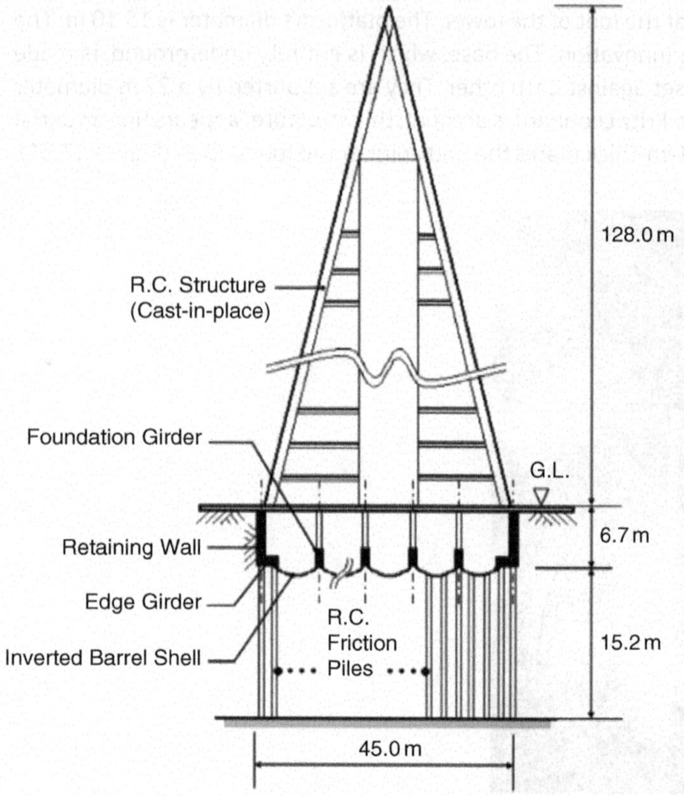

Figure 13.32 Section of building and foundation (Enriquez and Fierro 1963).

13.9 Questions and Problems

13.9.1 Explanatory

1 What are the advantages of shells in comparison to plain foundations?

2 Name some of the applications of conical and inverted dome foundations.

3 How does meridional compression change along the cone side?

4 What are the structural properties of Hypar?

5 What are the differences between synclastic and anticlastic shells?

6 How do pseudo stresses work in structural analysis?

13.9.2 Conceptual

1 How are shell foundations more cost-effective in comparison to plain foundations?

2 What are the differences between the membrane, bending, and ultimate strength theories?

3 How do shell contact pressure components change in friction and frictionless soils?

4 What is the effect of core soil on bearing capacity and settlement?

5 Where in the conical foundation has the highest thickness? Why?

6 Compare shell foundations with other common foundations regarding sustainability aspects.

7 What is the impact of increasing the base length to rise ratio of the shell on bearing capacity and settlement?

8 Compare different types of shell foundations with shallow foundations regarding geotechnical design requirements.

13.9.3 Problems

1 Calculate the area of the amount of steel required at the top of the conical foundation for a 450 mm diameter column carrying the load of 1000 KN if the safe bearing capacity of the soil is 60 KN/m^2. Consider a safety factor of 1.5, concrete $f_c' = 20$ MPa, and steel $f_y = 415$ MPa.

2 Calculate the hoop tension at a distance of the column base from the apex for a column with a 0.6 m diameter applying 1700 KN if the safe bearing capacity of the soil is 100 KN/m^2.

3 Calculate the total compression at the top of the conical foundation for a 500 mm diameter column carrying a load of 1500 KN if the safe bearing capacity of the soil is 100 KN/m^2.

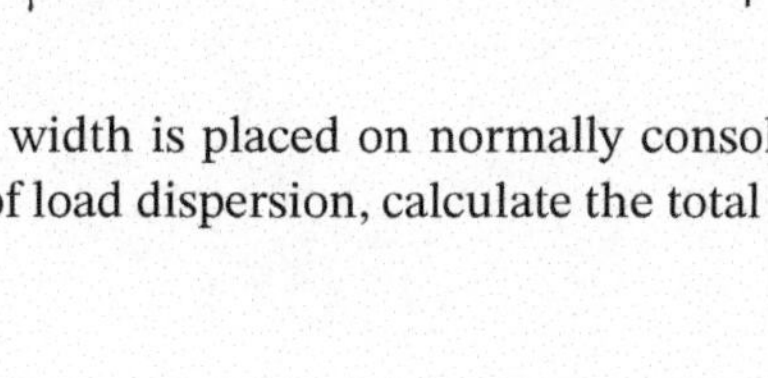

4 A Hypar shell foundation with a 5 m width is placed on normally consolidated clayey soil. The applied net load is assumed to be 750 kN. For a 30° area of load dispersion, calculate the total consolidation settlement of clayey layers.

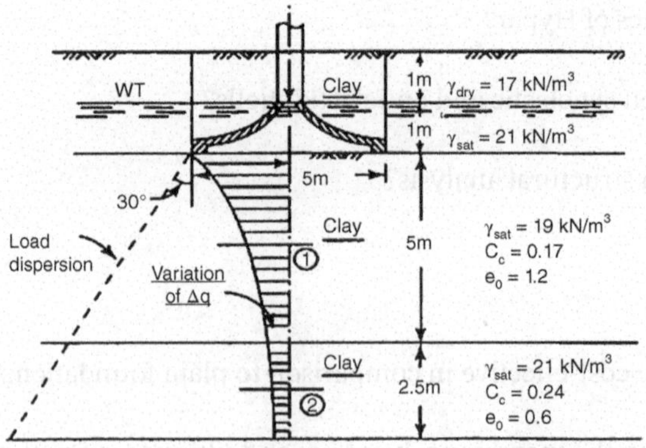

References

Abdel-Rahman, M. (1996). Geotechnical behavior of shell foundations. PhD. Thesis, Concordia University.

ACI Committee 318 (2008). Building Code Requirements for Structural Concrete (ACI 318-08). American Concrete Institute.

Agarwal, K.B & Gupta, R.N. (1983). Soil-structure interaction in shell foundations. Proc. Ill/ International Workshop on Soil-Structure Interaction, University of Roorkee, Nov–Dec. 1983, Vol. I, pp. 110–112.

Enriquez, R.R. and Fierro, A. (1963). A new project for Mexico city. *Civil Engineering* 33 (6): 36–38.

Feickert, K.A. (2022). Thin shell foundations: embodied carbon reduction through materially efficient geometry. MSc. Thesis. Department of Architecture, Massachusetts Institute of Technology, USA.

Jayakrishna Devaki, Y.M. (2002). Numerical studies on the geotechnical performance of shell foundations by the finite element method. M.Sc. Thesis, Indian Institute of Technology.

Huat, B.B.K., Mohammed, T.A., Abang Ali, A.A.A., and Abdullah, A.A. (2007). Numerical and field study on triangular shell footing for low rise building. *International Journal of Engineering Technology* 4 (2): 194–204.

Kidder, F. E. (1905). *Building, Construction and Superintendence.* W. T. Comstock, Chapter III, pp. 87–126.

Kurian, N.P. (2006). *Shell Foundations: Geometry, Analysis, Design and Construction.* Alpha Science International, Limited.

Leonhardt, F. (1967). The present position of reinforced concrete tower design. Bulletin of the IASS(29).

Leonhardt, F. (1956). *Der Stuttgarter Fernsehturm.* Ernst.

Murray, L. (2023). Sydney Opera House. https://www.britannica.com/topic/Sydney-Opera-House.

Ramaswamy, G.S. (1968). *Design and Construction of Concrete Shell Roofs.* McGraw-Hill.

Ramaswamy, G.S. & Keshava Rao, M.N. (1961). The membrane theory applied to hyperbolic paraboloid shells. *The Indian Concrete Journal*, 35, 5, 156–162 and 168–171.

Reiff, A. and Burkhardt, B. (2020). Repairs of the Stuttgart Television Tower. *ICOMOS–Hefte des Deutschen Nationalkomitees* 73: 52–58.

Rinaldi, R. (2012). Inverted shell foundation performance in soil. PhD. Thesis. Concordia University.

Szechy, C. (1965). The influence of the shape of contact surface upon the bearing capacity and settlement of strip foundations. Proceedings of the 5th Symposium of the Civil and Hydraulic Engineering Department, Indian Institute of Science, Bangalore, Vol. 3, pp. C2/1–7.

Tadayon, A., Eslami, A., and Fakharian, K. (2022). Load–displacement behavior of semi-deep shell foundations on sand by experimental study. *International Journal of Civil Engineering* 20 (7): 839–855.

Topic D

Deep Foundations

Section VI

Single Piles and Drilled Shafts

14

Types and Installation

Chapter Highlight

- Applications
- Materials and Geometry Types
- Installation and Construction Technics
- Driven Piles
  - Timbe
  - Steel
  - Precast Concrete and Composite
- Drilled Shafts (Bored Piles)
  - Wet and Dry Construction
  - Enlarged Base and Post-Grouted
- Special Piles
  - Continuous Flight Auger Piles (CFA)
  - Drilled Displacement Piles (DDP)
  - Helical Piles
  - Self-Expanded Piles
  - Mini and Micro-Piles
- Case Study
  - Drilled Shafts Behavior in Caliche Layers
  - Piling and Rigid Intrusions Combination
  - FHWA Source Manual for Drilled Shafts Performance
  - Installation Effect via Physical Modeling (FCV-AUT)

14.1 Deep Foundations Applications

Deep foundations, as vertical or slightly sloped structural elements embedded in the soil, are designed to transfer the load from upper structures to deeper and more competent layers of soil. They are typically used when shallow foundations are not sufficient to support the loads or when the upper soil layers are weak or unstable. Deep foundations can be constructed using various materials such as wood, concrete, steel, or a combination of these materials. One key consideration when using deep foundations is the increased cost compared to shallow foundations. The construction process for deep foundations involves more extensive excavation, specialized equipment, and additional materials, leading to higher expenses. However, in certain projects, ensuring structural safety and stability outweighs the economic concerns, making deep foundations a necessary choice. Deep foundations offer several advantages. They can provide higher load-bearing capacity and resistance to settlement compared to shallow foundations. They are also effective in mitigating the effects of poor soil conditions, such as weak or expansive soils. Additionally, deep foundations can be used in areas with high water tables or in regions prone to seismic activity, where stability and performance are critical. The selection of deep foundations over shallow foundations

Advanced Foundation Engineering: Principles, Performance and Prospects, First Edition. Moses Karakouzian and Abolfazl Eslami.

is determined by site-specific factors, such as soil conditions, structural requirements, and project constraints. Detailed geotechnical investigations and engineering analyses are conducted to assess the soil properties and determine the most suitable type of deep foundation for a given project.

Deep foundation systems are recommended in at least one of the following conditions:

- When the upper soil layers have low-bearing capacity, they are unable to support the load transferred by the superstructure. Soil layers with higher resistance are located at greater depths, necessitating the use of deep foundations. Even if mat foundations are used, adequate-bearing capacity cannot provided by the surface layers.
- Scouring is a common problem in bridge foundations or areas near beaches, where structures can be eroded by water and sediment. Deep foundations provide a solution to this problem.
- When large concentrated loads need to be transferred from the structure to the soil, shallow foundations or mats may not be able to withstand these forces, requiring the use of deep foundations.
- When high groundwater levels or artesian pressure in the soil make it impractical to construct shallow foundations, necessitating the use of deep foundations.
- Increasing the soil's stiffness beneath machine foundations helps control foundation vibrations and maintain the system's natural frequency.
- Deep foundations are required to resist tensile or overturning forces, particularly for tall structures where stability is crucial.
- Lateral and earthquake forces need to be restrained and controlled, requiring the use of deep foundations.
- Deep foundations are employed to mitigate landslides, enhance slope stability, and provide support against ground motion.
- When structures require sufficient pullout capacity and external stability under combined vertical, horizontal, and moment (VHM) loading conditions, deep foundations are essential.
- Implementing barriers within deep foundations helps control seepage and prevent unwanted water flow.
- Deep foundations can be used to enhance the capacity of existing shallow foundations by intrusion or confinement using deep-seated elements.

Deep Foundations are generally categorized into three different types:

- Piles (displacement deep foundations)
- Drilled Shafts (replacement deep foundations)
- Special deep foundations

Piles are driven into the ground using methods such as hammering, vibrating, or bottom driving. Drilled shafts, also known as bored deep foundations, are created by excavating a hole and filling it with concrete, often with the addition of a reinforcement cage. Special deep foundations refer to those that do not fit into the categories of piles or drilled shafts and encompass various specialized techniques and systems.

Deep foundations are installed and constructed in a variety of forms and conditions. A summary of different subgroups based on some factors is presented in Table 14.1.

The selection, construction, and application of deep foundations are influenced by various factors, including the type of structure, the magnitude of the loads, the availability of equipment and materials, and the geotechnical conditions of the site. In this context, in situ testing methods, particularly cone penetration testing (CPT & CPTu), play a crucial role (Eslami et al. 2020).

14.2 Piles Materials and Cross-Sections

Piles are available in a variety of shapes and configurations, including circular, square, H-shaped, triangular, polygonal, longitudinal, cylindrical, and conical cross-sections. The ends of piles can be either open or closed, depending on the specific design requirements. Due to the diverse range of pile characteristics, it is challenging to categorize them comprehensively. However, piles are often classified based on factors such as installation methods, geometric shapes, soil displacement during installation, materials, bearing capacity, performance, and intended use.

Regarding geometric shapes, the longitudinal sections of piles can have a cylindrical or conical shape. Conical piles can further be classified into two types: those with a steeply tapered gradient and those with a gradually tapered gradient. As

Table 14.1 Deep foundations types based on different factors (CFEM 2006 / Canadian Geotechnical Society).

Factor	Subgroup
Installation	Driven, bored, jacked, cast in situ, jetted, excavated, augured
Displacement	Displacement, low–displacement, non-displacement
Material	Concrete, steel, wood
Function	Shaft bearing, toe bearing, combination
Capacity	High, moderate, low
Shape	Square, round, hexagonal, octagonal, H-section, Tapered
Environment	Land, marine, off-shore
Inclination	Vertical, battered
Length	Long, short
Structure	Bridge, building, platform, towers, and machinery

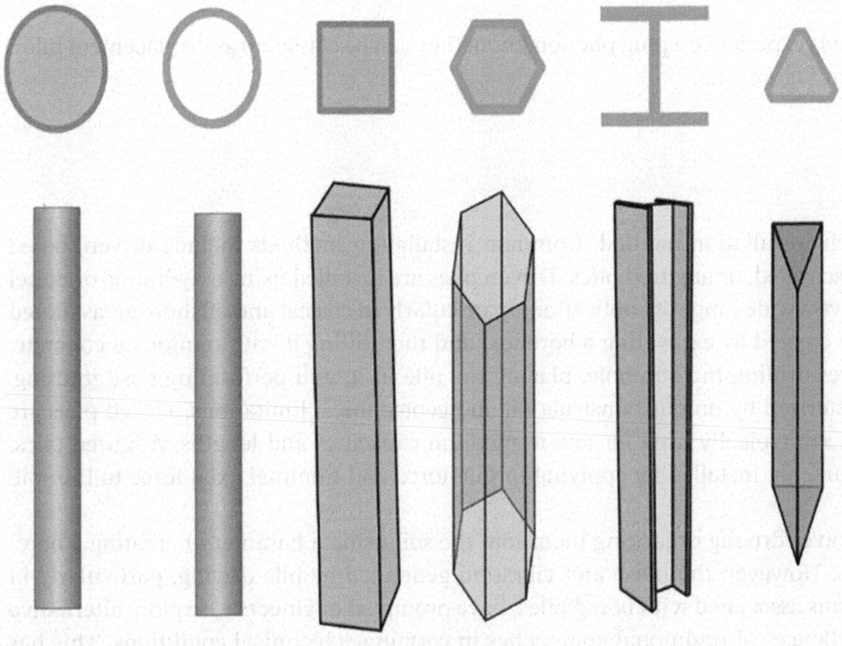

Figure 14.1 Illustration of pile longitudinal and cross-sections.

mentioned earlier, in terms of cross-sections, piles can have circular, square, hexagonal, octagonal, H-shaped, and even triangular shapes. They can be solid or hollow. Figure 14.1 provides examples of different longitudinal and cross-section configurations of piles.

The surface of the pile shaft can be smooth or rough, and the tip of the pile can have a conical shape or a broader area. In some cases, conical or wide shoes may be attached to the pile toe. Additionally, the toe of a deep foundation can be either open or closed. For instance, pipe piles can be used with either an open or closed-end, depending on the project requirements.

Piles can be classified into three categories based on how soil is affected during installation: displacement piles, low displacement piles, and non-displacement piles. Displacement piles involve radial compaction of the soil as they are installed, and this category includes closed steel piles. Non-displacement piles, on the other hand, involve creating a hole in the soil and then filling it with reinforced concrete. Piles with low displacement exhibit behavior that falls between these two categories. This category includes driven piles with open ends, drilled displacement piles, and H-shaped piles. It is worth noting

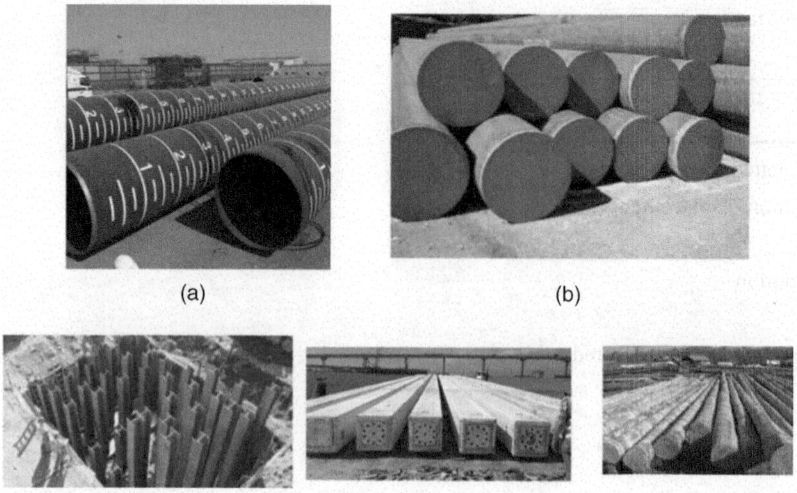

(a) (b) (c) (d) (e)

Figure 14.2 Typical kinds of driven piles: (a) open-end pipe pile (Nan steel), (b) closed-end pipe pile, (c) H pile (Structotag), (d) precast concrete pile, and (e) timber pile.

that in cases where driven piles with open ends experience a plug phenomenon, they can be classified as displacement piles. Typical types of driven piles are illustrated in Figure 14.2.

14.3 Pile Installation

Piles can also be categorized based on their installation method. Common installation methods include driven, bored (cast-in-place), jacked, cast-in situ, jetted, excavated, or augured piles. Driven piles are installed using a hydraulic or diesel hammer to drive them into the soil. They have a wide range of applications, particularly in coastal and offshore areas. Bored piles, also known as cast-in-place piles, are created by excavating a borehole and then filling it with reinforced concrete. The drilling and placement method involves drilling the borehole, placing the pile in it, and performing post-grouting. This method is well-suited for areas characterized by unique construction and geotechnical limitations. Jacked piles are installed by jacking them into the soil and are typically used for low to medium capacities and lengths. Augured piles, which have gained popularity in recent years, are installed by applying torque force and minimal axial force to the soil. They are primarily used in urban areas.

In situ construction of piles typically involves driving or jacking them into the soil using a hammer or creating a borehole and filling it with reinforced concrete. However, the noise and vibration generated by pile driving, particularly in urban areas, as well as the practical limitations associated with bored piles, have prompted engineers to explore alternative methods to address the drawbacks and challenges of traditional approaches in certain geotechnical conditions. This has led to the development of specific types of piles, such as continuous flight auger (CFA) piles, drilled-displacement piles (DDP), micro-piles, post-grouted piles, helical piles, and self-expanded piles. These innovative pile types offer improved performance and enhanced construction efficiency in various geotechnical settings.

In geotechnically favorable conditions and with appropriate equipment, driven piles are typically preferred due to their higher bearing capacity compared to other piles with similar geometry. However, challenges arise when driving these piles in certain soil types, and the associated noise and vibration can be problematic in urban environments, particularly in close proximity to sensitive structures like hospitals or historical buildings. Consequently, the use of special deep foundations has become more prevalent in urban areas to address these concerns. On the other hand, in cases where geotechnical conditions are suitable, particularly when a competent rock layer is present at a depth of up to 30 meters, bored piles remain widely used in engineering practice. These piles offer advantages such as the ability to be installed in larger diameters and greater lengths, resulting in higher bearing capacity. Bored piles are especially suitable in situations where challenging geotechnical conditions exist, such as collapsed layers or high groundwater levels.

The installation of driven piles is relatively straightforward in soft soils but can pose challenges in dense soils. Difficulties may arise when encountering a dense layer during the driving process. Several solutions can be employed to overcome these difficulties. One approach involves using a water jet device to change the soil structure beneath the pile before driving it. This method helps to ease the pile's penetration into the soil. Alternatively, a driving pile shoe can be utilized at the pile's tip to facilitate its movement through the soil. In granular soils, a higher impact typically results in an increased penetration rate. On the other hand, in stiff clays, a lower impact combined with a heavier driving force often yields a better penetration rate.

Piles made of timber, steel, pre-stressed concrete, precast concrete, or a combination of these materials are installed in the soil using diesel or hydraulic hammers. One advantage of the driving procedure is that it preloads the soil under the pile tip, which can enhance the pile's bearing capacity. However, there are some limitations associated with driven piles. One limitation is the high noise and vibration generated during installation. This can be problematic, particularly in urban environments with noise and vibration restrictions. The vibration caused by hammer impacts and the excessive noise can pose risks to adjacent structures, especially older buildings, and may result in structural failure. Noise pollution is another issue that needs to be addressed. Despite these limitations, driven piles are widely used in environments where noise and vibration are not significant concerns. They are particularly common for offshore structures, where open pipe piles are frequently employed.

14.3.1 Hammer Driving Equipment

The selection of a suitable hammer for pile installation is indeed a crucial aspect of successful pile installation. The choice of the hammer should aim to achieve the desired pile depths specified in the design while minimizing any potential damage to the piles. Additionally, a pile-driving system can provide valuable information for estimating the geotechnical-bearing capacity of the piles. By monitoring and analyzing the driving parameters such as hammer energy, blow count, and penetration resistance, a geotechnical engineer can assess the behavior of the piles during installation and make informed decisions regarding their capacity. Therefore, one of the primary challenges for a geotechnical engineer is to carefully evaluate the geotechnical conditions at the site and select a suitable hammer that is capable of effectively driving the piles to the required depths, considering factors such as soil type, pile characteristics, and project requirements. The goal is to achieve cost-effectiveness while ensuring the integrity and performance of the installed piles.

Pile driving systems consist of several essential components, including cranes, leaders, hammers, helmets, and cushions, all working together to facilitate the pile installation process and minimize direct impact on the pile. Figure 14.3 depicts

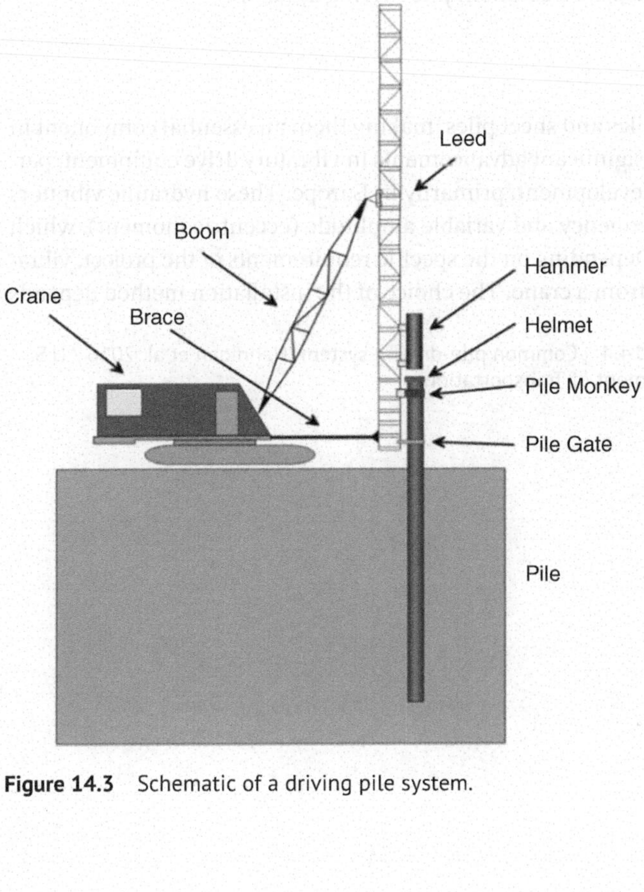

Figure 14.3 Schematic of a driving pile system.

a standard rammer system commonly used for pile driving. Among these components, the leader or vertical axis plays a crucial role as it guides the pile in the desired direction during installation. The leader serves as a framework where the pile, hammer, and other equipment are positioned. It provides stability and ensures accurate alignment of the pile during driving. The piles are typically loaded onto the leader using a pile gate, which securely holds them in place. In addition to the leader, other components, such as cranes, provide the necessary lifting capability, while hammers deliver the impact force to drive the pile into the ground. Helmets are used to transmit the hammer's energy to the pile, while cushions are employed to reduce the impact and protect the pile from damage during driving.

When installing piles in large groups in close proximity to each other, a careful evaluation of the ground stiffness and consideration of pile breaks are crucial. It is important to determine the proper order of pile installation to avoid any negative impact on the overall bearing capacity of the group. To achieve this, it is recommended to prioritize driving the piles in a sequence that moves outward from the center of the group. By starting from the center and gradually working towards the outer edges, the load distribution among the piles can be more effectively managed. This approach helps minimize any potential differential settlements and ensures that the group of piles functions collectively to support the desired load.

Successful pile installation requires careful coordination between various factors, including the pile type, cross-section, ground conditions, and driving method. One crucial aspect to consider is the selection of the appropriate hammer size based on the pile's characteristics and the desired driving depth or resistance.

The force exerted on the pile top by the driving equipment should be sufficient to overcome the resistance of the pile and the surrounding ground. However, it is important to carefully consider the weight and height of the hammer fall to avoid damaging the pile, especially in soft soils. When driving piles in soft soils, using a heavy hammer with a low fall height can help minimize the risk of pile damage. This combination allows for longer stress wavelengths and increased durability, reducing the likelihood of excessive stress concentrations 6on the pile head. By controlling the stress levels, the risk of damaging the pile during driving can be mitigated. It should be noted that the penetration of the pile will increase by increasing stress much more effectively than increasing the impact durability under challenging driving conditions.

While visible damage to piles during driving is typically observed near the pile head, it is important to recognize that the shaft and tip of the pile can also be susceptible to damage. Common types of damage that can occur during driving include buckling, crumbling, twisting, and deformation, as well as longitudinal cracking in steel piles and cutting, cracking, and spalling in concrete piles.

Problems at the pile tip can sometimes be detected through driving records. Easier penetration and high temporary density indicate damage to the lower part of the pile. Figure 14.4 depicts a common pile-driving system.

14.3.2 Vibratory Driving

Vibrators play a vital role in the installation and extraction of piles and sheet piles, making them an essential component in many construction processes. In recent years, there have been significant advancements in vibratory drive equipment, particularly hydraulic vibrators, with a focus on innovation and development, primarily in Europe. These hydraulic vibrators offer several innovative features, such as variable operating frequency and variable amplitude (eccentric moment), which provide versatility and adaptability for different applications. Depending on the specific requirements of the project, vibrators can be operated either by being led by leads or suspended from a crane. The choice of the installation method depends

Figure 14.4 Common pile-driving system (Hannigan et al. 2016 / U.S. Department of Transportation).

on factors such as the type of piles, site conditions, and project constraints. Modern hydraulic vibrators are equipped with advanced control systems that allow for precise control, monitoring, and recording of the vibration process throughout all stages of operation. This level of control and monitoring ensures the accuracy and efficiency of the installation or extraction process and provides valuable data for quality control and documentation purposes.

Vibrators have proven to be highly effective in various geotechnical applications, including pile driving, casing installation, and sheet pile driving. The use of vibratory driving techniques offers several advantages, such as high production rates and minimal environmental impact. However, it is crucial to consider project-specific requirements and factors, including geotechnical and hydrogeological conditions, and environmental considerations, when selecting a vibratory drive system. Failure to account for these factors can lead to issues related to environmental impacts, such as ground vibrations, which may cause disturbances or damage to surrounding structures or sensitive environments. The selection of the appropriate vibrator type and operating parameters should be based on a thorough understanding of the project requirements and geotechnical principles. Unfortunately, in some cases, foundation companies rely on mechanical engineers who may have limited knowledge of geotechnical principles. This can lead to suboptimal vibrator selection and operation. To ensure the proper selection and use of vibratory drives, it is essential to involve geotechnical engineers who possess the necessary expertise and understanding of the specific geotechnical challenges and considerations involved. By employing empirical principles and engaging knowledgeable geotechnical engineers, the risks associated with vibratory driving can be mitigated, and the selection of the most suitable vibrator type and operating parameters can be optimized. This approach ensures that the vibratory driving process is carried out efficiently, effectively, and with minimal environmental impact.

The following three factors must be taken into consideration while installing piles with vibrators:

- **Driveability:** To ensure effective performance, choose the proper vibrator system and operation.
- **Bearing Capacity:** Check to ensure that the required bearing capacity can be met.
- **Environmental Impact:** Reduce ground noise and/or vibrations (Massarsch et al. 2017).

Figure 14.5a presents an actual view of the installation of a pile via the vibration process. Figure 14.5b depicts a simplified model for toe force (F_z) vs. penetration (z).

14.3.3 Bottom Driving

The bottom driving technique is a method used for installing piles by driving a closed-ended thin-wall steel tube into the ground. The steel tube, typically with a length ranging between 2 and 6 meters, is driven into the soil using an internal drop

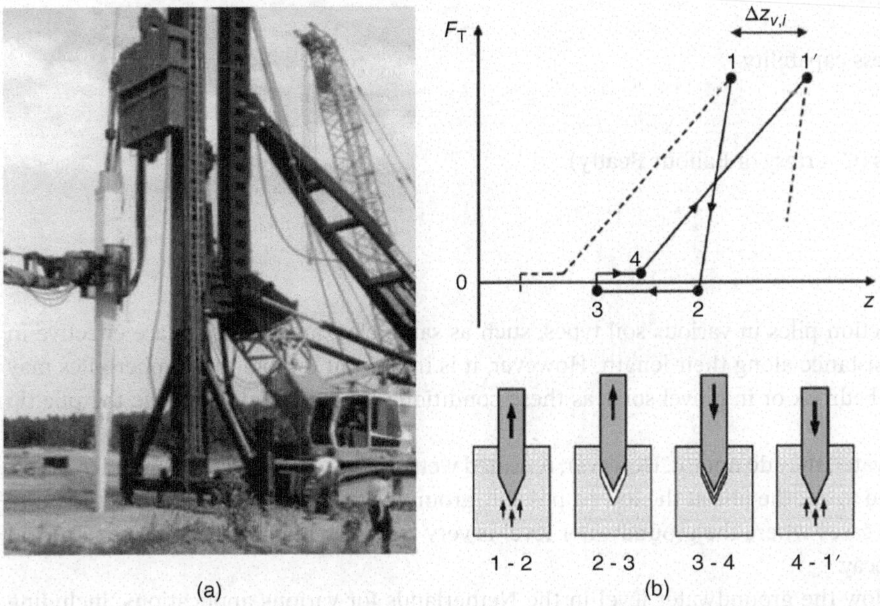

Figure 14.5 Vibratory driving: (a) actual case, (b) simplified model of pile toe–soil interaction during installation (Massarsch et al. 2017 / With permission of John Wiley & Sons).

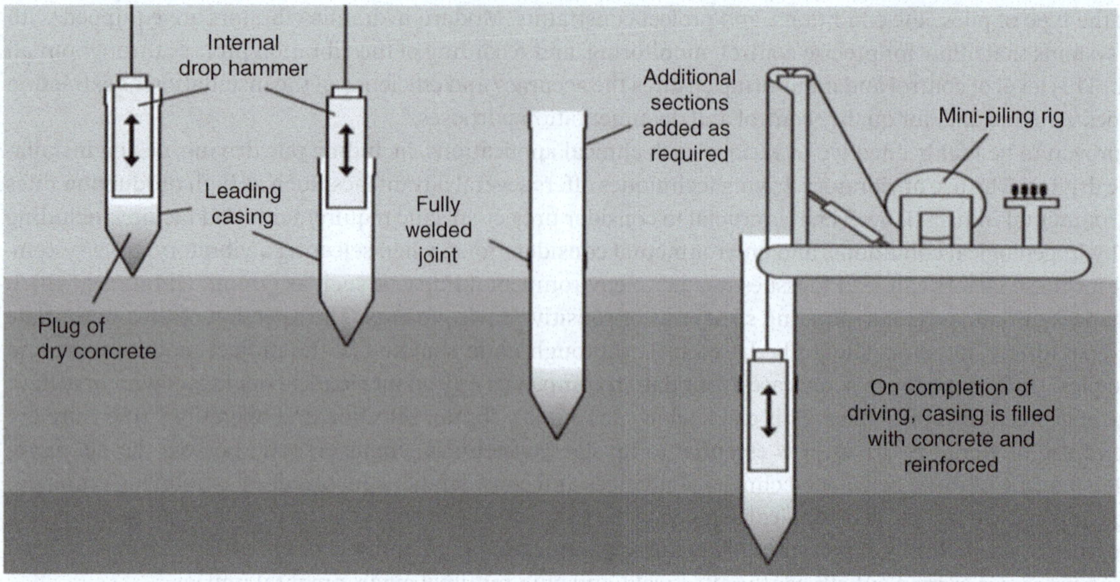

Figure 14.6 Installation sequence of bottom driving piles.

hammer. At the end of the tube, there is a dry concrete plug that helps facilitate the driving process. During the installation procedure, the individual steel tubes are joined together using full but nonstructural fillet welds. This ensures a continuous and secure connection between the tubes. As the driving continues, the tubes are driven to the predetermined length. Once the desired depth is reached, the steel tube is filled with high-slump concrete or grout. This filling process helps increase the overall strength and stability of the pile. After the filling, a single reinforcing bar or cage is inserted into the tube to provide additional structural reinforcement. Figure 14.6 illustrates the construction sequence of bottom driving piles, showing the step-by-step process of driving the steel tubes, filling them with concrete or grout, and inserting the reinforcing bar or cage.

The advantages of these piles are as follows:

- Small and light installation equipment
- Minimal site preparation is required
- Piles self-testing
- Limited headroom and restricted access capability
- Low levels of noise and vibration
- No debris to be removed
- Applicable in most ground conditions (Courtesy of Balfour Beatty)

14.3.4 Timber Piles

Timber piles are commonly used as friction piles in various soil types, such as sand, clay, and silt. They are effective in transferring loads through frictional resistance along their length. However, it is important to note that timber piles may not be suitable as end-bearing piles on bedrock or in gravel soils, as these conditions can potentially damage the pile tip during driving.

When timber piles are submerged in water, they do not rot. However, repeated wetting and drying cycles can lead to decay. To mitigate this issue, it is recommended to cut the pile at the lowest possible groundwater level and use concrete to build and extend the pile above that level. In cases where the groundwater level is very high, the pile head can be constructed below the groundwater level to avoid decay.

Timber piles are extensively used below the groundwater level in the Netherlands for various applications, including light structures, house construction, and roads. They are also popular in other countries, such as England, the United States, Australia, and the Scandinavian Islands.

14.3.5 H-shaped Steel Piles

H-shaped piles are commonly employed in situations where there are challenging or problematic soil layers, and their ends are embedded in a hard layer or bedrock. These piles can be utilized in both friction and end-bearing modes, depending on the specific project requirements. They are capable of supporting axial loads ranging from 350 to 1800 kN and can be used for a wide range of depths.

Proper connection of H-shaped piles is crucial to ensure their structural integrity. The connection points should be at least as strong as the pile section itself. Typically, H-shaped piles are connected through welding. The advantage of these piles is that their length can be extended as needed. When increasing the length of H-shaped piles, a full penetration weld of the web and flange is commonly employed. It is important to conduct thorough inspections of each weld, and in appropriate locations, additional welds should be tested using mechanical or radiographic methods to ensure the quality of the connections.

14.3.6 Pipe Steel Piles

These piles are driven in two forms, with closed and open ends, in the soil. Inside, they may be empty or filled with concrete. These piles may be used as end-bearing, friction, or combination piles. Due to the ability to increase and reduce these piles' lengths easily, they are typically used in long lengths.

14.3.6.1 Closed-end Pipe Piles

Closed-end pipe piles are available in various configurations, including jointless, single-jointed, or spiral designs. They are typically used in diameters ranging from 20 to 80 cm, although larger diameters may have open bottoms. These piles can serve as end-bearing piles, friction piles, or a combination of both. One of the advantages of closed-end pipe piles is their flexibility in terms of length. They can be easily adjusted to suit the specific requirements of a project, making them a preferred choice when different pile lengths are needed. The interior of these piles can be either empty or filled with concrete. In cases where the concrete inside the pile is subjected to tensile forces, such as uplift or lateral loads, or when the pile has an extended free length, reinforced concrete is necessary to provide the required strength and stability. The choice between using empty or concrete-filled piles, as well as the decision to reinforce the concrete or leave it unreinforced, depends on the specific project requirements, soil conditions, and structural considerations.

14.3.6.2 Open-End Pipe Piles

Open-end pipe piles are specifically employed in challenging driving conditions, such as difficult soil types or when deep penetration is required. These piles are available in diameters ranging from 20 to 400 cm. Larger-diameter piles, typically with diameters exceeding 90 cm, are referred to as large-diameter piles. The selection of large diameter piles is driven by their ability to withstand impacts from ships berthing, as well as considerations for overturning, seismic activity, and scouring effects. These piles are commonly used as fender systems in dolphin barriers and offshore platforms, providing protection and stability during berthing. They are also utilized as primary foundations for large river bridges and other structures that experience significant lateral loads. Open-end pipe piles are typically driven into hard sandy soils, or bedrock to ensure proper bearing capacity. Alternatively, they can function as friction piles in longer lengths, relying on the surrounding soil for support. During the driving process, it is possible for open-end piles to experience a phenomenon known as plugging. This occurs when soil fills the pile tip, effectively converting it into a closed-end pile. As a result, the tip resistance of the pile significantly increases, providing additional load-bearing capacity.

14.3.6.3 Composite Pipe Piles

In some cases, pipe piles, both open-ended and closed-ended ones, are filled with concrete after installation, aiming to enhance structural stiffness and capacity, particularly under lateral loading and corrosion prevention. These piles are recognized as composite pipe piles.

14.3.7 Precast Concrete Piles

Precast concrete piles are manufactured at specific lengths and then transported to the construction site. It is crucial to handle them with care during transportation to avoid subjecting the concrete to excessive bending or other forms of stress.

These piles are available in square, hexagonal, and octagonal cross-sections, either in empty or full form. As the dimensions of the piles increase, they are often manufactured in an empty form to reduce weight and facilitate transportation.

Precast piles can be pre-stressed by incorporating high-strength pre-stressed cables. These cables typically have a final resistance of about 1800 kN/m^2 (18,000 kg/m^2). In the manufacturing process, the pre-stressed cables, with strengths ranging from 900 to 1300 N/mm^2, are tensioned inside the pile mold and then filled with concrete. Once the concrete cures, the pre-stressed cables are cut, resulting in the development of compressive forces within the pile section.

One of the significant advantages of precast, pre-stressed piles is their enhanced durability compared to cast-in-place concrete piles. The constant pressure from the pre-stressed cables contributes to better resistance against corrosion and weathering, resulting in improved long-term performance. Unlike in situ concrete piles, pre-stressed concrete piles have inherent protection against corrosion in most cases, eliminating the need for additional corrosion treatment methods.

Pre-stressed concrete piles may be more susceptible to damage during driving in stiff soils and layers compared to precast concrete piles. This is because the application of pre-stressing force can reduce the compressive strength of the concrete. The reduction in compressive strength may result in higher vulnerability to cracking or spalling when encountering challenging soil conditions during installation.

To achieve longer lengths, precast concrete piles are often manufactured in shorter segments, typically up to 15 meters long, and then connected or patched together using suitable methods to form longer piles. On the other hand, pre-stressed concrete piles can be manufactured in longer lengths, usually up to 40 meters, by incorporating pre-stressed cables within the concrete during the casting process. They can also be connected or patched together using suitable methods to form longer piles.

However, one challenge with precast concrete piles is the difficulty of patching or cutting the ends after driving. Unlike in situ cast piles, precast piles have predetermined lengths and configurations, making it challenging to modify their length or shape once they are installed. Therefore, careful planning and accurate determination of pile lengths are crucial during the design and manufacturing stages to ensure proper alignment and compatibility with the project requirements.

Although the length of a precast concrete pile cannot be easily changed once it is manufactured, there are methods available to extend the length of precast piles if needed. One approach is to extend the pile through its head by adding additional concrete to the top of the pile. This extension is typically done using a concrete pour, but it requires a time delay to allow the initial sections of the pile to achieve sufficient strength before the extension is added. This method ensures a continuous structural connection between the existing pile and the extension. Alternatively, special mechanical pile fittings can be used to connect vertical sections of precast piles. These fittings are designed to provide a secure and reliable connection between the sections, ensuring structural continuity along the length of the pile. In some cases, epoxy connections may also be employed to join precast pile segments. Epoxy connections involve bonding the ends of the pile sections using epoxy resin, creating a strong and durable connection between the segments.

Another type of precast pile widely used in marine structures is the centrifuge method's concrete piles, also known as spun piles. These piles are constructed using the centrifugal method, as shown in Figure 14.7. The construction process begins by placing a reinforcement cage inside a mold. Then, concrete is poured into the mold, and the mold is closed. Special equipment is used to rotate the mold at high speeds. As a result of the centrifugal force acting on the concrete, it

(a) (b)

Figure 14.7 (a) Square precast concrete piles and (b) precast concrete piles made by the centrifugal method (Hannigan et al. 2016 / U.S. Department of Transportation).

takes the shape of a pipe, forming the concrete pile. During the rotation process, excess water is separated from the concrete, resulting in a concrete mixture with a low water-to-cement ratio (typically between 26% and 30%). This reduction in water content improves the concrete's durability, reduces permeability, and increases its strength against corrosion. As a result, these piles are particularly well-suited for use in high-corrosion environments. After the initial concrete setting, the pile is transferred to an Autoclave machine for further processing. The Autoclave provides high-quality and controlled curing conditions, ensuring the development of strong and durable concrete.

14.4 Drilled Shafts (Bored Piles): Applications and Construction

Bored piles, also known as drilled piles or cast-in-place piles, are a type of non-displacement pile that is created by drilling a borehole using an auger. Once the borehole is prepared, a reinforced concrete foundation is constructed inside the hole. The load-bearing capacity of bored piles is achieved through a combination of shaft friction along the pile and resistance at the pile tip. During the drilling process, the soil under the pile tip is disturbed. In order to mobilize the tip resistance, a significant amount of movement of the pile is required, which is more than the permissible displacement in some cases. Despite the potential challenges associated with drilling and the disturbance of soil under the pile tip, bored piles offer several advantages. They can be constructed to reach significant depths and are suitable for various ground conditions. Bored piles provide reliable load-bearing capacity and are commonly used in a wide range of construction projects, including high-rise buildings, bridges, and other structures that require deep foundations.

Construction of bored piles has its challenges and special considerations. Below is a list of the major challenges.

1) Bored piles are often required to be excavated to considerable depths, penetrating through various materials ranging from soft soils to very hard rock formations.
2) It is crucial to ensure the stability of the borehole in caving soils under groundwater conditions. Proper measures should be taken to prevent the collapse or sloughing of soil during the drilling process.
3) The borehole should be filled with reinforced concrete in a manner that provides adequate confinement and restraint to the surrounding layers and tip of the pile. This ensures efficient transfer of large lateral and axial loads to the underlying bearing layer.
4) A completed bored pile should be designed and constructed as an integral and efficient structural element capable of withstanding the imposed structural loads, including compressive, tensile, and bending forces.

Bored piles are most efficient when they are located in strong bearing layers, as these layers provide optimal support for the piles. In situations where driven piles cannot be installed due to various reasons, bored piles can be a suitable alternative. They are particularly useful in areas with dense soil that can resist scouring or in locations where rock is present beneath collapsible soil layers. Concrete piles are generally easy to install in dense and cohesive soils. However, in very soft clay or loose granular soils with significant water flow, the execution of concrete piles without the use of drilling mud can be challenging or even impossible. The presence of drilling mud helps stabilize the borehole and prevents collapse or excessive water flow during the installation of the piles. Figure 14.8 presents a general view of a drilled shaft's construction process.

One of the key advantages of utilizing bored piles is the ability of the engineer to inspect the underlying subsurface materials, which helps determine the nature of the bearing layer. This advantage allows for better planning and design of the piles. However, it is important to note that bored piles are sensitive to the sublayer conditions. The success of their implementation is greatly influenced by the specific site location and the characteristics of the subsoil. Factors such as the presence of water, the stability of the borehole, and the properties of the surrounding soil can affect the performance of bored piles.

Table 14.2 provides a summary of the advantages and limitations of bored piles, highlighting their benefits in terms of subsurface inspection but also acknowledging their sensitivity to sublayer conditions. It is crucial to consider these factors and conduct a thorough geotechnical analysis when considering the use of bored piles in a construction project.

14.4.1 Construction Methods

The implementation of bored piles can be done using three different methods: dry, casing, and wet methods. Among these, the dry method is considered the most favorable condition for the installation of bored piles. In the dry method, the borehole is drilled to the desired depth in soils and rocks above the groundwater level. Care is taken during the drilling process to

Figure 14.8 Drilled shafts construction: (a) boring a hole, (b) lifting and placing the reinforcement cage, (c) placing the concrete, and (d) finishing the top, including the connection to the structure (FHWA 2018 / U.S. Department of Transportation).

Table 14.2 Advantages and limitations of bored piles (Eslami et al. 2020).

Advantages	Shortcomings
Straightforward implementation in cohesive materials and most rock	Sensitivity of construction to groundwater or challenging drilling conditions
Appropriate for different ground conditions	Probable influence of construction method on the performance of drilled shafts
Visual assessment of bearing layer	No direct measurement of axial resistance through installation, similar to driven piles
Probable extremely high axial resistance	Complicated and costly high axial resistance load tests
Considerable bending strength	Careful construction of reinforced concrete members for structural integrity
Small footprint for single shaft foundation with no need for a pile cap	Single piles are statically indeterminate so they must have a high degree of reliability
Low noise and vibration, appropriate for urban areas and adjacent to existing structures	Experienced, capable contractor for construction, often performed as extraordinary work of subcontractors
Can penetrate below scour zone into a stable, scour-resistant formation	Inefficient in deep soft soils without appropriate bearing layers
easy adjustment to accommodate variable conditions encountered in production	Comprehensive site investigation comprising an evaluation of conditions affecting construction; the potential for differing site conditions to impact costs, and a schedule is required

prevent sudden collapse of the borehole. Sufficient time is allowed for the borehole to stabilize before proceeding with further construction activities. By using the dry method, the stability of the borehole can be maintained, allowing for efficient and accurate installation of the bored piles. This method is commonly employed when the subsoil conditions are suitable and do not pose significant risks of collapse or instability.

The stability of the borehole during the installation of bored piles depends on the characteristics of the soil and its layering. In cases where the soil profile consists of highly fissured or cracked clay, with thicker layers of silt or sand ranging from 2.5 to

5 cm, there is a risk of instability, scaling, or excessive seepage. The casing method is employed to address these challenges. In the casing method, a steel casing is inserted into the borehole to provide support and prevent the collapse of the unstable soil layers. The casing helps maintain the integrity of the borehole and prevents the entry of water or soil particles, thereby ensuring the stability of the well.

During the concreting of a bored pile, it is important to ensure proper water control within the well to maintain the integrity and quality of the concrete. Ideally, there should be no more than 7.5 cm of water on the bottom of the well, and the inward flow rate should not exceed 30 cm/h. If seepage occurs during the concreting process, steps should be taken to manage the water effectively. It is recommended to drain as much water as possible from the well using an in-well pump before discharging the concrete. This helps to reduce the water level and minimize the risk of excessive water mixing with the concrete. However, if the seepage exceeds the acceptable amount and the well becomes flooded with water, it may be necessary to switch to the wet method of concreting. In the wet method, the concrete is pumped into the well under controlled conditions to ensure proper placement and consolidation.

The dry well execution process is illustrated in Figure 14.9:

a) The drilling tool is used to create the well by removing soil or rock from the ground. It can be a drill bit attached to a drilling rig or other equipment suitable for the specific ground conditions.
b) After drilling the well, debris such as soil, rocks, or drilling mud may accumulate at the shaft base. It is important to clean and remove these debris to ensure a clear and stable base for the pile.
c) The reinforcement cage, consisting of steel bars or rebars, is inserted into the well. The cage provides structural reinforcement to the bored pile, enhancing its load-bearing capacity and overall strength.
d) The concrete is then poured into the well using a falling concrete shot or similar method. A guiding tool is used to direct the flow of concrete to the center of the well, ensuring proper placement and consolidation of the concrete around the reinforcement cage.

The casing method is employed in sites where there are collapsible soils or excessive deformation risks. It is also used to drill wells in water or permeable layers until they reach a stable dry layer unless there is a stable bearing layer present. The casing method typically requires the addition of water or drilling lubricants to facilitate the process.

Casing pipes, which are often made of steel, are inserted into the borehole during drilling. These casings provide temporary support and prevent the collapse of unstable soil layers. Once the concrete is poured and cured, the casings are then extracted from the bored piles. In certain cases, a permanent casing may be utilized as an integral part of the design of bored piles. This permanent casing serves as a structural mold or element, providing additional strength and stability to the pile. Figure 14.10 shows the schematic of the casing method.

The wet method is employed when soil conditions require the well to remain filled with drilling fluid during the drilling and construction processes. This method ensures the stability of the borehole and prevents the collapse of the well. The drilling fluid used can be either water or a specially designed slurry, depending on the stability of the well. There are two types of slurry commonly used in bored piles: mineral slurry and polymer slurry. Bentonite, a type of mineral slurry,

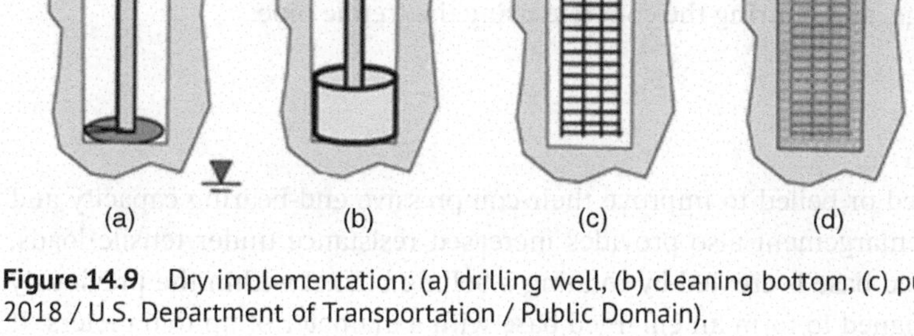

Figure 14.9 Dry implementation: (a) drilling well, (b) cleaning bottom, (c) putting reinforcement, and (d) pouring concrete (FHWA 2018 / U.S. Department of Transportation / Public Domain).

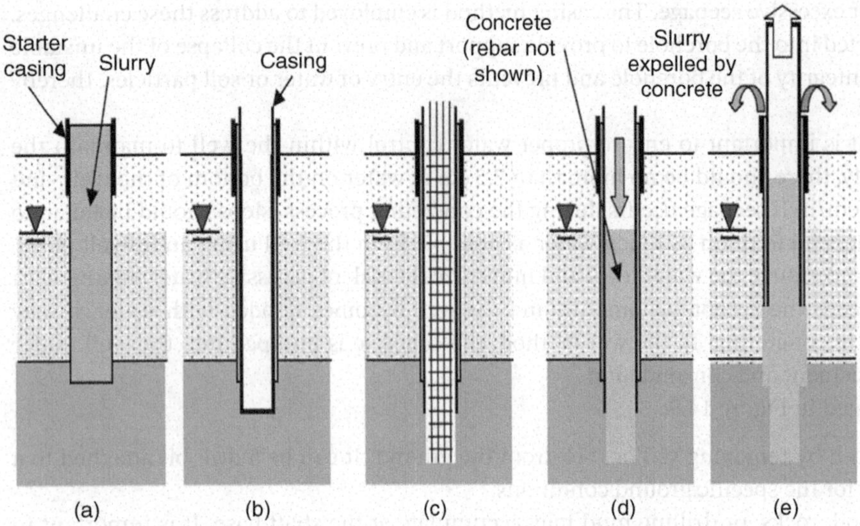

Figure 14.10 Stages of casing method implementation (FHWA 2018 / U.S. Department of Transportation / Public Domain).

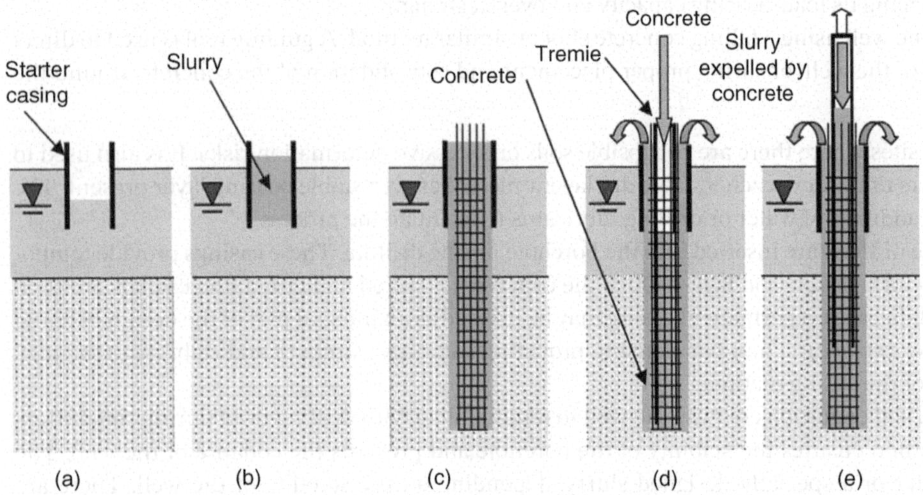

Figure 14.11 Stages of wet method implementation (FHWA 2018 / U.S. Department of Transportation / Public Domain).

is widely used in the construction of bored piles. It helps maintain the stability of the well by providing support to the surrounding soil.

The execution of the wet method involves several steps. First, a primary or surface casing is installed from the ground surface. This casing provides initial support and helps prevent the collapse of the well. Next, the reinforcement cage is placed inside the well to provide strength to the pile. Finally, the concrete is poured into the well using a tremie pipe. The tremie pipe ensures that the concrete is placed from the bottom of the well, displacing the drilling fluid or slurry as it fills the well. Figure 14.11 illustrates the typical execution process of the wet method for bored piles. It demonstrates the sequence of installing the casing, placing the reinforcement cage, and pouring the concrete using the tremie pipe.

14.4.2 Enlarged Base (Belled) Drilled Shafts

In certain cases, the base of drilled shafts is expanded or belled to improve their compressive end-bearing capacity and enhance the utilization of the toe resistance. This enlargement also provides increased resistance under tensile loads, improving the pullout capacity of the shaft. The belled base is created by rotating a belling bucket within the previously drilled straight-sided shaft. The belling bucket is designed to form an enlarged base with a diameter of up to 6 meters. It

(a) (b)

Figure 14.12 Enlarged base drilled shaft: (a) schematic view, (b) a tool for boring (Viggiani et al. 2012 / Spon Press).

is necessary to have a minimum shaft diameter of 800 mm to accommodate the belling bucket. However, it is important to note that this technique is suitable only for relatively stiff cohesive soils (Viggiani et al. 2012).

Figure 14.12a illustrates a schematic representation of a belled drilled shaft, showing the enlarged base at the bottom. Figure 14.12b shows the process of using a rotating belling bucket to create the enlarged base within the previously drilled shaft. The belling bucket is advanced into the soil while rotating, forming the desired enlargement. This technique allows for an increased bearing capacity and improved performance of the drilled shaft in specific soil conditions, enhancing its overall effectiveness as a deep foundation element.

14.4.3 Post-Grouting

In the 1960s, research and investigations began on enhancing the toe capacities of drilled shafts through a technique called post-grouting. This method involves injecting high-pressure grout beneath the toe of the drilled shaft after the concrete has cured. The aim of post-grouting is to densify the soil and debris that remain after the drilling process, effectively preloading the soil at the toe and improving its bearing capacity.

The process of post-grouting typically involves the following steps:

1) Installation of grout pipes within the conventional reinforcement cage, extending to the bottom of the shaft.
2) Once the concrete in the shaft has cured, high-pressure grout is injected through the grout pipes into the bottom of the shaft. The grout penetrates the surrounding soil, compacting it and increasing its density. This process helps to improve the end-bearing capacity of the shaft.

Studies conducted by Gouvenot and Gabiax in 1975 showed that post-grouting large-diameter shafts resulted in significantly enhanced ultimate load capacities, with load capacities up to three times greater in sands and clays. Since then, post-grouting methods have become widely adopted globally.

Figure 14.13 illustrates the concept of post-grouting, showing the injection of high-pressure grout into the toe of the drilled shaft. This process helps to improve the performance and load-bearing capabilities of the shaft, particularly in terms of service displacement limits. By densifying the soil at the toe, the shaft can effectively distribute and support the applied loads within the desired limits.

This post-grouting method is particularly suitable for projects in urban areas where vibrations caused by traditional pile driving methods are not tolerated. It is also beneficial in situations where the soil conditions would have necessitated extremely long drilled shaft lengths to achieve significant end-bearing contributions. By applying post-grouting, the overall efficiency and load-bearing capacity of the drilled shafts can be enhanced, making them a viable alternative in such challenging soil and construction conditions.

The performance of a drilled shaft in terms of end bearing and shaft friction is often limited due to certain factors that affect the development of bearing capacity. These factors lead to substantial displacements in the shaft, which are necessary to fully mobilize the bearing capacity, particularly in sands and clays. Three main mechanisms contribute to these displacements:

Figure 14.13 Equipment and method of grouting from the tip (FHWA 2018 / U.S. Department of Transportation).

1) **Strain Incompatibilities:** Strain incompatibilities usually occur between the end bearing and side friction, relevant to service displacement criteria. The ultimate side frictional component develops with relatively small shaft displacements compared to the required displacements to mobilize the ultimate end bearing. Development of the side friction component can be 50% of ultimate at displacements of approximately 0.2% of the shaft diameter (D) and completely developed within 0.5 to 1.0% D. On the other hand, end bearing component mobilization can be 50% mobilized at 2.0% D and completely mobilized within 10 to 15% D. The end bearing component needs 10 to 30 times more shaft displacement to mobilize the same percentage of its ultimate value than the side shear component. The side friction is strained well beyond its ultimate strength and is forced into a residual state by the time the end-bearing capacity is realized. Furthermore, the service load deflection
2) **Disturbance of the Shaft Tip Zone:** In cohesionless soils, the shaft toe zone is frequently disturbed during construction. This can occur due to soil stress relaxation caused by overburden excavation, inflow of groundwater due to insufficient hydrostatic head, or rapid removal of the excavation tool. The disturbance in the shaft toe zone is challenging to remove and requires displacements that exceed the allowable service limits to mobilize the end-bearing capacity. This issue is particularly prominent in less competent cohesionless soils, such as loose sands.
3) **Soft debris at the Excavation Bottom:** Construction techniques may leave soft debris at the bottom of the excavation. Factors contributing to this include inadequate shaft bottom cleanliness, nonuniform distribution of tip debris leading to a decreased shaft area bearing on the soil, excessive sand content in the drilling fluid, prolonged time for placing the reinforcement cage and concreting, and drilling fluid deposits at the bottom. The presence of soft debris at the toe of the shaft requires excessive displacements to mobilize the end bearing, which may not be evident in an otherwise clean excavation.

Overall, these factors result in the need for significant shaft displacements to fully develop the bearing capacity in drilled shafts, particularly in sands and clays. Addressing these challenges during the construction process is crucial to optimize the performance of drilled shafts and ensure their reliable load-bearing capacity.

According to Loehr et al. (FHWA 2018), two mechanisms, known as "pre-mobilization" and "ground improvement," contribute to enhancing the performance of drilled shafts across various ground conditions when post-grouting is implemented.

The pre-mobilization mechanism involves "stiffening" the load–settlement response of the shaft, enabling it to mobilize more resistance for a given settlement. While pre-mobilization itself does not increase the ultimate shaft resistance, it often increases the nominal design resistance by allowing for greater resistance mobilization at settlements corresponding to a chosen failure criterion. In other words, pre-mobilization improves the load-bearing behavior of the shaft, leading to improved performance.

The term "ground improvement mechanism" refers to beneficial changes in the ground below the shaft tip that increase the available tip resistance. These changes can be achieved through several means, such as densification of the ground beneath the tip, grout penetration into the ground, enlargement of the shaft tip, or a combination of these factors. These

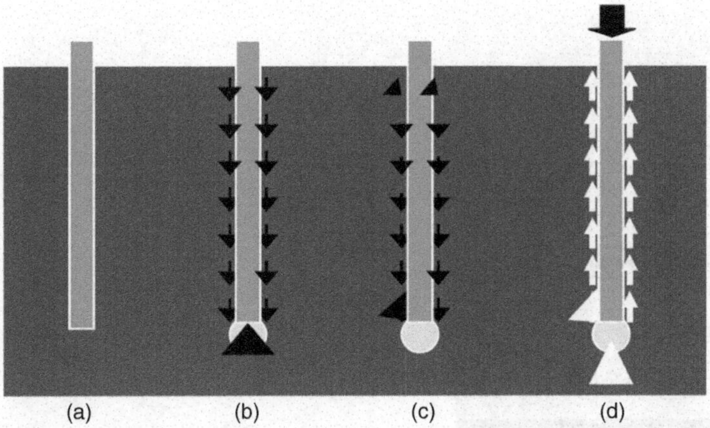

Figure 14.14 Post-grouting process: (a) shaft constructed, (b) grout pressure at tip, reacts against side shear, (c) some relaxation occurs, but some preload remains, and (d) structural loads on shaft engage side and tip resistance (FHWA 2018 / U.S. Department of Transportation / Public Domain).

alterations effectively enhance the capacity of the ground to provide resistance at the shaft tip, resulting in improved performance of the drilled shaft.

The post-grouting procedure, as shown in Figure 14.14, involves injecting grout into the shaft after the initial concrete curing. This process helps facilitate the pre-mobilization and ground improvement mechanisms, leading to enhanced load-bearing capacity and overall performance of the drilled shaft.

14.5 Special Piles

In recent years, the implementation of special piles has been on the rise due to the challenges associated with bored piles, especially in soft and problematic soils and high groundwater sites. Special piles offer several advantages, such as low noise and vibration during installation, the ability to be constructed without the need for massive equipment, and the potential for soil improvement. The list of special piles includes continuous flight auger piles (CFA), drilled displacement piles (DDP), helical piles, and self-expanded piles, along with post-grouting, micro, and mini piles. These special piles will be discussed in the following sections.

These piles have unique characteristics and exhibit different performance compared to conventional piles. The increasing popularity of special piles is driven by their ability to address unique and specific site requirements and provide reliable and cost-effective foundation solutions.

14.5.1 Continuous Flight Auger Piles (CFA)

CFA piles, also known as auger pressure grout (APG) or auger cast-in-place (ACIP) piles, were invented in the 1940s and have since become a popular choice for foundation construction. These piles are installed using a continuous flight auger, which drills into the soil and creates a well for the pile. The installation process begins with the continuous flight auger drilling into the ground under vertical force and torque. The soil excavated from the well is contained within the flight parts of the auger, ensuring the stability of the well. Once the desired depth is reached, the auger is partially lifted, leaving space for concrete or grout to be poured into the well. The soil within the flight parts and the concrete or grout inside the well prevent the well from collapsing. After the concrete or grout is placed, a reinforcement cage is typically installed in the upper portion of the pile. This reinforcement cage is usually extended to a depth of 10 to 15 meters, as this is where the bending stresses are higher. Below this depth, the reinforcement is often omitted, as the bending stresses are lower and the installation of the reinforcement becomes easier.

CFA piles have diameters ranging from 0.3 to 1 meter and lengths of 30 to 35 meters. The low displacement nature of the installation process, where the surrounding soils are moved radially, makes CFA piles a suitable choice when low displacement piles are required. These piles offer a cost-effective and efficient solution for various construction projects. Figure 14.15 shows the installation of CFA piles.

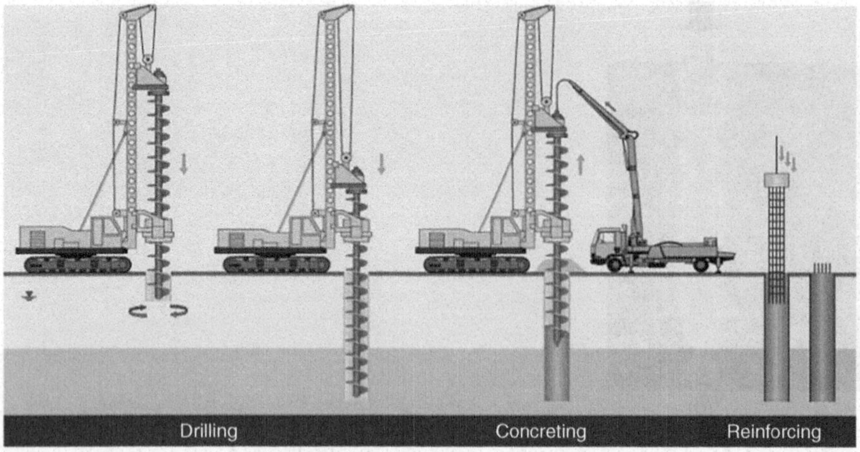

Figure 14.15 CFA pile installation process (Brown et al. 2007 / U.S. Department of Transportation / Public Domain).

The main distinction between CFA piles and bored piles lies in the method of installation and the limitations of the respective techniques. CFA piles do not require drilling fluids or casings to stabilize the well, resulting in a faster drilling speed compared to bored piles. However, CFA piles require higher torque during installation, making them more suitable for weaker rock layers and soils.

It is important to consider the specific geotechnical conditions of a site when selecting between CFA piles and bored piles, as well as the load requirements and project specifications. Consulting with a geotechnical engineer or foundation specialist is recommended to determine the most suitable pile type for a given project.

CFA piles are indeed well-suited for certain soil conditions and specific project requirements. The characteristics and conditions that make CFA piles suitable include:

1) **Medium to High-Density Clay:** These soils provide sufficient pile sleeve friction, allowing for significant bearing capacity. CFA piles can effectively transfer loads through the frictional resistance along the pile shaft in these soil types.
2) **Cement Sands and Weak Stones:** CFA piles can penetrate these types of soils effectively, making them suitable for projects where such soil conditions are encountered.
3) **Sedimentary Soils:** Silty and clay soils with a small amount of cohesion are favorable for CFA pile installation. These soils allow for fast and economical pile installation using CFA techniques.
4) **Medium to High-Density Silty Sand or Well-Graded Sand:** CFA piles can be used in these soil types, particularly if the groundwater level is low enough. These soils provide suitable conditions for efficient CFA pile installation.
5) **Speed of Installation:** CFA piles can be installed quickly, especially when favorable geotechnical conditions are present, and if the necessary equipment and resources are available at the project site.
6) **Large Number of Piles:** When a project requires a significant quantity of piles, CFA piles can be a suitable choice due to their high production rate. This can help expedite the construction process.
7) **Projects Executed Under a Roof:** Special equipment can be used to install CFA piles in confined spaces, such as projects carried out under a roof. If the geotechnical conditions allow, CFA piles can be more economical compared to other alternatives like micro-piles.
8) **Pile sleeve Friction Dominates:** In projects where the majority of the load can be carried through pile sleeve friction due to favorable geotechnical conditions and the absence of hard layers or stones in the soil, CFA piles can provide a cost-effective solution.

It is important to conduct a thorough geotechnical investigation and consult with a geotechnical engineer or foundation specialist to assess the suitability of CFA piles for a specific project based on the site conditions, load requirements, and other project considerations.

14.5.2 Drilled Displacement Piles (DDP)

Drilled displacement piles (DDP) are a type of deep foundation that is installed by drilling into the ground and displacing the soil rather than removing it. Unlike other types of piles, such as bored or driven piles, DDPs do not rely on soil removal

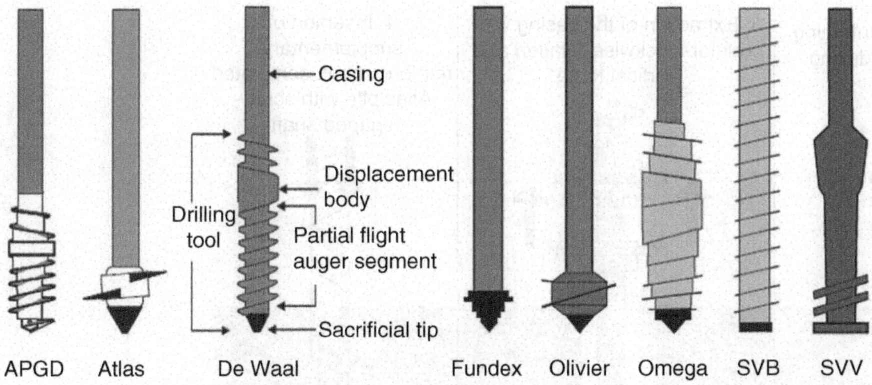

Figure 14.16 Types of drilling equipment of DDPs (adapted from De Cock and Imbo 1994).

for installation. DDPs have gained popularity due to their advantages, such as low noise and vibration, improved concrete corrosion resistance, and environmental compatibility.

The process of installing DDPs involves the use of specialized drilling equipment, which typically consists of a drilling tool with a sacrificial tip and a larger diameter displacement body or casing. The sacrificial tip is attached to the drilling tool and helps create a borehole by breaking and loosening the soil as it is rotated and advanced into the ground. As the drilling tool progresses, the displacement body or casing follows, displacing the surrounding soil radially outward. Once the desired depth is reached, the sacrificial tip is separated from the drilling tool, leaving behind a void space. This void is then filled with grout or concrete, which is typically pumped through the drilling tool. The grout or concrete fills the space and forms the body of the pile. The result is a pile that has displaced the surrounding soil and is fully encased within the grout or concrete. The displacement of the soil during installation provides several benefits. It densifies the soil surrounding the pile, increasing the side friction and improving load transfer capacity. It also helps improve the pile's resistance to axial and lateral loads. Additionally, the displacement process helps minimize soil disturbance, reducing the potential for settlement and improving the overall stability of the pile.

DDPs typically have diameters ranging from 270 to 610 mm and can reach lengths of up to 40 meters. The specific design and components of DDPs may vary depending on the manufacturer and the drilling equipment used. Additionally, DDPs are commonly used in a variety of soil conditions, including cohesive soils, non-cohesive soils, and mixed soil profiles. They are suitable for tensile, compressive and moment loads, and can be designed to meet specific project requirements.

There are many differences between the drilling equipment used for DDPs. In general, these pile drilling equipment include one or more of the components introduced below (Figure 14.16): The soil displacement body (a part with a higher diameter that moves the soil around the pile) is shown in Figure 14.16 of this section as a displacement body. The drilling section with a partial flight auger is also shown in Figure 14.16. A specially designed sacrificial tip is attached to the end of the digging tool.

It is important to note that the installation of DDPs requires specialized equipment and experienced contractors to ensure proper execution and achieve the desired performance. Various types of drilling equipment are available, such as Atlas, Berkl, Fundex, Oliver, DeWaal, Omega, and Bauer, each with its own installation techniques and equipment. Figure 14.16 shows the different drilling equipment provided by several companies.

14.5.2.1 Types of Drilled Displacement Piles

Drilled displacement piles (DDPs) can be classified into different categories based on their performance and pile shape after installation. The four main categories you mentioned are:

1) **DDPs with a Screw-Shaped Body:** Piles such as Atlas, Oliver, and Screwsol feature a screw-shaped body that is formed during the pile installation process. The drilling tool is rotated clockwise to create the screw-like path in the soil, and once the desired depth is reached, the drilling tool is extracted with counter-clockwise torque. These screw-shaped piles offer advantages in terms of their improved resistance against tensile loading compared to other types of piles. They can be used in a wide range of soil conditions, similar to driven piles. However, their installation in dense sandy soils may pose challenges. Figure 14.17 illustrates the steps involved in the execution of Atlas piles, serving as an example of the implementation process for screw-shaped drilled displacement piles.

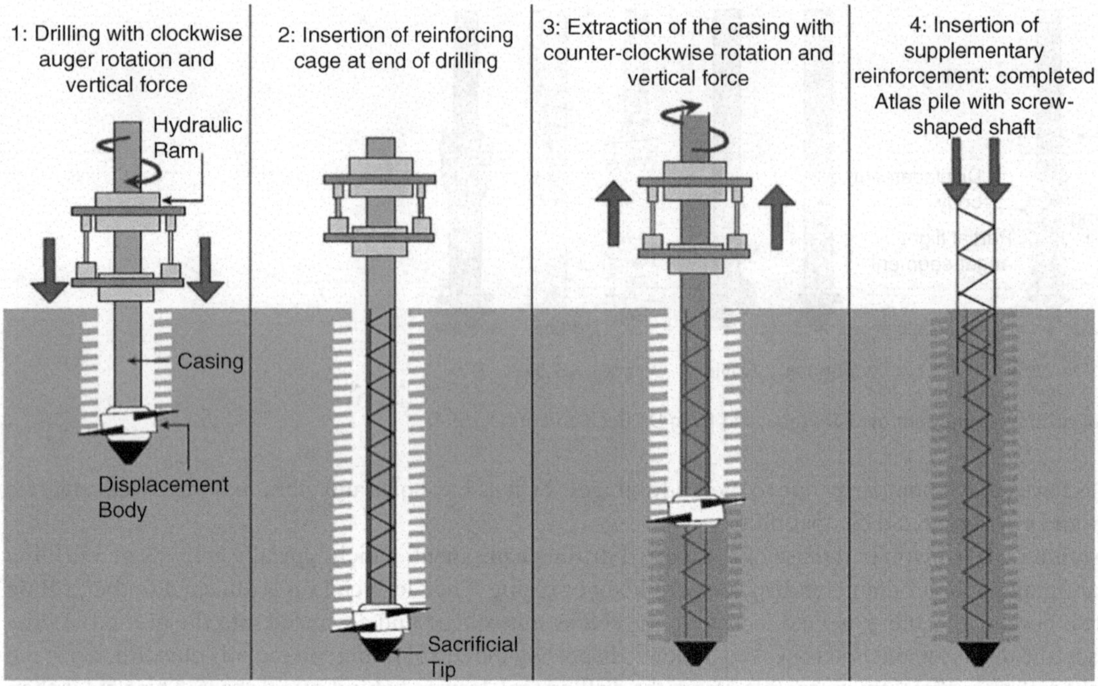

Figure 14.17 Installation procedures and equipment Atlas drilled displacement piles (Prezzi and Basu 2005).

2) **DDPs with Soil Displacement Body in Drilling Tools:** Piles in this category, such as DeWaal and Omega, have a specific soil displacement body integrated into the drilling tool. As the drilling tool advances, the displacement body helps to displace the soil and create a void for the grout or concrete to be injected. These piles take on a smooth, cylindrical shape after installation. One of the distinguishing features of this category is that there is no spoil generated during the installation process. As a result, these piles are considered displacement piles, as they displace the soil rather than removing it. This can be advantageous in certain ground conditions where spoil removal is challenging or undesirable. Figure 14.18 illustrates the steps involved in the installation of DeWaal piles, serving as an example of the implementation process for drilled displacement piles with soil displacement bodies. The installation process for other piles in this category would be similar. In Figure 14.19, different parts of these piles are shown.
3) **DDPs with Continuous Flight Augers:** This category includes piles like SVB, which utilize continuous flight augers during the drilling process. SVB piles are an evolution of the CFA model, but they have a larger diameter for the central shaft. The installation method for SVB piles is similar to that of CFA piles, where the auger is rotated into the ground to create the borehole, and then concrete or grout is injected through the hollow stem of the auger as it is slowly withdrawn. One key difference between SVB piles and CFA piles is the larger diameter of the central shaft in SVB piles. This results in a lower spoil generation during installation, as the soil is moved more radially rather than being carried up the auger flights. This radial displacement of soil contributes to increased bearing capacity for SVB piles compared to CFA piles.
4) **DDPs with Simple Cylindrical Drilling Tools:** Piles such as Fundex and steel drilled displacement piles fall into this category. They employ a simple cylindrical drilling tool for creating the borehole, with the soil displacement occurring primarily through the rotation and advancement of the tool. **Fundex** piles are characterized by a cylindrical casing with a tip at the bottom. The installation process involves rotating the casing into the ground while applying downward pressure to cut through the soil. The cylindrical casing helps to compress the soil radially and displace it to the sides as the pile is installed. Once the desired depth is reached, the casing is filled with concrete, and reinforcement can be added if necessary. The Fundex pile remains in the ground as a load-bearing element. **Drilled displacement steel** piles, on the other hand, are specifically designed piles that have gained popularity in North America over the past two decades. These piles consist of a cylindrical shaft with a specially designed tip. During installation, torque is applied to the pile, causing the cutting tip to move downward and compress the soil radially. The displacement of soil is completed by the cylindrical part of the pile. Once the pile is installed, it is filled with reinforced concrete, and the drilling tool remains in the ground. This differs from other drilled displacement piles, where the drilling tool is typically withdrawn after installation.

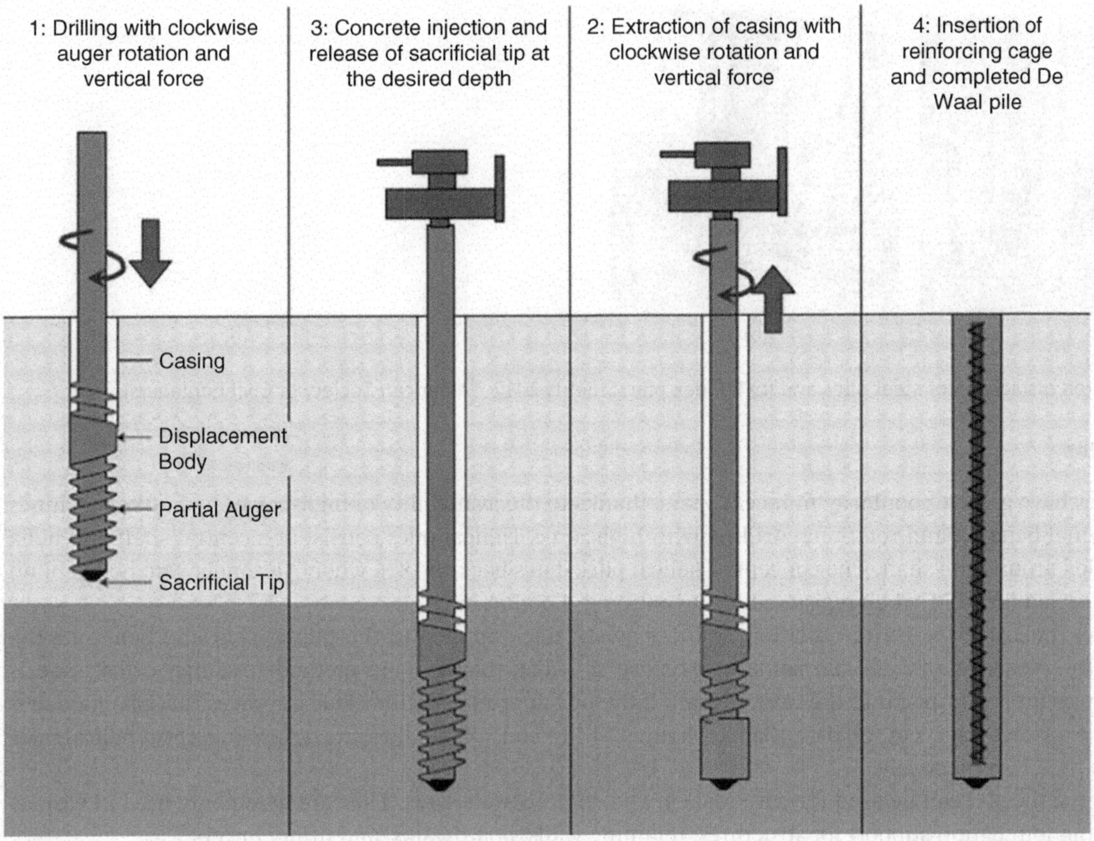

Figure 14.18 Installation procedures and equipment DeWaal drilled displacement piles (Prezzi and Basu 2005).

Figure 14.19 Different parts of drilled displacement piles with soil displacement body (Prezzi and Basu 2005).

Both Fundex piles and drilled displacement steel piles offer advantages such as low noise and vibration during installation, as well as the ability to displace soil without the need for spoils removal. These piles can be used in various soil conditions and provide efficient load-bearing capacity. Two Fundex piles and drilled displacement steel piles are among these categories, as shown in Figure 14.20.

Each category of DDPs offers specific advantages and may be more suitable for certain ground conditions or project requirements. It is important to consider the specific characteristics and performance of each type when selecting the appropriate DDP for a given project. Consulting with experienced geotechnical engineers and foundation specialists can help determine the most suitable DDP type for a particular application.

Figure 14.20 (a) Drilled displacement steel piles and (b) Fundex pile (Shields 2017 / American Society of Civil Engineers).

14.5.3 Helical Piles

Helical or screw piles have gained popularity in recent years, thanks to the availability of high-power hydraulic machines that can apply torsional bending and rotational drilling with high force. Helical piles consist of a central shaft and helix plates that are spiraled around the shaft. The concept of helical piles dates back to 1836 when Alexander Mitchell, a blind architect, first constructed them. Mitchell registered his invention in London in 1833.

During installation, helical piles use the rotation and downward force applied to the pile to split and penetrate the surrounding soil. However, this process can sometimes cause soil disturbance. One method to address this issue is post-grouting, where grout is pumped into the central shaft of the pile after installation. This improves the soil characteristics and reduces the adverse effects of soil disturbance. Figures 14.21a and 14.21b illustrate different parts of helical piles and installation equipment, respectively.

Helical piles are versatile and can be used for both tension and compression loads. They are commonly used in various applications, including foundation support for structures, retaining walls, boardwalks, and utility towers.

During the installation of helical piles, a standard method involves using a hydraulic torque motor to screw the pile into the soil. The helix plates on the pile shaft penetrate the soil as torque is applied to the pile's head. It is important to maintain regular control of the pile's verticality during installation to ensure proper alignment. In addition to the rotational force, a fixed axial force is applied while inserting the helical pile into the ground. The magnitude of this axial force is determined in such a way that the pile advances approximately 80% of the helix pitch with each full rotation. This helps ensure sufficient penetration and engagement of the helix plates into the soil, providing the necessary load-bearing capacity.

Helical piles have environmental advantages that have garnered attention toward their use in geotechnical works, such as lower material consumption in making them recoverable and requiring smaller equipment during installation, thus reducing carbon emissions. A plot of the number of U.S. patents filed regarding helical piles is shown in Figure 14.22.

Figure 14.21 Helical piles: (a) different parts and (b) typical installation equipment (Perko 2009 / With permission of John Wiley & Sons).

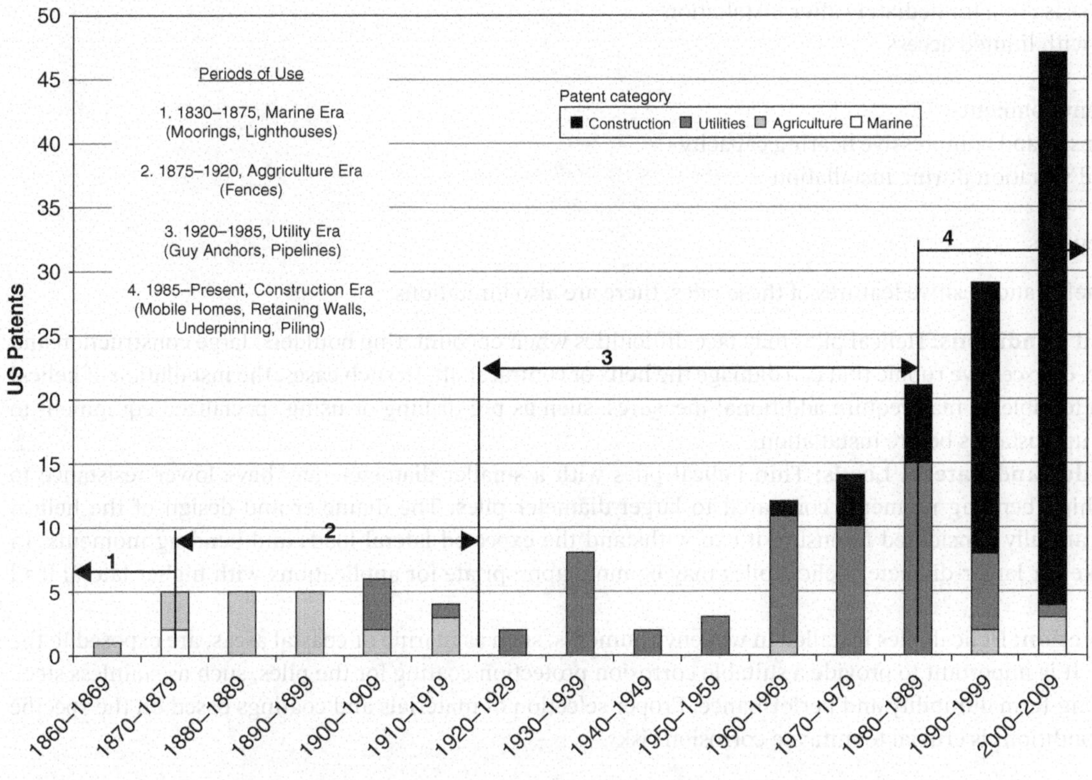

Figure 14.22 U.S. helical pile patents (Perko 2009 / With permission of John Wiley & Sons).

Figure 14.23 presents some typical applications of helical piles for offshore and inland projects. Helical piles were initially used as anchors, but they were also used to tolerate tensile loads in the soil, including transmission towers and pipelines. However, their use as compressive and lateral load structures has been expanded. Today, these piles can be exploited against tensile, compressive, and lateral loads, as well as overturning bending. The advantages of helical piles are:

- Easy to install
- Require minimal equipment
- Can be installed quickly

Figure 14.23 Typical applications of helical piles for offshore and inland projects.

- The loading process is done immediately after installation
- Suitable for areas with limited access
- Reusability
- Used in offshore environments
- The acceptable tensile and compressive bearing capacity
- Minimal noise and vibration during installation
- Affordable
- Environmental benefits

Even with the benefits and positive features of these piles, there are also limitations:

1) **Challenging Soil Conditions:** Helical piles may face difficulties when encountering boulders, large construction and demolition waste, or excessive rubble that can damage the helix or central shaft. In such cases, the installation of helical piles may not be feasible or may require additional measures, such as pre-drilling or using specialized equipment to remove or break up obstacles before installation.
2) **Thin Helical Piles and Lateral Loads:** Thin helical piles with a smaller diameter may have lower resistance to lateral loads or high bending moments compared to larger-diameter piles. The diameter and design of the helical pile need to be carefully considered to ensure it can withstand the expected lateral loads and bending moments. In some cases, thicker or larger-diameter helical piles may be more appropriate for applications with higher lateral load requirements.
3) **Corrosion Protection:** Helical piles installed in wet environments, such as marine or coastal areas, are exposed to the risk of corrosion. It is important to provide a suitable corrosion protection coating for the piles, such as stainless steel, to ensure their long-term durability and performance. Proper selection of materials and coatings based on the specific environmental conditions is crucial to mitigate corrosion risks.

14.5.4 Expander Body Piles

Bored cast in situ deep foundations with an expander body (EB) system is a method that aims to improve the geotechnical performance of the foundation while minimizing soil disturbance. The EB system involves the use of a bent steel tube, which is expanded by pressure grouting. This inflation causes the decompression of soil beneath the EB, compensated by a second grouting stage of the soil at the pile toe (Monteiro et al. 2021). The installation technique is illustrated in Figure 14.24.

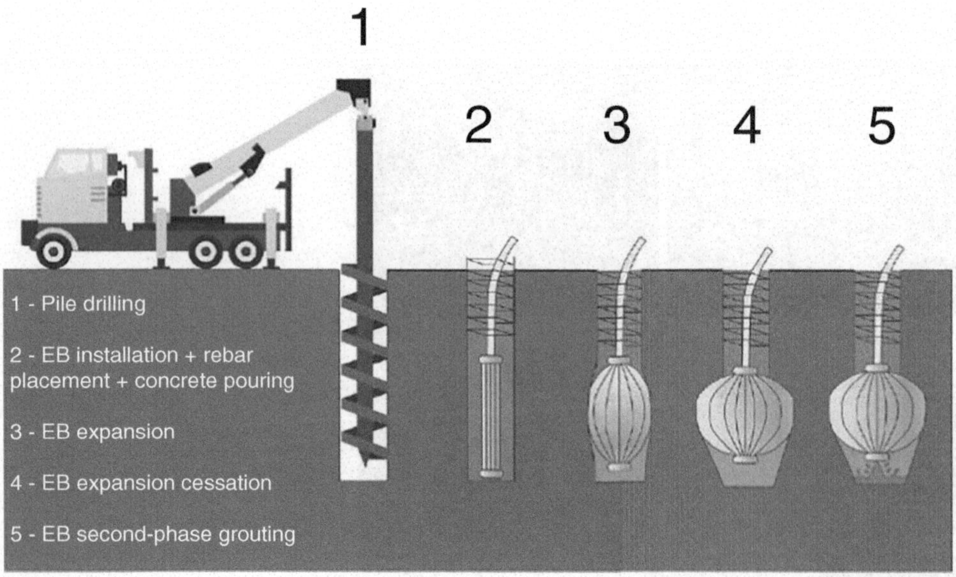

Figure 14.24 Bored cast-in situ piles equipped with the EB system installation technique (Monteiro et al. 2021 / SciELO – Scientific Electronic Library Online / CC BY 4.0).

The installation process of this system typically involves the following steps:

1) **Boring:** A borehole is first drilled into the ground using a suitable drilling technique, such as a drilling rig or auger. The diameter of the borehole is generally larger than the final desired diameter of the foundation.
2) **Expander Body Installation:** Once the borehole is drilled, the expander body, which is a bent steel tube, is inserted into the borehole. The expander body is positioned at the desired depth and orientation within the soil.
3) **Grouting:** The first grouting stage is carried out by injecting grout into the annular space between the expander body and the surrounding soil. The grout is typically a cementitious material that helps expand the expander body and exert pressure on the surrounding soil.
4) **Soil Decompression:** The expansion of the expander body causes the soil to be decompressed beneath the body. This decompression allows for increased bearing capacity and improved soil conditions around the foundation.
5) **Toe grouting:** After the soil decompression, a second grouting stage is performed at the pile toe. Grout is injected into the soil at the pile toe to further enhance the load-bearing capacity and stability of the foundation.

The EB system offers several advantages, including improved load-bearing capacity, reduced soil disturbance, and enhanced geotechnical performance. It is commonly used in projects where deep foundations are required, such as high-rise buildings, bridges, and other structures with significant vertical loads.

14.5.4.1 Wing Pile (WP)

The wing pile is made of a square-shaped shaft. The pile consists of three parts: body, helix, and joints, as well as the locking section. Figure 14.25 displays the wing pile in closed and open modes.

This pile model utilizes torque to penetrate the ground, and once it reaches the desired depth, the penetration is halted by torque. Subsequently, the helix is opened using a compressive force or a jack. This method minimizes soil disturbance during installation, making it more efficient compared to helical piles.

14.5.4.2 Self-Expanded Pile (SEP)

Helical piles are used nowadays due to their convenient installation and need for less equipment. However, disturbed soil during installation occurs due to helix penetration, which consequently reduces the tensile bearing capacity, and this is one limitation that various researchers have tried to overcome. Post-grouting is one method that has been considered to overcome this limitation. However, the impossibility of grouting in some soils led to the creation of a new species of helical piles called self-expanded piles (SEP), in which piles have a self-expanded helix (Shojaei et al. 2019). In these piles, the helix is closed during installation. After reaching the deployment depth, they are opened by applying a particular mechanism, thus reducing the amount of undisturbed soil during installation and increasing its tensile bearing capacity.

The self-expanded pile (SE) model consists of a cylinder that houses the helix internally, as depicted in Figure 14.26. The pile is initially driven into the ground using a clockwise torque force. Once it reaches the desired depth, the upper part of the pile is subjected to counter-clockwise torque, causing the blades to open and complete the deployment process.

14.5.5 Mini- and Micro-Piles

Micro-piles are generally known as bored piles with small diameters between 90 and 300 mm, and driven piles with diameters smaller than 150 mm. They can have a bearing capacity of between 5 and 100 tons when inserted by the pressure

Figure 14.25 Wing pile: (a) expanded mode and (b) closed mode (Shojaei et al. 2019 / with permission of Taylor & Francis Group).

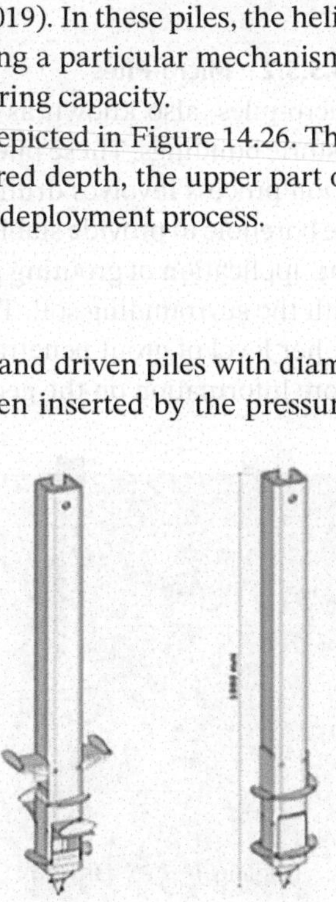

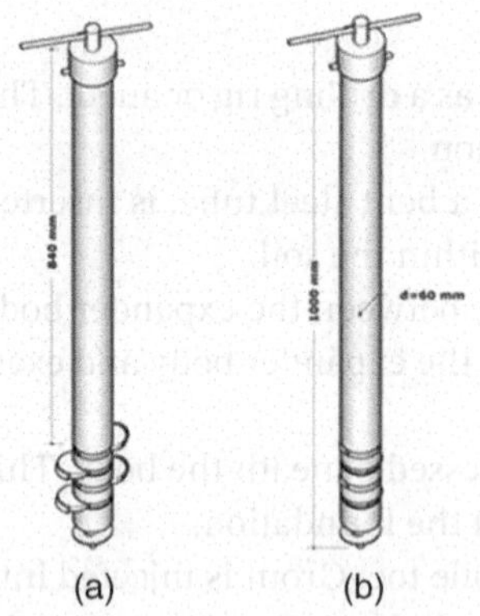

Figure 14.26 Self-expanded pile SE model: (a) open mode and (b) closed mode (Shojaei et al. 2019 / with permission of Taylor & Francis Group).

grouting method. Mini piles, bored piles with diameters between 200 and 600 mm, and driven piles with diameters of 300 mm are used when more bearing capacity is necessary, but there is limited access, and only smaller equipment than bored and driven pile equipment can be used.

14.5.5.1 Mini Piles

A pile with dimensions ranging from 200 to 600 mm is commonly referred to as a mini pile. This term encompasses various types of piles, including driven piles, bored piles, CFA piles, drilled displacement piles, and helical piles within this size range. The main advantage of mini piles is their suitability for limited access conditions, thanks to the use of lighter equipment compared to larger piles. In some cases, mini piles can be installed within height restrictions of up to 2 meters. Despite their smaller size, these piles offer cost savings and can still withstand significant loads.

These piles are used when one of the following conditions exists:

- Access to the project is limited, and using conventional installation equipment is impossible.
- Situations where piles need to be installed close to existing structures
- When the ground conditions are not suitable for drilling
- Risk of soil liquefaction during pile driving
- Repairing existing foundations.

14.5.5.2 Micro-Piles

Micro-piles, also known as mini piles, were developed in Italy in the 1950s primarily for the purpose of underpinning historic buildings. These piles are characterized by their smaller diameters, typically ranging from 90 to 300 mm. The installation process involves drilling a borehole into the soil to the desired depth. A perforated steel casing is then inserted into the borehole to provide stability. Inside the casing, reinforcement is placed, and concrete or grout is pumped into the casing. The application of grouting pressure ensures that the grout penetrates through the perforations in the casing and interacts with the surrounding soil. This grouting process helps improve the soil's properties, particularly in granular soils where a higher level of grout penetration can be achieved. Figure 14.27 shows the steps to install a micro-pile. Chapter 9 presents more information on the geotechnical design of micro-piles.

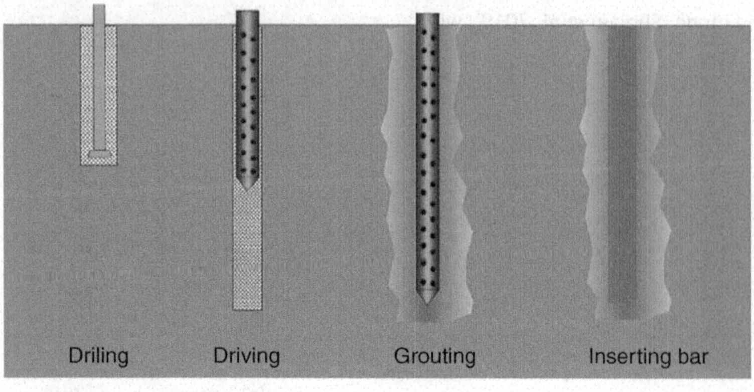

Figure 14.27 Micro-pile implementation steps.

Figure 14.28 Schematic of using micro-pile for soil stabilization.

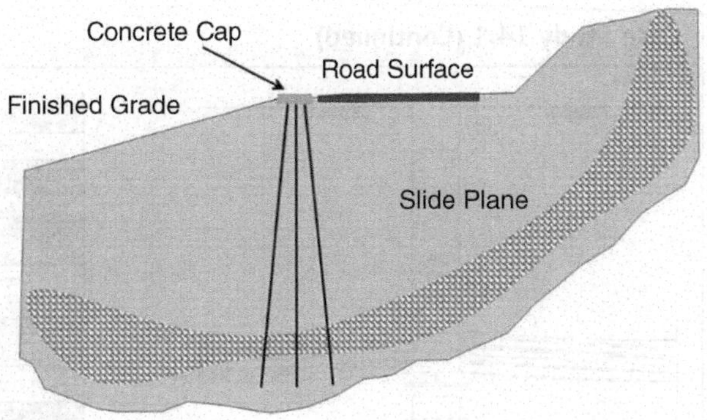

These piles can be used in most highways or bridges, as well as for soil stabilization or improvement, as shown in Figure 14.28. These piles can also be used in environments where access is restricted, such as improving the foundations of old buildings or any other environment that has access restrictions. Some applications of micro-piles include:

Micro-piles have a wide range of applications and can be used in various scenarios for soil stabilization, foundation improvement, and structural support, as shown in Figure 14.28. Some common applications of micro-piles include:

1) **Improving Old Building Foundations:** Micro-piles can be used to stabilize and strengthen the foundations of existing buildings, especially in cases where the original foundations are inadequate or have deteriorated over time.
2) **New Building Foundations:** Micro-piles can be utilized as an alternative foundation system for new construction projects, especially in areas with challenging soil conditions or limited access.
3) **Slope Stabilization:** Micro-piles can be installed to reinforce slopes and prevent soil erosion or slope failures. They provide additional stability to slopes and help mitigate the risk of landslides.
4) **Controlling Foundation Settlement:** Micro-piles can be used to control and minimize the settlement of foundations, ensuring the stability and structural integrity of the supported structures.
5) **Enhancing Bearing Capacity:** Micro-piles can increase the bearing capacity of foundations, allowing them to support higher loads. They can be designed to withstand compressive, tensile, and lateral loads, providing additional strength and stability to the foundation system.

Case Study 14.1

Performance of Drilled Shafts in Caliche Layers

Caliche is a calcium carbonate deposit that is cemented and only found in dry areas (Karakouzian et al. 2015). Because of the extremely strong bonding that prevents slippage between the caliche layers and the shaft concrete, the shaft-caliche system functions as a single unit. The conventional design and analysis procedure of drilled shaft foundations is changed by the presence of caliche layers in the soil profile, so analyzing and designing drilled shafts in caliche may not be compatible with the typical design methodologies used to evaluate ultimate skin friction. A novel method for designing drilled shafts in soil profiles with intermediate caliche layers was presented by Karakouzian et al. (2015). This method makes the assumption that caliche is a continuous plate with a significant bending capacity.

According to the findings of the load tests, the caliche layers and the shaft may function as a continuous plate attached monolithically to the shaft, as indicated in Figure 14.29a. One of the most recent Las Vegas hotel construction projects had been chosen for an O-cell test. At various depths, the project site had layers of thick, competent caliche. Based on the boring data the upper caliche layer is constant over the site. In most borings, this layer is at least 10 feet deep, while the upper caliche zone normally has a 1.5- to 2-foot soil layer. Figure 14.29b depicts the boring log for the O-cell load test and the test setup.

(Continued)

Case Study 14.1 (Continued)

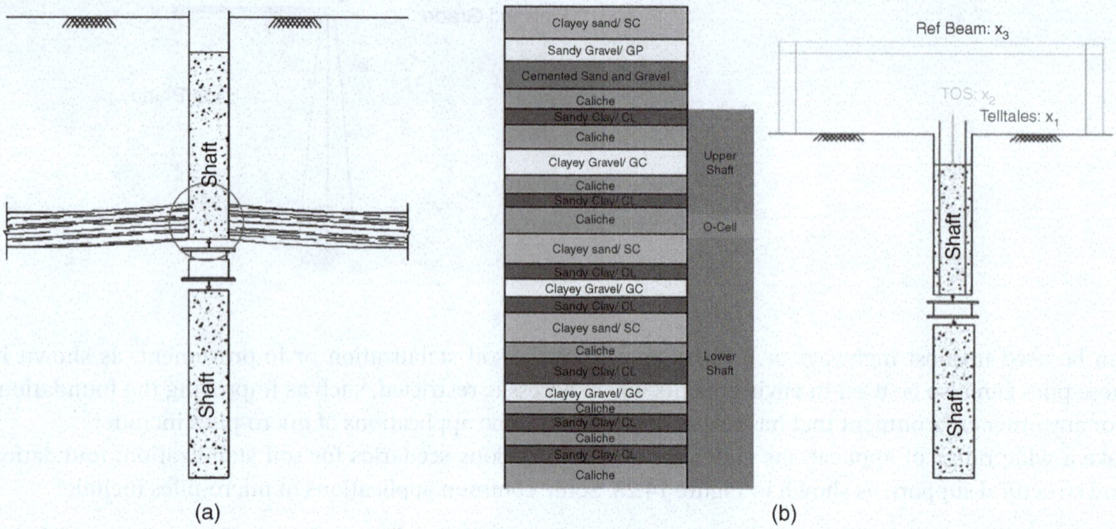

Figure 14.29 (a) Monolith behavior of deep foundation system and (b) O-cell load test and associated boring log (Karakouzian et al. 2015 / With permission of American Society of Civil Engineers).

According to the findings of the numerical study, the maximum slippage that can occur between the shaft and soil layers is less than what is necessary to provide the ultimate skin resistance values. The interface between the shaft and caliche layers exhibits even more of a slippage reduction, as illustrated in Figure 14.30, indicating the monolithic behavior of the shaft-caliche system.

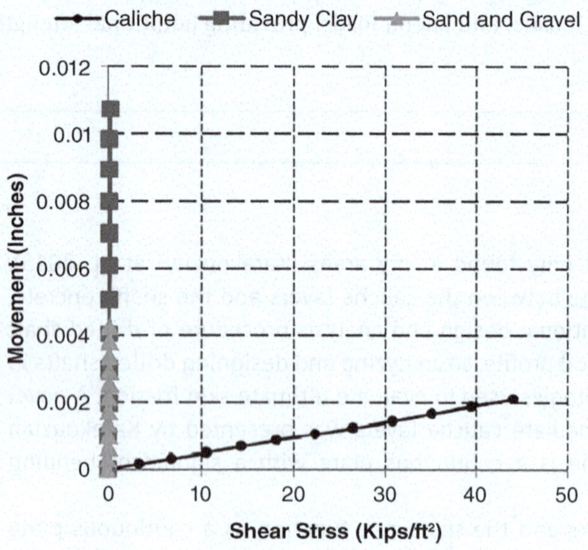

Figure 14.30 Comparison of caliche, sand and gravel, and sandy clay (Karakouzian et al. 2015 / With permission of American Society of Civil Engineers).

The Caliche layers and shaft are strongly bonded, which prevents slipping. Therefore, the analysis and design of drilled shafts in caliche may not be compatible with conventional design methodologies based on ultimate skin friction. The slippage values for sandy clay and sand and gravel demonstrate that the soil layers are not sufficiently mobilized to produce ultimate skin resistance. Additionally, the slippage value decreases even more at the interference between the caliche layers and the shaft, demonstrating monolithic behavior of the caliche layers and the shaft. According to the research, caliche layers create an elastic system by holding onto the shaft and lowering the slippage values. It is necessary to conduct additional research to analyze the Osterberg test. The method currently utilized to analyze the findings of O-cell tests may need to be modified because of the monolith behavior of drilled shafts in caliche.

Case Study 14.2

Piling and Rigid Intrusions Combination

A number of LPG and oil-storage tanks are installed on the man-made island of Mikagehama, which is located in Kobe's port region (Ishihara and Cubrinovski 2004). The oil storage tank, designated as Tank TA72, is the focus of this study. It is situated on the island's western region, roughly 20 meters inland from the revetment line. The tank has a diameter of around 15 meters.

Figure 14.31a depicts the tank's cross-section and its foundation, whereas Figure 14.31b depicts the foundation plan view. The 69 pre-stressed, high-strength precast concrete piles measuring 23 meters in length and 45 cm in diameter support the tank. The piles are embedded in a 50 cm thick concrete slab at the top. Sand compaction piles had been erected around the tank's perimeter, down to a depth of roughly 15 meters, in order to densify and strengthen the foundation soil. Figure 14.31b depicts the compaction piles' dimensions and placement.

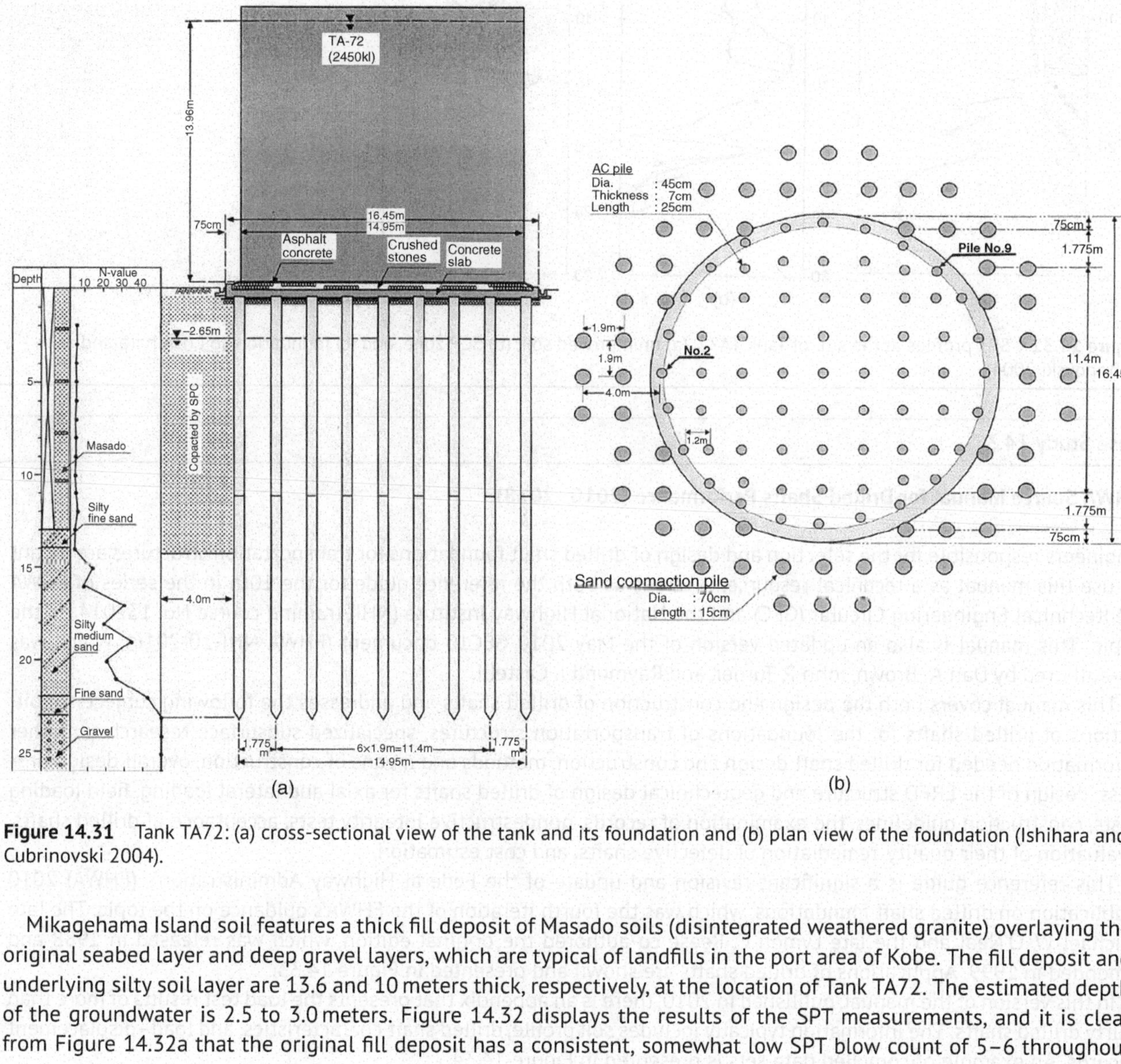

Figure 14.31 Tank TA72: (a) cross-sectional view of the tank and its foundation and (b) plan view of the foundation (Ishihara and Cubrinovski 2004).

Mikagehama Island soil features a thick fill deposit of Masado soils (disintegrated weathered granite) overlaying the original seabed layer and deep gravel layers, which are typical of landfills in the port area of Kobe. The fill deposit and underlying silty soil layer are 13.6 and 10 meters thick, respectively, at the location of Tank TA72. The estimated depth of the groundwater is 2.5 to 3.0 meters. Figure 14.32 displays the results of the SPT measurements, and it is clear from Figure 14.32a that the original fill deposit has a consistent, somewhat low SPT blow count of 5–6 throughout

(Continued)

Case Study 14.2 (Continued)

the depth. With the exception of the top 1–2 meters of the layer, where very low blow counts have been reported, the SPT resistance in the silty layer is typically in the range of 20 to 35 blow counts. The deep gravel layer's *N* value is 50. Standard penetration tests have also been carried out in the foundation soil (the soil between the concrete piles) and the sand compaction pile zone (the soil between the compaction piles). As illustrated in Figures 14.32b and 14.32c, the SPT resistance of the Masado layer greatly increased as a result of the ground improvement, reaching values on the order of 10–30 within the zone of the sand compaction pile and 20–40 blow counts in the foundation soil.

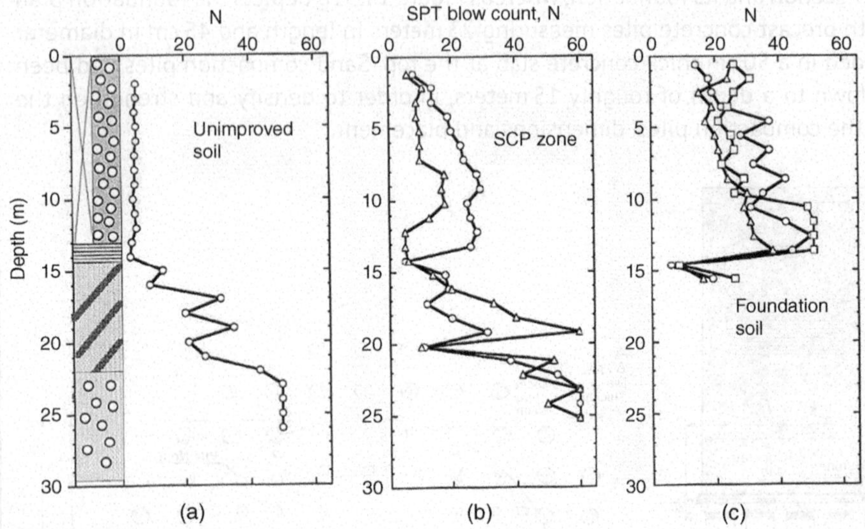

Figure 14.32 SPT profiles at the site of Tank TA72: (a) unimproved soil, (b) SCP zone, and (c) foundation soil (Ishihara and Cubrinovski 2004).

Case Study 14.3

FHWA Source Manual for Drilled Shafts Performance (2010–2018)

Engineers responsible for the selection and design of drilled shaft foundations for transportation structures are meant to use this manual as a technical resource. It serves as both the reference guide for the 10th in the series of FHWA Geotechnical Engineering Circular (GEC) and the National Highway Institute (NHI) training course No. 132014 on the topic. This manual is also an updated version of the May 2010 GEC10 document (FHWA-NHI-10-2016), which was co-authored by Dan A. Brown, John P. Turner, and Raymond J. Castelli.

This manual covers both the design and construction of drilled shafts and addresses the following subjects: applications of drilled shafts for the foundations of transportation structures, specialized subsurface research to gather information needed for drilled shaft design and construction, methods and means of construction, overall design process, design of the LRFD structure and geotechnical design of drilled shafts for axial and lateral loading, field loading tests, construction guidelines, the examination of records, nondestructive integrity tests, acceptance of drilled shafts, evaluation of their quality, remediation of defective shafts, and cost estimation.

This reference guide is a significant revision and update of the Federal Highway Administration's (FHWA) 2010 publication on drilled shaft foundations, which was the fourth iteration of the FHWA's guidance on the topic. The late Michael W. O'Neal and the late Lymon C. Reese co-authored the original edition, which was released in 1988 and amended in 1999. Applications of drilled shafts are shown and presented in Figure 14.33.

In this version of the manual, published in 2010, there is an appendix that presents the load test results of more than thirty drilled shafts. The information typically includes soil profile, drilled shaft characteristics, and load–displacement records. An example of compiled data sets is presented in Figure 14.34.

Figure 14.33 Various applications of drilled shafts: (a) group of drilled shafts for large loads, (b) drilled shafts for individual column support over water, (c) drilled shaft secant wall, and (d) drilled shafts used to support earth retaining structures (FHWA 2018 / With permission of U.S. Department of Transportation).

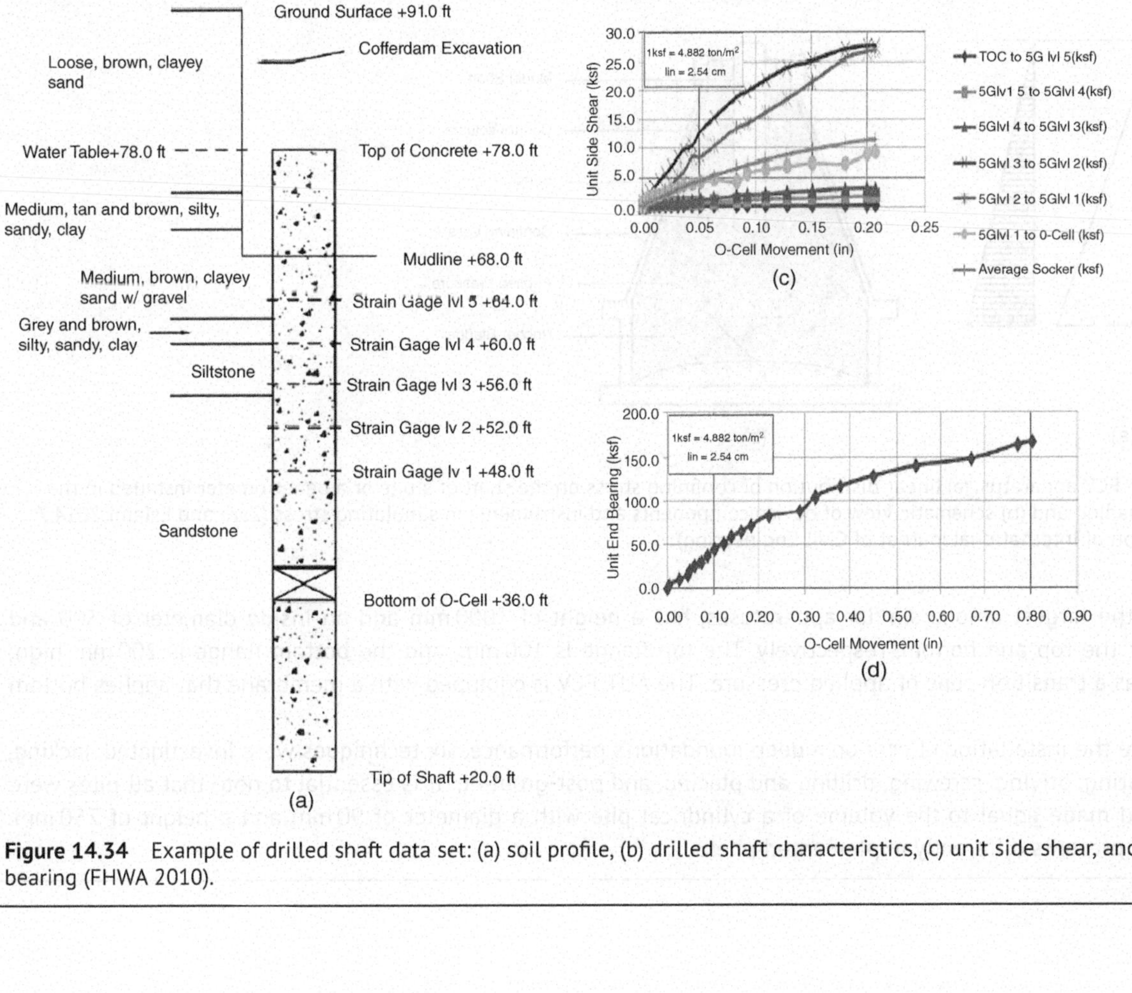

Figure 14.34 Example of drilled shaft data set: (a) soil profile, (b) drilled shaft characteristics, (c) unit side shear, and (d) unit end bearing (FHWA 2010).

(Continued)

Case Study 14.3 (Continued)

Shaft Diameter (in)	Shaft Length (ft)	Shaft Length Below O-cell (ft)	Shaft Length Above O-cell (ft)	O-cell Size (in)	Max Net Load (kips)	Max Deflection Up (in)
48.0	58.0	16.0	42.0	34.0	4394.00	0.21

(b)

Figure 14.34 *(Continued)*

Case Study 14.4

Installation Effect via Physical Modeling (FCV-AUT)

FCV (frustum confining vessel) has been developed for the physical modeling of deep foundations and penetrometers (Zare and Eslami 2014; Zarrabi and Eslami 2016). An FCV combines the advantages of centrifuges and CCs in one apparatus (Zarrabi and Eslami 2016). This device is a frustum chamber that applies steady pressure to its bottom and creates linear stress distribution along its vertical central line. The confining stress level in soil mass has been controlled by applying pressure to the device's bottom. Schematic profiles of FCV and an idealized distribution of stresses within a control volume are illustrated in Figure 14.35. The purpose of FCV is to produce, within the control volume, a state of stresses similar to those typically found in the field, while simultaneously controlling the stress levels.

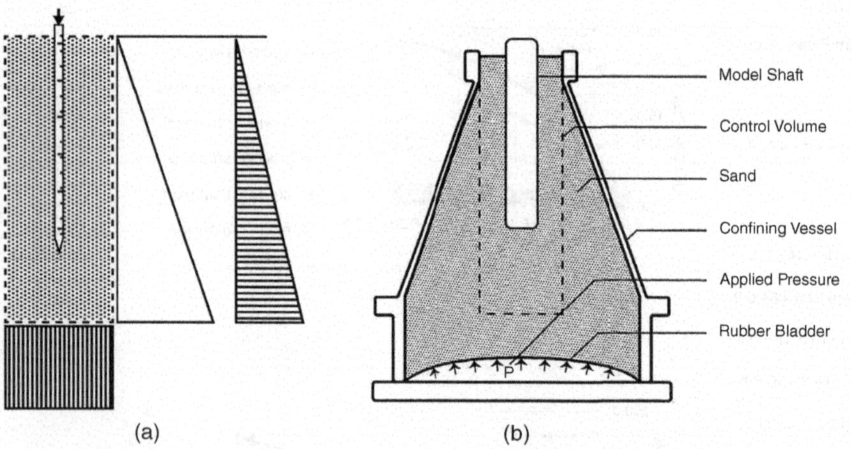

Figure 14.35 FCV apparatus: (a) linear distribution of confining stress on the shaft of a pile or a penetrometer installed in the sand before loading and (b) schematic view of device components and instruments in simulating stress (Zare and Eslami 2014 / With permission of International Journal of Civil Engineering).

FCV-AUT, the largest among similar apparatuses, has a height of 1000 mm and an inside diameter of 300 and 1300 mm for the top and bottom, respectively. The top flange is 100 mm, and the bottom flange is 200 mm high; it functions as a transition zone of applied pressure. The AUT FCV is equipped with a membrane that applies bottom pressure.

To evaluate the installation effects on a deep foundation's performance, six techniques were investigated: jacking, drilling, grouting, driving, screwing, drilling and placing, and post-grouting. It is essential to note that all piles were designed and made equal to the volume of a cylindrical pile with a diameter of 90 mm and a height of 750 mm. Figure 14.36 presents differently implemented piles.

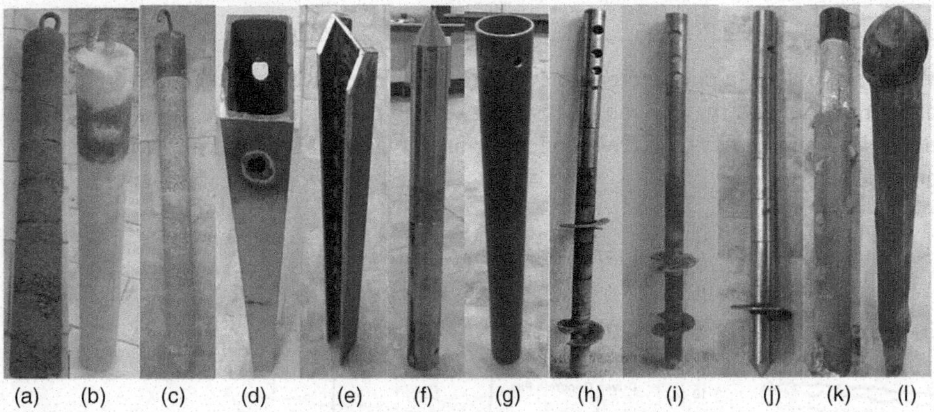

Figure 14.36 Various piles implemented in the FCV-AUT: (a) drilled shaft, (b) precast-in-place pile, (c) precast concrete pile, (d) box-shaped steel pile, (e) H-shaped steel pile, (f) closed-ended steel pile, (g) open-ended steel pile, (h) helical pile with two helixes ($S/D = 3$), (i) helical pile with two helixes ($S/D = 2$), (j) helical pile with single helix, (k) shaft post-grouted precast-in-place pile, and (l) tip post-grouted drilled shaft (Zarrabi and Eslami 2016 / American Society of Civil Engineers).

Figure 14.37 depicts six different installation methods that were used in the existing study.

Figure 14.37 Installation methods of model piles in the FCV-AUT: (a) drilling and placing, (b) drilling and grouting, (c) jacking, (d) driving, (e) screwing, and (f) post-grouting (Zarrabi and Eslami 2016 / American Society of Civil Engineers).

The load–displacement curves of the driven and jacked piles are shown in Figure 14.38. In compression load tests, jacked piles had noticeably higher axial capacity values than those of conventionally driven piles. The results show that driven and jacked piles can display different behaviors regarding pile geometries, so the performance of tubular and other kinds of piles was not nearly the same among piles with different geometries. This fact is also valid in

(Continued)

Case Study 14.4 (Continued)

tension load testing, as the axial capacity of jacked piles increased compared with the driven piles. In tension loading, jacked piles showed greater maximum capacities than driven piles. However, the behavior of driven piles was stiffer at the initial pile head displacements. This indicates that driven piles within service settlements can bear greater or at least equal loads than jacked piles.

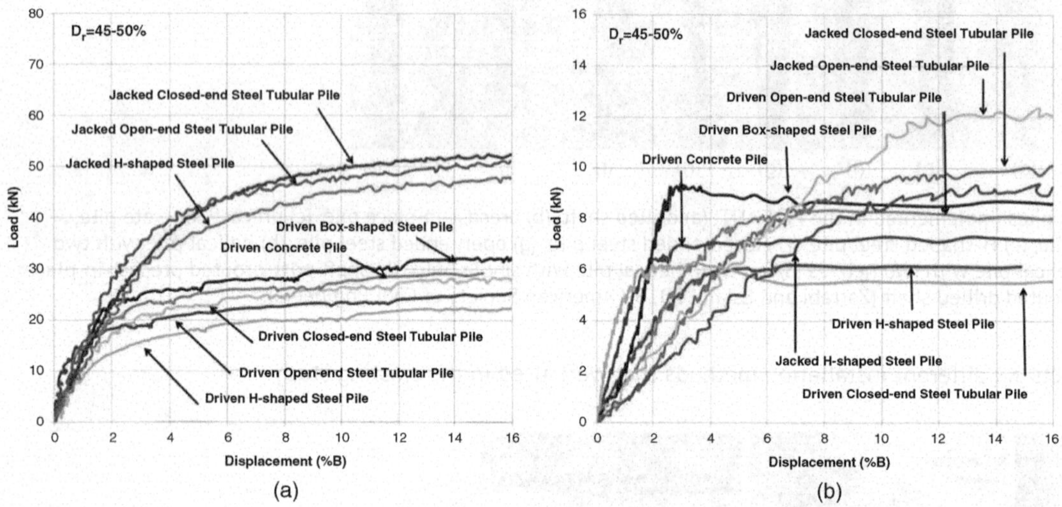

Figure 14.38 Load–displacement curves of jacked and driven piles: (a) compression and (b) tension (Zarrabi and Eslami 2016).

14.6 Major Points

1) Deep foundations are slender column elements designed to transfer the load from upper structures to soil-bearing layers at greater depths. They are used when the upper layers of soil are unable to support the structural loads effectively. Deep foundations are employed to enhance safety against overturning and to provide resistance against lateral forces and slip.
2) Deep foundations are classified based on several factors, including the installation method, cross-section, soil displacement during installation, materials, bearing capacity, performance, and purpose of use. Classification based on appearance, installation method, and soil displacement is particularly important in categorizing deep foundations.
3) Piles are classified based on the installation method, including driving, cast-in-place, jacking, placement, and augured techniques. Piles can be constructed using materials such as wood, steel, concrete, or a combination of these materials (composite piles). They are installed by driving or jacking using hydraulic or diesel hammers, vibratory methods, or bottom-driving techniques. The use of piles offers advantages such as improving the soil beneath the pile tip and increasing the bearing capacity. However, one limitation of piles is the noise and vibration generated during driving, especially in urban environments. In coastal and offshore areas, closed-end steel pipe piles with larger diameters are commonly employed.
4) Drilled shafts, also known as bored piles, are constructed by excavating a well or borehole and then filling it with reinforced concrete. One of their advantages is their capability to be installed in dense soils and weak rock layers. However, they face challenges when dealing with collapsing layers during the construction process. They are also difficult to execute in high groundwater environments, and their bearing capacity depends on the specific implementation conditions. Drilled shafts are commonly used as caissons in the foundations of bridges or docks, typically in larger diameters ranging from 2 to 5 meters. Techniques such as enlarging the base and post-grouting can be employed to enhance the geotechnical performance of drilled shafts.
5) Due to limitations in the implementation of bored and driven piles, as well as advancements in technology, engineers have developed new methods for pile installation. These innovative methods fall under the category of special piles, which have been specifically designed to overcome the limitations associated with bored and driven piles. Special piles

are utilized to implement deep foundations in projects that face construction or economic constraints. These piles offer alternative solutions to meet the specific requirements of such projects.

6) CFA (continuous flight auger) piles were developed to address the challenges faced in bored pile execution. These piles are installed using equipment that consists of a continuous flight auger, which is rotated into the soil by applying torque force. While CFA piles offer several advantages, such as faster installation compared to bored piles, their implementation in dense and rocky layers can be challenging. Therefore, CFA piles are typically more suitable for medium-density and weak rock layers. However, they are well-suited for use in weak rock layers due to their efficient installation process.
7) Drilled displacement piles, similar to CFA piles, involve the use of drilling tools that are installed in the soil by applying torque and a small axial force. Once the desired tip deployment depth is reached, the sacrificial tip is detached from the drilling tool. Concrete is then injected into the well, the drilling tool is extracted, and a reinforcement cage is inserted into the well. These piles are preferred in urban environments due to their low vibration and noise levels during installation. Additionally, their implementation in compressible soil layers improves geotechnical conditions and enhances the bearing capacity of the foundation. As a result, the use of drilled displacement piles is increasing in many countries as they offer efficient and effective solutions for various construction projects.
8) Helical piles consist of a shaft with one or more helixes that are installed into the soil using torque force. These piles generate both rotational and axial forces in the soil. Similar to drilled displacement piles, helical piles offer the advantage of minimal vibration and noise during installation. Helical piles are particularly valued in urban environments and for light structures due to their nondisruptive installation process. They can be installed in a variety of soil conditions, including cohesive soils, loose sands, and silts. The helical shape of the piles allows for efficient load transfer and enhanced stability. These piles are highly regarded for their versatility, ease of installation, and cost-effectiveness. They have found extensive applications in residential, commercial, and industrial projects where noise and vibration control is crucial or where access is limited.

14.7 Questions and Problems

14.7.1 Explanatory

1 Define the application as well as the benefits and shortcomings of the driven piles below:
- Timber Piles
- Steel Piles
- Precast Concrete Piles

2 Explain the reason for choosing the appropriate pile type to support heavy structures underlain by bedrock from 25 to 40 meters in depth.

3 Compare the benefits and shortcomings and describe the dissimilarities in constructing bored and self-expanded piles.

4 Explain the dissimilarities in the construction of CFA piles and DD piles. Give instances of soils in which each type would be the most appropriate option.

5 What are the effects of densifying the loose sands as a result of pile driving on the load-bearing capacity of such piles?

6 What is the difference between mini piles and micro-piles?

7 Explain the most suitable types of soil for the dry and wet drilled shaft construction methods.

8 What are the reasons for using piles rather than other types of foundations in marine projects?

9 What kind of piles are used for the improvement and repair of foundations?

10 For which conditions are deep foundation systems recommended?

11 List some typical applications of drilled shafts in civil engineering projects. Name and discuss other efficient solutions for each case.

12 Compare and contrast two different methods of pile installation, vibration, and bottom driving. Discuss the benefits and limitations of each technique.

13 What are the benefits of an enlarged base for drilled shafts in comparison with typical ones? What are other suggestions for enhancing the geotechnical performance of drilled shafts?

14.7.2 Conceptual

1 List and explain the results and relevance of deep foundation geometry and soil conditions on bearing capacity.

2 What are typical solutions to overcome soil disturbance during helical pile installation?

3 Illustrate and explain piles that are applicable in case of high pullout capacity requirements.

4 Consider Case Study 14.4 and discuss the effect of the installation procedure on pile behavior under compressive and tensile loads.

5 Compare and contrast the FCV apparatus with other common physical modeling tools in geotechnical engineering, particularly in regard to application for deep foundation testing.

6 Discuss the effect of caliche layers based on Case Study 14.1.

7 Consider Case Study 14.2 and explain the necessity of the hybrid application of deep foundations and rigid intrusions in challenging conditions. What are the advantages and disadvantages of this category of foundation systems?

8 How can the assessment of two major types of deep foundations contribute to the realization of sustainable development concepts?

9 Explain the mechanism of geotechnical performance enhancement of drilled shafts after post-grouting.

10 What are the effects of driving piles in regard to existing soil improvement? List the main influential factors.

References

Brown, D.A., Dapp, S., Thompson, R., and Lazarte, C. (2007). *Design and construction of continuous flight auger piles.* Federal Highway Administration Office of Technology Application, Office of Engineering/Bridge Division.

CGS, CFEM: Canadian foundation engineering manual. (2006). CGS Richmond, BC, Canada.

De Cock, F., and Imbo, R. (1994). Atlas screw pile: a vibration-free, total displacement, cast-in-place pile. Transportation Research Record, 49–62.

Eslami, A., Moshfeghi, S., Molaabasi, H., and Eslami, M. (2020). Piezocone and Cone Penetration Test (CPTu and CPT). In: *Applications in Foundation Engineering*, 1ste. Elsevier.

FHWA. (2010). Drilled Shafts: Construction Procedures and Design Methods. Publication No. FHWA-NHI 10-016.

FHWA. (2018). Drilled Shafts: Construction Procedures and Design Methods. Publication No. FHWA-NHI 18-024.

Hannigan, P.J., Rausche, F., Likins, G. et al. (2016). *Design and Construction of Driven Pile Foundations–Volume I.* National Highway Institute.

Ishihara, K. and Cubrinovski, M. (2004). Case studies of pile foundations undergoing lateral spreading in liquefied deposits. *International Conference on Case Histories in Geotechnical Engineering* 5.

Karakouzian, M., Afsharhasani, R., and Kluzniak, B. (2015). Elastic Analysis of Drilled Shaft Foundations in Soil Profiles with Intermediate Caliche Layers. IFCEE 2015 - ASCE 2015.

Massarsch, K.R., Fellenius, B.H., and Bodare, A. (2017). *Fundamentals of Vibratory Driving of Piles and Sheet Piles*, 1–16. Berlin: Geotechnik, Ernst & Sohn Verlag.

Monteiro, F.F., Da Cunha, R.P., De Aguiar, M.F.P., and Silva, C.M. (2021). Settlement of bored piles with expander body system in lateritic soils. *REM–International Engineering Journal* 74 (3): 309–318.

Perko, H.A. (2009). Helical piles: a practical guide to design and installation. In: *Helical Piles: A Practical Guide to Design and Installation*. John Wiley & Sons.

Prezzi, M., and Basu, P. (2005). Overview of construction and design of auger cast-in-place and drilled displacement piles. DFI's 30th annual conference on deep foundations. Chicago, U.S.A.

Shields, C. (2017). Fundamentals of Torque Down Piles. Geo-Institute Dinner Presentation, p. 63.

Shojaei, E., Eslami, A., and Ganjian, N. (2019). Self-expanded piles: a new approach to unconventional piles development. *Marine Georesources & Geotechnology*. https://doi.org/10.1080/1064119X.2019.1679924.

Viggiani, C., Mandolini, A., and Russo, G. (2012). *Piles and Pile Foundations*. London: Spon Press.

Zare, M. and Eslami, A. (2014). Study of deep foundations performance by frustum confining vessel (FCV). *International Journal of Civil Engineering* 12 (4): 271–280.

Zarrabi, M. and Eslami, A. (2016). Behaviour of piles under different installation effects by physical modeling. *International Journal of Geomechanics* 16 (5): 04016014.

15

Geotechnical Design

Chapter Highlight

- **Deep Foundation Design Strategy**
- **Load Transfer Mechanism**
- **Bearing Capacity: Static Analysis**
  - Toe Resistance
  - Shaft Resistance
- **Load–Displacement**
- **Settlement**
- **Resistance Distribution and Residual Loads**
- **Helical Piles Capacity**
- **Pullout Capacity of Deep Foundations**
- **Post-Installation Capacity Variations**
- **Case Studies**
  - Pile Prediction Symposium (Finno 1989)
  - Unified Design Approach (Fellenius 1984, 2004; Hai and Fellenius 2014)

15.1 Deep Foundation Design

Deep foundations play a crucial role in a wide range of projects, whether they are located in nearshore, onshore, or offshore environments. These foundations become especially significant when dealing with relatively soft to medium compressible deltaic deposits or when faced with complex loading conditions involving lateral and overturning forces. Extensive research, along with years of practical experience, has shed light on the importance of deep foundation systems and has paved the way for a knowledge-based approach to their analysis and design. Despite the advancements in pile engineering, there are still certain simplifications, conservative practices, and uncertainties that need to be addressed (Eslami and Fellenius 1997). Additionally, the equipment used for installing and executing different types of piles is a critical factor in the successful implementation of these foundation systems. Therefore, pile engineering is a comprehensive and time-consuming profession that requires careful design and construction.

The initial step in the design of deep foundations involves studying project constraints, assessing potential events like earthquakes and floods, considering design requirements, accounting for environmental factors, and estimating foundation loads. The second step includes site visits, conducting geotechnical investigations, performing subsurface explorations, and conducting laboratory testing to gather essential information for subsequent design steps. Based on the collected data, the geotechnical designer selects the appropriate foundation system for the superstructure, taking into account site conditions and economic considerations. If deep foundations are required, a decision is made among various types of systems, such as drilled shafts, driven piles, helical piles, micro-piles, or drilled displacement piles, considering cost, schedule, and constructability. Once the deep foundation system is finalized, a design subsurface profile is established for each foundation location, considering the variability of soil properties inherent in site conditions. The diameter and embedment depth of the

Advanced Foundation Engineering: Principles, Performance and Prospects, First Edition. Moses Karakouzian and Abolfazl Eslami.

piles, their arrangement in the plan, and the spacing between them are determined to achieve optimal pile group-bearing capacity. The structural design of the piles and their connection to the column or cap is then detailed. Consideration is given to construction methods, including the installation plan and potential construction challenges. Load testing is often conducted to obtain valuable information for site-specific evaluation of the design, which influences the selection of safety factors or resistance factors. Procedures for observation, inspection, and documentation of construction activities should be defined, with effective communication between inspectors and the designer to address any unusual conditions or problems in an efficient manner.

- Documentation plays a crucial role in the construction of deep foundations. A final report is often prepared to document the foundation construction process, any remediation measures taken, and the final as-built conditions of the foundation. This documentation serves as a valuable resource for future projects that may involve the rehabilitation, modification, or replacement of the structure. By documenting the construction and testing procedures, valuable lessons can be learned from each project. This documentation serves as a reference for future projects that may encounter similar circumstances. It allows engineers and stakeholders to analyze the successes, challenges, and best practices of the project, facilitating continuous improvement in the design, construction, and management of deep foundations. Furthermore, the documentation serves as a historical record of the project, providing a comprehensive account of the construction process and ensuring that important details are preserved for future reference. This information is particularly valuable when conducting inspections, assessments, or modifications to the foundation in the future. Overall, proper documentation of the construction and testing of deep foundations ensures transparency, accountability, and knowledge transfer, contributing to the ongoing improvement of construction practices and the successful execution of future projects.

15.1.1 Deep Foundations Application Strategy

The general strategy for deep foundations application is as follows:

- Determining the magnitude and combination of loads on the superstructure, including vertical, horizontal, and moments.
- Evaluating the need for deep foundations based on geotechnical site investigations, subsurface conditions, and economic requirements.
- Selecting the appropriate type of deep foundation, such as piles, drilled shafts, or special foundations, considering factors like material, section, vertical or inclined installation, and equipment requirements.
- Determining the optimal embedment depth based on geotechnical parameters, imposed loads, and available equipment capabilities.
- Configuring the arrangement of deep foundations in the plan, considering both individual and group performance, and their interaction with the raft or superstructure.
- Calculating the bearing capacity of the deep foundations using static analysis methods.
- Refining the estimated bearing capacity through in situ tests or dynamic analysis techniques.
- Controlling and assessing the settlement behavior of the foundation system.
- Conducting structural design of the deep foundation elements and their connection to the cap or column.
- Performing practical evaluation of the foundation's bearing capacity through static or dynamic load tests to validate the construction procedure, ensure construction quality, and optimize the safety factor and overall project economy.
- Considering post-installation aspects such as post-grouting, soil setup and relaxation, and monitoring fluctuations in ground conditions and groundwater levels.

15.2 Load Transfer Mechanism

The mechanisms of load transfer from a deep foundation to the surrounding ground are depicted in Figure 15.1. When a deep foundation, such as a pile, is subjected to an axial compression load (Q_{ult}) applied at the top, the load is transferred to the underlying soil both along the shaft and the toe of the pile. This distinguishes deep foundations from shallow foundations, as the load transfer in shallow foundations primarily occurs through the base.

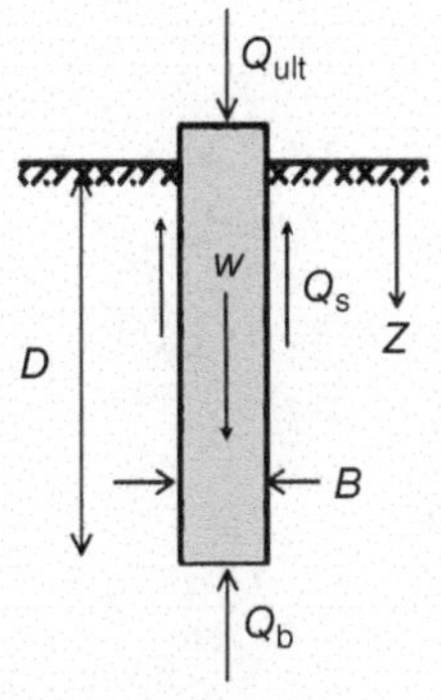

Figure 15.1 Load transfer mechanism under axial compression loading.

The axial pile-bearing capacity is determined by the following equation.

$$Q = Q_b + Q_S \tag{15.1}$$

$$Q_b = A_b \cdot q_b \tag{15.2}$$

$$Q_S = \int_0^L q_{si} P dx \tag{15.3}$$

In the above equations, Q, Q_b, and Q_s are the load at the head of the pile, toe-bearing capacity, and shaft-bearing capacity. A_b = pile base area, q_b = unit toe resistance, q_{si} = unit shaft resistance, P = pile perimeter, and L = pile embedment length.

Figure 15.2a provides an overview of the relationship between downward displacement and axial resistance. The resistance can be divided into three components: side resistance, tip resistance, and total resistance. Side resistance refers to the shear resistance along the shaft, tip resistance is the resistance at the pile's tip, and total resistance is the combined effect of both side and tip resistance. In Figure 15.2b, the distribution of axial load with depth (z) is depicted for different levels of displacement. As the axial load on the pile shaft increases from zero, the shaft starts to displace downward, resulting in the mobilization of side resistance (shown as point A). The load is transferred to the surrounding soil or rock, leading to a decreasing load with increasing depth, as indicated by the dashed curve in Figure 15.2b. At this stage, the load is primarily supported by side resistance, and the load transmitted to the base of the pile may be relatively small. As the load continues to increase, the full side resistance is mobilized (shown as point B), typically occurring at a displacement of around 1% to 2% of the pile diameter. The base of the pile must then withstand further load increments until the maximum base and combined resistances are reached (shown as point C). The displacement required to mobilize the maximum base resistance can vary between 3% and 15% of the pile diameter, depending on the soil conditions.

In Figure 15.3, six different axial loading modes of a pile are presented. The upper portion of six piles labeled A through F are depicted, with their toes located at the elevation of plane Z–Z. Pile A, or mode A, represents the behavior of a pile subjected to a push load applied at the pile head. As shown on the left side of the pile, this loading results in a downward deformation of the soil layers and an upward-directed shaft resistance. The pile is under compression, which theoretically leads to a Poisson's ratio effect. This effect causes an increase in the pile diameter, resulting in an increase in the earth pressure coefficient K_s, as indicated. In addition to Poisson's ratio effect, the positive shaft resistance transfers the load from the pile to the surrounding soil. As a result, the effective overburden stress in the soil increases, leading to an increase in the shaft resistance.

Mode B represents a pile subjected to pull load applied at the pile head. This type of loading causes an upward deformation of the soil layers and results in a negative shaft resistance. In this mode, there is a decrease in lateral pressure and a decrease in effective stress. Mode B, also known as the uplift loading mode, is often considered representative of a drag load mode. However, in contrast to the pile in the drag load mode, mode C, pile B is under tension rather than compression.

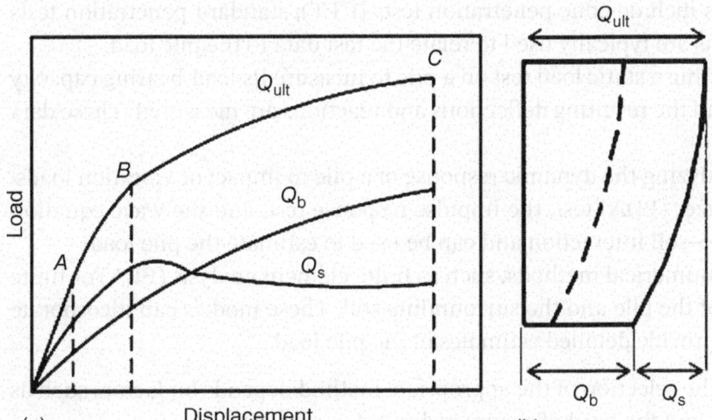

Figure 15.2 Pile axial load transfer. (a) relationship between downward displacement and axial resistance, (b), the distribution of axial load with depth for different levels of displacement.

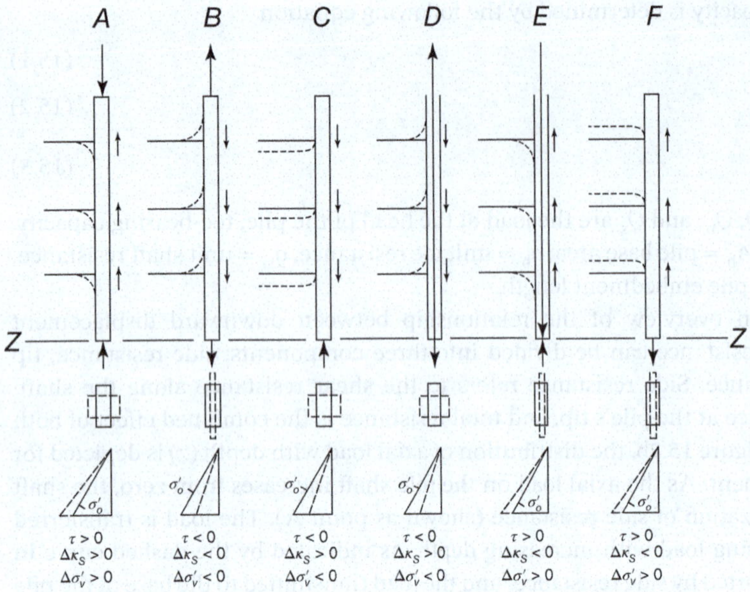

Figure 15.3 Behavior modes of a pile subjected to different loading conditions (adapted from Fellenius 1984).

Therefore, attempting to predict the drag load based on static uplift testing might underestimate its magnitude. To accurately study the drag load, an uplift test should be conducted in the arrangement shown in mode D. In this mode, the pull load is applied at the pile toe, making all three aspects similar to the drag load mode. Mode E represents a push test where the load is applied to the pile toe. This test aims to simulate the behavior of pile F, which is influenced by swelling soil above plane *Z*–*Z*.

15.3 Bearing Capacity: Static Analysis

Various methods have been developed to estimate the total load on a pile, *Q*, which consists of the base resistance, Q_b, and the shaft resistance, Q_S. These methods can be categorized into five main groups based on their simplifying assumptions and empirical approaches:

1) **Static Analyses:** These methods involve using simplified mathematical models and analytical solutions to estimate the load on the pile. They often rely on simplified soil–structure interaction models and assumptions about soil behavior.
2) **In situ Test-Based Methods:** These methods utilize data obtained from in situ tests, either directly or indirectly, to estimate the load on the pile. Examples of such tests include cone penetration tests (CPT), standard penetration tests (SPT), and pressuremeter tests. Empirical correlations are typically used to relate the test data to the pile load.
3) **Pile Static Load Test:** This method involves conducting a static load test on a pile to measure its load-bearing capacity directly. The pile is subjected to incremental loads, and the resulting deflections and reactions are measured. These data are then used to estimate the load on the pile.
4) **Pile Dynamic Load Test:** This method involves analyzing the dynamic response of a pile to impact or vibration loads. It includes techniques such as the pile driving analyzer (PDA) test, the impulse response test, and the wave equation analysis. These methods provide insights into the pile–soil interaction and can be used to estimate the pile load.
5) **Numerical Modeling:** This method involves using numerical methods, such as finite element analysis (FEA) or finite difference analysis (FDA), to simulate the behavior of the pile and the surrounding soil. These models can incorporate complex soil properties and boundary conditions to provide detailed estimates of the pile load.

Each method has its advantages and limitations, and the selection of the appropriate method depends on factors such as the site conditions, available data, project requirements, and the level of accuracy desired.

In this chapter, only the static analysis methods will be discussed. The other methods will be covered in subsequent chapters. In the following two subsections, static analyses will be presented to estimate the toe capacity and the shaft resistance.

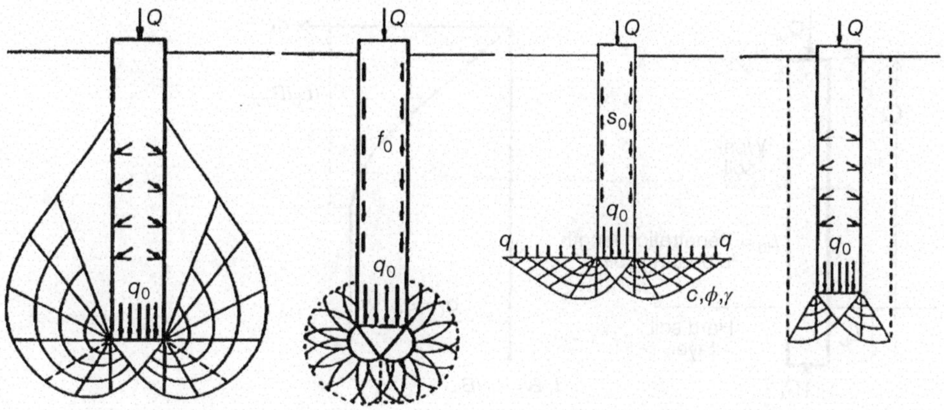

Figure 15.4 Load failure mechanism around the pile toe (adapted from Vesic 1977).

15.3.1 Toe Capacity (Q_b)

The general equation for estimating pile unit toe-bearing capacity is similar to the Terzaghi equation for shallow foundations as following

$$q_{\text{ult}} = cN_C^* + \overline{q}N_q^* + \frac{1}{2}B\gamma N_\gamma^* \tag{15.4}$$

where C is soil cohesion, γ is soil unit weight, $\overline{q} = \gamma D$ is effective vertical stress at the toe, and B and D are the pile width/diameter and embedment length, respectively. N_C^*, N_q^*, and N_γ^* are dimensionless bearing capacity factors for the pile foundations.

The dimensionless bearing capacity factors are the main difference between shallow and deep foundations. Different researchers investigated the load transfer mechanism and failure patterns around the pile toe is provided in Figure 15.4 and developed their proposed equations for toe-bearing capacity according to the Terzaghi equation.

The dimensionless bearing capacity factors differ among these methods. Since the embedment depth of piles is much more than their width, the third term in Equation 15.4 is neglected. Moreover, for cohesionless soils, the first term is omitted. Even if the first and third terms were considered in the above equation, the second term, i.e., the depth factor, is dominant. Hence, the unit toe-bearing capacity can be rewritten as follows:

For cohesive soils: $q_{\text{ult}} = cN_C^* + \overline{q}N_q^*$ (15.5)

For cohesionless soils: $q_{\text{ult}} = \overline{q}N_q^*$ (15.6)

Accordingly, the general equation for estimating toe-bearing capacity is

$$Q_t = A_t\left(cN_C^* + \overline{q}N_q^*\right) \tag{15.7}$$

It should be noted that the pile installation method, such as driven, drilled, or CFA piles, as well as factors such as pile material, soil variability, drainage conditions, and other parameters, can significantly influence the estimation of toe-bearing capacity. Several methods have been developed to estimate the toe-bearing capacity, including the Meyerhof, Vesic, Janbu, and Kulhawy methods.

15.3.1.1 The Meyerhof (1976) Method

Meyerhof's method assumes that the toe-bearing capacity of a pile driven into a sand layer increases with the embedment length in the layer and reaches a maximum at a critical depth ratio ($L_b/B = (L_b/B)_{\text{cr}}$). In a homogeneous soil layer, L_b is equal to the embedment length (L) (Figure 15.5). As shown in Figure 15.5, when the pile is penetrated into the bearing layer, the value of L_b is less than L. For depth ratios greater than the critical depth ratio, the unit toe-bearing capacity remains constant, as shown in Figure 15.5. Please see Table 15.3 for the details of bearing capacity determination.

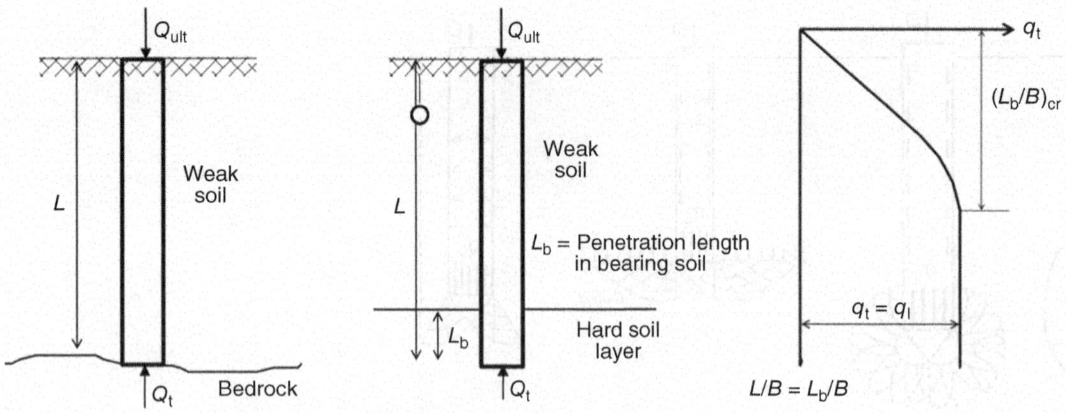

Figure 15.5 Toe-bearing capacity variation with depth.

15.3.1.2 The Vesic (1977) Method

Vesic (1977) derived a solution based on the pressure required to expand a cylindrical or spherical cavity at the level of the base of the pile and carried out a large-scale experimental study. He related the rigidity index to the bearing capacity factors. Please see Table 15.3 for the details of bearing capacity determination.

15.3.1.3 The Janbu (1976) Method

Janbu (1976) proposed an equation similar to Equation 15.7. However, the bearing capacity factors were different, as presented in Table 15.2. Please see Table 15.3 for the details of bearing capacity determination.

15.3.1.4 The Kulhawy (1991) Method

Kulhawy (1991) developed his method for drilled shafts with circular sections. The diameter of drilled shafts varies between 1 to 3 m and a maximum of 6 m with embedment lengths even more than 76 m. The performance of drilled shafts depends on the execution method and should be considered in analyzing and designing. He related the combination of rigidity, shape, and depth factors to the bearing capacity factors as the rigidity index. Kulhawy (1991) proposed the below equation for the estimation of unit toe-bearing capacity.

$$q_{ult} = CN_c\xi_{cs}\xi_{cd}\xi_{cr} + \overline{q}N_q\xi_{qs}\xi_{qd}\xi_{qr} + \frac{1}{2}\gamma BN_\gamma\xi_{\gamma s}\xi_{\gamma d}\xi_{\gamma r} \tag{15.8}$$

N_c and N_q are nondimensional bearing capacity factors and are estimated according to Prandtl and Reissner's proposed equations. N_γ is estimated according to the Vesic equation.

Shape correction factors are dependent on the soil internal friction angle. Depth correction factors are dependent on the soil's internal friction angle and depth-to-diameter ratio (D/B). Depth correction factors are applicable for shallow and deep foundations since D/B is a continuous function. For D/B of more than four, the value of ξ_{qd} remains constant. The rigidity correction factor is dependent on the soil's internal friction angle and I_{rr}. These equations can be classified according to the soil type and loading rate for drained and undrained conditions. Please refer to Tables 15.1 and 15.2 to determine the correction factors for the Kulhawy method. Additionally, please see Table 15.3 for the details of bearing capacity determination.

If the piles are installed in coarse-grained soils or the long-term toe-bearing capacity of piles installed in fine-grained soils is essential, the drained condition is considered to estimate toe-bearing capacity.

When saturated fine-grained soils are loaded quickly, and there is not enough time for dissipation of excess pore water pressure, it is called the undrained condition. The total stress approach or $\phi = 0$ is considered in this condition, and soil strength is dependent on undrained shear strength, C_u.

Table 15.3 summarizes the equations for these different methods.

15.3.2 Shaft Resistance (Q_S)

Pile unit shaft or side resistance, represented by q_s, refers to the frictional stress developed between the soil and the pile along the length of the pile's embedment. The value of q_s is influenced by several factors, such as the material of the pile, the

Table 15.1 Kulhawy bearing capacity correction factors for circular piles in drained condition.

Correction factor	Sign	Equation
Shape	ξ_{cs}	$1 + (N_q/N_c)$
	$\xi_{\gamma s}$	0.6
	ξ_{qs}	$1 + \tan\varphi$
Depth	ξ_{cd}	$\xi_{qd} - [(1 - \xi_{qd})/(N_c \tan\varphi)]$
	$\xi_{\gamma r}$	1
	ξ_{qd}	$1 + 2\tan\varphi(1 - \sin\varphi)^2[(\pi/180)\tan^{-1}(D/B)]$
Rigidity	ξ_{cr}	$\xi_{qr} - [(1 - \xi_{qr})/(N_c \tan\varphi)] \le 1$
	$\xi_{\gamma r}$	ξ_{qr}
	ξ_{qr}	$\exp\{[-3.8\tan\varphi] + [3.07\sin\varphi)(\log_{10}2I_{rr})/(1 + \sin\varphi)]\} \le 1$

Table 15.2 Kulhawy bearing capacity correction factors for circular piles in undrained condition.

Correction Factor	Sign	Value
Shape	ξ_{cs}	1.2
	ξ_{qs}	1
Depth	ξ_{cd}	$1 + 0.33[(\pi/180)\tan^{-1}(D/B)]$
	ξ_{qd}	1
Rigidity	ξ_{cr}	$0.44 + 0.6\log_{10}I_{rr} \le 1$
	ξ_{qr}	1

type of soil surrounding the pile shaft, the drainage condition, the installation method, and the extent of soil displacement. Figure 15.6 illustrates the behavior of shear strength during loading in both drained and undrained conditions.

15.3.2.1 Effective Stress Approach; Beta Method

The effective friction forces acting on the pile shaft depend significantly on the active forces between soil and pile, as presented in Figure 15.6.

$$\tau = r_s = \mu.\sigma'_h = \mu.k.\sigma'_v \tag{15.9}$$

where τ is the effective shear strength, r_s is the unit shaft resistance, $\mu = \tan\varphi_f$ is the coefficient of friction between pile and soil, and φ_f is the pile–soil interface friction angle, σ'_h is effective horizontal stress, σ'_v is effective vertical stress, and k is the earth pressure coefficient. Tables 15.4 and 15.5 present the variation of φ_f and k for driven piles.

Salgado (2008) presented graphs for estimating the ratio of K/K_0 for driven piles according to their induced displacements in cohesionless soils with various relative densities (Figure 15.7). He proposed to be cautious about using design values greater than 1.5 for low displacement piles or greater than 2.5 for high displacement piles. Accordingly, Equation 15.9 can be rewritten as

$$r_s = \mu.K.\sigma'_v = K_0\sigma'_v\left(\frac{K}{K_0}\right)\tan\left[\varphi'\left(\frac{\varphi_f}{\varphi'}\right)\right] \tag{15.10}$$

According to the difficulties in evaluation of φ_f and K, a single factor, i.e., β was introduced as below (Burland 1973):

$$\beta = K_0\left(\frac{K}{K_0}\right)\tan\left[\varphi'\left(\frac{\varphi_f}{\varphi'}\right)\right] \tag{15.11}$$

And accordingly, Equation 15.9 is $r_s = \beta\sigma'_v$ and this method of estimating shaft friction is known as the beta method. Table 15.6 shows the beta value estimations in different soil types.

Table 15.3 Summary of methods for predicting pile toe-bearing capacity.

Method	**Condition**	**Q_t**	**Commentary**
Meyerhof (1976)	Cohesive soil, undrained	$Q_t = A_t\left(C_u N_C^*\right) = 9A_t C_u$	$\varphi = 0$ C_u: the undrained soil shear strength below the pile toe
	Cohesionless soils, drained	$Q_t = A_t\left(\overline{q}N_q^*\right) < A_t q_L$ $q_L(\text{kN/m}^2) = 50N_q^* \tan\varphi$	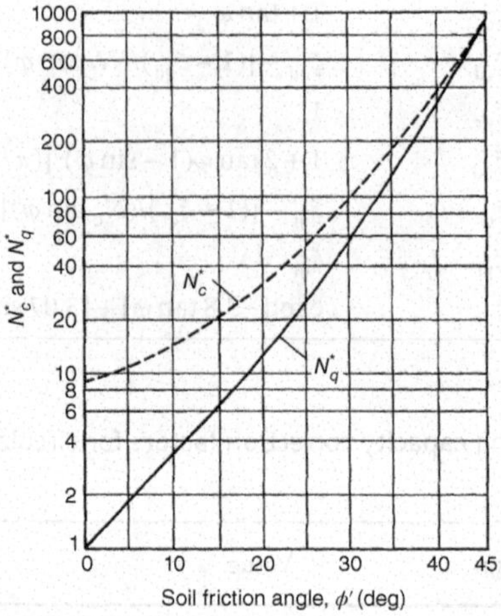
Vesic (1977)		$Q_t = A_t q_t = A_t\left(cN_C^* + \sigma_0' N_q^*\right)$	$\sigma_0' = \left(\frac{1+2K_0}{3}\right)\overline{q}$ $\overline{q}$ is the effective vertical stress at the pile toe $K_0 = 1 - \sin\varphi.$ $N_q^* = \frac{3}{3-\sin\varphi}\left\{\exp\left[\left(\frac{\pi}{2}-\varphi\right)\tan\varphi\right]\tan^2\left(45+\frac{\varphi}{2}\right)I_{rr}^{\frac{1.333\sin\varphi}{1+\sin\varphi}}\right\}$ $I_{rr} = \frac{I_r}{1+\xi_v I_r}$ $I_r = \frac{E_s}{2(1+\mu_S)(C+\overline{q}\tan\varphi)} = \frac{G_s}{C+\overline{q}\tan\varphi}$ In undrained soil condition or dense soil layer, $\xi_v = 0$ and $I_{rr} = I_r$. $N_c^* = \left(N_q^* - 1\right)\cot\varphi$ If $\varphi = 0$ then $N_c^* = \frac{4}{3}(LN(I_{rr})+1) + \frac{\pi}{2} + 1$
Janbu (1976)		$Q_t = A_t q_t = A_t\left(cN_C^* + \sigma_0' N_q^*\right)$	$N_q^* = (\tan\varphi + \sqrt{1+\tan^2\varphi}).e^{2\eta'\tan\varphi}$ $N_c^* = \left(N_q^* - 1\right)\cot\varphi$
Kulhawy (1991)	Drained	$q_{ult} = \overline{q}N_q\xi_{qs}\xi_{qd}\xi_{qr}$ $+\frac{1}{2}\gamma BN_\gamma\xi_{\gamma s}\xi_{\gamma d}\xi_{\gamma r}$	$\overline{\gamma}$ is the average effective unit weight of soil from depth D to $D+B$, $\overline{q}$ effective vertical stress at depth D and ϕ and I_{rr} are averaged from depth D to $D+B$. $I_r = \frac{G_d}{\overline{C}+\overline{q}_0\tan\varphi}$ G_d is the drained shear modulus, $\overline{q}_0$ is average effective vertical stress from D to $D+B$ Other parameters are defined in Table 14.2.
	Undrained	$q_{ult} = 5.14C_u\xi_{cs}\xi_{cd}\xi_{cr}$ $+\overline{q}\xi_{qs}\xi_{qd}\xi_{qr}$	$\varphi = 0 \rightarrow N_C = 5.14, N_q = 1$ and $N_\gamma = 0$ $I_r = \frac{G_u}{C_u} = \frac{E_u}{2(1+v_u)C_u}$ E_u and G_u are undrained Young and shear modulus and v_u is undrained Poisson ratio and is equal to 0.5 for saturated cohesive soils in rapid loading. Other parameters are defined in Table 14.3.

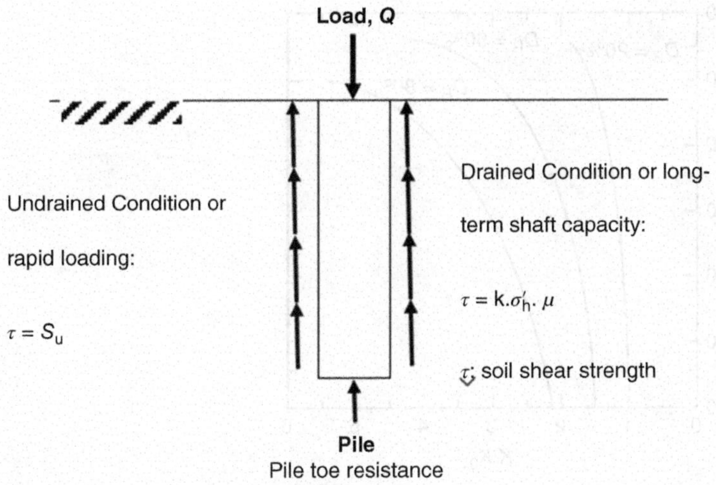

Figure 15.6 The active shear strength along the pile length under drained and undrained conditions.

Table 15.4 Approximate ratios of coefficient of lateral earth pressure for driven piles.

Method of construction	K/K_0
Driven pile-small displacement	1–1.4
Driven pile-large displacement	1–1.8

Table 15.5 Approximate ϕ_f/ϕ' values for the interface between soil and driven piles (Das 2016 / Cengage Learning).

Driven pile type	ϕ_f/ϕ'
Concrete	0.8–1
Rough steel-step-taper pile	0.7–0.9
Smooth steel-pipe pile or H pile	0.5–0.7
Timber	0.8–0.9

15.3.2.2 Total Stress Approach; Alpha Method

As depicted in Figure 15.6, total stress analysis is implemented in cohesive soils for estimating pile shaft resistance based on undrained shear strength. As presented in Equation 15.12 in this method, an adhesion factor, i.e., α, multiplied by undrained shear strength resulting in unit shaft resistance

$$r_s = \alpha S_u \tag{15.12}$$

where r_s is unit shaft resistance, α is adhesion factor, and S_u is undrained shear strength.

Randolph and Murphy (1985) proposed to implement the shear strength, the over-consolidation ratio, and the initial stress rate for estimation of α and developed the following equations:

$$\alpha = \frac{\left(S_u/\sigma'_Z\right)_{NC}^{0.5}}{\left(S_u/\sigma'_Z\right)^{0.5}} \text{ if } S_u/\sigma'_Z \leq 1 \tag{15.13}$$

$$\alpha = \frac{\left(S_u/\sigma'_Z\right)_{NC}^{0.5}}{\left(S_u/\sigma'_Z\right)^{0.25}} \text{ if } S_u/\sigma'_Z > 1 \tag{15.14}$$

where S_u/σ'_Z is the ratio of undrained shear strength to effective vertical stress, and NC denotes normally consolidated soils.

Table 15.7 summarizes different static analysis methods.

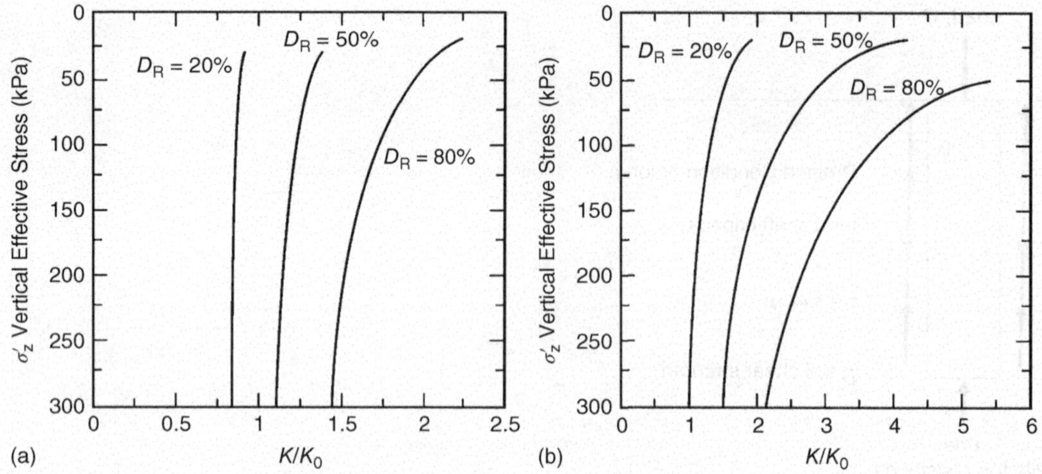

Figure 15.7 The variation of K/K_0 according to the soil relative density and effective vertical stress: (a) low displacement piles and (b) high displacement piles (Salgado 2008).

Table 15.6 Estimation of the β values in different soils (CFEM 1992).

Soil type	Soil friction angle	β
Clay	25–30	0.23–0.4
Silt	28–34	0.27–0.5
Sand	32–40	0.3–0.8
Gravel	35–45	0.35–0.8

15.4 Code-Based Methods

Several codes and standards have been developed and implemented worldwide for the geotechnical design of deep foundations. In this context, two selected approaches are the Unified approach from the Canadian Foundation Engineering Manual (CFEM) and the API approach from the American Petroleum Institute Guideline. These approaches provide guidance and recommendations for the design of deep foundations, taking into account various factors such as soil conditions, loading conditions, and structural requirements.

15.4.1 The Unified Pile Design

According to the various methods for determining r_s and r_t and the presence of different relations for estimating N_q, the Unified pile design method was developed as follows:

$$r_t = N_t.\sigma'_{Z=D_f} \tag{15.15}$$

$$r_s = \beta\sigma'_z \tag{15.16}$$

where $\sigma'_{Z=D_f}$ is the effective vertical stress at depth $Z = D_f$. The values of β and N_t for driven and drilled piles are given in Table 15.8.

15.4.2 API Method (2011)

This code presents equations for toe and shaft capacity in cohesive and cohesionless soils separately. This code is primarily intended for analyzing and determining the bearing capacity of deep foundations used in offshore structures. The deep

Table 15.7 The summary of static analyses methods for predicting axial pile-bearing capacity.

Soil type	**Unit toe resistance, r_t**		**Unit shaft resistance, r_s**
Fine-grained soils	Effective stress approach $r_t = N_q \sigma'_{z=D_f}$	Total stress approach $r_t = N_c S_u$ S_u: Undrained soil shear strength ($^{kg}/_{cm^2}$) $N_c = 9 - 12$ $N_c = 6\left[1 + 0.2\left(\frac{D}{B}\right)\right]$ $r_t \leq 40\,^{kg}/_{cm^2}$ $N_c = 6.5\ S_u = 0.25$ $N_c = 8.0\ S_u = 50$ $N_c = 9\ S_u \geq 1$	API (1984) $r_s = \alpha S_u$ $\alpha = 1 - \frac{(S_u - 0.25)}{0.7} 0.25 \leq S_u \leq 0.7$ $\alpha = 1\ S_u \leq 0.25$ $\alpha = 0.5\ S_u \geq 0.7$ For drilled shafts: $\alpha = 1\ S_u \leq 0.35$ $\alpha = 0.5\ S_u \geq 0.8$
Coarse-grained soils	Effective stress approach $r_t = N_q \sigma'_{z=D_f}$ $N_q = 2N_{q,\text{shallow}}$ $N_{q,\text{shallow}}$: N_q for shallow foundations $r_t \leq 120$ kg/cm^2 For over-consolidated soils $r_t = CN_c + qN_q$ C: soil cohesion.		McClelland (1974) $r_s = \beta \sigma'_z$ $\beta = 0.15 - 0.35$ Compressive $\beta = 0.1 - 0.24$ Tension Meyerhof (1976) Driven Piles $\beta = 0.44\ \varphi = 28$ $\beta = 0.75\ \varphi = 35$ $\beta = 1.2\ \varphi = 37$ $r_s \leq 1\,^{kg}/_{cm^2}$ Drilled Piles $\beta = 0.1\ \varphi = 28$ $\beta = 0.2\ \varphi = 35$ $\beta = 0.35\ \varphi = 37$ Burland (1973) For over-consolidated soils: $r_s = C + \beta \sigma'_z$ $\beta = (1 - \sin\varphi)\tan\varphi(\text{OCR})^{0.5}$ $\beta = 0.5 - 0.8$ Poulos (1989) $r_s = \beta \sigma'_z\ \beta = 0.5 - 0.8$ $0.6 \leq r_s \leq 1\,^{kg}/_{cm^2}$

foundations commonly employed in this context are hollow steel pipe piles, which are installed into the seabed or ocean floor using different driving systems and hammers. The calculations for bearing capacity take into account the weight of the soil inside the pile and the uplift load resulting from hydrostatic forces. The code provides separate equations for calculating the toe capacity and shaft capacity in cohesive soils and cohesionless soils.

15.4.2.1 Cohesive Soils

For calculating the shaft resistance of driven piles in cohesive soils, Equation 15.12 is used, similar to the Alpha method, but the adhesion factor is calculated as follows:

$$\text{For } \Psi \leq 1 \rightarrow \alpha = 0.5\Psi^{-0.5} \quad (15.17)$$

$$\text{For } \Psi > 1 \rightarrow \alpha = 0.5\Psi^{-0.25} \quad (15.18)$$

with the constraint that $\alpha \leq 1$.

Table 15.8 The values of β and N_t for piles and drilled shafts in the Unified design method.

Soil type	Soil friction angle	Drilled piles		Driven piles	
		β	N_t	β	N_t
Clay	25–30	0.25–0.32	3–10	0.25–0.32	3–10
Silt	28–34	0.2–0.3	10–30	0.3–0.5	20–40
Loose sand		0.2–0.4	20–30	0.3–0.8	30–80
Medium sand	32–42	0.3–0.5	30–60	0.6–1	50–120
Dense sand		0.4–0.6	50–100	0.8–1.2	100–120
Gravel	35–45	0.4–0.7	80–150	0.8–1.5	150–350

Where, at depth, z;

$$\Psi = \frac{S_u}{p'_0(z)} \tag{15.19}$$

$p'_0(z)$ = effective vertical stress at depth z.
The tip resistance is calculated via:

$$r_t = 9S_u \tag{15.20}$$

where S_u is the undrained shear strength of soil at the pile toe depth.

The shaft resistance acts on both the inner and outer sides of the pile. The total compressive resistance is the sum of the inner and outer shaft resistance and the end bearing on the pile wall annulus or the end bearing of the plug, whichever is the lesser. For piles considered to be plugged, the bearing pressure may be assumed to act over the entire pile cross-section. For unplugged piles, the bearing pressure acts on the pile wall annulus only. That a pile is considered plugged or unplugged should be based on static calculations. A pile can be driven in an unplugged condition but behaves as plugged under static loads.

In layered soils, shaft friction must be calculated via the alpha method. The toe capacity for piles tipped in cohesive layers with adjacent weaker layers may be taken as given in Equation 15.20 if the two requirements below are provided. Otherwise, some modification of the end-bearing is usually necessary.

(a) the pile achieves penetration of two or three diameters or more into the layer in question; and
(b) the tip is approximately three diameters or more above the bottom of the layer to prevent punching failure

15.4.2.2 Cohesionless Soils

Equations similar to what was introduced for the unified method are used for shaft and toe resistance. Relevant amounts of β and N_q that are used in place of N_t are presented in Table 15.9.

For full displacement piles (i.e., closed-ended or fully plugged open-ended piles), values of β may be assumed to be 25% higher than those given in Table 15.8. For long piles, r_s, does not necessarily increase linearly with the overburden stress. In those cases, it may be appropriate to limit shaft resistance to the given values in Table 15.8. A similar recommendation is also available for toe resistance.

Considerations for plugged and unplugged conditions are as similar as for cohesive soils. For soils that do not fall within the ranges of relative density and soil description presented in Table 15.8 or for materials with unusually weak grains or compressible structure, the given amounts are not necessarily appropriate for the selection of design parameters. Very loose soils or soils containing large amounts of mica or volcanic grains are an instance in this regard, which may require special laboratory or field tests for design parameter selection. Of particular importance are sands containing calcium carbonate, which are found extensively in vast areas of the oceans.

Table 15.9 Design parameters for cohesionless siliceous soil (API 2011).

Relative density[a)]	Soil description	β	Limiting shaft friction values (kPa)	N_q	Limiting end bearing values (MPa)
Very loose	Sand	Not applicable[c)]	Not applicable[c)]	Not applicable[c)]	Not applicable[c)]
Loose	Sand				
Loose	Sand–silt[b)]				
Medium dense	Silt				
Dense	Silt				
Medium dense	Sand–silt[b)]	0.29	67	12	3
Medium dense	Sand	0.37	81	20	5
Dense	Sand–silt[b)]				
Dense	Sand	0.46	96	40	10
Very dense	Sand–silt[b)]				
Very dense	Sand	0.56	115	50	12

Note: The listed parameters are intended as guidelines only. Other values may be justified in cases where detailed information, such as CPT records, strength tests on high-quality samples, model tests, or pile driving performance, is available.

a) The definitions for the relative density percentage description are as follows: very loose, 0–15; loose, 15–35; medium dense, 35–65; dense, 65–85; very dense, 85–100.

b) Sand–silt includes those soils with significant fractions of both sand and silt. Strength values generally increase with increasing sand fractions and decrease with increasing silt fractions.

c) Design parameters given in previous editions for these soil/relative density combinations may be unconservative. Hence, it is recommended to use CPT-based methods.

Example 15.1 An open-ended steel pile with a length of 19 meters, an outer diameter of 80 cm, and a thickness of 30 mm is driven in two different sites. Determine the ultimate capacity in the following conditions:

(a) Compressive loads, sand, unified method, in case of formation of the plug.
(b) Compressive loads, sand, β method, in case of no plug.
(c) Compressive loads, clay, API method, in case of formation of the plug.
(d) Tensile loads, sand and clay, API, and β method, in case of no plug.

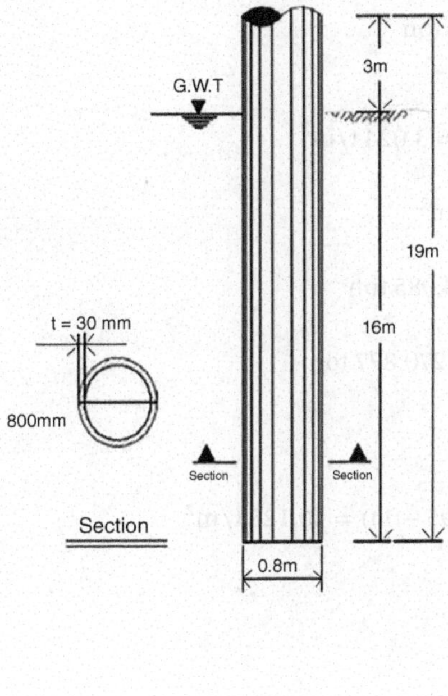

Sand:

$$K_c = 0.8, M = 0.75, \phi = 35°, N_t = 35, \gamma_{sat} = 1.9\ t/m^3$$

Clay:

$$S_u = 0.22\sigma'_v, N_c = 12, \gamma_{sat} = 1.95\ t/m^3$$

In the case of the formation of the plug, the problem is similar to having a closed-end pile. No shaft resistance is mobilized in the inner side of the pile, and the base area is calculated assuming a complete circle. The issue will become more evident via the following solution.

(a)
Toe resistance:

$$q_t = N_t \times \sigma'_v = 35 \times (16 \times (1.9 - 1)) = 504\ t/m^2$$

$$A_t = \left(\frac{\pi}{4}\right) \times 0.8^2 = 0.503\ m^2$$

$$Q_t = q_t \times A_t = 504 \times 0.503 = 253.512\ ton$$

Shaft resistance:

$$\beta = K_c \times M \times \tan\phi = 0.8 \times 0.75 \times \tan 35 = 0.42$$

$$q_s = \beta \times \sigma'_v = 0.42 \times (8 \times (1.9 - 1)) = 3.024\ t/m^2$$

$$A_s = \pi \times 0.8 \times 16 = 40.212\ m^2$$

$$Q_s = q_s \times A_s = 3.024 \times 40.212 = 121.601\ ton$$

$$Q_u = Q_t + Q_s = 253.512 + 121.601 = 375.113\ ton$$

(b)
Toe resistance:

$$q_t = N_t \times \sigma'_v = 35 \times (16 \times (1.9 - 1)) = 504\ t/m^2$$

$$A_t = \left(\frac{\pi}{4}\right) \times (0.8^2 - 0.74^2) = 0.073\ m^2$$

$$Q_t = q_t \times A_t = 504 \times 0.073 = 36.792\ ton$$

Shaft resistance:

$$q_s = \beta \times \sigma'_v = 0.42 \times (8 \times (1.9 - 1)) = 3.024\ t/m^2$$

$$A_s = \pi \times (0.8 + 0.74) \times 16 = 77.409\ m^2$$

$$Q_s = q_s \times A_s = 3.024 \times 77.409 = 234.085\ ton$$

$$Q_u = Q_t + Q_s = 36.792 + 234.085 = 270.877\ ton$$

(c)
Toe resistance:

$$q_t = N_c \times S_u = 12 \times 0.22 \times (16 \times (1.95 - 1)) = 40.128\ t/m^2$$

$$A_t = \left(\frac{\pi}{4}\right) \times 0.8^2 = 0.503\ \text{m}^2$$

$$Q_t = q_t \times A_t = 40.128 \times 0.503 = 20.184\ \text{ton}$$

Shaft resistance:

$$S_{u-Z=8m} = 0.22 \times 8 \times (1.95 - 1) = 1.672\ \text{t/m}^2 = 0.167\ \text{kg/cm}^2$$

$$S_u = 0.167\ \text{kg/cm}^2 \leq 0.25\ \text{kg/cm}^2 \rightarrow \alpha = 1$$

$$q_s = \alpha \times S_u = 1 \times 1.672 = 1.672\ \text{t/m}^2$$

$$A_s = \pi \times 0.8 \times 16 = 40.212\ \text{m}^2$$

$$Q_s = q_s \times A_s = 1.672 \times 40.212 = 67.234\ \text{ton}$$

$$Q_u = Q_t + Q_s = 20.184 + 67.234 = 87.418\ \text{ton}$$

(d)
The resistance of the pile against tensile loads comprises the shaft resistance and the weight of the pile. In this problem, the weight of the piles is not considered conservatively.

Sand:

$$T_u = Q_s$$

$$q_s = \beta \times \sigma'_v = 0.42 \times (8 \times (1.9 - 1)) = 3.024\ \text{t/m}^2$$

$$A_s = \pi \times (0.8 + 0.74) \times 16 = 77.409\ \text{m}^2$$

$$T_u = Q_s = q_s \times A_s = 3.024 \times 77.409 = 234.085\ \text{ton}$$

Clay:

$$T_u = Q_s$$

$$S_{u-Z=8m} = 0.22 \times 8 \times (1.95 - 1) = 1.672\ \text{t/m}^2 = 0.167\ \text{kg/cm}^2$$

$$S_u = 0.167\ \text{kg/cm}^2 \leq 0.25\ \text{kg/cm}^2 \rightarrow \alpha = 1$$

$$q_s = \alpha \times S_u = 1 \times 1.672 = 1.672\ \text{t/m}^2$$

$$A_s = \pi \times (0.8 + 0.74) \times 16 = 77.409\ \text{m}^2$$

$$T_u = Q_s = q_s \times A_s = 1.672 \times 77.409 = 129.428\ \text{ton}$$

Example 15.2 For the given pile, determine the allowable axial-bearing capacity if the safety factor is 3.

(a) Implementing the Unified design method
(b) If the values of β and N_t are increased by 30%, and the embedment depth remains constant, how does the pile diameter change?

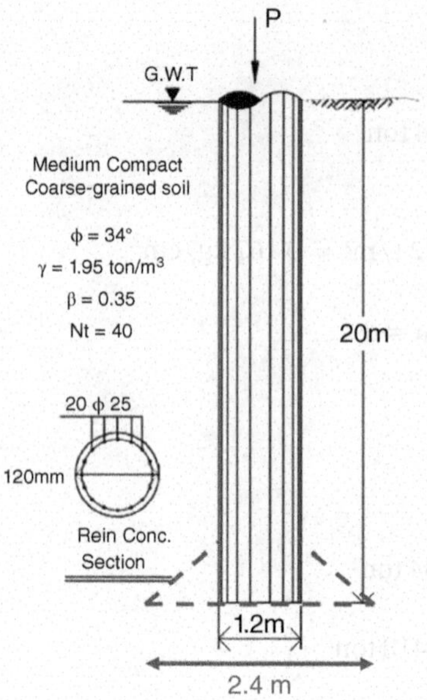

Answer:

$\beta = 0.35$ and $N_t = 40$:

Z	u_0	σ_v (ton/m^2)	σ'_v (ton/m^2)	r_s (ton/m^2)	R_s (ton)
0	0	0	0	0	0
1	1	1.95	0.95	0.33	1.25
2	2	3.9	1.9	0.67	2.51
3	3	5.85	2.85	1.00	3.76
4	4	7.8	3.8	1.33	5.01
5	5	9.75	4.75	1.66	6.27
6	6	11.7	5.7	2.00	7.52
7	7	13.65	6.65	2.33	8.77
8	8	15.6	7.6	2.66	10.03
9	9	17.55	8.55	2.99	11.28
10	10	19.5	9.5	3.33	12.53
11	11	21.45	10.45	3.66	13.79
12	12	23.4	11.4	3.99	15.04
13	13	25.35	12.35	4.32	16.30
14	14	27.3	13.3	4.66	17.55
15	15	29.25	14.25	4.99	18.80
16	16	31.2	15.2	5.32	20.06
17	17	33.15	16.15	5.65	21.31
18	18	35.1	17.1	5.99	22.56
19	19	37.05	18.05	6.32	23.82
20	20	39	19	6.65	25.07
$R_t = 40*19*\pi*1.2^2/4 = 859.5$ ton				$R_s =$	263.2 ton

$R = R_t + R_s = 859.5 + 263.2 = 1122.8$ ton and $R_{all} = 1122.8/3 = 374.3$ ton

(b) $\beta = 0.455$ and $N_t = 52$:

z	u_0	σ_v (ton/m^2)	σ'_v (ton/m^2)	r_s (ton/m^2)
0	0	0	0	0
1	1	1.95	0.95	0.43225
2	2	3.9	1.9	0.8645
3	3	5.85	2.85	1.29675
4	4	7.8	3.8	1.729
5	5	9.75	4.75	2.16125
6	6	11.7	5.7	2.5935
7	7	13.65	6.65	3.02575
8	8	15.6	7.6	3.458
9	9	17.55	8.55	3.89025
10	10	19.5	9.5	4.3225
11	11	21.45	10.45	4.75475
12	12	23.4	11.4	5.187
13	13	25.35	12.35	5.61925
14	14	27.3	13.3	6.0515
15	15	29.25	14.25	6.48375
16	16	31.2	15.2	6.916
17	17	33.15	16.15	7.34825
18	18	35.1	17.1	7.7805
19	19	37.05	18.05	8.21275
20	20	39	19	8.645

$$R = R_s + R_t = 8.645 \times \pi \times D \times 20 + 19 \times \pi \times {}^{D^2}\!/_4 = 1122.8 \text{ ton} \Rightarrow D = 1.03 \text{ m}$$

15.5 Load–Displacement of Piles (*P*–Δ)

The pile static load test is widely regarded as one of the most reliable methods for determining the axial-bearing capacity of a pile. It can be conducted during the design phase using test piles or during the construction phase using production piles to ensure the quality of the construction. The number of tested piles depends on factors such as the size, importance, and environmental considerations of the project, which are determined by a geotechnical engineering consulting company. While static pile loading tests provide accurate results, they are often costly and time-consuming. Therefore, a limited number of static pile load tests are typically performed in each project based on its significance. To overcome this limitation, several methods have been developed for predicting the load transfer mechanism in piles.

The resistance of a pile to the load applied in a static load test depends on the relative movement between the pile and the soil. During a head-down static loading test, the upper regions of the soil profile are the first to engage in resistance. The shaft resistance progressively engages from the pile head and moves down the pile. The pile toe does not receive any load until all the soil along the pile shaft has been engaged. Until that point, the movement of the pile head corresponds to the accumulated elastic shortenings of the pile. The resistance versus movement curves for shaft resistance are referred to as *t*–*z* functions, while the curves for toe resistance are called *q*–*z* functions.

15.5.1 The *t*–*z* Method

Kraft et al. (1981) introduced the *t*–*z* method as a means to develop pile-load displacement diagrams. This method takes into account the various components involved, including side friction, toe bearing, and elastic compression, as well as their

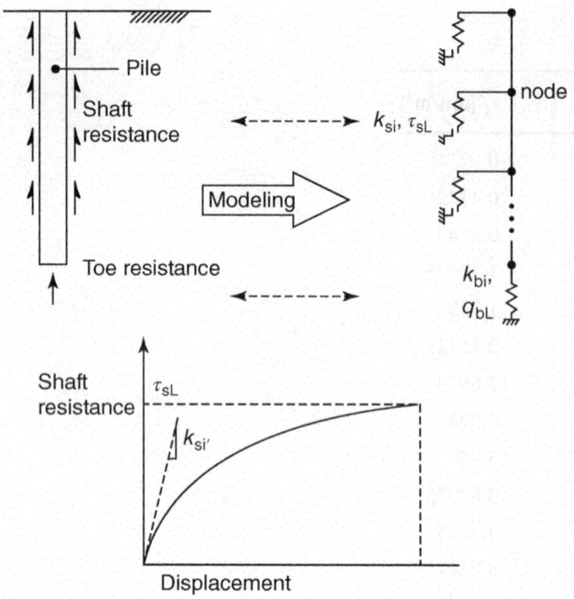

Figure 15.8 Numerical models for the *t–z* method.

Table 15.10 Different developed *t–z* functions.

Method	*t–z* formula	Comments
Ratio function	$r = r_{trg}\left(\dfrac{z}{z_{trg}}\right)^{\theta}$	r: shaft or toe resistance r_{trg}: resistance at a specific location
Hyperbolic function	$r = \dfrac{z}{C_1 z + C_2}\dfrac{1}{r_{trg}}$	Z: soil displacement z_{trg}: soil displacement at a specific location
Exponential function	$r = r_{inf}(1 - e^{-bz})$	r_{inf}: shaft or toe resistant at infinite displacement
The 80%-Hansen function,	$r = \dfrac{\sqrt{z}}{C_1 z + C_2}\dfrac{1}{r_{trg}}$	C_1 and C_2: coefficients (Fellenius 2015) a, b, and c: coefficients estimated as below
The Zhang strain-softening function	$r = \dfrac{z\,(a + cz)}{(a + bz)^2}$	$b = \dfrac{1}{2r_{peak}} - \dfrac{a}{z_{peak}}$ $c = \dfrac{1}{4r_{peak}} - \dfrac{a}{z_{peak}}$

interaction. The *t–z* method divides the pile into discrete segments, with each segment characterized by its cross-sectional area (A), modulus of elasticity (E), and length. The side friction resistance between each segment and the surrounding soil is represented by a nonlinear spring, with the load–displacement behavior defined by a *t–z* curve. Similarly, the toe bearing is modeled using a nonlinear spring defined by a *q–z* curve (as shown in Figure 15.8). The *t–z* curves depict the progressive shear strength corresponding to the vertical displacements along the pile shaft during axial loading. Several parameters, such as depth, friction angle between the pile and soil, pile installation method, pile length and diameter, pile axial stiffness and resistance, and the distribution of rigidity along the pile shaft, influence the shape of the *t–z* curves.

Several researchers have developed different models for predicting the *t–z* and *q–z* curves. These equations can be categorized into five groups, including (a) the ratio function, (b) the hyperbolic function, (c) the exponential function, (d) the 80%-Hansen function, and (e) the Zhang strain-softening function. Table 15.10 provides the proposed *t–z* formulations. Figure 15.9 compares the performance of these various predictive functions.

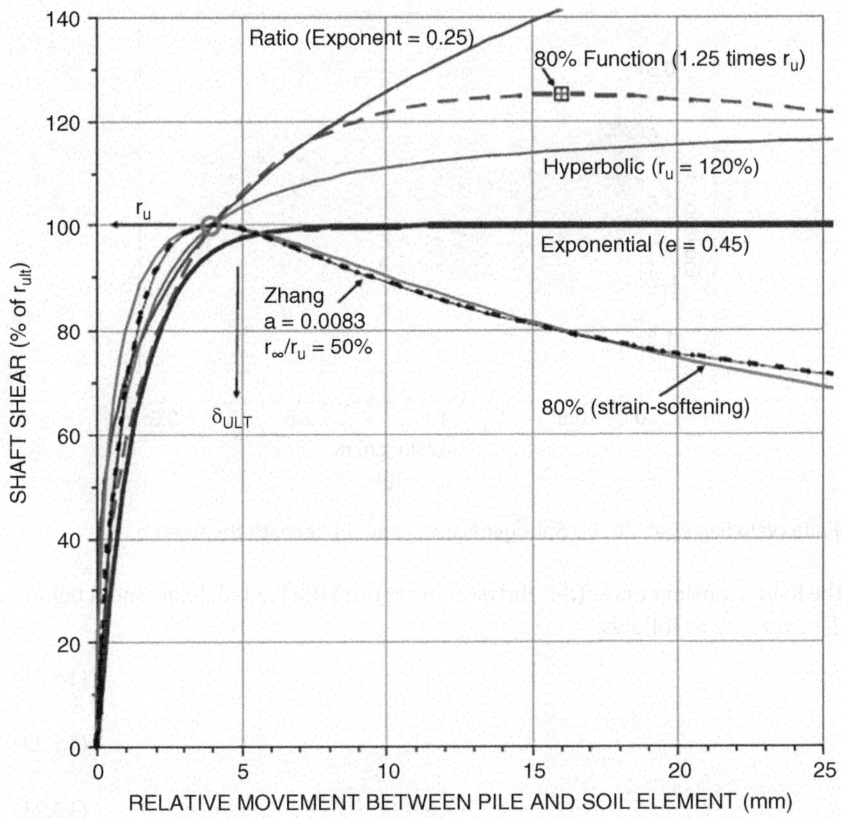

Figure 15.9 The performance of different proposed *t–z* functions (Fellenius 2013 / Canadian Science Publishing).

15.5.2 CPT-Based Load–Displacement Behavior of Foundations (Valikhah et al. 2019)

Valikhah et al. (2019) proposed a new analytical–numerical method to estimate the bearing capacity and axial load–displacement behavior of driven piles in granular soils using CPT records. They used the method of stress characteristics to analyze the stress field below and around the pile and, in effect, the failure mechanism. This failure mechanism has been then used by implementing the kinematic approach of the limit analysis to compute the displacement field. This procedure is employed in a step-wise manner to gradually calculate the stress and displacement field as the pile is assumed to penetrate into the ground. In their proposed method, the mobilization of the friction angle is linked to the gradual increase in shear strains in the field, which is done by using the CPT results, which are both continuous and reliable compared to standard laboratory tests often conducted on disturbed samples at discrete intervals (Eslami and Fellenius 1997). Hence, the step-wise procedure is expected to give rise to a complete load–displacement behavior of driven piles shown in a practical case.

To develop their analytical–numerical method, Valikhah et al. (2019) compiled a database of case histories consisting of 165 full-scale pile load tests. The database included comprehensive information on the soil type as well as the results of cone penetration test (CPT) soundings conducted near the pile locations. It is important to note that all the piles in the investigated database belonged to the "driven pile" type. The majority of the cases involved sand soils, while some cases involved silt and mixed soils.

The procedure of utilizing CPT-based stress and strain fields for predicting load–displacement behavior is not limited to driven piles but has also been applied to shallow foundations. This approach was previously introduced by Veiskarami et al. (2014, 2015) as a method for estimating the bearing capacity of shallow foundations. The schematic diagram in Figure 15.10 depicts the stress fields around the pile and beneath the footing, as determined through numerical analysis.

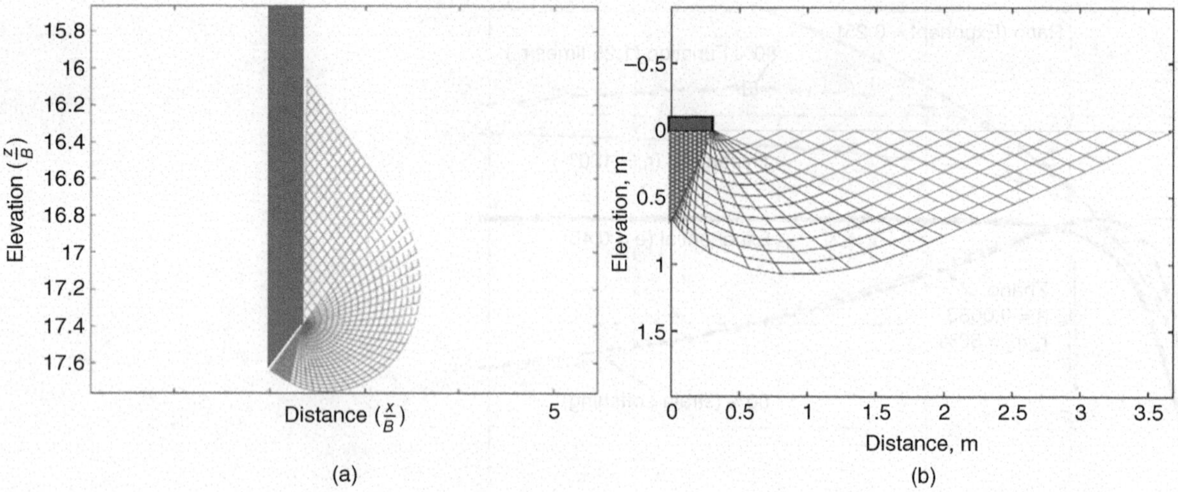

Figure 15.10 Stress field formed: (a) around the pile (Valikhah et al. 2019 / Springer Nature) and (b) beneath the footing.

The direct CPT-based equations to predict the load–transfer curves (t–z and q–z curves) and the load–displacement behavior of driven piles proposed by Valikhah et al. (2019) are as follows:

$$q = \frac{z}{a_t + b_t z} \tag{15.21}$$

$$t = \frac{z}{a_s + b_s z} \tag{15.22}$$

$$P = \frac{\Delta}{a + b\Delta} \tag{15.23}$$

In these equations, q is the effective overburden pressure of soil at the pile tip, t is shear stress at the pile shaft, z is the vertical movement of the pile, P is the axial pile load, and Δ is the movement of the pile head.

Valikhah et al. (2019), with a study on the properties of the collected database, proposed the relations for a and b parameters for pile tip and shaft based on CPT results as follows:

$$a_t = 0.0009 \ln \frac{1}{\left(\frac{q_c}{\sigma_0}\right)} + 0.0067 \tag{15.24}$$

$$b_t = \frac{0.002}{\sqrt{\sigma_0}} \tag{15.25}$$

$$a_s = 0.0017e^{\left(\frac{97.7}{f_s}\right)} \tag{15.26}$$

$$b_s = 0.0242 \left(\frac{1}{f_s}\right)^{0.038} \tag{15.27}$$

$$a = \frac{0.006}{\left(\frac{q_c}{\sigma_0}\right)^{0.4}} \tag{15.28}$$

$$b = \frac{0.089 R_f}{B} \tag{15.29}$$

Inspection of the results revealed that these parameters are functions of the ratio q_c/σ_0, σ_0, f_s, R_f and B, where σ_0 is the initial soil pressure at the depth at which these curves are required. $R_f = \frac{f_s}{q_c}$ is friction ratio, and B is pile width or diameter. This dependency seems to be logical as the stress level effect cannot be disregarded. Therefore, if the soil conditions and pile properties are provided, the load–transfer curves and load–displacement behavior of piles could be predicted using the abovementioned relations.

Example 15.3 The steel closed-ended pipe pile 355 mm wide and 9 m long is investigated. The pile is under 1 MN loading (800 kN dead load and 200 kN live load). Predict the load–displacement behavior based on the given CPT profile.

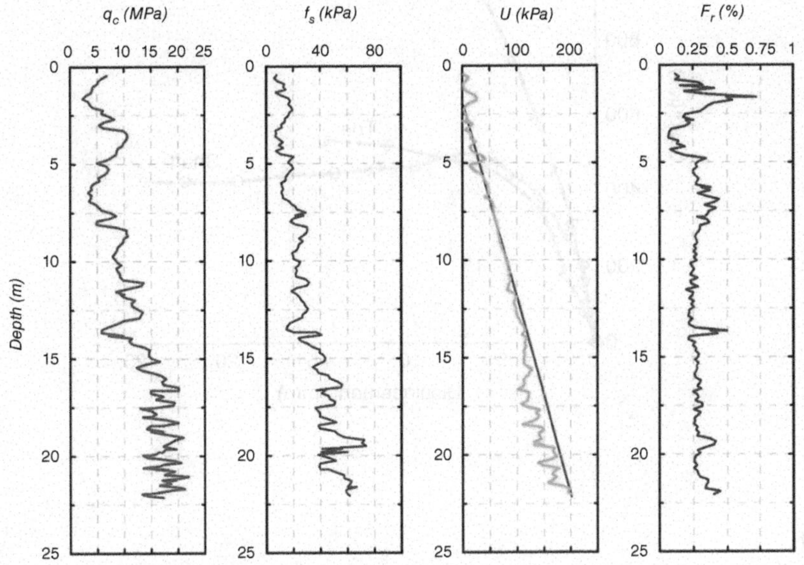

The step-by-step procedure to predict the load–transfer curves and load–displacement of the pile using the analytical–numerical approach proposed by Valikhah et al. (2019) is as below:

The geometric mean of q_c in the influenced zone (4 b bellow and 8 b above the pile tip): 7 MPa

The geometric mean of f_s along the pile length: 25 kPa

Unit weight from Mayne (2001) Equation: $\gamma_t = 12 + 1.5(\ln(f_s + 1)) = 16.89$ kPa

The effective stress at the pile tip level: $\sigma_0 = 2 \times 16.89 + 7 \times 6.89 = 82.01$ kPa

$$a_t = 0.0009 \ln \frac{1}{\left(\frac{q_c}{\sigma_0}\right)} + 0.0067 = 0.0009 \ln \frac{1}{\left(\frac{7000}{82}\right)} + 0.0067 = 0.0027$$

$$b_t = \frac{0.002}{\sqrt{\sigma_0}} = \frac{0.002}{\sqrt{82}} = 0.00022$$

$$a_s = 0.0017e^{\left(\frac{97.7}{f_s}\right)} = 0.0017e^{\left(\frac{97.7}{25}\right)} = 0.085$$

$$b_s = 0.0242\left(\frac{1}{f_s}\right)^{0.038} = 0.0242\left(\frac{1}{25}\right)^{0.038} = 0.021$$

$$\text{Friction ratio: } R_f = \frac{f_s}{q_c} = \frac{25}{7000} = 0.0036$$

$$a = \frac{0.006}{\left(\frac{q_c}{\sigma_0}\right)^{0.4}} = \frac{0.006}{\left(\frac{7000}{82}\right)^{0.4}} = 0.001$$

$$b = \frac{0.089R_f}{B} = \frac{0.089 \times 0.0036}{0.355} = 0.0009$$

Using the obtained parameters and the Equations (15.20) to (15.23), $t - z$, $q - z$, and $P - \Delta$ curves. The results obtained by the proposed approach are shown.

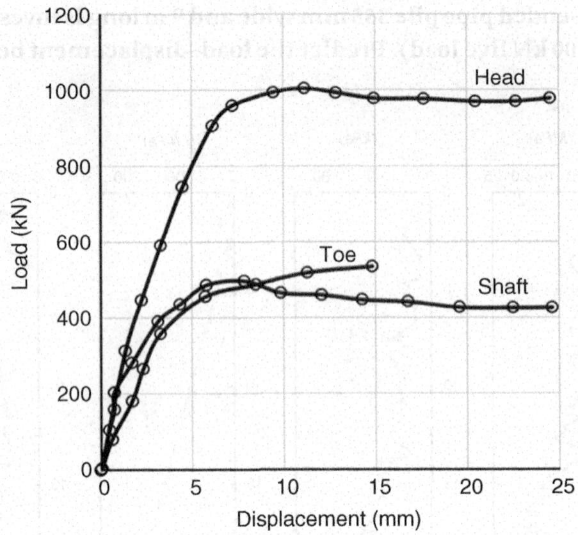

15.6 Settlement

Piles are typically designed to withstand axial loads. When a pile is subjected to a compressive axial load, it undergoes downward displacement, activating the shear strength of the surrounding soil. This load transfer mechanism results in an increase in effective stress within the soil, leading to an increase in compressive force within the pile and an increase in the lateral pressure coefficient of the soil. It should be noted that a significant amount of displacement is required at the pile tip to activate the pile toe resistance, which is considerably higher than the displacement required to mobilize friction resistance. Moreover, larger diameter piles require a greater amount of displacement compared to smaller diameter piles. Additionally, the displacement requirement for driven piles is generally lower compared to drilled piles. This is because driven piles compact the subsoil during installation, simulating the preloading of the soil. On the other hand, drilled shafts create a smear zone and result in reduced soil compaction due to the drilling process.

Studies have indicated that the displacement necessary to fully activate the ultimate toe resistance of driven piles typically falls within the range of 3 to 10% of the pile diameter (Fellenius 1989). However, it is important to note that the load–displacement relationship is not linear. For instance, when the displacement reaches half of the value required to activate the toe resistance fully, more than half of the total toe resistance will already be mobilized. In the case of drilled shafts, it is challenging to establish a specific range or estimate the required displacement needed to activate the toe resistance due to the variability associated with the drilling process.

Static analysis methods are implemented to evaluate the load–displacement response at the pile head. The required displacement to mobilize shaft and toe resistances and the elastic compression of the pile contribute to the pile's load–displacement behavior, as presented in Equation 15.30.

$$S_e = (S_e)_1 + (S_e)_2 + (S_e)_3 \tag{15.30}$$

where S_{e1}, S_{e2}, and S_{e3} are elastic settlements of the pile, the settlement required to mobilize shaft resistance and the elastic settlement required to mobilize the toe resistance, respectively.

$$(S_e)_1 = \frac{(Q_{wp} + \xi Q_{ws})}{A_p E_p} \tag{15.31}$$

where

Q_{wp} = load carried at the pile point under working load condition
Q_{ws} = load carried by side resistance under working load condition
A_p = the area of cross-section of the pile
E_p = the modulus of elasticity of the pile material

Table 15.11 Typical values of C_p (Das 2016 / Cengage Learning).

Type of soil	Driven pile	Bored pile
Sand (dense to loose)	0.02–0.04	0.09–0.18
Clay (stiff to soft)	0.02–0.03	0.03–0.06
Silt (dense to loose)	0.03–0.05	0.09–0.12

The magnitude of ξ varies between 0.5 and 0.67 and will depend on the nature of the distribution of the unit side resistance along the pile shaft

$$(S_e)_2 = \frac{Q_{ws}}{pL}\frac{D}{E_s}\left(1-\mu_s^2\right) I_{ws} \tag{15.32}$$

where P is the perimeter of the pile, L is the embedded length of the pile, and I_{ws} is the influence factor.

$$I_{ws} = 2 + 0.35\sqrt{\frac{L}{D}} \tag{15.33}$$

Vesic (1977) also proposed a simple empirical relation as follows:

$$(S_e)_2 = \frac{Q_{ws}C_s}{Lq_p} \tag{15.34}$$

where C_s is an empirical constant, $C_s = (0.93 + 0.16\sqrt{L/D})C_p$. The values of C_p are estimated from Table 15.11.

$$(S_e)_3 = \frac{q_{wp}D}{E_s}\left(1-\mu_s^2\right) I_{wp} \tag{15.35}$$

where

D = the width or diameter of the pile
Q_{wp} = point load per unit area at the pile point (Q_{wp}/A_p)
E_s = the modulus of elasticity of soil at or below the pile point
μ_s = Poisson's ratio of soil
I_{wp} = an influencing factor equal to 0.85

Vesic (1977) proposed a semiempirical method for obtaining the magnitude of the settlement of S_{e2} as below

$$(S_e)_3 = \frac{Q_{wp}C_p}{Dq_p} \tag{15.36}$$

where q_p is the ultimate point resistance of the pile and C_p is an empirical coefficient

15.6.1 Negative Skin Friction

Relative movements between the pile shaft and the surrounding soil can result in either negative skin friction or shaft resistance. Negative skin friction refers to the shear forces acting along the pile surface induced by the soil. As a result, each pile reaches an equilibrium between the applied loads, including the loads on the pile head, the drag force caused by negative skin friction in the upper portion of the pile, and the shaft and toe resistances in the lower portion of the pile. The location where the forces are in equilibrium is referred to as the neutral plane. The neutral plane determines the depth at which the shear stresses along the pile transition from negative skin friction to positive skin friction, and no relative displacement occurs between the pile and the soil (Figure 15.11).

15.7 Resistance Distribution; Residual Stress

There are different theories about how the toe and shaft resistances are distributed along with the pile embedment depth. However, the estimation of bearing capacity using basic equations shows that the shaft and toe resistances of the piles are a function of the effective stress and increase with increasing depth (pile length).

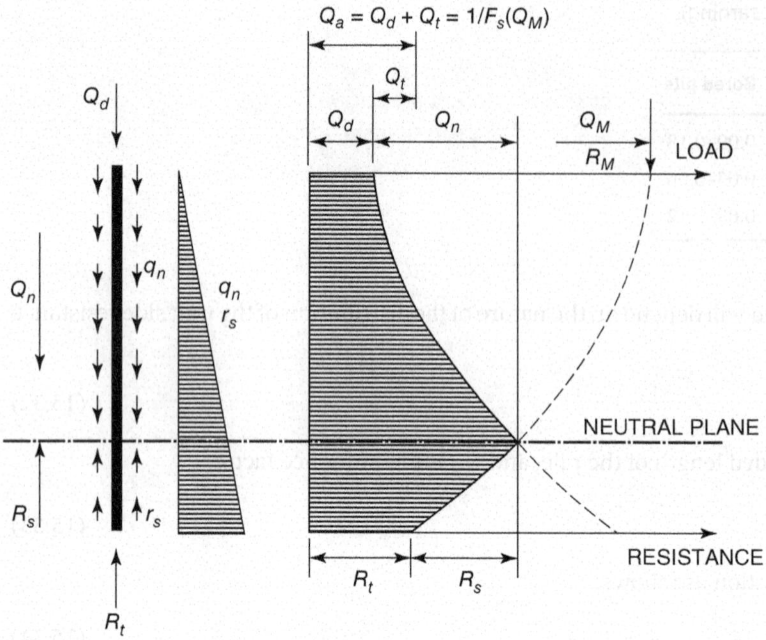

Figure 15.11 The location of the neutral plane.

Some researchers believe that after the depth referred to as the critical depth, the shear strength along the pile shaft does not increase linearly with the depth, and it remains constant (Nordlund 1963; Vesic 1967; Meyerhof 1976). According to the critical depth theory, the ratio of critical depth to pile diameter (L_c/d) is a function of the internal friction angle of the soil. Therefore, as the angle of internal friction of the soil increases, so does the critical depth ratio.

On the other hand, some researchers believe that the pile shaft and toe resistances after the critical depth have neither a constant value nor a linear trend. Instead, the distribution of pile resistance with depth is nonlinear (Fellenius and Altaee 1995). Furthermore, the soil's internal friction angle will not increase as the critical depth ratio increases but may be variable or even decreasing (Fellenius 2015). Figure 15.12 illustrates how the toe and shaft resistances are distributed along with the pile depth.

Indeed, reconsolidating the soil around the pile after its installation results in the settlement of soil layers around the pile. Accordingly, negative skin friction will be developed along the pile shaft, which, in turn, creates a compressive force in a pile called "residual compression." The residual force generated at the top of the driven pile is balanced by force due to the positive friction generated at the bottom of the driven pile and the tip resistance. Therefore, incorrect interpretations are obtained if the residual force is not considered in the pile static loading test results.

Figure 15.13a illustrates the Load transfer mechanism (axial force variation along with the pile depth) with and without considering the residual force. Six different forces were applied to the pile head. Figure 15.13a shows that the frictional resistance at the top of the pile is at its maximum, and its value decreases with increasing depth, and at the pile toe level, it reaches zero. The latter means that the axial force in a pile remains constant from a certain depth down (i.e., critical depth) due to the small frictional resistance. Therefore, this was considered as a reason for considering critical depth. Figure 15.13b shows that the assumption of critical depth is meaningless if the residual force is considered in the analysis. It is noteworthy that the small effect of removing the residual force in the analysis is to reduce the contribution of frictional resistance and increase the contribution of tip resistance. Fellenius and Altaee (1995) argued that neglecting the presence of residual loads in full-scale pile load tests leads to regular load distribution. This load distribution appears linear below some particular depth, while residual loads existed prior to pile loading and analysis.

Eslami et al. (2017) applied different analytical methods comprising static analysis, unified method, API approach, and CPT direct correlations to determine the bearing capacity of piles. As depicted in Figure 15.14, Comparisons demonstrated that bearing capacity at the critical depth is neither linear-constant nor linear as suggested by unified codes. Generally, implementing direct and indirect CPT-based method calculations and comparing them with measured records revealed

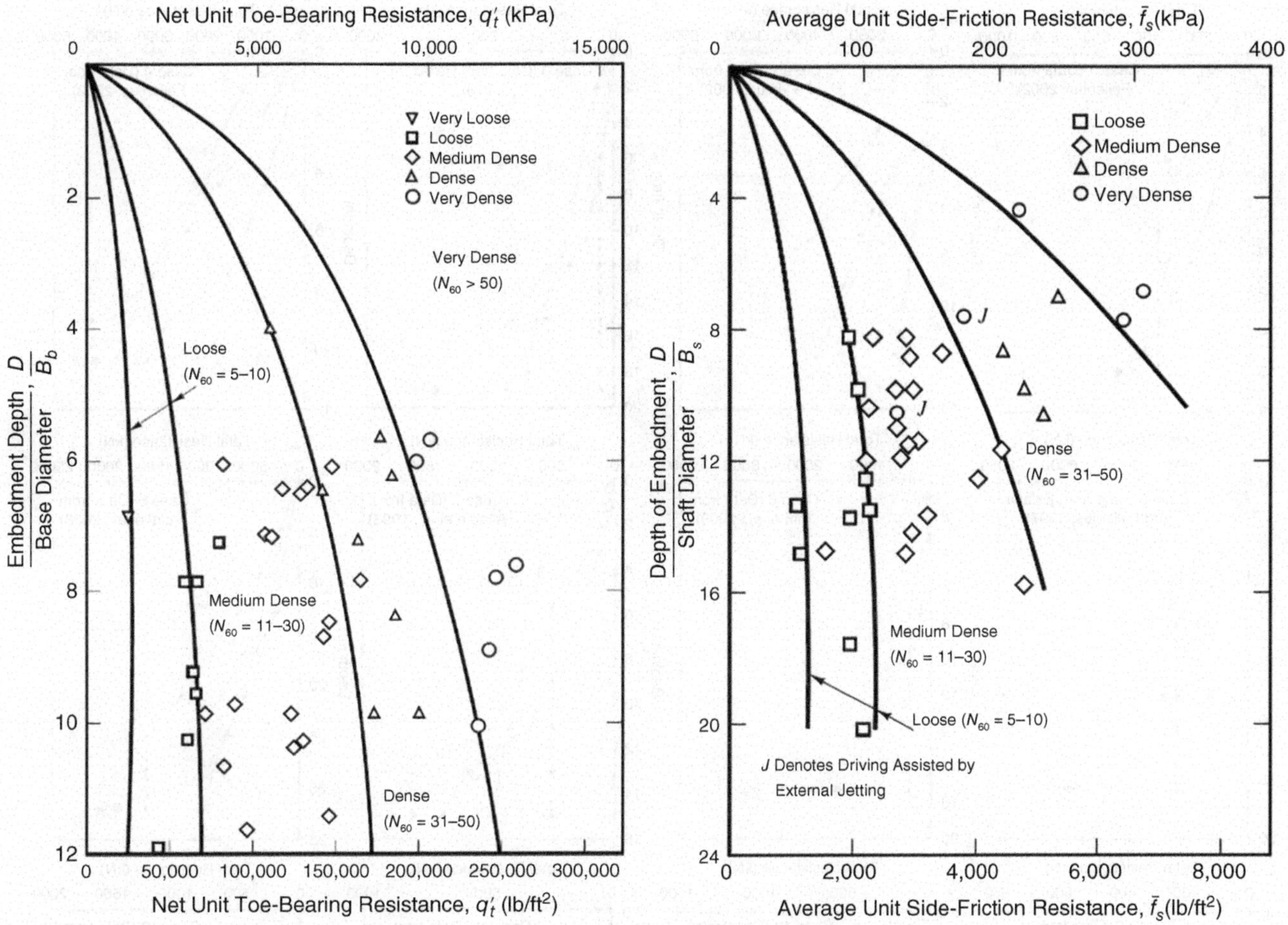

Figure 15.12 The distribution of toe and shaft resistances with pile embedment depth (Coduto et al. 2016).

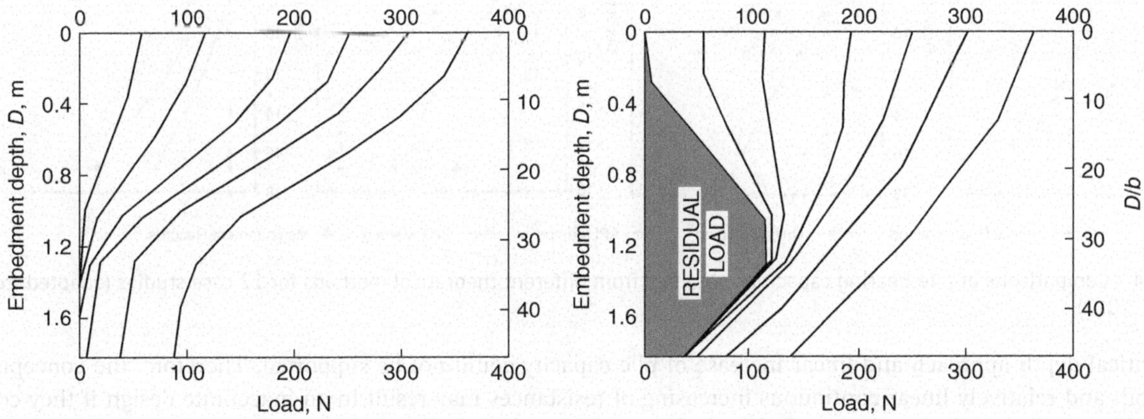

Figure 15.13 Pile axial load distribution: (a) ignoring residual loads, (b) considering residual loads (adapted from Hanna and Tan 1973).

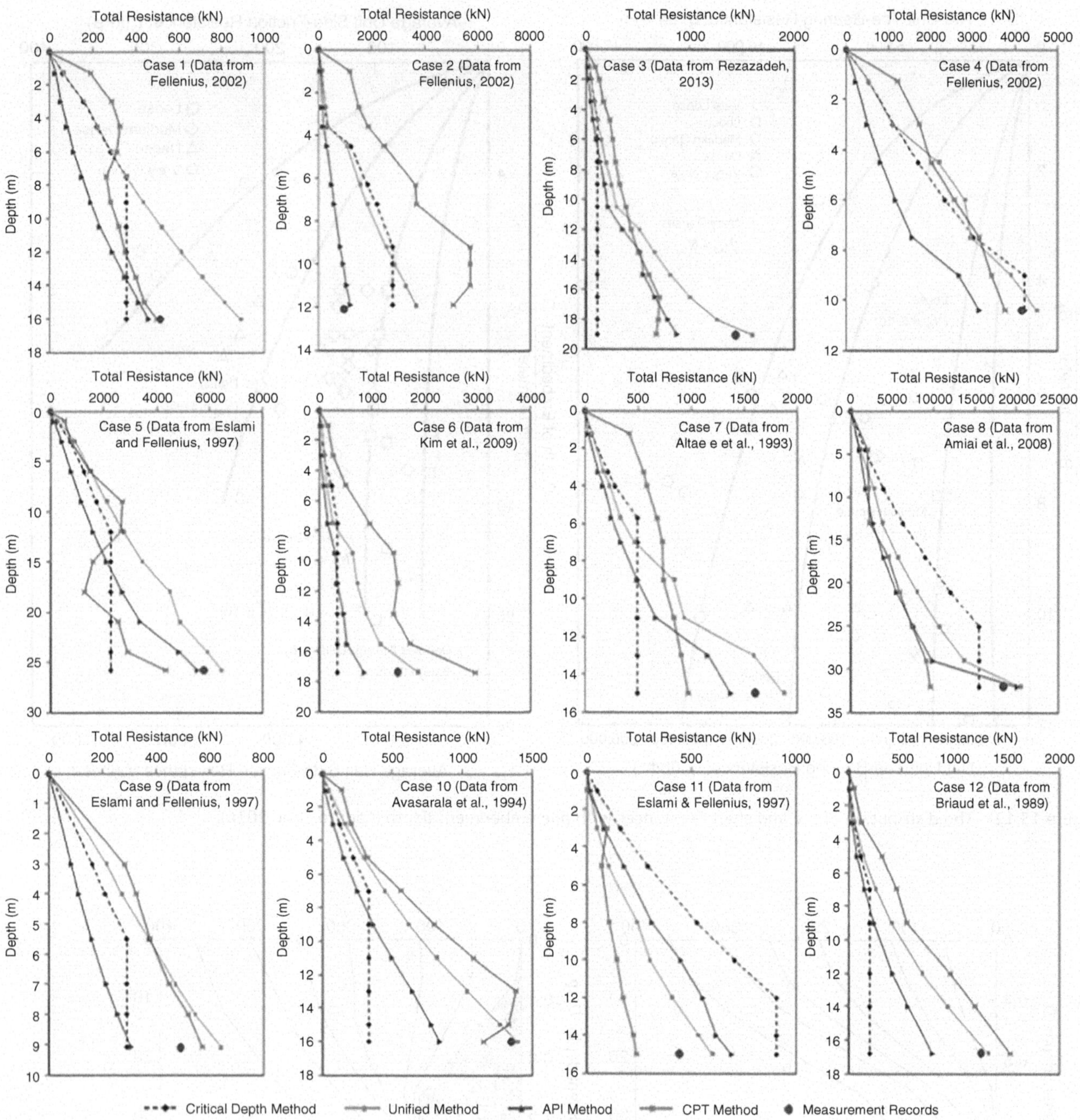

Figure 15.14 Comparisons of pile-bearing capacities obtained from different theoretical methods for 12 case studies (adapted from Eslami et al. 2017).

that the critical depth approach and linear increase of pile capacity could not be supported. Therefore, the concepts of critical depth and relatively linear continuous increasing of resistances may result in an inaccurate design if they come solely into capacity determination, especially for long piles.

15.8 Helical Piles Capacity

The use of helical piles has become a popular technique among geotechnical engineers for improving problematic soils due to their efficient installation and optimal construction material (Sakr 2015; Motamedinia et al. 2019). Figure 15.15 illustrates two common failure modes observed in helical piles: individual and cylindrical failure. In individual failure mode, the load

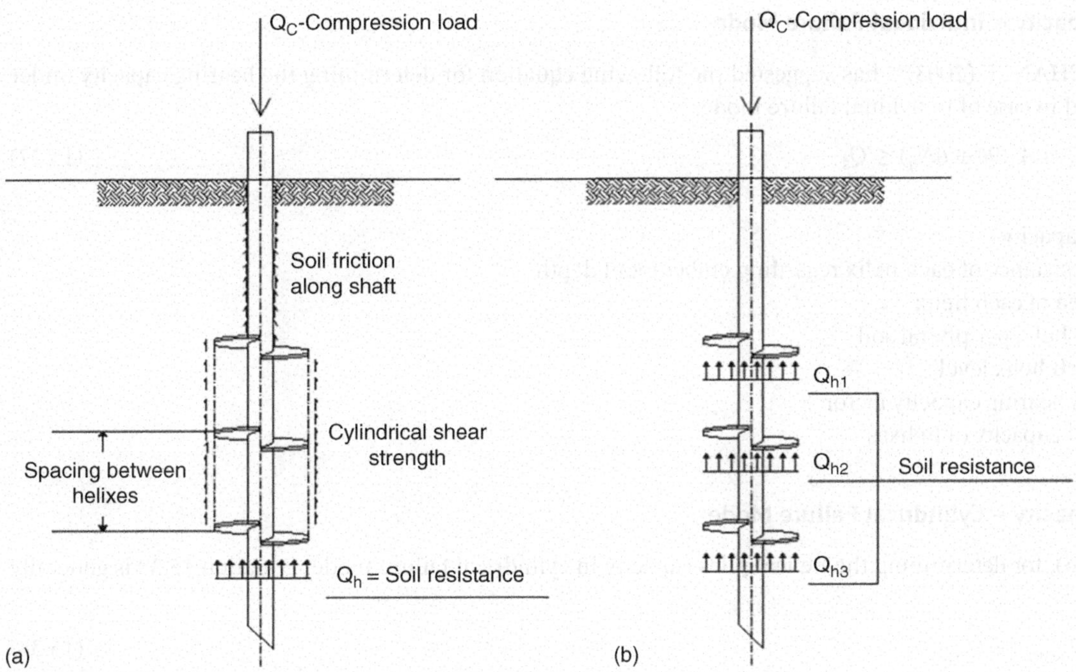

Figure 15.15 Types of failure: (a) cylindrical, (b) individual (Arabameri and Eslami 2021 / Springer Nature).

is transferred through the helix using a bearing mechanism. On the other hand, in cylindrical failure mode, the load is transmitted to the soil through the frictional interaction between the cylindrical surface around the helixes and the bearing strength of the lower helix (Perko 2009).

In the individual failure mechanism, the spacing between helical plates is significant enough that each plate experiences bearing failure independently without any interaction or interference with adjacent helixes. On the other hand, in the cylindrical failure mechanism, the soil is trapped between adjacent helixes, leading to the development of a cylindrical shearing surface between the upper and lower helixes.

The optimal distance between helixes is determined by achieving the same bearing capacity for the pile under both the individual and cylindrical failure mechanisms. This can be achieved when the distance between helixes, expressed as a ratio to the helix diameter, is increased. With a larger distance between helixes, the interacting stress bubbles are farther apart, allowing each helix to behave independently. On the other hand, reducing the distance between helixes brings the stress bubbles closer together, leading to their interaction and the development of a cylindrical failure mechanism. Sakr (2009) estimated that the optimal distance between helixes should be approximately three times the average diameter of the helixes to ensure the desired performance and load transfer mechanism. This distance ensures that the helixes operate under the individual failure mode, maximizing their bearing capacity and minimizing interaction between adjacent helixes.

The installation of helical piles involves a spiral motion of the helixes, which leads to soil disturbance in the surrounding area. This disturbance results in vertical and radial displacement of the soil during installation, causing changes in the geotechnical parameters and bearing capacity of the soil beneath the helix. Various factors influence the soil disturbance, including the diameter, pitch, geometry, number, and spacing of the helixes, as well as the soil type, compaction, penetration rate, and applied crowd load. Studies by Tsuha et al. (2012), Spagnoli and de Hollanda Cavalcanti Tsuha (2020), Khazaei and Eslami (2016a,b), and Sharma et al. (2020) have investigated the effects of these parameters on soil disturbance during helical pile penetration. Perko (2009) specifically examined the impact of pitch number, spacing, geometry, and diameter of the helixes on installation and soil disturbance. The research findings indicate that the influence of these parameters on pile capacity depends on the number of helixes, their diameter, and the relative density of the soil. As a general rule, the compression capacity of the pile increases with an increase in the number of helixes, their diameter, and the relative density of the sand. This leads to a steeper pile load–displacement curve. However, it should be noted that the increased passage of more helixes, particularly in compacted sandy soils, can result in greater soil disturbance near the upper regions of the pile. This soil disturbance reduces the strength characteristics of the soil and subsequently affects the load-bearing capacity of the pile (Perko 2009).

15.8.1 Bearing Capacity – Individual Failure Mode

American company CHANCE (2003)** has suggested the following equation for determining the bearing capacity under compressive axial load in case of individual failure mode:

$$Q_t = \sum Q_h; Q_h = A_h(9c + qN_q) \leq Q_s \quad (15.37)$$

where

Q_t = total bearing capacity
Q_h = end bearing resistance of each helix regarding embedment depth
A_h = the contact area of each helix
C = cohesion of the helix peripheral soil
q = surcharge in each helix level
N_q = dimensionless bearing capacity factor
Q_s = structural limit capacity of helixes

15.8.2 Bearing Capacity – Cylindrical Failure Mode

Based on CFEM (2006), for determining the helical piles capacity in cylindrical failure mode, Equation 15.38 is generally applied:

$$Q_c = Q_{helix} + Q_{bearing} + Q_{shaft} \quad (15.38)$$

where;

Q_{helix} = shaft resistance of failure cylinder
$Q_{bearing}$ = end bearing of the lowest helix
Q_{shaft} = side capacity of the central shaft
In particular, for cohesive soils:

$$Q_{helix} = S_f \cdot (\pi \cdot D \cdot L_C) \cdot S_u \quad (15.39)$$

$$Q_{bearing} = A_H \cdot S_u \cdot N_C \quad (15.40)$$

$$Q_{shaft} = \pi \cdot d \cdot H_{eff} \cdot \alpha \cdot S_u \quad (15.41)$$

where

S_f = the index of helixes spacing ratio, i.e., the ratio of helix spacing to diameter (S/D), which is calculated through:

$$\text{For } \frac{S}{D} < 1.5 \rightarrow S_f = 1 \quad (15.42)$$

$$\text{For } 1.5 \leq \frac{S}{D} \leq 3.5 \rightarrow S_f = 0.683 + 0.069\left(3.5 - \frac{S}{D}\right) \quad (15.43)$$

$$\text{For } 3.5 < \frac{S}{D} \leq 4.6 \rightarrow S_f = 0.700 + 0.148\left(4.6 - \frac{S}{D}\right) \quad (15.44)$$

D = diameter of helixes (or an average of diameters in conical helical piles)
L_C = the space between the shallowest and deepest helix
S_u = undrained shear strength of soil
N_c = dimensionless bearing capacity factor
d = diameter of the central shaft
H_{eff} = effective height, equal to the difference between the first helix embedment depth and its diameter ($H_1 - D_1$)
The following equations are applicable for homogenous cohesionless soils and in the case of groundwater level at the ground surface:

$$Q_{helix} - \frac{1}{2} \cdot \pi \cdot D_a \cdot \gamma' \cdot (H_3^2 - H_1^2) \cdot K_h \cdot \tan\phi \quad (15.45)$$

$$Q_{bearing} = \gamma' \cdot H \cdot A_H \cdot N_q \quad (15.46)$$

$$Q_{shaft} = \frac{1}{2} \cdot P_S \cdot H_{eff}^2 \cdot \gamma' \cdot K_h \cdot \tan\phi \quad (15.47)$$

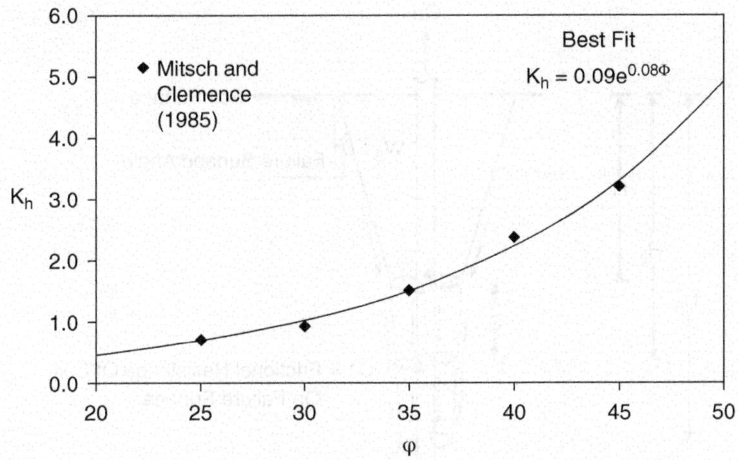

Figure 15.16 Relation of K_h factor and soil internal friction angle (φ).

where

D_a = average diameter of helixes
H_i = embedment depth of each helix
K_h = lateral earth pressure under compressive loading (Figure 15.16)
ϕ = internal friction angle of soil
P_S = perimeter of the central shaft

Example 15.4 Estimate the bearing capacity of the given helical pile according to the cylindrical failure pattern.

$d = 73$ mm
$D = 305$ mm
$S/D = 2$
$H = 5$ m
$S_u(\text{kg/cm}^2) = 0.25 + 0.05Z(\text{m})$
$\alpha = 0.84$

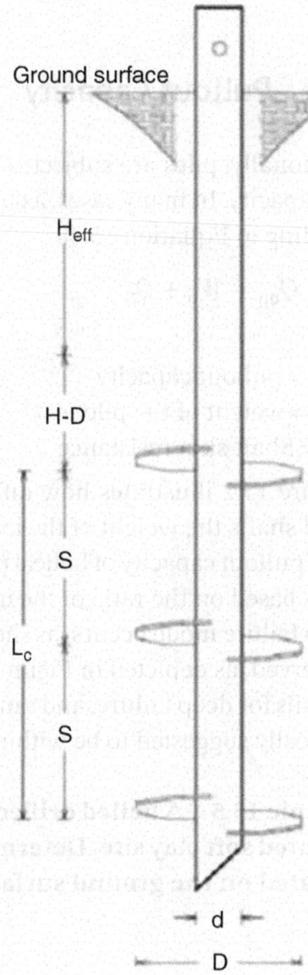

$$\text{Helix area} = \pi/4 \times 0.305^2 = 0.073\ \text{m}^2$$

$$\text{Shaft area} = \pi/4 \times 0.073^2 = 0.00418\ \text{m}^2$$

$$Q_c = Q_{helix} + Q_{bearing} + Q_{shaft}$$

$$Q_{helix} = s_f(\pi D L_c)C_u$$

$$s_f = 0.683 + 0.069(3.5 - 2) = 0.7865$$

$$z = 5\ \text{m} \rightarrow C_u = 0.25 + 0.05 \times 5 = 0.5\ \text{kg/cm}^2$$

$$z = 6.22\ \text{m} \rightarrow C_u = 0.25 + 0.05 \times 6.22 = 0.561\ \text{kg/cm}^2$$

$$\Rightarrow (C_u)_{avg} = 5.31\ \text{ton/m}^2$$

$$Q_{helix} = 0.7865 \times \pi \times 0.305 \times (4 \times 0.305) \times 5.305 = 4.88\ \text{ton}$$

$$Q_{bearing} = 0.073 \times 9 \times 5.61 = 3.69\ \text{ton}$$

$$Q_{shaft} = \pi D H_{eff} \alpha C_u$$

$$H_{eff} = 5 - 0.305 = 4.695$$

$$z = 0\ \text{m} \rightarrow C_u = 0.25 + 0.05 \times 0 = 0.25\ \text{kg/cm}^2$$

$$z = 4.695\ \text{m} \rightarrow C_u = 0.25 + 0.05 \times 4.695 = 0.48\ \text{kg/cm}^2$$

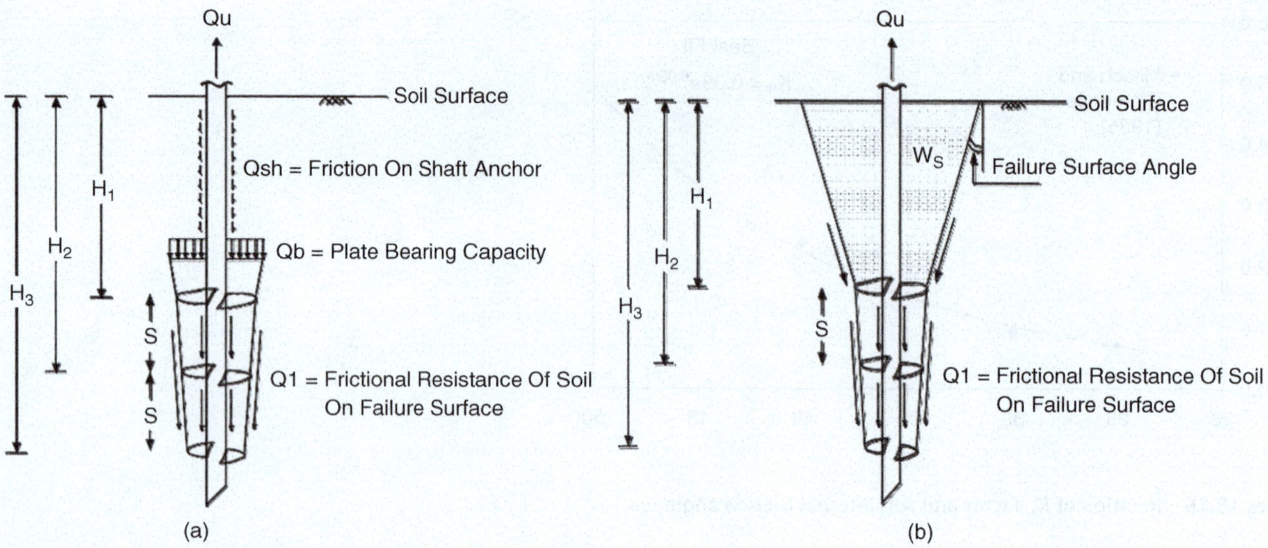

Figure 15.17 (a) Proposed failure mode for multi-helix anchors in clay and silt (deep mode), (b) failure surface for multi-helix anchors in sand (shallow mode) (Salem and Hussein 2017 / Salem and Hussein).

$\Rightarrow (C_u)_{avg} = 3.65\ ^{ton}/_{m^2}$

$Q_{shaft} = \pi \times 0.073 \times 4.695 \times 0.84 \times 3.65 = 3.3$ ton

$Q_c = Q_{helix} + Q_{bearing} + Q_{shaft} = 3.3 + 3.69 + 4.88 = 11.87$ ton

15.9 Pullout Capacity

Occasionally, piles are subjected to upward loads. Both the weights of the pile and its shaft friction contribute to pullout load capacity. In many cases, axial and lateral loads are applied simultaneously. The pullout capacity of the pile is estimated according to Equation 15.48.

$$Q_{up} = W_p + Q_S \tag{15.48}$$

where

Q_{up} = pullout capacity
W_p = weight of the pile
Q_S = Shaft skin resistance

Figure 15.2 illustrates how different bearing components are activated during loading. For the enlarged base (belled) drilled shafts the weight of the involved soil is also considered in uplift capacity.

The pullout capacity of helical piles has been extensively investigated by researchers, who have identified two main failure modes based on the ratio of the upper helix embedment depth to the helix diameter (H_1/D). When H_1/D is greater than 5, a deep failure mode occurs, as shown in Figure 15.17a. On the other hand, for H_1/D values below 5, a shallow failure mode is observed, as depicted in Figure 15.18b. These failure modes are more predominant in fine-grained soils such as silty and clay soils for deep failure, and sandy soils for shallow failure. The angle of the failure surface, as illustrated in Figure 15.17b, is typically suggested to be within the range of $\varphi/4$ to $\varphi/2$, where φ represents the soil friction angle.

Example 15.5 A belled drilled shaft (enlarged base) with a total length of 15 meter is implemented in a uniform saturated soft clay site. Determine the ultimate pullout capacity according to the figure and the given data. GWT is located on the ground surface.

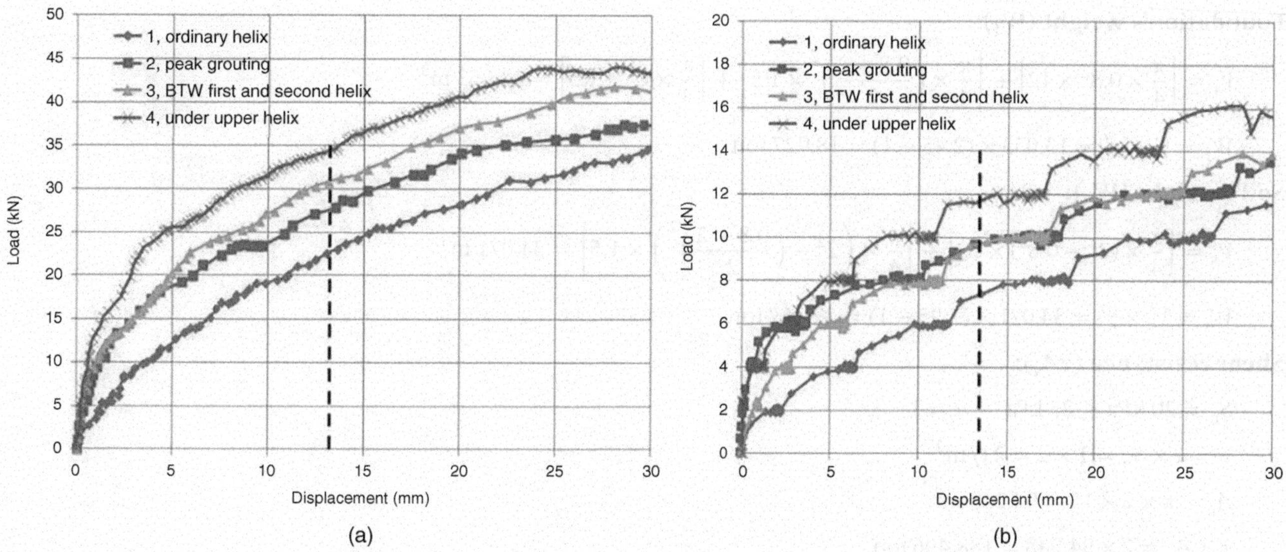

Figure 15.18 Post-grouted helical model piles performance: (a) compression tests, (b) tension tests (Khazaei and Eslami 2016a,b / Conscientia Beam).

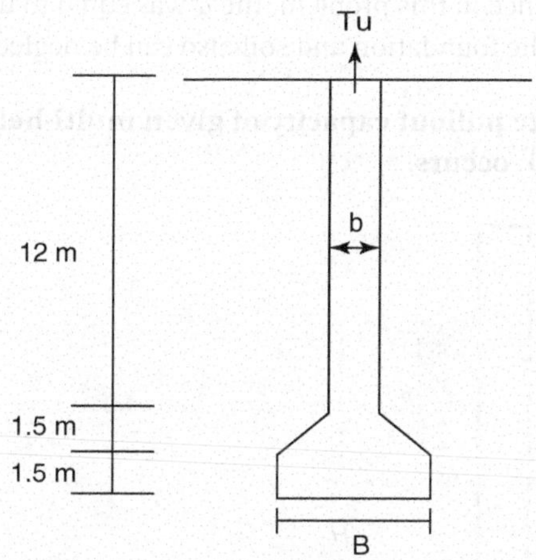

$$b = 80\ \text{cm}, B = 200\ \text{cm}, S_u = 20\ \text{kPa}, \gamma_{sat} = 1.95\frac{\text{t}}{\text{m}^3}, \gamma_{con} = 2.45\ \text{t/m}^3$$

The resistant components against uplift loads are depicted below. The resistant zone of shear forces around the foundation is assumed to be conservative and straightforward in the form of a cylinder.

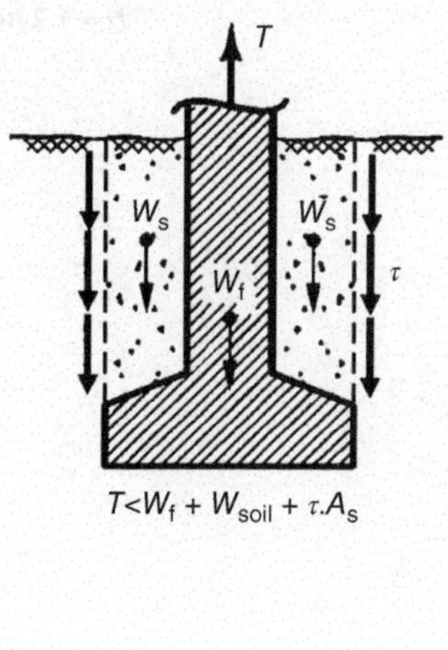

Foundation's weight (W_f):

$$V_f = \left[\frac{\pi}{4} \times 0.8^2 \times 12\right] + \left[\frac{\pi}{4} \times \left(\frac{0.8+2}{2}\right)^2 \times 1.5\right] + \left[\frac{\pi}{4} \times 2^2 \times 1.5\right] = 13.053\ \text{m}^3$$

$$W_f = V_f \times \gamma' = 13.053 \times (2.45 - 1) = 18.927\ \text{ton}$$

Soil's weight (W_s):

$$V_s = \left[\frac{\pi}{4} \times (2^2 - 0.8^2) \times 12\right] + \left[\frac{\pi}{4} \times \left(2^2 - \left(\frac{0.8+2}{2}\right)^2\right) \times 1.5\right] = 34.071\ \text{m}^3$$

$$W_s = V_s \times \gamma' = 34.071 \times (1.95 - 1) = 32.367\ \text{ton}$$

Shear resistance (τA_s):

$$S_u = 20\ \text{kPa} \leq 25\ \text{kPa} \rightarrow \alpha = 1$$

$$\tau = \alpha \times S_u = 1 \times 2 = 2\ \text{t/m}^2$$

$$A_s = \pi \times 2 \times 15 = 94.248\ \text{m}^2$$

$$\tau \times A_s = 2 \times 94.248 = 188.496\ \text{ton}$$

$$T_u = W_f + W_s + \tau A_s = 13.053 + 34.071 + 188.496 = 235.62\ \text{ton}$$

It should be noticed that in similar problems, usually, the shear resistance between two surfaces of soil is S_u, however αS_u can be used instead conservatively. Since in this problem, the α was equal to unity, there was no difference. The shear resistance between the lowest section of the foundation and soil also can be neglected considering conservatism.

Example 15.6 Determine the ultimate pullout capacity of given multi-helix helical pile. Assume that shallow failure mode with a failure angle of 10° occurs.

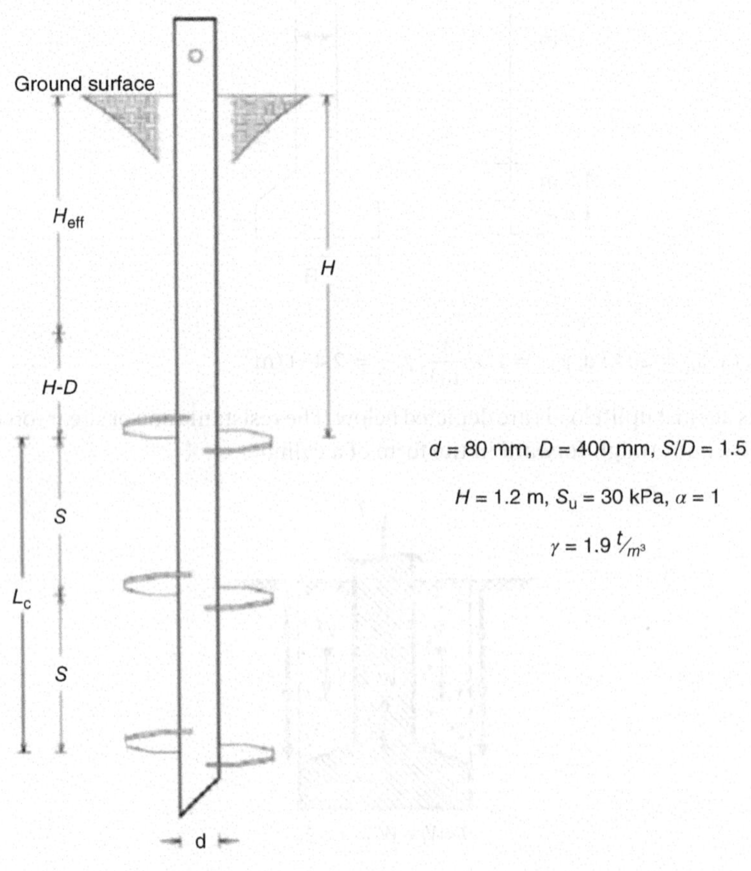

$$\frac{H_1}{D} = \frac{1.2}{0.4} = 3 < 5 \rightarrow \text{Shallow failure mode}$$

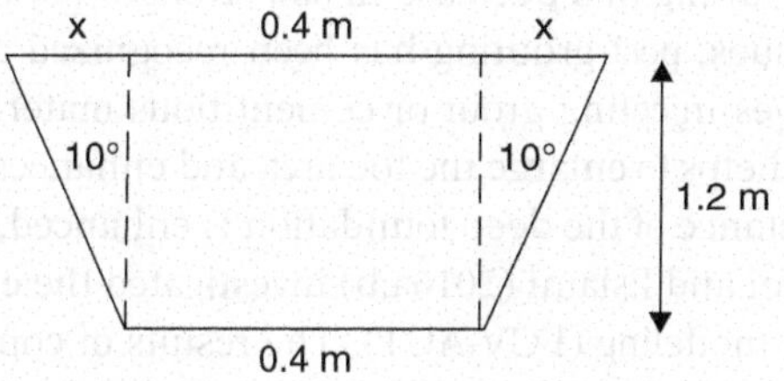

Upper wedge

$$\tan 10 = \frac{x}{1.2} \rightarrow x = 0.21\text{m}$$

Bigger diameter: $D_2 = 0.4 + (2 \times 0.21) = 0.82$ m

Weight of the soil in the wedge

$$v_{\text{soil}} = \left(\frac{\pi}{4} . D_{\text{avg}}{}^2 . H_1\right) - \frac{\pi d^2 H_1}{4}$$

$$D_{\text{avg}} = 0.61 = \frac{0.82 + 0.4}{2}$$

$$v_{\text{soil}} = \frac{\pi}{4} \times 1.2(0.61^2 - 0.08^2) = 0.345\text{m}^3$$

$$\omega_{\text{soil1}} = (\gamma - \gamma_\omega) \times v = (1.9 - 1) \times 0.345 = 0.311\text{ton}$$

Shaft resistance of the wedge

$$R_{\text{S1}} = S_{\text{u}} \times \pi \times D_{\text{avg}} \times H_1 = 30 \times \pi \times 0.61 \times 1.2 = 69 \text{ KN} = 6.9 \text{ ton}$$

Cylinder of Helixes

Weight of the soil in the cylinder

$$v_{\text{soil}} = \frac{\pi}{4} \times D^2(2s) - \frac{\pi}{4} d^2(2s)$$

$$v_{\text{soil}} = \frac{\pi}{4}(2 \times 0.6)(0.4^2 - 0.08^2) = 0.145 \text{ m}^3$$

$$\omega_{\text{soil2}} = (\gamma - \gamma_\omega) \times v = (1.9 - 1) \times 0.145 = 0.131 \text{ ton}$$

Shaft resistance of the cylinder

$$\frac{S}{D} = 1.5$$

$$S_{\text{f}} = 0.683 + 0.069(3.5 - 1.5) = 0.821$$

$$R_{\text{S2}} = S_{\text{u}} \times \pi \times D \times 2S \times S_{\text{f}} = 30 \times \pi \times 0.4 \times 2 \times 0.6 \times 0.821 = 37.141 \text{ KN} = 3.7 \text{ ton}$$

Total pullout capacity

$$T_{\text{u}} = (\omega_{\text{soil1}} + \omega_{\text{soil2}}) + (R_{\text{S1}} + R_{\text{S2}}) = 0.311 + 0.131 + 6.9 + 3.7 = 11 \text{ ton}$$

15.10 Post-Installation Capacity Variations

Variations in deep foundation resistance can occur due to various factors following installation or construction. Among these factors, four significant influences have been identified and discussed: post-grouting, soil setup and relaxation, the combined application of deep foundation and ground modification techniques, and fluctuations in-ground and groundwater levels.

15.10.1 Post-Grouting

Construction of drilled shafts can sometimes encounter challenges such as debris flow in the base and defects in the shaft, which can lead to a reduction in the estimated toe and shaft resistance. Additionally, the installation of piles can disturb the surrounding soil, resulting in a decrease in soil strength parameters and a subsequent decrease in geotechnical capacity. To address these issues, post-grouting has been recognized as a compensation solution and is widely used in practice. Post-grouting involves injecting grout or cementitious materials into the surrounding soil and can be applied to drilled shafts. This process helps to enlarge the toe area and enhances the interaction between the shaft and the surrounding soil. As a result, the resistance of the deep foundation is enhanced, as discussed in Chapter 14.

A study by Khazaei and Eslami (2016a,b) investigated the effect of post-grouting on the behavior of model-scale helical piles using physical modeling (FCV-AUT). The results of compression and tension loading tests, shown in Figure 15.18, indicate an improvement of approximately 25 to 50% in both tension and compression states.

15.10.2 Soil Setup and Relaxation

Variation of resistance with time has been recognized as one of the most essential issues in driven piles. This issue is well-studied, and it is understood that regarding the soil type, both soil setup and relaxation may happen. Soil setup and relaxation result in an increase and decrease in the pile-bearing capacity, respectively. Based on reported cases, the setup has occurred more often than relaxation. The generation of excess pore water pressure during pile driving and its dissipation with time, as well as aging, are two main reasons that are stated for setup phenomena.

Fakharian et al. (2013) have reported the piling practice of a 32-hectare utility plant located in the southwest of Iran near the Persian Gulf. About 5000 spun piles with an outer diameter of 45 cm were driven down to embedment depths ranging between 14 to 22 m, to support 12 water tanks. Pile dynamic tests (PDA) and static load tests were carried out on 30 test piles and 221 construction piles. The general soil profile consists of a very soft to stiff silty clay, with an average depth of 15 m, overlain by medium dense to dense sand, with 4 to 8 m thickness. Based on shaft capacities from dynamic analysis, β factor versus depth is back-calculated. As shown in Figure 15.19, the β factor for the restrike condition, which is driving again after a while is appreciably greater than those for the EOD condition or at the end of the initial drive, representing the occurrence of soil setup.

15.10.3 Combination of Deep Foundation and Ground Modification

Ground modification techniques can be effectively used in conjunction with deep foundations to provide additional support and address specific geotechnical challenges. One such technique is massive grouting, which involves injecting grout into the soil to improve its strength and stability. This can enhance the capacity of the shaft and toe of deep foundations, increasing their load-bearing capacity. Another technique is the installation of micro-piles, which are small-diameter piles that are drilled and grouted into the ground. Micro-piles can be used to reinforce the soil and improve the overall stability of the deep foundation system. They can also be employed to mitigate liquefaction potential in seismic areas and enhance the

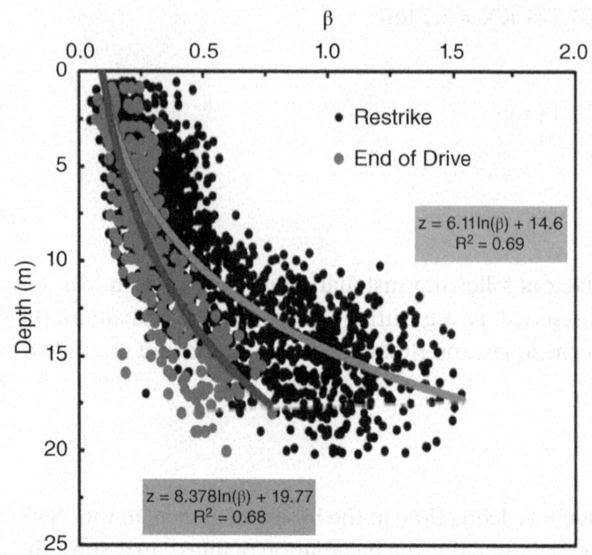

Figure 15.19 β factor variation with depth at EOD and restrike (Fakharian et al. 2013).

resistance of the foundation against lateral loads. By combining deep foundations with ground modification techniques, engineers can achieve economic and sustainable solutions that address both technical and practical requirements. These hybrid approaches provide enhanced performance and ensure the long-term stability and safety of structures. Additionally, they align with the principles of sustainable development by optimizing the use of resources and minimizing environmental impact.

In the study conducted by Ishihara and Cubrinovski (2004), a combination of deep foundations and ground improvement techniques was utilized for the foundation system of an oil storage tank located on an artificial island in Japan. The deep foundations consisted of round pre-stressed precast concrete piles with a diameter of 45 cm and a length of 23 m. These piles were interconnected by a concrete slab with a thickness of 50 cm. To enhance the foundation soil, sand compaction columns were installed around the perimeter of the tank. These columns aimed to densify and strengthen the existing soil. The original soil profile comprised a thick layer of disintegrated weathered granite fill overlaying the original seabed and deep gravel layers. The effectiveness of this combined approach was demonstrated through in situ testing, which showed improvements in the soil parameters and an enhancement in the overall geotechnical performance of the foundation system. By utilizing deep foundations in conjunction with ground improvement techniques, the project successfully addressed the specific challenges posed by the existing soil conditions and ensured the stability and long-term performance of the oil storage tank.

Maeda et al. (2006) conducted a comprehensive review on the composite application of piles and the deep mixing (DM) method for bridge foundations. This approach is particularly beneficial in cases where the bridge foundation is situated on thick soft clay or sand prone to liquefaction. Typically, a large number of piles are required to ensure the desired geotechnical performance in terms of displacement and strength. To address the limitations of the existing soil and improve its resistance, the deep mixing (DM) method, also known as deep soil mixing (DSM), is employed. This technique involves mechanically mixing a cementitious binder or other stabilizing agents with the in situ soil to enhance its properties. By utilizing the deep mixing method, the horizontal displacement can be effectively restrained, which subsequently reduces the number of required piles. This optimization leads to cost savings for the overall project. The combination of piles and deep mixing offers a synergistic approach that capitalizes on the advantages of both techniques. It enables the foundation to overcome the challenges associated with soft and liquefiable soils while optimizing the use of resources and achieving a more cost-effective design.

Courtesy of Sham-e Consulting Engineering Co., 2022 presented a case study involving a hybrid foundation system for a 35-story twin tower located near the coastal line of the Caspian Sea. The site's soil profile primarily consists of marine alluvial sand, occasionally intermixed with layers of silt and clay. To address the challenges posed by the extreme structural loads, nonuniform settlements, high groundwater level, and soil type, a combination of drilled shafts, raft foundations, and micro-piles was proposed and implemented. The drilled shafts, with diameters ranging from 1 to 1.2 m, were constructed to depths of 25 and 30 m. These deep foundations provide the necessary load-bearing capacity and stability required for the twin tower. To further enhance the foundation system and mitigate potential subsurface instability, micro-piles were incorporated as a complementary solution. The micro-piles, with a diameter of 10 cm and length of 12 m, were strategically designed to address specific localized issues and ensure uniform settlements across the site. By employing this hybrid foundation system, the project aimed to achieve a comprehensive and robust solution that considers the site's challenging soil conditions and structural requirements. The combination of drilled shafts, raft foundation, and micro-piles offers a synergistic approach to ensure the stability, integrity, and long-term performance of the twin tower in a complex geotechnical environment.

15.10.4 Fluctuation in Ground and Groundwater Level

As previously mentioned, the occurrence of scouring and degradation in deep foundations is a significant concern, particularly in cohesionless soils. Scouring refers to the loss of soil grains due to the flow of water, which can undermine the stability and load-bearing capacity of deep foundations. Figure 15.20 illustrates an example of scouring and degradation observed in a bridge project supported by deep foundations. In addition to scouring, fluctuations in groundwater level can also have a significant impact on the effective stress and, consequently, the shaft and toe resistance of deep foundations. Groundwater level fluctuations can be caused by various factors such as precipitation, water pumping, or freeze-and-thaw cycles. These fluctuations can induce changes in the soil's mechanical properties, affecting the overall performance and stability of deep foundations. Recognizing the importance of groundwater level fluctuations, the Federal Highway Administration (FHWA 2006) manual for drilled shafts emphasizes the need to consider and address this issue in the design and

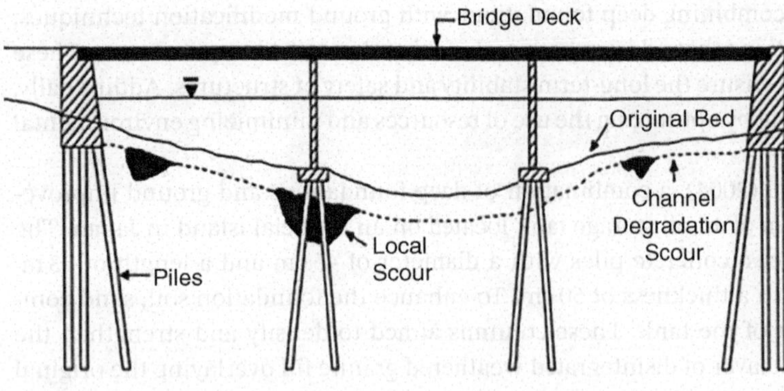

Figure 15.20 Bridge supported on piles affected by scouring.

construction of deep foundations. Proper measures, such as appropriate drainage systems and groundwater control techniques, may be employed to mitigate the adverse effects of groundwater level fluctuations and maintain the integrity and performance of deep foundations.

Case Study 15.1

Predicted and Observed Axial Behavior of Piles: Results of a Pile Prediction Symposium (Finno 1989)

The pile capacity event conducted by Finno (1989) at the Lakefill site on the Evanston Campus of Northwestern University was an important endeavor in foundation engineering. The event focused on conducting axial load tests on four piles, and it attracted significant interest from participating predictors and event attendees. A total of 23 predictors were involved in the event, contributing to the knowledge and understanding of pile capacity.

To gather comprehensive information about the site, a thorough soil investigation was conducted. The investigation included various testing methods such as standard penetration testing with split-spoon sampling and tube sampling, laboratory classification and testing of soil samples, *in situ* vane shear testing, static-cone penetration testing, pressuremeter testing, and dilatometer testing. Based on the soil investigation, two primary soil layers were identified at the site. The upper layer was approximately 7 m (23 feet) thick and consisted of sand. Beneath the sand layer was a thick layer of clay, into which the test piles were installed to a depth of 8 m (27 feet). The phreatic elevation, representing the groundwater level, was reported to be 4.5 m (15 feet) below the ground surface, and the pore pressure was hydrostatically distributed. Figure 15.21 illustrates the test piles and reaction frames used in the axial load testing. These components were essential for conducting the load tests and evaluating the pile capacity under the given soil conditions.

The pile capacity testing conducted by Finno (1989) involved four test piles, two of which were driven piles, and the other two were bored piles. The specifications of the test piles are as follows:

1) **Driven Pipe Pile:**
   - Diameter: 450 mm (18 in)
   - Closed-toe pile with a 9.5 mm (0.375 in) wall
   - Toe-plate diameter: 480 mm (19 in)
   - Nominal cross-sectional steel area: 134 cm^2 (20.8 in^2)
   - Instrumentation: Vibrating-wire strain gages at the toe and Embedment Depths 3, 5, and 11 m (10, 23, and 36 feet)
   - Additional instrumentation: Telltales anchored at four depths
   - Protected by one C8-18.75 channel with a cross-sectional area of 35.5 cm^2 (5.5 in^2)
2) **Driven H-Pile:**
   - Type: 355HP120 (14HP73) pile
   - Nominal cross-sectional steel area: 138 cm^2 (21.4 in^2)
   - Two C8-18.75 channels to protect telltales, each with a cross-sectional area of 35.5 cm^2 (5.5 in^2)

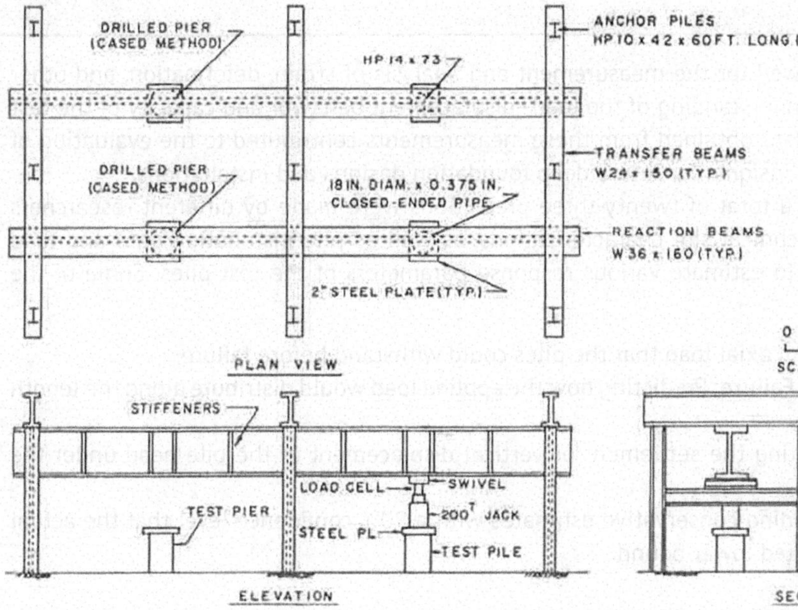

Figure 15.21 Test piles and reaction frames (Finno 1989).

- Instrumentation: Vibrating-wire strain gages at the toe and Embedment Depths 3, 5, and 11 m (10, 23, and 36 feet)
- Additional instrumentation: Telltales anchored at four depths

3) **Bored Pile (Installed through Bentonite Slurry):**
   - Diameter: 450 mm (18 in)
   - Instrumentation: Similar to the driven piles (vibrating-wire strain gages and telltales)
4) **Bored Pile (Installed with a 24-in Casing):**
   - Diameter: 450 mm (18 in)
   - Installed to a depth of 9.4 m (31 feet) using a 24-in casing
   - Instrumentation: Similar to the other bored pile (vibrating-wire strain gages and telltales)

All four test piles were installed to an embedment depth of 15.2 m (50 feet). The instrumentation, including vibrating-wire strain gages and telltales, was crucial for monitoring the behavior and performance of the piles during the axial load tests. Figure 15.22 provides detailed illustrations of the test piles, highlighting their specific features and instrumentation.

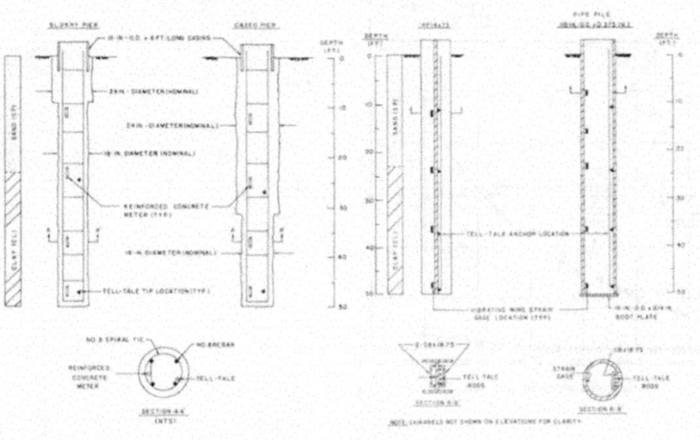

Figure 15.22 Details of test piles (Finno 1989).

(Continued)

Case Study 15.1 (Continued)

The comprehensive instrumentation allowed for the measurement and analysis of strain, deformation, and other relevant parameters, enabling a thorough understanding of the load–displacement behavior and capacity of the test piles under the given soil conditions. The data obtained from these measurements contributed to the evaluation of the pile performance and provided valuable insights for future deep foundation designs and installations.

In the study conducted by Finno (1989), a total of twenty-three predictions were made by different researchers based on the in situ, laboratory, and geotechnical site characterizations, as well as pile installation data and load testing schedule. These predictions aimed to estimate various response parameters of the test piles. Some of the predicted responses included:

1) **Axial Capacity:** Estimation of the maximum axial load that the piles could withstand before failure.
2) **Load Distribution along the Pile Length at Failure:** Predicting how the applied load would distribute along the length of each pile at the point of failure.
3) **Axial Load-Pile Head Settlement:** Estimating the settlement or vertical displacement of the pile head under the applied axial load.
4) **Lower Bound Estimates of Capacity:** Providing conservative estimates with a 90% confidence level that the actual capacity would be higher than the estimated lower bound.

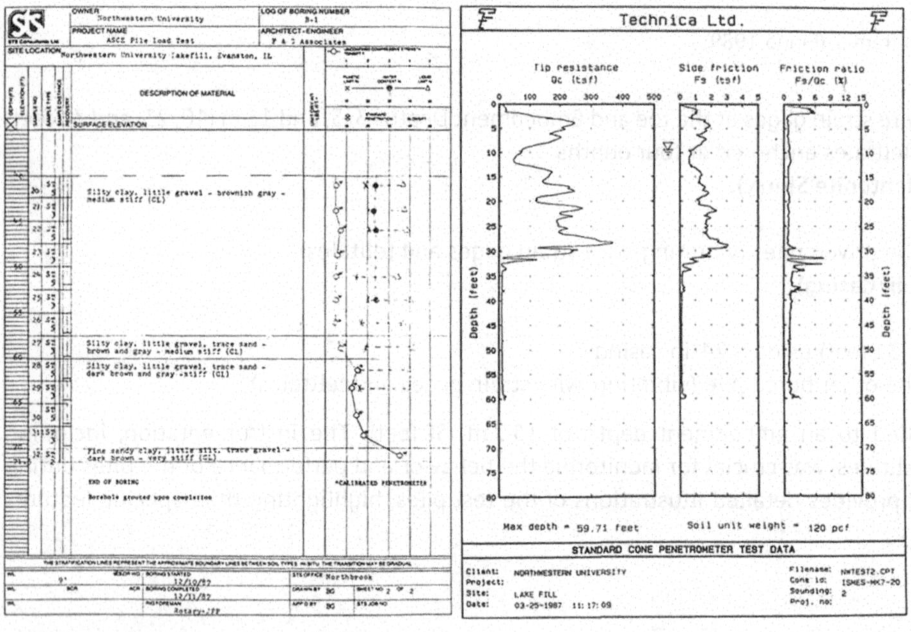

DEPTH (meters)	DEPTH (feet)	Su (tsf)	Su (psf)	REMOLDED Su (tsf)	REMOLDED Su (psf)	SENSITIVITY
9.0	29.5	0.33	660	0.10	192	3.4
10.0	32.8	NO READING, POSSIBLE COBBLE OR GRAVEL				
10.5	34.4	0.41	820	0.21	420	2.0
11.5	37.7	0.33	660	0.15	300	2.2
12.5	41.0	0.37	740	0.19	380	1.9
13.5	44.3	0.40	800	0.21	420	1.9
14.5	47.6	0.41	820	0.21	420	2.0
15.5	50.8	0.38	760	0.16	320	2.4
16.5	54.1	0.43	860	0.19	380	2.3
17.5	57.4	0.34	680	0.21	420	1.6
18.5	60.7	0.78	1560	0.45	900	1.7
19.5	64.0	0.71	1420	0.44	880	1.6
[illegible]	[illegible]	1.79	3580	0.63	1260	2.8

POSSIBLE GRAVEL AT 66 FEET

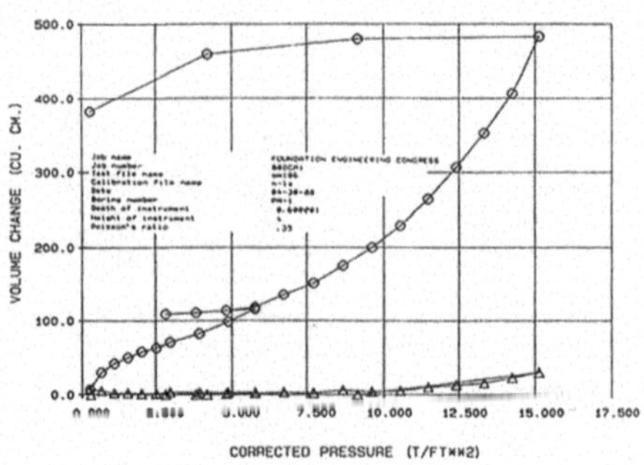

Figure 15.23 Summary of in situ test results (Finno 1989).

5) **Residual Load:** Predicting the residual load that would remain in the piles after the completion of the load testing.
6) **Pore Pressure Generation and Dissipation:** Estimation of the pore pressure changes within the soil during and after the load testing.

The predictions were based on the available data and the methodologies and judgments employed by the researchers. Each researcher utilized their own approach and analysis techniques to estimate the pile axial-bearing capacities. The in situ test results, which provide valuable information about the soil properties and behavior, are summarized in Figure 15.23, aiding in the interpretation and validation of the predictions made by the researchers.

Figure 15.24 shows the axial load–settlement response of piles, and Figure 15.25 provides a typical axial load distribution in two weeks.

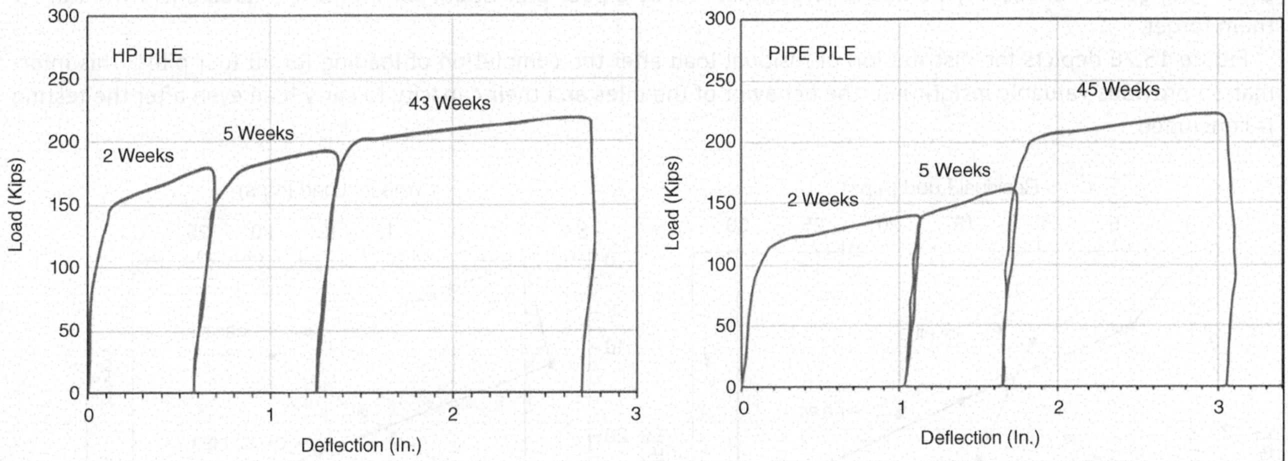

Figure 15.24 Axial load–deflection response of driven piles (Finno 1989).

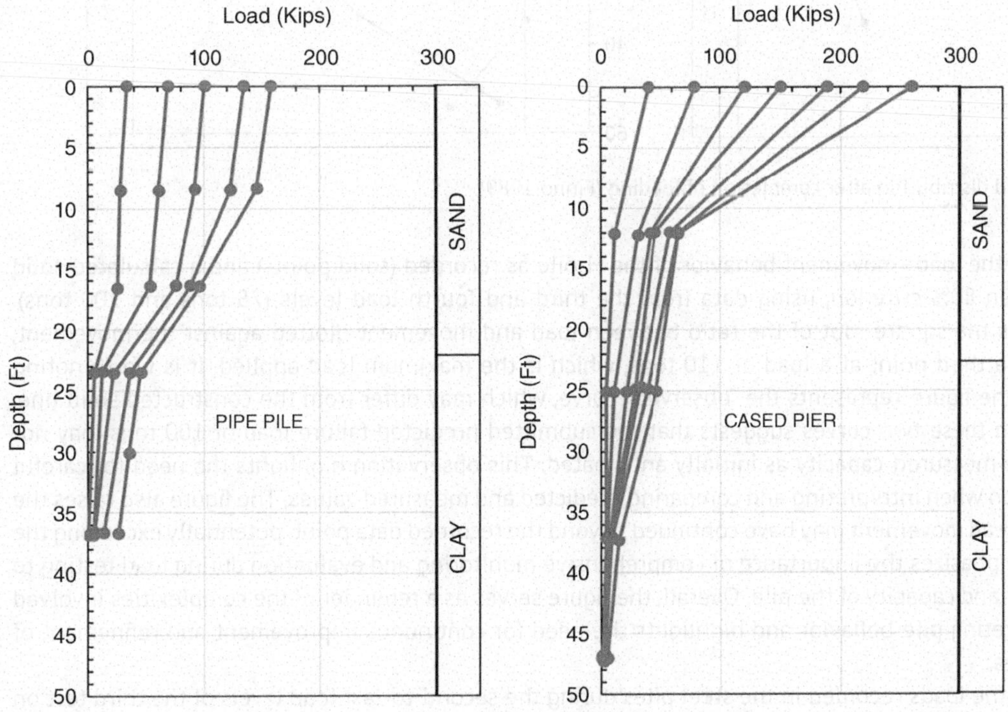

Figure 15.25 Typical load distribution in two-week tests (Finno 1989).

(Continued)

Case Study 15.1 (Continued)

The primary takeaway from the prediction exercise is not the comparison of individual predictions with the measured values but rather the assessment of the collective or average prediction in relation to the measurements. This evaluation provides insight into the predictive ability of the profession as a whole. It is not about identifying who is exceptionally accurate or significantly off in their predictions. Instead, the focus should be on recognizing those individuals who actively contribute to the profession through their participation and willingness to make predictions.

In essence, the question should be: "Who is taking the initiative to make predictions and contribute to the advancement of the profession?" It is more important to foster an environment where individuals are encouraged to make predictions, even if they may not always be accurate. It is through this active engagement that progress is made. As the saying goes, "I expect my friends to forget that I made a poor prediction, but if I make a good one, I will not let them forget."

Figure 15.26 depicts the distribution of residual load after the completion of loading for all four piles. This information provides valuable insight into the behavior of the piles and their capacity to carry load even after the testing is concluded.

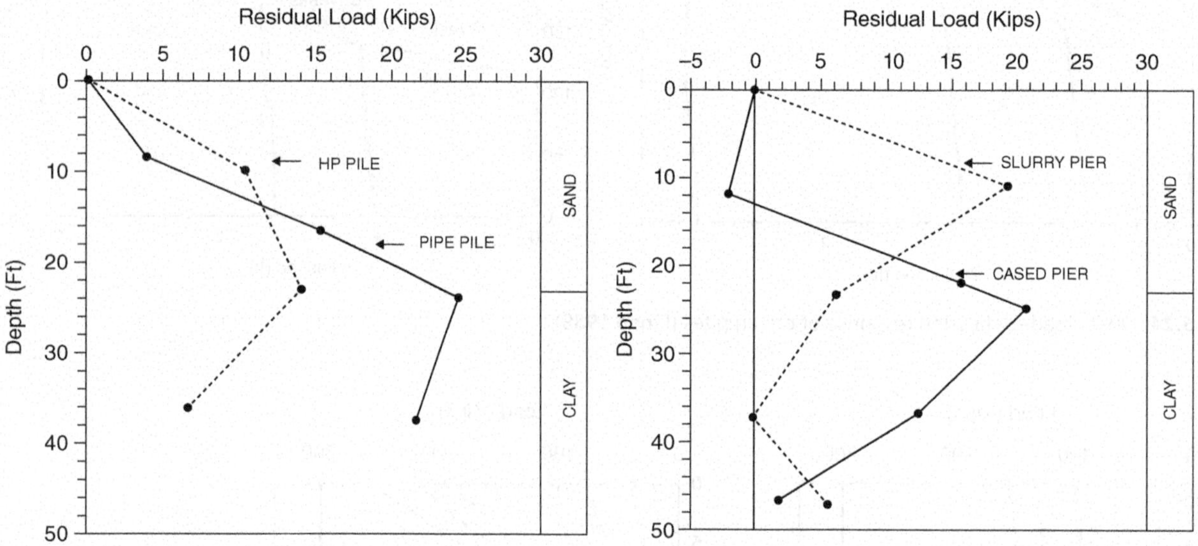

Figure 15.26 Residual load distribution after completion of loading (Finno 1989).

Figure 15.27 presents the load–movement behavior of the H-pile as recorded (solid points) and a calculated solid line based on the Hansen 80%-criterion, using data from the third and fourth load levels (75 tons and 100 tons). The solid line represents the square root of the ratio between load and movement plotted against the movement, extrapolated to include a third point at a load of 110 tons, which is the maximum load applied. It is worth noting that the dashed line in the figure represents the "observed" curve, which may differ from the constructed solid line. The discrepancy between these two curves suggests that the submitted predicted failure load of 100 tons may not align as closely with the measured capacity as initially anticipated. This observation highlights the need for careful analysis and consideration when interpreting and comparing predicted and measured values. The figure also raises the possibility that the pile head movement may have continued beyond the recorded data point, potentially exceeding the load of 110 tons. This emphasizes the importance of comprehensive monitoring and evaluation during load testing to capture the full behavior and capacity of the pile. Overall, the figure serves as a reminder of the complexities involved in predicting and interpreting pile behavior and highlights the need for continuous improvement and refinement of prediction methodologies.

Figure 15.28a depicts the loads recorded in the steel piles during the second-to-last load levels of the third test on the H-pile and pipe pile. The zero load reference corresponds to the readings taken at the beginning of the third test. It is important to note that the plotted data exclude any residual loads in the pile. Additionally, a continuous line is shown in the figure, representing the load distribution predicted discussed in the previous paragraphs. In Figure 15.28b,

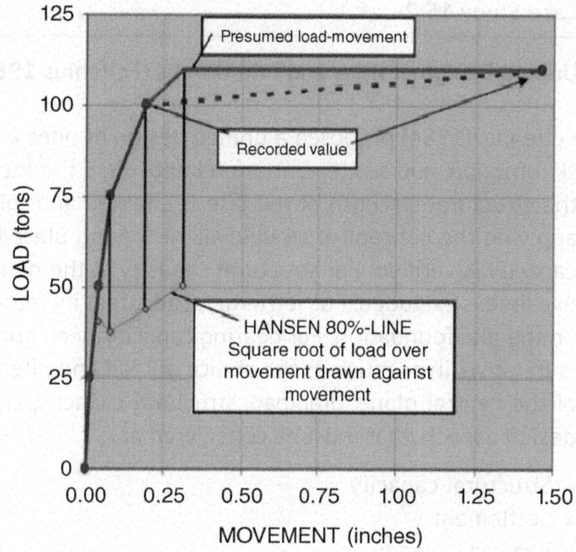

Figure 15.27 Presumed pile-head load–movement curve with correction construed through the Brinch Hansen 80%-line (Fellenius 1991a).

the load distribution in the pile during the third test is shown, with the zero load reference being the readings taken at the beginning of the first test. This means that the plotted data include the residual loads recorded in the pile during the two preceding tests, as indicated by the dashed curves. However, it is important to highlight that the residual loads induced during the driving and reconsolidation period before the start of the first test are not known and are therefore not included in the figure. The continuous line in the figure represents my predicted load distribution, taking into account all residual loads. This prediction relies entirely on the magnitude and distribution of the residual loads induced prior to the start of the static testing. These figures demonstrate the importance of considering and accounting for residual loads in pile testing and analysis. The accuracy of load distribution predictions heavily relies on understanding and quantifying the effects of residual loads induced during various stages of pile installation and testing.

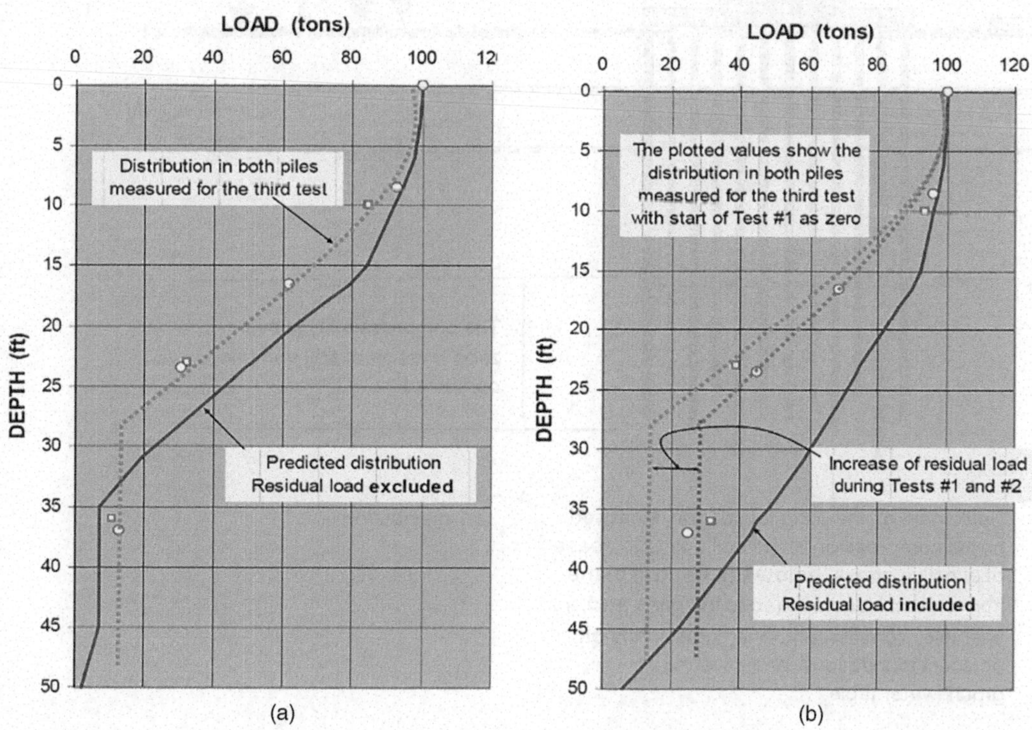

Figure 15.28 Loads during the third tests as calculated from observed strain data: (a) residual loads excluded, (b) including residual loads (Fellenius 1991b).

Case Study 15.2

Unified Design of Piles and Pile Groups (Fellenius 1984, 2004; Hai and Fellenius 2014)

Fellenius (1984) proposed a unified design of piles and pile groups wherein capacity, residual compression, negative skin friction, and settlement are related. First, the location of the neutral plane is determined. Then, the adequacy of the structural strength of the pile is checked and followed by an analysis of the settlement of the pile foundation, applying the concept of an equivalent footing placed at the neutral plane. Finally, the adequacy of the pile-bearing capacity is verified. For structural capacity at the neutral plane, dead load and dragload are considered together, but live load is excluded. For settlement, all stress increases in the soil are considered, not just that of the dead load acting on the pile foundation. For bearing capacity, dead and live loads are considered, but dragload is excluded. The design is Iterative, in as much as the choice of load and pile length will have an interactive influence on all aspects: location of the neutral plane, dragload, structural capacity, and settlement, as well as bearing capacity (Fellenius 1984). The design aspects that must be considered are:

- Structural capacity
- Settlement
- Bearing capacity
- Special considerations realizing negative skin friction and dragload

As presented in Section 15.6.1, the neural plane is located where the negative skin friction changes to positive shaft resistance. In other words, the neutral plane is determined by the requirement that the sum of the applied dead load plus the dragload is in equilibrium with the sum of the positive shaft and toe resistance, which will be discussed in more detail in Chapter 18. Figure 15.29 depicts the general principles of settlement analysis for a pile group.

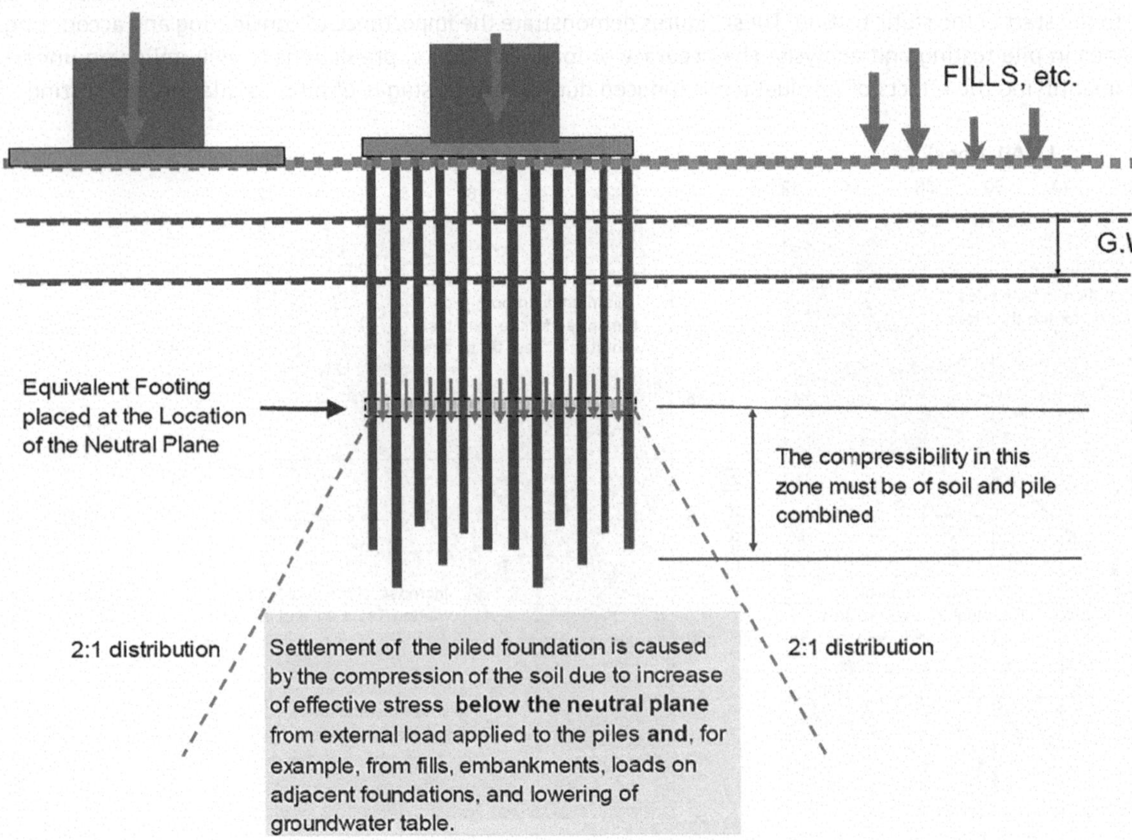

Figure 15.29 Principle of analysis of settlement for a pile group (Fellenius 2004 / American Society of Civil Engineers).

The design of a piled foundation should take into account three key aspects: capacity, dragload, and settlement. However, many design recommendations and guidelines tend to focus primarily on the capacity aspect, neglecting the other two aspects. This often leads practicing engineers to make decisions that may be overly conservative, unsafe, or unnecessarily costly. Some designers, who acknowledge the presence of settlement and the potential for dragload on the piles, attempt to address these factors by either adding the dragload to the external loads or by reducing the pile capacity by the amount of the dragload before calculating the allowable load using a factor of safety. While this approach may help reduce downdrag by lowering the depth of the neutral plane, it is not always cost-effective and may not fully consider all the relevant factors involved. In some cases, designers may attempt to mitigate negative skin friction by applying a bitumen coat to the piles, without realizing that this can also reduce the overall capacity of the piles. It is important for designers to recognize the significance of dragload and settlement in piled foundation design and to adopt comprehensive approaches that consider all three aspects holistically. By properly considering capacity, dragload, and settlement, designers can make more informed and effective decisions that result in optimized and cost-efficient designs (Fellenius 2004).

The residual load in a pile can be attributed to two main factors: the driving process and the reconsolidation of the surrounding soil. During pile driving, a portion of the load is introduced into the pile due to the impact and penetration of the pile into the ground. This initial load is known as the driving-induced residual load. Additionally, after the pile is driven, the surrounding soil undergoes a reconsolidation process. The disturbance caused by the pile driving leads to changes in the soil structure and stress distribution. As the soil gradually readjusts and reconsolidates, it exerts an additional load on the pile, known as the reconsolidation-induced residual load. It is important to note that residual load is not limited to driven piles in sand. Even for bored piles, where the installation method involves less disturbance to the surrounding soil, there can still be residual load due to reconsolidation effects. Figure 15.30 provides an illustration of the load distribution in a 0.9 m diameter bored pile installed in sand. The figure demonstrates how the residual load is distributed along the length of the pile, indicating the presence of both driving-induced and reconsolidation-induced residual loads.

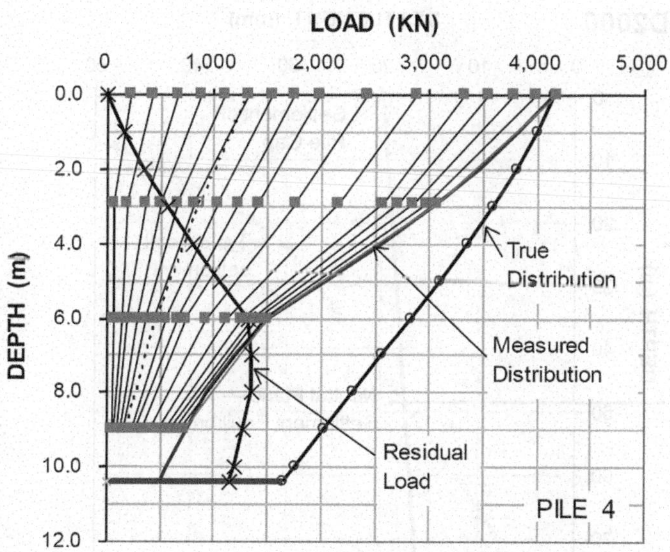

Figure 15.30 Measured load increase from the start of a static loading test in sand with evaluated distributions of residual and true load (Fellenius 2004 / American Society of Civil Engineers).

Nguyen and Fellenius (2014) conducted static loading tests using single-level bidirectional cells on two bored piles constructed at the Binh Loi Bridge in Ho Chi Minh City, Vietnam. The site consists of soft clay and silt layers near the surface, extending to depths of 11 m and 15 m, respectively. Below these layers, there is a compact silty sand layer that extends to a depth of approximately 55 m, followed by dense to very dense silty sand. The test program involved two bored piles with diameters of 2.0 m and 1.5 m, and depths of 73.5 m and 73.0 m, respectively. The piles were instrumented with single-level bidirectional cells to measure the load and displacement response during the static

(Continued)

Case Study 15.2 (Continued)

loading tests. Figure 15.31 presents the standard penetration test (SPT) records and the soil profile at the Binh Loi Bridge site. The SPT records provide information about the soil's resistance to penetration and help characterize the

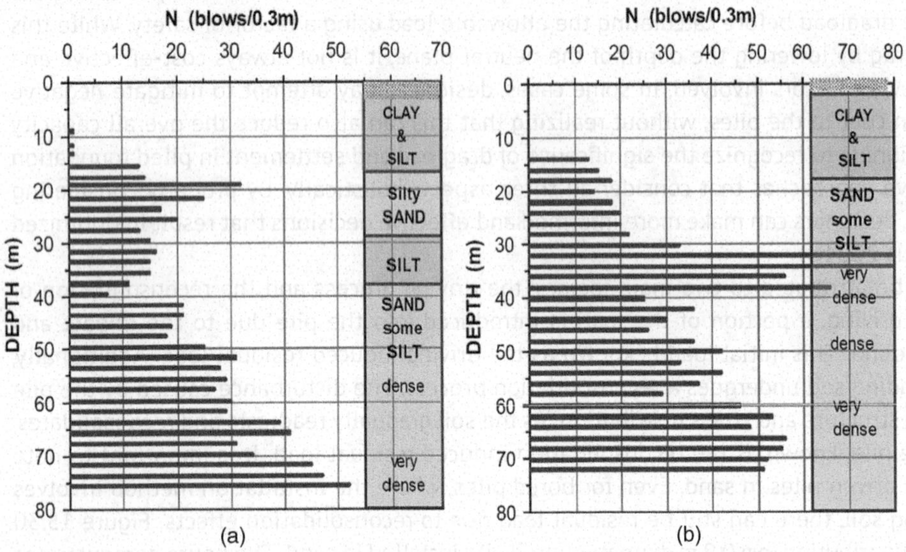

Figure 15.31 SPT N-indices at drilled shafts: (a) $D = 2$ m, (b) $D = 1.5$ m (Nguyen and Fellenius 2014 / American Society of Civil Engineers).

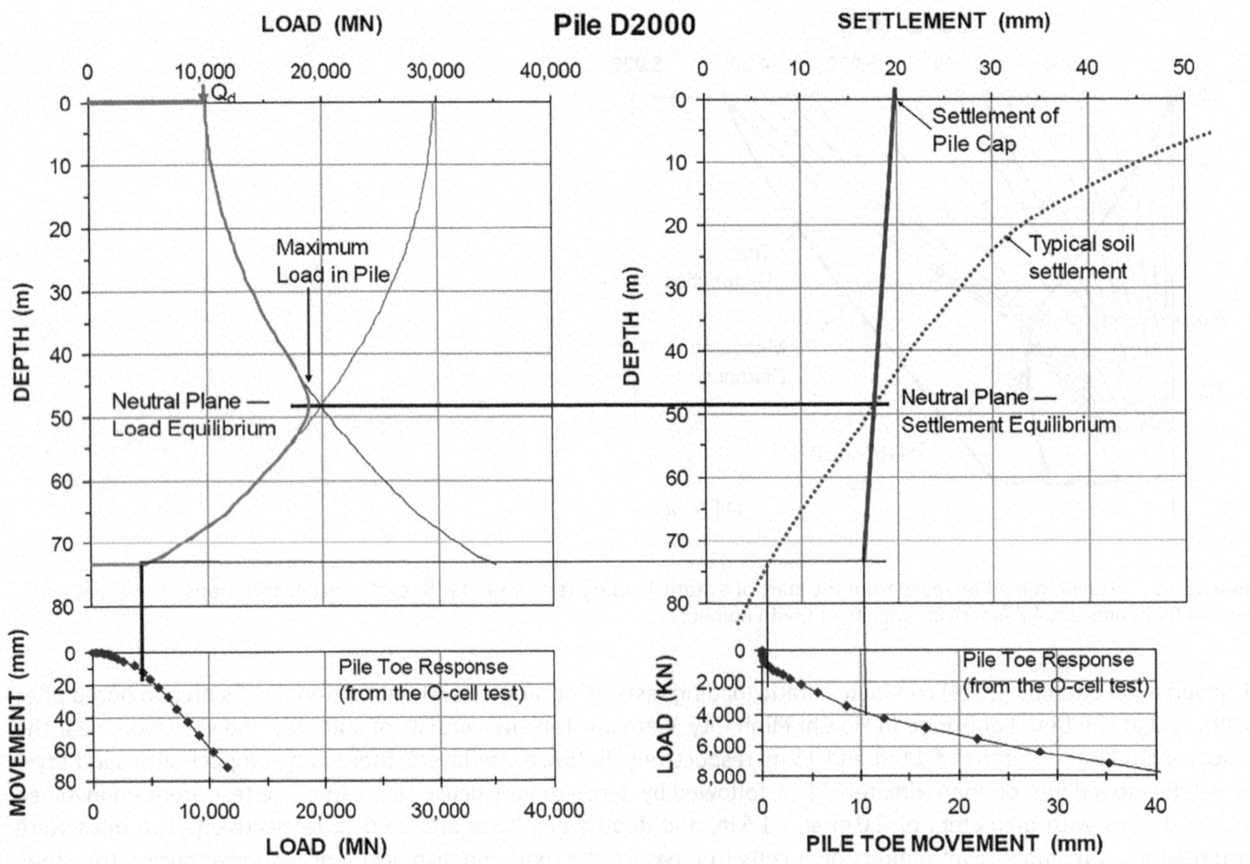

Figure 15.32 Load and settlement compilation for the design of a drilled shaft with a diameter of 2 m (Nguyen and Fellenius 2014 / American Society of Civil Engineers).

soil layers present at different depths. The soil profile indicates the layering and composition of the subsurface soil strata encountered at the site.

In Figure 15.32, the results of the loading test are presented, focusing on the pile toe response and settlement of the piled foundation. This analysis provides valuable insights into the behavior of the foundation system under load and can inform the design process. The figure illustrates the settlement behavior of the tested deep foundation, showing the relationship between applied load and resulting settlement. By examining the settlement characteristics, engineers can assess the performance and stability of the piled foundation and make informed design decisions.

15.11 Major Points

1) Deep foundations play a crucial role in various projects across different environments, including nearshore, onshore, and offshore. While pile engineering has advanced significantly with extensive research and knowledge-based approaches, there are still simplifications, conservativeness, and uncertainty concerns that need to be addressed. Due to the complexity of the field, pile engineering requires careful design, and construction.
2) The process of applying deep foundations involves determining the combination and magnitude of superstructure loads, evaluating the need for deep foundations, selecting the appropriate type and characteristics of deep foundations, determining the embedment depth and configuration, and conducting static analysis to calculate the bearing capacity. Refinement of the design is then done based on in situ tests and dynamic analysis. Settlement assessment, structural design requirements, and performance evaluation through load tests are also important considerations. Post-installation monitoring is essential.
3) Piles transfer loads to the soil through their shafts and toes. As axial load is applied to the pile shaft, it displaces downward, mobilizing side resistance in shear. The load transfer to the surrounding soil or rock results in a decrease in load with depth. Full-side resistance is typically mobilized at a displacement of approximately 1% to 2% of the pile diameter. The base of the pile must resist further load increases until the maximum base and combined resistances are reached. The displacement required to mobilize the maximum base resistance varies between 3% and 15% of the pile diameter depending on environmental conditions.
4) Various methods have been developed to estimate the shaft and toe resistance components of piles. However, these methods often rely on simplifying assumptions and empirical approaches. Different failure patterns, such as those proposed by Meyerhof, Vesic, Janbu, and Kulhawy, have been considered for pile toe resistance. The effective friction forces acting on the pile shaft depend on the active forces between the soil and pile, considering total and effective stress approaches.
5) While static pile loading tests provide more accurate results compared to other prediction methods, they can be costly and time-consuming. As a result, a limited number of static pile load tests are typically conducted in each project, usually for critical structures. Therefore, predicting the load transfer mechanism in piles through t–z (load–settlement) and q–z (load–stress) curves is crucial. Researchers have developed various models for predicting these curves.
6) Static analysis methods are used to evaluate the load–displacement response at the pile head. The required displacement to mobilize shaft and toe resistances includes the elastic settlement of the pile, the settlement required to mobilize shaft resistance, and the elastic settlement required to mobilize toe resistance. Relative movements between pile shafts and the soil can result in either negative skin friction or shaft resistance. Negative skin friction refers to the shear acting along the pile surface induced by the soil.
7) There are different theories regarding the distribution of toe and shaft resistances along the pile embedment depth. Some researchers suggest that after a certain depth known as the critical depth, the shear strength along the pile shaft does not increase linearly but remains constant. On the other hand, some researchers propose that the shaft and toe resistances exhibit nonlinear distribution after the critical depth, neither having a constant value nor a linear trend.
8) Contemporary issues in piling engineering, such as the application of helical piles, have gained significant attention. Efforts have been made to improve the performance of piles, including their immediate loading capability, ease of installation, resistance to uplift and bearing loads, and reduction of negative environmental impacts. Helical piles, in particular, have become popular among geotechnical engineers due to their fast installation and optimal construction material.

9) Several factors can lead to variations in deep foundation capacity after installation or construction. These factors include post-grouting, soil setup and relaxation, the combined application of deep foundation and ground modification practices, as well as fluctuations in-ground and groundwater levels. These factors need to be considered to ensure the long-term stability and performance of deep foundations.
10) A unified design approach has been proposed for piles and pile groups, which considers capacity, residual compression, negative skin friction, and settlement as interconnected factors. The design process is iterative, as the choice of load and pile length influences the location of the neutral plane, dragload, structural capacity, settlement, and bearing capacity. This comprehensive design approach aims to optimize the performance and reliability of deep foundations.

15.12 Questions and Problems

15.12.1 Explanatory

1 What is negative skin friction?

2 What are the different modes of failure for helical piles?

3 Which parameters influence the beta factor?

4 While are piles commonly considered for offshore structures?

5 What do you propose for reducing safety factors according to the dynamic and static pile load testing?

6 What are the pros and cons of static analyses for predicting axial pile-bearing capacity? To overcome their disadvantages, what do you suggest?

7 Explain the general strategy for deep foundations applications.

8 Illustrate how to determine the pullout capacity of helical piles.

9 List and discuss some factors that may affect the capacity after installation.

10 How fluctuation in groundwater can change the shaft and toe capacity of deep foundations?

15.12.2 Conceptual

1 How do the N_c and N_q coefficients vary with depth?

2 Is the following sentence true? Why?
- The movement necessary for mobilizing shaft resistance is smaller than the movement necessary for mobilizing toe resistance.

3 Compare and contrast two different failure modes of helical piles under pullout loading.

4 Discuss the differences, similarities, and main causes of soil setup and relaxation.

5 Based on Case Study 15.1, list and explain the main factors that are influential in the load–displacement response of deep foundations. How predictors can overcome these issues and predict the response as accurately as possible?

6 What are the advantages of applying the unified pile design instead of typical codes and standards recommendations? Name the main results and relevant in this regard. Define your answer based on what you learned from Case Study 15.2.

15.12.3 Problems

$\varphi = 33°$, $E_s = 35$ MPa, $\gamma_{sat} = 20$ kN/m³, $D_r = 80\%$ MPa, $\nu = 0.3$, $\zeta_v = 0$

1 A drilled shaft with a diameter of 80 cm and embedment depth of 15 m is constructed in uniform medium-dense sand. The groundwater level is at the ground surface. Other soil parameters are given below. Determine the toe capacity via the following methods:
a) Meyerhof (1976)
b) Vesic' (1977)
c) Kulhawy (1991)

2 An open-ended steel pile with a length of 19 m, an outer diameter of 80 cm, and a thickness of 30 mm is driven in two different sites. Determine the ultimate capacity in the following conditions:
a) Compressive loads, sand, unified method, in case of formation of the plug.
b) Compressive loads, sand, β method, in case of no plug.
c) Compressive loads, clay, API method, in case of formation of the plug.
d) Tensile loads, sand and clay, API, and β method, in case of no plug.

Sand:

$$K_c = 0.75, M = 0.8, \phi = 35°, \quad N_t = 35, \gamma_{sat} = 2\,\text{t/m}^3$$

Clay:

$$S_u = 0.25\sigma'_v, N_c = 11, \gamma_{sat} = 1.9\,\text{t/m}^3$$

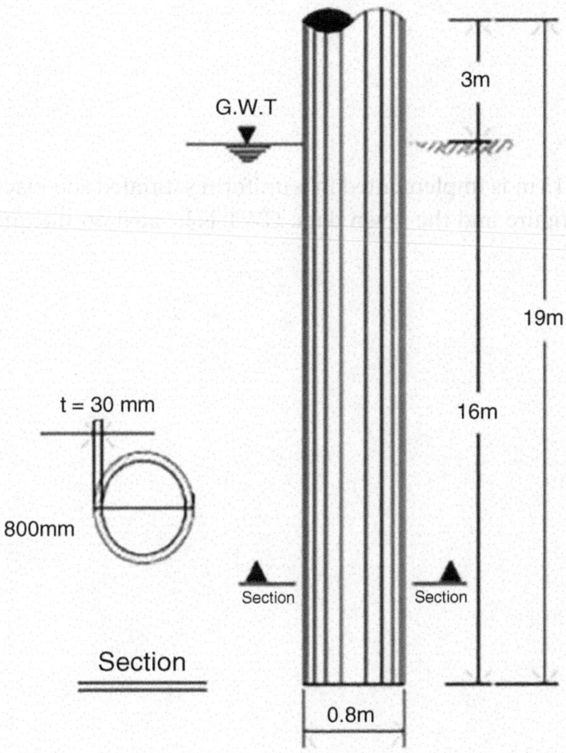

3 Determine the bearing capacity of given helical piles under compressive and pullout loading.
$d = 80$ mm, $D = 450$ mm, $S/D = 1.5$, $H = 1.5$ m, $S_u = 25$ kPa, $\alpha = 1$

$$\gamma_s = 2\,\text{t/m}^3$$

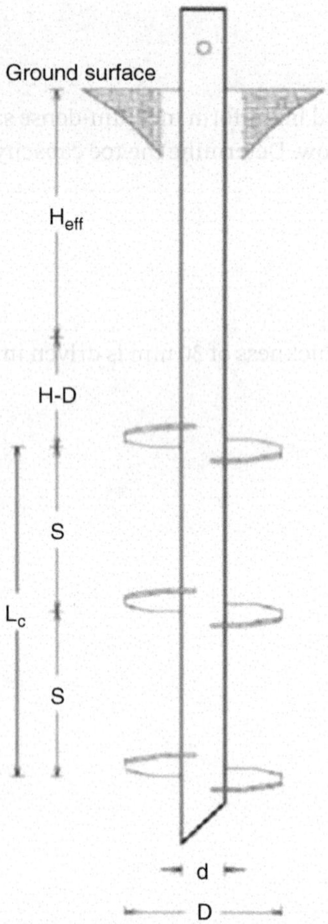

4 A belled drilled shaft (enlarged base) with a total length of 15 m is implemented in a uniform saturated soft clay site. Determine the ultimate pullout capacity according to the figure and the given data. GWT is located on the ground surface.

$$b = 75\text{ cm}, B = 220\text{ cm}, S_u = 30\text{ kPa}$$

$$\gamma_{sat} = 1.9\frac{t}{m^3}, \gamma_{con} = 2.4\text{ t/m}^3$$

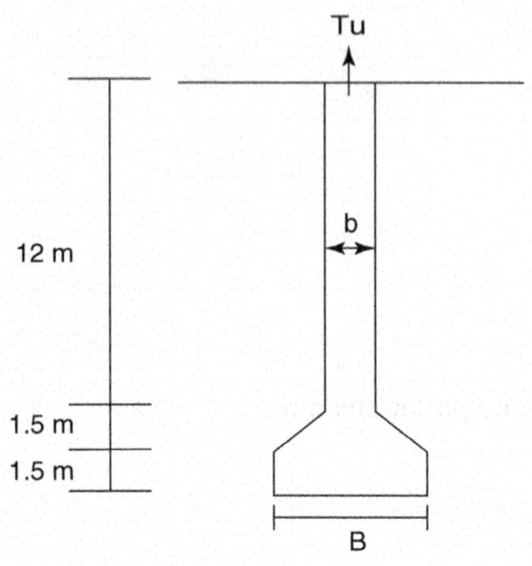

References

API (1984). *API-RP-2GEO Geotechnical and Foundation Design Considerations*. Washington: American Petroleum Institute.

API (2011). *API-RP-2GEO Geotechnical and Foundation Design Considerations*. Washington: American Petroleum Institute.

Arabameri, M. and Eslami, A. (2021). Microstructure and strength effect on bearing capacity of helical piles installed in Golestan loess. *International Journal of Civil Engineering* 1–18.

Burland, J. (1973). Shaft friction of piles in clay—a simple fundamental approach. *Publication of: Ground Engineering/UK/* 6 (3).

Canadian Geotechnical Society (2006). *Canadian Foundation Engineering Manual*, 4e. Richmond, BC, Canada: BiTech Publishing Ltd.

CFEM (1992). *Canadian Foundation Engineering Manual*, 3e, 512. Vancouver: Canadian Geotechnical Society, BiTech. Publishers.

Coduto, D.P., Kitch, W.A., and Yeung, M.C.R. (2016). *Foundation design: principles and practices*. USA: Prentice-Hall.

Das, B. (2016). Principles of Foundation Engineering 8th Ed. Instructor, 201605.

Eslami, A. and Fellenius, B.H. (1997). Pile capacity by direct CPT and CPTu methods applied to 102 case histories. *Canadian Geotechnical Journal* 34 (6): 886–904.

Eslami, A., Valikhah, F., Veiskarami, M., and Salehi, M. (2017). CPT-based investigation for pile toe and shaft resistances distribution. *Geotechnical and Geological Engineering* 35 (6): 2891–2905.

Fakharian, K., Attar, I.H., Sarrafzadeh, A., & Haddad H. (2013). Contributing factors on soil setup and the effects on pile design parameters. Proceedings of the 18th International Conference on Soil Mechanics and Geotechnical Engineering, Paris.

Fellenius, B. H. (1984). Unified Design of Piles and Pile Groups. Transportation Research Record 1169.

Fellenius, B.H. (1989). Prediction of pile capacity. Proceedings of ASCE, Geotechnical Engineering Division, the 1989 Foundation Engineering Congress, Symposium on Predicted and Observed Behavior of Piles, R. J. Finno, Editor, ASCE Geotechnical Special Publication No. 23, pp. 293–302.

Fellenius, B.H. (1991a). Results of foundation engineering congress pile loading tests. *ASCE, Journal of Geotechnical Engineering* 117 (1): 188–191.

Fellenius, B.H. (1991b). Summary of pile capacity predictions and comparison with observed behavior. *ASCE Journal of Geotechnical Engineering* 117 (1): 192–195.

Finno, R.J. (1989). *Predicted and Observed Axial Behavior of Piles: Results of a Pile Prediction Symposium*. ASCE.

Fellenius, B.H. and Altaee, A. (1995). Critical depth: how it came into being and why it does not exist. *Proceedings of the ICE-Geotechnical Engineering* 113 (2): 107–111.

Fellenius, B.H. (2004). Unified Design of Piled Foundations with Emphasis on Settlement Analysis. Paper to ASCE Geotechnical Special Publication (GSP 125).

Fellenius, B.H. (2013). Simplified non-linear approach for single pile settlement analysis. *Discussion, Canadian Geotechnical Journal* 50 (6): 685–687.

Fellenius, B.H. (2015). Basics of Foundation Design, Electronic Edition. www.Fellenius.net, 432 p.

FHWA (2006). *Design and Construction of Driven Pile Foundations-Volume II*. Federal Highway Administration FHWA-NHI-05-043.

Hai, N. M. and Fellenius, B. H. (2014). O-Cell tests on two 70 m long bored piles in Vietnam. In: From soil behavior fundamentals to innovations in geotechnical engineering: honoring Roy E. Olson (pp. 482–496).

Hanna, T.H. and Tan, R.H.S. (1973). Prediction of residual driving stresses in piles. *Canadian Geotechnical Journal* 10 (3): 311–339.

Ishihara, K. & Cubrinovski, M. (2004). Case Studies of Pile Foundations Undergoing Lateral Spreading in Liquefied Deposits. International Conference on Case Histories in Geotechnical Engineering. 5.

Janbu, N. (1976). March. Static bearing capacity of friction piles. In Sechste Europaeische Konferenz Fuer Bodenmechanik Und Grundbau (Vol. 1).

Khazaei, J. and Eslami, A. (2016a). Geotechnical behavior of helical piles via physical modeling by Frustum Confining Vessel (FCV). *International Journal of Geography and Geology* 5 (9): 167–181. https://doi.org/10.18488/journal.10/2016.5.9/10.9.167.181.

Khazaei, J. and Eslami, A. (2016b). Post-grouted helical piles behavior via physical modeling by FCV. *Marine Georesources & Geotechnology*.

Kraft, L.M., Ray, R.P., and Kagawa, T. (1981). Theoretical t-z curves. *Journal of Geotechnical and Geoenvironmental Engineering*, ASCE 107 (11): 1543–1560.

Kulhawy, F.H. (1991). *Drilled Shaft Foundations, Foundation Engineering Handbook*, Chap. 14, 2e, 537–552. New York: Van Nostrand Reinhold.

Maeda, Y., Omine, K., Ochiai, H., Furuki, H., Naemura, S., Okochi, Y., Ichikawa, A. (2006). Improvement of horizontal bearing capacity by composite ground foundation method in soft ground. In: Proc. of the 22nd US-Japan Bridge Engineering Workshop, Seattle, WA, USA.

Mayne, P.W. (2001). Stress-strain-strength-flow parameters from enhanced in-situ tests. *Proceedings of International Conference on In-Situ Measurement of Soil Properties and Case Histories*, Bali, Indonesia, pp. 27-48.

McClelland, B. (1974). Design of deep penetration piles for ocean structures. *Journal, Geotechnical Engineering Division, ASCE* 100 (GT7): 705–747.

Meyerhof, G.G. (1976). Bearing capacity and settlement of pile foundations. *Journal of the Geotechnical Engineering Division* 102 (3): 197–228.

Motamedinia, H., Hataf, N., and Habibagahi, G. (2019). A study on failure surface of helical anchors in the sand by PIV/DIC technique. *International Journal of Civil Engineering* 17 (12): 1813–1827.

Nguyen, M.H., & Fellenius, B.H. (2014). Bidirectional-Cell Tests on Two 70 m Long Bored Piles in Vietnam. GeoInstitute Geo Congress, February 23-26, ASCE, Geotechnical Special Publication, GSP 233, pp. 482–496.

Nordlund, R.L. (1963). Bearing capacity of piles in cohesionless soils. *Journal of the Soil Mechanics and Foundations Division* 89 (3): 1–35.

Perko, H.A. (2009). Helical Piles: a practical guide to design and installation. In: *Helical Piles: A Practical Guide to Design and Installation*. John Wiley & Sons.

Poulos, H.G. (1989). Pile behaviour—theory and application. *Géotechnique* 39 (3): 365–415.

Randolph, M.F. and Murphy, B. (1985). Shaft capacity of driven piles in clay. In: . Offshore Technology Conference.

Salem, T.N. and Hussein, M.M. (2017). Axial tensile capacity of helical piles from field tests and numerical study. *Port-Said Engineering Research Journal*.

Salgado, R. (2008). *The Engineering of Foundations*, vol. 888. New York: McGraw Hill.

Sakr, M. (2015). Retracted: relationship between installation torque and axial capacities of helical piles in cohesionless soils. *Journal of Performance of Constructed Facilities* 29 (6): 04014173. https://doi.org/10.1061/(ASCE)CF.1943-5509.0000621.

Sakr, M. (2009). Performance of helical piles in oil sand. *Canadian Geotechnical Journal* 46 (9): 1046–1061. https://doi.org/10.1139/T09-044.

Sharma, P., Rawat, S., and Gupta, A.K. (2020). Horizontal pullout behavior of novel open-ended pipe helical soil nail in frictional soil. *International Journal of Civil Engineering* 1–16. https://doi.org/10.1007/s40999-020-00535-2.

Sham-e Consulting Engineering Co. (2022). Technical report of site investigation and foundation design for Babolsar Twin towers.

Spagnoli, G. and de Hollanda Cavalcanti Tsuha, C. (2020). A review on the behavior of helical piles as a potential offshore foundation system. *Marine Georesources & Geotechnology* 1–24. https://doi.org/10.1080/1064119X.2020.1729905.

Tsuha, C.H.C., Aoki, N., Rault, G. et al. (2012). Evaluation of the efficiencies of helical anchor plates in the sand by centrifuge model tests. *Canadian Geotechnical Journal*. https://doi.org/10.1139/T2012-064.

Valikhah, F., Eslami, A., and Veiskarami, M. (2019). Load–displacement behavior of driven piles in sand using CPT-based stress and strain fields. *International Journal of Civil Engineering* 17: 1879–1893.

Veiskarami, M., Kumar, J., and Valikhah, F. (2014). Effect of the flow rule on the bearing capacity of strip foundations on sand by the upper-bound limit analysis and slip lines. *International Journal of Geomechanics* 14 (3): 04014008.

Veiskarami, M., Kumar, J., and Valikhah, F. (2015). A practical procedure to estimate the bearing capacity of footings on sand–application to 87 case studies. *Iranian Journal of Science and Technology Transactions of Civil Engineering* 39 (C2+): 469–483.

Vesic, A.S. (1967). *A study of the bearing capacity of deep foundations*.

Vesic, A.S. (1977). Design of pile foundations. NCHRP synthesis of highway practice, (42).

16

In Situ Penetration Testing Applications

Chapter Highlights

- **The Importance of In situ Testing in Pile Design**
- **Role and Significancy of Scale Effect in Axial-Bearing Capacity**
  - Embedment depth
  - Influence zone
  - Data production processing and averaging
  - Diameter
  - Nonhomogeneous condition
  - Penetration rate and failure mechanism
- **In Situ-Based Methods for Predicting Axial Pile-Bearing Capacity**
- **Model Uncertainty of In Situ-Based Predictive Methods**
- **Comments on Current Methods**
- **Case Histories**
  - Driven Piles, The Urmia Lake, Iran (Eslami et al. 2011)
  - Similarity of CPT records in two different locations and its advantages in predicting axial pile-bearing capacity (Eslami et al. 2020)
  - Performance-based assessment of different predictive pile-bearing capacity approaches (Heidarie Golafzani et al. 2020a,b)
  - Special Piles: Drilled Displacement Piles (Moshfeghi and Eslami 2019)

16.1 Introduction

As a complementary tool to static analysis for determining pile capacity, in situ testing methods have gained increased popularity in recent years. This can be attributed to advancements in in situ testing instruments, improved understanding of actual soil behavior, and the recognition of limitations in conventional laboratory testing (Wroth 1988). Several in situ-based methods have been developed for pile design and estimating bearing capacity, including the standard penetration test (SPT), cone penetration test (CPT), pressure meter test (PMT), vane shear test (VST), and more. However, this chapter focuses on SPT and CPT-based methods due to their widespread usage and availability among geotechnical engineers. Other tests, such as the PMT, are discussed in relevant chapters, such as Chapter 17, which specifically covers design concepts for lateral loading.

The SPT remains the most commonly used in situ test in geotechnical investigations. However, it has certain limitations and challenges regarding interpretation and repeatability. In contrast, the CPT is a simpler, faster, and relatively more economical test that provides continuous records with depth. Its results can be interpreted empirically and analytically. Consequently, the CPT is increasingly preferred for pile analysis.

Advanced Foundation Engineering: Principles, Performance and Prospects, First Edition. Moses Karakouzian and Abolfazl Eslami.

Due to its similarity to a pile and the advantages it offers as a geotechnical investigation tool, the CPT has been extensively used for estimating pile capacity. During site investigations, particularly in challenging underground conditions, CPT soundings provide continuous depth profiles, which are valuable for bearing capacity estimation.

By using the cone penetrometer as a model pile, the measured cone resistance and sleeve friction can be utilized to estimate unit toe and shaft resistances, respectively. The CPT allows for the development of nearly continuous profiles with depth, eliminating the need for obtaining representative undisturbed samples and determining soil parameters through conventional laboratory testing. Additionally, there is no requirement for estimating intermediate parameters such as shaft coefficient (K_s) and toe coefficient (N_q) for pile capacity estimation.

Both static and dynamic loading tests offer significant advantages and limitations, and they should be considered in the interpretation and analysis of pile behavior. During the initial design stage, piling configuration, including the type, length, diameter, group pattern, and installation, is determined and evaluated using static analysis through analytical approaches or in situ testing methods.

It is important to note that static analysis approaches, despite their limitations, are integral and essential steps in piled foundation design. They cannot be fully replaced by dynamic approaches or static loading tests. Even when pile load testing is economically justifiable, static analysis approaches are still utilized for the initial estimation of pile capacity to determine approximate pile characteristics.

16.2 Scale Effect in Correlations Between Pile, Foundation, and CPT

The cone penetrometer test (CPT) and piles exhibit certain similarities, but there are important differences that must be considered when using CPT data to estimate pile capacity. The cone point resistance and sleeve friction measured during CPT are not directly equivalent to the base and shaft capacities of a pile due to differences in geometry, critical depth, strain level, penetration rate, and soil failure mechanisms (Takesue et al. 1998; Fleming et al. 2009; Eslami et al. 2013).

In some direct methods that account for scale effects, the focus has been on geometric parameters such as pile diameter and embedment depth. However, other factors such as penetration rate, strain level, non-homogeneity of the soil, ultimate pile capacity, and friction fatigue also play crucial roles in relating sleeve friction (f_s) and cone resistance (q_c) to the unit shaft capacity of the pile (r_s).

Figure 16.1 illustrates the differences between piles and CPT in terms of material, penetration rate, and dimensions. Piles have limited displacement, preventing the mobilization of ultimate soil strength and the occurrence of soil failure (Fleming et al. 2009). On the other hand, CPT can induce soil failure due to its larger displacements. In essence, the penetration mechanisms of a pile and CPT are different, with CPT causing significant soil displacements while pile penetration is very small. Additionally, differences in materials contribute to variations in rigidity and roughness between piles and the cone penetrometer.

16.3 CPT and Pile Scale Effect for Toe Capacity

A cone penetration test (CPT)-based design methods often establish a direct correlation between the unit toe capacity of a pile, r_t, and the cone resistance, q_c. This approach is based on the similarities in geometry and vertical penetration between piles and CPT. Researchers such as De Beer (1963) and Eslami and Fellenius (1997) have proposed considering the unit toe

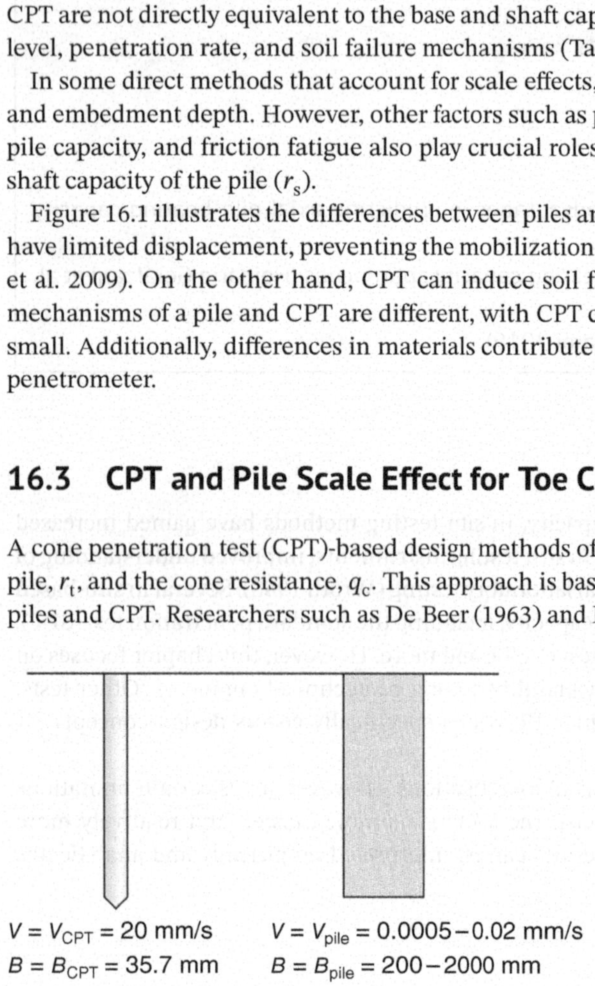

Figure 16.1 Schematic view of pile and CPT differences in penetration rates, dimensions, and shape.

capacity equal to the cone resistance. However, there are differing opinions among researchers regarding the equivalence of unit toe resistance and cone resistance. Meyerhof (1983), Gui and Bolton (1998), and Jardine et al. (2005) argue that the unit toe resistance is not necessarily equal to the cone resistance.

Some aspects regarding the difference between a pile and CPT are as follows:

1) Embedment depth
2) Influence zone
3) Data production processing and averaging
4) Diameter
5) Nonhomogeneous condition
6) Penetration rate and failure mechanism

16.3.1 Embedment Depth

Several studies, including De Beer (1963), Meyerhof (1983), and Gui and Bolton (1998), have highlighted that the toe capacity of a pile or the cone resistance of a penetrometer is not fully mobilized upon initial penetration into a new layer. Full mobilization of toe capacity occurs when the pile or penetrometer has advanced sufficiently into the target layer. The embedment depth, which represents the depth at which the pile or penetrometer has fully mobilized its capacity, is influenced by the diameter of the pile. It is commonly expressed as a multiple of the pile diameter.

Nottingham (1975) conducted a theoretical and analytical analysis using conventional bearing capacity theory to compare the length of the failure surface zone for different foundation embedments. The results, depicted in Figure 16.2, demonstrate that as the embedment depth increases from shallow to deep, the failure mechanism undergoes a transformation from shallow failure to deep failure. Consequently, the failure zone extends, and ultimately, the ultimate capacity of the foundation increases.

As a pile has a greater diameter than a CPT instrument, a deeper embedment from the ground surface or into a hard layer is required to mobilize that layer's "full" strength. Until sufficient penetration is achieved, the unit toe capacity (r_t) will be less than the cone resistance (q_c) because the previous layer's influence is still present at the pile tip. This mechanism is depicted in Figure 16.3, highlighting the difference in behavior between a pile and a cone penetrometer. Additionally, the L/D ratio (length-to-diameter ratio) of a cone penetrometer is typically higher than that of a pile, resulting in a higher ratio of shaft capacity (q_s) to base capacity (q_b). This ratio (q_s/q_b) is taken into consideration in pile design to account for these differences.

16.3.2 Influence Zone

To estimate the unit toe resistance of a large scale pile, it is necessary to determine the failure zone around the pile toe based on the average cone point resistance obtained from small-scale cone penetrometer tests. However, there is no specific evidence to confirm whether the failure is localized punching or general shear failure.

Experimental studies by Meyerhof (1956, 1976) suggest that a penetration distance into a soil layer approximately ten times the pile diameter is required to fully mobilize the ultimate unit toe resistance of the bearing layer. De Beer (1963) found a similar depth requirement for developing resistance in penetrometers in sand. Experimental and numerical work

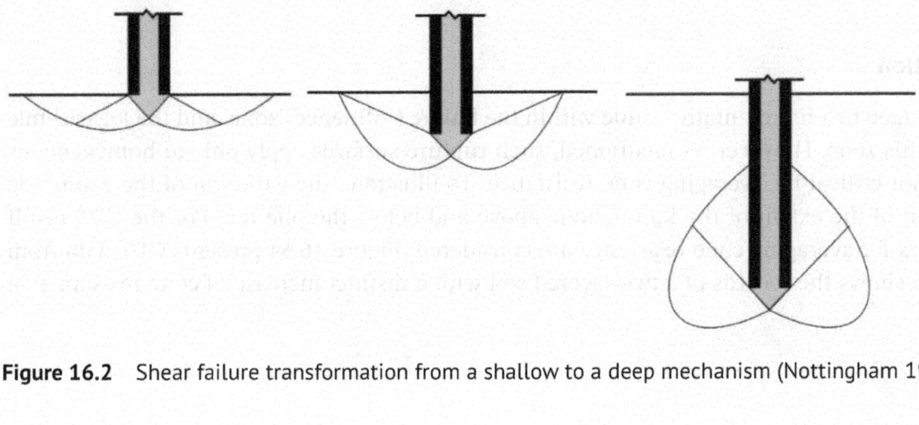

Figure 16.2 Shear failure transformation from a shallow to a deep mechanism (Nottingham 1975).

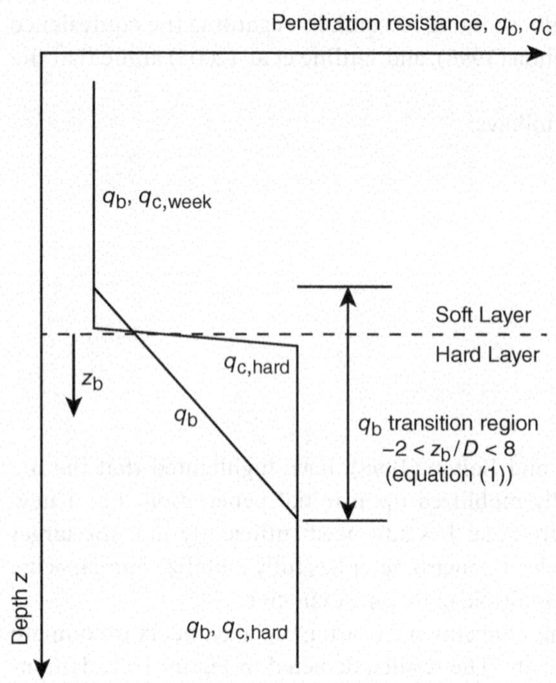

Figure 16.3 Partial embedment reduction factor on base resistance (White and Bolton 2005 / With permission of Institution of Civil Engineers).

by Altaee et al. (1993) on two concrete piles in medium-dense sand revealed that the zone influencing the toe resistance extends approximately 5*b* below and 5*b* above the pile toe, where *b* is the pile diameter.

Ghionna et al. (1993) demonstrated that both displacement and non-displacement piles achieve an ultimate unit penetration resistance value similar to the cone point unit resistance value. The slight difference observed is attributed to the variation in apex angles between the soil cone formed beneath a pile and the apex angle of the cone penetrometer. However, the ratio of penetration required to reach ultimate resistance is higher for non-displacement piles compared to displacement piles.

Salgado (1995) conducted tests on dry sand in calibration chambers and found that the penetration resistance for a penetrometer with a specific apex angle is independent of the penetrometer diameter. He also observed that the difference in penetration resistance between a flat-ended penetrometer and a standard cone penetrometer (with an apex angle of 60°) is approximately 15% for dense sands under moderately low confining stress, and much smaller for loose sands under high confining stress.

According to Eslami's (1996) experimental and analytical study, a logarithmic spiral can be used to represent the failure zone in homogeneous soil, as shown in Figure 16.4. The failure mechanism is characterized as a general shear failure, where the rupture surfaces extend toward the body of the pile or penetrometer, similar to the concepts proposed by Meyerhof (1951) and De Beer (1963).

The deepest point of the rupture surface beneath the pile toe can be determined by maximizing the projection of the radius of the logarithmic spiral on the vertical axis. This provides an estimation of the extent of the failure zone around the pile toe.

16.3.3 Nonhomogeneous Condition

The cone resistance needs to be averaged to a representative value within the failure (influence) zone, and the logarithmic spiral is but one way to determine this zone. However, as mentioned, such rupture surfaces apply only to homogeneous soil, and the height of the zone is not critical for averaging cone resistance. To illustrate, the variation of the geometric average can be studied as a function of the extent of the failure zone above and below the pile toe. For the CPT result shown in Figure 16.5, different zones for averaging cone resistance are considered. Figure 16.5a presents CPT data from homogeneous soil, and Figure 16.5b shows the records of a two-layered soil with a distinct increase of cone resistance in

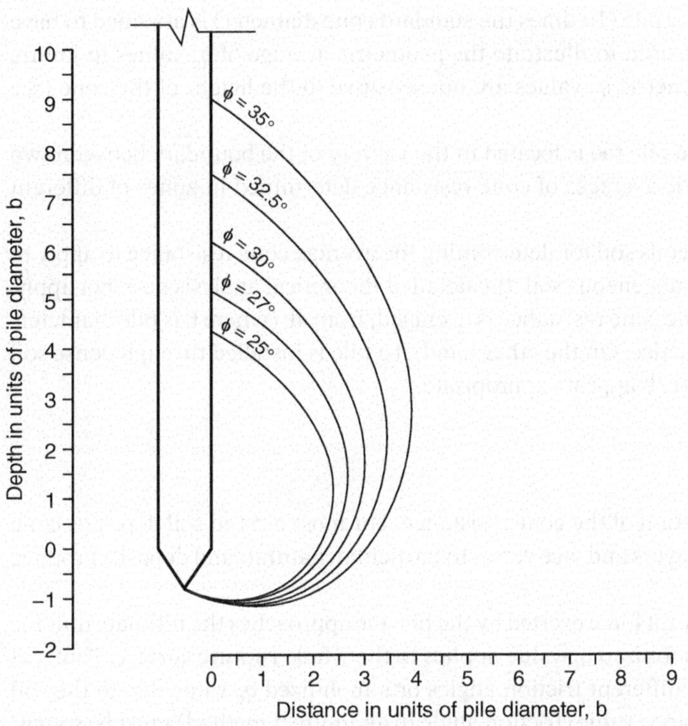

Figure 16.4 Rupture surfaces around pile toe for different soils (Eslami 1996 / with permission of University of Ottawa Press).

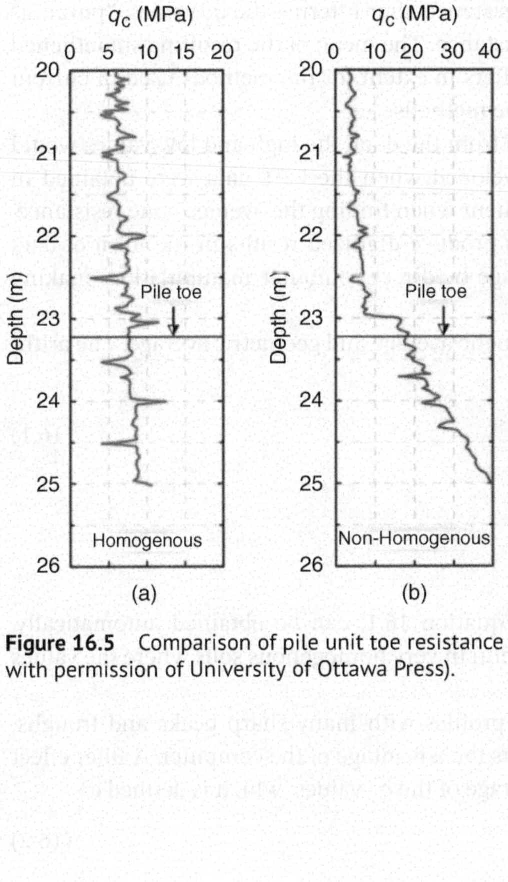

Geometric Average of Cone Resistance For Different zones ($b = 350$ mm), MPa

Zone	Homogenous	Non-Homogenous
2b–2b	7.66	16.13
2b–4b	7.43	11.79
2b–6b	7.09	9.00
2b–8b	6.93	8.06
4b–2b	7.89	19.92
4b–4b	7.65	14.89
4b–6b	7.38	11.97
4b–8b	7.15	10.11

Figure 16.5 Comparison of pile unit toe resistance for different zones: (a) Homogeneous and (b) Nonhomogeneous (Eslami 1996 / with permission of University of Ottawa Press).

the lower layer (nonhomogeneous soil). A 350 mm diameter pile (10 times the standard cone diameter) is assumed to have been installed a small distance into the lower layer and is used to illustrate the geometric average of q_c values in failure zones of different heights. For homogeneous soil, the geometric q_c values are not sensitive to the height of the zone (see Figure 16.5a).

In contrast, for a pile in nonhomogeneous soil, when the pile toe is located in the vicinity of the boundary between two layers, a significant difference exists between the geometric averages of cone resistance determined in zones of different heights.

As a result, the extent of the influence zone in homogeneous soil for determining the average cone resistance to apply to pile unit toe resistance is not critical. In the case of nonhomogeneous soil, the detailed theoretical analysis does not apply. When a pile is installed through weak soil into dense soil, the cone resistance is averaged, from 4b (where b is pile diameter) below to 8b above the pile toe, which is used by current practice. On the other hand, if a pile is installed through dense soil into weak soil, decreasing the upper limit height from 8b to 2b appears appropriate.

16.3.4 Data Processing and Averaging

The variation of soil characteristics is reflected in the variation of the cone resistance. For most natural soil deposits, cone resistance ranges from soft or loose layers to hard or dense layers and vice versa. In particular, natural sand deposits produce q_c profiles with many sharp peaks and troughs.

Along a specific, well-defined rupture surface, when the unit force exerted by the pile toe approaches the ultimate unit toe resistance, no single friction angle or single fraction of a mobilized q_c-value applies to the whole rupture surface. Points at the rupture surface, located at different depths, may have different friction angles or a mobilized q_c value due to the soil variation and confining stress. A representative cone resistance value (friction angle in an indirect method) must be sought to obtain cone data for pile design, especially for unit toe resistance. It is important to note that, the larger the pile diameter, the greater the extent of rupture surface below and above the pile toe. Therefore, a procedure must be established to find the representative q_c-value related to the unit toe resistance, and the value must be a function of the pile diameter.

All current direct CPT methods apply a correlation factor to the cone resistance after filtering the q_c-values, "purging" the peak and trough values from the data to determine the pile unit toe resistance. The mean of the resulting smoothened curve, the q_c-average, is then determined over a particular zone, which differs in extent for the methods used in current practice. The correlation factor to apply to the q_c average differs between the methods.

Filtering the cone data is necessary because, were a true mean produced from the data, the high and low values would disproportionately influence the average. The filtering approach was developed when the CPT data were obtained in hard-copy diagrams only, and it brings about considerable leeway for judgment when finding the average cone resistance. However, subjective filtering is no longer necessary because current tests produce digitized results in the form of data tables. This makes the records easily accessible for determining the average by direct computer manipulation, making working with hard-copy diagrams redundant.

Two types of cone resistance averaging methods can be considered: arithmetic average and geometric average. The arithmetic average is defined as:

$$q_{ca} = \frac{q_{c1} + q_{c2} + \cdots q_{cn}}{n}, \tag{16.1}$$

where

q_{ca} = arithmetic average of values which range from qc1 through q_{cn}
n = number of considered values

Having the CPT data in the computer, the average q_c, according to Equation 16.1, can be obtained automatically. However, without first excluding peaks and troughs, this average is only useful in very homogenous soils where the values are uniform.

In contrast, the natural variability of many sand deposits produces qc profiles with many sharp peaks and troughs. Filtering is therefore necessary in most cases, and if done manually, it offsets the advantage of the computer. A filter effect can be achieved directly, however, by instead calculating the geometric average of the q_c-values, which is defined as

$$q_{cg} = \sqrt[n]{q_{c1} \cdot q_{c2} \cdot \cdots \cdot q_{cn}}, \tag{16.2}$$

where

q_{cg} = geometric average of values that range from q_{c1} to q_{cn}

The bias in the arithmetic mean arises from the influence of the absolute magnitude instead of ratios of variations (Kennedy and Neville 1986). For example, the arithmetic average of the numbers 0.5 and 2.0 is 1.25, and the geometric average is 1.00. If the numbers are 0.33 and 3.00 instead, the arithmetic average becomes 1.65, while the geometric average is still 1.00. For a data set of the numbers 0.33, 0.50, 2.00, and 3.00, the arithmetic and geometric averages are 1.46 and 1.00, respectively.

Assume that a set of values is as follows: 5, 5, 2, 5, 25, 5, 6, 1, 6, 6, 30, and 6, where the dominant values lie between 5 and 6. The arithmetic and geometric averages are 8.50 and 5.71, respectively. The result shows that the geometric average is closer to the dominant values, as opposed to the arithmetic average, which is not representative of the dominant range. Thus, by taking the geometric average of q_c-values in a zone in the vicinity of the pile toe, a filtered representative value is obtained unaffected by the operator's judgment and is, therefore, repeatable.

16.3.5 Diameter

Given the larger diameter, piles require larger penetration into a layer to mobilize the full capacity, while CPT, with a smaller diameter, reaches its highest resistance values instantly as it penetrates a soil layer, as shown in Figure 16.6.

Meyerhof (1983) and Jardine et al. (2005) suggested that the pile unit toe capacity significantly decreases as the pile diameter increases, in a way that a penetrometer diameter (B_{CPT} = 36 mm) has considerably higher unit toe resistance than a pile installed in the same soil. In this regard, using a database of 29 closed-end piles with diameters ranging from 0.1 to 0.9 m, Chow (1997) stated that the pile-to-cone unit resistance ratio is invariably less than unity, and this ratio decreases according to the following equation (Axelsson 2000):

$$\frac{r_t}{q_c} = 1 - 0.5 \log\left(\frac{B}{B_{CPT}}\right) \tag{16.3}$$

Eslami and Fellenius (1997) proposed the factor, C_t, in their method to correlate the pile unit toe resistance with effective cone resistance. They suggested the C_t value of unity for most cases. Actually, a one-to-one relation exists between q_c and q_b (r_t). This relation is true while other factors affecting scale are realized.

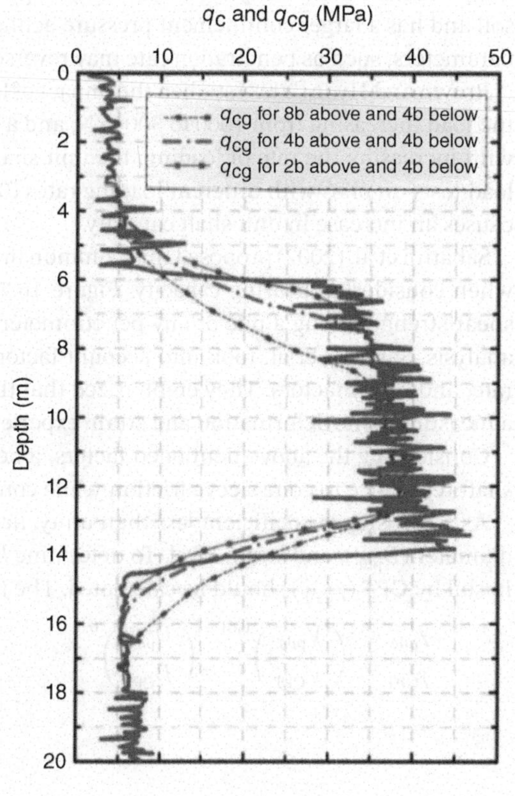

Figure 16.6 Comparison of cone resistance and calculated geometric average for a dense soil layer laid between loose layers (Eslami and Fellenius 1997 / With permission of National Research Council Canada).

White and Bolton (2005) reassembled the Chow (1997) database to examine the relationship between CPT resistance and toe capacity. They concluded that, as long as the modification of embedment depth is applied and the loading test is performed according to the defined settlement criterion with residual stress modifications, the scale effect in terms of diameter would be negligible. In such conditions, the unit toe resistance of the pile is approximately 90% of the cone resistance.

16.3.6 Penetration Rate and Mechanism

Generally, the CPT is considered in drained and undrained conditions in sands and clays, respectively (Campanella and Robertson 1981). However, the lower penetration rate during the pile static loading test provides sufficient time for the pore pressure to dissipate. Consequently, it appears to have fully drained conditions in the sand and relatively drained conditions in clays around the pile during a static load test.

Al-Mhaidib (1999) performed static compression loading tests on a model pile with a diameter of 30 mm under various loading rates of 0.01, 0.05, 0.1, 0.5, and 1 mm/min. This highlighted the effect of loading rates on the compressive capacity of piles in a way that increasing the loading rate leads to an increase in pile capacity.

16.4 Scale Effect in Estimation of Shaft Capacity

The studies on the effect of pile length on pile shaft capacity show that unit shaft capacity increases with depth and is proportional to the effective overburden stress. According to conventional interpretations, above a certain depth, called "critical depth," shaft capacity increases with the effective stress, and below the critical depth, the shaft capacity remains constant (Meyerhof 1976). Also, De Beer (1963) found a reduction in unit shaft capacity with depth and attributed it to the arching that takes place close to the pile toe. This arching results in decreased vertical stress at the pile toe level and leads to a reduced unit shaft capacity close to the pile toe. However, investigations demonstrate that pile shaft and toe capacity do not increase linearly with depth, and they do not reach a limiting value after a certain depth. (Randolph et al. 1994).

The effect of diameter of the shaft on the shaft capacity has not been considered in any of the direct CPT methods except for the Schmertmann (1978), Nottingham (1975), and UWA (2005) methods. It is expected that, for two piles with the same soil profile and the same penetration rate, the pile with the larger diameter induces larger displacements in the surrounding soil and has a larger confinement pressure acting on the pile shaft, resulting in a larger pile shaft capacity. However, other parameters, such as penetration rate may reverse this trend.

Brown and Hyde (2008) studied the impact of loading rates on the unit shaft capacity with two tests: a rapid load test, with the load increasing from 1000 to 3000 kN, and a load test with a constant displacement rate (0.01 mm/s). They stated that, with increasing the rate of loading, the unit shaft capacity increases. Moreover, Al-Mhaidib (1999), by conducting tensile load tests on piles with different loading rates (0.01, 0.05, 0.1, 0.5, and 1 mm/s), remarked that increasing the loading rate causes an increase in unit shaft capacity.

Sabatini et al. (2002) proposed that common in situ tests, such as SPT and CPT, fall into the category of large strain analysis when considering bearing capacity. Figure 16.7a and b demonstrate this concept. They argued that the mobilization of shear strength along a pile or any penetrometer is influenced by the level of shear strain experienced in the soil. In their analysis, Sabatini et al. took into account factors such as deviator stress, strain, variations in shear modulus with strain rate, and safety factors. They emphasized that the shear strength mobilization is related to the shear strain level, which is a measure of the deformation and strain experienced by the soil, as depicted in Figure 16.7c.

Considering the abovementioned factors, a new approach was proposed by Eslami et al. (2013) that correlates pile unit shaft resistance to cone sleeve friction while considering scale effects.

As a first step, a coefficient less than unity, named k, must be determined based on the pile penetration rate (V_{pile}), pile diameter (D_{pile}), and strain level. To determine k, the ratio of induced soil shear strain by pile (γ_{pile}) to induce shear strain in soil by CPT (γ_{CPT}) should be calculated. The final formula derived after calibration is as follows:

$$\frac{\gamma_{\text{pile}}}{\gamma_{\text{CPT}}} = \left(\frac{V_{\text{pile}}}{V_{\text{CPT}}}\right)^{0.61} \cdot \left(\frac{D_{\text{pile}}}{D_{\text{CPT}}}\right)^{0.5} \tag{16.4}$$

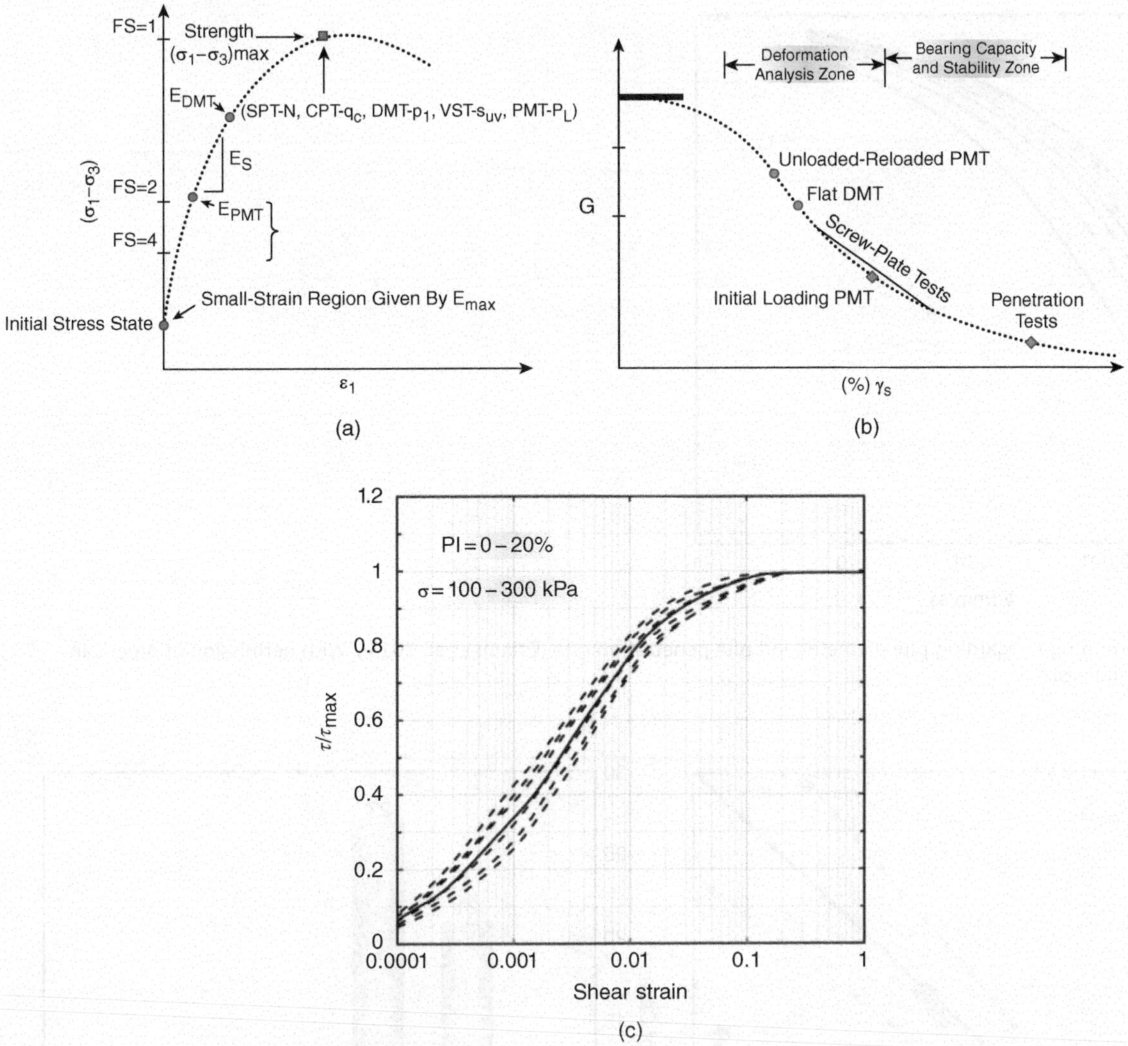

Figure 16.7 Stress–strain–strength curves for different in situ tests: (a) strength measured by in situ tests at the peak of the stress–strain curve, (b) variation of shear modulus with strain level, and (c) variation of shear stress with shear strain (Sabatini et al. 2002).

where h is the embedment length, and D_{CPT} and V_{CPT} are CPT diameter and the rate of penetration, respectively. It should be mentioned that in this formula, it is assumed that $h_{CPT} = h_{pile}$.

Considering $\gamma_{CPT} = 0.3$, $D_{CPT} = 35$ mm, and $V_{CPT} = 20$ mm/s, induced shear strain in soil by pile can be expressed as Equation 16.5:

$$\gamma_{pile} = 0.0255\,(V_{pile})^{0.61} \cdot (D_{pile})^{0.5} \tag{16.5}$$

where V_{pile} is the penetration rate of the pile (mm/s) and D_{pile} is the pile diameter (mm).

Now, assuming $\frac{\tau}{\tau_{max}} = k$ and using Ishibashi and Zhang's Chart (1993) for determining $\frac{\tau}{\tau_{max}}$ from shear strain, a new graph is proposed from which k can be obtained, as presented in Figure 16.8. After calculating k, depending on the pile diameter and the penetration rate, pile unit shaft capacity, r_s, can be calculated from Equation 16.6:

$$r_s = k \cdot f_s \tag{16.6}$$

As depicted in Figure 16.9, for the investigated database of Eslami et al. (2013), unit pile shaft resistance values are invariably less than the cone sleeve friction for almost all the cases. A comparison of the load distribution along the pile is also delineated in Figure 16.10.

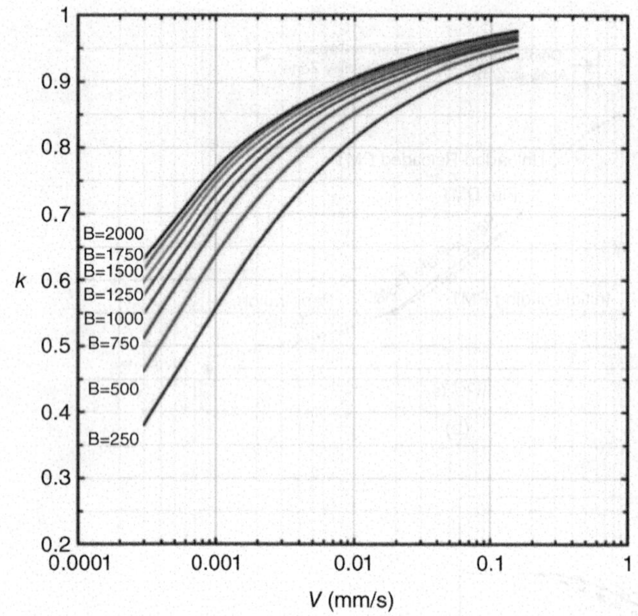

Figure 16.8 Determining k regarding pile diameter and pile penetration rate (Eslami et al. 2013 / With permission of American Society of Civil Engineers).

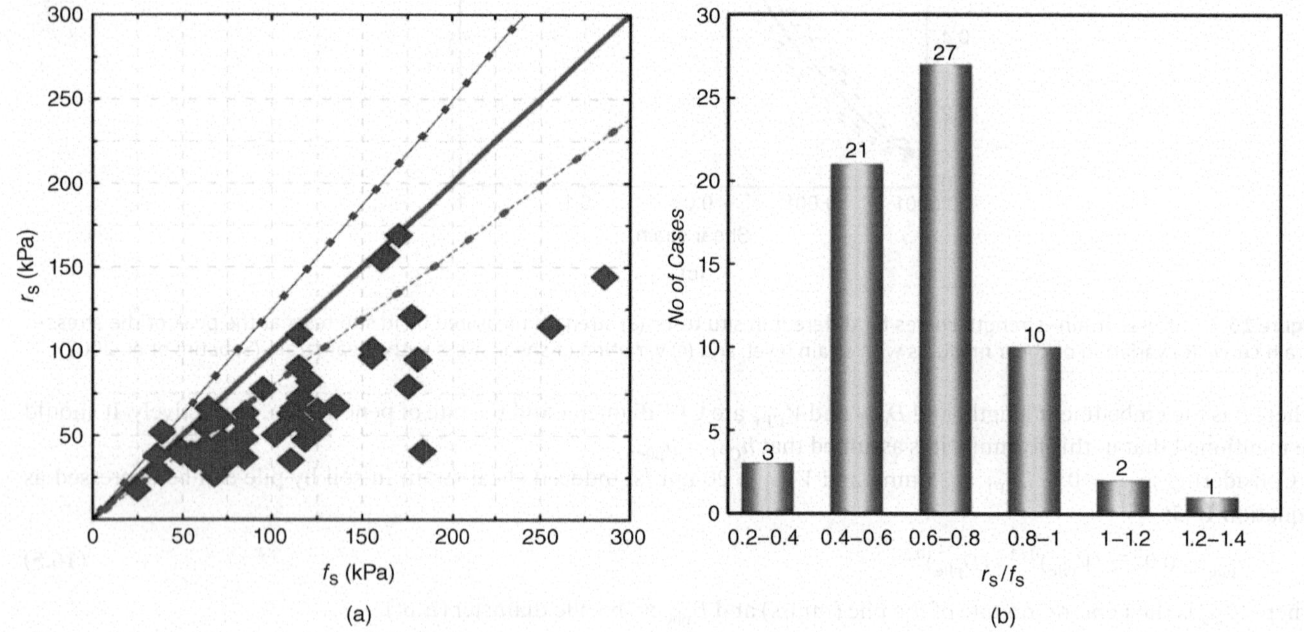

Figure 16.9 (a) Comparison of r_s and f_s, (b) distribution of r_s/f_s values for Eslami et al. (2013) database (Eslami et al. 2013 / With permission of American Society of Civil Engineers).

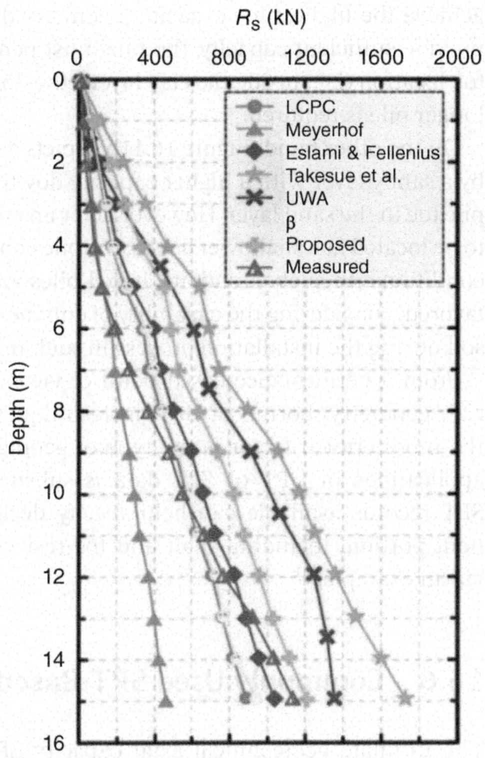

Figure 16.10 An example of shaft capacity distribution along with the pile by CPT-based methods and the measured values (Eslami et al. 2013 / With permission of American Society of Civil Engineers).

16.5 CPT or SPT Records for Pile Installation

Example 16.1 According to Figure 16.11a, down to 7.5 m, the soil type is sand (high q_c and low R_f) and can be considered as a bearing layer, while below 7.5 m, the soil type is clay with relatively low tip resistance and high friction ratio, as indicated in the q_c profile. Thus, as a first choice, it is more appropriate to locate the pile toe somewhere within the first 7.5 m to

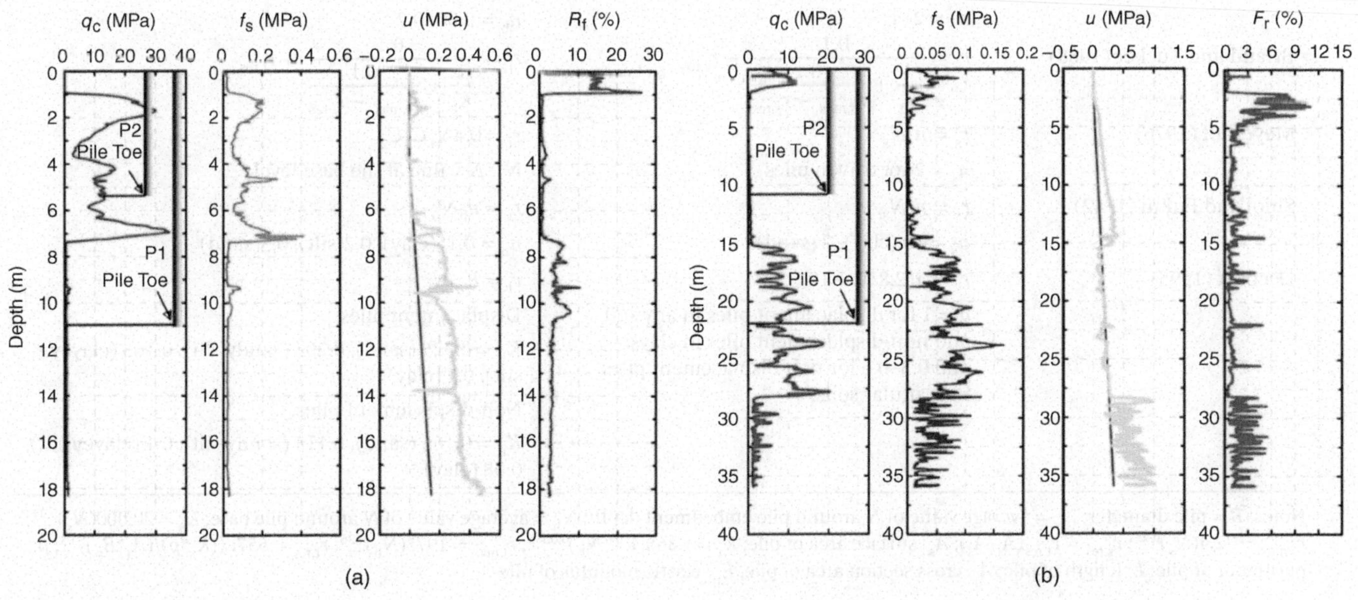

Figure 16.11 CPT profiles and pile alternatives: (a) NWU Site (CPT data from Finno 1989) and (b) UBC site (CPT data from Campanella et al. 1989).

achieve the highest toe capacity (geometry dictation). However, if the placement of the pile toe in this stratum does not provide sufficient capacity, the pile must penetrate inside the clay layer. In this case, given the low resistance at the pile toe location (i.e., inside the clay layer), the dominant bearing capacity of the pile would be from the shaft capacity where a longer pile is required.

On the other hand, Figure 16.11b depicts a clayey soil layer with limited resistance within the first 15 m depth underlain by a sandy layer with a higher capacity down to 28 m. In this case, to have optimal toe capacity, it is desirable to locate the pile toe in the sand layer. However, if for any reason (e.g., less required capacity and insufficient equipment) the pile toe has to be located at a shallower depth, the pile construction type must be opted in a way that also accompany improving the soil conditions. In such situations, bored piles are not recommended. Instead, driven or drilled displacement piles are more favored, considering the capability of enhancing soil conditions by vibration or displacements imposed on the surrounding soil during the installation process in such methods.

From a performance-based point of view, installing deep foundations depends on external conditions such as material, geometry, boring or driving techniques or equipment, environmental issues, and adjacent structures. However, the more critical factor is the type of geomaterial classified as soft, medium, or hard subsurface soils. One of the first applications of CPT or SPT data is subsurface profiling and soil behavior classification. Therefore, having CPT or SPT records available can help specify design aspects such as pile geometry, construction type, and bearing component portion, including shaft and toe resistance. This concept is illustrated in CPT profiles, as shown in Figure 16.11 via an example.

16.6 Commonly Used SPT-Based Methods for Pile Axial-Bearing Capacity

The ultimate geotechnical axial capacity of a single pile can be estimated using correlations based on the standard penetration test (SPT) results, as shown in Table 16.1. The SPT is widely used in situ tests due to its simplicity and cost-effectiveness, leading to the development of various correlations that utilize SPT data for geotechnical analysis

Table 16.1 Summary of common SPT-based methods for evaluation of axial pile capacity.

Method/references	Pile unit side resistance (kPa)	Pile unit end bearing (MPa)
Bazaraa and Kurkur (1986)	$r_s = n_s N_s$ $n_s = 2\text{–}4$	$r_t = n_b N_b$ $n_b = 0.06\text{–}0.2$
Briaud and Tucker (1988)	$r_s = \dfrac{0.1}{\dfrac{1}{K_s} + \dfrac{0.1}{r_{s,max} + r_{s,res}}} - r_{s,res}$	$r_t = \dfrac{0.1}{\dfrac{1}{K_t} + \dfrac{0.1}{r_{t,max} - r_{t,res}}} + r_{t,res}$
Meyerhof (1976)	$r_s = n_s N_s$ $n_s = 2$ for driven piles	$r_t = 0.4 N_1 C_1 C_2$ N_1: N value at the base level
Shioi and Fukui (1982)	$r_s = n_s N_s$ $n_s = 10$ (clay), 2 (sand)	$r_t = n_b N_b$ $n_b = 0.15$(clay), 0.2(silt), 0.3(sand)
Decourt (1995)	$r_s = \alpha(2.8N_s + 10)$ $\alpha = 1$ for displacement piles in any soil and non-displacement piles in clays, and 0.5–0.6 for non-displacement piles in granular soils.	$r_t = K_b\ N_b$ Displacement piles: $K_b = 0.325$ (sand), 0.205 (sandy silt), 0.165 (clayey silt), 0.1 (clay) Non-displacement piles: $K_b = 0.165$ (sand), 0.115 (sandy silt), 0.1 (clayey silt), 0.08 (clay)

Notes: B = pile diameter; N_s = average value of N around pile embedment depth; N_b = average value of N around pile base; $K_s = 20{,}000(N_s)^{0.27}$; $r_{s,max} = 22.4(N_s)^{0.29}$; $r_{s,res} = r_{t,res}(A_t/A_s)$; A_s: surface area of pile; $K_t = 1{,}868{,}400(N_b)^{0.0065}$; $r_{t,max} = 1975(N_b)^{0.36}$; $r_{t,res} = 557L((K_s{*}p)/(A_t{*}E_p))^{0.5}$; p: perimeter of pile; L: length of pile; A_t: cross section area of pile, E_p: elastic modulus of pile.

Table 16.2 A and B factors for SPT-based methods (Budhu 2011; Eslami 2013).

References	*B*	*A*	Deep foundation type	Soil type
Shioi and Fukui (1982)	5	0.15	Bored	Fine-grained
Decourt (1982)	3.3	0.15		
Decourt (1982)	Mix of coarse and fine	Clay, 0.12	Driven	
	3.3	Silt, 0.2		
	3.3	Sand, 0.25		
Shioi and Fukui (1982)	10	Sand, 0.1–0.3		
Shioi and Fukui (1982)	5	0.1–0.15	Bored	Coarse-grained
Yamashita et al. (1987)	$r_s \leq 150$ kPa	$r_t < 3$ MPa		
Decourt (1982)	3.3 $r_s \leq 170$ kPa	0.15 $r_t \leq 7.5$ MPa		
Meyerhof (1976)	1	0.12		
Meyerhof (1956, 1976)	2	0.45	Driven	
Decourt (1982)	3.3	0.4		
Shioi and Fukui (1982)	10	0.3		
Shariatmadari et al. (2008)	3.65	0.385		

and design. Accordingly, a simple method was developed to estimate the SPT N value from CPT results as follows (Lunne et al. 1997):

$$I_C = \left((3.47 - \log Q_t)^2 + (\log F_r + 1.22)^2\right) \tag{16.7}$$

$$\frac{\frac{q_c}{P_a}}{N_{60}} = 8.5\left(1 - \frac{I_C}{4.6}\right), \tag{16.8}$$

where Q_t is the normalized cone penetration resistance (dimensionless) and is considered as $Q_t = \frac{q_t - \sigma_{v0}}{\sigma'_{v0}}$. F_r is the normalized friction ratio expressed in percent ($F_r = \frac{f_s}{q_t - \sigma_{v0}} \times 100$). I_c is the Robertson soil type index, and P_a is the reference stress = 100 kPa. Jefferies and Davies (1993) suggested that the mentioned method provides a better estimate for the SPT N values than the actual SPT. Table 16.1 presents a summary of common SPT-based methods.

A general trend has been suggested for SPT records application for deep foundations axial bearing capacity as follows:

$$r_s = BN \tag{16.9}$$

$$r_t = AN \tag{16.10}$$

where

r_s = shaft unit resistance (kPa)
B = proportional index between shaft unit resistance and the average value of N along the shaft
r_t = toe unit resistance (MPa)
A = proportional index between toe unit resistance and the average value of N around the toe

Some of the proposed values for A and B factors by different researchers have been encapsulated in Table 16.2.

Example 16.2 The SPT records for a site are provided below. A drilled shaft with two embedment lengths of 10 and 16 m and a diameter of 1 m was constructed on this site. Determine the allowed axial bearing capacity via the Bazaraa and Kurkur (1986) and Meyerhof (1976) methods.

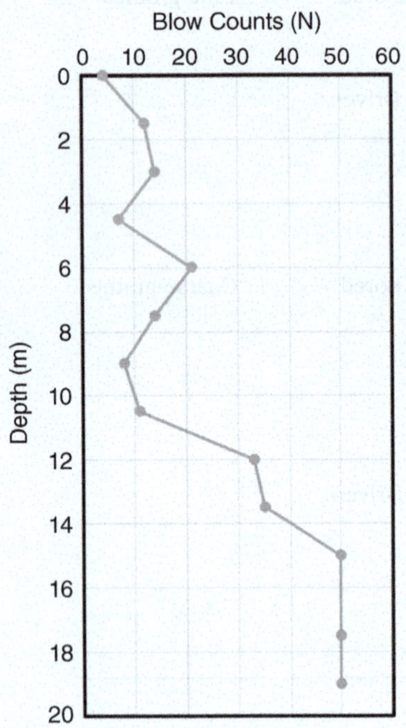

Depth (m)	Blow Counts (*N*)
0	4
1.5	12
3	14
4.5	7
6	21
7.5	14
9	8
10.5	11
12	33
13.5	35
15	50
17.5	50
19	50

$D_f = 10$ m

Meyerhof (1976)

Shaft Resistance:

$$N_{\text{ave-shaft}} = \frac{4+12+14+7+21+14+8}{7} = 11.4$$

$$B = 1$$

$$r_s = BN_{\text{ave}} = 1 \times 11.4 = 11.4 \text{ kPa}$$

$$A_s = \pi DL = \pi(1)(10) = 31.4 \text{ m}^2$$

$$R_s = r_s A_s = 11.4 \times 31.4 = 358 \text{ kN} = 35.8 \text{ tons}$$

Tip Resistance:

$$A = 0.12$$

$$N_{\text{ave-base}} = \frac{8+11}{2} = 9.5$$

$$r_t = AN_{\text{base}} = 0.12 \times 9.5 = 1.1 \text{ MPa}$$

$$A_t = \frac{\pi D^2}{4} = \frac{\pi \times 1^2}{4} = 0.79 \text{ m}^2$$

$$R_t = r_t A_t = 1100 \times 0.79 = 869 \text{ kN} = 86.9 \text{ tons}$$

Total Bearing Capacity:

$$R_u = R_t + R_s = 86.9 + 35.8 = 122.7 \text{ tons}$$

$$R_{allowed} = \frac{122.7}{3} = 40.9 \text{ tons}$$

Bazaraa and Kurkur (1986)

Shaft Resistance:

$$B > 0.5 \text{ m} \Rightarrow n_s = 1.34B = 1.34$$

$$N_{ave-shaft} = \frac{4 + 12 + 14 + 7 + 21 + 14 + 8}{7} = 11.4$$

$$r_s = n_s N_{ave} = 1.34 \times 11.4 = 15.3 \text{ kPa}$$

$$A_s = \pi DL = \pi(1)(10) = 31.4 \text{ m}^2$$

$$R_s = r_s A_s = 15.3 \times 31.4 = 480.4 \text{ kN} = 48 \text{ tons}$$

Tip Resistance:

$$\text{Effect Zone} = [D - 3.75\text{B}, D + B] = [6.25 \text{ m}, 11 \text{ m}]$$

$$N_{ave-base} = \frac{14 + 8 + 11}{3} = 11$$

$$B > 0.5 \text{ m} \Rightarrow n_t = 270$$

$$r_t = n_t N_{ave-base} = 270 \times 11 = 2970 \text{ kPa}$$

$$A_t = \frac{\pi D^2}{4} = \frac{\pi \times 1^2}{4} = 0.79 \text{ m}^2$$

$$R_t = r_t A_t = 2970 \times 0.79 = 2346.3 \text{ kN} = 234.6 \text{ tons}$$

Total Bearing Capacity:

$$R_u = R_t + R_s = 234.6 + 48 = 282.6 \text{ tons}$$

$$R_{allowed} = \frac{282.6}{3} = 94.2 \text{ tons}$$

D_f = 16 m

Meyerhof (1976)

Shaft Resistance:

$$N_{ave-shaft} = \frac{4 + 12 + 14 + 7 + 21 + 14 + 8 + 11 + 33 + 35 + 50}{11} = 19$$

$$B = 1$$

$$r_s = BN_{ave} = 1 \times 19 = 19 \text{ kPa}$$

$$A_s = \pi DL = \pi(1)(16) = 50.3 \text{ m}^2$$

$$R_s = r_s A_s = 19 \times 50.3 = 955.7 \text{ kN} = 95.6 \text{ tons}$$

Tip Resistance:

$$A = 0.12$$

$$N_{\text{ave-base}} = \frac{50 + 50}{2} = 50$$

$$r_t = AN_{\text{base}} = 0.12 \times 50 = 6\text{ MPa}$$

$$A_t = \frac{\pi D^2}{4} = \frac{\pi \times 1^2}{4} = 0.79\text{ m}^2$$

$$R_t = r_t A_t = 6000 \times 0.79 = 4740\text{ kN} = 474\text{ tons}$$

Total Bearing Capacity:

$$R_u = R_t + R_s = 474 + 95.6 = 569.6\text{ tons}$$

$$R_{\text{allowed}} = \frac{569.6}{3} = 189.9\text{ tons}$$

Bazaraa and Kurkur (1986)

Shaft Resistance:

$$B > 0.5\text{ m} \Rightarrow n_s = 1.34B = 1.34$$

$$N_{\text{ave-shaft}} = \frac{4 + 12 + 14 + 7 + 21 + 14 + 8 + 11 + 33 + 35 + 50}{11} = 19$$

$$r_s = n_s N_{\text{ave}} = 1.34 \times 19 = 25.5\text{ kPa}$$

$$A_s = \pi DL = \pi(1)(16) = 50.3\text{ m}^2$$

$$R_s = r_s A_s = 25.5 \times 50.3 = 1282.7\text{ kN} = 128.3\text{ tons}$$

Tip Resistance:

$$\text{Effect Zone} = [D - 3.75\text{B}, D + B] = [12.25\text{ m}, 17\text{ m}]$$

$$N_{\text{ave-base}} = \frac{35 + 50}{2} = 42.5$$

$$B > 0.5\text{ m} \Rightarrow n_t = 270$$

$$r_t = n_t N_{\text{ave-base}} = 270 \times 42.5 = 11{,}475\text{ kPa}$$

$$A_t = \frac{\pi D^2}{4} = \frac{\pi \times 1^2}{4} = 0.79\text{ m}^2$$

$$R_t = r_t A_t = 11{,}475 \times 0.79 = 9065.3\text{ kN} = 906.5\text{ tons}$$

Total Bearing Capacity:

$$R_u = R_t + R_s = 906.5 + 128.3 = 1034.8\text{ tons}$$

$$R_{\text{allowed}} = \frac{1034.8}{3} = 344.9\text{ tons}$$

The capacity is increased significantly as a result of deeper embedment depth.

16.7 Commonly Used Direct CPT-Based Methods for Pile Axial-Bearing Capacity

Given the penetrometer analogy with the pile, it can be considered a model pile. Figure 16.12 illustrates how closely the CPT tip resistance and sleeve friction resemble the pile toe and shaft capacity components (i.e., r_t and r_s).

Over 30 approaches have been proposed for estimating the pile axial-bearing capacity by CPT records within the last 30 years. A summary of such methods has been gathered by Niazi and Mayne (2013). Table 16.3 demonstrates a list of these methods. Below, we discuss the more commonly used methods.

16.7.1 Nottingham (1975) and Schmertmann (1978)

The Schmertmann and Nottingham method is based on a summary of the work on the model and full-scale piles presented by Nottingham (1975). The unit toe resistance of a pile(r_t) in the sand and the clay is equal to the cone resistance average. The average q_c value is determined from the graphical representation of the CPT measurement in a zone ranging from 8B above (B is the pile diameter) and from 0.7B through 4B below the pile toe, where the actual value depends on the trend of q_c values (see Figure 16.13). This zone is defined by the failure pattern for the pile toe, which is assumed to follow a log-spiral shape, as initially introduced by Begemann (1963). An upper limit of 15 MPa is imposed for the unit toe resistance.

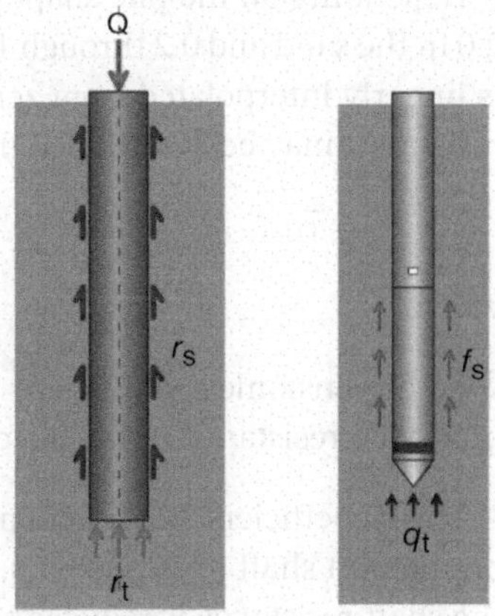

Figure 16.12 Analogy between pile and penetrometer.

Table 16.3 List of common CPT- and CPTu-based methods for pile-bearing capacity estimation.

No.	Method/references	No.	Method/references
1	Bogdanovic (1961)	17	UCD-05 (Gavin and Lehane 2005)
2	Begemann (1969)	18	ICP-05 (Jardine et al. 2005)
3	Meyerhof (1983)	19	UWA-05 (Lehane et al. 2005)
4	Aoki and Velloso (1975)	20	NGI-05 (Clausen et al. 2005)
5	Nottingham (1975), Schmertmann (1978)	21	Cambridge-05 (White and Bolton 2005)
6	Penpile (Clisby et al. 1978)	22	Togliani (2008)
7	Dutch (de Ruiter and Beringen 1979)	23	German (Kempfert and Becker 2010)
8	Philipponnat (1980)	24	UCD-11 (Igoe et al. 2010, 2011)
9	LCPC (Bustamante and Gianeselli 1982)	25	*V–K* (Van Dijk and Kolk 2011)
10	Cone-m (Tummay and Fakhroo 1982)	26	SEU (Cai et al. 2011, 2012)
11	Price and Wardle (1982)	27	HKU (Yu and Yang 2012)
12	Gwizdala (1984)	28	UWA-13 (Lehane et al. 2013)
13	Eslami and Fellenius (1997), Fellenius (2002)	29	Probabilistic Approach by Chen and Zhang (2013)
14	KTRI (Takesue et al. 1998)	30	Enhanced UniCone method (Niazi and Mayne 2016)
15	TCD-03 (Gavin and Lehane 2003)	31	Valikhah et al. (2018)
16	Fugro-05 (Kolk et al. 2005)	32	Data-based pile design approach (Eslami et al. 2020)

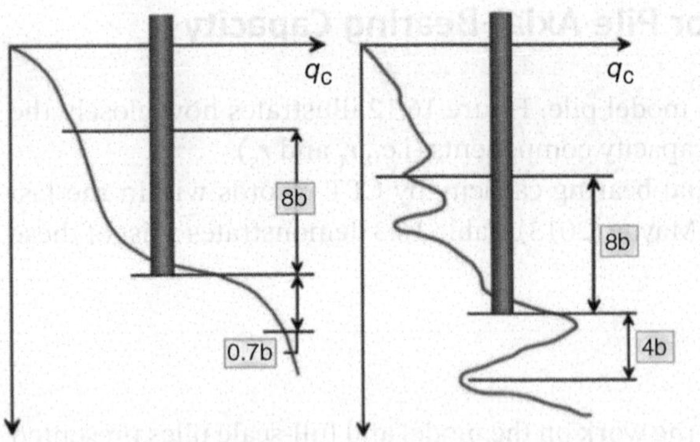

Figure 16.13 Application of CPT data for pile design (adapted from Schmertmann 1978).

The unit shaft resistance of a pile (r_s) is determined from either sleeve friction or cone resistance:

$$r_s = Kf_s \tag{16.11}$$

where K = dimensionless coefficient for clay defined as the reverse ratio in Figure 16.14.

Depending on the pile shape and material, cone type, and embedment ratio, the K-coefficient ranges from 0.8 through 2.0 in the sand and 0.2 through 1.25 in the clay. Above an embedment depth of eight pile diameters, the unit shaft resistance is linearly interpolated from zero at the ground surface to the value of Equation 16.11. In sand, but not in clay, the shaft resistance may be determined from the cone resistance instead:

$$r_s = \bar{c}q_c, \tag{16.12}$$

where

$\bar{c}$ = a dimensionless coefficient
q_c = cone resistance (total; uncorrected for pore pressure)

The $\bar{c}$-coefficient is a function of the pile type and ranges from 0.8% through 1.8%. An upper limit of 120 kPa is imposed on the unit shaft resistance (r_s), regardless of whether it is determined from Equation 16.11 or 16.12. For tension capacity, the shaft resistance is reduced to 70% of that determined by Equation 16.11 and Equation 16.12.

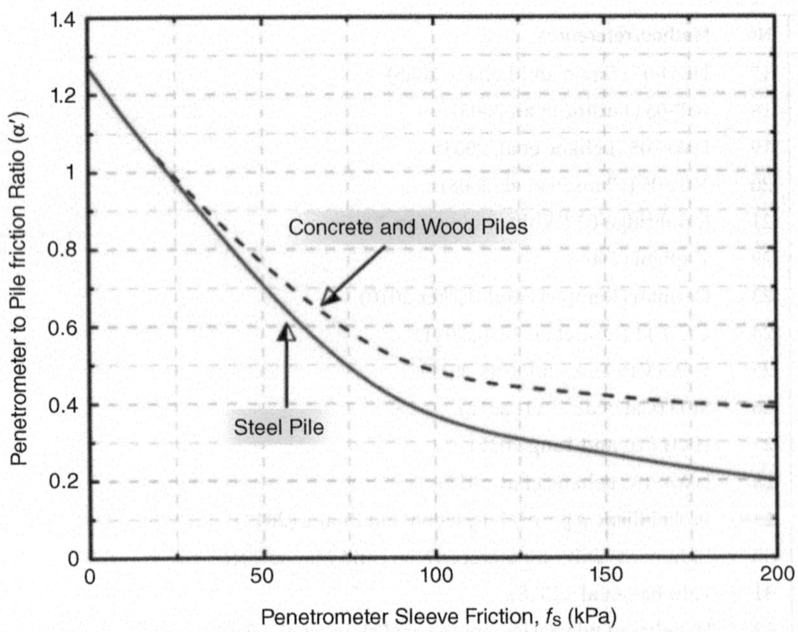

Figure 16.14 Design curves for pile shaft resistance in clay (adapted from Schmertmann 1978).

16.7.2 LCPC (Bustamante and Gianeselli 1982)

The Laboratoire Central des Ponts et Chausees (LCPC) method (also called the French method) is based on the experimental work of Bustamante and Gianeselli (1982) for the French Highway Department. Both the unit toe and unit shaft resistances are determined from the cone resistance, q_c. The sleeve friction (f_s) is not used. The unit toe resistance is determined as a range of 40% through 55% of the q_c, averaged within a zone of 1.5B above and 1.5B below the pile toe (see Table 16.4 and Figure 16.15).

The unit shaft resistance is determined from Equation 16.10 with the c-coefficient ranging from 0.5% through 3.0%, as governed by rules for the cone resistance magnitude, soil type, and pile type. Upper limits of the unit shaft resistance are imposed, ranging from 15 kPa through 120 kPa depending on soil type, pile type, and pile installation method (see Table 16.4).

16.7.3 Meyerhof (1976, 1983)

The Meyerhof method is based on theoretical and experimental studies of deep foundations in sand. The influence of the scale effect of piles and shallow penetration in dense sand strata has been considered for unit toe resistance. The unit toe

Table 16.4 Empirical coefficients for French method for pile capacity calculation using CPT (modified after Bustamante and Gianeselli 1982).

			Driven piles				Bored piles			
			K_s**		Upper limit of r_s (kPa)		K_s**		Upper limit of r_s (kPa)	
Nature of soil	q_c (MPa)	K_c*	A	B	A	B	A	B	A	B
Soft clay and mud	<1	0.5	30	30	15	15	30	30	15	15
Moderately compact clay	1–5	0.45	40	80	35 (80)	35 (80)	40	80	35 (80)	35 (80)
Compact to stiff clay and compact silt	>5	0.55	60	120	35 (80)	35 (80)	60	120	35 (80)	35 (80)
Silt and loose sand	<5	0.5	60	120	35	35	60	150	35	35
Moderately compact sand and gravel	5–12	0.5	100	200	80 (120)	80	100	200	80	35 (80)
Compact to very compact sand and gravel	>12	0.4	150	200	120 (150)	120	150	300	120 (150)	80 (120)

CATEGORY:
A – Driven precast piles, prestressed tubular piles, and Jacked concrete piles
B – Driven metal piles and jacked metal piles
*$r_t = K_c\ q_{c,s}$
($q_{c,s}$ = average q_c values)
**$r_s = q_c/K_s$
Note: Bracket value for r_s: apply to careful execution and minimum disturbance of soil due to disturbance.

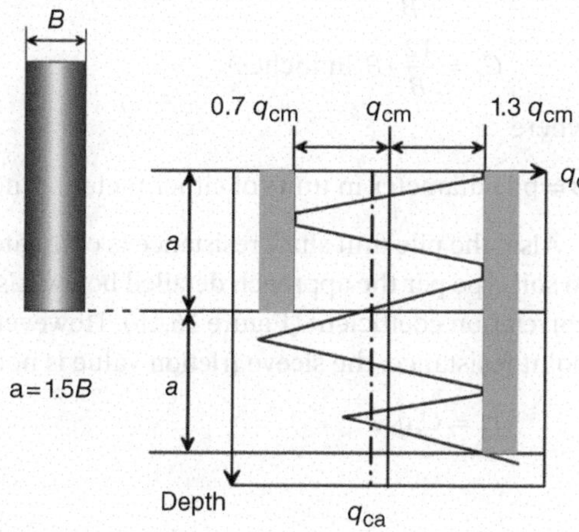

Figure 16.15 LCPC CPT method for determining equivalent cone resistance at pile toe (adapted from Bustamante and Gianeselli 1982).

resistance in the sand is determined from cone resistance through an empirical relation as follows:

$$r_t = q_{c,a} C_1 C_1 \tag{16.13}$$

where

$q_{c,a}$ = arithmetic average of q_c values in a zone ranging from "1*b*" below through "4*b*" above pile toe
$C_1 = \left[{}^{(b+0.5)}/_{2B}\right]^n$; modification factor for scale effect when $b > 0.5$, otherwise $C_1 = 1$
$C_2 = {}^{D_b}/_{10B}$; modification factor for penetration into dense strata when $Db < 10b$, otherwise $C_2 = 1$
B = pile diameter (m)
n = an index; 1 for loose sand, 2 for medium dense sand, and 3 for dense sand
D_b = embedment of the pile (m) in dense sand strata

The unit toe resistance of a bored pile is reduced to 30% of that determined from Equation 16.13.

The unit shaft of pile resistance is determined from either Equation 16.9 with $K = 1$ or Equation 16.12 with $C = 0.5\%$. For bored piles, 70% and 50% reduction factors are applied to the calculated values.

16.7.4 UniCone (Eslami and Fellenius 1997)

The UniCone method (Eslami and Fellenius 1997) is based on CPTu data but can also make use of CPT data. The method is developed from an extensive database of CPT and CPTu tests coupled with tests for axial pile capacity.

The data are unfiltered in the Eslami and Fellenius CPTu method (in contrast to other methods). Instead, a potentially disproportionate influence of odd "peaks and troughs" is reduced by employing the geometric average of the cone stress, as opposed to the arithmetic average used by the CPT methods. Furthermore, the cone stress is transferred to the "effective" cone stress (q_E) by subtracting the measured pore pressure (u_2) from the measured total cone stress (corrected for pore pressure acting against the shoulder). The pile unit toe resistance is the geometric average of the "effective" cone stress over an influence zone that depends on the soil layering.

When a pile is installed through weak soil into dense soil, the average is determined over an influence zone extending from 4*B* below the pile toe through a height of 8*B* above the pile toe. On the other hand, when a pile is installed through dense soil into weaker soil, the average is determined over an influence zone extending from 4B below the pile toe through a height of 2*B* above the pile toe. The relation is given in the following equation:

$$r_t = C_t q_{Eg}, \tag{16.14}$$

where C_t is the toe correlation coefficient and q_{Eg} is the geometric average of the cone point resistance over the influence zone after correction for pore pressure on the shoulder and adjustment to "effective" stress. The toe correlation coefficient (C_t) also called the toe adjustment factor, is a function of the pile size (toe diameter). The larger the pile diameter, the larger the movement required to mobilize the toe resistance.

Therefore, the "usable" pile toe capacity diminishes with increasing pile toe diameter. For pile diameters larger than about 0.4 m, the adjustment factor should be determined by the relation given in Equation 16.15:

$$C_t = \frac{1}{3B} \ (B \text{ in meter}) \tag{16.15a}$$

$$C_t = \frac{12}{B} \ (B \text{ in inches}) \tag{16.15b}$$

where

B = pile diameter in units of either meter or inch

Also, the pile unit shaft resistance is correlated to the average "effective" cone point stress with a modification according to soil type per the approach detailed below. Using cone stress and sleeve friction, the soil profiling chart determines the C_s correlation coefficient (Figure 16.16). However, because the sleeve friction is a more variable measurement than the cone point resistance, the sleeve friction value is not applied directly.

$$r_s = C_s q_E, \tag{16.16}$$

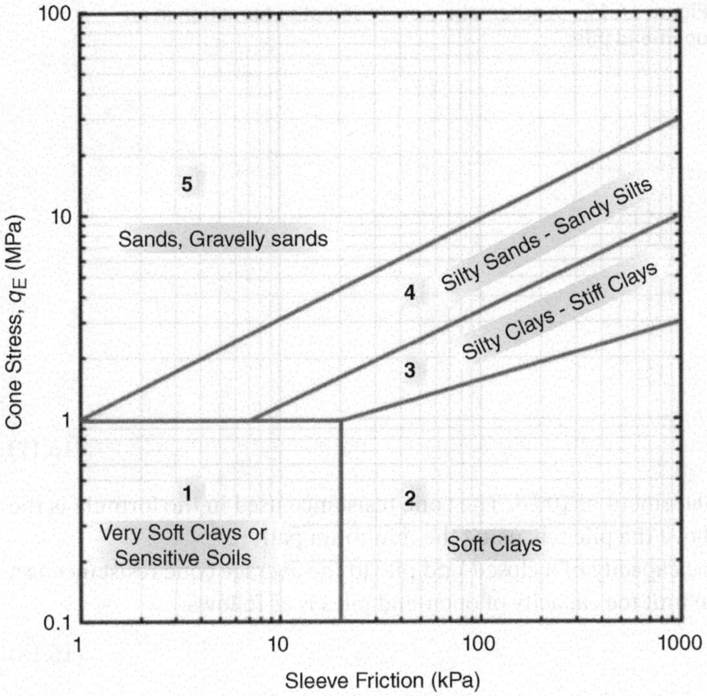

Figure 16.16 The Eslami and Fellenius (1997) chart for soil classification (Eslami and Fellenius 1997 / With permission of National Research Council Canada).

Table 16.5 Shaft coefficient correlation, C_s (Eslami and Fellenius 1997 / With permission of National Research Council Canada).

Soil Type	C_s
Soft sensitive soils	8.0%
Clay	5.0%
Stiff clay and a mixture of clay and silt	2.5%
Mixture of silt and sand	1.0%
Sand	0.4%

where

r_s = pile unit shaft resistance
C_s = shaft correlation coefficient, which is a function of soil type determined from the soil profiling chart (Table 16.5)
q_E = cone point resistance after correction for pore pressure on the cone shoulder and adjustment to "effective" stress

16.7.5 UWA (Lehane et al. 2005)

This approach is only applicable for driven piles in sand. The main difference between this approach and other methods is the need to consider the effect of partial plugging (Figure 16.17). In this regard, the incremental filling ratio (IFR) is introduced. The IFR is the variation in soil plug thickness per pile penetration into the soil.

Figure 16.17 demonstrates a schematic view of soil plug formation in an open-end pile. In a fully plugged condition, IFR is zero, and IFR equals unity in an unplugged condition. For intermediate partial plugging conditions, IFR varies between

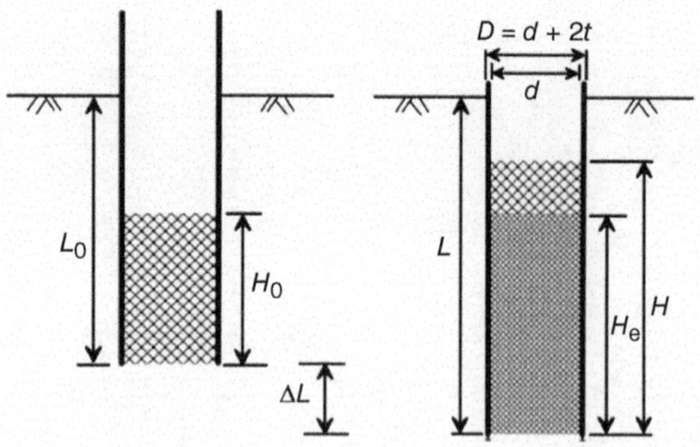

Figure 16.17 A schematic view of soil plug formation in an open-end pile.

0 and 1.0.

$$\mathrm{IFR} = \frac{H - H_0}{L - L_0} \quad (16.17)$$

The pile capacity is defined as the load at pile toe displacement of 10%B. The cone resistance used in the formula is the average of the cone resistance within 4B below and 8B above the pile toe, using the minimum path rule.

According to the UWA method, the ratio of the unit toe capacity of a closed-end pile to the average cone resistance is a constant value equal to 0.6. The proposed formula for the unit toe capacity of open-end piles is as follows:

$$\frac{q_{b0.1}}{q_{c,avg}} = 0.15 + 0.45A_{rb,eff} \quad (16.18)$$

$$A_{rb,eff} = 1 - \mathrm{FFR}\left(\frac{B_i^2}{B^2}\right) \quad (16.19)$$

In the above equation, FFR is the Final Filling Ratio, which is the average of IFR over the last 3B of pile penetration. For open-end piles, it is necessary to use the effective diameter for defining the averaging zone.

$$B_{eff} = B\sqrt{A_{rb,eff}} \quad (16.20)$$

As mentioned previously, the unit toe resistance of a pile depends on FFR. Since measuring this parameter is not common in practice, the following Equation is proposed for estimating FFR:

$$\mathrm{FFR} \approx \min\left[1, \left(\frac{B_i(m)}{1.5(m)}\right)^{0.2}\right] \quad (16.21)$$

The unit shaft resistance of a pile can be calculated using Equation 16.22:

$$r_s = \frac{f_t}{f_c}\left[0.03q_c A_{rs,eff}^{0.3}\left[\max\left(\frac{h}{B}, 2\right)\right]^{-0.5} + \Delta\sigma'_{rd}\right]\tan\delta_f \quad (16.22)$$

$$A_{rs,eff} = 1 - \mathrm{IFR}\left(\frac{B_i}{B}\right)^2 \quad (16.23)$$

where $\frac{f_t}{f_c} = 1.0$ in compression, 0.75 in tension, h = distance from pile toe, q_c = cone resistance at a given depth, $\Delta\sigma'_{rd}$ = effective radial stress caused by dilation during loading as defined by Equation 16.20, and δ_f = friction angle between soil and pile.

The average IFR over the last 20B of pile penetration can be estimated by Equation 16.24:

$$\mathrm{IFR}_{mean} \approx \min\left[1, \left(\frac{B_i(m)}{1.5(m)}\right)^{0.2}\right] \quad (16.24)$$

The following equation defines the effective radial stress due to the dilation effect during loading:

$$\Delta\sigma'_{rd} = \frac{4G}{B}\Delta r \quad (16.25)$$

$$G \approx 185 q_c q_{c1N}^{-0.7} \quad (16.26)$$

$$q_{c1N} = \frac{(q_c/p_{ref})}{(\sigma'_{v0}/p_{ref})^{0.5}} \quad (16.27)$$

16.7.6 Imperial College Pile ICP

The ICP approach (Jardine et al. 2005/2013) provides separate equations for sand and clay. In sands, this method is applicable for driven piles for which the SML (slow maintained load) test is conducted around 10 days after installation. Friction fatigue and the pile tip condition are considered in terms of h/r^* and A_r, respectively. Changes in radial stress, the interface friction angle, and the direction of loading are taken into account in this method.

The pile capacity is defined as the capacity mobilized at the pile head settlement equal to 10% of the pile diameter. In the unit toe resistance formula, the averaging zone is 1.5, and the pile diameter above and below the pile tip for the cone resistance. The following equation calculates the unit toe resistance for closed-end piles:

$$\frac{q_{b0.1}}{q_{c,avg}} = \max\left[1 - 0.5\log\left(\frac{B}{B_{CPT}}\right), 0.3\right] \tag{16.28}$$

For open-end piles, two conditions, plugged and unplugged, are considered. To calculate the toe capacity of an open-end pile, it is necessary to first define whether the pile is plugged or unplugged, then use the corresponding formula for toe capacity calculation. If one of the following inequations is satisfied, the pile is plugged. Otherwise, it is unplugged.

$$B_i < 0.02(D_r - 30) \text{ or} \tag{16.29a}$$

$$B_i < 0.083\left(\frac{q_{c,avg}}{p_{ref}}\right)B_{CPT} \tag{16.29b}$$

where

D_r = relative density in percent.

In the case of the unplugged condition, the pile unit toe capacity is calculated by the following equation:

$$\frac{q_{b0.1}}{q_{c,avg}} = \max\left[0.5 - 0.25\log\left(\frac{B}{B_{CPT}}\right), 0.15, A_r\right] \tag{16.30}$$

where A_r is the effective area ratio defined as Equation 16.31:

$$A_r = 1 - \left(\frac{B_i^2}{B^2}\right) \tag{16.31}$$

For plugged piles, the unit toe resistance is determined as

$$\frac{q_{b0.1}}{q_{c,avg}} = A_r \tag{16.32}$$

The mobilized unit shaft friction can be estimated by Equation 16.33:

$$r_s = a\left[0.029bq_c\left(\frac{\sigma'_{v0}}{p_{ref}}\right)^{0.13}\left[\max\left(\frac{h}{R^*}, 8\right)\right]^{-0.38} + \Delta\sigma'_{rd}\right]\tan\delta_f, \tag{16.33}$$

where a is 0.9 for open-end piles in tension and 1.0 for other conditions, and b is 0.8 for piles in tension and 1.0 for piles in compression. Other parameters in this equation are similar to Equation 16.22. Also, for open-end piles, the effective pile radius (R^*) is used in calculations, which is determined as

$$R^* = \sqrt{R^2 - R_i^2} \tag{16.34}$$

16.7.7 German

The German method, proposed by Kempfert and Becker (2010), is based on loading tests on roughly 1000 case histories of bored, precast concrete, pipe, screw cast-in-place piles, and micro-piles. According to this method, the pile's unit shaft and toe resistance in non-cohesive soil is calculated in cone resistance. In cohesive soils, the pile capacity is calculated using undrained shear strength (s_u) and is not directly correlated to the CPT measurements. This method is presented in the form of r_s and r_t curves in terms of q_c or s_u.

As depicted in Figure 16.18, the bearing capacity for each of the five pile type categories is determined by the proposed intervals within two upper and lower boundaries:

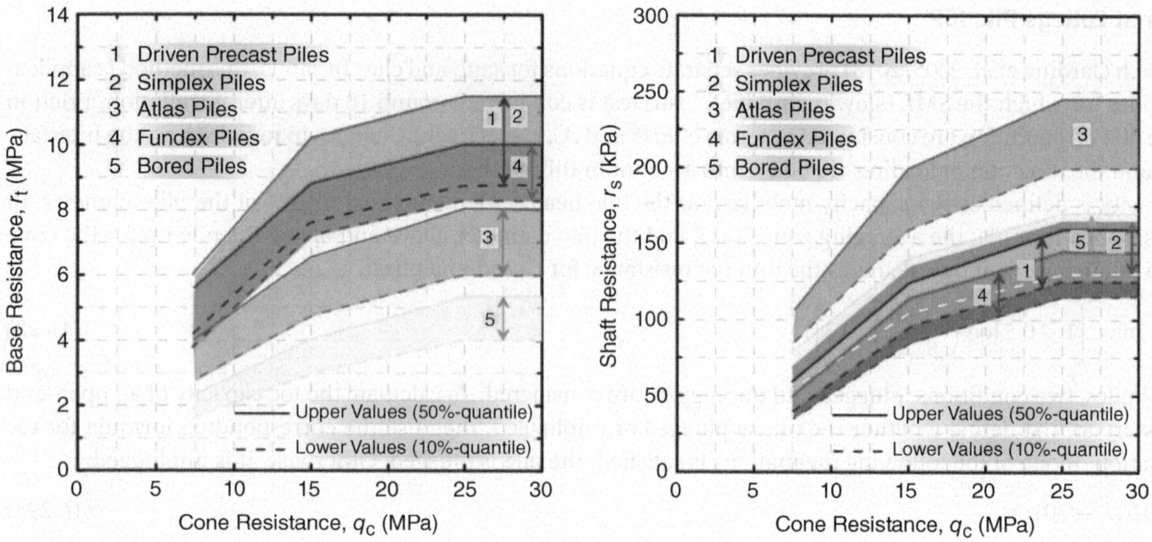

Figure 16.18 Upper and lower boundaries for pile capacity calculation according to the German method (Kempfert and Becker 2010 / With permission of American Society of Civil Engineers).

16.7.8 Enhanced UniCone (Niazi and Mayne 2016)

This method is a modified version of the Eslami and Fellenius (1997) method, using a database of 153 pile load tests introduced by Niazi and Mayne (2016). The modifications are offered to derive coefficients C_{se} for the shaft component and C_{te} for the toe component of the axial capacity from CPTu data. Also, they present a step-by-step methodology to use the method via a flowchart (Figure 16.19).

In order to estimate Q_{tn}, F_r, and I_c, it is necessary to apply a trial-and-error procedure to find "n," as stated below (Zhang et al. 2002):

Step 1: Assume $n = 1$ and calculate Q_{tn}, F_r, and I_c.
Step 2: If $I_c \leq 1.64$, $n = 0.5$.
If $1.64 < I_c < 3.30$, $n = (I_c - 1.64)*0.3 + 0.5$.
If $3.30 \leq I_c$, $n = 1$.
Iterate until the change in n, $\Delta n < 0.01$.
If $\sigma'_{v0} > 300$ kPa, let $n = 1$ for all soils

16.7.9 Summary of the CPT-Based Methods

Fifteen methods are presented in this section, a summary of which is presented in Table 16.6.

16.8 Comments on the Methods

1) A few methods, such as those developed in the '70s and '80s, do not consider the more accurate measurements achievable by CPTu since piezocone was generally not available during those times.
2) While the recommendations are specified to soil type (clay and sand) for a few methods, none of them, except for UniCone (Eslami and Fellenius 1997) and enhanced UniCone (Niazi and Mayne 2016), include a means for identifying the soil type from CPT data. Instead, the coefficients' soil profile relies on information from conventional boring and sampling as well as laboratory testing, which may not be fully relevant to the CPT data.
3) The CPT-based methods include random smoothing and filtering of the CPT data, eliminating peaks and troughs that expose the results to considerable subjective operator influence.
4) The cone resistance (total resistance) has not been corrected for the pore pressure on the cone shoulder, and therefore, the data behind the methods include errors that are smaller for the sand and larger for the clay. This matter (i.e., penetration pore pressure, u_2) is realized by UniCone (Eslami and Fellenius 1997).

Table 16.6 Summary of CPT- and CPTu-based methods.

No.	Method/references	Pile unit side resistance (r_s)	Pile unit end bearing (r_t)
1	Nottingham (1975) Schmertmann (1978)	$r_s = C_s q_c$ $r_s = K f_s$ $C_s = 0.8 - 1.8\%$ $K = 0.8 - 2(\text{sand})$	$r_t = C_t q_{ca}$
2	Meyerhof (1976)	$r_s = k f_s$ $k = 1$ $r_s = \bar{c} q_c$ $\bar{c} = 0.5\%$	$r_t = q_{c,a} c_1 c_2$ $c_1 = \left(\frac{B + 0.5}{2B}\right)^n$ $c_2 = \frac{D_b}{10B}$ D_b = bearing embedment depth n = 1 (loose), 2 (medium dense), 3 (dense)
3	LCPC (Bustamante and Gianeselli 1982)	$r_s = \frac{1}{k_s} q_c$ $k_s = 30 - 150$	$r_t = k_b q_{eq}$ $k_b = 0.4 - 0.55$
4	Tummay and Fakhroo (1982)	$r_s = K f_s$ $K = 0.5 + 9.5 e^{-90 f_s}$	$r_t = C_t q_{ca}$
5	Price and Wardle (1982)	$r_s = k_s \cdot f_s \leq 120$ kPa $k_s = 0.53$ for driven piles, 0.62 for jacked piles, and 0.49 for bored piles.	$r_t = k_b \cdot q_{ca} \leq 15$ MPa $k_b = 0.35$ for driven piles, 0.35 for jacked piles
6	Penpile (Clisby et al. 1978)	$r_s = \frac{f_{sa}}{1.5 + 14.47 f_{sa}}$ f_{sa} = average cone sleeve friction in MPa.	$r_t = 0.25 q_{ca}$ (in clay) $r_t = 0.125 q_{ca}$ (in sand)
7	Unicone (Eslami and Fellenius 1997)	$r_s = c_{se} \times q_E$ $q_E = q_t - u_2$ $c_{se} = 0.3\% - 8\%$	$r_t = c_{te} \times q_{Eg}$ $q_{cg} = (q_{c1} \times q_{c2} \times q_{c3} \times \ldots \times q_{cn})^{\frac{1}{n}}$ $c_{te} = 1$
8	Takesue et al. (1998)	$\Delta u < 300\text{ kPa} \Rightarrow {}^{r_s}/_{f_s} = \frac{\Delta u}{1250} + 0.76$ $300 < \Delta u < 1200\text{ kPa} \Rightarrow {}^{r_s}/_{f_s} = \frac{\Delta u}{200} - 0.5$	—
9	UWA method (Lehane et al. 2005)	$r_s = \frac{f_t}{f_c}\left[0.03 q_c A_{rs,eff}^{0.3}\left[\max\left(\frac{h}{D}, 2\right)\right]^{-0.5} + \Delta\sigma'_{rd}\right]\tan\delta_f$ $A_{rs,eff} = 1 - \text{IFR}\left(\frac{B_i}{B}\right)^2$ $\text{IFR}_{mean} \approx \min\left[1, \left(\frac{B_i(m)}{1.5(m)}\right)^{0.2}\right]$	$\frac{r_{t0.1}}{q_{c,avg}} = 0.15 + 0.45 A_{rb,eff}$ $A_{rb,eff} = 1 - \text{FFR}\left(\frac{B_i^2}{B^2}\right)$ $\text{FFR} \approx \min\left[1, \left(\frac{B_i(m)}{1.5(m)}\right)^{0.2}\right]$
10	Fugro method (Kolk et al. 2005)	Compression Loading: $h/R^* \geq 4: r_s = 0.08 q_c \left(\frac{\sigma'_{v0}}{p_{ref}}\right)^{0.05}\left(\frac{h}{R^*}\right)^{-0.90}$ $h/R^* \leq 4: r_s = 0.08 q_c \left(\frac{\sigma'_{v0}}{p_{ref}}\right)^{0.05}(4)^{-0.90}\left(\frac{h}{4R^*}\right)$ Tension Loading: $r_s = 0.045 q_c \left(\frac{\sigma'_{v0}}{p_{ref}}\right)^{0.15}\left(\max\left(\frac{h}{R^*}, 4\right)\right)^{-0.85}$	$r_{t0.1} = 8.5 q_{c,avg}\left(\frac{p_{ref}}{q_{c,avg}}\right)^{0.5} A_r^{0.25}$ $A_r = 1 - \left(\frac{B_i^2}{B^2}\right)$

(Continued)

Table 16.6 (Continued)

No.	Method/references	Pile unit side resistance (r_s)	Pile unit end bearing (r_t)
11	ICP method (Jardine et al. 2005)	$r_s =$ $a\left[0.029bq_c\left(\frac{\sigma'_{v0}}{p_{ref}}\right)^{0.13}\left[\max\left(\frac{h}{R^*},8\right)\right]^{-0.38}+\Delta\sigma'_{rd}\right]\tan\delta_f$ $a = 0.9$ (OE piles in tension), 1.0 (all other cases) $b = 0.8$ (tension), 1.0 (compression) δ_f measured or estimated as fctn(d_{50})	$\frac{r_{t0.1}}{q_{c,avg}} = \max\left[1-0.5\log\left(\frac{B}{B_{CPT}}\right),0.3\right]$ The pile is fully plugged if: $B_i < 0.02(D_r - 30)$ *or* $B_i < 0.083\left(\frac{q_{c,avg}}{p_{ref}}\right)B_{CPT}$ Fully Plugged: $\frac{r_{t0.1}}{q_{c,avg}} =$ $\max\left[0.5-0.25\log\left(\frac{B}{B_{CPT}}\right),0.15,A_r\right]$ Coring: $\frac{q_{b0.1}}{q_{c,avg}} = A_r$
12	NGI method (Clausen et al. 2005)	$r_s = \left(\frac{z}{Dp_{ref}F_{D_r}F_{sig}F_{tip}F_{load}F_{mat}}\right) \geq 0.1\sigma'_{v0}$ $F_{D_r} = 2.1(D_r - 0.1)^{1.7}$ $F_{sig} = \left(\frac{\sigma'_{v0}}{p_{p_a}}\right)^{0.25}$ $F_{tip} = 1.0$ (driven OE), 1.6 (driven CE) $F_{load} = 1.0$ (tension), 1.3 (compression) $F_{mat} = 1.0$ (steel), 1.2 (concrete)	Closed-end pile: $\frac{r_{t0.1}}{q_{c,tip}} = \frac{0.8}{1+D_r^2}$ Open-end pile Plugged: $\frac{r_{t0.1}}{q_{c,tip}} = \frac{0.7}{1+3D_r^2}$ Unplugged: $r_{t0.1} = r_{t,ann}A_r + r_{t,plug}(1-A_r)$ $r_{t,ann} = q_{c,tip}$, $r_{t,plug} = \frac{12r_{s,avg}L}{\pi D_i}$ $r_{t0.1} = \min(r_{t0.1,plugged}, r_{t0.1,unplugged})$
13	Togliani (2008)	$r_s = k_1\, q_t^{0.5}$ $k_1 = \beta\left\{1.2\left(0.8+\frac{R_f}{8}\right)\right\}$ for $R_f < 1$ $k_1 = \beta\{1.1(0.4+Ln(R_f))\}$ for $R_f \geq 2$ $k_1 = \beta\left\{1.2\left(0.8+\frac{R_f}{8}\right)+1.1(0.4+Ln(R_f))\right\}/2$ *for* $1 \leq R_f < 2$ $B = f$ (pile installation type)	$r_t = k_3 q_t$ $k_3 = \lambda + [0.01(L_{pile}/d_{toe})]$ λ is considered to be 0.2 for driven piles and 0.1 for bored piles; L_{pile}/d_{toe} is pile slenderness ratio
14	German method (Kempfert and Becker 2010)	Defined based on the proposed chart of $q_c - r_s$	Defined based on the proposed chart of $q_c - r_t$
15	Enhanced UniCone (Niazi and Mayne 2016)	According to the chart in Figure 16.19	According to the chart in Figure 16.19

Note: D = pile embedment depth, B = pile diameter, B_i = pile inner diameter, R = pile radius, R_i = pile inner radius, h = height above pile tip, $B_{CPT} = 0.036\ m$, $p_{ref} = 100$ kPa, $R^* = \sqrt{R^2 - R_i^2}$, $A_r = 1 - \left(\frac{D_i^2}{D^2}\right)$, $\Delta\sigma'_{rd} = \frac{2G}{R^*}\Delta y$; $\Delta y \approx 2\,R_a \approx 0.02$ mm, $G = 185q_c q_{c1N}^{-0.7}$ or $G = q_c[0.0203 + 0.00125q_{c1N} - 1.216e^{-6}q_{c1N}^2]^{-1}$, $q_{c1N} = \frac{(q_c/p_{ref})}{(\sigma'_{v0}/p_{ref})^{0.5}}$, and $D_r = 0.4\left[\ln\frac{q_{c,tip}}{22(\sigma'_{v0}p_{ref})^{0.5}}\right]$.

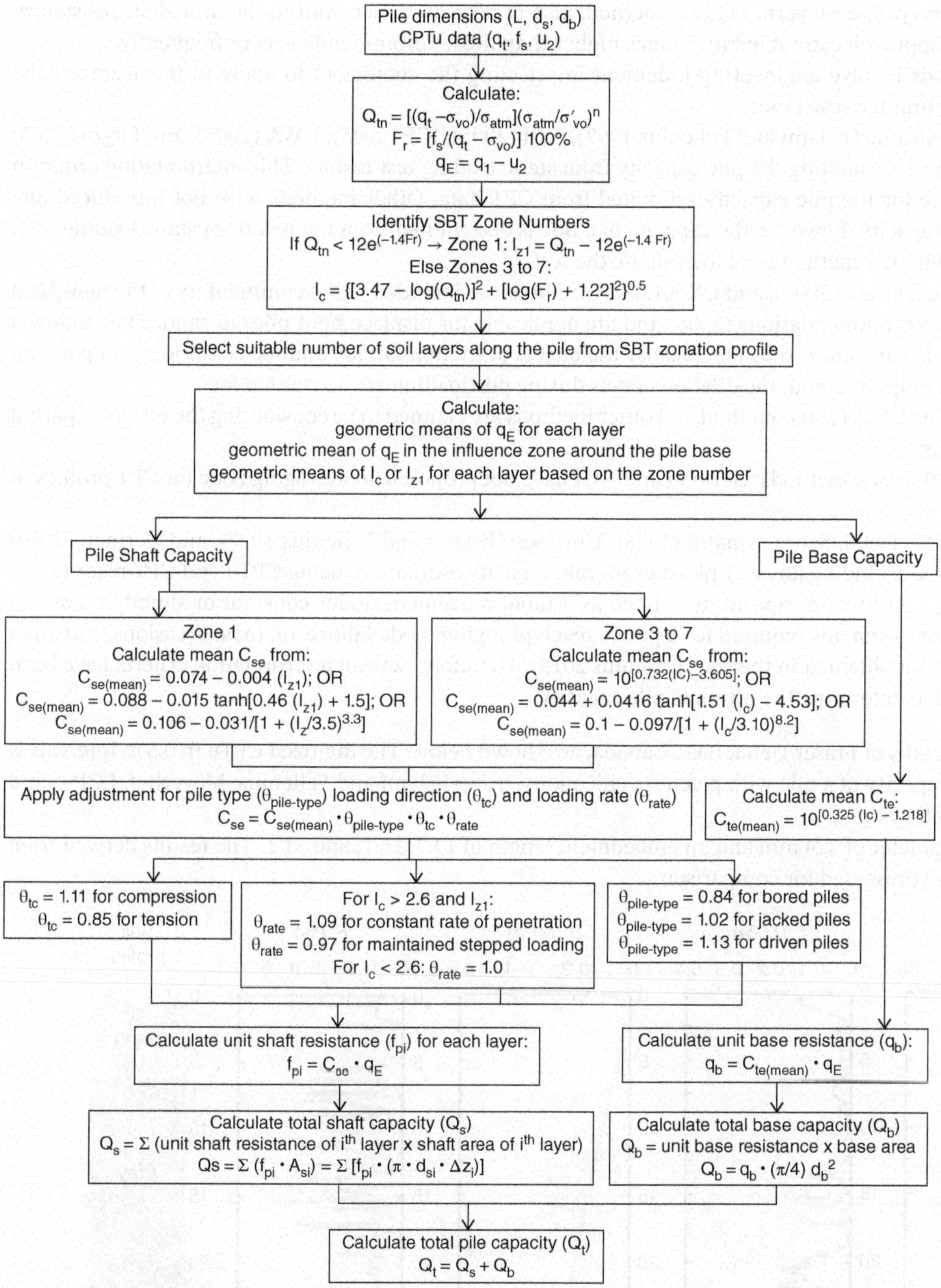

Figure 16.19 Modified UniCone flowchart for evaluating axial pile capacity based on CPTu records (Niazi and Mayne 2016 / with permission of Elsevier).

5) In the French method, the extent of the zone above and below the pile toe in which the cone resistance is averaged appears to be too limited. As considered in the Schmertmann method, particularly if the soil strength decreases below the pile toe, the soil average must include the conditions over a depth larger than 1.5B in the distance below the pile toe.
6) The methods employ total stress values, whereas effective stress governs pile capacity in the long term.
7) Some of the methods are locally developed; that is, they are based on limited types of piles and soils, such as Schmertmann (1978).
8) The upper limit resistance imposed on the unit toe resistance in the Schmertmann and European methods is not reasonable in very dense sands where values of pile unit toe resistance (r_t) higher than 15 MPa frequently occur. All methods

developed before 1990, except the Meyerhof (1976) method, also impose an upper limit to the unit shaft resistance, which an effective stress approach cannot justify. Values higher than those recommended occur frequently.

9) Most of the direct methods involve engineering judgment in selecting the coefficient to apply to the average cone resistance to arrive at the unit toe resistance.
10) Some methods such as UniCone (Eslami and Fellenius 1997), NGI (2005), ICP (2005), UWA (2005), and Fugro (2005) specify a certain criterion for evaluating the pile capacity from static loading test results. This interpretation criterion can be used as a reference for the pile capacity estimated from CPT data. Other methods have not introduced any criteria for ultimate pile capacity; however, the capacity of a pile is determined from the results of static loading tests and varies considerably with the method used to evaluate the test.
11) The NGI (2005), ICP (2005), Fugro (2005), and UWA (2005) methods are included in the commentary of the new 22nd edition of the API RP 2A Recommendations (2006) and are applicable for displacement piles in sand. They follow a relatively similar format. For instance, they all consider the effects of friction fatigue and toe conditions in open-end piles. Also, except for the Fugro method, the dilation effects during pile loading are accounted for.
12) The distinction between the UWA (2005) method and other methods are claimed to be considering the effects of partial plugging in open-end piles.
13) Unlike the majority of CPT-based methods, German and NGI have not proposed an averaging zone for CPT profiles in toe capacity calculation.
14) Methods such as Meyerhof (1983), Schmertmann (1978), UniCone (Eslami and Fellenius 1997), and German (2010) involve routine procedures, excluding any complexities for pile capacity estimation, using CPTu and CPT records.
15) In a static pile load test, the ultimate capacity is defined as a rapid movement under constant or slightly increased load. Because of the large movements required for a pile to reach plunging mode failure, on most occasions, a distinct plunging ultimate load is not obtained in the test (Fellenius 2015). Therefore, several interpretation criteria have been proposed and are applied to determine the ultimate pile capacity.

Example 16.3 The CPTu results of Fraser Delta, B.C., Canada, are shown below. The digitized CPTu in 0.5 m intervals is also presented. The bearing capacity of a pile with given specifications, using Eslami and Fellenius, Meyerhof, LCPC, and Schmertmann, would be as follows:

The pile is driven with a diameter of 324 mm and an embedment length of 13.7, 16.1, and 31.1. The results derived from the static pile load test are also presented for comparison.

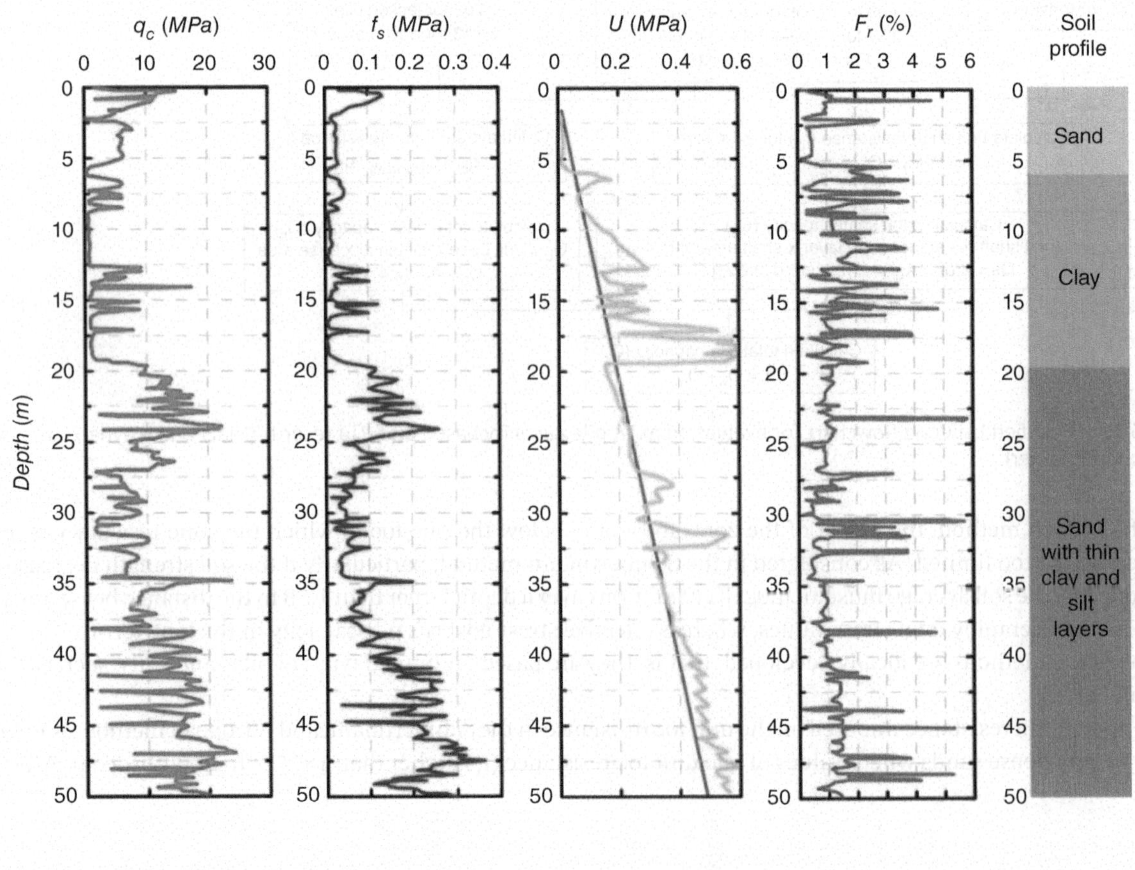

a) Unit shaft resistance for the pile with a length of 31.1 m.
The geometric mean of cone resistance from surface to pile toe:

According to the CPTu profile, two layers are considered: one from the depth of 0–17 m and below.

$$R_s = C_s . q_E$$

In soft clay (0–17.0 m): $C_s = 0.08$, $\overline{q}_E = 0.1895$ MPa
In sand (17.0–31.0 m): $C_s = 0.003$, $\overline{q}_E = 9.017$ MPa

$$r_{1s} = 0.185 \times 0.08 = 0.0152 \text{ MPa}$$

$$r_{2s} = 9.017 \times 0.003 = 0.027 \text{ MPa}$$

$$R_s = r_s \times A_s$$

$$R_{1s} = 0.0152 \times \pi \times 0.324 \times 17 = 0.263 \text{ MN} = 263 \text{ kN}$$

$$R_{2s} = 0.027 \times \pi \times 0.324 \times (31.1 - 17) = 0.387 \text{ MN} = 387 \text{ kN}$$

$$R_s = R_{1s} + R_{2s} = 263 + 387 = 650 \text{ kN}$$

b) Unit toe resistance:
the averaging zone is $8B$ above and $4B$ below the pile toe. Considering the pile embedment depth, the averaging zone depth is from 29.5 m to 33.5 m.

$$r_t = C_t \times \overline{q}_{E(geo)}$$

$$\overline{q}_{E(geo)} = 1.587 \text{ MPa}$$

$$r_t = 1 \times 1.587 = 1.587$$

$$R_t = A_t \times r_t = \frac{\pi \times 0.324^2}{4} \times 1.587 = 0.131 \text{ MN} = 131 \text{ kN}$$

$$R = R_t + R_s = 1005 + 131 = 1136 \text{ kN}$$

The summary of the results for other methods and pile embedment depths is as follows:

Pile no.	Bearing capacity (kN)				
	Schmertmann	LCPC	Meyerhof	UniCone	Measured capacity
1	141	217	–	244	290
2	431	633	–	630	630
3	825	1053	1001	1136	1100

Example 16.4 The bearing capacity of a closed-end driven steel pipe pile with the length and diameter of 22.5 m and 273 mm by 10 of the presented methods is as follows:

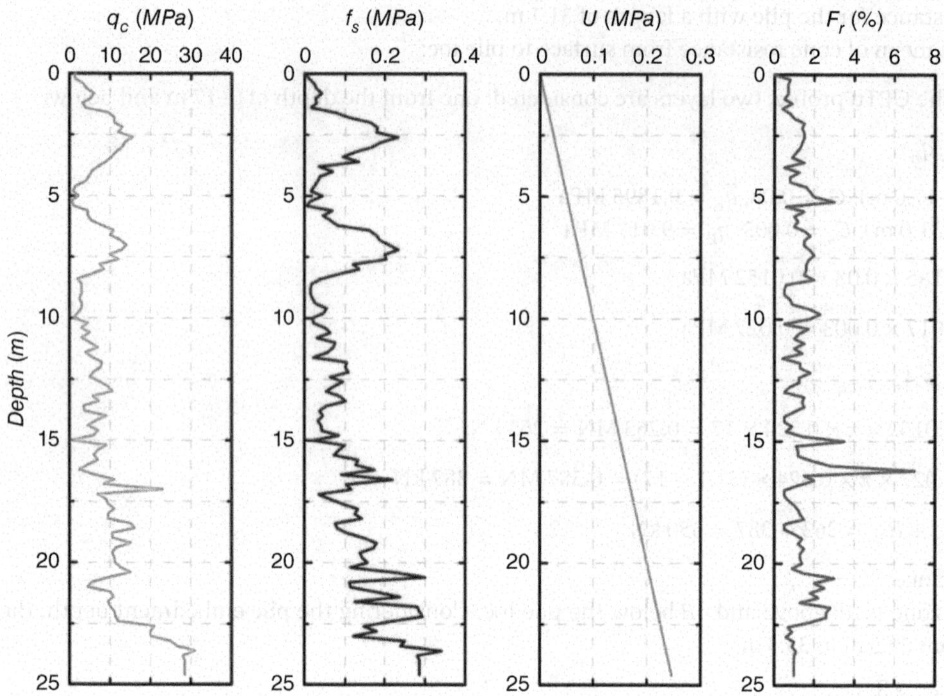

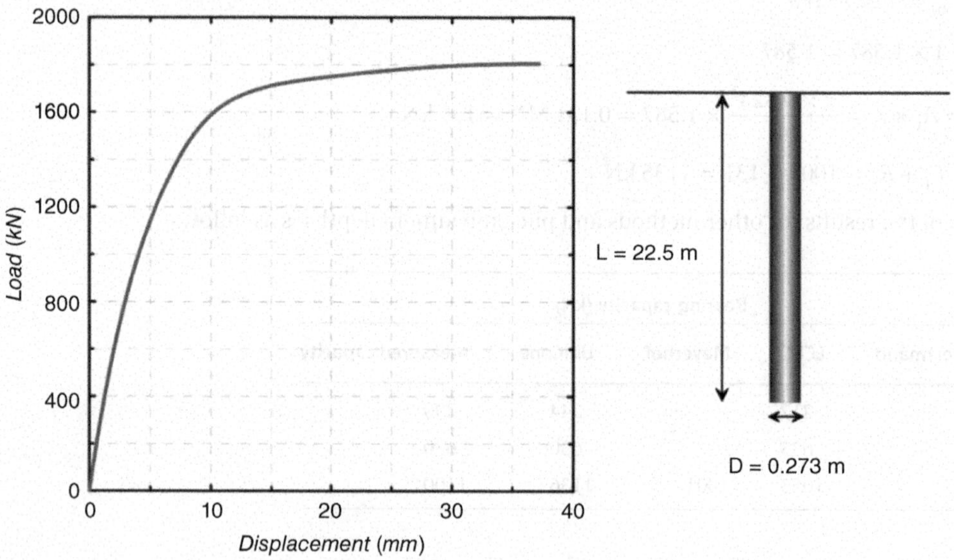

Results of pile capacity estimations using CPT-based direct methods are encapsulated below.

Method	Bearing Capacity (kN)
Brinch Hansen 80% criterion	1620
Eslami and Fellenius (1997)	1713
LCPC	1161
Meyerhof	1451
Schmertmann	2078
UWA	1379
NGI	1410
Fugro	1162
ICP	1438
German-Upper Bound	1836
German-Lower Bound	1291
German-Average	1564

Example 16.5 The bearing capacity of a square concrete pile with the characteristics presented below is as follows:

Case ID		134-1-FITTJA C
References		Axelsson (1998)
Location		Sweden
Shape		Square
Material		Concrete
Installation		Driven
Embedment length, D (m)		19.0
Diameter, B (mm)		235
Cross Sectional Area, At (m^2)		0.055
Perimeter (m)		0.940
GWL (m)		1.0

Bearing capacity from dynamic load test (CAPWAP)		
1 day after installation	Toe capacity, R_t (kN)	349
	Shaft capacity, R_s (kN)	387
	Total capacity, R_u (kN)	736
72 days after installation	Toe capacity, R_t (kN)	319
	Shaft capacity, R_s (kN)	1122
	Total capacity, R_u (kN)	1441

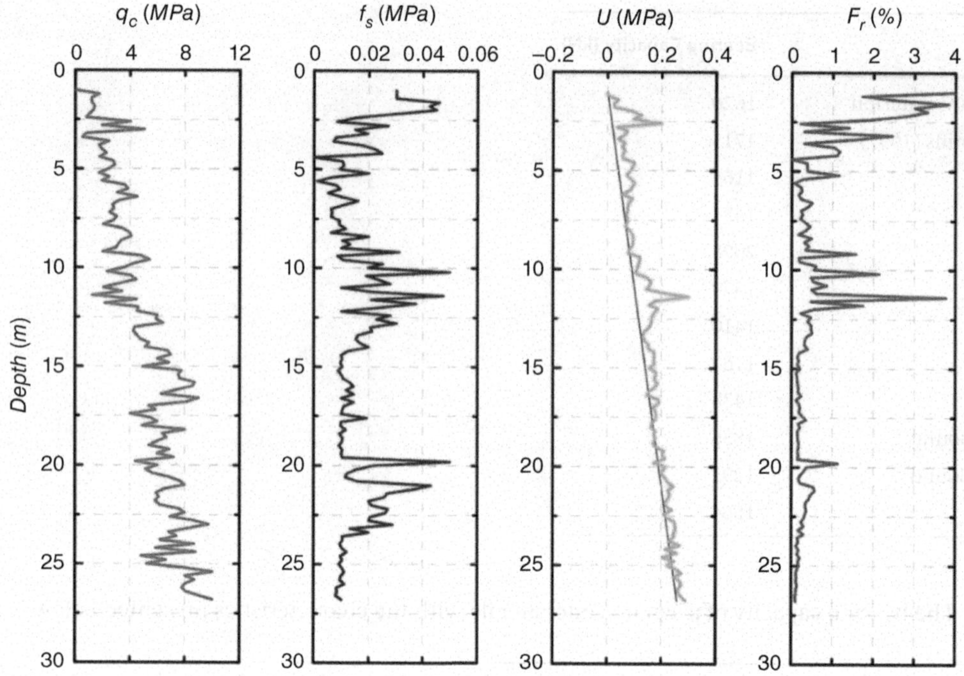

a) Toe capacity:
calculating the average cone resistance using the minimum path rule.

$$q_{c,avg\ (0.7B\ \text{below pile tip})} = 5.38\ \text{MPa}$$

$$q_{c,avg\ (4B\ \text{below pile tip})} = 5.96\ \text{MPa}$$

$$\rightarrow q_{c,avg\ (0.7B\ \text{bpt})} < q_{c,avg\ (4B\ \text{bpt})}$$

$$q_{c,min\ (0.7B\ \text{below pile tip})} = 5.38\ \text{MPa}$$

$$q_{c,avg\ (8B\ \text{above pile tip})} = 4.96\ \text{MPa}$$

$$\rightarrow q_{c,avg} = \frac{\frac{5.38+5.38}{2} + 4.96}{2} = 5.17\ \text{MPa}$$

The pile is closed-end $\rightarrow \frac{q_{b0.1}}{q_{c,avg}} = 0.6 \rightarrow q_{b0.1} = 0.6 \times 5.17 = 3.102\ \text{MPa} = 3102\ \text{kPa}$

$$Q_b = 3102 \times 0.055 = 171\ \text{kN}$$

b) Shaft capacity:

$$A_{rs,eff} = 1, B_{eff} = B = 0.265\ \text{m}, \Delta r = 0.02\ \text{mm}, \tan\delta_f = 0.52$$

Calculation of the pile capacity using the UWA method is as follows:

Z (m)	h/B	q_{c1N}	G (MPa)	$\Delta\sigma'_{rd}$ (kPa)	r_s (kPa)	ΔZ (m)	P (m)	ΔR_s (kN)	R_s (kN)
1.00	76.596	12,364.7	0.01	0.00	0.07	0	0.940	0.00	0.00
1.40	74.894	63.37	12.19	4.15	4.33	0.40	0.940	1.63	1.63
1.80	73.191	52.46	16.28	5.54	5.46	0.40	0.940	2.05	3.68
2.20	71.489	32.03	17.20	5.86	4.99	0.40	0.940	1.88	5.56
2.60	69.787	105.87	28.43	9.68	12.55	0.40	0.940	4.72	10.27

Z (m)	h/B	q_{c1N}	G (MPa)	$\Delta\sigma'_{rd}$ (kPa)	r_s (kPa)	ΔZ (m)	P (m)	ΔR_S (kN)	R_S (kN)
3.00	68.085	118.73	32.90	11.20	15.36	0.40	0.940	5.78	16.05
3.40	66.383	11.36	17.82	6.07	4.17	0.40	0.940	1.57	17.62
3.80	64.681	34.34	26.83	9.13	8.10	0.40	0.940	3.05	20.67
4.20	62.979	42.38	30.55	10.40	9.89	0.40	0.940	3.72	24.38
4.60	61.277	48.60	33.76	11.49	11.50	0.40	0.940	4.32	28.71
5.00	59.574	40.62	33.72	11.48	10.91	0.40	0.940	4.10	32.81
5.40	57.872	39.30	35.02	11.92	11.28	0.40	0.940	4.24	37.05
5.80	56.170	55.13	40.49	13.78	14.73	0.40	0.940	5.54	42.59
6.20	54.468	44.53	39.53	13.46	13.45	0.40	0.940	5.06	47.65
6.60	52.766	44.84	41.10	13.99	14.13	0.40	0.940	5.31	52.96
7.00	51.064	38.53	40.66	13.84	13.39	0.40	0.940	5.04	58.00
7.40	49.362	39.71	42.37	14.42	14.21	0.40	0.940	5.34	63.34
7.80	47.660	25.99	38.46	13.09	11.41	0.40	0.940	4.29	67.63
8.20	45.957	48.94	47.85	16.29	17.55	0.40	0.940	6.60	74.23
8.60	44.255	40.74	46.53	15.84	16.15	0.40	0.940	6.07	80.30
9.00	42.553	35.10	45.65	15.54	15.22	0.40	0.940	5.72	86.03
9.40	40.851	55.20	53.58	18.24	21.22	0.40	0.940	7.98	94.01
9.80	39.149	48.00	52.59	17.90	19.98	0.40	0.940	7.51	101.52
10.20	37.447	25.60	44.53	15.16	13.84	0.40	0.940	5.20	106.72
10.60	35.745	47.81	54.87	18.68	21.33	0.40	0.940	8.02	114.74
11.00	34.043	21.84	44.27	15.07	13.39	0.40	0.940	5.04	119.78
11.40	32.340	13.00	38.64	13.15	10.30	0.40	0.940	3.87	123.65
11.80	30.638	21.81	45.99	15.65	14.22	0.40	0.940	5.35	129.00
12.20	28.936	44.98	58.19	19.81	23.42	0.40	0.940	8.81	137.80
12.60	27.234	57.69	63.80	21.72	28.95	0.40	0.940	10.88	148.69
13.00	25.532	41.89	58.95	20.07	23.90	0.40	0.940	8.99	157.68
13.40	23.830	42.44	60.16	20.48	25.01	0.40	0.940	9.40	167.08
13.80	22.128	50.12	64.25	21.87	29.25	0.40	0.940	11.00	178.07
14.20	20.426	61.82	69.49	23.66	35.60	0.40	0.940	13.39	191.46
14.60	18.723	54.07	67.76	23.07	33.60	0.40	0.940	12.63	204.09
15.00	17.021	43.92	64.59	21.99	30.11	0.40	0.940	11.32	215.41
15.40	15.319	66.09	74.05	25.21	43.14	0.40	0.940	16.22	231.63
15.80	13.617	75.55	78.14	26.60	50.75	0.40	0.940	19.08	250.72
16.20	11.915	62.93	74.97	25.52	46.59	0.40	0.940	17.52	268.23
16.60	10.213	75.98	80.36	27.36	58.24	0.40	0.940	21.90	290.13
17.00	8.511	44.78	69.45	23.64	41.07	0.40	0.940	15.44	305.57
17.40	6.809	33.07	64.20	21.86	35.42	0.40	0.940	13.32	318.89
17.80	5.106	48.65	72.96	24.84	54.27	0.40	0.940	20.41	339.30
18.20	3.404	64.35	80.28	27.33	81.99	0.40	0.940	30.83	370.13
18.60	1.702	52.89	76.57	26.07	87.08	0.40	0.940	32.74	402.87
19.00	0.000	42.29	72.41	24.65	72.27	0.40	0.940	27.18	430.05

A summary of the bearing capacity values from CPT-based methods is presented in the below table.

	UWA-05	NGI-05	ICP-05	German method		
				Upper bound	Lower bound	Average value
R_s (kN)	430	408	419	519	266	392
R_t (kN)	171	223	234	266	181	223
R_u (kN)	601	631	653	784	447	615

16.9 Case Histories

Case History 16.1

Urmia Lake Causeway (Eslami et al. 2011)

This project aimed to establish a connection between the cities of Tabriz and Urmia in northwestern Iran. The construction site chosen was the narrowest part of the lake, requiring fill of a 13.5 km stretch from both the east and west coast sides. However, due to the presence of thick layers of fine and sensitive soils, the occurrence of sliding in the middle section, and environmental concerns, it was not feasible to continue with the fill for the remaining path. As an alternative solution, a causeway was designed to bridge the two filled sections. To gather crucial information about the soil parameters and soil profiling along the causeway path, eight piezocone tests, also known as CPTu tests, were conducted. These tests provide valuable data for determining the properties of the soil layers. The geographical location of the site can be seen in Figure 16.20. Figure 16.21 displays an example of the CPTu profiles recorded at the site. The soil type was classified using the charts developed by Robertson and Campanella (1989) as well as Eslami and Fellenius (1997). According to these charts, the predominant soil composition at the site consists of 90% sensitive soft clayey soils, clay, and stiff clay, while the remaining 10% comprises silt, sandy silt, and sand–gravel.

Figure 16.20 The geographical location of the causeway site (Eslami et al. 2011 / Elsevier / CC BY 4.0).

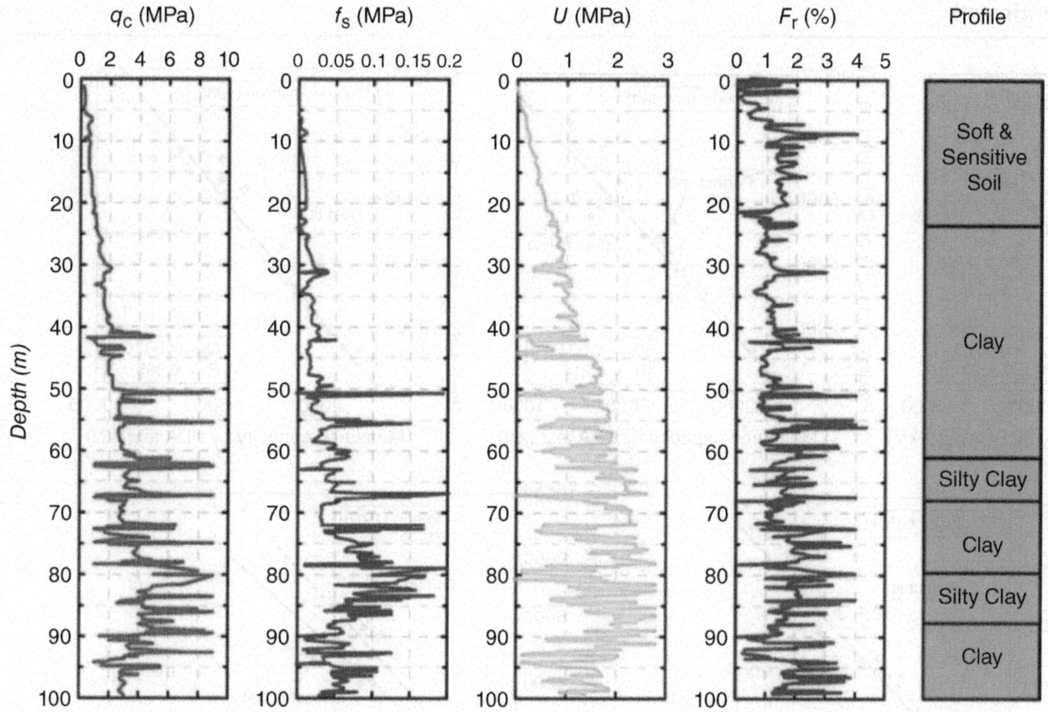

Figure 16.21 A typical CPTu profile of the Urmia site (Eslami et al. 2011 / Elsevier / CC BY 4.0).

For the Urmia Lake Causeway project, a total of 10 open-ended steel pipe piles with constant sections along their lengths were driven into the site. The pile diameters ranged from 685 to 913 mm, and the embedment depths varied from 41 to 85 m. To assess the capacity of these circular steel driven piles, four direct cone penetration test (CPT) methods and one CPTu method were evaluated. The cone tests were conducted using an electrical piezocone, which provides valuable data on the soil properties and behavior. By analyzing the CPT and CPTu data, the capacity of the piles could be estimated. The average embedment depth for the piles in this project was 70 m. During the different design stages of the project, it was found that CPTu and dynamic load tests were suitable tools for rapid and optimal pile design and verification. The pile-bearing capacity derived from the dynamic load tests ranged from 4700 to 16,500 kN, indicating the varying load-bearing capabilities of the driven piles.

The assessment of various CPT and CPTu-based methods, along with static and dynamic load tests, resulted in a predicted to measured capacity ratio ranging from 0.92 to 1.08. This indicates the satisfactory accuracy of the investigated methods in estimating the pile capacity. However, it should be noted that correction/calibration factors ranging from 0.5 to 1.3 were applied to the Dutch, LCPC, and Schmertmann methods to obtain this range. The correction factors were necessary because these methods were developed based on the mechanical cone, which does not consider the effect of pore pressure on their predictions. Given the significant values of pore pressures observed in the site, it was important to apply these correction factors to improve the accuracy of the predictions. Figure 16.22 provides a comparison of the performance of these methods. Overall, the precise records obtained from CPTu testing can be considered a valuable tool for evaluating pile capacity, especially in soft and loose deposits, particularly in marine environments where pore pressures play a significant role.

(Continued)

Case History 16.1 (Continued)

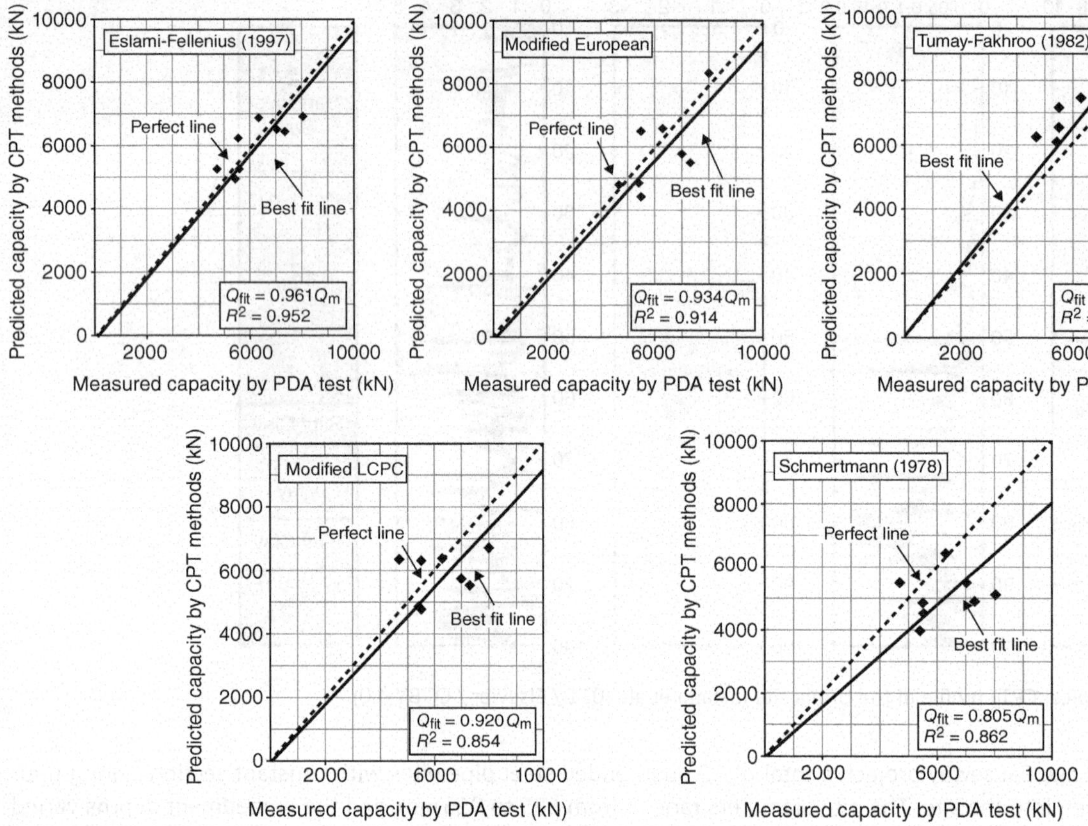

Figure 16.22 Comparing the performance of different predictive CPT-based methods (Eslami et al. 2011 / Elsevier / CC BY 4.0).

Case History 16.2

Similarity of CPT records in two different locations and its advantages in predicting axial pile-bearing capacity (Eslami et al. 2020)

The study by Eslami et al. (2020) demonstrated the potential benefits of using similar site conditions in different locations to inform the selection of suitable predictive methods for pile design. In this case, let's consider a driven pile that is planned to be installed in loose to medium-coarse sand in Norway. The pile has a length of 15.5 m and a diameter of 280 mm. The AUT: GEO-CPT&Pile database (2016) provided information on another pile installed in Sweden under similar site conditions. Figure 16.23 presents the CPT records for both sites and the results of static pile load tests for the selected and design piles.

In the study, three different CPT-based methods, namely the Dutch method (de Ruiter and Beringen, 1979), the Meyerhof method (1983), and the UniCone method (Eslami and Fellenius 1997), were evaluated for their performance in predicting pile capacity. Figure 16.24 illustrates the performance of these methods, with lower and upper threshold values set at 0.8 and 1.2 for the ratio of measured to predicted capacity. Based on the comparison, it can be observed that the Meyerhof (1983) and the UniCone method (Eslami and Fellenius 1997) exhibit good performance, as their predictions are close to each other and fall within the defined threshold values. These methods are considered

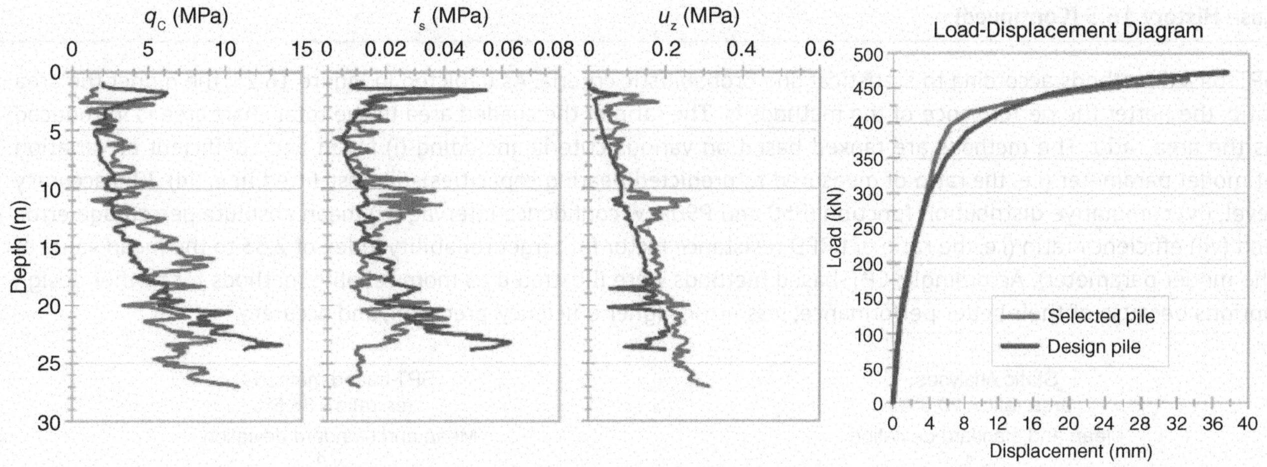

Figure 16.23 The CPTu records and the static pile load test result for the design and selected piles (Eslami et al. 2020).

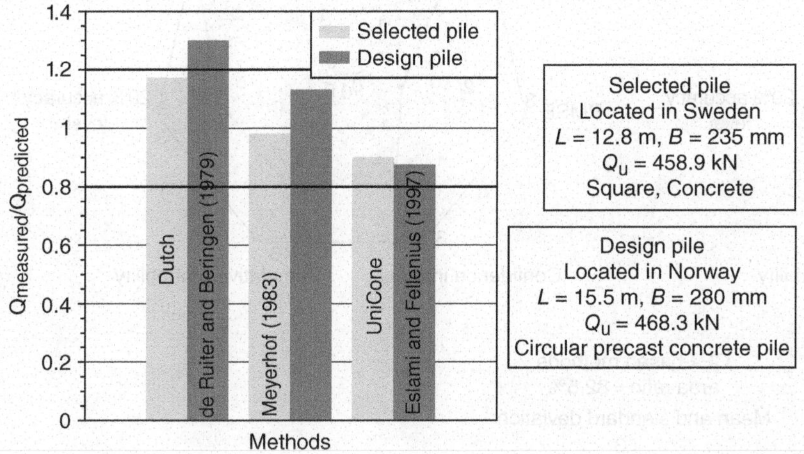

Figure 16.24 The variation of model parameter for the selected and designed piles (Eslami et al. 2020).

well-performing, as their predictions align well with the measured capacities for both the selected and designed piles. On the other hand, the Dutch method (de Ruiter and Beringen 1979) tends to provide predictions that are closer to the upper threshold value. While it still falls within the acceptable range, it shows a slightly higher level of uncertainty compared to the other methods. The results highlight the importance of considering similar site conditions and pile characteristics when selecting predictive methods. The performance of the methods can vary depending on the specific soil conditions and pile parameters. By utilizing the information from similar site conditions and pile characteristics, engineers can make more informed decisions and choose the most appropriate predictive method for pile design.

Case History 16.3

Performance-based assessment of different predictive pile-bearing capacity approaches (Heidarie Golafzani et al. 2020a,b)

Heidarie Golafzani et al. (2020a,b) compiled a database of 60 driven piles and demonstrated that among three different approaches for estimating axial pile-bearing capacity, CPT-based methods perform better than Static Analyses and

(Continued)

Case History 16.3 (Continued)

SPT-based methods according to statistical and probabilistic criteria. As depicted in Figure 16.25, the higher the area ratio, the better the performance of the methods is. The ratio of the shaded area to the total chart area is introduced as the area ratio. The methods are ranked based on various criteria, including (i) mean and coefficient of variation of model parameter (i.e. the ratio of measured to predicted bearing capacities), (ii) best-fitted line, (iii) 20% accuracy level, (iv) cumulative distribution function (P50 and P90), (v) confidence interval, (vi) mean absolute percentage error, and (vii) efficiency ratio (i.e. the ratio of LRFD resistance factor for target reliability index of 2.33 to the mean value of the model parameter). Accordingly, CPT-based methods were illustrated as more reliable methods for further design options because of their better performance, less error, higher efficiency, precision, and accuracy.

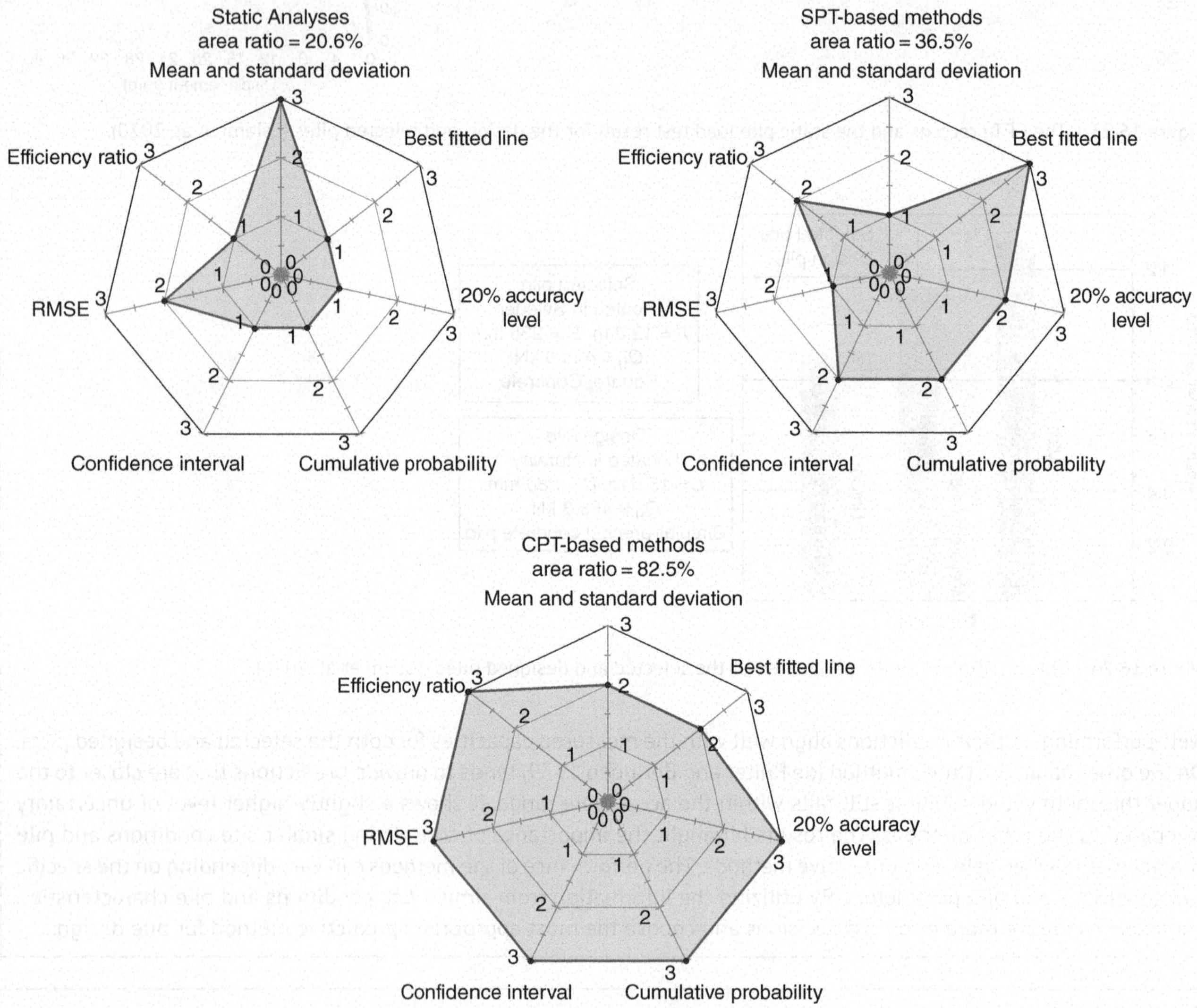

Figure 16.25 Assessment of the three common axial pile-bearing capacity prediction approaches (Heidarie Golafzani et al. 2020a, b / with permission of Taylor & Francis Group).

Case History 16.4

Special Piles-Drilled Displacement Piles (Moshfeghi and Eslami 2018)

Moshfeghi and Eslami (2019) compiled a database including 65 records of static load tests on drilled displacement piles (DDPs) in addition to the adjacent CPT profiles. Statistical reliability-based assessments, focusing on soil–pile specifications, have been conducted for six current CPT-based methods of determining the piles' bearing capacity. Figure 16.26 presents the frequency of different piles in this database.

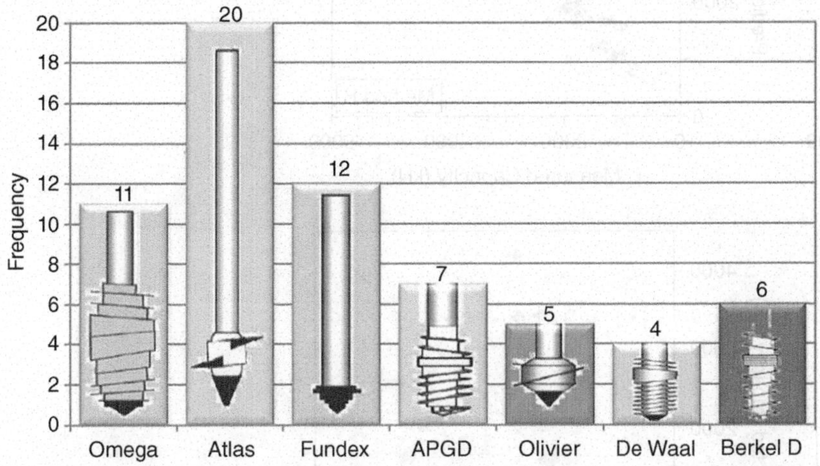

Figure 16.26 Database characterization based on pile type (Moshfeghi and Eslami 2019 / with permission of Taylor & Francis Group).

The primary approach of determining the bearing capacity of DDPs is in situ-based methods, including standard penetration test (SPT), CPT as well as pressure meter test. Six methods are selected in this study to estimate the bearing capacity of these piles. Two of them are exclusively for DDPs, i.e., method A (NeSmith 2002; Brettmann and NeSmith 2005) and method B (Bustamante and Gianeselli 1993, 1998) and the other four are commonly used methods of determining pile-bearing capacity, i.e., the UniCone method (Eslami and Fellenius 1997), the UWA method (Lehane et al. 2005), the Togliani method (Togliani 2008) and the German method (Kempfert and Becker 2010). Figure 16.27 compares the performance of these methods in predicting the bearing capacity of these piles.

Except for method B, all of the investigated methods generally underestimated the capacity of DDP (displacement driven piles), with the exception of Berkel displacement piles. Among the methods, method A performed the best for Atlas piles, while Eslami and Fellenius's (1997) method showed good performance for Omega and Fundex piles. Method B, on the other hand, performed well for Olivier and De Waal piles. Togliani (2008) method showed good estimation for APGD and Berkel displacement piles, which are types of displacement piles. It is important to note that the more commonly used CPT-based methods, such as Togliani (2008) and Eslami and Fellenius (1997), which were not specifically developed or calibrated for DDP, showed promising potential and provided reasonable estimations. These methods demonstrated reliability across various categories and can be considered in the design of DDP. Overall, the findings suggest that Togliani (2008) and Eslami and Fellenius's (1997) methods, among others, can be valuable tools for estimating the capacity of DDP. However, it is essential to exercise caution and consider site-specific conditions and project requirements when selecting and applying these methods.

(Continued)

Case History 16.4 (Continued)

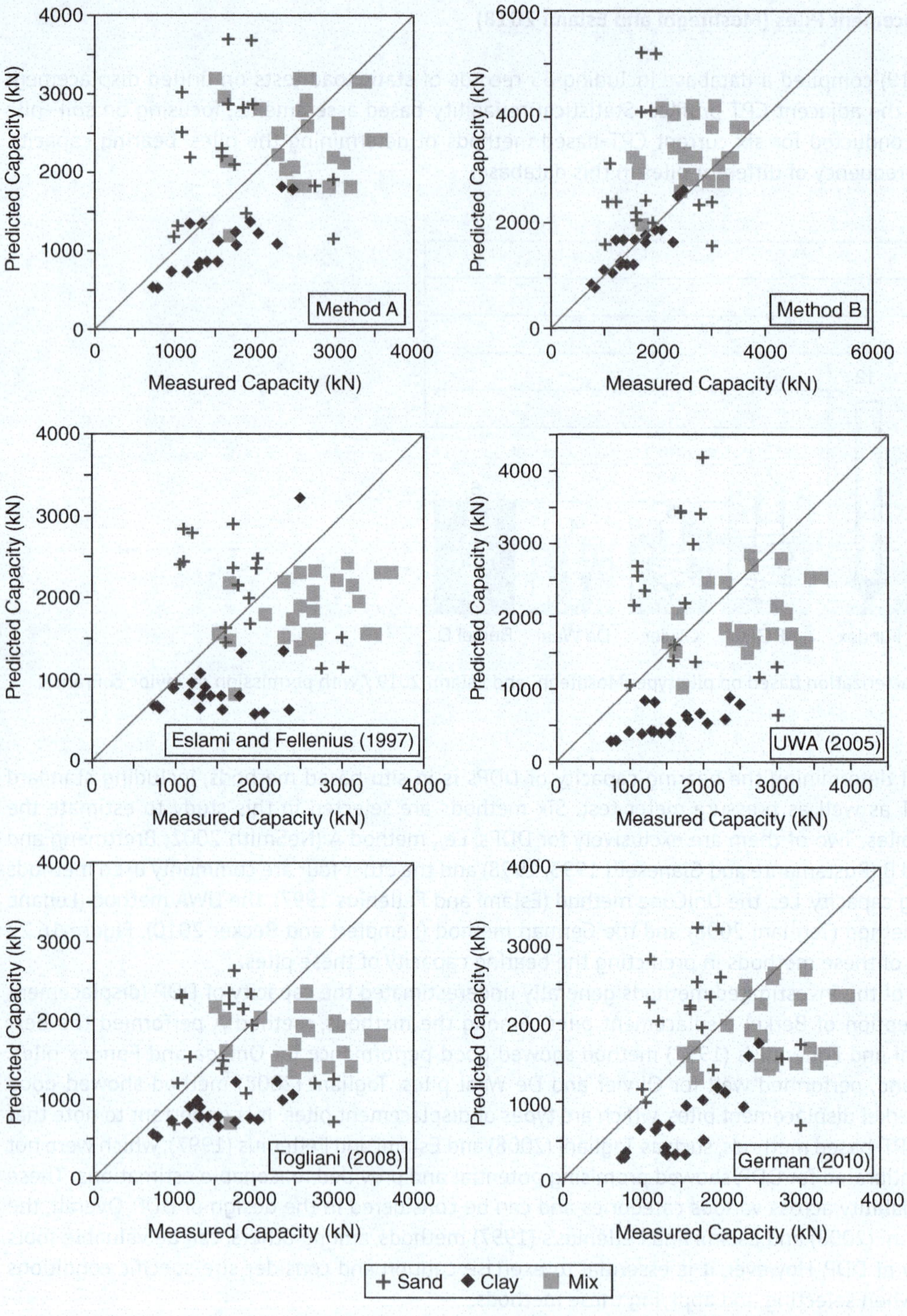

Figure 16.27 The performance of CPT-based methods based on soil type (Moshfeghi and Eslami 2019 / with permission of Taylor & Francis Group).

16.10 Major Points

1) In situ tests, such as SPT and CPT, provide more accurate and reliable records for geotechnical purposes, especially in pile geotechnical design, as they are performed under existing stresses and boundary conditions in the field.
2) Two main approaches exist for utilizing in situ test records in determining the bearing capacity of deep foundations: direct and indirect. In the direct approach, test records are directly used, often with some modification factors, to determine the bearing capacity. The indirect approach involves estimating soil shear strength through correlations and applying static analysis methods. Direct approaches are more commonly used and recommended.
3) CPT, acting as a model pile, offers continuous records with depth and the ability to incorporate various sensors. It can be used to determine pile capacity by considering scale effects and applying corrections for differences in geometry of the pile versus the CPT model and differences in displacements of the pile vs. the CPT model.
4) Important factors in understanding scale effects in toe capacity estimation include embedment depth, influence zone, data processing and averaging, diameter, nonhomogeneous conditions, penetration rate, and failure mechanism. Penetration rate (velocity), diameter, and strain level are particularly important in predicting shaft capacity.
5) While CPT-based design methods directly correlate the pile toe unit capacity with the cone resistance, it should be noted that unit toe resistance is not necessarily equal to cone resistance due to differences in factors such as embedment depth, influence zone, data processing and averaging, diameter, nonhomogeneous conditions, penetration rate, and failure mechanism.
6) The unit shaft capacity of piles is influenced by the effective overburden stress. Above the critical depth, shaft capacity increases with depth, while below this depth, it remains constant. Other factors, such as diameter, loading, and penetration rates, have also been investigated for their effects on shaft capacity.
7) In situ testing methods, such as CPT and SPT, are valuable in predicting axial pile-bearing capacity more precisely and reliably. Various common SPT and CPT-based methods have been developed and employed for this purpose.
8) Different CPT-based methods can be categorized and assessed based on various aspects, including consideration of excess pore water pressure, averaging approaches, the definition of toe failure zones, and the provision of specific and relevant soil behavior charts.
9) Each method is developed based on local data availability and specific site conditions. Different assumptions and models are applied for validation. Therefore, it is important to employ site-specific design and use methods carefully and conservatively.
10) While SPT or CPT provides insight into the inherent variability of soils, different SPT and CPT-based methods have been developed based on various assumptions and simplifications. Consequently, the estimation of axial pile capacity often yields a wide range of predictions, emphasizing the need to select an appropriate method for optimal and performance-based pile design.

16.11 Questions and Problems

16.11.1 Explanatory

1 What are the important parameters of the difference between unit toe and cone resistance?

2 How do the parameters below affect the shaft capacity?
a) Penetration rate
b) Effective overburden stress
c) Pile diameter

3 Why are the cone resistance records averaged around the pile toe?

4 Explain the role of scale effects in developing CPT-based methods. List the important factors.

5 Explain how the similarity of CPT records in two different locations can be advantageous in predicting axial pile-bearing capacity based on what you learned from Case Study 16.2.

6 Compare and contrast direct and indirect approaches in the application of in situ test records for designating bearing capacity.

7 Illustrate the significance of in situ testing in determining the embedment depth of deep foundations.

8 Discuss the reasons for higher safety factors in SPT-based methods compared to the safety factors in CPT-based ones.

16.11.2 Conceptual

1 Which approach is more reliable: SPT-based methods or CPT-based methods?

2 How does the inherent soil variability influence the reliability of in situ-based methods?

3 Why do various CPT-based methods lead to a wide range of predictions?

4 For Case Study 16.1, define the role of deep foundations as well as in situ testing in recognizing and dealing with problematic soils.

5 Compare and contrast different predictive pile-bearing capacity approaches, including CPT-based, SPT-based, and static analysis realizing performance-based assessment. Additionally, what are the reasons for dissimilarities based on the results of Case Study 16.3.

6 For Case Study 16.4, discuss the efficiency of CPT-based methods for special piles, in particular drilled displacement ones.

7 Explain the effect of considering excess porewater pressure in CPT-based methods, in particular for soft and challenging soils.

8 How can bearing capacity and cavity expansion concepts be justified through CPT records?

9 What are the probable solutions to consider strength and stiffness loss due to liquefaction through CPT and CPTu records?

10 Discuss the significance and role of considering complete CPT records (q_c, f_s, and u_2) instead of a few of them in developing direct methods.

16.11.3 Problems

1 A square concrete pile with an embedment depth of 15.2 m and a width of 0.305 m is driven in VA, USA. The variation of CPT and SPT records with depth are given below. Predict the axial-bearing capacity of this pile according to the following methods:

a) UniCone (Eslami and Fellenius 1997)
b) Schmertmann (1978)
c) LCPC (Bustamante and Gianeselli 1982)
d) Bazaraa and Kurkur (1986)
e) Briaud and Tucker (1988)
f) Decourt (1995)
g) Compare the results with the given static pile load test

h)

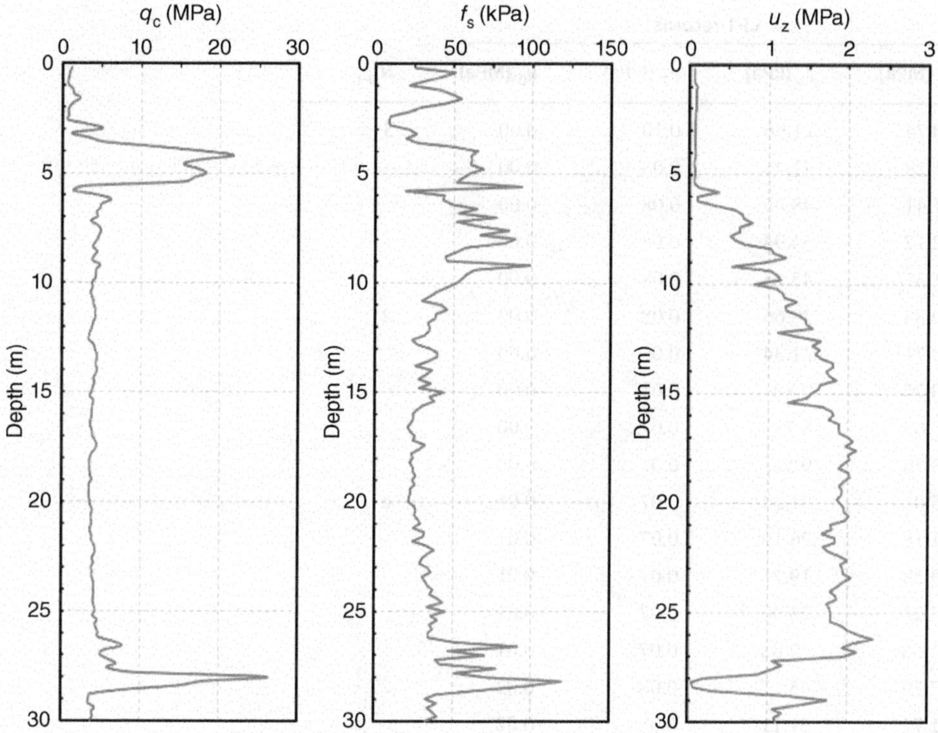

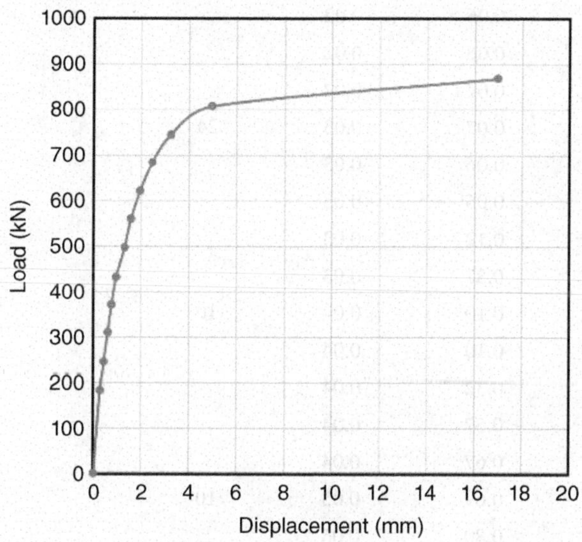

Depth (m)	CPT records				N_{SPT}
	q_c (MPa)	f_s (kPa)	u_2 (MPa)	u_0 (MPa)	
0.20	1.16	25.84	0.06	0.00	
0.40	0.92	50.79	0.06	0.00	
0.60	0.82	42.33	0.06	0.00	
0.80	0.69	31.66	0.06	0.00	

Depth (m)	CPT records				N_{SPT}
	q_c (MPa)	f_s (kPa)	u_2 (MPa)	u_0 (MPa)	
1.00	0.74	21.96	0.10	0.00	3
1.20	1.28	41.45	0.08	0.00	
1.40	1.41	48.85	0.08	0.00	
1.60	2.17	54.94	0.08	0.00	
1.80	1.37	45.15	0.08	0.00	
2.00	0.88	28.66	0.08	0.00	2
2.20	0.71	21.34	0.07	0.00	
2.40	0.76	9.88	0.08	0.00	
2.60	0.69	8.73	0.07	0.00	
2.80	3.36	9.52	0.07	0.00	
3.00	5.02	21.34	0.07	0.01	6
3.20	1.18	26.19	0.07	0.01	
3.40	3.38	19.75	0.07	0.01	
3.60	5.25	26.90	0.07	0.01	
3.80	11.56	49.82	0.07	0.01	
4.00	17.70	65.17	0.08	0.02	27
4.20	21.71	61.11	0.08	0.02	
4.40	17.36	61.55	0.06	0.02	
4.60	15.34	60.76	0.06	0.02	
4.80	17.13	69.14	0.07	0.02	
5.00	18.24	60.00	0.07	0.03	24
5.20	16.77	56.08	0.06	0.03	
5.40	15.39	51.94	0.06	0.03	
5.60	2.80	93.56	0.12	0.03	
5.80	1.34	19.49	0.37	0.03	
6.00	4.29	44.09	0.10	0.04	10
6.20	6.12	43.12	0.10	0.04	
6.40	4.98	48.94	0.32	0.04	
6.60	4.77	64.81	0.57	0.04	
6.80	4.31	53.17	0.67	0.04	
7.00	4.44	77.43	0.69	0.05	10
7.20	3.61	51.85	0.80	0.05	
7.40	4.43	65.96	0.64	0.05	
7.60	4.86	85.19	0.64	0.05	
7.80	4.71	66.93	0.52	0.05	
8.00	4.48	89.42	0.57	0.05	9
8.20	4.14	82.36	0.69	0.06	
8.40	3.87	57.58	1.02	0.06	
8.60	4.39	51.76	1.06	0.06	
8.80	3.99	45.00	1.21	0.06	
9.00	3.81	46.03	0.98	0.06	10
9.20	4.75	98.77	0.54	0.07	
9.40	4.31	63.49	1.03	0.07	

	CPT records				
Depth (m)	q_c (MPa)	f_s (kPa)	u_2 (MPa)	u_0 (MPa)	N_{SPT}
9.60	4.25	58.38	1.10	0.07	
9.80	4.20	55.73	1.15	0.07	
10.00	3.76	51.94	0.82	0.07	8
10.20	3.85	45.50	1.09	0.08	
10.40	3.62	39.24	1.17	0.08	
10.60	3.68	35.27	1.19	0.08	
10.80	3.89	30.00	1.34	0.08	
11.00	3.97	42.50	1.20	0.08	9
11.20	4.16	46.03	1.22	0.09	
11.40	4.14	39.15	1.37	0.09	
11.60	4.18	38.27	1.46	0.09	
11.80	3.95	33.77	1.47	0.09	
12.00	3.80	34.83	1.55	0.09	8
12.20	3.61	34.57	1.11	0.10	
12.40	3.41	31.48	1.36	0.10	
12.60	3.34	23.81	1.65	0.10	
12.80	3.91	27.16	1.55	0.10	
13.00	4.12	35.54	1.64	0.10	10
13.20	4.27	40.04	1.59	0.11	
13.40	4.44	39.86	1.66	0.11	
13.60	4.31	33.86	1.81	0.11	
13.80	4.27	25.84	1.82	0.11	
14.00	4.43	36.60	1.73	0.11	8
14.20	4.25	32.72	1.77	0.12	
14.40	3.97	27.95	1.85	0.12	
14.60	4.08	35.80	1.65	0.12	
14.80	3.87	28.31	1.62	0.12	
15.00	3.78	44.36	1.53	0.12	9
15.20	3.51	35.19	1.47	0.13	
15.40	3.97	38.10	1.24	0.13	
15.60	3.85	30.78	1.68	0.13	
15.80	3.80	27.07	1.80	0.13	
16.00	3.78	24.51	1.82	0.13	8
16.20	3.70	25.04	1.76	0.14	
16.40	3.61	23.46	1.89	0.14	
16.60	3.62	21.16	1.89	0.14	
16.80	3.81	23.02	1.89	0.14	
17.00	3.85	25.93	2.00	0.14	10
17.20	4.16	24.60	2.06	0.15	
17.40	4.39	32.19	1.97	0.15	
17.60	4.25	28.66	2.10	0.15	
17.80	3.80	23.72	2.00	0.15	
18.00	3.57	24.43	2.02	0.15	8

Depth (m)	CPT records				N_{SPT}
	q_c (MPa)	f_s (kPa)	u_2 (MPa)	u_0 (MPa)	
18.20	3.49	22.66	1.95	0.15	
18.40	3.51	24.60	1.88	0.16	
18.60	3.51	24.43	2.01	0.16	
18.80	3.59	26.28	2.02	0.16	
19.00	3.76	24.25	1.95	0.16	8
19.20	3.47	26.01	1.89	0.16	
19.40	3.49	22.75	1.93	0.17	
19.60	3.43	23.19	1.92	0.17	
19.80	3.45	22.13	1.86	0.17	
20.00	3.57	21.69	1.91	0.17	9
20.20	3.62	25.31	1.96	0.17	
20.40	3.85	26.54	1.96	0.18	
20.60	3.72	25.57	2.02	0.18	
20.80	3.68	27.60	1.97	0.18	
21.00	3.80	29.37	1.79	0.18	9
21.20	3.61	25.04	1.81	0.18	
21.40	3.62	32.19	1.68	0.19	
21.60	3.80	31.04	1.84	0.19	
21.80	3.78	24.51	1.85	0.19	
22.00	4.10	33.33	1.69	0.19	10
22.20	4.22	38.27	1.76	0.19	
22.40	4.14	33.51	1.90	0.20	
22.60	3.85	30.34	2.00	0.20	
22.80	4.02	29.19	1.91	0.20	
23.00	3.78	34.22	1.94	0.20	9
23.20	3.87	36.33	1.95	0.20	
23.40	3.81	30.95	2.04	0.21	
23.60	4.01	32.54	1.94	0.21	
23.80	4.08	34.13	1.87	0.21	
24.00	4.06	37.21	1.78	0.21	10
24.20	4.12	35.45	1.85	0.21	
24.40	4.02	39.95	1.87	0.22	
24.60	4.01	44.00	1.74	0.22	
24.80	4.16	33.95	1.78	0.22	
25.00	4.29	45.50	1.79	0.22	11
25.20	3.97	36.16	1.80	0.22	
25.40	4.54	40.39	1.80	0.23	
25.60	4.25	35.45	1.94	0.23	
25.80	4.44	37.48	2.01	0.23	
26.00	4.65	34.66	2.14	0.23	15
26.20	4.65	34.66	2.32	0.23	
26.40	6.58	51.15	2.07	0.24	

CPT records					
Depth (m)	q_c (MPa)	f_s (kPa)	u_2 (MPa)	u_0 (MPa)	N_{SPT}
26.60	7.65	91.89	1.94	0.24	
26.80	5.30	47.35	2.11	0.24	
27.00	4.92	70.63	1.95	0.24	16
27.20	5.95	39.33	1.06	0.24	
27.40	6.96	42.06	1.17	0.24	
27.60	5.84	77.78	1.04	0.25	
27.80	8.91	51.59	0.89	0.25	
28.00	26.09	73.02	0.20	0.25	31
28.20	18.25	119.58	0.05	0.25	
28.40	14.90	95.86	0.07	0.25	
28.60	7.82	68.69	0.32	0.26	
28.80	3.68	39.77	1.22	0.26	
29.00	3.49	32.72	1.75	0.26	11
29.20	3.41	40.21	1.42	0.26	
29.40	4.22	34.39	1.06	0.26	
29.60	3.85	37.83	1.18	0.27	
29.80	3.99	40.56	1.08	0.27	
30.00	3.66	32.72	1.10	0.27	10

Displacement (mm)	Load (kN)
0.00	0.00
0.27	182.02
0.43	244.94
0.59	310.11
0.75	370.79
0.94	431.46
1.29	496.63
1.56	559.55
1.94	620.23
2.46	683.15
3.25	743.82
4.96	806.74
16.96	867.42

2 The SPT records for a site are provided below. A square concrete pile was driven on this site. Its embedment length is 15.24 m, and its width is 305 mm. Determine the axial-bearing capacity of this pile according to the methods of Bazaraa and Kurkur (1986), Briaud and Tucker (1988), and Decourt (1995). The static pile load test showed that the ultimate pile capacity is about 872 kN.

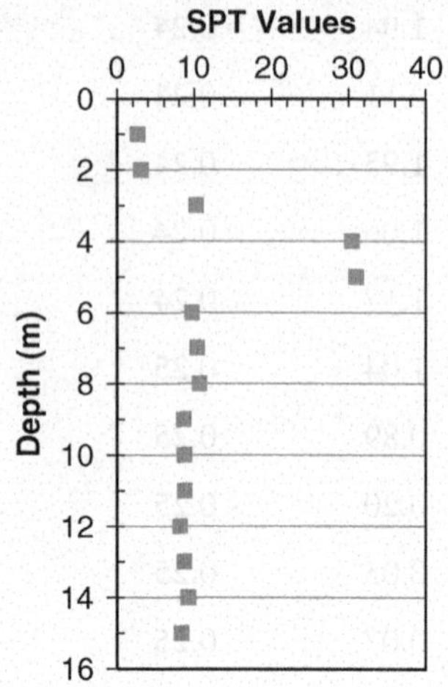

3 The records of SPT & CPT tests in a piling site are given below. The installed piles are closed-end steel pipe piles with a diameter of 324 mm and an embedded length of 14 m. The existing soil type is recognized as sand-based on-site investigation. Determine the axial-bearing capacity using the four methods listed below and compare the results with the load–displacement diagram.

SPT-based:

- Meyerhof (1976)
- Shariatmadari et al. (2008)

CPT-based:

- Eslami and Fellenius (1997)
- German (2010)

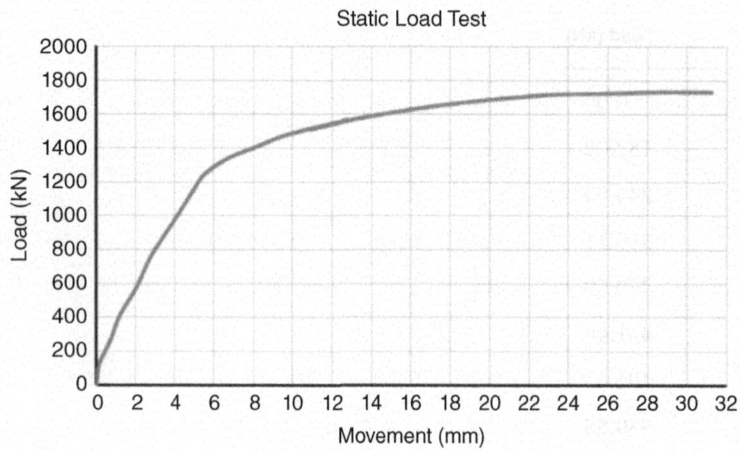

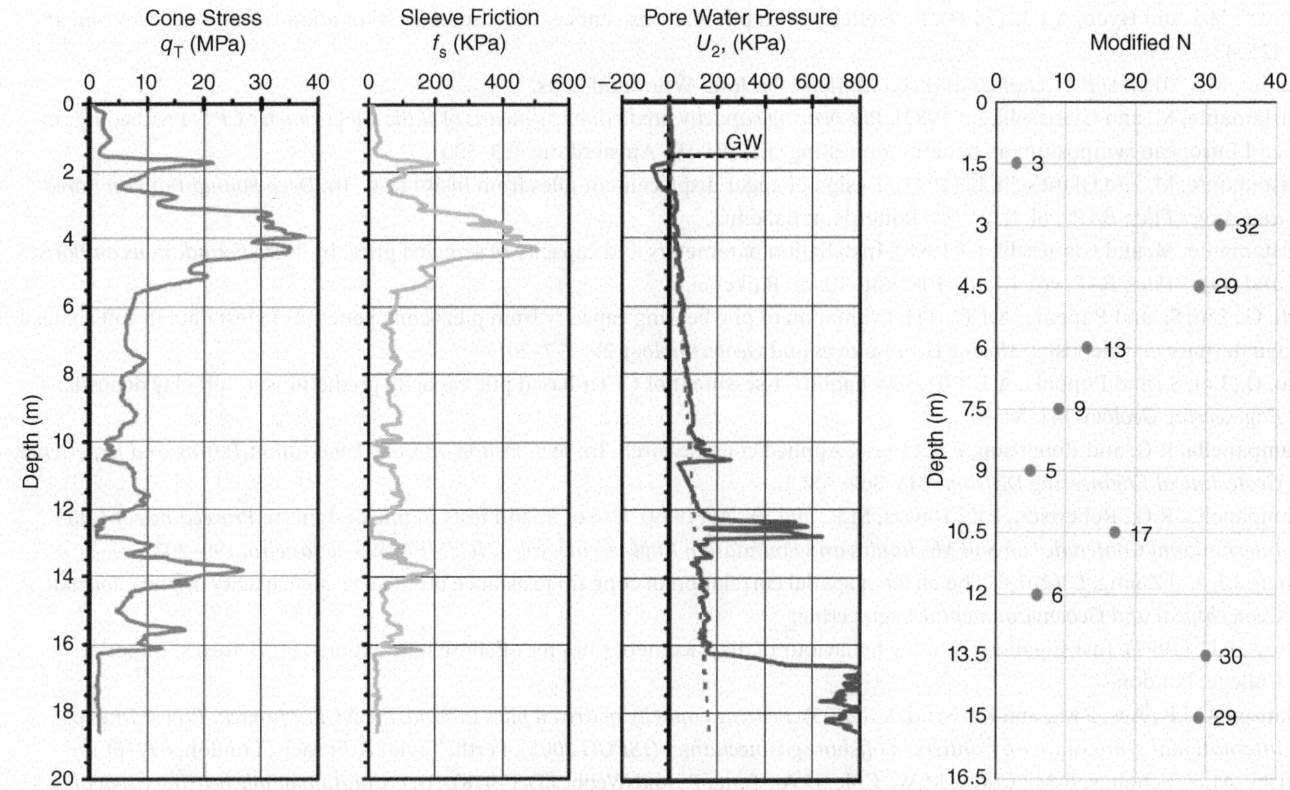

References

AL-Mhaidib, A.I. (1999). Loading rate effect on piles in clay from laboratory tests. *Journal of King Saud University* 13: 39–55.

Altaee, A., Evgin, E., and Fellenius, B.H. (1993). Axial load transfer for piles in sand. III: Load transfer for piles in sand and the critical depth. *Canadian Geotechnical Journal* 30 (3): 455–463.

Aoki, N. and Velloso, D.A. (1975). *An approximate method to estimate the bearing capacity of piles*. Proceedings of 5th Pan-American Conference of soil mechanics and foundation engineering, Buenos Aires, 367–376.

AUT: GEO-CPT&Pile database (2016). https://civil.aut.ac.ir/content/.

Axelsson, G. (1998). Long-term increase in shaft capacity of driven piles in sand. *Fourth International Conference on Case Histories in Geotechnical Engineering*, St. Louis, Missouri, USA.

Axelsson, G. (2000). Long-term set-up of driven piles in sand. Ph.D. Thesis. Royal Institute of Technology, Stockholm. in partial fulfillment of the requirements for the degree of Doctor of Philosophy.

Bazaraa, A.R. and Kurkur, M.M. (1986). N-values used to predict settlements of piles in Egypt. In Use of In Situ Tests in Geotechnical Engineering (462–474). ASCE.

Begemann, H.K.S.P. (1963). The use of the static penetrometer in Holland. *New Zealand Engineering* 18 (2): 41.

Begemann, H.K.S.Ph. (1969). The Dutch static penetration test with the adhesion jacket cone. Delft Soil Mechanics and Delft Hydraulics Laboratory, XIII(1).

Bogdanovic L (1961). The use of penetration tests for determining the bearing capacity of piles. *Proceedings of the 5th International Conference on Soil Mechanics and Foundation Engineering*, Paris 2, p 17.

Brettmann, T. and NeSmith, W. (2005). Advances in auger pressure grouted piles: design, construction and testing. In: *GSP 129. Advances In Deep Foundations: Construction, Design, And Testing-In Memory of Michael W. O'Neill* (ed. C. Vipulanandan and F. Townsend), 290–307. ASCE.

Briaud, J.L. and Tucker, L.M. (1988). Measured and predicted axial capacity of 98 piles. *Journal of Geotechnical Engineering* 114 (9): 984–1001.

Brown, M.J. and Hyde, A.F.L. (2008). Rate effects from pile shaft resistance measurements. *Canadian Geotechnical Journal* 45: 425–431.

Budhu, M. (2011). *Soil Mechanics and Foundations*, 3e. John Wiley and Sons.

Bustamante, M. and Gianeselli, L. (1982). *Pile bearing capacity predictions by means of static penetrometer CPT*. Proceedings of 2nd European symposium on penetration testing (ESOPT-II), Amsterdam, 493–500.

Bustamante, M. and Gianeselli, L. (1993). Design of auger displacement piles from insitu tests. In: *Deep Foundations on Bored and Auger Piles, BAP*, vol. II, 21–34. Rotterdam: Balkema.

Bustamante, M. and Gianeselli, L. (1998). Installation parameters and capacity of screwed piles. In: *Deep Foundations on Bored and Auger Piles, BAP*, vol. III, 95–108. Rotterdam: Balkema.

Cai, G., Liu, S., and Puppala, A.J. (2011). Evaluation of pile bearing capacity from piezocone penetration test data in soft Jiangsu Quaternary clay deposits. *Marine Georesources and Geotechnology* 29: 177–201.

Cai, G., Liu, S., and Puppala, A.J. (2012). Reliability assessment of CPTu-based pile capacity predictions in soft clay deposits. *Engineering Geology* 141: 84–91.

Campanella, R.G. and Robertson, P.K. (1981). Applied cone research. In: *Symposium on Cone Penetration Testing and Experience. Geotechnical Engineering Division*, 343–362. ASCE.

Campanella, R.G., Robertson, P.K., Davies, M.P., and SY, A. (1989). Use of in-situ tests in pile design. In: *Proceedings of 12th International Conference on Soil Mechanics and Foundation Engineering, vol. 1. ICSMFE, Rio de Janeiro*, 199–203.

Chen, J.J. and Zhang, L. (2013). The effect of spatial correlation of cone tip resistance on the bearing capacity of piles. *Journal of Geotechnical and Geoenvironmental Engineering*.

Chow, F.C. (1997). Investigations into the behaviour of displacement piles for offshore foundations. Ph.D. thesis. Imperial College, London.

Clausen, C.J.F., Aas, P.M., and Karlsrud, K. (2005). *Bearing capacity of driven piles in sand, the NGI approach. Proceedings of international symposium on frontiers in offshore geomechanics (ISFOG 2005)*, Perth, Taylor & Francis, London, 677–681.

Clisby, M.B., Scholtes, R.M., Corey, M.W., Cole, H.A., Teng, P., and Webb, J.D. (1978). *An evaluation of pile bearing capacities*, Vol. I, Final Report, Mississippi State Highway Department.

De Beer, E.E. (1963). The scale effect in the transposition of the results of deep sounding tests on the ultimate bearing capacity of piles and caisson foundations. *Géotechnique* 13 (1): 39–75.

De Ruiter, J. and Beringen, F.L. (1979). Pile foundations for large North Sea structures. *Marine Geotechnology and Georesources* 3 (3): 267–314.

Decourt, L. (1995). *Prediction of Load Settlement Relationships for Foundations on the Basis of the SPT-T*. Ciclo de Conferencias Inter. "Leonardo Zeevaert", UNAM, Mexico, 85–104.

Eslami, A. (1996). *Bearing capacity of piles from cone penetrometer test data*. Ph.D. Thesis, University of Ottawa, Department of Civil Engineering, 516p.

Eslami, A. (2013). *Foundation Engineering, Design and Construction*, 2e. Building and Housing Research Center.

Eslami, A. and Fellenius, B.H. (1997). Pile capacity by direct CPT and CPTu methods applied to 102 case histories. *Canadian Geotechnical Journal* 34 (6): 880–898.

Eslami, A. and Heidarie Golafzani, S. (2020). Relevant data-based approach upon reliable safety factor for pile axial capacity. *Marine Georesources & Geotechnology* 1–14.

Eslami, A., Aflaki, E., and Hosseini, B. (2011). Evaluating CPT and CPTu based pile bearing capacity estimation methods using Urmiyeh Lake Causeway piling records. *Scientia Iranica* 18 (5): 1009–1019.

Eslami, A., Lotfi, S., and Eslami, M. (2013). *Pile shaft capacity from cone penetration test (CPT) records, considering scale effects.* Proceedings of the 38th Annual Conference on Deep Foundations, Deep Foundation Institute (DFI), Phoenix AZ, USA.

Eslami, A., Heidarie Golafzani, S., and Moshfeghi, S. (2020). CPT and pile database for performance-based design of pile axial bearing capacity. In Proceedings of the 45th Annual Conference on Deep Foundations, Deep Foundation Institute (DFI), Oxon Hill, Maryland, October 13–16, 12p.

Fellenius, B.H. (2002). Excerpt from Chapter 6 of the red book: direct methods for estimating pile capacity, in Background to UniCone. http://www.fellenius.net, accessed May 22, 2010.

Fellenius, B.H. (2015). *Basics of Foundation Design*, Electronice. 432pp. www. Fellenius.net.

Finno, R.J. (1989). *Predicted and Observed Axial Behavior of Piles: Results of a Pile Prediction Symposium (Geotechnical Special Publication No 23)*. American Society of Civil Engineers.

Fleming, K., Weltman, A., Randolph, M., and Elson, K. (2009). *Piling Engineering*, 3e. New York: CRC Press.

Gavin, K.G. and Lehane, B.M. (2003). The shaft capacity of pipe piles in sand. *Canadian Geotechnical Journal* 40: 36–45.

Gavin, K. and Lehane, B.M. (2005). Estimating the end bearing resistance of pipe piles in sand using the final filling ratio. In: *Proceedings of the International Symposium on Offshore Geotechnics, Perth, Australia, 19–21 September 2005* (ed. S. Gourvenec and M. Cassidy), 717–723. London: Taylor and Francis Group.

Ghionna, V.N., Jamiolkowsi, M., Lancellotta, R., and Pedroni, S. (1993). Base capacity of bored piles in sands from in-situ tests. Proceeding of 2nd International Geotechnic Seminar on Deep Foundations on Bored and Auger Piles, VanImpe, ed., Balkema, Rotterdam, The Netherlands, 67–74.

Gui, M.W. and Bolton, M.D. (1998). *Geometry and scale effects in CPT and pile design*. Proceedings of the 1st International Conference on Site Characterization, Atlanta, 1063–1068.

Gwizdala, K. (1984). *Large diameter bored piles in non-cohesive soils*. Swedish Geotechnical Institute, Report No. 26, Linkoping, Sweden.

Heidarie Golafzani, S., Jamshidi Chenari, R., and Eslami, A. (2020a). Reliability based assessment of axial pile bearing capacity: static analysis, SPT and CPT-based methods, Georisk: Assessment and Management of Risk for Engineered Systems and Geohazards, 10.1080/17499518.2019.1628281.

Heidarie Golafzani, S., Eslami, A., and Jamshidi Chenari, R. (2020b). Probabilistic assessment of model uncertainty for prediction of pile foundation bearing capacity; static analysis, SPT and CPT-based methods. *Geotechnical and Geological Engineering* 1–19.

Igoe, D.J.P., Gavin, K.G., and O'Kelly, B.C. (2010). *Field tests using an instrumented model pile in sand*. Proceedings of physical modeling in geotechnics, Taylor and Francis Group, 775–780.

Igoe, D.J.P., Gavin, K.G., & O'Kelly, B.C. (2011). Shaft capacity of open-ended piles in sand. *Journal of Geotechnical and Geoenvironmental Engineering* (ASCE), 137(10), 903–913.

Ishibashi, I. and Zhang, X. (1993). Unified dynamic shear module and damping ratios of sand and clay. *Soils and Foundation* 33 (1): 182–191.

Jardine, R.J., Chow, F.C., Overy, R., and Standing, J. (2005). *ICP Design Methods for Driven Piles in Sands and Clays*, 105. London: Thomas Telford Publishing.

Jefferies, M. and Davies, M. (1993). Use of CPTu to estimate equivalent SPT N60. ASTM. *Geotechnical Testing Journal* 16 (4): 458e468.

Kempfert, H.G. and Becker, P. (2010). Axial pile resistance of different pile types based on empirical values. In: *Deep Foundations and Geotechnical In-Situ Testing*, 149–154.

Kennedy, J.B. and Neville, A.M. (1986). *Basic Statistical Methods for Engineering and Scientists*, 519. New York: Harper and Row.

Kolk, H. J., Baaijens, A.E., and Senders, M. (2005). *Design criteria for pipe piles in silica sands*. Proceedings of international symposium on frontiers in offshore geomechanics (ISFOG 2005), Perth, Taylor and Francis Group, London, 711–716.

Lehane, B. M., Schneider, J. A., and Xu, X. (2005). *The UWA-05 method for prediction of the axial capacity of driven piles in sand*. Proceedings of international symposium on frontiers in offshore geomechanics (ISFOG 2005), Perth, Taylor and Francis, London, 683–689.

Lehane, B.M., Li, Y., and Williams, R. (2013). Shaft capacity of displacement piles in clay using the cone penetration test. *Journal of Geotechnical and Geoenvironmental Engineering* 139 (2): 253–266.

Lunne, T., Robertson, P.K., and Powell, J.J. (1997). *Cone Penetration Test in Geotechnical Practice*. Blackie Academic & Professional.

Meyerhof, G.G. (1951). The ultimate bearing capacity of foundations. *Géotechnique* 2 (4): 301–332.

Meyerhof, G.G. (1956). Penetration tests and bearing capacity of cohesionless soils. *Journal of Soil Mechanics and Foundation Engineering Division* 82 (SM1): 1–12.

Meyerhof, G.G. (1976). Bearing capacity and settlement of ASCE. *Journal of Geotechnical Engineering* 102 (GT3): 195–228.

Meyerhof, G.G. (1983). Scale effects of pile capacity. *Journal of Geotechnical Engineering* 108 (GT3): 195–228.

Moshfeghi, S. and Eslami, A. (2018). Reliability-based assessment of drilled displacement piles bearing capacity using CPT records. *Marine Georesources and Geotechnology* 1–14.

Moshfeghi, S. and Eslami, A. (2019). Reliability-based assessment of drilled displacement piles bearing capacity using CPT records. *Marine Georesources & Geotechnology* 37 (1): 67–80.

NeSmith, W. M. (2002). Static capacity analysis of augered, pressure-injected displacement piles. *Deep Foundations 2002: An International Perspective on Theory, Design, Construction, and Performance* (pp. 1174–1186).

Niazi, F.S. and Mayne, P.W. (2013). Cone penetration test based direct methods for evaluating the static axial capacity of single piles. *Geotechnical and Geological Engineering* 31 (4): 979–1009.

Niazi, F.S. and Mayne, P.W. (2016). CPTu-based enhanced UniCone method for pile capacity. *Engineering Geology* 212: 21–34.

Nottingham, L.C. (1975). Use of quasi-static friction cone penetrometer data to predict load capacity of displacement piles. Ph. D. Thesis, Dept. of Civil Engineering, University of Florida, 553p.

Philipponnat, G. (1980). Méthode pratique de calcul d'un pieu isolé, à l'aide du pénétromètre statique. *Revue Francaise de* Geotechnique(10): 55–64.

Price, G. and Wardle, I. (1982). A comparison between cone penetration test results and the performance of small diameter instrumented piles in stiff clay. *Proceedings of the 2nd European Symposium on Penetration Testing*, Amsterdam.

Randolph, M.F., Dolwin, J., and Beck, R. (1994). Design of driven piles in sand. *Géotechnique* 44 (3): 427–448.

Sabatini, P.J., Bachus, R.C., Mayne, P.W., Schneider, J.A., and Zettler, T.E. (2002). Geotechnical engineering circular no. 5: evaluation of soil and rock properties (No. FHWA-IF-02-034).

Salgado, R. (1995). Design of piles in sands based on CPT results. Proceeding of 10th Pan-Am. Conference of Soil Mechanics and Foundations. Engrg., Guadalajara, 3, 1261–1274.

Schmertmann, J.H. (1978). Guidelines for cone test, performance, and design. Federal Highway Administration, Report FHWA TS 78209, Washington, 145pp.

Shariatmadari, N., Eslami, A., and Karimpour-Fard, M. (2008). Bearing capacity of driven piles in sands from SPT–applied to 60 case histories. *Iranian Journal of Science & Technology, Transaction B, Engineering* 32 (B2): 125–140.

Shioi, Y. and Fukui, J. (1982). Application of N-value to design of foundations in Japan. *Proceeding of the Second European Symposium on Penetration Testing* 1 (1982): 159–116.

Takesue, K, Sasao, H., and Matsumoto, T. (1998). Correlation between ultimate pile skin friction and CPT data. Geotechnical Site Characterization (2). Balkema, Rotterdam, 1177–1182.

Togliani, G. (2008). Pile capacity prediction for in-situ tests. Proceedings of geotechnical and geophysical site characterization, Taylor and Francis Group, London, 1187–1192.

Tummay, M. T., and Fakhroo, M. (1982). Friction pile capacity in cohesive soils using electric quasi-static penetration tests, Louisiana Department of Transportation and Development Research and Development Section, Baton Rouge, La. Interim Research Rep. No. 1.

Valikhah, F., Eslami, A., and Veiskarami, M. (2018). Load-displacement Behavior of Driven Piles in Sand Using CPT-Based Stress and Strain Fields. *International Journal of Civil Engineering*. https://doi.org/10.1007/s40999-018-0388-7.

Van Dijk, B., and Kolk, H. (2011). CPT-based design method for axial capacity of offshore piles in clays. Proceedings of frontiers in offshore geotechnics II. Taylor and Francis Group, London, 555–560.

White, D.J. and Bolton, M.D. (2005). Comparing CPT and pile base resistance in the sand. *Proceedings of the Institution of Civil Engineers-Geotechnical Engineering* 158 (1): 3–14.

Wroth, C.P. (1988). Penetration testing-a more rigorous approach to interpretation. In: *Proceedings of the 1st International Symposium on Penetration Testing, ISOPT-1*, vol. 1, 303e311. Disney World.

Yu, F. and Yang, J. (2012). Base capacity of open-ended steel pipe piles in sand. *Journal of Geotechnical and Geoenvironmental Engineering (ASCE)* 138 (9): 1116–1128.

Zhang, G., Robertson, P.K., and Brachman, R.W.I. (2002). Estimating liquefaction induced ground settlements from CPT for level ground. *Canadian Geotechnical Journal* 39 (5): 1168e1180.

17

Lateral Loading and Structural Aspect

Chapter Highlights

This chapter is divided into two parts:

Part I – Evaluation of soil–pile interaction in lateral loading conditions – Sections 17.1–17.4

- Background of Lateral Loading
  - Performance and Boundary Conditions
- Types of Foundations for Lateral Load Resistance
- Analyses Methods
  - Critical Length
  - Broms Method
  - Nondimensional Method
- *p–y* Approach
  - *p–y* Curves for Different Soil Types
  - *p–y* Curves Directly from In Situ Tests

Part II – Structural design considerations of piles – Sections 17.5–17.8

- Structural Design Aspects
  - Structural Problems and Protection
  - Design Criteria
- Structural Considerations of Piles
- Piles Design Under Combined Loading
- Case Studies
  - Oil-Storage Tank Foundation, Japan
  - Single Piles Adjacent Excavation, China
  - Two Case Histories of Load Tests on Piles or Drilled Shafts Under Lateral Load, USA
  - Three Laterally Loaded Piles with Different Cross Sections, USA
  - Two Bridge Foundations, Costa Rica
  - Fourteen Case Histories in Soil Liquefied, Japan and Nesw Zealand

17.1 Background of Lateral Loading

The loads acting on piles can be divided into two categories:

- **Axial Loads:** These loads act along the pile axis and result in compressive or tensile stresses. In certain situations, the primary design load is the force of gravity acting on a structure. This gravity load is transmitted to the foundation through columns and bearing walls, causing vertical piles to experience predominantly axial loading.
- **Lateral Loads (or Transverse Loads):** These loads act perpendicular to the pile axis, generating shear and moments.

Advanced Foundation Engineering: Principles, Performance and Prospects, First Edition. Moses Karakouzian and Abolfazl Eslami.

17.1.1 Source of Lateral Loads

In many cases, the presence of applied lateral loads, either independently or in combination with axial loads, must be taken into account in foundation design. These lateral loads can arise from various sources, such as:

- Earth pressures, which occur on the back of retaining walls
- Water pressures, which act on structures like weirs or hydraulic structures
- Lateral thrust from arched structures, even when gravity loads are the only applied loads
- Wind forces
- Earthquakes
- Berthing and impact loads from ships on piers or harbor structures
- Downhill movements of earth slopes, including mudflows and lateral spread
- Vehicle acceleration and braking forces on bridges
- Eccentric or inclined loads on columns
- Ocean wave forces on offshore or harbor structures
- River current forces on bridge piers
- Cable forces on electrical transmission or railway support towers
- Cable anchors for suspension bridges
- Stresses induced by temperature changes in the structure

These lateral loads can be static, cyclic, or dynamic, presenting unique challenges for foundation engineers. The ultimate limit state (ULS) and serviceability limit state (SLS) are crucial considerations in the design process.

17.1.2 Performance and Boundary Conditions

Flexible piles and rigid piles are two categories used to describe the behavior of piles under lateral loads.

1) **Flexible Piles:** Flexible piles, also known as long piles, are characterized by their relatively large length compared to their diameter. These piles are able to undergo significant deflection and bending under load. The flexural stiffness (EI) of the pile material plays a significant role in the pile's behavior. The pile shaft can experience distributed bending moments along its length, and the pile can deform in a curved shape. Flexible piles are capable of redistributing loads and providing resistance against lateral forces.
2) **Rigid Piles:** Rigid piles, also known as short piles, are characterized by their limited length compared to their diameter. These piles behave as rigid members and do not exhibit significant bending or deflection under load. The flexural stiffness (EI) of the pile material is relatively high compared to the applied loads, resulting in minimal deformation. Rigid piles primarily resist loads through axial compression or tension and transmit them directly to the underlying soil or rock without significant bending or rotation.

It is important to note that the distinction between flexible and rigid piles is not based on a specific length threshold but rather on the relative behavior of the pile in response to the applied loads. The specific categorization of a pile as flexible or rigid depends on various factors, such as the pile length, diameter, material properties, and the magnitude and distribution of the applied loads.

Piles transmit applied axial loads to the ground through two mechanisms: toe bearing and side friction (Figure 17.1a). While a shallow foundation's upward (tensile) load capacity is limited to its weight, piles utilize both the weight and the side friction (acting in the opposite direction), making them more effective in resisting upward loads (Figure 17.1b). When subjected to lateral loading, piles rely on flexural stiffness and lateral soil resistance (Figure 17.1c).

The boundary conditions at the pile heads play a crucial role in analyzing laterally loaded pile problems. Three types of boundary conditions are commonly considered:

- **Free-Head Pile:** In this case, the lateral load may act at or above-ground level, and the pile head is free to rotate without any restraint.
- **Fixed-Head Pile:** A fixed-head pile is free to move laterally but is completely prevented from rotating.
- **Partially Restrained-Head Pile:** A pile with a partially restrained head is allowed to move and rotate under certain restraints. This type of boundary condition is often encountered in offshore drilling platforms and similar structures.

The key to the solution of laterally loaded pile problems lies in determining the value of the modulus of subgrade reaction with respect to depth along the pile.

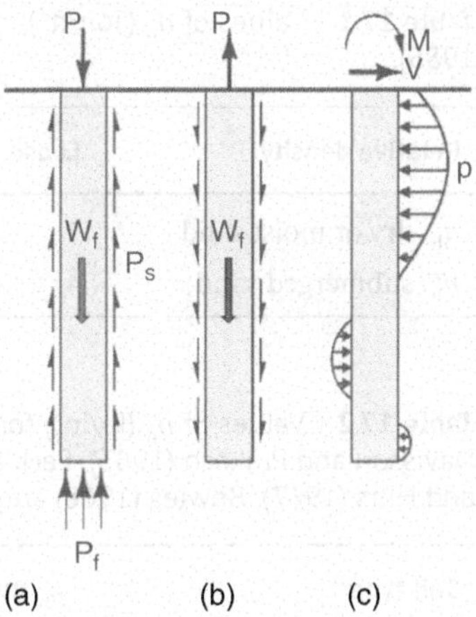

Figure 17.1 Transfer of structural loads from a pile foundation into the ground: (a) axial downward (compressive) loads, (b) axial upward (tensile) loads, and (c) lateral loads.

17.1.3 Modulus of Horizontal Subgrade Reaction

Figure 17.2a depicts a vertical pile experiencing a lateral load P_t at the ground level. The figure also illustrates the deflection profile of the pile and the corresponding curves representing the soil reaction. In this context, the lateral elastic modulus of the soil, E_s, at any point x beneath the surface along the pile is defined according to Winkler's hypothesis.

$$E_s = -\frac{P}{y} \tag{17.1}$$

y is the pile lateral deflection, and p is the soil horizontal reaction expressed as force per unit length of the pile. As the load P_t at the top of the pile increases, the lateral deflection y and the corresponding soil horizontal reaction, p, increase. The relationship between p and y can be established along the pile at depths $x = x_1$, $x = x_2$, etc.

Figure 17.2b displays a family of p–y curves plotted in the appropriate quadrant. The curves in the second and fourth quadrants indicate that the soil resistance, p, opposes the lateral deflection, y. While the p–y curves in Figure 17.2b serve as illustrations, they are representative of many families of curves where the initial stiffness and maximum resistance increase with depth.

A typical p–y curve is presented in Figure 17.2c, exhibiting strong nonlinearity. The curve transitions from an initial stiffness, E_{si}, to an ultimate resistance, P_u. It is important to note that the soil modulus, E_s, is not a constant and varies with deflection. Many factors influence the value of E_s. The most important factors are the pile's width or diameter. The soil properties, the nature and magnitude of loading, and The flexural stiffness EI of the pile material.

The nonlinear force-deformation behavior depicted by the p–y curves at different depths implies that the lateral soil modulus, E_s, is a function of both the depth, x, and the lateral deflection, y. However, when considering a specific load level, E_s can be approximated as a function of only the depth, x. The actual variation of E_s with depth can take various

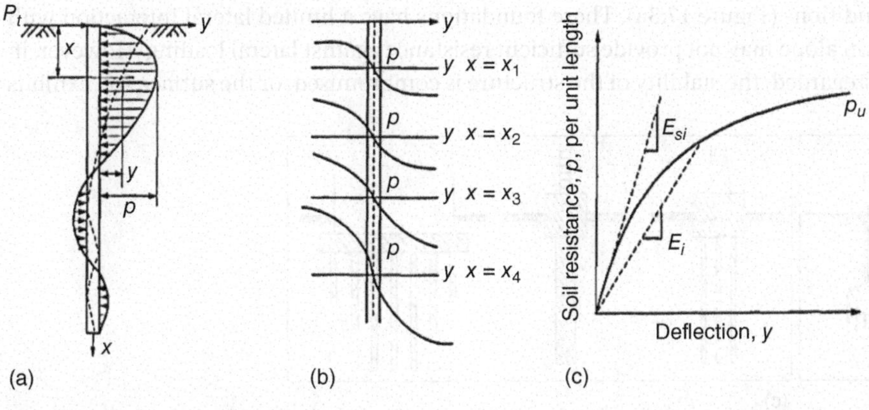

Figure 17.2 The concept of p–y curves: (a) laterally loaded pile, (b) family of p–y curves, (c) characteristic shape of a curve (Murthy 2007/CBS Publishers and Distributers).

Table 17.1 Values of n_h (Ton/ft^3) for sand (after Terzaghi 1955).

Relative density	Loose	Medium	Dense
n_h, dry or moist sand	7	21	56
n_h, submerged sand	4	14	34

Table 17.2 Values of n_h (lb/in^3) for cohesive soils (Reese and Matloc (1956), Davisson and Prakash (1963), Peck and Davisson (1962), Davisson (1970), Wilson and Hills (1967), Bowles (1968) are reproduced from Poulos & Davis (1980)).

Soil type	n_h (lb/in^3)	References
Soft NC clay	0.6–12.7	Reese and Matloc (1956)
	1.0–2.0	Davisson and Prakash (1963)
NC organic clay	0.4–1.0	Peck and Davisson (1962)
	0.4–3.0	Davisson (1970)
Peat	0.2	Davisson (1970)
	0.1–0.4	Wilson and Hills (1967)
Loess	29–40	Bowles (1968)

forms depending on the soil properties and the inclination of the pile. The most popular and useful form of soil modulus variation is:

$$E_s = n_h x^n \tag{17.2}$$

n_h is termed as *the coefficient of soil modulus variation*. The value of the power n depends upon the pile's type of soil and batter. It may be noted here that the n_h values proposed by the investigators are not constants. It varies with the soil properties (density for granular materials and shear strength for clay soils), the pile's width or diameter, the pile's material flexural stiffness, and the pile's deflection. It is a complex phenomenon that does not depend on a single factor. The values of n_h by Terzaghi and others are given in Tables 17.1 and 17.2.

17.2 Foundation Resistance Against Lateral Loading

Engineers often opt for shallow foundations when the lateral load on the structure is low or negligible, or when the structure is not subjected to significant seismic conditions (Figure 17.3a). These foundations have a limited lateral interaction with the surrounding soil, and their wall friction alone may not provide sufficient resistance against lateral loading. However, in cases where the lateral load cannot be disregarded, the stability of the structure is compromised, or the surface soil exhibits

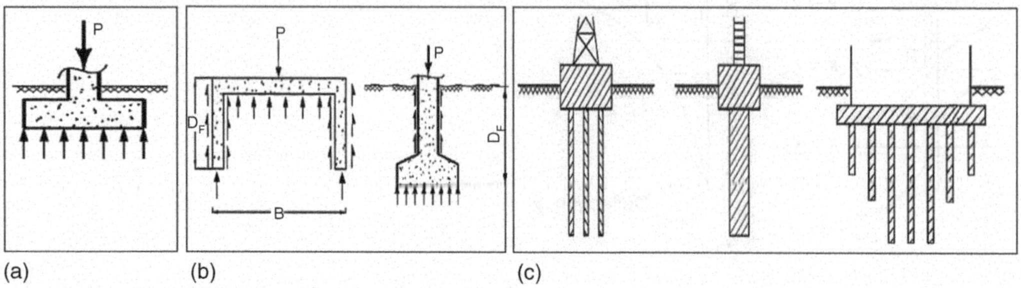

Figure 17.3 (a) Shallow foundation, (b) two types of semi-deep foundations, (c) types of deep foundations (Eslami 2013).

settlement and bearing capacity issues, shallow foundations are not suitable. In such situations, semi-deep foundations like skirted foundations, well foundations, bucket foundations, caissons, and box foundations are employed. These foundations offer greater wall friction and passive pressure compared to shallow foundations, while still being more practical and cost-effective than deep foundations due to their increased surface area in contact with the soil (Figure 17.3b). If none of the aforementioned foundation types suffice, deep foundations such as single piles, pile groups, and PRFs (Pile-Raft Foundations) are ultimately utilized as a final resort (Figure 17.3c).

Vertical piles are highly efficient in supporting axial loads but are less effective in resisting lateral loads. To address this, one approach to withstand horizontal structural loads is by installing some of the piles at an inclined angle with respect to the vertical axis. These piles, known as battered piles or raker piles, are inclined from the vertical line. In this configuration, both the vertical and horizontal loads are transmitted from a pile cap to a group of piles. Battered piles are used for several reasons:

1) **Lateral Load Resistance:** Battered piles are particularly effective in resisting lateral loads. By installing piles at an angle from the vertical, the pile's capacity to resist lateral forces is enhanced. The inclined orientation allows the piles to better distribute and transfer the applied lateral loads to the surrounding soil. This is especially beneficial in situations where the lateral loads are significant, such as in waterfront structures subject to ocean currents, wind loads, or seismic forces.
2) **Space Constraints:** Battered piles can be advantageous in situations where there are space limitations or site constraints. By installing piles at an angle, they can be accommodated within a smaller footprint or in areas where vertical piles may not be feasible due to obstructions or subsurface conditions. This provides flexibility in the design and allows for efficient use of the available space.
3) **Structural Efficiency:** Battered piles can contribute to the overall structural efficiency of a foundation system. By incorporating an arrangement of both vertical and battered piles, the load-carrying capacity of the foundation can be optimized. The vertical piles primarily carry axial loads, while the battered piles are more efficient in resisting lateral forces. This combination allows for a more balanced and efficient distribution of loads within the foundation system.
4) **Soil Conditions:** Battered piles can be advantageous in certain soil conditions. In soils where the lateral resistance provided by the surrounding soil is limited, the inclined orientation of battered piles can compensate for the lower soil capacity. This is particularly relevant in soils with low shear strength or weak cohesion, where the use of vertical piles alone may not provide sufficient lateral load resistance.
5) **Structural Stability:** Battered piles can enhance the stability and performance of a structure, especially under large and static lateral loading conditions. The inclined orientation of the piles can provide additional resistance against horizontal movements induced by earthquakes or impact loads. This improves the overall stability and resilience of the structure against lateral forces.
6) Overall, battered piles offer an effective solution for addressing lateral load requirements, space limitations, soil conditions, and structural stability considerations. Their utilization can lead to more efficient and reliable foundation designs in various engineering applications.

Although battered piles have generally performed well, there are two main situations where problems can arise:

- Offshore structures, such as drilling platforms, are subjected to significant lateral loads from ocean currents, wind, and seismic forces. However, it is not practical to drive piles at a large enough angle to accommodate these loads.
- Battered pile foundations create a very stiff system, which is suitable for static loads but can pose challenges for soil/structure when dynamic loads, such as earthquakes or impact loads from ships, are applied.

To overcome these challenges and ensure practical and cost-effective foundation systems, vertical piles are preferred to battered piles.

When a lateral load and moment are applied to the head of a vertical pile, the behavior of the pile can be described as that of a long flexible pile. This behavior is illustrated in Figure 17.4. The lateral load primarily affects the upper soil zone surrounding the pile. In this zone, the mobilization of frictional forces is insufficient, and the passive pressure is negligible. As a result, the lateral displacement of the pile is relatively large in this upper zone compared to the lower regions. This behavior can be attributed to the relatively lower stiffness of the upper soil layers compared to the deeper soil layers. The upper soil zone experiences greater deformation under the lateral load, leading to a larger lateral displacement. In contrast, the deeper soil layers provide more resistance to lateral movement due to their higher stiffness and the mobilization of frictional forces.

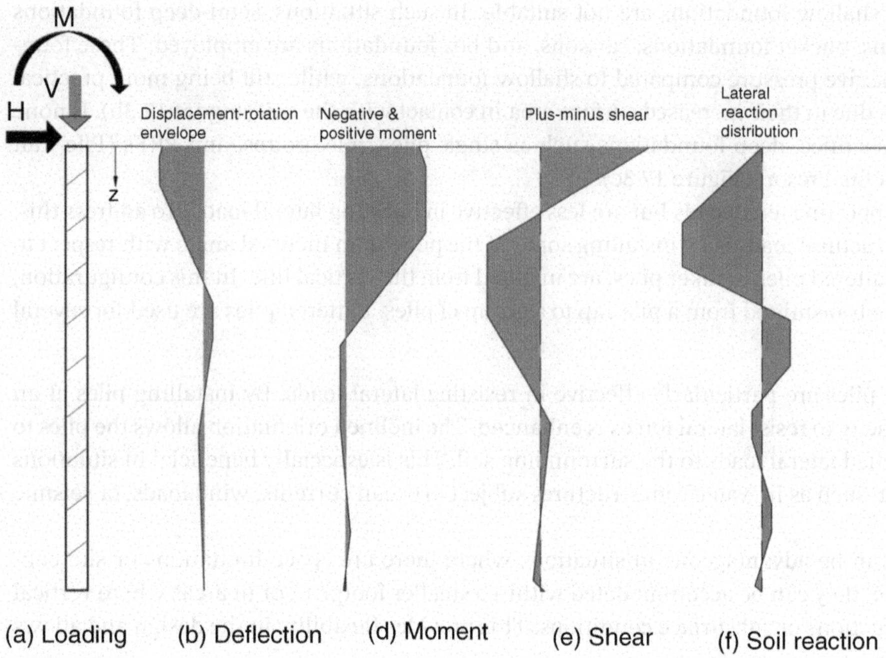

Figure 17.4 A complete solution to a laterally loaded pile problem (Coduto 2016).

17.3 Analysis of Piles with Lateral Loads

17.3.1 Critical Length

For a uniform pile with a given bending stiffness (E_pI_p), there is a critical length (L_c) beyond which the pile behaves under a lateral load as if it were infinitely long and can be termed a "*flexible*" pile. The expressions for the critical lengths are given in the following:

$$L_c = 4\sqrt[4]{\frac{E_pI_p}{k_hD}} = 4R \tag{17.3}$$

$$L_c = 3.5\sqrt[5]{\frac{E_pI_p}{n_h}} = 3.5T \tag{17.4}$$

For constant soil modulus with depth (e.g., stiff over-consolidated clays), pile stiffness factor, $R = \sqrt[4]{\frac{E_pI_p}{k_hD}}$ (in units of length), where E_pI_p is the bending stiffness of the pile, D is the width of the pile, and k_h is the coefficient of horizontal subgrade reaction.

For soil, modulus increases linearly with depth (e.g., normally consolidated clay and granular soils), pile stiffness factor, $T = \sqrt[5]{\frac{E_pI_p}{n_h}}$, where n_h is the constant of horizontal subgrade reaction. The criteria for behavior as a short (rigid) pile or as a long (flexible) pile are presented in Table 17.3.

Table 17.3 Behavior of piles (Tomlinson and Woodward 2015/Taylor & Francis Group).

	Soil modulus	
Pile type	**Linearly increasing**	**Constant**
Flexible piles	$L \geq 4T$	$L \geq 3.5R$
Rigid piles	$L \leq 2T$	$L \leq 2R$

The rigidity or flexibility of piles is dependent on the following factors:

- Pile width
- Pile embedment length
- Pile material
- The type of loading imposing the pile
- Direction of loads
- Constant horizontal subgrade reaction
- Elastic modulus of the pile
- Moment of inertia of the pile
- The soil lateral stiffness
- Pile cap thickness

Many methods are available to solve the laterally loaded pile problem. Each has its own merits and demerits. Some of the methods that are discussed in this Chapter are:

- Broms Method – Section 17.3.2
- Brinch Hansen Method – Section 17.3.3
- Evan-Duncan Method – Section 17.3.4
- Elastic continuum Method – Section 17.3.5
- Nondimensional Method – Section 17.3.6
- Lateral Pressure Method – Displacement Approach (*p–y* Curves) – Section 17.4

17.3.2 Broms Method

Broms (1964) produced simplified solutions for the ultimate lateral load capacity of piles by considering both the ultimate strength of the bearing ground and the yield stress of the pile material. For simplicity, the Broms solutions are presented separately for cohesive soils and cohesionless soil.

17.3.2.1 Rigid Piles (Short Piles) – Cohesive Soils

Broms uses a simplified distribution of soil reaction for cohesive soils, as shown in Figure 17.5. Additionally, the horizontal bearing capacity and maximum bending moment for short piles in cohesive soils are shown in Figure 17.6 (Broms 1964).

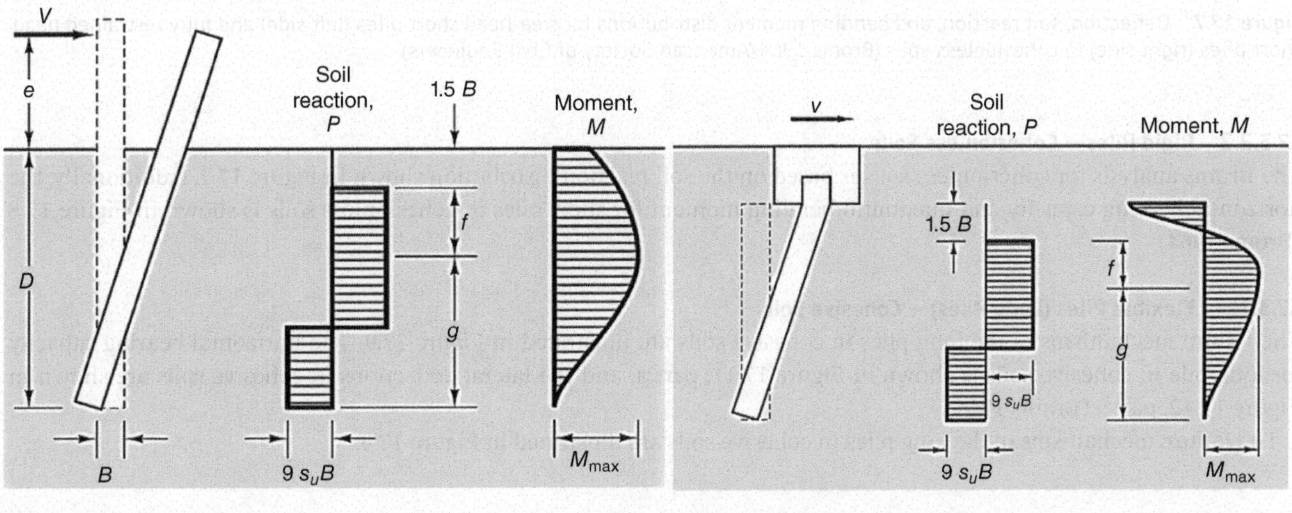

Figure 17.5 Deflection, soil reaction and bending moment distributions for free-head short piles (left side) and fully restrained head short piles (right side) in cohesive soil (Broms 1964/American Society of Civil Engineers).

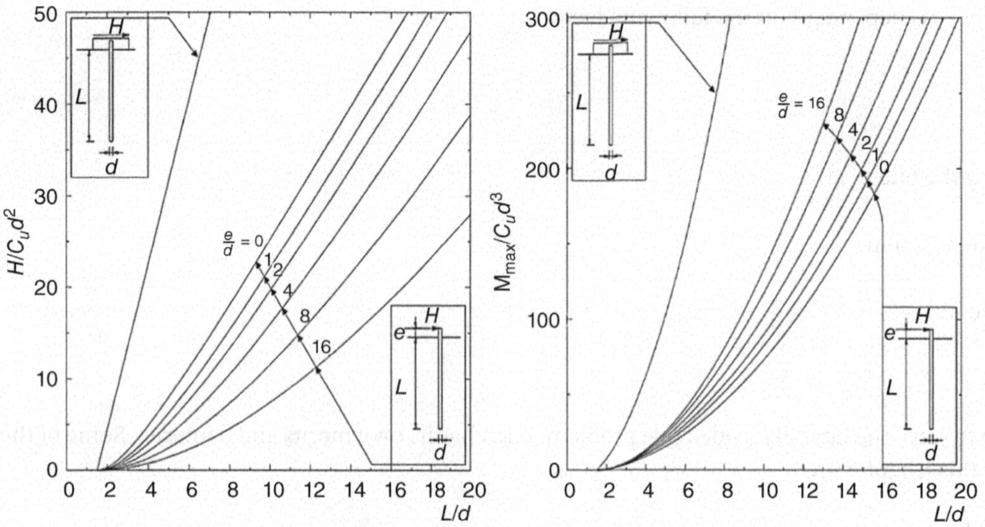

Figure 17.6 Horizontal bearing capacity (left side) and maximum bending moment (right side) for short piles in cohesive soils (Broms 1964/American Society of Civil Engineers).

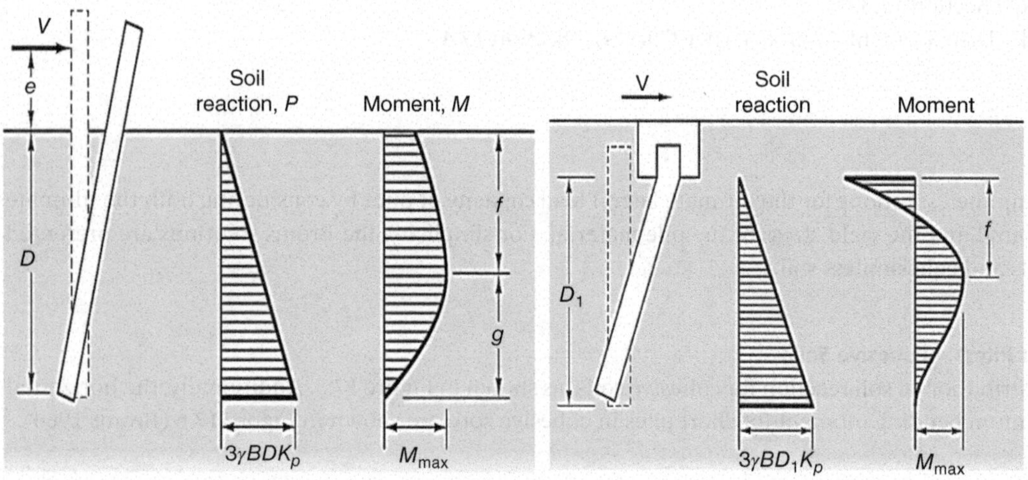

Figure 17.7 Deflection, soil reaction, and bending moment distributions for free-head short piles (left side) and fully restrained head short piles (right side) in cohesionless soils (Broms 1964/American Society of Civil Engineers).

17.3.2.2 Rigid Piles – Cohesionless Soils

The Broms analysis for cohesionless soils is based on the soil reaction distribution shown in Figure 17.7. Additionally, the horizontal bearing capacity and maximum bending moment for short piles in cohesionless soils is shown in Figure 17.8 (Broms 1964).

17.3.2.3 Flexible Piles (Long Piles) – Cohesive Soils

The failure mechanisms of the long piles in cohesive soils are illustrated in Figure 17.9. The horizontal bearing capacity for long pile in cohesive soils is shown in Figure 17.11, part a, and the lateral deflections in cohesive soils are shown in Figure 17.12, part a (Broms 1964).

The failure mechanisms of the long piles in cohesive soils are illustrated in Figure 17.9.

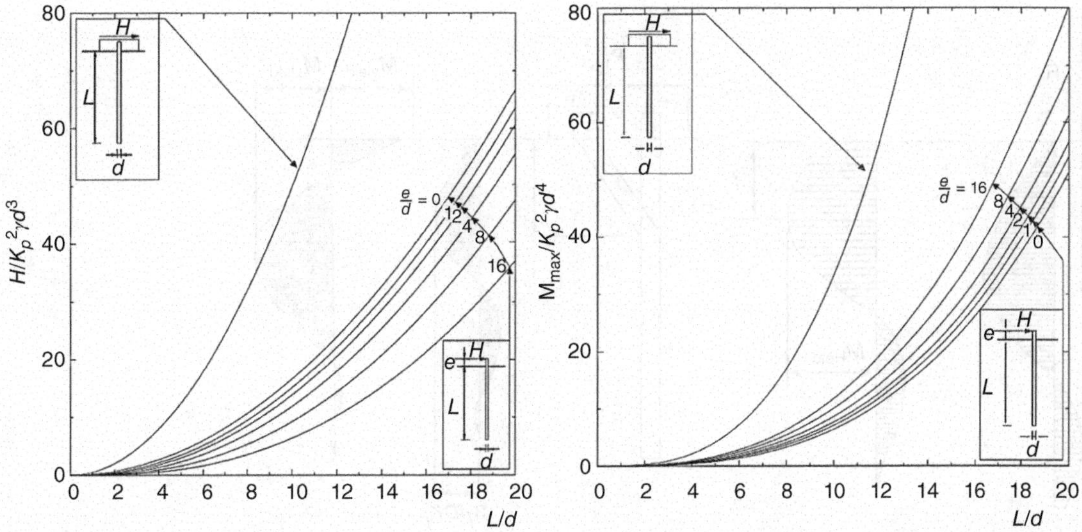

Figure 17.8 Horizontal bearing capacity (left side) and maximum bending moment (right side) for short piles in cohesionless soils (Broms 1964/American Society of Civil Engineers).

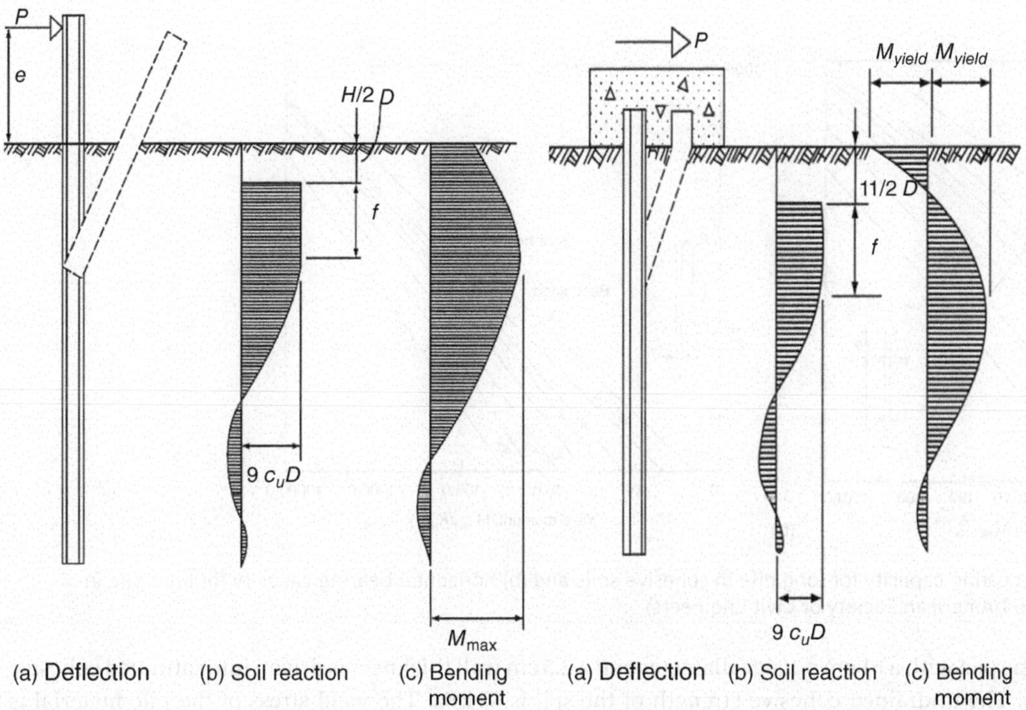

Figure 17.9 The failure mechanism for free-head and fully restrained head long piles in cohesive soil (Broms 1964/American Society of Civil Engineers).

17.3.2.4 Flexible Piles (Long Piles) – Cohesionless Soils

The failure mechanisms of the long piles in cohesionless soils are illustrated in Figure 17.10. Horizontal bearing capacity for long pile in cohesionless soils are shown in Figure 17.11, part b, and the lateral deflections in cohesionless soils are shown in Figure 17.12, part b (Broms 1964).

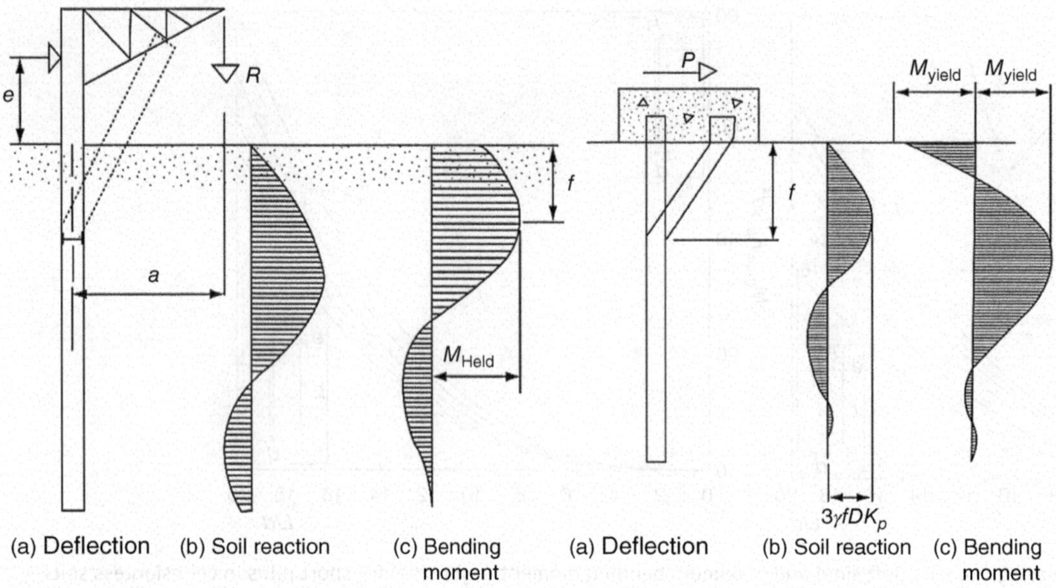

Figure 17.10 Failure mechanism for free-head and fully restrained head long piles in cohesionless soil (Broms 1964/American Society of Civil Engineers).

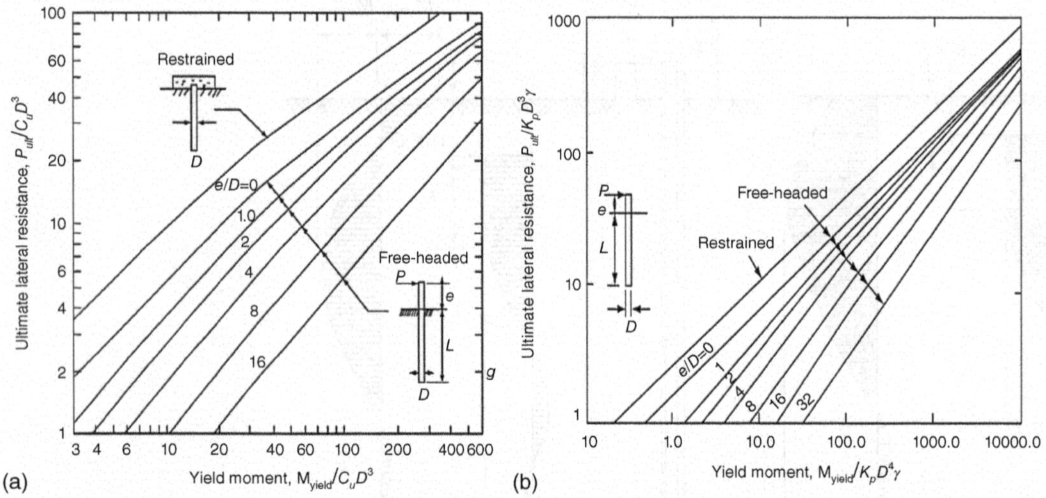

Figure 17.11 (a) Horizontal bearing capacity for long pile in cohesive soils and (b) horizontal bearing capacity for long pile in cohesionless soils (Broms 1964/American Society of Civil Engineers).

Example 17.1 A steel pipe pile with a 61 cm outside diameter and a 2.5 cm wall thickness is driven into saturated cohesive soil up to a depth of 20 m. The undrained cohesive strength of the soil is 85 kPa. The yield stress of the pile material is 2800 kg/cm^2.

a) Calculate the ultimate lateral resistance of the pile using the Broms method with the load applied at ground level. (Assume the yield resistance of the pile section is equal to $1.3 f_y Z$.)
b) If the given pile is restrained against rotation, calculate the ultimate lateral resistance P_u.

Part a

The pile is considered long. Use Figure 17.11a to obtain the ultimate lateral resistance P_u of the pile.

The nondimensional yield moment $= \frac{M_y}{C_u D^3}$

$$Z = \text{section modulus} = \frac{1}{\mathrm{I}} = \frac{\pi(D_0^4 - D_i^4)}{64R} = \frac{\pi(61^4 - 56^4)}{64 \times 30.5} = 6452.6\ \text{cm}^3$$

$$M_y = 1.3 \times 2800 \times 6452.6 = 23.487 \times 10^6\ \text{kg cm}$$

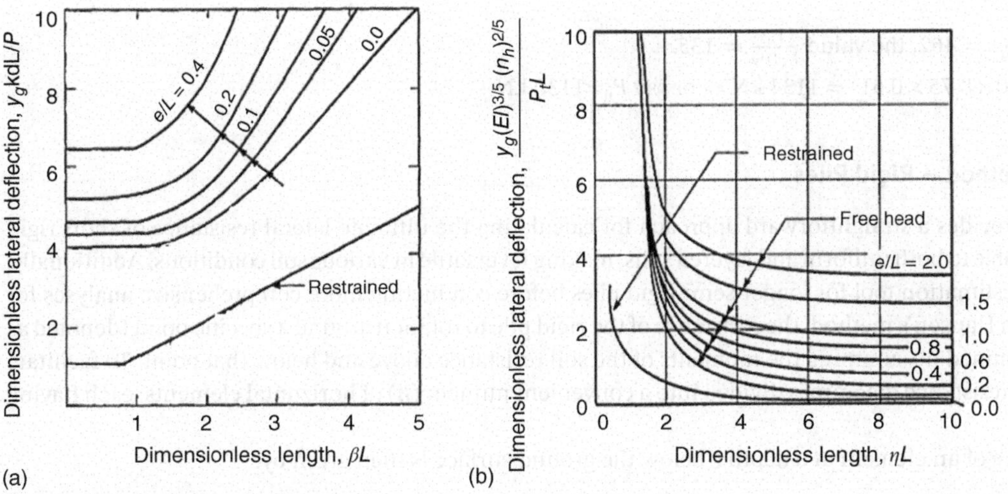

Figure 17.12 (a) Lateral deflections in cohesive soils and (b) lateral deflections in cohesionless soils (Broms 1964/American Society of Civil Engineers).

$$\frac{M_y}{C_u D^3} = \frac{23.487 \times 10^6}{0.85 \times 61^3} = 122$$

From Figure 17.11a for $e/D = 0$, $\frac{M_y}{C_u D^3} = 122 \Rightarrow \frac{P_u}{C_u D^2} \approx 35$
$P_u = 35 C_u D^2 = 35 \times 85 \times 0.612 = 1107$ kN $\Rightarrow$ or say P_u=1100 kN

Part b

$$\frac{M_y}{C_u D^3} = \frac{23.487 \times 10^6}{0.85 \times 61^3} = 122$$

From Figure 17.11a for the restrained pile, $\frac{M_y}{C_u D^3} = 122 \Rightarrow \frac{P_u}{C_u D^2} \approx 50$
$P_u = 50 C_u D^2 = \frac{50}{35} \times 1107$ kN $= 1581$ kN $\Rightarrow$ or say $P_u = 1550$ kN

Example 17.2 A steel pipe pile of 61 cm, with an inner diameter of 56 cm is driven into medium-dense sand under submerged conditions, which has a relative density of 60% and an angle of internal friction of 38°. Assume that the yield resistance of the pile section is the same as the given in Example 17.1. The submerged until weight of the soil $\gamma_b = 8.75$ kN/m^3. Using the Broms Method,

a) Compute the ultimate lateral resistance of the pile.
b) If the pile is restrained, what is the ultimate lateral resistance of the pile?

Part a

Nondimensional yield moment $= \frac{M_y}{K_p D^4 \gamma}$

$$K_p = \tan^2(45 + \phi/2) = \tan^2 64° = 4.20$$

$$M_y = 23.487 \times 10^6 \text{ kg cm}$$

$$\gamma = 8.75 \text{ kN/m}^3 \approx 8.75 \times 10^{-4} \text{ kg/cm}^3$$

$$D = 61 \text{ cm}$$

$$\frac{M_y}{K_p D^4 \gamma} = \frac{23.487 \times 10^6}{4.20 \times 61^4 \times 8.75 \times 10^{-4}} = 462$$

From Figure 17.11b, for $\frac{M_y}{K_p D^4 \gamma} = 462$, for $e/D = 0$, we have $\frac{P_u}{K_p D^3 \gamma} = 80$
$P_u = 80\ K_p \gamma D^3 = 80 \times 4.20 \times 8.75 \times 0.61^3 = 667$ kN $\Rightarrow$ or say $P_u = 660$ kN

Part b

From Figure 17.11b, for $\frac{M_y}{K_p D^4 \gamma} = 462$, the value $\frac{P_u}{K_p D^3 \gamma} = 135$.

$P_u = 135\ K_p \gamma D^3 = 135 \times 4.20 \times 8.75 \times 0.61^3 = 1134$ kN $\Rightarrow$ or say P_u=1130 kN

17.3.3 Brinch Hansen Method – Rigid Piles

Brinch Hansen's method provides a straightforward approach for calculating the ultimate lateral resistance of short rigid piles. This method is applicable to both uniform and layered soils, making it versatile in various soil conditions. Additionally, it can be used as an initial estimation tool for longer semi-rigid piles before conducting more comprehensive analyses for long-slender piles. In Brinch Hansen's method, the resistance of the rigid pile to rotation around a specific point (denoted as X in Figure 17.13a) is determined by summing the moments of the soil resistance above and below that point. To facilitate the calculation, the passive resistance diagram is divided into a convenient number (n) of horizontal elements, each having a depth of L/n.

The unit passive resistance of an element at a depth z below the ground surface is then given by:

$$p_z = p_{0z} K_{qz} + c K_{cz} \tag{17.5}$$

where p_{0z} is the effective overburden pressure at depth z, c is the cohesion of the soil at depth z, and K_{qz} and K_{cz} are the passive pressure coefficients for the frictional and cohesive components, respectively, at depth z. Brinch Hansen has established values of K_q and K_c in relation to the depth z and the width of pile B in the direction of rotation, as shown in Figure 17.14.

The total passive resistance on each horizontal element is $p_z \times \frac{L}{n} \times B$ and, by taking moments about the point of application of the horizontal load:

$$\sum M = \sum_{z=0}^{z=x} p_z \frac{L}{n}(e+z)B - \sum_{z=x}^{z=L} p_z \frac{L}{n}(e+z)B \tag{17.6}$$

The point of rotation at depth x is correctly chosen when $\sum M = 0$, that is, when the passive resistance of the soil above the point of rotation balances that below it. Point X is thus determined by the process of trial and adjustment. If the head of the pile carries a moment M instead of a horizontal force, the moment can be replaced by a horizontal force H at a distance e above the ground surface where M is equal to $H \times e$. Where the head of the pile is fixed against rotation, the equivalent height e_1 above-ground level of a force H acting on a pile with a free head is given by:

$$e_1 = \frac{1}{2}(e + z_f) \tag{17.7}$$

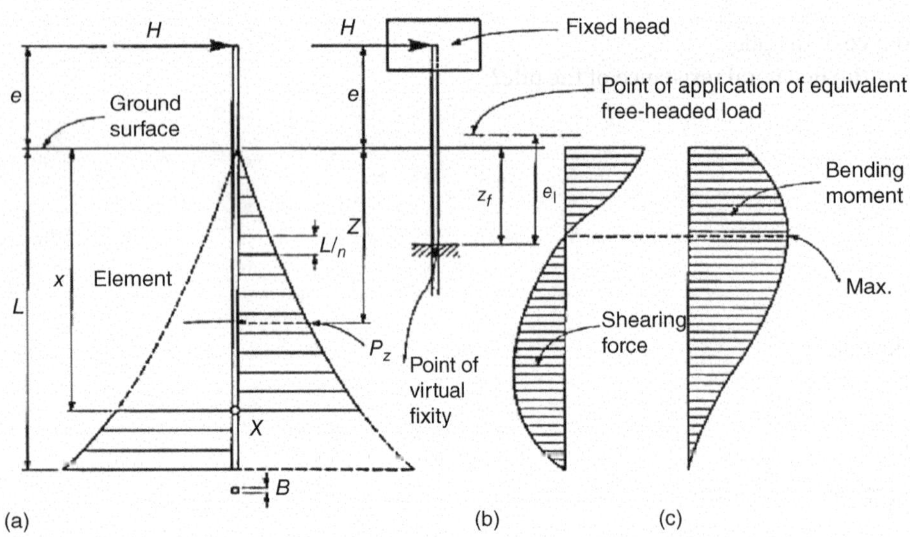

Figure 17.13 Brinch Hansen's method for calculating ultimate lateral resistance of short piles: (a) soil reactions, (b) shearing force diagram, and (c) bending moment diagram (Brinch Hansen 1961).

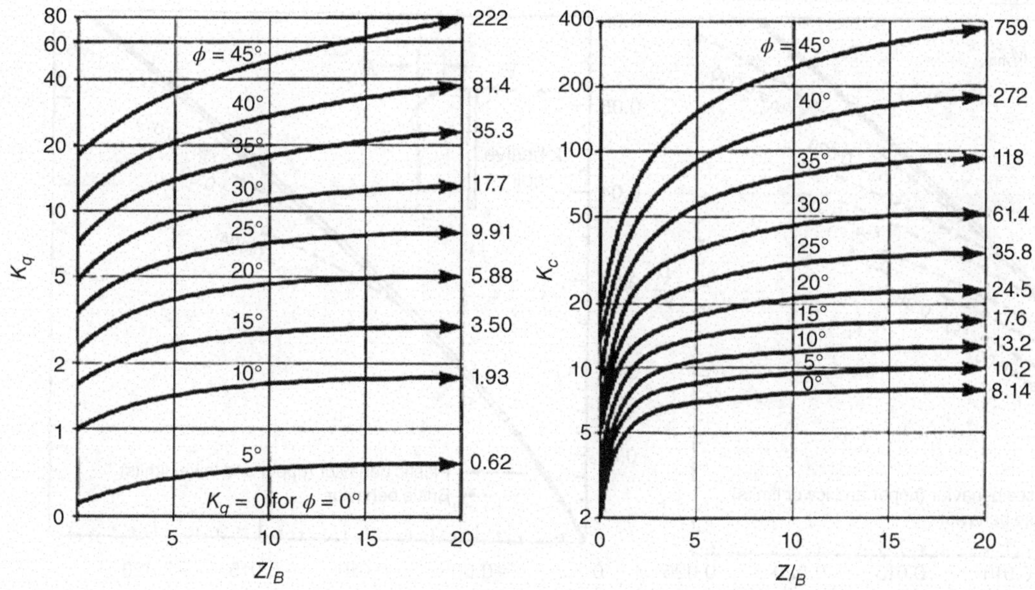

Figure 17.14 Brinch Hansen's coefficients K_q and K_c (Brinch Hansen 1961).

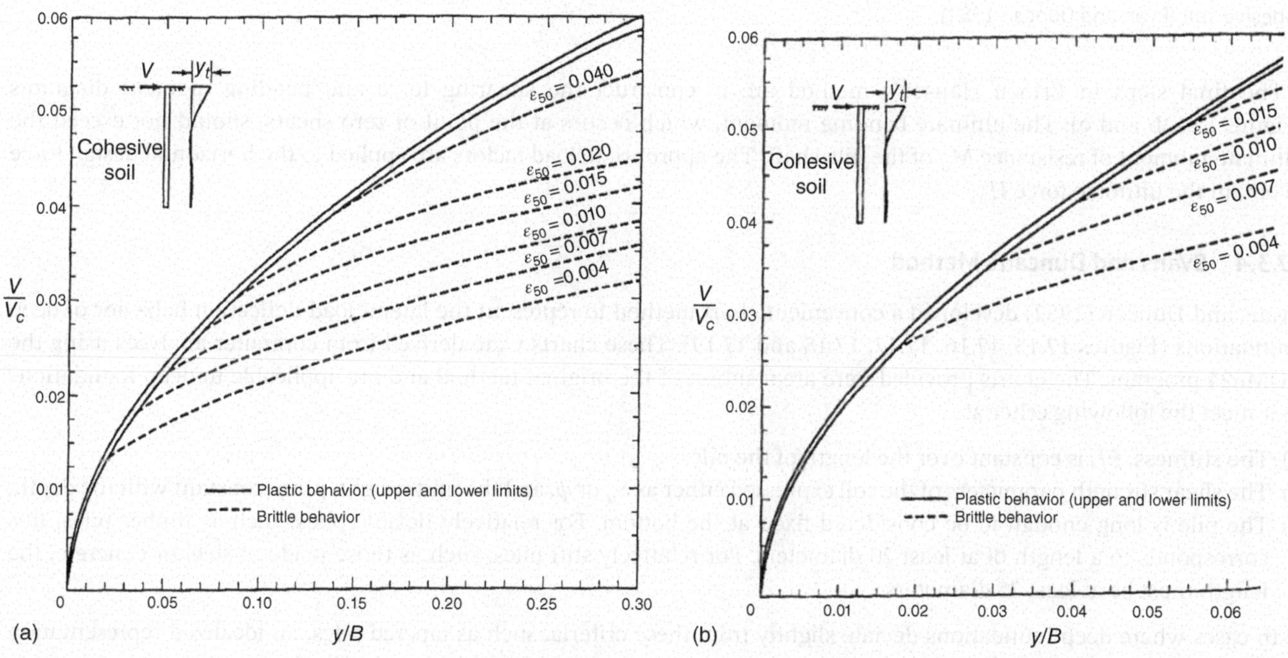

Figure 17.15 (a) Shear load vs. lateral deflection curves in cohesive soil for free-head conditions and (b) restrained-head conditions (Evan and Duncan 1982).

where e is the height from the ground surface to the point of application of the load at the fixed head of the pile (Figure 17.13a), and z_f is the depth from the ground surface to the point of virtual fixity. The depth z_f is not known at this stage, but for practical design purposes, it can be taken as 1.5 m for compact granular soil or stiff clay (below the zone of soil shrinkage in the latter case) and 3 m for soft clay or silt.

Having obtained the depth to the center of rotation from Equation 17.7, the ultimate lateral resistance of the pile to the horizontal force H_u can be obtained by taking moments about the point of rotation, When,

$$H_u(e+x) = \sum_{z=0}^{z=x} p_z \frac{L}{n}(x-z)B + \sum_{z=x}^{z=x+L} p_z \frac{L}{n}(z-x)B \tag{17.8}$$

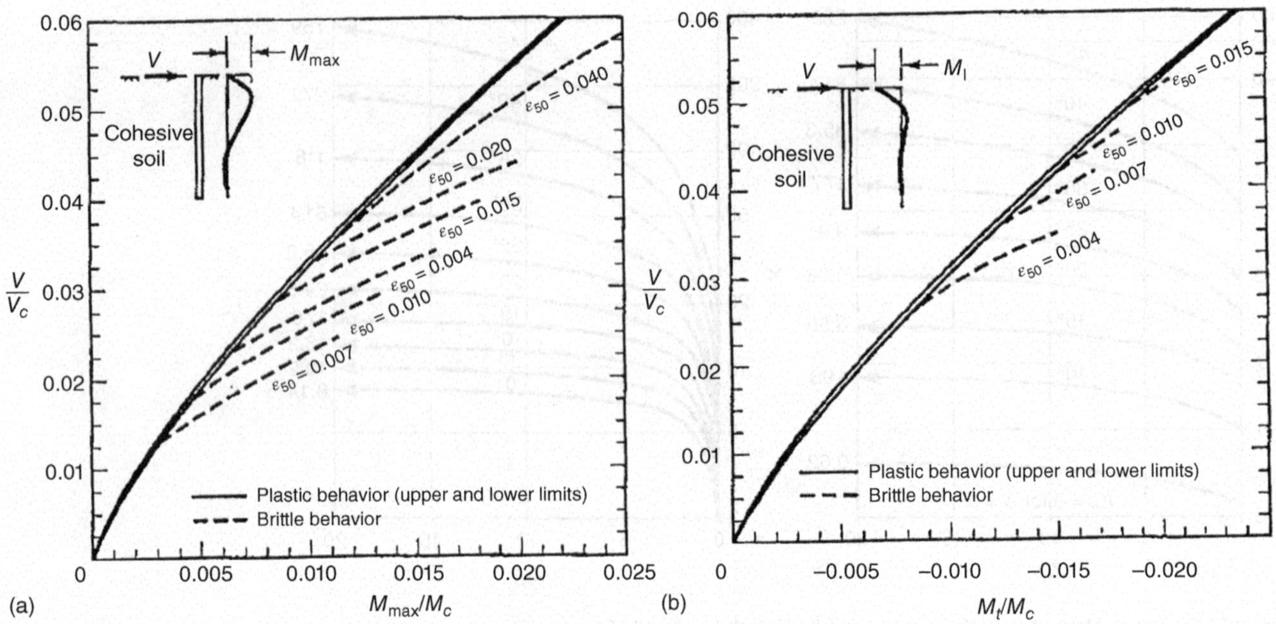

Figure 17.16 (a) Shear load vs. maximum moment curves for free-head conditions and (b) restrained-head conditions (right side) in cohesive soil (Evan and Duncan 1982).

The final steps in Brinch Hansen's method are to construct the shearing force and bending moment diagrams (Figure 17.13b and c). The ultimate bending moment, which occurs at the point of zero shears, should not exceed the ultimate moment of resistance M_u of the pile shaft. The appropriate load factors are applied to the horizontal design force to obtain the ultimate force H_u.

17.3.4 Evans and Duncan's Method

Evans and Duncan (1982) developed a convenient chart method to represent the lateral load-deflection behavior of deep foundations (Figures 17.15, 17.16, 17.17, 17.18 and 17.19). These charts were derived from computer analyses using the COM624 program. The charts provided here are a subset of the original method and are applicable to deep foundations that meet the following criteria:

1) The stiffness, EI, is constant over the length of the pile.
2) The shear strength parameters of the soil expressed either as S_u or ϕ, and the unit weight, γ, are constant with the depth.
3) The pile is long enough to be considered fixed at the bottom. For relatively flexible piles, such as timber piles, this corresponds to a length of at least 20 diameters. For relatively stiff piles, such as those made of steel or concrete, the length must be at least 35 diameters.

In cases where deep foundations deviate slightly from these criteria, such as tapered piles, an idealized representation can be obtained by averaging the values of EI, S_u, ϕ, or γ from the ground surface to a depth of 8 pile diameters.

Evans and Duncan defined the characteristic shear load, V_c, and characteristic moment load, M_c, as follows:

$$V_c = \lambda B^2 E R_I \left(\frac{\sigma_p}{E R_I}\right)^m (\varepsilon_{50})^n \tag{17.9}$$

$$M_c = \lambda B^3 E R_I \left(\frac{\sigma_p}{E R_I}\right)^m (\varepsilon_{50})^n \tag{17.10}$$

$$R_I = \frac{I}{\pi B^4/64} \tag{17.11}$$

For solid circular cross-sections, R_I is equal to 1 and for square cross-sections, R_I is equal to 1.7. For plastic clay and sand, λ is equal to 1, and for brittle clay, λ is equal to $(0.14)^n$.

For cohesive soils σ_p is equal to $4.2S_u$ and for cohesionless soils σ_p is equal to $2C_{p\phi}\gamma B\tan^2\left(45 + \frac{\phi}{2}\right)$.

Table 17.4 Values of exponents *m* and *n* (Evans and Duncan 1982).

	For V_c		For M_c	
Soil Type	*m*	*n*	*m*	*n*
Cohesive	0.683	−0.22	0.46	−0.15
Cohesionless	0.57	−0.22	0.40	−0.15

where:

V_c = characteristic shear load
M_c = characteristic moment load
λ = a dimensionless parameter dependent on the soil's stress–strain behavior
B = diameter of the foundation
E = modulus of elasticity of foundation
= 29,000,000 lb/in^2 (200,000 MPa) for steel
= $15{,}200\ \sigma_r(\acute{f}_c/\sigma_r)^{0.5}$ for concrete
= 1,600,000 lb/in^2 (11,000 MPa) for Southern Pine or Douglas Fir
$\acute{f}_c$ = 28-day compressive strength of concrete
σ_r = reference stress = 14 lb/in^2 = 0.10 MPa
R_I = moment of inertia ratio (dimensionless)
σ_p = representative passive pressure of soil
ε_{50} = strain at which 50% of the soil strength is mobilized
m, n = exponents from Table 17.4
I = moment of inertia of foundation
= $\pi B^4/64$ for solid circular cross-sections
= $B^4/12$ for square cross-sections
S_u = undrained shear strength of soil from the ground surface to a depth of 8 pile diameters
ϕ = friction angle of soil (deg) from the ground surface to a depth of 8 pile diameters
$C_{p\phi}$ = passive pressure factor = $\phi/10$
γ = unit weight of soil from the ground surface to a depth of 8 pile diameters. If the groundwater table is in this zone, use a weighted average of γ and γ_b, where γ_b is the buoyant unit weight in the zone below the groundwater table.

The value of ε_{50} could be determined from triaxial compression tests. Typically, $\varepsilon_{50} = 0.01$ for clays and $\varepsilon_{50} = 0.002$ for medium-dense sands that contain little or no mica.

Example 17.3 A 20 k (~9 tons) shear load will be applied to a 12-inch (~30 cm) square, 60 ft (18.29 m) long resistance-head concrete pile. The soil is sand with $\phi = 36°$ and $\gamma = 120\ \text{lb/ft}^3$ (19.22 kN/m^3). The groundwater table is at a depth of 40 ft (12.19 m). The pile is made of concrete with a 28-day compressive strength of 6000 lb/in^2 (41,383 kPa). Compute the lateral deflection at the top of this pile and the maximum moment using Evans and Duncan's Method.

Use units of pounds and inches. $\lambda = 1.00$; $R_I = 1.70$; $\varepsilon_{50} = 0.002$

$$C_{p\phi} = \frac{\phi}{10} = \frac{36}{10} = 3.6$$

$$\sigma_p = 2C_{p\phi}\gamma B\tan^2\left(45 + \frac{\phi}{2}\right) = (2)(3.6)\left(\frac{120}{12^3}\right)(12)\tan^2\left(45 + \frac{36}{2}\right) = 23.1\frac{\text{lb}}{\text{in}^2}(159.3\ \text{kPa})$$

$$E = 15{,}200\sigma_r(\acute{f}_c/\sigma_r)^{0.5} = (15{,}200)(14)(6000/14)^{0.5} = 4400\ \text{k/in}^2(3.0347\times10^7\text{kPa})$$

Using Equation 17.9:

$$V_c = \lambda B^2 E R_I\left(\frac{\sigma_p}{ER_I}\right)^m(\varepsilon_{50})^n = (1.00)(12)^2(4{,}400{,}000)(1.70)\left(\frac{23.1}{(4{,}400{,}000)(1.70)}\right)^{0.57}(0.002)^{-0.22} = 3{,}056{,}000\ \text{lb}\ (13.6\ \text{MN})$$

$$V/V_c = 20{,}000/3{,}056{,}000 = 0.0065$$

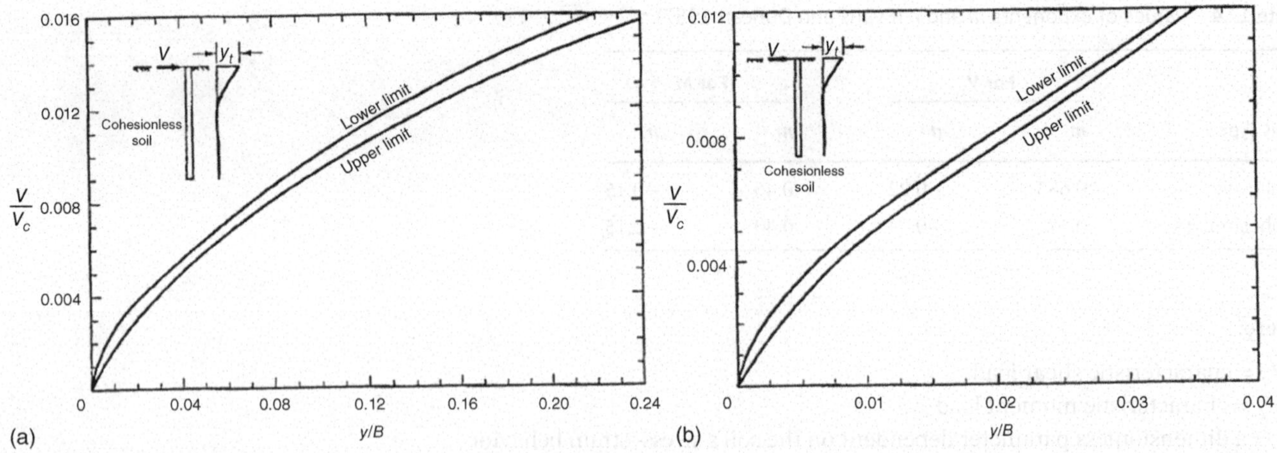

Figure 17.17 Shear load vs. lateral deflection curves in cohesionless soil for (a) free-head conditions, (b) restrained-head conditions (Evan and Duncan 1982).

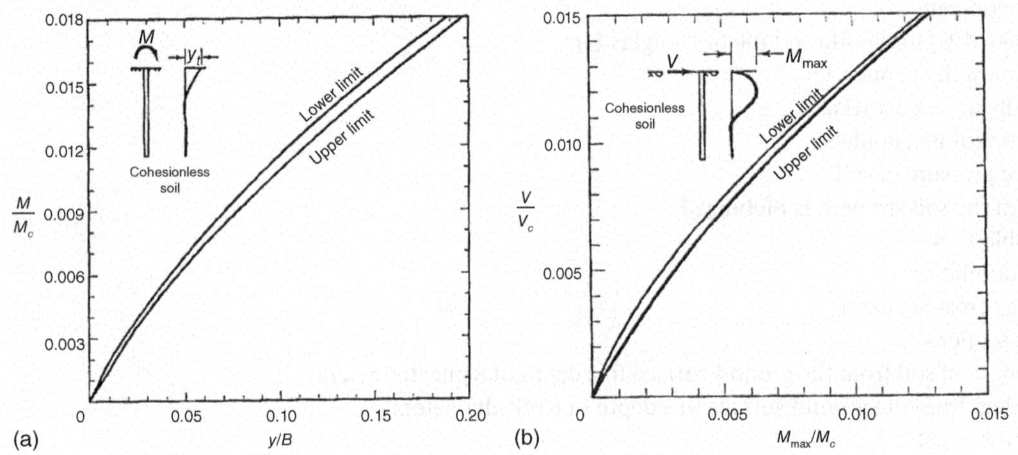

Figure 17.18 (a) Moment load vs. lateral deflection curves for free-head conditions, (b) shear load vs. maximum moment curves for free-head condition in cohesionless soil (Evan and Duncan 1982).

From Figure 17.17b: $y_t/B = 0.0150$

$\Rightarrow y_t = (0.0150)(12) = 0.18$ in (4.57 mm)

Using Equation 17.10:

$$M_c = \lambda B^3 E R_I \left(\frac{\sigma_p}{E R_I}\right)^m (\varepsilon_{50})^n = (1.00)(12)^3(4{,}400{,}000)(1.70)\left(\frac{23.1}{(4{,}400{,}000)(1.70)}\right)^{0.40}(0.002)^{-0.15}$$

$$\Rightarrow M_c = 205{,}200{,}000 \text{ in} - \text{lb (2319.28 kN m)}$$

From Figure 17.18b: $M_{max}/M_c = 0.0041$

$$\Rightarrow M_{max} = (0.0041)(205{,}200{,}000) = 841{,}000 \text{ in} - \text{lb (95.054 kN m)}$$

17.3.5 Elastic Continuum Method

Poulos and Davis (1980) developed an analysis method that considers the soil as an elastic continuum. This approach considers the influence of stresses and forces acting at different points within the soil on the displacements at a given point. The method presented in this section can be used to model the behavior of piles under strain conditions beyond the elastic range. In situations where lateral forces on piles are relatively low, and the piles themselves are rigid, the pile head movements generally fall within the elastic range. In such cases, it is convenient to use the elastic continuum model to

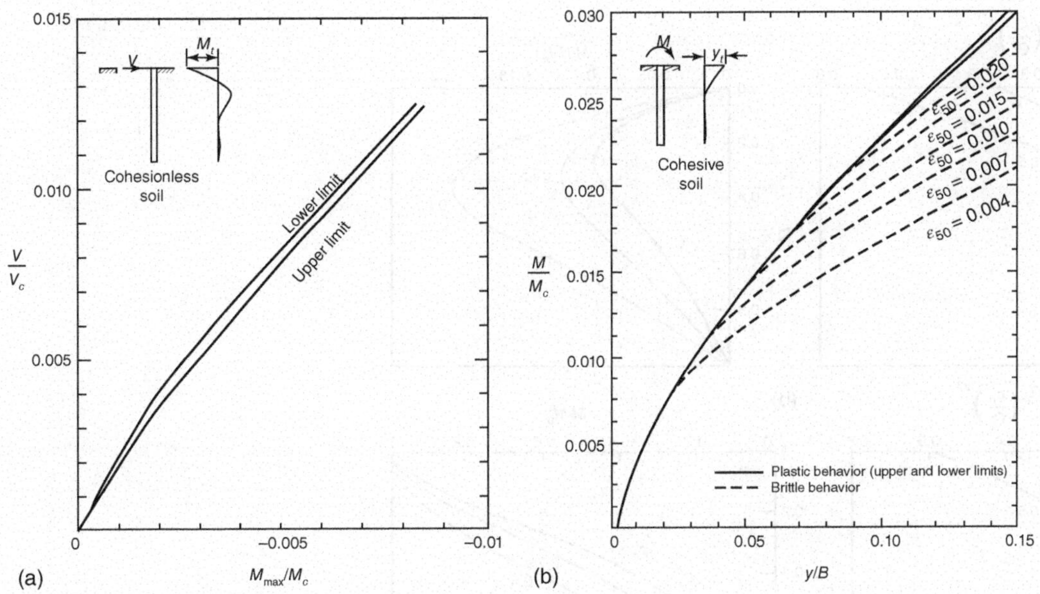

Figure 17.19 (a) Shear load vs. maximum moment curves for the restrained-head conditions in cohesionless soil, (b) moment load vs. lateral deflection curves for free-head conditions in cohesive soil (Evan and Duncan 1982).

calculate deflections and bending moments. To establish relationships between pile deflections and bending moments with depth for lateral force and moment loading, Randolph employed finite element analyses. These relationships are depicted in Figure 17.20. The following notation applies to the parameters in this Figure:

y_0 = lateral displacement at the ground surface
z = depth below ground level
H_0 = lateral load applied at the ground surface
M = bending moment in the pile
M_0 = bending moment at the ground surface
r_0 = radius of the pile
$\acute{E}_p$= effective young's modulus of a solid circular pile of radius r_0
G_c = characteristic modulus of the soil, that is, the average value of G^* over depths less than l_c

$$G^* = G\left(1 + \frac{3}{4}\upsilon\right)$$

G = shear modulus of the soil
υ = Poisson's ratio
l_c = critical length of the pile
$l_c = 2r_0(\acute{E}_p/G^*)^{2/7}$ for homogeneous soil
$l_c = 2r_0(\acute{E}_p/m^* r_0)^{2/9}$ for soil increasing linearly in stiffness by the depth
$m^* = m\left(1 + \frac{3}{4}\upsilon\right)$
$m = \frac{G}{z}$, where G varies with depth as $G = mz$
ρ_c = homogeneity factor = $\frac{G^* \text{ at } l_c/4}{G^* \text{ at } l_c/2}$

17.3.6 Nondimensional Method

The nondimensional method for analyzing laterally loaded piles presented in this section is based on the paper published by Reese and Matlock (1956). In the case of an applied lateral load P_t and moment M_t, the solution for deflections of the elastic curve will include the relative stiffness factor and be expressed as:

$$y = f_y(x, T, L, E_s, EI, P_t, M_t) \tag{17.12}$$

where:

$T = \left(\frac{EI}{n_h}\right)^{\frac{1}{5}}$ is the Stiffness Factor

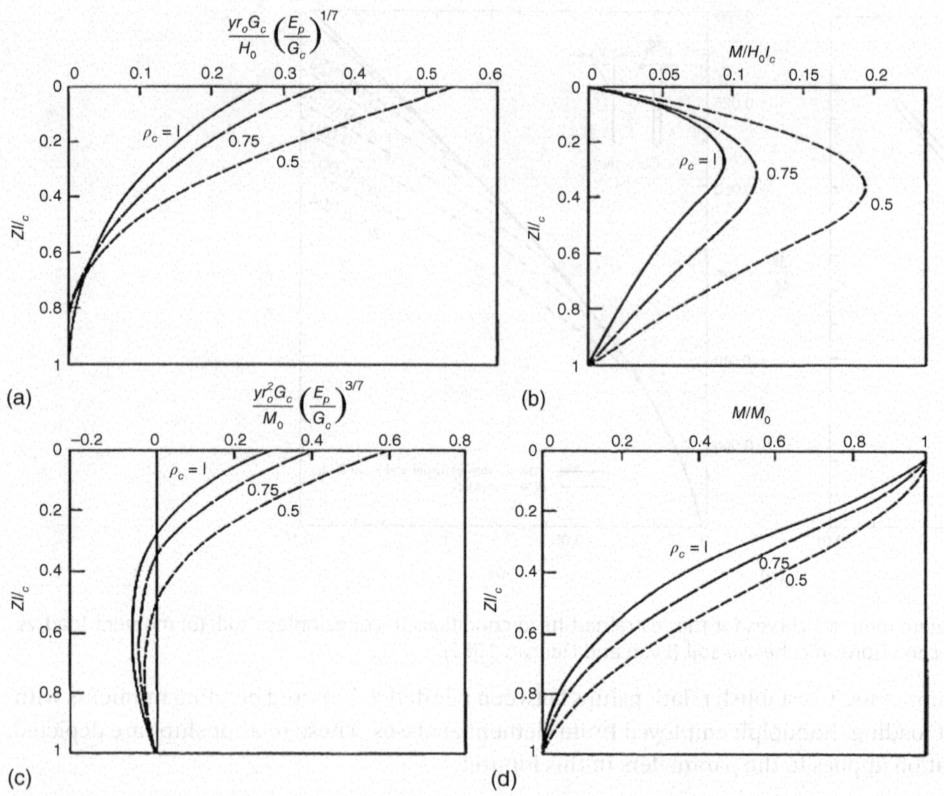

Figure 17.20 Generalized curves giving deflected pile shape and bending moment profile for lateral force and bending moment applied to pile head (after Randolph): (a) deflected pile shape for lateral force loading, (b) bending moment profile for lateral force loading, (c) Deflected pile shape for moment loading, and (d) Bending moment profile for moment loading.

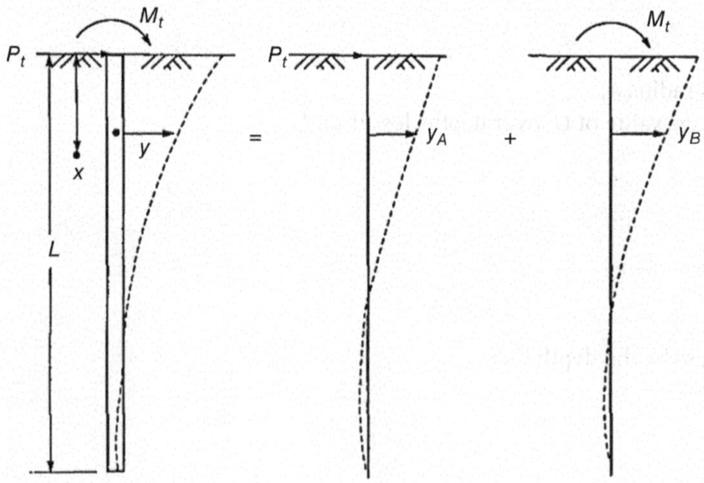

Figure 17.21 Principle of superposition for the deflection of laterally loaded piles (Murthy 2007/CBS Publishers and Distributers).

Other boundary conditions can be substituted for P_t and M_t. The assumption of elastic behavior is introduced for the pile, and if deflections remain small relative to the pile dimensions, the principle of superposition may be employed. Thus, the effects of an imposed lateral load P_t and an imposed moment M_t, if considered independently, give rise to deflection.

$$y = y_A + y_B \tag{17.13}$$

where, y_A is the deflection due to P_t, and y_B is the deflection caused by the moment M_t (Figure 17.21). The solution for two cases of deflection may be expressed in the following forms.

$$\text{For case A: } \frac{y_A}{P_t} = f_A(x, T, L, E_s, EI) \tag{17.14}$$

$$\text{For case B: } \frac{y_B}{M_t} = f_B(x, T, L, E_s, EI) \tag{17.15}$$

where, f_A and f_B represent two different functions of the same terms. Each case has six terms and two dimensions (force and length). There are, therefore, four independent nondimensional groups that can be formed as follows:

$$\text{For case A: } \frac{y_A EI}{P_t T^3}, \frac{x}{T}, \frac{L}{T}, \frac{E_s T^4}{EI} \tag{17.16}$$

$$\text{For case B: } \frac{y_B EI}{M_t T^2}, \frac{x}{T}, \frac{L}{T}, \frac{E_s T^4}{EI} \tag{17.17}$$

To satisfy the conditions of similarity, each of these groups must be equal for both the model and prototype. A group of nondimensional parameters may be defined with the same numerical value for any model and prototype. They are:

$$\text{Depth coefficient, } z = \frac{x}{T} \tag{17.18}$$

$$\text{Maximum depth coefficient, } z_{max} = \frac{L}{T} \tag{17.19}$$

$$\text{Soil modulus function, } \phi(z) = \frac{E_s T^4}{EI} \tag{17.20}$$

$$\text{Case A, deflection coefficient, } A_y = \frac{y_A EI}{P_t T^3} \tag{17.21}$$

$$\text{Case B, deflection coefficient, } B_y = \frac{y_B EI}{M_t T^2} \tag{17.22}$$

By carrying out dimensional analysis in the same way as for deflection, the solutions for slope, moment, shear, and soil reaction may be expressed as:

$$\text{Slope, } S = S_A + S_B = \left[\frac{P_t T^2}{EI}\right] A_s + \left[\frac{M_t T}{EI}\right] B_s \tag{17.23}$$

$$\text{Moment, } M = M_A + M_B = [P_t T]A_m + [M_t]B_m \tag{17.24}$$

$$\text{Shear, } V = V_A + V_B = [P_t]A_v + \left[\frac{M_t}{T}\right] B_v \tag{17.25}$$

$$\text{Soil reaction, } p = p_A + p_B = \left[\frac{P_t}{T}\right] A_p + \left[\frac{M_t}{T^2}\right] B_p \tag{17.26}$$

The corresponding A and B coefficients may be obtained from Table 17.5. However, often the deflection and slope at ground level are required.

Example 17.4 A steel pipe pile with 61 cm outer diameter and a wall thickness of 2.5 cm is driven into loose sand (D_r = 30%) under submerged conditions to a depth of 20 m. The submerged unit weight of the soil is 8.75 kN/m³, and the angle of internal friction is 33°. The *EI* value of the pile is 4.35 × 1011 kg-cm². Compute the ground line deflection of the pile with a nondimensional method in three situations that are given below (n_h = 6 MN/m³) using the nondimensional method.

a) A pile under a lateral load of 268 kN at ground level under free head conditions,
b) A lateral load at a height of 2 m above-ground level,
c) A pile fixed against rotation.

Part a

$$P_t = 268\ \text{kN}, M_t = 0\ \text{kN.m}$$

$$y_g = y_A + y_B = A_y \frac{P_t T^3}{EI} + B_y \frac{M_t T^2}{EI}$$

Table 17.5 The A and B coefficients, as obtained by Reese and Matlock for long vertical piles.

z	A_y	A_s	A_m	A_v	A_p
0.0	2.435	−1.623	0.000	1.000	0.000
0.1	2.273	−1.618	0.100	0.989	−0.277
0.2	2.112	−1.603	0.198	0.966	−0.422
0.3	1.952	−1.578	0.291	0.906	−0.586
0.4	1.796	−1.545	0.379	0.840	−0.718
0.5	1.644	−1.503	0.459	0.764	−0.822
0.6	1.496	−1.454	0.532	0.677	−0.897
0.7	1.353	−1.397	0.595	0.585	−0.947
0.8	1.216	−1.335	0.649	0.489	−0.973
0.9	1.086	−1.268	0.693	0.392	−0.977
1.0	0.962	−1.197	0.727	0.295	−0.962
1.2	0.738	−1.047	0.767	0.109	−0.885
1.4	0.544	−0.893	0.772	−0.056	−0.761
1.6	0.381	−0.741	0.746	−0.193	−0.609
1.8	0.247	−0.596	0.696	−0.298	−0.445
2.0	0.142	−0.464	0.628	−0.371	−0.283
3.0	−0.075	−0.040	0.225	−0.349	0.226
4.0	−0.050	0.052	0.000	−0.016	0.201
5.0	−0.009	0.025	−0.033	0.013	0.046
z	B_y	B_s	B_m	B_v	B_p
0.0	1.623	−1.750	1.000	0.000	0.000
0.1	1.453	−1.650	1.000	−0.007	−0.145
0.2	1.293	−1.550	0.999	−0.028	−0.259
0.3	1.143	−1.450	0.994	−0.058	−0.343
0.4	1.003	−1.351	0.987	−0.095	−0.401
0.5	0.873	−1.253	0.976	−0.137	−0.436
0.6	0.752	−1.156	0.960	−0.181	−0.451
0.7	0.642	−1.061	0.939	−0.226	−0.449
0.8	0.540	−0.968	0.914	−0.270	−0.432
0.9	0.448	−0.878	0.885	−0.312	−0.403
1.0	0.364	−0.792	0.852	−0.350	−0.364
1.2	0.223	−0.629	0.775	−0.414	−0.268
1.4	0.112	−0.482	0.668	−0.456	−0.157
1.6	0.029	−0.354	0.594	−0.477	−0.047
1.8	−0.030	−0.245	0.498	−0.476	0.054
2.0	−0.070	−0.155	0.404	−0.456	0.140
3.0	−0.089	0.057	0.059	−0.0213	0.268
4.0	−0.028	0.049	0.042	0.017	0.112
5.0	0.000	0.011	0.026	0.029	−0.002

$$z = 0 \rightarrow y_g = 2.435\frac{P_tT^3}{EI} + 1.623\frac{M_tT^2}{EI} = 2.435\frac{P_tT^3}{EI}$$

$$T = \left(\frac{EI}{n_h}\right)^{\frac{1}{5}} = \left(\frac{4.35 \times 10^2}{6}\right)^{\frac{1}{5}} = 2.3554\ \text{m}$$

$$y_g = 2.435\frac{P_tT^3}{EI} = 2.435\left(\frac{268 \times 2.3554^3}{4.35 \times 10^2}\right) = 0.0196\ \text{m} = 1.96\ \text{cm}$$

Part b

$$P_t = 268\ \text{kN}, M_t = 2 \times 0.268 = 0.536\ \text{MN.m}$$

$$y_g = y_A + y_B = A_y\frac{P_tT^3}{EI} + B_y\frac{M_tT^2}{EI}$$

$$z = 0 \rightarrow y_g = 2.435\frac{P_tT^3}{EI} + 1.623\frac{M_tT^2}{EI}$$

$$T = \left(\frac{EI}{n_h}\right)^{\frac{1}{5}} = \left(\frac{4.35 \times 10^2}{6}\right)^{\frac{1}{5}} = 2.3554\ \text{m}$$

$$y_g = 2.435\left(\frac{268 \times 2.3554^3}{4.35 \times 10^2}\right) + 1.62\frac{0.536 \times 2.3554^2}{4.35 \times 10^2} = 0.0196 + 0.0111 = 0.0307\ \text{m} = 3.07\ \text{cm}$$

Part c

$$z = 0 \rightarrow S_g = 0$$

$$\frac{M_g}{P_gT} = -\left(\frac{A_s}{B_s}\right)_{z=0} = -\left(\frac{-1.623}{-1.75}\right) = -0.93$$

$$C_m = A_m - 0.93B_m$$

$$z = 0 \rightarrow C_m = 0 - 0.93(1) = -0.93$$

$$y_g = 0.93\frac{P_tT^3}{EI} = \frac{0.93}{2.435} \times 1.96 = 0.75\ \text{cm}$$

17.4 Lateral Pressure – Displacement Approach (*p–y* Curves)

The p–y method is a soil–structure interaction analysis technique that employs a series of nonlinear springs to model the behavior of the foundation. Dunnavant and O'Neill (1989)introduced a new stiff clay criterion for this method based on laboratory findings at the University of Houston. In the p–y method, the foundation is represented using a two-dimensional finite difference analysis, as outlined by Reese and Van Impe (2001). The foundation is divided into n intervals, with a node located at the end of each interval. The soil is modeled as a series of nonlinear springs at each node. These springs are treated as uncoupled, meaning that each spring acts independently of the others. Additionally, appropriate boundary conditions, as discussed earlier, must be applied to capture the behavior of the system accurately. By inputting the necessary information and incrementally applying the structural loads, the analysis software can determine the static equilibrium condition and calculate the shear, moment, and lateral deflection at each interval along the foundation.

The exact shape of the p–y curve at a particular point on a pile depends on many factors, including:

- Soil properties
- Type of loading (i.e., short-term static, sustained static, repeated, cyclic, or dynamic)
- Pile diameter and cross-sectional shape
- Coefficient of friction between the pile and the soil
- Depth below the ground surface
- Pile construction methods
- Group interaction effects

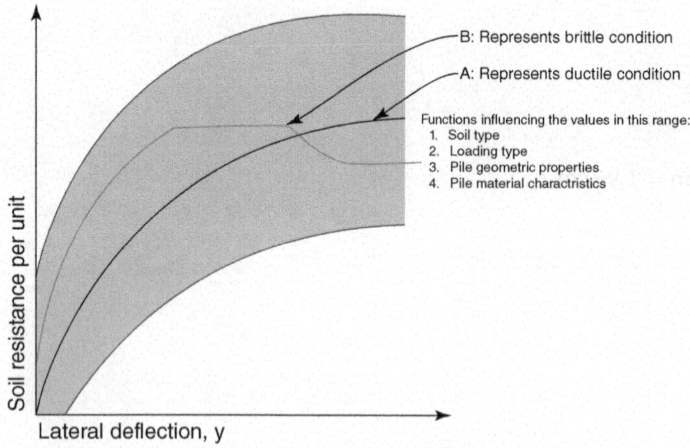

Figure 17.22 Typical p–y curves (Coduto et al. 2016/with permission of Pearson Education).

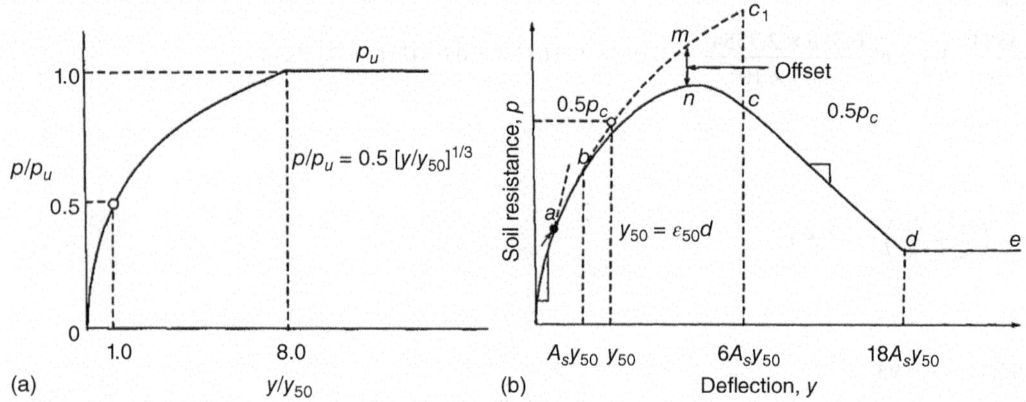

Figure 17.23 Characteristic shapes of the p–y curves for: (a) soft clay below the water table and (b) stiff clay below the water table (after Reese et al. 1974).

The individual influences of these various factors are not well established, so it has been necessary to develop p–y curves empirically by back-calculating them from static load tests and combining this data with analytical models based on soil mechanics to develop a comprehensive understanding of the soil–structure interaction. There are different types of p–y curves with distinct behaviors. Ductile curves, represented by Curve A in Figure 17.22, exhibit an increase in resistance up to a maximum value, p_u, at a certain deflection, and maintain this resistance even at larger deflections. On the other hand, brittle or strain-softening curves, shown as Curve B in Figure 17.22, initially reach a maximum value of resistance p_u but then decrease with further deflection. Soft clays under static loading and sands typically exhibit ductile curves. However, some clays, particularly stiff ones or when subjected to repeated or dynamic loading, can exhibit brittle curves. Brittle curves are concerning as they have the potential to cause significant foundation movements, posing challenges to structural stability.

17.4.1 *p–y* Curves for Different Soil Types (Reese et al. 1970–1975)

Reese and his colleagues have developed methods for constructing p–y curves for both clay and sand, which are valuable in predicting the nonlinear behavior of laterally loaded piles. These curves can be utilized with the difference equation method or the nondimensional solutions. According to their research, the p–y curves are independent of the shape and stiffness of the pile and represent the deformation of a specific vertical area of the soil, unaffected by the loading above and below it. Although this assumption is not strictly true, Reese et al. (1975) have confirmed through their experimental investigations that the soil reaction at a point primarily depends on the pile deflection at that specific point, rather than the pile deflections above and below it.

The following types of clays are considered when constructing p–y curves.

17.4.1.1 p–y Curves for Clays

17.4.1.1.1 p–y Curves for Soft Clays Below Water Table This procedure was developed from the tests on fully instrumented flexible pipe piles subjected to short-term static and cyclic loading. Correlations were made with the results of field and laboratory tests on "Undisturbed" soil samples obtained from the test sites. The ultimate soil resistance, P_u can be obtained by using the equation:

$$P_u = N_p c_x d \tag{17.27}$$

Where N_p is normalized ultimate soil resistance, c_x is undrained shear strength at depth x, and d is pile width or diameter. The value of N_p is a function of depth below the ground surface. The value of N_p increases with depth until it reaches some limiting value, which remains constant at greater depths. For shallow depths, Equation 17.27 may be written as:

$$P_u = \left(3 + \frac{\acute{\gamma} x}{c_x} + \frac{Jx}{d}\right) c_x d \tag{17.28}$$

and of deeper depths:

$$P_u = 9 c_x d \tag{17.29}$$

Where P_u is the ultimate soil resistance per unit length of pile, $\acute{\gamma}$ is the average effective unit weight of soil from the ground surface to the p–y curve, x is the depth from the ground surface to the p–y curve, and J is 0.50 for all practical purposes. To obtain the shape of the p–y curve, a mathematical expression that fitted with the experimental p–y curves was selected. Matlock selected the following equation:

$$\frac{P}{P_u} = 0.5\left(\frac{y}{y_{50}}\right)^{0.33} \tag{17.30}$$

Where, $y_{50} = 2.5b\varepsilon_{50}$ and ε_{50} is the strain at 50% of the maximum stress difference, determined from a UU triaxial compression test.

Figure 17.23a shows a characteristic shape of the p–y curve for soft clay below the water table.

17.4.1.1.2 p–y Curves for Stiff Clays Below Water Table For stiff clays, Reese et al. considered that the clay close to the ground surface fails in a wedge shape, and the soil moves up. At a depth greater than a critical depth, the soil moves around the pie. In both cases, the movement of the pile should be able to cause soil failure. The two theoretical expressions they developed are:

$$P_{c1} = 2c_a d + \acute{\gamma} dx + 2.83 c_a x \tag{17.31}$$

$$P_{c2} = 11cd \tag{17.32}$$

Where P_{c1} is ultimate soil resistance near the surface, P_{c2} is ultimate soil resistance well below the ground surface, c_a is average shear strength over the depth x, $\acute{\gamma}$ is submerged unit weight, and d is pile diameter or width.

Figure 17.23b shows a characteristic shape of the p–y curve for a stiff clay below the water table.

17.4.1.1.3 p–y Curves for Stiff Clays above Water Table Reese and Welch (1975) proposed criteria for predicting the behavior of flexible piles in stiff clays above the groundwater table. The following procedure is adopted for constructing *p–y* curves.

1) Obtain the value of c, γ, and d as before. Obtain the values of ε_{50} from stress–strain curves.
2) Compute P_u from Equations 17.31 and 17.32. The smaller of the two is used.
3) Compute the deflection y_{50}.
4) Points describing the *p–y* curves may be computed from the relationship $\frac{P}{P_u} = 0.5\left(\frac{y}{y_{50}}\right)^{0.25}$.
5) Beyond $y = 16y_{50}$, $P = P_u$ for values y.

Figure 17.24a shows the *p–y* curve for static loading in stiff clays above the groundwater table.

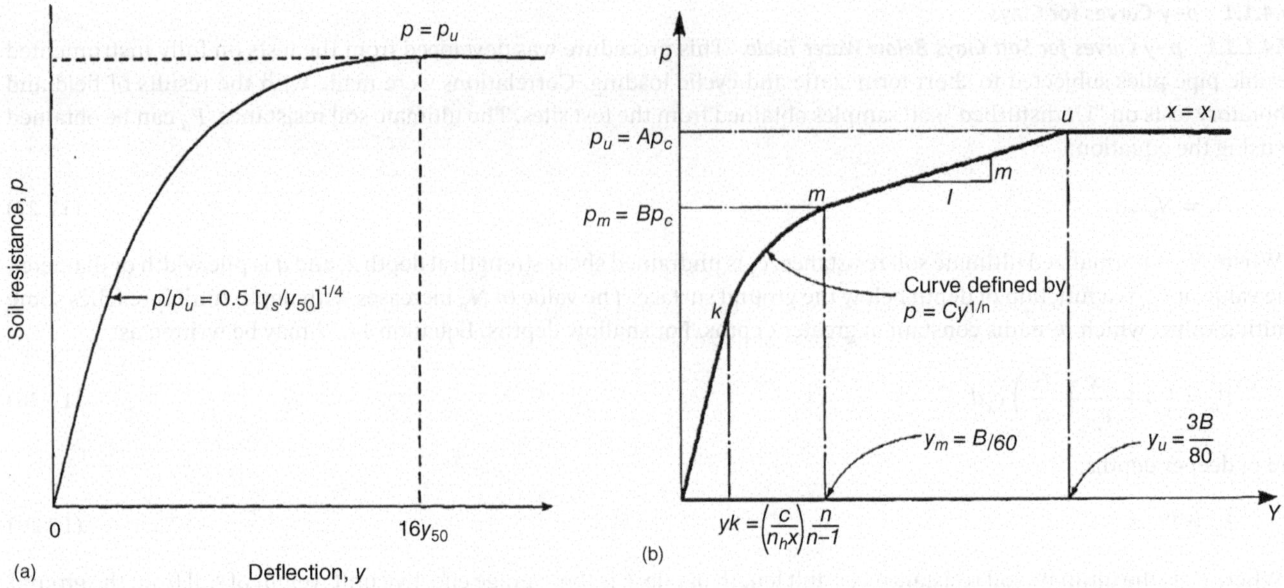

Figure 17.24 Characteristic shapes of the p–y curves for: (a) stiff clay above the water table and (b) sand (after Reese et al. 1974).

17.4.1.2 *p–y* Curves for Sand

As established by Reese et al., the shape of a p–y curve for a pile in the sand is shown in Figure 17.24b. It is in the form of a three-part curve up to the stage of the ultimate failure p_u.

The procedure for obtaining the shape of the curve and the trial and adjustment process using various assumed values of the subgrade modulus variation coefficient n_h to obtain the stiffness factor T are more complex than those described above for piles in normally consolidated clays. The values of n_h are obtained by plotting the soil modulus E_u against the depth, but the trial line is vertical, passing through the plotted points, again with weight given to depths of 0.5R or less.

Example 17.5 A 400 mm diameter pile is embedded into a saturated medium clay that has an average unit weight of 15.0 kN/m³ and an undrained shear strength of 30 kPa. The groundwater table is near the ground surface. Using the Matlock method, develop the static p–y curve at a depth of 5 m ($\varepsilon_{50} = 0.01$)

None of the available methods specifically address medium clays, but the Matlock method (for soft clay) is the closest match and thus is an appropriate choice.

$$\acute{\gamma} = 15.0 - 9.8 = 5.2\ \text{kN/m}^3$$

$$y_{50} = 2.5b\varepsilon_{50} = 2.5(400)(0.01) = 10\ \text{mm}$$

$$P_u = \left(3 + \frac{\acute{\gamma}x}{c_x} + \frac{Jx}{d}\right)c_x d = \left(3 + \left(\frac{5.2\ \text{kN/m}^3}{30\ \text{kPa}}\right)(5\ \text{m}) + \left(\frac{0.25}{0.4\ \text{m}}\right)(5\ \text{m})\right)(30\ \text{kPa})(0.4\ \text{m}) = 84\ \text{kN/m}$$

$$P_u = 9c_x d$$

$$P_u = 9(30\ \text{kPa})(0.4\ \text{m}) = 108\frac{\text{kN}}{\text{m}}$$

Use 84 kN/m

$$y_{50} = 2.5b\varepsilon_{50} = (2.5)(0.01)(400) = 10\ \text{mm}$$

$$\frac{P}{P_u} = 0.5\left(\frac{y}{y_{50}}\right)^{0.33} \leq 1$$

$$P = 0.5\left(\frac{y}{10\ \text{mm}}\right)^{0.33}(84\ \text{kN/m}) \leq 84\ \text{kN/m}$$

The resulting curve is shown in Figure 17.25.

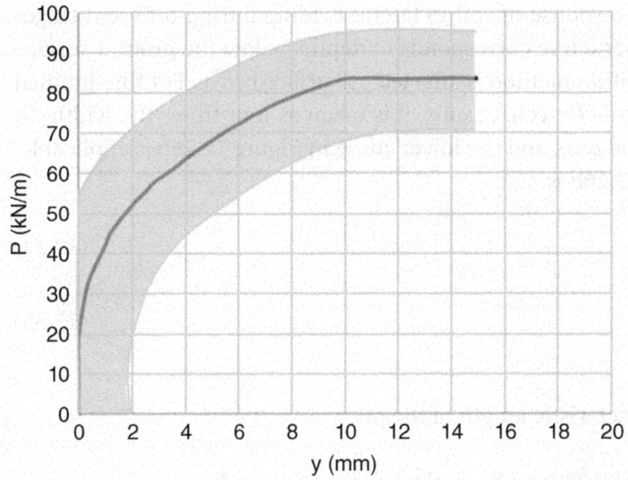

Figure 17.25 Computed p–y curve for example 17.5 (Coduto 2016/with permission of Pearson Education).

17.4.2 *p–y* Curves Directly from In situ Tests

17.4.2.1 Application of the Pressuremeter Test to Establish *p–y* Curves

The pressuremeter test conducted in a borehole is highly suitable for determining p–y curves for laterally loaded piles. In the initial phase of the test, there is a linear correlation between the applied pressure and the corresponding volume change, which represents the radial expansion of the borehole walls. At the creeping pressure, p_f, the pressure–volume relationship transitions into a nonlinear behavior, indicating the plastic yielding of the soil. Finally, at the limit pressure, p_l, the volume increases rapidly without a significant increase in pressure, resulting in the horizontal portion of the p–y curve. Figure 17.26a shows a typical pressure/volume change curve. The equation is given by Baguelin (1978):

$$E_m = 2.66 V_m \frac{\Delta_p}{\Delta_v} \tag{17.33}$$

Where:

$\Delta_p = p_f - p_{0m}$
$\Delta_v = V_f - V_{0m}$
$\frac{\Delta_p}{\Delta_v}$ = slope of the curve between V_0 and V_f
V_m = midpoint volume

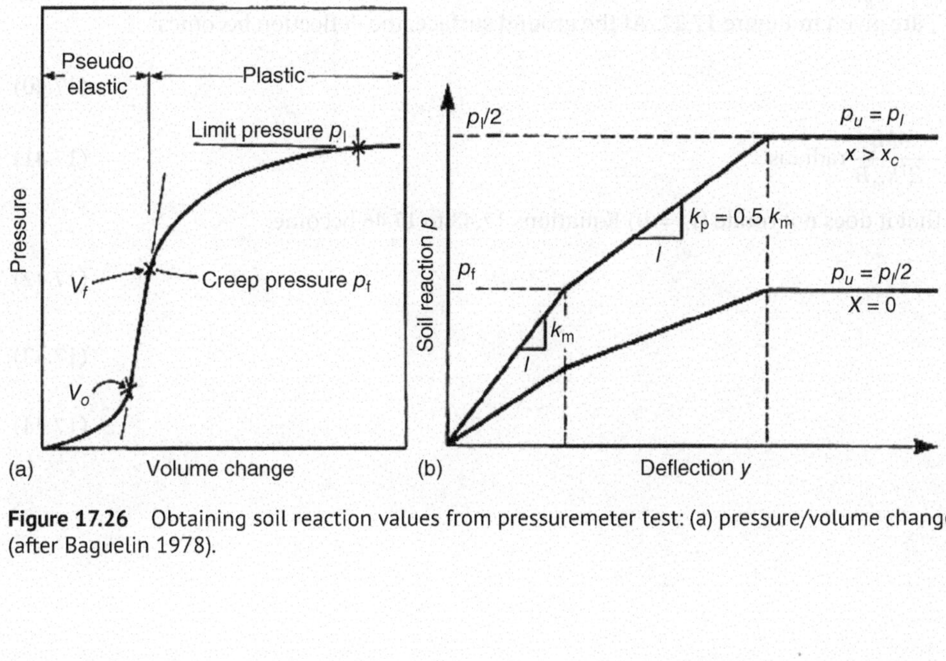

Figure 17.26 Obtaining soil reaction values from pressuremeter test: (a) pressure/volume change curve, (b) design reaction curve (after Baguelin 1978).

Baguelin (1978) provide two sets of curves that illustrate the response of soil to lateral loading during different stages of the pressuremeter test, as depicted in Figure 17.26b. The upper curve corresponds to depths below the ground surface equal to or greater than the critical depth, x_c, at which the calculation method is affected by surface heave. For fine-grained soils, x_c is typically considered as twice the width of the pile, while for coarse soils, it is taken as four times the width. In the presence of a pile cap, where surface heave is not present, x_c is zero, and the lower curve in Figure 17.26b is applicable. The value of the coefficient of subgrade reaction, k_m, in Figure 17.26b is:

$$\frac{1}{k_m} = \frac{2}{9E_m} B_0 \left(\frac{B}{B_0} \cdot 2.65 \right)^{\alpha} + \frac{\alpha B}{6E_m} \quad \text{for B > 600 mm} \tag{17.34}$$

$$\frac{1}{k_m} = \frac{B}{E_m} \left(\frac{4(2.65)^{\alpha} + 3\alpha}{18} \right) \quad \text{for B < 600 mm} \tag{17.35}$$

Where

E_m = mean value of the pressuremeter modulus over the characteristic length of the pile.
B_0 = pile width
α = rheological factor varying from 1.0 to 0.5 for clays, 0.67 and 0.33 for silts, and 0.5 to 0.33 for sand
Between the ground surface and the critical depth, X_c, the value of k_m, should be reduced by the coefficient λ_z, given by:

$$\lambda_z = \frac{1 + \frac{X}{X_c}}{2} \tag{17.36}$$

In a homogeneous soil, a simplified approach involves assuming no lateral soil reaction between the ground surface and a depth equivalent to 0.5–0.75 times the pile width (B), and then utilizing the full reaction provided by the upper curve in Figure 17.26b. Baguelin (1978) provide the following equations for calculating deflections, bending moments, and shears at any depth (z) below the ground surface, under the condition of a constant value of the pressuremeter modulus with depth:

$$\text{Deflection } y(z) = \frac{2H}{Rk_m B} F_1 + \frac{2M_t}{R^2 k_m B} F_4 \tag{17.37}$$

$$\text{Moment } M(z) = \frac{H.R.F_3}{R} + M_t F_2 \tag{17.38}$$

$$\text{Shear } T_z = HF_4 - \frac{2M_t}{R} F_3 \tag{17.39}$$

Where

R = stiffness coefficient (Baguelin refers to this as the transfer length, l_0)
H = horizontal load applied to the pile head
M_t = bending moment at the pile head
z = dimensionless coefficient equal to $\frac{X}{R}$
Values of the coefficients F_1 to F_4 are given in Figure 17.27. At the ground surface, the deflection becomes:

$$y_0 = \frac{2H}{Rk_m B} + \frac{2M_t}{R^2 k_m B} \tag{17.40}$$

$$\text{and slope} = \acute{y}_0 = -\frac{2H}{R^2 k_m B} - \frac{4M_t}{R^1 k_m B} \text{ radians} \tag{17.41}$$

If the head of the pile is fixed so that it does not rotate ($y_0 = 0$) Equations 17.43 to 17.46 become:

$$y(z) = \frac{H}{Rk_m B} F_2 \tag{17.42}$$

$$M(z) = -\frac{HR}{2} F_4 \tag{17.43}$$

$$T(z) = H \cdot F_3 \tag{17.44}$$

Figure 17.27 Values of the coefficients $F_1 - F_4$ (after Baguelin 1978).

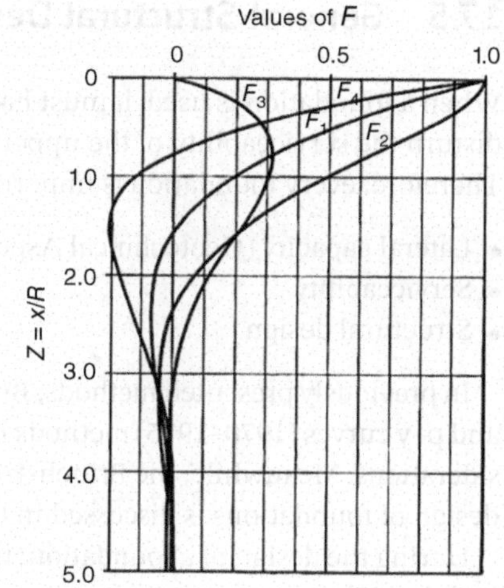

$$y_0 = \frac{H}{Rk_m B} \tag{17.45}$$

$$M_t = -\frac{HR}{2} \tag{17.46}$$

To draw the pile load/deflection curve, the deflections corresponding to soil reactions equal to the creeping pressure, p_f, and the limit pressure, p_l, are calculated from the relationship $p = k_m y$. The lateral pile loads follow Equations 17.37, 17.40, 17.42, or 17.45. For soil reactions between the limit pressure and creeping pressure, the k_m value is halved, as shown in Figure 17.26.

17.4.2.2 Application of the Cone Penetration Test (CPT) to Establish *p–y* Curves

Truong and Lehane (2014) proposed numerically derived CPT-based formulations for p–y curves in soft clays, while Suryasentana and Lehane (2014b) focused on sands. In their study on soft clay (Truong and Lehane 2014), they performed a numerical analysis using Plaxis 3D, a 3D finite element analysis (FEA) program. They considered six models of a pile with a length-to-diameter ratio of 11.8, and a lateral load was applied to the pile at a distance of 1.5 times the pile diameter above the ground surface. The regression analyses of all cases resulted in the following two equations:

$$\frac{P}{P_u} = \tanh\left[(0.26I_r + 3.98)\left(\frac{y}{D}\right)^{0.85}\left(\frac{Z}{D}\right)^{-0.5}\right] \quad \text{for: } 0 < \frac{Z}{D} < 3 \tag{17.47}$$

$$\frac{P}{P_u} = \tanh\left[(0.15I_r + 2.3)\left(\frac{y}{D}\right)^{0.85}\right] \quad \text{for: } \frac{Z}{D} \geq 3 \tag{17.48}$$

According to their work, the p–y curves can be directly derived from CPT, q_{net} profiles, but an additional assumption will be made to determine the value of I_r, which is defined in the following equation:

$$I_r = \frac{1}{3}\left(\frac{E}{q_{net}}\right)N_{kt} = \frac{1}{3\varepsilon_c} \tag{17.49}$$

As a part of their work for sands, Suryasentana and Lehane (2014b) proposed the following equation:

$$\frac{P}{\gamma z D} = 2.4\left(\frac{q_c}{\gamma z}\right)^{0.67}\left(\frac{z}{D}\right)^{0.75}\left(1 - \exp\left(\frac{z}{D}\right)^{-1.2}\left(\frac{y}{D}\right)^{0.89}\right) \tag{17.50}$$

Initial regression analyses of the collated p and y data yielded a format comparable to that proposed by Novello (1999), which was simply estimated from back-analyses of a small number of lateral pile tests. The best-fitted obtained equation was:

$$\frac{P}{\gamma z D} = 4.2\left(\frac{q_c}{\gamma z}\right)^{0.68}\left(\frac{y}{D}\right)^{0.56} = 4.2(q_c)^{0.68}(z)^{0.32}(y)^{0.56}(\gamma)^{0.32}(D)^{0.44} \tag{17.51}$$

17.5 General Structural Design Aspects

When a foundation is used, it must have a controlled amount of displacement in addition to bearing the load that does not disturb the serviceability of the upper structure. On the other hand, the foundation must be durable enough and not fail. Therefore, every foundation is important from three aspects during design:

- Lateral capacity (Geotechnical Aspect)
- Serviceability
- Structural design

In previously presented methods, Broms (1964), Evan-Duncan (1982), Elastic Continuum (1981), nondimensional (1956), and p–y curves (1970–1975) methods have considered two aspects. These aspects are lateral capacity and serviceability considerations. Meanwhile, the Brinch-Hansen method only obtains the lateral capacity. Therefore, the aspect of the structural design of foundations is discussed in this section.

During the design of a foundation, three important aspects are taken into consideration to ensure its proper performance and durability:

1) **Geotechnical Aspect:** Lateral Capacity: In addition to resisting vertical (axial) loads, the foundation must have sufficient lateral capacity to resist applied lateral loads, such as those caused by wind, earthquakes, or other lateral forces. Various methods, such as Broms (1964), Evans-Duncan (1982), Elastic Continuum (1981), and p–y curves (1970–1975), address the lateral capacity of the foundation.
2) **Serviceability:** The foundation must not cause excessive displacements or settlements that could compromise the serviceability of the upper structure. Excessive movements may lead to cracking or other forms of damage to the structure. The lateral deflection and settlement also need to be controlled within acceptable limits to ensure serviceability.
3) **Structural Design:** The foundation design should also consider the structural aspect of the foundation itself. It should be able to withstand the applied loads and remain stable without failing. The foundation's structural design ensures that it has sufficient strength and stiffness to bear the loads without experiencing excessive deformations or failures.

By addressing all three aspects in the foundation design process, engineers can create a well-designed foundation that meets both geotechnical and structural requirements while ensuring the serviceability of the overall structure.

17.5.1 Importance of Pile Structural Aspects

Correctly designing piles to withstand stresses is crucial for their successful installation and long-term performance as supporting elements in a foundation structure. When piles are driven as preformed elements, stresses can occur during their installation process. These stresses arise as the piles are lifted from their fabrication bed and pitched in the piling rig. As they are driven into the ground, piles experience compressive and sometimes tensile stresses. Additionally, if they deviate from their intended alignment, they may suffer bending stresses.

Regardless of the type of piles, they can also be subjected to bending stresses due to lateral and eccentric loading. Eccentric loading may be designed deliberately or may occur when the pile heads deviate from their intended positions. Bending moments near the pile heads can result from a differential settlement between adjacent piles or pile groups, especially if there are distortions in the pile caps or connecting beams.

In pile design, it is essential to account for various factors that can affect the working stresses. These factors include unseen breakage that may happen during pile driving, potential imperfections in concrete cast in situ, and the long-term effects of corrosion or biological decay. Considering these factors ensures that the working stresses adopted for the piles are appropriate and lead to a safe and reliable foundation system.

In addition to guidance on structural design and detailing, matters of relevance to the design of piled foundations in EC2 (BS EN 1992-1-1: 2004) include the following:

1) Dimensional tolerances of cast-in-place piles
2) Partial factors for the ultimate limit state of the material
3) The influence of soil–structure interaction caused by differential settlement

4) Strength classes of concrete and reinforcement cover for various exposure conditions
5) Slenderness and effective lengths of isolated members
6) Punching shear and reinforcement in pile caps
7) Limits for crack widths and
8) Minimum reinforcement for bored piles

17.5.2 Structural Problems and Protection

Damage to piles during driving can be visible near the pile head, but it is important to recognize that the shaft and toe of the pile may also suffer damage. The damage to a pile section or casing during driving can take various forms, including buckling, crumbling, twisting, distortion, and longitudinal cracking of steel, as well as shattering, shearing, cracking, and spalling of concrete.

Several factors can contribute to pile damage during driving, such as overdriving caused by an unsuitable combination of hammer weight, drop, or misalignment of the pile. When the hammer delivers blows to the pile head, it can lead to eccentric stresses. To minimize the risk of damage, it is crucial to ensure that the hammer blows are directed along the axis of the pile. However, the pile head should be allowed to move slightly inside the driving helmet to prevent the transmission of excessive torsion or bending forces.

To avoid pile damage during driving, proper driving techniques and equipment must be employed. Engineers and construction crews must carefully consider the hammer characteristics, pile material, alignment, and driving process to prevent overdriving or misalignment, and to ensure the piles are installed safely and effectively. Regular inspections during the driving process can help identify any potential damage, allowing for timely corrective actions to be taken. Structural defects can also occur due to liquefaction, as depicted in Figure 17.28.

Excessive compressive stress is a common cause of failure at the pile head, and it is essential to carefully consider the ground conditions and driving process to prevent such failures. Tensile stresses can also arise during driving, particularly when there is low ground resistance or when hammer rebound occurs due to hard packing and a light hammer. In such situations, the reflection of compressive waves at a free end can lead to tensile stresses. Moreover, the transition from a dense stratum to weaker materials during driving can cause damage to the pile. Driving the pile too quickly through this transition can induce tensile stresses, especially if the weaker soil cannot provide adequate support. When the head of a steel pile sustains severe damage, cutting it back and welding an extension may be necessary to restore its integrity and performance. This repair process allows the damaged section to be removed and replaced with a new extension, ensuring the pile's structural capacity is maintained. Consequently, it is crucial to control the driving process to avoid excessive stress and potential damage. Proper monitoring and supervision during pile driving, along with adherence to appropriate safety measures and engineering guidelines, are essential to avoid failure and ensure the successful installation of piles as reliable supporting members in foundation structures. Figure 17.29 presents the effect of liquefaction on pile performance and relevant defects. Table 17.6 encapsulates possible defects in displacement piles due to driving.

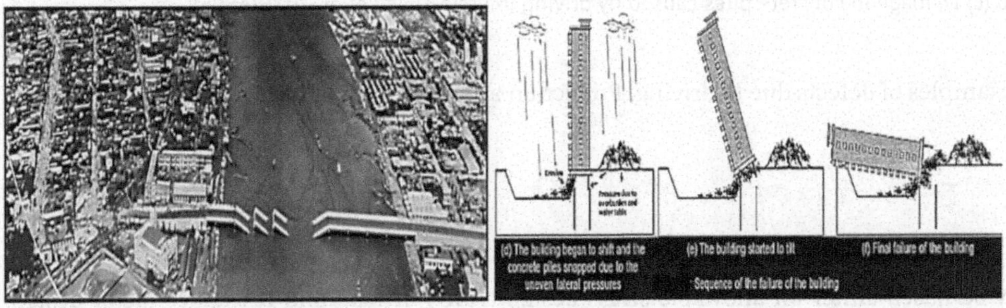

(a) (b)

Figure 17.28 (a) Showa Bridge collapse during 1964 Niigata earthquake (Bhattacharya et al. 2014/with permission of Elsevier), (b) a rare foundation failure of a flat building in Shanghai, China (https://www.academia.edu/39172880/A_Rare_Foundation_Failure_of_a_Flat_Building_in_Shanghai_risk_engineering_assignment).

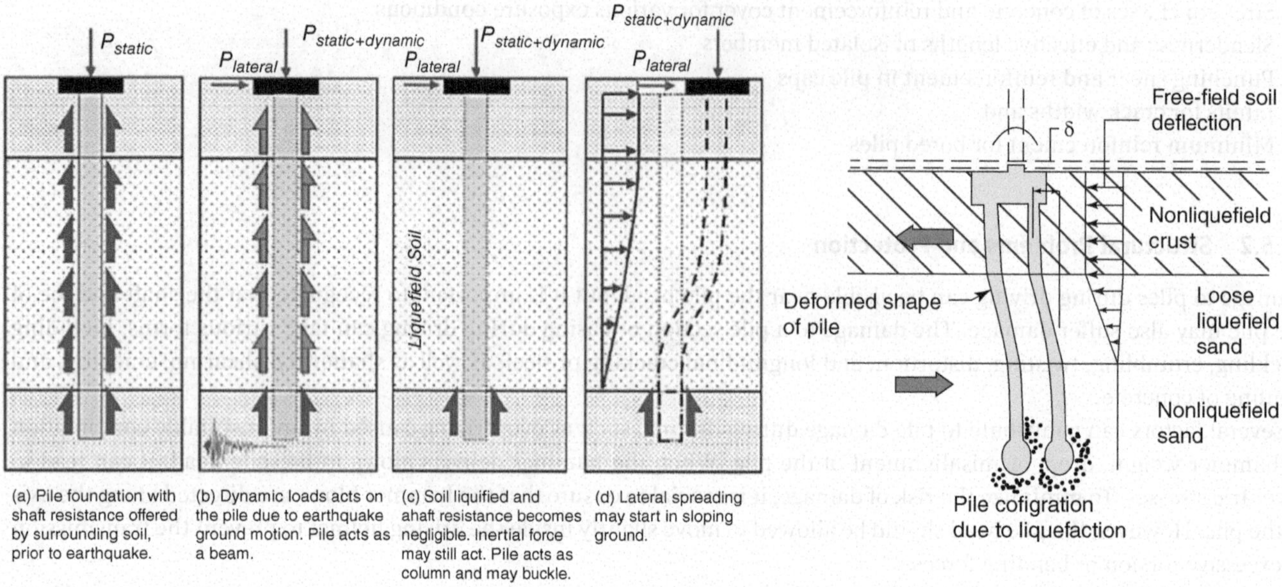

(a) (b)

Figure 17.29 (a) Stages of loading for pile foundation in liquefiable soil (adopted from Bhattacharya, 2003), (b) pile configuration due to liquefaction (Finn and Thavaraj 2001).

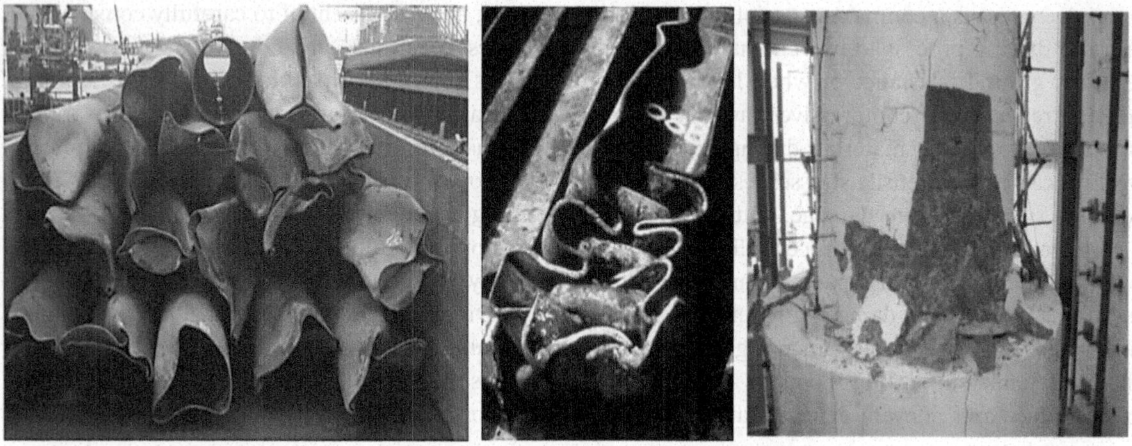

(a) (b) (c)

Figure 17.30 (a) Tip buckling of pipe piles from the port basin in Rotterdam, (b) tip buckling of H piles (Randolph 2018/The University of Western Australia), and (c) Damage in concrete piles caused by driving (Murcia-Delso et al. 2016/ASCE).

Figure 17.30 illustrates some examples of defects due to driving. Protection approach and damages relevant to corrosion is depicted in Figure 17.31.

17.5.3 Design Criteria

The pile's structural design may be based either on allowable stress design (ASD) or load and resistance factor design (LRFD), as permitted by the applicable code. The International Building Code (ICC 2012) defines allowable stresses in piles in terms of ASD, even when the superstructure is designed using LRFD (although some LRFD provisions may apply). The American Association of State Highway and Transportation Officials (AASHTO) bridge code (AASHTO 2012) uses LRFD.

Table 17.6 Possible defects in displacement piles caused by driving (CEDD and CEDB 2006).

Pile type	Problems	Possible causes
Steel piles	Damaged pile top (head) (e.g., buckling, longitudinal cracking, distortion)	(a) Unsuitable hammer weight (b) Incorrect use of dollies, helmets, and packing (c) Rough cutting of pile ends (d) Overdriving
	Damaged pile shaft (e.g., twisting, crumpling, bending)	(a) Unsuitable hammer weight (b) Inadequate directional control of driving (c) Overdriving (d) Obstructions
	Collapse of tubular piles	Insufficient thickness
	Damaged pile toe (e.g., buckling, crumpling)	(a) Overdriving (b) Obstructions (c) Difficulty in-toeing into rock
	Base plate rising relative to the casing, loss of plugs or shoes in cased piles	(a) Poor welding (b) Overdriving (c) Incorrect use of concrete plugs
Concrete piles	Damaged pile head (e.g., shattering, cracking, spalling of concrete)	(a) Unsuitable reinforcement details (b) Insufficient reinforcement (c) Poor quality concrete (d) Excessive concrete cover (e) Unsuitable hammer weight (f) Incorrect use of dollies, helmets, packing (g) Overdriving
	Damaged pile shaft (e.g., fracture, cracking, spalling of concrete)	(a) Excessive restraint on piles during driving (b) Unsuitable hammer weight (c) Poor quality concrete (d) Excessive or incorrect concrete cover (e) Obstructions (f) Overdriving (g) Incorrect distribution of driving stresses from the use of incorrect dollies, helmets, or packing
	Damaged pile toe (e.g., collapsing, cracking, spalling of concrete)	(a) Overdriving (b) Poor-quality concrete (c) Insufficient reinforcement (d) Inadequate or incorrect concrete cover (e) Obstructions (f) Absence of rock shoe where required

17.5.3.1 ASD

For steel and timber piles subject only to axial compressive loading, the allowable compressive load is:

$$P_a = f_a A \tag{17.52}$$

Moreover, the design must stratify the following criterion:

$$P \leq P_a \tag{17.53}$$

When flexural stresses are also present, the design must satisfy an interaction criterion, such as:

$$\frac{P}{f_a A} \pm \frac{M}{f_b S} \leq 1 \tag{17.54}$$

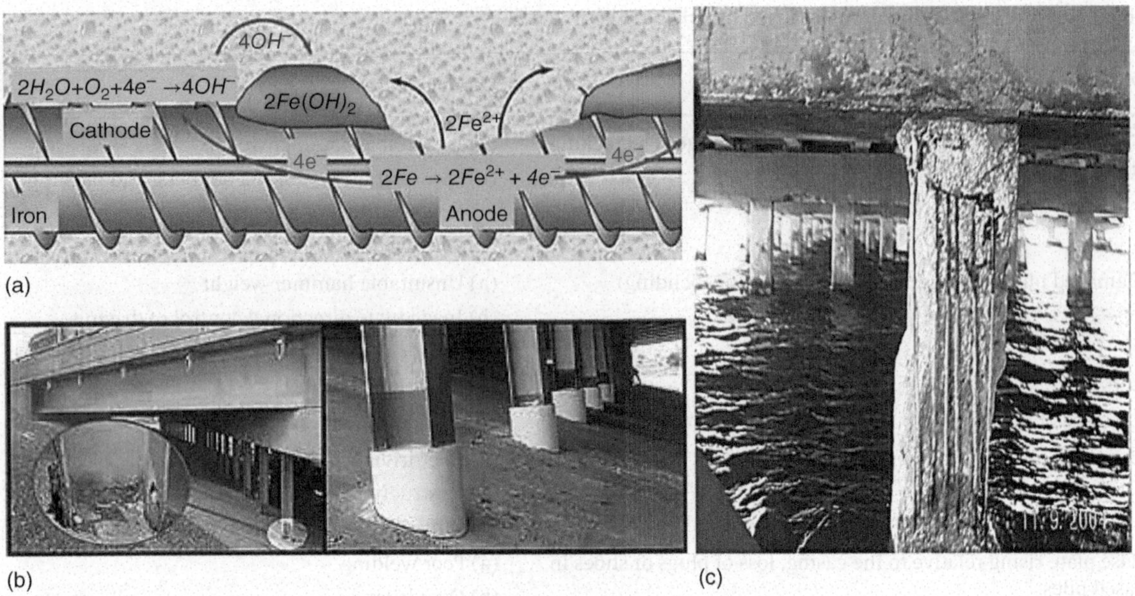

Figure 17.31 (a) Anode and cathode in corrosion microcell on reinforcement (PCA 2007), (b) corrosion and protection on steel piles, and (c) damage to the concrete pile (taken from Sen and Mullins, 2007).

Where:

P = Axial compressive stress on the pile from axial load
M = External moment
P_a = Allowable compressive stress of the pile
f_a = Allowable compressive stress of the pile
f_b = Allowable flexural stress of the pile
A = Cross-sectional area of the pile
S = Elastic section modulus

Note that Equation 17.54 demonstrates how the axial load capacity of a pile is reduced when flexural stresses are added. Specific codes often have more detailed interaction criteria intended for specific structural materials. For piles subject to axial tension loading with or without flexural loading, substitute P_{up} into Equations 17.52–17.54 and use the tensile values of f_a and f_b.

Because this is an ASD analysis, the values of P and M should be based on the unfactored loads computed using similar equations presented in the governing code. The value of P or P_{up} becomes smaller with depth as some axial load shifts to the soil due to side friction. However, for simplicity, structural engineers usually neglect any load transfer due to side friction and use the P or P_{up} applied by the structure at the top of the pile. The value of M can increase or decrease with depth and can be obtained by developing a moment diagram or determining the maximum moment.

For analysis purposes, neglect any interaction between the shear and the axial or moment loads. For an ASD analysis, the shear force, V, must not exceed the allowable shear force, V_a:

$$V_a = f_v A \tag{17.55}$$

$$V < V_a \tag{17.56}$$

Where,

f_v = allowable shear stress

The shear force, V, is greatest at the top of the pile and varies with depth as determined by a shear diagram.

17.5.3.2 LRFD

When using LRFD, the structural ULS requirement for piles subject only to axial compression or axial tension is:

$$P_u \leq \phi P_n \tag{17.57}$$

Where:

P_u = factored demand
ϕ = resistance factor
P_n = nominal load capacity

For shear V and moment M:

$$V_u \leq \phi V_n \tag{17.58}$$

$$M_u \leq \phi M_n \tag{17.59}$$

Methods for determining P_n, M_n, and V_n, and the interaction between axial and flexural loads depend on the structural material and the governing code. As with ASD, Equations 17.57 to 17.58 must be satisfied across the entire pile length.

17.6 Structural Considerations of Piles and Drilled Shafts

17.6.1 Drilled Shaft and Auger Piles

Drilled shafts are typically constructed with concrete that has a 28-day compressive strength $\acute{f}_c$ of 24 to 35 MPa (3,500–5,000 lb/in^2). The maximum aggregate size for shafts constructed using the dry method may be no more than 1/5 of the clearance between the rebars. However, in order to provide for proper concrete flow during placement when using tremie concrete, Brown (2010) recommend a maximum aggregate size of no more than 1/10 of the clearance between the rebars.

Auger piles are normally constructed with grout having a design 28-day compressive strength $\acute{f}_c$ of 24 to 35 MPa (3,500–5,000 lb/in^2). Various additives are generally used to obtain the needed fluidity to facilitate steel placement. Reinforcing steel with f_y = 420 MPa (60 k/in^2) is most commonly specified for both drilled shafts and auger piles.

ACI specifies a minimum concrete cover of 75 mm (3 in) between reinforcement and soil. AASHTO specifies a minimum of 50 to 75 mm (2–3 in) depending on the circumstances. However, Brown (2010) recommend the following larger values for large-diameter drilled shafts:

- 75 mm (3.0 in) for shaft diameter $\leq$ 900 mm (3 ft)
- 100 mm (4.0 in) for 900 mm (3 ft) $<$ shaft diameter $<$ 1,500 mm (5 ft)
- 150 mm (6.0 in) for shaft diameter $\geq$ 1,500 mm (5.0 ft)

The transverse reinforcement is typically placed outside the longitudinal reinforcement, and the required cover for the transverse steel is 12 mm (0.5 in) less than the above values.

17.6.2 Prestressed Concrete Piles

Prestressed concrete piles are subject to a number of code requirements, which include various detailing and seismic design provisions. They are also subject to the general code requirements for prestressed concrete. Typically, the effective prestress f_{pc} is about 4.8 MPa (700 lb/in^2), but values as high as 8.3 MPa (1,200 lb/in^2) have been used. The concrete strength $\acute{f}_c$ is usually between 35 and 70 MPa (5,000–10,000 lb/in^2). Figure 17.32 shows different prestressed concrete piles.

Filling empty steel pipe piles with concrete is a common practice to enhance their structural capacity and provide additional benefits. By infilling the pile with concrete after driving, the compressive and flexural design load capacities of the pile increase significantly. The concrete core adds to the section's moment of inertia, making it more resistant to bending and compression forces. Moreover, the concrete infilling also offers corrosion protection to the interior of the steel pipe, extending the service life of the pile and reducing maintenance requirements. This is particularly advantageous when the pile is exposed to harsh environments or corrosive substances.

However, it is essential to note that concrete infilling generally does not improve the geotechnical downward load capacity of the pile. Even open-end pipe piles tend to become fully plugged with soil particles and debris during driving or over time, limiting their ability to transfer additional loads to the underlying soil through the open end. The load transfer mechanism in these cases primarily relies on the shaft friction and end-bearing capacity of the pile. On the other hand, the geotechnical uplift capacity of the pile may increase slightly due to the added weight of the concrete infilling. The increased self-weight contributes to the resistance against upward forces or uplift that the pile may experience.

Figure 17.33 shows schematic views of concrete-filled piles as well as the installation procedure and pile shoe.

(a) (b) (c)

Figure 17.32 (a) Pretensioned spun concrete piles (MKH building material), (b) prestressed concrete piles (PCM construction material), and (c) circular and rectangular cross sections for prestressed concrete (Belarbi et al. 2017).

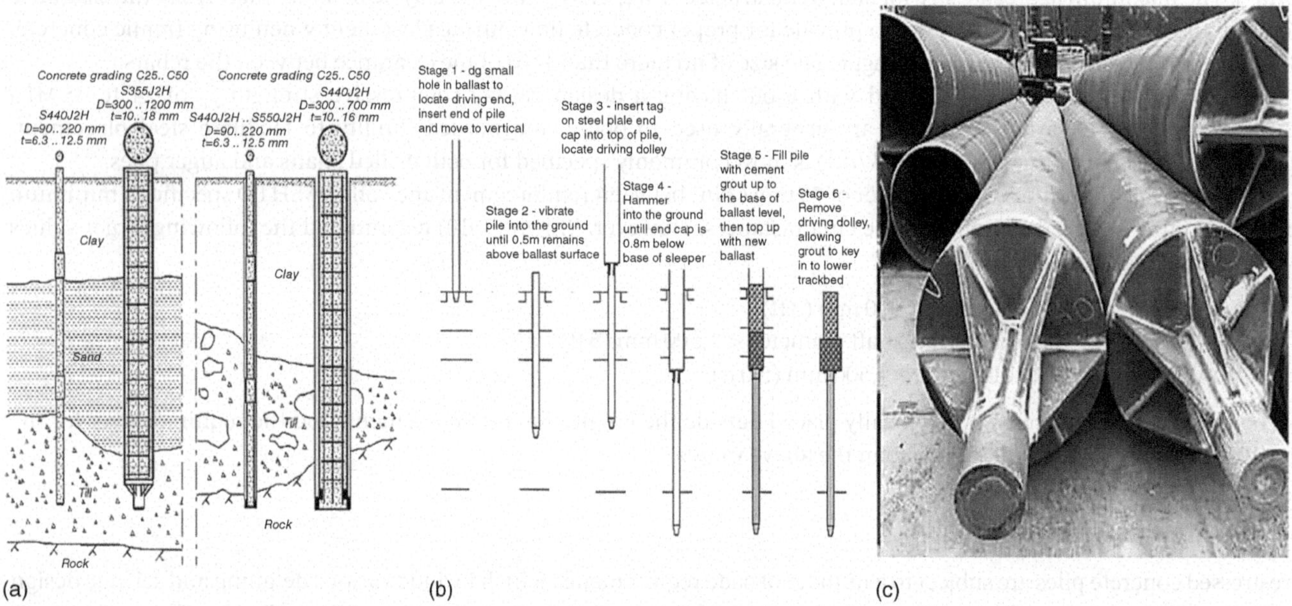

(a) (b) (c)

Figure 17.33 (a) Concrete-filled circular steel tubes (semantic scholar), (b) installation-procedure-for-concrete filled tubular-steel-piles (scientific diagram), and (c) steel pipe piles with pile shoe (SSAB).

17.6.3 Steel Piles

Steel pipe piles have a constant moment of inertia, I, as well as constant section modulus S, regardless of the direction of the lateral load. However, the I and S for H-piles depend on the direction of the load relative to the web of the H-pile. If the design engineer does not specify the as-built web orientation, then the I and S of the weak (Y–Y) axis of H-piles must be used. However, if the orientation is specified on the design drawings, the stiffer X–X axis properties must be used. Figure 17.34 depicts steel piles and their driving procedure.

For many years, ASTM A36 steel with a yield strength f_y of 250 MPa (36,000 lb/in^2) was commonly used for steel piles. However, most steel piles now are made of higher strength steel, such as ASTM A572 Grade 50 or ASTM A690, with a yield strength f_y of 345 MPa (50 k/in^2)). There are also pipe piles made of even higher strength steel with $f_y = 420$ MPa (60 k/in^2)) available. The selection of steel grade for steel piles is crucial in ensuring their structural capacity and performance.

The notation f_y is often used to signify the yield strength of structural steel and steel rebars. For consistency, we will use f_y throughout the remainder of this chapter.

(a) (b)

Figure 17.34 (a) Steel piles (the constructor), (b) steel driving piling (DFI piling).

When using allowable stress design (ASD) with the International Building Code (IBC) design requirements, the tensile and compressive stresses in steel piles under design loads are limited to allowable values. The allowable tensile stress, f_a, is limited to $0.35 f_y$, not to exceed 110 MPa (16 k/in^2). However, if the design is supported by a geotechnical investigation and a static load test, the allowable tensile stress can be increased to $0.50 f_y$, not to exceed 220 MPa (32 k/in^2). This increase is still conservative compared to the $0.60 f_y$ to $0.66 f_y$ permitted in the superstructure design, reflecting the cautious approach commonly taken in geotechnical and foundation engineering to ensure safety and reliability.

For LRFD, P_n, M_n, and V_n are the same as for members of the same material in the superstructure. However, the resistance factors are lower to reflect a more conservative design. The AASHTO resistance factors for steel piles are:

For axial resistance of piles in compression and subject to damage due to severe driving conditions where the use of a pile tip is necessary:

- H piles: $\phi = 0.50$
- Pipe piles: $\phi = 0.60$

For axial resistance of piles in compression under good driving conditions where the use of a pile tip is not necessary:

- H piles: $\phi = 0.60$
- Pipe piles: $\phi = 0.70$

For combined axial and flexural resistance of undamaged piles:

- The axial capacity of H piles: $\phi = 0.70$
- The axial capacity of pipe piles: $\phi = 0.80$
- Flexural capacity: $\phi = 1.00$

Example 17.6 A 12.75 in (32.3 cm), diameter steel pipe pile with a 0.500 in (1.27 cm), wall thickness is subject to the following ASD demands: $P = 200$ k (~90 tons), $V = 50$ k (22.7 tons), $M = 250$ in-k (2880 ton-m). Using ASD, the IBC allowable stresses, and $f_y = 50$ k/in^2 (35,154 ton/m^2). determine if this section is adequate. The driving conditions will be checked using a wave equation analysis, and the requirements of IBC 1810.3.2.8 will be satisfied ($S = 929$ cm^3)

$$f_a = f_b = 0.50 f_y = (0.50)(50{,}000) = 25{,}000\ \text{lb/in}^2 (17{,}577\ \text{ton/m}^2)$$

$$A = \frac{\pi}{4}(12.75^2 - 11.75^2) = 9.82\ \text{in}^2\ (59.87\text{cm}^2)$$

$$\frac{P}{F_z A} + \frac{M}{F_b S} = \frac{200{,}000}{(25{,}000)(9.82)} + \frac{250{,}000}{(25{,}000)(56.7)} = 0.99 < 1.00\ \text{OK}$$

$$f_v = 0.40 f_y = (0.40)(50{,}000) = 20{,}000\ \text{lb/in}^2\ (14{,}061.6\ \text{ton/m}^2)$$

$$V_a = \frac{f_v A}{2} = \frac{(20{,}000)(9.82)}{2} = 98{,}200\ \text{lb} = 44.54\ \text{tons} \geq 22.7\ \text{tons OK}$$

Therefore, the structural design is satisfactory.

17.7 Piles and Drilled Shafts Design Under Combine Loading (VMH)

17.7.1 Drilled Shafts and Auger Piles

- **Axial Loads**

For drilled shafts and auger piles subjected to axial compression with no shear or moment, the ASD allowable loads per IBC (2012) are:

$$P_a = 0.3f'_c A_c + 0.50 f_y A_s \tag{17.60}$$

$$P_{up,a} = 0.50 f_y A_s \tag{17.61}$$

The tensile and compressive stresses in the steel are limited to 165 MPa (24,000 lb/in²) and 220 MPa (32,000 lb/in²), respectively. When a permanent casing is used, the first coefficient in Equation 17.69 is increased to 0.33, and the area of the casing may be considered as part of the steel area. For LRFD using AASHTO:

$$P_n = \beta[0.85f'_c(A_g - A_s) + A_s f_y] \tag{17.62}$$

$$P_{up,a} = f_y A_s \tag{17.63}$$

Where
 P_a = allowable axial compressive load
 $P_{up,a}$ = allowable axial tension load
 β = 0.85 for spiral reinforcement or 0.80 for tie reinforcement in compression; 0.90 in tension
 f'_c = specified 28-day compressive strength of concrete
 f_y = specified yield strength of reinforcing steel
 A_g = gross cross-sectional area
 A_s = cross-sectional area of longitudinal steel

The AASHTO ϕ is 0.75, except in certain seismic loading where a value of 0.90 may be used.

- **Combined Axial and Flexural Loading**

Similar to prestressed concrete piles, drilled shafts, and auger piles are subjected to combined axial and flexural loading. As a result, their design and analysis involve considering various code requirements to ensure their structural capacity and performance.

To represent the combined axial and flexural loading capacity of drilled shafts and auger piles, engineers often use an interaction diagram. The interaction diagram is a graphical representation that shows the allowable combinations of axial load and bending moment that the pile can sustain while satisfying the design criteria and safety factors specified in the applicable building codes or design standards. The interaction diagram is essential in understanding the behavior of the pile under different loading conditions. It helps engineers identify the safe operating range for the pile's axial and bending loads, as well as the limitations on load combinations to prevent failure or excessive deformations. By considering the interaction between axial and flexural loads, the design of drilled shafts and auger piles can be optimized to ensure their structural integrity and efficiency. Overall, the use of interaction diagrams is a valuable tool in the design and analysis of drilled shafts and auger piles, enabling engineers to meet code requirements and ensure the safe and reliable performance of these deep foundation elements.

17.7.2 Prestressed Concrete Piles

- **Axial Loads**

For prestressed concrete piles subjected only to axial compression, IBC (2012), PCI (2004), and ACI (2012) specify an allowable structural load using ASD:

$$f_a = 0.33f'_c - 0.27f_{pc} \tag{17.64}$$

For prestressed concrete piles subjected only to axial tension, PCI (2004) recommends:
 Permanent and repetitive loads: $f_a = 0$

Transient loads: $\begin{cases} f_a = 0.5\sqrt{\acute{f}_c} & \text{SI} \\ f_a = 6\sqrt{\acute{f}_c} & \text{English} \end{cases}$

ACI (2012) recommends: $f_a = 0.5 f_y$

Where

f_a = allowable tensile or compressive stress due to normal load (MPa, lb/in^2)

$\acute{f}_c$ = 28-day compressive strength of concrete (MPa, lb/in^2)

f_{pc} = effective prestress on the gross cross-section (MPa, lb/in^2)

Careful detailing between the pile and the pile cap is needed to properly transfer tension loads.

- **Combined Axial and Flexural Loading**

PCI (2004) recommends the following ASD capacities for prestressed concrete piles subjected to combined axial and flexural loads:

Compression with all loads applied: $f_a + f_b = 0.60\acute{f}_c$

Compression with dead loads only: $f_a + f_b = 0.40\acute{f}_c - f_{pc}$

Compression with live load plus half of the dead load: $f_a + f_b = 0.40\acute{f}_c$

Tension in the normal environment: $\begin{cases} f_a + f_b = 0.5\sqrt{\acute{f}_c} & \text{SI} \\ f_a + f_b = 6\sqrt{\acute{f}_c} & \text{English} \end{cases}$

And half of these allowable tension values in a corrosive environment. Where:

f_a = allowable tensile or compressive stress due to flexural load

Table 17.7 presents octagonal and round prestressed concrete piles based on PCI (2010).

17.7.3 Concrete-Filled Steel Pipe Piles

When empty steel pipe piles do not offer sufficient structural capacity on their own, engineers often resort to filling them with concrete to enhance their load-bearing capabilities. This process involves placing concrete inside the steel pipe pile

Table 17.7 Standard Octagonal and Round Prestressed Concrete Piles (adapted from PCI 2010).

Section properties								Allowable concentric service load (tons) $\acute{f}_c$ (psi)			
Size (in)	Core Diameter (in)	Area A_g (in^2)	Weight (Lb/ft)	Moment of Inertia (in^4)	Elastic Section Modulus (in^3)	Radius of Gyration (in)	Perimeter (ft)	5,000	6,000	8,000	10,000
Square piles											
12	Solid	144	150	1,728	288	3.46	4.00	105	129	176	224
14	Solid	196	204	3,201	457	4.04	4.67	143	175	240	305
16	Solid	256	267	5,461	683	4.62	5.33	187	229	314	398
18	Solid	324	338	8,748	972	5.20	6.00	236	290	397	504
20	Solid	400	417	13,333	1,333	5.77	6.67	292	358	490	622
20	11	305	318	12,615	1,262	6.43	6.67	222	273	373	474
24	Solid	576	600	27,648	2,304	6.93	8.00	420	515	705	896
24	12	463	482	26,630	2,219	7.58	8.00	338	414	567	720
24	14	422	439	25,762	2,147	7.81	8.00	308	377	517	656
24	15	399	415	25,163	2,097	7.94	8.00	291	357	488	621
30	18	646	672	62,347	4,157	9.82	10.00	471	578	791	1,005
36	18	1,042	1,085	134,815	7,490	11.38	12.00	761	933	1,276	1,621

after it has been driven into the ground. The addition of concrete significantly improves the compressive and flexural design load capacities of the pile. The concrete acts as additional material that contributes to the overall strength of the pile, enabling it to support higher loads and resist bending moments more effectively. Furthermore, filling the steel pipe piles with concrete provides some corrosion protection to the interior of the pipe. The concrete acts as a barrier, shielding the steel from corrosive elements in the soil or surrounding environment, thereby extending the pile's service life.

However, it is essential to note that the concrete infilling generally does not enhance the geotechnical downward load capacity of the pile significantly. Even in the case of open-end pipe piles, they eventually become fully plugged with soil, which limits any potential increase in geotechnical downward load capacity. On the other hand, the geotechnical uplift capacity of the pile experiences a slight improvement due to the added weight of the concrete. The increased weight contributes to the resisting force against upward loads, making the pile more stable under uplift conditions.

When using ASD, the IBC allowable axial compressive stress on the gross cross-section for concrete-filled steel pipe piles with no flexural stresses is:

$$P_a = 0.33\acute{f}_c A_c + 0.50 f_y A_s \tag{17.65}$$

Only the steel resists uplift, so:

$$P_{up,a} = 0.50 f_y A_s \tag{17.66}$$

Where

A_c = cross-sectional area of concrete
A_s = cross-sectional area of steel
f_y = yield strength of steel

If the conditions for using higher allowable stress in steel, as described above, are not satisfied, the coefficient for the steel term in Equations 17.65 and 17.66 should be reduced to 0.35.

The stress analysis becomes more complex when flexural stresses are also present because concrete acts primarily in compression. In such cases, only a portion of the concrete area may be considered, and the neutral axis does not necessarily pass through the center of the pile. Unfortunately, evaluating flexural stiffness and capacity poses challenges due to the differing methodologies presented by AISC, ACI, and AASHTO (Roeder et al. 2010). For example, the AISC formulation for flexural stiffness is:

$$(EI)_{eff} = E_s I_s + C_3 E_c I_c \tag{17.67}$$

$$C_3 = 0.6 + 2\left(\frac{A_s}{A_s + A_c}\right) \leq 0.9 \tag{17.68}$$

Where

$(EI)_{eff}$ = effective flexural stiffness of the composite section
E_s, E_c = elastic modulus of the steel and concrete
I_s, I_c = moment of inertia of the steel and concrete

The concrete is placed after the pile is driven, so driving stresses must be evaluated using only the steel area. The allowable driving stresses are the same as for regular steel pipe piles.

Example 17.7 A 60-ft (18.29 m) long, 14-in (35.5 cm). square prestressed concrete pile is to be made of concrete with $\acute{f}_c$ = 6,000 lb/in² (4218.5 ton/m²). It is to be hoisted to the pile driver by picking it up at a single point located 0.293 L (5.36 m) from one end. This is the distance that produces the smallest moments in the pile. Determine whether this arrangement satisfies the handling stress requirements (Coduto et al., 2016).

Per Table 17.7, the weight of this pile is 204 lb/ft (303.6 kg/m). Multiplying by 1.5 (as recommended by PCI) gives 306 lb/ft (455.4 kg/m). Using the principles of statics, the maximum moment in the pile is 47,300 ft-lb (6.54 ton-m), as shown in Figure 17.36. The tensile stress controls, so that;

$$f_b = 5\sqrt{\acute{f}_c} = 5\sqrt{6{,}000} = 387\ \text{lb/in}^2\ (272.1\ \text{ton/m}^2)$$

$$\frac{P}{F_z A} + \frac{M}{F_b S} = 0 + \frac{(47{,}300)(12\text{in/ft})}{(387)(457)} = 3.2 > 1\ \text{Not Acceptable}$$

The flexural stresses are too high, so another pickup arrangement must be used. Try using two pickup points, which are optimally located at a distance of 0.207 L from each end of the pile.

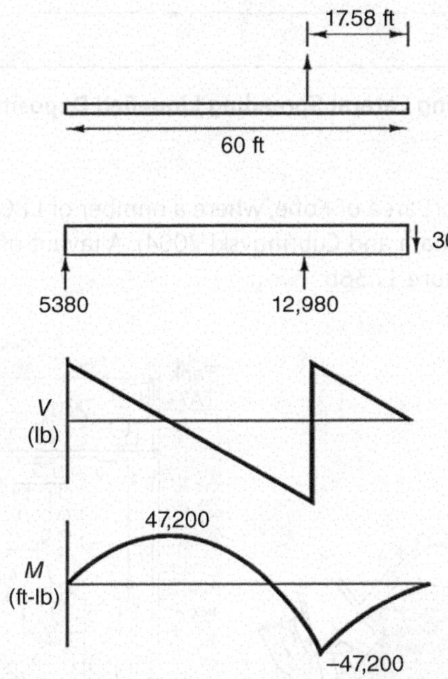

17.8 Shear Piles

Shear piles are reinforced concrete cylindrical piles that extend through a landslide and are anchored at their lower end in stable soils or bedrock to provide lateral bearing resistance near the base of the moving ground (see Figure 17.35). The design of the reinforcement steel is determined by the maximum bending moment developed in the pile.

Shear piles can be arranged in a line to form a wall that supports critical infrastructure, such as a road or railroad, within a landslide. Some of the benefits of using shear piles include:

1) Lateral deflection is usually not a concern, allowing for flexible anchoring near the top.
2) The top of the pile can be buried below the ground surface, making it invisible after construction or easily covered by a structure, playing surface, road, etc.

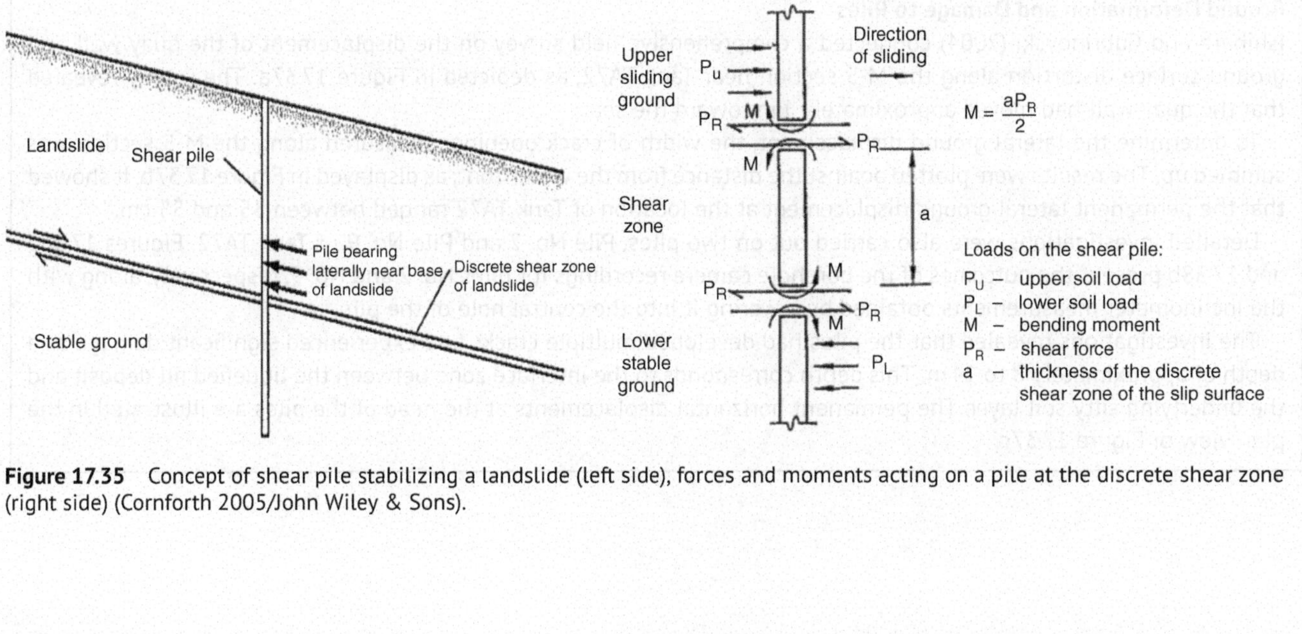

Figure 17.35 Concept of shear pile stabilizing a landslide (left side), forces and moments acting on a pile at the discrete shear zone (right side) (Cornforth 2005/John Wiley & Sons).

3) Installation locations can be selected with greater flexibility.
4) Individual shear piles offer more passive resistance per pile compared to a group of piles.

Case Study 17.1

Case Studies of Pile Foundations Undergoing Lateral Spreading Liquefied Deposits

Mikagehama Tank

Mikagehama is an artificial island in the port area of Kobe, where a number of LPG and oil-storage tanks are installed. Its location is shown in Figure 17.36a (Ishihara and Cubrinovski 2004). A layout of the tanks and auxiliary facilities in the Mikagehama tank farm is shown in Figure 17.36b.

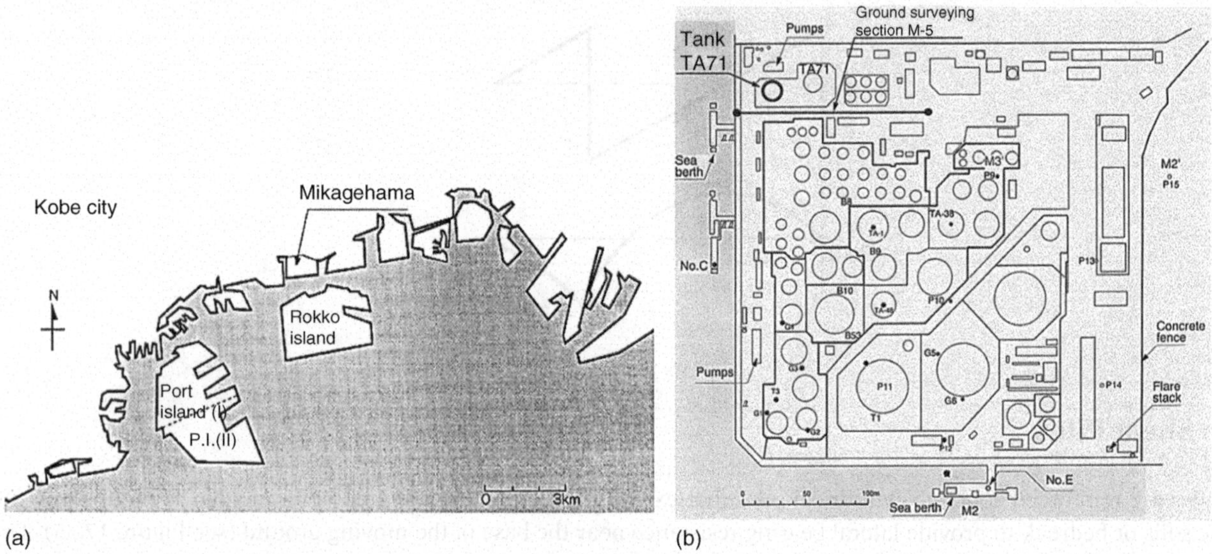

Figure 17.36 (a) Overview of reclaimed lands in the port area of Kobe, (b) location of oil tank TA72 at Mikagehama Island (Ishihara and Cubrinovski 2004).

The conditions of the soil profile and the numbers related to SPT field test results have already been briefly stated in Chapter 14. In this chapter, lateral displacements and the performance of the support piles of Tank TA72 will be investigated.

Ground Deformation and Damage to Piles

Ishihara and Cubrinovski (2004) conducted a comprehensive field survey on the displacement of the quay wall and ground surface distortion along the M-5 section near Tank TA72, as depicted in Figure 17.37a. The survey revealed that the quay wall had moved approximately 1 m toward the sea.

To determine the lateral ground displacement, the width of crack openings measured along the M-5 section was summed up. The results were plotted against the distance from the waterfront, as displayed in Figure 17.37b. It showed that the permanent lateral ground displacement at the location of Tank TA72 ranged between 35 and 55 cm.

Detailed investigations were also carried out on two piles, Pile No. 2 and Pile No. 9 of Tank TA72. Figures 17.38a and 17.38b present the outcomes of the borehole camera recordings for piles No. 2 and No. 9, respectively, along with the inclinometer measurements obtained by lowering it into the central hole of the pile.

The investigations revealed that the piles had developed multiple cracks and experienced significant damage at a depth of approximately 8 to 14 m. This depth corresponds to the interface zone between the liquefied fill deposit and the underlying silty soil layer. The permanent horizontal displacements at the head of the piles are illustrated in the plan view of Figure 17.37c.

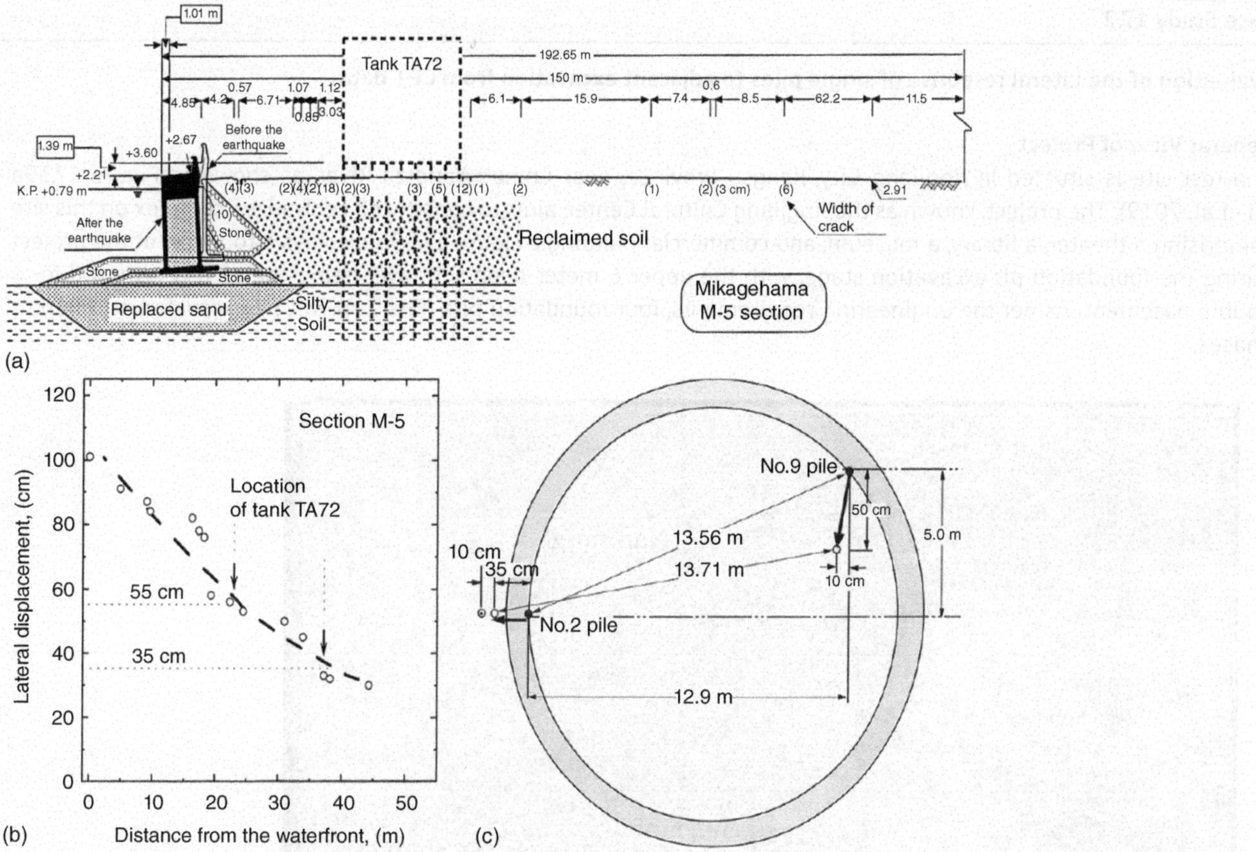

Figure 17.37 (a) Detailed profile of the quay wall movement and ground distortion in the backfills at Section M-5, (b) lateral ground displacement versus distance from the waterfront along Section M-5, and (c) permanent lateral displacements of Pile No. 2 and Pile No. 9 at the head of the pile (Ishihara and Cubrinovski 2004).

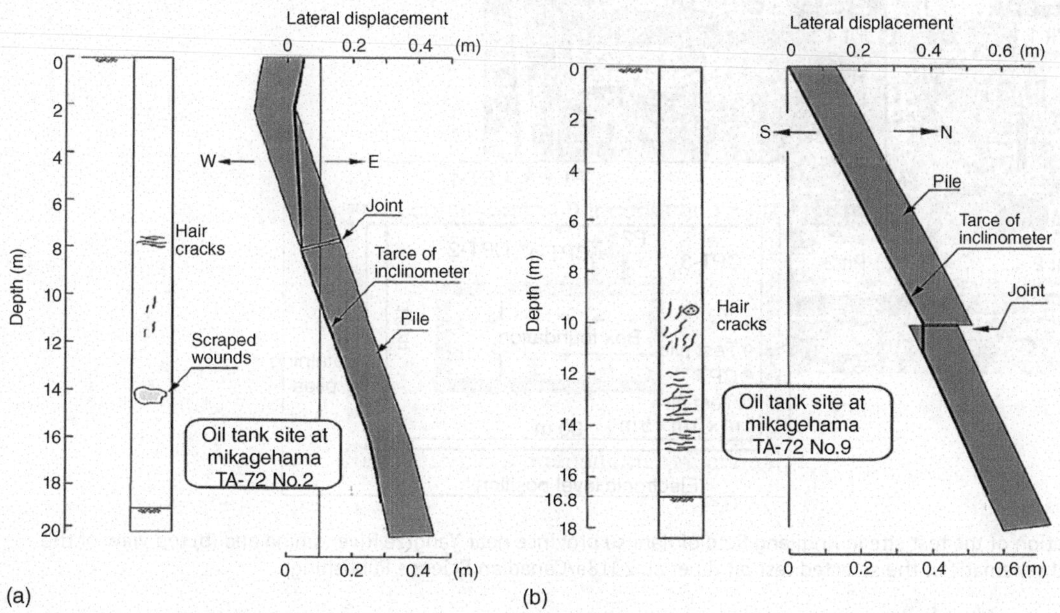

Figure 17.38 Lateral displacement and observed cracks on the inside wall of (a) Pile No. 2 and (b) Pile No. 9 (Ishihara and Cubrinovski 2004).

Case Study 17.2

Evaluation of the lateral response of single piles to adjacent excavation from CPT data

General View of Project

The test site is situated in Jingjiang City, Jiangsu Province, near China's Yangtze River, as shown in Figure 17.39a (Li et al. 2019). The project, known as the Jingjiang Cultural Center, aims to construct a centralized complex on this site comprising a theater, a library, a museum, and commercial buildings. Figure 17.39b displays a top view of the project during the foundation pit excavation stage, with the upper 6-meter-deep soil layer being completely removed for a double basement. As per the engineering requirements, four foundation pits (Pits 1-4) will be excavated at different phases.

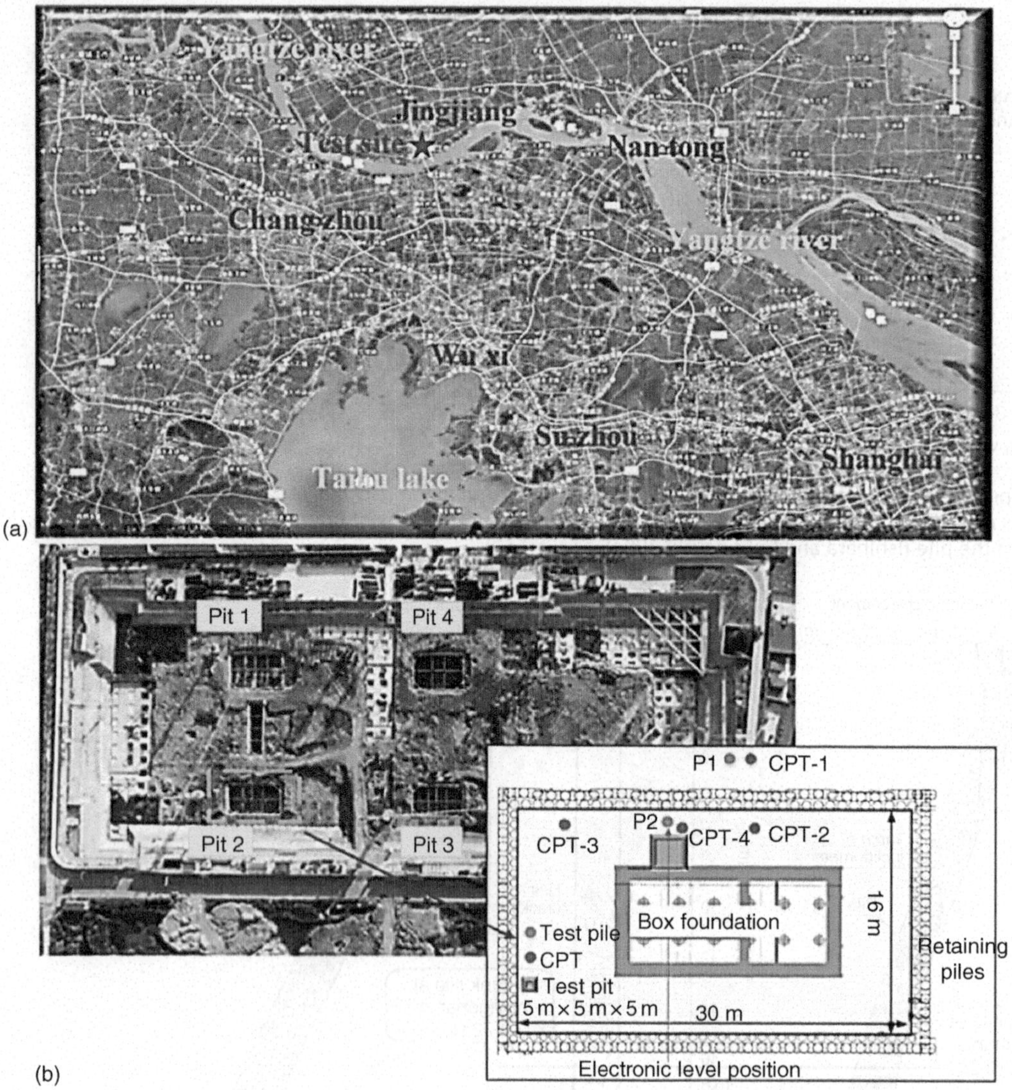

Figure 17.39 (a) Location of the test site in Jingjiang field of Jiangsu province near Yangtze River, China and (b) top view of the excavation project and schematic of the selected test pit (Li et al. 2018a/Canadian Science Publishing).

Soil Type and Material of Piles

A schematic of the cross-section of the excavation and the bored pile is shown in Figure 17.40a. The bored pile in question has a diameter of 1 meter and an effective embedded length of 30 meters, with C30 grade concrete used. The concrete cover was 50 mm. For steel reinforcement, it consists of 16 #25 main bars, evenly spaced around the circumference, along with #8 spiral links at a pitch of 100 mm. The reinforcement extends to the full length of the pile. The pile characterization is summarized in Table 17.8.

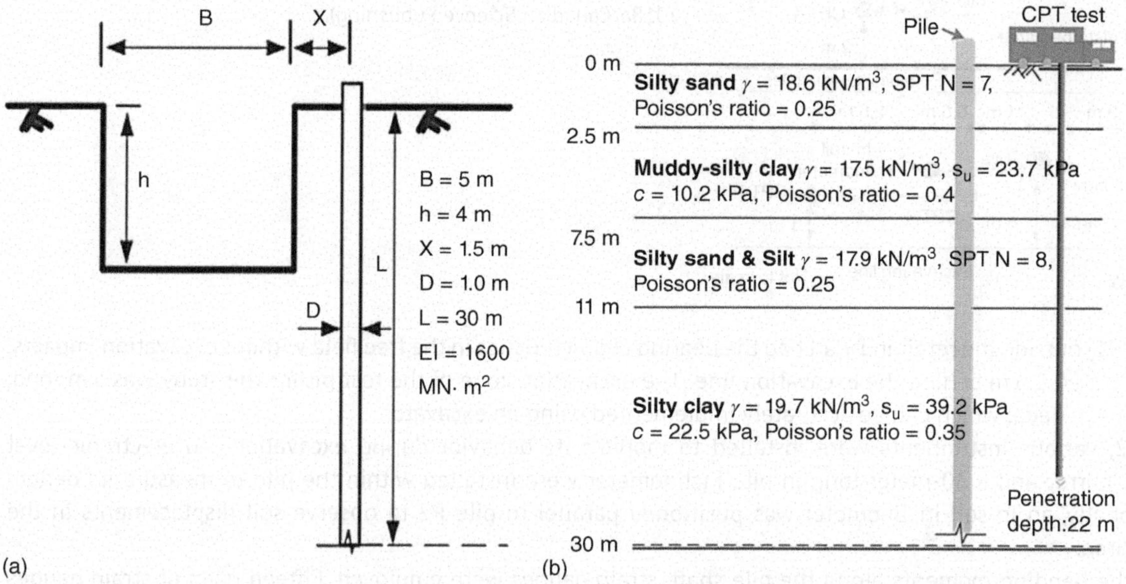

Figure 17.40 Schematic of: (a) cross-section of excavation, (b) site geotechnical characteristics (Li et al. 2018a/Canadian Science Publishing).

Table 17.8 Test pile parameters (Li et al. 2018a/Canadian Science Publishing).

Pile type	Effective pile length (L)	Diameter (D)	Flexural rigidity (EI)	Steel reinforcement	Reinforcement ratio (ρ)
Bored pile	30 m	1000 mm	0.16×10^7 kN.m^2	16 #25	0.95%
	Elastic modulus of reinforcement E_r	Concrete grade	Protective thickness d_p	Elastic modulus of concrete E_c	
	2.0×10^8 kN/m^2	C30	50 mm	3.0×10^7 kN/m^2	

To determine the geotechnical properties of the site, in situ, and laboratory tests were conducted. The soil characterization is shown in Figure 17.40b. The subsoil profile along the effective depth of the pile consists of four main layers:

1) Upper silty sand with a thickness of 2.5 meters
2) Middle muddy–silty clay with a thickness of 5 meters
3) Middle silty sand and silt with a thickness of 3.5 meters
4) Bottom silty clay with a thickness of 17 meters

It should be noted that the actual pile length exceeds 30 meters, with a sand-socketing layer present below the silty clay layer. These strata are commonly found in the Yangtze River basin in China. The presence of muddy–silty or silty clay layers can result in significant lateral movements during soil excavation, leading to substantial bending moments and deflections in the existing piles.

(Continued)

Case Study 17.2 (Continued)

Location of CPT Tests and Instrumentation

Figure 17.41 shows the measurement points with the locations of the test piles and test pit excavation corresponding to Figure 17.40b. Figure 17.41 also shows a series of cone penetration tests (CPTs) to examine the changes in soil stress and pile response before and after the test pit excavation.

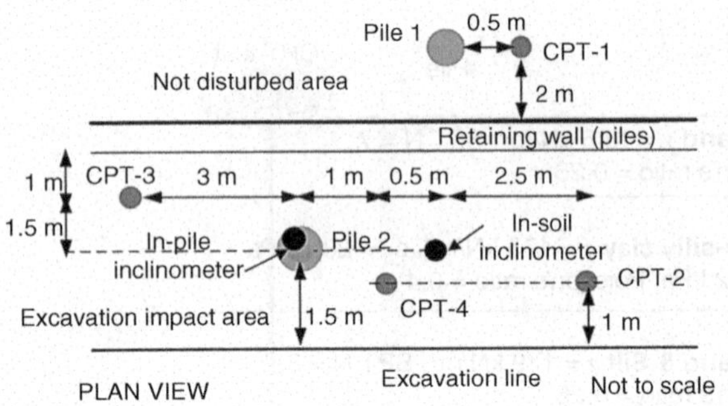

Figure 17.41 Plan layout of test piles and CPTs (Li et al. 2018a/Canadian Science Publishing).

Test pile 1 (P1) outside the retaining wall had the bearing characteristics in the free field without excavation impacts. Pile P2 was located 1.5 m behind the excavation line. The excavation zone of the test pit for the study was 5 m long, 5 m wide, and 4 m deep, which can be accurately implemented using an excavator.

For pile P2, various instruments were installed to monitor its behavior during excavation. An electronic-level displacement mirror and a 30-meter-long in-pile inclinometer were installed within the pile to measure its deflections. Additionally, an in-soil inclinometer was positioned parallel to pile P2 to observe soil displacements in the surrounding area.

To assess the bending moments along the pile shaft, strain gauges were employed. Fifteen pairs of strain gauges were distributed on both the front and rear sides of the upper pile, with vertical intervals of 1 meter between each pair. Furthermore, seven pairs of strain gauges were placed along the lower half of the pile, with vertical intervals of 2 meters between each pair. These strain gauges were utilized to measure bending moments at different levels of the pile, providing valuable data for assessing its structural response during excavation.

Results of CPT Tests

Figure 17.42a displays the CPT q_c profiles measured at various locations before excavation, along with the corresponding soil behavior type determined using the Robertson et al. (1986) classification method. This classification method helps categorize the soil based on its behavior in response to the cone resistance measurements.

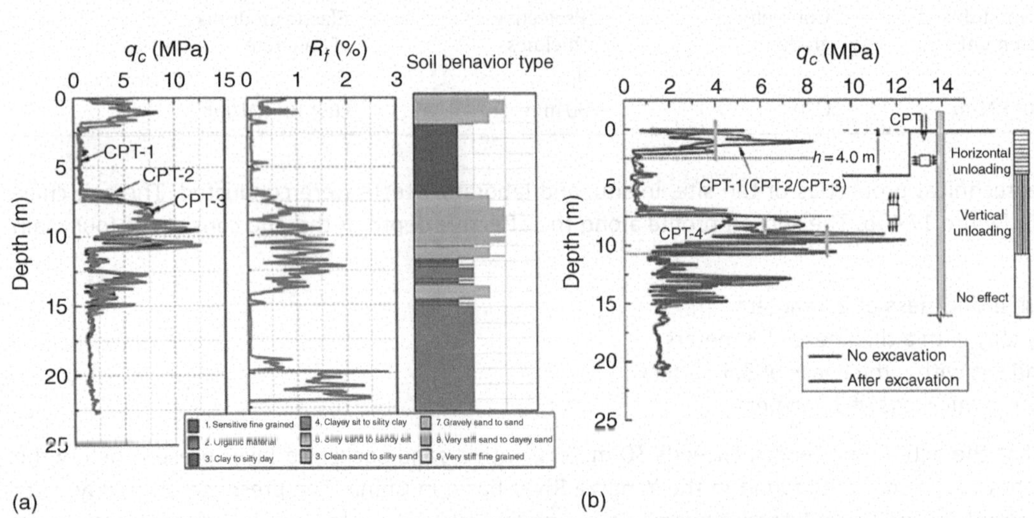

Figure 17.42 (a) Results of cone tests before the excavation, (b) comparison of CPT q_c profiles before and after the test pit excavation (Li et al. 2018a/Canadian Science Publishing).

Figure 17.42b presents a comparison of the soil resistance profiles of CPT-1 and CPT- 4 relative to the 4.0-meter-deep test pit excavation. The excavation process has disturbed the soil stress, leading to significant changes in the cone resistance profiles. The alterations in the CPT q_c profiles are attributed to the different soil stress paths and lateral movements induced by the excavation. These changes are essential to consider when analyzing the behavior and stability of the soil and pile foundations in the construction area.

For this study, the CPT-based p–y curves by Li et al. (2018b) for soft clays and Suryasentana and Lehane (2014a) for sands were used, which were directly generated from CPT data.

Lateral Response after Excavation

In Figures 17.43a and 17.43b, the p–y curves are depicted during excavation and for load–displacement behavior, respectively. These curves were computed using PYGMY software and were compared to the measured results for the two test piles.

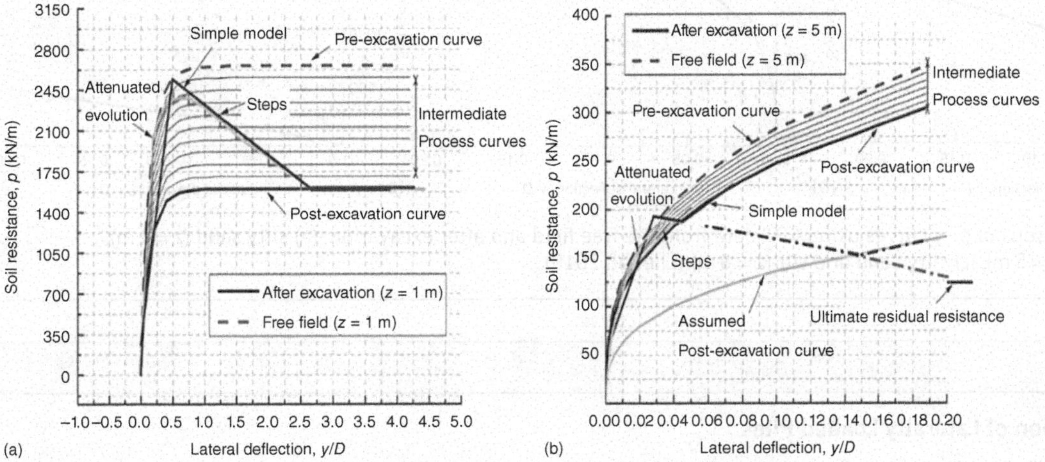

Figure 17.43 Attenuated evolution of the p–y curves undergoing excavation unloading effect in: (a) silty sand, (b) muddy–silty clay (Li et al. 2019).

Figures 17.44a and 17.44b show a simplified model for the attenuated p–y evolution based on the pre-excavation and post-excavation p–y curves. Additionally, the measured and predicted load–displacement curves are presented using p–y curves from the free field (before excavation) and after excavation, respectively. This model helps in understanding the changes in soil behavior and pile response due to the excavation process, providing valuable insights for the analysis and design of the foundation system in the context of soil disturbance during construction.

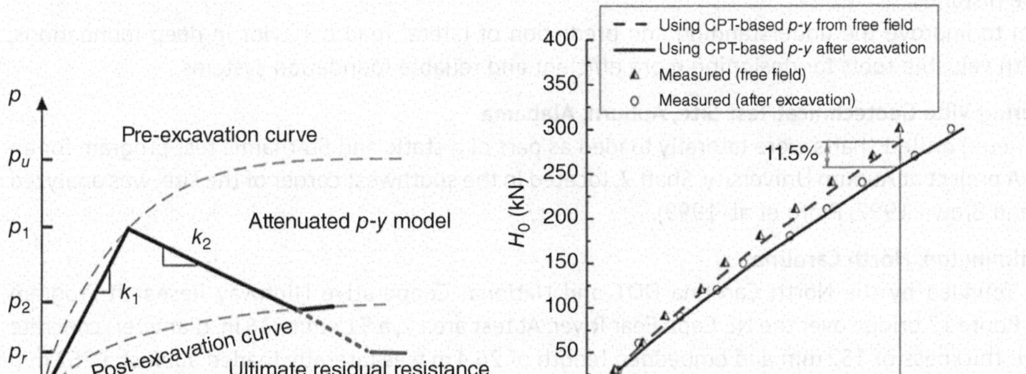

Figure 17.44 (a) Simple model for the attenuated p–y evolution based on pre-excavation and post-excavation p–y curves, (b) measured and predicted load–displacement curves using p–y curves from the free field and after excavation (Li et al. 2019).

(Continued)

Case Study 17.2 (Continued)

The excavation unloading effects on the p–y curves associated with the lateral pile response are well reflected. The post-excavation p–y curves of silty sand (z = 1 m), muddy–silty clay (z = 5 m), and silty sand and silt (z = 8 m) in Figure 17.45 have different degrees of decline compared to the p–y curves before excavation.

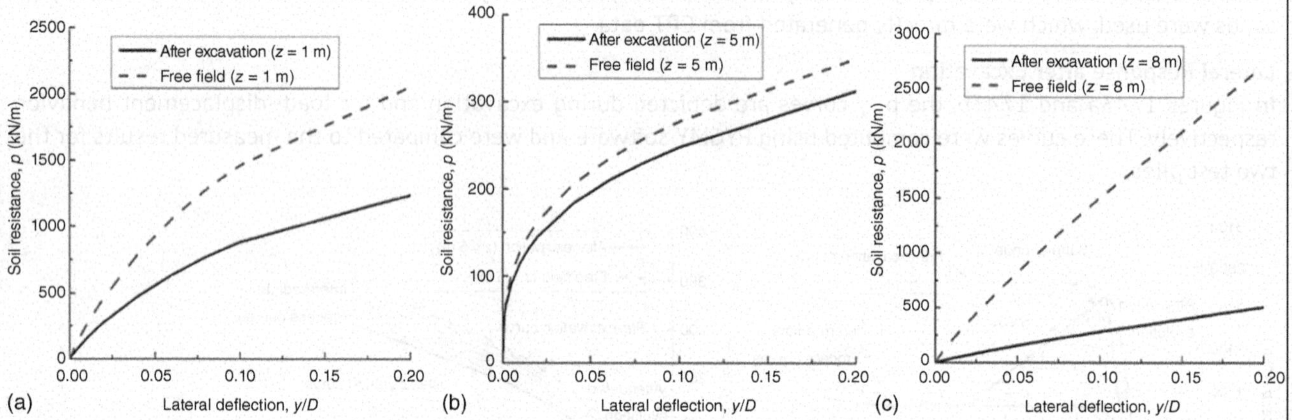

Figure 17.45 Comparisons of p–y curves using CPT data from the free field and after excavation: (a) silty sand (z = 1 m), (b) muddy–silty clay (z = 5 m), (c) silty sand and silt (z = 8 m) (Li et al. 2019).

Case Study 17.3

Case History Evaluation of Laterally Loaded Piles

General View of Project

In this section, two case histories of load tests on piles or drilled shafts under lateral loads are presented (Anderson et al. 2003). The first approach utilized a correlative method, where soil parameters obtained from in situ tests, such as standard penetration test (SPT) and cone penetration test (CPT), were used as input values for standard p–y curves. These curves are commonly used in design software to estimate the lateral load resistance of deep foundations.

The second approach involved calculating p–y curves directly from stress-deformation data measured in dilatometer (DMT) and cone pressuremeter tests. These tests provide valuable information about the soil's behavior and response to lateral loading, enabling a more accurate and customized determination of p–y curves for the specific soil conditions encountered in the case histories.

Both approaches aim to improve the understanding and prediction of lateral load behavior in deep foundations, providing engineers with valuable tools for designing more efficient and reliable foundation systems.

Case History 17.3.1: Spring Villa Geotechnical Test Site, Auburn, Alabama

Six 915 mm (36 in. diameter) drilled shafts were laterally loaded as part of a static and Statnamic test program for an Alabama DOT and FHWA project at Auburn University. Shaft 2, located in the southwest corner of the site, was analyzed for this study (Vinson and Brown 1997; Pinto et al. 1999).

Case History 17.3.2: Wilmington, North Carolina

The test program was founded by the North Carolina DOT and National Cooperative Highway Research Program (NCHRP) for a new U.S. Route 17 bridge over the NE Cape Fear River. At test area 2, a 915 mm (36 in. diameter) concrete cylinder pile with a wall thickness of 152 mm and embedded length of 26.4 m was laterally loaded against a 762 mm (30 in.) square prestressed concrete pile embedded 27.6 m.

Pile Properties in Case Histories

A summary of piles properties is presented in Table 17.9.

Table 17.9 Summary of pile properties (Anderson et al. 2003/American Society of Civil Engineers).

Case history	Pile length (m)	Embedded depth (m)	Width (mm)	Modulus E (GPa)
Spring Villa Auburn	12	11.7	915	27.2
Wilmington, pipe pile	34	26.4	915	30.4
Wilmington, prestressed concrete pile	34	27.6	760	30.4

Soil Properties in Case Histories

Tables 17.10 and 17.11 encapsulate soil properties in Auburn and Wilmington, respectively.

Table 17.10 Summary of soil properties in Auburn case history (Anderson et al. 2003/American Society of Civil Engineers).

Case history	Soil type	Elevation (m)	SPT N_{60}	CPT q_c (kPa)	CPT S_u (kPa)	DMT E_{ed} (kPa)	DMT S_u (kPa)
Auburn, Alabama	Clay	0 to −2.0	Table 17.10 Summary of soil properties in Auburn case history (Anderson et al. 2003) 26			16,470	73
	Clay	0 to −4.0		2494	163.8		
	Clay	−2 to −6.5	13			11,910	63
	Clay	−6.5 to −9.5	19			14,650	51
	Silty clay	−4.0 to −8.0		4044	262.2		
	Clay	−9.5 to −15.0	14				
	Clay–silt	−8.0 to −12.0		5388	346.4		
	Clay–silt	−12.0 to −15.0		4650	293.6		

Table 17.11 Summary of soil properties in Wilmington case history (Anderson et al. 2003/American Society of Civil Engineers).

Case histories	Soil type	Elevation (m)	SPT N_{60}	CPT q_c (kPa)	CPT S_u (kPa)	CPT ϕ (deg)	DMT E_d (kPa)	DMT S_u (kPa)	DMT ϕ (deg)
Wilmington	Clay	−1.8 to −2.8	2						
	Clay	−2.8 to −7.3	10	400		23	2,880	12	
	Sand	−7.3 to −8.8	19	3,600	30		330	11	
	Sand	−8.8 to −15.5	28	15,200	41		27,700		35
	Sand	−15.5 to −37.5	48		44		47,000		35

Results and Discussion

The comparison between the measured and predicted results in Figures 17.46 and 17.47 shows varying degrees of success. In the Auburn case (clays above the water table), all four in situ test methods (SPT, CPT, DMT, and PMT) were accurately predicted. At the Wilmington site (submerged clay), three in situ methods were satisfactorily predicted, but only the DMT showed satisfactory results. Since there is no standardized "Davisson" type capacity to evaluate failure for piles under lateral loads, the evaluation of the predictions' accuracy was done differently. The comparison was made by plotting the "measured load" versus the "predicted" deformation at different percentages (25%, 50%, 75%, and 100%) of the applied lateral test load. An R-value was then defined as the ratio of the two methods being evaluated. This R-value provides a measure of the "goodness" of predictions, giving engineers a useful metric to assess the reliability of their design calculations for lateral load behavior in deep foundations.

(Continued)

Case Study 17.3 (Continued)

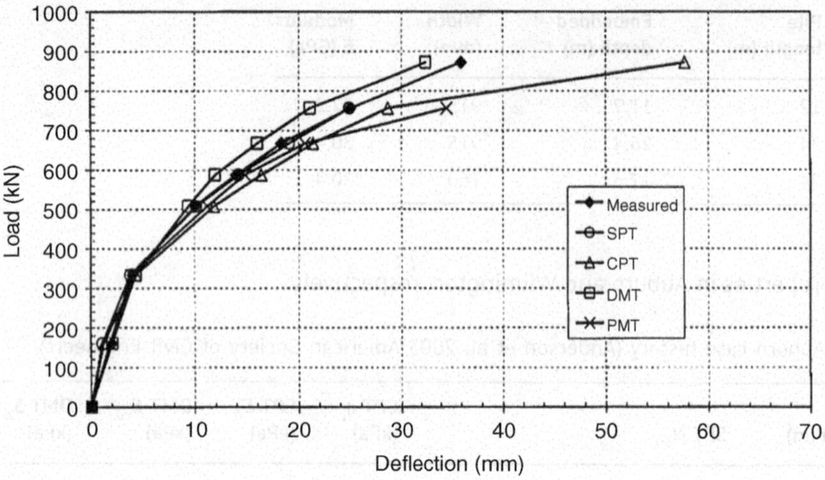

Figure 17.46 Comparison between measured and predicted lateral deflections for Auburn test site (Anderson et al. 2003/American Society of Civil Engineers).

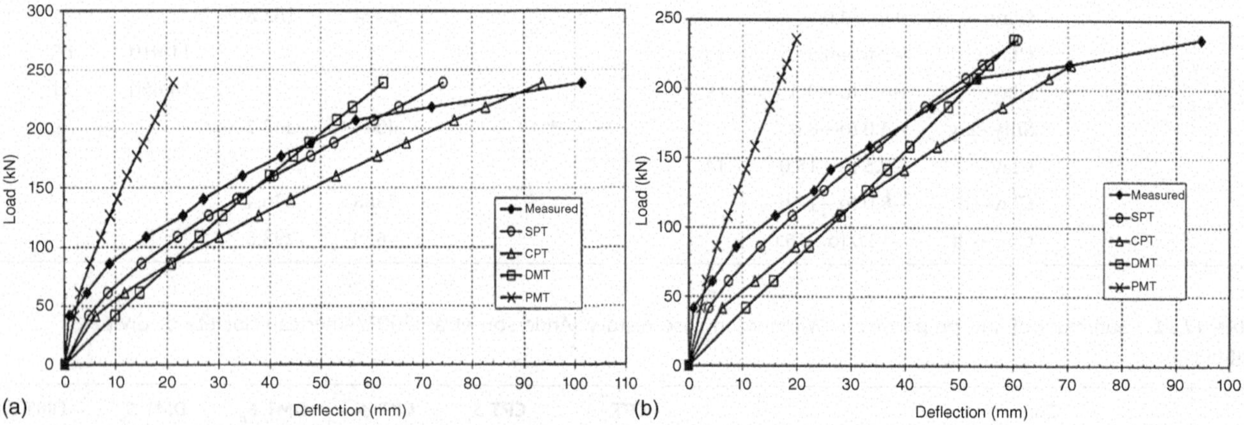

Figure 17.47 Comparison between measured and predicted lateral deflections for Wilmington test site: (a) 762 mm prestressed concrete pile and (b) 915 mm pipe pile (Anderson et al. 2003/American Society of Civil Engineers).

A summary of *R*-values for lateral load case histories is given in Table 17.12.

$$\text{R value} = \frac{\text{Measured deflection}}{\text{Predicted deflection}} \times 100\%$$

Table 17.12 summary of *R*-values for lateral load case histories (Anderson et al. 2003/American Society of Civil Engineers).

	Loading level			
Test	**25%**	**50%**	**75%**	**100%**
SPT	77.15%	78.35%	82.56%	95.25%
CPT	78.23%	81.11%	90.48%	109.81%
DMT	92.67%	120.13%	139.76%	160.49%
PMT	108.05%	151.28%	177.79%	230.54%

The average predictions based on SPT correlations were found to be conservative for all loading levels. In other words, the predicted results tended to be on the safe side, with the actual behavior likely to be better than predicted. On average, CPT predictions were mostly conservative with *R*-values ranging between 78% and 90%, except at the maximum load where the predictions were slightly overestimated (*R*-value ≈ 110%). The DMT-derived p–y curves showed accurate predictions at low lateral loads, which is expected since the DMT membrane only deforms 1 mm into the soil, making it more sensitive to small displacements. For PMT-derived p–y curves, accurate predictions were obtained for both "sands" and "clays" where pore pressures were not expected to be significant. However, in clays below the groundwater table, rate effects may cause the predictions to be unconservative (too stiff).

Case Study 17.4

Influence of Flexural Stiffness on the Developed Soil Reactions of Three Laterally Loaded Piles

This case study presents the results and analyses of three full-scale lateral load tests conducted on test piles installed at a well-characterized sandy site (Fernandez et al. 2010). The primary objective of this study is to evaluate the influence of the pile bending stiffness on the soil reactions acting around the laterally loaded pile. The three test piles were subjected to similar testing conditions, allowing for a comprehensive comparison and assessment of their behavior.

Hampton, Virginia Case History

In 2003, Pando conducted a comprehensive test pile program in Hampton, Virginia, to study the feasibility of using composite piles for bridge foundations. As part of this research, the author installed three test piles near the Route 351 bridge crossing of the Hampton River in the Chesapeake Bay area of Virginia. To investigate the test site, two hollow stem auger borings, four cone penetrometer probes, and one dilatometer probe were utilized. Figure 17.48 displays the SPT N-values along with descriptions of the materials encountered in the borings. Additionally, the tip resistance and sleeve friction from the four-cone penetrometer test (CPT) probes are shown in Figure 17.48.

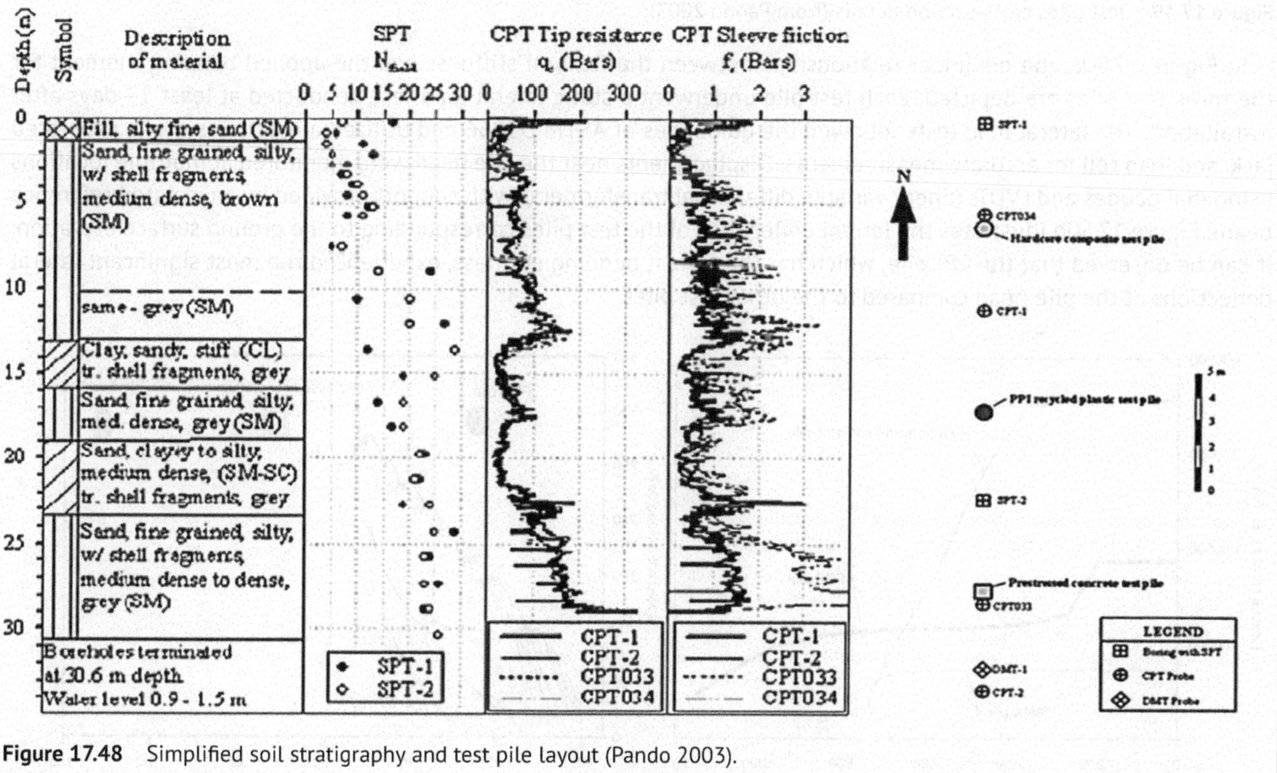

Figure 17.48 Simplified soil stratigraphy and test pile layout (Pando 2003).

(Continued)

Case Study 17.4 (Continued)

Soil Conditions on Site

At the test site, the general soil stratigraphy consisted of a 1.0-meter-thick layer of silty fine sand fill on top, followed by loose to medium-dense silty fine sand extending to a depth of 13.0 meters. Below the silty fine sand, there was a layer of stiff sandy clay reaching a depth of 15.5 meters. Further down, at a depth of 30.6 meters, medium-dense to dense silty and clayey sand was encountered. The groundwater level, controlled by the adjacent Hampton River, was observed to be 0.9 to 1.5 meters (3 to 5 feet) below the ground surface during the field testing.

Pile Conditions

In this case history, three test piles were used, each with different construction materials and cross-section details. The test piles included a prestressed concrete test pile (PSC), a concrete-filled FRP tubular composite pile, and a plastic matrix composite pile (PPI). Figure 17.49 shows the cross-section details of these three test piles. The PSC test pile has a square cross-section with dimensions of 610 mm on each side. It is 18 meters long and has an initial flexural stiffness (*EI*) of 355,610 kN.m^2. The FRP test pile has a circular cross-section with a diameter of 622 mm and a length of 18.3 meters. Its initial flexural stiffness is 186,510 kN.m^2. The PPI test pile also has a circular cross-section with a diameter of 592 mm and a length of 18.3 meters. Its initial flexural stiffness is 71,705 kN.m^2.

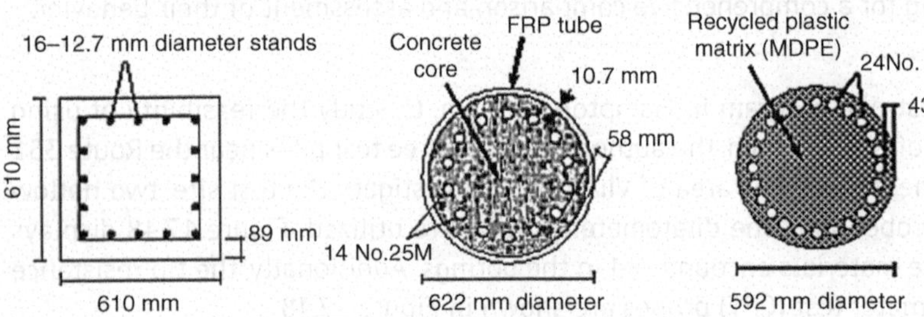

Figure 17.49 Test piles cross-section details (from Pando 2003).

In Figure 17.50a, the nonlinear relationships between the flexural stiffness and the applied bending moment for the three test piles are depicted. Each test pile underwent a static lateral load test, conducted at least 14 days after installation. The lateral load tests followed the guidelines of ASTM D3966 and utilized a reaction system, calibrated jack, and load cell for accurate measurements. Displacements near the pile head were monitored at multiple locations using dial gauges and LVDTs (linear variable differential transformers), with support provided by an isolated reference beam. Figure 17.50b illustrates the lateral deflections of the test piles corresponding to the ground surface elevation. It can be observed that the PPI pile, which has the lowest bending stiffness, experienced the most significant lateral deflections at the pile head compared to the other test piles.

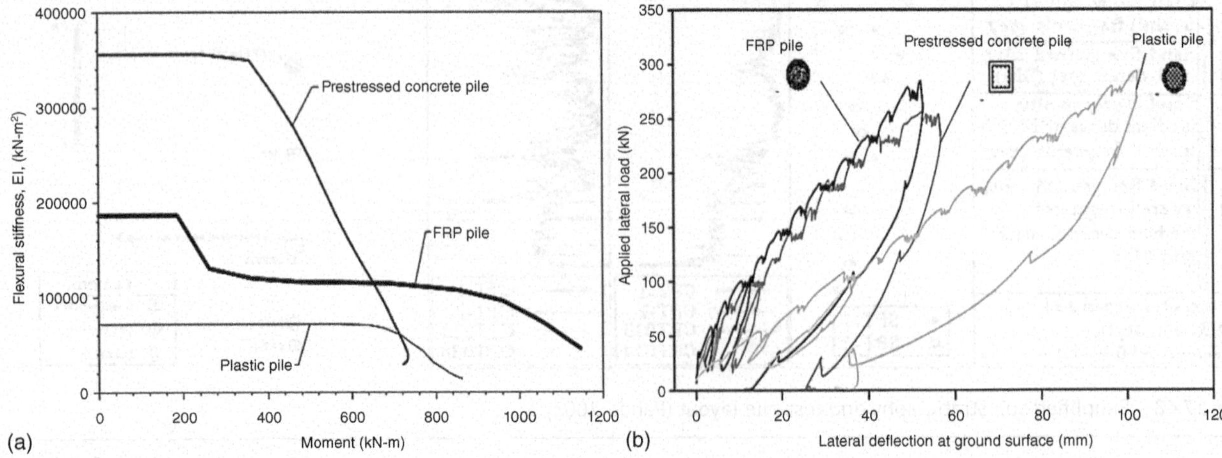

Figure 17.50 (a) Flexural stiffness as a function of the applied moment and (b) measured lateral deflection at the ground surface for the three test piles (Pando 2003).

Analyses with LPILE and SWM Software

Soil reactions were computed using LPILE and SWM 6.1. Soil layers and properties used in the analyses with LPILE and SWM are presented in Table 17.13. Also, selected analytical results obtained using both approaches and for the high level of applied load are shown in Figures 17.51a–c.

Table 17.13 Soil layers and properties used in the analyses with LPILE and SWM (Fernandez et al. 2010/American Society of Civil Engineers).

Test pile	Layer thick (m)	Soil type	γ (kN/m^3)	ϕ (deg)	K (MN/m^3)	ε_{50} (strain)
PSC	0.91	Sand	19.8	31 (31)	2.8	0.04
	1.5	Sand	10	30 (27)	1.7	0.07
	7.5	Sand	10	33 (33)	1.7	0.04
	10	Sand	11	35 (35)	1.7	Default
FRP	1.2	Sand	19.8	34 (33)	9.1	Default
	7.2	Sand	10	33 (33)	5.5	Default
	11.6	Sand	11	35 (34)	5.5	Default
PPI	0.91	Sand	19.8	33 (31)	3.6	Default
	1.5	Sand	10	30 (27)	2.2	Default
	7.5	Sand	10	33 (33)	2.2	Default
	10	Sand	11	35 (35)	2.2	Default

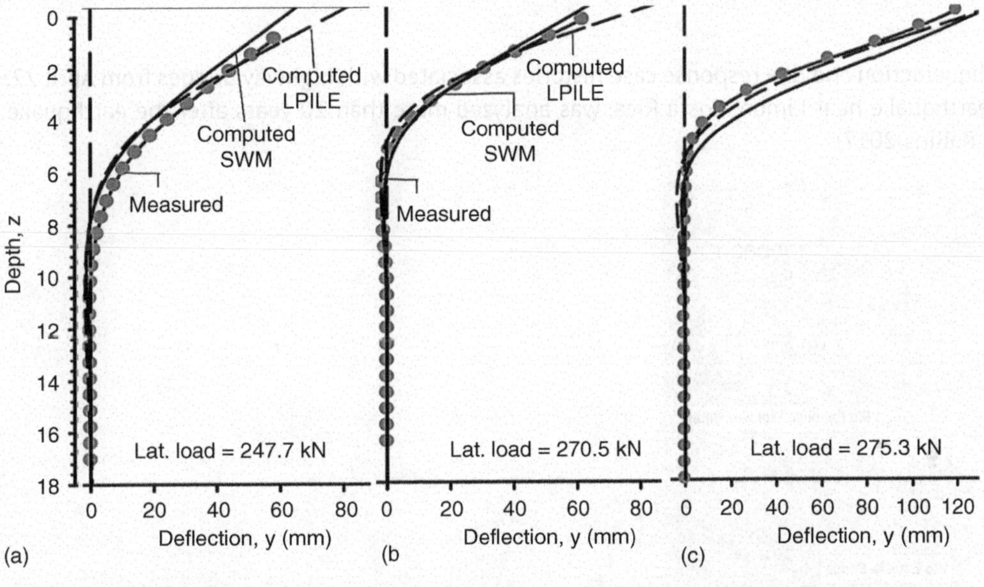

Figure 17.51 Computed best fits versus measured lateral displacement profiles for the: (a) PSC test pile, (b) FRP test pile, and (c) PPI plastic test pile (Fernandez et al. 2010/American Society of Civil Engineers).

In general, both analytical approaches provide a reliable agreement between the measured and computed deflection profiles for the different test piles and different levels of lateral loading. The soil reactions are discussed below.

The p–y curves obtained from the LPILE and SWM analyses are presented in Figures 17.52a and 17.52b. It can be observed that the LPILE analyses resulted in different p–y curves for all three test piles, showing variations in both stiffness and ultimate resistance values. Surprisingly, the PSC pile yielded the lowest p–y curve, while the FRP test pile exhibited the highest curve. This outcome was unexpected, as the tendency was to obtain stiffer p–y curves for piles with higher flexural stiffness (EI).

(Continued)

Case Study 17.4 (Continued)

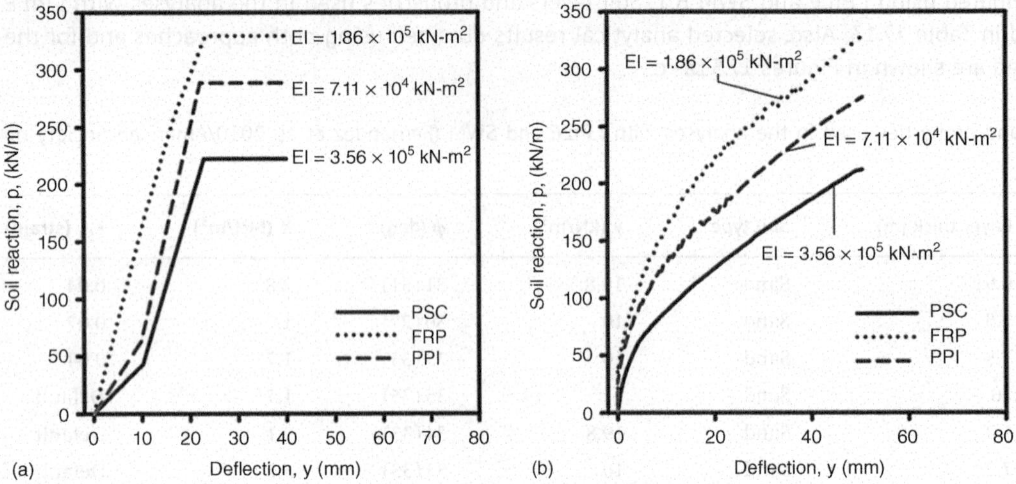

Figure 17.52 Self-generated p–y curves for the three test piles at 2.91 m depth: (a) LPILE p–y curves and (b) SWM p–y curves (Fernandez et al. 2010/American Society of Civil Engineers).

Case Study 17.5

Lateral Spread Displacement and Bridge Foundation Case Histories from the 1991 Magnitude 7.6 Earthquake near Limon, Costa Rica

This section presents two liquefaction and pile response case histories associated with highway bridges from April 22, 1991. 7.6 magnitude (M) earthquake near Limon, Costa Rica, was analyzed more than 20 years after the earthquake (Figure 17.53) (Franke and Rollins 2017).

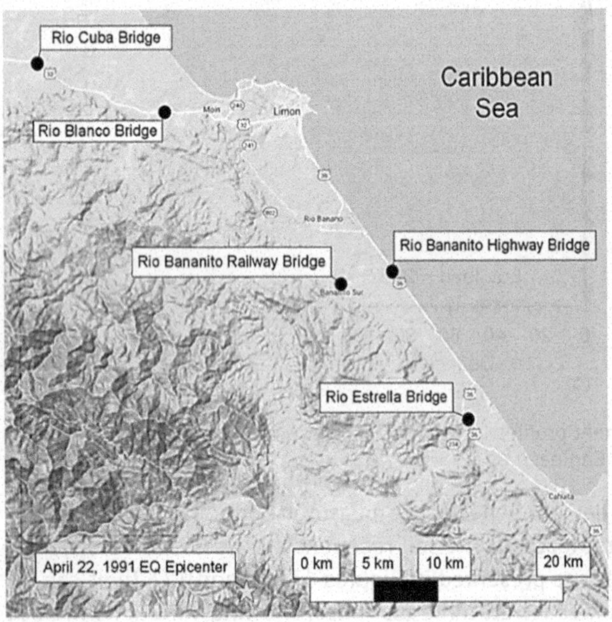

Figure 17.53 Site vicinity map of the five bridges reconnoitered and analyzed in this study and their locations relative to the April 1991 Earthquake Epicenter (star) (© 2016, Google LLC) (Franke and Rollins 2017).

Earthquake Description

On April 22, 1991, at 15:57 local time, a powerful earthquake with a moment magnitude of 7.6 struck the Limón Province of Costa Rica, near the Caribbean Sea. The epicenter was recorded at approximately 9.67° North latitude and 83.07° West longitude, with a focal depth of about 12.9 km. This devastating earthquake caused significant damage and loss of life, with 53 people killed and 193 others injured. The impact of the earthquake on the region was severe, as it led to the disruption of around 30% of the highway pavement and railways. The damage resulted from various effects, including fissures, scarps, and settlements caused by soil liquefaction (Priestley et al. 1991).

Case Study 1: Rio Bananito Railway Bridge

The Rio Bananito Railway Bridge is a historic single-span truss bridge, believed to have been constructed sometime before 1890. The bridge is a steel through-truss design and spans approximately 50 meters in length. Figure 17.54 provides both plan and profile views of the bridge and its foundation. The foundation of the bridge consists of four large caissons, each with an elliptical shape, measuring 1.46 meters across the minor axis and 2.16 meters across the major axis. These caissons were built using a 12-mm thick steel shell, comprised of narrow sheet piles that were interconnected and then filled with concrete to provide stability and strength.

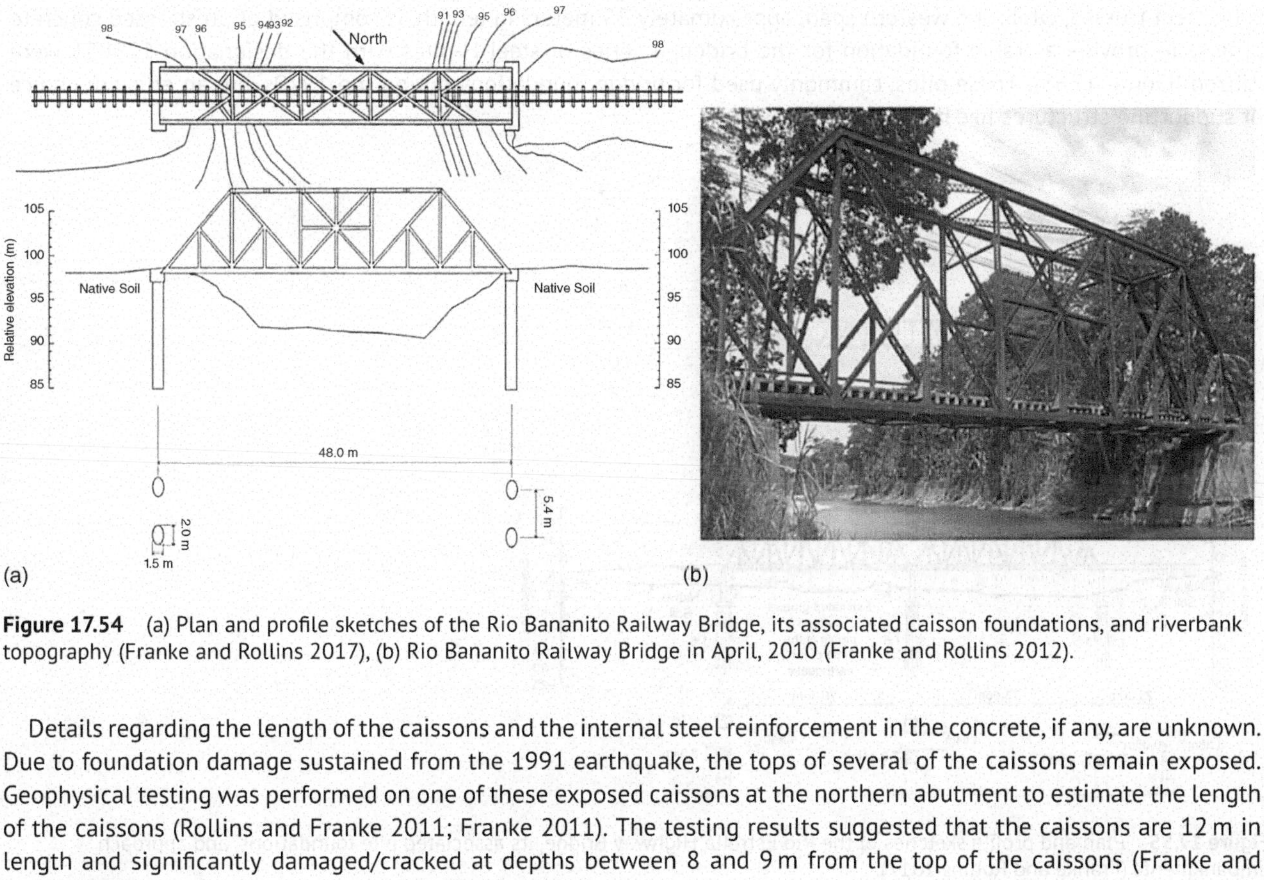

Figure 17.54 (a) Plan and profile sketches of the Rio Bananito Railway Bridge, its associated caisson foundations, and riverbank topography (Franke and Rollins 2017), (b) Rio Bananito Railway Bridge in April, 2010 (Franke and Rollins 2012).

Details regarding the length of the caissons and the internal steel reinforcement in the concrete, if any, are unknown. Due to foundation damage sustained from the 1991 earthquake, the tops of several of the caissons remain exposed. Geophysical testing was performed on one of these exposed caissons at the northern abutment to estimate the length of the caissons (Rollins and Franke 2011; Franke 2011). The testing results suggested that the caissons are 12 m in length and significantly damaged/cracked at depths between 8 and 9 m from the top of the caissons (Franke and Rollins 2012).

Soil Profile

Table 17.14 presents the representative soil profile at the northern abutment of the bridge.

(Continued)

Case Study 17.5 (Continued)

Table 17.14 Representative Soil Profile at the Northern Abutment of the Rio Bananito Railway Bridge (Franke and Rollins 2017).

Thickness (m)	USCS	Avg $(N_1)_{60}$	γ (kN/m³)	Fines (%)	D_{50} (mm)	LL (%)	PI (%)	Moisture (%)	ϕ (Deg)	D_r (%)	C_u (kPa)
6.0	CH	4	18.9	95	—	61	31	46	0	—	27.0
1.4	SM	12	18.1	23	0.4	32	5	27	33	45	—
3.7	SC/SM	27	18.8	21	2.0	34	10	19	37	67	—
2.8	SC/SM	20	18.8	19	2.4	36	13	16	35	58	—

Case Study 2: Rio Estrella Highway Bridge

The Rio Estrella Highway Bridge is a significant transportation structure constructed in 1971. It consists of three spans supporting two lanes of traffic. The eastern and central spans, each about 75 meters long, are constructed using steel trusses, while the western span, approximately 25 meters in length, is comprised of prestressed concrete girders. To provide a stable foundation for the bridge, a series of steel H-piles with the designation 12BP53 were utilized (Figure 17.55). These piles, commonly used for bridge foundations during the 1970s, are an efficient choice for supporting structures like highway bridges.

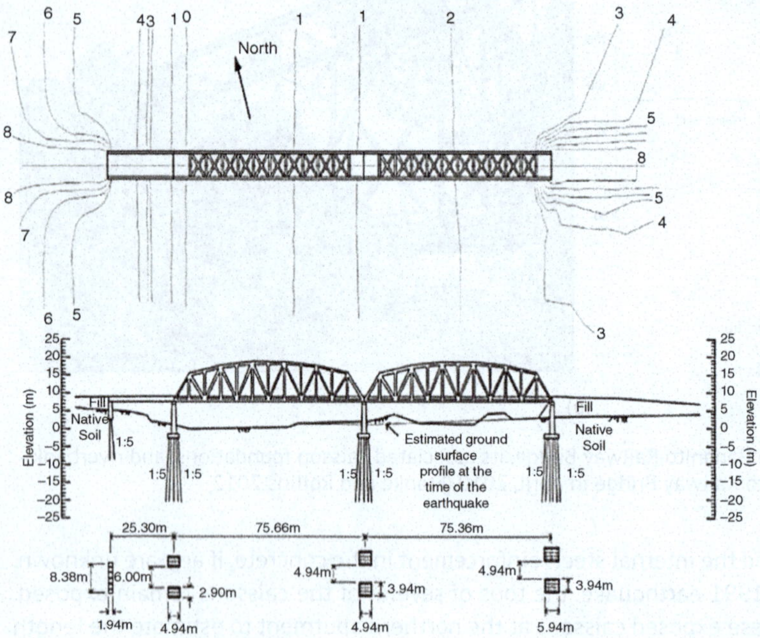

Figure 17.55 Plan and profile sketches of the Rio Estrella Highway Bridge, its associated pile foundations, and approach embankments (Franke and Rollins 2017).

Additional details regarding pile spacing and batter are shown in Figure 17.55.

Soil Profile

Table 17.15 presents the representative soil profile at the eastern abutment of the bridge.

Table 17.15 Representative Soil Profile at the Eastern Abutment of the Rio Estrella Bridge. (Franke and Rollins 2017).

Thickness (m)	USCS	Avg $(N_1)_{60}$	γ (kN/m^3)	Fines (%)	D_{50} (mm)	LL (%)	PI (%)	Moisture (%)	ϕ (Deg)	D_r (%)	C_u (kPa)
6.0	SM	—	18.5	25	—	NP	NP	9	35	62	—
3.0	ML/SM	9	18.1	52	0.1	46	8	25	32	39	—
1.6	MH	4	18.8	97	—	53	18	46	0	—	19.0
2.8	GW	14	18.1	3	5.0	—	—	14	33	48	—
2.7	SW-SM	14	18.1	7	1.0	NP	NP	16	33	48	—
0.9	SM	37	18.3	27	0.2	NP	NP	24	40	77	—
2.3	SM	25	18.3	32	0.2	NP	NP	24	37	65	—
7.1	SM	32	18.3	43	0.2	NP	NP	24	39	73	—

Abutments Properties

The Rio Bananito Railway Bridge is supported by four large elliptically-shaped caissons, each measuring 1.46-by-2.16 meters across the major axes. These caissons are constructed with a 12-mm thick steel shell made up of narrow sheet piles, which are connected and then filled with concrete. While specific details regarding the caissons' length and internal steel reinforcement in the concrete are not available, it can be assumed that they are designed to provide a stable foundation for the railway bridge.

The Rio Estrella Highway Bridge's foundation comprises 12BP53 steel H-piles, which were commonly used for bridge foundations in the 1970s. These piles are similar in properties to the 12HP53 H-piles used today. The bridge's western abutment is supported by ten piles (two rows of five), each approximately 17.2 meters in length. The western bent of the bridge is founded on two 2.90 × 4.94 meters pile caps, with each supported by 15 piles (three rows of five) that are 20 meters in length. The eastern bent, on the other hand, is founded on two 3.94 × 4.94 meters pile caps, each supported by 20 piles (four rows of five) that are 20 meters in length. Additionally, the eastern abutment of the bridge appears to be designed for potential conversion into a bent abutment if needed. This abutment is supported by two 3.96 × 5.96 meters pile caps, each supported by 24 piles (four rows of six) that are 20 meters in length.

Table 17.16 presents pile stiffness parameters used in the kinematic pile response analysis.

Table 17.16 Pile Stiffness Parameters Used in the Kinematic Pile Response Analysis (Franke and Rollins 2017).

Bridge	Pile head behavior	EI per pile (MN.m^2)	Number of piles/ caissons	EI for pile cap (MN.m^2)	EI for equiv. single pile (MN. m^2)	P-multiplier	Rotational stiffness (MN.m/rad)
Rio Bananito Railway	Free-head	13,312	2	N/A	26,624	2.0	N/A
Rio Estrella	Fixed-head	0.04024	48	4,190,900	1,931.52	26.2	67.274

(Continued)

Case Study 17.5 (Continued)

Observed Damage in Case Studies

The Rio Bananito Railway Bridge experienced significant liquefaction-related issues during the earthquake, leading to the deformation of the ground parallel to the river bank. Liquefaction features such as fissures and sand boils were observed, indicating ground movement in the direction of the river. The lateral spread displacements at both the north and south bridge abutments exerted considerable force on the four caissons supporting the bridge, causing them to move toward the river.

The northern abutment was particularly affected, with the east and west caissons displacing by 4.3 and 5.7 meters, respectively, and rotating by 26° and 37°, respectively. Similarly, the southern abutment experienced displacement, with the east and west caissons moving 2.8 and 1.9 meters, respectively, and rotating 19° and 7°, respectively. The movement of the caissons resulted in three of the four caissons losing contact with the bridge. As a consequence, the bridge itself slumped into the surrounding soil and tilted to the east by nearly 15°, as shown in Figure 17.56.

(a) (b)

Figure 17.56 (a) Rio Bananito Railway Bridge after Limon earthquake (Courtesy of Earthquake Engineering Research Institute, 1991), (b) Floods damaged a bridge over the La Estrella river along route 36 in Talamanca Canton, Limón province, Costa Rica, July 2021 (CNE / Government of Costa Rica).

In the Rio Estrella bridge, the earthquake caused significant damage to the eastern and western spans of the steel truss bridge. The eastern span collapsed from its support at the east abutment, fracturing the two bottom cords adjacent to the central slab pier. Meanwhile, the western span fell from its support on the central pier but remained supported at the west pier. The pre-tensioned reinforced concrete section at the west end of the bridge remained intact during the earthquake. Near the eastern abutment of the bridge, lateral spread displacement of up to 2 meters in the direction of the Rio Estrella was measured by Youd (1993).

The computed kinematic pile response results for equivalent single piles are shown in Figures 17.57 and 17.58 for the case histories. These results provide information on the displacement (in meters), bending moment (in kN.m), shear force (in kN), and curvature (in degrees) profiles of the piles during the earthquake.

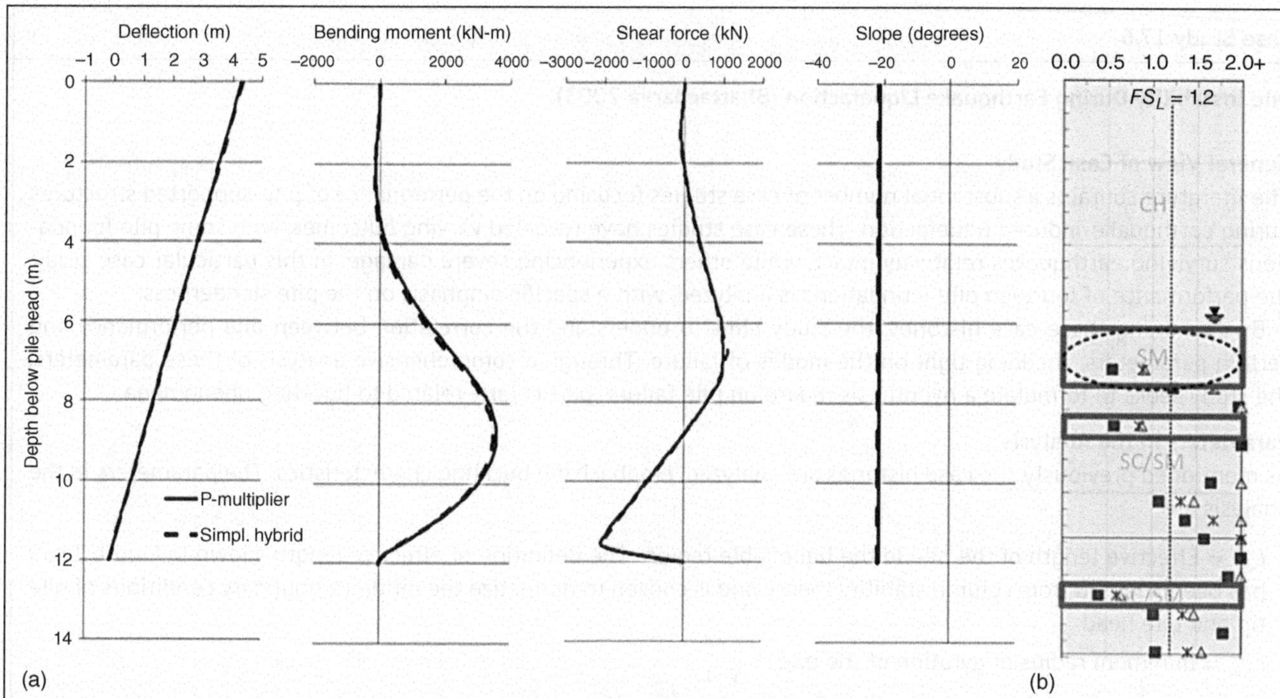

Figure 17.57 (a) Kinematic caisson response results per average caisson at the northern abutment of the Rio Bananito Railway Bridge, (b) deterministic liquefaction triggering results for Rio Bananito Railway bridge (Franke and Rollins 2017).

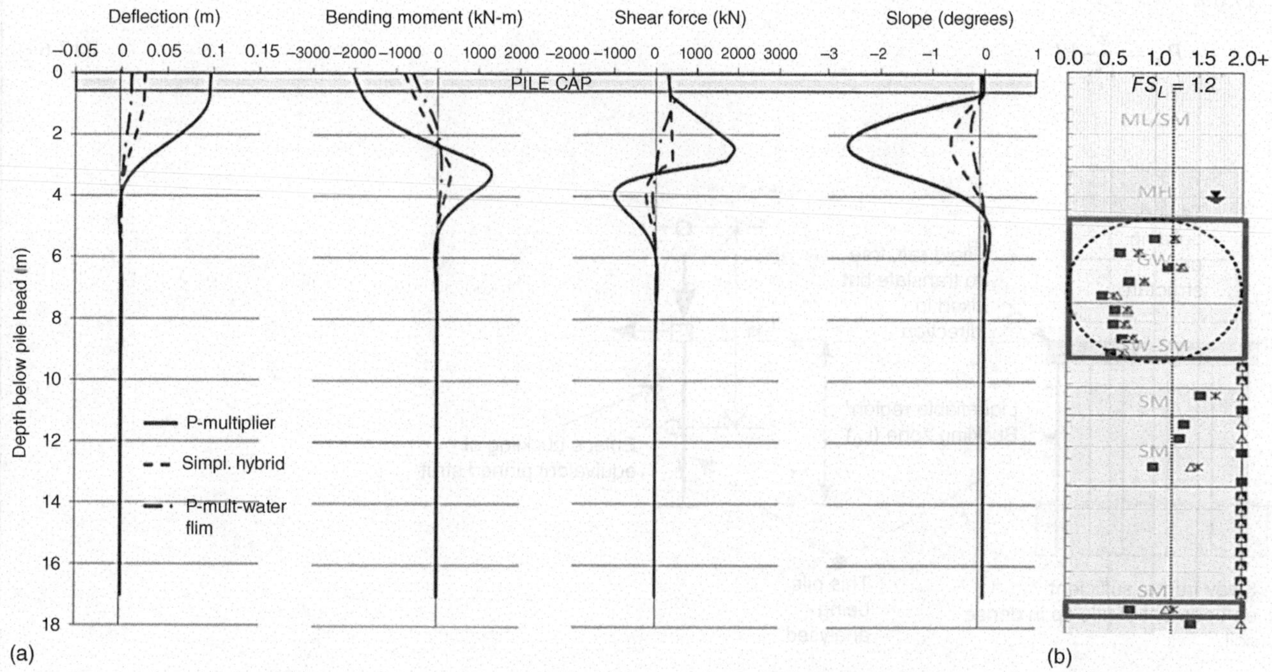

Figure 17.58 (a) Kinematic caisson response results per average caisson at the eastern abutment of the Rio Estrella Bridge, (b) deterministic liquefaction triggering results for Rio Estrella Bridge (Franke and Rollins 2017).

Case Study 17.6

Pile Instability During Earthquake Liquefaction (Bhattacharya 2003).

General View of Case Study:

The literature contains a substantial number of case studies focusing on the performance of pile-supported structures during earthquake-induced liquefaction. These case studies have revealed varying outcomes, with some pile foundations surviving earthquakes relatively intact, while others experiencing severe damage. In this particular case study, the performance of fourteen pile foundations is analyzed, with a specific emphasis on the pile slenderness.

By examining these case histories, the study aims to understand the correlation between pile performance and certain parameters, shedding light on the modes of failure. Through a comprehensive analysis of these parameters, the study seeks to formulate a hypothesis regarding pile failure, particularly related to buckling phenomena.

Parameters in the Analysis

As mentioned previously, the case histories are analyzed, emphasizing buckling characteristics. The parameters in the analysis are:

- L_{eff} = Effective length of the pile in the liquefiable region. The definition of effective length shown in Figure 17.59 has been adapted from column stability theory and is chosen to normalize the different boundary conditions of pile tip and pile head.
- r_{min} = minimum radius of gyration of the pile = $\sqrt{\frac{I}{A}}$
- Slenderness ratio of the pile in the liquefiable region, $\frac{L_{eff}}{r_{min}}$
- Based on conventional design procedures, there is an allowable load on pile P, with no allowance for liquefaction, which can also be called a conventionally allowable load in a pile.
- Euler's elastic critical load of the pile (P_{cr}) calculated from the well-known buckling formula as shown by Equation 17.69.

$$P_{cr} = \frac{\pi^2}{L_{eff}^2} EI \tag{17.69}$$

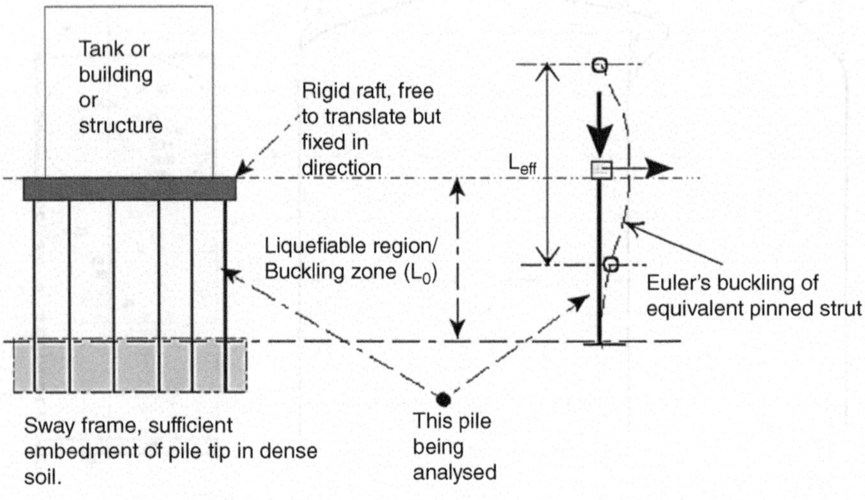

Figure 17.59 Effective length for a tank type structure (Bhattacharya 2003/Cambridge University Press).

- Axial stress σ in the pile, calculated by dividing P by the cross-sectional area of the pile, A. The value at failure is denoted as σ_f.

Reported Case Histories of Pile Foundation Performance during Earthquakes
This section of the chapter describes the fourteen case histories studied and analyzed. Out of the fifteen case histories, six were found to survive the earthquake, and the remaining eight either collapsed or suffered severe damage. Table 17.17 gives a summary of piles performance in case histories.

Analysis and Discussion
Figure 17.60a depicts a plot of the effective length of piles situated in a liquefiable zone, plotted against the minimum radius (r_{min}) of the pile section. A critical line is also displayed, representing a slenderness ratio (L_{eff}/r_{min}) of 50. This line serves as a crucial point of differentiation in structural engineering, distinguishing between "long" and "short" columns. Columns with slenderness ratios below 50 are classified as "short" columns and are expected to fail due to crushing. On the other hand, columns with slenderness ratios above 50 are considered "long" columns and are anticipated to fail due to buckling instability.

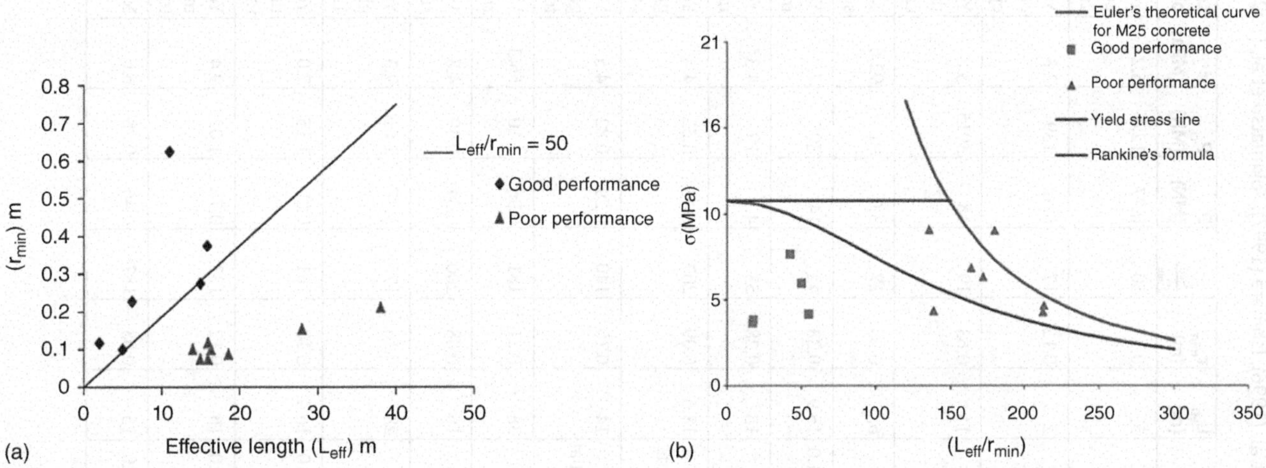

Figure 17.60 (a) L_{eff} versus r_{min} for piles studied, (b) plot of concrete pile performance mentioned in Table 17.19 (Bhattacharya 2003/Cambridge University Press).

Figure 17.60b shows the plot of 12 concrete piles mentioned in Table 17.17. The piles are assumed to be of M25 grade concrete (BS 8110), with a characteristic strength of 25 MPa. In the plot, three well-defined lines are drawn. These are:

(I) Yield stress line ($\sigma_y = 11.2$ MPa) taken as the design crushing value,
(II) Euler's curve for σ_{cr}, given by Equation 17.70, which is the elastic stability limit obtained by substituting the value of I into Equation 17.69 and noting that $\sigma_{cr}$is the critical stress given by (P_{cr}/A).

$$\sigma_{cr} = \frac{P_{cr}}{A} = \frac{\pi^2}{\left(\frac{L_{eff}}{r_{min}}\right)^2}E \tag{17.70}$$

(III) A curve for σ_f drawn using Rankine's formula (1866) shown by Equation 17.71. This design curve mediates the transition between strength and stability. Many other similar curves, such as Perry-Robertson (1886, 1925), can equally be used.

$$\frac{1}{\sigma_f} = \frac{1}{\sigma_{cr}} + \frac{1}{\sigma_y} \tag{17.71}$$

σ_y is yield stress of the material and σ_{cr} is the elastic critical stress as calculated by Equation 17.70, leading to an estimate of the combined failure stress σ_f.

Figure 17.61 plots the allowable load (P) against the Euler's buckling load (P_{cr}) for the failed piles. It may be observed that the piles that failed had a P/P_{cr} ratio between 0.5 and 1. On the other hand, Table 17.17 demonstrates that the piles that survived the earthquake had a (P/P_{cr}) ratio below 0.1.

(Continued)

Case Study 17.6 (Continued)

Table 17.17 Summary of pile performances. (Youd (1993), Hamada (1992), Tokimatsu et al. (1996), Ishihara (1997), Tokimatsu et al. (1998) are reproduced from Bhattacharya (2003) / Cambridge University Press).

S1 no.	Case History and Ref.	L[a] (m)	L_0[b] (m)	Pile Sec./type	Framing action/β[c]	L_{eff} (m)	r_{min} (m)	$\frac{L_{eff}}{r_{min}}$	P (MN)	P_{cr} (MN)	σ (MPa)	Lateral spreading observed?	Performance
1	10-storey Hokuriku building, 1964 Niigata earthquake, Hamada (1992)	12	5	0.4m dia. RCC	Large piled raft with basement. 1	5	0.1	50	0.77	12.4	6.2	Yes, the nearby ground moved by 2m.	Good
2	Landing bridge, 1987 Edgecumbe earthquake, Berrill et al (2001)	9	4	0.4m square PSC	Raked piles, no sway frame 0.5	2	0.12	17	0.62	139	3.8	Yes, the ground cracked and sand ejected.	Good
3	14-story building in an American park, 1995 Kobe earthquake, Tokimatsu et al. (1996)	33	11	2.5m dia. RCC	Large pile group and large pile dia. 1.0	11	0.63	19	18	3915	3.8	Yes, quay walls on the west, south, and east moved.	Good
4	Kobe Shimim Hospital, 1995 Kobe earthquake, Soga (1997	30	6.2	0.66m dia. Steel tube	Large piled raft with basement. 1.0	6.2	0.23	27	3.0	91	92.6	No, the ground subsided	Good
5	Hanshin expressway pier, 1995 Kobe earthquake, Ishihara (1997)	41	15.9	1.5m dia. RCC	Small group (22 piles) 1.0	15.9	0.38	42	14	305	7.9	Yes, the ground moved by 0.62m.	Good
6	LPG tank 101, Kobe earthquake, Ishihara (1997)	27	15	1.1m dia. RCC	Large piled raft 1.0	15	0.28	53	4.1	79	4.3	Yes, the ground moved by 0.7m.	Good
7	N.H.K building, 1964 Niigata earthquake, Hamada (1992)	12	9.3	0.35m dia. RCC	Groups tied by flexible beams, Less embedment at pile tip. 2.0	18.6	0.09	207	0.43	0.52	4.4	Yes, the nearby ground moved by 2m.	Poor
8	NFCH building, 1964 Niigata earthquake, Hamada (1992	9	7	0.35m dia. RCC hollow	Groups tied by flexible beam, Less embedment at pile tip 2.0	14	0.10	140	0.29	0.82	4.5	Yes, the nearby ground moved by 1 to 2m.	Poor
9	Showa bridge, 1964 Niigata earthquake, Hamada (1992)	25	19	0.6m dia. Steel tube	A single row of piles 2.0	38	0.21	181	0.96	1.10	56.3	Yes, the width of the river decreased.	Poor
10	Yachiyo Bridge, 1964 Niigata earthquake, Hamada (1992)	11	8	0.3m dia. RCC	Isolated footing 2.0	16	0.08	200	0.34	0.39	4.8	Yes, the width of the river decreased.	Poor
11	Gaiko Ware House, 1983 Chubu earthquake, Hamada (1992	18	14	0.6m dia. PSC hollow	Isolated footing 2.0	28	0.16	175	1.47	1.61	9.3	Yes, the nearby ground moved by 1.5 m.	Poor
12	4-storey firehouse, 1995 Kobe earthquake, Tokimatsu et al (1996)	30	16.4	0.4m dia. PSC	Groups tied by beam. 1.0	16.4	0.10	161	0.89	1.15	7.0	Yes, the building moved and tilted toward the sea.	Poor
13	3-story building at Fukae, 1995 Kobe earthquake, Tokimatsu et al. (1998)	20	16	0.4m dia. PSC hollow	Groups tied by beam. 1.0	16	0.12	133	0.72	1.02	9.4	Yes, the building moved and tilted toward the sea.	Poor
14	LPG tank 106,107 –1995 Kobe earthquake, Ishihara (1997)	20	15	0.3m dia. RCC hollow	Groups tied by beam. 1.0	15	0.08	187	0.46	0.38	6.6	No, ground subsided.	Poor

a) L = Length of the pile
b) L_0 = Length of pile in liquefiable region/buckling zone
c) β = Factor for estimating effective length. $L_{eff} = \beta L_0$

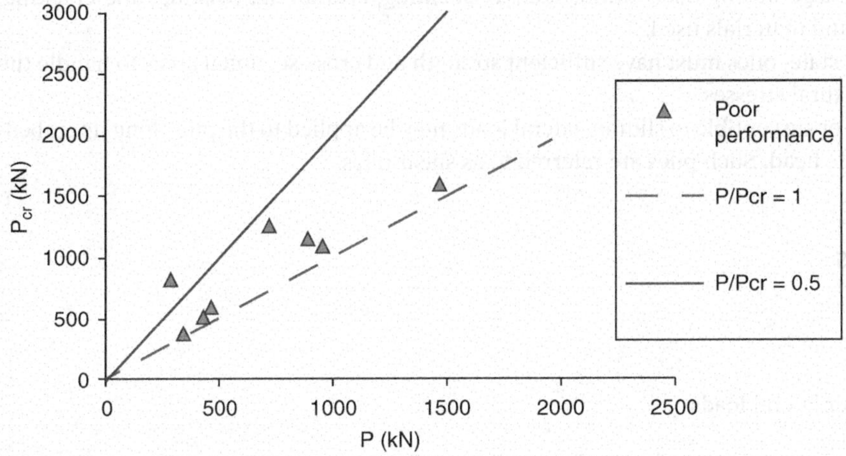

Figure 17.61 Plot of the poor performance of piled foundations (Bhattacharya 2003/Cambridge University Press).

The analysis of the case histories reveals a clear correlation between pile performance during liquefaction and buckling parameters. The behavior of short columns, which fail in crushing, and long columns, which fail in buckling, is analogous to the failure modes observed in pile foundations during liquefaction-induced earthquakes. In liquefied soils, the lateral support provided to the pile by the surrounding soil is diminished or completely removed due to the loss of soil strength during the earthquake. This loss of lateral support significantly impacts the behavior of the pile, making it more susceptible to buckling instability, similar to above-ground long columns failing without lateral support.

17.9 Major Points

1) The use of deep foundations allows for effective resistance against lateral loads and moment loads, especially compared to shallow and semi-deep foundations. Deep foundations can mobilize wall friction forces and utilize the lateral stiffness of the soil, contributing to their favorable performance under lateral loads.
2) During lateral load analyses, engineers consider different boundary conditions at the head of the pile, including the free-head, restrained-head, and partially-restrained-head conditions.
3) The behavior of a pile is classified as either a long flexible pile or a short rigid pile, depending on its length and flexural stiffness. When a pile is greater than a particular length, the length loses its significance. The behavior of the pile will not be affected if the length is greater than this particular length. Such piles are called *long flexible piles*. In the case of short piles, the flexural stiffness *EI* of the material of the pile loses its significance. The pile behaves as a rigid member and rotates as a unit.
4) The subgrade modulus variation coefficient (n_h) values used in the analysis are not constants and depend on various factors, such as soil properties, pile width or diameter, pile material's flexural stiffness, and pile deflection.
5) The rigidity or flexibility of a pile is influenced by several factors, including pile width, embedment length, material type, loading conditions, elastic modulus of the pile, and others.
6) The p–y method employs nonlinear p–y curves to model lateral soil pressure acting on the foundation and uses finite difference analysis to calculate deflections, shears, and moments.
7) Design p–y curves are derived from empirical data obtained through full-scale lateral load tests or in situ testing methods like PMT and CPT. They use common soil properties, such as undrained shear strength, as input parameters.
8) In foundation design, three critical aspects must be considered: lateral capacity, structural design, and serviceability issues.
9) The structural design of piles can be based on either allowable stress design (ASD) or load and resistance factor design (LRFD), as specified by applicable codes. Additionally, driven piles must be designed to withstand transportation, handling, and driving loads.

10) Piles may sustain various types of damage during installation, such as buckling, crumbling, twisting, and cracking, depending on the driving conditions and materials used.
11) To satisfy the structural ultimate limit state, piles must have sufficient strength and cross-sectional areas to handle the resulting compressive, tensile, and flexural stresses.
12) In cases where the ground is liquefied or susceptible to sliding, lateral loads may be applied to the pile along its embedment length instead of solely at the pile head. Such piles are referred to as shear piles.

17.10 Questions and Problems

17.10.1 Explanatory

1 Explain the applications of piles under lateral loads.

2 List the types of piles for lateral load resistance and compare them with one another.

3 Explain what the critical length represents.

4 List the factors that are influential on the flexibility or rigidity of piles.

5 Explain four manual analysis methods for piles under lateral loads.

6 List the factors that influence the shape of the p–y curves and name methods that can be used to establish them.

7 Explain how to establish p–y curves using CPT test data.

8 List 4 items of possible defects in displacement piles (Steel Piles) caused by driving.

9 Explain structural considerations of the drilled shaft and auger piles.

10 List 4 cases of problems with driven cast-in-place piles caused by groundwater.

17.10.2 Conceptual

1 Discuss the important boundary conditions in lateral load piles.

2 Define the role of the modulus of subgrade reaction in the deflection of the pile in-depth and discuss which issues can influence the modulus of subgrade reaction.

3 Explain how the load is transferred to soils in Battered Piles.

4 Evaluate the effect of the following parameters on the rigidity of piles:
a) Embedment length of the pile
b) Horizontal subgrade reaction
c) Pile cross-section
d) Pile material

5 Analyze how the coefficients A_y, A_s, A_m, A_v, and A_p are calculated.

6 Evaluate the effect of the following parameters on the shape of the p–y curves.
a) Soil properties
b) Pile cross-section

c) Coefficient of friction between the pile and the soil
d) Depth below the ground surface

7 Explain why there is a difference between the shape of p–y curves in stiff clays above and below the groundwater table.

8 Which test better agrees with the practical on-site results on the laterally loaded piles? PMT or CPT? Why?

9 Explain the difference between lateral load pile and shear pile. Describe the behavioral mechanism of each one.

10 Compare the structural considerations of steel piles with pre-stressed concrete piles. Investigate the effect of the shape of the cross-section of a pile on structural considerations of Steel piles.

11 Consider Case Studies 17.1 and 17.3 and explain the significant points learned through this case study.

12 Consider Case Studies 17.5 and 17.6 and explain the significant points learned through this case study.

13 Explain how to recognize the good or poor structural performance of a deep foundation under earthquake conditions (refer to Case Study 17.6).

17.10.3 Problems

1 A 12.75 in. (32.4 cm) diameter steel pipe pile is driven into a saturated medium clay with $S_u = 900\ \mathrm{lb/ft^2}$ (43.11 kPa) and $\acute{\gamma} = 80\ \mathrm{lb/ft^3}$ (12.57 kN/m^3). This pile is embedded to a depth great enough to be considered a long pile. Using a spreadsheet, using Reese to develop a family of p–y curves at depths of 5, 10, 15, and 20 ft. Plot all four curves on the same diagram. Given: $\varepsilon_{50} = 0.02$.

2 A steel pile 900 mm circular in the section is driven into saturated clay up to a depth of 12 m. The clay is in a submerged state. A lateral load of 40 tons and a vertical load of 60 tons is applied to the pile at 2 m above the ground level. Compute the lateral deflection of the pile at ground level. Given: $n_h = 15\ \mathrm{MN/m^3}$, $k_s = 7500\ \mathrm{kN/m^3}$, $E = 200$ GPa.

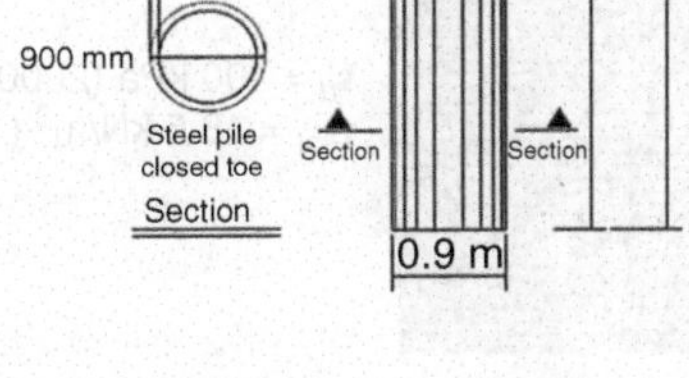

3 A vertical bored and cast-in-place pile 900 mm in diameter is installed to a depth of 6 m in a stiff over-consolidated clay ($C_u = 120\ \mathrm{kN/m^2}$, $\acute{c} = 10\ \mathrm{kN/m^2}$, $\acute{\phi} = 25°$). Find the maximum permanent horizontal load that can be applied at a point 4 m above-ground level using Broms and Evan-Duncan. Also, find the maximum applied load if the lateral deflection of the pile at ground level is limited to not more than 25 mm. Given that the soil modulus is 7.5 MN/m^2, the elastic modulus of concrete is $26 \times 10^3\ \mathrm{MN/m^2}$, and the moment of inertia of the pile is $0.0491 \times (900)^4\ \mathrm{mm^4}$, $\acute{\gamma} = 15\ \mathrm{MN/m^3}$ and $k_s = 7500\ \mathrm{kN/m^3}$.

4 Pressuremeter tests made at intervals of depth in a highly weathered weak broken siltstone gave the following parameters:

- Pressuremeter modulus = E_m = 30 MN/m^2
- Limit pressure = P_l = 1.8 MN/m^2
- Creeping pressure = P_f = 0.8 MN/m^2

The above values were reasonably constant with depth. Draw the deflection curve for a horizontal load applied to the head of a 750 mm pile at the ground surface up to the ultimate load, then obtain the deflection for a horizontal load of half the ultimate.

5 Develop an interaction diagram (a plot of allowable moment load versus allowable compressive load) for an HP 12 × 84 pile made of A572 steel (f_y = 50 k/in^2 = 344.86 MPa). Assume bending occurs along the weak axis. The required data are given below.

Section	Weight (lb/ft)	Area (in^2)	Depth (in)	Width (in)	I_x (in^4)	I_y (in^4)	Z_x (in^3)	Z_y (in^3)	S_x (in^3)	S_y (in^3)	ASD Allowable axial load (k)
HP 12 × 84	84.0	24.6	12.3	12.3	650	213	120	53.2	106	34.6	615

6 A 12.75-in-diameter steel pipe pile with a 0.50 in wall thickness is subject to the following ASD demands: P = 200 k (90.72 tons), V = 50 k (22.68 tons), M = 250 in-k (2.88 ton-m). Using ASD with the IBC allowable stresses and f_y = 50 k/in2 (344.86 MPa), determine if this section is adequate. The driving conditions will be checked using a wave equation analysis, and the requirements of IBC 1810.3.2.8 will be satisfied (V_a = 50,000 lb (22.68 tons), ϕ_v = 0.4).

7 A square precast concrete pile of sides 0.45 m (18 in.) will be driven 15 m (49 ft) into the soil strata shown in the bottom figure. Estimate the allowable load capacity for a safety factor of 2. If the lateral load is equal to 25% of the bearing capacity at the pile head, calculate the characteristic shear load and moment load using Evans and Duncan's method. Given: 28-day compressive strength is equal to 6000 lb/in^2, N_c = 12, ε_{50} = 0.01, δ = 0.75 φ.

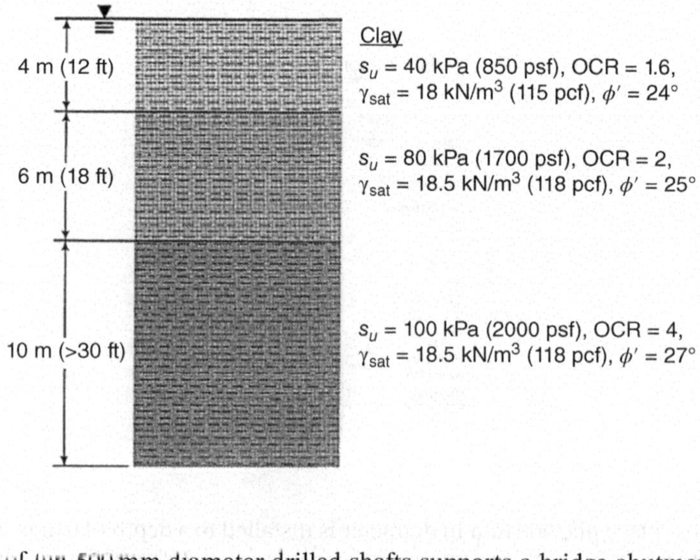

8 A pile group consisting of ten 500 mm diameter drilled shafts supports a bridge abutment. The pile spacing in both directions is 2 m, and the length of the pile group is 17 m. The soil profile consists of fine sand 2 m thick (γ_{sat} = 19 kN/m^3) above-medium sand (γ_{sat} = 20 kN/m^3). Groundwater is located 2 m below the ground surface. If the lateral load is 4 MN, compute the deflection of the pile head with the Elastic Continuum Method. The soil profile is homogenous. Given: $\acute{E}_P$ = 4400 ksi, υ = 0.25 and G = 5 MN.

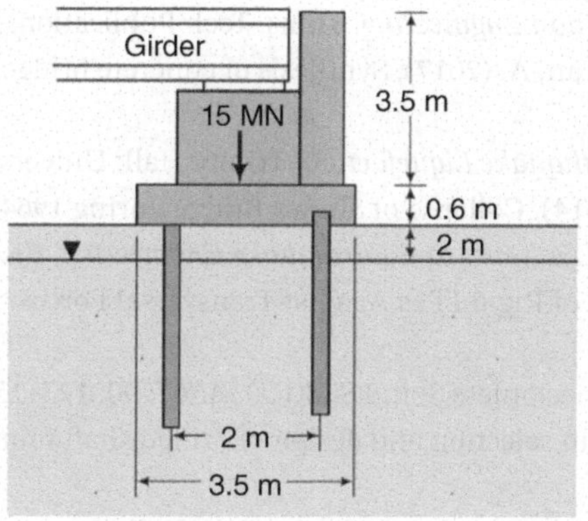

9 Plot a p–y curve with CPT results given in the figure below. Use the Novello Method (1999) to solve this problem. The pile length is 60 ft (18.29 m). The length of the pile is equal to its embedment depth. The diameter of the pile is 500 mm. Calculate soil-specific weight from the equation below:

$$\gamma = 26 - \frac{14}{1 + (0.5\log(f_s + 1))^2}$$

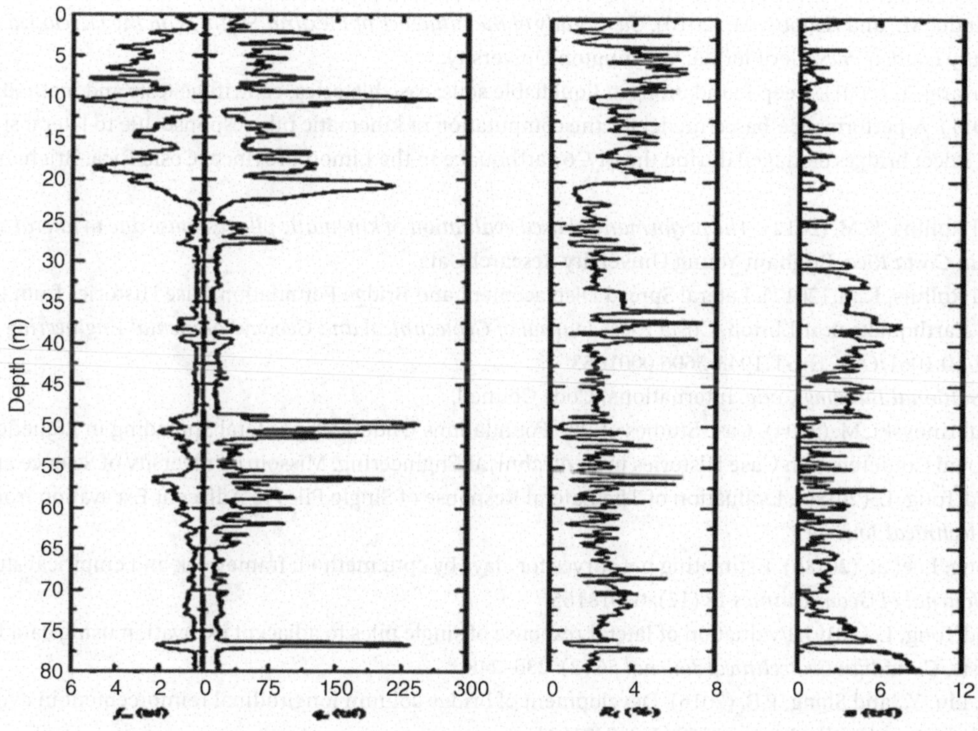

References

ACI (2012). *Guide to Design, Manufacture, and Installation of Concrete Piles*, Publication 543 R-12. American Concrete Institute.

American Association of State Highway and Transportation Officials (AASHTO) (2012). *AASHTO LRFD Bridge Design Specifications*, 6e, 5-15–5-17. AASHTO.

Anderson, J., Townsend, F., and Grajales, B. (2003). Case History Evaluation of Laterally Loaded Piles. *Journal of Geotechnical and Geo-environmental Engineering.*

Baguelin, F. (1978). *The Pressuremeter and Foundation Engineering*. Trans. Tech Publications.

Belarbi, A., Dawood, M., Bowman, M., and Mirmiran, A. (2017). Synthesis of concrete bridge piles prestressed with CFRP systems.

Bhattacharya, S. (2003). *Pile Instability During Earthquake Liquefaction*. Trinity Hall: University of Cambridge.

Bhattacharya, S., Tokimatsu, K., Goda, K. et al. (2014). Collapse of Showa Bridge during 1964 Niigata earthquake: A quantitative reappraisal on the failure mechanisms. In: *Soil Dynamics and Earthquake Engineering*. Elsevier.

Brinch Hansen, J. (1961). The Ultimate Resistance of Rigid Piles Against Transversal Forces. Danish Geotechnical Institute Bulletin, No.12, Copenhagen, 5–9.

Broms, B. (1964). Lateral Resistance of Piles in Cohesionless Soil. J.S.M.F.D. *ASCE* 90: 123–156.

Brown, D. (2010). Constructability considerations in selection and design of drilled shaft foundations for bridges. *Transportation Research Record* 2202 (1): 3–9.

Coduto, D.P. (2016). *Foundation Design, Principles and Practices*, 3rde. Prentice Hall.

Coduto, D.P., Kitch, W.A., and Yeung, M.R. (2016). *Foundation Design: Principles and Practices*, 3e. Upper Saddle River, NJ: Pearson Education.

Cornforth, D.H. (2005). *Landslides in Practice*. John Wiley & Sons.

Dunnavant, W. and O'Neill, M.W. (1989). Experimental p-y Model for Submerged Stiff Clay. *J. Geotech. Engineering* 1989 (115): 95–114.

Eslami, A. (2013). *Foundation Engineering, Design and Construction*, 2e. Building and Housing Research Center.

Evans, L. T. Jr., and Duncan, J. M. (1982). "Simplified analysis of laterally loaded piles." Rep. No. UCB/GT/82-04, University of California, Berkeley, CA.

Fernandez, S., Pando, M., and Ashour, M. (2010). *Case Study of the Influence of Flexural Stiffness on the Developed Soil Reactions of Three Laterally Loaded Piles*. GeoFlorida: Washington University.

Finn, W. and Thavaraj, T. (2001). Deep foundations in liquefiable soils: case histories, centrifuge tests and methods of analysis.

Franke, K. W. (2011). A performance-based model for the computation of kinematic pile response due to lateral spread and its application on select bridges damaged during the M7. 6 earthquake in the Limon Province, Costa Rica. Brigham Young University.

Franke, K.W. and Rollins, K.M. (2012). *The performance-based evaluation of kinematic pile response due to lateral spread at an historic bridge in Costa Rica*. Brigham Young University ResearchGate.

Franke, K.W. and Rollins, K.M. (2017). Lateral Spread Displacement and Bridge Foundation Case Histories from the 1991 Magnitude 7.6 Earthquake near Limon, Costa Rica. *Journal of Geotechnical and Geoenvironmental Engineering* 143 (6): https://doi.org/10.1061/(ASCE)GT.1943-5606.0001653.

ICC (2012). *International Building Code*. International Code Council.

Ishihara, K. & Cubrinovski, M. (2004). Case Studies of Pile Foundations Undergoing Lateral Spreading in Liquefied Deposits. 50th International Conference on Case Histories in Geotechnical Engineering. Missouri University of Science and Technology.

Li, H., Liu, S., and Tong, L. (2018a). Evaluation of The Lateral Response of Single Piles to Adjacent Excavation from CPT Data. *Canadian Geotechnical Journal*.

Li, H., Liu, S., Tong, L. et al. (2018b). Estimating p-y curves for clays by cptu method: framework and empirical study. *International Journal of Geomechanics* 18 (12): 04018165.

Li, H., Liu, S., and Tong, L. (2019). Evaluation of lateral response of single piles to adjacent excavation using data from cone penetration tests. *Canadian Geotechnical Journal* 56 (2): 236–248.

Murcia-Delso, J., Liu, Y., and Shing, P.B. (2016). Development of bridge column longitudinal reinforcement in oversized pile shafts. *Journal of Structural Engineering* 142 (11): 04016114.

Murthy, V.N.S. (2007). *Advanced Foundation Engineering*, 1ste. CBS Publishers and Distributers.

Novello, E. A. (1999). From static to cyclic p-y data in calcareous sediments Engineering for calcareous sediments 1, 17–24.

Pando, M. A. (2003). A laboratory and field study of composite piles for bridge substructures.

PCA (2007). *Corrosion of Embedded Metals*. Portland Cement Association.

PCI (2004). *Precast Prestressed Concrete Piles*. Precast Prestressed Concrete Institute.

PCI (2010). *PCI Design Handbook*, 7e. Precast/Prestressed Concrete Institute.

Pinto, P.L., Anderson, B., and Townsend, F.C. (1999). Comparison of horizontal load transfer curves for laterally loaded piles from strain gages and slope inclinometer: a case study. *ASTM Special Technical Publication* 1358: 3–15.

Poulos, H.G. and Davis, E.H. (1980). *Pile foundation analysis and design*, vol. 397. New York: Wiley.

Priestley, M.J., Singh, J.P., Youd, T.L., and Rollins, K.M. (1991). Costa Rica Earthquake of April 22, 1991, reconnaissance report. *Earthquake Spectra* 7 (1B): 79–90.

Randolph, M.F. (2018). *Potential Damage to Steel Pipe Piles During Installation. Fugro Chair in Geotechnics, Centre for Offshore Foundation Systems*. The University of Western Australia.

Reese, L.C. and Van Impe, W.F. (2001). *Single Piles and Pile Group Under Lateral Loading*, 463. Rotterdam: A. A Balkema.

Reese, L.C. and Welch, R.C. (1975). Lateral loading of deep foundations in stiff clay. *Journal of Geotechnical Engineering Division, ASCE* 101 (7): 633–649.

Reese, L. C., Cox, W. R., and Koop, F. D. (1974). Analysis of laterally loaded piles in sand. *Offshore Technology Conference*.

Reese, L.C., Cox, W.R., and Koop, F.D. (1975). Field testing and analysis of laterally loaded piles in stiff clay. *Proceedings, VII Annual Offshore Technology Conference*, Houston, TX, II(OTC 2312), pp. 672–690.

Roeder, C.W., Lehman, D.E., and Bishop, E. (2010). Strength and stiffness of circular concrete-filled tubes. *ASCE Journal of Structural Engineering* 136 (12): 1545–1553.

Robertson, P.K., Campanella, R.G., Gillespie, D., and Greig, J. (1986). Use of Piezometer Cone data. In-Situ'86 Use of In-situ testing in Geotechnical Engineering, GSP 6, ASCE, Reston, VA, Specialty Publication, SM 92, pp. 1263–1280.

Rollins, K. and Franke, K. (2011). *Subsurface Investigation and Analysis of Bridges in Costa Rica Damaged by Lateral Spreading to Evaluate and Improve Design Procedures for Kinematic Loading of Piles*. Golden, CO: USGS.

Sen, R. and Mullins, G. (2007). Application of FRP composites for underwater piles repair. *Composites Part B: Engineering* 38 (5–6): 751–758.

Suryasentana, S. and Lehane, B.M. (2014a). *Verification of numerically derived CPT based p-y curves for piles in sand*. The University of Western Australia.

Suryasentana, S.K. and Lehane, B.M. (2014b). Numerical derivation of CPT-based p–y curves for piles in sand. *Géotechnique* 64 (3): 186–194.

Terzaghi, K.V. (1955). Evaluation of coefficient of subgrade reaction. *Géotechnique* 5 (4): 297–326.

Tomlinson, M. and Woodward, J. (2015). *Pile Design and Construction Practice*, 60the. USA and Canada: Taylor & Francis.

Truong, P. and Lehane, B. (2014). Numerically derived CPT-based p-y curves for a soft clay modeled as an elastic perfectly plastic material. *3rd International Symposium on Cone Penetration Testing*.

Vinson, J. and Brown, D. (1997). Site characterization of the Spring Villa geotechnical test site and a comparison of strength and stiffness parameters for a Piedmont residual soil. Highway Research Center, Report IR-97, 4.

Section VII

Combined Loading and Group Actions

18

Combined Loading and Pile Group

Chapter Highlights

- **General Introduction to Pile Groups**
- **Insight on Applied Pressure and Displacement Relation**
- **Pile and Pile Group under Combined Loading**
- **Load Distribution in a Pile Group**
  - Rigid Pile Cap
  - Flexible Pile Cap
- **Capacity under Lateral Loads**
- **Structural Design of Pile Cap**
  - Sectional Method
  - Strut and Tie Method
- **Estimation of Settlement**
  - Neutral Plane Approach
- **Case Studies**
  - Three Oil Tanks Foundation, Belgium (Van Impe et al. 2013, 2015)
  - Heave of a Railway Bridge, Spain (Alonso and Raomn 2013; Alonso et al. 2015)
  - Static Response of a Group of 13 Piles Tested Simultaneously, Bolivia (Fellenius et al. 2019)
  - Actual Full-scale Pile Group Lateral Loading Tests, Japan (Teramoto et al. 2018)
  - Full-Scale Load Tests on Drilled Shaft Groups, China (Dai et al. 2012)

18.1 Introduction

Structures are rarely found on a single pile. Typically, a column or foundation part is supported by a minimum of three piles to address alignment issues and unintended eccentricity. Piled foundations consist of piles that strengthen the soil and transfer the applied load to a deeper and stronger soil layer. These piles are arranged in various geometric designs, such as squares, rectangles, circles, and octagons, with a pile center-to-center spacing (s) of at least two times the diameter or width of the pile (B). Examples of different geometric patterns are shown in Figure 18.1a.

A reinforced concrete block, known as the pile cap, is commonly used to cap the pile group and make the piles work together as a unit. The pile cap may or may not be in direct contact with the ground. When in contact with the ground, a portion of the load is directly transferred to the ground. Figure 18.1b,c illustrate two different pile cap conditions in pile groups.

Analyzing the behavior of a pile group foundation is a complex task that involves considering the interaction between the soil and the structure. Several factors influence the behavior of a pile group, including:

1) Pile installation methods: Whether the piles are bored or driven into the ground will affect their behavior.
2) Load transfer mode: The load can be transferred through shaft resistance or toe resistance of the piles or both.

Advanced Foundation Engineering: Principles, Performance and Prospects, First Edition. Moses Karakouzian and Abolfazl Eslami.

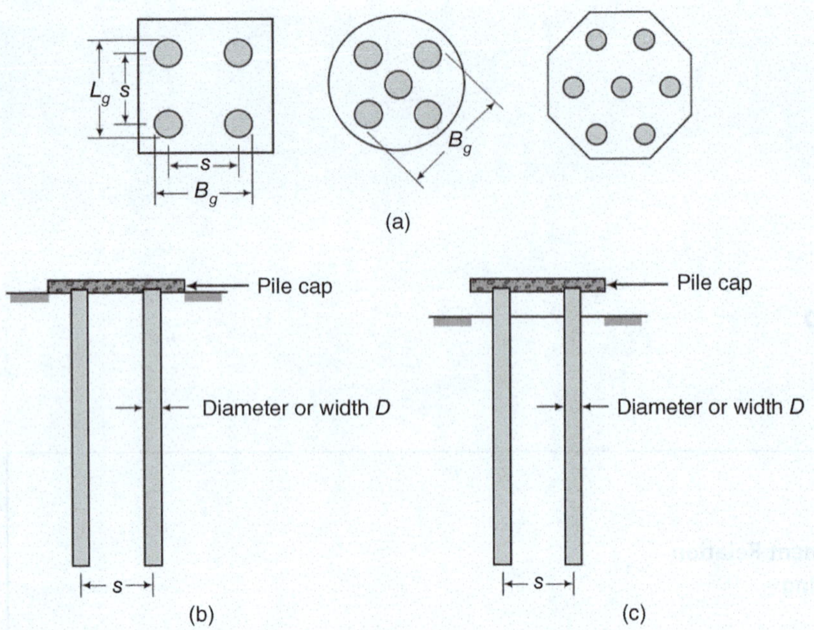

Figure 18.1 Pile groups: (a) example of different geometrical arrangements of pile groups; pile cap, (b) in contact with the ground, (c) not in contact with the ground (Budho 2011).

Table 18.1 Summary of example cases characteristics.

Case ID	Case description	$q_{avg.}/P_a$	Δ (mm)	B (m)	Δ/B	References
Case 1.a	Hypothetical Foundations	0.5	2.25	2.5	0.009	–
Case 1.b		2	2.25	2.5	0.009	
Case 2	Babolsar Twin Towers	6.4	70	30	0.0023	Sham-e Consulting Engineering Co. (2016)
Case 3	Messeturm building	5.44	144	58.8	0.002	Hemsley (2000)
Case 4	Third Millennium Bridge	4.8	20	22.4	0.0009	Fellenius and Ochoa (2016)

3) Subsurface geomaterials: The nature of the geomaterials in which the piles are constructed will influence pile group performance.
4) Pile group three-dimensional geometry: The arrangement and layout of the piles in three-dimensional space play a crucial role.
5) Presence of load-bearing pile cap: The presence or absence of a load-bearing pile cap can impact the load distribution.
6) Relative stiffness: The relative stiffness of the structure, the piles, and the ground will affect the overall response of the pile group.
7) Number, length, diameter, arrangement, and spacing of piles: The layout and arrangement of the individual piles within the group will influence the group's behavior.
8) Sequence of pile installation: The order in which the piles are installed can affect their load-sharing behavior.
9) Soil type: The properties of the surrounding soil, such as its strength and stiffness, are important factors.
10) Elapsed time since pile installation: The behavior of the pile group may change over time after the piles are installed.
11) Direction of applied load: The direction from which the load is applied can affect how the pile group responds.

Considering these factors and their interactions is crucial in accurately analyzing the behavior and performance of a pile group foundation.

Advancements in computational capabilities have significantly improved the analysis and design of pile groups. Traditional methods, often based on "rules of thumb" and empirical observations, have given way to more rational and

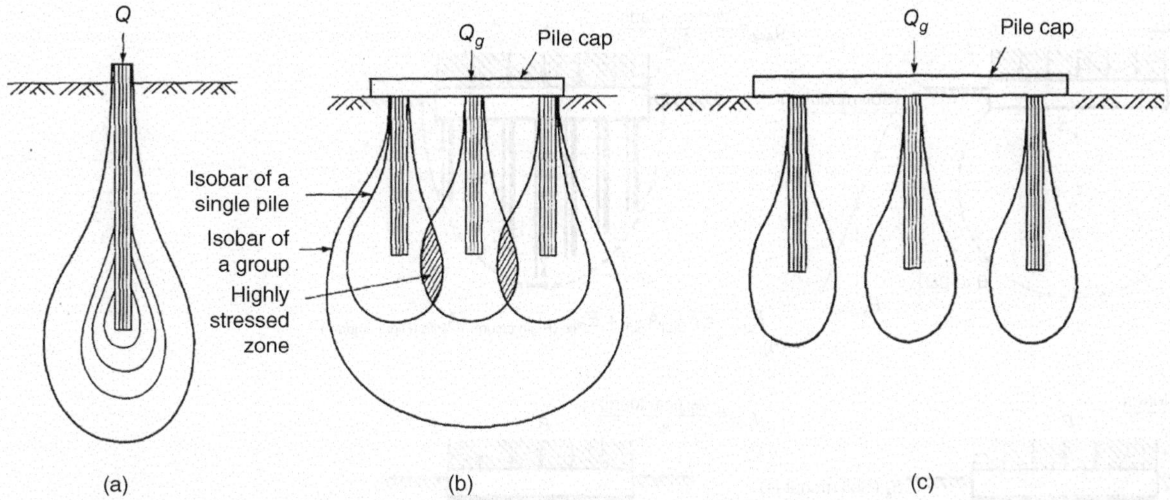

Figure 18.2 Stress bulbs: (a) single pile, (b) closely spaced pile group, (c) far apart pile group (Murthy 2007 / CBS Publishers and Distributers).

rigorous design approaches. With sophisticated numerical analysis tools and enhanced computing power, engineers can now examine typical pile groups, including those with vertical and battered piles, under complex loading conditions more accurately and efficiently. Parametric studies, which involve systematically varying different parameters to understand their effects on pile group behavior, can now be conducted more affordably and effectively. These studies allow engineers to explore various design scenarios and optimize the performance of pile groups in different soil and loading conditions.

18.1.1 Pile Spacing

In general, the proximity of adjacent piles diminishes a pile's efficiency, particularly in terms of stiffness. To address this issue, it is recommended to use fewer, more widely spread piles, emphasizing the role of piles as reinforcement.

Figure 18.2a illustrates the pressure isobars of a single pile with load Q acting on the top. When piles are grouped, the pressure isobars of adjacent piles may overlap, as shown in Figure 18.2b. In zones with overlapping pressures, the soil experiences extreme stress. If there is sufficient overlap, either the soil fails or the pile group settles severely, as the combined pressure bulb extends to a substantial depth below the pile base. To avoid overlap, spacing the piles widely apart is necessary, as depicted in Figure 18.2c. However, it is essential to note that using large spacings is not always advisable, as it results in a larger pile cap, increasing the overall cost of the foundation.

The spacing between piles within a group is determined by a variety of factors, including the following:

1) Stresses from adjacent piles overlap,
2) Foundation cost,
3) Pile group efficiency.

The spacing between piles in a pile group is influenced by the installation technique and the type of soil. When piles are driven, soil displacement can lead to increased overlapping stresses. In loose sandy soils where soil displacement compacts the soil between the piles, the piles can be spaced closer together. However, in saturated clay or silty soils where the displaced soil does not compact, the soil between the piles may shift upward, lifting the pile cap. To avoid this issue in such soils, larger spacings between piles are required.

When piles are cast in situ, the soils close to the piles are not subjected to the same degree of stress, allowing for smaller spacings. However, using closer spacing should only be considered after a thorough analysis of its effect on settlement and bearing capacity. Special attention should be paid to adjacent piles at varying depths to investigate the effects of load transfer and soil deformation from upper-level piles to lower-level ones. It is also essential for the designer to specify a pile installation schedule within the group to provide the largest possible space between installed shafts and newly concreted shafts. This careful consideration of pile spacing ensures the stability and effectiveness of the pile group foundation.

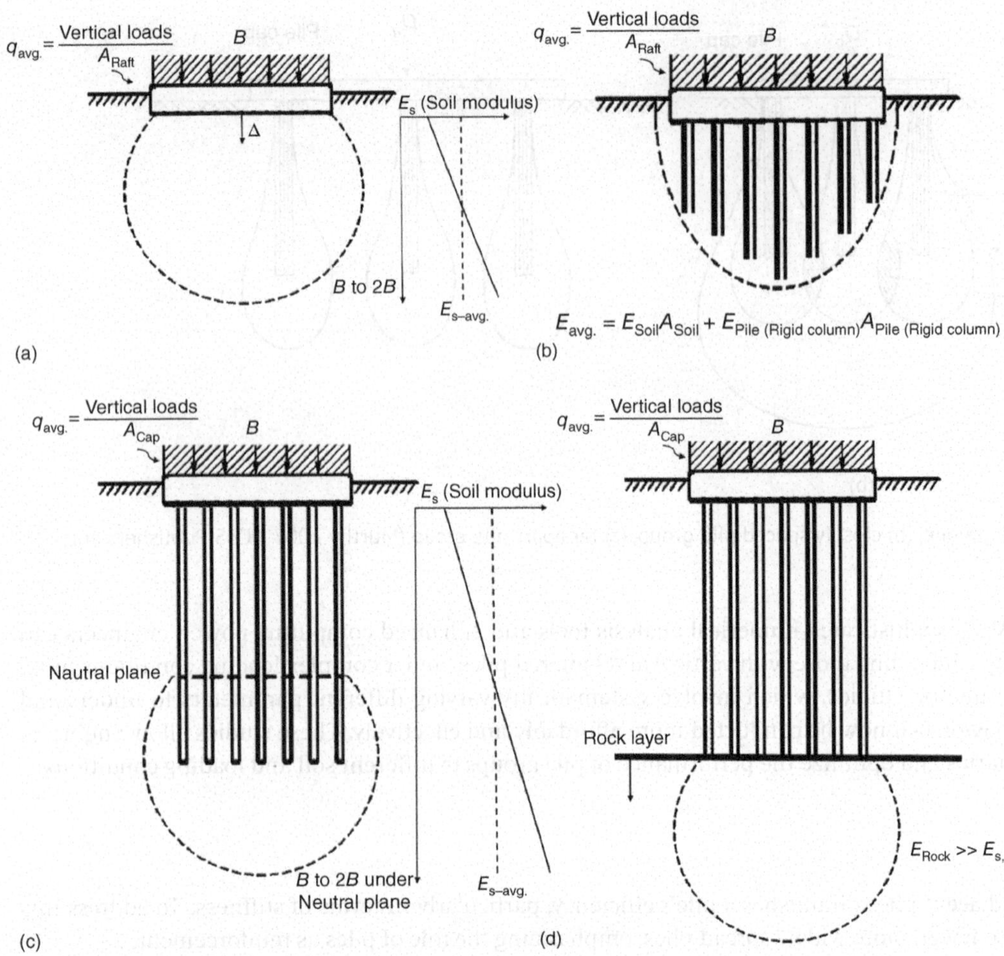

Figure 18.3 Schematic stress bulb for settlement calculation: (a) unpiled raft, (b) piled raft (or raft combined with rigid intrusions), (c) pile group implemented in uniform soil, and (d) pile group underlain by rock.

18.2 Insight on Applied Pressure and Displacement Relation

Engineers usually consider a depth ranging from B to 2B for stress-influencing zones in settlement calculations. Figure 18.3 illustrates the schematic stress bulbs induced in the underlying geomaterials due to applied loads. Figures 18.3a to 18.3d depict a mat without piles, a mat (or raft) combined with piles or rigid columns, a pile group implemented in uniform soil, and a pile group underlain on rock, respectively.

Figure 18.4a provides an overview of the relationship between the projected pressure and the displacement ratio in medium-dense or medium-strength soils, considering typical ranges for applied pressure and displacement ratio in actual projects. The parameters used in Figure 18.4a have been introduced and illustrated in Figure 18.3. On the other hand, Figure 18.4b illustrates the general condition of bearing capacity and settlement, along with usual instances relevant to and compatible with each zone. Several example cases have been used to demonstrate the application of Figure 18.4a. It is important to note that Figure 18.4a presents a general approach and can be further improved or modified through additional investigations.

18.3 Pile and Pile Group under Combined Loading

Structures supported on piles often encounter not only vertical loads but also lateral and moment loads. Lateral loads can arise from various sources, such as wind, traffic, seismic events, waves, docking ships, and soil pressures. Additionally,

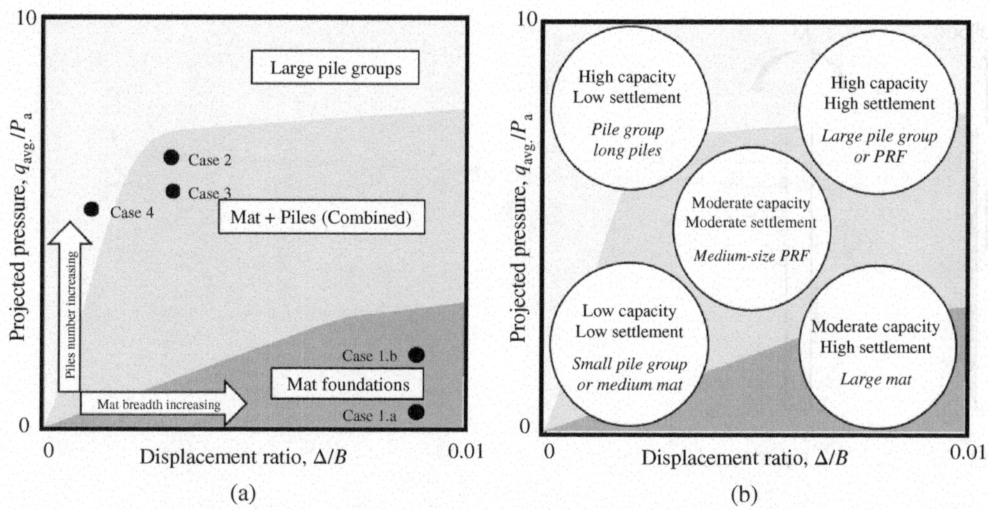

Figure 18.4 (a) Schematic $q_{avg.}/P_a$-Δ/B curves for mats, combined foundations and large pile groups, (b) overall condition of bearing capacity and settlement along with common instances in each zone.

moments can be induced by the eccentricity of the vertical force, the connection between the superstructure and the piles, or the application of lateral forces on the pile relative to the ground surface. Figure 18.5 illustrates the load transfer mechanism and the deflection behavior of single piles under various load conditions.

Figures 18.5d to 18.5f clearly show that non-axial loads can lead to significant deflections of the pile head. When subjected to combined loading, the individual performance of single piles may raise concerns in terms of geotechnical, structural, and serviceability aspects. However, when piles are grouped together with pile caps and interconnected soil between them, they function as a unified block foundation, effectively transferring a combination of loads to the underlying soil in a safe manner. Figure 18.6a illustrates a pile group comprising both battered and vertical piles subjected to combined loading. Figure 18.6.b demonstrates how a pile group operates as a block foundation.

18.4 Load Distribution in a Pile Group

Pile groups typically encounter a combination of loads, including vertical and horizontal forces, as well as moments in one or two directions. It is crucial to meticulously determine the axial loads on each pile within the group, considering both geotechnical and structural requirements. In this context, the rigidity of the pile cap assumes a vital role. The pile cap's stiffness significantly influences the load distribution among the individual piles. It is essential to ensure that each pile effectively carries its share of the vertical loads and moments to maintain the foundation's stability and structural integrity. Improper load distribution could lead to uneven settlements and potential overloading of specific piles, jeopardizing the overall performance of the foundation.

To accurately determine the axial loads on each pile, engineers must analyze the behavior of the pile group under different loading conditions, accounting for the interaction between the piles, soil, and pile cap. Advanced analytical methods, such as finite element analysis, may be employed to achieve precise load distribution based on the pile cap's rigidity and the applied loads. Considering the pile cap's stiffness, the foundation can be designed to effectively withstand the combined loads, ensuring the safety and reliability of the structure.

18.4.1 Rigid Pile Cap

When using a rigid pile cap, it is common to perform an uncoupled analysis of the axial and lateral pile reactions. This approach assumes that the pile–soil interaction mechanisms are independent of the loading direction, whether axial or lateral. The spacing between the piles allows for balancing the structure's moments by developing increments (both positive and negative) to the axial force exerted on each pile. In this analysis, the axial loads on individual piles are adjusted to account for the lateral loads and moments acting on the structure. The stiffness of the pile cap ensures that the loads

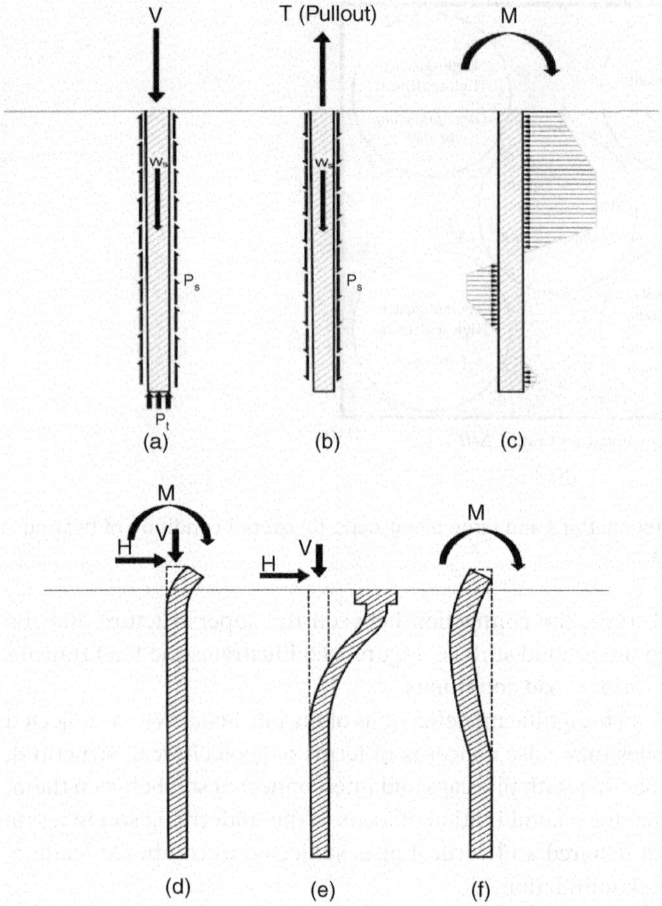

Figure 18.5 Load transfer mechanism of piles: (a) compressive loads, (b) tensile loads, (c) lateral loads. Different pile head deflection under uniaxial loads and head connection type: (d) free-head, (e) fully restrained head, and (f) pure rotation.

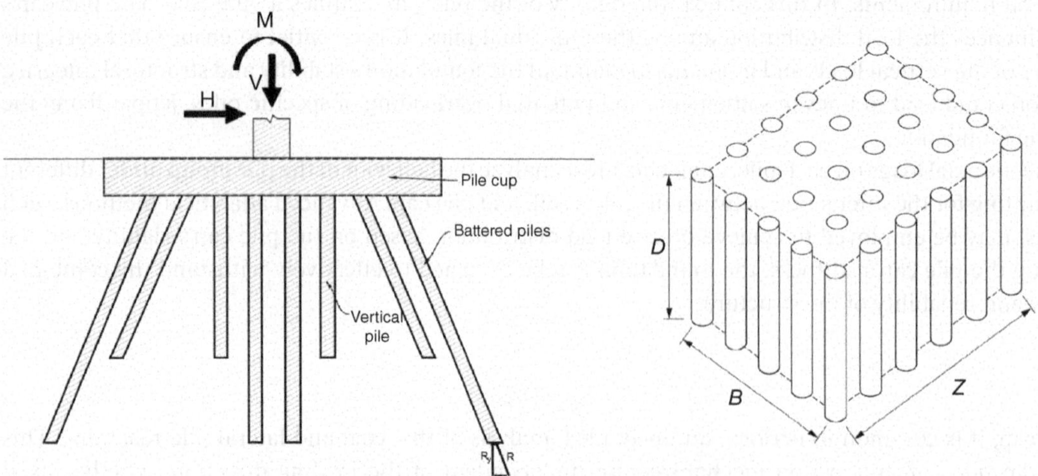

Figure 18.6 (a) Battered piles being used in combination with vertical piles to resist combined vertical and horizontal loads, (b) pile group acting as block foundation (Hannigan et al. 2016; Tomlinson and Woodward 2014).

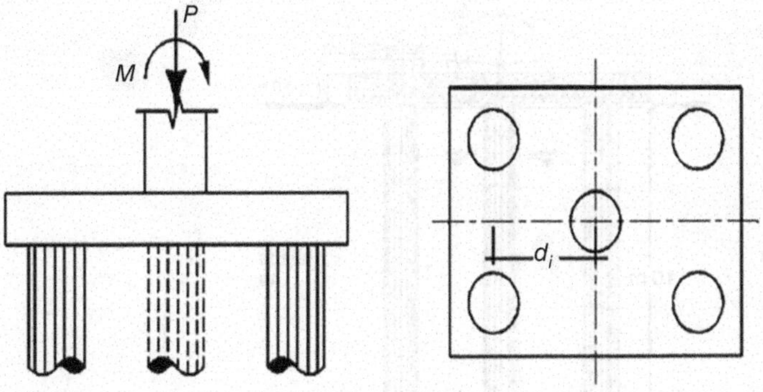

Figure 18.7 Axial force designation of a pile group (Eslami 2013).

are distributed appropriately among the piles to achieve equilibrium and maintain the stability of the foundation. By considering the pile cap's rigidity and the interaction between the piles and the surrounding soil, engineers can accurately determine the distribution of axial loads and ensure that the foundation can effectively resist axial and lateral loads. This uncoupled approach simplifies the analysis while still providing reliable results for the design of pile groups under combined loading conditions.

$$\sigma = \frac{P}{nA} \pm \frac{Md_i}{I} \tag{18.1}$$

Where:
σ = the stress in each pile
A = the cross-section of piles
n = the number of piles
d_i = Distance of each pile to the center of gravity
I = Moment of inertia of piles based on neutral axis (flexural axis)

$$I = \sum I_0 + \sum Ad_i^2 \tag{18.2}$$

Neglecting I_0 (the inertia moment of each pile concerning the passed axis through the gravity center of each pile parallel to the main axis):

$$I = \sum Ad_i^2 \tag{18.3}$$

By replacing Equation 18.3 in Equation 18.1:

$$\sigma = \frac{P}{nA} \pm \frac{Md_i}{\sum Ad_i^2} \tag{18.4}$$

By multiplying Equation 18.4 in A and considering $F = \sigma \times A$, the axial force of each pile will be:

$$F = \frac{P}{n} \pm \frac{Md_i}{\sum d_i^2} \tag{18.5}$$

Plus is used for compressive loads, and negative is used for tensile loads.

When horizontal loads are applied to the pile group, the load distribution is not uniform, which is a result of the pile–soil–pile interaction process discussed in Section 18.6.

Due to the development of a complicated load transfer system, the response of a group of piles can differ from that of a single isolated pile, resulting in the "group effect." This issue will be discussed in detail in the following sections.

Example 18.1 Consider the given pile group, assuming the pile cap as rigid and ultimate loads as: M = 60 ton.m, P = 360 tons and H = 180 tons, determine the maximum and minimum stress and axial load of piles. Control the compressive stress in piles section ($\sigma_{allowed}$ = 15 MPa)

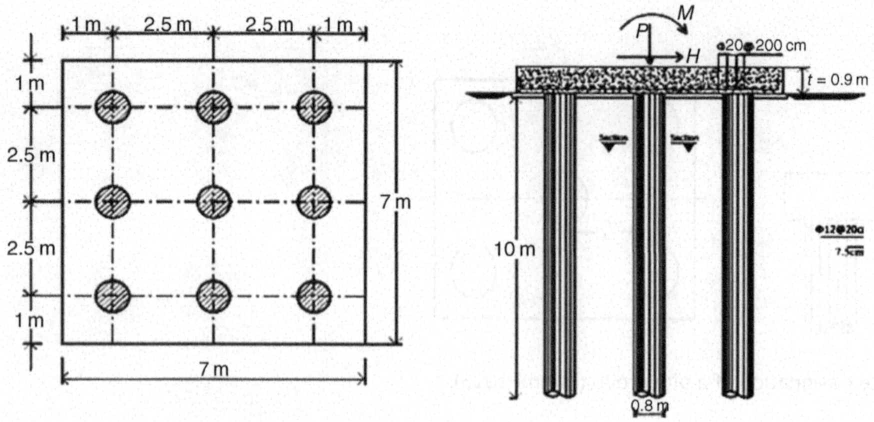

Stress:

$$\text{Maximum Stress: } \sigma = \frac{P}{nA} + \frac{Md_i}{\sum Ad_i^2} = \frac{360}{9 \times \frac{\pi \times 0.8^2}{4}} + \frac{60 \times 2.5}{6 \times \frac{\pi \times 0.8^2}{4} \times 2.5^2} = 87.5\ \frac{ton}{m^2}$$

$$\text{Minimum Stress: } \sigma = \frac{P}{nA} - \frac{Md_i}{\sum Ad_i^2} = \frac{360}{9 \times \frac{\pi \times 0.8^2}{4}} - \frac{60 \times 2.5}{6 \times \frac{\pi \times 0.8^2}{4} \times 2.5^2} = 71.6\ \frac{ton}{m^2}$$

$$\sigma_{max} = 87.5\ \frac{ton}{m^2} < \sigma_{allowed} = 15\ MPa = 1500\ \frac{ton}{m^2} \rightarrow O.K.$$

Force:

$$\text{Maximum Force: } F = \sigma \times A = 87.5 \times \frac{\pi \times 0.8^2}{4} = 44\ ton$$

$$\text{Minimum Force: } F = \sigma \times A = 71.6 \times \frac{\pi \times 0.8^2}{4} = 36\ ton$$

18.4.2 Flexible Pile Cap

In pile foundations, the rigidity of the pile cap is a crucial factor in ensuring the proper distribution of loads among the piles. If the cap is considered rigid enough, all the piles will be equally loaded. However, in cases where the cap is flexible, the piles closest to the loading point may become overloaded compared to the other piles in the group.

The thickness of the pile cap is a critical design element in pile–cap foundations. A thin cap acts as a flexible slab, leading to uneven load distribution, which is contrary to the principles of pile foundation construction. On the other hand, a thick cap functions as a rigid slab, evenly distributing the load on the piles and effectively resisting bending moments and punching shear. However, the additional weight of a thick cap may impose extra stress on the piles. Thus, the ideal cap should be thick enough to provide rigidity but also flexible enough to distribute the load uniformly without adding excessive weight.

In conventional approaches to pile cap design, shear is determined by assuming that each pile reacts independently to the external load. The shear is then calculated as the total load applied to the cap divided by the number of piles. However, making such assumptions about flexible caps may lead to an unsafe or overdesigned foundation.

El Hammouli (2021) conducted a study to investigate the influence of pile cap thickness and its rigidity on load distribution within a pile group. The main findings from the research are as follows:

- The pile cap exhibits flexibility up to a certain thickness, beyond which it transforms into a rigid slab, leading to nearly uniform load distribution among the piles.
- The stiffness of the pile cap improves as the spacing between piles decreases, especially for larger caps.
- The behavior of the pile cap can vary based on the soil's strength; it may act rigid on strong soil and flexible on weak soil.
- The stiffness of the pile cap increases as the diameter of the piles increases, resulting in a larger contact area between the piles and the cap, effectively reducing settlement.
- The rigidity of the pile cap increases proportionally with the length of the piles, similar to the frictional resistance of the piles.

- Among the parameters studied, the spacing between piles does not significantly impact load sharing between the cap and the pile group. However, load sharing increases linearly with both pile diameter and length.

These conclusions shed light on the importance of pile cap thickness and its rigidity in influencing the load distribution within a pile group. Understanding these factors can lead to more efficient and reliable pile foundation designs.

18.4.3 Comments on Load Distribution in Pile Groups

The interaction between piles in a group can lead to nonuniform load distributions, particularly in groups with uniformly loaded piles. When a rigid pile cap is used to distribute the load, the outer piles tend to bear a greater proportion of the applied load compared to the center piles. This phenomenon has been observed in typical load distributions for small pile groups, as reported by Poulos and Davis (1980), where corner piles may hold up to three to four times the load of central piles. Similarly, Cooke et al. (1981) demonstrated that in a large group of 351 piles, piles near the corners took two to three times the load as the middle ones. Engineers are unsure of how to mitigate such load distributions.

Engineers often face uncertainty in dealing with such nonuniform load distributions in pile groups. This uneven distribution of loads among the piles can have significant implications for the overall performance and safety of the foundation system. Therefore, it is essential for engineers to carefully consider and account for the effects of pile interaction and load distribution when designing pile foundations to ensure their stability and effectiveness under various loading conditions. Innovative design approaches and advanced analysis techniques are being explored to better address this issue and optimize the performance of pile groups under complex loading scenarios.

Numerous options exist:

1) Extend the length of the outer piles to ensure their stability under increased loads.
2) Abandon the center piles as they add little to the pile group's stiffness.
3) Lengthen the central piles to support a greater portion of the load, thus balancing the load distribution.

Upon closer examination, the first solution of longer outer piles transferring additional load to the foundation's borders may indeed result in increased bending moments in the pile cap or increased differential settlements. This could potentially lead to uneven settlement and differential movements of the foundation, which can be detrimental to the overall stability and performance of the structure.

Similarly, the second approach of concentrating the structure's weight at the foundation's borders may not always be practical or feasible, as the distribution of loads in a building or structure is often complex and cannot be easily concentrated at specific points.

The third alternative, which involves finding a viable balance between pile lengths and load distribution, seems to be a more promising solution. This approach requires careful engineering analysis and consideration of various factors, such as pile spacing, pile stiffness, and soil properties, to ensure that the load is adequately distributed among the piles and that the foundation can effectively support the structure's weight without excessive settlements or differential movements.

Assuming that the loading from the superstructure is sufficiently uniform, the foundation must be capable of uniformly transmitting the load to the earth. In an idealistic situation, bending moments in the pile cap and differential settlements are reduced, which often requires more support from piles in the center of the foundation than towards the periphery (to counteract the natural tendency of a uniformly loaded region to form a dish shape in the center). This strategy becomes especially appealing when the higher soil layers are reasonably competent, and the pile cap may directly transfer some of the load to the ground.

In many circumstances, the stiffness of the structure above the pile cap will be sufficient to minimize concerns about high loads in the peripheral piles. A straightforward technique based on equally distributed piles will suffice, even if it is not the most efficient. The design should guarantee that the pile material is not overstressed, and an adequate allowance is made for the possibility of induced high bending moments at the pile cap's edges and corners.

18.5 Bearing Capacity Under Axial Loads

The interaction between piles of a group and the surrounding soil is complicated, and the bearing capacity of the group may differ from the total of the individual piles' bearing capacities. Figure 18.2 illustrates the rising soil loading due to the overlapping of soil stresses caused by each pile. As implied by the same Figure, the settlement of a pile group is also greater than the settlement of a single pile given to the same axial force as one of the piles due to the increased volume of soil subjected to significant stress increments.

The pile group bearing capacity, Q_{ug}, is given by:

$$Q_{ug} = E_g n Q_u \tag{18.6}$$

where,

Q_{ug} = the pile group bearing capacity
E_g = the group efficiency factor
n = the number of piles
Q_u = the axial-bearing capacity of a single pile

The group efficiency factor may be less than or more than one based on various factors. (Coduto et al. 2016):

(i) the number, length, diameter, arrangement, and spacing of piles;
(ii) the relative contributions of shaft and toe resistances to the ultimate bearing capacity;
(iii) the sequence of pile installation, mainly in displacement piles;
(iv) the soil type;
(v) the interaction between the pile cap and the soil;
(vi) the direction of the applied load (tension or compression).

Due to a relative lack of experimental data, validating theoretical models used to predict the group efficiency factor is challenging (Fernandes 2020). However, several trends are readily apparent:

(i) in loose sand, the group efficiency of displacement piles is typically greater than one (Lo 1967; Vesić 1969), owing to the increased normal effective stress on the shaft and the increased compaction of the soil around the piles driven in a group (Fleming et al. 2009);
(ii) due to soil remolding, the group efficiency of displacement piles installed in clayey soils is typically less than one. Another point to consider is the group bearing capacity's time dependence;
(iii) There are even fewer experimental data for non-displacement piles; hence, according to some writers (Das 2004; Reese et al. 2005), the group efficiency of bored piles should be slightly less than 1 for pile spacing smaller than 3B.

Figure 18.8 illustrates the variation in group efficiency for a three-group pile in sand (Kishida and Meyerhof 1965). Group efficiency, in this context, refers to the ratio of the ultimate bearing capacity of the pile group to the sum of the individual piles' ultimate bearing capacities. For loose and medium sands, the group efficiency can be more than unity, primarily due to the densification of the sand surrounding the piles. On the other hand, in the case of dense sands, they can dilate and become looser during pile driving, leading to a group efficiency that is less than unity. When the piles are spaced far apart, the group efficiency tends to be close to unity, regardless of the relative density of the sand (Das and Sivakugan 2019).

In close-spaced pile groups, an alternative failure mechanism can be observed, as depicted in Figure 18.9. For such groups installed in clay under undrained conditions, the bearing capacity is defined as the total of the lateral (skin) resistances

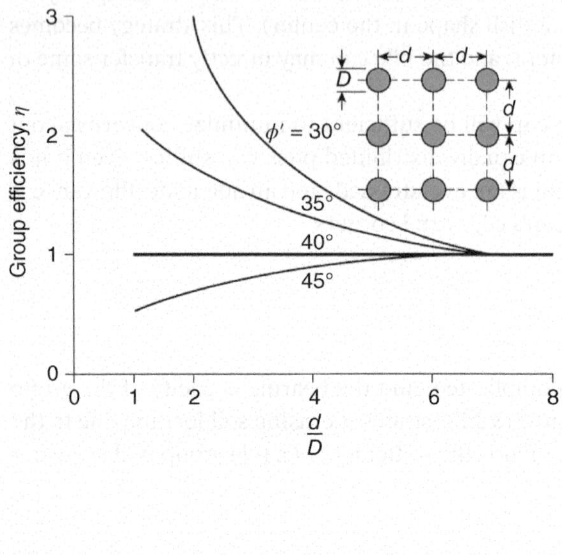

Figure 18.8 Variation of pile groups efficiency in sand (adapted from Kishida and Meyerhof 1965).

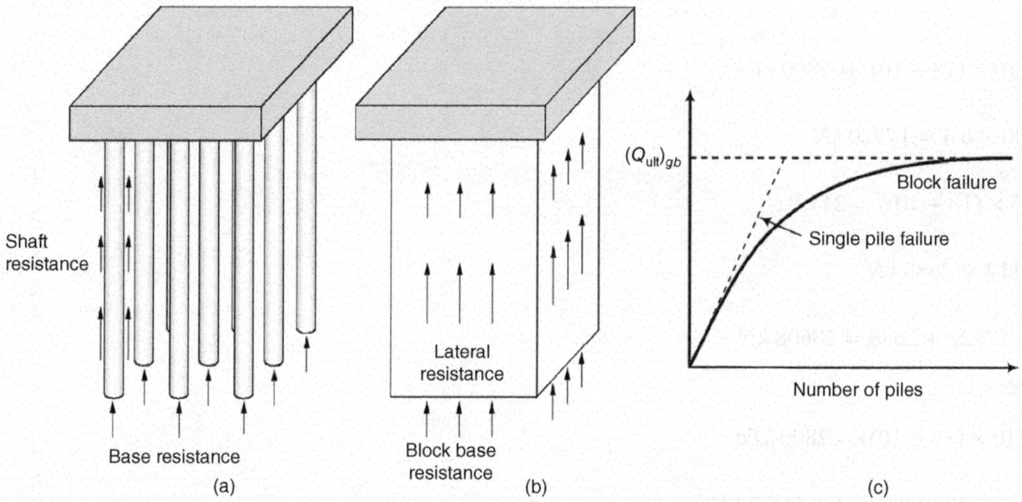

Figure 18.9 Pile groups failure modes: (a) single pile failure, (b) block failure (Reese et al. 2005), and (c) bearing capacity variation based on pile numbers (Budho 2011).

along the block perimeter, with the base resistance calculated using the area corresponding to the block's base. Figure 18.9c illustrates the bearing capacity variation based on the number of piles for two different failure modes in such close-spaced pile groups.

Example 18.2 A group of piles, comprising 8 driven steel piles (H-Piles) is used to construct an open-type wharf. The dimensions and positions of piles are as follows. Designate the ultimate axial-bearing capacity of the pile group in two different conditions of underlying soil, assuming that no plug has been formed:

a) Sand: $\gamma_{sat} = 18\ \frac{kN}{m^3}, \beta = 0.6, N_t = 35$
b) Clay: $S_u(kPa) = 20 + Z\ (m), \alpha = 1, N_C = 9$

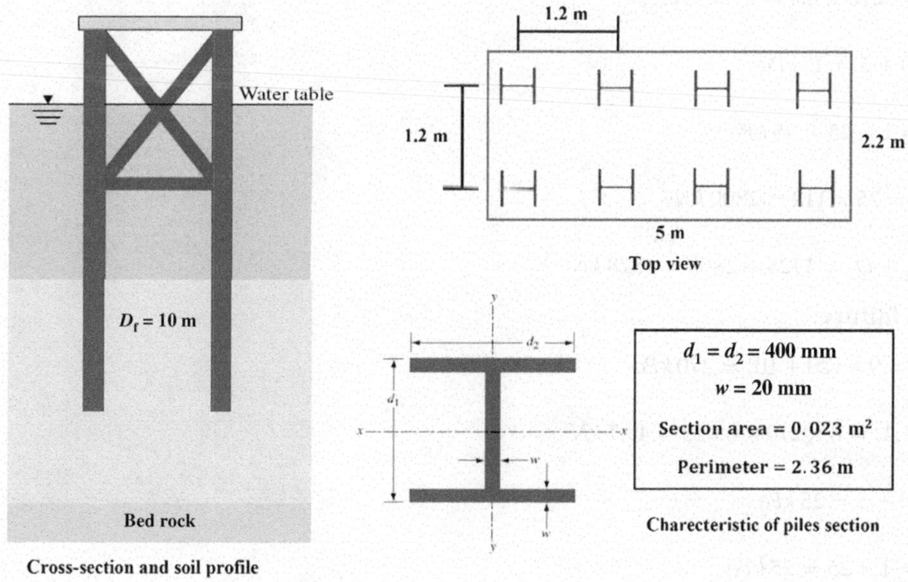

Group dimensions:

$$B_g = 1.2 + 0.4 = 1.6\,m, L_g = (3 \times 1.2) + 0.4 = 4\,m$$

$$A_{shaft} = 2 \times (B_g + L_g) \times D_f = 2 \times (1.6 + 4) \times 10 = 112\,m, A_{base} = B_g \times L_g = 1.6 \times 4 = 6.4\,m^2$$

Sand, block failure:

$$r_t = N_t \times \sigma'_v = 35 \times (10 \times (18 - 10)) = 2800\ kPa$$

$$Q_t = r_t \times A_{base} = 2800 \times 6.4 = 17920\ kN$$

$$r_s = \beta \times \sigma'_v = 0.6 \times (5 \times (18 - 10)) = 24\ kPa$$

$$Q_s = r_s \times A_s = 24 \times 112 = 2688\ kN$$

$$Q_{u-Block} = Q_t + Q_s = 17920 + 2688 = 20608\ kN$$

Sand, individual failure:

$$r_t = N_t \times \sigma'_v = 35 \times (10 \times (18 - 10)) = 2800\ kPa$$

$$Q_t = 8 \times r_t \times A_{base} = 8 \times 2800 \times 0.023 = 515.2\ kN$$

$$r_s = \beta \times \sigma'_v = 0.6 \times (5 \times (18 - 10)) = 24\ kPa$$

$$Q_s = 8 \times r_s \times A_s = 8 \times 24 \times (2.36 \times 10) = 4531.2\ kN$$

$$Q_{u-Individual} = Q_t + Q_s = 515.2 + 4531.2 = 5046.4\ kN$$

$$Q_{u-Sand} = min\{Q_{u-Block} = 20608\ kN, Q_{u-Individual} = 5046.4\ kN\} = 5046.4\ kN$$

Clay, block failure:

$$r_t = N_c \times S_u = 9 \times (20 + 10) = 270\ kPa$$

$$Q_t = r_t \times A_t = 270 \times 6.4 = 1728\ kN$$

$$S_{u-Z=5m} = 20 + 5 = 25\ kPa$$

$$r_s = \alpha \times S_u = 1 \times 25 = 25\ kPa$$

$$Q_s = r_s \times A_s = 25 \times 112 = 2800\ kN$$

$$Q_{u-Block} = Q_t + Q_s = 1728 + 2800 = 4528\ kN$$

Clay, individual failure:

$$r_t = N_c \times S_u = 9 \times (20 + 10) = 270\ kPa$$

$$Q_t = 8 \times r_t \times A_t = 8 \times 270 \times 0.023 = 49.7\ kN$$

$$S_{u-Z=5m} = 20 + 5 = 25\ kPa$$

$$r_s = \alpha \times S_u = 1 \times 25 = 25\ kPa$$

$$Q_s = 8 \times r_s \times A_s = 8 \times 25 \times (2.36 \times 10) = 4720\ kN$$

$$Q_{u-Block} = Q_t + Q_s = 49.7 + 4720) = 4769.7\ kN$$

$$Q_{u-Clay} = min\{Q_{u-Block} = 4528\ kN, Q_{u-Individual} = 4769.7\ kN\} = 4528\ kN$$

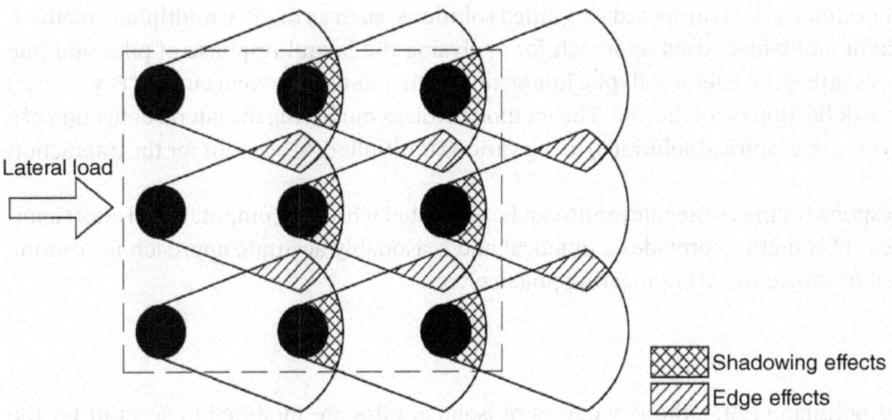

Figure 18.10 Shadowing and edge effects in a laterally loaded pile group (Walsh 2005).

Soil Type (Analyze Type)	Block Failure (kN)	Individual Failure (kN)
Sand (ESA*) not defined	20608	5046.4
Clay (TSA**) not defined	4528	4769.7

*ESA: Effective Stress Analysis
**ESA: Total Stress Analysis

When a pile group consisting of n piles is subjected to a lateral force exerted at the cap level, the following takes place:

- the lateral displacement of the pile head on all piles in the group is equal to the cap displacement;
- when subjected to a lateral load corresponding to uniform load distribution, the group's lateral displacement is higher than the displacement of a single isolated pile with identical boundary conditions at the head;
- the load distribution provided by the piles is not uniform and is dependent on the location of each pile.

When a pile group is subjected to lateral loading, each individual pile pushes against the adjacent soil, creating localized shear zones. As the load increases, these shear zones expand to mobilize the required soil reaction, leading to overlapping effects between adjacent piles. This overlapping effect, known as the shadowing effect, reduces the soil response mobilization in the intermediate rows of piles. Additionally, the overlapping effect can also occur between adjacent piles in the same row, known as the edge effect. As a result, the piles positioned in the corners of the group often play a significant role in the overall lateral load balancing. Figure 18.10 illustrates these overlapping and shadowing effects between piles in a pile group under lateral loading.

As depicted in Figure 18.10, the reaction generated by the peripheral corner piles in a pile group is only slightly less than the reaction provided by an isolated pile under identical conditions. However, due to the shadowing and edge effects, the piles positioned within the group experience substantially smaller reactivity. Despite all piles experiencing the same head displacement, the soil reaction is site-dependent, leading to greater shear forces on piles in the external rows compared to piles in the interior rows.

Experimental evidence provided by Brown et al. (1988) and Rollins et al. (1998) supports the existence of this interaction problem. The effect becomes more pronounced in pile groups with numerous piles. Neglecting this effect may result in a considerable overestimation of the mobilized internal forces on periphery piles during the design process. Therefore, considering the interaction effects in pile groups is essential to accurately analyze and design pile foundations for lateral loading conditions.

A detailed understanding of the complex phenomena involved in geotechnical problems, including the behavior of pile groups under lateral loading, requires a method based on a continuous formulation. However, these approaches, which account for the three-dimensional nature of the problem and the nonlinearity of the soil, can be computationally costly and may not be practical for routine design support.

To address this challenge, Brown and Bollman (1993) proposed simplified solutions, such as the P–y multipliers method. The P–y multipliers method is an efficient and widely used approach for analyzing the lateral response of piles and pile groups. It simplifies the problem by representing the lateral soil–pile interaction with one-dimensional curves (P–y curves) that relate the lateral load (P) to the lateral deflection (y) of the pile. The method involves modifying the lateral deflection of a single isolated pile (obtained from analytical or empirical solutions) using various multipliers to account for the interaction effects between piles in a group.

By using P–y multipliers, the lateral response of the entire pile group can be estimated with less computational effort compared to full three-dimensional analyses. This method provides a practical and reasonably accurate approach for routine design support, making it a valuable tool in geotechnical engineering practice.

18.5.1 P–y Curves Method

In the method proposed by Brown and Bollman (1993), the P–y curves of isolated piles are modified to account for the pile–soil–pile interaction effects in a pile group. The modification is done as a function of the position of each pile within the group. Figure 18.11 illustrates this concept. This approach is applied as follows:

- For a single pile, use a P–y curve that is suitable for the soil and pile properties;
- Applying a P-multiplier, P_m, to the chosen P–y curve results in the generation of a revised P–y curve, as seen in Figure 18.5. P_m values of 0.8 for the lead row, 0.4 for the second row, and 0.3 for the third and subsequent rows are suggested with $s = 3B$ (B is the diameter of the pile). P_m values of 1.0 are recommended for the lead row, 0.85 for the second row, and 0.7 for the third and subsequent rows when $s = 5B$. Linear interpolation may be used for intermediate pile spacing;
- Create a numerical model (for example, using the finite element technique) in which the piles are represented as beam elements connected by a rigid cap, and the updated P–y curves are assumed as functions of the pile position relative to the load direction. (see Figure 18.11).

Because all group piles typically have the same structural configuration, pile design is determined by the shear forces and bending moments mobilized in the important row, i.e., the front row, as seen in Figure 18.11.

18.6 Structural Design of Pile Cap

For the design of pile caps, engineers commonly employ one of two methods: the sectional approach for shallow pile caps and the strut-and-tie (STM) method for deep pile caps. Both approaches utilize the same procedure to determine the cap thickness and are designed to satisfy the requirements for one-way and two-way shear. However, in both cases, shear is typically determined by assuming the reaction of each pile to the external load, which is calculated by dividing the total load applied to the cap by the number of piles. It is important to note that this assumption holds true for rigid caps.

For flexible caps, the situation is different. In such cases, the piles interact with the cap and the surrounding soil in a more complex manner, leading to varying load distributions among the piles. The traditional approach of equally distributing loads to each pile may not accurately represent the actual behavior of the system. In these instances, a more detailed analysis considering the flexibility of the cap and pile–soil–pile interaction may be necessary to assess the load distribution and shear forces accurately.

Engineers must carefully consider the cap's rigidity and the interaction between the piles and the cap to ensure a reliable design. When dealing with flexible caps, more advanced techniques like the modified p–y curves method or other numerical methods may be required to model the behavior of the pile group accurately under lateral loads. While the sectional approach and STM method are effective for rigid caps and meet shear requirements, flexible caps demand special attention and may require more sophisticated analysis techniques to accurately predict shear forces and ensure the structural integrity of the foundation.

18.6.1 Sectional Method

The method described is based on flexural theory and treats the pile cap as a shallow foundation. It models the piles and pile caps as single foundations, simulating the load from the superstructure as similar to a column. Instead of considering a distributed reaction of soil beneath the entire pile cap, individual reactions from each pile in the group are taken into

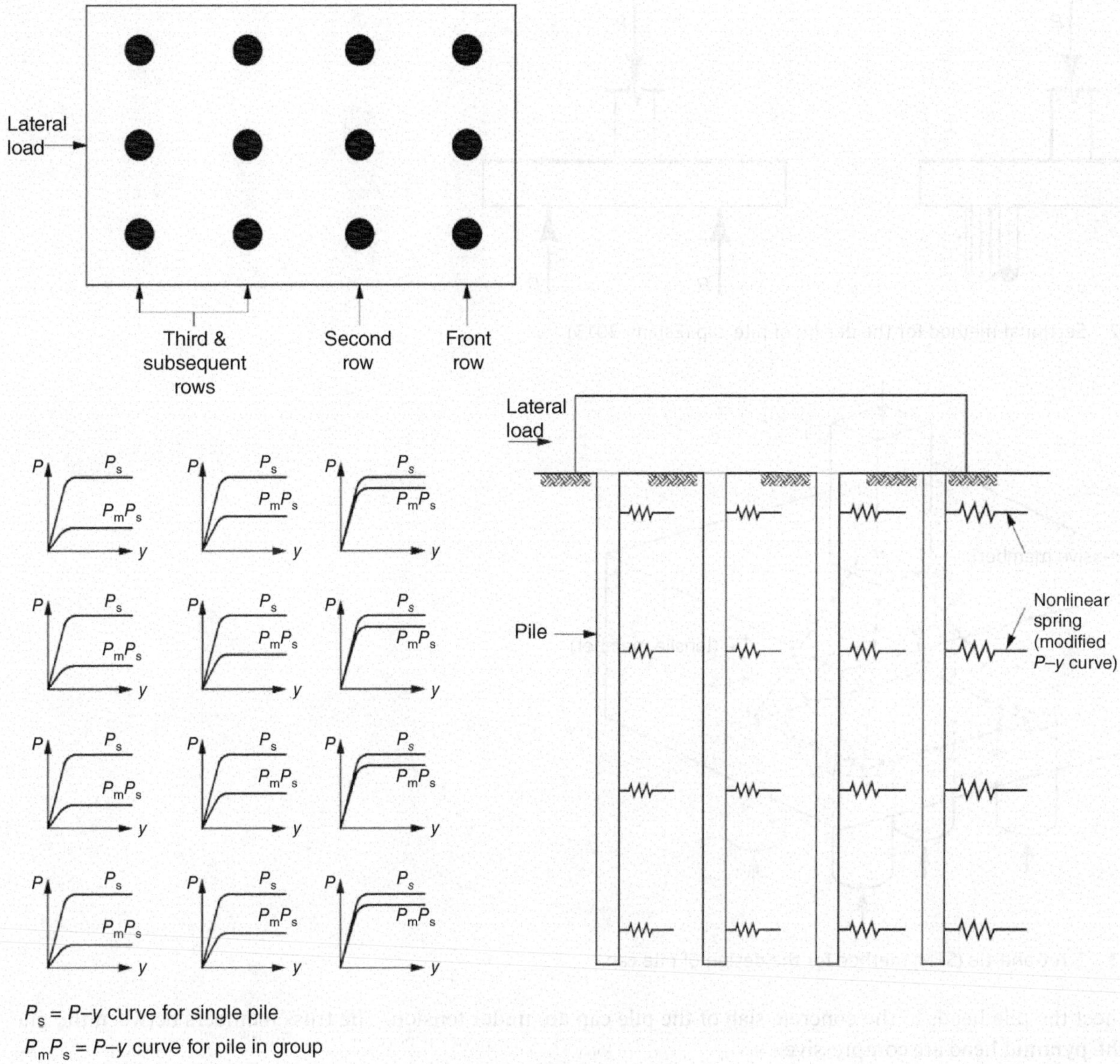

Figure 18.11 P-multiplier idea for lateral analysis (Hannigan et al. 2016).

account. The critical section for moment calculation is located adjacent to the column, wall, or pillar. To determine the thickness of the pile cap, punching shear is checked at a distance of d/2 around the pillars and the contact area between piles and pile caps (d is the effective depth of the pile cap section). Additionally, one-directional shear is checked at a distance of d from the column or pillar.

Figure 18.12 provides a schematic view of this method, illustrating how the piles and pile cap are treated as a single unit, allowing for a more straightforward analysis of the pile group's response to applied loads from the superstructure. By focusing on individual pile reactions and considering specific critical sections, engineers can accurately design the thickness and reinforcement of the pile cap to ensure its structural integrity and capacity to transmit loads effectively to the underlying soil.

18.6.2 Strut-and-Tie (STM) Method

In this method, the piles and pile cap are assumed to be a three-dimensional truss in which the compressive loads of superstructure are imposed on the pick of the pyramid (see Figure 18.13), and the piles support this truss. The truss members

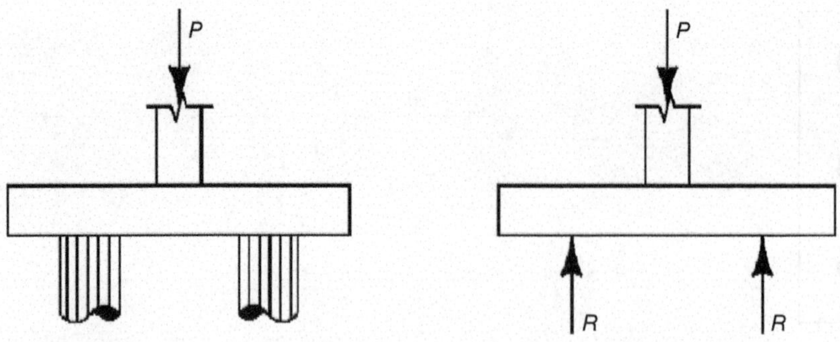

Figure 18.12 Sectional method for the design of pile cap (Eslami 2013).

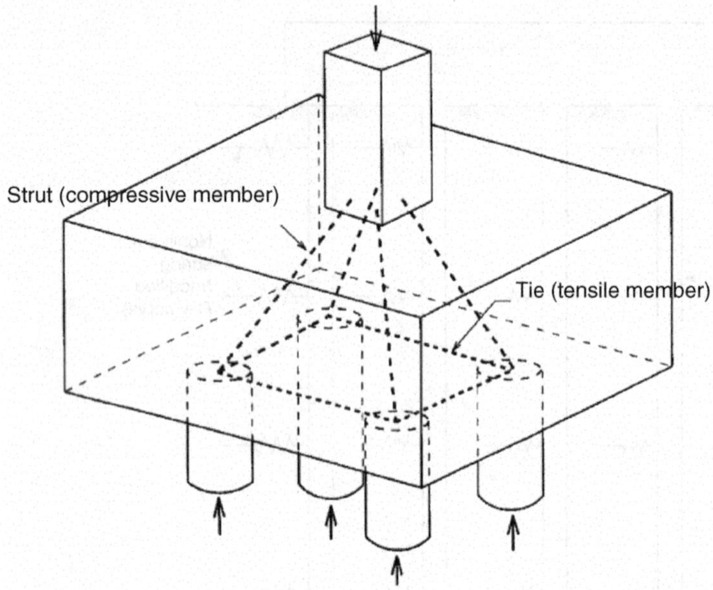

Figure 18.13 Strut-and-tie (STM) method for the design of pile cap.

which connect the pile heads to the concrete slab of the pile cap are under tension. The truss members between the pile head and the pyramid head are compressive.

In the design of pile caps and other concrete structures, concrete is primarily used to bear compressive loads, while steel reinforcement is employed to provide tensile strength.

18.6.3 General Guides in Design (Bowles 1997)

The following guidelines can be used for designing pile caps:

- Reinforcement of pile caps is required for both positive and negative bending forces. Ensure that concrete close to the earth has a minimum cover of 70 mm. When piles extend about 70 mm into the cap, the bottom cap reinforcement should extend 75 mm above the pile top to prevent concrete cracking at the pile head.
- Pile caps should extend at least 150 mm, preferably 250 mm, beyond the outer face of external piles. When piles extend more than 75 mm into the cap, the bottom rebars should wrap around the pile to prevent the cap from being split by pile head moments and shears.
- Pile heads should extend at least 300 mm into the cap, and the minimum thickness of the pile caps above the pile heads should be maintained.

- If the piles are exposed to tension stresses, certain tension shear connections should be installed on the pile heads.

It is recommended to obtain pile cap moments and shears using finite element analysis (FEM) or a computer program for design purposes. When dealing with battered piles or additional degrees of freedom in the load, a specific program must be employed to produce an accurate pile cap solution.

18.7 Settlement of Pile Group

Depending on their spacing, piles in a group tend to interact. The closer the spacing, the more interaction, and settlement.

18.7.1 Equivalent Footing Method

An equivalent footing method is a simplified approach for determining the settlement of a pile group by treating the group as a single footing. The location and dimensions of the equivalent footing are determined based on how the load is transferred.

For granular soils, the settlement of the equivalent footing can be estimated using the elastic approach, while for clays, it can be calculated using the consolidation theory. The settlement at the pile's tip is obtained by adding the settlement of the equivalent footing and the elastic compression of the pile's length above the footing.

The equivalent footing method is commonly used to estimate the settlement of pile groups with simple geometry in routine calculations. However, it may not properly consider the effects of pile spacing or interaction effects. The application of stiffness correction factors accounts for the relative stiffness between the structure and the foundation. It is essential to use this method cautiously when analyzing groups of piles with complex geometry, different pile lengths, or under completely nonuniform loading conditions. A more comprehensive analysis may be necessary for such cases.

Figure 18.14 presents the Terzaghi and Peck model, which is based on the assumption of a wide equivalent footing located at a depth of 2D/3 from the pile's bottom, with a slope of 2 to 1, to distribute service loads. The model calculates the pile group settlement based on this approach.

18.7.2 United Method Based on Neutral Plane

In the unified method proposed by CFEM (1992) and Fellenius (1989), the load distribution form for pile group settlement calculations is referred to as the neutral plane. It represents the point where the settlements of the surrounding soil and the pile are equal. Above this neutral plane, the soil settles more than the pile, and below it, the pile settles more than the soil

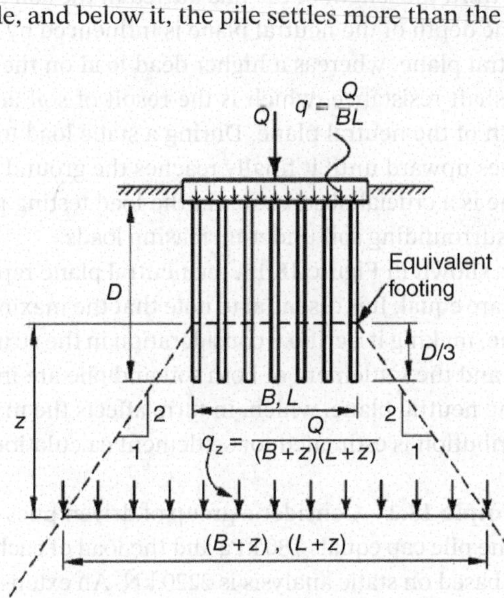

Figure 18.14 Terzaghi and Peck proposed a method for a pile group settlement.

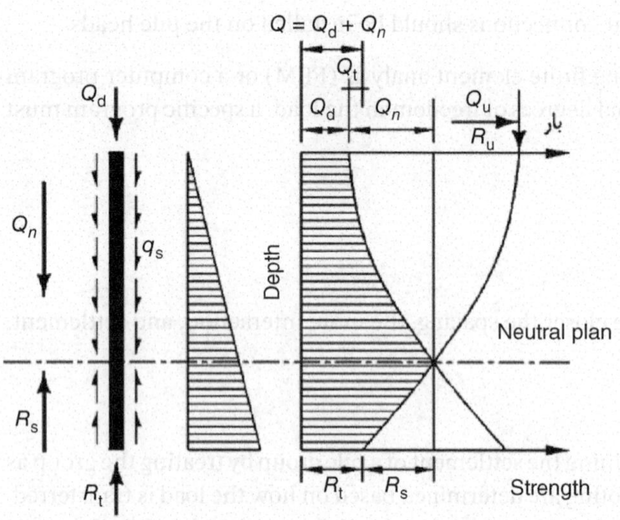

Figure 18.15 Determination of neutral plane (Fellenius 1989).

(Figure 18.15). This concept helps in understanding the distribution of settlements within the pile group and is essential for accurate settlement analysis and design considerations.

To justify the concept of the neutral plane, it is necessary to analyze the relative displacement between the pile and the surrounding soil. When the pile moves relative to the soil, positive friction is mobilized, while negative friction is activated when the soil pulls the pile down (Figure 18.15). For equilibrium of forces, the resultant of all static forces must be zero. At the top of the pile, negative friction creates a force that changes from zero to its maximum value at a certain depth called the equilibrium depth. Below this depth, the axial load of the pile decreases due to the positive frictional resistance and the end resistance. In each pile, the applied dead loads and the negative friction are in equilibrium with the positive frictional resistance and the toe resistance. The neutral plane is the depth at which the shear stress on the pile wall changes from negative to positive, and the relative displacement between the pile and the soil is zero.

The location of the neutral plane is usually lower than the middle of the pile. In the special case where the bottom of the pile is on a rock mass, the neutral plane is located at the pile's toe. For a floating pile based solely on frictional resistance, in a homogeneous soil with a linear increase in frictional resistance, the neutral plane is typically placed at a point about one-third the length of the pile buried in the soil above the pile's end.

The depth of the neutral plane is influenced by several factors. A higher toe resistance results in a deeper location of the neutral plane, whereas a higher dead load on the pile head leads to a lower placement of the neutral plane. Additionally, the shaft resistance, which is the result of soil and pile shaft interaction, along with the pile length, also influences the depth of the neutral plane. During a static load test on a pile, with a gradual increase in load, the neutral plane gradually moves upward until it finally reaches the ground surface at the moment of failure. This upward movement of the neutral plane is a crucial observation in the load testing process, as it provides valuable insights into the behavior of the pile and the surrounding soil under increasing loads.

As shown in Figure 18.16, the neutral plane represents the level at which the settlement of the surrounding soil and the pile are equal. It is essential to note that the maximum structural loads in the piles coincide with the location of the neutral plane, making it a critical consideration in the structural design requirements. The load distribution mechanism within the pile and the settlement of both soil and pile are interrelated. Changes in the applied loads cause variations in the position of the neutral plane, which, in turn, affects the maximum load distribution along the pile and alters the way constant load distribution is estimated for settlement calculations.

Example 18.3 Consider a group of driven piles (D=0.45m) under a steel tank of water. The constant dead load imposed on the pile cap equals 180 kPa and the load of each pile is equal and will be 1200 kN. The ultimate-bearing capacity of each pile based on static analysis is 2220 kN. An extensive fill with the height of 2 m and $\gamma = 22\ \frac{kN}{m^3}$ is constructed in a wide area around the pile cap, which is circular in shape with a diameter of 42 m. Assuming the failure mode as individual and sand layer is extended to the depth of 60 m.

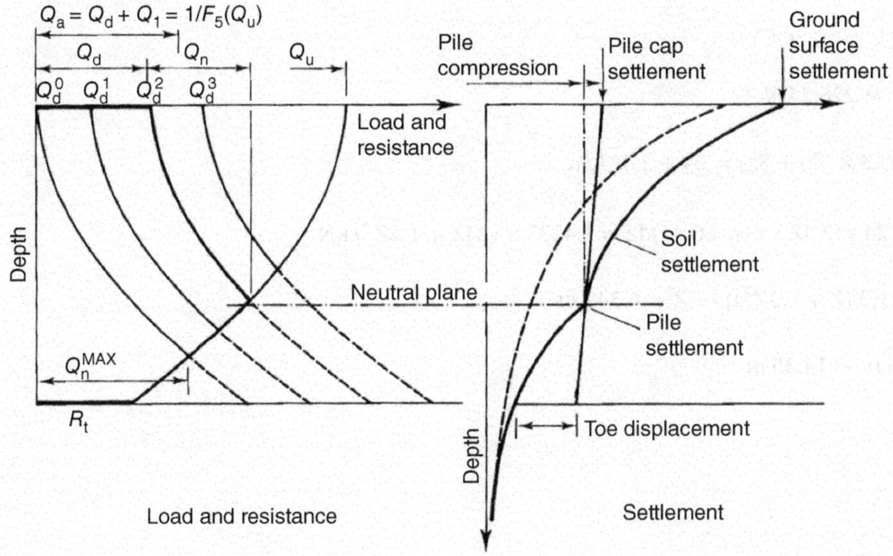

Figure 18.16 Settlement dependence on the neutral plane location (Fellenius 1989).

a) Determine the depth of neutral plane
b) Plot the axial load of piles
c) Estimate the settlement of pile group

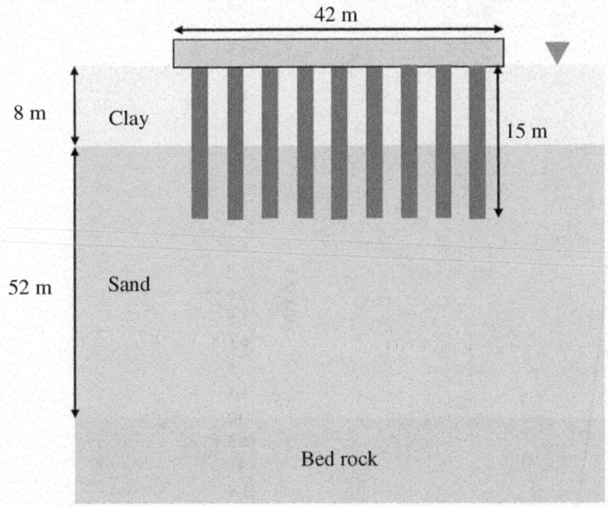

Clay: $\gamma_{sat} = 20\ \frac{kN}{m^3}$, $S_u = 20\ kPa$, $e_0 = 0.75$, $C_c = 0.22$, $C_r = 0.04$, **OCR = 8, 6, 4, 2 (2 m sublayer from top)**

Sand: $\gamma_{sat} = 18\ \frac{kN}{m^3}$, $\beta = 0.3$, $N_t = 30$, $m = 250$, $j = 0.5$

a) Determining the depth of the neutral plane

$$Q_d + Q_n = Q_{ult} - Q_n$$

$$Q_n = \frac{Q_{ult} - Q_d}{2} = \frac{2220 - 1200}{2} = 510\ kN$$

$$Clay:\ q_{s(z=0-8m)} = \alpha \times S_u = 1 \times 20 = 20\ kPa$$

$$A_s = \pi \times 0.45 \times 8 = 11.31\ m^2$$

$$Q_{s(z=0-8m)} = q_s \times A_s = 20 \times 11.31 = 226.2\ kN$$

$$Sand:\ q_{s(avg,\ z=8-15m)} = \beta \times \sigma'_0 = 0.3 \times (80 + 8z) = 24 + 2.4z\ kPa$$

$$Q_{s(z=8-15m)} = \int_0^z q_s \times A_s = \int_0^z ((24 + 2.4Z) \times \pi \times 0.45)dz = 1.4137 \times (24Z + 1.2Z^2)\ kN$$

$$Q_n = 510\ kN = 226.2 + 1.4137 \times [(24Z + 1.2Z^2)] \rightarrow Z = 6.349\ m$$

$$Z_{nutral-plane} = 8 + 6.349 = 14.349\ m \approx 14.35\ m$$

b) Axial load of each pile

$$Z < Z_{neutral-plane}:\ Q_n = Q_d + Q_s$$

$$Z > Z_{neutral-plane}:\ Q_n = Q_{ult} - Q_s$$

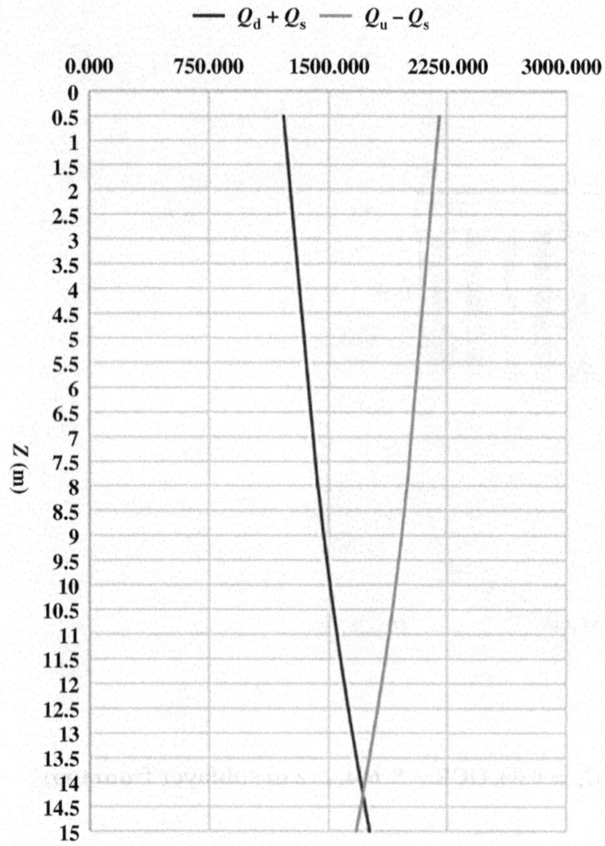

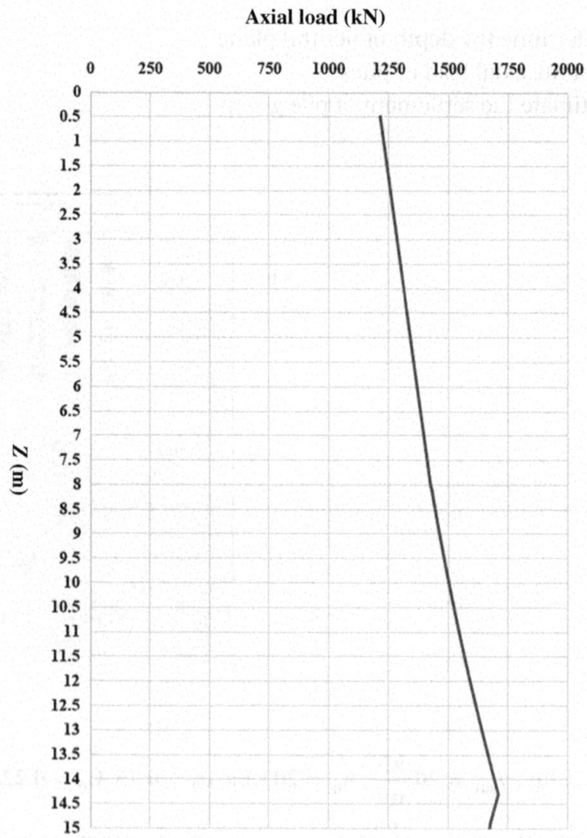

c) Estimation of settlement

$$\varepsilon = \frac{1}{5m}\left(\sqrt{\sigma'_1} - \sqrt{\sigma'_0}\right)$$

$$\Delta q_{Service} = q_{Service}\left(\frac{D}{D+Z}\right)^2 = 180\left(\frac{42}{42+Z}\right)^2,\ \Delta q_{Fill} = 2 \times 22 = 44\ kPa$$

Calculations are done in 2 m layers, a sample calculation for one layer is provided below:

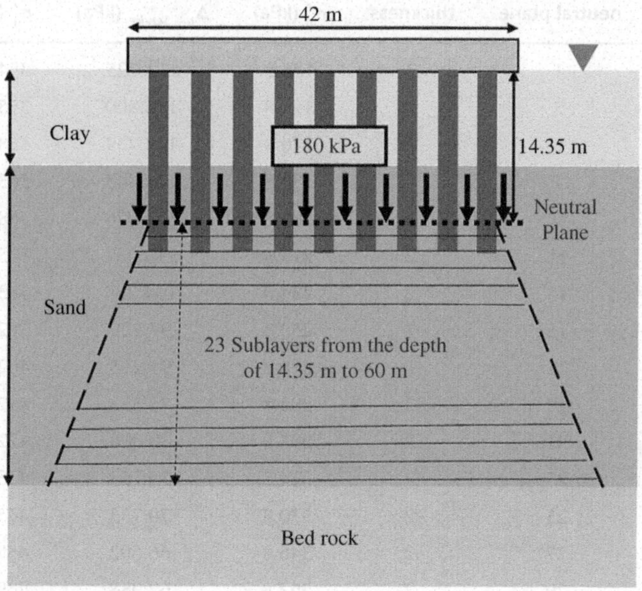

$$\text{Layer 1: } 14.35\text{ m to }16.35\text{ m: } \sigma_0' = \sum \gamma' \times z = 8 \times (20-10) + (15.35-8) \times (18-10) = 138.8\text{ kPa}$$

$$\Delta q_{Service} = 180\left(\frac{42}{42+1}\right)^2 = 171.725\ kPa$$

$$\sigma_1' = \sigma_0' + \Delta q_{Service} + \Delta q_{Fill} = 138.8 + 171.725 + 44 = 354.525\ kPa$$

$$\varepsilon = \frac{1}{5m}\left(\sqrt{\sigma_1'} - \sqrt{\sigma_0'}\right) = \frac{1}{5 \times 250}(\sqrt{354.525} - \sqrt{138.8}) = 0.00564$$

$$s = \varepsilon \times \Delta h = 0.00564 \times 2 = 0.01127m = 11.27\ mm$$

The distribution of stresses is as illustrated below:

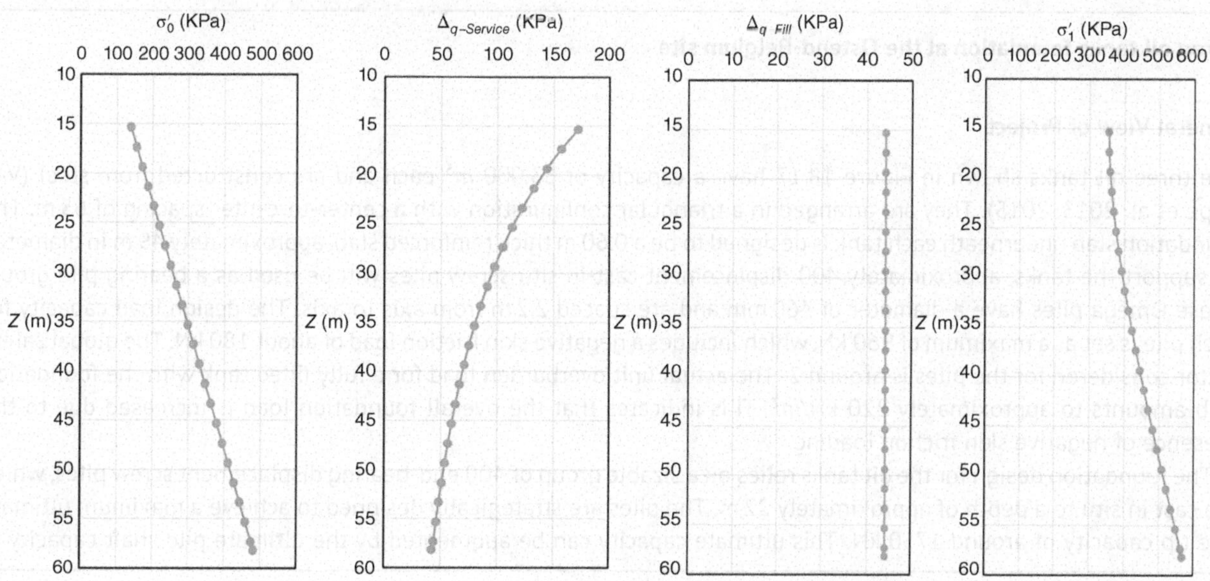

Layer No.	Mid. from the ground surface	Mid. from the neutral plane	Layer thickness	σ'_0 (kPa)	$\Delta_{q-Service}$ (kPa)	σ'_1 (kPa)	Strain	Settlement (mm)
1	15.35	1	2	138.8	171.725	354.525	0.00564	11.276
2	17.35	3	2	154.8	156.800	355.600	0.00513	10.265
3	19.35	5	2	170.8	143.739	358.539	0.00469	9.386
4	21.35	7	2	186.8	132.245	363.045	0.00431	8.618
5	23.35	9	2	202.8	122.076	368.876	0.00397	7.945
6	25.35	11	2	218.8	113.037	375.837	0.00368	7.351
7	27.35	13	2	234.8	104.965	383.765	0.00341	6.827
8	29.35	15	2	250.8	97.729	392.529	0.00318	6.361
9	31.35	17	2	266.8	91.215	402.015	0.00297	5.946
10	33.35	19	2	282.8	85.332	412.132	0.00279	5.575
11	35.35	21	2	298.8	80.000	422.800	0.00262	5.242
12	37.35	23	2	314.8	75.153	433.953	0.00247	4.942
13	39.35	25	2	330.8	70.733	445.533	0.00234	4.672
14	41.35	27	2	346.8	66.692	457.492	0.00221	4.426
15	43.35	29	2	362.8	62.988	469.788	0.00210	4.204
16	45.35	31	2	378.8	59.583	482.383	0.00200	4.001
17	47.35	33	2	394.8	56.448	495.248	0.00191	3.815
18	49.35	35	2	410.8	53.554	508.354	0.00182	3.646
19	51.35	37	2	426.8	50.876	521.676	0.00174	3.490
20	53.35	39	2	442.8	48.395	535.195	0.00167	3.346
21	55.35	41	2	458.8	46.091	548.891	0.00161	3.214
22	57.35	43	2	474.8	43.947	562.747	0.00155	3.092
23	58.175	44.825	1.65	481.4	42.119	567.519	0.00151	2.710
Total Settlement (mm)								130

Case Study 18.1

Three oil tanks foundation at the Ostend-Belgium site

General View of Project

The three oil tanks shown in Figure 18.17 have a capacity of 33,000 m^3 each and are constructed from steel (Van Impe et al. 2013, 2015). They are arranged in a triangular configuration with a center-to-center spacing of 65 m. The foundation slab underneath each tank is designed to be a 0.60 m thick reinforced slab, approximately 49 m in diameter. To support the tanks, approximately 400 displacement cast-in situ screw piles will be used as a bearing pile group. These Omega piles have a diameter of 460 mm and are spaced 2.2 m from axis to axis. The design load capacity for each pile is set at a maximum of 960 kN, which includes a negative skin friction load of about 180 kN. The global safety factor considered for the piles is around 2. The actual unit overburden load for a fully filled tank with the foundation slab amounts to approximately 220 kN/m^2. This indicates that the overall foundation load is increased due to the presence of negative skin friction loading.

The foundation design for the oil tanks relies on a sizable group of 400 end-bearing displacement screw piles, which are cast in situ to a depth of approximately 22 m. The piles are strategically designed to achieve a maximum ultimate pile tip capacity of around 1750 kN. This ultimate capacity can be augmented by the ultimate pile shaft capacity in

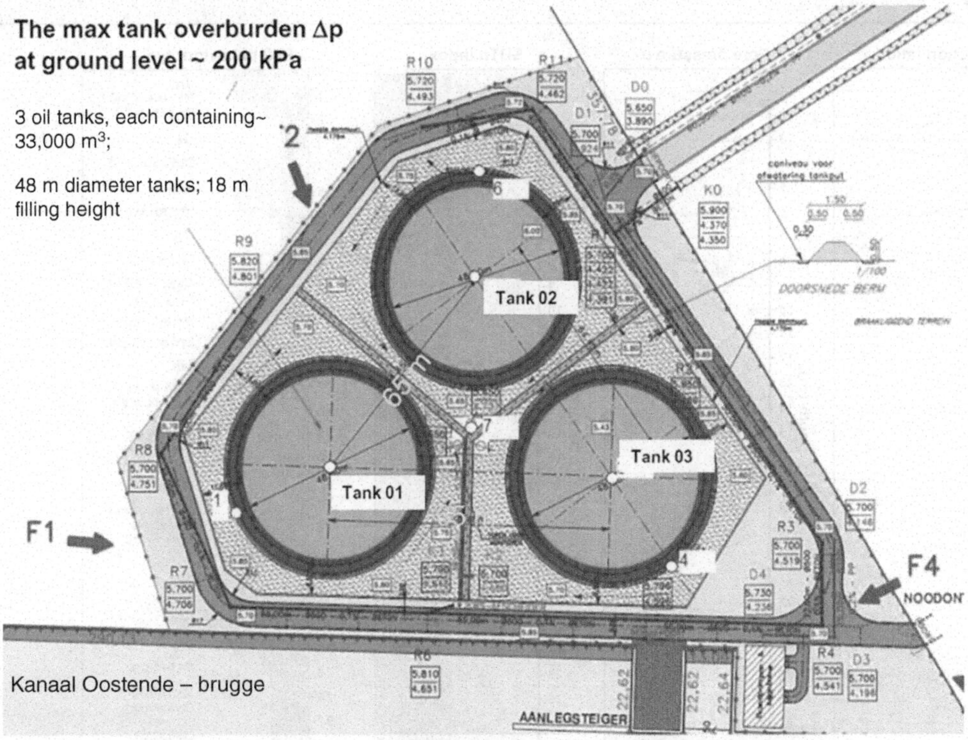

Figure 18.17 General view of oil tanks at Ostend – Belgium (Van Impe et al. 2013).

the approximately 3.5 m thick bearing sand layer above the pile tip, which adds approximately 200 kN of additional capacity. As a result, the total ultimate pile capacity reaches around 1950 kN, thereby ensuring that the allowable capacity of 960 kN is well within the safe range. This design approach guarantees a robust and reliable foundation for the oil tanks, capable of supporting their considerable load requirements.

Soil Conditions

At the Ostend oil tank site, the soil conditions up to a depth of approximately 12 to 15 m are classified as heterogeneous fill. The entire area seems to have undergone excavation and hydraulic refilling over several decades before being repurposed as an oil tank facility.

Between depths of around 12 to 15 m and 18 m, the first natural bearing layer consists of tertiary sand known as the Tielt formation or the "Kortemark" sand package. This sand layer is about 4 to 5 m thick. Below the sand layer, a much thicker tertiary OC clay formation, referred to as the "Kortrijk" formation of the OC clay of Western Flanders, is found, extending to a depth of about 80 to 100 m. This clay exhibits some resemblance to London clay.

The bearing sand layer, with a depth of about 4 to 5 m, shows CPT cone resistances ranging around 20 MPa, and in some instances, exceeding 30 MPa. In contrast, the slightly OC clay layer beneath gradually changes its cone resistance with depth, ranging from about 2.5 to 5 MPa in the top section. The results of the CPTu tests conducted on the site are presented in Figure 18.18

Test Piles Installation and Axial-Bearing Capacity Estimation

To optimize the design procedure, a fully instrumented test pile (Figure 18.19) was constructed and subjected to loading until a pile base settlement of 10% of the pile base diameter was achieved. This settlement corresponds to the required deformation at the "failure load" for a soil displacement pile in Belgian practice. It is assumed that the groundwater table is located approximately 0.5 m below the natural ground level.

Based on the capacity forecast at the foot of the test pile (21.56 m depth), the total ultimate pile tip capacity is estimated to be 2168 kN, with a total ultimate shaft capacity of 1943 kN, resulting in a total ultimate pile capacity of

(Continued)

Case Study 18.1 (Continued)

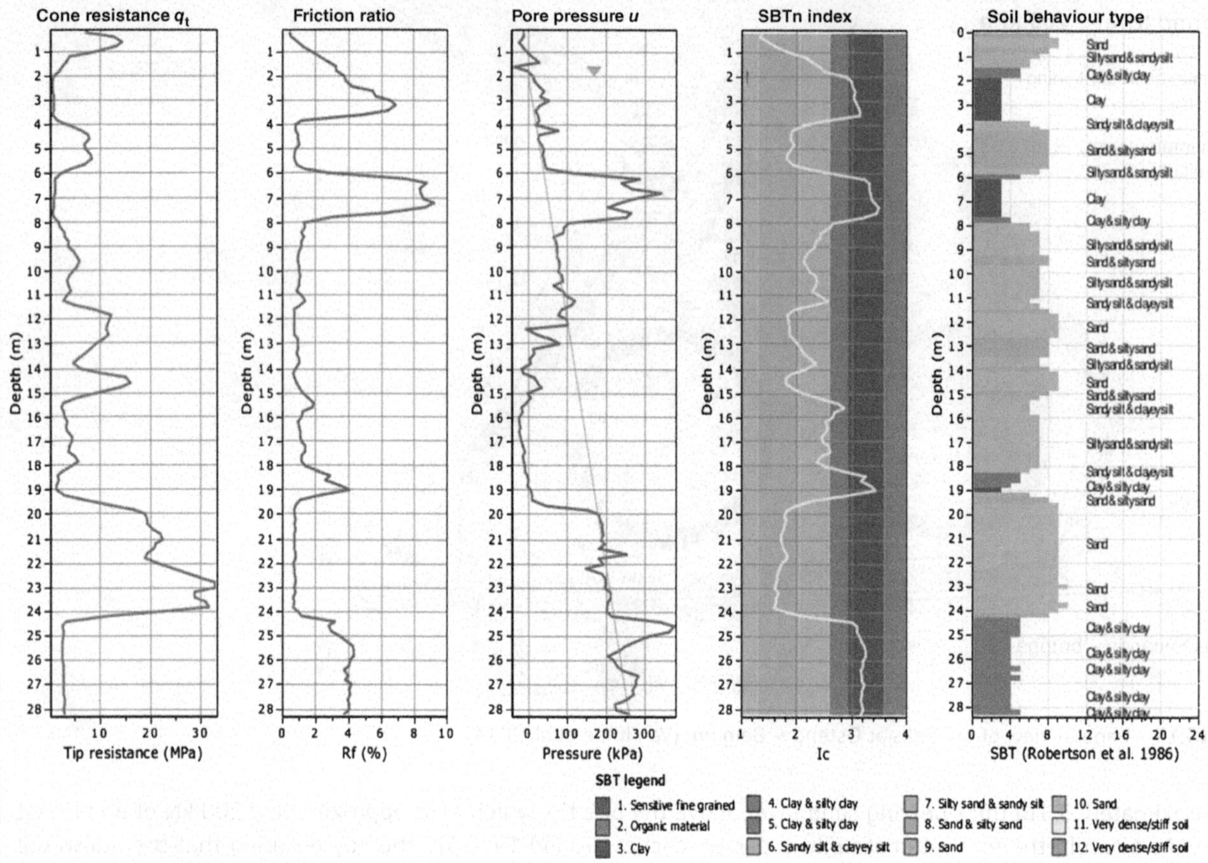

Figure 18.18 Relevant CPTu log in Ostend (Van Impe et al. 2013).

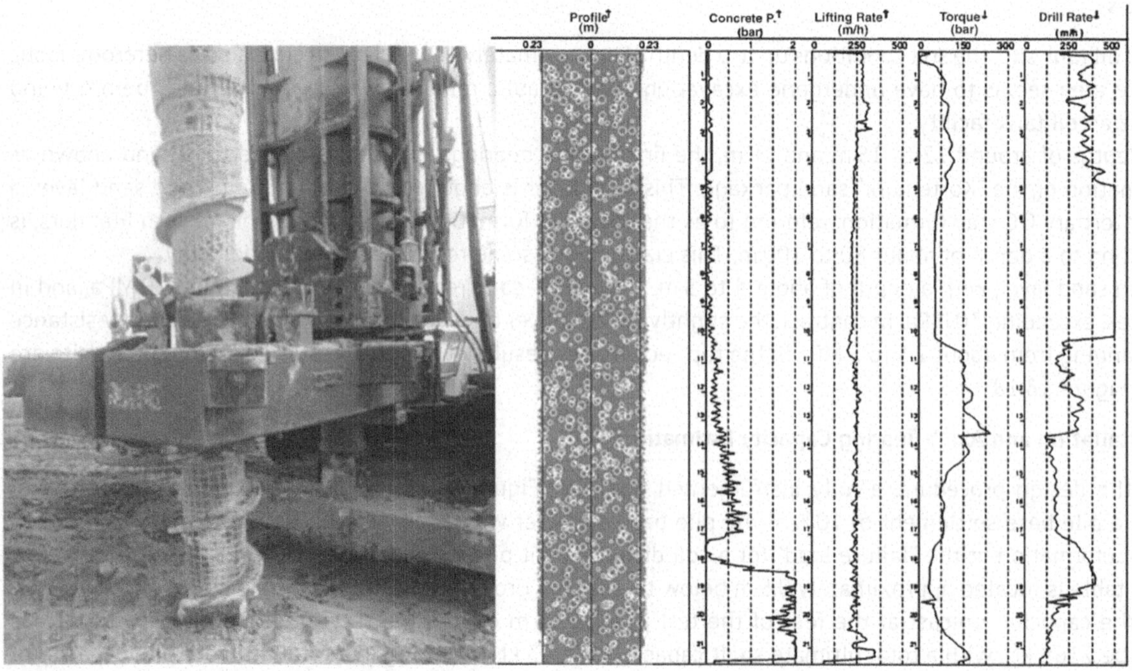

Figure 18.19 Test pile installation parameters (Van Impe et al. 2013).

Q_{ult} ~ 4.1 MN. However, caution should be exercised regarding the previously indicated pile shaft capacity estimate using CPT, as pile installation in this type of soil, characterized as heterogeneous fill material, is highly susceptible to reconsolidation and negative skin friction effects. The contribution of the top 11 m of the shaft capacity to the CPT result is questionable and should be disregarded.

Pile Group Settlement, from the Hydro–Load Testing

In April 2013, the hydro-testing of the three tanks was conducted. Each tank was filled to a height of approximately 18 m, and the filling process for one tank took about three days. After filling, the water level in the tanks was maintained steadily for about four days, followed by a three-day emptying operation. The sequence of tank filling was as follows: Tank 01 was filled first, followed by Tank 02 with some overlap in time, and finally, after emptying Tank 02, Tank 03 was filled. Figure 18.20 illustrates the loading pattern history, along with the measured mean settlements.

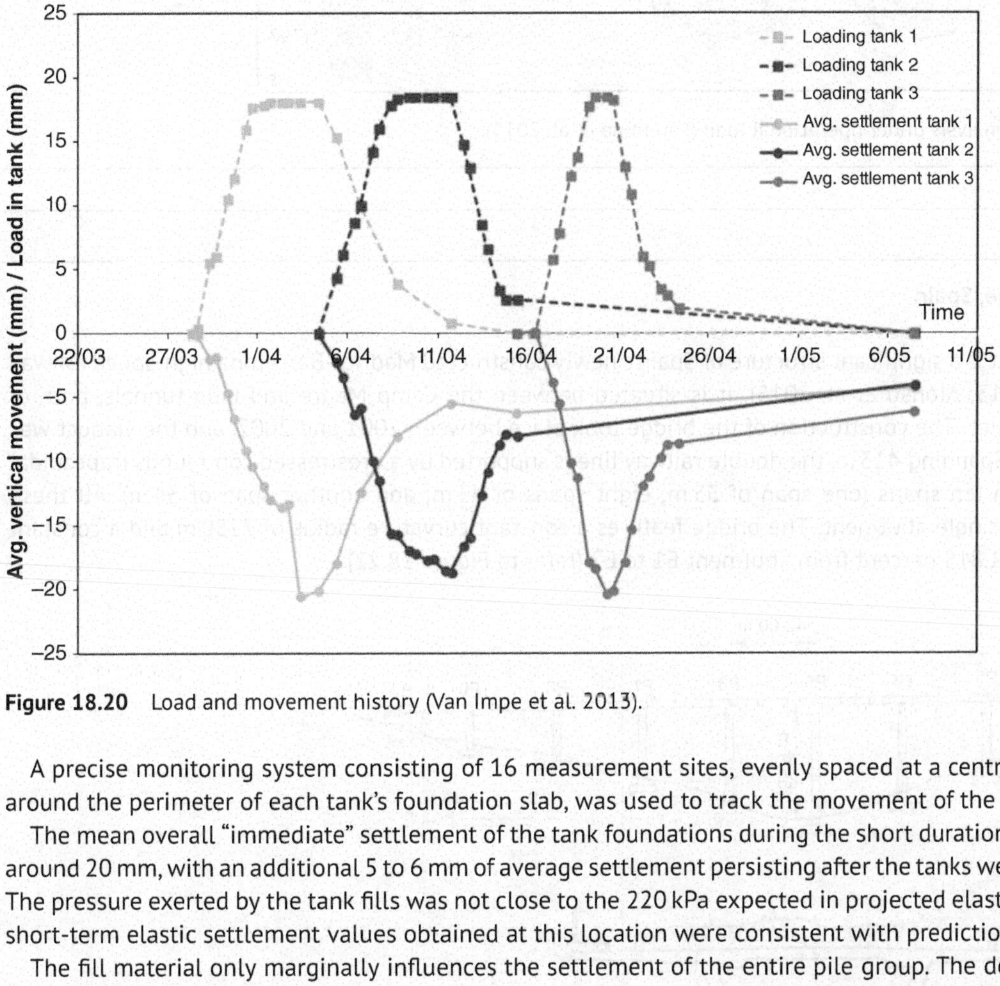

Figure 18.20 Load and movement history (Van Impe et al. 2013).

A precise monitoring system consisting of 16 measurement sites, evenly spaced at a central angle of 22.5 degrees around the perimeter of each tank's foundation slab, was used to track the movement of the foundation slabs.

The mean overall "immediate" settlement of the tank foundations during the short duration of hydro testing ranged around 20 mm, with an additional 5 to 6 mm of average settlement persisting after the tanks were completely unloaded. The pressure exerted by the tank fills was not close to the 220 kPa expected in projected elastic settlements. However, short-term elastic settlement values obtained at this location were consistent with predictions.

The fill material only marginally influences the settlement of the entire pile group. The dense sand layer, the silty clay, and, in the long run, the thick OC clay layer all contribute significantly to the settlements. To establish the best match between prediction and measurement, a single value of the consolidation coefficient (c_v) for the silty clay was determined using this compressibility parameter in combination with the immediate compressibility parameters acquired from the quick hydro test. Fitting was performed using the average settlement's values.

A basic three-dimensional elastic stress model was used to make the prediction. The model takes into account the minor difference in thickness of the sand layer underneath the various pile groups (slightly lower thickness below tank 3). The tanks' final average settlement varies between 87 and 90 mm (See Figure 18.21). The tank's centers settle to a depth of 132 to 136 mm. The long-term tilt is between 19 and 21 mm.

(Continued)

Case Study 18.1 (Continued)

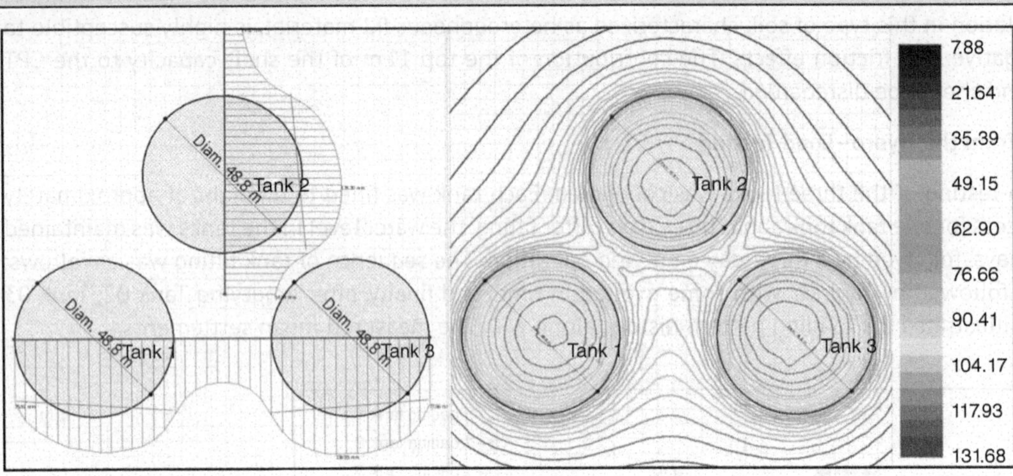

Figure 18.21 Settlement analysis under operational load (Van Impe et al. 2015).

Case Study 18.2

Heave of a Railway Bridge, Spain

The Pont de Candí viaduct is a significant structure of Spain's newly constructed Madrid–Barcelona high-speed railway (Alosnso and Ramon 2013; Alonso et al. 2015). It is situated between the Camp Magre and Lilla tunnels, both of which faced swelling issues. The construction of the bridge took place between 2001 and 2002, and the viaduct was completed in July 2002. Spanning 413 m, the double railway line is supported by a prestressed continuous trapezoidal concrete box girder with ten spans (one span of 35 m; eight spans of 43 m; and another span of 34 m). All these spans are anchored in a single abutment. The bridge features a constant curvature radius of 7250 m and a constant longitudinal gradient of 1.815 percent from abutment E1 to E2 (refer to Figure 18.22).

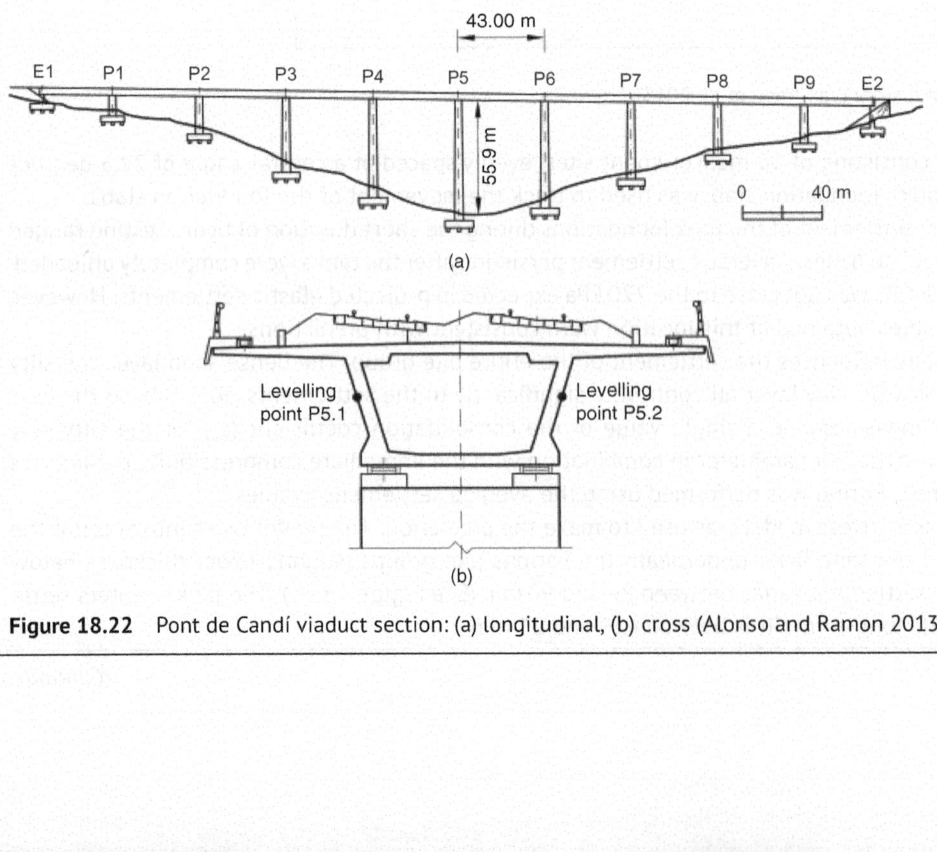

Figure 18.22 Pont de Candí viaduct section: (a) longitudinal, (b) cross (Alonso and Ramon 2013).

Long pillars (P1–P9) support the top deck; their height ranges from 11 m to a maximum of 55.9 m in the valley's middle (pillar P5). Each bridge pillar is supported by three drilled piles with a diameter of 1.65 m and an average length of 20 m, as seen in Figure 18.23 longitudinal profile. The pile cap thickness is 3 m for pillars P1, P2, and P9, and 3.5 m for pillars P3–P8.

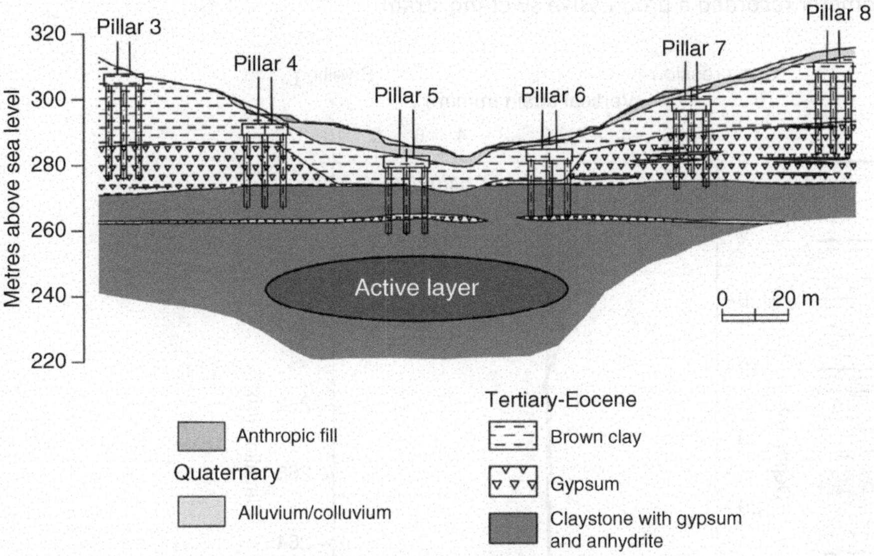

Figure 18.23 Geological profile along bridge axis (Alonso et al. 2015).

After the construction of the railway lines, the railway administration conducted systematic leveling and observed a continuous increase in vertical displacements in the central viaduct pillars, particularly in pillars P5 and P6. Figure 18.24 presents heave profiles showing the heave pattern at two different times, with the most significant heave observed in the middle pillars. The heave was accumulating at a rate of approximately 5 to 10 mm per month. The cause of this unusual behavior was initially unknown. The huge, deep foundations of the pillars experienced only minor vertical displacements caused by a superficial swelling layer except for P5 and P6. The piles were embedded or socketed in the hard layer of anhydrite claystone, providing substantial resistance to the swelling stresses along the pile shafts.

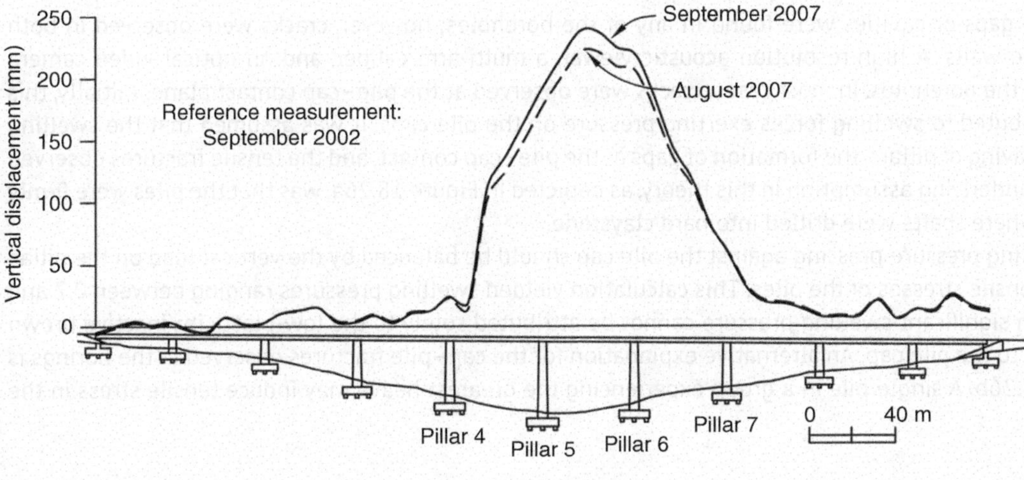

Figure 18.24 Heave profiles (Alonso et al. 2015).

(Continued)

Case Study 18.2 (Continued)

Extensometer measurements indicate the presence of a 10–12 m thick active layer under the pile's tip. Figure 18.25 illustrates the geometry of the pile group beneath pillar 5 and the strains measured using a sliding micrometer (Kovári and Amstad 1982). No deformations were observed along the length of the pile's shaft. However, at depths greater than 3 m from the pile's tip, the micrometer recorded a progressive swelling strain.

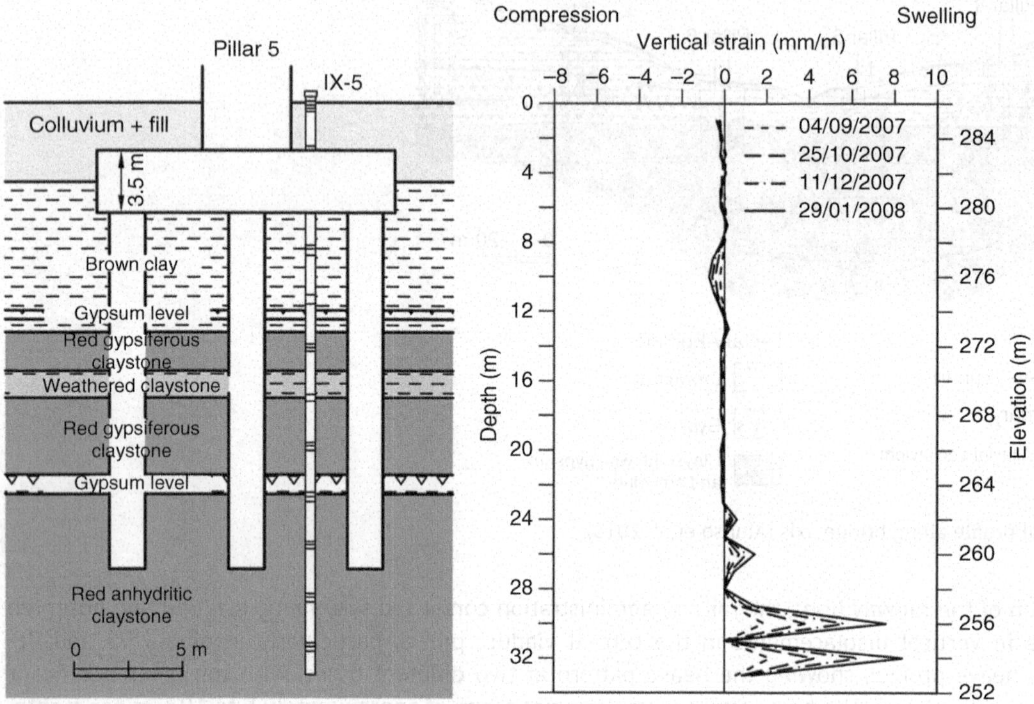

Figure 18.25 Measured strains (Alonso et al. 2015).

Vertical boreholes were drilled along the entire length of a specific pile, allowing for the retrieval of concrete cores. Additional boreholes were also made a few meters below the pile tops into the natural earth. In pillars P4, P5, P6, and P7, twelve such boreholes were drilled (three per pillar).

Upon inspection, no gaps or cavities were found in any of the boreholes; however, cracks were observed in both the cores and borehole walls. A high-resolution acoustic viewer, a multi-arm caliper, and an optical video camera were utilized to assess the boreholes. In most cases, cracks were observed at the pile–cap contact plane. Initially, this phenomenon was attributed to swelling forces exerting pressure on the pile caps. It was assumed that the swelling pressures led to the heaving of pillars, the formation of gaps at the pile–cap contact, and the tensile fractures observed in the pile shafts. The underlying assumption in this theory, as depicted in Figure 18.26a, was that the piles were firmly embedded at depths where shafts were drilled into hard claystone.

The calculated swelling pressure pressing against the pile cap should be balanced by the vertical load on the pillar, thus overcoming the tensile stresses of the piles. This calculation yielded swelling pressures ranging between 0.7 and 0.8 MPa. However, such significant swelling pressure cannot be attributed solely to the low-plasticity, inactive brown clay layer located next to the pile cap. An alternative explanation for the cap–pile fractures observed in the borings is presented in Figure 18.26b. A single pile in a group experiencing the greatest heave may induce tensile stress in the other piles.

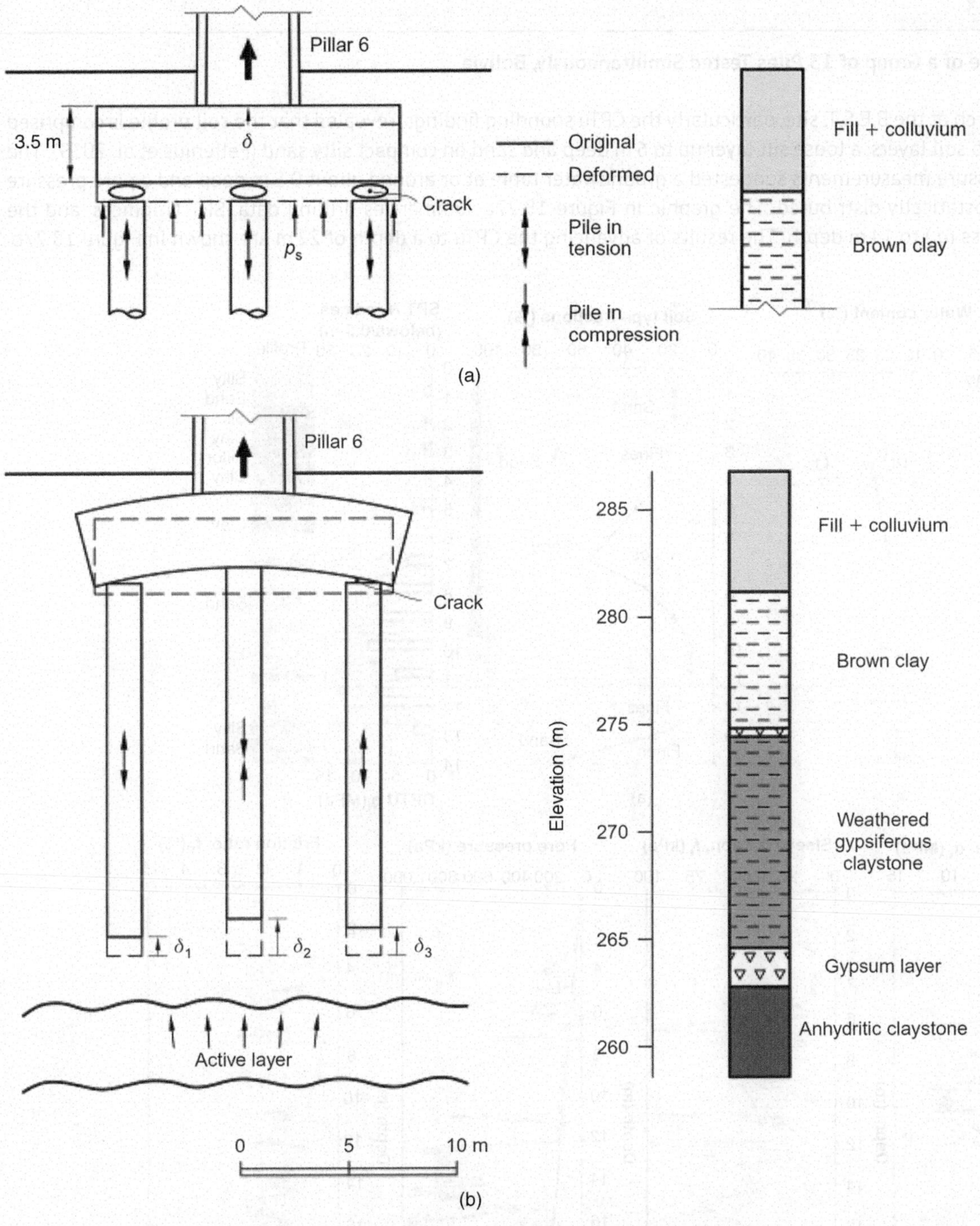

Figure 18.26 Two alternative interpretations of foundation–heave interaction: (a) swelling pressure against the pile cap, (b) piles pushed upwards because of heave at the deep active zone (Alonso and Ramon 2013).

Originally, the boreholes made through the piles were filled with drilling fluid. Upon emptying, it was noted that the rate of water level recovery was faster in piles belonging to pillars P5 and P6 (those positioned at the bottom of the valley and experiencing the most heave) compared to piles in pillars P4 and P7. These observations could be attributed to the more severe pile cracking in pillars P5 and P6. Furthermore, the increased permeability of the soil/rock around the piles in pillars P5 and P6 explains the varying rates of water level recovery.

Case Study 18.3

Static Response of a Group of 13 Piles Tested Simultaneously, Bolivia

The site research at the B.E.S.T. site, particularly the CPTu sounding findings, revealed that the soil profile is comprised basically of two soil layers: a loose silt layer up to 6 m deep and sand on compact silty sand (Fellenius et al. 2019). The CPTu pore pressure measurements suggested a groundwater table at or around about 0.5 m deep and a pore pressure that was hydrostatically distributed. The graphic in Figure 18.27a summarizes drilling data, SPT N-indices, and the CPTu cone stress (q_t) to 14 m depth. The results of advancing the CPTu to a depth of 22 m are shown in Figure 18.27b.

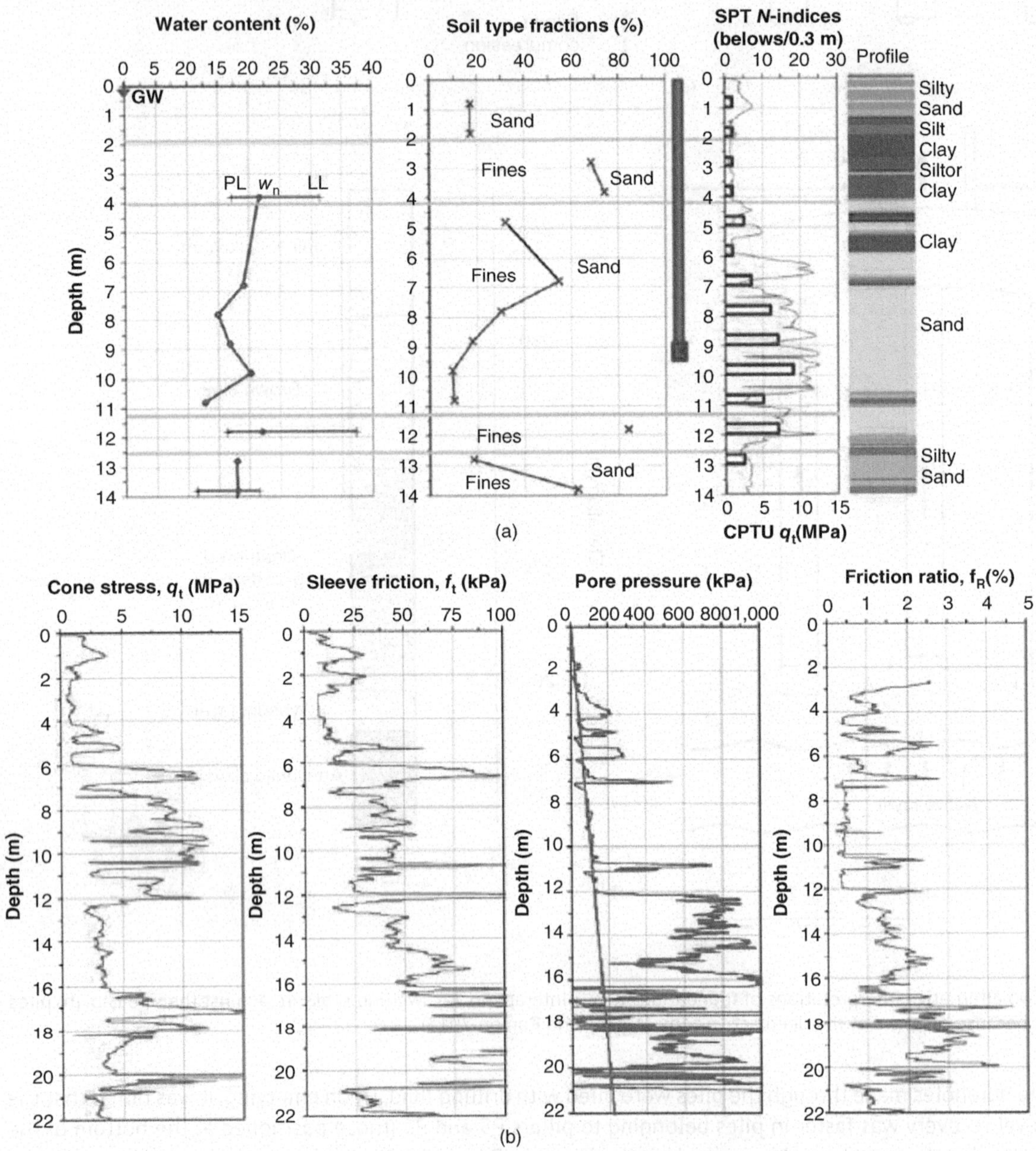

Figure 18.27 (a) Soil profile from BH-E2, (b) CPTu sounding diagram, CPTu E2 (Fellenius et al. 2019).

The pile group consisted of thirteen 300-mm-diameter pressure-grouted full displacement piles (FDP) installed to a depth of 9.5 m. As a reference for the group test, a single pile, identical to the group piles in all respects, was placed 5.0 m away from the closest group pile and 3.5 m away from the nearest response pile. Each pile was equipped with a bidirectional cell (BD), which included a 200 mm high and 80 mm wide hydraulic jack located in each pile at a depth of 8.3 m (the lower end of the BD). All BDs in the group piles were interconnected to a common pump, ensuring that an equal load was applied to all BDs. Immediately under the BD, the piles were equipped with a 1.2-m long expander body (EBI612), which was expanded (pressurized with grout) during the shaft's construction. The soil under the EBI was post-grouted twenty-four hours after the EB expansion to recompress the earth beneath the EBI. In most piles, post-grouting introduced little or no grout.

The piles were installed in the arrangement illustrated in Figure 18.28 at a three-pile-shaft diameter, with a center-to-center spacing of 0.90 m. The spacing equates to an 11% overall footprint ratio (nominal total pile area over group footprint area, excluding the EBI area).

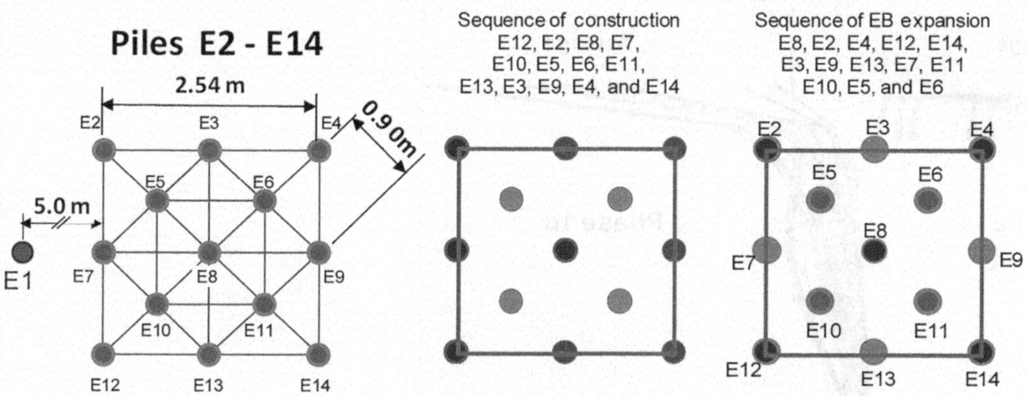

Figure 18.28 Pile locations and sequence of construction and EB expansion (Fellenius et al. 2019).

Static loading tests were performed on both single piles and the group piles by activating the BD cells, which propelled the pile shafts upwards, while the EB supplied the necessary response force. All BD cells in the group piles (Piles E2–E14) were connected to the same pump, ensuring equal loading for all group piles. The pile heads were unbounded, and their individual movements were recorded. The overall layout is illustrated in Figure 18.29a, showing the positions of the beams and response piles, while Figure 18.29b provides a photograph of the system used to record the individual pile motions. The measurements involved monitoring ground surface movements at ten locations around the perimeter and inside the group's core. The testing procedure was the same for both the single pile, Pile E1, and the group piles (Piles E2–E14). It included applying equal load increments of 50 kN per pile every ten minutes, with continuous monitoring of the pile head's movements until they reached excessive levels.

Figure 18.29 (a) General arrangement, (b) frame for recording pile movements (Fellenius et al. 2019).

(Continued)

Case Study 18.3 (Continued)

During the test, the pump experienced a leak, causing it to lose pressure and become incapable of sustaining the load. As a result, the BDs were unloaded to nearly zero pressure and secured to prevent further hydraulic pressure loss. The pump was subsequently repaired, and the reloading process was performed using the same load and movement reference points as in the initial Phase 1b. The load responses of Piles E2–E14 are presented in Figure 18.30. These 13 piles exhibited similar behavior to the single pile (Pile E1), with a gradual increase in load and then a decrease after a few millimeters of movement. However, the ultimate load per pile was approximately 70% of that observed for Pile E1, with values of 430 kN compared to 600 kN.

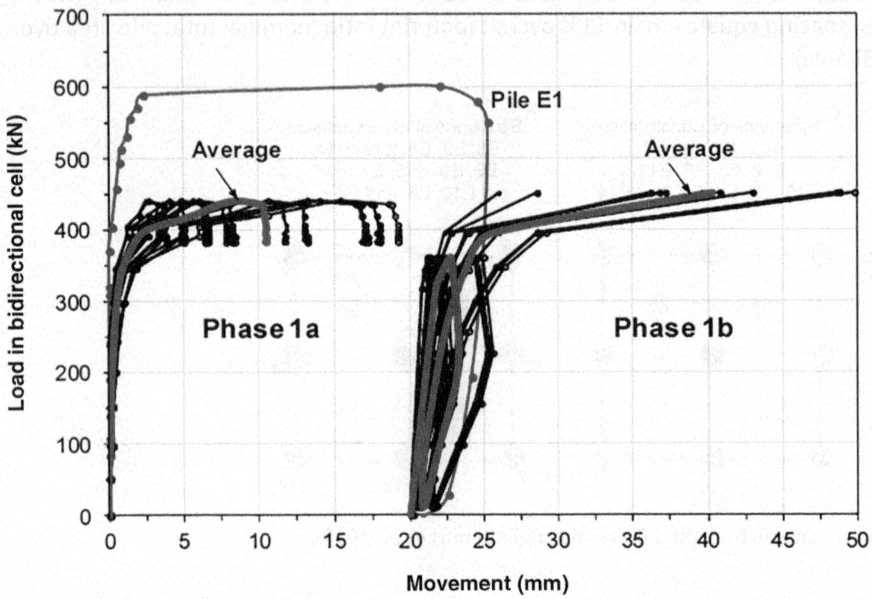

Figure 18.30 Pile E1 and Piles E2–E14. Pile head upward load-movements (Fellenius et al. 2019).

Case Study 18.4

Evaluation of ultimate behavior of actual large-scale pile group foundation by in situ lateral loading tests and numerical analysis, Japan

The damaged L.N.G. (liquefied natural gas) tank used for the loading test is visible in Figure 18.31a, while its foundation is shown in Figure 18.35b (Teramoto et al. 2018). The external view of the specimen and the reaction piles for loading can be seen in Figure 18.35c. This metal tank has a capacity of up to 45,000 m^3 and is supported by a 48 m diameter reinforced concrete slab and 496 steel pipe piles. Further details of the piles are illustrated in Figure 18.32a. The slab is 0.8 m thick and rests 0.8 m above the ground surface.

The piles are 23.5 m in length, 406.4 mm in diameter, with a 12.7 mm thickness top section and a 9.5 mm thickness bottom section. The land surrounding the tank's perimeter is improved using sand compaction piles, enhancing its N-value to approximately 15.

Each pile is an open-toe end-supported pile with an N-value of about 50 for the bearing ground at the pile tip. The top section of the pile is reinforced concrete to a depth of 6 m, and the upper 0.8 m above the ground and 0.3 m below the ground is covered with a protective concrete layer that is rectangular and 150 mm thick.

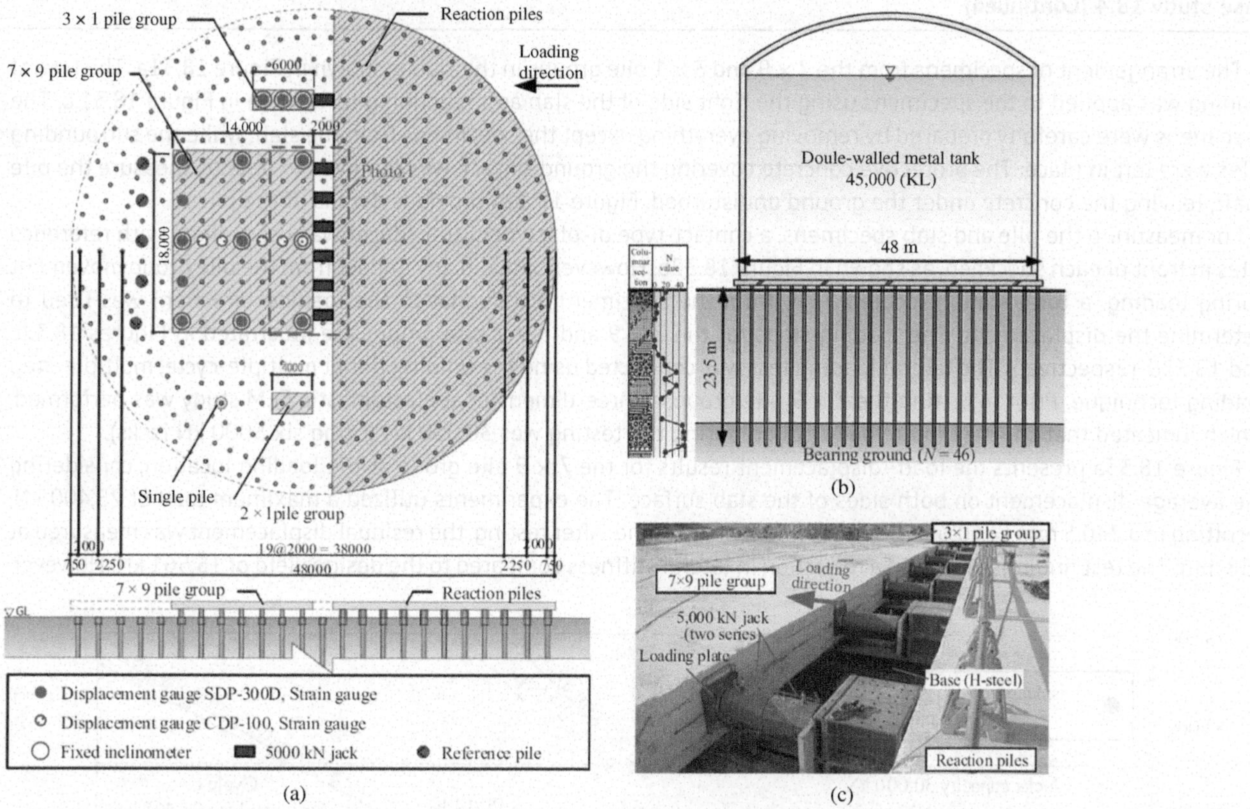

Figure 18.31 (a) Elevated view of the tank, (b) specimen and measurement arrangement, and (c) externals of the specimen and reaction piles (Teramoto et al. 2018 / Tomonari Niimura / Tomihiro Akutsu / Makoto Kimura / Elsevier B.V.).

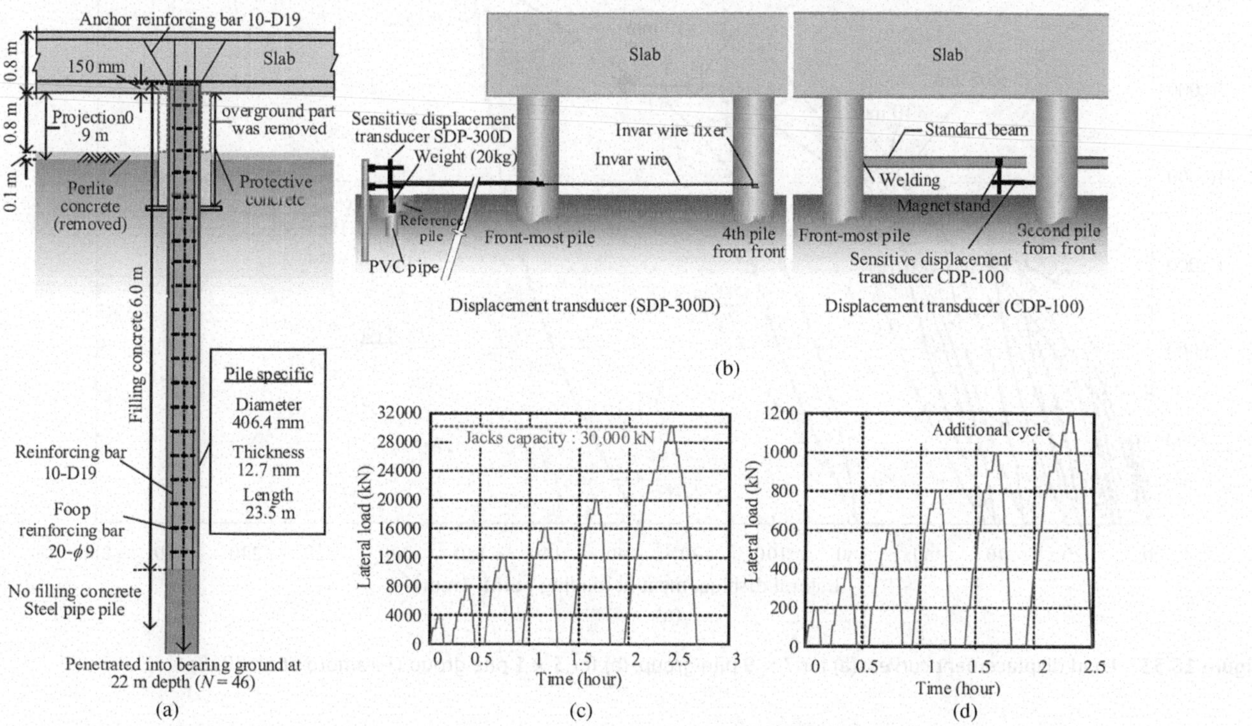

Figure 18.32 (a) Pile details, (b) measurement setup, (c) loading stages of 7×9 pile group, and (d) loading stages of 3×1 pile group (Teramoto et al. 2018).

(Continued)

Case Study 18.4 (Continued)

The arrangement of specimens from the 7 × 9 and 3 × 1 pile groups in the slab is shown in Figure 18.31a. The lateral loading was applied to the specimens using the right side of the slab as a reaction pile, as seen in Figure 18.31.c. The specimens were carefully prepared by removing everything except the specimens from the slab, while the surrounding piles were left in place. The protective concrete covering the ground surface was removed to directly measure the pile shaft, leaving the concrete under the ground undisturbed. Figure 18.32a provides details of the piles.

For measuring the pile and slab specimens, a contact-type displacement transducer was used, along with reference piles in front of each specimen, as shown in Figure 18.32b. However, to avoid any impact from the pile group movement during loading, a total station located 25 m from the specimen (orthogonal to the loading direction) was used to determine the displacement. The loading steps for the 7 × 9 and 3 × 1 pile groups are illustrated in Figures 18.32c and 18.32d, respectively. The lateral loading test was conducted using the unidirectional multiple-cycle, multiple-step loading technique. Prior to testing the 7 × 9 pile group, a three-dimensional elastoplastic FEM study was performed, which indicated that the maximum load applied during the testing was 30,000 kN (using six 5000 kN jacks).

Figure 18.33a presents the load–displacement results for the 7 × 9 pile group at the loading location, considering the average displacement on both sides of the slab surface. The experiments utilized a maximum load of 25,400 kN, resulting in a 240.5 mm (= 60% × d) displacement at that time. After testing, the residual displacement was measured at 158 mm. The test findings indicated a reduction in lateral stiffness compared to the design yield of 15,561 kN. However,

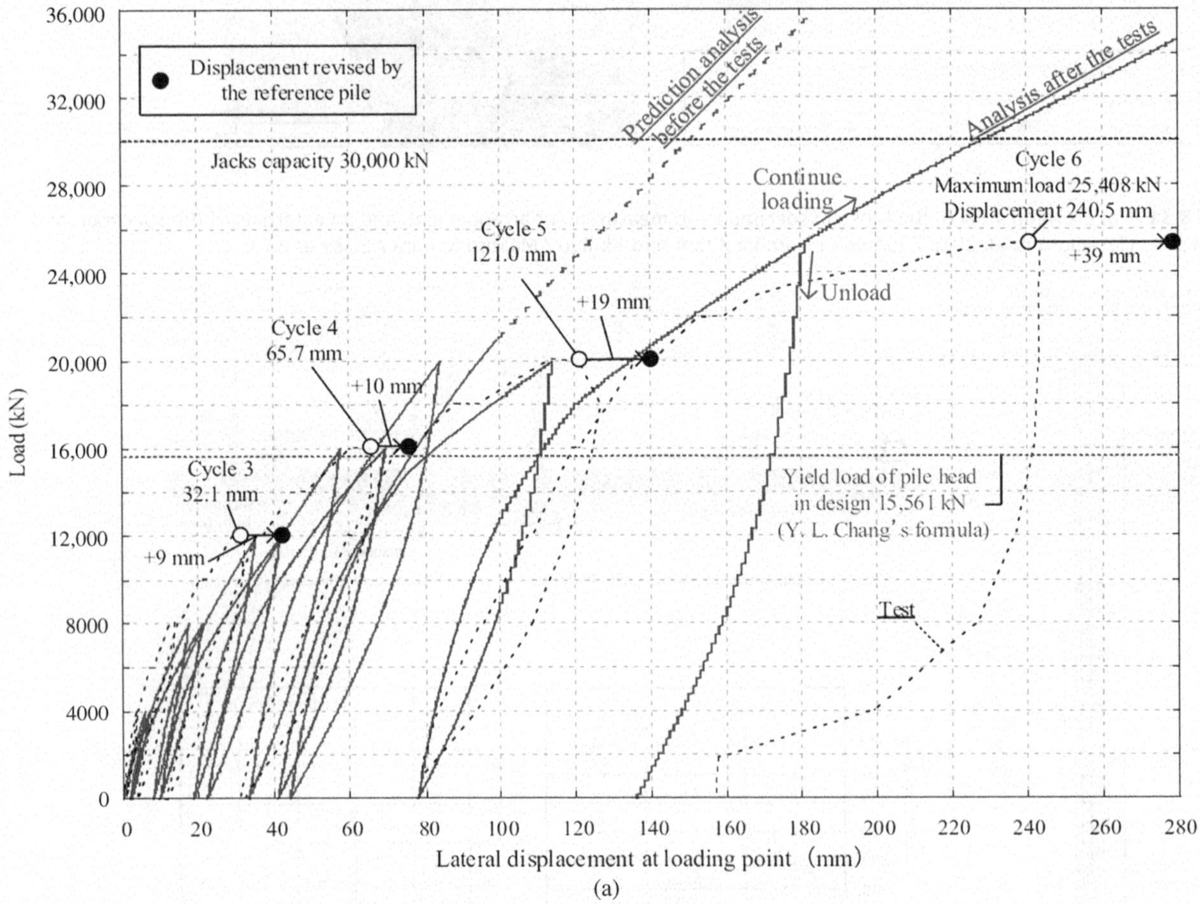

Figure 18.33 Load displacement curves: (a) for 7 × 9 pile group, (b) for 3 × 1 pile group (Teramoto et al. 2018).

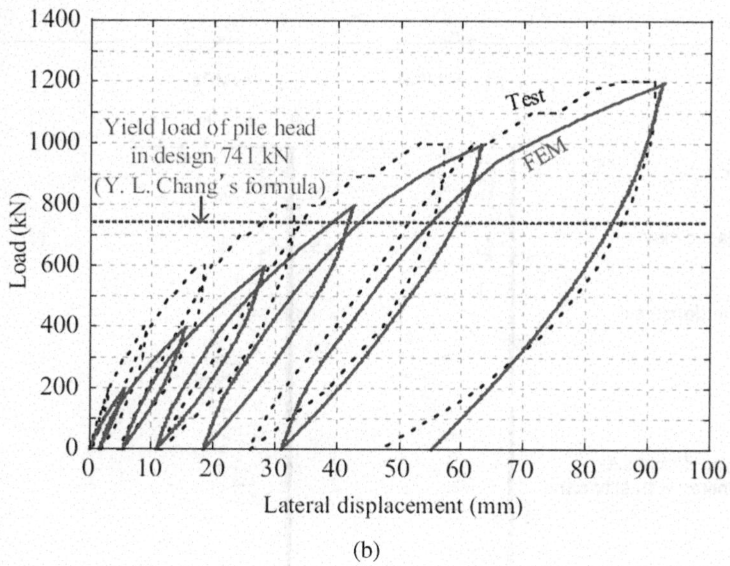

Figure 18.33 *(Continued)*

the ductile deformation continued to progress, and the load was steadily increased without any brittle failure. Similarly, Figure 18.33b illustrates the load–displacement results for the 3 × 1 pile group at the loading position. The initial stiffness from the test results was found to be lower than that of the 7 × 9 pile group, with an accuracy of approximately 60 mm (= 15% × d). These results suggest that the FEM simulation may have underestimated the soil parameter, as the lateral stiffness in the early phase is more influenced by the ground stiffness rather than the pile stiffness.

Case Study 18.5

Full-Scale Load Tests on Drilled Shaft Groups, China

Pile group interactions with closely spaced piles can be highly complex, and there is a lack of experimental data from full-scale bored pile groups subjected to loading (Dai et al. 2012). However, Dai et al. (2012) reported the results of axial static load tests conducted on both full-scale instrumented pile groups and single piles. The objective of these load experiments was to investigate the influence of pile length, spacing, and number on the load response of pile groups. The experiments involved varying the number of piles in a group, their spacing, composition, and length. Each pile had a diameter of 400 mm.

The test site in Jiangning, Nanjing, China, was situated on level ground within the QinHuai River floodplain. An auger boring (BH) was conducted at the site, reaching a depth of 29.50 m. The boring revealed a uniform and thick layer of soft clay extending from a depth of 17 m and likely beyond. For piles with a length of 24 m, the auger boring depth was 11B deeper than the test pile base (B = 400 mm, the pile diameter). To obtain a continuous record of soil resistance with depth, static cone penetration tests (CPTs) were performed close to the BH. The achieved CPT profiles are shown in Figure 18.34. However, pore-water pressure measurement was not feasible with the employed cone, as CPTu is not commonly used in China. The subsurface profile revealed several silt and clay layers, with the thickness of each layer determined based on CPT data. Soil samples were collected using split-spoon samplers, and the groundwater level was encountered at a depth of 2.60 m.

(Continued)

Case Study 18.5 (Continued)

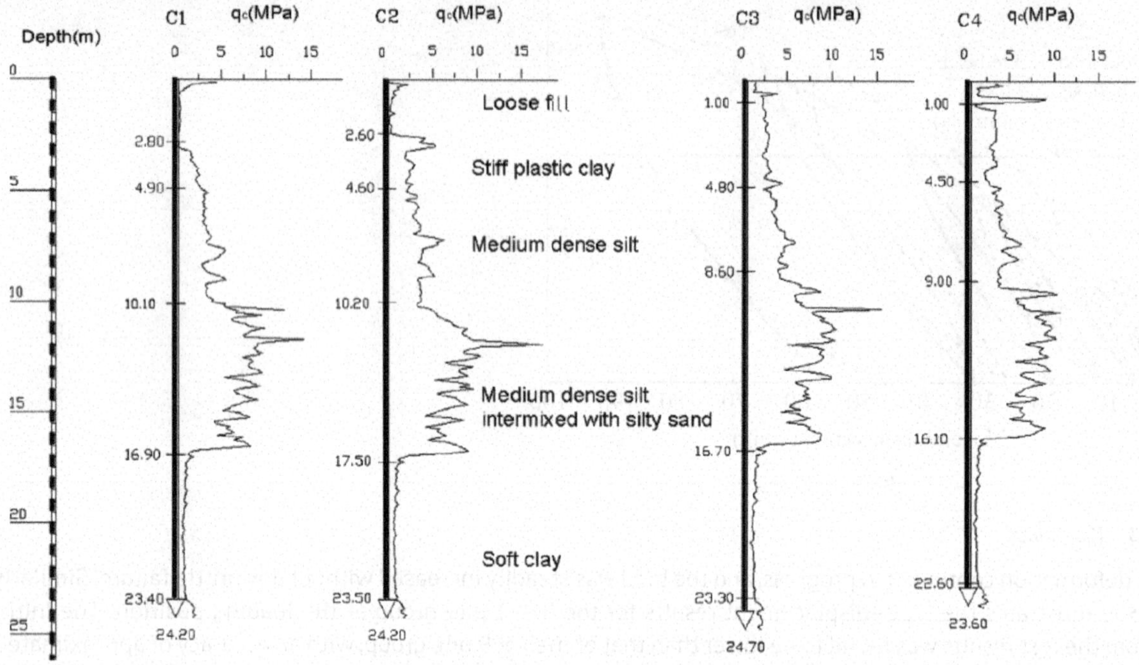

Figure 18.34 CPT profiles (Dai et al. 2012 / With permission of Canadian Science Publishing).

Two-, four-, and nine-pile groups with pile lengths of 20 and 24 m were evaluated in this study. Load transfer and load–settlement curves were generated for both isolated piles and individual instrumented piles within the groups. The interaction coefficients for each pile in the group were back-calculated from the measured data. The study showed that the interaction coefficients depend on settlement, group size, and pile proximity, challenging the common assumption of elastic behavior in calculations. The piles were installed using the slurry method, protruding 0.1 m above the ground and drilled to depths of 20 or 24 m. A reinforced concrete cap of 0.4, 0.8, or 1.2 m in thickness was poured over the pile groups and single piles. A layout plan for the test piles and pile groups is shown in Figure 18.35. The pile caps were practically considered rigid as they rested on the ground. The pile spacing in groups QZ2, QZ4, and QZ9 was 2.5 times the pile diameter (B), while it was 3.0B in groups QZ2L, QZ4L, and QZ9L. The layout plan of the test single and group deep foundations is shown in Figure 18.35.

The load–settlement curves for single piles and pile groups with two, four, and nine piles are displayed in Figure 18.36. Although the settlements of each pile in the group were not measured separately, it may be presumed that all of the piles have the same settlement because the pile cap is rigid.

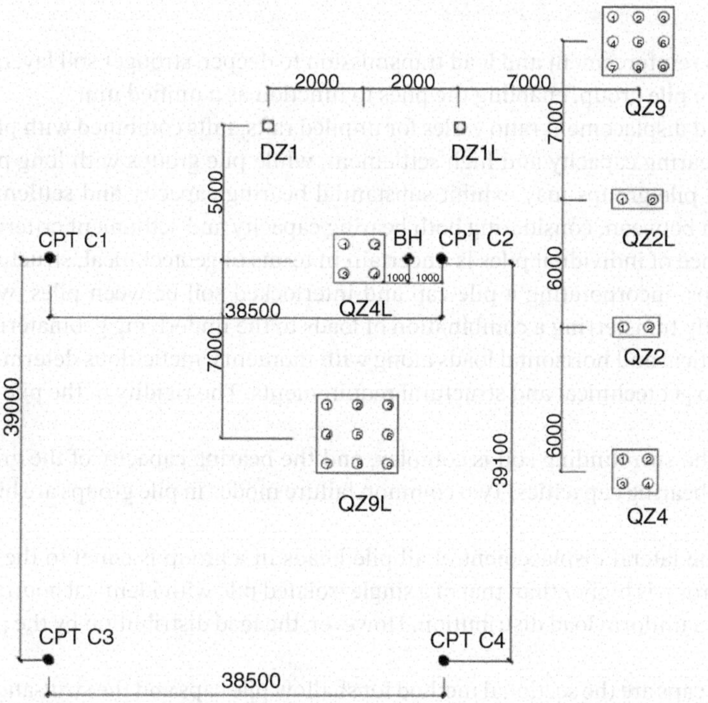

Figure 18.35 Layout plan of test foundations and CPTs (dimensions in mm) (Dai et al. 2012 / With permission of Canadian Science Publishing).

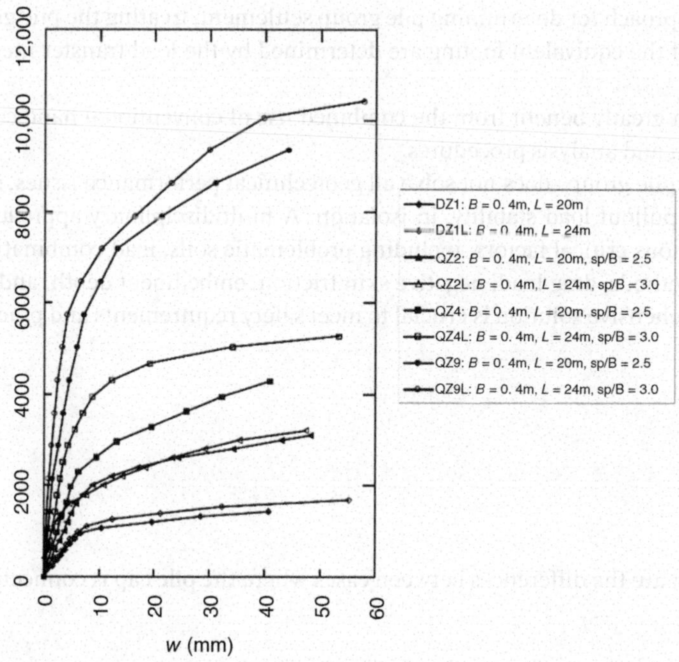

Figure 18.36 Load–displacement for single and group foundations. sp, spacing of pile group (Dai et al. 2012 / With permission of Canadian Science Publishing).

18.8 Major Points

1) Piled foundations serve the dual purpose of soil reinforcement and load transmission to deeper, stronger soil layers. A reinforced concrete pile cap typically crowns the pile group, enabling the piles to function as a unified unit.
2) The relationship between projected pressure and displacement ratio varies for unpiled rafts, rafts combined with piles, and pile groups. Large mats exhibit medium bearing capacity and high settlement, while pile groups with long piles show high capacity and low settlement. Large pile groups may exhibit substantial bearing capacity and settlement simultaneously. Medium-sized piled rafts fall in between, considering both bearing capacity and settlement criteria.
3) In the case of combined loading, the performance of individual piles is uncertain in terms of geotechnical, structural, and serviceability aspects. However, pile groups, incorporating a pile cap and interlocked soil between piles, work together as a unique block foundation, efficiently transferring a combination of loads to the underlying geomaterial.
4) When a pile group is subjected to combined vertical and horizontal loads along with moments, meticulous determination of each pile's axial loads is necessary due to geotechnical and structural requirements. The rigidity of the pile cap plays a critical role in this regard.
5) The interaction between piles in a group and the surrounding soil is complex, and the bearing capacity of the group may differ from the sum of the individual piles' bearing capacities. Two common failure modes in pile groups are block failure and individual pile failure.
6) Under lateral forces applied at the cap level, the lateral displacement of all pile heads in a group is equal to the cap displacement. The lateral displacement of the group is higher than that of a single isolated pile with identical boundary conditions when subjected to lateral loads with a uniform load distribution. However, the load distribution by the piles is nonuniform and site-dependent.
7) Two frequently used methods for designing pile caps are the sectional method for shallow pile caps and the strut-and-tie (STM) method for deep pile caps. Both approaches use the same procedure to determine the cap thickness, and both satisfy the requirements for one-way and two-way shear. However, in both circumstances, the shear is computed by assuming the reaction of each pile to the external load, which is calculated by dividing the total load applied to the cap by the number of piles. This issue is not valid for flexible caps.
8) Piles in a group tend to interact depending on their spacing. Closer spacing leads to more interaction and settlement. An equivalent footing method is a streamlined approach for determining pile group settlement, treating the pile group as a single equal footing. The location and size of the equivalent footing are determined by the load transfer method and are defined by the neutral plane depth.
9) Pile group issues in geotechnical engineering can greatly benefit from the combined use of conventional hand calculations and computer-based simulations in design and analysis procedures.
10) It is essential to recognize that the application of pile groups does not solve all geotechnical performance issues, such as bearing capacity, settlement, and lateral and pullout load stability, in isolation. A multidisciplinary approach to design and analysis is necessary, considering various critical factors, including problematic soils, load combinations, ground modification techniques, construction methods, drag load, negative skin friction, embedment depth, and pile cap area and thickness. An integrated and comprehensive solution is crucial to meet safety requirements and practical considerations.

18.9 Questions and Problems

18.9.1 Explanatory

1 What are the pile caps and their functions? What are the differences between cases where the pile cap is connected to the ground surface and cases where it is not?

2 List the factors that influence the behavior of a pile group.

3 Explain the effects of pile spacing on the behavior of a pile group.

4 Define the roles of pile cap thickness in load distribution in a pile group.

5 Which issues can influence the group efficiency factor?

6 Explain and compare two different failure modes of pile groups under axial loads.

7 Illustrate and define the shadowing effect in pile groups under lateral loads.

8 What are the steps to apply the P–y curve method in pile groups?

9 Explain and compare the sectional method and STM method in the structural design of pile caps.

10 Illustrate and define the concept of equivalent foundation and neutral plane in settlement estimation of pile groups.

11 Explain why pile groups are more favorable than single piles in extraordinary load combinations.

18.9.2 Conceptual

1 Compare the isobars and settlement of a single pile, a pile group with closely spaced piles, and a group with piles far apart.

2 Validate the equations of load distribution in the case of rigid pile caps through basic theories.

3 Define the roles of pile cap thickness in load distribution in a pile group.

4 Evaluate the effect of the following parameters on the depth of the neutral plane:
(a) dead load magnitude
(b) piles length
(c) toe capacity of the pile
(d) shaft friction of pile
(e) internal friction angle of soil in a sandy deposit
(f) undrained shear strength of soil in a fine-grained deposit

5 Evaluate the settlement of the given pile group in Example 18.3 in two following conditions:
(a) Only raft as the foundation
(b) Removing the fill adjacent to the pile cap

6 Consider case 18.1, and explain the significant points that this case study highlights.

7 Consider case 18.2. Explain the effects of heave on pile group behavior, and illustrate two alternatives regarding foundation-heave interaction.

8 Explain and illustrate the differences between the response of a single pile and a group pile based on case 18.3.

9 Compare and contrast the load–displacement diagrams of single and group drilled shafts, considering the results of Case Study 18.5.

10 Consider buildings with given characteristics and assume a height of 3.3 m for each story. Consider each building in three soil conditions: weak, medium, and hard. Calculate the amount of settlement and discuss on resultant $\frac{\Delta}{B}$ in each case through the presented figure. Use $s = k\frac{qB}{E}$ in your calculations assuming $k = 0.4$.

Building ID	Height/Breadth	Number of stories
Building 1	2	30
Building 2	3	60
Building 3	4	90

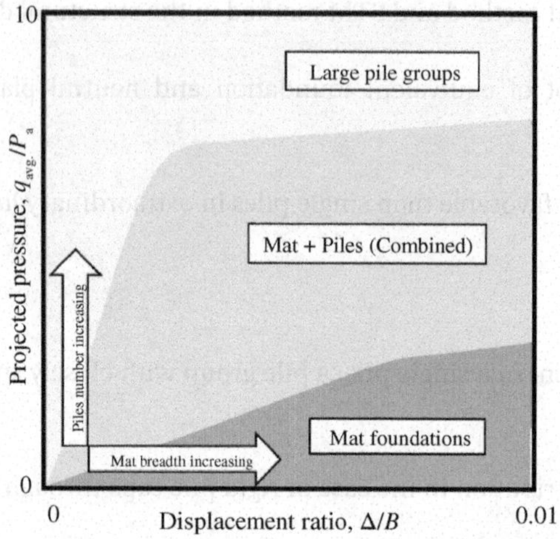

11 Compare and discuss the five zones introduced in the figure below regarding bearing capacity and settlement.

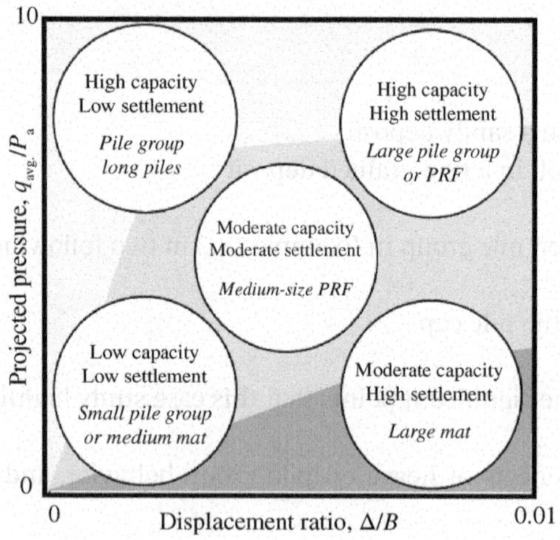

18.9.3 Problems

1 A pile group comprising nine driven piles with a diameter of 0.4 m is implemented in a two-layered soil with the following characteristics. Determine the ultimate bearing capacity of the pile group under axial loads in two different total and effective stress analysis conditions. The first 8 m of soil is soft clay, and the remaining pile length is embedded in stiff clay.

$$Soft\ Clay:\ S_u = 20\,kPa, \varphi'_{cs} = 30°, \gamma = 17\,\frac{kN}{m^3}, \gamma_{sat} = 18\,\frac{kN}{m^3}, OCR = 1$$

$$Stiff\ Clay:\ S_u = 90\,kPa, \varphi'_{cs} = 28°, \gamma = 17.5\,\frac{kN}{m^3}, \gamma_{sat} = 18.5\,\frac{kN}{m^3}, OCR = 5$$

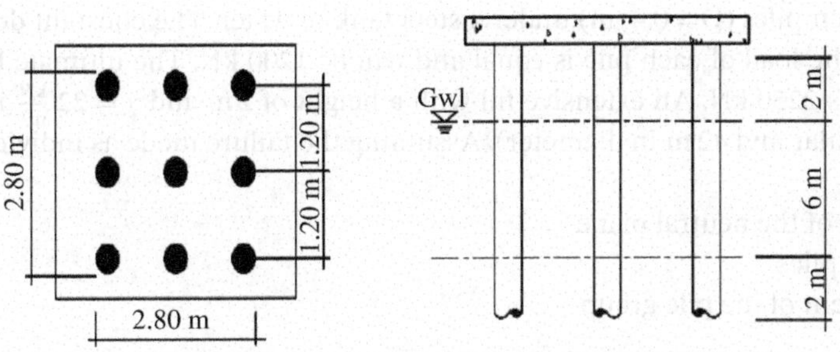

2 Consider the given pile group, assuming the pile cap is rigid and the ultimate loads as M = 90 ton.m, P = 450 tons, and H = 210 tons, determine the maximum and minimum stress and axial load of piles. Control the compressive stress in the piles section - ($\sigma_{allowed}$ = 15 MPa).

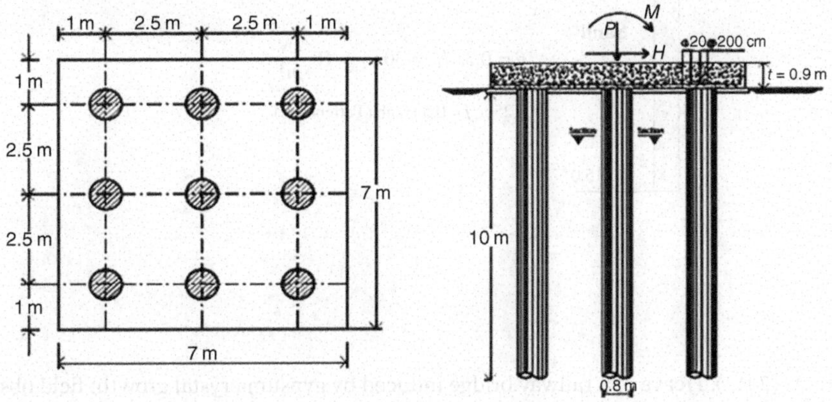

3 A group of piles, comprising 8 driven steel piles (H-Piles) is used to construct an open-type wharf. The dimensions and positions of the piles are as follows. Designate the ultimate axial-bearing capacity of the pile group in two different conditions of underlying soil, assuming that no plug has been formed:

(a) Sand: $\gamma_{sat} = 20\ \frac{kN}{m^3}$, $\beta = 0.5$, $N_t = 35$

(b) Clay: S_u(kPa) = 25 kPa, $\alpha = 1$, $N_C = 10$

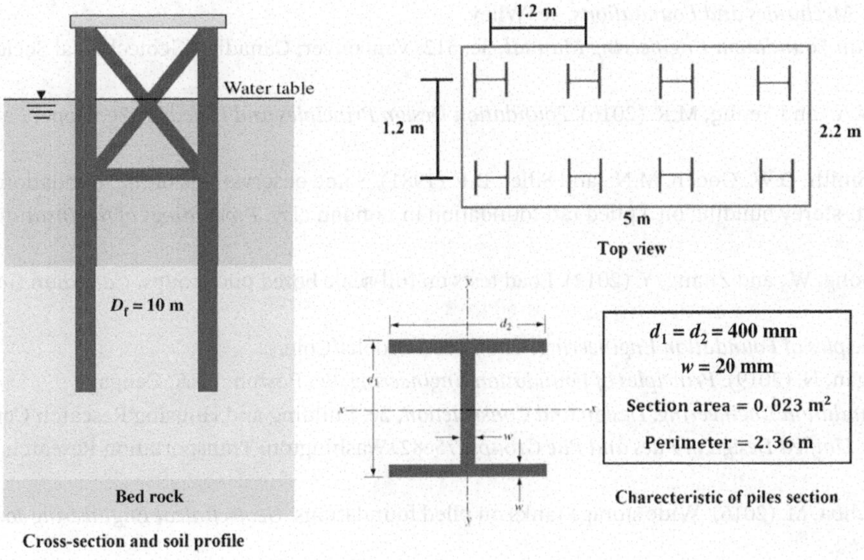

4 Consider a group of driven piles (D = 0.45m) under a steel tank of water. The constant dead load imposed on the pile cap equals 200 kPa and the load of each pile is equal and will be 1200 kN. The ultimate bearing capacity of each pile based on static analysis is 2250 kN. An extensive fill with a height of 2 m and $\gamma = 22\ \frac{kN}{m^3}$ is constructed in a wide area around the pile cap (circular and 42 m in diameter). Assuming the failure mode as individual and the thickness of the sand layer equal to 52 m.

(a) Determine the depth of the neutral plane
(b) Plot the axial load of piles
(c) Estimate the settlement of the pile group

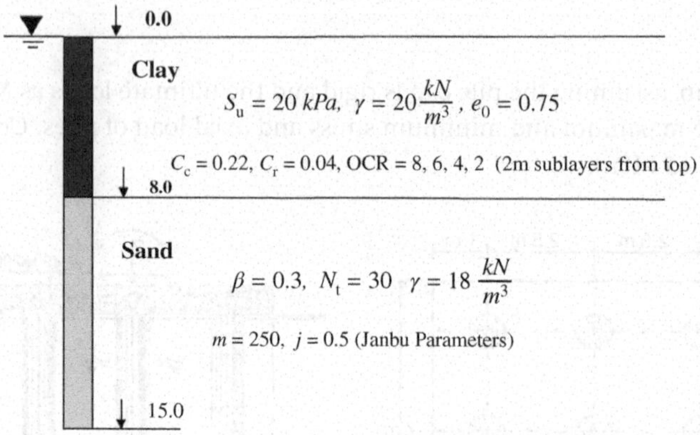

References

Alonso, E.E. and Ramon, A. (2013). Heave of a railway bridge induced by gypsum crystal growth: field observations. *Geotechnique* 63 (9): 707–719.

Alonso, E.E., Sauter, S., and Ramon, A. (2015). Pile groups under deep expansion: a case history. *Canadian Geotechnical Journal* 52 (8): 1111–1121.

Bowles, J.E. (1997). *Foundation Analysis and design*, 5e. McGraw-Hill.

Brown, D.A. and Bollman, H.T. (1993). Pile-supported bridge foundations designed for impact loading. *Transportation Research Record* 1331: 87–91.

Brown, D.A., Morrison, C., and Reese, L.C. (1988). Lateral load behavior of pile group in sand. *Journal of Geotechnical Engineering, ASCE* 114 (11): 1261–1276.

Budho, M. (2011). *Soil Mechanics and Foundations*, 3e. Wiley.

CFEM (1992). *Canadian Foundation Engineering Manual*, 3e, 512. Vancouver: Canadian Geotechnical Society, BiTech. Publishers.

Coduto, D.P., Kitch, W.A., and Yeung, M.R. (2016). *Foundation Design Principles and Practices*, 3e. Upper Saddle River, NJ: Prentice Hall, Inc.

Cooke, R.W., Bryden Smith, D.W., Gooch, M.N., and Sillet, D.F. (1981). Some observations of the foundation loading and settlement of a multi-storey building on a piled raft foundation in London clay. *Proceedings of the Institution of Civil Engineers* 107 (1): 433–460.

Dai, G., Salgado, R., Gong, W., and Zhang, Y. (2012). Load tests on full-scale bored pile groups. *Canadian Geotechnical Journal* 49: 1293–1308.

Das, B.M. (2004). *Principles of Foundation Engineering*. Thompson Brooks/Cole.

Das, B.M. and Sivakugan, N. (2019). *Principles of Foundation Engineering*, 9e. Boston, USA: Cengage.

Eslami, A. (2013). *Foundation Engineering, Design and Construction*, 2e. Building and Housing Research Center.

Fellenius, B.H. (1989). *Unified Design of Piles and Pile Groups*, 75e82. Washington: Transportation Research Board. TRB Record 1169.

Fellenius, B.H. and Ochoa, M. (2016). Wide storage tanks on piled foundations. *Geotechnical Engineering Journal of the SEAGS & AGSSEA* 47 (1): 50–62.

Fellenius, B.H., Mario Terceros, H., Mario Terceros, A. et al. (2019). Static response of a group of 13 piles tested simultaneously. In: *Proceedings of XVI Pan-American Conference on Soil Mechanics and Geotechnical Engineering (XVI PCSMGE)* (ed. N.P. López-Acosta et al.), Mexico, Cancun (17–20 November).

Fernandes, M.M. (2020). *Analysis and Design of Geotechnical Structures*. CRC Press eBooks. https://doi.org/10.1201/9780429398452.

Fleming, K., Weltman, A., Randolph, M., and Elson, K. (2009). *Piling Engineering*, 3e. Taylor & Francis.

Hammouli, S.E., Hanna, A., Lin, L., and Khalifa, M. (2021). Role of cap thickness in pile-cap foundation. *International Journal of Civil Infrastructure*.

Hannigan, P. J., Rausche, F., Likins, G. E., et al. (2016). Design and Construction of Driven Pile Foundations. Publication No. FHWA-NHI-16-009, vol. I. Federal Highway Administration, U.S. Department of Transportation.

Hemsley, J.A. (2000). *Design Applications of Raft Foundations*, 1e. Thomas Telford.

Kishida, H. and Meyerhof, G.G. (1965). Bearing capacity of pile groups under eccentric loads in sand. *Proceedings, Sixth International Conference on Soil Mechanics and Foundation Engineering*, Montreal (8–15 September), vol. 2, 270–274. Toronto: University of Toronto Press.

Kovári, K. and Amstad, C. (1982). A new method of measuring deformations in diaphragm walls and piles. *Géotechnique* 32 (4): 402–406.

Lo, M.B. (1967). Discussion to paper by B. O. Beredugo. *Canadian Geotechnical Journal* 4 (3): 353–356.

Murthy, V.N.S. (2007). *Advanced Foundation Engineering*, 1e. CBS Publishers and Distributers.

Poulos, H.G. and Davis, E.H. (1980). *Pile Foundation Analysis and Design*. New York: Wiley.

Reese, L.C., Isenhower, W., and Wang, S. (2005). *Analysis and Design of Shallow and Deep Foundations*. John Wiley & Sons.

Rollins, K.M., Peterson, K.T., and Weaver, T.J. (1998). Lateral load behavior of full-scale pile group in clay. *Journal of Geotechnical and Geoenvironmental Engineering, ASCE* 124 (6): 468–478.

Sham-e Consulting Engineering Co. (2016). Technical report of site investigation and foundation design for Babolsar Twin towers.

Teramoto, S., Niimura, T., Akutsu, T., and Kimura, M. (2018). Evaluation of ultimate behavior of actual large-scale pile group foundation by in-situ lateral loading tests and numerical analysis. *Soils and Foundations* 58 (4): 819–837.

Tomlinson, M. and Woodward, J. (2014). *Pile Design and Construction Practice*, 6e. Taylor & Francis.

Van Impe, P., Van Impe, W.F., and Seminck, L. (2013). Discussion of an instrumented screw pile load test and connected pile group load settlement behavior. *Journal of Geo-Engineering Sciences* 1 (1): 13–36.

Van Impe, P., Van Impe, W.F., Manzotti, A. and Seminck, L. (2015). Load-settlement behaviour of three pile groups: a case study. *Proceedings of 16th European Conference on Soil Mechanics and Geotechnical Engineering*, Edinburgh, UK.

Vesić, A.S. (1969). Experiments with instrumented pile groups in sand. In: *Performance of Deep Foundations* (ed. R. Lundgren and E. D'Appolonia), 177–222.

Walsh, J.M. (2005). *Full Scale Lateral Load Test of a 3x5 Pile Group in Sand*. Department of Civil and Environmental Engineering, Brigham Young University.

19

Piled Raft Foundations

Chapter Highlight

- **Introduction to Piled Raft Foundations**
- **Principles**
  - Analysis Process
  - GASP (Geotechnical Analysis of Strip with Piles, Poulos 1991)
  - GARP (Geotechnical Analysis of Raft with Piles, Poulos 1991)
  - Numerical Analysis (2D & 3D)
  - Design Requirements
  - Design Philosophies
  - Favorable & Unfavorable Conditions in Application
- **Geotechnical Performance**
  - Bearing Capacity
  - Vertical Load
  - Lateral Load
  - Flexural Loads (Moments)
  - Load–Displacement Behavior
  - Settlement
  - Load Distribution
- **Special Considerations**
  - Connected & Nonconnected
  - Cushion Application
  - Hybrid Intrusions and Piles PRFs
  - Geothermally Activated Foundation Systems
- **Case Studies**
  - Messe Turm Tower, Frankfurt, Germany (Hemsley 2000)
  - Full-Scale Load Tests on PRFs, Brazil (Garcia and Albuquerque 2019)
  - PRFs under Combined Loading (VMH) (Chanda et al. 2021)
  - Dynamic Response of Nonconnected Piled Raft Foundation Using Centrifuge Modeling (Baziar et al. 2018)

19.1 Introduction

Pile groups have traditionally been designed with a relatively high safety factor, with the bearing capacity being the primary design criterion. The arrangement of piles in the group is intended to support the loads from the superstructure. However, the connection "cap," which is typically a raft (or mat) in direct contact with the soil, is seldom taken into account for its contribution to the overall performance of the pile group.

Advanced Foundation Engineering: Principles, Performance and Prospects, First Edition. Moses Karakouzian and Abolfazl Eslami.

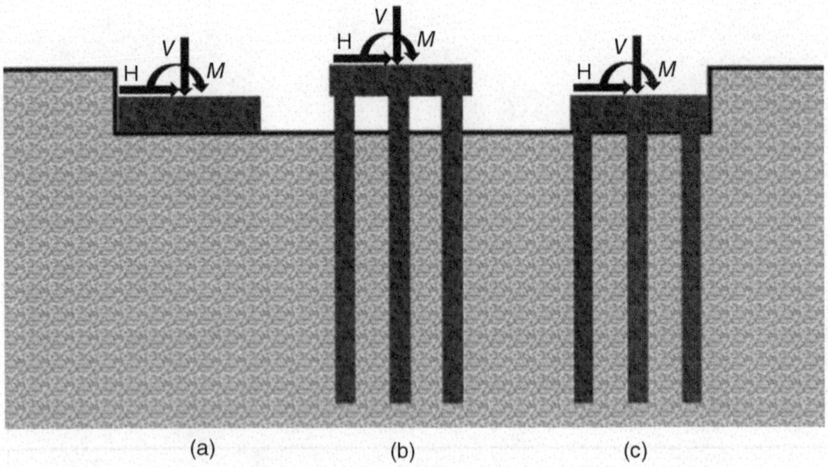

Figure 19.1 (a) Raft, (b) pile group, and (c) piled raft (after Mandolini 2003).

Pile performance enhancement has gained recognition as one of the most effective solutions for increasing bearing capacity and reducing settlement in foundation systems. The hybrid foundation, also known as the piled-raft foundation (PRF) (Eslami et al. 2008; Veiskarami et al. 2007), combines the benefits of both piles and a raft to support the loads. In PRF, the design procedure differs from conventional foundation design, and the loads are assumed to be borne by both the raft and the piles. The piles in a PRF can carry loads up to levels comparable to a single pile's bearing capacity or even higher (Eslami et al. 2008; Veiskarami et al. 2007). Figure 19.1 provides a schematic view of a raft, a pile group, and a piled raft foundation. In challenging loading conditions, such as combined vertical, moment, and horizontal loading, the performance of raft foundation alone is often not adequate in terms of bearing capacity, settlement, and stability. Pile groups are commonly preferred in such situations. However, PRFs with their comprehensive and interactive response to loads emerge as a more favorable option to address these challenges effectively.

The piled raft foundation offers a highly economical approach to reduce both total and differential settlements compared to conventional foundation solutions. In recent advancements, piles have often been used not primarily for overall stability but as settlement reducers. In such cases, the raft itself possesses sufficient bearing capacity, and the piles are strategically incorporated to control or minimize the displacements of the foundation system, rather than bearing the majority of the loads (Reul and Randolph 2004; Randolph 1994). This innovative design approach provides an efficient and cost-effective solution to mitigate settlement-related challenges in foundation engineering.

In the traditional design approach, the pile head is typically connected to or embedded in the raft to establish a rigid connection. In such cases, a significant number of piles are used, and each pile bears a relatively low load, allowing for a high safety margin against geotechnical or structural failure. The piles are usually designed to resist horizontal forces effectively through well-designed connections, supported by the large number of piles. However, in settlement-reducing pile designs that serve as structural members, a different approach can be adopted. A lower safety factor may be applied to the geotechnical capacity of these piles, with the raft alone providing sufficient support (Eslami et al. 2008; Reul and Randolph 2004). Yet, when these piles are structurally connected to the raft, as in conventional construction, high axial stress may be induced in the relatively small number of piles. As a result, the load-carrying capacity of settlement-reducing piles might be controlled by their structural capacity rather than their geotechnical capacity. Additionally, sparsely arranged piles under a raft may not offer adequate resistance against lateral loads. To address these challenges, considering these piles as stiffeners for the soil offers a viable alternative (Wong et al. 2004). By adopting this approach, the piles can efficiently improve the overall performance of the piled raft foundation system, ensuring better resistance against lateral loads and optimizing the load-carrying capacity of the settlement-reducing piles. This integrated design strategy enhances the reliability and effectiveness of the piled raft foundation.

Figure 19.2 showcases the evolution of high-rise foundation systems in Frankfurt am Main, Germany, where numerous buildings in recent decades have been constructed on the challenging Frankfurt Clay deposit. The illustration reveals a general trend wherein the foundation systems shift from simple rafts to more sophisticated solutions like piled raft foundations as the height of structures increases.

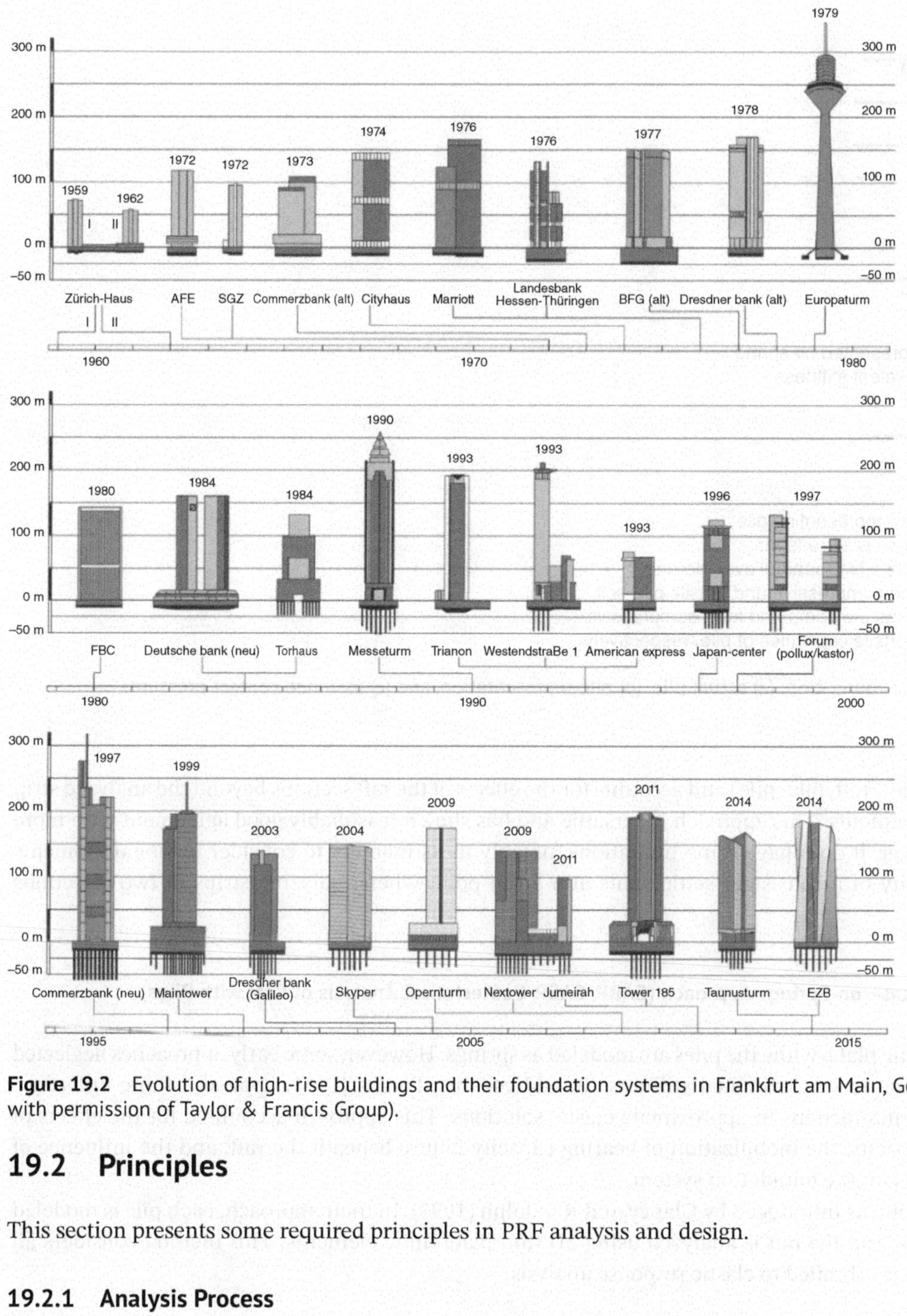

Figure 19.2 Evolution of high-rise buildings and their foundation systems in Frankfurt am Main, Germany (Katzenbach et al. 2017 / with permission of Taylor & Francis Group).

19.2 Principles

This section presents some required principles in PRF analysis and design.

19.2.1 Analysis Process

Two broad classes of detailed analysis methods will be considered:

- Approximate computer-based methods:
  - GASP (Geotechnical Analysis of Strip with Piles)
  - GARP (Geotechnical Analysis of Raft with Piles)
- more rigorous computer-based methods.

19.2.1.1 Method Employing a Strip-on-Springs Approach (GASP, Geotechnical Analysis of Strip with Piles, Poulos 1991)

Poulos (1991) proposed a method for analyzing piled raft foundations, as shown in Figure 19.3. In this approach, a strip section represents a portion of the raft, and the piles are modeled as springs. The method considers all four components of

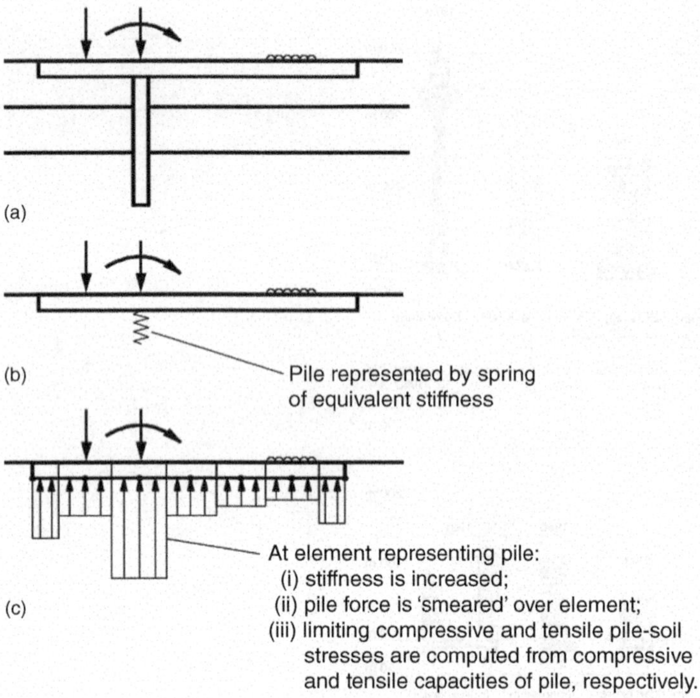

Figure 19.3 Modelling of piled strip foundation: (a) actual pile, (b) pile representation, and (c) assumed contact pressures (Poulos 1991).

interaction (pile–raft, raft–pile, raft–raft, pile–pile) and accounts for the effects of the raft sections beyond the analyzed strip by computing free-field soil movements. This approach is versatile and has shown reasonably good agreement with more comprehensive analyses. However, it does have some limitations, notably in its inability to consider torsional moments within the raft and the possibility of inconsistent settlements at a given point when analyzing strips in two directions through that point.

19.2.1.2 Method Employing a Plate-on-Springs Approach (GARP, GASP, Geotechnical Analysis of Raft with Piles, Poulos 1991)

The raft is represented by an elastic plate, while the piles are modeled as springs. However, some early approaches neglected certain interaction effects, leading to overestimated stiffness. To address this, Poulos (1994) applied a finite difference method and considered various interactions in approximate elastic solutions. This approach accounted for the effects of piles reaching their ultimate capacity, the mobilization of bearing capacity failure beneath the raft, and the influence of free-field vertical soil movements on the foundation system.

Another more rigorous concept was introduced by Clancy and Randolph (1993). In their approach, each pile is modeled as a series of rod finite elements, and the raft is analyzed using 2D thin-plate finite elements. This method considers all interaction components, though it is limited to elastic response analysis.

19.2.1.3 More Rigorous Numerical Methods of Analysis

These methods offer more detailed modeling of the various components in piled raft foundations compared to previous approaches. They heavily rely on computer analysis and often utilize specialized software. Two main numerical methods employed are the boundary element method and the finite element method, and sometimes a combination of both.

When it comes to modeling real-world problems, 3D finite element methods are often considered the most powerful tool, at least from an analysis perspective (although the challenge of assigning appropriate parameters still remains).

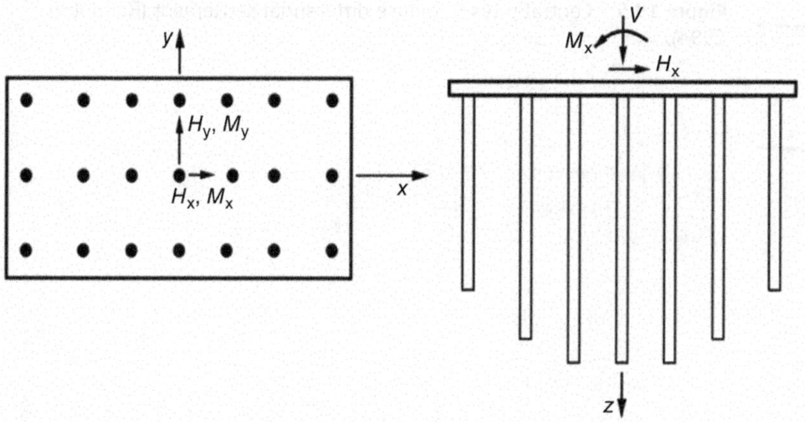

Figure 19.4 Piled raft foundation system (Hemsley 2000 / With permission of Thomas Telford Ltd.).

19.2.2 Design Requirements

Figure 19.4 depicts a PRF and the overall imposed forces and moments. The aspects that are required to be considered in the design are as follows:

- ultimate geotechnical capacity under vertical, lateral, and moment loading,
- overall settlement and stiffness,
- differential settlements and angular rotations,
- lateral movements and stiffness,
- structural design of both rafts and piles (Hemsley 2000).

The design criterion for total and differential settlements, as well as lateral deformations, is to ensure that the maximum or differential movement does not exceed the allowable limit. This limit varies depending on the type of structure being considered.

The structural design requires an estimation of the following:

- bending moments and shear forces within the raft,
- axial and lateral loads, as well as bending moments in the piles.

A comprehensive design approach can address all the mentioned aspects in a single coherent analysis. While complex 3D numerical simulations provide detailed analyses, relatively simple methods should also be capable of preliminary design and serve as a control for computer-based design methods.

19.2.3 Design Philosophies

Three distinct design approaches can be identified for piled rafts:

- Conventional
- Creep piling
- Differential settlement control (Randolph 1994)

19.2.3.1 Conventional

In the conventional approach, the foundation is primarily designed as a pile group, with evenly spaced piles distributed throughout the entire foundation area. The load is directly transferred from the pile cap to the ground. The main advantage of this approach is a reduction in the total number of piles required, as only approximately 60–75% of the total structural load is borne by the piles.

An example of the conventional approach is presented in a case history by Smith and Randolph (1990). In this case, the piles were 0.8 m in diameter and 20 m in length. The total structural load of 1144 MN was carried by 280 piles, resulting in

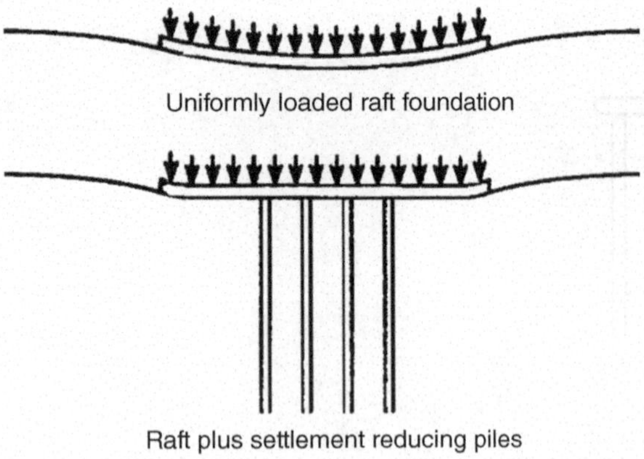

Figure 19.5 Central piles to reduce differential settlement (Randolph 1994).

an estimated average working load per pile of around 3 MN, considering that over 25% of the load was transferred from the pile cap to the ground.

19.2.3.2 Creep Piling

The two principles behind this approach (for relatively soft cohesive soils) are:

- each pile is designed to operate at a working load at which significant creep starts to occur (about 70–80% of ultimate bearing capacity),
- adequate piles are implemented to decrease the net contact pressure between the raft and soil lower than the pre-consolidation pressure of the clay.

The foundation is primarily designed as a raft, but the total settlement is reduced through the inclusion of piles, which are evenly distributed beneath the raft. The working load of each pile is chosen as the creep load to prevent high loads from developing in piles located at the edges of the foundation, enabling more accurate determination of bending moments in the raft.

This approach has also been studied for noncohesive soil, and it has been observed that the transfer of load from the cap results in an increase in the supporting piles' capacity due to the effective rise in stress. The piles mainly function under their enhanced creep load, and the foundation behavior can be analyzed using the conventional methods described earlier.

19.2.3.3 Differential Settlement Control

Two previously mentioned approaches apply a uniform piles distribution, to limit settlements to allowed limits. Consequently, the differential settlements will be decreased. However, a more direct method is to design the pile support in a way that minimizes differential settlements at the outset without essentially a considerable decrease in the average settlement.

Figure 19.5 depicts schematically the principles behind the design of piles to decrease differential settlements. Presuming that the structural load is approximately uniformly distributed beneath the building, then there will be a tendency for an unpiled raft to "dish" in the center. Adding some piles in the central region, which may be loaded near their ultimate capacity, will decrease that tendency, and therefore minimize differential settlements.

Figure 19.6 depicts the load–displacement performance of piled rafts designed according to the first two strategies. Curve 0 indicates the behavior of the raft alone, in which excessive settlement occurs at the design load. Curve 1 represents the conventional design philosophy, for which the behavior of the pile-raft system is governed primarily by the pile group behavior and which may be largely linear at the design load. In this case, the piles carry the great majority of the load. Curve 2 represents the case of creep piling, where the piles perform at a lower safety factor, but due to fewer piles, the raft bears more load than Curve 1. Curve 3 presents the strategy piles application as settlement reducers and utilizing the full capacity at the design load. Consequently, the load–displacement relation may be nonlinear at the design load, but the overall foundation system has an adequate margin of safety, and the settlement criterion is satisfied. So, the design presented by Curve 3 is likely to be significantly more economical than the designs presented by Curves 1 and 2.

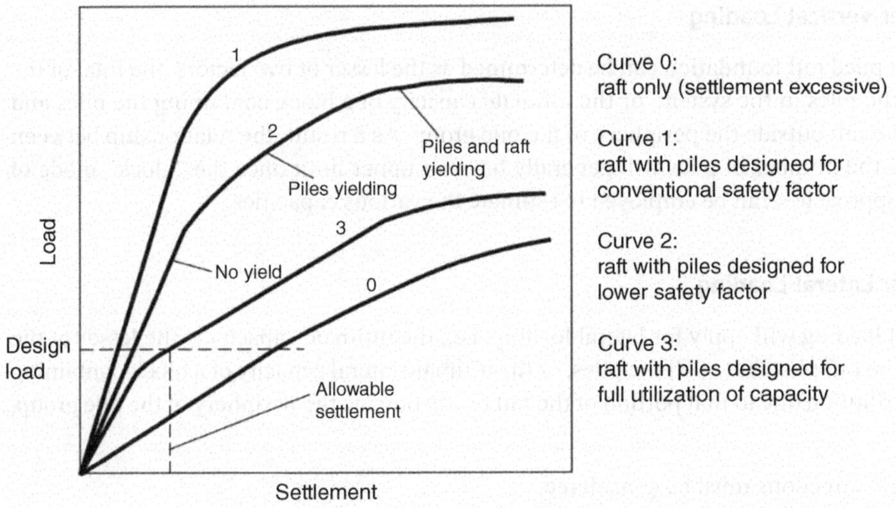

Figure 19.6 Load–settlement curves for piled rafts according to various design philosophies (Hemsley 2000 / Thomas Telford Ltd.).

19.2.4 Favorable and Unfavorable Conditions in Application

The main advantages of applying PRFs are:

- As piles are not required to bear all the load, there is the potential for considerable savings in the foundation's cost.
- Piles can be positioned strategically under the raft so that differential settlements can be limited.
- Variety in length and/or diameter of piles as well as raft thickness can be applied aiming for optimized design.
- Piles can be designed to bear a load approaching (or equal to) their ultimate geotechnical load, provided that the raft can develop a sufficient proportion of the required ultimate load capacity (Hemsley 2000).

The most effective application of piled rafts occurs when the raft itself can provide adequate load capacity. However, the settlement and/or differential settlements of the raft alone most probably exceed the allowable settlement values. Poulos (2001) examined some idealized soil profiles and found that soil profiles consisting of relatively stiff clays as well as relatively dense sands or weak rocks may be favorable. In both circumstances, the raft can provide a considerable part of the required load capacity and can also contribute to the foundation stiffness, particularly after the complete mobilization of pile capacity (Mandolini et al. 2005; Viggiani et al. 2012).

It has also been observed that the performance of a piled raft foundation can be optimized by selecting appropriate locations for the piles beneath the raft. Generally, the piles should be concentrated in the most heavily loaded parts, while the number of piles can be decreased, or even eliminated, in less heavily loaded parts (Horikoshi and Randolph 1998).

There are soil profiles where PRF may not provide much if any, advantage over a conventional piled foundation, including:

- Very soft clays at or near the surface of the raft, where the raft can contribute only a relatively small part of the required ultimate load capacity.
- Subjected to long-term consolidation settlement; in which, the soil may tend to lose contact with the raft and transfer all the load to the piles.
- Subjected to expansive movements; in which, the soil movements will lead to an increase in contact pressures on the raft and consequently, the development of tensile forces in the piles (Hemsley 2000).

19.3 Geotechnical Performance

This section presents geotechnical aspects of piled raft foundations performance, especially bearing capacity and load–displacement issues.

19.3.1 Geotechnical Capacity Under Vertical Loading

The ultimate geotechnical capacity of a piled raft foundation can be determined as the lesser of two factors: the total of the ultimate capacities of the raft plus all the piles in the system, or the ultimate capacity of a block containing the piles and raft, along with that of the portion of the raft outside the periphery of the pile group. As a result, the relationship between the ultimate geotechnical capacity and the number of piles will generally have an upper limit once the "block" mode of failure develops. Conventional design approaches can be employed to estimate the various capacities.

19.3.2 Geotechnical Capacity Under Lateral Loading

The same approach as used for vertical loading will apply for lateral loading, i.e., the ultimate capacity is the lesser of the sum of the ultimate lateral capacity of the raft plus that of all the piles, or the ultimate lateral capacity of a block containing the piles, raft and the soil, plus the contribution due to that portion of the raft or cap outside the periphery of the pile group. The following points need to be noted:

- the response in both orthogonal lateral directions must be considered;
- the ultimate lateral capacity of the raft may comprise both shear resistance at the underside of the raft and passive resistance of the embedded portion of the raft;
- for the ultimate lateral capacity of the piles, both "short-pile" (lateral failure of the supporting soil) and "long-pile" (yield or failure of the pile itself) modes of failure must be considered;
- for the ultimate lateral capacity of the pile–soil–raft block, it will generally be adequate to consider only the "short-pile" failure of the block.

The general form of the relationship between ultimate lateral capacity and the number of piles will be similar to that for vertical loading, with an upper limit being the "block" capacity of the group. As with vertical loading, the various ultimate capacities can be assessed from conventional foundation design procedures.

19.3.3 Geotechnical Capacity Under Moment Loading

The ultimate moment capacity of the piled raft can be estimated approximately as the lesser of:
(a) the ultimate moment capacity of the raft (M_{ur}) and the individual piles (M_{up});
(b) the ultimate moment capacity of a block containing the piles, raft, and soil (M_{ub}).
The ultimate moment capacity of the raft can be estimated through:

$$\frac{M_{ur}}{M_m} = \frac{27}{4}\frac{V}{V_u}\left[1 - \left(\frac{V}{V_u}\right)^{\frac{1}{2}}\right] \tag{19.1}$$

where,
M_m = maximum possible moment that soil can support
V = applied vertical load
V_u = ultimate centric load on the raft when no moment is applied
Considering loading in the x-direction only, for a rectangular raft, the maximum moment M_m in the x-direction can be expressed as

$$M_m = \frac{P_{ur}BL^2}{8} \tag{19.2}$$

where:
P_{ur} = ultimate bearing capacity beneath the raft
B = width of the raft
L = length of the raft
The ultimate moment contributed by the piles can be estimated from

$$M_{up} \approx \sum_{i=1}^{n_p} P_{uui}|x_i| \tag{19.3}$$

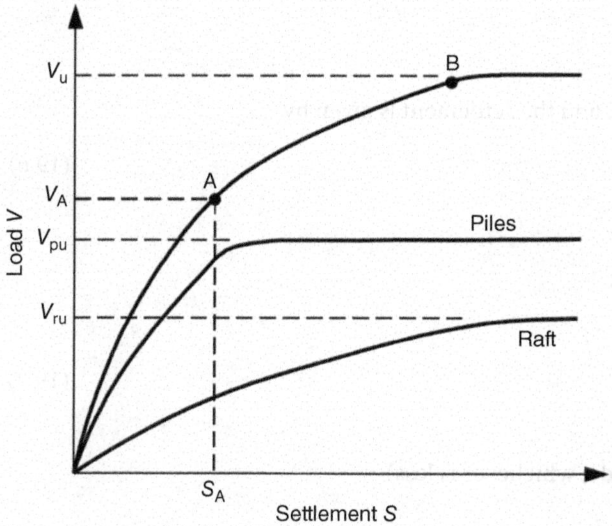

Figure 19.7 Construction of load–settlement curve for piled raft (Hemsley 2000 / With permission of Thomas Telford Ltd.).

where,

P_{uui} = ultimate uplift capacity of typical pile i
$|x_i|$ = absolute distance of pile i from the centre of gravity of the group
n_p = number of piles

The ultimate moment capacity of the block can be estimated (conservatively) from the theory for the "short-pile" failure of a rigid pile subjected to moment loading. Poulos and Davis (1980) give the solution for ultimate moment capacity M_{uB} (if no horizontal force is acting) as

$$M_{uB} = \alpha_B \overline{p}_u B_B D_B^2 \tag{19.4}$$

where,

B_B = width of block perpendicular to the direction of loading
D_B = depth of the block
$\overline{p}_u$ = average ultimate lateral resistance of soil along the block
α_B = factor depending on the distribution of ultimate lateral pressure with depth
= 0.25 for constant p_u with depth
= 0.2 for linearly increasing p_u with depth from zero at the surface.

19.3.4 Estimation of Load–Settlement Behavior

Several simplified analyses have been developed to estimate the load–settlement behavior of a piled raft. The presented method is a combination and extension of Poulos and Davis (1980) and Randolph (1994) methods. The following aspects are included:

- estimation of the load sharing between the raft and the piles, using the approximate solution of Randolph (1994);
- hyperbolic load–deflection relationships for the piles and the raft, thus providing a more realistic overall load–settlement response for the piled raft system than the original tri-linear approach of Poulos and Davis (1980).

Figure 19.7 shows the load–settlement relationship for the piled raft. Point A represents the point at which the pile capacity is fully mobilized when the total vertical applied load is V_A. Up to that point, both the piles and the raft share the load, and the settlement (S) can be expressed as

$$S = \frac{V}{K_{pr}} \tag{19.5}$$

where,

V = vertical applied load

K_{pr} = axial stiffness of piled raft system

Beyond point A, the additional load must be carried by the raft, and the settlement is given by

$$S = \frac{V_A}{K_{pr}} + \frac{V - V_A}{K_r} \tag{19.6}$$

where,

V_A = applied load at which pile capacity is mobilised

K_r = axial stiffness of raft

The load V_A can be estimated from

$$V_A = \frac{V_{pu}}{\beta_p} \tag{19.7}$$

where,

V_{pu} = ultimate capacity of piles (single pile or block failure mode, whichever is less)

β_p = proportion of load carried by piles

Use can be made of the approximate expressions described by Randolph (1994) for K_{pr} in Equation (19.5) and β_p in Equation (19.7), namely

$$K_{pr} = XK_p \tag{19.8}$$

where K_p denotes the stiffness of the pile group alone and, for fairly large numbers of piles:

$$X \approx \frac{1 - 0.6\left(\frac{K_r}{K_p}\right)}{1 - 0.64\left(\frac{K_r}{K_p}\right)} \tag{19.9}$$

$$\beta_p = 1/(1 + \alpha) \tag{19.10}$$

$$\alpha \approx \frac{0.2}{1 - 0.8\left(\frac{K_r}{K_p}\right)}\left(\frac{K_r}{K_p}\right) \tag{19.11}$$

If it is assumed that the pile and raft load–settlement relationships are hyperbolic, then the secant stiffnesses of the piles (K_p) and the raft (K_r) can be expressed as

$$K_p = K_{pi}\left(1 - \frac{R_{fp}V_p}{V_{pu}}\right) \tag{19.12}$$

$$K_r = K_{ri}\left(1 - \frac{R_{fr}V_r}{V_{ru}}\right) \tag{19.13}$$

where,

K_{pi} = initial tangent stiffness of pile group

R_{fp} = hyperbolic factor for pile group

V_p = load carried by piles

V_{pu} = ultimate capacity of piles

K_{ri} = initial tangent stiffness of raft

R_{fr} = hyperbolic factor for raft

V_r = load carried by raft

V_{ru} = ultimate capacity of raft

The load carried by the piles is given by

$$V_p = \beta_p V \leq V_{pu} \tag{19.14}$$

and the load carried by the raft is

$$V_r = V - V_p \tag{19.15}$$

where V denotes the total vertical applied load.

Substituting Equations (19.7)–(19.15) in Equations (19.5) and (19.6), the following expressions are obtained for the load–settlement relationship of the piled raft system.

For $V \leq V_A$:

$$S = \frac{V}{XK_{pi}\left(1 - \frac{R_{fp}\beta_p V}{V_{pu}}\right)} \tag{19.16}$$

For $V > V_A$:

$$S = S_A + \frac{V - V_A}{K_{ri}\left[1 - R_{fr}\frac{(V - V_{pu})}{V_{ru}}\right]} \tag{19.17}$$

where

$$S_A = \frac{\frac{V_{pu}}{\beta_p}}{XK_{pi}(1 - R_{fp})} \tag{19.18}$$

Equations (19.16)–(19.18) provide the means for estimating the average load–settlement relationship for the piled raft. Because K_r and K_p will vary with the applied load level, the parameters X and β_p will also generally vary. Thus, it may be necessary to carry out an iterative or incremental analysis, commencing with the initial stiffnesses K_{ri} and K_{pi}.

19.3.5 Immediate and Final Settlements

As proposed by Poulos and Davis (1980), it is preferable to calculate the consolidation settlement from the difference between the elastic total final and consolidation settlements and add this to the immediate settlement computed from a nonlinear analysis. The overall total final settlement is calculated via the:

$$S_{TF} = \frac{V}{K_u} + V\left(\frac{1}{K'_e} - \frac{1}{K_{ue}}\right) \tag{19.19}$$

where,

V = applied vertical load

K_u = undrained foundation stiffness (from nonlinear analysis)

K_{ue} = undrained foundation stiffness (from elastic undrained analysis)

K'_e = drained foundation stiffness (from elastic drained analysis)

The procedure outlined above will generally provide an estimate of the average load–settlement behavior of the piled raft foundation. However, the differential settlements within the foundation will depend largely on the distribution of applied loads, the arrangement of the piles, and the relative rigidity of the raft (Hemsley 2000).

19.3.6 Estimation of Pile Loads

An estimation of the largest compressive and tensile forces in the piles is essential for the structural design. A first estimate of the axial forces in the piles can be made through the following equations. The axial force P_i in any pile i is:

$$P_i = \frac{V\beta_p}{n_p} + \frac{M_x^* x_i}{I_y} + \frac{M_y^* y_i}{I_x} \tag{19.20}$$

with

$$M_x^* = \frac{M_x - \frac{M_y I_{xy}}{I_x}}{1 - \frac{I_{xy}^2}{I_x I_y}}, \quad M_y^* = \frac{M_y - \frac{M_x I_{xy}}{I_y}}{1 - \frac{I_{xy}^2}{I_x I_y}} \tag{19.21}$$

where,

V = total vertical load acting at the centroid of the foundation

n_p = number of piles in the group

M_x, M_y = moments about the centroid of the pile group in the direction of x- and y-axes

β_p = proportion of load carried by piles

I_x, I_y = moment of inertia of pile group for x- and y-axes
I_{xy} = product of inertia of pile group about the centroid
x_i, y_i = distance of pile i from y- and x-axes
M_x^*, M_y^* = effective moments in x- and y-directions, respectively, taking the symmetry of pile layout into account.
For a symmetrical pile group layout, $I_{xy} = 0$ and $M_x^* = M_x$, $M_y^* = M_y$. Therefore, Equation (19.20) reduces to

$$P_i = \frac{V\beta_p}{n_p} \pm \frac{M_x x_i}{\sum_{i=1}^{n_p} x_i^2} \pm \frac{M_y y_i}{\sum_{i=1}^{n_p} y_i^2} \tag{19.22}$$

The reviewed approach inherently assumes that the raft is rigid, the pile heads are pinned to the raft, i.e., no moment transfer occurs from the raft to the piles and piles are vertical.

19.3.7 Estimation of Raft Moments and Shears

The contact pressures below the raft balance only the load carried by the raft, while the piles bear the remaining load. Another approach is to consider the piled raft as a series of piled strip foundations, with the behavior of each piled strip obtained either by assuming the strip as rigid or, preferably, using solutions for a strip on an elastic foundation, with the piles treated as supports (or negative loads). This allows for an estimation of the torsional moments in the raft, which requires a more refined analysis where the raft is treated as a plate (Hemsley 2000).

Example 19.1 It is required to assess the adequacy of the given piled raft foundation, realizing the following design criteria (Hemsley 2000):

(a) a minimum overall factor of safety of 2.5 against bearing capacity, overturning, and lateral failure for the ultimate load case;
(b) a maximum long-term average settlement of 50 mm

All 9 precast concrete piles are of 0.6 m diameter, each driven to a penetration depth of 15 m. The average ultimate shaft friction is assumed to be 60 kPa in compression and 42 kPa in tension, while the ultimate end-bearing capacity is 900 kPa. For each pile, the ultimate axial capacities are then calculated to be 1.95 MN in compression and 1.20 MN in tension.

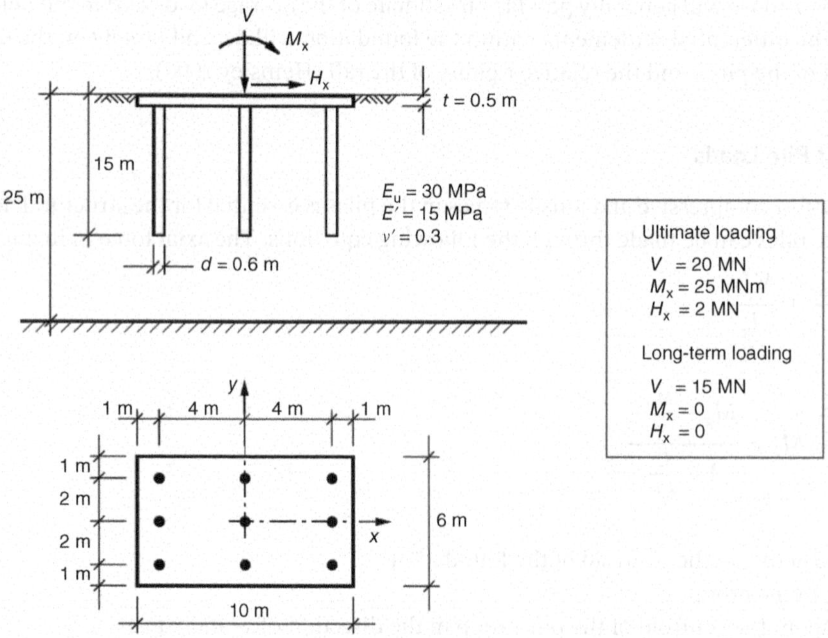

Vertical load capacity

Raft

$$q_u = 6c_u = 0.6\ \text{MPa}$$

$$P_{\text{raft}} = 10 \times 6 \times 0.6 = 36\ \text{MN}$$

Singular failure mode: Raft + Piles

$$P_{\text{total}} = 36 + (9 \times 1.95) = 53.55\ \text{MN}$$

Block failure mode:

$$2 \times (8.6 + 4.6) \times 0.1 \times 15 + 8.6 \times 4.6 \times 0.9 + (10 \times 6 - 8.6 \times 4.6) \times 0.6 = 87.46\ \text{MN}$$

Vertical capacity = min {53.55, 87.46} = 53.55 MN

F.S. = $\frac{53.55}{20} = 2.67 > 2.5$ O.K.

Moment capacity

Soil: $M_{\text{m}} = \frac{P_{\text{ur}}BL^2}{8} = 0.6 \times 6 \times \frac{10^2}{8} = 45\ \text{MNm}$

Raft: $M_{\text{ur}} = 45 \times \frac{27}{4} \times \frac{20}{53.55}\left(1 - \sqrt{\frac{20}{53.55}}\right) = 44.1\ \text{MNm}$

Piles: $M_{\text{up}} = \sum_{i=1}^{9} P_{\text{uui}}|x_i| = 1.2 \times (3 \times 4 + 3 \times 4 + 3 \times 0) = 28.8\ \text{MNm}$

Total: 44.1 + 28.8 = 72.9 MNm

A check must now be made of the moment capacity of the block. For the block, the length is 2.5 times the width, so that the average ultimate lateral pressure along the block, $\overline{P_{\text{u}}}$, is approximately

$$4.5 \times 0.1 = 0.45\ \text{MPa}$$

$$M_{\text{uB}} = 0.25 \times 0.45 \times 6 \times 15^2 = 151.9\ \text{MNm}$$

Lateral capacity = min {72.9, 151.9} = 53.55 MN

F.S. = $\frac{72.9}{25} = 2.92 > 2.5$ O.K.

Lateral load capacity

The ultimate lateral load capacity of single piles is computed using Broms' solutions (1964), assuming fixed pile heads. For short-pile failure, the theory gives an ultimate lateral load capacity of 7.6 MN per pile. Considering long-pile failure and taking the yield moment of the pile itself to be 0.45 MN m, the ultimate lateral load is found to be 0.61 MN, which is the critical value. For the nine piles, the total lateral load capacity is 5.49 MN, which is found to be less than the corresponding value for the block.

F.S. = $\frac{5.49}{2} = 2.74 > 2.5$ O.K.

Total settlement

Undrained: $K_{ri} = 420\ \text{MN/m}$

Drained: $K_{ri} = 169\ \text{MN/m}$

Undrained: $K_{\text{pi}} = 651\ \text{MN/m}$

Drained: $K_{\text{pi}} = 366\ \text{MN/m}$

$$K_{pri} = X\,K_{\text{pi}}$$

Undrained:

$$X = \frac{1-0.6\left(\frac{420}{651}\right)}{1-0.64\left(\frac{420}{651}\right)} = 1.044$$

$$K_{\text{ue}} = 1.044 \times 651 = 680\ \text{MN/m}$$

Drained:

$$X = \frac{1-0.6\left(\frac{169}{366}\right)}{1-0.64\left(\frac{169}{366}\right)} = 1.026$$

$$\acute{K}_e = 1.026 \times 366 = 375\ \text{MN/m}$$

The immediate settlement based on purely elastic analysis, under a load of 15 MN is 31 mm. It will be assumed that the final consolidation settlement (S_{CF}) can be computed as the difference between the total final and immediate settlements.

$$S_{CF} = \frac{V}{\acute{K}_e} - \frac{V}{K_{ue}} = 15\left(\frac{1}{375} - \frac{1}{680}\right) = 0.0179\ \text{m}$$

Total settlement:

$0.0311 + 0.0179 = 0.049\ \text{m} = 49\ \text{mm} < 50\ \text{mm}$ O.K.

More detail and calculations are available in the main resource.

19.4 Special Considerations

The current section is devoted to special considerations in the design and construction of piled raft foundations. These considerations include the design of connected and nonconnected piled rafts, the utilization of cushions, the hybrid application of ground modification techniques in conjunction with deep foundations under a raft foundation, and geothermally activated PRFs.

19.4.1 Connected and Nonconnected PRFs

The bearing and deformation aspects of a piled raft foundation or combined piled raft foundation (CPRF) are determined via interactions between the bearing elements and the existing subsoil, as depicted in Figure 19.8. The raft resistance $R_{raft,\,k\,(S)}$ is the integration of the soil contact pressure beneath the foundation. The total resistance, $R_{tot,\,k\,(S)}$ is delivered through the addition of raft resistance and piles resistance, $\sum R_{pile,\,k\,(S)}$. The resistance of a single pile is comprised of skin and base resistance. The existing interactions in a CPRF are as follows:

- Pile–soil
- Pile–pile
- Raft–soil
- Pile–raft

Piled raft foundations are generally classified into two different setups based on whether the piles are connected or nonconnected to the raft. The second one, recently becoming more popular, is recognized as disconnected piled rafts. In this configuration, the gap between the pile and raft can be filled with a selected suitable material, chosen on a case-specific basis. In such cases, the piles perform as soil reinforcement, and the loads are transferred indirectly from the superstructure to the piles. Since piles are not purely structural members in these cases, they can be designed with lower factors of safety, especially when a reduction in settlement to achieve service limit states is required. The intrinsic complexity of nonconnected piled rafts makes predicting their mechanical response challenging (Tradigo et al. 2015). Figure 19.9 depicts the interaction between soil and structural members for both connected and nonconnected piled rafts.

Eslami et al. (2011) conducted a study on the performance of optimized piled raft foundations (PRFs) with both connected and nonconnected piles. The research involved 2D and 3D finite element analyses of three case histories, including a 12-story residential building in Iran, a 39-story twin tower in Indonesia, and the Messeturm tower in Germany. Various influential parameters were evaluated, such as pile spacing, embedment depth, piling configuration, and raft thickness, with the aim of optimizing the design. The results demonstrated that the optimal design is achieved by concentrating the piles in the center of the raft and minimizing the total length of the piles. Furthermore, the implementation of nonconnected piled raft foundation systems led to a significant reduction in settlement and internal bending moments. In nonconnected piled rafts, the axial load on the piles is decreased. In such cases, settlement is controlled by the piles, while the bearing capacity is provided by the raft. Unlike connected cases, the piles have a lesser contribution to the load-carrying system (Eslami et al. 2011).

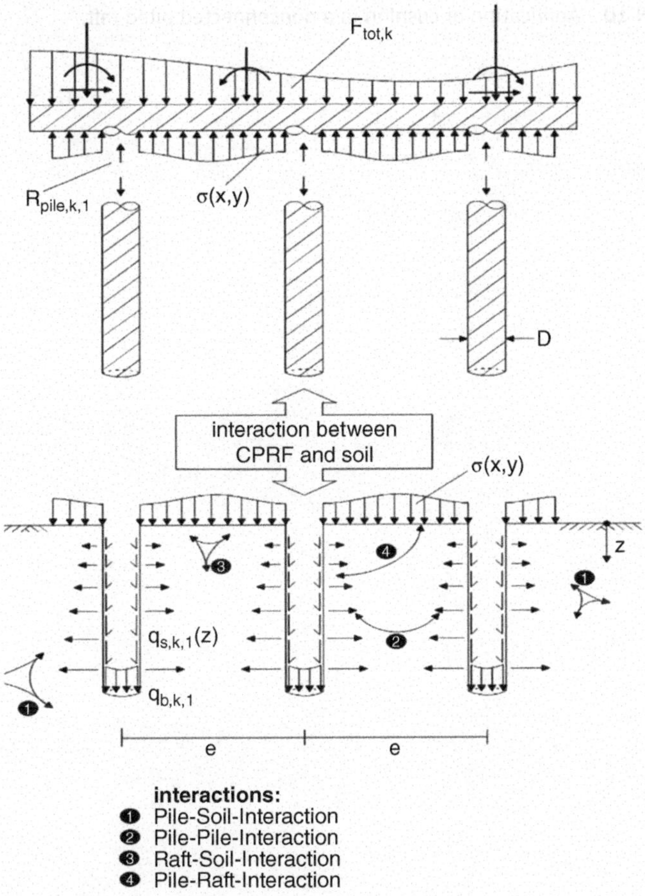

Figure 19.8 Interactions in a PRF (Katzenbach et al. 1998, 2020).

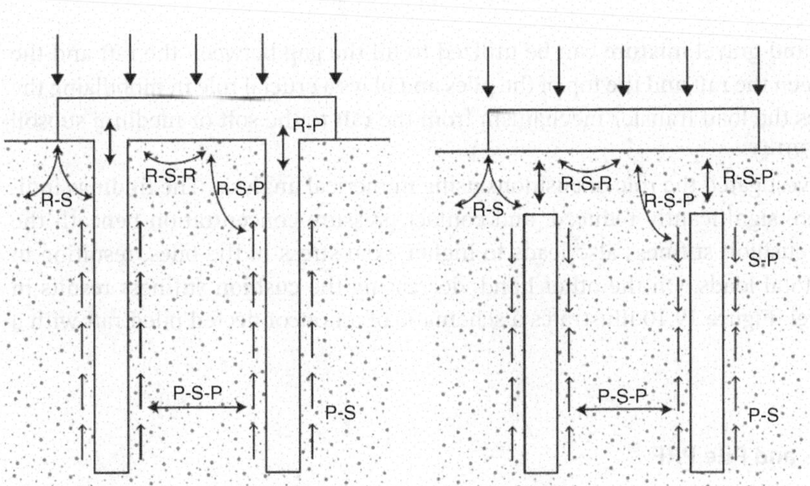

Figure 19.9 Soil and structural members interaction (R = raft; P = pile; S = soil): (a) connected, (b) disconnected (nonconnected) (Tradigo et al. 2015 / with permission of Elsevier).

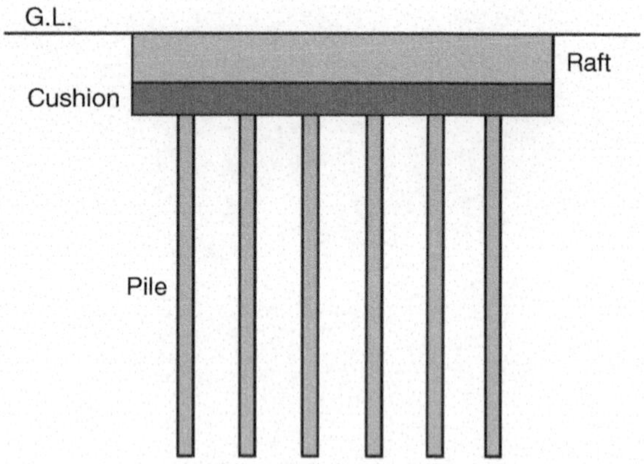

Figure 19.10 Application of cushion in a nonconnected piled raft.

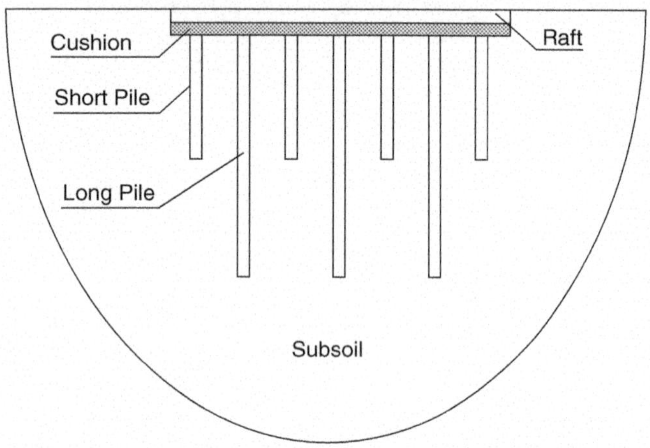

Figure 19.11 Hybrid soil improvement intrusion and pile PRF (Liang et al. 2003 / with permission of Elsevier).

19.4.2 Cushion Application

In nonconnected PRFs, a cushion made of a sand-gravel mixture can be utilized to fill the gap between the raft and the piles. This cushion is compacted in layers between the raft and the top of the piles and plays a crucial role in mobilizing the subsoil-bearing capacity. Moreover, it facilitates the load transfer mechanism from the raft to the soft or medium subsoil (Eslami and Salehi Malekshah 2011; Ata et al. 2014).

Eslami and Salehi Malekshah (2011) have investigated the role of cushions using numerical analysis. The findings indicate that with a stiffer cushion, settlements are significantly reduced, and contact pressure concentration beneath the raft at pile locations increases. Increasing the cushion stiffness also leads to higher axial stress in the piles, resulting in a higher ratio of the sum of pile loads to the total loads. On the other hand, decreasing the cushion stiffness results in the neutral plane being located at a deeper level. Figure 19.10 illustrates a schematic of a nonconnected piled raft with a cushion.

19.4.3 Hybrid Soil Improvement Intrusions and Pile PRF

Similar to Chapter 18's discussion on pile groups, a piled raft foundation can be created using a combination of rigid or flexible soil improvement methods and piles. Liang et al. (2003) proposed a hybrid approach where short piles made of flexible materials like soil–cement or sand–gravel are used to improve the bearing capacity of the shallow existing soil. On the other hand, long piles composed of rigid materials like reinforced concrete are embedded in the bearing stratum to reduce settlement. Additionally, a cushion is employed to enhance the overall performance of the foundation system. Figure 19.11 illustrates a schematic view of the reviewed hybrid piled raft foundation.

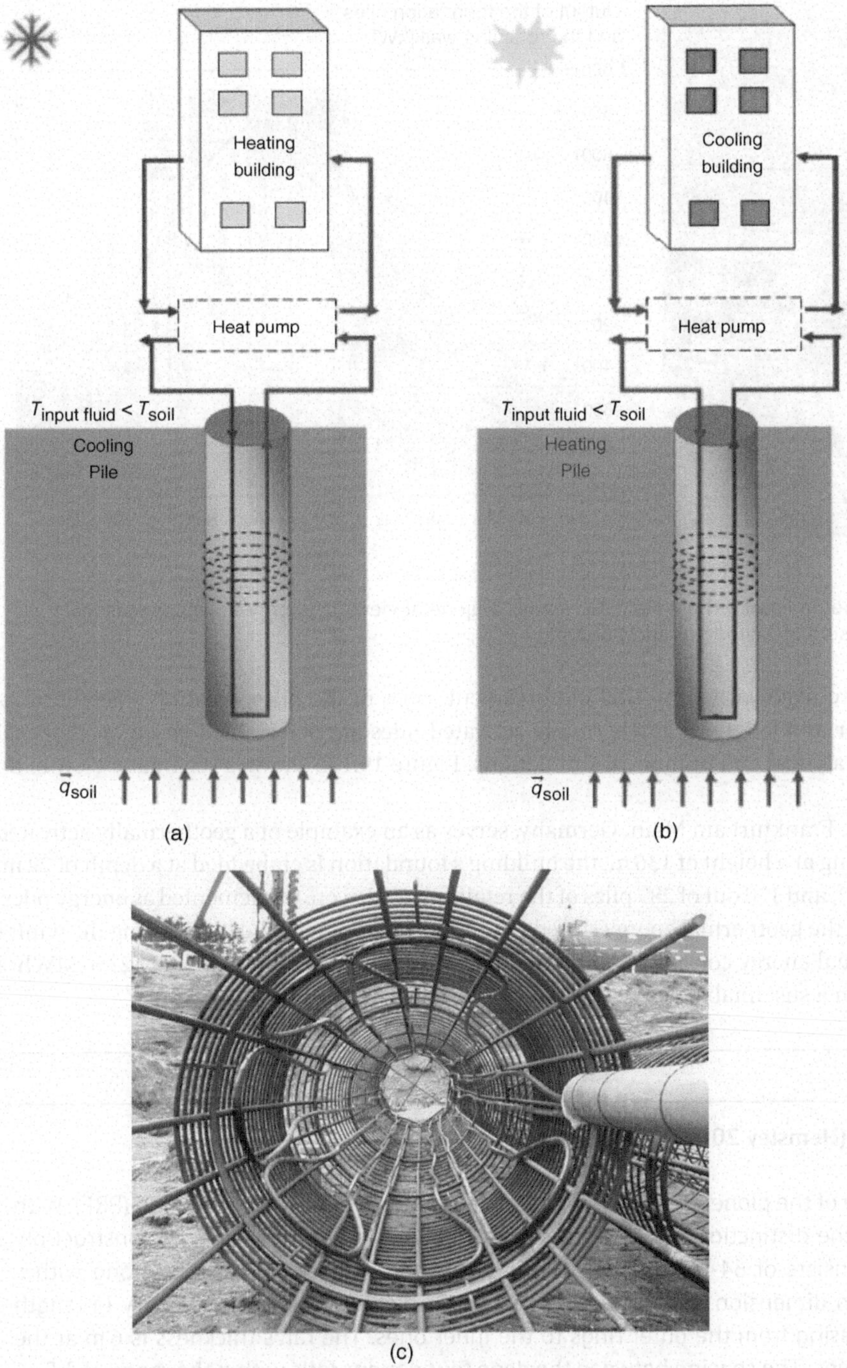

Figure 19.12 (a) Energy extraction. (b) Energy storage (M.E. Suryatriyastuti et al. 2012/with permission of Elsevier). (c) Absorber pipes fitted to the reinforcement cage of a large diameter bored energy pile (Heinz Brandl 2016 / with permission of Elsevier).

19.4.4 Geothermally Activated PRFs

The application of geothermal energy for building temperature regulation offers an environmentally friendly and sustainable approach. Geothermal energy utilizes the subsoil and groundwater to extract and store thermal energy. This system can be used for both heating and cooling the building, depending on the temperature difference between the outside and the subsoil. Figure 19.12a illustrates the general process of a geothermally activated piled raft foundation (PRF).

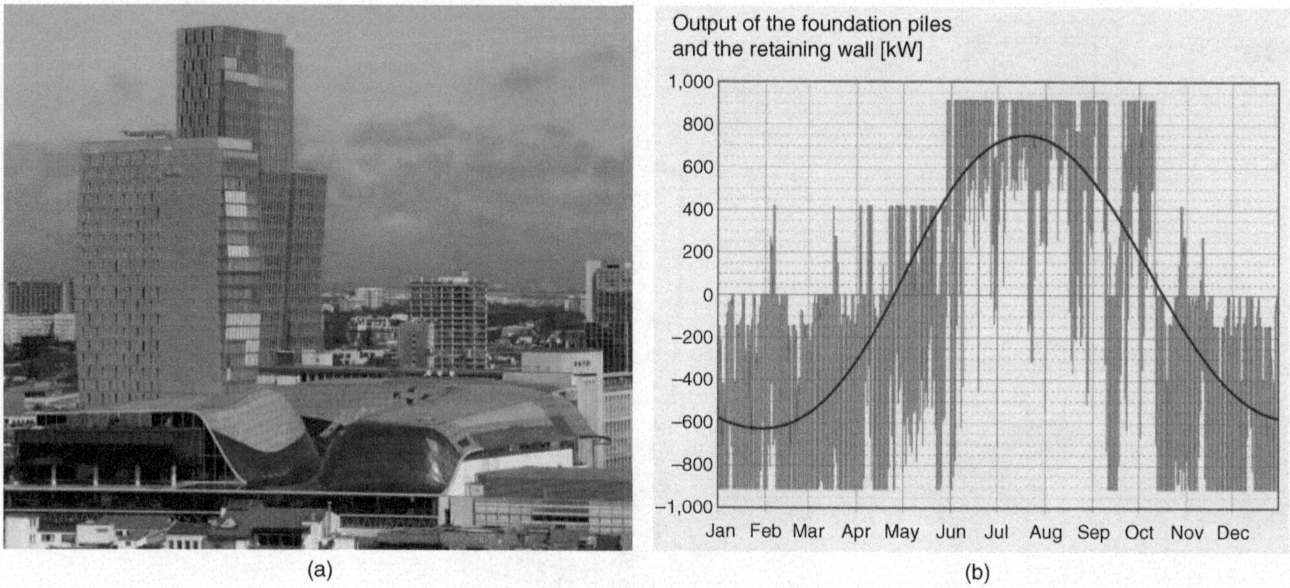

Figure 19.13 High-rise building PalaisQuartier in Frankfurt am Main, Germany: (a) general view, (b) energy-balanced seasonal storage (Katzenbach et al. 2020 / with permission of Taylor & Francis Group).

In this system, heat exchanger tubes are installed within the reinforcement cages of the piles, creating a geothermal-activated foundation system. The number and length of geothermally activated piles are determined based on practical experience and calculations, including analytical and numerical simulations. Figure 19.12b shows a schematic view of an energy pile head.

The high-rise building PalaisQuartier in Frankfurt am Main, Germany, serves as an example of a geothermally activated foundation system (Figure 19.13a). Standing at a height of 136 m, the building's foundation is embedded at a depth of 22 m. Among the foundation piles, 252 out of 302, and 130 out of 289 piles of the retaining wall, were implemented as energy piles. This innovative system effectively utilizes the geothermal energy stored in the subsoil to provide heating during the winter and cooling during the summer. The annual energy consumption for heating and cooling are approximately 2350 MWh/a and 2410 MWh/a, respectively, resulting in a sustainable and efficient energy balance (Figure 19.13b).

Case Study 19.1

Messe Turm Tower, Frankfurt, Germany (Hemsley 2000; Randolph and Reul 2019)

The Messe Turm Tower stands out as one of the pioneering structures supported by the piled raft foundation (PRF). With an impressive height of 256 m, it held the distinction of being the tallest building in Europe during its construction. The foundation of this iconic tower consists of 64-drilled shafts arranged in three concentric circles, along with a square-shaped raft measuring 58.8 m in dimension. The drilled shafts have a diameter of 1.3 m and vary in length between 26.9, 30.9, and 34.9 m, progressing from the outer rings to the inner ones. The raft's thickness is 6 m at the center and tapers down to 3 m at the edges. The spacing between the deep foundations falls within the range of 3.5–6 times their diameter. Figure 19.14 provides a detailed illustration of the foundation along with the instrumentation used for monitoring. The subsoil in the vicinity consists of fill and Quaternary sand and gravel to a depth of 10 m below ground level, underlain by the Frankfurt clay with a minimum thickness of 60 m.

Extensive instrumentation has been employed to monitor the performance of the foundation. This comprehensive set of measurements includes monitoring the rotation and settlement of the foundation, as well as sub-surface settlement. Additionally, the loads acting on the pile heads and the load distribution along the length of the piles are closely monitored to ensure the foundation's stability and structural integrity.

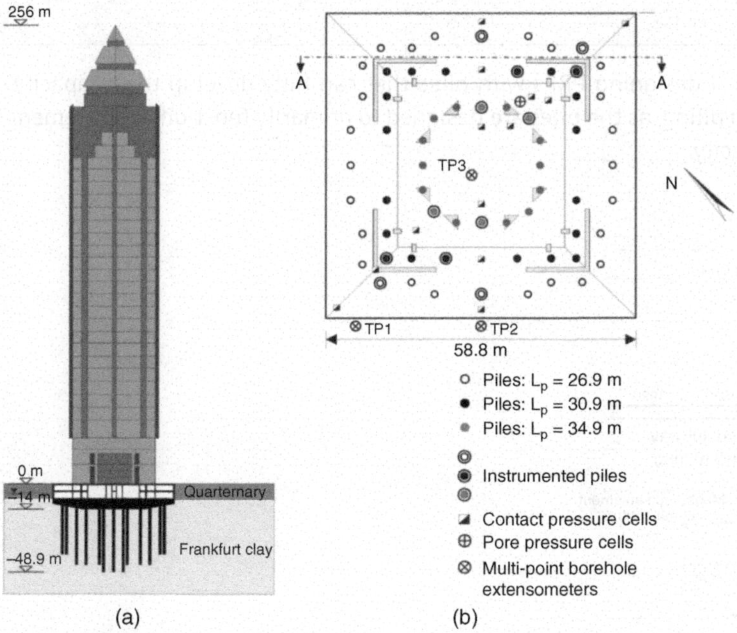

Figure 19.14 Details of Messe Turm Tower foundation: (a) cross-section of A-A, (b) ground plan of the raft (Randolph and Reul 2019).

The groundwater level is typically situated at a depth of 4.5–5 m below the ground surface. However, during the construction of a subway tunnel with a station located 47 m east of the Tower, the groundwater level had to be drawn down more than 12 m. As a result, the groundwater level near the Messe Turm Tower decreased by about 10 m. This significant drawdown led to a reduction of the uplift force on the raft, amounting to 278 MN. Figure 19.15 illustrates the variation in groundwater level and the average measured resistance for the inner, middle, and outer drilled shafts. Notably, the groundwater drawdown and rise correspond with increases and decreases in the loads experienced by the drilled shafts, respectively.

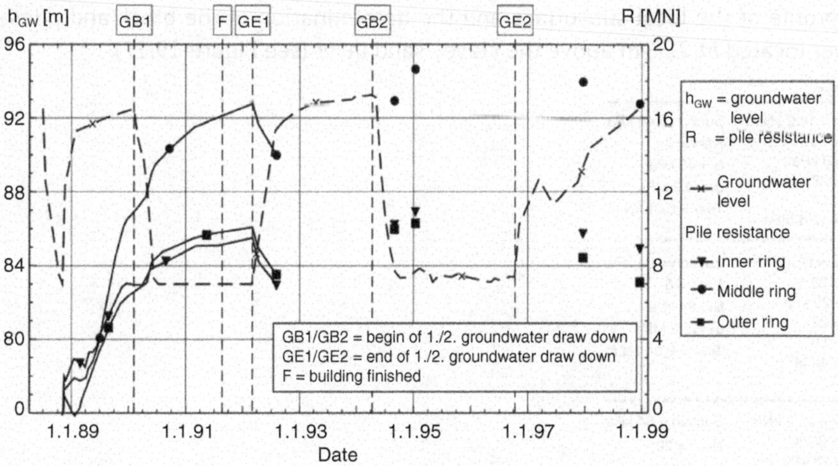

Figure 19.15 Measured variation of groundwater level and pile resistance with time (Randolph and Reul 2019).

The measured time-settlement of the tower is depicted in Figure 19.16. Over the course of approximately seven years since the beginning of construction, the total building settlement reached about 115 mm by the end of 1995. Interestingly, the predicted time-settlement behavior aligns well with the measured settlement amounts, as demonstrated in Figure 19.16.

(Continued)

Case Study 19.1 (Continued)

The reviewed project highlights the feasibility of designing PRFs with piles that can fully develop their capacity. This approach allows for economic optimization in piling, as the piles are designed to primarily function as settlement reducers while operating at or near their full capacity.

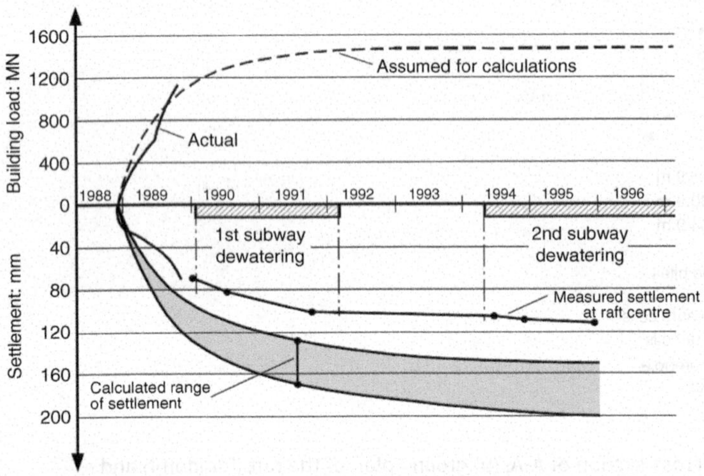

Figure 19.16 Calculated and measured settlements for the Tower (Hemsley 2000 / With permission of Thomas Telford Ltd.).

Case Study 19.2

Full-Scale Load Tests on Piled Rafts, Brazil (Garcia and Albuquerque 2019)

The full-scale tests were conducted at Unicamp Experimental Site II, SP, Brazil. The average parameters for the soil deposits were determined using SPT and CPT tests. The results of the subsurface characterization enabled the development of the geological-geotechnical profile of the local subsurface and the determination of the block and piles positions, grounded on the sandy silt layer located at 2.35 m above the clayey sand layer (see Figure 19.17).

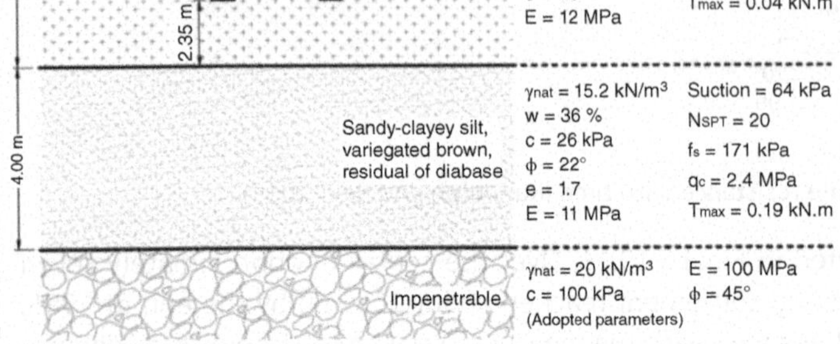

Figure 19.17 Soil profile and position of the PRFs (Garcia and Albuquerque 2019 / SciELO – Scientific Electronic Library Online / CC BY 4.0).

Sets of blocks consisting of one-, two-, three-, and four-drilled shafts with a diameter of 0.25 m and a length of 5 m were constructed and tested. All deep foundations were instrumented with strain gauges installed on the top and the tip of the piles. The spacing between the piles was set at five times their diameter to ensure a larger contact area between the block and the ground. The geometry of the tested piled foundations and the placement of the piles are shown in Figure 19.18.

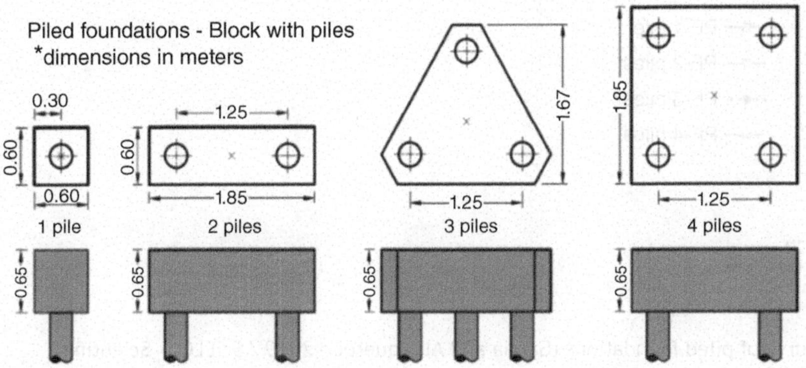

Figure 19.18 Plan view and elevation details of piled foundations (Garcia and Albuquerque 2019 / SciELO – Scientific Electronic Library Online / CC BY 4.0).

The piled foundations were constructed by excavating trenches in the soil mass, and then the sides of the blocks were carefully excavated to free them from lateral contact with the soil (see Figure 19.19).

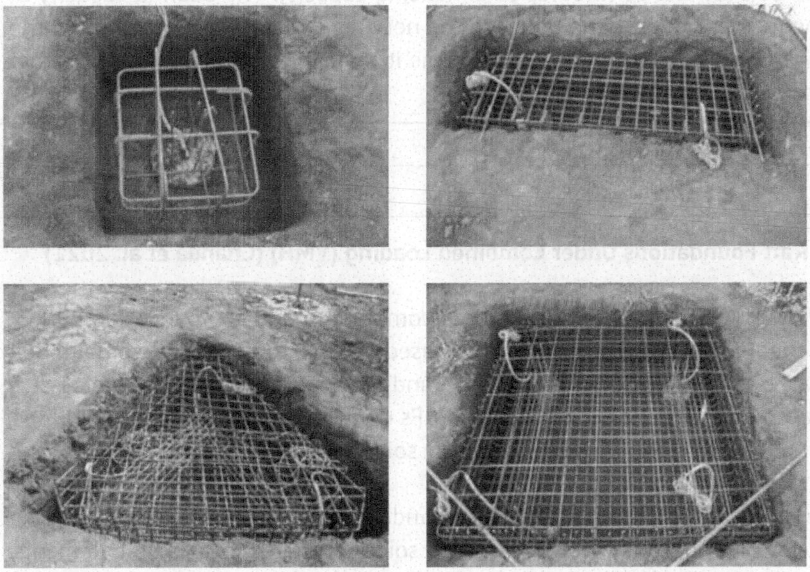

Figure 19.19 Details of the piled foundations construction (Garcia and Albuquerque 2019 / Latin American Journal of Solids and Structures / CC BY 4.0).

The foundations were subjected to a slow-maintained load test (SMLT) based on NBR 12.131. The foundation components were loaded up to a displacement that characterized a continuous displacement until reaching the ultimate load, with values exceeding 40 mm. The load–settlement curves obtained from the tests are illustrated in Figure 19.20.

(Continued)

Case Study 19.2 (Continued)

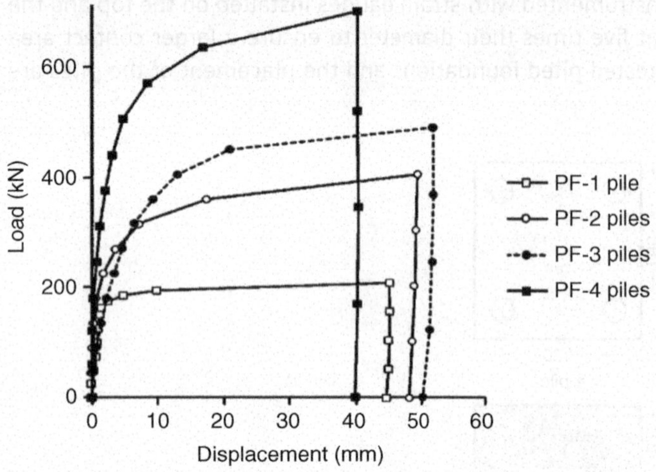

Figure 19.20 Results for load–displacement curve of piled foundations (Garcia and Albuquerque 2019 / SciELO – Scientific Electronic Library Online / Public Domain CC BY 4.0).

The load distribution between the foundation elements, both deep and shallow, that compose the piled raft was obtained through instrumentation. The bearing capacity contribution of the shallow element varied in the range of 15–27%, while the load borne by piles remained between 73% and 85%.

For PRFs composed of piles with a spacing of five times their diameter, the contribution of contact was around 21%, indicating that even for porous soils, block–soil contact contributes to enhancing capacity. The bearing capacity of piled rafts increased concomitantly with the increase in the ratio between the net area of soil–block contact and the total foundation plan area (A_{net}/A_{total}). However, this cannot be generalized, as it relies on the number of piles that form the group and the resistance parameters of the soil mass.

Case Study 19.3

Lateral Capacity-Based Envelopes of Piled Raft Foundations Under Combined Loading (VMH) (Chanda et al. 2021)

When combined vertical (*V*), horizontal (*H*), and moment (*M*) loads act on a combined piled raft foundation (CPRF), the traditional design philosophy may not adequately consider the capacity based on such loading. This combination of loads arises due to the action of wind, waves, earth pressure, earthquakes, and gravity loads. To address this issue, Chanda et al. (2021) conducted a comprehensive study on the response of CPRFs under combined loading conditions, utilizing both 3D finite element analysis and experimental investigation on soil deposits composed of both sand and clay.

The numerical simulations were performed using PLAXIS 3D, with the foundation system positioned below the ground surface. The study considered soft clay and Agartala loose sand as the soil deposits. The structural elements, including the piles, raft, and superstructure, were composed of concrete. Figure 19.21 illustrates the foundation system used in the study.

A 3×3 pile group with a length-to-diameter ratio (*L/D*) of 33.33 and center-to-center spacing equal to 5D was utilized for experimental validation. The model piles are rigidly connected via a raft with a thickness of 20 mm. The experiment was conducted in a cylindrical container with lateral and vertical boundaries measuring 1 meter. The model piles and rafts used for the experiment are illustrated in Figure 19.22.

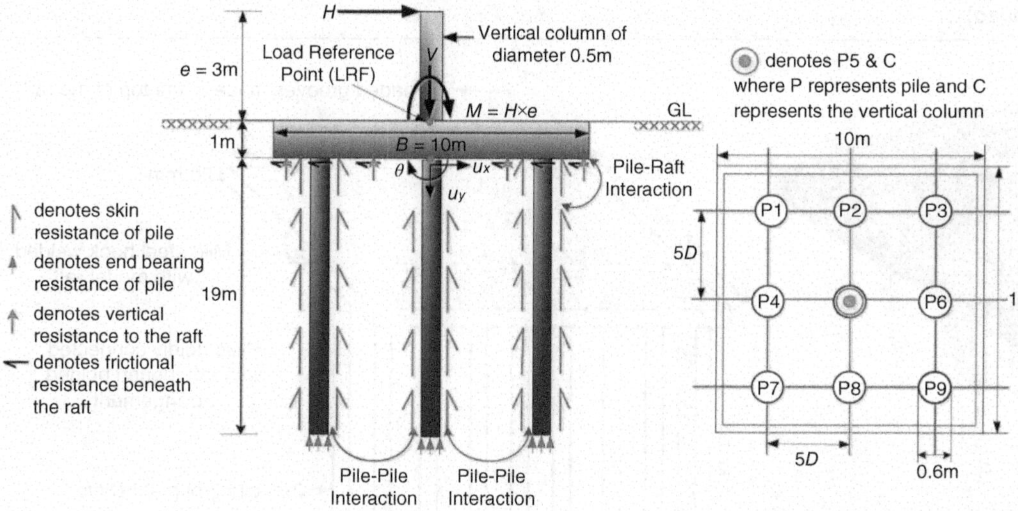

Figure 19.21 Schematic view of the applied CPRF: (a) front view, (b) top view (Chanda et al. 2021 / With permission of American Society of Civil Engineers).

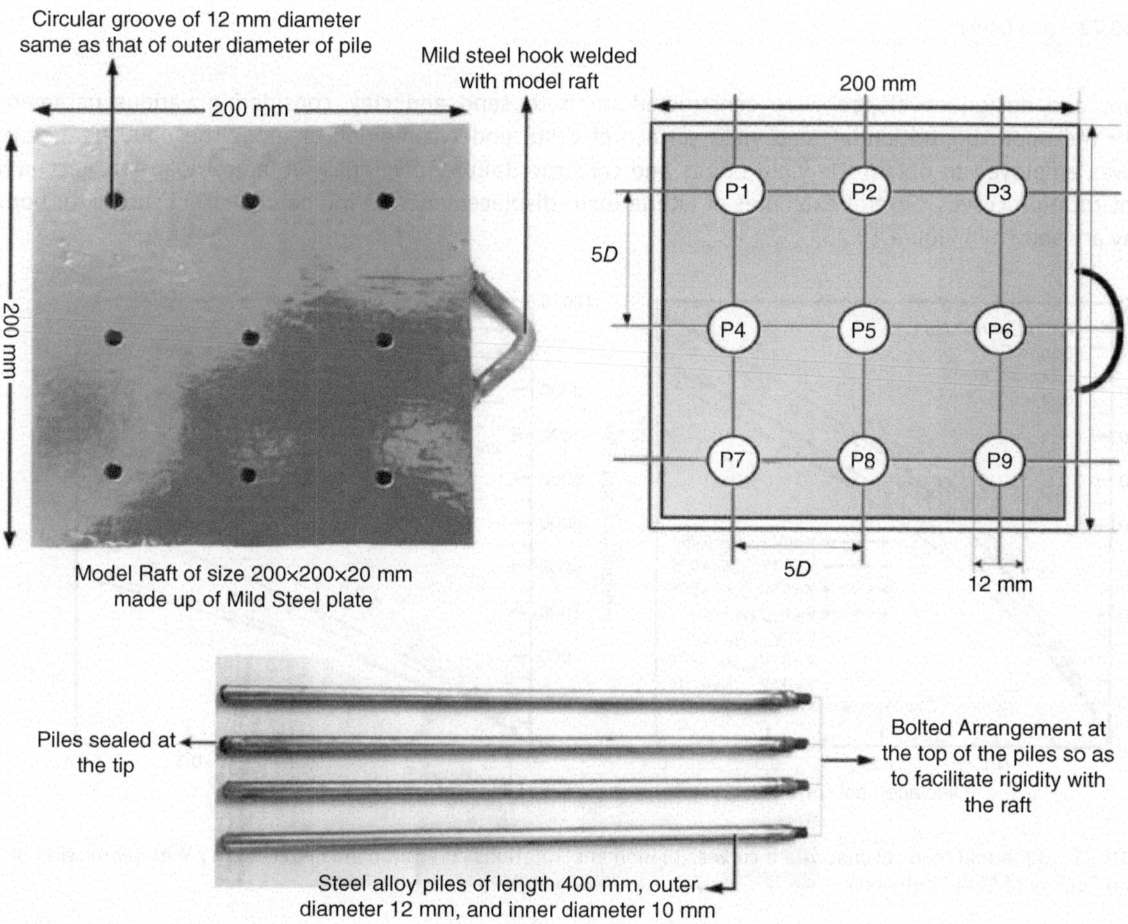

Figure 19.22 Model pile and raft used for the experiment (Chanda et al. 2021 / American Society of Civil Engineers).

(Continued)

Case Study 19.3 (Continued)

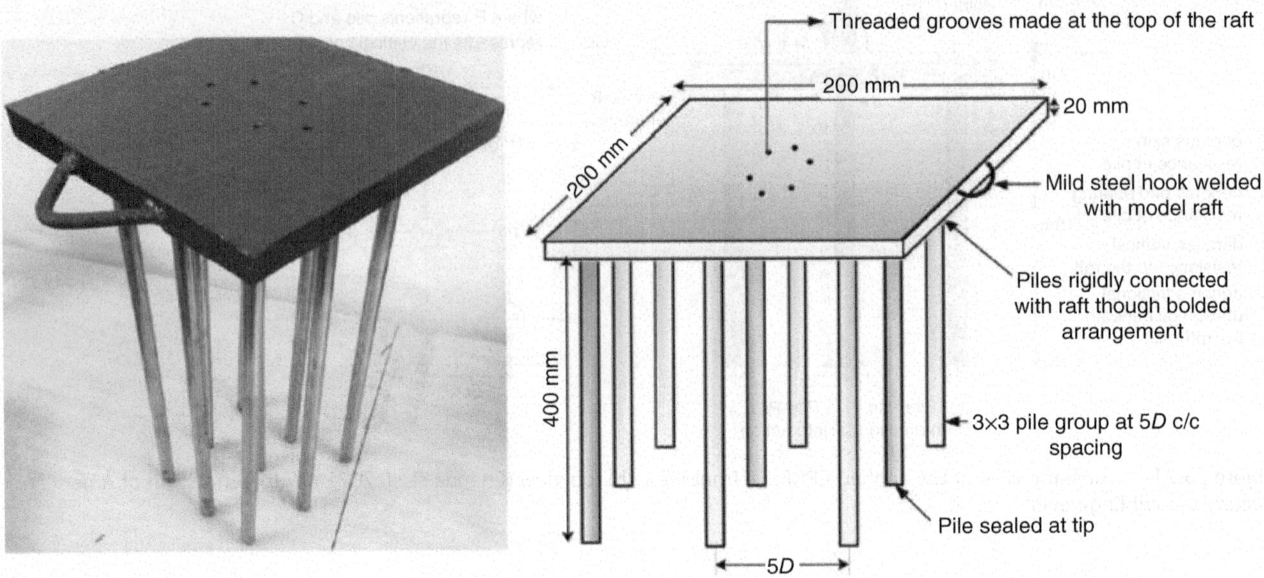

Figure 19.22 *(Continued)*

Failure and design envelopes were constructed for both sand and clay, considering various parameters, to evaluate the upper bound lateral load yield surface of CPRF under combined loading. The double tangent technique was employed to obtain the yield points and construct failure envelopes on lateral load–displacement and moment-rotation curves. Several examples of lateral load–displacement and moment-rotation curves for both sand and clay are shown in Figure 19.23.

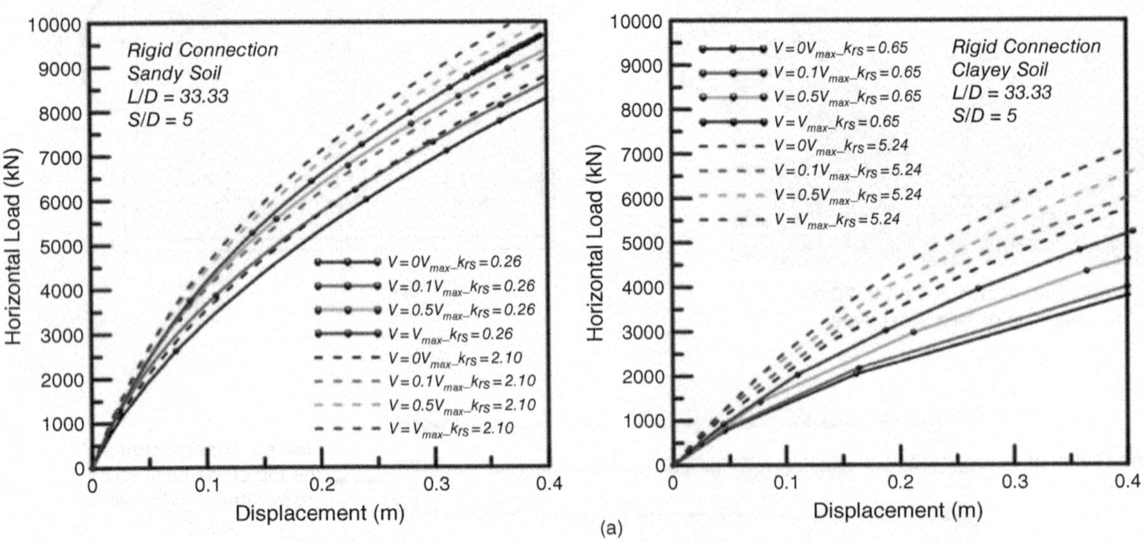

Figure 19.23 (a) lateral load–displacement curves, (b) moment–rotations curves (Chanda et al. 2021 / With permission of American Society of Civil Engineers).

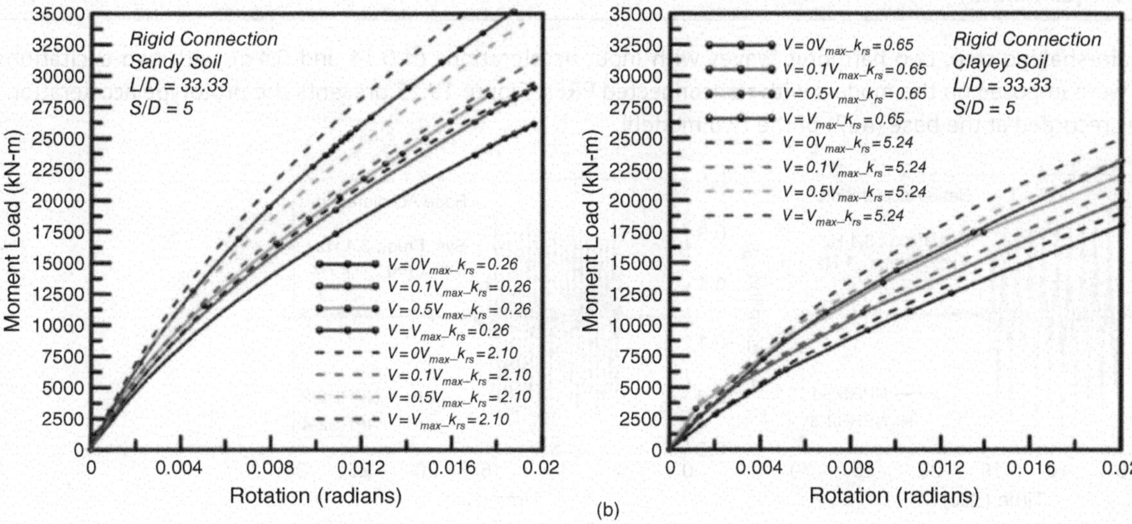

Figure 19.23 *(Continued)*

Case Study 19.4

Dynamic Response of Nonconnected Piled Raft Foundation Using Centrifuge Modeling (Baziar et al. 2018)

Several centrifuge model tests were conducted to study the behavior of two different superstructures supported by a nonconnected PRF (NPRS) in dry sand. The tests were performed using a beam centrifuge at the National Central University (NCU) of Taiwan at 50 times gravity (50 g). A laminar container with dimensions of 711 × 356 × 353 mm (length × width × height) was used to minimize wave reflections from the boundaries. The superstructures were subjected to sinusoidal waves with two amplitudes (0.14 and 0.4 g) and two motion frequencies (1 and 2 Hz), applied via a shaker installed on the base of the centrifuge. A cushion layer was included to resist horizontal movement during shaking.

The PRF consisted of a raft positioned on a 1.5-meter thick gravel layer (in prototype scale) placed on the sand. Four piles were installed in the sandy soil, creating a gap between the raft and piles, resulting in a nonconnected piled raft configuration, as depicted in Figure 19.24.

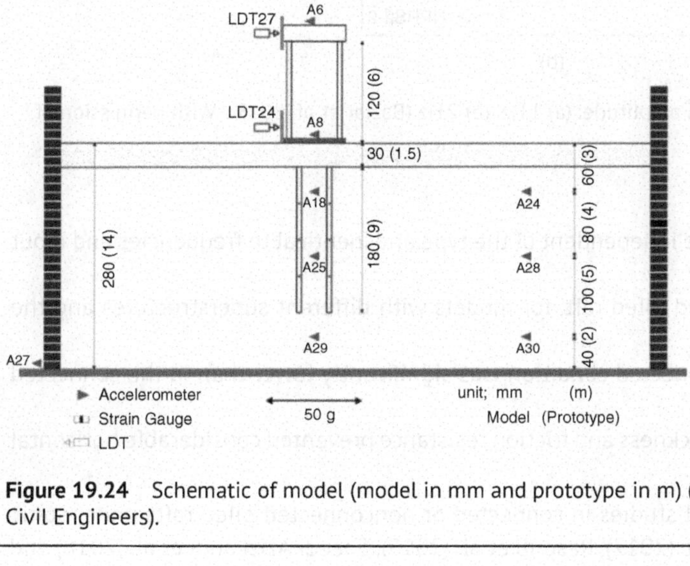

Figure 19.24 Schematic of model (model in mm and prototype in m) (Baziar et al. 2018 / With permission of American Society of Civil Engineers).

(Continued)

Case Study 19.4 (Continued)

For the main-shaking step, two harmonic waves with input accelerations of 0.14 and 0.4 g), with two excitation frequencies, were imposed on the models with nonconnected PRFs. Figure 19.25 presents the prototype acceleration time histories recorded at the base (A27) of the two models.

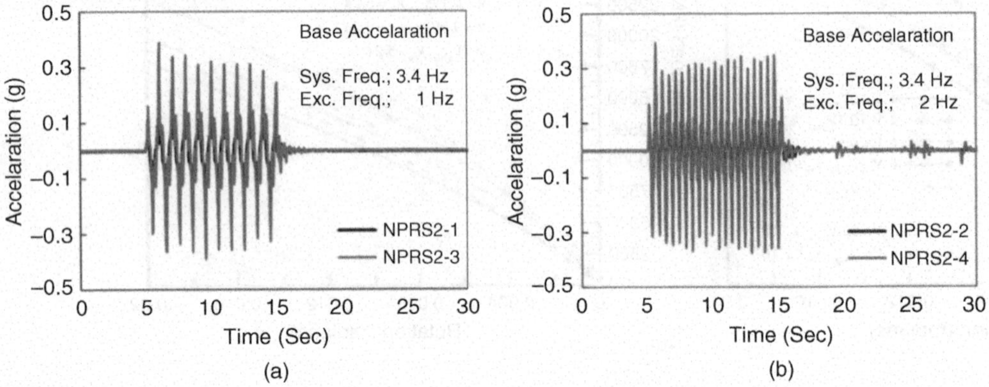

Figure 19.25 Prototype acceleration time histories: (a) 1 Hz, (b) 2 Hz (Baziar et al. 2018 / With permission of American Society of Civil Engineers).

Three strain gauges were installed to measure the axial force along pile P3, and the results are shown in Figure 19.26. At low dynamic loading (0.14 g) with both excitation frequencies, the axial forces for both superstructures were similar. When the acceleration amplitude was increased from 0.14 to 0.4 g, there were only minor increases in the axial forces. Therefore, the type of superstructure appears to have an insignificant effect on the axial force.

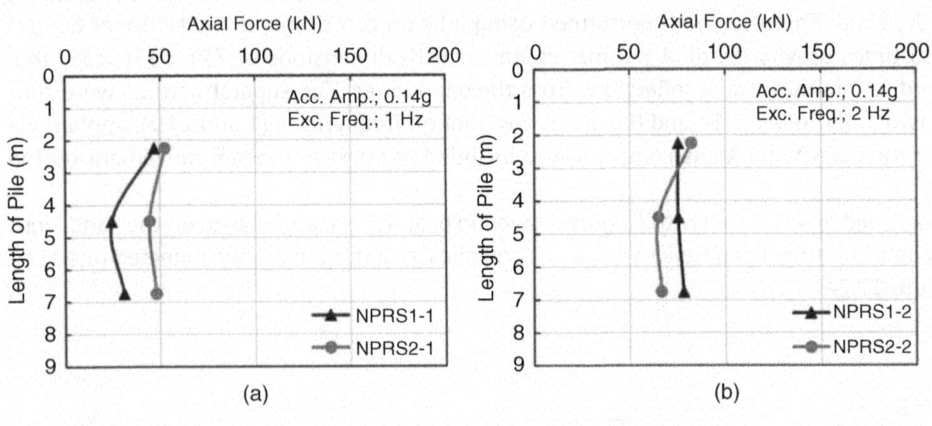

Figure 19.26 Prototype axial force with 0.14 g acceleration amplitude: (a) 1 Hz, (b) 2 Hz (Baziar et al. 2018 / With permission of American Society of Civil Engineers).

The main findings of the study are as follows:

- Soil amplifications for different superstructures were independent of the type and identical to frequencies and input characteristics of the piled raft.
- The internal forces of the piles in the nonconnected piled raft, for models with different superstructures and the same test loading conditions, were identical.
- The bending moment along the piles in the nonconnected condition was significantly lower than in the connected condition.
- The presence of a suitable cushion with sufficient thickness and friction resistance prevented considerable horizontal displacements.

Some other noticeable numerical and experimental studies in connected or nonconnected piled rafts under static and dynamic loads are Baziar et al. (2009), Eslami et al. (2012), Rasouli et al. (2015), Saeedi Azizkandi et al. (2018) and Akbari et al. (2021).

19.5 Major Points

1 One of the most effective methods for increasing the bearing capacity of a foundation is pile enhancement, known as the hybrid foundation or piled raft foundation (PRF).

2 Piled raft foundations allow for the reduction of settlements and differential settlements in a cost-effective manner compared to traditional foundation concepts. In modern PRF design, piles are typically utilized to act as settlement reducers rather than being solely responsible for the overall stability of the foundation.

3 Among the main advantages of adopting a piled raft foundation are substantial cost savings in foundations, control of differential settlements, and the ability to optimize the foundation design by varying pile length, location, and raft thickness.

4 There are two broad categories of detailed analysis methods for PRFs: approximate computer-based methods, including the strip-on-springs approach (GASP), and the plate-on-springs approach (GARP). More rigorous methods involve comprehensive computer analysis and modeling of all PRF components.

5 The design requirements for piled raft foundations encompass the ultimate geotechnical capacity under vertical, lateral, and moment loading, overall settlement and stiffness, differential settlements and angular rotations, lateral movements, and stiffness, as well as structural design considerations for both the raft and piles.

6 Three distinct design philosophies are identified for piled rafts: (a) the "conventional" approach, where the foundation is primarily designed as a pile group with regular spacing; (b) the "creep piling" approach, where the foundation is designed as a raft with reduced total settlement due to the inclusion of piles uniformly distributed beneath the raft; and (c) the "differential settlement control" approach, which aims to minimize differential settlements without significantly reducing average settlement.

7 Favorable soil profiles for PRFs typically consist of relatively stiff clays or relatively dense sands and weak rocks.

8 There are some soil profiles in which piled rafts may not provide significant advantages over conventional piled foundations. These include profiles with very soft clays at or near the surface of the raft, those subjected to long-term consolidation settlement, and those prone to expansive (upward) movements.

9 Special considerations must be considered in the design and construction of piled raft foundations, such as connected and nonconnected piled rafts, utilization of cushions, hybrid application of ground modification intrusions, deep foundations under a raft foundation, and geothermally activated PRFs.

10 Designing and analyzing piled raft foundations can be inherently challenging due to complex interactions between soil, raft, and pile. It requires a comprehensive understanding of geotechnical and foundation engineering principles, making routine and closed-form calculations inadequate in most cases. Rigorous advanced computer-based analysis and simulations with powerful modeling tools are recommended for accurate analysis and design.

19.6 Questions

19.6.1 Explanatory

1 Compare and contrast pile group and piled raft regarding construction and load transfer mechanism.

2 List and explain the main advantages of PRF applications.

3 What are the similarities and differences between GASP and GARP analysis approaches?

4 Name major design requirements of piled raft foundations.

5 Define three different philosophies in the design of PRFs and compare them.

6 Explain favorable and unfavorable conditions for piles raft application.

7 Discuss the general issues that must be controlled to realize geotechnical performance.

8 Illustrate the steps and aspects to estimate the load–displacement behavior of a PRF.

9 Compare and contrast connected and nonconnected piled raft foundations based on geotechnical and practical aspects.

10 Explain the role of cushion in the performance of PRFs.

19.6.2 Conceptual

1 Discuss the role of existing problematic soils in the application of piled raft foundations and the special consideration that must be taken in this regard.

2 Explain the effect of using a cushion on the load transfer mechanism and consequently the load carried by piles.

3 Define the influence of fluctuation in groundwater level on piled raft foundations performance based on Case Study 19.1, the Messe Turm Tower.

4 Consider the load–displacement diagram in Case Study 19.2, resulting from full-scale load testing. Discuss the reasons for differences in stiffness and capacity of different foundations.

5 Upon what is presented in Case Study 19.3, how can combined loading (VMH) affect the capacity of a piled raft?

6 What are your suggestions for designing PRFs in a more economical and environment-friendly way?

7 How can optimized design be achieved? Discuss the crucial factors and their major effect on performance realizing both technical and practical aspects.

8 Based on the results of Case Study 19.4, compare and contrast the performance of nonconnected PRFs under static and dynamic loadings.

9 Explain the reason for the fact that the edge piles are usually shorter than central ones in a PRF.

10 Define the role of raft thickness in a PRF and illustrate the reasons and outcomes of variation in raft thickness based on what you learned from Case Study 19.1, Messe Turm Tower.

11 Discuss the effects of a geothermally activated system on the geotechnical performance of piled raft foundations.

References

Akbari, A., Eslami, A., and Nikookar, M. (2021). Influence of soil stiffness on the response of piled raft foundations under earthquake loading. *Transportation Infrastructure Geotechnology*.

Ata, A., Badrawi, E., and Nabil, M. (2014). Numerical analysis of unconnected piled raft with cushion. *Ain Shams Engineering Journal*.

Baziar, M.H., Ghorbani, A., and Katzenbach, R. (2009). Small-scale model test and three-dimensional analysis of pile-raft foundation on medium-dense sand. *International Journal of Civil Engineering* 7 (3): 170–175.

Baziar, M.H., Rafiee, F., Lee, C.J., and Saeedi Azizkandi, A. (2018). Effect of superstructure on the dynamic response of non-connected piled raft foundation using centrifuge modeling. *International Journal of Geomechanics* 18 (10).

Brandl, H. (2016). Geothermal geotechnics for urban undergrounds. *Procedia Engineering* 165: 747–764.

Clancy, P. and Randolph, M.F. (1993). An approximate analysis procedure for piled raft foundations. *International Journal of Numerical and Analytical Methods in Geomechanics* 17 (12): 849–869.

Chanda, D., Nath, U., Saha, R., and Haldar, S. (2021). Development of lateral capacity-based envelopes of piled raft foundation under combined V-M-H loading. *International Journal of Geomechanics.*

Eslami, A., Veiskarami, M., and Eslami, M.M. (2008). Piled-raft foundation (PRF) optimization design with connected and disconnected piles, Proceedings of the 33rd Annual and 11th Int'l Conference on Deep Foundations, October 15–17 2008, Deep Foundations Institute (DFI), New York, NY, U.S.A. 201–211.

Eslami, M.M., Aminikhah, A., and Ahmadi, M.M. (2011). A comparative study on pile group and piled raft foundations (PRF) behavior under seismic loading. *Computational Methods in Civil Engineering* 2 (2): 185–199.

Eslami, A., Veiskarami, M., and Eslami, M.M. (2012). Study on optimized piled-raft foundations (PRF) performance with connected and non-connected piles- three case histories. *International Journal of Civil Engineering* 10 (2): 100–111.

Eslami, A. and Salehi Malekshah, S. (2011). Analysis of non-connected piled raft foundation (NCPRF) with cushion by finite element method. *Computational Methods in Civil Engineering.* 2: 153–168.

Garcia, J.R. and Albuquerque, P.J.R. (2019). Analysis of the contribution of the block-soil contact in piled foundations. *Latin American Journal of Solids and Structures* 16 (6).

Hemsley, J.A. (2000). *Design applications of raft foundations.* London: Thomas Telford Ltd.

Horikoshi, K. and Randolph, M.F. (1998). A contribution to optimum design of piled rafts. *Geotechnique* 48 (3): 301–317.

Liang, F.Y., Chen, L.Z., and Shi, X.G. (2003). Numerical analysis of composite piled raft with cushion subjected to vertical load. *Computers and Geotechnics* 30: 443–453.

Katzenbach, R., Arslan, U., Moormann, C., and Reul, O. (1998). Piled raft foundation – interaction between piles and raft. Darmstadt geotechnics. *Darmstadt University of Technology* 4: 279–296.

Katzenbach, R., Leppla, S., and Choudhury, D. (2017). *Foundation Systems for High-Rise Structures*, 1e. CRC Press, Taylor & Francis Group.

Katzenbach, R., Leppla, S., and Alzaylaie, M. (2020). Sustainable foundation systems of high-rise buildings. *Research Bulletin of Center for Excellence in Smart Construction* 2: 14–16.

Mandolini, A. (2003). Design of piled raft foundations: practice and development. In BAP IV, International Geotechnical Seminar on Deep Foundations on Bored and Auger Piles. Vol. 1. pp. 59–80.

Mandolini, A., Russo, G., and Viggiani, C. (2005). Pile foundations: experimental investigations, analysis and design. Proceedings of the 16th International Conference on Soil Mechanics and Geotechnical Engineering, Osaka, Vol. 1, pp. 177–213.

Poulos, H.G. and Davis, E.H. (1980). *Pile foundation analysis and design.* New York: John Wiley.

Poulos, H.G. (1991). Analysis of piled strip foundation. In: *Comp. Methods & Advances in Geomech*, vol. 1 (ed. G. Beer et al.), 183–191. Rotterdam: Balkema.

Poulos, H.G. (1994). An approximate numerical analysis of pile–raft interaction. *International Journal of Numerical and Analytical Methods in Geomechanics* 18: 73–92.

Poulos, H.G. (2001). Piled raft foundations – design and applications. *Geotechnique* 51 (2): 95–113.

Randolph, M.F. (1994). Design methods for pile groups and piled rafts. State-of-the-Art Report, 13th Int. Conf. Soil Mech. Foundn Engng, New Delhi, 5, 61–82.

Rasouli, H., Saeedi Azizkandi, A., Baziar, M.H. et al. (2015). Centrifuge modeling of non-connected piled raft system. *International Journal of Civil Engineering* 13 (2): 114–123.

Reul, O., and Randolph, M.F. (2004). Design strategies for piled rafts subjected to nonuniform vertical loading. *Journal of Geotechnical and Geoenvironmental Engineering*, ASCE, 130 (1), 1–13.

Randolph, M.F. and Reul, O. (2019). Practical approaches for design of pile groups and piled rafts. 4th Bolivian International Conference on Deep Foundations.

Saeedi Azizkandi, A., Baziar, M.H., and Fallah Yeznabad, A. (2018). 3D dynamic finite element analyses and 1 g shaking table tests on seismic performance of connected and nonconnected piled raft foundations. *KSCE Journal of Civil Engineering* 22: 1750–1762.

Smith, D.M.A. and Randolph, M.F. (1990). Piled raft foundations – a case history. Proc. Conf. on Deep Found. Practice, Singapore, 237–245.

Suryatriyastuti, M.E., Mroueh, H., and Burlon, S. (2012). Understanding the temperature-induced mechanical behaviour of energy pile foundations. Renewable and Sustainable Energy Reviews. 16 (5): 3344–3354.

Tradigo, F., Pisano, F., and di Prisco, C. (2015). On the use of embedded pile elements for the numerical analysis of disconnected piled rafts. *Computers and Geotechnics* 72: 89–99.

Veiskarami, M., Eslami, A., Ranjbar, M.M., and Riyazi, T. (2007). Geotechnical Interaction of Mat Foundation and Pile Group, Two Case Studies, Esteghlal Journal of Engineering, Isfahan University of Technology, September 2007, Vol. 36, No. 1, pp. 93–107.

Viggiani, C., Mandolini, A., and Russo, G. (2012). *Piles and Pile Foundations*. London: Spon Press.

Wong, I.H., Chang, M.F., and Cao, X.D. (2004). Behavior of model rafts resting on pile-reinforced sand. *Journal of Geotechnical Engineering*, ASCE, 130(2), 129–138.

20

Performance Under Extreme Conditions

Chapter Highlight

- **Deep Foundations for Slope Stability**
  - Soil Slopes and Failure Modes
  - Prevention and Remediation Approaches
  - Shear Pile Walls
  - Practical Case Studies
    - Washington Park Station Slide (Cornforth 2005)
    - The Goat Lick Slide (Cornforth 2005)
    - Dashtehgan Landslide (Ekramirad et al. 2012)
- **Downdrag Forces**
  - Potential for Mobilization
  - Neutral Plane Concept
  - Determinant Factors
  - Soil Improvement Effects
- **Cyclic Loading and Its Effects on Deep Foundation performance**
  - Variables and Characterization
  - Effects on Foundations
  - Resistance Degradation
    - Axial
    - Lateral
  - Cyclic Transversal Loading
- **Earthquake and Liquefaction Conditions**
- **Scour**

20.1 Introduction

In certain instances, deep foundations may be used at or subjected to exceptional conditions that necessitate more rigorous analysis and design considerations. In this chapter, five different extraordinary conditions are discussed. These five conditions are: (i) deep foundations for slope stability; (ii) downdrag forces; (iii) cyclic loading and its effects on deep foundation performance; (iv) earthquake and liquefaction conditions; and (v) scour.

20.2 Deep Foundations for Slope Stability

Using deep foundations is among the typical methods for controlling and providing stability of natural and artificial slopes.

Advanced Foundation Engineering: Principles, Performance and Prospects, First Edition. Moses Karakouzian and Abolfazl Eslami.

20.2.1 Introduction

Landslides represent a formidable force of nature, a topic of paramount significance, and a chapter in the ever-evolving story of our planet. These powerful geological events have left indelible marks throughout history, causing catastrophic economic losses and tragically claiming countless lives. While the term "landslide" is commonly used to describe a range of slope movements, both natural and human-induced, it can be somewhat inadequate since not all of these movements fit the classic image of earth or rock sliding down a hillside. Geomorphologists employ the term "mass wasting" to encompass these diverse phenomena, although it offers only a marginal improvement in precision. Landslides take on various forms, each exhibiting its unique behavior and underlying mechanisms, as depicted in Figure 20.1. Following is a partial list of the types of landslide movements.

- Falls
- Topples
- Slides (rotational and translational)
- Lateral spreads
- Flows
- Composite or combination type

Landslides, in all their complexity, are not easily defined by a single set of characteristics. Their dimensions and geometry encompass a spectrum of attributes, including length, width, depth, and slope.

Landslide shear surfaces can take various forms, each requiring a specific analytical approach. Common shapes include single planes (or slightly curved surfaces), typically found in steep areas and relatively easy to analyze; circular arcs, prevalent in uniform strata like fill and soft clays, often in smaller landslides; and double and triple wedges, seen in medium to large translational landslides, requiring more complex numerical or graphical analysis. The introduction of computer software has revolutionized landslide analysis, enabling engineers to perform precise and efficient calculations, and providing a powerful tool to address the diverse shapes and complexities found in landslides.

To effectively model landslides for stability analysis, it is crucial to accurately define the shape of shear surfaces. The cornerstone of a reliable assessment lies in capturing the geometry of a landslide. This assessment is performed by obtaining data from boreholes in the area of the landslide. Borehole sampling is a very effective method for studying landslides since it provides a direct means of observation. Ideally, most boreholes should be strategically placed along the maximum slope line, near the mid-width of the landslide. This placement creates a robust model of the landslide form, offering valuable insights for subsequent analysis. While some practitioners scatter boreholes across the entire landslide area to cover every inch, this can hinder the creation of a dependable cross-sectional view needed for comprehensive stability analysis. Therefore, the top priority should always be establishing a well-placed line of boreholes. Additional boreholes can be considered as necessary, taking into account factors such as the landslide's shape and project budget constraints.

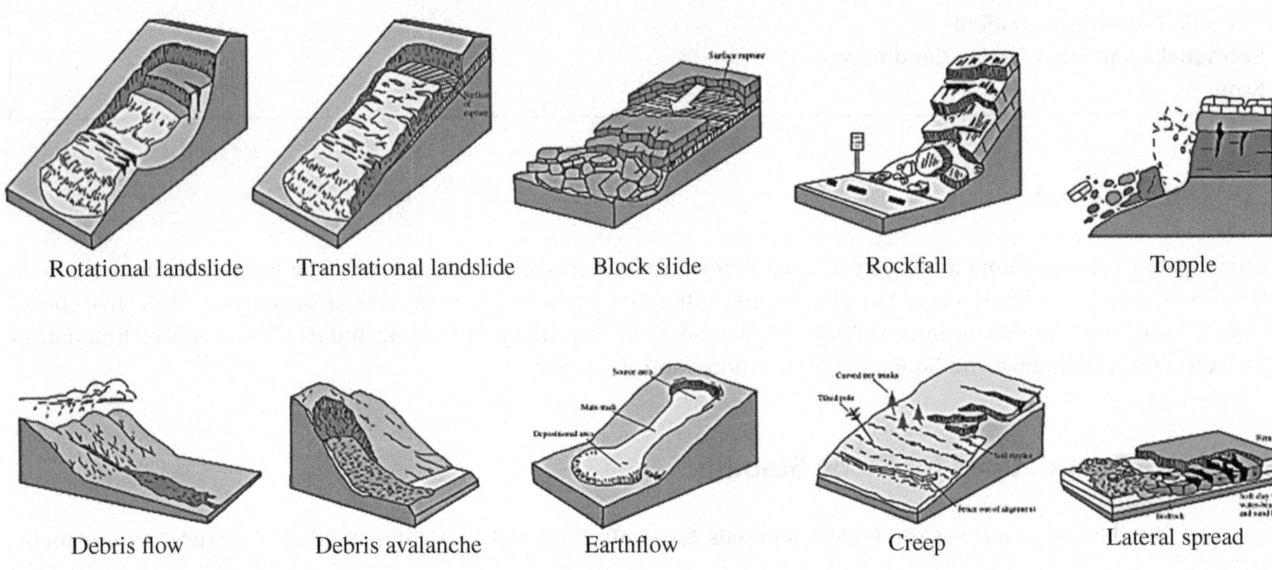

Figure 20.1 Schematic illustration of the major types of landslide movement (Papadopoulos and Loupasakis 2014).

Slope stability analysis relies on the factor of safety to measure stability, with equilibrium principles as its foundation. Achieving precision in this analysis is challenging due to soil property variability. Back-analysis, a critical procedure, estimates shear strength in problematic layers, offering advantages over lab and in situ testing. Even when the factor of safety is known, back-analysis introduces uncertainties related to factors like the failure surface and slope geometry.

Among the various methods for slope analysis, back analysis stands out as the most effective approach for comprehending the underlying causes of slope instability and determining the preventive and remediation measures that could have averted landslides. It serves as a valuable tool not only for identifying the causes of landslides but also for designing effective preventive measures to mitigate the risks associated with slope instability.

Back-analysis is more effective than traditional lab testing for estimating shear strength, though it does not eliminate uncertainties; instead, it presents new challenges. Factors affecting back-calculated shear strength include various material properties, slope geometry, phreatic surface, porewater pressures, rainfall effects, failure surface location, and tension cracks. The process involves understanding subsurface conditions, defining representative cross-sections, determining the failure surface, selecting a stability method, adjusting shear strength to approach unity, comparing with lab results, validating with empirical correlations, investigating discrepancies, and iteratively refining the process for alignment.

20.2.2 Prevention and Remediation Methods

Preventing landslides becomes of utmost importance in regions where the likelihood of rapid slope failure is high, and the risk of damage and injury is substantial. From the looming threat of flow slides to earthquake-induced landslides and the formidable force of rockslides, proactive measures can help mitigate the potential devastation.

Remediation methods can enhance general slope stability and avert landslides. Some of these remediation methods are mentioned in Table 20.1 (Eslami 2013). Effectively tackling landslides requires understanding their specific triggers and determining precise remediation levels for future stability. The landslide's cause shapes the appropriate remedy, extending beyond correction to include an added safety margin. Factors like historical weather patterns, such as heavy rainfall or storms, must be considered in the remediation design. In regions with a history of these weather events, adjusting for a higher groundwater level is essential for increased resilience against potential future challenges. The additional safety margin is more than just good practice; it is a critical component in transforming how we approach the analysis of landslide issues, especially when assessing stability.

Recognizing that safety is at the precipice of compromise when a landslide is imminent (indicated by FS = 1), allows for the precise determination of the slope geometry. Instruments like inclinometers are used to measure slope steepness, and pressure gauges are employed to assess underground water levels at the failure surface. When accurate data is fed into the back analysis of the landslide at the point of failure, it facilitates the creation of an exceptionally precise model. This level of precision empowers us to conduct comprehensive parametric analyses, exploring a wide range of potential scenarios. What distinguishes landslide remediation design is the integration of remedial and back analyses. These two approaches are not isolated; they entail comparative studies on the same cross sections. Furthermore, any uncertainties or errors in the assumptions made during the back analysis seamlessly carry over into the remedial analysis. This merging of information and methodology significantly enhances the likelihood of achieving stability. Consequently, in landslide remediation, a lower factor of safety is not only permissible but often advisable when compared to general slope stability design.

20.2.3 Shear Pile Walls

Slope stability is essential for preventing landslides and preserving the integrity of landscapes and structures. One effective method to enhance the stability of slopes is the use of shear piles. Shear piles have the ability to transfer lateral loads from weak or compressible soil layers to deeper, more competent strata, thus providing passive resistance, mitigating lateral movements, and ensuring stability. The advantages of using piles for slope stability enhancement include:

- **Resistance to Lateral Deflection:** Piles are highly resistant to lateral deflection, reducing the need for anchoring near the top of the slope.
- **Invisibility:** Piles can be installed with their tops buried below ground or concealed by structures, roads, or other surfaces, making them visually unobtrusive.
- **Installation Flexibility:** Engineers have flexibility in selecting the precise location for pile installation.
- **Individual Load Capacity:** Individual piles can provide substantial passive resistance, often more than a group of piles combined.

Table 20.1 Remediation methods and types for slopes.

Remediation method	Types	Schematic view
Geometry modification	1. Height reduction 2. Slope reduction 3. Removing the unstable slope fragments	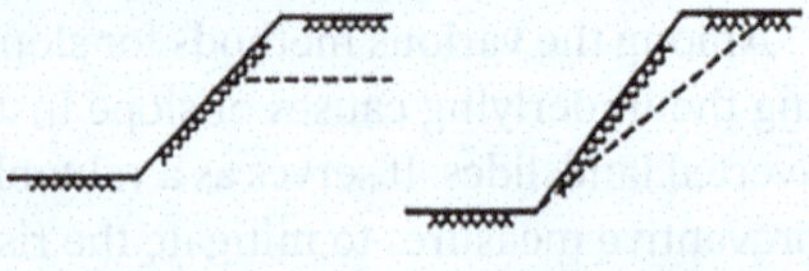 Height reduction Slope reduction
Hydraulic modification	1. Surface drainage 2. Subsurface drainage 3. Impermeable covers	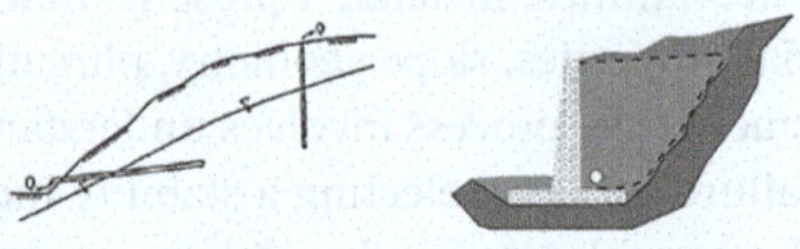 Vertical and horizontal drainage Drainage behand the wall
Shear strength improvement	1. Mechanical modifications (DDC or explosion) 2. Physical modification 3. Chemical modification (admixtures and grouting)	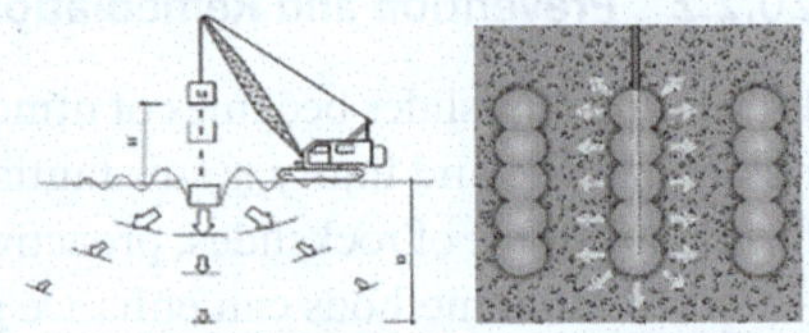 Dynamic compaction Grouting
Reinforcement	1. Micro-piles 2. Soil nails and anchorage systems 3. Shear pile walls	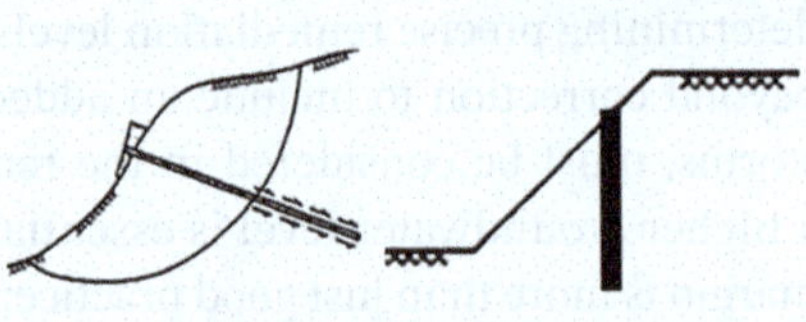 Soil-nail and anchorage Shear pile wall
Retaining structures	1. Gravity wall 2. Gabion structures 3. Sloped Wall	 Gabion wall Sloped wall

Concrete shear pile walls, as shown in Figure 20.2, offer a formidable solution for landslide stabilization. These piles, constructed from heavily reinforced concrete, simplify construction without requiring specialized subcontractors. Their unique design allows discreet integration within the landslide, minimizing their visual impact compared to alternative methods. Shear piles can be aligned to create a supporting wall for critical infrastructure like roads or railroads near landslide-prone locations.

The reinforced concrete shear piles traverse the landslide's sliding surface, firmly anchoring into stable soil or bedrock below. They can stand independently or be tied back with ground anchors. Figure 20.3 shows these two conditions.

An advantage is their top-down construction method of shear piles, which is ideal for landslide remediation. Construction starts at ground level, with excavation on the downhill side only, if necessary, after erecting the pile wall. Tying the piles together with a heavily reinforced grade beam enhances their wall-like function. While shear piles may entail higher costs, they excel in scenarios requiring partial stabilization, particularly for safeguarding vital infrastructure in challenging access areas. Reinforcement can be achieved through steel H-beams or a cage of reinforcing bars, allowing flexibility in design based on bending stress considerations. Typically, shear piles are cast in place, with drilling facilitated by equipment like helical augers or bucket augers. They can penetrate hard bedrock or stable ground, sometimes requiring larger coring barrels for very hard rock. Temporary casing is necessary when drilling through caving or cohesionless soils below the

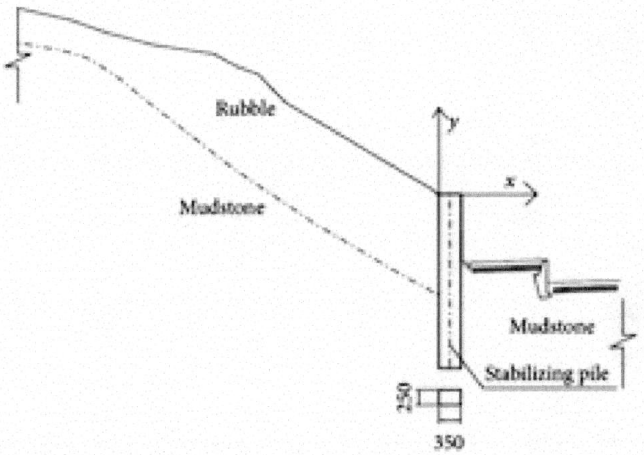

Figure 20.2 Shear pile wall section in the landslide (Cornforth 2005/John Wiley & Sons).

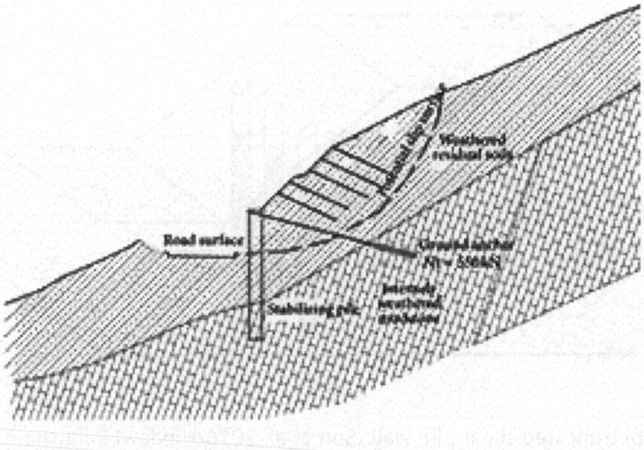

A pile retaining wall with prestressed tie-back anchors (design case I).

Figure 20.3 Tied-back shear pile wall: (a) cantilevered shear pile and (b) tied-back shear pile wall (Cornforth 2005/John Wiley & Sons).

water table. The casing can be installed by reaming the hole or driving into the ground with care to avoid damaging nearby piles. Thin-wall casing, though not a primary sliding resistance component, contributes to stability. An alternative involves precast concrete segments for pile formation. In this method, a large-diameter auger drills the hole, and pipe segments provide casing. Alternatively, soil can be excavated from inside the pipe, and it is hammered into the ground to case the hole. Precast concrete pipe construction is typically reserved for large-diameter piles. Once the pipe reaches full depth, reinforcement and mass concrete are added, creating the structural pile.

20.2.4 Shear Piles Construction and Performance

The configuration of these reinforced concrete cylinder piles (as shown in Figure 20.4), aligned in rows to create a wall, offers several variations, including:

- **Tangent Pile Wall:** Piles arranged side by side in a straight formation.
- **Secant Pile Wall:** Piles are positioned side by side but intersect the adjacent piles on each side.
- **Staggered Pile Wall:** Piles placed side by side in a staggered alignment.
- **Spaced Pile Wall:** Gaps between adjacent piles filled with soil or weak reinforced concrete.

All these configurations are viable for stabilizing landslides, although it is crucial to note that the forces they need to withstand are often much greater than the Coulomb or Rankine pressures experienced by walls built in initially stable

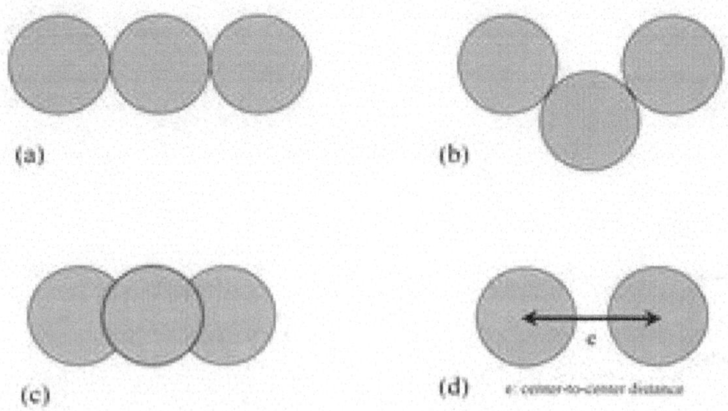

Figure 20.4 Shear pile wall configurations: (a) tangent, (b) secant, (c) staggered, and (d) spaced (Cornforth 2005 Cornforth, 2005/John Wiley & Sons).

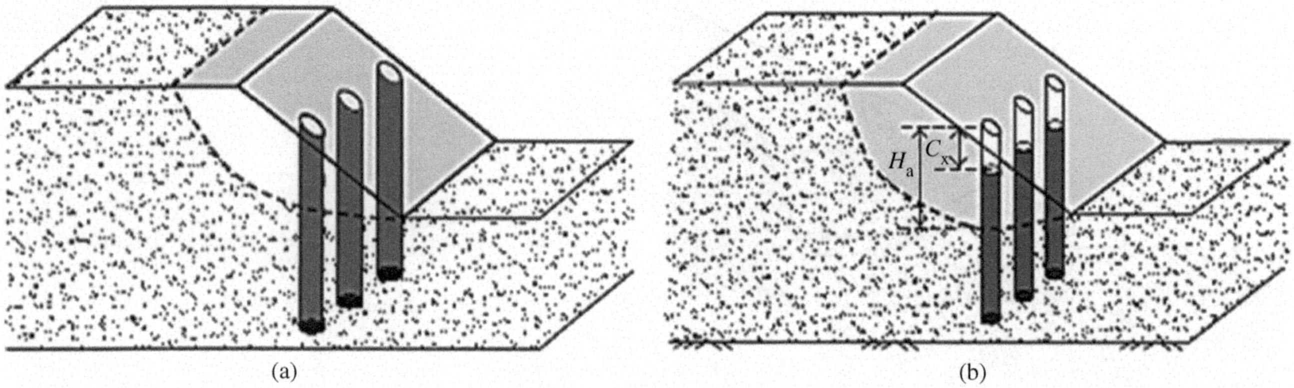

Figure 20.5 Shear pile wall types: (a) conventional shear pile and (b) truncated shear pile wall (Sun et al. 2016/Hindawi Publishing Corporation/Public Domain CC BY 4.0).

ground. As a result, these piles must be closely spaced and heavily reinforced. The shear pile walls can have full-length (conventional type) or truncated types, which will be elaborated upon in the following section, as illustrated in Figure 20.5. They can also be constructed with or without ground anchoring, referred to as a cantilever shear pile wall.

A truncated shear pile wall is a specialized engineering solution used to reinforce and stabilize slopes susceptible to landslides or instability. It involves the installation of piles with shortened upper portions to resist shear forces and maintain the stability of the slope while minimizing visual impact and potentially reducing costs. The design and implementation of truncated shear pile walls require careful consideration of site-specific conditions and engineering requirements. Designing a truncated shear pile wall involves assessing factors such as soil properties, slope angle, expected loads, and the desired safety factor. The depth of truncation and pile spacing are crucial parameters to determine during the design phase.

20.2.5 Case Studies

Case Study 20.1

The Ancient Landslide Challenge: Washington Park Station Slide (Cornforth 2005)

The Washington Park Station Slide case history begins with a monumental feat – the excavation into a terrain steeped in ancient landslides (Cornforth 2005). As the excavation work commenced, a subtle yet concerning development came to light – shear movements at the depths of the ancient landslide slip plane, hidden beneath the earth's surface. These movements underwent continuous scrutiny via an in-place inclinometer, seamlessly integrated into an automated data acquisition system. The inclinometers detected shear movements on the old landslide surface as excavation continued. This was exacerbated by the lower strength of the ancient slip surface compared to the surrounding soil. Triaxial tests on the soil above showed a strength of $c = 660$ psf and $\varphi = 33°$, while the ancient slip surface had much lower strength values ($c_r = °0$ and $\varphi_r = 7.9°$). Deep-seated movements along the ancient slip plane, especially in the east pit excavation, raised concerns. The base of the excavation was only 10 ft above the ancient slip plane, leading geotechnical engineers to conclude there was an unacceptable risk of a localized landslide behind a 115-foot-long north-cut slope in the east pit excavation. A meticulous examination, comparing shear movement trends with excavation rates and winter storm data, unequivocally established that the excavation was the primary instigator of these movements.

At the mid-depth of the excavation, a pivotal decision was made to temporarily suspend operations due to mounting apprehensions about a potential landslide cascading into the excavation site. This precautionary measure, termed the "prevention move," marked a significant juncture of the excavation operations.

Responding to the potential instability, an innovative solution emerged – the shear pile wall. This ingenious construction not only provided support for the temporary cut slope but also became an integral component of the structure's foundation. These piles reached deep into the ancient slip plane, securely anchoring themselves in the bedrock, ultimately evolving into a cost-effective solution. Figure 20.6 displays an illustration depicting the implementation of shear piles at Washington Park Station.

Figure 20.6 Washington Park Station Slide implementing the shear piles (Cornforth 2005/John Wiley & Sons).

Case Study 20.2

Dashtehgan Landslide

This landslide took place in the central region of Iran, situated on the outskirts of the Alborz Mountains and in proximity to the Caspian Sea, within an area known as Dashtehgan Rudbar (Ekramirad et al. 2012). The predominant soil composition in this region consists of altered shales, as opposed to unaltered shales or non-shale materials like limestone and granite. Furthermore, the area experiences significant weathering and is considered one of the most seismically active regions in Iran, elevating the risk of landslides.

The trigger for this event was an engineering oversight during the expansion of the Rasht–Qazvin highway, where soil was removed from the base of the slope. This excavation led to soil instability. Subsequently, structures situated

(Continued)

Case Study 20.2 (Continued)

higher up the slope, including agricultural lands, olive orchards, and high-pressure electrical towers, collapsed onto the construction path of the highway. Various factors contributed to the acceleration of this failure, including the introduction of moisture to the soil and increasing its weight, as well as making the upper surface of the bedrock slippery. In a matter of minutes, approximately one and a half million cubic meters of soil succumbed to the landslide, resulting in substantial financial losses. With proper precautions and remediation methods, this catastrophe could have been averted. Figure 20.7a–c provide a visual depiction of the Dashtehgan region and relevant details regarding the landslide and soil profile.

Figure 20.7 View of the Dashtehgan landslide: (a) two general views, (b) landslide location near the power transmission tower, and (c) simulated critical failure surface and profiling (Ekramirad et al. 2012/Sharif Journal of Civil Engineering/CC BY 4.0).

A preliminary analysis of the mass stability was conducted both before and after the excavation under natural soil moisture conditions, and the obtained safety factor indicated stability of the soil mass. Under the conditions that arose after the landslide occurred, a back analysis following the excavation from the slope toe was performed to calculate soil parameters under highly moist conditions and at the threshold of instability. The objective of this back analysis was to identify suitable solutions for stabilizing the slope and preventing landslides. Furthermore, by comparing the laboratory test results of soil in unsaturated conditions and some of the saturation parameters of the soils, it can be stated that the stability of the soil mass in dry conditions was expected, which is consistent with what occurred at the site. However, after rainfall and water infiltration into the soil mass, especially in the fine-grained portion, soil erosion took place. In this study, a series of back analysis procedures took place with different assumptions based on the real geotechnical data of the Dashtehgan landslide. One is mentioned as an example of the procedure.

In response to unsaturated soil conditions, a comprehensive back-analysis of the excavated section was conducted by Ekramirad et al. (2012). Equipped with geotechnical data and a clear understanding of the excavation site's location, a thorough section back-analysis was performed. (See Figure 20.8 for a numerical modeling of the critical shear surface of the slope before excavation with unsaturated soil in PLAXIS software.) Despite the relatively small thickness of the cut section compared to the landslide mass, it measured at least 30–50 cm. Consequently, we applied three distinct levels of required sliding through block landslide simulations on the cut section, leveraging the precise location and initial soil parameters. A back-analysis was then carried out to calculate the factor of safety regarding mass stability. Even after significant alterations to soil parameters within the cut section (as shown in Figure 20.8), the factor of safety against mass stability consistently remained at approximately 1.209, highlighting system stability under unsaturated conditions.

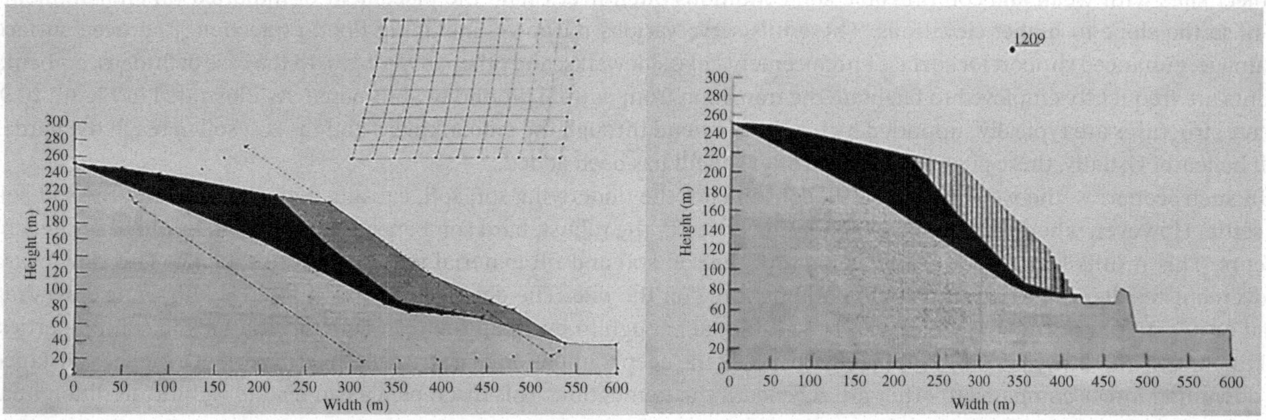

Figure 20.8 Numerical modeling before Dashtehgan landslide occurrence with unsaturated soil in PLAXIS software (Ekramirad et al. 2012).

The back-analysis procedure was repeated for various conditions, including post-excavation with saturated soil. The results indicated a factor of safety greater than 1 (FS = 1.090), confirming system stability even in saturated conditions. However, it was observed that in the case of a circular shear surface, the factor of safety against stability was significantly less than 1 under saturated conditions, suggesting that rainfall infiltration into the soil mass, especially in the fine-grained section, could be a primary cause of the landslide. Therefore, implementing drainage measures to prevent water infiltration into the mass is essential among all available solutions.

Furthermore, based on stability analysis conducted after slope excavation, it became evident that the removal of the weight of the slope toe was one of the most significant contributing factors to the landslide occurrence.

Proposed Remediation Methods: Before commencing any construction operations for building the highway, several proposed solutions have been considered:

1) Excavation and grading on the top of the soil slope above the neutral axis.
2) Construction of a restrained retaining wall at the base of the soil slope, extending through the landslide area. Options include soldier piles and combinations of pile walls and anchors.
3) Building reinforced concrete retaining walls at the base of the slope and anchoring them to piles installed at the top of the slope.

After conducting analyses using PLAXIS software, it was observed that the best stability method is the use of a restrained retaining wall. This option excels in terms of execution, technical considerations, and economic factors.

20.3 Downdrag Forces

Subsurface soil conditions play a crucial role in determining the type and characteristics of the foundation system. Problematic soils, often referred to as challenging soils, are the main culprit to a range of geotechnical issues related to the performance of foundations, such as bearing failure, excessive total and differential settlements, instability, liquefaction, erosion potential, and water seepage. Examples of such problematic soils include collapsible soils, loess, expansive soils, dispersive clays, organic soils, peats, colluvium, talus, degradable materials, soft and sensitive clays, liquefiable sands, uncontrolled fill, and recycled materials (Coduto et al. 2016; Han 2015; Holtz et al. 2023).

Deep foundations may encounter performance issues when constructed in problematic soils, particularly in soft, highly compressible and sensitive clays, and fills. Deep foundations constructed in these soils may encounter, in addition to the forces from the superstructure, downdrag forces as these soils consolidate and compress with time. This section will focus on the evaluation of the downdrag forces imposed on the pile.

20.3.1 Introduction

Project sites with weak and compressible soil conditions often necessitate the placement of imported fill embankments to raise the slope to higher elevations. These fills serve various purposes, including flood protection, improved surface drainage, enhanced support for surface enhancements like sidewalks, and other advantages. In the case of bridges, embankments are frequently employed to facilitate the transition from natural ground to abutments. As illustrated in Figure 20.9, heavy structures are typically supported by piles that extend through the embankment and the soft soil to reach the harder soil beneath. Usually, these piles are installed after the fill has been added.

In such scenarios, the weight of the infill consolidates the underlying soft soil, causing both the infill and the soft soil to settle. However, when the pile is adequately anchored in the robust, hard soil beneath the soft soil, minimal settlement occurs. This results in a downward movement of the soft soil and fill material with respect to the pile. This downward movement results in a downward frictional drag force on the pile. The downward friction force is known as downdrag load or negative skin friction, which can be substantial enough to overload the pile. Due to variations in soil properties, fill thickness, and other factors, certain areas beneath the deep foundation of a structure may experience higher resistance to downdrag forces compared to other areas beneath the foundation. This discrepancy can lead to significant differential settlement within the structure.

Downdrag loads can also manifest in other scenarios where the ground moves downward relative to the top of the pile. For instance, a lowering of the water table, even without the presence of an embankment, can induce settlement due to the associated increase in effective stress. Downdrag may also occur during earthquakes, particularly in loose sand, where seismic forces induce ground settlement (Coduto et al. 2016).

The process for calculating downdrag loads are as same as presented in Chapter 15 for shaft resistance, except the loads act downward instead of upward.

At a project site, the geotechnical properties of natural deposits of different soil layers can vary. Accordingly, the downdrag forces on piles will vary as the properties of the soil layers vary at the site. Several combinations of alternating layers of soft (weak) and stiff (strong) soils can be found in nature. The interaction of the downdrag forces with the pile will be different with each different combination. The three most common scenarios of combinations of layers encountered are as follows: (i) the most extreme scenario is where the weak and compressible deposit extends to the full depth of the pile influence

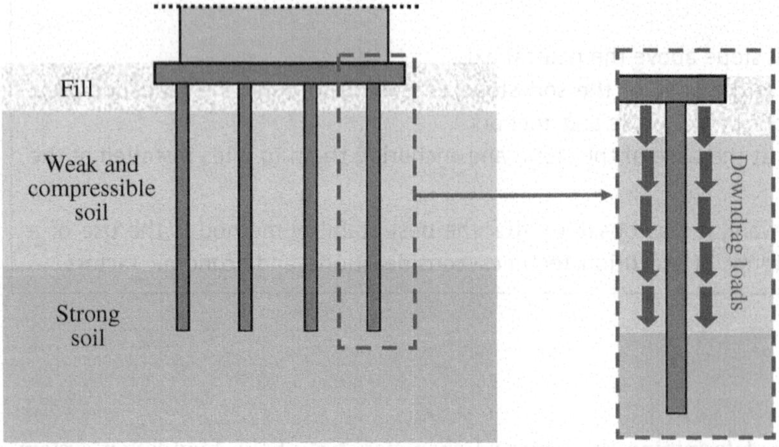

Figure 20.9 Schematic of downdrag forces on piles.

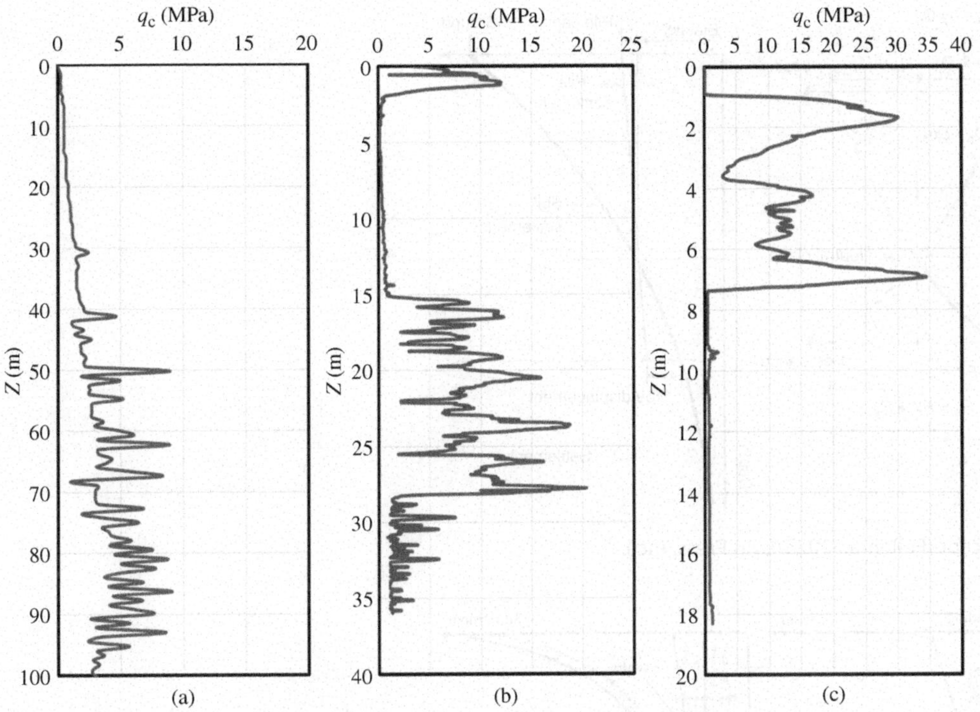

Figure 20.10 Typical CPTu profile: (a) Urmia Lake Causeway (Eslami et al. 2019/with permission of Taylor & Francis Group), (b) UBC research pile site (adapted from Campanella and Sy 1991), and (c) Evanston pile prediction symposium (Finno 1989/American Society of Civil Engineers).

zone, as reported in the Urmia Lake Causeway Project in Case Study 4.1. A typical CPTu record of this site is presented in Figure 20.10a. (ii) Another scenario is when a weak soil layer is located on a strong soil layer. In this situation, the pile toe is embedded in the strong layer while the shaft is in the weak layer. One such case is reported by Campanella and Sy (1991) in the University of British Columbia (UBC) pile research site located in Vancouver, Canada. A typical CPTu record of this site is presented in Figure 20.10b. (iii) Another scenario is when a strong soil layer is located on a weak soil layer. In this situation, the pile toe is embedded in the weak layer while the shaft is in the strong layer. Such a case was reported by Finno (1989) at the Pile Prediction Symposium held at the Lakefill site on the Evanston Campus of Northwestern University in conjunction with the 1989 Foundation Engineering Congress. A typical CPTu record of this site is presented in Figure 20.10c.

20.3.2 Performance of Piles with Downdrag Forces: Neutral Plane Concept

In cases where drag forces are present, the negative friction generated at the top of the pile creates a drag force that starts from zero and reaches its maximum value at a specific depth known as the equilibrium depth. This point signifies the transition from negative to positive frictional stress within the pile. Below this equilibrium depth, which is referred to as the neutral plane, the axial load on the pile diminishes due to the combined positive frictional resistance and the ultimate tip resistance. Each pile is in equilibrium between the combined downward force of the applied dead load from the superstructure, the negative friction force (downdrag), and the upward counterbalancing force due to the combined positive frictional resistance and the pile tip resistance. This equilibrium spans the length of the pile and corresponds to the depth of the neutral plane, where the frictional shear stress along the pile wall shifts from negative to positive, and the relative displacement between the pile and the ground becomes negligible. Figure 20.11 provides a visual representation of the concept of the neutral plane (Fellenius 2023).

The settlement of the pile group can be estimated using the equivalent foundation approach, where the load is transferred to the depth of the neutral plane, and then the settlement is calculated from this depth similar to the raft foundation. In other words, settlement is caused by soil compression due to increased effective stress below the neutral plane due to external loads. Based on the neutral plane concept, the decrease in the toe and the shaft resistance along with the increase in the negative side friction leads to a decrease in the neutral plane, so a larger amount of settlement is expected.

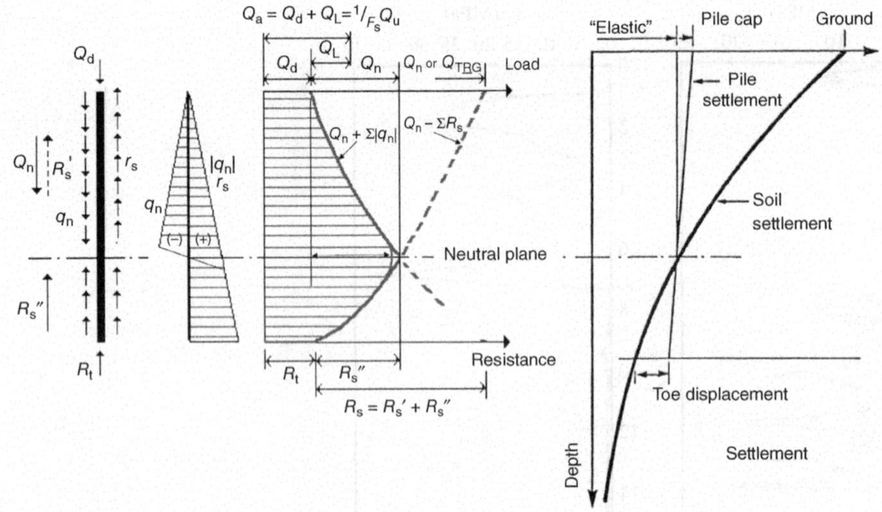

Figure 20.11 Neutral plane concept (Fellenius 2023/Lulu Press, Inc.).

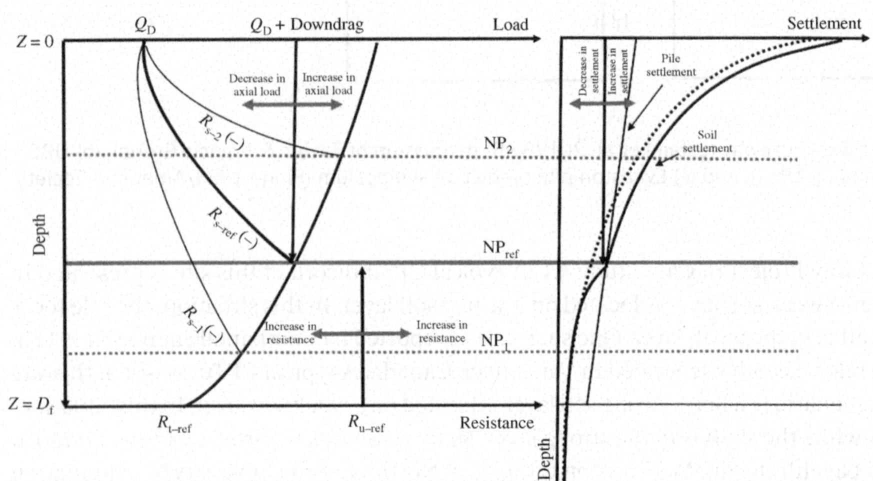

Figure 20.12 Schematic view of variation in neutral plane depth by altering the negative skin friction.

Negative surface friction and drag loads are particularly pronounced in super-soft and sensitive soils with a predisposition for collapse, as exemplified by Lake Urmia sediments (Case Study 4.1). As depicted in Figure 20.12, when negative surface friction decreases (from the reference condition to condition 1), holding other parameters constant, the neutral plane shifts, resulting in reduced axial loads and correspondingly lesser settlement in the structure. Conversely, in scenarios involving sensitive soft soils, increasing negative surface friction (from reference conditions to condition 2) leads to an upward shift of the neutral plane. Consequently, higher axial loads and increased settlement are anticipated. This situation presents a more complex and challenging design scenario, demanding more stringent considerations.

Bearing capacity, settlement, and axial load are intricately interrelated and operate iteratively, unified by the concept of the neutral plane. Within this concept, the occurrence of collapse in sensitive soils leads to an increase in tensile load while simultaneously reducing shaft and tip resistance. While the resistance may remain constant, the applied load rises as the downdrag load diminishes, consequently reducing the factor of safety. Generally, in soils prone to collapse, the distance between the neutral surface and the ground decreases, resulting in reduced bearing capacity and increased settlement. When all other factors remain constant, an elevation in negative surface friction brings about a significant upsurge in the axial load experienced by the structure. This factor must be duly accounted for in the design process.

Figure 20.13 provides an alternative perspective on the impact of sensitive soils through the neutral plane. Settlement in soft clays is predominantly attributed to consolidation settlements, primarily influenced by the thickness and stiffness of

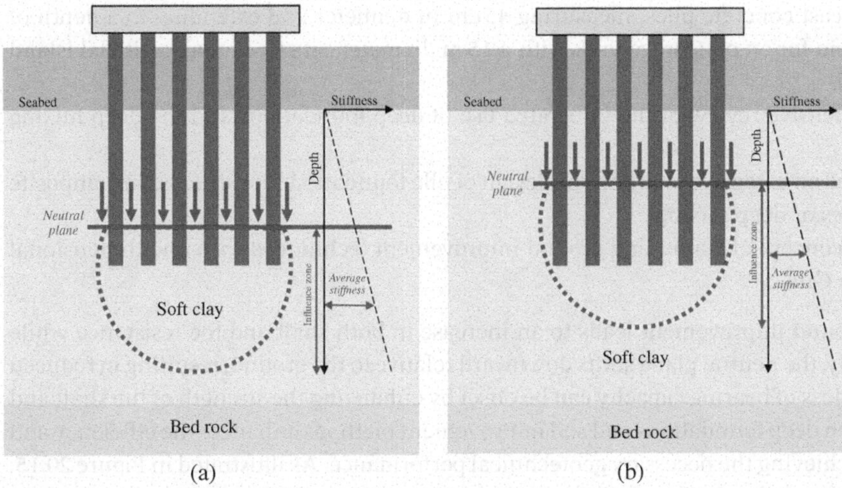

Figure 20.13 Schematic alteration of neutral plane location and relevant settlement: (a) undisturbed and (b) disturbed.

the deposit. An increase in the depth of the neutral plane results in an expanded thickness of the soft clay layer susceptible to settlement, coupled with a reduction in stiffness. The diminished stiffness is attributed to deposits at lower depths, which are typically younger and less compacted compared to deeper layers, as highlighted in the q_t profile.

20.3.3 Solutions to Overcome Dragload

Soil improvement is acknowledged as an alternative approach to complement deep foundations. Various techniques, including compaction, deep mixing, and micro-piles, are often employed in conjunction with deep foundations. The objective is to enhance shaft and toe capacity, mitigate liquefaction risks, and bolster lateral capacity. These composite foundation solutions typically align with economic and sustainable development goals while meeting geotechnical and construction requirements. Figure 20.14a and b illustrate a combined application scheme for the foundations of a reservoir (one quarter) and a river bridge, respectively. The foundation of the reservoir is a combination of rigid intrusions and driven piles. The deep foundation of the bridge is partially constructed in a zone of extensively treated and improved soil.

The application of various soil improvement techniques enhances the geotechnical properties of the existing subsoil, resulting in increased bearing capacity and reduced settlement. Furthermore, soil improvement augments the modulus of elasticity, signifying improved stiffness and reduced compressibility, ultimately leading to a reduction in settlement. The literature has several examples of deep foundations constructed in improved soil. Select cases are listed below:

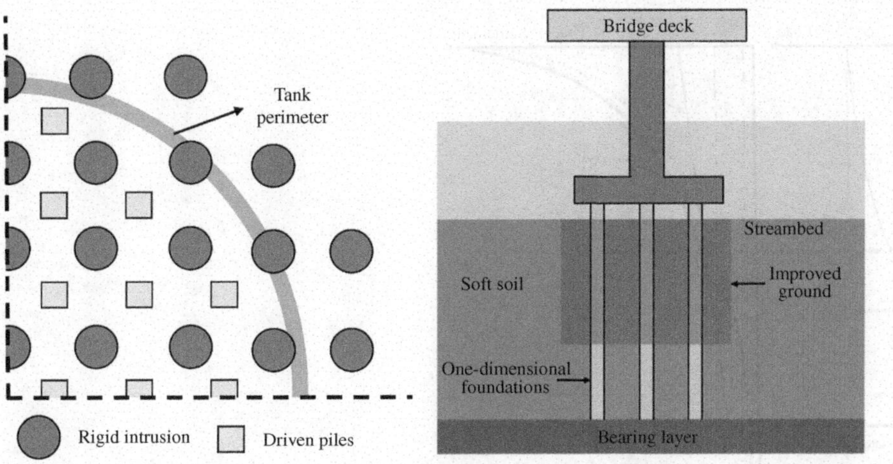

Figure 20.14 Schematic view of combined application of ground improvement and deep foundations (a) columnar approach for a tank and (b) massive approach for a riverine bridge.

- A combination of round pre-stressed precast concrete piles, measuring 45 cm in diameter and extending to a depth of 23 m, is employed as the foundation system for an oil storage tank with a 15 m diameter, situated on an artificial island in Japan (Ishihara and Cubrinovski 2004).
- Maeda et al. (2006) conducted a comprehensive review of the integrated use of deep foundations and the deep mixing (DM) method for bridge foundations.
- Tomisawa and Miura (2007) have presented an extensive study on the design of pile foundations constructed in composite ground improved by sand compaction piles or deep mixing.
- Asadi et al. (2016) have documented the concept of combining ground improvement techniques with one-dimensional foundations for a causeway in the Persian Gulf.

Adopting the neutral plane approach, ground improvement leads to an increase in both shaft and toe resistance while reducing negative skin friction. Consequently, the neutral plane shifts downward relative to the ground, resulting in reduced settlement, as depicted in Figure 20.15. The loss of bearing capacity can be offset by enhancing the strength of the shaft and toe resistance. Generally, the synergy between deep foundations and soil improvement methods enhances the efficiency and cost-effectiveness of the design, while still achieving the necessary geotechnical performance. As illustrated in Figure 20.15, although the axial structural load increases due to the shallower position of the neutral plane, this is not a concern for deep foundations owing to their high material strength. The prioritization of increased strength and reduced settlement takes precedence over a slight rise in axial structural loads.

Coduto et al. (2016) have suggested the following solutions for overcoming the excessive downdrag loads:

- Applying a bituminous coating to the pile to reduce the friction coefficient. This method proves highly effective, unless the pile encounters an abrasive substrate like sand that could potentially scrape off the bituminous coating.
- Driving the pile into place before backfilling and optionally covering it with a bitumen mulch. Exposed parts of the pile can be shielded with a lubricated plastic sheet or another low-friction material, after which a backfill is placed around the pile.
- Employing pre-drilling with a bit larger than the pile, followed by optional partial filling of the hole with bentonite, and then driving the pile into the prepared hole.
- Utilizing a pile head with a larger diameter than the pile itself, which allows for the excavation of a larger hole during pile driving. Alternatively, installing a ring at the appropriate depth on the pile achieves a similar effect, focusing the effect at the tip of the installed pile.
- Employing an open steel pipe pile driven through consolidating soil, followed by the removal of the plug. Subsequently, a smaller diameter bearing pile is driven through the pipe and into the underlying support layers. This method isolates the inner pile from downdrag loads and is sometimes referred to as an isolation casing.
- Accelerating settlement through additional backfilling or other techniques, and then proceeding to install the foundation after the settlement process is complete.

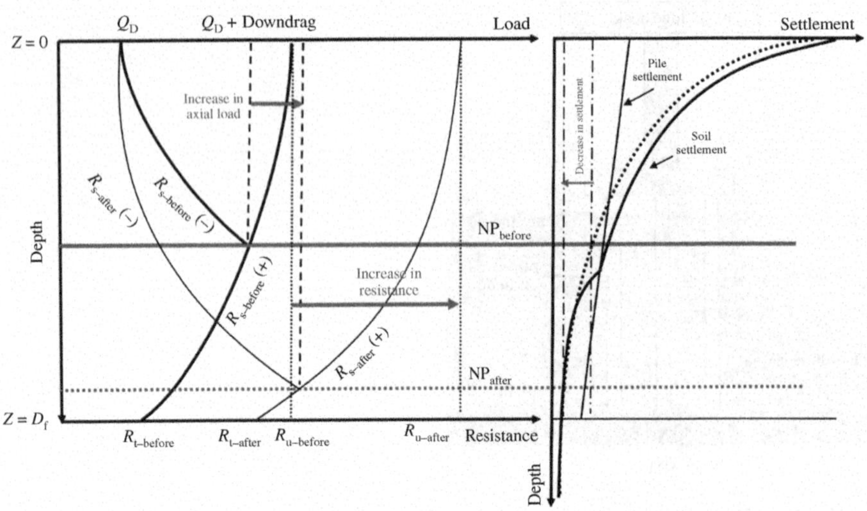

Figure 20.15 Effect of soil improvement on geotechnical performance based on neutral plane.

20.4 Cyclic Loading and Its Effects on Deep Foundations Performance

20.4.1 Background

Structures are exposed to various types of loading, including static and variable loading. The term "static" pertains to constant loading, primarily stemming from gravitational forces. The term "variable" refers to loadings that fluctuate and are caused by the environment or industrial and anthropogenic activities. Variable loadings generated from environmental sources include loads induced by wind, earthquakes, tides, swells, and potentially from changes in temperature, pressure, phreatic zone, etc. Variable loadings generated from industrial and anthropogenic activities include loads induced by operating machines, presses, rail, road, air traffic, materials handling machinery (forklifts, cranes, etc.), or by filling/emptying of tanks. Within the category of variable loads, some may be characterized as "cyclic," denoting actions that vary with a distinct repetitive nature, featuring a certain level of regularity in both amplitude and recurrence periods.

The common regulations – (Eurocode 7 2004) and NF P 94-262 for piles – alert engineers to the potential effects of cyclic loading on foundation soils, including degradation of the bearing capacity of the foundations, amplifying movements and internal forces, or altering the stiffness of the materials. Also, the cyclic loading effect must be considered in design via cyclic *P–y* curves, as recommended by API (2014). However, it must be admitted that there are no guides to help geotechnicians assess the severity of the phenomena likely to be encountered or how to take their effects into account.

Cyclic stresses lead to a complex response of soils and foundations. Multiple phenomena are intuitively related to the effect of cycles, such as:

- Degradation of the shear strength, which can reduce the bearing capacity;
- Generation of excess pore pressures, which reduces the effective stresses and can lead to the phenomenon of liquefaction;
- Variation of shear strength of clay soils with the shearing rate; and
- Creep or "fatigue" of soils and interfaces over a large number of cycles, can increase long-term displacements (Puech and Garnier 2017).

The ideal case of cyclic loading has a constant period and amplitude, which is called regular loading, and it is easy to characterize the loading experience utilizing the following values (illustrated in Figure 20.16):

- Q_m: mean load or mean component of the cyclic load;
- Q_c: cyclic component or half-amplitude of the cyclic load;
- Q_{min}: minimum cyclic load ($Q_{min} = Q_m - Q_c$);
- Q_{max}: maximum cyclic load ($Q_{max} = Q_m + Q_c$);
- T: period of the cycles ($T = 1/f$, where f = frequency of the cycles);
- N: number of cycles.

In the real world, cyclic loading seldom exhibits a consistent amplitude or fixed period. Consequently, there is a need for a procedure to transform the actual, unpredictable loading into an idealized, regular loading pattern. Cycle counting

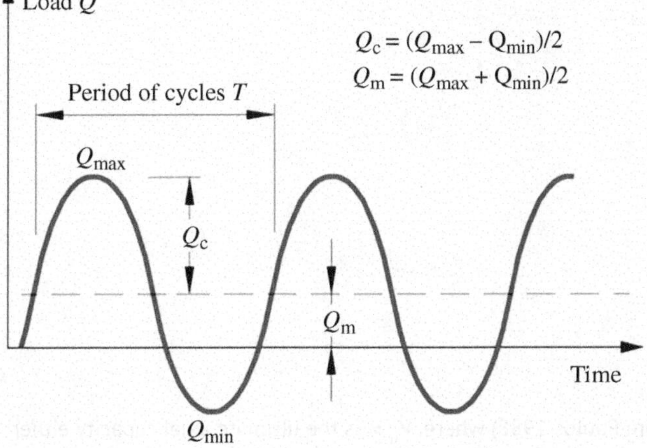

Figure 20.16 Regular cyclic loading definitions.

methods, stemming from "rainflow" analysis, are extensively employed to convert histograms of actual loads into a series of cycles with uniform amplitude and frequency (Polák and Vašek 1992). This transformation ensures that the material experiences equivalent damage as it would under real-world conditions.

20.4.2 Cyclic Stability Diagram

Poulos (1988) introduced the concept of the cyclic stability diagram for a pile, which has proven to be a valuable tool for analyzing the holistic response of a relatively rigid pile subjected to axial cyclic forces and assessing its performance based on the combination of these forces and the number of cycles (Poulos 1988). Figure 20.17 illustrates the underlying principle of the cyclic stability diagram in the context of axial loading.

On the horizontal axis, the normalized mean component of the cyclic loading is V_m/V_u; and on the vertical axis: the normalized amplitude of the cyclic loading V_c/V_u. The sides of the triangle joining the points where V_m/V_u and V_c/V_u are equal to 1 are the points of static failure (monotonic loading for which $V_m/V_u + V_c/V_u = 1$).

The diagram is dissymmetrical in relation to the vertical axis, because the compressive capacity is greater than the tension capacity ($V_{uc} > V_{ut}$). In its simplest form, the diagram (Figure 20.17) can be broken down into three zones defined by combinations of the couple (V_m/V_u; V_c/V_u):

- Zone A permits the application of an extensive number of cycles, often reaching into the thousands or tens of thousands, without inducing significant irreversible displacements in the pile. This zone is considered cyclically stable, as the pile's capacity remains unaffected by the repeated cycles.
- In Zone C, cyclic loading leads to irreversible displacements that inevitably culminate in failure after a relatively small number of cycles, typically just a few or a few tens. Zone C is classified as cyclically unstable, as the pile's capacity is significantly compromised by the cyclic loading.
- Zone B exhibits intermediate behavior, where a limited number of cycles, usually around a hundred, produce pile displacements without an immediate progression toward failure. If cycles persist, these displacements may either tend toward apparent stabilization or ultimately lead to complete pile failure after a high number of cycles, ranging from a few hundred to several thousand. This latter phase is known as the metastable zone. In Zone B, the pile's capacity experiences an impact from the sequence of cycles.

It is important to be aware that the dimensions of each zone in Figure 20.17 are contingent upon the specific soil type and the type of pile under consideration. Under certain favorable circumstances, such as those found in normally consolidated or slightly over-consolidated clays, it has been feasible to enhance the insights derived from the stability diagram by generating isovalue curves for the number of cycles required for failure.

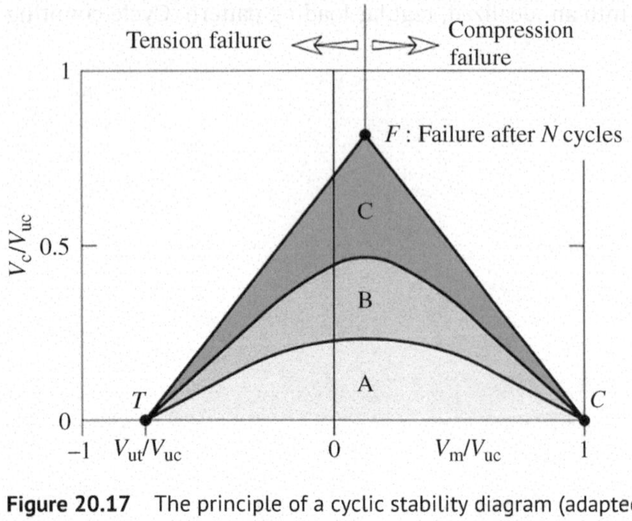

Figure 20.17 The principle of a cyclic stability diagram (adapted from Poulos 1988) where, V_u = is the ultimate axial capacity either in compression or tension (pullout); V_{uc} = is the ultimate monotonic compression resistance; V_{ut} = is the ultimate monotonic tension resistance (pull out capacity); V_c = is the cyclic load; V_m = is the mean load.

20.4.3 Degradation of Pile Foundations Under Axial Cyclic Loading

Cyclic degradation manifests itself within soils, particularly at the soil–pile interfaces, and influences the overall response of the pile to both axial and lateral loading.

Under axial cyclic loading, the overall response of the soil–pile system is an outcome of integrating local responses along the length of the pile. The axial compressibility of the pile, denoted as K_a, plays a role in generating these mechanisms, as depicted in Figure 20.18.

Figure 20.18a shows an infinitely rigid pile within the soil. The soil is characterized by constant mechanical properties throughout its depth (τus constant). The application of axial cyclic loading at the pile head results in:

- Constant cyclic displacements along the entire length of the pile, and
- Constant relative (normalized) cyclic shear stress levels at all points.

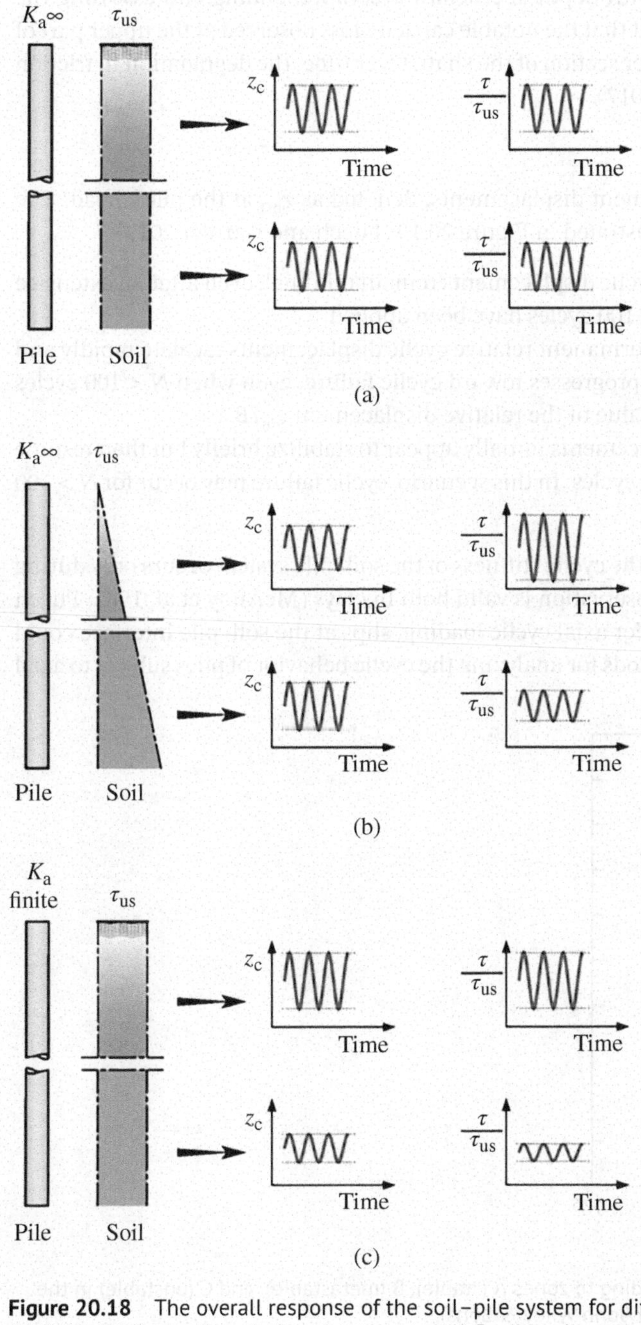

Figure 20.18 The overall response of the soil–pile system for different configurations: (a) rigid pile; constant shaft resistance, (b) rigid pile; increasing shaft resistance, and (c) flexible pile; constant shaft resistance (Puech and Garnier 2017/John Wiley & Sons).

In Figure 20.18b, the same assumption of an infinitely rigid pile is made, but this time within the soil, where mechanical properties of the soil increase linearly with depth. When cycles are applied at the pile head, we observe:

- Uniform cyclic displacements along the pile's length, and
- A decrease in relative cyclic shear stress levels with increasing depth.

Finally, in Figure 20.18c, we consider a compressible pile within soil characterized by constant mechanical properties throughout its depth (τ_{us} constant). The application of cycles at the pile head results in:

- Noticeable cyclic displacements at the pile head, gradually diminishing with depth,
- A concurrent decrease in relative cyclic shear stress levels with increasing depth.

Actually, piles are relatively compressible, and soil typically exhibits improved mechanical properties with depth. Understanding the phenomenon of decreasing cyclic degradation with depth is crucial for comprehending and assessing the overall response of a soil–pile system. Consequently, it is evident that the notable capacity loss observed at the upper part of the shaft necessitates the redistribution of loads toward the lower section of the shaft. Over time, the degradation in friction tends to propagate toward the pile's base (Puech and Garnier 2017).

20.4.3.1 Pile Head Response

Various tests were conducted to assess the evolution of permanent displacements, denoted as z_m, at the pile's head. The measurements revealed three distinct types of behaviors, as illustrated in Figure 20.19 (Puech and Garnier 2017).

A: Stable Response: In this scenario, the permanent relative cyclic displacements remain minimal, even after an extensive number of cycles. They stabilize quickly, typically when $N < 100$ cycles have been applied.

C: Unstable Response: In cases of an unstable response, the permanent relative cyclic displacements escalate rapidly and continue to increase at an accelerated pace. The pile swiftly progresses toward cyclic failure, even when $N < 100$ cycles have been applied. This failure can be defined by a specific value of the relative displacement z_m/B.

B: Metastable Response: The permanent relative cyclic displacements initially appear to stabilize briefly but then resume an upward trajectory, accelerating beyond approximately 100 cycles. In this scenario, cyclic failure may occur for $N > 100$ cycles.

Several researchers have shown that a significant decrease in the cyclic stiffness of the soil–pile system occurs only during the last few cycles before the general failure of the pile. This observation is valid both in clays (McAnoy et al. 1982; Puech 2013) and sands (Jardine and Standing 2012; Puech 2013). Under axial cyclic loading, slips at the soil–pile interface could be the main cause of the degradation of pile response. The methods for analyzing the cyclic behavior of piles subject to axial

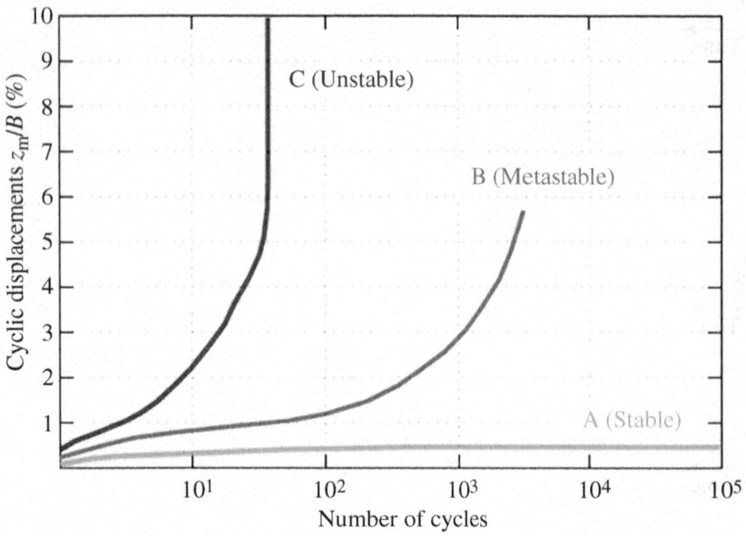

Figure 20.19 Typical pile head permanent displacements corresponding to zones A (stable), B (metastable), and C (unstable) in the cyclic stability diagram. $z_m = 1/2(z_{max} + z_{min})$ (Puech and Garnier 2017/John Wiley & Sons).

loading, rarely consider the contribution of the variations in cyclic stiffness. Atkins (2000) published a report on the cyclic degradation of offshore piles.

Field cyclic tests on driven piles in sandy soils have been conducted and reported by Lehane (1992), Chow (1997), and Jardine (1991, 1994). Tsuha et al. (2012) have reported laboratory model-scale tests aimed at simulating these field pile conditions. Airey et al. (1992) studied pile friction degradation using shearbox tests. The cyclic behavior of pile groups has been assessed by Al-Douri and Poulos (1994). DeJong et al. (2003) investigated interface load transfer degradation through cyclic loading. Sand-steel direct shear tests were performed by Mortara et al. (2007) to examine cyclic shear stress degradation and post-cyclic behavior. Rattley et al. (2017) reported laboratory test predictions of the cyclic axial resistance of a pile installed in the North Sea. Rimoy et al. (2022) conducted research on the stability and load–displacement of driven piles under axial cyclic loading.

20.4.4 Pile Foundations Degradation Under Lateral Cyclic Loading

Under the influence of cyclic lateral loading, the initial signs of degradation in the lateral reactions of the soil are evident in the displacement of the pile's head and the maximum moment experienced by the pile, both of which increase with the number of cycles. Numerous studies have validated this observation, and some studies have even derived equations that describe the relationship between pile head displacement and the number of cycles N.

Given the multitude of factors influencing soil–pile displacement, the distribution of horizontal stresses that develop within the soil is intricate. To simplify the modeling of soil reactions, it is common practice to consider their resultant horizontal force, denoted as H in Figure 20.20, applied at the pile head. The average stress across the width, B, of the pile, still in the same direction, is represented as $p = H/B$, as depicted in Figure 20.20.

Examining the successive values of the resultant reaction P during the cycles enables us to study the evolution of soil–pile interaction. However, determining this reaction experimentally is a challenging task. An example of static P–y curves for a 12 m deep pile embedded in dense sand and subjected to a load $H = 960$ kN at its head is illustrated in Figure 20.21. Figure 20.21 also illustrates that under static loading conditions, the mobilized reactions increase with depth (a reasonable behavior for this soil type). It is also noteworthy that the lateral displacement, y, and the corresponding reaction, P, change sign at approximately the same depth, aligning with the pivot or rotation point of the pile (around 5 m in the example provided here).

20.4.5 Piles Under Cyclic Transversal Loading

The content presented here is a compilation from various studies conducted in France and elsewhere, including recent work conducted within the SOLCYP project, the French acronym for "piles under cyclic solicitations" (Puech and Garnier 2017). In these experiments, piles were typically subjected to a series of identical cycles characterized by the following parameters:

- The number of cycles N and the frequency (or period) of the cyclic loading;
- The mean value of the applied lateral load H_m; and
- The half-amplitude of the cyclic component H_c.

The load applied to the pile head oscillates during the cycles between $H_{max} = H_m + H_c$ and $H_{min} = H_m - H_c$. The sign of the ratio H_{min}/H_{max} indicates whether the loading is two-way (if negative) or one-way (if positive or zero). H_{max} is often normalized by the limiting resistance to static loading, H_{lim}, the definition of which can vary depending on the publication consulted. Specific aspects of pile response to horizontal loading were obtained from these experiments, which must be taken into consideration in the design of such structures (Puech and Garnier 2017).

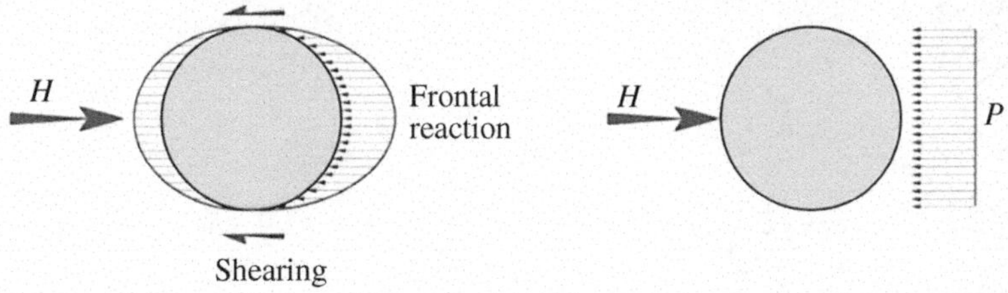

Figure 20.20 Real distribution of soil reactions and simplified model (Puech and Garnier 2017/John Wiley & Sons).

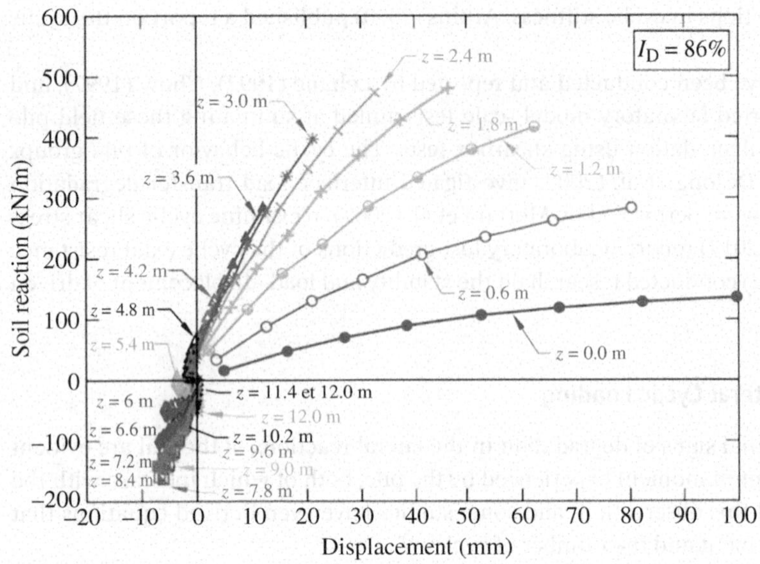

Figure 20.21 Static *P–y* curves at different depths (Rosquoët 2004).

20.4.5.1 Critical Role of Surface Layers

The response of the soil mass surrounding the pile differs under axial and lateral loading. While axial loading can mobilize the entire soil mass, lateral loading, applied at the head of the pile, primarily affects the layers of soil in direct contact with the upper part of the pile. Design methods are naturally influenced by this specific aspect of soil–pile interaction when subjected to lateral loads.

Moreover, specific mechanisms occur in the immediate vicinity of the soil surface, typically within a depth of approximately 2*B* or 3*B* (*B* being the diameter of the pile), depending on the source (Baguelin and Jézéquel 1972; Broms 1964; Merrill and Onge 1974; Randolph and Houlsby 1984). In this region, soil displacements are not purely horizontal; they also exhibit a vertical component. A bulge may form in front of the pile, and a gap may appear behind it due to soil detachment, especially if the soil possesses some degree of cohesion (refer to Figure 20.22).

Within each cycle, a gap forms around the pile. In the presence of a water layer on the soil's surface, as is often the case with offshore piles, water infiltrates this gap and contributes to soil erosion, as depicted in Figure 20.23. The reactions taking place within this extremely shallow zone are considerably more complex, primarily due to the aforementioned factors. Consequently, modeling them accurately can be challenging, and these reactions may not strictly adhere to the same principles observed in the deeper soil layers (Puech and Garnier 2017).

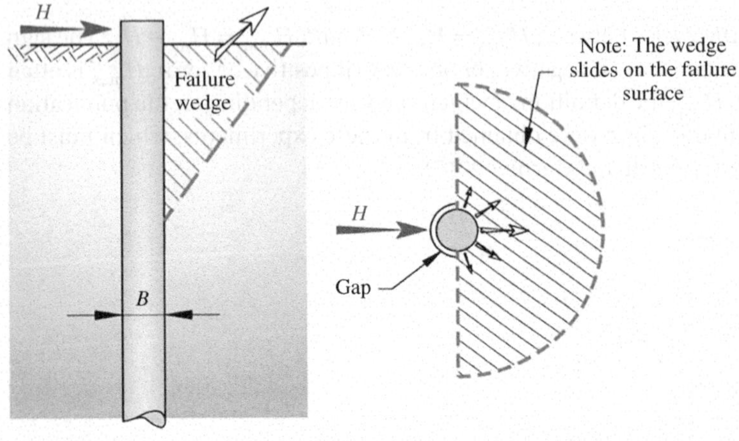

Figure 20.22 Mechanisms of soil deformation around the pile (Murff and Hamilton 1993/American Society of Civil Engineers).

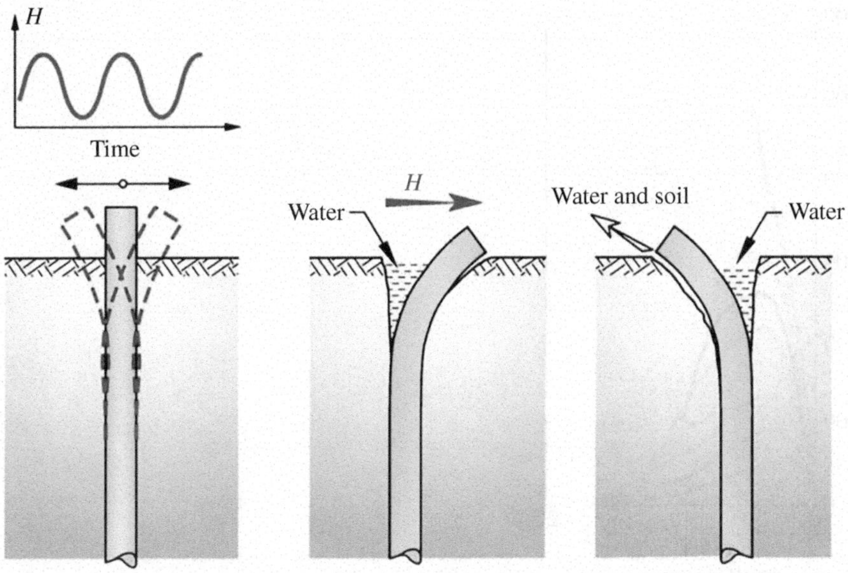

Figure 20.23 Repeated phenomena of detachment and erosion under cyclic loading (after Bea 1979).

20.4.5.2 Cycles and the *P–y* Reaction Curves

Reaction curves form the foundation of contemporary design methods, establishing the connection between horizontal soil stress (p) and lateral pile displacement (y) at a specific depth. Typically expressed as functions $P = f(y)$, with $P = p \cdot B$, determining *P–y* curves experimentally has remained a formidable challenge, even under the simplest conditions of monotonic loading. Efforts have been made, albeit with limited success, to measure soil reactions directly and precisely using pressure sensors. For instance (Baguelin and Jézéquel 1972) attempted such measurements in tests conducted on piles at the Plancoët site, while (Mayne et al. 1995; Prasad and Chari 1999) made similar attempts in experiments involving tank models.

Nonetheless, a handful of rare studies have managed to investigate the impact of cyclic loading on *P–y* reaction curves. Figure 20.24, for instance, illustrates the *P–y* curves obtained by Reese et al. (1974) in a stiff clay under both monotonic and cyclic loads. When cyclic loading is applied, a reduction in reaction can be observed for a given pile displacement.

Several proposals have been put forward to incorporate cycles into relatively straightforward calculations. These methods aim to derive cyclic reaction curves from static *P–y* curves and involve two approaches:

- **Reducing the Ultimate Strength:** This approach, suggested by Reese and Van Impe (2000) and endorsed by the API, is depicted in Figure 20.25.
- **Increasing the Stiffness:** Peralta (2010) suggests modifying the static *P–y* curves by applying the same coefficient to the displacement y as observed for the pile head displacement.

The two methods are then compared by introducing a degradation of the soil response, either across the entire pile penetration depth (which is highly unrealistic) or solely in the surface layers (to a depth of $z = 2.265B$ as per API recommendations). However, the results have thus far yielded inconclusive findings.

These approaches were also applied to stiff clays but have not found practical application (Reese and Welch 1975). It is important to note that the two methods are not equivalent, as the second method (increasing lateral displacement y) does not alter the plastic plateau, unlike the first method (reducing the reaction P). At present, the API recommendations are the only practical method for generating cyclic *P–y* curves, but they do not account for the number of cycles or their characteristics. These recommendations are primarily focused on simulating the overall effects of waves generated by a reference storm.

The effect of cycles on the soil reactions can be divided into two zones. In the upper layers, where the lateral pile displacement under maximum load is significant, the reactions decrease with the cycles for a given pile displacement. However, this degradation of soil reaction due to cyclic loading is not observed at greater depths because the lateral pile displacement remains small. In the deeper layers, the points on the *P–y* diagram represent a shift in soil–pile interaction, but they still align with the static reaction curve, with an increase in reaction P compensating for the decrease in reactions observed in the upper layers. Figures 20.26 (for sand) and 20.27 (for saturated clay) illustrate these observations for piles (Puech and Garnier 2017).

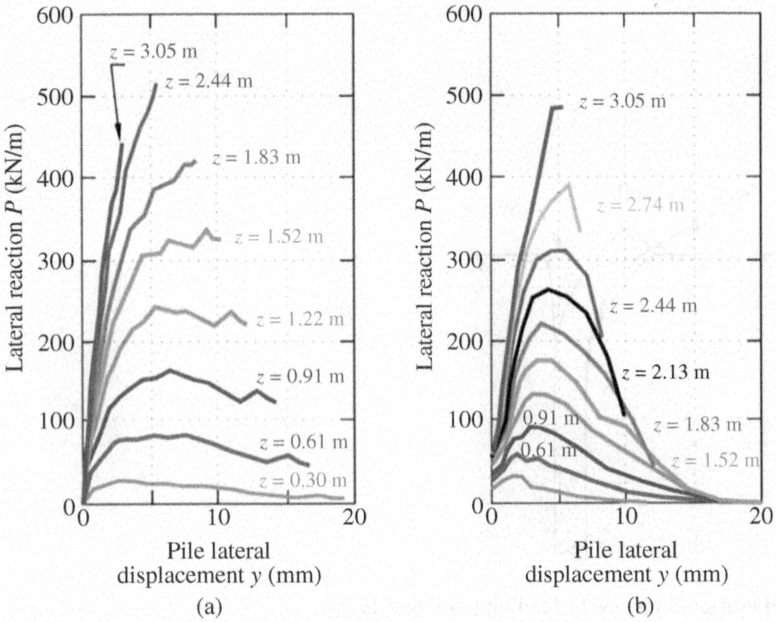

Figure 20.24 Experimental P–y curves in a stiff OC clay: (a) monotonic loading and (b) cyclic loading (P–y envelop curves) (Reese et al. 1974/American Society of Civil Engineers).

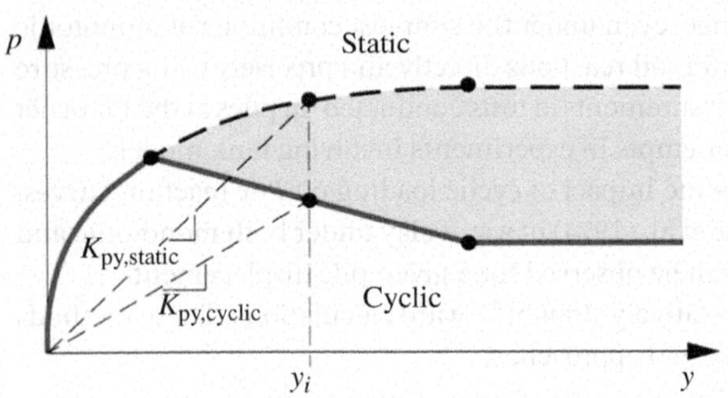

Figure 20.25 Principle of construction of cyclic P–y curves (adapted from Reese and Van Impe 2000).

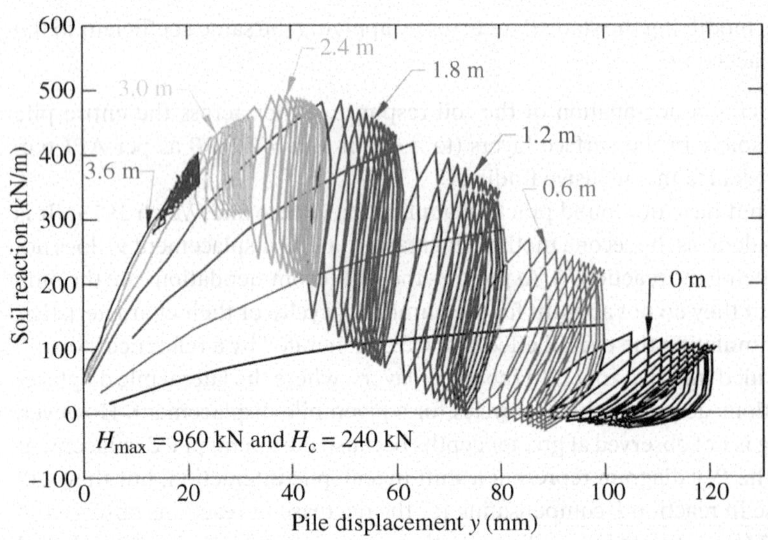

Figure 20.26 Example of cyclic P–y curves at different depths for H_{max} = 960 kN and H_c = 240 kN in dry sand (Rosquoët 2004).

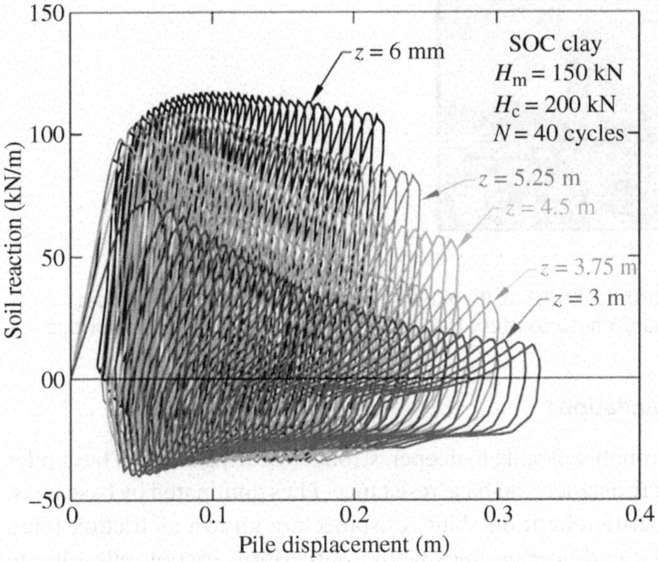

Figure 20.27 Reaction curves under 40 cycles in SOC clay at $H_m = 150$ kN and $H_c = 200$ kN (Khemakhem 2012).

20.5 Earthquake and Liquefaction

20.5.1 Failure of Structures During Earthquakes

Damage to structures caused by earthquakes is an inevitable occurrence, often resulting from a combination of structural deficiencies, foundation failures, or a combination of both. The failures depicted in Figure 20.28a and b stem from foundation failures, where the condition of the soil supporting the foundation plays a pivotal role in these incidents.

During an earthquake, the behavior of the underlying soil that supports a foundation is a critical factor in determining the foundation's performance. Broadly, there are two types of ground responses that can lead to damage in structures (Dobry and Iai 2000). In one scenario, the soil experiences failure, often through liquefaction, as seen in the 1995 Kobe earthquake. In the other scenario, the soil amplifies the ground motion, as illustrated in the 1989 Loma Prieta earthquake in California. Figure 20.29 depicts instances of pile-supported structures collapsing when situated on or passing through liquefiable soils during earthquakes. The objective is to investigate the failure mechanisms associated with these types of foundations during seismic events (Bhattacharya et al. 2003).

(a) (b)

Figure 20.28 (a) a high-rise building during the 1995 Kobe earthquake, after U.C. Berkeley (1995), and (b) tilting of a building during the 1999 Kocaeli earthquake (Turkey), after EERI (1999) (Bhattacharya et al. 2003/Cambridge University Press).

Figure 20.29 (a) Piled "Million Dollar" bridge after 1964 Alaska earthquake (USA), (b) piled "Showa Bridge" after 1964 Niigata earthquake (Japan), (c) piled tanks after 1995 Kobe earthquake (Japan), photo courtesy NISEE (Bhattacharya et al. 2003/Cambridge University Press).

20.5.2 Earthquake Loading and Performance of Pile Foundations

Pile foundations are primarily used to transfer axial loads through soft soils to deeper, stronger-bearing strata. These piles operate through two distinct load transfer mechanisms: shaft resistance and base resistance. Piles dominated by base resistance are referred to as end-bearing piles, while those primarily reliant on shaft resistance are known as friction piles. In regions prone to earthquakes, there is an expectation that end-bearing piles would outperform friction piles due to their end restraints. However, numerous instances of damage to end-bearing piles and pile-supported structures have been observed in major earthquakes. By gaining insights into the failure mechanisms of end-bearing piles in liquefiable soils, a deeper understanding can be developed, resulting in more reliable and sustainable designs for earthquake-prone areas.

The fundamental assumption here is that the load applied to the pile at its head is distributed between the sum of shaft resistance and end bearing. The actual values of end bearing and shaft resistances that come into play will depend on the displacements experienced by the pile when subjected to the applied working loads. Consequently, geotechnical engineers must possess a thorough understanding of the displacements incurred by a pile under working loads. Furthermore, the contemporary shift toward performance-based design necessitates that geotechnical engineers be capable of estimating these displacements across various load combinations to establish performance criteria effectively.

Dynamic loading from earthquakes will be superimposed upon the existing working loads that act on the pile. Consequently, when designing piles in earthquake-prone regions, it is crucial to establish a clear distinction regarding the anticipated performance of these piles. In some cases, it may be a design requirement for the pile to experience only minor displacements during a more probable but moderate earthquake. In other words, the pile must meet serviceability requirements and perform satisfactorily. In accordance with seismic design guidelines such as those provided by Burcharth et al. (2001) and the more recent ISO 23469 guidelines, this type of earthquake motion may be designated as a level 1 (L1) earthquake. For L1 earthquakes, there is a 50% probability of displacement occurring during the structure's design life. Conversely, during less probable but more severe earthquakes, the pile may be permitted to undergo greater displacement, but it must still function effectively to prevent the total collapse of the superstructure. This earthquake motion is categorized as a level 2 (L2) earthquake, where there is a 10% probability of displacement during the structure's design life. On average, an L1 earthquake occurs approximately every 72 years, while an L2 earthquake has an average return period of around 475 years.

Performance guidelines for any geotechnical structure can be established by considering an acceptable level of damage, following the PIANC guidelines. These guidelines were developed with the design philosophy centered on safeguarding human life and property. They also take into account the structure's capacity to offer emergency services in the aftermath of an earthquake and aim to prevent any potential release of hazardous materials, a concern that may be especially relevant to harbor structures (Madabhushi et al. 2009).

20.5.3 Current Understanding of Pile Failure and Design Methods

To the best of our knowledge, soil liquefaction entails the loss of shear strength in the soil, causing it to transform into a flowing state, and pulling along any overlying, non-liquefied crust. These soil layers carry the pile with them, resulting in a bending failure commonly referred to as lateral spreading failure. The prevailing failure mechanism suggests that the soil exerts a pushing force on the pile, as indicated by ground surface deformations observed near piled foundations.

The Japanese highway code of practice (JRA 1996) has embraced this understanding of pile failure. The code advises practicing engineers to design piles against bending failure, assuming that the non-liquefied crust provides passive earth pressure to the pile, while the liquefied soil itself exerts a drag force equal to 30% of the total overburden pressure. Other codes, such as the USA code (NEHRP 2000) and Eurocode 8, part 5 (EC8 1998), also emphasize the bending strength of the pile (Bhattacharya et al. 2003).

20.5.4 Liquefaction

Under earthquake loading, loosely deposited sandy soil layers undergo reorganization, leading to a denser arrangement of soil grains and causing volumetric contraction. In fully saturated soil, this contractive behavior transitions into a rise in pore water pressure since there isn't enough time for pore water to escape from the soil layers. When the excess pore water pressure increases to match the effective stress in the soil, it is considered that the soil layers have experienced complete liquefaction. Conversely, if the excess pore water pressures induced by earthquake loading remain limited, the soil may still soften, and this condition is referred to as partial liquefaction.

The foundations of civil engineering structures are susceptible to settlement or rotation when underlying soil layers undergo liquefaction. Among these structures, pile foundations can be particularly vulnerable to soil liquefaction. If the soil beneath the pile's base liquefies, it reduces the base capacity, potentially resulting in excessive settlements of the pile. Conversely, in situations where soil liquefaction occurs to a limited depth, such as in relatively small magnitude earthquakes, the soil surrounding the pile shaft may liquefy, causing a loss of shaft friction. This loss can lead to an increase in the base load of the pile, which, in turn, may result in elevated settlement. Furthermore, in addition to these scenarios, pile foundations are also at risk when subjected to laterally spreading ground.

Due to the occurrence of full or partial soil liquefaction within soil layers, lateral spreading of sloping ground can take place. When excess pore pressures build up within a sloping layer, the slope's ability to withstand static shear stress may diminish, resulting in the initiation of flow. Pile foundations passing through such laterally spreading ground are consequently exposed to lateral kinematic loads. Typically, if the soil layer has completely liquefied, the lateral forces exerted on the piles by the flowing ground may be relatively modest. However, if a stiff non-liquefied soil layer exists above the liquefied layer, this upper layer may also undergo lateral spreading. These non-liquefied layers can exert substantial lateral loads on pile foundations, as they possess the capacity to generate significant passive pressures against the piles or pile cap (Madabhushi et al. 2009).

20.5.4.1 Liquefaction Susceptibility

The assessment of soil liquefaction potential is crucial, and various methods have been employed for this purpose, including laboratory testing and field testing. Laboratory testing primarily focuses on evaluating soil susceptibility based on its grain-size distribution, while field testing involves techniques such as the standard penetration test (SPT), cone penetration test (CPT), and shear wave velocity measurements within the soil deposit.

In 1970, H. Tsuchida proposed boundary curves based on grain size distribution to identify soils resistant to liquefaction. These curves were originally published in Japanese by the Port and Harbour Research Institute (Tsuchida 1970). Translations of these curves into English, obtained from NRC (1985), are presented in Figure 20.30. Notably, Fraction E sand, used in the subsequent centrifuge tests to be discussed, falls within the boundaries applicable to most liquefiable soils, as indicated by the red line in Figure 20.30.

Several empirical correlations exist that relate earthquake intensity, SPT or CPT values of the soil, and the cyclic stress ratio required for liquefaction. These correlations are based on extensive field performance data collected from past earthquakes. Figure 20.31 illustrates a recent correlation developed by Seed et al. (1985) using data from American, Japanese, and Chinese sources for an earthquake with a magnitude of 7.5 on the Richter scale. The correlation depicted in Figure 20.30 is utilized to analyze the case histories of pile failures presented later (Bhattacharya et al. 2003).

20.5.4.2 Liquefaction Effect on *P–y* Curves

The effect of liquefaction on the construction of a *P–y* curve primarily can be addressed through variations in the soil properties and behavior. Cone penetration testing (CPT and CPTu) is among the most versatile tools in this regard. Liquefaction can impact the construction of a *P–y* curve through the following items:

- **Soil Strength Reduction:** Liquefaction can significantly reduce the strength and shear resistance of the soil. The CPT data collected from liquefiable soils may indicate lower values of soil strength parameters, such as friction angle and cohesion, compared to non-liquefiable soils. This reduced strength needs to be considered when constructing the *P–y* curve to accurately represent the soil's response.

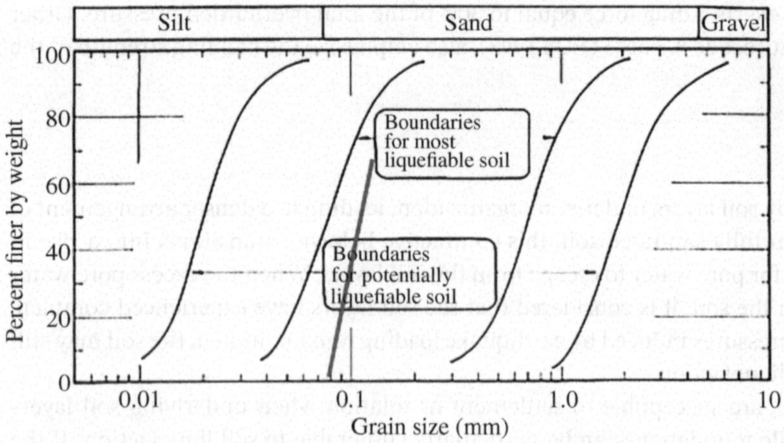

Figure 20.30 Limits in the grading curves separating liquefiable and non-liquefiable soils (after NRC 1985); The red line denotes the grading curve of fraction *E* used in the centrifuge tests (Bhattacharya et al. 2003).

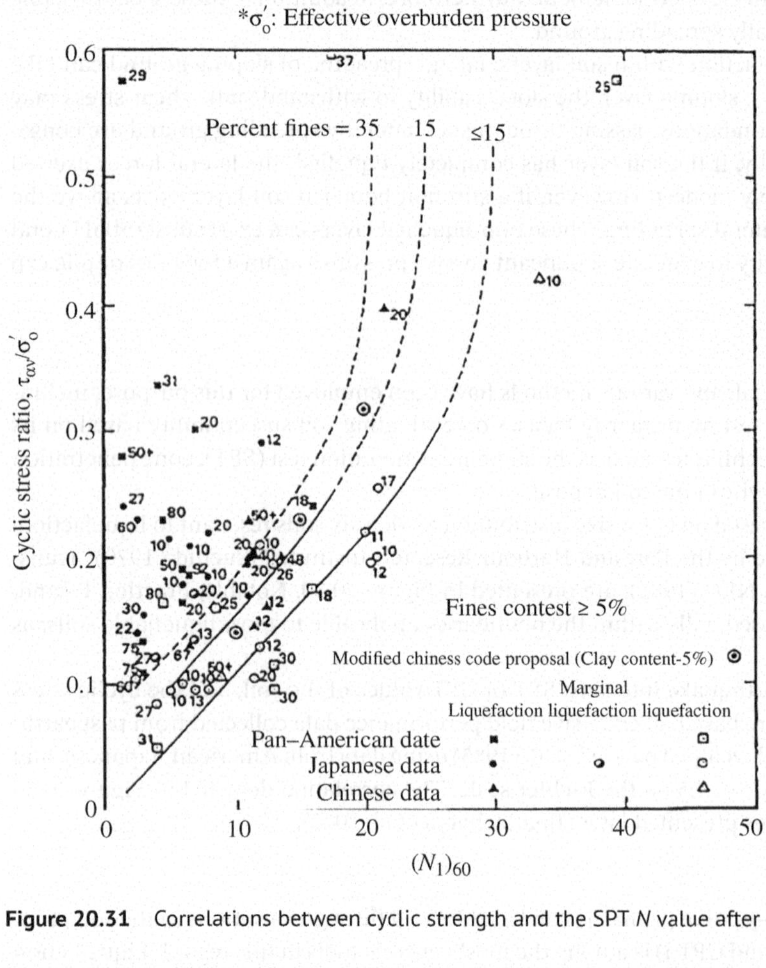

Figure 20.31 Correlations between cyclic strength and the SPT *N* value after Seed et al. (1985) (Tsuchida 1970).

- **Change in Soil Stiffness:** Liquefied soils experience a substantial decrease in stiffness during liquefaction. The CPT data obtained from liquefied layers may indicate lower values of soil stiffness parameters, such as shear wave velocity (V_s) or soil modulus (E). Incorporating the reduced stiffness in the *P*–*y* curve construction is crucial to capture the soil's actual behavior during lateral loading.
- **Porewater Pressure Generation:** During liquefaction, excess pore water pressure builds up in the soil. This excess pressure affects the effective stress and can lead to a decrease in the lateral soil resistance against pile deflection. The

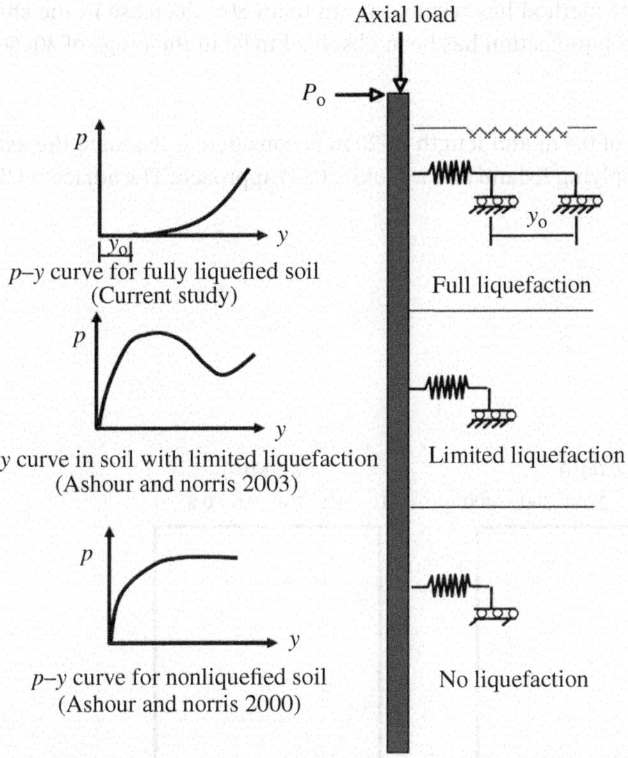

Figure 20.32 *P–y* curve for liquefied and non-liquefied sands (Ashour and Ardalan 2012/Canadian Science Publishing).

P–y curve construction should consider the impact of excess pore water pressure on the lateral soil response, which may require specialized analysis or empirical methods.

- **Lateral Soil Displacement:** Liquefaction-induced ground shaking can cause significant lateral displacement of the soil. The displacement profile can vary depending on factors such as soil layering and ground conditions. The *P–y* curve should be constructed considering the potential for lateral soil movement during liquefaction to accurately capture the pile–soil interaction and estimate lateral pile deflection.

It is important to note that accurately modeling and incorporating the effects of liquefaction in the construction of a *P–y* curve can be challenging. The behavior of liquefied soils is complex, and additional site-specific investigations, such as field testing or laboratory testing on liquefied soil samples, may be necessary for more accurate assessments.

Both full- and model-scale load tests have been recognized as major tools in the characterization of single and group piles performance in post-liquefaction conditions. However, significant uncertainty is still existing in the response of pile and liquefied soil as well as relevant *P–y* curves.

Due to the complexity of the *P–y* curve in liquefied sand and its unusual pattern, as illustrated in Figure 20.32, the undrained stress-strain of the soil should be applied in an actual soil–pile interaction model to evaluate the relevant *P–y* curves. In this regard, Ashour and Ardalan (2012) have suggested a technic to assess the laterally loaded piles response and associated *P–y* curves under full liquefaction conditions, accounting for water pressure fluctuation adjacent to the pile. Also, a constitutive stress-strain model has been established. A schematic view of the suggested *P–y* curve in the reviewed study is depicted in Figure 20.38, in which P_0, is the pile-head lateral load and y_0 is the pile deflection within the lateral spread phase.

20.5.4.3 Capacity of Deep Foundations During Liquefaction (Eslami and Maleki 2023)

The reduction in soil strength occurs during both full and limited liquefaction. In such conditions, the lateral support of the deep foundation from the soil is lost and therefore a considerable drop in shaft and toe capacity is observed during the liquefaction. Eslami and Maleki (2023) have assessed this phenomenon realizing the increase in excess porewater pressure (Δu) through some relevant parameters, including S_t, B_q, and u_2/q_t which have been introduced previously in Chapter 4.

The suggested approach recommends multiplying S_t to the shaft resistance, calculated before liquefaction, aiming to modify it. A similar approach is applied for the toe resistance, in which one of the decreasing indexes of $1 - B_q$ or $1 - u_2/q_t$

is used. Based on limited research work, utilizing the current method has resulted in up to an 80% decrease in the shaft and up to 20% loss of toe resistance. Total capacity loss due to liquefaction has been observed to be in the range of 30–50% through the current method.

Example 20.1 A closed-end pipe pile with outer diameter of 0.5 m and length of 20 m is considered. Evaluate the axial capacity prior and post-liquefaction via suggested methods applying Eslami and Maleki (2023) approach. The adjacent CPT profile is also given.

- Eslami and Fellenius (1997)
- ICP (Jardine et al. 2005)
- Fugro (Kolk et al. 2005)
- UWA (Lehane et al. 2005)
- German (Kempfert and Becker 2010)

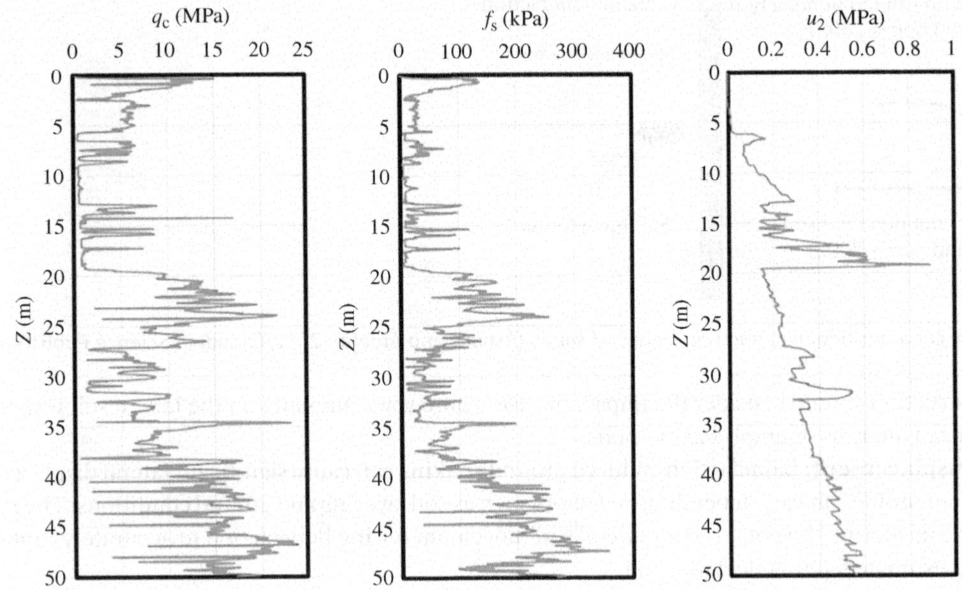

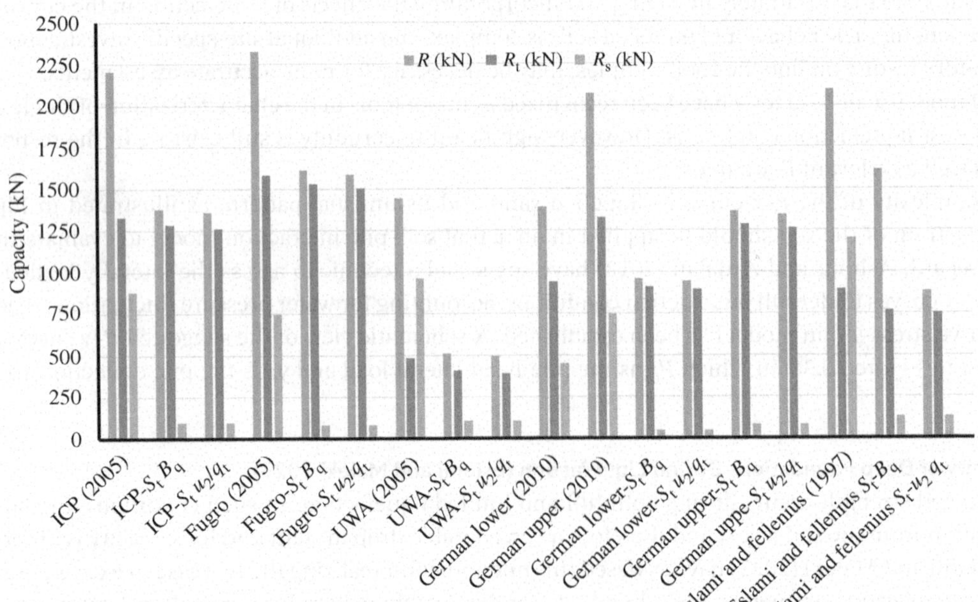

Method	*R* (kN)	R_t (kN)	R_s (kN)
ICP	2206.76	1327.58	879.18
ICP – S_tB_q	1379.18	1283.76	95.43
ICP – $S_t\ u_2/q_t$	1357.57	1262.14	95.43
Fugro	2326.18	1581.50	744.67
Fugro – S_tB_q	1609.95	1529.30	80.65
Fugro – $S_t\ u_2/q_t$	1584.20	1503.56	80.65
UWA	1440.48	481.60	958.88
UWA – S_tB_q	510.36	406.51	103.85
UWA – $S_t\ u_2/q_t$	494.46	390.62	103.85
German Lower	1388.83	938.69	450.14
German Upper	2071.19	1326.49	744.70
German Lower – S_tB_q	956.45	907.70	48.75
German Lower – $S_t\ u_2/q_t$	941.17	892.42	48.75
German Upper – S_tB_q	1363.36	1282.71	80.65
German Upper – $S_t\ u_2/q_t$	1341.76	1261.11	80.65
Eslami and Fellenius	2093.17	893.32	1199.85
Eslami and Fellenius S_tB_q	1609.95	764.80	130.30
Eslami and Fellenius $S_t\ u_2/q_t$	882.13	751.82	130.30

20.5.5 Pile Failure Theories in Areas of Seismic Liquefaction

The failure of piled foundations has been a recurrent issue in recent powerful earthquakes. In many instances, the failure of end-bearing piles in liquefiable regions during earthquakes is primarily attributed to the consequences of liquefaction-induced lateral spreading (Finn and Thavaraj 2001; Ishihara 1997; Hamada 1992b; Tokimatsu et al. 1996; Tokimatsu and Asaka 1998). The observed downslope deformation of the ground surface adjacent to the pile foundation lends support to this explanation. These theories regarding pile failure treat the pile as a beam element and posit that the lateral loads stemming from inertia and slope movement induce bending failure in the pile (Bhattacharya et al. 2003).

The concept of lateral spreading was initially introduced as a potential cause of pile failure during liquefaction in a report published by the National Research Council in 1985. This report asserts that "lateral spreading" is responsible for more damage during earthquakes than any other form of liquefaction-induced ground failure. This mechanism has gained acceptance as the leading explanation for pile failure in numerous earthquake events. Notably, the Japanese Code of Practice (JRA 1996) is the sole code that has incorporated this understanding of pile failure. Concerning pile–soil interaction, the current mechanism described by JRA posits that the soil pushes the pile, as depicted in Figure 20.35, ultimately leading to bending failure. This section elaborates on the theory of pile failure based on lateral spreading, as elucidated by Ishihara (1997) and Tokimatsu and Asaka (1998).

20.5.5.1 Ishihara's (1997) Concept of Pile Failure

In his Terzaghi Lecture at the Hamburg ICSMFE in 1997, Ishihara introduced the seismic loading on piles by elucidating the concepts of the "top-down effect" and "bottom-up effect," which are described as follows:

- At the initiation of seismic shaking, the inertia forces generated by the superstructure are transmitted to the top of the pile and subsequently to the underlying soil. Ishihara posits that, during the primary shaking, sandy soil deposits experience minimal softening due to liquefaction, and the relative displacement between the piles and the ground remains limited. Nevertheless, he theorizes that if the ground motion reaches a sufficiently high level, the induced bending moment in the piles could surpass a critical value, potentially leading to pile failure. Since this load originates from the inertia force of the superstructure, it is termed the "top-down effect." Ishihara concludes that observed pile failures in the upper sections following an earthquake can be attributed to this effect.

- Ishihara (1997) also states, "It has been known that the onset of liquefaction occurs approximately at the same time as the instant when peak acceleration is reached in the course of seismic loading, which often has an irregular time history." Consequently, the softened ground starts to experience horizontal movement on sloping terrain shortly after the onset of liquefaction. In this scenario, lateral forces are applied to the pile embedded in the ground, resulting in deformation of the pile in the direction of the slope. Ishihara assumes that seismic motion has already passed its peak, and subsequent shaking may persist with reduced intensity. Therefore, the inertia force transmitted from the superstructure becomes negligible. Under such loading conditions, the maximum induced bending moment in the pile may not occur near the pile head but at a lower depth within the pile, a phenomenon referred to as the "bottom-up effect" (Ishihara 1997).

20.5.5.2 Failure Theory Based on Tokimatsu et al. (1998)

In their work, Tokimatsu et al. (1998) provided a schematic representation of soil–pile structure interaction in liquefiable soil, as depicted in Figure 20.36. Their analysis is based on the following assumptions:

1) In the initial stage, denoted as Stage I in Figure 20.36, before the development of pore water pressure, the dominant influence on the pile is the inertia force generated by the superstructure.
2) As pore pressure in the liquefied soil increases, kinematic forces from the moving soil come into play, marking the transition to Stage II in Figure 20.33.
3) Toward the conclusion of the seismic shaking event, kinematic forces become more prominent and exert a significant impact on pile behavior, especially when permanent displacements occur in laterally spreading soil (Tokimatsu and Asaka 1998).

20.5.6 Review of Design Methods in the Major Codes of Practice

This section serves two main objectives:

1) To examine the design approaches for pile foundations in regions prone to seismic liquefaction, as outlined in prominent codes of practice, including the Japanese Highway Code of Practice (JRA 1996), NEHRP (USA code), and Eurocode 8. Particular emphasis will be placed on the underlying principles and assumptions that inform these design methods.
2) To provide a historical overview of the evolution of the Japanese Code of Practice (JRA 1996). This specific code is selected for discussion because it uniquely includes guidelines for the design of pile foundations in soil subjected to lateral spreading effects.

20.5.6.1 Development of Japanese Code of Practice (1972–1996)

Several bridges, including the Showa and Yachiyo bridges, suffered damage during the 1964 Niigata earthquake due to soil liquefaction. In response to these past incidents and the valuable insights gained, the "Seismic coefficient method" was introduced in the Highway Bridge Specification (JRA 1972) to address liquefaction effects, as noted by Yasuda and Berrill (2000). Subsequent revisions were made in 1980, leading to the emergence of an alternative approach known as the "Seismic deformation method." Following the extensive damage experienced by piled bridges in the aftermath of the 1995 Kobe earthquake, a comprehensive revision of the Highway Bridge Specification was undertaken, as detailed in Kawashima (2000). This updated edition incorporated a novel approach with provisions for assessing lateral spreading. This section provides an overview of the development of this code (Bhattacharya et al. 2003).

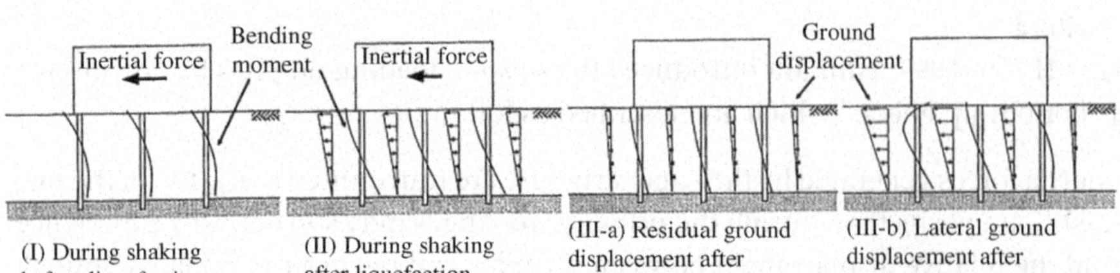

Figure 20.33 Schematic diagram showing the pile failure (after Tokimatsu et al. 1998/with permission of Elsevier).

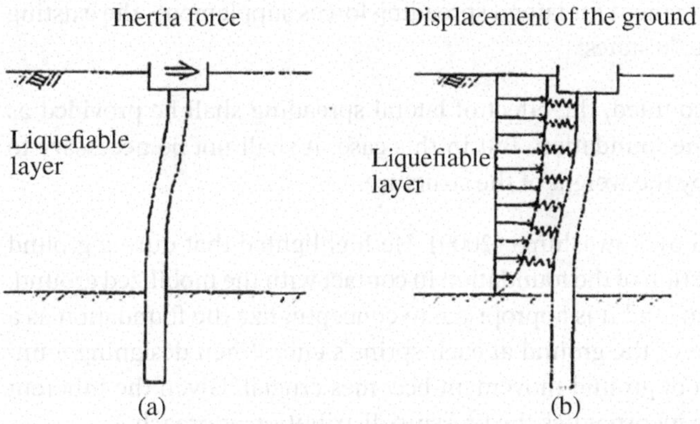

Figure 20.34 Two types of methods for static analysis: (a) seismic coefficient method and (b) seismic deformation method (Yasuda and Berrill 2000/Taylor & Francis Group (GeoEng conference held in Melbourne in 2000)).

20.5.6.2 Japanese Highway Bridge Specification (JRA 1971)

In the seismic coefficient method, the inertia force generated by earthquake shaking on the superstructure is transferred to the pile head, as illustrated in Figure 20.34a. It is assumed that in liquefied soil, the soil spring coefficient is zero due to the assumption that liquefied soil offers no resistance. Consequently, pile design entails consideration of the maximum bending moment occurring at the interface between the liquefiable and non-liquefiable hard layer. This design approach was adopted for all pile foundations in Japan from 1972 to 1980 (Bhattacharya et al. 2003). Another common approach is seismic deformation method as depicted in Figure 20.34b.

20.5.6.3 Japanese Highway Bridge Specification (JRA 1980)

Yasuda and Berrill (2000) documented that the Japanese engineers expressed concerns that the JRA (1972) approach resulted in excessive design requirements. This concern arose because it necessitated a large number of piles, prompting researchers like Iwasaki (1981) to seek methods for quantifying the resistance of liquefied soil by estimating the reduction in the bearing capacity of pile foundations due to liquefaction. To predict the decrease in bearing capacity caused by an increase in excess pore water pressure, model plate load tests were conducted. Shaking table tests were also undertaken to establish the relationship between pore water pressure (Lu) and the safety factor against liquefaction (F_L). By combining the results of these two types of tests, the relationship between F_L and D_E was determined, as presented in Table 20.2. This relationship was subsequently adopted in the Highway Bridge Specification JRA (1980) (Bhattacharya et al. 2003).

20.5.6.4 Japanese Highway Bridge Specification (JRA 1996)

The Japanese design specifications for highway bridges underwent a significant revision in response to the extensive bridge damage caused by the 1995 Kobe earthquake, primarily attributed to liquefaction-induced lateral spreading. Consequently, guidelines were introduced to consider the forces resulting from ground movement induced by liquefaction. The JRA 1996 code recommends that practicing engineers assess pile designs for bending failure, taking into account pressure distribution. This distribution was developed through a retrospective analysis of some piled bridge foundations along the Hanshin

Table 20.2 Reduction factor introduced in JRA (1980).

F_L (Factor of safety against liquefaction)	Depths (m)	D_E (Reduction factor)
$F_L \leq 0.6$	$0 < X \leq 10$	0
	$10 < X \leq 20$	1/3
$0.6 < F_L \leq 0.8$	$0 < X \leq 10$	1/3
	$10 < X \leq 20$	2/3
$0.8 < F_L \leq 1.0$	$0 < X \leq 10$	2/3
	$10 < X \leq 20$	1

Expressway that had experienced less severe damage. This assessment for lateral spreading forces supplements the existing requirements related to inertia. Nevertheless, the code on page 78 states:

> In a case where the effects of lateral spreading are accounted, the effect of lateral spreading shall be provided as horizontal force to study the seismic performance of the foundation. But in this case, it shall not be necessary to simultaneously account for the inertia force produced by the weight of the structure.

The rationale behind this design philosophy was elucidated by Kawashima (2000). He highlighted that during ground movement, the forces exerted by the moving ground act on a section of the foundation in contact with the mobilized ground. He posited that this is fundamentally a force-based mechanism, and it is appropriate to conceptualize the foundation as a structure supported by soil springs, prescribing the movement of the ground at each spring's end. When designing foundations based on such an analytical model, precise prediction of ground movement becomes crucial. Given the inherent difficulty in evaluating maximum ground movement, the code incorporates the pressure distribution approach.

Dynamic centrifuge tests were conducted to measure the earth pressure acting on a piled foundation behind a retaining wall during and after earthquake loading (Sato et al. 2001). The authors concluded that the JRA (1996) code tends to overestimate lateral pressure in both liquefiable and non-liquefiable layers. Similar findings were observed in the centrifuge test results reported by Dobry and Abdoun (2001), which showed pressures of a similar order of magnitude to those predicted by the JRA (1996) code.

This code has not been immune to criticism, as indicated by Haigh (2002). His centrifuge results demonstrated that JRA (1996) underestimates the lateral pressure distribution during the peak transient phase but provides a reasonable prediction for the post-earthquake residual flow. He also made the following comment.

> This also seems a strange design philosophy to adopt, as in almost all other cases, horizontal stresses are taken to be some multiple of vertical effective stresses (i.e., a K value) plus pore pressure, rather than a multiple of total stresses.
>
> (Bhattacharya et al. 2003)

20.5.6.5 Eurocode 8 (Part 5)

Eurocode 8 – Part 5 (EC8 2000), addressing geotechnical aspects of earthquake engineering, incorporates provisions relevant to the design of pile foundations in seismic regions. This code also provides a methodology for assessing liquefaction potential at a site, employing a modified version of the Seed and Idriss type chart that relates the cyclic resistance ratio (CRR) to corrected SPT-N values. The comprehensive design of piles under static loading, including axial and lateral loading, is covered by Eurocode 7 (1997) for geotechnical design. It should be noted that Eurocode 8 does not exclusively focus on the design of piles passing through liquefiable soil deposits (Madabhushi et al. 2009).

Eurocode advises designers to account for the design of piles against bending resulting from inertia and kinematic forces arising from the deformation of the surrounding soil. It further states:

> Piles shall be designed to remain elastic. When this is not feasible, the sections of the potential plastic hinging must be designed according to the rules of Part 1-3 of Eurocode 8.
>
> (Bhattacharya et al. 2003)

Eurocode 8 (Part 5) also states:

> "Potential plastic hinging shall be assumed for:
>
> - A region of $2d$ from the pile cap
> - A region of $\pm\ 2d$ from any interface between two layers with markedly different shear stiffness (ratio of shear moduli > 6)
>
> Where d denotes the pile diameter. Such region shall be ductile, using proper confining reinforcements."

20.5.6.6 National Earthquake Hazards Reduction Program (NEHRP) (2000) Code

Figure 20.35 shows an illustration from the NEHRP code of practice. The code on page 203 notes:

> If an unloaded pile was placed in the soil, it would be forced to bend similar to a pile supporting a building. The primary requirement is stability, and this is best provided by piles that can support their loads while still conforming to the ground motions and, hence the need for ductility.

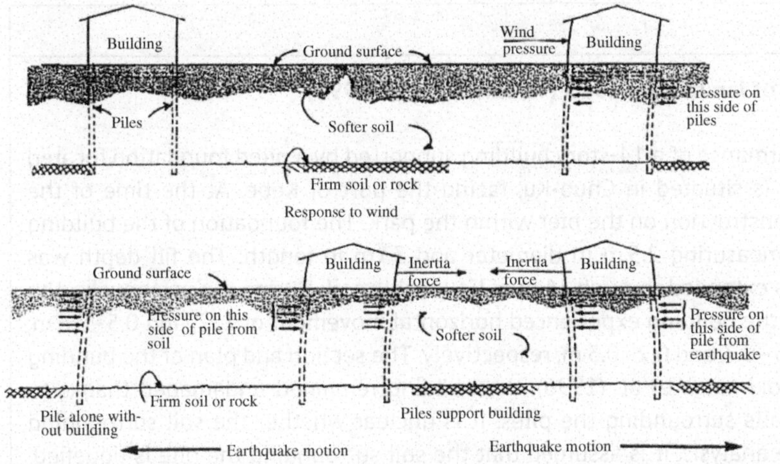

Figure 20.35 Response to earthquake, NEHRP (2000) code of practice (adapted from Hawkins and Ghosh 2000).

20.5.7 Case Histories of Pile Foundation Performance During Earthquakes

20.5.7.1 Excellent Performances of Pile Foundations

Case Study 20.3

10-Story Hokuriku Building During the 1964 Niigata Earthquake (Hamada 1992b)

This case study highlights the remarkable performance of a 10-story building, including a basement floor, supported by reinforced concrete piles. These piles have a diameter of 400 mm and a length of 12 m. The piles pass through a maximum of 5 m in liquefied soil layer and are embedded 7 m in dense soil (Hamada 1992b). The tops of the piles are integrally cast with the basement slab. The tips of the piles can be assumed to be fixed at the dense sand layer, while the top heads can be assumed to be free to translate laterally but anchored in direction.

Following the earthquake, no damage was reported to the superstructure. Notably, no cracks appeared on the walls, beams, or columns, and there was no structural inclination observed. The ground surface in close proximity to the building shifted by approximately 2 m (Hamada 1992b). Therefore, the building demonstrated strong performance despite severe lateral loadings. The schematic diagram of the building and the soil profile can be seen in Figure 20.36.

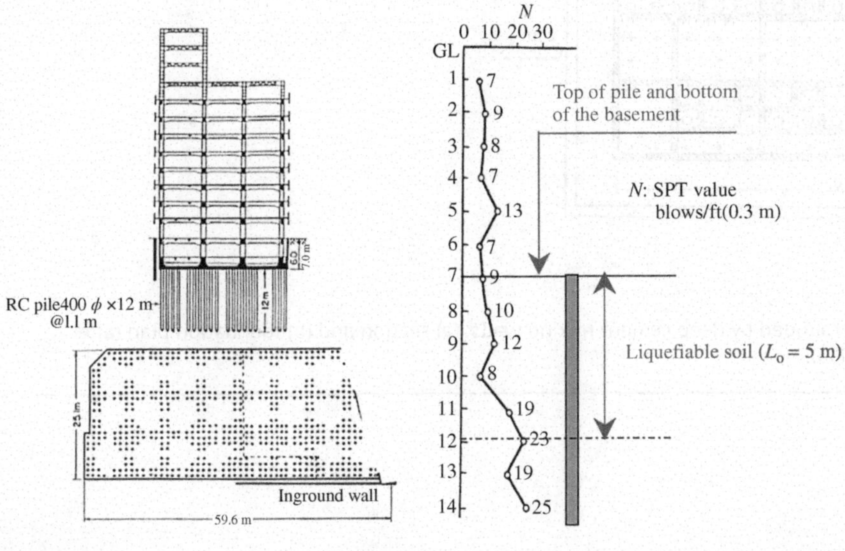

Figure 20.36 The structural arrangement of the 10-story piled building along with the soil profile (after Hamada 1992a; Bhattacharya et al. 2003).

Case Study 20.4

14-Story Building During the Kobe Earthquake of January 17, 1995 (Tokimatsu et al. 1996)

This case study highlights the remarkable performance of a 14-story building supported by a piled foundation located near a park (Tokimatsu et al. 1996). The park is situated in Chuo-Ku, facing the port of Kobe. At the time of the earthquake, this 14-story building was under construction on the pier within the park. The foundation of the building consists of cast-in-place concrete piles, each measuring 2.5 m in diameter and 33 m in length. The fill depth was 12.2 m, and the water table was 1.2 m. The piles extended through 11 m of liquefiable soil. During the earthquake, the quay wall on the west, south, and east sides of the building experienced horizontal movement of 1, 2, and 0.5–0.6 m, along with settlement ranging from 0.4–0.6, 0.5–0.7, and 0.2–0.3 m, respectively. The section and plan of the building are illustrated in Figure 20.37. According to Tokimatsu et al. (1996), the building remained undamaged thanks to the presence of deep cement mixing (DCM) walls surrounding the piles. It is unclear whether the soil surrounding the pile liquefied during the earthquake. In the analysis, it is assumed that the soil surrounding the pile is liquefied. Figure 20.40 shows a section of the foundation and the plan of the foundation system.

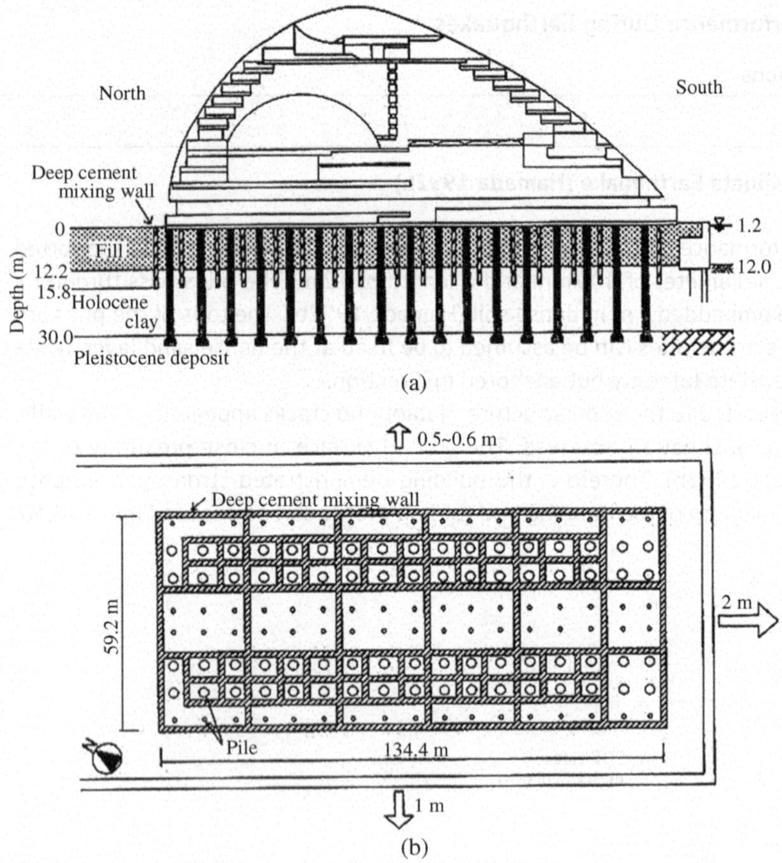

Figure 20.37 Building founded on piles surrounded by deep cement mixing walls: (a) section and (b) foundation plan (after Tokimatsu et al. 1996; Bhattacharya et al. 2003).

20.5.7.2 Poor Performance of Pile Foundations

Case Study 20.5

N.H.K. Building During the 1964 Niigata Earthquake (Hamada 1992b)

This case study illustrates the failure of piles in the N.H.K. building, a four-story R.C. building supported by reinforced concrete piles. These piles have a diameter of 350 mm and a length ranging from 11 to 12 m. Following the earthquake, an investigation was conducted on 74 piles, all of which exhibited similar damage patterns (Hamada 1992b). The piles failed at two distinct positions, specifically 2.5–3.5 m from the upper end and 2.0–3.0 m from the bottom, as depicted in Figure 20.38. While the exact location of the water table is not documented in Hamada's report, it is assumed to be at a depth of 1.7 m below ground level. Based on the soil investigation report, the liquefied layer is most likely the upper 11 m. Consequently, the pile was found at a depth of 1–2 m within a non-liquefiable layer. Thus, the portion of the pile within the liquefiable soil is approximately 9.3 m (11 m to 1.7 m) (Bhattacharya et al. 2003).

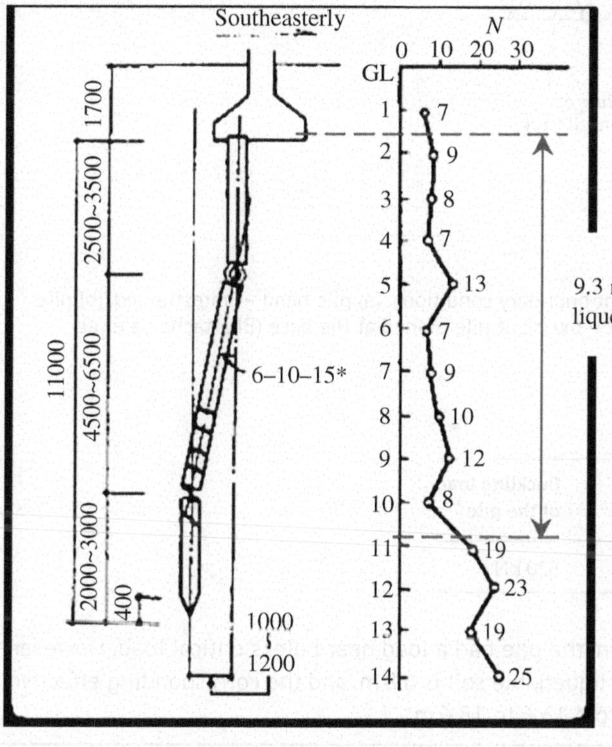

Figure 20.38 Failed piles of N.H.K building (after Hamada 1992b; Bhattacharya et al. 2003).

From Figure 20.41, it may be noted that the pile base rotated 10°–15°, which suggests that it was pinned. Figures 20.39a and b depict unrestrained and free to translate but restrained against rotation conditions for the pile head, respectively. Following Figure 20.39c, the effective length of the pile is twice the length of the pile in liquefiable soil. Table 20.3 quotes the parameters.

(Continued)

Case Study 20.5 (Continued)

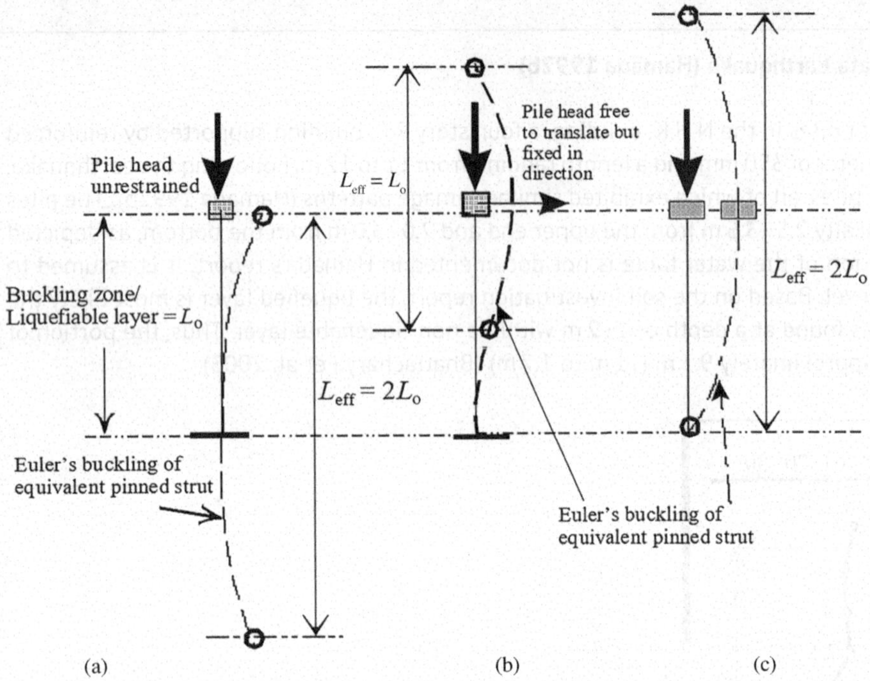

Figure 20.39 Concept of the effective length of the pile for different boundary conditions: (a) pile head – unrestrained, (b) pile head – free to translate but restrained against rotation, (c) pile head – fixed but pile pinned at the base (Bhattacharya et al. 2003/Cambridge University Press).

Table 20.3 Parameters of the pile, Case 3.

L_{eff} of the pile in liquefiable zone	r_{min} of the pile	Slenderness ratio	Conventionally allowable load	Buckling load of the pile
18.6 m	0.09 m	207	430 kN	520 kN

Thus, it can be seen in Figure 20.42 that during liquefaction, the pile had a load near Euler's critical load. However, if the water table is 4.2 m deep, then the length of the pile in liquefiable soil is 6.8 m, and the corresponding effective length would be 13.6 m. The effective length thus can vary from 13.6 to 18.6 m.

Case Study 20.6

Failure of Showa Bridge During the 1964 Niigata Earthquake (Takata et al. 1965)

Figure 20.40 illustrates the failure of the Showa Bridge, a 10-span structure spanning 307 m. Figure 20.41 provides additional insight, indicating that the piles extend through 10 m of potentially liquefiable soil and rest upon a stable non-liquefiable soil layer measuring 6 m in depth. The length of the pile within the liquefiable zone amounts to 10 m, while the portion in contact with air or water spans 9 m. Consequently, the length of the pile within the unsupported zone during liquefaction reaches 19 m. Based on the buckled shape observed, it is evident that the pile featured fixed-free boundary conditions, thereby rendering the effective length twice the length within the unsupported zone.

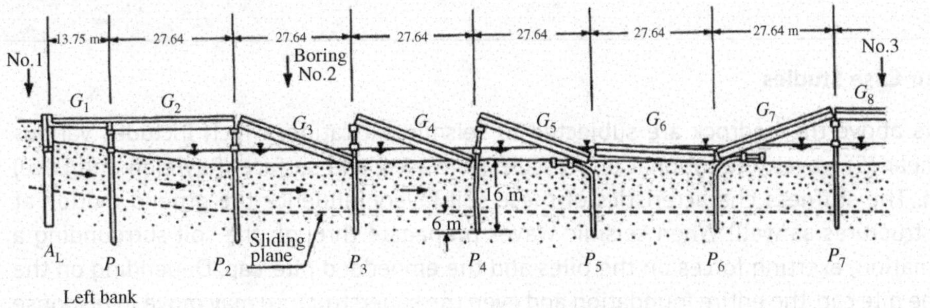

Figure 20.40 Schematic view of the girders fall-off in Showa bridge (Takata et al. 1965).

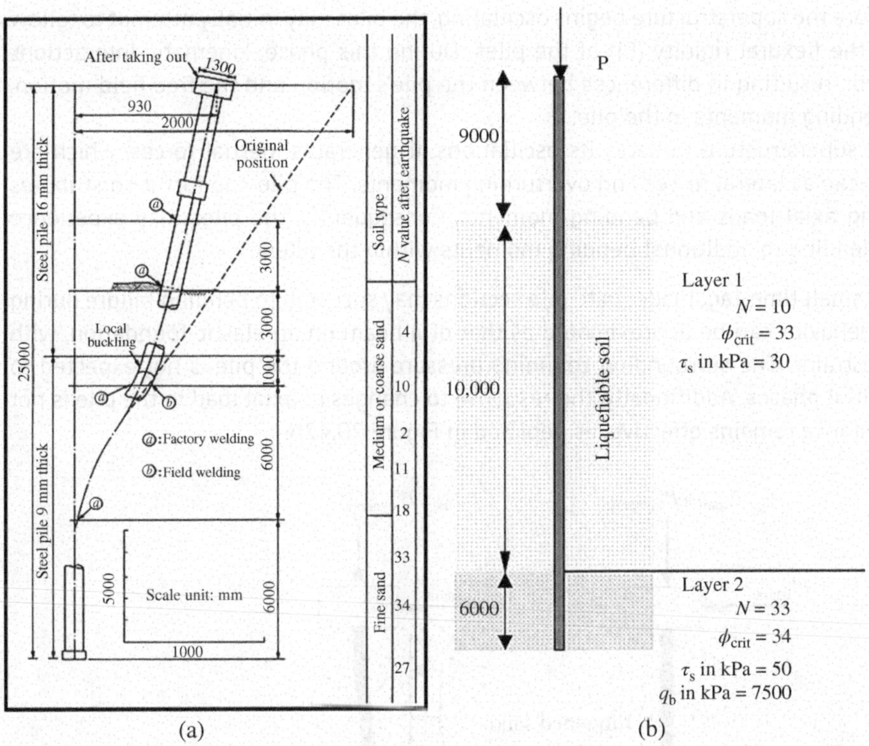

Figure 20.41 Piles of Showa bridge: (a) post-earthquake recovery (after Fukuoka 1966) and (b) schematic diagram of the pile and the soil profile (Bhattacharya et al. 2003/Cambridge University Press).

Structural Properties of Pile

Effective length of the pile (L_{eff}) = 2 × 19 = 38 m

Moment of inertia of the pile (I) = $\frac{\pi}{64}(0.609^4 - 0.591^4)\ \text{m}^4 = 7.63 \times 10^{-4}\ \text{m}^4$

Buckling load of pile = $\frac{\pi^2}{L_{eff}^2}\text{EI} = \frac{\pi^2}{38^2\text{m}^2} \times 210\ \text{GPa} \times 7.63 \times 10^{-4}\ \text{m}^4 = 1095\ \text{kN}$

Structural engineers generally demand a load factor of at least two against linear elastic buckling to allow for eccentricities and stiffness reduction due to yielding.

L_{eff} of the pile in liquefiable zone	r_{min} of the pile	Slenderness ratio	Conventionally allowable load	Buckling load of the pile
38 m	0.21 m	181	965 kN	1095 kN

(Continued)

Case Study 20.6 (Continued)

Comments Based on the Above Four Case Studies

During earthquakes, the soil layers above the bedrock are subjected to seismic excitation, which includes various types of incident waves such as shear (*S*) waves, dilatational or pressure (*P*) waves, and surface (Rayleigh and Love) waves, resulting in ground motion. The stiffness characteristics of these soil layers influence the ground motion at a particular site, impacting piled structures as well. When seismic waves propagate through the soil surrounding a pile, the soil layers undergo deformation, exerting forces on the piles and the embedded pile-cap. Depending on the rigidity of the superstructure and the pile cap, the entire foundation and even the superstructure may move in response to this soil–structure interaction. Consequently, the piles may experience two distinct phases of initial soil–structure interaction:

1) **Pre-superstructure Oscillation:** Before the superstructure begins oscillating, the piles may initially attempt to follow the soil motion, contingent upon the flexural rigidity (EI) of the piles. During this phase, kinematic interactions between the soil and pile may occur, resulting in differences between the pile's motion and the free field motion. These discrepancies can induce bending moments in the pile.
2) **Superstructure Oscillation:** As the superstructure initiates its oscillations, it generates inertial forces, which are transmitted to the pile via the pile-cap as lateral forces and overturning moments. The pile-cap further distributes these forces to the piles as varying axial loads and bending moments. Consequently, the piles may experience additional axial and lateral loads, leading to additional bending moments within the piles.

These two effects occur with only a small time-lag. Inadequate pile sections may succumb to bending failure during this process. At this stage, the pile's behavior can be approximated as that of a beam on an elastic foundation, with the soil providing sufficient lateral restraint. The surrounding confining pressure around the pile is not expected to decrease significantly during these initial phases. Additionally, the response to changes in axial load in the pile is not anticipated to be severe, as shaft resistance remains effective, as depicted in Figure 20.42b.

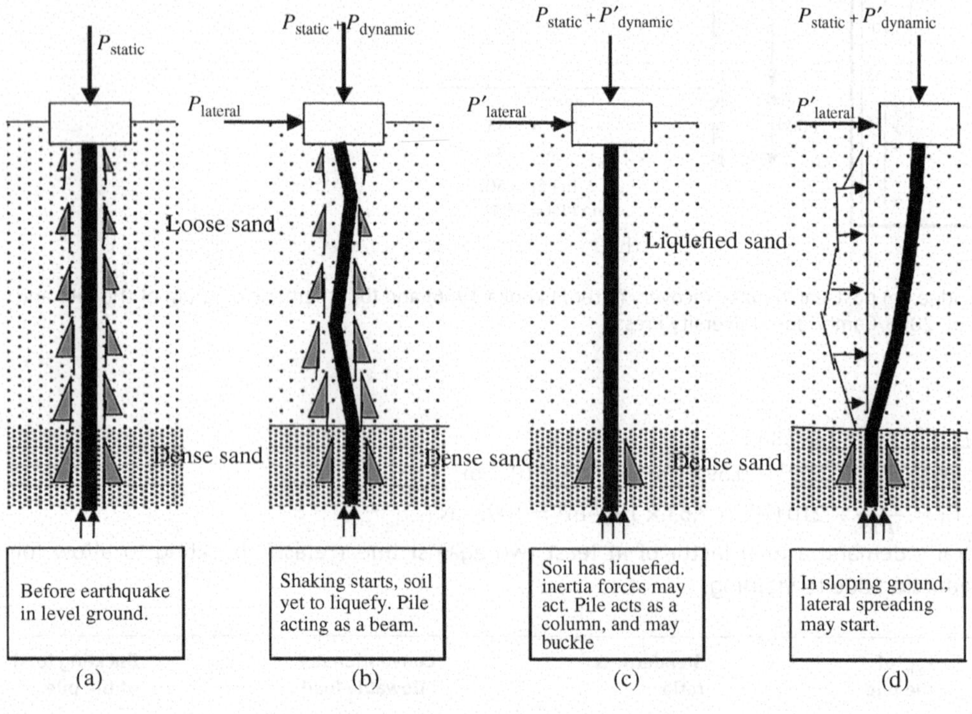

Figure 20.42 The time history of loading in the proposed failure mechanism; (a) Before earthquake in level ground, (b) Shaking starts, soil yet to liquefy. Pile acting as a beem, (c) Soil has liquefied. Inertia forces may act. Pile acts as a column, and may buckle, (d) In sloping ground, lateral spreading may start (Bhattacharya et al. 2004/Cambridge University Press).

However, in loose saturated sandy soil, as the seismic shaking persists, pore pressure will build up, and the soil may undergo liquefaction. The onset of liquefaction triggers distinctive changes in the pile's stress state:

- **Loss of Shaft Resistance:** The pile loses its shaft resistance within the liquefied layer, and axial loads are transferred downward to activate additional base resistance. If the base capacity is exceeded, settlement failure can occur.
- **Reduction in Soil Stiffness:** The liquefied soil loses its stiffness, causing the pile to act as an unsupported column. Piles with high slenderness ratios become susceptible to axial instability, potentially leading to pile buckling failure. This risk is heightened by lateral disturbing forces and the degradation of bending stiffness due to the onset of plastic yielding.

In sloping ground conditions, even if the pile withstands the aforementioned load conditions, it may encounter additional drag loads due to lateral soil spreading. In such scenarios, the pile's behavior may resemble that of a beam column (a column subjected to lateral loads), as illustrated in Figure 20.42d.

20.5.7.3 Summary

Several case histories of pile foundation performance have been analyzed, shedding light on the potential role of buckling as a failure mechanism. However, it is important to acknowledge the presence of numerous uncertainties associated with the parameters used in these case studies. Notably, the parameter rmin is devoid of uncertainty. Conversely, parameters like L_{eff} and P_{cr} rely on the accurate estimation of the liquefiable soil depth, which introduces an element of uncertainty. Additionally, the actual load acting on the pile, denoted as P, can deviate significantly from the allowable load determined through conventional procedures.

While buckling appears to be a plausible failure mechanism, it is crucial to consider that when a pile tends to buckle laterally, it must also engage in shearing the soil in front of it. This shearing action has the potential to cause dilation in the sheared zone, resulting in suctions and an increase in the stiffness of the soil–pile system. Consequently, the likelihood of buckling becoming a critical issue may be mitigated.

To gain a more comprehensive understanding and validation of the buckling mechanism, controlled physical modeling techniques such as dynamic centrifuge tests can be employed. These experiments allow for a more controlled exploration of pile behavior under seismic conditions and can help verify the potential for buckling as a significant failure mode.

20.5.8 Steps of the Design Method

This section of the chapter will outline the algorithm of the proposed design process.

Step 1: Use inclined piles to take any horizontal components of static load, e.g., retaining structures. Establish static axial loads and minimum eccentricity on all piles.

Step 2: From the site investigation data, identify any non-liquefiable crust, e.g., clay layer, that might be mobilized to translate laterally when $r_u = 1$ beneath it. From the slope of the ground and seasonal variation of the groundwater table, identify whether ground spreading is possible.

Step 3: Identify the depth of any liquefiable layer, e.g., loose- or medium-dense sands or silts. Based on the pile material and the liquefiable layer's depth, choose a pile section. Estimate lateral pressures on the piles in these zones. Ignore skin friction and lateral support in these zones. Assume that the piled structure does not move, and that the crust does move, creating passive earth pressures on walls and ultimate lateral pressures on piles. Calculate shear force H.

Step 4: For the chosen pile diameter, estimate the length of the pile required beneath the liquefiable zone to carry the axial load. Ignore the skin friction in the liquefiable layer. From economic considerations, choose the number of piles and the length. However, a minimum depth of embedment is necessary in the hard layer for the fixity of the pile, i.e., to create moment restraint.

Step 5: Estimate the depth of fixity of the pile in the non-liquefiable hard layer based on the standard conventional procedure for partially embedded pile, e.g., Davisson and Robertson (1965). Create a pile cap with full moment fixity at the pile heads wherever possible. Estimate the equivalent length of the pile for Euler's buckling considering the restraints offered by the pile cap and the zone of embedment beneath the liquefiable soil layer. Keep the slenderness ratio of the pile in the unsupported layer below 50.

Step 6: Assume possible collapse mechanisms, and estimate the lateral load required to form the mechanisms. The reduction of plastic moment capacity due to axial load is to be taken into consideration. Calculate the safety factor for forming plastic collapse mechanisms against lateral loads. If the pile section is inadequate, increase the plastic moment capacity of the section. The options are using higher-grade concrete or increasing the thickness of the steel pile section. If that is not feasible, a tubular steel pile filled with concrete can be used, or the diameter of the pile may be increased (Bhattacharya et al. 2003).

Example 20.2 An eight-story building is considered as shown below to explain the proposed design method. It is founded on liquefiable soil with a depth of 10 m and underlying hard soil with a modulus of 40 MN/m^3. Grade of concrete = 25 MPa. In this example, the lateral loads on the pile are estimated using (JRA 1996). Detailed calculations are omitted in the interests of brevity.

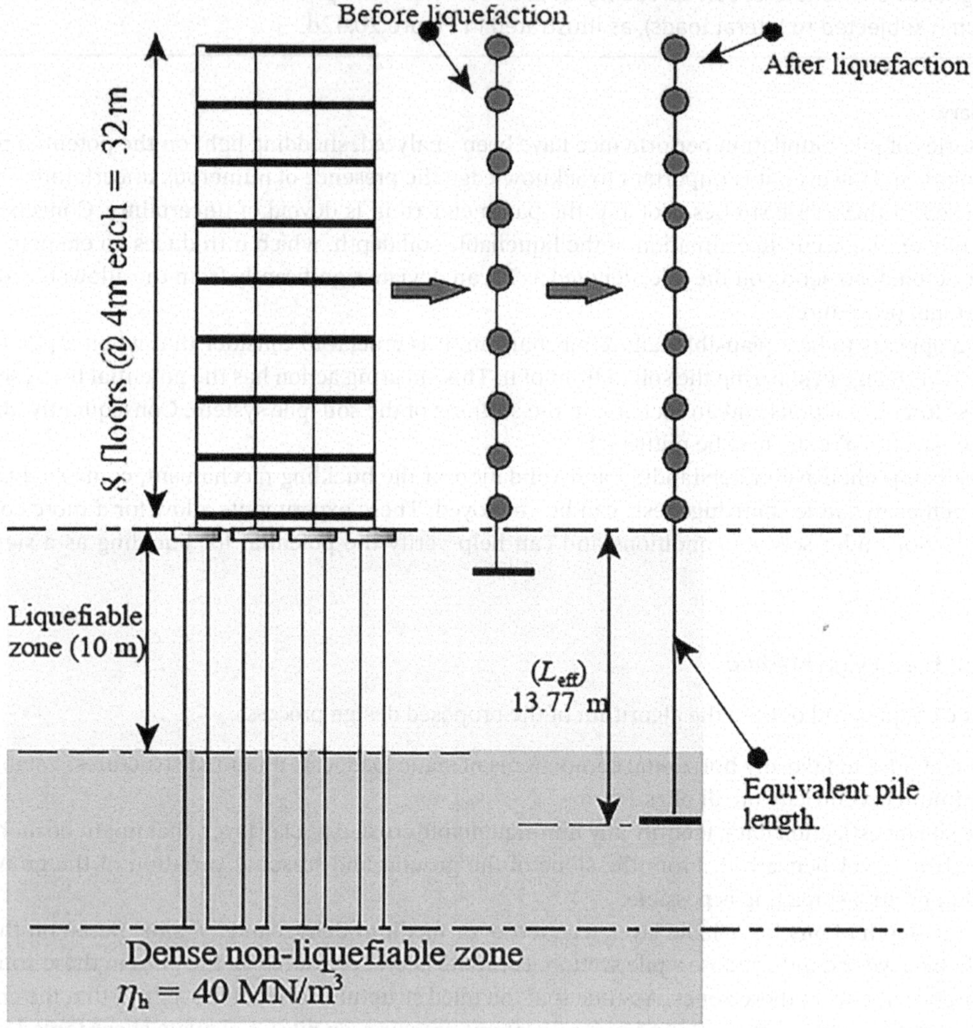

Design Loads

The following data are extracted from detailed structural analysis and design calculations. The typical axial load at each column base equals 1800 kN (dead load + live load + appropriate dynamic earthquake loads), and the typical lateral inertia load at each column base is 50 kN (from base shear) for the design earthquake.

Stability Considerations

The initial guess of the pile diameter = 1.1 m

Depth of fixity = 3.77 m.

Unsupported length = (10 + 3.77) m = 13.77 m
Effective length = 13.77 m (From the boundary conditions, Figure 20.42)
Slenderness ratio = 49.7 (less than 50, SAFE against buckling).

Length of the Pile for Serviceability Limit State

Here the shaft resistance in the top 10 m is ignored. For the type of soil the

- End bearing capacity of the pile is given by = 2450 D^2 kN, where D = dia. of the pile
- Shaft capacity in Layer II (non-liquefiable layer) = (248 L) kN, where L = length of the pile in non-liquefiable hard layer.

Choosing 1.1 m dia. pile:
End bearing resistance = 2960 kN and shaft resistance:

- For 6 m embedment in hard layer = 1488 kN
- For 10 m embedment in hard layer = 2480 kN
- For 15 m embedment in hard layer = 3720 kN

Depth of fixity = 3.77 m. Allowing a "Factor of Safety" of 1.5 the minimum embedment = 6 m.
Choosing 1.1 m dia., 20 m of total pile length. It gives ultimate load of (2480 + 2960) kN = 5440 kN. Allowable load allows a safety factor of 3 = 5440/3 = 1813 kN.
Thus, one pile (1.1 m) is required per column.

Check Against Mechanism Formation

For M25 grade concrete, assuming yield strength in bending = 11.2 MPa (0.446 f_{ck})
M_P of the section of the pile in absence of axial load = $\left(\frac{1.1^3}{6}\right) \times 11.2$ MPa = 2485 kN.m
P = 1800 kN and P_Y = 15,918 kN (assuming yield strength in compression = 16.75 MPa)
M'_p = available plastic moment capacity = 96% of MP = 2385 kN.m
Considering a single pile–pile cap connection and assuming there is no non-liquefied crust, the lateral load required to form the mechanism is estimated below.

$$\text{Collapse Load } (H_{\text{collapse}}) = 6\left(\frac{M'_p}{L_{\text{eff}}}\right) = 1039 \text{ kN}$$

The lateral spreading forces should not exceed the collapse load (1039 kN)
Calculation of Lateral Spreading Forces Following JRA 1996
For 10 m of liquefiable layer: Lateral pressure at the bottom of the pile = 30% of overburden pressure = 0.3 × 10 m × 18 kN/m³ = 54 kPa.
Total load = 297 kN
Thus, factor of safety against residual lateral spreading loads = (1039/297) = 3.49
For seasonal variation of water table the depth of non-liquefied crust can be estimated and thus the corresponding factor of safety.

20.6 Scour

20.6.1 Background and Definitions

Bridge scour refers to the erosion-induced loss of soil around bridge supports due to the flow of water. The total scour experienced at a highway crossing encompasses three primary components:

1) **Long-term Aggradation and Degradation of the Riverbed:** Aggradation signifies the gradual accumulation of sediment along the longitudinal profile of a channel bed, while degradation entails the progressive lowering of the channel bed due to natural erosion. These processes occur independently of the presence of the bridge.
2) **Contraction Scour at the Bridge:** Contraction refers to the narrowing of the channel caused by factors such as bridge construction. Contraction scour results from the increased flow velocity stemming from the channel's constriction, leading to the removal of material from the bed and banks across a substantial portion of the channel width.

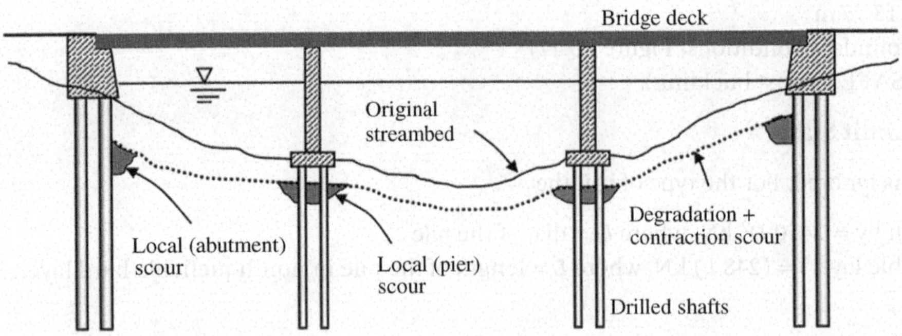

Figure 20.43 Components of scour affecting bridge support on deep foundations (FHWA 2010/U.S. Department of Transportation/Public Domain).

3) **Local Scour at Piers or Abutments:** Local scouring arises from the erosion of material in the vicinity of obstacles obstructing water flow. It occurs due to the localized acceleration of flow and the generation of vortices caused by these obstructions, predominantly at bridge piers and abutments.

Total scour encompasses the cumulative impact of long-term degradation, contraction scour, and local scour. Figure 20.43 visually illustrates the scour components that can affect bridges supported by deep foundations. Scour effects that necessitate consideration in drilled shaft design encompass:

1) Alterations in subsurface stress.
2) Reductions in embedment leading to variations in geotechnical resistances.
3) Potential modifications in the response of the structure to changes in the performance of the foundation.

It is noteworthy that channel migration, though not categorized as a specific scour component, denotes the lateral movement of an alluvial river channel across its floodplain. This phenomenon primarily results from the interplay between bank scour (erosion) and point bar deposition over time. Channel migration can have comparable scour effects on foundations, entailing either soil removal or buildup at pier and abutment locations (NHI Course 2018).

20.6.2 Design Philosophy for Scour

The Federal Highway Administration (FHWA) document titled "Evaluating Scour at Bridges" (Richardson and Davis 2001), often referred to as HEC-18, provides practical methodologies for scour analysis and general bridge design procedures to address the risk of scour. The following paragraphs outline the fundamental approach to bridge scour as outlined in HEC-18, aligning with AASHTO specifications.

For new bridge foundations, it is imperative to design them with the capacity to withstand the effects of scour resulting from hydraulic conditions associated with floods larger than the intended design flood. The selection of the hydraulic design flood is a risk-based decision, considering factors such as the bridge's importance, safety, reliability, and potential consequences of failure. Once the frequency of the hydraulic design flood is determined, both the minimum scour design flood and the scour design check flood are chosen based on Table 20.4. In this table, Q_D represents the hydraulic event (flood) with a return interval equal to D in years (e.g., Q_{25} signifies the flood with a 25-year return interval). It is noteworthy that both scour floods exceed the magnitude of the hydraulic design flood.

Table 20.4 Flood frequencies for hydraulic design, scour design, and scour design check floods (based on Table 2.1, HEC-18).

Hydraulic design flood frequency, Q_D	Scour design flood frequency, Q_D	Scour design check flood frequency, Q_D
Q_{10}	Q_{25}	Q_{50}
Q_{25}	Q_{50}	Q_{100}
Q_{50}	Q_{100}	Q_{200}
Q_{100}	Q_{200}	Q_{500}

This approach adheres to the principle that bridge design should account for a higher level of scour than the hydraulic design flood because exceeding the latter, which has a reasonably high likelihood over the bridge's service life, could result in more extensive scour and potential bridge failure. Moreover, designing for a higher scour level than the hydraulic design flood ensures redundancy in the bridge's performance after the hydraulic event transpires.

Bridges must be designed to withstand the worst conditions arising from the scour design flood without failure. Consequently, bridge foundations are thoroughly examined to ascertain that they can withstand scour resulting from the scour design check flood, an extreme event. It is important to emphasize that the scour check flood is an exceptional occurrence, whereas the scour design flood is not. The scour from the scour design flood is considered for all strength and service limit states for which the bridge is designed.

The approach to designing for scour is founded on several key principles:

1) **Interdisciplinary Design Team:** The foundation design process should involve a collaborative team of engineers possessing expertise in hydraulic, geotechnical, and structural engineering.
2) **Hydraulic Studies:** Comprehensive hydraulic studies of bridge sites are an integral part of the design process. These studies should encompass two primary aspects: (i) optimizing the sizing of the bridge waterway opening to minimize adverse impacts on upstream and downstream landowners while ensuring secure foundations, and (ii) conducting thorough scour calculations at the bridge foundations. The scope of the hydraulic analysis should be tailored to match the complexity of the situation, the significance of the highway, and the potential consequences of failure.
3) **Consideration of Existing Formulas:** When utilizing existing formulas to estimate scour, it is crucial to acknowledge the limitations and gaps in our current knowledge. The interdisciplinary team should exercise engineering judgment by comparing the results obtained from scour calculations with available hydrologic and hydraulic data, as well as on-site conditions, to arrive at a sound and prudent design. This data should encompass the following factors.
   a) The performance of existing structures during previous flood events
   b) The impacts of flood discharge regulation and control
   c) Hydrologic characteristics and historical flood patterns of the stream and similar watercourses
   d) The presence of redundant structural elements in the bridge design
   e) Historical trends in channel migration

It is essential to recognize that while occasional damage to highway approaches resulting from rare floods can be swiftly repaired to restore traffic service, a bridge that collapses or sustains major structural damage due to scour can pose substantial safety risks to motorists.

Additionally, it can lead to significant societal consequences and economic losses over an extended period. Beyond the expenses incurred by Departments of Transportation (DOTs) for bridge replacement or repair, detour construction, and maintenance, there can be substantial costs to communities and entire regions, including extended travel times, inconvenience, and lost business opportunities. Consequently, it is justifiable to establish a higher hydraulic standard for bridge foundation design to withstand scour compared to the typical requirements for sizing bridge approaches (NHI Course 2018).

20.6.3 Analysis and Prediction of Scour

The comprehensive approach to addressing scour at bridge sites comprises three distinct stages of assessment and analysis:

1) **Stream Stability and Geomorphic Assessment:** This initial stage involves conducting a stream stability and geomorphic assessment in accordance with the guidelines outlined in the publication "Stream Stability at Highway Structures," commonly referred to as HEC-20 (Lagasse et al. 2012).
2) **Scour Analysis and Foundation Design:** The second stage encompasses a detailed scour analysis following the procedures outlined in HEC-18 (Arneson et al. 2012). This stage also involves the design of foundations and is particularly pertinent to drilled shaft design. Figure 20.44 illustrates a flowchart depicting the scour analysis process. In cases where existing bridges are identified as critically susceptible to scour, it may be necessary to develop a new bridge design. Alternatively, the process may proceed to Stage 3.
3) **Development of Countermeasures Plan:** The final stage revolves around devising a comprehensive plan of action aimed at implementing countermeasures to address both bridge scour and stream instability. It is important to note that this stage primarily pertains to existing bridges in need of mitigation measures. It is not applicable to new bridge designs and is not directly relevant to drilled shaft design considerations.

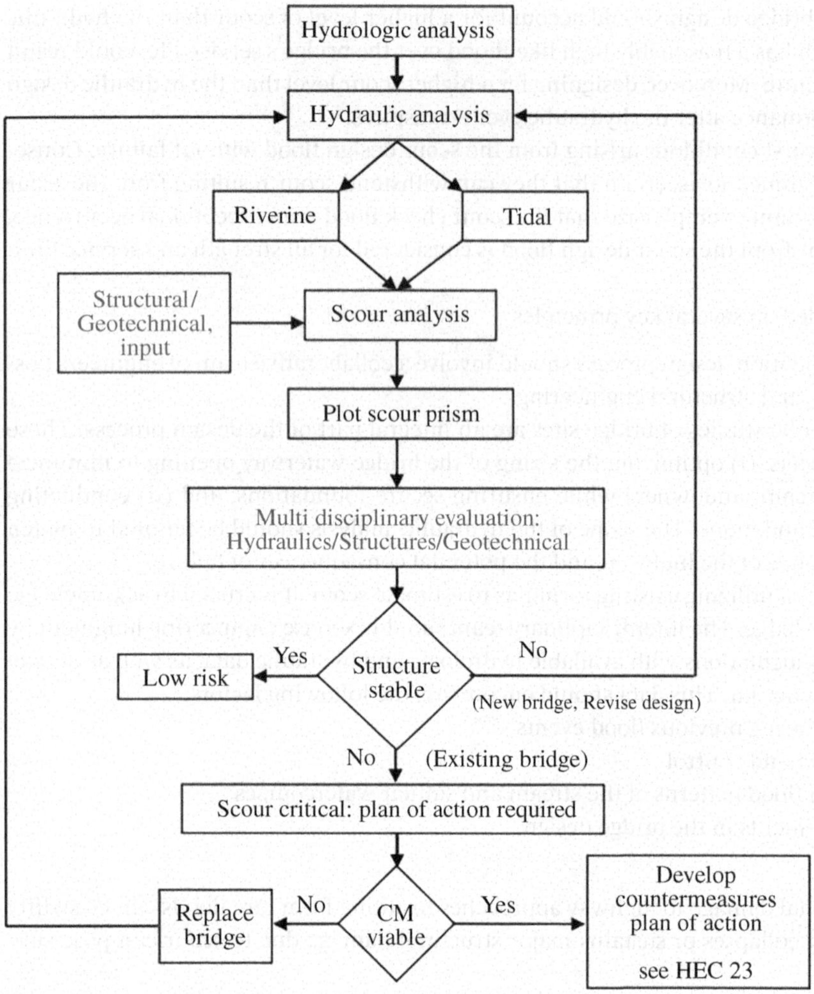

Figure 20.44 Flow chart for hydrology, hydraulics, and scour analysis by HEC-18 (adapted from Arneson et al. 2012).

By systematically progressing through these three stages, transportation agencies and engineering teams can effectively evaluate, analyze, and address scour-related challenges at bridge sites, ensuring the safety and integrity of critical infrastructure.

Figure 20.44 underscores the imperative collaboration among hydraulic, geotechnical, structural, and civil engineers in addressing scour-related challenges. While hydraulic and hydrologic analyses are typically spearheaded by hydraulic specialists, they necessitate valuable insights from structural and geotechnical experts. Geotechnical information crucial at this stage encompasses subsurface profiles, derived from borings and geophysical tests, classification of geomaterials using the Unified Soil Classification System, and an assessment of subsurface unit erodibility. Structural contributions encompass critical details such as the quantity, placement, and geometry of bridge piers and abutments. Civil engineers play a pivotal role by providing data on station limits for approach embankments, as well as the size and location of any hydraulic structures integrated into these embankments. The development of a scour prism for each pier and abutment, based on scour analysis, is paramount.

The multidisciplinary evaluation of the bridge is a collaborative effort, encompassing geotechnical considerations related to strength, serviceability, and extreme event limit states of deep foundations. If this evaluation confirms bridge stability, the trial design is deemed low risk and thus acceptable. However, should issues arise, modifications can be made to the design, or countermeasures may be implemented to mitigate scour effects. These countermeasures might involve lengthening the bridge, adjusting bridge spans, or revising the number, size, and arrangement of drilled shafts within each foundation. In general, foundations are designed to withstand predicted scour, rather than relying solely on auxiliary measures like rip-rap armor protection, as the long-term effectiveness of such measures is often unreliable.

The detailed methods and equations for calculating total scour, comprising degradation, contraction scour, and local scour, are comprehensively outlined in HEC-18 and are not reiterated here. Of particular relevance to drilled shaft design is the scour prism, a graphical representation quantifying the total volume of material removed by scour. This serves as a crucial output arising from the scour analysis (NHI Course 2018).

20.6.4 Application to Drilled Shaft Design

AASHTO Specifications (2017) require the evaluation of bridge foundations for two scour conditions: (i) the design flood scour condition is assumed for foundation strength and service limit state evaluations, and (ii) the check flood scour condition is assumed for extreme limit state evaluations. The commentary under Article 10.5.5.2.1 states:

> Scour design for the design flood must satisfy the requirement that the factored foundation resistance after scour is greater than the factored load determined with the scoured soil removed. The resistance factors will be those used in the Strength Limit State, without scour.

Accordingly, the resistance factors listed in Table 8.4 (NHI Course 2018) are utilized for the analysis of strength limit states in foundations subjected to scour from the scour design flood. In the evaluation of the Extreme Event II limit state, the resistance factors provided in this table for Extreme Event II are applied in conjunction with scour resulting from the scour design check flood.

For every assessment of strength or service limit states, it is assumed that all material within the scour prism above the total scour line has been removed and is no longer available for axial or lateral support. The removal of soil also leads to changes in subsurface stress, potentially affecting the foundation's resistance to both axial and lateral loads. The extent of stress changes is contingent upon whether scour occurs over a wide area (degradation and contraction scour) or a confined area (local scour). In cases where local scour is confined to a small region around the shaft, the streambed associated with degradation and contraction scour exerts a more substantial influence on the stress conditions along the embedded length of the drilled shaft.

Figure 20.45 illustrates a single drilled shaft at bridge support, depicting the scour prism, as indicated by the shaded region, resulting from scour analysis. The initial streambed elevation is represented by Point A. Subsequently, the streambed is lowered to Point B, based on the forecasted degradation and contraction scour. The analysis predicts a local scour hole around the pier, with the depth of the local scour hole, denoted as y_s, computed following methods outlined in HEC-18. For practical applications, it is recommended to use a top width value (w) of 2.0 times the local scour depth on each side of the pier (Arneson et al. 2012).

Vertical stress at any depth along the shaft can be estimated as follows.

- At the top of the embedded length of the shaft (Point C), the vertical stress is considered zero.
- At a depth of embedment equal to or greater than 1.5 y_s, assume that the vertical stress is primarily influenced by the streambed elevation at Point B.
- Assume a linear variation in vertical stress over the depth of 1.5 y_s. In other words, the effect of local scour on stress diminishes linearly over a depth equal to 1.5 times the scour hole depth, y_s.

20.6.5 Effects of Scour on Axial Resistance

Scour conditions are incorporated into the analysis of side resistance in cohesionless soils using the beta method, as introduced in Chapter 15.

Considering that vertical effective stress σ'_v is one of the crucial parameters affecting side resistance, it is important to account for changes in vertical stress along the section of the drilled shaft that is embedded in cohesionless layers. This concept is illustrated in Figure 20.45.

In cohesive soils, the α-method is typically used to calculate nominal side resistance. For these soils, changes in effective overburden stress resulting from scour are assumed to have no significant impact on the undrained shear strength and, consequently, the unit side resistance remains unchanged. Similarly, the effects of scour are presumed to have no substantial effect on the strength and nominal axial resistances provided by rock.

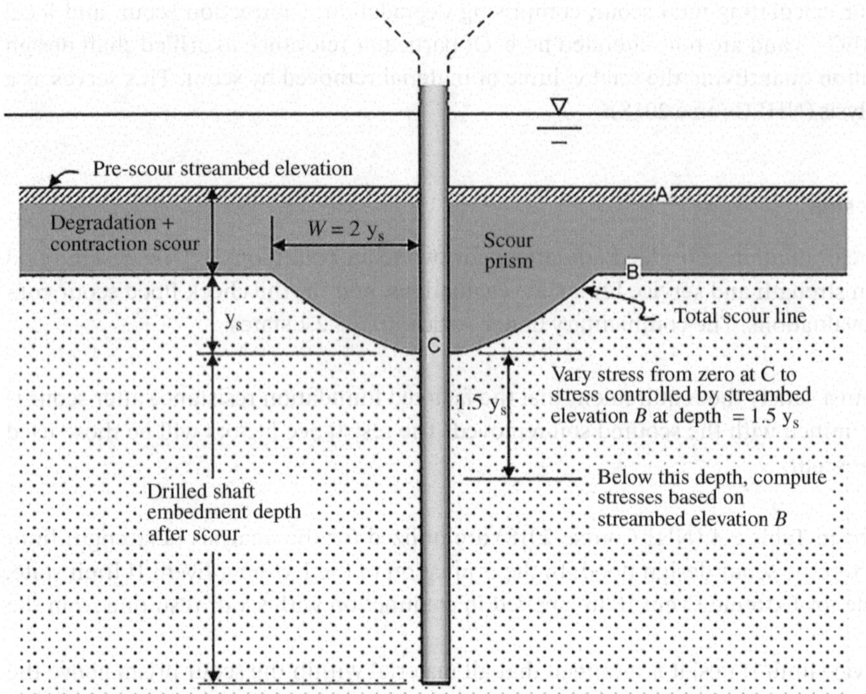

Figure 20.45 Illustration of scour prism and effects on drilled shaft.

Additional issues to be considered by foundation designers in connection with scour include:

- When designing drilled shafts, it is essential to account for column action, particularly in cases where the unsupported shaft length increases due to scour.
- In certain situations, local scour holes at piers and abutments may overlap, leading to an indeterminate scour depth that could be deeper. For practical applications, it is recommended to use a top width value of 2.0 times the local scour depth on each pier side when calculating local scour dimensions.
- To minimize obstruction to flood flows and reduce the risk of local scour, it is advisable to position the top of a shaft-supported footing or cap below the streambed at a depth equal to the estimated long-term degradation and contraction scour depth. In cases where drilled shafts may be exposed to erosion from river or tidal currents, even lower footing elevations may be desirable. However, for deep water situations, it might be more cost-effective to place the pile cap at or above the mudline or waterline and design the foundation accordingly.
- Stub abutments situated within the embankment should have foundations supported by drilled shafts or piles extending below the thalweg elevation, which includes accounting for long-term degradation and contraction scour in the bridge waterway. This ensures structural integrity in case the thalweg shifts and the bed material around the abutment foundations scours down to the thalweg elevation. Thalweg refers to the lowest point of the streambed.
- When estimating contraction and local scour, it is essential to consider the foundation cap (if present) and any pier protection system attached to or surrounding the pier.

20.6.6 Laterally Loaded Pile Behavior Under Scour Conditions (Considering the Stress History)

Scour can lead to the removal of soil around bridge foundations, potentially diminishing the capacity and safety of the bridge. Many bridge failures have been linked to scour events, underscoring the importance of evaluating the impact of scour on bridge stability. Typically, assessing the performance of bridge piles under scoured conditions involves the straightforward removal of soil around the piles, with possible alterations in the stress history of the remaining soil often disregarded. However, in reality, as scour removes soil from above the remaining soil, the latter undergoes an unloading process that increases the over-consolidation ratios.

In this study, we took into account the stress history of soft clay when analyzing laterally loaded piles using the *P–y* method. To accommodate the stress history effect, we modified the conventional *P–y* curves for the soft clay to assess the undrained responses of laterally loaded piles at various scour depths. To validate our approach, we conducted a case study based on Matlock's work from 1970, comparing computed results when considering and when ignoring the stress history effect. Our findings revealed that accounting for the stress history effect led to a reduction in the ultimate soil resistance ranging from 21.8% to 32.7%. Disregarding the stress history of the soft clay resulted in an unconservative design for laterally loaded piles under scour conditions.

This study delved into the effects of scour on the responses of laterally loaded piles in soft clays with horizontal mudlines, all while considering changes in the stress history of the remaining soils. The conclusions drawn from our analyses are as follows:

1) Scour had a negligible impact on altering the effective unit weight of the remaining soil post-scour. Consequently, when accounting for the stress history effect, changes in the effective unit weight can be safely disregarded. However, we observed that alterations in the undrained shear strength of the soil were substantial and played a crucial role in the variations in responses of laterally loaded piles, depending on whether the stress history effect was considered or neglected. The over-consolidation ratio (OCR) emerged as the primary influencing factor affecting changes in undrained shear strength.
2) The stress history effect, particularly when the scour depth equaled 5 and 10 times the pile diameter, resulted in a notable decrease in ultimate soil resistance – ranging from 21.8% to 32.7% – compared to scenarios where stress history effects were disregarded. This reduction in soil resistance translated into increases of 13–15% in lateral pile-head deflections and 5–7% in maximum bending moments within the pile.
3) Neglecting the stress history of the soft clay led to an analysis that was overly optimistic for laterally loaded piles under scour conditions.

Typically, conventional scour analyses for pile foundations involve simply excavating soils around the foundation, without considering changes in soil properties and the stress state of the remaining soils after scour. This approach may be overly optimistic for pile foundations in soft clay, tending to underestimate both lateral deflections and bending moment demands. We strongly recommend that when conducting a scour analysis in soft clay settings, practitioners take into account the implications of stress history effects and soil property changes. This paper outlines an approach to adapt existing *P–y* curves to incorporate both soil property changes and stress history effects, providing valuable guidance to engineers dealing with pile foundations in soft clay susceptible to scour (Lin et al. 2014).

20.6.7 Scour Effects on the Lateral Behavior of a Large-Diameter Monopile in Soft Clay

This paper presents a numerical investigation aimed at exploring the scour effect on the lateral behavior of monopiles. It utilizes an advanced hypoplastic model that takes into account the influence of stress history on both soil stiffness and strength. Additionally, the suitability of the *P–y* method proposed by Lin et al. (2014), which incorporates the stress history effect, is evaluated. From this study, the following conclusions can be derived:

1) Scour has a significant impact on increasing the over-consolidation ratio and reducing the undrained shear strength of the remaining soil. This leads to substantial differences in pile behavior when considering or neglecting the stress history effect.
2) As the scour depth increases from 0.2 *D* to 0.5 *D*, the consideration of the stress history effect results in a maximum reduction of 30.1–39.8% in ultimate soil resistance and a reduction of 20.7–25.8% in the initial stiffness of the *P–y* curves. These reductions translate to a 13–49% increase in lateral pile-head deflection and a 1–2% increase in maximum bending moments within the pile. Neglecting the stress history effect leads to an analysis that is overly optimistic for laterally loaded piles under scour conditions.
3) The failure mechanism of the large-diameter monopile comprises two components: wedge failure at a shallow depth and rational soil flow at a deeper level. Disregarding the stress history effect results in an underestimation of the width and depth of the wedge failure zone, while overestimating the location of the rotational soil flow zone.
4) The modified *P–y* curves proposed by Lin et al. (2014) for considering the stress history effect tend to overestimate the percentage reduction in ultimate soil resistance. Consequently, the Lin *P–y* method is likely to overestimate the percentage difference in pile-head deflection between scenarios considering and ignoring the stress history effect (He et al. 2019).

20.6.8 Case Studies

Case Study 20.7

Railway Bridge Pier Repair Work in Columbus, Ohio, United States (Takefumi and Takayuki 2018)

One of the rail bridge foundations in the Columbus, Ohio, area required immediate repair due to scouring around its base (Takefumi and Takayuki 2018). Initially, the pier foundation relied on timber piles with a concrete footer. The repair project involved (i) installing a sheet pile cofferdam around the foundation; (ii) driving micro-piles around the original footer; and (iii) constructing a larger concrete encasement footer. Detailed information can be found in Figures 20.46 and 20.47. Figure 20.48 illustrates the soil conditions, complete with *N*-values and the horizontal placement of the 25-foot (7.5 m) long sheet piles.

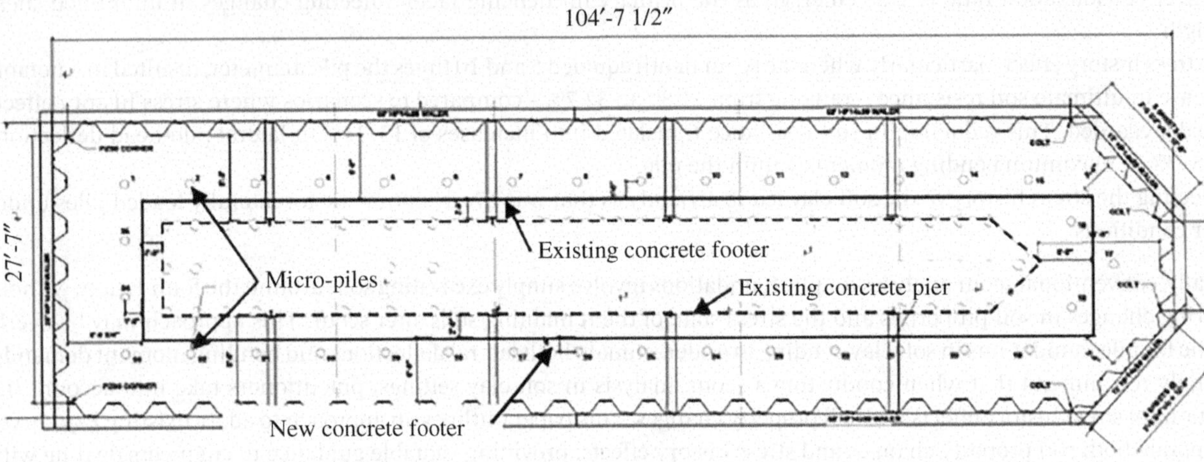

Figure 20.46 Plan view of cofferdam (Takefumi and Takayuki 2018/American Society of Civil Engineers).

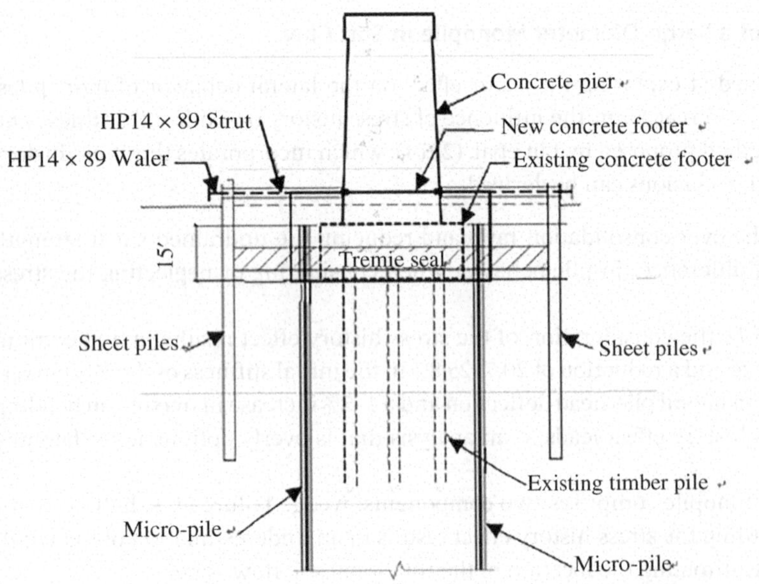

Figure 20.47 Cross-section of the cofferdam and pier foundation (Takefumi and Takayuki 2018/American Society of Civil Engineers).

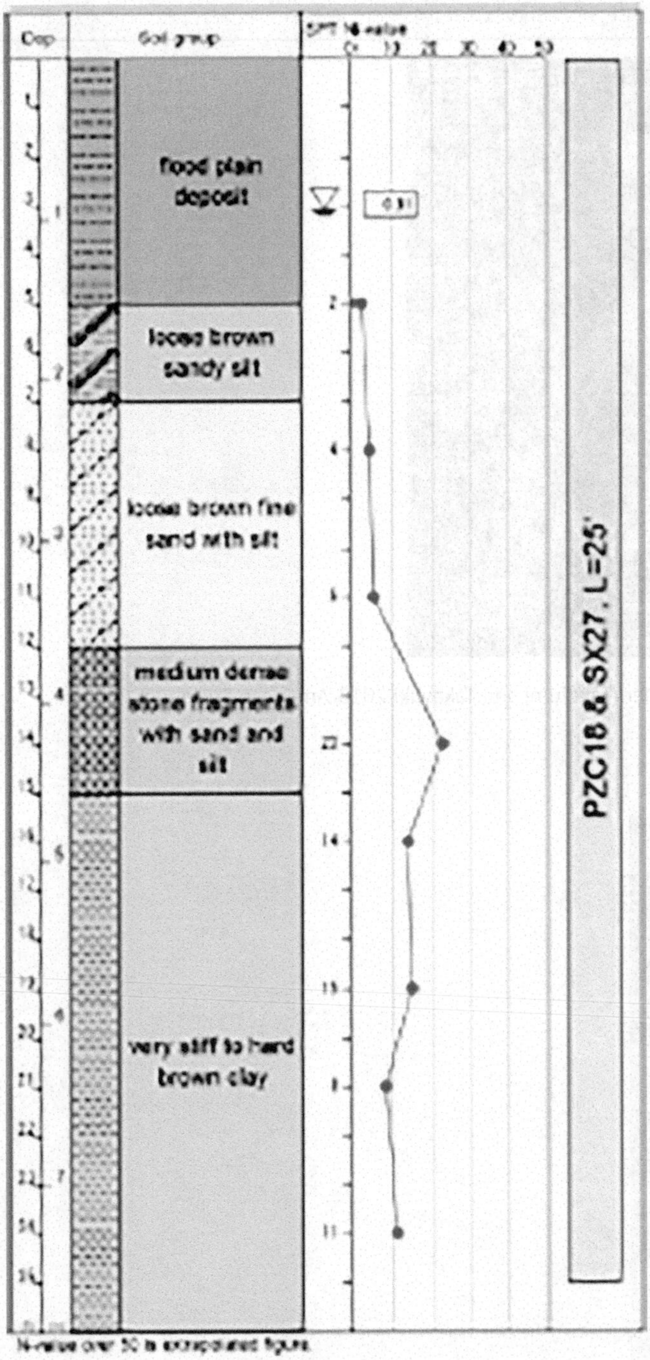

Figure 20.48 Soil conditions (Takefumi and Takayuki 2018/American Society of Civil Engineers).

To facilitate the movement of sheet pile sections, a small track loader equipped with suitable attachments was employed to navigate within the limited headroom, as depicted in Figure 20.49. The intricate process of splicing sheet piles beneath the girder is showcased in Figure 20.50. Finally, Figure 20.51 offers a view of the finished sheet pile cofferdam, featuring the concrete pier, the existing concrete footer, and a series of new micro-piles with steel bar reinforcement on top. Remarkably, the cofferdam was erected swiftly, without compromising the integrity of the weakened pier foundation.

(Continued)

Case Study 20.7 (Continued)

Figure 20.49 Cofferdam sheet pile press-in work under rail girder (Takefumi and Takayuki 2018/American Society of Civil Engineers).

Figure 20.50 Splicing sheet piles by welding (Takefumi and Takayuki 2018/American Society of Civil Engineers).

Figure 20.51 Sheet pile cofferdam and micro-piles for larger footer (Takefumi and Takayuki 2018/American Society of Civil Engineers).

Case Study 20.8

Scour Repairs on Bridge on Road 71N Over Portage Diversion Channel, Manitoba, Canada (Takefumi and Takayuki 2018)

The bridge on Road 71N, spanning the Portage Diversion Channel, is a two-lane structure with a length of 127.9 m, comprising five spans and being simply supported. It was originally constructed in 1967. The substructure of this bridge includes reinforced cast-in-place concrete abutments founded on timber piles, as well as reinforced cast-in-place concrete piers that rest upon steel H-piles. During the flood of 2011, the bridge foundations experienced the intense force of turbulent flow as water entered the diversion channel (Figure 20.52). The volume of inflow significantly exceeded the channel's designed capacity, resulting in substantial scouring of the piers. This scouring exposed the steel H-pile foundations at several pier locations, with depths of up to 6 m.

Despite the considerable extent of pier scouring, a comprehensive assessment determined that the overall substructure remained in good condition. A structural analysis of the existing foundations was conducted to ascertain whether additional piling was necessary to maintain the current level of service. This analysis relied on existing subsurface stratigraphy and pile driving records. The examination concluded that the existing foundations possessed adequate capacity to support the current loads and to accommodate future increases in ice loading for an expanded diversion channel.

The design solution involved the creation of new reinforced cast-in-place concrete pier footings, extending to a depth of 3.0 m. These footings would encapsulate and connect the two existing 900 mm deep pier footing components, ensuring overall structural integrity. The connection between the new and existing concrete utilized dowels to effectively transfer loads to the new foundations. Additionally, erosion protection measures around the bridge piers were implemented, featuring the placement of Class 600 and 900 stone ripraps. Figures 20.53–20.56 show the progression of the repair and the completion of the project.

(Continued)

Case Study 20.8 (Continued)

Figure 20.52 Scouring beneath bridge piers (Takefumi and Takayuki 2018/American Society of Civil Engineers).

Figure 20.53 Excavation complete beneath pier (Takefumi and Takayuki 2018/American Society of Civil Engineers).

Figure 20.54 Reinforcing steel for new pier footing.

Figure 20.55 Completed pier footing.

Figure 20.56 Completed bridge scour repair works.

Case Study 20.9

Please see the Case Study 26.10 in Chapter 26

Case Study 26.10: Emergency Bridge Pier Foundation Repair with Pressed-in Piles: Kumakami River Bridge Foundation Repair, Fukuoka Prefecture, Japan (Takefumi and Takayuki 2018)

20.7 Major Points

1) **Challenging and Extreme Conditions Impact on Foundations:** The performance of foundations can be significantly affected by challenging and extreme conditions. These conditions encompass various factors, including soil quality, superstructure loads, environmental conditions, construction constraints, serviceability requirements, and natural or artificial disasters.
2) **Cyclic Loading, Liquefaction, and Scouring:** The current chapter addresses three critical phenomena: cyclic loading, liquefaction, and scouring. Cyclic loading and liquefaction often arise due to seismic activity, leading to reduced soil strength and stiffness. Scouring directly erodes soil around and beneath foundations, compromising their stability.
3) **Pile Failures Stem from Uncertainty:** Pile failures are often attributed to uncertainties in extreme loads and geotechnical aspects. To enhance reliability, efforts should focus on improving knowledge in relevant fields. Extreme loads affecting pile foundations include cyclic loading, earthquakes, liquefaction, and scour.
4) **Pile Design for Various Needs:** Piles are employed for diverse purposes such as slope stability control and mitigating downdrag. Cyclic loading, liquefaction, and scouring are some critical considerations in pile foundation design, especially in earthquake-prone areas where soil strength and stiffness can degrade.
5) **Landslides and Remediation:** Landslides present significant geological events with potential for catastrophic consequences. Understanding landslide behavior is crucial for prevention and safety. Various types of landslides, early warning signs, and effective remediation methods are discussed.
6) **Shear Pile Walls for Stabilization:** Shear pile walls constructed from reinforced concrete offer a practical solution for landslide stabilization. These piles traverse the sliding surface and anchor into the stable ground, providing an effective means of addressing landslide-related challenges.
7) **Negative Shaft Friction (Downdrag):** In certain construction scenarios, soil settlement relative to the pile can result in negative shaft friction, known as downdrag. This issue is especially significant when piles penetrate soft clay layers into harder bearing layers, impacting both ultimate and serviceability states.
8) **Complex Interaction in Pile Settlement:** The relationship between bearing capacity, settlement, and axial load is intricate and interconnected, based on the neutral plane concept. Factors such as soil collapse, negative skin friction, and changes in axial load must be considered during design.
9) **Soil Improvement as a Solution:** Soil improvement serves as a complementary approach to deep foundations, reducing negative skin friction and dragload while increasing shaft and toe resistance. The minor increase in structural axial load is typically manageable and secondary to the enhancement of bearing capacity and settlement performance.
10) **Three Approaches for Design:** Designers can choose from three approaches of varying complexity: "GLOBAL," "LOCAL," and "FEM" methods. These methods necessitate the collection of specific parameters, particularly related to the soil's response to cyclic stresses and degradation laws, for quantifying soil–pile interface friction loss or lateral reactions.
11) **Transforming Real-World Loading:** Real-world cyclic loading is characterized by varying amplitudes and periods. Design efforts involve converting this random loading into an idealized, regular loading pattern. Cycle counting methods are commonly used to create idealized cycles that mimic real loads, causing similar material damage.
12) **Soil–Pile Degradation under Cyclic Loading:** Soil–pile degradation is a crucial factor influencing pile response, particularly under axial and lateral cyclic loading. Extensive studies and tests have been conducted to predict pile head response and maximum bending moments. Another vital aspect is deriving cyclic P–y curves from static P–y reaction curves.
13) **Cyclic Pile Response and Soil Behavior:** Understanding soil–pile degradation under axial and lateral cyclic loading is crucial. Studies and tests have been conducted to predict pile head response and maximum moments. Additionally, the derivation of cyclic P–y curves from static P–y reaction curves is essential for accurate design.

14) **Liquefaction and Soil Behavior:** Soil liquefaction involves the loss of shear strength in soil layers, causing them to flow and drag the pile, resulting in bending failures attributed to lateral spreading. Distinguishing soil liquefaction susceptibility is essential for designing specific solutions.
15) **Landslides and Remediation:** Landslides present significant geological events with potential for catastrophic consequences. Understanding landslide behavior is crucial for prevention and safety. Various types of landslides, early warning signs, and effective remediation methods are discussed.
16) **Scour Impact on Bridges:** Scour predominantly affects bridges, with total scour comprising long-term degradation, contraction scour, and local scour. A multidisciplinary team of engineers in hydraulic, geotechnical, and structural design must evaluate scour at bridge sites and design foundations accordingly.

20.8 Questions and Problems

20.8.1 Explanatory

1 Name and explain each kind of extreme loadings and conditions on piles.

2 What is the role of back-analysis in estimating shear strength for landslide understanding, and why is it considered a powerful tool despite its associated uncertainties?

3 How does the text describe the relationship between landslide prevention and hazard mapping, and why is this relationship significant in areas prone to slope failure?

4 Explain the concept of "permanent stability" in the context of landslide remediation, and why it is crucial to account for various factors when considering future conditions.

5 What is the method of transforming real-world cyclic load to equivalent load?

6 What are the three types of pile head responses due to cyclic axial loading?

7 How does an earthquake cause damage to structures?

8 Describe conditions where the downdrag force is mobilized.

9 List some measures that are taken to mitigate the consequences of downdrag force.

10 What is bridge scour, and how can it be predicted?

20.8.2 Conceptual

1 How do the principles of force and moment equilibrium apply to slope stability calculations, and why is it challenging to achieve a high degree of precision in these calculations?

2 What are the key differences between shear piles and truncated shear piles in terms of design and application, and how do these differences impact their effectiveness in slope stabilization?

3 Conceptually, why is it essential for engineers to strategically position boreholes along the line of the maximum slope when conducting site investigations for landslide remediation, and what insights does this positioning provide for stability analysis?

4 Explain the difference between axial and lateral cyclic loading effects on soil–pile response.

5 Explain the effect of cycles on *P–y* reaction curves.

6 Explain the role of laboratory and in situ tests in designing piles for cyclic loading.

7 What are the differences between the API approach and the SOLCYP approach, for cyclic design of pile foundations?

8 Define the pile failure mechanism due to liquefaction and list some suggestions to provide more resistance against liquefaction failure.

9 Compare and contrast various determinant factors on the depth of the neutral plane.

10 List the lessons that can be learned via the case histories explained in this chapter.

20.8.3 Problems

1 Consider the procedure introduced in Example 20.1 for a closed-end pipe pile with an outer diameter of 0.6 m and length of 30 m. Evaluate the axial capacity prior to and post-liquefaction via suggested methods. The adjacent CPT profile is also given. These methods have been introduced in Chapter 16 and were used in Example 20.1.

- Eslami and Fellenius (1997)
- ICP (Jardine et al. 2005)
- Fugro (Kolk et al. 2005)
- UWA (Lehane et al. 2005)
- German (Kempfert and Becker 2010)

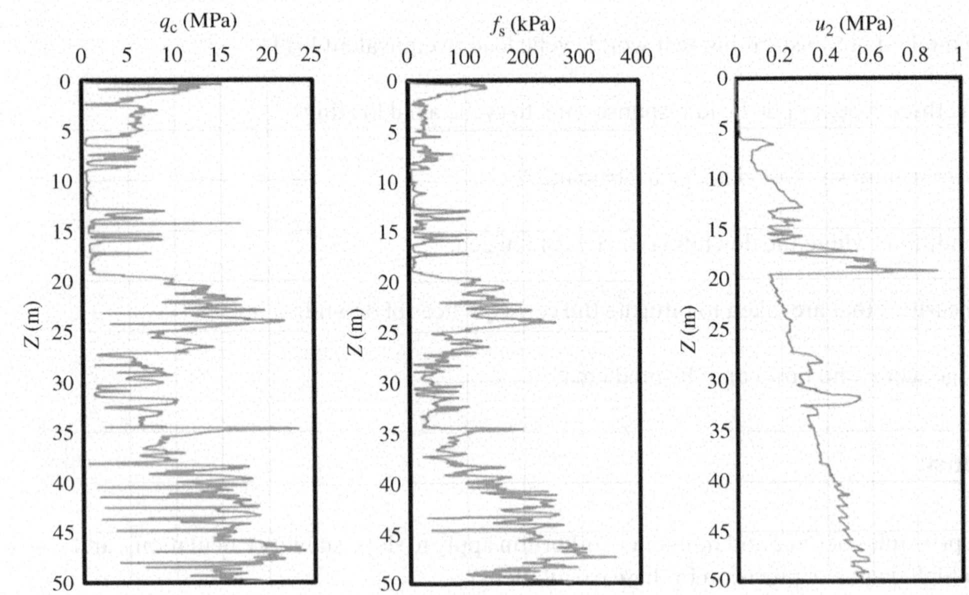

2 A belled bored pile (enlarged base pile) with a total length of 15 m is implemented in a uniform saturated soft clay site. According to the figure and the given data, determine the ultimate uplift capacity before and after scour. GWT is located on the ground surface (scour depth is 3 m).

$$b = 80\text{ cm}, B = 200\text{ cm}, S_u = 20\text{ kPa}, \gamma_{sat} = 1.95\text{ t/m}^3, \gamma_{con} = 2.45\text{ t/m}^3$$

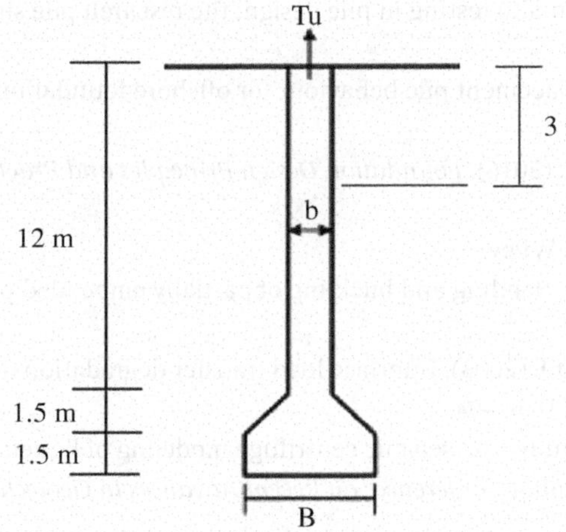

References

AASHTO (2017). *Load and Resistance Factor Design Bridge Design Specifications*, 8e. Washington D.C: AASHTO.

Airey, D.W., Al-Douri, R.H., and Poulos, H.G. (1992). Estimation of pile friction degradation from shearbox tests. *Geotechnical Testing Journal* 15 (4): 388–392.

Al-Douri, R. and Poulos, H.G. (1994). Cyclic behaviour of pile groups in calcareous sediments. *Soils and Foundations* 34 (2): 49–59.

American Petroleum Institute (2014). API: recommended practice 2GEO-geotechnical and foundation design considerations, American Petroleum Institute, Version October 2014.

Arneson, L.A., Zevenbergen, L.W., Lagasse, P.F., and Clopper, P.E. (2012). *Evaluating Scour at Bridges*. National Highway Institute (US).

Asadi, F., Eslami, A., and Valikhah, F. (2016). Ground improvement and foundation practice for Persian Gulf Bridge (Causeway); Bandar Abbass Harbour – Qeshm Island. *Marine Georesources and Geotechnology*. https://doi.org/10.1080/1064119X.2016.1213774.

Ashour, M. and Ardalan, H. (2012). P–y curve and lateral response of piles in fully liquefied sands. *Canadian Geotechnical Journal* 49: 633–650.

Association, Japan Road (1971). Specifications for earthquake resistant design of highway bridges. January.

Association, Japanese Road (1996). *Specification for Highway Bridges, Part V*. Tokyo, Japan: Seismic Design.

Atkins Consultants Ltd (2000). Cyclic degradation of offshore piles. HSE Offshore Technology Report OTO 2000 013. London: Health and Safety Executive.

Baguelin, F. and Jézéquel, J.F. (1972). Etude Experimentale Du Comportement De Pieux Sollicites Horizontalement. Ann. Itbtp-Serie Sols Et Fondations-91, 297.

Bea, R.G. (1979). Performance of dynamically loaded pile foundations. *Proceedings of the 2nd International Conference BOSS'79*, Paper 68, vol. 3, 728–746. Paper: P1979-4 Proceedings. Woodward-Clyde Consultants, USA.

Bhattacharya, S., Madabhushi, S.P.G., and Bolton, M.D. (2003). *Pile Instability During Earthquake Liquefaction*. Cambridge, UK: University of Cambridge.

Bhattacharya, S., Madabhushi, S.P.G., and Bolton, M.D. (2004). An alternative mechanism of pile failure in liquefiable deposits during earthquakes. *Geotechnique* 54 (3): 203–213.

Broms, B.B. (1964). Lateral resistance of piles in cohesive soils. *Journal of the Soil Mechanics and Foundations Division* 90 (2): 27–63.

Burcharth, H.F., Bernal, A., Blazquez, R. et al. (2001). *Seismic Design Guidelines for Port Structures*. PIANC/International Navigation Association.

Campanella, D. and Sy, A. (1991). Use on in-situ testing in pile design; the research pile site as UBC. *Geotechnical News* 9 (2): 42–45.

Chow, F.C. (1997). Investigations into displacement pile behaviour for offshore foundations. PhD thesis. University of London, Imperial College.

Coduto, D.P., Kitch, W.A., and Yeung, M.R. (2016). *Foundation Design Principles and Practices*, 3e. Upper Saddle River, NJ: Prentice Hall, Inc.

Cornforth, D. (2005). *Landslide in Practice.* Wiley.

Davisson, M.T. and Robertson, K.E. (1965). Bending and buckling of partially embedded piles. *Proceedings of 6th ICSMFE* 2: 243–246.

DeJong, J.T., Randolph, M.F., and White, D.J. (2003). Interface load transfer degradation during cyclic loading: a microscale investigation. *Soils and Foundations* 43 (4): 91–94.

Dobry, R. and Abdoun, T. (2001). Recent studies on seismic centrifuge modeling of liquefaction and its effect on deep foundations. *Proceedings of 4th International Conferences on Recent Advances in Geotechnical Earthquake Engineering and Soil Dynamics*, San Diego, USA (24–29 May).

Dobry, R. & Iai, S. (2000). Recent developments in the understanding of earthquake site response and associated seismic code implementation. *ISRM International Symposium.*

EC8 (2000). *Design Provisions for Earthquake Resistance of Structures, Part 5: Foundations, Retaining Structures and Geotechnical Aspects.* Brussels: European Committee for Standardization.

Ekramirad, S.A., Eslami, A., and Razaghi, J. (2012). *Huge Landslide of Dashtehgan Rudbar; Causes and Stabilizations Approach.* Sharif Civil Engineering (In Persian).

Eslami, A. (2013). *Foundation Engineering, Design, and Construction*, 2e. Tehran: Building and Housing Research center.

Eslami, A. and Fellenius, B.H. (1997). Pile capacity by direct CPT and CPTu methods applied to 102 case histories. *Canadian Geotechnical Journal* 34 (6): 886–904.

Eslami, A. and Maleki, M. (2023). Piles performance in liquefied soils; bearing capacity and stability issues. *Journal of Structural & Construction Engineering* (Under Publication).

Eslami, A., Akbarimehr, D., Aflaki, E., and Hajitaheriha, M.M. (2019). Geotechnical site characterization of the Lake Urmia super-soft sediments using laboratory and CPTu records. *Marine Georesources and Geotechnology* 38 (10): 1223–1234.

Fellenius, B.H. (2023). Basics of foundation design. Electronic Edition. www.Fellenius.net (accessed 6 September 2023).

FHWA (2010). Drilled shafts: construction procedures and design methods. Publication No. FHWA-NHI 10-016.

Finn, W.D., & Thavaraj, T. (2001). Deep foundations in liquefiable soils: case histories, centrifuge tests and methods of analysis. *Proceedings of 4th International Conferences on Recent Advances in Geotechnical Earthquake Engineering and Soil Dynamics*, San Diego, USA (24–29 May).

Finno, R.J. (1989). *Predicted and Observed Axial Behavior of Piles: Results of a Pile Prediction Symposium* (Geotechnical Special Publication No 23). American Society of Civil Engineers.

Fukuoka, M. (1966). Damage to civil engineering structures. *Soils and Foundations, Tokyo, Japan* 6 (2): 45–52.

Haigh, S.K. (2002). *Effects of Earthquake-Induced Liquefaction on Pile Foundations in Sloping Ground*. The University of Cambridge.

Hamada, M (1992a). Large ground deformations and their effects on lifelines: 1964 Niigata earthquake. Case Studies of liquefaction and lifelines performance during past earthquake", Technical Report NCEER-92-0001, Volume-1, Japanese case studies, National Centre for Earthquake Engineering Research, Buffalo, NY.

Hamada, M. (1992b). Large ground deformations and their effects on lifeline: 1964 Niigata Earthquake. *Proceedings, Case Studies of Liquefaction and Lifeline Performance During Past Earthquakes*, Japanese Case Studies, Japan, 1.

Han, J. (2015). *Principles and Practices of Ground Improvement*. Wiley.

Hawkins, N.M. and Ghosh, S.K. (2000). Proposed revisions to 1997 NEHRP recommended provisions for seismic regulations for precast concrete structures. Part 2-seismic-force-resisting systems. *PCI Journal* 45 (5): 34–44.

He, B., Lai, Y., Wang, L. et al. (2019). Scour effects on the lateral behavior of a large-diameter monopile in soft clay: role of stress history. *Journal of Marine Science and Engineering* 7 (6): 170.

Holtz, R.D., Kovacs, W.D., and Sheahan, T.C. (2023). *An Introduction to Geotechnical Engineering*, 3e. Pearson.

Ishihara, K. (1997). Terzaghi oration: geotechnical aspects of the 1995 Kobe earthquake. *Proceedings of The International Conference on Soil Mechanics and Foundation Engineering-International Society for Soil Mechanics and Foundation Engineering*, Hamburg, Germany (6–12 September 1997), vol. 4, 2047–2074. ISSMFE Society.

Ishihara, K. and Cubrinovski, M. (2004). Case studies of pile foundations undergoing lateral spreading in liquefied deposits. *International Conference on Case Histories in Geotechnical Engineering*, New York, USA (13–17 April 2004), p. 5.

Iwasaki, T. (1981). Dynamic soil-structure interaction with emphasis on geotechnical engineering aspects. *Tshuchi-to-Kiso, JGS* 29 (9): 7–10. (In Japanese).

Jardine, R.J. (1991). The cyclic behaviour of offshore piles. In: *The Cyclic Loading of Soils* (ed. O.'.R. Brown). Glasgow: Blackie and Son.

Jardine, R.J. (1994). Offshore pile design for cyclic loading: North Sea clays. HSE Offshore Technology Report, OTN 94 157.85.

Jardine, R.J. and Standing, J.R. (2012). Field axial cyclic loading experiments on piles driven in sand. *Soils and Foundations* 52 (4): 723–736.

Jardine, J., Chow, F., Overy, R., and Standing, J. (2005). *ICP Design Method for Driven Piles in Sands and Clays*. London: Thomas Telford Publishing Ltd., 105pp.

JRA (1980, 1972). Japanese Road Association, Specification for Highway Bridges, Part V, Seismic Design.

Kawashima, K. (2000). Seismic design and retrofit of bridges. *Bulletin of the New Zealand Society for Earthquake Engineering* 33 (3): 265–285.

Kempfert, H.G., Becker, P., 2010. Axial pile resistance of different pile types based on empirical values. *Proceedings of Geo-Shanghai 2010 Deep Foundations and Geotechnical In-Situ Testing (GSP 205)*. ASCE, Reston, VA, pp. 149-154.

Khemakhem, M. (2012). *Etude expérimentale de la réponse aux charges latérales monotones et cycliques d'un pieu foré dans l'argile*. Ecole centrale de Nantes.

Kolk, H., Baaijens, A., and Senders, M. (2005). Design criteria for pipe piles in silica sands. In: *Proceedings of International Symposium on Frontiers in Offshore Geomechanics (ISFOG 2005), Perth*, 711–716. London: Taylor and Francis Group.

Lagasse, P.F., Zevenbergen, L.W., Spitz, W., and Arneson, L.A. (2012). *Stream Stability at Highway Structures*. The United States: Federal Highway Administration. Office of Bridge Technology.

Lehane, B.M. (1992). Experimental investigations of pile behaviour using instrumented field piles. PhD thesis. University of London, Imperial College.

Lehane, B.M., Scheider, J.A., and Xu, X. (2005). CPT Based Design of Driven Piles in Sand for Offshore Structures. Internal Rep. No. GEO: 05345. University of Western Australia, Perth, Australia.

Lin, C., Han, J., Bennett, C., and Parsons, R.L. (2014). Behavior of laterally loaded piles under scour conditions considering the stress history of undrained soft clay. *Journal of Geotechnical and Geoenvironmental Engineering* 140 (6): 6014005.

Madabhushi, G., Haigh, S., and Knappett, J. (2009). *Design of Pile Foundations in Liquefiable Soils*. World Scientific.

Maeda, Y., Omine, K., Ochiai, H., et al. (2006). Improvement of horizontal bearing capacity by composite ground foundation method in soft ground. *Proc. of the 22nd US-Japan Bridge Engineering Workshop*, Seattle, WA, USA (23–28 October 2006).

Mayne, P.W., Kulhawy, F.H., and Trautmann, C.H. (1995). Laboratory modeling of laterally-loaded drilled shafts in clay. *Journal of Geotechnical Engineering* 121 (12): 827–835.

McAnoy, R.P.L., Cashman, A.C., and Purvis, D. (1982). Cyclic tensile testing of a pile in glacial till. *2nd Int. Conf. Num. Meth. in Offshore Piling*, 257–291.

Merrill, E.F. and Onge, G.H.S. (1974). American Petroleum Institute recommended practice for squirrel-cage induction motors. *IEEE Transactions on Industry Applications* 4: 477–484.

Mortara, G., Mangiola, A., and Ghionna, V.N. (2007). Cyclic shear stress degradation and post-cyclic behaviour from sand–steel interface direct shear tests. *Canadian Geotechnical Journal* 44 (7): 739–752.

Murff, J.D. and Hamilton, J.M. (1993). P-ultimate for undrained analysis of laterally loaded piles. *Journal of Geotechnical Engineering* 119 (1): 91–107.

National Research Council (NRC) (1985). *Liquefaction of Soils During Earthquakes*, Report No. CETS-EE-001, Committee on Earthquake Engineering. Washington, DC: National Academy Press.

NEHRP (2000). NEHRP recommended provisions for seismic regulations for new buildings and other structures 2000 edition, part 1: Provisions, Report no. FEMA368, Building Seismic Safety Council for the Federal Emergency Management Agency, Washington, DC, USA.

NHI Course (2018). NHI Course No. 132014 drilled shafts: construction procedures and design methods. 132014.

Papadopoulos, V. and Loupasakis, K. (2014). *Soil Mechanics and Foundation Elements*, 1–22. Athens: NTUA.

Peralta, P.K. (2010). *Investigations on the Behavior of Large Diameter Piles under LongTerm Lateral Cyclic Loading in Cohesionless Soil*. Hannover: Leibniz University.

Polák, J. and Vašek, A. (1992). Analysis of the cyclic stress-strain response in variable amplitude loading using the rain-flow method. In: *The Rain-Flow Method in Fatigue* (ed. Y. Mukarami), 123–131. Butterworth-Heinemann ISBN 9780750605045.

Poulos, H.G. (1988). Cyclic stability diagram for axially loaded piles. *Journal of Geotechnical Engineering* 114 (8): 877–895. https://doi.org/10.1061/(ASCE)0733-9410(1988)114:8(877).

Prasad, Y.V.S.N. and Chari, T.R. (1999). Lateral capacity of model rigid piles in cohesionless soils. *Soils and Foundations* 39 (2): 21–29.

Puech, A. (2013). Advances in axial cyclic pile design: contribution of the SOLCYP project. *Proceedings TC 209 Workshop*, 18 ICSMGE, 45–57. Paris, France: Presses des Ponts.

Puech, A. and Garnier, J. (2017). Design of Piles Under Cyclic Loading: SOLCYP Recommendations. John Wiley & Sons, Inc. https://doi.org/10.1002/9781119469018.

Randolph, M.F. and Houlsby, G.T. (1984). The limiting pressure on a circular pile loaded laterally in cohesive soil. *Geotechnique* 34 (4): 613–623.

Rattley, M.J., Costa, L., Jardine, R.J., and Cleverly, W. (2017). Laboratory test predictions of the cyclic axial resistance of a pile driven in North Sea soils. *Offshore Site Investigation Geotechnics 8th International Conference Proceeding* (12–14 September), vol. 2, 636–643. London, UK: Society for Underwater Technology.

Reese, L.C. and Van Impe, W.F. (2000). *Single Piles and Pile Groups Under Lateral Loading*. CRC Press.

Reese, L.C. and Welch, R.C. (1975). Lateral loading of deep foundations in stiff clay. *Journal of the Geotechnical Engineering Division* 101 (7): 633–649.

Reese, L.C., Cox, W.R., and Koop, F.D. (1974). Field testing and analysis of laterally loaded piles in stiff clay. In: *7th Offshore Technology Conference (OTC), Paper OTC 2 312*, 671–690. Houston, Texas.

Richardson, E.V. and Davis, S.R. (2001). *Evaluating Scour at Bridges*. The United States of America: Federal Highway Administration. Office of Bridge Technology.

Rimoy, S., Silva, I.M., and Jardine, R.J. (2022). Stability and load-displacement behavior of axially cyclic loaded displacement piles in sands. *Canadian Geotechnical Journal* 59 (8): 1358–1372.

Rosquoët, F. (2004). Pieux sous charge latérale cyclique. Nantes.

Sato, M., Ogasawara, M., and Tazoh, T. (2001). Reproduction of lateral ground displacements and lateral-flow earth pressures acting on pile foundations using centrifuge modeling. *Proceedings of 4th International Conferences on Recent Advances in Geotechnical Earthquake Engineering and Soil Dynamics*, San Diego, USA (24–29 May).

Seed, H.B., Tokimatsu, K., Harder, L.F., and Chung, R.M. (1985). Influence of SPT procedures in soil liquefaction evaluations. *Journal of Geotechnical Engineering, ASCE* 3 (12): 1425–1445.

Sun, S.-W., Zhao, F., and Zhang, K. (2016). Stability of slopes reinforced with truncated piles. *Advances in Materials Science and Engineering* 2016: 1–16.

Takata, T., Tada, Y., Toshida, I., and Kuribayashi, E. (1965). Damage to bridges in Niigata earthquake. Report No, 125(5), Public Works Research Institute (in Japanese) – Japan.

Takefumi, T. and Takayuki, S. (2018). Emergency bridge pier foundation repair with pressed-in piles. *Proceedings, IFCEE 2018*, Orlando, USA, 224–237. doi: https://doi.org/10.1061/9780784481608.022.

Tokimatsu, K. and Asaka, Y. (1998). Effects of liquefaction-induced ground displacements on pile performance in the 1995 Hyogoken-Nambu earthquake. *Soils and Foundations* 38: 163–177.

Tokimatsu, K., Mizuno, H., and Kakurai, M. (1996). Building damage associated with geotechnical problems. *Soils and Foundations* 36: 219–234.

Tokimatsu K., Oh-oka Hiroshi, Satake, K., Shamoto Y. and Asaka Y (1998). "Effects of Lateral ground movements on failure patterns of piles in the 1995 Hyogoken-Nambu earthquake". *Proceedings of a speciality conference, Geotechnical Earthquake Engineering and Soil Dynamics III*, ASCE Geotechnical Special publication No 75, pp. 1175–1186.

Tomisawa, K. and Miura, S. (2007). Mechanical behavior of pile foundation constructed in composite ground and its evaluation. *Soils and Foundations* 47 (5): 961–972.

Tsuchida, H. (1970). Prediction and countermeasure against the liquefaction in sand deposits. *Abstract of the Seminar in the Port and Harbor Research Institute*, 31–333.

Tsuha, C.H.C., Foray, P.Y., Jardine, R.J. et al. (2012). Behaviour of displacement piles in sand under cyclic axial loading. *Soils and Foundations* 52 (3): 393–410.

Yasuda, S. and Berrill, J.B. (2000). Observations of the earthquake response of foundations in soil profiles containing saturated sands. *ISRM International Symposium*, Melbourne, Australia (November 19–24).

Topic E

Complementary Issues

Section VIII

Testing and Numerical Approaches

21

Full Scale Load Testing and Inspections

Chapter Highlight

- **Shallow Foundations Testing**
  - Bench-Scale Testing
  - Plate-Load Testing
  - Full-Scale Field Testing
- **Deep Foundation Testing**
  - Pile Testing
    - Static Load Testing
    - Dynamic Load Testing
    - Rapid Load Testing
  - Drilled Piers/Shafts Testing
    - Static Testing: Conventional and Osterberg
    - Dynamic Testing: Statnamic and Hammer Drop
- **Hybrid Foundations Testing**
- **Rigid Intrusion Testing (RI) and Micro-piles**
- **Integrity and Soundness Inspection**
- **Case Studies**

21.1 Introduction

Foundations play a pivotal role in providing essential support for structures, ensuring their stability, safety, and durability. To ensure that the foundation provides stability, safety, and durability, it is crucial that the engineer perform testing and inspection to evaluate the soundness of the foundation design and construction practices. This chapter aims to provide a comprehensive overview of testing and inspection methods for various types of foundations, including shallow, deep, hybrid, and rigid intrusion foundations. Additionally, the chapter includes a section on integrity and soundness inspection of existing foundations.

Shallow Foundations Testing: Shallow foundations are commonly used for low-rise structures and are located near the ground surface. Bench-scale testing involves testing soil samples in a laboratory to determine their strength and deformation characteristics. Plate-load testing involves applying a load to a steel plate and measuring the settlement of the soil. Full-scale field testing involves loading the foundation with a known weight or pressure and measuring the resulting settlement.

Deep Foundation Testing: Deep foundations are used for high-rise structures. Pile testing involves testing the capacity and integrity of individual piles using static, dynamic, or rapid load testing methods. Drilled piers/shaft testing involves testing the capacity and integrity of individual shafts using static or dynamic testing methods, such as the conventional or Osterberg tests, and Statnamic or hammer drop tests.

Advanced Foundation Engineering: Principles, Performance and Prospects, First Edition. Moses Karakouzian and Abolfazl Eslami.

Hybrid Foundations Testing: Hybrid foundations are a combination of shallow and deep foundations and are commonly used for complex structures or sites with challenging soil conditions. The testing methods for hybrid foundations depend on the type of foundation used and typically involve a combination of shallow and deep foundation testing methods.

Rigid Intrusion and Micro-pile Testing: Rigid intrusion testing is used to test the capacity and integrity of rigid inclusions, such as stone columns or concrete piles, that are used to strengthen weak soils. Micro-piles are small-diameter piles that are used to support structures with limited access or low headroom sites. Testing methods for micro-piles involve both static and dynamic load testing (Bowles 1996).

Integrity and Soundness Inspection: In situations where the health of existing foundations of existing structures needs to be evaluated, integrity and soundness inspection methods are utilized to identify any potential issues or defects. These methods include visual inspection, nondestructive testing, and load testing.

21.2 Shallow Foundations Testing

21.2.1 Bench-Scale Testing

Generally, bench-scale tests involve conducting experiments on a reduced scale compared to the actual size. Inherent advantages of employing bench-scale tests are cost effectiveness, reduced time requirement, rapid data acquisition, high level of accuracy of results, capability of swift adjustments to processes and materials. Moreover, these tests serve as invaluable tools for product innovation, aiding in design refinement and optimal production parameter determination.

An example of bench-scale shallow foundation testing is shown in Figure 21.1 (Safarzadeh Gendeshmin and Aminfar 2020) and Figure 21.2 (Jiu et al. 2022). This downscaled loading apparatus consists of a metal tank, featuring one side crafted from sturdy glass measuring 20 mm in thickness. This glass face facilitates visual observation and particle velocity measurement to understand soil liquefaction mechanisms beneath the foundation. The external surface of the glass face has small circular markers, forming reference points within a 100 by 100-mm grid. An accompanying piezometer is strategically positioned on one of the lateral sides to monitor and regulate water levels. Five miniature 2100 strain gauges are also affixed to the tank's lateral surface for assessing non-saturated soil behavior during loading.

For loading procedures, a custom metal frame or holder is devised to mount the tank. This holder is equipped with a loading system with a capacity of 5 tons. A force measurement mechanism and a displacement measurement system are incorporated into the arrangement.

The loading system of this apparatus features an 850-mm hydraulic jack designed to be affixed on fixed rails of the holder frame. This hydraulic jack is adaptable in both horizontal directions. To exert force, a hydraulic control valve and an electrically controlled hydraulic unit are employed. The hydraulic unit assembly, force sensor, displacement sensor, and two negative pressure sensors, linked to two of the strain gauges, are integrated into a comprehensive control and data collection system. This integrated system, marked as "1," could be connected to a computer through software to facilitate simultaneous soil loading commands and data acquisition. To accommodate water absorption and soil drainage within the tank, a discharge valve is thoughtfully incorporated at the lower part of the apparatus.

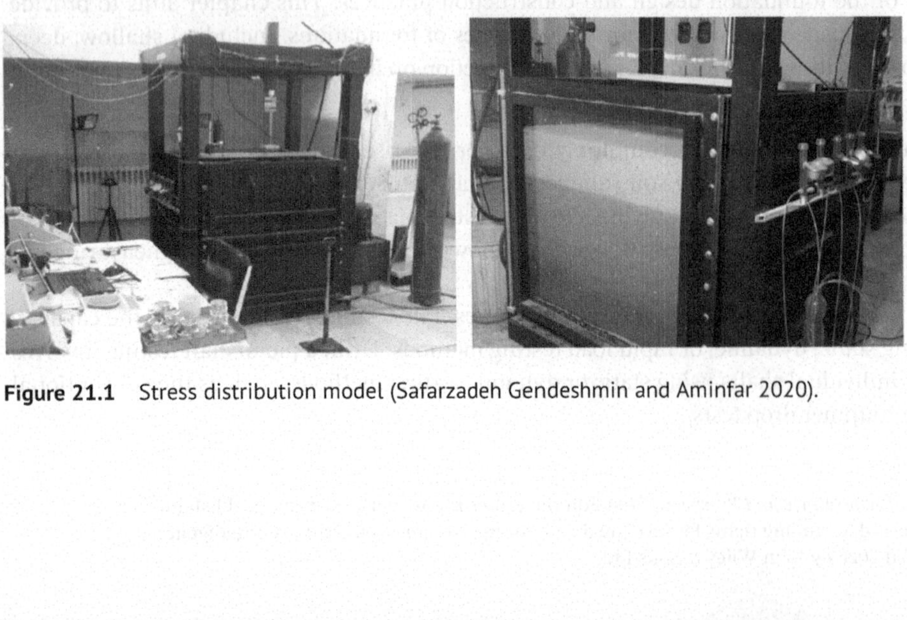

Figure 21.1 Stress distribution model (Safarzadeh Gendeshmin and Aminfar 2020).

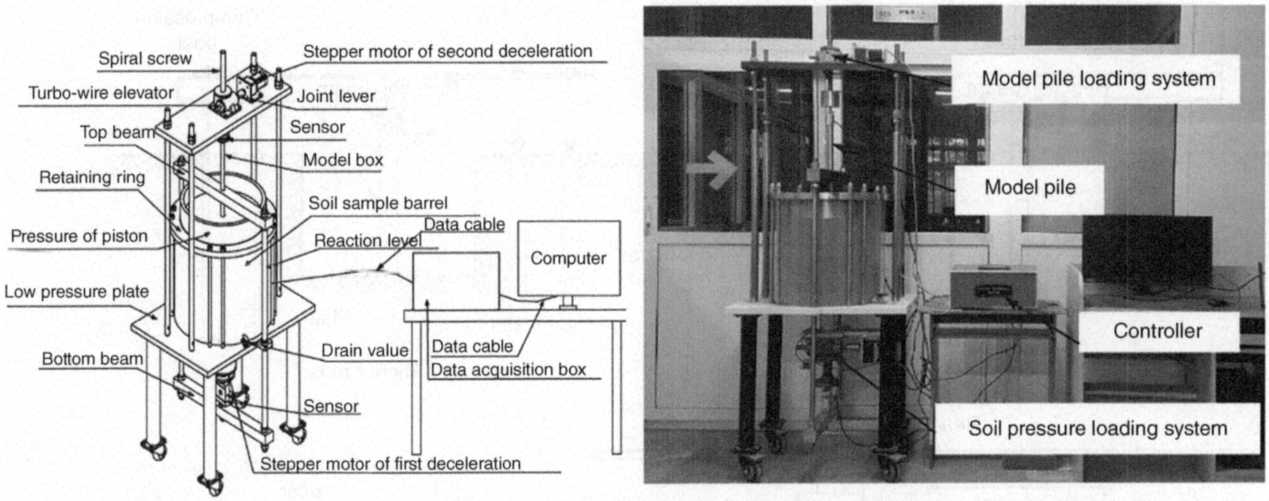

Figure 21.2 Schematic diagram and photograph of actual testing device (Jiu et al. 2022/Hindawi).

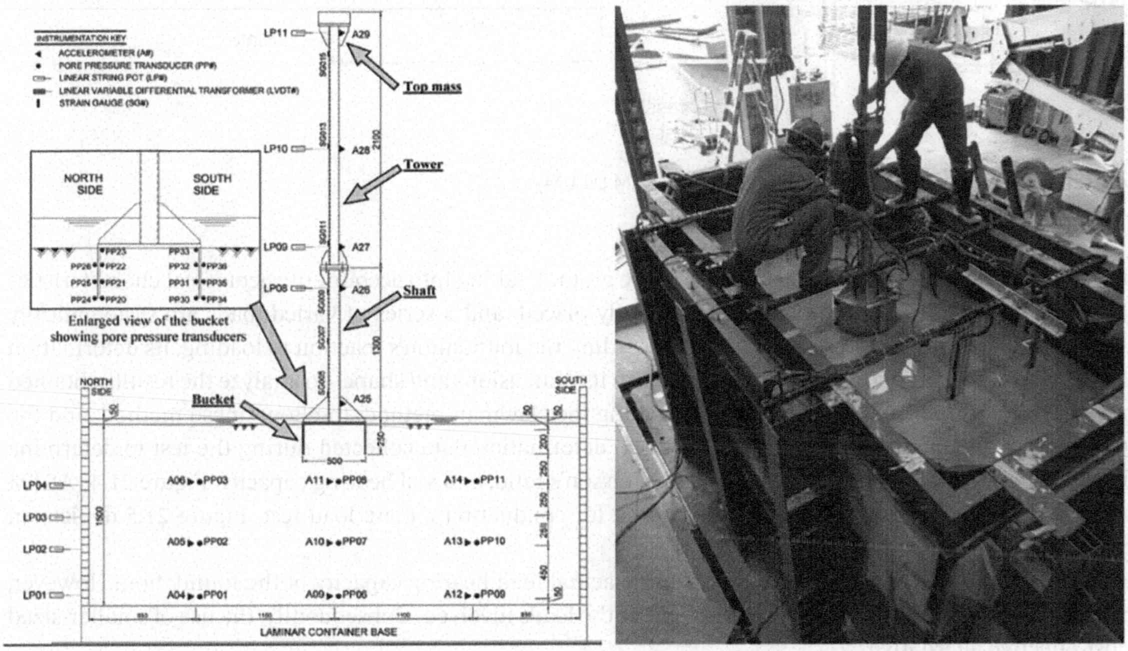

Figure 21.3 Schematic diagram and physical test deployment of the bucket and the attached shaft in the saturated soil (Zayed et al. 2023/Elsevier B.V.).

Another example of a bench-scale loading device is shown in Figure 21.2 (Jiu et al. 2022).

Another type of bench-scale testing is the shake table. Shake tables are used to test the response of structures and foundations to seismic (earthquake). An example of the use of the shake table is shown in Figure 21.3 (Zayed et al. 2023), where a scale model of a bucket foundation of an offshore wind turbine is tested for seismic response.

21.2.2 Plate Load Testing

Plate load testing is a widely used method for evaluating the behavior and performance of shallow foundations under both static and dynamic loading conditions. The procedure for conducting this test involves several essential steps. Initially,

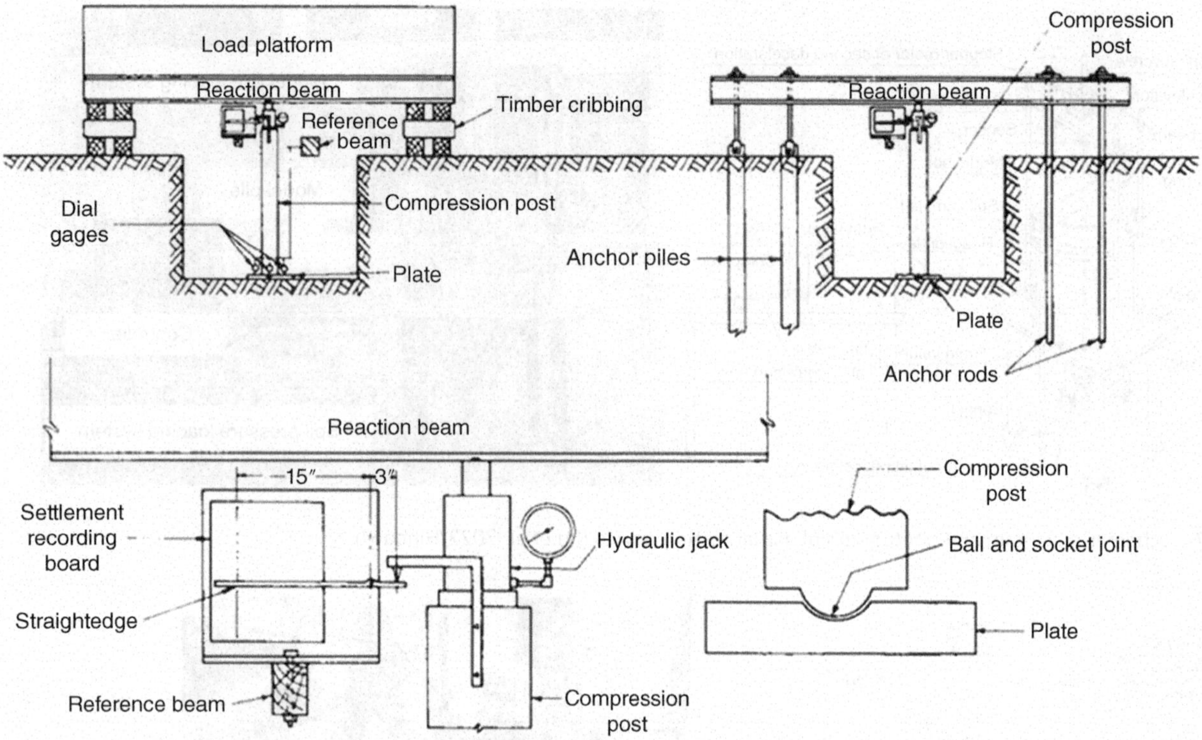

Figure 21.4 Typical setup for conducting static plate load tests (ASTM D1194).

the foundation under examination is carefully positioned on the ground, taking into account the geological characteristics of the site. Subsequently, a machine loading plate is strategically placed, and a series of varied loads are systematically applied to the foundation. The test generates valuable data regarding the foundation's reaction to loading, its deformation characteristics, the distribution of pressure, and any alterations in its dimensions and shape. To analyze the results obtained from the plate load test, various methods are employed, including the Newman method, the Boussinesq method, and the Parkinson method. These methods utilize load distribution and deformation data collected during the test to determine essential soil properties such as the modulus of soil elasticity, Poisson's ratio, and soil bearing capacity. Figure 21.4 (ASTM D1194) provides a visual representation of a typical setup used for conducting a plate load test. Figure 21.5 depicts an actual field plate load test.

Full-scale plate load tests can also be performed to obtain the actual load-bearing capacity of the foundation. However, full-scale tests are costly due to the substantial size and weight of the loads involved. Consequently, the use of smaller-sized plates is a more cost-effective alternative.

To conduct the plate load test, the ASTM D1194 standard is typically followed. The test involves an initial estimation of the soil's load-bearing capacity. This estimated load is then divided into several sections, typically four or five, which are successively applied to the soil. After each load application, the system is left undisturbed for one hour, allowing the necessary soil settlement to occur in response to the applied load. The resulting settlement measurements against the corresponding applied loads are recorded at each stage of the test. At the conclusion of the test, the data is used to generate a graph, illustrated in Figure 21.6 (Baca 2020), which aids in the assessment of the foundation's load-bearing capacity.

The load-bearing capacity can be determined from the plate load test graph by utilizing the two tangents drawn at the start and end segments, as demonstrated in Figure 21.6. The plate load test is applicable to nearly all soil types, whether at the surface or at deeper levels. The test proves highly valuable and effective in scenarios involving hard soils, soft rocks, and coarse sands where conventional testing methods such as SPT and CPT are impractical. Nonetheless, the plate load test does have its limitations:

Figure 21.5 Setup of field plate load tests (Çevik 2022/Eskişehir Osmangazi University).

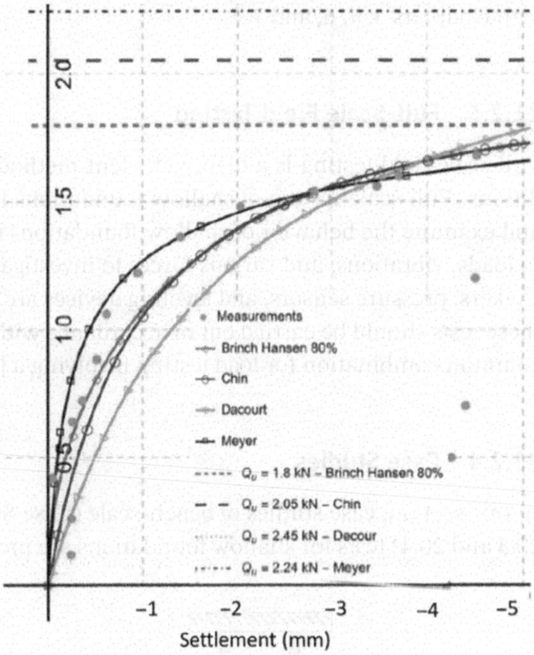

Figure 21.6 Static load tests graph is plotted (Baca 2020/IOP Publishing/Public Domain CC BY 3.0).

- In many instances, the loading system may not possess the capacity to induce soil yielding.
- Conducting the plate load test at significant depths or below the groundwater table can be unfeasible.
- The implementation cost of the plate loading test method is substantial, making it primarily only cost-effective for major projects.
- Soil sampling cannot be performed within this testing method.
- The depth of the soil bed influenced by the plate load test loading is considerably shallower than the actual structure's depth, potentially leading to test result inaccuracies.
- Compliance with building code regulations may not be achievable if the test is conducted within wells or galleries.
- Due to the utilization of dead loads or heavy machinery to provide reaction forces, a level terrain is necessary for the test.
- Plate load tests in soils undergoing consolidation settlements will yield erroneous results (Mokhberi et al. 2019).

21.2.2.1 Analysis of Plate Load Tests

- **Load-Bearing Capacity:** The load-bearing capacity of saturated clay soils (internal friction is zero) is independent of the dimensions of the shallow foundation, because the term that has the width of the footing in the Terzaghi and Hansen equation for bearing capacity is zero. Therefore, the load-bearing capacity determined by the plate load test is identical to the load-bearing capacity of the actual size shallow footing. However, this only applies to saturated clays. For other soils that have friction and cohesion values, the load-bearing capacity of the actual footing is determined by multiplying the bearing capacity obtained from the plate test by the ratio of the width of the actual shallow footing to the width of the test plate.
- **Settlements:** The settlement of the actual footing can be derived from the settlements of the plate load test. Various equations have been proposed to calculate the settlement of the actual shallow footing.
- **Soil Shear Strength:** Although the main and important method for obtaining shear strength is the triaxial test, we can also use the plate load test to estimate the shear strength in unsaturated cohesive soils. In this case, the load should not be applied step by step, and it is necessary to apply the load uniformly to the plate.
- **Soil Reaction Modulus (K_s):** The reaction modulus of the soil is the normal stress applied to the soil surface that causes a unit settlement. The results of the plate load tests are used to determine the reaction modulus of the soil.
- Applications and scaleup concepts of plate load tests (PLTs) for shallow foundations have been presented in more detail in Chapters 3, 4, 6, and 7.

21.2.3 Full-Scale Field Testing

Full-scale field testing is a highly efficient method for investigating the behavior of shallow foundations under real conditions. Full-scale testing of shallow foundations involves a series of precise tests and large-scale simulations to evaluate and examine the behavior of shallow foundations under real conditions. In these tests, shallow foundations are subjected to loads, vibrations, and various forces to investigate their behavior. Advanced and powerful equipment and tools such as shakers, pressure sensors, and leveling devices are required to perform full-scale shallow foundation testing. Additionally, these tests should be carried out in accordance with valid standards and acceptable testing methods. Figure 21.7 presents a common combination for load testing involving a jack and a bearing plate.

21.2.4 Case Studies

In this section, case studies of bench-scale (Case Study 20.1), plate loading (Case Study 20.2), and full-scale (Case Studies 20.3 and 20.4) tests for shallow foundations are presented.

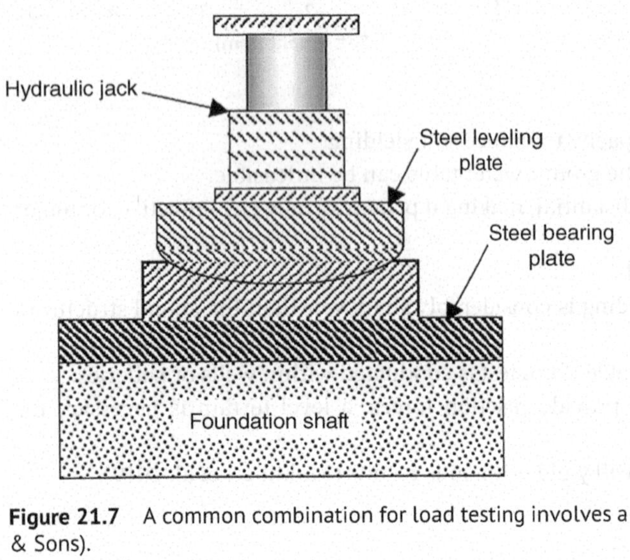

Figure 21.7 A common combination for load testing involves a jack and a bearing/leveling plate (Hertlein and Davis 2007/John Wiley & Sons).

Case Study 21.1

Bench-Scale Testing of Shallow Foundations on Reinforced Soil

This case study investigates the pressure and settlement behavior of a soil bed reinforced with bitumen-coated geogrids as a substitute for geosynthetics in shallow foundation projects (Akbar et al. 2022). The study includes a comparative analysis of the behavior of prestressed and non-prestressed geogrids with different tensile strengths. Figure 21.8 depicts model steel tanks. According to the test results, as presented in Figure 21.9, the bitumen prestressing force improves the soil bed's resistance to pressure and reduces settlement caused by cracking in the geosynthetic-reinforced soil (Akbar et al. 2022).

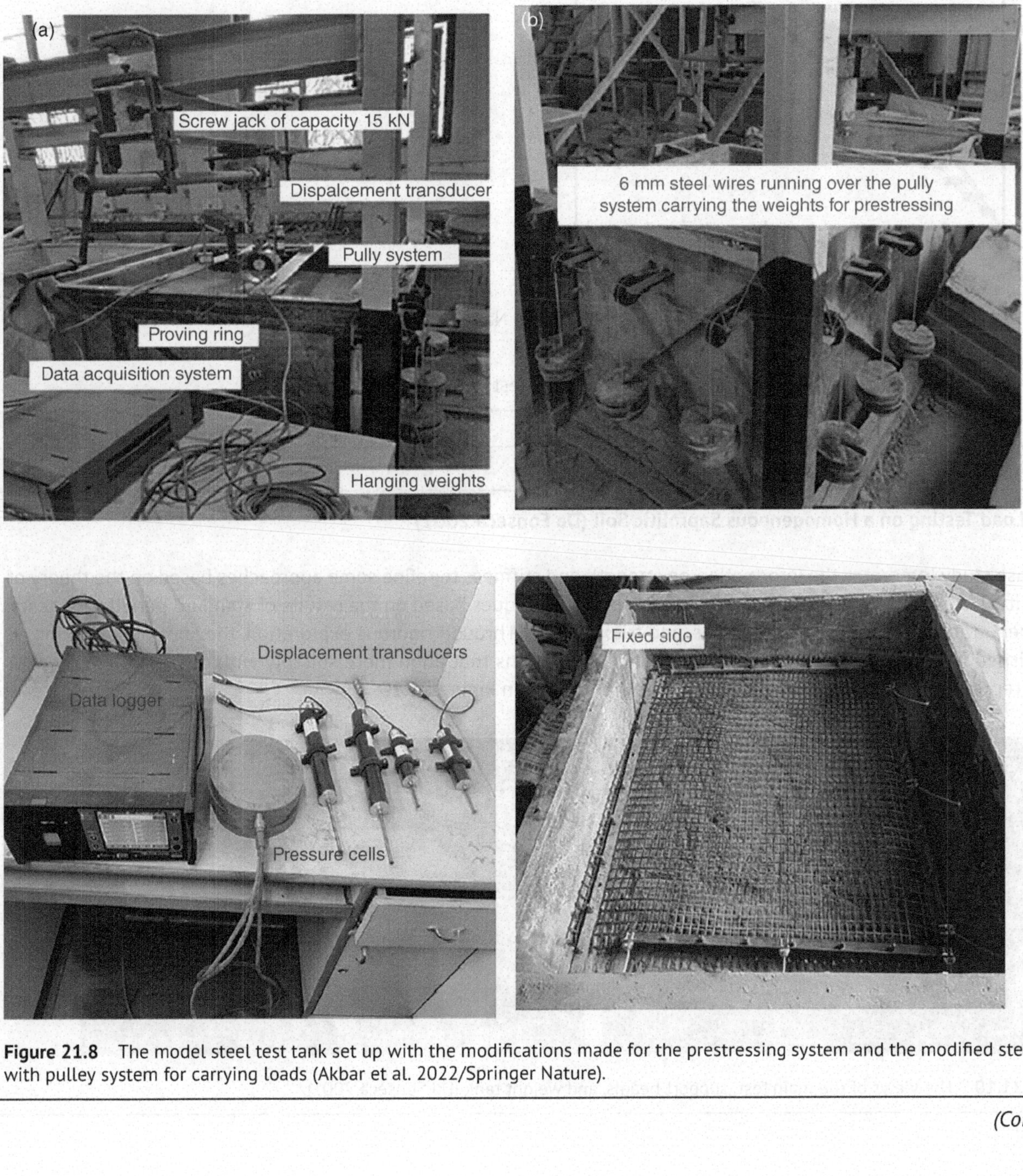

Figure 21.8 The model steel test tank set up with the modifications made for the prestressing system and the modified steel tank with pulley system for carrying loads (Akbar et al. 2022/Springer Nature).

(Continued)

Case Study 21.1 (Continued)

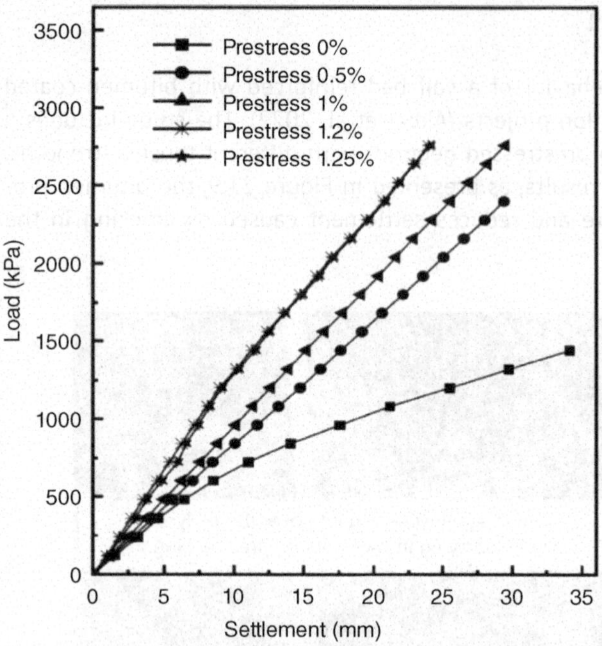

Figure 21.9 The load–settlement graph (Akbar et al. 2022/Springer Nature).

The diagram of the displacement load resulting from this test is shown in Figure 21.9.

Case Study 21.2

Plate-Load Testing on a Homogeneous Saprolitic Soil (Da Fonseca 2001)

This case study integrates the information on strength and stiffness to refine some approaches based on the theory of elasticity. The study chiefly centered on semi-empirical techniques based on the results of standard penetration tests, cone penetration tests, plate loading tests, and triaxial tests. Through rigorous exploration, they not only scrutinized established methods but also introduced potential adaptations that align more suitably with observed behaviors. A visual representation of the primary experiment is portrayed in Figure 21.10.

Figure 21.10 The views of the main test, support beams, and weight tank (Da Fonseca 2001).

The load–settlement curve resulting from this test is shown in Figure 21.11.

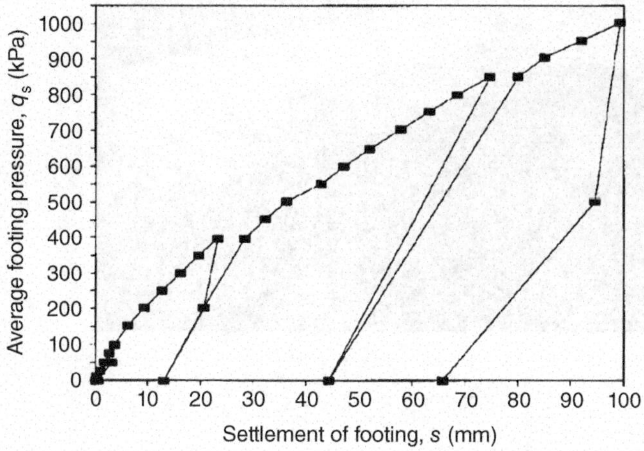

Figure 21.11 The load–settlement curve (Da Fonseca 2001).

Case Study 21.3

Testing of Shallow Foundations on a Backfill of Tire-Derived Aggregate (Mahgoub and El Naggar 2020)

This case study outlines a comprehensive study conducted in Antrim, Nova Scotia, Canada, aimed at investigating the behavior of shallow foundations constructed atop a backfill consisting of tire-derived aggregate (TDA) materials. The study involved the creation and evaluation of three distinct square concrete footings, each featuring different thicknesses of TDA layers. Corresponding field tests were devised to gauge various aspects, including the loads exerted on the concrete footing, the extent of footing settlement, and the stresses transmitted within the replacement layers. The native soil composition at the site primarily comprised glacial till combined with sand and gravel. Characterization efforts encompassed Atterberg limit tests and sieve analysis. To establish the shear strength parameters of the till layer at the proposed foundation level, triaxial tests were carried out. These tests revealed undrained shear strength values (c_u) ranging from 100 to 140 kPa, while the drained shear strength parameters yielded a cohesion of 22 kPa and an internal friction angle of 20°. Figure 21.12 shows the steps for the field test procedures (pouring the concrete, aligning the loading truck, applying the load, and installing the membrane sheet). Figure 21.13 shows the load–settlement curve obtained from the test.

Figure 21.12 Test procedures: (a) pouring the concrete, (b) aligning the loading truck, (c) applying the load, and (d) installation of the membrane sheet (Mahgoub and El Naggar 2020/with permission of Elsevier).

(Continued)

Case Study 21.3 (Continued)

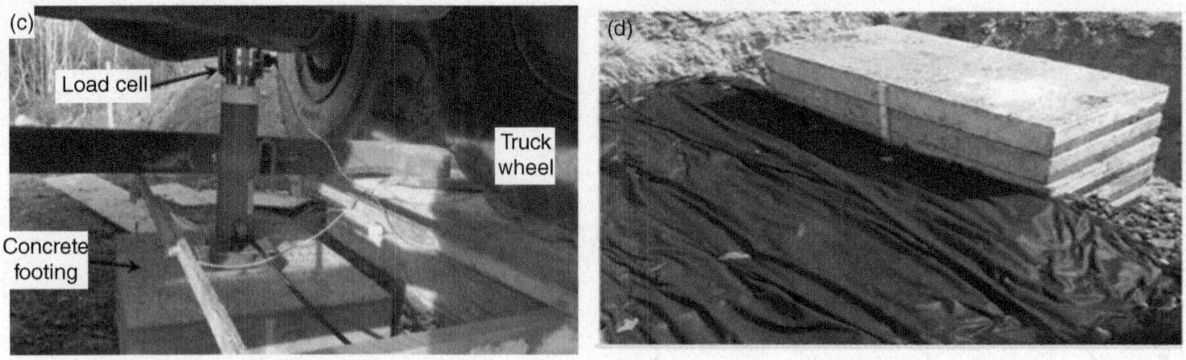

Figure 21.12 *(Continued)*

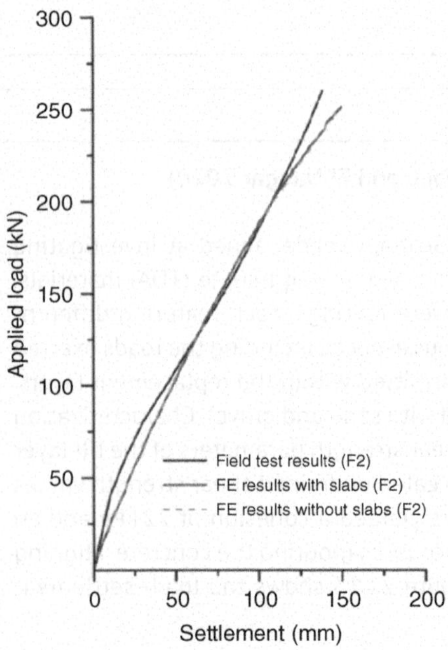

Figure 21.13 Load settlement curve (Mahgoub and El Naggar 2020/with permission of Elsevier).

The load settlement curve resulting from this test is shown in Figure 21.13.

21.3 Deep Foundation Testing

21.3.1 Pile Testing

In cases where the use of shallow foundations is not feasible due to excessive settlement of loose soil or environmental constraints, deep foundations, such as piles or caissons, can serve as an alternative solution. The load transfer mechanism via piles involves a combination of frictional resistance and end bearing, with the load being transferred to the soil. The designer determines the appropriate type and length of the pile through the utilization of data obtained from either in situ or laboratory tests. The load-bearing capacity of the pile at the site is assessed through a variety of pile loading tests, including static and dynamic pile load tests.

Figure 21.14 Images of static loading of the pile ((a) GRL Engineers, Inc).

21.3.1.1 Static Load Testing

One approach to determining the load-bearing capacity of piles is to conduct pile loading tests, with one such test being the static pile load test. After conducting the pile loading test and obtaining the necessary resistance data, the initial design is modified as needed to meet the required specifications. It is preferable to conduct pile loading tests during the design phase, but they are typically carried out during the construction phase. The general nature of pile loading tests is straightforward, involving the application of a significant force, either discretely or continuously, to the pile head through a mechanism that varies depending on available access, methods, and equipment, while recording the resulting vertical displacement. The results and data are then plotted on a load–displacement curve for each level of applied load, and the final load-bearing capacity of the pile is predicted and tested according to various available methods (Das 2010). Photos depicting static load tests of piles are shown in Figure 21.14.

Static load testing on piles is conducted by directly measuring the displacement of the pile head in response to the load applied at the pile head. It is the most fundamental type of load testing for piles and is considered the basis for pile performance evaluation. This test is typically performed with a load ranging from 100 to 12,000 kN. Static load testing may be applied in one of the following modes: compression, lateral, and tension (e.g., uplift).

In static load testing, the load is usually applied through a jack that acts on a reaction beam, which is supported by an anchor system or pulled against a reaction mass (dead weight). The anchoring system may consist of cable anchors or reaction piles attached to the ground to provide the necessary tensile resistance. The nominal test load is usually applied in several increments according to the relevant code or customized testing specifications for a particular project. The load applied at each increment is held constant for a specified period of time or until the rate of pile displacement is less than a nominal value. Static load testing methods can be implemented for all types of piles, on land and in water. This test can be performed on both full-scale piles and model piles specifically made for load testing. Model piles are deliberately loaded until they fail, while full-scale piles are only subjected to a limited amount of load to ensure that they perform satisfactorily under the operational or design load. Additionally, an extra load is applied to determine the reserve capacity of the pile. Views of lateral and tensile/uplift static load tests are shown in Figure 21.15a and b. Examples of compressive static load tests are presented in Figure 21.16.

The loads for static pile tests can be applied in several methods. These methods include:

- Rapid loading
- Slow loading
- Incremental loading with a constant rate
- Stepwise loading at equal time intervals
- Loading with a constant settlement rate
- Stepwise loading with constant settlements
- Cyclic loading

Figure 21.15 (a) Lateral static load test and (b) tensile/uplift static load (Geo Dynamics Engineers LLP/https://www.geodynamics.net/conventional-pile-load-tests/ / last accessed on August 20, 2024.).

Figure 21.16 Images of different formations of pile compressive static loading ((a) Total Foundations/http://totalfoundation.in/other-techniques.php/last accessed on August 20, 2024. (b) Battista et al. 2016/Emerald Publishing Limited.).

There are several methods to interpret the results of static pile tests. Some of these methods include:

- Hansen Brinch method (Hansen 1963)
- DeBeer yield limit method (DeBeer 1968)
- Failure load Chin–Kondner extrapolation method (Chin 1970; Kondner 1963)
- Davisson offset limit method (Davisson 1972)
- Mazurkiewicz method (Mazurkiewicz 1972)
- Butler and Hoy method (Butler and Hoy 1977)
- Maximum curvature point method (Shen and Niu 1991)
- Decourt Method (Decourt 1999)

Figure 21.17 illustrates the application of some of the mentioned methods on a load–displacement curve. More details and discussion on different methods can be found in Fellenius (1980), Fellenius (1990), and Fellenius (2001).

In crucial projects like the one at hand, it becomes imperative to ensure the performance and load-carrying capacity of the designed piles. To achieve this assurance, on-site static load testing is conducted either during or after the design phase. This process serves to validate the load-carrying capacity of the piles, thereby ensuring the overall success and safety of the project. As an example, Houseinzadeh and Alielahi (2018) conducted a study involving static load tests on the piles used for an access bridge near the Persian Gulf. This cable-stayed bridge spans approximately 2230 m in length and features bored concrete piles, each accompanied by its respective pile cap. The bored piles come in varying diameters, ranging from

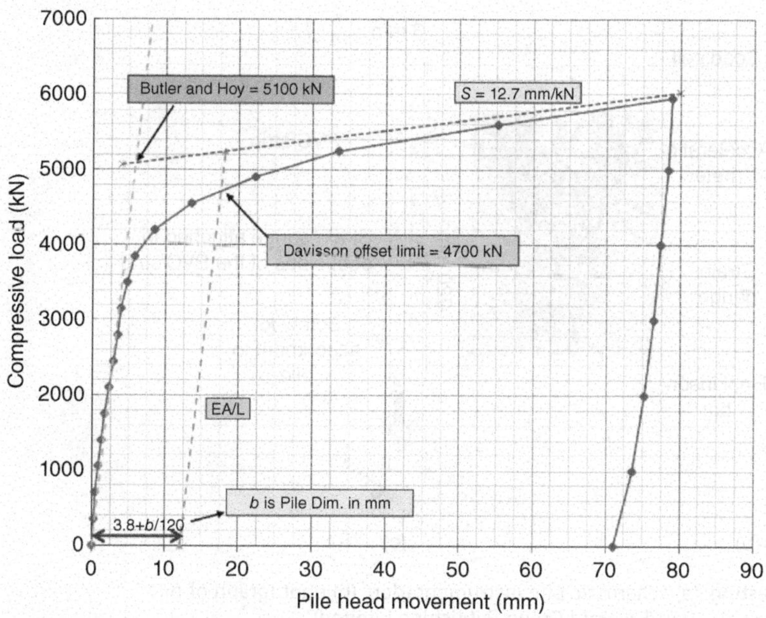

Figure 21.17 An example of a static load test graph for a pile (Fellenius 1975/American Society of Civil Engineers).

1.2 m, with depths spanning from 25 to 32 m. The specific dimensions of the piles were determined based on thorough geotechnical studies and soil condition testing within the project area.

21.3.1.2 Dynamic Load Testing

Although static load testing is a reliable method for determining load-carrying capacity, it can be expensive and time-consuming. Dynamic load testing, on the other hand, is a more cost-effective and faster alternative. It allows for testing a greater number of piles while saving time and money. In coastal structures where piles are installed in water, conducting static load tests can be practically impossible, making dynamic load testing the only viable option. The philosophy of dynamic load testing is based on measuring dynamic force and velocity values resulting from the impact of a hammer on the pile head. All dynamic relationships in determining the load-carrying capacity of a pile are derived from force–momentum relationships (Heins et al. 2020). Figure 21.18 presents some details of dynamic load testing.

The following factors are influential in dynamic load testing:

- Efficiency of the hammer
- Weight of the hammer
- Drop the height of the hammer
- Penetration per blow
- Elastic compression of the cushion
- Elastic compression of the pile
- Elastic compression of the soil
- Pile weight

Types of dynamic load testing for piles include:

- Dynamic pile testing with the impact of drop weight
- Statnamic testing
- Integrity testing
- Rapid load testing

The steps involved in conducting the dynamic load test are:

- Instrumentation of the pile
- Application of the dynamic load – dropping the hammer, vibration
- Acquisition of data – pile penetration, wave propagation
- Analysis and interpretation of data

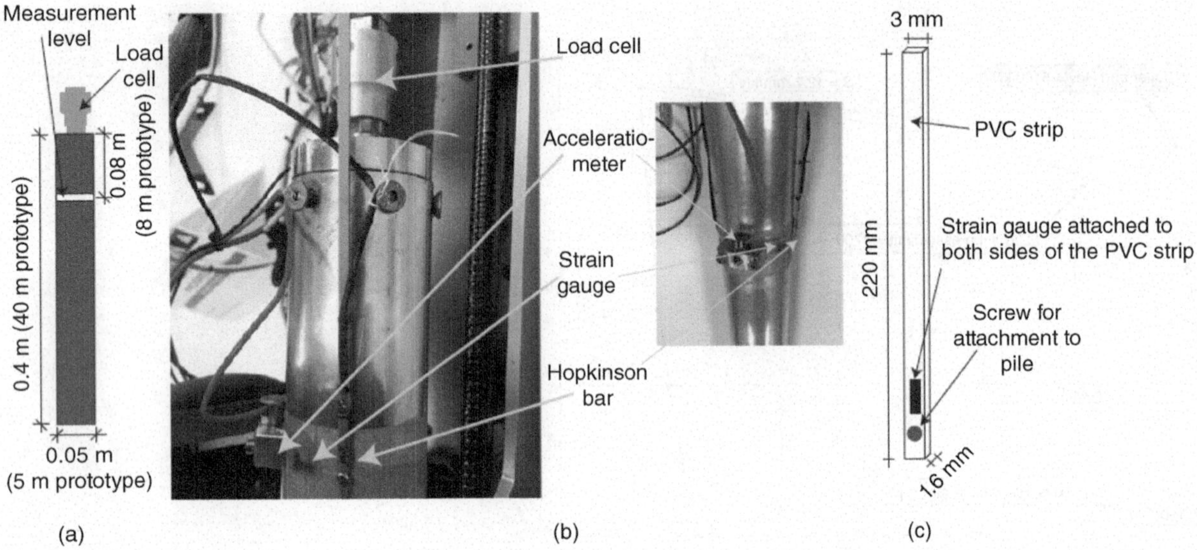

Figure 21.18 Displaying images of the dynamic load testing: (a) schematic pile instrumentation, (b) photograph of pile instrumentation, and (c) sketch of Hopkinson bar (Heins et al. 2020/Emerald Group Publishing Limited).

21.3.1.3 Acquisition and Analysis of Dynamic Load Testing Data

A strain gauge and an accelerometer are affixed to the pile for data collection purposes. The strain gauge enables the calculation of strain, which, in conjunction with the pile's area and modulus of elasticity, facilitates the determination of the force applied to the pile. Simultaneously, the accelerometer serves to compute wave velocity and the resultant force within the pile, derived from the wave's acceleration. It is crucial that the forces acquired from these two instruments align and are interrelated. The data gathered by these instruments are then transmitted to a computer for analysis, with the outcomes presented to the operator (in accordance with ASTM D4945-12) as illustrated in Figure 21.19. The computer captures the wave generated by the impact of the pile and records the extent of the pile's penetration into the soil.

Figure 21.19 Dynamic testing and instrumentation (Alvarez et al. 2006/American Society of Civil Engineers).

The Pile Driving Analyzer® (PDA) system is the most widely employed system for dynamic load testing and pile driving monitoring. During the test, the PDA system collects crucial information concerning the velocity and stress induced within the pile. The PDA system assesses the capacity of the tested pile, as well as evaluates shaft integrity, driving stresses, and hammer energy when monitoring installation.

Part of the PDA system is the CAse Pile Wave Analysis Program (CAPWAP®). CAPWAP is a widely employed tool for the analysis of data collected during dynamic pile testing. CAPWAP operates as an iterative curve-fitting technique, aligning the pile's response, as determined by a wave equation model, with the measured response from the actual pile subjected to a single hammer blow. The pile model is composed of continuous segments, representing the overall resistance of the embedded section of the pile through a series of springs (static resistance) and dashpots (dynamic resistance). The testing process begins by recording the forces and accelerations acting on the actual pile during the initial impact. This data is captured using a strain gauge and accelerometer affixed at the pile head. These recorded measurements, along with reasonable estimates of soil resistance, seismic activity (quake), and damping parameters, serve as inputs for the pile model. Subsequently, the force-time signal at the pile head is computed using the model and compared against the measured force-time signal. The soil–resistance distribution, seismic activity, and damping parameters are iteratively adjusted until a close alignment is achieved between the measured and calculated signals. Figure 21.20 (GZA GeoEnvironmental, Inc. 1999) provides an illustrative example, demonstrating a comparison between a measured and calculated force signal from a test pile. Upon achieving an acceptable alignment, the analysis furnishes an estimate of the ultimate static capacity, details on the distribution of soil resistance along the pile, and values for seismic activity and damping parameters.

By employing the CAPWAP and the data extracted, it becomes feasible to calculate the pile's load-carrying capacity. This analysis also allows for an examination of various factors, such as the hammer's performance, cushion efficiency, alignment of the hammer with the pile, and hammer efficiency through the study of wave motion. Additionally, monitoring stress within the pile due to wave motion helps identify potential issues related to pile damage and failure during driving. Furthermore, this approach enables the determination of crack length and depth in concrete piles and identifies the location of buckling in steel piles.

For a comprehensive understanding of dynamic testing fundamentals, including CAPWAP, readers are referred to the following document:

- ASTM (1996a), "Standard Test Method for High-Strain Dynamic Testing of Piles;" D4945-89, 1996 Annual Book of ASTM Standards, ASTM International, West Conshohocken, PA.
- FHWA (1998), Design and Construction of Driven Foundations, Report No. FHWA-HI-97-013, Washington, DC.

21.3.1.4 Advantages and Disadvantages of Pile Dynamic Analysis (PDA) System Load Test

Conducting a PDA system load test on full-sized piles and under real-world conditions yields precise results. Furthermore, dynamic testing is faster and more cost-effective compared to static testing. It also permits the examination of soil resistance during both initial driving and restrikes resulting from hammering. One limitation of the PDA test is that if the hammer

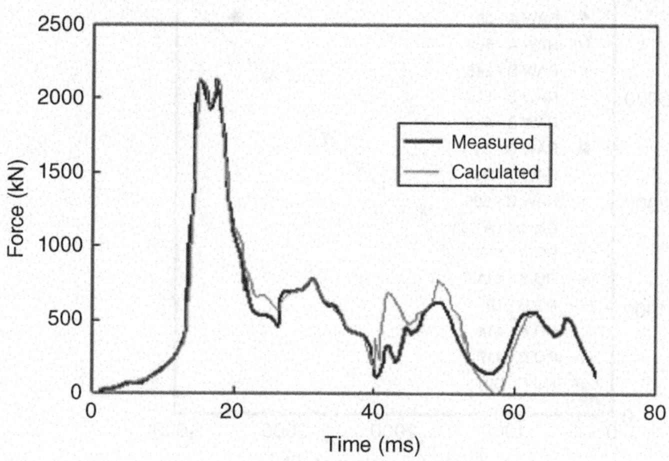

Figure 21.20 Comparison between a measured and calculated force signal from a test pile using CAPWAP® signal matching (GZA GeoEnvironmental, Inc. 1999).

energy is insufficient, the pile resistance may not activate, making it challenging to predict the pile's load-carrying capacity (ASTM D4945-12).

Information that can be gleaned from PDA load testing includes:

- Utilizing the Case method to ascertain load-carrying capacity
- Assessing pile driving performance and energy transfer to the pile
- Determining stress levels induced in the pile during driving
- Monitoring the pile's health and damage throughout the driving process
- Providing input parameters for wave equation analysis.

In 2013, Hosseinzadeh and Fakharian (2013) conducted a study at the Fajr Petrochemicals site utilizing dynamic load testing. An essential aspect of these tests is the determination of β during and after pile driving. Figures 21.21 and 21.22 illustrate the data sourced from the Fajr Petrochemicals site concerning the values of β (used in calculating shaft friction

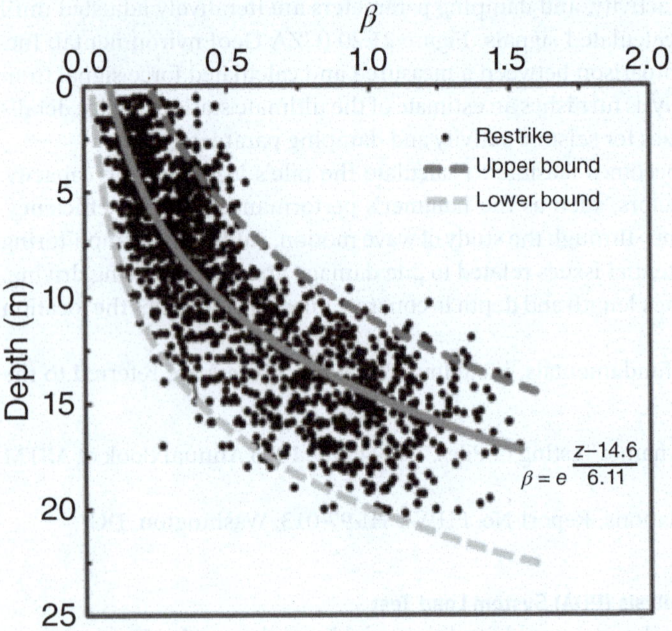

Figure 21.21 Fajr petrochemical site data for beta values (Hosseinzadeh and Fakharian 2013).

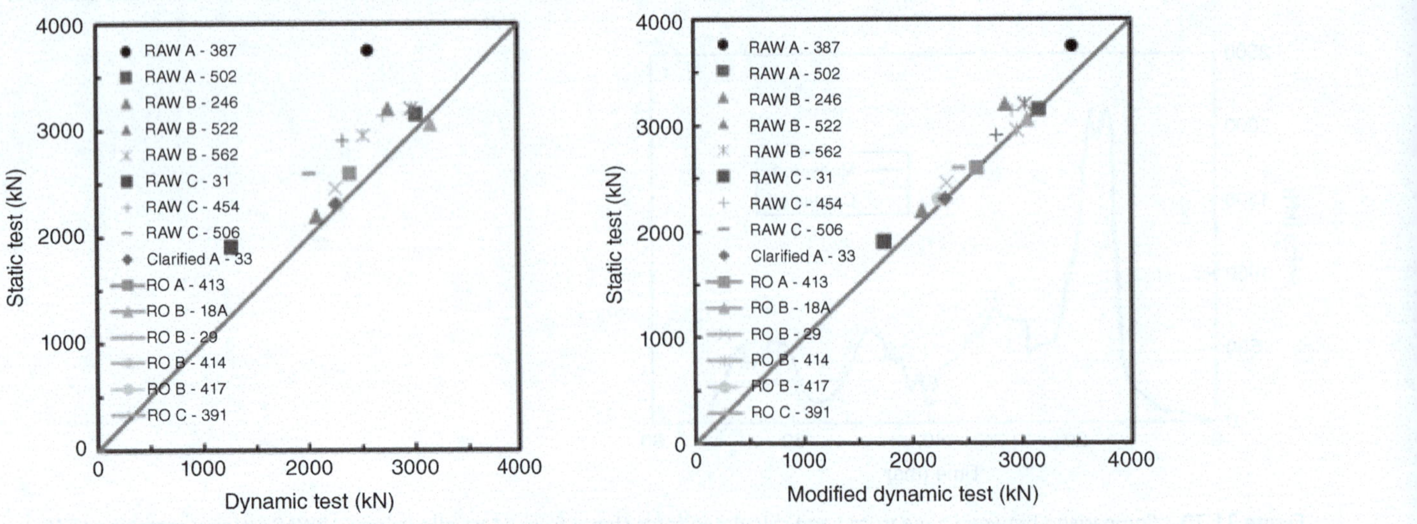

Figure 21.22 Correlation between static and dynamic tests on the Fajr Petrochemicals site (Hosseinzadeh and Fakharian 2013).

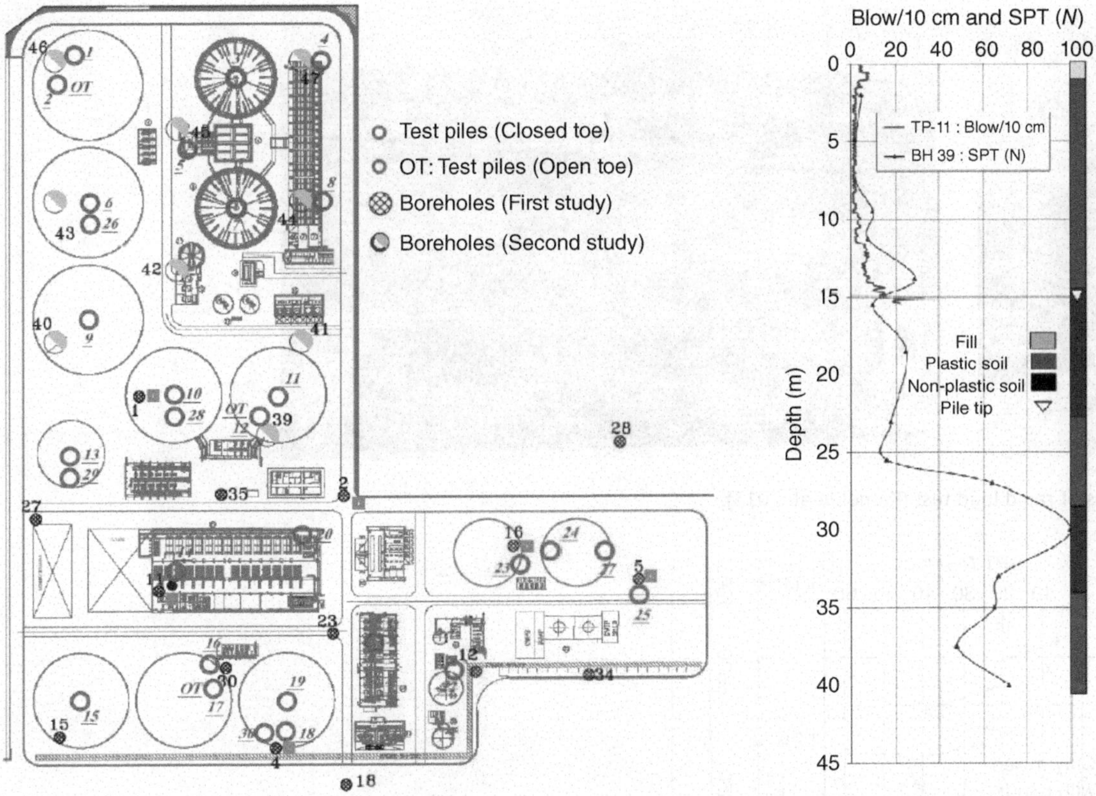

Figure 21.23 Plan view of water unit structures, boreholes, and test pile locations, Fajr II (Hosseinzadeh and Fakharian 2013).

in saturated clays) and the correlation between static and dynamic tests, respectively. Figure 21.23 provides a plan view of water unit structures, boreholes, and test pile locations at the site.

21.3.1.5 Rapid Load Testing

The main alternative to static and dynamic tests is rapid load testing. Usually, the diameter and depth of current buried piles are higher compared to driven piles, and it is difficult to activate their resistance with a pile-driving hammer. However, rapid load testing can activate the pile resistance. To activate the pile resistance, it is necessary to use very large masses for falling onto the pile, which significantly increases the potential for pile damage (Holscher and van Tol 2008). Images of the rapid load test are given in Figure 21.24.

The testing location was in the Okayama area, Japan. The soil profile and layer conditions are depicted by SPT *N*-values, as shown in Figure 21.25. Weathered rocks, characterized by *N*-values exceeding 50, are located below a depth of 20 m (Kamei et al. 2013). Figure 21.26 illustrates the relationship between the load–displacement curve and soil resistance, as well as ULP (ultimate limit pile) resistance.

21.3.2 Drilled Piers/Shafts Testing

21.3.2.1 Static Testing: Conventional

Conventional testing refers to the use of traditional kentledge methods to determine reaction loads. Kentledge, a traditional technique, involves the use of heavy weights (such as concrete blocks) to generate a reaction load on the structure. When compared to modern methods for determining reaction loads, traditional kentledge methods have several drawbacks, including the requirement for heavy equipment, time-consuming installation, transportation and relocation, and the high associated costs. Therefore, new methods like impact testing, rapid load testing, and nondestructive testing techniques such as seismic testing and ultrasonic measurement are employed to ascertain reaction loads in structures (Yen and Lee 2000). Figure 21.27 illustrates a schematic of a static axial compression load test using traditional "kentledge" reaction load, while Figure 21.28 provides a comparison between Statnamic (dynamic testing) and traditional kentledge methods for shafts of the same size.

Figure 21.24 Images of rapid load test (Kamei et al. 2013).

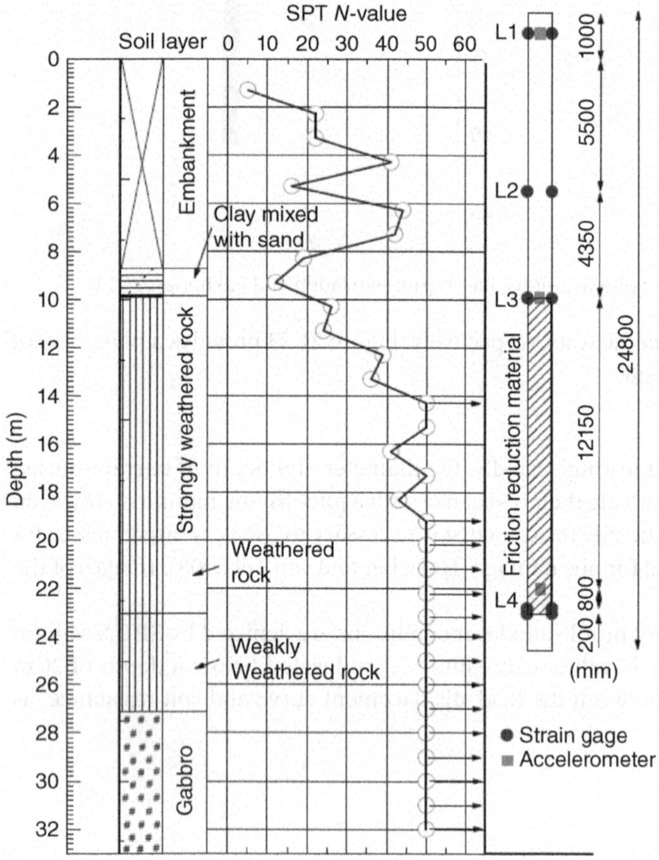

Figure 21.25 Profiles of soil layers and SPT *N*-values, together with the test pile No. 1 (Kamei et al. 2013).

21.3.2.2 Osterberg Testing

The bidirectional load test method for bored piles was revolutionized by Osterberg (1984) through the introduction of the O-cell. The O-cell is a high-capacity hydraulic jack placed inside the shaft of a pile to test its load-carrying capacity. By applying hydraulic force, the jack exerts force in two directions, upward and downward. This device, when internally pressurized, applies equal upward and downward loads, enabling separate determination of side shear and end-bearing capacities. The Osterberg cell (O-cell) bi-directional testing method offers a cost-effective means of conducting high-capacity static load tests of drilled piers that would otherwise be prohibitively expensive or technically

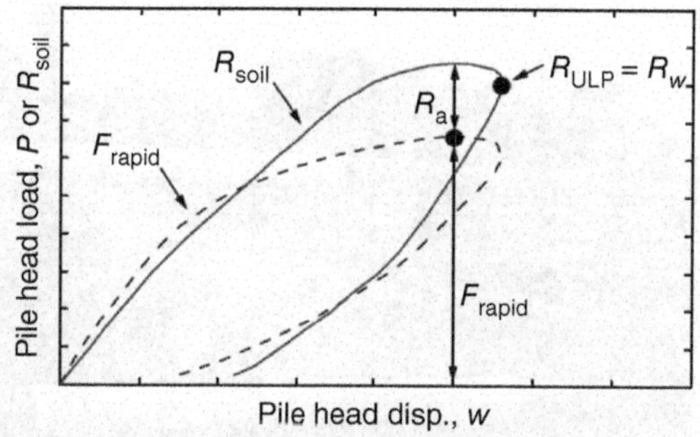

Figure 21.26 Relationship between load–displacement curve and soil resistance and ULP resistance (Kamei et al. 2013).

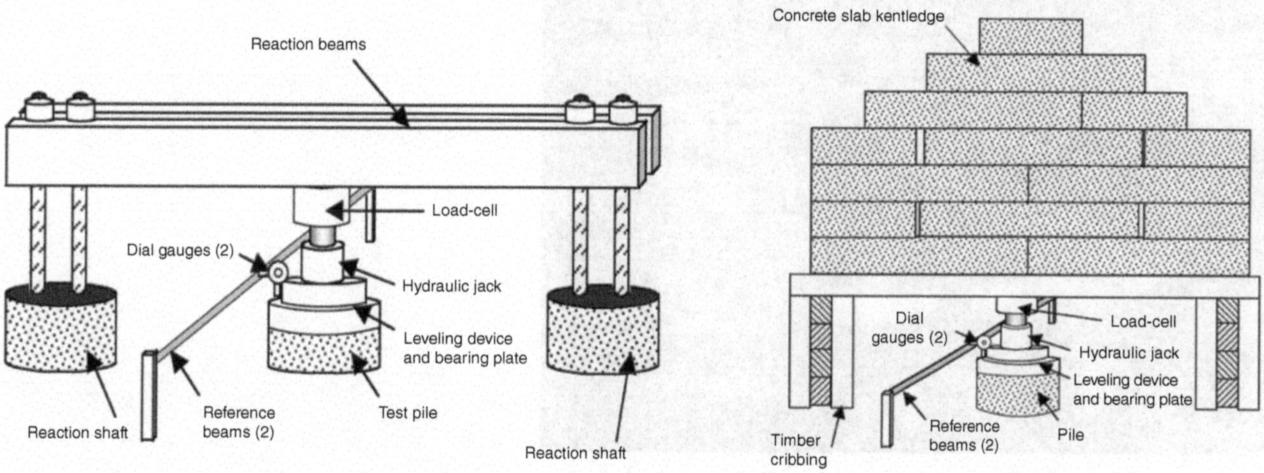

Figure 21.27 Schematic of a static axial compression load test with a traditional "kentledge" reaction and with reaction shafts (Yen and Lee 2000).

Figure 21.28 Comparison of Statnamic and a traditional kentledge for the same-sized shafts (Yen and Lee 2000).

challenging. The key innovation lies in a specially designed hydraulic jack (O-cell assembly) that is welded directly into the pile at a predetermined location, as depicted in Figure 21.29 (Caltrans 1998). Figure 21.30 (Salman 2021), illustrates the bidirectional load test: (a) steel cage reinforcement, (b) bidirectional load cell (Salman 2021). The O-cell is hydraulically pressurized from the surface, simultaneously loading both the pile section above and below the O-cell. This internal loading approach ensures that the pile component above the O-cell acts as a reaction for loading the

Figure 21.29 O-cell installation in a drilled shaft (Caltrans 1998/State of California/Public Domain).

(a) (b)

Figure 21.30 Illustration of the bidirectional load test: (a) steel cage reinforcement and (b) bidirectional load cell (Salman 2021/The University of Texas at Arlington).

pile component below it and vice versa. The upward force is resisted by the wall friction, and the downward force is resisted by the pile's bearing capacity, or the combination of the pile's bearing capacity and wall friction, depending on the O-cell's location. With the O-cell testing method, engineers can statically load test full-scale piles to near their ultimate capacity in place. During testing, electronic sensors measure the displacement of both pile sections, allowing the O-cell to assess end bearing and skin friction separately, thus maximizing the information obtained. This method allows for load testing to be performed with greater ease, economy, and safety and provides more information about

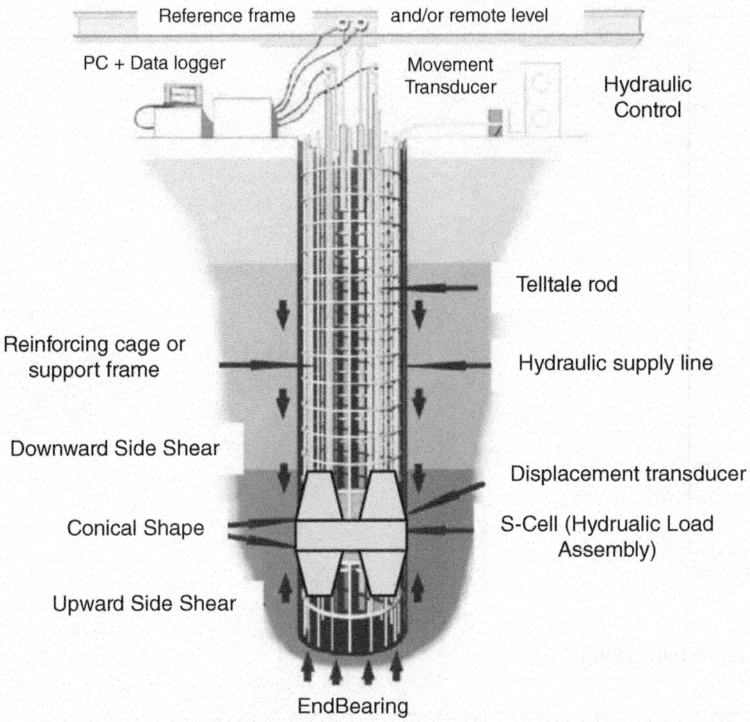

Figure 21.31 Typical Osterberg cell test instrumentation (Patidar et al. 2020).

the behavior of the pile. The use of Osterberg cell methods has led to a reduction in the conservative design of drilled piers and improvements in construction methods, leading to increased use of deep drilled pier foundations in structures (Fellenius et al. 1999). Figure 21.31 (Schmertmann et al. 1988) shows the different components of an Osterberg cell instrumentation setup.

The O-cell method not only enhances safety but also saves time and resources due to reduced preparation efforts for testing. It has become the preferred method for static load testing of bored piles and auger cast-in-place piles. The O-cell is positioned between two steel plates, and the reinforcement cage is tack-welded to these steel plates for ease of handling. During the load test, these tack-welds break, causing the two sections to be loaded in opposite directions. This method facilitates the measurement of side resistance and end-bearing capacity of the drilled shaft, as demonstrated in Figure 21.32.

To interpret the results of a bidirectional load test, they are typically converted to results from a head-down test. The equivalent top–down curve, reflecting the measured side resistance and end-bearing curve, is determined, as shown in Figure 21.33.

21.3.3 Dynamic Testing

21.3.3.1 Statnamic Testing

The Statnamic test is a recently developed rapid load testing method for piles. Statnamic device involves propelling a reaction mass onto the pile head using the high pressure of explosion of gas or solid fuel with a high expansion rate. Figure 21.34 (Garner 2007) shows the basic components of the Statnamic test. The explosion pressure exerts a downward force on the pile in accordance with Newton's third law and is equal in magnitude and opposite in direction to the mass's reaction. The maximum acceleration of the reaction mass generated during loading is approximately 5 g. Consequently, the weight of the reaction mass is roughly 20% of the intended maximum load. One of the notable advantages of Statnamic is its high loading capacity, which can reach up to 60 MN. The results obtained from the Statnamic test are analyzed by subtracting the inertial and damping components to determine the equivalent static load-carrying capacity of the pile. This method has led to highly accurate results in predicting the load-carrying capacity of piles in coarse-grained soils but is less accurate in determining the load-carrying capacity in fine-grained soils. Figure 21.35 shows (a) a schematic view indicating different stages of Statnamic testing and testing and (b) equipment for conducting the Statnamic pile test (Hoefsloot 2010).

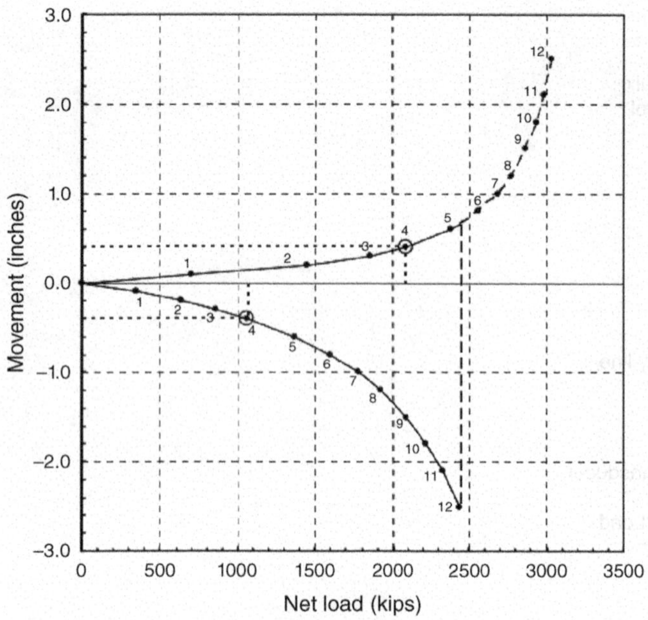

Figure 21.32 Typical bidirectional load test results after (Osterberg 1998).

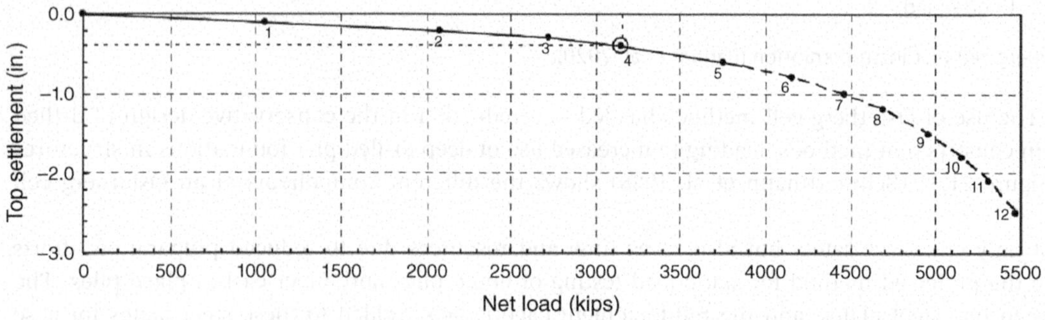

Figure 21.33 Equivalent settlement curve for top-down loading (adapted from Osterberg 1998) .

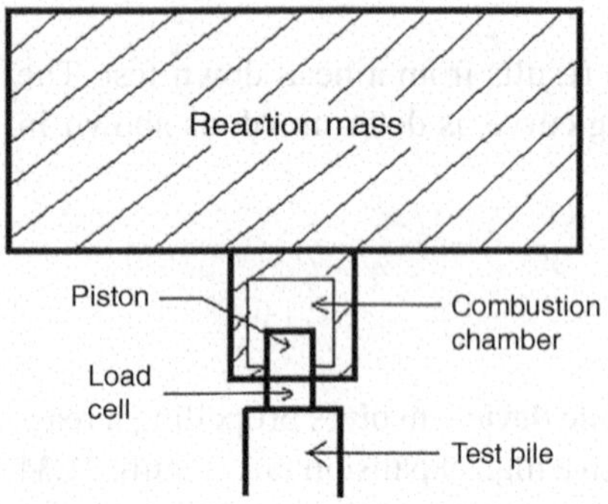

Figure 21.34 Components of Statnamic test (Garner 2007).

Figure 21.36 shows Statnamic testing performed in a river (Haque et al. 2018). It is important to mention that performing repeated loading with the Statnamic device can be challenging.

Statnamic loading has a duration of approximately 120 ms, which is significantly longer than the natural period of even the stiffest pile. This extended duration minimizes stress wave effects, allowing the pile to be treated as a rigid body, and conventional static analytical methods can be applied. It is important to note that while Statnamic load tests have demonstrated negligible rate effects for piles in very stiff soils and those end-bearing in rock, rate effects for piles in soft soils have been relatively substantial and have had a significant impact on load–displacement behavior (Swaroopa et al. 2012). The Unloading Point model (Weaver and Rollins 2009), depicted in Figure 21.37, serves as a straightforward analytical method for determining the static resistance observed during a Statnamic test. Additionally, the Unloading-Point model enables the quantification of rate effects that may occur during such a test.

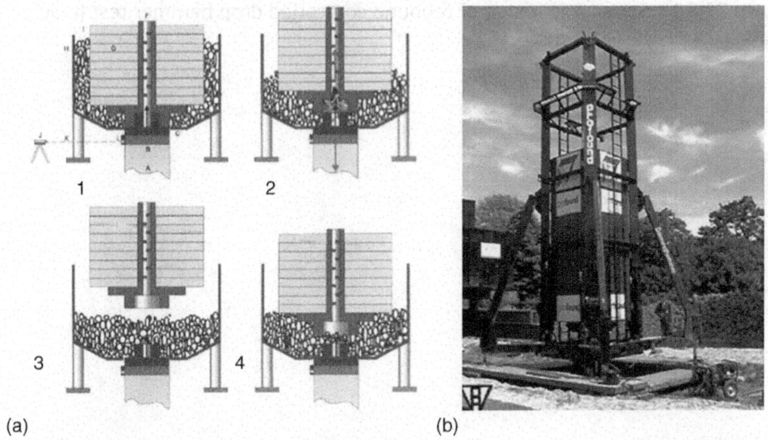

Figure 21.35 Statnamic pile test (SNPT): (a) schematic view indicating different stages of testing and (b) equipment for conducting the Statnamic pile test (Hoefsloot 2010/Fugro).

Figure 21.36 Rigolets pass bridge – Statnamic test of the pile in the river (Haque et al. 2018/SAGE Publications, Inc.).

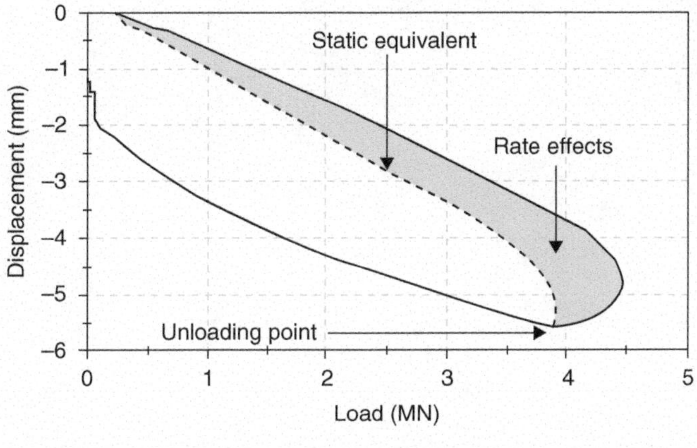

Figure 21.37 The unloading point model (Swaroopa et al. 2012).

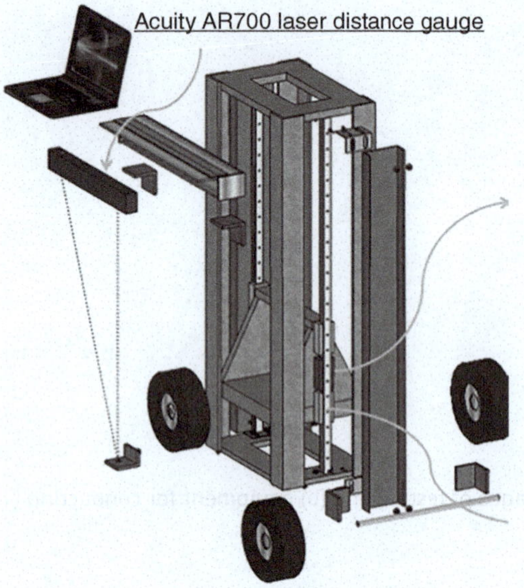

Figure 21.38 Schematic image of rebound controlled drop hammer test (Liao et al. 2018).

21.3.3.2 Hammer Drop Testing

Hammer drop testing is a fast-loading test method used to investigate the behavior and load-bearing capacity of piles and deep foundations. In this test, a heavy mass is placed at a high speed on the head of the pile or foundation to apply a rapid loading as an impact load. Then, by monitoring the movement of piles or foundations, their behavior under rapid loading is investigated. This method is very popular in the construction industry due to its simplicity and speed (Liao et al. 2018). Figure 21.38 depicts a schematic view of the rebound-controlled drop hammer test. A schematic image of the drop hammer test device is given in Figure 21.39 (Elgamal 2020). Figure 21.40 shows a schematic of typical results of force and velocity from the drop hammer test from which the pile capacity is derived (Elgamal 2020).

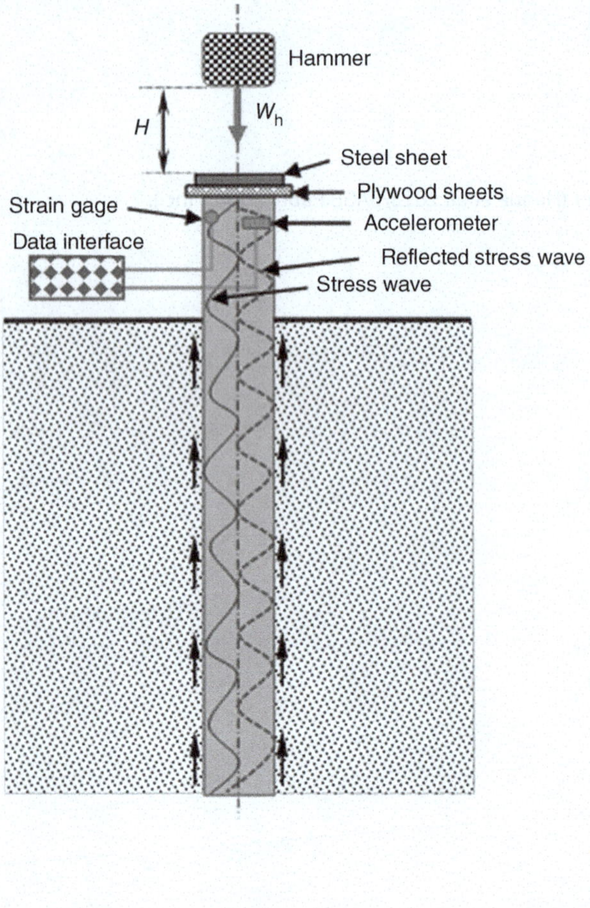

Figure 21.39 Schematic sketch showing high strain dynamic testing of the drilled shaft (Elgamal 2020).

Figure 21.40 Typical force and velocity schematic (Elgamal 2020).

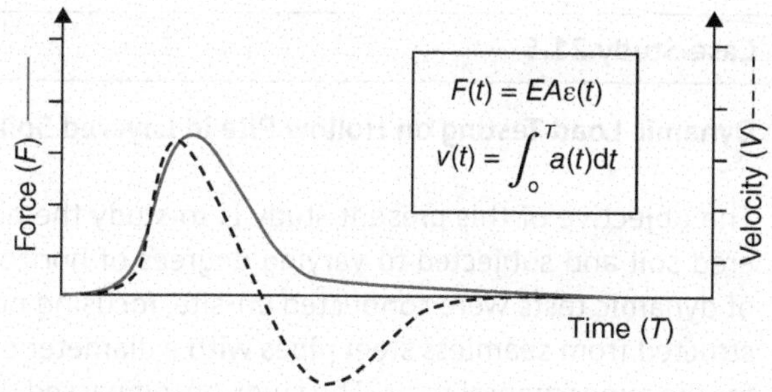

21.3.4 Case Studies

Further case studies on deep foundation have been presented in this section.

Case Study 21.4

SLT on Open-Ended Pipe Pile (Han et al. 2019)

This research executed a full-scale static load test on a double-wall, open-ended pipe pile with a diameter of 660 mm, which was driven into a gravelly sand to a depth of 30.48 m. The pile was instrumented with a combination of electrical resistance, and vibrating-wire strain gauges to measure the outer and inner shaft resistances and the annulus resistance during a static load test. The measurements taken during the test provided data on the load–settlement curve, load–transfer curves, and a comprehensive profile detailing the unit limit shaft resistance. Two distinct methods, both based on cone penetration tests, were employed to estimate the pile's resistance. The findings indicated that while both methods capably approximated unit shaft resistance in sandy layers characterized by low gravel content, they notably overestimated the unit shaft resistance activated within soils characterized by high gravel content. Furthermore, the estimated base resistances, as determined by both methods, are closely aligned with those observed during the static load test, offering validation for their accuracy (Han et al. 2019).

Pile head settlement and load transfer curves at the ultimate load level are shown in the diagrams in Figure 21.41.

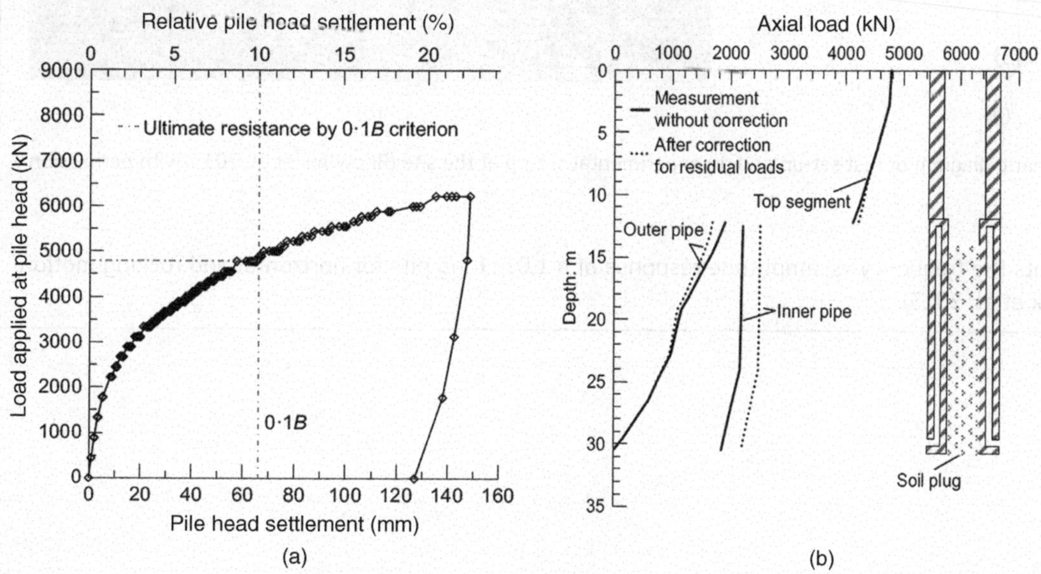

Figure 21.41 (a) Pile head settlement and (b) load transfer curves at the ultimate load level (Han et al. 2019/Institution of Civil Engineers).

Case Study 21.5

Dynamic Load Testing on Hollow Pile in Layered Soil (Bhowmik et al. 2013)

The objective of this present study is to study the nonlinear behavior exhibited by a hollow pile situated within layered soil and subjected to varying degrees of horizontal dynamic load. In order to realize this investigation, a series of dynamic tests were conducted on-site, focusing on the behavior of hollow, closed-end steel piles. These piles, constructed from seamless steel pipes with a diameter of 100 mm and a wall thickness of 5 mm, were the chosen subjects for the experimental trials. The study encompassed three distinct *L/d* ratios (10, 15, and 20) to facilitate comprehensive analysis. A conical steel shoe was affixed to the pile's tip to enhance its performance. To mitigate resistance during the pile driving phase, preliminary auger boring was executed using a 75 mm auger bit. Subsequently, the 100 mm diameter piles were meticulously driven into the soil utilizing a 65 kg drop hammer, with each drop from a height of roughly 300 mm, all carried out through the augered borehole. The driving process was meticulously overseen using a tripod stand to ensure the pile's vertical alignment. Post-driving, a resting period of approximately one month was allocated to facilitate optimal bonding between the pile and the soil. This period was deemed essential due to the potential soil disturbance caused during the driving operation. Allowing for soil settlement and the solidification of the bond was vital to ensure the structural integrity of the pile-soil system (Bhowmik et al. 2013). A visual representation of the testing arrangement can be observed in Figure 21.42.

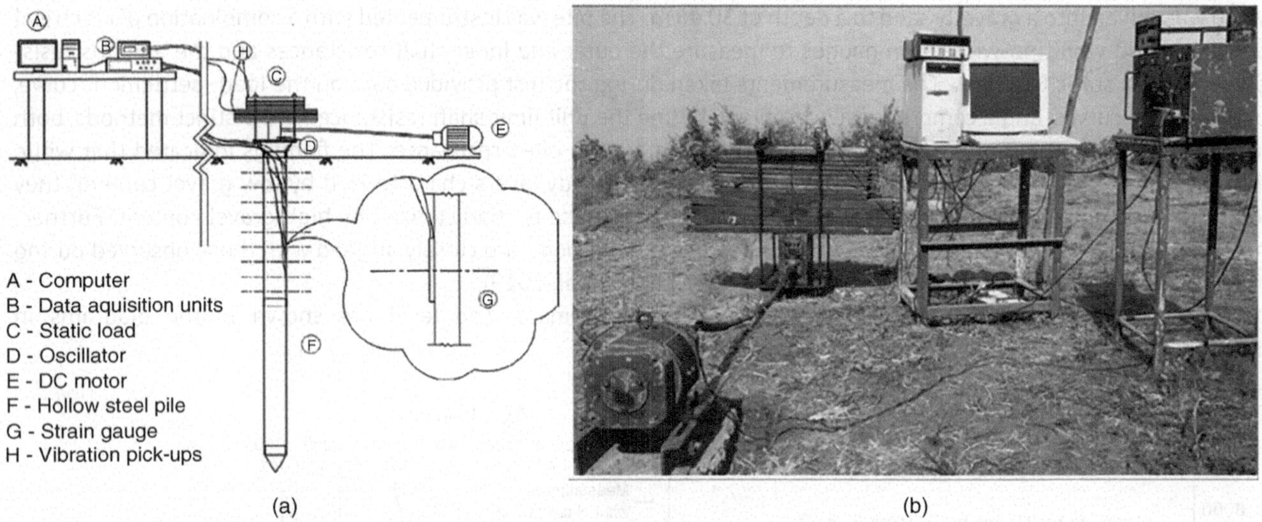

Figure 21.42 (a) Schematic diagram of test set-up and (b) experimental set-up at the site (Bhowmik et al. 2013/with permission of Elsevier).

Figure 21.43 presents the frequency vs. amplitude response of a 1.0 m long pile for horizontal and rocking motion, respectively (Bhowmik et al. 2013).

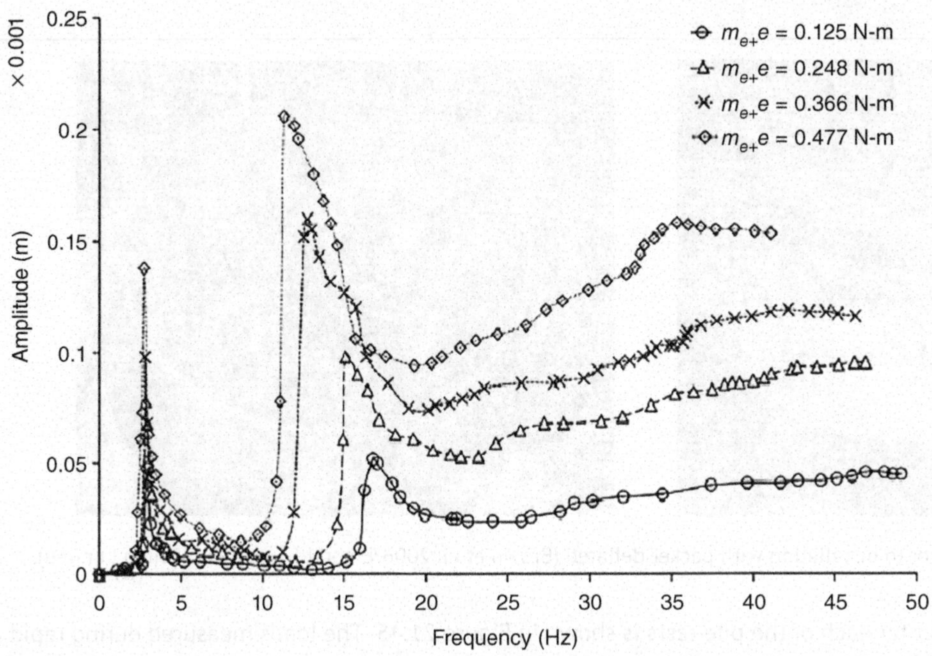

Figure 21.43 Frequency vs. horizontal and rocking amplitude for *L*/*d*¼10 (Bhowmik et al. 2013/with permission of Elsevier).

Case Study 21.6

Rapid Load Testing in Clays (Brown et al. 2006)

The utilization of rapid load testing methodologies for piled foundations is increasingly gaining prominence, particularly with the Statnamic test experiencing growing adoption in the United Kingdom. However, prevalent methods of analysis tend to overestimate pile capacities for clays, with discrepancies reaching up to 50%. In order to probe the behavior of rapid load pile testing within clay formations comprehensively, a thorough investigation was undertaken, featuring a full-scale pile subjected to testing. This pile's performance was contrasted against the constant rate of penetration and maintained load static tests. To decode the complexities, nonlinear viscous parameters derived from laboratory models and element tests were introduced to effectively emulate the rate-dependent shear resistance of clay. The investigation involved a comparative assessment of shaft frictions, which were obtained through strain-gauged reinforcement within the pile, with shear strains and stresses originating from accelerations within the surrounding soil. This rigorous approach aimed to elucidate the intricate load transfer mechanisms governing the rapid loading of a pile within clay formations. For deeper insight, the study conducted by Brown et al. in 2006 examined these dynamics in detail. A visual depiction of the pile-tip load cell before installation, with the packer in a deflated state, is presented in Figure 21.44.

(Continued)

Case Study 21.6 (Continued)

Figure 21.44 Pile-tip load cell prior to installation with packer deflated (Brown et al. 2006/Emerald Group Publishing Limited).

The load–settlement behavior for each of the pile tests is shown in Figure 21.45. The loads measured during rapid load testing do not start at zero because, prior to testing, the 18-t reaction mass was lowered onto the pile.

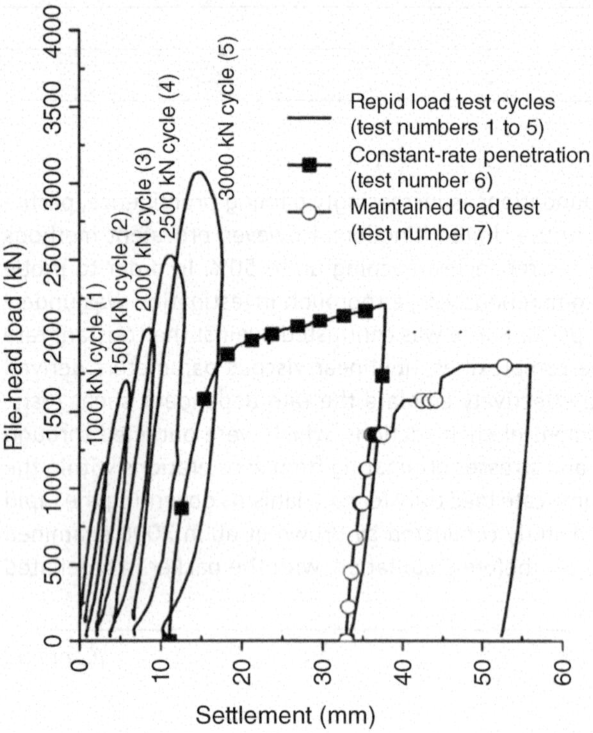

Figure 21.45 Pile load–settlement history from different load-testing methods (Brown et al. 2006/Emerald Publishing Ltd).

Case Study 21.7

Osterberg Testing on an Instrumented Pile (Pelecanos et al. 2017)

This paper introduces a compelling case study centered around an Osterberg-cell test performed on a meticulously instrumented pile situated in London. The pile's instrumentation included an array of cutting-edge tools encompassing distributed optical fiber sensors, strain gauges, displacement transducers, and extensometers. The data gleaned from the optical fiber sensors distinctly revealed the progressive development of shaft friction in direct correlation with applied loads. Intriguingly, an unexpected slip between the pile and the soil emerged within two specific soil layers, a phenomenon ascribed to the formation of a filter cake stemming from an extended construction period. To unravel the complexities of pile-soil slip and its crucial role in predicting vertical displacements, the researchers embarked on finite-element analyses. These analyses served as a platform for comprehending the underlying mechanisms of pile-soil slip and its implications. The study was further enriched by insights into the behavior of the founding chalk layer. The work conducted by Pelecanos et al. in 2017 brought to light these noteworthy findings. A visual depiction detailing the geometry of the test pile and elucidation of the employed instrumentation can be observed in Figure 21.46.

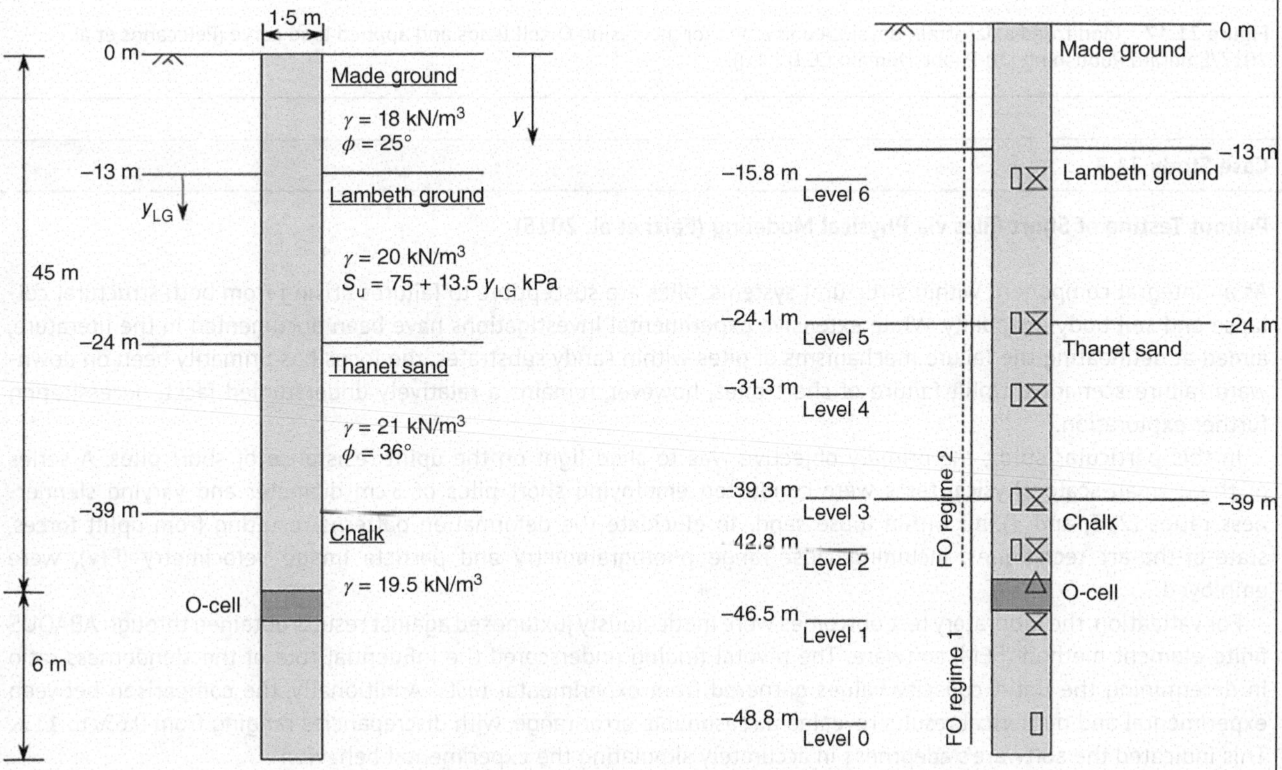

Figure 21.46 Geometry of the test pile and description of instrumentation (Pelecanos et al. 2017/Emerald Publishing Ltd/Public Domain CC BY 4.0).

Figure 21.47 shows the monitored axial mechanical strain during the load test, which has been compensated for temperature change (Pelecanos et al. 2017).

(Continued)

Case Study 21.7 (Continued)

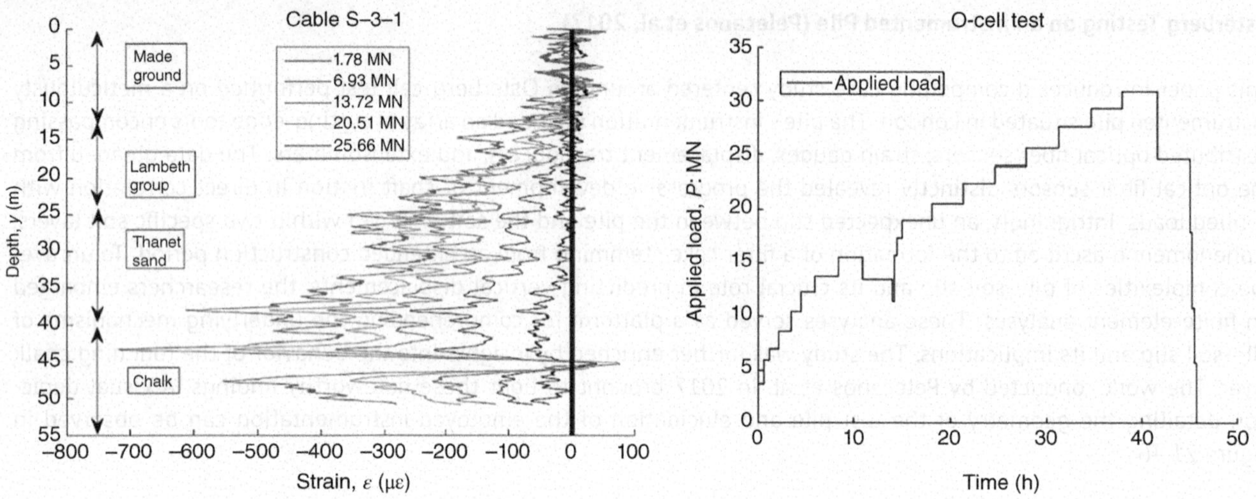

Figure 21.47 Monitored axial strain developed in a pile for increasing O-cell loads and applied load curve (Pelecanos et al. 2017/Emerald Publishing Ltd/Public Domain CC BY 4.0).

Case Study 21.8

Pullout Testing of Short Piles via Physical Modeling (Faizi et al. 2015)

As an integral component within structural systems, piles are susceptible to failures arising from both structural collapse and soil body instability. While extensive experimental investigations have been documented in the literature, aimed at delineating the failure mechanisms of piles within sandy substrates, the focus has primarily been on downward failure scenarios. Uplift failure of short piles, however, remains a relatively understudied facet, necessitating further exploration.

In this particular study, the primary objective was to shed light on the uplift resistance of short piles. A series of three small-scale physical tests were conducted employing short piles of 5 cm diameter and varying slenderness ratios (2, 3, and 4), all within loose sand. To elucidate the deformation patterns resulting from uplift forces, state-of-the-art techniques, including close-range photogrammetry and particle image velocimetry (PIV), were employed.

For validation, the laboratory test outcomes were meticulously juxtaposed against results obtained through ABAQUS finite element method (FEM) software. The pivotal finding underscored the influential role of the slenderness ratio in determining the uplift capacity values garnered from experimental tests. Additionally, the comparison between experimental and numerical results revealed a reasonable error range, with discrepancies ranging from 0.6% to 11%. This indicated the software's adeptness in accurately simulating the experimental behavior.

Both PIV and numerical techniques converged to reveal a congruent failure zone. Photogrammetric corrections were diligently applied to counteract distortions introduced by the camera lens and external camera-control plane movements. Ensuring measurement accuracy, the image scale was verified using control target markers. The ensuing data underwent meticulous processing, thereby facilitating the evaluation of sand movement during the pullout tests involving the semi-circular pile.

Visual aids enhance understanding. Figure 21.48 provides a comparative depiction of the maximum deformations observed during the complete removal of piles, specifically focusing on cases with a slenderness ratio of 4. Moreover, Figure 21.49 illustrates the uplift load–displacement curve of circular short pile sections derived from both experimental and numerical methodologies (Faizi et al. 2015).

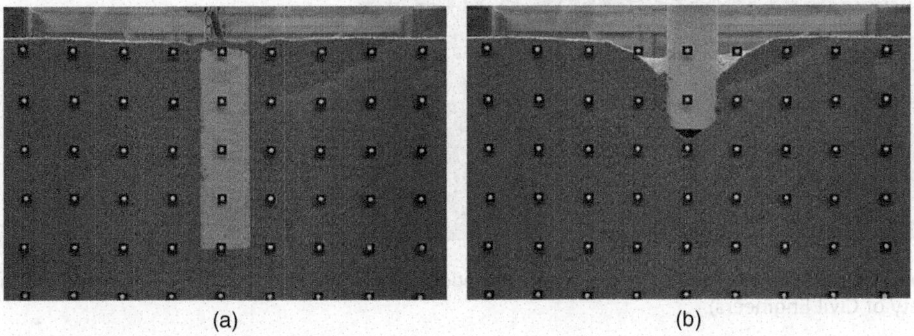

Figure 21.48 Schematic steps of failure mechanism for pile with $L/d = 4$ using PIV method: (a) before starting the test and (b) after failure (Faizi et al. 2015/with permission of Elsevier).

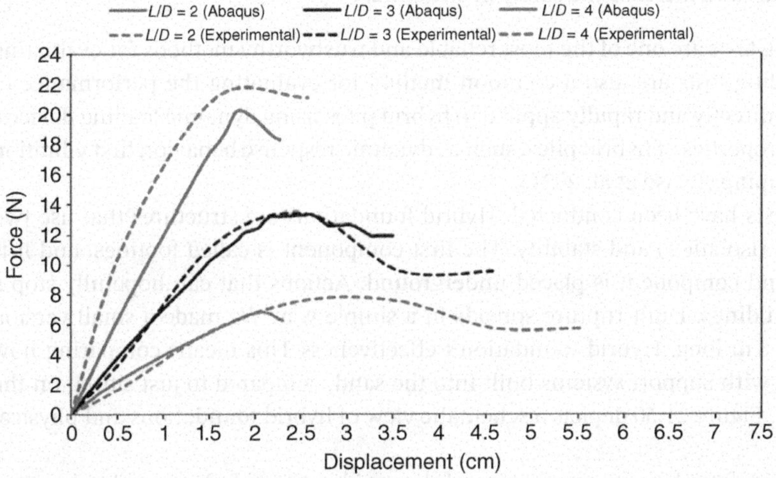

Figure 21.49 Load displacement curve for short piles obtained by the experimental and numerical techniques (Faizi et al. 2015/with permission of Elsevier).

21.4 Hybrid Foundations Testing

The tests used to investigate hybrid piles may differ from those used for other types of foundations. These differences are usually due to the unique characteristics of hybrid piles, which are formed by combining two types of traditional piles and monopile foundations. For example, one of the main differences that may exist is that hybrid piles usually have different mechanical properties that require specific and suitable testing. Additionally, due to the differences in the structure and shape of hybrid piles, some of the tests used for other foundations, such as static or dynamic loading tests, require specific adjustments and modifications to investigate hybrid piles. Therefore, specific and appropriate tests should be designed and conducted to evaluate and investigate hybrid piles, taking into account their unique characteristics and features (Russo et al. 2021). To investigate hybrid piles, there are specific tests that can be used as methods to evaluate and examine the characteristics and performance of this type of deep foundation. Some of these tests are mentioned below:

- **Static Loading Test:** In this test, a static load is applied to hybrid piles, and their behavior and load-bearing capacity under static loads are investigated. Static loading devices are usually used to conduct this test, which is used as a method to investigate the mechanical specifications of hybrid piles and determine their load-bearing capacity.
- **Dynamic Loading Test:** In this test, impact loads are applied to hybrid piles using dynamic hammers, and their behavior and load-bearing capacity under dynamic loads are investigated. This test is used as a method to investigate the tolerance of hybrid piles against earthquake vibrations and dynamic loads.

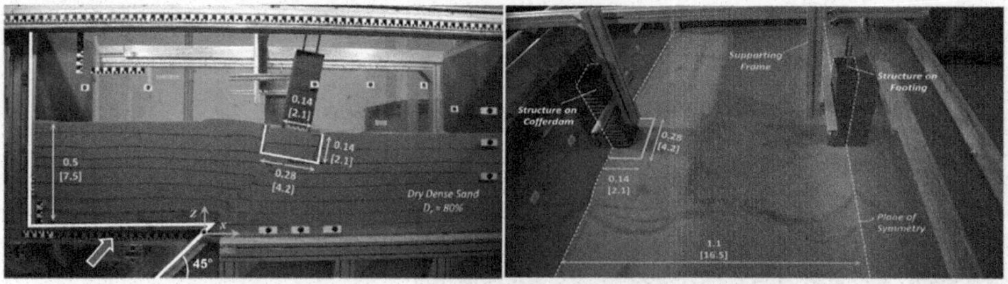

Figure 21.50 Schematic image of hybrid foundations and photos of the physical model in the split box indicating key dimensions in meters (Loli et al. 2018/American Society of Civil Engineers).

- **Seismic Restraint Test:** In this test, hybrid piles are subjected to seismic loads using special devices, and their behavior and load-bearing capacity against earthquake vibrations are investigated. This test is used as a method to evaluate the ability of hybrid piles to withstand seismic loads and increase the safety of structures.

Previous studies have shown that static loading tests are one of the most reliable and trustworthy methods for evaluating the performance of hybrid piles. Dynamic loading tests are also a common method for evaluating the performance of hybrid piles. In this method, dynamic loading is directly and rapidly applied to hybrid piles using dynamic loading devices. This method is used to determine the dynamic properties of hybrid piles, such as dynamic response behavior, first vibration frequency, and changes in pile stiffness and damping (Russo et al. 2021).

In the sample study below, physical model tests have been conducted. Hybrid foundations are structures that use two components to provide both seismic protection (isolation) and stability. The first component is called footings, and it is designed to rock during earthquakes. The second component is placed underground. Actions that can hopefully stop a break in the ground and prevent damage to buildings. Fault rupture spreads in a simple way. We made a small version of dense dry sand using a special box that was 3 m long. Hybrid foundation's effectiveness This means comparing how well a structure like a building sits on the sand with support systems built into the sand, compared to just sitting on the sand without any extra support (Loli et al. 2018). Figure 21.50 depicts a schematic view of hybrid foundations and physical modeling.

The evaluation of the performance of hybrid foundations requires careful consideration of these and other factors to ensure that the foundation system is designed and constructed to meet the specific requirements of the project.

There are many examples of successful hybrid foundation projects around the world. Here are a few examples (Poulos 2011):

1) **Petronas Twin Towers, Kuala Lumpur, Malaysia:** The Petronas Twin Towers, completed in 1998, are the world's tallest twin towers. The towers are supported by a hybrid foundation system that includes both raft foundations and piles. The foundation system was designed to resist the high wind and seismic forces in the region, and it has performed well since construction.
2) **Burj Khalifa, Dubai, UAE:** The Burj Khalifa, completed in 2010, is the world's tallest building. The building is supported by a hybrid foundation system that includes both raft foundations and piles. The foundation system was designed to resist the high wind and seismic forces in the region, and it has performed well since construction (see Figure 21.51).
3) **Shard London Bridge, London, UK:** The Shard London Bridge, completed in 2012, is one of the tallest buildings in Europe. The building is supported by a hybrid foundation system that includes both raft foundations and piles. The foundation system was designed to accommodate the complex geology of the site, which includes layers of soft clay and dense sand.
4) **Sydney Opera House, Sydney, Australia:** The Sydney Opera House, completed in 1973, is one of the world's most iconic buildings. The building is supported by a hybrid foundation system that includes both raft foundations and piles. The foundation system was designed to accommodate the complex geology of the site, which includes layers of sandstone and shale.

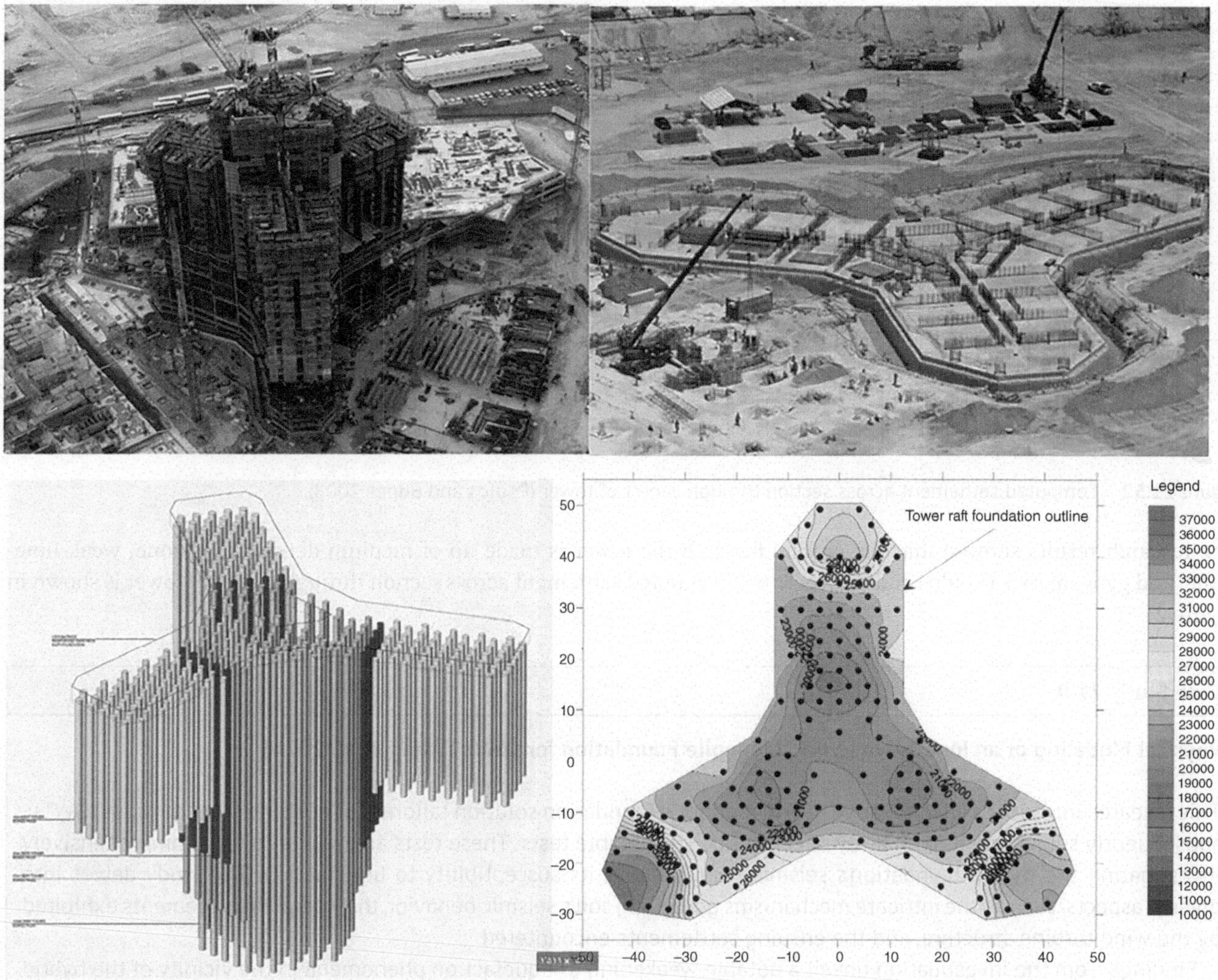

Figure 21.51 Images of the implementation of the Burj Khalifa hybrid foundation in Dubai (Poulos and Bunce 2008/Missouri University Of Science And Technology).

5) **Taipei 101, Taipei, Taiwan:** Taipei 101, completed in 2004, was the world's tallest building until it was surpassed by the Burj Khalifa in 2010. The building is supported by a hybrid foundation system that includes both raft foundations and piles. The foundation system was designed to resist the high wind and seismic forces in the region, and it has performed well since construction.

These projects demonstrate the versatility and effectiveness of hybrid foundation systems in supporting some of the world's most iconic and challenging structures. In the construction of Burj Khalifa, shallow foundation elements were used to support the tower columns. Full-scale tests were conducted to examine the behavior of these shallow foundation elements under real conditions. In these tests, the shallow foundation elements were subjected to loads and different forces, and their behavior was examined. Based on the test results, the design of the shallow foundation elements in Burj Khalifa was optimized, leading to increased safety and useful life of the building. Constructing a structure as magnificent as Burj Khalifa in desert geography brings about numerous engineering challenges, and one of the most important challenges is designing the foundation or the building's footing. If the foundation is not large and deep enough, there is a risk of gradual settlement and sinking of the tower in the loose desert soil. To investigate the characteristics and properties of the soil at the site of Burj Khalifa's construction, engineers created 33 deep boreholes using various techniques.

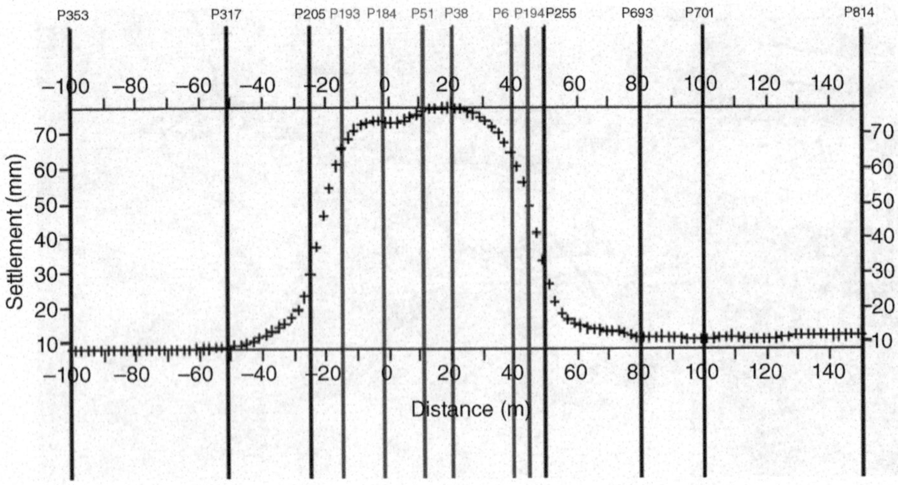

Figure 21.52 Computed settlement across section through center of tower (Poulos and Bunce 2008).

The research results showed that the ground beneath the tower is made up of medium-density sandstone, weak limestone, and gypsum rock (Poulos and Bunce 2008). Computed settlement across section through centre of tower is shown in Figure 21.52.

Case Study 21.9

Physical Modeling of an Innovative Hybrid Monopile Foundation for OWTs (Wang et al. 2020)

This research introduces an innovative hybrid monopile foundation solution tailored for offshore wind turbines (OWTs), subsequently subjecting it to an array of centrifuge shake table tests. These tests are geared toward comprehensively investigating the hybrid foundation's seismic response and its susceptibility to liquefaction. The study delves into diverse aspects, such as the intricate mechanisms governing soil's seismic behavior, the lateral displacements exhibited by the wind turbine structure, and the ensuing settlements encountered.

Findings from the investigation unveil a notable weakening of liquefaction phenomena in the vicinity of the hybrid monopile foundation, setting it apart from conventional single-pile systems. This innovative approach brings about augmented lateral stability, bolstering the overall system's robustness. However, it is noteworthy that wind turbines employing the hybrid monopile foundation exhibit increased settlement levels during seismic events, attributable to both soil–structure interaction and static-bearing induced soil shearing effects.

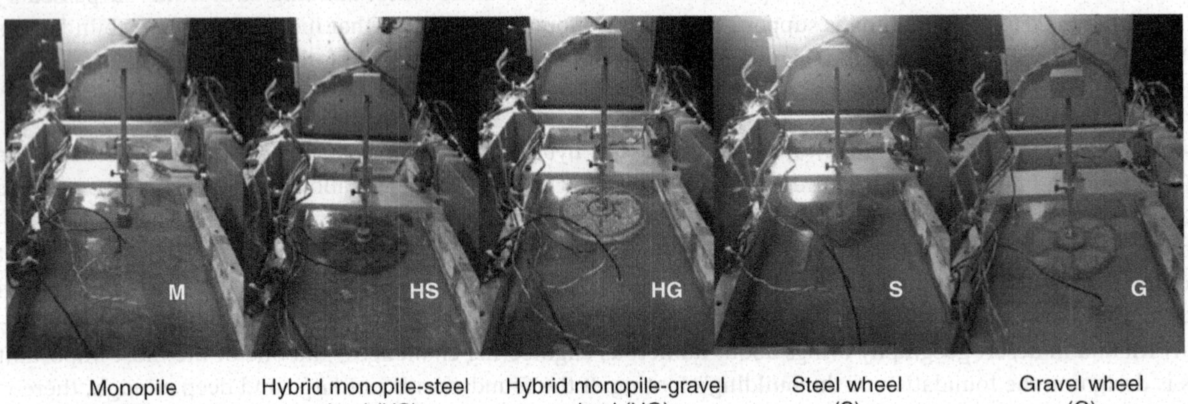

Figure 21.53 Images of the test models (Wang et al. 2020/with permission of Elsevier).

The study is dedicated to an in-depth exploration of the seismic dynamics of hybrid monopile foundations for OWTs within the context of earthquake occurrences. The insights garnered through this study serve as valuable references for practical design implementations (Wang et al. 2020).

Five distinct models of offshore wind turbines were fabricated to represent diverse foundation configurations. These models encompassed a monopile foundation (M), a hybrid monopile-steel wheel foundation (HS), a hybrid monopile-gravel wheel foundation (HG), a steel wheel foundation (S), and a gravel wheel foundation (G), as visually illustrated in Figure 21.53.

Lateral displacements of the five models are monitored by an LVDT fixed at 4.5 m 303 above the ground surface. Time histories of lateral displacements are plotted together in Figure 21.54.

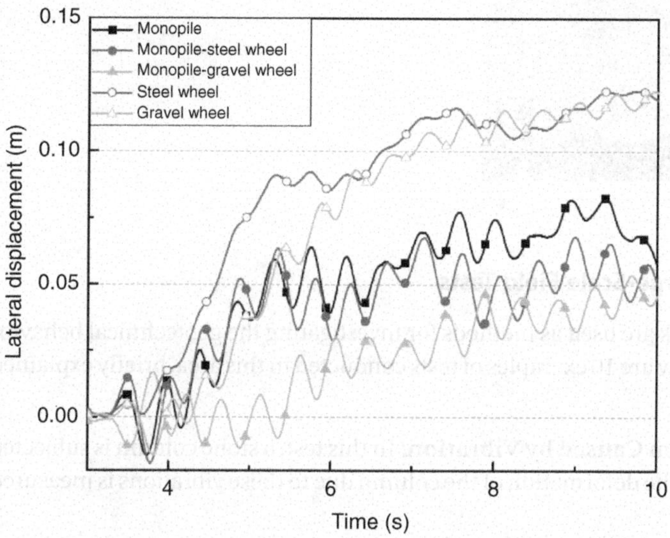

Figure 21.54 Time history of lateral displacement for dry tests (Wang et al. 2020/with permission of Elsevier).

21.5 Rigid Intrusion And Micro-pile Testing

The load capacity test on stone column improved soil is one of the methods used to assess the strength of improved soil. This test involves covering the surface of the improved soil with stone columns and applying a load either at an incline or directly. Initially, stone columns are placed at a specific depth in the improved soil using a drilling machine. Then, using a mixer device, the improved soil with stone columns is placed on the surface of the soil as the base for the test load. In this test, the load is gradually increased on the surface, and as long as the improved soil with stone columns does not fail, the applied load is increased. By considering the load applied at the moment of failure, the load-bearing capacity of the improved soil with stone columns can be calculated. This test is used as one of the methods to assess the strength of improved soil and can be an effective method for improving and increasing the strength of improved soil with stone columns in construction and civil engineering projects (Kumar and Kumar 2019; McCabe et al. 2007). Figure 21.55 depicts typical testing on rigid intrusions.

21.5.1 ASTM Standards for Testing of Stone Columns

To perform different tests on stone columns, using various standards that align with the type of test and its implementation conditions is recommended. Some of these standards include,

- ASTM D1143/D1143M-07 standard for conducting compressive load testing on stone columns.
- ASTM D3966-90 standard for conducting bending load testing on stone columns.
- ASTM D4428-10 standard for conducting shear testing on stone columns.
- ASTM D5731-08 standard for conducting elastic modulus testing on stone columns.
- ASTM D6026-14 standard for conducting deformation testing on stone columns.

Figure 21.55 Rigid intrusion testing (RI) (hodim/Adobestock).

21.5.2 Types of Small-Scale Laboratory Tests and Large-Scale Field Tests

Small-scale physical modeling (Figure 21.56) and field tests are used as methods for investigating the geotechnical behavior of rocks and analyzing the stability of rock structures. Below are 10 examples of tests conducted in this area, briefly explained (Fakhroo and Al-Mahaidi 2019):

- **Testing of Stone Columns by Analyzing Deformation Caused by Vibration:** In this test, a stone column is subjected to various pressure vibrations using a seismograph, and the deformation of the column due to these vibrations is measured using digital cameras.
- **Modeling Test of the Behavior of a Stone Column Under Dynamic Loading:** In this test, a stone column is subjected to dynamic loading from impact pressure, and the behavior of the column due to this loading is measured using a set of pressure and acceleration sensors.
- **Testing of Stone Columns Using Nondestructive Equipment:** In this test, a stone column is examined using nondestructive equipment such as magnetic imaging and ultrasonic waves to study the internal structure of the stone.
- **Scaling Test on Rock Samples:** In this test, rock samples are produced in small sizes, such as millimeters, and their behavior under different loads is examined.
- **Testing of Stone Columns Using Nano-equipment:** In this test, a stone column is examined using nano-equipment to study the effects of nanomaterials on the behavior of the column.

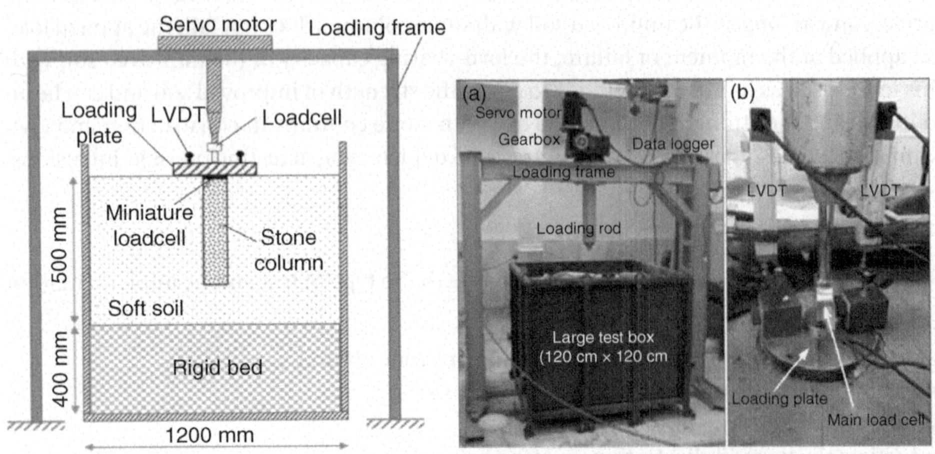

Figure 21.56 Schematic images of small-scale modeling of stone column implementation (Mehrannia et al. 2018/Springer Nature).

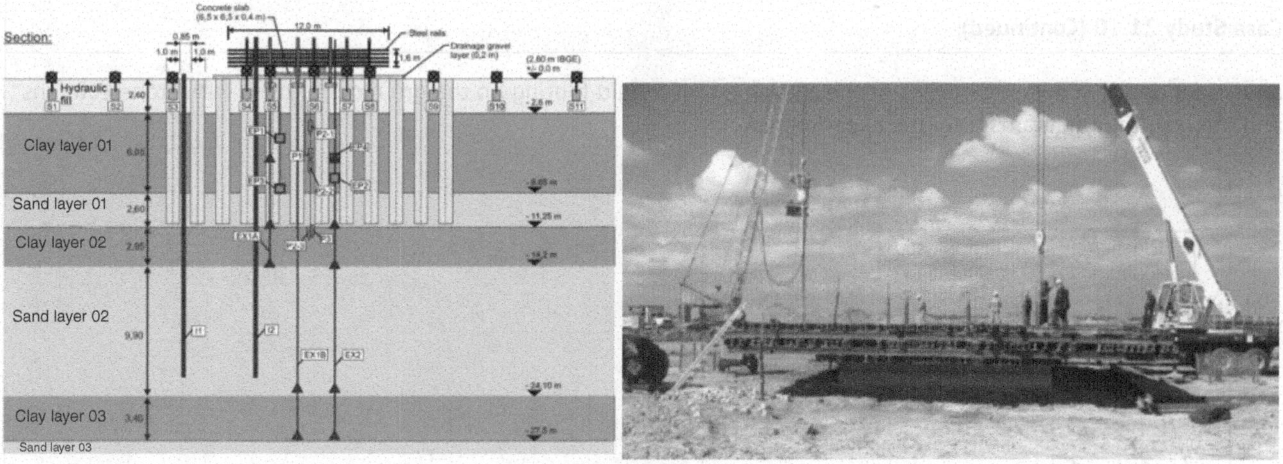

Figure 21.57 Schematic images of field-scale modeling of stone column implementation (Almeida et al. 2014).

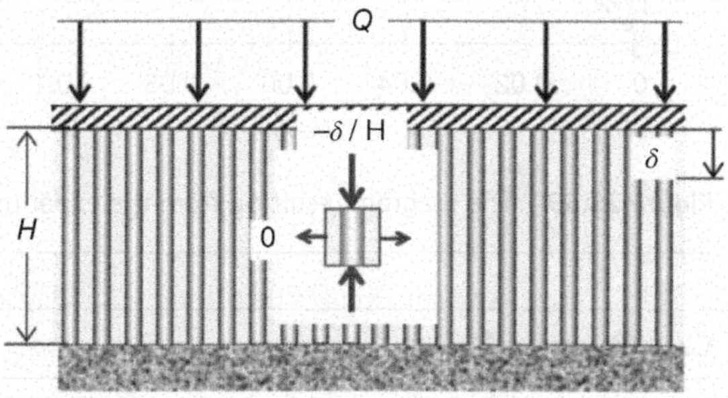

Figure 21.58 Schematic images of the settlement of a rigid footing on the ground improved with stone columns (Abdelkrim and de Buhan 2007).

- **Testing of Stone Columns Using Radiographic Equipment:** In this test, a stone column is examined using radiographic equipment to study the internal structure of the column and control the quality of the stone.
- **Testing of Stone Columns Using Optical Equipment:** In this test, a stone column is examined using optical equipment to study its deformation and behavior under different loads.
- **Testing of Stone Columns Using Image Analysis:** In this test, a stone column is examined using image analysis to study its deformation and behavior under different loads.
- **Testing of Stone Columns Using Electron Microscopy:** In this test, a stone column is examined using electron microscopy to study its internal structure and analyze its stability under different loads.
- **Testing of Stone Columns Using Spectral Analysis:** In this test, a stone column is examined using spectral analysis to study its properties and analyze its stability under different loads.

The schematic images shown in Figure 21.57 represent field-scale modeling of a stone column conducted in the laboratory. Figure 21.58 shows schematic images of field-scale modeling of stone column implementation (Almeida et al. 2014).

Case Study 21.10

The Settlement of a Rigid Footing on Ground Improved with Stone Columns (Abdelkrim and de Buhan 2007)

Abdelkrim and de Buhan (2007) presented a cohesive soil layer without any friction angle, which was reinforced with a network of stone columns to demonstrate their analysis and the effect of considering elastoplastic conditions. They evaluated a reinforced analysis and plotted a chart for the reinforced soil, as well as two elastoplastic analyses for unreinforced and reinforced soils. As observed, for the same load magnitude, the predicted settlement in the elastic analysis is about half of the elastoplastic condition, and designing with an elastic method is illogical. Figures 21.58

(Continued)

Case Study 21.10 (Continued)

and 21.59 depict schematic images of the settlement of a rigid footing on the ground improved with stone columns and the resultant diagrams from the experiment.

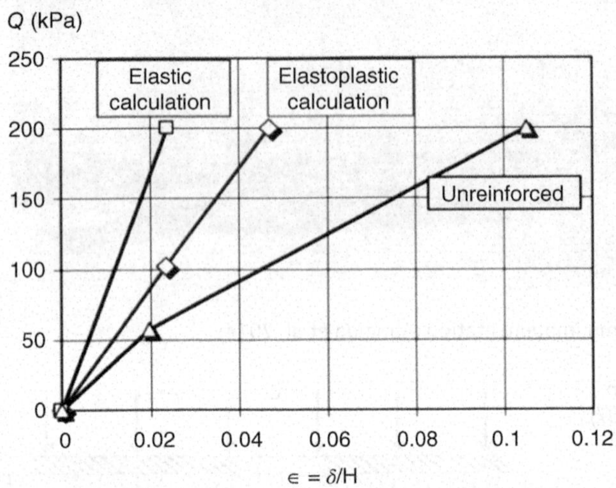

Figure 21.59 The diagram resulting from the experiment (Abdelkrim and de Buhan 2007).

Case Study 21.11

Oil Tanks Footing on Ground Improved with Stone Columns (Raju and Daramalinggam 2012)

Raju and Daramalinggam (2012) investigated a gas tank retrofitting project using stone columns. The project aimed to construct two LNG gas tanks with a diameter of 84 m on loose and semi-compacted silty soil in India. The two tanks were filled with liquid gas up to a height of approximately 35 meters and exerted a pressure of about 230 kPa on the tank floor. The high risk of liquefaction and excessive settlements in the sandy soil were among the concerns of this project. An image of the mentioned tanks is shown in Figure 21.60.

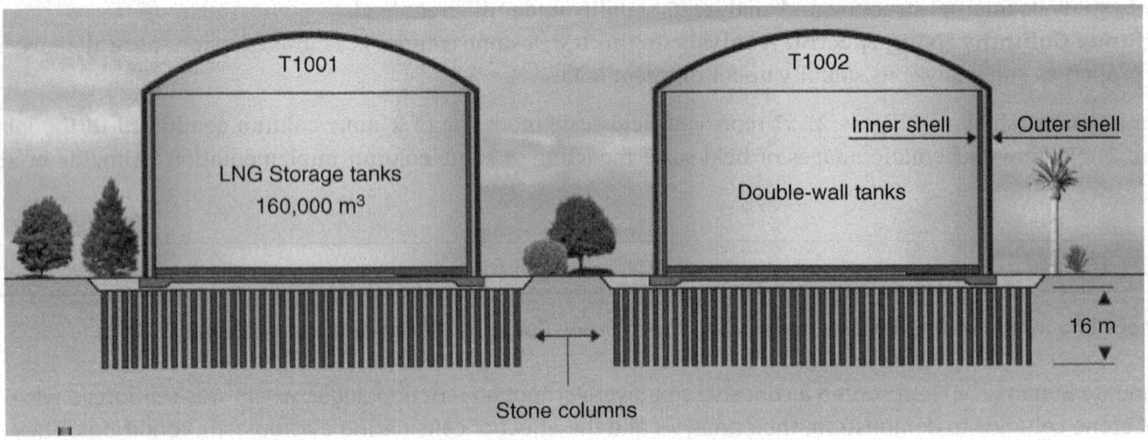

Figure 21.60 Images of oil tanks placed on stone columns (Raju and Daramalinggam 2012).

To estimate the necessary consolidation settlement, it is essential to study the stress distribution inside the stone column-reinforced soil. Moreover, vertical stresses in both stone and the cohesive soil around it play a role in stability analysis. By constructing embankments or piles on reinforced soil with stone columns, a significant stress concentration occurs in the stone column and a reduction in stress in the surrounding soil, with less stiffness than the column. This stress concentration occurs because the vertical settlement is approximately the same in the stone column and the surrounding soil, as the stiffness of the column is greater than that of the cohesive or non-cohesive soil around it (Raju and Daramalinggam 2012).

Case Study 21.12

Water Storage Tanks on Stone Columns (Bhushan et al. 2004)

At the site of two 190-ft (57.9 m) diameter, 40-ft (12.2 m) high, 8 million-gallon (30,300 m^3) circular steel water storage tanks, ground improvement was performed through the installation of stone columns to mitigate liquefaction and lateral spreading potential, as well as a surcharge program aimed at reducing post-construction settlements. Settlements observed during the surcharge program ranged between 9 and 15 in. (225 and 375 mm), while post-construction settlement during the hydrotest was approximately 1.2 in. (305 mm) (Bhushan et al. 2004).

21.5.3 Micro-piles Tests

Considering the purpose of micro-pile implementation and a performance-based approach to it, various quality control tests can be defined for this method, some of which include (Hesham and Abdelrahman 2006):

- Compression load test
- Tension load test
- Lateral load test

To ensure that the designed micro-piles are capable of withstanding the intended load with a high degree of confidence throughout their life span, field load tests are performed on micro-piles. The purpose of quality control tests is to reach

Figure 21.61 Field test images of micro-piles and load settlement curves (Elsiragy 2021/European Journal of Engineering and Technology Research).

a maximum load, axial displacement, and creep threshold predetermined in advance (see Figure 21.61). These tests are performed in four stages (Elsiragy 2021):

- **Final Test:** Loading continues until soil movement and slip between soil and the pile are observed.
- **Verification Test:** It is conducted to verify the design method for implementation and the ability to provide the predetermined resistance.
- **Proof Test:** Usually conducted on 5% of the micro-piles, it involves gradually increasing the load until reaching a percentage of the predicted load in the design.
- **Creep Test:** It is conducted to measure the amount of displacement under a constant load and at specified time intervals. It should be noted that the maximum displacement in each stage is 2 mm, and exceeding that is not acceptable.

Case Study 21.13

Testing on Pile and Micro-pile Groups

In project carried out in Iran, specifically in Mazandaran province, micro-piles were used to reinforce the foundation of blocks F and G of the Narenjestan commercial-administrative complex. The execution of these micro-piles is shown in Figure 21.62.

Figure 21.62 Field test images of micro-piles in the Narenjestan project in Mazandaran province (Segara et al. 2002), Courtesy of Samanpey Co. 2017.

The largest implementation of micro-piles in the world is related to the Jiaozhou Bay Bridge in China. With a length of 55 km, this bridge is the world's largest sea bridge, and micro-piles were used to construct its foundations. Each of the bridge's pillars is 102 m long, weighs 100,000 tons, and more than 15,000 micro-piles were used to support them. The project was carried out by the China State Construction Engineering Corporation (CSCEC). The bridge connects three major cities in China through a 34-mi (approximately 54.5-km) highway. It has six lanes for traffic and four tunnels, one of which passes under the water (Ji et al. 2012).

Case Study 21.14

Study on Hollow Bar Micro-piles (Bayesteh and Sabermahani 2018)

The preference for self-drilled hollow bar micro-piles over the traditional solid bar system is on the rise in soil improvement projects due to the associated reduction in construction costs and time. Despite their increasing popularity, the load-deformation behavior of hollow bar micro-piles remains relatively underexplored. Consequently, the impact of installation techniques and the characteristics of the surrounding soil on their bond strength remains unverified. This current study aims to address these gaps by introducing three distinct grouting methods. These methods were employed to scrutinize the load–displacement behavior of hollow-bar micro-piles within both clay and gravel environments. A comprehensive field study was undertaken, involving the installation of 21 full-scale tests on hollow bar micro-piles at two different sites. Time-dependent grouting techniques were utilized to assess the strength of the bond between grout and ground. The observed load–displacement behavior revealed several key insights. Notably, the induced bond diameter of the hollow bar micro-piles exceeded theoretical calculations, while their estimated bond strength surpassed the values recommended by prevailing codes. Moreover, micro-piles subjected to higher drilling and injection durations and sequences exhibited enhanced elastic behavior. The study conducted by Bayesteh and Sabermahani in 2018 successfully sheds light on these dynamics. Further elaboration, including details about the drilling equipment and the micro-pile testing procedure at the site, is visualized in Figure 21.63.

Figure 21.64 illustrates the load–displacement compression test curves for the micro-piles at both locations using grouting Method 3. The data presented is for a four-cycle compression test.

The load–displacement curves for compression testing on the 9- and 15-m micro-piles are presented in Figure 21.65. Regardless of the proposed grouting method, the primary focus of this graph is to investigate the influence of soil type on the performance of hollow-bar micro-piles. The movement of micro-piles at Site 1 under 300 kN of compressive loading was about five to six times greater than that of micro-piles located at Site 2.

The load–displacement curves pertaining to the compression testing of both 9 and 15-m micro-piles are depicted in Figure 21.65. The central objective of this illustration revolves around comprehending the impact of soil type on the efficacy of hollow-bar micro-piles, irrespective of the selected grouting approach.

A noteworthy observation arising from this graph lies in the dissimilarities between the micro-pile behaviors at Site 1 and Site 2. Under a compressive load of 300 kN, the movement exhibited by micro-piles situated at Site 1 was approximately five to six times more pronounced compared to their counterparts positioned at Site 2.

(Continued)

Case Study 21.14 (Continued)

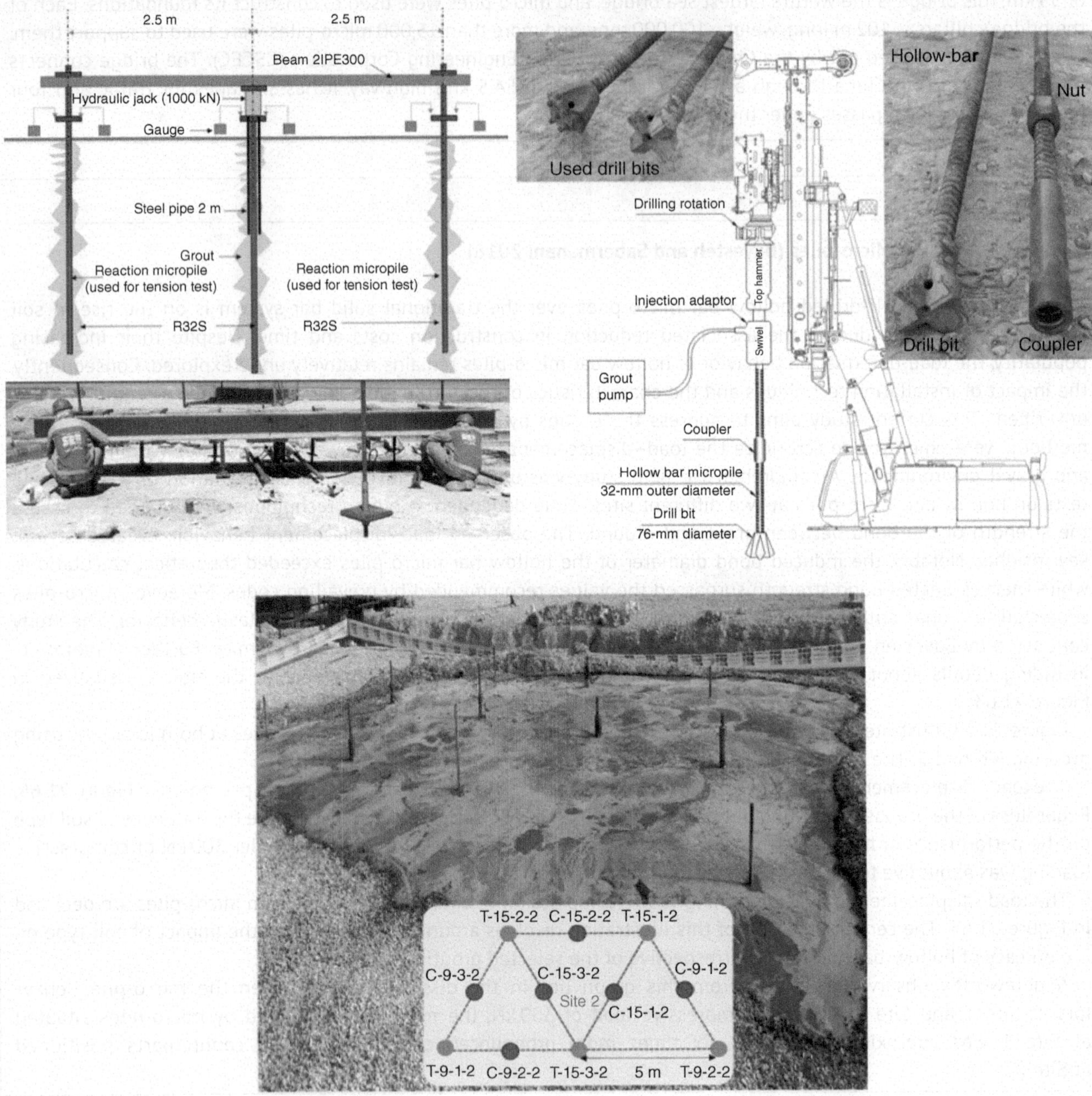

Figure 21.63 The details of drilling equipment and micro-piles testing at site (Bayesteh and Sabermahani 2018/American Society of Civil Engineers).

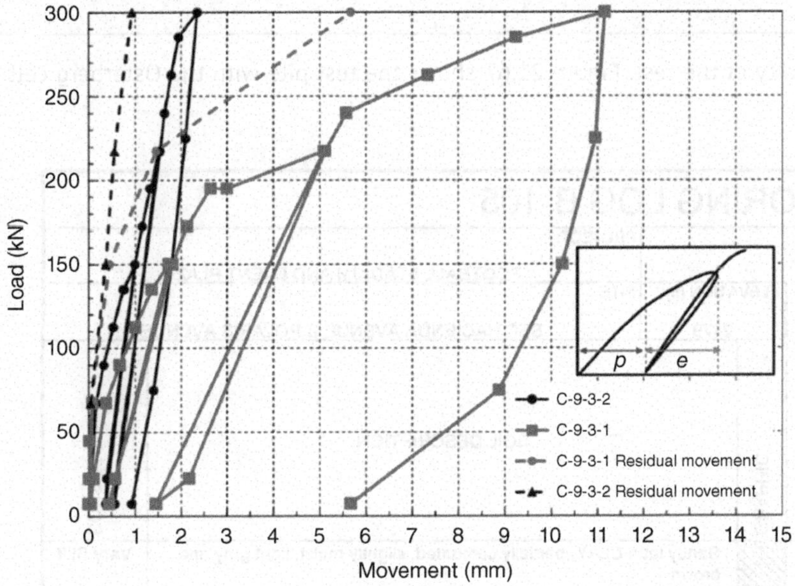

Figure 21.64 Load–displacement compression test for both soil types (Bayesteh and Sabermahani 2018/American Society of Civil Engineers).

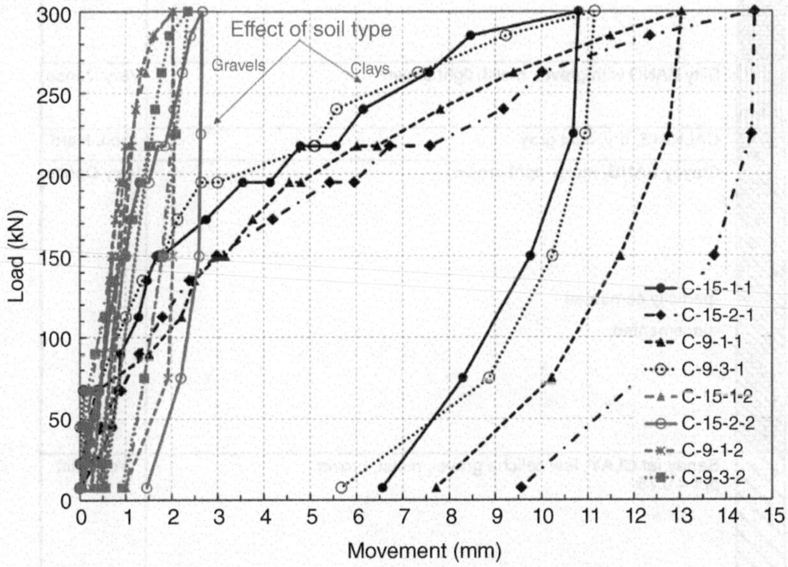

Figure 21.65 Load–displacement curves for all grouting methods at both sites (Bayesteh and Sabermahani 2018/American Society of Civil Engineers).

Case Study 21.15

Osterberg Test Results at Allegiant Stadium, Las Vegas, Nevada, USA (Report on CFA Pile Load Testing, Osterberg Method, Clark County Project Records 2018)

An Osterberg cell (O-cell) test on the nominal 24-in. diameter and 66.2-foot-deep test pile was performed. Subsurface conditions at the test pile location consist primarily of sandy clays and layers of caliche. Figure 21.66 shows the

(Continued)

Case Study 21.15 (Continued)

borehole of the soil stratigraphy in the vicinity of the test. Figure 21.67 shows the test pile with the Osterberg cell and its location in the pile.

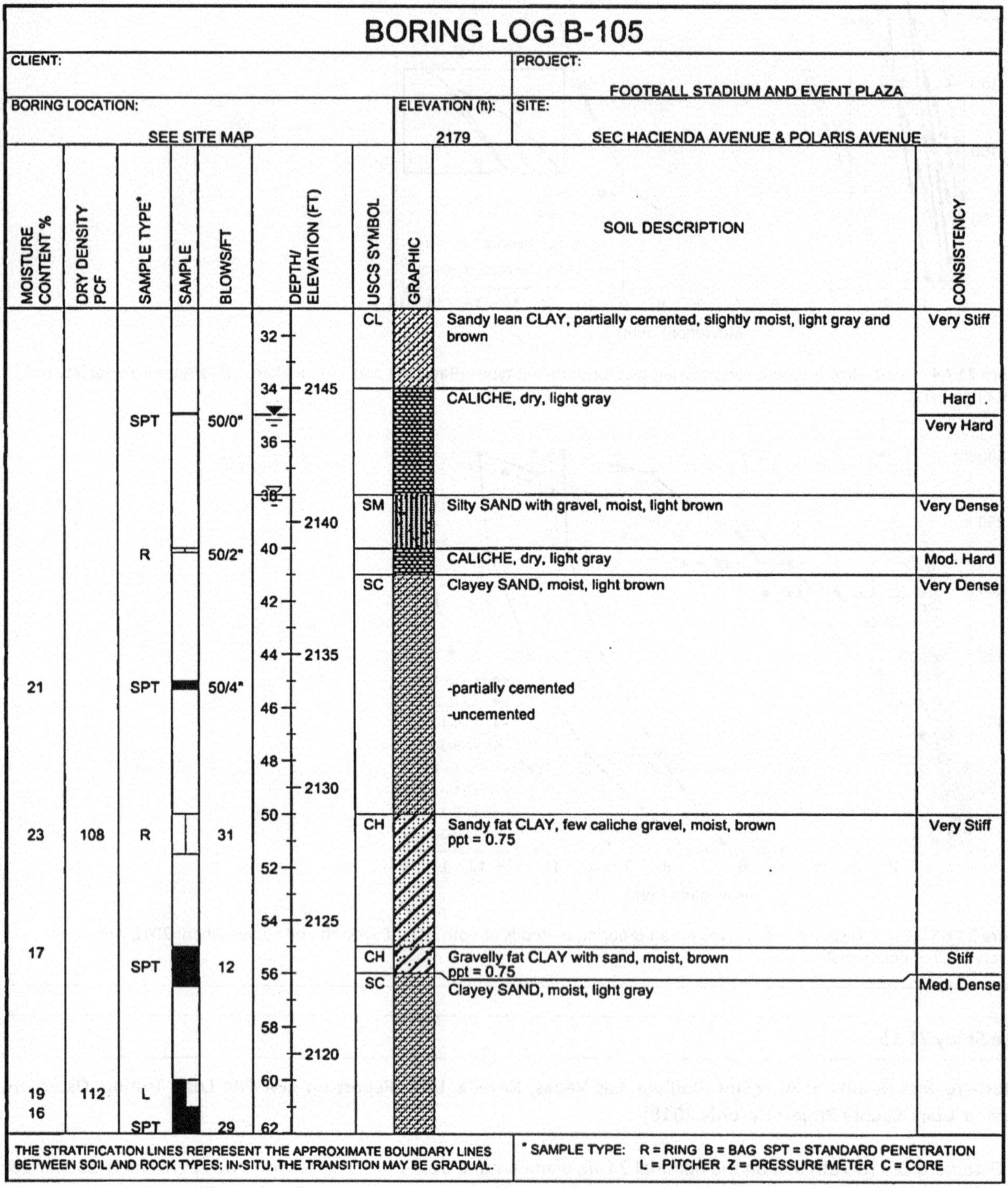

Figure 21.66 Borehole the soil stratigraphy in the vicinity of the test (Report on CFA pile load testing, Osterberg method, Clark County Project Records 2018).

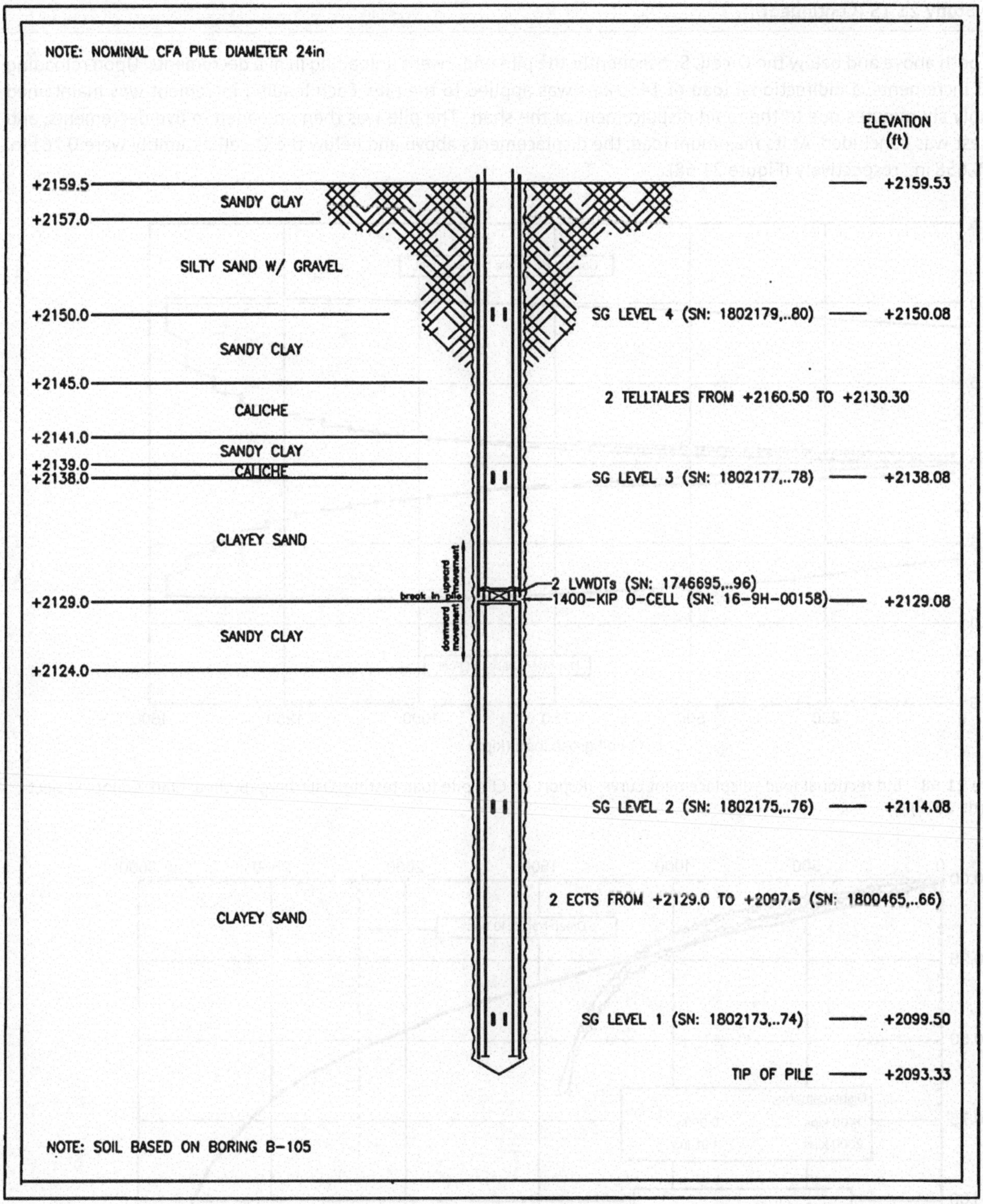

Figure 21.67 The test pile with the Osterberg cell and its location in the pile (Report on CFA pile load testing, Osterberg method, Clark County Project Records 2018).

In the conducted test, a 16-in. diameter O-cell, positioned 35.8 ft above the pile tip, was utilized. The O-cell underwent pressurization in 18 nearly equal increments, resulting in a maximum bidirectional load of 884 kips applied to the

(Continued)

Case Study 21.15 (Continued)

pile, both above and below the O-cell. Subsequently, the pile underwent unloading in five decrements. Upon reloading in 16 increments, a bidirectional load of 1430 kips was applied to the pile. Each loading increment was maintained for only six minutes due to the rapid displacement of the shaft. The pile was then unloaded in five decrements, and the test was concluded. At its maximum load, the displacements above and below the O-cell assembly were 0.761 in. and 0.658 in., respectively (Figure 21.68).

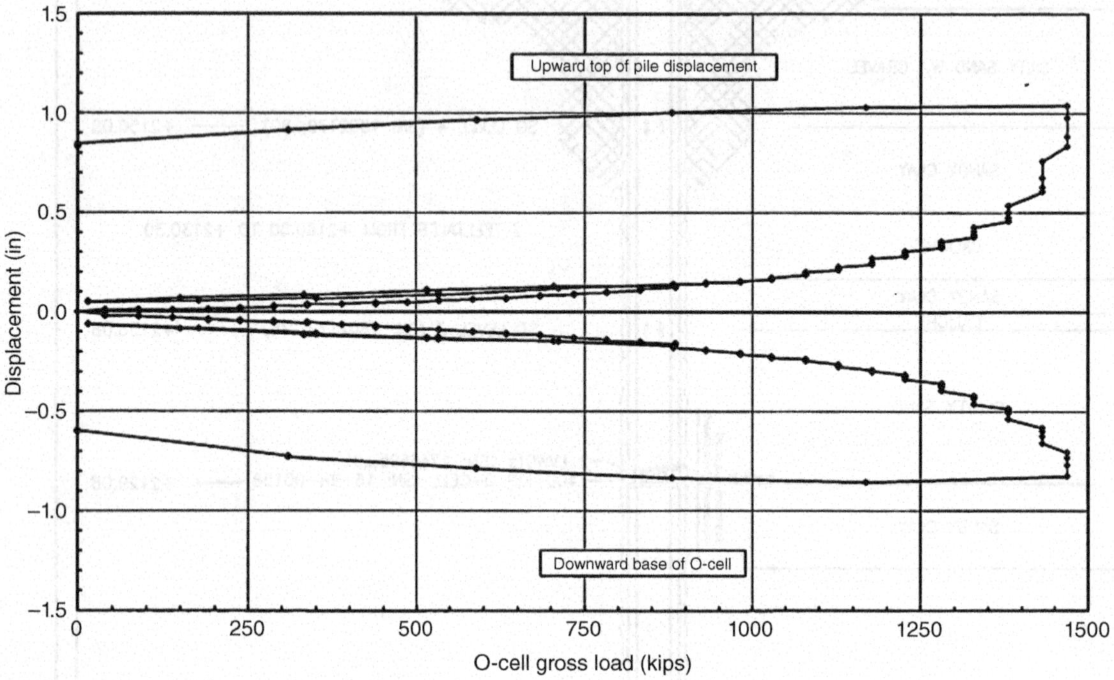

Figure 21.68 Bidirectional load–displacement curve. (Report on CFA pile load testing, Osterberg method, Clark County Project Records 2018).

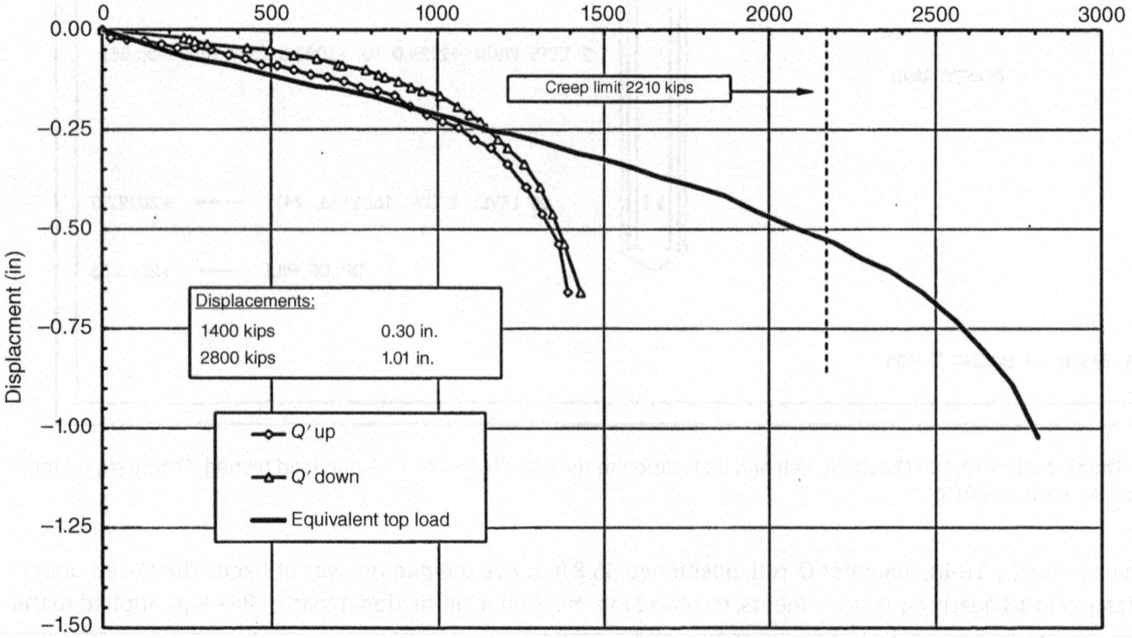

Figure 21.69 Equivalent top load curve. (Report on CFA pile load testing, Osterberg method, Clark County Project Records 2018).

Equivalent top load curves for the test pile were constructed. For a top loading of 1400 kips, the adjusted test data indicated an approximate displacement of 0.30 in. Similarly, for a top loading of 2800 kips, the adjusted test data suggested a displacement of approximately 1.01 in. Unit side shear data calculated from strain gages revealed a maximum mobilized average net side shear of 12.0 ksf. Interestingly, the load distribution calculated from strain gages indicated no load was transmitted to the end bearing at the aforementioned maximum downward displacement (Figure 21.69).

21.6 Integrity and Soundness Inspection

Nondestructive evaluation tests are used to assess the condition of structural members in deep foundations, such as piles. The quality of soil layers, sufficient load-bearing capacity, and settlement within the desired range must be analyzed by load testing. Load testing on piles can have destructive and irreversible effects on the structural components, and despite these challenges, the performance of the pile may deteriorate, and its load-bearing capacity may decrease after testing. The quality and consistency of executed piles in construction projects are of great importance. Evaluating the health of cast-in-place piles and their behavior in terms of consistency during and after construction is one of the most significant challenges in geotechnical studies. Issues such as breakage, bulging, reduction in cross-sectional area (narrowing), nonuniformity of concrete, expansion of the cross-sectional area, etc., have led to the use of nondestructive control testing to the control consistency and health of executed piles. Among these tests, the PIT test is used in numerous projects due to its ease of execution and low cost (Wu et al. 2015).

21.6.1 Health and Integrity Testing of Piles

The execution of in situ piles always raises questions about the quality and consistency of the final product because, unlike precast piles, there is no direct inspection of the pile in place after installation. In this type of pile, which is formed simultaneously with drilling inside the soil, naturally, more ambiguities arise regarding the actual quality and dimensions of the pile structure. Of course, initial controls such as sampling of concrete before discharge into the borehole or recording the poured concrete volume for the pile are performed in most pile projects. However, the in situ pile execution process is faced with difficulties that affect the quality of their implementation and make their inspection necessary. Various nondestructive methods are used to control the length or quality of deep foundation concrete. All of these methods have their own unique advantages and limitations. The use of the pile integrity testing method is a suitable way to determine the quality of the executed pile in terms of its material and consistency. Figure 21.70 illustrates some of the defects that can be detected by the PIT test in in situ piles. Some of these problems include (Zandbabarasi et al. 2022):

- Slippage or collapse of the borehole wall and the creation of a narrow zone in the cross-section of the pile
- Failure to reach the final drilling length due to errors in the execution team or noncompliance with guidelines
- Improper use or lack of use of tremie pipe for concrete pouring
- Interruption or cessation during drilling and concrete pouring operations

21.6.2 PIT Test

In this test, a light handheld hammer weighing 0.5–10 kg hits the head of the pile, creating a compressive stress wave along the pile. The primary wave is reflected back toward the head of the pile wherever it encounters a change in the dimensional and mechanical properties of the pile structure or the impedance. The reflections of this compressive stress wave at the head of the pile are measured by an accelerometer sensor and are recorded after initial processing in the device's memory. The accelerometer sensor is temporarily attached to the head of the pile using an adhesive. For this purpose, inappropriate concrete must be completely removed, and smooth surfaces must be created for the sensor to connect and for the hammer to strike. These tests are performed using the PIT device and are analyzed by the PIT-W software (Wu et al. 2015). Figure 21.71 shows a schematic diagram of performing the PIT test.

Figure 21.70 Examples of images of defects present in piles (Zandbabarasi et al. 2022).

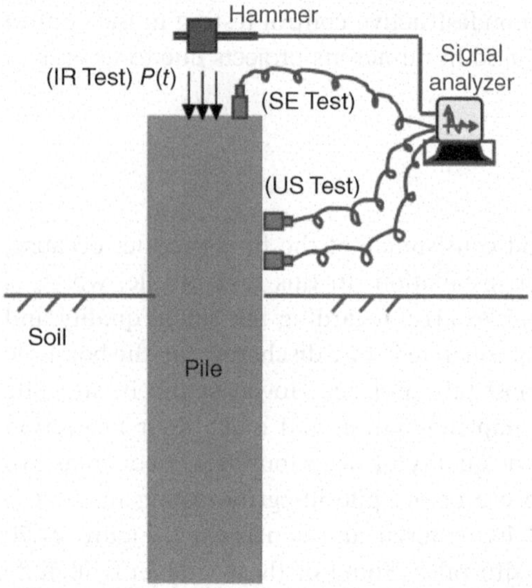

Figure 21.71 Schematic diagram of performing PIT test (Wu et al. 2015/NDT.net/Public Domain CC BY 4.0).

21.6.3 The Classification of PIT Test Results

Pile integrity is a comprehensive quality indicator that reflects the relative changes in the size of the pile cross-section, density, and continuity of the pile materials. Evaluating the integrity of a pile is a systematic and comprehensive assessment. In 2011, Webster and Raveshahi classified in situ and driven piles based on the analysis of wave reflections and the resulting waveform shape, using the results of the PIT test. Based on the results of these studies, piles were classified into seven groups in terms of health and integrity, or the presence of minor or major defects (Webster et al. 2011).

21.6.4 Interpretation of PIT Test Results

When a hammer strikes the top of a pile, a momentary compression is created at the top of the pile, causing a particle to move downward on the surface of the pile with a velocity, V. This compression generates a compressive stress wave that is propagated downward along the pile. The velocity of this stress wave is a function of the properties and material of the pile.

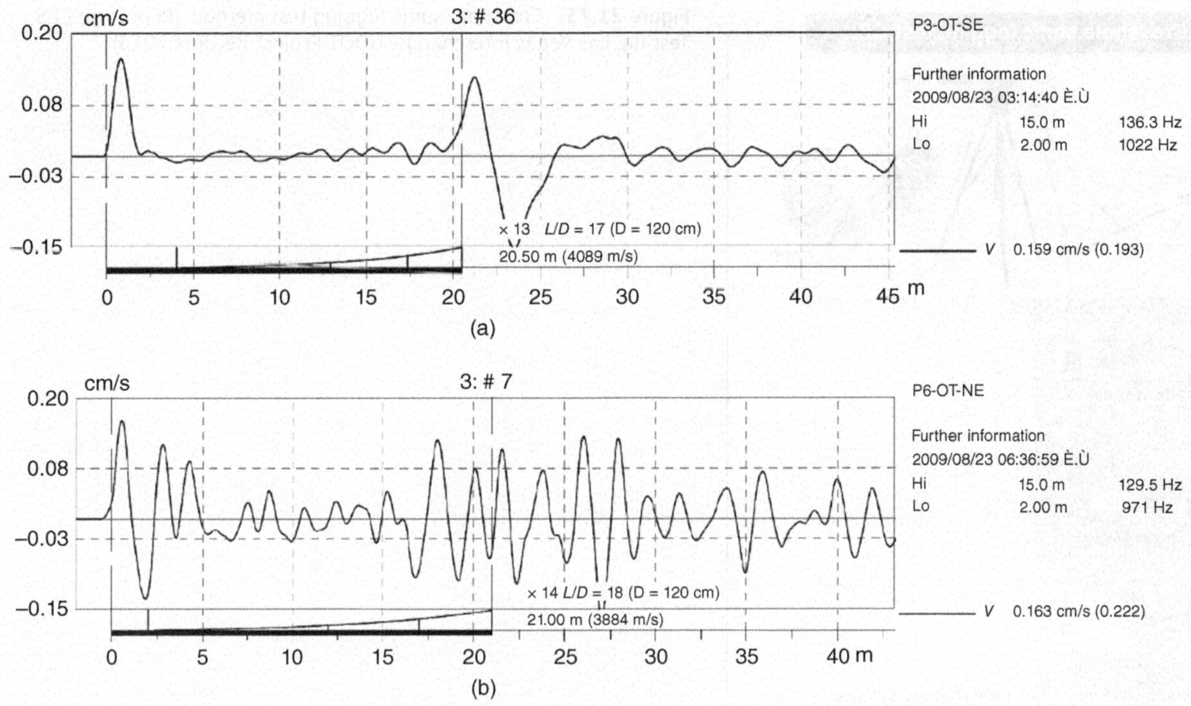

Figure 21.72 (a) Healthy pile graph and (b) defective pile graph (Fakharian and Hosseinzadeh 2010).

In steel piles, the velocity of this wave is 5130 m/s, while in concrete, it ranges between 3000 and 4500 m/s. The velocity records and reflections generated from the pile's tip or areas of discontinuity are displayed graphically. The damaged area is estimated based on the velocity of the wave in the pile and the width of the damaged region, considering the size of the reflection generated (Webster et al. 2011).

The time required for the reflections caused by impedance changes to return to the top of the pile as a reflection depends on the distance from the top of the pile to the point of wave reflection and the velocity of the wave. Therefore, with the constant velocity of the wave along the pile, time can be converted to length or distance. Figure 21.72 shows (a) a healthy pile graph and (b) a defective pile graph (Fakharian and Hosseinzadeh 2010).

Two methods can be used (Fakharian and Hosseinzadeh 2010):

- Suppose the exact length of the pile is known, and therefore, the velocity of the wave can be calculated from the time of the wave's round trip to the pile's tip and upward reflection.
- If the length of the pile is not known, the velocity of the wave should be assumed based on the material of the pile, and then the length of the pile should be calculated based on the time it takes for the wave to travel to the reflection point at the pile's tip.

21.6.5 CSL Testing

The CSL (crosshole sonic logging) test is a crucial quality assurance method utilized for evaluating drilled shaft foundations and concrete slurry walls. Sonic integrity testing is performed according to the ASTM 6760 Standard, "Integrity testing of concrete deep foundations by ultrasonic crosshole testing." During construction, access tubes made of materials such as PVC or steel are integrated into the concrete, or coreholes are created to facilitate logging, as illustrated in Figure 21.73.

21.6.5.1 CSL Testing Procedure

In a CSL test, an ultrasonic pulse is transmitted through the concrete between paired source and receiver probes inside water-filled tubes or holes. Subsequently, these probes are retracted to the surface while measuring the depth

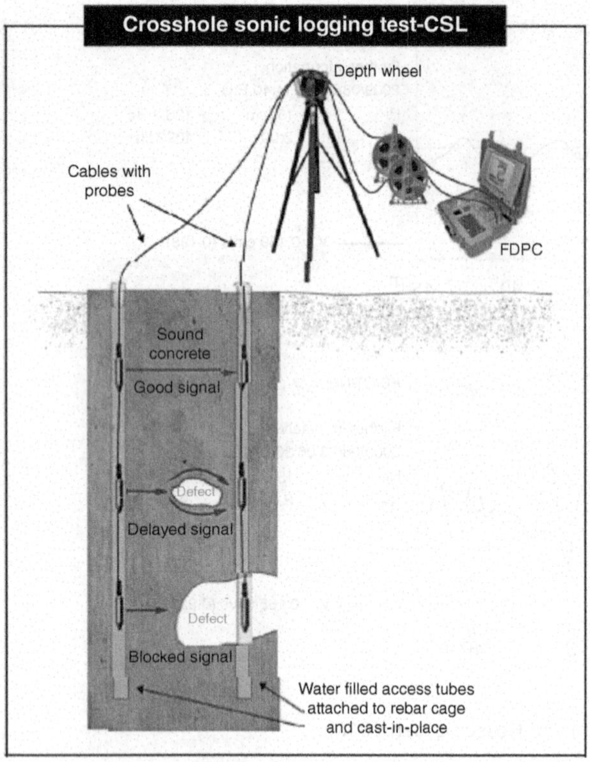

Figure 21.73 Crosshole sonic logging test method. (Report on CLS Testing, Las Vegas Interchange, NDOT Project Records 2018).

using a measurement wheel, allowing the CSL method to assess the concrete quality, The CSL method cannot identify bulbs (diameter increases) due to test geometry. However, it is effective in detecting significant necking or caving, as these generally intrude into the signal paths between the specific pair of tubes or holes (Tinkey and Olson 2007).

In the CSL test, the source is stimulated by a high-voltage pulse every 2.2 in. or 5.5 cm. Simultaneously, the receiver response and measurement depth are recorded for each measurement. To reduce noise, the receiver response is electronically filtered around the receiver's resonant frequency. Data from the receiver probe are recorded and processed using a PC-based sonic logging system (Sheinin et al. 2012).

21.6.5.2 Analysis of CSL Data

Analyzing concrete integrity involves measuring wave travel times between the source and receiver, calculating corresponding wave velocities, and evaluating receiver response energies. Prolonged travel times and slower velocities indicate irregularities in the concrete between the tubes, assuming good bonding between tubes and concrete. Complete signal loss indicates a significant defect in the concrete between one or more tube pairs.

Positive results display consistent pulse arrival times and reasonable compression wave velocities for concrete. Defects such as contaminated or weak concrete and soil intrusions cause delayed arrivals (slower velocity) or no arrivals within the defect zone. The signal energy level serves as a secondary indicator, with low energy indicating poorer concrete quality when the arrival time is delayed (not in the case of a good arrival time). Concrete wave velocity increases over time, especially during the initial curing days. Angled source and receiver CSL logs (with vertical offset) improve the understanding of the defect nature and locations identified in the initial same-elevation source and receiver logs. Additionally, single-hole sonic logs (SSL) provide information about the vertical extent of a defect zone for a specific tube and validate tube-to-concrete coupling in case of an anomaly in a CSL log.

Favorable outcomes are characterized by uniform pulse arrival times aligning with concrete-appropriate compression wave speeds. However, issues like impurities, feeble concrete, or soil infiltration cause delays (slower speed) or absence of signals within the flawed area. The signal's energy level is a secondary gauge of concrete quality, where diminished energy signifies subpar concrete. Concrete's wave speed heightens over time, especially in initial curing days.

21.6.5.3 CSL Rating Criteria

The CSL rating criteria identify abrupt increases in signal arrival times corresponding to decreases in the average signal velocity between the test probes. These changes stem from conditions like increased signal path length around a flaw, decreased signal velocity passing through lower velocity materials (weaker concrete, honeycomb, or contaminated concrete), and deterioration of the bond between the access tube and the concrete. Tube-concrete bond deterioration is rare and mainly found in upper shaft portions or in shafts with PVC access tubes. These criteria rely on the percentage reduction of signal velocity through the flawed area vs. the velocity through adjacent sound material. This approach ensures consistent material measurement, which is crucial if tube pairs are unevenly spaced from shaft top to bottom. While CSL criteria draw from experience with ultrasonic pulse velocity measurements, calculated signal velocities in CSL tests are affected by delays from slower water and PVC tube materials. This delay results in artificially lower signal velocity for paths less than 24 in. However, the percentage change in signal velocity between good and flawed material remains accurate despite inherent signal delays. Here are the general rating criteria for CSL results:

- **Good (G):** No signal distortion and a decrease in signal velocity of 10% or less indicate good quality concrete.
- **Questionable (Q):** Minor signal distortion and lower signal amplitude with a 10–20% decrease in signal velocity. Indicates minor contamination or intrusion and/or questionable quality concrete. Anomalies with 10–15% velocity reductions have revealed sound or flawed concrete at different sites.
- **Poor (P):** Severe signal distortion and significantly lower signal amplitude with a 20% or more decrease in signal velocity. Indicates water slurry contamination, soil intrusion, and/or poor-quality concrete.
- **No Signal (NS):** No signal received. Likely due to soil intrusion or other severe defects absorbing the signal (assuming good tube-concrete bonding). If PVC tubes are used or if measurements are near the shaft top, tube-concrete bonding is suspect.
- **Water (W):** Measured signal velocity of nominally $V = 4800$ to 5000 fps. Indicates water intrusion or a water-filled gravel intrusion with minimal fines.

There is no direct correlation between low-strain velocity and high-strain strength in concrete. However, the modulus is proportional to the square of velocity, and using correlations between modulus and compressive strength (ACI 228.1 R-89, In-Place Methods for Determination of Strength of Concrete), strength can be inferred from velocity. According to the relationship outlined in the American Concrete Institute (ACI) reference, strength is proportional to the square of modulus, which translates to strength being proportional to velocity raised to the power of 4. Hence, a 10% reduction in velocity theoretically signifies a material strength of approximately 65% of normal, while a 20% reduction results in a material strength of about 40% of normal.

Case Study 21.16

CSL Testing Results at Las Vegas Interchange, Las Vegas, Nevada, USA (Report on CLS Testing, Las Vegas Interchange, NDOT Project Records, 2018)

The following are the results of two CSL integrity tests of concrete in two deep foundations. The first test resulted in a Good (G) rating (Figures 21.74 and 21.75). The second test resulted in a poor (P) rating (Figures 21.76 and 21.77). The presented sonic logs show the continuous measurements of energy levels and arrival times along the length of the piles. In these GSL tests, 42 kilohertz (kHz) transducers with a 1.25-in. outer diameter (OD) were utilized.

(Continued)

Case Study 21.16 (Continued)

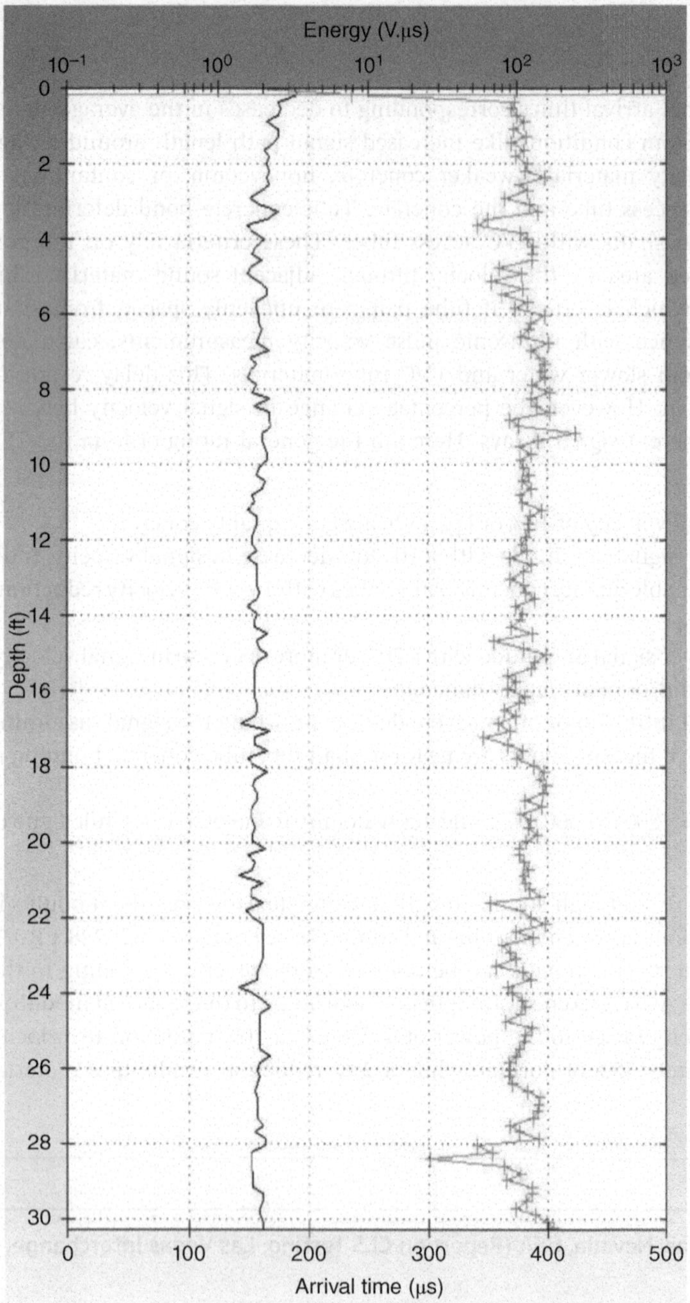

Figure 21.74 Energy (red) and Arrival Time (blue), Good (G) Rating of CSL Testing Results at Las Vegas Interchange, Las Vegas, Nevada, USA (Report on CLS Testing, Las Vegas Interchange, NDOT Project Records 2018).

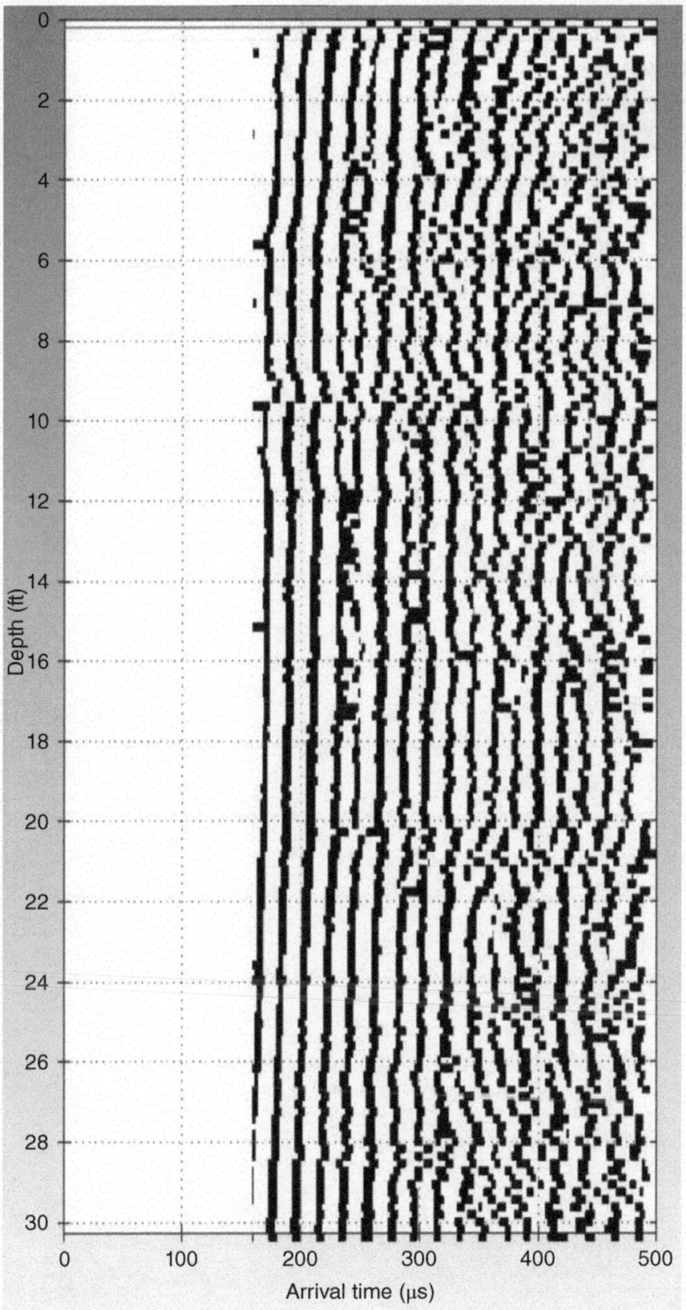

Figure 21.75 Arrival times of successive pulses, good (G) rating of CSL testing results at Las Vegas Interchange, Las Vegas, Nevada, USA (Report on CLS Testing, Las Vegas Interchange, NDOT Project Records 2018).

(Continued)

Case Study 21.16 (Continued)

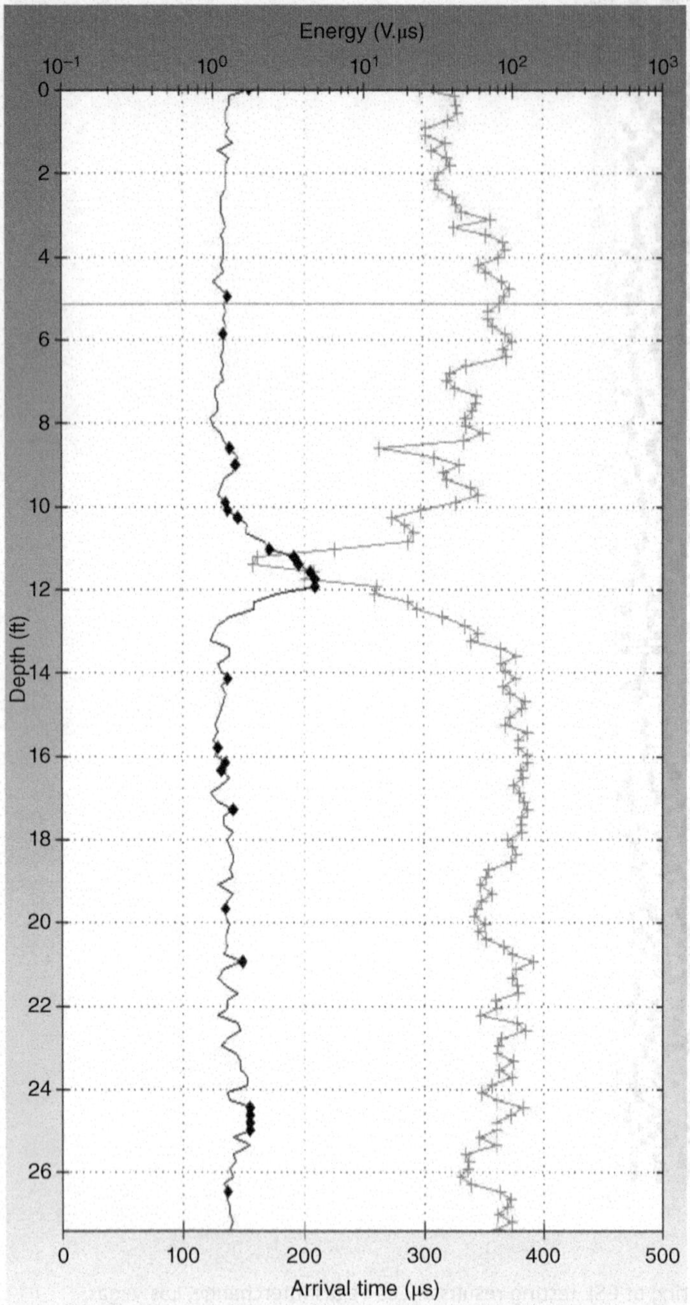

Figure 21.76 Energy (red) and arrival time (blue), poor (P) rating of CSL testing results at Las Vegas Interchange, Las Vegas, Nevada, USA (Report on CLS Testing, Las Vegas Interchange, NDOT Project Records 2018).

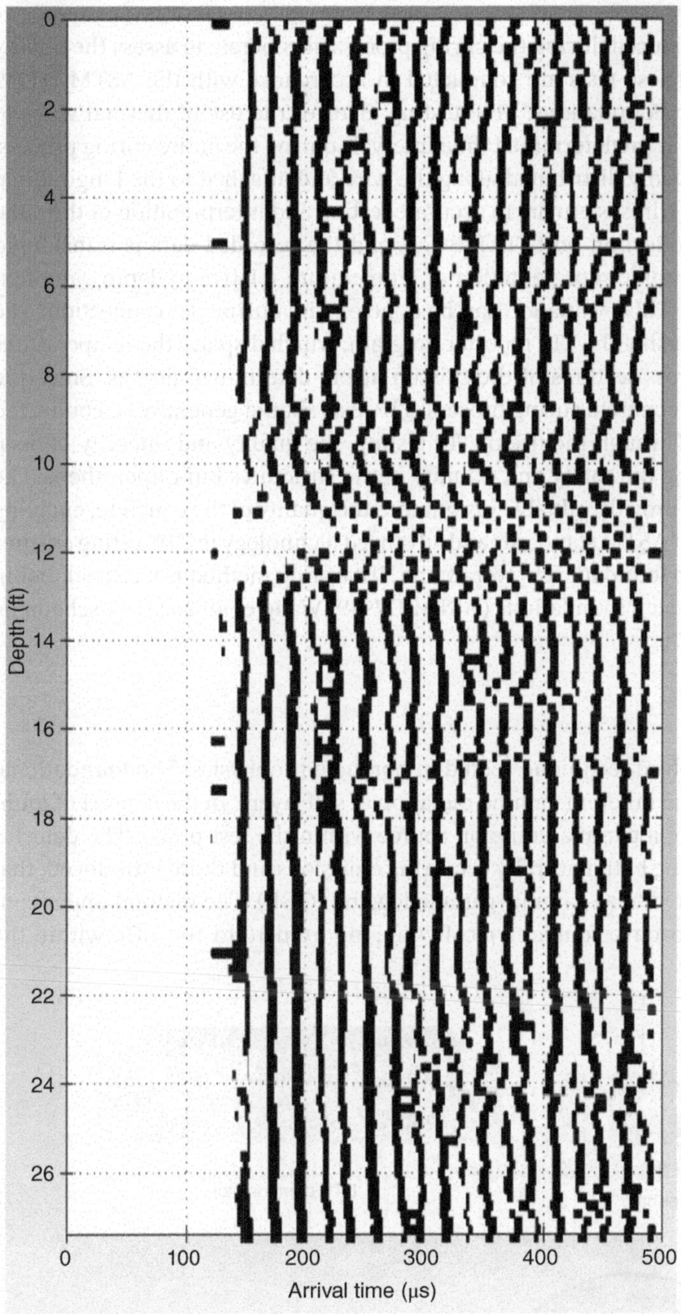

Figure 21.77 Arrival times of successive pulses, poor (P) rating of CSL testing results at Las Vegas Interchange, Las Vegas, Nevada, USA (Report on CLS Testing, Las Vegas Interchange, NDOT Project Records 2018).

21.6.6 TIP Testing

Thermal integrity profiling tests (TIP) utilize the heat produced during the curing process of concrete to assess the quality and integrity of the concrete inside deep foundations. These tests are conducted in accordance with the ASTM D7949 standard, "Thermal Integrity Profiling of Concrete Deep Foundations." TIP testing involves the use of thermal wires to measure the temperature of the concrete inside a caisson at predetermined time intervals during the entire curing process. The wires contain digital temperature sensors that are evenly distributed along the wire and attached to the longitudinal bars of the reinforcement cage. Proper installation of the wires is crucial to accurate testing and interpretation of the data. Each thermal wire is connected to a continuous data collection port (TAP BOX), and the recorded data is transmitted to the TIP main unit. A qualitative assessment of the graphic representation of temperature relative to depth can often indicate if the caisson has any integrity issues or defects. This evaluation can be done on-site during data collection. The TIP system data is uploaded to a computer and analyzed using the TIP reporter program, which displays the temperatures measured over time according to depth and presents cross-sections of the caisson at pre-determined depths. Since the testing is based on the temperature generated during the concrete curing process, TIP tests should generally be conducted between 4 and 72 hours after the concrete installation. Through the use of TIP testing, the quality and integrity of deep concrete foundations can be accurately assessed, ensuring the safety and stability of the structures built upon them. The precise measurement of temperature and analysis of data provide valuable insight into the quality of the concrete, enabling early detection of any potential issues or defects. The use of ASTM standards and advanced technology in TIP testing ensures its accuracy, reliability, and usefulness in evaluating deep foundations. Overall, the TIP testing method is a crucial tool in ensuring the safety and durability of structures built on deep foundations (ASTM D7949; Wang et al. 2023). A schematic diagram of the TIP testing setup is depicted in Figure 21.78.

21.6.7 GL Testing

Gamma Logging of concrete foundation shafts is an evolved technique rooted in borehole geophysics. The foundational approach, known as passive gamma logging, quantifies the inherent gamma radiation of soil layers. In the context of foundation assessment, this method extends by incorporating a gamma emission source within the test probe. The detector then gauges the composite gamma radiation, encompassing both naturally occurring emissions and those introduced, that the probe captures. Therefore, it is more accurately referred to as gamma–gamma logging (GGL). The method under consideration involves the installation of a low-energy radioactive source, typically ranging from 10 to 100 mCi, within the

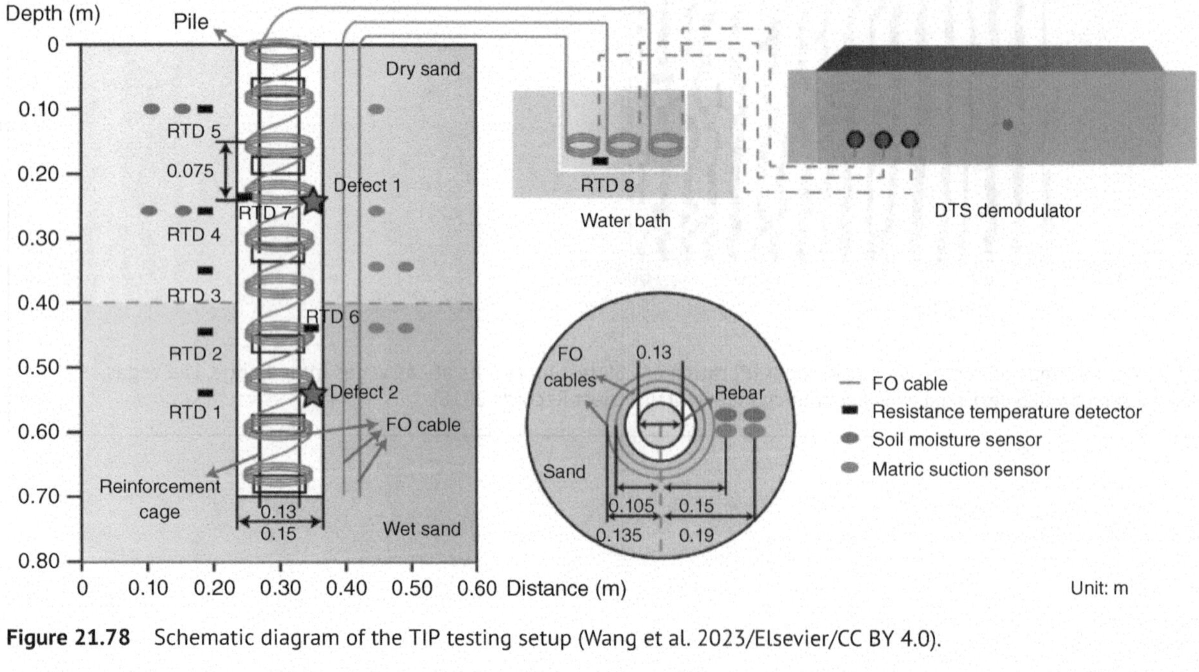

Figure 21.78 Schematic diagram of the TIP testing setup (Wang et al. 2023/Elsevier/CC BY 4.0).

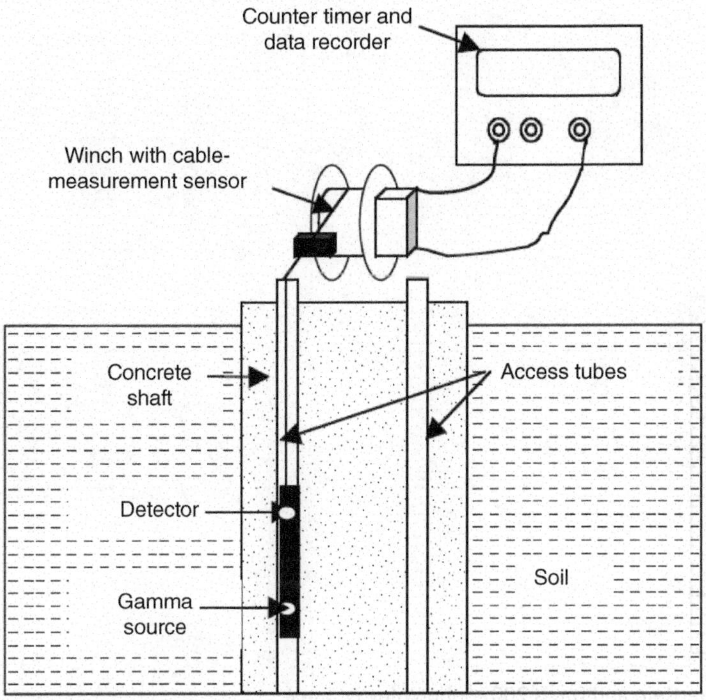

Figure 21.79 Schematic of the gamma–gamma log test set-up (Hertlein and Davis 2007/John Wiley & Sons).

test probe. This source emits gamma radiation into the concrete material surrounding it. The incident particles emanating from a source are subject to absorption, refraction, or reflection by the proximate material via the physical mechanism commonly referred to as "Compton scattering." A number of particles that have undergone refraction or reflection ultimately arrive at a radiation detector, also known as a scintillometer, situated within the test probe. These particles are sequentially enumerated as they transit through the detector. These particles are commonly referred to as "backscatter" in the academic literature. The number of particles detected by the scintillometer is influenced by the density of concrete, owing to its propensity to absorb a proportionate number of radiated particles in relation to its density. This phenomenon underscores the significance of concrete density in the context of scintillometry. The statement posits that GGL, as a method of evaluation, proves efficacious in determining the uniformity of concrete. This assertion was corroborated in a previous study conducted by Preiss and Caiserman in 1975 (Hertlein and Davis 2007). Figure 21.79 depicts a schematic of the gamma–gamma log test setup.

21.6.8 Parallel Seismic Testing

The present study concerns a data-acquisition system. The hydrophone is deployed to the base of the borehole, while the adjacent structure is in close proximity. The superior edge of the cylinder is subjected to a percussive blow with a hammer. The generation of an acoustic pulse is initiated by. The impact delivered by the hammer serves as the initiation of the data-acquisition cycle. The hydrophone signal is a crucial component in underwater acoustic monitoring, as it provides valuable information regarding the source, direction, and intensity of acoustic emissions within aquatic environments. The time at which the acoustic wave reaches the hydrophone is accurately captured and documented. The outcome is yet to be ascertained. The presence of external sounds interferes with the ability of the individual to effectively receive auditory information; then, the communication process may be impeded. This may lead to errors or misunderstandings in the transmission of information. Therefore, it is important to minimize background noise as much as possible, and to ensure that any important auditory information is conveyed clearly and effectively. The inbound signal may experience attenuation resulting from varied factors, such as the distance traveled or soil conditions. Consequently, the hammer impulse may be affected. The procedure involves iteratively performing a series of repetitions, followed by aggregating (i.e., summation

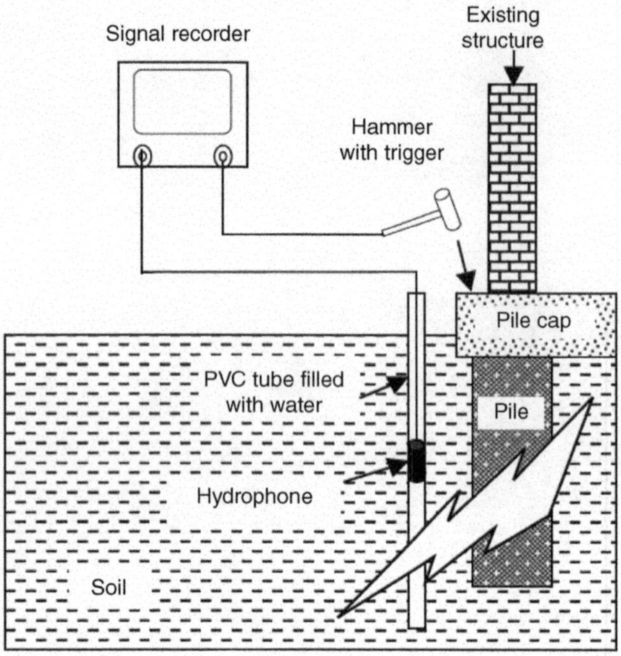

Figure 21.80 Schematic of the parallel seismic test set-up (Hertlein and Davis 2007/John Wiley & Sons).

and subsequent averaging) the resultant signals. The aim is to minimize the occurrence of unsystematic disturbances in the data under analysis, while augmenting the consistent and predictable component of the signal. “The hydrophone is a highly sensitive underwater micro-phone device used for detecting and recording sound waves in aquatic environments.” In order to examine the potential relationship between the variable under investigation and the observed outcome, a systematic approach is employed whereby the variable is increased in a uniform manner. The experimental procedure involves conducting the test at each incremental level of the variable, as illustrated in Figure 21.80 (Hertlein and Davis 2007).

Case Study 21.17

PIT Testing on 378 Piles (Zandbabarasi et al. 2022)

In 2020, Mr. Hasan and his team conducted PIT testing on 378 piles in a project in Bangladesh to classify and assess the quality of the installed piles. The results of the study classified the piles into six groups, with 56% of the installed piles falling into Group 1, which represents piles with no defects or issues (AA grade), and approximately 30% of them having cracks or thinning in cross-section and being placed in Group PFX. The velocity of the shear wave was reported to be around 3000 m/s. Similar research was conducted in Nigeria in 2018 using PIT testing, which found that about 78% of the tested piles were classified as completely healthy (AA grade). Another laboratory field study was conducted in 2022 by Ibrahim Zand Babarisi and his colleagues to examine the quality of concrete piles in Jareez using the nondestructive PIT testing method on the Tehran-Isfahan high-speed train site. A schematic diagram of the implementation of the PIT test and its plotted graph is shown in Figure 21.81 (Zandbabarasi et al. 2022).

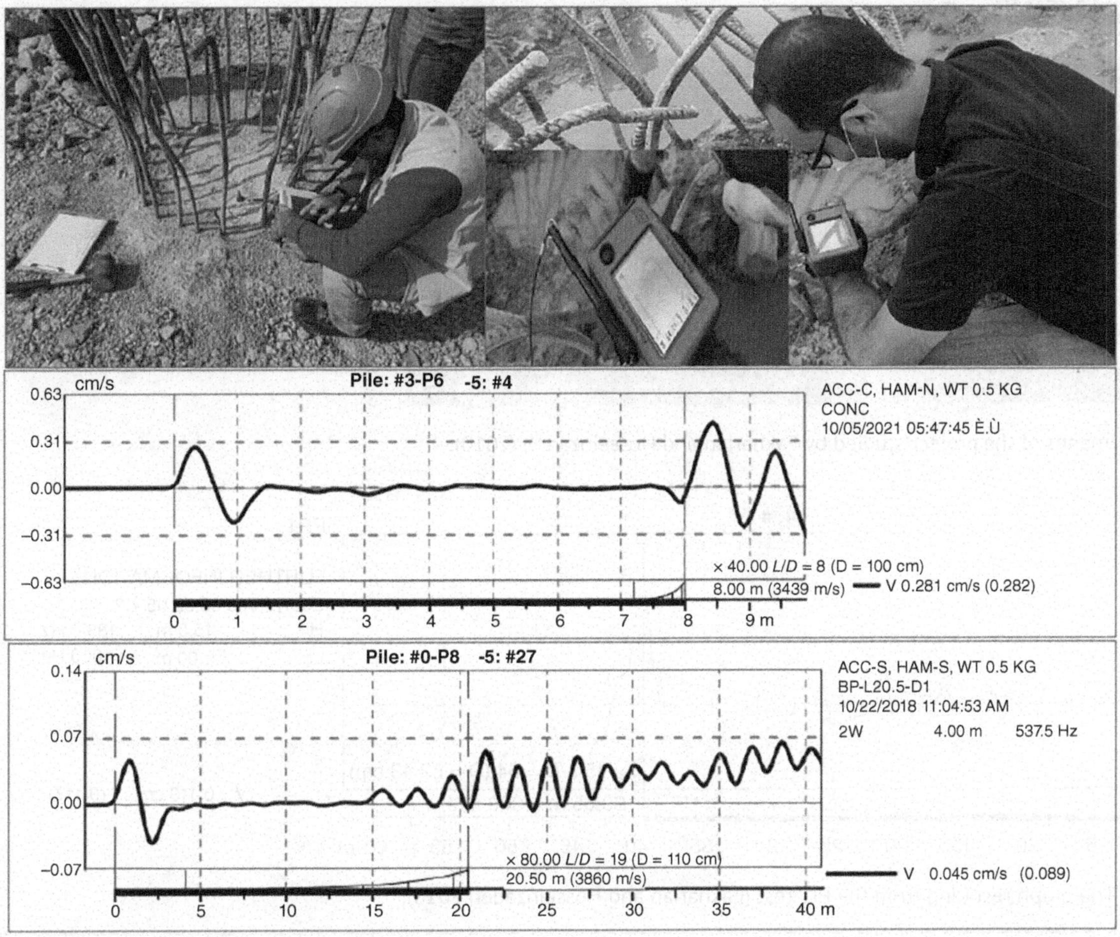

Figure 21.81 A schematic diagram of the implementation of the PIT test and its plotted graph (Zandbabarasi et al. 2022).

Case Study 21.18

PIT on Installed Piles (Fakharian and Hosseinzadeh 2010)

Fakharian and Hosseinzadeh conducted a study in 2010 on more than 30 cases of installed piles in Bandar Imam Khomeini. Quays 10 and 14 are located adjacent to each other and were constructed in the mid-1970s with an approximate length of 5000 m and a width of 55 m. The diameter of the piles is 800 mm. The images of the case study and its diagram are shown in Figures 21.82 and 21.83.

(Continued)

Case Study 21.18 (Continued)

Figure 21.82 Images of the project studied by Fakharian and Hosseinzadeh (2010).

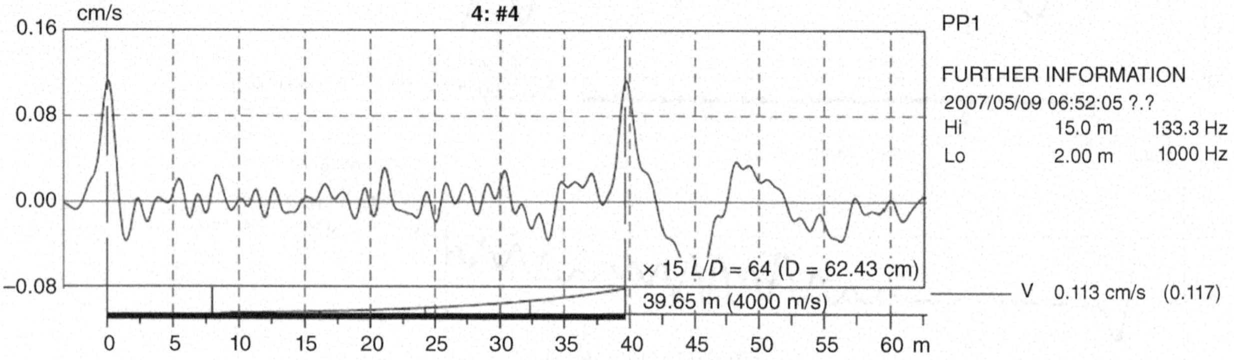

Figure 21.83 The graph resulting from the PIT test (Fakharian and Hosseinzadeh 2010).

21.7 International Prediction Symposium Incorporating Comprehensive Pile Load Tests (Fellenius 2017)

As presented in Chapter 3, Case History 3.3, the goal of the Bolivian Experimental Site for Testing (B.E.S.T.) is to give detailed and complete information on the geomaterial and geotechnical conditions of a specific location in which a variety of pile tests have been conducted. The research done in the field was shared at the 3rd International Conference on Deep Foundations in Bolivia (C.F.P.B).

The first 10–20 m of deposit is composed of NC clays, silts, sands in different combinations, and thicknesses. The groundwater table at the site varies based on the season between the ground surface and about 0.5 m depth. Different types of single piles and one pile group are being constructed and tested. Figure 21.84 shows the pile locations, and Table 21.1 shows a summary of the different piles and tests. Figure 21.85 presents the equipment for a 450 mm diameter full displacement pile (FDP).

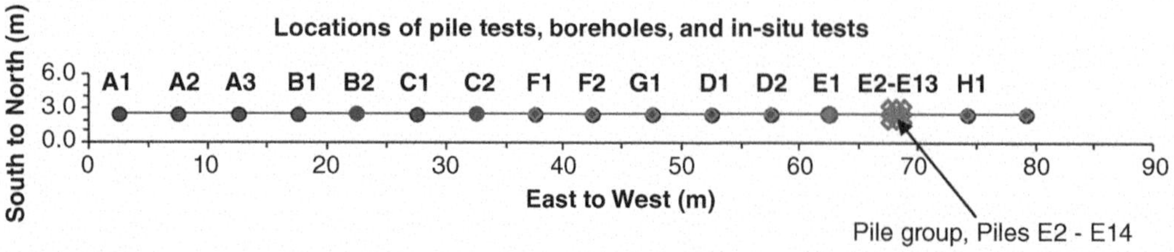

Figure 21.84 The location of test piles (Fellenius 2017/with permission of Bengt Fellenius).

Table 21.1 Pile and test summary (Fellenius 2017).

PILE ID	PILE TYPE	PILE DIAMETER (mm)	ATTACHED DEVICE	TEST and SEQUENCE	THERMAL INTEGRITY PROFILER (TIP)	CROSS-HOLE CHECK pvc or steel pipe	GAGE TYPES and LEVELS	
							VW	RESISTIVE
A-1	Drilled with slurry	620	EB (800)	BD+HD+DT	2 wires		L1, L2, L3	L1, L2, L3
A-2			TB + SB	BD+HD+DT			L1, L2, L3	L1, L2, L3
A-3*				HD+DT				L1, L2, L3
B-1	CFA	450	EB (600)	BD+HD+DT			L1, L2, L3	L1, L2, L3
B-2*				HD+DT				L1, L2, L3
C-1	FDP	450	EB (600)	BD+HD+DT			L1, L2, L3	L1, L2, L3
C-2*				HD+DT	2 wires			L1, L2, L3
D-1	Self boring Micropile	150	EB (500)	HD			L1, L2, L3	L1, L2, L3
D-2				HD				L1, L2, L3
E-1*	FDP	220	EB (300)	BD+HD			L1, L2, L3	L1, L2, L3
E-2 to E-14			EB (300)	BD+HD			L1, L3	L2
F-1	Drilled with slurry	450	EB (400)	BD+HD+DT	2 wires	1 pvc	L1, L2, L3	L1, L2, L3
F-2		600	EB (600)	BD+HD+DT	4 wires	3 pvc	L1, L2, L3	L1, L2, L3
F-3		1,200			6 wires	5 pvc		L1, L2, L3
G-1	Helical	300		HD			L1, L2, L3	L1, L2, L3
DC1200-1	Bored Pile with retrievable casing	1,200			6 wires	5 steel	L1, L2, L3	L1, L2, L3
DC620-1		620			4 wires		L1, L2, L3	L1, L2, L3
DC620-2		620			4 wires	3 steel	L1, L2, L3	L1, L2, L3
DC620-3		620			4 wires	3 pvc	L1, L2, L3	L1, L2, L3
CFA450-1	CFA	450			2 wires		L1, L2, L3	L1, L2, L3
CFA450-2		450			2 wires	1 pvc	L1, L2, L3	L1, L2, L3
FDP450-1	FDP	450			2 wires		L1, L2, L3	L1, L2, L3
FDP360-1	FDP	360			2 wires		L1, L2, L3	L1, L2, L3

NOTES:

Length: All piles intended for static testing will be installed with pile toe at 9.5 m below grade
For pile with EBI unit, the depth is measured to bottom of the EBI before expansion

* PREDICTION PILE
BD: BIDIRECTIONAL STATIC LOADING TEST
HD: HEAD-DOWN STATIC LOADING TEST
SG: INSTRUMENTED WITH STRAIN GAGES
DT: DYNAMIC TEST
(dark shading) PILES FOR LOADING AND INTEGRITY TESTS
(light shading) PILES FOR INTEGRITY TESTS ONLY

EB: EXPANDER BASE
TB: TOE BOX
SB: SHAFT BOX

L: GAGE LEVEL
L1 at 2.0 m depth
L2 at 5.0 m depth
L3 at 7.5 m depth

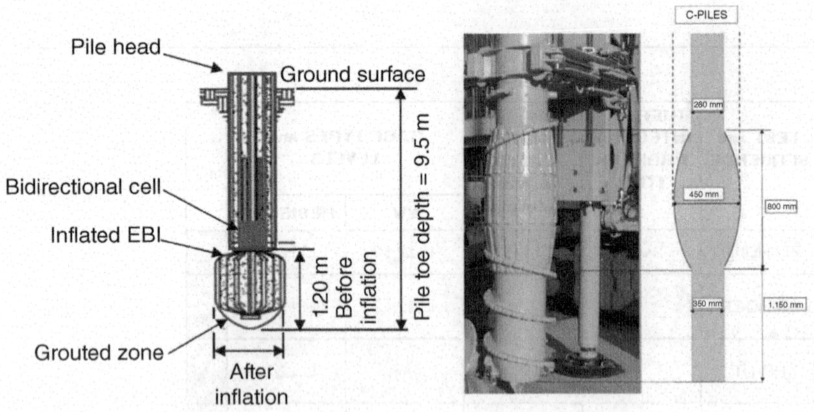

Figure 21.85 Equipment and geometry of the 450 mm-full displacement pile (C2) (Fellenius 2017).

The geotechnical conditions of the site were investigated using conventional in situ and laboratory methods. At each single pile location and at the location of two piles of the group (E-piles), the following in situ tests were performed:

- SPT (standard penetration test)
- SCPTu (seismic piezocone penetration test with pore water pressure measurement)
- SDMT (seismic dilatometer test)
- PMT (pressuremeter test)
- SASW and REMI geophysical tests

Each borehole and field test are located at a 0.8-m radial distance from the test pile. When possible, water content and grain size distribution were determined for samples.

The results of laboratory and SPT investigations of borehole BH-B2 are shown in Figure 21.86. The cone penetrometer soundings with pore water pressure measurements and dilatometer test results for piles C1 are presented in Figure 21.87a and b.

In this event, 71 engineers and researchers participated and implemented various predictive methods and software. Figure 21.88 presents a wide range of predicted load–displacement diagrams for the investigated piles. The deviation of

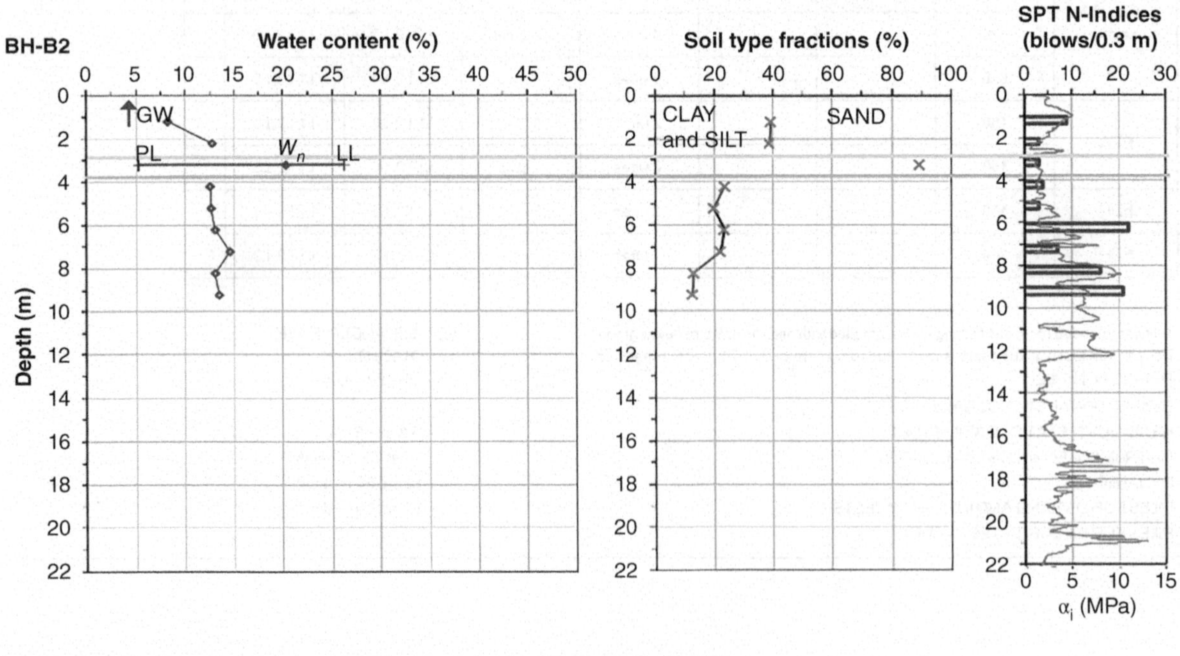

Figure 21.86 Borehole data and SPT N values for BH-B2 (Fellenius 2017).

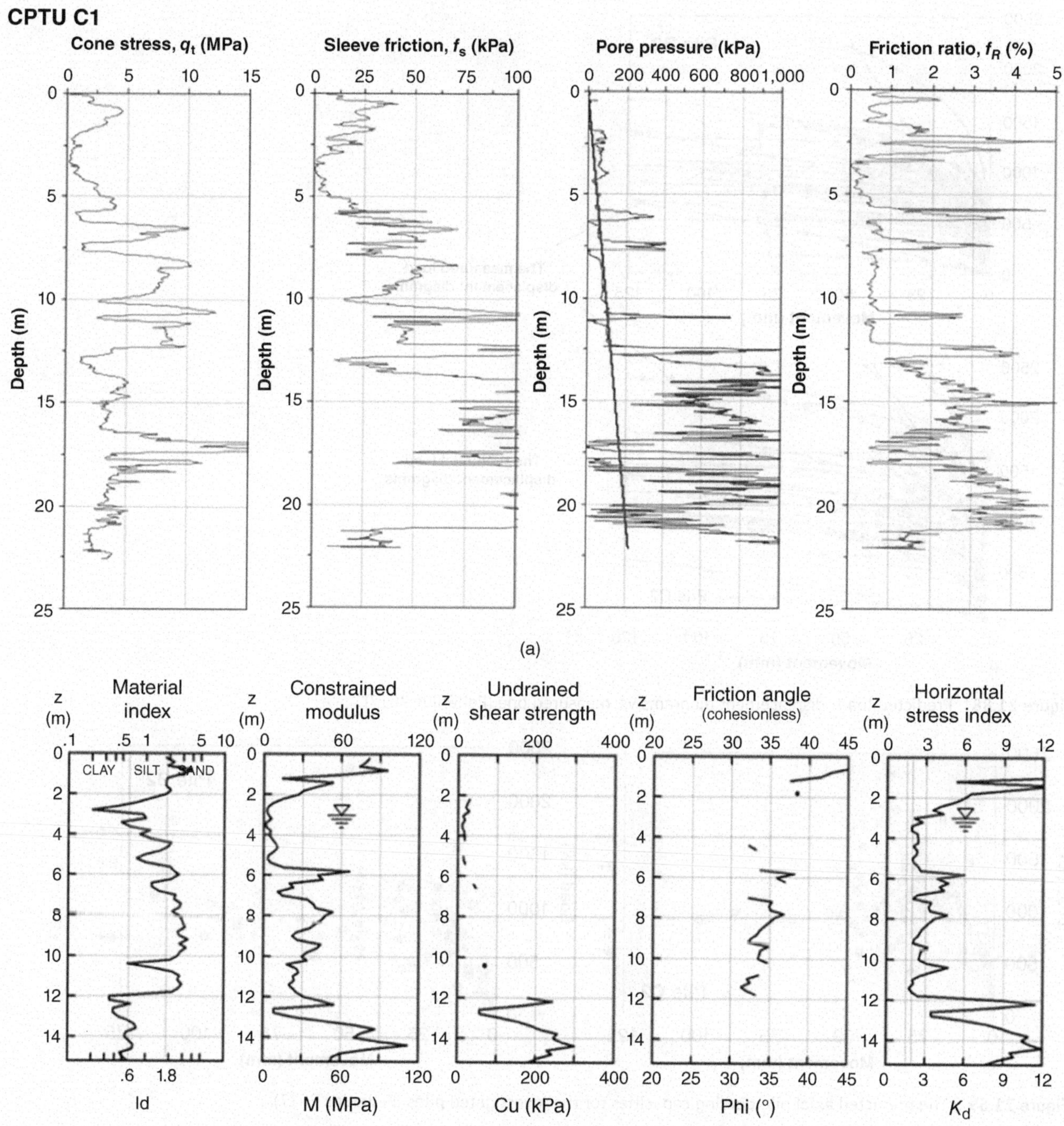

Figure 21.87 In situ tests records for BH-C1: (a) CPTu and (b) DMT (Fellenius 2017).

the proposed load–displacement diagrams from the measured one illustrates the effect of latent uncertainty sources on axial pile design.

Furthermore, various interpretation methods led to a broad range of predictions, as presented in Figure 21.89. Therefore, the predicted bearing capacities were determined in this figure as hollow red circles. Also, Figure 21.90 elucidates where the predicted axial bearing capacity and its corresponding displacement lay in the measured load–displacement diagram.

Moreover, Figure 21.91 provides the resistance distribution and the development of shaft resistance along with the pile embedment depth according to the predictions of shaft and toe bearing capacities for B2 and C2.

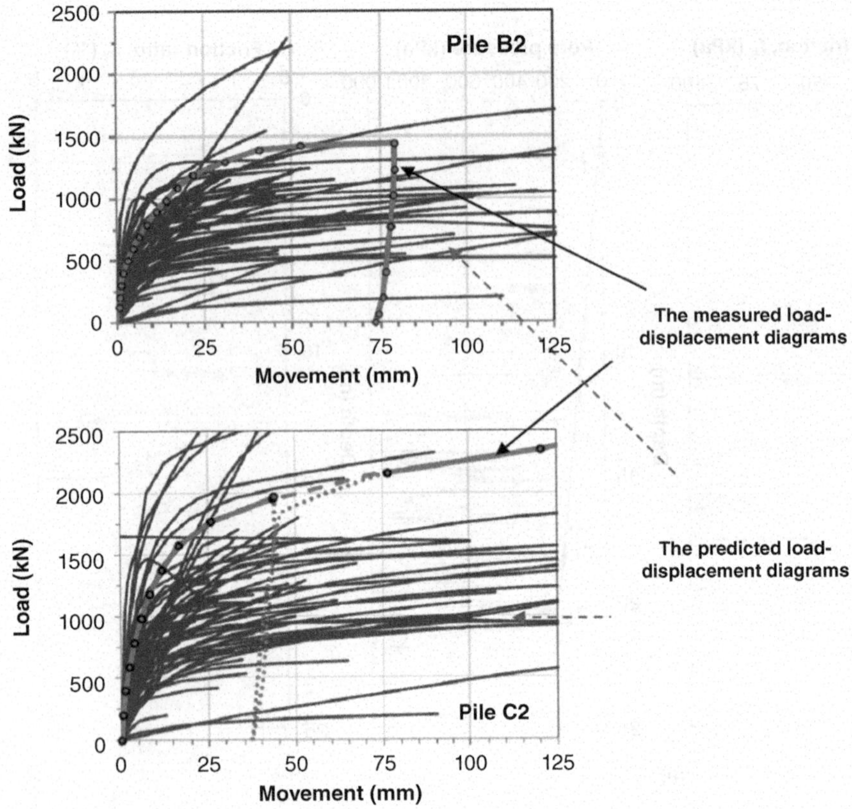

Figure 21.88 Predicted load–displacement diagrams vs. measured one (Fellenius 2017).

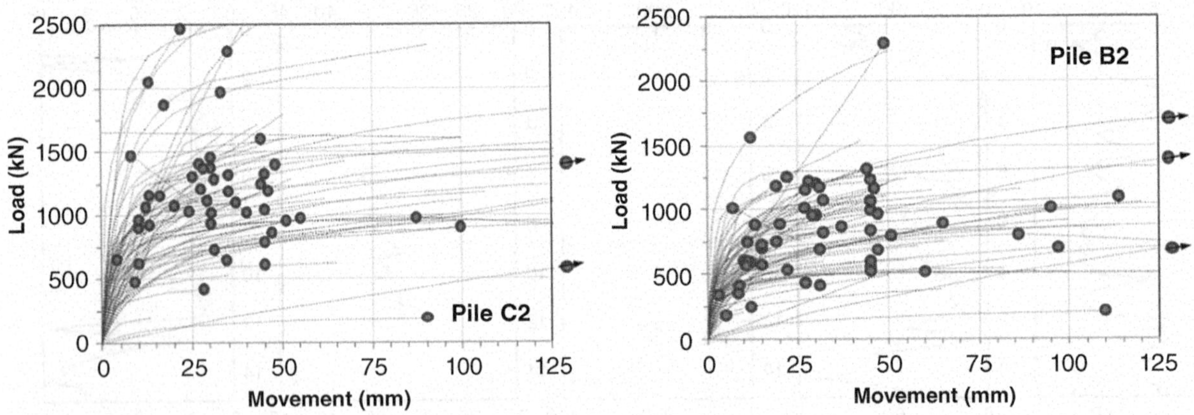

Figure 21.89 The predicted axial pile bearing capacities for the investigated piles (Fellenius 2017).

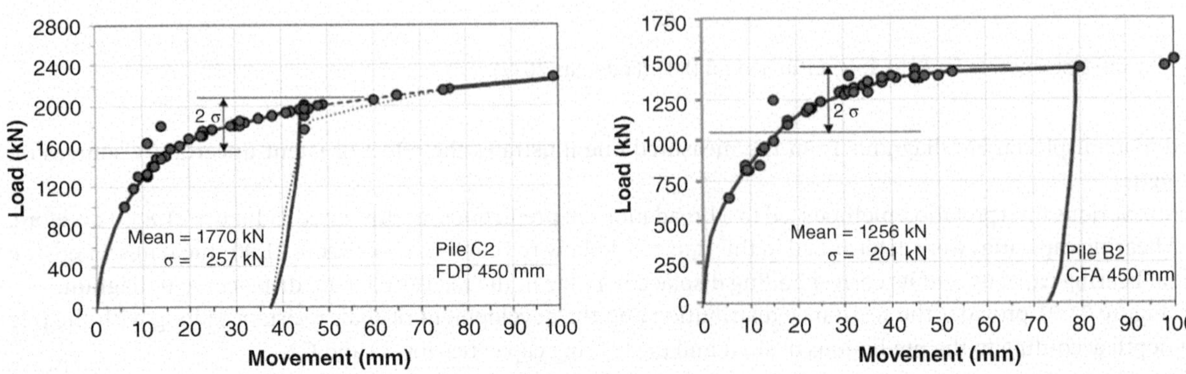

Figure 21.90 The variation of predicted axial bearing capacity on measured load–displacement diagrams (Fellenius 2017).

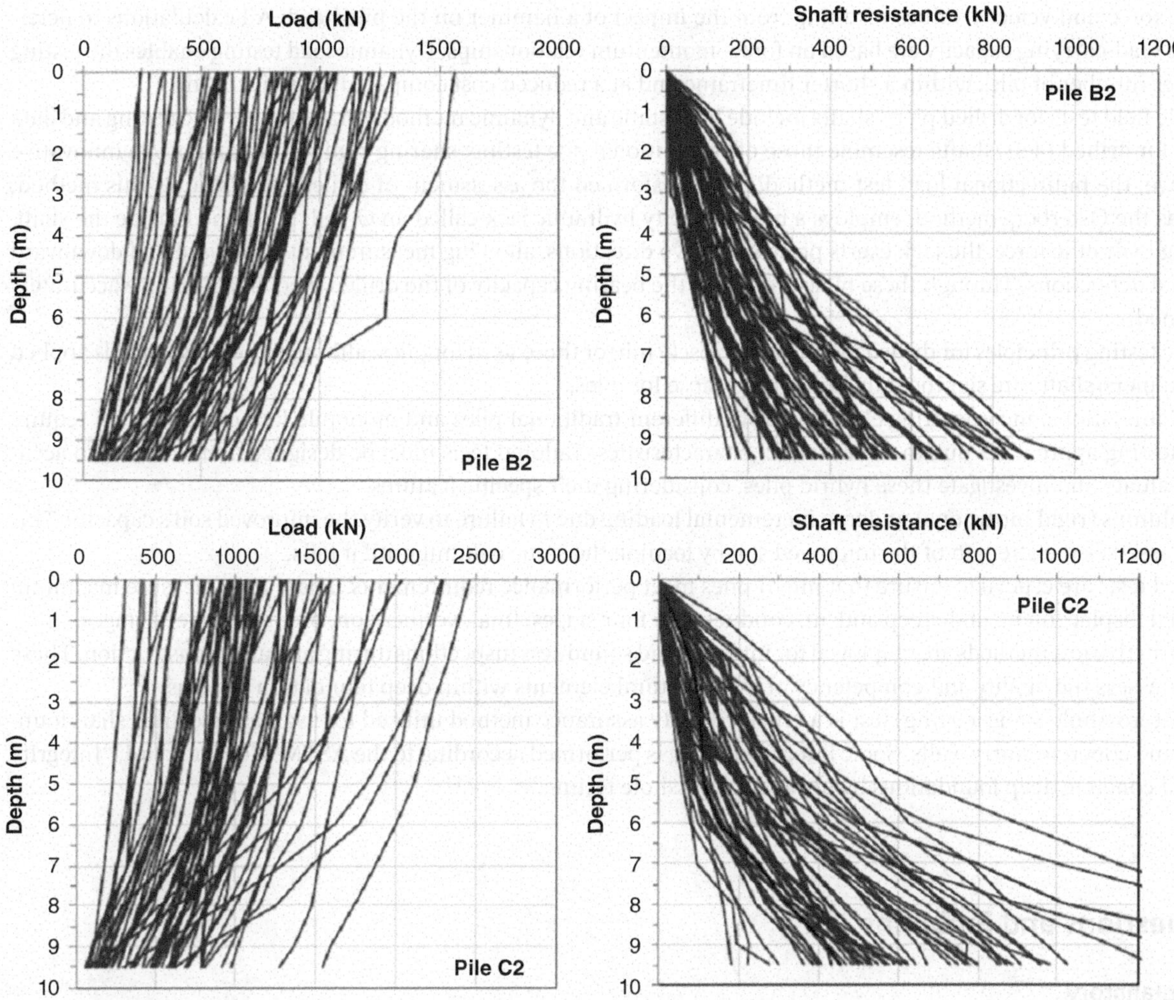

Figure 21.91 The resistance and load distribution along the pile shaft for the investigated piles (Fellenius 2017).

21.8 Major Points

1) For any important project, the foundation engineers need to perform appropriate testing to confirm their design of the foundation system. Depending on the foundation type bench scale and/or full-field testing are strongly recommended. Additionally, the foundation engineer needs to perform integrity testing of the constructed foundation system to ensure that there are no defects in the foundation system.
2) In the realm of shallow foundations, bench-scale tests entail experiments conducted on a reduced scale compared to the actual size. The advantages of employing these tests are manifold: they are cost-effective, require less time, enable rapid data acquisition, yield highly accurate results, and allow for swift adjustments to processes and materials. Also, for shallow foundations, plate load testing stands out as a widely used method for assessing the behavior and performance of shallow foundations under both static and dynamic loading conditions. On the other end of the spectrum, full-scale testing of shallow foundations involves a series of precise tests and large-scale simulations, providing a comprehensive evaluation of the behavior of shallow foundations under real-world conditions.
3) Full-scale field testing is essential to determine the load-bearing capacity of piles and typically involves two types of tests: static and dynamic. In static tests, incremental static loads are applied to the pile head, and settlements are observed for each load increment. The load-settlement data collected is then used to determine the pile's load-bearing capacity through various methods. Although static load testing is reliable, it can be both expensive and time-consuming. An alternative, more cost-effective, and efficient approach is dynamic load testing. This method entails measuring

dynamic force and velocity values resulting from the impact of a hammer on the pile head. All calculations to determine the load-carrying capacity are based on force–momentum relationships. Dynamic load testing enables the testing of a larger number of piles within a shorter timeframe and at a reduced cost compared to static testing.

4) Full-scale field tests for drilled piers/shafts include both static and dynamic methods. Traditional static testing and data analysis for drilled piers/shafts resemble those of conventional pile testing, sharing similar limitations. An innovative alternative, the bidirectional load test method, has transformed the assessment of drilled piers/shafts. This method, known as the Osterberg method, employs a high-capacity hydraulic jack called an O-cell, positioned inside the shaft. Applying hydraulic force, the jack exerts pressure in two directions, allowing measurement of upward and downward forces and deflections. Through these measurements, the bearing capacity of the drilled piers/shafts can be accurately determined.
5) Dynamic testing principles for drilled piers/shafts closely mirror those used for piles, although the impact loads applied to drilled piers/shafts are significantly larger than those for piles.
6) Hybrid foundations, incorporating elements from different traditional piles and monopile foundations, often require unique testing approaches due to their distinct characteristics. Tailored tests must be designed and executed to accurately evaluate and investigate these hybrid piles, considering their specific features.
7) Stone columns (rigid intrusions) undergo incremental loading due to failure to verify the improved soil's capacity. This method assesses the strength of the improved soil by loading the stone column until it fails.
8) Field load tests are crucial to ensure that micro-piles meet performance requirements. These tests measure maximum load, axial displacement, and creep and are conducted in four stages: final, verification, proof, and creep stages.
9) Nondestructive test methods are employed for integrity and soundness inspections during and after construction. These methods assess the quality and competency of the structural elements within deep foundation systems.
10) The CSL (crosshole sonic logging) test is a crucial quality assurance method utilized for evaluating drilled shaft foundations and concrete slurry walls. Sonic integrity testing is performed according to the ASTM 6760 Standard, "Integrity testing of concrete deep foundations by ultrasonic crosshole testing."

21.9 Questions and Problems

21.9.1 Explanatory

1 Explain the difference between Bench-Scale tests and field tests.

2 What is the reason for using small-scale tests in geotechnical engineering?

3 Explain the plate load test and name the advantages and disadvantages of this test.

4 Why field tests are used in geotechnical engineering, and what is the importance of these types of tests?

5 Compare the static and dynamic tests of piles with each other and explain the advantages and disadvantages of each.

6 Explain the rapid load test and state in which type of piles it is better to use this test.

7 Explain Osterberg testing by drawing the schematic form of the test and writing its main difference from common tests.

8 Explain the mechanism of the Statnamic test.

9 Explain rigid intrusion (RI) testing and name the various methods of its implementation.

10 Name possible tests on micro-piles and explain them.

21.9.2 Conceptual

1 Why can't the results of small-scale tests be used as a design basis in the field?

2 Compare the types of load testing according to the ASTM standards method with each other for shallow foundations.

3 What is the innovation of the Osterberg test method?

4 Compare and contrast static and dynamic load tests for piles regarding application, results, and interpretation.

5 Explain the significance of the load–displacement diagram in the engineering of foundation as the major record.

6 Driving a pile is realized as a testing process. How it can be interpreted and applied?

7 What are the aspects of drilled shaft construction upon concepts of foundations testing?

8 What are the similarities and differences in load testing of shallow and deep foundations? How testing concepts can be extended for semi-deep foundations.

9 Discuss different interpretation methods for pile load–displacement realizing strain level. How does this issue affect the ultimate loads? Which in situ tests are compatible with high-strain capacity?

10 List and define lessons learned from the pile prediction symposium reported by Fellenius (2017).

References

Abdelkrim, M. and de Buhan, P. (2007). An elastoplastic homogenization procedure for predicting the settlement of a foundation on a soil reinforced by columns. *European Journal of Mechanics-A/Solids* 26 (4): 736–757.

ACI 228.1 R-89 (1989). *In-Place Methods for Determination of Strength of Concrete*. Detroit: American Concrete Institute.

Akbar, A., Bhat, J.A., and Mir, B.A. (2022). Laboratory study on pressure settlement behavior of shallow foundation on bitumen coated geogrid reinforced soil bed with varying percentage of prestress. *International Journal of Pavement Research and Technology* 16 (3): 544–562.

Almeida, M., Lima, B., Riccio, M. et al. (2014). Stone columns field test: monitoring data and numerical analyses. *Geotechnical Engineering Journal of the SEAGS & AGSSEA* 45 (1): 1–12.

Alvarez, C., Zuckerman, B., and Lemke, J. (2006). Dynamic pile analysis using CAPWAP and multiple sensors. *GeoCongress 2006: Geotechnical Engineering in the Information Technology Age*, Atlantia, Georgia (26 February–1 March 2006), pp. 1–5. American Society of Civil Engineers.

ASTM (1996a). *"Standard Test Method for High-Strain Dynamic Testing of Piles"; D4945-89, 1996 Annual Book of ASTM Standards*. West Conshohocken, PA: ASTM International.

ASTM D1143/D1143M-07 (2018). *Standard Test Methods for Deep Foundations Under Static Axial Compressive Load*. West Conshohocken, PA: ASTM International.

ASTM D1194-18 (2018). *Standard Test Method for Bearing Capacity of Soil for Static Load and Spread Footings*. West Conshohocken, PA: ASTM International.

ASTM D3966-20 (2020). *Standard Test Methods for Deep Foundations Under Lateral Load*. West Conshohocken, PA: ASTM International.

ASTM D4945-12 (2012). *Standard Test Method for High-Strain Dynamic Testing of Deep Foundations*. West Conshohocken, PA: ASTM International.

Baca, M. (2020). Static load test interpretation methods of pile base capacity approximation for design of wind turbine deep foundation. *Journal of Physics: Conference Series* 1614: 012020.

Battista, N., Kechavarzi, C., Seo, H., et al. (2016). Distributed fibre optic sensors for measuring strain and temperature of cast-in-situ concrete test piles. *Proceedings of the International Conference on Smart Infrastructure and Construction (ICSIC)*, Cambridge, UK, 27–29.

Bayesteh, H. and Sabermahani, M. (2018). Full-scale field study on effect of grouting methods on bond strength of hollow-bar micropiles. *Journal of Geotechnical and Geoenvironmental Engineering* 144 (12): 04018091.

Bhowmik, D., Baidya, D.K., and Dasgupta, S.P. (2013). A numerical and experimental study of hollow steel pile in layered soil subjected to lateral dynamic loading. *Soil Dynamics and Earthquake Engineering* 53: 119–129.

Bhushan, K., Dhingra, A., Scheyhing, C., and Zhai, E. (2004). Ground improvement by stone columns and surcharge at a tank site. *Proceedings of the International Conference on Case Histories in Geotechnical Engineering*, New York, New York (13–17 April 2004), vol. 1, pp. 19–30. University of Missouri—Rolla.

Bowles, J.E. (1996). *Foundation Analysis and Design*. McGraw-Hill Education.

Brown, M.J., Hyde, A.F.L., and Anderson, W.F. (2006). Analysis of a rapid load test on an instrumented bore pile in clay. *Géotechnique* 56 (9): 627–638.

Butler, H.D. and Hoy, H.E. (1977). *Users Manual for the Texas Quick-Load Method for Foundation Load Testing*. Washington: Federal Highway Administration, Office of Development.

Caltrans (1998). Benicia-Martinez bridge retrofit. http://www.dot.ca.gov/hq/esc/tollbridge/Ben-Mar/0440U4/Photos/0440U4ProjectPhotos01.html (accessed 2 July 2023).

Çevik, S. (2022). Evaluation of plate load test results: a case study in Ankara. *Eskişehir Osmangazi Üniversitesi Mühendislik ve Mimarlık Fakültesi Dergisi* 30 (2): 253–259. https://doi.org/10.31796/ogummf.1055962.

Chin, F.K. (1970). Estimation of the ultimate load of piles not carried to failure. *Proceedings of the 2nd Southeast Asian Conference on Soil Engineering*, Singapore (11–15 June), pp. 81–90. Asian Institute of Technology.

Das, B.M. (2010). *Principles of Geotechnical Engineering*. Cengage Learning.

Davisson, M.T. (1972). High capacity piles. *Proceedings of Lecture Series on Innovations in Foundation Construction*, Illinois Section, Chicago (January to May 1972), 81–112. American Society of Civil Engineers, ASCE.

DeBeer, E.E. (1968). Proefondervindlijke bijdrage tot de studie van het grensdraag vermogen van zand onder funderingen op staal. Tijdshift der Openbar Verken van Belgie, No. 6, 1967 and No. 4, 5, and 6, 1968.

Decourt, L. (1999). Behavior of foundations under working load conditions. *Proceedings of the 11th Pan-American Conference on Soil Mechanics and Geotechnical Engineering*, Foz DoIguassu, Brazil (August 1999), vol. 4, 453–488. ISSMGE (International Society for Soil Mechanics and Geotechnical Engineering).

Elgamal, A. (2020). Empirical approach to predict static capacity of drilled shafts based on high strain dynamic pile test results. *5th World Congress on Civil, Structural, and Environmental Engineering (CSEE'20)*, Lisbon, Portugal Virtual Conference – October 2020, Paper No. ICGRE 146. https://doi.org/10.11159/icgre20.146.

Elsiragy, M.N. (2021). Field comparative investigation of loading test on micro-piles installed with different technique – (Case study). *European Journal of Engineering and Technology Research* 6 (4): 88–93.

Faizi, K., Armaghani, D.J., Sohaei, H. et al. (2015). Deformation model of sand around short piles under pullout test. *Measurement* 63: 110–119.

Fakharian, K. and Hosseinzadeh Attar, I. (2010). Application of PIT integrity test in estimating the length of 10 and 14-meter piles in the Imam Khomeini port's structures. *Proceedings of the 9th International Conference on Coasts, Ports and Marine Structures*, Tehran (28–30 November 2012).

Fakhroo, R. and Al-Mahaidi, R. (2019). Non-destructive testing of grouted stone columns. *Journal of Rock Mechanics and Geotechnical Engineering* 11 (2): 244–256.

Fellenius, B.H. (1975). Test loading of piles. Methods, interpretation, and new proof testing procedure. *Journal of the Geo-technical Engineering Division* 101 (GT9): 855–869.

Fellenius, B.H. (1980). The analysis of results from routine pile loading tests. *Ground Engineering* 13 (6): 19–31.

Fellenius, B.H. (1990). *Guidelines for the Interpretation and Analysis of the Static Loading Test*, 1e. Sparta, NJ: Continuing Education Committee, Deep Foundation Institute.

Fellenius, B.H. (2001). What capacity value to choose from the results of a static load test fulcrum, Winter, Deep Foundation Institute, 19–22.

Fellenius, B.H. (2017). Report on the B.E.S.T. prediction survey of the 3rd CBFP event. *Proceedings of the 3rd Bolivian International Conference on Deep Foundations*, Santa Cruz de la Sierra, Bolivia (27–29 April 2017), vol. 3, 7–25. Madison, WI, USA: Omnipress.

Fellenius, B.H., Altaee, A., Kulesza, R., and Hayes, J. (1999). O-cell testing and FE analysis of 28-m-deep barrette in Manila, Philippines. *Journal of Geotechnical and Geoenvironmental Engineering* 125 (7): 566–575.

FHWA (1998). Design and construction of driven foundations. Report No. FHWA-HI-97-013, Washington, DC.

Fonseca, A.V.d. (2001). Load tests on residual soil and settlement prediction on shallow foundation. *Journal of Geotechnical and Geoenvironmental Engineering* 127 (10): 869–883.

Garner, M.P. (2007). Loading rate effects on axial pile capacity in clays. Theses and Dissertations. 1434. https://scholarsarchive.byu.edu/etd/1434, BYU Scholars Archive Citation.

GZA GeoEnvironmental, Inc. (1999). Letter to Slattery/Interbeton/J.F. White/Perini Regarding Dynamic Pile Testing Results, CA/T Contract C09A4, Approach No. 5, July 26, 1999.

Han, F., Ganju, E., Prezzi, M. et al. (2019). Axial resistance of open-ended pipe pile driven in gravelly sand. *Géotechnique* 70 (2): 138–152.

Hansen, J.B. (1963). Discussion on hyperbolic stress-strain response: cohesive soils. *Journal for Soil Mechanics and Foundation Engineering* 89, SM4: 241–242.

Haque, M.N., Abu-Farsakh, M.Y., Nickel, C. et al. (2018). A load-testing program on large-diameter (66-inch) open-ended and PSC-instrumented test piles to evaluate design parameters and pile setup. *Transportation Research Record* 2672 (52): 291–306. https://doi.org/10.1177/0361198118776781.

Heins, E., Bienen, B., Randolph, M.F., and Grabe, J. (2020). Effect of installation method on static and dynamic load test response for piles in sand. *International Journal of Physical Modelling in Geotechnics* 20 (1): 1–23.

Hertlein, B. and Davis, A. (2007). *Nondestructive Testing of Deep Foundations*. Chichester, England: STS Consultants Ltd and Construction Technology Labs Inc; Wiley.

Hesham, E.N.M. and Abdelrahman, E.M. (2006). Performance-based quality control tests of micropiles. *Journal of Geotechnical and Geoenvironmental Engineering* 132 (9): 1112–1122.

Hoefsloot, F. (2010). *Statnamic Pile Load Testing*. The Netherlands: Fugro.

Holscher, P. and van Tol, F.A. (2008). *Rapid Load Testing on Piles*. IOS Press.

Hosseinzadeh, M. and Fakharian, K. (2013). Dynamic load testing of piles in fajr petrochemicals site: a case study. *International Journal of Geosciences* 4 (7): 1133–1139.

Houseinzadeh, H and Alielahi, H. (2018). Experiences from pile Static load Tests; a case study of an access bridge in Persian Gulf Project. *Proceeding of International Conference on Civil Engineering, Architecture and Urban Management*, Tehran, Iran (1 March 2018) (In Persian).

Ji, Z., Liu, H., and Han, J. (2012). Micropile foundation design and construction for the Jiaozhou Bay Bridge in China. *Journal of Geotechnical and Geoenvironmental Engineering* 138 (2): 221–230.

Jiu, Y.Z., Zhang, Z., and Zhang, X.Y. (2022). Experimental investigation on the vertical bearing characteristics of jacked piles in saturated silt foundations under excavation. *Advances in Civil Engineering* 2022: 9460701. 12 pages.

Justason, M., Janes, M., Middendorp, P., and Mullins, A. (2000). Statnamic load testing using water as reaction mass. In: *The 6th International Conference on the Application of Stress Wave Theory to Piles*. Sao Paulo, Brazil.

Kamei, S., Takano, K., and Fujita, T. (2013). Comparison of static load test and rapid load test on steel pipe piles in two sites. *Soils and Foundations* 53 (3): 375–383.

Kondner, R.L. (1963). Hyperbolic stress-strain response: cohesive soils. *Journal for Soil Mechanics and Foundation Engineering* 89 (SM1): 115–143.

Kumar, A. and Kumar, P. (2019). Load capacity test on stone column improved soil. *Journal of Materials and Environmental Science* 10 (2): 171–176.

Liao, Y., Zhou, Y., and Qin, P. (2018). Multiple reference impact testing for bridge assessment with drop hammer. *Shock and Vibration* 2018: 1–17.

Loli, M., Kourkoulis, R., and Gazetas, G. (2018). Physical and numerical modeling of hybrid foundations to mitigate seismic fault rupture effects. 144 (11): 04018083.

Mahgoub, A. and El Naggar, H. (2020). Shallow foundations on lightweight TDA backfill: field tests and 3D numerical modelling. *Computers and Geotechnics* 126: 103761.

Mazurkiewicz, B.K. (1972). Test loading of piles according to Polish regulations. Royal Swedish Academy of Engineering Sciences, Commission on Pile Research, Report No. 35, pp. 20, Stockholm.

McCabe, B.A., McNeill, J.A., and Black, J.A. (2007). Ground Improvement using the Vibro-stone Column Technique. The Institute of Engineers of Ireland, Technical Publication.

Mehrannia, N., Kalantary, F., and Ganjian, N. (2018). Experimental study on soil improvement with stone columns and granular blankets. *Journal of Central South University* 25 (4): 866–878.

Mokhberi, E., Ghorbani, A., and Mohebbi, M. (2019). Plate load testing: a review of analysis methods and results. *Bulletin of the Iranian Association of Engineering Geologists* 11 (3): 1–16.

Naranjestan Complex. (2017). Samanpey. https://samanpey.com/projects/naranjestan-commercial-complex-black-fg?lang=en.

Osterberg, J.O. (1984). A new simplified method for load testing drilled shafts. *Found Drill* 23 (6), pub. Association of Drilled Shaft Contractors, ADSC: 9–11.

Osterberg, J.O. (1998). The Osterberg load test method for bored and driven piles: The first ten years. *Paper presented at the Proc., 7th Int. Conf. and Exhibition on Piling and Deep Foundations,* Vienna, Austria (15–17 June 1998), 1.28.

Patidar, V.K., Kaur, S., and Dindorkar, N. (2020). Bi-directional Static Pile Load Test by S-Cell Method. In: *Advances in Geotechnical and Transportation Engineering*, Lecture Notes in Civil Engineering, vol. 71 (ed. S. Saride, B. Umashankar, and D. Avirneni). Singapore: Springer. https://doi.org/10.1007/978-981-15-3662-5_2.

Pelecanos, L., Soga, K., Chung, M.P.M. et al. (2017). Distributed fibre-optic monitoring of an Osterberg-cell pile test in London. *Géotechnique Letters* 7: 152–160.

Poulos, H.G. (ed.) (2011). *Foundations: Analysis and Design*, vol. 2. Wiley.

Poulos, H. and Bunce, G. (2008) Foundation design for the Burj Khalifa, Dubai – the World's tallest building. *Proceedings of the 6th International Conference on Case Histories in Geotechnical Engineering*, Arlington, VA (11–16 August 2008).

Raju, V.R. and Daramalinggam, J. (2012). Ground improvement: principles and applications in Asia. *Proceedings of the Institution of Civil Engineers-Ground Improvement* 165 (2): 65–76.

Report on CFA Pile Load Testing, Osterberg Method (2018). LT-A – raiders stadium production piles, Clark County Project Records, Las Vegas, NV (LT-1839-1), Las Vegas, Nevada, USA.

Report on CLS Testing, Las Vegas Interchange, NDOT Project Records (2018).

Russo, G., Marone, G., and Di Girolamo, L. (2021). Hybrid energy piles as a smart and sustainable foundation. *Journal of Human, Earth, and Future* 2 (3): 306–322.

Safarzadeh Gendeshmin, Z. and Aminfar, M.H. (2020). Study the behavior of shallow footings subjected to vertical static loads in unsaturated granular soils. PhD thesis. University of Tabriz.

Salman, H. (2021). Utilizing load tests and site investigation data to predict design load capacities for drilled shafts in different geologic formations. The University of Texas at Arlington ProQuest Dissertations Publishing, 2021. 28826570. Available from ProQuest One Academic. (2596687372).

Samanpey. (2017). https://samanpey.com/projects/naranjestan-commercial-complex-black-fg?lang=en.

Schmertmann, J.H., Hayes, J.A., Molnit, T., and Osterberg, L.O. (1988). O-cell testings case histories demonstrate the importance of bored pile (drilled shaft) construction technique. *International Conference on Case Histories in Geotechnical Engineering,* St. Louis, Missouri (31 May–4 June 1988). 7. https://scholarsmine.mst.edu/icchge/4icchge/4icchge-session00/7.

Sheinin, V.I., Dzagov, A.M., Blokhin, D.I. et al. (2012). Use of ultrasonic-test data for quality and strength evaluation of concrete. *Soil Mechanics and Foundation Engineering* 49: 124–131. https://doi.org/10.1007/s11204-012-9179-3.

Shen, B.H. and Niu, D.S. (1991). A new method for determining the yield load of piles. *Proceedings of the Fourth International Conference on Piling and Deep Foundations*, Deep Foundation Institute, Stresa (7–12 April), vol. 1, 531–534. Balkema Publishers.

Swaroopa, S., Gokhale, C.S., and Mungi, M.S. (2012). Statnamic test on piles. *IJCA Proceedings on Emerging Technology Trends on Advanced Engineering Research* – 2012 ICETT(4):38-44.

Tinkey, Y. and Olson, L.D. (2007) Turning crosshole sonic logs into crosshole velocity tomograms. *Proceedings of GeoDenver – 2007*, Issues in Deep Foundations. https://doi.org/10.1061/40902(2221)4.

Wang, X., Zeng, X., Li, X., and Li, J. (2020). Liquefaction characteristics of offshore wind turbine with hybrid monopile foundation via centrifuge modelling. *Soil Dynamics and Earthquake Engineering* 145: 2358–2372.

Wang, J., Zhu, H., Tan, D. et al. (2023). Thermal integrity profiling of cast-in-situ piles in sand using fiber-optic distributed temperature sensing. *Journal of Rock Mechanics and Geotechnical Engineering* 15 (12): 3244–3255.

Weaver, T.J. and Rollins, K.M. (2009). Reduction factor for the unloading point method at clay soil sites. *Journal of Geotechnical and Geoenvironmental Engineering* 136 (4): https://doi.org/10.1061/(ASCE)GT.1943-5606.0000262.

Webster, K., Rausche, F., and Webster, S. (2011). Pile and shaft integrity test results classification, mitigation, acceptance, and/or rejection. *Transportation Research Board 90th Annual Meeting*, Washington DC, United States (23-27 January).

Wu, S., Lai, J., Yang, B.H., and Cheng, C.F. (2015). Integrity testing of model piles with pile cap. *International Symposium Non-Destructive Testing in Civil Engineering (NDT-CE)*, Berlin, Germany (15–17 September).

Yen, T.F. and Lee, V.S. (2000). *Bridge Engineering Handbook, Second Edition: Construction and Maintenance*. CRC Press.

Zandbabarasi, E., Ghasemi Tabar, A.H., and Abdollahi, A. (2022). Evaluation of quality of concrete piles in Jareez using non-destructive PIT-Test method (Case study: Tehran-Isfahan high-speed train). *Proceedings of the 20th Concrete Day Conference and 14th National Concrete Conference*, Tehran (8 October 2022).

Zayed, M., Kim, K., Prabhakaran, A., and Elgamal, A. (2023). Shake table testing and computational framework for seismic response of utility-scale bucket foundation offshore wind turbines. *Soil Dynamics and Earthquake Engineering* 171: 107939.

22
Numerical Analysis

Chapter Highlight

- Introduction
- Preliminaries and Basic Concepts
  - What Is Numerical Modeling?
  - Analytical Solutions vs. Numerical Methods
  - Geotechnical Problem Classification
  - Significance of Numerical Modeling in Foundation Problems
  - Geometrical Considerations of the Model in Foundation Problems
  - Boundary Conditions
  - Constitutive Models for Soil
  - 2D vs. 3D Models
- Numerical Analysis in Geotechnical Engineering
  - Analysis Planning
  - Geometry
  - Boundary Conditions
  - Structural Elements
  - Meshing
  - Failure Criteria and Material Properties
  - Loading
  - Initial Conditions
  - Analysis Stages
  - Outputs
- Different Numerical Methods in Foundation Engineering
  - Finite Difference Method (FDM)
  - Finite Element Method (FEM)
  - Finite Element Limit Analysis (FELA)
  - Boundary Element Method (BEM)
  - Discrete Element Method (DEM)
  - Material Point Method (MPM)
  - Kinematic Element Method (KEM)
- Common Software in Foundation Engineering
  - Applications, Advantages, and Limitations
  - Numerical Modeling of Special Issues in Foundation Engineering

Advanced Foundation Engineering: Principles, Performance and Prospects, First Edition. Moses Karakouzian and Abolfazl Eslami.

22.1 Introduction

One important issue in geotechnical engineering is using numerical methods in analysis and design. This issue is particularly important in complex geometries and loading conditions. To use numerical methods to solve problems related to geotechnical engineering, it is necessary to obtain information such as the geometric specifications of the foundation, the applied loading, and the results of geotechnical tests (laboratory and in situ tests). With this information, modeling can be done, and the output can be used to design the desired structure. Numerical methods are mainly available in the form of commercial software, which can be used easily as a result.

This chapter examines the application of numerical methods in geotechnical engineering, and the most common engineering software will also be discussed. Finally, with examples and case studies of specific topics in geotechnical engineering, the use of numerical methods in solving these problems will be demonstrated.

22.2 Preliminaries and Basic Concepts

22.2.1 What Is Numerical Modeling?

Numerical modeling refers to the use of mathematical algorithms and computer simulations to study and predict the behavior of complex systems or processes. This approach is commonly used in many fields, including physics, engineering, finance, and environmental science. The process of numerical modeling typically involves the following steps:

- Formulating a mathematical model that describes the system or process of interest.
- Discretizing the model into a set of equations that can be solved using numerical methods.
- Implementing the numerical solution algorithm on a computer, often using specialized software.
- Running simulations to generate numerical results that can be analyzed and interpreted.

Numerical modeling has many advantages over analytical or experimental approaches, including the ability to study complex systems that are difficult or impossible to observe directly, and the ability to rapidly test and evaluate different scenarios or hypotheses. However, numerical modeling also has limitations, such as the need for accurate data inputs and assumptions and the risk of errors or biases in the model implementation.

In geotechnical engineering, numerical modeling is an essential tool to simulate and predict the behavior of soils, rocks, and structures. It involves using mathematical and computational methods to solve complex geotechnical problems such as analyzing the stability of slopes, evaluating the settlement of foundations, and designing retaining structures.

Numerical modeling in geotechnical engineering typically involves the following steps:

- **Defining the Problem:** This involves specifying the site conditions, soil properties, and the type of analysis to be performed.
- **Creating a Numerical Model:** This involves creating a computer model of the problem using appropriate software, such as finite elements or finite difference methods. The model should accurately represent the geometry, boundary conditions, and the problem's material properties.
- **Analyzing the Model:** This involves running the model and analyzing the results, such as stresses, strains, displacements, and velocities, to evaluate the behavior of the soil or structure under different loading conditions.
- **Validating the Model:** This involves comparing the results of the numerical model with field observations or laboratory tests to ensure that the model accurately represents the problem's real-world behavior.
- **Implementation of the Results:** Based on the results of the numerical model, recommendations can be made to design or modify the structure, select appropriate materials, and determine appropriate safety factors.

Some of the advantages of numerical modeling in geotechnical engineering include:

- Providing insights into complex geotechnical problems that may be difficult or impossible to analyze using traditional methods.
- Testing different scenarios and design options before construction can reduce costs and improve safety.
- Providing a more comprehensive understanding of the behavior of soils and structures can lead to improved design and construction practices.
- Evaluating the long-term performance of structures and the assessment of their durability and service life.

Overall, numerical modeling is a powerful tool that can aid in the analysis, design, and construction of geotechnical projects and can ultimately improve the safety, reliability, and sustainability of infrastructure.

22.2.2 Analytical Solutions vs. Numerical Methods

Numerical methods and analytical solutions are two different approaches for solving mathematical problems, particularly in the field of science and engineering.

An analytical solution commonly involves finding a closed-form expression for a mathematical problem, usually by using mathematical formulas or equations. This solution may be exact and can provide insights into the behavior of the system being studied. Analytical solutions can often be found for simple mathematical problems, but for complex systems, it may not be possible to find an analytical solution.

On the other hand, numerical methods involve using algorithms and computational techniques to obtain approximate solutions to a problem. Numerical methods rely on discretization of the problem domain and the use of iterative procedures to approximate the solution. Numerical methods are particularly useful when it is not possible to obtain an analytical solution or when the problem is too complex to solve analytically.

Differential equations play a significant role in geotechnical engineering, as many of the phenomena studied in this field are governed by differential equations. Some important differential equations are:

- **Consolidation Differential Equation:** This equation describes the rate of consolidation of a soil layer due to the dissipation of excess pore water pressure.
- **Seepage Differential Equation:** This equation governs the flow of water through porous media, such as soil and rock.
- **Elasticity Equation:** This equation describes the behavior of soil and rock under elastic deformation, which is important in foundation design and slope stability analysis.
- **Plasticity Equation:** This equation describes the behavior of soil and rock under plastic deformation, which is important in the analysis of landslides and soil liquefaction.
- **Wave Equation:** This equation is used to model the propagation of seismic waves through soil and rock, which is important in the design of earthquake-resistant structures.

Generally, under simple initial and boundary conditions and considering many simplifications, analytical formulations are commonly solvable to reach results such as settlement, bearing capacities, failure surfaces, etc.; however, the obtained results may not be reliable. In many cases, engineers cannot use these simplifications because this makes the actual behavior and response of the structure against the loads to be different from the answers resulting from the simplification. This issue causes the design to not be done correctly, and as a result, the safety and stability of earthen structures such as tunnels, slopes, foundations, and pits are endangered. Hence, simplification is not an efficient tool to solve the governing equations in geotechnical problems analytically. In many cases, the problem should be solved with minimal assumptions and simplifications and sometimes without any simplification at all.

22.2.3 Geotechnical Problem Classification

Geotechnical problems related to soil can be broadly classified into three categories:

- **Soil Stability Problems:** Some examples of soil stability problems include slope stability analysis, retaining wall design, and foundation design. Slope stability analysis involves assessing the stability of slopes under the influence of gravity and external loads. Retaining wall design involves designing walls to retain soil and prevent slope failure. Foundation design involves designing structures that transfer loads to the soil.
- **Soil Mechanics Problems:** Some examples of soil mechanics problems include soil classification, soil compaction, and shear soil strength. Soil classification involves categorizing soil based on its physical properties. Soil compaction involves the densification of soil through mechanical means. Shear soil strength involves measuring the soil's ability to resist deformation under shear stresses.
- **Groundwater Problems:** Some examples of groundwater problems include seepage analysis, well design, and dewatering. Seepage analysis involves assessing water flow through porous media such as soil and rock. Well design involves designing wells to extract or inject water from or into the ground. Dewatering involves removing groundwater from an excavation site to facilitate construction.

Due to the high complexity of geometric conditions, constitutive models, and loading in many cases, most of these problems cannot be solved by traditional methods. Therefore, solving some of the geotechnical problems requires an understanding of the properties of soil and rock, as well as the use of mathematical and computational tools such as numerical methods. Geotechnical engineers use this knowledge and these tools to design safe and effective structures and foundations.

22.2.4 Significance of Numerical Modeling in Foundation Problems

Numerical modeling plays a significant role in foundation engineering. It is used to simulate the behavior of soil and foundation systems under different loading conditions and to evaluate the performance of various foundation types. Some of the key benefits of numerical modeling in foundation engineering include:

- **Improved Understanding of Soil Behavior:** Numerical modeling allows engineers to better understand the complex behavior of soil and rock, which can vary widely depending on factors such as soil type, moisture content, and stress history. By modeling the soil and foundation system, engineers can gain insight into how the system will behave under different conditions and identify potential failure modes.
- **Optimization of Foundation Design:** Numerical modeling allows engineers to optimize the design of foundations to meet specific performance criteria. By evaluating different foundation types and configurations in a virtual environment, engineers can identify the most effective and cost-efficient solutions for a given project.
- **Reduction in Construction Costs:** Numerical modeling can help reduce construction costs by identifying potential issues before construction begins. By simulating the behavior of the soil and foundation system, engineers can identify potential issues such as settlement or lateral displacement and develop solutions to address these issues before construction begins.
- **Improved Safety:** Numerical modeling can improve safety by allowing engineers to evaluate the performance of a foundation system under extreme loading conditions such as earthquakes or high winds. This information can be used to design foundations that can withstand these loads and protect people and property.

22.2.4.1 Application in Shallow Foundations

Numerical modeling is an important tool to analyze the behavior of shallow foundations. Shallow foundations are typically used for structures that have relatively low loads, such as typical low-rise residential buildings and small commercial structures. Numerical modeling can be used to evaluate a variety of factors that can affect the behavior of shallow foundations, including:

- **Soil Properties:** The behavior of a shallow foundation is heavily influenced by the properties of the underlying soil, such as its strength, stiffness, and density. Numerical modeling can be used to simulate the behavior of the soil under different loading conditions and to evaluate the effects of soil variability on the foundation's performance.
- **Foundation Geometry:** The geometry of a shallow foundation, including its size, shape, and depth, can affect its behavior. Numerical modeling can be used to evaluate the effects of different foundation geometries on factors such as settlement, bearing capacity, and lateral displacement.
- **Loading Conditions:** The behavior of a shallow foundation is also influenced by the loading conditions it is subjected to such as the weight of the structure and any additional loads from wind, seismic activity, or other sources. Numerical modeling can be used to simulate the effects of these loads on the foundation and to evaluate the potential for failure under different loading conditions.
- **Soil–Structure Interaction:** The behavior of a shallow foundation is also affected by the interaction between the soil and the structure it supports. Numerical modeling can be used to simulate this interaction and to evaluate the effects of different soil–structure interfaces on the foundation's performance.

22.2.4.2 Application in Deep Foundations

Numerical modeling is a critical tool for analyzing and designing deep foundations such as single piles, pile groups, drilled shafts, etc. Numerical modeling can be used to evaluate a variety of factors that can affect the behavior of pile foundations and pile groups, including:

- **Soil–Pile Interaction:** The behavior of pile foundations is heavily influenced by the interaction between the pile and the surrounding soil. Numerical modeling can be used to simulate this interaction and to evaluate the effects of factors such as pile size, shape, and material properties on the performance of the foundation.

- **Loading Conditions:** The behavior of pile foundations is also influenced by the loading conditions they are subjected to such as the weight of the structure and any additional loads from wind, seismic activity, or other sources. Numerical modeling can be used to simulate the effects of these loads on the pile foundation and to evaluate the potential for failure under different loading conditions.
- **Pile Group Interaction:** When multiple piles are used to support a structure, the interaction between the piles can affect the performance of the foundation. Numerical modeling can be used to simulate this interaction and to evaluate the effects of factors such as pile spacing, pile arrangement, and pile cap design on the pile group's behavior.
- **Soil Variability:** The performance of pile foundations can be affected by variability in the soil properties. Numerical modeling can be used to simulate the behavior of the soil under different loading conditions and to evaluate the effects of soil variability on the performance of the pile foundation.

In summary, numerical modeling is a powerful tool that allows engineers to better understand the behavior of soil and foundation systems, optimize foundation design, reduce construction costs, and improve safety. It is an essential component of modern foundation engineering practice.

22.2.5 Geometrical Considerations of the Model in Foundation Problems

The geometry of a numerical model in foundation problems should be representative of the actual foundation system being analyzed. The model should include all relevant features and boundaries that may affect the behavior of the foundation.

- **Foundation Shape and Size:** The foundation shape and size should be accurately represented in the model. The geometry of the foundation can significantly affect its behavior and the stresses and strains that it experiences.
- **The Soil Profile:** The soil profile should be accurately represented in the model. This includes the different soil layers, their thickness, and their properties, such as soil type, stiffness, and strength. The soil profile can affect the foundation's behavior and its response to loading.
- **Adjacent Structures:** Any adjacent structures or soil layers that may affect the behavior of the foundation should be included in the model. This can include adjacent buildings, retaining walls, or soil layers with varying properties.

22.2.5.1 The Effects of Shape and Size

- **Bearing Capacity:** The shape and size of a foundation can affect its bearing capacity. A wider foundation will typically have a higher bearing capacity than a narrower foundation with the same depth, as the wider foundation can distribute the load over a larger area. Similarly, a foundation with a larger depth will typically have a higher bearing capacity than a shallower foundation. In addition, the shape of the foundation can affect the distribution of stresses in the soil, which can affect the bearing capacity.
- **Settlement:** The shape and size of a foundation can also affect its settlement behavior. A wider foundation will typically have a lower settlement than a narrower foundation with the same depth, as the wider foundation can distribute the load over a larger area. Similarly, a foundation with a larger depth will typically have a lower settlement than a shallower foundation. The shape of the foundation can also affect the distribution of stresses in the soil, which can affect settlement.
- **Lateral Resistance:** The shape and size of a foundation can also affect its resistance to lateral loads such as wind or earthquake loads. A wider and deeper foundation will typically have a higher resistance to lateral loads than a narrower and shallower foundation. The shape of the foundation can also affect the distribution of lateral stresses in the soil, which can affect lateral resistance.
- **Soil–Structure Interaction:** The shape and size of a foundation can also affect the interaction between the foundation and the surrounding soil. A larger and wider foundation will typically interact with a larger volume of soil than a smaller and narrower foundation. The shape of the foundation can also affect the distribution of stresses in the soil, which can affect soil–structure interaction behavior.

22.2.5.2 The Effects of Soil Profile

- **Bearing Capacity:** The bearing capacity of a shallow foundation depends on the soil's strength and stiffness. A soil layer with high strength and stiffness will provide a higher bearing capacity than a layer with low strength and stiffness for identical foundation dimensions. Therefore, the soil profile must be accurately characterized to determine the bearing capacity of the foundation.

- **Settlement:** The settlement of a shallow foundation is affected by the soil stiffness and compressibility. A soft or compressible soil layer will cause a higher settlement than a stiff or less compressible soil layer. Therefore, the soil profile must be accurately characterized to determine the foundation's potential settlement.
- **Lateral Earth Pressure:** The lateral earth pressure acting on a shallow foundation depends on the soil type and slope of the soil layers. A soil layer with a high angle of internal friction will provide a higher lateral earth pressure than a layer with a low angle of internal friction. Therefore, the soil profile must be accurately characterized to determine the lateral earth pressure acting on the foundation.
- **Slope Stability:** The stability of the slope surrounding a shallow foundation depends on the soil strength, stiffness, and angle of internal friction. A soil layer with a low angle of internal friction can result in slope instability, which can affect the behavior of the foundation. Therefore, the soil profile must be accurately characterized to determine the potential for slope instability.

22.2.5.3 The Effects of Adjacent Structure

Adjacent structures can have a significant effect on the behavior of shallow foundations. When a shallow foundation is located near an adjacent structure, there are several potential effects that may occur:

- **Soil Movement:** The presence of an adjacent structure can cause soil movement, such as settlement or lateral movement, which can affect the behavior of the foundation. This can be caused by excavation activities or changes in groundwater conditions, among other factors.
- **Soil Stress Redistribution:** The presence of an adjacent structure can cause stress redistribution in the soil, which can affect the behavior of the foundation. This can be caused by the weight of the adjacent structure, changes in the soil moisture content, or the presence of underground utilities, among other factors.
- **Soil–Structure Interaction:** The presence of an adjacent structure can cause soil–structure interaction effects, which can affect the behavior of the foundation. This can be caused by the interaction between the foundation and the adjacent structure, such as through the transfer of loads or the sharing of a common foundation.
- **Vibration and Dynamic Effects:** The presence of an adjacent structure can cause vibration and dynamic effects, which can affect the behavior of the foundation. This can be caused by machinery or traffic near the foundation, or by the presence of adjacent structures with different natural frequencies.

22.2.6 The Definition of Boundary Conditions

The boundary conditions in a shallow foundation system's numerical model depend on the problem being modeled and the behavior of the soil and foundation at the edges of the computational domain. In general, there are two types of boundary conditions that can be used in numerical models for shallow foundations: displacement boundary conditions and load boundary conditions.

22.2.6.1 Displacement Boundary Conditions

Displacement boundary conditions are used to specify the displacement of the soil or foundation at the edges of the computational domain. These boundary conditions are typically used when the edges of the computational domain are fixed or partially fixed. For example, a displacement boundary condition might be used to specify that the foundation cannot move vertically or that the soil cannot move horizontally.

22.2.6.2 Load Boundary Conditions

Load boundary conditions are used to specify the loads acting on the soil or foundation at the edges of the computational domain. These boundary conditions are typically used when the edges of the computational domain are free or partially free. For example, a load boundary condition might be used to specify the weight of the soil acting on the foundation or the pressure exerted by a retaining wall on the soil.

In addition to these boundary conditions, some numerical models for shallow foundations may also use interface elements to represent the interaction between the soil and the foundation. Interface elements are used to model the behavior of the soil and foundation at the interface between them. They can be used to capture the effects of soil–structure interaction on the behavior of the foundation system. Additional discussion on boundary conditions is presented in Section 22.3.3.

22.2.7 Constitutive Models for Soil

A constitutive model for soil is a mathematical model that describes the soil's stress–strain behavior. The soil's behavior under loading is complex and can be influenced by many factors, including soil type, moisture content, density, and stress history. As a result, there are many different constitutive models for soil that have been developed to capture the soil's behavior under different conditions.

22.2.7.1 Mohr–Coulomb

The Mohr–Coulomb constitutive model for soil and rock is a simple model that is widely used in geotechnical engineering to describe the soil's stress–strain behavior. The model is based on the principles of Coulomb's law of friction and assumes that the soil and rock's shear strength is governed by the angle of internal friction and the effective stress acting on the soil.

The Mohr–Coulomb model describes the soil and rock's stress–strain behavior using two parameters: the angle of internal friction (ϕ) and the cohesion (c). The angle of internal friction is a measure of the soil and rock's resistance to sliding along a failure plane, while the cohesion represents the soil's strength due to interparticle forces.

The Mohr–Coulomb model can be used to predict the soil and rock's failure under various loading conditions, including triaxial compression, triaxial extension, and direct shear. The model is particularly useful for analyzing the stability of slopes and retaining walls, as well as the bearing capacity of shallow foundations.

However, it is important to note that the Mohr–Coulomb model is a simplified model that does not consider the effects of soil and rock structure and plasticity. In situations where these effects are important, more advanced constitutive models, such as the modified Cam-Clay model, may be required.

22.2.7.2 Drucker–Prager

The Drucker–Prager (DP) constitutive model is a popular elastoplastic model used to describe the soil's stress–strain behavior. It is an extension of the Mohr–Coulomb model that accounts for the effects of stress path and strain rate on the behavior of soil. The DP model describes the stress–strain behavior of soil using four parameters: the cohesion (c), the angle of internal friction (φ), the bulk modulus (K), and the plasticity parameter (η). The cohesion and angle of internal friction have the same meaning as in the Mohr–Coulomb model, while the bulk modulus represents the soil's incompressibility, and the plasticity parameter governs the soil's plastic deformation.

The DP model assumes that the soil behaves as an elastic material up to a certain point, after which it undergoes plastic deformation. The yield surface of the DP model is a nonlinear surface that depends on the current state of stress and strain rate. The model assumes that the yield surface is a function of the effective mean stress (p) and the deviatoric stress (q), defined as:

$$p = (\sigma_1 + \sigma_2 + \sigma_3)/3 - (1 - \sin(\varphi))/(3K)^*(\sigma_1 + \sigma_2 + \sigma_3) \tag{22.1}$$

$$q = \left((\sigma_1 - \sigma_2)^2 + (\sigma_2 - \sigma_3)^2 + (\sigma_3 - \sigma_1)^2\right)/2 \tag{22.2}$$

where

σ_1, σ_2, and σ_3 = the principal stresses.

The DP model also includes a hardening rule that governs the increase in strength with plastic deformation. The hardening rule is defined by the plasticity parameter (η), which is related to the amount of plastic deformation. As the plastic deformation increases, the yield surface shifts outward, representing an increase in strength.

The DP model can be used to predict the behavior of soil under various loading conditions, including triaxial compression and extension, as well as cyclic loading. It is particularly useful for analyzing the stability of slopes and retaining walls, as well as the bearing capacity of shallow foundations.

22.2.7.3 Cam-Clay

The Cam-Clay model is a widely used elastoplastic constitutive model for soil that considers the effects of soil structure and plasticity. It is a mathematical model that describes the soil's stress–strain behavior under various loading conditions, including compression, extension, and shear.

The Cam-Clay model assumes that the soil is composed of a network of interconnected particles that can slide and deform relative to one another. The model consists of two components: an elastic component that describes the response of the soil under small strains and a plastic component that accounts for the soil's behavior under large strains.

The model is defined by three parameters: the pre-consolidation pressure (σ_p), the compression index (C_c), and the swelling index (C_s). The pre-consolidation pressure measures the soil's initial state of stress, while the compression and swelling indices describe the soil's response to compression and swelling, respectively.

The Cam-Clay model assumes that the soil undergoes two types of deformation: volume change and shear deformation. The model uses a yield surface to describe the onset of plastic deformation. The yield surface is defined by the effective mean stress (p) and the deviatoric stress (q), which are calculated as follows:

$$p = (\sigma_1 + \sigma_2 + \sigma_3)/3 \tag{22.3}$$

$$q = \left[(\sigma_1 - \sigma_2)^2 + (\sigma_2 - \sigma_3)^2 + (\sigma_3 - \sigma_1)^2\right]/2 \tag{22.4}$$

The Cam-Clay model also includes a hardening rule that accounts for the soil's response to plastic deformation. As the soil undergoes plastic deformation, the yield surface shifts outward, representing an increase in strength.

The Cam-Clay model also includes a hardening rule that accounts for the soil's response to plastic deformation. As the soil undergoes plastic deformation, the yield surface shifts outward, representing an increase in strength.

The Cam-Clay model is particularly useful for analyzing the behavior of clay soils, which are known to exhibit significant plasticity and sensitivity to changes in stress and strain. The model can be used to predict the behavior of soil under a wide range of loading conditions, including those encountered in geotechnical engineering applications such as slope stability analysis, foundation design, and embankment design.

22.2.7.4 Modified Cam-Clay

The modified Cam-Clay (MCC) model is a variation of the Cam-Clay model that is often used to describe the stress–strain behavior of soils with significant plasticity and sensitivity to changes in stress and strain. The MCC model includes modifications to the original Cam-Clay model that improve its ability to predict the soil's behavior under complex loading conditions.

Like the Cam-Clay model, the MCC model assumes that the soil is composed of a network of interconnected particles that can slide and deform relative to one another. The model includes both an elastic component and a plastic component, and is defined by several parameters, including the pre-consolidation pressure (σ_p), the compression index (C_c), and the swelling index (C_s).

The MCC model includes several modifications to the original Cam-Clay model, including the addition of a state parameter (K) and the use of a modified yield function. The state parameter measures the soil's current state of plastic deformation, while the modified yield function accounts for the effect of the state parameter on the soil's response to stress and strain.

The MCC model assumes that the soil undergoes two types of deformation: volume change and shear deformation. The model uses a modified yield surface to describe the onset of plastic deformation. The yield surface is defined by the effective mean stress (p) and the deviatoric stress (q), which are calculated similarly to the original Cam-Clay model.

The MCC model also includes a hardening rule that accounts for the soil's response to plastic deformation. As the soil undergoes plastic deformation, the yield surface shifts outward, representing an increase in strength.

22.2.8 2D vs. 3D Models

Due to their reduced degrees of freedom, 2D models have shorter execution times than their 3D counterparts. Additionally, analyzing simulation outcomes is simpler with 2D models, as results are easier to visualize. As a result, it is often preferable to conduct 2D simulations when feasible. However, 2D models have limitations when it comes to modeling three-dimensional physical phenomena. These constraints must be considered when formulating and interpreting 2D models. Nevertheless, creating simplified 2D models to verify hypotheses about phenomena or numerical algorithms is recommended at any stage of the modeling process, if possible. Even with a 3D software package, a 2D "slice" can be generated as an initial modeling step.

22.3 Numerical Analysis in Geotechnical Engineering

Nowadays, numerical methods are used to solve a wide range of problems related to geotechnical and foundation engineering (i.e., the design of shallow and deep foundations, the design of soil and rock slopes, and excavations). Since numerical methods are mostly used in the form of software, the generalities of their use for analyzing and designing the intended problems are similar. The following examines the general process of these modellings and the challenges that may arise at each stage of numerical analysis and design.

22.3.1 Analysis Planning

Because solving problems using numerical methods usually requires significantly more time and money compared to traditional methods, choosing a numerical method for engineering problems requires a convincing and logical justification. Merely having various capabilities in numerical methods for modeling problems does not necessarily lead to more accurate results, as there may be a higher margin of error due to the complexity of software and the diversity of input parameters for modeling engineering problems. Traditional design methods are usually faster and less expensive, but they often involve many simplifying assumptions and provide limited and approximate results. However, in many specific geotechnical engineering problems, there may be no traditional and simple method for analysis. In such cases, the accuracy and details provided by numerical analysis can lead to significant savings during construction. For example, when it is necessary to consider the following factors, numerical analysis may be required instead of traditional analysis:

- Complex soil behavior (such as nonlinear stiffness, hardening soil, inhomogeneity, and creep) as well as behavioral changes due to specific operations (such as soil improvement or rehabilitation)
- Complex hydraulic conditions
- Unusual geometries and complex loading conditions
- Soil–structure interaction and internal structural forces in complex structures and interaction with adjacent structures
- The effect of construction sequence and method
- Use of observational approaches in design
- Time effects (such as creep and consolidation)
- The necessity of conducting back-analysis of field tests or monitored projects.

Before starting to build a numerical model, the objectives of the numerical analysis must be defined. For example, the goal of a numerical model may be to demonstrate the sufficient safety of a geotechnical structure against failure, the absence of serious damage to the structure due to its small movements, or the amount of water flowing into the reservoir of a dam. Given the different objectives for each of these cases, the modeling approach in each of them also differs.

To model an accurate geotechnical numerical model, comprehensive information about the project site's history and current status is needed. To gather this information, it is crucial to examine the observed characteristics of samples taken from the site. Then, depending on the desired results from each analysis, the required geotechnical parameters must be obtained by conducting specialized tests. When interpreting the results of laboratory tests and field studies, an understanding of the site's geological history and the mechanisms that make up the soil layers is essential. The number of residual stresses and the stress path that the soil has experienced at the site have significant effects on the ground's behavior. This may affect both the input parameters of the model and the need for simulation in the numerical model-building stages.

For example, if existing structures or infrastructure are at the site, their geometric details and loading should be determined to be included in the numerical model. Furthermore, any uncertainties in the interpretation of the ground conditions and soil or rock parameters in the area should be examined. Performing sensitivity analysis to evaluate the impact of uncertainties on numerical model outputs is also necessary. It should be noted that regular communication with those who conduct field studies helps to judge uncertainties.

In summary, any information required for constructing a structure is also necessary when modeling that structure numerically. The minimum information required for numerical modeling includes the following items:

- Structural plots for defining the geometry of the numerical model
- Structural analysis results for determining the loads imposed on the ground
- Allowable values for displacement and deformation of the structure and ground
- Sequence (stages) of constructing different parts of the structure for software simulation
- Construction schedule for identifying time intervals between construction stages and making decisions about drainage conditions that govern the model.

22.3.2 Geometry

Choosing a 2D or 3D numerical model is an important decision in the stages of constructing a numerical model, as it has a significant impact on the time required to build and analyze the model. Generally, analyses should always be done in 3D unless there are logical reasons to show that the problem can be analyzed in 2D. To model a fundamentally 3D problem in 2D, one dimension of the problem must be much longer than the other two dimensions, provided that the material properties and geometry are constant along the longer dimension. In other words, if a uniform 2D section is created along the longer dimension from any arbitrary location, 2D modeling can be used instead of 3D modeling. For example, for a long strip foundation, retaining wall, tunnel, or long earth dam, if the above conditions are met, 2D modeling can be used. Such problems are called plane strain problems. In this case, strains and displacements only occur within the plane, and zero is assumed for the third dimension (i.e., perpendicular to the plane). In some special cases, such as circular pile and single circular cross-sectional shafts subjected to axial loads and with symmetry in geometry, loading, and material properties with respect to their central axis, 2D modeling can be used instead of 3D modeling. Such problems are called axisymmetric problems.

In a very small percentage of 2D models, one dimension is much smaller than the other dimensions (unlike the assumption in plane strain problems, where one dimension is assumed to be longer than the other two). In these problems, known as plane stress problems, two-dimensional modeling can be used instead of three-dimensional modeling. Analysis of a very thin steel sheet is an example of this type of problem.

When none of the above conditions are met, three-dimensional modeling should be used. Figure 22.1 shows examples of problems in geotechnical engineering and the appropriate analysis method.

Geotechnical engineering problems occur in a field that extends infinitely in size. However, when constructing a numerical model of geotechnical engineering problems, considering suitable dimensions for the model from the perspective of

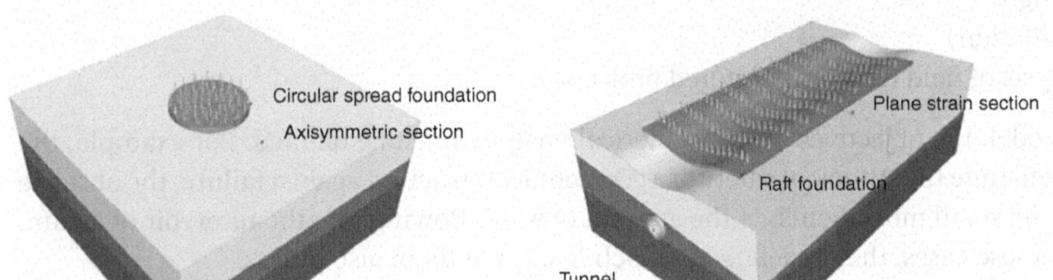

Figure 22.1 Examples of analysis types in geotechnical engineering (Lees 2016 / Emerald Publishing Limited).

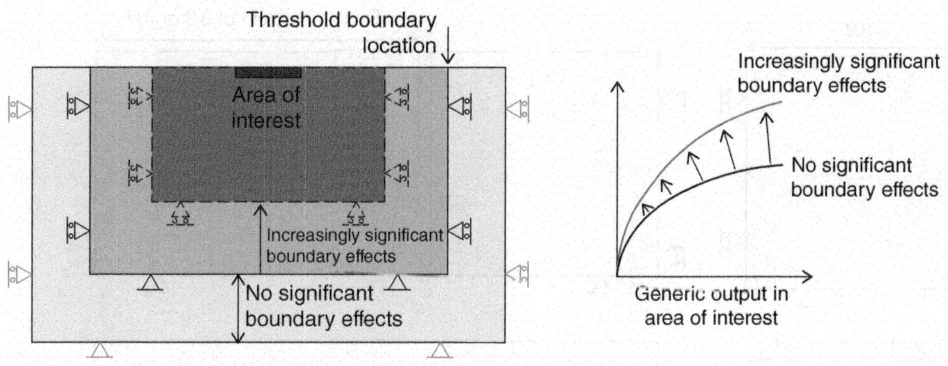

Figure 22.2 The effect of model dimensions on the results (Lees 2016 / Emerald Publishing Limited).

the time required for analysis and the accuracy of the results obtained is of great importance. Generally, the larger the size of the model, the less the effect of the model boundaries on the results and the greater the accuracy of the results, but on the other hand, the analysis time also increases significantly. The dimensions of the model must be considered in such a way as to include all possible failure areas around the problem area. Usually, the boundaries extend more in soft material layers and less in hard materials. Also, the bottom boundary of the model can usually be closer to the area of interest (location of the applied load, location of the retaining wall, proximity to the excavation, etc.). Still, the lateral boundaries usually need to be farther away from the area of interest. This is because the soil's hardness and strength usually increase with depth. Therefore, the effects of the area of interest in the model are less important at depth than at the lateral boundaries. To determine the most suitable dimensions for the model, which leads to results with adequate accuracy and does not unnecessarily increase the analysis time, various values must be considered for the dimensions of the model, and the best value for the dimensions of the model should be selected based on the results obtained for each model. First, a model of the problem must be created and analyzed, as shown in Figure 22.2. Then, by changing the dimensions of the model (moving the boundaries), gradually increasing or decreasing the size of the model, and examining the key outputs, the appropriate geometry of the model and its final boundaries can be determined. Finally, the boundary position where the boundary effects are no longer significant and the dimensions of the model have no effect on the analysis is selected as the model boundary.

As an initial recommendation for modeling surface foundations, excavation, and tunneling, the model dimensions can be considered according to the recommendation in Figure 22.3. However, it is necessary to refine the model dimensions with more detailed analyses for each specific project.

22.3.3 Boundary Conditions

After determining the model dimensions, suitable boundary conditions must be defined for the boundaries. Due to the infinite continuation of the earth, it is often impossible to consider a specific boundary for the area around engineering problems located on the ground. However, in numerical models, due to the limitation of model dimensions and the inability to model the entire surrounding area of the problem, it is necessary to consider suitable boundary conditions for the model to achieve balance and convergence toward a final solution. Usually, there is no horizontal displacement at the lateral boundaries of the model. However, in the vertical direction, there is displacement due to ground settlement. Therefore, in numerical modeling, the movement of the vertical boundaries is limited, but they are allowed to move in the vertical direction. These conditions, which are present in most engineering problems, are called standard boundary conditions.

In dynamic analyses, first, the static boundary conditions mentioned above must be assigned to the model, then the dynamic boundary conditions should be defined. This is because the system is in a static state before being subjected to dynamic loading, such as an earthquake. In dynamic loading conditions, when the force of seismic waves hits the lateral boundaries, it is repeatedly reflected and remains inside the model. However, this is not the case, and the earthquake passes through the desired location in the ground after being created. To address this issue and create realistic dynamic conditions for the numerical model, there are three solutions, namely:

- The dimensions of the boundaries should be considered large enough so that the waves, after colliding with the boundaries and reflecting in the model, are sufficiently attenuated before reaching the desired structure. The drawback of this method is that the analysis time increases due to the model becoming large.

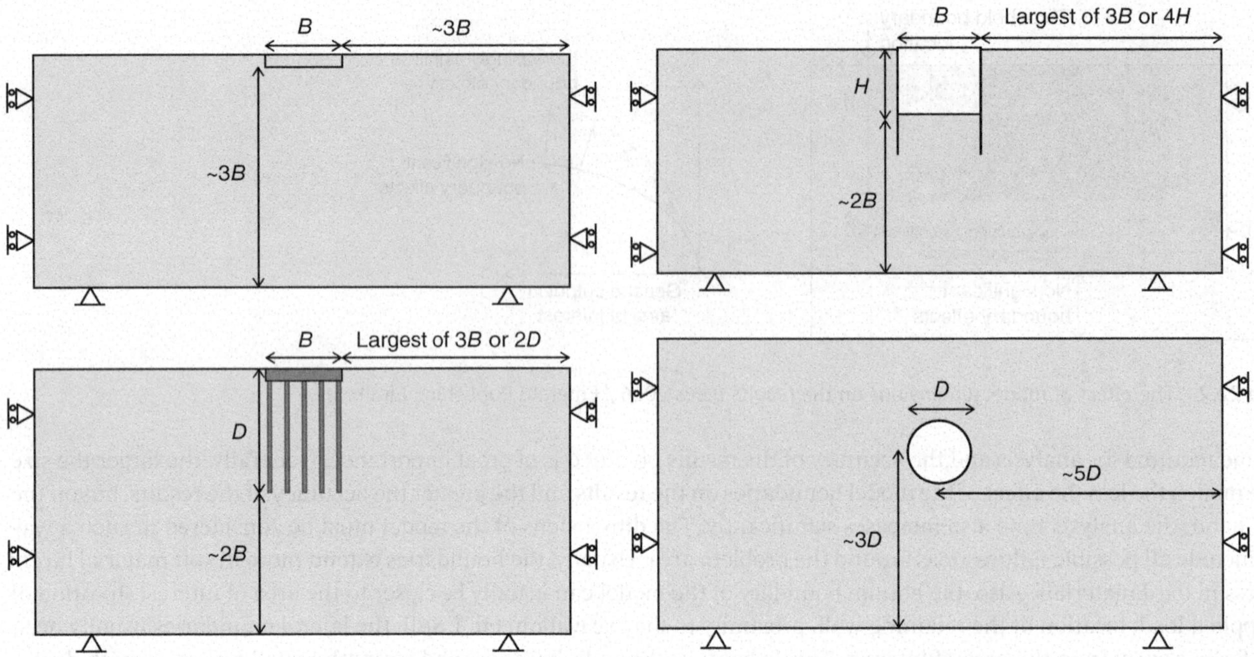

Figure 22.3 Initial proposed dimensions for numerical models (Lees 2016 / Emerald Publishing Limited).

- Elements should be used at the boundaries that absorb earthquake waves after colliding with them and do not reflect them back into the model. Such boundaries are called absorbent boundaries.
- The use of free boundaries allows waves to pass through them without reflecting them back into the model.

In general, free boundaries perform better than absorbent boundaries, especially when the wave is oblique. For example, under waves with an angle greater than 30° to the horizon, absorbent boundaries do not perform well, while free boundaries do.

The important point is that in dynamic loading caused by earthquakes, the lower boundary of the model should not be absorbent, whereas in dynamic loading, such as the vibration of machinery, the lower boundary should be absorbent. In the first case, because the distribution of earthquake forces is from the bottom of the model, if the lower boundary is also absorbent, it absorbs the earthquake force and prevents its propagation toward the top of the model. However, in the second case, the propagation of the dynamic load is from the surface of the model, and therefore, all boundaries should be absorbent.

22.3.4 Structural Elements

There are various types of elements for modeling geotechnical structures. Understanding the behavior of each type of these elements is important for selecting the most appropriate ones in modeling. For modeling structures with significant volume, such as piles, diaphragm walls, soldier pile walls, large-diameter drilled shafts, and tunnel linings, three-dimensional volumetric elements, two-dimensional surface elements, or one-dimensional linear elements can be used. On the other hand, linear elements are more suitable for modeling structures with small volumes, such as retaining walls.

Another type of element used in numerical models is the interface element. This element is used at the junction between the ground and the structure in faults and rock discontinuities, and even at locations that contain a change in the type of soil layer, where there is potential for relative movement. The interface is usually modeled using thin continuous elements with suitable stiffness and resistance. Geometrically, the thickness of these elements is equal to zero because the nodes of the elements on each side of an interface are aligned with each other due to the contact between the structure and the ground. However, a virtual thickness needs to be considered so that numerical distortions do not occur during the calculation of elastic deformations. If the thickness of the interface elements is too large, large elastic deformations occur. The appropriate range for this thickness is usually between 1% and 10% of the interface element's length, but it is better to determine the optimal value through parameter analysis. As for the properties used for the interface surface

Table 22.1 Reduction coefficient of resistance for interface element (Lees 2016 / Emerald Publishing Limited).

Interaction sand/steel	$= R_{\text{inter}} \approx 0.6\text{–}0.7$
Interaction clay/steel	$= R_{\text{inter}} \approx 0.5$
Interaction sand/concrete	$= R_{\text{inter}} \approx 1.0\text{–}0.8$
Interaction clay/concrete	$= R_{\text{inter}} \approx 1.0\text{–}0.7$
Interaction soil/geogrid (grouted body) (interface is not necessary)	$= R_{\text{inter}} \approx 1.0$

element, a reduction coefficient (R_{inter}) is usually used. Table 22.1 provides some recommended values for this reduction coefficient.

22.3.5 Meshing

Selecting an appropriate size and type of mesh, just like the geometric dimensions of the model, has a significant impact on the numerical analysis results. Meshes have different shapes, including triangular and quadrilateral (in 2D) and tetrahedral and hexahedral (in 3D). The advantages of triangular (2D) and tetrahedral (3D) elements over quadrilateral (2D) and hexahedral (3D) elements are that they can be easily applied to complex geometric shapes and are less susceptible to errors caused by distortion. In numerical software, meshes are applied either automatically or manually. Although automatic meshing is simpler and faster, it cannot be relied upon without careful and sufficient examination. The resulting mesh from the automatic meshing process must be carefully examined before executing the model. While implementing a very fine mesh with small elements throughout the model yields the most accurate results, it requires a long computational time. The most logical meshing method is to use smaller elements in important areas of the model (locations of stiffness changes, discontinuities, around the foundation and piles, excavation boundaries, around the retaining wall, around the tunnel, etc.) and larger elements in areas where stresses and strains are more uniform. This increases the analysis speed without significantly reducing the accuracy of the results.

Usually, trial and error are used to determine the mesh's optimal size. To do this, the size of the elements must be reduced in several stages so that the results become independent of the size of the elements. For example, in determining the load-bearing capacity of a pile using software, it is necessary to start the analysis with a specific mesh size and gradually reduce the mesh size. Finally, a graph of the changes in load-bearing capacity vs. mesh size is plotted. It is obvious that the final and appropriate size of the mesh for the model is the size at which no change in bearing capacity is observed for values smaller than it. Figure 22.4 shows an example of the selection process for the optimal size mesh in a numerical model to determine the load-carrying capacity of a surface pile on a rock foundation.

22.3.6 Failure Criteria and Material Properties

In numerical analysis, the relationships between force and stress, displacement and strain, and stress and strain are considered in the form of constitutive models. The most basic behavioral model used in numerical analysis is the linear elastic model, in which materials are generally assumed to be homogeneous and isotropic. This model is not able to predict plastic behavior and is not suitable for modeling load-bearing conditions. Two parameters, the elastic modulus and Poisson's ratio of the materials, are required for this behavioral model, which can be obtained from laboratory test results or some field tests such as SPT. This model is used to model the behavior of concrete materials (if plastic behavior is not considered) and sometimes healthy rocks without cracks, fissures, and weathering. Also, one well-known behavioral model in geotechnical engineering, the Mohr–Coulomb model, works similarly to the linear elastic model in the elastic part, but if the stress conditions reach the yield point, it can simulate complete plastic behavior. This model considers the stiffness of soil masses to be constant and independent of stress, and the soil settles or swells with this constant stiffness. If the soil is subjected to loading in the plastic range, the soil will undergo deformation with a slope equal to the loading slope, which this behavioral model is unable to predict. To simulate complete plastic behavior, this model requires the definition of two soil parameters, internal friction angle (ϕ), and cohesion (c). If $\phi = 0$, the Mohr–Coulomb criterion is equivalent to the Tresca criterion, and if $\phi = 90$, it is equivalent to the Rankine model. Also, this constitutive model is not suitable

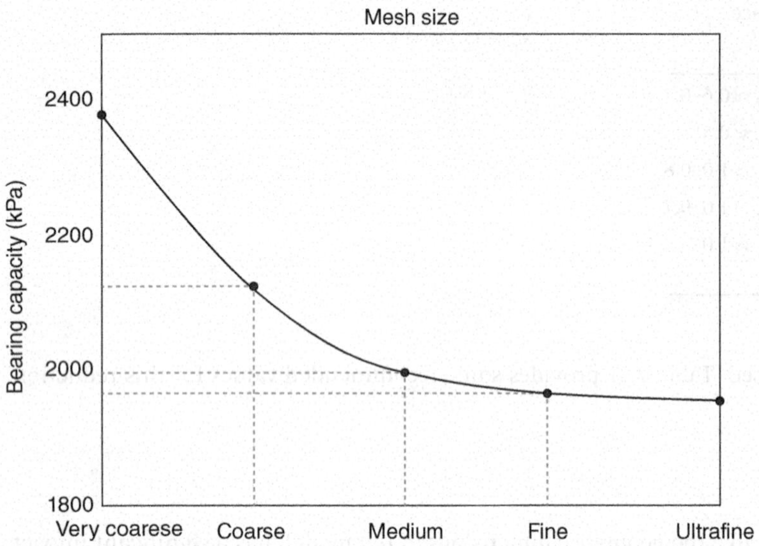

Figure 22.4 An example of the trial-and-error process for selecting the appropriate dimensions for the mesh.

for modeling cyclic problems. In addition to these two common models, various other behavioral models are applicable in the numerical analysis of geotechnical engineering problems, including the transverse isotropic elastic model, hardening soil, hardening soil with small strain, Cam-Clay, soft soil, soft soil creep, jointed rock, Hoek–Brown, Drucker–Prager, and ubiquitous-joint, among others.

After selecting an appropriate behavioral model based on the conditions of the problem, the type of analysis and the nature of the materials, it is necessary to define the properties of the materials. These properties include all mechanical and resistance characteristics for materials (such as soil, rocks, concrete, metals, interface elements, etc.) whose values are obtained from laboratory or field tests. Additionally, drainage conditions are also determined based on the type of materials and the loading type and duration.

22.3.7 Loading

Another important part of numerical analysis is structural loading. The speed at which the load is applied must be carefully selected so that it corresponds to the nature of the actual load on the structure. In numerical analysis, applied loads are generally static, quasi-static, or dynamic. Static loading is applied in various forms, such as point, linear, and surface loading. In quasi-static loading, earthquake force is applied as an equivalent static force in two horizontal and vertical directions to the model. Dynamic loading is also introduced by applying a complete earthquake acceleration, velocity, or displacement history. Dynamic loading is generally used in analyzing loads resulting from earthquakes, explosions, and machine foundation vibrations.

22.3.8 Initial Conditions

The first stage of any geotechnical numerical analysis is establishing initial stresses in the model. Every soil naturally undergoes stress caused by weight and the pressure of existing water and upper soil layers before any structure is built on it. These stresses are not necessarily constant at every depth and layer of soil, and their value varies especially near the ground surface. Properly establishing initial stresses in the numerical model has a significant impact on the final analysis results. Any changes in stress caused by the construction history or groundwater fluctuations must be simulated in later stages of the modeling process. In special cases such as laboratory test simulations, initial stresses (caused by the weight of the sample) are negligible compared to the applied stresses during the test. The sample can be assumed to have zero density and initial stresses in such cases. Then, the applied stresses on the sample during the real test can be applied in the numerical analysis software.

In general, in important geotechnical resources, K_0 (the ratio of horizontal effective stress to vertical stress) is referred to as an important parameter in establishing initial stresses. In numerical software, the definition of the K_0 parameter enables the establishment of initial stresses in the model. The vertical stresses are caused by the weight of the soil layers, and the horizontal stresses are obtained by multiplying the vertical stresses by the K_0 value. It should be noted that in conditions where the material and geotechnical parameters of the soil are variable, the value of K_0 should be defined separately for each type of soil.

22.3.9 Analysis Stages

One of the most important benefits of using numerical methods in analyzing geotechnical engineering problems is the ability to perform a step-by-step analysis to simulate the construction process of engineering projects. After applying initial stress conditions, the loading stages can be applied to the model step by step and in separate phases. For example, when modeling a stabilized excavation using a soldier pile system, after establishing the initial stress conditions, the first excavation stage to a certain depth is modeled. Depending on the time interval between the excavation of the first stage and the installation of the first row of piles, it may be necessary to release some or all of the stresses caused by excavation. Then, in the next stage, the first row of piles is modeled with an appropriate distance between them. The above two stages are then sequentially modeled in subsequent separate phases to complete the modeling of the entire excavation. Finally, the entire analysis is performed step by step by running the model. The results obtained from such analysis have significant adaptations to reality, which cannot be easily provided in nonnumerical analyses.

22.3.10 Outputs

After running the numerical model, its outputs are observed. The complete outputs obtained from a numerical analysis are voluminous. However, usually the simplest outputs that are controlled as part of an initial review for each stage of the numerical analysis include the following:

- Examination of deformed meshes using realistic and exaggerated scales to ensure logical control of deformations.
- Examination of the pressure contour of a borehole to ensure the accuracy of boundary conditions for groundwater flow and drainage conditions.
- Control of stress and displacement vector outputs to investigate their compatibility with applied loads.
- Control of normal and shear strain contours to check their logical behavior and help highlight problematic areas in the numerical model.

After completing the analysis using numerical software, the required outputs are obtained. The necessary items can be extracted using the outputs. These items include stress–strain diagrams and changes in flexural anchor and shear forces generated in shallow and deep foundations, axial forces generated in retaining and guard structures, diagrams of changes in shear stress and horizontal displacement over time in dynamic analyses, and so on.

22.4 Different Numerical Methods in Foundation Engineering

Numerical methods are widely used for solving problems related to geotechnical engineering. The most important and common methods include the finite element method (FEM) and the finite difference method (FDM). Of course, other numerical methods are also used for problem-solving, which this section explains. Each of these methods has specific assumptions that distinguish it from other methods.

22.4.1 Finite Difference Method

22.4.1.1 Introduction

The finite difference method (FDM) is a numerical technique used in geotechnical engineering to solve differential equations that describe the behavior of soil and rock. FDM involves dividing a continuous domain into a finite number of discrete elements or cells, and approximating the differential equations using finite difference equations. These equations are then solved numerically to obtain an approximate solution to the original differential equation.

FDM is commonly used in geotechnical engineering for solving problems related to soil mechanics, such as slope stability, consolidation, and seepage analysis. The method is particularly useful for problems that are too complex to be solved analytically, or for problems that involve irregular geometries or nonlinear behavior.

In slope stability analysis, FDM can be used to model the distribution of stresses and strains within a slope, and to evaluate the stability of the slope under different loading conditions. In consolidation analysis, FDM can be used to simulate the time-dependent behavior of a soil deposit undergoing consolidation due to changes in pore water pressure. In seepage analysis, FDM can be used to model the flow of water through a porous medium, and to evaluate the potential for soil erosion or piping.

One of FDM's advantages is that it is relatively easy to implement and can be used to solve problems with irregular geometries or complex boundary conditions. However, its accuracy depends on the resolution of the grid used to divide the domain, and it can be computationally expensive for large or complex problems. Therefore, FDM is typically used in conjunction with other numerical techniques, such as the finite element method or boundary element method, to obtain more accurate and efficient solutions.

22.4.1.2 Formulating and Solving Procedures in FDM

FDM involves dividing the domain of the differential equation into a grid of discrete points, and approximating the derivatives at each point using the values at neighboring points. The resulting system of equations can be solved using matrix methods. Here are the general steps for formulating and solving a problem using the finite difference method. The following steps are general and can be applied to various differential equations. However, the specific details of the procedure may vary depending on the problem being solved.

- **Discretize the Domain:** Divide the domain of the differential equation into a grid of discrete points. The spacing between points is called the "grid spacing" or "mesh size" and is denoted by Δx.
- **Approximate the Derivatives:** Approximate the derivatives in the differential equation at each point using the values at neighboring points. This is done using finite difference approximations, which express the derivative in terms of differences between neighboring grid points. For example, the first derivative of a function $f(x)$ can be approximated as:

$$f'(x) \approx (f(x+h) - f(x))/\Delta x \text{ (forward difference)} \tag{22.5}$$

$$f'(x) \approx (f(x) - f(x-h))/\Delta x \text{ (backward difference)} \tag{22.6}$$

$$f'(x) \approx (f(x+h) - f(x-h))/(2\Delta x) \text{ (centered difference)} \tag{22.7}$$

Higher-order derivatives can be approximated using similar techniques.

- **Discretize the Differential Equation:** Replace the derivatives in the differential equation with their finite difference approximations. This results in a system of equations that can be solved for the values of the function at each grid point.
- **Boundary Conditions:** Apply the boundary conditions to the system of equations. These conditions specify the values of the function at the boundaries of the domain.
- **Solve the System of Equations:** Solve the resulting system of equations using matrix methods. This involves setting up a matrix equation, where the coefficients of the matrix are determined by the finite difference approximations and the boundary conditions. The solution to the matrix equation gives the values of the function at each grid point.
- **Analyze the Solution:** Once the system of equations has been solved, the resulting values can be analyzed to determine the behavior of the function. This may involve computing quantities such as the maximum and minimum values, the rate of change, or the integral of the function over the domain.

22.4.1.3 Application of FDM in Geotechnical Engineering

One of the main applications of FDM in geotechnical engineering is in the analysis of soil–structure interaction. FDM can be used to model the complex behavior of soil and structures under different loading conditions, including static, dynamic, and seismic loads. The method can be used to study the deformation and stress distribution in soils and structures, as well as their interaction.

Another important application of FDM in geotechnical engineering is in the analysis of consolidation. FDM can be used to simulate the process of soil consolidation, which occurs when water is squeezed out of the soil due to the application of load. The method can be used to predict the rate and extent of consolidation, and to analyze the effect of different factors such as soil type, loading conditions, and drainage conditions.

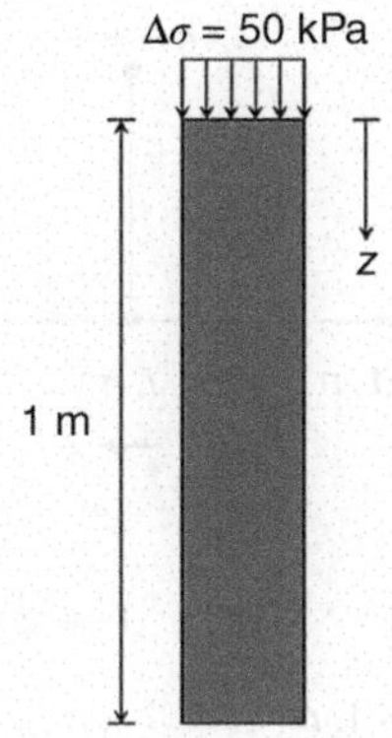

Figure 22.5 Soil layer under surcharge load (GitHub, Inc. / www. Github.com, 2022 / last accessed November 17, 2023).

FDM can also be used to analyze seepage in soils. Seepage analysis is important in many geotechnical engineering applications, such as the design of dams, levees, and groundwater remediation systems. FDM can be used to model the flow of water through soils and structures, and to predict the rate and direction of water movement under different conditions.

Example 22.1 ***One-dimensional Consolidation Simulation Using FDM Coding in Python***

(a) Explanation:
A soil layer with a thickness of 1 m is experiencing a surcharge load of 50 kPa, as depicted in Figure 22.5. The soil layer is drained at both its upper and lower boundaries, implying that the additional pore pressure can disperse through both limits. The soil has a coefficient of consolidation of 2×10^{-6} m^2/s. Determine the consolidation process (excessive pore pressure changes) by passing time.
The equation that describes the process of one-dimensional consolidation through partial differentiation is provided as follows.

$$\frac{\partial u}{\partial t} = C_v \frac{\partial^2 u}{\partial z^2} \tag{22.8}$$

$$C_v = \frac{k}{m_v \gamma_w} \tag{22.9}$$

The symbol u represents the additional water pressure within pores, t refers to the duration of the process, z denotes the depth, C_v the consolidation coefficient, k the coefficient of permeability, m_v the coefficient of volumetric compressibility, and γ_w the unit weight of water.

(b) Formulation
Explicit Method
In the explicit method, the solution at a point in space and time is calculated based on the values of the solution and its derivatives at previous time steps and neighboring points in space. The calculation involves using a finite difference approximation to discretize the PDE and then applying a formula that expresses the solution at the current time step in terms of the solution and its derivatives at previous time steps.
The explicit method is called "explicit" because the formula used to calculate the solution at each time step involves only known quantities from previous steps, and does not involve solving any additional equations. This makes the method relatively simple to implement and computationally efficient. However, the explicit method has a stability restriction that limits the size of the time step that can be used. If the time step is too large, the solution may become unstable, and the method may fail to converge. The formulation for Equation 22.8 using the explicit method (Figure 22.6) is expressed as:

$$\frac{u_i^{n+1} - u_i^n}{\Delta t} - C_v \frac{u_{i+1}^n - 2u_i^n - u_{i-1}^n}{\Delta z^2} = 0 \tag{22.10}$$

Implicit Method
In the context of finite difference methods for solving partial differential equations (PDEs), the implicit method is a numerical technique that involves solving a set of equations at each time step to obtain the solution at the next step. In contrast to the explicit method, the solution at the current time step is not explicitly expressed in terms of the solution at

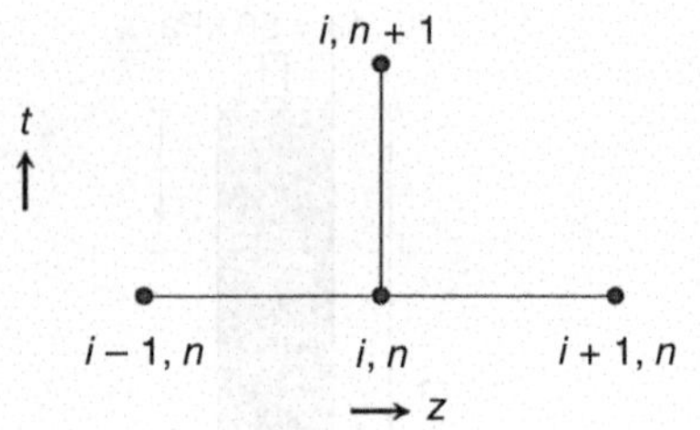

Figure 22.6 Explicit arrangement (GitHub, Inc. / www. Github.com, 2022 / last accessed November 17, 2023).

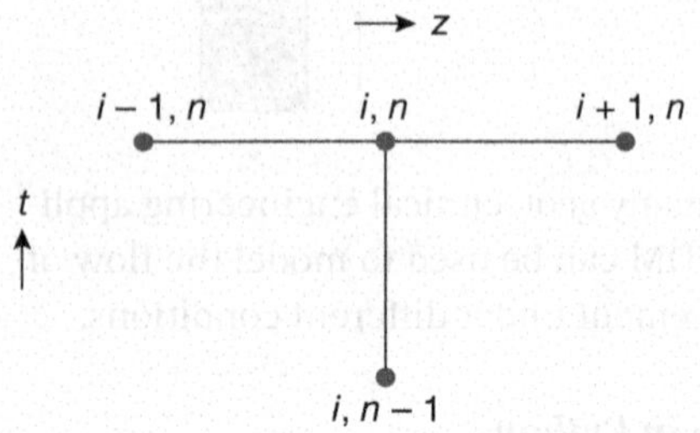

Figure 22.7 Implicit arrangement (GitHub, Inc. / www. Github.com, 2022 / last accessed November 17, 2023).

previous steps. Instead, the implicit method involves solving a system of equations that includes the unknown solution values at the current time step, as well as the known values at previous steps.

The advantage of the implicit method is that it is unconditionally stable, meaning that it can use larger time steps than the explicit method without becoming unstable. This makes it more suitable for solving PDEs with stiff terms, which require small time steps to remain stable in the explicit method. However, the implicit method is more computationally intensive and requires solving a system of equations at each time step, which can be more challenging than the explicit method. The formulation for Equation (22.8) using the implicit method (Figure 22.7) is expressed as:

$$\frac{u_i^n - u_i^{n-1}}{\Delta t} - C_v \frac{u_{i+1}^n - 2u_i^n - u_{i-1}^n}{\Delta z^2} = 0 \tag{22.11}$$

(c) Coding in Python

```
import numpy as np
import matplotlib.pyplot as plt
import pylab as py
py.rcParams['figure.figsize'] = 4.5, 6.5
# Soil layer properties
H = 1.0         # Depth of soiil layer in m
cv = 2e-6       # Coefficient of consolidation in m^2/s
# Spatial discretization
N = 10                    # Number of elements of 1D mesh (N+1 nodes)
z = np.linspace(0, H, N+1)
dz = z[1] - z[0]   # Node spacing for a uniform discretization
# Temporal discretization (explicit method)
dt_max = dz**2 / 2.0 / cv
#print('Time step should be less than or equal to %'), dt_max
dt = 1.0 * dt_max
# Calculate \kappa = cv * dt / dz^2
k = cv * dt / dz**2
# Total time for simulation (t) and number of time steps (Nt)
t = 86400.0 * 2
Nt = int(t / dt) + 1
# Create array for storing pore pressures
u = np.zeros(N+1)                   # unknown u at new time level
u_1 = np.zeros(N+1)                   # u at the previous time level
```

```
# Set initial condition
for i in range(0, N+1):
    u_1[i] = 50.0
# Compute the pore pressure for subsequent time steps
for n in range(0, Nt):
    # Compute pore pressure at inner nodes
    for i in range(1, N):
u[i] = u_1[i] + k * (u_1[i-1] - 2 * u_1[i] + u_1[i+1])
        # Insert boundary conditions
        u[0] = 0.0
        #u[N] = 0.0

# Update u_1 for next time step
u_1[:] = u
# Dimensionless time
T = cv * n * dt / (H/2)**2
# Plot results
# if n > 0 and n % (N/2) == 0:
if T > 0.0 and T % 0.1 == 0:
plt.plot(u, z, label='T =' + str(T))
plt.gca().invert_yaxis()
plt.gca().xaxis.tick_top()
plt.gca().set_xlabel('X LABEL')
plt.gca().xaxis.set_label_position('top')
plt.xlabel('u [kPa]')
plt.ylabel('z [m]')
plt.legend(loc='best')
plt.savefig('Explicit_1D_N-' + str(N) + '.eps')
plt.show()
```

(d) Results
Figure 22.8 shows the variations of the pore pressure with time (0.1 day, 0.2 day, 0.4 day, and 0.8 day). *N* is number of nodes in *z* direction.

22.4.1.4 Application of FDM in Shallow Foundations

To apply the FDM for modeling shallow foundations, the following steps can be followed:

- **Define the Domain of the Problem:** The domain of the problem should be defined by specifying the boundary conditions and the geometry of the foundation.
- **Discretize the Domain:** Divide the domain into a grid of cells, and assign values of the soil properties to each cell.
- **Approximate the Partial Derivatives:** Approximate the partial derivatives in the governing equations using finite difference approximations.
- **Formulate the Algebraic Equations:** Substitute the approximated partial derivatives into the governing equations to obtain a set of algebraic equations.
- **Solve the Algebraic Equations:** Solve the set of algebraic equations using iterative methods to obtain the solution.
- **Analyze the Results:** Analyze the results to understand the behavior of the shallow foundation under different loading and soil conditions.

Example 22.2 ***Load–Settlement Curve of a Shallow Foundation on Cohesive Frictionless Soil***
In this example, FLAC3D software, working based on explicit finite difference approximation to time derivatives, is used to model the problem. Figure 22.9 shows the geometry of the problem. Table 22.2 provides the soil properties per 2023 FLAC3D user manual.

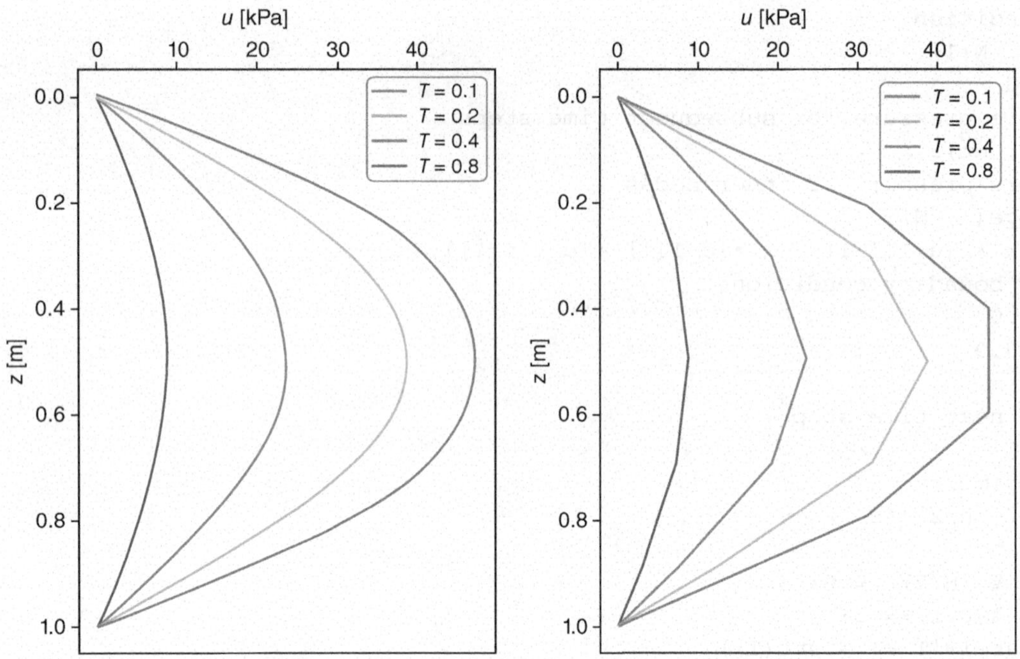

Figure 22.8 Pore pressure variations with time left: $N = 10$, right: $N = 80$ (www.Github.com 2022).

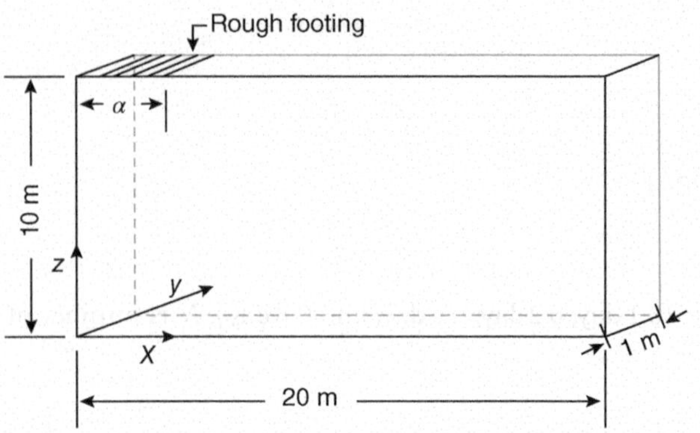

Figure 22.9 The geometry of the problem (adapted from Itasca Consulting Group, Inc. 2023).

Table 22.2 Soil properties (adapted from Itasca Consulting Group, Inc. 2023).

Shear modulus (G)	Bulk modulus (K)	Cohesion (c)	Friction angle (ϕ)	Dilation angle (ψ)
0.1 GPa	0.2 GPa	0.1 MPa	0°	0°

In this problem, the numerical simulation assumes half-symmetry and plane-strain conditions. Figure 22.10 shows the selection of a coordinate system. The strip footing area has a half-width value. The following figure illustrates the boundary conditions that are applied to this domain. The displacement of the rough footing is constrained in the y-direction, and a velocity is imposed in the negative z-direction to simulate the footing load.

The domain is divided into a single layer of 520 zones that are arranged in a graded pattern, as depicted in Figure 22.11. The contact nodes (soil and footing) are subjected to a velocity of 0.5×10^{-5} m/step for a total of 25,000 calculation steps.

Figure 22.12 shows the load–displacement curve that corresponds to the numerical simulation, where the normalized average footing pressure is the load (p/c), and the magnitude of the normalized vertical displacement is represented by the variable "disp" (u_z/a).

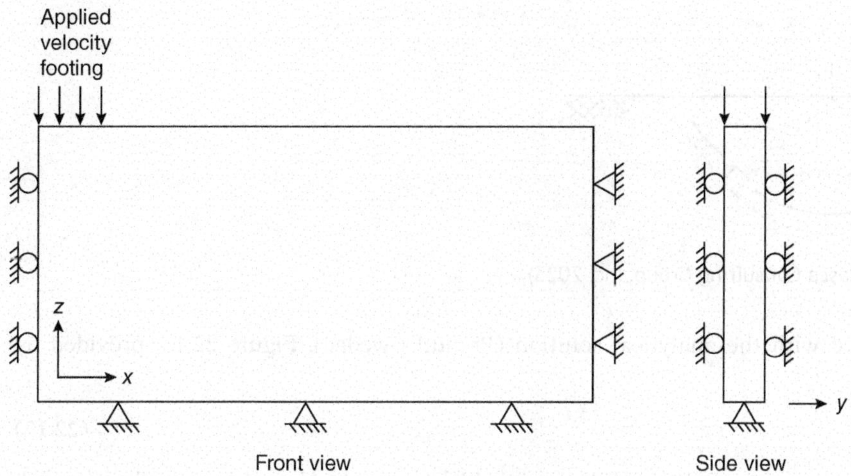

Figure 22.10 Boundaries and coordinate system for numerical models (adapted from Itasca Consulting Group, Inc. 2023).

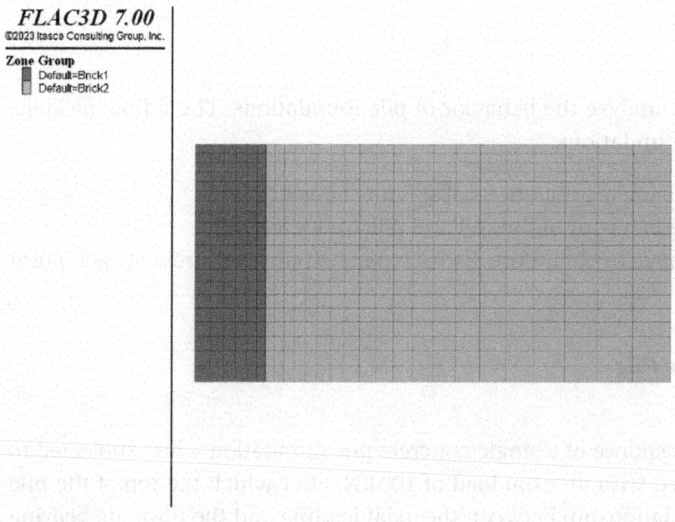

Figure 22.11 FLAC3D model and meshing (adapted from Itasca Consulting Group, Inc. 2023).

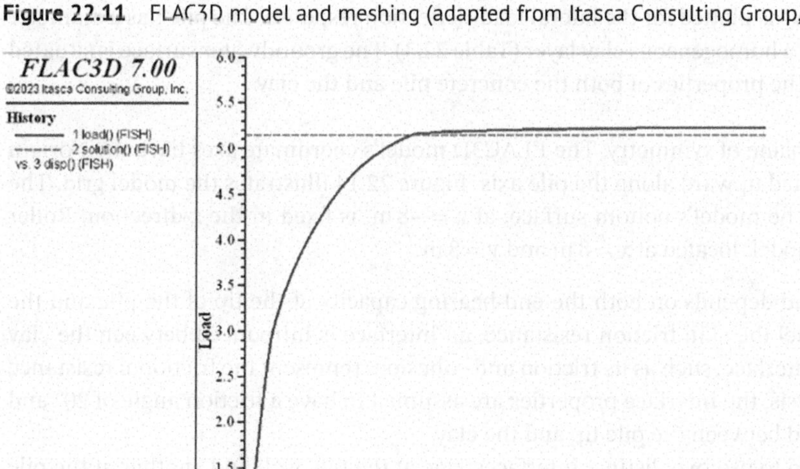

Figure 22.12 Load–displacement curve (adapted from Itasca Consulting Group, Inc. 2023).

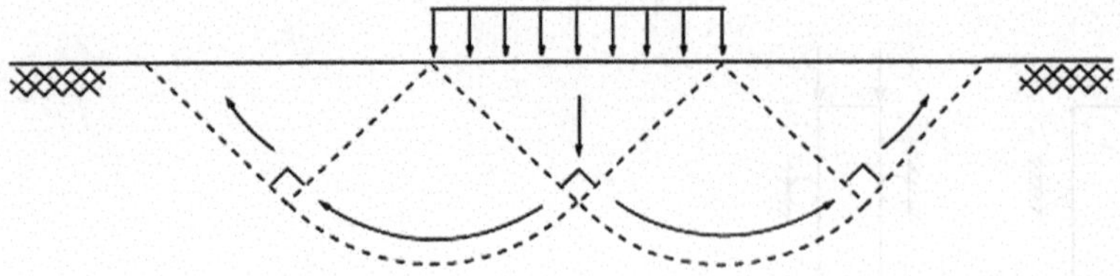

Figure 22.13 Prandtl's wedge (adapted from Itasca Consulting Group, Inc. 2023).

The numerical result has been compared with the analytical solution (Prandtl's wedge), Figure 22.13, provided by Terzaghi and Peck (1967) as:

$$q = (2 + \pi) \cdot c \tag{22.12}$$

The variable q denotes the average footing pressure at failure, while the variable c represents the cohesion of the material. The numerical value of the bearing capacity, q, is 523.0 kPa, and the relative error is 1.72% when compared to the analytical value of 514.2 kPa.

22.4.1.5 Application of FDM in Pile Foundations

FDM can be applied to model the soil–pile interaction and analyze the behavior of pile foundations. The following steps are typically followed in a finite difference analysis of pile foundations:

- Determine the stress and strain at each point in the grid using the equations of soil mechanics.
- Apply the boundary conditions to the grid to simulate the behavior of the soil and pile under load.
- Solve the system of equations using finite difference methods to obtain the displacement, strain, and stress at each point in the grid.

Example 22.3 ***Axial and Lateral Loading of a Concrete Pile***

(a) Explanations:
This problem involves calculating the load-deflection response of a single concrete pile foundation when subjected to both axial and lateral loading. Initially, the pile is loaded with an axial load of 100 kN, after which the top of the pile is horizontally displaced to a maximum of 4 cm. The relationship between the axial loading and the ultimate bearing capacity of the pile is determined, and a curve is plotted for the lateral load-deflection response. The pile has a diameter of 0.6 m, a length of 5 m, and is embedded in a homogeneous clay layer (Table 22.3). The groundwater surface is situated at a depth of 5.5 m. Table 22.3 summarizes the properties of both the concrete pile and the clay.

(b) Modeling in FLAC3D
In this analysis, the pile axis has a vertical plane of symmetry. The FLAC3D model's coordinate axes have their origin at the top of the pile, and the z-axis is oriented upward along the pile axis. Figure 22.14 illustrates the model grid. The top surface of the model, at $z = 0$, is free. The model's bottom surface, at $z = -8$ m, is fixed in the z-direction. Roller boundaries are applied to the sides of the model, located at $x = 8$ m and $y = 8$ m.

The ability of a pile to withstand an axial load depends on both the end-bearing capacity at the tip of the pile and the skin friction resistance along the shaft. To model the skin friction resistance, an interface is introduced between the clay and the walls of the pile. The properties of the interface, such as its friction and cohesion, represent the frictional resistance between the concrete and the clay. In this analysis, the interface properties are assumed to have a friction angle of 20° and a cohesion of 30 kPa. Another interface is placed between the pile tip and the clay.

To achieve the best results, it is recommended to use two distinct interfaces: one at the pile wall and another at the pile base in order to ensure that the grid points shared by both interfaces are also separated.

Before simulating the pile installation, the model is first brought to a state of equilibrium under gravitational loading. After this, a horizontal water table is created at a depth of $z = -5.5$ m, and the zones below the water table are assigned the wet density of the clay material. In the next stage of analysis, the model is brought to an equilibrium state after the pile has been installed. To model the installation, the properties of the pile zones are changed from those representing the

Table 22.3 Pile-soil system properties (adapted from Itasca Consulting Group, Inc. 2023).

	Concrete pile	Clay
Dry density	2500 kg/m^3	1230 kg/m^3
Wet density	—	1550 kg/m^3
Elastic properties		
Young's modulus	25.0 GPa	100.0 MPa
Poisson's ratio	0.20	0.30
Bulk modulus	13.9 GPa	83.33 MPa
Shear modulus	10.4 GPa	38.46 MPa
Strength properties		
Cohesion	—	30 kPa
Friction angle	—	0.0

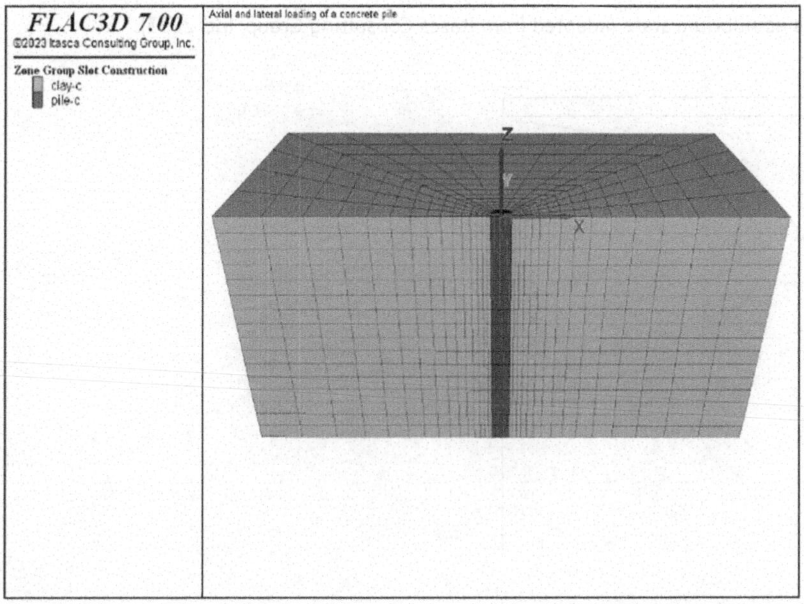

Figure 22.14 The pile and soil griding in FLAC3D (Itasca Consulting Group, Inc. 2023).

clay material to those representing the concrete pile material. Figure 22.15 shows the vertical stress distribution at this equilibrium state, which includes the weight of the pile.

To determine the ultimate bearing capacity of the pile in the axial direction, a vertical velocity is applied at the top of the pile using a "ramp" boundary condition. This linear increase from zero to the desired value is useful in this case due to the significant difference in stiffness between the concrete and clay, resulting in varying natural periods of the model. Since the concrete's high stiffness controls the critical timestep, many timesteps are necessary to propagate loading through the model. If the velocity is abruptly applied, the initial inertial effects will dominate, making it harder to identify the system's steady-state response.

The table ramp is employed to gradually increase the velocity applied to the pile top grid points. To calculate the axial stress at the top of the pile, the FISH function vert_load is used, which also saves the value as a history. To improve efficiency, the grid points on the cap surface are saved as a map in the symbol cap. Figure 22.16 depicts a graph of axial stress vs. axial displacement at the pile top when the velocity is increased gradually from 0.0 to $5*10^{-8}$ over 30,000 steps. Notably, only a minor amount of oscillation is observed initially in the response.

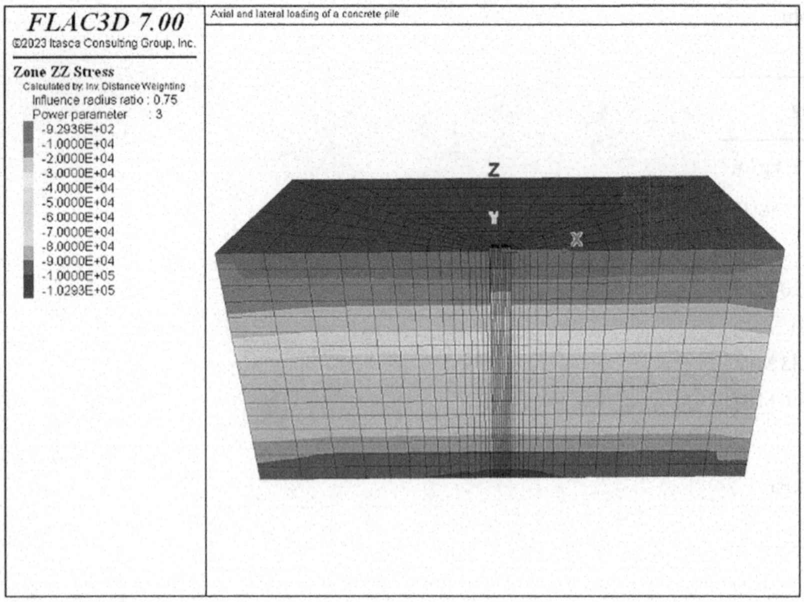

Figure 22.15 The vertical stress distribution at this equilibrium state (adapted from Itasca Consulting Group, Inc. 2023).

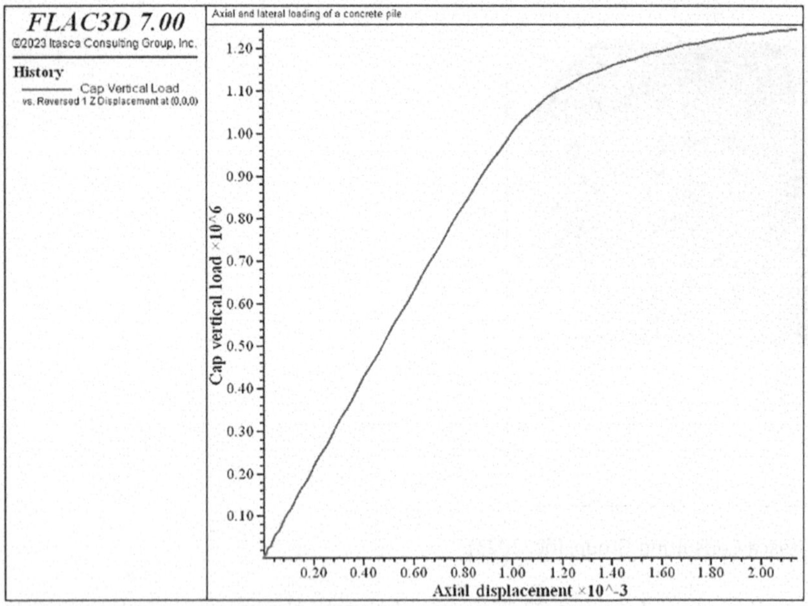

Figure 22.16 Load–displacement curve for single pile (adapted from Itasca Consulting Group, Inc. 2023).

For this stage of the analysis, a combined damping approach is used since it is more effective in removing kinetic energy from the model for the given loading condition. In this case, most of the grid points will not experience a change in the velocity component's sign due to the applied velocity loading. Hence, local damping, which relies on velocity sign-changes, is not effective.

Figure 22.16 displays that the pile's ultimate bearing capacity is around 1.1 MPa. This capacity is determined by both the skin friction resistance and the pile-tip bearing capacity. When the interface cohesion is set to zero, the pile tip is the only resistance, and the ultimate capacity is calculated to be about 0.3 MPa. To determine the appropriate values for interface strength properties, it is suggested to compare the calculated pile capacity to the field test results.

To analyze the response to an axial loading of 100 kN, the analysis is performed again starting from the initial gravitational loading state. In order to apply this loading, an axial stress of 0.354 MPa is applied to the top of the pile. This value is about three times smaller than the ultimate axial capacity. As a result of this loading, the pile top moves downward by 0.33 mm.

Once the model reaches equilibrium for the 100 kN axial loading, a lateral deflection of 4 cm is induced at the top of the pile by applying a horizontal velocity. Unlike the previous loading, a ramp is not used for this stage as the inertial effects are negligible for this type of loading. A horizontal velocity of $1*10^{-7}$ is applied to the pile cap to achieve the deflection.

Figure 22.17 displays the relationship between the lateral force and deflection of the pile's top, indicating that a lateral load of around 300 kN is needed to cause 4 cm of deflection. Figure 22.18 illustrates the displacement of the pile and the soil's plastic state at this point in the analysis.

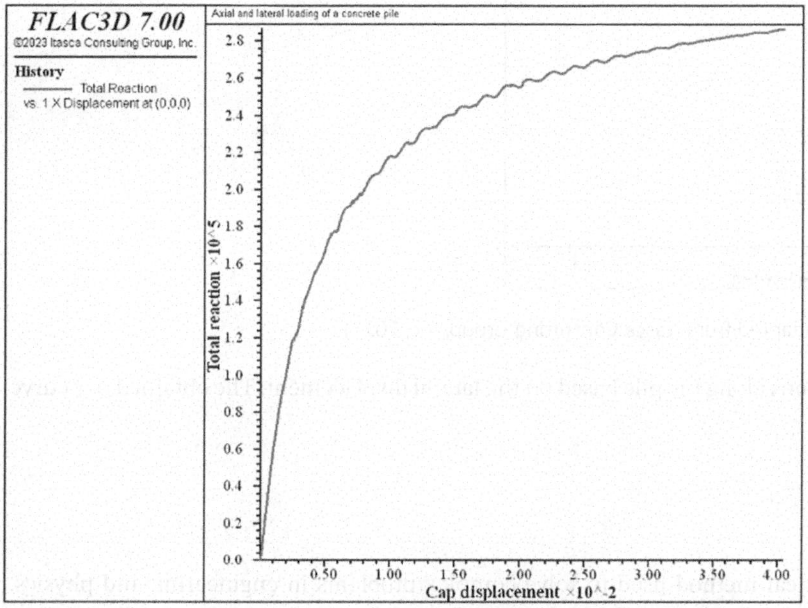

Figure 22.17 The relationship between the lateral force and deflection of the pile's top (adapted from Itasca Consulting Group, Inc. 2023).

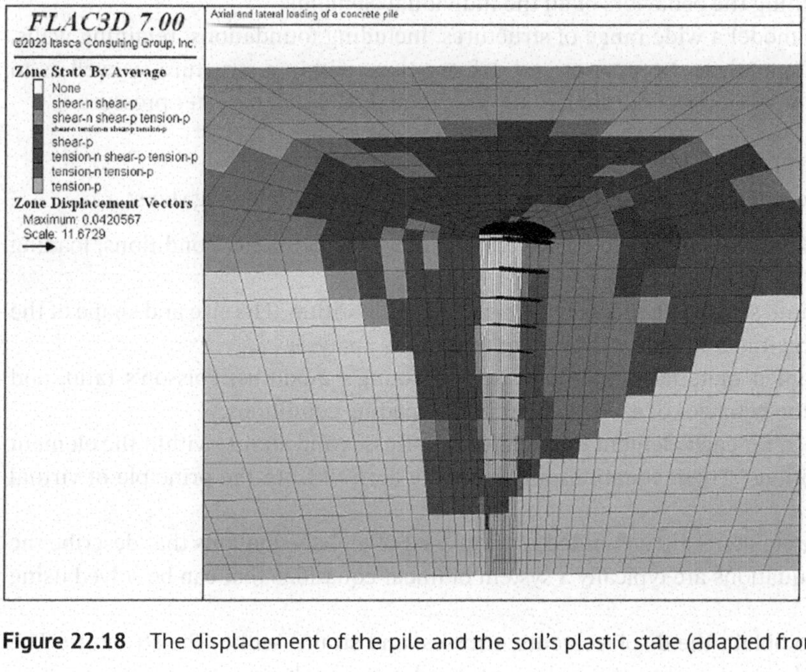

Figure 22.18 The displacement of the pile and the soil's plastic state (adapted from Itasca Consulting Group, Inc. 2023).

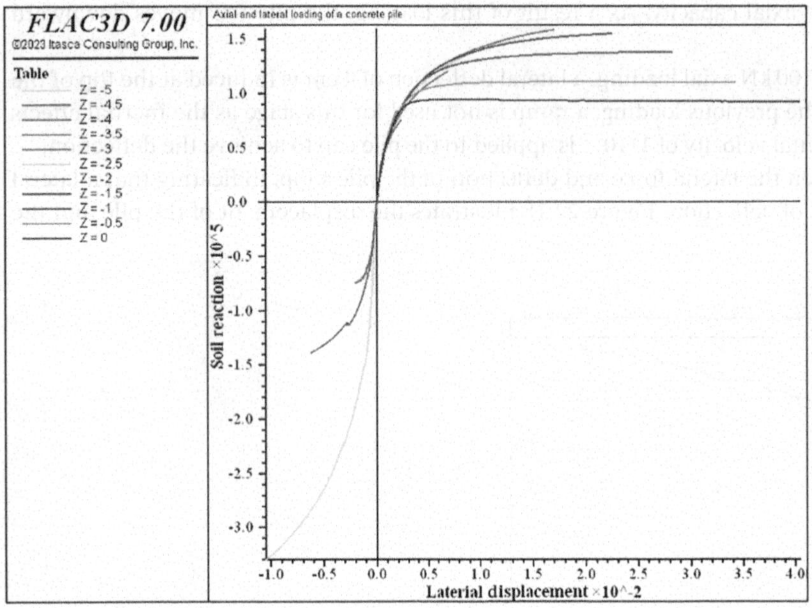

Figure 22.19 The *p–y* curve at different heights (adapted from Itasca Consulting Group, Inc. 2023).

A FISH function is utilized to track soil reactions along the pile based on the lateral displacement. The obtained *p–y* curve is shown in Figure 22.19.

22.4.2 Finite Element Method (FEM)

22.4.2.1 Introduction

The finite element method (FEM) is a numerical method used to solve complex problems in engineering and physics. In geotechnical engineering, FEM is used to model and analyze the behavior of soil and rock structures under various loading conditions.

FEM divides a complex structure into a mesh of smaller, simpler elements, which are interconnected at their nodes. The behavior of each element is then analyzed using mathematical equations and boundary conditions, and the overall behavior of the structure is calculated by combining the behaviors of all the individual elements.

In geotechnical engineering, FEM is used to model a wide range of structures, including foundations, retaining walls, tunnels, and slopes. The method can be used to analyze the stresses and deformations within a structure, as well as to predict the behavior of the structure under various loading conditions such as earthquakes, wind, or water pressure.

22.4.2.2 Formulating and Solving Procedures in FEM

The formulation and solving procedure in finite element method (FEM) generally involves the following steps:

- **Problem Definition:** The first step is to define the problem to be solved, including the boundary conditions, loading conditions, and material properties.
- **Mesh Generation:** The structure is divided into smaller elements using a meshing algorithm. The size and shape of the elements depend on the complexity of the structure and the accuracy required in the analysis.
- **Element Properties:** Each element is assigned material properties, such as Young's modulus, Poisson's ratio, and density. The properties are used to calculate the behavior of each element under loading conditions.
- **Element Equations:** Equations are developed for each element that relates the stresses and strains within the element to the material properties and loading conditions. These equations are generally derived from the principle of virtual work or energy minimization.
- **Global Equations:** The equations for all the elements are combined to form a set of global equations that describe the behavior of the entire structure. The global equations are typically a system of linear equations that can be solved using matrix algebra.
- **Solution:** The global equations are solved to obtain the displacements, stresses, and strains within the structure. The accuracy of the solution depends on the number and quality of the elements used in the analysis.

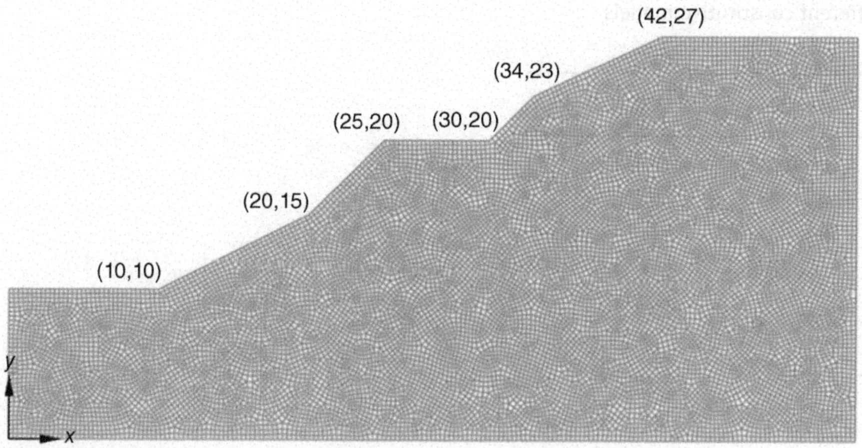

Figure 22.20 Terraced slope geometry and meshing (Karami et al. 2023a / Springer Nature).

- **Post-processing:** The results are post-processed to generate visualizations of the structure's behavior, such as stress and strain distributions, displacement plots, and animations.

Overall, FEM is a powerful and versatile method for solving complex engineering problems, and its effectiveness depends on the accuracy of the input data, the quality of the mesh, and the expertise of the analyst.

22.4.2.3 Applications of FEM in Geotechnical Engineering

Some specific applications of FEM in geotechnical engineering include:

- **Analysis of Soil–Structure Interaction:** FEM can be used to model the interaction between the soil and a structure, such as a foundation, retaining wall, or tunnel. This helps to understand the stresses and deformations in the soil and the structure, and to design the appropriate foundations and support systems.
- **Slope Stability Analysis:** FEM can be used to model the stability of slopes under different loading and boundary conditions. The analysis helps to identify potential failure modes and to design appropriate stabilization measures.
- **Seepage Analysis:** FEM can be used to model the flow of water through soils and to analyze the seepage characteristics of dams, embankments, and other geotechnical structures. This helps to design appropriate drainage and seepage control systems.
- **Ground Improvement Analysis:** FEM can be used to model the effects of various ground improvement techniques such as soil stabilization, grouting, and pile installation. This helps to evaluate the effectiveness of these techniques and to design appropriate solutions.
- **Tunneling and Excavation Analysis:** FEM can be used to model the behavior of soil and rock around tunnels and excavations. This helps to understand the deformation and stress distribution in the soil and to design appropriate support systems.

Example 22.4 ***Terraced Slope Stability Simulation Using FEM***

Explanation: Terraced slopes (also multilevel and multitier slopes) are groups of slopes constructed at different elevations. The following terraced slope with the given geometry (Figure 22.20) is modeled in ABAQUS software. The mechanical properties of the soil for different constitutive models are given in Table 22.4. To analyze this slope, extended cam clay is selected as the constitutive model for the soil. After modeling and running the numerical model in ABAQUS software working based on FEM, the results of failure surfaces as well as safety factors are provided in the figure. As shown in Figure 22.21, five shear zones emerge on the terraced slope. The safety factors calculated using the FEM are nearly identical to LEM (the limit equilibrium method) results.

22.4.2.4 Applications of FEM in Shallow Foundations with Examples

Finite element method (FEM) can be used to model shallow foundations, which are commonly used in building and civil engineering projects. Shallow foundations are usually constructed on or near the surface of the soil and are designed to spread the load of the structure over a larger area to prevent excessive settlement or bearing capacity failure.

Table 22.4 Soil mechanical properties for different constitutive models (Karami et al. 2023b / Springer Nature).

Item	Parameter	Value
General	k (m/s)	$3.63*10^{-6}$
	e (−)	0.606
	γ_t (kN/m³)	17.9
	γ_d (kN/m³)	16.5
	G_s (−)	2.67
Elasticity	κ (−)	$C_s/2.3 = 0.007$
	ν (−)	0.28
Extended Cam-Clay plasticity	λ (−)	$C_c/2.3 = 0.07$
	M (−)	1.12
	a (MPa)	0.22
	χ (−)	1
	K (−)	0.788
Modified Drucker–Prager/cap plasticity	d (kPa)	66.03
	B	48.28°
	R (−)	1
	ε_{vol}^{pl} (−)	0
	α (−)	0.05
	K (−)	0.79
Mohr–Coulomb plasticity	c (kPa)	8.820
	ϕ	28.2°
	ψ	0.1°
	E (MPa)	30

The following steps can be taken to model a shallow foundation using FEM:

- **Geometry and Mesh Generation:** The geometry of the foundation and the surrounding soil are created using CAD software. The geometry is then meshed using a finite element mesh generator. The mesh size and element type should be selected based on the size and shape of the foundation and the soil properties.
- **Material Properties:** The soil properties, including the density, Young's modulus, Poisson's ratio, and shear strength, are assigned to the elements in the mesh. The soil can be modeled as either linear or nonlinear depending on the soil type and the loading conditions.
- **Boundary Conditions:** Boundary conditions are applied to the model to simulate the actual conditions under which the foundation is constructed. The foundation is usually subjected to vertical and horizontal loads, and the soil is subjected to vertical stress due to the weight of the foundation and any overlying structures. The interface between the foundation and soil is also modeled to consider the interaction between the two.
- **Analysis:** The analysis is performed by solving the global equations derived from the element equations. The analysis can be performed using commercial FEM software or by coding the equations in a programming language. The solution provides the displacements, stresses, and strains in the soil and the foundation.
- **Results and Interpretation:** The results can be visualized using post-processing techniques, including contour plots, stress and strain distributions, and animations. The results can be interpreted to evaluate the performance of the foundation under different loading conditions and to optimize the foundation designs.

Example 22.5 ***Bearing Capacity of a Strip Footing Under Inclined Load***
This instance illustrates a solution for the equilibrium limit of a layer of sand loaded by a rigid strip footing with inclined loading. The behavior of the sand is simulated using the cap model with modified Drucker–Prager parameters matched to

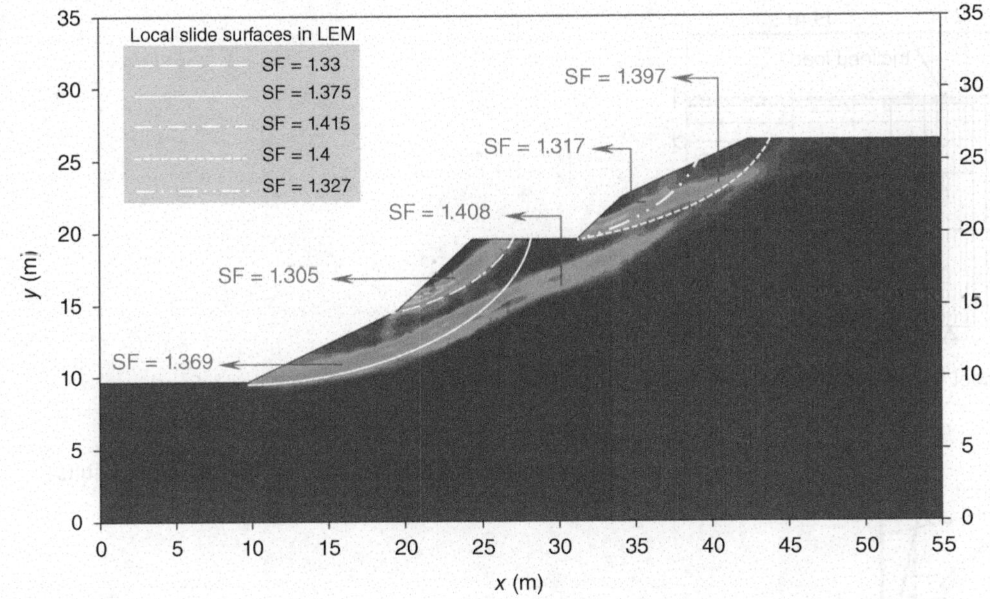

Figure 22.21 Formation of shear zones and calculated safety factors (SF); comparison of the FEM results with LEM (Karami et al. 2023a / Springer Nature).

Table 22.5 Cap model parameters for sand (Helwany 2007 / John Wiley & Sons).

General		**Plasticity**	
ρ (kg/m^3)	1923	d	10^{-5}
e_0	1.5	β (°)	44.56 (plane strain)
Elasticity			55.7 (3-D)
E (MPA)	182	R	0.4
ν	0.28	Initial yield	0.0
		α	0.05
		K	1

the classical Mohr–Coulomb yield model parameters. The sand's elastic parameters and cap model parameters are the same as those used for Ottawa sand in Table 22.5. The concrete foundation is linear elastic with a Young's modulus of 1435 MPa and a Poisson ratio of 0.2. The finite element model used for the analysis is displayed in Figure 22.22 and covers the entire sand layer and foundation due to the system's asymmetry caused by the inclined load. The strip foundation is in perfect contact with the soil and uses reduced-integration, bilinear, and plane strain quadrilateral elements. The base of the sand layer is fixed in all directions, and all vertical boundaries are fixed horizontally but free vertically. The mesh is finer near the foundation since it is the zone of stress concentration, but no mesh convergence studies have been conducted. The dimensions of the sand layer are selected to minimize the boundary effect on foundation behavior.

At the start of the analysis, gravity loads are applied to the sand layer, which is crucial in determining the initial stresses in all soil elements. Since the soil behavior is stress-dependent, the cap model used in this analysis takes this important fact into consideration. During this step, the "geostatic" command is used to ensure that equilibrium is maintained within the sand layer. This option guarantees that the initial stress condition in any element within the sand layer falls within the initial yield surface of the cap model. Next, a point load with horizontal and vertical components is applied to the center of the foundation, inclined at a 15° angle from the vertical. The load–displacement relationship for the 2-m-wide strip footing is established to determine its bearing capacity, as shown in Figure 22.23. The bearing capacity of the footing can be determined from the point where the initial and final straight-line portions of the curve meet. The finite element analysis predicts a bearing capacity of approximately 675 kPa, slightly smaller than the 738 kPa Meyerhof's equation predicts. This difference is because Meyerhof's equation assumes that the soil is a rigid-perfectly plastic material that fails abruptly when

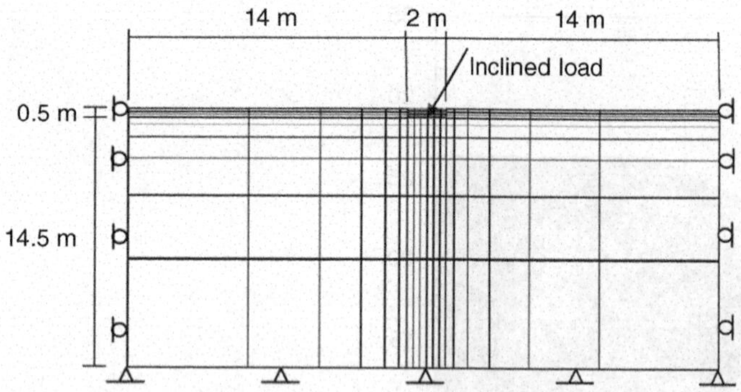

Figure 22.22 A finite element model including boundaries, geometry, and meshing (Helwany 2007 / John Wiley & Sons).

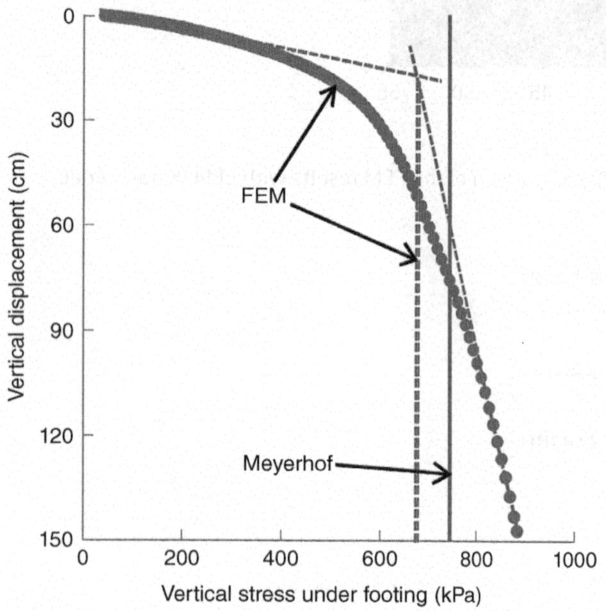

Figure 22.23 Comparison between the FEM and Meyerhof results (Helwany 2007 / John Wiley & Sons).

the bearing capacity of the soil is reached. In contrast, the present finite element analysis assumes that the soil is an elasto-plastic material that deforms under applied loads, and the yielding can occur in a progressive manner due to the nature of the finite element formulation. As a result, yielding in one element can cause the next element to yield, leading to a gradual and progressive slip surface formation.

Example 22.6 ***Bearing Capacity of a Square Foundation (3D Modeling)***
In this example, a limit equilibrium solution is presented for a layer of Ottawa sand that is loaded by a rigid and perfectly rough square footing. To simulate the behavior of the Ottawa sand, the cap model, which is a modified Drucker–Prager model with a cap, is used with parameters matching the classical Mohr Coulomb yield model parameters c' and φ'. The model is analyzed in three dimensions and is shown in Figure 22.24. The sand layer has a depth of 50 m and a plan of 100 m × 100 m, while the loaded area is 3 m × 3 m. To take advantage of symmetry, only one-fourth of the sand layer and the loaded area are considered. The loaded area represents a foundation that has perfect contact with the soil. The sand layer is modeled using reduced-integration, eight-node linear brick elements. The base of the sand layer is fixed in all directions, and all vertical boundaries are fixed in the horizontal direction but free in the vertical direction. Figure 22.24 shows the finite element mesh used in the analysis. Notably, the mesh is finer in the vicinity of the foundation, as this is the zone of stress concentration. Although no mesh convergence studies have been performed, the dimensions of the sand layer are chosen to minimize the boundary effect on foundation behavior.

In this instance, we establish the relationship between the load applied and the displacement for the 3 m × 3 m footing. The bearing capacity of the footing can be determined from the load–displacement curve. Two different numerical

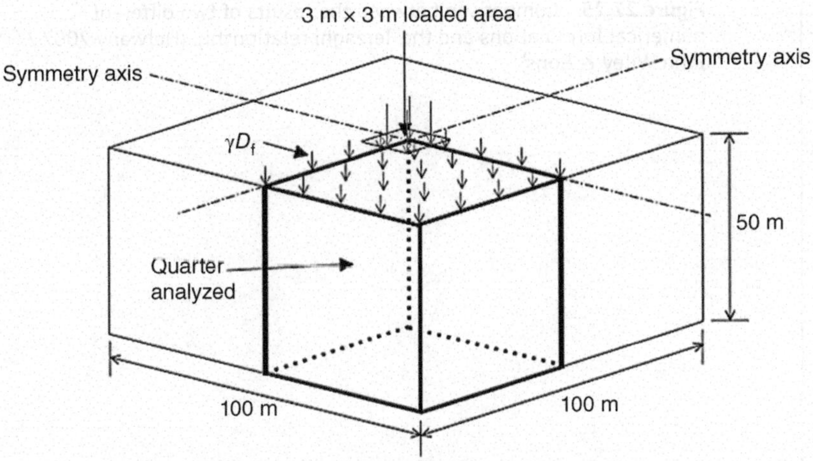

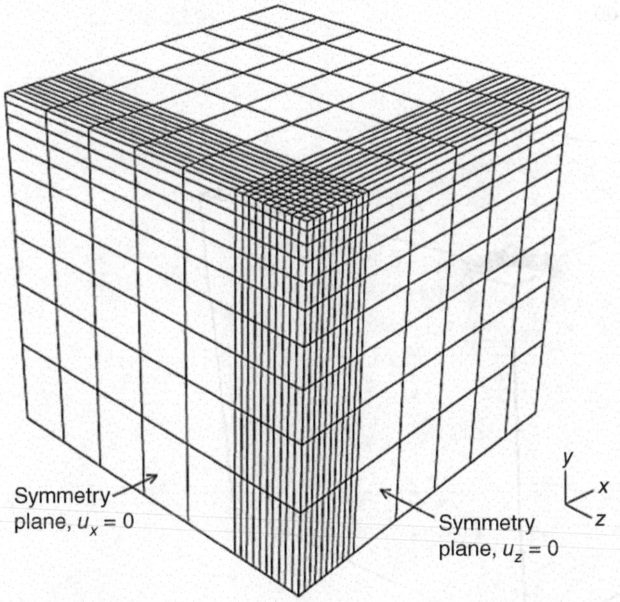

Figure 22.24 Left: geometry of problem; right: numerical model with meshing (Helwany 2007 / John Wiley & Sons).

formulations are employed in the analysis: (i) static analysis and (ii) dynamic-explicit analysis with slow loading. Static analysis is typically utilized for problems with statically applied loads, like the present case. The dynamic-explicit option can also be used to examine the problem; however, this must be done cautiously. The model must not be excited, so the loads should be applied very slowly. It should be noted that explicit analyses use small time increments to ensure stability, making them computationally expensive.

At the start of the analysis, the sand layer is subjected to both gravity loads and surcharge loads, which are crucial as they dictate the initial stress conditions in all soil elements. The behavior of the soil is influenced by stress and the cap model utilized in this case accounts for this fundamental fact. In both analyses, a constant downward velocity boundary condition is implemented to apply the foundation load with a speed of 6 cm/s for 10 seconds.

Figure 22.25 illustrates the pressure–settlement curve obtained from the static analysis and dynamic–explicit analysis. The two analyses show good agreement up to the point of failure. They both predict a bearing capacity of about 1350 kPa. For comparison, the figure also displays the bearing capacity of 1474 kPa predicted by Terzaghi's equation. It should be noted that the finite element prediction of the bearing capacity is slightly lower than Terzaghi's prediction. There are several reasons why there is a difference, but the main cause is that Terzaghi's equation assumes that the soil is a rigid-perfectly plastic substance that fails suddenly when it reaches the bearing capacity of the soil. On the other hand, the finite element analysis currently used assumes that the soil is an elastoplastic material with hardening, which means that it deforms

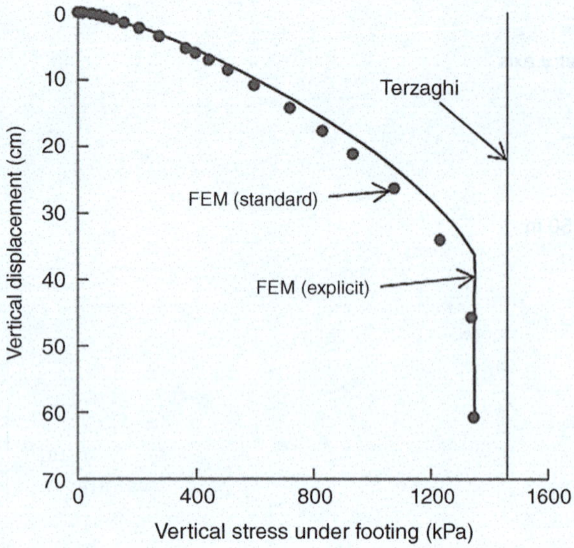

Figure 22.25 Comparison between the results of two different numerical formulations and the Terzaghi relationship (Helwany 2007 / John Wiley & Sons).

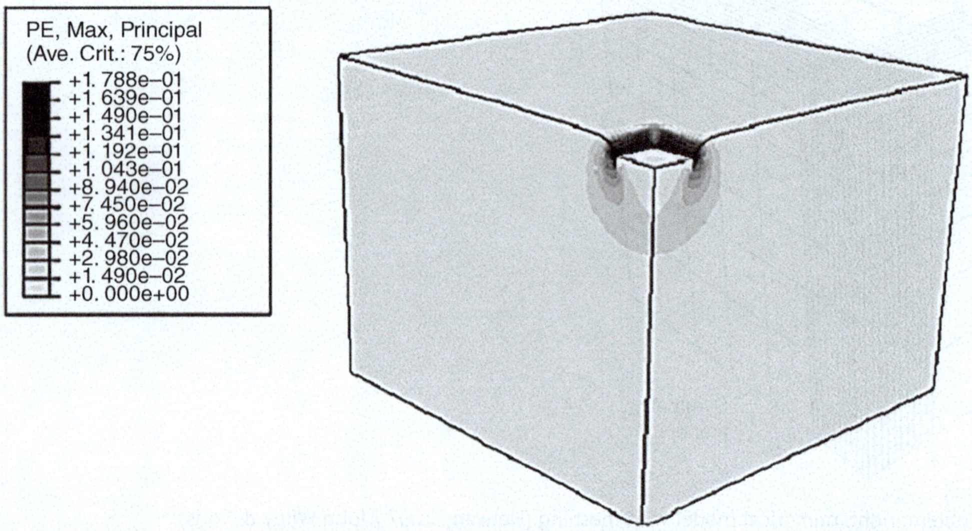

Figure 22.26 The shear strain contours at failure load (Helwany 2007 / John Wiley & Sons).

under applied loads. Additionally, the nature of the finite element formulation allows for a progressive yielding of the soil, as elements can yield gradually and progressively, causing adjacent elements to yield. Figure 22.26 shows the shear strains contours at failure load, and Figure 22.27 presents the formation of vertical stresses at failure load.

22.4.2.5 Application of FEM in Deep Foundations

Soil–pile interaction is an important aspect to consider in the design of foundation systems such as piles used to support structures. The Finite Element Method (FEM) can be used to model soil–pile interaction, which involves the behavior of soil and pile under loading conditions. The following steps can be taken to model soil–pile interaction using FEM:

- **Geometry and Mesh Generation:** The geometry of the pile and the surrounding soil are created using CAD software. The geometry is then meshed using a finite element mesh generator. The mesh size and element type should be selected based on the size and shape of the pile and the soil properties.
- **Material Properties:** The soil and pile properties, including the density, Young's modulus, Poisson's ratio, and shear strength, are assigned to the elements in the mesh. The soil can be modeled as either linear or nonlinear depending on the soil type and the loading conditions. The pile can also be modeled as a linear or nonlinear material depending on its material properties.

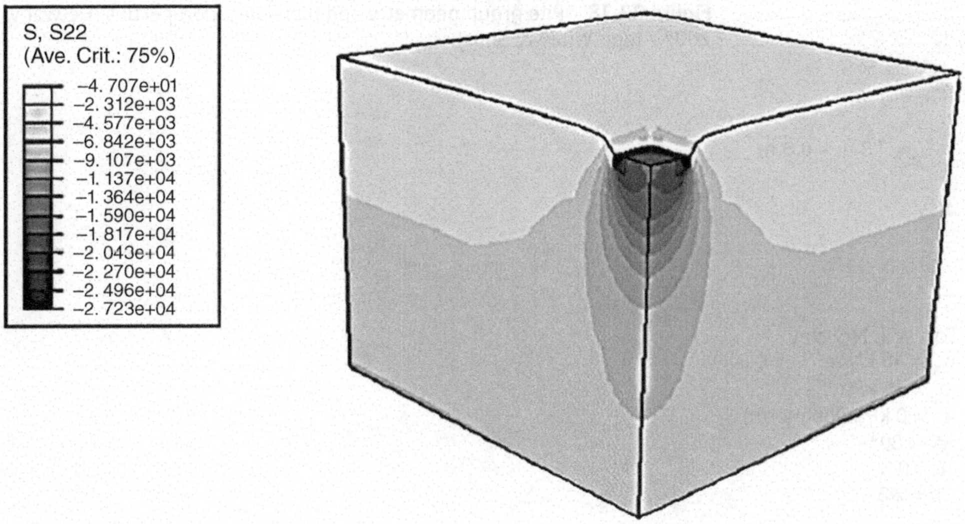

Figure 22.27 The formation of vertical stresses at failure load (Helwany 2007 / John Wiley & Sons).

- **Boundary Conditions:** Boundary conditions are applied to the model to simulate the actual conditions under which the pile is constructed. The pile is usually subjected to vertical and lateral loads, and the soil is subjected to vertical stress due to the weight of the pile and any overlying structures. The interface between the pile and soil is also modeled to consider the interaction between the two.
- **Analysis:** The analysis is performed by solving the global equations derived from the element equations. The analysis can be performed using commercial FEM software or by coding the equations in a programming language. The solution provides the displacements, stresses, and strains in the soil and the pile.
- **Results and Interpretation:** The results can be visualized using post-processing techniques, including contour plots, stress and strain distributions, and animations. The results can be interpreted to evaluate the performance of the pile under different loading conditions and to optimize the pile design.

Example 22.7 ***Pile Group Bearing Capacity Simulation Using FEM***

Consider a group of four concrete piles with a square cross-section of $0.6 \times 0.6\,\text{m}^2$ placed as illustrated in Figure 22.28. These piles have an embedded length of $L = 9.15\,\text{m}$ and are positioned in a uniform, thick layer of saturated clay with an average undrained shear strength of $c_u = 14\,\text{kPa}$ and a friction angle of $\varphi' = 30°$. The clay layer below the pile base has an undrained shear strength of $c_u = 21.3\,\text{kPa}$ and a friction angle of $\varphi' = 30°$. We will determine the total load capacity of the pile group using the finite element method in ABAQUS software for both undrained loading conditions and drained loading conditions.

(a) Undrained loading conditions

This example aims to find a solution using limit equilibrium for a thick clay layer loaded under undrained conditions by a pile group that includes four concrete piles, each of which is 9.15 m long and has a square cross section of 0.6 m-by-0.6 m.

The analysis of pile group problems is conducted in three dimensions. In this case, the pile group is arranged in a square configuration, which allows for taking advantage of the symmetry when generating the finite element mesh of the pile group and the surrounding soil. As a result, only one-fourth of the geometry is analyzed, as depicted in Figure 22.29. The finite element mesh for this section is shown in Figure 22.30. This simplification is not applicable to pile groups that are subjected to lateral loading since those problems are not symmetrical, and the entire group should be analyzed together.

In any soil-pile system, it is necessary to use interface elements that can simulate the frictional interaction between the soil and the surface of the pile. However, when modeling a pile group using finite element analysis, it is not possible to enforce a particular mode of failure, such as a single-pile failure mechanism or block failure mechanism. The type of failure mechanism that will occur is dependent on several factors, such as the problem configuration, properties of the piles and surrounding soil, and the loading conditions (drained or undrained). For this analysis, it is assumed that

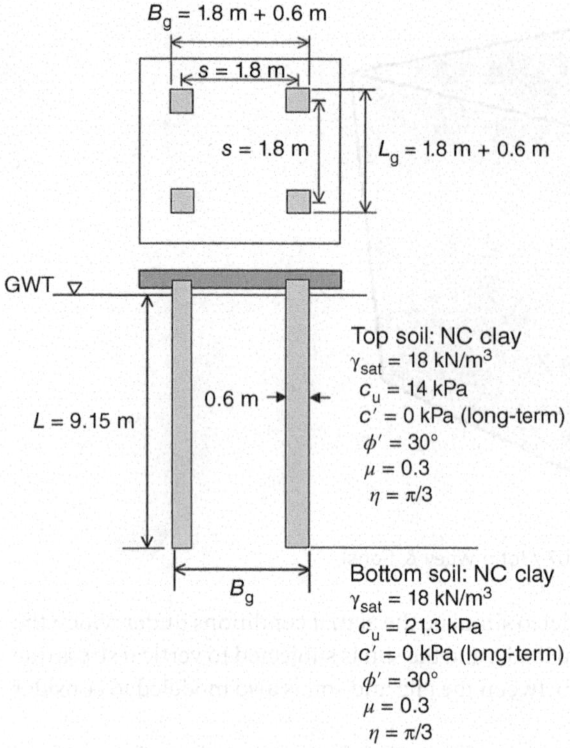

Figure 22.28 Pile group geometry and mechanical properties (Helwany 2007 / John Wiley & Sons).

the piles are already embedded in the soil before the pile loads are applied. Any excess pore water pressures caused by pile driving are also assumed to have dissipated completely before application.

The finite element mesh used in this analysis consists of two parts: the concrete pile and the surrounding soil, with interface elements between them. The mesh takes advantage of symmetry in two planes, as depicted in Figure 22.30. The mesh has dimensions of 15 m in the x-direction, 15 m in the y-direction, and 15.15 m in the z-direction. When choosing mesh dimensions, it is important to ensure that the boundaries do not affect the solution. A generally accepted rule is that the length and width of the mesh should be greater than 30 times the diameter of the pile, while the height should be greater than twice the embedded length of the pile. However, it should be noted that the present mesh is used only for illustration purposes and does not conform to these values.

The simulation assumes that the pile is initially in full contact with the soil. To model the frictional interaction between the pile and the soil, penalty-type interface elements are utilized with a friction coefficient of 0.3. These elements can accurately represent the Coulomb-type frictional interaction between the soil and the surface of the pile that is in contact.

Linear brick elements with eight nodes are utilized for the pile, and pore fluid/stress elements with eight nodes are used for the soil. The brick elements have reduced integration and do not account for pore water pressure. The pore fluid/stress elements are trilinear displacement and pore pressure elements with reduced integration. The clay layer base is fixed in all directions. Two planes of symmetry are present, one allowing sliding in the y and z directions but not the x-direction, and the second allowing sliding in the x and z directions but not the y-direction. The mesh is denser near the pile to capture the concentrated stress zone. No convergence studies have been conducted.

The clay's elastic response is modeled as linear and isotropic, with a Young's modulus of 68.9 × 103 kPa and a Poisson's ratio of 0.3. The cap model is used to simulate the drained and undrained behavior of the clay with the parameters $d = 0$ and $\varphi = 50.2°$, which were calibrated to match the Mohr–Coulomb failure criterion parameters $c' = 0$ and $\varphi' = 30°$. The estimated undrained shear strength of the clay layer using this model is $c_u \approx 14$ kPa at the pile midheight and $c_u \approx 21$ kPa at the pile tip, which is consistent with the undrained shear strength values of the top and bottom clay layers shown in Figure 22.29.

The cap eccentricity parameter, denoted by R, is set to 0.4. The initial position of the cap, which indicates the initial consolidation of the specimen, is $\varepsilon_{vol}^{pl}(0) = 0.0$, and the cap hardening curve is assumed to be a straight line passing

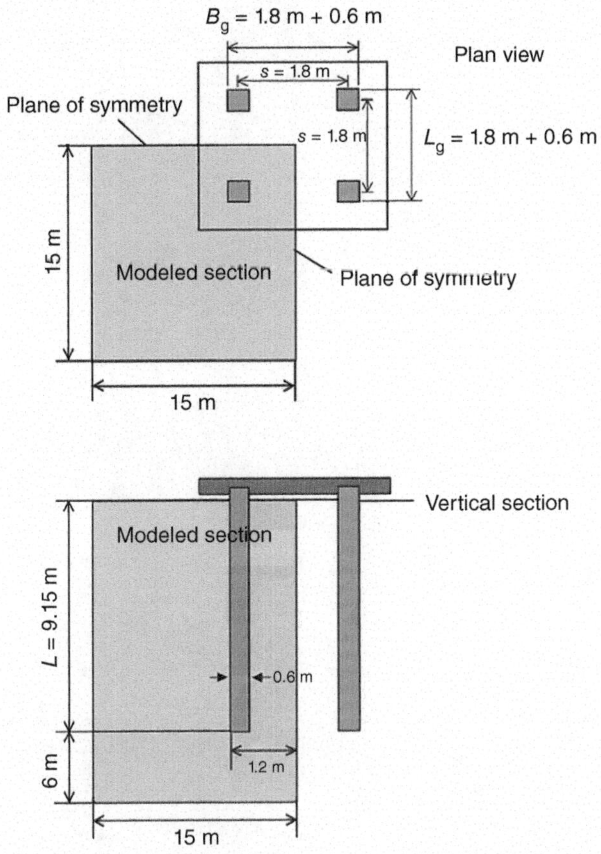

Figure 22.29 Geometry and simplification of that (Helwany 2007 / John Wiley & Sons).

through two points: (p' = 0.57 kPa, ε_{vol}^{pl} = 0.0) and (p' = 103 kPa, ε_{vol}^{pl} = 0.0032). A transition surface parameter $\alpha = 0.1$ is also assumed. The input data required for this consolidation analysis include the initial void ratio, $e_0 = 1.5$, and the vertical and horizontal effective stress profiles of the clay layer. In this study, the initial horizontal effective stress is 50% of the vertical effective stress.

Solving the problem is divided into two steps. In the first step, the effective self-weight of the clay layer is applied using the "body-force" option, where the groundwater table coincides with the ground surface. The clay layer is assumed to be elastoplastic, and the cap model is used to analyze the ability of the clay layer to withstand the end-bearing and skin-shear stresses induced by the pile. The "geostatic" command is used during step 1 to ensure that equilibrium is satisfied within the clay layer, and it verifies that the initial stress condition in any element within the clay layer falls within the initial yield surface of the cap model. This model, along with others like the Cam clay model and the Mohr–Coulomb model, can detect failure within the clay layer.

During step 2 of the analysis, a consolidation analysis is performed, and the pile is loaded using a vertical displacement boundary condition. This causes the top surface of the pile to move downward by 30 cm at a constant rate of 0.003 cm/s. This loading rate is sufficient to cause undrained soil behavior. Figure 22.31 shows the excess pore water pressure in a soil element directly below the pile tip, which indicates a significant increase in the pore water pressure during pile loading, suggesting undrained loading.

Figure 22.32 displays the pile group capacity vs. settlement curve that was obtained from the finite element analysis. As the pile load is increased, the settlement also increases. When the pile load reaches approximately 3000 kN, a pile group settlement of around 4 cm is observed. Shortly after this point, the pile moves downwards at an even faster rate, indicating that the load capacity of the pile has been reached. The figure also includes the pile group capacity predicted by the α-method using both the single-pile failure mechanism and the block failure mechanism. The predicted pile group capacity using the single-pile failure mechanism is 1504 kN, while the predicted capacity using the block failure mechanism is 2336 kN. It is important to note that the pile group capacity predicted

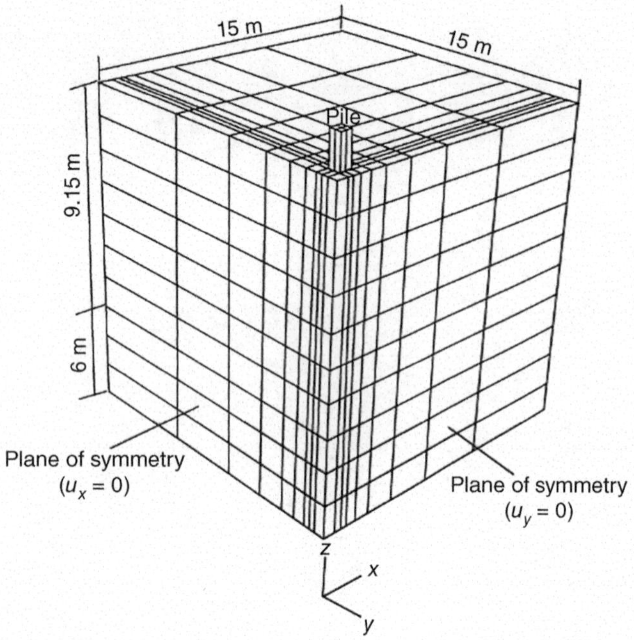

Figure 22.30 The finite element mesh (Helwany 2007 / John Wiley & Sons).

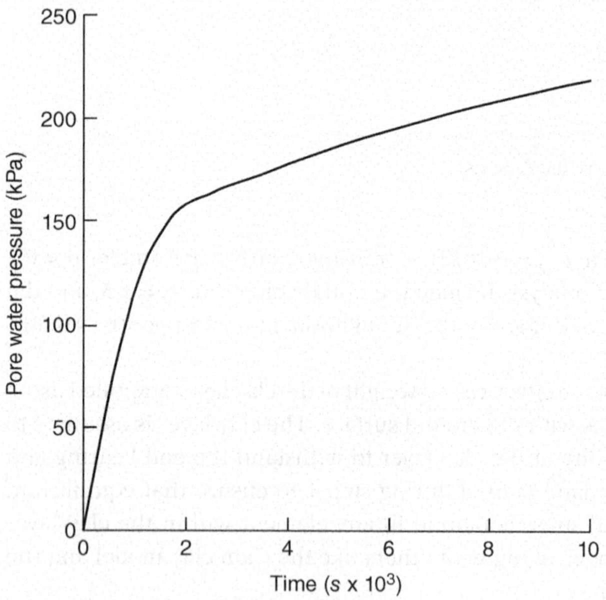

Figure 22.31 The development of excess pore water pressure in a soil element positioned directly beneath the tip of a pile during undrained loading (Helwany 2007 / John Wiley & Sons).

by the finite element analysis is higher than the bearing capacity predicted by the α-method using either failure mechanism.

(b) Drained loading condition

The setup for this problem is the same as the one described in part (a), including the geometry, boundary conditions, and materials. The finite element mesh used is also the same, and the material parameters used are identical. Like part (a), the problem is run in two steps, with the effective self-weight of the clay layer being applied in step 1 in the same way.

In this case, the main difference from part (a) is the loading rate applied in step 2. The loading rate is low because a drained condition is required. Additionally, the top surface of the clay layer is made permeable to allow excess pore

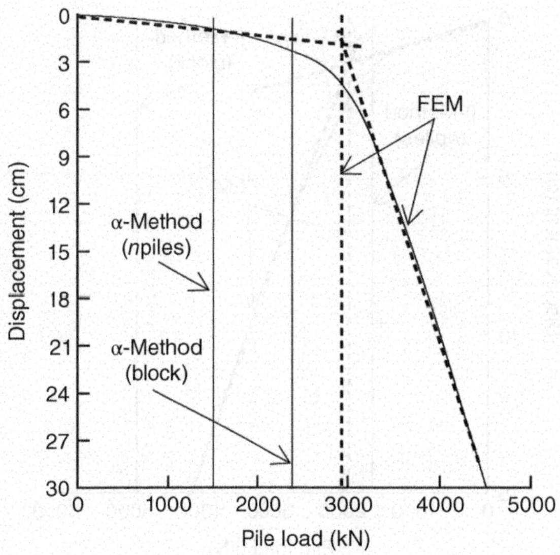

Figure 22.32 Load–settlement curve and a comparison of α-method and FEM results (Helwany 2007 / John Wiley & Sons).

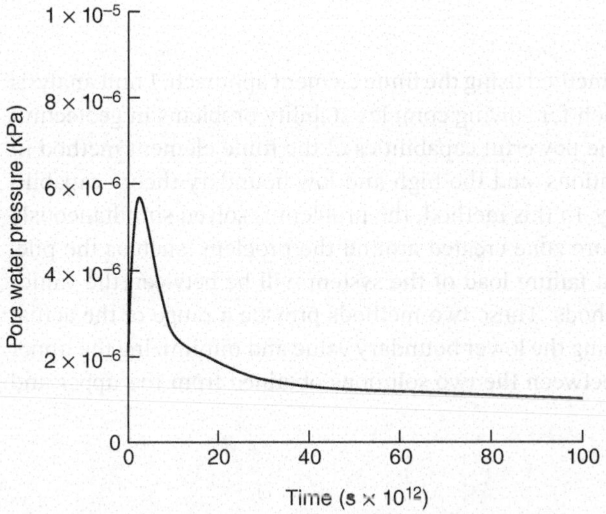

Figure 22.33 The development of excess pore water pressure in a soil element positioned directly beneath the tip of a pile during drained loading (Helwany 2007 / John Wiley & Sons).

water pressure to dissipate. To achieve a drained condition, the pile load is applied at a very slow and constant rate of 3×10^{-13} cm/s in step 2 using a vertical displacement boundary condition to move the top surface of the pile downward by 30 cm. This slow loading rate ensures that the soil behaves in a drained manner. Figure 22.33 confirms that there is essentially no excess pore water pressure in an element located immediately below the pile tip during loading, which indicates that it is a drained loading condition. The problem geometry, boundary conditions, and materials used in this part are identical to those used in part (a), and the finite element mesh and material parameters are also the same.

Figure 22.34 shows the plot of pile group load vs. settlement obtained from the finite element analysis. As the load is increased, the settlement increases gradually up to around 2000 kN, after which the pile group settlement rapidly increases, indicating that the pile group's load capacity has been exceeded. Figure 22.34 also displays the pile group capacity of 1578 kN, predicted by the β-method using the single-pile failure mechanism, for comparison. The β-method with the block failure mechanism predicts a pile group capacity of 4830 kN, and this prediction is also shown in the figure. The pile group load capacity predicted by the finite element analysis is slightly higher than that predicted by the α-method with the single-pile failure mechanism, but much lower than the load capacity predicted by the α-method with the block failure mechanism.

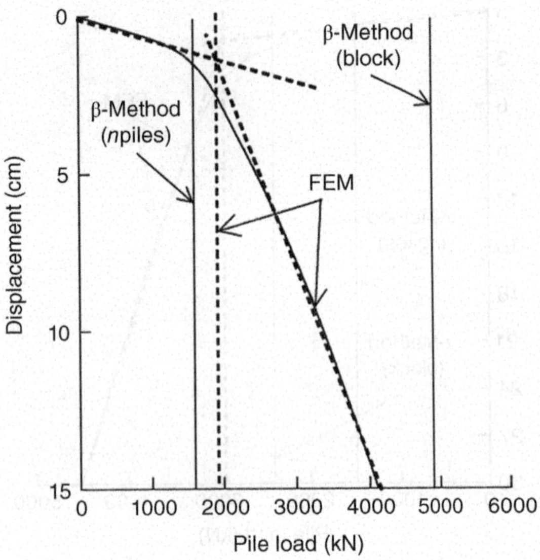

Figure 22.34 Load–settlement curve and a comparison of β-method and FEM results (Helwany 2007 / John Wiley & Sons).

22.4.3 Finite Element Limit Analysis (FELA)

This method is based on solving the equations in the limit analysis method using the finite element approach. Limit analysis based on the finite element method (FELA) is a powerful approach for solving complex stability problems in geotechnical engineering. This combined technique simultaneously uses the powerful capabilities of the finite element method in modeling complex soil layers, loads, and various boundary conditions, and the high and low boundary theory capabilities of limit analysis to achieve solutions with maximum accuracy. In this method, the problem is solved simultaneously using both upper and lower-bound methods, and the plastic failure zone created around the problem (such as the pile, retaining wall, excavation, etc.) is obtained. Therefore, the actual failure load of the system will be between the values obtained from the upper bound (UB) and lower bound (LB) methods. These two methods provide a range of the actual solution (accurate failure load), which can be reduced by maximizing the lower boundary value and minimizing the upper boundary value (Figure 22.35). Because the actual solution lies between the two solutions obtained from the upper and

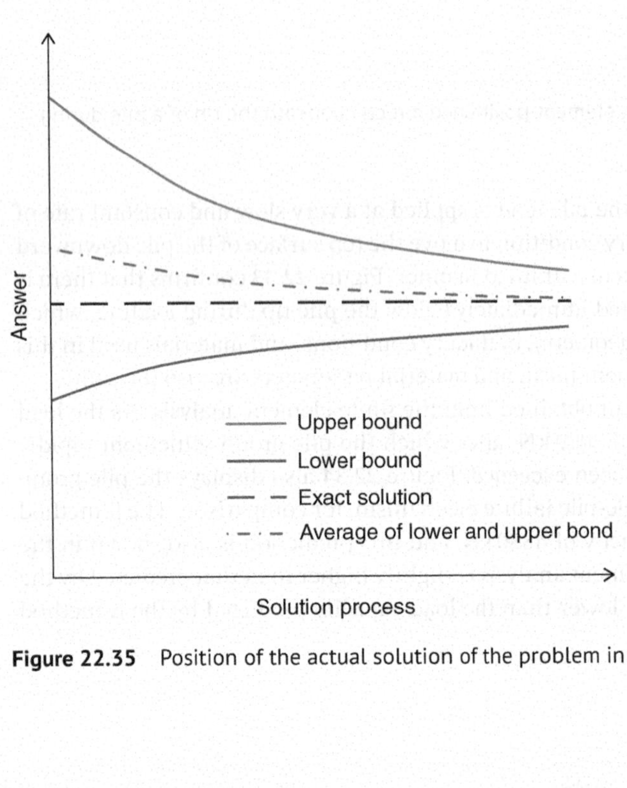

Figure 22.35 Position of the actual solution of the problem in FELA analysis (Ebrahimi and Khazaei 2019 / Springer Nature).

lower boundary methods, the result is taken as the mean of the upper and lower boundaries with appropriate accuracy (Equation 22.13).

$$\text{Result} \cong \frac{\text{UB} + \text{LB}}{2} \tag{22.13}$$

22.4.4 Boundary Element Method (BEM)

The boundary element method is another form of the finite element method. In this method, instead of modeling the entire domain, its boundaries are discretized. As a result, the dimensionality of the problem is reduced by one order. For example, a three-dimensional problem is reduced to a two-dimensional problem, and a two-dimensional problem is reduced to a one-dimensional problem. Therefore, the boundary element method is suited to solve many three-dimensional and two-dimensional problems that other methods cannot.

22.4.5 Discrete Element Method (DEM)

This method was first introduced as a computational method to analyze rock mechanic problems, in which the rock is considered as a collection of separate blocks. In this method, soil or rock particles are modeled as circular or elliptical disks. These particles only interact with each other at their contact points through normal and shear springs and dampers. Also, tangential forces have no effect on normal forces. Since the particles in the DEM method move and interact based on simple physical laws, the theory of this method is generally much simpler than other numerical methods. In this method, by calculating the resultant forces and moments acting on each particle according to the force-displacement law, acceleration is calculated according to Newton's second law, and by integrating acceleration twice over a short time interval, the new position of each particle is determined, and these steps are repeated until the desired state is reached. In this approach, due to the small deformation of the particles compared to the overall deformation of the sample, the particles are assumed to be rigid.

22.4.6 Material Point Method (MPM)

With the continuous development of MPM for geotechnical applications, it has been identified that this method has strong potential in analyzing problems associated with large soil displacements (such as the installation of marine piles, driven piles, and installation of soil anchors that require significant soil displacement for activation). Considering that commonly used methods such as finite elements still have limitations in considering large deformations and material flow, MPM overcomes this limitation and enables previously impossible analyses.

22.4.7 Kinematic Element Method (KEM)

This method is proposed for solving geotechnical stability problems, and it can be used to obtain failure loads in various pile engineering problems (such as pile bearing capacity). The KEM method essentially consists of three parts, including kinematic analysis, static analysis, and optimization process. In this method, a suitable and relatively simple failure mechanism is considered according to Figure 22.36, which occurs under the failure system. This mechanism itself consists of several solid and movable blocks called kinematic elements. A unit vertical displacement is applied to the solid element

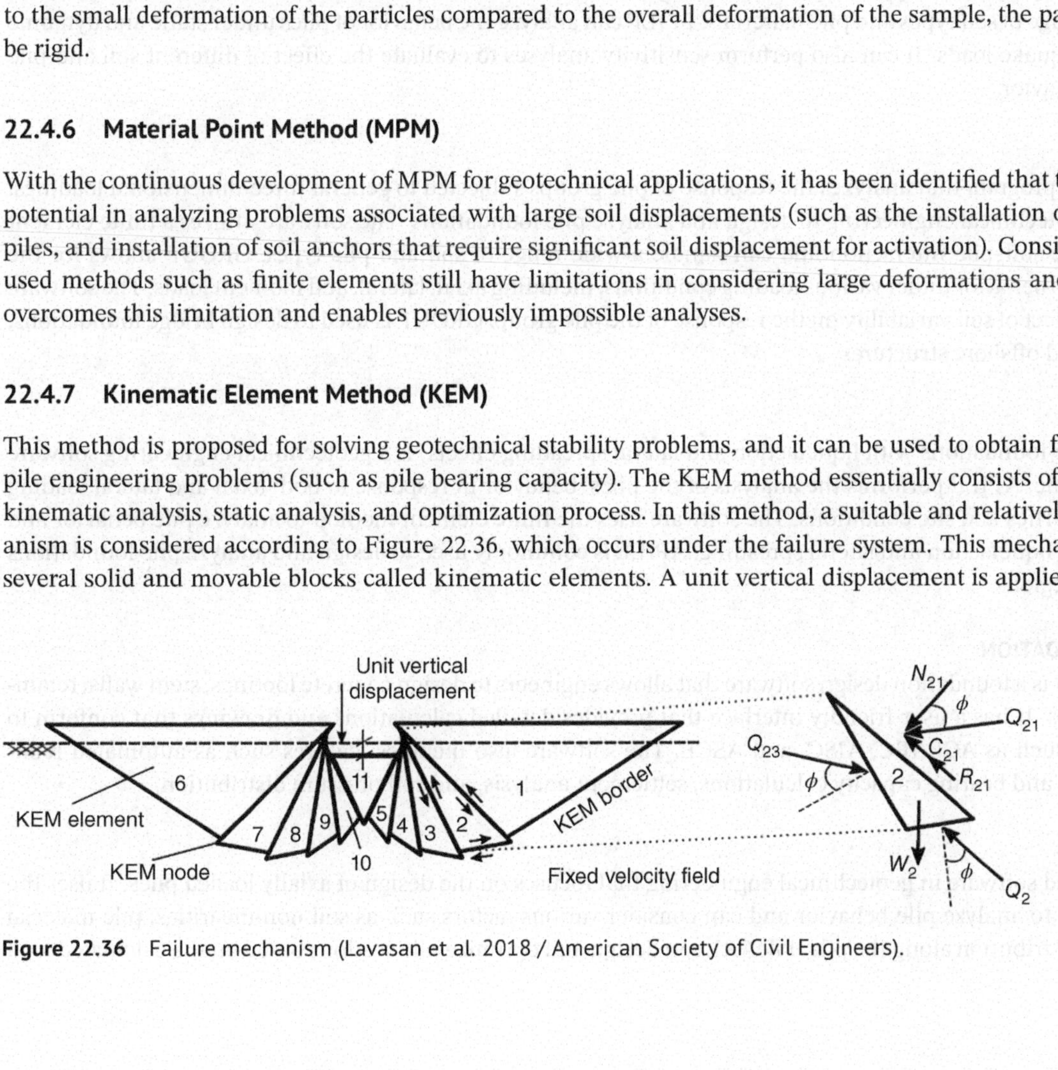

Figure 22.36 Failure mechanism (Lavasan et al. 2018 / American Society of Civil Engineers).

under the pile, which leads to the displacement of the remaining elements along their sliding boundaries. The final shape of the failure mechanism and the forces applied to the elements are obtained through the optimization process.

22.5 Common Software in Foundation Engineering

Most numerical methods are used in commercial software. In this section, the most important numerical software, along with its major applications, strengths, and limitations in geotechnical engineering are introduced.

22.5.1 Design Group

22.5.1.1 SAFE

SAFE is a software program developed by CSI (Computers and Structures, Inc.) to analyze and design foundation systems, including mat foundations, pile foundations, and drilled pier foundations. It is commonly used in geotechnical and structural engineering to design building foundations. SAFE can be used to analyze both static and dynamic loads on the foundation system, and it includes features that model soil–structure interaction and consider the effects of soil settlements on the foundation performance. The software also supports various design codes, including ACI, ASCE, and AISC.

22.5.1.2 LPILE

LPILE is a widely used software program in geotechnical engineering to analyze single piles and pile groups under lateral and vertical loading. It is based on the *p*–*y* method, which allows the user to model the soil–pile interaction using nonlinear soil response curves (*p*–*y* curves) that can be generated based on soil properties and pile dimensions.

The program can analyze various types of piles, including drilled shafts, driven piles, and precast concrete piles, and it can consider a wide range of soil types and pile materials. LPILE can analyze the behavior of piles under static and dynamic loads, including earthquake loads. It can also perform sensitivity analyses to evaluate the effect of different soil and pile parameters on pile behavior.

22.5.1.3 GROUP

GROUP is a computer program that analyzes the response of pile groups subjected to general three-dimensional loadings. It is widely used in geotechnical engineering to design and analyze pile foundations. The software utilizes a finite element approach to model the soil–pile interaction and can handle a wide range of soil and pile types. GROUP allows for the evaluation of pile group behavior under various loading conditions, including axial, lateral, and moment loads. The software can also analyze the effect of soil variability on the response of the pile group. GROUP is used to design bridge foundations, high-rise buildings, and offshore structures.

22.5.1.4 APILE

APILE (analysis of pile foundations with liquefaction and lateral spreading effects) is a geotechnical engineering software developed by Ensoft, Inc. APILE performs the analysis of the piles' behavior in response to both axial and lateral loading based on the soil properties and site conditions. The software uses the finite element method to analyze pile behavior and considers the effects of liquefaction and lateral spreading. APILE is commonly used to design and analyze pile foundations in earthquake-prone regions.

22.5.1.5 ASDIP FOUNDATION

ASDIP FOUNDATION is a foundation design software that allows engineers to design concrete footings, stem walls, retaining walls, and pile caps. It has a user-friendly interface that provides detailed calculations and drawings that conform to various design codes such as ACI, IBC, AISC, and ASCE. The software also includes features such as automated rebar detailing, soil pressure and bearing capacity calculations, settlement analysis, and soil pressure distribution.

22.5.1.6 PileAXL

PileAXL is a specialized software in geotechnical engineering that focuses on the design of axially loaded piles. It uses the finite element method to analyze pile behavior and can consider various factors such as soil nonlinearities, pile material properties, and load distribution along the pile. PileAXL can design and optimize pile sections based on the calculated load capacities and can also analyze the behavior of existing piles. The software is commonly used by geotechnical engineers to design and analyze piles for various structures such as bridges, buildings, and offshore platforms.

22.5.1.7 ALLPile

ALLPile is a geotechnical engineering software specifically designed to analyze and design-driven pile foundations. It is used to analyze both single piles and pile groups subjected to vertical, lateral, and overturning loads. ALLPile can also be used to design pile caps and evaluate the capacity of existing piles. The software uses various methods, including the American Petroleum Institute (API) method, to calculate pile capacities and settlement. It can also perform dynamic analysis using the case and wave equation analysis methods. ALLPile allows users to model complex soil conditions and offers various options for defining soil and pile properties.

22.5.2 Analysis Group

22.5.2.1 DIANA

DIANA (displacement analyzer) is a finite element software commonly used in geotechnical engineering to analyze and design complex soil–structure interaction problems. It can handle a wide range of geotechnical problems such as foundation design, slope stability analysis, excavation design, and retaining wall analysis. DIANA allows for nonlinear analyses, which are necessary to accurately model soil–structure interaction behavior. It is also equipped with various material models to simulate different soil behaviors. The software provides users with detailed graphical outputs and reports to help interpret and analyze the results.

22.5.2.2 Midas GTS NX

Midas GTS NX is a geotechnical and tunneling analysis software that allows for the analysis of various geotechnical problems such as soil–structure interaction, rock mechanics, groundwater seepage, and more. It uses finite element analysis to model complex geotechnical problems and offers a range of analysis capabilities, including static and dynamic analysis, slope stability analysis, and seismic analysis. The software can also be used to design various geotechnical structures such as foundations, retaining walls, and tunnels. Additionally, Midas GTS NX supports various international design codes such as ACI, AISC, Eurocode, and more.

22.5.2.3 FLAC 3D/2D

FLAC (Fast Lagrangian Analysis of Continua) is a software package developed by Itasca Consulting Group that is widely used in geotechnical engineering to analyze the behavior of soil, rock, and other materials under complex loading conditions. FLAC is a finite difference numerical modeling tool that can simulate the behavior of soils, rocks, and other materials using a wide range of constitutive models, including linear elastic, Mohr–Coulomb, Hoek–Brown, and others. FLAC can be used to model various geotechnical problems, such as soil–structure interaction, slope stability, seepage analysis, and underground excavations.

22.5.2.4 PLAXIS 3D/2D

PLAXIS is a well-known software in geotechnical engineering for finite element analysis of soil, rock, and structural interaction. It is used to analyze, design, and verify various geotechnical structures, such as excavations, foundations, tunnels, embankments, and retaining walls. The software uses advanced material models, including both linear and nonlinear behavior, to simulate various geotechnical problems accurately. The program provides a user-friendly interface, various customization options, and extensive post-processing capabilities to help the user visualize the results in a clear and effective manner. It is widely used in academia, consulting, and construction industries for geotechnical engineering design and analysis.

22.5.2.5 ABAQUS

ABAQUS is a powerful finite element analysis software used in various fields including geotechnical engineering. It can be used to model and analyze complex soil–structure interaction problems, such as deep excavations, tunnels, foundations, retaining structures, and slope stability. It offers a range of constitutive models for soil and rock, including nonlinear and time-dependent behavior, as well as advanced analysis capabilities such as dynamic and thermal analysis. ABAQUS also provides pre-processing and post-processing tools for easy model creation, visualization, and interpretation of results.

22.5.2.6 GeoStudio

GeoStudio is a suite of software applications for geotechnical engineering and geoenvironmental modeling developed by GEO-SLOPE International. It includes several modules such as SEEP/W for groundwater flow and seepage analysis, SLOPE/W for slope stability analysis, SIGMA/W for stress and deformation analysis, TEMP/W for thermal analysis, CTRAN/W for contaminant transport analysis, and QUAD/DS for dynamic analysis. GeoStudio is widely used in the geotechnical engineering industry for various applications, including soil–structure interaction analysis, foundation design, retaining wall design, and landfill design.

22.6 Numerical Modeling of Special Issues in Foundation Engineering

In foundation engineering, there may be special conditions in which using existing classical methods for analysis can cause significant errors. Soil in projects rarely conforms to common assumptions in homogeneous and isotropic states. Additionally, problems in the underground such as ground slippage, liquefaction, water seepage, and problematic layers present the pile designer with numerous problems in using the analytical and empirical solutions available in engineering references and standards. To overcome these problems, using numerical methods is useful and appropriate. One of the subtopics of special conditions in pile engineering is the existence of complex geometries or the proximity of multiple structures/geotechnical problems that may not have any easy analytical or empirical solutions. These conditions include:

1) Analysis of the effects of adjacent piles (on soil or rock)
2) Calculation of bearing capacity in the presence of voids (such as tunnels) under piles
3) Existence of piles adjacent to excavations
4) Presence of faults adjacent to geotechnical structures
5) Drawing of settlement contours of piles under combined loads
6) Issues with various geometric complexities and loading on geotechnical structures.

Using numerical methods and engineering software developed based on them is a suitable solution to deal with the above cases. In the following section, several examples of these methods will be presented.

Example 22.8 ***Bearing Capacity of Two Adjacent Footings on Rock Masses***
Two adjacent strip footings with a width of $B = 1$ m and edge-to-edge spacing of S are located on a rocky ground with Mohr–Coulomb failure parameters, as shown below. Neglecting any surcharge loads, calculate the bearing capacity of these two footings for S/B ratios of 1 and 1.2.

$$\sigma_{ci} = 10\text{ MPa}, D = 0, \text{GSI} = 30, m_i = 7, \gamma_{rock} = 21\text{ kN/m}^3$$

Table 22.6 Summarizes of the information used for the performed modeling. (Shamloo and Imani, 2021)

Solution:

Because the piles are strip-shaped, a two-dimensional numerical analysis using the plane strain method has been performed. Using the results obtained from the PHASE 2 software, the stress–settlement curve for a point in the rock mass located below the center of the pile was plotted. Then, the load-bearing capacity was obtained using this curve. In addition,

Table 22.6 Summary of the information used for the performed modeling. (Shamloo and Imani 2021 / with permission of Elsevier).

Software	Analysis method	Loading	Failure criterion	Mesh size	Boundary conditions	Dimensions of geometry in terms of foundation width *B*
PHASE2	FEM	Uniformly at a speed of $2*10^{-5}$ m/s.	Hoek–Brown	0.1*B*	Lateral boundary: closed in horizontal direction and open in vertical direction. Bottom boundary: closed in both directions.	7*B* from the edge of the foundation 5*B* from the bottom of the foundation

Table 22.7 The load-bearing capacity calculated using the upper bound limit analysis method compared to the results using the curve (Shamloo and Imani 2021 / with permission of Elsevier).

Method	q_u (MPa), considering $S/B = 1.2$	q_u (MPa), considering $S/B = 1$
FEM (PHASE 2 software)	2.233	2.948
Upper Bound	2.284	2.502

Table 22.8 Soil properties (Lavasan et al. 2016 / Springer Nature).

Dry unit weight	Angle of internal friction	Cohesion	Dilation angle	Elastic modulus	Poisson's ratio
γ_d (kN/m^3)	φ (°)	c (kN/m^2)	ψ (°)	E_s (kN/m^2)	υ_s
17.2	18	40	0	35,000	0.3

Table 22.9 Comparison of ultimate bearing capacity results of the rigid strip footing.

FEM	Hansen and Vesic	Terzaghi	Method
573.45	576.89	662.57	q_{ult} (kN/m^2)
0	0.6	15.5	Difference with numerical method (%)

the load-bearing capacity was also calculated using the upper bound limit analysis method proposed by Shamlou and Imani (2021), and the results are compared in Table 22.7. These two methods have yielded relatively close results.

Example 22.9 ***Effect of Tunneling on the Bearing Capacity of a Strip Footing on Soil***
According to Figure 22.35, a tunnel with a diameter of $d = 6$ m has been excavated at a depth (distance from the tunnel ceiling to the ground surface) of $C = $ d in the soil with the characteristics specified in Table 22.8. A solid strip footing with a width of $B = 0.25d$ is located exactly along the centerline of the tunnel on the ground surface. The final bearing capacity of the strip footing should be estimated in the following two conditions: (i) without the presence of the tunnel using (a) Terzaghi's equation, (b) Vesic and Hansen's equation, and (c) the output of finite element modeling; and then (ii) calculating the final bearing capacity in the case of the excavated tunnel (without lining) using the finite element method.

Solving Part (1) of the Problem:

PLAXIS 2D software has been used to calculate the bearing capacity of the strip pile using the finite element method. The complete Mohr–Coulomb elastic-plastic behavior model has been assumed for the soil with the assumption of the non-associated flow rule. Also, according to the problem conditions, the strip pile is assumed to be a solid type with a rough contact surface on the ground and without any surcharge pressure on the ground surface. The modeling is done symmetrically according to Figure 22.37 with the assumption of constrained horizontal displacement for the vertical boundaries of the model and a hinged support assumption for the model's base points. However, in this stage, the tunnel has not been modeled according to the problem statement. Triangular 15-node elements have been used, and to increase the accuracy of the results obtained, the dimensions of the mesh have been chosen to be finer in the area between the strip pile and the tunnel.

Figure 22.38 shows the load–settlement curve obtained from numerical modeling for the case without the tunnel and the bearing capacity obtained from this curve. In Table 22.9, the results obtained from the three methods mentioned in parts (a), (b), and (c) are compared.

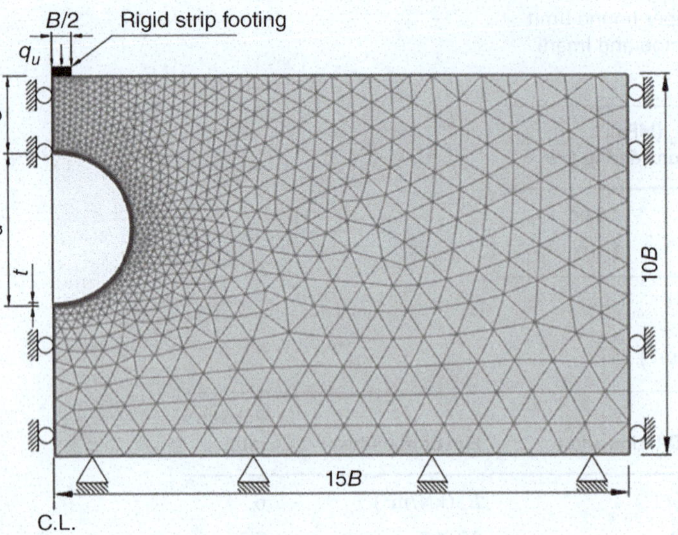

Figure 22.37 Schematic view of meshing, boundary conditions, and representation of circular tunnel and strip footing.

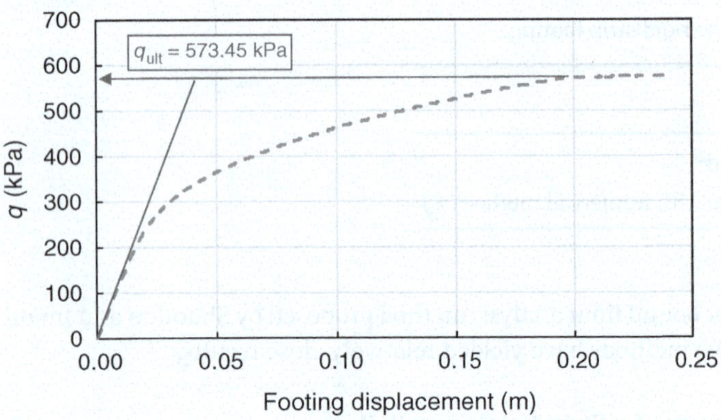

Figure 22.38 Output results of the load–settlement curve under the rigid strip foundation without the excavated tunnel.

Solving Part (2) of the Problem:

In Figure 22.39, the load–settlement curve obtained from the finite element analysis for the point below the pile in an excavated but not lined tunnel is shown. The ultimate bearing capacity of the strip footing will be equal to:

$$q_{ult} = 447.31\ \text{kN/m}^2$$

The results of this analysis indicate a 22% reduction in the load-bearing capacity due to tunnel excavation. In Figure 22.40, the soil settlement lines under the foundation are shown for both tunneling and non-tunneling conditions.

Example 22.10 ***Bearing Capacity of a Strip Footing Close to an Excavation***

According to Figure 22.41, a strip footing with $B = 1$ m width is located at a distance of $S = 1.5B$ from a vertical excavation of depth $H = 6B$. The soil properties are given in Table 22.10. An estimation of the ultimate bearing capacity of the strip footing is required in the following two cases: (i) without the presence of the excavation wall using (a) Terzaghi's equation, (b) Vesic and Hansen's equation, and (c) the output of the finite element numerical modeling, and then, (ii) in the presence of the excavation wall using the finite element numerical method.

Solving Part (1) of the Problem

To calculate the bearing capacity of the strip footing using the finite element method, PLAXIS 2D software has been used. The complete Mohr–Coulomb elastic-plastic behavior model has been used for the soil with the assumption of a

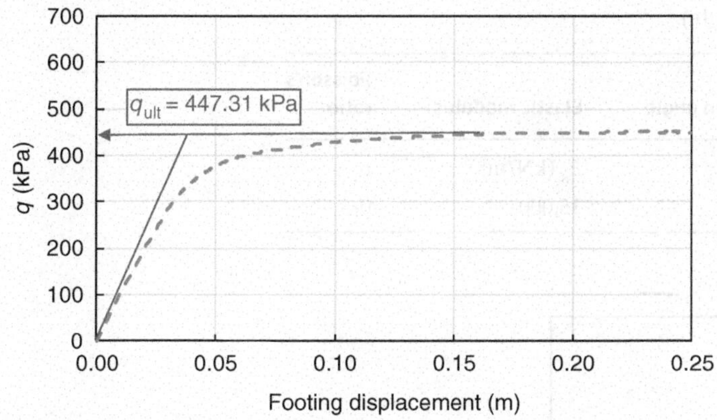

Figure 22.39 Output results of the load–settlement curve under the solid strip foundation in an excavated tunnel without covering, obtained from the finite element analysis.

Figure 22.40 Failure of soil under the footing in (a) the absence of a tunnel and (b) tunneling without lining.

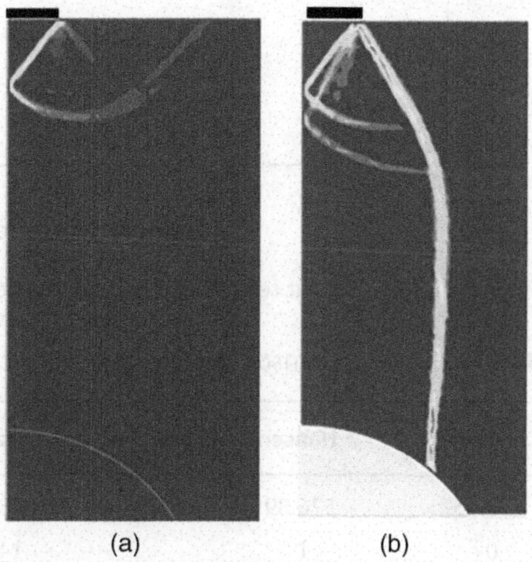

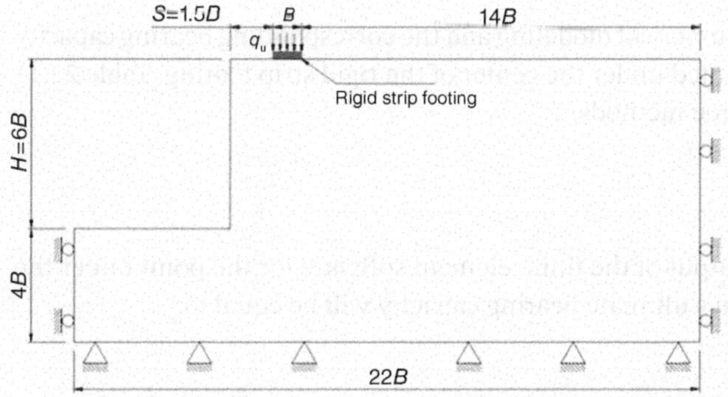

Figure 22.41 Schematic view of the mesh configuration, boundary conditions, and the strip footing and excavation wall display.

non-associated flow rule. Additionally, according to the problem conditions, the strip footing is assumed to be rigid with a rough contact surface located on the ground and without any surcharge pressure on the ground surface. The modeling is performed symmetrically, assuming that the horizontal displacements are constrained for the vertical boundaries of the model and with a hinged support assumption for the model's floor points. The numerical model dimensions are shown in Figure 22.41. Fifteen-node triangular elements have been used, and to increase the accuracy of the obtained solutions, the dimensions of the mesh have been selected finer in the area between the strip footing and the tunnel.

Table 22.10 Soil properties (Lavasan et al. 2016 / Springer Nature).

Dry unit weight	Angle of internal friction	Cohesion	Dilation angle	Elastic modulus	Poisson's ratio
γ_d (kN/m^3)	φ (°)	c (kN/m^2)	ψ (°)	E_s (kN/m^2)	υ_s
17/2	18	40	0	35,000	0/3

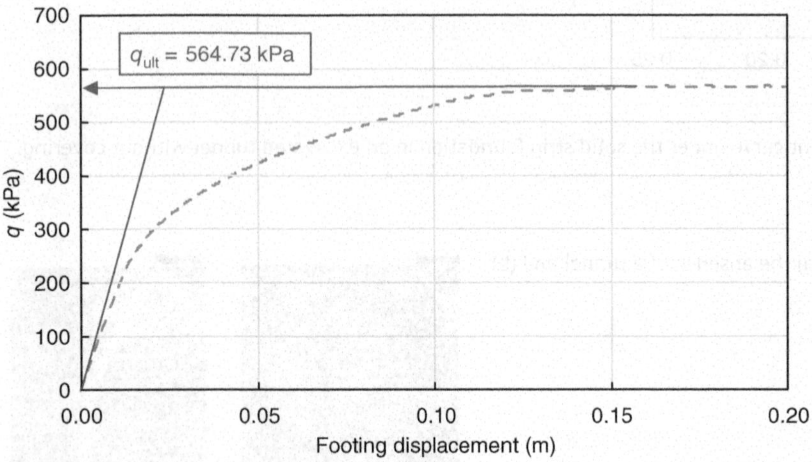

Figure 22.42 Output results of the load–settlement curve under the solid strip footing in the absence of an excavation wall.

Table 22.11 Comparison of ultimate bearing capacity results for strip foundation.

FEM	Hansen and Vesic	Terzaghi	Method
573.45	576.89	648.38	q_{ult} (kN/m^2)
0	1	14.9	Difference with numerical method (%)

Figure 22.42 shows the load–settlement curve obtained from numerical modeling and the corresponding bearing capacity for the no-excavation case. This curve is plotted for the point located under the center of the rigid strip footing. Table 22.11 presents a comparison of the bearing capacity obtained from three methods.

Solving Part (2) of the Problem

In Figure 22.43, the load–settlement curve obtained from the output of the finite element software for the point under the pile in the presence of excavation is shown. The strip foundation's ultimate bearing capacity will be equal to:

$$q_{\text{ult}} = 273.85\ \text{kN/m}^2$$

The results of this analysis indicate more than a 51% reduction in bearing capacity due to excavation near the pile. In Figure 22.44, the soil settlement curves are shown for both the presence and absence of excavation.

Example 22.11 ***Pile Group Bearing Capacity Subjected to Cyclic Lateral Load Modeling Using ABAQUS Software***
In this example, the vertical and lateral capacity of a steel pile group in loose sand after subjecting lateral cyclic loading is investigated. Figure 22.45 gives the geometry of the laboratory model. Also, the mechanical properties of soil and pile dimensions are provided in Table 22.12. First, the pile group are applied a cyclic lateral loading then the vertical and lateral capacity of the pile group is evaluated. Finally, the experimental FEM and BEM (Boundary element method) results are compared.

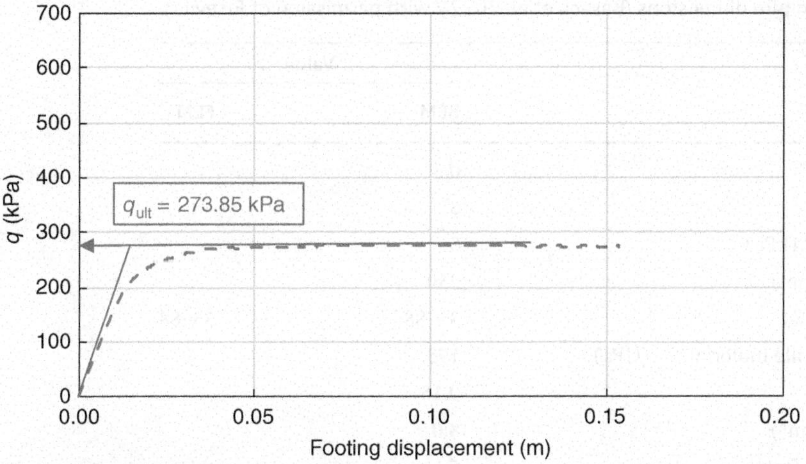

Figure 22.43 The output results of the strip foundation's load–settlement curve under the excavation wall.

Figure 22.44 Soil settlement curves under the footing in: (a) absence of excavation and (b) presence of excavation.

Figure 22.45 The laboratory model's geometry (Basack et al. 2022 / with permission of Elsevier).

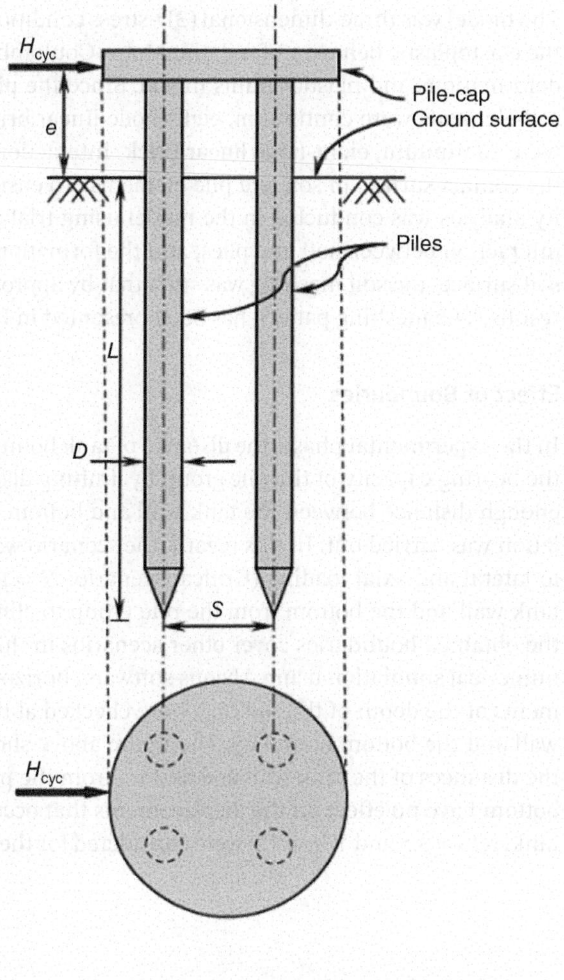

Table 22.12 The mechanical properties of soil and pile dimensions (Basack et al. 2022 / with permission of Elsevier).

Item	Parameter	Value	
		BEM	FEM
Sand	Poisson's ratio, μ_s	0.3	
	Dilation angle, ψ (°)	3	
	Effective cohesion, c' (kPa)	0	
	Layer thickness, h (mm)	250	
	Relative density, D_r (%)	35–88	35–88
Piles	Young's modulus of pile material, E_p (GPa)	195	
	Poisson's ratio, μ_p	0.15	
	Unit weight, γ_p (kN/m^3)	8.0	
	External diameter, D (mm)	5	
	Embedded length, L (mm)	90	
	Dimension of pile cap:		
	Diameter (mm); Thickness (mm)	50; 10	
	Load eccentricity, e (mm)	35	
Cyclic loading	Number of cycles (N)	100–1000	
	Frequency, f (Hz)	0.833–1.583	
	Amplitude, A (%)	30–70	

Finite Element Modeling Procedure

The model was three-dimensional (3D-stress conditions) and consisted of two parts: the soil and the pile group. Considering the elastoplastic behavior of soil, the Mohr–Coulomb failure criterion was implemented as a constitutive model to analyze deformations and plastic strains in soil. Since the piles are steel, the pile group was modeled as an elastic material. The soil elements were continuum, eight-node linear brick reduced integration with hourglass control, and the pile elements were continuum, eight-node linear brick. Interaction between soil piles was defined by tangential and normal stiffness at the contact surface of soil and pile elements. To ensure convergence of the finite element equations, a mesh size sensitivity analysis was conducted in the model using trial-and-error procedures to reach the best mesh size density. Due to the interaction between soil and piles, and the formation of basin-like depressions around the pile group in the vicinity of the soil surface, the soil meshing was such that by approaching the piles, the mesh size gradually decreased to achieve precise results. The meshing pattern has been presented in Figure 22.46.

Effect of Boundaries

In the experimental phase, the distance of tank boundaries (the chamber that contains the soil) from the piles' body affects the bearing capacity of the pile group by limiting displacements around the body and the tip of the piles. In order to place enough distance between the tank wall and bottom from the pile group in the experimental program, a numerical simulation was carried out. In this regard, the scenario was selected in which the soil is going to be disturbed considerably due to lateral and axial loading (Critical scenario: $N = 1000$, $f = 1.583$ Hz, $A = 70\%$). By designating enough distance for the tank wall and the bottom from the pile group to eliminate the boundaries' effects on the results for this critical scenario, the obtained boundaries cover other scenarios in this study as well. By defining semi-infinite elements for boundaries in numerical simulation using Abaqus software, horizontal displacements at the surface of the soil (δ_H) and vertical displacements at the depth of the soil (δ_V) were checked at the end of the critical scenario to obtain the optimum distance for the wall and the bottom boundary. The figure above shows the influence of tank boundaries on displacements. X and Y are the distances of the tank wall and bottom from the pile group, respectively. For $X/S \geq 2.7$ and $Y/L \geq 0.7$, the tank wall and bottom have no effect on the displacements that occurred around the pile group. To have conservative dimensions for the tank, $X/S \approx 3.5$ and $Y/L \approx 1.8$ were considered for the experiment. In this way, X and Y values are equal to 52.5 and 160 mm,

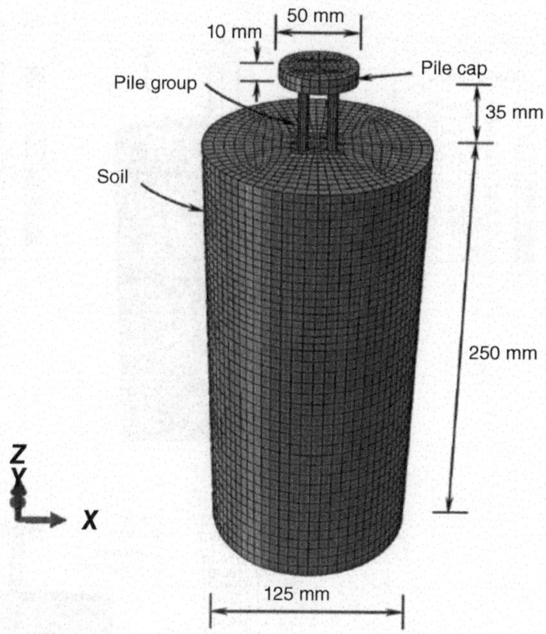

Figure 22.46 The meshing pattern of the piles-soil system (Basack et al. 2022 / with permission of Elsevier).

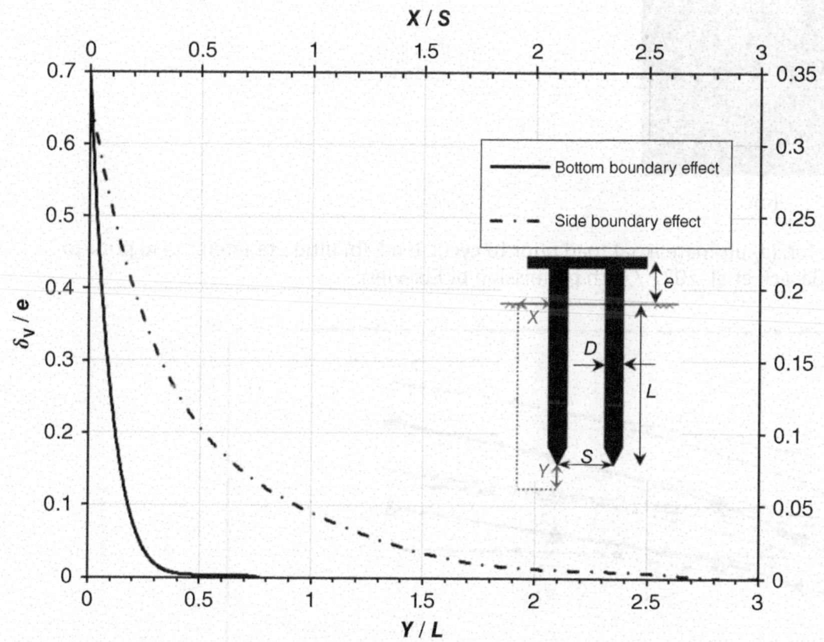

Figure 22.47 Boundary distance validation in experimental using numerical evaluation (Basack et al. 2022 / with permission of Elsevier).

respectively. As a result, due to symmetry in the problem, the diameter and height of the tank are equal to 125 and 250 mm, respectively. Figure 22.47 presents the domain of the boundaries that have a minimum effect on the pile displacement response.

Figure 22.48 shows the numerical results using Abaqus. The experimental and numerical load-displacement responses of the pile group are presented in Figure 22.49. The curves are hyperbolic, and the ultimate capacities of the pile group have been estimated using the double-tangent method. Accordingly, the experimental ultimate axial and lateral pile capacities are found to be 140 and 161 N, while the relevant values in the BEM and FEM analyses have been 160, 162, and 188,

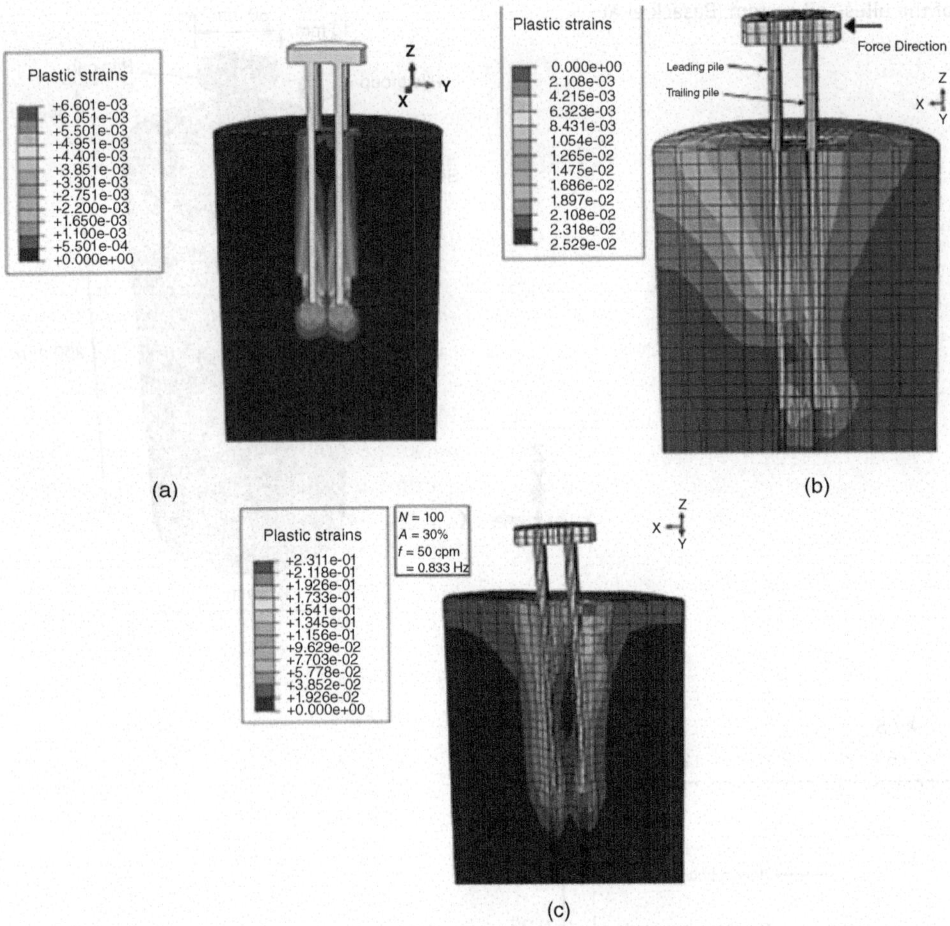

Figure 22.48 Contours of plastic strains at failure for: (a) ultimate axial load prior to cyclic load, (b) ultimate lateral load prior to cyclic load, and (c) ultimate post-cyclic axial load (Basack et al. 2022 / with permission of Elsevier).

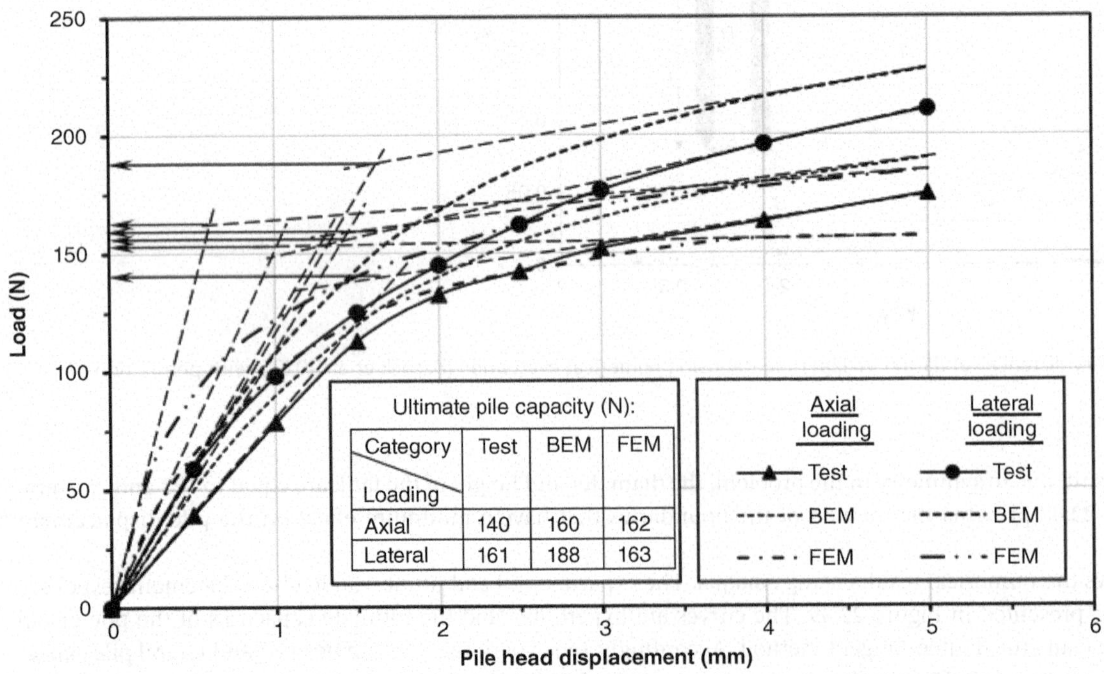

Figure 22.49 Load–settlement curves for vertical and lateral capacities in different methods (Basack et al. 2022).

163 N, respectively; the deviations are 14.3%, 15.7%, 16.8%, and 1.2%, respectively. Thus, the axial capacities obtained by the BEM and FEM analyses are close to the test data, while for lateral capacities, only the FEM result is close to the experimental value.

Example 22.12 ***Vertical Bearing Capacity of Helical Piles***

The objective of this study was to conduct a series of small laboratory tests on model screw piles, which were placed in a bed of soft clayey soil, to examine their load–settlement characteristics under axial loads. Based on the observed data, the ultimate capacities of the model piles were evaluated for different pile diameters and embedded lengths. In addition, a 3D finite element model (3D FEM) was developed using ABAQUS to analyze the stress distributions along the threaded pile surface and failure patterns. The numerical results were compared to the test results and found to be in reasonable agreement. The numerical model was then used to further investigate the stress distributions and failure patterns.

To analyze the axial load–settlement response of helical piles, a finite element model (FEM) was employed with the ABAQUS software. The model was developed by creating a three-dimensional representation of a single helical pile in soil, based on the experimental setup illustrated in Figure 22.50a–d. Elastic and elastoplastic properties were assigned to the pile and soil (Tables 22.13 and 22.14, respectively). The Mohr–Coulomb failure criterion was used to determine soil failure during analysis. Geometric boundary conditions were maintained by fixing the soil sides and the bottom along the X, Y, and Z directions. Interface elements, including tangential and normal stiffnesses, were used to define the interaction between the pile and soil. The pile and soil were meshed using C3D10 elements (10-node tetrahedral elements). The pile was slowly inserted into the soil using a displacement function, and quasi-static analysis was employed to avoid dynamic effects in the model.

The experimental and numerical results are presented herein with adequate analyses and interpretations. In the experimental program, the helical pile is slowly inserted into the test bed by applying torque on the pile head. In the numerical modeling, on the other hand, it is assumed that the pile is already embedded into the soil, for simplification. Thus, in order to approach the experimental results, the effect of the installation is considered using the calibration of the contact parameters between the soil and the pile, such as tangential behavior, normal behavior, and cohesive behavior in the software.

Figure 22.51 presents the load–settlement responses of the three helical piles as obtained from both tests and the FEM analysis. The results show that a logarithmic best-fit curve was the most appropriate with R_2 values ranging from 0.969 to 0.998. The ultimate capacity values were determined from the relevant best-fit curves using the double tangent method. The experimental and numerical curves were quite similar with an average deviation of less than 10%.

Example 22.13 ***Three-dimensional Finite Element Analysis of the Interaction Between Tunneling and Pile Foundations***

This example investigates the interaction between a shallow tunnel and a nearby single pile. The tunnel has a diameter of $D = 7.5$ m and a cover depth of $H = 2.5D$. The lining of the tunnel is expected to behave in a linear-elastic manner with Young's modulus of $E = 35{,}000$ MPa and a Poisson's ratio of $\nu = 0.25$. The lining possesses axial and bending stiffnesses of $K_{sn} = 4978$ MPa and $K_{sf} = 7.4$ MNm^2, respectively. The pile is 22.5 m long and 1 m wide, and it is positioned relative to the tunnel by a distance Δ_{tp} (the distance between the pile axis and the tunnel center) and $i_{tp} = +0.5D$ (the pile tip is located below the horizontal axis of the tunnel). The pile is anticipated to behave in a linear-elastic fashion, with a Young's modulus of $E = 23{,}500$ MPa and a Poisson's ratio of $\nu = 0.25$. The pile possesses axial and bending stiffnesses of EA = 23,500 MN and EI = 1960 MNm^2, respectively. The soil is supposed to behave according to an elastic perfectly-plastic constitutive relationship that is based on the Mohr–Coulomb criterion, with a Young's modulus of $E = 30$ MPa, a Poisson's ratio of $\nu = 0.3$, a cohesion of $C = 5$ kPa, a friction angle of $\varphi = 27$, and a dilatancy angle of $\psi = 5$. Table 22.15 provides a summary of the characteristics of the soil, lining, and pile materials.

Figure 22.52 displays the finite element mesh that was utilized for numerical simulation. The mesh is composed of 3111 20-node isoparametric hexahedral elements and 14,300 nodes. The soil mass is constrained to expand up to $12D$ in the lateral axis (x), $7D$ in the longitudinal axis (y), and $5.5D$ in the vertical axis (z).

The numerical modeling is executed in two stages. The initial stage involves the application of an axial load of $Q = 1000$ kN to the pile head. The subsequent stage involves the construction of the tunnel, which is accomplished in 12 successive steps.

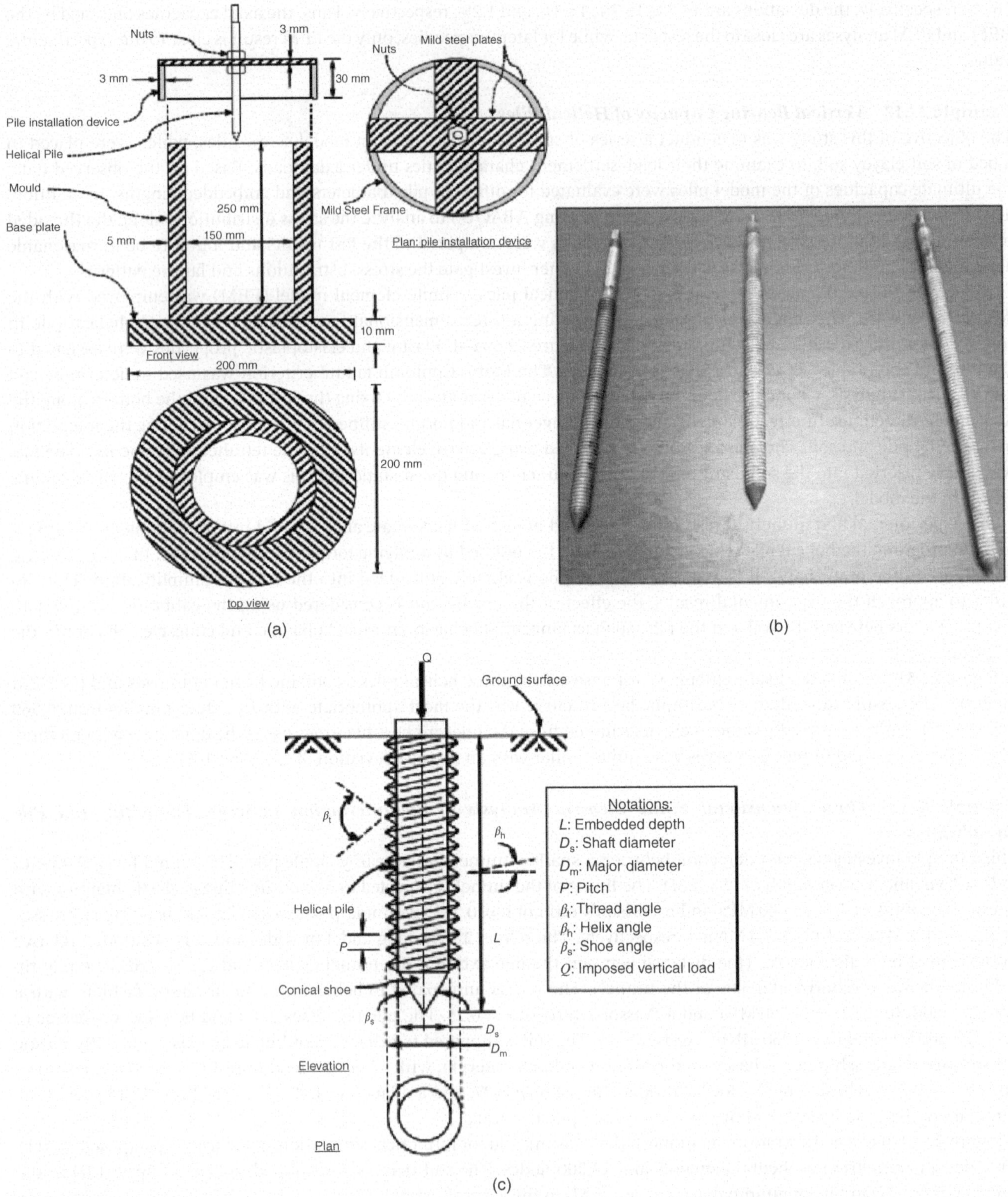

Figure 22.50 (a) Experimental setup, (b) helical piles, (c) definitions of helical pile, and (d) numerical model of pile and soil (Karami et al. 2023b / Springer Nature).

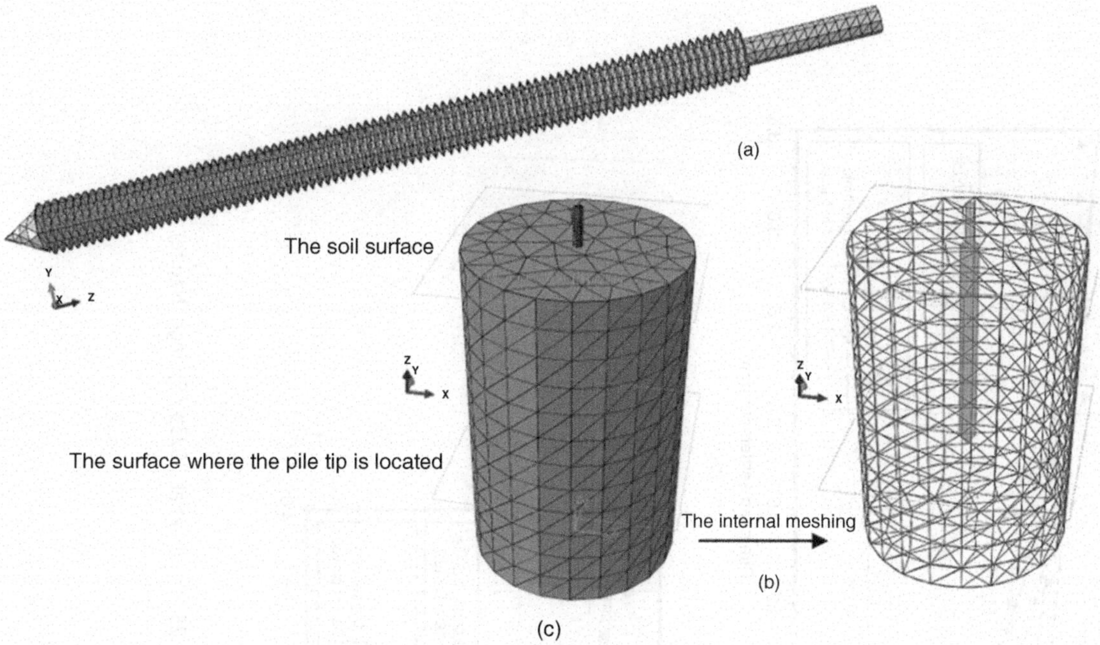

Figure 22.50 (*Continued*)

Table 22.13 Soil properties (Karami et al. 2023b / Springer Nature).

Material	Parameter		Value
Soil	Particle size distribution	Clay	61%
		Silt	33%
		Sand	5.5%
		Gravel	0.5%
	Atterberg limit	Liquid limit	46%
		Plastic limit	17%
	Specific gravity of solid particles		2.8
	Standard proctor compaction test	Optimum moisture content	18.75%
		Maximum dry density	19.1 kN/m^3
	Experimental moisture content		40%
	Unconfined undrained triaxial test	c	21 kPa
		ϕ	15°

Figure 22.53 displays how the position of the pile tip relative to the tunnel horizontal axis (measured by the i_{tp} parameter) impacts the axial force distribution. The axial force distribution is substantially influenced by i_{tp}. When the pile tip is positioned above the tunnel horizontal axis, a continuous increase in axial force is evident due to the downward movement of soil adjacent to the pile. On the other hand, when the pile tip is situated below the tunnel horizontal axis ($i_{tp} > 0$), an increase in axial force up to a maximum N_{max} (axial force) is observed, followed by a decrease caused by the upward movement of the soil situated beneath the tunnel horizontal axis. The bending moment is also greatly affected by the position of the pile tip relative to the tunnel axis. When the pile tip is above the tunnel axis, a low bending moment is induced by tunneling compared to that produced when the pile tip is located below the tunnel horizontal axis. The maximum bending moment occurs when $i_{tp} = 0.5$.

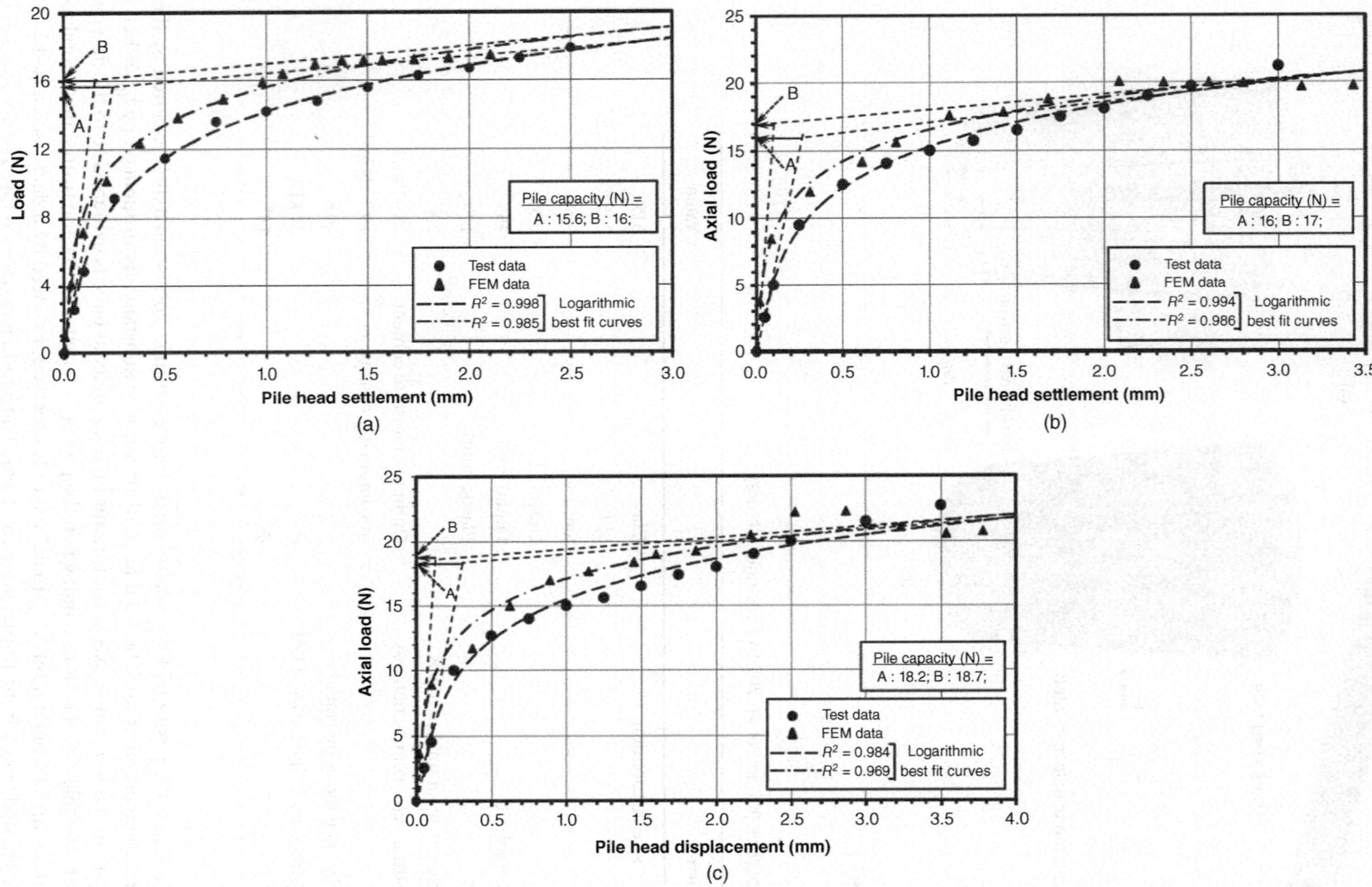

Figure 22.51 Axial load–settlement response of helical piles with L/D_m ratio of: (a) 13.33, (b) 12.14, and (c) 11.25 (Karami et al. 2023b / Springer Nature).

Table 22.14 Pile properties (Karami et al. 2023 / Springer Nature).

	D_m (mm)	D_s (mm)	L (mm)	p (mm)	β_t (mm)	β_s (mm)
Geometrical parameters	12	10.5	160	1.75	60	60
	14	12.5	170	2.0	60	60
	16	14.5	180	2.0	60	60
Strength	Young's modulus (GPa)					889
	Ultimate tensile strength (MPa)					725

Table 22.15 The characteristics of the soil, lining, and pile materials (Mroueh and Shahrour 2002 / John Wiley & Sons).

Material	E (MPa)	v (–)	C (kPa)	φ (°)	Ψ (°)
Soil (sandy)	30	0.3	5	27	5
Liner (concrete)	35, 000	0.25	—	—	—
Pile (concrete)	23, 500	0.25	—	—	—

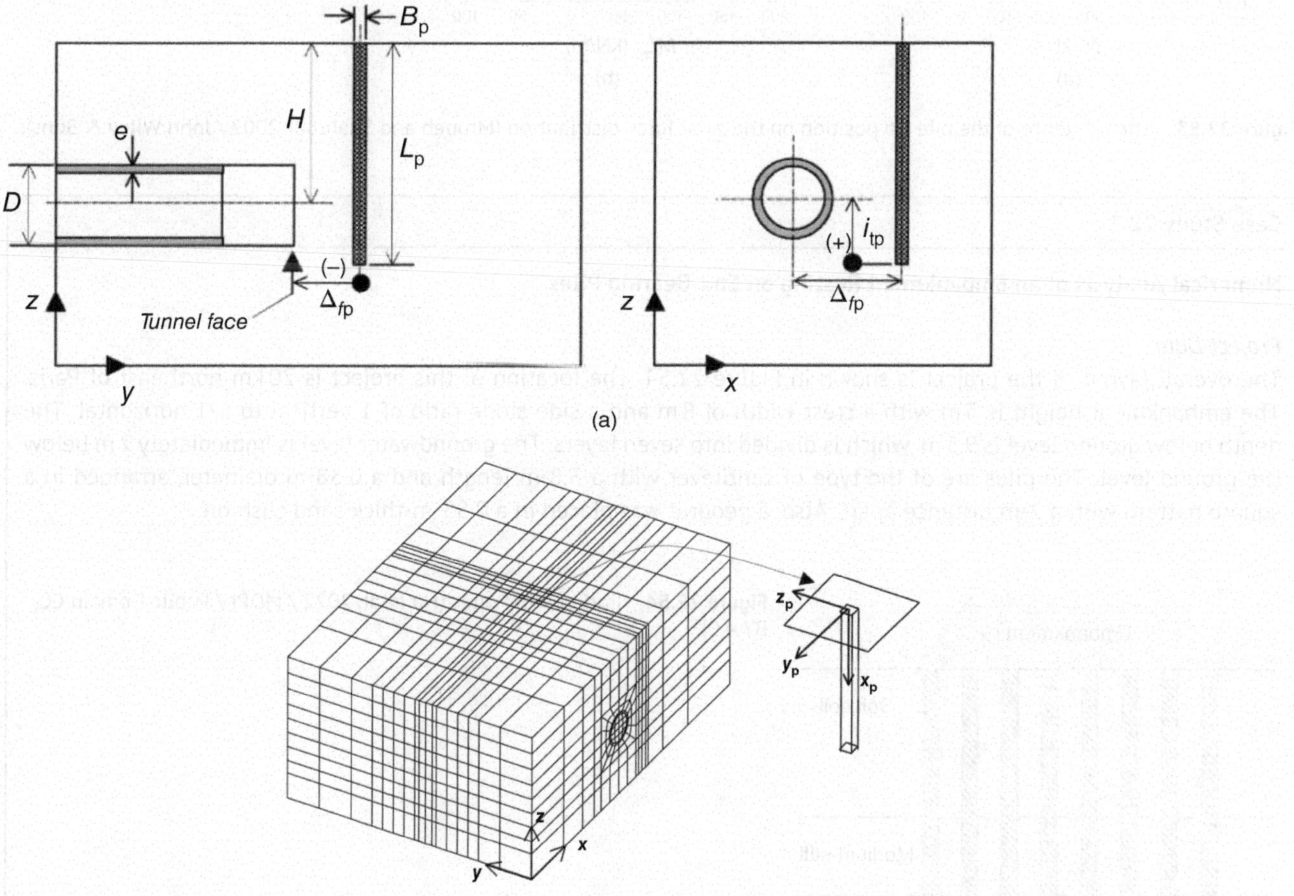

Figure 22.52 (a) Definitions of symbols and (b) FEM meshing pattern (Mroueh and Shahrour 2002 / John Wiley & Sons).

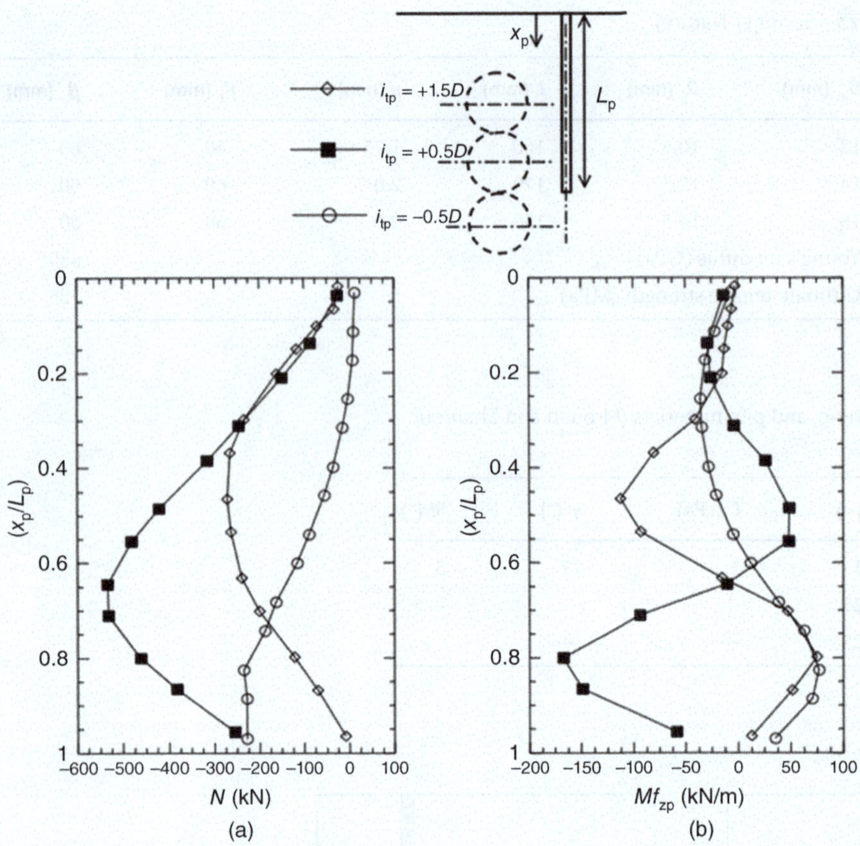

Figure 22.53 The influence of the pile tip position on the axial force distribution (Mroueh and Shahrour 2002 / John Wiley & Sons).

Case Study 22.1

Numerical Analysis of an Embankment Resting on End-Bearing Piles

Project Data

The overall layout of the project is shown in Figure 22.54. The location of this project is 20 km northeast of Paris. The embankment height is 5 m with a crest width of 8 m and a side slope ratio of 1 vertical to 5/1 horizontal. The depth below ground level is 9.5 m, which is divided into seven layers. The groundwater level is immediately 2 m below the ground level. The piles are of the type of cantilever with a 3.8-m length and a 0.38-m diameter, arranged in a square pattern with a 2-m distance apart. Also, a geogrid was placed in a 0.55-m-thick sand cushion.

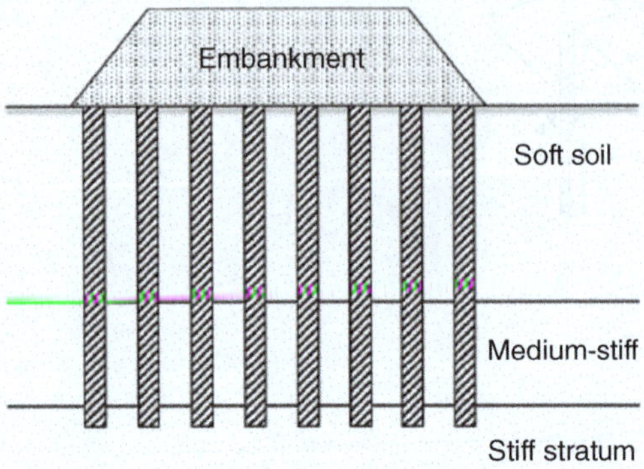

Figure 22.54 End-bearing piles (Hu et al. 2022 / MDPI / Public Domain CC BY 4.0).

Modeling

The 3D modeling was performed using Abaqus software. Figure 22.55 shows the finite element model used in the analysis. The geotechnical parameters of each sublayer are listed in Table 22.16. Isotropic geogrid with a tensile strength of 500 kN/m was considered. The information of the subgrade, embankment fill materials, and piles are also presented in Table 22.17.

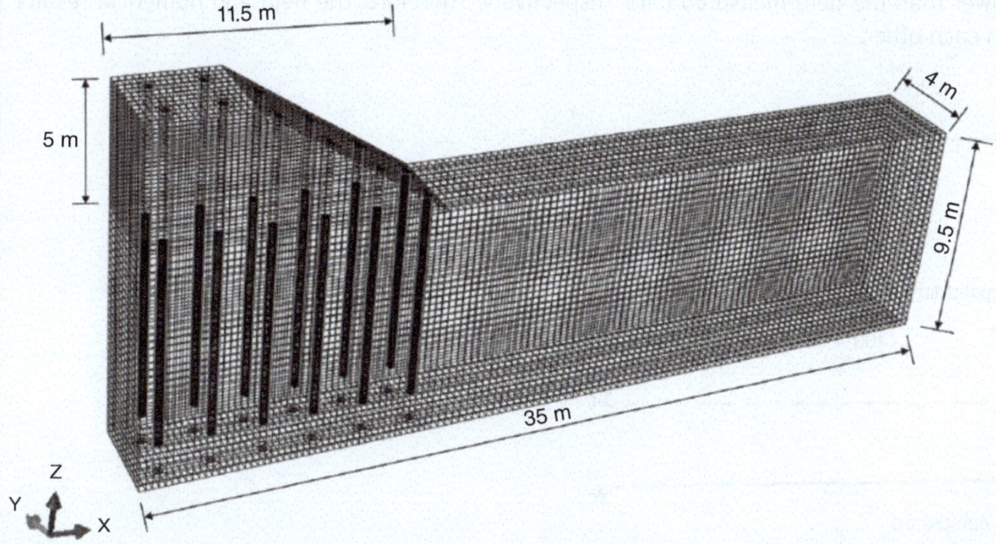

Figure 22.55 Three-dimensional finite element model of end-bearing piles (Hu et al. 2022 / MDPI / Public Domain CC BY 4.0).

Table 22.16 Geotechnical parameters of each soil layer (Hu et al. 2022 / MDPI / Public Domain CC BY 4.0).

Soil layer	Material	(kN/m³)γ	′(°)ϕ	e_0	υ	λ	κ	k (m/s)
S1	Clay backfill	20	30.6	0.75	0.35	0.092	0.014	—
S2	Clay backfill	20	30.6	1.3	0.36	0.191	0.029	—
S3	Beige clayey	14	30.6	2	0.37	0.308	0.046	8.8×10^{-8}
S4	Gray sand lightly	19	30.6	0.7	0.35	0.071	0.011	1.0×10^{-6}
S5	Plastic clay	20.5	27	0.9	0.32	0.116	0.017	6.1×10^{-7}
S6	Sandy clay	20.5	27	0.8	0.32	0.088	0.013	6.1×10^{-7}
S7	Gravel	21	34	0.75	0.31	0.027	0.004	6.6×10^{-5}

Table 22.17 Mechanical properties of subgrade, embankment fill, and piles (Hu et al. 2022 / MDPI / Public Domain CC BY 4.0).

Material	(kN/m³)γ	*E* (MPa)	υ	c′ (kPa)	′ (°)ϕ
Embankment	18.5	20	0.3	10	30
Gravel (cushion)	20	70	0.3	60	36
Pile	24	20,000	0.2	18,000	—

The above tables include γ, specific weight, ν, Poisson's ratio, λ, logarithmic plastic bulk modulus, κ, logarithmic bulk modulus, c', effective cohesion, ϕ', effective friction angle, k, permeability coefficient, e^0, initial porosity ratio, and E, elastic modulus.

(Continued)

Case Study 22.1 (Continued)

Analysis of Results

Considering that the final settlement was estimated in the field measurement, the results of the finite element analysis were compared with the field-measured settlement. As Figure 22.56 shows, the final settlement of the top surface of the pile in the finite element results was 31 mm, and the surface below it (under the soil) was 61.2 mm, which was 10.7% higher and 6.2% lower than the field-measured data, respectively. Therefore, the field and numerical results had good agreements with each other.

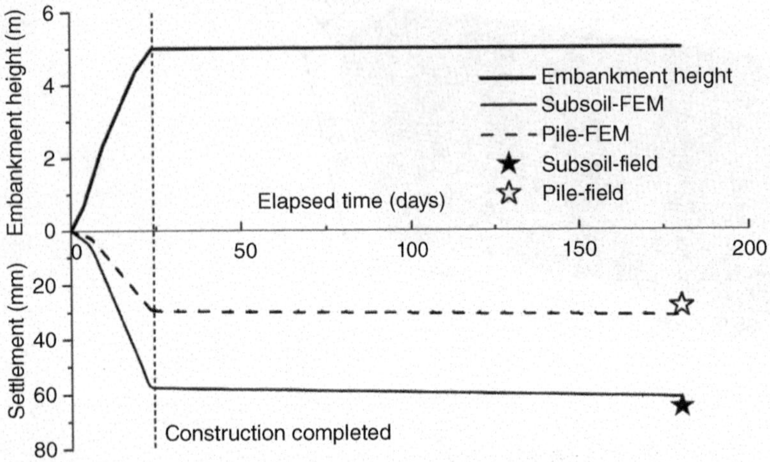

Figure 22.56 Height of embankment vs. time and settlement vs. time measured at the top of the pile and below the soil surface (Hu et al. 2022 / MDPI / Public Domain CC BY 4.0).

Figure 22.57 also shows the variation of vertical stresses applied to the center of the soil layer surrounded by four adjacent piles vs. the elapsed time for the FEM analysis, field data, and the main numerical results presented by Rowe and Liu (2015). The numerical analysis results have a very good agreement with the field data. With these comparisons, the accuracy of the numerical analysis is also confirmed.

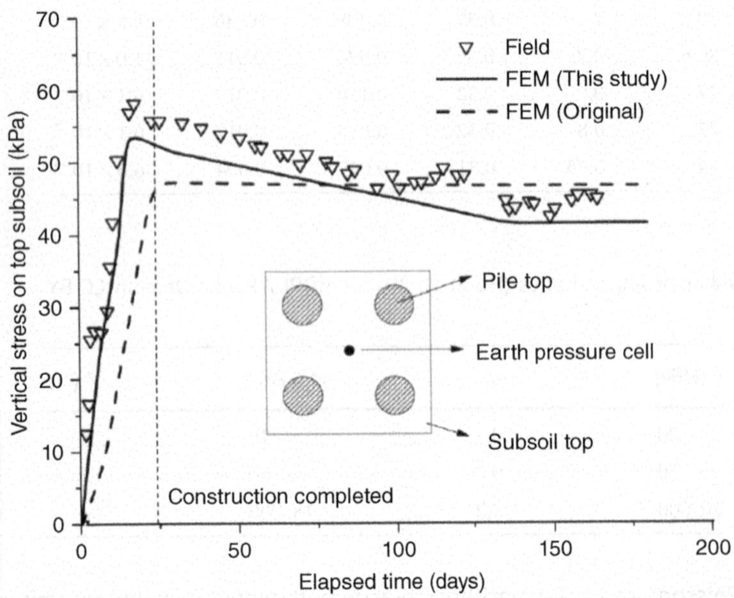

Figure 22.57 Vertical stress on the bottom surface of the soil vs. time (Hu et al. 2022 / MDPI / Public Domain CC BY 4.0).

Case Study 22.2

Numerical Analysis of a Retaining Wall Supporting a Steep Slope

A high embankment supported by a combination retaining structure (CRS) was constructed near the Lanai expressway tunnel between Jiaou and Libo in Guizhou province, China. The slope has an angle of approximately 55° with the horizon after construction. Figure 22.58 shows a view of the project, and Figure 22.59 shows a schematic of the constructed combination structure. The layers from the surface to the bottom include a granular soil layer, a highly weathered limestone layer, and a very thick limestone layer with moderate weathering. The granular soil layer is approximately 0–6 m thick and consists of 70% rock and 30% silt. The highly weathered limestone layer is classified as soft and jointed rock and is approximately 0.9–3.2 m thick.

Figure 22.58 Site conditions for the elevated embankment supported by CRS: (a) before embankment construction and (b) after embankment construction (Xie et al. 2020 / American Society of Civil Engineers).

Figure 22.59 Schematic of the composite structure built on a steep slope (Xie et al. 2020 / American Society of Civil Engineers).

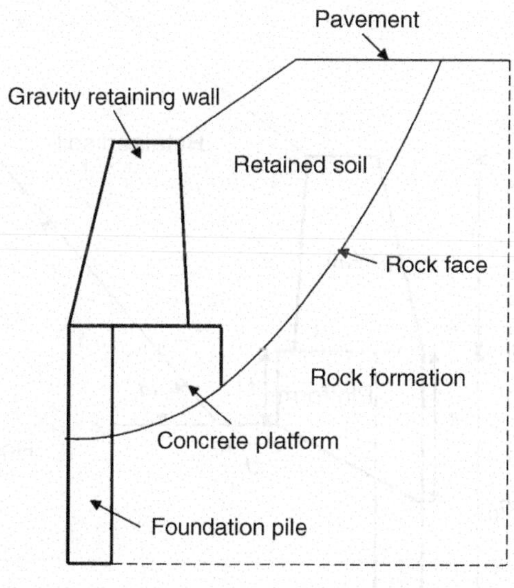

Problem Data

Due to the steep slope in this location, a special composite construction method was chosen and implemented. The sandy layer was removed before constructing the CRS. Six rectangular piles with the same cross-sectional area (2.4 m × 1.8 m) but different lengths (variable from 9 to 16 m) were installed in the rock formation with varying thicknesses (variable from 5.5 to 8 m). The distance between the center of two adjacent piles is 6 m. A concrete platform (width of 7.5 m) was constructed to connect the upper console sections of the piles, which extended from the surface of the moderately weathered limestone layer to the top of the piles. A gravity retaining wall with a top width of 3.5 m and a bottom width of 6.5 m was built on the platform. The length of the wall is 36 m, and its height is 10 m (Figure 22.60).

(Continued)

Case Study 22.2 (Continued)

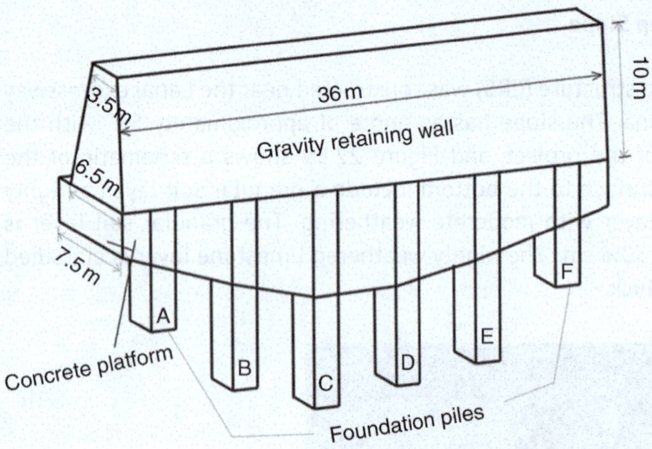

Figure 22.60 Side view of the problem configuration (Xie et al. 2020 / American Society of Civil Engineers).

The piles are made of reinforced concrete with a yield moment of 3.98*104 kN/m. Figure 22.61 shows the cross-sectional geometry of the embankment. There are five variables to determine the cross-sectional geometry, including pile length (h_p), angle of rock surface slope (ω), internal slab thickness (l), external slab thickness (s), and horizontal distance between the bottom of the internal slab edge and the slope facing the rock (d). Table 22.18 shows the values of these five variables for six cross-sections of the project.

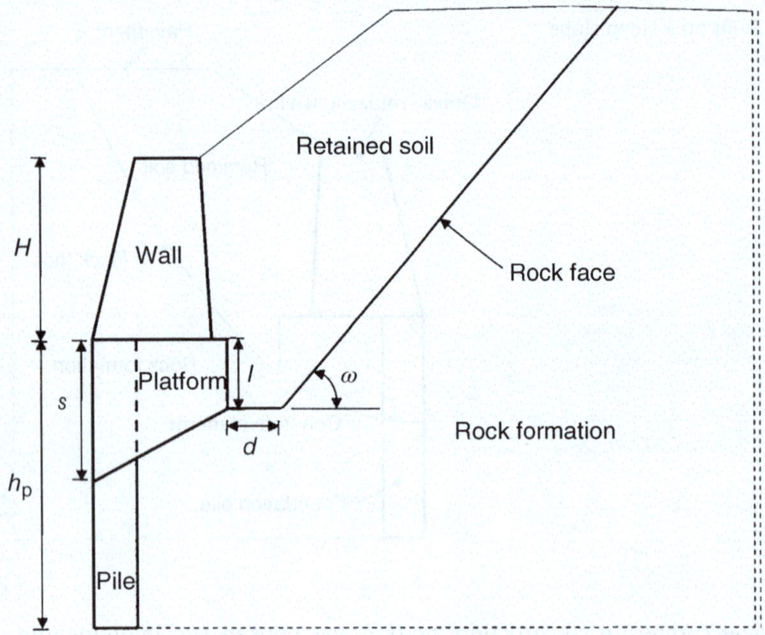

Figure 22.61 Cross-sectional view of the desired embankment (Xie et al. 2020 / American Society of Civil Engineers).

Numerical Modeling

Numerical simulation has been performed using the FLAC 3D software. The mesh configuration for section C is shown in Figure 22.62. The mechanical behavior of the rock formation, anchors, platform, and retaining wall was elastic. Table 22.19 presents the input parameters for the retaining wall, anchors, platform, retaining soil, and rock formation.

Table 22.18 Geometric information for six cross sections along the project (Xie et al. 2020 / American Society of Civil Engineers).

	Section					
Parameter	**A**	**B**	**C**	**D**	**E**	**F**
ω (°)	55	54	52	51	54	55
d (m)	0	2	3	1	0	0
l (m)	2	4	5	4	3	0
s (m)	3.5	7	8	7	6	2.5
h_p (m)	9	14	16	15	14	9

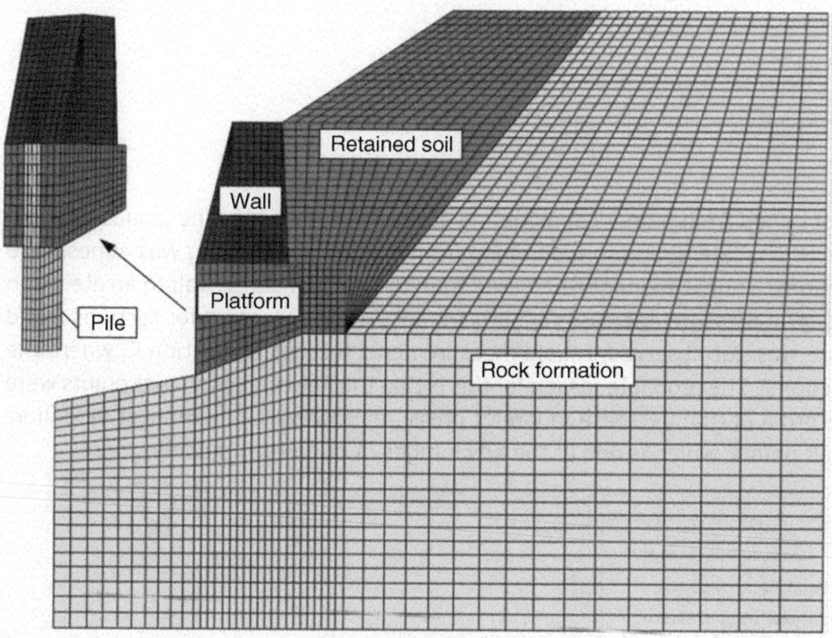

Figure 22.62 The simulated model for embankment stabilization using the combined method ($\omega = 52°$, $l = 5$ m, $d = 3$ m, $s = 8$ m) (Xie et al. 2020 / American Society of Civil Engineers).

Table 22.19 Properties of materials used in numerical modeling (Xie et al. 2020 / American Society of Civil Engineers).

Parameters	**Rock formation (elastic)**	**Pile (elastic)**	**Platform (elastic)**	**Wall (elastic)**	**Retained soil (Mohr–Coulomb)**
Unit weight (kN/m^3)	20.5	25	24	24	19.8
Cohesion (kPa)	—	—	—	—	0
Friction angle (°)	—	—	—	—	40
Modulus of elasticity (MPa)	11,000	30,000	22,000	21,000	59
Poisson's ratio	0.26	0.2	0.2	0.2	0.3

(Continued)

Case Study 22.2 (Continued)

For the common section element, the value of the friction angle for the joint has also been considered based on various references and available recommendations equal to the values in Table 22.20.

Table 22.20 Different values of friction angles for common surfaces (Xie et al. 2020 / American Society of Civil Engineers).

Angle (degree)	Element of common interface
28	Between the soil and the retaining wall. $\delta_1 = 0.7\varphi$
38	Between soil and rock $\delta_1 = 0.95\varphi$
35	Between the concrete slab and the concrete retaining wall. $\delta_1 = 0.7\varphi$
31	Between concrete slab and rock
39	Between pile and rock

Results

Figure 22.63 shows the lateral pressure obtained from numerical modeling and field monitoring. The results of these two methods have considerable agreement. The lateral pressure obtained from numerical modeling was almost zero at the bottom of the wall for all cross-sections, while it suddenly increased from the bottom of the wall to an elevation of 2 m. From an elevation of 2–4 m, the lateral pressure decreases slightly in most sections (except for sections A and F). However, a decrease in lateral pressure was not observed in field measurements (except for section C, where the decrease was greater than in the other sections). The probable reason for this is that the field measurement points were installed at 2-m intervals and did not record a partial decrease in lateral pressure. However, in numerical modeling, it is possible to predict the proximity of all points, which is one of the advantages of numerical modeling.

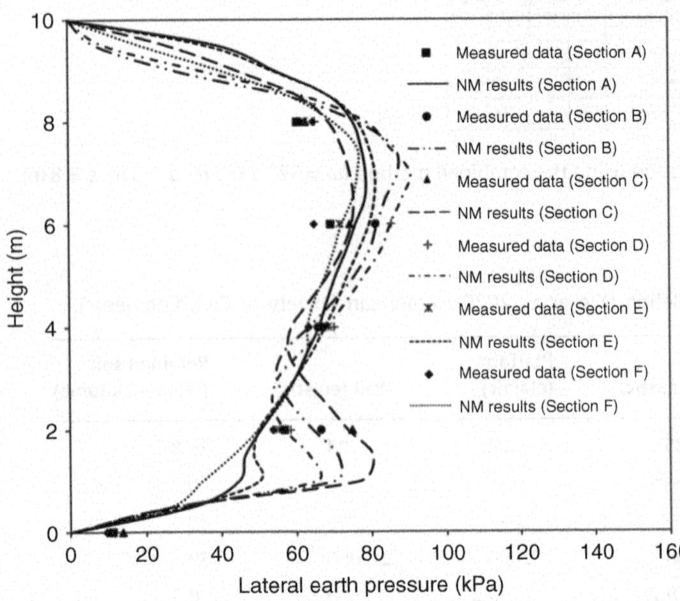

Figure 22.63 Comparison of lateral earth pressure obtained from field monitoring and numerical modeling (Xie et al. 2020 / American Society of Civil Engineers).

In addition, in the numerical model, many required data such as maximum shear strain in the soil, lateral and vertical displacement of the soil, changes in lateral pressure due to the displacement of the barrier wall (Figure 22.64), and many other cases can be observed.

Figure 22.64 Lateral pressure changes due to displacement of the wall for section C (Xie et al. 2020 / American Society of Civil Engineers).

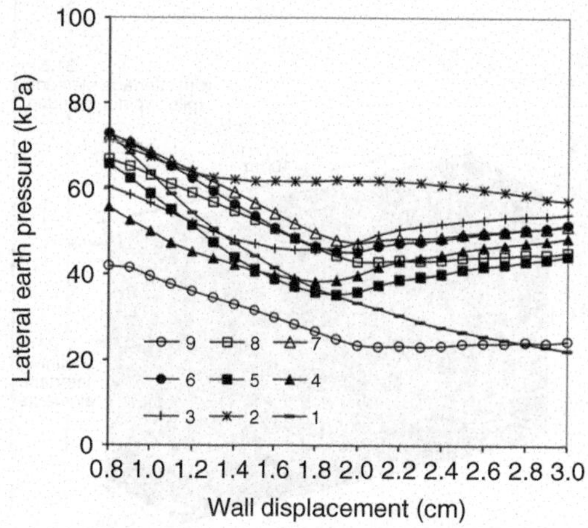

Case Study 22.3

Finite Element Analysis of Tunnel–Soil–Pile Interaction

Pile foundations located close to newly constructed tunnels may experience significant additional loads. In order to accurately calculate the bending and axial stresses induced by tunneling, it is crucial to predict the magnitude and shape of ground displacements that lead to curvature changes. However, the finite element simulation of tunneling by removing initial stress-state forces tends to predict incorrect ground displacement profiles and, as a result, incorrect forces in pile foundations. To address this challenge, this paper presents a simple and effective displacement-controlled model (DCM) that can be used to predict the effects of tunnel excavation on adjacent pile foundations. The DCM works by simulating tunneling through the application of displacements to the tunnel boundary. The paper proposes a method to determine the magnitude and direction of tunnel boundary displacements based on convergence patterns observed in field and centrifuge test results. Back analyses of numerous greenfield tunnel case histories using the DCM show that there is good agreement between the computed displacement profiles and the field/test data. The suitability of the DCM in modeling tunnel–soil–pile interaction problems is demonstrated through back analysis of a centrifuge test and a field case study. Figure 22.65 indicates the 3D numerical model built in ABAQUS software.

Results

Figure 22.66 compares the obtained numerical results with the field data. The maximum induced bending moment for pile P_1, as computed and presented in Figure 22.66a, is observed to occur after the excavation of the SB tunnel. This result is consistent with the fact that the curvature of pile P1 decreases due to reversing the lateral soil movements following the excavation of the second NB tunnel. Meanwhile, the induced axial force for both piles increases after the excavation of the NB tunnel as expected, due to soil settlement that contributes to the development of soil–pile interfacial friction. Figure 22.66b shows the induced axial force for pile P2, along with the measured maximum bending moment and axial forces, which are in good agreement.

(Continued)

Case Study 22.3 (Continued)

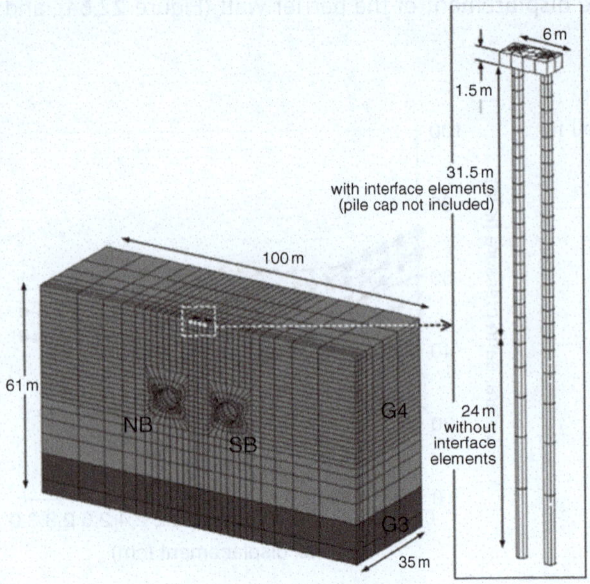

Figure 22.65 FEM model of the problem (Cheng et al. 2007 / with permission of Elsevier).

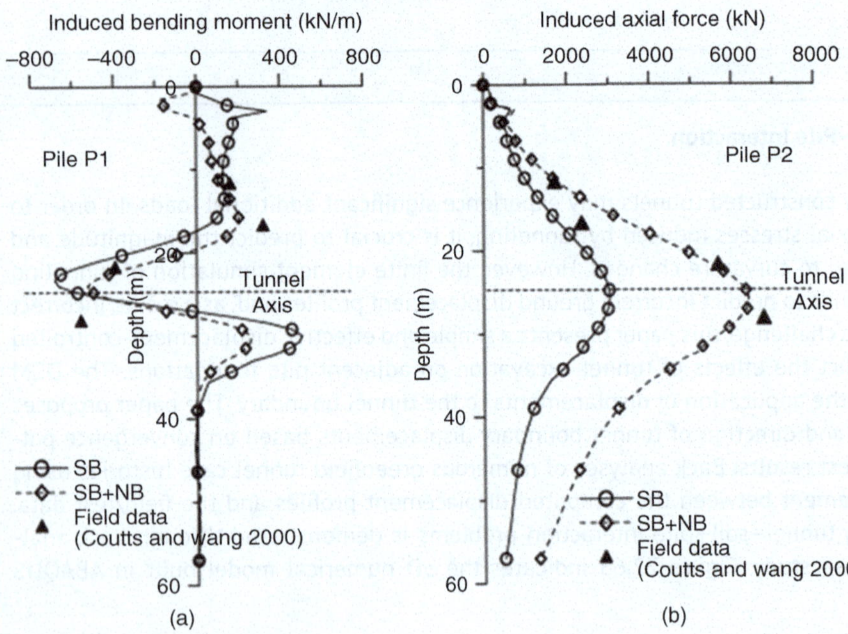

Figure 22.66 Induced (a) bending moment – P1 and (b) axial force – P2 with NB and SB tunnel excavation (Cheng et al. 2007 / with permission of Elsevier).

Case Study 22.4

Bearing Capacity of Shallow Foundations on Unsaturated Soils

The study investigates the ultimate bearing capacity of shallow foundations on unsaturated soils by considering the suction-dependent effective stress. The research presents two approaches: the first is an analytical solution that extends Vesic's solution for saturated soils to incorporate the effect of matric suction on ultimate bearing capacity. The solution introduces a modification factor as a nonlinear function of matric suction to adjust the cohesion-dependent component in the bearing capacity equation. The second approach involves integrating the unsaturated effective stress state with the suction-dependent cohesion into a 3D finite-difference code. This approach considers the variation of matric suction with soil depth and the dependency of saturation degree on soil suction. Four experimental data series from physical plate load tests were used to validate both approaches with water retention curves used as input parameters. The results (Figure 22.67) indicate a good agreement between analytical, numerical, and experimental data, particularly for sandy soils. Table 22.21 shows physical modeling and in situ test properties.

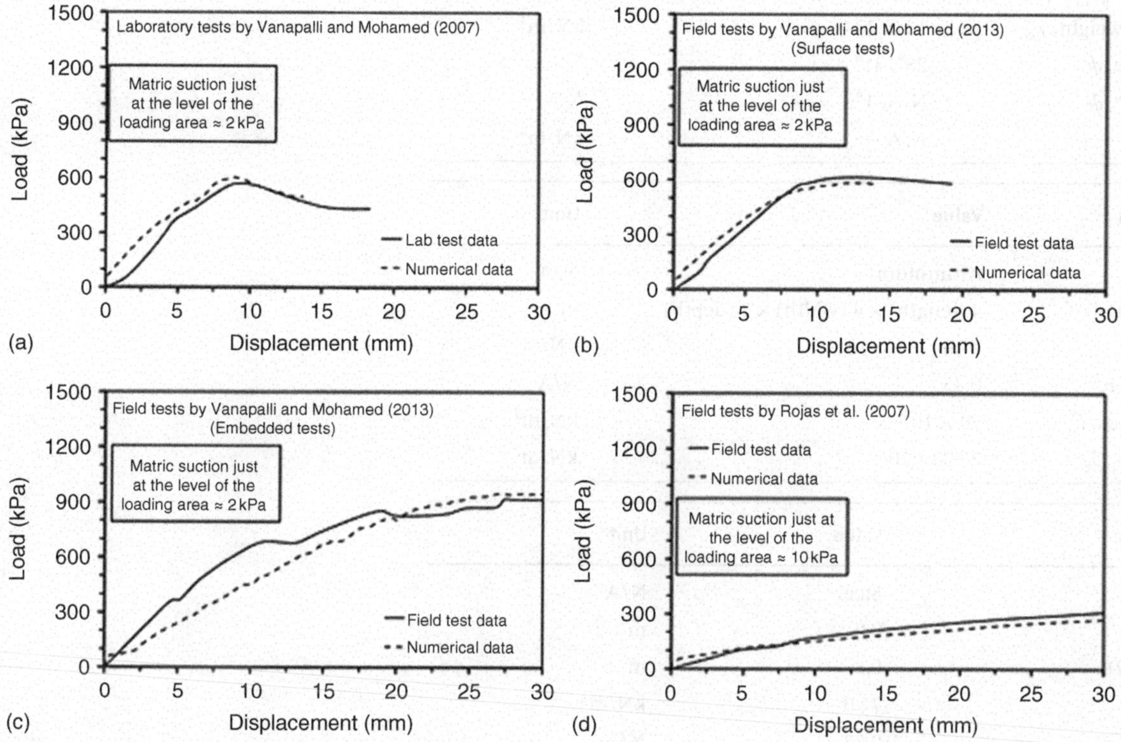

Figure 22.67 Validation of the load–displacement curves in numerical modeling and field tests in four different scenarios: (a) test 1, (b) test 2, (c) test 3, and (d) test 4 (refer to Table 22.21) (Akbari et al. 2020 / American Society of Civil Engineers).

Table 22.21 Soil and in situ test properties (Akbari et al. 2020 / American Society of Civil Engineers).

Property	Test #1	Test #2	Test #3	Test #4
Shape and size of the loading area	Square10 × 10 cm^2	Square15 × 15 cm^2	Square15 × 15 cm^2	CircleD ¼31 cm
Depth of testing (m)	0	0	0.15	0
Soil type	Sand	Sand	Sand	Clay
Soil density (kN/m^3)	16	16	16	16
Soil suctions below the loading area, ψ, (kN/m^3)	0, 2, 4, 6	0, 2, 4, 6	0, 2, 6	0, 10, 48, 56, 60, 63
Soil cohesion c' (kPa)	0.6	0.6	0.6	0.6
Soil friction angle, φ' (°)	35.3	35.3	35.3	26

(Continued)

Case Study 22.4 (Continued)

Table 22.22 Properties of pile, raft, superstructure, and soil (Chanda et al. 2020 / with permission of Elsevier).

Soil parameters	Value	Unit
Soil condition	Loose Sand, Medium Sand	N/A
Young's modulus, E_s	7,000, 13,000	kN/m^2
Poisson's ratio, υ_s	0.3	N/A
Saturated unit weight, γ_{sat}	18.0, 20.0	kN/m^3
Frictional angle, ϕ	25°, 31°	°
Dilatancy angle, ϕ	N/A, 1°	°
Cohesion, c	N/A	kN/m^2

Raft parameters	Value	Unit
Material	Aluminum	N/A
Raft dimension	4 (length) × 4 (width) × 1 (depth)	m
Unit weight, γ	27.0	kN/m^3
Poisson's ratio, υ_r	0.35	N/A
Young's modulus, E_r	70×10^6	kN/m^2
Shear modulus, G_r	2.593×10^7	kN/m^2

Pile parameters	Value	Unit
Material	Steel	N/A
Pile length, L	9.0	m
Pile diameter, D	0.5	m
Unit weight, γ	77.0	kN/m^3
Poisson's ratio, υ_p	0.33	N/A
Young's modulus, E_p	20×10^7	kN/m^2
Cross sectional area, A_p	0.196	m^2
Moment of inertia, I_p	0.003068	m^4

Vertical column parameters	Value	Unit
Material	Steel	N/A
Column diameter, D	0.5	m
Column length, L	3.0	m
Unit weight, γ	77.0	kN/m^3
Poisson's ratio, υ_c	0.33	N/A
Young's modulus, E_c	20×10^7	kN/m^2
Cross sectional area, A_c	0.196	m^2
Moment of inertia, I_c	0.003068	m^4

Case Study 22.5

Behavior of Piled Raft Foundation in Sand Subjected to Combined VMH Loading

The current research emphasizes how the interplay of combined vertical, moment, and horizontal loads affect the stationary behavior of a piled raft foundation that is placed in uniform sandy soil. The study includes a thorough analysis of various factors that have an impact, such as altering the relative stiffness of the raft (k_{rs}), changing the foundation's length-to-width ratio (L/B), adjusting the spacing in relation to the pile diameter (S/D), modifying the way the pile heads are connected to the raft, and varying the ratio of Young's modulus between the piles and the soil (E_p/E_s). The material properties for the soil, raft foundation, pile, and structural elements are presented in Table 22.22. This investigation aims to understand how these factors influence the response of the pile-raft system under static vertical, moment, and horizontal interactions. Furthermore, the research also provides two-dimensional design charts that can be used to estimate lateral load, considering the changes in these parameters (Figure 22.68).

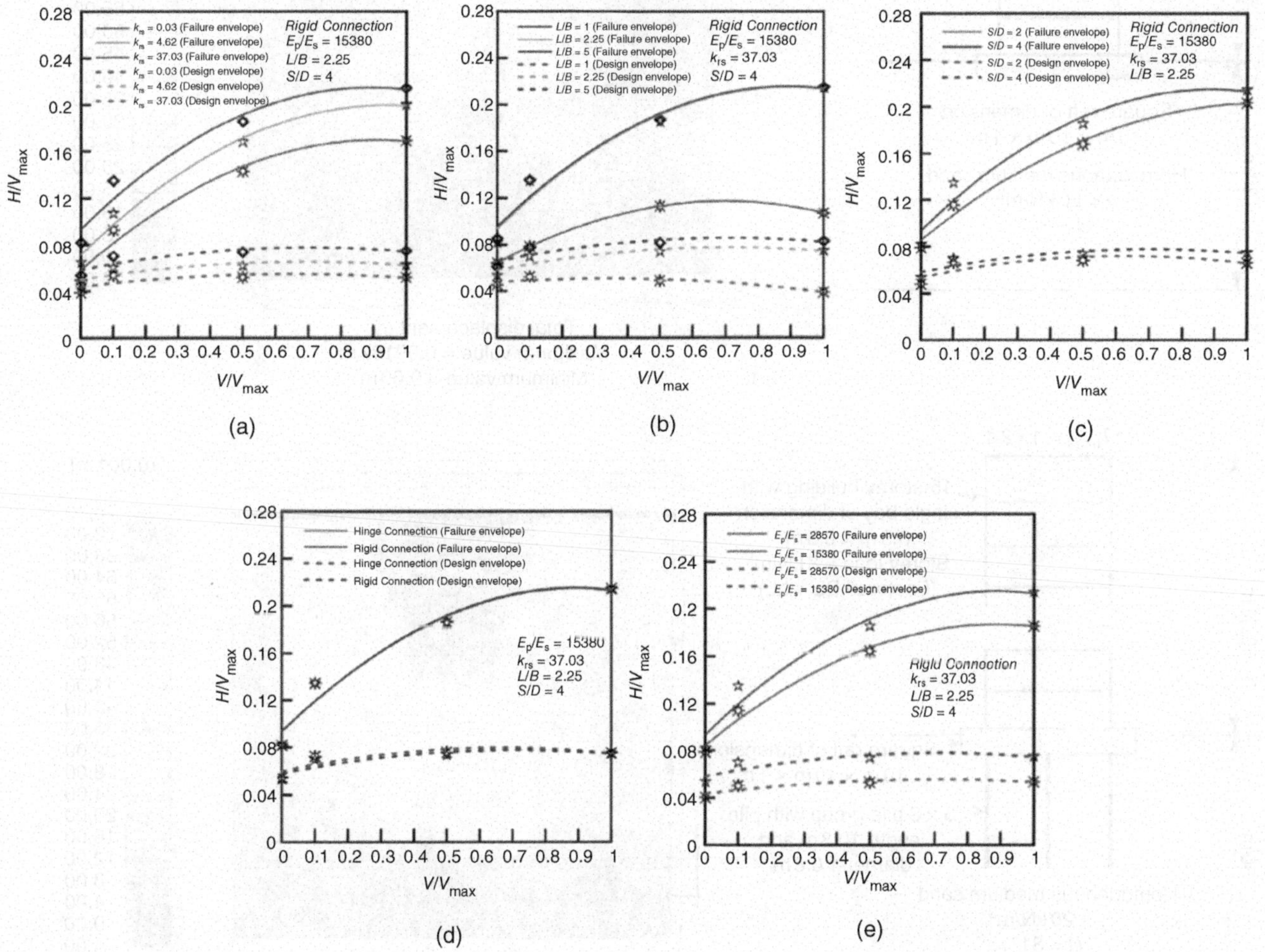

Figure 22.68 Lateral failure and design envelopes for piled raft foundations considering different parametric variations, viz.: (a) k_{rs}, (b) L/B ratio, (c) S/D ratio, (d) pile head connection, and (e) E_p/E_s ratio (Chanda et al. 2020 / with permission of Elsevier).

(Continued)

Case Study 22.5 (Continued)

After a series of parametric studies and simulations lateral failure and design envelopes for piled raft foundations considering different parametric variations have been presented (Figure 22.68). Then, as a case study, this study focuses on the behavior of a 15-story building supported by a piled raft foundation in medium sandy soil. The building

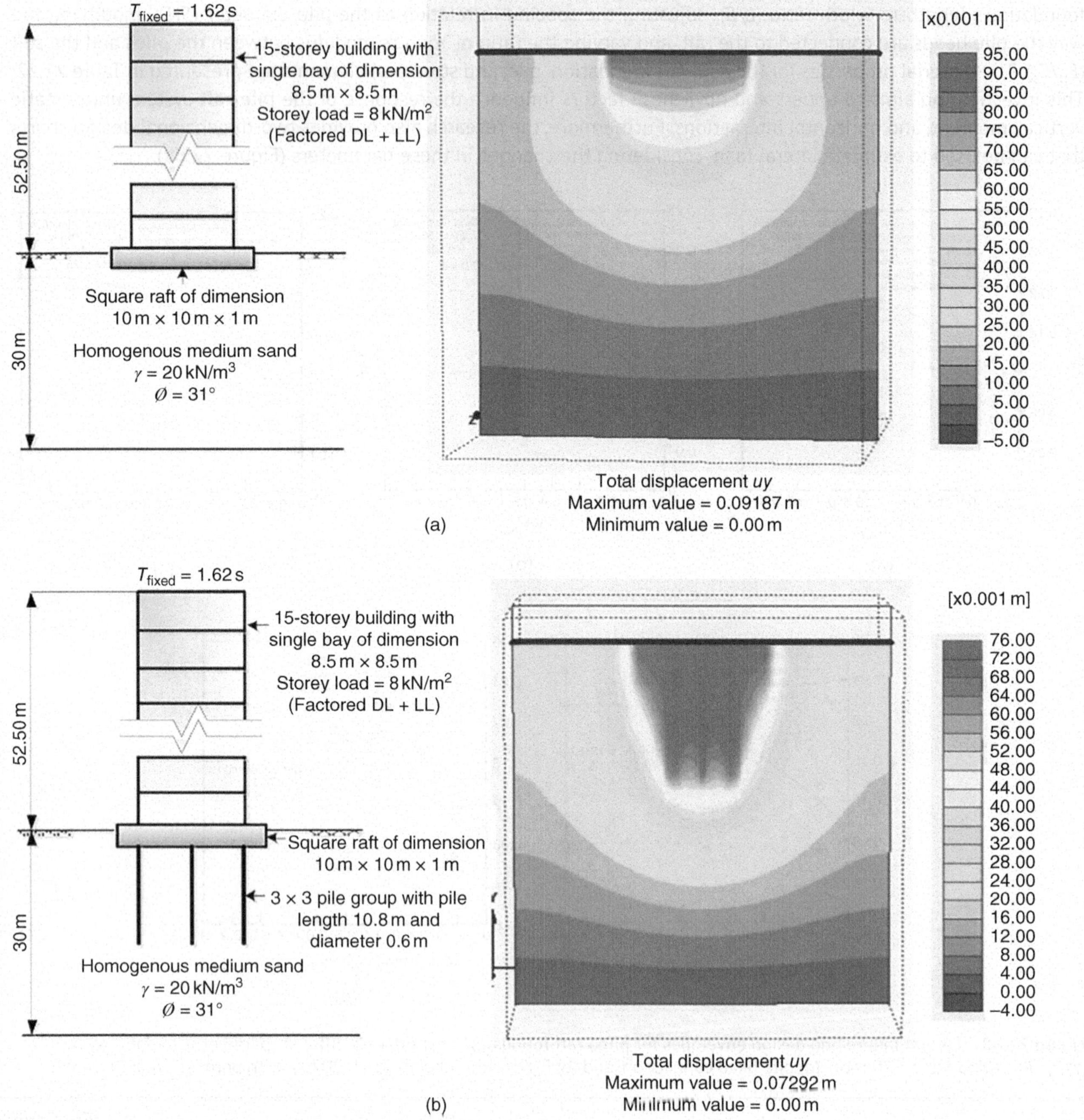

Figure 22.69 Settlement contour of a case study representative 15 storey building structure supported on: (a) raft foundation and (b) piled raft foundation under application of design vertical load (Chanda et al. 2020 / with permission of Elsevier).

dimensions are 8.5 m × 8.5 m per floor, with each floor being 3.5 m high. The initial design is conducted under gravity loading at 8.0 kN/m^2 per floor (including Dead Load and Live Load factors), resulting in a total vertical load of 9248 kN due to building and roof weight. The raft dimensions are 10 m × 10 m × 1 m. The raft demonstrates a satisfactory safety factor (FS > 3) against shear failure under vertical loads. However, the calculated raft settlement under this loading is 91.87 mm, exceeding the allowable 75 mm limit by Bowels (1997). To mitigate this, a 3 × 3 pile group, each pile being 10.8 m long and 0.6 m in diameter, is suggested, resulting in a 72.92 mm settlement reduction (about 20.63%). The ultimate vertical load capacity (V_{max}) of this 3 × 3 piled raft is computed at 67,000 kN using PLAXIS 3D (2008).

The subsequent analysis focuses on lateral load-carrying capacity. The design lateral capacity under no applied lateral load (V/V_{max} = 0) is determined as 3630 kN, allowing 30 mm lateral displacement. The proposed dimensionless envelope approach agrees closely with PLAXIS 3D results (discrepancy of 7%). However, when considering combined vertical, moment, and horizontal loading, the actual lateral capacity (H) of the 3 × 3 pile group is 4490 kN at V/V_{max} = 0.14, about 24% higher than that under purely vertical loading. This implies a conservative design for lateral capacity under different loading conditions, influenced by the safety factor with respect to vertical loading.

To achieve an economical design, a redesign of the 3 × 3 piled raft foundation is carried out, aiming for a target lateral capacity of 3630 kN (under V/V_{max} = 0) and a settlement limit of 75 mm. Multiple iterations lead to a revised 3 × 2 pile configuration, maintaining the same pile length and diameter but adjusting the inter-pile spacing to five times the diameter. Settlement marginally increases from 72.92 to 73.90 mm, and the lateral design capacity reaches 4000 kN for the 3 × 2 piled raft foundation under V/V_{max} = 0.14, surpassing the 3 × 3 configuration's capacity under purely vertical loading.

Overall, the study's revised design strategy reduces pile numbers while maintaining economic efficiency and foundation stability (Figures 22.69–22.71).

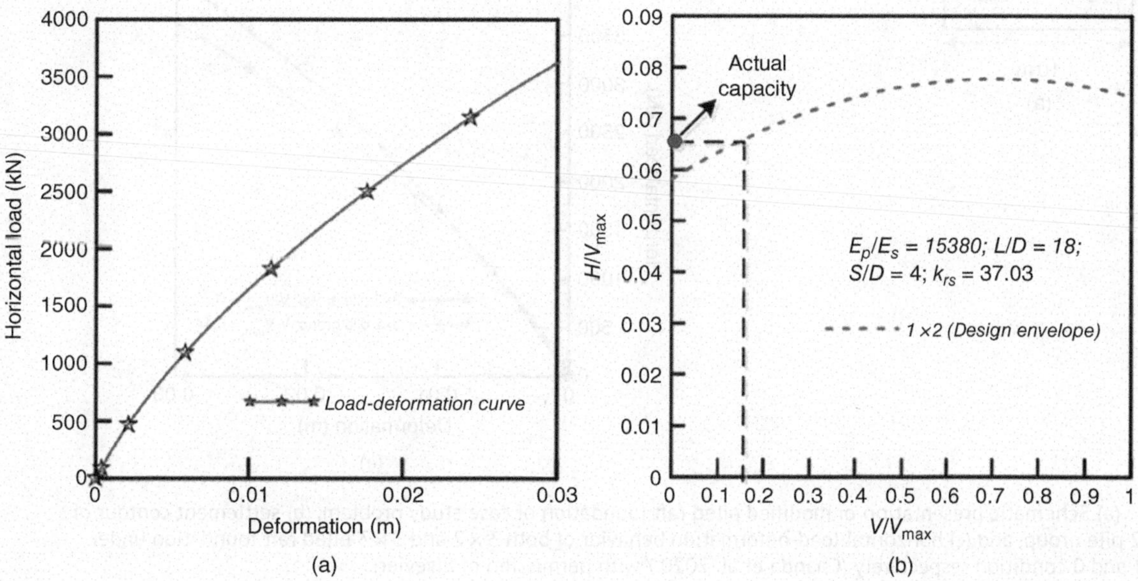

Figure 22.70 (a) Horizontal load deformation of case study piled raft system under sole lateral loading and (b) Actual design lateral capacity calculation for case study foundation using design envelope proposed in the present study (Chanda et al. 2020 / with permission of Elsevier).

(Continued)

Case Study 22.5 (Continued)

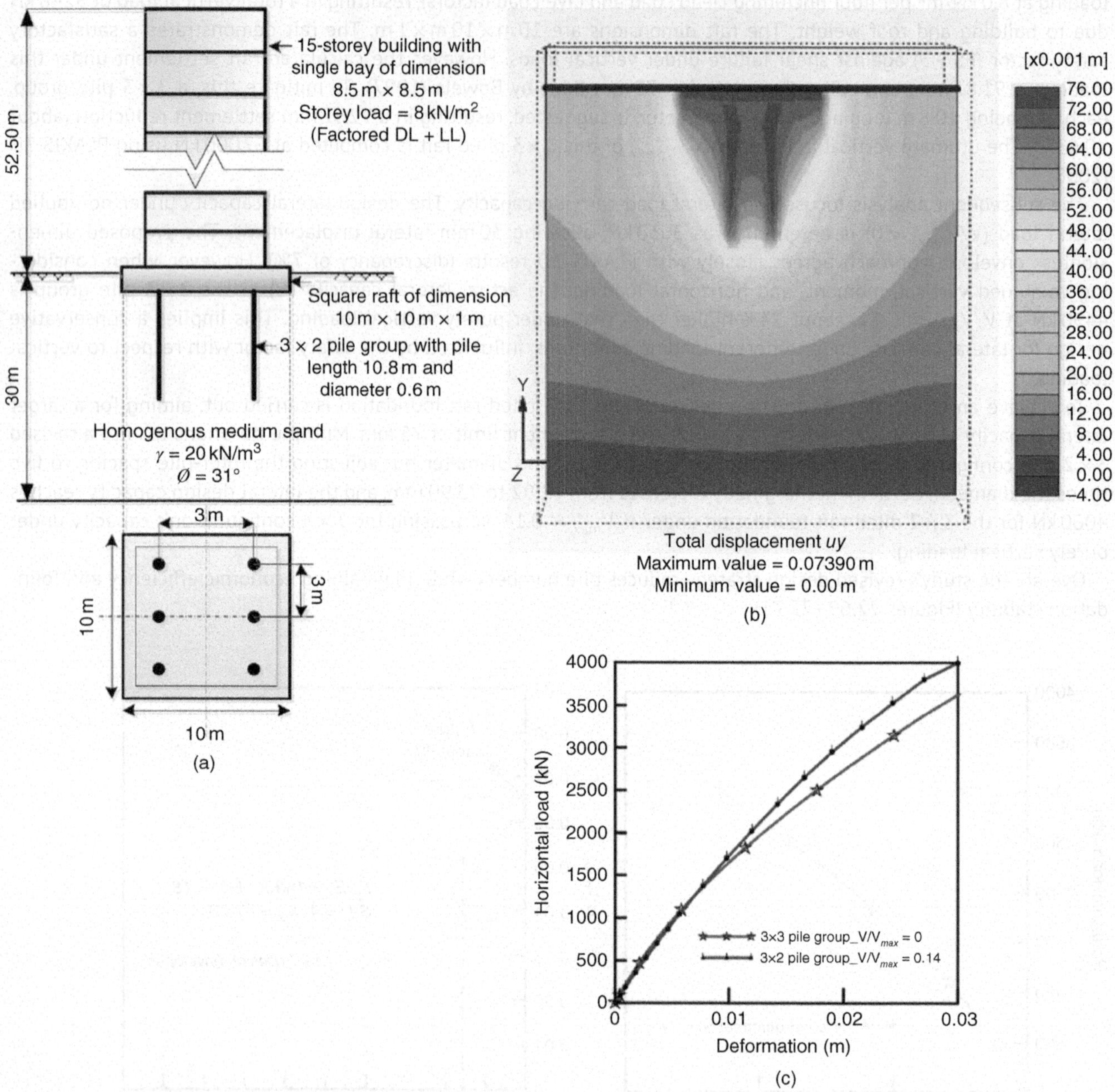

Figure 22.71 (a) Schematic presentation of modified piled raft foundation of case study problem, (b) settlement contour of modified 3 × 2 pile group, and (c) horizontal load-deformation behavior of both 3 × 2 and 3 × 3 piled raft foundation under V/V_{max} = 0.14 and 0 condition respectively (Chanda et al. 2020 / with permission of Elsevier).

Case Study 22.6

Analysis of a Caliche Stiffened Pile Foundation

While the methodologies for designing deep foundations are predominantly tailored for uniform soil arrangements, the existence of densely cemented strata could result in the miscalculation of load resistance and an overestimation of settlement in pile foundations. This investigation introduces an innovative strategy that employs proficient caliche layers affixed to both the upper and lower segments of a continuous flight auger (CFA) pile. This amalgamation creates a fresh composite foundation system termed as the caliche stiffened pile (CSP). The primary aim is to efficiently optimize the necessary pile length through an economical approach, all while keeping the inherent soil properties unaffected. The settlement behaviors of the CSP foundation, implemented for a tall building, were vigilantly tracked, and thorough full-scale experiments were carried out to quantify the pile's load-bearing capacity. Finite element retrospective analyses were conducted to counteract any detrimental effects arising from disturbances in soil samples during settlement calculations. Furthermore, a reverse calculation was performed involving a test embankment fill to ascertain the stiffness parameters of the soil. This simulation emulated an unforeseen external load applied to the structure. By employing finite element analysis and a parametric exploration, the effects of the CSP on managing pile foundation settlement and optimizing the required pile length were comprehensively explored. The proposed CSP foundation exhibits the capability to significantly diminish settlement in CFA piles when confronted with caliche

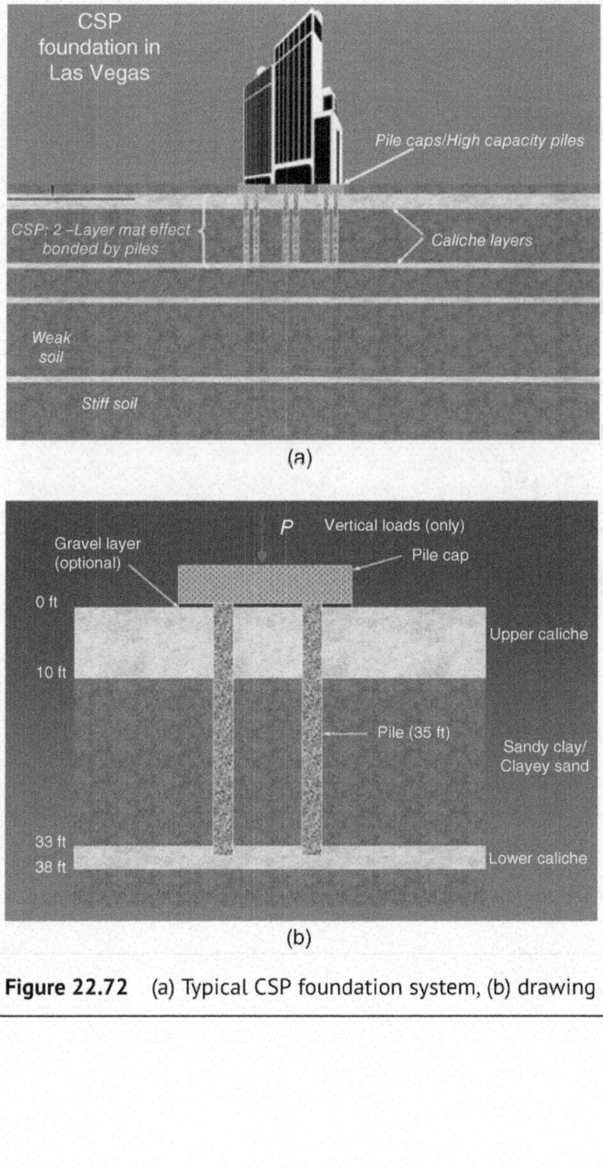

Figure 22.72 (a) Typical CSP foundation system, (b) drawing of single CSP cap foundation system (Stone 2009 / UNLV).

(Continued)

Case Study 22.6 (Continued)

layers possessing a thickness equal to or exceeding the pile diameter at both the CFA pile head and toe, precisely where the CSP components are positioned.

This research introduced the CSP foundation system, which combines the settlement-reducing attributes of a pile foundation with the stiffness effects akin to plate structures from nearby caliche layers near the surface (Figure 22.72). Through the utilization of finite element analyses, comprehensive field tests, and laboratory experiments involving the settlement response of an individual pile in a stratified soil containing caliche, it was noted that the behavior of the suggested foundation closely resembles the load transfer mechanisms observed in mat foundations supported by piles.

The CSP foundation was employed for a prominent Las Vegas Strip hotel, standing at 190 m tall with 51 floors (Figure 22.73). The hotel's circular arc-shaped plan measures 213 m in length and 24 m in width. The soil profile includes upper and lower caliche layers overlying a soil deposit. The building, constructed with reinforced concrete

(a)

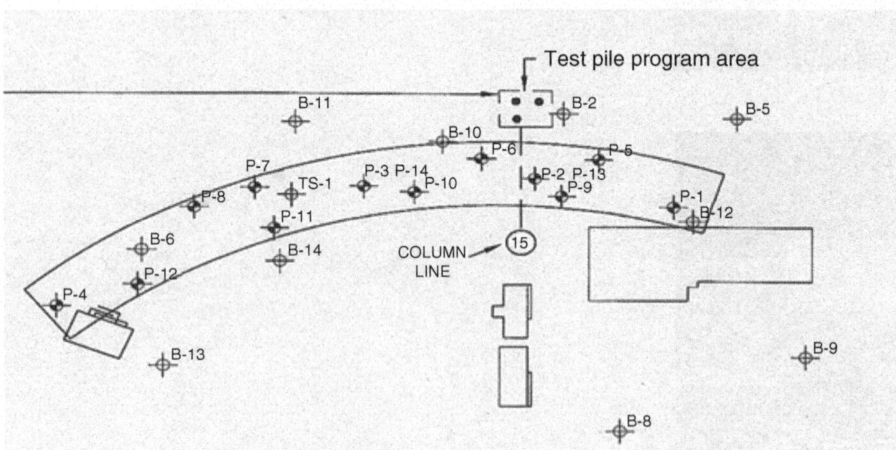

(b)

Figure 22.73 (a) Photograph of case study hotel tower during construction (Mikerussell / Wikimedia Commons / Public domain). (b) Case study hotel tower boring plan (Stone 2009 / UNLV).

and post-tensioned floor slabs, experienced combined dead and live loads ranging from 22,240 to 40,000 kN for typical column lines, translating to a vertical contact pressure of 464 kPa for each $3.2 \times 7.4\,m^2$ tributary area. The hotel has two elevator cores and two isolated shear walls, with dead plus live contact pressure loads in the elevator cores at approximately 670 kPa.

Caliche core sample tests encompassed UCS and consolidated undrained (CU) triaxial tests, obtained through a California-modified ring sampler. UCS results indicated an average unconfined compressive strength of 61.8 MPa for the caliche layers. Saturated CU triaxial tests applied a 96.5 kPa confining pressure with a constant rate of strain (CRS) of 0.04, revealing average modulus of elasticity (E_c) and Poisson ratio (v_c) values of 14.36 GPa and 0.32, respectively.

The field tests involved predrilling excavations during pile installation to identify caliche properties and thicknesses, especially for 300 pile positions. The upper caliche layer, located around an elevation of 625–630 m, exhibited a depth of 3 m or more in most borings. A 0.45 to 0.6 m soil layer typically lies within this upper caliche zone. A second layer existed at 618.4 m elevation, below which the presence of caliche zones became more sporadic. Soil profiles above 61 m depth consisted of sands and gravels, with deeper soils being predominantly fine-grained clayey sands. Standard penetration data (SPT) indicated denser and stiffer soils in the upper 15 m, possibly due to surface over-consolidation.

Two full-scale pile load tests were conducted using the bidirectional Osterberg cell method, measuring ultimate axial capacity at 2224 kN for a pile tip displacement of 5% of the shaft diameter. Shear transfer mechanisms in soil and caliche layers were recorded through load–displacement responses, with maximum interface shear capacity reached at displacement of 0.51 to 1.02 cm. Another series of top-down load tests were performed to measure failure parameters of the upper caliche deposit and the soil immediately below it, along with load distribution and pile settlement at a design load of 14,946 kN.

A test fill embankment was constructed to simulate additional loads from an adjacent structure, aiming to determine in situ stiffness parameters for accurate evaluation of tower settlement (Figure 22.74). Settlement plates were placed in the embankment and monitored during fill placement, resulting in an average maximum settlement of 4.57 cm. Settlements were measured in the tower after constructing the first-floor columns, revealing higher settlements in the south core attributed to varying caliche thicknesses. The maximum recorded settlement in the south core was nearly 3 cm higher than in the north core, increasing by 26% after four months due to both dead and live loads. The south core settlement was mainly influenced by dead and live loads, while the north core settlement was primarily affected by dead loads.

Figure 22.74 Test fill embankment (Stone et al. 2023 / Canadian Science Publishing).

(Continued)

Case Study 22.6 (Continued)

A 3D FEM analysis using PLAXIS 3D was conducted to evaluate settlements over the entire building area capturing piles, upper and lower caliche, and interaction between the pile and soil. The 3D model includes each pile that was installed in the building foundation (300 piles); the ultimate skin friction properties of the piles may vary based on their installed location beneath the mat foundation. The pile caps and core mats are simulated using structural plate elements. All pile tops are rigidly connected to pile caps or core mats. The finite element mesh, structural elements, and embedded piles are shown in Figure 22.75.

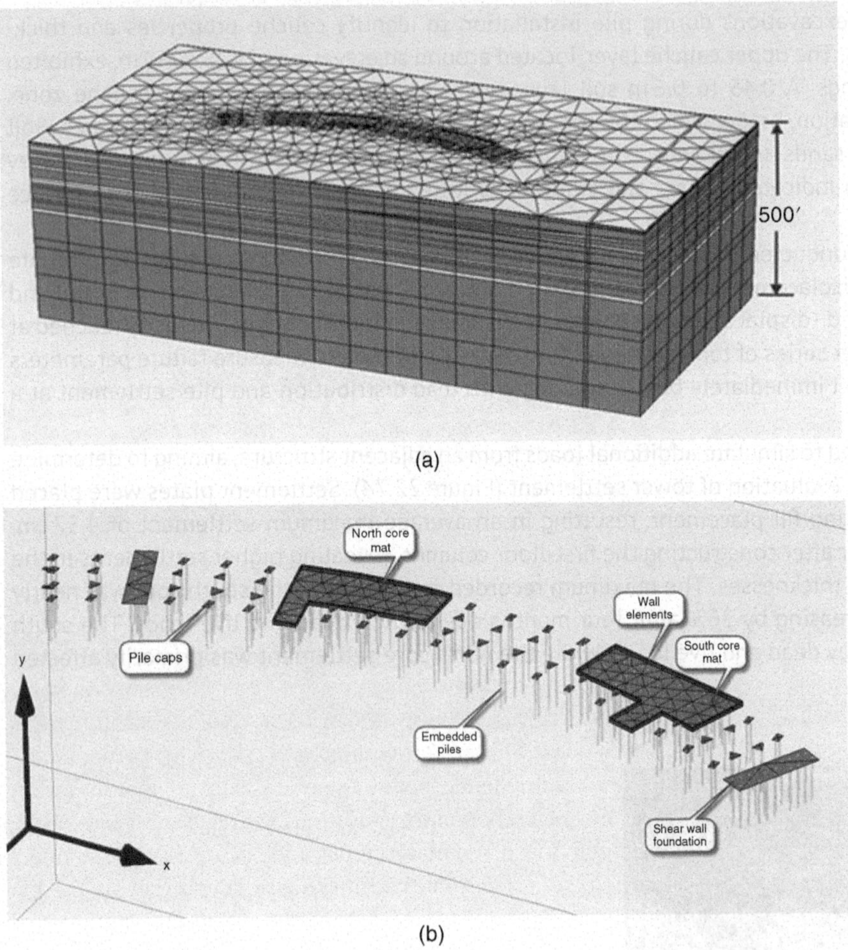

Figure 22.75 (a) 3D model mesh for settlement analysis, side view. (b) Structural elements in the modeling (Stone 2009 / UNLV).

Figure 22.76 shows the settlements based on the numerical analysis of the 3D model. Figure 22.77 compares the predicted settlements based on the 3D model to the actual settlement measurements.

The study highlights the advantages of the CSP foundation in soils with caliche layers, contrasting with other foundation solutions or ground improvements suited for homogeneous soils without caliche. Parametric and finite element analyses support the CSP foundation's minimal settlement and shortest required pile length, outperforming cases with

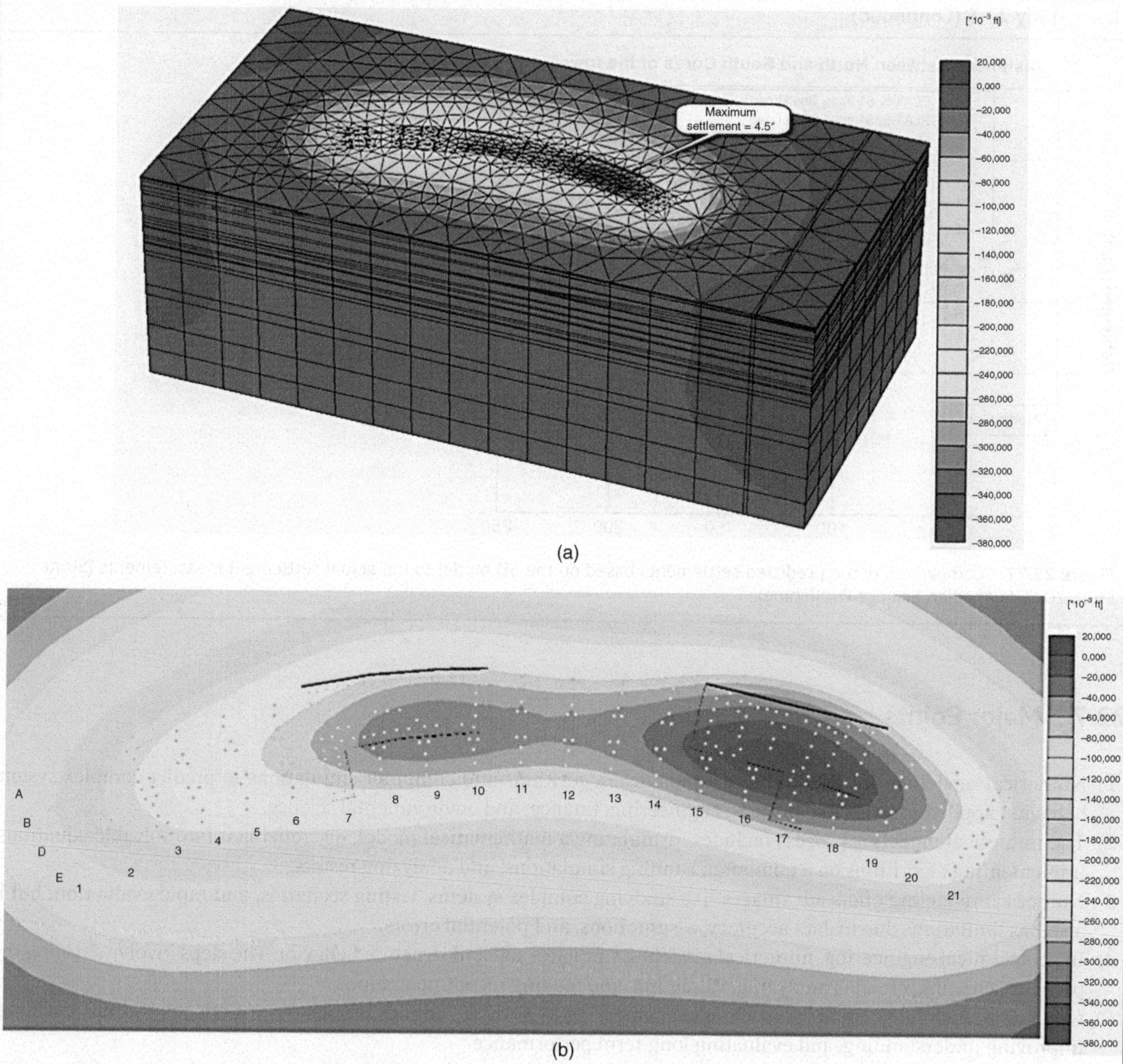

Figure 22.76 Model settlement contours for $D + L$ load case: (a) side view and (b) plan view (Stone 2009 / UNLV).

caliche layers at the head, toe, or absent. The method's applicability lies in the accessibility of caliche layers near the surface in Southern Nevada, wherein attaching pile tips and tops to caliche layers can control settlement. Soil layers between these caliche deposits contribute to settlement control, while the soil below the pile tip has limited influence. The study's data is based on the analyzed case study, emphasizing the need for additional data on group pile foundation behavior and enhanced soil stratigraphy modeling using more borehole information about caliche properties and variations.

(Continued)

Case Study 22.6 (Continued)

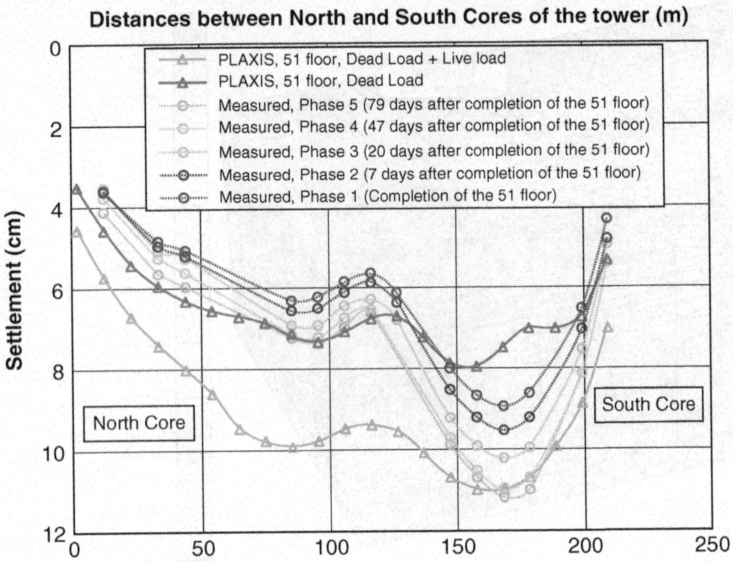

Figure 22.77 Comparison of the predicted settlements based on the 3D model to the actual settlement measurements (Stone et al. 2023 / Canadian Science Publishing).

22.7 Major Points

1) Numerical modeling involves using mathematical algorithms and computer simulations to predict complex system behavior, applied in fields like physics, engineering, finance, and environmental science.
2) The numerical modeling process includes formulating a mathematical model, discretizing it into solvable equations, implementing algorithms on a computer, running simulations, and analyzing results.
3) Numerical modeling offers advantages like studying complex systems, testing scenarios, and rapid evaluation, but it also has limitations due to data accuracy, assumptions, and potential errors.
4) In geotechnical engineering, numerical modeling simulates soil and structure behavior. The steps involve defining the problem, creating a model, analyzing, validating, and making recommendations.
5) Numerical modeling benefits geotechnical engineering by solving complex problems, cost-effective scenario testing, improving understanding, and evaluating long-term performance.
6) Analytical solutions find closed-form expressions for problems, while numerical methods use algorithms for approximations. Differential equations play a significant role in geotechnical engineering.
7) Geotechnical problems are classified into soil stability, soil mechanics, and groundwater problems. Numerical modeling is vital due to complex geometric and loading conditions.
8) Numerical modeling is crucial for foundation engineering, offering insights into soil behavior, optimizing design, reducing costs, and enhancing safety.
9) Numerical modeling aids in analyzing the behavior of shallow and pile foundations, considering soil properties, geometry, loading, and soil–structure interaction.
10) Accurate geometric representation and boundary conditions are essential in numerical modeling for shallow foundations. Displacement and load boundary conditions are used, and interface elements can model soil–structure interaction.
11) Different constitutive models describe soil behavior. Mohr–Coulomb model uses friction and cohesion, Drucker–Prager accounts for stress path, Cam-Clay models soil as a network of particles, and Modified Cam-Clay refines Cam-Clay for plasticity.
12) 2D models are computationally efficient and easier to visualize but may not capture full 3D phenomena. Considerations for using 2D models include simplified hypotheses and algorithm verification.

13) FDM (finite difference method) is used to solve differential equations describing soil and rock behavior. It involves discretizing a domain into elements and approximating differential equations using finite difference equations. FDM is particularly useful for complex problems and irregular geometries.
14) FDM is commonly applied in soil mechanics, such as slope stability, consolidation, and seepage analysis. It models stress, strains, and water flow, making it valuable for understanding behavior under different conditions.
15) FEM (finite element method) divides complex structures into interconnected elements, analyzing their behavior using equations and boundary conditions. It's widely used in geotechnical engineering for understanding stresses, deformations, and behaviors under various loads.
16) Applications of FEM in geotechnical engineering involve analyzing soil–structure interaction, slope stability, seepage, ground improvement, tunneling, and excavation. It helps design foundations, support systems, and drainage control.
17) In addition to FDM and FEM, finite element limit analysis (FELA), boundary element method (BEM), discrete element method (DEM), material point method (MPM), and kinematic element method (KEM), each offering unique ways to solve geotechnical problems with varying complexities and limitations.

22.8 Questions and Problems

22.8.1 Explanatory

1 What is numerical modeling?

2 What are the differences between analytical solutions and numerical methods?

3 How are geotechnical problems classified?

4 Why is numerical modeling important in foundation problems?

5 What geometrical considerations are important in foundation problems?

6 What are constitutive models for soil?

7 What are the implications of using 2D vs. 3D models?

8 What are the common numerical methods used in geotechnical engineering, and why are they important?

9 How does the finite difference method (FDM) work, and what are its advantages and limitations?

10 What are the key applications of the finite difference method (FDM) in geotechnical engineering?

11 How does the finite element method (FEM) work, and what are its steps in problem formulation and solution?

12 What are the specific applications of finite element method (FEM) in geotechnical engineering?

13 How does the finite element limit analysis (FELA) method combine finite element and limit analysis techniques, and why is it useful?

14 How does the discrete element method (DEM) work, and what types of problems is it well-suited for?

22.8.2 Conceptual

1 What are shallow and deep foundations, and why is numerical modeling crucial for understanding their behavior?

2 How does the process of numerical modeling differ for shallow and deep foundations?

3 What factors influence the selection of appropriate numerical methods for modeling shallow and deep foundations?

4 How does the finite difference method (FDM) or finite element method (FEM) apply to the analysis of shallow and deep foundations?

5 What are the main challenges when modeling soil–pile interaction using numerical methods for deep foundations?

6 What are the key differences in modeling soil–structure interaction for shallow and deep foundations?

7 How does the numerical modeling of soil–pile interaction help optimize the design and performance of deep foundations?

8 What are some limitations and uncertainties associated with numerical modeling of foundation behavior in geotechnical engineering?

9 How does the material point method (MPM) address the challenges of modeling large soil displacements in deep foundation scenarios?

10 What role do boundary conditions play in the accuracy and reliability of numerical models for both shallow and deep foundations?

References

Akbari Garakani, A., Sadeghi, H., Saheb, S., and Lamei, A. (2020). Bearing capacity of shallow foundations on unsaturated soils: analytical approach with 3D numerical simulations and experimental validations. *International Journal of Geomechanics* 20 (3): 04019181.

Basack, S., Karami, M., and Karakouzian, M. (2022). Pile–soil interaction under cyclic lateral load in loose sand: experimental and numerical evaluations. *Soil Dynamics and Earthquake Engineering* 162: 107439.

Bowels, J.E. (1997). *Foundation Analysis and Design*. The McGraw Hill Companies.

Chanda, D., Saha, R., and Haldar, S. (2020). Experimental investigation of piled raft foundation under combined vertical, lateral and moment loads. *Construction in Geotechnical Engineering: Proceedings of IGC 2018* (pp. 325–338).

Cheng, C.Y., Dasari, G.R., Chow, Y.K., and Leung, C.F. (2007). Finite element analysis of tunnel–soil–pile interaction using displacement-controlled model. *Tunnelling and Underground Space Technology* 22 (4): 450–466.

Ebrahimi, K. and Khazaei, J. (2019). Investigation of bearing capacity and failure pattern in shell foundations by FELA method. *Geotechnical and Geological Engineering* 37: 3523–3534.

Helwany, S. (2007). *Applied Soil Mechanics with ABAQUS Applications*. Wiley.

Hu, S., Zhuang, Y., Wu, Y. et al. (2022). Numerical study of bearing capacity of the pile-supported embankments for the flexible floating, rigid floating and end-bearing piles. *Symmetry* 2022 (14): 1981.

Itasca Consulting Group, Inc. (2023). *FLAC3D — Fast Lagrangian Analysis of Continua in Three-Dimensions, Ver. 9.0.* Minneapolis: Itasca.

Karami, M., Nazari-Sharabian, M., Bristow, J., and Karakouzian, M. (2023a). Evaluation of the stability of terraced slopes in clayey gravel soil using a novel numerical technique. *Frontiers of Structural and Civil Engineering* 1–16.

Karami, M., Basack, S., Das, G. et al. (2023b). Experimental and numerical evaluations of load settlement response of helical piles in soft clay. *Marine Georesources and Geotechnology journal* 1–18. https://doi.org/10.1080/1064119X.2023.2254790.

Lavasan, A.A., Talsaz, A., Ghazavi, A., and Schanz, T. (2016). Behavior of shallow strip footing on twin voids. *Geotechnical and Geological Engineering* 34: 1791–1805.

Lavasan, A.A., Ghazavi, M., Blumenthal, A.V., and Schanz, T. (2018). Bearing capacity of interfering strip footings. *Journal of Geotechnical and Geoenvironmental Engineering* 144 (3): 04018003.

Lees, A. (2016). *Geotechnical Finite Element Analysis. A Practical Guide*. ICE.

Mroueh, H. and Shahrour, I. (2002). Three-dimensional finite element analysis of the interaction between tunneling and pile foundations. *International Journal for Numerical and Analytical Methods in Geomechanics* 26 (3): 217–230.

Rowe, R.K. and Liu, K.W. (2015). Three-dimensional finite element modelling of a full-scale geosynthetic-reinforced, pile-supported embankment. *Canadian Geotechnical Journal.* 52: 2041–2054.

Shamloo, S. and Imani, M. (2021). Upper bound solution for the bearing capacity of two adjacent footings on rock masses. *Computers and Geotechnics* 129 (2021): 103855.

Stone Jr, R.C. (2009). Analysis of a caliche stiffened pile foundation. Doctoral dissertation. University of Nevada, Las Vegas, USA.

Stone, R.C., Farhangi, V., Fatahi, B., and Karakouzian, M. (2023). A novel short pile foundation system bonded to highly cemented layers for settlement control. *Canadian Geotechnical Journal* 60 (9): 1332–1351.

Terzaghi, K. and Peck, R.B. (1967). *Soil mechanics in engineering practice*, 2e. New York, N.Y: John Wiley and Sons, Inc.

Xie, M., Zheng, J., Zhang, R. et al. (2020). Performance of a combined retaining wall structure supporting a high embankment on a steep slope: case study. *International Journal of Geomechanics* 20 (6): 05020002.

Section IX

Trends and Prospects

23

Uncertainty and Reliability

Chapter Highlights

- Uncertainty Sources in Geotechnical Engineering
- Random Analysis Techniques in Geotechnical Engineering
- Methods for Determining the Reliability Index:
  - First Order Second Moment Method (FOSM)
  - First Order Reliability Method (FORM)
  - Monte Carlo Simulation
- Case Histories
  - Associated Uncertainties Affecting the Babolsar Bridge Failure (Application of MCS-Eslami et al. 2016)
  - Model Uncertainty Appraisal of Various Methods for Predicting Axial Pile-Bearing Capacity (Heidarie Golafzani et al. 2020a,b)
  - Assessment of Model Uncertainty for Predictive Axial Pile-Bearing Capacity Approaches (Heidarie Golafzani et al. 2020a,b)
  - The Reliability Assessment of Different Design Cases for the James Bay Dikes (Baecher and Christian 2003)

23.1 Introduction

Geotechnical design, especially in foundation engineering, involves a combination of site investigations, design models, codes, and computations, as depicted in Figure 23.1. Engineers are tasked with making decisions based on prevailing conditions to achieve an optimum design that aligns with the project's significance.

In conventional design approaches, engineers often rely on material properties, such as friction angles, soil undrained shear strength, or the uniaxial compressive resistance of concrete, as fixed parameters with specific values. However, a closer examination of borehole data, SPT or CPT records, reveals the inherent uncertainty in soil properties. These uncertainties are not restricted solely to material properties or simplified constitutive models; even the estimation of material characteristics introduces numerous uncertainties in the design process. Consequently, discrepancies between measured and predicted design values are frequently observed in this field.

The uncertainty in geotechnical design can be broadly categorized into two main types. The first category arises from the inherent random nature of materials and is a fundamental aspect that cannot be eliminated. However, it can be better described and accounted for by conducting more tests or utilizing extensive databases. Parameters like undrained shear strength fall into this category, as their values vary spatially and over time. Shear strength is often estimated based on a limited number of samples and tests, and then applied for design at different locations from where the samples were obtained. The inherent uncertainty represents the natural variability of soil characteristics from one point to another, resulting from complex processes such as sedimentation, weathering, stress history, and time.

The second category of uncertainty arises from our inadequate and limited knowledge of a system or process. This type of uncertainty can generally be classified into three main types: model uncertainty, transformation uncertainty, and

Advanced Foundation Engineering: Principles, Performance and Prospects, First Edition. Moses Karakouzian and Abolfazl Eslami.

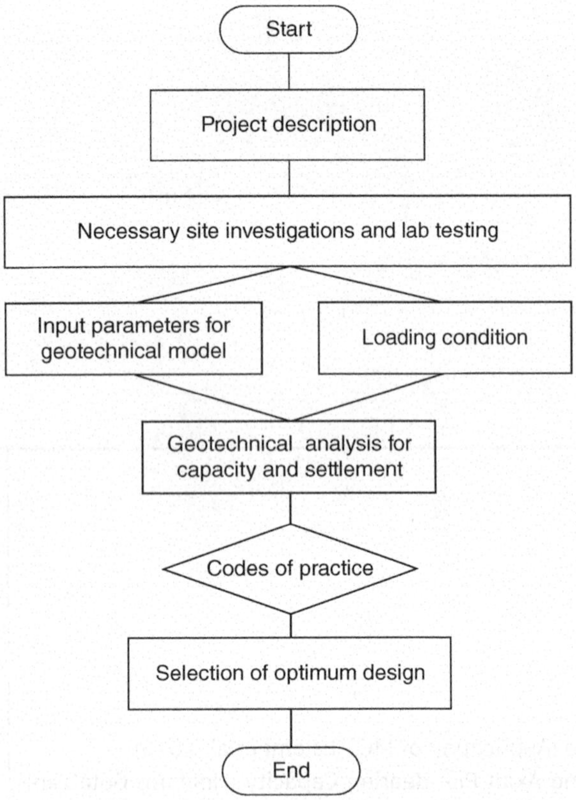

Figure 23.1 Foundation design algorithm.

measurement/random noise error. These uncertainties stem from the challenges in accurately representing and predicting the behavior of complex geotechnical systems, the conversion of data from one form to another (e.g., laboratory to field conditions), and errors in measurements or random fluctuations in data. Addressing these uncertainties is crucial for robust and reliable geotechnical design.

Measurement errors in geotechnical design arise from various sources, including equipment limitations, operator proficiency, and random variations during testing. This category encompasses empirical methods or models that incorporate laboratory or field measurements to derive soil design characteristics. Increasing the quantity and quality of experiments or using better models can help reduce this type of uncertainty. Statistical uncertainty, resulting from a limited amount of information, can be mitigated by testing more samples. However, uncertainties caused by the sampling or measurement process cannot be eliminated merely by increasing the number of samples.

Estimating live and dynamic loads, such as earthquakes, winds, floods, etc., introduces significant uncertainty into the design process. On the other hand, uncertainties associated with static loads can often be neglected compared to uncertainties in material parameters. Moreover, it is important to recognize that under dynamic loads, the material parameters may exhibit significant variations, which may not be evident under static conditions. For instance, cyclic loading can lead to a decrease in the shear strength of the soil, introducing greater uncertainty in the performance of this geotechnical parameter. Addressing these uncertainties and considering the variability of material behavior under different loading conditions is crucial for accurate and reliable geotechnical design.

23.2 Uncertainty Sources in Geotechnical Engineering

Soil properties can vary significantly with both depth and location. For example, let us take cone penetration testing (CPT) measurements. CPT with continuous recording outputs provides geotechnical engineers with valuable insights into inherent soil variability. Figure 23.2 illustrates a typical CPT sounding, showing variations in soil properties with depth. In conventional design methods, the "trend," which represents the average value of recorded parameters, is often used for design purposes. However, to better express soil's spatial variability, the use of random field and probabilistic approaches

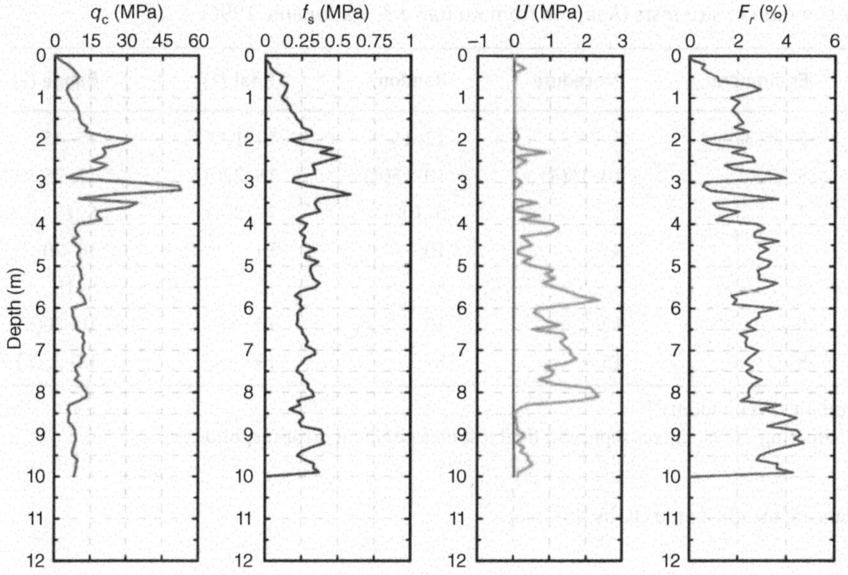

Figure 23.2 An example of CPT records and the variation of soil properties among depths (Eslami 1996 / with permission of University of Ottawa Press).

becomes an essential and advantageous tool. These methods consider the stochastic nature of soil properties, enabling engineers to better assess and account for uncertainties in their design processes.

To consider soil as a random field, it is necessary to have the trend or average value as well as the coefficient of variation (COV) and correlation length for that particular soil characteristic. The COV is determined as the ratio of standard deviation to data mean. The correlation length is the length at which soil parameters are strongly correlated. It can be calculated in several ways, such as the variance reduction method, direct integration of sample auto-correlation function, Bartlett's limit method, and auto-correlation model fitting (Kenarsari et al. 2011).

As depicted in Figure 23.3, a property denoted as "3" can be divided into two components. The first component is represented by a straight line known as the trend $t(z)$. By subtracting the trend value from the overall data, the remaining portions display variations around the zero line, referred to as the residual component or $w(z)$. This residual component serves as an indicator of soil spatial variability and may also encompass other accompanying uncertainties, which will be further discussed in the subsequent paragraphs (Phoon and Kulhawy 1999).

The statistical uncertainty in parameter estimation arises from the limited data available and the techniques used for estimation. Measurement uncertainty, on the other hand, can be attributed to three primary sources: (i) equipment error, (ii) errors caused by operators, and (iii) the effects of random tests during measurements. The magnitude of this type of

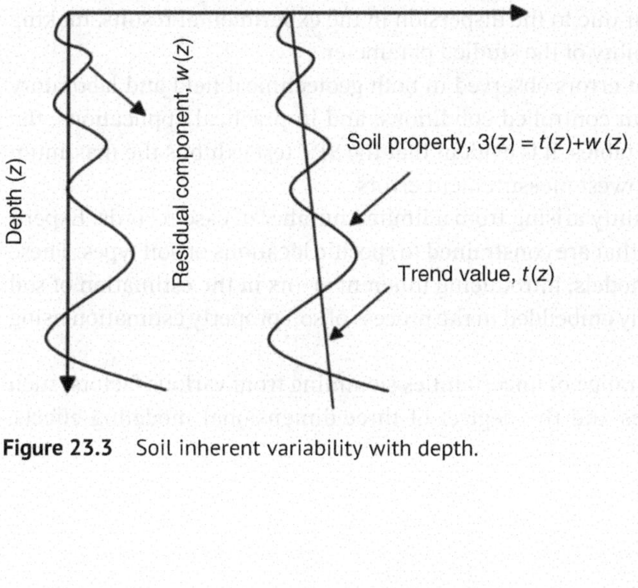

Figure 23.3 Soil inherent variability with depth.

Table 23.1 Summary of measurement error for common in situ tests (Adapted from Kulhawy & Trautmann, 1996).

Test	Equipment	Procedure	Random	Total (1)	Range (2)
Standard penetration test (SPT)	5–75(3)	5–75(3)	12–15	14–100(3)	15–45
Mechanical cone penetration test (MCPT)	5	10–15(4)	10–15(4)	15–22(4)	15–25
Electric cone penetration test (ECPT)	3	5	5–10(4)	7–12(4)	5–15
Vane shear test (VST)	5	8	10	14	10–20
Dilatometer test (DMT)	5	5	8	11	5–15
Pressuremeter test, pre-bored (PMT)	5	12	10	16	10–20(5)
Self-boring pressuremeter test (SBPMT)	8	15	8	19	15–25(5)

COV(Total) = $[\text{COV(Equipment)}^2 + \text{COV(Procedure)}^2 + \text{COV(Random)}^2]^{0.5}$
Because of limited data and judgment involved in estimating COVs, ranges represent field test measurement error magnitudes.
Best to worst case scenarios, respectively, for SPT.
Tip and side resistance, respectively, for CPT.
Results may likely differ for P_0, P_f,and P_L, but the data are insufficient to clarify this issue.

Table 23.2 Summary of total measurement error of some laboratory tests (Look 2007 Adapted from Look, 2007).

			Coefficient of variation (%)	
Test	Property	Soil type	Range	Mean
Atterberg tests	Plasticity index	Fine-grained	5–51	24
Triaxial compression	Effective angle of friction	Clay, silt	7–56	24
Direct shear	Shear strength, C_u	Clay, silt	19–20	20
Triaxial compression	Shear strength, C_u	Clay, silt	8–38	19
Direct shear	Effective angle of friction	Sand	13–14	14
Direct shear	Effective angle of friction	Clay	6–22	14
Direct shear	Effective angle of friction	Clay, silt	3–29	13
Atterberg tests	Plastic limit	Fine-grained	7–18	10
Triaxial compression	Effective angle of friction	Sand, silt	2–22	8
Atterberg tests	Liquid limit	Fine-grained	3–11	7
Unit weight	Density	Fine-grained	1–2	1

uncertainty can be assessed through the manufacturer's information about the equipment and the proficiency of the operators during laboratory and field tests. Random test errors occur due to the dispersion in the experimental results, making it challenging to distinguish them from the inherent soil variability of the studied parameter.

Tables 23.1 and 23.2 provide a summary of the measurement errors observed in both geotechnical field and laboratory tests. It is important to note that these results are derived from controlled conditions, and in practical applications, the measurement error is expected to be even higher. Based on the tables, it is evident that the SPT test exhibits the maximum measurement error, while the DMT and ECPT tests show the lowest measurement errors.

Transformation error is another significant source of uncertainty arising from a limited number of case records. Experimental correlations are typically developed based on databases that are constrained to specific locations or soil types. These correlations involve various assumptions and computational models, introducing inherent errors in the estimation of soil properties. Consequently, there is a level of uncertainty implicitly embedded in the process of soil property estimation using these correlations.

Model errors in geotechnical engineering encompass a wide range of uncertainties stemming from various factors, such as the simplification of equilibrium and deformation analyses and the neglect of three-dimensional modeling effects.

Model uncertainty is quantified as the ratio of the real quantity to the predicted quantity using a model. The mean and coefficient of variation are used to characterize the uncertainty of the model. The discrepancy between the mean value of the unit number signifies a systematic error in the model, whereas the standard deviation reflects the variability in the model's predicted values.

23.3 Random Analysis Techniques in Geotechnical Engineering

Random analyses offer effective tools for addressing the variability of soil properties, enabling the development of rational algorithms to estimate soil design parameters based on probabilities and assess the associated design risks. Depending on the level and type of reliability analysis, these analyses yield one or more of the following outcomes:

- Probability of failure (or the likelihood of undesirable performance)
- Reliability Index (β)
- The most probable combination of parameters leading to failure
- Sensitivity of the results to changes in the parameters

The appropriate target reliability index for each geotechnical design should be determined based on the significance of the supported structure. In geotechnical engineering, numerous sources of uncertainties, both quantified and unquantified, exist. Consequently, relying solely on a safety factor greater than one may not adequately ensure safety objectives.

Various reliability methods, including first and second-order methods, have been developed in structural engineering. However, due to the inherent complexity of considering uncertainty in geotechnical designs, it is preferable not to directly apply these methods. It is essential to familiarize geotechnical engineers with the intricacies of addressing uncertainties in their designs.

In structural engineering, reliability-based design has been extensively studied, mainly because constitutive models for materials like steel and concrete are linear, and these materials are manufactured under controlled conditions. However, geotechnical engineering presents distinct challenges. Soil properties exhibit inherent spatial variability, and their characteristics are influenced by statistical uncertainty, measurement errors, and transformation uncertainty. Moreover, the constitutive models used in geotechnical engineering are nonlinear and more complex.

Despite these complexities, reliability methods remain valuable tools for investigating geotechnical engineering issues. Commonly used applications include determining the probability of failure and assessing the reliability index of proposed geotechnical designs. Embracing these methods empowers engineers to better account for uncertainties and enhance the safety and performance of geotechnical structures.

23.4 Methods for Determining the Reliability Index

For each geotechnical problem, a performance function, also known as the limit state function or margin of safety (M), is defined. This function relies on random variables for resistance (R) and loading (Q). To fully characterize these variables, parameters like mean, standard deviation, and scale of fluctuation (SOF) are determined using relevant databases. Additionally, each random variable is associated with a specific distribution function, such as normal, beta, lognormal, etc. With the identified distribution functions, the performance function takes on its own distribution. According to the central limit theorem, this function tends to converge toward a normal distribution. This transformation into a normal distribution is advantageous, as it simplifies the application of reliability methods and facilitates the assessment of uncertainties in geotechnical designs.

According to the performance function, failure occurs when $M < 0$, as illustrated in Figure 23.4 and defined by Equation 23.1. However, accurately determining the probability of failure for a nonlinear performance function requires calculating multiple integrals, which can be a complex task. To address this challenge, various methods have been developed for determining the probability of failure, or more precisely, the reliability index in geotechnical designs. Some of these methods include the first order second moment method (FOSM), the first order reliability method (FORM), the second-order reliability method (SORM), the point estimation method (PEM), and the Monte Carlo simulation (MCS) (Ang and Tang 2007; Baecher and Christian 2003; Harr 1987). These techniques offer valuable tools for assessing the

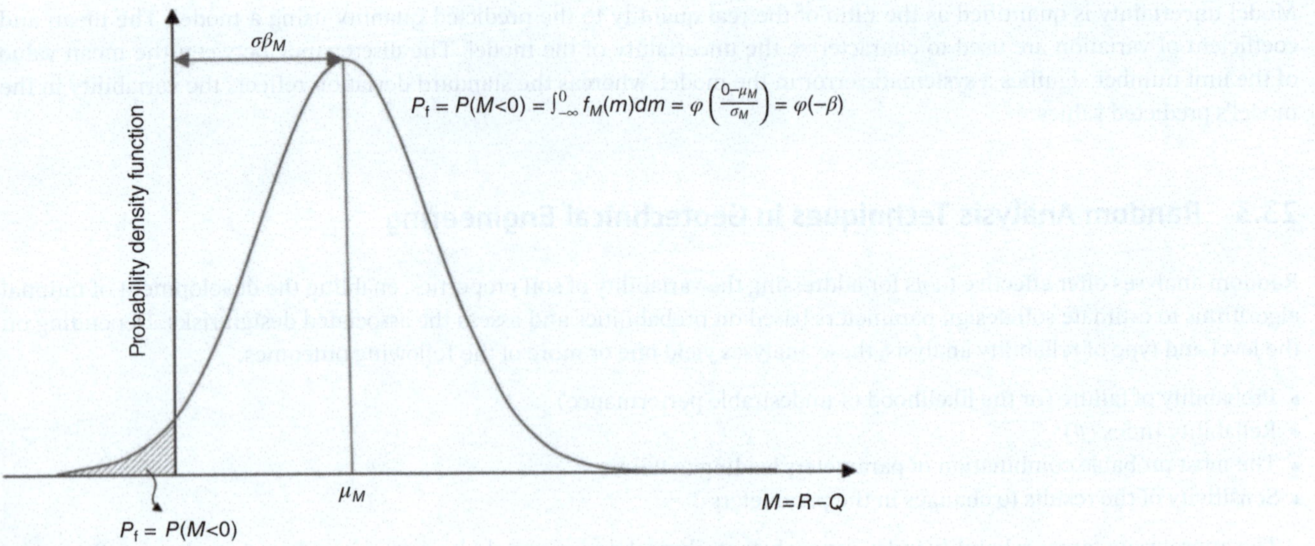

Figure 23.4 Definition of the probability of failure and reliability index (β).

reliability and risk associated with geotechnical designs, considering the complexities and uncertainties inherent in such engineering problems.

$$P_f = \iiint_{M<0}^{1} \cdots \int f_{xy...z}(X, Y, \ldots, Z)\mathrm{d}x\mathrm{d}y \ldots \mathrm{d}z \tag{23.1}$$

where,

x = the vector of the input parameters of the problem

f_x = the joint probability density function of the random variables

M = the limit state function, which can be defined in different forms and represents safety

23.4.1 First Order Second Moment Method

This approach, known as the first-order second-moment method (FOSM), is based on the Taylor expansion theory. It is referred to as "first-order" because it considers only the first-order terms in the Taylor expansion, approximating the non-linear function as a linear function near the mean value. Additionally, it is called the "second moment" method because it provides estimates of the variance, i.e., the second moment of the function.

In the FOSM, the reliability index is expressed as a function of the random input parameters. The method calculates the mean and standard deviation of the function based on the mean and standard deviation, as well as the correlation coefficient of the input parameters.

Geometrically, the reliability index represents the shortest distance between the mean, or the most probable point, and the failure point, as illustrated in Figure 23.4. In this case, Φ denotes the standard normal distribution function, and β denotes the reliability index, which is calculated as the ratio of the mean to the standard deviation of the performance function, as presented in Equation 23.2.

$$\beta = \frac{\mu_M}{\sigma_M} \tag{23.2}$$

Let the performance function be $f(X)$, and its expansion around the mean is

$$f(X) = f(\mu_X) + (X - \mu_X)\frac{\mathrm{d}f}{\mathrm{d}x} + \ldots \tag{23.3}$$

Hence, the expectation of a nonlinear function of a single random variable is

$$E[f(X)] \approx E\left[f(\mu_X) + (X - \mu_X)\left.\frac{\mathrm{d}f}{\mathrm{d}x}\right|_{X=\mu_X} \right] = f(\mu_X) \tag{23.4}$$

Table 23.3 Reliability index for mechanically equal performance functions.

Performance function (M)	Reliability index (β)
$R - Q$	$\dfrac{\mu_R - \mu_Q}{\sqrt{\sigma_R^2 + \sigma_Q^2}}$
$(R/Q) - 1$	$\dfrac{\mu_R - \mu_Q}{\mu_R\left(\sqrt{\frac{\sigma_R^2}{\mu_R^2} + \frac{\sigma_Q^2}{\mu_Q^2}}\right)}$
$\ln(R/Q)$	$\dfrac{\ln\left(\frac{\mu_R}{\mu_Q}\right)}{\sqrt{\frac{\sigma_R^2}{\mu_R^2} + \frac{\sigma_Q^2}{\mu_Q^2}}}$

Moreover, its variance is

$$\text{Var}[f(X)] \approx \text{var}\left[f(\mu_X) + (X - \mu_X)\frac{df}{dx}\bigg|_{X=\mu_X}\right] = \left(\frac{df}{dx}\right)^2 \text{Var}[X] \tag{23.5}$$

Table 23.3 presents the reliability indexes for three different performance functions in the outward direction. In this table, R and Q represent resistance and load, respectively. Although these functions are mechanically equivalent, their first derivatives differ due to their distinct expressions. Consequently, the reliability index for each performance function varies, as shown in Table 23.3, highlighting a significant limitation of this method. Furthermore, utilizing first derivatives for nonlinear functions does not result in a precise approximation of the reliability index, leading to potential inaccuracies.

Example 23.1 If the performance function for the two random variables R (i.e., resistance) and Q (i.e., load) is considered as $M = \ln(R/Q)$, What is the FOSM reliability index?

$$M = \ln\left(\frac{R}{Q}\right) \Rightarrow E[M] = \ln(\mu_R) - \ln(\mu_Q)$$

Since $E[f(x, y)] = f(\mu_x, \mu_y)$,

$$E[M] = E\left[\ln\left(\frac{R}{Q}\right)\right] = \ln\left(\frac{\mu_R}{\mu_Q}\right) = \ln(\mu_R) - \ln(\mu_Q)$$

$$\Rightarrow E[M] = \ln(\mu_R) - \ln(\mu_Q)$$

$$\text{Var}[f(x,y)] = \text{Var}[f(\mu_x, \mu_y)] + \text{Var}\left[(x - \mu_x)\frac{\partial f}{\partial x}\bigg|\mu_x\right] + \text{Var}\left[(y - \mu_y)\frac{\partial f}{\partial y}\bigg|\mu_y\right]$$

$$= 0 + \left(\frac{\partial f}{\partial x}\bigg|\mu_x\right)^2 \text{Var}[x] + \left(\frac{\partial f}{\partial y}\bigg|\mu_y\right)^2 \text{Var}[y]$$

$$\Rightarrow \text{Var}[M] = \text{Var}\left[\ln\left(\frac{R}{Q}\right)\right] = \left(\frac{\partial M}{\partial R}\bigg|\mu_R\right)^2 \text{Var}[R] + \left(\frac{\partial M}{\partial Q}\bigg|\mu_Q\right)^2 \text{Var}[Q]$$

$$= \left(\frac{\frac{1}{\mu_Q}}{\frac{\mu_R}{\mu_Q}}\right)^2 \text{Var}[R] + \left(\frac{\frac{-\mu_R}{\mu_Q^2}}{\frac{\mu_R}{\mu_Q}}\right)^2 \text{Var}[Q] = \frac{\text{Var}[R]}{\mu_R^2} + \frac{\text{Var}[Q]}{\mu_Q^2}$$

$$\Rightarrow \mathrm{Var}[M] = \frac{\mathrm{Var}[R]}{\mu_R^2} + \frac{\mathrm{Var}[Q]}{\mu_Q^2}$$

$$\beta = \frac{\mu_M}{\sigma_M} = \frac{\ln\left(\frac{\mu_R}{\mu_Q}\right)}{\sqrt{\frac{\sigma_R^2}{\mu_R^2} + \frac{\sigma_Q^2}{\mu_Q^2}}}$$

23.4.2 First Order Reliability Method (FORM)

To address the limitations of the FOSM approach, as discussed earlier, the first order reliability method (FORM) analysis was introduced. However, the initial iteration algorithm of FORM, proposed by Hasofer and Lind (1974), had some shortcomings due to its underlying assumptions (Baecher and Christian 2003). To overcome these issues, Rackwitz and Fiessler (1978) developed an improved iteration process (Haldar and Mahadevan 2000). For estimating the reliability index, the steps are considered as follows (Kim et al. 2005):

Step 1: Define the limit state function g in terms of the random variables.
Step 2: Assume an initial value for the reliability index β. Any value of β can be assumed.
Step 3: Assume the initial values of the design points (dp). The initial design points can be assumed to be at the mean values of the random variables.
Step 4: Compute the mean and standard deviation at the design point of the equivalent normal distribution for the random variables that are non-normal.
Step 5: Compute the partial derivatives evaluated at the design points.

$$\mathrm{pder} = \left(\frac{\partial g}{\partial x}\right)_{x=\mathrm{dp}} \tag{23.6}$$

Step 6: Compute the direction cosines (α) at the design points.

$$\alpha = \left[\frac{\left(\frac{\partial g}{\partial x}\right) * \sigma_x^N}{\sqrt{\Sigma_i^n\left(\left(\frac{\partial g}{\partial x}\right) * \sigma_x^N\right)^2}}\right] \tag{23.7}$$

Step 7: Compute the new values for the design points as

$$\mathrm{dp} = \mu_x^N - \alpha * \beta * \sigma_x^N \tag{23.8}$$

Repeat steps 4 through 7 until the direction cosines (α) converge to the specified tolerance value of 0.005.
Step 8: Once αs converge, the new design points can be expressed in terms of β as the unknown parameter. These new design points must satisfy the limit state function. Substitute the random variables in the limit state function with these new design points and solve g for β.
Repeat steps 3 through 8 until β converges to the tolerance value of 0.001.

Example 23.2 What is the FORM reliability index for a vertical cut with the given conditions as below? The random variables are uncorrelated and normally distributed.

The height of the cut (m)	Cohesion (kPa)	Unit weight (kN/m^3)
10	$\mu_c = 100$ kPa, $\sigma_c = 30$ kPa	$\mu_\gamma = 100$ kPa, $\sigma_\gamma = 30$ kPa

Step 1: The limit state function is $g = c\text{-}\gamma H/4$
Step 2: Assume $\beta = 3$.
Step 3: The initial values of the design points are considered as their mean values.

		1st iteration
Step 1	$g = c - \gamma H/4$	
Step 2	β	3
Steps 3 and 4	dp_c	100
		30
	dp_γ	20
		2
Step 5	∂_g/∂_c	1
	$\partial_g/\partial_\gamma$	−2.5
Step 6	α_c	0.9864
	α_γ	−0.1644
Step 7	dp_c	11.22
	dp_γ	20.99

Since the performance function is linear and random variables are normally distributed, iterations are limited to 1. Since αs are convergent, the new design point for cohesion and unit weight random variables are expressed in terms of β and substituted in the performance function as follows (step 8:):

$$(100 - 0.986 * \beta * 30) - \frac{10}{4}(20 + 0.164 * \beta * 2) = 0$$

$$\Rightarrow \beta = 1.67$$

If the performance function is considered as $g = 4c/(\gamma H) - 1$, then the reliability index is calculated as follows:

		1st iteration	2nd iteration	3rd iteration	4th iteration
Step 1	$g = 4c/(\gamma H) - 1$				
Step 2	β	3	3	3	3
Step 3	dp_c	$\mu = 100$	100	100	100
		$\sigma = 30$	30	30	30
Step 4	dp_γ	$\mu = 20$	20	20	20
		$\sigma = 2$	2	2	2
Step 5	$\partial_g/\partial_c = 4/(\gamma H)$	0.02	0.01827	0.01974	0.0198
	$\partial_g/\partial_\gamma = -4c/(\gamma^2 H)$	−0.1	−0.01219	−0.00983	−0.00985
Step 6	α_c	0.9487	0.9990	0.9994	0.9995
	α_γ	−0.3162	−0.0445	−0.0332	−0.0332
Step 7	dp_c	14.6185	10.0890	10.0495	10.0495
	dp_γ	21.8974	20.2668	20.1990	20.1989
		$\lvert\alpha i - \alpha i + 1\rvert =$	0.0503	0.00044	6.287E-07
			0.2718	0.01129	1.895E-05

After the convergence of α values:

$$\text{Step 8:}\ \frac{4(100 - 30 * 0.999\beta)}{10(20 - 2 * (-0.0332) * \beta)} - 1 = 0 \Rightarrow \beta = 1.66.$$

23.4.3 Monte Carlo Simulation (MCS)

The Monte Carlo method is a highly versatile and powerful simulation tool applicable to a wide range of problems. Its simplicity and completeness in simulations often lead to reasonably accurate results. One of its key advantages is the flexibility to handle various probabilistic distributions without creating high estimates, and it can effectively model correlations between variables.

In this method, the frequency of negative events for the performance function, based on the actual number of realizations, serves as an indicator for calculating the probability of failure. In contrast, the first and second-order methods assume a normal distribution for the performance function to estimate the probability of failure.

Optimizing the number of simulations is crucial for obtaining accurate estimates of statistical moments of the joint distribution function, which in turn provides valuable information such as failure probability. The Monte Carlo method offers great potential for reliably analyzing complex engineering scenarios.

In this method, the value of the reliability index is calculated according to Equation 23.9:

$$R = 1 - \frac{N_f}{N} \tag{23.9}$$

where N is the total number of simulations performed, and N_f is the number of failed simulations. In this way, the probability of failure is computed as follows:

$$P_f = \frac{N_f}{N} \tag{23.10}$$

Example 23.3 Estimate a simple integral of $\int x^2 dx$ in MCS.

One can run the following MATLAB code:

```
sum=0;
% "n" shows the dimension of the matrix, which contains generated random numbers.
% "k" shows the number of iterations, which is considered to estimate this integral.
n=100;
k=100;
for kk=1:k
hit=0;
miss=0;
X=rand(n);
Y=rand(n);
  for i=1:n
    for j=1:n
        W(i,j)=X(i,j)^2;
      if Y(i,j)<W(i,j) ;
           hit=hit+1;
      else
           miss=miss+1;
        end
    end
end
           integration=hit/(n*n);
           sum=sum+integration;
end
% the number, called "result," is the mean of iterations which is done above and is
also the final result.
result=sum/k
```

Example 23.4 What is the probability of failure for a shallow foundation given below?

Square footing: footing width = 2 m; embedment depth: 1 m, q_{all} = 300 kPa, γ = 18 kPa, c' and tan ϕ' are independent random variables.

Random variable	Mean	COV	PDF
c' (kPa)	5	1	Normal
tan ϕ'	0.58	0.09	Normal

File Home Insert Page Layout Formulas Data Review View Help Foxit PDF Tell me what you want to do

B993 =NORMINV(RAND(), 5, 1)

	A	B	C	D	E	F	G	H	I	J	K	L	M	N	O
1	No.	c'	tanϕ'	ϕ'	Nq	Nc	Nγ	ics	icd	iqs	iqd	iγs	iγd	qult	# Fail
993	992	7.55	0.66	0.58	27.05	39.65	36.85	1.68	1.20	1.66	1.13	0.60	1.00	1916.99	0.00
994	993	5.33	0.55	0.50	15.90	27.22	18.50	1.58	1.20	1.55	1.15	0.60	1.00	984.14	0.00
995	994	4.74	0.53	0.49	14.57	25.62	16.49	1.57	1.20	1.53	1.15	0.60	1.00	868.17	0.00
996	995	4.69	0.75	0.64	41.93	54.67	64.27	1.77	1.20	1.75	1.12	0.60	1.00	2716.63	0.00
997	996	5.36	0.58	0.53	18.59	30.36	22.71	1.61	1.20	1.58	1.14	0.60	1.00	1164.68	0.00
998	997	3.87	0.41	0.39	8.03	17.18	7.40	1.47	1.20	1.41	1.16	0.60	1.00	432.96	0.00
999	998	3.13	0.57	0.52	17.91	29.57	21.63	1.61	1.20	1.57	1.15	0.60	1.00	992.33	0.00
1000	999	5.06	0.47	0.44	10.70	20.78	10.93	1.52	1.20	1.47	1.16	0.60	1.00	635.77	0.00
1001	1000	6.38	0.58	0.53	18.59	30.36	22.71	1.61	1.20	1.58	1.14	0.60	1.00	1224.78	0.00
1002															
1003	Mean	5.00	0.58												5.00
1004	ST.Dev	1.00	0.09												
1005															
1006															
1007		c'	tanϕ'	ϕ'	Nq	Nc	Nγ	ics	icd	iqs	iqd	iγs	iγd	qult	Pf (%)
1008		5.00	0.58	0.52	20.51	32.07	26.57	1.62	1.20	1.58	1.14	0.60	1.00	1284.37	0.50
1009															

Example 23.5 The bearing capacity of 30 piles was predicted by four A, B, C, and D methods and compared to measured values, as provided in the table. Assess the performance of these methods via (a) mean and COV of the model parameter; (b) best-fitted line; (c) 20% accuracy level; and (d) P_{50} and P_{90}–P_{50}.

The predicted and measured bearing capacity values for 30 piles.

	Predicted (kN)				
Pile	**Method A**	**Method B**	**Method C**	**Method D**	**Measured (kN)**
1	1,280.1	767.2	1,839.4	1,545.3	2,150
2	1,686.1	1,324.9	4,189.3	1,595.3	1,050
3	1,067.4	809.6	847.5	996.6	1,550
4	1,568.7	960.6	2,403.6	1,140	2,618
5	1,727.4	1,130.2	2,253.4	1,941.6	2,445
6	1,107.3	523.7	1,666.7	998	855
7	1,146.1	1,079.3	1,197.8	1,905.5	1,750
8	1,242.3	648	1,919	1,099.3	1,527
9	1,979.8	1,140.3	3,929.6	1,400.3	1,742.9
10	1,979.8	1,140.3	3,929.6	1,400.3	1,249.1
11	2,740.1	2,403	2,602.9	6,713.5	3,798.1

Pile	Predicted (kN) Method A	Method B	Method C	Method D	Measured (kN)
12	2,949.5	2,272.7	5,080	6,748.5	3,291
13	3,530	2,836.1	39,57.2	9,200.5	3,400
14	657.5	313.2	750.6	600.3	842
15	549.8	559.6	278.7	450.8	805.3
16	34,772.8	41,645.4	67,208.5	10,0621.4	35,897.9
17	18,373.5	57,134.9	35,499	80,361.5	32,000
18	3,993	2,802	3,300.4	2,858.1	4,200
19	1,596	1,241	1,373.8	940.6	1,300
20	5,145	4,475	6,718.9	9,344	4,700
21	3,971	3,984	6,718.9	9,344	3,800
22	3,971	3,984	6,718.9	9,344	3,700
23	933	903	871.4	1,045.7	997.1
24	934	902	909.6	1,087.6	1,013
25	4,790	3,152	4,154	4,131.4	4,322.5
26	2,516	1,896	1,755.3	2,408.6	2,500
27	4,343	3,402	4,415.1	4,174.6	4,460
28	244	217	171.4	266.9	301.3
29	630	633	836.8	650.4	636
30	1,136	1,053	1,226.9	1,420.5	1,100

The performance of four predictive methods.

Methods	A: Mean (Q_m/Q_p)	COV	R_1	B: Best fitted line (Q_{fit}/Q_m)	R^2	R_2	C: 20% accuracy level	R_3	D: P_{50}	$P_{90}-P_{50}$	R_4	Final rank
Method A	1.13	0.27	3	0.81	0.92	1	52	1	1.04	0.50	3	8
Method B	1.45	0.40	4	1.39	0.93	2	29	4	1.34	1.04	4	14
Method C	1.02	0.52	2	1.51	0.92	3	29.8	3	0.98	0.51	1	9
Method D	1.00	0.47	1	2.58	0.97	4	30.1	2	0.97	0.51	2	9

23.5 Case Studies

In this section, different aspects of uncertainty in geotechnical engineering will be investigated as follows. The first example is about the loading uncertainty effect on a relic bridge in Babolsar, Iran, elaborated by Eslami et al. (2016). The second is about model uncertainty of various axial pile-bearing capacity predictive methods (Heidarie Golafzani et al. 2020a, b).

Case Study 23.1

Associated Uncertainties Affecting Babolsar Bridge Failure (Application of MCS)

In 2016, Eslami et al. conducted a study on a historic pedestrian bridge situated in Babolsar, northern Iran, to explore the impact of loading uncertainties in its design (Eslami et al. 2016). This bridge had encountered damage from a flood in 1964, resulting from geotechnical, hydraulic, and structural factors. Its structural components comprised reinforced concrete, featuring four river piers and five spans with hinge supports of approximately equal lengths. Each of the four piers was supported by a shallow foundation, measuring 1.4 m in width and 4 m in length, with an initial embedment depth of 2 m before the flood.

The bridge's overall dimensions included a width of 2 m and a length of 101 m. The flood in 1964 led to increased water levels and the formation of whirl water streams, causing scouring around the foundation. Consequently, settlement occurred in the bridge's piers, leading to subsidence in its deck supports. The evolution of this historical bridge over time is illustrated in Figure 23.5.

Figure 23.5 Babolsar Pedestrian Bridge over time: (a) original bridge after the 1964 flood, (b) damaged bridge, and (c) retrofitted bridge (Eslami et al. 2016 / with permission of Elsevier).

As time passed, the settlement of the historic city bridge increased, leading to its eventual damage. The bridge was analyzed under three loading conditions: (i) before the occurrence of the flood, (ii) during the flood event, and (iii) after the flood. In deterministic analyses, failure was assumed to occur when the safety factor (FS) was less than two. The safety factor represents the ratio of ultimate bearing capacity to the applied vertical load on the footing area. Figure 23.6 presents the results of the deterministic analyses, where the symbols B, D, and A represent "before,"

(Continued)

Case Study 23.1 (Continued)

"during," and "after" flood conditions, respectively. TSA and ESA stand for total stress analysis and effective stress analysis, respectively. *G* and *L* refer to general shear failure and local/punching shear failure, respectively.

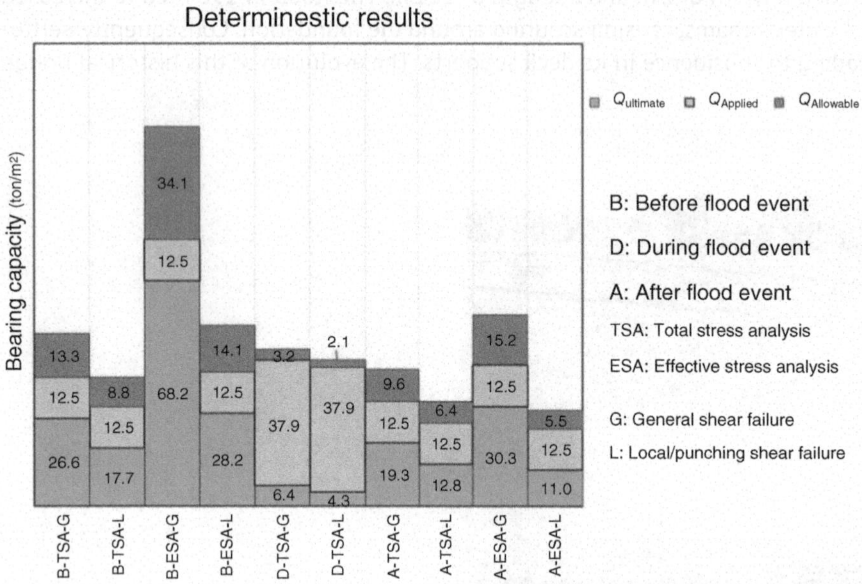

Figure 23.6 Results of deterministic solutions (Eslami et al. 2016 / with permission of Elsevier).

Under the "before flood" loading condition, the total stress analysis showed a probability of local shear/punching failure, indicating that the bridge did not meet the capacity requirements. However, in other cases, the bridge remained safe. During the "during flood" conditions, changes in loading and foundation conditions resulted in unsatisfactory bridge performance, with q_{ult} (ultimate bearing capacity) being significantly lower than q_{app} (applied stress) in all cases. For the "after flood" conditions, the analyses assumed that no failure had occurred, and only the embedment depth was reduced to zero due to the flood. The results indicated that with the reduction in embedment depth, at least three out of four cases showed failure occurrence.

In reliability-based analyses, soil shear strength parameters, i.e., c (cohesion) and ϕ (friction angle), were considered as random variables, distributed normally under two conditions: uncorrelated and correlated. The safety factor, FS, was defined as the ratio of ultimate bearing capacity to the applied stress. The probability of failure, P_f, was calculated as the ratio of the frequency of FS being less than unity to the total number of FS.

In Table 23.4, the results of the reliability analysis are presented. A comparison of the ultimate bearing capacity for both methods indicates that the calculated bearing capacity values are similar. However, the reliability analysis reveals that there is always a probability of failure for all cases. The probability of failure before the flood is lower than the probability after the flood conditions, which is attributed to the reduction in foundation embedment depth during the flood. Notably, during the flood, the probabilities of failure for both cases were 100%, indicating the significant impact of extreme loading and the reduction in foundation-bearing capacity on the structural integrity of the bridge.

Due to the lateral applied pressures induced by the flood, eccentric loading becomes inevitable. When the eccentricity exceeds certain limits, tension in the foundation base (uplift) occurs. Since soil has almost zero tensile strength, a portion of the footing will not effectively transmit loads to the soil. To avoid this, the footing should be designed to satisfy the conditions $e_B < B/6$ and $e_L < L/6$ (where B and L are footing width and length, respectively; e_B and e_L are eccentricity in these directions). In the case of the Babolsar Bridge, changing the bearing width from 2 to 3 m results in an eccentricity (i.e., e_L) change from 1.67 to 2.5 m. To meet the requirement of $6e_L < L$, the footing requires a minimum length of 10 to 15 m. Compared to the present length, it is evident that the bridge would collapse during a flood event.

Table 23.4 MCS results compared to deterministic results (Eslami et al. 2016 / with permission of Elsevier).

Condition			P_f (%)	FS		q_{ult}		q_{ult}
			MCS	Mean	SD	Mean	SD	(Deterministic)
Before flood	TSA	General shear failure	1.68	2.13	0.53	26.63	6.66	26.56
		Local shear failure	11.83	1.42	0.36	17.74	4.44	17.68
	ESA	General shear failure	0	5.70	1.66	71.17	20.77	68.18
		Local shear failure	0.06	2.18	0.45	27.29	5.6	28.18
During flood	TSA	General shear failure	100	0.07	0.02	2.7	0.67	6.37
		Local shear failure	100	0.05	0.01	1.8	0.45	4.25
After flood	TSA	General shear failure	8.1	1.54	0.39	19.26	4.8	19.25
		Local shear failure	45.8	1.03	0.26	12.8	3.2	12.8
	ESA	General shear failure	0.25	3.3	1.08	41.67	13.55	30.3
		Local shear failure	27.27	1.2	0.32	15.05	4.06	11.01

q_{ult} is in ton/m^2.

These investigations highlight the importance of thorough site investigations, soil laboratory testing, and the consideration of loading combinations, particularly in extreme events. The deterministic and reliability analyses clearly indicate that a shallow foundation is not suitable in this case. Therefore, to withstand the effects of flood occurrences or other dynamic loads, the pier foundation of this bridge should be constructed deeper, or at the very least semi-deep, to provide the necessary stability for the bridge superstructure.

Case Study 23.2

Model Uncertainty Appraisal of Various Methods for Predicting Axial Pile-Bearing Capacity

A comprehensive database has been compiled from the results of 60 full-scale pile loading tests characterized by penetrometer soundings conducted in the proximity of pile locations (Heidarie Golafzani et al. 2020a, b). The case records were obtained from three reputable sources: Eslami and Fellenius (1997) contributed 13 cases, Niazi (2014) added 30 cases, and the AUT-CPT and pile database released by Moshfeghi and Eslami (2016) provided 17 cases. The compiled database includes data from 13 countries spanning four different continents. The geographical distribution of the cases is as follows: North America contributes 26 cases (43.4%), Europe provides 17 cases (28.3%), Asia accounts for 15 cases (25%), and there are two cases (3.3%) from Australia. Among the individual countries, the maximum number of cases (16) is from the USA, followed by ten cases from Canada, seven from Norway, and five from Japan. The pile characteristics in the database vary significantly, with pile lengths ranging from 6 to 79 m and diameters spanning from 114 to 1500 mm. The shape and material of the piles also differ, including round, square, and H piles made from either steel or concrete. For further details and a comprehensive summary of the case records, please refer to the work by Heidarie Golafzani et al. (2020a).

Pile Load Test Failure Criteria Interpretation

The selected piles underwent axial load testing, either under top-down compression or top-up tension, with the exception of AZDEGAN (TP14, TP22, and TP28), for which pile driving analyzer (PDA) tests were conducted. Approximately 95% of the load test results include a load–displacement diagram, while the remaining 5% contain only the ultimate bearing capacity value. To ensure consistency and account for model uncertainties, a standardized load test interpretation procedure was employed to derive a single "measured" quantity from each test (Phoon and Kulhawy 2005). Among the six available failure criteria, three are commonly used: the Davisson offset limit (Davisson, 1972), the Brinch Hansen 80% criterion (Brinch Hansen 1963), and the Chin–Kondner extrapolation (Chin 1978). The Davisson

(Continued)

Case Study 23.2 (Continued)

limit typically reports loads in the lower part of the load–displacement diagram. Conversely, the Chin–Kondner criterion extrapolates the static pile load test diagram and assumes that the measured capacity is even higher than the maximum applied load during the test.

On the other hand, the Brinch-Hansen 80% criterion generally corresponds with the intuitively perceived plunging failure of the pile. The so-called Hansen 80% criterion yields the most consistent results with the least scatter, making it the more appropriate criterion to define the reference load in such analyses (Fellenius and Tech 2001; Moshfeghi and Eslami 2018). In cases where the load from Hansen (1963) exceeded the maximum applied load, the maximum applied load was considered the ultimate measured capacity. The pile capacities in the database range from 75 to 36,000 kN, with an average capacity of 2771 kN.

Model Uncertainty

Predictive methods incorporate various assumptions and bases, impacting their performance due to model uncertainty. Additionally, inherent soil variability is directly reflected in the measurements when CPT records are used to predict pile-bearing capacity. Studies by Jamshidi Chenari and Kamyab Farahbakhsh (2015) and Jamshidi Chenari et al. (2018) have clearly distinguished the stationary and nonstationary aspects of inherent soil variability using CPT records. Moreover, recent studies (e.g., Paikowsky et al. 2004; Allen 2005; Heidarie Golafzani et al. 2020a) have suggested that there is no need to account for error induced by inherent soil spatial variability or measurement error due to the accuracy and size of the CPT database.

As a result, the model and transformation uncertainties associated with the used predictive models significantly influence the accuracy and efficiency of a method (Phoon and Kulhawy 2005). To quantify these uncertainties, a parameter termed the model parameter is defined as the ratio of the measured to predicted pile axial capacity ($\lambda_R = Q_m/Q_p$). Figure 23.7 illustrates the scatter predictions against the measured pile-bearing capacities. Since the resistance bias factor does not attain negative values, it is assumed to follow a log-normal distribution.

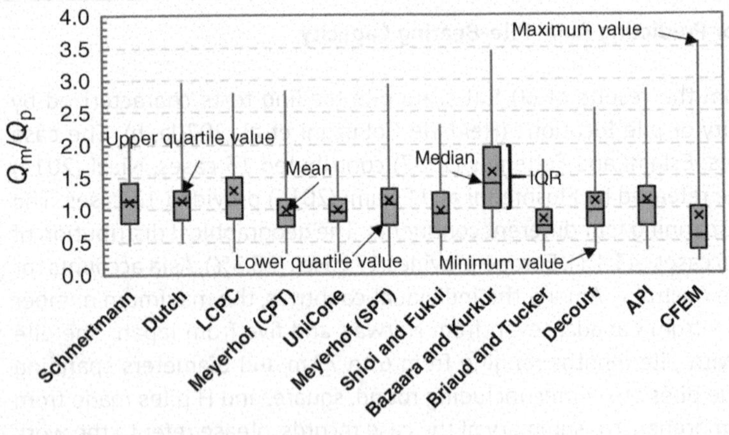

Figure 23.7 Box plot for resistance bias factor (Q_m/Q_p) from different predictive methods (Heidarie Golafzani et al. 2020b).

Statistical Assessment of Model Parameter

This classification involves assessing the mean and standard deviation of the model parameter, as well as comparing the best-fit line and root mean square error (RMSE). The comparison of the mean and standard deviation of the model parameters for different methods reveals the accuracy and precision of each method. A value closer to 1 for the mean (μ) and closer to 0 for the standard deviation (σ) indicates a method's proximity to an ideal case ($\mu_{\lambda R} = 1, \sigma_{\lambda R} = 0$). Additionally, the best-fit line examines the accuracy of the methods, and the closer Q_m/Q_p is to 1, the better the performance of the method. The scatter of the data and the best-fit line criteria are presented in Figure 23.8. To monitor model errors associated with each method, the root mean square error (RMSE) is used, and a lower RMSE value indicates a better method. The results of the statistical criteria are presented in Table 23.5.

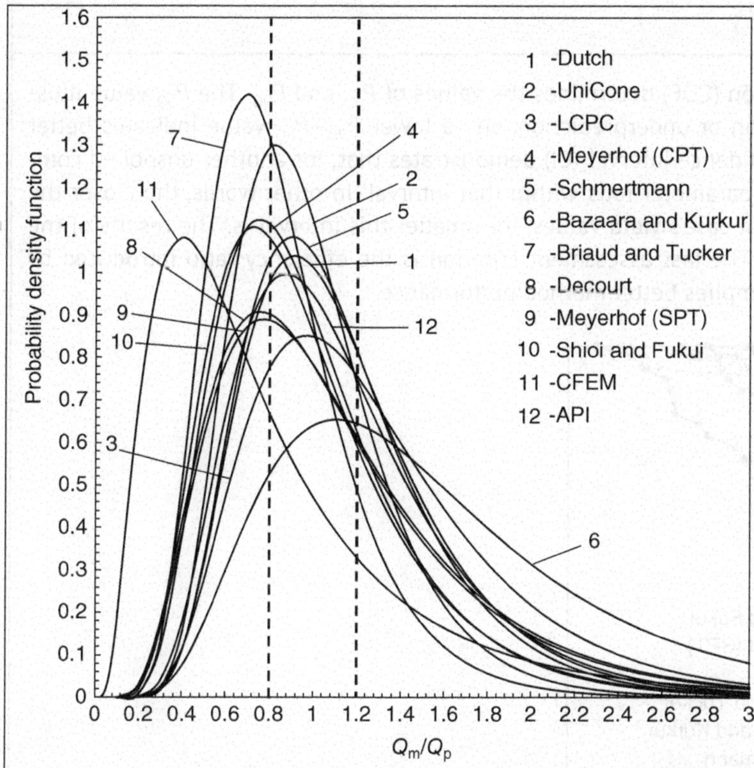

Figure 23.8 Log-normal distribution of Q_m/Q_p for different methods (Heidarie Golafzani et al. 2020b).

Table 23.5 Statistical assessment of the model parameter.

	Mean and standard deviation			Best fit line		
	μ	σ	COV	Q_{fit}/Q_m	R^2	RMSE
Schmertmann (CPT)	1.12	0.39	0.35	1.18	0.98	1,499
Dutch (CPT)	1.14	0.47	0.41	1.23	0.98	1,866
LCPC (CPT)	1.29	0.6	0.47	1.38	0.93	3,401
Meyerhof (CPT)	0.99	0.44	0.44	0.98	0.92	1,714
UniCone (CPT)	0.99	0.33	0.33	0.82	0.92	1,864
Meyerhof (SPT)	1.13	0.59	0.52	1.48	0.92	4,196
Shioi and Fukui (SPT)	0.98	0.42	0.43	2.51	0.96	10,572
Bazaraa and Kurkur (SPT)	1.57	0.78	0.5	1.2	0.95	2,087
Briaud and Tucker (SPT)	0.85	0.34	0.4	1.03	0.98	876
Decourt (SPT)	1.12	0.61	0.54	2.09	0.96	7,722
API (static)	1.14	0.5	0.44	1.17	0.98	1,486
CFEM (static)	0.87	0.76	0.86	1.81	0.92	6,250

Probabilistic Assessment of the Model Parameter

This approach investigates the model parameter through its distribution function. The 20% accuracy level is an efficient indicator of the model performance, assessing the precision of the model parameter around unity. A larger area between $0.8 < Q_m/Q_p = 1.2$ indicates better method performance, implying higher accuracy and precision. Figure 23.8 provides a more comprehensive performance evaluation of the different methods by superimposing a log-normal density function on the model parameter (i.e., the measured ratio to predicted pile axial capacity data).

(Continued)

Case Study 23.2 (Continued)

Additionally, the cumulative distribution function (CDF) determines the values of P_{50} and P_{90}. The P_{50} value illustrates the tendency of a method to overprediction or underprediction, and a lower $P_{90}-P_{50}$ value indicates better precision for each method (Figure 23.9). The confidence interval (CI) demonstrates that, for another unspoiled compiled data bank/sample, the mean of the model parameter falls within that interval. In other words, the closer the method's statistical characteristics are to the ideal case's yield values, the smaller this interval is. The results of the probabilistic criteria are provided in Table 23.6. The last assessment criterion is the efficiency ratio introduced by Paikowsky et al. (2004). A higher efficiency ratio implies better method performance.

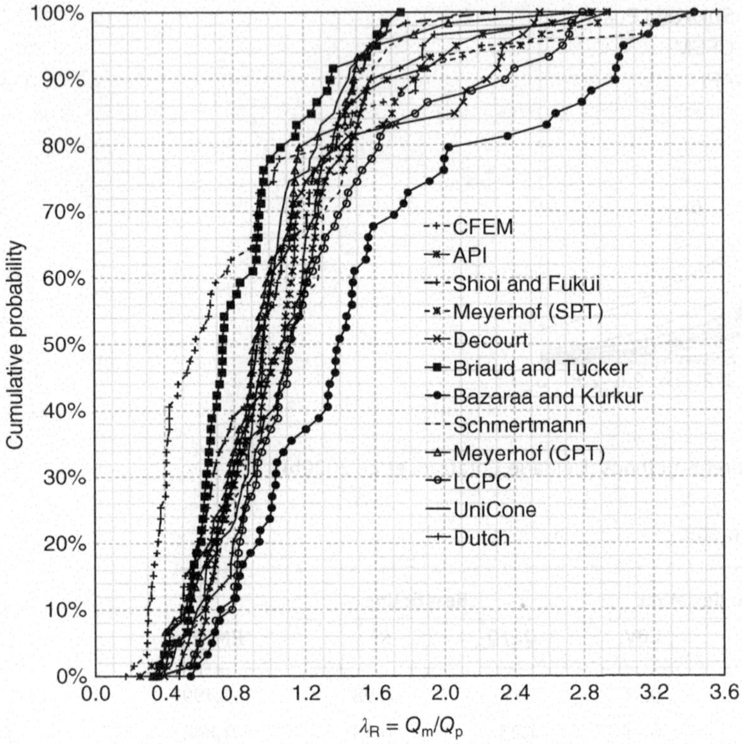

Figure 23.9 Cumulative distribution function; comparison of P_{50} and $P_{90}-P_{50}$ (Heidarie Golafzani et al. 2020b).

Table 23.6 Probabilistic assessment of the model parameter (Heidarie Golafzani et al. 2020b).

Methods	20% accuracy level	Cumulative distribution function			Confidence interval
		P_{50}	P_{90}	$P_{90}-P_{50}$	
Schmertmann (CPT)	0.4	1.13	1.59	0.46	0.1
Dutch (CPT)	0.39	1.13	1.66	0.53	0.12
LCPC (CPT)	0.33	1.12	2.38	1.26	0.16
Meyerhof (CPT)	0.36	0.92	1.49	0.57	0.11
UniCone (CPT)	0.43	0.96	1.46	0.5	0.08
Meyerhof (SPT)	0.31	1.09	1.85	0.76	0.15
Shioi and Fukui (SPT)	0.34	0.93	1.51	0.58	0.11
Bazaraa and Kurkur (SPT)	0.25	1.4	2.99	1.59	0.2
Briaud and Tucker (SPT)	0.35	0.73	1.36	0.63	0.09
Decourt (SPT)	0.31	0.97	2.28	1.31	0.16
API (static)	0.37	1.09	1.79	0.7	0.13
CFEM (static)	0.19	0.57	1.86	1.29	0.2

Figure 23.10 presents a comparison of different methods' performance using radar charts, assessing statistical, probabilistic, and reliability-based criteria. In this study, scores on each criterion range from 12 to 1, with the method showing the best performance receiving a score of 12 and the poorest performance receiving a score of 1. A larger shaded area on the radar chart indicates a method that outperforms the others.

UniCone appears to have a better performance, possibly due to several factors, including: (i) incorporation of excess pore water pressure to predict the long-term behavior of piles; (ii) utilization of a CPT-based soil classification system for improved estimation of shaft resistance; (iii) correction of the cone stress for the pore water pressure acting on the cone shoulder; and (iv) consideration of the failure zone for estimating the unit toe resistance. These features contribute to the superior performance of the UniCone method compared to others.

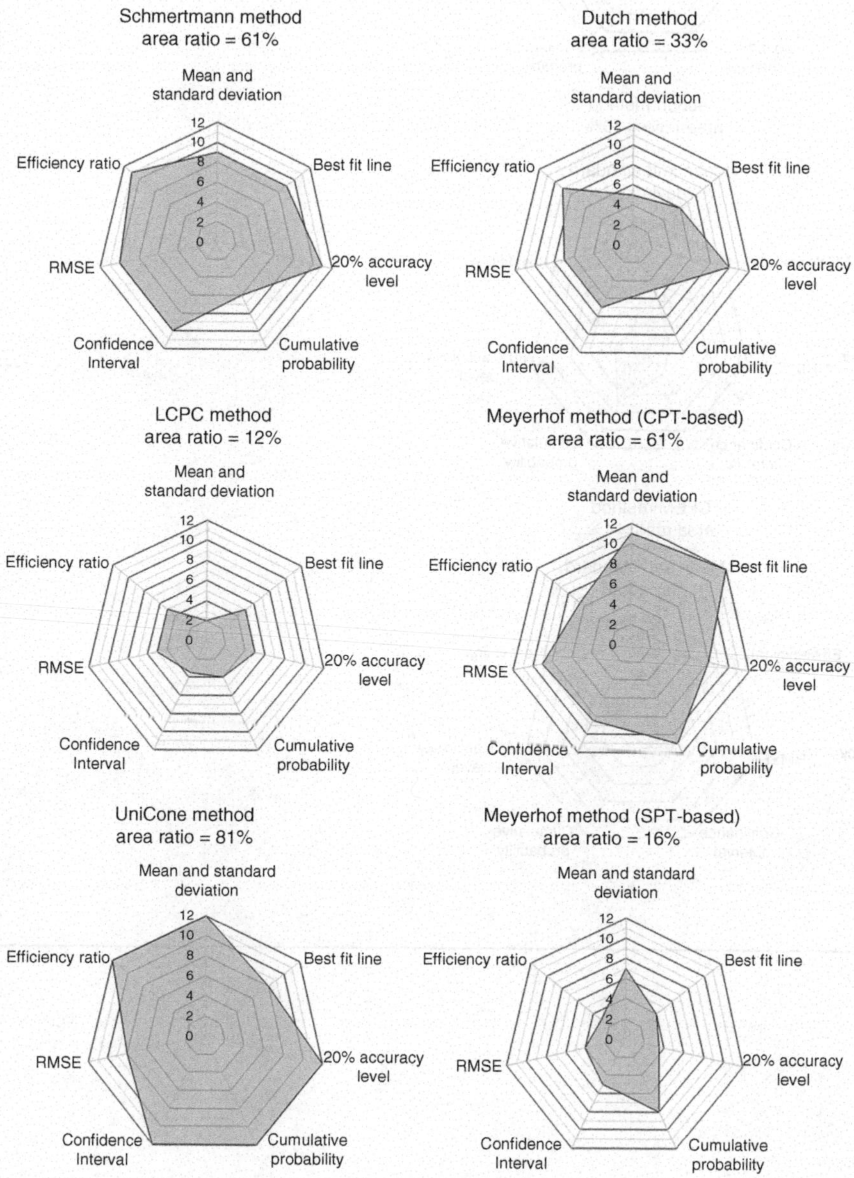

Figure 23.10 Assessment of different methods based on the investigated criteria (Heidarie Golafzani et al. 2020b).

(Continued)

Case Study 23.2 (Continued)

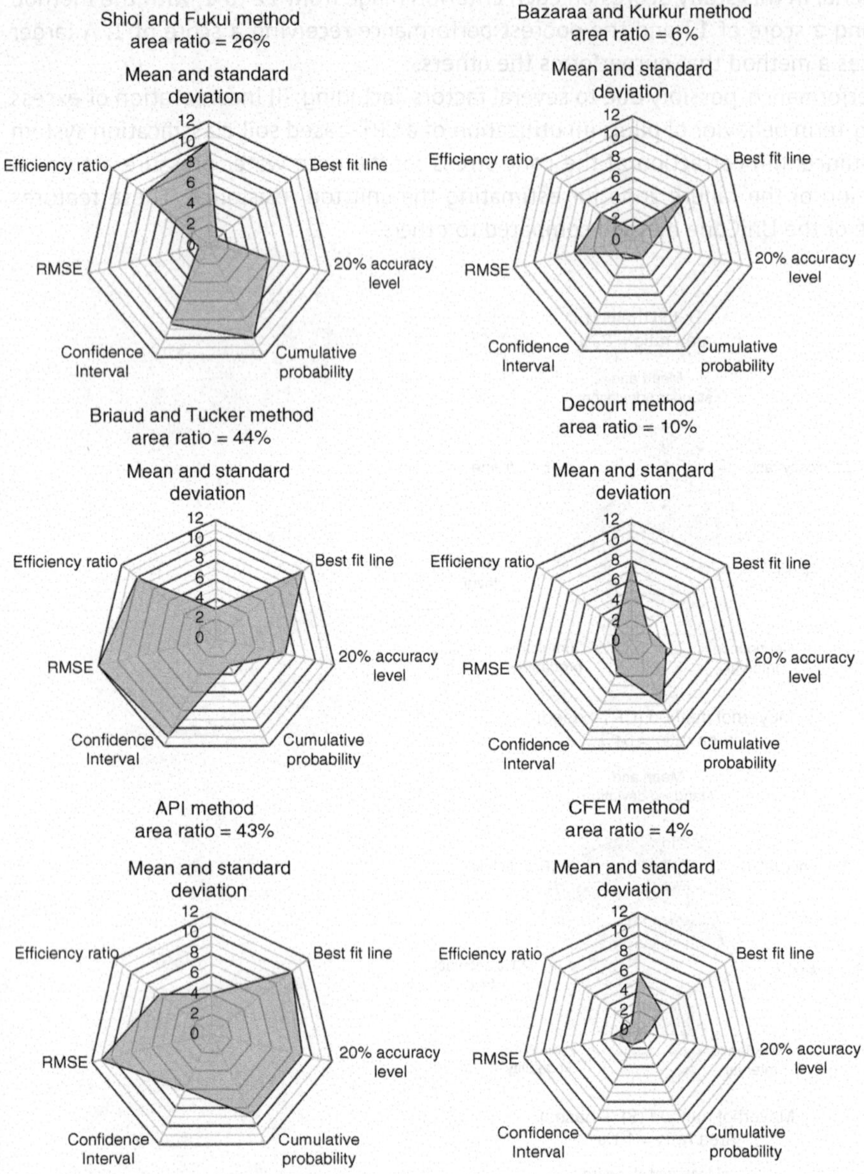

Figure 23.10 *(Continued)*

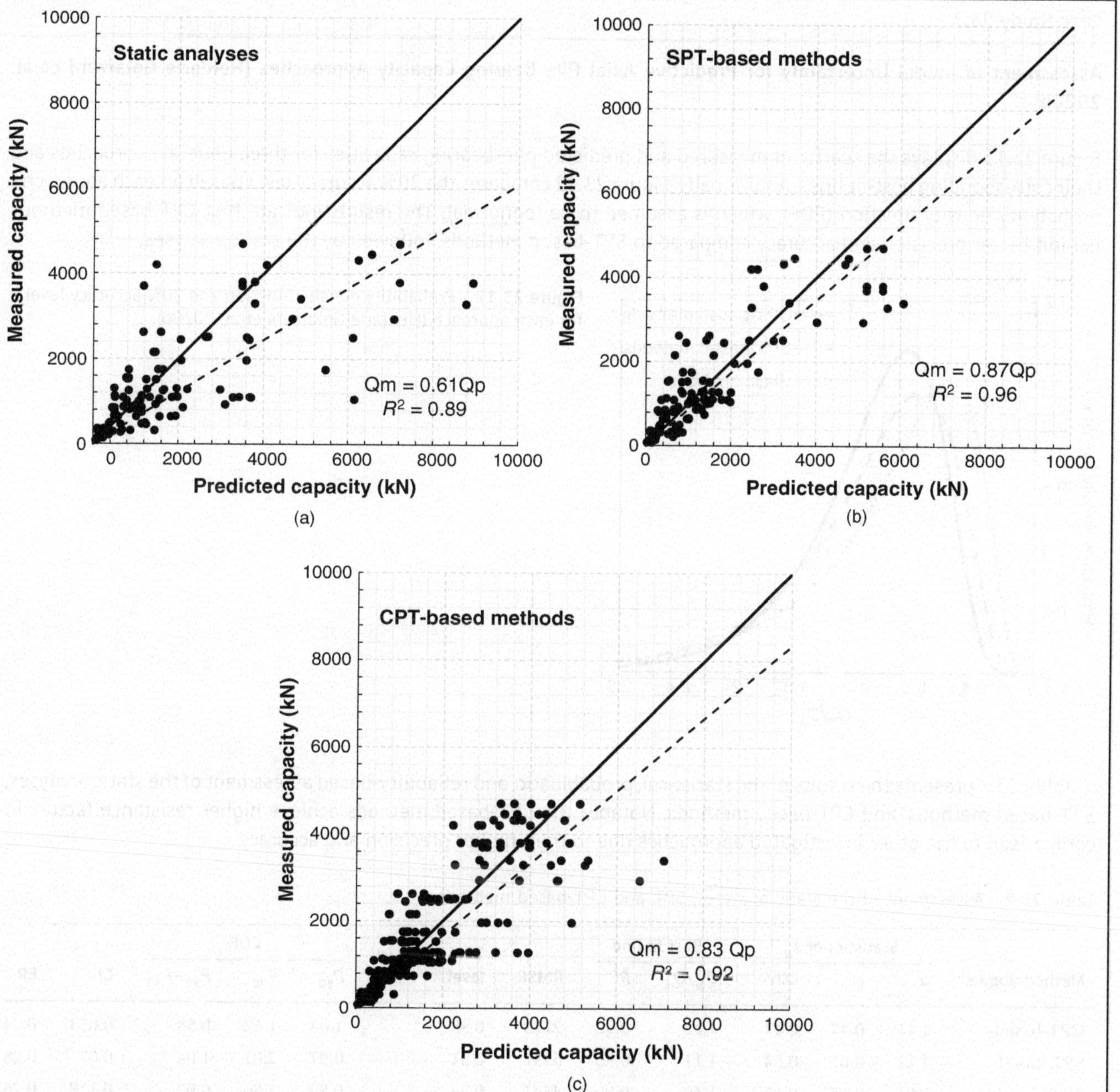

Figure 23.11 Predicted vs. measured pile bearing capacity for three general approaches: (a) Static analysis method, (b) SPT-based method, and (c) CPT-based method.

Case Study 23.3

Assessment of Model Uncertainty for Predictive Axial Pile-Bearing Capacity Approaches (Heidarie Golafzani et al. 2020b)

Figure 23.11 displays the scatter of measured and predicted pile-bearing capacities for three general approaches and their corresponding best-fit lines. Additionally, Figure 23.12 compares the 20% accuracy level based on each approach's probability density function (PDF), which is assumed to be lognormal. The results indicate that CPT-based methods exhibit better precision and accuracy compared to SPT-based methods, followed by the static analyses.

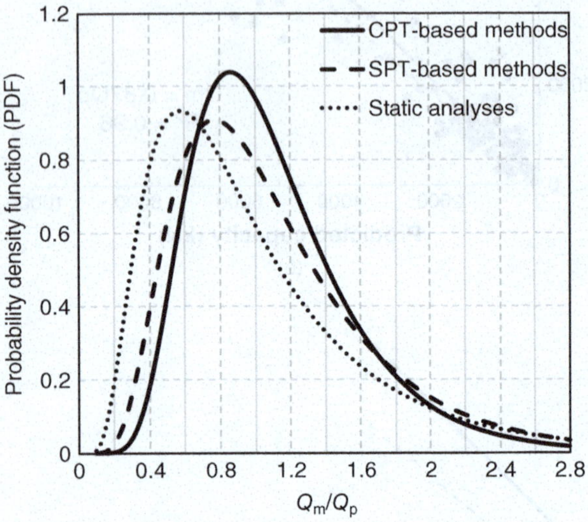

Figure 23.12 Probability density function and 20% accuracy level for each approach (Heidarie Golafzani et al. 2020b).

Table 23.7 presents the results of the statistical, probabilistic, and reliability-based assessment of the static analyses, SPT-based methods, and CPT-based methods. Notably, the CPT-based methods achieve higher resistance factors in comparison to the other investigated approaches due to their higher precision and accuracy.

Table 23.7 Assessment of the static analyses, SPT, and CPT-based methods.

Methodologies	Statistics of λ_R			Best fit line		RMSE	20% accuracy level	CDF			CI	ER
	μ	σ	COV	Q_{fit}/Q_m	R^2			P_{50}	P_{90}	$P_{90}-P_{50}$		
CPT-based	1.11	0.47	0.42	1.12	0.92	2178	0.38	1.04	1.62	0.58	0.053	0.44
SPT-based	1.13	0.61	0.54	1.11	0.96	6231	0.31	0.97	2.01	1.04	0.07	0.28
Static analyses	1.01	0.65	0.65	1.49	0.89	4543	0.24	0.89	1.86	0.97	0.118	0.26

Figure 23.13 presents a comprehensive comparison of all assessment criteria. As previously mentioned, a higher area ratio indicates better performance among the approaches. Figures 23.11–23.13 clearly illustrate the higher reliability of direct CPT-based methods compared to SPT-based methods, static analyses, or indirect methods. The use of continuous CPT soundings, recorded under actual field stresses and boundary conditions, provides more reliable and repeatable input data for CPT-based methods. These methods establish a direct correlation between pile unit toe and shaft resistance with q_c and f_s readings, employing coefficients primarily dependent on soil and pile types. As a result, direct CPT-based methods exhibit lower uncertainty in predicting pile capacity. Table 23.7 provides a detailed comparison of each approach based on the introduced criteria.

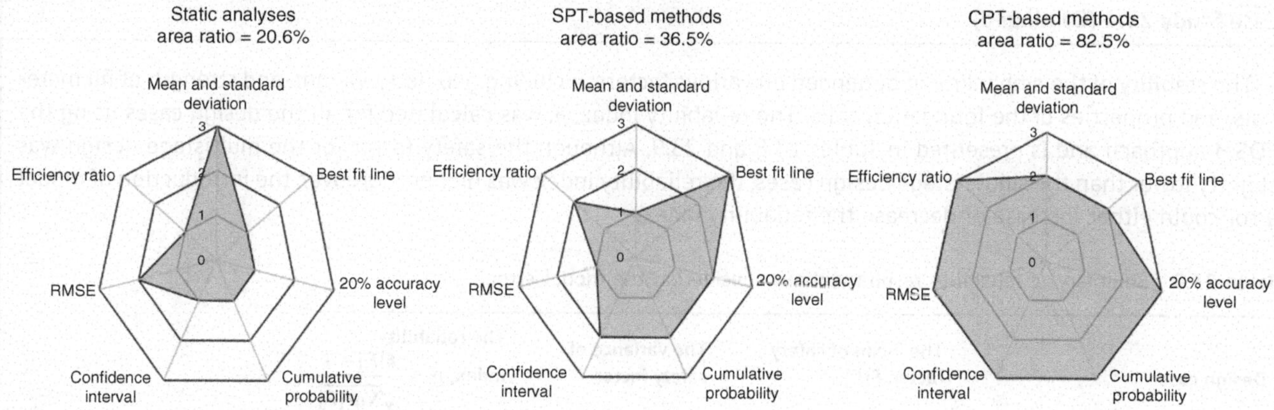

Figure 23.13 Assessment of the three common universal axial pile-bearing capacity prediction approaches (Heidarie Golafzani et al. 2020b).

While SPT is a widely used, cost-effective, and simple in situ test, the NSPT values are recorded at discontinuous intervals, influenced by various factors, and require correction. Consequently, SPT-based methods lack the same level of reliability as direct CPT-based methods. However, due to its in situ nature, SPT is preferred over static analyses, as evident in Figures 23.11–23.13.

Static analyses rely on different empirical correlations to estimate geo-parameters and incorporate them into simplified geotechnical models. Furthermore, these geo-parameters are treated as deterministic in these methods, while SPT and CPT records demonstrate the variability of soil properties in both horizontal and vertical directions. Ignoring the inherent soil variability leads to a higher level of error associated with the estimation of pile axial-bearing capacity in static analyses.

Case Study 23.4

The Reliability Assessment of Different Design Cases for the James Bay Dikes (Baecher and Christian 2003)

During the James Bay hydroelectric project, the construction of approximately 50 km of dikes on soft and sensitive clays posed several challenges. Strengthening the foundation soil during construction, selecting appropriate safety factors, and evaluating strength properties led to diverse professional opinions. To address these issues and ensure slope stability for dike heights ranging from 6 to 23 m, an international committee was formed to reach a consensus. Two main design approaches were proposed: single-stage and multistage construction of the typical dike. The objective of the analysis was to understand the relative safety of different designs, gain insight into the influence of various parameters, and establish consistent criteria for preliminary designs.

For the single-stage cases, embankments were constructed either to a height of 6 m without berms or to a height of 12 m with one berm. Stability analyses for these cases utilized undrained shear strength from vane tests and the simplified Bishop circular method. In contrast, the multistage construction involved vertical drains, construction up to 12 m, subsidence to 80% of primary consolidation, and subsequent building up to 23 m. Consolidated-undrained shear tests, combined with stress histories and the Morgenstern–Price method of slices, were employed for stability analyses. The simplified soil profile for the investigated site consisted of a 4 m crust layer, followed by 8 m of highly sensitive marine clay, and further underlain by 6.5 m of lacustrine clay, with a total clay thickness of 18.5 m.

In the single-stage cases, there were uncertainties about the crust thickness, depth to the till, unit weight, and friction angle of the fill due to the extensive length of the embankment and limited exploration program at each site. On the other hand, the multistage case had a more limited length of embankment, but an extensive exploration and instrumentation program. The major uncertainty in the multistage design was related to the properties of the fill and the partially consolidated clay.

(Continued)

Case Study 23.4 (Continued)

The stability of the embankment depended on various factors, including geometry, weight, and strength of fill materials, and properties of the foundation soils. The reliability index, β, was calculated for all the design cases using the FOSM approach and is presented in Tables 23.8 and 23.9. Although the safety factor for the multistage design was slightly lower than the single-stage design cases, the reliability index was higher. Moreover, the introduction of model error could either increase or decrease the reliability index.

Table 23.8 Summary of reliability results without considering the model error.

Design case	The mean of safety factor, $E[F]$	The variance of safety factor	The reliability index, $\beta = \frac{E[F]-1}{\sqrt{\text{Var}[F]}}$
Single stage dike, $H = 6$ m	1.5	0.074	1.84
Single stage dike, $H = 12$ m	1.453	0.029	2.66
Multistage dike, $H = 23$ m	1.427	0.0134	3.69

Table 23.9 Summary of reliability results considering the model error.

Design case	The mean of safety factor, $E[F]$	The variance of safety factor	The reliability index, $\beta = \frac{E[F]-1}{\sqrt{\text{Var}[F]}}$
Single stage dike, $H = 6$ m	1.575	0.086	1.96
Single stage dike, $H = 12$ m	1.526	0.041	2.60
Multistage dike, $H = 23$ m	1.498	0.0244	3.19

The reliability index holds greater significance than the safety factor in stability analysis. Selecting an appropriate value for the safety factor can be challenging, and the inherent soil variability plays a crucial role in uncertainty assessment for slope and embankment stability analyses. In the context of the James Bay dikes, there was insufficient information available to consider the correlation effect. However, not accounting for correlation effects was conservative in this instance, as its inclusion would have reduced the probability of failure and increased the reliability index. It is important to emphasize that these reliability analyses provide estimates of the probability of failure, not its absolute value, necessitating more comprehensive probabilistic risk analysis. Nevertheless, for most practical cases, calculating the relative probability of failure is adequate for parametric studies.

23.6 Major Points

1) Geotechnical engineering is accompanied by uncertainty sources, including soil's inherent variability, measurement error, transformation uncertainty, and model error. Random analysis techniques have been developed to quantify these uncertainty sources.
2) Instead of implementing a global factor of safety or an ASD approach for geotechnical designs, a reliability index has been introduced to quantify these uncertainties. The higher the reliability index, the lower the probability of failure.
3) The performance function and the joint distribution function are important for determining the reliability index. Various numerical methods have been developed to estimate the probability of failure, including FOSM, FORM, MCS, and Modified FOSM.
4) Reliability analysis reveals that failure occurrence is not improbable in different cases, enabling engineers to make informed decisions compared to deterministic methods. While a factor of safety is often used in deterministic

approaches to mitigate the risk of failure, it may not be sufficient, as demonstrated by the case of the Babolsar Pedestrian Bridge failure during a flood due to excessive displacements in the deck or bridge supports. Reliability assessment has shown that an increase in the negative correlation coefficient slightly affects mean FS and q_{ult} but decreases the standard deviation, resulting in increased homogeneity.

5) Among various predictive methods, in situ-based methods have shown better predictions for the axial-bearing capacity of piles. In situ tests are performed under existing field stresses and boundary conditions, making them more reliable and repeatable compared to SPT, the most commonly used in situ test, which may lack reliability.
6) Static analyses rely on empirical correlations to estimate geo-parameters for simplified geotechnical models. However, treating these geo-parameters as deterministic in static methods ignores the variation of soil properties in both horizontal and vertical directions, leading to higher-level errors in estimating pile axial-bearing capacity. The efficiency ratio proves to be a more appropriate factor for better understanding the reliability assessment.
7) Understanding the physical meaning of the safety factor is difficult, and selecting an appropriate value is fraught with difficulty in geotechnical engineering. However, the reliability index is more meaningful in the measure of stability than the safety factor. Furthermore, these reliability analyses lead to an estimation of the probability of failure, not its absolute value, requiring more elaborated probabilistic risk analysis.

23.7 Questions and Problems

23.7.1 Explanatory

1 Which method has a better estimation of axial pile-bearing capacity: in situ-based methods or static analysis?

2 What do the statistical and probabilistic criteria measure for the assessment of model parameters? Moreover, how do they work?

3 How does the probability of failure change if the random variables in Example 23.2 are negatively or positively correlated?

4 What is the difference between correlation length and the scale of fluctuation?

5 If R and Q are correlated random variables, and the performance function is $M = R - Q$, what is the FOSM reliability index?

6 Why do various predictive methods for estimating axial pile-bearing capacity lead to a wide range of predictions?

7 Why were methods such as FOSM, FORM, or MCS developed to estimate the reliability index?

8 What is the target reliability index? Furthermore, how is it defined?

23.7.2 Conceptual

1 What is the role of engineering judgment in geotechnical designs?

2 How do uncertainty sources affect geotechnical designs?

3 Why does FOSM result in an inaccurate prediction of the reliability index compared to FORM or MCS?

4 What is the shortcoming of the Hasofer and Lind (1974) method in determining the FORM reliability index?

5 How is the suitable number of simulations determined in MCS?

6 As illustrated in Figure 23.13, explain why CPT-based methods outperformed for the gathered databank.

7 Why do Baecher and Christian (2003) believe the reliability index is more meaningful than the safety factor?

8 Regarding Figure 23.8, which method outperforms and why?

9 Compare the performance of the predictive methods shown in Figure 23.7 via their IQR.

23.7.3 Problems

1 As presented in Table 23.3, R and Q are independent random variables:
(a) If $M = (R/Q) - 1$, prove that $\beta = \frac{\mu_R - \mu_Q}{\mu_R\left(\sqrt{\frac{\sigma_R^2}{\mu_R^2} + \frac{\sigma_Q^2}{\mu_Q^2}}\right)}$.
(b) If $M = R - Q$, prove that $\beta = \frac{\mu_R - \mu_Q}{\sqrt{\sigma_R^2 + \sigma_Q^2}}$.
(c) If R and Q are correlated and dependent random variables, how do the above reliability index formulas change?

2 For Example 23.3, how does the probability of failure change if the random variables are changed as below?
(a) If $\text{St.Dev}_c = 3$ and $\text{St.Dev}_{\tan\phi} = 0.3$
(b) If $\text{St.Dev}_c = 0.5$ and $\text{St.Dev}_{\tan\phi} = 0.09$

References

Allen, T.M. (2005). *Development of Geotechnical Resistance Factors and Downdrag Load Factors for LRFD Foundation Strength Limit State Design: Reference Manual*. US Department of Transportation, Federal Highway Administration, National Highway Institute.

American Petroleum Institute. (2000). Recommended practice for planning, designing, and constructing fixed offshore platforms–working stress design. API RP A, 2.

Ang, A.H. and Tang, W.H. (2007). *Probability Concepts in Engineering Planning: Emphasis on Applications to Civil and Environmental Engineering*. John Wiley and Sons.

Baecher, G.B. and Christian, J.T. (2003). *Reliability and Statistics in Geotechnical Engineering*. John Wiley & Sons.

Bazaraa, A.R. and Kurkur, M.M. (1986). N-values used to predict settlements of piles in Egypt. In: *Use of In Situ Tests in Geotechnical Engineering*, 462–474. ASCE.

Briaud, J.L. and Tucker, L.M. (1988). Measured and predicted axial capacity of 98 piles. *Journal of Geotechnical Engineering* 114 (9): 984–1001.

Brinch Hansen, J. (1963). Discussion of hyperbolic stress-strain response: cohesive soil by Robert L. Kondner. *Journal for Soil Mechanics and Foundation Engineering* 89 (4): 241–242.

CFEM (2006). *Canadian Foundation Engineering Manual*. Richmond, B.C. Canadian Geotechnical Society (CGS).

Chin, F.K. (1978). Diagnosis of pile condition. *Geotechnical Engineering* 9: 85–104.

Davisson, M.T. (1972). High capacity piles. *Proceedings, Soil Mechanics Lecture Series on Innovations in Foundation Construction* (January to May 1972), 81–112. Chicago: ASCE, Illinois Section.

Decourt, L. (1995). Prediction of load settlement relationships for foundations on the Basis of the SPT-T. *Ciclo de Conferencias Inter. Leonardo Zeevaert*, UNAM, Mexico, 85–104.

Eslami, A. (1996). Bearing capacity of piles from cone penetrometer test data. Doctoral dissertation, Ph D thesis. University of Ottawa, Department of Civil Engineering.

Eslami, A. and Fellenius, B.H. (1997). Pile capacity by direct CPT and CPTu methods applied to 102 case histories. *Canadian Geotechnical Journal* 34: 886–904.

Eslami, A., Golafzani, S.H., and Jamshidi Chenari, R. (2016). Assessment of Babolsar concrete pedestrian bridge failure for 1964 flood event and retrofitting practice. *Engineering Failure Analysis* 68: 101–112.

Fellenius, B.H. and Tech, P. (2001). *What Capacity Value to Choose from the Results a Static Loading Test*. New Jersey, NJ: Fulcrum, Deep Foundation Institute.

Haldar, A. and Mahadevan, S. (2000). *Probability, Reliability, and Statistical Methods in Engineering Design*. John Wiley.

Hansen, J.B. (1963). Discussion of hyperbolic stress-strain response: cohesive soil. *Journal of Soil Mechancis and Foundations Division, ASCE* 89 (4): 241–242.

Harr, M.E., 1987. *Reliability-based Design in Civil Engineering*. (No Title).

Hasofer, A.M. and Lind, N.C. (1974). Exact and invariant second-moment code format. *Journal of the Engineering Mechanics Division* 100 (1): 111–121.

Heidarie Golafzani, S., Jamshidi Chenari, R., and Eslami, A. (2020a). Reliability based assessment of axial pile bearing capacity: static analysis, SPT and CPT-based methods. *Georisk: Assessment and Management of Risk for Engineered Systems and Geohazards* 14 (3): 216–230. https://doi.org/10.1080/17499518.2019.1628281.

Heidarie Golafzani, S., Eslami, A., and Jamshidi Chenari, R. (2020b). Probabilistic assessment of model uncertainty for prediction of pile foundation bearing capacity; static analysis, SPT, and CPT-based methods. *Geotechnical and Geological Engineering* 38: 5023–5041.

Jamshidi Chenari, R. and Kamyab Farahbakhsh, H. (2015). Generating non-stationary random fields of auto-correlated, normally distributed CPT profile by matrix decomposition method. *Georisk: Assessment and Management of Risk for Engineered Systems and Geohazards* 9 (2): 96–108.

Jamshidi Chenari, R., Kamyab Farahbakhsh, H., Heidarie Golafzani, S., and Eslami, A. (2018). Non-stationary realization of CPT data: considering lithological and inherent heterogeneity. *Georisk: Assessment and Management of Risk for Engineered Systems and Geohazards* 12 (4): 1–14.

Kenarsari, E., Oloomi, R., Jamshidi Chenari, R. and Eslami, A. (2011). Effect of vertical heterogeneity in soil strength on pile bearing capacity prediction from CPT data. *Proceedings of the 36th Annual Conference on Deep Foundations*, Boston, MA.

Kim, K.J., Rahman, M.S., Gabr, M.A. et al. (2005). Reliability based calibration of resistance factors for axial capacity of driven piles. In: *Advances in Deep Foundations*, 1–12.

Kulhawy, F.H. and Trautmann, C.H. (1996, August). Estimation of in-situ test uncertainty. In: *Uncertainty in The Geologic Environment: From Theory to Practice*, 269–286. ASCE.

Look, B.G. (2007). *Handbook of Geotechnical Investigation and Design Tables*. Taylor & Francis.

Meyerhof, G.G. (1976). Bearing capacity and settlement of pile foundations. *Journal of Geotechnical and Geoenvironmental Engineering* 102, no. ASCE 11962: 197–228.

Meyerhof, G.G. (1983). Scale effects of ultimate pile capacity. *Journal of Geotechnical Engineering* 109: 797–806.

Moshfeghi, S. and Eslami, A. (2016). Study on pile ultimate capacity criteria and CPT-based direct methods. *International Journal of Geotechnical Engineering* 12 (1): 1–12.

Moshfeghi, S. and Eslami, A. (2018). Failure analysis of CPT-based direct methods for axial capacity of driven piles in sand. *Georisk: Assessment and Management of Risk for Engineered Systems and Geohazards*. https://doi.org/10.1080/17499518.2018.1478105.

Niazi, F.S. (2014). Static axial pile foundation response using seismic piezocone data. Doctoral dissertation. Georgia Institute of Technology.

Paikowsky, S.G., Birgisson, B., McVay, M., et al. (2004). Load and resistance factor design for deep foundations, *NCHRP Report 507*. Washington DC: Transportation Research Board of the National Academies.

Phoon, K.K. and Kulhawy, F.H. (1999). Characterization of geotechnical variability. *Canadian Geotechnical Journal* 36: 612–624.

Phoon, K.K. and Kulhawy, F.H. (2005). Characterization of model uncertainties for laterally loaded rigid drilled shafts. *Geotechnique* 55 (1): 45–54.

Rackwitz, R. and Fiessler, B. (1978). Structural reliability under combined random load sequences. *Computers and Structures* 9 (5): 489–494.

Schmertmann, J.H. (1978). Guidelines for cone penetration test: performance and design *(No. FHWA-TS-78-209)*. The United States: Federal Highway Administration.

Shioi, Y. and Fukui, J. (1982). Application of N-value to design of foundations in Japan. In: *Penetration Testing*, 1e, vol. 1. *Proceeding of the Second European Symposium on Penetration Testing*, Amsterdam (24–27 May 1982) (ed. A. Verruijt). Routledge.

24

LRFD Approach and Some Codes Review

Chapter Highlight

- Load and Resistance Factor Design (LRFD)
- Methods for Determining the Resistance Factor in LRFD:
  - First Order Second Moment Method (FOSM)
  - First Order Reliability Method (FORM)
  - Monte Carlo Simulation
  - Modified First Order Second Moment Method (FOSM)
- Axial Bearing Capacities of Piles in Some Codes
- LRFD Approach for Piles in Some Codes
- Case Studies
  - Calibration of Resistance Factor Regarding Different Approaches (Heidarie Golafzani et al. 2020b)
  - The Resistance Factor for Different Piles (Paikowsky et al. 2004)

24.1 Introduction

In the previous chapter, we explored various uncertainty sources in geotechnical engineering and their impact on geotechnical designs. It became evident that relying solely on the allowable stress design (ASD) approach and using a lumped factor of safety may not lead to cost-effective and optimal designs, considering the latent uncertainties. In conventional deterministic solutions, geotechnical parameters like undrained shear strength were assumed to be constant within each layer, introducing additional uncertainties into designs.

To address these concerns and achieve consistency between the design of superstructures and substructures, the load and resistance factor design (LRFD) approach was developed. In LRFD, the contributing parameters, including loads and resistance, are treated as random variables, and probability theory is utilized to quantify encountered risks through the reliability index (Eslami and Heidarie Golafzani 2021).

As shown in Figure 24.1, the ASD approach relies on a misleading safety factor, defined as the ratio of mean measured values of resistance ($\overline{R}$) to mean measured values of load ($\overline{Q}$), promoting the concept of safe design. However, the shaded area highlights that there is always a probability of failure due to uncertainties inherent in both load and resistance. The solid and dashed lines for the resistance (R) both show two different probability density functions for the random variable of the resistance (R). The dashed line emphasizes that the COV is higher in comparison to the solid line, resulting in a greater shaded area and consequently, a greater probability of failure. Consequently, accurately determining the statistics for load and resistance becomes crucial.

In this chapter, we will delve into the LRFD approach and its implementation in geotechnical designs, with a particular focus on geotechnical pile design. By adopting LRFD, engineers can better account for uncertainties and develop more reliable and efficient designs.

Advanced Foundation Engineering: Principles, Performance and Prospects, First Edition. Moses Karakouzian and Abolfazl Eslami.

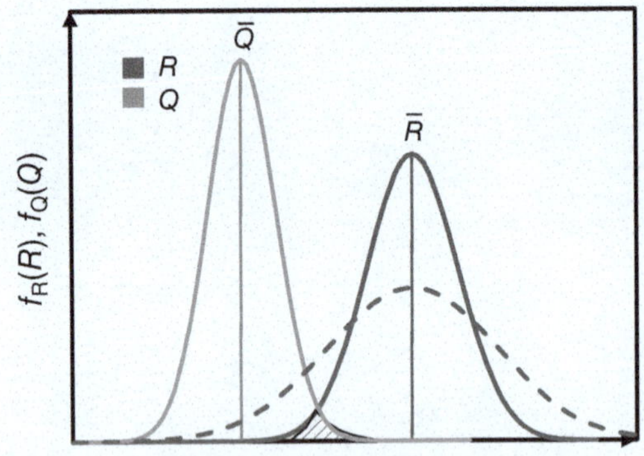

Figure 24.1 Schematic diagram for load and resistance distribution (Eslami & Heidarie Golafzani, 2021 / with permission of Taylor & Francis Group).

24.2 Load and Resistance Factor Design

As mentioned earlier, in ASD, the design loads or Q, including actual forces considered to be applied to the structure, are compared to resistance or strength, i.e., R via a factor called the factor of safety or FS:

$$Q \leq Q_{\text{all}} = \frac{R}{\text{FS}} = \frac{Q_{\text{ult}}}{\text{FS}} \tag{24.1}$$

where Q is the design load, Q_{all} is the allowable design load, R is the resistance of the structure, and Q_{ult} is the ultimate resistance, e.g., geotechnical pile resistance.

Despite ASD, in LRFD, the variability in loads and resistances are considered via partial safety factors implemented separately to load effects and resistance. The uncertainty in loads and resistance leads to a load factor, i.e., γ and a resistance factor, i.e., φ, respectively. Therefore, the formulation in LRFD for the strength limit state is as follows (Whithiam et al. 1997):

$$R_r = \varphi R_n \geq \eta \sum \gamma_i Q_i \tag{24.2}$$

where R_r is factored resistance, R_n is ultimate resistance, Q_i is the force or load effect, φ is the resistance factor, γ_i is load factor, and $\eta \geq 0.95$ is a dimensionless load modifier taken from the structural code. The strength limit states ensure structural safety during the structure design life and in relation to piles it involves the ultimate bearing capacity of the soil (Petek et al. 2016; NCHRP, 507).

The service limit states provide limits on stress, deformation, and cracking under expected working loads. Accordingly, the formulation in LRFD for the serviceability state is as follows:

$$\varphi \delta_n \geq \eta \sum \gamma_i \delta_i \tag{24.3}$$

where δ_i is the estimated displacement, δ_n is the tolerable displacement.

Geotechnical projects are designed based on their importance and expected lifetime serviceability, and this is accomplished by setting a target reliability index (β_{Target}) in the load and resistance factor design (LRFD) approach. The resistance factor (φ) is then calibrated accordingly to achieve this desired reliability index. The resistance factor is typically less than one and is multiplied by the resisting component. On the other hand, load factors are greater than one to account for potential load uncertainties and increase the load effects. As a result, in the LRFD approach, it is crucial that the reduced resistance is greater than the increased load effects to ensure a reliable and safe design.

Various methods have been developed for calibrating the resistance factor (Heidarie Golafzani et al. 2020a, b), considering the specific requirements of each geotechnical project. These calibration methods help engineers achieve the desired reliability level and ensure the overall stability and performance of the designed structure.

24.2.1 Calibration to Allowable Stress Design

Calibration by fitting to ASD is a benchmark in geotechnical designs. Also, when insufficient statistical data is available to perform a calibration by optimization, fitting to ASD is used. It is presented as follows:

$$\text{FS} = \frac{R}{\sum Q_i} \tag{24.4}$$

$$\varphi R_n = \sum \gamma_i Q_i \tag{24.5}$$

$$\varphi = \frac{\sum \gamma_i Q_i}{\text{FS} \sum Q_i} \tag{24.6}$$

If only the dead and live loads are considered (AASHTO load case I), the above equation is as following

$$\varphi = \frac{\gamma_D Q_D + \gamma_L Q_L}{\text{FS}(Q_D + Q_L)} \tag{24.7}$$

$$\varphi = \frac{\gamma_D {}^{Q_D}/_{Q_L} + \gamma_L}{\text{FS}\left({}^{Q_D}/_{Q_L} + 1\right)} \tag{24.8}$$

Example 24.1 According to calibration to ASD, compute the resistance factor for FS = 3 and $Q_D/Q_L = 1$ to 3 regarding the probabilistic characteristics of the random variables assumed to be those used by AASHTO (Nowak 1999).

AASHTO considers the probabilistic characteristics of the random variables as follows. Also, it assumes that they are log-normally distributed and mutually independent.

$\gamma_L = 1.75, \lambda_{Q_L} = 1.15, \text{COV}_{Q_L} = 0.18$

$\gamma_D = 1.25, \lambda_{Q_D} = 1.08, \text{COV}_{Q_D} = 0.13$

$$\text{if FS} = 3 \text{ and } {}^{Q_D}/_{Q_L} = 1, \text{then } \varphi = \frac{\gamma_D {}^{Q_D}/_{Q_L} + \gamma_L}{\text{FS}\left({}^{Q_D}/_{Q_L} + 1\right)} = \frac{1.25 * 1 + 1.75}{3 * (1 + 1)} = 0.5$$

$$\text{if FS} = 3 \text{ and } {}^{Q_D}/_{Q_L} = 2, \text{then } \varphi = \frac{\gamma_D {}^{Q_D}/_{Q_L} + \gamma_L}{\text{FS}\left({}^{Q_D}/_{Q_L} + 1\right)} = \frac{1.25 * 2 + 1.75}{3 * (2 + 1)} = 0.4722$$

$$\text{if FS} = 3 \text{ and } {}^{Q_D}/_{Q_L} = 3, \text{then } \varphi = \frac{\gamma_D {}^{Q_D}/_{Q_L} + \gamma_L}{\text{FS}\left({}^{Q_D}/_{Q_L} + 1\right)} = \frac{1.25 * 3 + 1.75}{3 * (3 + 1)} = 0.4583$$

24.2.2 Calibration Using the FOSM Approach

Calibrating using the FOSM approach requires considering a safety margin function, i.e., $g(R, Q)$. Since parameters attain positive values in geotechnical engineering, it is assumed that they (i.e., R and Q) are log-normally distributed and independent variables. Therefore

$$g(R, Q) = \ln(R) - \ln(Q) = 0 \tag{24.9}$$

Failure happens when $g(R, Q) < 0$ or $\ln(R/Q) < 0$. From the previous chapter, we know that the reliability index is calculated from the below equation (FHWA 2001);

$$\beta = \frac{\ln\left[\left(\frac{\overline{R}}{\overline{Q}}\right)\sqrt{\frac{1 + \text{COV}_Q^2}{1 + \text{COV}_R^2}}\right]}{\sqrt{\ln\left[\left(1 + \text{COV}_R^2\right)\left(1 + \text{COV}_Q^2\right)\right]}} \tag{24.10}$$

where $\overline{R}$ and $\overline{Q}$ are the mean values of resistance (R) and load (Q), respectively, and COV_R and COV_Q are the coefficients of variation for resistance and load. The value of β can be related to the traditional allowable stress design (ASD) by a factor of safety. In cases where only dead and live loads are considered (AASHTO Load Case I), the reliability index can be expressed as follows (FHWA 2001; McVay et al. 2000):

$$\beta = \frac{\ln\left[\frac{\lambda_R \text{FS}\left(\frac{Q_D}{Q_L} + 1\right)}{\lambda_{Q_D}\frac{Q_D}{Q_L} + \lambda_{Q_L}}\sqrt{\frac{1 + \text{COV}_{Q_D}^2 + \text{COV}_{Q_L}^2}{1 + \text{COV}_R^2}}\right]}{\sqrt{\ln\left[\left(1 + \text{COV}_R^2\right)\left(1 + \text{COV}_{Q_D}^2 + \text{COV}_{Q_L}^2\right)\right]}} \tag{24.11}$$

In which, λ_R, λ_{Q_D}, and λ_{Q_L} are the bias factors for the resistance, dead and live loads, respectively; COV_R, COV_Q, and COV_{Q_L} are the coefficients of variation for the resistance, dead and live loads, respectively; and Q_D/Q_L is the ratio of means of dead and live loads.

Accordingly, the resistance factor calibrated using the FOSM approach is attained from the following equation:

$$\varphi = \frac{\dfrac{\lambda_R \sum \gamma_i Q_i}{\lambda_Q}\sqrt{\dfrac{1+\mathrm{COV}_Q^2}{1+\mathrm{COV}_R^2}}}{\exp\left\{\beta_{\text{Target}}\sqrt{\ln\left[\left(1+\mathrm{COV}_Q^2\right)\left(1+\mathrm{COV}_R^2\right)\right]}\right\}} \tag{24.12}$$

where λ_R and λ_Q are bias resistance and load factors, respectively, and COV_R and COV_Q are coefficients of variation for resistance and load, respectively. γ_Q is the load factor. Equation 24.4 highlights how the prediction of axial pile bearing capacity by different methods affects design efficiency. Bias resistance/load factors represent the mean ratio of the measured to predicted resistance/load. In geotechnical engineering, the nature of imposed loads is less uncertain than the resisting component.

When dead and live loads are considered separately, the above equation becomes (Barker et al. 1991; Withiam et al. 1997)

$$\varphi = \frac{\lambda_R\left(\dfrac{\gamma_D Q_D}{Q_L}+\gamma_L\right)\sqrt{\dfrac{1+\mathrm{COV}_{Q_D}^2+\mathrm{COV}_{Q_L}^2}{1+\mathrm{COV}_R^2}}}{\left(\lambda_{Q_D}\dfrac{Q_D}{Q_L}+\lambda_{Q_L}\right)\exp\left\{\beta_{\text{Target}}\sqrt{\ln\left[\left(1+\mathrm{COV}_{Q_D}^2+\mathrm{COV}_{Q_L}^2\right)\left(1+\mathrm{COV}_R^2\right)\right]}\right\}} \tag{24.13}$$

Example 24.2 According to calibration to FOSM, Compute the resistance factor for $\lambda_R = 1.14$, $\mathrm{COV}_R = 0.47$, $\beta_{\text{Target}} = 3$ and $Q_D/Q_L = 1$ to 3 regarding the probabilistic characteristics of the random variables assumed to be those used by AASHTO (Nowak 1999).

By substituting the parameters into Equation 24.13,

$$\gamma_L = 1.75,\ \lambda_{Q_L} = 1.15,\ \mathrm{COV}_{Q_L} = 0.18$$

$$\gamma_D = 1.25,\ \lambda_{Q_D} = 1.08,\ \mathrm{COV}_{Q_D} = 0.13$$

λ_R	COV_R	β_{Target}	$Q_D/Q_L = 1$	$Q_D/Q_L = 2$	$Q_D/Q_L = 3$
1.14	0.47	3	$\varphi = 0.319$	$\varphi = 0.305$	$\varphi = 0.297$

24.2.3 Calibration Using the FORM Approach

Although FOSM is a simple method for evaluating the reliability index, it has some shortcomings, particularly when dealing with nonlinear performance functions. Linearizing the performance function at the mean values of the random variables can introduce errors in such cases. Additionally, for mechanically equivalent performance functions, the FOSM reliability index may not be unique, as different functional representations of the performance function can lead to different mean estimates and first derivatives.

In comparison, FORM is more versatile and overcomes the shortcomings of FOSM, as addressed earlier. The iteration algorithm introduced by Rackwitz and Fiessler (1978) facilitates the computing process. In FORM, random variables are considered to be independent and log-normally distributed with a nonlinear limit state function. The steps for calculating the FORM-based reliability index were reviewed in Chapter 23.

Regarding the calibration of the FORM resistance factor, it follows a similar iterative process as the basic algorithm of FORM reliability analysis. In this case, the performance function is defined by Kim et al. (2005):

$$g = \ln\left(\frac{\lambda_R\left(\gamma_{QD}\frac{Q_D}{Q_L}+\gamma_{QL}\right)}{\phi\left(\lambda_{QD}\frac{Q_D}{Q_L}+\lambda_{QL}\right)}\right) \tag{24.14}$$

Example 24.3 According to calibration to FORM, Compute the resistance factor for $\lambda_R = 1.14$, $\mathrm{COV}_R = 0.47$, $\beta_{\text{Target}} = 3$ and $Q_D/Q_L = 1$ to 3 regarding the probabilistic characteristics of the random variables assumed to be those used by AASHTO (Nowak 1999).

The basic algorithm of the FORM for resistance factor calibration is like that of the respective reliability analysis presented in Chapter 23. The limit state function for FORM is presented in Equation 24.14. A spreadsheet can be used following the iteration algorithm to facilitate the computing processes of the resistance factor.

$$\gamma_L = 1.75, \lambda_{Q_L} = 1.15, \mathrm{COV}_{Q_L} = 0.18$$

$$\gamma_D = 1.25, \lambda_{Q_D} = 1.08, \mathrm{COV}_{Q_D} = 0.13$$

For the given assumption, the resistance factor is as follows:

The FORM calibration of the resistance factor

λ_R	COV_R	β_{Target}	$Q_D/Q_L = 3$
1.14	0.47	3	$\varphi = 0.3293$

step 1		g()=LN(()/())													
												miuBFR=	1.14	miuLNBFR	0.03
	β	0.1		1.79	3.00	3.00	3.00	3.00	3.00			COVBFR=	0.47	sigLNBFR	0.45
step 3															
	BFR*	1.14	0.977	0.272	0.280	0.280	0.280	0.280	0.280			miuBFD=	1.08	miuLNBFD	0.07
	BFD*	1.08	1.074	1.123	1.161	1.162	1.162	1.162	1.162	gammaD	1.25	COVBFD=	0.13	sigLNBFD	0.13
	BFL*	1.15	1.134	1.168	1.194	1.194	1.194	1.194	1.194	gamma L	1.75				
step 4												miuBFL=	1.15	miuLNBFL	0.12
	miuNBFR	1.026	1.030	0.634	0.645	0.645	0.645	0.645	0.645			COVBFL=	0.18	sigLNBFL	0.18
	sigNBFR	0.509	0.436	0.121	0.125	0.125	0.125	0.125	0.125						
	miuNBFD	1.071	1.071	1.070	1.067	1.067	1.067	1.067	1.067		QD/QL=	3			
	sigNBFD	0.140	0.139	0.145	0.150	0.150	0.150	0.150	0.150		φ =	0.3293			
	miuNBFL	1.132	1.132	1.131	1.130	1.130	1.130	1.130	1.130						
	sigNBFL	0.205	0.202	0.209	0.213	0.213	0.213	0.213	0.213						
	∂g/∂BFR	0.877	1.024	3.682	3.571	3.569	3.569	3.569	3.569						
	∂g/∂BFD	-0.683	-0.689	-0.661	-0.641	-0.641	-0.641	-0.641	-0.641						
	∂g/∂BFL	-0.228	-0.230	-0.220	-0.214	-0.214	-0.214	-0.214	-0.214						
	αBFR	0.973	0.973	0.973	0.973	0.973	0.973	0.973	0.973						
	αBFD	-0.208	-0.209	-0.209	-0.210	-0.210	-0.210	-0.210	-0.210						
	αBFL	-0.102	-0.101	-0.100	-0.099	-0.099	-0.099	-0.099	-0.099						
step 7			ok	ok	ok	ok	ok	ok	ok						
step 8	β		1.787	3.000	3.000	3.000	3.000	3.000	3.000						
step 9				Continue	ok	ok	ok	ok	ok						

The iteration algorithm for the FORM calibration of the resistance factor

24.2.4 Calibration Using Monte Carlo Simulation (MCS) Approach

In the MCS method, random variables are generated based on their distribution functions and incorporated into the limit state function. The probability of failure, and subsequently the reliability index, is determined by calculating the ratio of the number of limit state functions that are less than zero to the total number of generated random variables. By adjusting the resistance factor, the desired level of reliability can be achieved. The limit state function utilized in this analysis, which is mechanically equivalent to Equation 24.14, is as follows (Kwak et al. 2010):

$$g = \left(\lambda_R/\varphi\right)\left[\gamma_{Q_D} Q_D/Q_L + \gamma_{Q_L}\right] - \left[\lambda_{Q_D} Q_D/Q_L + \lambda_{Q_L}\right] \tag{24.15}$$

Example 24.4 According to calibration to MCS, Compute the resistance factor for $\lambda_R = 1.14$, $\mathrm{COV}_R = 0.47$, $\beta_{\mathrm{Target}} = 2.5$ and $Q_D/Q_L = 1$ to 3 regarding the probabilistic characteristics of the random variables assumed to be those used by AASHTO (Nowak 1999).

$$\gamma_L = 1.75, \lambda_{Q_L} = 1.15, \mathrm{COV}_{Q_L} = 0.18$$
$$\gamma_D = 1.25, \lambda_{Q_D} = 1.08, \mathrm{COV}_{Q_D} = 0.13$$

The MCS calibration of the resistance factor

λ_R	COV_R	β_{Target}	$Q_D/Q_L = 3$
1.14	0.47	2.5	$\varphi = 0.4152$

X1	X2	X3	φ	QD/QL	sigma ln r	miu ln r		Qdi	Qli	g			
0.378884	0.893258	-0.66836	0.4152	3	0.446753	1.516295	5.395486	1.202277	0.334827	3.858381	0		
1.420224	-0.68374	0.401423					8.591597	0.980263	0.405306	7.206027	0	count g<0	635
0.689046	-1.24577	1.500484					6.197393	0.911474	0.493192	4.792727	0		
1.809144	-1.03671	-0.50011					10.22193	0.936479	0.34504	8.940415	0	Pf	0.00635
-0.05551	0.18831	0.582398					4.44373	1.097415	0.418618	2.927696	0		
-2.7317	-2.22757	-0.26343					1.344335	0.802685	0.359935	0.181715	0		
-0.24646	1.392861	0.011829					4.080381	1.282606	0.378068	2.419707	0		
-1.63357	-1.06282	1.379409					2.195679	0.933319	0.482644	0.779716	0	beta	2.5
1.696903	-0.72809	1.009969					9.722007	0.974651	0.451832	8.295524	0		
-0.99155	-0.18074	-0.07515					2.925066	1.046219	0.372242	1.506605	0		
0.153227	-0.10986	0.312517					4.878072	1.055862	0.398923	3.423287	0	**Input Variables**	
-1.07797	-1.7633	-1.43021					2.814284	0.852408	0.29224	1.669636	0	miu	1.14
-0.12158	-0.51539	-0.17755					4.314486	1.001861	0.365497	2.947128	0	cov	0.47
1.628282	-0.88674	0.08701					9.428485	0.954838	0.383178	8.090469	0		
0.530764	0.634821	-1.42954					5.774291	1.162719	0.292275	4.319298	0		
0.075719	-0.7499	-0.50173					4.712049	0.971903	0.34494	3.395206	0		
-0.36848	0.431599	1.944268					3.863901	1.132529	0.533866	2.197507	0		
0.153386	-0.35441	-0.5926					4.878417	1.022959	0.339388	3.51607	0		
1.217838	0.260918	-1.55139					7.84886	1.107779	0.285984	6.455096	0		
-0.27454	-0.34123	-1.44482					4.029504	1.024706	0.291478	2.71332	0		

The iteration algorithm for the MCS calibration of the resistance factor

24.2.5 Calibration Using Modified FOSM Approach

Styler (2006) introduced a modification to the FOSM approach by correcting the coefficient of variation for load (COV_Q). In this enhanced version of FOSM, COV_Q, which is the sum of the coefficients of variation for dead and live loads, is replaced with higher-order terms, leading to a more accurate resistance factor than the original version. This method not only requires significantly less computational effort but also yields results that closely align with those obtained from the FORM approach (Hu et al. 2012). The modified FOSM equation for the calculation of the resistance factor is as follows:

$$\varphi = \frac{\lambda_R\left(\gamma_{Q_D}\frac{Q_D}{Q_L}+\gamma_{Q_L}\right)\sqrt{\dfrac{1+\dfrac{\frac{Q_D^2}{Q_L^2}(\lambda_{Q_D})^2COV_{Q_D}^2+(\lambda_{Q_L})^2COV_{Q_L}^2}{\frac{Q_D^2}{Q_L^2}(\lambda_{Q_D})^2+2\frac{Q_D}{Q_L}\lambda_{Q_D}\lambda_{Q_L}+(\lambda_{Q_L})^2}}{1+COV_R^2}}}{\left(\lambda_{Q_D}\frac{Q_D}{Q_L}+\lambda_{Q_L}\right)\exp\left(\beta_T\sqrt{LN\left[(1+COV_R^2)\left(1+\dfrac{\frac{Q_D^2}{Q_L^2}(\lambda_{Q_D})^2COV_{Q_D}^2+(\lambda_{Q_L})^2COV_{Q_L}^2}{\frac{Q_D^2}{Q_L^2}(\lambda_{Q_D})^2+2\frac{Q_D}{Q_L}\lambda_{Q_D}\lambda_{Q_L}+(\lambda_{Q_L})^2}\right)\right]}\right)} \tag{24.16}$$

Example 24.5 According to calibration to Modified FOSM, Compute the resistance factor for $\lambda_R = 1.14$, $COV_R = 0.47$, $\beta_{Target} = 2.5$ and 3, $Q_D/Q_L = 1$ to 3 regarding the probabilistic characteristics of the random variables assumed to be those used by AASHTO (Nowak 1999).

$$\gamma_L = 1.75,\ \lambda_{Q_L} = 1.15,\ COV_{Q_L} = 0.18$$

$$\gamma_D = 1.25,\ \lambda_{Q_D} = 1.08,\ COV_{Q_D} = 0.13$$

λ_R	COV_R	β_{Target}	$Q_D/Q_L = 3$
1.14	0.47	2.5	$\varphi = 0.4120$
		3	$\varphi = 0.3280$

24.3 Axial Pile Bearing Capacity of Piles in Codes

The axial bearing capacity of piles is determined by considering two main components: tip resistance and shaft resistance, which are estimated using various methods with different assumptions and empirical approaches. These methods can be broadly categorized into five groups:

1) Static analyses.
2) In situ test-based methods (direct or indirect).
3) Pile static load test.
4) Pile dynamic load test (including pile dynamic analysis and testing).
5) Numerical modeling.

Static analyses are the initial and commonly used approaches to estimate pile-bearing capacity. These methods rely on bearing capacity and/or cavity expansion theories and involve the utilization of soil shear strength parameters, which introduce significant uncertainties.

In situ, test-based methods provide an alternative approach for estimating pile-bearing capacity. These methods directly or indirectly use in situ test data to estimate pile capacity, thereby reducing some of the uncertainties associated with soil shear strength parameters.

Performing a full-scale pile-loading test is another effective way to reduce uncertainties in estimating pile-bearing capacity. However, the practicality and cost-effectiveness of conducting such tests depend on the scale of the project.

Table 24.1 presents a summary of different methods implemented in various design codes. The selection of an appropriate method depends on project-specific factors such as available data, project scale, and budget constraints.

24.3.1 Design of Piles Regarding Eurocode 7

Eurocode 7 considers only three methods for designing pile foundations, including: (i) design by static and dynamic load tests, (ii) design by calculation, and (iii) design by observation.

Static pile load test has a great role in Eurocode 7 for pile designs as their primary design method or providing validity to designs based on dynamic load tests or calculations. Eurocode 7 considers two types of piles, working piles as permanent piles and trial piles installed before the finalization of the design and in the same condition as the working piles. Pile load tests should be performed when there is uncertainty and no reliable experience about the pile installation and its ultimate bearing capacity. It is expected that the load tests allow the delineation of load–displacement behavior of piles, creep, rebound, and ultimate failure load. However, this code does not specify the number of tested piles and leaves this decision to engineering judgment. Furthermore, the pile load test should be performed after excess pore water pressure dissipation and soil set-up leading to the desired strength.

Table 24.1 Components of axial pile bearing capacities in common codes.

No.	Method	Unit toe resistance	Unit shaft resistance
1	Unified	$r_t = N_t \cdot \sigma'_{Z=D_f}$	$r_s = \beta \cdot \sigma'_Z$
2	API (2000)	Cohesive soils $r_t = 9S_u$ S_u is undrained shear strength around tip	$r_s = \alpha S_u \begin{cases} \alpha = 0.5\psi^{-0.5} & \psi \le 1 \\ \alpha = 0.5\psi^{-0.25} & \psi > 1 \end{cases}$ $\psi = \frac{S_u}{P'_0}$ P'_0 is effective vertical stress and S_u is the undrained shear strength
		Cohesionless soils $r_t = N_q \sigma'_z$	$r_s = KP'_0 \tan\delta$
3	NAVFAC	Cohesive soils $R_b = 9S_u \cdot A_b$	$r_s = \alpha S_u$
		Cohesionless soils $N_c = (N_q - 1)\cot\phi'$ $q_b = (\sigma'_v)_b N_q + c'_b N_c$	$\beta = K(z) \cdot \tan\delta(z)$

Moreover, this code allows for determining the compressive resistance of a pile according to dynamic load tests calibrated by static load tests performed in the same manner, including similar pile dimensions and ground conditions.

The primary preference of Eurocode 7 is performing pile load tests rather than calculation. The calculation approach emphasizes that the ultimate bearing capacity is the sum of a pile's characteristic base (R_{bk}) and shaft resistances (R_{sk}) obtained from an alternative procedure, i.e., R_{ck} or the total characteristic compressive resistance.

For piles subjected to compression, Eurocode 7 requires the design compressive action, including the self-weight of the pile, F_{cd}, to be less than or equal to the design bearing resistance, R_{cd} or $F_{cd} \leq R_{cd}$.

Table 24.2 Resistance factors for deep foundations (Eurocode 7).

Parameter		Model factor	Resistance factors R1	R2	R3	R4
Base resistance (R_b)	γ_b			1.1	1.0	
driven pile			1.0			1.3
bored pile			1.25			1.6
CFA pile			1.1			1.45
Shaft resistance (R_s)	γ_s		1.0	1.1	1.0	1.3
Total resistance (R_c)	γ_t			1.1	1.0	
driven pile			1.0			1.3
bored pile			1.15			1.5
CFA pile			1.1			1.4
Tensile resistance (R_{st})	γ_{st}		1.25	1.15	1.1	1.6
Calculation model	γ_{Rd}					

Table 24.3 Limit state, load case, and load combination.

Limit state and load case	Load combination
Strength I	Basic load combination related to normal vehicular use of the bridge without wind.
Strength II	Load combination relating to the use of the bridge by owner-specified special vehicles, evaluation permit vehicles, or both, without wind.
Strength III	Load combination for bridge exposed to wind velocity exceeding 55 mph without live loads.
Strength IV	Load combination for very high dead load to live load force effect ratios. (Typical for bridge spans greater than 250 ft.)
Strength V	Load combination for normal vehicular use of the bridge with a wind velocity of 55 mph.
Extreme Event I	Load combinations including earthquake.
Extreme Event II	Load combinations relating to ice load, collision by vessels and vehicles, and certain hydraulic events with a reduced live load other than that which is part of the vehicular collision load.
Service I	Load combinations relating to the normal operational use of the bridge with a 55 mph wind and all loads taken at their nominal values.
Service II	Load combinations intended to control the yielding of steel structures and slip of slip-critical connections due to vehicular live load.
Service III	Load combinations relating to tension in prestressed concrete superstructures with the objective of crack control and to principal tension in the webs of segmental concrete girders.
Service IV	Load combinations intended to control tension in prestressed concrete columns with the objective of crack control.
Fatigue I	Fatigue and fracture load combination related to infinite load-induced fatigue life.
Fatigue II	Fatigue and fracture load combination related to finite load-induced fatigue life.

After AASHTO, LRFD (2014).

The relationship between R_{cd} and R_{ck} is as follows:

$$R_{cd} = \frac{R_{sk}}{\gamma_s} + \frac{R_{bk}}{\gamma_b} \text{ or } R_{cd} = \frac{R_{tk}}{\gamma_t} = \frac{R_{sk} + R_{bk}}{\gamma_t} \tag{24.17}$$

where γ_s, γ_b, and γ_t are partial factors on the shaft, and base resistances when designing pile by calculation regarding in situ and laboratory testing and total characteristic resistance, R_{tk} when the toe and shaft components are not determined separately. Table 24.2 provides these partial factors for different sets: R1–R4 to resistance.

Verified by static pile load tests, traditional design is based on geotechnical calculations. Trial piles are tested to failure to determine the ultimate axial pile capacity with certainty. In this case, a factor of safety of 2 is implemented in designs, and when no pile load tests are performed, the value of 3 is assigned to the factor of safety.

24.3.2 Design of Piles Regarding AASHTO

The AASHTO LRFD Bridge Design Specifications (AASHTO, LRFD 2014) define four limit states for the design of driven piles: strength, service, extreme event, and fatigue. These limit states result in up to thirteen load combinations, as shown in Table 24.3.

However, only certain limit states are commonly considered for the design of driven pile foundations, including strength I, strength IV, extreme event I (earthquake), extreme event II (involving ice, vessel collision, blast, or vehicle collision), and service I. Other load combinations are typically not applicable to foundation design.

Table 24.4 provides the load combinations and partial load factors for these speciic limit states used in the design of driven piles. According to AASHTO, LRFD (2014), the designer must ensure that the summation of factored force efects on the

Table 24.4 Load combinations and load factors.

Load combination limit state	DC DD DW EHEV ES EL PS CR SH	LL IM CE BR PL LS	WA	WS	WL	FR	TU	TG	SE	EQ[a]	BL[a]	IC[a]	CT[a]	CV[a]
Strength I	γ_p	1.75	1.00	—	—	1.00	0.50/1.20	γ_{TG}	γ_{SE}	—	—	—	—	—
Strength II	γ_p	1.35	1.00	—	—	1.00	0.50/1.20	γ_{TG}	γ_{SE}	—	—	—	—	—
Strength III	γ_p	—	1.00	1.40	—	1.00	0.50/1.20	γ_{TG}	γ_{SE}	—	—	—	—	—
Strength IV	γ_p	—	1.00	—	1.00	1.00	0.50/1.20	—	—	—	—	—	—	—
Strength V	γ_p	1.35	1.00	0.40	—	1.00	0.50/1.20	γ_{TG}	γ_{SE}	—	—	—	—	—
Extreme Event I	γ_p	γ_{EQ}	1.00	—	—	1.00	—	—	—	1.00	—	—	—	—
Extreme Event II	γ_p	0.50	1.00	—	—	1.00	—	—	—	—	1.00	1.00	1.00	1.00
Service I	1.00	1.00	1.00	0.30	1.00	1.00	1.00/1.20	γ_{TG}	γ_{SE}	—	—	—	—	—
Service II	1.00	1.30	1.00	—	—	1.00	1.00/1.20	—	—	—	—	—	—	—
Service III	1.00	0.80	1.00	—	—	1.00	1.00/1.20	γ_{TG}	γ_{SE}	—	—	—	—	—
Service IV	1.00	—	1.00	0.70	—	1.00	1.00/1.20	—	1.00	—	—	—	—	—
Fatigue I -[b]LL, IM and CE	—	1.50	—	—	—	—	—	—	—	—	—	—	—	—
Fatigue II -[b]LL, IM and CE	—	0.75	—	—	—	—	—	—	—	—	—	—	—	—

The maximum and minimum values of γ_p are 1.25 and 0.9, respectively, for dead load components and attachments.
PL, pedestrian live load; LS, live load surcharge; WA, water load and steam pressure; WS, wind load on structure; WL, wind on live load; FR, friction load; TU, force effect due to uniform temp.; TG, force effect due to temp. gradient; SE, force effect due to settlement; EQ, earthquake load; BL, blast load; IC, ice load; CT, vehicular collision force; CV, vessel collision force; DC, dead load components and attachments; DD, downdrag; DW, wearing surface and utilities; EH, horizontal earth pressure; EV, vertical earth pressure; ES, earth surcharge; EL, locked-in stress; PS, secondary forces from post-tensioning; CR, force effects due to creep; SH, force effects due to shrinkage; LL, vehicular live load; IM, vehicular dynamic load allowance; CE, vehicular centrifugal force; BR, vehicular braking force.
a) Use one of these at a time
b) Load factors only applied to LL, IM, and CE
After AASHTO, LRFD (2014).

Table 24.5 Summary of resistance factors for LRFD design of drilled shaft foundations (Brown et al. 2018).

Limit state	Component of resistance	Geomaterial	Equation, method	Resistance factor, ϕ
Strength I through Strength V	Overturning of individual elastic shaft; head free to rotate	All geomaterials	*p–y* method pushover analysis	0.67
Geotechnical lateral resistance	Overturning of single row, retaining wall or abutment; head free to rotate	All geomaterials	*p–y* pushover analysis	0.67
	Pushover of elastic shaft within multiple-row group, w/moment connection to cap	All geomaterials	*p–y* pushover analysis	0.80
Strength I through Strength V	Side resistance in compression/uplift	Cohesionless soil	Beta method	0.55/0.45
Geotechnical axial resistance		Cohesive soil	Alpha method	0.45/0.35
		Rock		0.50 / 0.40
		Cohesive IGM	Modified alpha method	0.60/0.50
	Base resistance in compression	Cohesionless soil		0.50
		Cohesive soil		0.40
		Rock		0.50
	Static compressive resistance from load tests	All geomaterials		0.70
	Static uplift resistance from load tests	All geomaterials		0.60
	Group block failure	Cohesive soil		0.55
	Group uplift resistance	Cohesive and cohesionless soil		0.45
Strength I through Strength V;	Axial compression			0.75
	Combined axial and flexure			0.75–0.90
Structural resistance of reinforced concrete	Shear			0.90
Service I	All cases, all geomaterials			1.00
Extreme Events I and II	Axial geotechnical uplift resistance	All geomaterials	Methods cited above for strength limit states	0.80
	Geotechnical lateral resistance	All geomaterials	*p–y* method pushover analysis; Chapter 9	0.80
	All other cases	All geomaterials	Methods cited above for strength limit states	1.00

foundation is less than or equal to the summation of factored nominal geotechnical resistances for all relevant geotechnical limit states. This requirement ensures the safety and stability of the pile foundation under different loading conditions. The summary of resistance factors for drilled shafts in different limit states is presented in Table 24.5.

$$\sum \varphi_i R_{\mathrm{ni}} \geq \sum \gamma_i Q_i \quad \text{for all applicable geotechnical limit states} \tag{24.18}$$

and

$$\sum \varphi_{\mathrm{str}} P_{\mathrm{ni}} \geq \sum \gamma_i Q_i \quad \text{for all applicable structural limit states} \tag{24.19}$$

where Q_i is the force effects, γ_i is the load factor for that force effect, R_{ni} is the nominal geotechnical resistance, φ_i is its resistance factor, and P_{ni} and φ_{str} are the nominal structural resistance and its resistance factor, respectively.

According to this code, the limits considered for the design of driven piles are (i) strength limit states and (ii) extreme event limit states for axial compression resistance of a single pile and other geotechnical and structural limit states.

24.4 Case Studies

This section is dedicated to comparing the calibrated resistance factors via different approaches and their variation for different pile types installed in various soil types.

Case Study 24.1

Calibration of Resistance Factor Regarding Different Approaches (Heidarie Golafzani et al. 2020a, b)

Heidarie Golafzani et al. (2020a, b) compiled a comprehensive database consisting of 60 driven piles installed in various soil types. They evaluated the performance of 12 predictive methods in estimating axial pile-bearing capacities. Figure 24.2 illustrates the difference between the resistance factors obtained using the FORM and FOSM approaches. The results indicate that the CPT-based methods exhibit higher resistance factors compared to the SPT-based methods and static analyses, respectively.

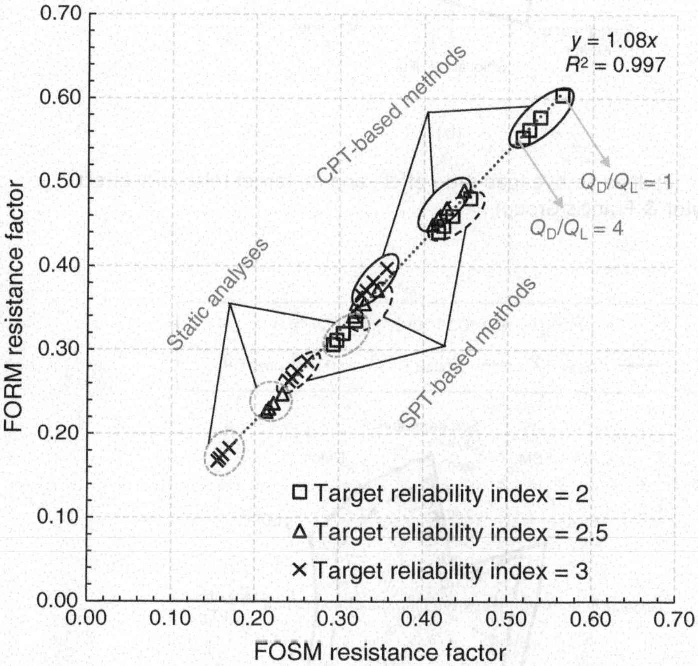

Figure 24.2 Resistance factors for different dead to live load ratios and target reliabilities (Eslami & Heidarie Golafzani, 2021 / with permission of Taylor & Francis Group).

Furthermore, the markers in each bundle represent different dead/live load ratios, specifically 1, 2, 3, and 4. Lower dead/live load ratios appear to result in higher resistance factors while maintaining a constant target reliability level. However, the resistance factors demonstrate limited sensitivity to changes in this ratio.

Figures 24.3a and 24.4a present the variation of FOSM and FORM resistance factors for each method, providing additional evidence that CPT-based methods such as Bazaraa and Kurkur (1986), Schmertmann (1978), UniCone of Eslami and Fellenius (1997), and LCPC of Bustamante and Gianeselli (1982) yield higher resistance factors compared to other methods, with CFEM (2006) showing lesser resistance factors. Moreover, Figures 24.3b and 24.4b demonstrate the sensitivity of resistance factors to the target reliability index. As the target reliability index increases to achieve a lower probability of failure, the resistance factors decrease accordingly. The results from Figure 24.2 further validate that the resistance factor obtained from the FORM method is generally greater than that obtained from the FOSM approach for the current dataset. Figures 24.3 and 24.4 consistently support this finding, emphasizing the difference between the two approaches.

(Continued)

Case Study 24.1 (Continued)

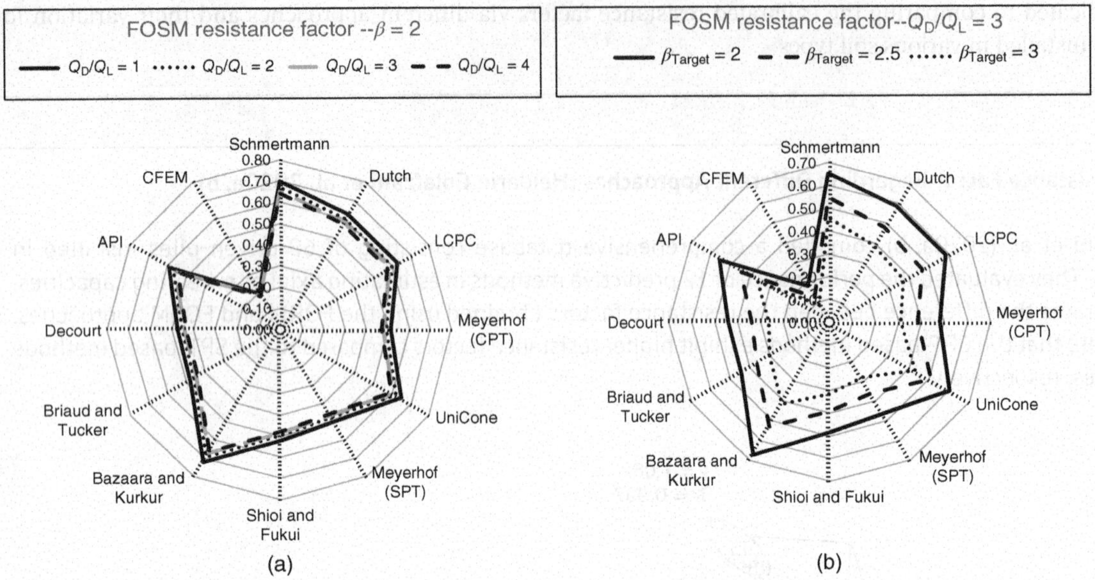

Figure 24.3 FOSM resistance factors for different methods: (a) dead-to-live load ratio effect and (b) target reliability effect (Eslami & Heidarie Golafzani, 2021 / with permission of Taylor & Francis Group).

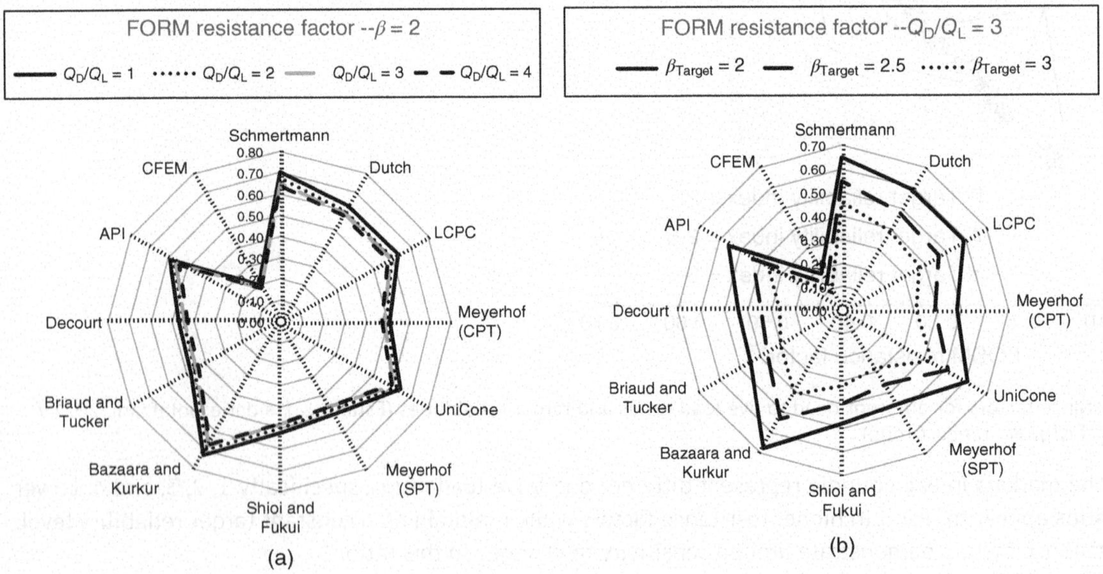

Figure 24.4 FORM resistance factors for different methods: (a) dead-to-live load ratio effect and (b) target reliability effect (Eslami & Heidarie Golafzani, 2021 / with permission of Taylor & Francis Group).

McVay et al. (2000) introduced a novel parameter, the ratio of resistance factor to resistance bias factor (i.e., $\phi/\lambda R$), which characterizes the contribution of measured capacity in a design for a specific structural reliability level, aligning with the reliability index. A higher $\phi/\lambda R$ value indicates greater reliability and efficiency in design. Figure 24.5 displays this efficiency ratio for different methods, considering a target reliability index of 3.

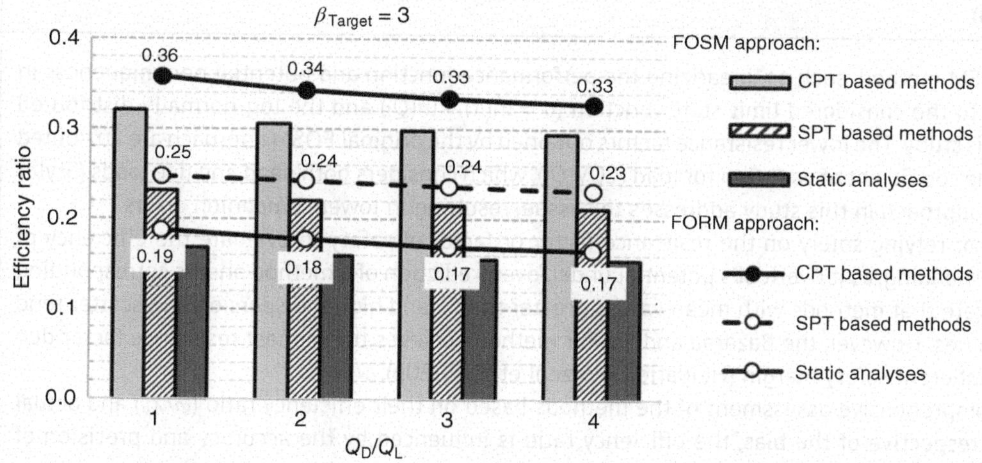

Figure 24.5 Efficiency ratio for different reliability approaches and groups of methods (Eslami & Heidarie Golafzani, 2021 / with permission of Taylor & Francis Group).

The results indicate that both the FOSM and FORM approaches show higher efficiency ratios for the CPT-based methods compared to the SPT-based methods and static analyses. Additionally, the efficiency ratios obtained from the FORM approach are slightly greater than those from the FOSM approach. This implies that the CPT-based methods perform better in terms of reliability and efficiency when assessing axial pile-bearing capacities, and the FORM approach generally provides slightly superior results compared to the FOSM approach.

Figure 24.6 presents a comparison of the resistance factor (ϕ) obtained by different methods. While different target reliabilities influence the resistance factor, uncertainties originating from the model also significantly impact the results. The FOSM resistance factors in Figure 24.6 are lower compared to the other three methods, which produce statistically equivalent factors. However, the modified FOSM approach performs better, as it yields resistance factors that are nearly identical to those obtained by the FORM and MCS methods.

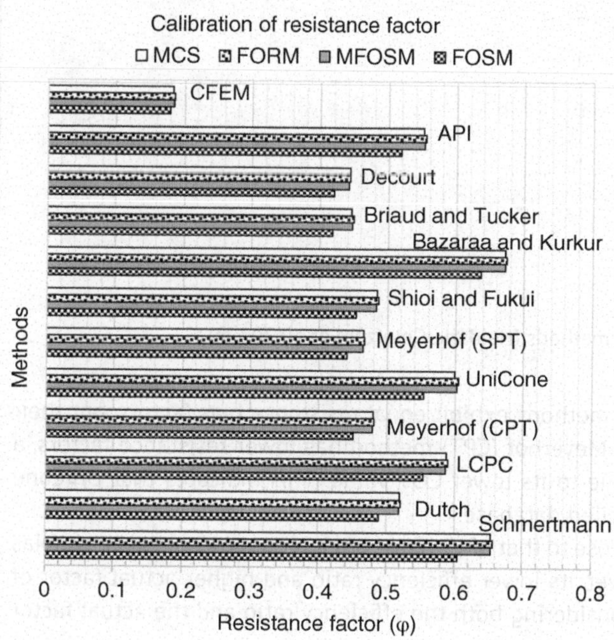

Figure 24.6 Calibration of resistance factor for $\beta_{\text{Target}} = 2$ and $Q_D/Q_L = 3$ (Heidarie Golafzani et al., 2021 / with permission of Springer Nature).

(Continued)

Case Study 24.1 (Continued)

The limitations of the FOSM method, such as linearizing the performance function and potential nonuniqueness in certain cases, do not apply to the considered limit state function ($g = \ln(R) - \ln(Q)$) and the log-normally distributed random variables used in this study. The lower resistance factors obtained by the original FOSM approach are attributed to the method of deriving the coefficient of variation for load (COV(Q)), which considers both dead and live loads (Styler 2006). The modified FOSM approach in this study addresses this issue, resulting in lower companion errors.

It is important to note that relying solely on the resistance factor or factor of safety to evaluate the efficiency of different methods can be misleading, as it overlooks potential under/overprediction of a method and its corresponding COV. Statistical results indicate that methods with mean values greater than 1 and higher dispersion and scatter tend to yield lower resistance values. However, the Bazaraa and Kurkur method achieves the highest resistance factor due to its underprediction and inherent safety margin (Heidarie Golafzani et al. 2020b).

Figure 24.7 provides a comprehensive assessment of the methods based on their efficiency ratio (φ/λ_R) and actual factor of safety (FS $\times \lambda_R$). Irrespective of the bias, the efficiency ratio is influenced by the accuracy and precision of a method (i.e., lower COV value and bias closer to unity). Methods that are more accurate and precise tend to have higher efficiency ratios and lower actual factors of safety.

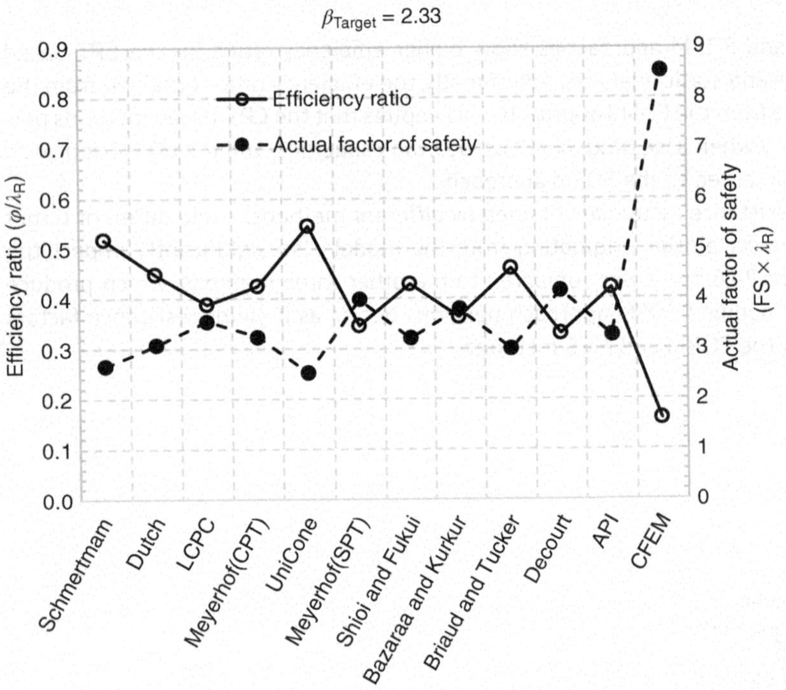

Figure 24.7 Efficiency ratio and actual factor of safety for different methods (Heidarie Golafzani et al. 2020b).

For example, both the UniCone and Meyerhof (CPT-based) methods exhibit equal resistance bias factors, but their COVs differ. Comparing these methods, it is evident that the Meyerhof (CPT) method has lower resistance factors, a higher efficiency ratio, and a lower actual factor of safety due to its lower COV value. This indicates that UniCone outperforms the Meyerhof (CPT) method for the current compiled database.

Additionally, the Schmertmann method, with a COV value close to that of UniCone but a higher mean resistance bias factor, yields greater resistance factors than UniCone. However, its lower efficiency ratio and higher actual factor of safety compared to UniCone emphasize the importance of considering both the efficiency ratio and the actual factor of safety for sound engineering judgment.

Case Study 24.2

The Resistance Factor for Different Piles (Paikowsky et al. 2004)

Paikowsky et al. (2004) emphasized the significance of large databases in providing a quantitative assessment of pile capacity evaluation methods during design and construction. Compiling such databases makes it possible to identify the controlling parameters of capacity evaluation methods that require calibration. This information is essential for examining design methodologies and testing the validity and cost-effectiveness of factors of safety used by working stress design (WSD) and for developing alternative methodologies like load and resistance factor design (LRFD).

LRFD offers a more suitable design methodology for geotechnical applications compared to WSD. The ability to determine design factors while quantifying their impact is a powerful tool in engineering. The development of load and resistance factors through reliability-based calibration and databases is a crucial step toward objectively quantifying LRFD parameters. This approach facilitates meaningful utilization of LRFD principles (in contrast to parameter fitting in WSD) and lays the groundwork for future developments that can further rationalize design and lead to more cost-effective construction.

The NCHRP Report 507 by Paikowsky et al. (2004) based on analyses of various databases provides recommendations for resistance and efficiency factors for static and dynamic analyses of different pile types constructed in various soils.

24.5 Major Points

1) In the ASD approach, the factor of safety, defined as the ratio of mean measured values of resistance ($\overline{R}$) to mean measured values of load ($\overline{Q}$), is commonly used to achieve safe design. However, it is essential to recognize that there is always a probability of failure due to uncertainties inherent in both load and resistance. Therefore, accurately determining statistics for load and resistance is of utmost importance.
2) Unlike the ASD approach, in LRFD, the variability in loads and resistances is explicitly considered through the application of separate partial safety factors for load effects (γ) and resistance (φ). By accounting for uncertainties in loads and resistance separately, LRFD provides a more comprehensive and rigorous design framework.
3) Various methods have been developed to calibrate the resistance factor (φ) in LRFD, including: (i) calibration based on the allowable stress design approach, (ii) calibration using the first order second moment (FOSM) method, (iii) calibration using the first order reliability method (FORM), (iv) calibration using the Monte Carlo simulation (MCS) method, and (v) calibration using the modified FOSM approach. These methods play a crucial role in achieving accurate and efficient design in geotechnical engineering.

24.6 Questions and Problems

24.6.1 Explanatory

1 What is the priority of LRFD compared to ASD in geotechnical designs?

2 Compare different resistance factor calibration approaches.

24.6.2 Conceptual

1 According to Figure 24.2, how does the variation of the target reliability index affect the resistance factor variation?

2 Regarding Figures 24.3 and 24.4, how does the variation of Q_D/Q_L affect the resistance factor?

3 What is the priority of the efficiency ratio in comparison to the resistance factor for performance assessment of predictive methods for axial pile bearing capacity?

4 Figure 24.7 provides the variation of efficiency ratio and actual factor of safety for different predictive methods. Explain why the higher the efficiency ratio, the less the actual factor of safety.

5 Compare the resistance factor for static and dynamic analyses of driven piles regarding Table 24.5.

References

AASHTO, LRFD (2014). *LRFD Highway Bridges Specifications, SI Units*, 7the. Washington, DC: American Association of State Highway and Transportation Officials.

American Petroleum Institute. (2000). Recommended practice for planning, designing, and constructing fixed offshore platforms–working stress design. API RP A, 2.

Barker, R.M., Duncan, J.M., Rojiani, K.B. et al. (1991). *NCHRP Report 343: Manuals for the Design of Bridge Foundations.* Washington, DC: Transportation Research Board, National Research Council.

Bazaraa, A.R. and Kurkur, M.M. (1986). N-values used to predict settlements of piles in Egypt. In: *Use of In Situ Tests in Geotechnical Engineering*, 462–474. ASCE.

Brown, D.A., Turner, J.P., Castelli, R.J., and Loehr, E. (2018). *Drilled Shafts: Construction Procedures and Design Methods* (No. FHWA-NHI-18-024). National Highway Institute (US).

Bustamante, M. and Gianeselli, L. (1982). Pile bearing capacity prediction by means of static penetrometer CPT. *Proceedings of the 2nd European Symposium on Penetration Testing*, Amsterdam (24–27 May 1982), 493–500.

CFEM (2006). *Canadian Foundation Engineering Manual.* Richmond, B.C: Canadian Geotechnical Society (CGS).

Eslami, A. and Fellenius, B.H. (1997). Pile capacity by direct CPT and CPTu methods applied to 102 case histories. *Canadian Geotechnical Journal* 34: 886–904.

Eslami, A. and Heidarie Golafzani, S. (2021). Relevant data-based approach upon reliable safety factor for pile axial capacity. *Marine Georesources and Geotechnology* 39 (11): 1373–1386.

FHWA (2001). *Load and Resistance Factor Design (LRFD) for Highway Bridge Substructures.* Washington DC: FHWA.

Heidarie Golafzani, S., Jamshidi Chenari, R., and Eslami, A. (2020a). Reliability based assessment of axial pile bearing capacity: static analysis, SPT and CPT-based methods. *Georisk: Assessment and Management of Risk for Engineered Systems and Geohazards* 14 (3): 216–230. https://doi.org/10.1080/17499518.2019.1628281.

Heidarie Golafzani, S., Eslami, A., and Jamshidi Chenari, R. (2020b). Probabilistic assessment of model uncertainty for prediction of pile foundation bearing capacity; static analysis, SPT, and CPT-based methods. *Geotechnical and Geological Engineering* 38: 5023–5041.

Hu, Z., McVay, M., Bloomquist, D. et al. (2012). New ultimate pile capacity prediction method based on cone penetration test (CPT). *Canadian Geotechnical Journal* 49 (8): 961–967.

Kim, K.J., Rahman, M.S., Gabr, M.A. et al. (2005). Reliability based calibration of resistance factors for axial capacity of driven piles. In: *Advances in Deep Foundations* (ed. C. Vipulanandan and F.C. Townsend), 1–12. Austin, TX: ASCE.

Kwak, K., Kim, K.J., Huh, J. et al. (2010). Reliability-based calibration of resistance factors for static bearing capacity of driven steel pipe piles. *Canadian Geotechnical Journal* 47 (5): 528–538.

McVay, M.C., Birgisson, B., Zhang, L. et al. (2000). Load and resistance factor design (LRFD) for driven piles using dynamic methods – a Florida perspective. *Geotechnical Testing Journal* 23 (1): 55–56.

Meyerhof, G.G. (1976). Bearing capacity and settlement of pile foundations. *Journal of Geotechnical and Geoenvironmental Engineering* 102: 197–228. no. ASCE 11962.

Meyerhof, G.G. (1983). Scale effects of ultimate pile capacity. *Journal of Geotechnical Engineering* 109: 797–806.

Nowak, A.S. (1999). Calibration of LRFD bridge design code (No. Project C12-33 FY'88-'92).

Paikowsky, S.G., Birgisson, B., McVay, M., et al. (2004). Load and resistance factor design for deep foundations, *NCHRP Report 507*. Washington DC: Transportation Research Board of the National Academies.

Petek, K., Mitchell, R., and Ellis, H. (2016). *FHWA Deep Foundation Load Test Database Version 2.0 User Manual* (No. FHWA-HRT-17-034). United States. Federal Highway Administration.

Rackwitz, R. and Fiessler, B. (1978). Structural reliability under combined random load sequences. *Computers and Structures* 9 (5): 489–494.

Schmertmann, J.H. (1978). Guidelines for cone penetration test: performance and design (*No. FHWA-TS-78-209*). The United States: Federal Highway Administration.

Styler, M.A. (2006). Development and implementation of the DIGGS format to perform LRFD resistance factor calibration of driven concrete piles in Florida. Doctoral dissertation. University of Florida.

Withiam, J., Voytko, E., Barker, R., et al. (1997). Load and resistance factor design (LRFD) of highway bridge substructures. Washington, DC: FHWA Publication no. HI-98-032, July.

Topic F

Recent Advances and Developments

Section X

Sustainable Development

25
Data Mining-Based Design

Chapter Highlight

- Role of the Databases in Geotechnical Engineering
- The Application of CPT and Soil Databases
- The Implementation of CPT and Pile Databases
  - First AUT: Geo-CPT&Pile Database
- Relevant Data-based Pile Design Approach
- Case Studies
  - Piles Installed in Sandy Soils
  - Piles Installed in Clayey Soils
- Reliability Appraisal of CPT-based Methods for Axial Pile-Bearing Capacity
- Multi-Criteria Decision-Making Models for Optimized Selection of Methods

25.1 Introduction

Databases play a crucial role in geotechnical engineering, as they provide organized collections of data, making it easier to access and retrieve information as needed. Given the empirical nature of geotechnical engineering, databases are valuable tools that enable engineers to approach optimum designs efficiently in terms of both timing and cost. Numerous geotechnical databases have been developed, including those introduced by Long et al. (1999), Kalavar and Ealy (2000), Roling et al. (2011), and Anderson et al. (2007), among others. These databases contribute significantly to enhancing the knowledge and understanding of geotechnical behavior, leading to more informed and reliable engineering decisions.

Several existing databases in the geotechnical literature are related to these areas:

- In situ testing
- Liquefaction records
- Soil improvement performance
- Retaining structure performance and ground deformations due to excavation
- Slope failure aspects
- Substructure damage
- Soil behavior classification
- Shallow foundations
- Pile axial loading tests
- Pile lateral loading tests

Databases in geotechnical engineering serve two main purposes: research and practical applications. These two categories are interconnected, as advancements in research lead to improvements in practical engineering practices.

In research, geotechnical databases offer a valuable resource for evaluating existing design methods and identifying their strengths and weaknesses under various conditions. They play a crucial role in the development of new design methods, such as settlement, bearing capacity, and stability analyses, by providing data for calibration and validation of these

Advanced Foundation Engineering: Principles, Performance and Prospects, First Edition. Moses Karakouzian and Abolfazl Eslami.

approaches. Additionally, assessing the performance of new methods through case records is essential to gain a deeper understanding of their capabilities.

Geotechnical databases also play a significant role in modifying existing methods by back-analyzing the behavior of geotechnical elements, such as foundation systems and excavation retaining structures. This enables engineers to obtain more accurate design procedures by learning from observed behaviors and incorporating statistical and reliability-based approaches. Such insights obtained from databases help in better understanding and dealing with various scenarios encountered in practical engineering projects.

In geotechnical design procedures, two conditions may arise: (i) when sufficient site investigations are conducted due to low variability at a specific site, providing abundant information on site conditions such as cone penetration testing (CPT) records and pile load test results; and (ii) when economic constraints or project limitations prevent comprehensive in situ testing, especially in sites with high inherent variability. In both cases, geotechnical databases can play a crucial role when combined with performance-based approaches. These databases offer the opportunity to replicate data, such as CPT and pile load–displacement records, similar to those encountered on the project site. By leveraging available data, the need for time-consuming or expensive loading tests can be reduced, leading to improved project efficiency.

In this chapter, we focus on typical databases in geotechnical engineering, including CPT results, with a specific emphasis on pile and CPT data. The chapter also provides examples of studies conducted using pile and CPT databases, showcasing their significance in geotechnical engineering practices.

25.2 Soil Parameters and In Situ Testing Databases

Several empirical correlations have been established between in situ testing records and other geotechnical parameters of geo-materials, such as soil relative density, unit weight, soil modulus, over-consolidation ratio, and more. These correlations were previously discussed and referenced in Chapter 4. For instance, Mayne and Illingworth (2010) compiled a database to estimate soil shear wave velocity and subsequently correlated it with CPT measurements to estimate soil unit weight. The database comprised data from 44 sites, encompassing a wide range of soil types, including sand, silt, clay, and tilt.

Eslami and Mohammadi (2016) gathered a database of CPTu records combined with laboratory test results from various regions worldwide. This database was used to validate their approach for estimating soil shear strength parameters in drained conditions. The records were collected from eight sites, primarily near the shoreline (marine and sea).

CPT records can also be utilized to monitor and assess ground improvement in problematic soils. Farhadi Nasl (2013) utilized a database to investigate the feasibility of soil behavior classification charts in identifying problematic soils and to illustrate changes in soil behavior after ground modification.

Another database related to the behavior of problematic soils was compiled by Eslami et al. (2017). This database was used to develop a new triangular chart that serves as a better indicator of problematic soil behavior. It included case records of cone penetrometer tests carried out in 54 sites with problematic soils.

In the Netherlands, Lengkeek et al. (2018) analyzed a database comprising 11 sites to propose a correlation for estimating soil unit weight.

Similarly, other researchers, such as Eslaamizaad and Robertson (1997), Mayne (2006), and Chanmee et al. (2017), have established correlations between CPT data and various soil parameters, including shear modulus, shear wave velocity, and over-consolidation ratio, respectively. These databases and correlations contribute significantly to advancing geotechnical engineering practices and understanding soil behavior in different conditions.

25.3 Shallow Foundation and In Situ Testing Databases

Chapter 4 of this book presents common geotechnical parameters used in foundation engineering and various correlations that relate in situ test records to the required design parameters for foundations. These correlations have been developed based on data gathered from geotechnical databases. However, the application of these databases is not limited to these correlations, and several other studies have utilized these databases to advance geotechnical knowledge and design methods. Some of these studies are elaborated below.

Eslaamizaad and Robertson (1996) gathered the results of nine load tests of full-size shallow foundations on lightly cemented, medium-dense quartz sands, and the CPT records were performed near these footings. Five of them were square footings with a width of 1–3 m and an embedment depth of 0.8 m. The rest were circular footings with a diameter of about 2 m and an embedment depth of 2.3 m. They performed regression analysis to investigate the effect of shape, size, and depth of footing on the predicted bearing capacity. Subsequently, they proposed a new approach for predicting the bearing capacity of shallow foundations according to the available CPT records.

Briaud and Gibbens (1999) conducted full-scale load tests on five square spread footings with widths ranging from 1 to 3 m and an embedment depth of 0.75 m. These footings were placed in medium-dense, fairly uniform, silty silica sand. They evaluated six bearing capacity methods and twelve settlement methods to assess their accuracy in predicting bearing capacity and settlement. The study also examined various factors, including settlement measurements, load test results, design correlations, creep settlement, zone of influence under the footing, mode of deformation of the soil mass, and volume change observations.

Eslami and Gholami (2006) compiled a database of 28 footings with full-scale loading test results and closely matched CPT records near the foundations. They assessed the performance of various methods in predicting the bearing capacity of shallow foundations. In situ-based methods were explored as a solution for predicting shallow foundations' bearing capacities, as they address issues like providing representative undisturbed samples, estimating required soil parameters according to site and stress conditions, and offering records with depth. Additionally, they proposed a new CPT-based method for predicting the bearing capacity of shallow foundations.

Valikhah and Eslami (2019) compiled a database of 46 cases of shallow foundations along with their adjacent CPT records. The settlements for these shallow foundations ranged from 25 to 260 mm, approximately 2–10% of the foundation width. They proposed a new analytical CPT-based approach founded on the Junbu nonlinear stress–strain method to estimate foundation settlement. The efficiency of the suggested method was compared to other methods, such as Schmertmann (1978), Berardi and Lancellotta (1991), and Malekdoost and Eslami (2011), and the new approach demonstrated higher efficiency and accuracy than the others.

These studies demonstrate the significant contributions of geotechnical databases in advancing design methodologies and understanding soil behavior in various practical engineering scenarios.

25.4 Deep Foundations and In Situ Testing Databases

Several well-established geotechnical databases have been presented and reviewed in the literature, focusing on determining pile capacity through in situ testing. These databases have significantly contributed to understanding the performance of current methods.

Briaud and Tucker (1988) conducted an extensive evaluation of 13 methods for estimating the settlement and bearing capacity of piles using standard penetration test (SPT), cone penetration test (CPT), and pressuremeter test (PMT) based on a comprehensive database of 98 case records. The database consists of 64 square concrete piles with widths ranging from 36 to 46 cm, 27 H piles, and 7-drilled shafts with diameters from 30 to 41 cm. The pile embedment lengths vary between 3 and 25 m, with an average of 12.2 m. The pile capacities range from 307 to 2890 kN, with an average of 1213 kN.

In a similar context, Alsamman (1995) compiled a comprehensive database of 95 full-scale drilled shaft load tests from 29 sites across eight countries, with a primary focus on sites in Germany and the United States. The main objectives of this database were twofold: to evaluate the performance of five methods for bearing capacity estimation and to develop a new procedure for determining capacity using CPT data. The database encompasses 48 pile load tests conducted in cohesionless soils, 16 load tests in cohesive soils, and 31 load tests in mixed soils. The diameters of the drilled shafts vary between 30 and 213 cm, and the shaft lengths range from 4.6 to 42 m. The CPTs were primarily performed using mechanical cones, and the data contain valuable information on cone tip resistance variation with depth. Ultimate loads were defined at a shaft displacement of 5% of the diameter, along with the elastic compression of the shaft. In cases where the shafts were subjected to tension, an axial load at 12.7 mm of displacement was used to define the ultimate load.

The Eslami and Fellenius (1997) database comprises 102 case records from 40 sites across 13 countries, with a primary focus on the United States. The sites were characterized by sediments of clay, silt, and sand. The majority of the CPT cases included in the database were obtained using electrical cones, with most of the CPT measurements at a vertical spacing of 300 mm or smaller. The piles in the database mostly had square or round cross-sections, and they were constructed using steel and concrete materials. The installation method for all but 10 of the piles was driving. The pile embedment lengths ranged from 5 to 67 m, with pile diameters ranging from 200 to 900 mm, and pile capacities varying from 80 to 8000 kN. Fourteen of the case records were instrumented, allowing the determination of toe and shaft resistances separately. This extensive database served as the basis for evaluating and comparing six different methods for determining pile-bearing capacity using both CPT and CPTu data.

Van Impe (1999) conducted a survey in Waddinxween, The Netherlands, with the objective of elucidating the impact of different installation techniques on the behavior of pile foundations. The study involved testing a total of 55 piles, classified into six types, including 15 slender precast piles, 10 steel H-beams, 9 vibro-piles, 10 vibrex-piles, 8 CFA piles, and three prestressed concrete sheet piles. Through this investigation, the study aimed to gain insights into how each installation technique influenced the performance and behavior of the respective pile types, contributing valuable data to the understanding of pile foundation behavior under different construction methods.

Abu-Farsakh and Titi (2004) conducted a study where they compiled the results of static load tests on 35 square precast prestressed concrete (PPC) piles to assess the suitability of eight CPT-based methods for determining bearing capacity. The piles' embedment lengths varied from 9 to 38 m, with pile diameters ranging from 356 to 762 mm. Among the 35 piles tested, 26 were driven in clay soils, while the other 9 were driven in layered soils. The ultimate pile capacity from the load test was determined using the Butler–Hoy tangent method (1977), providing valuable insights into the performance of CPT-based methods for estimating the bearing capacity of square precast prestressed concrete piles under different soil conditions.

The UWA Database (Schneider et al. 2008) comprises 77 case records of impact-driven piles and was specifically developed to evaluate seven design methods applicable to siliceous sands. All CPT data were accurately digitized to a depth interval of 0.1 m or smaller, providing detailed information about the subsurface soil conditions. The database includes a diverse range of sands with varying relative densities. The diameters of the piles in this database are predominantly less than 80 cm, with pile lengths ranging from 5 to 80 m, most of which fall between 10 and 20 m. Pile capacities are predominantly less than 5000 kN. In this database, pile capacity in compression was determined by measuring the load at the displacement of 10% of the pile diameter, reaching the plunging failure point. In cases where plunging failure was not achieved, Chin's hyperbolic extrapolation (1978) was employed to estimate the pile capacity. Pile capacity in tension was defined as the maximum uplift load minus the pile weight, offering valuable insights into the behavior of impact-driven piles in siliceous sands and contributing to the evaluation of design methods for such geotechnical conditions.

Shariatmadari et al. (2008) conducted a comprehensive compilation of a database comprising 60 driven piles along with their corresponding standard penetration test (SPT) results. The database encompasses 43 full-scale pile load tests and 17 dynamic tests that underwent CAPWAP (Case Pile Wave Analysis Program) analysis. These valuable case histories were meticulously gathered from 18 different sources, encompassing data from 26 sites spanning 7 countries. The piles in the database vary in embedment length from 8.5 to 34.2 m, with pile diameters ranging from 246 to 1424 mm, while the pile capacities range from 520 to 6450 kN. Most of the piles are constructed with steel and possess a round cross-section. The majority of the loading tests performed on the piles were compressive, and in 11 cases, pull-out tests were conducted. The primary objective of this database compilation was to propose a novel SPT-based method for predicting the axial-bearing capacity of piles. To assess its performance, the method was compared with five other SPT-based methods, namely Shioi and Fukui (1982), Bazaraa and Kurkur (1986), Briaud and Tucker (1988), Meyerhof (1976), and Aoki and De'Alencar (1975). The results demonstrated that the newly proposed method exhibited an acceptable level of error with low variance compared to the other investigated methods. The Meyerhof (1976) and Briaud and Tucker (1988) methods achieved second and third place, respectively, in terms of performance among the evaluated methods. This database and the corresponding analysis contribute significantly to the understanding and prediction of axial-bearing capacity for driven piles based on SPT data.

Van Dijk and Kolk's (2011) database comprises 33 case records of driven piles conducted at 15 different locations, all of which were installed in clay soils. The database includes results from cone penetration tests (CPTs) and the equilibrium in situ pore pressure, determined at 0.1 m depth intervals. The diameters of the piles in the database range from 102 to 812 mm, while the pile lengths vary from 3.6 to 71.4 m. Among the 33 case records, 18 piles were subjected to compression tests, and 15 were subjected to tension tests. The time intervals between pile installation and load tests in the database range from 6 hours to 134 days. The ultimate capacity of a pile in this database was defined as the maximum resistance measured during the test, corrected for the effective weight of the pile, which accounts for the total weight of the pile minus the weight of the displaced soil. This database provides valuable insights into the behavior of driven piles in clay soils under varying conditions.

The Hassani (2011) database consists of 70 case records obtained from 16 sources, which report data from 18 sites in 10 countries. The soils at these sites include sediments of soft and stiff clay, silt, silty sand, and sand. The majority of the CPT cases in this database were obtained using an electrical cone, and all of the CPT measurements were taken at a vertical spacing of 25–300 mm. Most of the piles in this database had a round cross-section, and the pile materials were primarily steel and concrete. With the exception of two piles, all others were installed by driving. The pile embedment lengths in the database range from 8.2 to 75 m, while the pile diameters vary from 270 to 813 mm. The pile capacities in the database varied from 485 to 6860 kN. The Hassani database was utilized for training the group method of data handling (GMDH) type neural networks to model the effects of effective cone point resistance and cone sleeve friction on pile unit shaft resistance.

Eslami et al. (2011) utilized 13 case records of driven pipe piles from four marine environment sites in Iran. The soils at these sites mainly comprised soft, sensitive, and stiff clay. All cases had available results from the electrical piezo penetrometer. The database included pile embedment lengths ranging from 41 to 85 m, pile diameters varying from 685 to 913 mm, and pile capacities ranging from 4700 to 16,500 kN. The researchers employed this database to assess the performance of five CPT and CPTu-based methods specifically tailored for Iran's marine environments. The predicted bearing capacities obtained from these methods were then compared to the static capacity derived from both static and dynamic load tests, serving as the reference test.

Gwizdala and Krasinski (2013) compiled a database of bridge structures founded on driven displacement Vibro and Franki piles. They introduced a novel methodology to derive the load–displacement diagram of piles using CPT results. Subsequently, the design assumptions and computational results were rigorously verified through static load tests on the piles.

The ZJU-ICL (2015) database was jointly developed by Yang et al. (2015), in collaboration with Zhejiang University and the Imperial College London. This database comprises 115 records of driven piles in sand and is publicly accessible, providing valuable resources for researchers in the field.

Niazi and Mayne (2016) utilized a comprehensive database of piles and piezocone test results to present an improved version of the UniCone method – a CPTu-based technique used to determine the axial-bearing capacity of piles. The database includes 153 case records of piles installed using various methods, such as driven, bored, cast-in situ, and jacked techniques. The piles were constructed in both sandy and clayey soils. Notably, all the case records in the database contain essential measurements of excess pore pressure, allowing for a more accurate and reliable assessment of pile behavior and performance.

The development, calibration, and validation of CPT-based methods have predominantly relied on the use of CPT and pile databases. For instance, the Schmertmann (1978) and Nottingham (1975) methods were formulated based on a synthesis of research conducted by Nottingham (1975) on both model and full-scale piles. The De Ruiter and Beringen (1979) method, also known as the European method, draws from the valuable insights gained from offshore construction projects in the North Sea. Likewise, the LCPC method builds upon the experimental work conducted by Bustamante and Gianeselli (1982) for the French Highway Department, providing a solid foundation for its calibration and validation. Similarly, the Tumay and Fakhroo (1982) method finds its roots in an experimental study involving various piles in clay soils in Louisiana, USA. These databases have significantly contributed to the robustness and reliability of the CPT-based methods, enabling engineers to make informed decisions in geotechnical engineering projects.

A summary of the characteristics these CPT and piling databases is presented in Table 25.1.

Table 25.1 Summary of CPT and pile databases in geotechnical literature.

No.	Database	Number of records	Pile installation technique	Pile surrounding soil
1	Schmertmann (1978)	More than 50	Driven	Sand, clay
2	Bustamante and Gianeselli (1982)	More than 200	Driven, bored	Sand, clay
3	Tumay and Fakhroo (1982)	40	Driven	Clay
4	Briaud and Tucker (1988)	98	Driven, bored	Clay, sand
5	Alsamman (1995)	95	Drilled shaft	Clay, sand
6	Eslami and Fellenius (1997)	102	Driven, bored	Clay, sand
7	Van Impe (1999)	55	Precast piles, steel H-beams, vibro-piles, vibrex-piles, CFA, prestressed concrete sheet piles	Clay, sand
8	Abu-Farsakh and Titi (2004)	35	Prestressed driven	Clay
9	UWA (Lehane et al. 2005)	77	Driven	Sand
10	Hassani (2011)	70	Driven, bored	Clay, sand
11	Van Dijk and Kolk (2011)	33	Circular driven	Clay
12	Eslami et al. (2011)	13	Driven	Clay
13	Gwizdala and Krasinski (2013)	6	Driven displacement Vibro and Franki piles	Clay, sand
14	ZJU-ICL (2015)	115	Driven	Sand
15	Niazi and Mayne (2016)	153	Driven, bored cast in situ, jacked	Clay, sand

Case Study 25.1

AUT: Geo-CPT&Pile Database

Establishment

Considering the primary goal of evaluating various methods for determining static pile capacity based on cone and piezocone penetration tests, the AUT: Geo-CPT&Pile database has been meticulously compiled from reputable and well-documented geotechnical engineering sources (Eslami et al. 2019a). Initially introduced in 2015 as the "AUT-CPT&Pile database" (Moshfeghi et al. 2015), this database comprises 466 records of pile loading tests along with the corresponding results of a cone or piezocone penetration tests conducted in the vicinity of the pile locations (AUT: GEO-CPT&Pile Database 2016).

Updates and Accessibility

The AUT: Geo-CPT&Pile database has recently undergone significant expansion, now encompassing a total of 600 records. These case records have been meticulously collected from 68 sources spanning 24 countries, with a notable concentration of cases from the United States. The database upgrade reflects the continuous efforts to enhance its comprehensiveness and representativeness in geotechnical engineering research. The distribution of the case records is depicted in Figure 25.1.

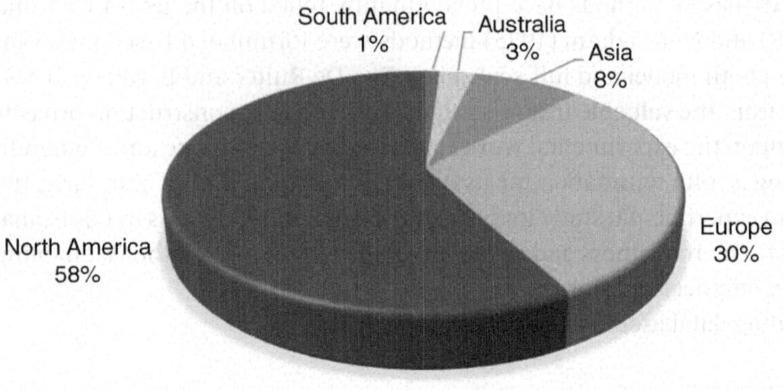

Figure 25.1 Distribution of case records of AUT: Geo-CPT&Pile database.

The AUT: Geo-CPT&Pile database comprises case records that feature both axial load tests and CPT records. CPT logs have been meticulously digitized at various depth intervals, including 0.05, 0.1, 0.2, 0.3, and occasionally 0.5 and 1 m, utilizing the GetData Graph Digitizer 2.24 software. Notably, 46% of the CPT results include measurements of cone tip resistance (q_c) and sleeve friction (f_s), while 29% solely contain q_c measurements. Additionally, 21% of the records include data for q_c, f_s, and pore pressure (u_2 or u_3), and 4% present measurements of q_c in combination with u_2 or u_3. Regarding the site soils, a wide array of clayey, silty, and sandy deposits are included, ensuring the representation of diverse geotechnical conditions.

The load test records feature head-down static load tests in both compression and tension, static O-cell tests, dynamic load tests (PDA and CAPWAP), and Statnamic load tests. Approximately 96% of the load test results provide comprehensive load–displacement diagrams, offering valuable insights into pile behavior. For the remaining 4% of cases, only the ultimate capacity is reported.

Case records in the AUT: Geo-CPT&Pile database consist of various pile types, including driven piles, auger cast piles, drilled displacement piles, continuous flight auger piles (CFA), driven cast-in situ piles (DCIS), post-grouted piles, jacked piles, vibro-driven piles, and helical piles. These piles are made from materials such as steel, concrete, composite (steel and concrete), and CFG (cement fly ash gravel). The shapes of the piles vary and include round, square, pipe, triangular, octagonal, H, X, and helical designs. The embedment lengths of the piles in the database range from 3.0 to 100 m, with the majority being less than 50 m. The pile diameters span from 50 to 2500 mm, with an average diameter of 450 mm. The pile capacities in the database range from 50 to 36,000 kN, with an average capacity of 1620 kN.

Figure 25.2 shows the distribution of case records in the AUT: Geo-CPT&Pile database.

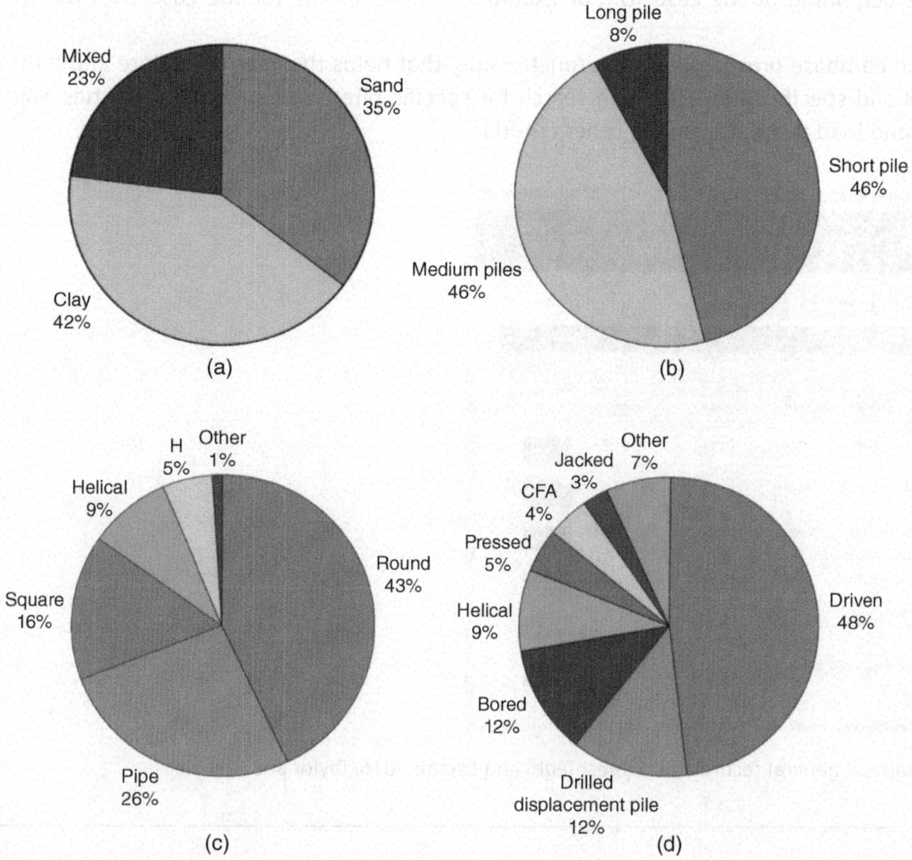

Figure 25.2 Distribution of AUT: Geo-CPT&Pile database case records based on: (a) soil types, (b) pile length, (c) shape, and (d) installation procedure.

Organization

The AUT: Geo-CPT&Pile database has been efficiently organized using Microsoft Access software, with different sections dedicated to essential information, CPT data, pile characteristics, and data sources. Additionally, a convenient search section is available to retrieve data with specific characteristics.

The General Records Form: The form in Figure 25.3 serves as the initial section of the database, encompassing a comprehensive list of all the records. Each record is assigned a unique identification name, enabling easy differentiation between its source, CPT profiles, and pile-loading test details. This form also provides a concise summary of each record's key characteristics. For more in-depth information on a particular record, users can access the "Details" button located at the end of each row, facilitating a more detailed view of the specific data.

CPT Data: This section of the database provides the CPT profiles and digitized results for each case. The data are conveniently presented in MS Excel files, which you can access by simply clicking on the provided preview link.

Piles' Characteristics: This section of the database provides information about pile geometry and installation type, along with the results of pile loading tests presented in both diagram and digitized formats. The database includes data on the time intervals between pile installation and performing the load tests, as well as the separated shaft and toe capacities, if available. This part is organized into subsections based on pile installation types for easy navigation. As with the "CPT results" section, you can access the MS Excel files containing the pile load test results by clicking on the provided preview links.

(Continued)

Case Study 25.1 (Continued)

Information Sources: This section of the database provides the original sources of the data for more detailed reference. Each case is assigned a unique identification name, and the name of the reference file for each case starts with a three-digit number at the beginning of the case ID. For example, the source file for the case with the ID "064-SANDPOINT" is "064."

Search Form: This section of the database provides a search functionality that helps the user to explore and filter the data based on various factors and specifications. One can search for specific pile types and characteristics, soil types, and the availability of CPT and load test data, among other criteria.

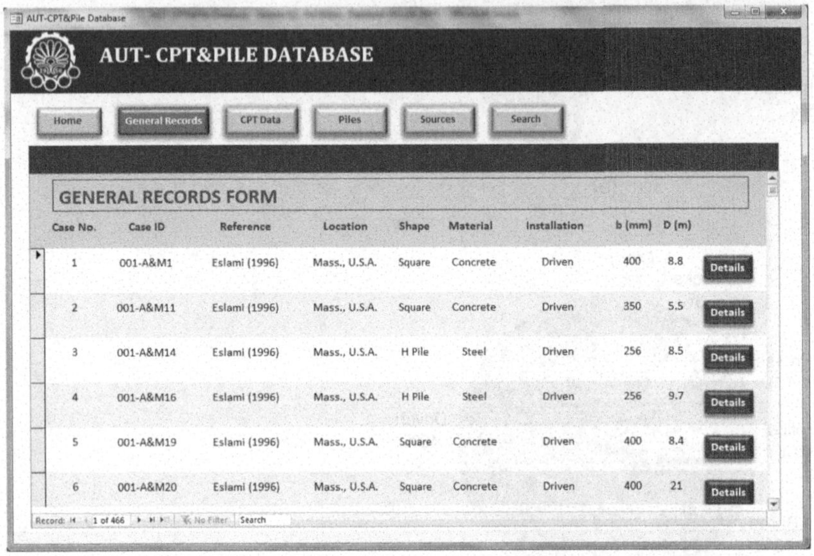

Figure 25.3 AUT: Geo-CPT&Pile database: general records form (Moshfeghi and Eslami 2016/Taylor and Francis).

25.5 New Approach for Pile Design Upon Cone Penetration Test Databases

Cone penetration test (CPT) records are considered reliable with relatively few measurement errors, offering valuable insights into the inherent soil variability at the investigated site. These records play a significant role in estimating soil shear strength parameters and contribute to determining both pile unit side and toe resistances. Although CPT-based methods have demonstrated their superiority over other approaches, they have been developed and calibrated using diverse assumptions, simplifications, and compiled databases. It is essential to note that the methods used for interpreting pile load tests (PLT) and the time interval between pile installation and conducting PLT can significantly influence the performance of predictive methods. Therefore, careful consideration and validation of these factors are crucial to ensuring accurate and dependable results.

The impact of uncertainty sources on the performance of predictive methods varies depending on the site conditions. Different compiled databases may exhibit varying accuracy and precision of the predictive methods. To mitigate the effects of these inherent uncertainties, a data-based approach is proposed to predict axial pile-bearing capacity. CPT and pile databases are considered valuable sources of big indirect data (BID) that hold potential for improving pile design (Phoon 2020). The proposed data-based approach offers a platform for selecting appropriate methods tailored to the specific site and project characteristics. By assessing the performance of predictive methods for similar sites and piles in the database using statistical, probabilistic, and reliability-based criteria, geotechnical engineers gain insights into the involvement of uncertainty sources. This analysis helps understand how both known and unknown uncertainties influence the predicted axial-bearing capacity among different methods and approaches. Among the various prevalent CPT-based methods, the proposed approach serves as an algorithm for decision-making, enabling the identification of the best-suited predictive methods for a particular site location (Heidarie Golafzani and Eslami 2021). As a result, a higher resistance factor is implemented in the load and resistance factor design (LRFD) approach. Furthermore, value engineering aims to enhance functionality and quality while optimizing costs. Therefore, geotechnical data-based pile design leads to a more reliable and cost-effective approach, taking into account latent uncertainties in the design process. This approach fosters a more robust and efficient design that aligns with the specific challenges and complexities of each project.

The data-based pile design procedure comprises several essential steps. First, a database containing CPT and pile load test records is compiled. Subsequently, a thorough assessment of the site's prevailing conditions is carried out. Depending on the project's significance, financial constraints, and other factors, one of the following three conditions may apply:

1) A comprehensive static or dynamic loading test plan is deemed necessary; or
2) Performing the minimum number of tests is sufficient; or
3) Testing is not feasible due to limitations.

Incorporating databases becomes especially crucial in cases of minimum to no testing conditions. By identifying similar cases with comparable geology to the project or gathering information from adjacent projects, the inherent soil variability at the investigated site can be somewhat reduced and provide valuable insights for the design process.

After conducting a thorough search for similar site conditions in the database and selecting pertinent data and cases, the predictive methods for axial pile-bearing capacity are evaluated using two main criteria:

a) Statistical and probabilistic analysis to assess the variability and uncertainty in the predictions; and
b) Reliability, risk, and efficiency considerations to make informed and intelligent selections for the design procedure of the site.

The first category of criteria is focused on assessing precision, accuracy, and estimated error in the predictive methods. The aim is to understand the variability and uncertainty associated with the predictions. The second group of criteria includes the efficiency ratio, which is the ratio of the LRFD resistance factor calibrated for a target reliability index to the mean resistance bias factor. Additionally, the wasted capacity index is considered to evaluate the effectiveness of these methods in the prediction process. Selecting appropriate methods based on these criteria enhances efficiency and ensures better project performance.

Subsequently, for the LRFD approach, load factors are determined in accordance with local codes and available information. The resistance factor for the target reliability index is then calculated. As the design methods are rationally selected, the resistance factor leads to an optimum site-specific design. The algorithm for data-based pile design is illustrated in Figure 25.4, showcasing the systematic process for making informed decisions and optimizing pile designs.

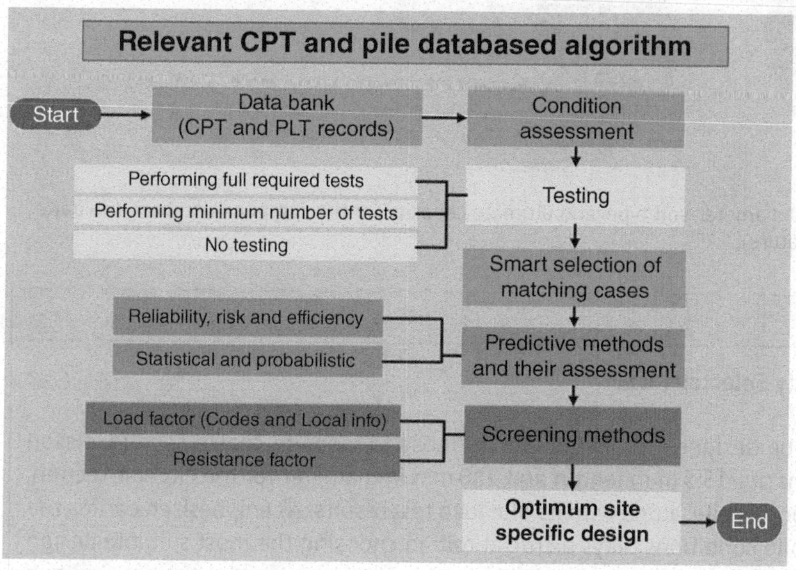

Figure 25.4 Relevant data-based pile design algorithm (Heidarie Golafzani and Eslami 2021).

25.6 Implementations of Proposed Procedure

Heidarie Golafzani et al. (2020a,b) compiled a database consisting of 60 driven piles, along with pile load tests and the corresponding CPT records obtained from the AUT: Geo-CPT&Pile database (Moshfeghi and Eslami 2016). The cases were collected from 13 countries spread across four different continents, with a majority from North America (44%), followed by Europe (28%), Asia (25%), and Australia (3%). The site soils in the database were categorized as 45% clayey soil, 32% sandy soil, and approximately 23% mixed soil. The distribution of soil types, pile lengths, and pile diameters for both steel and concrete piles is illustrated in Figure 25.5. This compiled database was then utilized to apply the data-based pile design algorithm in two case studies, demonstrating the practical implementation and effectiveness of the proposed approach.

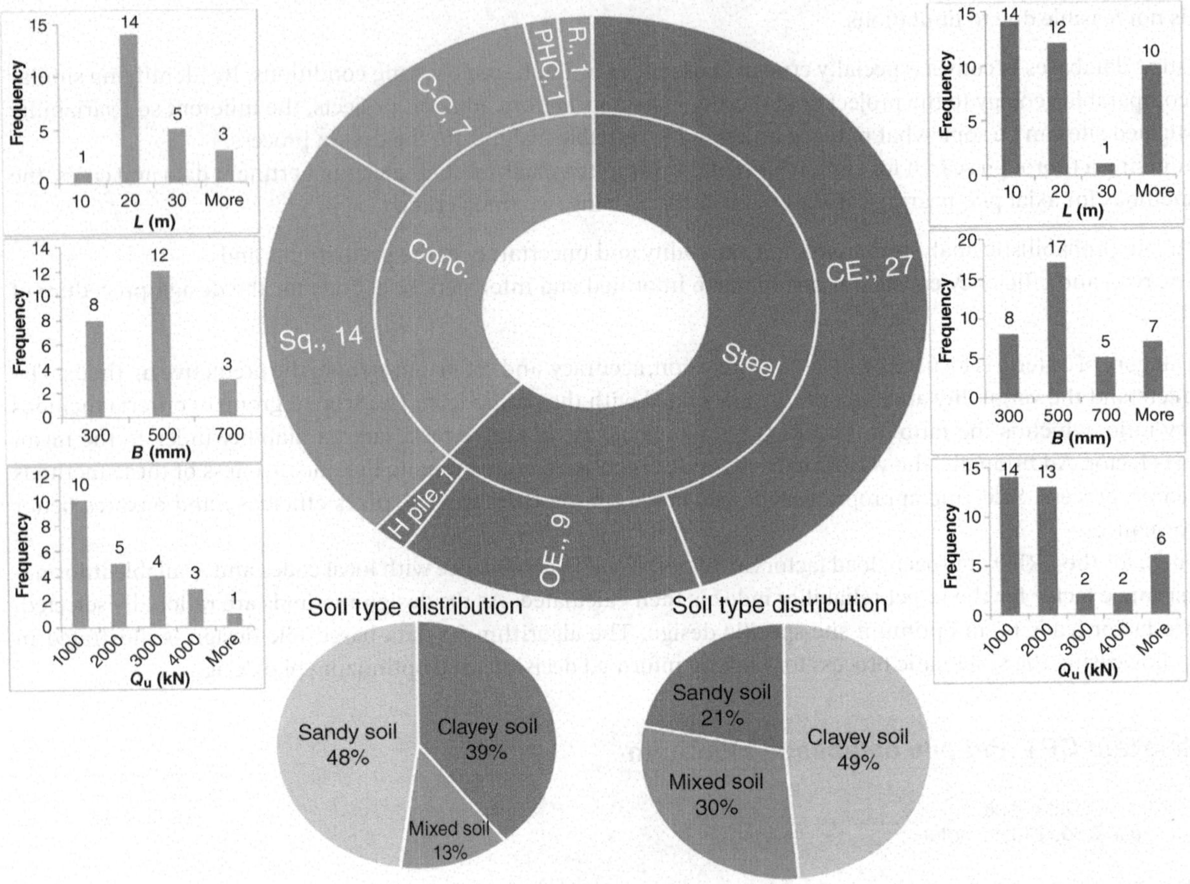

Figure 25.5 Distribution of pile material, length and diameter, soil type, and ultimate capacity for the compiled database (Heidarie Golafzani et al. 2022 / With permission of Springer nature).

Case Study 25.2

Comparing the Bearing Capacity of Two Smartly Selected Piles

A project in Norway requires a driven pile to be designed for installation in loose to medium-coarse sand (Axelsson 2000; Lunne et al. 2003). The pile's dimensions are 15.5 m in length and 280 mm in diameter for the circular section. Figure 25.6 shows the available CPT records for this pile, along with its pile load test results. As emphasized earlier, the intelligent selection of a site with matching pile conditions plays a crucial role in choosing the most suitable design methods for this specific pile. It involves finding a pile or piles in the database that closely resembles the given pile in terms of CPT records, including length, diameter, and shape of the pile, as well as the installation method and loadin g

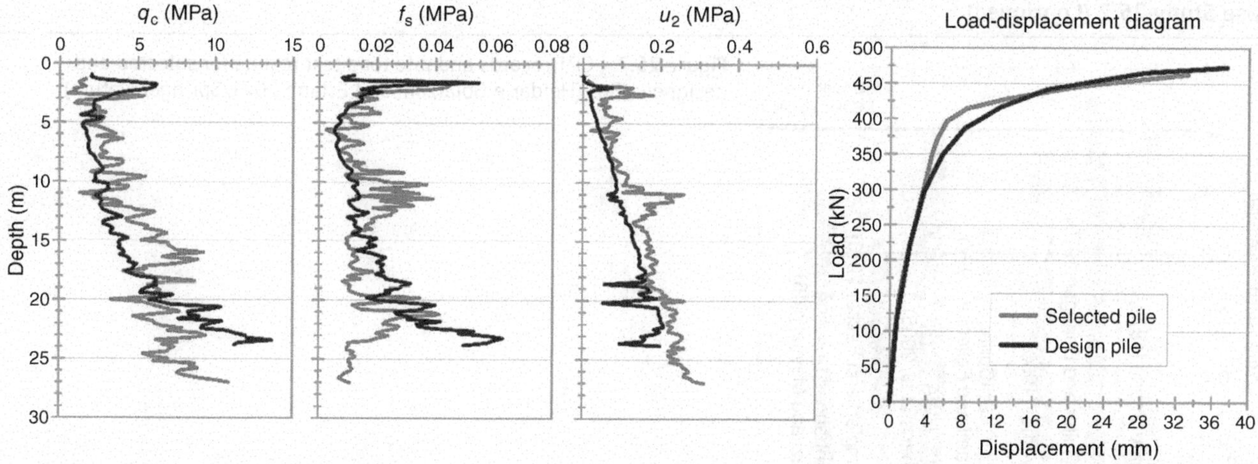

Figure 25.6 CPT records and pile load test results for selected and design pile (Heidarie Golafzani and Eslami 2021 / With permission of Springer Nature).

conditions. After evaluating 27 sites with sandy and mixed soils, one particular site in Sweden is identified as the most suitable match for the investigated 15.5 m-long pile. The site consists of a soil layer comprising sand, underlain by silty sand and gravelly sand layers. Table 25.2 presents the relevant characteristics of the selected piles for this site. To facilitate the design process, Figure 25.6 provides the CPT records and pile load test results for the chosen pile from the database.

Table 25.2 Summary of investigated piles from AUT: Geo-CPT&Pile database for the case study NO.1.

Source	No.	Case ID	Location	Shape	Material	*L* (m)	*B* (mm)	Loading mode	Q_u (kN)
Axelsson (2000)	1	Site ID 18-FSB 1 (Selected pile)	Sweden	Square	Concrete	12.8	235	Compression	458.9
Lunne et al. (2003)	2	Site ID 30-HSD E4 (Design pile)	Norway	Circular	Concrete	15.5	280	Compression	468.3

The performance of different predictive methods for the two investigated piles is depicted in Figure 25.7. Acceptable prediction values are defined based on two threshold values for the model parameter, Q_m/Q_p, namely 0.8 and 1.2 (Titi and Abu-Farsakh 1999). Methods that demonstrate good performance and fall within these upper and lower threshold values are UniCone (Eslami and Fellenius 1997), Meyerhof (1983), German (Kemfert and Becker 2010), and enhanced UniCone (Niazi and Mayne 2016). Their predictions are close together and lie between upper and lower threshold values. On the other hand, the predictions of Schmertmann (1978), Dutch (De Ruiter and Beringen 1979), and UWA (Lehane et al. 2005) methods for the design pile slightly exceed the upper threshold value. However, the predictions for the selected pile lie between the upper and lower threshold values. Comparing the predictions for these two piles

(Continued)

Case Study 25.2 (Continued)

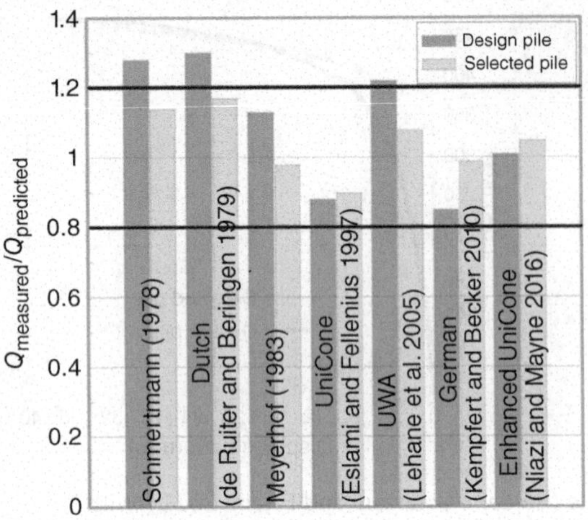

Figure 25.7 CPT records and pile load test results for selected and designed piles (Heidarie Golafzani and Eslami 2021/Springer Nature).

highlights the importance of considering similar site conditions and pile characteristics in the selection and adoption of predictive methods. When the project conditions closely resemble the conditions of the selected pile in the database, it leads to better decisions and more accurate predictions. This emphasizes the significance of utilizing a comprehensive and data-based approach to ensure optimal pile design for specific site conditions.

Case Study 25.3

Comparing the Axial-Bearing Capacity of Piles Installed in Clayey Soils

Ten circular steel piles, each with a diameter of 813 mm and an average embedment length of 70 m, were driven into the soft and sensitive clay subsoil of Urmiyeh Lake as part of a 1500 m-long causeway construction (Eslami et al. 2011). The subsoil consists of 100 m of soft and sensitive clay, as shown in Figure 25.8. To estimate the axial pile-bearing capacity for these piles, five different methods were investigated: Schmertmann (1978), Dutch (De Ruiter

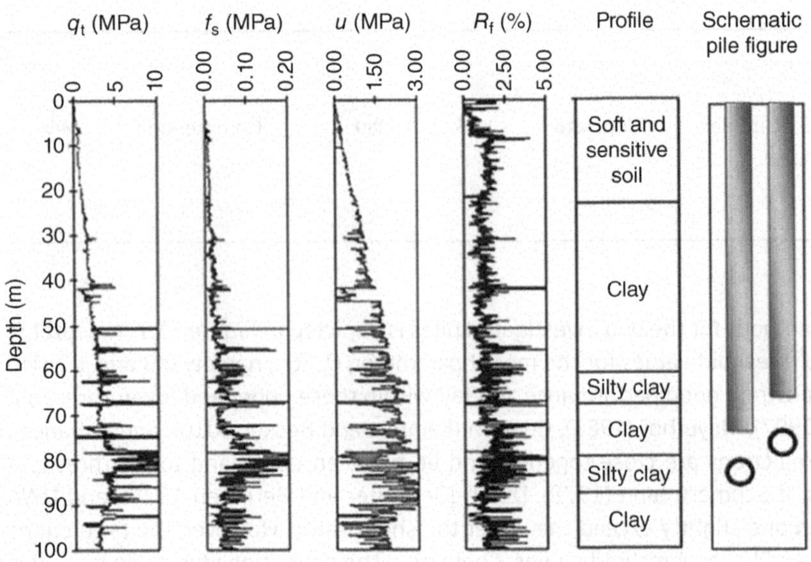

Figure 25.8 CPTu logs and soil profile for one of the boreholes in Urmiyeh Lake (Eslami et al. 2011/Elsevier/CC BY 4.0).

and Beringen 1979), LCPC (Bustamante and Gianeselli 1982), Tumay and Fakhroo (1982), and Eslami and Fellenius (1997). Among these methods, the Eslami and Fellenius (1997) method provided the most accurate estimation of the axial pile-bearing capacity. The model parameters for the other four methods varied within the range of 0.81–1.08. Based on these findings, the Eslami and Fellenius (1997) method is recommended for accurate estimation of the axial pile-bearing capacity in similar soft clay subsoil conditions.

The database investigated by Heidarie Golafzani and Eslami (2021) predominantly comprises clayey soil, accounting for approximately 45% of the cases. The performance assessment of 12 different methods for piles installed in clayey soils revealed that methods such as UniCone (Eslami and Fellenius 1997), Meyerhof (1983), Schmertmann (1978), and Dutch (De Ruiter and Beringen, 1979) exhibit good predictive capabilities. The comparison of the model parameter for piles installed in clayey soil is presented in Figure 25.9, showcasing the results obtained from various investigated CPT-based methods.

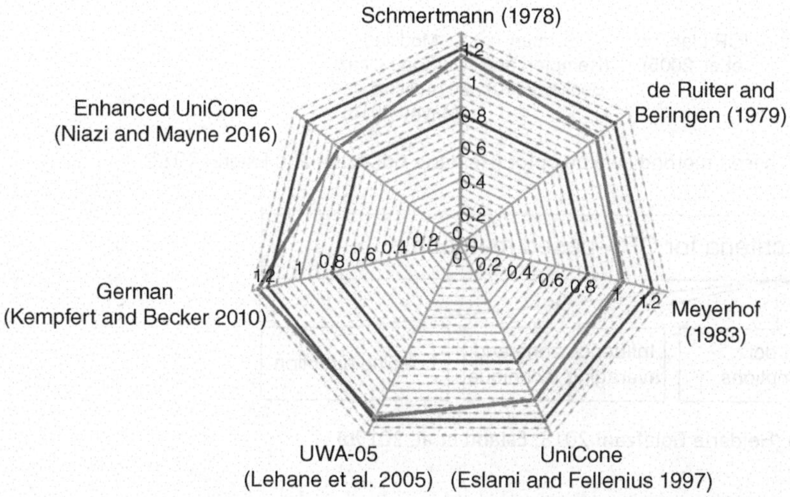

Figure 25.9 Model parameter of various methods for piles installed in clayey soils (Heidarie Golafzani and Eslami 2021).

25.7 Appraisal of Predictive CPT-Based Methods

Heidarie Golafzani and Eslami (2021)assessed the performance of six selected CPT-based methods developed over time from 1982 to 2016 in predicting the axial pile-bearing capacity. These methods apply different CPTu records and formulas to estimate unit side and end resistances separately (Niazi 2014). Figure 25.10 compares the box plot for these methods and provides information about the scatter of data, their mean (i.e., the cross inside the box), median (i.e., the line inside the box), the maximum and minimum of the model parameter, upper and lower quartile values, and the interquartile range (IQR). The IQR shows the middle 50% of data and is a measure of statistical dispersion. The closer the mean to unity, the lower the IQR and the smaller the scatter of data, which indicates the methods' satisfactory performance.

Model parameter, or the ratio of measured to predicted bearing capacity, is a valuable means of tracking the performances of various methods in predicting the axial-bearing capacity of piles. Various assessment criteria have been defined for statistical and probabilistic appraisal of predictive methods (Heidarie Golafzani et al. 2020b). These criteria measure the accuracy, precision, and error of a specific method.

25.7.1 Reliability-Based Criteria

Although statistical and probabilistic criteria shed light on the performance of the CPT-based methods, their performances are heavily dependent on the investigated database. Moreover, these methods are developed from various databases differing from one another in the gathered piles, the installation procedure and environment, the cone penetration test type, pile loading tests, and considering the effect of pile setup. These differences and further external factors bring about various formulas

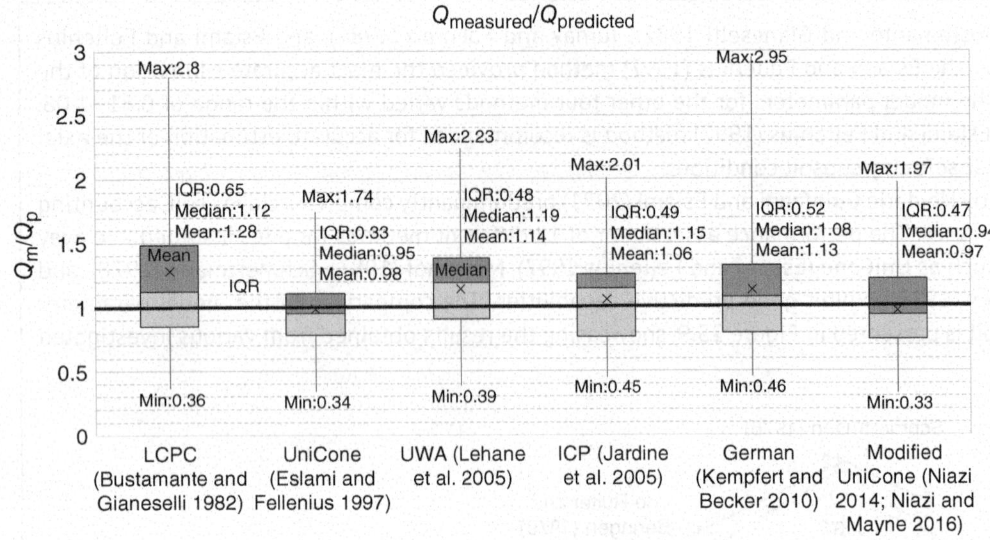

Figure 25.10 The performance of investigated CPT-based methods via box plot (Heidarie Golafzani and Eslami 2021).

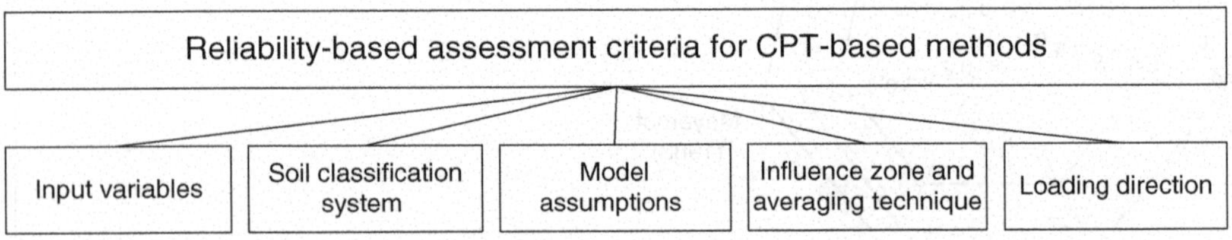

Figure 25.11 Reliability-based assessment criteria (Heidarie Golafzani 2018; Eslami et al. 2019b).

for predicting axial pile capacity, as provided in Eslami et al. (2019a,b) grouped the influencing factors into 11 categories. Five of these assessment criteria are considered for reliability-based evaluation in this study, as depicted in Figure 25.11. Other factors are data processing, failure criterion, pile installation method, time frame (i.e., the time elapsed between pile installation and pile load test delineating the effect of pile setup), friction fatigue, and data production- cone type.

25.7.2 Input Variables

In situ, tests such as CPT are valuable since the soil is tested under actual environmental conditions and stresses. Due to the similarity between pile and CPT, various correlations have been developed between CPT records and pile toe and side unit resistances. Formerly, these correlations were developed based on qc and further qc and fs since these records were recorded during the penetration test. Accompanying more various sensors during testing, i.e., measuring hydrostatic and excess pore water pressure or shear wave velocity, other correlations were developed based on the recorded data. The CPT and CPTu-based methods can be categorized according to their input variables as follows:

Depending solely on q_c, the first group includes LCPC (Bustamante and Gianeselli 1982), ICP-05 (Jardine et al. 2005), UWA-05 (Lehane et al. 2005), and German (Kempfert and Becker 2010) methods. The second group utilizes q_c, f_s, and u_2 to estimate pile axial-bearing capacity and covers UniCone (Eslami and Fellenius 1997) and Modified UniCone (Niazi 2014; Niazi and Mayne 2016).

This classification necessitates merging the ratios of measured to predicted capacities for various methods in one group. According to this classification, the performance of the methods is illustrated in Figure 25.12. The bias resistance factor is estimated for each group individually, and the efficiency ratio is calculated for the investigated methods, as presented in Table 25.3. The second group methods attain lower IQR and range values, a closer mean to unity, and lower COV than the

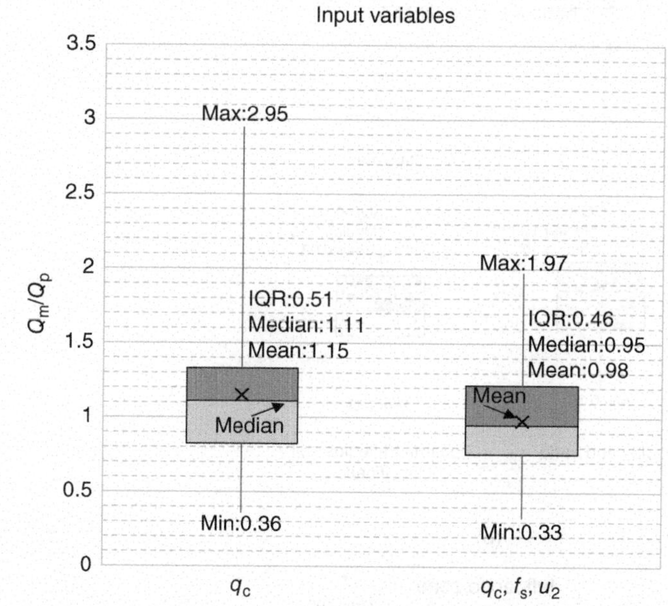

Figure 25.12 The efficiency ratio for CPT-based methods classified according to "input variables" (Heidarie Golafzani and Eslami 2021).

Table 25.3 Characteristics of classified groups according to "input variables" criterion; $\beta_{Target} = 2.33$, $Q_D/Q_L = 3$.

	1 variable	3 variables
μ_{Q_m/Q_p}	1.15	0.98
$COV\mu_{Q_m/Q_p}$	0.4	0.33
Range	2.59	1.65
ER‡	0.48	0.54

ER‡: Efficiency ratio.

first group. Although both groups attained similar resistance factors, the second group possessed a higher efficiency ratio due to its lower scatter and higher precision and accuracy.

25.7.3 Model Assumptions

The second criterion implemented to categorize the methods is the model assumptions. The investigated methods were developed based on various simplifications and assumptions. Having a look through the methods, one can divide them into four major groups. The first group consists of methods utilizing empirical correlations relating CPT q_c and f_s parameters to pile unit side and toe resistances. LCPC (Bustamante and Gianeselli 1982) belongs to this group. The second group was developed based on the total stress approach, especially in fine-grained soils. These methods apply undrained shear strength of soils either at the pile tip or along the pile embedment depth to estimate pile-bearing capacity. The German (Kempfert and Becker 2010) method is in this group. The third group employs the effective stress approach (β method) indirectly. These methods implement CPT records to estimate shear strength parameters and then apply these geotechnical parameters for predicting the axial pile-bearing capacity. The ICP-05 (Jardine et al. 2005) and UWA-05 (Lehane et al. 2005) methods come into this category. The fourth group is based on the direct effective stress method. These methods directly implement CPT records and the effective stress approach to predict the axial-bearing capacity of the investigated piles. UniCone (Eslami

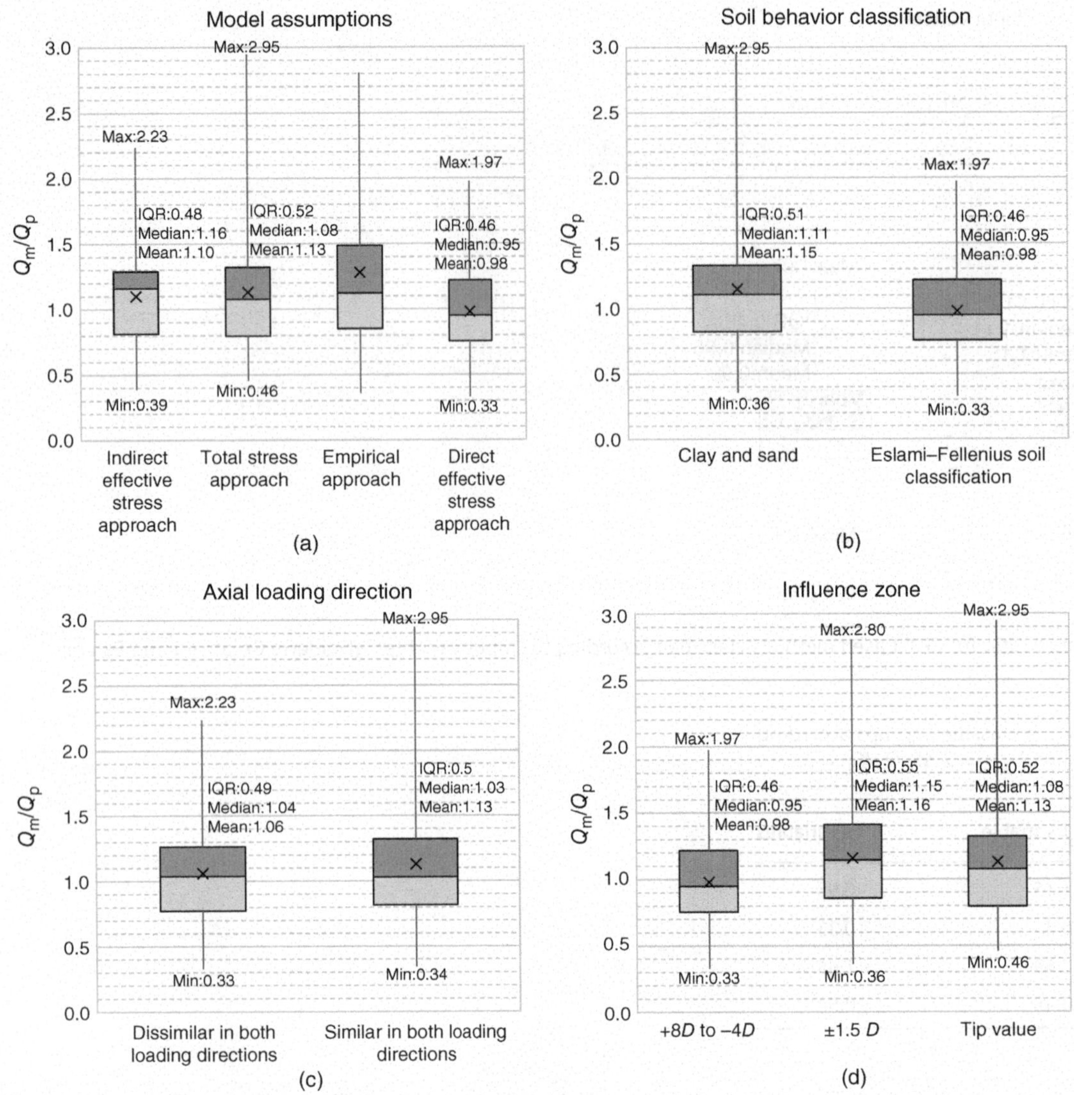

Figure 25.13 The efficiency ratio for CPT-based methods classified according to: (a) model assumptions, (b) soil behavior classification, (c) axial loading direction, and (d) influence zone axial loading direction (Heidarie Golafzani and Eslami 2021).

and Fellenius 1997) and Modified UniCone (Niazi 2014; Niazi and Mayne 2016) fit into this category. Figure 25.13a and Table 25.4 demonstrate that methods utilizing the direct effective stress approach perform better than other groups.

25.7.4 Soil Behavior Classification (SBC)

Pile shaft capacity depends on acting friction forces on the sides, which are related to the soil type (Eslami et al. 2020a). Regarding the implemented soil classification systems, methods were divided into two groups. The methods in the first group consider a broad classification of soils like sand and clay, while the second group utilizes more specific soil behavior classification based on CPT records. Except for UniCone (Eslami and Fellenius 1997) and Modified UniCone (Niazi 2014; Niazi and Mayne 2016), the remaining methods belong to the first group. The UniCone method, by implementing the Eslami and Fellenius soil behavior classification (SBC) chart, considers five different soil types along with the pile embedment depth and around the pile toe. This method suggested five constant coefficients (i.e., C_{se}) for estimating unit shaft resistance according to the soil type. Modified UniCone (Niazi 2014; Niazi and Mayne 2016) considers a modified Eslami and Fellenius soil classification system, classifying soils into 11 different soil types and proposing 11 C_{se}s as side correlation coefficients for different soil types. Figure 25.13b and Table 25.4 portray the efficiency of implementing different SBCs on the performance of the methods.

Table 25.4 Reliability-based criteria characteristics for predictive methods according to $\beta_{Target} = 2.33$, $Q_D/Q_L = 3$.

		μ_{Q_m/Q_p}	$COV\mu_{Q_m/Q_p}$	Range	ER‡
Model assumptions	Indirect effective stress approach	1.10	0.35	1.85	0.52
	Total stress approach	1.13	0.42	2.49	0.44
	Empirical approach	1.28	0.44	2.44	0.42
	Direct effective stress approach	0.98	0.33	1.65	0.54
Soil behavior classification	Clay and sand	1.15	0.4	2.59	0.46
	Eslami–Fellenius classification	0.98	0.33	1.65	0.54
Axial loading direction	Distinction between axial loading directions	1.06	0.35	1.91	0.46
	No distinction between axial loading directions	1.13	0.42	2.60	0.44
Influence zone	+8D to −4D (geometric mean)	0.98	0.33	1.65	0.54
	±1.5D	1.16	0.4	2.44	0.46
	Tip value	1.13	0.42	2.49	0.44

ER‡: Efficiency ratio.

Typical pile-bearing capacity components are shaft and toe resistance in compression loading and shaft resistance under tension for regular piles. Some CPT-based methods, including the LCPC (Bustamante and Gianesselli 1982), UniCone (Eslami and Fellenius 1997), and German (Kempfert and Becker 2010), suggested considering the proposed relations for prediction of shaft capacity of regular piles under tension loading. However, some methods encompassing the ICP (Jardine et al. 2005), UWA (Lehane et al. 2005), and Modified UniCone (Niazi 2014; Niazi and Mayne 2016) distinguished between the shaft capacity under compression and tension loadings. Figure 25.13c and Table 25.4 compare the efficiency ratio and scatter of model parameters for these two groups.

25.7.5 Influence Zone

Methods consider various failure patterns around the pile toe to estimate unit toe-bearing capacity. Accordingly, the influence zone around the pile tip varies among different methods. Generally, methods are divided into four categories. The first and second groups consider a log spiral failure pattern around the pile toe and average the qc records from 8*D* above to 4*D* below and 4*D* above to 1*D* below the pile toe, respectively. The third group proposed implementing the cavity expansion theory to average the q_c records 1.5*D* above and below the pile tip, and the fourth group considers no failure pattern around the pile base, where *D* is defined as the pile diameter. Moreover, the averaging techniques for the records lying between the considered zone above and below the pile tip are different. Figure 25.13d and Table 25.4 illustrate the efficiency ratio and dispersion of the model parameter for various categories.

Figure 25.14 compares the area ratio, i.e., the shaded to the total area ratio for CPT-based methods through the first set of criteria. Higher values of area ratio indicate better performance of the methods. Various CPT-based methods were developed regarding dissimilar databases differing in the number of piles, pile installation method, the investigated site location(s), static pile load test interpretation, and cone type. These differences and the background of each method, such as model assumptions, are influential parameters and control the associated efficiency and error in axial pile-bearing capacity predictions. The more differences between the initial and investigated databases, the more errors and the less the predictive methods' effectiveness.

The second set of criteria was selected according to the fundamental distinctions among CPT-based methods. Accordingly, five criteria were introduced, and methods were fitted into subcategories of each criterion. The model parameters of 62 investigated piles were gathered for methods belonging to each subcategory. Then, the efficiency ratios were calculated. Figure 25.15 illustrates the performance of the methods for the introduced reliability-based criteria via radar charts.

Methods possessing the following characteristics perform better than other investigated methods for this database, as depicted in Figures 25.12 and 25.13. As presented in Figure 25.12, the first criterion, i.e., input parameters, implementing q_c, f_s, and u_2, results in higher efficiency ratios. Figure 25.16a demonstrates that applying the effective stress approach for direct

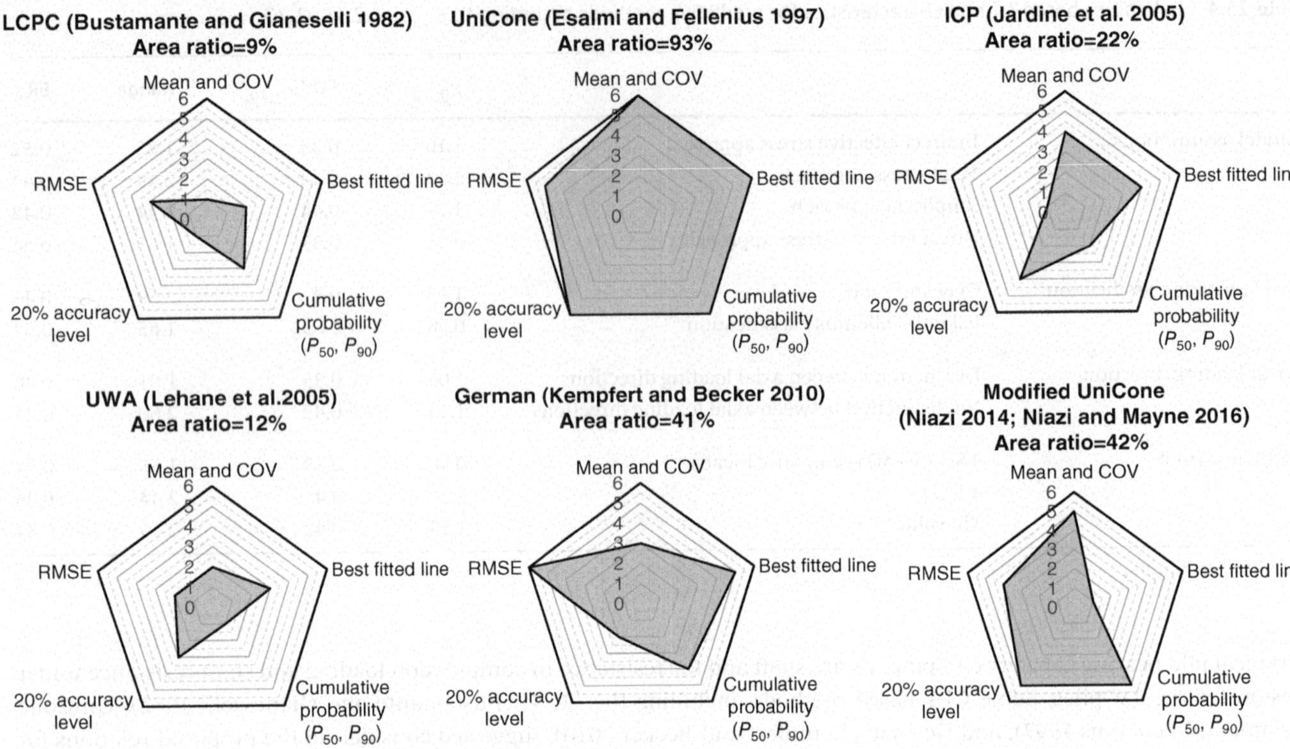

Figure 25.14 The performance of CPT-based methods according to statistical and probabilistic criteria (Heidarie Golafzani and Eslami 2021).

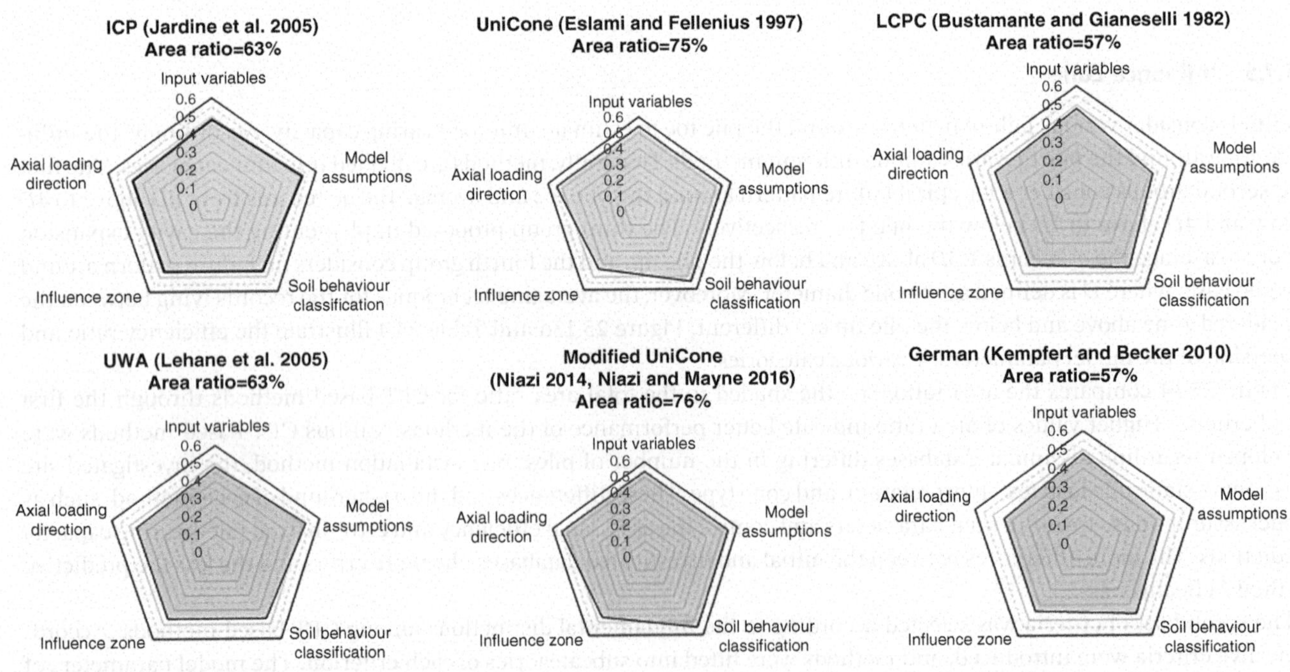

Figure 25.15 The performance of CPT-based methods according to reliability-based criteria (Heidarie Golafzani and Eslami 2021).

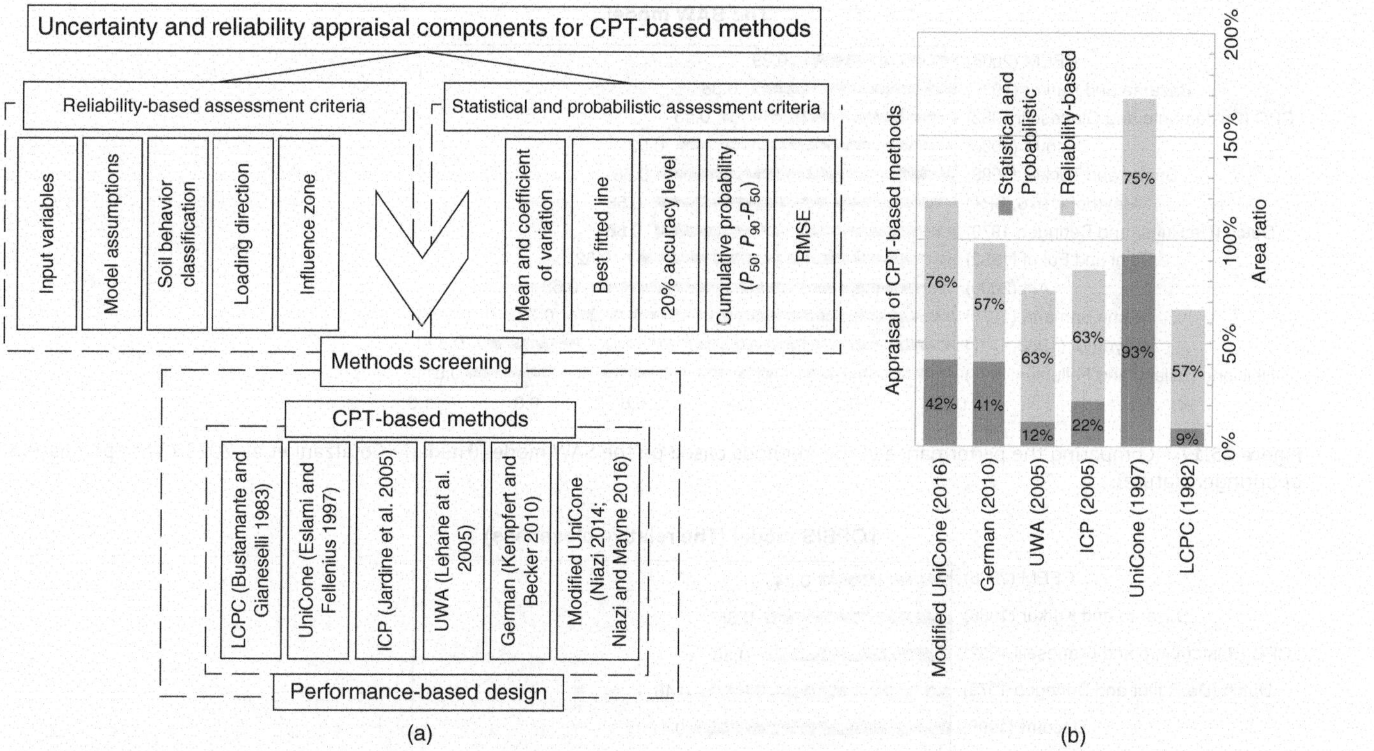

Figure 25.16 Methods screening: (a) appraisal components, (b) the performance of CPT-based methods via investigated criteria (Heidarie Golafzani and Eslami 2021).

CPT-based methods increases the efficiency ratio among other classifications. Figure 25.13b illustrates methods considering a more complicated SBC for estimating shaft capacity, e.g., the Eslami and Fellenius chart, which has better predictions. Figure 25.13c confirms that considering the loading direction in shaft capacity prediction is more reliable. Figure 25.13d shows the log-spiral failure pattern, from 8D above to 4D below the pile toe, which deals with minor uncertainties.

Assessing the performance of CPT-based methods is one of the steps in the site-specific design algorithm proposed by Eslami et al. (2020b). According to this algorithm, it is necessary to find matching available site conditions and pile characteristics in the compiled database for a particular site condition. The compatible methods and design approaches can be selected in the next step by implementing statistical, probabilistic, and reliability-based criteria. Hence, geotechnical data-based pile design is accompanied by a cost-effective design and more reliability regarding value engineering. Indeed, data mining provides invaluable information and enhances the geotechnical engineer to evaluate the performance of the predictive methods. Figure 25.16 delineates the uncertainty and reliability appraisal components. Accordingly, the geotechnical engineer will select the prioritized methods for design purposes according to the limitations and prevailing environmental conditions to make a better decision.

25.8 Multi-Criteria Decision-Making Models for Optimized Selection of Methods

Heidarie Golafzani et al. (2022) compiled a database of 60 driven piles installed in various soil types. To compare the performance of predictive methods in estimating the axial pile-bearing capacity, they selected 12 methods, including 5 CPT-based methods, 5 SPT-based methods, and 2 static analyses. Previously, when choosing various methods, some criteria were considered, and the methods were ranked according to these criteria and their summation led to a final rank for prioritizing one method over the others. However, MCDM models enhance the designer to arrive at a more conscious decision. Accordingly, they opted for seven criteria and implemented (i) simple additive weighting (SAW), (ii) technique for order preference by similarity to ideal solution (TOPSIS), and (iii) Elimination Et Choix Traduisant la REalite' (ELECTRE) models for assessing the performance of the predictive methods. Figures 25.17 and 25.18 delineate the performance of the SAW AND TOPSIS models. Furthermore, the ELECTRE model by pairwise comparison reveals the precedence of one method over the others.

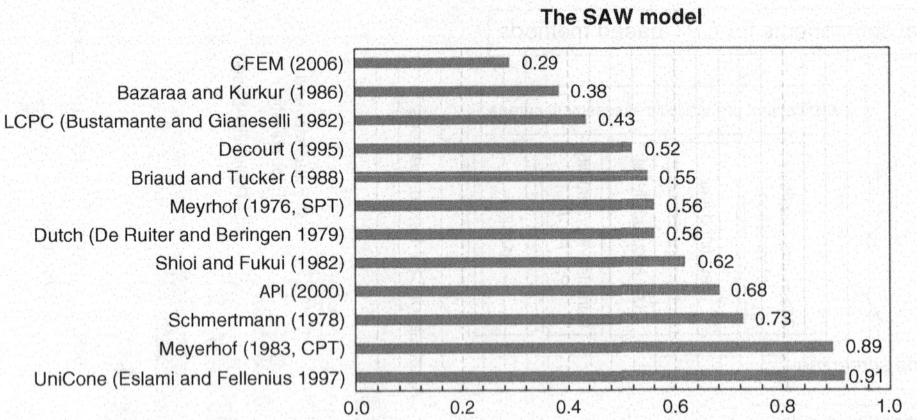

Figure 25.17 Comparing the performance of the methods based on the SAW model (Heidarie Golafzani et al. 2022 / With permission of Springer Nature).

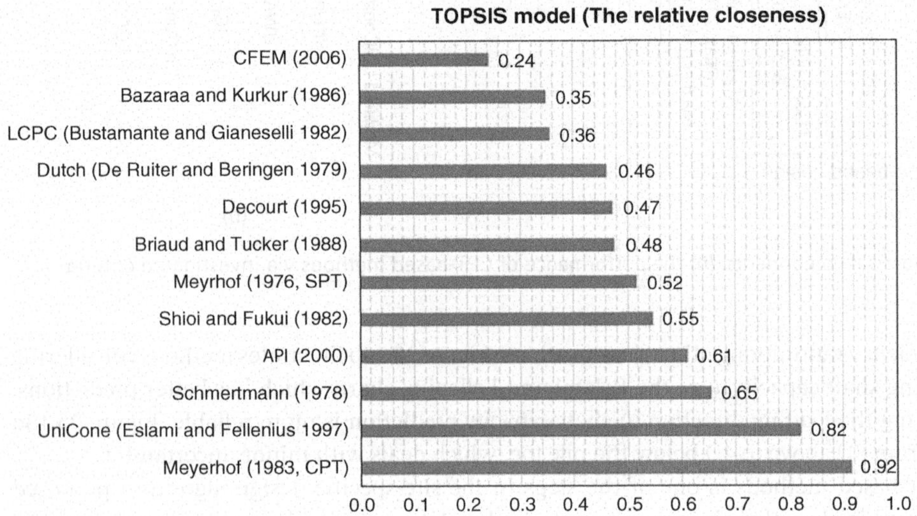

Figure 25.18 Comparing the performance of the methods based on the TOPSIS model (Heidarie Golafzani et al. 2022/ With permission of Springer Nature).

According to these results, the methods of Meyerhof (1983) and UniCone (Eslami and Fellenius 1997) outweigh the others, followed by the Schmertmann (1978) and API (2000) methods.

Additionally, there are various selection strategies known as aggregate methods, one of which is the Copeland method. This approach takes into account not only the number of times a method is ranked higher than others (i.e., $\sum C$) but also the number of times it is ranked lower ($\sum R$). Consequently, the difference between these two numbers determines the ranking of the methods. A larger difference value indicates a better performance for the method, as depicted in Figure 25.19.

25.9 Major Points

1) Databases are essential collections of organized data that facilitate easy access and information retrieval when needed, proving to be valuable tools in geotechnical engineering due to their empirical nature. These databases help achieve optimal designs in terms of timing and cost.
2) Various empirical correlations exist between CPT data and other crucial geotechnical parameters like soil-relative density, unit weight, soil modulus, over-consolidation ratio, and more. The CPT records serve as valuable indicators to monitor and assess ground improvement in problematic soils.
3) Several well-established geotechnical databases in the literature, including piling and cone penetration testing records, have been presented and reviewed. These databases have been extensively utilized to study different aspects of determining pile capacity through in situ tests.

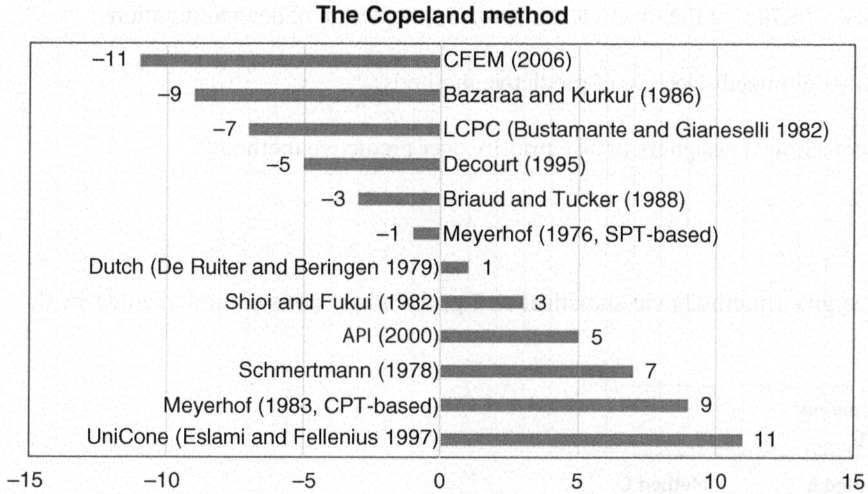

Figure 25.19 The Copeland for determining the priority among predictive methods based on the decision-making models (Heidarie Golafzani et al. 2022 / With permission of Springer Nature).

4) The AUT: Geo-CPT&Pile database is primarily focused on evaluating different approaches to determine static pile capacity based on cone and piezocone penetration tests. Compiled from reliable and documented geotechnical engineering sources, the database comprises 600 records of pile loading tests and corresponding cone or piezocone penetration test results. The case records originate from 68 sources spanning 24 countries, with a majority of cases from the United States.
5) The proposed data-based approach offers a robust platform for selecting appropriate methods tailored to specific site and project characteristics. By assessing the performance of predictive methods using statistical, probabilistic, and reliability-based criteria on similar sites and piles in the database, valuable insights are gained into how uncertainty sources influence the predicted axial pile-bearing capacity across different methods and approaches. This approach serves as an algorithm to make informed decisions, identifying the best-suited predictive methods for specific site locations. As a result, the LRFD approach incorporates a higher resistance factor, leading to a more reliable and cost-effective geotechnical data-based pile design, accounting for latent uncertainties.

25.10 Questions and Problems

25.10.1 Explanatory

1 What are the main applications of databases, including in situ test records in geotechnical engineering?

2 Why are in situ tests important in geotechnical engineering?

3 How do geotechnical databases affect embedded uncertainty sources in projects?

4 What are the pros and cons of a relevant data-based pile design approach?

5 What are the main steps for the relevant data-based pile design approach?

25.10.2 Conceptual

1 Compare the CPT-based methods from a reliability-based design criteria point of view regarding Figure 25.15.

2 What is the priority of a relevant data-based pile design approach compared to other common methods?

3 What is the role of geotechnical databases, including the in situ test records, in the design of deep foundations?

4 What is the role of MCDM models in the optimized selection of predictive methods?

5 How do the selection strategies help geotechnical designers to take priority over predictive methods?

25.10.3 Problem

1 Compare the performance of the three given methods via statistical and probabilistic criteria implemented in this chapter.

	Measured to predicted bearing capacity ratio; Q_m/Q_p		
Number	**Method A**	**Method B**	**Method C**
1	1.30	1.59	1.10
2	0.96	1.20	0.79
3	1.25	1.45	1.09
4	0.63	0.74	0.55
5	0.99	1.29	0.81
6	1.03	1.18	0.91
7	0.68	0.78	0.60
8	1.17	1.30	1.07
9	1.32	2.08	0.97
10	1.32	3.55	1.06
11	1.23	2.02	0.89
12	1.32	2.10	0.96
13	1.14	1.63	0.87
14	1.00	1.32	0.80
15	0.65	0.95	0.49
16	1.63	1.98	1.39
17	2.04	2.37	1.79
18	0.86	1.01	0.75
19	1.13	1.24	1.05
20	1.10	1.32	0.94
21	1.09	1.23	0.97
22	1.19	1.48	1.00
23	0.59	0.68	0.52
24	0.46	0.53	0.40
25	0.58	0.67	0.51

References

Abu-Farsakh, M.Y. and Titi, H.H. (2004). Assessment of direct cone penetration test methods for predicting the ultimate capacity of friction driven piles. *Journal of Geotechnical and Geoenvironmental Engineering* 130 (9): 935–944.

Alsamman, O.M. (1995). The use of CPT for calculating axial capacity of drilled shafts. Doctoral dissertation. University of Illinois at Urbana-Champaign.

Anderson, B.J., Townsend, F.C., and Rahelison, L. (2007). Load testing and settlement prediction of shallow foundation. *Journal of Geotechnical and Geoenvironmental Engineering* 133 (12): 1494–1502.

Aoki, N. and De Alencar, D. (1975). An approximate method to estimate the bearing capacity of piles. *Proceedings, 5th Pan-American Conf. of Soil Mechanics and Foundation Engineering, Buenos Aires* 1: 367–376.

API (2000). Recommended practice for planning, designing, and constructing fixed offshore platforms-working stress design. API RP A, 2.

AUT: GEO-CPT&Pile Database (2016). https://civil.aut.ac.ir/content/ (accessed 28 December 2021).

Axelsson, G. (2000). Long-term increase in shaft capacity of non-cohesive soils. PhD thesis. Royal Institute of Technology, Division of Soil and Rock Mechanics, Stockholm, 194 p.

Bazaraa, A.R. and Kurkur, M.M. (1986). N-values used to predict settlements of piles in Egypt. In: *Use of In Situ Tests in Geotechnical Engineering*, 462–474. ASCE.

Berardi, R. and Lancellotta, R. (1991). Stiffness of granular soils from field performance. *Geotechnique* 41 (1): 149–157.

Briaud, J.L. and Gibbens, R. (1999). Behavior of five large spread footings in sand. *Journal of Geotechnical and Geoenvironmental Engineering* 125 (9): 787–796.

Briaud, J.L. and Tucker, L.M. (1988). Measured and predicted axial capacity of 98 piles. *Journal of Geotechnical Engineering* 114 (9): 984–1001.

Bustamante, M. and Gianeselli, L. (1982). Pile bearing capacity prediction by means of static penetrometer CPT. *Proceedings of the 2nd European Symposium on Penetration Testing*, Amsterdam (24–27 May 1982). A.A. Balkema, pp. 493–500.

Chanmee, N., Chai, J., Hino, T., and Wang, J. (2017). Methods for evaluating over consolidation ratio from piezocone sounding results. *Underground Space* 2 (3): 182 e194.

De Ruiter, J. and Beringen, F. (1979). Pile foundations for Large North Sea structures. *Marine Georesources and Geotechnology* 3 (3): 267–314.

Eslaamizaad, S. and Robertson, P.K. (1996). Cone penetration test to evaluate bearing capacity of foundation in sands. *Proceedings of the 49th Canadian Geotechnical Conference*, St. John's, Newfoundland, Canada (23–25 September 1996). The Canadian Geotechnical Society, pp. 429–438.

Eslaamizaad, S. and Robertson, P.K. (1997). Evaluation of settlement of footings on sand from seismic in-situ tests. *Proceedings of the 50th Canadian Geotechnical Conference*, Ottawa, Ontario, vol. 2, pp. 755 e764.

Eslami, A. and Fellenius, B.H. (1997). Pile capacity by direct CPT and CPTu methods applied to 102 case histories. *Canadian Geotechnical Journal* 34: 886–904.

Eslami, A. and Gholami, A.M. (2006). Analytical model for the ultimate bearing capacity of foundations from cone resistance. *Scientia Iranica* 13 (3): 223–233.

Eslami, A. and Mohammadi, A. (2016). Drained soil shear strength parameters from CPTu data for marine deposits by analytical model. *Ships and Offshore Structures* 11 (8): 913–925.

Eslami, A., Aflaki, E., and Hosseini, B. (2011). Evaluating CPT and CPTu based pile bearing capacity estimation methods using Urmiyeh Lake causeway piling records. *Scientia Iranica* 18 (5): 1009–1019.

Eslami, A., Alimirzaei, M., Aflaki, E., and Molaabasi, H. (2017). Deltaic soil behavior classification using CPTu records – proposed approach and applied to fifty-four case histories. *Marine Georesources and Geotechnology* 35 (1): 62–79.

Eslami, A., Moshfeghi, S., Heidarie, S., and Valikhah, F. (2019a). AUT:Geo-CPT&Pile database updates and implementations for pile geotechnical design. *Geotechnical Engineering Journal of the SEAGS and AGSSEA* 50 (2).

Eslami, A., Moshfeghi, S., MolaAbasi, H., and Eslami, M.M. (2019b). *Piezocone and Cone Penetration Test (CPTu and CPT) Applications in Foundation Engineering*. Butterworth-Heinemann.

Eslami, A., Lotfi, S., Infante, J.A. et al. (2020a). Pile shaft capacity from cone penetration test records considering scale effects. *International Journal of Geomechanics* 20 (7): 04020073.

Eslami, A., Heidarie Golafzani, S., and Moshfeghi, S. (2020b). CPT and pile database for performance-based design of pile axial bearing capacity. *Proceedings of the 45th Annual Conference on Deep Foundations*, Deep Foundation Institute (DFI), Oxon Hill, Maryland (13–16 October), 12p.

Farhadi Nasl, F. (2013). Behavioral Classification of Problematic Soils by CPT and CPTu for Improvement Application. MSc. thesis. Amirkabir University of Technology.

Gwizdala, K. and Krasinski, A. (2013). Bearing capacity of displacement piles in layered soils with highly diverse strength parameters. *Proceedings of the 18th International Conference on Soil Mechanics and Geotechnical Engineering*. Paper No. 2519.

Hassani, A. (2011). Pile Bearing Capacity Prediction Based on CPT Records Using GMDH Type Neural Network. MSC thesis. Amirkabir University of Technology.

Heidarie Golafzani, S. (2018). Application of reliability method in processing of CPT data for determination of pile axial capacity. PhD thesis. Department of Civil Engineering, University of Guilan, Guilan, Iran.

Heidarie Golafzani, S. and Eslami, A. (2021). Uncertainty and reliability appraisal of CPT-based methods for axial pile bearing capacity. *Proceedings of the 46th Annual Conference on Deep Foundations, Deep Foundation Institute (DFI)*, Las Vegas, Nevada (12–15 October 2021), 12pp, Presented by Dr. Abolfazl Eslami.

Heidarie Golafzani, S., Jamshidi Chenari, R., and Eslami, A. (2020a). Reliability based assessment of axial pile bearing capacity: static analysis, SPT and CPT-based methods. *Georisk: Assessment and Management of Risk for Engineered Systems and Geohazards* 14 (3): 1–15.

Heidarie Golafzani, S., Eslami, A., and Jamshidi Chenari, R. (2020b). Probabilistic assessment of model uncertainty for prediction of pile foundation bearing capacity; static analysis, SPT and CPT-based methods. *Geotechnical and Geological Engineering* 38: 1–19.

Heidarie Golafzani, S., Eslami, A., Jamshidi Chenari, R., and Hamed Saghaian, M. (2022). Optimized selection of axial pile bearing capacity predictive methods based on multi-criteria decision-making models and database approach. *Soft Computing Journal*. https://doi.org/10.1007/s00500-021-06583-7.

Jardine, R., Chow, F., Overy, R., and Standing, J. (2005). *ICP Design Methods for Driven Piles in Sands and Clays*. Thomas Telford Ltd.

Kalavar, S. and Ealy, C. (2000). FHWA Deep Foundation Load Test Database. *Geotechnical Special Publication* 192.

Kempfert, H.G. and Becker, P. (2010). Axial pile resistance of different pile types based on empirical values. *Proceedings of Geo-Shanghai*, Shanghai, China (3–5 June 2010), pp. 149–154.

Lehane, B., Schneider, J., and Xu, X. (2005). The UWA-05 method for prediction of axial capacity of driven piles in sand. *Frontiers in Offshore Geotechnics: ISFOG* 683–689.

Lengkeek, H.J., de Greef, J., Joosten, S., et al. (2018). CPT based unit weight estimation extended to soft organic soils and peat. *Cone Penetration Testing 2018: Proceedings of the 4th International Symposium on Cone Penetration Testing (CPT'18)*, Delft, The Netherlands (21–22 June 2018). CRC Press, p. 389.

Long, J., Bozkurt, D., Kerrigan, J., and Wysockey, M. (1999). Value of methods for predicting axial pile capacity. *Transportation Research Record: Journal of the Transportation Research Board* 1663: 57–63.

Lunne, T., Long, M., and Forsberg, C.F. (2003). Characterisation and engineering properties of Onsøy clay. *Characterisation and Engineering Properties of Natural Soils (Proceedings Singapore Workshop)*, pp. 395–428. Rotterdam: Balkema.

Malekdoost, M. and Eslami, A. (2011). Application of CPT data for estimating foundations settlement. Sharif Civil. *Engineering Journal* 27 (2): 75 e85. No. 1.

Mayne, P.W. (2006). Interrelationships of DMT and CPT readings in soft clays. *Flat Dilatometer Testing* 231–236.

Mayne, P.W. and Illingworth, F., 2010. Direct CPT method for footing response in sands using a database approach. *Proceedings, 2nd International Symposium on Cone Penetration Testing* (CPT'10), 3, Huntington Beach, California, pp. 315 e322.

Meyerhof, G.G. (1976). Bearing capacity and settlement of pile foundations. *Journal of Geotechnical and Geoenvironmental Engineering* 102, no. ASCE 11962.

Meyerhof, G.G. (1983). Scale effects of ultimate pile capacity. *Journal of Geotechnical Engineering* 109: 797–806.

Moshfeghi, S. and Eslami, A. (2016). Study on pile ultimate capacity criteria and CPT-based direct methods. *International Journal of Geotechnical Engineering* 1–12.

Moshfeghi, S., Eslami, A., and Hosseini, S.M.M. (2015, October). AUT-CPT&Pile database for piling performance using CPT and CPTu records. In: *Proceedings of the 40th Annual Conference on Deep Foundations*, 12–15. Oakland, CA, Deep Foundation Institute (DFI).

Niazi, F.S. (2014). Static axial pile foundation response using seismic piezocone data. Doctoral dissertation. Georgia Institute of Technology.

Niazi, F.S. and Mayne, P.W. (2016). CPTu-based enhanced UniCone method for pile capacity. *Engineering Geology* 212: 21–34.

Nottingham, L.C. (1975). *Use of quasi-static friction cone penetrometer data: To Predict Load Capacity of Displacement Piles*. University of Florida.

Phoon, K.K. (2020). The story of statistics in geotechnical engineering. *Georisk: Assessment and Management of Risk for Engineered Systems and Geohazards* 14 (1): 3–25.

Roling, M.J., Sritharan, S., and Suleiman, M.T. (2011). Introduction to PILOT database and establishment of LRFD resistance factors for the construction control of driven steel H-piles. *Journal of Bridge Engineering* 16 (6): 728 e738.

Schmertmann, J.H. (1978). Guidelines for cone penetration test: performance and design (*No. FHWA-TS-78-209*). The United States: Federal Highway Administration.

Schneider, J.A., Xu, X., and Lehane, B.M. (2008). Database assessment of CPT-based design methods for axial capacity of driven piles in siliceous sands. *Journal of Geotechnical and Geoenvironmental Engineering* 134 (9): 1227–1244.

Shariatmadari, N., Eslami, A., and Karimpour Fard, M. (2008). Bearing capacity of driven piles in sands from SPT – applied to 60 case histories. *Iranian Journal of Science and Technology, Transaction B: Engineering* 32 (2): 125–140.

Shioi, Y. and Fukui, J. (1982). Application of N-value to design of foundations in Japan. *Proceeding of the Second European Symposium on Penetration Testing* 1 (1982): 159–116.

Titi, H.H. and Abu-Farsakh, M.Y. (1999). *Evaluation of Bearing Capacity of Piles from Cone Penetration Test Data*. Louisiana, USA: Louisiana Transportation Research Center.

Tumay, M.T. and Fakhroo, M. (1982). Friction pile capacity prediction in cohesive soils using electric quasi- static penetration tests. Interim Research Report No. 1. Baton Rouge, LA: Louisiana Department of Transportation and Development, Research and Development Section.

Valikhah, F. and Eslami, A. (2019). CPT-based non-linear stress–strain approach for evaluating foundation settlement: analytical and numerical analysis. *Arabian Journal for Science and Engineering* 44 (10): 8819–8834.

Van Dijk, B. and Kolk, H. (2011). CPT-based design method for axial capacity of offshore piles in clays. *Proceedings of Frontiers in Offshore Geotechnics II*. Taylor and Francis Group, London, pp. 555–560.

Van Impe, W.F. (1999). International exercise on piled rafts-Sweden examples. International Society of Soil Mechanics and Geotechnical Engineering. ITC-18 Committee, WVI/HDC/WIV.3.5, Gent University.

Yang, Z.X., Jardine, R.J., Guo, W.B., and Chow, F. (2015). A new and openly accessible database of tests on piles driven in sands. *Geotechnique Letters* 5 (1): 12–20.

26
Foundation Damages: Recognition and Repair

Chapter Highlight

- Insight on Foundation Damages
- Geotechnical Aspects
  - Bearing Capacity
  - Settlement
  - Stability
- Structural and Construction Faults
- Durability and Aging Effects
- Neighboring and Circumferential Complications
- Damage Assessment
- Foundation Restore and Retrofitting
  - Compensation Grouting
  - Foundation Underpinning
  - Structural Retrofitting
- Case Histories
  - Pisa Tower Tilting Due to Problematic Soil (Salgado 2008; Coduto et al. 2016)
  - Subsidence Due to Gypsum Dissolution, Ripon (North Yorkshire) in England Restoration (Cooper 1998)
  - Santos Adjacent Buildings
  - Relevelling of a Seven-Story Concrete Building that Settled in Clayey Sand (Ni and Cheng 2010)
  - Babolsar Concrete Pedestrian Bridge Retrofitting Practice (Eslami et al. 2016)
  - Restoration and Strengthening of the Cathedral Tower Foundation. Theresa of Avila (Santrač et al. 2021)
  - A Study on Increasing Building Story by Foundation Strengthening (Zumrawi and Aldawi 2018)
  - Micro-pile Underpinning of Mandalay Bay Hotel and Casino (Richards and Kardofilis 2006)
  - Performance of Buildings over Liquifiable Ground in Adapazari, Turkey (Sancio et al. 2004)
  - Emergency Bridge Pier Foundation Repair with Pressed-in Piles: Kumakami River Bridge Foundation Repair, Fukuoka Prefecture, Japan (Takefumi and Takayuki 2018)

26.1 Insight on Foundation Damages

Foundation damages represent one of the primary concerns for both inland and onshore structures, including buildings, silos, bridges, and storages. As the critical interface between the structure and the ground for load transfer, the foundation's soundness plays a vital role in the overall performance and longevity of the structure. Any failure in the foundation can disrupt the structure's function and efficiency, leading to potential structural instability and safety issues.

Advanced Foundation Engineering: Principles, Performance and Prospects, First Edition. Moses Karakouzian and Abolfazl Eslami.

Various factors can contribute to foundation damage, such as the presence of problematic soils under the foundation, unusual loads applied to the structure, poor construction practices, failure to account for weak or compressible deposits, changes in environmental conditions (e.g., deep excavation nearby), and the effects of adjacent buildings, tunneling, groundwater fluctuations, scouring, rock dissolution, and more. Addressing these potential causes and ensuring a robust foundation design and construction are essential for maintaining the structural integrity and safety of buildings and infrastructure.

Signs of foundation damage, particularly nonuniform settlements, can manifest in various ways, including cracks in interior and exterior elements of the building such as walls, damage to water and sewage transmission lines, gaps in doors and windows, difficulty in opening and closing doors and windows, misalignment of slabs and floors, and cracks in ceilings or floors. Detecting any of these warning signs is crucial, as they may indicate more severe damage to the building and its structure.

The causes of foundation damage can be categorized into geotechnical, structural, and construction-related factors. This chapter will discuss the different types of foundation damage and their origins. Additionally, methods for restoring and repairing the foundation will be explained to ensure the stability and longevity of the structure. Addressing foundation issues promptly is essential to prevent further deterioration and maintain the overall integrity of the building.

26.2 Geotechnical Aspects

Geotechnical issues can pose significant problems for foundations, leading to volume changes such as settlement and heave, as well as issues with bearing capacity and stability. One critical geotechnical concern that can have a lasting impact on projects is the presence of problematic soil at the project site, which can greatly affect the performance and integrity of the structure. Addressing these geotechnical defects is essential to ensure the long-term success and safety of the construction project.

26.2.1 Bearing Capacity

Soil failure beneath the foundation, caused by vertical and lateral displacement, is categorized as a bearing capacity failure. Shallow foundation failures can occur due to poor bearing capacity or when the applied load exceeds the design capacity. These failures are classified as general, local, or punching failures, each corresponding to different shear failure mechanisms (Kimmerling 2002).

For deep foundations, weak bearing capacity can result from various factors, including low pile tip resistance due to implementation in a weak or unsuitable layer, insufficient shaft resistance, or inadequate lateral resistance around the pile

(a) (b)

Figure 26.1 Bearing capacity failure of: (a) silo foundation (Kimmerling 2002) and (b) shallow foundation (Prakash 2007).

or drilled shafts. Figure 26.1 illustrates the collapse of soils due to bearing capacity weakness. Addressing these issues is crucial to prevent foundation failures and ensure the stability and longevity of the structures.

26.2.2 Settlement and Distortion

All structures can be settled due to loading. The critical factor in building settlement is the nonuniform settlement formed due to local movement in the structure.

Where foundation settlement takes place, a building may endure tilt and distortion. Although some structural distortion may occur, the building rotates as a rigid block for chimneys and high-rise buildings with a relatively large height-to-length ratio. The tilt of structures caused by nonuniform settlements for masonry towers and chimneys is dangerous and can even lead to the collapse of the structure. Distortion due to differential settlement in low-rise buildings is more important than tilt and makes more concern. Tilt is unlikely to be so severe that the entire building is in danger of overturning, as typically explained in Chapter 6. Since the use of deep foundations for low-rise buildings on poor ground is not economically justified, raft foundations are usually used for such lightly loaded buildings. Differential settlement in structures with sufficiently rigid foundations causes the entire building to tilt as a "rigid" body and prevents significant distortion of the building and any consequential cracking of the walls (Charles and Skinner 2004). Also, intermediate approaches such as ground modifications or semi-deep foundations can be considered.

Foundation displacement in the form of settlement is usually defined as differential settlement (Δs), in which the amount of vertical movement of one part of the foundation is more than another. A heave can cause the same type of vertical displacement, but settlement is more common than a heave. Six types of foundation settlements are defined in Table 26.1 and are further illustrated in Figure 26.2.

The results of unacceptable ground displacements are classified as follows:

- **Aesthetics:** Apply a negative effect on the appearance of the building.
- **Serviceability:** Some functions of the building or services such as drains, gas and water pipes, opening and closing of doors and windows, are disrupted.
- **Stability:** There is a risk of collapse of the building or some part of it.

26.2.3 Stability

26.2.3.1 Due to Loads

Combined loading (VMH), composed of vertical load (V), horizontal load (H), and overturning moments (M), is considered one of the most influential factors leading to the instability of structures, causing rotation, and horizontal and vertical

Table 26.1 Types of foundation settlement (Charles and Skinner 2004 / Emerald Publishing Limited).

Type	Settlement	Tilt	Distortion	Settlement (s)
i	Uniform	None	None	$s_A = s_B = s_C$
ii	Nonuniform	Uniform	None	$s_C - s_A = (L_{AC}/L_{AB})(s_B - s_A)$
iii	Nonuniform	None	Sagging	$s_A = s_C$ $s_A < s_B$
iv	Nonuniform	None	Hogging	$s_A = s_C$ $s_A > s_B$
v	Nonuniform	Nonuniform	Sagging	$s_A \neq s_C$ $s_C - s_A = (L_{AC}/L_{AB})(s_B - s_A)$
vi	Nonuniform	Nonuniform	Hogging	$s_A \neq s_C$ $s_C - s_A = (L_{AC}/L_{AB})(s_B - s_A)$

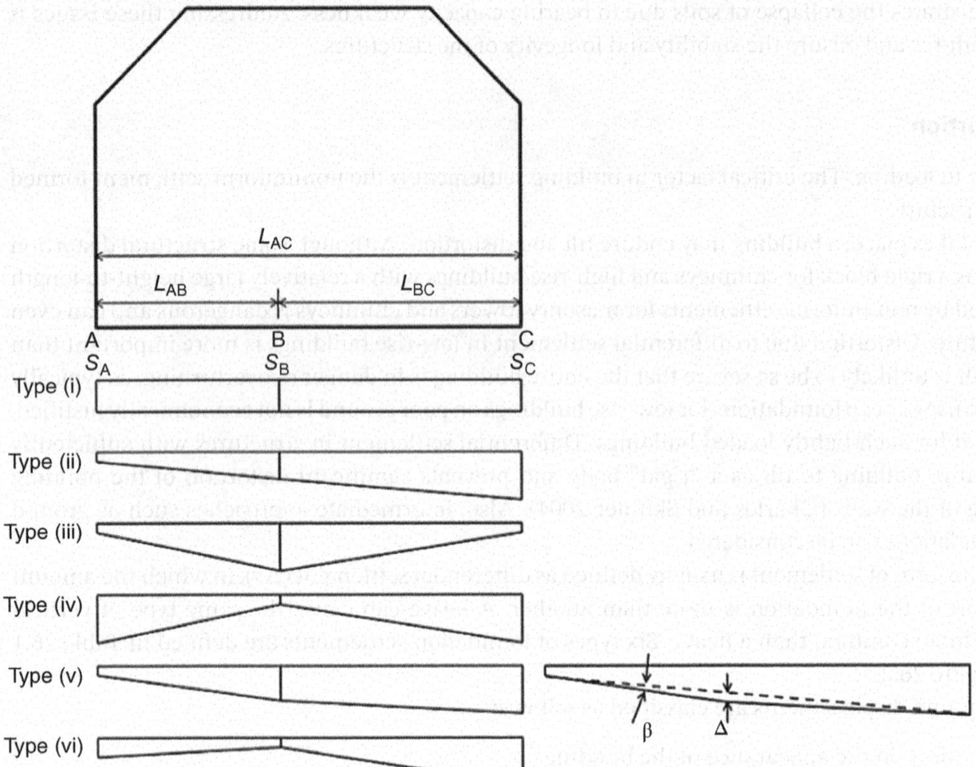

Figure 26.2 Types of foundation settlement (Charles and Skinner 2004 / Emerald Publishing Limited).

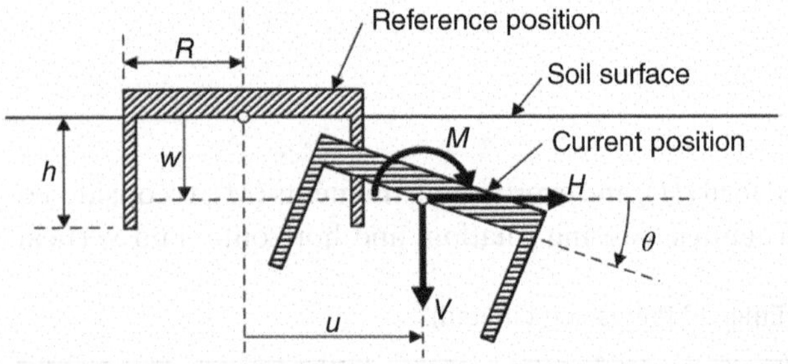

Figure 26.3 Combined loads and displacement systems (Villalobos et al. 2010).

movements. Combined loadings typically result from gravity loads, wind loads, earthquake loads, as well as wave loads in marine sites. Various load and displacement systems have been used in previous investigations. Figure 26.3 depicts one of the typical combined load and displacement systems.

26.2.3.2 Due to Soil Conditions

Soils are materials formed by weathering and the erosion process of rocks. However, changes in the water content of certain soils can pose significant challenges for construction projects. These challenges include swelling, dispersing, and collapse (Rezaei et al. 2012). The presence of problematic soils in many regions around the world makes foundation construction extremely challenging. For instance, expansive or collapsible soils can cause high differential movements in structures due to excessive heave or settlement (Das 2014). Changes in water content in expansive soils lead to alterations in their volume, potentially damaging structures built on such soil. In dispersive soils, particles move through the soil with water flow, creating conduits that can harm the structural integrity of buildings. Figure 26.4 illustrates some building damages caused by problematic soils.

Figure 26.4 Damage to buildings caused by problematic soils.

Case Study 26.1

Pisa Tower Tilting Due to Problematic Soil (Coduto et al. 2016; Salgado 2008)

The Tower of Pisa, a renowned tourist attraction located in the city of Pisa, Tuscany, Italy, stands tall at 54 m and weighs 142,000 kN. It consists of a tall ground story, six loggias (open galleries), and a belfry at the top. The foundation is a spread foundation in the form of a hollow cylinder, with an outer diameter of 19.48 m and an inner diameter of 4.47 m. It is embedded approximately 3 m into the soil.

Situated on top of 300 m of sediments deposited by the Arno River and the sea, the tower's foundation rests on about 9 m of silty/sandy soil deposited by the river, followed by approximately 30 m of marine clay. Due to the differing compressibility of the silt layer on the south side of the tower compared to the north side, the settlement developed faster on the south side, resulting in the tower's present inclination (Salgado 2008).

Throughout the twentieth century, the tower experienced a gradual tilt at a rate of about six arc seconds per year. By the end of the century, the total tilt reached approximately 5.5° to the south, meaning the top of the tower structure was 5.2 m (17.0 ft) off plumb. The average bearing pressure under the tower is 497 kPa (10,400 psf), but the tilting caused the weight to act eccentrically on the foundation, leading to nonuniform bearing pressure. At the end of the century, it ranged from 62 to 930 kPa (1300–19,600 psf). The tower was evidently on the verge of collapse, and even a minor earthquake could potentially topple it. Consequently, it became apparent that remedial measures were necessary.

In 1990, the interior of the tower was closed to visitors due to concerns about its stability. In 1993, a temporary stabilization measure was implemented by placing about 600 tons of lead ingots on the north side of the tower. However, this approach only slightly decreased the tilt, and further efforts were needed to address the issue.

In 1995, engineers installed a concrete ring around the foundation and began drilling tie-down anchors through the ring into the dense sand stratum located approximately 40 ft deep. While this caused some compression of the underlying soils and a slight reduction in tilt, the construction of the anchors unexpectedly increased the tower's tilt by about 1.5 mm in a single night, equivalent to a year's worth of normal movement. Consequently, this method was abandoned, and additional lead ingots were added to the north side as a temporary solution.

The collapse of a tower in nearby Assisi after an earthquake in 1997 prompted renewed efforts to stabilize the Tower of Pisa. The overseeing committee approved the use of the soil extraction method, or under-excavation, to decrease the tilt from 5.5° to 5.0°, aiming to return the tower to a position similar to 300 years ago. The goal was not to make the tower perfectly plumb.

Between February 2000 and February 2001, a careful soil extraction process was carried out underneath the north side of the tower by drilling 41 diagonal borings. The process involved removing only small amounts of soil at a time, and the tower's movement was closely monitored throughout. As a result of this effort, the tower's tilt was reduced by 0.5°. Further progress was made between February 2001 and September 2008, achieving an additional reduction of 150 arc seconds of tilt. However, the rate of tilt reduction has since slowed to less than 0.2 arc seconds per year. Engineers estimate that this remediation will ensure the tower's stability for another 200 years (Coduto et al. 2016).

26.3 Issues Related to Structural Design and Construction Errors

Foundations commonly experience two major types of failures due to loading: flexural (bending) and shearing. Additionally, foundations experience various issues, including deterioration of the foundation material and construction defects.

Among the challenges in the interaction between concrete structures and the underlying ground, the punching shear resistance of reinforced concrete ground-supported slabs represents a significant category. The spatial failure mechanism in this case is influenced by critical parameters such as the geometric dimensions of the structure, properties of the concrete materials, applied loads, and the soil properties below. Punching shear failure occurs when the applied load causes a portion of the foundation to punch through and separate from the rest of the footing. On inclined failure surfaces, this type of failure combines both tension and shear. Figure 26.5 illustrates the punching shear model at the ultimate limit state.

In most cases, shallow foundations such as mats, slabs, or even pile caps experience punching failure due to the combination of existing loads and relatively low thickness or depth. Flexural failure in the foundation can be attributed to factors such as excessive loads applied to the foundation and improper design and execution of longitudinal lower bars, as explained in Chapter 7.

Moreover, structural damage related to the foundation may also result from weaknesses in concrete performance, which can be caused by improper execution and design of concrete, inadequate coverage of steel rebar, concrete texture bubbling, and unfavorable conditions during concreting or mixing different materials with concrete.

During construction, structures may encounter adverse conditions that were not identified beforehand, faulty design, or poor quality of work, as shown in Figure 26.6. Some examples of these issues include (Dadarya 2007):

- Foundations supported on problematic soil without appropriate soil improvement.
- Deficient construction techniques or the use of substandard materials.
- Foundation designs that are too complex or challenging to construct correctly.
- Foundations that do not perform as anticipated.

Damages and defects related to the foundation can include the following:

- Incorrect dimensions and location during the execution of the foundation or pile may require assessing the efficiency of the foundation structure.
- Damage to the pile during installation.
- Defects in the foundation's concrete structure during construction, such as soil entering the concrete, the appearance of bubbles due to improper water mixing, or a reduction in concrete efficiency. Additionally, inadvertent use of an improper mixing design with poor compressive strength, lack of proper concrete cover on steel rebars, or foundation failure due to excessive loading after construction can also be a concern.
- Leaching of the concrete due to the inflow of water through joints or voids in the subsurface.

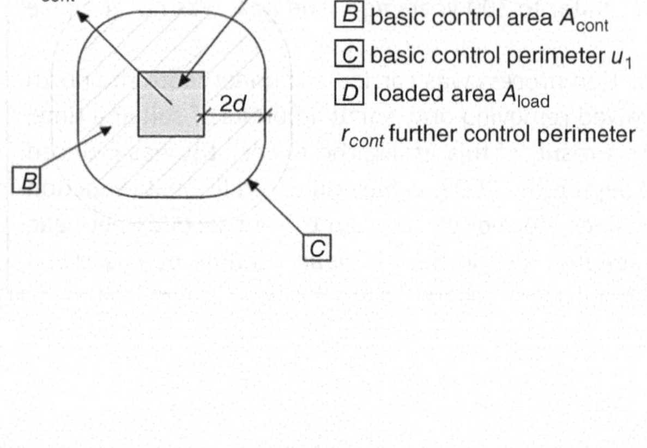

Figure 26.5 Punching shear verification model at the ultimate limit state (Sucharda et al. 2018).

Figure 26.6 A few damages due to poor constructions.

- Major cracks in the concrete pile can affect its quality and reliability. These cracks may result from improper construction practices or damage during transportation and installation.

26.4 Durability and Aging Effects

In certain cases, foundation damage may occur during the life span of a structure due to environmental factors. Such failures can be categorized as either internal or external. Internal factors encompass damages to structural elements like steel rebar and concrete. These damages can arise from concrete deterioration (physical, biological, and chemical), reinforcement degradation, soil erosion and scour, rock dissolution, sulfate attack, and corrosion of steel rebar. Additionally, older masonry and stone structures may require retrofitting of their foundation structures due to aging, chemical attack, erosion, and corrosion. Figure 26.7 illustrates damages observed in masonry structures.

Figure 26.7 The damage to masonry structures (D'Ayala and Paganoni 2011 / Frontiers Media S.A.).

26.4.1 Distortion and Cracks

Deformations resulting from foundation movements can generate strains within the building, potentially leading to structural damage. Identifying the exact type of damage caused by these strains is challenging, even if the magnitude of the strain is determined. When a building undergoes distortion, visible damage manifests as cracks in brick walls and doors/windows jamming. Cracks are the primary indicators of structural damage in buildings. External factors like temperature, humidity, chemical reactions, or building deformations can contribute to crack formation (Korff 2009). The characteristics of cracks due to the deformation of the building are:

- Cracks are typically tapered, starting small at one end and widening towards the other.
- Cracks can be observed on both the interior and exterior of the building.
- Cracks may extend below and above ground level.
- The location and direction of the cracks are directly related to the deformation mode (hogging or sagging). Additional forms of damage may include broken windows and jammed doors due to distorted window and door openings, sloping floors, and tilting walls.

Cracks tend to emerge at the weakest points of the structure, commonly found around door and window openings, between walls and floors or skirting boards, and between walls and ceilings (Shabha and Kuhwald 1995). These cracks typically follow two main directions: vertical, leading towards the center of the building, or conversely, outward, extending towards opposite corners, as illustrated in Figure 26.8. Settlement cracks are often relatively large and isolated.

The presence of cracks can have adverse effects on the building's structural integrity, causing doors and windows to shift out of the plumb line and become distorted. As the damage progresses, they may begin to stick, making it increasingly difficult to open and close them. Moreover, exterior walls might buckle and deviate from the original alignment, leading to severe cases of roof damage, such as dislodged tiles and roof failure (Shabha and Kuhwald 1995).

Reinforced concrete structural damage can be classified into two primary categories: concrete damage and reinforcement damage. These failures can exert a substantial impact on the longevity and load-bearing capacity of foundations.

26.4.2 Concrete Damages

Numerous detrimental factors have the potential to compromise the integrity of a steel-reinforced concrete structure over its life span. These influences can emerge from the environment or be linked to the structure's usage. They can be categorized into groups encompassing physical, biological, and chemical factors (Bonić et al. 2015).

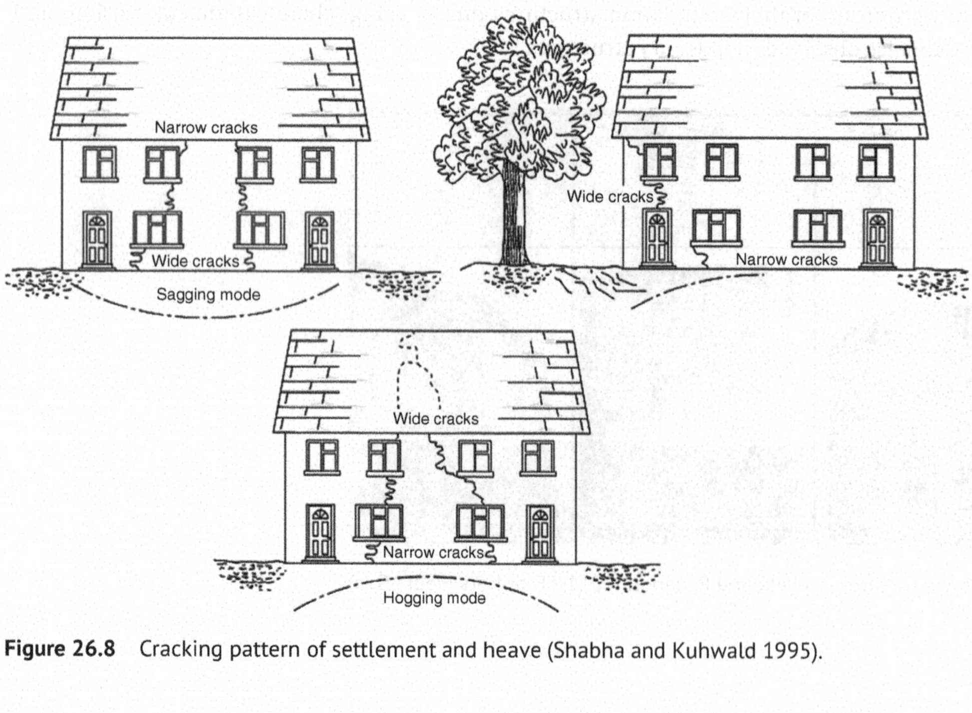

Figure 26.8 Cracking pattern of settlement and heave (Shabha and Kuhwald 1995).

26.4.2.1 Physical Damages

Frost action stands out as a primary contributor to the physical deterioration of concrete structures. The presence of water within concrete pores, coupled with freezing at low temperatures, exerts significant pressure on the material. Repeated occurrences of this phenomenon can result in severe concrete damage, a particularly prevalent issue in colder regions.

Several strategies can be employed to counteract the detrimental effects of frost. These encompass careful selection of foundation depth, incorporation of a sand layer beneath foundations, implementation of effective drainage systems, and establishment of robust moisture and water insulation for below-ground portions of buildings. These measures work collectively to mitigate the impact of frost on foundations (Bonić et al. 2015).

26.4.2.2 Biological Damages

The expansion of cracks in the foundation due to the root system of trees and the effect of vegetation is known as biological effects (Bonić et al. 2015).

26.4.2.3 Chemical Effects

Degradation of concrete can be caused by chemical effects as a result of chemical reactions of aggressive compounds from the structural environment and the components of the cement aggregate themselves. Their destructive effects cause concrete corrosion. This type of deterioration can be divided into aggregate expansion, salt weathering, sulfate attack, and carbonation (Bonić et al. 2015).

26.4.3 Reinforcement Damages

Corrosion stands as a prominent challenge that undermines the durability of reinforced concrete (RC) structures. The process of corrosion gives rise to brown discolorations on the concrete surface, accompanied by the development of cracks along the reinforcing bars. This occurrence is particularly pronounced in the corners of elements exposed to the infiltration of aggressive substances into the concrete matrix. As illustrated in Figure 26.9, within concrete, the corrosion of steel results in the formation of cracks due to the expansion of rust, which occupies a larger volume than the steel itself. This expansion facilitates the deeper penetration of aggressive materials (Bonić et al. 2015).

The consequences of corrosion extend to the initiation of cracking and scaling in concrete. The expansion of corrosion products can lead to these defects, while concurrently causing a reduction in the cross-sectional area of the steel reinforcement. This combination of effects imperils the safety and functionality of RC structures. Notably, a corrosion level exceeding a 15% mass loss has been shown to notably compromise the ductility and plastic deformation capacity of tensioned bars (Zhou et al. 2015).

26.4.4 Soil Erosion and Scour

In regions characterized by high-intensity rainfall, the likelihood of soil surface erosion caused by flowing water is considerable. If the topsoil is loose and the structure is situated on a sloped surface, the vulnerability to surface erosion is further exacerbated. Drawing from local insights, it becomes essential to gauge the depth of erosion and subsequently position the foundation beneath these estimated levels. Erosion effectively erodes the foundational support, often leading to settlement in the underlying structure.

Cajka and Manasek (2005) highlight that in loose sandy soils, post-flood scenarios can result in the displacement of soil particles by the force of flowing water. These consequences can ultimately manifest as subsidence within the foundation.

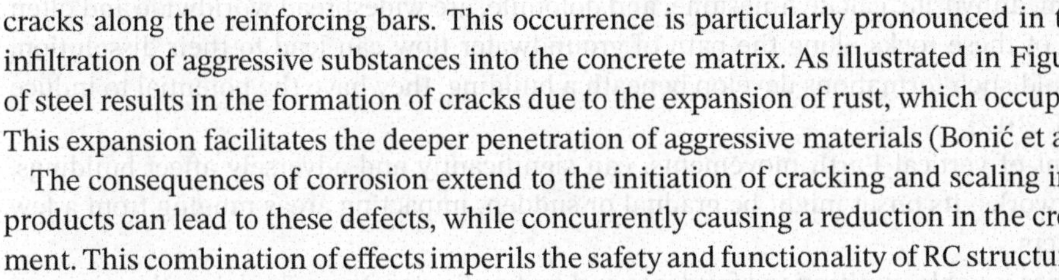

Figure 26.9 Development of corrosion of a reinforcement bar in concrete (Bonić et al. 2015 / with permission of Elsevier).

Their study also suggests that water velocities exceeding 0.6–1.2 m/s can induce scouring in gravel soils along the Bečva River in the Czech Republic. Such scouring can compromise the load-bearing capacity of foundation piles. Consequently, it is advisable to employ a greater number of piles with smaller dimensions, as the collective failure probability in a pile group is lower compared to that of a single pile. Nonetheless, it is important to acknowledge that smaller piles might be susceptible to damage at their heads due to factors like buckling and lateral forces arising from water flow (Toll et al. 2012).

- **Internal Erosion**
  Fine particles can undergo displacement from the interstitial gaps between larger grains, a phenomenon often referred to as suffusion or piping, particularly in the presence of high-velocity groundwater. This process leads to a loosening of the soil skeleton, rendering it more vulnerable to potential collapse. Moreover, when a layer of fine-grained soil interfaces with a layer of coarse-grained soil, fine particles can be transported into the pores of the coarser soil. This occurrence causes a reduction in the permeability of the coarse-grained layer, while concurrently triggering an expansion of void spaces within the fine-grained soil. This intricate mechanism is commonly known as contact erosion (Herle et al. 2010).
- **Erosion of Soil Surface (Scour)**
  The gradual erosion of soil surface layers due to the action of flowing water is termed scouring, resulting in the deepening of the affected area (Terzaghi et al. 1996). Scouring ranks among the three primary culprits behind bridge failures. Astonishingly, statistics indicate that as much as 60% of all bridge failures can be attributed to scouring (Herle et al. 2010). Scouring has the potential to extract a considerable volume of soil beneath a foundation, a process that can lead to deformations and cracks in the superstructure. The outcome could be an uneven settlement or even a complete structural collapse.

26.4.5 Rock Dissolution Caused Settlement

Soluble rock formations, such as gypsum, anhydrite, calcite limestone, and dolomite, are widespread worldwide and often occur at shallow depths. The presence of these rocks along the path of groundwater flow can lead to their dissolution, ultimately resulting in subsidence. Should such formations develop beneath a building, they have the potential to induce substantial structural damage (Toll et al. 2012).

Subsidence, constituting a component of vertical Earth movements, can significantly and adversely affect buildings, infrastructure, and communication networks. Its onset might be gradual or sudden, impacting areas ranging from a few square meters to several square kilometers.

The existence of soluble deposits carries notable implications for dams and hydraulic structures. In dams, the presence of soluble layers can lead to a gradual dissolution of these rocks, causing cavities to expand as water flows through joints and fissures due to the hydraulic gradient. This process leads to increased water seepage, potentially culminating in the collapse of the dam foundation.

A notable instance described by Gonzalez-Nicieza et al. (2008) involves a gypsiferous ground foundation failure in Oviedo, Spain. In this case, buildings exhibited a significant crack, measuring approximately 10 cm in width. The aftermath encompassed damage to 362 apartments, rendering them uninhabitable and necessitating the rehousing of 300 families, thus inducing substantial social and economic repercussions (Toll et al. 2012).

Case Study 26.2

Subsidence Due to the Dissolution of Gypsum, Ripon (North Yorkshire) in England (Cooper 1998)

In England, the town of Ripon in North Yorkshire has faced notable cases of subsidence attributed to gypsum dissolution, as illustrated in Figure 26.10. Over the past 150 years, Ripon has witnessed no fewer than 30 significant instances of subsidence, with the potential for further occurrences. Typically, the formed cavities span between 10 and 30 m in diameter and extend to depths of around 20 m. Towards the eastern part of the city, these dimensions can expand to up to 80 m in diameter, accompanied by a depth of 30 m. The formation of these cavities is abrupt and often transpires without any preceding signs or warnings.

Figure 26.10 The subsidence that occurred on July 28, 1979, behind the houses on Magdalens Road, Ripon (Cooper 1998 / The Geological Society of London).

26.5 Neighboring and Circumferential Complications

External factors contribute to heightened settlement within the foundation or the emergence of uneven settlement patterns. These effects encompass the impact of the following elements:

- Deep Excavation in Proximity to the Structure
- Settlement Arising from Tunneling Activities
- Consequences of Liquefaction and Lateral Spreading
- Challenges Presented by Nearby and Surrounding Construction
- Fluctuations in Groundwater Levels
- Ramifications of Flooding Incidents

26.5.1 Deep Excavation in the Vicinity of the Structure

Within urban areas, the necessity for constructing deep basements arises due to spatial constraints. As depicted in Figure 26.11, excavation processes can induce additional settlements in neighboring structures. Therefore, it becomes

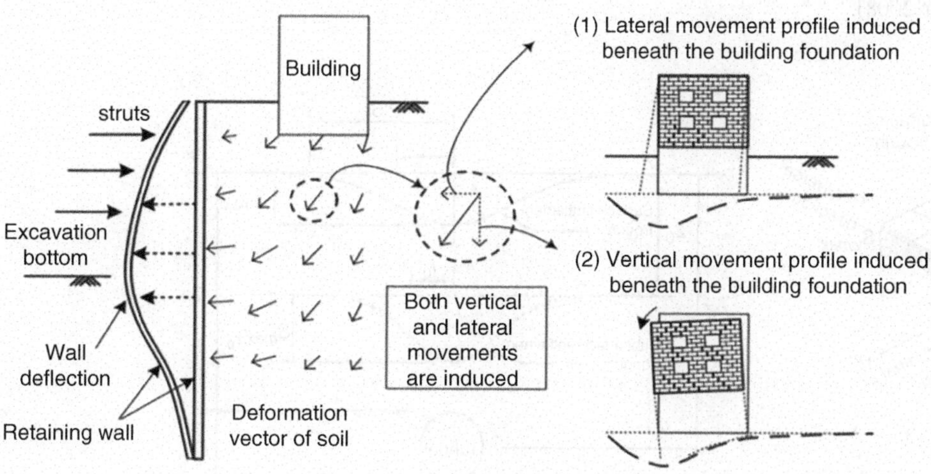

Figure 26.11 Schematic diagram of excavation effects (Schuster et al. 2009 / American Society of Civil Engineers).

imperative to rigorously manage the extent of settlement stemming from these operations to safeguard surrounding constructions from potential harm. Numerous factors, including soil and structural properties, construction sequencing, workmanship, and excavation geometry, collectively influence the magnitude of soil displacement.

The evaluation of building vulnerability is conducted to gauge damage extents and identify vulnerable structures. This evaluation encompasses the following steps: (i) determination of displacements in an unaltered state, (ii) application of these displacements to the building, (iii) assessment of potential damage, and (iv) formulation of necessary remedial measures if required (Korff 2009).

In a study by Tan et al. (2016), the mechanisms behind building settlements in proximity to excavations were investigated. Their research indicated that buildings supported by short piles, similarly situated on pile foundations, are susceptible to damage due to settlements resulting from adjacent excavations.

Ground movements associated with deep excavations stem from a variety of sources or triggers, and their prediction can be undertaken on a holistic basis for all construction phases or on a per-stage basis, as indicated by Korff (2009):

- Vibrations leading to densification resulting from the installation of walls and other structural components.
- Displacements of structural elements due to excavation activities.
- Consolidation effects brought about by the decrease in groundwater levels.

26.5.2 Tunneling-Induced Settlement

Tunneling within challenging ground conditions induces alterations in soil structure, which in turn trigger fluctuations in pore pressure, resulting in a reduction of soil stiffness and strength (Gonga et al. 2020). Typically, modest ground movements occur towards the tunnel during tunneling operations due to stress relief prior to the installation of the lining (Standing and Burland 2006). Consequently, the accurate prediction and management of settlements stemming from tunneling assume a paramount role, particularly in soft ground conditions, where these settlements can jeopardize serviceability and induce damage to existing structures (Gonga et al. 2020).

In a broader sense, tunneling operations can yield horizontal and vertical displacements in structures. The vertical displacements manifest as transverse shifts and longitudinal settlement troughs, as depicted in Figure 26.12. The magnitude and intensity of settlements are contingent on various factors, including the vertical and horizontal distances from the foundation of the structure. The influence of these factors on the stability of the structure during tunnel excavation has been subject to exploration (Obaid Abbas 2020).

The effective range of the tunnel on the structures depends on the trough parameter. That parameter is Trough (*i*), which indicates the distance from the points of penetration to the centerline of the tunnel (Wu and Lee 2003). The relations for the trough parameter are shown in Table 26.2.

Horizontal tensile strains can cause structural damage, and therefore, a prediction of horizontal movement is required. Figure 26.13 depicts the relation between the settlement trough, horizontal movements, and horizontal strains occurring at ground level (Dimmock and Mair 2008).

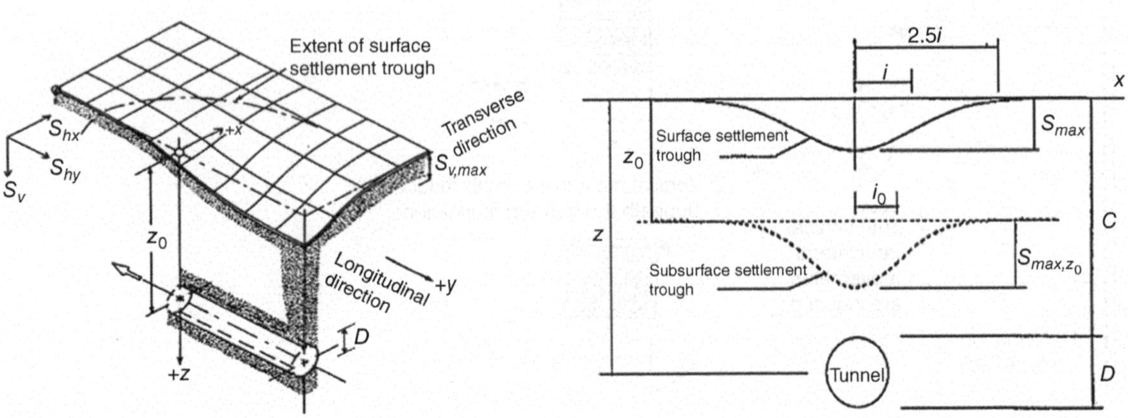

Figure 26.12 Tunneling-induced settlement trough (Attewell et al. 1986) and surface and subsurface settlement troughs (Wu and Lee 2003).

Table 26.2 Relation for trough parameter.

Source/References	Parameter	Remarks
O'Reilly and New (1982)	$i = 0.43z_0 + 1.1$	Z_0 (buried depth of depth)
O'Reilly and New (1982) (Simplified method)	$i = Kz_0$	K = An empirical constant ranging from 0.4 to 0.7 for cohesive soils and from 0.2 to 0.3 for granular soils
Clough and Schmidt (1981)	$\frac{2i}{D} = \left(\frac{z}{D}\right)^{0.8}$	Z (depth of tunnel center), D (tunnel diameter)

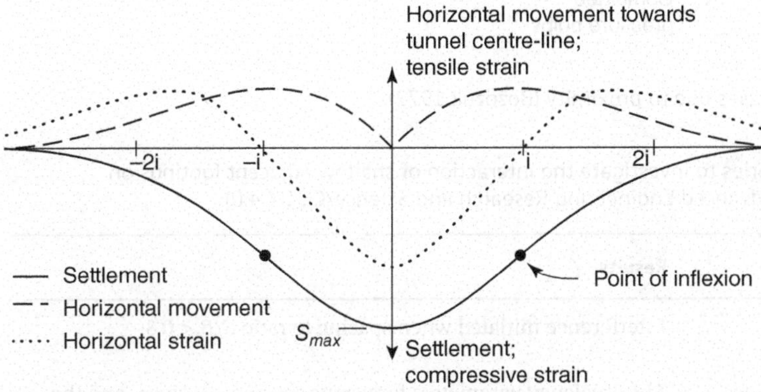

Figure 26.13 Transverse settlement trough (Dimmock and Mair 2008).

26.5.3 Liquefaction and Lateral Spreading

Subsequent to earthquakes, the settlement and inclination of shallow foundations resting on saturated sandy/silty soils have been frequently observed on a global scale, often attributed to liquefaction or softening of the underlying soil. The seismic activity leads to a reduction in effective stresses, consequently diminishing soil stiffness and shear strength. This, in turn, facilitates the settlement of structures founded on shallow footings. When excess pore pressure generated during earthquakes reaches a critical level, triggering full liquefaction beneath the foundation, the result can be substantial deformations with profound implications for the structures above (Bertalot et al. 2012). Liquefaction has been the predominant cause of damage to shallow foundations in previous seismic events, primarily due to the compromised bearing capacity it entails (Bahadori et al. 2020). A comprehensive discussion and elucidation of liquefaction effects can be found in Chapter 20.

26.5.4 Adjacent and Circumferential Construction Complications

When two buildings are constructed in proximity to each other, the foundations of the buildings are adjacent to each other. Due to the proximity of the foundations, the pressure bubbles generated in the soil beneath the foundations of the two buildings overlap, as shown in Figure 26.14. This situation causes larger settlements on the overlapping bubble side of the buildings compared to the opposite side. Additionally, when pressure bubbles interfere, they can cause the buildings to rotate and tilt.

The interference of two closely spaced legs that are loaded simultaneously with equal loads has been investigated experimentally by researchers. Most of these studies conclude that reducing the distance between two adjacent silos increases the final bearing capacity of each in comparison with an isolated footing due to the inverted arch developed within the soil between the two footings (Salamatpoor et al. 2019).

The load-deformation interference of two footings resting on a cohesionless soil medium has been studied by several researchers, as shown in Table 26.3.

Amir (1992) predicted the load–displacement and load–tilt characteristics of neighbored footings on clay. He noticed that the interaction started to occur at an $S/B = 4$, reaching a maximum interference at an $S/B = 1.5$. With any further reduction in spacing, the footing started to act as one block with a width equal to $2B$ (Alwalan 2018).

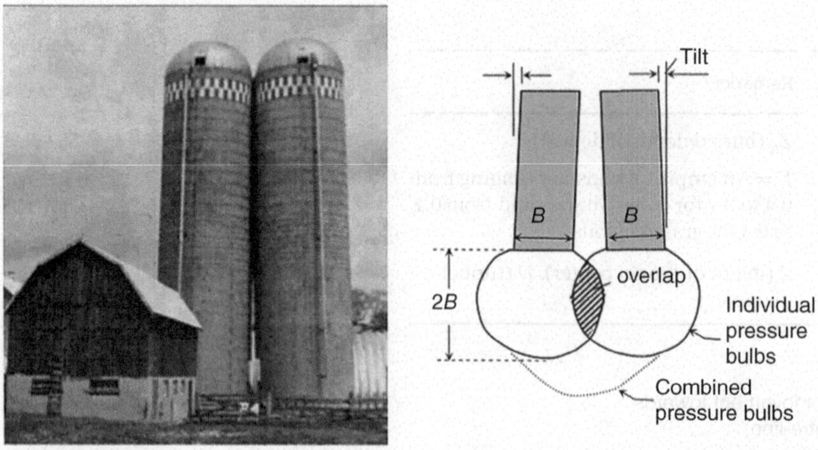

Figure 26.14 Leaning silos as a result of overlapping stresses due to proximity (Bozozuk 1972).

Table 26.3 An overview of the study done in the laboratories to investigate the interaction of shallow adjacent footings on cohesionless soil (Alwalan 2018/International Journal of Advanced Engineering Research and Science/CC BY 4.0).

No.	Names of the researchers	Type of soil	Results
1	Selvadurai and Rabbaa (1983)	Ottawa sand	Interference initiated when spacing to ratio $S/B < 0.3$.
2	Graham (1984)	Ottawa and silica sand	The efficiency parameters for versus spacing are given, and the interaction is dependent on the soil friction angle.
3	Lee and Eun (2009)	Sand	Conducted a circular plate test in the field. The soil's failure stress beneath the neighbored footing is higher than isolated footing; however, larger settlements occur beneath the neighbored footing.
4	Srinivasan and Ghosh (2011)	Dry, dense homogeneous sand	They tested circular footings in the lab using scaled model experiments. Efficiency factors (ξ) are found to be maximum at $S/B = 0.5$.
5	Reddy et al. (2012)	Medium dense sand	Models of square and circular footings were tested. On the sand, the closer the footings were, the better the foundations responded in terms of settlement and ultimate bearing capacity; however, between $B \leq S \leq 6B$, increasing settlements were noted.
6	Srinivasan and Ghosh (2013)	Two layers of sand (weak layer underline by strong layer)	The bearing capacity and the developed settlement at failure declined when the depth of the weak upper layer was increased. Efficiency factors (ξ) are found to be maximum at $S/B = 0.5$.

Various scenarios describe the overlapping of pressure zones, as demonstrated in Figure 26.15. In the scenario depicted in Figure 26.15a, the greater distance between the two structures results in the pressure overlap zone not aligning beneath either structure. Consequently, both structures rotate towards the area of overlap. In the scenarios described in Figure 26.15b and c, the two buildings are very close to each other. The soil beneath the new building is not previously loaded, but the soil beneath the old building is previously loaded. In Figure 26.15b scenario, the new low-rise building experiences rotation. In Figure 26.15c scenario, due to the increased height of the new building, cracks manifest on the walls of the old building.

26.5.5 Groundwater Fluctuation Defects

Variations in groundwater levels wield a significant impact on the ultimate load-bearing capacity and settlement tendencies of shallow foundations. The origins of these fluctuations can be attributed to diverse factors such as seasonal shifts, leaky pipes, dewatering processes, and soil desiccation.

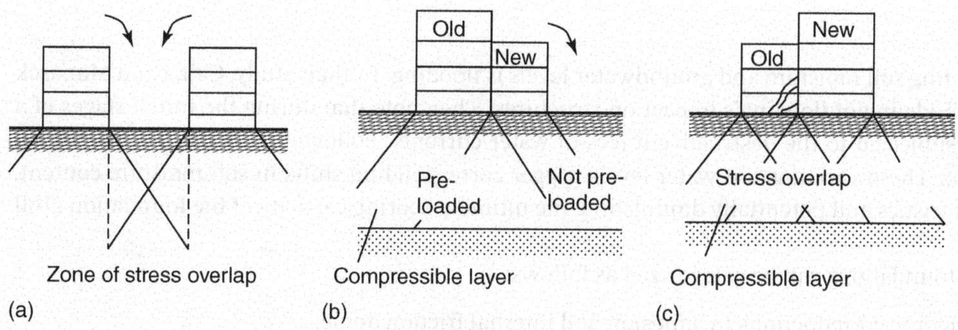

Figure 26.15 Stress zones: (a) same old and new buildings, (b) new building's height lower than the old building, and (c) old building's height lower than a new building (Marseh et al. 2013).

Alterations in water levels due to seasonal changes, including floods or heavy rainfall, can trigger supplementary settlements in shallow foundations. In cases where groundwater level shifts exceed tolerable settlement thresholds, these changes can induce substantial settlements that endanger the structural integrity.

Lowering the groundwater level is often necessary to facilitate drilling and excavation requirements. However, this drainage can lead to unintended consequences such as additional settlement in adjacent buildings, pipeline rupture, subsidence, or tilting of structures (Wu et al. 2019).

Over the years, it has been recognized that the process of soil desiccation caused by vegetation root systems, followed by subsequent swelling, can result in damage to shallow foundations and road pavements situated on clay soils (Blight 2009a,b). Figure 26.16 visually presents the settlement attributed to root-induced desiccation.

The influence of tree roots on foundation performance manifests in two distinct ways:

- Roots growing beneath foundations might experience an increase in cross-sectional dimensions, thereby exerting upward pressure on the foundation.
- Adjacent foundation-bound roots tend to absorb a significant portion of moisture from the underlying soil, consequently inducing shrinkage or settlement.

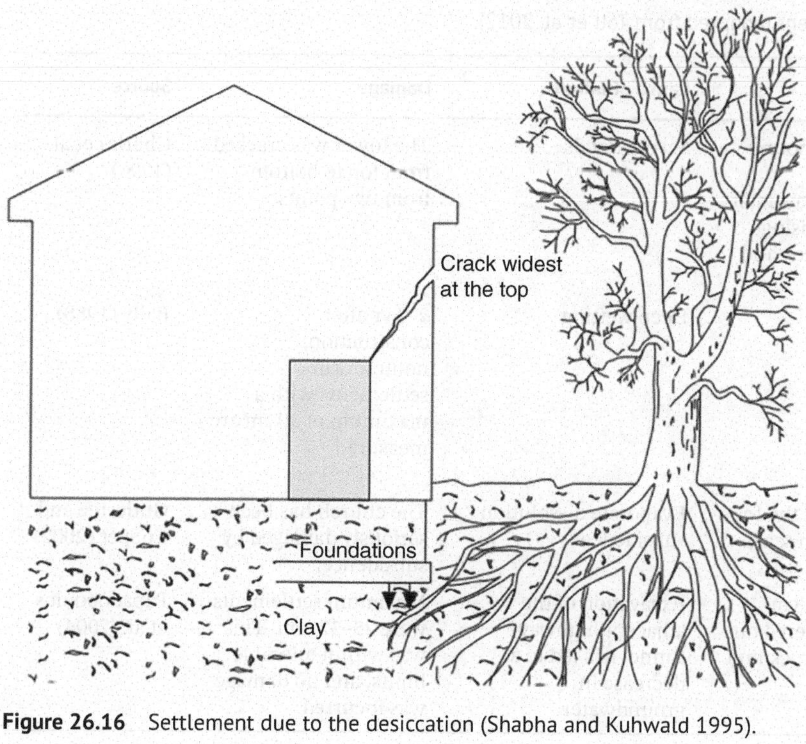

Figure 26.16 Settlement due to the desiccation (Shabha and Kuhwald 1995).

26.5.6 Flooding Effects

One of the events capable of elevating soil moisture and groundwater levels is flooding. In their study, Cajka and Manasek (2005) delve into the numerical modeling of flooding's impact on structures. They note that during the initial stages of a flood, the building's structure is subjected to the destructive forces of water currents. Following a flood, the underlying soil becomes saturated with water. These variations in water levels trigger corresponding shifts in soil moisture content, consequently altering soil characteristics and potentially diminishing the ultimate bearing capacity of the foundation (Toll et al. 2012).

The repercussions of floods on foundations can be categorized as follows:

- Alterations in soil properties encompass reductions in adhesion and internal friction angle.
- Decreased foundation stability due to soil erosion beneath the foundation and diminished friction between the soil and foundation base, as illustrated in Figure 26.17.

Figure 26.17 Loss of buried depth of the foundation due to the occurrence of scouring.

Table 26.4 Cases associated with collapse settlement (adapted from Toll et al. 2012).

Structure	Location	Soil type	Cause of failure	Damage	Source
A stone tower	York, UK	15 m high mound made up of horizontal layers of a fill comprising stones, gravel, and clay	Severe floods	The tower was cracked from top to bottom from two points	Charles et al. (1996)
11-story housing consists of two independent and similar structures with the dimensions of a 22 × 16-m plan	Romania	Loess	Deep wetting	2–3 yr after construction, nonuniform settlements with a maximum of 30 cm are measured	Bally (1988)
San Pedro de Los Francos church	Calatayud, NE, Spain	Situation at the foot of a gypsum scarp	Evaporate dissolution subsidence	The church has been seriously damaged by subsidence.	Gutierrez and Cooper (2002)
11-story reinforced concrete building founded on 42 belled caissons to 10–11 m depth	Boston, USA	Hard clay 18 m thick is covered on glacial bedrock and shale	Excavation of the adjacent highway tunnel has led to a decrease in groundwater	Maximum settlements were 25–32 mm. This was within tolerable limits, and no damage was incurred	Papadopoulos et al. (2004)

- Reduction in bearing capacity caused by elevated groundwater levels and a decrease in effective soil-specific gravity.
- Additional horizontal pressures on structures leading to potential instability.
- Introduction of unexpected moments to the foundation, thereby elevating the likelihood of overturning.
- Excessive uplift forces exerted on structures.
- Deterioration of foundation materials, particularly in the context of wooden piles.

The manifestation of these effects can vary based on local conditions, the predominant influences, and the severity of their consequences, leading to potential variations in the simultaneous appearance of these effects.

Table 26.4 shows the number of case studies whose foundations have been damaged due to various environmental factors.

Case Study 26.3

The Leaning Towers of Santos, Brazil

Buildings in Santos, Brazil, were built on shallow foundations until 1968 due to a lack of constraints on the type of foundation used in high-rise constructions. Due to differential settlements, about a hundred structures are now noticeably tilted, Figure 26.18. According to studies, there are three main layers in this area, each with its geotechnical features. The first layer contains sand and intermediate sand mixed with marine clay; the next layer is marine clay, followed by rock and strong residual soils. Studies showed two main factors induced differential settlement and tilting. The first is a change in soil qualities in the higher sand layer, and the second is adjacency, which is known as an aggravating factor in this scenario. Many of the structures in Santos were built next to each other, causing additional settlement and tilting.

Figure 26.18 The leaning towers of Santos (Diego Grandi / Alamy Stock Photo).

26.6 Damage Assessment

26.6.1 Assessment of Potential Damage to Buildings (Fok et al. 2012)

Figure 26.19 proposes a framework for assessing the risk of damage to adjacent constructions due to excavation and tunneling.

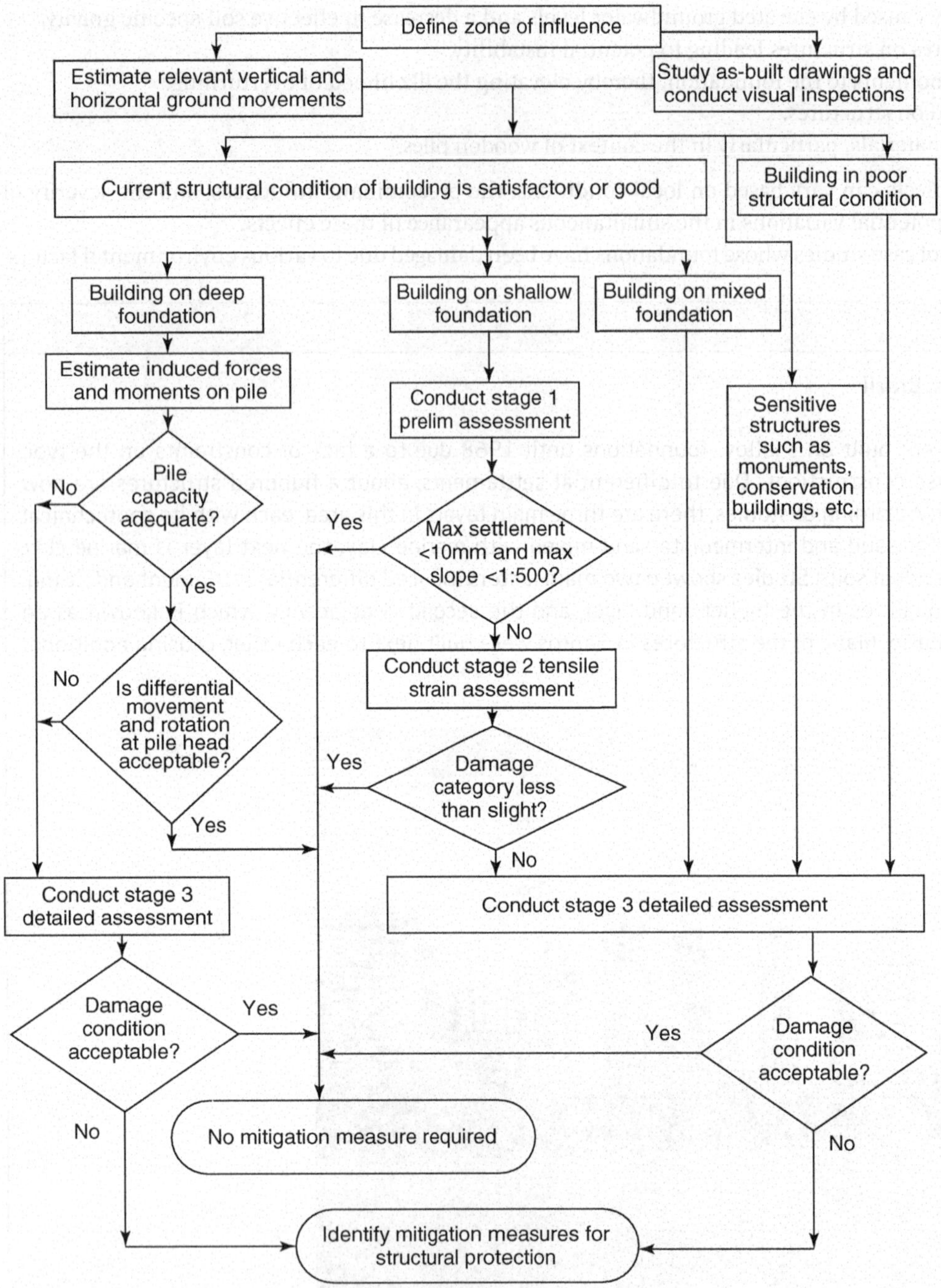

Figure 26.19 Framework for assessing the risk of building damage (Desmond TEO).

26.6.2 Zone of Influence (Fok et al. 2012)

The initial step in the process involves defining the construction's zone of influence. Identifying this zone enables the recognition of buildings susceptible to potential threats. Consequently, all structures situated within the zone of influence can be subjected to evaluations for potential damage. When specific details regarding construction methodologies are available, the zone of influence should serve as the foundation for excavation activity assessments. This can be achieved by examination of previously completed case studies, with validation through numerical modeling.

According to Clough and O'Rourke (1990), the extent of the zone of influence spans up to three times the excavation's depth behind the excavation support wall. However, in cases where a substantial layer of soft soil lies beneath the final excavation level, the determination of the zone of influence aligns with the depth of the soft soil rather than the excavation

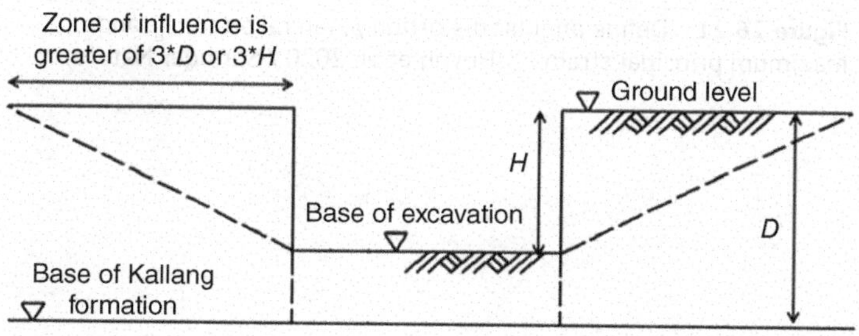

Figure 26.20 Zone of influence for excavations without far-field consolidation effects (Desmond TEO).

depth. Consequently, the zone of influence in such scenarios would encompass three times the depth of the underlying soft soils, as illustrated in Figure 26.20.

26.6.3 Risk Assessment (Fok et al. 2012)

A phased analytical strategy for assessing potential building damage is depicted in Figure 26.19. The buildings are categorized into the following groups based on as-built structural data and visual inspection surveys: (i) those constructed atop shallow foundations, (ii) those erected on deep foundations, (iii) those featuring mixed foundations, (iv) structures confronting unfavorable structural conditions, and (v) buildings situated either on or in proximity to sensitive grounds. The methodology for buildings on shallow foundations closely adheres to the framework outlined by Dimmock and Mair (2008) and encompasses three stages:

- **Stage 1 Preliminary Assessment:** Buildings with a settlement zone measuring less than 10 mm and a slope not exceeding 1:500 can be assessed to carry a negligible risk of damage.
- **Stage 2 Assessment using Tensile Strain Method:** The maximum tensile strain induced in the building is estimated by representing the structure as a beam and calculating the maximum tensile strain through the deflection ratio from the settlement profile and the horizontal strain from the horizontal displacement profile. This calculated maximum tensile strain is then juxtaposed with the critical tensile strain associated with different types of structural damage. It is important to note that this method assumes a building's inherent stiffness is absent, resulting in a conservative assumption. A real-world building would possess stiffness that would reduce both deflection ratios and horizontal strains.
- **Stage 3 Detailed Evaluation:** For buildings assessed with a moderate or higher risk of damage, a comprehensive evaluation is mandated. This evaluation encompasses various facets, including scrutiny of the building's structural particulars, consideration of excavation sequence and methodology, and integration of soil–structure interaction effects. Moreover, the impact of the building's stiffness is investigated. When warranted, this detailed evaluation might extend to analyzing additional stresses borne by structural components due to externally generated distortions and movements. This stage could also encompass ensuring that the building's serviceability limit states remain unbreached, involving a structural analysis of critical elements.

Should the potential for damage remain unacceptable even after the detailed evaluation (i.e., moderate or greater risk), protective measures are employed to safeguard the buildings. These measures encompass the following:

1) **Internal Structural Measures:** These pertain to actions taken within the excavation during construction to mitigate ground movements at their source. Measures might involve augmenting wall thickness, reinforcing strutting supports, and optimizing construction sequencing.
2) **External Structural Treatments:** This category involves methods like underpinning and structural reinforcement, which enhance the adjacent structure's capacity to endure, adapt to, or accommodate ground movements.
3) **Ground Treatment Techniques:** Encompassing a range of techniques, this approach seeks to curtail or modify induced ground movements by manipulating the ground's engineering response.

The chosen protective measures depend on the assessment outcomes and the severity of the potential damage.

26.6.4 Assessment of Damage Level on the Adjacent Building (Huynh et al. 2020)

A set of criteria has been established to evaluate the potential and severity of damage to neighboring buildings stemming from deep excavations. The primary parameters for assessment encompass angular distortion (β), lateral extension

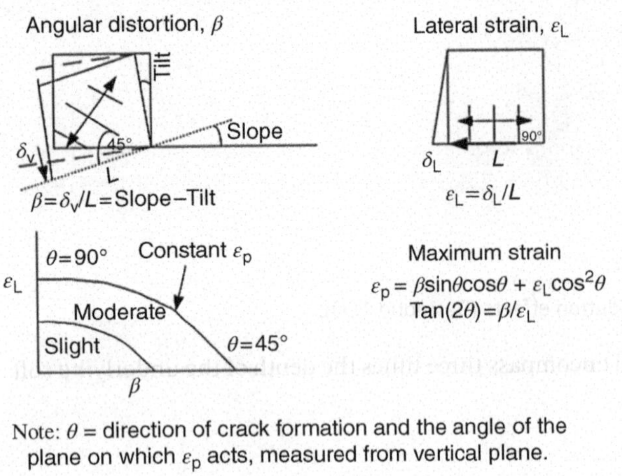

Figure 26.21 Define angular distortion β, lateral strain ε_L, and maximum principal strain ε_p (Huynh et al. 2020 / Springer Nature).

strain (ε_L), and principal strain (ε_p). Angular distortion (β) pertains to diagonal cracks observed in affected buildings. Lateral extension strain (ε_L) captures the deformation resulting from lateral ground displacement, calculated utilizing Equation 26.1, and the extension of a building foundation sideways, normalized by its length. This parameter (ε_L) holds relevance for assessing vertical cracks in impacted buildings. Principal strain (ε_p) is approximated as the maximum strain on a building, combining the effects of β and ε_L. The intricate details of this determination are illustrated in Figure 26.21. Often, the highest principal strain (ε_p) is compared to critical strain thresholds established for categorizing building damage severity. In isolated exceptional instances, particularly for buildings with shorter lengths, angular distortion (β) predominantly triggers damage, with the influence of lateral strain being marginal.

$$\beta = \frac{\delta_V}{L} = \text{Slope}(S) - \text{Tilt}(T) \tag{26.1}$$

$$\varepsilon_L = \frac{\delta_L}{L} \tag{26.2}$$

$$\varepsilon_p = \varepsilon_L \cos \theta_{max}^2 + \beta \sin \theta_{max} \cos \theta_{max} \tag{26.3}$$

In the equations presented above, the symbols are:

- L is the length of the building.
- δ_v is the induced angular distortion.
- δ_L is the induced lateral extension.
- β is the angular distortion.
- ε_L is the lateral strain.
- θ_{max} is the angle of crack formation, measured from the vertical plane.

26.6.4.1 Damage Potential Index Method

The introduction of the Damage Potential Index (DPI) offers a method to evaluate the likelihood of damage to neighboring structures. The DPI serves as an adjustment to the maximum principal tensile strain ε_p, which is calculated using Equation 26.4 (Schuster et al. 2009).

$$\text{DPI} = \frac{\varepsilon_p}{\frac{1}{200}} \times 100 \tag{26.4}$$

As shown in Table 26.5, Schuster et al. (2009) identified six building damage levels associated with DPI values.

26.6.4.2 Burland (1977) Method to Assess Damages Level on Building

Burland (1977) introduced a classification system encompassing six categories of building damage, denoted from 0 to 5. These categories incrementally denote greater severity and are determined by factors such as fracture width and the limiting

Table 26.5 Levels of building damage according to DPI.

	Level of building damage caused by excavation	DPI Sagging $\frac{d}{H_c} \leq 1.4$	DPI Hogging $\frac{d}{H_c} \geq 1.4$	Remedial measures
1	Negligible to very slight	0–15	0–10	Such levels of damage are regarded bearable, and no plan to protect adjacent buildings is required.
2	Very slight to slight	15–25	10–20	
3	Slight to moderate	25–35	20–30	The damage to adjacent buildings may be too much to tolerate. During the design phase, a protective plan may be required. If this is not done, significant care must be taken to monitor the building while it is being constructed.
4	Moderate	35–60	30–50	These damage levels (levels 4–6) are unacceptable. The excavation design should be re-evaluated and possibly revised. Alternatively, an appropriate protection strategy for adjacent structures must be implemented.
5	Severe	60–85	50–80	
6	Very severe	>85	>80	

d is the distance between the structure and the excavation.
H_c is the excavation depth.

tensile strain of structures. The details are outlined in Table 26.6. Burland devised this approach through the aggregation and examination of data from construction sites, which enabled the identification and categorization of building damage.

- **Aesthetic Damage:** This pertains to minor cracks observed in building structures that are easily remedied.
- **Loss of Efficiency or Serviceability:** Instances where building components experience a loss of effectiveness or functionality, such as doors and windows becoming jammed, and pipelines rupturing. Although the building's structural integrity remains unimpaired, the compromised serviceability may bear commercial and economic implications for both the structure and its operations.
- **Excessive Cracks or Deformations in Load-Bearing Structures:** This results in structural damage, which can lead to either partial or complete collapse. Such structural issues might at times be partially concealed, with whitewash and plaster serving as effective indicators of fracture extension.

26.6.5 Effect of Building Stiffness

The stiffness of a building plays a pivotal role in dictating its response to movements induced by excavation. Buildings possessing high stiffness tend to pivot and shift as a solid entity, thereby encountering diminished tensile strain (Fok et al. 2012). When buildings exhibit behavior resembling a rigid box, the deviation of the walls (α_w) from vertical alignment corresponds to the tilt of the foundations (α_f), as depicted in Figure 26.22. However, in practical scenarios, this ideal alignment is improbable due to the fact that significant tilting of the foundation triggers some degree of distortion within the structure itself (Charles and Skinner 2004).

26.7 Foundation Restoration and Retrofitting

Addressing foundation defects requires the implementation of various solutions either individually or in combination. Generally, the repairing and alteration of foundations can be accomplished through the following methods:

- **Soil Reinforcement (Geotechnical Improvement):** This approach aims to fortify and enhance the soil beneath the foundation to reduce settlements. This approach includes several techniques to improve the soil, such as compensation grouting and underpinning using piles, micro-piles, and helical piles.

Table 26.6 The relationship between the degree of damage and the limiting tensile strain (after Boscardin and Cording 1989) was established by Burland et al. (1997) (Huynh et al. 2020).

Category of damage	Degree of severity	Description of typical damage	Crack width (mm)	Limiting tensile strain (%)
0 aesthetic	Negligible	Hairline cracks	<0.1	0–0.05
1 aesthetic	Very slight	Regular decorating can be used to treat fine cracks. Damage is generally limited to the interior wall finishes. Close inspection may reveal some cracks in external brickwork or masonry.	<0.1	0.05–0.075
2 aesthetic	Slight	Cracks are easily repaired. Some redecorating is likely required. Recurrent cracks can be hidden with the appropriate lining. Externally, cracks may be apparent, and repointing may be required to ensure weather tightness. Doors and windows may stick lightly	<0.5	0.075–0.15
3 aesthetic/ functional	Moderate	The fissures will need to be opened up and repaired by a mason. Repointing of external brickwork and possibly a small amount of brickwork to be replaced. Doors and windows are sticking. Service pipes may fracture. Weather-tightness often impaired	5–15 (many cracks with width > 3 mm)	0.15–0.3
4 functional/ serviceability	Severe	Sections of walls are being broken out and replaced, particularly around doors and windows. The frames of the windows and doors were deformed. The floor is considerably sloped. Walls leaning or bulging noticeably Some loss of bearing in beams. Service pipes have been disrupted. Extensive repair work is involved.	15–25 (but depend on the number of cracks)	>0.3
5 structural	Very severe	Damage to the structure required a huge repair job, including partial or total reconstruction. Beams start to lose their bearing. Walls lean badly and require shoring. Windows are distorted and broken. Instability issue	>25 (but depend on the number of cracks)	

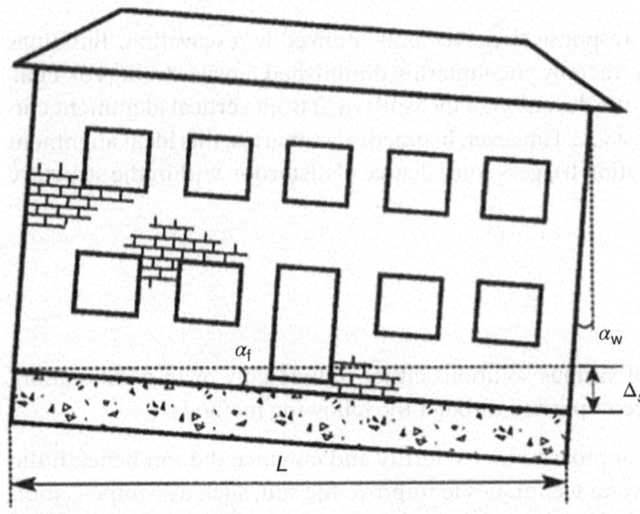

Figure 26.22 A building on a stiff raft foundation that has undergone differential settlement (Charles and Skinner 2004 / Emerald Publishing Limited).

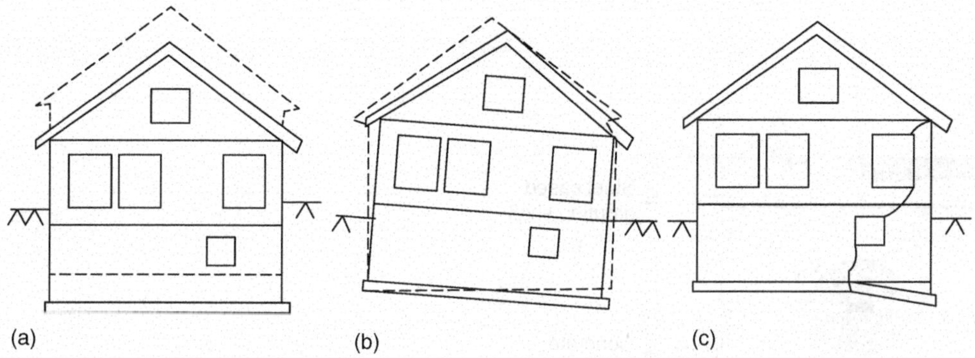

Figure 26.23 Types of building settlements: (a) uniform, (b) tilting, and (c) differential.

- **Structural Reinforcement of Existing Foundations:** This approach includes expanding the dimensions of the foundation, installing ties to the existing foundation, enhancing the bending and shear resistance of the foundation, and increasing the capacity of existing piles.
- **Load Reduction:** If practical, one approach can be to reduce the building loads on the foundation.

These methods individually or collectively contribute to the enhancement and preservation of foundation integrity, ensuring stable and durable structures.

26.7.1 Relevelling

One of the paramount challenges in civil engineering, with significant implications for the stability, safety, and functionality of structures, pertains to the tilting of buildings due to foundation issues. Addressing tilted structures and compromised foundations involves two primary approaches: either the structure can be demolished and reconstructed, or the existing structure can undergo repairs. The decision-making process for selecting between these options is frequently governed by cost considerations. Notably, when dealing with historically valuable structures, restoration often emerges as the preferred course of action, as preserving the architectural heritage outweighs demolition.

Over the past four decades, the domain of civil engineering has devoted considerable attention to the design and construction procedures for rectifying tilted structures. However, the absence of a standardized design framework in this field renders the restoration process a complex technical challenge. Attaining a secure and cost-effective restoration design mandates the possession of specialized engineering expertise, which can only be cultivated through extensive practical and theoretical immersion within the geotechnical and structural engineering domains (Wen et al. 2020).

In a broader context, building settlements can be categorized into three primary types: uniform, tilted, and differential, as depicted in Figure 26.23. Uniform settlements typically pose minimal threats to the structure's stability and functionality. Conversely, the primary concerns revolve around differential settlements and tilting. For nonuniform settlements, diverse techniques are employed to restore the structure to its original level. These methods include grouting, lowering the elevated portion, slab elevation using hydraulic jacks, and underpinning via techniques like piling, micro-piling, or helical piling.

26.7.1.1 Lightening the Weight of the Existing Building

One strategy for mitigating and curbing the expansion of foundation settlements involves reducing the overall weight of the building. By diminishing the dead weight of the structure, the stresses exerted by the foundation on the underlying strata are alleviated, resulting in controlled and reduced settlement rates. However, for historic buildings where safeguarding the architectural essence is imperative, the approach of lightening the building's load might not be a suitable solution. In such cases, alternative methods, as detailed below, should be pursued.

26.7.1.2 Compensation Grouting

Compensation grouting broadly can be described as the injection of fluid or semi-fluid material into the ground, increasing the local volume at the injection point. According to the European regulation EN 12715, grouting for the purpose of

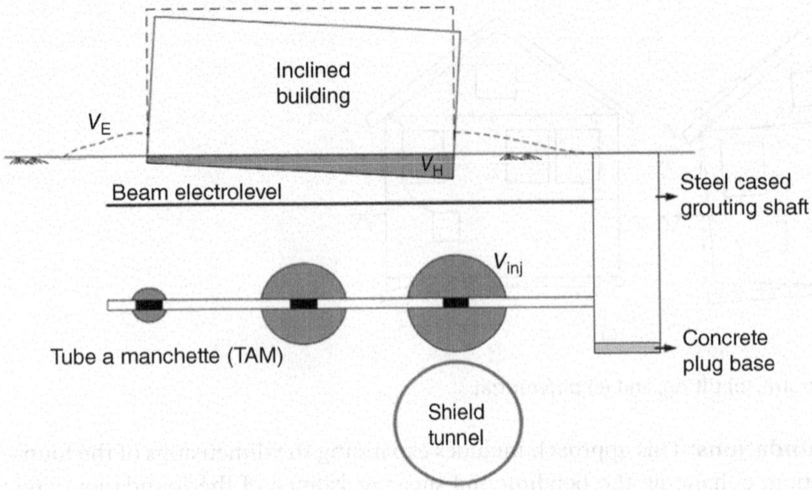

Figure 26.24 Conceptual diagram of typical compensation grouting (Ni and Cheng 2010).

soil displacement like compaction grouting and hydro-fracture grouting, is known as compensatory grouting. Figure 26.24 shows the conceptual diagram of typical compensation grouting.

Jet Grouting: Jet grouting is one of the most common soil remediation and stabilization methods. This method can be used for various purposes, such as increasing the bearing capacity and reducing the settlement of new and existing foundations, supporting underground excavations, and creating water cut-offs for dams (Modoni et al. 2006).

Fracture Grouting: Fracture grouting, also referred to as compensation grouting, involves the injection of grout under pressure to create fractures within soil or solid rock. This process yields densely compacted soil and rock masses. Fracture grouting serves as a ground improvement technique applied when other methods, particularly permeation grouting, are unsuitable or infeasible.

Compaction Grouting: Compaction grouting has been commonly used for correcting differential settlements, underpinning slabs and foundations, liquefaction mitigation and prevention of seismic problems, compensation of superficial settlements due to tunnel excavations in soft or loose soils, for improving foundations of underground pipelines and culverts, and for preventing the formation of sinkholes in dam foundations (Tunçdemir 2007).

This method can be used to treat and improve many geotechnical problems as well as repair and strengthen many structures and infrastructures in different soil conditions. This method can be used to compact and improve a wide range of soil types, a sampling of which is graphically illustrated in Figure 26.25 (Szynakiewicz 2016).

In the course of continuous injection, the soil mass experiences a sustained compaction due to the recurring hydraulic fracturing phenomenon. This leads to lateral displacement of the soil mass beneath the foundation. If the slurry fails to induce lateral soil movement, vertical displacement occurs, initiating the process of lifting the structure. This phase is termed the structural lift stage.

26.7.1.3 Compensation Grouting Effectiveness

The efficiency of compensatory injection is determined by the amount of soil swelling caused by a particular volume of slurry injection (Soga et al. 2004). Grouting efficiency, η, is defined as the ratio of the soil heaved volume, V_E, to the injected volume of grout, V_{inj}, Equation 26.5 (Ni and Cheng 2010).

$$\text{Grouting efficiency;}\quad \eta = \frac{V_E}{V_{inj}} \tag{26.5}$$

The grout efficiency is usually smaller than one due to fluid loss during injection, grout escape from the designated area along the discontinuities and fractures, and soil settlement by the dissipation of excess pore pressures generated during grout injection. The injection efficiency in clay depends on the OCR; with increasing OCR, injection efficiency also increases (Ni and Cheng 2010).

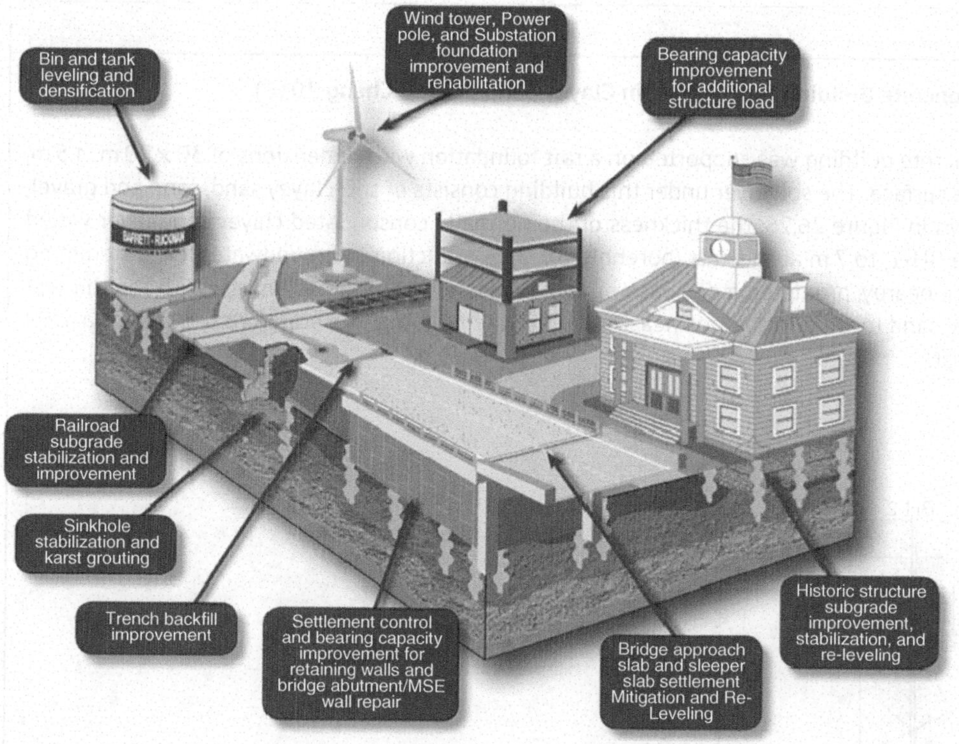

Figure 26.25 Typical compaction grouting solutions (Szynakiewicz 2016).

The efficiency of grout injection can, in some instances, be negative. This scenario is particularly evident in soils that are typically normally or slightly consolidated. In such cases, the injection does not lead to repairs but rather contributes to supplementary settlements due to soil consolidation. As the injected slurry attains stability within the injection sites, the consolidated mass amplifies pressure, reinforces the material, and subsequently diminishes the efficacy of the injected slurry (Ni and Cheng 2010).

The advantages of relevelling the foundation by grouting are:

- No need for removal and replacement.
- Efficient for preserving existing structures.
- Convenient accessibility and feasibility in restricted spaces.
- Minimal vibration impact on nearby buildings.
- No need to construct any attachment to the foundation, such as a pile.
- Repeatable implementation procedure.

Disadvantages of relevelling the foundation by injection method are:

- Challenges in estimating the requisite slurry volume.
- Potential toxicity and negative impact on groundwater due to specific chemical slurries.
- Difficulties in regulating the injection rate to prevent damage to adjacent structures. It is mandatory to have specialized and experienced contractors be engaged for this work, and to use instruments with high operating precision.

26.7.1.4 Lifting Foundation with Hydraulic Jacks

Employing jacks for lifting and repositioning displaced foundations is another common approach for foundation leveling. Injectable materials can be utilized to fill the void between the soil mass and the foundation. In this technique, to manage the displacement of the repositioned foundation before restoring it to its original level, methods such as micro-piles, helical piles, or conventional piles can be employed beneath the foundation to supply the reaction force required by the jack. Further discussion on this will follow.

Case Study 26.4

Relevelling of a Seven-Story Concrete Building that Settled in Clayey Sand (Ni and Cheng 2010)

The seven-story reinforced concrete building was supported on a raft foundation with dimensions of 30 × 20 m, 4.5 m, and a depth of 4.5 m below the surface. The soil layer under this building consists of soft clayey sand, sand and gravel, and siltstone bedrock. As shown in Figure 26.26, the thickness of the normally consolidated clayey sand layer varied from 1 m at the west (borehole BH-1) to 7 m at the east (borehole BH-2). A reduction in groundwater level, stemming from extensive excavation work nearby, prompted a decrease in pore pressure within the soil mass. With the thickest stratum of compressible clayey sand located in the northeast corner of the building, the structure underwent a 1/99 inclination toward the northeast.

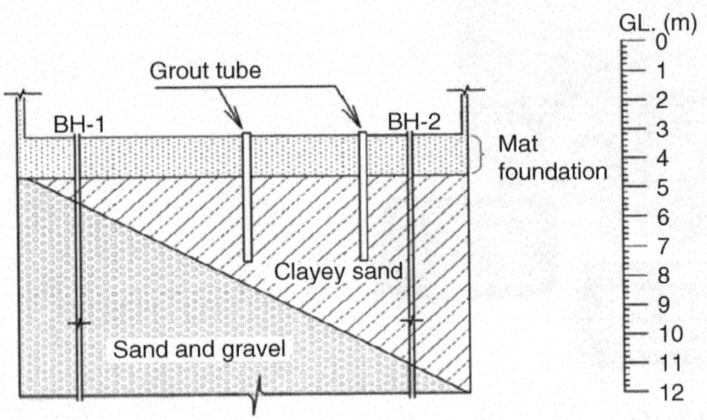

Figure 26.26 Subsurface profile, building basement, and injection pipe (Ni and Cheng 2010).

Fracture grouting was chosen as the preferred grouting method. Grouting pipes were strategically positioned in the soil, maintaining a 2-m distance from the foundation floor and a depth of 2.5 m. The injection of quick-setting grouts was then initiated.

Across two grouting phases, a total of approximately 460 m^3 of slurry was injected, resulting in the elevation of the northeast corner of the building by 32.37 cm. The initial grouting phase aimed to stabilize the clayey sand layer beneath the mat foundation. The grouted clayey sand layer, in conjunction with the underlying sand and gravel layer, facilitated the required reaction for the subsequent grouting stage. The ultimate compensation efficiency reached 27%. Refer to Figure 26.27 for a visual representation of the grouting's performance.

Figure 26.27 Performance of grouting: (a) before fracture grouting and (b) after fracture grouting (Ni and Cheng 2010 / ASCE).

Case Study 26.5

Babolsar Concrete Pedestrian Bridge Retrofitting Practice (Eslami et al. 2016)

The concrete bridge stands as the oldest pedestrian crossing in Babolsar, Iran, situated along the picturesque Caspian Sea shoreline. In 1964, the bridge encountered the effects of local scouring around its piers, brought about by river flooding. This resulted in localized and uneven settlement beneath the three piers of the middle spans, inducing the rotation of the piers and causing localized deformation in the deck's support points. The damage sustained by the bridge can be attributed to a combination of geotechnical, structural, and hydraulic factors.

The process of reconstructing and repairing the bridge encompasses several stages, including strengthening the foundation framework, releveling the structure, and raising the bridge deck followed by necessary reparations. Two methods were considered for strengthening: (i) installation of vibratory metal piles and (ii) installation of micro-piles. Given the specific conditions and available resources in the region, the decision was made to proceed with the installation of vibratory metal piles.

Here are the steps taken for repairing the bridge foundation.

The initial phase involved:

1) Erecting a scaffold around the piers to provide the necessary workspace for the jacking operation.
2) Employing beam supports at the joint location of the deck to maintain the deck's alignment at the center of the span.
3) Conducting all jacking operations while the deck is in place.

The second phase involved:

1) Demolishing the concrete seat of the deck on both sides of the pier to create a suitable space for the jacks.
2) Installing and elevating the deck to the designated height.

The deck was raised by 400 mm from its initial balanced position in anticipation of potential future settlements and to establish a suitable arch within the bridge's structure. Due to the significant settlements of the piers, achieving deck equilibrium required a two-month-long process. Figure 26.28 illustrates the progressive elevation and upgrade of the deck using the metal plates, implemented in a series of steps to reach the intended level. Figure 26.29a depicts the bridge prior to the renovation, while 26.30b showcases the bridge's appearance after the retrofitting procedure.

Figure 26.28 Deconstruction of the pier and balancing the deck to the required height (Eslami et al. 2016 / with permission of Elsevier).

(Continued)

Case Study 26.5 (Continued)

(a)

(b)

Figure 26.29 Babolsar concrete pedestrian bridge: (a) preceding renovation in 2012 and (b) three years afterward, in 2015 (Eslami et al. 2016 / with permission of Elsevier).

26.7.2 Underpinning with Helical (Screw) Piles and Brackets

This method of elevating buildings is the most prevalent and widely employed technique for building restoration. This approach involves raising a portion or the entirety of a building using jacks. To facilitate the building's elevation, a reaction force is indispensable (the jack necessitates a location to generate a reaction force to lift the structure). This reactive force can be established by linking the foundation to spiral piles.

For underpinning with helical piles, the initial step involves excavating a small area to expose the base of the existing foundation. Subsequently, a helical pile is installed utilizing a hydraulic torque motor to generate the requisite load-bearing capacity at the specified design depth. A foundation "L" bracket is then affixed to the pile's shaft and the existing foundation to create a structural connection between them, facilitating load transfer (Lutenegger and Kemper 2008).

Advantages of repairing foundations with helical underpinning systems:

- Rapid and cost-effective installation.
- Minimal construction-induced vibrations.
- Minimal disruption to the existing structure.
- Ease of construction in confined space, Figure 26.30b.
- Immediate loading capability, leading to reduced project duration and costs.
- Built-in quality control (installation torque to capacity correlation).
- Flexibility of geometry.
- Installation in high groundwater conditions.

(a)

(b)

Figure 26.30 (a) Typical L-style foundation bracket and (b) installation of a helical pile adjacent to the existing drain line (Lutenegger 2013 / Missouri University of Science and Technology).

26.7.3 Underpinning with Micro-piles

Micro-piles have been used for underpinning since the 1980s. Because of their fast installation, compatibility with confined physical locations, and high bearing capacity, micro-piles are commonly utilized for structural support and in situ reinforcing, Figure 26.31 (Wen et al. 2020).

Figure 26.32 depicts the installation technique of micro-piles. As shown in Figure 26.32a, the installation process involves excavating an expanded working area around the foundation to expose the sidewalls of the lower step of the foundation. In cases of elevated groundwater levels, a surface dewatering method is employed to dry the working area. Subsequently, as depicted in Figures 26.32b and c, two steel brackets per micro-pile are positioned on the sides of the foundation's lower step. These steel brackets serve a dual purpose: (i) they contribute to the reaction frame that furnishes the necessary driving force during the micro-pile installation via electric hydraulic jacks (refer to Figure 26.32d), and (ii) they are integrated into the correction system that aligns the inclined tower (Wen et al. 2020).

In certain cases, micro-piles can also serve as anchors to secure a shaft into an underlying formation, as demonstrated in Figure 26.33. To achieve this, a micro-pile is inserted into the drilled shaft and extended to the desired depth within the bed. This method enhances the structural integrity of the pile. For successful implementation, access to the top of the shaft is necessary (Brown and Powell 2013).

In situations where existing structures require support or underpinning, other deep foundation elements such as driven piles or drilled shafts are often not feasible or practical. This can be due to physical constraints, geological considerations, cost, or a combination of various factors. In such scenarios, pin piles offer a cost-effective solution and deliver the following advantages (Pearlman 2000):

- The equipment is compact yet robust, enabling operation in areas with limited accessibility.

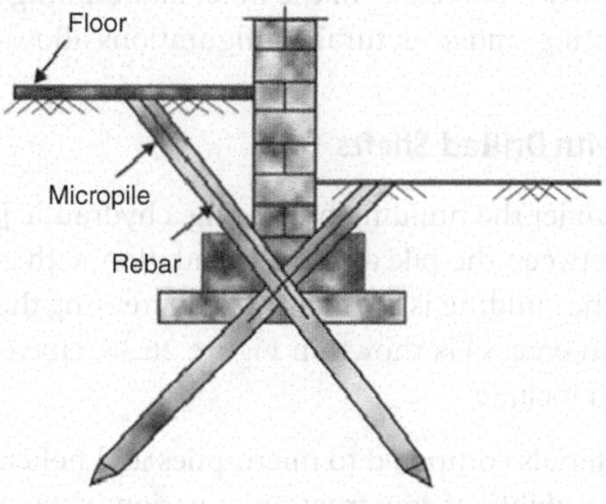

Figure 26.31 Strengthening foundation by micro-piles (Bhattacharjee 2018) (Systab).

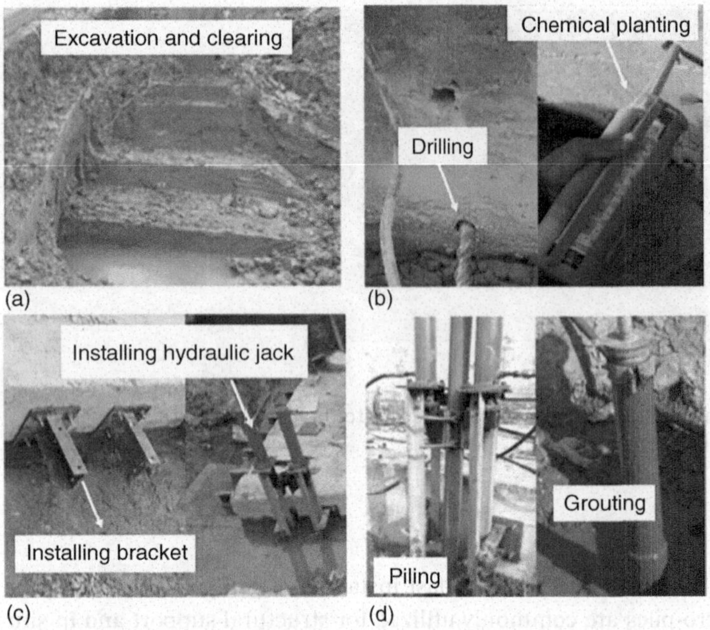

Figure 26.32 Construction processes of the grouted steel-pipe micro-pile: (a) excavation and clearing, (b) drilling and chemical planting, (c) installing hydraulic jack and bracket, (d) piling and grouting (Wen et al. 2020) (Lei Wen / Gangqiang Kong / Hossam Abuel-Naga / Qingsong Li / Zhendong Zhang / American Society of Civil Engineers).

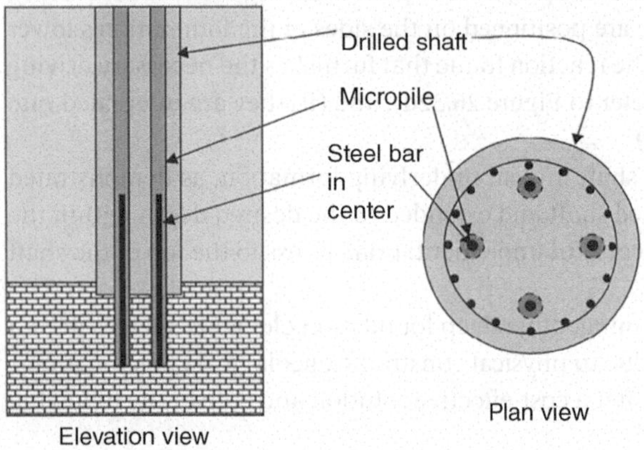

Figure 26.33 Micro-piles drilled through an existing drilled shaft are being used (Brown and Powell 2013).

- The drilling process is quiet, generating minimal vibration, and resulting in negligible ground disturbance around each pile installation.
- Drill spoil can be efficiently removed from the hole, maintaining a clean operational environment.
- A range of drilling, grouting, and structural configurations allows for tailored and optimized solutions.

26.7.4 Underpinning with Drilled Shafts

The pile can be installed under the building by placing a hydraulic jack on the top of the pile created under the foundation and replacing the space between the pile and the foundation with concrete and composite parts, Figure 26.34. In this way, the settlement created in the building is compensated, increasing the foundation system's load-bearing capacity. The underpinning by pile installation process is shown in Figure 26.35. However, it is important to note some limitations associated with pile execution, which include:

- A higher volume of materials compared to micro-piles and helical piles.
- Difficult access and infeasibility of construction in various subsurface conditions.
- Longer installation duration.

Figure 26.34 Underpinning by pile and releveling foundation with hydraulic jacks.

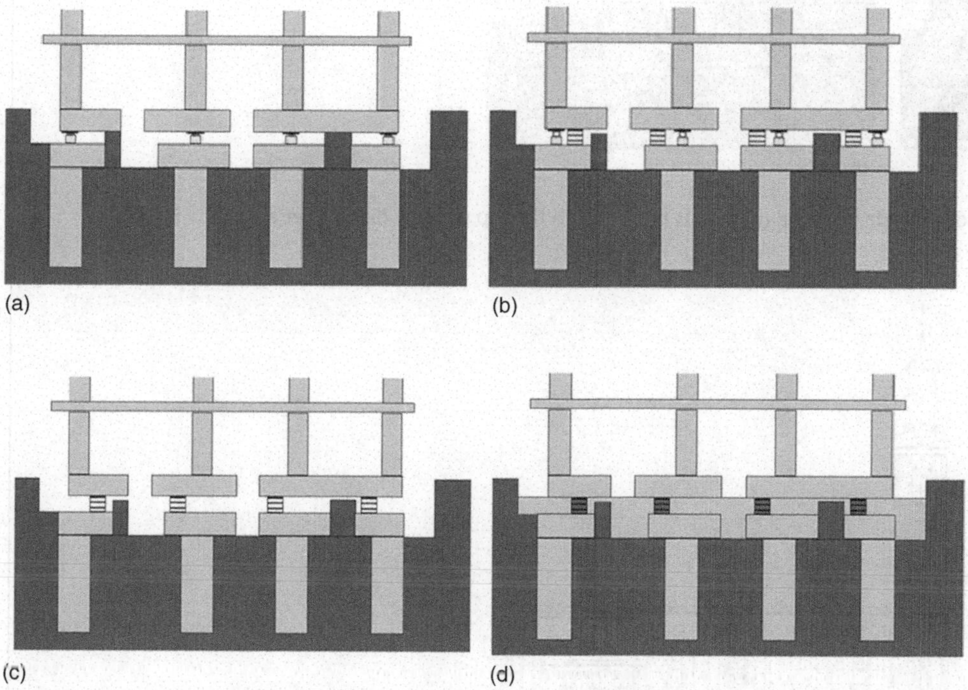

Figure 26.35 Steps of pile execution in relevelling foundation: (a) execution of piles and modification of building assembly with hydraulic jacks, (b) placement of composite parts between the pile and the modified foundation, (c) removal of the hydraulic jack, and (d) placing concrete in the space between the pile and the foundation.

Case Study 26.6

Repair and Foundation Strengthening of the Towers of the Cathedral of St. Theresa of Avila in Subotica (Santrač et al. 2021)

The Cathedral of St. Theresa of Avila in Subotica was constructed between 1773 and 1798. It boasts a total length of 61 m, a nave width of 21 m, a ridge height of 28 m, and a towering spire standing at 54 m. Historical records indicate that cracks emerged on the front façade a few years after the completion of construction. These cracks gradually expanded in size over time. When restoration efforts commenced in 2015, the notable cracks had widened to approximately 15 cm. Figures 26.36 and 26.37 depict the comprehensive mapping of all cracks and defects. Figure 26.37 illustrates

(Continued)

Case Study 26.6 (Continued)

the substantial cracks that developed on the frontal facade. Presently, these cracks have attained a maximum width of approximately 19 cm, spanning the entire thickness of the wall.

Figure 26.36 (a) Cracks in the front facade wall and (b) details of cracks on the top window (Santrač et al. 2021 / MDPI).

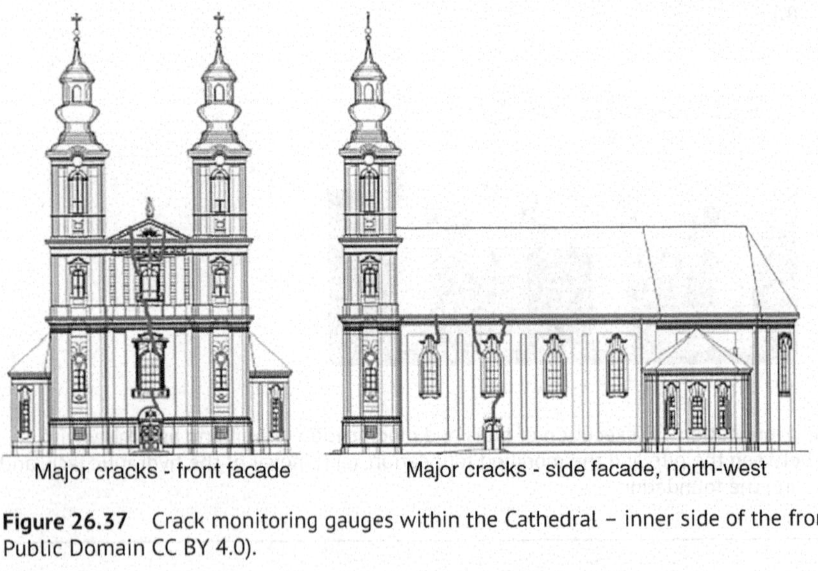

Figure 26.37 Crack monitoring gauges within the Cathedral – inner side of the front facade wall (Santrač et al. 2021 / MDPI / Public Domain CC BY 4.0).

Soil Profile

The soil profile can be divided into three distinct layers. The uppermost layer extends from the ground surface down to a depth of 9 m and comprises low to medium plasticity soft clay. Following this, the second layer, approximately 2 m thick, consists of sand. Subsequently, the third layer of clay reappears at a depth ranging from 18 to 20 m, as depicted in Figure 26.38.

The groundwater depth is at 3.2 to 4.2 m from the ground level around the Cathedral.

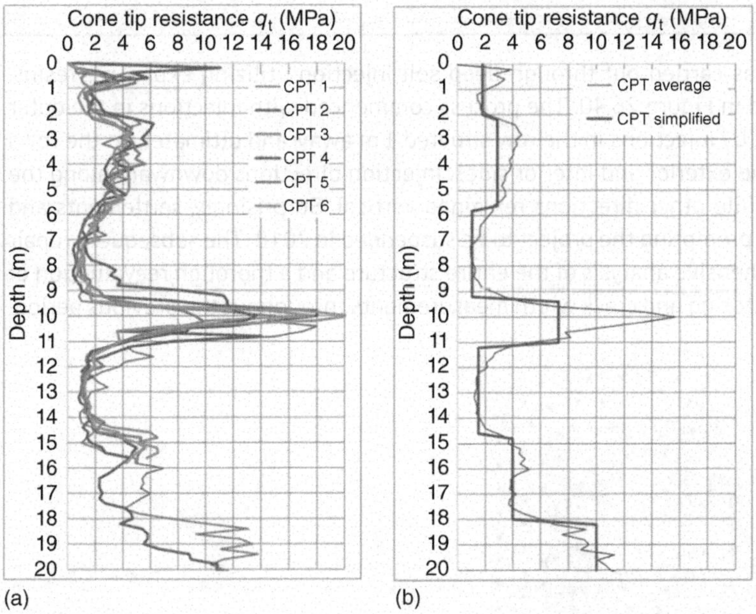

Figure 26.38 (a) All penetration tests next to the towers and (b) the average of all tests (Santrač et al. 2021 / MDPI / Public Domain CC BY 4.0).

Cause of Damages

The primary contributor to the settlement issue in the Cathedral Towers is the imposition of an excessive load by the structure onto the compressible clay, resulting in substantial consolidation settlement. Another contributing factor is the application of eccentric loads on the foundation, leading to the tilting of the towers. The critical location of the center of gravity renders the structure susceptible and unstable in response to relatively minor horizontal forces.

Foundation Rehabilitation Methods

a) In 2017, the first modern attempt to restore the towers was initiated by inserting piles beneath the foundations, as shown in Figure 26.39. Each pile comprised multiple steel cylindrical segments with a diameter of 323 mm and a length of 500 mm. Hydraulic jacks were employed to install the steel cylinders into the ground below. The initial steel cylinder of each pile featured a conical tip. During the installation of 11 piles (not in numerical order), settlements intensified with each new pile installed, and new cracks emerged. The south tower's inclination increased by 0.4 mm/m, the north tower by 0.6 mm/m, and both towers inclined towards the nave by 0.4 mm/m. The overall crack width expanded by approximately 4 cm, escalating from 15 to 19 cm. Consequently, after the installation of the 11 piles, the work was halted, and piles numbered 2, 5, 6, 11, and 14 remained uninstalled.

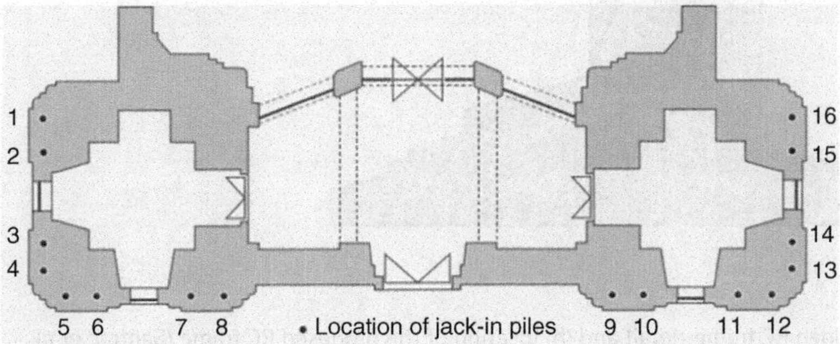

Figure 26.39 The layout of the piles beneath the tower foundations (Santrač et al. 2021 / MDPI / Public Domain CC BY 4.0).

(Continued)

Case Study 26.6 (Continued)

b) In the second attempt, the reinforcement was carried out through deep soil injection utilizing expansive resins. The injection points' positioning is illustrated in Figure 26.40. The process commenced with injections in the outer row, positioned at a distance of 2 m, followed by injections in the row situated 1 m away, and ultimately in the rows immediately adjacent to the tower wall on the exterior and interior sides. Injection directions downward along the wall were at an 8° angle toward the wall, while other directions remained vertical. Surprisingly, settlements and damages intensified during these operations, prompting the project to be suspended in 2018. The subsequent repair approach was formulated based on a comprehensive analysis of the entire structure and a thorough reevaluation of data collected from ongoing settlement monitoring and crack width measurements throughout the previous period.

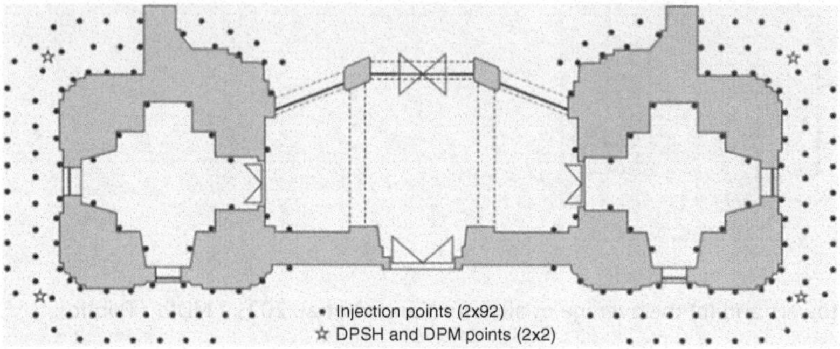

Figure 26.40 The location of injection locations around the tower's foundations (Santrač et al. 2021 / MDPI / Public Domain CC BY 4.0).

c) This series of observations and analyses converge to the conclusion that the most effective rehabilitation strategy does not lie within the realm of soil modification but rather hinges on mitigating the weight of the towers – in essence, alleviating the ground pressure. While it is true that, due to the phenomenon of creep, complete cessation of settlement is not achievable even with weight reduction, substantial deceleration can be achieved. The envisioned solution entails a partial deconstruction of the towers, their interconnection, and linkage to the Cathedral nave through a robust reinforced concrete framework. Subsequently, the towers would be reconstructed employing lighter materials, while preserving their existing appearance. This approach holds the potential to proactively address the issue of cracks over the long term, significantly elongating the serviceable life of the structure, as depicted in Figures 26.41 and 26.42.

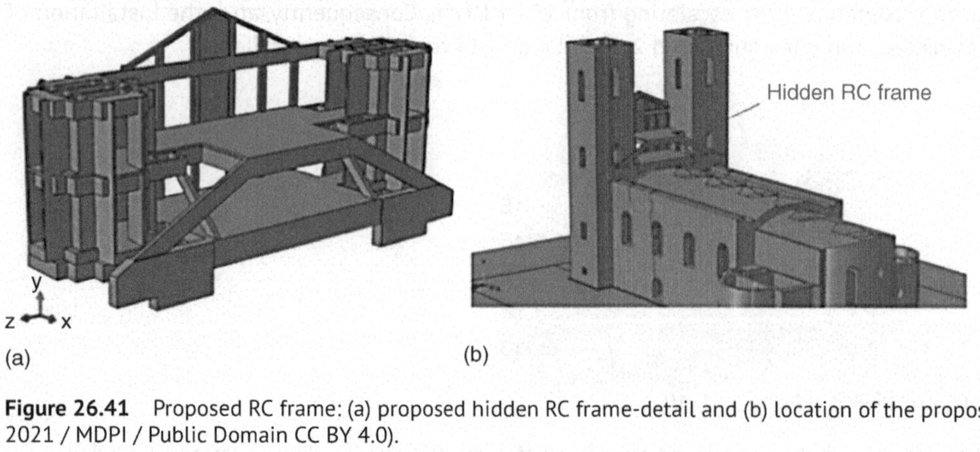

Figure 26.41 Proposed RC frame: (a) proposed hidden RC frame-detail and (b) location of the proposed RC frame (Santrač et al. 2021 / MDPI / Public Domain CC BY 4.0).

Figure 26.42 Upper sections of the towers and front façade-proposed total reconstruction (Santrač et al. 2021 / MDPI).

Comments on Selection Criteria Foundation Strengthening Methods

The decision regarding the selection of foundation-strengthening techniques is influenced by a range of criteria, as outlined by Zakharov in 2015:

- Geotechnical and hydrogeological conditions.
- Current state and original design of the foundation.
- Characteristics of the strengthening process.
- Capabilities and requirements of the building.

The following are the selection criteria for foundation-strengthening methods:

- Feasibility of executing the process under load, without altering foundation dimensions.
- The simplicity and rapidity of the implementation.
- Ability to swiftly transfer live loads post-strengthening.
- Suitability for constrained working spaces.
- Degree of mechanization achievable.
- Presence or absence of perceptible structural appearance alterations post-reconstruction.
- Viability of utilizing the structure during the reconstruction phase.
- Significance of factoring in usage patterns and preserving the building's architectural integrity, especially when dealing with historic structures.

26.7.5 Retrofitting and Strengthening

Numerous existing buildings and structures have been constructed using concrete. However, a considerable portion of these buildings and structures prove incapable of withstanding lateral or vertical loads, as stipulated by current code requirements. This inadequacy may stem from poorly executed construction or the presence of damaged building components. Moreover, the various features of a building may degrade, sustain damage, or simply deteriorate over time (Elsamny et al. 2020).

The reinforcement of existing buildings serves various purposes, including altering building functions, rectifying and rejuvenating structures, and increasing the number of floors (Zumrawi and Aldawi 2018). Undertaking repairs and enhancements on foundations constitutes a financially demanding and intricate construction endeavor. This approach is deemed necessary in the following scenarios:

- Degradation of concrete elements due to the advanced age of the building.
- Excessive settling of the foundation due to subpar soil conditions.
- Structural harm to the foundation is caused by seismic overloading.

Figure 26.43 Planting the steel rebars (Horse Construction / https://www.horseen.com/solution/control-construction-quality-for-planting-rebars / last accessed 23, November, 2023).

- Amplified seismic loads due to revisions in building codes or strengthening initiatives.
- Requirement for supplementary foundation support for additional floors.

Addressing the repair and reinforcement of a foundation structure can be accomplished through several methods, such as fortifying the existing foundation structure, introducing a fresh foundation structure, or modifying the soil to enhance foundation support. Strengthening the existing footing requires increasing the dimensions of the footing to increase the bearing area of the foundation, adding a tie to existing foundations, and underpinning measures.

26.7.5.1 Implanting Steel Rebars

Planting steel rebars is essential to establish a dependable bond among steel bars, beams, walls, columns, and concrete slabs. This practice serves to mitigate potential harm to the existing structure while simultaneously enhancing its load-bearing capacity. In the case of columns, retrofitting measures involve the incorporation of steel rebars to bolster their structural integrity. Moreover, the utilization of steel rebar planting facilitates the connection between a new foundation and the existing one. A visual representation of the process of implanting steel rebars is illustrated in Figure 26.43.

26.7.5.2 Foundation Sectional Enlargement

By increasing the dimensions, the contact surface on the foundation can be increased, and the applied stresses on the foundation can be reduced, leading to increased bearing capacity, decreased soil subsidence with increasing dimensions, and subsequent reduction of the existing stress. Increasing the dimensions can be done alone or by increasing the dimensions of the foundation and the column attached to it.

Case Study 26.7

A Study on Strengthening of Building Foundation for Story Extension (University of Khartoum) (Zumrawi and Aldawi 2018)

One of the simplest ways to strengthen concrete members is to enlarge sections with additional reinforced or plain concrete. Its purpose is to increase the thickness of the member. However, when introducing new concrete onto preexisting concrete, several crucial factors necessitate careful attention. The existing surface must undergo roughening and thorough cleaning to establish a suitable bond. In some cases, the application of epoxy adhesive can offer added efficacy in fortifying the bond. Should the shear forces at the joint interface prove substantial, the inclusion of additional anchors in the form of bolts or stirrups might be essential.

To increase the dimensions of a shallow foundation, the following steps are taken:

1) As shown in Figure 26.44a, excavating the area surrounding the footing to the appropriate dimensions.

2) Creating surface abrasions on the concrete surface at the top and the vertical sides of the foundation.
3) Drilling small holes on the surface and sides of the footing and filling with the holes epoxy grout. Then, as illustrated in Figure 26.44b and c, inserting dowel bars into the drilled holes.
4) Pouring additional plain fresh concrete of 10 cm thickness in the extended footing area before arranging the reinforcement bars surrounding the footing, as shown in Figure 26.44d and e.
5) Cleaning foundation contact surfaces.
6) Installing additional bars for new reinforced concrete according to plans.
7) Pouring fresh concrete with proper compaction to cover the enlarged footing area, as shown in Figure 26.44f.
8) Refilling with suitable materials to the desired level.
9) A polymeric binder was used on the exposed surface of the base to ensure the connection of fresh concrete with old concrete.

Figure 26.44 Foundation sectional enlargement: (a) excavation around the footing, (b) drilled holes, (c) inserted dowels bars, (d) plain concrete in the extended area, (e) reinforcement bars arranged surrounding the footing, and (f) fresh concrete cast in the enlarged area of the footing (Zumrawi and Aldawi 2018) (University Amar Telidji – Laghouat).

One of the methods to increase the dimensions of the foundation is to simultaneously increase the dimensions of the foundation and the column attached to it, which is shown in Figure 26.45.

Figure 26.45 Increasing the dimensions of the foundation and the column attached to it (Bhattacharjee 2018).

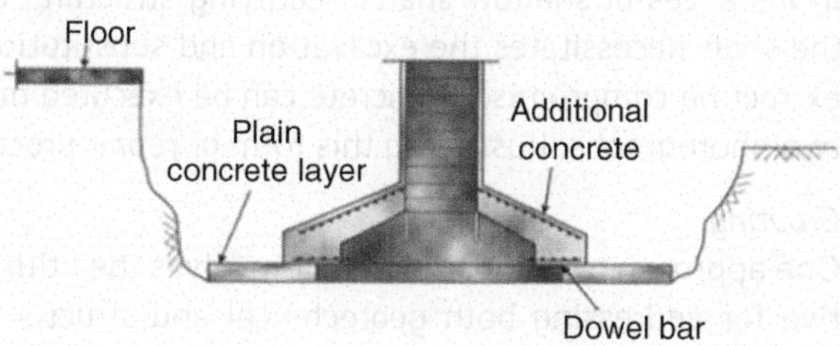

(Continued)

Case Study 26.7 (Continued)

Add the Tie to the Existing Foundation

Enhancing foundation stability to prevent slippage involves the strategy of consolidating the foundations by using ties to connect different foundations to each other thus enhancing the resistance against lateral loads. Various approaches can be adopted to establish an improved linkage between the fresh concrete overlay and the preexisting surface. One such method involves the roughening and cleansing of the existing surface. Alternatively, mechanical joints like nails can be utilized to ensure a more robust connection.

Strengthen Foundation by Underpinning

One of the methods of strengthening the foundation is underpinning, which can be used alone or by increasing the foundation's dimensions. The underpinning method was presented in Sections 26.7.2–26.7.4.

Structural Enhancement of Piles

To strengthen the structure of piles, two methods can be used, such as (Brown and Powell 2013):

- Jacketing the pile with a layer of steel and concrete.
- Removing part of the damaged pile, such as drilling inside it and replacing it with a structural element such as sheath and high-strength concrete. To increase the geotechnical strength of the shaft, the central element can be extended to depths below the bottom of the shaft, which increases the shaft's lateral and vertical bearing capacity, as shown in Figure 26.46.
- Drilling holes in the shaft and grouting. Next installing additional steel rebar, high-strength bars, or strands that can also be used to improve structural integrity, as shown in Figure 26.46 with micro-pile repair.

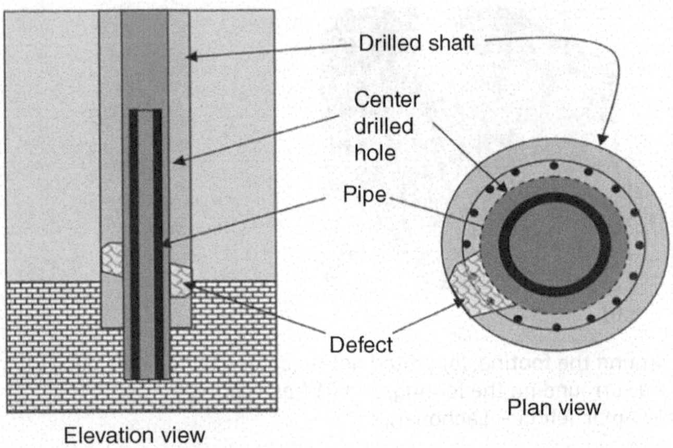

Figure 26.46 Use of a steel section cast into a drilled shaft to form a structural bridge (Brown and Powell 2013).

Excavation and Replacement of Shallow Shafts (Brown and Powell 2013)

In instances of shallow shafts, rectifying structural deficiencies such as weakened concrete in the upper section of the shaft necessitates the excavation and substitution of new structural components. In this scenario, the process of extracting compromised concrete can be executed manually using impact tools or rotary tools. Figure 26.47 presents two photographs illustrating this form of repair procedure.

Grouting

One approach to foundation repair involves the utilization of the grout injection method. This technique proves effective for addressing both geotechnical and structural anomalies. Injection procedures can be employed at various locations, including the base of the shaft or beneath the foundation to enhance geotechnical stability, within the structural element to rectify foundation structural issues, or around the shaft to establish a connection with the soil or provide a protective cover against corrosion.

(a) (b)

Figure 26.47 Excavation and replacement of damaged concrete with (a) a drill within the cage and (b) hand methods (State of California/Public Domain).

Grouting Within the Shaft (Brown and Powell 2013)

Grouting serves as a valuable technique for repairing a compromised concrete section within a shaft. The removal of low-strength material often poses the most challenging aspect of this process. To address this, the use of high-pressure jets is employed to eliminate weakened materials. Following the completion of water jetting to clear the hole of cuttings and debris, it is advisable to thoroughly drain or vacuum the access hole. This ensures optimal conditions for the effective infusion of grouting material to fill cavities. This method is also effective in both enhancing and rectifying concrete areas that contain soil inclusions or exhibit low-strength characteristics. It generally suffices to reinstate the structural integrity to levels that meet project requirements.

Grouting Around the Perimeter of the Shaft (Brown and Powell 2013)

In cases where there is an insufficient cover on the reinforcing bars of the shaft, or there are voids between the exterior of the shaft and the surrounding soil, it is advisable to consider grout injection around the shaft's perimeter. It is worth noting that grouting around the shaft can start and end at specified heights along the shaft based on the targeted treatment zone, without requiring full-length extension along the entire shaft. Grouting can also be used to stabilize an area surrounding a shaft in preparing it for internal hydro demolition and subsequent grouting.

However, it is important to recognize that partial or patch repairs in a concrete structure offer only a limited resolution to the issue and are suitable only for minor damages. Corrosion inhibitors, which encompass chemical compound admixtures or surface applications to cured concrete, fall within the realm of patch solutions. These protective coatings establish an impermeable barrier over the concrete, effectively obstructing the ingress of chloride ions (James et al. 2019).

Case Study 26.8

Micro-pile Underpinning of Mandalay Bay Hotel and Casino (Richards and Kardofilis 2006)

Mandalay Bay Hotel and Casino, located in Las Vegas, Nevada, is a 3700-room hotel and casino complex spanning 43 stories. The structure features three radial wings extending from the central elevator core, each positioned at 120° intervals (Figure 26.48). The central elevator core is designed to house 28 vertical cars and is enclosed within four substantial shear wall enclosures. The estimated total combined dead and live loads exerted on the central core amounts to 250,000 kips.

The foundation of the complex is a 10-foot-thick mat, and the central core is situated approximately 20 ft below ground level. Subsequent to the completion of the reinforced concrete framework, concerns arose regarding unacceptable vertical movements, particularly differential shifts between the tower core and the wings. The central core

(Continued)

Case Study 26.8 (Continued)

Figure 26.48 Overall elevation view of Mandalay Bay in 1998 (Richards and Kardofilis 2006 / Nicholson Construction Company).

was subsiding at a rate of approximately ½ to ¾ inches per week, while the wings exhibited a slower rate of sinking, exacerbating the disparity in column bay distortions.

Geotechnical Conditions

The valley of Las Vegas consists of very deep interbedded layers of alluvial sands, silts, and clay (Vanderpool et al. 1998). Some layers of caliche and cemented sands and gravels exist at various elevations. Most large structures are built on mats foundations that gain support from the caliche layers. At the Mandalay Bay site, caliche layers several feet in thickness were identified at 20, 25, 65, and 85 ft below the surface. Harder layers of partially to fully cemented materials were also identified at depths of 185–200 ft below the mat and another at 300 ft below the mat. Under the mat foundation, the soil sandwiched between the caliche layers was predominantly clayey sands and sandy clays. To obtain a sunken resort feel at Mandalay Bay, the shallow caliche layers were at least partially removed.

Retrofitting of the Foundation System

The approach was to drill and install micro-piles through holes cored into the mat and not bonded in the mat, so that the piles could be jacked into the ground and maintain the building at a desired level. Then, the structural beam supports would be installed to act as permanent attachments and jacking frames, as shown in Figure 26.49. The entire system had the capacity to lift the center of the tower if that proved to be necessary. In order to support the center core, a layout consisting of 536 micro-piles (Pin Piles) was developed. The resulting system was designed to support the core as if it was one very large pile cap. All the micro-piles used to support the hotel core were 200 feet deep, were fully bonded with grout to the various soil and caliche layers and were isolated from the mat. The decision to drill 200 ft was based on a fairly substantial caliche layer being encountered at the depth in a preliminary methods hole, and subsequent boring also often encountered a similar layer.

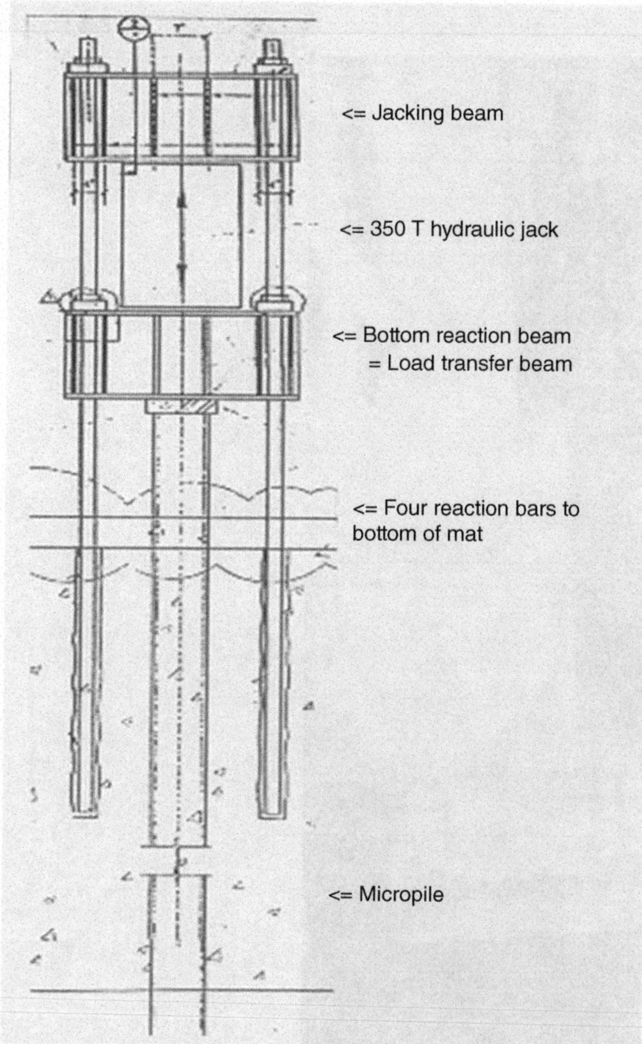

Figure 26.49 Load transfer and jacking frame (Richards and Kardofilis 2006).

Each micro-pile consisted of 7-in OD 80 ksi pipe sections drilled down with threaded flush jointed 10-foot long pieces. External flushing with water and polymer additives ensured that the heavier soil particles were lifted around the annulus of the pile and ejected. The external flush drilling also provided very rapid advancement of the casing. A preliminary test hole was drilled outside the work area, and cuttings were captured to prove this method would not cause a major loss of ground. Once the casing reached the desired bottom elevation, a flexible tremie pipe was lowered inside the pipe casing to the bottom of the hole. Neat cement grout, mixed in high shear colloidal mixers, with a water-to-cement ratio of 0.45 and a super-plasticizing admixture was then pumped through the tremie pipe. Full external grouting was performed by attaching the drill head back on the casing, and pumping additional grout through the head of the drill until it was observed to be flowing up from around the outside of the pile (Figure 26.50).

(Continued)

Case Study 26.8 (Continued)

Figure 26.50 Externally flushing with grout (Richards and Kardofilis 2006 / Nicholson Construction Company).

A 10-foot-long PVC pipe was inserted over the top of the pile into the mat to act as a bond breaker. To connect the micro-piles to the mat foundation each pile was constructed with its own reaction frame/transfer beam assembly and its own 350-tons jack (Figure 26.49). Four steel reaction bars were grouted into the mat around each pile. A small beam was placed on top of the pile, followed by the hydraulic jack and then another beam. The four reaction bars were attached to the top beam with large nuts. When the jack was extended, the force pushed down against the pile and up against the mat. When the desired load was reached, nuts were tightened down on the bottom beam to lock in the load on the pile.

Each pile was cyclically load tested (Figure 26.51) with its own jack to 600 kips and finally locked off at a nominal load of 50 kips. This testing also tested the grouting of the reaction bars into the mat. During retrofitting and installation of the micro-piles the building was continuously monitored using level survey and tiltmeters to verify that retrofitting was improving the settlement rate. No lifting of the core was attempted since the differential settlement between the core and wings was continually improving and since the floors were being leveled continuously during construction and attachment of the adjacent structures.

Figure 26.51 Overhead view of pile attachment frames (Richards and Kardofilis 2006 / Nicholson Construction Company).

Long Term Performance
A long-term continuous settlement monitoring system was installed after the completion of the retrofitting project. The monitoring system indicated that the settlements of the building halted right after the completion of the retrofitting. To date, the building has not experienced any settlements.

Case Study 26.9

Performance of Buildings over Liquifiable Ground in Adapazari, Turkey (Sancio et al. 2004)

The city of Adapazari in Turkey bore the brunt of the extensive building damage and loss of life resulting from the August 17, 1999, 7.4 Magnitude Kocaeli earthquake (Bray and Stewart 2000). The seismic impact triggered the liquefaction of shallow silt deposits prevalent in the region, leading to the collapse of numerous buildings due to ground failure. This failure was characterized by buildings undergoing relative vertical displacement into the ground, tilting, and lateral shifting. The incidence of structural damage was closely linked to the occurrence of ground failure.

(Continued)

Case Study 26.9 (Continued)

Typical Building Foundation

The prevalent construction in the area comprised 3 to 6-story reinforced concrete buildings, designed with a beam-column system and generally lacking shear walls. Foundations of these structures typically featured a reinforced concrete mat of 30–40 cm thickness, reinforced further with 30 cm wide and 100–120 cm deep concrete grade beams, spaced between 4 and 6 m in both directions. The voids between adjacent grade beams were backfilled with compacted soil, subsequently covered by a thin concrete floor slab. This arrangement resulted in a highly rigid and robust mat foundation. The ability of these foundations to absorb the seismic impact is often credited for preventing significant structural damage, allowing the buildings to tilt, translate, or undergo differential downward movement while maintaining their integrity.

Figure 26.52 Examples of ground failure-induced building damage in Adapazari after the August 17, 1999 Kocaeli earthquake: (a) vertical displacement, (b) vertical displacement with a ground heave, (c) vertical displacement with significant tilt, and (d) bearing capacity failure (Sancio et al. 2002 / with permission of Elsevier).

Mechanism of Building Settlements

In the aftermath of the Kocaeli earthquake, the predominant mechanism of building settlement in Adapazari was attributed to the lateral spreading of soil directly beneath the buildings. This phenomenon resulted from a temporary loss of bearing capacity caused by seismically induced pore water pressures. Swift and substantial deformations were observed, followed by additional vertical displacement prompted by the gradual dissipation of excess pore water pressures once the ground shaking ceased. In instances where the building exhibited a slender configuration (high H/B ratio) and lacked lateral constraints, it could undergo catastrophic bearing capacity-related failure, leading to excessive tilting or even complete collapse.

Forms of Ground Failure-Related Building Damage

During the reconnaissance missions conducted post-earthquake, various forms of building damage arising from ground failure were documented, some of which are visually represented in Figures 26.52 and 26.52, serving to illustrate the ensuing descriptions:

Uniform Vertical Displacement: Numerous structures in Adapazari experienced sinking into the ground, often without pronounced tilting, as depicted in Figure 26.52a. Instances of ground heave were also observed, as evidenced in Figure 26.52b.

Vertical Displacement with Tilt: Certain buildings encountered nonuniform vertical deformation, resulting in the condemnation of the structures despite their lack of structural damage, exemplified by Figure 26.52c. Buildings tipping over, shown in Figure 26.52d, were primarily noticed in slender structures lacking lateral support, characterized by a high building height (H) to width (B) ratio.

Lateral Translation: Several buildings exhibited lateral translation over soil that had undergone liquefaction directly beneath their foundations. Figure 26.53 illustrates two instances of this phenomenon. In the first scenario, the building shifted laterally by 31 cm, creating a noticeable gap between the structure and the adjacent sidewalk. In the second example, the building's translation amounted to around 110 cm, directed towards an open alley, resulting in the mobilization of a wedge of soil. Additionally, this building underwent a perpendicular translation of 50–55 cm.

Figure 26.53 Examples of lateral displacement of structures on mat foundations in Adapazari after the 1999 Kocaeli earthquake. Note the development of a passive resistance wedge in the photograph on the right (Sancio et al. 2002 / with permission of Elsevier).

Case Study 26.10

Emergency Bridge Pier Foundation Repair with Pressed-in Piles: Kumakami River Bridge Foundation Repair, Fukuoka Prefecture, Japan (Takefumi and Takayuki 2018)

This case study presents the repairs of one of the pier foundations of the Kumakami River Bridge on the JR Kushu Railway Company's Kyudai Line in Fukuoka Prefecture, Japan. In July 2012, the area suffered significant scouring and settlement due to flooding caused by a typhoon. As a result, the railway service was temporarily suspended until interim repair work could be undertaken. Following the resumption of service, train speeds on the bridge were kept low, awaiting permanent repair. Figure 26.54 illustrates a passing train on the bridge, with the circled area denoting the pier foundation that underwent temporary repair.

Figure 26.54 Damaged pier foundation (in circle) of Kumakami River Bridge (Takefumi and Takayuki 2018 / American Society of Civil Engineers).

The original foot protection of the pier was lost due to scouring. The soil conditions at the pier site primarily consisted of boulder-mixed gravelly sand, with a 3.3-foot (1.0 m) thick sand layer located at 6.6 ft (2.0 m) beneath the ground surface. *N*-values indicated substantial strength beyond 60 at 10 ft (3.0 m) below the surface, as depicted in Figure 26.55. Concerns arose for the railway company due to safety considerations during repair work. These concerns stemmed from the proximity of the work to the girder, which bore frequent live rail traffic, as well as the sensitive foundation. The overhead clearance beneath the girder measured approximately 13 ft (4.0 m), as shown in Figure 26.56. The proposed permanent repair solution entailed constructing a 25-foot × 16.5-foot (7.6 m × 5.4 m) cofferdam using 23.6-in. (600 mm) diameter and 16-foot (4.8 m) long pipe piles, as depicted in Figure 26.57. The interior of the cofferdam would then be filled with reinforced concrete.

Given the need to install pipe piles within boulder-mixed gravelly sand beneath the girder, a Gyropress™ piling equipment was selected for the task. For more information regarding Gyropress Method™, please refer to Hirata and Matsui (2016). This piling method commences by applying hydraulic force to initiate the installation of the first few piles. This is achieved through the use of the equipment's own weight and additional counterweight. Once this initial stage is completed, the equipment secures itself onto the previously installed piles and advances autonomously as the pile installation progresses. Notably, this method stands out for its environmentally friendly approach, as it does not rely on vibratory or percussive force for pile installation.

Figure 26.55 Soil conditions and location of pipe piles (Takefumi and Takayuki 2018 / American Society of Civil Engineers).

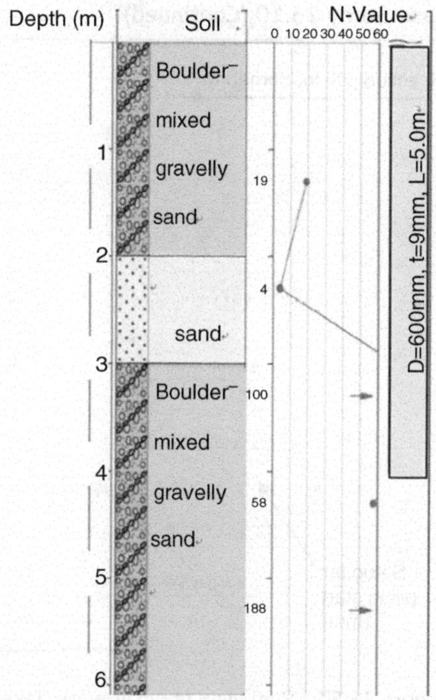

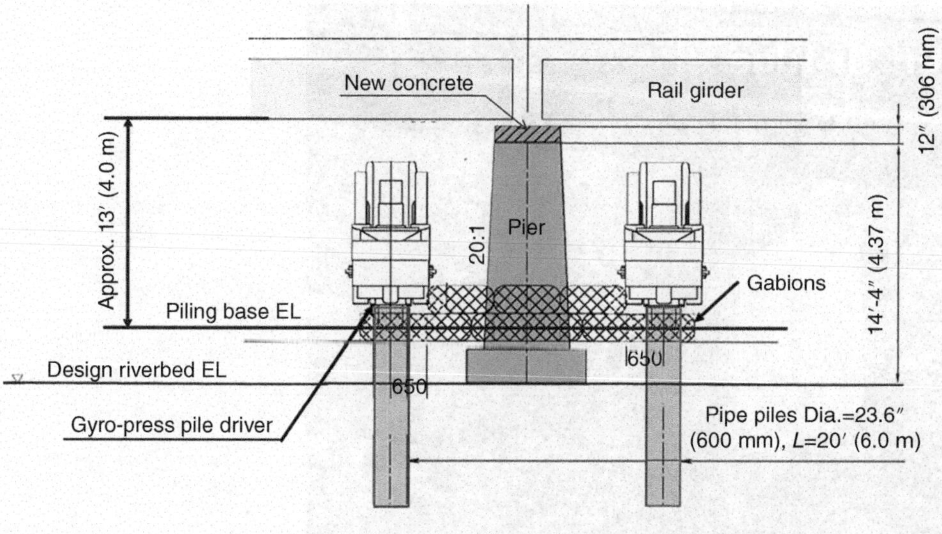

Figure 26.56 Cross-section of repair work (Takefumi and Takayuki 2018 / American Society of Civil Engineers).

Under the girder, eight piles were spliced twice each due to limited headroom, while those outside the girder were driven as complete units. Reaction sheet piles and reaction pipe piles, as shown in Figure 26.57, were essential for the Gyropress equipment to initiate pile installation and make directional changes, respectively. Temporary foot protection gabions had been positioned, as illustrated in Figure 26.58. Subsequently, after partially removing the gabions as part of the preparations for permanent work, the Gyropress equipment expediently pressed the piles into place to establish the cofferdam, as depicted in Figure 26.59.

(Continued)

Case Study 26.10 (Continued)

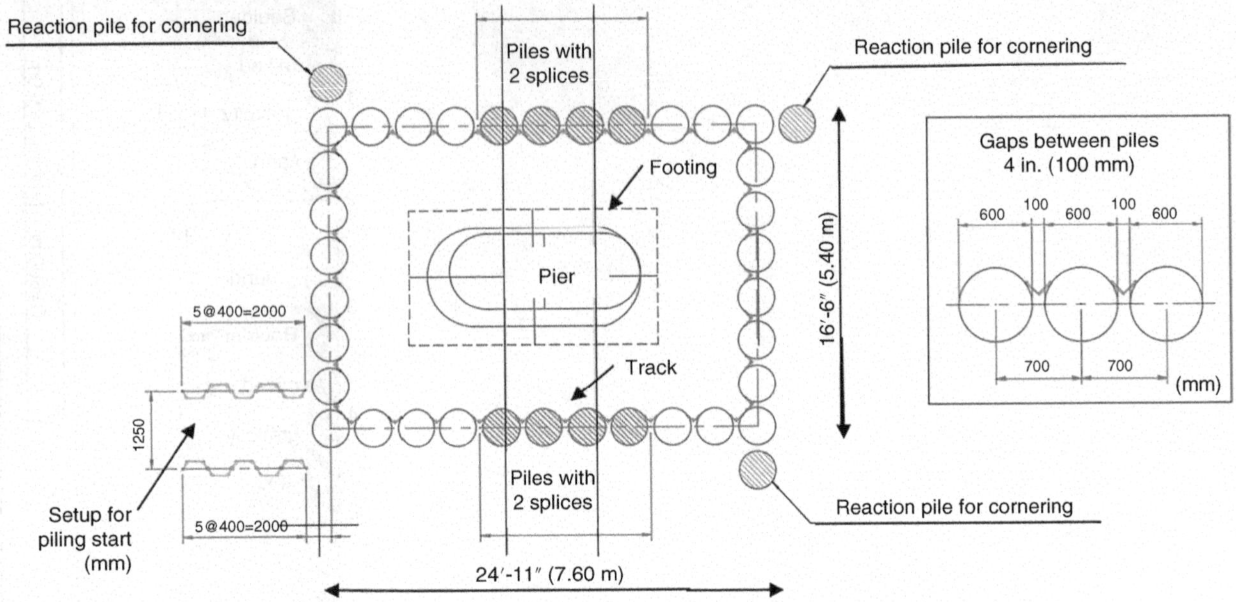

Figure 26.57 Plan view of repair work (Takefumi and Takayuki 2018 / American Society of Civil Engineers).

Figure 26.58 Pier before Cofferdam Construction (Takefumi and Takayuki 2018 / American Society of Civil Engineers).

Considering that the pier had settled and slightly tilted due to scouring, inclinometers were employed to closely monitor the movement during construction. Figure 26.60 presents the monitoring results, indicating a minor 0.045° tilt occurred along the bridge axis, while nearly negligible tilt was observed in the normal direction. Importantly, rail traffic remained uninterrupted during the piling activity, as shown in Figure 26.61, enabling the construction to be efficiently completed within a relatively short timeframe.

Figure 26.59 Tubular piles being driven by Gyropress piling equipment (Takefumi and Takayuki 2018 / American Society of Civil Engineers).

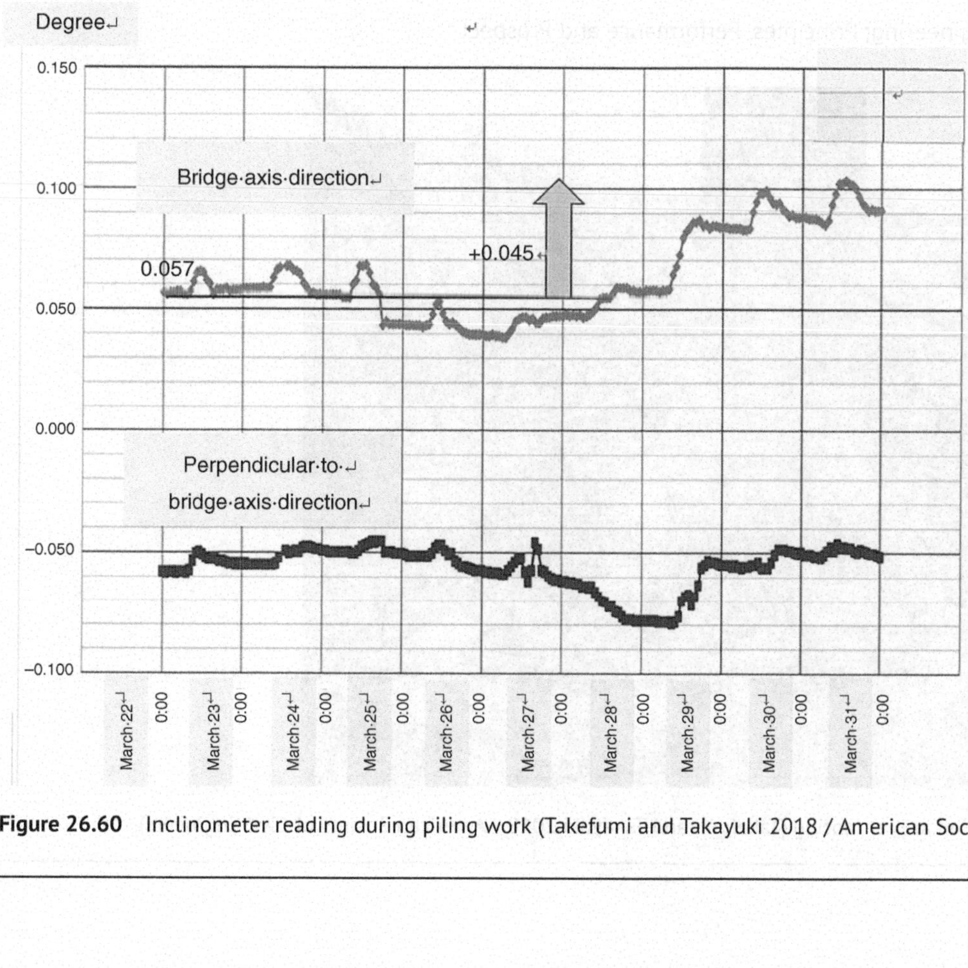

Figure 26.60 Inclinometer reading during piling work (Takefumi and Takayuki 2018 / American Society of Civil Engineers).

(Continued)

Case Study 26.10 (Continued)

Figure 26.61 Gyropress piling continued under the passing train (Takefumi and Takayuki 2018 / American Society of Civil Engineers).

Advanced Foundation Engineering; Principles, Performance and Prospect

Figure 26.62 Re-bar installation at pier footing (Takefumi and Takayuki 2018 / American Society of Civil Engineers).

To bridge the 4-in. (100 mm) gap between the piles, a pressed-in steel angle bar was welded to both sides of the piles. Subsequently, the piles were cored and filled with concrete to enhance stability. Rebars were introduced inside the cofferdam, as illustrated in Figure 26.62, followed by the casting of concrete to create a larger and more robust foundation compared to its previous state, as seen in Figure 26.63.

Figure 26.63 Concrete placed (Takefumi and Takayuki 2018).

26.8 Major Points

1) Damages to structures usually occur due to reduced or loss of foundation-bearing capacity, excessive settlement, and tilt that can cause overturning structures. However, defects internal to the foundation can be caused by bending and shearing weaknesses due to poor construction, improper foundation dimensions, and material deterioration of the structural elements during the life cycle. Generally, these damages can be categorized into geotechnical, structural, and construction causes.
2) A bearing capacity failure occurs when the shear strength of the soil beneath the foundation is exceeded. General shear, local shear, and punched shear are the three types of bearing capacity failures.
3) Settlement in a building is divided into three categories: uniform, tilted, and differential. Differential settlement and tilting are the causes of most structural damages. The main sign of settlement issues following the construction of the building is the occurrence of cracks in various elements such as floors and ceilings, door frames, and windows.
4) Factors that cause internal damage to the structural elements of a foundation (steel rebar and concrete), can be physical, biological, or chemical in nature such as sulfate attack, and rebar corrosion.
5) Factors external to the foundation can cause increased uniform and differential settlements. These factors include deep excavation or tunneling in the vicinity of the structure, liquefaction of soil due to earthquakes, groundwater fluctuation, and flooding.
6) Addressing foundation defects necessitates employing various solutions, either individually or in combination. Broadly speaking, foundation repair and reconstruction can be accomplished through one of the following methods:

soil reinforcement (geotechnical enhancement), structural reinforcement of existing foundations, and alleviating foundation loads by lightening the existing building.

7) Geotechnical improvement aims to strengthen and enhance the soil beneath the foundation or rectify settlements. This can be achieved through underpinning techniques like compensation grouting, foundation underpinning using drilled shafts, micro-piles, and helical piles.
8) Structural enhancement encompasses strategies such as enlarging foundation dimensions, introducing ties to the existing foundation, reinforcing the foundation against bending and shearing, and augmenting the resistance of existing piles.
9) Jacks are utilized for leveling foundations, while injectable materials can fill the gap between the soil mass and the foundation. This method involves elevating a section or the entirety of a building using jacks. The reaction force required for lifting the building necessitates a connection between the foundation and micro-pile, helical pile, or pile to provide the necessary reaction force for lifting the structure.
10) There are methods for retrofitting and strengthening that include implanting steel rebars and enlarging the dimensions of the foundation.

26.9 Questions and Problems

26.9.1 Explanatory

1 Categorize the types of foundation damages and briefly describe each one.

2 What are the signs of foundation damage?

3 Describe the types of foundation settlements and explain what type causes the most problems for the structure.

4 Explain the effect of adjacent buildings on the behavior of foundations.

5 How do trees damage the foundation?

6 What are the effects of floods on foundation subsoil and foundation structures?

7 Explain the effect of steel rebar corrosion on foundation performance.

8 What is the grouting efficiency, and how does it change with OCR?

9 What are the advantages of repairing foundations with helical underpinning systems?

10 Describe methods of strengthening and repairing existing foundations.

11 Describe the techniques employed for safeguarding reinforced concrete foundations against corrosion and corrosive materials, and outline the types of structures that are most susceptible to experiencing corrosion.

26.9.2 Conceptual

1 According to the proposed methods for repairing settled foundations, what is the best repair method in terms of environmental compatibility and convenience?

2 With respect to the vulnerability of bridges to floods, what are the strategies that increase the stability of these structures against floods?

3 Considering the Babolsar concrete pedestrian bridge as a successful example, what are your suggestions for engineers in similar projects? Explain the lessons and experiences that can be gained through this case history.

4 Describe the Tower of Pisa project. What were the causes of failure? What are your recommendations to solve the problem?

5 Explain three different failure conditions of two adjacent buildings regarding the difference in their height. Suggest suitable methods for failure prevention in each condition.

6 The bridge is damaged due to flood. Which geotechnical parameters caused this damage?

7 What are the main causes of nonuniform settlement?

8 What are the similarities and differences in the application of micro-piles and helical piles in restoration practice?

9 Consider Case Study 26.7 and discuss the effectiveness of the strategies used considering technical and practical aspects.

10 Discuss the role of implanting steel rebars in enhancing the geotechnical and structural capacity of existing foundations.

References

Abbas, H.O. (2020). Influence of tunnel excavation on tower foundation settlement constructed on sandy soil. *Transportation Infrastructure Geotechnology.* 7: 651–663. https://doi.org/10.1007/s40515-020-00114-x.

Alwalan, M.F. (2018). Interaction of closely spaced shallow foundations on sands and clays: a review. *International Journal of Advanced Engineering Research and Science (IJAERS)* 5 (9): 101–110. https://doi.org/10.22161/ijaers.5.9.11.

Amir, A. (1992). Interference effect on the behaviour of footings. A thesis for PhD. University of Roorkee – 267-667, India.

Attewell, P. B., Yeates, J., and Selby, A. R. (1986). Soil movements induced by tunnelling and their effects on pipelines and structures. Blackie; New York, 325 P.

Bahadori, H., Motamedi, H., Hasheminezhad, A., and Motamed, R. (2020). Shaking table tests on shallow foundations over geocomposite and geogrid-reinforced liquefiable soils. *Soil Dynamics and Earthquake Engineering* 128: 105896.

Bally, R.J. (1988). Some specific problems of wetted loessial soils in civil engineering. *Engineering Geology* 25 (2-4): 303–324.

Bertalot, D., Brennan, A., Knappett, J., et al. (2012). Use of centrifuge modelling to improve lessons learned from earthquake case histories. *Eurofuge 2012*, Delft, The Netherlands (23–24 April 2012).

Bhattacharjee, J. (2018). Deterioration of concrete structures along with case studies in India. *Proceedings of the Institution of Civil Engineers - Forensic Engineering* 171 (2): 80–90. https://doi.org/10.1680/jfoen.17.00010.

Blight, G.E. (2009a). *Geotechnical Engineering for Mine Waste Storage Facilities.* CRC Press.

Blight, G. (2009b). Desiccation of soil by vegetation and potential interaction with buildings-a field study. *Journal of the South African Institution of Civil Engineering= Joernaal van die Suid-Afrikaanse Instituut van Siviele Ingenieurswese* 51 (2): 20–29.

Bonić, Z., Ćurčić, G.T., Davidović, N., and Savić, J. (2015). Damage of concrete and reinforcement of reinforced-concrete foundations caused by environmental effects. *Procedia Engineering* 117: 411–418. https://doi.org/10.1016/j.proeng.2015.08.187.

Boscardin, M.D. and Cording, E.J. (1989). Building response to excavation-induced settlement. *Journal of Geotechnical Engineering* 115 (1): 1–21.

Bozozuk, M. (1972). Foundation failure of the Vanleek Hill Tower silo. *Performance of Earth and Earth Supported Structures* 22: 885–902.

Bray, J.D. and Stewart, J.P. (2000). Damage patterns and foundation performance in Adapazari, in Kocaeli, Turkey Earthquake of August 17, 1999 Reconnaissance Report. *Earthquake Spectra* 16 (Suppl. A): 163–189.

Brown, M.J. and Powell, J.J.M. (2013). Comparison of rapid load test analysis techniques in clay soils. *ASCE Journal of Geotechnical & Geoenvironmental Engineering* 139 (1): 152–161. https://doi.org/10.1061/(ASCE)GT.1943-5606.0000733.

Burland, J.B., Broms, B.B., and de Mello, V.F.B. (1977). Behavior of foundations and structures on soft ground. *Proceedings of the 9th International Conference on Soil Mechanics and Foundation Engineering (ICSMFE)*, Tokyo (10–15 July 1977), 495–546. Japanese Geotechnical Society.

Burland, J.B. (1997). Theme lecture: Subsidence due to tunnelling and its effects on buildings. *Proceedings of the International Conference on Soil Mechanics and Foundation Engineering* 14 (4): 2399–2402.

Cajka, R. and Manasek, P. (2005). Building structures in danger of flooding. In: *In Proceedings IABSE Conference on Role of Structural Engineers Towards Reduction of Powerty*. New Delhi, India.

Charles, J.A. and Skinner, H.D. (2004). Settlement and tilt of low-rise buildings. *Proceedings of the Institution of Civil Engineers-Geotechnical Engineering* 157 (2): 65–75. https://doi.org/10.1680/geng.2004.157.2.65.

Charles, J., Watts, K., and BRE. (1996). The assessment of the collapse potential of fills and its significance for building on fill. *Proceedings of the Institution of Civil Engineers-Geotechnical Engineering* 119 (1): 15–28.

Clough, G.W. and O'Rourke, T.D. (1990). *Construction Induced Movements of In situ Walls Design and Performance of Earth Retaining Structures Geotechnical Special Publication No. 25 R* (ed. C. Lambe and L.A. Hansen). New York, NY: ASCE.

Clough, G. W., & Schmidt, B. (1981). Design and performance of excavations and tunnels in soft clay. *Developments in Geotechnical Engineering* 567–634. https://doi.org/10.1016/b978-0-444-41784-8.50011-3.

Coduto, D.P., Kitch, W.A., and Yeung, M.C.R. (2016). *Foundation Design Principles and Practices*, 3e. USA: Prentice Hall ISBN 13: 9780133411898.

Cooper, A. (1998). Subsidence hazards caused by the dissolution of Permian gypsum in England: geology, investigation and remediation. *Geological Society, London, Engineering Geology Special Publications* 15 (1): 265–275.

D'Ayala, D.F. and Paganoni, S. (2011). Assessment and analysis of damage in L'Aquila historic city centre after 6th April 2009. *Bulletin of Earthquake Engineering* 9: 81–104.

Dadarya, A.K. (2007). *Foundation Settlement in Buildings (Causes & Prevention)*. Gwalior, India: Centre for Advanced Maintenance Technology, Ministry of Railways, Government of India.

Das, B.M. (2014). *Principles of Foundation Engineering*, 8e. Global Engineering: Timothy L. Anderson ISBN-13: 978-1-305-08155-0.

Dimmock, P.S. and Mair, R.J. (2008). Effect of building stiffness on tunnelling-induced ground movement. *Tunnelling and Underground Space Technology* 23 (4): 438–450.

Elsamny, M.K., Elbatal, S.A., Abo-Alanwar, M.M., and Abdel-Mohsen, A.M. (2020). Retrofitting of damaged RC columns using spiral stirrups. *New York Science Journal* 13 (3): 16–38. https://doi.org/10.7537/marsnys130320.02.

Eslami, A., Golafzani, S.H., and Chenari, R.J. (2016). Assessment of Babolsar Concrete Pedestrian Bridge Failure for 1964 flood event and retrofitting practice. *Engineer Failure Analysis* 68: 101–112. https://doi.org/10.1016/j.engfailanal.2016.05.025.

Fok, P., Neo, B.H., Goh, K.H., and Wen, D. (2012). Assessing the impact of excavation-induced movements on adjacent buildings. *The IES Journal Part A: Civil and Structural Engineering* 5 (3): 195–203. https://doi.org/10.1080/19373260.2012.696444.

Gonga, C., Ding, W., and Xie, D. (2020). Twin EPB tunneling-induced deformation and assessment of a historical masonry building on Shanghai soft clay. *Tunneling and Underground Space Technology.* 98: 103300. https://doi.org/10.1016/j.tust.2020.103300.

González-Nicieza, C., Álvarez-Fernández, M.I., Menéndez-Díaz, A. et al. (2008). Failure analysis of concrete sleepers in heavy haul railway tracks. *Engineering Failure Analysis* 15 (1–2): 90–117.

Graham, J., Raymond, G. P., and Suppiah, A. (1984). Bearing capacity of three closely-spaced footings on sand. *Geotechnique* 34 (2): 173–181.

Gutiérrez, F. and Cooper, A.H. (2002). Evaporite dissolution subsidence in the historical city of Calatayud, Spain: damage appraisal and prevention. *Natural Hazards* 25: 259–288.

Herle, I., Herbstová, V., Kupka, M., and Kolymbas, D. (2010). Geotechnical problems of cultural heritage due to floods. *Journal of Performance of Constructed Facilities* 24 (5): 446–451.

Hirata, H. and Matsui, N. (2016). Expanding applications of the Gyropress Method™, Nippon steel and Sumitomo metal Technical Report No. 113, December 2016. Japan: Nippon Steel Corporation.

Huynh, Q.T., Lai, V.Q., Boonyatee, T., and Keawsawasvong, S. (2020). Behavior of a deep excavation and damages on adjacent buildings: a case study in Vietnam. *Transportation Infrastructure Geotechnology.* https://doi.org/10.1007/s40515-020-00142-7.

James, A., Bazarchi, E., Chiniforush, A.A. et al. (2019). Rebar corrosion detection, protection, and rehabilitation of reinforced concrete structures in coastal environments: a review. *Construction and Building Materials* 224: 1026–1039. https://doi.org/10.1016/j.conbuildmat.2019.07.250.

Kimmerling, R. (2002). Geotechnical engineering circular No. 6 shallow foundations by Federal Highway Administration. Lulu.com; null edition.

Korff, M. (2009). *Deformations and Damage to Buildings Adjacent to Deep Excavations in Soft Soils*. Delft Cluster.

Lee, J. and Eun, J. (2009). Estimation of bearing capacity for multiple footings in sand. *Computers and Geotechnics* 36 (6): 1000–1008.

Lutenegger, A.J. (2013). Underpinning a residential structure on uncontrolled fill with helical helical-piles. *International Conference on Case Histories in Geotechnical Engineering* (29 April–4 May 2013), 36. Wheeling (Chicago), Illinois: Missouri University of Science and Technology.

Lutenegger, A.J. and Kemper, J.H. (2008). Preservation of historic structures using Screw-Pile foundations. In: *Structural Analysis Of Historic Construction: Preserving Safety And Significance, Vols 1 And 2*.

Marseh, B.P., Heidarpour, B., Hosseinnezhad, A., and Soumehsaraei, B.G. (2013). Effects of adjacent building construction: a case study. *Journal of Basic and Applied Scientific Research* 3 (1): 370–375.

Modoni, G., Croce, P., and Mongiovì, L. (2006). Theoretical modelling of jet grouting. *Géotechnique* 56 (5): 335–347.

Ni, J.C. and Cheng, W.-C. (2010). Monitoring and modeling grout efficiency of lifting structure in soft clay. *International Journal of Geomechanics* 10 (6): 223–229.

O'reilly, M. P., & New, B. (1982). Settlements above tunnels in the United Kingdom-their magnitude and prediction. *Proceedings of Tunnelling '82 Symposium*, London, 173–181.

Papadopoulos, A., Dusseault, S., and Bobrow, D. (2004). Dewatering effects on structures adjacent to central artery/tunnel project. *Geotechnical Engineering for Transportation Projects* (pp. 1524–1533).

Pearlman, S. L. (2000). Pin piles for structural underpinning. *Proceedings of the 25th Annual Deep Foundations Institute Meeting*

Prakash, S. (2007). *Soil Dynamics*. USA: McGraw-Hill.

Reddy, E.S., Borzooei, S., and Reddy, G. (2012). Interference between adjacent footings on sand. *International Journal of Advanced Engineering Research and Studies* 1 (4): 95–98.

Rezaei, M., Ajalloeian, R., and Ghafoori, M. (2012). Geotechnical properties of problematic soils emphasis on collapsible cases. *International Journal of Geosciences* 3 (1): 105–110.

Richards, T.D. and Kardofilis, D. (2006). Underpinning of the Mandalay Bay Hotel and Casino. *54th Annual Geotechnical Engineering Conference*, St. Paul, Minnesota (17 February 2006).

Salamatpoor, S., Jafarian, Y., and Hajiannia, A. (2019). Bearing capacity and uneven settlement of consecutively constructed adjacent footings rested on saturated sand using model tests. *International Journal of Civil Engineering*. 17 (6): 737–749. https://doi.org/10.1007/s40999-018-0295-y.

Salgado, R. (2008). *The Engineering of Foundations*. USA: McGraw-Hill.

Sancio, R.B., Bray, J.D., Stewart, J.P. et al. (2002). Correlation between ground failure and soil conditions in Adapazari, Turkey. *Soil Dynamics and Earthquake Engineering* 22: 1093–1102.

Sancio, R., Bray, J.D., Durgunoglu, T., and Onalp, A. (2004). Performance of buildings over liquefiable ground in Adapazari, Turkey. *Proceedings, 13th World Conference on Earthquake Engineering*. Vancouver, Canada: Canadian Association for Earthquake Engineering.

Santrač, P., Grković, S., Kukaras, D. et al. (2021). Case study-an extreme example of soil-structure interaction and the damage caused by works on foundation strengthening. *Applied Sciences* 11 (11): 5201. https://doi.org/10.3390/app11115201.

Schuster, M., Kung, G.T.C., Juang, C.H., and Hashash, Y.M.A. (2009). Simplified model for evaluating damage potential of buildings adjacent to a braced excavation. *Journal of Geotechnical and Geoenvironmental Engineering* 135 (12): 1823–1835. https://doi.org/10.1061/(ASCE)GT.1943-5606.0000161.

Selvadurai, A. P. S., & Rabbaa, S. (1983). Some experimental studies concerning the contact stresses beneath interfering rigid strip foundation resting on a granular stratum. *Canadian Geotechnical Journal* 20: 385–386.

Shabha, G. and Kuhwald, K. (1995). Subsidence and the associated problems with reference to low-risehousing. *Structural Survey* 13 (3): 28–35.

Soga, K., Au, S., Jafari, M., and Bolton, M. (2004). Laboratory investigation of multiple grout injections into clay. *Géotechnique* 54 (2): 81–90.

Srinivasan, V. and Ghosh, P. (2011). Interaction problem of circular footings on homogeneous soil deposit. Proceedings of *Indian Geotechnical Conference* (pp. 823–826).

Srinivasan, V. and Ghosh, P. (2013). Experimental investigation on interaction problem of two nearby circular footings on layered cohesionless soil. *Geomechanics and Geoengineering* 8 (2): 97–106.

Standing, J. and Burland, J. (2006). Unexpected tunnelling volume losses in the Westminster area, London. *Géotechnique* 56 (1): 11–26.

Sucharda, O., Smirakova, M., Vaskova, J. et al. (2018). Punching shear failure of concrete ground supported slab. *International Journal of Concrete Structures and Materials* 12 (1): 36.

Szynakiewicz, T. (2016). Compaction grouting for ground improvement and structure rehabilitation. *Rocky Mountain Geo-Conference* (pp. 23–31).

Takefumi, T. and Takayuki, S. (2018). Emergency bridge pier foundation repair with pressed-in piles. *Proceedings, IFCEE*, 224–237. https://doi.org/10.1061/9780784481608.022.

Tan, Y., Huang, R., Kang, Z., and Bin, W. (2016). Covered semi-top-down excavation of subway station surrounded by closely spaced buildings in downtown Shanghai: building response. *Journal of Performance of Constructed Facilities* 30 (6): 04016040.

Terzaghi, K., Peck, R.B., and Mesri, G. (1996). *Soil Mechanics in Engineering Practice*. New York: John Wiley and Sons.

Toll, D. G., Abedin, Z., Buma, J. et al. (2012). The impact of changes in the water table and soil moisture on the structural stability of buildings and foundation systems: a systematic review (CEE10-005 (SR90)) (Technical report-collaboration for environmental evidence). Durham: Durham University Library.

Tuncdemir, F. (2007). Theoretical and practical aspects of compaction grouting. *Teknik Dergi* 18 (1): 4069–4080.

Vanderpool, W.E., Doehring, M., and Palmer, S.D. (1998). Geotechnical investigation report, project Y, Las Vegas, Nevada. 24 November 1998.

Villalobos, F.A., Byrne, B.W., and Houlsby, G.T. (2010). Model testing of suction caissons in clay subjected to vertical loading. *Applied Ocean Research* 32 (4): 414–424.

Wen, L., Kong, G., Li, Q., and Zhang, Z. (2020). Field tests on axial behavior of grouted steel pipe micropiles in marine soft clay. *International Journal of Geomechanics* 20 (6): 06020006.

Wu, B. and Lee, C. (2003). Ground movements and collapse mechanisms induced by tunneling in clayey soil. *International Journal of Physical Modelling in Geotechnics* 3 (4): 15–29.

Wu, Y.X., Lyu, H.M., Han, J., and Shen, S.L. (2019). Dewatering–induced building settlement around a deep excavation in soft deposit in Tianjin, China. *Journal of Geotechnical and Geoenvironmental Engineering* 145 (5): 05019003. https://doi.org/10.1061/(ASCE)GT.1943-5606.0002045.

Zhou, Y., Gencturk, B., Willam, K., and Attar, A. (2015). Carbonation-induced and chloride-induced corrosion in reinforced concrete structures. *Journal of Material in Civil Engineering* 27 (9): 04014245. https://doi.org/10.1061/(ASCE)MT.1943-5533.0001209.

Zumrawi, M.M. and Aldawi, H.K. (2018). A study on strengthening of building foundation for storey extension. *Journal of Building Materials and Structures* 5 (2): 218–226.

27

Foundation Selection: Performance, Sustainability, and Value Engineering/Optimization

Chapter Highlight

- Overview of Foundation Alternatives
- General Selection Criteria
- Multidisciplinary Aspects in Engineering of Foundations
- Determinant Factors for Foundations Selection
- Project-Based Foundation Selection
- Sustainability Applications in Foundation Engineering
- Reuse of Existing Foundations After Demolition of the Structures
- Holistic Approach to Foundation Systems
- Value Engineering and Optimization

27.1 Overview of Foundations

The role of foundations as a crucial intermediary for transferring superstructure loads to the ground is indispensable throughout a structure's entire life span (Fakher 2022; Salgado 2008). Therefore, a comprehensive understanding of foundation types, construction methods, and performance is imperative. Traditionally, foundations are broadly categorized as shallow and deep foundations based on their embedment depth relative to their dimensions. In an effort to address some limitations associated with major foundation categories and to incorporate recent advancements in civil engineering, two intermediate solutions have been introduced: shallow foundations on treated grounds and semi-deep foundations (Eslami et al. 2020). Figure 27.1 illustrates various foundation categories concerning their construction processes. The sequence of adopting foundation alternatives, considering embedment depth, is depicted in Figure 27.1.

Shallow foundations are widely utilized due to their cost-effectiveness and ease of construction. They efficiently transfer superstructure loads to the underlying soil strata at a relatively shallow depth. However, in certain scenarios, intermediate options are disregarded, and when limitations of shallow foundations are encountered, deep foundations become the final resort. Deep foundations penetrate surface problematic deposits and effectively transfer loads to deeper layers. Nonetheless, challenges related to cost and installation may limit their widespread application.

Drawing inspiration from structural analysis concepts, foundations are categorized into one-dimensional (linear), two-dimensional (planar), and three-dimensional (volumetric) forms. Furthermore, considering the mechanism for load transfer, one-dimensional foundations are conceptualized as vector-act foundations. Planar foundations, on the other hand, are further classified into section-act and surface-act foundations. Additionally, three-dimensional foundations are identified as block-act (hybrid) foundations. Detailed specifications for each category of existing and innovative foundations are presented in Chapters 1 and 2. Also, more quantitative details can be found in Ebrahimipour et al. (2023) as well as Ebrahimipour and Eslami (2024). Figure 27.2 provides illustrative examples of these major foundations, considering subsurface conditions, form, and load transfer mechanisms.

Advanced Foundation Engineering: Principles, Performance and Prospects, First Edition. Moses Karakouzian and Abolfazl Eslami.

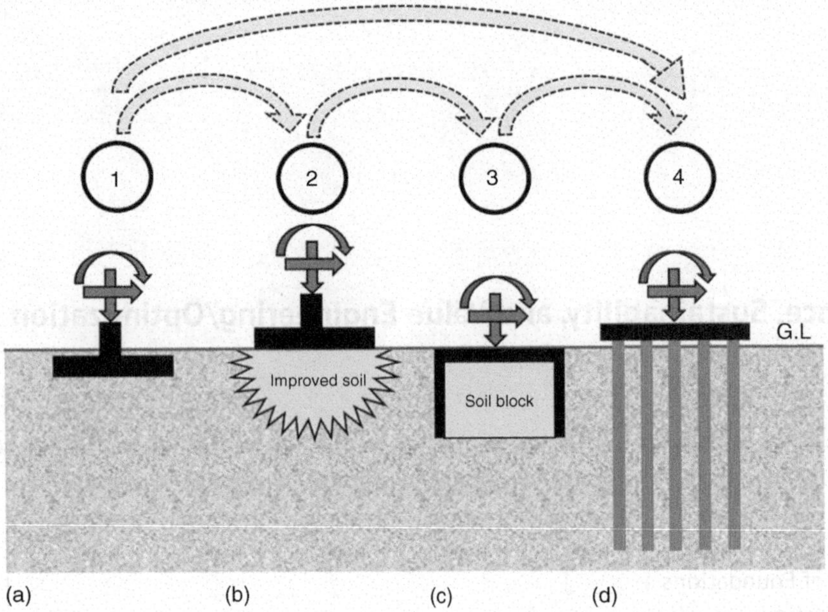

Figure 27.1 Priority of foundations construction: (a) shallow, (b) shallow on improved soil, (c) semi-deep, and (d) deep.

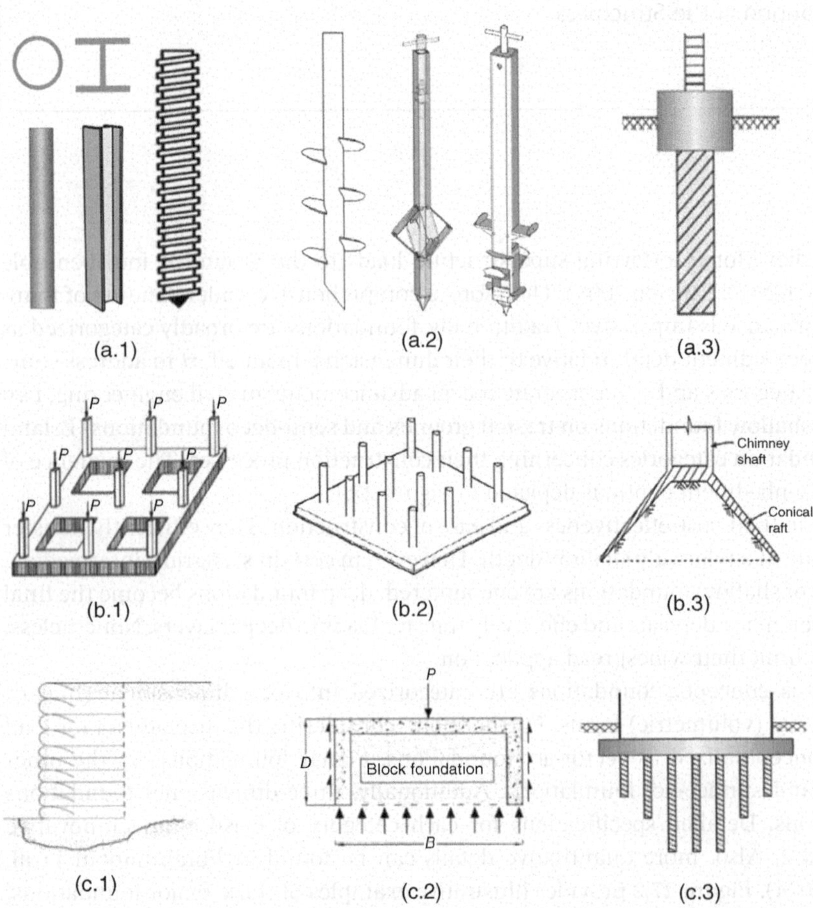

Figure 27.2 Typical examples of foundation systems based on form: (a) 1D: (1) driven and drilled displacement piles, (2) helical (Fateh et al. 2017) and self-expanded piles (Shojaei et al. 2019), (3) single shaft (Eslami et al. 2020 / with permission of Elsevier), (b) 2D: (1) grid foundation (Coduto et al. 2016), (2) raft foundation, (3) inverted cone shell foundations (Kurian 2006), (c) 3D: (1) geotextile reinforced soil, (Das 2014 / Cengage Learning), (2) bucket foundation (Rezazadeh and Eslami 2018), (3) PRF (Eslami et al. 2020 / with permission of Elsevier).

Shallow foundations, recognized globally as a viable and cost-effective option, are among the most widely used types of foundations. Their various types and construction methods are detailed in Chapter 5. Chapters 6 and 7 delve into the fundamental aspects of these foundations, analyzing their design and behavior concerning geotechnical and structural considerations. However, these foundations have limitations, especially in cases of weak soil conditions, large loads, and stability concerns.

To overcome these limitations, ground modification serves as a supplementary and complementary approach for enhancing problematic in situ soil conditions while optimizing foundation dimensions and material consumption. Chapter 8 focuses on introducing diverse soil improvement techniques and their implementation under different conditions. This approach addresses several geomaterial issues, including strength, stiffness, and serviceability requirements. Chapter 9 evaluates and quantifies the performance of foundations on stabilized grounds.

Semi-deep foundations, a recent development, are now available in various types and geometries for inland, nearshore, and offshore projects. Chapter 10 discusses the major types and relevant construction techniques. In Chapter 11, special emphasis is placed on the design principles and behavior of hybrid and skirted foundations, particularly for offshore sites. Floating foundations, incorporating the concept of compensating for excavated soil, are extensively detailed in Chapter 12. Additionally, Chapter 13 introduces and evaluates shell foundations, a significant type of semi-deep foundation.

This comprehensive exploration of foundation types and their respective methodologies provides a valuable resource for understanding the complexities and nuances of foundation engineering.

Deep foundations, which effectively transfer imposed loads to deep strata, find extensive application in areas characterized by challenging surface deposits, as well as sites susceptible to scour, liquefaction, and severe earthquakes. Chapter 14 provides a detailed exploration of the types and installation methods of deep foundations, serving as a pivotal resource for their selection, construction, and performance assessment.

Chapters 15 and 16 are dedicated to the design of individual deep foundations under axial loads, employing static analysis, and in situ testing applications, respectively. Penetrative in-situ tests are recognized as model piles (Shirani et al. (2023) as discussed in Chapter 16. Chapter 17 comprehensively covers their behavior under lateral loads and the fundamental principles of structural design for deep foundations.

In scenarios involving exceptionally high superstructure loads or a combination thereof, deep foundations are utilized in groups or combinations. Pile groups and pile raft foundations (PRFs) emerge as dominant alternatives, forming an interactive soil-foundation system suitable for substantial structures on challenging soils. Chapters 18 and 19 delve into the essential principles of design and practical implementation of pile groups and PRFs, offering a sequential and comprehensive understanding of these interactive foundation solutions.

27.2 Multidisciplinary Approach

As highlighted in Chapter 1, foundation engineering stands as a multidisciplinary discipline, recognized for its interactive nature among structural, constructional, and geotechnical engineering. Consequently, considerations for foundation selection are generally addressed from three key aspects of foundation design: structural, constructional, and geotechnical dimensions. This concept, along with the pivotal factors associated with each aspect, is illustrated in Figure 27.3. Furthermore, it is imperative to acknowledge novel criteria and effective approaches, such as sustainable design and value engineering, as integral requirements in the foundation selection process. These complementary considerations play a crucial role in enhancing the overall efficiency and viability of foundation engineering solutions.

27.3 Determinant Factors for Foundations Selection

Recognizing the array of foundation alternatives available, it is crucial to take into account several influential factors before selecting the ideal foundation. These factors include:

1) **Superstructure Geometry in Plan:** The design and size of the superstructure are essential considerations in the foundation selection procedure. The nature and magnitude of loads vary in this regard and must be realized in preliminary and final decisions.
2) **Superstructure Height:** Apart from the size of the structure in the plan, the height is also a critical parameter. Requirements for a tower and a building with a few stories are completely different.

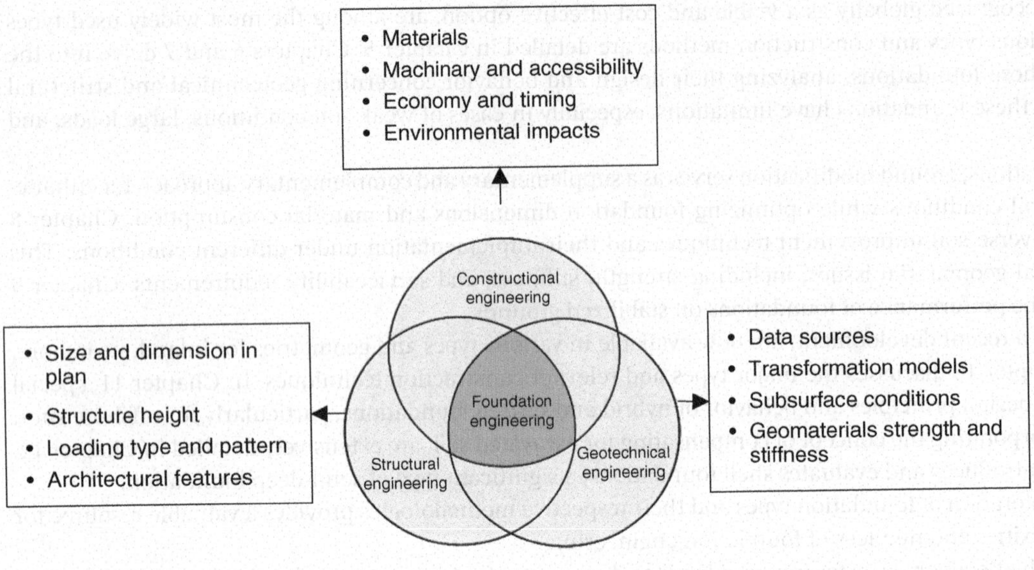

Figure 27.3 Foundation engineering: interactive practice of structural, constructional, and geotechnical engineering.

3) **Loading Pattern and Combination:** Imposed loads on a particular building may vary concerning type and magnitude as well as pattern and combination. Combined loading is highly influential in the design and analysis of foundations. In such situations, more rigorous considerations for foundation selection are mandatory.
4) **Building Importance Level:** The importance of the building, i.e., the purpose of use, for example, residential, commercial, and hospital must be recognized and realized in foundation selection. Requirements and considerations for more important structures are more rigorous than ordinary ones.
5) **Architectural Features (Facet):** Architectural features are another important factor that can dictate the foundation type. Unique architectural designs, such as basements or large open-concept designs, must be considered to designate a suitable foundation alternative to support and satisfy specific designs.
6) **Subsurface Condition:** The existing soil type and condition in the construction highly affect the foundation performance and consequently dictate its type and dimensions. Soils have a variety of important technical properties, including bearing capacity, drainage potential, and deformation characteristics. Clayey soils are highly prone to shrink and swell with alteration in moisture content. On the other hand, sands in case of low relative density and high groundwater table may liquefy under earthquake loading.
7) **Building Codes and Regulations:** Building codes and regulations are essential to certify that the project does not exceed safety standards. The minimum requirements for construction comprising substructure is set by local authorities or municipalities. It is highly important to prioritize compliance in case of foundation selection since failure to comply with standards may result in expensive postponement or legal issues.
8) **Materials:** As presented in Chapter 5, a variety of materials comprising masonry, wood, rock, steel as well as plain and reinforced concrete can be used for foundation construction. Selecting the most appropriate material realizing the technical aspects along with local availability and cost are among the most important steps in appropriate foundation design and selection.
9) **Construction Expertise (Operators):** A variety of levels and types of skills and expertise are needed for different foundation construction. The proper performance of some substructure systems is highly dependent on the correct installation procedure. Therefore, specialized and knowledge-based human sources are mandatory. It is essential to consider the availability of expert local contractors prior to selecting and designing proposed foundations.
10) **Machinery:** Both machinery and human sources have a significant role in the correct construction of a foundation. Prior to selecting the foundation alternative the availability of the required machinery must be controlled.
11) **Groundwater Table Situation:** The level of subsurface water is highly influential on the performance of foundations considering bearing capacity, settlement, and general stability issues. The influence zone for each geotechnical criterion must be controlled in this regard. Moreover, a high groundwater table dictates special conditions during installation and makes the inspection and maintenance process more challenging. Overall, the temporary and constant water level must be recognized and considered carefully in the design and selection processes.

12) **Geohazards:** Major geohazards, including flood, earthquake, scouring, and liquefaction are highly influential in the design and selection of a foundation system. These issues must be considered in the preliminary stages of design to prevent detrimental consequences.
13) **Accessibility and Installation Ease:** Understanding installation requirements and challenges along with different foundation feasibility is of high importance. Foundations with complex implementation processes are unsuitable for locations with limited access, which must be considered in preliminary analysis. For instance, driven piles, are not a proper option for populated zones due to excessive noise and vibrations. Drilled shafts or helical piles are more appropriate foundation alternatives. Overhead and vertical clearances are other issues that must be recognized.
14) **Budget:** Generally, construction practice is in conjunction with meticulous economic planning. In this regard, realizing the foundation system is one of the most critical structural elements, a significant section of the budget must be devoted. The required budget for each foundation alternative is different, therefore it must be determined how much can be expended. The role of the foundation in providing safety and stability of the structure is undeniable; thus, balancing the budget and technical requirements is crucial in the foundation selection procedure.
15) **Cost-effectiveness:** Apart from considering technical and durability issues, the foundation option must be controlled to fit with the available budget. The foundation costs include not only the initial cost of implementation but also preservation and repair costs after construction during the life span of the superstructure. Balancing the foundation's total cost and performance-based criteria leads to an economically wise option.
16) **Construction Timeline:** The construction timeline is another important factor that must be realized in foundation selection. Opting for a foundation alternative compatible with tight deadlines is essential since the required time for installation varies among different types. In this regard, precast or cast-in-place deep foundations can be named and compared.
17) **Climate and Weather Conditions:** The climate and weather conditions of the proposed area for construction are influential on foundation selection. For instance, in zones with a high potential for earthquakes, choosing and designing foundations resistant to seismic loads is required. Another example is for areas that have a high water content or frequent floods, in which foundations must be designed in a way that drains the water properly along with water-resistant features.
18) **Environmental Impact**: The focus on sustainability and environmental concerns has been increasing continuously in recent decades. For instance, foundations built from concrete have a significant carbon footprint resulting from cement production procedures. On the other side, foundations constructed by consuming recycled or reclaimed materials are more environmentally friendly options.
19) **Existing Utilities**: Existing facilities and utilities in the project site can affect the foundation type and design. Since they may be sensitive to excavation, vibration or noise, rigorous considerations must be satisfied. Moreover, space for construction may be limited due to existing utilities resulting in particular foundation alternatives.
20) **Adjacent Structures and Construction:** Similar to existing facilities and utilities in the project site, adjacent structures are influential on the proposed foundation design and selection. Some foundation options are not available based on neighboring conditions. Furthermore, future adjacent construction affects the behavior of existing foundations in regard to nonuniform settlement and tilting. In this regard, foundations must be selected and designed in a way that the least failure occurs due to adjacent structures.
21) **Maintenance Requirements:** The maintenance requirements must be considered prior to selecting the foundation type. Some foundation types need regular inspection and preservation for efficient performance and durability. Resealing of raft foundation on the ground can be named in this regard.
22) **Energy Efficiency:** Foundations play a significant role in the overall energy consumption of the project. More insulated foundations are typically more energy-efficient. Realizing the increasing attention to energy efficiency worldwide, more energy-efficient foundation options can result in less carbon footprint and consequently optimization in heating and cooling costs.
23) **Durability:** Durability is recognized as a key feature for foundations. Issues such as resistance against settlement and cracking must be considered in the case of selecting appropriate foundation alternatives. Foundations that have a proven history of long-term durability and efficient behavior in each area are the most preferred and recommended ones. Solution and corrosion are two major detrimental phenomena that must be realized to be prevented through compatible selection and design.
24) **Project Function:** The personal preferences of the project owner(s) can be influential in the foundation selection process. Some issues like personal lifestyle, design preferences, and overall vision for the building are among the factors

that may need to be covered. Future building plans can affect the foundation type. For example, if the basement is proposed to be used as a storage of car parking, box foundations are a suitable foundation option; otherwise, a typical raft foundation on the ground can be implemented as a cost-effective option in case minimum technical requirements are satisfied. Rotatory floors and locating pools in the middle and top stories are examples of relevant issues.

27.4 Project-Based Foundation Selection

The general type of a project plays a significant role in determining the most suitable and efficient foundation alternative, as common choices are often derived from well-documented past projects. This section outlines some major project types and their corresponding foundation options.

27.4.1 Storage Tanks

There is no one-size-fits-all solution when it comes to selecting the ideal tank foundation. Each situation demands a unique evaluation based on its specific requirements. Chapter 2, Case Study 1.6, illustrates an actual case study involving a storage tank supported on stabilized soil. Case Study 3.2, also delves into the site investigation process for an onshore tank farm. Generally, foundations for storage tanks can be categorized into the following types.

- **Earth Foundations:** Earth foundations are recognized as the most cost-effective foundation alternative for storage tanks. They also have the benefit of straightforward construction practice. Figure 27.4a depicts an example of earth foundations.
- **Earth Foundation with Concrete or Granular Ringwall:** A widely applied foundation alternative for tanks is an earth foundation surrounded by a concrete ring wall. The concrete ring foundation is implemented beneath the shell to support its heavy loads (Figure 27.4b). In some cases, crushed stone or gravel can be applied instead of the concrete.
- **Reinforced Concrete Slab Foundation:** A reinforced concrete slab foundation is typically applied for small tanks rather than large ones. They are used in case of the need to distribute the loads over an area larger than the tank area or a rigid uniform foundation (Figure 27.4c).
- **Piled Foundations:** In particular cases where tanks or facilities are so sensitive to deformation or in case of problematic deposits existing in surface layers, piled foundations are implemented. Figure 27.4d depicts a usual view of piled as support of a tank. One related record has been presented in Case Study 18.1.

27.4.2 Buildings

Buildings are the most commonly employed structures across various applications, ranging from residential to commercial spaces such as offices and hotels. Different standards exist to classify buildings based on their height. Ordinary buildings

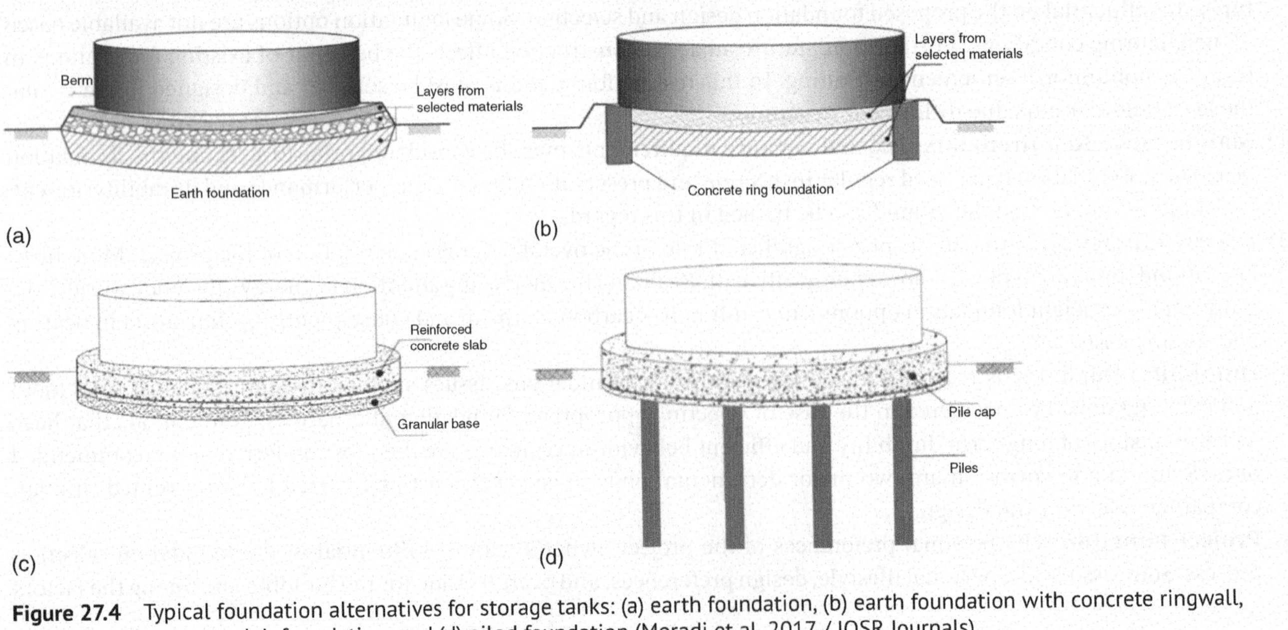

Figure 27.4 Typical foundation alternatives for storage tanks: (a) earth foundation, (b) earth foundation with concrete ringwall, (c) reinforced concrete slab foundation, and (d) piled foundation (Moradi et al. 2017 / IOSR Journals).

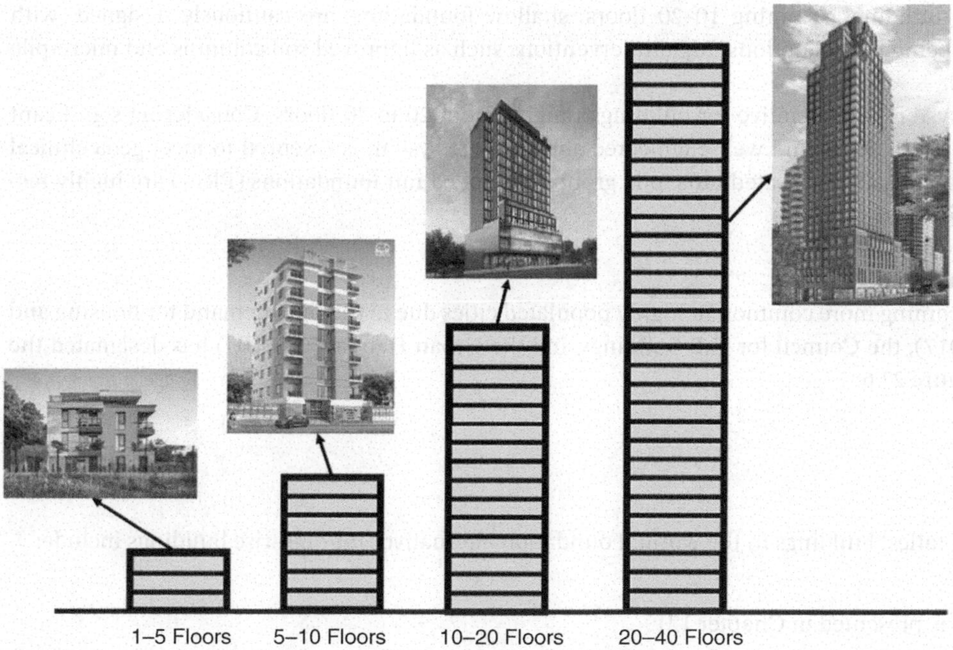

Figure 27.5 Schematic view of buildings classes based on height.

with just a few stories are typically supported on standard spread footings. However, as the number of stories increases, the demand for more intricate and comprehensive substructure systems becomes apparent.

27.4.2.1 Ordinary Buildings

He et al. (2019) have classified buildings into six types: low-rise, multistory, middle-rise, small high-rise, high-rise, and ultra-high-rise based on the Code for Design of Civil Buildings of China (2005), in which small high-rise and high-rise have been initiated from "High-rise buildings" of 30–100 m (Table 27.1).

Following the introduced classification by He et al. (2019), four classes have been considered for further discussion on their appropriate foundation alternatives. These classes are composed of low-rise (1–5 floors), middle-rise (5–10 floors), small high-rise (10–20 floors) and high-rise (20–40 floors). Buildings that have more than 40 floors are discussed separately as tall buildings. Figure 27.5 depicts a schematic view of four classes of foundations regarding height apart from tall buildings (more than 40 floors).

Low-rise buildings, ranging from 1 to 5 floors, are commonly supported by a single foundation, occasionally supplemented with ties if necessary. Strip foundations in both directions and slab foundations are also frequently utilized for these structures.

For middle-rise buildings, approximately 5–10 to 10 stories high, box and floating foundations find extensive application. Ground improvement techniques are often incorporated as additional measures to enhance substructure systems. In challenging soil conditions, deep foundations are employed to ensure the necessary technical performance.

Table 27.1 Buildings categorization based on the height (He et al. 2019 / MDPI / Public Domain CC BY 4.0).

Classification	Floor	Height (m)
Low-rise building	1–3	≤9
Multistory building	4–6	9–21
Middle-rise building	7–9	21–30
Small high-rise building	10–16	30–50
High-rise building	17–40	50–100
Ultra-high-rise building	>40	>100

In the case of small high-rise buildings spanning 10–20 floors, shallow foundations are cautiously designed, with semi-deep and deep foundations being typical options. Rigid interventions such as improved soil columns and micro-pile elements are also viable solutions.

A comprehensive foundation system is imperative for buildings ranging from 20 to 40 floors. Considering significant overturning loads and moments, an intricate and well-engineered substructure system is essential to meet geotechnical requirements and ensure serviceability. Compensated rafts, pile groups, and piled raft foundations (PRFs) are highly recommended and commonly used in such scenarios.

27.4.2.2 Tall or High-Rise Buildings

Tall or high-rise buildings are becoming more common in highly populated cities due to the high demand for housing and office space. Based on Poulos (2017), the Council for Tall Buildings and the Urban Habitat (CTBUH) has designated the following criteria as shown in Figure 27.6:

- **200–300 m:** tall buildings
- **300–600 m:** super-tall buildings
- **>600 m:** mega-tall buildings

Figure 27.7 depicts some of the tallest buildings in the world. Foundation alternatives for high-rise buildings include:

- Shallow raft foundation
- Compensated raft foundation (as presented in Chapter 12)
- Pile or barrette foundation
- Piled raft foundation (as Case Study 19.1 in Chapter 19)
- Compensated piled raft foundation (Poulos 2017)

Case Studies 2.2, 2.4, and 2.5 in Chapter 2 are devoted to the assessment of high-rise building foundations. The interactive performance of PRF and basement walls as the substructure of the Shanghai megatall building are presented in Case Study 7.2.

Mandolini (2003) provides a useful chart to assist in the selection of an appropriate foundation type, considering the conventional range of foundation options. This chart is shown in Figure 27.8, and contains four regions:

- Region I, for which a raft foundation may be relevant. In this region, the computed factor of safety against failure may be large, and the estimated maximum settlement may be less than the specified value (8 cm in this case).
- Region II, in which a "small" piled raft may be appropriate. In such cases, the settlement criteria may be satisfied but the factor of safety is too low. Thus, the design of a foundation in this category is governed by load capacity considerations.

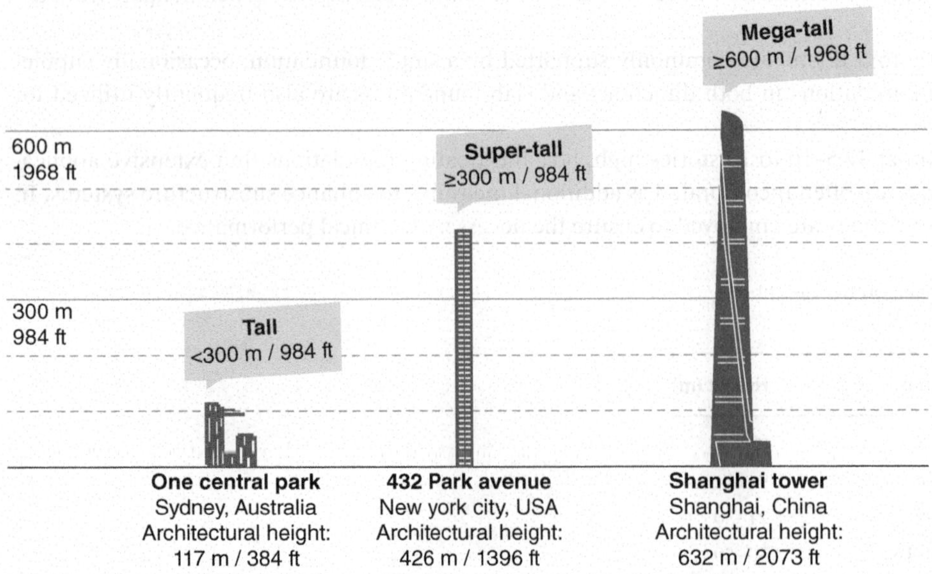

Figure 27.6 Classification of tall buildings upon height (CTBUH).

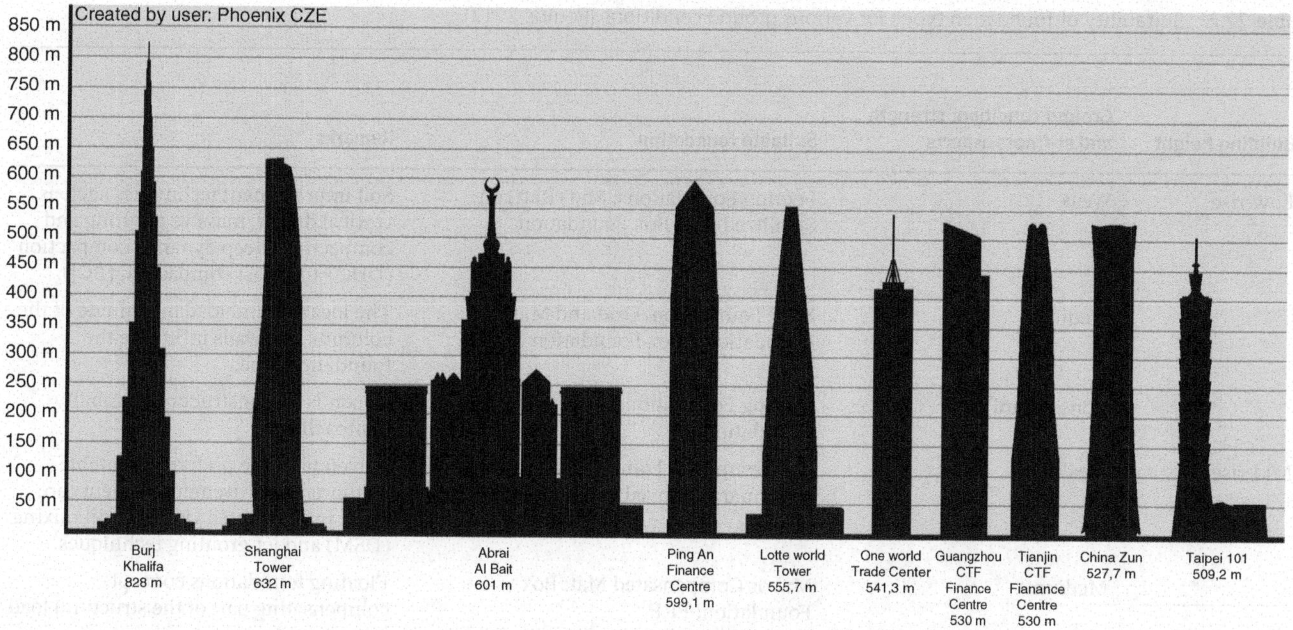

Figure 27.7 Schematic view of some of the tallest buildings in the world (Phoenix CZE / Wikipedia Commons / https://en.wikipedia.org/wiki/List_of_tallest_buildings#/media/File:Tallest_buildings_2022.png / last accessed November 17, 2023 / CC BY 4.0).

Figure 27.8 Example of chart for selection of foundation type (Mandolini 2003 / Millpress Science Publishers). Targets: minimum factor of safety (FS) = 3, maximum settlement = 8 cm.

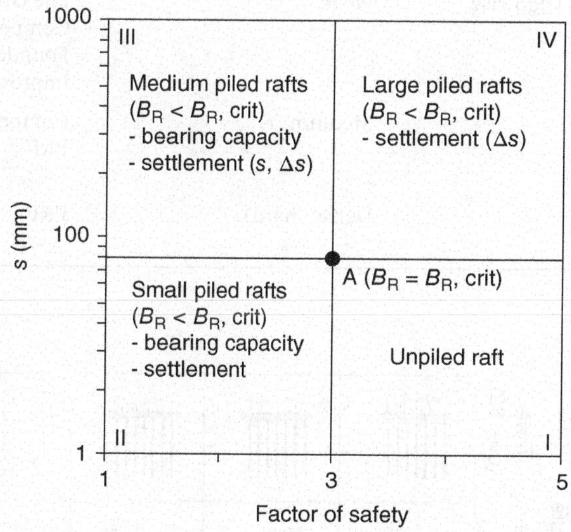

- Region III, in which both the settlement and the capacity criteria require attention.
- Region IV, where the foundation design is controlled by settlement, rather than load capacity, considerations. It is Region IV, in which many high-rise building foundations lie and for which a piled raft foundation may be a suitable option.

It is Region IV, in which many high-rise building foundations lie and for which a piled raft foundation may be a suitable option.

Foundation systems for high-rise structures have been evaluated and presented byKatzenbach et al. (2017). A general trend to which foundation systems and types may be suitable to adopt for buildings based on their height on various ground profiles is given in Table 27.2. It should be emphasized that this table is a general algorithm only and that a proper decision on the type of foundation requires careful consideration of the factors set out in the current chapter, such as construction techniques and equipment, urban and rural location, accessibility and environmental requirements.

Eslami et al. (2025) have introduced a method to recommend suitable foundation systems for buildings under typical conditions. They have categorized existing soil strength levels from very soft to hard and buildings into different classes, from first class (1–20 floors) to the highest class (more than 80 floors). An overview of their approach is outlined in Figure 27.9.

Table 27.2 Suitability of foundation types for various ground conditions (Poulos 2017).

Building height	Ground condition; strength and stiffness aspects	Suitable foundation	Remarks
Low-rise	Weak	Footing Foundations; Mat (Raft) on Stabilized Soil; Box Foundation	Soil improvement techniques such as vertical drains, massive grouting and compaction (deep dynamic compaction (DDC) and blast compaction [BC]).
	Medium	Shell Foundation; Grid and Mat Foundations; Box Foundation	The location and load magnitude of the columns and walls influence the foundation type.
	Dense (hard)	Footing Foundations; Grid and Mat Foundations	Depends on construction feasibility approaches.
Mid-rise	Weak	Pile group; Box Foundations; raft on columnar improved soils	Driven piles are preferred. Suitable columnar improvement elements are those implemented via deep soil mixing (DSM) and jet-grouting techniques.
	Medium	Mat or Compensated Mat; Box Foundation; PRF	Floating foundations concept, compensating part of the structural load is more preferred.
	Dense (hard)	Mat or Compensated Mat; PRF	Drilled shafts are commonly used as deep foundations.
High-rise	Weak	Pile Group; Hybrid Foundation 1: Compensated Raft + Pile Group; Hybrid Foundation 2: Pile Group + Soil improvement	Driven piles are commonly used as deep foundations.
	Medium	Compensated Mat; Shell Foundation; PRF	The feasibility of shell foundations in terms of construction aspects must be considered.
	Dense (hard)	PRF	Drilled shafts are commonly used as deep foundations.

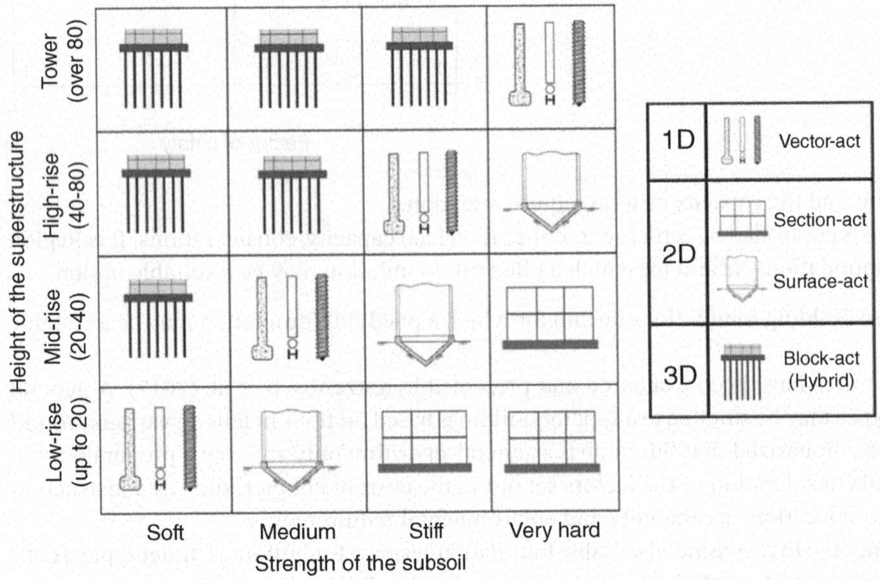

Figure 27.9 Conceptual framework for determination of recommended foundation in different cases regarding (Eslami et al. 2025).

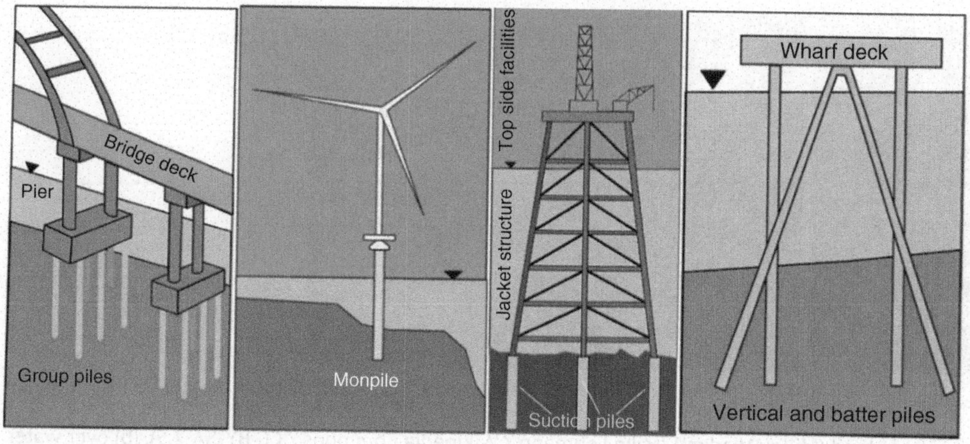

Figure 27.10 Typical applications of piles for structures in marine environments (Eslami and Heidarie 2020 / with permission of Taylor & Francis Group).

27.4.3 Coastal, Onshore, and Offshore Structures

In recent decades, global population growth has shifted the focus of developers toward nearshore and offshore sites in recent decades. These developments are driven by the urgent need for new food and energy sources, as well as spaces for residential and recreational purposes. In offshore, onshore, and nearshore locations, deep foundations, especially driven piles, have emerged as the most preferred foundation solutions. Their widespread adoption can be attributed to their efficient geotechnical performance, construction feasibility, and optimal material usage. Figure 27.10 illustrates various typical applications of deep foundations in marine projects.

27.4.3.1 Bridges

Bridge foundations are exceptionally critical, as they bear the entire load of the bridge along with the traffic loads it carries. These foundations are constructed under piers or abutments and rest upon the underlying soil or rock. The loads transferred by the foundations to the underlying soil must not cause soil shear failure or compromise the superstructure's settlement. This section discusses bridge foundations, the construction process, and various types of bridge foundations.

The foundation constitutes a vital structural element of bridges, serving as the primary component of the bridge structure. It receives loads from the piers or abutments and transfers them to the soil. The purpose of the foundation is to distribute the bridge load over a large bearing area and ensure the bridge's stability against settlement and lateral movement. The foundation is designed to be sufficiently deep to prevent the effects of scouring (Ahmed 2021). Chapter 18 includes Case Study 18.2, which is dedicated to analyzing the performance of a bridge supported by pile groups in problematic soil conditions.

Two major categories of bridges are:

- In Land (Figure 27.11a)
- Over Water (Figure 27.11b)

Different types of bridge foundations are employed based on the nature and depth of the soil profile:

- **Raft Foundation:** This type is ideal for bridges situated on soft clay and silt, where hard soil is not available within a reasonable depth (1.5–2.5 m). Raft foundations are best suited for areas with low allowable bearing capacity, heavy bridge loads, and the possibility of unequal settlement.
- **Pile Foundation:** Pile foundations are suitable for areas with very soft soil and a lack of hard strata at a reasonable depth below the riverbed. They are well suited for locations where heavy scouring of the riverbed is expected, and the foundation needs to bear concentrated loads. Pile foundations are also chosen when providing a raft or grillage foundation would be significantly more expensive.
- **Well Foundation:** Well foundations are appropriate when good soil is available approximately 3–4 m below the riverbed level, particularly in sandy soil beds. They are suitable for soft soils or sandy beds, where the river water's high velocity could lead to heavy scouring.

(a) (b)

Figure 27.11 View of typical bridges: (a) in land (Millau Viaduct) (Mike Lehmann / Wikipedia Commons / CC BY-SA 2.5), (b) over water (Obinna 2023 / Structville Integrated Services Limited).

- **Caisson Foundation:** Caisson foundations are employed when a hard stratum is present near the riverbed, but the water depth is excessive. This option is chosen when it is not economically feasible to exclude water from a dry bed for sinking wells to create a suitable foundation (Team 2023).

27.4.3.2 In Land Bridges

For inland bridges featuring short to medium spans (up to 50 m) and resting on favorable subsurface conditions with standard pier heights, shallow or semi-deep foundations stand out as the preferred choice. However, when confronted with significant loads and challenging soil conditions, such as those found in viaduct bridges, the adoption of deep foundations becomes imperative.

Extensive research on bridge elements' performance has been conducted by Prof. Rahai and colleagues, as documented in Rahai (2004), Hedjazi et al. (2007), and Saberi et al. (2016). A critical consideration in bridge foundation design is the potential for rocking isolation. Sharifzadeh Asli et al. (2018) have delved into the impact of soil reinforcement on the rocking isolation potential of high-rise bridge foundations. Figure 27.12 visually portrays the disparities between the behaviors of traditional designs and those based on rocking isolation principles concerning both structural and foundation elements.

27.4.3.3 Over Water Bridges

In addition to the factors discussed in Section 27.3, several critical considerations must be taken into account for selecting an appropriate foundation for an overwater bridge:

- **Conditions of Exposure and Water Depths:** Bridges spanning vast estuaries or bays are exposed to harsh environments due to winds and waves. To combat this, prefabricated massive components that can be quickly sunk onto a prepared bed or piled platform are preferred. This method enhances construction efficiency while minimizing damage risks to partially completed structures, which might otherwise be susceptible to adverse weather conditions.

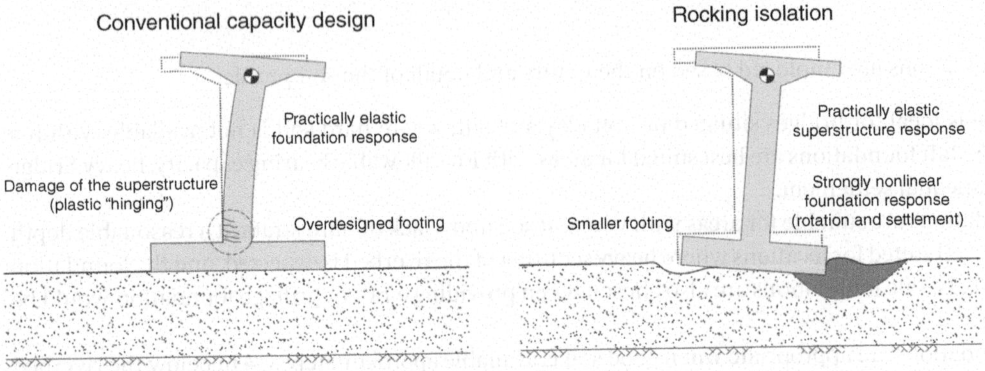

Figure 27.12 Differences between structure and foundation behaviors in the traditional design as opposed to the design based on rocking isolation (Anastasopoulos et al. 2012 / with permission of Elsevier).

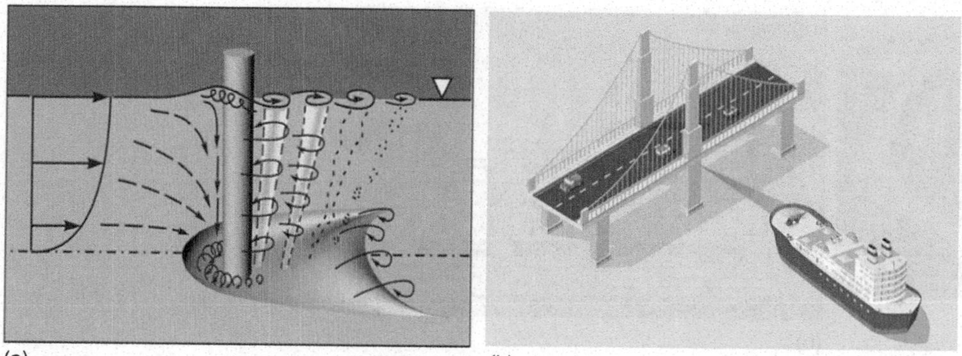

(a) (b)

Figure 27.13 (a) Scouring in bridge piers, (b) schematic of ship collision (Obinna 2023 / Structville Integrated Services Limited).

- **Water Currents and Tides:** River currents and tidal streams can lead to scour holes and erosion at the bed level, exerting drag forces on piers or piles. Scouring is particularly crucial in areas near cofferdams, where eddies caused by factors like partially driven sheet piles might intensify erosion risks. Implementing measures such as mattressing can be necessary under restricted flow conditions to prevent erosion during the caisson's descent in the final few meters of water. Figure 27.13a illustrates the scouring process around bridge piers.
- **Ship Collision:** Preventing ship collisions with bridge piers is vital, not only within designated navigation channels but also in areas where ships may drift outside the channel. Rapid changes in tides or shifts in deep-water channels in large estuaries can put multiple bridge piers at risk. Implementing protective measures to reduce collision risks can significantly impact foundation costs. Figure 27.13b provides a schematic view of a ship collision scenario.
- **Earthquakes:** Earthquakes present significant challenges for deep-water piles due to the combination of forces acting on the bridge superstructure and the pier body, leading to strong overturning moments at the base level. Circular structures may be necessary to counteract forces acting in various directions, including vertically. Ground shaking can result in the liquefaction of loose to medium-dense granular soils, making piled foundations or ground treatment essential to densify loose soil deposits and provide stability for the pier (Obinna 2023).

27.4.3.4 Coastal Structures

Coastal structures hold immense global importance, serving the marine, riverine, transportation, and tourism industries. Constructing coastal protection structures is crucial to shielding harbors and other installations from erosive sea waves, as depicted in Figure 27.14a. Seawalls, vertical barriers parallel to shorelines (Figure 27.14b), serve as protective structures. As illustrated in Figure 27.14c, Berthing facilities offer essential support to ships and facilitate seamless transitions between ships and land transportation. Dolphins, distinctive marine structures extending above water level and detached from the shore, are created by driving piles into the seabed or riverbed. They provide a platform or fixing point, as shown in Figure 27.14d. Deep foundations, particularly driven piles, are extensively utilized in constructing these vital coastal structures.

27.4.3.5 Offshore Structures

Offshore environments, especially those with soft to medium deposits, intricate combinations of loads, and inherent spatial variability in engineering parameters, pose significant challenges. These conditions, coupled with stringent environmental regulations and construction complexities, create exceptionally demanding situations. Addressing these challenges necessitates the use of appropriate foundation types, emphasizing efficiency and sufficiency. In Case Study 6.3 provides a comparative overview of offshore loads and foundations in contrast to inland settings. Chapter 10 introduces and reviews major foundations applied in offshore sites. For visual clarity, Figure 27.15 showcases several offshore structural systems and their corresponding support mechanisms.

Spudcans, featured in Chapter 10, are specially crafted foundations for independent-leg jack-up platforms. These structures, resembling large inverted cones, are characterized by a circular plan and a shallow conical underside with a prominent spigot. In sizable jack-ups, spudcans often exceed 20 m in diameter, displaying varied shapes across different manufacturers and rigs.

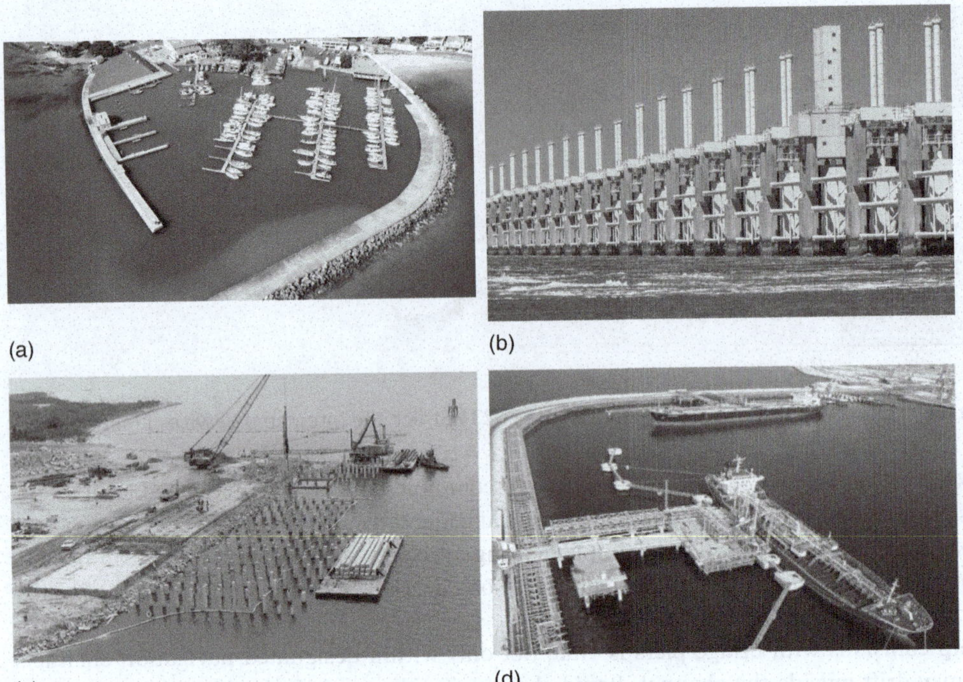

Figure 27.14 Typical coastal structures: (a) breakwater structure (Brandon / Adobe Stock Photos), (b) seawall (allard1 / Adobe Stock Photos), (c) open berthing facility (JillFine / Wikimedia Commons), and (d) dolphin structure.

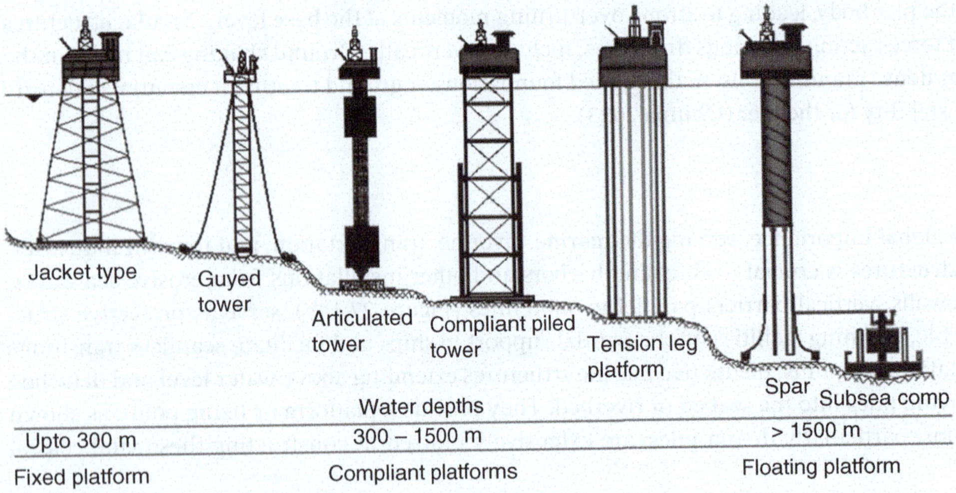

Figure 27.15 Typical offshore structures and their support system (Hasan et al. 2010).

27.4.4 Wind Turbines and Transmission Utilities

Transmission and cell towers, highlighted in Case Study 10.1, serve as tall structures made primarily of steel, designed to support overhead power lines. Despite their lightweight nature concerning gravity loads, their substantial height creates significant overturning forces. Consequently, employing deep foundations becomes crucial to ensure stability. Deep foundations are the standard choice, but in cases of challenging accessibility and transportation, semi-deep foundations like belled well foundations offer a viable alternative. These foundations can effectively bear and transfer both compressive and pullout loads. Figure 27.16 provides visual examples, showcasing various cell towers and a typical transmission tower supported on belled well foundations.

Installing offshore wind turbines (OWTs) presents unique challenges due to their specific geometry and the diverse loads they encounter. Managing combined loads, including vertical and horizontal forces along with overturning moments, is a primary focus for offshore structures. Figure 27.17 provides a visual representation of the sources and equivalent loads acting on an offshore wind turbine.

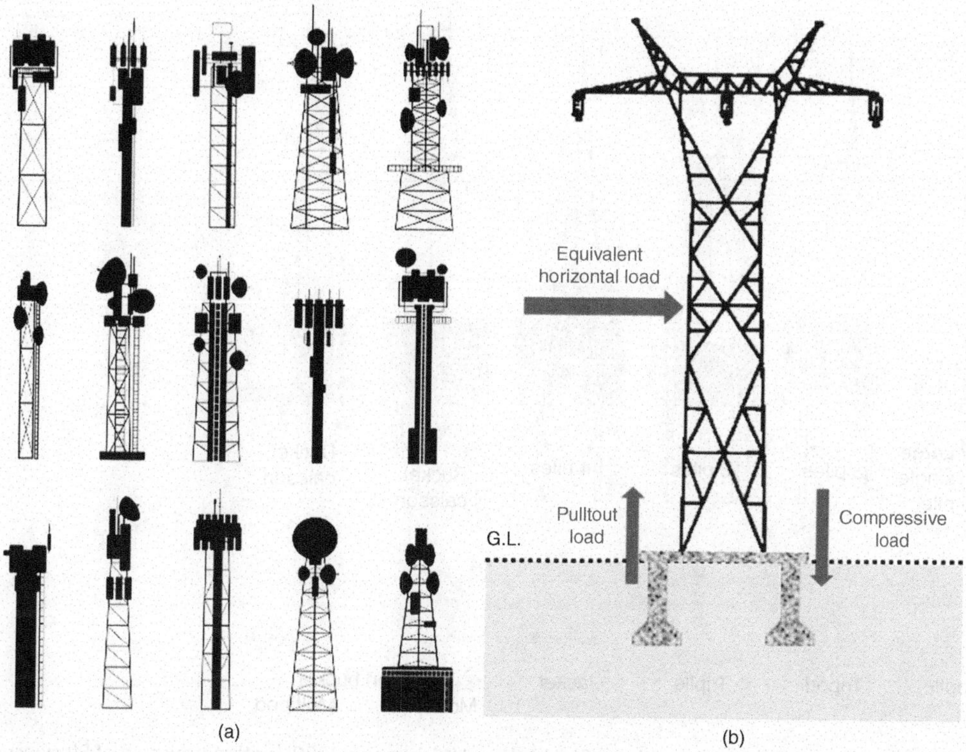

Figure 27.16 (a) Various cell towers (Freepik 2021 / Freepik Company S.L.), (b) typical transmission tower supported on belled well foundations.

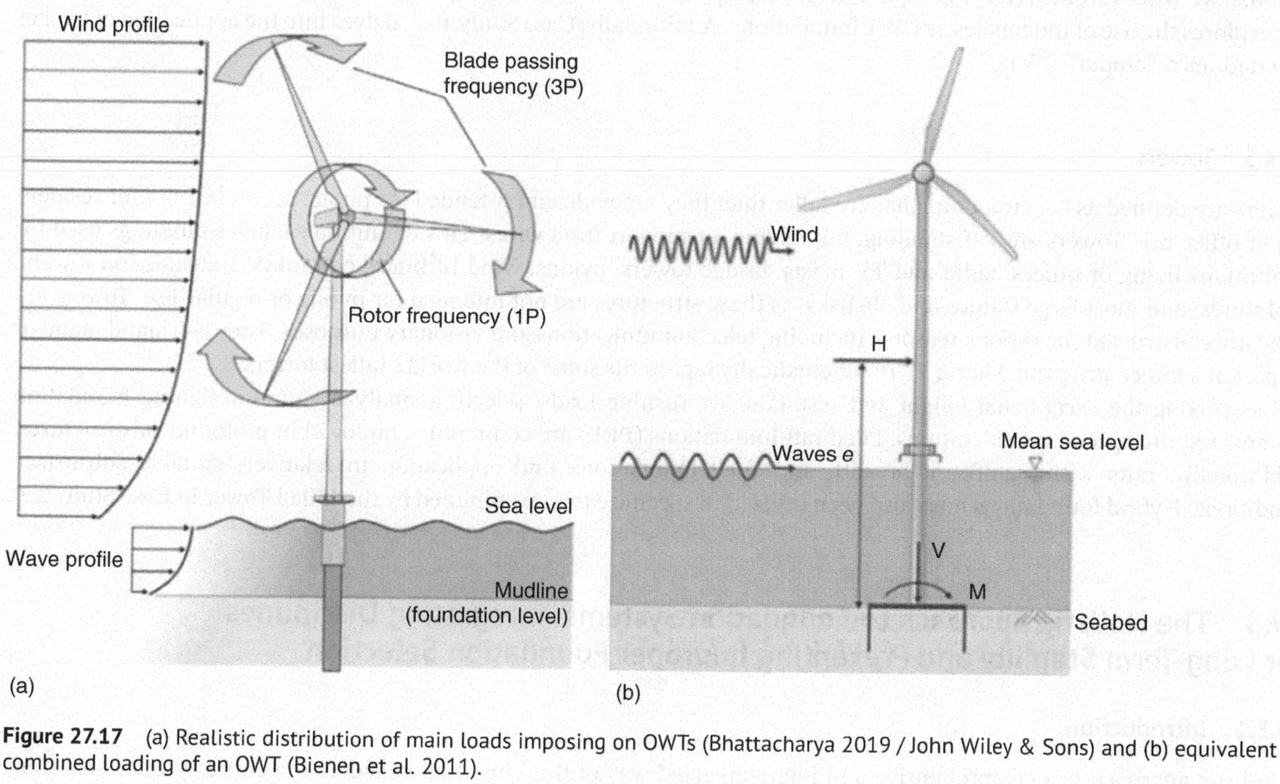

Figure 27.17 (a) Realistic distribution of main loads imposing on OWTs (Bhattacharya 2019 / John Wiley & Sons) and (b) equivalent combined loading of an OWT (Bienen et al. 2011).

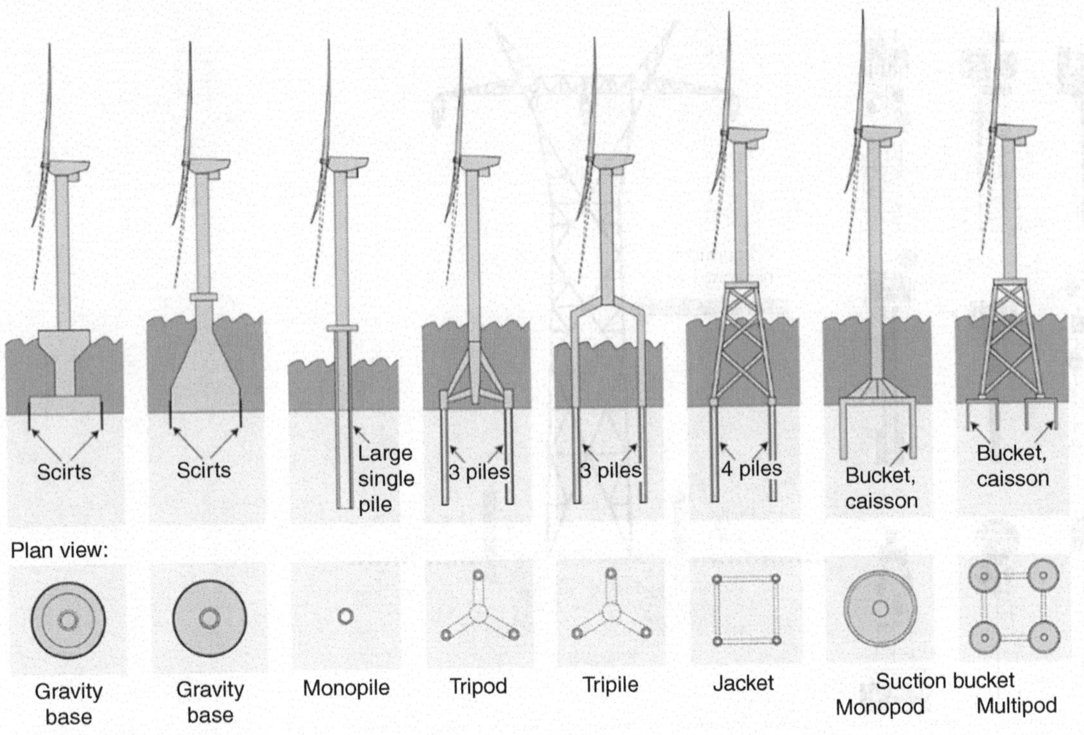

Figure 27.18 Typical foundation alternatives for offshore wind turbines (Staubach and Wichtmann 2020 / with permission of Elsevier).

Considering deposit characteristics, load requirements, and factors from Section 27.3, a range of foundations are utilized for offshore wind turbines (OWTs). Figure 27.18 displays several common foundation alternatives for OWTs. Case Study 2.1, explores the use of monopoles as OWT foundations. Additionally, Case Study 10.3, delves into the application of bucket foundations to support OWTs.

27.4.5 Towers

Towers are defined as constructions that are taller than they are wide and intended for public access but not for residential or office use. Towers are self-standing, not relying on support from wires. This definition excludes buildings used for continuous living or offices, radio and TV masts, bridge towers, pylons, wind turbines, chimneys, transmission towers, sculptures, and most large statues and obelisks, as these structures are not intended for public or regular use. Towers are typically constructed for various reasons, including telecommunications and visionary purposes. They can stand alone or be part of a larger structure. Figure 27.19 schematically represents some of the world's tallest towers.

Recognizing the exceptional height and resultant overturning loads, selecting, analyzing, and designing foundation systems require special considerations. Piled raft foundations (PRF) are commonly employed in profound substructures. Additionally, rafts with significant breadth and shell foundations find application in relatively suitable subsurface conditions. Hybrid foundations have also been utilized in specific cases, exemplified by the Milad Tower in Case Study 2.3.

27.5 The Holistic Approach to Foundation Systems: Integrating Disciplines for Long-Term Stability and Preventing Improper Foundation Selection

27.5.1 Introduction

A holistic approach is a comprehensive and interconnected way of thinking that considers the entire system, including all its components, elements, and interrelationships, rather than focusing solely on individual parts. It emphasizes the understanding that every aspect of a system is intricately linked and influences the functioning of the whole. Taking a

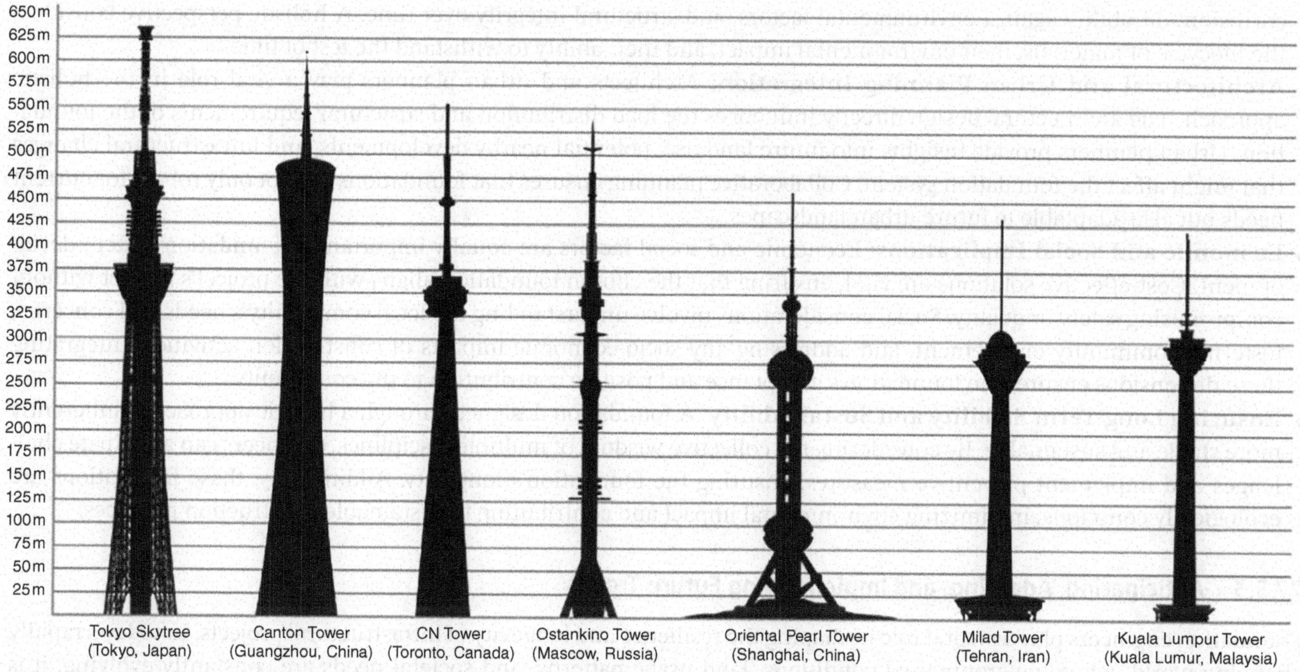

Figure 27.19 Schematic view of some of the tallest towers in the world (Timsdad / Wikipedia Commons / https://commons.wikimedia.org/wiki/File:Tallest_towers_in_the_world.svg / CC BY 4.0).

holistic approach, one examines the complexities and interdependencies within a system, recognizing that changes in one area can have ripple effects throughout the entire system. This approach encourages a broader perspective, integrating diverse elements and disciplines, to gain a comprehensive understanding and make well-informed decisions that account for the intricate web of connections within the system. Whether applied in healthcare, environmental science, engineering, or other fields, a holistic approach fosters a more nuanced understanding of complex issues, leading to more effective and sustainable solutions. As our understanding of construction and environmental dynamics deepens, it becomes increasingly clear that a holistic approach is essential in developing foundation systems that stand the test of time. This section discusses the importance of integrating various disciplines to create robust foundation systems that ensure long-term stability, safety, and sustainability.

27.5.2 Understanding the Holistic Approach for Foundations

A holistic approach to foundation systems transcends traditional boundaries, involving collaboration between architects, civil engineers, geologists, environmental scientists, economists, urban planners, and sociologists. By synthesizing knowledge from these diverse fields, a comprehensive understanding of the site's geological, environmental, and social context emerges. This interdisciplinary synergy forms the cornerstone of a robust foundation design.

- **Multidisciplinary Collaboration:** The holistic approach to foundation system design involves collaboration among various disciplines, such as structural engineering, geotechnical engineering, architecture, environmental science, and even urban planning. Engineers, architects, and geologists work together to understand the geological conditions, structural requirements, and environmental impact of the proposed structure. This collaborative effort ensures that the foundation system is tailored to the specific needs and challenges of the site.
- **Considering Geological and Environmental Factors:** Geological and environmental considerations are pivotal in foundation design. Geologists assess soil composition, stability, and seismic activity, providing crucial insights into the ground's behavior under stress. Environmental scientists study factors like groundwater levels, climate patterns, and potential pollution, all of which impact soil properties. By integrating these insights, engineers can develop foundations tailored to the specific needs of the site, ensuring resilience against natural disasters and environmental changes.
- **Ensuring Proper Materials Selection and Compatibility:** Materials science plays a pivotal role in foundation construction. Engineers must select materials compatible with the geological conditions of the site, ensuring resistance to

corrosion, durability against environmental factors, and structural integrity over time. A holistic perspective considers the lifecycle of materials, their environmental impact, and their ability to withstand the test of time.

- **Architectural and Urban Planning Integration:** Architects and urban planners play a vital role in the holistic approach. The architectural design directly influences the load distribution and structural requirements of the foundation. Urban planners provide insights into future land use, potential nearby developments, and infrastructural changes that might affect the foundation system. Collaborative planning ensures that foundations are not only robust for current needs but also adaptable to future urban landscapes.
- **Economic and Social Implications:** Economic and social factors are equally important in foundation system development. Cost-effective solutions are vital, ensuring that the chosen foundation aligns with the project's budget without compromising safety or quality. Social considerations involve understanding the local community's needs and concerns, fostering community engagement, and addressing any socio-economic impacts of construction activities. Integrating these dimensions ensures the foundation's acceptance and positive contribution to the community.
- **Ensuring Long-Term Stability and Sustainability:** A foundation designed through a holistic approach is inherently more stable and sustainable. By considering the collective wisdom of multiple disciplines, engineers can anticipate challenges and implement preventive measures, ensuring the foundation's longevity. Additionally, these foundations are ecologically conscious, minimizing environmental impact and contributing to sustainable construction practices.

27.5.3 Anticipating, Adopting, and Implementing Future Trends

Foundation engineers play a pivotal role in shaping the resilience and longevity of infrastructural projects. In today's rapidly changing world, where environmental conditions, land usage patterns, and societal needs are constantly evolving, it is imperative for foundation engineers to adopt a forward-thinking approach. The foundation engineer needs to understand the significance of selecting foundation types with the anticipation of potential future changes in the environment, use of the facility, and other dynamic factors.

- **Adaptability to Environmental Changes:** Environmental changes, including climate change, rising sea levels, and increased instances of extreme weather events, pose significant challenges to infrastructural stability. Foundation engineers must anticipate these changes and select foundation types that can withstand future environmental stresses. Foundations designed to adapt to varying soil conditions, water levels, and seismic activity ensure the longevity of structures amidst changing environmental dynamics.
- **Flexibility for Evolving Land Use:** As urban landscapes continue to evolve, the usage of land can transform dramatically over time. Foundation engineers must consider potential changes in land-use, such as converting industrial areas into residential spaces or vice versa. Foundation systems that offer flexibility in supporting different types of structures allow for seamless transitions, accommodating the evolving needs of communities and economies.
- **Technological Advancements and Innovation:** The rapid advancement of technology impacts how buildings are constructed and utilized. Foundation engineers must stay abreast of innovative construction techniques, materials, and foundation systems. Embracing cutting-edge technologies such as smart sensors, self-healing materials, and advanced geotechnical analysis tools enables the development of foundations capable of adapting to future technological innovations.
- **Sustainable Practices for Long-Term Viability:** Sustainability is the core of future-oriented foundation engineering. Engineers are increasingly focusing on sustainable foundation solutions, incorporating eco-friendly materials, energy-efficient construction methods, and renewable energy sources. Sustainable foundations not only reduce the environmental impact but also enhance the long-term viability of infrastructural projects, aligning with the global push for green and resilient infrastructure.
- **Community Engagement and Future Needs:** Foundation engineers should engage with local communities to understand their needs and aspirations. By incorporating community input into foundation design, engineers can create foundations that align with future societal requirements. For instance, foundations supporting community centers can be designed to accommodate additional floors or spaces as the community grows, ensuring the facility's relevance and usability in the years to come.

27.5.4 The Perils of Ignoring the Holistic Approach in Foundation System Design and Construction

Ignoring the holistic nature of this process, which integrates geological, environmental, structural, and social considerations, can have dire consequences. The following are potential repercussions of overlooking the comprehensive approach to designing and constructing foundation systems of structures.

- **Structural Instability and Failures:** Neglecting the holistic approach can lead to foundation systems that are ill-suited for the site's geological conditions. Without a thorough understanding of the soil composition and bearing capacity, foundations might lack the necessary stability. This oversight can result in structural failures, compromising the safety of inhabitants and leading to catastrophic consequences.
- **Environmental Degradation:** Inadequate consideration of the environmental impact during foundation construction can lead to soil erosion, groundwater contamination, and disruption of local ecosystems. Ignoring sustainable practices can harm the natural environment, diminish biodiversity, and affect the delicate balance of local ecosystems. Contamination of groundwater due to construction chemicals and improper waste disposal can have long-term environmental consequences.
- **Increased Maintenance and Repair Costs:** Foundations that do not consider the holistic approach often require frequent repairs and maintenance due to premature wear and tear. Structural elements strained beyond their capacity due to inadequate foundation design may lead to recurring issues, necessitating costly repairs. Ignoring the long-term implications of foundation systems can burden property owners with substantial maintenance costs over time.
- **Decreased Property Value and Marketability:** Structures with unstable foundations are often perceived as liabilities in the real estate market. Potential buyers and investors are hesitant to invest in properties with a history of foundation problems. Consequently, the market value of such properties decreases significantly, leading to financial losses for property owners. Ignoring the holistic approach jeopardizes the marketability and desirability of structures.
- **Legal and Liability Issues:** Ignoring essential factors in foundation design and construction may lead to legal disputes and liabilities. Property owners and construction professionals can face legal repercussions if structural failures result in injuries, property damage, or financial losses. Legal battles arising from negligence in foundation construction can tarnish reputations and result in severe financial penalties.
- **Community Disruptions:** Structural failures due to poorly designed foundation systems can cause significant disruptions within communities. Evacuations, road closures, and the demolition of unstable structures disrupt the daily lives of residents. These events can lead to public outcry, erode trust in local authorities, and create social unrest within affected communities.

27.5.5 Improper Foundation Selection

Dr. Ralph B. Peck was a highly influential civil engineer known for his significant contributions to geotechnical engineering. Dr. Peck was a Professor of Foundation Engineering at the University of Illinois at Urbana-Champaign. Throughout his career, Dr. Peck emphasized the importance of understanding soil behavior and the interaction between soil and structures. He was a key figure in advancing the field of geotechnical engineering, particularly in the areas of soil mechanics, foundation engineering, and geotechnical site characterization. Dr. Peck believed that failures, collapses, or catastrophes in civil engineering projects rarely occur due to a single isolated cause, but rather stem from a combination of factors that engineers may have overlooked or not fully understood. Dr. Peck always emphasized the importance of considering various interrelated factors in any engineering project. These factors could range from geological conditions and structural design to material properties and construction techniques. By acknowledging this complexity, engineers are reminded of the need for a holistic perspective, ensuring that they carefully assess all aspects of a project to prevent oversights and anticipate potential challenges. Dr. Peck's insight highlights the necessity of a comprehensive and multidisciplinary approach, where engineers collaborate across disciplines to account for the intricate interplay of factors and ensure the safety and stability of structures.

One notable example of improper foundation selection is the Millennium Tower in San Francisco, which has been gradually sinking and tilting since its completion in 2008. The tilting of the Millennium Tower in San Francisco was primarily caused by the excessive settlement of its foundation. The building's foundation was constructed on layers of soft soil, which led to uneven settling. The issue was attributed to the construction of adjacent buildings causing the Millennium Tower's foundation to settle unevenly. The settlement and tilt of the tower have been studied comprehensively by Stewart et al. (2023). This situation emphasizes the importance of considering the broader context and potential interactions with neighboring structures during the design and construction of foundations. Several factors contributed to this issue, including:

- **Adjacent Construction:** The construction of the nearby Transbay Transit Center involved significant excavation and dewatering, which likely disturbed the soil and influenced the settlement of Millennium Tower.
- **Groundwater Pumping:** Construction projects in the vicinity involved pumping groundwater, which can lead to soil compaction and settlement in the surrounding area.
- **Inadequate Foundation Design:** The foundation was designed to rest on dense sands and mud, but the presence of softer, compressible clay layers beneath the building was not accounted for in the original design.
- **Liquefaction Risk:** San Francisco is prone to liquefaction during earthquakes, where saturated soil temporarily loses strength and behaves like a liquid. While the tower's foundation was designed to withstand liquefaction, the settlement still occurred.

These factors, combined with the building's considerable height and weight, contributed to its tilting issue. As a result, the incident highlighted the importance of thorough geotechnical investigations, proper foundation design, and considering the potential impacts of neighboring construction activities to prevent such problems in the future. Among several other documented examples of case histories on improper foundation selection are: Two-grain silos (kissing silos) in Canada (Bozozuk 1972); El Sagrario Church and Metropolitan Cathedral in Mexico (Gonzalez et al. 1997); The leaning Tower of Pisa in Italy (Salgado 2008); Fargo Grain Elevator in the USA (Coduto et al. 2016); and Underpinning of the Mandalay Bay Hotel and Casino (Richards and Kardofilis 2006).

27.5.6 Concluding Remarks

Improper selection of foundations for structures can have devastating consequences, both in terms of safety and financial impact. Inadequate understanding of soil conditions, incorrect assessment of loads, or failure to consider environmental factors can lead to structural instability, compromising the safety and longevity of the entire building. Subsidence, settlement, and even structural failure are potential outcomes of a mismatched foundation choice, jeopardizing the structural integrity and posing risks to occupants and passersby. Subsidence, cracks, and structural failures can lead to costly repairs, disruptions, and legal liabilities. Furthermore, an unsuitable foundation choice can compromise the building's functionality and aesthetics, affecting its overall value and usability. The types of foundation damages, their causes, and restoration and repair techniques are discussed in Chapter 26. Failures have also been discussed in Chapter 26.

To mitigate these risks, a thorough understanding of the site, comprehensive soil analysis, and expert engineering consultation are indispensable. Making informed decisions based on accurate data and professional expertise is not just a matter of compliance; it is an essential investment in the longevity, safety, and sustainability of any construction project. Properly chosen foundations form the bedrock of a secure, resilient, and enduring built environment.

Utilizing a holistic approach to foundation systems results in comprehensive, forward-thinking design. By tapping insights from geology, environmental science, architecture, urban planning, economics, and social sciences, engineers can create foundations that not only support structures but also bolster communities. The integration of disciplines ensures that foundation systems are not just strong for today but resilient for the challenges of tomorrow, embodying the essence of sustainable and enduring construction practices. Embracing this holistic perspective is not just a choice; it is an imperative, paving the way for safer, stable, and sustainable structures in our ever-evolving world.

In a world characterized by rapid change and uncertainty, foundation engineers serve as the vanguards of sustainable infrastructure. By anticipating potential future changes in the environment, land usage, technology, and societal needs, foundation engineers can design resilient, adaptable, and eco-friendly foundation systems. Embracing a forward-thinking mindset not only ensures the stability of structures but also contributes significantly to the creation of a sustainable and harmonious built environment for generations to come.

The consequences of ignoring the holistic approach to foundation system design and construction are far-reaching and impactful. Structural instability, environmental degradation, increased maintenance costs, decreased property value, legal issues, and community disruptions are just a few of the potential outcomes. It is imperative for engineers, architects, and construction professionals to adopt a comprehensive and integrated approach, considering geological, environmental, social, and structural factors. Embracing a holistic perspective ensures the creation of stable, sustainable, and safe structures, benefiting both present and future generations.

27.6 Sustainability Applications in Foundation Engineering

27.6.1 Introduction

Sustainability, in general terms, refers to the ability to meet the needs of the present without compromising the ability of future generations to meet their own needs. It embodies the concept of balancing economic, social, and environmental considerations to create a harmonious and enduring coexistence between humanity and the natural world (Figure 27.20, UN 2005). Sustainable practices aim to conserve resources, minimize environmental impact, and promote social equity, fostering a stable and equitable society for current and future generations. Sustainability involves responsible resource management, energy efficiency, waste reduction, conservation of biodiversity, and equitable social development. It is not merely a goal but a holistic approach that permeates various aspects of human life, including business, education, urban planning, and policymaking, to ensure a better quality of life for all while preserving the planet for the long term.

An adaptive system possesses the innate ability to gradually and organically adjust to its surroundings, demonstrating flexibility in navigating complex situations. These traits are crucial in assessing an engineered system's resilience in a constantly changing socio-economic landscape, ensuring its ability to maintain its identity and purpose without causing environmental harm. Sustainable engineering aims to harmonize engineered systems with both natural and human-made environments, preserving their functionality and societal benefits. Striking a balance between the engineered system's efficiency, economic viability, environmental impact, and social equity is vital. This delicate equilibrium is represented in Figure 27.21, emphasizing sustainability as a dynamic interplay between engineering design, economy, environment, and equity within the global context.

Some of the major implementations of sustainability applications in foundation engineering can be listed as follows:

- Recycling construction and demolition waste (C&DW)
- Reuse of existing foundations
- Furnace slag-based cement grout (Ortega et al. 2014)
- Glass granules for soil stabilization (Umu 2023)
- Biopolymer-based materials for soil improvement (Chang et al. 2020)
- Geosynthetics-based solutions (Dąbrowska et al. 2023).
- Energy piles

In the following Subsections of Section 27.6, the following three topics will be presented: (i) Construction and Demolition Waste (C&DW), (ii) Recycling construction and demolition waste (C&DW), and (iii) Reuse of existing foundations after the demolition of the structure.

Figure 27.20 Pillars of sustainable development (United Nations 2005).

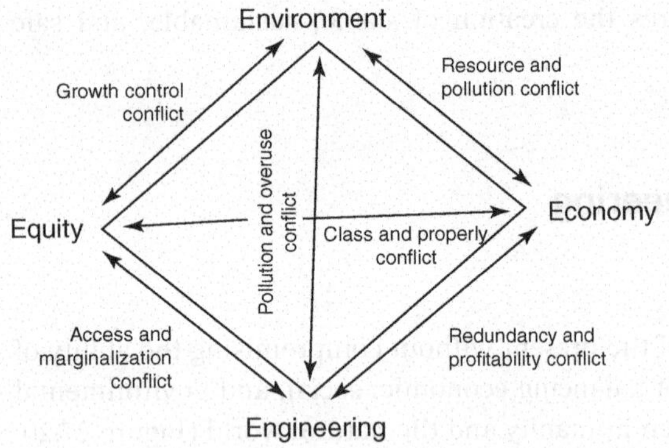

Figure 27.21 The four E's of sustainability in engineering projects (Basu et al. 2013).

27.6.2 Construction and Demolition Waste

Construction waste, stemming from various activities like public works, demolition, and excavation, constitutes a significant environmental concern (Gunnell et al. 2009). The European Union alone generates approximately 850 million tons of demolition waste annually, constituting 31% of total waste production (Solís-Guzmán et al. 2009). Global statistics indicate that in 2012, municipal and construction waste amounted to 3.5 million tons per day, with a per capita production rate of 1.19 kg daily for the urban population. Predictions suggest an increase to 6.96 million tons daily by 2025, considering a global urban population of 4.287 million, resulting in a per capita production rate of 1.42 kg daily. Managing construction waste is crucial, with the U.S. Environmental Protection Agency collaborating extensively with the construction industry to enhance operations, reduce waste, promote recycling, and gather accurate data on waste characteristics (Señas et al. 2016). Natural disasters like earthquakes and floods exacerbate the problem, causing significant damage to urban structures and generating substantial waste volumes. Instances like the Taiwan earthquake in 1999, which led to the destruction of 53,000 residential units, and the recent Turkey earthquake resulting in the demolition of over 190,000 residential units, highlight the pressing global issue of construction waste management (Asif et al. 2007).

27.6.2.1 The Process of Recycling with Complete Material Separation

Recycling construction waste has become a widespread practice in the construction industry, playing a vital role in addressing environmental concerns. Inadequate management of construction waste poses significant future environmental challenges, especially in rapidly urbanizing cities experiencing economic development. The recycling process, as depicted in Figure 27.22, although effective, can be time-consuming and expensive, especially when dealing with waste depots

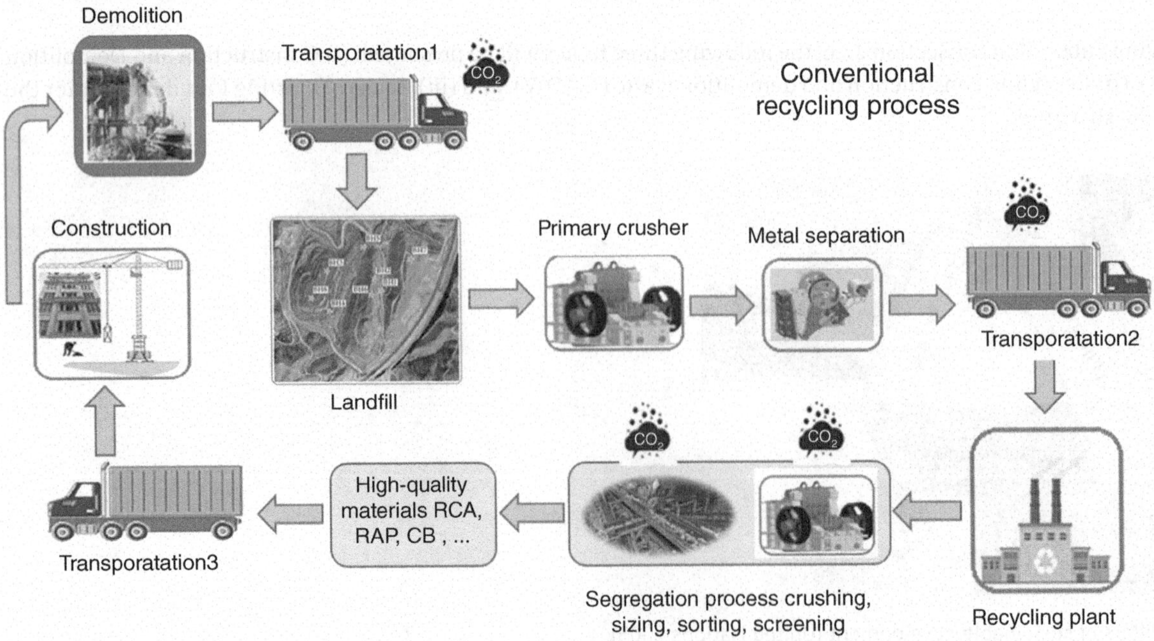

Figure 27.22 The conventional segregating recycling process of C&Dw (Sani et al. 2019 / with permission of Elsevier).

Figure 27.23 Crushing and screening of concrete waste elements (Courtesy Keller Polska).

Figure 27.24 Storage plant of crushed concrete waste at a construction site (Courtesy Keller Polska).

in urban centers. Figure 27.23 illustrates the crushing and screening of concrete waste, while Figure 27.24 showcases a storage plant for crushed concrete waste at a construction site (Kawalec et al. 2017).

27.6.2.2 Typical Engineering Application

Utilizing recycled construction and demolition (C&D) materials in engineering applications and geo-environmental infrastructure offers significant economic and environmental advantages. These materials have demonstrated their worth in various applications, such as concrete production, road infrastructure, paving, and asphalt production, proving their economic viability. Often, construction waste is neglected and left in landfills or along roadsides due to insufficient waste management control in many cities. However, focusing on these materials can lead to substantial cost savings by reducing energy consumption and waste management expenses. Currently, implementing engineering projects using recycled C&D materials can significantly decrease project implementation costs. Studies have shown that construction waste from sites, obtained through the demolition of old structures and previously left as debris, can serve as an effective substitute for natural materials. Recycling construction waste not only preserves natural resources but also helps prevent future environmental challenges. Several applications of recycled C&D materials are presented in the following:

- **Landfills, Slopes, and Road Construction:** Recycling C&D materials can be applied at various stages, serving multiple purposes. These materials find utility in covering layers in landfills, filling embankments, constructing village roads,

Figure 27.25 Photographs of the back wall-filling near the Faraz town area in the Karaj north highway using nonselected C&D waste reinforced by geogrids: (a) the Faraz town geogrid wall, (b) geogrid layers, (c) nonselected C&D wastes with geogrids, (d) the geogrid wall blockage, (e) keeping all heavy equipment at least 0.9 m away from the back of the blocks, (f) geogrid wall block execution; (g) the final concreting of the geogrid wall, and (h) the geogrid wall area executed with C&D (Eslami et al. 2022 / Springer Nature).

and forming base and subbase layers in roads and asphalt production. They also function effectively in sidewalks and geotechnical applications like pipe bedding, low-stress structures, and filling materials for retaining walls. Combining these recycled materials with high-quality natural resources enhances their geotechnical applications, preventing floods and aiding water infiltration in slopes. A significant advantage of construction waste recycling lies in its role as filling material, enhancing land and natural resources, reducing primary and environmental costs, transportation expenses, and energy consumption.

- **C&D Materials in Geosynthetic Reinforced Infrastructures:** Recycled construction waste materials can reinforce road construction through geogrids, exhibiting notably higher shear resistance at interfaces compared to unreinforced samples. A pilot project (Figure 27.25) showcased the use of geogrid-reinforced waste materials in cost-effective, low-energy road embankment walls, minimizing land ownership costs. The strategy involved crushing materials without separation and combining them with geogrids, offering a desirable alternative to natural materials in road and filling projects (Eslami et al. 2022).
- **Recycled Materials in Concrete Production:** Integrating construction waste with natural aggregates in concrete production, including structural, fill, and lightweight concrete, proves to be economically efficient and ecologically beneficial. While regulations for recycled waste in concrete products are limited, understanding the properties of waste materials is crucial. Through comprehensive examinations, recycled concrete derived from construction waste can be effectively utilized in road construction projects, pedestrian walkways, piles, and urban endeavors.
- **Construction Waste in Pile Foundations:** Traditional recycling processes for high-quality segregated aggregates entail significant energy consumption, rendering them costly (Figure 27.22). Conversely, a nonsegregated construction waste recycling scheme, depicted in Figure 27.26, markedly reduces energy consumption. Establishing specific regulations for low-energy consumption of recycled materials from nonsegregated construction waste becomes pivotal to boost recycling rates. This approach involves crushing materials separately and subsequently mixing them thoroughly to create a homogenous mixture.

Utilizing these lower-quality materials proves beneficial in embankments, rural roads, pedestrian pathways, and pile foundations. Additionally, when blended with higher-quality counterparts, they find applications in pipe coating, ground enhancement, pavements, new asphalt and concrete products, foundations, and subbase materials. A key advantage lies in increasing recycling rates and significantly curbing costs related to recycling, environmental impact, transportation, and landfill disposal. The proposed approach, illustrated in Figure 27.26, involves employing nonsegregated construction waste, eliminating separation costs and reducing energy consumption. While resulting in lower-quality materials compared to conventional processes, these materials can be effectively utilized in various applications, including pile foundations, as depicted in Figure 27.27.

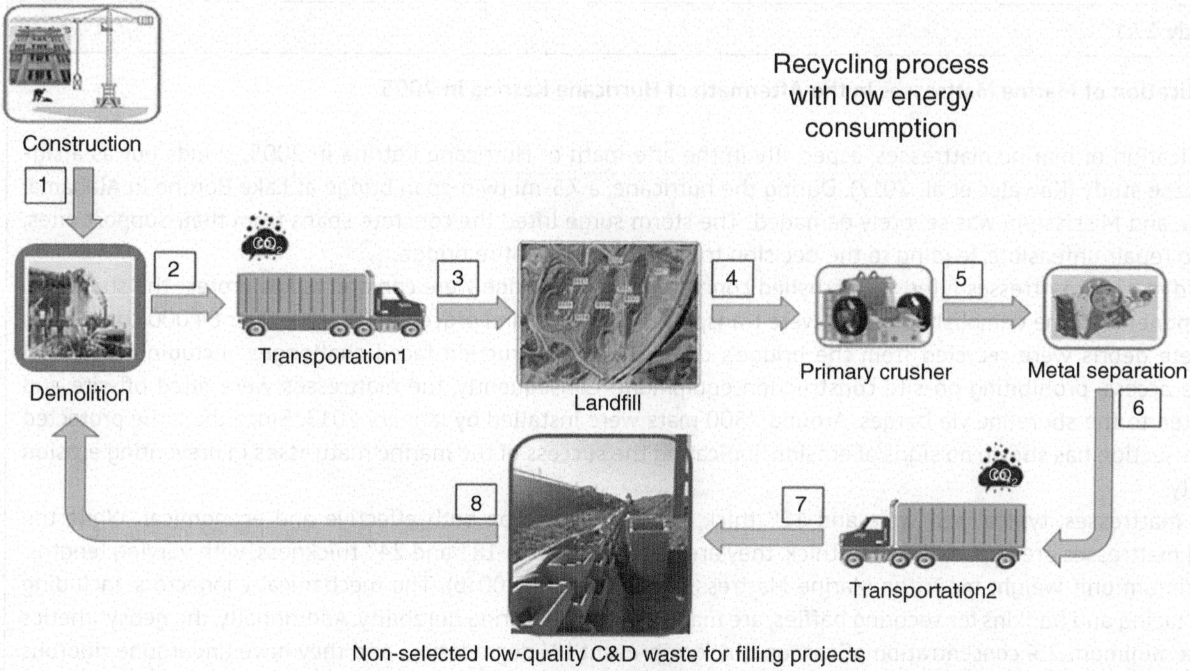

Figure 27.26 Low-energy consumption non-segregating recycling process scheme for the use of waste in piles (Nasiri et al. 2023 / Springer Nature).

Figure 27.27 Application of recycled materials in pile foundations (Nasiri et al. 2023 / Springer Nature).

27.6.3 Use of Construction and Demolition Waste (C&DW) in Foundations

Problematic soils pose significant challenges in engineering projects, especially soft and expansive clays, which can jeopardize infrastructure stability if not properly addressed. Stone columns have long been a reliable method for improving soil-bearing capacity and reducing settlements, with a history dating back to the 1960s in European countries. However, the extensive use of natural stones in these columns raises sustainability concerns, given the environmental impact of mining and quarrying. In the pursuit of eco-friendly alternatives, various waste materials like shredded tire wastes, coconut shells, fly ash chips, and crushed concrete have been explored. Among these, concrete debris emerges as a promising candidate, especially in countries like India, where construction and demolition waste abound. The present study delves into the innovative approach of partially replacing crushed stones in stone columns with concrete debris. Through detailed investigations and finite element modeling, this study aims to explore the efficacy of this sustainable alternative, shedding light on its performance and feasibility in real-world engineering applications. By addressing both engineering challenges and environmental concerns, this research contributes significantly to the evolution of sustainable foundation practices.

Case Study 27.1

The Application of Marine Mattresses in the Aftermath of Hurricane Katrina in 2005

The application of marine mattresses, especially in the aftermath of Hurricane Katrina in 2005, stands out as a significant case study (Kawalec et al. 2017). During the hurricane, a 7.5-mi twin-span bridge at Lake Borgne in Alabama, Louisiana, and Mississippi was severely damaged. The storm surge lifted the concrete spans from their support piles, rendering repair unfeasible, leading to the decision to dismantle the entire bridge.

Geogrid marine mattresses filled with crushed concrete from the bridge were constructed to protect the shoreline. The components of the demolished bridge were transported via barge and processed on-site. Over 84,000 cubic yards of concrete debris were recycled from the bridge's demolition. Construction faced challenges, including restricted shoreline access, prohibiting on-site construction equipment. Consequently, the mattresses were filled off-site and transported to the shoreline via barges. Around 7500 mats were installed by January 2013. Since then, the protected shoreline section has shown no signs of erosion, indicating the success of the marine mattresses in preventing erosion effectively.

These mattresses, typically 5′ wide and 12″ thick, have proven to be both effective and economical. While the standard mattresses are 5′ wide and 12″ thick, they are also available in 18″ and 24″ thickness, with varying lengths. The maximum unit weight for Triton Marine Mattress is 7.87 tons (15,700 lb). The mechanical connectors, including braid for lacing and bodkins for securing baffles, are made of HDPE, ensuring durability. Additionally, the geosynthetics contain a minimum 2% concentration of carbon black to prevent UV degradation, and they have undergone rigorous independent tests for UV resistance. This case study underscores the practicality and efficiency of geogrid marine mattresses in shoreline protection, emphasizing their resilience and cost-effectiveness. Figures 27.28–27.30 show the marine mattresses filled on site with crushed concrete, the marine mattresses installed from the barge, and the view of completed protection marine mattresses, respectively (Kawalec et al. 2017, courtesy Keller Polska).

Figure 27.28 Marine mattresses filled on site with crushed concrete (Tensar International).

Figure 27.29 Marine mattresses installed from barge (Tensar International).

Figure 27.30 View on completed protection Marine mattress (Tensar International).

Case Study 27.2

The Utilization of Crushed Concrete in Soil Improvement Beneath a New 2 km Railway Line (Kawalec et al. 2017)

The utilization of crushed concrete in soil improvement beneath a new 2 km railway line in Poland proved to be a challenging yet innovative endeavor. Facing weak soil conditions, the project demanded a method that ensured stability and cost-effectiveness. Figure 27.31 shows the geotechnical cross section in the longitudinal axis (Kawalec et al. 2017).

(Continued)

Case Study 27.2 (Continued)

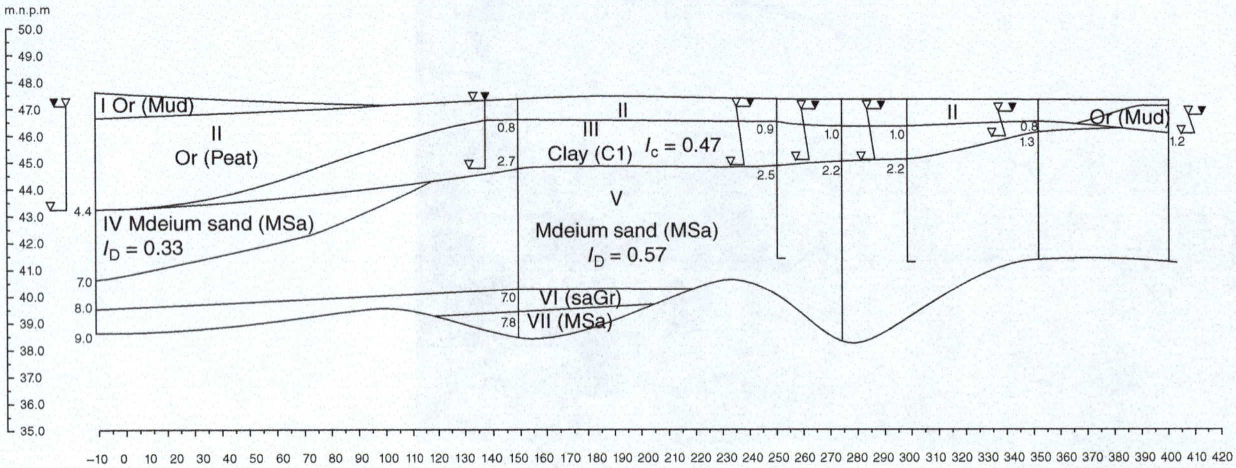

Figure 27.31 Geotechnical cross section in the longitudinal axis (Kawalec et al. 2017, Courtesy Keller Polska / IOP Publishing / Public Domain CC BY 4.0).

After an in-depth geological analysis, the dynamic replacement method was selected. This approach involved driving granular material, including crushed concrete, into the ground using a heavy barrel-shaped rammer. The energy from each blow compacted the material, creating columns that significantly enhanced soil-bearing capacity and stiffness. Figure 27.32 shows the process of column formation during the execution of dynamic replacement (Kawalec et al. 2017).

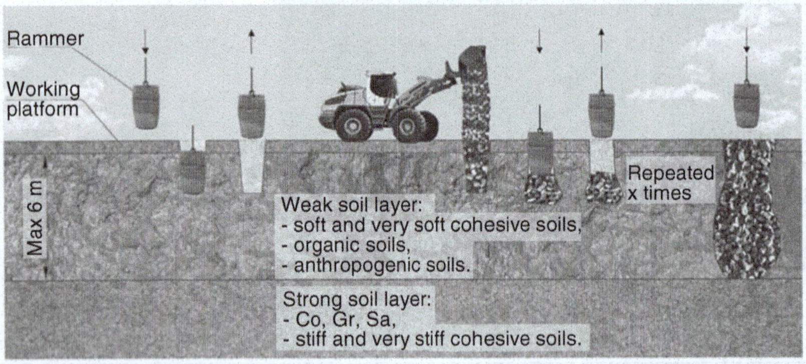

Figure 27.32 Process of column formation during the execution of dynamic replacement (Kawalec et al. 2017, courtesy Keller Polska).

The dynamic replacement process commenced with the formation of a crater by releasing the rammer from a considerable height. Subsequently, the crater was backfilled with granular material, and the rammer delivered blows to drive the material to the desired depth. The process was repeated, controlling the drop height to regulate energy, ensuring optimal compaction. The resulting columns had diameters ranging from 1.8 to 2.4 m and were situated under the transmitted loads in a triangular pattern with a 3.0 m span between columns. Figures 27.33 and 27.34 show the dynamic replacement by the heavy crane with a barrel-like shape rammer and the crushed concrete with sand used for the column construction, respectively (Kawalec et al. 2017).

Figure 27.33 The dynamic replacement: heavy crane with barrel-like shape rammer (IOP Publishing / CC by 3.0).

Figure 27.34 The dynamic replacement: crushed concrete with sand for column construction (IOP Publishing / CC by 3.0).

Quality checks, including static load plate tests, verified the effectiveness of the columns. The primary deformation modulus of the columns reached up to 26 MPa in most cases, with minimal settlement observed three months after completion. These results aligned with earlier numerical modeling and validated the success of the dynamic replacement method in stabilizing weak soils beneath the railway line.

Case Study 27.3

Effect of Replacing Crushed Stone in Stone Columns by Waste Material on Soil Improvement (Salim et al. 2020)

Stone columns have proven to be very effective in enhancing clayey soil-bearing capacity. An experimental laboratory study at the University of Technology, Baghdad, explored the feasibility of replacing crushed stone in stone columns by

(Continued)

Case Study 27.3 (Continued)

demolition concrete waste material to enhance clayey soil-bearing capacity. The concrete waste used in this research had a density and grain size distribution identical to that of the crushed stone. Varying concrete waste percentages by weight (25%, 50%, 75%, 100%) replacing the crushed stone were examined. The findings revealed a 119% bearing capacity increase in the clayey soil was achieved using crushed stone. However, a 155% bearing capacity increase of the clayey soil was achieved using 100% crushed concrete waste.

Case Study 27.4

Reuse of Flood Damaged Roadway Debris (Unpublished personal exploration 2023)

A community of approximately 150 residents faced unprecedented devastation when Tropical Storm Hilary unleashed torrential rainfall, bringing 8 in. of water to Mt. Charleston, Nevada. The event, which occurred in the early hours of August 21, 2023, led to a historic flood, with some areas receiving their entire yearly rainfall within just 48 hours, as reported by the U.S. Forest Service. The aftermath revealed extensive damage, with the floodwaters wreaking havoc on crucial infrastructure, particularly miles of the two-lane access road. Figure 27.35 shows an example of the damage to the roadway.

Figure 27.35 Example of the damage to the roadway (The Nevada Department of Transportation).

In response, the Nevada Department of Transportation (NDOT) swiftly assessed the road damage and initiated a relentless, round-the-clock effort to restore the affected sections of State Route 156 (Lee Canyon Road) and SR-157 (Kyle Canyon Road). Their ambitious goal was to complete the restoration work by early October, aiming to provide the residents and the community with functional roads as swiftly as possible. The damaged roadway and asphalt pavement debris were collected and placed in open spaces along the road (Figures 27.36 and 27.37). Since the debris was conveniently available, it was used as fill to enhance the existing slopes (Figures 27.38 and 27.39).

Figure 27.36 An example of the collected debris (Moses Karakouzian [Book Author]).

Figure 27.37 An example of the collected debris (Moses Karakouzian [Book Author]).

(Continued)

Case Study 27.4 (Continued)

Figure 27.38 Example of debris used as fill to enhance the existing slopes (Moses Karakouzian [Book Author]).

Figure 27.39 Example of debris used as fill to enhance the existing slopes (Moses Karakouzian [Book Author]).

Case Study 27.5

Construction and Demolition Waste as Valuable Resources for Geosynthetic-Encased Stone Columns (Anita et al. 2022)

The study investigates the potential use of mixed construction and demolition waste (CDW) in geosynthetic-encased stone columns. The investigation was performed in three steps. First, a detailed geotechnical analysis, including the physical, mechanical, and hydraulic properties of CDW, was conducted. Figure 27.40 shows the processed CDW

Figure 27.40 Processed CDW used in the study (Anita et al. 2022 / American Society of Civil Engineers).

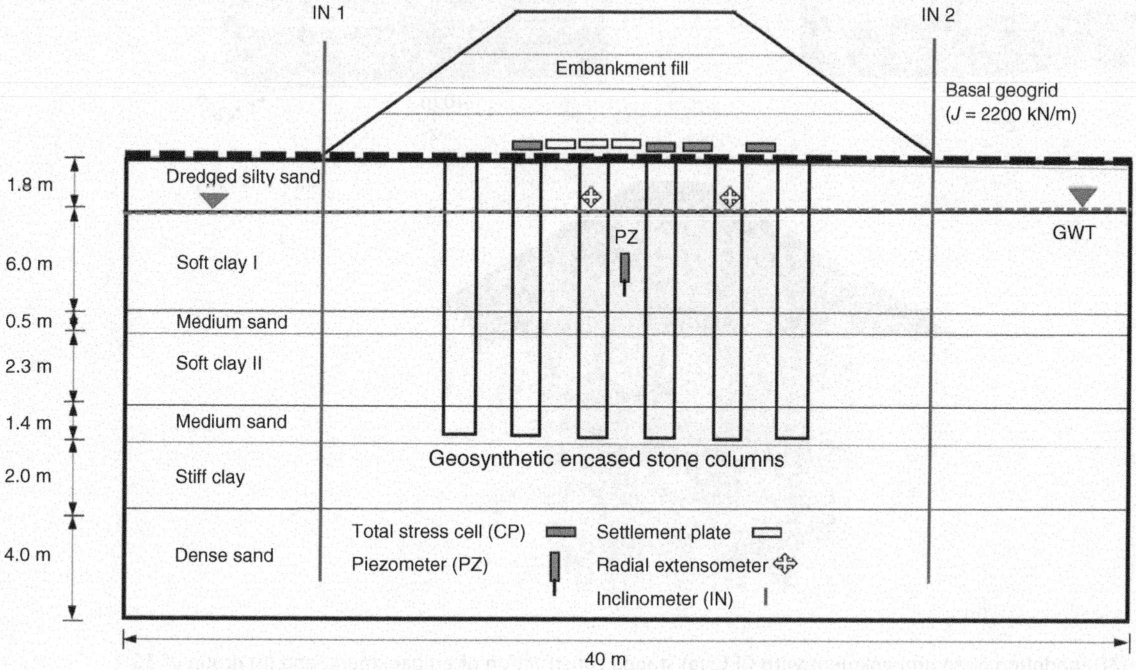

Figure 27.41 Soil profiles adopted for the numerical modeling (Anita et al. 2022 / American Society of Civil Engineers).

(Continued)

Case Study 27.5 (Continued)

used in the study. Next, a 3D Plaxis numerical model of a field study by Almeida et al. (2015) on the behavior of 36 geotextile-encased granular columns supporting test embankment on soft deposit and constructed in several stages was developed and verified (Figure 27.41). Figure 27.42 shows the staged construction of the field test embankment, and the group of 36 geosynthetic-encased stone columns. Figure 27.43 shows the verification results of the numerical model. Finally, the performance of 36 geosynthetic-encased stone columns constructed using CDW (CDW-GECs) was explored through the verified 3D numerical model. Figure 27.44 presents the comparison of the performance of the 36 geosynthetic-encased stone columns using CSA-GECs and CDW-GECs. The results indicate that the performance of the columns using CSA-GECs and CDW-GECs are almost identical.

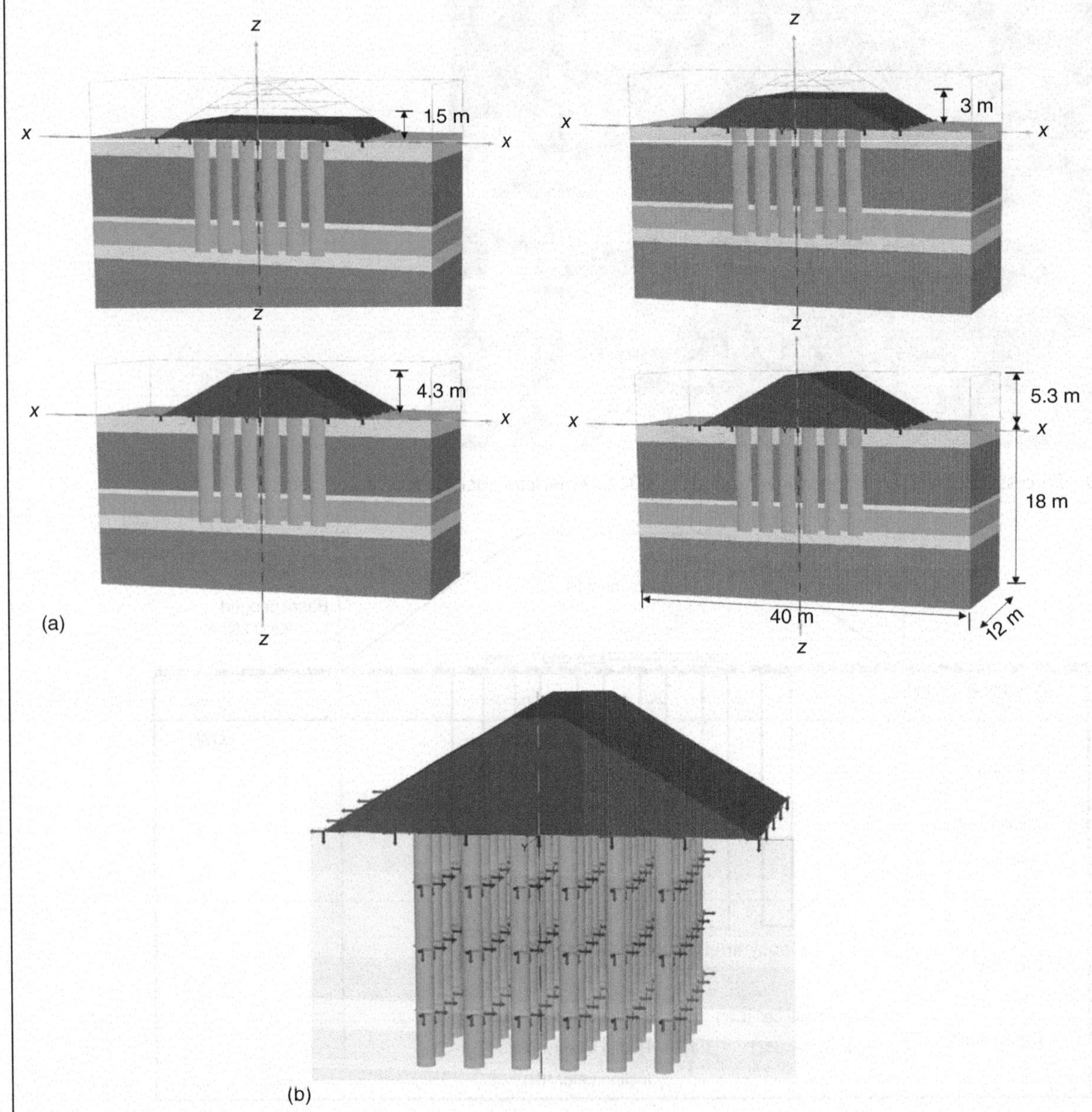

Figure 27.42 3D modeling of an embankment with GEC: (a) staged construction of embankment, and (b) group of 36 geosynthetic-encased stone columns (Anita et al. 2022 / American Society of Civil Engineers).

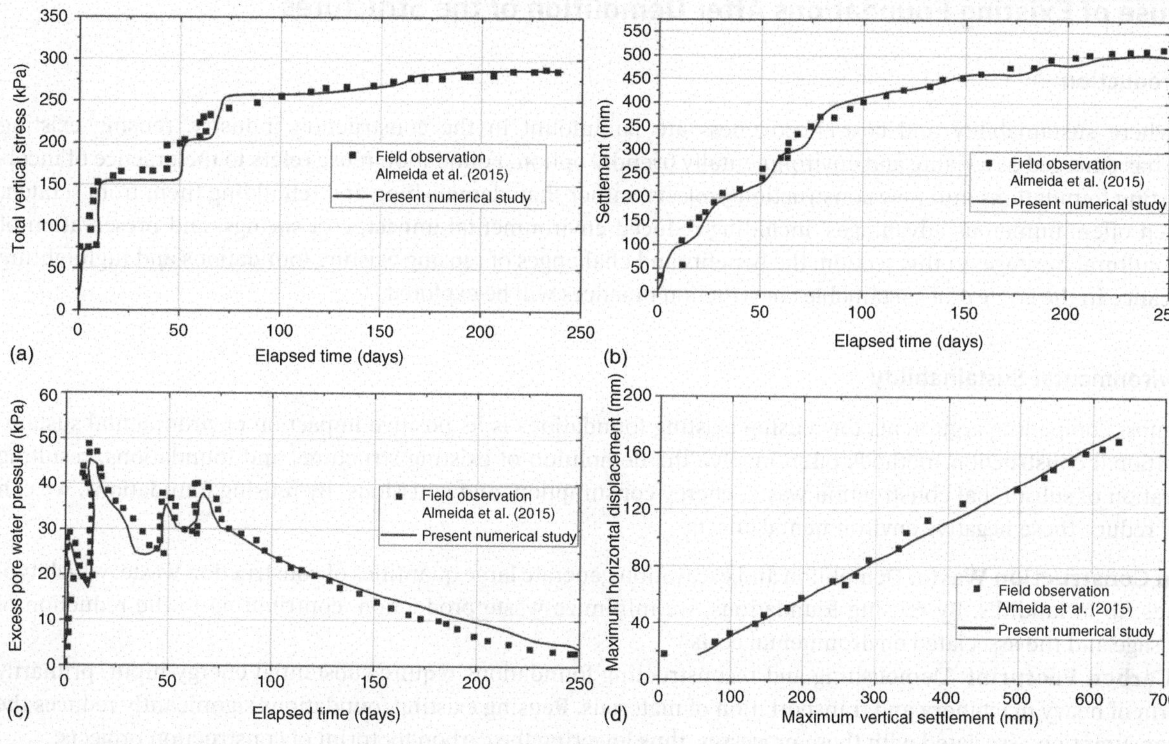

Figure 27.43 Verification of the numerical model: (a) variation of total vertical stress with time, (b) variation of settlement with time, (c) variation of excess hydrostatic pressure with time, and (d) variation of maximum horizontal displacement with maximum vertical settlement (Anita et al. 2022 / American Society of Civil Engineers).

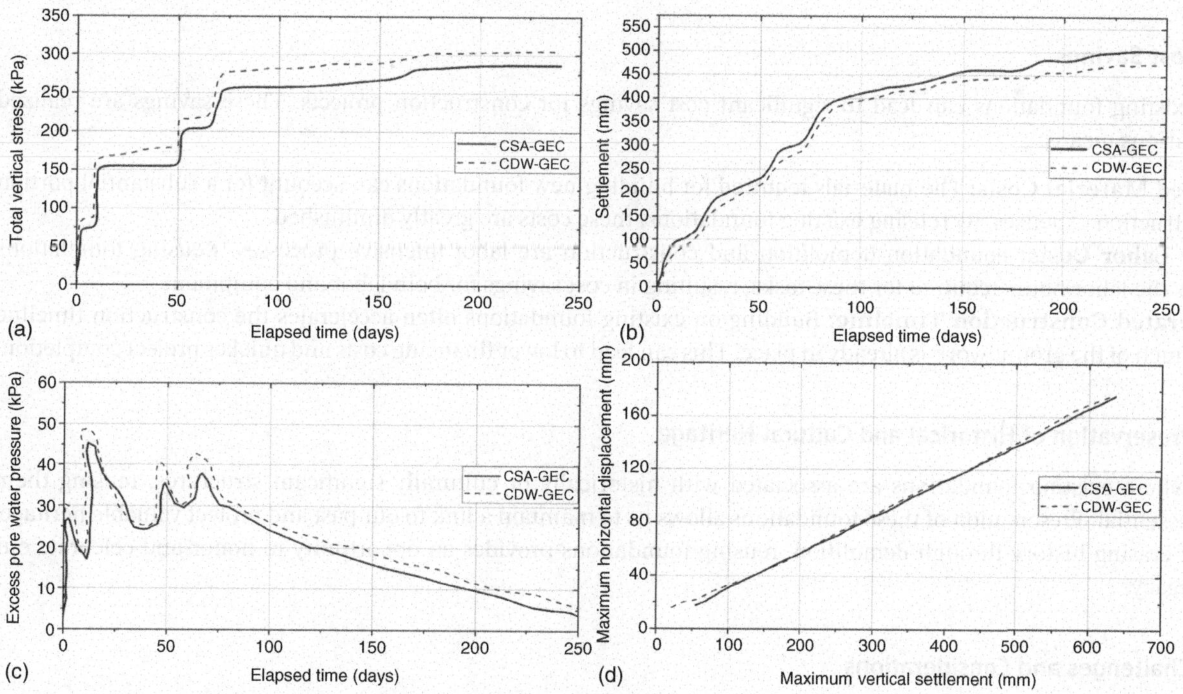

Figure 27.44 Comparison of performance of CSA-GECs and CDW-GECs: (a) variation of total vertical stress with time, (b) variation of settlement with time, (c) variation of excess hydrostatic pressure with time, and (d) variation of maximum horizontal displacement with maximum vertical settlement (Anita et al. 2022 / American Society of Civil Engineers).

27.7 Reuse of Existing Foundations After Demolition of the Structures

27.7.1 Introduction

In an era where sustainability and cost-effectiveness are paramount in the construction industry, reusing existing foundations has emerged as a viable and environmentally friendly option. Foundation reuse refers to the practice of incorporating existing foundations into new construction projects, rather than demolishing and rebuilding them from scratch. This approach offers numerous advantages, including reduced environmental impact, cost savings, and preservation of historical or cultural heritage. In this section, the benefits and challenges of reusing existing foundations and highlighting their significance in the context of sustainable construction practices will be explored.

27.7.2 Environmental Sustainability

One of the most compelling arguments for reusing existing foundations is its positive impact on environmental sustainability. Traditional construction methods often involve the demolition of existing structures and foundations, resulting in the generation of substantial construction waste, energy consumption, and emissions. By reusing foundations, we can significantly reduce these negative environmental effects.

a) **Reduced Construction Waste:** Demolition and excavation generate large quantities of construction waste, which typically ends up in landfills. By reusing foundations, we minimize waste production, contributing to the reduction of landfill usage and the associated environmental costs.
b) **Lower Carbon Footprint:** Demolishing and reconstructing foundations require substantial energy input, primarily in the form of heavy machinery and transportation of materials. Reusing existing foundations significantly reduces the energy consumption associated with these processes, thus lowering the carbon footprint of construction projects.
c) **Preservation of Natural Resources:** Constructing new foundations often involves extracting and consuming raw materials such as concrete, steel, and aggregates. Foundation reuse minimizes the need for these resource-intensive materials, conserving natural resources and preserving the environment.

27.7.3 Cost Savings

Reusing existing foundations can lead to significant cost savings for construction projects. These savings are realized through various means:

a) **Reduced Material Costs:** The materials required for building new foundations can account for a substantial portion of construction expenses. By reusing existing foundations, these costs are greatly diminished.
b) **Lower Labor Costs:** Foundation demolition and construction are labor-intensive processes. Reusing foundations reduces the labor hours required for these tasks, resulting in cost savings for both labor and equipment.
c) **Accelerated Construction Timeline:** Building on existing foundations often accelerates the construction timeline since much of the groundwork is already in place. This can lead to lower financing costs and quicker project completion.

27.7.4 Preservation of Historical and Cultural Heritage

In cases where existing foundations are associated with historically or culturally significant structures, reusing them becomes essential. Preservation of these foundations allows us to maintain a link to our past and protect valuable heritage. Instead of erasing history through demolition, reusing foundations provides an opportunity to honor and celebrate our cultural legacy.

27.7.5 Challenges and Considerations

While the advantages of reusing existing foundations are evident, it is crucial to acknowledge the challenges and considerations involved:

a) **Structural Integrity:** Assessing the structural integrity of existing foundations is paramount. Engineers must thoroughly evaluate the condition of the foundation to ensure it can safely support the new structure.

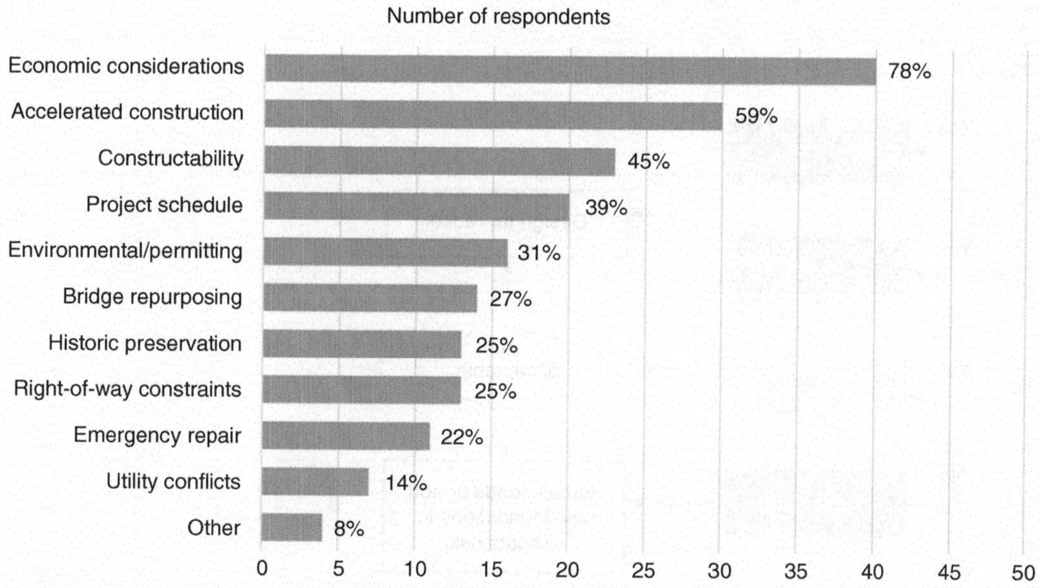

Figure 27.45 Motivations for foundation reuse (Boeckmann and Loehr 2017).

b) **Compatibility:** The existing foundation's design and load-bearing capacity must be compatible with the requirements of the new construction. Modifications or enhancements may be necessary.
c) **Regulatory Compliance:** Construction regulations and codes may vary by location and change over time. Compliance with current standards is essential when reusing existing foundations to ensure safety and legality.

Figure 27.45 presents different motivations regarding foundation reuse.

27.7.6 Risk Assessment Process for Foundation Reuse

In the CIRIA publication C6532, the concept of "reuse load factors" is introduced to establish a connection between the demand placed on foundations and their permissible capacity, enabling the quantification of the potential for reuse. The existing foundation capacity is determined through one of the following methods: the initial capacity recorded in historical archives, the capacity assessed based on the as-built foundation's geometry at a later stage, or the actual load to which the foundations have been proven to perform satisfactorily. These defined criteria can be employed to categorize the level of risk associated with foundation reuse (Figure 27.46).

27.7.7 Overview of the Process Involved in Reusing Foundations

Following are the activities the foundation engineer needs to incorporate in executing a project that involves the reuse of existing foundations (Agrawal et al. 2018; Boeckman et al. 2017).

Site Investigation: Before considering foundation reuse, a thorough site investigation is essential. This includes soil testing, geotechnical studies, and structural assessments of the existing foundations. The goal is to assess the condition and capacity of the foundations and the suitability of the site for the new construction.

Structural Assessment: Engineers need to evaluate the structural integrity of the existing foundations. They assess whether the foundations can support the intended load and meet the safety requirements for the new structure. If the foundations are found to be inadequate, they may require reinforcement or modification.

Compatibility Analysis: The design of the new structure needs to be compatible with the existing foundations. This involves ensuring that the size, shape, and configuration of the reused foundations match the requirements of the new construction. Modifications may be necessary to accommodate any differences.

Load Testing: Load testing may be performed to verify the capacity and stability of the existing foundations. This testing can involve applying simulated loads to the foundation to determine its performance. Results from load testing help engineers make decisions regarding reuse.

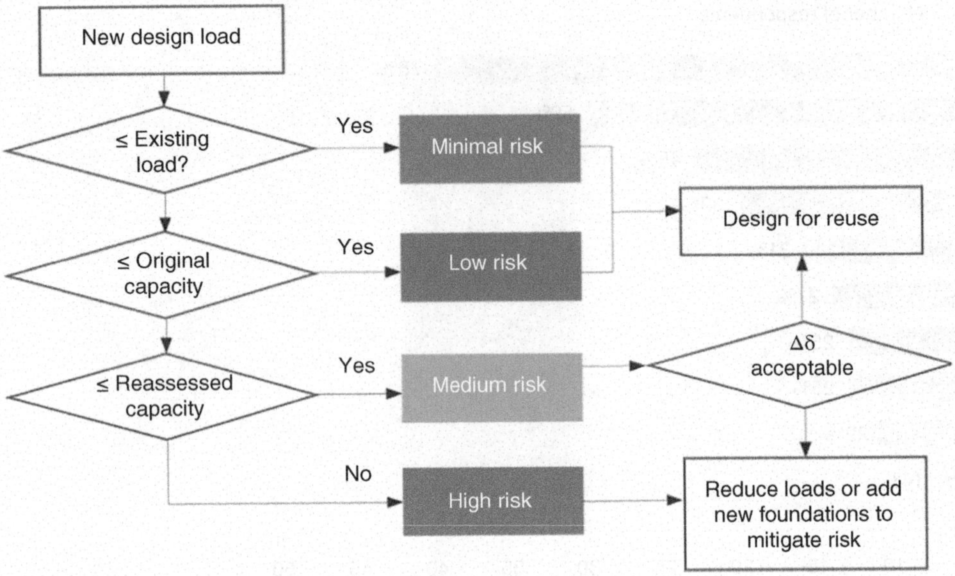

Figure 27.46 Risk assessment process for foundation reuse (Tayler 2020).

Foundation Modification: In some cases, existing foundations may need modifications or repairs to ensure they meet the new project's requirements. This could involve underpinning, strengthening, or other structural enhancements.

Environmental and Regulatory Considerations: Reusing foundations may have environmental and regulatory implications. Engineers and project managers must comply with local building codes, environmental regulations, and permits related to foundation reuse.

Foundation Excavation and Preparation: If modifications are required, part of the existing foundation may need to be exposed and prepared for construction. This could involve excavation, removal of old materials, and installation of new elements as necessary.

New Construction: After the existing foundations are assessed, modified, or prepared, the new construction can proceed. The new structure is built on or connected to the reused foundations.

Quality Assurance and Monitoring: Continuous monitoring and quality assurance measures are essential during the construction process to ensure that the reused foundations perform as intended. This includes inspections, testing, and verification of the foundations' structural integrity.

Documentation and Records: It is critical to maintain comprehensive records of the entire foundation reuse process, including site investigations, assessments, modifications, and any testing conducted. This documentation helps in future construction projects and can be useful for regulatory compliance.

27.7.8 Foundation Reuse in Bridge Construction

When evaluating the replacement of an existing bridge due to its deteriorated or outdated superstructure, it is important to recognize that the bridge's foundation might still have significant functional value. Utilizing these existing foundations in the process of replacing or expanding the bridge can offer substantial cost and time advantages compared to constructing entirely new components. The potential time savings associated with reusing foundations can mitigate disruptions to traffic flow, a primary objective of accelerated bridge construction (ABC), while the cost savings can enhance the economic feasibility and sustainability of an ABC bridge replacement project. However, it is worth noting that existing foundations may possess uncertain material properties, dimensions, or specific details that introduce uncertainties when considering reuse. Unlike a newly constructed foundation, an existing one may have experienced damage, may lack sufficient load-bearing capacity, and may have a limited remaining service life due to deterioration. This part delves into the examination of these concerns, potential measures to reinforce foundations, and innovative strategies for optimizing load-bearing capabilities.

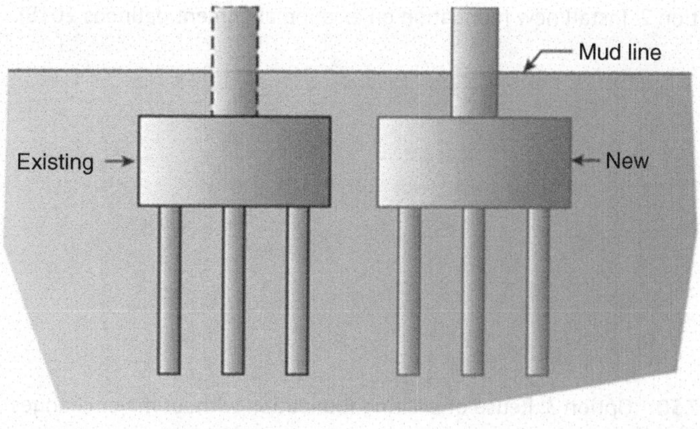

Figure 27.47 Option 1: Install new foundation on new alignment (Jalinoos 2018).

Options 1 up to 4 provide a visual representation of the available foundation reuse choices during the reconstruction of a bridge superstructure (Davis et al. 2019).

27.7.8.1 Option 1

Installing a new foundation on a new alignment typically refers to the process of constructing a fresh foundation for a building or structure at a location or alignment different from the previous one (Figure 27.47).

Case Study 27.6

Example of Option 1 Hurricane Bridge Deck Replacement, Missouri (Axtell et al. 2014)

This bridge originally opened in the 1930s, and the truss elements needed to be replaced (Figure 27.48). The foundation elements seemed to be in good shape, and the Missouri DOT performed an investigation on the existing foundation elements to discover if they could be reused in a new superstructure. The tests did not find any major issues, but ironically, the winning bid fell into the option 1 category because it was 1% cheaper than the suggested option 3 plans. In this case, reusing the elements of the bridge was certainly possible and would have saved material while maintaining its historical appearance.

Figure 27.48 Hurricane bridge deck replacement, Missouri: Option 1 (Axtell et al. 2014; American Bridge Company).

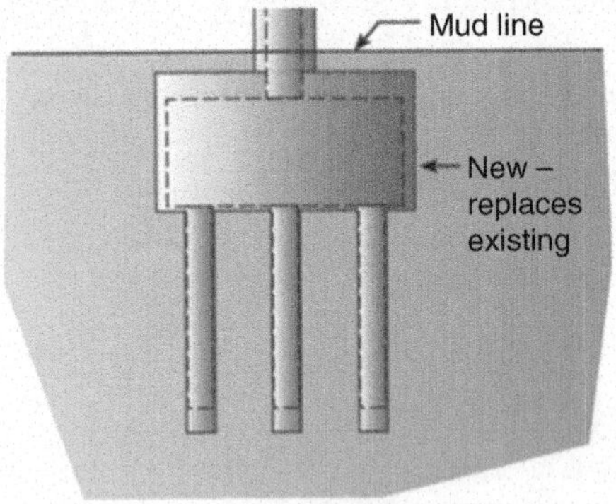

Figure 27.49 Option 2: Install new foundation on existing alignment (Jalinoos 2018).

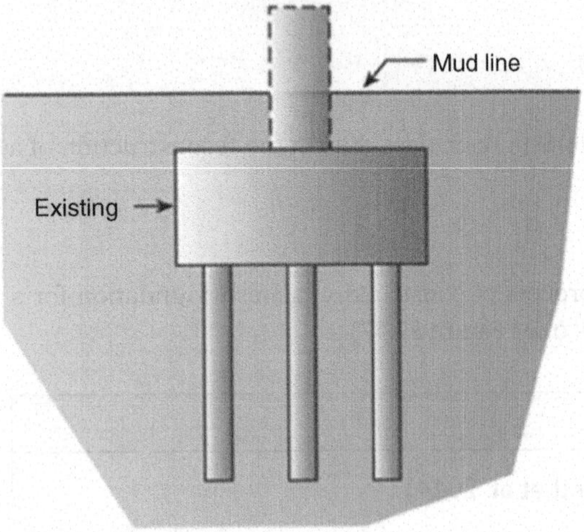

Figure 27.50 Option 3: Reuse of exciting foundation without major changes (Jalinoos 2018).

27.7.8.2 Option 2

Installing a new foundation on an existing alignment means constructing a fresh foundation for a building or structure at the same location or alignment where a previous foundation or structure once stood (Figure 27.49).

27.7.8.3 Option 3

The reuse of an existing foundation without major changes refers to the practice of retaining and utilizing an already existing foundation when constructing a new building or structure that has similar or compatible requirements. This approach is often chosen when the existing foundation is structurally sound and meets the needs of the new project, thus saving time and cost compared to building an entirely new foundation (Figure 27.50).

27.7.8.4 Option 4

The reuse of an existing foundation after strengthening involves the process of assessing, reinforcing, and upgrading an older or weakened foundation to make it suitable for supporting a new building or structure. This approach is often taken when the existing foundation can be retrofitted but requires additional support to meet the requirements of the new project (Figure 27.51).

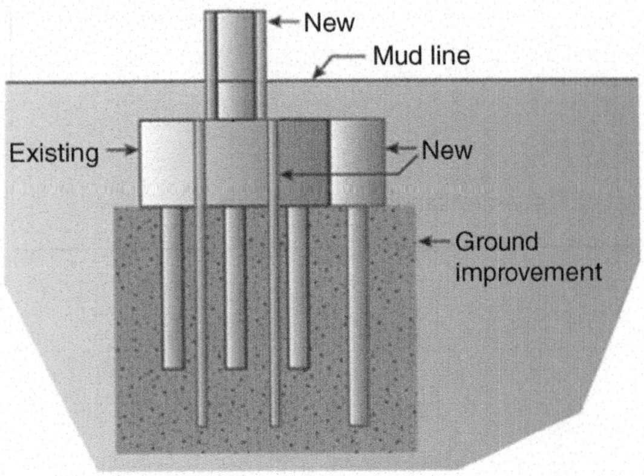

Figure 27.51 Option 4: Reuse of exciting foundation after strengthening (Jalinoos 2018).

Following cases are examples of Option 4 implementations.

Case Study 27.7

Example 1 of Option 4: Huey P. Long Bridge, Louisiana (Agrawal et al. 2018)

Originally built in 1935, the Huey P. Bridge had narrow lanes, highways without shoulders, and had been operating at capacity for some time. In order to ameliorate these problems, the bridge was widened using parallel trusses along the outside of the travel lanes. This necessary work, however, increased the deadweight on the supporting foundations and main-span piers, so additional testing needed to be done in order to verify if the foundations could support the weight without additional settlement. The analysis estimates "that the bearing capacity of the foundation was substantially higher than the original 0.53 MPa … and that additional proposed loading would lead to an additional 2.5–3.8 cm … of settlement" (Davis et al. 2019). The concrete in the existing pier columns was discovered to be stronger than what the design had anticipated, and the below-water portions of the pier were also investigated and appeared to be intact.

Case Study 27.8

Example 2 of Option 4: Milton Madison Bridge, Indiana and Kentucky (Agrawal et al. 2018)

This bridge was originally finished in 1928, and its superstructure was in poor condition, having no shoulders and lanes that were not wide enough to meet modern standards (Figure 27.52). In this case, option 4 was chosen over option 1 because it offered an estimated savings of 50 million dollars. The bridge's superstructure contained 4 piers founded on pneumatic caissons, which showed signs of aging due to "cracking on the pier faces, delaminated sections, lift

Figure 27.52 Milton Madison Bridge prior to renovation (Agrawal et al. 2018) (Gary / Adobe stock).

(Continued)

Case Study 27.8 (Continued)

lines from the original construction, and areas of paste erosion" (Davis et al. 2019). The development team conducted various tests, including "vertical core holes, sampling (pier and caisson concrete, underlying soil), and wireline logging performed from the bridge deck, and sample extraction from the bridge faces" (Davis et al. 2019), and 4 of the piers were determined to be salvageable. The improvements performed prevented those working on the project from needing to construct additional piers or foundation elements.

Case Study 27.9

Example 3 of Option 4: Mississippi River Bridge, Ontario, Canada (Agrawal et al. 2018)

The Mississagi River Bridge, located in Iron Bridge, Ontario, Canada, serves as a critical crossing point for a provincial highway, spanning the Mississagi River, with no other major river crossings for miles in either direction (Figure 27.53). Originally built in 1943, the bridge deck had suffered from corrosion and deterioration, leaving it in a generally poor state. Given the unavailability of the adjacent Right-of-Way (ROW) for acquisition, a solution was needed that preserved the existing alignment while making use of the existing bridge piers. These piers had been initially constructed by driving steel sheet piling into the riverbed, excavating soil to achieve the desired bearing surface, and filling the cofferdam with concrete. It was initially believed that all piers were founded directly on bedrock. However, subsequent investigation, involving the drilling of vertical core holes through the piers into the underlying soil and bedrock, revealed that three of the original piers had been founded on soil rather than bedrock. Additionally, the steel sheet piling had deteriorated, a fact observed through visual and ultrasonic inspections. The existing scour, or erosion of the riverbed, was also documented through visual observations. In some areas, the existing scour had eroded below the footing bottom, raising concerns that the soil-supported piers might become unstable as the cofferdams continued to deteriorate. To address these challenges, the foundation was strengthened through a process of underpinning, involving the installation of micro-piles drilled through the existing footings and the underlying soil until they reached bedrock. An advantage of this approach was that it allowed the underpinning of the existing foundation without necessitating a complete closure of the bridge, as the micro-piles were installed from the existing bridge deck.

Figure 27.53 Mississippi River Bridge, Ontario, Canada, Option 4. Micro-pile installation through the existing bridge deck (Agrawal et al. 2018).

27.7.8.5 Concluding Remarks on Foundation Reuse in Bridge Construction

Options 1 and 2 show the construction of new foundations and are classified as foundation reconstruction. On the other hand, Options 3 and 4 involve the reuse of existing foundations, either in whole or in part, and fall under the category of foundation reuse. It is important to note that these options may not be mutually exclusive, as a single replacement project could encompass several abutments and piers, with some being preserved and others replaced.

27.7.9 Reuse of Existing Piles in Building Reconstruction

The following cases illustrate the reuse of existing foundations in different situations.

Case Study 27.10

Pile Reuse in Tokyo (Tsubakihara 2005)

Two reinforced concrete buildings, dating back to 1972, stood on the construction site of a proposed 10-story building in Chiyoda-ku, central Tokyo. The northern building, comprising four aboveground and two below-ground stories, was supported by large-diameter cast-in-place concrete bell piles. In contrast, the southern building, with six aboveground and one below-ground stories, was supported by cast-in-place concrete piles. The new structure, rising 48 m high with three underground levels, covered a floor area of approximately 37 m by 50 m (Figure 27.54). In the southern part, the foundation utilized a raft foundation design. In the northern section, existing cast-in-place concrete bell piles were repurposed as part of the foundation. These original piles were initially designed to bear only vertical loads, lacking full consideration for seismic lateral forces in their initial design. To address this, eight new cast-in-place concrete bell piles were integrated with the existing 14 piles, as depicted in Figure 27.55. Both the existing and new piles had shaft diameters ranging from 2.4 to 3.9 m and toe diameters between 2.6 and 4.1 m. To accommodate the eccentricity of the piles and columns, the existing and new piles were connected to a 1.4-m thick reinforced concrete mat slab.

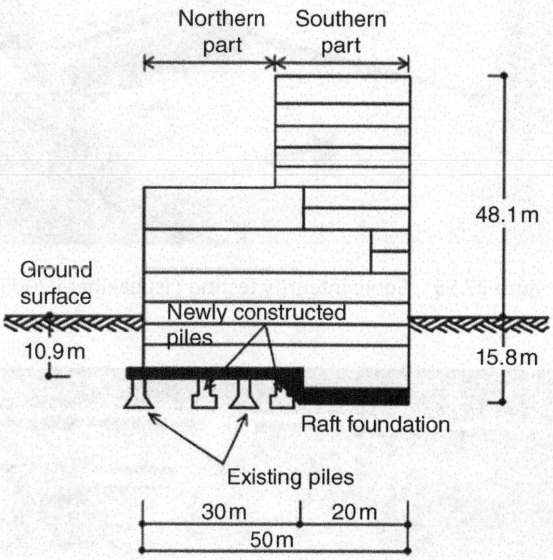

Figure 27.54 Schematic of the new building (Tsubakihara 2005).

Ensuring the quality and durability of pile materials is essential when considering the reuse of existing piles over their life span. To maintain the integrity of these pile materials, a series of tests were performed. To assess piles' structural integrity and length, sonic integrity testing using a low-strain method was conducted for all 14 existing piles (Figure 27.56). During these tests, no significant defects were detected in the pile bodies. To validate the accuracy of the pile length estimates obtained from the test results, three existing piles whose lengths were confirmed through core-borings of the pile concrete were selected (Figure 27.57).

(Continued)

Case Study 27.10 (Continued)

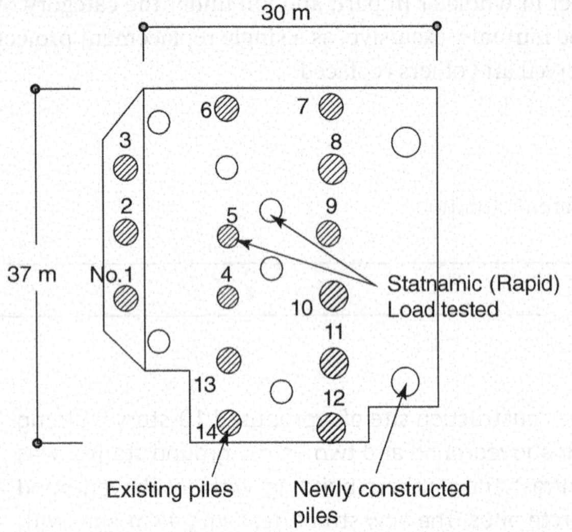

Figure 27.55 Layout of existing and newly constructed piles (Tsubakihara 2005).

Figure 27.56 Sonic integrity testing (Tsubakihara 2005).

Figure 27.57 Concrete cores of a pile (Tsubakihara 2005).

To assess the durability of the pile materials, we conducted concrete carbonation testing, compressive testing of concrete, and tensile testing of reinforcing bars. In the carbonation testing, the extent of carbonation was evaluated by applying a mixture consisting of 1% phenolphthalein and 99% ethyl alcohol onto the top surface of six selected piles (Figure 27.58). As these areas exhibited a color change to red, it was determined that the concrete in the piles had not undergone carbonation.

Figure 27.58 Carbonation testing of pile concrete (Tsubakihara 2005).

The rapid loading tests were performed on the second basement floor of the old building, following the demolition of its superstructure. The dimensions of the two test piles can be observed in Figure 27.59.

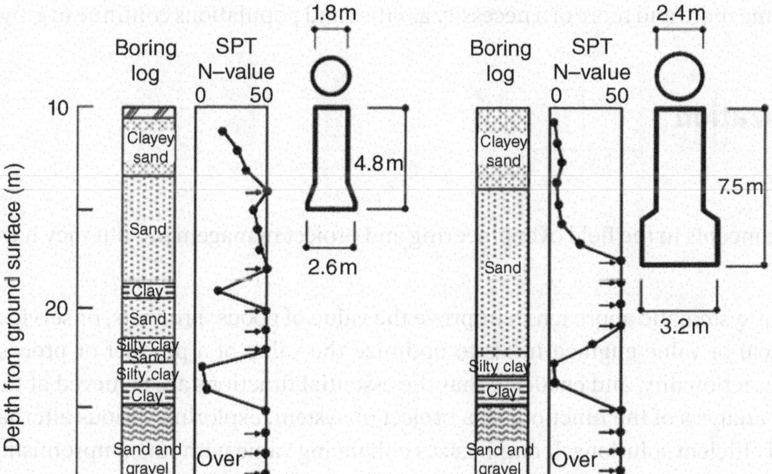

Figure 27.59 Soil conditions and schematics of load-tested piles (Tsubakihara 2005).

Figure 27.60 illustrates the static behavior relationship between the unit toe resistance and the ratio of pile head settlement to the pile toe diameter. The unit toe resistance was determined by dividing the pile head load by the cross-sectional area at the pile toe. The analysis led to the conclusion that the settlement stiffness per unit cross-sectional area at the pile toe for the existing pile was nearly identical to that of the new pile. Furthermore, the variation between the settlement stiffness calculated in the design phase and that derived from the rapid loading tests fell within a 25% margin, thereby confirming the validity of the design value.

(Continued)

Case Study 27.10 (Continued)

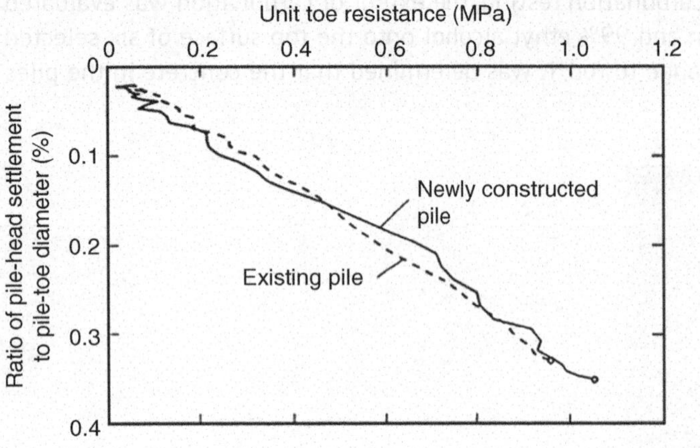

Figure 27.60 Relationship between unit toe resistance (Tsubakihara 2005).

27.7.9.1 Concluding Remarks

Despite the vastly differing nature of these examples, reusing existing foundations almost always resulted in economic, temporal, and environmental gains. Of course, these projects illustrate that each situation must be evaluated on a case-by-case basis and done so carefully. Engineering judgment will decide if a foundation can be reused and how, but if done properly, the development team, the people using these structures, and even the surrounding nature stand to benefit. Given the current issues of lacking ample space for new buildings in many cities, as well as the toll such large projects can have upon the environment, sustainability is more relevant than ever. Understanding this topic and how to properly reuse foundations will thus result in innumerable benefits and become more and more of a necessity as cities and populations continue to grow.

27.8 Value Engineering and Optimization

27.8.1 Introduction

Value engineering and optimization are related concepts in the field of engineering and project management, but they have distinct focuses and approaches.

Value Engineering: Value engineering (VE) is a systematic approach to improve the value of goods, products, or services by evaluating their functions. The primary goal of value engineering is to optimize the value of a product or process by identifying unnecessary costs, improving functionality, and ensuring that the essential functions are achieved at the lowest possible cost. VE involves a structured analysis of the functions of a project or system, exploring various alternatives, and selecting the most cost-effective and efficient solutions. It emphasizes enhancing value without compromising quality or performance.

Optimization: Optimization, on the other hand, refers to the process of making something as effective or functional as possible. In engineering, optimization involves finding the best solution to a problem within a defined set of constraints. It aims to maximize or minimize specific objectives, such as minimizing costs, maximizing efficiency, or maximizing performance. Optimization techniques use mathematical algorithms and models to search for the most optimal solution, considering various factors and constraints.

Key Differences:

1) **Focus:** Value engineering primarily focuses on enhancing value by analyzing functions and costs, while optimization concentrates on finding the best solution to a specific problem, considering various objectives and constraints.

2) **Approach:** Value engineering employs systematic methods to analyze functions, identify alternatives, and select the most cost-effective solution. Optimization uses mathematical algorithms and modeling techniques to search for the best solution based on defined objectives and constraints.
3) **Objectives:** Value engineering aims to improve value by reducing unnecessary costs and enhancing essential functions. Optimization aims to maximize or minimize specific objectives, such as minimizing costs, maximizing efficiency, or maximizing performance.

In summary, value engineering is a broader approach that encompasses the evaluation of functions and costs to enhance overall value, while optimization is a specific technique used to find the best solution based on defined objectives and constraints. Optimization can be one of the tools used within the value engineering process to achieve the desired objectives.

The first of the following two case studies is an example of value engineering applied to a slope stabilization project. The second case study, is the application of an optimization techniques for a pile foundation system.

Case Study 27.11

Value Engineering of Driven H-Piles for Slope Stability on the Missouri River (Thompson et al. 2012)

To enhance rail traffic capacity, a rail line owner expanded an existing main line near Gasconade, MO, by adding an adjacent track along the Missouri River. The new track was positioned on the existing railroad embankment between the existing track and the river, necessitating stability improvements for the embankment slopes (Figure 27.61). The initial design incorporated permanent soldier pile retaining walls in certain areas and jet-grout "shear pins" in others to ensure slope stability (Figure 27.62). However, due to challenges related to site access, the contractor proposed using driven HP 310 × 79 (HP12 × 53) piles as an alternative to the jet-grouted columns, a solution deemed "equal or better." Figures 27.63 and 27.64 display typical analysis outcomes for the H-pile and jet-grout column stabilization alternatives. In each case, computed safety factors are presented for potential sliding surfaces, encompassing either both sets of tracks (referred to as "double track") or solely the new set of tracks adjacent to the river (referred to as "single track").

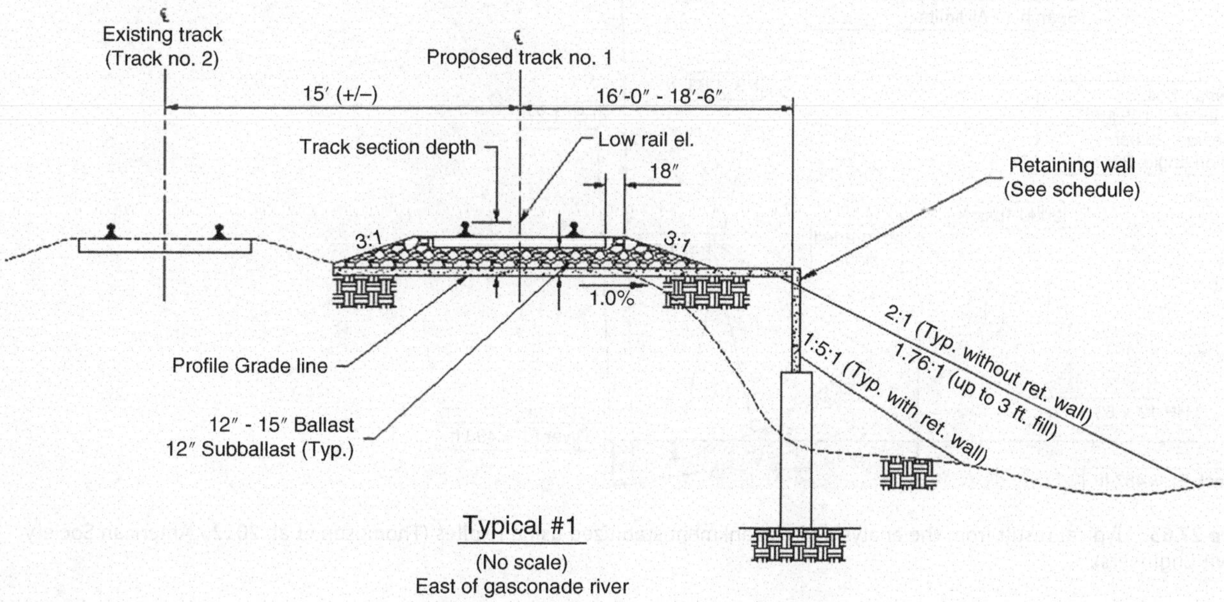

Figure 27.61 Typical cross-section of project alignment (Thompson et al. 2012 / American Society of Civil Engineers).

(Continued)

Case Study 27.11 (Continued)

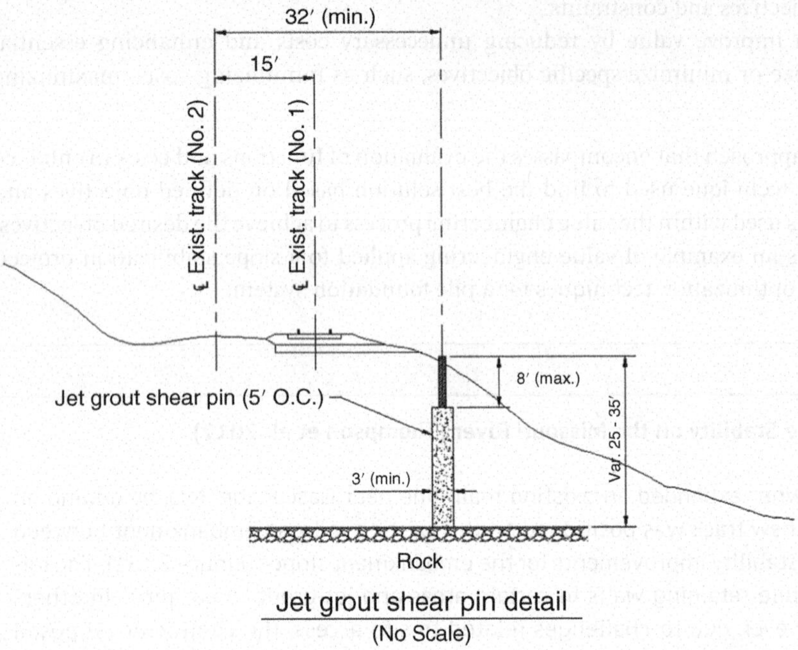

Figure 27.62 Proposed jet-grout shear pin detail (Thompson et al. 2012 / American Society of Civil Engineers).

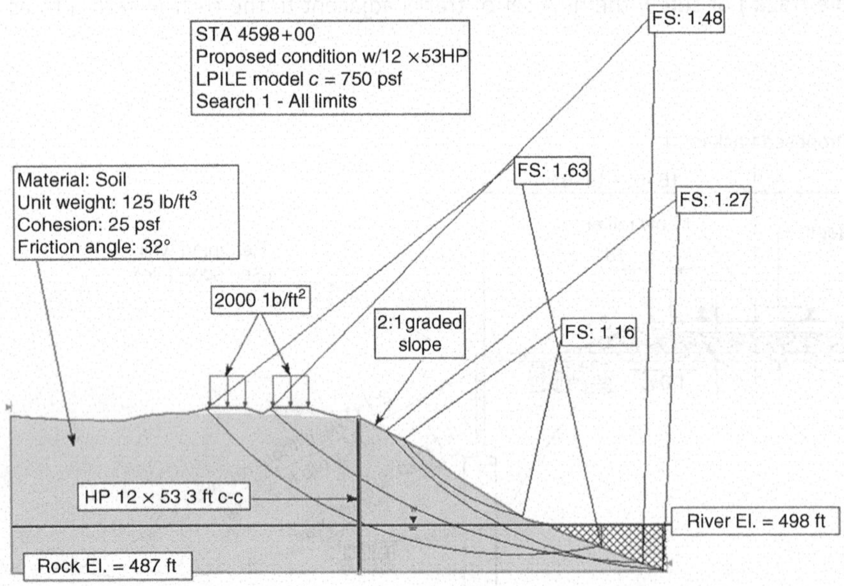

Figure 27.63 Typical result from the analysis of embankment stabilized using H-piles (Thompson et al. 2012 / American Society of Civil Engineers).

Table 27.3 presents the outcomes of stability analyses, displaying computed factors of safety for the initial design employing jet-grout columns and the suggested alternative utilizing steel H-piles at key cross-sections. For each stabilization option, factors of safety are provided for potential sliding surfaces covering both sets of tracks ("double track") and exclusively the new set of tracks closest to the river ("single track").

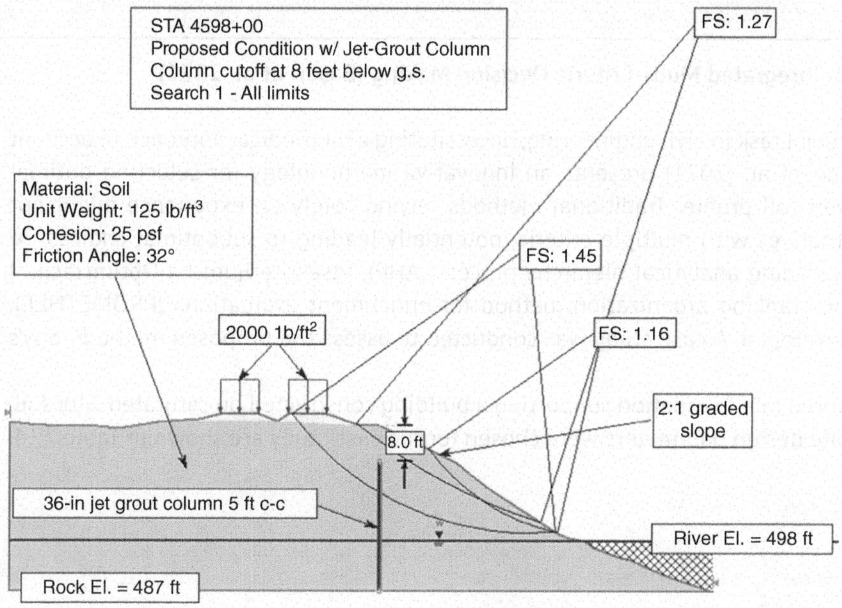

Figure 27.64 Typical result from analysis of embankment stabilized using jet-grout columns (Thompson et al. 2012 / American Society of Civil Engineers).

Table 27.3 Summary of calculated factors of safety (Thompson et al. 2012 / American Society of Civil Engineers).

	Jet grout column		HP12x53 H-pile	
Cross-section station	**Double track[a)]**	**Single track[b)]**	**Double track[a)]**	**Single track[b)]**
4546+00	1.5	1.4	1.5	1.4
4598+00	1.4	1.3	1.5	1.4
4606+00	1.5	1.4	1.5	1.4
4619+00	1.5	1.3	1.5	1.3

a) Double track – potential sliding surfaces encompassing both sets of tracks.
b) Single track – potential sliding surfaces encompassing only the new set of tracks (closest to the river).

This case study outlines the analyses conducted to compare the proposed use of driven H-piles for slope stabilization instead of the initially planned jet-grout columns. The findings demonstrated that utilizing HP 310x79 (HP12x53) A572 Grade 50 steel H-piles, spaced 0.9 m (3 ft) apart near the embankment crest and driven to bedrock, would offer stabilization comparable to or better than the jet-grouted columns specified in the original design. Consequently, the owner approved the proposal, opting for the driven piles as an “equal or better” alternative to the jet-grout columns. Pile installation commenced in September 2007 and concluded in late October 2007, covering approximately 10,400 m (34,000 ft) of piles. This approach resulted in cost savings exceeding US $275,000 compared to the jet-grout shear pin option. Additionally, the installation of driven piles was completed in about half the anticipated time for jet grout shear pin installation. The owner further benefited from cost and schedule savings by eliminating the need for temporary work areas required for staging jet grout equipment.

Case Study 27.12

Selection of Pile Foundation Systems: An Integrated Multi-Criteria Decision Making (Başarı et al. 2021)

Choosing the right pile foundation is a crucial task in civil engineering, necessitating a methodical approach to account for multiple criteria. This study by Başarı et al. (2021) presents an innovative methodology for selecting optimal pile foundation designs based on a given soil profile. Traditional methods relying solely on experience might not suffice when evaluating numerous alternatives with multiple criteria, potentially leading to suboptimal choices. To address this, an integrated approach leveraging analytical hierarchy process (AHP), Vlsekriterijumska Optimizacija I Kompromisno Resenje (VIKOR)/preference ranking organization method for enrichment evaluations (PROMETHEE), and finite element analysis (FEM) was developed. A case study was conducted to assess the proposed methodology's effectiveness.

The study focused on a hypothetical bored pile foundation supporting a building constructed on saturated silty soil, as shown in Figure 27.65. The levels of pile design parameters were chosen for the case study are shown in Table 27.4.

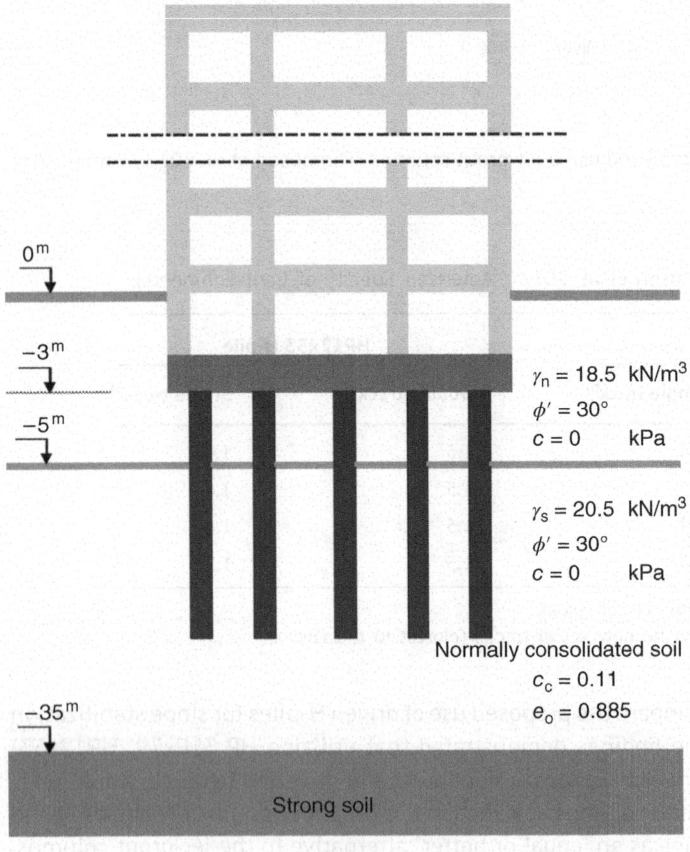

Figure 27.65 Hypothetical building, pile foundation, and soil profile (Başarı et al. 2021).

The following criteria were identified for the optimization, and the pairwise comparison matrix and weights of criteria were developed, as shown in Table 27.5. All feasible 48 design alternatives are shown in Table 27.6.

- **Settlement (S):** Total vertical settlement displacement of the foundation or structures.
- **Safety Factor of Bearing Capacity (SFBC):** SFBC is a ratio of the ultimate bearing capacity and the foundation loads.
- **Ease of Installation (EI):** This parameter describes the relative ease of installation of the support structure on site, in terms of equipment needed and its availability, manpower and time.

Table 27.4 The levels of pile design parameters (Başarı et al. 2021 / Dicle University Journal of Engineering / Public Domain CC BY 4.0).

Description	Unit	Value
Diameter (*D*)	m	0.45
		0.65
		0.80
		1.00
Length (*L*)	m	15
		20
		25
		30
Distance between piles (DBP)	m	2D
		3D
		4D

Table 27.5 The pairwise comparison matrix and weights of criteria (Başarı et al. 2021 / Dicle University Journal of Engineering / Public Domain CC BY 4.0).

	S	SFBC	EI	LDR	C
S	1	5	2	3	0.5
SFBC	0.2	1	0.5	0.5	0.2
EI	0.5	2	1	2	0.33
LDR	0.33	2	0.5	1	0.25
C	2	5	3	4	1
Weights	0.27	0.06	0.15	0.1	0.42

- **Length–Diameter Ratio (LDR):** It is the ratio of pile length and the pile diameter. Length–diameter ratio of piles is widely used for pile analyses.
- **Cost (C):** Unit prices cover the costs of labor, material, and construction.

The final ranking based on VIKOR and PROMETHEE II result is obtained by averaging the ranks as shown in Table 27.7.

Table 27.6 Feasible alternatives (Başarı et al. 2021 / Dicle University Journal of Engineering / Public Domain CC BY 4.0).

Alternative No.	*D* (m)	*L* (m)	Number of pile (#)	*S* (cm)	SFBC	EI	LDR	*C* (£)
1	0.5	15	99	8.41	2.79	1.7	33.3	234,027
2	0.5	20	99	6.02	4.7	2.1	44.4	312,036
3	0.5	25	99	4.25	5.94	2.5	55.6	390,045
4	0.5	30	99	2.9	7.18	3	66.7	468,054
5	0.7	15	48	8.81	2.86	1.3	23.1	195,305
6	0.7	20	48	6.32	3.95	1.6	30.8	279,751
7	0.7	25	48	4.47	5.04	1.9	38.5	349,689
8	0.7	30	48	3.06	6.13	2.2	46.2	419,627

(Continued)

(Continued)

Case Study 27.12 (Continued)

Table 27.6 (Continued)

Alternative No.	*D* (m)	*L* (m)	Number of pile (#)	*S* (cm)	SFBC	EI	LDR	*C* (£)
9	0.8	15	35	8.41	2.81	1.2	18.8	188,764
10	0.8	20	35	6	3.94	1.5	25	275,199
11	0.8	25	35	4.23	5.07	1.8	31.3	343,999
12	0.8	30	35	2.89	6.2	2	37.5	412,798
13	1	15	20	9.94	2.18	1	15	162,759
14	1	20	20	7.14	3.13	1.2	20	241,016
15	1	25	20	5.09	4.08	1.4	25	301,270
16	1	30	20	3.55	5.03	1.6	30	361,524
17	0.5	15	165	8.48	5.76	2.5	33.3	390,045
18	0.5	20	165	6.07	7.83	3.3	44.4	520,059
19	0.5	25	165	4.28	9.9	4	55.6	650,074
20	0.5	30	165	2.92	12	4.7	66.7	780,089
21	0.7	15	80	8.75	4.77	1.9	23.1	325,509
22	0.7	20	80	6.27	6.59	2.4	30.8	466,252
23	0.7	25	80	4.44	8.41	2.9	38.5	582,815
24	0.7	30	80	3.05	10.2	3.4	46.2	699,378
25	0.8	15	54	8.78	4.33	1.7	18.8	291,237
26	0.8	20	54	6.29	6.08	2.1	25	424,593
27	0.8	25	54	4.44	7.82	2.5	31.3	530,741
28	0.8	30	54	3.05	9.56	2.9	37.5	636,889
29	1	15	35	9.08	3.82	1.4	15	284,828
30	1	20	35	6.5	5.48	1.8	20	421,778
31	1	25	35	4.61	7.14	2.1	25	527,223
32	1	30	35	3.17	8.8	2.4	30	632,667
33	0.5	15	374	8.08	13.1	5.2	33.3	884,101
34	0.5	20	374	5.77	17.8	6.8	44.4	1,178,801
35	0.5	25	374	4.05	22.4	8.4	55.6	1,473,502
36	0.5	30	374	2.76	27.1	10	66.7	1,768,202
37	0.7	15	180	8.22	10.7	3.8	23.1	732,395
38	0.7	20	180	5.88	14.8	4.9	30.8	1,049,067
39	0.7	25	180	4.14	18.9	6	38.5	1,311,334
40	0.7	30	180	2.81	23	7.1	46.2	1,573,601
41	0.8	15	117	8.41	9.39	3.1	18.8	631,013
42	0.8	20	117	6	13.2	4	25	919,951
43	0.8	25	117	4.23	16.9	4.9	31.3	1,149,938
44	0.8	30	117	2.89	20.7	5.8	37.5	1,379,926
45	1	15	80	8.14	8.74	2.7	15	651,037
46	1	20	80	5.83	12.5	3.5	20	964,065
47	1	25	80	4.1	16.3	4.2	25	1,205,081
48	1	30	80	2.78	20.1	5	30	1,446,097

Table 27.7 Final ranking (Başarı et al. 2021 / Dicle University Journal of Engineering / Public Domain CC BY 4.0).

Alternative No.	PROMETHEE II rank	VIKOR rank	Average rank	Final rank
16	1	1	1	1
12	2	2	2	2
11	3	3	3	3
8	5	4	4.5	4
15	4	6	5	5
7	6	5	5.5	6

The results of the numerical analysis confirmed the effectiveness of the proposed methodology in choosing the most suitable pile foundation for a given soil profile (Table 27.8). While promising, further research is needed to expand the methodology's applicability across diverse soil profiles in a generalized manner. This integrated approach offers a systematic foundation selection process, empowering decision-makers to incorporate their preferences effectively.

Table 27.8 Settlement and safety factor values for the best five alternatives (Başarı et al. 2021 / Dicle University Journal of Engineering / Public Domain CC BY 4.0).

	Settlements (cm)		Safety factor of bearing capacity	
Alternative No	Conventional method	FEM analyses	Conventional method	FEM analyses
16	3.55	3.962	5.03	5.43
12	2.89	3.96	6.2	6.55
11	4.23	4.897	5.07	6.03
8	3.06	3.9	6.13	6.36
15	5.09	4.738	4.08	5.7

In this unique study, VIKOR and PROMETHEE methods were introduced for the first time in the context of selecting the best pile foundation, marking a significant advancement. The methodology not only considers pile design parameters but also integrates pile layout considerations, setting it apart from previous multi criteria decision-making (MCDM)-based pile selection studies.

27.8.1.1 Concluding Remarks

In the realm of foundation engineering, the principles of value engineering and optimization are paramount in guiding the selection of foundations for structures. Value engineering emphasizes maximizing the value of every component, ensuring optimal performance while minimizing costs. By meticulously assessing various factors such as soil conditions, loads, and structural requirements, engineers can strategize the most efficient and cost-effective foundation systems. Optimization techniques further refine these solutions and fine-tune designs to achieve the perfect balance between functionality, durability, and affordability. By embracing these methodologies, construction projects can not only stand on strong foundations but also thrive in terms of efficiency, sustainability, and economic viability, marking a significant stride toward a more robust and enduring built environment.

27.9 Major Points

1) Foundations play a significant in transferring loads from the superstructure to the substructure. A comprehensive knowledge of foundation types and behavior is essential for selecting the most appropriate alternative for a project.

2) Typically, foundations are classified as shallow and deep based on the embedment depth ratio. Shallow foundations on improved soil and semi-deep foundations have been introduced as supplementary solutions.
3) Drawing upon structural analysis principles and considering their geometrical forms, foundations are categorized into one-dimensional (linear), two-dimensional (planar), and three-dimensional (volumetric) structures. Additionally, in terms of load transfer mechanisms, foundations can be classified as vector-act, section-act, surface-act, and block-act (hybrid) systems.
4) Foundation engineering embodies a multidisciplinary approach, functioning as an interactive practice within the realms of structural, constructional, and geotechnical engineering. Consequently, the selection of foundations is typically categorized into three key aspects: structural, constructional, and geotechnical considerations.
5) Prior to choosing the suitable foundation, various influential factors need to be considered, including superstructure design and size, architectural features, building importance level, subsurface deposit type, climate and weather conditions, budget constraints, adherence to building codes and regulations, construction timeline, maintenance needs, energy efficiency, long-term durability, construction expertise, intended future use, personal preferences, accessibility, ease of installation, environmental impact, cost-effectiveness, groundwater table levels, existing utilities, and proximity to adjacent structures.
6) The selection of appropriate and efficient foundation options often relies on the specific project type. This choice is facilitated by the availability of typical alternatives documented in previous cases. Major project categories include storage tanks, buildings, coastal and offshore structures, wind turbines, transmission utilities, and towers.
7) Foundations designed for storage tanks typically fall into several categories, including earth foundations, earth foundations with concrete or granular ring walls, reinforced concrete slab foundations, and piled foundations.
8) Low-rise buildings, typically ranging from 1 to 5 floors, are usually supported by single foundations, strip foundations in double directions, and slab foundations. Middle-rise buildings, spanning from 5 to 10 to 10-story structures, commonly employ box and floating foundations in conjunction with ground improvement techniques. For small high-rise buildings with 10–20 floors, semi-deep and deep foundations are the norm. In more challenging conditions, comprehensive foundation systems like compensated rafts, pile groups, and piled raft foundations (PRFs) become essential for buildings with stories ranging from 20 to 40.
9) Common bridge foundations include raft foundations, pile foundations, well foundations, and caisson foundations, selected based on the nature and depth of the site. Design and analysis for bridges, whether on land or over water, must consider specific requirements, especially for high-rise piers dealing with problematic deposits and substantial loads. Additionally, for onshore and offshore structures, special considerations are essential, addressing challenges related to installation, challenging deposits, and complex loading conditions.
10) In the case of telecommunication, cell, and power towers, managing overturning loads is paramount. Hence, deeper foundations are preferable. In favorable subsurface conditions, shallow or semi-deep foundations (such as shell foundations) can be employed, provided there is a meticulous analysis of external stability.
11) A holistic approach is a comprehensive and interconnected way of thinking that considers the entire system, including all its components, elements, and interrelationships, rather than focusing solely on individual parts. It emphasizes the understanding that every aspect of a system is intricately linked and influences the functioning of the whole.
12) A holistic approach to foundation systems transcends traditional boundaries, involving collaboration between architects, civil engineers, geologists, environmental scientists, economists, urban planners, and sociologists. By synthesizing knowledge from these diverse fields, a comprehensive understanding of the site's geological, environmental, and social context emerges. This interdisciplinary synergy forms the cornerstone of a robust foundation design.
13) Incorrect foundation selection can lead to substructure damage, disrupting the entire supported system. It is crucial to have adequate data about problematic soils, structural loads, construction methods, adjacent constructions, and geohazards to prevent foundation damage. The causes of damage due to inappropriate foundation selection can be categorized as geotechnical, structural, and construction-related issues.
14) Value engineering is a systematic approach for enhancing the value of a project at the lowest cost, including applying less expensive and more available substitute materials and methods along with preserving functionality. Implementing the principles of value engineering in foundation engineering leads to cost-effective foundation solutions in regard to quantity and quality.
15) Sustainable development is developing in a way that the current needs of people are satisfied without interfering with the capability of future generations to meet their own. The three pillars of sustainable development are people, planet, and profit.

16) Recycling demolition waste in building foundations is a crucial step toward sustainable construction practices. Recycling these materials for foundations minimizes the need for virgin resources, curbing environmental degradation. Additionally, it lowers construction costs, making projects more economically viable. Moreover, recycling demolition waste promotes circular economy principles, where materials are reused and repurposed, contributing to a greener and more responsible construction industry.
17) Reusing existing building foundations is paramount for sustainable construction practices. It reduces construction waste, minimizes environmental impact, and conserves valuable resources. Additionally, reusing foundations can significantly cut down construction costs and time, promoting economic efficiency in building projects. Moreover, repurposing existing foundations preserves the historical and architectural heritage of structures, contributing to the preservation of cultural identity and heritage. Ultimately, it promotes a more eco-friendly and economically viable approach to construction, aligning with the principles of sustainable development.

27.10 Questions and Problems

27.10.1 Explanatory

1 List major characteristics of the main and intermediate foundation types in regard to the ratio of depth to breadth.

2 Define how structural concepts are incorporated with foundation engineering and name novel foundation categories in this regard.

3 What are the three main disciplines that form foundation engineering? Explain the role of each discipline.

4 Name the main factors that are influential in the foundation selection procedure.

5 Illustrate different common foundation alternatives for storage tanks.

6 Determine the sources of combined loading in an offshore wind turbine and explain how this issue can affect the implemented foundation.

7 Explain how improper foundation selection can affect the performance of the whole structure. Name the main issues that must be recognized well during the foundation selection process.

8 Name and explain principles of value engineering and its common application in the engineering of foundations.

9 Introduce C&DWs and their potential application in civil engineering. Use three concepts of energy, economy, and environment in your response.

10 Compare and contrast similarities and differences between offshore and onshore super and substructures.

11 Discuss the ways demolition waste can be used in foundation construction.

12 What factors need to be considered for reusing existing foundations after the demolition of the structure?

27.10.2 Conceptual

1 Compare the foundation selection procedure for an ordinary residential house and a tower.

2 What are the benefits and disadvantages of driven piles and drilled shafts realizing the factors presented in the current chapter?

3 Discuss the performance of different foundation options for a storage tank in regard to geotechnical and constructional issues.

4 Compare and contrast various suggested substructure systems for an offshore wind turbine.

5 How can adjacent construction influence the foundation selection of new buildings? Extend your response via geotechnical requirements of foundation design.

6 What are the mandatory characteristics of foundations that are implemented for towers based on technical and constructional aspects? Compare the suitable foundation options for soft and stiff deposits.

7 Define the role of value engineering in the foundation selection process. Expand your answers using actual examples and case studies.

8 Explain the significance of compensated foundation regarding geotechnical, structural, and constructional aspects.

9 Elaborate and illustrate foundation systems elements and roles for different structures.

10 Based on what you learned in the current chapter, define the role of isolation rocking in the performance of bridge foundations.

11 Discuss how to balance the competing sustainability factors in a construction project.

12 Discuss the relationship, if any, between satisfying sustainability requirements and the process of value engineering.

13 Discuss the risks associated with the reuse of existing foundations, and how to mitigate them.

14 Are the differences between value engineering and optimization? If yes, please discuss these differences.

References

Agrawal, A.K., Jalinoos, F., Davis, N.T. et al. (2018). *Foundation Reuse for Highway Bridges.* FHWA-HIF-18-055. Washington, DC: Federal Highway Administration.

Ahmed, S. (2021). Bridge foundation | 7 types of bridge foundation | bridge foundation construction process | Civilstring. CivilString. https://civilstring.com/bridge-foundation-and-types-of-bridge-foundation/ (accessed 21 September 2023).

Almeida, M.S.S., Hosseinpour, I., Riccio, M., and Alexiew, D. (2015). Behavior of geotextile-encased granular columns supporting test embankment on soft deposit. *Journal of Geotechnical and Geoenvironmental Engineering* 141 (3): 04014116. https://doi.org/10.1061/(ASCE)GT.1943-5606.0001256.

Anastasopoulos, I., Kourkoulis, R., Gelagoti, F., and Papadopoulos, E. (2012). Rocking response of SDOF systems on shallow improved sand: an experimental study. *Soil Dynamics and Earthquake Engineering* 40: 15–33.

Anita, A., Karthika, S., and Divya, V. (2022). Construction and demolition waste as valuable resources for geosynthetic-encased stone columns. *Journal of Hazardous, Toxic, and Radioactive Waste* 27 (2): https://doi.org/10.1061/JHTRBP.HZENG-1175.

Asif, M., Muneer, T., and Kelley, R. (2007). Life cycle assessment: a case study of a dwelling home in Scotland. *Building and Environment* 42: 1391–1394.

Axtell, P.J., Siegel, T.C., and Brown, D. (2014). Sustainability and reuse of foundations for Hurricane Deck bridge. In: *Proc., 39th Annual Conf. on Deep Foundations.* Hawthorne, NJ: Deep Foundation Institute.

Başarı, E., Eski, Ö., Araz, U., and Turan, M.E. (2021). Selection of pile foundation systems: an integrated multi-criteria decision making. *DUJE* 12 (1): 133–145.

Basu, D., Misra, A., Puppala, A.J., and Chittoori, C.S. (2013). Sustainability in geotechnical engineering. *In the Proceedings of the 18th International Conference on Soil Mechanics and Geotechnical Engineering*, Paris (2–6 September).

Bhattacharya, S. (2019). *Design of Foundations for Offshore Wind Turbines*. Wiley.

Bienen, B., Rausch, L., Gaudin, C., et al. (2011). Numerical study of the combined load capacity of a hybrid foundation. *21st International Offshore and Polar Engineering Conference*, Maui, Hawaii, USA (19–24 June 2011), vol. 8, 556–562.

Boeckmann, A. and Loehr, J.E. (2017). *Current Practices and Guidelines for the Reuse of Bridge Foundations*. NCHRP Synthesis Report 505, 118. Washington, D.C.: Transportation Research Board of the National Academies.

Bozozuk, M. (1972). Foundation failure of the Vankleek Hill tower silo. *Proceedings of the ASCE Specialty Conference on Performance of Earth and Earth-Supported Structures*, Purdue University, Lafayette, Indiana, USA (11–14 June 1972), 1, 2, pp. 885–902, 1972-07-01. ASCE.

Chang, I., Lee, M., Tran, T.P.A. et al. (2020). Review on biopolymer-based soil treatment (BPST) technology in geotechnical engineering practices. *Transportation Geotechnics* 24: 100385.

China National Standards (2005). *Code for Design of Civil*. Building (GB50352-2005). Beijing, China: China Architecture & Building Press.

Coduto, D.P., Kitch, W.A., and Yeung, M.R. (2016). *Foundation Design Principles and Practices*, 3e. Upper Saddle River, NJ: Prentice Hall, Inc.

Dąbrowska, J., Kiersnowska, A., Zięba, Z., and Trach, Y. (2023). Sustainability of geosynthetics-based solutions. *Environments* 10 (4): 64.

Das, B.M. (2014). *Principles of Foundation Engineering*, 8e. Cengage Learning.

Davis, N.T., Hoomaan, E., Agrawal, A.K. et al. (2019). Foundation reuse in accelerated bridge construction. *Journal of Bridge Engineering* 24 (10): 05019010.

Ebrahimipour, A., & Eslami, A. (2024). Analytical study of piles behavior for marine challenging substructures. *Ocean Engineering (Print)*, 292, 116514. PP 1–16. https://doi.org/10.1016/j.oceaneng.2023.116514.

Ebrahimipour, A., Eslami, A., Imani, M., & Imam, R. (2023). Foundation systems performance; technical and sustainability aspects. *In the Proceedings of the 13th International Congress on Civil Engineering*, Tehran, Iran (17–19 October).

Eslami, A. and Heidarie Golafzani, S. (2020). Relevant data-based approach upon reliable safety factor for pile axial capacity. *Marine Georesources and Geotechnology* 39 (11): 1373–1386.

Eslami, A., Moshfeghi, S., Molaabasi, H., and Eslami, M. (2020). *Piezocone and Cone Penetration Test (CPTu and CPT) Applications in Foundation Engineering*, 1e. Elsevier.

Eslami, A., Nasiri, A., and Fahimifar, A. (2022). Improving the geotechnical properties of recycled non-selected construction waste using geogrid reinforcement. *International Journal of Civil Engineering* 21: 1–20.

Eslami, A., Ebrahimipour, A., Imani, M., Imam, R., and Mo, P. (2025). Form and load transfer aspects of foundation systems: Case-based implementation and adaptation for buildings. Deep Underground Science and Engineering. https://doi.org/10.1002/dug2.12146.

Fakher, A. (2022). *Advanced Foundation Engineering*, 2e. Iran: University of Tehran Press (In Persian).

Fateh, A.M.A., Eslami, A., and Fahimifar, A. (2017). Direct CPT and CPTu methods for determining bearing capacity of helical piles. *Marine Georesources and Geotechnology* 35 (2): 193–207.

Freepik (2021). Internet technology set of cartoons about fifth generation of mobile communications Free Vector (2021, June 3). Freepik. https://www.freepik.com/free-photos-vectors/mobile-tower-icon/12 (accessed 21 September 2023).

Gonzalez, E.T., Villa, S.E., and Shelley, E.O. (1997). Options for correcting differential settlements in Mexico City's Metropolitan Cathedral. *Proceedings of 14th International Conference on Soil Mechanics and Foundation Engineering* (Hamburg) (6–12 September).

Gunnell, K., Du Plessis, C., and Gibberd, J. (2009). *Green Building in South Africa: Emerging Trends*. Department of Environmental Affairs and Tourism (DEAT).

Hasan, S.D., Islam, N., and Moin, K. (2010). A review of fixed offshore platforms under earthquake forces. *Structural Engineering and Mechanics* 35 (4): 479–491.

He, S., Wang, X., Dong, J. et al. (2019). Three-dimensional urban expansion analysis of valley-type cities: a case study of Chengguan District, Lanzhou, China. *Sustainability* 11 (20): 5663.

Hedjazi, S., Rahai, A., and Sennah, K. (2007). Evaluation of creep effects on the time-dependent deflections and stresses in prestressed concrete bridges. *Bridge Structures* 3 (2): 119–132.

Jalinoos, F. (2018). Foundation reuse: an option for Bridge Reconstruction Projects. Federal Highway Administration Office of Infrastructure R&D. *ASPIRE Summer* 2018: 36–37.

Katzenbach, R., Leppla, S., and Choudhury, D. (2017). *Foundation Systems for High-Rise Structures*, 1e. CRC Press, Taylor & Francis Group.

Kawalec, J., Kwiecien, S., Pilipenko, A., and Rybak, J. (2017). Application of crushed concrete in geotechnical engineering–selected issues. *IOP Conference Series: Earth and Environmental Science* 95: 022057.

Kurian, N.P. (2006). *Shell Foundations: Geometry, Analysis, Design, and Construction*. Alpha Science International Limited.

Mandolini, A. (2003). Design of piled rafts foundation: practice and development. In: *Deep Foundations on Bored and Auger Piles* (ed. W.F. van Impe), 59–80. Rotterdam: Millpress.

Moradi, A., Mahmoudzadeh, A., and Rahim Safavi, Y. (2017). Geotechnical aspects of storage tank foundation types. *Journal of Mechanical and Civil Engineering* 14 (1): 43–49.

Nasiri, A., Eslami, A., and Fahimifar, A. (2023). Assessment of construction and demolition waste depot materials for filling. *Transportation Infrastructure Geotechnology* 11: 1–27. Springer.

Obinna, U. (2023). *Foundation for Bridges Over Water – Structville*. Structville. https://structville.com/foundation-for-bridges-over-water.

Ortega, J.M., Pastor, J.L., Albaladejo, A. et al. (2014). Durability and compressive strength of blast furnace slag-based cement grout for special geotechnical applications. *Materiales De Construccion* 64 (313): e003.

Poulos, H.G. (ed.) (2017). *Tall Building Foundation Design*. Taylor & Francis Group.

Rahai, A. (2004). Effect of earthquake vertical motion on RC bridge piers. *Proceedings of 13th World Conference on Earthquake Engineering*, Vancouver, BC, Canada (1–6 August).

Rezazadeh, S. and Eslami, A. (2018). Bearing capacity of semi-deep skirted foundations on clay using stress characteristics and finite element analyses. *Marine Georesources & Geotechnology* 36 (6): 625–639.

Richards, T.D. and Kardofilis, D. (2006). Underpinning of the Mandalay Bay Hotel and Casino, *54th Annual Geotechnical Engineering Conference*, St. Paul, Minnesota (17 February 2006).

Saberi, M.R., Rahai, A., Sanayei, M., and Vogel, R.M. (2016). Steel Bridge service life prediction using Bootstrap method. *International Journal of Civil Engineering* 15: 51–61.

Salgado, R. (2008). *The Engineering of Foundations*, 1e. McGraw-Hill Education.

Salim, N.M., Hasan, S., and Al-Soudany, K. (2020). Effect of replacing crushed stone in stone columns by waste material on soil improvement ratio. *Proceedings of International Structural Engineering and Construction* 7: 2–7.

Sani, A.K., Singh, R.M., Amis, T., and Cavarretta, I. (2019). A review on the performance of geothermal energy pile foundation, its design process and applications. *Renewable and Sustainable Energy Reviews* 106: 54–78.

Señas, L., Priano, C., and Marfil, S. (2016). Influence of recycled aggregates on properties of self-consolidating concretes. *Construction and Building Materials* 113: 498–505.

Sharifzadeh Asli, M., Hosseini, S.M.M.M., and Jahanirad, A. (2018). Effect of soil reinforcement on rocking isolation potential of high-rise bridge foundations. *Soil Dynamics and Earthquake Engineering* 115: 142–155.

Shirani, S.S., Eslami, A., Ebrahimipour, A., and Karakouzian, M. (2023). Dominant factors in MiniCone, CPT and pile correlations: a data-based approach. *Deep Underground Science and Engineering* 2 (4): 346–358. https://doi.org/10.1002/dug2.12055.

Shojaei, E., Eslami, A., and Ganjian, N. (2019). Self-expanded piles: a new approach to unconventional. *Marine Georesources and Geotechnology* 39 (1): 115–128.

Solís-Guzmán, J., Marrero, M., Montes-Delgado, M.V., and Ramírez-de-Arellano, A. (2009). A Spanish model for quantification and management of construction waste. *Waste Management* 29: 2542–2548.

Staubach, P. and Wichtmann, T. (2020). Long-term deformations of monopile foundations for offshore wind turbines studied with a high-cycle accumulation model. *Computers and Geotechnics* 124: 103553.

Stewart, J.P., Wagner, N., Murphy, D. et al. (2023). Foundation settlement and tilt of Millennium Tower in San Francisco, California. *Journal of Geotechnical and Geoenvironmental Engineering* 149 (6): https://doi.org/10.1061/jggefk.gteng-10244.

Team, D. (2023). Types of Bridge Foundation. Daily Civil. https://dailycivil.com/types-bridge-foundations-suitability/.

Thompson, R.W. III, Hill, J.R., and Loehr, J.E. (2012). Case history: value engineering of driven H-piles for slope stability on the Missouri River. *Contemporary Topics in Deep Foundations* 207–214. https://doi.org/10.1061/41021(335)26.

Tsubakihara, Y. (2005). Reuse of existing piles in building reconstruction and its environmental effects. *Proceedings of the 2005 World Sustainable Building Conference in Tokyo (SB05Tokyo).*

Umu, S.U. (2023). Assessment of sustainable expanded glass granules for enhancing shallow soil stabilization and dynamic behaviour of clay through resonant column tests. *Engineering Science and Technology, an International Journal* 42: 101415.

United Nations General Assembly (2005). 2005 World Summit Outcome. United Nations, *Report A/60/1*.

Subject Index

Advanced Foundation Engineering: Principles, Performance and Prospects, First Edition. Moses Karakouzian and Abolfazl Eslami.

c

e

f

Author Index

Advanced Foundation Engineering: Principles, Performance and Prospects, First Edition. Moses Karakouzian and Abolfazl Eslami.

b

e

f

i

j

n

o

p

q

r

S